Hindu and Arabian Period 200 B.C. to 1250 A.D.
Hindu-Arabic numeral system; Arab absorption of Hindu arithmetic and Greek geometry

Modern Period (Early) 1450 A.D. to 1800 A.D.
Logarithms; modern number theory; analytic geometry; calculus; the exploitation of the calculus

Period of Transmission 1250 A.D. to 1500 A.D.
Learning preserved by Arabs slowly transmitted to Western Europe

0 TO 500 A.D.

Chinese begin to use negative numbers.

Zero symbol is invented.

Hypatia studies number theory, geometry, and astronomy; her death in Alexandria is followed by decline of Alexandria as center of learning.

Jesus Christ's teachings establish new religion.

Under Emperor Claudius, Romans conquer Britain.

Roman Empire is divided into east and west.

Visigoths sack Rome.

Christianity becomes official religion of Roman Empire.

500 A.D. TO 1000 A.D.

al-Khowarizmi composes key book on algebra and Hindu numerals.

Early computing algorithms are developed.

Omar Khayyam creates geometric solutions of cubic equations and calendric problems.

Justinian's legal code is instituted.

Hegira of Muhammad takes place.

Chinese invent compass, gunpowder, and printing.

Charlemagne is crowned Holy Roman Emperor.

1000 A.D. TO 1500 A.D.

Fibonacci's Liber Abaci advocates Hindu-Arabic numeral system, which supplants Roman system.

Jerusalem is captured in First Crusade.

Genghis Khan rules.

Marco Polo travels through the East.

Universities are established at Bologna, Paris, Oxford, and Cambridge.

Bubonic plague kills one-fourth of Europe's population.

Printing with movable type is invented; rise of humanism occurs.

Columbus discovers New World.

1500 A.D. TO 1700 A.D.

Francois Viète simplifies algebraic notation.

Galileo Galilei applies math to experiments with falling bodies.

John Napier invents logarithms.

René Descartes and Blaise Pascal unify algebra and geometry.

Pierre de Fermat develops modern number theory.

Fermat and Pascal help lay foundations for theory of probability.

Isaac Newton and Gottfried Leibniz independently discover calculus.

Bernoulli family makes numerous contributions in analysis.

Newton's Principia Mathematica has enormous impact throughout Europe.

Johannes Kepler describes laws governing planetary movement (important to geometry and astronomy).

Works of da Vinci, Michelangelo, Raphael, Titian, and others mark the High Renaissance in Italy.

Protestant Reformation begins with Martin Luther's ninety-five theses.

Nicolas Copernicus attacks theory of geocentric universe.

Elizabeth accedes the throne. Sir Francis Drake defeats Spanish Armada.

William Shakespeare's plays are published.

Galileo invents telescope.

Mayflower lands at Plymouth Rock.

Newton formulates laws of gravity.

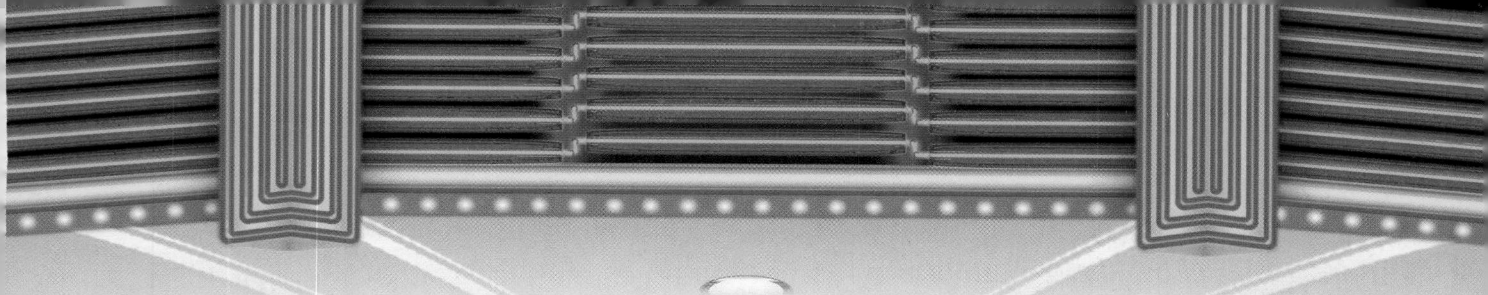

MATHEMATICAL IDEAS

EXPANDED ELEVENTH EDITION

CHARLES D. MILLER

VERN E. HEEREN
American River College

JOHN HORNSBY
University of New Orleans

AND

MARGARET L. MORROW
Plattsburgh State University of New York
for the chapter on Graph Theory

JILL VAN NEWENHIZEN
Lake Forest College
for the chapter on Voting and Apportionment

PEARSON
Addison Wesley

Boston San Francisco New York
London Toronto Sydney Tokyo Singapore Madrid
Mexico City Munich Paris Cape Town Hong Kong Montreal

Publisher: Greg Tobin
Executive Editor: Anne Kelly
Project Editor: Joanne Ha
Associate Project Editor: Elizabeth Bernardi
Assistant Editor: Ashley O'Shaughnessy
Senior Managing Editor: Karen Wernholm
Senior Production Supervisor: Peggy McMahon
Senior Cover Designer: Barbara T. Atkinson
Interior Design: Henry Rachlin
Photo Researcher: Beth Anderson
Media Producer: Sharon Smith
Software Development: Janet Szykowny and Mary Durnwald
Senior Marketing Manager: Becky Anderson
Marketing Coordinator: Maureen McLaughlin
Senior Author Support / Technology Specialist: Joe Vetere
Rights and Permissions Advisor: Shannon Barbe and Dana Weightman
Senior Manufacturing Buyer: Carol Melville
Text Design, Production Coordination, Composition, and Illustrations: Progressive Information Technologies

Cover photo: Fortuna Theatre, Photographer Ron Stroud / Masterfile

For permission to use copyrighted material, grateful acknowledgment has been made to the copyright holders on page C1 in the back of the book, which is hereby made part of this copyright page.

The Library of Congress has already cataloged the Expanded Student Edition as follows:

Library of Congress Cataloging-in-Publication Data
Miller, Charles David, 1942–1986
Mathematical ideas.—11th ed., expanded ed. / Charles D. Miller, Vern E. Heeren, John Hornsby. p. cm.
Includes index.
ISBN 0-321-36146-6-Expanded
1. Mathematics—Textbooks. I. Heeren, Vern E. II. Hornsby, E. John. III. Title.

QA39.3.M55 2007b
510—dc22 2005045896

ISBN-13 978-0-321-36146-2 ISBN-10 0-321-36146-6
5 6 7 8 9 10—DOW—10 09 08

To my classmates in the 1967 graduating class of Catholic High School of Pointe Coupee, New Roads, Louisiana: Eleanor André, Mary Lynn Brumfield, Marilyn Cazayoux, Greg Chustz, Suzanne Chustz, Fellman Chutz, Alexis Cotten, Gay Dabadie, Cathie Ducote, Johnny Forbes, Bonnie Garrett, Gregory Langlois, Hilary Langlois, Paul Lorio, Garrett Olinde, Lynne Olinde, J. D. Patin, Loretta Ramagos, Kackie Smith, Buddy Vosburg, Allen Wells—In memory of Shootie Gosserand and Tippy Hurst—And to the coolest teacher and the most influential priest ever: Sister Margaret Maggio (Stephen) and Fr. Jerome Dugas. I love you all.

JOHNNY

To all my math students, over these many years.

V. E. H.

CONTENTS

PREFACE

After ten editions and nearly forty years, *Mathematical Ideas* continues to be one of the premier textbooks in liberal arts mathematics education. We are proud to present the eleventh edition of a text that offers non-physical science students a practical coverage that connects mathematics to the world around them. It is a flexible book that has evolved alongside changing trends but remains steadfast to its original objectives.

For the first time, this book features a theme that spans its entire contents. Movies and television have become entrenched in our society and appeal to a broad range of interests. With this in mind, we have rewritten every chapter opener with reference to a popular movie or television show, including discussion of a scene that deals with the mathematics covered in the chapter. The margin notes, long a popular hallmark of the book, have been updated to include similar references as well. These references are indicated with a movie camera icon . We hope that users of this edition will enjoy visiting Hollywood while learning mathematics.

Mathematical Ideas is written with a variety of students in mind. It is well suited for several courses, including those geared toward the aforementioned liberal arts audience and survey courses in mathematics, finite mathematics, and mathematics for prospective and in-service elementary and middle-school teachers. Numerous topics are included for a two-term course, yet the variety of topics and flexibility of sequence makes the text suitable for shorter courses as well. Our main objectives continue to be comprehensive coverage, appropriate organization, clear exposition, an abundance of examples, and well-planned exercise sets with numerous applications.

Overview of Chapters

- **Chapter 1 (The Art of Problem Solving)** introduces the student to inductive reasoning, pattern recognition, and problem-solving techniques. Many of the new problems are taken from the popular monthly calendars found in the NCTM publication *Mathematics Teacher.*

- **Chapter 2 (The Basic Concepts of Set Theory) and Chapter 3 (Introduction to Logic)** give brief overviews of set theory and elementary logic. Instructors wishing to do so may cover Chapter 3 before Chapter 2.

- **Chapter 4 (Numeration and Mathematical Systems)** covers various types of numeration systems and group theory, as well as clock arithmetic and modular number systems.

- **Chapter 5 (Number Theory)** presents an introduction to topics such as prime and composite numbers, the Fibonacci sequence, and magic squares. There is updated information on new developments in the field of prime numbers. New to this edition is an extension on modern crypotography.

- **Chapter 6 (The Real Numbers and Their Representations)** introduces some of the basic concepts of real numbers, their various forms of representation, and operations of arithmetic with them.

- **Chapter 7 (The Basic Concepts of Algebra) and Chapter 8 (Graphs, Functions, and Systems of Equations and Inequalities)** offer numerous new applications that help form the core of the text's algebra component.
- **Chapter 9 (Geometry)** covers the standard topics of elementary plane geometry, a section on transformational geometry, an extension on constructions, non-Euclidean geometry, and material on chaos and fractals.
- **Chapter 10 (Trigonometry)** includes angles in standard position, right angle trigonometry, and the laws of sines and cosines.
- **Chapter 11 (Counting Methods)** focuses on elementary counting techniques, in preparation for the chapter to follow.
- **Chapter 12 (Probability)** covers the basics of probability, odds, and expected value.
- **Chapter 13 (Statistics)** has been revised to include new data in examples and exercises.
- **Chapter 14 (Personal Financial Management)** provides the student with the basics of the mathematics of finance as applied to inflation, consumer debt, and house buying. The chapter includes a section on investing, with emphasis on stocks, bonds, and mutual funds. Examples and exercises have been updated to reflect current interest rates and investment returns.

The following chapters are available in the Expanded Edition of this text:

- **Chapter 15 (Graph Theory)** covers the basic concepts of graph theory and its applications. Material on graph coloring is new to this edition.
- **Chapter 16 (Voting and Apportionment)** deals with issues in voting methods and apportionment of votes, topics which have become increasingly popular in liberal arts mathematics courses.

Course Outline Considerations

For the most part, the chapters in the text are independent and may be covered in the order chosen by the instructor. The few exceptions are as follows: Chapter 6 contains some material dependent on the ideas found in Chapter 5; Chapter 6 should be covered before Chapter 7 if student background so dictates; Chapters 7 and 8 form an algebraic "package" and should be covered in sequential order; a thorough coverage of Chapter 12 depends on knowledge of Chapter 11 material, although probability can be covered without teaching extensive counting methods by avoiding the more difficult exercises; and the latter part of Chapter 13, on inferential statistics, depends on an understanding of probability (Chapter 12).

Features of the Eleventh Edition

New: Chapter Openers In keeping with the Hollywood theme of this edition, all chapter openers have been rewritten to address a scene or situation from a popular movie or a television series. Some openers illustrate the correct use of mathematics, while others address how mathematics is misused. In the latter case, we subscribe to the premise that we can all learn from the mistakes of others. Some openers (e.g., Chapters 1 and 9) include a problem statement that the reader is asked to solve. We hope that you enjoy reading these chapter openers as much as we have enjoyed preparing them.

Enhanced: Varied Exercise Sets We continue to present a variety of exercises that integrate drill, conceptual, and applied problems. The text contains a wealth of exercises to provide students with opportunities to practice, apply, connect, and extend the mathematical skills they are learning. We have updated the exercises that focus on real-life data and have retained their titles for easy identification. Several chapters are enriched with new applications, particularly Chapters 7, 8, 10, and 14. We continue to use graphs, tables, and charts when appropriate. Many of the graphs use a style similar to that seen by students in today's print and electronic media.

Enhanced: Margin Notes This popular feature is a hallmark of this text and has been retained and updated where appropriate. These notes are interspersed throughout the text and deal with various subjects such as lives of mathematicians, historical vignettes, philatelic and numismatic reproductions, anecdotes on mathematics textbooks of the past, newspaper and magazine articles, and current research in mathematics. Completely new Hollywood-related margin notes have been included as well.

Collaborative Investigations The importance of cooperative learning is addressed in this end-of-chapter feature.

Problem-Solving Hints Special paragraphs labeled "Problem-Solving Hint" relate the discussion of problem-solving strategies to techniques that have been presented earlier.

Optional Graphing Technology We continue to provide sample graphing calculator screens (generated by a TI-83/84 Plus calculator) to show how technology can be used to support results found analytically. It is not essential, however, that a student have a graphing calculator to study from this text; *the technology component is optional.*

Flexibility Some topics in the first six chapters require a basic knowledge of equation solving. Depending on the background of the students, the instructor may omit topics that require this skill. On the other hand, the two algebra chapters (Chapters 7 and 8) provide an excellent overview of algebra, and because of the flexibility of the text, they may be covered at almost any time.

Art Program The text continues to feature a full-color design. Color is used for instructional emphasis in text discussions, examples, graphs, and figures. New and striking photos have been incorporated to enhance applications and provide visual appeal.

For Further Thought These entries encourage students to share amongst themselves their reasoning processes in order to gain a deeper understanding of key mathematical concepts.

New: Example Titles The numerous, carefully selected examples that illustrate concepts and skills are now titled so that students can see at a glance the topic under consideration. They prepare students for the exercises that follow.

Updated: Emphasis on Real Data in the Form of Graphs, Charts, and Tables We continue to use up-to-date information from magazines, newspapers, and the Internet to create real applications that are relevant and meaningful.

Chapter Tests Each chapter concludes with a chapter test so that students can check their mastery of the material.

MEDIA GUIDE

MathXL®

MathXL is a powerful online homework, tutorial, and assessment system that accompanies this Addison-Wesley textbook. With MathXL, instructors can create, edit, and assign online homework and tests using algorithmically generated exercises correlated at the objective level to the text. Instructors can also create and assign their own online exercises and import TestGen tests for added flexibility. All student work is tracked in MathXL's online gradebook. Students can take chapter tests in MathXL and receive personalized study plans based on their test results. The study plan diagnoses weaknesses and links students directly to tutorial exercises for the objectives they need to study and on which they need to be retested. Students can also access supplemental animations and video clips directly from selected exercises. MathXL is available to qualified adopters. For more information, visit our Web site at www.mathxl.com or contact your local sales representative.

MathXL® Tutorials on CD ISBN: 0-321-36972-6

This interactive tutorial CD-ROM provides algorithmically generated practice exercises that are correlated at the objective level to the exercises in the textbook. Every practice exercise is accompanied by an example and a guided solution designed to involve students in the solution process. Selected exercises may also include a video clip to help students visualize concepts. The software provides helpful feedback for incorrect answers and can generate printed summaries of students' progress.

MyMathLab

MyMathLab is a series of text-specific, easily customizable online courses for Addison-Wesley textbooks in mathematics and statistics. MyMathLab is powered by CourseCompass™—Pearson Education's online teaching and learning environment—and by MathXL—our online homework, tutorial, and assessment system. MyMathLab gives instructors the tools needed to deliver all or a portion of their course online, whether students are in a lab setting or working from home. MyMathLab provides a rich and flexible set of course materials, featuring free-response exercises that are algorithmically generated for unlimited practice and mastery. Students can also use online tools, such as video lectures, animations, and a multimedia textbook, to independently improve their understanding and performance. Instructors can use MyMathLab's homework and test managers to select and assign online exercises correlated directly to the textbook, and they can also create and assign their own online exercises and import TestGen tests for added flexibility. MyMathLab's online gradebook—designed specifically for mathematics and statistics—automatically tracks students' homework and test results and gives the instructor control over how to calculate final grades. Instructors can also add off-line (paper-and-pencil) grades to the gradebook. MyMathLab is available to qualified adopters. For more information, visit our Web site at www.mymathlab.com or contact your local sales representative.

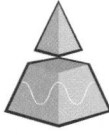

InterAct Math Tutorial Web site www.interactmath.com

Get practice and tutorial help online! This interactive tutorial Web site provides algorithmically generated practice exercises that correlate directly to the exercises in the textbook. Students can retry an exercise as many times as they like, with new values each time, for unlimited practice and mastery. Every exercise is accompanied by an interactive guided solution that provides helpful feedback for incorrect answers, and students can also view a worked-out sample problem that steps them through an exercise similar to the one they are working on.

Video Lectures on CD with Optional Captioning

In this comprehensive video series, an engaging team of instructors provide chapter- and section-based instruction on every topic in the textbook. These lessons present key concepts and show students how to work exercises, providing extra instruction for students who have missed a class or who are in need of a little extra help. The lectures are available on CD-ROM, for purchase with the text at minimal cost. Affordable and portable for students, this series makes it easy and convenient for students to watch the videos from a computer at home or on campus.

SUPPLEMENTS to accompany MATHEMATICAL IDEAS
Eleventh Edition • Expanded Eleventh Edition

STUDENT SUPPLEMENTS

Student's Study Guide and Solutions Manual

- By Emmett Larson, *Brevard Community College*
- This manual provides solutions to the odd-numbered exercises in the exercise sets, the Extensions, and the Appendix exercises, as well as solutions for all the Chapter Test exercises. Chapter summaries review key points in the text, providing extra examples, and enumerate major topic objectives.

ISBN 0-321-36971-8

Video Lectures on CD with Optional Captions

- This is a complete set of digitized videos for student use at home or on campus, making it ideal for distance learning or supplemental instruction.

ISBN 0-321-36954-8

Addison-Wesley Math Tutor Center

- The Tutor Center provides tutoring through a registration number that can be packaged with a new textbook or purchased separately. The Tutor Center is staffed by qualified college mathematics instructors who provide students with tutoring on examples and odd-numbered exercises from the textbook. It is accessible via toll-free telephone, toll-free fax, e-mail, and the Internet (www.aw-bc.com/tutorcenter).

INSTRUCTOR SUPPLEMENTS

NEW! Annotated Instructor's Edition

- This special edition of the text provides answers next to text exercises for quick reference, where possible. The remaining answers are found in the answer section.

ISBN 0-321-36147-4

Instructor's Solutions Manual

- By Emmett Larson, *Brevard Community College*
- This manual contains solutions to all end-of-section exercises, Extension, Chapter Test, and Appendix exercises.

ISBN 0-321-36970-X

Instructor's Testing Manual

- This manual contains four tests for each chapter of the text. Answer keys are included.

ISBN 0-321-36969-6

TestGen®

- TestGen enables instructors to build, edit, print, and administer tests using a computerized bank of questions developed to cover all text objectives. The software is available on a dual-platform Windows/ Macintosh CD-ROM.

ISBN 0-321-36966-1

NEW! Insider's Guide to Teaching with Mathematical Ideas, 11e

- The Insider's Guide includes resources to help faculty with course preparation and classroom management. It provides helpful teaching tips correlated to each section of the text, as well as general teaching advice.

ISBN 0-321-49090-8

PowerPoint Lecture Presentation

- The PowerPoint classroom presentation slides are geared specifically to sequence this textbook. They are available within MyMathLab or at www.aw-bc. com/irc.

NEW! Adjunct Support Center

- The Center offers consultation on suggested syllabi, helpful tips on using the textbook support package, assistance with content, and advice on classroom strategies. It is available Sunday through Thursday evenings from 5 P.M. to midnight EST; telephone: 1-800-435-4084; e-mail: adjunctsupport @aw.com; fax: 1-877-262-9774.

ACKNOWLEDGMENTS

We wish to thank the following reviewers for their helpful comments and suggestions for this and previous editions of the text. (Reviewers of the eleventh edition are noted with an asterisk.)

H. Achepohl, *College of DuPage*
Shahrokh Ahmadi, *Northern Virginia Community College*
Richard Andrews, *Florida A&M University*
Cindy Anfinson, *Palomar College*
Elaine Barber, *Germanna Community College*
Anna Baumgartner, *Carthage College*
James E. Beamer, *Northeastern State University*
Elliot Benjamin, *Unity College*
Jaime Bestard, *Barry University*
Joyce Blair, *Belmont University*
Gus Brar, *Delaware County Community College*
Roger L. Brown, *Davenport College*
Douglas Burke, *Malcolm X College*
John Busovicki, *Indiana University of Pennsylvania*
Ann Cascarelle, *St. Petersburg Junior College*
Kenneth Chapman, *St. Petersburg Junior College*
Gordon M. Clarke, *University of the Incarnate Word*
M. Marsha Cupitt, *Durham Technical Community College*
James Curry, *American River College*
*Rosemary Danaher, *Sacred Heart University*
Ken Davis, *Mesa State College*
Nancy Davis, *Brunswick Community College*
George DeRise, *Thomas Nelson Community College*
Catherine Dermott, *Hudson Valley Community College*
*Greg Dietrich, *Florida Community College at Jacksonville*
Diana C. Dwan, *Yavapai College*
Laura Dyer, *Belleville Area College*
Jan Eardley, *Barat College*
Joe Eitel, *Folsom College*
Azin Enshai, *American River College*
Gayle Farmer, *Northeastern State University*
Michael Farndale, *Waldorf College*
Gordon Feathers, *Passaic County Community College*
Thomas Flohr, *New River Community College*
Bill Fulton, *Black Hawk College—East*
Anne Gardner, *Wenatchee Valley College*
Donald Goral, *Northern Virginia Community College*
Glen Granzow, *Idaho State University*
Larry Green, *Lake Tahoe Community College*
Arthur D. Grissinger, *Lock Haven University*
Don Hancock, *Pepperdine University*
Denis Hanson, *University of Regina*
Marilyn Hasty, *Southern Illinois University*
Shelby L. Hawthorne, *Thomas Nelson Community College*
Jeff Heiking, *St. Petersburg Junior College*
*Laura Hillerbrand, *Broward Community College*
*Jacqueline Jensen, *Sam Houston State University*
Emanuel Jinich, *Endicott College*
*Frank Juric, *Brevard Community College-Palm Bay*
Karla Karstens, *University of Vermont*
Hilary Kight, *Wesleyan College*
Barbara J. Kniepkamp, *Southern Illinois University at Edwardsville*
Suda Kunyosying, *Shepherd College*
*Yu-Ju Kuo, *Indiana University of Pennsylvania*
Pam Lamb, *J. Sargeant Reynolds Community College*
John W. Legge, *Pikeville College*
*John Lattanzio, *Indiana University of Pennsylvania*
Leo Lusk, *Gulf Coast Community College*
Sherrie Lutsch, *Northwest Indian College*
Rhonda Macleod, *Florida State University*
Andrew Markoe, *Rider University*
Darlene Marnich, *Point Park College*
Victoria Martinez, *Okaloosa Walton Community College*
Chris Mason, *Community College of Vermont*

Mark Maxwell, *Maryville University*
Carol McCarron, *Harrisburg Area Community College*
Delois McCormick, *Germanna Community College*
Daisy McCoy, *Lyndon State College*
Cynthia McGinnis, *Okaloosa Walton Community College*
Vena McGrath, *Davenport College*
Robert Moyer, *Fort Valley State University*
Shai Neumann, *Brevard Community College*
*Vladimir Nikiforov, *University of Memphis*
Barbara Nienstedt, *Gloucester County College*
Chaitanya Nigam, *Gateway Community-Technical College*
Jean Okumura, *Windward Community College*
Bob Phillips, *Mesabi Range Community College*
Kathy Pinchback, *University of Memphis*
Priscilla Putman, *New Jersey City University*
Scott C. Radtke, *Davenport College*
John Reily, *Montclair State University*
Beth Reynolds, *Mater Dei College*
Shirley I. Robertson, *High Point University*
Andrew M. Rockett, *CW Post Campus of Long Island University*
Kathleen Rodak, *St. Mary's College of Ave Maria University*
*Abby Roscum, *Marshalltown Community College*
D. Schraeder, *McLennan Community College*
Wilfred Schulte, *Cosumnes River College*
Melinda Schulteis, *Concordia University*
Gary D. Shaffer, *Allegany College of Maryland*
*Doug Shaw, *University of North Iowa*
Jane Sinibaldi, *York College of Pennsylvania*
Larry Smith, *Peninsula College*
Marguerite Smith, *Merced College*
Charlene D. Snow, *Lower Columbia College*
H. Jeannette Stephens, *Whatcom Community College*
Suzanne J. Stock, *Oakton Community College*
Dian Thom, *McKendree College*
Claude C. Thompson, *Hollins University*
Mark Tom, *College of the Sequoias*
Ida Umphers, *University of Arkansas at Little Rock*
Karen Villarreal, *University of New Orleans*
Wayne Wanamaker, *Central Florida Community College*
David Wasilewski, *Luzerne County Community College*
William Watkins, *California State University, Northridge*
Susan Williford, *Columbia State Community College*
Tom Witten, *Southwest Virginia Community College*
Fred Worth, *Henderson State University*
Rob Wylie, *Carl Albert State College*
Henry Wyzinski, *Indiana University Northwest*

A project of this magnitude cannot be accomplished without the help of many other dedicated individuals. Anne Kelly served as sponsoring editor for this edition. Jeff Houck of Progressive Publishing Alternatives provided excellent production supervision. Ashley O'Shaughnessy, Greg Tobin, Barbara Atkinson, Becky Anderson, Peggy McMahon, Beth Anderson, and Joanne Ha of Addison-Wesley gave us their unwavering support. Terry McGinnis provided her usual excellent behind-the-scenes guidance. Thanks go to Dr. Margaret L. Morrow of Plattsburgh State University and Dr. Jill Van Newenhizen of Lake Forest College, who wrote the material on graph theory and voting/apportionment, respectively, for the Expanded Edition. Perian Herring, Cheryl Davids, Patricia Nelson, and Alicia Gordon did an outstanding job of accuracy- and answer-checking, and Becky Troutman provided the *Index of Applications*. And finally, we thank our loyal users over the past four decades for making this book one of the most successful in its market.

Vern E. Heeren
John Hornsby

1

THE ART OF PROBLEM SOLVING

The 1995 movie *Die Hard: With a Vengeance* is the third in a series of action films starring Bruce Willis as New York Detective John McClane. In this film, McClane is tormented by villain Simon Gruber (Jeremy Irons) who plants bombs around the city and poses riddles and puzzles for disarming them. In one situation, as McClane and store owner Zeus Carver (Samuel L. Jackson) open a briefcase containing a timer connected to a bomb near a park fountain, Simon relates the following riddle by telephone and gives them 5 minutes to solve it:

On the fountain there should be two jugs. Do you see them? A 5-gallon and a 3-gallon. Fill one of the jugs with exactly 4 gallons of water, and place it on the scale, and the timer will stop. You must be precise. One ounce more or less will result in detonation.

McClane and Carver were able to solve the riddle and defuse the bomb. Can you solve it? Variations of this problem have been around for many years. The answer is on page 26.

1.1 Solving Problems by Inductive Reasoning

Characteristics of Inductive and Deductive Reasoning • Pitfalls of Inductive Reasoning

The **Moscow papyrus,** which dates back to about 1850 B.C., provides an example of **inductive reasoning** by the early Egyptian mathematicians. Problem 14 in the document reads:

> *You are given a truncated pyramid of 6 for the vertical height by 4 on the base by 2 on the top. You are to square this 4, result 16. You are to double 4, result 8. You are to square 2, result 4. You are to add the 16, the 8, and the 4, result 28. You are to take one-third of 6, result 2. You are to take 28 twice, result 56. See, it is 56. You will find it right.*

What does all this mean? A *frustum* of a pyramid is that part of the pyramid remaining after its top has been cut off by a plane parallel to the base of the pyramid. The formula for finding the volume of the frustum of a pyramid with a square base is

$$V = \frac{1}{3}h(b^2 + bB + B^2),$$

where b is the area of the upper base, B is the area of the lower base, and h is the height (or altitude). The writer of the problem is giving a method of determining the volume of the frustum of a pyramid with square bases on the top and bottom, with bottom base side of length 4, top base side of length 2, and height equal to 6.

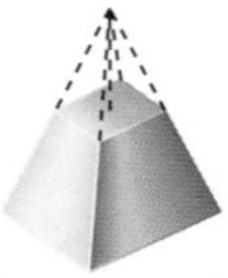

A truncated pyramid, or frustum of a pyramid

Characteristics of Inductive and Deductive Reasoning

The development of mathematics can be traced to the Egyptian and Babylonian cultures (3000 B.C.–A.D. 260) as a necessity for problem solving. To solve a problem or perform an operation, a cookbook-like recipe was given, and it was performed repeatedly to solve similar problems. During the classical Greek period (600 B.C.–A.D. 450), general concepts were applied to specific problems, resulting in a structured, logical development of mathematics.

By observing that a specific method worked for a certain type of problem, the Babylonians and the Egyptians concluded that the same method would work for any similar type of problem. Such a conclusion is called a *conjecture*. A **conjecture** is an educated guess based on repeated observations of a particular process or pattern. The method of reasoning we have just described is called *inductive reasoning*.

Inductive Reasoning

Inductive reasoning is characterized by drawing a general conclusion (making a conjecture) from repeated observations of specific examples. The conjecture may or may not be true.

In testing a conjecture obtained by inductive reasoning, it takes only one example that does not work in order to prove the conjecture false. Such an example is called a **counterexample.**

Inductive reasoning provides a powerful method of drawing conclusions, but there is no assurance that the observed conjecture will always be true. For this reason, mathematicians are reluctant to accept a conjecture as an absolute truth until it is formally proved using methods of *deductive reasoning*. Deductive reasoning characterized the development and approach of Greek mathematics, as seen in the works of Euclid, Pythagoras, Archimedes, and others.

Deductive Reasoning

Deductive reasoning is characterized by applying general principles to specific examples.

We now look at examples of these two types of reasoning. In this chapter, we often refer to the **natural** or **counting numbers:**

$$1, 2, 3, \ldots .$$ Natural (counting) numbers

(↑ Ellipsis points)

The three dots (*ellipsis points*) indicate that the numbers continue indefinitely in the pattern that has been established. The most probable rule for continuing this pattern is "add 1 to the previous number," and this is indeed the rule that we follow.

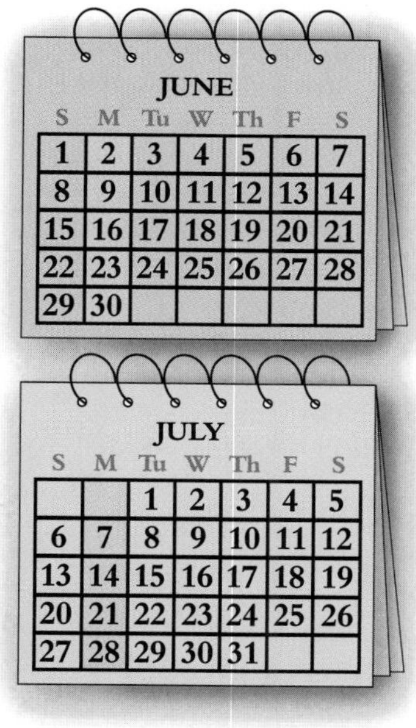

FIGURE 1

Now consider the following list of natural numbers: 2, 9, 16, 23, 30. What is the next number of this list? What is the pattern? After studying the numbers, we might see that $2 + 7 = 9$, and $9 + 7 = 16$. Do we add 16 and 7 to get 23? Do we add 23 and 7 to get 30? Yes; any number in the given list can be found by adding 7 to the preceding number, so the next number in the list should be $30 + 7 = 37$.

We set out to find the "next number" by reasoning from observation of the numbers in the list. We may have jumped from these observations to the general statement that any number in the list is 7 more than the preceding number. This is an example of *inductive reasoning*.

By using inductive reasoning, we concluded that 37 was the next number in the list. But this is wrong. We've been tricked into drawing an incorrect conclusion. The person making up the list has another answer in mind. The list of numbers

$$2, 9, 16, 23, 30$$

actually gives the dates of Mondays in June if June 1 falls on a Sunday. The next Monday after June 30 is July 7. With this pattern, the list continues as

$$2, 9, 16, 23, 30, 7, 14, 21, 28, \ldots .$$

See the calendar in Figure 1. The process used to obtain the rule "add 7" in the preceding list reveals one main flaw of inductive reasoning. We can never be sure that what is true in a specific case will be true in general. Inductive reasoning does not guarantee a true result, but it does provide a means of making a conjecture.

Throughout this book, we use *exponents* to represent repeated multiplication. For example, in the expression 4^3 the exponent is 3:

$$4^3 = 4 \cdot 4 \cdot 4 = 64. \quad \text{4 is used as a factor 3 times.}$$

Exponential Expression

If a is a number and n is a counting number (1, 2, 3, . . .), then the exponential expression a^n is defined as

$$a^n = \underbrace{a \cdot a \cdot a \cdot \ldots \cdot a}_{n \text{ factors of } a}.$$

The number a is the *base* and n is the exponent.

With deductive reasoning, we use general statements and apply them to specific situations. For example, one of the best-known rules in mathematics is the Pythagorean theorem: In any right triangle, the sum of the squares of the legs (shorter sides) is equal to the square of the hypotenuse (longest side). Thus, if we know that the lengths of the shorter sides are 3 inches and 4 inches, we can find the length of the longest side. Let h represent the longest side.

$$3^2 + 4^2 = h^2 \quad \text{Pythagorean theorem}$$
$$9 + 16 = h^2 \quad 3^2 = 3 \cdot 3 = 9;\ 4^2 = 4 \cdot 4 = 16$$
$$25 = h^2 \quad \text{Add.}$$
$$5 = h \quad \text{The positive square root of 25 is 5.}$$

Thus, the longest side measures 5 inches. We used the general rule (the Pythagorean theorem) and applied it to the specific situation.

Reasoning through a problem usually requires certain *premises*. A **premise** can be an assumption, law, rule, widely held idea, or observation. Then reason inductively or deductively from the premises to obtain a **conclusion.** The premises and conclusion make up a **logical argument.**

EXAMPLE 1 Identifying Premises and Conclusions

Identify each premise and the conclusion in each of the following arguments. Then tell whether each argument is an example of inductive or deductive reasoning.

(a) Our house is made of brick. Both of my next-door neighbors have brick houses. Therefore, all houses in our neighborhood are made of brick.

(b) All word processors will type the symbol @. I have a word processor. I can type the symbol @.

(c) Today is Monday. Tomorrow will be Tuesday.

SOLUTION

(a) The premises are "Our house is made of brick" and "Both of my next-door neighbors have brick houses." The conclusion is "Therefore, all houses in our neighborhood are made of brick." Because the reasoning goes from specific examples to a general statement, the argument is an example of inductive reasoning (although it may very well have a false conclusion).

(b) Here, the premises are "All word processors will type the symbol @" and "I have a word processor." The conclusion is "I can type the symbol @." This reasoning goes from general to specific, so deductive reasoning was used.

(c) There is only one premise here, "Today is Monday." The conclusion is "Tomorrow will be Tuesday." The fact that Tuesday immediately follows Monday is being used, even though this fact is not explicitly stated. Because the conclusion comes from general facts that apply to this special case, deductive reasoning was used. ■

The earlier calendar example illustrated how inductive reasoning may, at times, lead to false conclusions. However, in many cases, inductive reasoning does provide correct results if we look for the most *probable* answer.

In the 2003 movie *A Wrinkle in Time*, young Calvin O'Keefe, played by Gregory Smith, is challenged to identify a particular sequence of numbers. He correctly identifies it as the **Fibonacci sequence.**

EXAMPLE 2 Predicting the Next Number in a Sequence

Use inductive reasoning to determine the *probable* next number in each list below.

(a) 5, 9, 13, 17, 21, 25 **(b)** 1, 1, 2, 3, 5, 8, 13, 21 **(c)** 2, 4, 8, 16, 32

SOLUTION

(a) Each number in the list is obtained by adding 4 to the previous number. The probable next number is $25 + 4 = 29$.

(b) Beginning with the third number in the list, 2, each number is obtained by adding the two previous numbers in the list. That is,

$$1 + 1 = 2, \qquad 1 + 2 = 3, \qquad 2 + 3 = 5,$$

and so on. The probable next number in the list is $13 + 21 = 34$. (These are the first few terms of the famous *Fibonacci sequence*.)

(c) It appears here that to obtain each number after the first, we must double the previous number. Therefore, the most probable next number is $32 \times 2 = 64$. ■

Inductive reasoning often can be used to predict an answer in a list of similarly constructed computation exercises, as shown in the next example.

EXAMPLE 3 Predicting the Product of Two Numbers

$37 \times 3 = 111$
$37 \times 6 = 222$
$37 \times 9 = 333$
$37 \times 12 = 444$

Consider the list of equations in the margin. Use the list to predict the next multiplication fact in the list.

SOLUTION

In each case, the left side of the equation has two factors, the first 37 and the second a multiple of 3, beginning with 3. The product (answer) in each case consists of three digits, all the same, beginning with 111 for 37×3. For this pattern to continue, the next multiplication fact would be $37 \times 15 = 555$, which is indeed true. (*Note:* You might want to investigate what occurs after 30 is reached for the right-hand factor, and make conjectures based on those products.) ■

TABLE 1

Number of Points	Number of Regions
1	1
2	2
3	4
4	8
5	16

Pitfalls of Inductive Reasoning There are pitfalls associated with inductive reasoning. A classic example involves the maximum number of regions formed when chords are constructed in a circle. When two points on a circle are joined with a line segment, a *chord* is formed. Locate a single point on a circle. Because no chords are formed, a single interior region is formed. See Figure 2(a). Locate two points and draw a chord. Two interior regions are formed, as shown in Figure 2(b). Continue this pattern. Locate three points, and draw all possible chords. Four interior regions are formed, as shown in Figure 2(c). Four points yield 8 regions and five points yield 16 regions. See Figures 2(d) and 2(e).

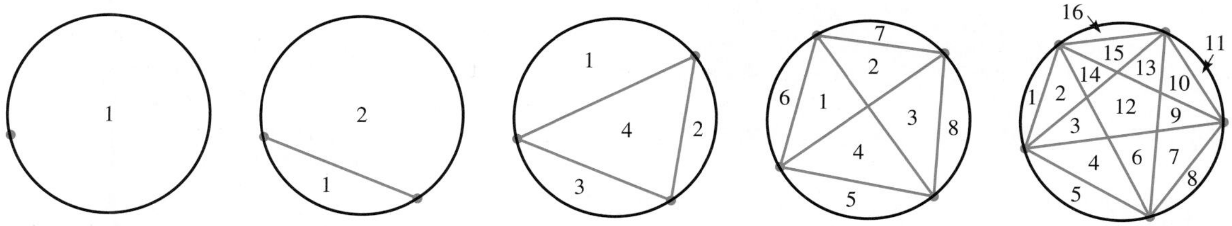

(a) (b) (c) (d) (e)

FIGURE 2

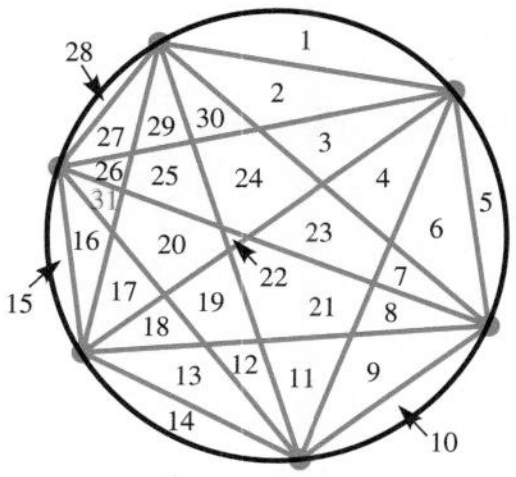

FIGURE 3

The results of the preceding observations are summarized in Table 1 in the margin. The pattern formed in the column headed "Number of Regions" is the same one we saw in Example 2(c), where we predicted that the next number would be 64. It seems here that for each additional point on the circle, the number of regions doubles. A reasonable inductive conjecture would be that for six points, 32 regions would be formed. But as Figure 3 indicates, there are only 31 regions! The pattern of doubling ends when the sixth point is considered. Adding a seventh point would yield 57 regions. The numbers obtained here are

1, 2, 4, 8, 16, 31, 57.

For n points on the circle, the number of regions is given by the formula

$$\frac{n^4 - 6n^3 + 23n^2 - 18n + 24}{24}.^*$$

We can use a graphing calculator to construct a table of values that indicates the number of regions for various numbers of points. Using X rather than n, we can define Y_1 using the expression above (see Figure 4(a) on the next page). Then, creating a table of values, as in Figure 4(b), we see how many regions (indicated by Y_1) there are for any number of points (X).

As indicated earlier, not until a general relationship is proved can one be sure about a conjecture because one counterexample is always sufficient to make the conjecture false.

For Further Thought

Inductive Reasoning Anecdote

The following anecdote concerning inductive reasoning appears in the first volume of the *In Mathematical Circles* series by Howard Eves (PWS-KENT Publishing Company).

A scientist had two large jars before him on the laboratory table. The jar on his left contained 100 fleas; the jar on his right was empty. The scientist carefully lifted a flea from the jar on the left, placed the flea on the table between the two jars, stepped back, and in a loud voice said, "Jump." The flea jumped and was put in the jar on the right. A second flea was carefully lifted from the jar on the left and placed on the table between the two jars. Again the scientist stepped back and in a loud voice said, "Jump." The flea jumped and was put in the jar on the right. In the same manner, the scientist treated each of the 100 fleas in the jar on the left, and each flea jumped as ordered. The two jars were then interchanged and the experiment continued with a slight difference. This time the scientist carefully lifted a flea from the jar on the left, yanked off its hind legs, placed the flea on the table between the jars, stepped back, and in a loud voice said, "Jump." The flea did not jump, and was put in the jar on the right. A second flea was carefully lifted from the jar on the left, its hind legs yanked off, and then placed on the table between the two jars. Again the scientist stepped back and in a loud voice said, "Jump." The flea did not jump, and was put in the jar on the right. In this manner, the scientist treated each of the 100 fleas in the jar on the left, and in no case did a flea jump when ordered. So the scientist recorded in his notebook the following induction: "A flea, if its hind legs are yanked off, cannot hear."

For Group Discussion or Individual Investigation

Discuss or research examples from advertising on television, in newspapers, magazines, etc., that lead consumers to draw incorrect conclusions.

*For more information on this and other similar patterns, see "Counting Pizza Pieces and Other Combinatorial Problems," by Eugene Maier, in the January 1988 issue of *Mathematics Teacher,* pp. 22–26.

Note the careful use of parentheses.

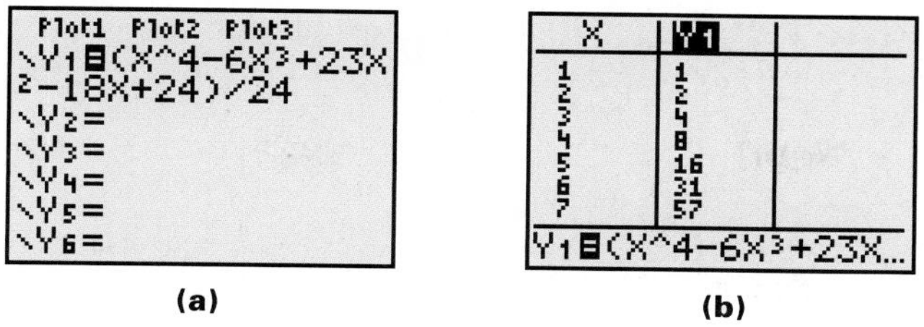

FIGURE 4

1.1 Exercises

In Exercises 1–12, determine whether the reasoning is an example of deductive or inductive reasoning.

1. If the mechanic says that it will take seven days to repair your car, then it will actually take ten days. The mechanic says, "I figure it'll take a week to fix it, ma'am." Then you can expect it to be ready ten days from now.

2. If you take your medicine, you'll feel a lot better. You take your medicine. Therefore, you'll feel a lot better.

3. It has rained every day for the past nine days, and it is raining today as well. So it will also rain tomorrow.

4. Marin's first three children were boys. If she has another baby, it will be a boy.

5. Finley had 95 baseball cards. His mom gave him 20 more for his birthday. Therefore, he now has 115 of them.

6. If the same number is subtracted from both sides of a true equation, the new equation is also true. I know that $9 + 18 = 27$. Therefore, $(9 + 18) - 12 = 27 - 12$.

7. If you build it, they will come. You build it. Therefore, they will come.

8. All men are mortal. Socrates is a man. Therefore, Socrates is mortal.

9. It is a fact that every student who ever attended Geekville University was accepted into graduate school. Because I am attending Geekville, I can expect to be accepted to graduate school, too.

10. For the past 53 years, a rare plant has bloomed in Columbia each summer, alternating between yellow and green flowers. Last summer, it bloomed with green flowers, so this summer it will bloom with yellow flowers.

11. In the sequence 5, 10, 15, 20, . . . , the most probable next number is 25.

12. Carrie Underwood's last four single releases have reached the Top Ten country list, so her current release will also reach the Top Ten.

13. Discuss the differences between inductive and deductive reasoning. Give an example of each.

14. Give an example of faulty inductive reasoning.

Determine the most probable next term in each list of numbers.

15. 6, 9, 12, 15, 18

16. 13, 18, 23, 28, 33

17. 3, 12, 48, 192, 768

18. 32, 16, 8, 4, 2

19. 3, 6, 9, 15, 24, 39

20. $\frac{1}{3}, \frac{3}{5}, \frac{5}{7}, \frac{7}{9}, \frac{9}{11}$

21. $\frac{1}{2}, \frac{3}{4}, \frac{5}{6}, \frac{7}{8}, \frac{9}{10}$

22. 1, 4, 9, 16, 25

23. 1, 8, 27, 64, 125

24. 2, 6, 12, 20, 30, 42

25. 4, 7, 12, 19, 28, 39

26. −1, 2, −3, 4, −5, 6

27. 5, 3, 5, 5, 3, 5, 5, 5, 3, 5, 5, 5, 5, 3, 5, 5, 5, 5

28. 8, 2, 8, 2, 2, 8, 2, 2, 2, 8, 2, 2, 2, 2, 8, 2, 2, 2, 2

29. Construct a list of numbers similar to those in Exercise 15 such that the most probable next number in the list is 60.

30. Construct a list of numbers similar to those in Exercise 26 such that the most probable next number in the list is 9.

In Exercises 31–42, a list of equations is given. Use the list and inductive reasoning to predict the next equation, and then verify your conjecture.

31.
$$\begin{aligned}(9 \times 9) + 7 &= 88\\ (98 \times 9) + 6 &= 888\\ (987 \times 9) + 5 &= 8888\\ (9876 \times 9) + 4 &= 88{,}888\end{aligned}$$

32.
$$\begin{aligned}(1 \times 9) + 2 &= 11\\ (12 \times 9) + 3 &= 111\\ (123 \times 9) + 4 &= 1111\\ (1234 \times 9) + 5 &= 11{,}111\end{aligned}$$

33.
$$\begin{aligned}3367 \times 3 &= 10{,}101\\ 3367 \times 6 &= 20{,}202\\ 3367 \times 9 &= 30{,}303\\ 3367 \times 12 &= 40{,}404\end{aligned}$$

34.
$$\begin{aligned}15873 \times 7 &= 111{,}111\\ 15873 \times 14 &= 222{,}222\\ 15873 \times 21 &= 333{,}333\\ 15873 \times 28 &= 444{,}444\end{aligned}$$

35.
$$\begin{aligned}34 \times 34 &= 1156\\ 334 \times 334 &= 111{,}556\\ 3334 \times 3334 &= 11{,}115{,}556\end{aligned}$$

36.
$$\begin{aligned}11 \times 11 &= 121\\ 111 \times 111 &= 12{,}321\\ 1111 \times 1111 &= 1{,}234{,}321\end{aligned}$$

37.
$$\begin{aligned}3 &= \frac{3(2)}{2}\\ 3 + 6 &= \frac{6(3)}{2}\\ 3 + 6 + 9 &= \frac{9(4)}{2}\\ 3 + 6 + 9 + 12 &= \frac{12(5)}{2}\end{aligned}$$

38.
$$\begin{aligned}2 &= 4 - 2\\ 2 + 4 &= 8 - 2\\ 2 + 4 + 8 &= 16 - 2\\ 2 + 4 + 8 + 16 &= 32 - 2\end{aligned}$$

39.
$$\begin{aligned}5(6) &= 6(6 - 1)\\ 5(6) + 5(36) &= 6(36 - 1)\\ 5(6) + 5(36) + 5(216) &= 6(216 - 1)\\ 5(6) + 5(36) + 5(216) + 5(1296) &= 6(1296 - 1)\end{aligned}$$

40.
$$\begin{aligned}3 &= \frac{3(3 - 1)}{2}\\ 3 + 9 &= \frac{3(9 - 1)}{2}\\ 3 + 9 + 27 &= \frac{3(27 - 1)}{2}\\ 3 + 9 + 27 + 81 &= \frac{3(81 - 1)}{2}\end{aligned}$$

41.
$$\begin{aligned}\frac{1}{2} &= 1 - \frac{1}{2}\\ \frac{1}{2} + \frac{1}{4} &= 1 - \frac{1}{4}\\ \frac{1}{2} + \frac{1}{4} + \frac{1}{8} &= 1 - \frac{1}{8}\\ \frac{1}{2} + \frac{1}{4} + \frac{1}{8} + \frac{1}{16} &= 1 - \frac{1}{16}\end{aligned}$$

42.
$$\begin{aligned}\frac{1}{1 \cdot 2} &= \frac{1}{2}\\ \frac{1}{1 \cdot 2} + \frac{1}{2 \cdot 3} &= \frac{2}{3}\\ \frac{1}{1 \cdot 2} + \frac{1}{2 \cdot 3} + \frac{1}{3 \cdot 4} &= \frac{3}{4}\\ \frac{1}{1 \cdot 2} + \frac{1}{2 \cdot 3} + \frac{1}{3 \cdot 4} + \frac{1}{4 \cdot 5} &= \frac{4}{5}\end{aligned}$$

A story is often told about how the great mathematician Carl Friedrich Gauss (1777–1855) at a very young age was told by his teacher to find the sum of the first 100 *counting numbers. While his classmates toiled at the problem, Carl simply wrote down a single number and handed it in to his teacher. His answer was correct. When asked how he did it, the young Carl explained that he observed that there were* 50 *pairs of numbers that each added up to* 101. *(See below.) So the sum of all the numbers must be* $50 \times 101 = 5050$.

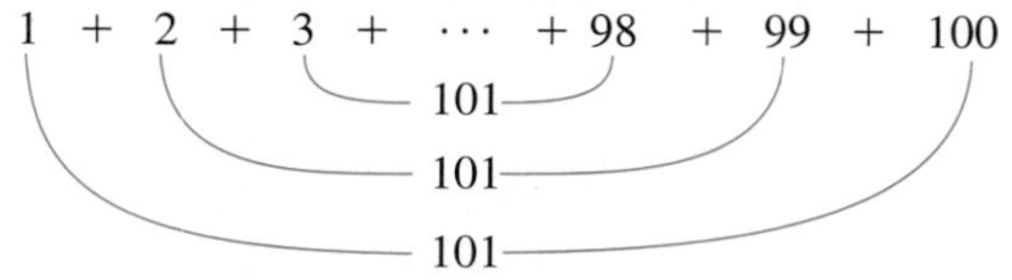

50 sums of 101 $= 50 \times 101 = 5050$

Use the method of Gauss to find each sum.

43. $1 + 2 + 3 + \cdots + 200$

44. $1 + 2 + 3 + \cdots + 400$

45. $1 + 2 + 3 + \cdots + 800$

46. $1 + 2 + 3 + \cdots + 2000$

47. Modify the procedure of Gauss to find the sum $1 + 2 + 3 + \cdots + 175$.

48. Explain in your own words how the procedure of Gauss can be modified to find the sum $1 + 2 + 3 + \cdots + n$, where n is an odd natural number. (When an odd natural number is divided by 2, it leaves a remainder of 1.)

49. Modify the procedure of Gauss to find the sum $2 + 4 + 6 + \cdots + 100$.

50. Use the result of Exercise 49 to find the sum $4 + 8 + 12 + \cdots + 200$.

51. Find a pattern in the following figures and use inductive reasoning to predict the next figure.

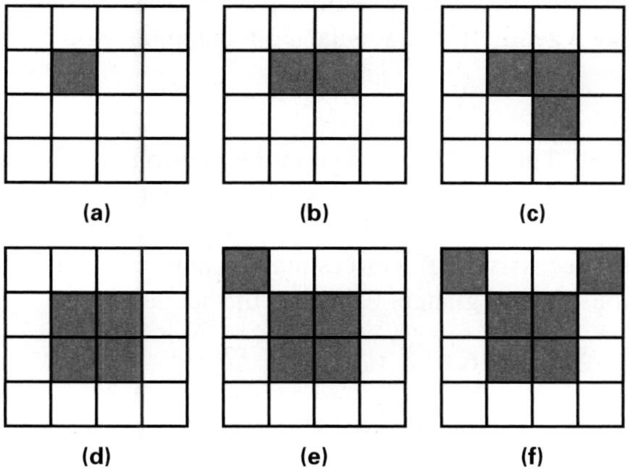

52. Consider the following table.

0	2	2	2	0	0	0	0	0
0	2	4	6	4	2	0	0	0
0	2	6	12	14	12	6	2	0
0	2	8	20	32	38	32	20	8

Find a pattern and predict the next row of the table.

53. What is the most probable next number in this list? 12, 1, 1, 1, 2, 1, 3 (*Hint:* Think about a clock.)

54. What is the next term in this list? O, T, T, F, F, S, S, E, N, T (*Hint:* Think about words and their relationship to numbers.)

55. **(a)** Choose any three-digit number with all different digits. Now reverse the digits, and subtract the smaller from the larger. Record your result. Choose another three-digit number and repeat this process. Do this as many times as it takes for you to see a pattern in the different results you obtain. (*Hint:* What is the middle digit? What is the sum of the first and third digits?)

(b) Write an explanation of this pattern. You may want to use this exercise as a "number trick" to amuse your friends.

56. Choose any number, and follow these steps.

(a) Multiply by 2.
(b) Add 6.
(c) Divide by 2.
(d) Subtract the number you started with.
(e) Record your result.

Repeat the process, except in Step (b), add 8. Record your final result. Repeat the process once more, except in Step (b), add 10. Record your final result.

(f) Observe what you have done; use inductive reasoning to explain how to predict the final result. You may want to use this exercise as a "number trick" to amuse your friends.

57. Complete the following.

$142{,}857 \times 1 =$ ______
$142{,}857 \times 2 =$ ______
$142{,}857 \times 3 =$ ______
$142{,}857 \times 4 =$ ______
$142{,}857 \times 5 =$ ______
$142{,}857 \times 6 =$ ______

What pattern exists in the successive answers? Now multiply 142,857 by 7 to obtain an interesting result.

58. Complete the following.

$$12{,}345{,}679 \times 9 = \underline{\qquad\qquad}$$
$$12{,}345{,}679 \times 18 = \underline{\qquad\qquad}$$
$$12{,}345{,}679 \times 27 = \underline{\qquad\qquad}$$

By what number would you have to multiply 12,345,679 to get an answer of 888,888,888?

59. Refer to Figures 2(b)–(e) and 3. Instead of counting interior regions of the circle, count the chords formed. Use inductive reasoning to predict the number of chords that would be formed if seven points were used.

60. The following number trick can be performed on one of your friends. It was provided by Dr. George DeRise of Thomas Nelson Community College.

(a) Ask your friend to write down his or her age. (Only whole numbers are allowed.)
(b) Multiply the number by 4.
(c) Add 10.
(d) Multiply by 25.
(e) Subtract the number of days in a non-leap year.
(f) Add the amount of change (less than a dollar, in cents) in his or her pocket.
(g) Ask your friend for the final answer.

If you add 115 to the answer, the first two digits are the friend's age, and the last two give the amount of change.

61. Explain how a toddler might use inductive reasoning to decide on something that will be of benefit to him or her.

62. Discuss one example of inductive reasoning that you have used recently in your life. Test your premises and your conjecture. Did your conclusion ultimately prove to be true or false?

1.2 An Application of Inductive Reasoning: Number Patterns

Number Sequences • Successive Differences • Number Patterns and Sum Formulas • Figurate Numbers

Number Sequences

An ordered list of numbers such as

$$3, 9, 15, 21, 27, \ldots,$$

is called a *sequence*. A **number sequence** is a list of numbers having a first number, a second number, a third number, and so on, called the **terms** of the sequence. The sequences in Examples 2(a) and 2(c) in the previous section are called *arithmetic* and *geometric sequences,* respectively. An **arithmetic sequence** has a common *difference* between successive terms, while a **geometric sequence** has a common *ratio* between successive terms.

Successive Differences

The sequences seen in the previous section were usually simple enough for us to make an obvious conjecture about the next term. However, some sequences may provide more difficulty in making such a conjecture, and often the **method of successive differences** may be applied to determine the next term if it is not obvious at first glance. Consider the sequence

$$2, 6, 22, 56, 114, \ldots.$$

Because the next term is not obvious, subtract the first term from the second term, the second from the third, the third from the fourth, and so on.

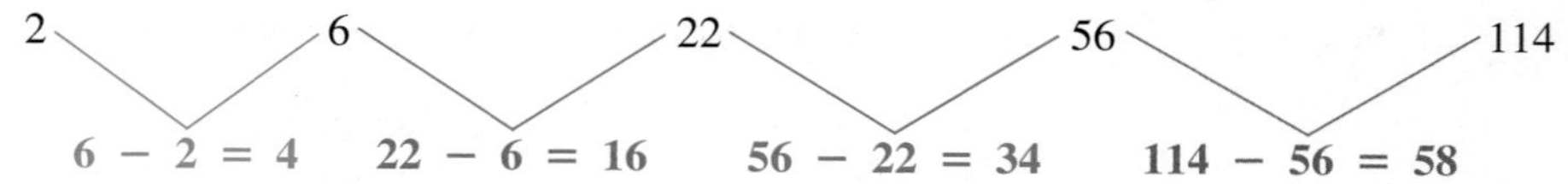

Now repeat the process with the sequence 4, 16, 34, 58 and continue repeating until the difference is a constant value, as shown in line (4):

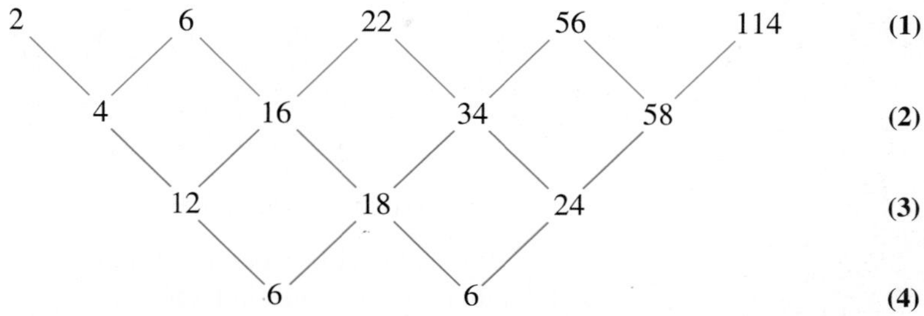

Once a line of constant values is obtained, simply work "backward" by adding until the desired term of the given sequence is obtained. Thus, for this pattern to continue, another 6 should appear in line (4), meaning that the next term in line (3) would have to be 24 + 6 = 30. The next term in line (2) would be 58 + 30 = 88. Finally, the next term in the given sequence would be 114 + 88 = **202**. The final scheme of numbers is shown below.

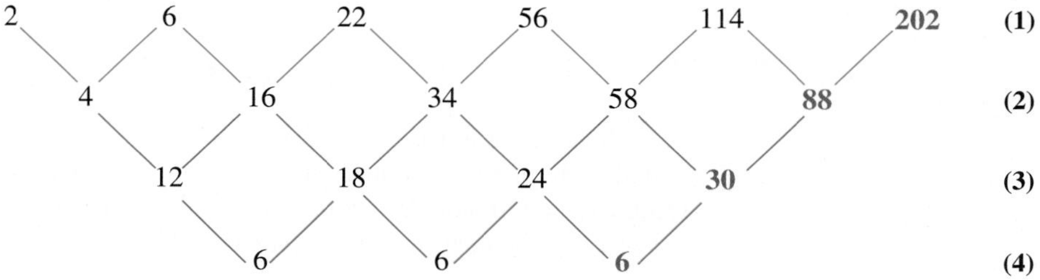

EXAMPLE 1 Using Successive Differences

Use the method of successive differences to determine the next number in each sequence.

(a) 14, 22, 32, 44, . . . **(b)** 5, 15, 37, 77, 141, . . .

SOLUTION

(a) Using the scheme described above, obtain the following:

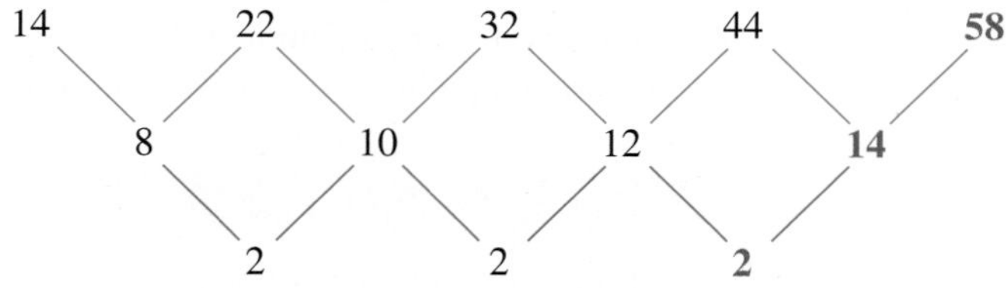

Once the row of 2s was obtained and extended, we were able to get 12 + 2 = 14, and 44 + 14 = 58, as shown above. The next number in the sequence is **58**.

(b) Proceeding as before, obtain the following diagram.

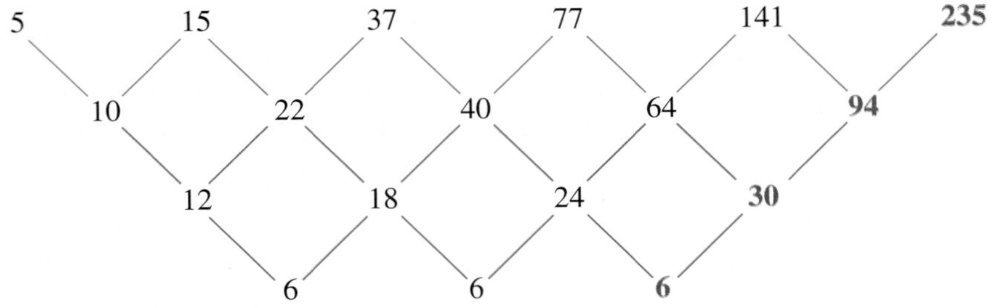

The next number in the sequence is **235**. ■

The method of successive differences will not always work. For example, try it on the Fibonacci sequence in Example 2(b) of Section 1.1 and see what happens!

Number Patterns and Sum Formulas Mathematics features a seemingly endless variety of number patterns. Observe the following pattern:

$$\begin{aligned} 1 &= 1^2 \\ 1 + 3 &= 2^2 \\ 1 + 3 + 5 &= 3^2 \\ 1 + 3 + 5 + 7 &= 4^2 \\ 1 + 3 + 5 + 7 + 9 &= 5^2. \end{aligned}$$

In each case, the left side of the equation is the indicated sum of the consecutive odd counting numbers beginning with 1, and the right side is the square of the number of terms on the left side. You should verify this in each case. Inductive reasoning would suggest that the next line in this pattern is

$$1 + 3 + 5 + 7 + 9 + 11 = 6^2.$$

Evaluating each side shows that each side simplifies to 36.

We cannot conclude that this pattern will continue indefinitely, because observation of a finite number of examples does not guarantee that the pattern will continue. However, mathematicians have proved that this pattern does indeed continue indefinitely, using a method of proof called *mathematical induction*. (See any standard college algebra text.)

Any even counting number may be written in the form $2k$, where k is a counting number. It follows that the kth odd counting number is written $2k - 1$. For example, the third odd counting number, 5, can be written $2(3) - 1$. Using these ideas, we can write the result obtained above as follows.

Sum of the First *n* Odd Counting Numbers

If n is any counting number, then

$$\mathbf{1 + 3 + 5 + \cdots + (2n - 1) = n^2.}$$

EXAMPLE 2 Predicting the Next Equation in a List

In each of the following, several equations are given illustrating a suspected number pattern. Determine what the next equation would be, and verify that it is indeed a true statement.

(a)

$$\begin{aligned} 1^2 &= 1^3 \\ (1+2)^2 &= 1^3 + 2^3 \\ (1+2+3)^2 &= 1^3 + 2^3 + 3^3 \\ (1+2+3+4)^2 &= 1^3 + 2^3 + 3^3 + 4^3 \end{aligned}$$

(b)

$$\begin{aligned} 1 &= 1^3 \\ 3+5 &= 2^3 \\ 7+9+11 &= 3^3 \\ 13+15+17+19 &= 4^3 \end{aligned}$$

(c)

$$\begin{aligned} 1 &= \frac{1 \cdot 2}{2} \\ 1+2 &= \frac{2 \cdot 3}{2} \\ 1+2+3 &= \frac{3 \cdot 4}{2} \\ 1+2+3+4 &= \frac{4 \cdot 5}{2} \end{aligned}$$

SOLUTION

(a) The left side of each equation is the square of the sum of the first n counting numbers, while the right side is the sum of their cubes. The next equation in the pattern would be

$$(1+2+3+4+5)^2 = 1^3 + 2^3 + 3^3 + 4^3 + 5^3.$$

Each side simplifies to 225, so the pattern is true for this equation.

(b) The left sides of the equations contain the sum of odd counting numbers, starting with the first (1) in the first equation, the second and third (3 and 5) in the second equation, the fourth, fifth, and sixth (7, 9, and 11) in the third equation, and so on. The right side contains the cube (third power) of the number of terms on the left side in each case. Following this pattern, the next equation would be

$$21 + 23 + 25 + 27 + 29 = 5^3,$$

which can be verified by computation.

(c) The left side of each equation gives the indicated sum of the first n counting numbers, and the right side is always of the form

$$\frac{n(n+1)}{2}.$$

For the pattern to continue, the next equation would be

$$1+2+3+4+5 = \frac{5 \cdot 6}{2}.$$

Because each side simplifies to 15, the pattern is true for this equation. ■

The patterns established in Examples 2(a) and 2(c) can be written as follows.

Special Sum Formulas

For any counting number n,

$$(1 + 2 + 3 + \cdots + n)^2 = 1^3 + 2^3 + 3^3 + \cdots + n^3$$

and $$1 + 2 + 3 + \cdots + n = \frac{n(n+1)}{2}.$$

The second formula given is a generalization of the method first explained preceding Exercise 43 in the previous section, relating the story of young Carl Gauss. We can provide a general, deductive argument showing how this equation is obtained. Suppose that we let S represent the sum $1 + 2 + 3 + \cdots + n$. This sum can also be written as $S = n + (n-1) + (n-2) + \cdots + 1$. Now write these two equations as follows.

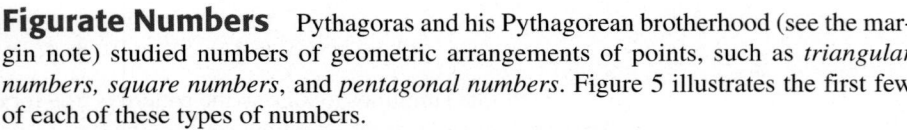

$$\begin{aligned} S &= 1 + 2 + 3 + \cdots + n \\ S &= n + (n-1) + (n-2) + \cdots + 1 \\ \hline 2S &= (n+1) + (n+1) + (n+1) + \cdots + (n+1) && \text{Add the corresponding sides.} \\ 2S &= n(n+1) && \text{There are } n \text{ terms of } n+1. \\ S &= \frac{n(n+1)}{2} && \text{Divide both sides by 2.} \end{aligned}$$

We can now apply deductive reasoning to find the sum of the first n counting numbers for any given value of n.

In the 1959 Disney animation *Donald in Mathmagic Land*, Donald Duck travels back in time to meet the Greek mathematician **Pythagoras** (c. 540 B.C.), who with his fellow mathematicians formed the Pythagorean brotherhood. The brotherhood devoted its time to the study of mathematics and music.

Figurate Numbers Pythagoras and his Pythagorean brotherhood (see the margin note) studied numbers of geometric arrangements of points, such as *triangular numbers, square numbers*, and *pentagonal numbers*. Figure 5 illustrates the first few of each of these types of numbers.

The *figurate numbers* possess numerous interesting patterns. Every square number greater than 1 is the sum of two consecutive triangular numbers. (For example, $9 = 3 + 6$ and $25 = 10 + 15$.) Every pentagonal number can be represented as the sum of a square number and a triangular number. (For example, $5 = 4 + 1$ and $12 = 9 + 3$.)

In the expression T_n, n is called a **subscript.** T_n is read **"T sub *n*,"** and it represents the triangular number in the nth position in the sequence. For example,

$$T_1 = 1, \quad T_2 = 3, \quad T_3 = 6, \quad \text{and} \quad T_4 = 10.$$

S_n and P_n represent the nth square and pentagonal numbers, respectively.

Formulas for Triangular, Square, and Pentagonal Numbers

For any natural number n,

the nth triangular number is given by $\mathbf{T}_n = \dfrac{n(n+1)}{2}$,

the nth square number is given by $\mathbf{S}_n = n^2$, and

the nth pentagonal number is given by $\mathbf{P}_n = \dfrac{n(3n-1)}{2}$.

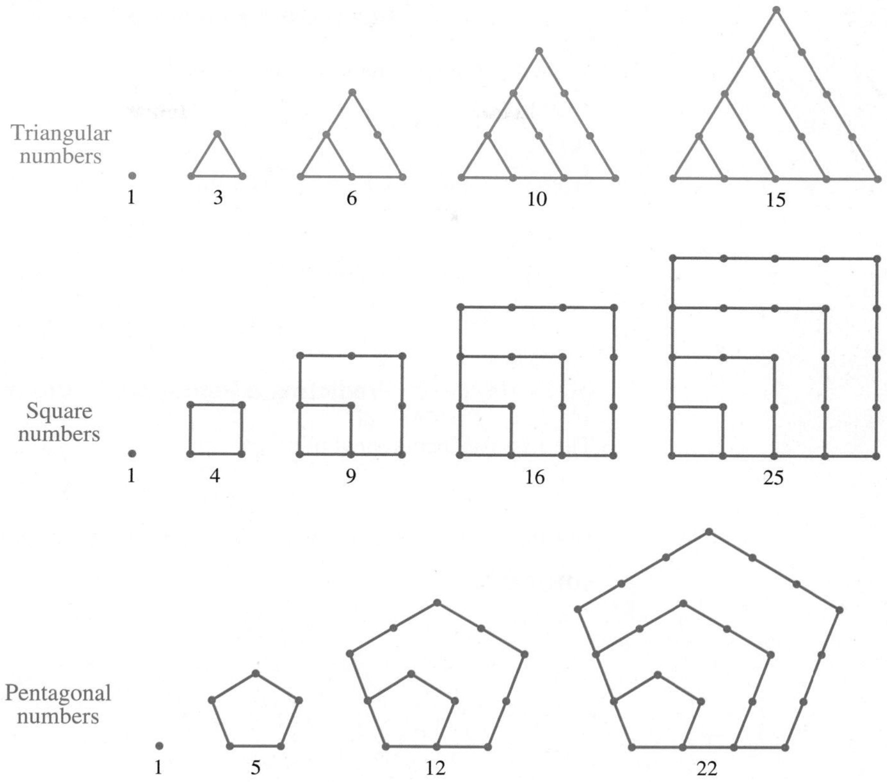

FIGURE 5

EXAMPLE 3 Using the Formulas for Figurate Numbers

Use the formulas to find each of the following.

(a) seventh triangular number
(b) twelfth square number
(c) sixth pentagonal number

SOLUTION

(a) $T_7 = \frac{n(n+1)}{2} = \frac{7(7+1)}{2} = \frac{7(8)}{2} = \frac{56}{2} = 28$ Formula for a triangular number, $n = 7$

(b) $S_{12} = n^2 = 12^2 = 144$ Formula for a square number, $n = 12$

$12^2 = 12 \cdot 12$

(c) $P_6 = \frac{n(3n-1)}{2} = \frac{6[3(6)-1]}{2} = \frac{6(18-1)}{2} = \frac{6(17)}{2} = 51$

Inside the brackets, multiply first and then subtract.

EXAMPLE 4 Illustrating a Figurate Number Relationship

Show that the sixth pentagonal number is equal to 3 times the fifth triangular number, plus 6.

SOLUTION

From Example 3(c), $P_6 = 51$. The fifth triangular number is 15. Thus,

$$51 = 3(15) + 6 = 45 + 6 = 51.$$

The general relationship examined in Example 4 can be written as follows.

$$P_n = 3 \cdot T_{n-1} + n \quad (n \geq 2)$$

EXAMPLE 5 Predicting a Pentagonal Number

The first five pentagonal numbers are

1, 5, 12, 22, 35.

Use the method of successive differences to predict the sixth pentagonal number.

SOLUTION

1		5		12		22		35		51
	4		7		10		13		16	
		3		3		3		3		

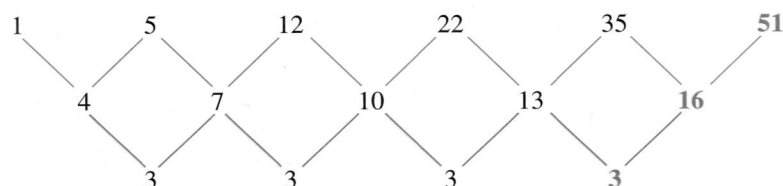

After the second line of successive differences, we work backward to find that the sixth pentagonal number is 51, which was also found in Example 3(c).

For Further Thought

Kaprekar Numbers

Take any three-digit number whose digits are not all the same. Arrange the digits in decreasing order, and then arrange them in increasing order. Now subtract. Repeat the process, using a 0 if necessary in the event that the difference consists of only two digits. For example, suppose that we choose a number whose digits are 1, 4, and 8, such as 841.

841	963	954
−148	−369	−459
693	594	495

Notice that we have obtained the number 495, and the process will lead to 495 again. The number 495 is called a **Kaprekar number.** The number 495 will eventually always be generated if this process is applied to such a three-digit number.

For Group Discussion or Individual Investigation

1. Apply the process of Kaprekar to a two-digit number, in which the digits are not the same. (Interpret 9 as 09 if necessary.) Compare the results. What seems to be true?
2. Repeat the process for four digits, comparing results after several steps. What conjecture can be made for this situation?

1.2 Exercises

Use the method of successive differences to determine the next number in each sequence.

1. 1, 4, 11, 22, 37, 56, . . .

2. 3, 14, 31, 54, 83, 118, . . .

3. 6, 20, 50, 102, 182, 296, . . .

4. 1, 11, 35, 79, 149, 251, . . .

5. 0, 12, 72, 240, 600, 1260, 2352, . . .

6. 2, 57, 220, 575, 1230, 2317, . . .

7. 5, 34, 243, 1022, 3121, 7770, 16799, . . .

8. 3, 19, 165, 771, 2503, 6483, 14409, . . .

9. Refer to Figures 2 and 3 in Section 1.1. The method of successive differences can be applied to the sequence of interior regions,

1, 2, 4, 8, 16, 31,

to find the number of regions determined by seven points on the circle. What is the next term in this sequence? How many regions would be determined by eight points? Verify this using the formula given at the end of that section.

10. Suppose that the expression $n^2 + 3n + 1$ determines the nth term in a sequence. That is, to find the first term, let $n = 1$; to find the second term, let $n = 2$, and so on.

(a) Find the first four terms of the sequence.

(b) Use the method of successive differences to predict the fifth term of the sequence.

(c) Find the fifth term by letting $n = 5$ in the expression $n^2 + 3n + 1$. Does your result agree with the one you found in part (b)?

In Exercises 11–20, several equations are given illustrating a suspected number pattern. Determine what the next equation would be, and verify that it is indeed a true statement.

11. $(1 \times 9) - 1 = 8$
$(21 \times 9) - 1 = 188$
$(321 \times 9) - 1 = 2888$

12. $(1 \times 8) + 1 = 9$
$(12 \times 8) + 2 = 98$
$(123 \times 8) + 3 = 987$

13. $999{,}999 \times 2 = 1{,}999{,}998$
$999{,}999 \times 3 = 2{,}999{,}997$

14. $101 \times 101 = 10{,}201$
$10{,}101 \times 10{,}101 = 102{,}030{,}201$

15. $3^2 - 1^2 = 2^3$
$6^2 - 3^2 = 3^3$
$10^2 - 6^2 = 4^3$
$15^2 - 10^2 = 5^3$

16. $1 = 1^2$
$1 + 2 + 1 = 2^2$
$1 + 2 + 3 + 2 + 1 = 3^2$
$1 + 2 + 3 + 4 + 3 + 2 + 1 = 4^2$

17. $2^2 - 1^2 = 2 + 1$
$3^2 - 2^2 = 3 + 2$
$4^2 - 3^2 = 4 + 3$

18. $1^2 + 1 = 2^2 - 2$
$2^2 + 2 = 3^2 - 3$
$3^2 + 3 = 4^2 - 4$

19. $1 = 1 \times 1$
$1 + 5 = 2 \times 3$
$1 + 5 + 9 = 3 \times 5$

20. $1 + 2 = 3$
$4 + 5 + 6 = 7 + 8$
$9 + 10 + 11 + 12 = 13 + 14 + 15$

Use the formula $S = \dfrac{n(n+1)}{2}$ *to find each sum.*

21. $1 + 2 + 3 + \cdots + 300$

22. $1 + 2 + 3 + \cdots + 500$

23. $1 + 2 + 3 + \cdots + 675$

24. $1 + 2 + 3 + \cdots + 825$

Use the formula $S = n^2$ *to find each sum. (Hint: To find* n*, add 1 to the last term and divide by 2.)*

25. $1 + 3 + 5 + \cdots + 101$

26. $1 + 3 + 5 + \cdots + 49$

27. $1 + 3 + 5 + \cdots + 999$

28. $1 + 3 + 5 + \cdots + 301$

29. Use the formula for finding the sum

$$1 + 2 + 3 + \cdots + n$$

to discover a formula for finding the sum

$$2 + 4 + 6 + \cdots + 2n.$$

30. State in your own words the following formula discussed in this section:

$$(1 + 2 + 3 + \cdots + n)^2 = 1^3 + 2^3 + 3^3 + \cdots + n^3.$$

31. Explain how the following diagram geometrically illustrates the formula $1 + 3 + 5 + 7 + 9 = 5^2$.

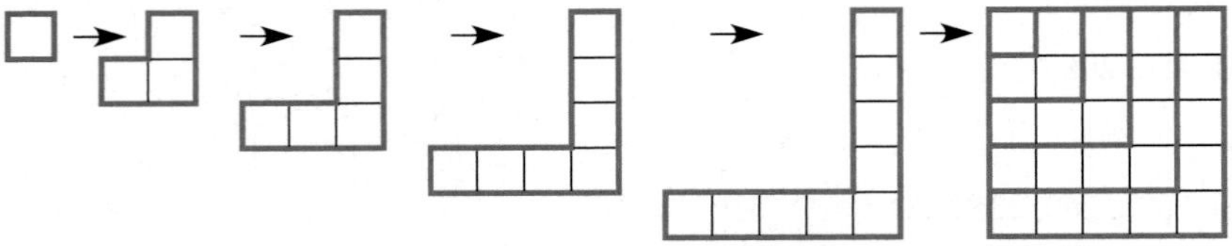

32. Explain how the following diagram geometrically illustrates the formula $1 + 2 + 3 + 4 = \frac{4 \times 5}{2}$.

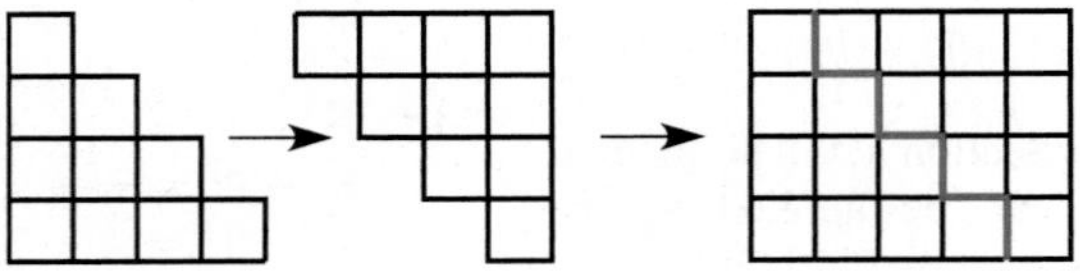

33. Use patterns to complete the table below.

Figurate Number	1st	2nd	3rd	4th	5th	6th	7th	8th
Triangular	1	3	6	10	15	21		
Square	1	4	9	16	25			
Pentagonal	1	5	12	22				
Hexagonal	1	6	15					
Heptagonal	1	7						
Octagonal	1							

34. The first five triangular, square, and pentagonal numbers may be obtained using sums of terms of sequences, as shown below.

Triangular	Square	Pentagonal
$1 = 1$	$1 = 1$	$1 = 1$
$3 = 1 + 2$	$4 = 1 + 3$	$5 = 1 + 4$
$6 = 1 + 2 + 3$	$9 = 1 + 3 + 5$	$12 = 1 + 4 + 7$
$10 = 1 + 2 + 3 + 4$	$16 = 1 + 3 + 5 + 7$	$22 = 1 + 4 + 7 + 10$
$15 = 1 + 2 + 3 + 4 + 5$	$25 = 1 + 3 + 5 + 7 + 9$	$35 = 1 + 4 + 7 + 10 + 13$

Notice the successive differences of the added terms on the right sides of the equations. The next type of figurate number is the **hexagonal** number. (A hexagon has six sides.) Use the patterns above to predict the first five hexagonal numbers.

35. Eight times any triangular number, plus 1, is a square number. Show that this is true for the first four triangular numbers.

36. Divide the first triangular number by 3 and record the remainder. Divide the second triangular number by 3 and record the remainder. Repeat this procedure several more times. Do you notice a pattern?

37. Repeat Exercise 36, but instead use square numbers and divide by 4. What pattern is determined?

38. Exercises 36 and 37 are specific cases of the following: In general, when the numbers in the sequence of n-agonal numbers are divided by n, the sequence of remainders obtained is a repeating sequence. Verify this for $n = 5$ and $n = 6$.

39. Every square number can be written as the sum of two triangular numbers. For example, $16 = 6 + 10$. This can be represented geometrically by dividing a square array of dots with a line as shown.

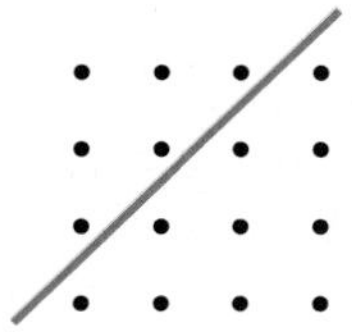

The triangular arrangement above the line represents 6, the one below the line represents 10, and the whole arrangement represents 16. Show how the square numbers 25 and 36 may likewise be geometrically represented as the sum of two triangular numbers.

40. A fraction is in *lowest terms* if the greatest common factor of its numerator and its denominator is 1. For example, $\frac{3}{8}$ is in lowest terms, but $\frac{4}{12}$ is not.

(a) For $n = 2$ to $n = 8$, form the fractions

$$\frac{n\text{th square number}}{(n + 1)\text{th square number}}$$

(b) Repeat part (a), but use triangular numbers instead.

(c) Use inductive reasoning to make a conjecture based on your results from parts (a) and (b), observing whether the fractions are in lowest terms.

41. Complete the following table.

n	2	3	4	5	6	7	8
A Square of n							
B (Square of n) + n							
C One-half of Row B entry							
D (Row A entry) − n							
E One-half of Row D entry							

Use your results to answer the following, using inductive reasoning.

(a) What kind of figurate number is obtained when you find the average of n^2 and n? (See Row C.)

(b) If you square n and then subtract n from the result, and then divide by 2, what kind of figurate number is obtained? (See Row E.)

42. Find the least integer N greater than 1 such that two different figurate numbers equal N. What are they?

In addition to the formulas for T_n, S_n, and P_n, the following formulas are true for ***hexagonal*** *numbers* (H), ***heptagonal*** *numbers* (Hp), *and* ***octagonal*** *numbers* (O):

$$H_n = \frac{n(4n - 2)}{2}, \quad Hp_n = \frac{n(5n - 3)}{2}, \quad O_n = \frac{n(6n - 4)}{2}.$$

Use these formulas to find each of the following.

43. the sixteenth square number

44. the eleventh triangular number

45. the ninth pentagonal number

46. the seventh hexagonal number

47. the tenth heptagonal number

48. the twelfth octagonal number

49. Observe the formulas given for H_n, Hp_n, and O_n, and use patterns and inductive reasoning to predict the formula for N_n, the nth **nonagonal** number. (A nonagon has nine sides.) Then use the fact that the sixth nonagonal number is 111 to further confirm your conjecture.

50. Use the result of Exercise 49 to find the tenth nonagonal number.

Use inductive reasoning to answer each question.

51. If you add two consecutive triangular numbers, what kind of figurate number do you get?

52. If you add the squares of two consecutive triangular numbers, what kind of figurate number do you get?

53. Square a triangular number. Square the next triangular number. Subtract the smaller result from the larger. What kind of number do you get?

54. Choose a value of n greater than or equal to 2. Find T_{n-1}, multiply it by 3, and add n. What kind of figurate number do you get?

1.3 Strategies for Problem Solving

A General Problem-Solving Method • Using a Table or Chart • Working Backward • Using Trial and Error • Guessing and Checking • Considering a Similar Simpler Problem • Drawing a Sketch • Using Common Sense

George Polya, author of the classic *How to Solve it*, died at the age of 97 on September 7, 1985. A native of Budapest, Hungary, he was once asked why there were so many good mathematicians to come out of Hungary at the turn of the century. He theorized that it was because mathematics is the cheapest science. It does not require any expensive equipment, only pencil and paper. He authored or coauthored more than 250 papers in many languages, wrote a number of books, and was a brilliant lecturer and teacher. Yet, interestingly enough, he never learned to drive a car.

A General Problem-Solving Method

In the first two sections of this chapter we stressed the importance of pattern recognition and the use of inductive reasoning in solving problems. There are other useful approaches. These ideas are used throughout the text.

Probably the most famous study of problem-solving techniques was developed by George Polya (1888–1985), among whose many publications was the modern classic *How to Solve It*. In this book, Polya proposed a four-step method for problem solving.

Polya's Four-Step Method for Problem Solving

Step 1 **Understand the problem.** You cannot solve a problem if you do not understand what you are asked to find. The problem must be read and analyzed carefully. You may need to read it several times. After you have done so, ask yourself, "What must I find?"

Step 2 **Devise a plan.** There are many ways to attack a problem. Decide what plan is appropriate for the particular problem you are solving.

Step 3 **Carry out the plan.** Once you know how to approach the problem, carry out your plan. You may run into "dead ends" and unforeseen roadblocks, but be persistent.

Step 4 **Look back and check.** Check your answer to see that it is reasonable. Does it satisfy the conditions of the problem? Have you answered all the questions the problem asks? Can you solve the problem a different way and come up with the same answer?

In Step 2 of Polya's problem-solving method, we are told to devise a plan. Here are some strategies that may prove useful.

Problem-Solving Strategies

Make a table or a chart.
Look for a pattern.
Solve a similar simpler problem.
Draw a sketch.
Use inductive reasoning.
Write an equation and solve it.
If a formula applies, use it.
Work backward.
Guess and check.
Use trial and error.
Use common sense.
Look for a "catch" if an answer seems too obvious or impossible.

A particular problem solution may involve one or more of the strategies listed here, and you should try to be creative in your problem-solving techniques. The examples that follow illustrate some of these strategies.

Fibonacci (1170–1250) discovered the sequence named after him in a problem on rabbits. Fibonacci (son of Bonaccio) is one of several names for Leonardo of Pisa. His father managed a warehouse in present-day Bougie (or Bejaia), in Algeria. Thus it was that Leonardo Pisano studied with a Moorish teacher and learned the "Indian" numbers that the Moors and other Moslems brought with them in their westward drive.

Fibonacci wrote books on algebra, geometry, and trigonometry.

Using a Table or Chart

EXAMPLE 1 Solving Fibonacci's Rabbit Problem

A man put a pair of rabbits in a cage. During the first month the rabbits produced no offspring but each month thereafter produced one new pair of rabbits. If each new pair thus produced reproduces in the same manner, how many pairs of rabbits will there be at the end of 1 year? (This problem is a famous one in the history of mathematics and first appeared in *Liber Abaci,* a book written by the Italian mathematician Leonardo Pisano (also known as Fibonacci) in the year 1202.)

SOLUTION

We apply Polya's method.

Step 1 **Understand the problem.** After several readings, we can reword the problem as follows:

How many pairs of rabbits will the man have at the end of one year if he starts with one pair, and they reproduce this way: During the first month of life, each pair produces no new rabbits, but each month thereafter each pair produces one new pair?

Step 2 **Devise a plan.** Because there is a definite pattern to how the rabbits will reproduce, we can construct Table 2. Once the table is completed, the final entry in the final column is our answer.

TABLE 2

Month	Number of Pairs at Start	Number of New Pairs Produced	Number of Pairs at End of Month
1st			
2nd			
3rd			
4th			
5th			
6th			
7th			
8th			
9th			
10th			
11th			
12th			

The answer will go here.

Step 3 **Carry out the plan.** At the start of the first month, there is only one pair of rabbits. No new pairs are produced during the first month, so there is $1 + 0 = 1$ pair present at the end of the first month. This pattern continues. In the table, we

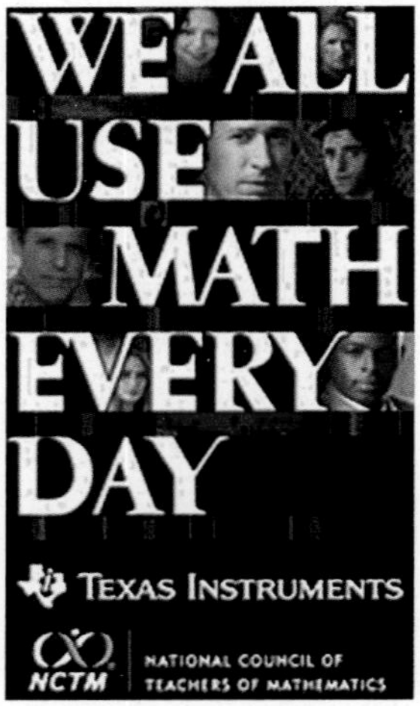

On January 23, 2005, the CBS television network presented the first episode of *NUMB3RS*, a show focusing on how mathematics is used in solving crimes. David Krumholtz plays Charlie Eppes, a brilliant mathematician who assists his FBI agent brother (Rob Morrow). In the first-season episode "Sabotage" (2/25/2005), one of the agents admits that she "never saw how math relates to the real world," and Charlie uses the **Fibonacci sequence** and its relationship to nature to enlighten her.

add the number in the first column of numbers to the number in the second column to get the number in the third.

Month	Number of Pairs at Start	+	Number of New Pairs Produced	=	Number of Pairs at End of Month	
1st	1		0		1	1 + 0 = 1
2nd	1		1		2	1 + 1 = 2
3rd	2		1		3	2 + 1 = 3
4th	3		2		5	.
5th	5		3		8	.
6th	8		5		13	.
7th	13		8		21	.
8th	21		13		34	.
9th	34		21		55	.
10th	55		34		89	.
11th	89		55		144	.
12th	144		89		**233**	144 + 89 = 233

The answer is the final entry.

There will be 233 pairs of rabbits at the end of one year.

Step 4 **Look back and check.** This problem can be checked by going back and making sure that we have interpreted it correctly, which we have. Double-check the arithmetic. We have answered the question posed by the problem, so the problem is solved.

The sequence shown in color in the table in Example 1 is the Fibonacci sequence, mentioned in Example 1 of the previous section. In the remaining examples of this section, we use Polya's process but do not list the steps specifically as we did in Example 1.

Working Backward

EXAMPLE 2 Determining a Wager at the Track

Rob Zwettler goes to the racetrack with his buddies on a weekly basis. One week he tripled his money, but then lost \$12. He took his money back the next week, doubled it, but then lost \$40. The following week he tried again, taking his money back with him. He quadrupled it, and then played well enough to take that much home with him, a total of \$224. How much did he start with the first week?

SOLUTION

This problem asks us to find Rob's starting amount, given information about his winnings and losses. We also know his final amount. The method of working backward can be applied quite easily.

Because his final amount was \$224 and this represents four times the amount he started with on the third week, we *divide* \$224 by 4 to find that he started the third week with \$56. Before he lost \$40 the second week, he had this \$56 plus the \$40 he lost, giving him \$96. This represented double what he started with, so he started with \$96 *divided by* 2, or \$48, the second week. Repeating this process once more for the first week, before his \$12 loss he had

$$\$48 + \$12 = \$60,$$

Augustus De Morgan was an English mathematician and philosopher, who served as professor at the University of London. He wrote numerous books, one of which was *A Budget of Paradoxes*. His work in set theory and logic led to laws that bear his name and are covered in other chapters. He died in the same year as Charles Babbage.

which represents triple what he started with. Therefore, he started with

$$\$60 \div 3 = \$20. \quad \text{Answer}$$

To check, observe the following equations that depict winnings and losses:

First week: $(3 \times \$20) - \$12 = \$60 - \$12 = \$48$

Second week: $(2 \times \$48) - \$40 = \$96 - \$40 = \$56$

Third week: $(4 \times \$56) = \$224$. His final amount

Using Trial and Error

Recall that $5^2 = 5 \cdot 5 = 25$, that is, 5 squared is 25. Thus, 25 is called a **perfect square.** Other perfect squares include

$$1, \quad 4, \quad 9, \quad 16, \quad 36, \quad \text{and so on.} \quad \text{Perfect squares}$$

The next example uses the idea of perfect square.

EXAMPLE 3 Finding Augustus De Morgan's Birth Year

The mathematician Augustus De Morgan lived in the nineteenth century. He made the following statement: "I was x years old in the year x^2." In what year was he born?

SOLUTION

We must find the year of De Morgan's birth. The problem tells us that he lived in the nineteenth century, which is another way of saying that he lived during the 1800s. One year of his life was a perfect square, so we must find a number between 1800 and 1900 that is a perfect square. Use trial and error.

$$42^2 = 1764$$
$$43^2 = 1849 \quad \text{1849 is between 1800 and 1900.}$$
$$44^2 = 1936$$

The only natural number whose square is between 1800 and 1900 is 43, since $43^2 = 1849$. Therefore, De Morgan was 43 years old in 1849. The final step in solving the problem is to subtract 43 from 1849 to find the year of his birth:

$$1849 - 43 = 1806. \quad \text{He was born in 1806.}$$

Although the following suggestion for a check may seem unorthodox, it works: Look up De Morgan's birth date in a book dealing with mathematics history, such as *An Introduction to the History of Mathematics*, Sixth Edition, by Howard W. Eves.

Guessing and Checking

As mentioned above, $5^2 = 25$. The inverse (opposite) of squaring a number is called taking the **square root**. We indicate the positive square root using a **radical sign** $\sqrt{\ }$. Thus, $\sqrt{25} = 5$. Also,

$$\sqrt{4} = 2, \quad \sqrt{9} = 3, \quad \sqrt{16} = 4, \quad \text{and so on.} \quad \text{Square roots}$$

The next problem deals with a square root, and dates back to Hindu mathematics, circa 850.

EXAMPLE 4 Finding the Number of Camels

One-fourth of a herd of camels was seen in the forest; twice the square root of that herd had gone to the mountain slopes; and 3 times 5 camels remained on the riverbank. What is the numerical measure of that herd of camels?

SOLUTION

The numerical measure of a herd of camels must be a counting number. Because the problem mentions "one-fourth of a herd" and "the square root of that herd," the number of camels must be both a multiple of 4 and a perfect square, so that only whole numbers are used. The least counting number that satisfies both conditions is 4. We write an equation where x represents the numerical measure of the herd, and then substitute 4 for x to see if it is a solution.

$$\underbrace{\frac{1}{4}x}_{\text{One-fourth of the herd}} + \underbrace{2\sqrt{x}}_{\text{Twice the square root of that herd}} + \underbrace{3 \cdot 5}_{\text{3 times 5 camels}} \underbrace{=}_{=} \underbrace{x}_{\text{The numerical measure of the herd.}}$$

$$\frac{1}{4}(4) + 2\sqrt{4} + 3 \cdot 5 = 4 \qquad \text{Let } x = 4.$$

$$1 + 4 + 15 \stackrel{?}{=} 4 \qquad \sqrt{4} = 2$$

$$20 \neq 4$$

Because 4 is not the solution, try 16, the next perfect square that is a multiple of 4.

$$\frac{1}{4}(16) + 2\sqrt{16} + 3 \cdot 5 = 16 \qquad \text{Let } x = 16.$$

$$4 + 8 + 15 \stackrel{?}{=} 16 \qquad \sqrt{16} = 4$$

$$27 \neq 16$$

Because 16 is not a solution, try 36.

$$\frac{1}{4}(36) + 2\sqrt{36} + 3 \cdot 5 = 36 \qquad \text{Let } x = 36.$$

$$9 + 12 + 15 \stackrel{?}{=} 36 \qquad \sqrt{36} = 6$$

$$36 = 36$$

We see that 36 is the numerical measure of the herd. Check in the words of the problem: "One-fourth of 36, plus twice the square root of 36, plus 3 times 5" gives 9 plus 12 plus 15, which equals 36. (Algebra shows that 36 is the *only* correct answer.) ■

Considering a Similar Simpler Problem

EXAMPLE 5 Finding the Units Digit of a Power

The digit farthest to the right in a counting number is called the *ones* or *units* digit, because it tells how many ones are contained in the number when grouping by tens is considered. What is the ones (or units) digit in 2^{4000}?

The 1952 film *Hans Christian Andersen* features Danny Kaye as the Danish writer of fairy tales. In a scene outside a schoolhouse, he sings a song to an inchworm: "Inchworm, inchworm, measuring the marigolds, you and your arithmetic, you'll probably go far." Following the scene, students in the schoolhouse are heard singing arithmetic facts:

Two and two are four,
Four and four are eight,
Eight and eight are sixteen,
Sixteen and sixteen are
thirty-two.

Their answers are all **powers of 2.**

SOLUTION

Recall that 2^{4000} means that 2 is used as a factor 4000 times:

$$2^{4000} = \underbrace{2 \times 2 \times 2 \times \ldots \times 2.}_{\text{4000 factors}}$$

Certainly, we are not expected to evaluate this number. To answer the question, we examine some smaller powers of 2 and then look for a pattern. We start with the exponent 1 and look at the first twelve powers of 2.

$$\begin{array}{lll} 2^1 = 2 & 2^5 = 32 & 2^9 = 512 \\ 2^2 = 4 & 2^6 = 64 & 2^{10} = 1024 \\ 2^3 = 8 & 2^7 = 128 & 2^{11} = 2048 \\ 2^4 = 16 & 2^8 = 256 & 2^{12} = 4096 \end{array}$$

Notice that in each of the four rows above, the ones digit is the same. The final row, which contains the exponents 4, 8, and 12, has the ones digit 6. Each of these exponents is divisible by 4, and because 4000 is divisible by 4, we can use inductive reasoning to predict that the units digit in 2^{4000} is 6.

(*Note*: The units digit for any other power can be found if we divide the exponent by 4 and consider the remainder. Then compare the result to the list of powers above. For example, to find the units digit of 2^{543}, divide 543 by 4 to get a quotient of 135 and a remainder of 3. The units digit is the same as that of 2^3, which is 8.)

Drawing a Sketch

EXAMPLE 6 Connecting the Dots

FIGURE 6

An array of nine dots is arranged in a 3×3 square, as shown in Figure 6. Is it possible to join the dots with exactly four straight line segments if you are not allowed to pick up your pencil from the paper and may not trace over a segment that has already been drawn? If so, show how.

SOLUTION

Figure 7 shows three attempts. In each case, something is wrong. In the first sketch, one dot is not joined. In the second, the figure cannot be drawn without picking up your pencil from the paper or tracing over a line that has already been drawn. In the third figure, all dots have been joined, but you have used five line segments as well as retraced over the figure.

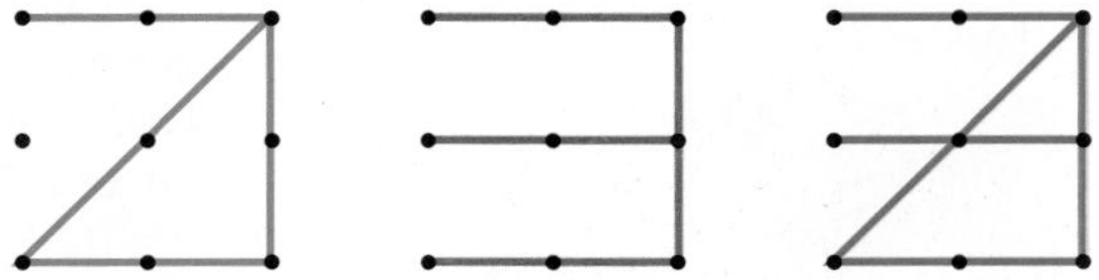

FIGURE 7

The conditions of the problem can be satisfied, as shown in Figure 8. We "went outside of the box," which was not prohibited by the conditions of the problem. This is an example of creative thinking—we used a strategy that is usually not considered at first.

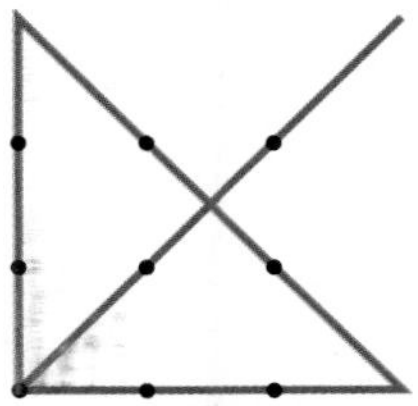

FIGURE 8

In *Die Hard: With a Vengeance* (see the Chapter Opener), Simon taunts McClane with a riddle that has its origins in Egyptian mathematics.

As I was going to St. Ives,
I met a man with seven wives.
Every wife had seven sacks,
Every sack had seven cats,
Every cat had seven kittens.
Kittens, cats, sacks and wives,
How many were going to St. Ives?

"My phone number is 555 and the answer. Call me in 30 seconds or die."

By calling 555-0001, he was able to contact Simon. Do you see why 1 is the answer to this riddle? (Use **common sense.**)

Using Common Sense The final example falls into a category of problems that involve a "catch." Some of these problems seem too easy or perhaps impossible at first because we tend to overlook an obvious situation. Look carefully at the use of language in such problems. And, of course, never forget to use common sense.

EXAMPLE 7 Determining Coin Denominations

Two currently minted United States coins together have a total value of \$1.05. One is not a dollar. What are the two coins?

SOLUTION

Our initial reaction might be, "The only way to have two such coins with a total of \$1.05 is to have a nickel and a dollar, but the problem says that one of them is not a dollar." This statement is indeed true. What we must realize here is that the one that is not a dollar is the nickel, and the *other* coin is a dollar! So the two coins are a dollar and a nickel. ■

Solution to the Chapter Opener Problem This is one way to do it: With both jugs empty, fill the 3-gallon jug and pour its contents into the 5-gallon jug. Then fill the 3-gallon jug again, and pour it into the 5-gallon jug until the latter is filled. There is now $(3 + 3) - 5 = 1$ gallon in the 3-gallon jug. Empty the 5-gallon jug, and pour the 1 gallon of water from the 3-gallon jug into the 5-gallon jug. Finally, fill the 3-gallon jug and pour all of it into the 5-gallon jug, resulting in $1 + 3 = 4$ gallons in the 5-gallon jug.

(*Note*: There is another way to solve this problem. See if you can discover the alternative solution.)

For Further Thought

A Brain Teaser

Various forms of the following problem have been around for many years.

In Farmer Jack's will, Jack bequeathed $\frac{1}{2}$ of his horses to his son Johnny, $\frac{1}{3}$ to his daughter Linda, and $\frac{1}{9}$ to his son Jeff. Jack had 17 horses, so how were they to comply with the terms of the will? Certainly, horses cannot be divided up into fractions. Their attorney, Garbarino, came to their rescue, and was able to execute the will to the satisfaction of all. How did she do it?

Here is the solution:

Garbarino added one of her horses to the 17, giving a total of 18. Johnny received $\frac{1}{2}$ of 18, or 9, Linda received $\frac{1}{3}$ of 18, or 6, and Jeff received $\frac{1}{9}$ of 18, or 2. That accounted for a total of $9 + 6 + 2 = 17$ horses. Then Garbarino took back her horse, and everyone was happy.

For Group Discussion or Individual Investigation

Make up a similar problem involving fractions. Check your work.

1.3 Exercises

One of the most popular features in the journal Mathematics Teacher, *published by the National Council of Teachers of Mathematics, is the monthly calendar, which provides an interesting, unusual, or challenging problem for each day of the month. Problems are contributed by the editors of the journal, teachers, and students, and the contributors are cited in each issue. Exercises 1–35 are problems chosen from these calendars over the past several years, with the day, month, and year for the problem indicated. The authors want to thank the many contributors for permission to use these problems.*

Use the various problem-solving strategies to solve each problem. In many cases there is more than one possible approach, so be creative.

1. ***Catwoman's Cats*** If you ask Batman's nemesis, Catwoman, how many cats she has, she answers with a riddle: "Five-sixths of my cats plus seven." How many cats does Catwoman have? (April 20, 2003)

2. ***Pencil Collection*** Bob gave four-fifths of his pencils to Barbara, then he gave two-thirds of the remaining pencils to Bonnie. If he ended up with ten pencils for himself, with how many did he start? (October 12, 2003)

3. ***Adding Gasoline*** The gasoline gauge on a van initially read $\frac{1}{8}$ full. When 15 gallons were added to the tank, the gauge read $\frac{3}{4}$ full. How many more gallons are needed to fill the tank? (November 25, 2004)

4. ***Gasoline Tank Capacity*** When 6 gallons of gasoline are put into a car's tank, the indicator goes from $\frac{1}{4}$ of a tank to $\frac{5}{8}$. What is the total capacity of the gasoline tank? (February 21, 2004)

5. ***Number Pattern*** What is the relationship between the rows of numbers?

18,	38,	24,	46,	42
8,	24,	8,	24,	8

(May 26, 2005)

6. ***Unknown Number*** The number in an unshaded square is obtained by adding the numbers connected with it from the row above. (The 11 is one such number.) What is the value of x? (August 9, 2004)

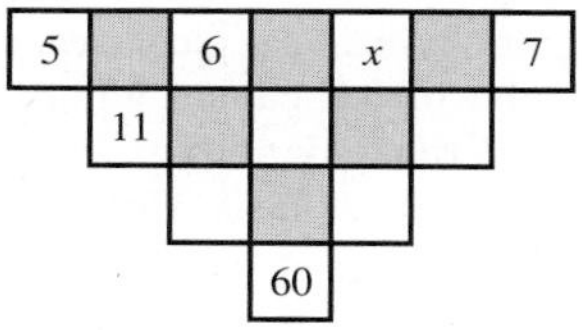

7. ***Locking Boxes*** You and I each have one lock and a corresponding key. I want to mail you a box with a ring in it, but any box that is not locked will be emptied before it reaches its recipient. How can I safely send you the ring? (Note that you and I each have keys to our own lock but not to the other lock.) (May 4, 2004)

8. ***Woodchuck Chucking Wood*** Nine woodchucks can chuck eight pieces of wood in 3 hours. How much wood can a woodchuck chuck in 1 hour? (May 24, 2004)

9. ***Number in a Sequence*** In the sequence 16, 80, 48, 64, A, B, C, D, each term beyond the second term is the arithmetic mean (average) of the two previous terms. What is the value of D? (April 26, 2004)

10. ***Unknown Number*** Cindy was asked by her teacher to subtract 3 from a certain number and then divide the result by 9. Instead, she subtracted 9 and then divided the result by 3, giving an answer of 43. What would her answer have been if she had worked the problem correctly? (September 3, 2004)

11. ***Labeling Boxes*** You are working in a store that has been very careless with the stock. Three boxes of socks are each incorrectly labeled. The labels say *red socks, green socks*, and *red and green socks*. How can you relabel the boxes correctly by taking only one sock out of one box, without looking inside the boxes? (October 22, 2001)

12. ***Vertical Symmetry in States' Names*** (If a vertical line is drawn through the center of a figure and the left and right sides are reflections of each other across this line, the figure is said to have vertical symmetry.) When spelled with all capital letters, each letter in HAWAII has vertical symmetry. Find the name of a state whose letters all have vertical and horizontal symmetry. (September 11, 2001)

13. ***Sum of Hidden Dots on Dice*** Three dice with faces numbered 1 through 6 are stacked as shown. Seven of the eighteen faces are visible, leaving eleven faces hidden on the back, on the bottom, and between dice. The total number of dots not visible in this view is ________.

A. 21
B. 22
C. 31
D. 41
E. 53
(September 17, 2001)

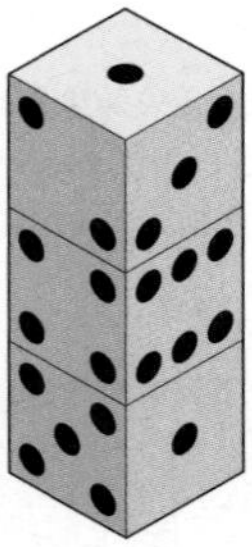

14. ***Mr. Green's Age*** At his birthday party, Mr. Green would not directly tell how old he was. He said, "If you add the year of my birth to this year, subtract the year of my tenth birthday and the year of my fiftieth birthday, and then add my present age, the result is eighty." How old was Mr Green? (December 14, 1997)

15. ***Unfolding and Folding a Box*** An unfolded box is shown below.

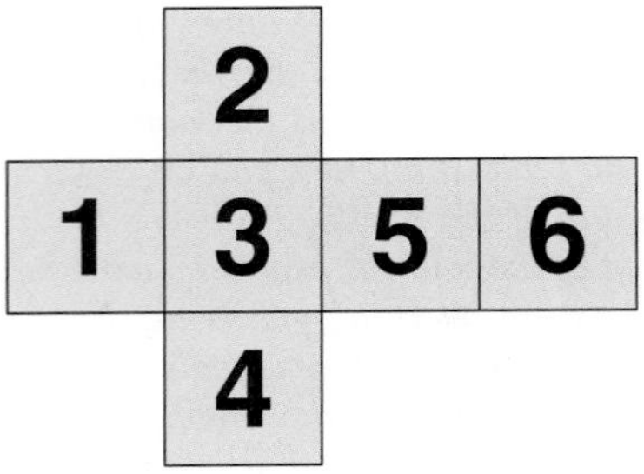

Which figure shows the box folded up? (November 7, 2001)

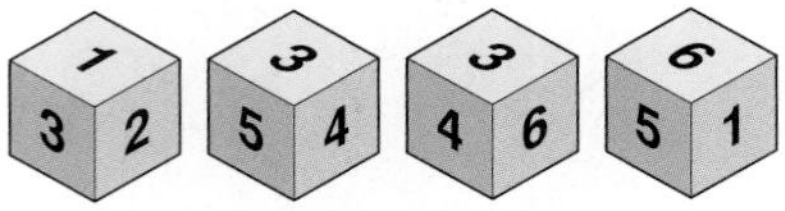

A B C D

16. ***Age of the Bus Driver*** Today is your first day driving a city bus. When you leave downtown, you have twenty-three passengers. At the first stop, three people exit and five people get on the bus. At the second stop, eleven people exit and eight people get on the bus. At the third stop, five people exit and ten people get on. How old is the bus driver? (April 1, 2002)

17. ***Matching Triangles and Squares*** How can you connect each square with the triangle that has the same number? Lines cannot cross, enter a square or triangle, or go outside the diagram. (October 15, 1999)

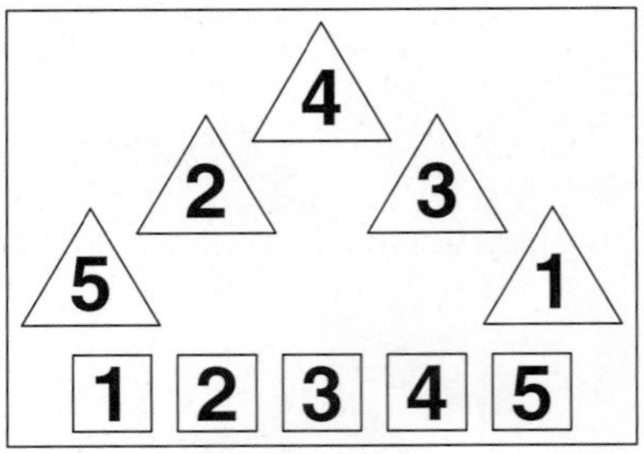

18. ***Ticktacktoe Strategy*** You and a friend are playing ticktacktoe, where three in a row loses. (See the next page.) You are O. If you want to win, what must your next move be? (October 21, 2001)

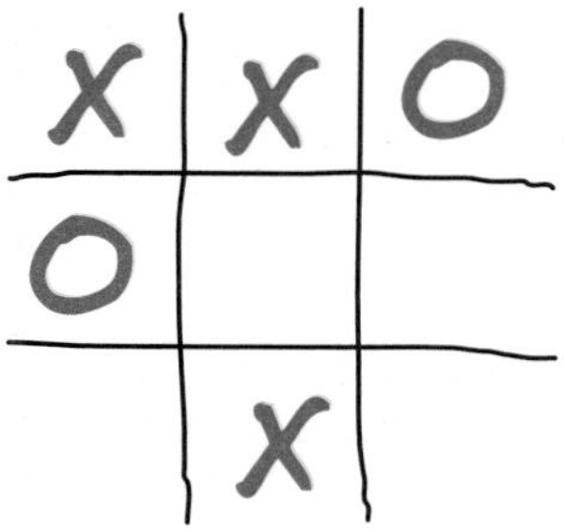

19. ***Forming Perfect Square Sums*** How must one place the integers from 1 to 15 in each of the spaces below in such a way that no number is repeated and the sum of the numbers in any two consecutive spaces is a perfect square? (November 11, 2001)

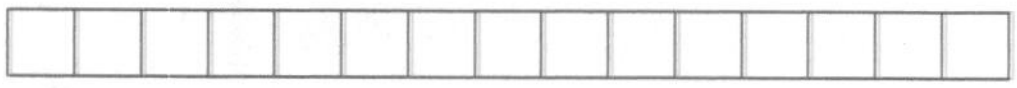

20. ***How Old?*** Pat and Chris have the same birthday. Pat is twice as old as Chris was when Pat was as old as Chris is now. If Pat is now 24 years old, how old is Chris? (December 3, 2001)

21. ***Difference Triangle*** Balls numbered 1 through 6 are arranged in a *difference triangle*. Note that in any row, the difference between the larger and the smaller of two successive balls is the number of the ball that appears below them. Arrange balls numbered 1 through 10 in a *difference triangle*. (May 6, 1998)

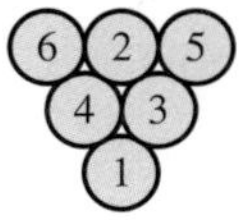

22. ***Clock Face*** By drawing two straight lines, divide the face of a clock into three regions such that the numbers in the regions have the same total. (October 28, 1998)

23. ***Alphametric*** If a, b, and c are digits for which

$$\begin{array}{r} 7\ a\ 2 \\ -4\ 8\ b \\ \hline c\ 7\ 3, \end{array}$$

then $a + b + c =$ ______.

A. 14 **B.** 15 **C.** 16 **D.** 17 **E.** 18
(September 22, 1999)

24. ***Perfect Square*** Only one of these numbers is a perfect square. Which one is it? (October 8, 1997)

329476 389372 964328
326047 724203

25. ***Sleeping on the Way to Grandma's House*** While traveling to his grandmother's for Christmas, George fell asleep halfway through the journey. When he awoke, he still had to travel half the distance that he had traveled while sleeping. For what part of the entire journey had he been asleep? (December 25, 1998)

26. ***Counting Puzzle (Rectangles)*** How many rectangles of any size are in the figure shown? (September 10, 2001)

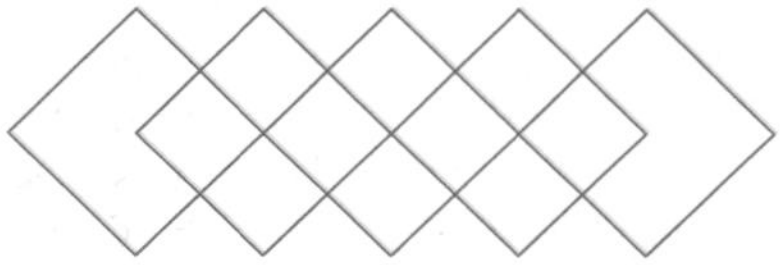

27. ***Buckets of Water*** You have brought two unmarked buckets to a stream. The buckets hold 7 gallons and 3 gallons of water, respectively. How can you obtain exactly 5 gallons of water to take home? (October 19, 1997)

28. ***Collecting Acorns*** Chipper and Dalie collected thirty-two acorns on Monday and stored them with their acorn supply. After Chipper fell asleep, Dalie ate half the acorns. This pattern continued through Friday night, with thirty-two acorns being added and half the supply being eaten. On Saturday morning, Chipper counted the acorns and found that they had only thirty-five. How many acorns had they started with on Monday morning? (March 12, 1997)

29. ***Counting Puzzle (Rectangles)*** How many rectangles are in the figure? (March 27, 1997)

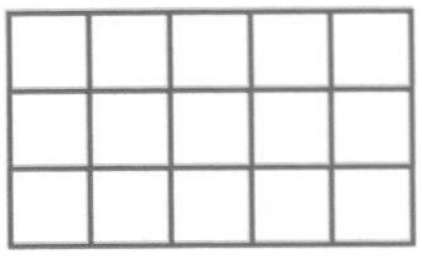

30. ***Digit Puzzle*** Place each of the digits 1, 2, 3, 4, 5, 6, 7, and 8 in separate boxes so that boxes that share common corners do not contain successive digits. (November 29, 1997)

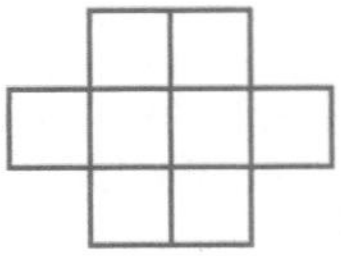

31. ***Palindromic Number*** (*Note:* A *palindromic* number is a number whose digits read the same left to right as right to left. For example, 383, 12321, and 9876789 are palindromic.) The odometer of the family car read 15951 when the driver noticed that the number was palindromic. "Curious," said the driver to herself. "It will be a long time before that happens again." But 2 hours later, the odometer showed a new palindromic number. (*Author's note:* Assume it was the next possible one.) How fast was the car driving in those 2 hours? (December 26, 1998)

32. ***Exchange Rate*** An island has no currency; it instead has the following exchange rate:

50 bananas = 20 coconuts
30 coconuts = 12 fish
100 fish = 1 hammock

How many bananas equal 1 hammock? (April 16, 1998)

33. ***Final Digits of a Power of 7*** What are the final two digits of 7^{1997}? (November 29, 1997)

34. ***Brightness of a Clock Display*** If a digital clock is the only light in an otherwise totally dark room, when will the room be darkest? Brightest? (May 1, 1996)

35. ***Value of Coins*** Which is worth more, a kilogram of \$10 gold pieces or half a kilogram of \$20 gold pieces? (March 20, 1995)

36. ***Units Digit of a Power of 3*** If you raise 3 to the 324th power, what is the units digit of the result?

37. ***Units Digit of a Power of 7*** What is the units digit in 7^{491}?

38. ***Money Spent at a Bazaar*** Ashley O'Shaughnessy bought a book for \$10 and then spent half her remaining money on a train ticket. She then bought lunch for \$4 and spent half her remaining money at a bazaar. She left the bazaar with \$8. How much money did she start with?

39. ***Unknown Number*** I am thinking of a positive number. If I square it, double the result, take half of that result, and then add 12, I get 37. What is my number?

40. ***Frog Climbing up a Well*** A frog is at the bottom of a 20-foot well. Each day it crawls up 4 feet, but each night it slips back 3 feet. After how many days will the frog reach the top of the well?

41. ***Matching Socks*** A drawer contains 20 black socks and 20 white socks. If the light is off and you reach into the drawer to get your socks, what is the minimum number of socks you must pull out in order to be sure that you have a matching pair?

42. ***Counting Puzzle (Squares)*** How many squares are in the following figure?

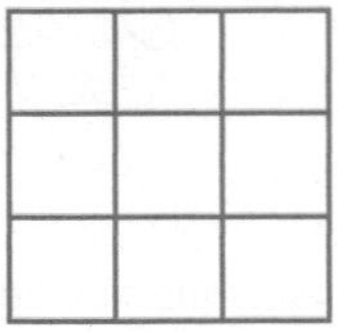

43. ***Counting Puzzle (Triangles)*** How many triangles are in the following figure?

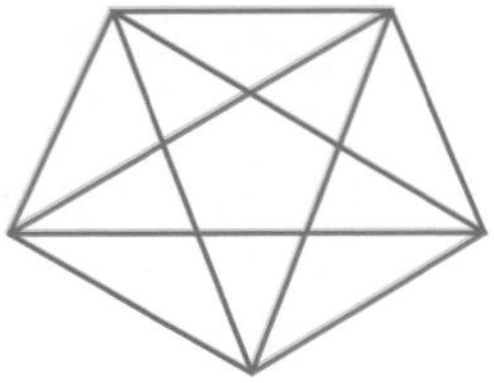

44. ***Children in a Circle*** Some children are standing in a circular arrangement. They are evenly spaced and marked in numerical order. The fourth child is standing directly opposite the twelfth child. How many children are there in the circle?

45. ***Perfect Number*** A *perfect number* is a counting number that is equal to the sum of all its counting number divisors except itself. For example, 28 is a perfect number because its divisors other than itself are 1, 2, 4, 7, and 14, and $1 + 2 + 4 + 7 + 14 = 28$. What is the least perfect number?

46. ***Naming Children*** Becky's mother has three daughters. She named her first daughter Penny and her second daughter Nichole. What did she name her third daughter?

47. ***Growth of a Lily Pad*** A lily pad grows so that each day it doubles its size. On the twentieth day of its life, it completely covers a pond. On what day was the pond half covered?

48. ***Interesting Property of a Sentence*** Comment on an interesting property of this sentence: "A man, a plan, a canal, Panama." (*Hint:* See Exercise 31.)

49. ***High School Graduation Year of Author*** One of the authors of this book graduated from high school in the year that satisfies these conditions: (1) The sum of the digits is 23; (2) The hundreds digit is 3 more than the tens digit; (3) No digit is an 8. In what year did he graduate?

50. ***Relative Heights*** Donna is taller than David but shorter than Bill. Dan is shorter than Bob. What is the first letter in the name of the tallest person?

51. ***Adam and Eve's Assets*** Eve said to Adam, "If you give me one dollar, then we will have the same amount of money." Adam then replied, "Eve, if you give me one dollar, I will have double the amount of money you are left with." How much does each have?

52. ***Missing Digits Puzzle*** In the addition problem at the top of the next column, some digits are missing as indicated by the blanks. If the problem is done correctly, what is the sum of the missing digits?

$$\begin{array}{rrrr} & \_\_ & 3 & 5 \\ & 8 & \_\_ & 6 \\ + & 1 & 4 & \_\_ \\ \hline \_\_ & 4 & 0 & 8 \end{array}$$

53. ***Missing Digits Puzzle*** Fill in the blanks so that the multiplication problem below uses all digits 0, 1, 2, 3, . . . , 9 exactly once, and is correctly worked.

$$\begin{array}{rrrrr} & & \_\_ & 0 & 2 \\ \times & & & 3 & \_\_ \\ \hline \_\_ & 5, & \_\_ & \_\_ & \_\_ \end{array}$$

54. ***Magic Square*** A *magic square* is a square array of numbers that has the property that the sum of the numbers in any row, column, or diagonal is the same. Fill in the square below so that it becomes a magic square, and all digits 1, 2, 3, . . . , 9 are used exactly once.

6		8
	5	
		4

55. ***Magic Square*** Refer to Exercise 54. Complete the magic square below so that all counting numbers 1, 2, 3, . . . ,16 are used exactly once, and the sum in each row, column, or diagonal is 34.

6			9
	15		14
11		10	
16		13	

56. ***Paying for a Mint*** Brian Altobello has an unlimited number of cents (pennies), nickels, and dimes. In how many different ways can he pay 15¢ for a chocolate mint? (For example, one way is 1 dime and 5 pennies.)

57. ***Pitches in a Baseball Game*** What is the minimum number of pitches that a baseball player who pitches a complete game can make in a regulation 9-inning baseball game?

58. *Weighing Coins* You have eight coins. Seven are genuine and one is a fake, which weighs a little less than the other seven. You have a balance scale, which you may use only three times. Tell how to locate the bad coin in three weighings. (Then show how to detect the bad coin in only *two* weighings.)

59. *Geometry Puzzle* When the diagram shown is folded to form a cube, what letter is opposite the face marked Z?

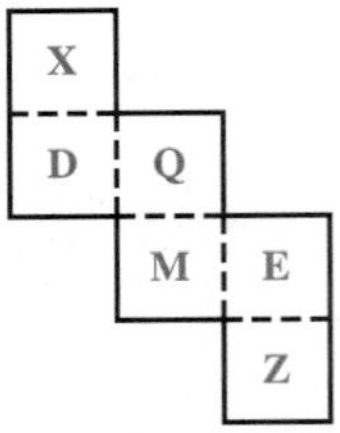

60. *Picture Puzzle* Draw a square in the following figure so that no two cats share the same region.

61. *Geometry Puzzle* Draw the following figure without picking up your pencil from the paper and without tracing over a line you have already drawn.

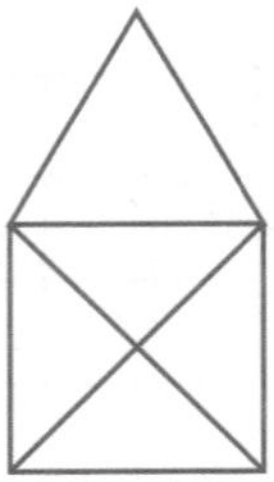

62. *Geometry Puzzle* Repeat Exercise 61 for this figure.

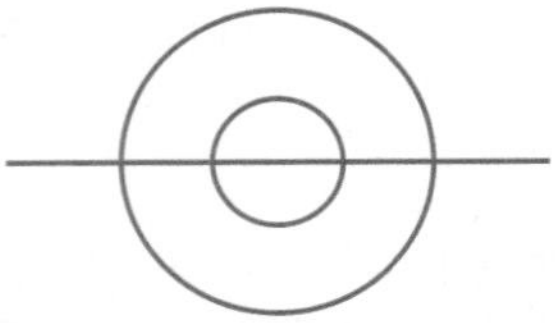

63. *Decimal Digit* What is the 100th digit in the decimal representation for $\frac{1}{7}$?

64. *Books on a Shelf* Volumes 1 and 2 of *The Complete Works of Wally Smart* are standing in numerical order from left to right on your bookshelf. Volume 1 has 450 pages and Volume 2 has 475 pages. Excluding the covers, how many pages are between page 1 of Volume 1 and page 475 of Volume 2?

65. *Oh Brother!* The brother of the chief executive officer (CEO) of a major industrial firm died. The man who died had no brother. How is this possible?

66. *Teenager's Age* A teenager's age increased by 2 gives a perfect square. Her age decreased by 10 gives the square root of that perfect square. She is 5 years older than her brother. How old is her brother?

67. *Ages* James, Dan, Jessica, and Cathy form a pair of married couples. Their ages are 36, 31, 30, and 29. Jessica is married to the oldest person in the group. James is older than Jessica but younger than Cathy. Who is married to whom, and what are their ages?

68. *Making Change* In how many different ways can you make change for a half dollar using currently minted U.S. coins, if cents (pennies) are not allowed?

69. *Days in a Month* Some months have 30 days and some have 31 days. How many months have 28 days?

70. *Dirt in a Hole* How much dirt is there in a cubical hole, 6 feet on each side?

71. *Fibonacci Property* Refer to Example 1, and observe the sequence of numbers in color. Choose any four successive terms. Multiply the first one chosen by the fourth; then multiply the two middle terms. Repeat this process. What do you notice when the two products are compared?

72. *Palindromic Greeting* The first man introduced himself to the first woman with a brief "palindromic" greeting. What was the greeting? (*Hint:* See Exercises 31, 48, and 51.)

73. *Geometry Puzzle* What is the maximum number of small squares in which we may place crosses (×) and not have any row, column, or diagonal completely filled with crosses? Illustrate your answer.

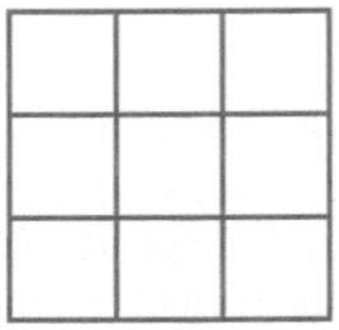

74. ***Determining Operations*** Place one of the arithmetic operations +, −, ×, or ÷ between each pair of successive numbers on the left side of this equation to make it true. Any operation may be used more than once or not at all. Use parentheses as necessary.

$$1 \quad 2 \quad 3 \quad 4 \quad 5 \quad 6 \quad 7 \quad 8 \quad 9 = 100$$

1.4 Calculating, Estimating, and Reading Graphs

Calculation • Estimation • Interpretation of Graphs

The photograph shows the **Sharp Elsimate EL-330M,** a typical four-function calculator.

Since the introduction of hand-held calculators in the early 1970s, the methods of everyday arithmetic have been drastically altered. One of the first consumer models available was the Texas Instruments SR-10, which sold for nearly \$150 in 1973. It could perform the four operations of arithmetic and take square roots, but could do very little more.

Calculation

The search for easier ways to calculate and compute has culminated in the development of hand-held calculators and computers. This text assumes that all students have access to calculators, allowing them to spend more time on the conceptual nature of mathematics and less time on computation with paper and pencil. For the general population, a calculator that performs the operations of arithmetic and a few other functions is sufficient. These are known as **four-function calculators.** Students who take higher mathematics courses (engineers, for example) usually need the added power of **scientific calculators. Graphing calculators,** which actually plot graphs on small screens, are also available. Remember the following.

> Always refer to your owner's manual if you need assistance in performing an operation with your calculator. If you need further help, ask your instructor or another student who is using the same model.

Graphing calculators have become the standard in the world of advanced hand-held calculators. One of the main advantages of a graphing calculator is that both the information the user inputs into the calculator and the result generated by that calculator can be viewed on the same screen. In this way, the user can verify that the information entered into the calculator is correct. Although it is not necessary to have a graphing calculator to study the material presented in this text, we occasionally include graphing calculator screens to support results obtained or to provide supplemental information.*

The screens that follow illustrate some common entries and operations.

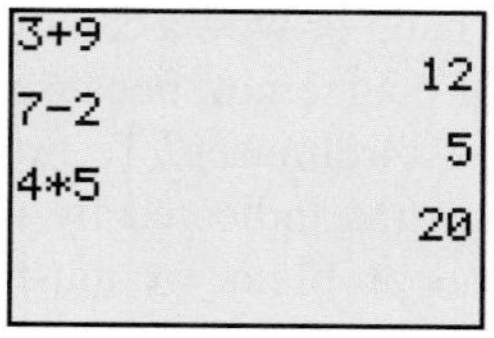

$3 + 9 = 12$
$7 - 2 = 5$
$4 \times 5 = 20$

A

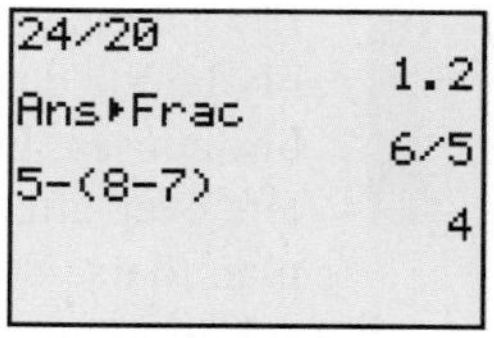

$\frac{24}{20} = 1.2$
$1.2 = \frac{6}{5}$
$5 - (8 - 7) = 4$

B

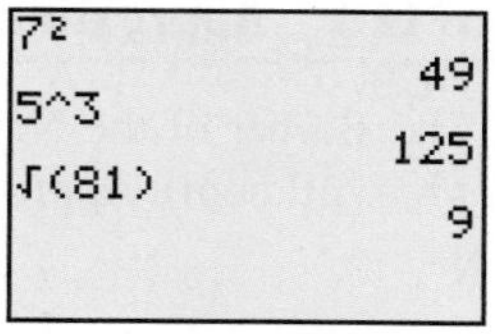

$7^2 = 49$
$5^3 = 125$
$\sqrt{81} = 9$

C

*Because they are the most popular models of graphing calculators, we include screens generated by TI-83 Plus and TI-84 Plus models from Texas Instruments.

Any calculator (particularly a graphing calculator) consists of two components: the electronic "box" and the owner's manual that explains how to use it. The **TI-84 Plus** graphing calculator is shown.

Screen A illustrates how two numbers can be added, subtracted, or multiplied. Screen B shows how two numbers can be divided, how the decimal quotient (stored in the memory cell Ans) can be converted into a fraction, and how parentheses can be used in a computation. Screen C shows how a number can be squared, how it can be cubed, and how its square root can be taken.

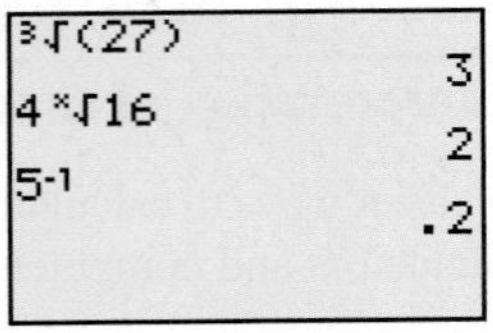

$\sqrt[3]{27} = 3$

$\sqrt[4]{16} = 2$

5^{-1} (or $\frac{1}{5}$) $= .2$

D

$\pi \approx 3.141592654$

5! (or $1 \times 2 \times 3 \times 4 \times 5$) $= 120$

$6{,}265{,}804 \times 8{,}980{,}591 \approx 5.627062301 \times 10^{13}$

$\approx$ indicates "is approximately equal to"

E

Screen D shows how other roots (cube root and fourth root) can be found, and how the reciprocal of a number can be found using -1 as an exponent. Screen E shows how π can be accessed with its own special key, how a *factorial* (as indicated by !) can be found and how a result might be displayed in *scientific notation.* (The "E13" following 5.627062301 means that this number is multiplied by 10^{13}. This answer is still only an approximation, because the product $6{,}265{,}804 \times 8{,}980{,}591$ contains more digits than the calculator can display.)

Estimation Although calculators can make life easier when it comes to computations, many times we need only estimate an answer to a problem, and in these cases a calculator may not be necessary or appropriate.

EXAMPLE 1 Estimating an Appropriate Number of Birdhouses

A birdhouse for swallows can accommodate up to 8 nests. How many birdhouses would be necessary to accommodate 58 nests?

SOLUTION

If we divide 58 by 8 either by hand or with a calculator, we get 7.25. Can this possibly be the desired number? Of course not, because we cannot consider fractions of birdhouses. Do we need 7 or 8 birdhouses? To provide nesting space for the nests left over after the 7 birdhouses (as indicated by the decimal fraction), we should plan to use 8 birdhouses. In this problem, we must round our answer *up* to the next counting number. ■

EXAMPLE 2 Approximating Average Number of Yards per Carry

In 2004, Tiki Barber of the New York Giants carried the football 322 times for 1518 yards (*Source:* nfl.com). Approximate his average number of yards per carry.

SOLUTION

Because we are are asked only to find Tiki's approximate average, we can say that he carried about 300 times for about 1500 yards, and his average was about $\frac{1500}{300} = 5$ yards per carry. (A calculator shows that his average to the nearest tenth was 4.7 yards per carry. Verify this.) ■

EXAMPLE 3 Comparing Proportions of Workers by Age Groups

In a recent year, there were approximately 127,000 males in the 25–29-year age bracket working on farms. This represented part of the total of 238,000 farm workers in that age bracket. Of the 331,000 farm workers in the 40–44-year age bracket, 160,000 were males. Without using a calculator, determine which age bracket had a larger proportion of males.

SOLUTION

Here, it is best to think in terms of thousands instead of dealing with all the zeros. First, let us analyze the age bracket 25–29 years. Because there were a total of 238 thousand workers, of which 127 thousand were males, there were $238 - 127 = 111$ thousand female workers. Here, more than half of the workers were males. In the 40–44-year age bracket, of the 331 thousand workers, there were 160 thousand males, giving $331 - 160 = 171$ thousand females, meaning fewer than half were males. A comparison, then, shows that the 25–29-year age bracket had the larger proportion of males.

Interpretation of Graphs

Using graphs is an efficient means of transmitting information in a concise way. Any issue of the newspaper *USA Today* will verify this. *Circle graphs* or *pie charts, bar graphs*, and *line graphs* are the most common.

A **circle graph** or **pie chart** is used to give a pictorial representation of data. A circle is used to indicate the total of all the categories represented. The circle is divided into sectors, or wedges (like pieces of a pie), whose sizes show the relative magnitudes of the categories. The sum of all the fractional parts must be 1 (for 1 whole circle).

EXAMPLE 4 Interpreting Information in a Circle Graph

Use the circle graph in Figure 9 to determine how much of the amount spent for a \$3.50 gallon of gasoline in California goes to refinery margin and to crude oil cost.

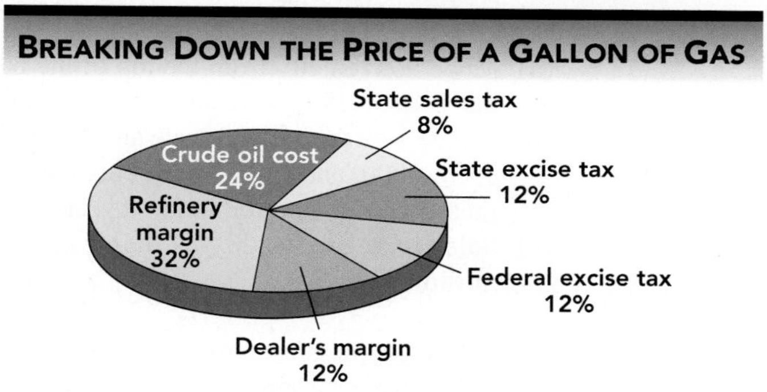

Source: California Energy Commission.

FIGURE 9

SOLUTION

The sectors are sized to match how the price is divided. For example, most of the price (32%) goes to the refinery, while the least portion (8%) goes for state sales tax. As expected, the percents total 100%. If the price of gasoline is \$3.50 per gallon,

Refinery margin: $\$3.50 \times \underbrace{.32}_{\text{32\% converted to a decimal}} = \$1.12$ Crude oil cost: $\$3.50 \times \underbrace{.24}_{\text{24\% converted to a decimal}} = \$.84.$

A **bar graph** is used to show comparisons. We illustrate with a bar graph where we must estimate the heights of the bars.

EXAMPLE 5 Interpreting Information in a Bar Graph

The bar graph in Figure 10 shows sales in millions of dollars for CarMax Auto Superstores, Inc. The graph compares sales for 5 years.

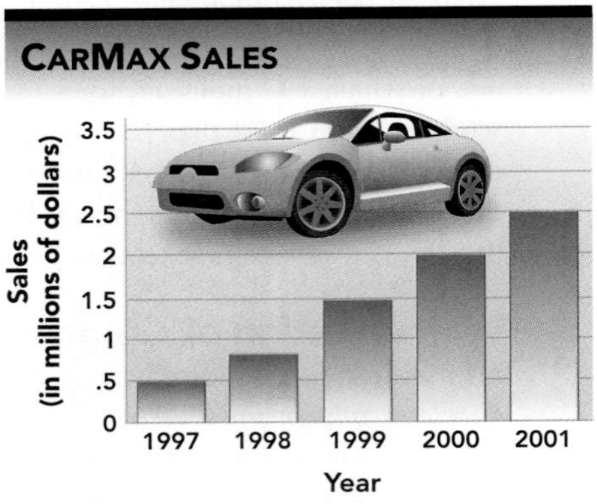

Source: Circuit City CarMax Group.

FIGURE 10

(a) Estimate sales in 1998.
(b) In what years were sales greater than $1 million?
(c) As the years progress, describe the change in sales.

SOLUTION

(a) Move horizontally from the top of the bar for 1998 to the scale on the left to see that sales in 1998 were about $.8 million.
(b) Locate 1 on the vertical scale and follow the line across to the right. Three years—1999, 2000, and 2001—have bars that extend above the line for 1, so sales were greater than $1 million in those years.
(c) Sales increase steadily as the years progress, from about $.5 million to $2.5 million. ■

A **line graph** is used to show changes or trends in data over time. To form a line graph, we connect a series of points representing data with line segments.

EXAMPLE 6 Interpreting Information in a Line Graph

The line graph in Figure 11 shows average prices for all types of gasoline in the U.S. for the years 1999 through 2004.

(a) In which years shown did the average price decrease from the previous year?
(b) What was the general trend in gasoline prices from 1999 to 2004?
(c) Estimate the average prices for 2002 and 2004. About how much did gasoline price rise from 2002 to 2004?

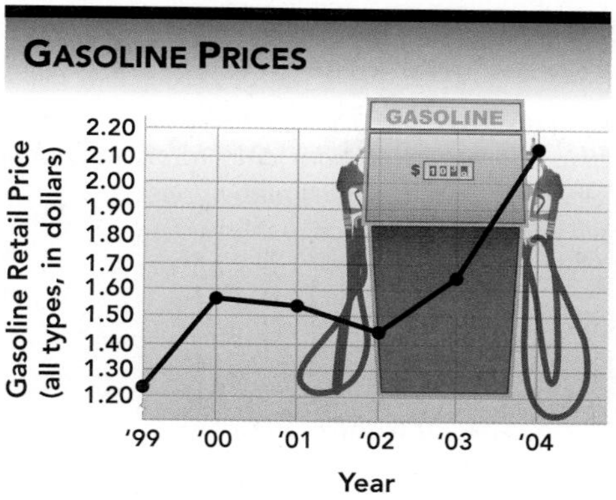

Source: Energy Information Administration.

FIGURE 11

SOLUTION

(a) The line segments joining the points for the years 2000, 2001, and 2002 fall from left to right. This indicates that average prices decreased from the previous years in 2001 and 2002.

(b) Although the prices fell in 2001 and 2002, the general trend is that prices rose from 1999 to 2004, as indicated by the overall rise of the line graph from left to right.

(c) It appears that in 2002 the average price was about \$1.44 and in 2004 about \$2.12. Thus, the price rose about $\$2.12 - \$1.44 = \$.68$ per gallon.

For Further Thought

Are You "Numerate"?

Letter is to *number* as *literacy* is to *numeracy*. Much has been written about how important it is that the general population be "numerate." The essay "Quantity" by James T. Fey in *On the Shoulders of Giants: New Approaches to Numeracy* contains this description of an approach to numeracy.

> Given the fundamental role of quantitative reasoning in applications of mathematics as well as the innate human attraction to numbers, it is not surprising that number concepts and skills form the core of school mathematics. In the earliest grades all children start on a mathematical path designed to develop computational procedures of arithmetic together with corresponding conceptual understanding that is required to solve quantitative problems and make informed decisions. Children learn many ways to describe quantitative data and relationships using numerical, graphic, and symbolic representations; to plan arithmetic and algebraic operations and to execute those plans using effective procedures; and to interpret quantitative information, to draw inferences, and to test the conclusions for reasonableness.

For Group Discussion or Individual Investigation

With calculator in hand, fill in the boxes with the digits 3, 4, 5, 6, 7, or 8, using each digit at most once. See how close you can come to the "goal number." You are allowed 1 minute per round. Good luck!

Round I $\square \times \square\square\square\square = 30{,}000$

Round II $\square \times \square\square\square\square = 40{,}000$

Round III $\square \times \square\square\square\square = 50{,}000$

Round IV $\square\square \times \square\square\square\square = 30{,}000$

Round V $\square\square \times \square\square\square\square = 60{,}000$

1.4 EXERCISES

Exercises 1–20 are designed to give you practice in learning how to do some basic operations on your calculator. Perform the indicated operations and give as many digits in your answer as shown on your calculator display. (The number of displayed digits may vary depending on the model used.)

1. $39.7 + (8.2 - 4.1)$

2. $2.8 \times (3.2 - 1.1)$

3. $\sqrt{5.56440921}$

4. $\sqrt{37.38711025}$

5. $\sqrt[3]{418.508992}$

6. $\sqrt[3]{700.227072}$

7. 2.67^2

8. 3.49^3

9. 5.76^5

10. 1.486^6

11. $\dfrac{14.32 - 8.1}{2 \times 3.11}$

12. $\dfrac{12.3 + 18.276}{3 \times 1.04}$

13. $\sqrt[5]{1.35}$

14. $\sqrt[6]{3.21}$

15. $\dfrac{\pi}{\sqrt{2}}$

16. $\dfrac{2\pi}{\sqrt{3}}$

17. $\sqrt[4]{\dfrac{2143}{22}}$

18. $\dfrac{12{,}345{,}679 \times 72}{\sqrt[3]{27}}$

19. $\dfrac{\sqrt{2}}{\sqrt[3]{6}}$

20. $\dfrac{\sqrt[3]{12}}{\sqrt{3}}$

21. Choose any number consisting of five digits. Multiply it by 9 on your calculator. Now add the digits in the answer. If the sum is more than 9, add the digits of this sum, and repeat until the sum is less than 10. Your answer will always be 9. Repeat the exercise with a number consisting of six digits. Does the same result hold?

22. Use your calculator to *square* the following two-digit numbers ending in 5: 15, 25, 35, 45, 55, 65, 75, 85. Write down your results, and examine the pattern that develops. Then use inductive reasoning to predict the value of 95^2. Write an explanation of how you can mentally square a two-digit number ending in 5.

By examining several similar computation problems and their answers obtained on a calculator, we can use inductive reasoning to make conjectures about certain rules, laws, properties, and definitions in mathematics. Perform each calculation and observe the answers. Then fill in the blank with the appropriate response. (Justification of these results will be discussed later in the book.)

23. $(-3) \times (-8)$; $(-5) \times (-4)$; $(-2.7) \times (-4.3)$
Multiplying a negative number by another negative number gives a ____________ product.
(negative/positive)

24. $5 \times (-4)$; -3×8; $2.7 \times (-4.3)$
Multiplying a negative number by a positive number gives a ____________ product.
(negative/positive)

25. 5.6^0; π^0; 2^0; 120^0; $.5^0$
Raising a nonzero number to the power 0 gives a result of ______.

26. 1^2; 1^3; 1^{-3}; 1^0; 1^{13}
Raising 1 to any power gives a result of ______.

27. $1/7$; $1/(-9)$; $1/3$; $1/(-8)$
The sign of the reciprocal of a number is ____________ the sign of the number.
(the same as/different from)

28. $5/0$; $9/0$; $\pi/0$; $-3/0$; $0/0$
Dividing a number by 0 gives a(n) ____________ on a calculator.

29. $0/8$; $0/2$; $0/(-3)$; $0/\pi$
Zero divided by a nonzero number gives a quotient of ______.

30. $(-3) \times (-4) \times (-5)$; $(-3) \times (-4) \times (-5) \times (-6) \times (-7)$; $(-3) \times (-4) \times (-5) \times (-6) \times (-7) \times (-8) \times (-9)$
Multiplying an *odd* number of negative numbers gives a ____________ product.
(positive/negative)

31. $(-3) \times (-4)$; $(-3) \times (-4) \times (-5) \times (-6)$; $(-3) \times (-4) \times (-5) \times (-6) \times (-7) \times (-8)$
Multiplying an *even* number of negative numbers gives a ________ product.
(positive/negative)

32. $\sqrt{-3}$; $\sqrt{-5}$; $\sqrt{-6}$; $\sqrt{-10}$
Taking the square root of a negative number gives a(n) ____________ on a calculator.

33. Find the decimal representation of $1/6$ on your calculator. Following the decimal point will be a 1 and a string of 6s. The final digit will be a 7 if your calculator *rounds off* or a 6 if it *truncates*. Which kind of calculator do you have?

34. Choose any three-digit number and enter the digits into a calculator. Then enter them again to get a six-digit number. Divide this six-digit number by 7. Divide the result by 13. Divide the result by 11. What is your answer? Explain why this happens.

35. Choose any digit except 0. Multiply it by 429. Now multiply the result by 259. What is your answer? Explain why this happens.

36. Choose two natural numbers. Add 1 to the second and divide by the first to get a third. Add 1 to the third and divide by the second to get a fourth. Add 1 to the fourth and divide by the third to get a fifth. Continue this process until you discover a pattern. What is the pattern?

When a four-function or scientific calculator (not a graphing calculator, however) is turned upside down, the digits in the display correspond to letters of the English alphabet as follows:

$0 \leftrightarrow$ O	$3 \leftrightarrow$ E	$7 \leftrightarrow$ L
$1 \leftrightarrow$ I	$4 \leftrightarrow$ h	$8 \leftrightarrow$ B
$2 \leftrightarrow$ Z	$5 \leftrightarrow$ S	$9 \leftrightarrow$ G.

Perform the indicated calculation on a four-function or scientific calculator. Then turn your calculator upside down to read the word that belongs in the blank in the accompanying sentence.

37. $(100 \div 20) \times 14{,}215{,}469$
I filled my tank with gasoline from the __________ station.

38. $\dfrac{10 \times 10{,}609}{\sqrt{4}}$
"It's got to be the ________."

39. $60^2 - \dfrac{368}{4}$
The electronics manufacturer _______ produces the Wave Radio.

40. $187^2 + \sqrt{1600}$
Have you ever read *Mother* _________ nursery rhymes?

41. Make up your own exercise similar to Exercises 37–40.

42. Displayed digits on some calculators show some or all of the parts in the pattern as in the figure at the top of the next column. For the digits 0 through 9:
(a) Which part is used most frequently?
(b) Which part is used the least?

(c) Which digit uses the most parts?
(d) Which digit uses the fewest parts?

Give an appropriate counting number answer to each question in Exercises 43–46. (Find the least counting number that will work.)

43. ***Pages to Store Trading Cards*** A plastic page designed to hold trading cards will hold up to 9 cards. How many pages will be needed to store 563 cards?

44. ***Drawers for Videocassettes*** A sliding drawer designed to hold videocassettes has 20 compartments. If Chris wants to house his collection of 408 Disney videotapes, how many such drawers will he need?

45. ***Containers for African Violets*** A gardener wants to fertilize 800 African violets. Each container of fertilizer will supply up to 60 plants. How many containers will she need to do the job?

46. ***Fifth-Grade Teachers Needed*** False River Academy has 155 fifth-grade students. The principal, Butch LeBeau, has decided that each fifth-grade teacher should have a maximum of 24 students. How many fifth-grade teachers does he need?

In Exercises 47–52, use estimation to determine the choice closest to the correct answer.

47. ***Price per Acre of Land*** To build a "millennium clock" on Mount Washington in Nevada that would tick once each year, chime once each century, and last at least 10,000 years, the nonprofit Long Now Foundation

purchased 80 acres of land for $140,000. Which one of the following is the closest estimate to the price per acre?
A. $1000 **B.** $2000 **C.** $4000 **D.** $11,200

48. ***Time of a Round Trip*** The distance from Seattle, Washington, to Sprinfield, Missouri, is 2009 miles. About how many hours would a roundtrip from Seattle to Springfield and back take a bus that averages 50 miles per hour for the entire trip?
A. 60 **B.** 70 **C.** 80 **D.** 90

49. ***People per Square Mile*** Buffalo County in Nebraska has a population of 40,249 and covers 968 square miles. About how many people per square mile live in Buffalo County?
A. 40 **B.** 400 **C.** 4000 **D.** 40,000

50. ***Revolutions of Mercury*** The planet Mercury takes 88.0 Earth days to revolve around the sun once. Pluto takes 90,824.2 days to do the same. When Pluto has revolved around the sun once, about how many times will Mercury have revolved around the sun?
A. 100,000 **B.** 10,000 **C.** 1000 **D.** 100

51. ***Rushing Average*** In 2004, Muhsin Muhammad of the Carolina Panthers caught 93 passes for 1405 yards. His approximate number of yards gained per catch was ________.
A. $\frac{9}{14}$ **B.** 140 **C.** 1.4 **D.** 14

52. ***Area of the Sistine Chapel*** The Sistine Chapel in Vatican City measures 40.5 meters by 13.5 meters.

Which is the closest approximation to its area?
A. 110 meters **B.** 55 meters
C. 110 square meters **D.** 600 square meters

The circle graph at the top of the next column shows the approximate percent of immigrants admitted into the United States during the 1990s. Use the graph to answer the questions in Exercises 53–56.

53. What percent of the immigrants were from the "Other" group of countries?

U.S. Immigrants by Region of Birth

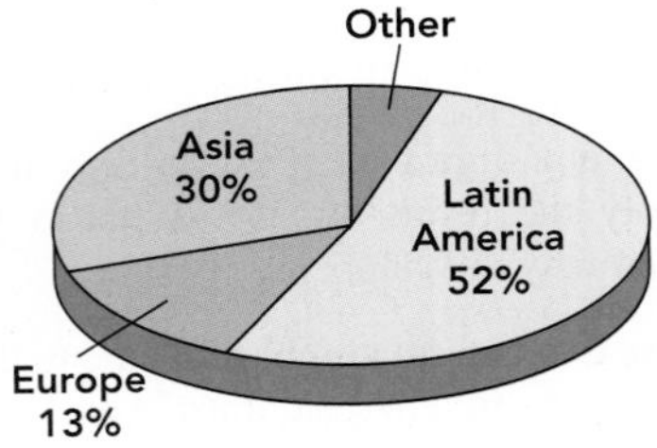

Source: U.S. Bureau of the Census.

54. What percent of the immigrants were not from Asia?

55. In a group of 2,000,000 immigrants, how many would you expect to be from Europe?

56. In a group of 4,000,000 immigrants, how many more would there be from Latin America than all the other regions combined?

The bar graph shows the amount of personal savings, in billions of dollars, accumulated during the years 1997 through 2001. Use the graph to answer the questions in Exercises 57–60.

Personal Savings in the United States

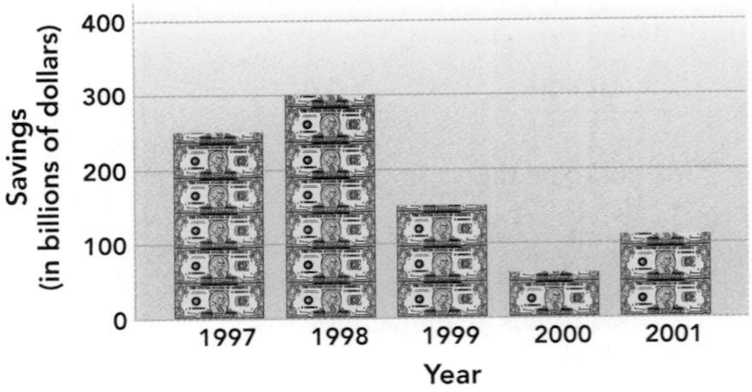

Source: U.S. Bureau of Economic Analysis.

57. Which year had the greatest amount of savings? Which had the least?

58. Which years had amounts greater than $200 billion?

59. Approximately how much was the amount for 1997? for 1998?

60. Approximately how much more was saved in 1998 than 1997?

The line graph indicates that current projections for Medicare funding will not cover its costs unless the program changes. Use the graph to answer the questions in Exercises 61–64.

61. Which is the only period in which Medicare funds are predicted to increase?

62. By approximately how much will the funds decrease between the years 2005 and 2006?

63. How do the amounts for 2004 and 2007 compare?

64. In which year will funds first show a deficit?

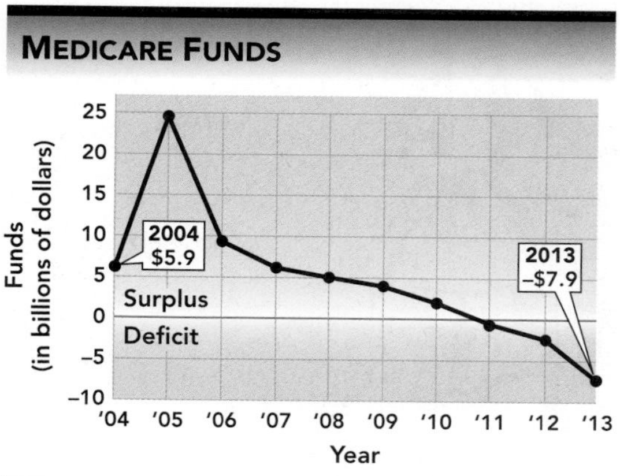

Source: Centers for Medicare and Medicaid Services.

EXTENSION
Using Writing to Learn About Mathematics

Research has indicated that the ability to express mathematical observations in writing can serve as a positive force in one's continued development as a mathematics student. The implementation of writing in the mathematics class can use several approaches.

Journals One way of using writing in mathematics is to keep a journal in which you spend a few minutes explaining what happened in class that day. The journal entries may be general or specific, depending on the topic covered, the degree to which you understand the topic, your interest level at the time, and so on. Journal entries are usually written in informal language, and are often an effective means of communicating to yourself, your classmates, and your instructor what feelings, perceptions, and concerns you are having at the time.

Learning Logs Although journal entries are for the most part unstructured writings in which the student's thoughts are allowed to roam freely, entries in learning logs are typically more structured. An instructor may pose a specific question for a student to answer in a learning log. In this text, we intersperse writing exercises in each exercise set that are appropriate for answering in a learning log. For example, consider Exercise 13 in the exercise set for the opening section in this chapter.

Discuss the differences between inductive and deductive reasoning. Give an example of each.

(continued)

Mathematical writing takes many forms. One of the most famous author/mathematicians was **Charles Dodgson** (1832–1898), who used the pen name **Lewis Carroll.**

Dodgson was a mathematics lecturer at Oxford University in England. Queen Victoria told Dodgson how much she enjoyed *Alice's Adventures in Wonderland* and how much she wanted to read his next book; he is said to have sent her *Symbolic Logic*, his most famous mathematical work.

The *Alice* books made Carroll famous. Late in life, however, Dodgson shunned attention and denied that he and Carroll were the same person, even though he gave away hundreds of signed copies to children and children's hospitals.

Here is a possible response to this exercise.

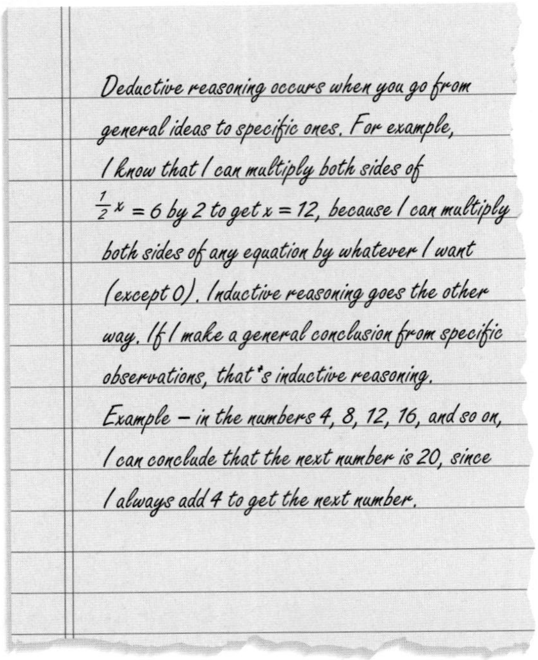

Deductive reasoning occurs when you go from general ideas to specific ones. For example, I know that I can multiply both sides of $\frac{1}{2}x = 6$ by 2 to get $x = 12$, because I can multiply both sides of any equation by whatever I want (except 0). Inductive reasoning goes the other way. If I make a general conclusion from specific observations, that's inductive reasoning. Example – in the numbers 4, 8, 12, 16, and so on, I can conclude that the next number is 20, since I always add 4 to get the next number.

Reports on Articles from Mathematics Publications The motto "Publish or perish" has long been around, implying that a scholar in pursuit of an academic position must publish in a journal in his or her field. There are numerous journals that publish papers in mathematics research and/or mathematics education. In Activity 3 at the end of this section, we provide some suggestions of articles that have appeared within the last few years. A report on such an article can help you understand what mathematicians do and what ideas mathematics teachers use to convey concepts to their students.

Term Papers Professors in mathematics survey courses are, in increasing numbers, requiring short term papers of their students. In this way, you can become aware of the plethora of books and articles on mathematics and mathematicians, many written specifically for the layperson. In Activities 5 and 6 at the end of this section, we provide a list of possible term paper topics.

EXTENSION ACTIVITIES

Rather than include a typical exercise set, we list some suggested activities in which writing can be used to enhance awareness and learning of mathematics.

Activity 1 Keep a journal. After each class, write for a few minutes on your perceptions about the class, the topics covered, or whatever you feel is appropriate. You may want to use the following guidelines.

Journal Writing*

1. *WHO should write in your journal?* You should.
2. *WHAT should you write in your journal?* New words, ideas, formulas, or concepts; profound thoughts; wonderings, musings, problems to solve; reflections on the class; questions—both answerable and unanswerable; writing ideas
3. *WHEN should you write in your journal?* After class each day; as you are preparing, reading, or studying for class; anytime an insight or question hits you.
4. *WHERE should you write in your journal?* Anywhere—so keep it with you when possible.
5. *WHY should you write in your journal?* It will help you record ideas that you might otherwise forget. It will be worthwhile for you to read later on so that you can note your growth. It will facilitate your learning, problem solving, writing, reading, and discussion in class.
6. *HOW should you write in your journal?* In wonderful, long, flowing sentences with perfect punctuation and perfect spelling and in perfect handwriting; or in single words that express your ideas, in short phrases, in sketches, in numbers, in maps, in diagrams, in sentences. (You may even prefer to organize your journal entries on your desktop, notebook, or palmtop computer.)

Activity 2 Keep a learning log, answering at least one writing exercise from each exercise set covered in your class syllabus. Ask your teacher for suggestions of other types of specific writing assignments. For example, you might want to choose a numbered example from a section in the text and write your own solution to the problem, or comment on the method that the authors use to solve the problem. Don't be afraid to be critical of the method used in the text.

Activity 3 The National Council of Teachers of Mathematics publishes journals in mathematics education: *Teaching Children Mathematics* (formerly called *Arithmetic Teacher*) and *Mathematics Teacher* are two of them. These journals can be found in the periodicals section of most college and university libraries. We have chosen several recent articles in each of these journals. There are thousands of other articles from which to choose. Write a short report on one of these articles according to guidelines specified by your instructor.

From *Mathematics Teacher*

2001

Johnson, Craig M. "Functions of Number Theory in Music." Vol. 94, No. 8, November 2001, p. 700.

Lightner, James E. "Mathematics Didn't Just Happen." Vol. 94, No. 9, December 2001, p. 780.

*"Journal Writing" from "No Time for Writing in Your Class?" by Margaret E. McIntosh in *Mathematics Teacher,* September 1991, p. 431. Reprinted by permission.

(continued)

McNeill, Sheila A. "The Mayan Zeros." Vol. 94, No. 7, October 2001, p. 590.

Socha, Susan. "Less Is Sometimes More." Vol. 94, No. 6, September 2001, p. 450.

2002

Houser, Don. "Roots in Music." Vol. 95, No. 1, January 2002, p. 16.

Howe, Roger. "Hermione Granger's Solution." Vol. 95, No. 2, February 2002, p. 86.

Kolpas, Sidney J. "Let Your Fingers Do the Multiplying." Vol. 95, No. 4, April 2002, p. 246.

Van Dresar, Vickie J. "Opening Young Minds to Closure Properties." Vol. 95, No. 5, May 2002, p. 326.

2003

McDaniel, Michael. "Not Just Another Theorem: A Cultural and Historical Event." Vol. 96, No. 4, April 2003, p. 282.

Nelson, Joanne E., Margaret Coffey, and Edie Huffman. "Stop This Runaway Truck, Please." Vol. 96, No. 8, November 2003, p. 548.

Roberts, David L., and Angela L. E. Walmsley. "The Original New Math: Storytelling versus History." Vol. 96, No. 7, October 2003, p. 468.

Yoshinobu, Stan T. "Mathematics, Politics, and Greenhouse Gas Intensity: An Example of Using Polya's Problem-Solving Strategy." Vol. 96, No. 9, December 2003, p. 646.

2004

Devaney, Robert L. "Fractal Patterns and Chaos Games." Vol. 98, No. 4, November 2004, p. 228.

Francis, Richard L. "New Worlds to Conquer." Vol. 98, No. 3, October 2004, p. 166.

Hansen, Will. "War and Pieces." Vol. 98, No. 2, September 2004, p. 70.

Mahoney, John F. "How Many Votes Are Needed to Be Elected President?" Vol. 98, No. 3, October 2004, p. 154.

From *Teaching Children Mathematics*

2001

Karp, Karen S., and E. Todd Brown. "Geo-Dolls: Traveling in a Mathematical World." Vol. 8, No. 3, November 2001, p. 132.

Randolph, Tamela D., and Helene J. Sherman. "Alternative Algorithms: Increasing Options, Reducing Errors." Vol. 7, No. 8, April 2001, p. 480.

Sun, Wei, and Joanne Y. Zhang. "Teaching Addition and Subtraction Facts: A Chinese Perspective." Vol. 8, No. 1, September 2001, p. 28.

Whitenack, Joy W., et. al. "Second Graders Circumvent Addition and Subtraction Difficulties." Vol. 8, No. 4, December 2001, p. 228.

2002

Agosto, Melinda. "Cool Mathematics for Kids." Vol. 8, No. 7, March 2002, p. 397.

Huniker, DeAnn. "Calculators as Learning Tools for Young Children's Explorations of Number." Vol. 8, No. 6, February 2002, p. 316.

Strutchens, Marilyn E. "Multicultural Literature as a Context for Problem Solving: Children and Parents Learning Together." Vol. 8, No. 8, April 2002, p. 448.

Whitin, David J. "The Potentials and Pitfalls of Integrating Literature into the Mathematics Program." Vol. 8, No. 9, May 2002, p. 503.

2003

Arvold, Bridget, Gina Stone, and Lynn Carter. "What Do You Get When You Cross a Math Professor and a Body Builder?" Vol. 9, No. 7, March 2003, p. 408.

Edelson, R. Jill, and Gretchen L. Johnson. "Integrating Music and Mathematics in the Elementary Classroom." Vol. 9, No. 8, April 2003, p. 474.

Phillips, Linda J. "When Flash Cards Are Not Enough." Vol. 9, No. 6, February 2003, p. 358.

Uy, Frederick L. "The Chinese Numeration System and Place Value." Vol. 9, No. 5, January 2003, p. 243.

2004

Anthony, Glenda J., and Margaret A. Walshaw. "Zero: A 'None' Number?" Vol. 11, No. 1, August 2004, p. 38.

Buschman, Larry. "Teaching Problem Solving in Mathematics." Vol. 10, No. 6, February 2004, p. 302.

Joram, Elana, Christina Hartman, and Paul R. Trafton. "'As People Get Older, They Get Taller': An Integrated Unit on Measurement, Linear Relationships, and Data Analysis." Vol. 10, No. 7, March 2004, p. 344.

Mann, Rebecca L. "Balancing Act: The Truth Behind the Equals Sign." Vol. 11, No. 2, September 2004, p. 65.

Activity 4 One of the most popular mathematical films of all time is *Donald in Mathmagic Land*, produced by Disney in 1959. Spend an entertaining half-hour watching this film, and write a report on it according to the guidelines of your instructor.

Activity 5 Write a report according to the guidelines of your instructor on one of the following mathematicians, philosophers, and scientists.

Abel, N.
Agnesi, M. G.
Agnesi, M. T.
Al-Khowârizmi
Apollonius
Archimedes
Aristotle
Babbage, C.
Bernoulli, Jakob
Bernoulli, Johann
Cantor, G.
Cardano, G.
Copernicus, N.
De Morgan, A.
Descartes, R.
Euler, L.
Fermat, P.
Fibonacci (Leonardo of Pisa)
Galileo (Galileo Galilei)
Galois, E.
Gauss, C.
Hilbert, D.
Kepler, J.
Kronecker, L.
Lagrange, J.
Leibniz, G.
L'Hospital, G.
Lobachevsky, N.
Mandelbrot, B.
Napier, J.
Nash, J.
Newton, I.
Noether, E.
Pascal, B.
Plato
Polya, G.
Pythagoras
Ramanujan, S.
Riemann, G.
Russell, B.
Somerville, M.
Tartaglia, N.
Whitehead, A.
Wiles, A.

(continued)

Activity 6 Write a term paper on one of the following topics in mathematics according to the guidelines of your instructor.

Babylonian mathematics
Egyptian mathematics
The origin of zero
Plimpton 322
The Rhind papyrus
Origins of the Pythagorean theorem
The regular (Platonic) solids
The Pythagorean brotherhood
The Golden Ratio (Golden Section)
The three famous construction problems of the Greeks
The history of the approximations of π
Euclid and his "Elements"
Early Chinese mathematics
Early Hindu mathematics
Origin of the word *algebra*
Magic squares
Figurate numbers
The Fibonacci sequence
The Cardano/Tartaglia controversy
Historical methods of computation (logarithms, the abacus, Napier's rods, the slide rule, etc.)
Pascal's triangle
The origins of probability theory
Women in mathematics
Mathematical paradoxes
Unsolved problems in mathematics
The four color theorem
The proof of Fermat's Last Theorem
The search for large primes
Fractal geometry
The co-inventors of calculus
The role of the computer in the study of mathematics
Mathematics and music
Police mathematics
The origins of complex numbers
Goldbach's conjecture
The use of the Internet in mathematics education
The development of graphing calculators
Mathematics education reform movement
Multicultural mathematics
The Riemann Hypothesis

Activity 7 Investigate a computer program that focuses on teaching children elementary mathematics and write a critical review of it as if you were writing for a journal that contains software reviews of educational material. Be sure to address the higher-level thinking skills in addition to drill and practice.

Activity 8 The following Web sites provide a fascinating list of mathematics-related topics. Go to one of them, choose a topic that interests you, and report on it, according to the guidelines of your instructor.

www.mathworld.wolfram.com

www.world.std.com/~reinhold/mathmovies.html

www.mcs.surrey.ac.uk/Personal/R.Knott/

www.dir.yahoo.com/Science/Mathematics/

www.cut-the-knot.com/

www.ics.uci.edu/~eppstein/recmath.html

Activity 9 A theme of mathematics-related scenes in movies and television is found throughout this book. Prepare a report on one or more such scenes, and determine whether the mathematics involved is correct or incorrect. If correct, show why; if incorrect, find the correct answer. The Website

www.world.std.com/~reinhold/mathmovies.html

provides a wealth of information on mathematics in the movies.

Activity 10 The longest running animated television show is *The Simpsons*, having begun in 1989. The Website

www.simpsonsmath.com

explores the occurrence of mathematics in the episodes on a season-by-season basis. Watch several episodes and elaborate on the mathematics found in them.

COLLABORATIVE INVESTIGATION

Discovering Patterns in Pascal's Triangle

One fascinating array of numbers, *Pascal's triangle,* consists of rows of numbers, each of which contains one more entry than the one before. The first five rows are shown here.

$$
\begin{array}{ccccccccc}
 & & & & 1 & & & & \\
 & & & 1 & & 1 & & & \\
 & & 1 & & 2 & & 1 & & \\
 & 1 & & 3 & & 3 & & 1 & \\
1 & & 4 & & 6 & & 4 & & 1
\end{array}
$$

To discover some of its patterns, divide the class into groups of four students each. Within each group designate one student as A, one as B, one as C, and one as D. Then perform the following activities in order.

1. Discuss among group members some of the properties of the triangle that are obvious from observing the first five rows shown.

2. It is fairly obvious that each row begins and ends with 1. Discover a method whereby the other entries in a row can be determined from the entries in the row immediately above it. (*Hint*: In the fifth row, $6 = 3 + 3$.) Then, as a group, find the next four rows of the triangle, and have each member prepare his or her own copy of the entire first nine rows for later reference.

3. Now each student in the group will investigate a particular property of the triangle. In some cases, a calculator will be helpful. All students should begin working at the same time. (A discussion follows.)
   Student A: Find the sum of the entries in each row. Notice the pattern that emerges. Now write the tenth row of the triangle.
   Student B: Investigate the successive differences in the diagonals from upper left to lower right. For example, in the diagonal that begins 1, 2, 3, 4, . . . , the successive differences are all 1; in the diagonal that begins 1, 3, 6, . . . , the successive differences are 2, 3, 4, and so on. Do this up through the diagonal that begins 1, 6, 21, ...
   Student C: Find the values of the first five powers of the number 11, starting with 11^0 (recall $11^0 = 1$).

Student D: Arrange the nine rows of the triangle with all rows "flush left," and then draw lightly dashed arrows as shown:

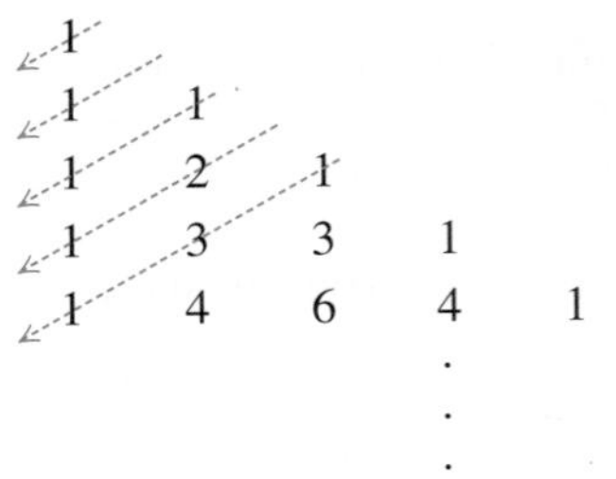

and so on. Then add along the diagonals. Write these sums in order from left to right.

4. After all students have concluded their individual investigations in Item 3, return to a group discussion.
   (a) Have student A report the result found in Item 3, and then make a prediction concerning the sum of the entries in the tenth row.
   (b) Have student B report the successive differences discovered in the diagonals. Then have all students in the group investigate the successive differences in the diagonal that begins 1, 7, 28. . . . (It may be necessary to write a few more rows of the triangle.)
   (c) Have student C report the relationship between the powers of 11 found, and then determine the value of 11^5. Why does the pattern not continue here?
   (d) Have student D report the sequence of numbers found. Then, as a group, predict what the next sum will be by observing the pattern in the sequence. Confirm your prediction by actual computation.
5. Choose a representative from each group to report to the entire class the observations made throughout this investigation.
6. Find a reference to Pascal's triangle using a search engine of the Internet and prepare a report on the reference.

CHAPTER 1 TEST

In Exercises 1 and 2, decide whether the reasoning involved is an example of inductive or deductive reasoning.

1. Jane Fleming is a sales representative for a publishing company. For the past 14 years, she has exceeded her annual sales goal, primarily by selling mathematics textbooks. Therefore, she will also exceed her annual sales goal this year.
2. For all natural numbers n, n^2 is also a natural number. 101 is a natural number. Therefore, 101^2 is a natural number.
3. What are the fourth and fifth numbers in this sequence?

   1, 4, 27, _____, _____, 46656, . . .

   (From *Mathematics Teacher* monthly calendar, April 25, 1994)
4. Use the list of equations and inductive reasoning to predict the next equation, and then verify your conjecture.

$$65{,}359{,}477{,}124{,}183 \times 17 = 1{,}111{,}111{,}111{,}111{,}111$$
$$65{,}359{,}477{,}124{,}183 \times 34 = 2{,}222{,}222{,}222{,}222{,}222$$
$$65{,}359{,}477{,}124{,}183 \times 51 = 3{,}333{,}333{,}333{,}333{,}333$$

5. Use the method of successive differences to find the next term in the sequence

   3, 11, 31, 69, 131, 223, ...
6. Find the sum $1 + 2 + 3 + \cdots + 250$.
7. Consider the following equations, where the left side of each is an octagonal number.

$$1 = 1$$
$$8 = 1 + 7$$
$$21 = 1 + 7 + 13$$
$$40 = 1 + 7 + 13 + 19$$

   Use the pattern established on the right sides to predict the next octagonal number. What is the next equation in the list?

8. Use the result of Exercise 7 and the method of successive differences to find the first eight octagonal numbers. Then divide each by 4 and record the remainder. What is the pattern obtained?

9. Describe the pattern used to obtain the terms of the Fibonacci sequence 1, 1, 2, 3, 5, 8, 13, 21, ...

Use problem-solving strategies to solve each problem, taken from the date indicated in the monthly calendar of Mathematics Teacher.

10. ***Building a Fraction*** Each of the four digits 2, 4, 6, and 9 is placed in one of the boxes to form a fraction. The numerator and the denominator are both two-digit whole numbers. What is the smallest value of all the common fractions that can be formed? Express your answer as a common fraction. (November 17, 2004)

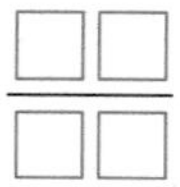

11. ***Units Digit of a Power of 9*** What is the units digit (ones digit) in the decimal representation of 9^{1997}? (January 27, 1997)

12. ***Counting Puzzle (Triangles)*** How many triangles are in this figure? (January 6, 2000)

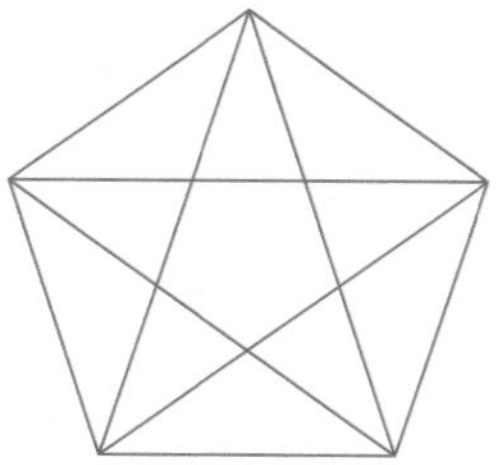

13. ***Devising a Correct Addition Problem*** Can you put the digits 1 through 9, each used once, in the boxes of the problem below to make an addition problem that has carrying and that is correct? If so, find a solution. If not, explain why no solution exists. (April 10, 2002)

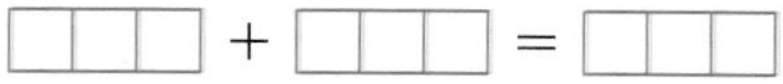

14. ***Missing Pages in a Newspaper*** A sixty-page newspaper, which consists of only one section, has the sheet with page 7 missing. What other pages are missing? (February 6, 1998)

15. ***Units Digit of a Sum*** Find the units digit (ones digit) of the decimal numeral representing the number $11^{11} + 14^{14} + 16^{16}$. (February 14, 1994)

16. Based on your knowledge of elementary arithmetic, describe the pattern that can be observed when the following operations are performed: 9×1, 9×2, $9 \times 3, \ldots, 9 \times 9$. (*Hint:* Add the digits in the answers. What do you notice?)

Use your calculator to evaluate each of the following. Give as many decimal places as the calculator displays.

17. $\sqrt{98.16}$

18. 3.25^3

19. ***Basketball Scoring Results*** During her NCAA women's basketball career, Seimone Augustus of LSU made 800 of her 1488 field goal attempts. This means that for every 15 attempts, she made approximately ________ of them.

A. 4 **B.** 8 **C.** 6 **D.** 2

20. ***Women in Mathematics*** The accompanying graph shows the number of women in mathematics or computer science professions during the past three decades.

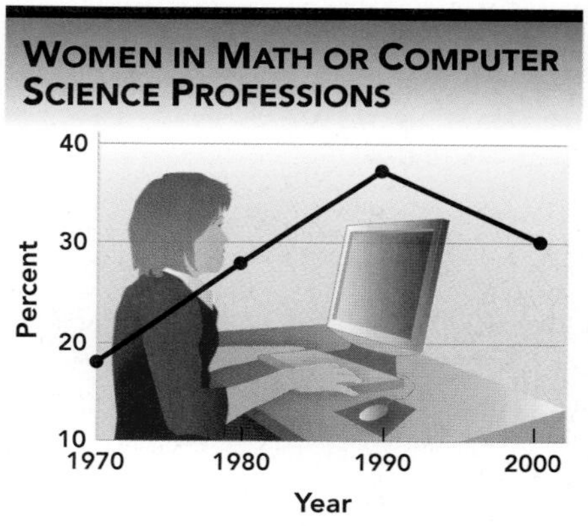

Source: U.S. Bureau of the Census and Bureau of Labor Statistics.

(a) In what decade (10-year period) did the percent of women in math or computer science professions decrease?

(b) When did the percent of women in math or computer science professions reach a maximum?

(c) In what year was the percent of women in math or computer science professions about 27%?

2

THE BASIC CONCEPTS OF SET THEORY

In the 1994 movie *I.Q.*, Meg Ryan plays Catherine Boyd, Alfred Einstein's brilliant niece, who is attracted to blue-collar worker Ed Walters (Tim Robbins). Ed pretends to be a physicist so that Catherine will not know his real background. In a charming scene, Catherine and Ed are standing a few feet apart while music is playing in the background.

ED: I think your uncle wants us to dance.

CATHERINE: Oh, now, don't be irrelevant, Ed. You can't get from there to here.

ED: Why not?

CATHERINE: Now don't tell me that a famous and brilliant scientist such as yourself doesn't know about Zeno's paradox.

ED: Remind me.

CATHERINE: You can't get from there to here because you always have to cover half the remaining distance, like from me to you. (Moving in increments of one-half) I have to cover half of it. Then, see, I still have half of that remaining, so I cover half that. I still have half of that left, so I cover half of that. Half of that . . . half of that . . . half of that . . . and since there are infinite halves left, I can't ever get there.

ED (reaching out and taking her in his arms and starting to dance): So how did that happen?

CATHERINE: I don't know.

Something in our human nature encourages us to collect things: baseball cards, Barbie dolls, coins, stamps, cars, and so on. A collection of objects is called a *set*, and this chapter deals with the mathematical aspects of sets. Prior to the twentieth century some ideas in set theory were considered *paradoxes* (wrong opinions). Zeno's paradox, as described by Catherine and seen in Exercises 51 and 52 of Section 2.5, has been around in several forms for thousands of years.

2.1 Symbols and Terminology

Designating Sets • Sets of Numbers and Cardinality • Finite and Infinite Sets • Equality of Sets

The basic ideas of set theory were developed by the German mathematician **Georg Cantor** (1845–1918) in about 1875. Cantor created a new field of theory and at the same time continued the long debate over infinity that began in ancient times. He developed counting by one-to-one correspondence to determine how many objects are contained in a set. Infinite sets differ from finite sets by not obeying the familiar law that the whole is greater than any of its parts.

Designating Sets

The human mind likes to create collections. Instead of seeing a group of five stars as five separate items, people tend to see them as one group of stars. The mind tries to find order and patterns. In mathematics this tendency to create collections is represented with the idea of a *set*. A **set** is a collection of objects. The objects belonging to the set are called the **elements,** or **members,** of the set.

Sets are designated using the following three methods: (1) *word description*, (2) the *listing method*, and (3) *set-builder notation*. A given set may be more conveniently denoted by one method rather than another, but most sets can be given in any of the three ways, as shown.

The set of even counting numbers less than 10	Word description
$\{2, 4, 6, 8\}$	Listing method
$\{x \mid x$ is an even counting number less than $10\}$	Set-builder notation

In the listing and set-builder notations, the braces at the beginning and ending indicate a set. Also, in the listing method, the commas between successive entries are essential. The set-builder notation above is read "the set of all x such that x is an even counting number less than 10."

Set-builder notation uses the algebraic idea of a *variable*. (Any symbol would do, but just as in other algebraic applications, the letter x is a common choice.) Before the vertical line we give the variable, which represents an element in general, and after the vertical line we state the criteria by which an element qualifies for membership in the set. By including *all* objects that meet the stated criteria, we generate (or build) the entire set.

Sets are commonly given names (usually capital letters). If E is selected as a name for the set of all letters of the English alphabet, then we can write

$$E = \{\text{a, b, c, d, e, f, g, h, i, j, k, l, m, n, o, p, q, r, s, t, u, v, w, x, y, z}\}.$$

The listing notation can often be shortened by establishing the pattern of elements included, and using ellipsis points to indicate a continuation of the pattern. Thus,

$$E = \{\text{a, b, c, d}, \ldots, \text{x, y, z}\} \quad \text{or} \quad E = \{\text{a, b, c, d, e}, \ldots, \text{z}\}.$$

The set containing no elements is called the **empty set,** or **null set.** The symbol $\emptyset$ is used to denote the empty set, so $\emptyset$ and $\{\ \}$ have the same meaning. We do *not* denote the empty set with the symbol $\{\emptyset\}$ because this notation represents a set with one element (that element being the empty set).

EXAMPLE 1 Listing Elements of Sets

Give a complete listing of all the elements of each of the following sets.

(a) the set of counting numbers between six and thirteen
(b) $\{5, 6, 7, \ldots, 13\}$
(c) $\{x \mid x$ is a counting number between 6 and $7\}$

SOLUTION

(a) This set can be denoted $\{7, 8, 9, 10, 11, 12\}$. (Notice that the word *between* excludes the endpoint values.)

(b) This set begins with the element 5, then 6, then 7, and so on, with each element obtained by adding 1 to the previous element in the list. This pattern stops at 13, so a complete listing is

$$\{5, 6, 7, 8, 9, 10, 11, 12, 13\}.$$

(c) There are no counting numbers between 6 and 7, and thus this is the empty set $\emptyset$. ■

For a set to be useful, it must be well defined. This means that if a particular set and some particular element are given, it must be possible to tell whether the element belongs to the set. For example, the preceding set E of the letters of the English alphabet is well defined. Given the letter q, we know that q is an element of E. Given the Greek letter θ (theta), we know that it is not an element of set E.

However, given the set C of all fat chickens, and a particular chicken, Hortense, it is not possible to say whether

Hortense is an element of C or Hortense is *not* an element of C.

The problem is the word "fat"; how fat is fat? Because we cannot necessarily decide whether a given chicken belongs to set C, set C is not well defined.

The letter q is an element of set E, where E is the set of all the letters of the English alphabet. To show this, $\in$ is used to replace the words "is an element of," or

$$\text{q} \in E,$$

which is read "q is an element of set E." The letter θ is not an element of E. To show this, $\in$ with a slash mark is used to replace the words "is not an element of," written

$$\theta \notin E.$$

This is read "θ is not an element of set E."

EXAMPLE 2 Applying the Symbol $\in$

Decide whether each statement is true or false.

(a) $3 \in \{1, 2, 5, 9, 13\}$

(b) $0 \in \{0, 1, 2, 3\}$

(c) $\frac{1}{5} \notin \left\{\frac{1}{3}, \frac{1}{4}, \frac{1}{6}\right\}$

SOLUTION

(a) Because 3 is *not* an element of the set $\{1, 2, 5, 9, 13\}$, the statement is false.

(b) Because 0 is indeed an element of the set $\{0, 1, 2, 3\}$, the statement is true.

(c) This statement says that $\frac{1}{5}$ is not an element of the set $\left\{\frac{1}{3}, \frac{1}{4}, \frac{1}{6}\right\}$, which is true. ■

Sets of Numbers and Cardinality Example 1 referred to counting numbers (or natural numbers), which were introduced in Section 1.1. Other important categories of numbers, which are used throughout the text, are summarized on the next page.

Sets of Numbers

Natural or Counting numbers $\{1, 2, 3, 4, \ldots\}$

Whole numbers $\{0, 1, 2, 3, 4, \ldots\}$

Integers $\{\ldots, -3, -2, -1, 0, 1, 2, 3, \ldots\}$

Rational numbers $\left\{\frac{p}{q} \mid p \text{ and } q \text{ are integers, and } q \neq 0\right\}$
(Some examples of rational numbers are $\frac{3}{5}$, $-\frac{7}{9}$, 5, and 0. Any rational number may be written as a terminating decimal number, like .25, or a repeating decimal number, like .666 . . . .)

Real numbers $\{x \mid x \text{ is a number that can be expressed as a decimal}\}$

Irrational numbers $\{x \mid x \text{ is a real number and } x \text{ cannot be expressed as a quotient of integers}\}$
(Some examples of irrational numbers are $\sqrt{2}$, $\sqrt[3]{4}$, and π. Decimal representations of irrational numbers never terminate and never repeat.)

The number of elements in a set is called the **cardinal number,** or **cardinality,** of the set. The symbol $\boldsymbol{n(A)}$, which is read **"n of A,"** represents the cardinal number of set A.

If elements are repeated in a set listing, they should not be counted more than once when determining the cardinal number of the set. For example, the set

$$B = \{1, 1, 2, 2, 3\}$$

has only three distinct elements, and so

$$n(B) = 3.$$

EXAMPLE 3 Finding Cardinal Numbers

Find the cardinal number of each set.

(a) $K = \{2, 4, 8, 16\}$ **(b)** $M = \{0\}$
(c) $R = \{4, 5, \ldots, 12, 13\}$ **(d)** $\emptyset$

SOLUTION

(a) Set K contains four elements, so the cardinal number of set K is 4, and $n(K) = 4$.
(b) Set M contains only one element, zero, so $n(M) = 1$.
(c) There are only four elements listed, but the ellipsis points indicate that there are other elements in the set. Counting them, we find that there are ten elements, so $n(R) = 10$.
(d) The empty set, $\emptyset$, contains no elements, and $n(\emptyset) = 0$.

A close-up of a camera lens shows the **infinity symbol,** ∞, defined as any distance greater than 1000 times the focal length of a lens.

The sign was invented by the mathematician John Wallis in 1655. Wallis used $1/\infty$ to represent an infinitely small quantity.

Finite and Infinite Sets If the cardinal number of a set is a particular whole number (0 or a counting number), as in all parts of Example 3, we call that set a **finite set.** Given enough time, we could finish counting all the elements of any finite set and arrive at its cardinal number. Some sets, however, are so large that we could never finish the counting process. The counting numbers themselves are such a set. Whenever a set is so large that its cardinal number is not found among the whole numbers, we call that set an **infinite set**. Infinite sets can be designated using the three methods already mentioned.

EXAMPLE 4 Designating an Infinite Set

Designate all odd counting numbers by the three common methods of set notation.

SOLUTION

The set of all odd counting numbers — Word description

$\{1, 3, 5, 7, 9, \ldots\}$ — Listing method

$\{x | x \text{ is an odd counting number}\}$ — Set-builder notation

Equality of Sets

Set Equality

Set A is **equal** to set B provided the following two conditions are met:

1. Every element of A is an element of B, and
2. Every element of B is an element of A.

Informally, two sets are equal if they contain exactly the same elements, regardless of order. For example,

$$\{a, b, c, d\} = \{a, c, d, b\}.$$ Both sets contain exactly the same elements.

Because repetition of elements in a set listing does not add new elements,

$$\{1, 0, 1, 2, 3, 3\} = \{0, 1, 2, 3\}$$ Both sets contain exactly the same elements.

EXAMPLE 5 Determining Whether Two Sets are Equal

Are $\{-4, 3, 2, 5\}$ and $\{-4, 0, 3, 2, 5\}$ equal sets?

SOLUTION

Every element of the first set is an element of the second; however, 0 is an element of the second and not the first. In other words, the sets do not contain exactly the same elements, so they are not equal: $\{-4, 3, 2, 5\} \neq \{-4, 0, 3, 2, 5\}$.

EXAMPLE 6 Determining Whether Two Sets are Equal

Decide whether each statement is *true* or *false*.

(a) $\{3\} = \{x | x \text{ is a counting number between 1 and 5}\}$

(b) $\{x | x \text{ is a negative natural number}\} = \{y | y \text{ is a number that is both rational and irrational}\}$

SOLUTION

(a) The set on the right contains *all* counting numbers between 1 and 5, namely 2, 3, and 4, while the set on the left contains *only* the number 3. Because the sets do not contain exactly the same elements, they are not equal. The statement is false.

(b) All natural numbers are positive, so the set on the left is $\emptyset$. By definition, if a number is rational, it cannot be irrational, so the set on the right is also $\emptyset$. Because each set is the empty set, the sets are equal. The statement is true.

2.1 EXERCISES

Match each set in Column I with the appropriate description in Column II.

I	II
1. $\{2, 4, 6, 8\}$	**A.** the set of all even integers
2. $\{x \mid x \text{ is an even integer greater than 4 and less than 6}\}$	**B.** the set of the five least positive integer powers of 2
3. $\{\ldots, -4, -3, -2, -1\}$	**C.** the set of even positive integers less than 10
4. $\{\ldots, -6, -4, -2, 0, 2, 4, 6, \ldots\}$	**D.** the set of all odd integers
5. $\{2, 4, 8, 16, 32\}$	**E.** the set of all negative integers
6. $\{\ldots, -5, -3, -1, 1, 3, 5, \ldots\}$	**F.** the set of odd positive integers less than 10
7. $\{2, 4, 6, 8, 10\}$	**G.** $\emptyset$
8. $\{1, 3, 5, 7, 9\}$	**H.** the set of the five least positive integer multiples of 2

List all the elements of each set. Use set notation and the listing method to describe the set.

9. the set of all counting numbers less than or equal to 6

10. the set of all whole numbers greater than 8 and less than 18

11. the set of all whole numbers not greater than 4

12. the set of all counting numbers between 4 and 14

13. $\{6, 7, 8, \ldots, 14\}$

14. $\{3, 6, 9, 12, \ldots, 30\}$

15. $\{-15, -13, -11, \ldots, -1\}$

16. $\{-4, -3, -2, \ldots, 4\}$

17. $\{2, 4, 8, \ldots, 256\}$

18. $\{90, 87, 84, \ldots, 69\}$

19. $\{x \mid x \text{ is an even whole number less than 11}\}$

20. $\{x \mid x \text{ is an odd integer between } -8 \text{ and } 7\}$

Denote each set by the listing method. There may be more than one correct answer.

21. the set of all counting numbers greater than 20

22. the set of all integers between -200 and 500

23. the set of Great Lakes

24. the set of U.S. presidents who served after Lyndon Johnson and before George W. Bush

25. $\{x \mid x \text{ is a positive multiple of 5}\}$

26. $\{x \mid x \text{ is a negative multiple of 6}\}$

27. $\{x \mid x \text{ is the reciprocal of a natural number}\}$

28. $\{x \mid x \text{ is a positive integer power of 4}\}$

Denote each set by set-builder notation, using x as the variable. There may be more than one correct answer.

29. the set of all rational numbers

30. the set of all even natural numbers

31. $\{1, 3, 5, \ldots, 75\}$

32. $\{35, 40, 45, \ldots, 95\}$

Identify each set as finite *or* infinite.

33. $\{2, 4, 6, \ldots, 32\}$

34. $\{6, 12, 18\}$

35. $\left\{\frac{1}{2}, \frac{2}{3}, \frac{3}{4}, \ldots\right\}$

36. $\{-10, -8, -6, \ldots\}$

37. $\{x \mid x$ is a natural number greater than $50\}$

38. $\{x \mid x$ is a natural number less than $50\}$

39. $\{x \mid x$ is a rational number$\}$

40. $\{x \mid x$ is a rational number between 0 and $1\}$

Find n(A) for each set.

41. $A = \{0, 1, 2, 3, 4, 5, 6, 7\}$

42. $A = \{-3, -1, 1, 3, 5, 7, 9\}$

43. $A = \{2, 4, 6, \ldots, 1000\}$

44. $A = \{0, 1, 2, 3, \ldots, 3000\}$

45. $A = \{\text{a, b, c}, \ldots, \text{z}\}$

46. $A = \{x \mid x$ is a vowel in the English alphabet$\}$

47. $A =$ the set of integers between -20 and 20

48. $A =$ the set of current U.S. senators

49. $A = \left\{\frac{1}{3}, \frac{2}{4}, \frac{3}{5}, \frac{4}{6}, \ldots, \frac{27}{29}, \frac{28}{30}\right\}$

50. $A = \left\{\frac{1}{2}, -\frac{1}{2}, \frac{1}{3}, -\frac{1}{3}, \ldots, \frac{1}{10}, -\frac{1}{10}\right\}$

51. Explain why it is acceptable to write the statement "x is a vowel in the English alphabet" in the set for Exercise 46, despite the fact that x is a consonant.

52. Explain how Exercise 49 can be answered without actually listing and then counting all the elements.

Identify each set as well defined *or* not well defined.

53. $\{x \mid x$ is a real number$\}$

54. $\{x \mid x$ is a negative number$\}$

55. $\{x \mid x$ is a good athlete$\}$

56. $\{x \mid x$ is a skillful typist$\}$

57. $\{x \mid x$ is a difficult course$\}$

58. $\{x \mid x$ is a counting number less than $2\}$

Fill each blank with either $\in$ or $\notin$ to make each statement true.

59. 5 _____ $\{2, 4, 5, 7\}$

60. 8 _____ $\{3, -2, 5, 7, 8\}$

61. -4 _____ $\{4, 7, 8, 12\}$

62. -12 _____ $\{3, 8, 12, 18\}$

63. 0 _____ $\{-2, 0, 5, 9\}$

64. 0 _____ $\{3, 4, 6, 8, 10\}$

65. $\{3\}$ _____ $\{2, 3, 4, 6\}$

66. $\{6\}$ _____ $\{2 + 1, 3 + 1, 4 + 1, 5 + 1, 6 + 1\}$

67. 8 _____ $\{11 - 2, 10 - 2, 9 - 2, 8 - 2\}$

68. The statement $3 \in \{9 - 6, 8 - 6, 7 - 6\}$ is true even though the *symbol* 3 does not appear in the set. Explain.

Write true *or* false *for each statement.*

69. $3 \in \{2, 5, 6, 8\}$

70. $6 \in \{-2, 5, 8, 9\}$

71. $\text{b} \in \{\text{h, c, d, a, b}\}$

72. $\text{m} \in \{\text{l, m, n, o, p}\}$

73. $9 \notin \{6, 3, 4, 8\}$

74. $2 \notin \{7, 6, 5, 4\}$

75. $\{\text{k, c, r, a}\} = \{\text{k, c, a, r}\}$

76. $\{\text{e, h, a, n}\} = \{\text{a, h, e, n}\}$

77. $\{5, 8, 9\} = \{5, 8, 9, 0\}$

78. $\{3, 7, 12, 14\} = \{3, 7, 12, 14, 0\}$

79. $\{4\} \in \{\{3\}, \{4\}, \{5\}\}$

80. $4 \in \{\{3\}, \{4\}, \{5\}\}$

81. $\{x \mid x$ is a natural number less than $3\} = \{1, 2\}$

82. $\{x \mid x$ is a natural number greater than $10\} = \{11, 12, 13, \ldots\}$

Write true *or* false *for each statement.*

Let $A = \{2, 4, 6, 8, 10, 12\}$, $B = \{2, 4, 8, 10\}$, *and* $C = \{4, 10, 12\}$.

83. $4 \in A$

84. $8 \in B$

85. $4 \notin C$

86. $8 \notin B$

87. Every element of C is also an element of A.

88. Every element of C is also an element of B.

89. This section opened with the statement, "The human mind likes to create collections." Why do you suppose this is so? In your explanation, use one or more particular "collections," mathematical or otherwise.

90. Explain the difference between a well defined set and a not well defined set. Give examples and use terms introduced in this section.

*Two sets are **equal** if they contain identical elements. However, two sets are **equivalent** if they contain the same number of elements (but not necessarily the same elements). For each condition, give an example or explain why it is impossible.*

91. two sets that are neither equal nor equivalent

92. two sets that are equal but not equivalent

93. two sets that are equivalent but not equal

94. two sets that are both equal and equivalent

95. ***Volumes of Stocks*** The table lists the ten most active stocks on the New York Stock Exchange in 2004.

MOST ACTIVE STOCKS ON NYSE

2004 Rank	Company name (symbol)	2004 share volume (in millions)
1.	Lucent Technologies, Inc. (LU)	5811.0
2.	Nortel Networks Corporation (NT)	4808.5
3.	Pfizer, Inc. (PFE)	4430.4
4.	General Electric Company (GE)	4118.5
5.	Motorola, Inc. (MOT)	2862.5
6.	Time Warner, Inc. (TWX)	2712.8
7.	Citigroup, Inc. (C)	2649.8
8.	Texas Instruments, Inc. (TXN)	2559.7
9.	EMC Corporation (EMC)	2506.7
10.	AT&T Wireless Services, Inc. (AWE)	2350.0

Source: www.info-please.com

(a) List the set of issues that had a share volume of at least 4118.5 million.

(b) List the set of issues that had a share volume of at most 4118.5 million.

96. ***Burning Calories*** Alexis Cotten is health conscious, but she does like a certain chocolate bar, each of which contains 220 calories. To burn off unwanted calories, Alexis participates in her favorite activities, shown below, in increments of 1 hour and never repeats a given activity on a given day.

Activity	Symbol	Calories Burned per Hour
Volleyball	v	160
Golf	g	260
Canoeing	c	340
Swimming	s	410
Running	r	680

(a) On Monday, Alexis has time for no more than two hours of activities. List all possible sets of activities that would burn off at least the number of calories obtained from three chocolate bars.

(b) Assume that Alexis can afford up to three hours of time for activities on Saturday. List all sets of activities that would burn off at least the number of calories in five chocolate bars.

2.2 Venn Diagrams and Subsets

Venn Diagrams • Complement of a Set • Subsets of a Set • Proper Subsets • Counting Subsets

Venn Diagrams In the statement of a problem, there is either a stated or implied **universe of discourse.** The universe of discourse includes all things under discussion at a given time. For example, in studying reactions to a proposal that a certain campus raise the minimum age of individuals to whom beer may be sold, the universe of discourse might be all the students at the school, the nearby members of the public, the board of trustees of the school, or perhaps all these groups of people.

In set theory, the universe of discourse is called the **universal set,** typically designated by the letter ***U*.** The universal set might change from problem to problem.

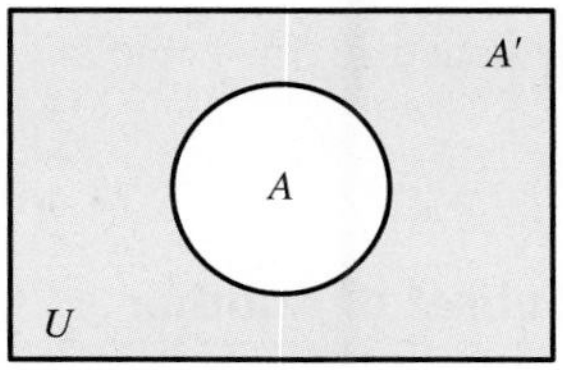

FIGURE 1

In set theory, we commonly use **Venn diagrams,** developed by the logician John Venn (1834–1923). In these diagrams, the universal set is represented by a rectangle, and other sets of interest within the universal set are depicted by circular regions. In the Venn diagram of Figure 1, the entire region bounded by the rectangle represents the universal set U, while the portion bounded by the circle represents set A.

Complement of a Set

The colored region inside U and outside the circle in Figure 1 is labeled A' (read **"*A* prime"**). This set, called the *complement* of A, contains all elements that are contained in U but not contained in A.

The Complement of a Set

For any set A within the universal set U, the **complement** of A, written A', is the set of elements of U that are not elements of A. That is,

$$A' = \{x \mid x \in U \text{ and } x \notin A\}.$$

EXAMPLE 1 Finding Complements

Find each of the following sets.

Let $U = \{a, b, c, d, e, f, g, h\}$, $M = \{a, b, e, f\}$, and $N = \{b, d, e, g, h\}$.

(a) M' **(b)** N'

SOLUTION

(a) Set M' contains all the elements of set U that are not in set M. Because set M contains the elements a, b, e, and f, these elements will be disqualified from belonging to set M', and consequently set M' will contain c, d, g, and h, or $M' = \{c, d, g, h\}$.

(b) Set N' contains all the elements of U that are not in set N, so $N' = \{a, c, f\}$.

Consider the complement of the universal set, U'. The set U' is found by selecting all the elements of U that do not belong to U. There are no such elements, so there can be no elements in set U'. This means that for any universal set U, $U' = \emptyset$.

Now consider the complement of the empty set, $\emptyset'$. Because $\emptyset' = \{x \mid x \in U$ and $x \notin \emptyset\}$ and set $\emptyset$ contains no elements, every member of the universal set U satisfies this description. Therefore, for any universal set U, $\emptyset' = U$.

Subsets of a Set

Suppose that we are given the universal set $U = \{1, 2, 3, 4, 5\}$, while $A = \{1, 2, 3\}$. Every element of set A is also an element of set U. Because of this, set A is called a *subset* of set U, written

$$A \subseteq U.$$

("A is not a subset of set U" would be written $A \nsubseteq U$.)

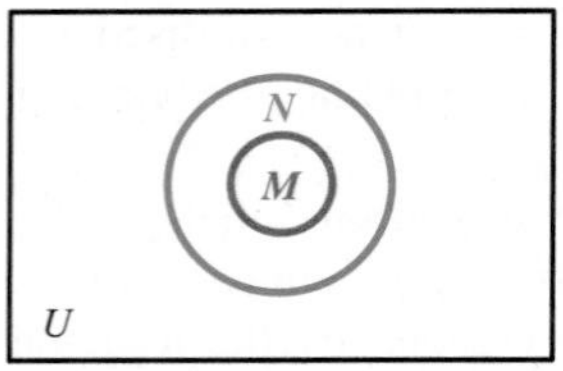

FIGURE 2

A Venn diagram showing that set M is a subset of set N is shown in Figure 2.

Subset of a Set

Set A is a **subset** of set B if every element of A is also an element of B. In symbols this is written $\boldsymbol{A \subseteq B}$.

EXAMPLE 2 Determining Whether One Set is a Subset of Another

Write $\subseteq$ or $\nsubseteq$ in each blank to make a true statement.

(a) $\{3, 4, 5, 6\}$ _____ $\{3, 4, 5, 6, 8\}$ **(b)** $\{1, 2, 6\}$ _____ $\{2, 4, 6, 8\}$
(c) $\{5, 6, 7, 8\}$ _____ $\{5, 6, 7, 8\}$

SOLUTION

(a) Because every element of $\{3, 4, 5, 6\}$ is also an element of $\{3, 4, 5, 6, 8\}$, the first set is a subset of the second, so $\subseteq$ goes in the blank:

$$\{3, 4, 5, 6\} \subseteq \{3, 4, 5, 6, 8\}.$$

(b) The element 1 belongs to $\{1, 2, 6\}$ but not to $\{2, 4, 6, 8\}$. Place $\nsubseteq$ in the blank.

(c) Every element of $\{5, 6, 7, 8\}$ is also an element of $\{5, 6, 7, 8\}$. Place $\subseteq$ in the blank. ■

As Example 2(c) suggests, every set is a subset of itself:

$$\boldsymbol{B \subseteq B}, \quad \textbf{for any set } \boldsymbol{B}.$$

The statement of set equality in Section 2.1 can be formally presented using subset terminology.

Set Equality (Alternative definition)

Suppose A and B are sets. Then $A = B$ if $A \subseteq B$ and $B \subseteq A$ are both true.

Proper Subsets When studying subsets of a set B, it is common to look at subsets other than set B itself. Suppose that $B = \{5, 6, 7, 8\}$ and $A = \{6, 7\}$. A is a subset of B, but A is not all of B; there is at least one element in B that is not in A. (Actually, in this case there are two such elements, 5 and 8.) In this situation, A is called a *proper subset* of B. To indicate that A is a proper subset of B, write $A \subset B$.

Notice the similarity of the subset symbols, $\subset$ and $\subseteq$, to the inequality symbols from algebra, $<$ and $\leq$.

Proper Subset of a Set

Set A is a **proper subset** of set B if $A \subseteq B$ and $A \neq B$. In symbols, this is written $\boldsymbol{A \subset B}$.

EXAMPLE 3 Determining Subset and Proper Subset Relationships

Decide whether $\subset$, $\subseteq$, or both could be placed in each blank to make a true statement.

(a) $\{5, 6, 7\}$ _____ $\{5, 6, 7, 8\}$ **(b)** $\{a, b, c\}$ _____ $\{a, b, c\}$

SOLUTION

(a) Every element of {5, 6, 7} is contained in {5, 6, 7, 8}, so $\subseteq$ could be placed in the blank. Also, the element 8 belongs to {5, 6, 7, 8} but not to {5, 6, 7}, making {5, 6, 7} a proper subset of {5, 6, 7, 8}. This means that $\subset$ could also be placed in the blank.

(b) The set {a, b, c} is a subset of {a, b, c}. Because the two sets are equal, {a, b, c} is not a proper subset of {a, b, c}. Only $\subseteq$ may be placed in the blank.

Set A is a subset of set B if every element of set A is also an element of set B. This definition can be reworded by saying that set A is a subset of set B if there are no elements of A that are not also elements of B. This second form of the definition shows that the empty set is a subset of any set, or

$$\emptyset \subseteq B, \quad \textbf{for any set } B.$$

This is true because it is not possible to find any elements of $\emptyset$ that are not also in B. (There are no elements in $\emptyset$.) The empty set $\emptyset$ is a proper subset of every set except itself:

$$\emptyset \subset B \quad \textbf{if } B \textbf{ is any set other than } \emptyset.$$

Every set (except $\emptyset$) has at least two subsets, $\emptyset$ and the set itself.

EXAMPLE 4 Listing All Subsets of a Set

Find all possible subsets of each set.

(a) {7, 8} **(b)** {a, b, c}

SOLUTION

(a) By trial and error, the set {7, 8} has four subsets: $\emptyset$, {7}, {8}, {7, 8}.

(b) Here, trial and error leads to eight subsets for {a, b, c}:

$\emptyset$, {a}, {b}, {c}, {a, b}, {a, c}, {b, c}, {a, b, c}.

Counting Subsets In Example 4, the subsets of {7, 8} and the subsets of {a, b, c} were found by trial and error. An alternative method involves drawing a **tree diagram,** a systematic way of listing all the subsets of a given set. Figures 3(a) and (b) show tree diagrams for {7, 8} and {a, b, c}.

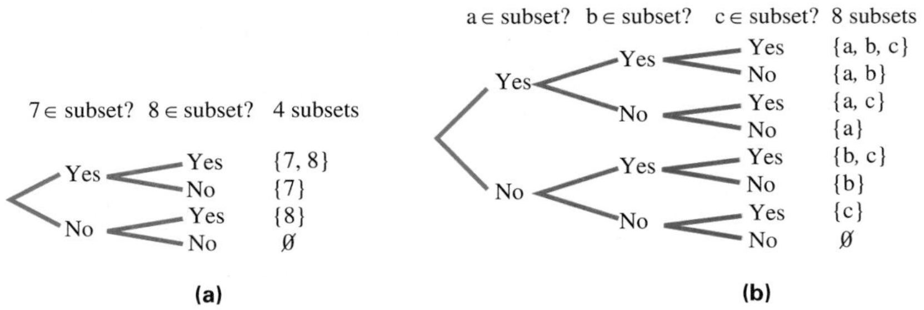

FIGURE 3

Powers of 2

$2^0 = 1$
$2^1 = 2$
$2^2 = 2 \times 2 = 4$
$2^3 = 2 \times 2 \times 2 = 8$
$2^4 = 2 \times 2 \times 2 \times 2 = 16$
$2^5 = 32$
$2^6 = 64$
$2^7 = 128$
$2^8 = 256$
$2^9 = 512$
$2^{10} = 1024$
$2^{11} = 2048$
$2^{12} = 4096$
$2^{15} = 32{,}768$
$2^{20} = 1{,}048{,}576$
$2^{25} = 33{,}554{,}432$
$2^{30} = 1{,}073{,}741{,}824$

In Example 4, we determined the number of subsets of a given set by making a list of all such subsets and then counting them. The tree diagram method also produced a list of all possible subsets. In many applications, we don't need to display all the subsets but simply determine how many there are. Furthermore, the trial and error method and the tree diagram method would both involve far too much work if the original set had a very large number of elements. For these reasons, it is desirable to have a formula for the number of subsets. To obtain such a formula, we use inductive reasoning. That is, we observe particular cases to try to discover a general pattern.

Begin with the set containing the least number of elements possible—the empty set. This set, ∅, has only one subset, ∅ itself. Next, a set with one element has only two subsets, itself and ∅. These facts, together with those obtained above for sets with two and three elements, are summarized here.

Number of elements	0	1	2	3
Number of subsets	1	2	4	8

This chart suggests that as the number of elements of the set increases by one, the number of subsets doubles. This suggests that the number of subsets in each case might be a power of 2. Every number in the second row of the chart is indeed a power of 2. Add this information to the chart.

Number of elements	0	1	2	3
Number of subsets	$1 = 2^0$	$2 = 2^1$	$4 = 2^2$	$8 = 2^3$

This chart shows that the number of elements in each case is the same as the exponent on the base 2. Inductive reasoning gives the following generalization.

Number of Subsets

The number of subsets of a set with n elements is $\mathbf{2^n}$.

Because the value 2^n includes the set itself, we must subtract 1 from this value to obtain the number of proper subsets of a set containing n elements.

Number of Proper Subsets

The number of proper subsets of a set with n elements is $\mathbf{2^n - 1}$.

As shown in Chapter 1, although inductive reasoning is a good way of *discovering* principles or arriving at a *conjecture,* it does not provide a proof that the conjecture is true in general. A proof must be provided by other means. The two formulas above are true, by observation, for $n = 0, 1, 2,$ or 3. (For a general proof, see Exercise 71 at the end of this section.)

EXAMPLE 5 Finding the Numbers of Subsets and Proper Subsets

Find the number of subsets and the number of proper subsets of each set.

(a) $\{3, 4, 5, 6, 7\}$ **(b)** $\{1, 2, 3, 4, 5, 9, 12, 14\}$

SOLUTION

(a) This set has 5 elements and $2^5 = 2 \cdot 2 \cdot 2 \cdot 2 \cdot 2 = 32$ subsets. Of these, $2^5 - 1 = 32 - 1 = 31$ are proper subsets.

(b) This set has 8 elements. There are $2^8 = 256$ subsets and 255 proper subsets. ■

2.2 EXERCISES

Match each set or sets in Column I with the appropriate description in Column II.

I	II
1. $\{p\}, \{q\}, \{p, q\}, \emptyset$	**A.** the complement of $\emptyset$
2. $\{p\}, \{q\}, \emptyset$	**B.** the proper subsets of $\{p, q\}$
3. $\{a, b\}$	**C.** the complement of $\{c, d\}$, if $U = \{a, b, c, d\}$
4. $\emptyset$	**D.** the complement of U
5. U	**E.** the complement of $\{b\}$, if $U = \{a, b\}$
6. $\{a\}$	**F.** the subsets of $\{p, q\}$

Insert $\subseteq$ *or* $\not\subseteq$ *in each blank so that the resulting statement is true.*

7. $\{-2, 0, 2\}$ _____ $\{-2, -1, 1, 2\}$

8. $\{M, W, F\}$ _____ $\{S, M, T, W, Th\}$

9. $\{2, 5\}$ _____ $\{0, 1, 5, 3, 4, 2\}$

10. $\{a, n, d\}$ _____ $\{r, a, n, d, y\}$

11. $\emptyset$ _____ $\{a, b, c, d, e\}$

12. $\emptyset$ _____ $\emptyset$

13. $\{-7, 4, 9\}$ _____ $\{x \mid x \text{ is an odd integer}\}$

14. $\left\{2, \frac{1}{3}, \frac{5}{9}\right\}$ _____ the set of rational numbers

Decide whether $\subset$, $\subseteq$, both, *or* neither *can be placed in each blank to make the statement true.*

15. $\{B, C, D\}$ _____ $\{B, C, D, F\}$

16. $\{red, blue, yellow\}$ _____ $\{yellow, blue, red\}$

17. $\{9, 1, 7, 3, 5\}$ _____ $\{1, 3, 5, 7, 9\}$

18. $\{S, M, T, W, Th\}$ _____ $\{M, W, Th, S\}$

19. $\emptyset$ _____ $\{0\}$

20. $\emptyset$ _____ $\emptyset$

21. $\{-1, 0, 1, 2, 3\}$ _____ $\{0, 1, 2, 3, 4\}$

22. $\left\{\frac{5}{6}, \frac{9}{8}\right\}$ _____ $\left\{\frac{6}{5}, \frac{8}{9}\right\}$

For Exercises 23–42, tell whether each statement is true *or* false.

Let $U = \{a, b, c, d, e, f, g\}$, $A = \{a, e\}$, $B = \{a, b, e, f, g\}$, $C = \{b, f, g\}$, *and* $D = \{d, e\}$.

23. $A \subset U$

24. $C \subset U$

25. $D \subseteq B$

26. $D \subseteq A$

27. $A \subset B$

28. $B \subseteq C$

29. $\emptyset \subset A$

30. $\emptyset \subseteq D$

31. $\emptyset \subseteq \emptyset$

32. $D \subset B$

33. $D \not\subseteq B$

34. $A \not\subseteq B$

35. There are exactly 6 subsets of C.

36. There are exactly 31 subsets of B.

37. There are exactly 3 subsets of A.

38. There are exactly 4 subsets of D.

39. There is exactly 1 subset of $\emptyset$.

40. There are exactly 127 proper subsets of U.

41. The Venn diagram below correctly represents the relationship among sets A, C, and U.

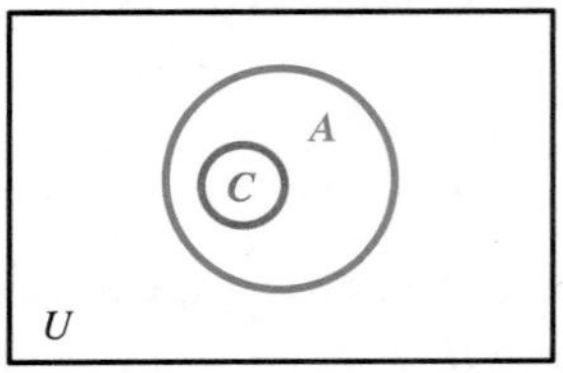

42. The Venn diagram below correctly represents the relationship among sets B, C, and U.

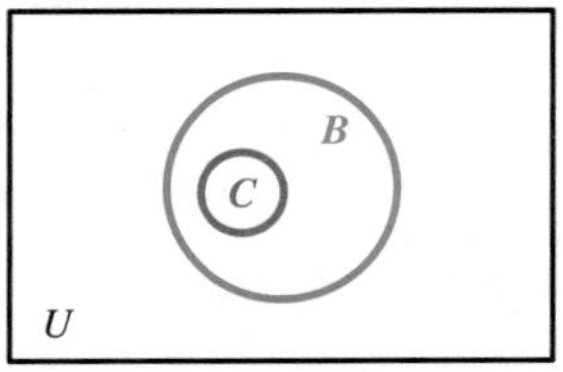

Find **(a)** *the number of subsets and* **(b)** *the number of proper subsets of each set.*

43. $\{1, 5, 10\}$

44. $\{8, 6, 4, 2\}$

45. {a, b, c, d, e, f}

46. the set of days of the week

47. $\{x \mid x$ is an odd integer between -7 and $4\}$

48. $\{x \mid x$ is an odd whole number less than $4\}$

Let $U = \{1, 2, 3, 4, 5, 6, 7, 8, 9, 10\}$ *and find the complement of each set.*

49. $\{1, 4, 6, 8\}$

50. $\{2, 5, 7, 9, 10\}$

51. $\{1, 3, 4, 5, 6, 7, 8, 9, 10\}$

52. $\{1, 2, 3, 4, 6, 7, 8, 9, 10\}$

53. $\emptyset$

54. U

Vacationing in Orlando, FL. *Terry McGinnis is planning to take her two sons to Orlando, FL, during their Thanksgiving vacation. In weighing her options concerning whether to fly or drive from their home in Iowa, she has listed the following characteristics.*

Fly to Orlando	Drive to Orlando
Higher cost	Lower cost
Educational	Educational
More time to see the sights	Less time to see the sights
Cannot visit relatives along the way	Can visit relatives along the way

Refer to these characteristics in Exercises 55–60.

55. Find the smallest universal set U that contains all listed characteristics of both options.

Let F represent the set of characteristics of the flying option and let D represent the set of characteristics of the driving option. Use the universal set from Exercise 55.

56. Give the set F'.

57. Give the set D'.

Find the set of elements common to both sets in Exercises 58–60.

58. F and D

59. F' and D'

60. F and D'

Meeting in a Hospitality Suite *Allen Wells, Bonnie Garrett, Cathie Ducote, David Bondy, and Eleanor André plan to meet at the hospitality suite after the CEO makes his speech at the January sales meeting of their publishing company. Denoting these five people by A, B, C, D, and E, list all the possible sets of this group in which the given number of them can gather.*

61. five people

62. four people

63. three people

64. two people

65. one person

66. no people

67. Find the total number of ways that members of this group can gather in the suite. (*Hint:* Find the total number of sets in your answers to Exercises 61–66.)

68. How does your answer in Exercise 67 compare with the number of subsets of a set of five elements? How can you interpret the answer to Exercise 67 in terms of subsets?

69. ***Selecting Bills from a Wallet*** Suppose that in your wallet you have the bills shown here.

(a) If you must select at least one bill, and you may select up to all of the bills, how many different sums of money could you make?

(b) In part (a), remove the condition "you must select at least one bill." Now, how many sums are possible?

70. ***Selecting Coins*** The photo shows a group of obsolete U.S. coins, consisting of one each of the penny, nickel, dime, quarter, and half dollar. Repeat Exercise 69, replacing "bill(s)" with "coin(s)."

71. In discovering the expression (2^n) for finding the number of subsets of a set with n elements, we observed that for the first few values of n, increasing the number of elements by one doubles the number of subsets. Here, you can prove the formula in general by showing that the same is true for any value of n. Assume set A has n elements and s subsets. Now add one additional element, say e, to the set A. (We now have a new set, say B, with $n + 1$ elements.) Divide the subsets of B into those that do not contain e and those that do.

(a) How many subsets of B do not contain e? (*Hint:* Each of these is a subset of the original set A.)

(b) How many subsets of B do contain e? (*Hint:* Each of these would be a subset of the original set A, with the element e inserted.)

(c) What is the total number of subsets of B?

(d) What do you conclude?

72. Explain why $\{\emptyset\}$ has $\emptyset$ as a subset and also has $\emptyset$ as an element.

2.3 Set Operations and Cartesian Products

Intersection of Sets • Union of Sets • Difference of Sets • Ordered Pairs • Cartesian Product of Sets • Venn Diagrams • De Morgan's Laws

Intersection of Sets Two candidates, Mary Lynn Brumfield and J.D. Patin, are running for a seat on the city council. A voter deciding for whom she should vote recalled the following campaign promises made by the candidates. Each promise is given a code letter.

Honest Mary Lynn Brumfield	Determined J.D. Patin
Spend less money, m	Spend less money, m
Emphasize traffic law enforcement, t	Crack down on crooked politicians, p
Increase service to suburban areas, s	Increase service to the city, c

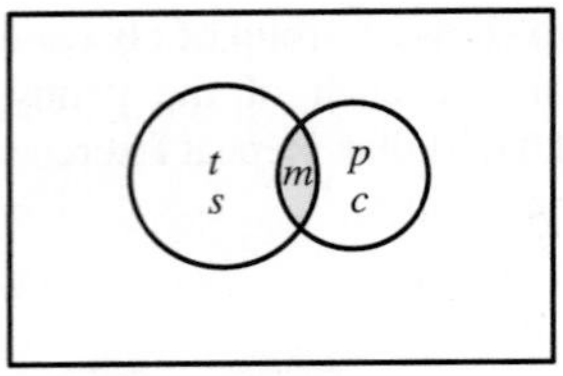

FIGURE 4

The only promise common to both candidates is promise m, to spend less money. Suppose we take each candidate's promises to be a set. The promises of Brumfield give the set $\{m, t, s\}$, while the promises of Patin give $\{m, p, c\}$. The only element common to both sets is m; this element belongs to the *intersection* of the two sets $\{m, t, s\}$ and $\{m, p, c\}$, as shown in color in the Venn diagram in Figure 4. In symbols,

$$\{m, t, s\} \cap \{m, p, c\} = \{m\},$$

where the cap-shaped symbol $\cap$ represents intersection. Notice that the intersection of two sets is itself a set.

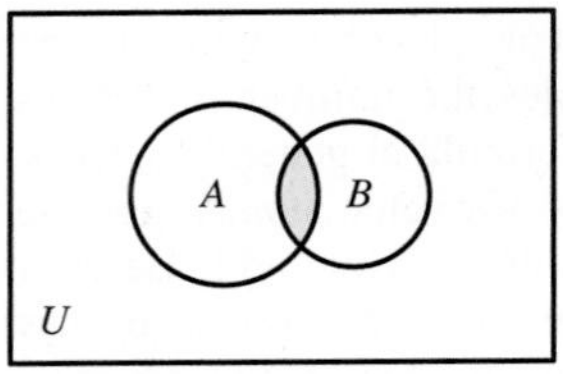

$A \cap B$

FIGURE 5

Intersection of Sets

The **intersection** of sets A and B, written $A \cap B$, is the set of elements common to both A and B, or

$$\boldsymbol{A \cap B = \{x \mid x \in A \text{ and } x \in B\}.}$$

Form the intersection of sets A and B by taking all the elements included in *both* sets, as shown in color in Figure 5.

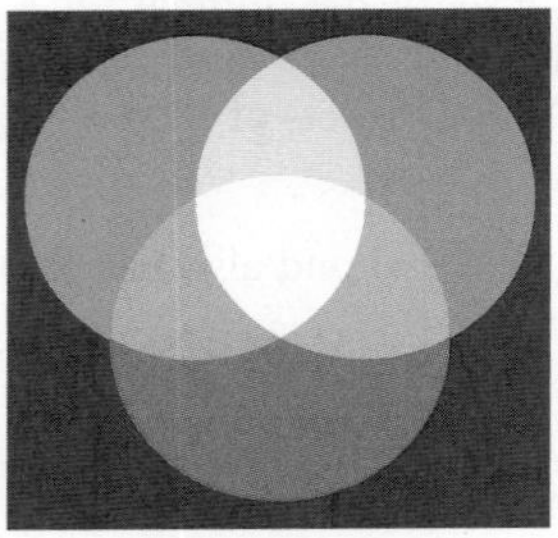

White light can be viewed as the intersection of the three primary colors.

EXAMPLE 1 Finding Intersections

Find each intersection.

(a) $\{3, 4, 5, 6, 7\} \cap \{4, 6, 8, 10\}$
(b) $\{9, 14, 25, 30\} \cap \{10, 17, 19, 38, 52\}$
(c) $\{5, 9, 11\} \cap \emptyset$

SOLUTION

(a) Because the elements common to both sets are 4 and 6,

$$\{3, 4, 5, 6, 7\} \cap \{4, 6, 8, 10\} = \{4, 6\}.$$

(b) These two sets have no elements in common, so

$$\{9, 14, 25, 30\} \cap \{10, 17, 19, 38, 52\} = \emptyset.$$

(c) There are no elements in $\emptyset$, so there can be no elements belonging to both $\{5, 9, 11\}$ and $\emptyset$. Because of this,

$$\{5, 9, 11\} \cap \emptyset = \emptyset.$$

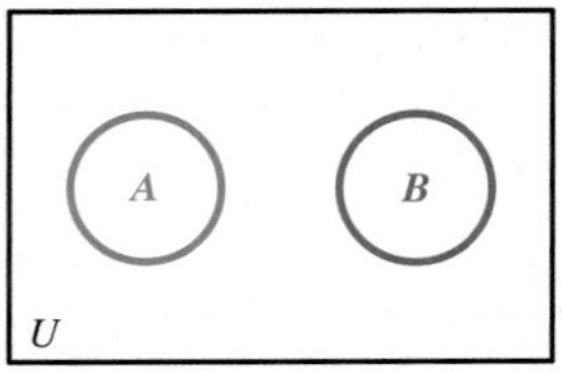

Disjoint sets

FIGURE 6

Examples 1(b) and 1(c) show two sets that have no elements in common. Sets with no elements in common are called **disjoint sets.** A set of dogs and a set of cats would be disjoint sets. In mathematical language, sets A and B are disjoint if $A \cap B = \emptyset$. Two disjoint sets A and B are shown in Figure 6.

Union of Sets We began this section with lists of campaign promises of two candidates running for city council. Suppose a pollster wants to summarize the types of promises made by candidates for the office. The pollster would need to study *all* the promises made by *either* candidate, or the set

$$\{m, t, s, p, c\},$$

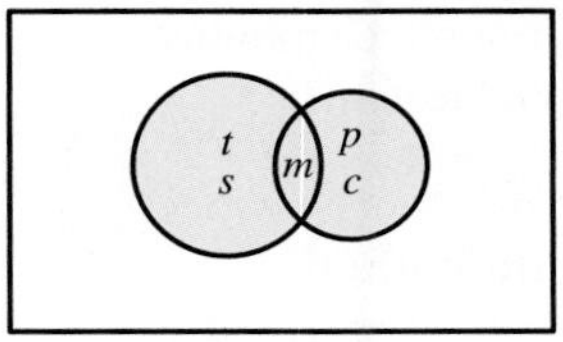

FIGURE 7

the *union* of the sets of promises made by the two candidates, as shown in color in the Venn diagram in Figure 7. In symbols,

$$\{m, t, s\} \cup \{m, p, c\} = \{m, t, s, p, c\},$$

where the cup-shaped symbol $\cup$ denotes set union. Be careful not to confuse this symbol with the universal set U. Again, the union of two sets is a set.

Union of Sets

The **union** of sets A and B, written $A \cup B$, is the set of all elements belonging to either of the sets, or

$$\mathbf{A \cup B = \{x \mid x \in A \text{ or } x \in B\}.}$$

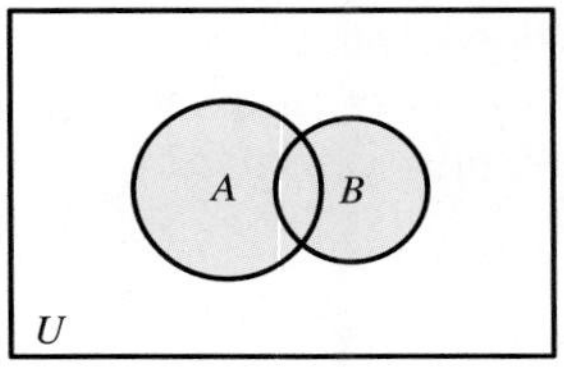

$A \cup B$

FIGURE 8

Form the union of sets A and B by taking all the elements of set A and then including the elements of set B that are not already listed. See Figure 8.

EXAMPLE 2 Finding Unions

Find each union.

(a) $\{2, 4, 6\} \cup \{4, 6, 8, 10, 12\}$
(b) {a, b, d, f, g, h} $\cup$ {c, f, g, h, k}
(c) $\{3, 4, 5\} \cup \emptyset$

SOLUTION

(a) Start by listing all the elements from the first set, 2, 4, and 6. Then list all the elements from the second set that are not in the first set, 8, 10, and 12. The union is made up of *all* these elements, written

$$\{2, 4, 6\} \cup \{4, 6, 8, 10, 12\} = \{2, 4, 6, 8, 10, 12\}.$$

(b) The union of these sets is

$$\{\text{a, b, d, f, g, h}\} \cup \{\text{c, f, g, h, k}\} = \{\text{a, b, c, d, f, g, h, k}\}.$$

(c) Because there are no elements in $\emptyset$, the union of $\{3, 4, 5\}$ and $\emptyset$ contains only the elements 3, 4, and 5, written

$$\{3, 4, 5\} \cup \emptyset = \{3, 4, 5\}.$$

For Further Thought

Comparing Properties

The arithmetic operations of addition and multiplication, when applied to numbers, have some familiar properties. *If a, b, and c are real numbers*, then the **commutative property of addition** says that the order of the numbers being added makes no difference: $a + b = b + a$. (Is there a **commutative property of multiplication**?) The **associative property of addition** says that when three numbers are added, the grouping used makes no difference: $(a + b) + c = a + (b + c)$. (Is there an **associative property of multiplication**?) The number 0 is called the **identity element for addition** since adding it to any number does not change that number: $a + 0 = a$. (What is the **identity element for multiplication**?) Finally, the **distributive property of multiplication over addition** says that $a(b + c) = ab + ac$. (Is there a distributive property of addition over multiplication?)

For Group Discussion or Individual Investigation

Now consider the operations of union and intersection, applied to sets. By recalling definitions, or by trying examples, answer the following questions.

1. Is set union commutative? How about set intersection?
2. Is set union associative? How about set intersection?
3. Is there an identity element for set union? If so, what is it? How about set intersection?
4. Is set intersection distributive over set union? Is set union distributive over set intersection?

Recall from the previous section that A' represents the *complement* of set A. Set A' is formed by taking all the elements of the universal set U that are not in A.

EXAMPLE 3 Finding Intersections and Unions of Complements

Find each set. Let

$$U = \{1, 2, 3, 4, 5, 6, 9\},\ A = \{1, 2, 3, 4\},\ B = \{2, 4, 6\},\ \text{and}\ C = \{1, 3, 6, 9\}.$$

(a) $A' \cap B$ **(b)** $B' \cup C'$ **(c)** $A \cap (B \cup C')$ **(d)** $(A' \cup C') \cap B'$

SOLUTION

(a) First identify the elements of set A', the elements of U that are not in set A:

$$A' = \{5, 6, 9\}.$$

Now, find $A' \cap B$, the set of elements belonging both to A' and to B:

$$A' \cap B = \{5, 6, 9\} \cap \{2, 4, 6\} = \{6\}.$$

(b) $B' \cup C' = \{1, 3, 5, 9\} \cup \{2, 4, 5\} = \{1, 2, 3, 4, 5, 9\}$.

(c) First find the set inside the parentheses:

$$B \cup C' = \{2, 4, 6\} \cup \{2, 4, 5\} = \{2, 4, 5, 6\}.$$

Now, find the intersection of this set with A.

$$\begin{aligned} A \cap (B \cup C') &= A \cap \{2, 4, 5, 6\} \\ &= \{1, 2, 3, 4\} \cap \{2, 4, 5, 6\} \\ &= \{2, 4\} \end{aligned}$$

(d) $A' = \{5, 6, 9\}$ and $C' = \{2, 4, 5\}$, so

$$A' \cup C' = \{5, 6, 9\} \cup \{2, 4, 5\} = \{2, 4, 5, 6, 9\}.$$

$B' = \{1, 3, 5, 9\}$, so

$$(A' \cup C') \cap B' = \{2, 4, 5, 6, 9\} \cap \{1, 3, 5, 9\} = \{5, 9\}.$$

It is often said that mathematics is a "language." As such, it has the advantage of concise symbolism. For example, the set $(A \cap B)' \cup C$ is less easily expressed in words. One attempt is the following: "The set of all elements that are not in both A and B, or are in C."

EXAMPLE 4 Describing Sets in Words

Describe each of the following sets in words.

(a) $A \cap (B \cup C')$ **(b)** $(A' \cup C') \cap B'$

SOLUTION

(a) This set might be described as "the set of all elements that are in A, and also are in B or not in C."

(b) One possibility is "the set of all elements that are not in A or not in C, and also are not in B."

Difference of Sets We now consider the *difference* of two sets. Suppose that $A = \{1, 2, 3, \ldots, 10\}$ and $B = \{2, 4, 6, 8, 10\}$. If the elements of B are excluded (or taken away) from A, the set $C = \{1, 3, 5, 7, 9\}$ is obtained. C is called the difference of sets A and B.

Difference of Sets

The **difference** of sets A and B, written $A - B$, is the set of all elements belonging to set A and not to set B, or

$$A - B = \{x \mid x \in A \text{ and } x \notin B\}.$$

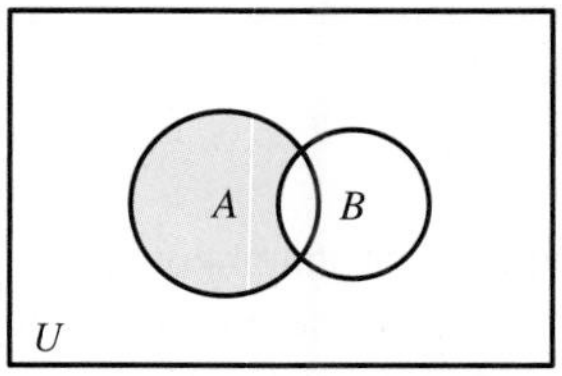

$A - B$

FIGURE 9

Because $x \notin B$ has the same meaning as $x \in B'$ the set difference $A - B$ can also be described as $\{x \mid x \in A \text{ and } x \in B'\}$, or $A \cap B'$. Figure 9 illustrates the idea of set difference. The region in color represents $A - B$.

EXAMPLE 5 Finding Set Differences

Find each set.

Let $U = \{1, 2, 3, 4, 5, 6, 7\}$, $A = \{1, 2, 3, 4, 5, 6\}$, $B = \{2, 3, 6\}$, and $C = \{3, 5, 7\}$.

(a) $A - B$ **(b)** $B - A$ **(c)** $(A - B) \cup C'$

SOLUTION

(a) Begin with set A and exclude any elements found also in set B. So,

$$A - B = \{1, 2, 3, 4, 5, 6\} - \{2, 3, 6\} = \{1, 4, 5\}.$$

(b) To be in $B - A$, an element must be in set B and not in set A. But all elements of B are also in A. Thus, $B - A = \emptyset$.

(c) From part (a), $A - B = \{1, 4, 5\}$. Also, $C' = \{1, 2, 4, 6\}$, so

$$(A - B) \cup C' = \{1, 2, 4, 5, 6\}.$$

The results in Examples 5(a) and 5(b) illustrate that, in general,

$$\mathbf{A - B \neq B - A.}$$

Ordered Pairs When writing a set that contains several elements, the order in which the elements appear is not relevant. For example, $\{1, 5\} = \{5, 1\}$. However, there are many instances in mathematics where, when two objects are paired, the order in which the objects are written is important. This leads to the idea of the *ordered pair.* When writing ordered pairs, use parentheses rather than braces, which are reserved for writing sets.

Ordered Pairs

In the **ordered pair** (a, b), a is called the **first component** and b is called the **second component.** In general, $(a, b) \neq (b, a)$.

Two ordered pairs (a, b) and (c, d) are **equal** provided that their first components are equal and their second components are equal; that is, $(a, b) = (c, d)$ if and only if $a = c$ and $b = d$.

EXAMPLE 6 Determining Equality of Sets and of Ordered Pairs

Decide whether each statement is *true* or *false*.

(a) $(3, 4) = (5 - 2, 1 + 3)$ **(b)** $\{3, 4\} \neq \{4, 3\}$ **(c)** $(7, 4) = (4, 7)$

SOLUTION

(a) Because $3 = 5 - 2$ and $4 = 1 + 3$, the ordered pairs are equal. The statement is true.

(b) Because these are sets and not ordered pairs, the order in which the elements are listed is not important. Because these sets are equal, the statement is false.

(c) These ordered pairs are not equal because they do not satisfy the requirements for equality of ordered pairs. The statement is false.

Cartesian Product of Sets A set may contain ordered pairs as elements. If A and B are sets, then each element of A can be paired with each element of B, and the results can be written as ordered pairs. The set of all such ordered pairs is called the *Cartesian product* of A and B, written $A \times B$ and read "A cross B." The name comes from that of the French mathematician René Descartes.

Cartesian Product of Sets

The **Cartesian product** of sets A and B, written $A \times B$, is

$$\boldsymbol{A \times B = \{(a, b) \mid a \in A \text{ and } b \in B\}.}$$

EXAMPLE 7 Finding Cartesian Products

Let $A = \{1, 5, 9\}$ and $B = \{6, 7\}$. Find each set.

(a) $A \times B$ **(b)** $B \times A$

SOLUTION

(a) Pair each element of A with each element of B. Write the results as ordered pairs, with the element of A written first and the element of B written second. Write as a set.

$$A \times B = \{(1, 6), (1, 7), (5, 6), (5, 7), (9, 6), (9, 7)\}$$

(b) Because B is listed first, this set will consist of ordered pairs that have their components interchanged when compared to those in part (a).

$$B \times A = \{(6, 1), (7, 1), (6, 5), (7, 5), (6, 9), (7, 9)\}$$

It should be noted that the order in which the ordered pairs themselves are listed is not important. For example, another way to write $B \times A$ in Example 7 would be

$$\{(6, 1), (6, 5), (6, 9), (7, 1), (7, 5), (7, 9)\}.$$

EXAMPLE 8 Finding the Cartesian Product of a Set with Itself

Let $A = \{1, 2, 3, 4, 5, 6\}$. Find $A \times A$.

SOLUTION

By pairing 1 with each element in the set, 2 with each element, and so on, we obtain the following set.

$$\begin{aligned} A \times A = \{&(1, 1), (1, 2), (1, 3), (1, 4), (1, 5), (1, 6),\\ &(2, 1), (2, 2), (2, 3), (2, 4), (2, 5), (2, 6),\\ &(3, 1), (3, 2), (3, 3), (3, 4), (3, 5), (3, 6),\\ &(4, 1), (4, 2), (4, 3), (4, 4), (4, 5), (4, 6),\\ &(5, 1), (5, 2), (5, 3), (5, 4), (5, 5), (5, 6),\\ &(6, 1), (6, 2), (6, 3), (6, 4), (6, 5), (6, 6)\} \end{aligned}$$

It is not unusual to take the Cartesian product of a set with itself, as in Example 8. The Cartesian product in Example 8 represents all possible results that are obtained when two distinguishable dice are rolled. This Cartesian product is important when studying certain problems in counting techniques and probability.

From Example 7 it can be seen that, in general, $\boldsymbol{A \times B \neq B \times A}$, because they do not contain exactly the same ordered pairs. However, each set contains the same number of elements, six. Furthermore, $n(A) = 3$, $n(B) = 2$, and $n(A \times B) = n(B \times A) = 6$. Because $3 \cdot 2 = 6$, one might conclude that the cardinal number of the Cartesian product of two sets is equal to the product of the cardinal numbers of the sets. In general, this conclusion is correct.

Cardinal Number of a Cartesian Product

If $n(A) = a$ and $n(B) = b$, then

$$\boldsymbol{n(A \times B) = n(B \times A) = n(A) \cdot n(B) = n(B) \cdot n(A) = ab = ba.}$$

EXAMPLE 9 Finding Cardinal Numbers of Cartesian Products

Find $n(A \times B)$ and $n(B \times A)$ from the given information.

(a) $A = \{\text{a, b, c, d, e, f, g}\}$ and $B = \{2, 4, 6\}$ **(b)** $n(A) = 24$ and $n(B) = 5$

SOLUTION

(a) Because $n(A) = 7$ and $n(B) = 3$, $n(A \times B)$ and $n(B \times A)$ both equal $7 \cdot 3$, or 21.

(b) $n(A \times B) = n(B \times A) = 24 \cdot 5 = 5 \cdot 24 = 120$ ■

Finding intersections, unions, differences, Cartesian products, and complements of sets are examples of *set operations.* An **operation** is a rule or procedure by which one or more objects are used to obtain another object. The most common operations on sets are summarized below, along with their Venn diagrams.

Set Operations

Let A and B be any sets, with U the universal set.

The **complement** of A, written A', is

$$\boldsymbol{A' = \{x \mid x \in U \text{ and } x \notin A\}.}$$

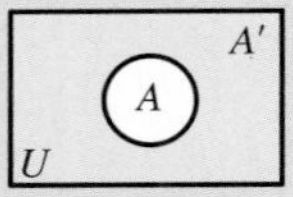

The **intersection** of A and B is

$$\boldsymbol{A \cap B = \{x \mid x \in A \text{ and } x \in B\}.}$$

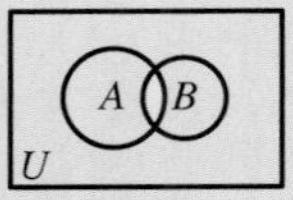

The **union** of A and B is

$$\boldsymbol{A \cup B = \{x \mid x \in A \text{ or } x \in B\}.}$$

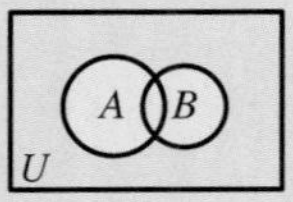

The **difference** of A and B is

$$\boldsymbol{A - B = \{x \mid x \in A \text{ and } x \notin B\}.}$$

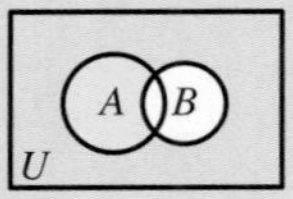

The **Cartesian product** of A and B is

$$\boldsymbol{A \times B = \{(x, y) \mid x \in A \text{ and } y \in B\}.}$$

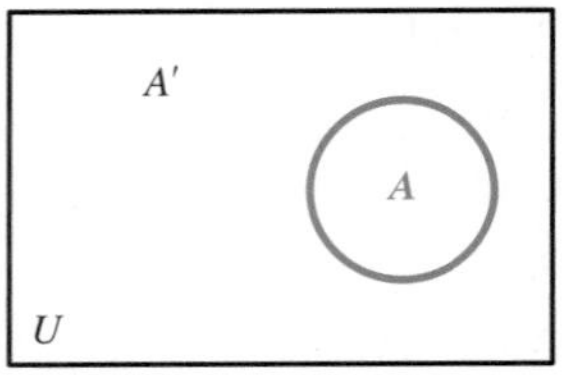

FIGURE 10

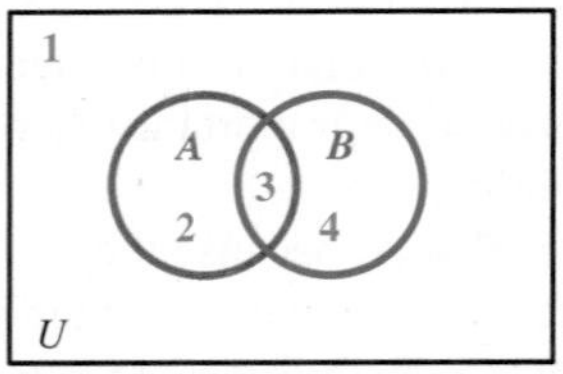

Numbering is arbitrary. The numbers indicate four regions, not cardinal numbers or elements.

FIGURE 11

Venn Diagrams

When dealing with a single set, we can use a Venn diagram as seen in Figure 10. The universal set U is divided into two regions, one representing set A and the other representing set A'. Two sets A and B within the universal set suggest a Venn diagram as seen in Figure 11, where the four resulting regions have been numbered to provide a convenient way to refer to them. (The numbering is arbitrary.) Region 1 includes those elements outside of both set A and set B. Region 2 includes the elements belonging to A but not to B. Region 3 includes those elements belonging to both A and B. How would you describe the elements of region 4?

EXAMPLE 10 Shading Venn Diagrams to Represent Sets

Draw a Venn diagram similar to Figure 11 and shade the region or regions representing the following sets.

(a) $A' \cap B$ **(b)** $A' \cup B'$

SOLUTION

(a) Refer to Figure 11. Set A' contains all the elements outside of set A—in other words, the elements in regions 1 and 4. Set B is made up of the elements in regions 3 and 4. The intersection of sets A' and B is made up of the elements in the region common to (1 and 4) and (3 and 4), which is region 4. Thus, $A' \cap B$ is represented by region 4, shown in color in Figure 12. This region can also be described as $B - A$.

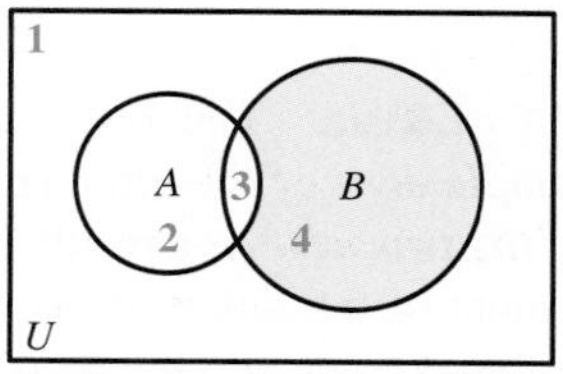

FIGURE 12

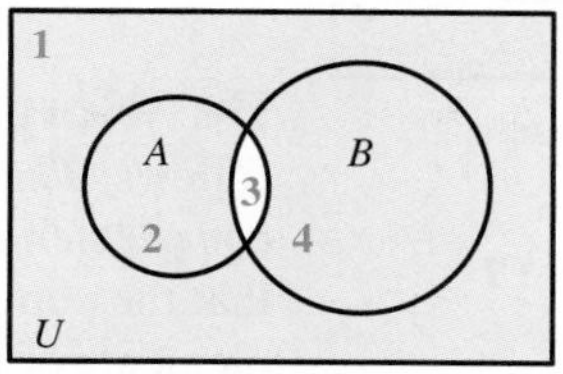

FIGURE 13

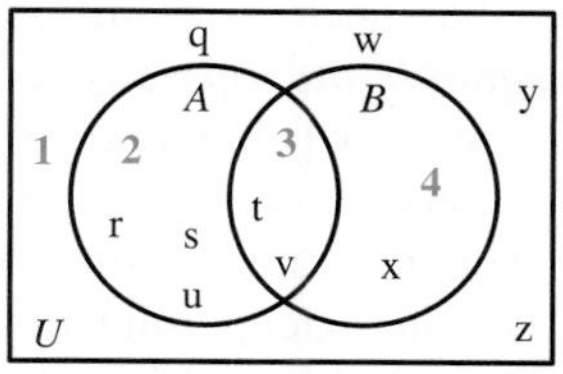

FIGURE 14

(b) Again, set A' is represented by regions 1 and 4, while B' is made up of regions 1 and 2. The union of A' and B', the set $A' \cup B'$, is made up of the elements belonging to the union of regions 1, 2, and 4, which are in color in Figure 13.

EXAMPLE 11 Locating Elements in a Venn Diagram

Place the elements of the sets in their proper locations in a Venn diagram.

Let $U = \{q, r, s, t, u, v, w, x, y, z\}$, $A = \{r, s, t, u, v\}$, and $B = \{t, v, x\}$.

SOLUTION

Because $A \cap B = \{t, v\}$, elements t and v are placed in region 3 in Figure 14. The remaining elements of A, that is r, s, and u, go in region 2. The figure shows the proper placement of all other elements.

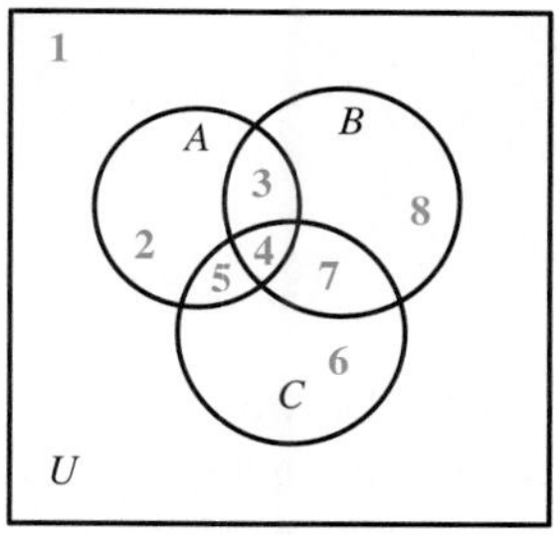

Numbering is arbitrary. The numbers indicate regions, not cardinal numbers or elements.

FIGURE 15

To include three sets A, B, and C in a universal set, draw a Venn diagram as in Figure 15, where again an arbitrary numbering of the regions is shown.

EXAMPLE 12 Shading a Set in a Venn Diagram

Shade the set $(A' \cap B') \cap C$ in a Venn diagram similar to the one in Figure 15.

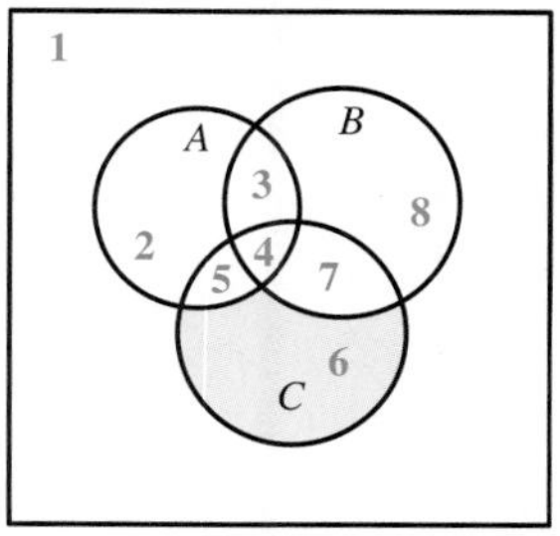

$(A' \cap B') \cap C$

FIGURE 16

SOLUTION

Work first inside the parentheses. As shown in Figure 16, set A' is made up of the regions outside set A, or regions 1, 6, 7, and 8. Set B' is made up of regions 1, 2, 5, and 6. The intersection of these sets is given by the overlap of regions 1, 6, 7, 8 and 1, 2, 5, 6, or regions 1 and 6. For the final Venn diagram, find the intersection of regions 1 and 6 with set C. As seen in Figure 16, set C is made up of regions 4, 5, 6, and 7. The overlap of regions 1, 6 and 4, 5, 6, 7 is region 6, the region in color in Figure 16.

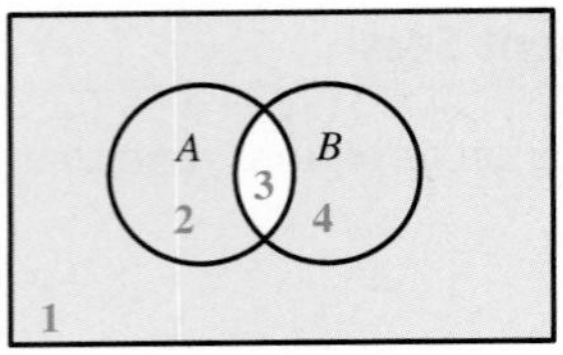

$(A \cap B)'$ is shaded.

(a)

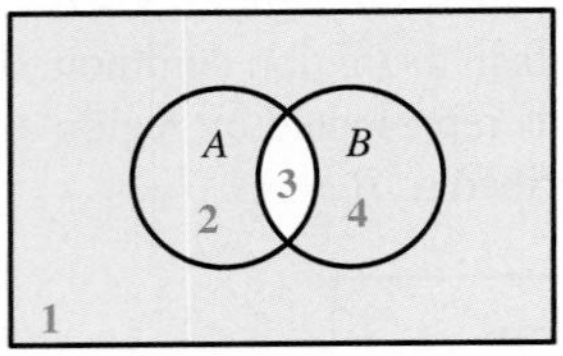

$A' \cup B'$ is shaded.

(b)

FIGURE 17

EXAMPLE 13 Verifying a Statement Using a Venn Diagram

Is the statement $(A \cap B)' = A' \cup B'$ true for every choice of sets A and B?

SOLUTION

To help decide, use the regions labeled in Figure 11. Set $A \cap B$ is made up of region 3, so that $(A \cap B)'$ is made up of regions 1, 2, and 4. These regions are in color in Figure 17(a).

To find a Venn diagram for set $A' \cup B'$, first check that A' is made up of regions 1 and 4, while set B' includes regions 1 and 2. Finally, $A' \cup B'$ is made up of regions 1 and 4, or 1 and 2, that is, regions 1, 2, and 4. These regions are in color in Figure 17(b).

The fact that the same regions are in color in both Venn diagrams suggests that

$$(A \cap B)' = A' \cup B'.$$

De Morgan's Laws The result of Example 13 can be stated in words as follows: ***The complement of the intersection of two sets is equal to the union of the complements of the two sets.*** As a result, it is natural to ask ourselves whether it is true that the complement of the *union* of two sets is equal to the *intersection* of the complements of the two sets (where the words "intersection" and "union" are substituted for each other). It turns out that this was investigated by the British logician Augustus De Morgan (1806–1871) and was found to be true. (See the margin note on page 23.) DeMorgan's two laws for sets follow.

De Morgan's Laws

For any sets A and B,

$$(A \cap B)' = A' \cup B' \quad \text{and} \quad (A \cup B)' = A' \cap B'.$$

The Venn diagrams in Figure 17 strongly suggest the truth of the first of De Morgan's laws. They provide a *conjecture*. Actual proofs of De Morgan's laws would require methods used in more advanced courses on set theory.

EXAMPLE 14 Describing Regions in Venn Diagrams Using Symbols

For the Venn diagrams, write a symbolic description of the region in color, using A, B, C, $\cap$, $\cup$, $-$, and $'$ as necessary.

(a)

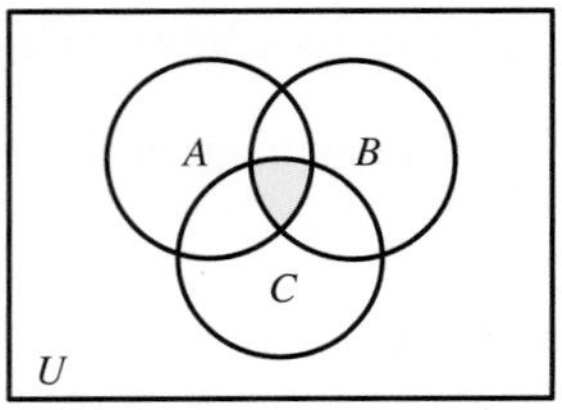

(b)

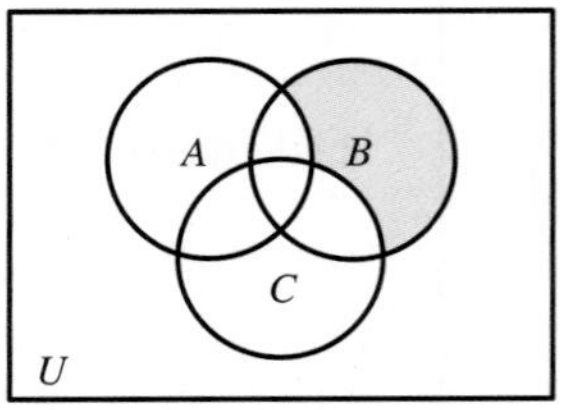

(c) Refer to the figure in part (b) and give two additional ways of describing the region in color.

SOLUTION

(a) The region in color belongs to all three sets, A and B and C. Therefore, the region corresponds to

$$A \cap B \cap C.$$

(b) The region in color is in set B and is not in A and is not in C. Because it is not in A, it is in A', and similarly it is in C'. The region is, therefore, in B and in A' and in C', and corresponds to

$$B \cap A' \cap C'.$$

(c) The region in color includes all of B, except for the regions belonging to either A or C. This suggests the idea of set difference. The region may be described as

$$B - (A \cup C), \quad \text{or equivalently,} \quad B \cap (A \cup C)'.$$

2.3 EXERCISES

Match each term in Column I with its proper designation in Column II. Assume that A and B are sets.

I

1. the intersection of A and B

2. the union of A and B

3. the difference of A and B

4. the complement of A

5. the Cartesian product of A and B

6. the difference of B and A

II

A. the set of elements in A that are not in B

B. the set of elements common to both A and B

C. the set of elements in the universe that are not in A

D. the set of elements in B that are not in A

E. the set of ordered pairs such that each first element is from A and each second element is from B, with every element of A paired with every element of B

F. the set of elements that are in A or in B or in both A and B

Perform the indicated operations.

Let $U = \{a, b, c, d, e, f, g\}$, $X = \{a, c, e, g\}$, $Y = \{a, b, c\}$, *and* $Z = \{b, c, d, e, f\}$.

7. $X \cap Y$

8. $X \cup Y$

9. $Y \cup Z$

10. $Y \cap Z$

11. $X \cup U$

12. $Y \cap U$

13. X'

14. Y'

15. $X' \cap Y'$

16. $X' \cap Z$

17. $X \cup (Y \cap Z)$

18. $Y \cap (X \cup Z)$

19. $(Y \cap Z') \cup X$

20. $(X' \cup Y') \cup Z$

21. $(Z \cup X')' \cap Y$

22. $(Y \cap X')' \cup Z'$

23. $X - Y$

24. $Y - X$

25. $X \cap (X - Y)$

26. $Y \cup (Y - X)$

27. $X' - Y$

28. $Y' - X$

29. $(X \cap Y') \cup (Y \cap X')$

30. $(X \cap Y') \cap (Y \cap X')$

Describe each set in words.

31. $A \cup (B' \cap C')$

32. $(A \cap B') \cup (B \cap A')$

33. $(C - B) \cup A$

34. $B \cap (A' - C)$

35. $(A - C) \cup (B - C)$

36. $(A' \cap B') \cup C'$

Adverse Effects of Alcohol and Tobacco *The table lists some common adverse effects of prolonged tobacco and alcohol use.*

Tobacco	Alcohol
Emphysema, e	Liver damage, l
Heart damage, h	Brain damage, b
Cancer, c	Heart damage, h

Let T be the set of listed effects of tobacco and A be the set of listed effects of alcohol. Find each set.

37. the smallest possible universal set U that includes all the effects listed

38. A'

39. T'

40. $T \cap A$

41. $T \cup A$

42. $T \cap A'$

Describe in words each set in Exercises 43–48.

Let U = the set of all tax returns,
A = the set of all tax returns with itemized deductions,
B = the set of all tax returns showing business income,
C = the set of all tax returns filed in 2005,
D = the set of all tax returns selected for audit.

Form 1040A
Department of the Treasury—Internal Revenue Service
U.S. Individual Income Tax Return (99) 2005
IRS Use Only—Do not write or staple in this space.
OMB No. 1545-0074
Label (See page 18.)
LABEL HERE
Your first name and initial | Last name | Your social security number
If a joint return, spouse's first name and initial | Last name | Spouse's social security number
Use the IRS label. Otherwise, please print or type.
Home address (number and street). If you have a P.O. box, see page 18. | Apt. no.
You **must** enter your SSN(s) above.
City, town or post office, state, and ZIP code. If you have a foreign address, see page 18.
Checking a box below will not change your tax or refund.
Presidential Election Campaign ▶ Check here if you, or your spouse if filing jointly, want $3 to go to this fund (see page 18) ▶ You Spouse
Filing status (Check only one box.)
1 Single
2 Married filing jointly (even if only one had income)
3 Married filing separately. Enter spouse's SSN above and full name here. ▶
4 Head of household (with qualifying person). (See page 19.) If the qualifying person is a child but not your dependent, enter this child's name here. ▶
5 Qualifying widow(er) with dependent child (see page 19)
Exemptions
6a Yourself. If someone can claim you as a dependent, **do not** check box 6a.
b Spouse
c Dependents: (1) First name Last name | (2) Dependent's social security number | (3) Dependent's relationship to you | (4) ✓ if qualifying child for child tax credit (see page 21)
If more than six dependents, see page 21.
Boxes checked on 6a and 6b
No. of children on 6c who: • lived with you • did not live with you due to divorce or separation (see page 22)
Dependents on 6c not entered above
d Total number of exemptions claimed.
Add numbers on lines above ▶
Income
Attach Form(s) W-2 here. Also attach Form(s) 1099-R if tax was withheld.
If you did not get a W-2, see page 24.
Enclose, but do not attach, any payment.
7 Wages, salaries, tips, etc. Attach Form(s) W-2. 7
8a **Taxable** interest. Attach Schedule 1 if required. 8a
b **Tax-exempt** interest. **Do not** include on line 8a. 8b
9a Ordinary dividends. Attach Schedule 1 if required. 9a
b Qualified dividends (see page 25). 9b
10 Capital gain distributions (see page 25). 10
11a IRA distributions. 11a | 11b Taxable amount (see page 25). 11b
12a Pensions and annuities. 12a | 12b Taxable amount (see page 26). 12b
13 Unemployment compensation and Alaska Permanent Fund dividends. 13
14a Social security benefits. 14a | 14b Taxable amount (see page 28). 14b
15 Add lines 7 through 14b (far right column). This is your **total income**. ▶ 15
Adjusted gross income
16 Educator expenses (see page 28). 16
17 IRA deduction (see page 28). 17
18 Student loan interest deduction (see page 31). 18
19 Tuition and fees deduction (see page 32). 19
20 Add lines 16 through 19. These are your **total adjustments**. 20
21 Subtract line 20 from line 15. This is your **adjusted gross income**. ▶ 21
For Disclosure, Privacy Act, and Paperwork Reduction Act Notice, see page 58. Cat. No. 11327A Form **1040A** (2005)

43. $B \cup C$

44. $A \cap D$

45. $C - A$

46. $D \cup A'$

47. $(A \cup B) - D$

48. $(C \cap A) \cap B'$

Assuming that A and B represent any two sets, identify each statement as either always true *or* not always true.

49. $A \subseteq (A \cup B)$

50. $A \subseteq (A \cap B)$

51. $(A \cap B) \subseteq A$

52. $(A \cup B) \subseteq A$

53. $n(A \cup B) = n(A) + n(B)$

54. $n(A \cup B) = n(A) + n(B) - n(A \cap B)$

For Exercises 55–60, use your results in parts (a) and (b) to answer part (c).

Let $U = \{1, 2, 3, 4, 5\}$, $X = \{1, 3, 5\}$, $Y = \{1, 2, 3\}$, *and* $Z = \{3, 4, 5\}$.

55. **(a)** Find $X \cup Y$.
(b) Find $Y \cup X$.
(c) State a conjecture.

56. **(a)** Find $X \cap Y$.
(b) Find $Y \cap X$.
(c) State a conjecture.

57. **(a)** Find $X \cup (Y \cup Z)$.
(b) Find $(X \cup Y) \cup Z$.
(c) State a conjecture.

58. **(a)** Find $X \cap (Y \cap Z)$.
(b) Find $(X \cap Y) \cap Z$.
(c) State a conjecture.

59. **(a)** Find $(X \cup Y)'$.
(b) Find $X' \cap Y'$.
(c) State a conjecture.

60. **(a)** Find $(X \cap Y)'$.
(b) Find $X' \cup Y'$.
(c) State a conjecture.

In Exercises 61 and 62, let set X equal the different letters in your last name.

61. Find $X \cup \emptyset$ and state a conjecture.

62. Find $X \cap \emptyset$ and state a conjecture.

Decide whether each statement is true or false.

63. $(3, 2) = (5 - 2, 1 + 1)$

64. $(10, 4) = (7 + 3, 5 - 1)$

65. $(6, 3) = (3, 6)$

66. $(2, 13) = (13, 2)$

67. $\{6, 3\} = \{3, 6\}$

68. $\{2, 13\} = \{13, 2\}$

69. $\{(1, 2), (3, 4)\} = \{(3, 4), (1, 2)\}$

70. $\{(5, 9), (4, 8), (4, 2)\} = \{(4, 8), (5, 9), (4, 2)\}$

Find $A \times B$ and $B \times A$, for A and B defined as follows.

71. $A = \{2, 8, 12\}$, $B = \{4, 9\}$

72. $A = \{3, 6, 9, 12\}$, $B = \{6, 8\}$

73. $A = \{d, o, g\}$, $B = \{p, i, g\}$

74. $A = \{b, l, u, e\}$, $B = \{r, e, d\}$

For the sets specified in Exercises 75–78, use the given information to find $n(A \times B)$ and $n(B \times A)$.

75. the sets in Exercise 71

76. the sets in Exercise 73

77. $n(A) = 35$ and $n(B) = 6$

78. $n(A) = 13$ and $n(B) = 5$

Find the cardinal number specified.

79. If $n(A \times B) = 72$ and $n(A) = 12$, find $n(B)$.

80. If $n(A \times B) = 300$ and $n(B) = 30$, find $n(A)$.

Place the elements of these sets in the proper locations on the given Venn diagram.

81. Let $U = \{a, b, c, d, e, f, g\}$,
$A = \{b, d, f, g\}$,
$B = \{a, b, d, e, g\}$.

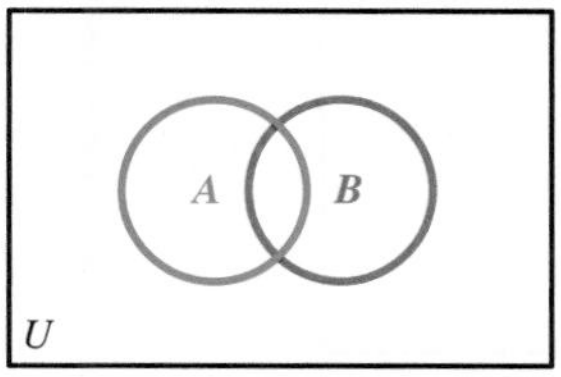

82. Let $U = \{5, 6, 7, 8, 9, 10, 11, 12, 13\}$,
$M = \{5, 8, 10, 11\}$,
$N = \{5, 6, 7, 9, 10\}$.

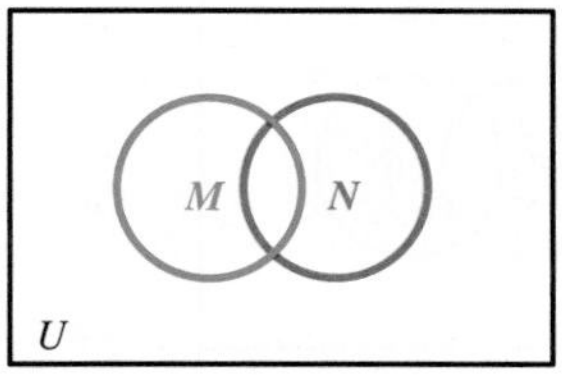

Use a Venn diagram similar to the one shown here to shade each set.

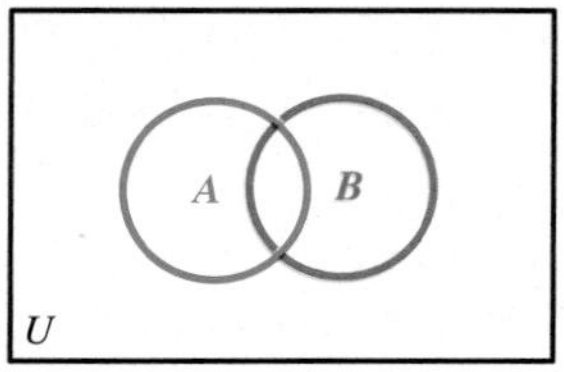

83. $B \cap A'$ **84.** $A \cup B$ **85.** $A' \cup B$

86. $A' \cap B'$ **87.** $B' \cup A$ **88.** $A' \cup A$

89. $B' \cap B$ **90.** $A \cap B'$ **91.** $B' \cup (A' \cap B')$

92. $(A \cap B) \cup B$ **93.** U' **94.** $\emptyset'$

95. Let $U = \{m, n, o, p, q, r, s, t, u, v, w\}$,
$A = \{m, n, p, q, r, t\}$,
$B = \{m, o, p, q, s, u\}$,
$C = \{m, o, p, r, s, t, u, v\}$.

Place the elements of these sets in the proper location on a Venn diagram similar to the one shown at the right.

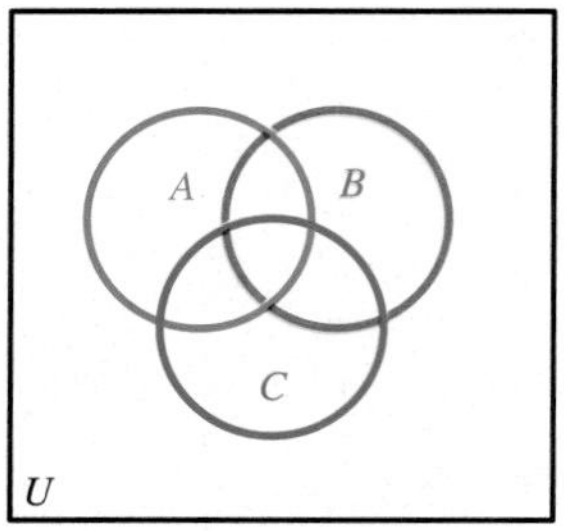

96. Let $U = \{1, 2, 3, 4, 5, 6, 7, 8, 9\}$,
$A = \{1, 3, 5, 7\}$,
$B = \{1, 3, 4, 6, 8\}$,
$C = \{1, 4, 5, 6, 7, 9\}$.

Place the elements of these sets in the proper location on a Venn diagram.

Use a Venn diagram to shade each set.

97. $(A \cap B) \cap C$ **98.** $(A \cap C') \cup B$ **99.** $(A \cap B) \cup C'$

100. $(A' \cap B) \cap C$ **101.** $(A' \cap B') \cap C$ **102.** $(A \cup B) \cup C$

103. $(A \cap B') \cup C$ **104.** $(A \cap C') \cap B$ **105.** $(A \cap B') \cap C'$

106. $(A' \cap B') \cup C$ **107.** $(A' \cap B') \cup C'$ **108.** $(A \cap B)' \cup C$

Write a description of each shaded area. Use the symbols A, B, C, ∩, ∪, −, and ′ as necessary. More than one answer may be possible.

109.

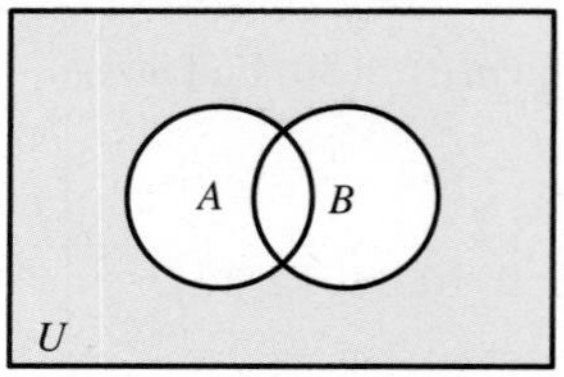

110.

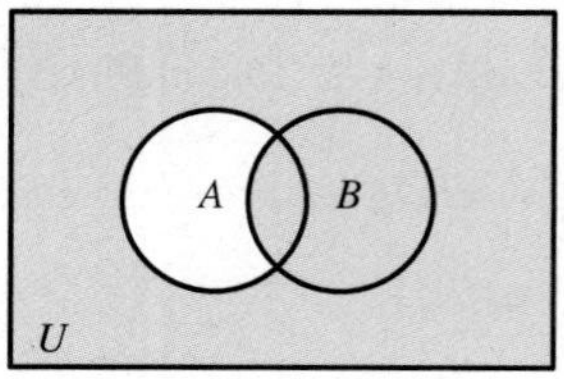

111.

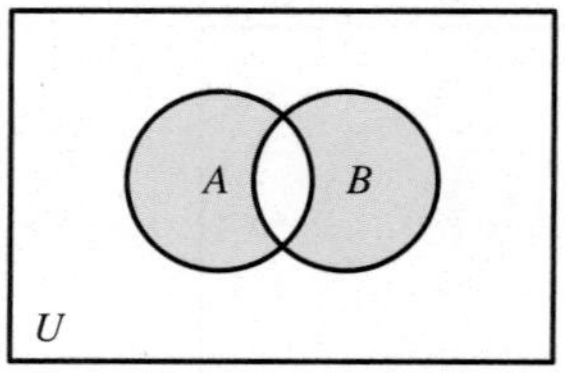

112.

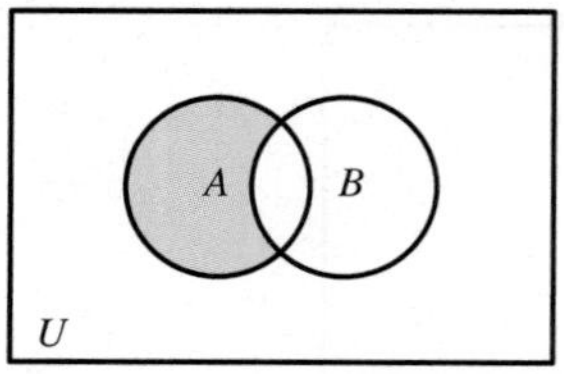

113.
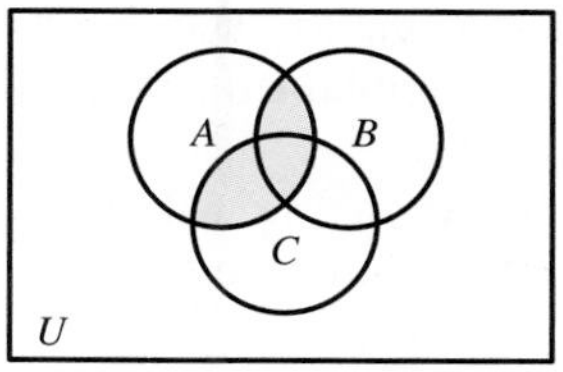

114.
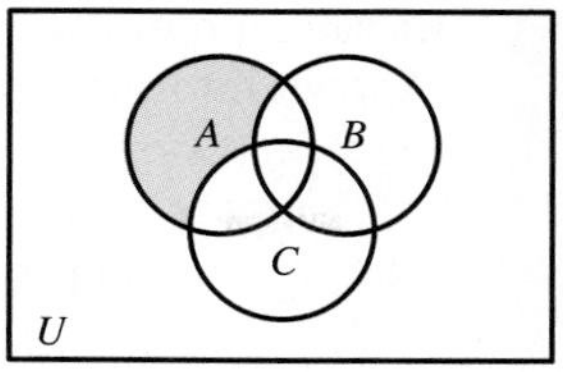

115.
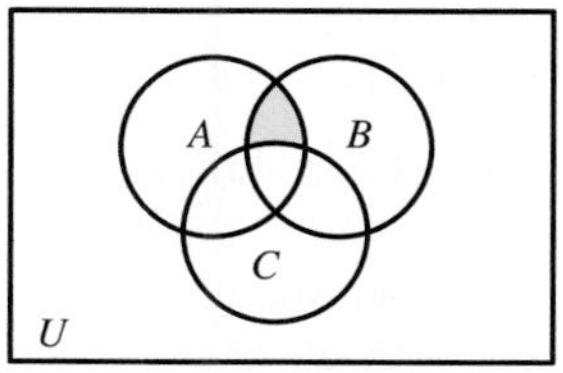

116.
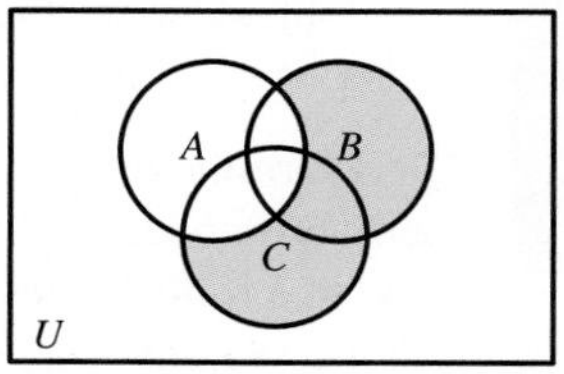

Suppose A and B are sets. Describe the conditions under which each statement would be true.

117. $A = A - B$

118. $A = B - A$

119. $A = A - \emptyset$

120. $A = \emptyset - A$

121. $A \cup \emptyset = \emptyset$

122. $A \cap \emptyset = \emptyset$

123. $A \cap \emptyset = A$

124. $A \cup \emptyset = A$

125. $A \cup A = \emptyset$

126. $A \cap A = \emptyset$

127. $A \cup B = A$

128. $A \cap B = B$

For Exercises 129–135, draw two appropriate Venn diagrams to decide whether the given statement is always true *or* not always true.

129. $A \cap A' = \emptyset$

130. $A \cup A' = U$

131. $(A \cap B) \subseteq A$

132. $(A \cup B) \subseteq A$

133. If $A \subseteq B$, then $A \cup B = A$.

134. If $A \subseteq B$, then $A \cap B = B$.

135. $(A \cup B)' = A' \cap B'$ (De Morgan's second law)

136. Give examples of how a language such as English, Spanish, Arabic, or Vietnamese can have an advantage over the symbolic language of mathematics.

137. If A and B are sets, is it necessarily true that $n(A - B) = n(A) - n(B)$?

138. If $Q = \{x \mid x \text{ is a rational number}\}$ and $H = \{x \mid x \text{ is an irrational number}\}$, describe each set.
(a) $Q \cup H$
(b) $Q \cap H$

2.4 Surveys and Cardinal Numbers

Surveys • Cardinal Number Formula

Surveys Problems involving sets of people (or other objects) sometimes require analyzing known information about certain subsets to obtain cardinal numbers of other subsets. In this section, we apply three problem-solving techniques

to such problems: Venn diagrams, cardinal number formulas, and tables. The "known information" is quite often (although not always) obtained by administering a survey.

Suppose a group of students on a college campus are questioned about some selected musical performers, and the following information is produced.

33 like Kenny Chesney.	15 like Kenny and Carrie.
32 like Beyoncé.	14 like Beyoncé and Carrie.
28 like Carrie Underwood.	5 like all three performers.
11 like Kenny and Beyoncé.	7 like none of these performers.

To determine the total number of students surveyed, we cannot just add the eight numbers above because there is some overlapping. For example, in Figure 18, the 33 students who like Kenny Chesney should not be positioned in region b but should be distributed among regions b, c, d, and e, in a way that is consistent with all of the given data. (Region b actually contains those students who like Kenny but do not like Beyoncé and do not like Carrie.)

Because, at the start, we do not know how to distribute the 33 who like Kenny, we look first for some more manageable data. The smallest total listed, the 5 students who like all three singers, can be placed in region d (the intersection of the three sets). The 7 who like none of the three must go into region a. Then, the 11 who like Kenny and Beyoncé must go into regions d and e. Because region d already contains 5 students, we must place $11 - 5 = 6$ in region e. Because 15 like Kenny and Carrie (regions c and d), we place $15 - 5 = 10$ in region c. Now that regions c, d, and e contain 10, 5, and 6 students respectively, region b receives $33 - 10 - 5 - 6 = 12$. By similar reasoning, all regions are assigned their correct numbers, as shown in Figure 19.

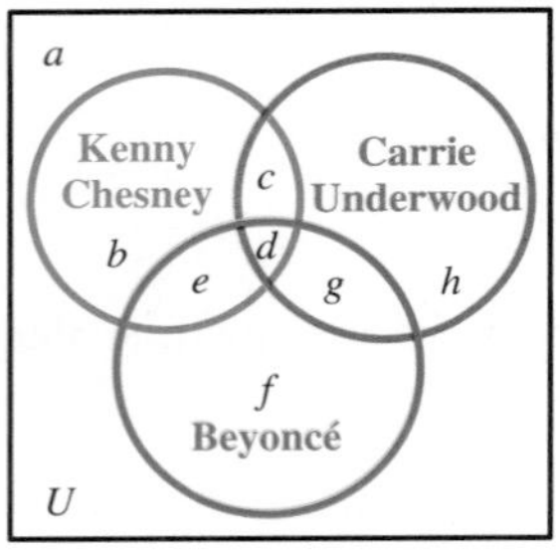

FIGURE 18

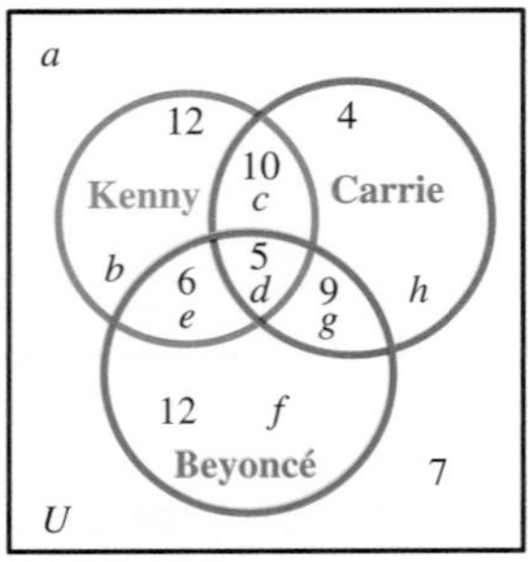

FIGURE 19

EXAMPLE 1 Analyzing a Survey

Using the survey data on student preferences for performers, as summarized in Figure 19, answer the following questions.

(a) How many students like Carrie Underwood only?
(b) How many students like exactly two performers?
(c) How many students were surveyed?

SOLUTION

(a) A student who likes Carrie only does not like Kenny and does not like Beyoncé. These students are inside the regions for Carrie and outside the regions for Kenny and Beyoncé. Region h is the appropriate region in Figure 19, and we see that four students like Carrie only.

(b) The students in regions c, e, and g like exactly two performers. The total number of such students is

$$10 + 6 + 9 = 25.$$

(c) Because each student surveyed has been placed in exactly one region of Figure 19, the total number surveyed is the sum of the numbers in all eight regions:

$$7 + 12 + 10 + 5 + 6 + 12 + 9 + 4 = 65.$$

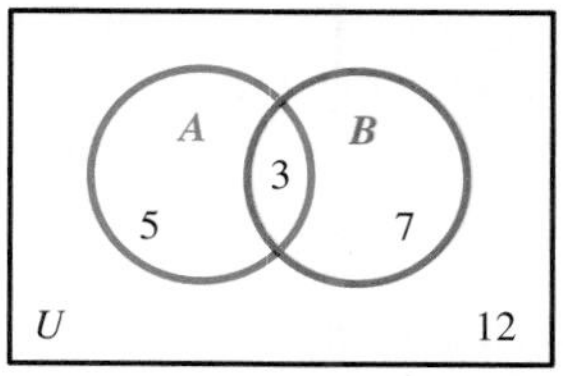

FIGURE 20

Cardinal Number Formula If the numbers shown in Figure 20 are the cardinal numbers of the individual regions, then $n(A) = 5 + 3 = 8$, $n(B) = 3 + 7 = 10$, $n(A \cap B) = 3$, and $n(A \cup B) = 5 + 3 + 7 = 15$. Notice that $n(A \cup B) = n(A) + n(B) - n(A \cap B)$ because $15 = 8 + 10 - 3$. This relationship is true for any two sets A and B.

Cardinal Number Formula

For any two sets A and B,

$$n(A \cup B) = n(A) + n(B) - n(A \cap B).$$

This formula can be rearranged to find any one of its four terms when the others are known.

EXAMPLE 2 Applying the Cardinal Number Formula

Find $n(A)$ if $n(A \cup B) = 22$, $n(A \cap B) = 8$, and $n(B) = 12$.

SOLUTION

The formula above can be rearranged. Thus,

$$\begin{aligned} n(A) &= n(A \cup B) - n(B) + n(A \cap B) \\ &= 22 - 12 + 8 \\ &= 18. \end{aligned}$$

Sometimes, even when information is presented as in Example 2, it is more convenient to fit that information into a Venn diagram as in Example 1.

EXAMPLE 3 Analyzing Data in a Report

Robert Hurst is a section chief for an electric utility company. Hurst recently submitted the following report to the management of the utility.

My section includes 100 employees, with

T = the set of employees who can cut tall trees,

P = the set of employees who can climb poles,

W = the set of employees who can splice wire.

$n(T) = 45$	$n(T \cap P) = 28$	$n(T \cap P \cap W) = 11$
$n(P) = 50$	$n(P \cap W) = 20$	$n(T' \cap P' \cap W') = 9$
$n(W) = 57$	$n(T \cap W) = 25$	

Is this a valid report? If not, why?

SOLUTION

The data supplied by Hurst are reflected in Figure 21. The sum of the numbers in the diagram gives the total number of employees in the section:

$$9 + 3 + 14 + 23 + 11 + 9 + 17 + 13 = 99.$$

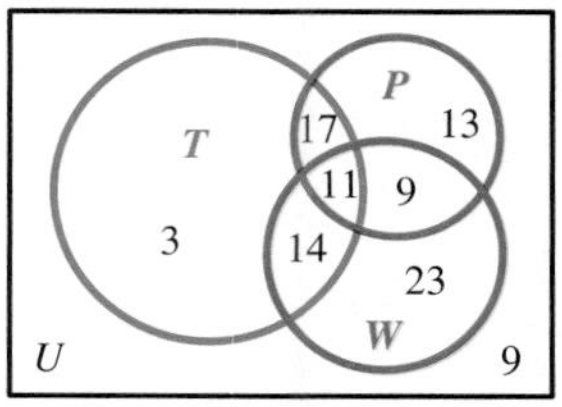

FIGURE 21

Hurst claimed to have 100 employees, but his data indicate only 99. The management decided that this error meant that Hurst did not qualify as section chief. He was reassigned as night-shift information operator at the North Pole. (The moral: Hurst should have taken this course.) ■

Sometimes information appears in a table rather than a Venn diagram, but the basic ideas of union and intersection still apply.

EXAMPLE 4 Analyzing Data in a Table

The officer in charge of the cafeteria on a North Carolina military base wanted to know if the beverage that enlisted men and women preferred with lunch depended on their ages. On a given day, she categorized her lunch patrons according to age and preferred beverage, recording the results in a table.

		Beverage			
		Cola (C)	**Iced Tea (I)**	**Sweet Tea (S)**	**Totals**
	18–25 (Y)	45	10	35	90
Age	**26–33 (M)**	20	25	30	75
	Over 33 (O)	5	30	20	55
	Totals	70	65	85	220

Using the letters in the table, find the number of people in each of the following sets.

(a) $Y \cap C$ **(b)** $O' \cup I$

SOLUTION

(a) The set Y includes all personnel represented across the top row of the table (90 in all), while C includes the 70 down the left column. The intersection of these two sets is just the upper left entry: 45 people.

(b) The set O' excludes the bottom row, so it includes the first and second rows. The set I includes the middle column only. The union of the two sets represents

$$45 + 10 + 35 + 20 + 25 + 30 + 30 = 195 \text{ people.}$$ ■

2.4 EXERCISES

Use the numerals representing cardinalities in the Venn diagrams to give the cardinality of each set specified.

1.

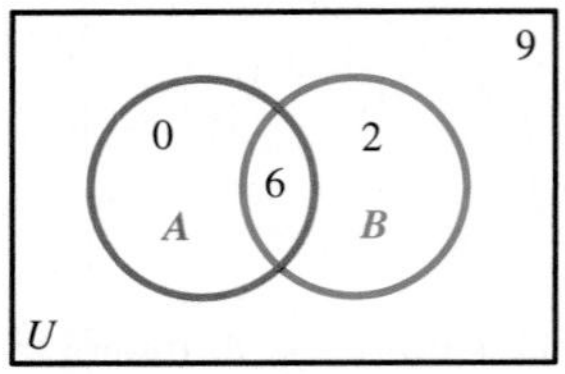

(a) $A \cap B$ **(b)** $A \cup B$ **(c)** $A \cap B'$
(d) $A' \cap B$ **(e)** $A' \cap B'$

2.

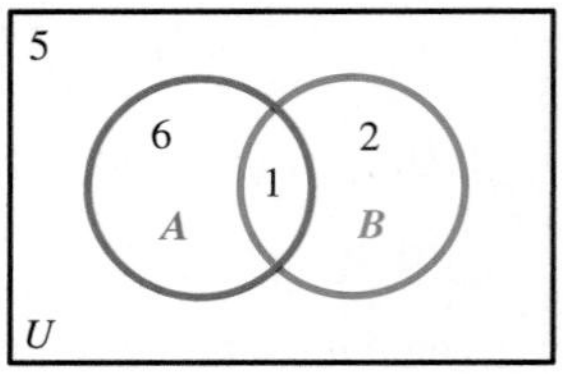

(a) $A \cap B$ **(b)** $A \cup B$ **(c)** $A \cap B'$
(d) $A' \cap B$ **(e)** $A' \cap B'$

3.

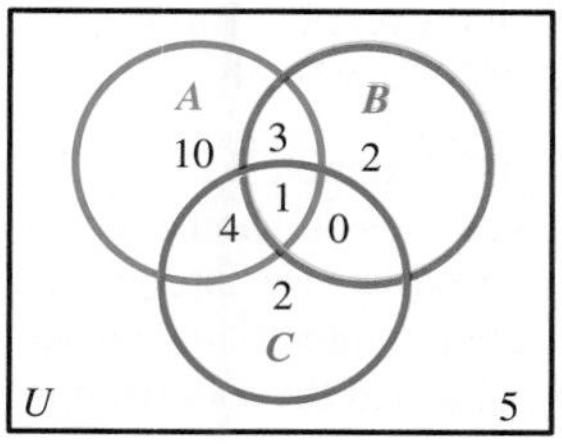

(a) $A \cap B \cap C$
(b) $A \cap B \cap C'$
(c) $A \cap B' \cap C$
(d) $A' \cap B \cap C$
(e) $A' \cap B' \cap C$
(f) $A \cap B' \cap C'$
(g) $A' \cap B \cap C'$
(h) $A' \cap B' \cap C'$

4.

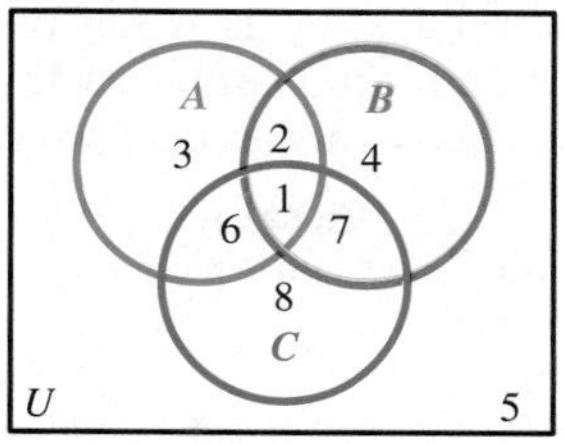

(a) $A \cap B \cap C$
(b) $A \cap B \cap C'$
(c) $A \cap B' \cap C$
(d) $A' \cap B \cap C$
(e) $A' \cap B' \cap C$
(f) $A \cap B' \cap C'$
(g) $A' \cap B \cap C'$
(h) $A' \cap B' \cap C'$

In Exercises 5–10, make use of an appropriate formula.

5. Find the value of $n(A \cup B)$ if $n(A) = 8$, $n(B) = 14$, and $n(A \cap B) = 5$.

6. Find the value of $n(A \cup B)$ if $n(A) = 16$, $n(B) = 28$, and $n(A \cap B) = 9$.

7. Find the value of $n(A \cap B)$ if $n(A) = 15$, $n(B) = 12$, and $n(A \cup B) = 25$.

8. Find the value of $n(A \cap B)$ if $n(A) = 20$, $n(B) = 14$, and $n(A \cup B) = 30$.

9. Find the value of $n(A)$ if $n(B) = 35$, $n(A \cap B) = 15$, and $n(A \cup B) = 55$.

10. Find the value of $n(B)$ if $n(A) = 20$, $n(A \cap B) = 6$, and $n(A \cup B) = 30$.

To prepare for the survey problems that follow later in this exercise set, draw an appropriate Venn diagram and use the given information to fill in the number of elements in each region.

11. $n(A) = 19$, $n(B) = 13$, $n(A \cup B) = 25$, $n(A') = 11$

12. $n(U) = 43$, $n(A) = 25$, $n(A \cap B) = 5$, $n(B') = 30$

13. $n(A') = 25$, $n(B) = 28$, $n(A' \cup B') = 40$, $n(A \cap B) = 10$

14. $n(A \cup B) = 15$, $n(A \cap B) = 8$, $n(A) = 13$, $n(A' \cup B') = 11$

15. $n(A) = 57$, $n(A \cap B) = 35$, $n(A \cup B) = 81$, $n(A \cap B \cap C) = 15$, $n(A \cap C) = 21$, $n(B \cap C) = 25$, $n(C) = 49$, $n(B') = 52$

16. $n(A) = 24$, $n(B) = 24$, $n(C) = 26$, $n(A \cap B) = 10$, $n(B \cap C) = 8$, $n(A \cap C) = 15$, $n(A \cap B \cap C) = 6$, $n(U) = 50$

17. $n(A) = 15$, $n(A \cap B \cap C) = 5$, $n(A \cap C) = 13$, $n(A \cap B') = 9$, $n(B \cap C) = 8$, $n(A' \cap B' \cap C') = 21$, $n(B \cap C') = 3$, $n(B \cup C) = 32$

18. $n(A \cap B) = 21$, $n(A \cap B \cap C) = 6$, $n(A \cap C) = 26$, $n(B \cap C) = 7$, $n(A \cap C') = 20$, $n(B \cap C') = 25$, $n(C) = 40$, $n(A' \cap B' \cap C') = 2$

Use Venn diagrams to work each problem.

19. ***Writing and Producing Music*** Joe Long writes and produces albums for musicians. Last year, he worked on 10 such projects.

He wrote and produced 2 projects.
He wrote a total of 5 projects.
He produced a total of 7 projects.

(a) How many projects did he write but not produce?
(b) How many projects did he produce but not write?

Joe Long, Bob Gaudio, Tommy DeVito, and Frankie Valli
The Four Seasons

20. ***Compact Disc Collection*** Paula Story is a fan of the music of Paul Simon and Art Garfunkel. In her collection of 22 compact discs, she has the following:

5 on which both Simon and Garfunkel sing
8 on which Simon sings
7 on which Garfunkel sings
12 on which neither Simon nor Garfunkel sings.

(a) How many of her compact discs feature only Paul Simon?
(b) How many of her compact discs feature only Art Garfunkel?
(c) How many feature at least one of these two artists?

21. ***Viewer Response to Movies*** Buddy Vosburg, a child psychologist, was planning a study of response to certain aspects of the movies *The Lion King*, *Shrek*, and *Finding Nemo*. Upon surveying a group of 55 children, he determined the following:

17 had seen *The Lion King*
17 had seen *Shrek*,
23 had seen *Finding Nemo*
6 had seen *The Lion King* and *Shrek*
8 had seen *The Lion King* and *Finding Nemo*
10 had seen *Shrek* and *Finding Nemo*
2 had seen all three of these movies.

How many children had seen:
(a) exactly two of these movies?
(b) exactly one of these movies?
(c) none of these movies?
(d) *The Lion King* but neither of the others?

22. ***Financial Aid for Students*** At the University of Louisiana, half of the 48 mathematics majors were receiving federal financial aid as follows:

5 had Pell Grants
14 participated in the College Work Study Program
4 had TOPS scholarships
2 had TOPS scholarships and participated in Work Study.

Those with Pell Grants had no other federal aid.

How many of the 48 math majors had:
(a) no federal aid?
(b) more than one of these three forms of aid?
(c) federal aid other than these three forms?
(d) a TOPS scholarship or Work Study?

23. ***Cooking Habits*** Robert Hurst (Example 3 in the text) was again reassigned, this time to the home economics department of the electric utility. He interviewed 140 people in a suburban shopping center to find out some of their cooking habits. He obtained the following results:

58 use microwave ovens
63 use electric ranges
58 use gas ranges
19 use microwave ovens and electric ranges
17 use microwave ovens and gas ranges
4 use both gas and electric ranges
1 uses all three
2 cook only with solar energy.

There is a job opening in Siberia. Should he be reassigned yet one more time?

24. ***Wine Tasting*** The following list shows the preferences of 102 people at a fraternity party:

99 like Spañada
96 like Ripple
99 like Boone's Farm Apple Wine
95 like Spañada and Ripple
94 like Ripple and Boone's
96 like Spañada and Boone's
93 like all three.

How many people like:
(a) none of the three?
(b) Spañada, but not Ripple?
(c) anything but Boone's Farm?
(d) only Ripple?
(e) exactly two of these wines?

25. ***Poultry on a Farm*** Old MacDonald surveyed her flock with the following results. She has:

9 fat red roosters	18 thin brown roosters
2 fat red hens	6 thin red roosters
26 fat roosters	5 thin red hens
37 fat chickens	7 thin brown hens.

Answer the following questions about the flock. (*Hint:* You need a Venn diagram with circles for fat (assuming that thin is not fat), for male (a rooster is a male; a hen is a female), and for red (assume that

brown and red are opposites in the chicken world).) How many chickens are:
(a) fat?
(b) red?
(c) male?
(d) fat, but not male?
(e) brown, but not fat?
(f) red and fat?

26. ***Student Goals*** Carol Britz, who sells college textbooks, interviewed freshmen on a community college campus to find out the main goals of today's students.

Let W = the set of those who want to be wealthy,
F = the set of those who want to raise a family,
E = the set of those who want to become experts in their fields.

Carol's findings are summarized here:

$n(W) = 160$ $\quad n(E \cap F) = 90$
$n(F) = 140$ $\quad n(W \cap F \cap E) = 80$
$n(E) = 130$ $\quad n(E') = 95$
$n(W \cap F) = 95$ $\quad n[(W \cup F \cup E)'] = 10.$

Find the total number of students interviewed.

27. ***Hospital Patient Symptoms*** Suzanne Chustz conducted a survey among 75 patients admitted to the cardiac unit of a Santa Fe hospital during a two-week period.

Let B = the set of patients with high blood pressure
C = the set of patients with high cholesterol levels,
S = the set of patients who smoke cigarettes.

Suzanne's data are as follows:

$n(B) = 47$ $\quad n(B \cap S) = 33$
$n(C) = 46$ $\quad n(B \cap C) = 31$
$n(S) = 52$ $\quad n(B \cap C \cap S) = 21$
$n[(B \cap C) \cup (B \cap S) \cup (C \cap S)] = 51.$

Find the number of these patients who:
(a) had either high blood pressure or high cholesterol levels, but not both
(b) had fewer than two of the indications listed
(c) were smokers but had neither high blood pressure nor high cholesterol levels
(d) did not have exactly two of the indications listed.

28. ***Song Themes*** It was once said that Country-Western songs emphasize three basic themes: love, prison, and trucks. A survey of the local Country-Western radio station produced the following data:

12 songs about a truck driver who is in love while in prison
13 about a prisoner in love
28 about a person in love
18 about a truck driver in love
3 about a truck driver in prison who is not in love
2 about people in prison who are not in love and do not drive trucks
8 about people who are out of prison, are not in love, and do not drive trucks
16 about truck drivers who are not in prison.

(a) How many songs were surveyed?

Find the number of songs about:
(b) truck drivers
(c) prisoners
(d) truck drivers in prison
(e) people not in prison
(f) people not in love.

29. The figure below shows U divided into 16 regions by four sets, A, B, C, and D. Find the numbers of the regions belonging to each set.
(a) $A \cap B \cap C \cap D$
(b) $A \cup B \cup C \cup D$
(c) $(A \cap B) \cup (C \cap D)$
(d) $(A' \cap B') \cap (C \cup D)$

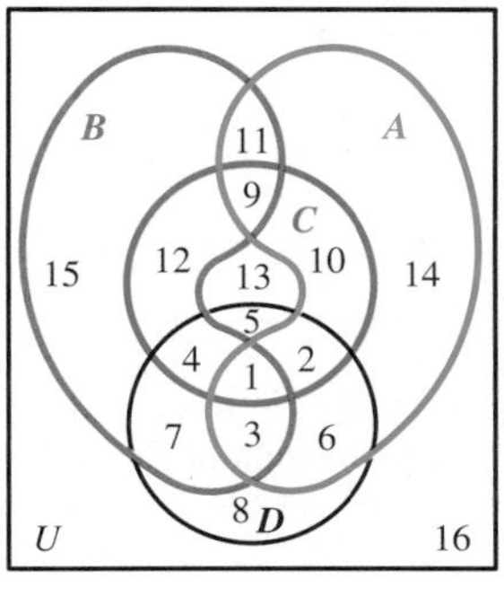

30. ***Sports Viewing Habits*** A survey of 130 television viewers revealed the following facts:

52 watch football
56 watch basketball
62 watch tennis
60 watch golf
21 watch football and basketball
19 watch football and tennis
22 watch basketball and tennis
27 watch football and golf
30 watch basketball and golf
21 watch tennis and golf
3 watch football, basketball, and tennis
15 watch football, basketball, and golf
10 watch football, tennis, and golf
10 watch basketball, tennis, and golf
3 watch all four of these sports
5 don't watch any of these four sports.

Use a diagram like the one in Exercise 29 to answer the following questions.

(a) How many of these viewers watch football, basketball, and tennis, but not golf?
(b) How many watch exactly one of these four sports?
(c) How many watch exactly two of these four sports?

Solve each problem.

31. ***Basketball Positions*** Dwaine Tomlinson runs a basketball program in California. On the first day of the season, 60 young women showed up and were categorized by age level and by preferred basketball position, as shown in the following table.

		Position			
		Guard (G)	Forward (F)	Center (N)	Totals
Age	Junior High (J)	9	6	4	19
	Senior High (S)	12	5	9	26
	College (C)	5	8	2	15
	Totals	26	19	15	60

Using the set labels (letters) in the table, find the number of players in each of the following sets.

(a) $J \cap G$ **(b)** $S \cap N$ **(c)** $N \cup (S \cap F)$
(d) $S' \cap (G \cup N)$ **(e)** $(S \cap N') \cup (C \cap G')$
(f) $N' \cap (S' \cap C')$

32. ***Army Housing*** A study of U.S. Army housing trends categorized personnel as commissioned officers (C), warrant officers (W), or enlisted (E), and categorized their living facilities as on-base (B), rented off-base (R), or owned off-base (O). One survey yielded the following data.

		Facilities			
		B	R	O	Totals
Personnel	C	12	29	54	95
	W	4	5	6	15
	E	374	71	285	730
	Totals	390	105	345	840

Find the number of personnel in each of the following sets.

(a) $W \cap O$ **(b)** $C \cup B$
(c) $R' \cup W'$ **(d)** $(C \cup W) \cap (B \cup R)$
(e) $(C \cap B) \cup (E \cap O)$ **(f)** $B \cap (W \cup R)'$

33. Could the information of Example 4 have been presented in a Venn diagram similar to those in Examples 1 and 3? If so, construct such a diagram. Otherwise, explain the essential difference of Example 4.

34. Explain how a cardinal number formula can be derived for the case where *three* sets occur. Specifically, give a formula relating $n(A \cup B \cup C)$ to $n(A)$, $n(B)$, $n(C)$, $n(A \cap B)$, $n(A \cap C)$, $n(B \cap C)$, and $n(A \cap B \cap C)$. Illustrate with a Venn diagram.

2.5 Infinite Sets and Their Cardinalities

One-to-One Correspondence and Equivalent Sets • The Cardinal Number $\aleph_0$ • Infinite Sets • Sets That Are Not Countable

The word **paradox** in Greek originally meant "wrong opinion" as opposed to orthodox, which meant "right opinion." Over the years, the word came to mean self-contradiction. An example is the statement "This sentence is false." By assuming it is true, we get a contradiction; likewise, by assuming it is false, we get a contradiction. Thus, it's a paradox.

Before the twentieth century it was considered a paradox that any set could be placed into one-to-one correspondence with a proper subset of itself. This paradox, called **Galileo's paradox** after the sixteenth-century mathematician and scientist **Galileo** (see the picture), is now explained by saying that the ability to make such a correspondence is how we distinguish infinite sets from finite sets. What is true for finite sets is not necessarily true for infinite sets.

One-to-One Correspondence and Equivalent Sets

As mentioned at the beginning of this chapter, most of the early work in set theory was done by Georg Cantor. He devoted much of his life to a study of the cardinal numbers of sets. Recall that the *cardinal number*, or *cardinality*, of a finite set is the number of elements that it contains. For example, the set $\{5, 9, 15\}$ contains 3 elements and has a cardinal number of 3. The cardinal number of $\emptyset$ is 0.

Cantor proved many results about the cardinal numbers of infinite sets. The proofs of Cantor are quite different from the type of proofs you may have seen in an algebra or geometry course. Because of the novelty of Cantor's methods, they were not quickly accepted by the mathematicians of his day. (In fact, some other aspects of Cantor's theory lead to paradoxes.) The results discussed here, however, are commonly accepted.

The idea of the cardinal number of an infinite set depends on the idea of one-to-one correspondence. For example, each of the sets $\{1, 2, 3, 4\}$ and $\{9, 10, 11, 12\}$ has four elements. Corresponding elements of the two sets could be paired off in the following manner (among many other ways):

$$\begin{array}{cccc} \{1, & 2, & 3, & 4\} \\ \updownarrow & \updownarrow & \updownarrow & \updownarrow \\ \{9, & 10, & 11, & 12\}. \end{array}$$

Such a pairing is a **one-to-one correspondence** between the two sets. The "one-to-one" refers to the fact that each element of the first set is paired with exactly one element of the second set and similarly each element of the second set is paired with exactly one element of the first set.

Two sets A and B which may be put in a one-to-one correspondence are said to be **equivalent.** Symbolically, this is written $\mathbf{A \sim B}$. The two sets shown above are equivalent but *not* equal.

The following correspondence between sets is not one-to-one because the elements 8 and 12 from the first set are both paired with the element 11 from the second set. These sets are not equivalent.

$$\begin{array}{ccc} \{1, & 8, & 12\} \\ \updownarrow & \searrow & \swarrow \\ \{6, & & 11\} \end{array}$$

It seems reasonable to say that if two non-empty sets have the same cardinal number, then a one-to-one correspondence can be established between the two sets. Also, if a one-to-one correspondence can be established between two sets, then the two sets must have the same cardinal number. These two facts are fundamental in discussing the cardinal numbers of infinite sets.

The Cardinal Number $\aleph_0$

The basic set used in discussing infinite sets is the set of counting numbers, $\{1, 2, 3, 4, 5, \ldots\}$. The set of counting numbers is said to have the infinite cardinal number $\aleph_0$ (the first Hebrew letter, aleph, with a zero subscript, read "aleph-null"). Think of $\aleph_0$ as being the "smallest" infinite cardinal number. To the question "How many counting numbers are there?", we answer "There are $\aleph_0$ of them."

From the discussion above, any set that can be placed in a one-to-one correspondence with the counting numbers will have the same cardinal number as the set of counting numbers, or $\aleph_0$. It turns out that many sets of numbers have cardinal number $\aleph_0$.

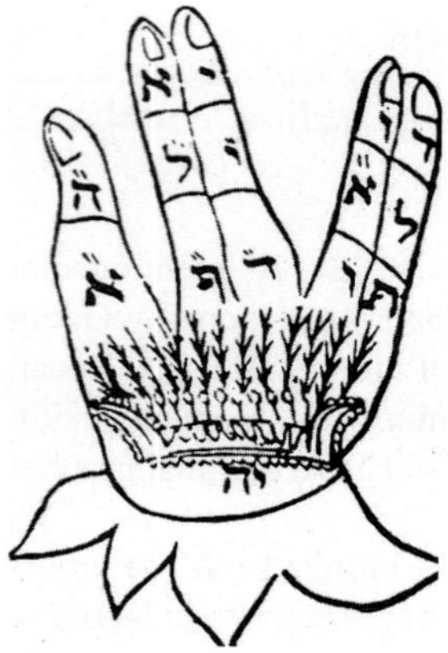

Aleph and other letters of the **Hebrew alphabet** are shown on a Kabbalistic diagram representing one of the ten emanations of God during Creation. Kabbalah, the ultramystical tradition within Judaism, arose in the fifth century and peaked in the sixteenth century in both Palestine and Poland.

Kabbalists believed that the Bible held mysteries that could be discovered in permutations, combinations, and anagrams of its very letters. They also "read" the numerical value of letters in a word by the technique called Gematria (from geometry?). This was possible because each letter in the aleph-bet has a numerical value (aleph = 1), and thus a numeration system exists. The letter Y stands for 10, so 15 should be YH (10 + 5). However, YH is a form of the Holy Name, so instead TW (9 + 6) is the symbol.

EXAMPLE 1 Showing that {0, 1, 2, 3, . . .} Has Cardinal Number $\aleph_0$

Verify that the set of whole numbers {0, 1, 2, 3, . . . } has cardinal number $\aleph_0$.

SOLUTION

All we really know about $\aleph_0$ is that it is the cardinal number of the set of counting numbers (by definition). To show that another set, such as the whole numbers, also has $\aleph_0$ as its cardinal number, we must show that set to be equivalent to the set of counting numbers. Equivalence is established by a one-to-one correspondence between the two sets. We exhibit such a correspondence, showing exactly how each counting number is paired with a unique whole number, as follows:

$$\begin{array}{cccccccccl} \{1, & 2, & 3, & 4, & 5, & 6, \ldots, & n, & \ldots\} & \text{Counting numbers} \\ \updownarrow & \updownarrow & \updownarrow & \updownarrow & \updownarrow & \updownarrow & \updownarrow & \updownarrow & \\ \{0, & 1, & 2, & 3, & 4, & 5, \ldots, & n-1, & \ldots\}. & \text{Whole numbers} \end{array}$$

The pairing of the counting number n with the whole number $n - 1$ continues indefinitely, with neither set containing any element not used up in the pairing process. Even though the set of whole numbers has an additional element (the number 0) compared to the set of counting numbers, the correspondence proves that both sets have the same cardinal number, $\aleph_0$.

Infinite Sets

The result in Example 1 shows that intuition is a poor guide for dealing with infinite sets. Intuitively, it would seem "obvious" that there are more whole numbers than counting numbers. However, because the sets can be placed in a one-to-one correspondence, the two sets have the same cardinal number.

The set {5, 6, 7} is a proper subset of the set {5, 6, 7, 8}, and there is no way to place these two sets in a one-to-one correspondence. However, the set of counting numbers is a proper subset of the set of whole numbers, and Example 1 showed that these two sets *can* be placed in a one-to-one correspondence. This important property is used in the formal definition of an infinite set.

Infinite Set

A set is **infinite** if it can be placed in a one-to-one correspondence with a proper subset of itself.

EXAMPLE 2 Showing that {. . . , −3, −2, −1, 0, 1, 2, 3, . . .} Has Cardinal Number $\aleph_0$

Verify that the set of integers {. . . ,−3, −2, −1, 0, 1, 2, 3, . . . } has cardinal number $\aleph_0$.

SOLUTION

A one-to-one correspondence can be set up between the set of integers and the set of counting numbers, as follows:

$$\begin{array}{ccccccccccc} \{1, & 2, & 3, & 4, & 5, & 6, & 7, & \ldots, & 2n, & 2n+1, & \ldots\} \\ \updownarrow & \updownarrow & \updownarrow & \updownarrow & \updownarrow & \updownarrow & \updownarrow & & \updownarrow & \updownarrow & \\ \{0, & 1, & -1, & 2, & -2, & 3, & -3, & \ldots, & n, & -n, & \ldots\}. \end{array}$$

Because of this one-to-one correspondence, the cardinal number of the set of integers is the same as the cardinal number of the set of counting numbers, $\aleph_0$.

The one-to-one correspondence of Example 2 proves that the set of integers is infinite; the set was placed in a one-to-one correspondence with a proper subset of itself.

As shown by Example 2, there are just as many integers as there are counting numbers. This result is not at all intuitive. However, the next result is even less intuitive. We know that there is an infinite number of fractions between any two counting numbers. For example, there is an infinite set of fractions $\left\{\frac{1}{2}, \frac{3}{4}, \frac{7}{8}, \frac{15}{16}, \frac{31}{32}, \ldots\right\}$ between the counting numbers 0 and 1. This should imply that there are "more" fractions than counting numbers. However, there are just as many fractions as counting numbers.

EXAMPLE 3 Showing that the Set of Rational Numbers Has Cardinal Number $\aleph_0$

Verify that the cardinal number of the set of rational numbers is $\aleph_0$.

SOLUTION

To show that the cardinal number of the set of rational numbers is $\aleph_0$, first show that a one-to-one correspondence may be set up between the set of nonnegative rational numbers and the counting numbers. This is done by the following ingenious scheme, devised by Georg Cantor. Look at Figure 22. The nonnegative rational numbers whose denominators are 1 are written in the first row; those whose denominators are 2 are written in the second row, and so on. Every nonnegative rational number appears in this list sooner or later. For example, $\frac{327}{189}$ is in row 189 and column 327.

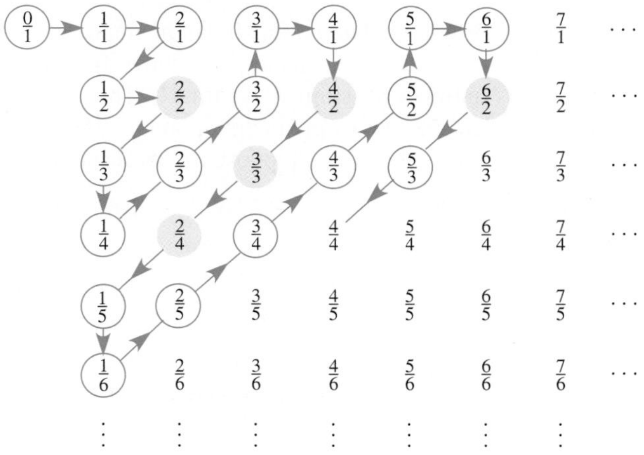

FIGURE 22

To set up a one-to-one correspondence between the set of nonnegative rationals and the set of counting numbers, follow the path drawn in Figure 22. Let $\frac{0}{1}$ correspond to 1, let $\frac{1}{1}$ correspond to 2, $\frac{2}{1}$ to 3, $\frac{1}{2}$ to 4 $\left(\text{skip } \frac{2}{2}, \text{ since } \frac{2}{2} = \frac{1}{1}\right)$, $\frac{1}{3}$ to 5, $\frac{1}{4}$ to 6, and so on. The numbers under the colored disks are omitted because they can be reduced to lower terms, and were thus included earlier in the listing.

This procedure sets up a one-to-one correspondence between the set of nonnegative rationals and the counting numbers, showing that both of these sets have the same cardinal number, $\aleph_0$. Now by using the method of Example 2, (i.e., letting each negative number follow its corresponding positive number), we can extend this correspondence to include negative rational numbers as well. Thus, the set of all rational numbers has cardinal number $\aleph_0$.

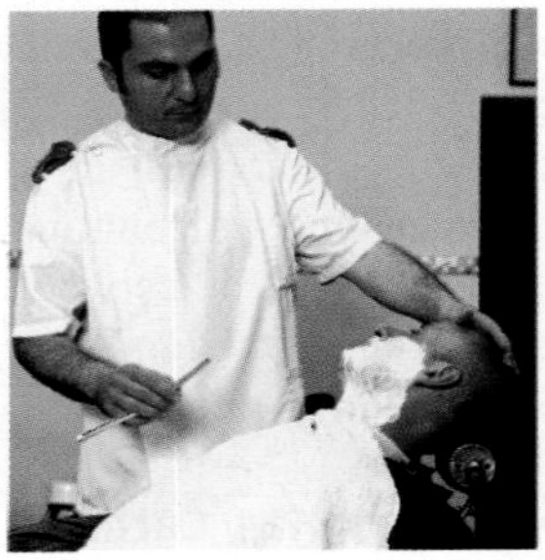

The Barber Paradox is a version of a paradox of set theory that Bertrand Russell proposed in the early twentieth century.

1. The men in a village are of two types: men who do not shave themselves and men who do.
2. The village barber shaves all men who do not shave themselves and he shaves only those men.

But who shaves the barber? The barber cannot shave himself. If he did, he would fall into the category of men who shave themselves. However, (2) above states that the barber does not shave such men.

So the barber does not shave himself. But then he falls into the category of men who do not shave themselves. According to (2), the barber shaves all of these men; hence, the barber shaves himself, too.

We find that the barber cannot shave himself, yet the barber does shave himself—a paradox.

A set is called **countable** if it is finite or if it has cardinal number $\aleph_0$. All the infinite sets of numbers discussed so far—the counting numbers, the whole numbers, the integers, and the rational numbers—are countable.

Sets That are Not Countable It would seem that every set is countable. However, the set of real numbers is not countable. That is, its cardinal number is not $\aleph_0$—in fact, it is greater than $\aleph_0$. The next example confirms this fact.

EXAMPLE 4 Showing that the Set of Real Numbers Does Not Have Cardinal Number $\aleph_0$

Verify that the set of all real numbers does not have cardinal number $\aleph_0$.

SOLUTION

There are two possibilities:

1. The set of real numbers has cardinal number $\aleph_0$.
2. The set of real numbers does not have cardinal number $\aleph_0$.

Assume for the time being that the first statement is true. If the first statement is true, then a one-to-one correspondence can be set up between the set of real numbers and the set of counting numbers. We do not know what sort of correspondence this might be, but assume it can be done.

In a later chapter, we show that every real number can be written as a decimal number (or simply "decimal"). Thus, in the one-to-one correspondence we are assuming, some decimal corresponds to the counting number 1, some decimal corresponds to 2, and so on. Suppose the correspondence is as follows:

$$1 \leftrightarrow .68458429006\ldots$$
$$2 \leftrightarrow .13479201038\ldots$$
$$3 \leftrightarrow .37291568341\ldots$$
$$4 \leftrightarrow .935223671611\ldots$$

and so on.

Assuming the existence of a one-to-one correspondence between the counting numbers and the real numbers means that every decimal is in the list above. Let's construct a new decimal K as follows. The first decimal in the above list has 6 as its first digit; let K start as $K = .4\ldots$. We picked 4 because $4 \neq 6$; we could have used any other digit except 6. Because the second digit of the second decimal in the list is 3, we let $K = .45\ldots$ (because $5 \neq 3$). The third digit of the third decimal is 2, so let $K = .457\ldots$ (because $7 \neq 2$). The fourth digit of the fourth decimal is 2, so let $K = .4573\ldots$ (because $3 \neq 2$). Continue defining K in this way.

Is K in the list that we assumed to contain all decimals? The first decimal in the list differs from K in at least the first position (K starts with 4, and the first decimal in the list starts with 6). The second decimal in the list differs from K in at least the second position, and the nth decimal in the list differs from K in at least the nth position. Every decimal in the list differs from K in at least one position, so that K cannot possibly be in the list. In summary:

We assume every decimal is in the list above.
The decimal K is not in the list.

Because these statements cannot both be true, the original assumption has led to a contradiction. This forces the acceptance of the only possible alternative to the original assumption: It is not possible to set up a one-to-one correspondence between the set of reals and the set of counting numbers. The cardinal number of the set of reals is not equal to $\aleph_0$. ■

The set of counting numbers is a proper subset of the set of real numbers. Because of this, it would seem reasonable to say that the cardinal number of the set of reals, commonly written c, is greater than $\aleph_0$. (The letter c here represents *continuum.*) Other, even larger, infinite cardinal numbers can be constructed. For example, the set of all subsets of the set of real numbers has a cardinal number larger than c. Continuing this process of finding cardinal numbers of sets of subsets, more and more, larger and larger infinite cardinal numbers are produced.

The six most important infinite sets of numbers were listed in an earlier section. All of them have been dealt with in this section, except the irrational numbers. The irrationals have decimal representations, so they are all included among the real numbers. Because the irrationals are a subset of the reals, you might guess that the irrationals have cardinal number $\aleph_0$, just like the rationals. However, because the union of the rationals and the irrationals is all the reals, that would imply that the cardinality of the union of two disjoint countable sets is c. But Example 2 showed that this is not the case. A better guess is that the cardinal number of the irrationals is c (the same as that of the reals). This is, in fact, true. The major infinite sets of numbers, with their cardinal numbers, are now summarized.

Zeno's paradox of the Tortoise and Achilles was given in its original form by Zeno of Elea. According to Math Academy Online, it "has inspired many writers and thinkers throughout the ages, notably Lewis Carroll and Douglas Hofstadter." More on Lewis Carroll can be found in Chapter 3.

In the original story, the Tortoise is able to convince Achilles (the Greek hero of Homer's *The Illiad*) that in a race, given a small head start, the Tortoise is always able to defeat Achilles. (See the Chapter Opener and Exercises 51 and 52 in this section.) The resolution of this paradox is discussed on the Website www.mathacademy.com.

Cardinal Numbers of Infinite Number Sets

Infinite Set	Cardinal Number
Natural or counting numbers	$\aleph_0$
Whole numbers	$\aleph_0$
Integers	$\aleph_0$
Rational numbers	$\aleph_0$
Irrational numbers	c
Real numbers	c

2.5 EXERCISES

Match each set in Column I with the set in Column II that has the same cardinality. Give the cardinal number.

I	II
1. {6}	**A.** $\{x \mid x \text{ is a rational number}\}$
2. {−16, 14, 3}	**B.** {26}
3. $\{x \mid x \text{ is a natural number}\}$	**C.** $\{x \mid x \text{ is an irrational number}\}$
4. $\{x \mid x \text{ is a real number}\}$	**D.** {x, y, z}
5. $\{x \mid x \text{ is an integer between 5 and 6}\}$	**E.** $\{x \mid x \text{ is a real number that satisfies } x^2 = 25\}$
6. $\{x \mid x \text{ is an integer that satisfies } x^2 = 100\}$	**F.** $\{x \mid x \text{ is an integer that is both even and odd}\}$

Place each pair of sets into a one-to-one correspondence, if possible.

7. {I, II, III} and {x, y, z}

8. {a, b, c, d} and {2, 4, 6}

9. {a, d, d, i, t, i, o, n} and {a, n, s, w, e, r}

10. {Reagan, Clinton, Bush} and {Nancy, Hillary, Laura}

Give the cardinal number of each set.

11. {a, b, c, d, . . . , k}

12. {9, 12, 15, . . . , 36}

13. ∅

14. {0}

15. {300, 400, 500, . . . }

16. {−35, −28, −21, . . . , 56}

17. $\left\{-\frac{1}{4}, -\frac{1}{8}, -\frac{1}{12}, \ldots\right\}$

18. $\{x \mid x \text{ is an even integer}\}$

19. $\{x \mid x \text{ is an odd counting number}\}$

20. {b, a, 1, 1, a, d}

21. {Jan, Feb, Mar, . . . , Dec}

22. {Alabama, Alaska, Arizona, . . . , Wisconsin, Wyoming}

23. Lew Lefton (www.math.gatech.edu/~llefton) has revised the old song "100 Bottles of Beer on the Wall" to illustrate a property of infinite cardinal numbers. Fill in the blank in the first line of Lefton's composition:

> $\aleph_0$ bottles of beer on the wall, $\aleph_0$ bottles of beer, take one down and pass it around, _____ bottles of beer on the wall.

24. Two one-to-one correspondences are considered "different" if some elements are paired differently in one than in the other. For example:

$$\begin{array}{ccc} \{a, & b, & c\} \\ \updownarrow & \updownarrow & \updownarrow \\ \{a, & b, & c\} \end{array} \quad \text{and} \quad \begin{array}{ccc} \{a, & b, & c\} \\ \updownarrow & \updownarrow & \updownarrow \\ \{c, & b, & a\} \end{array} \quad \text{are different,}$$

$$\text{while} \quad \begin{array}{ccc} \{a, & b, & c\} \\ \updownarrow & \updownarrow & \updownarrow \\ \{c, & a, & b\} \end{array} \quad \text{and} \quad \begin{array}{ccc} \{b, & c, & a\} \\ \updownarrow & \updownarrow & \updownarrow \\ \{a, & b, & c\} \end{array} \quad \text{are not.}$$

(a) How many *different* correspondences can be set up between the two sets {Jamie Foxx, Mike Myers, Madonna} and {Austin Powers, Ray Charles, Eva Peron}

(b) Which one of these correspondences pairs each person with the appropriate famous movie role?

Determine whether each pair of sets is equal, equivalent, both, *or* neither.

25. {u, v, w}, {v, u, w}

26. {48, 6}, {4, 86}

27. {X, Y, Z}, {x, y, z}

28. {top}, {pot}

29. $\{x \mid x$ is a positive real number$\}$, $\{x \mid x$ is a negative real number$\}$

30. $\{x \mid x$ is a positive rational number$\}$, $\{x \mid x$ is a negative real number$\}$

Show that each set has cardinal number $\aleph_0$ *by setting up a one-to-one correspondence between the given set and the set of counting numbers.*

31. the set of positive even integers

32. $\{-10, -20, -30, -40, \ldots\}$

33. {1,000,000, 2,000,000, 3,000,000, . . . }

34. the set of odd integers

35. $\{2, 4, 8, 16, 32, \ldots\}$ (*Hint*: $4 = 2^2$, $8 = 2^3$, $16 = 2^4$, and so on)

36. $\{-17, -22, -27, -32, \ldots\}$

In Exercises 37–40, identify the given statement as always true *or* not always true. *If* not always true, *give a counterexample.*

37. If A and B are infinite sets, then A is equivalent to B.

38. If set A is an infinite set and set B can be put in a one-to-one correspondence with a proper subset of A, then B must be infinite.

39. If A is an infinite set and A is not equivalent to the set of counting numbers, then $n(A) = c$.

40. If A and B are both countably infinite sets, then $n(A \cup B) = \aleph_0$.

Exercises 41 and 42 are geometric applications of the concept of infinity.

41. The set of real numbers can be represented by an infinite line, extending indefinitely in both directions. Each point on the line corresponds to a unique real number, and each real number corresponds to a unique point on the line.

(a) Use the figure below, where the line segment between 0 and 1 has been bent into a semicircle and positioned above the line, to prove that

$\{x \mid x$ is a real number between 0 and 1$\}$
is equivalent to $\{x \mid x$ is a real number$\}$.

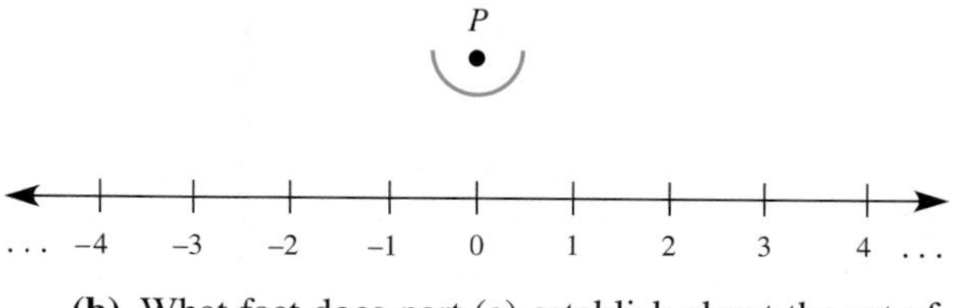

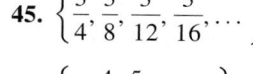

(b) What fact does part (a) establish about the set of real numbers?

42. Show that the two vertical line segments shown here both have the same number of points.

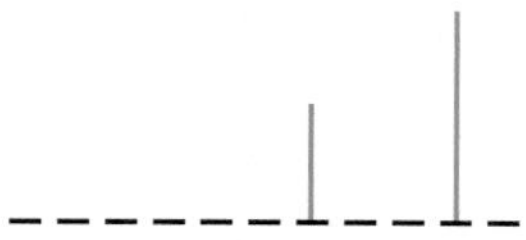

Show that each set can be placed in a one-to-one correspondence with a proper subset of itself to prove that each set is infinite.

43. $\{3, 6, 9, 12, \ldots\}$

44. $\{4, 7, 10, 13, 16, \ldots\}$

45. $\left\{\frac{3}{4}, \frac{3}{8}, \frac{3}{12}, \frac{3}{16}, \ldots\right\}$

46. $\left\{1, \frac{4}{3}, \frac{5}{3}, 2, \ldots\right\}$

47. $\left\{\frac{1}{9}, \frac{1}{18}, \frac{1}{27}, \frac{1}{36}, \ldots\right\}$

48. $\{-3, -5, -9, -17, \ldots\}$

49. Describe the distinction between *equal* and *equivalent* sets.

50. Explain how the correspondence suggested in Example 4 shows that the set of real numbers between 0 and 1 is not countable.

The Paradoxes of Zeno *The Chapter Opener discussed the scene in the movie* I.Q. *that deals with Zeno's paradox. Zeno was born about 496* B.C. *in southern Italy. Two forms of his paradox are given on the next page.*

What is your explanation for the following two examples of Zeno's paradoxes?

51. Achilles, if he starts out behind a tortoise, can never overtake the tortoise even if he runs faster.

Suppose Tortoise has a head start of one meter and goes one-tenth as fast as Achilles. When Achilles reaches the point where Tortoise started, Tortoise is then one-tenth meter ahead. When Achilles reaches *that* point, Tortoise is one-hundredth meter ahead. And so on. Achilles gets closer but can never catch up.

52. Motion itself cannot occur.

You cannot travel one meter until after you have first gone a half meter. But you cannot go a half meter until after you have first gone a quarter meter. And so on. Even the tiniest motion cannot occur because a tinier motion would have to occur first.

COLLABORATIVE INVESTIGATION

Surveying the Members of Your Class

This group activity is designed to determine the number of students present in your class without actually counting the members one by one. This will be accomplished by having each member of the class determine one particular set in which he or she belongs, and then finding the sum of the cardinal numbers of the subsets.

For this activity, we designate three sets: X, Y, and Z. Here are their descriptions:

X = {students in the class registered with the Republican party}

Y = {students in the class 24 years of age or younger}

Z = {students who have never been married}

Each student in the class will belong to one of the sets X, X', one of the sets Y, Y', and one of the sets Z, Z'. (Recall that the complement of a set consists of all elements in the universe (class) that are not in the set.)

As an example, suppose that a student is a 23-year-old divorced Democrat. The student belongs to the sets X', Y, and Z'. Joining these with intersection symbols, the set to which the student belongs is

$$X' \cap Y \cap Z'.$$

Now observe the Venn diagram that follows. The eight subsets are identified by lowercase letters (a) through (h). The final column in the following table will be completed when a survey is made. Each student should now determine to which set he or she belongs. (The student described earlier belongs to (g).)

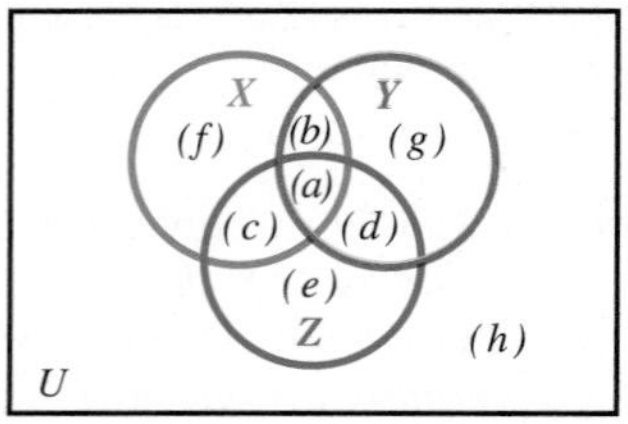

Region	Description in Terms of Set Notation	Number of Class Members in the Set
(a)	$X \cap Y \cap Z$	
(b)	$X \cap Y \cap Z'$	
(c)	$X \cap Y' \cap Z$	
(d)	$X' \cap Y \cap Z$	
(e)	$X' \cap Y' \cap Z$	
(f)	$X \cap Y' \cap Z'$	
(g)	$X' \cap Y \cap Z'$	
(h)	$X' \cap Y' \cap Z'$	

The instructor will now poll the class to see how many members are in each set. *Remember that each class member will belong to one and only one set.*

After the survey is made, find the sum of the numbers in the final column. They should add up to *exactly* the number of students present. Count the class members individually to verify this.

Topics for Discussion

1. Suppose that the final column entries do not add up to the total number of class members. What might have gone wrong?
2. Why can't a class member be a member of more than one of the eight subsets listed?

CHAPTER 2 TEST

In Exercises 1–14, let

$$U = \{a, b, c, d, e, f, g, h\}, \quad A = \{a, b, c, d\}, \quad B = \{b, e, a, d\}, \quad \textit{and} \quad C = \{a, e\}.$$

Find each set.

1. $A \cup C$

2. $B \cap A$

3. B'

4. $A - (B \cap C')$

Identify each statement as true *or* false.

5. $b \in A$

6. $C \subseteq A$

7. $B \subset (A \cup C)$

8. $c \notin C$

9. $n[(A \cup B) - C] = 4$

10. $\emptyset \subset C$

11. $A \cap B'$ is equivalent to $B \cap A'$

12. $(A \cup B)' = A' \cap B'$

Find each of the following.

13. $n(A \times C)$

14. the number of proper subsets of A

Give a word description for each set.

15. $\{-3, -1, 1, 3, 5, 7, 9\}$

16. {January, February, March, . . . , December}

Express each set in set-builder notation.

17. $\{-1, -2, -3, -4, \ldots\}$

18. $\{24, 32, 40, 48, \ldots, 88\}$

Place $\subset$, $\subseteq$, both, *or* neither *in each blank to make a true statement.*

19. $\emptyset$ ______ $\{x \mid x$ is a counting number between 20 and 21$\}$

20. $\{4, 9, 16\}$ ______ $\{4, 5, 6, 7, 8, 9, 10\}$

Shade each set in an appropriate Venn diagram.

21. $X \cup Y'$

22. $X' \cap Y'$

23. $(X \cup Y) - Z$

24. $[(X \cap Y) \cup (Y \cap Z) \cup (X \cap Z)] - (X \cap Y \cap Z)$

Facts About Inventions *The following table lists ten inventions, important directly or indirectly in our lives, together with other pertinent data.*

Invention	Date	Inventor	Nation
Adding machine	1642	Pascal	France
Barometer	1643	Torricelli	Italy
Electric razor	1917	Schick	U.S.
Fiber optics	1955	Kapany	England
Geiger counter	1913	Geiger	Germany
Pendulum clock	1657	Huygens	Holland
Radar	1940	Watson-Watt	Scotland
Telegraph	1837	Morse	U.S.
Thermometer	1593	Galileo	Italy
Zipper	1891	Judson	U.S.

Let U = the set of all ten inventions, A = the set of items invented in the United States, and T = the set of items invented in the twentieth century. List the elements of each set.

25. $A \cap T$

26. $(A \cup T)'$

27. $A - T'$

28. State De Morgan's laws for sets in words rather than symbols.

29. The numerals in the Venn diagram indicate the number of elements in each particular subset.

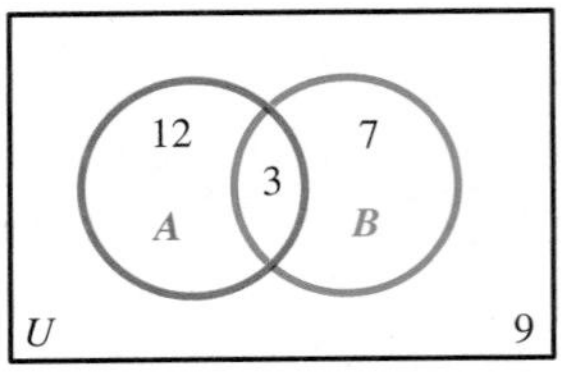

Determine the number of elements in each set.
(a) $A \cup B$ **(b)** $A \cap B'$ **(c)** $(A \cap B)'$

Financial Aid to College Students *In one recent year, financial aid available to college students in the United States was nearly \$30 billion. (Much of it went unclaimed, mostly because qualified students were not aware of it, did not know how to obtain or fill out the required applications, or did not feel the results would be worth their effort.) Three major sources of aid are government grants, private scholarships, and the colleges themselves.*

30. Marilyn Cazayoux, Financial Aid Director of a small private Southern college, surveyed the records of 100 sophomores and found the following:

49 receive government grants
55 receive private scholarships
43 receive aid from the college
23 receive government grants and private scholarships
18 receive government grants and aid from the college
28 receive private scholarships and aid from the college
8 receive help from all three sources.

How many of the students in the survey:
(a) have government grants only?
(b) have private scholarships but not government grants?
(c) receive financial aid from only one of these sources?
(d) receive aid from exactly two of these sources?
(e) receive no financial aid from any of these sources?
(f) receive no aid from the college or from the government?

3

INTRODUCTION TO LOGIC

The 1959 Oscar-nominated animated short *Donald in Mathmagic Land* was the first Disney cartoon televised in color. After nearly 50 years, it has proved to be a classic, rendering mathematical topics such as geometry, mathematics in music, games, and nature, and the amazing Golden Section in a way that anyone can understand. In one segment, Donald Duck, dressed as Alice from Lewis Carroll's *Through the Looking Glass*, is attacked by a "none-too-friendly group of chess pieces."

Logic (the subject of this chapter) and chess have been paired for centuries. Most scholars agree that chess dates back at least 1500 years, coming from Northern India and Afghanistan following trade routes through Persia. One does not have to have a high I.Q. to excel at chess. In fact, recent studies indicate that chess strategy might rely more on brain activity not usually associated with general intelligence. Good chess players rely on memory, imagination, determination, and inspiration. They are pattern thinkers that use long-established sets of consequences and probabilities resulting from countless hours of studying and playing. In the end, logic does not necessarily dictate the final outcome of any chess game, for if it did, humans would not stand a chance when playing faceless, number-crunching computers.

Sources: www.imdb.com, Walter A. Smart.

3.1 Statements and Quantifiers

Statements • Negations • Symbols • Quantifiers • Sets of Numbers

Gottfried Leibniz (1646–1716) was a wide-ranging philosopher and a universalist who tried to patch up Catholic–Protestant conflicts. He promoted cultural exchange between Europe and the East. Chinese ideograms led him to search for a universal symbolism. He was an early inventor of **symbolic logic.**

Statements This section introduces the study of *symbolic logic,* which uses letters to represent statements, and symbols for words such as *and, or, not.* One of the main applications of logic is in the study of the *truth value* (that is, the truth or falsity) of statements with many parts. The truth value of these statements depends on the components of which they are comprised.

Many kinds of sentences occur in ordinary language, including factual statements, opinions, commands, and questions. Symbolic logic discusses only the first type of sentence, the kind that involves facts. A **statement** is defined as a declarative sentence that is either true or false, but not both simultaneously. For example, both of the following are statements:

Electronic mail provides a means of communication.
$11 + 6 = 12.$
} Statements

Each one is either true or false. However, based on this definition, the following sentences are not statements:

Access the file.

Is this a great time, or what?

Luis Pujols is a better baseball player than Johnny Damon.

This sentence is false.

These sentences cannot be identified as being either true or false. The first sentence is a command, and the second is a question. The third is an opinion. "This sentence is false" is a paradox; if we assume it is true, then it is false, and if we assume it is false, then it is true.

A **compound statement** may be formed by combining two or more statements. The statements making up a compound statement are called **component statements.** Various **logical connectives,** or simply **connectives,** can be used in forming compound statements. Words such as *and, or, not,* and *if . . . then* are examples of connectives. (While a statement such as "Today is not Tuesday" does not consist of two component statements, for convenience it is considered compound, because its truth value is determined by noting the truth value of a different statement, "Today is Tuesday.")

EXAMPLE 1 Deciding Whether a Statement Is Compound

Decide whether each statement is compound.

(a) Shakespeare wrote sonnets, and the poem exhibits iambic pentameter.
(b) You can pay me now, or you can pay me later.
(c) If he said it, then it must be true.
(d) My pistol was made by Smith and Wesson.

SOLUTION

(a) This statement is compound, because it is made up of the component statements "Shakespeare wrote sonnets" and "the poem exhibits iambic pentameter." The connective is *and.*

(b) The connective here is *or.* The statement is compound.
(c) The connective here is *if . . . then,* discussed in more detail in a later section. The statement is compound.
(d) While the word "and" is used in this statement, it is not used as a *logical* connective, because it is part of the name of the manufacturer. The statement is not compound.

Negations

The sentence "Greg Chustz has a red truck" is a statement; the **negation** of this statement is "Greg Chustz does not have a red truck." The negation of a true statement is false, and the negation of a false statement is true.

EXAMPLE 2 Forming Negations

Form the negation of each statement.

(a) That state has a governor. **(b)** The sun is not a star.

SOLUTION

(a) To negate this statement, we introduce *not* into the sentence: "That state does not have a governor."
(b) The negation is "The sun is a star."

One way to detect incorrect negations is to check truth values. A negation must have the opposite truth value from the original statement.

The next example uses some of the inequality symbols in Table 1.

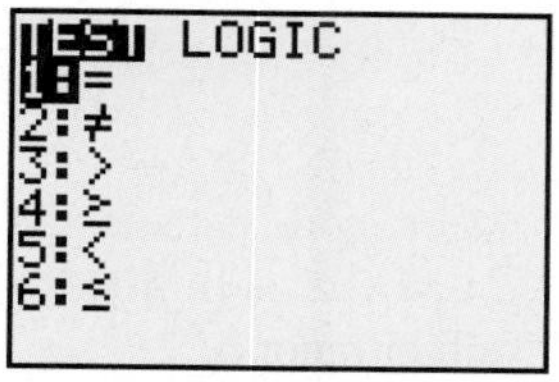

The TEST menu of the TI-83/84 Plus calculator allows the user to test the truth or falsity of statements involving $=$, $\neq$, $>$, $\geq$, $<$, and $\leq$. If a statement is true, it returns a 1; if false, it returns a 0.

TABLE 1

Symbolism	Meaning	Examples	
$a < b$	a is less than b	$4 < 9$	$\frac{1}{2} < \frac{3}{4}$
$a > b$	a is greater than b	$6 > 2$	$-5 > -11$
$a \leq b$	a is less than or equal to b	$8 \leq 10$	$3 \leq 3$
$a \geq b$	a is greater than or equal to b	$-2 \geq -3$	$-5 \geq -5$

EXAMPLE 3 Negating Inequalities

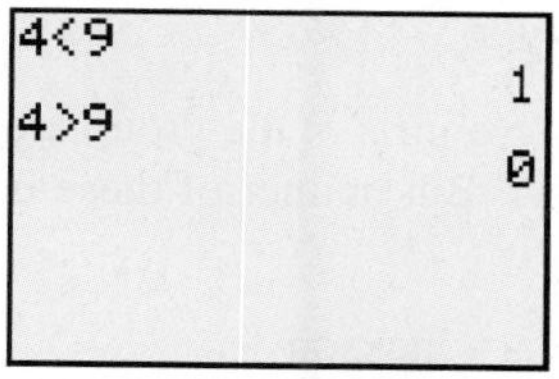

$4 < 9$ is true, as indicated by the 1. $4 > 9$ is false, as indicated by the 0.

Give a negation of each inequality. Do *not* use a slash symbol.

(a) $p < 9$ **(b)** $7x + 11y \geq 77$

SOLUTION

(a) The negation of "p is less than 9" is "p is *not* less than 9." Because we cannot use "not," which would require writing $p \not< 9$, phrase the negation as "p is greater than or equal to 9," or $p \geq 9$.
(b) The negation, with no slash, is $7x + 11y < 77$.

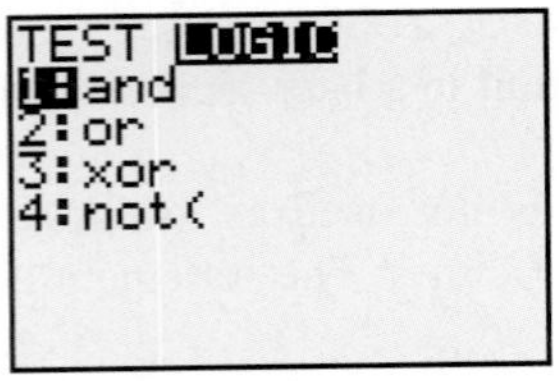

The LOGIC menu of the TI-83/84 Plus calculator allows the user to test truth or falsity of statements involving *and, or, exclusive or* (see Exercise 77 in the next section), and *not.*

Symbols To simplify work with logic, we use symbols. Statements are represented with letters, such as p, q, or r, while several symbols for connectives are shown in Table 2. The table also names the type of statement having the given connective.

TABLE 2

Connective	Symbol	Type of Statement
and	$\wedge$	Conjunction
or	$\vee$	Disjunction
not	$\sim$	Negation

The symbol $\sim$ represents the connective *not.* If p represents the statement "George W. Bush was president in 2005" then $\sim p$ represents "George W. Bush was not president in 2005."

Aristotle, the first to systematize the logic we use in everyday life, appears above in a detail from the painting *The School of Athens*, by Raphael. He is shown debating a point with his teacher **Plato.**

EXAMPLE 4 Translating from Symbols to Words

Let p represent "It is 80° today," and let q represent "It is Tuesday." Write each symbolic statement in words.

(a) $p \vee q$ **(b)** $\sim p \wedge q$ **(c)** $\sim(p \vee q)$ **(d)** $\sim(p \wedge q)$

SOLUTION

(a) From the table, $\vee$ symbolizes *or*; thus, $p \vee q$ represents

It is 80° today or it is Tuesday.

(b) It is not 80° today and it is Tuesday.
(c) It is not the case that it is 80° today or it is Tuesday.
(d) It is not the case that it is 80° today and it is Tuesday. ■

The statement in Example 4(c) usually is translated as "Neither p nor q."

Quantifiers The words *all, each, every,* and *no(ne)* are called **universal quantifiers,** while words and phrases such as *some*, *there exists*, and *(for) at least one* are called **existential quantifiers.** Quantifiers are used extensively in mathematics to indicate *how many* cases of a particular situation exist. Be careful when forming the negation of a statement involving quantifiers.

The negation of a statement must be false if the given statement is true and must be true if the given statement is false, in all possible cases. Consider the statement

All girls in the group are named Mary.

Many people would write the negation of this statement as "No girls in the group are named Mary" or "All girls in the group are not named Mary." But neither of these is correct. To see why, look at the three groups below:

Group I: Mary Lynn Brumfield, Mary Smith, Mary Jackson
Group II: Mary Johnson, Lynne Olinde, Margaret Westmoreland
Group III: Donna Garbarino, Paula Story, Rhonda Alessi, Kim Falgout.

These groups contain all possibilities that need to be considered. In Group I, *all* girls are named Mary; in Group II, *some* girls are named Mary (and some are not); in Group III, *no* girls are named Mary. Look at the truth values in Table 3 and keep in mind that "some" means "at least one (and possibly all)."

TABLE 3 Truth Value as Applied to:

	Group I	Group II	Group III	
(1) All girls in the group are named Mary. (Given)	T	F	F	Negation
(2) No girls in the group are named Mary. (Possible negation)	F	F	T	
(3) All girls in the group are not named Mary. (Possible negation)	F	F	T	
(4) Some girls in the group are not named Mary. (Possible negation)	F	T	T	Negation

The negation of the given statement (1) must have opposite truth values in *all* cases. It can be seen that statements (2) and (3) do not satisfy this condition (for Group II), but statement (4) does. It may be concluded that the correct negation for "All girls in the group are named Mary" is "Some girls in the group are not named Mary." Other ways of stating the negation are

Not all girls in the group are named Mary.

It is not the case that all girls in the group are named Mary.

At least one girl in the group is not named Mary.

Table 4 can be used to generalize the method of finding the negation of a statement involving quantifiers.

TABLE 4 Negations of Quantified Statements

Statement	Negation
All do.	Some do not. (Equivalently: Not all do.)
Some do.	None do. (Equivalently: All do not.)

The negation of the negation of a statement is simply the statement itself. For instance, the negations of the statements in the Negation column are simply the corresponding original statements in the Statement column. As an example, the negation of "Some do not" is "All do."

EXAMPLE 5 Forming Negations of Quantified Statements

Form the negation of each statement.

(a) Some cats have fleas. **(b)** Some cats do not have fleas.
(c) No cats have fleas.

SOLUTION

(a) Because *some* means "at least one," the statement "Some cats have fleas" is really the same as "At least one cat has fleas." The negation of this is "No cat has fleas."

(b) The statement "Some cats do not have fleas" claims that at least one cat, somewhere, does not have fleas. The negation of this is "All cats have fleas."

(c) The negation is "Some cats have fleas." ← Avoid the incorrect answer "All cats have fleas."

The 1997 film *Smilla's Sense of Snow* stars Julia Ormond as a brilliant young scientist who has been displaced from her beloved native Greenland. She has a passion for snow and mathematics. In a conversation, she speaks of her love of **numbers:**

To me, the number system is like human life. First you have the natural numbers, the ones that are whole and positive, like the numbers of a small child. Consciousness expands and a child discovers longing. Do you know the mathematical expression for longing? Negative numbers, the formalization of the feeling that you're missing something. Then the child discovers the in-between spaces, between stones, between people, between numbers, and that produces fractions. But it's like a kind of madness, because it doesn't even stop there. It never stops. There are numbers that we can't even begin to comprehend. Mathematics is a vast, open landscape. You head towards the horizon which is always receding, like Greenland.

Sets of Numbers Earlier we introduced sets of numbers that are studied in algebra, and they are repeated here.

Sets of Numbers

Natural or Counting numbers $\{1, 2, 3, 4, \ldots\}$

Whole numbers $\{0, 1, 2, 3, 4, \ldots\}$

Integers $\{\ldots, -3, -2, -1, 0, 1, 2, 3, \ldots\}$

Rational numbers $\left\{\frac{p}{q} \,\middle|\, p \text{ and } q \text{ are integers, and } q \neq 0\right\}$
(Some examples of rational numbers are $\frac{3}{5}$, $-\frac{7}{5}$, 5, and 0. Any rational number may be expressed as a terminating decimal number, such as .25 or a repeating decimal number, such as .666. . . .)

Real numbers $\{x \mid x \text{ is a number that can be written as a decimal}\}$

Irrational numbers $\{x \mid x \text{ is a real number and } x \text{ cannot be written as a quotient of integers}\}$
(Some examples of irrational numbers are $\sqrt{2}$, $\sqrt[3]{4}$, and π. Decimal representations of irrational numbers never terminate and never repeat.)

EXAMPLE 6 Deciding Whether Quantified Statements Are True or False

Decide whether each of the following statements about sets of numbers involving a quantifier is *true* or *false*.

(a) There exists a whole number that is not a natural number.
(b) Every integer is a natural number.
(c) Every natural number is a rational number.
(d) There exists an irrational number that is not real.

SOLUTION

(a) Because there is such a whole number (it is 0), this statement is true.

(b) This statement is false, because we can find at least one integer that is not a natural number. For example, -1 is an integer but is not a natural number.

(c) Because every natural number can be written as a fraction with denominator 1, this statement is true.

(d) In order to be an irrational number, a number must first be real. Therefore, because we cannot give an irrational number that is not real, this statement is false. (Had we been able to find at least one, the statement would have then been true.)

3.1 Exercises

Decide whether each is a statement or is not a statement.

1. September 11, 2001, was a Tuesday.
2. The ZIP code for Manistee, MI, is 49660.
3. Listen, my children, and you shall hear of the midnight ride of Paul Revere.
4. Yield to oncoming traffic.
5. $5 + 8 \neq 13$ and $4 - 3 = 12$
6. $5 + 8 \neq 12$ or $4 - 3 = 5$
7. Some numbers are negative.
8. James Garfield was president of the United States in 1881.
9. Accidents are the main cause of deaths of children under the age of 7.
10. *Shrek 2* was the top-grossing movie of 2004.

11. Where are you going today?
12. Behave yourself and sit down.
13. Kevin "Catfish" McCarthy once took a prolonged continuous shower for 340 hours, 40 minutes.
14. One gallon of milk weighs more than 4 pounds.

Decide whether each statement is compound.

15. I read the *Arizona Republic*, and I read the *Sacramento Bee*.
16. My brother got married in Amsterdam.
17. Tomorrow is Wednesday.
18. Mamie Zwettler is younger than 18 years of age, and so is her friend Emma Lister.
19. Jay Beckenstein's wife loves Ben and Jerry's ice cream.
20. The sign on the back of the car read "Alaska or bust!"
21. If Jane Fleming sells her quota, then Pam Snow will be happy.
22. If Tom is a politician, then Jack is a crook.

Write a negation for each statement.

23. Her aunt's name is Hildegard.
24. The flowers are to be watered.
25. Every dog has its day.
26. No rain fell in southern California today.
27. Some books are longer than this book.
28. All students present will get another chance.
29. No computer repairman can play poker.
30. Some people have all the luck.
31. Everybody loves somebody sometime.
32. Everyone loves a winner.

Give a negation of each inequality. Do not use a slash symbol.

33. $x > 12$
34. $x < -6$
35. $x \geq 5$
36. $x \leq 19$
37. Try to negate the sentence "The exact number of words in this sentence is ten" and see what happens. Explain the problem that arises.
38. Explain why the negation of "$x > 5$" is not "$x < 5$."

Let p represent the statement "She has green eyes" *and let q represent the statement* "He is 56 years old." *Translate each symbolic compound statement into words.*

39. $\sim p$
40. $\sim q$
41. $p \wedge q$
42. $p \vee q$
43. $\sim p \vee q$
44. $p \wedge \sim q$
45. $\sim p \vee \sim q$
46. $\sim p \wedge \sim q$
47. $\sim(\sim p \wedge q)$
48. $\sim(p \vee \sim q)$

Let p represent the statement "Chris collects DVDs" *and let q represent the statement* "Jack is an English major." *Convert each compound statement into symbols.*

49. Chris collects DVDs and Jack is not an English major.

50. Chris does not collect DVDs or Jack is not an English major.

51. Chris does not collect DVDs or Jack is an English major.

52. Jack is an English major and Chris does not collect DVDs.

53. Neither Chris collects DVDs nor Jack is an English major.

54. Either Jack is an English major or Chris collects DVDs, and it is not the case that both Jack is an English major and Chris collects DVDs.

55. Incorrect use of quantifiers often is heard in everyday language. Suppose you hear that a local electronics chain is having a 40% off sale, and the radio advertisement states "All items are not available in all stores." Do you think that, literally translated, the ad really means what it says? What do you think is really meant? Explain your answer.

56. Repeat Exercise 55 for the following: "All people don't have the time to devote to maintaining their vehicles properly."

Refer to the groups of art labeled A, B, *and* C, *and identify by letter the group or groups that are satisfied by the given statements involving quantifiers.*

A

B

C

57. All pictures have frames.

58. No picture has a frame.

59. At least one picture does not have a frame.

60. Not every picture has a frame.

61. At least one picture has a frame.

62. No picture does not have a frame.

63. All pictures do not have frames.

64. Not every picture does not have a frame.

Decide whether each statement in Exercises 65–74 involving a quantifier is true *or* false.

65. Every whole number is an integer.

66. Every natural number is an integer.

67. There exists a rational number that is not an integer.

68. There exists an integer that is not a natural number.

69. All rational numbers are real numbers.

70. All irrational numbers are real numbers.

71. Some rational numbers are not integers.

72. Some whole numbers are not rational numbers.

73. Each whole number is a positive number.

74. Each rational number is a positive number.

75. Explain the difference between the following statements:

All students did not pass the test.

Not all students passed the test.

76. The statement "For some real number x, $x^2 \geq 0$" is true. However, your friend does not understand why, because he claims that $x^2 \geq 0$ for *all* real numbers x (and not *some*). How would you explain his misconception to him?

77. Write the following statement using "every": There is no one here who has not done that at one time or another.

78. Only one of the following statements is true. Which one is it?

A. For some real number x, $x \not< 0$.

B. For all real numbers x, $x^3 > 0$.

C. For all real numbers x less than 0, x^2 is also less than 0.

D. For some real number x, $x^2 < 0$.

3.2 Truth Tables and Equivalent Statements

Conjunctions • Disjunctions • Negations • Mathematical Statements • Truth Tables • Alternative Method for Constructing Truth Tables • Equivalent Statements and De Morgan's Laws

Conjunctions The truth values of component statements are used to find the truth values of compound statements. To begin, let us decide on the truth values of the **conjunction** *p and q*, symbolized $p \wedge q$. In everyday language, the connective *and* implies the idea of "both." The statement

Monday immediately follows Sunday and March immediately follows February

is true, because each component statement is true. On the other hand, the statement

Monday immediately follows Sunday and March immediately follows January

is false, even though part of the statement (Monday immediately follows Sunday) is true. For the conjunction $p \wedge q$ to be true, both p and q must be true. This result is summarized by a table, called a **truth table,** which shows all four of the possible combinations of truth values for the conjunction *p and q*. The truth table for *conjunction* is shown here.

Truth Table for the Conjunction *p* and *q*

p and *q*

p	q	$p \wedge q$
T	T	T
T	F	F
F	T	F
F	F	F

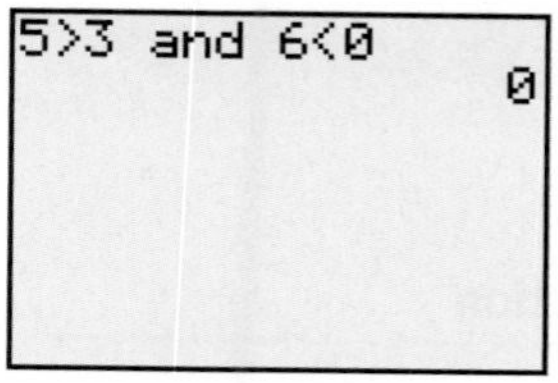

The calculator returns a "0" for $5 > 3$ *and* $6 < 0$, indicating that the statement is false.

EXAMPLE 1 Finding the Truth Value of a Conjunction

Let p represent "$5 > 3$" and let q represent "$6 < 0$." Find the truth value of $p \wedge q$.

SOLUTION

Here p is true and q is false. Looking in the second row of the conjunction truth table shows that $p \wedge q$ is false. ■

In some cases, the logical connective *but* is used in compound statements:

He wants to go to the mountains but she wants to go to the beach.

Here, *but* is used in place of *and* to give a different sort of emphasis to the statement. In such a case, we consider the statement as we would consider the conjunction using the word *and*. The truth table for the conjunction, given above, would apply.

Disjunctions In ordinary language, the word *or* can be ambiguous. The expression "this or that" can mean either "this or that or both," or "this or that but not both." For example, the statement

I will paint the wall or I will paint the ceiling

probably has the following meaning: "I will paint the wall or I will paint the ceiling or I will paint both." On the other hand, the statement

I will drive the Saturn or the BMW to the store

probably means "I will drive the Saturn, or I will drive the BMW, but I will not drive both."

The symbol $\vee$ normally represents the first *or* described. That is,

$$p \vee q \quad \text{means} \quad \text{"}p \text{ or } q \text{ or both."}$$

With this meaning of *or*, $p \vee q$ is called the *inclusive disjunction*, or just the **disjunction** of p and q. In everyday language, the disjunction implies the idea of "either." For example, the disjunction

I have a quarter or I have a dime

is true whenever I have either a quarter, a dime, or both. The only way this disjunction could be false would be if I had neither coin. A disjunction is false only if both component statements are false. The truth table for *disjunction* follows.

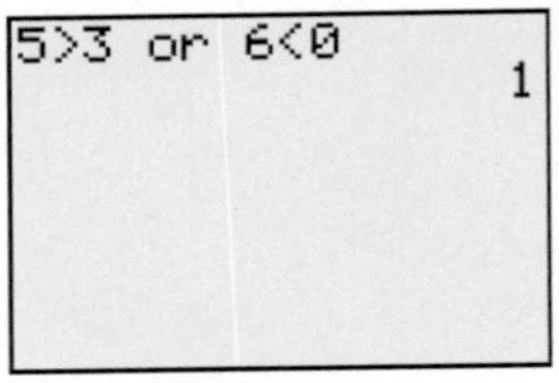

The calculator returns a "1" for $5 > 3$ *or* $6 < 0$, indicating that the statement is true.

Truth Table for the Disjunction *p* or *q*

p or *q*

p	q	$p \vee q$
T	T	T
T	F	T
F	T	T
F	F	F

TABLE 5

Statement	Reason That It Is True
$8 \geq 8$	$8 = 8$
$3 \geq 1$	$3 > 1$
$-5 \leq -3$	$-5 < -3$
$-4 \leq -4$	$-4 = -4$

EXAMPLE 2 Finding the Truth Value of a Disjunction

Let p represent "$5 > 3$" and let q represent "$6 < 0$." Find the truth value of $p \vee q$.

SOLUTION

Here, as in Example 1, p is true and q is false. The second row of the disjunction truth table shows that $p \vee q$ is true. ■

The symbol $\geq$ is read "is greater than or equal to," while $\leq$ is read "is less than or equal to." If a and b are real numbers, then $a \leq b$ is true if $a < b$ or $a = b$. Table 5 in the margin shows several statements and the reasons they are true.

Negations The **negation** of a statement p, symbolized $\sim p$, must have the opposite truth value from the statement p itself. This leads to the truth table for the negation, shown here.

Truth Table for the Negation not p

not p

p	$\sim p$
T	F
F	T

EXAMPLE 3 Finding the Truth Value of a Compound Statement

Suppose p is false, q is true, and r is false. What is the truth value of the compound statement $\sim p \wedge (q \vee \sim r)$?

SOLUTION

Here parentheses are used to group q and $\sim r$ together. Work first inside the parentheses. Because r is false, $\sim r$ will be true. Because $\sim r$ is true and q is true, find the truth value of $q \vee \sim r$ by looking in the first row of the *or* truth table. This row gives the result T. Because p is false, $\sim p$ is true, and the final truth value of $\sim p \wedge (q \vee \sim r)$ is found in the top row of the *and* truth table. From the *and* truth table, when $\sim p$ is true, and $q \vee \sim r$ is true, the statement $\sim p \wedge (q \vee \sim r)$ is true.

The preceding paragraph may be interpreted using a short-cut symbolic method. This method involves replacing the statements with their truth values, letting T represent a true statement and F represent a false statement:

$$\sim p \wedge (q \vee \sim r)$$
$$\sim \text{F} \wedge (\text{T} \vee \sim \text{F})$$
$$\text{T} \wedge (\text{T} \vee \text{T}) \qquad \sim\text{F gives T.}$$
$$\text{T} \wedge \text{T} \qquad \text{T} \vee \text{T gives T.}$$
$$\text{T.} \qquad \text{T} \wedge \text{T gives T.}$$

Work within parentheses first.

The T in the final row indicates that the compound statement is true. ■

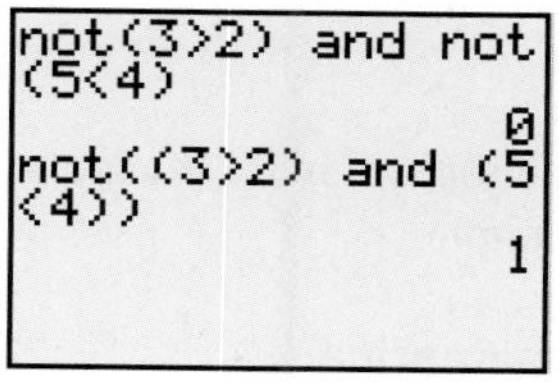

Example 4(a) explains why $\sim(3 > 2) \wedge [\sim(5 < 4)]$ is false. The calculator returns a 0. For a true statement such as $\sim[(3 > 2) \wedge (5 < 4)]$, it returns a 1.

Mathematical Statements We can use truth tables to determine the truth values of compound mathematical statements.

EXAMPLE 4 Deciding Whether a Compound Mathematical Statement Is True or False

Let p represent the statement $3 > 2$, q represent $5 < 4$, and r represent $3 < 8$. Decide whether each statement is *true* or *false*.

(a) $\sim p \wedge \sim q$ **(b)** $\sim(p \wedge q)$ **(c)** $(\sim p \wedge r) \vee (\sim q \wedge \sim p)$

SOLUTION

(a) Because p is true, $\sim p$ is false. By the *and* truth table, if one part of an "and" statement is false, the entire statement is false. This makes $\sim p \wedge \sim q$ false.

(b) For $\sim(p \wedge q)$, first work within the parentheses. Because p is true and q is false, $p \wedge q$ is false by the *and* truth table. Next, apply the negation. The negation of a false statement is true, making $\sim(p \wedge q)$ a true statement.

(c) Here p is true, q is false, and r is true. This makes $\sim p$ false and $\sim q$ true. By the *and* truth table, $\sim p \wedge r$ is false, and $\sim q \wedge \sim p$ is also false. Finally,

$$\begin{array}{ccc} (\sim p \wedge r) & \vee & (\sim q \wedge \sim p) \\ \downarrow & & \downarrow \\ \text{F} & \vee & \text{F}, \end{array}$$

which is false by the *or* truth table. (Alternatively, see Example 8(b).)

For Further Thought

Beauty or the Beast?

Raymond Smullyan is one of today's foremost writers of logic puzzles. This multitalented professor of mathematics and philosophy at City University of New York has written several books on recreational logic, including *The Lady or the Tiger?*, *What Is the Name of This Book?*, and *Alice in Puzzleland*. The title of the first of these is taken from the classic Frank Stockton short story, in which a prisoner must make a choice between two doors: behind one is a beautiful lady, and behind the other is a hungry tiger.

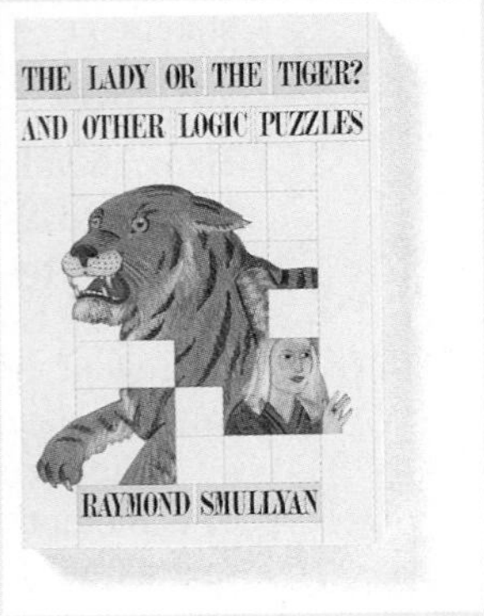

For Group Discussion or Individual Investigation

Smullyan proposes the following: What if each door has a sign, and the man knows that only one sign is true?

The sign on Door 1 reads:

IN THIS ROOM THERE IS A LADY AND IN THE OTHER ROOM THERE IS A TIGER.

The sign on Door 2 reads:

IN ONE OF THESE ROOMS THERE IS A LADY AND IN ONE OF THESE ROOMS THERE IS A TIGER.

With this information, the man is able to choose the correct door. Can you? (The answer is on page 110.)

When a quantifier is used with a conjunction or a disjunction, we must be careful in determining the truth value, as shown in the following example.

EXAMPLE 5 Deciding Whether a Quantified Mathematical Statement Is True or False

Decide whether each statement is *true* or *false*.

(a) For some real number x, $x < 5$ and $x > 2$.

(b) For every real number x, $x > 0$ or $x < 1$.

(c) For all real numbers x, $x^2 > 0$.

George Boole (1815–1864) grew up in poverty. His father, a London tradesman, gave him his first mathematics lessons and taught him to make optical instruments. Boole was largely self-educated. At 16 he worked in an elementary school and by age 20 had opened his own school. He studied mathematics in his spare time. He died of lung disease at age 49.

Boole's ideas have been used in the design of computers and telephone systems.

SOLUTION

(a) Replacing x with 3 (as an example) gives $3 < 5$ and $3 > 2$. Because both $3 < 5$ and $3 > 2$ are true statements, the given statement is true by the *and* truth table. (Remember: *some* means "at least one.")

(b) No matter which real number might be tried as a replacement for x, at least one of the statements $x > 0$ and $x < 1$ will be true. Because an "or" statement is true if one or both component statements are true, the entire statement as given is true.

(c) Because the quantifier is a universal quantifier, we need only find one case in which the inequality is false to make the entire statement false. Can we find a real number whose square is not positive (that is, not greater than 0)? Yes, we can—0 itself is a real number (and the *only* real number) whose square is not positive. Therefore, this statement is false.

Truth Tables In the preceding examples, the truth value for a given statement was found by going back to the basic truth tables. In the long run, it is easier to first create a complete truth table for the given statement itself. Then final truth values can be read directly from this table.

In this book we use the following standard format for listing the possible truth values in compound statements involving two component statements.

p	q	Compound Statement
T	T	
T	F	
F	T	
F	F	

EXAMPLE 6 Constructing a Truth Table

Consider the statement $(\sim p \wedge q) \vee \sim q$.

(a) Construct a truth table.

(b) Suppose both p and q are true. Find the truth value of this statement.

SOLUTION

(a) Begin by listing all possible combinations of truth values for p and q, as above. Then list the truth values of $\sim p$, which are the opposite of those of p.

p	q	$\sim p$
T	T	F
T	F	F
F	T	T
F	F	T

Use only the "$\sim p$" column and the "q" column, along with the *and* truth table, to find the truth values of $\sim p \wedge q$. List them in a separate column, as shown on the next page.

p	q	$\sim p$	$\sim p \wedge q$
T	T	F	F
T	F	F	F
F	T	T	T
F	F	T	F

Next include a column for $\sim q$.

p	q	$\sim p$	$\sim p \wedge q$	$\sim q$
T	T	F	F	F
T	F	F	F	T
F	T	T	T	F
F	F	T	F	T

Finally, make a column for the entire compound statement. To find the truth values, use *or* to combine $\sim p \wedge q$ with $\sim q$.

p	q	$\sim p$	$\sim p \wedge q$	$\sim q$	$(\sim p \wedge q) \vee \sim q$
T	T	F	F	F	F
T	F	F	F	T	T
F	T	T	T	F	T
F	F	T	F	T	T

(b) Look in the first row of the final truth table above, where both p and q have truth value T. Read across the row to find that the compound statement is false. ■

EXAMPLE 7 Constructing a Truth Table

Construct the truth table for $p \wedge (\sim p \vee \sim q)$.

SOLUTION

Proceed as shown.

p	q	$\sim p$	$\sim q$	$\sim p \vee \sim q$	$p \wedge (\sim p \vee \sim q)$
T	T	F	F	F	F
T	F	F	T	T	T
F	T	T	F	T	F
F	F	T	T	T	F

■

Answer to the Problem of *The Lady or the Tiger?*

The lady is behind Door 2. Suppose that the sign on Door 1 is true. Then the sign on Door 2 would also be true, but this is impossible. So the sign on Door 2 must be true, and the sign on Door 1 must be false. Because the sign on Door 1 says the lady is in Room 1, and this is false, the lady must be behind Door 2.

Emilie, Marquise du Châtelet (1706–1749) participated in the scientific activity of the generation after Newton and Leibniz. Educated in science, music, and literature, she was studying mathematics at the time (1733) she began a long intellectual relationship with the philosopher **François Voltaire** (1694–1778). She and Voltaire competed independently in 1738 for a prize offered by the French Academy on the subject of fire. Although du Châtelet did not win, her dissertation was published by the academy in 1744. During the last four years of her life she translated Newton's *Principia* from Latin into French—the only French translation to date.

If a compound statement involves three component statements p, q, and r, we will use the following standard format in setting up the truth table.

p	q	r	Compound Statement
T	T	T	
T	T	F	
T	F	T	
T	F	F	
F	T	T	
F	T	F	
F	F	T	
F	F	F	

EXAMPLE 8 Constructing a Truth Table

Consider the statement $(\sim p \wedge r) \vee (\sim q \wedge \sim p)$.

(a) Construct a truth table.

(b) Suppose p is true, q is false, and r is true. Find the truth value of this statement.

SOLUTION

(a) This statement has three component statements, p, q, and r. The truth table thus requires eight rows to list all possible combinations of truth values of p, q, and r. The final truth table, however, can be found in much the same way as the ones above.

p	q	r	$\sim p$	$\sim p \wedge r$	$\sim q$	$\sim q \wedge \sim p$	$(\sim p \wedge r) \vee (\sim q \wedge \sim p)$
T	T	T	F	F	F	F	F
T	T	F	F	F	F	F	F
T	F	T	F	F	T	F	F
T	F	F	F	F	T	F	F
F	T	T	T	T	F	F	T
F	T	F	T	F	F	F	F
F	F	T	T	T	T	T	T
F	F	F	T	F	T	T	T

(b) By the third row of the truth table in part (a), the compound statement is false. (This is an alternative method for working part (c) of Example 4.) ■

PROBLEM-SOLVING HINT One strategy for problem solving is to notice a pattern and use inductive reasoning. This strategy is applied in the next example.

TABLE 6

Number of Statements	Number of Rows
1	$2 = 2^1$
2	$4 = 2^2$
3	$8 = 2^3$

EXAMPLE 9 Using Inductive Reasoning

If n is a counting number, and a logical statement is composed of n component statements, how many rows will appear in the truth table for the compound statement?

SOLUTION

To answer this question, we examine some of the earlier truth tables in this section. The truth table for the negation has one statement and two rows. The truth tables for the conjunction and the disjunction have two component statements, and each has four rows. The truth table in Example 8(a) has three component statements and eight rows. Summarizing these in Table 6 seen in the margin reveals a pattern encountered earlier. Inductive reasoning leads us to the conjecture that if a logical statement is composed of n component statements, it will have 2^n rows. This can be proved using more advanced concepts.

The result of Example 9 is reminiscent of the formula for the number of subsets of a set having n elements.

Number of Rows in a Truth Table

A logical statement having n component statements will have 2^n rows in its truth table.

Alternative Method for Constructing Truth Tables

After making a reasonable number of truth tables, some people prefer the shortcut method shown in Example 10, which repeats Examples 6 and 8.

EXAMPLE 10 Constructing Truth Tables

Construct the truth table for each statement.

(a) $(\sim p \wedge q) \vee \sim q$ **(b)** $(\sim p \wedge r) \vee (\sim q \wedge \sim p)$

SOLUTION

(a) Start by inserting truth values for $\sim p$ and for q.

p	q	$(\sim p$	$\wedge$	$q)$	$\vee$	$\sim q$
T	T	F		T		
T	F	F		F		
F	T	T		T		
F	F	T		F		

Next, use the *and* truth table to obtain the truth values for $\sim p \wedge q$.

p	q	$(\sim p$	$\wedge$	$q)$	$\vee$	$\sim q$
T	T	F	F	T		
T	F	F	F	F		
F	T	T	T	T		
F	F	T	F	F		

Now disregard the two preliminary columns of truth values for $\sim p$ and for q, and insert truth values for $\sim q$. Finally, use the *or* truth table.

p	q	$(\sim p \wedge q)$	$\vee$	$\sim q$
T	T	F		F
T	F	F		T
F	T	T		F
F	F	F		T

p	q	$(\sim p \wedge q)$	$\vee$	$\sim q$
T	T	F	F	F
T	F	F	T	T
F	T	T	T	F
F	F	F	T	T

These steps can be summarized as follows.

p	q	$(\sim p$	$\wedge$	$q)$	$\vee$	$\sim p$
T	T	F	F	T	F	F
T	F	F	F	F	T	T
F	T	T	T	T	T	F
F	F	T	F	F	T	T
		①	②	①	④	③

The circled numbers indicate the order in which the various columns of the truth table were found.

(b) Work as follows.

p	q	r	$(\sim p$	$\wedge$	$r)$	$\vee$	$(\sim q$	$\wedge$	$\sim p)$
T	T	T	F	F	T	F	F	F	F
T	T	F	F	F	F	F	F	F	F
T	F	T	F	F	T	F	T	F	F
T	F	F	F	F	F	F	T	F	F
F	T	T	T	T	T	T	F	F	T
F	T	F	T	F	F	F	F	F	T
F	F	T	T	T	T	T	T	T	T
F	F	F	T	F	F	T	T	T	T
			①	②	①	⑤	③	④	③

The circled numbers indicate the order.

Equivalent Statements and De Morgan's Laws

One application of truth tables is to show that two statements are equivalent. Two statements are **equivalent** if they have the same truth value in *every* possible situation. The columns of each truth table that were the last to be completed will be the same for equivalent statements.

EXAMPLE 11 Deciding Whether Two Statements Are Equivalent

Are the following statements equivalent?

$$\sim p \wedge \sim q \quad \text{and} \quad \sim(p \vee q)$$

SOLUTION

Construct a truth table for each statement.

p	q	$\sim p \wedge \sim q$
T	T	F
T	F	F
F	T	F
F	F	T

p	q	$\sim(p \vee q)$
T	T	F
T	F	F
F	T	F
F	F	T

Because the truth values are the same in all cases, as shown in the columns in color, the statements $\sim p \wedge \sim q$ and $\sim(p \vee q)$ are equivalent. Equivalence is written with a three-bar symbol, $\equiv$. Using this symbol, $\sim p \wedge \sim q \equiv \sim(p \vee q)$.

In the same way, the statements $\sim p \vee \sim q$ and $\sim(p \wedge q)$ are equivalent. We call these equivalences *De Morgan's laws*.

De Morgan's Laws

For any statements p and q,

$$\boldsymbol{\sim(p \vee q) \equiv \sim p \wedge \sim q} \quad \text{and} \quad \boldsymbol{\sim(p \wedge q) \equiv \sim p \vee \sim q}.$$

(Compare the logic statements of De Morgan's laws with the set versions.) De Morgan's laws can be used to find the negations of certain compound statements.

EXAMPLE 12 Applying De Morgan's Laws

Find a negation of each statement by applying De Morgan's laws.

(a) I got an A or I got a B. **(b)** She won't try and he will succeed.
(c) $\sim p \vee (q \wedge \sim p)$

SOLUTION

(a) If p represents "I got an A" and q represents "I got a B," then the compound statement is symbolized $p \vee q$. The negation of $p \vee q$ is $\sim(p \vee q)$; by one of De Morgan's laws, this is equivalent to

$$\sim p \wedge \sim q,$$

or, in words,

I didn't get an A and I didn't get a B.

This negation is reasonable—the original statement says that I got either an A or a B; the negation says that I didn't get *either* grade.

(b) From one of De Morgan's laws, $\sim(p \wedge q) \equiv \sim p \vee \sim q$, so the negation becomes

She will try or he won't succeed.

(c) Negate both component statements and change $\vee$ to $\wedge$.

$$\sim[\sim p \vee (q \wedge \sim p)] \equiv p \wedge \sim(q \wedge \sim p)$$

Now apply De Morgan's law again.

$$\begin{aligned} p \wedge \sim(q \wedge \sim p) &\equiv p \wedge (\sim q \vee \sim(\sim p)) \\ &\equiv p \wedge (\sim q \vee p) \end{aligned}$$

A truth table will show that the statements

$$\sim p \vee (q \wedge \sim p) \quad \text{and} \quad p \wedge (\sim q \vee p)$$

are negations. ■

3.2 Exercises

Use the concepts introduced in this section to answer Exercises 1–6.

1. If q is false, what must be the truth value of the statement $(p \wedge \sim q) \wedge q$?

2. If q is true, what must be the truth value of the statement $q \vee (q \wedge \sim p)$?

3. If the statement $p \wedge q$ is true, and p is true, then q must be ________.

4. If the statement $p \vee q$ is false, and p is false, then q must be ________.

5. If $\sim(p \vee q)$ is true, what must be the truth values of the component statements?

6. If $\sim(p \wedge q)$ is false, what must be the truth values of the component statements?

Let p represent a false statement and let q represent a true statement. Find the truth value of the given compound statement.

7. $\sim p$

8. $\sim q$

9. $p \vee q$

10. $p \wedge q$

11. $p \vee \sim q$

12. $\sim p \wedge q$

13. $\sim p \vee \sim q$

14. $p \wedge \sim q$

15. $\sim(p \wedge \sim q)$

16. $\sim(\sim p \vee \sim q)$

17. $\sim[\sim p \wedge (\sim q \vee p)]$

18. $\sim[(\sim p \wedge \sim q) \vee \sim q]$

19. Is the statement $5 \geq 2$ a conjunction or a disjunction? Why?

20. Why is the statement $7 \geq 3$ true? Why is $9 \geq 9$ true?

Let p represent a true statement, and q and r represent false statements. Find the truth value of the given compound statement.

21. $(p \wedge r) \vee \sim q$

22. $(q \vee \sim r) \wedge p$

23. $p \wedge (q \vee r)$

24. $(\sim p \wedge q) \vee \sim r$

25. $\sim(p \wedge q) \wedge (r \vee \sim q)$

26. $(\sim r \wedge \sim q) \vee (\sim r \wedge q)$

27. $\sim[(\sim p \wedge q) \vee r]$

28. $\sim[r \vee (\sim q \wedge \sim p)]$

29. $\sim[\sim q \vee (r \wedge \sim p)]$

30. What is the only possible case in which the statement $(p \wedge \sim q) \wedge \sim r$ is true?

Let p represent the statement $15 < 8$, *let q represent the statement* $9 \not> 4$, *and let r represent the statement* $18 \leq 18$. *Find the truth value of the given compound statement.*

31. $p \wedge r$

32. $p \vee \sim q$

33. $\sim q \vee \sim r$

34. $\sim p \wedge \sim r$

35. $(p \wedge q) \vee r$

36. $\sim p \vee (\sim r \vee \sim q)$

37. $(\sim r \wedge q) \vee \sim p$

38. $\sim(p \vee \sim q) \vee \sim r$

Give the number of rows in the truth table for each compound statement.

39. $p \vee \sim r$

40. $p \wedge (r \wedge \sim s)$

41. $(\sim p \wedge q) \vee (\sim r \vee \sim s) \wedge r$

42. $[(p \vee q) \wedge (r \wedge s)] \wedge (t \vee \sim p)$

43. $[(\sim p \wedge \sim q) \wedge (\sim r \wedge s \wedge \sim t)] \wedge (\sim u \vee \sim v)$

44. $[(\sim p \wedge \sim q) \vee (\sim r \vee \sim s)] \vee [(\sim m \wedge \sim n) \wedge (u \wedge \sim v)]$

45. If the truth table for a certain compound statement has 128 rows, how many distinct component statements does it have?

46. Is it possible for the truth table of a compound statement to have exactly 54 rows? Why or why not?

Construct a truth table for each compound statement.

47. $\sim p \wedge q$

48. $\sim p \vee \sim q$

49. $\sim(p \wedge q)$

50. $p \vee \sim q$

51. $(q \vee \sim p) \vee \sim q$

52. $(p \wedge \sim q) \wedge p$

53. $\sim q \wedge (\sim p \vee q)$

54. $\sim p \vee (\sim q \wedge \sim p)$

55. $(p \vee \sim q) \wedge (p \wedge q)$

56. $(\sim p \wedge \sim q) \vee (\sim p \vee q)$

57. $(\sim p \wedge q) \wedge r$

58. $r \vee (p \wedge \sim q)$

59. $(\sim p \wedge \sim q) \vee (\sim r \vee \sim p)$

60. $(\sim r \vee \sim p) \wedge (\sim p \vee \sim q)$

61. $\sim(\sim p \wedge \sim q) \vee (\sim r \vee \sim s)$

62. $(\sim r \vee s) \wedge (\sim p \wedge q)$

Use one of De Morgan's laws to write the negation of each statement.

63. You can pay me now or you can pay me later.

64. I am not going or she is going.

65. It is summer and there is no snow.

66. $\frac{1}{2}$ is a positive number and -9 is less than zero.

67. I said yes but she said no.

68. Fellman Chutz tried to sell the wine, but he was unable to do so.

69. $5 - 1 = 4$ and $9 + 12 \neq 7$

70. $3 < 10$ or $7 \neq 2$

71. Dasher or Blitzen will lead Santa's sleigh next Christmas.

72. The lawyer and the client appeared in court.

Identify each statement as true *or* false.

73. For every real number x, $x < 13$ or $x > 6$.

74. For every real number x, $x > 9$ or $x < 9$.

75. For some integer n, $n \geq 4$ and $n \leq 4$.

76. There exists an integer n such that $n > 0$ and $n < 0$.

77. Complete the truth table for *exclusive disjunction*. The symbol $\underline{\vee}$ represents "one or the other is true, but not both."

p	q	$p \underline{\vee} q$
T	T	
T	F	
F	T	
F	F	

Exclusive disjunction

78. Attorneys sometimes use the phase "and/or." This phrase corresponds to which usage of the word *or*: inclusive or exclusive disjunction?

Decide whether each compound statement is true *or* false. *Remember that* $\underline{\vee}$ *is the* exclusive disjunction; *that is, assume* "either p or q is true, but not both."

79. $3 + 1 = 4 \ \underline{\vee}\ 2 + 5 = 7$

80. $3 + 1 = 4 \ \underline{\vee}\ 2 + 5 = 10$

81. $3 + 1 = 6 \ \underline{\vee}\ 2 + 5 = 7$

82. $3 + 1 = 12 \ \underline{\vee}\ 2 + 5 = 10$

3.3 The Conditional and Circuits

Conditionals • Negation of a Conditional • Circuits

Conditionals

"If you build it, he will come."
—The Voice in the movie *Field of Dreams*

Ray Kinsella, an Iowa farmer in the movie *Field of Dreams*, heard a voice from the sky. Ray interpreted it as a promise that if he would build a baseball field in his cornfield, then the ghost of Shoeless Joe Jackson (a baseball star in the early days of the twentieth century) would come to play on it. The promise came in the form of a conditional statement. A **conditional** statement is a compound statement that uses the connective *if . . . then.* For example, here are a few conditional statements:

If I read for too long, *then* I get a headache.

If looks could kill, *then* I would be dead.

If he doesn't get back soon, *then* you should go look for him.

In his April 21, 1989, five-star review of *Field of Dreams*, the *Chicago Sun-Times* movie critic Roger Ebert gave an explanation of why the movie has become an American classic.

There is a speech in this movie about baseball that is so simple and true that it is heartbreaking. And the whole attitude toward the players reflects that attitude. Why do they come back from the great beyond and play in this cornfield? Not to make any kind of vast, earthshattering statement, but simply to hit a few and field a few, and remind us of a good and innocent time.

In each of these conditional statements, the component coming after the word *if* gives a condition (but not necessarily the only condition) under which the statement coming after *then* will be true. For example, "If it is over 90°, then I'll go to the mountains" tells one possible condition under which I will go to the mountains—if the temperature is over 90°.

The conditional is written with an arrow, so "if p, then q" is symbolized

$$p \rightarrow q.$$

We read $p \rightarrow q$ as "p implies q" or "if p, then q." In the conditional $p \rightarrow q$, the statement p is the **antecedent,** while q is the **consequent.**

The conditional connective may not always be explicitly stated. That is, it may be "hidden" in an everyday expression. For example, the statement

Big girls don't cry

can be written in *if . . . then* form as

If you're a big girl, then you don't cry.

As another example, the statement

It is difficult to study when you are distracted

can be written

If you are distracted, then it is difficult to study.

In the quote from the movie *Field of Dreams*, the word "then" is not stated but understood to be there from the context of the statement. In that statement, "you build it" is the antecedent, and "he will come" is the consequent.

The importance of **symbols** was emphasized by the American philosopher-logician **Charles Sanders Peirce** (1839–1914), who asserted the nature of humans as symbol-using or sign-using organisms. Symbolic notation is half of mathematics, Bertrand Russell once said.

The conditional truth table is a little harder to define than the tables in the previous section. To see how to define the conditional truth table, let us analyze a statement made by a politician, Senator Shootie Gosserand:

If I am elected, then taxes will go down.

As before, there are four possible combinations of truth values for the two component statements. Let p represent "I am elected," and let q represent "Taxes will go down."

As we analyze the four possibilities, it is helpful to think in terms of the following: "Did Senator Gosserand lie?" If she lied, then the conditional statement is considered false; if she did not lie, then the conditional statement is considered true.

Possibility	Elected?	Taxes Go Down?	
1	Yes	Yes	p is T, q is T
2	Yes	No	p is T, q is F
3	No	Yes	p is F, q is T
4	No	No	p is F, q is F

The four possibilities are as follows:

1. In the first case assume that the senator was elected and taxes did go down (p is T, q is T). The senator told the truth, so place T in the first row of the truth table. (We do not claim that taxes went down *because* she was elected; it is possible that she had nothing to do with it at all.)
2. In the second case assume that the senator was elected and taxes did not go down (p is T, q is F). Then the senator did not tell the truth (that is, she lied). So we put F in the second row of the truth table.
3. In the third case assume that the senator was defeated, but taxes went down anyway (p is F, q is T). The senator did not lie; she only promised a tax reduction if she were elected. She said nothing about what would happen if she were not elected. In fact, her campaign promise gives no information about what would happen if she lost. Because we cannot say that the senator lied, place T in the third row of the truth table.
4. In the last case assume that the senator was defeated and taxes did not go down (p is F, q is F). We cannot blame her, because she only promised to reduce taxes if elected. Thus, T goes in the last row of the truth table.

The completed truth table for the conditional is defined as follows.

Truth Table for the Conditional If p, then q

If p, then q

p	q	$p \rightarrow q$
T	T	T
T	F	F
F	T	T
F	F	T

It must be emphasized that the use of the conditional connective in no way implies a cause-and-effect relationship. Any two statements may have an arrow placed between them to create a compound statement. For example,

If I pass mathematics, then the sun will rise the next day

is true, because the consequent is true. (See the special characteristics following Example 1.) There is, however, no cause-and-effect connection between my passing mathematics and the sun's rising. The sun will rise no matter what grade I get.

EXAMPLE 1 Finding the Truth Value of a Conditional

Given that p, q, and r are all false, find the truth value of the statement

$$(p \rightarrow \sim q) \rightarrow (\sim r \rightarrow q).$$

SOLUTION

Using the short-cut method explained in Example 3 of the previous section, we can replace p, q, and r with F (since each is false) and proceed as before, using the negation and conditional truth tables as necessary.

$$\begin{array}{ccc} (p \rightarrow \sim q) & \rightarrow & (\sim r \rightarrow q) \\ (\mathrm{F} \rightarrow \sim \mathrm{F}) & \rightarrow & (\sim \mathrm{F} \rightarrow \mathrm{F}) \\ (\mathrm{F} \rightarrow \mathrm{T}) & \rightarrow & (\mathrm{T} \rightarrow \mathrm{F}) \\ \mathrm{T} & \rightarrow & \mathrm{F} \\ & \mathrm{F} & \end{array}$$

Use the negation truth table.

Use the conditional truth table.

The statement $(p \rightarrow \sim q) \rightarrow (\sim r \rightarrow q)$ is false when p, q, and r are all false.

The following observations come from the truth table for $p \rightarrow q$.

Special Characteristics of Conditional Statements

1. $p \rightarrow q$ is false only when the antecedent is *true* and the consequent is *false*.
2. If the antecedent is *false*, then $p \rightarrow q$ is automatically *true*.
3. If the consequent is *true*, then $p \rightarrow q$ is automatically *true*.

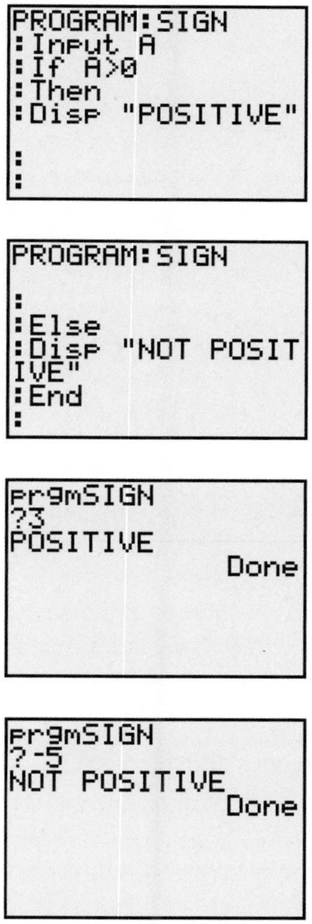

Conditional statements are useful in writing programs. The short program in the first two screens determines whether a number is positive. Notice the lines that begin with *If* and *Then.*

EXAMPLE 2 Determining Whether a Conditional Is True or False

Write *true* or *false* for each statement. Here T represents a true statement, and F represents a false statement.

(a) $\mathrm{T} \rightarrow (6 = 3)$ **(b)** $(5 < 2) \rightarrow \mathrm{F}$ **(c)** $(3 \neq 2 + 1) \rightarrow \mathrm{T}$

SOLUTION

(a) Because the antecedent is true, while the consequent, $6 = 3$, is false, the given statement is false by the first point mentioned above.

(b) The antecedent is false, so the given statement is true by the second observation.

(c) The consequent is true, making the statement true by the third characteristic of conditional statements.

Truth tables for compound statements involving conditionals are found using the techniques described in the previous section.

EXAMPLE 3 Constructing Truth Tables

Construct a truth table for each statement.

(a) $(\sim p \rightarrow \sim q) \rightarrow (\sim p \wedge q)$ **(b)** $(p \rightarrow q) \rightarrow (\sim p \vee q)$

SOLUTION

(a) First insert the truth values of $\sim p$ and of $\sim q$. Then find the truth values of $\sim p \rightarrow \sim q$.

p	q	$\sim p$	$\sim q$	$\sim p \rightarrow \sim q$
T	T	F	F	T
T	F	F	T	T
F	T	T	F	F
F	F	T	T	T

Next use $\sim p$ and q to find the truth values of $\sim p \wedge q$.

p	q	$\sim p$	$\sim q$	$\sim p \rightarrow \sim q$	$\sim p \wedge q$
T	T	F	F	T	F
T	F	F	T	T	F
F	T	T	F	F	T
F	F	T	T	T	F

Now find the truth values of $(\sim p \rightarrow \sim q) \rightarrow (\sim p \wedge q)$.

p	q	$\sim p$	$\sim q$	$\sim p \rightarrow \sim q$	$\sim p \wedge q$	$(\sim p \rightarrow \sim q) \rightarrow (\sim p \wedge q)$
T	T	F	F	T	F	F
T	F	F	T	T	F	F
F	T	T	F	F	T	T
F	F	T	T	T	F	F

(b) For $(p \rightarrow q) \rightarrow (\sim p \vee q)$, go through steps similar to the ones above.

p	q	$p \rightarrow q$	$\sim p$	$\sim p \vee q$	$(p \rightarrow q) \rightarrow (\sim p \vee q)$
T	T	T	F	T	T
T	F	F	F	F	T
F	T	T	T	T	T
F	F	T	T	T	T

As the truth table in Example 3(b) shows, the statement $(p \rightarrow q) \rightarrow (\sim p \vee q)$ is always true, no matter what the truth values of the components. Such a statement is called a **tautology.** Other examples of tautologies (as can be checked by forming truth tables) include $p \vee \sim p$, $p \rightarrow p$, $(\sim p \vee \sim q) \rightarrow \sim(q \wedge p)$, and so on. By the way, the truth tables in Example 3 also could have been found by the alternative method shown in the previous section.

Negation of a Conditional Suppose that someone makes the conditional statement

"If it rains, then I take my umbrella."

When will the person have lied to you? The only case in which you would have been misled is when it rains *and* the person does *not* take the umbrella. Letting p represent "it rains" and q represent "I take my umbrella," you might suspect that the symbolic statement

$$p \wedge \sim q$$

is a candidate for the negation of $p \rightarrow q$. That is,

$$\sim(p \rightarrow q) \equiv p \wedge \sim q.$$

This is indeed the case, as the next truth table indicates.

p	q	$p \rightarrow q$	$\sim(p \rightarrow q)$	$\sim q$	$p \wedge \sim q$
T	T	T	F	F	F
T	F	F	T	T	T
F	T	T	F	F	F
F	F	T	F	T	F

$\sim(p \rightarrow q)$ and $p \wedge \sim q$ columns: $\equiv$

Negation of $p \rightarrow q$

The negation of $p \rightarrow q$ is $p \wedge \sim q$.

Because

$$\sim(p \rightarrow q) \equiv p \wedge \sim q,$$

by negating each expression we have

$$\sim[\sim(p \rightarrow q)] \equiv \sim(p \wedge \sim q).$$

The left side of the above equivalence is $p \rightarrow q$, and one of De Morgan's laws can be applied to the right side:

$$p \rightarrow q \equiv \sim p \vee \sim(\sim q)$$
$$p \rightarrow q \equiv \sim p \vee q.$$

This final row indicates that a conditional may be written as a disjunction.

Writing a Conditional as an "or" Statement

$p \rightarrow q$ is equivalent to $\sim p \vee q$.

EXAMPLE 4 Determining Negations

Determine the negation of each statement.

(a) If you build it, he will come. **(b)** All dogs have fleas.

SOLUTION

Do not try to negate a conditional with another conditional.

(a) If b represents "you build it" and q represents "he will come," then the given statement can be symbolized by $b \rightarrow q$. The negation of $b \rightarrow q$, as shown earlier, is $b \wedge \sim q$, so the negation of the statement is

You build it and he will not come.

(b) First, we must restate the given statement in *if . . . then* form:

If it is a dog, then it has fleas.

Based on our earlier discussion, the negation is

It is a dog and it does not have fleas. ■

As seen in Example 4, the negation of a conditional statement is written as a conjunction.

EXAMPLE 5 Determining Statements Equivalent to Conditionals

Write each conditional as an equivalent statement without using *if . . . then.*

(a) If the Cubs win the pennant, then Gwen will be happy.
(b) If it's Borden's, it's got to be good.

SOLUTION

(a) Because the conditional $p \rightarrow q$ is equivalent to $\sim p \vee q$, let p represent "The Cubs win the pennant" and q represent "Gwen will be happy." Restate the conditional as

The Cubs do not win the pennant or Gwen will be happy.

(b) If p represents "it's Borden's" and if q represents "it's got to be good," the conditional may be restated as

It's not Borden's or it's got to be good. ■

FIGURE 1

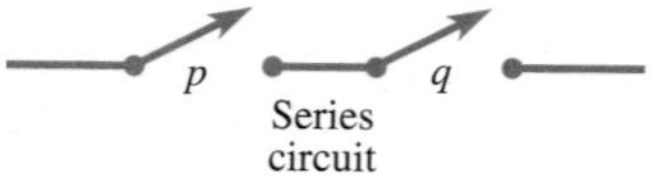

FIGURE 2

Circuits One of the first nonmathematical applications of symbolic logic was seen in the master's thesis of Claude Shannon in 1937. Shannon showed how logic could be used to design electrical circuits. His work was immediately used by computer designers. Then in the developmental stage, computers could be simplified and built for less money using the ideas of Shannon.

To see how Shannon's ideas work, look at the electrical switch shown in Figure 1. We assume that current will flow through this switch when it is closed and not when it is open.

Figure 2 shows two switches connected in *series;* in such a circuit, current will flow only when both switches are closed. Note how closely a series circuit

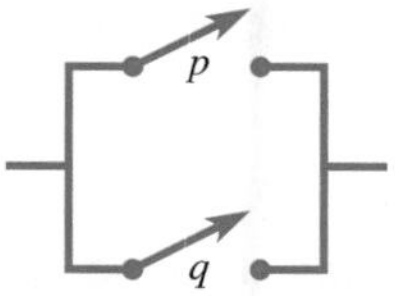

Parallel circuit

FIGURE 3

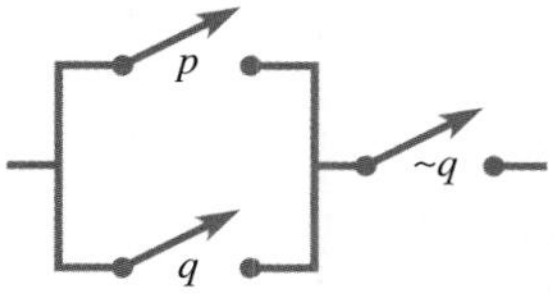

FIGURE 4

corresponds to the conjunction $p \wedge q$. We know that $p \wedge q$ is true only when both p and q are true.

A circuit corresponding to the disjunction $p \vee q$ can be found by drawing a *parallel* circuit, as in Figure 3. Here, current flows if either *p or q* is closed or if both *p and q* are closed.

The circuit in Figure 4 corresponds to the statement $(p \vee q) \wedge \sim q$, which is a compound statement involving both a conjunction and a disjunction.

Simplifying an electrical circuit depends on the idea of equivalent statements from Section 3.2. Recall that two statements are equivalent if they have the same truth table final column. The symbol $\equiv$ is used to indicate that the two statements are equivalent. Some equivalent statements are shown in the following box.

Equivalent Statements Used to Simplify Circuits

$p \vee (q \wedge r) \equiv (p \vee q) \wedge (p \vee r)$	$p \vee p \equiv p$
$p \wedge (q \vee r) \equiv (p \wedge q) \vee (p \wedge r)$	$p \wedge p \equiv p$
$p \rightarrow q \equiv \sim q \rightarrow \sim p$	$\sim(p \wedge q) \equiv \sim p \vee \sim q$
$p \rightarrow q \equiv \sim p \vee q$	$\sim(p \vee q) \equiv \sim p \wedge \sim q$

If T represents any true statement and F represents any false statement, then

$$p \vee \mathbf{T} \equiv \mathbf{T} \qquad p \vee \sim p \equiv \mathbf{T}$$
$$p \wedge \mathbf{F} \equiv \mathbf{F} \qquad p \wedge \sim p \equiv \mathbf{F}.$$

Circuits can be used as models of compound statements, with a closed switch corresponding to T, while an open switch corresponds to F. The method for simplifying circuits is explained in the following example.

EXAMPLE 6 Simplifying a Circuit

Simplify the circuit of Figure 5.

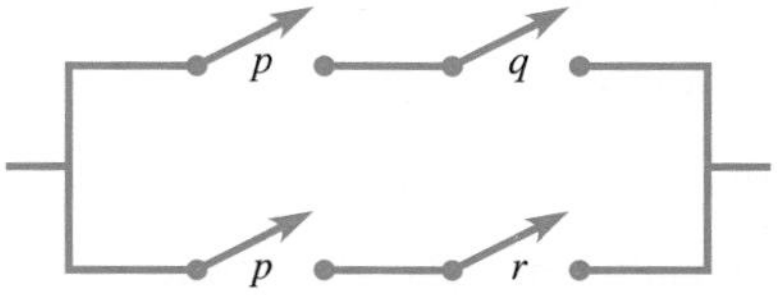

FIGURE 5

SOLUTION

At the top of Figure 5, p and q are connected in series, and at the bottom, p and r are connected in series. These are interpreted as the compound statements $p \wedge q$ and $p \wedge r$, respectively. These two conjunctions are connected in parallel, as indicated by the figure treated as a whole. Write the disjunction of the two conjunctions:

$$(p \wedge q) \vee (p \wedge r).$$

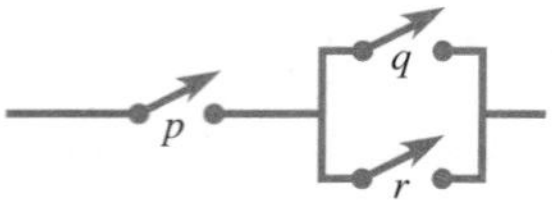

FIGURE 6

(Think of the two switches labeled "p" as being controlled by the same lever.) By one of the pairs of equivalent statements in the preceding box,

$$(p \wedge q) \vee (p \wedge r) \equiv p \wedge (q \vee r),$$

which has the circuit of Figure 6. This circuit is logically equivalent to the one in Figure 5, and yet it contains only three switches instead of four—which might well lead to a large savings in manufacturing costs.

EXAMPLE 7 Drawing a Circuit for a Conditional Statement

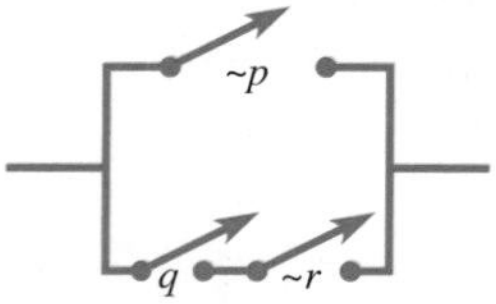

FIGURE 7

Draw a circuit for $p \rightarrow (q \wedge \sim r)$.

SOLUTION

From the list of equivalent statements in the box, $p \rightarrow q$ is equivalent to $\sim p \vee q$. This equivalence gives $p \rightarrow (q \wedge \sim r) \equiv \sim p \vee (q \wedge \sim r)$, which has the circuit diagram in Figure 7.

3.3 Exercises

Rewrite each statement using the if . . . then *connective. Rearrange the wording or add words as necessary.*

1. You can believe it if you see it on the Internet.

2. It must be alive if it is breathing.

3. Garrett Olinde's area code is 225.

4. Lorri Morgan visits Hawaii every summer.

5. All marines love boot camp.

6. Every picture tells a story.

7. No koalas live in Iowa.

8. No guinea pigs are scholars.

9. An opium eater cannot have self-command.

10. Running Bear loves Little White Dove.

Decide whether each statement is true *or* false.

11. If the consequent of a conditional statement is true, the conditional statement is true.

12. If the antecedent of a conditional statement is false, the conditional statement is true.

13. If p is true, then $\sim p \rightarrow (q \vee r)$ is true.

14. If q is true, then $(p \wedge q) \rightarrow q$ is true.

15. The statements "If it flies, then it's a bird" and "It does not fly or it's a bird" are logically equivalent.

16. The negation of "If pigs fly, I'll believe it" is "If pigs don't fly, I won't believe it."

17. Given that $\sim p$ is false and q is false, the conditional $p \rightarrow q$ is true.

18. Given that $\sim p$ is true and q is false, the conditional $p \rightarrow q$ is true.

19. In a few sentences, explain how to determine the truth value of a conditional statement.

20. Explain why the statement "If 3 = 5, then 4 = 6" is true.

Tell whether each conditional is true (T) *or* false (F).

21. $T \rightarrow (6 < 3)$
22. $F \rightarrow (4 \neq 7)$
23. $F \rightarrow (3 \neq 3)$
24. $(6 \geq 6) \rightarrow F$
25. $(4^2 \neq 16) \rightarrow (4 - 4 = 8)$
26. $(4 = 11 - 7) \rightarrow (8 > 0)$

Let s represent "She has a ferret for a pet," *let p represent* "he trains dogs," *and let m represent* "they raise alpacas." *Express each compound statement in words.*

27. $\sim m \rightarrow p$
28. $p \rightarrow \sim m$
29. $s \rightarrow (m \wedge p)$
30. $(s \wedge p) \rightarrow m$
31. $\sim p \rightarrow (\sim m \vee s)$
32. $(\sim s \vee \sim m) \rightarrow \sim p$

Let b represent "I ride my bike," *let r represent* "it rains," *and let p represent* "the concert is cancelled." *Write each compound statement in symbols.*

33. If I ride my bike, then the concert is cancelled.
34. If it rains, then I ride my bike.
35. If the concert is cancelled, then it does not rain.
36. If I do not ride my bike, then it does not rain.
37. The concert is cancelled, and if it rains then I do not ride my bike.
38. I ride my bike, or if the concert is cancelled then it rains.
39. It rains if the concert is cancelled.
40. I'll ride my bike if it doesn't rain.

Find the truth value of each statement. Assume that p and r are false, and q is true.

41. $\sim r \rightarrow q$
42. $\sim p \rightarrow \sim r$
43. $q \rightarrow p$
44. $\sim r \rightarrow p$
45. $p \rightarrow q$
46. $\sim q \rightarrow r$
47. $\sim p \rightarrow (q \wedge r)$
48. $(\sim r \vee p) \rightarrow p$
49. $\sim q \rightarrow (p \wedge r)$
50. $(\sim p \wedge \sim q) \rightarrow (p \wedge \sim r)$
51. $(p \rightarrow \sim q) \rightarrow (\sim p \wedge \sim r)$
52. $(p \rightarrow \sim q) \wedge (p \rightarrow r)$

53. Explain why, if we know that p is true, we also know that

$$[r \vee (p \vee s)] \rightarrow (p \vee q)$$

is true, even if we are not given the truth values of q, r, and s.

54. Construct a true statement involving a conditional, a conjunction, a disjunction, and a negation (not necessarily in that order), that consists of component statements p, q, and r, with all of these component statements false.

Construct a truth table for each statement. Identify any tautologies.

55. $\sim q \rightarrow p$
56. $p \rightarrow \sim q$
57. $(\sim p \rightarrow q) \rightarrow p$
58. $(\sim q \rightarrow \sim p) \rightarrow \sim q$
59. $(p \vee q) \rightarrow (q \vee p)$
60. $(p \wedge q) \rightarrow (p \vee q)$
61. $(\sim p \rightarrow \sim q) \rightarrow (p \wedge q)$
62. $r \rightarrow (p \wedge \sim q)$
63. $[(r \vee p) \wedge \sim q] \rightarrow p$
64. $[(r \wedge p) \wedge (p \wedge q)] \rightarrow p$
65. $(\sim r \rightarrow s) \vee (p \rightarrow \sim q)$
66. $(\sim p \wedge \sim q) \rightarrow (s \rightarrow r)$

67. What is the minimum number of Fs that must appear in the final column of a truth table for us to be assured that the statement is not a tautology?
68. If all truth values in the final column of a truth table are F, how can we easily transform the statement into a tautology?

Write the negation of each statement. Remember that the negation of $p \rightarrow q$ is $p \wedge \sim q$.

69. If that is an authentic Persian rug, I'll be surprised.

70. If Ella reaches that note, she will shatter glass.

71. If the English measures are not converted to metric measures, then the spacecraft will crash on the surface of Saturn.

72. If you say "I do," then you'll be happy for the rest of your life.

73. "If you want to be happy for the rest of your life, never make a pretty woman your wife." *Jimmy Soul*

74. If loving you is wrong, I don't want to be right.

Write each statement as an equivalent statement that does not use the if . . . then *connective. Remember that $p \rightarrow q$ is equivalent to $\sim p \vee q$.*

75. If you give your plants tender, loving care, they flourish.

76. If the check is in the mail, I'll be surprised.

77. If she doesn't, he will.

78. If I say "black", she says "white."

79. All residents of Oregon City are residents of Oregon.

80. All men were once boys.

Use truth tables to decide which of the pairs of statements are equivalent.

81. $p \rightarrow q$; $\sim p \vee q$

82. $\sim(p \rightarrow q)$; $p \wedge \sim q$

83. $p \rightarrow q$; $\sim q \rightarrow \sim p$

84. $q \rightarrow p$; $\sim p \rightarrow \sim q$

85. $p \rightarrow \sim q$; $\sim p \vee \sim q$

86. $p \rightarrow q$; $q \rightarrow p$

87. $p \wedge \sim q$; $\sim q \rightarrow \sim p$

88. $\sim p \wedge q$; $\sim p \rightarrow q$

89. $q \rightarrow \sim p$; $p \rightarrow \sim q$

90. Explain why the circuit shown will always have exactly one open switch. What does this circuit simplify to?

Write a logical statement representing each of the following circuits. Simplify each circuit when possible.

91.

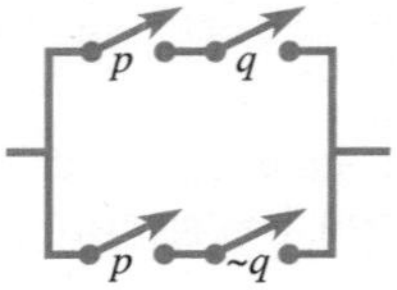

92.

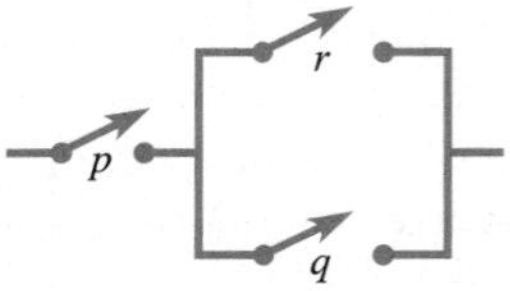

93.

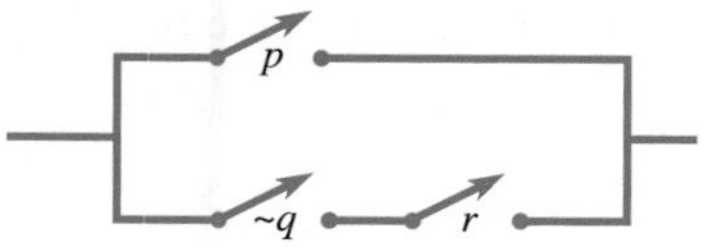

94.

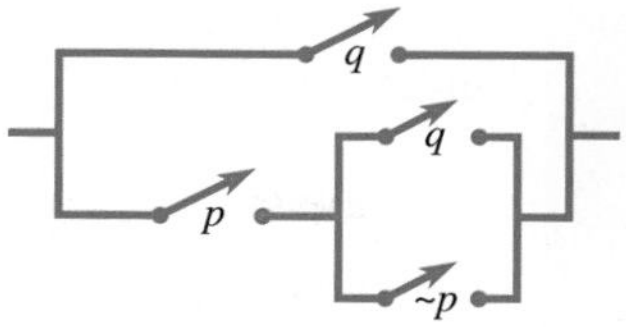

95.

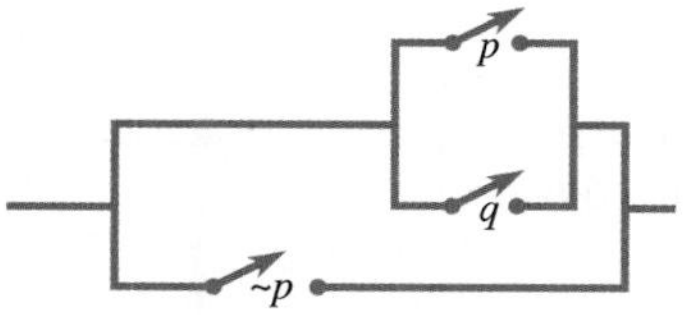

96.

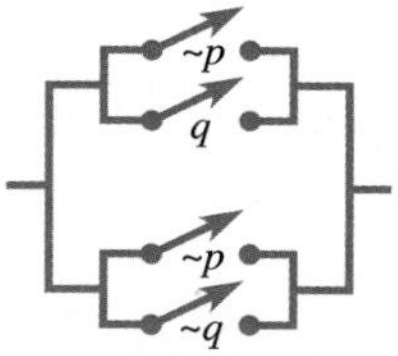

Draw circuits representing the following statements as they are given. Simplify if possible.

97. $p \wedge (q \vee \sim p)$

98. $(\sim p \wedge \sim q) \wedge \sim r$

99. $(p \vee q) \wedge (\sim p \wedge \sim q)$

100. $(\sim q \wedge \sim p) \vee (\sim p \vee q)$

101. $[(p \vee q) \wedge r] \wedge \sim p$

102. $[(\sim p \wedge \sim r) \vee \sim q] \wedge (\sim p \wedge r)$

103. $\sim q \rightarrow (\sim p \rightarrow q)$

104. $\sim p \rightarrow (\sim p \vee \sim q)$

105. Refer to Figures 5 and 6 in Example 6. Suppose the cost of the use of one switch for an hour is \$.06. By using the circuit in Figure 6 rather than the circuit in Figure 5, what is the savings for a year of 365 days, assuming that the circuit is in continuous use?

3.4 More on the Conditional

Converse, Inverse, and Contrapositive • Alternative Forms of "If *p*, then *q*" • Biconditionals • Summary of Truth Tables

Converse, Inverse, and Contrapositive

Many mathematical properties and theorems are stated in *if . . . then* form. Because of their usefulness, we expand our consideration of statements of the form $p \rightarrow q$. Any conditional statement is made up of an antecedent and a consequent. If they are interchanged, negated, or both, a new conditional statement is formed. Suppose that we begin with the given conditional statement

If you stay, then I go,

and interchange the antecedent ("you stay") and the consequent ("I go"). We obtain the new conditional statement

If I go, then you stay.

This new conditional is called the **converse** of the given statement.

Alfred North Whitehead (1861–1947) and Bertrand Russell worked together on *Principia Mathematica.* During that time, Whitehead was teaching mathematics at Cambridge University and had written *Universal Algebra.* In 1910 he went to the University of London, exploring not only the philosophical basis of science but also the "aims of education" (as he called one of his books). It was as a philosopher that he was invited to Harvard University in 1924. Whitehead died at the age of 86 in Cambridge, Massachusetts.

By negating both the antecedent and the consequent, we obtain the **inverse** of the given statement:

If you do not stay, then I do not go.

If the antecedent and the consequent are both interchanged *and* negated, the **contrapositive** of the given statement is formed:

If I do not go, then you do not stay.

These three related statements for the conditional $p \rightarrow q$ are summarized below. (Notice that the inverse is the contrapositive of the converse.)

Related Conditional Statements

Conditional Statement	$p \rightarrow q$	(If p, then q.)
Converse	$q \rightarrow p$	(If q, then p.)
Inverse	$\sim p \rightarrow \sim q$	(If not p, then not q.)
Contrapositive	$\sim q \rightarrow \sim p$	(If not q, then not p.)

EXAMPLE 1 Determining Related Conditional Statements

Given the conditional statement

If I live in Miami, then I live in Florida,

determine each of the following:

(a) the converse
(b) the inverse
(c) the contrapositive

SOLUTION

(a) Let p represent "I live in Miami" and q represent "I live in Florida." Then the given statement may be written $p \rightarrow q$. The converse, $q \rightarrow p$, is

If I live in Florida, then I live in Miami.

Notice that for this statement, the converse is not necessarily true, even though the given statement is true.

(b) The inverse of $p \rightarrow q$ is $\sim p \rightarrow \sim q$. For the given conditional statement, the inverse is

If I don't live in Miami, then I don't live in Florida,

which is again not necessarily true.

(c) The contrapositive, $\sim q \rightarrow \sim p$, is

If I don't live in Florida, then I don't live in Miami.

The contrapositive, like the given conditional statement, is true.

Bertrand Russell (1872–1970) was a student of Whitehead's before they wrote the *Principia*. Like his teacher, Russell turned toward philosophy. His works include a critique of Leibniz, analyses of mind and of matter, and a history of Western thought.

Russell became a public figure because of his involvement in social issues. Deeply aware of human loneliness, he was "passionately desirous of finding ways of diminishing this tragic isolation." During World War I he was an antiwar crusader, and he was imprisoned briefly. Again in the 1960s he championed peace. He wrote many books on social issues, winning the Nobel Prize for Literature in 1950.

Example 1 shows that the converse and inverse of a true statement need not be true. They *can* be true, but they need not be. The relationships between the truth values of the conditional statement, converse, inverse, and contrapositive are shown in the truth table that follows.

Equivalent (Conditional and Contrapositive)

Equivalent (Converse and Inverse)

		Conditional	Converse	Inverse	Contrapositive
p	q	$p \to q$	$q \to p$	$\sim p \to \sim q$	$\sim q \to \sim p$
T	T	T	T	T	T
T	F	F	T	T	F
F	T	T	F	F	T
F	F	T	T	T	T

As this truth table shows, a conditional statement and its contrapositive always have the same truth values, making it possible to replace any statement with its contrapositive without affecting the logical meaning. Also, the converse and inverse always have the same truth values.

This discussion is summarized as follows.

Equivalences

A conditional statement and its contrapositive are equivalent, and the converse and the inverse are equivalent.

EXAMPLE 2 Determining Related Conditional Statements

For the conditional statement $\sim p \to q$, write each of the following.

(a) the converse **(b)** the inverse **(c)** the contrapositive

SOLUTION

(a) The converse of $\sim p \to q$ is $q \to \sim p$.

(b) The inverse is $\sim(\sim p) \to \sim q$, which simplifies to $p \to \sim q$.

(c) The contrapositive is $\sim q \to \sim(\sim p)$, which simplifies to $\sim q \to p$. ■

In a speech during the 2004 presidential race, **John Kerry** made the following statement:

> *Mark my words. If I am elected president and there still has not been sufficient progress rapidly in these next months on these issues, then I will lead.*

This promise involves a conditional, a conjunction, and a negation.

Alternative Forms of "If *p*, then *q*"

The conditional statement "if p, then q" can be stated in several other ways in English. For example,

If you go to the shopping center, then you will find a place to park

can also be written

Going to the shopping center is *sufficient* for finding a place to park.

According to this statement, going to the shopping center is enough to guarantee finding a place to park. Going to other places, such as schools or office buildings, *might* also guarantee a place to park, but at least we *know* that going to the shopping center does.

Thus, $p \rightarrow q$ can be written "p is sufficient for q." Knowing that p has occurred is sufficient to guarantee that q will also occur. On the other hand,

Turning on the set is necessary for watching television (*)

has a different meaning. Here, we are saying that one condition that is necessary for watching television is that you turn on the set. This may not be enough; the set might be broken, for example. The statement labeled (*) could be written as

If you watch television, then you turned on the set.

As this example suggests, $p \rightarrow q$ is the same as "q is necessary for p." In other words, if q doesn't happen, then neither will p. Notice how this idea is closely related to the idea of equivalence between a conditional statement and its contrapositive.

Common Translations of $p \rightarrow q$

The conditional $p \rightarrow q$ can be translated in any of the following ways,

If p, then q.	p is sufficient for q.
If p, q.	q is necessary for p.
p implies q.	All p are q.
p only if q.	q if p.

The translation of $p \rightarrow q$ into these various word forms does not in any way depend on the truth or falsity of $p \rightarrow q$.

For example, the statement

If you are 18, then you can vote

can be written in any of the following alternative ways:

You can vote if you are 18.
You are 18 only if you can vote.
Being able to vote is necessary for you to be 18.
Being 18 is sufficient for being able to vote.
All 18-year-olds can vote.
Being 18 implies that you can vote.

EXAMPLE 3 Rewording Conditional Statements

Write each statement in the form "if p, then q."

(a) You'll be sorry if I go.
(b) Today is Friday only if yesterday was Thursday.
(c) All nurses wear white shoes.

SOLUTION

(a) If I go, then you'll be sorry.
(b) If today is Friday, then yesterday was Thursday.
(c) If you are a nurse, then you wear white shoes. ■

For Further Thought

A Word to the Wise Is Sufficient

How many times have you heard a wise saying like "A stitch in time saves nine," "A rolling stone gathers no moss," or "Birds of a feather flock together"? In many cases, such proverbial advice can be restated as a conditional in *if . . . then* form. For example, these three statements can be restated as follows:

"If you make a stitch in time, then it will save you nine (stitches)."

"If a stone rolls, then it gathers no moss."

"If they are birds of a feather, then they flock together."

For Group Discussion or Individual Investigation

1. Think of some wise sayings that have been around for a long time, and state them in *if . . . then* form.
2. You have probably heard the saying "All that glitters is not gold." Do you think that what is said here is actually what is meant? If not, restate it as you think it should be stated. (*Hint:* Write the original statement in *if . . . then* form.)

Principia Mathematica, the title chosen by Whitehead and Russell, was a deliberate reference to *Philosophiae naturalis principia mathematica*, or "mathematical principles of the philosophy of nature," Isaac Newton's epochal work of 1687. Newton's Principia pictured a kind of "clockwork universe" that ran via his Law of Gravitation. Newton independently invented the calculus, unaware that Leibniz had published his own formulation of it earlier. A controversy over their priority continued into the eighteenth century.

EXAMPLE 4 Translating from Words to Symbols

Let p represent "A triangle is equilateral," and let q represent "A triangle has three sides of equal length." Write each of the following in symbols.

(a) A triangle is equilateral if it has three sides of equal length.
(b) A triangle is equilateral only if it has three sides of equal length.

SOLUTION

(a) $q \rightarrow p$ **(b)** $p \rightarrow q$

Biconditionals The compound statement *p if and only if q* (often abbreviated *p iff q*) is called a **biconditional.** It is symbolized $p \leftrightarrow q$, and is interpreted as the conjunction of the two conditionals $p \rightarrow q$ and $q \rightarrow p$. Using symbols, this conjunction is written

$$(q \rightarrow p) \wedge (p \rightarrow q)$$

so that, by definition, $p \leftrightarrow q \equiv (q \rightarrow p) \wedge (p \rightarrow q)$.

The truth table for the biconditional $p \leftrightarrow q$ can be determined using this definition.

Truth Table for the Biconditional *p* if and only if *q*

p if and only if *q*

p	q	$p \leftrightarrow q$
T	T	T
T	F	F
F	T	F
F	F	T

From the truth table, we see that a biconditional is true when both component statements have the same truth value. It is false when they have different truth values.

EXAMPLE 5 Determining Whether Biconditionals Are True or False

Determine whether each biconditional statement is *true* or *false*.

(a) $6 + 9 = 15$ if and only if $12 + 4 = 16$
(b) $6 = 5$ if and only if $12 \neq 12$
(c) $5 + 2 = 10$ if and only if $17 + 19 = 36$

SOLUTION

(a) Both $6 + 9 = 15$ and $12 + 4 = 16$ are true. By the truth table for the biconditional, this biconditional is true.

(b) Both component statements are false, so by the last line of the truth table for the biconditional, this biconditional statement is true.

(c) Because the first component $(5 + 2 = 10)$ is false, and the second is true, this biconditional statement is false.

Summary of Truth Tables In this section and in the previous two sections, truth tables have been derived for several important types of compound statements. The summary that follows describes how these truth tables may be remembered.

Summary of Basic Truth Tables

1. $\sim p$, the **negation** of p, has truth value opposite of p.
2. $p \wedge q$, the **conjunction,** is true only when both p and q are true.
3. $p \vee q$, the **disjunction,** is false only when both p and q are false.
4. $p \rightarrow q$, the **conditional,** is false only when p is true and q is false.
5. $p \leftrightarrow q$, the **biconditional,** is true only when p and q have the same truth value.

3.4 EXERCISES

For each given conditional statement (or statement that can be written as a conditional), write **(a)** *the converse,* **(b)** *the inverse, and* **(c)** *the contrapositive in* if . . . then *form. In some of the exercises, it may be helpful to first restate the given statement in* if . . . then *form.*

1. If beauty were a minute, then you would be an hour.

2. If you lead, then I will follow.

3. If it ain't broke, don't fix it.

4. If I had a nickel for each time that happened, I would be rich.

5. Walking in front of a moving car is dangerous to your health.

6. Milk contains calcium.

7. Birds of a feather flock together.

8. A rolling stone gathers no moss.

9. If you build it, he will come.

10. Where there's smoke, there's fire.

11. $p \rightarrow \sim q$

12. $\sim p \rightarrow q$

13. $\sim p \rightarrow \sim q$

14. $\sim q \rightarrow \sim p$

15. $p \rightarrow (q \vee r)$ (*Hint:* Use one of De Morgan's laws as necessary.)

16. $(r \vee \sim q) \rightarrow p$ (*Hint:* Use one of De Morgan's laws as necessary.)

17. Discuss the equivalences that exist among a given conditional statement, its converse, its inverse, and its contrapositive.

18. State the contrapositive of "If the square of a natural number is even, then the natural number is even." The two statements must have the same truth value. Use several examples and inductive reasoning to decide whether both are true or both are false.

Write each statement in the form "if p, then q."

19. If it is muddy, I'll wear my galoshes.

20. If I finish studying, I'll go to the party.

21. "18 is positive" implies that $18 + 1$ is positive.

22. "Today is Tuesday" implies that yesterday was Monday.

23. All integers are rational numbers.

24. All whole numbers are integers.

25. Doing crossword puzzles is sufficient for driving me crazy.

26. Being in Baton Rouge is sufficient for being in Louisiana.

27. A day's growth of beard is necessary for Gerald Guidroz to shave.

28. Being an environmentalist is necessary for being elected.

29. I can go from Park Place to Baltic Avenue only if I pass GO.

30. The principal will hire more teachers only if the school board approves.

31. No whole numbers are not integers.

32. No integers are irrational numbers.

33. The Orioles will win the pennant when their pitching improves.

34. Rush will be a liberal when pigs fly.

35. A rectangle is a parallelogram with a right angle.

36. A parallelogram is a four-sided figure with opposite sides parallel.

37. A triangle with two sides of the same length is isosceles.

38. A square is a rectangle with two adjacent sides equal.

39. The square of a two-digit number whose units digit is 5 will end in 25.

40. An integer whose units digit is 0 or 5 is divisible by 5.

41. One of the following statements is not equivalent to all the others. Which one is it?
A. r only if s. **B.** r implies s.
C. If r, then s. **D.** r is necessary for s.

42. Many students have difficulty interpreting *necessary* and *sufficient*. Use the statement "Being in Quebec is sufficient for being in North America" to explain why "p is sufficient for q" translates as "if p, then q."

43. Use the statement "To be an integer, it is necessary that a number be rational" to explain why "p is necessary for q" translates as "if q, then p."

44. Explain why the statement "A week has eight days if and only if October has forty days" is true.

October

Sunday	Monday	Tuesday	Wednesday	Thursday	Friday	Saturday
1	2	3	4	5	6	7
8	9	10	11	12	13	14
15	16	17	18	19	20	21
22	23	24	25	26	27	28
29	30	31				

Identify each statement as true *or* false.

45. $5 = 9 - 4$ if and only if $8 + 2 = 10$.

46. $3 + 1 \neq 6$ if and only if $8 \neq 8$.

47. $8 + 7 \neq 15$ if and only if $3 \times 5 \neq 9$.

48. $6 \times 2 = 14$ if and only if $9 + 7 \neq 16$.

49. Bill Clinton was president if and only if Jimmy Carter was not president.

50. Burger King sells Big Macs if and only if Apple manufactures Ipods.

Two statements that can both be true about the same object are **consistent.** *For example,* "It is brown" *and* "It weighs 50 pounds" *are consistent statements. Statements that cannot both be true about the same object are called* **contrary;** "It is a Dodge" *and* "It is a Toyota" *are contrary. In Exercises 51–56, label each pair of statements as either* contrary *or* consistent.

51. Elvis is alive. Elvis is dead.

52. George W. Bush is a Democrat. George W. Bush is a Republican.

53. That animal has four legs. That same animal is a dog.

54. That book is nonfiction. That book costs more than $100.

55. This number is an integer. This same number is irrational.

56. This number is positive. This same number is a natural number.

57. Make up two statements that are consistent.

58. Make up two statements that are contrary.

3.5 Analyzing Arguments with Euler Diagrams

Logical Arguments • Arguments with Universal Quantifiers • Arguments with Existential Quantifiers

Leonhard Euler (1707–1783) won the academy prize and edged out du Châtelet and Voltaire. That was a minor achievement, as was the invention of "Euler circles" (which antedated Venn diagrams). Euler was the most prolific mathematician of his generation despite blindness that forced him to dictate from memory.

Logical Arguments

With inductive reasoning we observe patterns to solve problems. Now, in this section and the next, we study how deductive reasoning may be used to determine whether logical arguments are valid or invalid. A logical argument is made up of **premises** (assumptions, laws, rules, widely held ideas, or observations) and a **conclusion.** Together, the premises and the conclusion make up the argument. Also recall that *deductive* reasoning involves drawing specific conclusions from given general premises. When reasoning from the premises of an argument to obtain a conclusion, we want the argument to be valid.

Valid and Invalid Arguments

An argument is **valid** if the fact that all the premises are true forces the conclusion to be true. An argument that is not valid is **invalid.** It is called a **fallacy.**

It is very important to note that "valid" and "true" are not the same—an argument can be valid even though the conclusion is false. (See Example 4.)

Arguments with Universal Quantifiers

Several techniques can be used to check whether an argument is valid. One of these is the visual technique based on **Euler diagrams,** as shown in Examples 1–4.

EXAMPLE 1 Using an Euler Diagram to Determine Validity

Is the following argument valid?

All dogs are animals.
Puddles is a dog.
―――――――――――
Puddles is an animal.

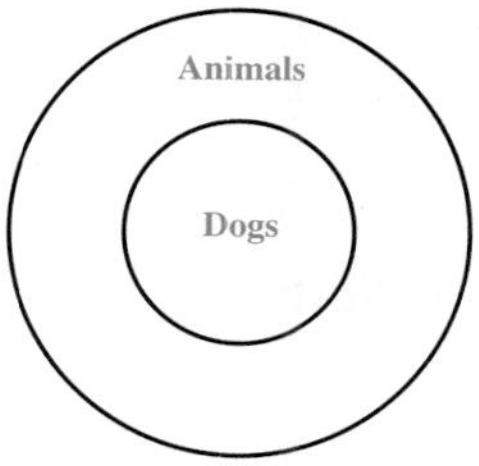

FIGURE 8

SOLUTION

Here we use the common method of placing one premise over another, with the conclusion below a line. To begin, draw regions to represent the first premise. One is the region for "animals." Because all dogs are animals, the region for "dogs" goes inside the region for "animals," as in Figure 8.

The second premise, "Puddles is a dog," suggests that "Puddles" would go inside the region representing "dogs." Let x represent "Puddles." Figure 9 shows that "Puddles" is also inside the region for "animals." If both premises are true, the conclusion that Puddles is an animal must be true also. The argument is valid. ■

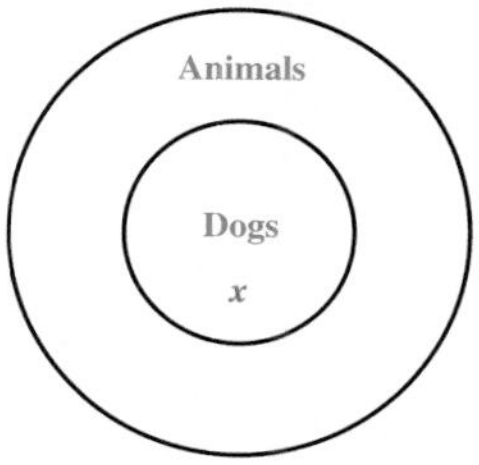

x represents Puddles.

FIGURE 9

EXAMPLE 2 Using an Euler Diagram to Determine Validity

Is the following argument valid?

All rainy days are cloudy.
Today is not cloudy.
―――――――――――
Today is not rainy.

SOLUTION

In Figure 10, the region for "rainy days" is drawn entirely inside the region for "cloudy days." Since "Today is *not* cloudy," place an x for "today" *outside* the region for "cloudy days." See Figure 11. Placing the x outside the region for "cloudy days" forces it also to be outside the region for "rainy days." Thus, if the first two premises are true, then it is also true that today is not rainy. The argument is valid.

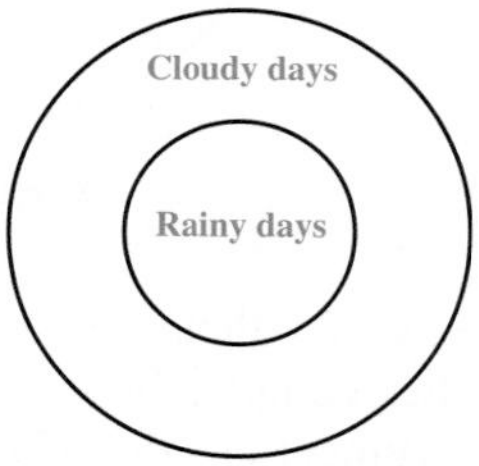

FIGURE 10

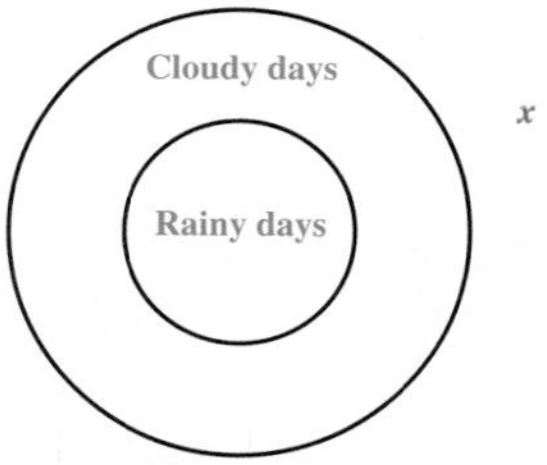

x represents today.

FIGURE 11 ■

EXAMPLE 3 Using an Euler Diagram to Determine Validity

Is the following argument valid?

All banana trees have green leaves.
That plant has green leaves.
―――――――――――
That plant is a banana tree.

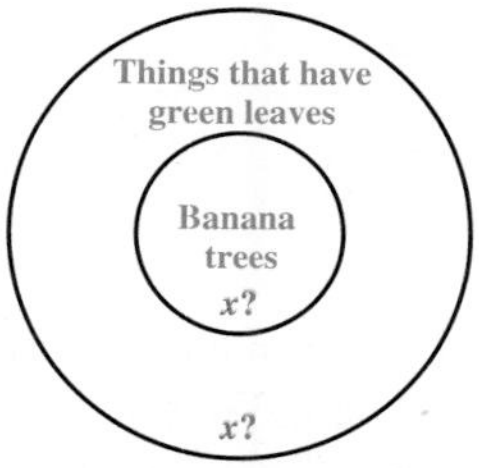

FIGURE 12

SOLUTION

The region for "banana trees" goes entirely inside the region for "things that have green leaves." See Figure 12. There is a choice for locating the x that represents "that plant." The x must go inside the region for "things that have green leaves," but can go either inside or outside the region for "banana trees." Even if the premises are true, we are not forced to accept the conclusion as true. This argument is invalid; it is a fallacy. ■

As mentioned earlier, the validity of an argument is not the same as the truth of its conclusion. The argument in Example 3 was invalid, but the conclusion "That plant is a banana tree" may or may not be true. We cannot be sure.

EXAMPLE 4 Using an Euler Diagram to Determine Validity

Is the following argument valid?

All expensive things are desirable.
All desirable things make you feel good.
All things that make you feel good make you live longer.
___
All expensive things make you live longer.

SOLUTION

A diagram for the argument is given in Figure 13.

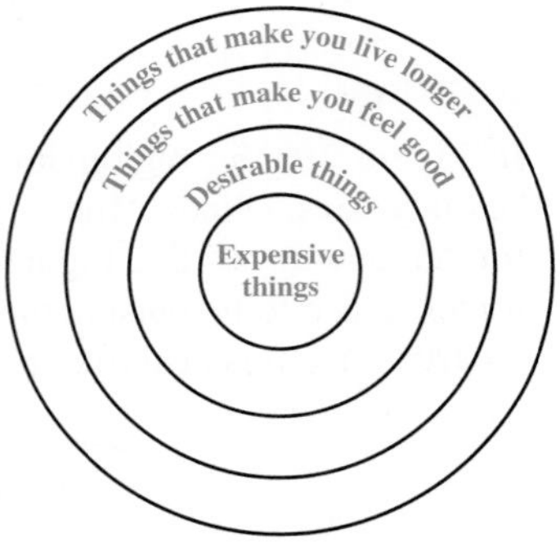

FIGURE 13

If each premise is true, then the conclusion must be true because the region for "expensive things" lies completely within the region for "things that make you live longer." Thus, the argument is valid. (This argument is an example of the fact that a *valid* argument need *not* have a true conclusion.)

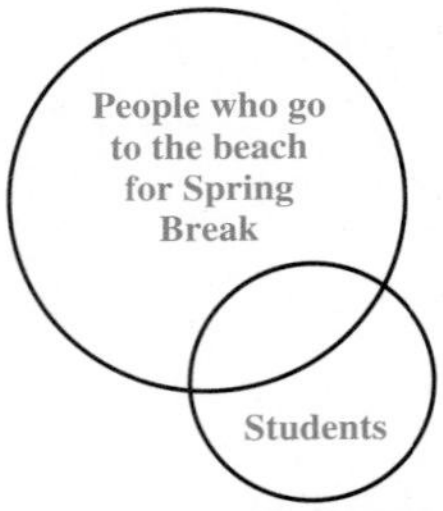

FIGURE 14

FIGURE 15

Arguments with Existential Quantifiers

EXAMPLE 5 Using an Euler Diagram to Determine Validity

Is the following argument valid?

Some students go to the beach for Spring Break.
I am a student.
___
I go to the beach for Spring Break.

SOLUTION

The first premise is sketched in Figure 14. As the sketch shows, some (but not necessarily *all*) students go to the beach. There are two possibilities for *I*, as shown in Figure 15.

One possibility is that *I* go to the beach; the other is that *I* don't. Since the truth of the premises does not force the conclusion to be true, the argument is invalid.

3.5 EXERCISES

Decide whether each argument is valid *or* invalid.

1. All amusement parks have thrill rides.
Great America is an amusement park.

Great America has thrill rides.

2. All disc jockeys play music.
Phlash Phelps is a disc jockey.

Phlash Phelps plays music.

3. All politicians lie, cheat, and steal.
That man lies, cheats, and steals.

That man is a politician.

4. All Southerners speak with an accent.
Bill Leonard speaks with an accent.

Bill Leonard is a Southerner.

5. All dogs love to bury bones.
Py does not love to bury bones.

Py is not a dog.

6. All handymen use cell phones.
Lee Guidroz does not use a cell phone.

Lee Guidroz is not a handyman.

7. All residents of Minnesota know how to live in freezing temperatures.
Wendy Rockswold knows how to live in freezing temperatures.

Wendy Rockswold lives in Minnesota.

8. All people who apply for a loan must pay for a title search.
Hilary Langlois paid for a title search.

Hilary Langlois applied for a loan.

9. Some dinosaurs were plant eaters.
Danny was a plant eater.

Danny was a dinosaur.

10. Some philosophers are absent minded.
Loretta Ramagos is a philosopher.

Loretta Ramagos is absent minded.

11. Some nurses wear blue uniforms.
Dee Boyle is a nurse.

Dee Boyle wears a blue uniform.

12. Some trucks have sound systems.
Some trucks have gun racks.

Some trucks with sound systems have gun racks.

13. Refer to Example 3. If the second premise and the conclusion were interchanged, would the argument then be valid?

14. Refer to Example 4. Give a different conclusion than the one given there so that the argument is still valid.

Construct a valid argument based on the Euler diagram shown.

15.

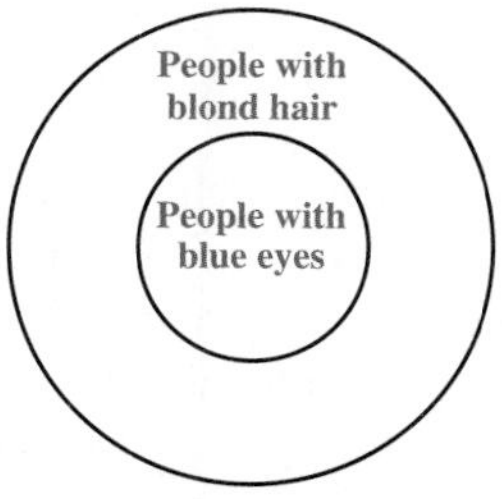

x represents Dinya Norris.

16.

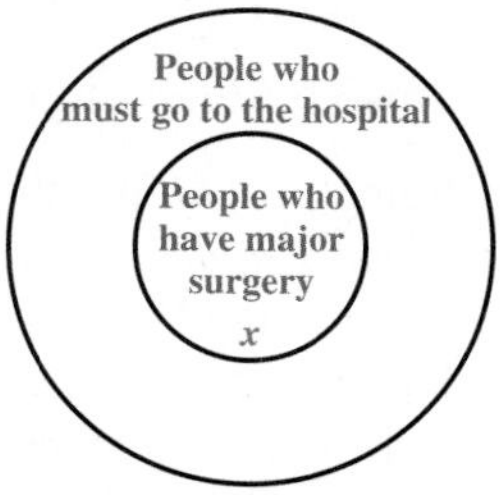

x represents Marty McDonald.

As mentioned in the text, an argument can have a true conclusion yet be invalid. In these exercises, each argument has a true conclusion. Identify each argument as valid *or* invalid.

17. All birds fly.
All planes fly.

A bird is not a plane.

18. All cars have tires.
All tires are rubber.

All cars have rubber.

19. All chickens have beaks.
All hens are chickens.
―――――――
All hens have beaks.

20. All chickens have beaks.
All birds have beaks.
―――――――
All chickens are birds.

21. Little Rock is northeast of Texarkana.
Little Rock is northeast of Austin.
―――――――
Texarkana is northeast of Austin.

22. Veracruz is south of Tampico.
Tampico is south of Monterrey.
―――――――
Veracruz is south of Monterrey.

23. No whole numbers are negative.
-4 is negative.
―――――――
-4 is not a whole number.

24. A scalene triangle has a longest side.
A scalene triangle has a largest angle.
―――――――
The largest angle in a scalene triangle is opposite the longest side.

In Exercises 25–30, the premises marked A, B, *and* C *are followed by several possible conclusions. Take each conclusion in turn, and check whether the resulting argument is* valid *or* invalid.

A. *All people who drive contribute to air pollution.*
B. *All people who contribute to air pollution make life a little worse.*
C. *Some people who live in a suburb make life a little worse.*

25. Some people who live in a suburb contribute to air pollution.

26. Some people who live in a suburb drive.

27. Suburban residents never drive.

28. Some people who contribute to air pollution live in a suburb.

29. Some people who make life a little worse live in a suburb.

30. All people who drive make life a little worse.

31. Find examples of arguments on television commercials. Check them for validity.

32. Find examples of arguments in magazine ads. Check them for validity.

EXTENSION

Logic Problems and Sudoku

• How to Solve Logic Problems • How to Solve Sudoku

Some people find that logic problems, which appear in periodicals such as *Official's Logic Problems, World-Class Logic Problems* and *England's Best Logic Problems* (both PennyPress), and *Logic Puzzles* (Dell), provide hours of enjoyment. They are based on deductive reasoning, and players answer questions based on clues given. The following explanation on solving such problems appeared in the May 2004 issue of *England's Best Logic Problems.*

How to Solve Logic Problems Solving logic problems is entertaining and challenging. All the information you need to solve a logic problem is given in the introduction and clues, and in illustrations, when provided. If you've never solved a logic problem before, our sample should help you get started. Fill in the Sample Solving

Chart as you follow our explanation. We use a "•" to signify "Yes" and an "X" to signify "No."

Sample Logic Problem

Five couples were married last week, each on a different weekday. From the information provided, determine the woman (one is Cathy) and man (one is Paul) who make up each couple, as well as the day on which each couple was married.

1. Anne was married on Monday, but not to Wally.
2. Stan's wedding was on Wednesday. Rob was married on Friday, but not to Ida.
3. Vern (who married Fran) was married the day after Eve.

Sample Solving Chart:

	PAUL	ROB	STAN	VERN	WALLY	MONDAY	TUESDAY	WEDNESDAY	THURSDAY	FRIDAY
ANNE										
CATHY										
EVE										
FRAN										
IDA										
MONDAY										
TUESDAY										
WEDNESDAY										
THURSDAY										
FRIDAY										

1

	PAUL	ROB	STAN	VERN	WALLY	MONDAY	TUESDAY	WEDNESDAY	THURSDAY	FRIDAY
ANNE		X	X		X	•	X	X	X	X
CATHY						X				
EVE						X				
FRAN						X				
IDA		X				X				X
MONDAY		X	X							
TUESDAY		X	X							
WEDNESDAY	X	X	•	X	X					
THURSDAY		X	X							
FRIDAY	X	•	X	X	X					

Explanation

Anne was married Mon. (1), so put a "•" at the intersection of Anne and Mon. Put "X"s in all the other days in Anne's row and all the other names in the Mon. column. (Whenever you establish a relationship, as we did here, be sure to place "X"s at the intersections of all relationships that become impossible as a result.) Anne wasn't married to Wally (1), so put an "X" at the intersection of Anne and Wally. Stan's wedding was Wed. (2), so put a "•" at the intersection of Stan and Wed. (Don't forget the "X"s.) Stan didn't marry Anne, who was married Mon., so put an "X" at the intersection of Anne and Stan. Rob was married Fri., but not to Ida (2), so put a "•" at the intersection of Rob and Fri., and "X"s at the intersections of Rob and Ida and Ida and Fri. Rob also didn't marry Anne, who was married Mon., so put an "X" at the intersection of Anne and Rob. Now your chart should look like chart 1.

Vern married Fran (3), so put a "•" at the intersection of Vern and Fran. This leaves Anne's only possible husband as Paul, so put a "•" at the intersection of Anne and Paul and Paul and Mon. Vern and Fran's wedding was the day after Eve's (3), which wasn't Mon. [Anne], so Vern's wasn't Tue. It must have been Thu. [see chart], so Eve's was Wed. (3). Put "•"s at the intersections of Vern and Thu., Fran and Thu., and Eve and Wed. Now your chart should look like chart 2.

(continued)

2

	PAUL	ROB	STAN	VERN	WALLY	MONDAY	TUESDAY	WEDNESDAY	THURSDAY	FRIDAY
ANNE	•	×	×	×	×	•	×	×	×	×
CATHY	×			×		×		×	×	
EVE	×			×		×	×	•	×	×
FRAN	×	×	×	•	×	×	×	×	•	×
IDA	×	×		×		×		×	×	×
MONDAY	•	×	×	×	×					
TUESDAY	×	×	×	×						
WEDNESDAY	×	×	•	×	×					
THURSDAY	×	×	×	•	×					
FRIDAY	×	•	×	×	×					

3

	PAUL	ROB	STAN	VERN	WALLY	MONDAY	TUESDAY	WEDNESDAY	THURSDAY	FRIDAY
ANNE	•	×	×	×	×	•	×	×	×	×
CATHY	×	•	×	×	×	×	×	×	×	•
EVE	×	×	•	×	×	×	×	•	×	×
FRAN	×	×	×	•	×	×	×	×	•	×
IDA	×	×	×	×	•	×	•	×	×	×
MONDAY	•	×	×	×	×					
TUESDAY	×	×	×	×	•					
WEDNESDAY	×	×	•	×	×					
THURSDAY	×	×	×	•	×					
FRIDAY	×	•	×	×	×					

The chart shows that Cathy was married Fri., Ida was married Tue., and Wally was married Tue. Ida married Wally, and Cathy's wedding was Fri., so she married Rob. After this information is filled in, Eve could only have married Stan. You've completed the puzzle, and your chart should now look like chart 3.

In summary: Anne and Paul, Mon.; Cathy and Rob, Fri.; Eve and Stan, Wed.; Fran and Vern, Thu.; Ida and Wally, Tue.

In some problems, it may be necessary to make a logical guess based on facts you've established. When you do, always look for clues or other facts that disprove it. If you find that your guess is incorrect, eliminate it as a possibility.

How to Solve Sudoku Sudoku is a simple game that has gained great popularity in the United States during the past few years. It is believed that the game originated as Number Place in the United States over 25 years ago, but gained in popularity only after it became a sensation in Japan, where it was renamed Sudoku, meaning "single number." (*Source*: *Sudoku #13*, 2005, Platinum Magazine Group.) Today it can be found in daily newspapers, on day-by-day calendars, and in periodical publications on newsstands.

There is only one rule in Sudoku: "Fill in the grid so that every row, every column, and every 3 × 3 box contains the digits 1 through 9." This involves scanning the given digits, marking up the grid, and analyzing. Here is a Sudoku in its original (given) form and in its final (solved) form.

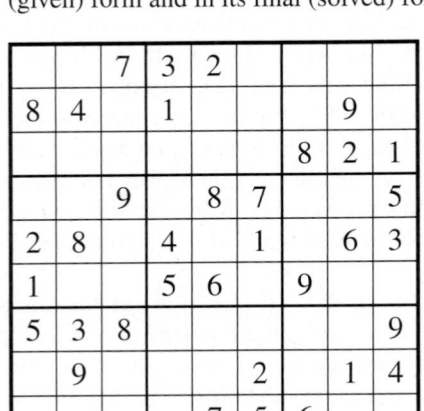

		7	3	2				
8	4		1				9	
						8	2	1
		9		8	7			5
2	8		4		1		6	3
1			5	6		9		
5	3	8						9
	9				2		1	4
				7	5	6		

Given Form

9	1	7	3	2	8	4	5	6
8	4	2	1	5	6	3	9	7
6	5	3	7	4	9	8	2	1
3	6	9	2	8	7	1	4	5
2	8	5	4	9	1	7	6	3
1	7	4	5	6	3	9	8	2
5	3	8	6	1	4	2	7	9
7	9	6	8	3	2	5	1	4
4	2	1	9	7	5	6	3	8

Solved Form

You can find Sudoku puzzles and solving strategies online at www.sudoku.org.

EXTENSION EXERCISES

Follow the guidelines to solve each logic problem, which appeared in the Spring 2004 issue of England's Best Logic Problems, *published by PennyPress.*

1. *A Moving Story* On the first day of her visit to Russia, British chess prodigy Queenie King, aged 14, played and defeated the adult champions from four different cities. From the clues below, can you work out the full name of her opponent in each game and the city from which he came?

	Boris	Ivan	Piotr	Yuri	Bishopnik	Knightovich	Pawnchev	Rookov	Corki	Gorki	Porki	Yorki
First												
Second												
Third												
Fourth												
Corki												
Gorki												
Porki												
Yorki												
Bishopnik												
Knightovich												
Pawnchev												
Rookov												

Clues:

(a) Boris was from the famous city of Gorki, while the other champions came from cities not so well known.

(b) Mr Rookov, Queenie's second opponent, was not the man from the city of Yorki, centre of the Russian chocolate industry.

(c) Piotr, who was the first Russian to play—and be beaten by—Queenie wasn't Mr Pawnchev.

(d) Mr Bishopnik played Queenie immediately after Ivan and immediately before the man from Corki.

2. *Dig This!* Professor Rosetta Stone, the eminent British archaeologist, is considering offers to participate in four different "digs" next year. From the clues below, can you work out the name of the fellow-archaeologist making each offer, which university he is from, what is to be excavated, and where it is?

	Universities								*Locations*			
	Arizona	Berlin	Miami	New York	Fort	Pyramid	Temple	Villa	China	Egypt	Peru	Scotland
Prof. Azimovic												
Prof. Katsouris												
Prof. Partington												
Prof. Voelkner												
Locations China												
Egypt												
Peru												
Scotland												
Fort												
Pyramid												
Temple												
Villa												

Clues:

(a) Professor Azimovic wants Rosetta to join him for an expedition into the wild highlands of Peru.

(b) The University of New York expedition is being mounted to excavate the site of a two thousand year old temple.

(c) Professor Katsouris isn't organising the expedition that is going to excavate an ancient fort.

(d) The site in the Takla Makan desert of China to be investigated by the University of Arizona expedition is not that of a villa.

(e) Professor Voelkner of the University of Berlin has no connection to the projected expedition to excavate a newly discovered pyramid in Egypt's Nile delta.

3. *Is That a Folk Song?* The traditional answer to the above question, when applied to something that may or may not be a folk song, is "Well, I never heard a cow sing it." None of the singers in this problem are cows, and they only sing folk songs. From the clues below, can you identify each male singer and his female partner, the name they perform under, and the type of folk songs they sing? (*Note*: The only vowels are A, E, I, O, and U.)

	Carol Dodds (Women)	Jane Kenny (Women)	Nancy O'Hara (Women)	Sue Rogers (Women)	Dirk and Daisy	Merlyns	Rose and Thorn	Starr Twins	American	English	Irish	Own compositions
Men Ben Ashby												
Hans Gruber												
Peter Owen												
Steven Thorp												
American												
English												
Irish												
Own compositions												
Dirk and Daisy												
Merlyns												
Rose and Thorn												
Starr Twins												

Clues:

(a) Although he was born in Munich, Hans Gruber and his partner only sing traditional English folk songs.

(b) Nancy O'Hara is the Rose half of Rose and Thorn.

(c) One of the men whose surname begins with a vowel sings with Carol Dodds, while the other performs traditional American material with partner Jane Kenny.

(d) Peter Owen and his partner—who resemble each other not at all—perform as the Starr Twins; Ben Ashby isn't one of the Merlyns.

(e) The male half of the duo who perform folk-type songs of their own composition has a first name with one more letter than that of Sue Rogers' partner.

4. *Monsieur le Duc* In the 1930s, the Duc de Bauch made a number of profitable business partnerships. The Duc went into business with four heiresses, each of whom had inherited a business from her father. From the clues given below, can you work out each heiress' name, where her family money came from, and the city and year in which she partnered with the Duc?

	Automobiles	Banking	Mining	Oil	Athens	Berlin	Monte Carlo	Paris	1932	1934	1936	1938
Drusilla Camden												
Horatia Hampton												
Mabelle Oakland												
Regina Stamford												
1932												
1934												
1936												
1938												
Athens												
Berlin												
Monte Carlo												
Paris												

Clues:

(a) Neither the banking heiress who partnered with the Duc de Bauch in 1938 nor Horatia Hampton, whose father was an oil millionaire, had ever been to Athens.

(b) One partnership was formed in Berlin in 1936.

(c) The Monte Carlo business arrangement was formed earlier than the one with Mabelle Oakland.

(d) The family of Regina Stamford, the Duc's 1934 partner, had no major involvement in the automobile industry.

(e) Drusilla Camden joined forces with the Duc in Paris.

Solve each Sudoku, which appeared in *Sudoku #13*, 2005, *Platinum Magazine Group*. (They are categorized according to difficulty level.)

5. ***Very Easy***

4	2	6	5	9	8	7	3	1
6	4		8				1	7
8				5	2	4		6
7	8	1	3					9
2	9						8	4
3					7	1	6	8
8		9	1	3				5
5	3				6		7	2
9	6			7		8	4	3

6. ***Very Easy***

		3				5	8	
8		2			7	6		
7	9		5	6			3	1
	7		3		1	9		
		8				7		
		5	9		2		6	
6	4			5	3		1	2
		1	4			3		9
	2	7				4		

7. ***Easy***

2				8			1	
			5	9		4		7
		9			2			
	4	7			8	1		6
6								3
5		1	6			2	4	
			8			3		
7		2		3	5			
	6			2				4

8. ***Easy***

			2	5			6	
7			6		4	5		8
	6	5			7			
		7				3		9
8								2
9		3				4		
			8			9	2	
2		8	9		5			7
	4			3	2			

9. ***Medium***

1						3	4	
5			8					
	6		2			8		5
4			7			6		
	7			8			2	
		5			3			1
9		1			7		6	
					1			4
	4	3						2

10. ***Medium***

5		6				7		8
	2		1		7		3	
7								4
		1	6	9	3	8		
		9	2	7	5	3		
1								6
	7		8		9		5	
6		3				4		9

11. ***Hard***

9		3		4				6
								5
	5	4			1			
1					3			8
		5				3		
3			6					7
			9			1	2	
8								
6				7		5		4

12. ***Hard***

7			8					
				5		9		6
	3		2			7		
			5				6	3
		4		7		1		
8	2				3			
		8			5		4	
9		2		1				
					8			1

3.6 Analyzing Arguments with Truth Tables

Truth Tables (Two Premises) • Valid and Invalid Argument Forms • Truth Tables (More Than Two Premises) • Arguments of Lewis Carroll

Truth Tables (Two Premises) In Section 3.5 we used Euler diagrams to test the validity of arguments. While Euler diagrams often work well for simple arguments, difficulties can develop with more complex ones, because Euler diagrams require a sketch showing every possible case. In complex arguments, it is hard to be sure that all cases have been considered.

In deciding whether to use Euler diagrams to test the validity of an argument, look for quantifiers such as "all," "some," or "no." These words often indicate arguments best tested by Euler diagrams. If these words are absent, it may be better to use truth tables to test the validity of an argument.

As an example of this method, consider the following argument:

If the floor is dirty, then I must mop it.
The floor is dirty.
―――――――――――――――
I must mop it.

To test the validity of this argument, we begin by identifying the *component* statements found in the argument. They are "the floor is dirty" and "I must mop it." We assign the letters p and q to represent these statements:

p represents "the floor is dirty";
q represents "I must mop it."

Now we write the two premises and the conclusion in symbols:

Premise 1: $p \rightarrow q$
Premise 2: p
Conclusion: q .

To decide if this argument is valid, we must determine whether the conjunction of both premises implies the conclusion for all possible cases of truth values for p and q. Therefore, write the conjunction of the premises as the antecedent of a conditional statement, and the conclusion as the consequent.

$$[(p \rightarrow q) \quad \wedge \quad p] \quad \rightarrow \quad q$$

$[(p \rightarrow q)$	$\wedge$	$p]$	$\rightarrow$	q
↑	↑	↑	↑	↑
premise	and	premise	implies	conclusion

Finally, construct the truth table for this conditional statement, as shown below.

p	q	$p \rightarrow q$	$(p \rightarrow q) \wedge p$	$[(p \rightarrow q) \wedge p] \rightarrow q$
T	T	T	T	T
T	F	F	F	T
F	T	T	F	T
F	F	T	F	T

Because the final column, shown in color, indicates that the conditional statement that represents the argument is true for all possible truth values of p and q, the statement is a tautology. Thus, the argument is valid.

The pattern of the argument in the floor-mopping example,

$$\begin{array}{l} p \rightarrow q \\ \underline{p\quad\;\;} \\ q \end{array} ,$$

is a common one, and is called **modus ponens,** or the *law of detachment*.

In summary, to test the validity of an argument using a truth table, follow the steps in the box.

Testing the Validity of an Argument with a Truth Table

Step 1 Assign a letter to represent each component statement in the argument.

Step 2 Express each premise and the conclusion symbolically.

Step 3 Form the symbolic statement of the entire argument by writing the *conjunction* of *all* the premises as the antecedent of a conditional statement, and the conclusion of the argument as the consequent.

Step 4 Complete the truth table for the conditional statement formed in Step 3 above. If it is a tautology, then the argument is valid; otherwise, it is invalid.

EXAMPLE 1 Using a Truth Table to Determine Validity

Determine whether the argument is *valid* or *invalid.*

If my check arrives in time, I'll register for the fall semester.
I've registered for the fall semester.
My check arrived in time.

SOLUTION

Let p represent "my check arrives (arrived) in time" and let q represent "I'll register (I've registered) for the fall semester." Using these symbols, the argument can be written in the form

$$\begin{array}{l} p \rightarrow q \\ \underline{q \quad\quad} \\ p \end{array} .$$

To test for validity, construct a truth table for the statement $[(p \rightarrow q) \wedge q] \rightarrow p$.

p	q	$p \rightarrow q$	$(p \rightarrow q) \wedge q$	$[(p \rightarrow q) \wedge q] \rightarrow p$
T	T	T	T	T
T	F	F	F	T
F	T	T	T	F
F	F	T	F	T

The third row of the final column of the truth table shows F, and this is enough to conclude that the argument is invalid.

If a conditional and its converse were logically equivalent, then an argument of the type found in Example 1 would be valid. Because a conditional and its converse are *not* equivalent, the argument is an example of what is sometimes called the **fallacy of the converse.**

EXAMPLE 2 Using a Truth Table to Determine Validity

Determine whether the argument is *valid* or *invalid.*

If a man could be in two places at one time, I'd be with you.
I am not with you.
A man can't be in two places at one time.

SOLUTION

If p represents "a man could be in two places at one time" and q represents "I'd be with you," the argument becomes

$$\begin{array}{l} p \rightarrow q \\ \underline{\sim q \quad\quad} \\ \sim p \end{array} .$$

The symbolic statement of the entire argument is

$$[(p \to q) \wedge \sim q] \to \sim p.$$

The truth table for this argument, shown below, indicates a tautology, and the argument is valid.

p	q	$p \to q$	$\sim q$	$(p \to q) \wedge \sim q$	$\sim p$	$[(p \to q) \wedge \sim q] \to \sim p$
T	T	T	F	F	F	T
T	F	F	T	F	F	T
F	T	T	F	F	T	T
F	F	T	T	T	T	T

The pattern of reasoning of this example is called **modus tollens,** or the *law of contraposition,* or *indirect reasoning.* ■

With reasoning similar to that used to name the fallacy of the converse, the fallacy

$$\begin{array}{l} p \to q \\ \underline{\sim p} \\ \sim q \end{array}$$

is called the **fallacy of the inverse.** An example of such a fallacy is "If it rains, I get wet. It doesn't rain. Therefore, I don't get wet."

EXAMPLE 3 Using a Truth Table to Determine Validity

Determine whether the argument is *valid* or *invalid.*

I'll buy a car or I'll take a vacation.
I won't buy a car.
____
I'll take a vacation.

SOLUTION

If p represents "I'll buy a car" and q represents "I'll take a vacation," the argument becomes

$$\begin{array}{l} p \vee q \\ \underline{\sim p} \\ q \end{array}.$$

We must set up a truth table for the statement $[(p \vee q) \wedge \sim p] \to q$.

p	q	$p \vee q$	$\sim p$	$(p \vee q) \wedge \sim p$	$[(p \vee q) \wedge \sim p] \to q$
T	T	T	F	F	T
T	F	T	F	F	T
F	T	T	T	T	T
F	F	F	T	F	T

The statement is a tautology and the argument is valid. Any argument of this form is valid by the law of **disjunctive syllogism.** ■

EXAMPLE 4 Using a Truth Table to Determine Validity

Determine whether the argument is *valid* or *invalid.*

If it squeaks, then I use WD-40.
If I use WD-40, then I must go to the hardware store.
____
If it squeaks, then I must go to the hardware store.

SOLUTION

Let p represent "it squeaks," let q represent "I use WD-40," and let r represent "I must go to the hardware store." The argument takes on the general form

$$\begin{array}{l} p \rightarrow q \\ \underline{q \rightarrow r} \\ p \rightarrow r. \end{array}$$

Make a truth table for the following statement:

$$[(p \rightarrow q) \wedge (q \rightarrow r)] \rightarrow (p \rightarrow r).$$

It will require eight rows.

p	q	r	$p \rightarrow q$	$q \rightarrow r$	$p \rightarrow r$	$(p \rightarrow q) \wedge (q \rightarrow r)$	$[(p \rightarrow q) \wedge (q \rightarrow r)] \rightarrow (p \rightarrow r)$
T	T	T	T	T	T	T	T
T	T	F	T	F	F	F	T
T	F	T	F	T	T	F	T
T	F	F	F	T	F	F	T
F	T	T	T	T	T	T	T
F	T	F	T	F	T	F	T
F	F	T	T	T	T	T	T
F	F	F	T	T	T	T	T

This argument is valid because the final statement is a tautology. The pattern of argument shown in this example is called **reasoning by transitivity,** or the *law of hypothetical syllogism.*

In a scene near the beginning of the 1974 film *Monty Python and the Holy Grail,* an amazing application of **poor logic** leads to the apparent demise of a supposed witch. A group of peasants have forced a young woman to wear a nose made of wood. The convoluted argument they make is this: Witches and wood are both burned, and because witches are made of wood, and wood floats, and ducks also float, if she weighs the same as a duck, then she is made of wood and therefore is a witch!

Valid and Invalid Argument Forms

A summary of the valid and invalid forms of argument presented so far follows.

Valid Argument Forms

Modus Ponens	Modus Tollens	Disjunctive Syllogism	Reasoning by Transitivity
$p \rightarrow q$	$p \rightarrow q$	$p \vee q$	$p \rightarrow q$
p	$\sim q$	$\sim p$	$q \rightarrow r$
q	$\sim p$	q	$p \rightarrow r$

Invalid Argument Forms (Fallacies)

Fallacy of the Converse	Fallacy of the Inverse
$p \rightarrow q$	$p \rightarrow q$
q	$\sim p$
p	$\sim q$

Truth Tables (More Than Two Premises) When an argument contains more than two premises, it is necessary to determine the truth values of the conjunction of *all* of them. Remember that if *at least one* premise in a conjunction of several premises is false, then the entire conjunction is false.

EXAMPLE 5 Using a Truth Table to Determine Validity

Determine whether the argument is *valid* or *invalid*.

If Eddie goes to town, then Mabel stays at home. If Mabel does not stay at home, then Rita will cook. Rita will not cook. Therefore, Eddie does not go to town.

SOLUTION

In an argument written in this manner, the premises are given first, and the conclusion is the statement that follows the word "Therefore." Let p represent "Eddie goes to town," let q represent "Mabel stays at home," and let r represent "Rita will cook."

$$\begin{array}{l} p \rightarrow q \\ \sim q \rightarrow r \\ \underline{\sim r} \\ \sim p \end{array}$$

To test validity, set up a truth table for the statement

$$[(p \rightarrow q) \wedge (\sim q \rightarrow r) \wedge \sim r] \rightarrow \sim p.$$

p	q	r	$p \rightarrow q$	$\sim q$	$\sim q \rightarrow r$	$\sim r$	$(p \rightarrow q) \wedge (\sim q \rightarrow r) \wedge \sim r$	$\sim p$	$[(p \rightarrow q) \wedge (\sim q \rightarrow r) \wedge \sim r] \rightarrow \sim p$
T	T	T	T	F	T	F	F	F	T
T	T	F	T	F	T	T	T	F	F
T	F	T	F	T	T	F	F	F	T
T	F	F	F	T	F	T	F	F	T
F	T	T	T	F	T	F	F	T	T
F	T	F	T	F	T	T	T	T	T
F	F	T	T	T	T	F	F	T	T
F	F	F	T	T	F	T	F	T	T

Because the final column does not contain all Ts, the statement is not a tautology. The argument is invalid.

Arguments of Lewis Carroll Consider the following poem, which has been around for many years.

For want of a nail, the shoe was lost.
For want of a shoe, the horse was lost.
For want of a horse, the rider was lost.
For want of a rider, the battle was lost.
For want of a battle, the war was lost.
Therefore, for want of a nail, the war was lost.

Each line of the poem may be written as an *if . . . then* statement. For example, the first line may be restated as "if a nail is lost, then the shoe is lost." The conclusion, "for want of a nail, the war was lost," follows from the premises, because repeated use of the law of transitivity applies. Arguments used by Lewis Carroll often take on a similar form. The next example comes from one of his works.

EXAMPLE 6 Supplying a Conclusion to Assure Validity

Tweedlogic "I know what you're thinking about," said Tweedledum, "but it isn't so, nohow." "Contrariwise," continued Tweedledee, "if it was so, it might be; and if it were so, it would be, but as it isn't, it ain't. That's logic."

Supply a conclusion that yields a valid argument for the following premises.

Babies are illogical.
Nobody is despised who can manage a crocodile.
Illogical persons are despised.

SOLUTION

First, write each premise in the form *if . . . then.*

If you are a baby, then you are illogical.
If you can manage a crocodile, then you are not despised.
If you are illogical, then you are despised.

Let p be "you are a baby," let q be "you are logical," let r be "you can manage a crocodile," and let s be "you are despised." With these letters, the statements can be written symbolically as

$$p \rightarrow \sim q$$
$$r \rightarrow \sim s$$
$$\sim q \rightarrow s.$$

Begin with any letter that appears only once. Here p appears only once. Using the contrapositive of $r \rightarrow \sim s$, which is $s \rightarrow \sim r$, rearrange the three statements as follows:

$$p \rightarrow \sim q$$
$$\sim q \rightarrow s$$
$$s \rightarrow \sim r.$$

From the three statements, repeated use of reasoning by transitivity gives the conclusion

$$p \rightarrow \sim r,$$

leading to a valid argument.

In words, the conclusion is "If you are a baby, then you cannot manage a crocodile," or, as Lewis Carroll would have written it, "Babies cannot manage crocodiles."

3.6 EXERCISES

Each argument is either valid by one of the forms of valid arguments discussed in this section, or it is a fallacy by one of the forms of invalid arguments discussed. (See the summary boxes.) Decide whether the argument is valid *or a* fallacy, *and give the form that applies.*

1. If Elton John comes to town, then I will go to the concert.
If I go to the concert, then I'll call in sick for work.
____
If Elton John comes to town, then I'll call in sick for work.

2. If you use binoculars, then you get a glimpse of the comet.
If you get a glimpse of the comet, then you'll be amazed.
____
If you use binoculars, then you'll be amazed.

3. If Kim Hobbs works hard enough, she will get a promotion.
Kim Hobbs works hard enough.
____
She gets a promotion.

4. If Johnny Forbes sells his quota, he'll get a bonus.
Johnny Forbes sells his quota.
____
He gets a bonus.

5. If he doesn't have to get up at 4:00 A.M., he's ecstatic.
He's ecstatic.
____
He doesn't have to get up at 4:00 A.M.

6. If she buys another pair of shoes, her closet will overflow.
Her closet will overflow.
____
She buys another pair of shoes.

7. If Kerry Wood pitches, the Cubs win.
The Cubs do not win.
____
Kerry Wood does not pitch.

8. If Nelson Dida plays, the opponent gets shut out.
The opponent does not get shut out.
____
Nelson Dida does not play.

9. "If we evolved a race of Isaac Newtons, that would not be progress." (quote from Aldous Huxley)
<u>We have not evolved a race of Isaac Newtons.</u>
That is progress.

10. "If I have seen farther than others, it is because I stood on the shoulders of giants." (quote from Sir Isaac Newton)
<u>I have not seen farther than others.</u>
I have not stood on the shoulders of giants.

11. She uses e-commerce or she pays by credit card.
<u>She does not pay by credit card.</u>
She uses e-commerce.

12. Mia kicks or Arnold pumps iron.
<u>Arnold does not pump iron.</u>
Mia kicks.

Use a truth table to determine whether the argument is valid *or* invalid.

13. $p \vee q$
$\underline{p}$
$\sim q$

14. $p \wedge \sim q$
$\underline{p}$
$\sim q$

15. $\sim p \rightarrow \sim q$
$\underline{q}$
p

16. $p \vee \sim q$
$\underline{p}$
$\sim q$

17. $p \rightarrow q$
$\underline{q \rightarrow p}$
$p \wedge q$

18. $\sim p \rightarrow q$
$\underline{p}$
$\sim q$

19. $p \rightarrow \sim q$
$\underline{q}$
$\sim p$

20. $p \rightarrow \sim q$
$\underline{\sim p}$
$\sim q$

21. $(\sim p \vee q) \wedge (\sim p \rightarrow q)$
$\underline{p}$
$\sim q$

22. $(p \rightarrow q) \wedge (q \rightarrow p)$
$\underline{p}$
$p \vee q$

23. $(\sim p \wedge r) \rightarrow (p \vee q)$
$\underline{\sim r \rightarrow p}$
$q \rightarrow r$

24. $(r \wedge p) \rightarrow (r \vee q)$
$\underline{q \wedge p}$
$r \vee p$

25. Earlier we showed how to analyze arguments using Euler diagrams. Refer to Example 4 in this section, restate each premise and the conclusion using a quantifier, and then draw an Euler diagram to illustrate the relationship.

26. Explain in a few sentences how to determine the statement for which a truth table will be constructed so that the arguments that follow in Exercises 27–36 can be analyzed for validity.

Determine whether each argument is valid *or* invalid.

27. Brian loves to watch movies. If Elayn likes to jog, then Brian does not love to watch movies. If Elayn does not like to jog, then Clay drives a school bus. Therefore, Clay drives a school bus.

28. If Hurricane Katrina hit that grove of trees, then the trees are devastated. People plant trees when disasters strike and the trees are not devastated. Therefore, if people plant trees when disasters strike, then Hurricane Katrina did not hit that grove of trees.

29. If the MP3 personal player craze continues, then downloading music will remain popular. American Girl dolls are favorites or downloading music will remain popular. American Girl dolls are not favorites. Therefore, the MP3 personal player craze does not continue.

30. Ashley Simpson sings or Ashton Kutcher is not a teen idol. If Ashton Kutcher is not a teen idol, then Fantasia does not win a Grammy. Fantasia wins a Grammy. Therefore, Ashley Simpson does not sing.

31. The Steelers will be in the playoffs if and only if Ben leads the league in passing. Bill coaches the Steelers or Ben leads the league in passing. Bill does not coach the Steelers. Therefore, the Steelers will not be in the playoffs.

32. If I've got you under my skin, then you are deep in the heart of me. If you are deep in the heart of me, then you are not really a part of me. You are deep in the heart of me or you are really a part of me. Therefore, if I've got you under my skin, then you are really a part of me.

33. If Dr. Hardy is a department chairman, then he lives in Atlanta. He lives in Atlanta and his first name is Larry. Therefore, if his first name is not Larry, then he is not a department chairman.

34. If I were your woman and you were my man, then I'd never stop loving you. I've stopped loving you. Therefore, I am not your woman or you are not my man.

35. All men are created equal. All people who are created equal are women. Therefore, all men are women.

36. All men are mortal. Socrates is a man. Therefore, Socrates is mortal.

37. Suppose that you ask a stranger for the time and you get the following response:

> "If I tell you the time, then we'll start chatting. If we start chatting, then you'll want to meet me at a truck stop. If we meet at a truck stop, then we'll discuss my family. If we discuss my family, then you'll find out that my daughter is available for marriage. If you find out that she is available for marriage, then you'll want to marry her. If you want to marry her, then my life will be miserable since I don't want my daughter married to some fool who can't afford a $10 watch."

Use reasoning by transitivity to draw a valid conclusion.

38. Calandra Davis made the following observation: "If I want to determine whether an argument leading to the statement

$$[(p \rightarrow q) \wedge \sim q] \rightarrow \sim p$$

is valid, I only need to consider the lines of the truth table which lead to T for the column headed $(p \rightarrow q) \wedge \sim q$." Calandra was very perceptive. Can you explain why her observation was correct?

In the arguments used by Lewis Carroll, it is helpful to restate a premise in if . . . then *form in order to more easily identify a valid conclusion. The following premises come from Lewis Carroll. Write each premise in* if . . . then *form.*

39. All my poultry are ducks.

40. None of your sons can do logic.

41. Guinea pigs are hopelessly ignorant of music.

42. No teetotalers are pawnbrokers.

43. No teachable kitten has green eyes.

44. Opium-eaters have no self-command.

45. I have not filed any of them that I can read.

46. All of them written on blue paper are filed.

Exercises 47–52 involve premises from Lewis Carroll. Write each premise in symbols, and then in the final part, give a conclusion that yields a valid argument.

47. Let p be "it is a duck," q be "it is my poultry," r be "one is an officer," and s be "one is willing to waltz."
- **(a)** No ducks are willing to waltz.
- **(b)** No officers ever decline to waltz.
- **(c)** All my poultry are ducks.
- **(d)** Give a conclusion that yields a valid argument.

48. Let p be "one is able to do logic," q be "one is fit to serve on a jury," r be "one is sane," and s be "he is your son."
- **(a)** Everyone who is sane can do logic.
- **(b)** No lunatics are fit to serve on a jury.
- **(c)** None of your sons can do logic.
- **(d)** Give a conclusion that yields a valid argument.

49. Let p be "one is honest," q be "one is a pawnbroker," r be "one is a promise-breaker," s be "one is trustworthy," t be "one is very communicative," and u be "one is a wine-drinker."
- **(a)** Promise-breakers are untrustworthy.
- **(b)** Wine-drinkers are very communicative.
- **(c)** A person who keeps a promise is honest.
- **(d)** No teetotalers are pawnbrokers. (*Hint:* Assume "teetotaler" is the opposite of "wine-drinker.")
- **(e)** One can always trust a very communicative person.
- **(f)** Give a conclusion that yields a valid argument.

50. Let p be "it is a guinea pig," q be "it is hopelessly ignorant of music," r be "it keeps silent while the *Moonlight Sonata* is being played," and s be "it appreciates Beethoven."
- **(a)** Nobody who really appreciates Beethoven fails to keep silent while the *Moonlight Sonata* is being played.
- **(b)** Guinea pigs are hopelessly ignorant of music.
- **(c)** No one who is hopelessly ignorant of music ever keeps silent while the *Moonlight Sonata* is being played.
- **(d)** Give a conclusion that yields a valid argument.

51. Let p be "it begins with 'Dear Sir'," q be "it is crossed," r be "it is dated," s be "it is filed," t be "it is in black ink," u be "it is in the third person," v be "I can read it," w be "it is on blue paper," x be "it is on one sheet," and y be "it is written by Brown."
- **(a)** All the dated letters are written on blue paper.
- **(b)** None of them are in black ink, except those that are written in the third person.
- **(c)** I have not filed any of them that I can read.
- **(d)** None of them that are written on one sheet are undated.
- **(e)** All of them that are not crossed are in black ink.
- **(f)** All of them written by Brown begin with "Dear Sir."
- **(g)** All of them written on blue paper are filed.
- **(h)** None of them written on more than one sheet are crossed.
- **(i)** None of them that begin with "Dear Sir" are written in the third person.
- **(j)** Give a conclusion that yields a valid argument.

52. Let p be "he is going to a party," q be "he brushes his hair," r be "he has self-command," s be "he looks fascinating," t be "he is an opium-eater," u be "he is tidy," and v be "he wears white kid gloves."
- **(a)** No one who is going to a party ever fails to brush his hair.
- **(b)** No one looks fascinating if he is untidy.
- **(c)** Opium-eaters have no self-command.
- **(d)** Everyone who has brushed his hair looks fascinating.
- **(e)** No one wears white kid gloves unless he is going to a party. (*Hint:* "a unless b" $\equiv \sim b \rightarrow a$.)
- **(f)** A man is always untidy if he has no self-command.
- **(g)** Give a conclusion that yields a valid argument.

COLLABORATIVE INVESTIGATION

Logic Problems and Sudoku Revisited

The logic problems and Sudoku in the Extension on pages 138–144 are fairly elementary, considering the complexity of some of the other problems found in the magazines mentioned. The problems here require more time and reasoning skills than the ones appearing in the Extension. They are taken from *England's Best Logic Problems*, May 2004, and *Sudoku #13*, 2005.

The class may wish to divide up into groups and see which group can solve these problems fastest.

EXERCISES

Note: As an exception to our usual style, answers to these Collaborative Investigation Exercises are given in the back of the book.

1. ***A Case of Foul Play*** At the end of a shelf on a bookcase is a pile of six murder novels published by a book club devoted to such works. From the clues given at the top of the next column, can you work out the titles and authors of the books numbered 1 to 6 in the stack, and work out the colour of its uniform style dust jacket? (*Note*: Women are Dahlia Dagger, Mary Hemlock, and Sandra Bludgeon, and men are Geoffrey Stringer, John Gunn, and Philip G Rott.)

	Death in Beijing	*The Final Case*	*A Killer Abroad*	*Lurking in the Shadows*	*Mayhem in Madagascar*	*Murder in the Sun*	Dahlia Dagger	Geoffrey Stringer	John Gunn	Mary Hemlock	Philip G Rott	Sandra Bludgeon	Black	Blue	Brown	Green	Red	Yellow
1																		
2																		
3																		
4																		
5																		
6																		
Black																		
Blue																		
Brown																		
Green																		
Red																		
Yellow																		
Dahlia Dagger																		
Geoffrey Stringer																		
John Gunn																		
Mary Hemlock																		
Philip G Rott																		
Sandra Bludgeon																		

Clues:

(a) *Murder in the Sun* is immediately below the novel by Mary Hemlock but somewhere higher in the pile than the book with the yellow dust jacket.

(b) Dahlia Dagger's contribution to the collection is entitled *Mayhem in Madagascar*; it is two below the green-covered book in the pile.

(c) The blue dust jacket belongs to the novel by Sandra Bludgeon, which occupies an even-numbered position on the shelf.

(d) The book with the red dust jacket is not *Lurking in the Shadows*.

(e) The brown dust jacket belongs to *A Killer Abroad*, which is by a female author.

(f) *The Final Case* occupies position 4 in the stack.

(g) The bottom book in the pile, which was not written by Geoffrey Stringer, has a black dust jacket.

(h) The author of the novel at the very top of the stack is John Gunn.

2. ***Very Hard Sudoku***

8		3	7				9	
		6	8					
2								6
	2		1	8				9
				3				
4				6	5		2	
7								3
					1	8		
	5				8	7		1

CHAPTER 3 TEST

Write a negation for each statement.

1. $6 - 3 = 3$

2. All men are created equal.

3. Some members of the class went on the field trip.

4. If that's the way you feel, then I will accept it.

5. She applied and got a FEMA trailer.

Let p represent "You will love me" *and let q represent* "I will love you." *Write each statement in symbols.*

6. If you won't love me, then I will love you.

7. I will love you if you will love me.

8. I won't love you if and only if you won't love me.

Using the same statements as for Exercises 6–8, write each of the following in words.

9. $\sim p \wedge q$

10. $\sim(p \vee \sim q)$

In each of the following, assume that p is true and that q and r are false. Find the truth value of each statement.

11. $\sim q \wedge \sim r$

12. $r \vee (p \wedge \sim q)$

13. $r \rightarrow (s \vee r)$ (The truth value of the statement s is unknown.)

14. $p \leftrightarrow (p \rightarrow q)$

15. Explain in your own words why, if p is a statement, the biconditional $p \leftrightarrow \sim p$ must be false.

16. State the necessary conditions for
(a) a conditional statement to be false.
(b) a conjunction to be true.
(c) a disjunction to be false.

Construct a truth table for each of the following.

17. $p \wedge (\sim p \vee q)$

18. $\sim(p \wedge q) \rightarrow (\sim p \vee \sim q)$

Decide whether each statement is true *or* false.

19. Some negative integers are whole numbers.

20. All irrational numbers are real numbers.

Write each conditional statement in if . . . then *form.*

21. All integers are rational numbers.

22. Being a rhombus is sufficient for a polygon to be a quadrilateral.

23. Being divisible by 3 is necessary for a number to be divisible by 9.

24. She digs dinosaur bones only if she is a paleontologist.

For each statement, write **(a)** *the converse,* **(b)** *the inverse, and* **(c)** *the contrapositive.*

25. If a picture paints a thousand words, the graph will help me understand it.

26. $\sim p \rightarrow (q \wedge r)$ (Use one of De Morgan's laws as necessary.)

27. Use an Euler diagram to determine whether the argument is *valid* or *invalid*.

All members of that athletic club save money.
Gregory Langlois is a member of that athletic club.
———
Gregory Langlois saves money.

28. Match each argument in parts (a) – (d) with the law that justifies its validity, or the fallacy of which it is an example, in choices A–F.
A. Modus ponens
B. Modus tollens
C. Reasoning by transitivity
D. Disjunctive syllogism
E. Fallacy of the converse
F. Fallacy of the inverse

(a) If he eats liver, then he'll eat anything.
He eats liver.
———
He'll eat anything.

(b) If you use your seat belt, you will be safer.
You don't use your seat belt.
___
You won't be safer.

(c) If I hear *Mr. Bojangles*, I think of her.
If I think of her, I smile.
___
If I hear *Mr. Bojangles*, I smile.

(d) She sings or she dances.
She does not sing.
___
She dances.

Use a truth table to determine whether each argument is valid *or* invalid.

29. If I write a check, it will bounce. If the bank guarantees it, then it does not bounce. The bank guarantees it. Therefore, I don't write a check.

30. $\sim p \rightarrow \sim q$
$q \rightarrow p$
___
$p \vee q$

4

NUMERATION AND MATHEMATICAL SYSTEMS

Bud Abbott and Lou Costello were probably the best-known comedy team in the United States during the 1940s and 1950s. They made nearly forty movies together, often re-creating their stage comedy acts on film. Their baseball routine "Who's on First," performed in *The Naughty Nineties* (1945), is an American classic and earned them entry into the National Baseball Hall of Fame.

In their 1941 film *In the Navy,* Seaman Pomeroy Watson (Costello) tries to convince Smokey Adams (Abbott) that he can feed seven sailors with a tray of twenty-eight doughnuts so that the sailors will each get thirteen doughnuts. In an amazing misuse of place value and arithmetic algorithms, he shows how 7 divided into 28 is 13. He then multiplies 13 by 7 to get 28, and finally adds 13 seven times to get 28. It is a routine that must be seen to be believed. While Costello's methods were done for laughs, there are algorithms that are unfamiliar to most students that do indeed yield correct answers. Some of them will be discussed in this chapter.

4.1 Historical Numeration Systems

Mathematical and Numeration Systems • Ancient Egyptian Numeration—Simple Grouping • Traditional Chinese Numeration—Multiplicative Grouping • Hindu-Arabic Numeration—Positional

Symbols designed to represent objects or ideas are among the oldest inventions of humans. These Indian symbols in Arizona are several hundred years old.

Mathematical and Numeration Systems

Earlier we introduced and studied the concept of a *set,* a collection of elements. A set, in itself, may have no particular structure. But when we introduce *ways of combining the elements* (called *operations*) and *ways of comparing the elements* (called *relations*), we obtain a **mathematical system.**

> **Mathematical System**
>
> A **mathematical system** is made up of three components:
>
> **1.** a set of elements;
> **2.** one or more operations for combining the elements;
> **3.** one or more relations for comparing the elements.

A familiar example of a mathematical system is the set of whole numbers {0, 1, 2, 3, . . . }, along with the operation of addition and the relation of equality.

Historically, the earliest mathematical system to be developed involved the set of counting numbers or at least a limited subset of the first few counting numbers. The various ways of symbolizing and working with the counting numbers are called **numeration systems.** The symbols of a numeration system are called **numerals.**

Numeration systems have developed over many millennia of human history. Ancient documents provide insight into methods used by the early Sumerian peoples, the Egyptians, the Babylonians, the Greeks, the Romans, the Chinese, the Hindus, and the Mayan people.

A practical method of keeping accounts by matching may have developed as humans established permanent settlements and began to grow crops and raise livestock. People might have kept track of the number of sheep in a flock by matching pebbles with the sheep, for example. The pebbles could then be kept as a record of the number of sheep.

Tally sticks like this one were used by the English in about 1400 A.D. to keep track of financial transactions. Each notch stands for one pound sterling.

A more efficient method is to keep a **tally stick.** With a tally stick, one notch or **tally** is made on a stick for each sheep. Tally sticks and tally marks have been found that appear to be many thousands of years old. Tally marks are still used today: for example, nine items are tallied by writing 𝍸 ||||.

Ancient Egyptian Numeration—Simple Grouping

Early matching and tallying led to the essential feature of all more advanced numeration systems, that of **grouping.** Grouping allows for less repetition of symbols and also makes numerals easier to interpret. Most historical systems, including our own, have used groups of ten, indicating that people commonly learn to count by using their fingers. The size of the groupings (again, usually ten) is called the **base** of the number system. Bases of five, twenty, sixty, and others have also been used historically.

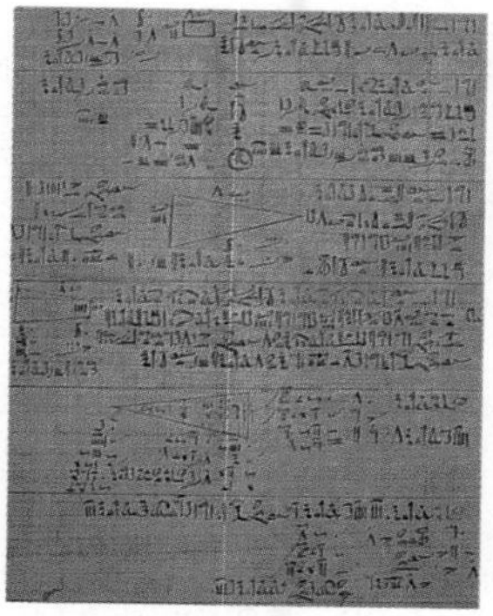

Much of our knowledge of **Egyptian mathematics** comes from the Rhind papyrus, from about 3800 years ago. A small portion of this papyrus, showing methods for finding the area of a triangle, is reproduced here.

The ancient Egyptian system is an example of a **simple grouping** system. It utilized ten as its base, and its various symbols are shown in Table 1. The symbol for 1 (|) is repeated, in a tally scheme, for 2, 3, and so on up to 9. A new symbol is introduced for 10 (∩), and that symbol is repeated for 20, 30, and so on, up to 90. This pattern enabled the Egyptians to express numbers up to 9,999,999 with just the seven symbols shown in the table.

TABLE 1 Early Egyptian Symbols

Number	Symbol	Description
1	𓏺	Stroke
10	𓎆	Heel bone
100	𓍢	Scroll
1000	𓆼	Lotus flower
10,000	𓂭	Pointing finger
100,000	𓆐	Burbot fish
1,000,000	𓁨	Astonished person

The numbers denoted by the seven Egyptian symbols are all *powers* of the base ten:

$$10^0 = 1, \quad 10^1 = 10, \quad 10^2 = 100, \quad 10^3 = 1000, \quad 10^4 = 10{,}000,$$
$$10^5 = 100{,}000, \quad \text{and} \quad 10^6 = 1{,}000{,}000.$$

These expressions, called *exponential expressions,* were first defined in Section 1.1. In the expression 10^4, for example, 10 is the *base* and 4 is the *exponent.* Recall that the exponent indicates the number of repeated multiples of the base.

EXAMPLE 1 Interpreting an Egyptian Numeral

Write in our system the number below.

𓆐𓆐𓆼𓆼𓆼𓆼𓆼𓍢𓍢𓍢𓍢𓎆𓎆𓎆𓎆𓎆𓎆𓎆𓎆𓎆𓏺𓏺𓏺𓏺𓏺𓏺𓏺

An Egyptian tomb painting shows scribes tallying the count of a grain harvest. **Egyptian mathematics** was oriented more to practicality than was Greek or Babylonian mathematics, although the Egyptians did have a formula for finding the volume of a certain portion of a pyramid.

SOLUTION

Refer to Table 1 for the values of the Egyptian symbols. Each 𓆐 represents 100,000. Therefore, two 𓆐s represent 2 · 100,000, or 200,000. Proceed as shown here.

two	𓆐	2 · 100,000 =	200,000
five	𓆼	5 · 1000 =	5000
four	𓍢	4 · 100 =	400
nine	𓎆	9 · 10 =	90
seven	𓏺	7 · 1 =	7
			205,497 ← Answer

Number	Symbol
1	I
5	V
10	X
50	L
100	C
500	D
1000	M

Roman numerals still appear today, mostly for decorative purposes: on clock faces, for chapter numbers in books, and so on. The system is essentially base ten, simple grouping, but with separate symbols for the intermediate values 5, 50, and 500, as shown above. If I is positioned left of V or X, it is subtracted rather than added. Likewise for X appearing left of L or C, and for C appearing left of D or M. Thus, for example, whereas CX denotes 110, XC denotes 90.

EXAMPLE 2 Creating an Egyptian Numeral

Write 376,248 in Egyptian symbols.

SOLUTION

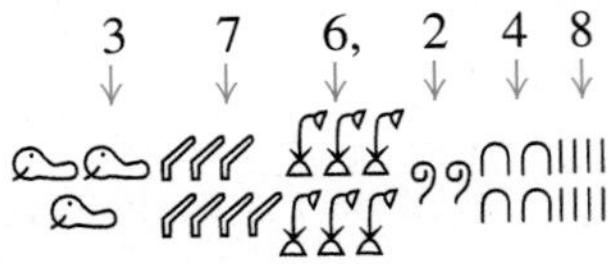

Refer to Table 1 as needed.

■

Notice that the position or order of the symbols makes no difference in a simple grouping system. Each of the numbers , , and 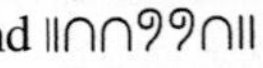 would be interpreted as 234. The most common order, however, is that shown in Examples 1 and 2, where like symbols are grouped together and groups of greater-valued symbols are positioned to the left.

A simple grouping system is well suited to the operations of addition and subtraction. For example, to add 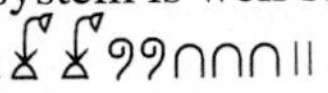 and 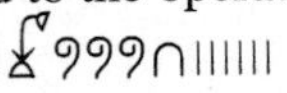 in the early Egyptian system, work as shown. Two Is plus six Is is equal to eight Is, and so on.

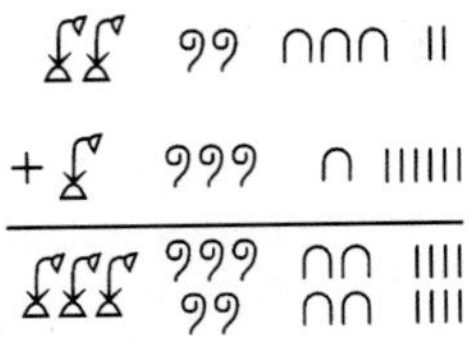

Sum:

While we used a + sign for convenience and drew a line under the numbers being added, the Egyptians did not do this.

Sometimes regrouping, or "carrying," is needed as in the example below. We get rid of ten heel bones from the tens group by placing an extra scroll in the hundreds group.

Sum:

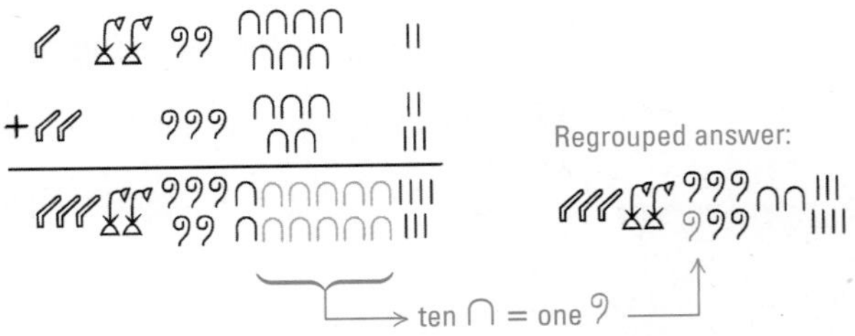

Subtraction is done in much the same way, as shown in the next example.

EXAMPLE 3 Subtracting Egyptian Numerals

Work each subtraction problem.

(a) 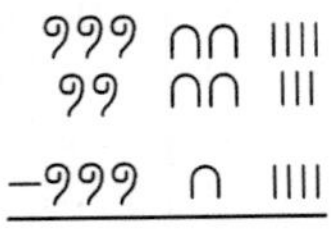

(b) 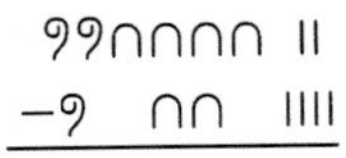

SOLUTION

(a) As with addition, work from right to left and subtract.

```
             999 ∩∩ ||||
              99 ∩∩ |||
            −999 ∩  ||||
            ─────────────
Difference:   99 ∩∩∩ |||
```

(b) To subtract four |s from two |s, "borrow" one heel bone, which is equivalent to ten |s. Finish the problem after writing ten additional |s on the right.

```
Regrouped:    99 ∩∩∩ ||||||      one ∩ = ten |s
                     ||||||
            −  9 ∩∩  ||||
            ─────────────
Difference:    9 ∩   ||||||||
```

■

A procedure such as those described above is called an **algorithm:** a rule or method for working a problem. The Egyptians used an interesting algorithm for multiplication that requires only an ability to add and to double numbers, as shown in Example 4. For convenience, this example uses our symbols rather than theirs.

Greek Numerals

1	α	60	ξ
2	β	70	o
3	γ	80	π
4	δ	90	φ
5	ϵ	100	ρ
6	ς	200	σ
7	ζ	300	τ
8	η	400	υ
9	θ	500	ϕ
10	ι	600	χ
20	κ	700	ψ
30	λ	800	ω
40	μ	900	ϡ
50	ν		

Classical Greeks used letters of their alphabet as numerical symbols. The base of the system was the number 10, and numbers 1 through 9 were symbolized by the first nine letters of the alphabet. Rather than using repetition or multiplication, they assigned nine more letters to multiples of 10 (through 90) and more letters to multiples of 100 (through 900). This is called a **ciphered system,** and it sufficed for small numbers. For example, 57 would be $\nu\zeta$; 573 would be $\phi o\gamma$; and 803 would be $\omega\gamma$. A small stroke was used with a units symbol for multiples of 1000 (up to 9000); thus 1000 would be $,\alpha$ or $'\alpha$. Often M would indicate tens of thousands (M for myriad = 10,000) with the multiples written above M.

EXAMPLE 4 Using the Egyptian Multiplication Algorithm

A rectangular room in an archaeological excavation measures 19 cubits by 70 cubits. (A cubit, based on the length of the forearm, from the elbow to the tip of the middle finger, was approximately 18 inches.) Find the area of the room.

SOLUTION

Multiply the width and length to find the area of a rectangle. Build two columns of numbers as shown below. Start the first column with 1, the second with 70. Each column is built downward by doubling the number above. Keep going until the first column contains numbers that can be added to equal 19. Then add the corresponding numbers from the second column.

	→ 1	70 ←	
	→ 2	140 ←	
$1 + 2 + 16 = 19$	4	280	$70 + 140 + 1120 = 1330$
	8	560	
	→ 16	1120 ←	

Thus $19 \cdot 70 = 1330$, and the area of the given room is 1330 square cubits. ■

Traditional Chinese Numeration—Multiplicative Grouping

Examples 1 through 3 above show that simple grouping, although an improvement over tallying, still requires considerable repetition of symbols. To denote 90, for example, the ancient Egyptian system must utilize nine ∩s: ∩∩∩∩∩∩∩∩∩. If an additional symbol (a "multiplier") was introduced to represent nine, say "9," then 90 could be denoted 9 ∩. All possible numbers of repetitions of powers of the base could be handled by introducing a separate multiplier symbol for each counting number less than the base.

TABLE 2

Number	Symbol
1	一
2	二
3	三
4	四
5	五
6	六
7	七
8	八
9	九
10	十
100	百
1000	千
0	零

Just such a system was developed many years ago in China. It was later adopted, for the most part, by the Japanese, with several versions occurring over the years. Here we show the predominant Chinese version, which used the symbols shown in Table 2. We call this type of system a **multiplicative grouping** system. In general, such a system would involve pairs of symbols, each pair containing a multiplier (with some counting number value less than the base) and then a power of the base. The Chinese numerals are read from top to bottom rather than from left to right.

Three features distinguish this system from a strictly pure multiplicative grouping system. First, the number of 1s is indicated using a single symbol rather than a pair. In effect, the multiplier $(1, 2, 3, \ldots, 9)$ is written but the power of the base (10^0) is not. (See Examples 5(a), (b), and (c).) Second, in the pair indicating 10s, if the multiplier is 1, then that multiplier is omitted. Just the symbol for 10 is written. (See Example 6(a).) Third, when a given power of the base is totally missing in a particular number, this omission is shown by the inclusion of the special zero symbol shown at the bottom of Table 2. (See Examples 5(b) and 6(b).) If two or more consecutive powers are missing, just one zero symbol serves to note the total omission. (See Example 5(c).) The omission of 1s and 10s, and any other powers occurring at the extreme bottom of a numeral need not be noted with a zero symbol. (See Example 5(d).) Note that, for clarification in the following examples, we have emphasized the grouping into pairs by spacing and by using braces. These features are *not* part of the actual numeral.

EXAMPLE 5 Interpreting Chinese Numerals

Interpret each Chinese numeral.

(a) 三
千
一
百
六
十
四

(b) 七
百
零
三

(c) 五
千
零
九

(d) 四
千
二
百

SOLUTION

(a)

三 千	$3 \cdot 1000 = 3000$
一 百	$1 \cdot 100 = 100$
六 十	$6 \cdot 10 = 60$
四	$4(\cdot 1) = 4$
Total:	3164

(b)

七 百	$7 \cdot 100 = 700$
零	$0(\cdot 10) = 00$
三	$3(\cdot 1) = 3$
Total:	703

(c)

五 千	$5 \cdot 1000 = 5000$
零	$0(\cdot 100) = 000$ $0(\cdot 10) = 00$
九	$9(\cdot 1) = 9$
Total:	5009

(d)

四 千	$4 \cdot 1000 = 4000$
二 百	$2 \cdot 100 = 200$
Total:	4200

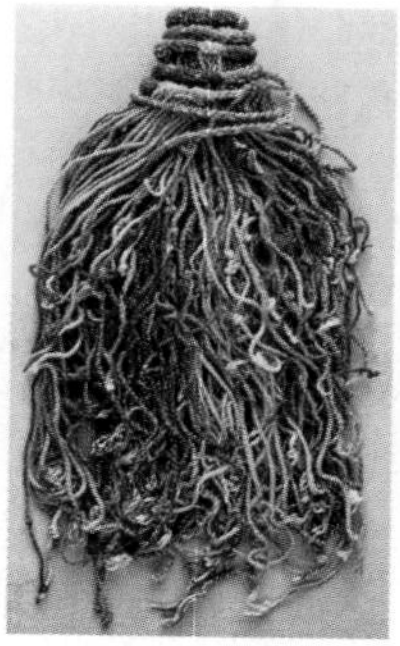

This photo is of a **quipu.** In *Ethnomathematics: A Multicultural View of Mathematical Ideas,* Marcia Ascher writes:

A quipu is an assemblage of colored knotted cotton cords. Used to construct bridges, in ceremonies, for tribute, and in every phase of the life cycle from birth to death, cotton cordage and cloth were of unparalleled importance in Inca culture. The colors of the cords, the way the cords are connected, the relative placement of the cords, the spaces between the cords, the types of knots on the individual cords, and the relative placement of the knots are all part of the logical-numerical recording.

EXAMPLE 6 Creating Chinese Numerals

Write a Chinese numeral for each number.

(a) 614 **(b)** 5090

SOLUTION

(a) The number 614 is made up of six 100s, one 10, and one 4, as depicted at the right.

$6 \cdot 100$: 六 百
$(1 \cdot)10$: 十
$4(\cdot 1)$: 四

(b) The number 5090 consists of five 1000s, no 100s, and nine 10s (no 1s).

$5 \cdot 1000$: 五 千
$(0 \cdot)100$: 零
$9(\cdot 10)$: 九 十

Hindu-Arabic Numeration—Positional A simple grouping system relies on repetition of symbols to denote the number of each power of the base. A multiplicative grouping system uses multipliers in place of repetition, which is more efficient. The ultimate in efficiency is attained with a **positional** system in which only multipliers are used. The various powers of the base require no separate symbols, because the power associated with each multiplier can be understood by the position that the multiplier occupies in the numeral. If the Chinese system had evolved into a positional system, then the numeral for 7482 could be written

七
四
八
二

rather than

七
千
四
百.
八
十
二

The lowest symbol is understood to represent two 1s (10^0), the next one up denotes eight 10s (10^1), then four 100s (10^2), and finally seven 1000s (10^3). Each symbol in a numeral now has both a *face value,* associated with that particular symbol (the multiplier value), and a *place value* (a power of the base), associated with the place, or position, occupied by the symbol.

Positional Numeration

In a positional numeral, each symbol (called a **digit**) conveys two things:

1. **face value**—the inherent value of the symbol
2. **place value**—the power of the base which is associated with the position that the digit occupies in the numeral.

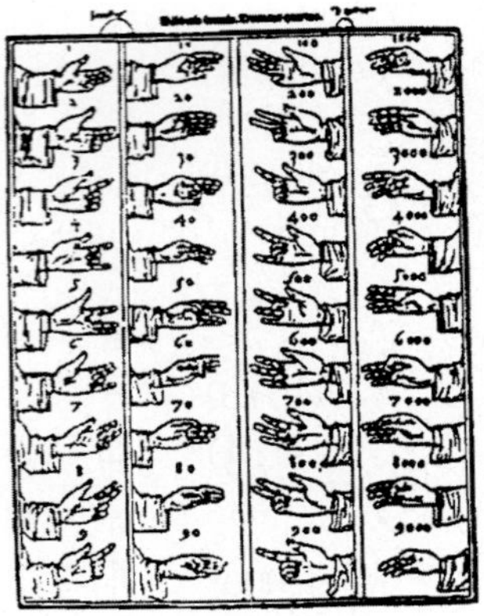

There is much evidence that early humans (in various cultures) used their fingers to represent numbers. As calculations became more complicated, finger reckoning, as shown in this sketch, became popular. The Romans became adept at this sort of calculating, carrying it to 10,000 or perhaps higher.

Number	Symbol
1	▼
10	‹

Babylonian numeration was positional, base sixty. But the face values within the positions were base ten simple grouping numerals, formed with the two symbols shown above. (These symbols resulted from the Babylonian method of writing on clay with a wedge-shaped stylus.) The numeral

‹‹▼▼▼‹‹‹‹▼

denotes 1421 ($23 \cdot 60 + 41 \cdot 1$).

The place values in a Hindu-Arabic numeral, from right to left, are 1, 10, 100, 1000, and so on. The three 4s in the number 46,424 all have the same face value but different place values. The first 4, on the left, denotes four 10,000s, the next one denotes four 100s, and the one on the right denotes four 1s. Place values (in base ten) are named as shown here.

Billions,	Hundred millions	Ten millions	Millions,	Hundred thousands	Ten thousands	Thousands,	Hundreds	Tens	Units	Decimal point.
8,	3	2	1,	4	5	6,	7	9	5	.

This numeral is read as eight billion, three hundred twenty-one million, four hundred fifty-six thousand, seven hundred ninety-five.

To work successfully, a positional system must have a symbol for zero to serve as a **placeholder** in case one or more powers of the base are not needed. Because of this requirement, some early numeration systems took a long time to evolve to a positional form, or never did. Although the traditional Chinese system does utilize a zero symbol, it never did incorporate all the features of a positional system, but remained essentially a multiplicative grouping system.

The one numeration system that did achieve the maximum efficiency of positional form is our own system, the **Hindu-Arabic** system. It was developed over many centuries. Its symbols have been traced to the Hindus of 200 B.C. They were picked up by the Arabs and eventually transmitted to Spain, where a late tenth-century version appeared like this:

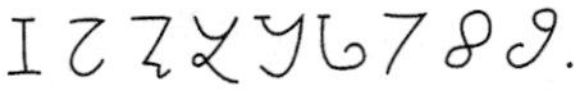

The earliest stages of the system evolved under the influence of navigational, trade, engineering, and military requirements. And in early modern times, the advance of astronomy and other sciences led to a structure well suited to fast and accurate computation. The purely positional form that the system finally assumed was introduced to the West by Leonardo Fibonacci of Pisa (1170–1250) early in the thirteenth century, but widespread acceptance of standardized symbols and form was not achieved until the invention of printing during the fifteenth century. Since that time, no better system of numeration has been devised, and the positional base ten Hindu-Arabic system is commonly used around the world today.

4.1 Exercises

Convert each Egyptian numeral to Hindu-Arabic form.

1. 

2. 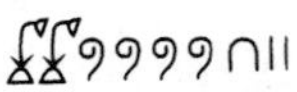

3. 

4. 

Convert each Hindu-Arabic numeral to Egyptian form.

5. 23,145 **6.** 427 **7.** 8,657,000 **8.** 306,090

Chapter 1 of the book of Numbers in the Bible describes a census of the draft-eligible men of Israel after Moses led them out of Egypt into the Desert of Sinai, about 1450 B.C. *Write an Egyptian numeral for the number of available men from each tribe listed.*

9. 59,300 from the tribe of Simeon

10. 46,500 from the tribe of Reuben

11. 74,600 from the tribe of Judah

12. 45,650 from the tribe of Gad

13. 62,700 from the tribe of Dan

14. 54,400 from the tribe of Issachar

Convert each Chinese numeral to Hindu-Arabic form.

15. 九
百
三
十
五

16. 二
百
四
十
六

17. 三
千
零
七

18. 四
千
九
百
零
二

Convert each Hindu-Arabic numeral to Chinese.

19. 960 **20.** 63 **21.** 7012 **22.** 2416

Though Chinese art forms began before written history, their highest development was achieved during four particular dynasties. Write traditional Chinese numerals for the beginning and ending dates of each dynasty listed.

23. Ming (1368 to 1644)

24. Sung (960 to 1279)

25. T'ang (618 to 907)

26. Han (202 B.C. to A.D. 220)

Work each addition or subtraction problem, using regrouping as necessary. Convert each answer to Hindu-Arabic form.

27. 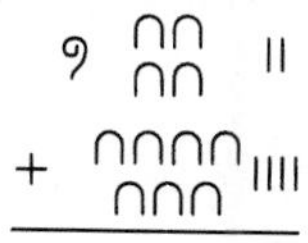

28. 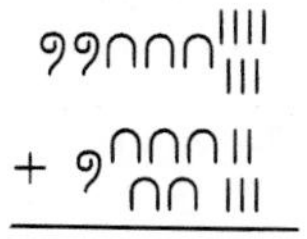

29. 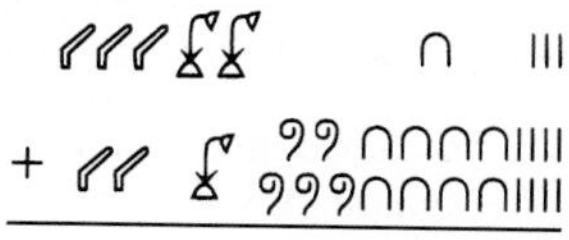

30. 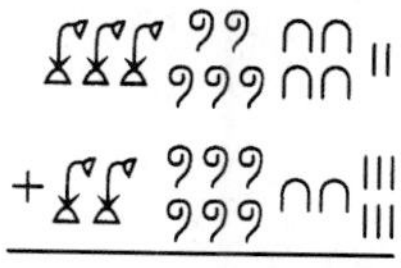

31. 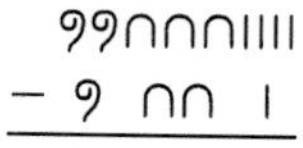

32. 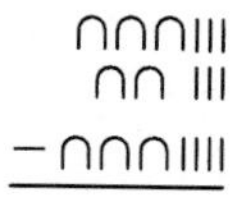

33. 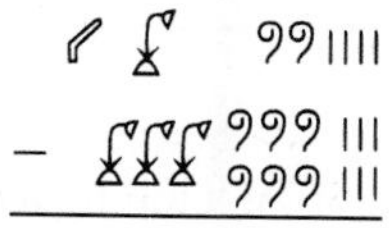

34. 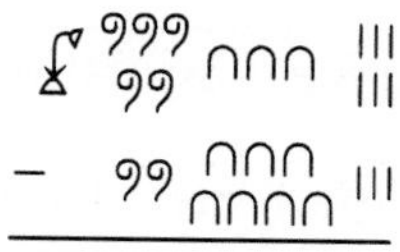

Use the Egyptian algorithm to find each product.

35. $26 \cdot 53$ **36.** $33 \cdot 81$ **37.** $58 \cdot 103$ **38.** $67 \cdot 115$

In Exercises 39 and 40, convert all numbers to Egyptian numerals. Multiply using the Egyptian algorithm, and add using the Egyptian symbols. Give the final answer using a Hindu-Arabic numeral.

39. ***Value of a Biblical Treasure*** The book of Ezra in the Bible describes the return of the exiles to Jerusalem. When they rebuilt the temple, the King of Persia gave them the following items: thirty golden basins, a thousand silver basins, four hundred ten silver bowls, and thirty golden bowls. Find the total value of this treasure, if each gold basin is worth 3000 shekels, each silver basin is worth 500 shekels, each silver bowl is worth 50 shekels, and each golden bowl is worth 400 shekels.

40. ***Total Bill for King Solomon*** King Solomon told the King of Tyre (now Lebanon) that Solomon needed the best cedar for his temple, and that he would "pay you for your men whatever sum you fix." Find the total bill to Solomon if the King of Tyre used the following numbers of men: 5500 tree cutters at two shekels per week each, for a total of seven weeks; 4600 sawers of wood at three shekels per week each, for a total of 32 weeks; and 900 sailors at one shekel per week each, for a total of 16 weeks.

Explain why each step would be an improvement in the development of numeration systems.

41. progressing from carrying groups of pebbles to making tally marks on a stick

42. progressing from tallying to simple grouping

43. progressing from simple grouping to multiplicative grouping

44. progressing from multiplicative grouping to positional numeration

Recall that the ancient Egyptian system described in this section was simple grouping, used a base of ten, and contained seven distinct symbols. The largest number expressible in that system is 9,999,999. Identify the largest number expressible in each of the following simple grouping systems. (In Exercises 49–52, d can be any counting number.)

45. base ten, five distinct symbols

46. base ten, ten distinct symbols

47. base five, five distinct symbols

48. base five, ten distinct symbols

49. base ten, d distinct symbols

50. base five, d distinct symbols

51. base seven, d distinct symbols

52. base b, d distinct symbols (where b is any counting number 2 or greater)

The Hindu-Arabic system is positional and uses ten as the base. Describe any advantages or disadvantages that may have resulted in each case.

53. Suppose the base had been larger, say twelve or twenty for example.

54. Suppose the base had been smaller, maybe eight or five.

4.2 Arithmetic in the Hindu-Arabic System

Expanded Form • Historical Calculation Devices

Expanded Form

The historical development of numeration culminated in positional systems. The most successful of these is the Hindu-Arabic system, which has base ten and, therefore, has place values that are powers of 10.

We now review exponential expressions, or powers (defined in Section 1.1), because they are the basis of expanded form in a positional system.

This Iranian stamp should remind us that counting on fingers (and toes) is an age-old practice. In fact, our word **digit,** referring to the numerals 0–9, comes from a Latin word for "finger" (or "toe"). Aristotle first noted the relationships between fingers and base ten in Greek numeration. Anthropologists go along with the notion. Some cultures, however, have used two, three, or four as number bases, for example, counting on the joints of the fingers or the spaces between them.

EXAMPLE 1 Evaluating Powers

Find each power.

(a) 10^3 **(b)** 7^2 **(c)** 5^4

SOLUTION

(a) $10^3 = 10 \cdot 10 \cdot 10 = 1000$
(10^3 is read "10 cubed," or "10 to the third power.")

(b) $7^2 = 7 \cdot 7 = 49$
(7^2 is read "7 squared," or "7 to the second power.")

(c) $5^4 = 5 \cdot 5 \cdot 5 \cdot 5 = 625$
(5^4 is read "5 to the fourth power.") ■

To simplify work with exponents, it is agreed that

$$a^0 = 1$$

for any nonzero number a. Thus, $7^0 = 1$, $52^0 = 1$, and so on. At the same time,

$$a^1 = a$$

for any number a. For example, $8^1 = 8$, and $25^1 = 25$. The exponent 1 is usually omitted.

By using exponents, numbers can be written in **expanded form** in which the value of the digit in each position is made clear. For example, write 924 in expanded form by thinking of 924 as nine 100s plus two 10s plus four 1s, or

$$\begin{aligned} 924 &= 900 + 20 + 4 \\ &= (9 \cdot 100) + (2 \cdot 10) + (4 \cdot 1) \\ &= (9 \cdot 10^2) + (2 \cdot 10^1) + (4 \cdot 10^0). \end{aligned}$$

$100 = 10^2$, $10 = 10^1$, and $1 = 10^0$.

EXAMPLE 2 Writing Numbers in Expanded Form

Write each number in expanded form.

(a) 1906 **(b)** 46,424

SOLUTION

(a) $1906 = (1 \cdot 10^3) + (9 \cdot 10^2) + (0 \cdot 10^1) + (6 \cdot 10^0)$
Because $0 \cdot 10^1 = 0$, this term could be omitted, but the form is clearer with it included.

(b) $46{,}424 = (4 \cdot 10^4) + (6 \cdot 10^3) + (4 \cdot 10^2) + (2 \cdot 10^1) + (4 \cdot 10^0)$ ■

EXAMPLE 3 Simplifying Expanded Numbers

Simplify each expansion.

(a) $(3 \cdot 10^5) + (2 \cdot 10^4) + (6 \cdot 10^3) + (8 \cdot 10^2) + (7 \cdot 10^1) + (9 \cdot 10^0)$
(b) $(2 \cdot 10^1) + (8 \cdot 10^0)$

SOLUTION

(a) $(3 \cdot 10^5) + (2 \cdot 10^4) + (6 \cdot 10^3) + (8 \cdot 10^2) + (7 \cdot 10^1) + (9 \cdot 10^0) = 326{,}879$
(b) $(2 \cdot 10^1) + (8 \cdot 10^0) = 28$ ■

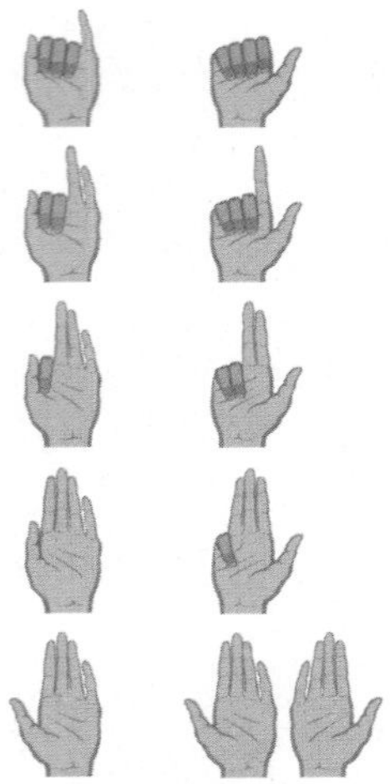

Finger Counting The first digits many people used for counting were their fingers. In Africa the Zulu used the method shown here to count to ten. They started on the left hand with palm up and fist closed. The Zulu finger positions for 1–5 are shown above on the left. The Zulu finger positions for 6–10 are shown on the right.

Expanded notation can be used to see why standard algorithms for addition and subtraction really work. The key idea behind these algorithms is based on the **distributive property,** which will be discussed more fully later in this chapter and also in Chapter 6. It can be written in one form as follows.

Distributive Property

For all real numbers *a, b,* and *c,*

$$(b \cdot a) + (c \cdot a) = (b + c) \cdot a.$$

For example,
$$(3 \cdot 10^4) + (2 \cdot 10^4) = (3 + 2) \cdot 10^4 = 5 \cdot 10^4.$$

EXAMPLE 4 Adding Expanded Forms

Use expanded notation to add 23 and 64.

SOLUTION

$$\begin{array}{rcl} 23 & = & (2 \cdot 10^1) + (3 \cdot 10^0) \\ +\ 64 & = & (6 \cdot 10^1) + (4 \cdot 10^0) \\ \hline & & (8 \cdot 10^1) + (7 \cdot 10^0) = 87 \end{array}$$ Sum

Subtraction works in much the same way.

EXAMPLE 5 Subtracting Expanded Forms

Use expanded notation to find $695 - 254$.

SOLUTION

$$\begin{array}{rcl} 695 & = & (6 \cdot 10^2) + (9 \cdot 10^1) + (5 \cdot 10^0) \\ -254 & = & (2 \cdot 10^2) + (5 \cdot 10^1) + (4 \cdot 10^0) \\ \hline & & (4 \cdot 10^2) + (4 \cdot 10^1) + (1 \cdot 10^0) = 441 \end{array}$$ Difference

Expanded notation and the distributive property can also be used to show how to solve addition problems that involve carrying and subtraction problems that involve borrowing.

EXAMPLE 6 Carrying in Expanded Form

Use expanded notation to add 75 and 48.

SOLUTION

$$\begin{array}{rcl} 75 & = & (7 \cdot 10^1) + (5 \cdot 10^0) \\ +\ 48 & = & (4 \cdot 10^1) + (8 \cdot 10^0) \\ \hline & & (11 \cdot 10^1) + (13 \cdot 10^0) \end{array}$$

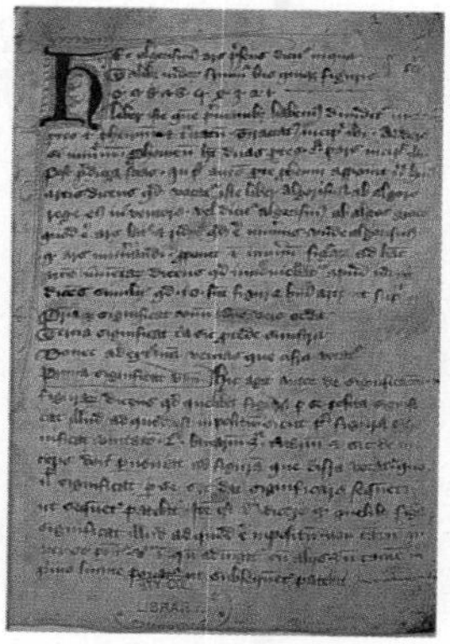

The ***Carmen de Algorismo*** (opening verses shown here) by Alexander de Villa Dei, thirteenth century, popularized the new art of "algorismus":

> *. . . from these twice five figures 0 9 8 7 6 5 4 3 2 1 of the Indians we benefit . . .*

The *Carmen* related that Algor, an Indian king, invented the art. But actually, "algorism" (or "algorithm") comes in a roundabout way from the name Muhammad ibn Musa al-Khorârizmi, an Arabian mathematician of the ninth century, whose arithmetic book was translated into Latin. Furthermore, this Muhammad's book on equations, *Hisab al-jabr w'almuqâbalah,* yielded the term "algebra" in a similar way.

Because the units position (10^0) has room for only one digit, $13 \cdot 10^0$ must be modified:

$$\begin{aligned} 13 \cdot 10^0 &= (10 \cdot 10^0) + (3 \cdot 10^0) \\ &= (1 \cdot 10^1) + (3 \cdot 10^0). \quad \text{Distributive property} \end{aligned}$$

In effect, the 1 from 13 moved to the left (carried) from the units position to the tens position. Now our sum is

$$\begin{aligned} &\underbrace{(11 \cdot 10^1) + (1 \cdot 10^1)} + (3 \cdot 10^0) \\ &= (12 \cdot 10^1) + (3 \cdot 10^0) && \text{Distributive property} \\ &= (10 \cdot 10^1) + (2 \cdot 10^1) + (3 \cdot 10^0) && \text{Modify } 12 \cdot 10^1. \\ &= (1 \cdot 10^2) + (2 \cdot 10^1) + (3 \cdot 10^0) \\ &= 123. && \text{Sum} \end{aligned}$$

EXAMPLE 7 Borrowing in Expanded Form

Use expanded notation to subtract 186 from 364.

SOLUTION

$$\begin{aligned} 364 &= (3 \cdot 10^2) + (6 \cdot 10^1) + (4 \cdot 10^0) \\ -186 &= (1 \cdot 10^2) + (8 \cdot 10^1) + (6 \cdot 10^0) \end{aligned}$$

Because, in the units position, we cannot subtract 6 from 4, we modify the top expansion as follows (the units position borrows from the tens position):

$$\begin{aligned} &(3 \cdot 10^2) + \underbrace{(6 \cdot 10^1) + (4 \cdot 10^0)} \\ &= (3 \cdot 10^2) + (5 \cdot 10^1) + (1 \cdot 10^1) + (4 \cdot 10^0) && \text{Distributive property} \\ &= (3 \cdot 10^2) + (5 \cdot 10^1) + \underbrace{(10 \cdot 10^0) + (4 \cdot 10^0)} \\ &= (3 \cdot 10^2) + (5 \cdot 10^1) + (14 \cdot 10^0) && \text{Distributive property} \end{aligned}$$

We can now subtract 6 from 14 in the units position, but cannot take 8 from 5 in the tens position, so we continue the modification, borrowing from the hundreds to the tens position.

$$\begin{aligned} &\underbrace{(3 \cdot 10^2)} + (5 \cdot 10^1) + (14 \cdot 10^0) \\ &= (2 \cdot 10^2) + (1 \cdot 10^2) + (5 \cdot 10^1) + (14 \cdot 10^0) && \text{Distributive property} \\ &= (2 \cdot 10^2) + \underbrace{(10 \cdot 10^1) + (5 \cdot 10^1)} + (14 \cdot 10^0) \\ &= (2 \cdot 10^2) + (15 \cdot 10^1) + (14 \cdot 10^0) && \text{Distributive property} \end{aligned}$$

Now we can complete the subtraction.

$$\begin{aligned} &(2 \cdot 10^2) + (15 \cdot 10^1) + (14 \cdot 10^0) \\ -\,&(1 \cdot 10^2) + (8 \cdot 10^1) + (6 \cdot 10^0) \\ \hline &(1 \cdot 10^2) + (7 \cdot 10^1) + (8 \cdot 10^0) = 178 && \text{Difference} \end{aligned}$$

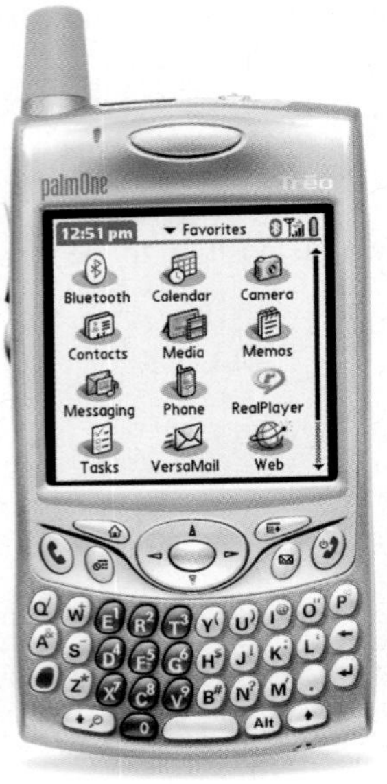

Palmtop computers are the latest in the development of calculating devices.

Examples 4 through 7 used expanded notation and the distributive property to clarify our usual addition and subtraction methods. In practice, our actual work for these four problems would appear as follows:

$$\begin{array}{r} 23 \\ +\ 64 \\ \hline 87 \end{array} \qquad \begin{array}{r} 695 \\ -\ 254 \\ \hline 441 \end{array} \qquad \begin{array}{r} {}^{1} \\ 75 \\ +\ 48 \\ \hline 123 \end{array} \qquad \begin{array}{r} {}^{2\ 15} \\ \not{3}\,\not{6}\,{}^{1}4 \\ -\ 1\ 8\ 6 \\ \hline 1\ 7\ 8. \end{array}$$

The procedures seen in this section also work for positional systems with bases other than ten.

Historical Calculation Devices

Because our numeration system is based on powers of ten, it is often called the **decimal system,** from the Latin word *decem,* meaning ten.* Over the years, many methods have been devised for speeding calculations in the decimal system. One of the oldest is the **abacus,** a device made with a series of rods with sliding beads and a dividing bar. Reading from right to left, the rods have values of 1, 10, 100, 1000, and so on. The bead above the bar has five times the value of those below. Beads moved *toward* the bar are in the "active" position, and those toward the frame are ignored. In our illustrations of abaci (plural form of abacus), such as in Figure 1, the activated beads are shown in black for emphasis.

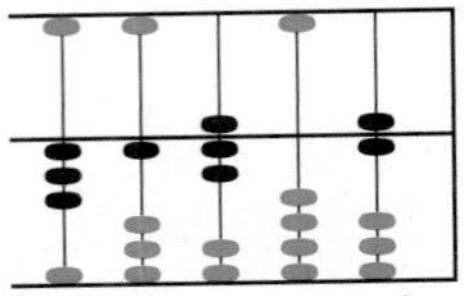

10^4 10^3 10^2 10^1 10^0

FIGURE 1

EXAMPLE 8 Reading an Abacus

What number is shown on the abacus in Figure 1?

SOLUTION

Find the number as follows:

$$(3 \cdot 10{,}000) + (1 \cdot 1000) + [(1 \cdot 500) + (2 \cdot 100)] + 0 \cdot 10 + [(1 \cdot 5) + (1 \cdot 1)]$$

$$= 30{,}000 + 1000 + 500 + 200 + 0 + 5 + 1$$

$$= 31{,}706.$$

Beads above the bar have five times the value.

Merry Math

The following two rhymes come from *Marmaduke Multiply's Merry Method of Making Minor Mathematicians,* a primer published in the late 1830s in Boston:

1. Twice 2 are 4. Pray hasten on before.
2. Five times 5 are 25. I thank my stars I'm yet alive.

As paper became more readily available, people gradually switched from devices like the abacus (though these still are commonly used in some areas) to paper-and-pencil methods of calculation. One early scheme, used both in India and Persia, was the **lattice method,** which arranged products of single digits into a diagonalized lattice, as shown in the following example.

**December* was the tenth month in an old form of the calendar. It is interesting to note that *decem* became *dix* in the French language; a ten-dollar bill, called a "dixie," was in use in New Orleans before the Civil War. "Dixie Land" was a nickname for that city before Dixie came to refer to all the Southern states, as in Daniel D. Emmett's song, written in 1859.

EXAMPLE 9 Using the Lattice Method for Products

Find the product 38 · 794 by the lattice method.

SOLUTION

Step 1 Write the problem, with one number at the side and one across the top.

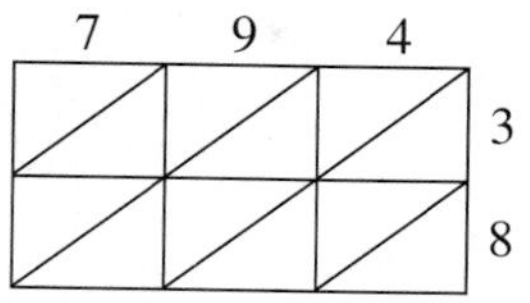

Step 2 Within the lattice, write the products of all pairs of digits from the top and side.

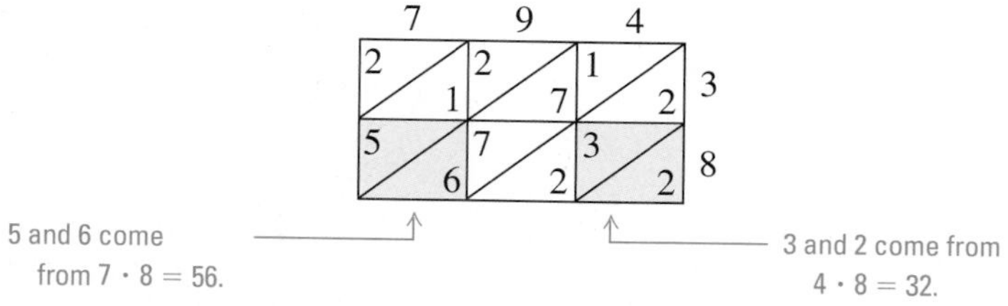

Step 3 Starting at the right of the lattice add diagonally, carrying as necessary.

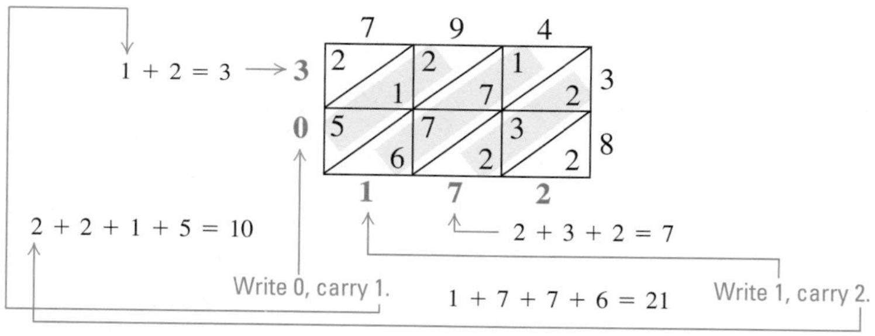

Step 4 Read the answer around the left side and bottom: 38 · 794 = **30,172.**

John Napier's most significant mathematical contribution, developed over a period of at least 20 years, was the concept of **logarithms,** which, among other things, allow multiplication and division to be accomplished with addition and subtraction. It was a great computational advantage given the state of mathematics at the time (1614).

Napier himself regarded his interest in mathematics as a recreation, his main involvements being political and religious. A supporter of John Knox and James I, he published a widely read anti-Catholic work that analyzed the Biblical book of Revelation. He concluded that the Pope was the Antichrist and that the Creator would end the world between 1688 and 1700. Napier was one of many who, over the years, have miscalculated the end of the world.

The Scottish mathematician John Napier (1550–1617) introduced a significant calculating tool called **Napier's rods,** or **Napier's bones.** Napier's invention, based on the lattice method of multiplication, is widely acknowledged as a very early forerunner of modern computers. It consisted of a set of strips, several for each digit 0 through 9, on which multiples of each digit appeared in a sort of lattice column. See Figure 2 on the next page.

An additional strip, called the *index,* could be laid beside any of the others to indicate the multiplier at each level. Napier's rods were used for mechanically multiplying, dividing, and taking square roots. Figure 3 on the next page shows how to multiply 2806 by 7. Select the rods for 2, 8, 0, and 6, placing them side by side. Then using the index, locate the level for a multiplier of 7. The resulting lattice, shown at the bottom of the figure, gives the product 19,642.

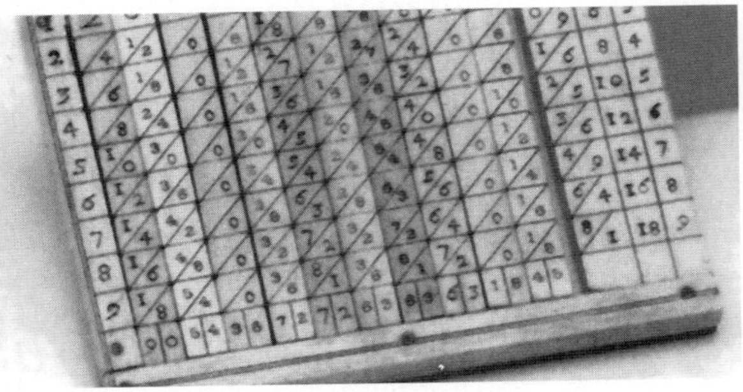

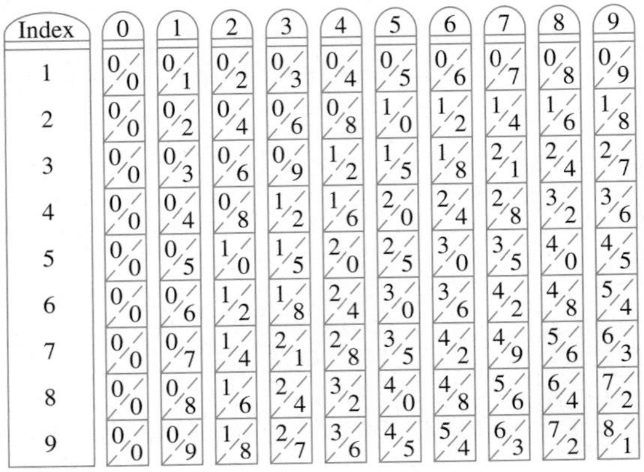

FIGURE 2

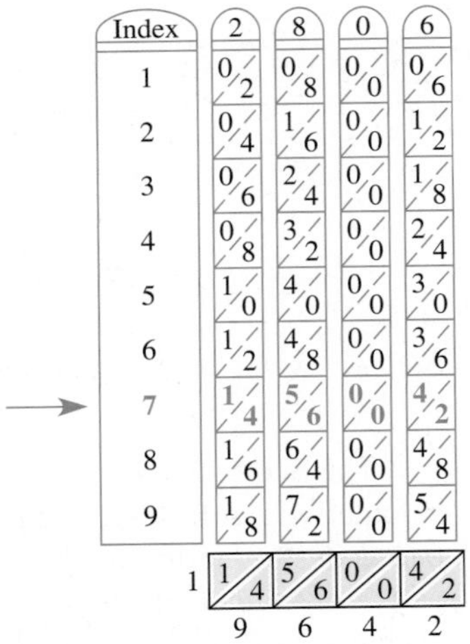

FIGURE 3

EXAMPLE 10 Multiplying with Napier's Rods

Use Napier's rods to find the product of 723 and 4198.

SOLUTION

We line up the rods for 4, 1, 9, and 8 next to the index as in Figure 4.

For a way to include a little magic with your calculations, check out http://trunks.secondfoundation.org/files/psychic.swf, and also http://digicc.com/fido.

The product $3 \cdot 4198$ is found as described in Example 9 and written at the bottom of the figure. Then $2 \cdot 4198$ is found similarly and written below, shifted one place to the left. (Why?) Finally, the product $7 \cdot 4198$ is written shifted two places to the left.

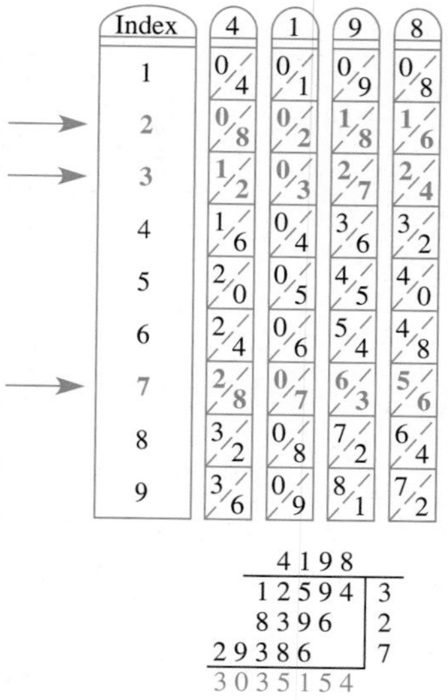

FIGURE 4

The final answer is found by addition to obtain $723 \cdot 4198 = 3{,}035{,}154$.

One other paper-and-pencil method of multiplication is the **Russian peasant method,** which is similar to the Egyptian method of doubling explained in the previous section. (In fact, both of these methods work, in effect, by expanding one of the numbers to be multiplied, but in base two rather than in base ten. Base two numerals are discussed in the next section.) To multiply 37 and 42 by the Russian peasant method, make two columns headed by 37 and 42. Form the first column by dividing 37 by 2 again and again, ignoring any remainders. Stop when 1 is obtained. Form the second column by doubling each number down the column.

Divide by 2, ignoring remainders. ↓	37	42	Double each number. ↓
	18	84	
	9	168	
	4	336	
	2	672	
	1	1344	

Now add up only the second column numbers that correspond to odd numbers in the first column. Omit those corresponding to even numbers in the first column.

→	**37**	**42**	←
	18	84	
Identify odd numbers. →	**9**	**168**	← Add these numbers.
	4	336	
	2	672	
→	**1**	**1344**	←

Finally, $37 \cdot 42 = 42 + 168 + 1344 = 1554$. ⟵ Answer

Most people use standard algorithms for adding and subtracting, carrying or borrowing when appropriate, as illustrated following Example 7. An interesting alternative is the **nines complement method** for subtracting. To use this method, we first agree that the nines complement of a digit n is $9 - n$. For example, the nines complement of 0 is 9, of 1 is 8, of 2 is 7, and so on, up to the nines complement of 9, which is 0.

To carry out the method, complete the following steps:

Step 1 Align the digits as in the standard subtraction algorithm.

Step 2 Add leading zeros, if necessary, in the subtrahend so that both numbers have the same number of digits.

Step 3 Replace each digit in the subtrahend with its nines complement, and then add.

Step 4 Finally, delete the leading digit (1), and add 1 to the remaining part of the sum.

EXAMPLE 11 Using the Nines Complement Method

Use the nines complement method to subtract 2803 − 647.

SOLUTION

Step 1	Step 2	Step 3	Step 4	
2803	2803	2803	2155	
− 647	− 0647	+ 9352	+ 1	
		12,155	2156	Difference

For Further Thought

Calculating on the Abacus

The abacus has been (and still is) used to perform rapid calculations. A simple example is adding 526 and 362. Start with 526 on the abacus:

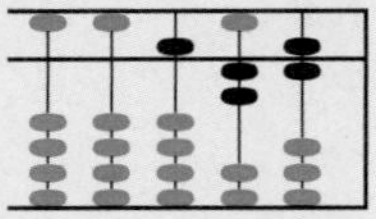

To add 362, start by "activating" an additional 2 on the 1s rod:

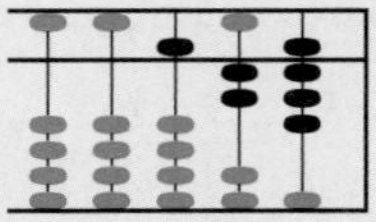

Next, activate an additional 6 on the 10s rod:

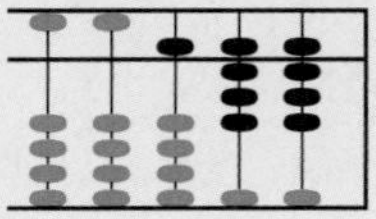

Finally, activate an additional 3 on the 100s rod:

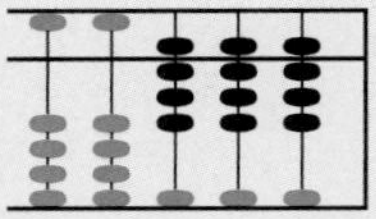

The sum, read from the abacus, is 888.

For problems where carrying or borrowing is required, it takes a little more thought and skill. Try to obtain an actual abacus (or, otherwise, make sketches) and practice some addition and subtraction problems until you can do them quickly.

For Group Discussion or Individual Investigation

1. Use an abacus to add: 13,728 + 61,455. Explain each step of your procedure.
2. Use an abacus to subtract: 6512 − 4816. Again, explain each step of your procedure.

4.2 EXERCISES

Write each number in expanded form.

1. 73

2. 925

3. 3774

4. 12,398

5. four thousand, nine hundred twenty-four

6. fifty-two thousand, one hundred eighteen

7. fourteen million, two hundred six thousand, forty

8. two hundred twelve million, eleven thousand, nine hundred sixteen

Simplify each expansion.

9. $(4 \cdot 10^1) + (2 \cdot 10^0)$

10. $(3 \cdot 10^2) + (5 \cdot 10^1) + (0 \cdot 10^0)$

11. $(6 \cdot 10^3) + (2 \cdot 10^2) + (0 \cdot 10^1) + (9 \cdot 10^0)$

12. $(5 \cdot 10^5) + (0 \cdot 10^4) + (3 \cdot 10^3) + (5 \cdot 10^2) + (6 \cdot 10^1) + (8 \cdot 10^0)$

13. $(7 \cdot 10^7) + (4 \cdot 10^5) + (1 \cdot 10^3) + (9 \cdot 10^0)$

14. $(3 \cdot 10^8) + (8 \cdot 10^7) + (2 \cdot 10^2) + (3 \cdot 10^0)$

In each of the following, add in expanded notation.

15. $54 + 35$

16. $782 + 413$

In each of the following, subtract in expanded notation.

17. $85 - 53$

18. $784 - 523$

Perform each addition using expanded notation.

19. $75 + 34$

20. $537 + 278$

21. $434 + 299$

22. $6755 + 4827$

Perform each subtraction using expanded notation.

23. $54 - 48$

24. $364 - 59$

25. $645 - 439$

26. $816 - 335$

Identify the number represented on each abacus.

27. 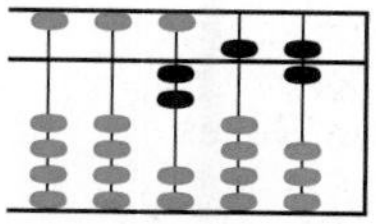

28. 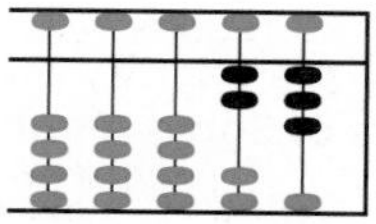

29. 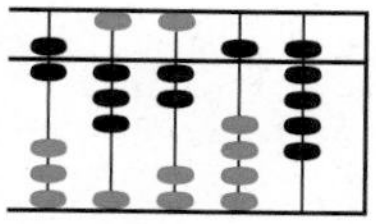

30. 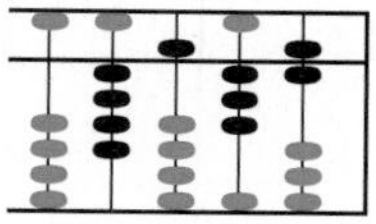

Sketch an abacus to show each number.

31. 38

32. 183

33. 2547

34. 70,163

Use the lattice method to find each product.

35. $65 \cdot 29$

36. $32 \cdot 741$

37. $525 \cdot 73$

38. $912 \cdot 483$

Refer to Example 10 where Napier's rods were used to find the product of 723 *and* 4198. *Then complete Exercises 39 and 40.*

39. Find the product of 723 and 4198 by completing the lattice process shown here.

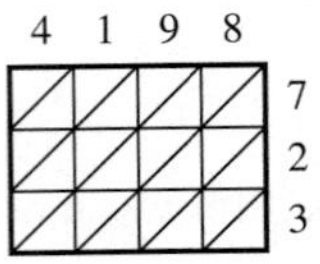

40. Explain how Napier's rods could have been used in Example 10 to set up one complete lattice product rather than adding three individual (shifted) lattice products. Illustrate with a sketch.

Use Napier's rods (Figure 2) to find each product.

41. 8 · 62 **42.** 32 · 73 **43.** 26 · 8354 **44.** 526 · 4863

Perform each subtraction using the nines complement method.

45. 283 − 41 **46.** 536 − 425 **47.** 50,000 − 199 **48.** 40,002 − 4846

Use the Russian peasant method to find each product.

49. 5 · 92 **50.** 41 · 53 **51.** 62 · 529 **52.** 145 · 63

4.3 Conversion Between Number Bases

General Base Conversions • Computer Mathematics

General Base Conversions Bases other than ten have occurred historically. For example, the ancient Babylonians used 60 as their base. The Mayan Indians of Central America and Mexico used 20. In this section we consider bases other than ten, but we use the familiar Hindu-Arabic symbols. We will consistently indicate bases other than ten with a spelled-out subscript, as in the numeral 43_{five}. Whenever a number appears without a subscript, it is to be assumed that the intended base is ten. Be careful how you read (or verbalize) numerals here. The numeral 43_{five} is read "four three base five." (Do *not* read it as "forty-three," as that terminology implies base ten and names a totally different number.)

For reference in doing number expansions and base conversions, Table 3 gives the first several powers of some numbers used as alternative bases in this section.

TABLE 3 Selected Powers of Some Alternative Number Bases

	Fourth Power	Third Power	Second Power	First Power	Zero Power
Base two	16	8	4	2	1
Base five	625	125	25	5	1
Base seven	2401	343	49	7	1
Base eight	4096	512	64	8	1
Base sixteen	65,536	4096	256	16	1

TABLE 4

Base Ten	Base Five
0	0
1	1
2	2
3	3
4	4
5	10
6	11
7	12
8	13
9	14
10	20
11	21
12	22
13	23
14	24
15	30
16	31
17	32
18	33
19	34
20	40
21	41
22	42
23	43
24	44
25	100
26	101
27	102
28	103
29	104
30	110

For example, the base two row of Table 3 indicates that

$$2^4 = 16, \quad 2^3 = 8, \quad 2^2 = 4, \; 2^1 = 2, \quad \text{and} \quad 2^0 = 1.$$

We begin with the base five system, which requires just five distinct symbols, 0, 1, 2, 3, and 4. Table 4 compares base five and decimal (base ten) numerals for the whole numbers 0 through 30. Notice that because only the symbols 0, 1, 2, 3, and 4 are used in base five, we must use two digits in base five when we get to 5_{ten}. The number 5_{ten} is expressed as one 5 and no 1s, that is as 10_{five}. Then 6_{ten} becomes 11_{five} (one 5 and one 1). Try to predict, without help, the smallest base ten numeral that would require three digits in base five. Is your answer consistent with Table 4? While base five uses fewer distinct symbols than base ten (an apparent advantage because there are few symbols to learn), it often requires more digits than base ten to denote the same number (a disadvantage because more symbols must be written).

You will find that in any base, if you denote the base "b," then the base itself will be 10_b, just as occurred in base five. For example,

$$7_{\text{ten}} = 10_{\text{seven}}, \quad 16_{\text{ten}} = 10_{\text{sixteen}}, \quad \text{and} \quad \text{so on.}$$

EXAMPLE 1 Converting from Base Five to Base Ten

Convert 1342_{five} to decimal form.

SOLUTION

Referring to the powers of five in Table 3, we see that this number has one 125, three 25s, four 5s, and two 1s, so

$$\begin{aligned} 1342_{\text{five}} &= (1 \cdot 125) + (3 \cdot 25) + (4 \cdot 5) + (2 \cdot 1) \\ &= 125 + 75 + 20 + 2 \\ &= 222. \end{aligned}$$

A shortcut for converting from base five to decimal form, which is *particularly useful when you use a calculator,* can be derived as follows:

$$\begin{aligned} 1342_{\text{five}} &= (1 \cdot 5^3) + (3 \cdot 5^2) + (4 \cdot 5) + 2 && \text{Factor 5 out of the three quantities in parentheses.} \\ &= ((1 \cdot 5^2) + (3 \cdot 5) + 4) \cdot 5 + 2 \\ &= (((1 \cdot 5) + 3) \cdot 5 + 4) \cdot 5 + 2. && \text{Factor 5 out of the two "inner" quantities.} \end{aligned}$$

The inner parentheses around $1 \cdot 5$ are not needed because the product would be automatically done before the 3 is added. Therefore, we can write

$$1342_{\text{five}} = ((1 \cdot 5 + 3) \cdot 5 + 4) \cdot 5 + 2.$$

This series of products and sums is easily done as an uninterrupted sequence of operations on a calculator, with no intermediate results written down. The same method works for converting to base ten from any other base.

This procedure is summarized as follows.

Calculator Shortcut for Base Conversion

To convert from another base to decimal form: Start with the first digit on the left and multiply by the base. Then add the next digit, multiply again by the base, and so on. The last step is to add the last digit on the right. Do *not* multiply it by the base.

Exactly how you accomplish these steps depends on the type of calculator you use. With some models, only the digits, the multiplications, and the additions need to be entered, in order. With others, you may need to press the [=] key following each addition of a digit. If you handle grouped expressions on your calculator by actually entering parentheses, then enter the expression just as illustrated above and in the following example. (The number of left parentheses to start with will be two fewer than the number of digits in the original numeral.)

The binary (base two) symbols of the *I Ching,* a 2000-year-old Chinese classic, permute into 8 elemental trigrams; 64 hexagrams are interpreted in casting oracles.

The basic symbol here is the ancient Chinese "**yin-yang,**" in which the black and the white enfold each other, each containing a part of the other. A kind of duality is conveyed between destructive (yin) and beneficial (yang) aspects. Leibniz (1646–1716) studied Chinese ideograms in search of a universal symbolic language and promoted East–West cultural contact. He saw parallels between the trigrams and his binary arithmetic.

Niels Bohr (1885–1962), famous Danish Nobel laureate in physics (atomic theory), adopted the yin-yang symbol in his coat of arms to depict his principle of *complementarity,* which he believed was fundamental to reality at the deepest levels. Bohr also pushed for East–West cooperation.

In its 1992 edition, *The World Book Dictionary* first judged "yin-yang" to have been used enough to become a permanent part of our ever changing language, assigning it the definition, "made up of opposites."

EXAMPLE 2 Using the Calculator Shortcut

Use the calculator shortcut to convert 244314_{five} to decimal form.

SOLUTION

$$\begin{aligned} 244314_{\text{five}} &= ((((2 \cdot 5 + 4) \cdot 5 + 4) \cdot 5 + 3) \cdot 5 + 1) \cdot 5 + 4 \\ &= 9334 \end{aligned}$$

Note the four left parentheses for a six-digit numeral.

EXAMPLE 3 Converting from Base Ten to Base Five

Convert 497 from decimal form to base five.

SOLUTION

The base five place values, starting from the right, are 1, 5, 25, 125, 625, and so on. Because 497 is between 125 and 625, it will require no 625s, but some 125s, as well as possibly some 25s, 5s, and 1s. Dividing 497 by 125 determines the proper number of 125s. The quotient is 3, with remainder 122. So we need three 125s. Next, the remainder, 122, is divided by 25 (the next place value) to find the proper number of 25s. The quotient is 4, with remainder 22, so we need four 25s. Dividing 22 by 5 yields 4, with remainder 2. So we need four 5s. Dividing 2 by 1 yields 2 (with remainder 0), so we need two 1s. Finally, we see that 497 consists of three 125s, four 25s, four 5s, and two 1s, so $497 = 3442_{\text{five}}$.

More concisely, this process can be written as follows.

$$\begin{aligned} 497 \div 125 &= 3 && \text{Remainder } 122 \\ 122 \div 25 &= 4 && \text{Remainder } 22 \\ 22 \div 5 &= 4 && \text{Remainder } 2 \\ 2 \div 1 &= 2 && \text{Remainder } 0 \\ 497 &= 3442_{\text{five}} \end{aligned}$$

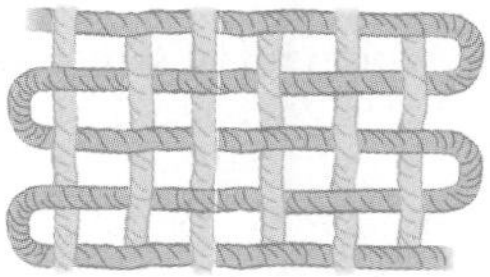

Woven fabric is a binary system of threads going lengthwise (warp threads—tan in the diagram above) and threads going crosswise (weft or woof). At any point in a fabric, either warp or weft is on top, and the variation creates the pattern.

Nineteenth-century looms for weaving operated using punched cards, "programmed" for pattern. The looms were set up with hooked needles, the hooks holding the warp. Where there were holes in cards, the needles moved, the warp lifted, and the weft passed under. Where no holes were, the warp did not lift, and the weft was on top. The system parallels the on–off system in calculators and computers. In fact, these looms were models in the development of modern calculating machinery.

Joseph Marie Jacquard (1752–1823) is credited with improving the mechanical loom so that mass production of fabric was feasible.

Check:

$$
\begin{aligned}
3442_{\text{five}} &= (3 \cdot 125) + (4 \cdot 25) + (4 \cdot 5) + (2 \cdot 1) \\
&= 375 + 100 + 20 + 2 \\
&= 497
\end{aligned}
$$

■

The calculator shortcut for converting from another base to decimal form involved repeated *multiplications* by the other base. (See Example 2.) A shortcut for converting from decimal form to another base makes use of repeated *divisions* by the other base. Just divide the original decimal numeral, and the resulting quotients in turn, by the desired base until the quotient 0 appears.

EXAMPLE 4 Using a Shortcut to Convert from Base Ten to Another Base

Repeat Example 3 using the shortcut just described.

SOLUTION

$$
\begin{array}{r|l}
\multicolumn{2}{r}{\qquad\qquad\qquad\text{Remainder}} \\
5 & \underline{497} \\
5 & \underline{\ 99} \quad \longleftarrow \quad 2 \\
5 & \underline{\ 19} \quad \longleftarrow \quad 4 \\
5 & \underline{\ \ 3} \quad \longleftarrow \quad 4 \\
 & \ \ 0 \quad \longleftarrow \quad 3
\end{array}
$$

Read the answer from the remainder column, reading from the bottom up:

$$497 = 3442_{\text{five}}.$$

■

To see why this shortcut works, notice the following:

The first division shows that four hundred ninety-seven 1s are equivalent to ninety-nine 5s and two 1s. (The two 1s are set aside and account for the last digit of the answer.)

The second division shows that ninety-nine 5s are equivalent to nineteen 25s and four 5s. (The four 5s account for the next digit of the answer.)

The third division shows that nineteen 25s are equivalent to three 125s and four 25s. (The four 25s account for the next digit of the answer.)

The fourth (and final) division shows that the three 125s are equivalent to no 625s and three 125s. The remainders, as they are obtained *from top to bottom,* give the number of 1s, then 5s, then 25s, then 125s.

The methods for converting between bases ten and five, including the shortcuts, can be adapted for conversions between base ten and any other base.

EXAMPLE 5 Converting from Base Seven to Base Ten

Convert 6343_{seven} to decimal form, by expanding in powers, and by using the calculator shortcut.

SOLUTION

$$\begin{aligned} 6343_{\text{seven}} &= (6 \cdot 7^3) + (3 \cdot 7^2) + (4 \cdot 7^1) + (3 \cdot 7^0) \\ &= (6 \cdot 343) + (3 \cdot 49) + (4 \cdot 7) + (3 \cdot 1) \\ &= 2058 + 147 + 28 + 3 \\ &= 2236 \end{aligned}$$

Calculator shortcut: $6343_{\text{seven}} = ((6 \cdot 7 + 3) \cdot 7 + 4) \cdot 7 + 3 = 2236.$

EXAMPLE 6 Converting from Base Ten to Base Seven

Convert 7508 to base seven.

SOLUTION

Divide 7508 by 7, then divide the resulting quotient by 7, until a quotient of 0 results.

$$\begin{array}{rl|c} & & \text{Remainder} \\ 7 \,\big|\, & 7508 & \\ 7 \,\big|\, & 1072 \leftarrow & 4 \\ 7 \,\big|\, & 153 \leftarrow & 1 \\ 7 \,\big|\, & 21 \leftarrow & 6 \\ 7 \,\big|\, & 3 \leftarrow & 0 \\ & 0 \leftarrow & 3 \end{array}$$

From the remainders, reading bottom to top, $7508 = 30614_{\text{seven}}$.

Because we are accustomed to doing arithmetic in base ten, most of us would handle conversions between arbitrary bases (where neither is ten) by going from the given base to base ten and then to the desired base, as illustrated in the next example.

EXAMPLE 7 Converting Between Two Bases Other Than Ten

Convert 3164_{seven} to base five.

SOLUTION

First convert to decimal form.

$$\begin{aligned} 3164_{\text{seven}} &= (3 \cdot 7^3) + (1 \cdot 7^2) + (6 \cdot 7^1) + (4 \cdot 7^0) \\ &= (3 \cdot 343) + (1 \cdot 49) + (6 \cdot 7) + (4 \cdot 1) \\ &= 1029 + 49 + 42 + 4 \\ &= \mathbf{1124} \end{aligned}$$

Next convert this decimal result to base five.

$$\begin{array}{rl|c} & & \text{Remainder} \\ 5 \,\big|\, & \mathbf{1124} & \\ 5 \,\big|\, & 224 \leftarrow & 4 \\ 5 \,\big|\, & 44 \leftarrow & 4 \\ 5 \,\big|\, & 8 \leftarrow & 4 \\ 5 \,\big|\, & 1 \leftarrow & 3 \\ & 0 \leftarrow & 1 \end{array}$$

From the remainders, $3164_{\text{seven}} = 13444_{\text{five}}$.

TABLE 5

Base Ten (decimal)	Base Two (binary)
0	0
1	1
2	10
3	11
4	100
5	101
6	110
7	111
8	1000
9	1001
10	1010
11	1011
12	1100
13	1101
14	1110
15	1111
16	10000
17	10001
18	10010
19	10011
20	10100

Computer Mathematics There are three alternative base systems that are most useful in computer applications. These are the **binary** (base two), **octal** (base eight), and **hexadecimal** (base sixteen) systems. Computers and handheld calculators actually use the binary system for their internal calculations because that system consists of only two symbols, 0 and 1. All numbers can then be represented by electronic "switches," of one kind or another, where "on" indicates 1 and "off" indicates 0. The octal and hexadecimal systems have been used extensively by programmers who work with internal computer codes and for communication between the CPU (central processing unit) and a printer or other output device.

The binary system is extreme in that it has only two available symbols (0 and 1); because of this, representing numbers in binary form requires more digits than in any other base. Table 5 shows the whole numbers up to 20 expressed in binary form.

Conversions between any of these three special base systems (binary, octal, and hexadecimal) and the decimal system can be done by the methods already discussed.

EXAMPLE 8 Converting from Binary to Decimal

Convert 110101_{two} to decimal form, by expanding in powers, and by using the calculator shortcut.

SOLUTION

$$\begin{aligned} 110101_{\text{two}} &= (1 \cdot 2^5) + (1 \cdot 2^4) + (0 \cdot 2^3) + (1 \cdot 2^2) + (0 \cdot 2^1) + (1 \cdot 2^0) \\ &= (1 \cdot 32) + (1 \cdot 16) + (0 \cdot 8) + (1 \cdot 4) + (0 \cdot 2) + (1 \cdot 1) \\ &= 32 + 16 + 0 + 4 + 0 + 1 \\ &= 53 \end{aligned}$$

Calculator shortcut:

$$\begin{aligned} 110101_{\text{two}} &= ((((1 \cdot 2 + 1) \cdot 2 + 0) \cdot 2 + 1) \cdot 2 + 0) \cdot 2 + 1 \\ &= 53. \end{aligned}$$

Note the four left parentheses for a six-digit numeral.

EXAMPLE 9 Converting from Decimal to Octal

Convert 9583 to octal form.

SOLUTION

Divide repeatedly by 8, writing the remainders at the side.

Division	Remainder
8 ⟌ 9 583	
8 ⟌ 1 197 ←	7
8 ⟌ 149 ←	5
8 ⟌ 18 ←	5
8 ⟌ 2 ←	2
0 ←	2

From the remainders, $9583 = 22557_{\text{eight}}$.

Trick or Tree? The octal number 31 is equal to the decimal number 25. This may be written as

31 OCT = 25 DEC

Does this mean that Halloween and Christmas fall on the same day of the year?

The hexadecimal system, having base 16, which is greater than 10, presents a new problem. Because distinct symbols are needed for all whole numbers from 0 up to one

Converting Calculators A number of scientific calculators are available that will convert between decimal, binary, octal, and hexadecimal, and will also do calculations directly in all of these separate modes.

less than the base, base sixteen requires more symbols than are normally used in our decimal system. Computer programmers commonly use the letters A, B, C, D, E, and F as hexadecimal digits for the numbers ten through fifteen, respectively.

EXAMPLE 10 Converting from Hexadecimal to Decimal

Convert $FA5_{\text{sixteen}}$ to decimal form.

SOLUTION

The hexadecimal digits F and A represent 15 and 10, respectively.

$$\begin{aligned} FA5_{\text{sixteen}} &= (15 \cdot 16^2) + (10 \cdot 16^1) + (5 \cdot 16^0) \\ &= 3840 + 160 + 5 \\ &= 4005 \end{aligned}$$

EXAMPLE 11 Converting from Decimal to Hexadecimal

Convert 748 from decimal form to hexadecimal form.

SOLUTION

Use repeated division by 16.

Division		Remainder		Hexadecimal notation
16 | 748				
16 | 46	←	12	←	C
16 | 2	←	14	←	E
0	←	2	←	2

From the remainders at the right, $748 = 2EC_{\text{sixteen}}$.

The decimal whole numbers 0 through 17 are shown in Table 6 on the next page along with their equivalents in the common computer-oriented bases (two, eight, and sixteen). Conversions among binary, octal, and hexadecimal systems can generally be accomplished by the shortcuts explained below.

The binary system is the natural one for internal computer workings because of its compatibility with the two-state electronic switches. It is very cumbersome, however, for human use, because so many digits occur even in the numerals for relatively small numbers. The octal and hexadecimal systems are the choices of computer programmers mainly because of their close relationship with the binary system. *Both eight and sixteen are powers of two.* When conversions involve one base that is a power of the other, there is a quick conversion shortcut available. For example, because $8 = 2^3$, every octal digit (0 through 7) can be expressed as a 3-digit binary numeral. See Table 7 on the next page.

EXAMPLE 12 Converting from Octal to Binary

Convert 473_{eight} to binary form.

SOLUTION

Replace each octal digit with its 3-digit binary equivalent. (Leading zeros can be omitted only when they occur in the leftmost group.) Then combine all the binary equivalents into a single binary numeral.

A T-shirt is currently being marketed with the following message printed on the front. "There are 10 kinds of people in the world: those who understand binary and those who don't." Do YOU understand this message?

$$\begin{array}{ccc} 4 & 7 & 3_{\text{eight}} \\ \downarrow & \downarrow & \downarrow \\ 100 & 111 & 011_{\text{two}} \end{array}$$

By this method, $473_{\text{eight}} = 100111011_{\text{two}}$. ■

TABLE 6 Some Decimal Equivalents in the Common Computer-Oriented Bases

Decimal (Base Ten)	Hexadecimal (Base Sixteen)	Octal (Base Eight)	Binary (Base Two)
0	0	0	0
1	1	1	1
2	2	2	10
3	3	3	11
4	4	4	100
5	5	5	101
6	6	6	110
7	7	7	111
8	8	10	1000
9	9	11	1001
10	A	12	1010
11	B	13	1011
12	C	14	1100
13	D	15	1101
14	E	16	1110
15	F	17	1111
16	10	20	10000
17	11	21	10001

TABLE 7

Octal	Binary
0	000
1	001
2	010
3	011
4	100
5	101
6	110
7	111

EXAMPLE 13 Converting from Binary to Octal

Convert 10011110_{two} to octal form.

SOLUTION

Start at the right and break the digits into groups of three. Then convert the groups to their octal equivalents.

$$\begin{array}{ccc} 10 & 011 & 110_{\text{two}} \\ \downarrow & \downarrow & \downarrow \\ 2 & 3 & 6_{\text{eight}} \end{array}$$

Finally, $10011110_{\text{two}} = 236_{\text{eight}}$. ■

TABLE 8

Hexadecimal	Binary
0	0000
1	0001
2	0010
3	0011
4	0100
5	0101
6	0110
7	0111
8	1000
9	1001
A	1010
B	1011
C	1100
D	1101
E	1110
F	1111

Because $16 = 2^4$, every hexadecimal digit can be equated to a 4-digit binary numeral (see Table 8), and conversions between binary and hexadecimal forms can be done in a manner similar to that used in Examples 12 and 13.

EXAMPLE 14 Converting from Hexadecimal to Binary

Convert $8B4F_{\text{sixteen}}$ to binary form.

SOLUTION

Each hexadecimal digit yields a 4-digit binary equivalent.

$$\begin{array}{cccc} 8 & B & 4 & F_{\text{sixteen}} \\ \downarrow & \downarrow & \downarrow & \downarrow \\ 1000 & 1011 & 0100 & 1111_{\text{two}} \end{array}$$

Combining these groups of digits, we see that

$$8B4F_{\text{sixteen}} = 1000101101001111_{\text{two}}.$$

Several games and tricks are based on the binary system. For example, Table 9 can be used to find the age of a person 31 years old or younger. The person need only tell you the columns that contain his or her age. For example, suppose Kellen Dawson says that her age appears in columns B, C, and D. To find her age, add the numbers from the top row of these columns:

$$\text{Kellen is } 2 + 4 + 8 = 14 \text{ years old.}$$

Do you see how this trick works? (See Exercises 69–72.)

TABLE 9

A	B	C	D	E
1	2	4	8	16
3	3	5	9	17
5	6	6	10	18
7	7	7	11	19
9	10	12	12	20
11	11	13	13	21
13	14	14	14	22
15	15	15	15	23
17	18	20	24	24
19	19	21	25	25
21	22	22	26	26
23	23	23	27	27
25	26	28	28	28
27	27	29	29	29
29	30	30	30	30
31	31	31	31	31

Several years ago, the Kellogg Company featured a **Magic Trick Age Detector** activity on specially marked packages of *Kellogg's*® *Rice Krispies*® cereal. The trick is simply an extension of the discussion in the text.

Kellogg's® *Rice Krispies*® and characters *Snap!*® *Crackle!*® and *Pop!*® are registered trademarks of Kellogg Company.

4.3 Exercises

List the first twenty counting numbers in each base.

1. seven (Only digits 0 through 6 are used in base seven.)

2. eight (Only digits 0 through 7 are used.)

3. nine (Only digits 0 through 8 are used.)

4. sixteen (The digits 0, 1, 2, . . . , 9, A, B, C, D, E, F are used in base sixteen.)

Write (in the same base) the counting numbers just before and just after the given number. (Do not convert to base ten.)

5. 14_{five}

6. 555_{six}

7. $\text{B6F}_{\text{sixteen}}$

8. 10111_{two}

Determine the number of distinct symbols needed in each of the following positional systems.

9. base three

10. base seven

11. base eleven

12. base sixteen

Determine, in each base, the least and greatest four-digit numbers and their decimal equivalents.

13. three

14. sixteen

Convert each number to decimal form by expanding in powers and by using the calculator shortcut.

15. 24_{five}

16. 62_{seven}

17. 1011_{two}

18. 35_{eight}

19. $\text{3BC}_{\text{sixteen}}$

20. 34432_{five}

21. 2366_{seven}

22. 101101110_{two}

23. 70266_{eight}

24. $\text{ABCD}_{\text{sixteen}}$

25. 2023_{four}

26. 6185_{nine}

27. 41533_{six}

28. 88703_{nine}

Convert each number from decimal form to the given base.

29. 86 to base five

30. 65 to base seven

31. 19 to base two

32. 935 to base eight

33. 147 to base sixteen

34. 2730 to base sixteen

35. 36401 to base five

36. 70893 to base seven

37. 586 to base two

38. 12888 to base eight

39. 8407 to base three

40. 11028 to base four

41. 9346 to base six

42. 99999 to base nine

Make each conversion as indicated.

43. 43_{five} to base seven

44. 27_{eight} to base five

45. 6748_{nine} to base four

46. $\text{C02}_{\text{sixteen}}$ to base seven

Convert each number from octal form to binary form.

47. 367_{eight}

48. 2406_{eight}

Convert each number from binary form to octal form.

49. 100110111_{two}

50. 11010111101_{two}

Make each conversion as indicated.

51. $DC_{sixteen}$ to binary

52. $F111_{sixteen}$ to binary

53. 101101_{two} to hexadecimal

54. 101111011101000_{two} to hexadecimal

Identify the greatest number from each list.

55. 42_{seven}, 37_{eight}, $1D_{sixteen}$

56. 1101110_{two}, 414_{five}, $6F_{sixteen}$

There is a theory that twelve would be a better base than ten for general use. This is mainly because twelve has more divisors (1, 2, 3, 4, 6, 12) *than ten* (1, 2, 5, 10), *which makes fractions easier in base twelve. The base twelve system is called the* **duodecimal system.** *In the decimal system we speak of a one, a ten, and a hundred (and so on); in the duodecimal system we say a* one, *a* dozen *(twelve), and a* gross *(twelve squared, or one hundred forty-four).*

57. Otis Taylor's clients ordered 9 gross, 10 dozen, and 11 copies of *The Minnie Minoso Story* during 2002. How many copies was that in base ten?

58. Which amount is larger: 3 gross, 6 dozen or 2 gross, 19 dozen?

One common method of converting symbols into binary digits for computer processing is called ASCII (American Standard Code of Information Interchange). The uppercase letters A *through* Z *are assigned the numbers* 65 *through* 90, *so* A *has binary code* 1000001 *and* Z *has code* 1011010. *Lowercase letters* a *through* z *have codes* 97 *through* 122 *(that is,* 1100001 *through* 1111010*). ASCII codes, as well as other numerical computer output, normally appear without commas.*

Write the binary code for each letter.

59. C

60. X

61. k

62. q

Break each code into groups of seven digits and write as letters.

63. 1001000100010110011001010000

64. 10000111001000101010110000111001011

Translate each word into an ASCII string of binary digits. (Be sure to distinguish uppercase and lowercase letters.)

65. New

66. Orleans

67. Explain why the octal and hexadecimal systems are convenient for people who code for computers.

68. There are thirty-seven counting numbers whose base eight numerals contain two digits but whose base three numerals contain four digits. Find the least and greatest of these numbers.

Refer to Table 9 for Exercises 69–72.

69. After observing the binary forms of the numbers 1–31, identify a common property of all Table 9 numbers in each of the following columns.

(a) Column A
(b) Column B
(c) Column C
(d) Column D
(e) Column E

70. Explain how the "trick" of Table 9 works.

71. How many columns would be needed for Table 9 to include all ages up to 63?

72. How many columns would be needed for Table 9 to include all numbers up to 127?

In our decimal system, we distinguish odd and even numbers by looking at their ones (or units) digits. If the ones digit is even (0, 2, 4, 6, *or* 8), *the number is even. If the ones digit is odd* (1, 3, 5, 7, *or* 9), *the number is odd. For Exercises 73–80, determine whether this same criterion works for numbers expressed in the given bases.*

73. two **74.** three **75.** four **76.** five

77. six **78.** seven **79.** eight **80.** nine

81. Consider all even bases. If the above criterion works for all, explain why. If not, find a criterion that does work for all even bases.

82. Consider all odd bases. If the above criterion works for all, explain why. If not, find a criterion that does work for all odd bases.

Determine whether the given base five numeral represents one that is divisible by five.

83. 3204_{five} **84.** 200_{five} **85.** 2310_{five} **86.** 342_{five}

Recall that conversions between binary and octal are simplified because eight is a power of 2: $8 = 2^3$. *(See Examples 12 and 13.) The same is true of conversions between binary and hexadecimal, because* $16 = 2^4$. *(See Example 14.)* Direct *conversion between octal and hexadecimal does not work the same way, because* 16 *is* not *a power of* 8. *Explain how to carry out each conversion without using base ten, and give an example.*

87. hexadecimal to octal **88.** octal to hexadecimal

Devise a method (similar to the one for conversions between binary, octal, and hexadecimal) for converting between base three and base nine, and use it to carry out each conversion.

89. 6504_{nine} to base three **90.** 81170_{nine} to base three

91. 212201221_{three} to base nine **92.** 200121021_{three} to base nine

4.4 Clock Arithmetic and Modular Systems

Finite Systems and Clock Arithmetic • Modular Systems

Finite Systems and Clock Arithmetic

At the beginning of this chapter we described a "mathematical system" as

1. a set of elements along with
2. one or more operations for combining those elements, and
3. one or more relations for comparing those elements.

FIGURE 5

The numeration systems studied in the first three sections mainly involved the set of whole numbers. The operations were mostly addition and multiplication, and the relation was that of equality. Because the set of whole numbers is infinite, that system is an **infinite mathematical system.** In this section, we consider some **finite mathematical systems,** based on finite sets.

The **12-hour clock system** is based on an ordinary clock face, except that 12 is replaced by 0 so that the finite set of the system is {0, 1, 2, 3, 4, 5, 6, 7, 8, 9, 10, 11}. (We will need just one hand on our clock.) See Figure 5.

As an operation for this clock system, addition is defined as follows: add by moving the hour hand in a *clockwise* direction. For example, to add 5 and 2 on a clock, first move the hand to 5, as in Figure 6. Then, to add 2, move the hand 2 more hours in a clockwise direction. The hand stops at 7, so

$$5 + 2 = 7.$$

FIGURE 6

This result agrees with traditional addition. However, the sum of two numbers from the 12-hour clock system is not always what might be expected, as the following example shows.

EXAMPLE 1 Finding Clock Sums by Hand Rotations

Find each sum in 12-hour clock arithmetic.

(a) $8 + 9$ **(b)** $11 + 3$

SOLUTION

(a) Move the hand to 8, as in Figure 7. Then advance the hand clockwise through 9 more hours. It stops at 5, so $8 + 9 = 5$.

(b) To find $11 + 3$, proceed as shown in Figure 8. Check that $11 + 3 = 2$.

FIGURE 7

Because there are infinitely many whole numbers, it is not possible to write a complete table of addition facts for that set. Such a table, to show the sum of every possible pair of whole numbers, would have infinite numbers of rows and columns, making it impossible to construct.

On the other hand, the 12-hour clock system uses only the whole numbers 0, 1, 2, 3, 4, 5, 6, 7, 8, 9, 10, and 11. In effect, the clock face serves to "reduce" the infinite set of whole numbers to the finite set {0, 1, 2, 3, 4, 5, 6, 7, 8, 9, 10, 11}. No matter how large a whole number results from additions (clockwise motions around the clock), the result is always equivalent, in this system, to one of the numbers 0 through 11. A table of all possible sums for this system requires only 12 rows and 12 columns. The 12-hour clock **addition table** is shown in Table 10 on the following page. The significance of the colored diagonal line will be discussed later.

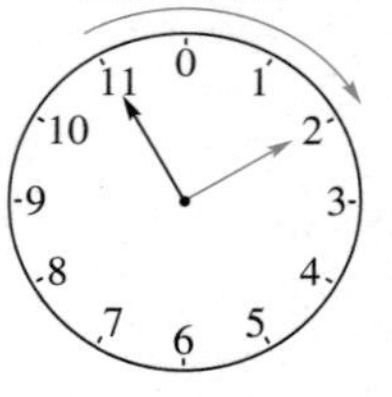

FIGURE 8

EXAMPLE 2 Finding Clock Sums by Addition Table

Use the 12-hour clock addition table to find each sum.

(a) $7 + 11$ **(b)** $11 + 1$

SOLUTION

(a) Rather than following rotations around the clock face, we simply refer to the table. Find 7 on the left of the addition table and 11 across the top. The intersection of the row headed 7 and the column headed 11 gives the number 6. Thus, $7 + 11 = 6$.

(b) Also from the table, $11 + 1 = 0$.

Mathematical systems are characterized by the properties they possess, specifically, the properties of their operations and relations. Five properties that many of the most commonly applied systems have are the *closure, commutative, associative, identity,* and *inverse* properties. We can check whether the 12-hour clock system has these properties.

Plagued by serious maritime mishaps linked to navigational difficulties, several European governments offered prizes for an effective method of determining longitude. The largest prize was 20,000 pounds (equivalent to several million dollars in today's currency) offered by the British Parliament in the Longitude Act of 1714. While famed scientists, academics, and politicians pursued an answer in the stars, **John Harrison,** a clock maker, set about to build a clock that could maintain accuracy at sea. This turned out to be the key, and Harrison's **Chronometer** eventually earned him the prize.

For a fascinating account of this drama and of Harrison's struggle to collect his prize money from the government, see the book *The Illustrated Longitude* by Dava Sobel and William J. H. Andrewes.

TABLE 10 12-Hour Clock Addition

+	0	1	2	3	4	5	6	7	8	9	10	11
0	0	1	2	3	4	5	6	7	8	9	10	11
1	1	2	3	4	5	6	7	8	9	10	11	0
2	2	3	4	5	6	7	8	9	10	11	0	1
3	3	4	5	6	7	8	9	10	11	0	1	2
4	4	5	6	7	8	9	10	11	0	1	2	3
5	5	6	7	8	9	10	11	0	1	2	3	4
6	6	7	8	9	10	11	0	1	2	3	4	5
7	7	8	9	10	11	0	1	2	3	4	5	6
8	8	9	10	11	0	1	2	3	4	5	6	7
9	9	10	11	0	1	2	3	4	5	6	7	8
10	10	11	0	1	2	3	4	5	6	7	8	9
11	11	0	1	2	3	4	5	6	7	8	9	10

Table 10 shows that the sum of two numbers on a clock face is always a number on the clock face. That is, if a and b are any clock numbers in the set of the system, then $a + b$ is also in the set of the system. Therefore, the system has the **closure property.** (The set of the system is *closed* under clock addition.)

Notice also that in this system $9 + 6$ and $6 + 9$ both yield 3. And the answers, the two 3s, are located in positions that are mirror images of one another with respect to the diagonal line shown. Observe another case—the results for $5 + 8$ and $8 + 5$ are also located symmetrically with respect to the diagonal line, and both results are 1. The entries throughout the entire table occur in equal, diagonally symmetric, pairs. This means that for any clock numbers a and b, $a + b = b + a$. The system, therefore, has the **commutative property.**

The next question is: When any three elements are combined in a given order, say $a + b + c$, does it matter whether the first and second or the second and third are associated initially? In other words, is it true that, for any elements a, b, and c in the 12-hour clock system, $(a + b) + c = a + (b + c)$?

EXAMPLE 3 Checking the Associative Property for Clock Addition

Is 12-hour clock addition associative?

SOLUTION

It would take lots of work to prove that the required relationship *always* holds. But a few examples should either disprove it (by revealing a *counterexample*—a case where it fails to hold), or should make it at least plausible. Using the clock numbers 4, 5, and 9, we see that

$$(4 + 5) + 9 = 9 + 9 = 6 \qquad 4 + (5 + 9) = 4 + 2 = 6.$$

Thus, $(4 + 5) + 9 = 4 + (5 + 9)$. Try another example:

$$(7 + 6) + 3 = 1 + 3 = 4 \qquad 7 + (6 + 3) = 7 + 9 = 4.$$

So $(7 + 6) + 3 = 7 + (6 + 3)$. Any other examples checked also will work. The 12-hour clock system therefore has the **associative property.**

Our next question is whether the clock face contains some element (number) that, when combined with any element (in either order), produces that same element. Such an element (call it e) would satisfy $a + e = a$ and $e + a = a$ for any element a of the system. Notice in Table 10 that $4 + 0 = 0 + 4 = 4$, $6 + 0 = 0 + 6 = 6$, and so on. The number 0 is the required *identity element.* The system has the **identity property.**

Generally, if a finite system has an identity element e, it can be located easily in the operation table. Check the body of Table 10 for a column that is identical to the column at the left side of the table. Because the column under 0 meets this requirement, $a + 0 = a$ holds for all elements a in the system. Thus, 0 is *possibly* the identity. Now locate 0 at the left of the table. Because the corresponding row is identical to the row at the top of the table, $0 + a = a$ also holds for all elements a, which is the other requirement of an identity element. Hence, 0 is *indeed* the identity.

FIGURE 9

Our 12-hour clock system can be expanded to include operations besides addition. For example, subtraction can be performed on a 12-hour clock. Subtraction may be interpreted on the clock face by a movement in the *counterclockwise* direction. For example, to perform the subtraction $2 - 5$, begin at 2 and move 5 hours counterclockwise, ending at 9, as shown in Figure 9. Therefore, in this system,

$$2 - 5 = 9.$$

In our usual system, subtraction may be checked by addition, and this is also the case in clock arithmetic. To check that $2 - 5 = 9$, simply add $9 + 5$, either by using rotation on the clock face or consulting the addition table. In either case, the result is 2, verifying the accuracy of this subtraction.

The *additive inverse,* $-a$, of an element a in clock arithmetic is the element that satisfies this statement: $a + (-a) = 0$ and $(-a) + a = 0$. Such an element, if it exists, can be determined either on the clock face or by using the addition table, as shown in the next example.

EXAMPLE 4 Finding Additive Inverses in Clock Arithmetic

Determine the additive inverse, if it exists, for each number in 12-hour clock arithmetic.

(a) 8 **(b)** 2

SOLUTION

(a) Use the clock face to solve the equation

$$8 + x = 0.$$

$$x = 4 \qquad \text{It is 4 hours from 8 to 0.}$$

The additive inverse of 8 is 4.

A **chess clock** or double clock is used to time chess, backgammon, and Scrabble games. Push one button, and that clock stops—the other begins simultaneously. When a player's allotted time for the game has expired, that player will lose if he or she has not made the required number of moves.

Mathematics and chess both involve structured relationships and demand logical thinking. Emanuel Lasker achieved mastery in both fields. He was best known as a World Chess Champion for 27 years, until 1921. Lasker also was famous in mathematical circles for his work concerning the theory of primary ideals, algebraic analogies of prime numbers. An important result, the Lasker-Noether theorem, bears his name along with that of Emmy Noether. Noether extended Lasker's work. Her father had been Lasker's Ph.D. advisor.

(b) Refer to the addition table to solve the equation

$$2 + x = 0.$$

$$x = 10$$ The row headed 2 has 0 in the column headed 10.

The additive inverse of 2 is 10.

The methods used in Example 4 may be used to verify that *every* element of the system has an additive inverse (also in the system). So the system has the **inverse property.**

A simpler way to verify the inverse property, if you have the table, is to make sure the identity element appears exactly once in each row, and that the pair of elements that produces it also produces it in the opposite order. (This last condition is automatically true if the commutative property holds for the system.) For example, note in Table 10 that row 3 contains one 0, under the 9, so $3 + 9 = 0$, and that row 9 contains 0, under the 3, so $9 + 3 = 0$ also. Therefore, 3 and 9 are inverses.

Table 11 lists all the elements and their additive inverses. Notice that one element, 6, is its own inverse for addition.

TABLE 11 Inverses for 12-Hour Clock Addition

Clock value a	0	1	2	3	4	5	6	7	8	9	10	11
Additive inverse $-a$	0	11	10	9	8	7	6	5	4	3	2	1

Using the additive inverse symbol, we can say that in clock arithmetic,

$$-5 = 7, \quad -11 = 1, \quad -10 = 2, \quad \text{and so on.}$$

We have now seen that the 12-hour clock system, with addition, has all five properties that we set out to check: closure, commutative, associative, identity, and inverse. Having discussed additive inverses, we can define subtraction formally. Notice that the definition is the same as for ordinary subtraction of whole numbers.

Subtraction on a Clock

If a and b are elements in clock arithmetic, then the **difference, $a - b$,** is defined as

$$a - b = a + (-b).$$

EXAMPLE 5 Finding Clock Differences

Find each difference.

(a) $8 - 5$ **(b)** $6 - 11$

SOLUTION

(a) $8 - 5 = 8 + (-5)$ Use the definition of subtraction.

$= 8 + 7$ The additive inverse of 5 is 7, from the table of inverses.

$= 3$

This result agrees with traditional arithmetic. Check by adding 5 and 3; the sum is 8.

(b) $6 - 11 = 6 + (-11)$
$= 6 + 1$ The additive inverse of 11 is 1.
$= 7$

Clock numbers can also be multiplied. For example,

$$5 \cdot 4 = 4 + 4 + 4 + 4 + 4 = 8. \quad \text{Add five 4s.}$$

EXAMPLE 6 Finding Clock Products

Find each product, using clock arithmetic.

(a) $6 \cdot 9$ **(b)** $3 \cdot 4$ **(c)** $6 \cdot 0$ **(d)** $0 \cdot 8$

SOLUTION

(a) $6 \cdot 9 = 9 + 9 + 9 + 9 + 9 + 9 = 6$ **(b)** $3 \cdot 4 = 4 + 4 + 4 = 0$
(c) $6 \cdot 0 = 0 + 0 + 0 + 0 + 0 + 0 = 0$ **(d)** $0 \cdot 8 = 0$

Some properties of the system of 12-hour clock numbers with the operation of multiplication will be investigated in Exercises 6–8.

Modular Systems We now expand the ideas of clock arithmetic to **modular systems** in general. Recall that 12-hour clock arithmetic was set up so that answers were always whole numbers less than 12. For example, $8 + 6 = 2$. The traditional sum, $8 + 6 = 14$, reflects the fact that moving the clock hand forward 8 hours from 0, and then forward another 6 hours, amounts to moving it forward 14 hours total. But because the final position of the clock is at 2, we see that 14 and 2 are, in a sense, equivalent. More formally, we say that 14 and 2 are **congruent modulo** 12 (or **congruent mod** 12), which is written

$$14 \equiv 2 \pmod{12} \quad \text{The sign } \equiv \text{ indicates congruence.}$$

By observing clock hand movements, you can also see that, for example,

$$26 \equiv 2 \pmod{12}, \qquad 38 \equiv 2 \pmod{12}, \qquad \text{and so on.}$$

In each case, the congruence is true because the difference of the two congruent numbers is a multiple of 12:

$$14 - 2 = 12 = 1 \cdot 12, \quad 26 - 2 = 24 = 2 \cdot 12, \quad 38 - 2 = 36 = 3 \cdot 12.$$

This suggests the following definition.

Congruence Modulo *m*

The integers a and b are **congruent modulo *m*** (where m is a natural number greater than 1 called the **modulus**) if and only if the difference $a - b$ is divisible by m. Symbolically, this congruence is written

$$a \equiv b \pmod{m}.$$

Because being divisible by m is the same as being a multiple of m, we can say that

$$a \equiv b \pmod{m} \text{ if and only if } a - b = km \text{ for some integer } k.$$

EXAMPLE 7 Checking the Truth of Modular Equations

Decide whether each statement is *true* or *false*.

(a) $16 \equiv 10 \pmod{2}$ **(b)** $49 \equiv 32 \pmod{5}$ **(c)** $30 \equiv 345 \pmod{7}$

SOLUTION

(a) The difference $16 - 10 = 6$ is divisible by 2, so $16 \equiv 10 \pmod{2}$ is true.

(b) The statement $49 \equiv 32 \pmod{5}$ is false, because $49 - 32 = 17$, which is not divisible by 5.

(c) The statement $30 \equiv 345 \pmod{7}$ is true, because $30 - 345 = -315$ is divisible by 7. (It doesn't matter if we find $30 - 345$ or $345 - 30$.)

There is another method of determining if two numbers, a and b, are congruent modulo m.

Criterion for Congruence

$a \equiv b \pmod{m}$ if and only if the same remainder is obtained when a and b are divided by m.

For example, we know that $27 \equiv 9 \pmod{6}$ because $27 - 9 = 18$, which is divisible by 6. Now, if 27 is divided by 6, the quotient is 4 and the remainder is 3. Also, if 9 is divided by 6, the quotient is 1 and the remainder is 3. According to the criterion above, $27 \equiv 9 \pmod{6}$ since both remainders are the same.

Addition, subtraction, and multiplication can be performed in any modular system just as with clock numbers. Because final answers should be whole numbers less than the modulus, we can first find an answer using ordinary arithmetic. Then, as long as the answer is nonnegative, simply divide it by the modulus and keep the remainder. This produces the smallest nonnegative integer that is congruent (modulo m) to the ordinary answer.

EXAMPLE 8 Performing Modular Arithmetic

Find each sum, difference, or product.

(a) $(9 + 14) \pmod{3}$ **(b)** $(27 - 5) \pmod{6}$ **(c)** $(50 + 34) \pmod{7}$
(d) $(8 \cdot 9) \pmod{10}$ **(e)** $(12 \cdot 10) \pmod{5}$

SOLUTION

(a) First add 9 and 14 to get 23. Then divide 23 by 3. The remainder is 2, so we obtain $23 \equiv 2 \pmod{3}$ and

$$(9 + 14) \equiv 2 \pmod{3}.$$

(b) $27 - 5 = 22$. Divide 22 by 6, obtaining 4 as a remainder:

$$(27 - 5) \equiv 4 \pmod{6}.$$

(c) $50 + 34 = 84$. When 84 is divided by 7, a remainder of 0 is found:

$$(50 + 34) \equiv 0 \pmod{7}.$$

(d) Since $8 \cdot 9 = 72$, and 72 leaves a remainder of 2 when divided by 10,

$$(8 \cdot 9) \equiv 2 \pmod{10}.$$

(e) $(12 \cdot 10) \pmod{5} = 120 \equiv 0 \pmod{5}$

PROBLEM-SOLVING HINT Modular systems can often be applied to questions involving cyclical changes. For example, our method of dividing time into weeks causes the days to repeatedly cycle through the same pattern of seven. Suppose today is Sunday and we want to know what day of the week it will be 45 days from now. Because we don't care how many weeks will pass between now and then, we can discard the largest whole number of weeks in 45 days and keep the remainder. (We are finding the smallest nonnegative integer that is congruent to 45 modulo 7.) Dividing 45 by 7 leaves remainder 3, so the desired day of the week is 3 days past Sunday, or *Wednesday.*

EXAMPLE 9 Using Modular Methods to Find the Day of the Week

If today is Thursday, November 12, and *next* year is a leap year, what day of the week will it be one year from today?

SOLUTION

A modulo 7 system applies here, but we need to know the number of days between today and one year from today. Today's date, November 12, is unimportant except that it shows we are later in the year than the end of February and therefore the next year (starting today) will contain 366 days. (This would not be so if today were, say, January 12.) Now dividing 366 by 7 produces 52 with remainder 2. Two days past Thursday is our answer. That is, one year from today will be a Saturday.

PROBLEM-SOLVING HINT A modular system (mod m) allows only a fixed set of remainder values, $0, 1, 2, \ldots, m - 1$. One practical approach to solving modular equations, at least when m is reasonably small, is to simply try all these integers. For each solution found in this way, others can be found by adding multiples of the modulus to it.

EXAMPLE 10 Solving Modular Equations

Solve each modular equation for whole number solutions.

(a) $(3 + x) \equiv 5 \pmod 7$ **(b)** $5x \equiv 4 \pmod 9$
(c) $6x \equiv 3 \pmod 8$ **(d)** $8x \equiv 8 \pmod 8$

SOLUTION

(a) Because dividing 5 by 7 yields remainder 5, the criterion for congruence is that the given equation is true only if dividing $3 + x$ by 7 also yields remainder 5. Try replacing x, in turn, by 0, 1, 2, 3, 4, 5, and 6.

$x = 0$: $(3 + 0) \equiv 5 \pmod 7$ is false. The remainder is 3.

$x = 1$: $(3 + 1) \equiv 5 \pmod 7$ is false. The remainder is 4.

$x = 2$: $(3 + 2) \equiv 5 \pmod 7$ is true. The remainder is 5.

Try $x = 3$, $x = 4$, $x = 5$, and $x = 6$ to see that none work. Of the integers from 0 through 6, only 2 is a solution of the equation $(3 + x) \equiv 5 \pmod 7$.

Because 2 is a solution, find other solutions to this mod 7 equation by repeatedly adding 7:

$$2 + 7 = 9, \quad 9 + 7 = 16, \quad 16 + 7 = 23, \quad \text{and so on.}$$

The set of all nonnegative solutions of $(3 + x) \equiv 5 \pmod 7$ is

$$\{2, 9, 16, 23, 30, 37, \ldots\}.$$

(b) Dividing 4 by 9 yields remainder 4. Because the modulus is 9, check the remainders when $5x$ is divided by 9 for $x = 0, 1, 2, 3, 4, 5, 6, 7,$ and 8.

$x = 0$: $5 \cdot 0 \equiv 4 \pmod 9$ is false. The remainder is 0.

$x = 1$: $5 \cdot 1 \equiv 4 \pmod 9$ is false. The remainder is 5.

Continue trying numbers. Only $x = 8$ works:

$5 \cdot 8 = 40 \equiv 4 \pmod 9$. The remainder is 4.

The set of all nonnegative solutions to the equation $5x \equiv 4 \pmod 9$ is

$$\{8, 8 + 9, 8 + 9 + 9, 8 + 9 + 9 + 9, \ldots\} \quad \text{or} \quad \{8, 17, 26, 35, 44, 53, \ldots\}.$$

(c) To solve $6x \equiv 3 \pmod 8$, try the numbers 0, 1, 2, 3, 4, 5, 6, and 7. None work. Therefore, the equation $6x \equiv 3 \pmod 8$ has no solutions. Write the set of all solutions as the empty set, $\emptyset$.

This result is reasonable because $6x$ will always be even, no matter which whole number is used for x. Because $6x$ is even and 3 is odd, the difference $6x - 3$ will be odd and therefore not divisible by 8.

(d) To solve $8x \equiv 8 \pmod 8$, trying the integers 0, 1, 2, 3, 4, 5, 6, and 7. *Any* replacement will work. The solution set is $\{0, 1, 2, 3, \ldots\}$.

Some problems can be solved by writing down two or more modular equations and finding their common solutions. The next example illustrates the process.

EXAMPLE 11 Finding the Number of Discs in a CD Collection

Julio wants to arrange his CD collection in equal size stacks, but after trying stacks of 4, stacks of 5, and stacks of 6, he finds that there is always 1 disc left over. Assuming Julio owns more than one CD, what is the least possible number of discs in his collection?

SOLUTION

The given information leads to three modular equations,

$$x \equiv 1 \pmod 4), \quad x \equiv 1 \pmod 5), \quad \text{and} \quad x \equiv 1 \pmod 6).$$

For the first equation, try $x = 0$, $x = 1$, $x = 2$, and $x = 3$. The value 1 works, as it does for the other two equations as well. So the solution sets are, respectively,

$$\{1, 5, 9, 13, 17, 21, 25, 29, 33, 37, 41, 45, 49, 53, 57, \mathbf{61}, 65, 69. \ldots\},$$
$$\{1, 6, 11, 16, 21, 26, 31, 36, 41, 46, 51, 56, \mathbf{61}, 66, 71, 76, \ldots\},$$
and
$$\{1, 7, 13, 19, 25, 31, 37, 43, 49, 55, \mathbf{61}, \ldots\}.$$

The least common solution greater than 1 is 61, so the least possible number of discs in the collection is 61. ■

EXAMPLE 12 Applying Congruences to a Construction Problem

A dry-wall contractor is ordering materials to finish a 17-foot-by-45-foot room. The wallboard panels come in 4-foot widths. Show that, after uncut panels are applied, all four walls will require additional partial strips of the same width.

SOLUTION

The width of any partial strip needed will be the remainder when the wall length is divided by 4 (the panel width). In terms of congruence, we must show that $17 \equiv 45 \pmod 4$. By the criterion for congruence, we see that this is true because both 17 and 45 give the same remainder (namely 1) when divided by 4. A 1-foot partial strip will be required for each wall. (In this case four 1-foot strips can be cut from a single panel, so there will be no waste.) ■

For Further Thought

A Card Trick

Many card "tricks" that have been around for years are really not illusions at all but are based on mathematical properties that allow anyone to do them with no special conjuring abilities. One of them is based on mod 14 arithmetic.

In this trick, suits play no role. Each card has a numerical value: 1 for ace, 2 for two, . . . , 11 for jack, 12 for queen, and 13 for king. The deck is shuffled and given to a spectator, who is instructed to place the deck of cards face up on a table and is told to follow the procedure described: The top card is removed from the deck and laid on the table with its face up. (We shall call it the "starter" card.) The starter card will be at the bottom of a pile. In order to form a pile, note

(continued)

the value of the starter card, and then add cards on top of it while counting up to 13. For example, if the starter card is a six, pile up seven cards on top of it. If it is a jack, add two cards to it, and so on.

When the first pile is completed, it is picked up and placed face down. The next card from the deck becomes the starter card for the next pile, and the process is repeated. This continues until all cards are used or until there are not enough cards to complete the last pile. Any cards that are left over are put aside, face down, for later use. We will refer to these as "leftovers."

The performer then requests that a spectator choose three piles at random. The remaining piles are added to the leftovers. The spectator is then instructed to turn over any two top cards from the piles. The performer is then able to determine the value of the third top card.

The secret to the trick is that the performer adds the values of the two top cards that were turned over, and then adds 10 to this sum. The performer then counts off this number of cards from the leftovers. The number of cards remaining in the leftovers is the value of the remaining top card!

For Group Discussion or Individual Investigation

1. Obtain a deck of playing cards and perform the "trick" as described above. (As with many activities, you'll find that doing it is simpler than describing it.) Does it work?
2. Explain why this procedure works. (If you want to see how someone else explained it, using modulo 14 arithmetic, see "An Old Card Trick Revisited," by Barry C. Felps, in the December 1976 issue of the journal *The Mathematics Teacher*.)

4.4 Exercises

Find each difference on the 12-hour clock.

1. $8 - 3$ **2.** $4 - 9$ **3.** $2 - 8$ **4.** $0 - 3$

5. Complete the 12-hour clock multiplication table below. You can use repeated addition and the addition table (for example, $3 \cdot 7 = 7 + 7 + 7 = 2 + 7 = 9$) or use mod 12 multiplication techniques, as in Example 8, parts (d) and (e).

·	0	1	2	3	4	5	6	7	8	9	10	11
0	0	0	0	0	0	0	0	0	0	0	0	0
1	0	1	2	3	4	5	6	7	8	9	10	11
2	0	2	4	6	8	10		2	4		8	
3	0	3	6	9	0	3	6			3	6	
4	0	4	8			8		4			4	8
5	0	5	10	3	8		6	11	4			
6	0	6	0		0	6	0	6		6		6
7	0	7	2	9				1			10	
8	0	8	4	0				8	4		8	4
9	0	9			0		6		0			
10	0	10	8			2						2
11	0	11										1

By referring to your table in Exercise 5, determine which properties hold for the system of 12-hour clock numbers with the operation of multiplication.

6. closure

7. commutative

8. identity (If so, what is the identity element?)

A 5-hour clock system utilizes the set {0, 1, 2, 3, 4}, and relates to the clock face shown here.

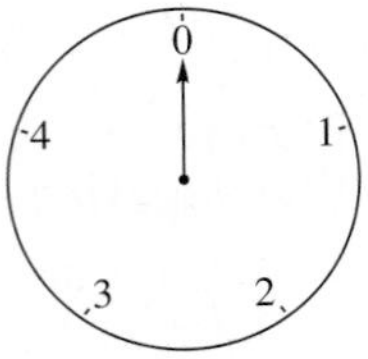

5-hour clock

9. Complete this 5-hour clock addition table.

+	0	1	2	3	4
0	0	1	2	3	4
1	1	2	3	4	
2	2	3	4		1
3	3	4			
4	4				3

Which properties are satisfied by the system of 5-hour clock numbers with the operation of addition?

10. closure

11. commutative

12. identity (If so, what is the identity element?)

13. inverse (If so, name the inverse of each element.)

14. Complete this 5-hour clock multiplication table.

·	0	1	2	3	4
0	0	0	0	0	0
1	0	1	2	3	4
2	0	2	4		
3	0	3			
4	0	4			

Determine which properties hold for the system of 5-hour clock numbers with the operation of multiplication.

15. closure

16. commutative

17. identity (If so, what is the identity element?)

In clock arithmetic, as in ordinary arithmetic, $a - b = d$ is true if and only if $b + d = a$. Similarly, $a \div b = q$ if and only if $b \cdot q = a$.

Use the idea above and your 5-hour clock multiplication table of Exercise 14 to find each quotient on a 5-hour clock.

18. $1 \div 3$

19. $3 \div 1$

20. $2 \div 3$

21. $3 \div 2$

22. Is division commutative on a 5-hour clock? Explain.

23. Is there an answer for $4 \div 0$ on a 5-hour clock? Find it or explain why not.

The military uses a 24-hour clock to avoid the problems of "A.M." and "P.M." For example, 1100 hours is 11 A.M., while 2100 hours is 9 P.M. (12 noon + 9 hours). In these designations, the last two digits represent minutes, and the digits before that represent hours. Find each sum in the 24-hour clock system.

24. 1400 + 500

25. 1300 + 1800

26. 0750 + 1630

27. 1545 + 0815

28. Explain how the following three statements can *all* be true. (*Hint:* Think of clocks.)

$$1145 + 1135 = 2280$$
$$1145 + 1135 = 1120$$
$$1145 + 1135 = 2320$$

Answer true *or* false *for each statement.*

29. $5 \equiv 19 \pmod 3$

30. $35 \equiv 8 \pmod 9$

31. $5445 \equiv 0 \pmod 3$

32. $7021 \equiv 4202 \pmod 6$

Work each modular arithmetic problem.

33. $(12 + 7)(\text{mod } 4)$

34. $(62 + 95)(\text{mod } 9)$

35. $(35 - 22)(\text{mod } 5)$

36. $(82 - 45)(\text{mod } 3)$

37. $(5 \cdot 8)(\text{mod } 3)$

38. $(32 \cdot 21)(\text{mod } 8)$

39. $[4 \cdot (13 + 6)](\text{mod } 11)$

40. $[(10 + 7) \cdot (5 + 3)](\text{mod } 10)$

41. The text described how to do arithmetic mod m when the ordinary answer comes out nonnegative. Explain what to do when the ordinary answer is negative.

Work each modular arithmetic problem.

42. $(3 - 27)(\text{mod } 5)$

43. $(16 - 60)(\text{mod } 7)$

44. $[(-8) \cdot 11] \pmod 3$

45. $[2 \cdot (-23)] \pmod 5$

In Exercises 46 and 47:

(a) *Complete the given addition table.*

(b) *Decide whether the closure, commutative, identity, and inverse properties are satisfied.*

(c) *If the inverse property is satisfied, give the inverse of each number.*

46. mod 4

+	0	1	2	3
0	0	1	2	3
1				
2				
3				

47. mod 7

+	0	1	2	3	4	5	6
0	0	1	2	3	4	5	6
1	1	2	3	4	5	6	
2							
3							
4							
5							
6							

In Exercises 48–51:

(a) *Complete the given multiplication table.*

(b) *Decide whether the closure, commutative, identity, and inverse properties are satisfied.*

(c) *Give the inverse of each nonzero number that has an inverse.*

48. mod 2

·	0	1
0	0	0
1	0	

49. mod 3

·	0	1	2
0	0	0	0
1	0	1	2
2	0	2	

50. mod 4

·	0	1	2	3
0	0	0	0	0
1	0	1	2	3
2	0	2		
3	0	3		

51. mod 9

·	0	1	2	3	4	5	6	7	8
0	0	0	0	0	0	0	0	0	0
1	0	1	2	3	4	5	6	7	8
2	0	2	4	6	8			5	
3	0	3	6	0		6		3	6
4	0	4	8		7		6		5
5	0	5	1		2		3	8	
6	0	6	3	0	6	3	0	6	3
7	0	7	5			8			2
8	0	8	7			4	3		1

52. Explain why a modular system containing the number 0 cannot satisfy the inverse property for multiplication.

Find all nonnegative solutions for each equation.

53. $x \equiv 3 \pmod 7$

54. $(2 + x) \equiv 7 \pmod 3$

55. $6x \equiv 2 \pmod 2$

56. $(5x - 3) \equiv 7 \pmod 4$

Solve each problem.

57. ***Odometer Readings*** For many years automobile odometers showed five whole number digits and a digit for tenths of a mile. For those odometers showing just five whole number digits, totals are recorded according to what modulus?

58. ***Distance Traveled by a Car*** If a car's five-digit whole number odometer shows a reading of 29,306, *in theory* how many miles might the car have traveled?

59. ***Determining Day of the Week*** Refer to Example 9 in the text. (Recall that *next* year is a leap year.) Assuming today was Thursday, January 12, answer the following questions.

(a) How many days would the next year (starting today) contain?

(b) What day of the week would occur one year from today?

60. ***Silver Spoon Collection*** Roxanna Parker has a collection of silver spoons from all over the world. She finds that she can arrange her spoons in sets of 7 with 6 left over, sets of 8 with 1 left over, or sets of 15 with 3 left over. If Roxanna has fewer than 200 spoons, how many are there?

61. ***Piles of Ticket Stubs*** Lawrence Rosenthal finds that whether he sorts his White Sox ticket stubs into piles of 10, piles of 15, or piles of 20, there are always 2 left over. What is the least number of stubs he could have (assuming he has more than 2)?

62. ***Determining a Range of Dates*** Assume again, as in Example 9, that *next* year is a leap year. If the next year (starting today) does *not* contain 366 days, what is the range of possible dates for today?

63. ***Flight Attendant Schedules*** Robin Strang and Kristyn Wasag, flight attendants for two different airlines, are close friends and like to get together as often as possible. Robin flies a 21-day schedule (including days off), which then repeats, while Kristyn has a repeating 30-day schedule. Both of their routines include layovers in Chicago, New Orleans, and San Francisco. The table below shows which days of each of their individual schedules they are in these cities. (Assume the first day of a cycle is day number 1.)

	Days in Chicago	Days in New Orleans	Days in San Francisco
Robin	1, 2, 8	5, 12	6, 18, 19
Kristyn	23, 29, 30	5, 6, 17	8, 10, 15, 20, 25

If today is July 1 and both are starting their schedules today (day 1), list the days during July and August that they will be able to see each other in each of the three cities.

The following formula can be used to find the day of the week on which a given year begins. Here y represents the year (which must be after* 1582, *when our current calendar began). First calculate*

$$a = y + [\![(y - 1)/4]\!] - [\![(y - 1)/100]\!] + [\![(y - 1)/400]\!],$$

where $[\![x]\!]$ *represents the greatest integer less than or equal to x. (For example,* $[\![9.2]\!] = 9$, *and* $[\![\pi]\!] = 3$.*) After finding a, find the smallest nonnegative integer b such that*

$$a \equiv b \pmod 7.$$

Then b gives the day of January 1, with $b = 0$ *representing Sunday,* $b = 1$ *Monday, and so on.*

Find the day of the week on which January 1 would occur in each year.

64. 1812 **65.** 1865

66. 2006 **67.** 2020

Some people believe that Friday the thirteenth is unlucky. The table below shows the months that will have a Friday the thirteenth if the first day of the year is known. A year is a leap year if it is divisible by* 4. *The only exception to this rule is that a century year* (1900, *for example) is a leap year only when it is divisible by* 400.

First Day of Year	Non-leap Year	Leap Year
Sunday	Jan., Oct.	Jan., April, July
Monday	April, July	Sept., Dec.
Tuesday	Sept., Dec.	June
Wednesday	June	March, Nov.
Thursday	Feb., March, Nov.	Feb., Aug.
Friday	August	May
Saturday	May	Oct.

Use the table to determine the months that have a Friday the thirteenth for each year.

68. 2007 **69.** 2008

70. 2009 **71.** 2200

72. Modular arithmetic can be used to create **residue designs.** For example, the designs (11, 3) and (65, 3) are shown here.

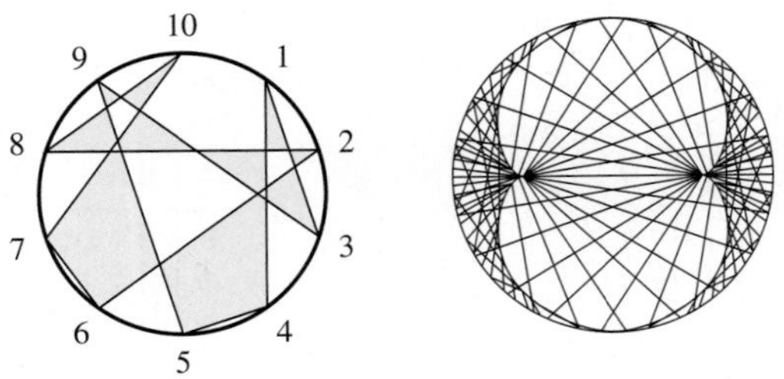

(11, 3) (65, 3)

To see how such designs are created, construct a new design, (11, 5), by proceeding as follows.

(a) Draw a circle and divide the circumference into 10 equal parts. Label the division points as 1, 2, 3, . . . , 10.

*Given in "An Aid to the Superstitious," by G. L. Ritter, S. R. Lowry, H. B. Woodruff, and T. L. Isenhour. *The Mathematics Teacher,* May 1977, pp. 456–457.

(b) Since $1 \cdot 5 \equiv 5 \pmod{11}$, connect 1 and 5. (We use 5 as a multiplier because we are making an (11, 5) design.)

(c) $2 \cdot 5 \equiv 10 \pmod{11}$
Therefore, connect 2 and _____.

(d) $3 \cdot 5 \equiv$ _____ (mod 11)
Connect 3 and _____.

(e) $4 \cdot 5 \equiv$ _____ (mod 11)
Connect 4 and _____.

(f) $5 \cdot 5 \equiv$ _____ (mod 11)
Connect 5 and _____.

(g) $6 \cdot 5 \equiv$ _____ (mod 11)
Connect 6 and _____.

(h) $7 \cdot 5 \equiv$ _____ (mod 11)
Connect 7 and _____.

(i) $8 \cdot 5 \equiv$ _____ (mod 11)
Connect 8 and _____.

(j) $9 \cdot 5 \equiv$ _____ (mod 11)
Connect 9 and _____.

(k) $10 \cdot 5 \equiv$ _____ (mod 11)
Connect 10 and _____.

(l) You might want to shade some of the regions you have found to make an interesting pattern. For more information, see "Residue Designs," by Phil Locke in *The Mathematics Teacher,* March 1972, pages 260–263.

Identification numbers are used in various ways for many kinds of different products. Books, for example, are assigned International Standard Book Numbers (ISBNs). Each ISBN is a ten-digit number. It includes a check digit, which is determined on the basis of modular arithmetic. The ISBN for one version of this book is*

0-321-36146-6.

The first digit, 0, identifies the book as being published in an English-speaking country. The next digits, 321, identify the publisher, while 36146 identifies this particular book. The final digit, 6, is a check digit. To find this check digit, start at the left and multiply the digits of the ISBN by 10, 9, 8, 7, 6, 5, 4, 3, and 2, respectively. Then add these products. For this book we get

$$(10 \cdot 0) + (9 \cdot 3) + (8 \cdot 2) + (7 \cdot 1) + (6 \cdot 3) + (5 \cdot 6) + (4 \cdot 1) + (3 \cdot 4) + (2 \cdot 6) = 126.$$

The check digit is the smallest number that must be added to this result to get a multiple of 11. Because 126 + 6 = 132, a multiple of 11, the check digit is 6. (It is possible to have a check "digit" of 10; the letter X is used instead of 10.)

When an order for this book is received, the ISBN is entered into a computer, and the check digit evaluated. If this result does not match the check digit on the order, the order will not be processed. Does each ISBN have the correct check digit?

73. 0-275-98341-2

74. 0-374-29288-7

Find the appropriate check digit for each ISBN. (Note: The positions of hyphens (or spaces) may vary (or there may be none), but this does not affect the determination of the check digit.)

75. *Man of the Century,* by Jonathan Kwitny, 0-8050-2688- _____

76. *Winning,* by Jack Welch, 0-06-075394- _____

77. *1776,* by David McCullough, 0-7432-2671- _____

78. *The Da Vinci Code,* by Dan Brown, 0-385-50420- _____

*For an interesting general discussion, see "The Mathematics of Identification Numbers," by Joseph A. Gallian in *The College Mathematics Journal,* May 1991, p. 194.

4.5 Properties of Mathematical Systems

An Abstract System • Closure Property • Commutative Property • Associative Property • Identity Property • Inverse Property • Distributive Property

An Abstract System Clock arithmetic and modular systems, discussed in the previous section, were built upon ordinary numbers and involved familiar operations such as addition, subtraction, multiplication, and division. We begin this section by presenting a more abstract system, where the elements and the operations have no implied mathematical significance. This way, we can concentrate on investigating the properties of the system without preconceived notions of what they may be.

TABLE 12

☆	*a*	*b*	*c*	*d*
a	*a*	*b*	*c*	*d*
b	*b*	*d*	*a*	*c*
c	*c*	*a*	*d*	*b*
d	*d*	*c*	*b*	*a*

To begin, we introduce a finite mathematical system made up of the set of elements $\{a, b, c, d\}$ and an operation we will write with the symbol ☆. We define the system in Table 12, an **operation table** that shows how operation ☆ combines any two elements from the set $\{a, b, c, d\}$. To use the table to find, say, $c ☆ d$, first locate c on the left, and d across the top. This row and column intersect at b, so that

$$c ☆ d = b.$$

As with clock arithmetic, the important properties we shall look for in this system are the following: *closure, commutative, associative, identity,* and *inverse.*

TABLE 13

☆	*a*	*b*	*c*	*d*
a	*a*	*b*	*c*	*d*
b	*b*	*d*	*a*	*c*
c	*c*	*a*	*d*	*b*
d	*d*	*c*	*b*	*a*

Closure Property For this system to be closed under the operation ☆, the answer to any possible combination of elements from the system must be in the set $\{a, b, c, d\}$. A glance at Table 12 shows that the answers in the body of the table are all elements of this set. This means that the system is closed. If an element other than *a, b, c,* or *d* had appeared in the body of the table, or if any position in the body of the table had contained no entry, the system would not have been closed.

Commutative Property In order for the system to have the commutative property, it must be true that $\Gamma ☆ \Delta = \Delta ☆ \Gamma$, where Γ and Δ stand for any elements from the set $\{a, b, c, d\}$. For example,

$$c ☆ d = b \quad \text{and} \quad d ☆ c = b, \quad \text{so} \quad c ☆ d = d ☆ c.$$

To see that the same is true for *all* choices of Γ and Δ, observe that Table 13 is symmetric with respect to the diagonal line shown. This "diagonal line test" establishes that ☆ is a commutative operation for this system.

Associative Property The system is associative if $(\Gamma ☆ \Delta) ☆ \Upsilon = \Gamma ☆ (\Delta ☆ \Upsilon)$, where Γ, Δ, and Υ represent any elements from the set $\{a, b, c, d\}$. There is no quick way to check a table for the associative property, as there is for the commutative property. All we can do is try some examples. Using the table that defines operation ☆,

$$(a ☆ d) ☆ b = d ☆ b = c, \quad \text{and} \quad a ☆ (d ☆ b) = a ☆ c = c,$$

so that $$(a ☆ d) ☆ b = a ☆ (d ☆ b).$$

In the same way, $$b ☆ (c ☆ d) = (b ☆ c) ☆ d.$$

In both these examples, changing the location of parentheses did not change the answers. Because the two examples worked, we suspect that the system is associative. We cannot be sure of this, however, unless every possible choice of three letters from the set is checked. (Although we have not completely verified it here, this system does, in fact, satisfy the associative property.)

Identity Property For the identity property to hold, there must be an element Δ from the set of the system such that $\Delta ☆ X = X$ and $X ☆ \Delta = X$, where X represents any element from the set $\{a, b, c, d\}$. We can see that a is such an element as follows. In Table 13, the column below a (at the top) is identical to the column at the left, and the row across from a (at the left) is identical to the row at the top. Therefore, a is in fact the identity element of the system. (It is shown in more advanced courses that if a system has an identity element, it has *only* one.)

Bernard Bolzano (1781–1848) was an early exponent of rigor and precision in mathematics. Many early results in such areas as calculus were produced by the masters in the field; these masters knew what they were doing and produced accurate results. However, their sloppy arguments caused trouble in the hands of the less gifted. The work of Bolzano and others helped put mathematics on a strong footing.

Inverse Property We found earlier that a is the identity element for the system using operation ☆. If there is an inverse in this system for, say, the element b, and if Δ represents the inverse of b, then

$$b \star \Delta = a \quad \text{and} \quad \Delta \star b = a \quad \text{(because } a \text{ is the identity element).}$$

Inspecting the table for operation ☆ shows that Δ can be replaced with c:

$$b \star c = a \quad \text{and} \quad c \star b = a.$$

So we see that c is the inverse of b.

We can inspect the table to see if every element of our system has an inverse in the system. We see (in Table 13) that the identity element a appears once in each row, and that, in each case, the pair of elements that produces a also produces it in the opposite order. Therefore, we conclude that the system satisfies the inverse property.

In summary, the mathematical system made up of the set $\{a, b, c, d\}$ and operation ☆ satisfies the closure, commutative, associative, identity, and inverse properties.

Potential Properties of a Single-Operation System

Here a, b, and c represent elements from the set of any system, and $\circ$ represents the operation of the system.

Closure The system is closed if for all elements a and b,

$$a \circ b$$

is in the set of the system.

Commutative The system has the commutative property if

$$a \circ b = b \circ a$$

for all elements a and b of the system.

Associative The system has the associative property if

$$(a \circ b) \circ c = a \circ (b \circ c)$$

for every choice of three elements a, b, and c of the system.

Identity The system has the identity property if there exists an identity element e (where e is in the set of the system) such that

$$a \circ e = a \quad \text{and} \quad e \circ a = a$$

for every element a of the system.

Inverse The system has the inverse property if, for every element a of the system, there is an element x in the system such that

$$a \circ x = e \quad \text{and} \quad x \circ a = e,$$

where e is the identity element of the system.

EXAMPLE 1 Identifying the Properties of a System

Table 14 on the next page defines a system consisting of the set $\{0, 1, 2, 3, 4, 5\}$ under an operation designated $\otimes$. Which properties above are satisfied by this system?

TABLE 14

$\otimes$	0	1	2	3	4	5
0	0	0	0	0	0	0
1	0	1	2	3	4	5
2	0	2	4	0	2	4
3	0	3	0	3	0	3
4	0	4	2	0	4	2
5	0	5	4	3	2	1

SOLUTION

All the numbers in the body of the table come from the set {0, 1, 2, 3, 4, 5}, so the system is closed. If we draw a line from upper left to lower right, we could fold the table along this line and have the corresponding elements match; the system has the commutative property.

To check for the associative property, try some examples:

$$2 \otimes (3 \otimes 5) = 2 \otimes 3 = 0 \quad \text{and} \quad (2 \otimes 3) \otimes 5 = 0 \otimes 5 = 0,$$

so that $$2 \otimes (3 \otimes 5) = (2 \otimes 3) \otimes 5.$$

Also, $$5 \otimes (4 \otimes 2) = (5 \otimes 4) \otimes 2.$$

Any other examples that we might try would also work. The system has the associative property.

Because the column at the left of the operation table is repeated under 1 in the body of the table, 1 is a candidate for the identity element in the system. To be sure that 1 is the identity element here, check that the row corresponding to 1 at the left is identical with the row at the top of the table. Since it is, 1 is indeed the identity element.

To find inverse elements, look for the identity element, 1, in the rows of the table. The identity element appears in the second row, $1 \otimes 1 = 1$; and in the bottom row, $5 \otimes 5 = 1$; so 1 and 5 both are their own inverses. There is no identity element in the rows opposite the numbers 0, 2, 3, and 4, so none of these elements has an inverse.

In summary, the system made up of the set {0, 1, 2, 3, 4, 5} under this operation $\otimes$ satisfies the closure, associative, commutative, and identity properties, but not the inverse property. ■

TABLE 15

$\boxtimes$	1	2	3	4	5	6
1	1	2	3	4	5	6
2	2	4	6	1	3	5
3	3	6	2	5	1	4
4	4	1	5	2	6	3
5	5	3	1	6	4	2
6	6	5	4	3	2	1

EXAMPLE 2 Identifying the Properties of a System

Table 15 defines a system consisting of the set of numbers {1, 2, 3, 4, 5, 6} under an operation designated $\boxtimes$. Which properties are satisfied by this system?

SOLUTION

Notice here that 0 is not an element of this system. This is perfectly legitimate. Because we are defining the system, we can include (or exclude) whatever we wish. Check that the system satisfies the closure, commutative, associative, and identity properties, with identity element 1. Let us now check for inverses. The element 1 is its own inverse, because $1 \boxtimes 1 = 1$. In row 2, the identity element 1 appears under the number 4, so $2 \boxtimes 4 = 1$ (and $4 \boxtimes 2 = 1$), with 2 and 4 inverses of each other. Also, 3 and 5 are inverses of each other, and 6 is its own inverse. Because each number in the set of the system has an inverse, the system satisfies the inverse property. ■

Distributive Property

When a mathematical system has two operations, rather than just one, we can look for the **distributive property.**

Distributive Property

Let ☆ and ∘ be two operations defined for elements in the same set. Then ☆ is distributive over ∘ if

$$a \star (b \circ c) = (a \star b) \circ (a \star c)$$

for every choice of elements *a*, *b*, and *c* from the set.

It is a well-known fact that multiplication is distributive over (or with respect to) addition on the set of real numbers. For example,

$$5 \cdot (8 + 3) = 5 \cdot 11 = 55 \quad \text{and} \quad 5 \cdot 8 + 5 \cdot 3 = 40 + 15 = 55,$$

so

$$5 \cdot (8 + 3) = 5 \cdot 8 + 5 \cdot 3.$$

EXAMPLE 3 Testing for the Distributive Property

Is addition distributive over multiplication on the set of whole numbers?

SOLUTION

To find out, replace ☆ with addition (+) and ∘ with multiplication (·) in the statement of the distributive property at the bottom of the previous page:

$$a + (b \cdot c) = (a + b) \cdot (a + c). \quad \text{Is this true in general?}$$

We need to find out whether this statement is true for *every* choice of three whole numbers that we might make. Try an example. If $a = 3$, $b = 4$, and $c = 5$,

$$a + (b \cdot c) = 3 + (4 \cdot 5) = 3 + 20 = 23,$$

while

$$(a + b) \cdot (a + c) = (3 + 4) \cdot (3 + 5) = 7 \cdot 8 = 56.$$

Since $23 \neq 56$, we have $3 + (4 \cdot 5) \neq (3 + 4) \cdot (3 + 5)$. This false result is a *counterexample* (an example showing that a general statement is false). This counterexample shows that addition is *not* distributive over multiplication on the whole numbers. ■

The final example illustrates how the distributive property may hold for an abstract finite system.

TABLE 16

☆	*a*	*b*	*c*	*d*	*e*
a	*a*	*a*	*a*	*a*	*a*
b	*a*	*b*	*c*	*d*	*e*
c	*a*	*c*	*e*	*b*	*d*
d	*a*	*d*	*b*	*e*	*c*
e	*a*	*e*	*d*	*c*	*b*

TABLE 17

∘	*a*	*b*	*c*	*d*	*e*
a	*a*	*b*	*c*	*d*	*e*
b	*b*	*c*	*d*	*e*	*a*
c	*c*	*d*	*e*	*a*	*b*
d	*d*	*e*	*a*	*b*	*c*
e	*e*	*a*	*b*	*c*	*d*

EXAMPLE 4 Testing for the Distributive Property

Suppose that the set $\{a, b, c, d, e\}$ has two operations ☆ and ∘ defined by Tables 16 and 17. The distributive property of ☆ with respect to ∘ holds in this system. Verify for the following case: $e \star (d \circ b) = (e \star d) \circ (e \star b)$.

SOLUTION

First evaluate the left side of the equation by using the tables.

$$\begin{aligned} e \star (d \circ b) &= e \star e && \text{Use the } \circ \text{ table.} \\ &= b && \text{Use the } \star \text{ table.} \end{aligned}$$

Now, evaluate the right side of the equation.

$$\begin{aligned} (e \star d) \circ (e \star b) &= c \circ e && \text{Use the } \star \text{ table twice.} \\ &= b && \text{Use the } \circ \text{ table.} \end{aligned}$$

Each time the final result is b; the distributive property is verified for this case. ■

4.5 Exercises

For each system in Exercises 1–10, decide which of the properties of single-operation systems are satisfied. If the identity property is satisfied, give the identity element. If the inverse property is satisfied, give the inverse of each element. If the identity property is satisfied but the inverse property is not, name the elements that have no inverses.

1. {1, 2}; operation $\otimes$

$\otimes$	1	2
1	1	2
2	2	1

2. {1, 2, 3, 4}; operation $\otimes$

$\otimes$	1	2	3	4
1	1	2	3	4
2	2	4	1	3
3	3	1	4	2
4	4	3	2	1

3. {1, 2, 3, 4, 5, 6, 7}; operation $\boxtimes$

$\boxtimes$	1	2	3	4	5	6	7
1	1	2	3	4	5	6	7
2	2	4	6	0	2	4	6
3	3	6	1	4	7	2	5
4	4	0	4	0	4	0	4
5	5	2	7	4	1	6	3
6	6	4	2	0	6	4	2
7	7	6	5	4	3	2	1

4. {1, 2, 3, 4, 5}; operation $\boxtimes$

$\boxtimes$	1	2	3	4	5
1	1	2	3	4	5
2	2	4	0	2	4
3	3	0	3	0	3
4	4	2	0	4	2
5	5	4	3	2	1

5. {1, 3, 5, 7, 9}; operation ☆

☆	1	3	5	7	9
1	1	3	5	7	9
3	3	9	5	1	7
5	5	5	5	5	5
7	7	1	5	9	3
9	9	7	5	3	1

6. {1, 3, 5, 7}; operation ☆

☆	1	3	5	7
1	1	3	5	7
3	3	1	7	5
5	5	7	1	3
7	7	5	3	1

7. {*A*, *B*, *F*}; operation *

*	*A*	*B*	*F*
A	*B*	*F*	*A*
B	*F*	*A*	*B*
F	*A*	*B*	*F*

8. {*m*, *n*, *p*}; operation ***J***

J	*m*	*n*	*p*
m	*n*	*p*	*n*
n	*p*	*m*	*n*
p	*n*	*n*	*m*

9. {*r*, *s*, *t*, *u*}; operation ***Z***

Z	*r*	*s*	*t*	*u*
r	*u*	*t*	*r*	*s*
s	*t*	*u*	*s*	*r*
t	*r*	*s*	*t*	*u*
u	*s*	*r*	*u*	*t*

10. {*A*, *J*, *T*, *U*}; operation #

#	*A*	*J*	*T*	*U*
A	*A*	*J*	*T*	*U*
J	*J*	*T*	*U*	*A*
T	*T*	*U*	*A*	*J*
U	*U*	*A*	*J*	*T*

The tables in the finite mathematical systems that we developed in this section can be obtained in a variety of ways. For example, let us begin with a square, as shown in the figure. Let the symbols a, b, c, and d be defined as shown in the figure.

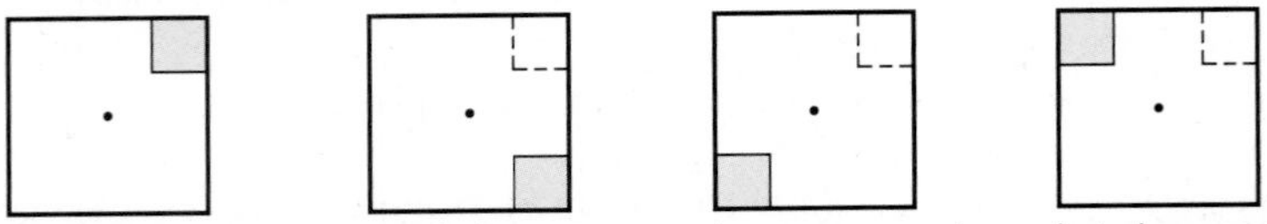

Let *a* represent zero rotation—leave the original square as is.

Let *b* represent rotation of 90° clockwise from original position.

Let *c* represent rotation of 180° clockwise from original position.

Let *d* represent rotation of 270° clockwise from original position.

Define an operation □ *for these letters as follows. To evaluate* $b \square c$, *for example, first perform b by rotating the square* 90°. *(See the figure.) Then perform operation c by rotating the square an additional* 180°. *The net result is the same as if we had performed d only. Thus,*

$$b \square c = d.$$

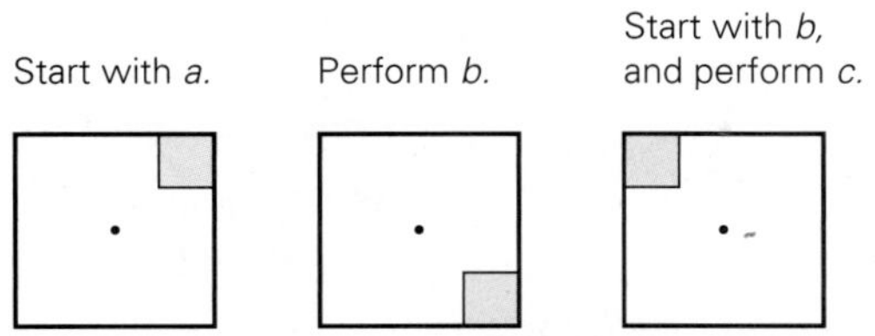
Start with *a*.
Perform *b*.
Start with *b*, and perform *c*.

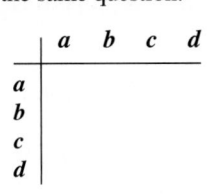

Use this method to find each of the following.

11. $b \square d$ **12.** $b \square b$ **13.** $d \square b$ **14.** $a \square b$

Solve each problem.

15. Complete the table at the right for the system of square rotations described above.

□	*a*	*b*	*c*	*d*
a	*a*	*b*	*c*	*d*
b	*b*	*c*		*a*
c	*c*		*a*	
d	*d*	*a*		

16. Which of the properties from this section are satisfied by this system?

17. Define a universal set U as the set of counting numbers. Form a new set that contains all possible subsets of U. This new set of subsets together with the operation of set intersection forms a mathematical system. Which of the properties listed in this section are satisfied by this system?

18. Replace the word "intersection" with the word "union" in Exercise 17; then answer the same question.

19. Complete the table at the right so that the result is *not* the same as operation □ of Exercise 15, but so that the five properties listed in this section still hold.

	a	*b*	*c*	*d*
a				
b				
c				
d				

Try examples to help you decide whether each operation, when applied to the integers, satisfies the distributive property.

20. subtraction with respect to multiplication

21. addition with respect to subtraction

22. subtraction with respect to addition

Recall that Example 3 provided a counterexample for the general statement

$$a + (b \cdot c) = (a + b) \cdot (a + c).$$

Thus, addition is not *distributive with respect to multiplication. Now work Exercises 23–26.*

23. Decide if the statement above is true for each of the following sets of values.
(a) $a = 2, b = -5, c = 4$
(b) $a = -7, b = 5, c = 3$
(c) $a = -8, b = 14, c = -5$
(d) $a = 1, b = 6, c = -6$

24. Find another set of a, b, and c values that make the statement true.

25. Under what general conditions will the statement above be true?

26. Explain why, regardless of the results in Exercises 23–25, addition is still *not* distributive with respect to multiplication.

27. Give the conditions under which each equation would be true.
(a) $a + (b - c) = (a + b) - (a + c)$
(b) $a - (b + c) = (a - b) + (a - c)$

28. **(a)** Find values of a, b and c such that

$$a - (b \cdot c) = (a - b) \cdot (a - c).$$

(b) Does this mean that subtraction is distributive with respect to multiplication? Explain.

Verify for the mathematical system of Example 4, defined by Tables 16 and 17, that the distributive property holds for each case.

29. $c \star (d \circ e) = (c \star d) \circ (c \star e)$

30. $a \star (a \circ b) = (a \star a) \circ (a \star b)$

31. $d \star (e \circ c) = (d \star e) \circ (d \star c)$

32. $b \star (b \circ b) = (b \star b) \circ (b \star b)$

Exercises 33 *and* 34 *are for students who have studied sets.*

33. Use Venn diagrams to show that the distributive property for union with respect to intersection holds for sets *A, B,* and *C.* That is,

$$A \cup (B \cap C) = (A \cup B) \cap (A \cup C).$$

34. Use Venn diagrams to show that *another* distributive property holds for sets *A, B,* and *C.* It is the distributive property of intersection with respect to union.

$$A \cap (B \cup C) = (A \cap B) \cup (A \cap C)$$

Exercises 35 *and* 36 *are for students who have studied logic.*

35. Use truth tables to show that the following distributive property holds:

$$p \vee (q \wedge r) \equiv (p \vee q) \wedge (p \vee r).$$

36. Use truth tables to show that *another* distributive property holds:

$$p \wedge (q \vee r) \equiv (p \wedge q) \vee (p \wedge r).$$

4.6 Groups

Groups • Symmetry Groups • Permutation Groups

Groups We have considered some mathematical systems, most of which have satisfied some or all of the closure, associative, commutative, identity, inverse, and distributive properties. Systems are commonly classified according to which properties they satisfy. One important category, when a single operation is considered, is the mathematical *group,* which we define here.

Group

A mathematical system is called a **group** if, under its operation, it satisfies the closure, associative, identity, and inverse properties.

Some sets of numbers, under certain operations, form groups. Others do not.

EXAMPLE 1 Checking the Group Properties

Does the set $\{-1, 1\}$ under the operation of multiplication form a group?

SOLUTION

Check the necessary four properties.

Closure The given system leads to the multiplication table below. All entries in the body of the table are either -1 or 1; the system is closed.

$\cdot$	**−1**	**1**
−1	1	−1
1	−1	1

Associative Try some examples:

$$-1 \cdot (-1 \cdot 1) = -1 \cdot (-1) = 1$$

and $$[-1 \cdot (-1)] \cdot 1 = 1 \cdot 1 = 1,$$

so $$-1 \cdot (-1 \cdot 1) = [-1 \cdot (-1)] \cdot 1.$$

Also $$1 \cdot [(-1) \cdot 1] = 1 \cdot (-1) = -1$$

and $$[1 \cdot (-1)] \cdot 1 = -1 \cdot 1 = -1,$$

so $$1 \cdot [(-1) \cdot 1] = [1 \cdot (-1)] \cdot 1.$$

Any other examples likewise will work. This system satisfies the associative property.

Identity The operation table for the system shows that the identity element is 1. (The column under 1 is identical to the column at the left, and the row to the right of 1 is identical to the row at the top.)

Inverse Check in the table that -1 is its own inverse, because $-1 \cdot (-1) = 1$ (the identity element); also, 1 is its own inverse.

All four of the properties are satisfied, so the system is a group.

Niels Henrik Abel (1802–1829) of Norway was identified in childhood as a mathematical genius but never received in his lifetime the professional recognition his work deserved.

At 16, influenced by a perceptive teacher, he read the works of Newton, Euler, and Lagrange. One of Abel's achievements was the demonstration that a general formula for solving fifth-degree equations does not exist. The quadratic formula (for equations of degree 2) is well known, and formulas do exist for solving third- and fourth-degree equations. Abel's accomplishment ended a search that had lasted for years.

In the study of abstract algebra, groups that have the commutative property are referred to as **abelian groups** in honor of Abel. He died of tuberculosis at age 27.

EXAMPLE 2 Checking the Group Properties

Does the set $\{-1, 1\}$ under the operation of addition form a group?

SOLUTION

The addition table below shows that closure is not satisfied, so there is no need to check further. The system is not a group.

+	**−1**	**1**
−1	−2	0
1	0	2

EXAMPLE 3 Checking the Group Properties

Does the set of integers $\{\ldots, -3, -2, -1, 0, 1, 2, 3, \ldots\}$ under the operation of addition form a group?

SOLUTION

Check the required properties.

Closure The sum of any two integers is an integer; the system is closed.

Associative Try some examples:

$$2 + (5 + 8) = 2 + 13 = 15$$

and $$(2 + 5) + 8 = 7 + 8 = 15,$$

so $$2 + (5 + 8) = (2 + 5) + 8.$$

Also $$-4 + (7 + 14) = -4 + 21 = 17$$

and $$(-4 + 7) + 14 = 3 + 14 = 17,$$

so $$-4 + (7 + 14) = (-4 + 7) + 14.$$

Apparently, addition of integers is associative.

Identity We know that $a + 0 = a$ and $0 + a = a$ for any integer a. The identity element for addition of integers is 0.

Inverse Given any integer a, its additive inverse, $-a$, is also an integer. For example, 5 and -5 are inverses. The system satisfies the inverse property.

Since all four properties are satisfied, this (infinite) system *is* a group.

Symmetry Groups Groups can be built upon sets of objects other than numbers. An example is the group of **symmetries of a square**, which we now develop. First, cut out a small square, and label it as shown in Figure 10.

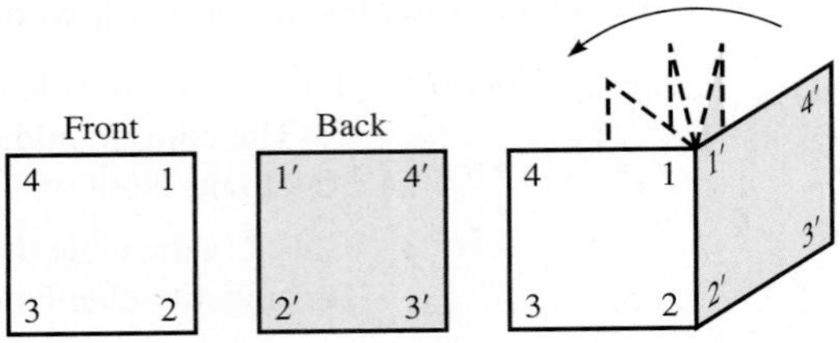

FIGURE 10

Make sure that 1 is in front of 1′, 2 is in front of 2′, 3 is in front of 3′, and 4 is in front of 4′. Let the letter M represent a clockwise rotation of 90° *about the center of the square* (marked with a dot in Figure 11). Let N represent a rotation of 180°, and so on. A list of the symmetries of a square is given in Figure 11.

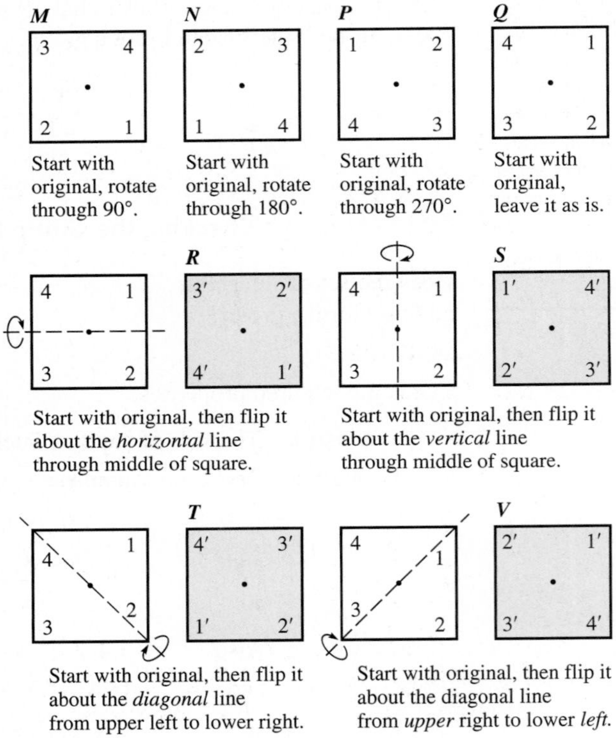

Symmetries of a square

FIGURE 11

Amalie ("Emmy") Noether (1882–1935) was an outstanding mathematician in the field of **abstract algebra.** She studied and worked in Germany at a time when it was very difficult for a woman to do so. At the University of Erlangen in 1900, Noether was one of only two women. Although she could attend classes, professors could and did deny her the right to take the exams for their courses. Not until 1904 was Noether allowed to officially register. She completed her doctorate four years later.

In 1916 Emmy Noether went to Göttingen to work with David Hilbert on the general theory of relativity. But even with Hilbert's backing and prestige, it was three years before the faculty voted to make Noether a *Privatdozent,* the lowest rank in the faculty. In 1922 Noether was made an unofficial professor (or assistant). She received no pay for this post, although she was given a small stipend to lecture in algebra.

Noether's area of interest was abstract algebra, particularly structures called rings and ideals. (Groups are structures, too, with different properties.) One special type of ring bears her name; she was the first to study its properties.

Évariste Galois (1811–1832), as a young Frenchman, agreed to fight a duel. He had been engaged in profound mathematical research for some time. Now, anticipating the possibility of his death, he summarized the essentials of his discoveries in a letter to a friend. The next day Galois was killed. He was not yet 21 years old when he died.

It was not until 1846 that Galois's theories were published. Mathematicians began to appreciate the importance of Galois's work, which centered on solving equations by using groups. Galois found a way to derive a group that corresponds to each equation. So-called **Galois groups** form an important part of modern abstract algebra.

A 2005 book by Mario Livio, titled *The Equation That Couldn't Be Solved: How Mathematical Genius Discovered the Language of Symmetry,* explores the story of Galois and Abel.

Combine symmetries as follows: Let NP represent N followed by P. Performing N and then P is the same as performing just M, so that $NP = M$. See Figure 12.

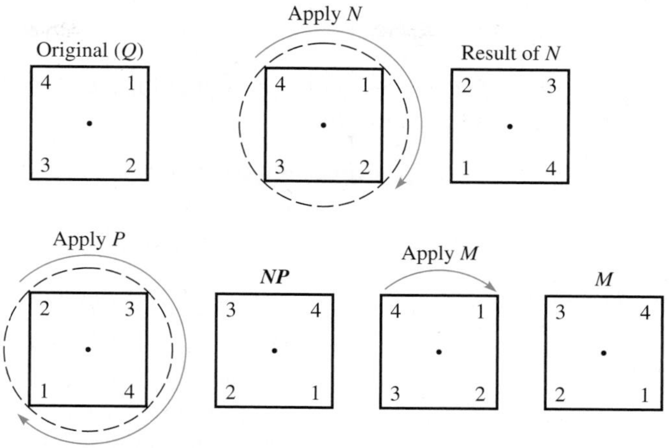

Think of N as advancing each corner two quarter turns clockwise. Thus 4 goes from upper left to lower right. To this result apply P, which advances each corner three quarter turns. Thus 2 goes from upper left to lower left. The result, NP, is the same as advancing each (original) corner one quarter turn, which M does. Thus $NP = M$.

FIGURE 12

EXAMPLE 4 Combining Symmetries of a Square

Find RT using the symmetries of Figure 11.

SOLUTION

First, perform R by flipping the square about a horizontal line through the middle. Then, perform T by flipping the result of R about a diagonal from upper left to lower right. The result of RT is the same as performing only M, so that $RT = M$. ■

The method in Example 4 can be used to complete Table 18 for combining the symmetries of a square.

TABLE 18

□	*M*	*N*	*P*	*Q*	*R*	*S*	*T*	*V*
M	N	P	Q	M	V	T	R	S
N	P	Q	M	N	S	R	V	T
P	Q	M	N	P	T	V	S	R
Q	M	N	P	Q	R	S	T	V
R	T	S	V	R	Q	N	M	P
S	V	R	T	S	N	Q	P	M
T	S	V	R	T	P	M	Q	N
V	R	T	S	V	M	P	N	Q

EXAMPLE 5 Verifying the Group Properties

Show that the system made up of the symmetries of a square is a group.

SOLUTION

For the system to be a group, it must satisfy the closure, associative, identity, and inverse properties.

Closure All the entries in the body of Table 18 come from the set $\{M, N, P, Q, R, S, T, V\}$. Thus, the system is closed.

Associative Try examples:

$$P(MT) = P(R) = T.$$

Also, $(PM)T = (Q)T = T,$

so that $P(MT) = (PM)T.$

Other similar examples also work. (See Exercises 25–28.) Thus, the system has the associative property.

Identity The column at the left in the table is repeated under Q. Check that Q is indeed the identity element.

Inverse In the first row, Q appears under P. Check that M and P are inverses of each other. In fact, every element in the system has an inverse. (See Exercises 29–34.)

Because all four properties are satisfied, the system is a group.

EXAMPLE 6 Verifying that a System Has a Subgroup

Form a mathematical system by using only the set $\{M, N, P, Q\}$ from the group of symmetries of a square. Is this new system a group?

SOLUTION

Table 19 for the elements $\{M, N, P, Q\}$ is just one corner of the table for the entire system. Verify that the system represented by this table satisfies all four properties and thus is a group. This new group is a *subgroup* of the original group of the symmetries of a square.

TABLE 19

□	*M*	*N*	*P*	*Q*
M	*N*	*P*	*Q*	*M*
N	*P*	*Q*	*M*	*N*
P	*Q*	*M*	*N*	*P*
Q	*M*	*N*	*P*	*Q*

Permutation Groups

A very useful example of a group comes from studying the arrangements, or permutations, of a list of numbers. Start with the symbols 1-2-3, in that order.

There are several ways in which the order could be changed—for example, 2-3-1. This rearrangement is written

1-2-3
2-3-1.

Much research in group theory has been devoted to "simple" groups (which are not at all simple). These groups are fundamental—other groups are built up from them. One of the largest of these simple groups has

$$2^{46} \cdot 3^{20} \cdot 5^{9} \cdot 7^{6} \cdot 11^{2} \cdot 13^{3} \cdot 17 \cdot 19 \cdot 23 \cdot 29 \cdot 31 \cdot 41 \cdot 47 \cdot 59 \cdot 71$$

different elements. This huge number of elements is why the group is called the **"monster group."** If this "monster group" contains too many elements, how about the "baby monster" simple group, with only

$$2^{41} \cdot 3^{13} \cdot 5^{6} \cdot 7^{2} \cdot 11 \cdot 13 \cdot 17 \cdot 19 \cdot 23 \cdot 31 \cdot 47$$

different elements? (For more information, see "The Classification of the Finite Simple Groups," by Michael Aschbacher in *The Mathematical Intelligencer,* Issue 2, 1981.)

Replace 1 with 2, replace 2 with 3, and replace 3 with 1. In the same way,

1-2-3
3-1-2

means replace 1 with 3, 2 with 1, and 3 with 2, while

1-2-3
3-2-1

says to replace 1 with 3, leave the 2 unchanged, and replace 3 with 1. All possible rearrangements of the symbols 1-2-3 are listed below where, for convenience, a name has been given to each rearrangement.

A^*: 1-2-3	B^*: 1-2-3	C^*: 1-2-3	D^*: 1-2-3	E^*: 1-2-3	F^*: 1-2-3
2-3-1	2-1-3	1-2-3	1-3-2	3-1-2	3-2-1

Two rearrangements can be combined as with the symmetries of a square; for example, the symbol B^*F^* means to first apply B^* to 1-2-3 and then apply F^* to the result. Rearrangement B^* changes 1-2-3 into 2-1-3. Then apply F^* to this result: 1 becomes 3, 2 is unchanged, and 3 becomes 1. In summary:

1-2-3
2-1-3 Rearrange according to B^*.
3 By F^*, 1 is replaced by 3.
2-3 Next, 2 remains unchanged.
2-3-1. As a last step, 3 changes into 1.

The net result of B^*F^* is to change 1-2-3 into 2-3-1, which is exactly what A^* does to 1-2-3. Therefore,

$$B^*F^* = A^*.$$

EXAMPLE 7 Combining Rearrangements

Find D^*E^*.

SOLUTION

Use the procedure described above.

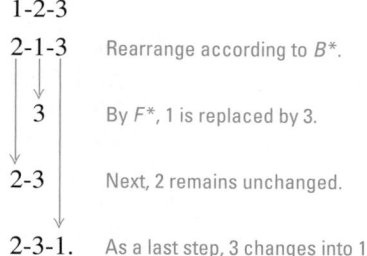

1-2-3
1-3-2 Rearrange according to D^*.
3 E^* replaces 1 with 3.
3 1 E^* replaces 2 with 1.
3-2-1 E^* replaces 3 with 2.

Elie-Joseph Cartan (1869–1951) did extensive work in **group theory.** His 1894 doctoral thesis completely categorized all finite groups known at the time. The classification of *all* finite groups took 150 years and culminated with the monster group (see page 215) constructed in 1980 by Robert Griess, Jr.

Throughout the last half of the twentieth century, physicists were able to apply the structure of some of these groups to improve their understanding of the basic particles of matter and their quantum interactions, thereby enabling them to formulate ever better theories of the fundamental forces of nature.

The result is that D^*E^* converts 1-2-3 into 3-2-1, as does F^*, so

$$D^*E^* = F^*.$$

As further examples, $A^*B^* = D^*$ and $F^*E^* = B^*$.

Once again, we see that we encountered a mathematical system: the set $\{A^*, B^*, C^*, D^*, E^*, F^*\}$ and the operation of the combination of two rearrangements. To see whether this system is a group, check the requirements.

Closure Combine any two rearrangements and the result is another rearrangement, so the system is closed.

Associative Try an example:

First $(B^*D^*)\,A^* = E^*A^* = C^*$.

Also $B^*(D^*\,A^*) = B^*\,B^* = C^*$,

so that $(B^*D^*)\,A^* = B^*(D^*A^*)$.

Because other examples will work out similarly, the system is associative.

Identity The identity element is C^*. If x is any rearrangement, then we have $xC^* = C^*x = x$.

Inverse Does each rearrangement have an inverse rearrangement? Begin with the basic order 1-2-3 and then apply, say B^*, resulting in 2-1-3. The inverse of B^* must convert this 2-1-3 back into 1-2-3, by changing 2 into 1 and 1 into 2. But B^* itself will do this. Hence, $B^*B^* = C^*$ and B^* is its own inverse. By the same process, E^* and A^* are inverses of each other. Also, each of C^*, D^*, and F^* is its own inverse.

Because all four requirements are satisfied, the system is a group. Rearrangements are also referred to as *permutations,* so this group is sometimes called the **permutation group on three symbols.** The total number of different permutations of a given number of symbols can be determined by techniques described in the chapter on counting methods.

4.6 EXERCISES

What is wrong with the way in which each question is stated?

1. Do the integers form a group?

2. Does multiplication satisfy all of the group properties?

Decide whether each system is a group. If not a group, identify all properties that are not satisfied. (Recall that any system failing to satisfy the identity property automatically fails to satisfy the inverse property also.) For the finite systems, it may help to construct tables. For infinite systems, try some examples to help you decide.

3. $\{0\}$; multiplication

4. $\{0\}$; addition

5. $\{0, 1\}$; addition

6. $\{0\}$; subtraction

7. $\{-1, 1\}$; division

8. $\{0, 1\}$; multiplication

9. $\{-1, 0, 1\}$; multiplication

10. $\{-1, 0, 1\}$; addition

11. integers; subtraction

12. integers; multiplication

13. odd integers; multiplication

14. counting numbers; addition

15. rational numbers; addition

16. even integers; addition

17. prime numbers; addition

18. nonzero rational numbers; multiplication

19. Explain why a *finite* group based on the operation of ordinary addition of numbers cannot contain the element 1.

20. Explain why a group based on the operation of ordinary addition of numbers *must* contain the element 0.

Exercises 21–34 apply to the system of symmetries of a square presented in the text. Find each combination.

21. RN

22. PR

23. TV

24. VP

Verify each statement.

25. $N(TR) = (NT)R$

26. $V(PS) = (VP)S$

27. $T(VN) = (TV)N$

28. $S(MR) = (SM)R$

Find the inverse of each element.

29. N

30. Q

31. R

32. S

33. T

34. V

A group that also satisfies the commutative property is called a **commutative group** *(or an* **abelian group,** *after Niels Henrik Abel). Determine whether each group is commutative.*

35. the group of symmetries of a square

36. the subgroup of Example 6

37. the integers under addition

38. the permutation group on three symbols

Give illustrations to support your answers for Exercises 39–42.

39. Produce a mathematical system with two operations which is a group under one operation but not a group under the other operation.

40. Explain what property is gained when the system of counting numbers is extended to the system of whole numbers.

41. Explain what property is gained when the system of whole numbers is extended to the system of integers.

42. Explain what property is gained when the system of integers is extended to the system of rational numbers.

Consider the following set of "actions" (A, B, C, and D) on three symbols (a, b, and c):

$$A: \begin{matrix} a & b & c \\ a & b & c \end{matrix} \qquad B: \begin{matrix} a & b & c \\ c & b & a \end{matrix} \qquad C: \begin{matrix} a & b & c \\ a & -b & c \end{matrix} \qquad D: \begin{matrix} a & b & c \\ c & -b & a \end{matrix}$$

(continued)

The resulting system is somewhat similar to the permutation group on three symbols discussed in the text, except that not all possible rearrangements of a, b, and c are included, and two of the actions involve sign changes. The operation of the system is the combination of actions as follows, for example, BD represents B followed by D:

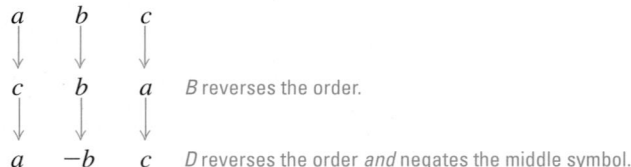

$$
\begin{array}{ccc l}
a & b & c & \\
\downarrow & \downarrow & \downarrow & \\
c & b & a & \text{B reverses the order.} \\
\downarrow & \downarrow & \downarrow & \\
a & -b & c & \text{D reverses the order \textit{and} negates the middle symbol.}
\end{array}
$$

The net result, from beginning to end, is simply to negate the middle symbol, which is exactly what C does. Therefore, BD = C.

43. Verify, for yourself, the entries shown in the following operation table and fill in the missing entries.

	A	***B***	***C***	***D***
A	*A*	*B*	*C*	*D*
B	*B*	*A*		*C*
C	*C*		*A*	
D	*D*			*A*

44. **(a)** Is there an identity element in this system?
(b) If so, what is it?

45. **(a)** Is closure satisfied by this system?
(b) Explain.

46. **(a)** Is this system commutative?
(b) Explain.

47. **(a)** Is the distributive property satisfied in this system?
(b) Explain.

48. **(a)** Assuming the associative property is satisfied, is the system a group?
(b) Explain.

COLLABORATIVE INVESTIGATION

A Perpetual Calendar Algorithm

In this chapter we examined some alternative algorithms for arithmetic computations. Algorithms appear in various places in mathematics and computer science, and in each case give us a specified method of carrying out a procedure that produces a desired result. The algorithm that follows allows us to find the day of the week on which a particular date occurred or will occur.

In applying the algorithm, you will need to know whether a particular year is a leap year. In general, if a year is divisible (evenly) by 4, it is a leap year. However, there are exceptions. Century years, such as 1800 and 1900, are not leap years, despite the fact that they are divisible by 4. Furthermore, as an exception to the exception, a century year that is divisible by 400 (such as the year 2000) is a leap year.

In groups of three to five students, read the algorithm and then work the Topics for Discussion.

The Algorithm

This algorithm requires several *key numbers.* Key numbers for the month, day, and century are determined by the following tables.

Month	Key
January	1 (0 if a leap year)
February	4 (3 if a leap year)
March	4
April	0
May	2
June	5
July	0
August	3
September	6
October	1
November	4
December	6

Day	Key
Saturday	0
Sunday	1
Monday	2
Tuesday	3
Wednesday	4
Thursday	5
Friday	6

Century	Key
1700s	4
1800s	2
1900s	0
2000s	6

The algorithm works as follows. (We use October 12, 1949, as an example.)

		Example
Step 1	Obtain the following five numbers.	
	1. The number formed by the last two digits of the year	49
	2. The number in Step 1, divided by 4, with the remainder ignored	12
	3. The month key (1 for October in our example)	1
	4. The day of the month (12 for October 12)	12
	5. The century key (0 for the 1900s)	0
Step 2	Add these five numbers.	74

Step 3 Divide the sum by 7, and retain the remainder. $\left(\frac{74}{7} = 10\text{, with remainder } 4\right)$

Step 4 Find this remainder in the day key table. (The number 4 implies that October 12, 1949 was a Wednesday.)

Topics for Discussion

1. Have each person in the group determine the day of the week on which he or she was born.

2. Among the group members, discuss whether the following poem applies. (This is all in good fun, of course.)

Monday's child is fair of face,
Tuesday's child is full of grace.
Wednesday's child is full of woe,
Thursday's child has far to go.
Friday's child is loving and giving,
Saturday's child works hard for a living.
But the child that is born on the Sabbath
day is bonny and good, happy and gay.

3. Determine the day of the week on which the following important historical events occurred.

(a) December 7, 1941 (the bombing of Pearl Harbor)
(b) November 22, 1963 (the assassination of John F. Kennedy)
(c) July 4, 1976 (the bicentennial of the United States)
(d) January 1, 2000 (the "dreaded" Y2K day)
(e) September 11, 2001 (the terrorist attacks on the United States)

CHAPTER 4 TEST

1. For the numeral 𓆼𓆼𓍢𓍢𓍢𓍢𓎆𓎆𓏺𓏺𓏺𓏺𓏺𓏺, identify the numeration system, and give the Hindu-Arabic equivalent.

2. Simplify:

$(7 \cdot 10^3) + (5 \cdot 10^2) + (6 \cdot 10^1) + (1 \cdot 10^0)$.

3. Write in expanded notation: 60,923.

Perform each operation using the alternative algorithm specified.

4. $37 \cdot 54$ (Russian peasant or Egyptian method)

5. $236 \cdot 94$ (Lattice method)

6. $21{,}325 - 8498$ (Nines complement method)

Convert each number to base ten.

7. 324_{five} **8.** 110010_{two} **9.** $\text{DEAF}_{\text{sixteen}}$

Convert as indicated.

10. 49 to base two

11. 2930 to base five

12. 10101110_{two} to base eight

13. Find all positive solutions for the equation

$$3x + 1 \equiv 2 \pmod 5.$$

14. Find each quantity on the 12-hour-clock.
(a) $8 + 9$ **(b)** $4 \cdot 10$
(c) -3 **(d)** $3 - 7$

15. Work each modular arithmetic problem.
(a) $(7 + 11) \pmod 5$ **(b)** $(8 \cdot 15) \pmod{17}$

16. ***Two-Dollar Bill Collection*** Alec has a collection of two-dollar bills which he can arrange in groups of 5 with 3 left over, groups of 7 with 6 left over, or groups of 10 with 8 left over. What is the least number of bills he could have?

Briefly explain each of the following.

17. the advantage of multiplicative grouping over simple grouping

18. the advantage, in a positional numeration system, of a smaller base over a larger base

19. the advantage, in a positional numeration system, of a larger base over a smaller base

Answer the questions in Exercises 20–22.

20. For addition of whole numbers, what is the identity element?

21. For multiplication of rational numbers, what is the inverse of 3?

22. For any whole numbers a, b, and c, $(a + b) + c = (b + a) + c$. What property does this illustrate?

Consider the mathematical system with the set $\{1, 3, 5, 7\}$ *and with the operation of multiplication modulo* 8.

23. Complete the operation table for the system.

∘	**1**	**3**	**5**	**7**
1	1	3	5	7
3	3			5
5	5		1	
7	7		3	

24. **(a)** Is there an identity element in this system?
(b) If so, what is it?

25. **(a)** Is closure satisfied by this system?
(b) Explain.

26. **(a)** Is this system commutative?
(b) Explain.

27. **(a)** Is the distributive property satisfied in this system?
(b) Explain.

28. **(a)** Assuming the associative property is satisfied, is the system a group?
(b) Explain.

5

NUMBER THEORY

The first episode of the animated series *The Simpsons* aired on December 17, 1989 and has since become a pop culture icon, providing humor, social commentary, and even lessons in mathematics. An annual treat for fans is the *Treehouse of Horror* episode, which airs near Halloween. In the sixth installment, this episode featured a segment titled *Homer 3D* in which a two-dimensional Homer Simpson became trapped in the third dimension. Computer graphics portrayed a three-dimensional coordinate system, the Parthenon (an example of the *Golden Ratio*), a cone, a black hole, and several mathematical equations. One such equation was

$$1782^{12} + 1841^{12} = 1922^{12}.$$

In Section 5.2 you will learn that ***this equation cannot be true.*** It is of the form of an equation of Fermat's Last Theorem, one of the most famous theorems in *number theory*, the topic of this chapter. The theorem was stated by the French mathematician Pierre de Fermat over 400 years ago, but was not proved until 1994. A graphing calculator such as the TI-83/84 Plus will indicate that the equation is true, but only because the calculator is unable to compute powers of this size. Can you explain why it can't be true? (*Hint:* Show that one side is an odd number and the other is an even number.) See page 237 for the complete answer.

5.1 Prime and Composite Numbers

Primes, Composites, and Divisibility • The Fundamental Theorem of Arithmetic • The Infinitude of Primes • The Search for Large Primes

Primes, Composites, and Divisibility

The famous German mathematician Carl Friedrich Gauss once remarked, "Mathematics is the Queen of Science, and number theory is the Queen of Mathematics." This chapter is centered around the study of number theory. **Number theory** is the branch of mathematics devoted to the study of the properties of the natural numbers. In earlier chapters we discussed the set of **natural numbers**, also called the **counting numbers** or the **positive integers:**

$$\{1, 2, 3, \ldots\}.$$

Number theory deals with the study of the properties of this set of numbers, and a key concept of number theory is the idea of *divisibility*. Informally, we say that one counting number is *divisible* by another if the operation of dividing the first by the second leaves a remainder 0. A formal definition follows.

Do not confuse $b \mid a$ with b/a. The expression $b \mid a$ denotes the *statement* "b divides a." For example, $3 \mid 12$ is a true statement, while $5 \mid 14$ is a false statement. On the other hand, b/a denotes the *operation* "b divided by a." For example, 28/4 yields the result 7.

Divisibility

The natural number a is **divisible** by the natural number b if there exists a natural number k such that $a = bk$. If b divides a, then we write $b \mid a$.

The ideas of **even** and **odd natural numbers** are based on the concept of divisibility. A natural number is even if it is divisible by 2 and odd if it is not. Every even number can be written in the form $2k$ (for some natural number k), while every odd number can be written in the form $2k + 1$. Another way to say the same thing: 2 divides every even number but fails to divide every odd number. (If a is even, then $2 \mid a$, whereas if a is odd, then $2 \nmid a$.)

Notice that if b divides a, then the quotient a/b or $\frac{a}{b}$ is a natural number. For example, 4 divides 20 because there exists a natural number k such that $20 = 4k$. The value of k here is 5, because $20 = 4 \cdot 5$. The natural number 20 is not divisible by 7, for example, since there is no natural number k satisfying $20 = 7k$. Alternatively, we think "20 divided by 7 gives quotient 2 with remainder 6" and since there is a nonzero remainder, divisibility does not hold. We write $7 \nmid 20$ to indicate that 7 does not divide 20.

If the natural number a is divisible by the natural number b, then b is a **factor** (or **divisor**) of a, and a is a **multiple** of b. For example, 5 is a factor of 30, and 30 is a multiple of 5. Also, 6 is a factor of 30, and 30 is a multiple of 6. The number 30 equals $6 \cdot 5$; this product $6 \cdot 5$ is called a **factorization** of 30. Other factorizations of 30 include $3 \cdot 10$, $2 \cdot 15$, $1 \cdot 30$, and $2 \cdot 3 \cdot 5$.

EXAMPLE 1 Checking Divisibility

Decide whether the first number is divisible by the second.

(a) 45; 9 **(b)** 60; 7 **(c)** 19; 19 **(d)** 26; 1

SOLUTION

(a) Is there a natural number k that satisfies $45 = 9k$? The answer is yes, because $45 = 9 \cdot 5$, and 5 is a natural number. Therefore, 9 divides 45, written $9 \mid 45$.

(b) Because the quotient $60 \div 7$ is not a natural number, 60 is not divisible by 7, written $7 \nmid 60$.

For any natural number a, it is true that $a \mid a$, and also that $1 \mid a$.

(c) The quotient $19 \div 19$ is the natural number 1, so 19 is divisible by 19. (In fact, any natural number is divisible by itself.)

(d) The quotient $26 \div 1$ is the natural number 26, so 26 is divisible by 1. (In fact, any natural number is divisible by 1.)

EXAMPLE 2 Finding Factors

Find all the natural number factors of each number.

(a) 36 **(b)** 50 **(c)** 11

SOLUTION

(a) To find the factors of 36, try to divide 36 by 1, 2, 3, 4, 5, 6, and so on. Doing this gives the following list of natural number factors of 36:

1, 2, 3, 4, 6, 9, 12, 18, and 36.

(b) The factors of 50 are 1, 2, 5, 10, 25, and 50.

(c) The only natural number factors of 11 are 11 and 1.

Like the number 19 in Example 1(c), the number 11 has only two natural number factors, itself and 1. Such a natural number is called a *prime number*.

The 1997 movie *Contact*, based on the Carl Sagan novel of the same name, portrays Jodie Foster as scientist Ellie Arroway. After years of searching, Ellie makes contact with intelligent life in outer space. Her contact is verified after receiving radio signals that indicate **prime numbers:** 2, 3, 5, 7, 11, and so on. Her superiors are not convinced, asking why the aliens don't just speak English. Ellie's response:

Well, maybe because 70% of the planet speaks other languages. Mathematics is the only true universal language, Senator. It's no coincidence that they're using primes . . . Prime numbers—that would be integers that are divisible only by themselves and 1.

Prime and Composite Numbers

A natural number greater than 1 that has only itself and 1 as factors is called a **prime number.** A natural number greater than 1 that is not prime is called **composite.**

Mathematicians agree that the natural number 1 is neither prime nor composite. The following alternative definition of a prime number clarifies that 1 is not a prime.

Alternative Definition of a Prime Number

A **prime number** is a natural number that has *exactly* two different natural number factors.

There is a systematic method for identifying prime numbers in a list of numbers: 2, 3, . . . , n. The method, known as the **Sieve of Eratosthenes,** is named after the Greek geographer, poet, astronomer, and mathematician, who lived from about 276 to 192 B.C. To construct such a sieve, list all the natural numbers from 2 through some given natural number n, such as 100. The number 2 is prime, but all other multiples of 2 (4, 6, 8, 10, and so on) are composite. Circle the prime 2, and cross out all other multiples of 2. The next number not crossed out and not circled is 3, the next prime. Circle the 3, and cross out all other multiples of 3 (6, 9, 12, 15, and so on) that are not already crossed out. Circle the next prime, 5, and cross out all other multiples of 5 not already crossed out. Continue this process for all primes less than or equal to the square root of the last number in the list. For this list, we may stop with 7, because the next prime, 11, is greater than the square root of 100, which is 10. At this stage, simply circle all remaining numbers that are not crossed out.

Table 1 shows the Sieve of Eratosthenes for 2, 3, 4, . . . , 100, identifying the 25 primes in that range. Theoretically, such a sieve can be constructed for any value of n.

TABLE 1 Sieve of Eratosthenes

	(2)	(3)	~~4~~	(5)	~~6~~	(7)	~~8~~	~~9~~	~~10~~	(11)	~~12~~	(13)	~~14~~
~~15~~	~~16~~	(17)	~~18~~	(19)	~~20~~	~~21~~	~~22~~	(23)	~~24~~	~~25~~	~~26~~	~~27~~	~~28~~
(29)	~~30~~	(31)	~~32~~	~~33~~	~~34~~	~~35~~	~~36~~	(37)	~~38~~	~~39~~	~~40~~	(41)	~~42~~
(43)	~~44~~	~~45~~	~~46~~	(47)	~~48~~	~~49~~	~~50~~	~~51~~	~~52~~	(53)	~~54~~	~~55~~	~~56~~
~~57~~	~~58~~	(59)	~~60~~	(61)	~~62~~	~~63~~	~~64~~	~~65~~	~~66~~	(67)	~~68~~	~~69~~	~~70~~
(71)	~~72~~	(73)	~~74~~	~~75~~	~~76~~	~~77~~	~~78~~	(79)	~~80~~	~~81~~	~~82~~	(83)	~~84~~
~~85~~	~~86~~	~~87~~	~~88~~	(89)	~~90~~	~~91~~	~~92~~	~~93~~	~~94~~	~~95~~	~~96~~	(97)	~~98~~
~~99~~	~~100~~												

EXAMPLE 3 Identifying Prime and Composite Numbers

Decide whether each number is prime or composite.

(a) 97 **(b)** 59,872 **(c)** 697

SOLUTION

(a) Because 97 is circled in Table 1, it is prime. If 97 had a smaller prime factor, 97 would have been crossed out as a multiple of that factor.

(b) The number 59,872 is even, so it is divisible by 2. It is composite. (There is only one even prime, the number 2 itself.)

(c) For 697 to be composite, there must be a number other than 697 and 1 that divides into it with remainder 0. Start by trying 2, and then 3. Neither works. There is no need to try 4. (If 4 divides with remainder 0 into a number, then 2 will also.) Try 5. There is no need to try 6 or any succeeding even number. (Why?) Try 7. Try 11. (Why not try 9?) Try 13. Keep trying numbers until one works, or until a number is tried whose square exceeds the given number, 697. Try 17:

$$697 \div 17 = 41.$$

The number 697 is composite: $697 = 17 \cdot 41.$ ■

How to Use Up Lots of Chalk

In 1903, the mathematician F. N. Cole presented before a meeting of the American Mathematical Society his discovery of a factorization of the number

$$2^{67} - 1.$$

He walked up to the chalkboard, raised 2 to the 67th power, and then subtracted 1. Then he moved over to another part of the board and multiplied out

$$193{,}707{,}721 \times 761{,}838{,}257{,}287.$$

The two calculations agreed, and Cole received a standing ovation for a presentation that did not include a single word.

An aid in determining whether a natural number is divisible by another natural number is called a **divisibility test.** Some simple divisibility tests exist for small natural numbers, and they are given in Table 2 on the next page. Divisibility tests for 7 and 11 are a bit involved, and they are discussed in the exercises for this section. Each test in the table is both a necessary and a sufficient condition. If the test statement is true, then divisibility occurs. If the test statement is not true, then divisibility does not occur.

Writing in the October 1, 1994 issue of *Science News*, Ivars Peterson gives a fascinating account of the recent discovery of a 75-year-old **factoring machine** ("Cranking Out Primes: Tracking Down a Long-lost Factoring Machine"). In 1989, Jeffrey Shallit of the University of Waterloo in Ontario came across an article in an obscure 1920 French journal, in which the author, Eugene Olivier Carissan, reported his invention of the factoring apparatus. Shallit and two colleagues embarked on a search for the machine. They contacted all telephone subscribers in France named Carissan and received a reply from Eugene Carissan's daughter. The machine was still in existence and in working condition, stored in a drawer at an astronomical observatory in Floirac, near Bordeaux.

Peterson explains in the article how the apparatus works. Using the machine, Carissan took just ten minutes to prove that 708,158,977 is a prime number, and he was able to factor a 13-digit number. While this cannot compare to what technology can accomplish today, it was a significant achievement for Carissan's day.

TABLE 2 Divisibility Tests

Divisible By	Test	Example
2	Number ends in 0, 2, 4, 6, or 8. (The last digit is even.)	9,489,994 ends in 4; it is divisible by 2.
3	Sum of the digits is divisible by 3.	897,432 is divisible by 3, since $8 + 9 + 7 + 4 + 3 + 2 = 33$ is divisible by 3.
4	Last two digits form a number divisible by 4.	7,693,432 is divisible by 4, since 32 is divisible by 4.
5	Number ends in 0 or 5.	890 and 7635 are divisible by 5.
6	Number is divisible by both 2 and 3.	27,342 is divisible by 6 since it is divisible by both 2 and 3.
8	Last three digits form a number divisible by 8.	1,437,816 is divisible by 8, since 816 is divisible by 8.
9	Sum of the digits is divisible by 9.	428,376,105 is divisible by 9 since sum of digits is 36, which is divisible by 9.
10	The last digit is 0.	897,463,940 is divisible by 10.
12	Number is divisible by both 4 and 3.	376,984,032 is divisible by 12.

EXAMPLE 4 Applying Divisibility Tests

In each case, decide whether the first number is divisible by the second.

(a) 2,984,094; 4 **(b)** 4,119,806,514; 9

SOLUTION

(a) The last two digits form the number 94. Since 94 is not divisible by 4, the given number is not divisible by 4.

(b) The sum of the digits is $4 + 1 + 1 + 9 + 8 + 0 + 6 + 5 + 1 + 4 = 39$, which is not divisible by 9. The given number is therefore not divisible by 9.

The Fundamental Theorem of Arithmetic A *composite* number can be thought of as "composed" of smaller factors. For example, 42 is composite: $42 = 6 \cdot 7$. If the smaller factors are all primes, then we have a *prime factorization*. For example, $42 = 2 \cdot 3 \cdot 7$. An important theorem in mathematics states that there is only one possible way to write the prime factorization of a given composite natural number. A form of this theorem was known to the ancient Greeks.*

*A theorem is a statement that can be proved true from other statements. For a proof of this theorem, see *What Is Mathematics?* by Richard Courant and Herbert Robbins (Oxford University Press, 1941), p. 23.

The following program, written by Charles W. Gantner and provided courtesy of Texas Instruments, can be used on the TI-83/84 Plus calculator to list all primes less than or equal to a given natural number *N*.

```
PROGRAM: PRIMES
: Disp "INPUT N ≥ 2"
: Disp "TO GET"
: Disp "PRIMES ≤ N"
: Input N
: 2→T
: Disp T
: 1→A
: Lbl 1
: A + 2→A
: 3→B
: If A > N
: Stop
: Lbl 2
: If B ≤ √(A)
: Goto 3
: Disp A
: Pause
: Goto 1
: Lbl 3
: If A/B ≤ int (A/B)
: Goto 1
: B + 2→B
: Goto 2
```

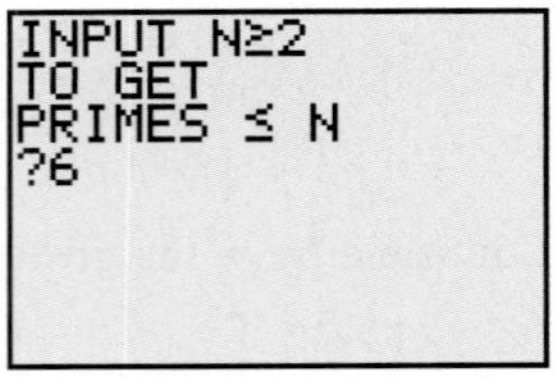

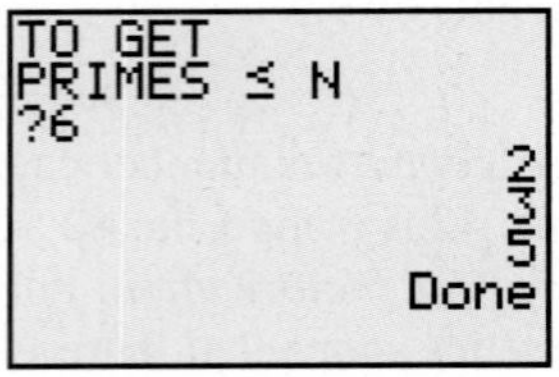

The display indicates that the primes less than or equal to 6 are 2, 3, and 5.

The Fundamental Theorem of Arithmetic

Every natural number can be expressed in one and only one way as a product of primes (if the order of the factors is disregarded). This unique product of primes is called the **prime factorization** of the natural number.

Because a prime natural number is not composed of smaller factors, its prime factorization is simply itself. For example, $17 = 17$ (or $17 = 1 \cdot 17$).

The following example shows two ways to factor a composite number into primes: (1) using a "factor tree" and (2) using repeated division.

EXAMPLE 5 Finding the Unique Prime Factorization of a Composite Number

Find the prime factorization of the number 504.

SOLUTION

The factor tree can start with $504 = 2 \cdot 252$, as shown below on the left. Then $252 = 2 \cdot 126$, and so on, until every branch of the tree ends with a prime. All the resulting prime factors are shown circled in the diagram.

Alternatively, the same factorization is obtained by repeated division by primes, as shown on the right. (In general, you would divide by the primes 2, 3, 5, 7, 11, and so on, each as many times as possible, until the answer is no longer composite.)

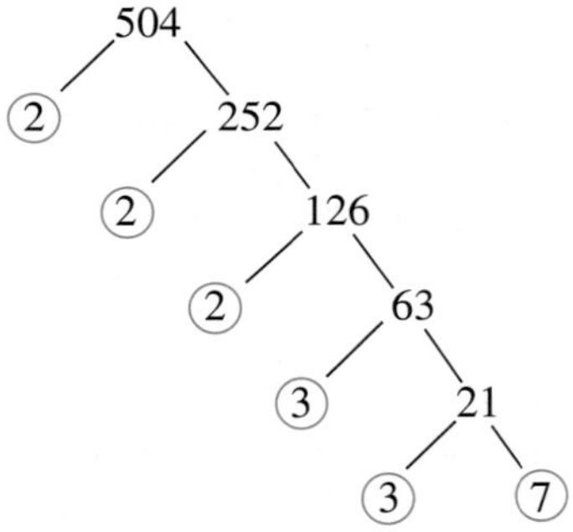

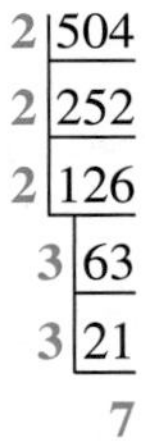

By either method, the prime factorization, in exponential form, is

$$504 = 2^3 \cdot 3^2 \cdot 7. \quad 2 \cdot 2 \cdot 2 = 2^3; 3 \cdot 3 = 3^2$$

The Infinitude of Primes

Mathematicians (amateur as well as professional) have sought for thousands of years to learn as much as possible about prime numbers. One important basic result was proved by Euclid around 300 B.C., namely that there are infinitely many primes. This means that no matter how large a prime we identify, there are always others even larger. Euclid's proof remains today as one of the most elegant proofs in all of mathematics. (An *elegant* mathematical proof is one that demonstrates the desired result in a most direct, concise manner. Mathematicians strive for elegance in their proofs.) It is called a **proof by contradiction.**

The Riemann Hypothesis is an insightful conjecture stated in the mid-1800s by Georg Friedrich Bernhard Riemann (profiled on page 601). It concerns how the prime numbers are distributed on the number line and is undoubtedly the most important unproven claim in all of mathematics. Thousands of other "theorems" are built upon the assumption of its truth. If it were ever disproved, those results would fall apart. A proof, on the other hand, may provide sufficient understanding of the primes to demolish public key cryptography, upon which rests the security of all Internet commerce (among other things).

Two excellent books on the Riemann Hypothesis, and the mathematicians at the forefront of its study over the years, both published in 2003, are *Prime Obsession*, by John Derbyshire, and *The Music of the Primes*, by Marcus du Sautoy.

A statement can be proved by contradiction as follows: We assume that the negation of the statement is true. The assumption that the negation is true is used to produce some sort of contradiction, or absurdity. The fact that the negation of the original statement leads to a contradiction means that the original statement must be true.

In order to better understand a particular part of the proof that there are infinitely many primes, it is helpful to examine the following argument.

Suppose that $M = 2 \cdot 3 \cdot 5 \cdot 7 + 1 = 211$. Now M is the product of the first four prime numbers, plus 1. If we divide 211 by each of the primes 2, 3, 5, and 7, the remainder is always 1.

$$\begin{array}{r} 105 \\ 2\overline{)211} \\ \underline{105} \\ 1 \end{array} \qquad \begin{array}{r} 70 \\ 3\overline{)211} \\ \underline{210} \\ 1 \end{array} \qquad \begin{array}{r} 42 \\ 5\overline{)211} \\ \underline{210} \\ 1 \end{array} \qquad \begin{array}{r} 30 \\ 7\overline{)211} \\ \underline{210} \\ 1 \end{array}$$

All remainders are 1.

So 211 is not divisible by any of the primes 2, 3, 5, and 7.

Now we are ready to prove that there are infinitely many primes. If it can be shown that *there is no largest prime number*, then there must be infinitely many primes.

THEOREM

Statement: There is no largest prime number.

Proof: Suppose that there is a largest prime number and call it P. Now form the number M such that

$$M = p_1 \cdot p_2 \cdot p_3 \cdots P + 1,$$

where $p_1, p_2, p_3, \ldots, P$ represent all the primes less than or equal to P. Now the number M must be either prime or composite.

1. Suppose that M is prime.
   M is obviously larger than P, so if M is prime, it is larger than the assumed largest prime P. We have reached a *contradiction*.
2. Suppose that M is composite.
   If M is composite, it must have a prime factor. But none of $p_1, p_2, p_3, \ldots, P$ are factors of M, because division by each will leave a remainder of 1. (Recall the above argument.) So if M has a prime factor, it must be greater than P. But this is a *contradiction*, because P is the assumed largest prime.

In either case 1 or 2, we reach a contradiction. The whole argument was based upon the assumption that a largest prime exists, but as this leads to contradictions, there must be no largest prime, or equivalently, there are infinitely many primes. ■

During the first season of the CBS television series *NUMB3RS,* the episode *Prime Suspect* was based on the premise that a mathematician was very close to proving the Riemann Hypothesis (which later proved to be erroneous). His daughter was kidnapped by people who were hoping to exchange her for the results in order to unlock financial security codes.

The Search for Large Primes Identifying larger and larger prime numbers and factoring large composite numbers into their prime components is of great practical importance today, because it is the basis of modern **cryptography systems,** or secret codes. Various codes have been used for centuries in military applications. Today the security of vast amounts of industrial, business, and personal data also depend upon the theory of prime numbers. See the Extension following Section 5.3.

At one time, $2^{11,213}-1$ was the largest known **Mersenne prime.** To honor its discovery, the Urbana, Illinois, post office used the cancellation picture above.

As mathematicians continue to search for larger and larger primes, a formula for generating all the primes would be nice (something similar, for example, to the formula $2n$, which generates all even counting numbers for $n = 1, 2, 3, \ldots$, or the formula n^2, which generates all the perfect squares). Numbers generated by the formula $M_n = 2^n - 1$ are called **Mersenne numbers** to honor the French monk Marin Mersenne (1588–1648). It was long known that a *composite* value of n would always generate a composite Mersenne number. (See Exercises 83–88.) And some early mathematicians believed (incorrectly) that a *prime* value of n would always generate a prime Mersenne number. That is, they believed that, starting with any known prime number n, one could always produce another, larger prime number $2^n - 1$. (Although "always generate a prime" is not the same as "generate all primes," at least it would have been an unending source of guaranteed primes.)

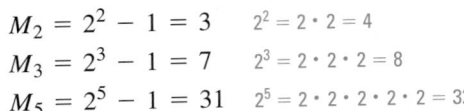

EXAMPLE 6 Finding Mersenne Numbers

Find each Mersenne number M_n for $n = 2, 3$, and 5.

SOLUTION

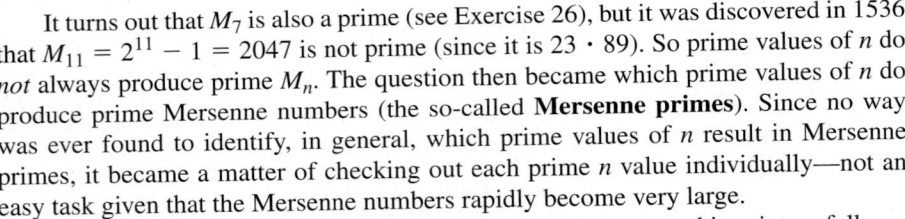

$$M_2 = 2^2 - 1 = 3 \qquad 2^2 = 2 \cdot 2 = 4$$
$$M_3 = 2^3 - 1 = 7 \qquad 2^3 = 2 \cdot 2 \cdot 2 = 8$$
$$M_5 = 2^5 - 1 = 31 \qquad 2^5 = 2 \cdot 2 \cdot 2 \cdot 2 \cdot 2 = 32$$

Note that all three values, 3, 7, and 31, are indeed primes.

Marin Mersenne (1588–1648), in his *Cogitata Physico-Mathematica* (1644), claimed that M_n was prime for $n = 2, 3, 5, 7, 13, 17, 19, 31, 67, 127$, and 257, and composite for all other prime numbers n less than 257. Other mathematicians at the time knew that Mersenne could not have actually tested all these values, but no one else could prove or disprove them either. It was more then 300 years later before all primes up to 257 were legitimately checked out, and Mersenne was finally revealed to have made five errors:

M_{61} is prime.
M_{67} is composite.
M_{89} is prime.
M_{107} is prime.
M_{257} is composite.

It turns out that M_7 is also a prime (see Exercise 26), but it was discovered in 1536 that $M_{11} = 2^{11} - 1 = 2047$ is not prime (since it is $23 \cdot 89$). So prime values of n do *not* always produce prime M_n. The question then became which prime values of n do produce prime Mersenne numbers (the so-called **Mersenne primes**). Since no way was ever found to identify, in general, which prime values of n result in Mersenne primes, it became a matter of checking out each prime n value individually—not an easy task given that the Mersenne numbers rapidly become very large.

We can summarize the discussion of Mersenne numbers, up to this point, as follows.

Mersenne Numbers and Mersenne Primes

For $n = 1, 2, 3, \ldots$, the **Mersenne numbers** are those generated by the formula

$$\mathbf{M_n = 2^n - 1.}$$

(1) If n is composite, then M_n is also composite.
(2) If n is prime, then M_n may be either prime or composite.

The prime values of M_n are called the **Mersenne primes.** Large primes being verified currently (though at a rather gradual pace) are commonly Mersenne primes.

The Mersenne prime search yielded results slowly. By about 1600, M_n had been verified as prime for all prime n up to 19 (except for 11, as mentioned above). The next one was M_{31}, verified by Euler in 1732. In 1876, French mathematician Edouard Lucas used a clever test he had developed to show that M_{127} (a 39-digit number) is prime. In the 1930s Lucas's method was further simplified by D. H. Lehmer, and the

Prime Does Pay On December 15, 2005, the 43rd known Mersenne prime was identified by Curtis Cooper and Steven Boone of Central Missouri State University, using a 2 gigahertz (upgraded to 3 gigahertz) Pentium 4 computer, as part of the GIMPS program. The Electronic Frontier Foundation has paid \$50,000 for record primes exceeding a million digits and offers \$100,000 for the first ten-million-digit prime. Because the 2005 record has 9,152,052 digits, it left the big prize still "up for grabs."

If you would like to join the **Great Internet Mersenne Prime Search** (or just find out more about it), check out www.mersenne.org.

testing of Mersenne numbers for primality has been done ever since with the Lucas–Lehmer test. In 1952 an early computer verified that M_{521}, M_{607}, M_{1279}, M_{2203}, and M_{2281} are primes.

Over the last half century, most new record-breaking primes have been identified by computer algorithms devised and implemented by mathematicians and programmers. In 1996, the **Great Internet Mersenne Prime Search (GIMPS)** was launched. Since then, many thousands of individuals have signed on, receiving free software and source code to run during slack time on personal computers throughout the world. The nine largest record Mersenne primes have been found by GIMPS.

As of early 2006, the record largest prime (discovered on December 15, 2005) was

$$M_{30,402,457} = 2^{30,402,457} - 1,$$

a number with 9,152,052 digits. It was the 43rd known Mersenne prime.

During the same general period that Mersenne was thinking about prime numbers, Pierre de Fermat (about 1601–1665) conjectured that the formula

$$2^{2^n} + 1$$

would always produce a prime, for any whole number value of n. Table 3 shows how this formula generates the first four **Fermat numbers,** which are all primes. The fifth Fermat number (from $n = 4$) is likewise prime. Fermat had verified these first five by around 1630. But the sixth Fermat number (from $n = 5$) turns out to be 4,294,967,297, which is *not* prime. (See Exercises 75 and 76.) To date, no more primes have been found among the Fermat numbers.

TABLE 3 The Generation of Fermat Numbers

n	2^n	2^{2^n}	$2^{2^n} + 1$
0	1	2	3
1	2	4	5
2	4	16	17
3	8	256	257

Of historical note are a couple of polynomial formulas that produce primes. (A *polynomial* in a given variable involves adding or subtracting integer multiples of whole number powers of the variable. Polynomials are among the most basic mathematical functions. They are discussed in Section 7.6.) In 1732, Leonhard Euler offered the formula $n^2 - n + 41$, which generates primes for n up to 40 and fails at $n = 41$. In 1879, E. B. Escott produced more primes with the formula $n^2 - 79n + 1601$, which first fails at $n = 80$.

EXAMPLE 7 Finding Numbers Using Euler's and Escott's Formulas

Find the first five numbers produced by each of the polynomial formulas of Euler and Escott.

SOLUTION

Table 4 on the next page shows the required numbers.

TABLE 4 A Few Polynomial-Generated Prime Numbers

n	Euler formula $n^2 - n + 41$	Escott formula $n^2 - 79n + 1601$
1	41	1523
2	43	1447
3	47	1373
4	53	1301
5	61	1231

All values found here are primes. (Use Table 1 to verify the Euler values.) ■

Actually, it is not hard to prove that there can be no polynomial that will consistently generate primes. More complicated mathematical formulas exist for generating primes, but none produced so far can be practically applied in a reasonable amount of time, even using the fastest computers.

5.1 EXERCISES

Decide whether each statement is true *or* false.

1. Every natural number is divisible by 1.

2. No natural number is both prime and composite.

3. There are no even prime numbers.

4. If n is a natural number and $9|n$, then $3|n$.

5. If n is a natural number and $5|n$, then $10|n$.

6. 1 is the least prime number.

7. Every natural number is both a factor and a multiple of itself.

8. If 16 divides a natural number, then 2, 4, and 8 must also divide that natural number.

9. The composite number 50 has exactly two prime factorizations.

10. The prime number 53 has exactly two natural number factors.

11. The number $2^{11} - 1$ is an example of a Mersenne prime.

12. As of early 2006, only five Fermat primes had ever been found.

Find all natural number factors of each number.

13. 12 **14.** 18

15. 20 **16.** 28

17. 120 **18.** 172

Use divisibility tests to decide whether the given number is divisible by each number.

(a) 2 **(b)** 3 **(c)** 4 **(d)** 5 **(e)** 6 **(f)** 8
(g) 9 **(h)** 10 **(i)** 12

19. 315 **20.** 630

21. 25,025 **22.** 45,815

23. 123,456,789 **24.** 987,654,321

25. **(a)** In constructing the Sieve of Eratosthenes for 2 through 100, we said that any composite in that range had to be a multiple of some prime less than or equal to 7 (since the next prime, 11, is greater than the square root of 100). Explain.

(b) To extend the Sieve of Eratosthenes to 200, what is the largest prime whose multiples would have to be considered?

(c) Complete this statement: In seeking prime factors of a given number, we need only consider all primes up to and including the ______ ____ of that number, since a prime factor greater than the ______ ____ can only occur if there is at least one other prime factor less than the ______ ____.

(d) Complete this statement: If no prime less than or equal to $\sqrt{n}$ divides n, then n is a ______ number.

26. (a) Continue the Sieve of Eratosthenes in Table 1 from 101 to 200 and list the primes between 100 and 200. How many are there?
(b) From your list in part (a), verify that the Mersenne number M_7 is indeed prime.

27. List two primes that are consecutive natural numbers. Can there be any others?

28. Can there be three primes that are consecutive natural numbers? Explain.

29. For a natural number to be divisible by both 2 and 5, what must be true about its last digit?

30. Consider the divisibility tests for 2, 4, and 8 (all powers of 2). Use inductive reasoning to predict the divisibility test for 16. Then, use the test to show that 456,882,320 is divisible by 16.

31. Redraw the factor tree of Example 5, assuming that you first observe that $504 = 12 \cdot 42$, then that $12 = 3 \cdot 4$ and $42 = 6 \cdot 7$. Complete the process and give the resulting prime factorization.

32. Explain how your result in Exercise 31 verifies the fundamental theorem of arithmetic.

Find the prime factorization of each composite number.

33. 240 **34.** 300

35. 360 **36.** 425

37. 663 **38.** 885

Here is a divisibility test for 7.

(a) *Double the last digit of the given number, and subtract this value from the given number with the last digit omitted.*
(b) *Repeat the process of part* **(a)** *as many times as necessary until the number obtained can easily be divided by 7.*
(c) *If the final number obtained is divisible by 7, then the given number also is divisible by 7. If the final number is not divisible by 7, then neither is the given number.*

Use this divisibility test to determine whether each number is divisible by 7.

39. 142,891 **40.** 409,311

41. 458,485 **42.** 287,824

Here is a divisibility test for 11.

(a) *Starting at the left of the given number, add together every other digit.*
(b) *Add together the remaining digits.*
(c) *Subtract the smaller of the two sums from the larger. (If they are the same, the difference is 0.)*
(d) *If the final number obtained is divisible by 11, then the given number also is divisible by 11. If the final number is not divisible by 11, then neither is the given number.*

Use this divisibility test to determine whether each number is divisible by 11.

43. 8,493,969 **44.** 847,667,942

45. 453,896,248 **46.** 552,749,913

47. Consider the divisibility test for the composite number 6, and make a conjecture for the divisibility test for the composite number 15.

48. Explain what is meant by a "proof by contradiction."

49. Give two factorizations of the number 75 that are not prime factorizations.

50. Explain, in general, when a factorization is a prime factorization.

Determine all possible digit replacements for x so that the first number is divisible by the second. For example, 37,58x is divisible by 2 if

$$x = 0, 2, 4, 6, \text{ or } 8.$$

51. 398,87x; 2 **52.** 2,45x,765; 3

53. 64,537,84x; 4 **54.** 2,143,89x; 5

55. 985,23x; 6 **56.** 7,643,24x; 8

57. 4,329,7x5; 9 **58.** 23,x54,470; 10

There is a method to determine the **number of divisors** *of a composite number. To do this, write the composite number in its prime factored form, using exponents. Add 1 to each exponent and multiply these numbers. Their product gives the number of divisors of the composite number. For example,*

$$24 = 2^3 \cdot 3 = 2^3 \cdot 3^1.$$

Now add 1 to each exponent:

$$3 + 1 = 4, 1 + 1 = 2.$$

Multiply $4 \cdot 2$ to get 8. There are 8 divisors of 24. (Because 24 is rather small, this can be verified easily. The divisors are 1, 2, 3, 4, 6, 8, 12, and 24, a total of eight as predicted.)

(continued)

Find the number of divisors of each composite number.

59. 48

60. 144

61. $2^8 \cdot 3^2$

62. $2^4 \cdot 3^4 \cdot 5^2$

Leap years occur when the year number is divisible by 4. An exception to this occurs when the year number is divisible by 100 (that is, it ends in two zeros). In such a case, the number must be divisible by 400 in order for the year to be a leap year. Determine which years are leap years.

63. 1776

64. 1894

65. 2400

66. 1800

67. Why is the following *not* a valid divisibility test for 8? "A number is divisible by 8 if it is divisible by both 4 and 2." Support your answer with an example.

68. Choose any three consecutive natural numbers, multiply them together, and divide the product by 6. Repeat this several times, using different choices of three consecutive numbers. Make a conjecture concerning the result.

69. Explain why the product of three consecutive natural numbers must be divisible by 6.

70. Choose any 6-digit number consisting of three digits followed by the same three digits in the same order (for example, 467,467). Divide by 13. Divide by 11. Divide by 7. What do you notice? Why do you think this happens?

71. Verify that Euler's polynomial prime-generating formula

$$n^2 - n + 41$$

fails to produce a prime for $n = 41$.

72. Evaluate Euler's polynomial formula for **(a)** $n = 42$, and **(b)** $n = 43$.

73. Choose the correct completion: For $n > 41$, Euler's formula produces a prime
A. never. **B.** sometimes. **C.** always.
(*Hint:* If no prime less than or equal to $\sqrt{n}$ divides n, then n is prime.)

74. Recall that Escott's formula, $n^2 - 79n + 1601$, fails to produce a prime for $n = 80$. Evaluate this formula for $n = 81$ and $n = 82$ and then complete the following statement: For $n > 80$, Escott's formula produces a prime
A. never. **B.** sometimes. **C.** always.

75. (a) Evaluate the Fermat number $2^{2^n} + 1$ for $n = 4$.
(b) In seeking possible prime factors of the Fermat number of part (a), what is the largest potential prime factor that one would have to try? (As stated in the text, this "fifth" Fermat number is in fact prime.)

76. (a) Verify the value given in the text for the "sixth" Fermat number (i.e., $2^{2^5} + 1$).
(b) Divide this Fermat number by 641. (Euler discovered this factorization in 1732, proving that the sixth Fermat number is not prime.)

77. Write a short report on the Great Internet Mersenne Prime Search (GIMPS).

78. Write a short report identifying the 40th, 41st, and 42nd known Mersenne primes and how, when, and by whom they were found.

79. The Mersenne margin note on page 228 cites a 1644 claim that was not totally resolved for some 300 years. Find out when, and by whom, Mersenne's five errors were demonstrated. (*Hint:* One was mentioned in the margin note on page 224.)

80. In Euclid's proof that there is no largest prime, we formed a number M by taking the product of primes and adding 1. Observe the pattern below.

$M = 2 + 1 = 3$	(3 is prime)
$M = 2 \cdot 3 + 1 = 7$	(7 is prime)
$M = 2 \cdot 3 \cdot 5 + 1 = 31$	(31 is prime)
$M = 2 \cdot 3 \cdot 5 \cdot 7 + 1 = 211$	(211 is prime)
$M = 2 \cdot 3 \cdot 5 \cdot 7 \cdot 11 + 1 = 2311$	(2311 is prime)

It seems as though this pattern will always yield a prime number. Now evaluate

$$M = 2 \cdot 3 \cdot 5 \cdot 7 \cdot 11 \cdot 13 + 1.$$

81. Is M prime or composite? If composite, give its prime factorization.

82. Explain in your own words the proof by Euclid that there is no largest prime.

The text stated that the Mersenne number M_n is composite whenever n is composite. Exercises 83–86 on the next page develop one way you can always find a factor of such a Mersenne number.

83. For the composite number $n = 6$, find

$$M_n = 2^n - 1.$$

84. Notice that $p = 3$ is a prime factor of $n = 6$. Find $2^p - 1$ for $p = 3$. Is $2^p - 1$ a factor of $2^n - 1$?

85. Complete this statement: If p is a prime factor of n, then _____ is a factor of the Mersenne number $2^n - 1$.

86. Find $M_n = 2^n - 1$ for $n = 10$.

87. Use the statement of Exercise 85 to find two distinct factors of M_{10}.

88. Do you think this procedure will always produce *prime* factors of M_n for composite n? (*Hint:* Consider $n = 22$ and its prime factor $p = 11$, and recall the statement following Example 6.) Explain.

5.2 Selected Topics from Number Theory

Perfect Numbers • Deficient and Abundant Numbers • Amicable (Friendly) Numbers • Goldbach's Conjecture • Twin Primes • Fermat's Last Theorem

The mathematician **Albert Wilansky,** when phoning his brother-in-law, Mr. Smith, noticed an interesting property concerning Smith's phone number (493–7775). The number 4,937,775 is composite, and its prime factorization is $3 \cdot 5 \cdot 5 \cdot 65{,}837$. When the digits of the phone number are added, the result, 42, is equal to the sum of the digits in the prime factors: $3 + 5 + 5 + 6 + 5 + 8 + 3 + 7 = 42$. Wilansky termed such a number a **Smith number.** In 1985 it was proved that there are infinitely many Smith numbers, but there still are many unanswered questions about them.

Perfect Numbers

In an earlier chapter we introduced figurate numbers, a topic investigated by the Pythagoreans. This group of Greek mathematicians and musicians held their meetings in secret, and were led by Pythagoras. In this section we examine some of the other special numbers that fascinated the Pythagoreans and are still studied by mathematicians today.

Divisors of a natural number were covered in Section 5.1. The **proper divisors** of a natural number include all divisors of the number except the number itself. For example, the proper divisors of 8 are 1, 2, and 4. (8 is *not* a proper divisor of 8.)

Perfect Numbers

A natural number is said to be **perfect** if it is equal to the sum of its proper divisors.

Is 8 perfect? No, because $1 + 2 + 4 = 7$, and $7 \neq 8$. The least perfect number is 6, because the proper divisors of 6 are 1, 2, and 3, and

$$1 + 2 + 3 = 6. \quad \text{6 is perfect.}$$

EXAMPLE 1 Verifying a Perfect Number

Show that 28 is a perfect number.

SOLUTION

The proper divisors of 28 are 1, 2, 4, 7, and 14. The sum of these is 28:

$$1 + 2 + 4 + 7 + 14 = 28.$$

By the definition, 28 is perfect.

The numbers 6 and 28 are the two least perfect numbers. The next two are 496 and 8128. The pattern of these first four perfect numbers led early writers to conjecture that

1. The nth perfect number contains exactly n digits.
2. The even perfect numbers end in the digits 6 and 8, alternately.

} Conjectures

(Exercises 41–43 will help you evaluate these conjectures.)

There still are many unanswered questions about perfect numbers. Euclid showed that if $2^n - 1$ is prime, then $2^{n-1}(2^n - 1)$ is perfect, and conversely. Because the prime values of $2^n - 1$ are the Mersenne primes (discussed in the previous section), this means that for every new Mersenne prime discovered, another perfect number is automatically revealed. (Hence, as of early 2006, there were also 43 known perfect numbers.) It is also known that all even perfect numbers must take the form $2^{n-1}(2^n - 1)$ and it is strongly suspected that no odd perfect numbers exist. (Any odd one would have at least eight different prime factors and would have at least 300 decimal digits.) Therefore, Euclid and the early Greeks most likely identified the form of all perfect numbers.

Deficient and Abundant Numbers Earlier we saw that 8 is not perfect because it is not equal to the sum of its proper divisors ($8 \neq 7$). Next we define two alternative categories for natural numbers that are *not* perfect.

A number is said to be a **weird number** if it is abundant without being equal to the sum of any set of its own proper divisors. For example, 70 is weird because it is abundant ($1 + 2 + 5 + 7 + 10 + 14 + 35 = 74 > 70$), but no set of the factors 1, 2, 5, 7, 10, 14, 35 adds up to 70.

Deficient and Abundant Numbers

A natural number is **deficient** if it is greater than the sum of its proper divisors. It is **abundant** if it is less than the sum of its proper divisors.

Based on this definition, a *deficient number* is one with proper divisors that add up to less than the number itself, while an *abundant number* is one with proper divisors that add up to more than the number itself. For example, because the proper divisors of 8 (1, 2, and 4) add up to 7, which is less than 8, the number 8 is deficient.

EXAMPLE 2 Identifying Deficient and Abundant Numbers

Decide whether each number is deficient or abundant.

(a) 12 **(b)** 10

SOLUTION

(a) The proper divisors of 12 are 1, 2, 3, 4, and 6. The sum of these divisors is 16. Because $16 > 12$, the number 12 is abundant.

(b) The proper divisors of 10 are 1, 2, and 5. Since $1 + 2 + 5 = 8$, and $8 < 10$, the number 10 is deficient.

Amicable (Friendly) Numbers Suppose that we add the proper divisors of 284:

$$1 + 2 + 4 + 71 + 142 = 220.$$

Their sum is 220. Now, add the proper divisors of 220:

$$1 + 2 + 4 + 5 + 10 + 11 + 20 + 22 + 44 + 55 + 110 = 284.$$

An extension of the idea of amicable numbers results in **sociable numbers.** In a chain of sociable numbers, the sum of the proper divisors of each number is the next number in the chain, and the sum of the proper divisors of the last number in the chain is the first number. Here is a 5-link chain of sociable numbers:

12,496
14,288
15,472
14,536
14,264.

The number 14,316 starts a 28-link chain of sociable numbers.

Notice that the sum of the proper divisors of 220 is 284, while the sum of the proper divisors of 284 is 220. Number pairs such as these are said to be *amicable*, or *friendly*.

Amicable or Friendly Numbers

The natural numbers a and b are **amicable,** or **friendly,** if the sum of the proper divisors of a is b, and the sum of the proper divisors of b is a.

The smallest pair of amicable numbers, 220 and 284, was known to the Pythagoreans, but it was not until more than 1000 years later that the next pair, 17,296 and 18,416, was discovered. Many more pairs were found over the next few decades, but it took a 16-year-old Italian boy named Nicolo Paganini to discover in the year 1866 that the pair of amicable numbers 1184 and 1210 had been overlooked for centuries!

Today, powerful computers continually extend the lists of known amicable pairs. The last time we checked, over ten million pairs were known. It still is unknown, however, if there are infinitely many such pairs. Finally, no one has found an amicable pair without prime factors in common, but the possibility of such a pair has not been eliminated.

A Dull Number? The Indian mathematician **Srinivasa Ramanujan** (1887–1920) developed many ideas in number theory. His friend and collaborator on occasion was G. H. Hardy, also a number theorist and professor at Cambridge University in England.

A story has been told about Ramanujan that illustrates his genius. Hardy once mentioned to Ramanujan that he had just taken a taxicab with a rather dull number: 1729. Ramanujan countered by saying that this number isn't dull at all; it is the smallest natural number that can be expressed as the sum of two cubes in two different ways:

$$1^3 + 12^3 = 1729$$
$$\text{and} \quad 9^3 + 10^3 = 1729.$$

Show that 85 can be written as the sum of two squares in two ways.

Goldbach's Conjecture One of the most famous unsolved problems in mathematics is Goldbach's conjecture. The mathematician Christian Goldbach (1690–1764) stated the following conjecture (guess).

Goldbach's Conjecture (Not Proved)

Every even number greater than 2 can be written as the sum of two prime numbers.

Examples: $8 = 5 + 3$

$10 = 5 + 5$ (or $10 = 7 + 3$)

Mathematicians have tried to prove the conjecture but have not been successful. However, the conjecture has been verified (as of early 2006) for numbers up to 2×10^{17}.

EXAMPLE 3 Expressing Numbers as Sums of Primes

Write each even number as the sum of two primes.

(a) 18 **(b)** 60

SOLUTION

(a) $18 = 5 + 13$. Another way of writing it is $7 + 11$. Notice that $1 + 17$ is *not* valid because by definition 1 is not a prime number.

(b) $60 = 7 + 53$. Can you find other ways? Why is $3 + 57$ not valid?

Mathematics professor Gregory Larkin, played by Jeff Bridges, woos colleague Rose Morgan (Barbra Streisand) in the 1996 film *The Mirror Has Two Faces*. Larkin's research and book focus on the **twin prime conjecture,** which he correctly states in a dinner scene. He is amazed that his nonmathematician friend actually understands what he is talking about.

Twin Primes Prime numbers that differ by 2 are called **twin primes.** Some twin primes are 3 and 5, 5 and 7, 11 and 13, and so on. Like Goldbach's conjecture, the following conjecture about twin primes has never been proved, although significant progress toward a proof was announced in 2005.

Twin Prime Conjecture (Not Proved)

There are infinitely many pairs of twin primes.

You may wish to verify that there are eight such pairs less than 100, using the Sieve of Eratosthenes in Table 1. As of early 2006, the largest known twin primes were

$$16{,}869{,}987{,}339{,}975 \cdot 2^{171{,}960} \pm 1.$$

Each contains 51,779 digits.

Recall from Section 5.1 that Euclid's proof of the infinitude of primes used numbers of the form $p_1 \cdot p_2 \cdot p_3 \cdots p_n + 1$, where all the ps are prime. It may seem that any such number must be prime, but that is not so. (See Exercise 80 of Section 5.1.) However, this form often does produce primes (as does the same form with the plus replaced by a minus). When *all* the primes up to p_n are included, the resulting numbers, if prime, are called **primorial primes.** They are denoted

$$p\# \pm 1.$$

For example, $5\# + 1 = 2 \cdot 3 \cdot 5 + 1 = 31$ is a primorial prime. (In late 2005, the largest known primorial prime was $392{,}113\# + 1$, a number with 169,966 digits.) The primorial primes are a popular place to look for twin primes.

Sophie Germain (1776–1831) studied at the École Polytechnique in Paris in a day when female students were not admitted. A **Sophie Germain prime** is an odd prime p for which $2p + 1$ also is prime. Lately, large Sophie Germain primes have been discovered at the rate of several per year. As of early 2006, the largest one known was $137{,}211{,}941{,}292{,}195 \cdot 2^{171{,}960} - 1$, which has 51,780 digits.

Source: www.utm.edu/research/primes

EXAMPLE 4 Verifying Twin Primes

Verify that the primorial formula $p\# \pm 1$ produces twin prime pairs for both **(a)** $p = 3$ and **(b)** $p = 5$.

SOLUTION

(a) $3\# \pm 1 = 2 \cdot 3 \pm 1 = 6 \pm 1 = 5$ and 7 Twin primes

Multiply, then add and subtract.

(b) $5\# \pm 1 = 2 \cdot 3 \cdot 5 \pm 1 = 30 \pm 1 = 29$ and 31 Twin primes

Fermat's Last Theorem In any right triangle with shorter sides a and b, and longest side (hypotenuse) c, the equation $a^2 + b^2 = c^2$ will hold true. This is the famous Pythagorean theorem. For example,

$$3^2 + 4^2 = 5^2 \quad a = 3, b = 4, c = 5$$
$$9 + 16 = 25$$
$$25 = 25.$$

It is known that there are infinitely many such triples (a, b, c) that satisfy the equation

$$a^2 + b^2 = c^2.$$

Is something similar true of the equation

$$a^n + b^n = c^n$$

for natural numbers $n \geq 3$? Pierre de Fermat, profiled in a margin note on page 241, thought that not only were there not infinitely many such triples, but that there were, in fact, none. He made the following claim in the 1600s.

Fermat's Last Theorem (Proved in the 1990s)

For *any* natural number $n \geq 3$, there are *no* triples (a, b, c) that satisfy the equation

$$\boldsymbol{a^n + b^n = c^n}.$$

Fermat wrote in the margin of a book that he had "a truly wonderful proof" for this, but that the margin was "too small to contain it." Did he indeed have a proof, or did he have an incorrect proof?

Whatever the case, Fermat's assertion was the object of some 350 years of attempts by mathematicians to provide a suitable proof. While it was verified for many specific cases (Fermat himself proved it for $n = 3$), a proof of the general case could not be found until the Princeton mathematician Andrew Wiles announced a proof in the spring of 1993. Although some flaws were discovered in his argument, Wiles was able, by the fall of 1994, to repair and even improve the proof.

There were probably about 100 mathematicians around the world qualified to understand the Wiles proof. Many of these examined and approved it. Today Fermat's Last Theorem finally is regarded by the mathematics community as officially proved.

The solution to the Chapter Opener problem is as follows.

The first term on the left side, 1782^{12}, must be an even number, because raising an even number to any power yields an even number. The second term on the left side, 1841^{12}, must be odd, because raising an odd number to any power yields an odd number (in this case, we know that it must have 1 as units digit as well). The sum on the left side must be odd, because even + odd = odd.

The right side must be an even number using the earlier reasoning. So the equation indicates that an odd number is equal to an even number, which is impossible.

EXAMPLE 5 Using a Theorem Proved by Fermat

One of the theorems legitimately proved by Fermat is as follows:

Every odd prime can be expressed as the difference of two squares in one and only one way.

Express each odd prime as the difference of two squares.

(a) 3 **(b)** 7

SOLUTION

(a) $3 = 4 - 1 = 2^2 - 1^2$

(b) $7 = 16 - 9 = 4^2 - 3^2$ ■

For Further Thought

Curious and Interesting

One of the most remarkable books on number theory is *The Penguin Dictionary of Curious and Interesting Numbers* (1986) by David Wells. This book contains fascinating numbers and their properties, including the following.

- There are only three sets of three digits that form prime numbers in all possible arrangements: {1, 1, 3}, {1, 9, 9}, {3, 3, 7}.
- Find the sum of the cubes of the digits of 136: $1^3 + 3^3 + 6^3 = 244$. Repeat the process with the digits of 244: $2^3 + 4^3 + 4^3 = 136$. We're back to where we started.
- 635,318,657 is the least number that can be expressed as the sum of two fourth powers in two ways:

$$635{,}318{,}657 = 59^4 + 158^4 = 133^4 + 134^4.$$

- The number 24,678,050 has an interesting property:

$$24{,}678{,}050 = 2^8 + 4^8 + 6^8 + 7^8 + 8^8 + 0^8 + 5^8 + 0^8.$$

- The number 54,748 has a similar interesting property:

$$54{,}748 = 5^5 + 4^5 + 7^5 + 4^5 + 8^5.$$

- The number 3435 has this property:

$$3435 = 3^3 + 4^4 + 3^3 + 5^5.$$

For anyone whose curiosity is piqued by such facts, this book is for you!

For Group Discussion or Individual Investigation

Have each student in the class choose a three-digit number that is a multiple of 3. Add the cubes of the digits. Repeat the process until the same number is obtained over and over. Then, have the students compare their results. What is curious and interesting about this process?

5.2 Exercises

Decide whether each statement in Exercises 1–10 is true *or* false.

1. There are infinitely many prime numbers.

2. The prime numbers 2 and 3 are twin primes.

3. There is no perfect number between 496 and 8128.

4. $2^n - 1$ is prime if and only if $2^{n-1}(2^n - 1)$ is perfect.

5. Any prime number must be deficient.

6. The equation $17 + 51 = 68$ verifies Goldbach's conjecture for the number 68.

7. There are more Mersenne primes known than there are perfect numbers.

8. The number 31 cannot be represented as the difference of two squares.

9. The number $2^6(2^7 - 1)$ is perfect.

10. A natural number greater than 1 will be one and only one of the following: perfect, deficient, or abundant.

11. The proper divisors of 496 are 1, 2, 4, 8, 16, 31, 62, 124, and 248. Use this information to verify that 496 is perfect.

12. The proper divisors of 8128 are 1, 2, 4, 8, 16, 32, 64, 127, 254, 508, 1016, 2032, and 4064. Use this information to verify that 8128 is perfect.

13. As mentioned in the text, when $2^n - 1$ is prime, $2^{n-1}(2^n - 1)$ is perfect. By letting $n = 2, 3, 5,$ and 7, we obtain the first four perfect numbers. Show that $2^n - 1$ is prime for $n = 13$, and then find the decimal digit representation for the fifth perfect number.

14. At the end of 2005, the largest known prime number was $2^{30{,}402{,}457} - 1$. Use the formula in Exercise 13 to

write an expression for the perfect number generated by this prime number.

15. It has been proved that the reciprocals of *all* the positive divisors of a perfect number have a sum of 2. Verify this for the perfect number 6.

16. Consider the following equations.

$$6 = 1 + 2 + 3$$
$$28 = 1 + 2 + 3 + 4 + 5 + 6 + 7$$

Show that a similar equation is valid for the third perfect number, 496.

Determine whether each number is abundant *or* deficient.

17. 36 **18.** 30

19. 75 **20.** 95

21. There are four abundant numbers between 1 and 25. Find them. (*Hint:* They are all even, and no prime number is abundant.)

22. Explain why a prime number must be deficient.

23. The first odd abundant number is 945. Its proper divisors are 1, 3, 5, 7, 9, 15, 21, 27, 35, 45, 63, 105, 135, 189, and 315. Use this information to verify that 945 is abundant.

24. Explain in your own words the terms *perfect number, abundant number,* and *deficient number.*

25. Nicolo Paganini's numbers 1184 and 1210 are amicable. The proper divisors of 1184 are 1, 2, 4, 8, 16, 32, 37, 74, 148, 296, and 592. The proper divisors of 1210 are 1, 2, 5, 10, 11, 22, 55, 110, 121, 242, and 605. Use the definition of amicable (friendly) numbers to show that they are indeed amicable.

26. An Arabian mathematician of the ninth century stated the following.

If the three numbers

$$x = 3 \cdot 2^{n-1} - 1,$$
$$y = 3 \cdot 2^{n} - 1,$$
and
$$z = 9 \cdot 2^{2n-1} - 1$$

are all prime and $n \geq 2$, then $2^n xy$ and $2^n z$ are amicable numbers.

(a) Use $n = 2$, and show that the result is the least pair of amicable numbers, namely 220 and 284.

(b) Use $n = 4$ to obtain another pair of amicable numbers.

Write each even number as the sum of two primes. (There may be more than one way to do this.)

27. 14 **28.** 22

29. 26 **30.** 32

31. Joseph Louis Lagrange (1736–1813) conjectured that every odd natural number greater than 5 can be written as a sum $a + 2b$, where a and b are both primes. Verify this for the odd natural number 11.

32. Another unproved conjecture in number theory states that every natural number multiple of 6 can be written as the difference of two primes. Verify this for 6, 12, and 18.

Find one pair of twin primes between the two numbers given.

33. 65, 80

34. 85, 105

35. 125, 140

While Pierre de Fermat probably is best known for his now famous "last theorem," he did provide proofs of many other theorems in number theory. Exercises 36–40 investigate some of these theorems.

36. If p is prime and the natural numbers a and p have no common factor except 1, then $a^{p-1} - 1$ is divisible by p.

(a) Verify this for $p = 5$ and $a = 3$.

(b) Verify this for $p = 7$ and $a = 2$.

37. Every odd prime can be expressed as the difference of two squares in one and only one way.

(a) Find this one way for the prime number 5.

(b) Find this one way for the prime number 11.

38. A prime number of the form $4k + 1$ can be represented as the sum of two squares.

(a) The prime number 5 satisfies the conditions of the theorem, with $k = 1$. Verify this theorem for 5.

(b) Verify this theorem for 13 (here, $k = 3$).

39. There is only one solution in natural numbers for $a^2 + 2 = b^3$, and it is $a = 5$, $b = 3$. Verify this solution.

40. There are only two solutions in integers for $a^2 + 4 = b^3$. One solution is $a = 2$, $b = 2$. Find the other solution.

The first four perfect numbers were identified in the text: 6, 28, 496, and 8128. The next two are 33,550,336 and 8,589,869,056. Use this information about perfect numbers to work Exercises 41–43.

41. Verify that each of these six perfect numbers ends in either 6 or 28. (In fact, this is true of all even perfect numbers.)

42. Is conjecture (1) in the text (that the nth perfect number contains exactly n digits) true or false? Explain.

43. Is conjecture (2) in the text (that the even perfect numbers end in the digits 6 and 8, alternately) true or false? Explain.

According to the Web site www.shyamsundergupta.com/amicable.htm, *a natural number is* happy *if the process of repeatedly summing the squares of its decimal digits finally ends in 1. For example, the least natural number (greater than 1) that is happy is 7, as shown here.*

$$7^2 = 49,\quad 4^2 + 9^2 = 97,\quad 9^2 + 7^2 = 130,$$
$$1^2 + 3^2 + 0^2 = 10,\quad 1^2 + 0^2 = 1.$$

An amicable pair is a **happy amicable pair** *if and only if both members of the pair are happy numbers. (The first 5000 amicable pairs include only 111 that are happy amicable pairs.) For each amicable pair, determine whether neither, one, or both of the members are happy, and whether the pair is a happy amicable pair.*

44. 220 and 284

45. 1184 and 1210

46. 10,572,550 and 10,854,650

47. 35,361,326 and 40,117,714

48. If the early Greeks knew the form of all even perfect numbers, namely $2^{n-1}(2^n - 1)$, then why did they not discover all the ones that are known today?

49. Explain why the primorial formula $p\# \pm 1$ does not result in a pair of twin primes for the prime value $p = 2$.

50. **(a)** What two numbers does the primorial formula produce for $p = 7$?
(b) Which, if either, of these numbers is prime?

51. Choose the correct completion: The primorial formula produces twin primes
A. never. **B.** sometimes. **C.** always.

See the margin note (on page 236) defining a Sophie Germain prime, and complete this table.

	p	$2p + 1$	Is p a Sophie Germain prime?
52.	2	____	____
53.	3	____	____
54.	5	____	____
55.	7	____	____
56.	11	____	____
57.	13	____	____

Factorial primes *are of the form $n! \pm 1$ for natural numbers n. (n! denotes "n factorial," the product of* all *natural numbers up to n, not just the primes as in the primorial primes. For example, $4! = 1 \cdot 2 \cdot 3 \cdot 4 = 24$.) As of early 2006, the largest verified factorial prime was 34,790! − 1, which has 142,891 digits. Find the missing entries in this table.*

	n	$n!$	$n! - 1$	$n! + 1$	Is $n! - 1$ prime?	Is $n! + 1$ prime?
	2	2	1	3	no	yes
58.	3	____	____	____	____	____
59.	4	____	____	____	____	____
60.	5	____	____	____	____	____

61. Explain why the factorial prime formula does not give twin primes for $n = 2$.

Based on the preceding table, complete each statement with one of the following: **A.** *never,* **B.** *sometimes, or* **C.** *always. When applied to particular values of n, the factorial formula $n! \pm 1$ produces*

62. no primes ____

63. exactly one prime ____

64. twin primes ____

5.3 Greatest Common Factor and Least Common Multiple

Greatest Common Factor • Least Common Multiple

Pierre de Fermat (about 1601–1665), a government official who did not interest himself in mathematics until he was past 30, devoted leisure time to its study. He was a worthy scholar, best known for his work in number theory. His other major contributions involved certain applications in geometry and his original work in probability.

Much of Fermat's best work survived only on loose sheets or jotted, without proof, in the margins of works that he read. Mathematicians of subsequent generations have not always had an easy time verifying some of those results, though their truth has generally not been doubted.

Greatest Common Factor

The **greatest common factor** is defined as follows.

Greatest Common Factor

The **greatest common factor (GCF)** of a group of natural numbers is the largest natural number that is a factor of all the numbers in the group.

Examples: 18 is the GCF of 36 and 54, because 18 is the largest natural number that divides both 36 and 54.

1 is the GCF of 7 and 16.

Greatest common factors can be found by using prime factorizations. To verify the GCF of 36 and 54, first write the prime factorization of each number (perhaps by using factor trees or repeated division):

$$36 = 2^2 \cdot 3^2 \quad \text{and} \quad 54 = 2^1 \cdot 3^3.$$

The GCF is the product of the primes common to the factorizations, with each prime raised to the power indicated by the *least* exponent that it has in any factorization. Here, the prime 2 has 1 as the smallest exponent (in $54 = 2^1 \cdot 3^3$), while the prime 3 has 2 as the least exponent (in $36 = 2^2 \cdot 3^2$). The GCF of 36 and 54 is

$$2^1 \cdot 3^2 = 2 \cdot 9 = 18,$$

as stated earlier. We summarize as follows.

Finding the Greatest Common Factor (Prime Factors Method)

Step 1 Write the prime factorization of each number.

Step 2 Choose all primes common to *all* factorizations, with each prime raised to the *least* exponent that it has in any factorization.

Step 3 Form the product of all the numbers in Step 2; this product is the greatest common factor.

EXAMPLE 1 Finding the Greatest Common Factor by the Prime Factors Method

Find the greatest common factor of 360 and 2700.

SOLUTION

Write the prime factorization of each number:

$$360 = 2^3 \cdot 3^2 \cdot 5 \quad \text{and} \quad 2700 = 2^2 \cdot 3^3 \cdot 5^2.$$

Now find the primes common to both factorizations, with each prime having as its exponent the *least* exponent from either product: 2^2, 3^2, 5. Then form the product of these numbers.

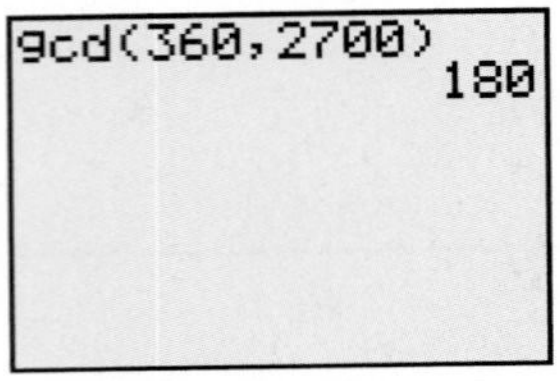

The calculator shows that the greatest common divisor (factor) of 360 and 2700 is 180. Compare with Example 1.

$$\text{GCF} = 2^2 \cdot 3^2 \cdot 5 = 180$$ Use the smallest exponents.

The greatest common factor of 360 and 2700 is 180. ■

EXAMPLE 2 Finding the Greatest Common Factor by the Prime Factors Method

Find the greatest common factor of 720, 1000, and 1800.

SOLUTION

Write the prime factorization for each number:

$$720 = 2^4 \cdot 3^2 \cdot 5, \qquad 1000 = 2^3 \cdot 5^3, \qquad \text{and} \qquad 1800 = 2^3 \cdot 3^2 \cdot 5^2.$$

Use the smallest exponent on each prime common to the factorizations:

$$\text{GCF} = 2^3 \cdot 5 = 40.$$

(The prime 3 is not used in the greatest common factor because it does not appear in the prime factorization of 1000.) ■

EXAMPLE 3 Finding the Greatest Common Factor by the Prime Factors Method

Find the greatest common factor of 80 and 63.

SOLUTION

Start with

$$80 = 2^4 \cdot 5 \quad \text{and} \quad 63 = 3^2 \cdot 7.$$

There are no primes in common here, so the GCF is 1. The number 1 is the largest number that will divide into both 80 and 63. ■

Two numbers, such as 80 and 63, with a greatest common factor of 1 are called **relatively prime numbers**—that is, they are prime *relative* to one another. (They have no common factors other than 1.)

Another method of finding the greatest common factor involves dividing the numbers by common prime factors.

Finding the Greatest Common Factor (Dividing by Prime Factors Method)

Step 1 Write the numbers in a row.

Step 2 Divide each of the numbers by a common prime factor. Try 2, then try 3, and so on.

Step 3 Divide the quotients by a common prime factor. Continue until no prime will divide into all the quotients.

Step 4 The product of the primes in Steps 2 and 3 is the greatest common factor.

EXAMPLE 4 Finding the Greatest Common Factor by Dividing by Prime Factors

Find the greatest common factor of 12, 18, and 60.

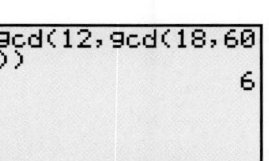

This screen uses the fact that gcd $(a, b, c) = \text{gcd}(a, \text{gcd}(b, c))$. Compare with Example 4.

SOLUTION

Write the numbers in a row and divide by 2.

$$\begin{array}{r|rrr} 2 & 12 & 18 & 60 \\ \hline & 6 & 9 & 30 \end{array}$$

The numbers 6, 9, and 30 are not all divisible by 2, but they are divisible by 3.

$$\begin{array}{r|rrr} 2 & 12 & 18 & 60 \\ \hline 3 & 6 & 9 & 30 \\ \hline & 2 & 3 & 10 \end{array}$$

No prime divides into 2, 3, and 10, so the greatest common factor of the numbers 12, 18, and 60 is given by the product of the primes on the left, 2 and 3.

$$\begin{array}{r|rrr} 2 & 12 & 18 & 60 \\ \hline 3 & 6 & 9 & 30 \\ \hline & 2 & 3 & 10 \end{array}$$

$2 \cdot 3 = 6$

The GCF of 12, 18, and 60 is 6.

There is yet another method of finding the greatest common factor of two numbers (but not more than two) that does not require factoring into primes or successively dividing by primes. It is called the **Euclidean algorithm,*** and it is illustrated in the next example.

EXAMPLE 5 Finding the Greatest Common Factor Using the Euclidean Algorithm

Use the Euclidean algorithm to find the greatest common factor of 90 and 168.

SOLUTION

Step 1 Begin by dividing the larger, 168, by the smaller, 90. Disregard the quotient, but note the remainder.

$$\begin{array}{r} 1 \\ 90\overline{)168} \\ \underline{90} \\ 78 \end{array}$$

Step 2 Divide the smaller of the two numbers by the remainder obtained in Step 1. Once again, note the remainder.

$$\begin{array}{r} 1 \\ 78\overline{)90} \\ \underline{78} \\ 12 \end{array}$$

Step 3 Continue dividing the successive remainders, as many times as necessary to obtain a remainder of 0.

$$\begin{array}{r} 6 \\ 12\overline{)78} \\ \underline{72} \\ 6 \end{array}$$

Greatest common factor

Step 4 The *last positive remainder* in this process is the greatest common factor of 90 and 168. It can be seen that their GCF is 6.

$$\begin{array}{r} 2 \\ 6\overline{)12} \\ \underline{12} \\ 0 \end{array}$$

*For a proof that this process does indeed give the greatest common factor, see *Elementary Introduction to Number Theory, Second Edition*, by Calvin T. Long, pp. 34–35.

The Euclidean algorithm is particularly useful if the two numbers are difficult to factor into primes. We summarize the algorithm here.

Finding the Greatest Common Factor (Euclidean Algorithm)

To find the greatest common factor of two unequal numbers, divide the larger by the smaller. Note the remainder, and divide the previous divisor by this remainder. Continue the process until a remainder of 0 is obtained. The greatest common factor is the last positive remainder obtained in this process.

Least Common Multiple Closely related to the idea of the greatest common factor is the concept of the *least common multiple*, which we define as follows.

Least Common Multiple

The **least common multiple (LCM)** of a group of natural numbers is the smallest natural number that is a multiple of all the numbers in the group.

Example: 30 is the LCM of 15 and 10 because 30 is the smallest number that appears in both sets of multiples.

Multiples of 15: {15, **30,** 45, 60, 75, 90, 105, . . . }
Multiples of 10: {10, 20, **30,** 40, 50, 60, 70, . . . }

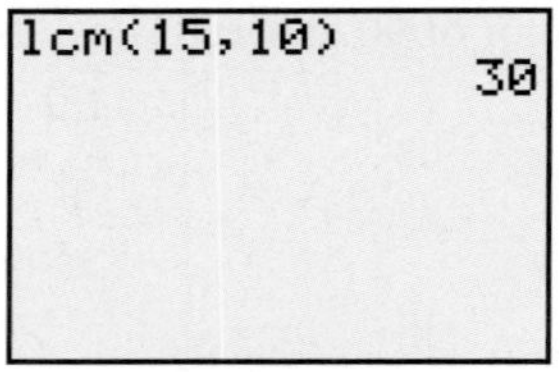

The least common multiple of 15 and 10 is 30.

The set of natural numbers that are multiples of *both* 15 and 10 form the set of *common multiples:*

$$\{30, 60, 90, 120, \ldots\}.$$

While there are infinitely many common multiples, the *least* common multiple is observed to be 30.

A method similar to the first one given for the greatest common factor may be used to find the least common multiple of a group of numbers.

Finding the Least Common Multiple (Prime Factors Method)

Step 1 Write the prime factorization of each number.

Step 2 Choose all primes belonging to *any* factorization, with each prime raised to the power indicated by the *largest* exponent that it has in any factorization.

Step 3 Form the product of all the numbers in Step 2; this product is the least common multiple.

EXAMPLE 6 Finding the Least Common Multiple by the Prime Factors Method

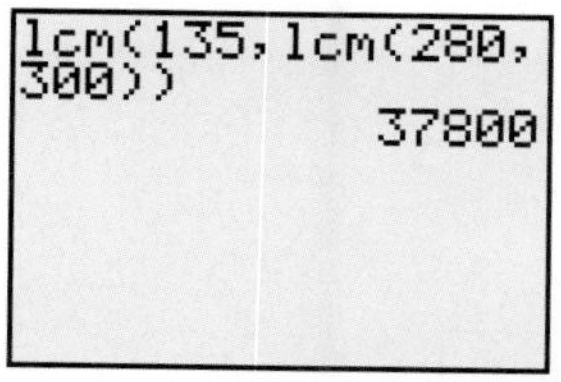

The least common multiple of 135, 280, and 300 is 37,800. Compare with Example 6.

Find the least common multiple of 135, 280, and 300.

SOLUTION

Write the prime factorizations:

$$135 = 3^3 \cdot 5, \quad 280 = 2^3 \cdot 5 \cdot 7, \quad \text{and} \quad 300 = 2^2 \cdot 3 \cdot 5^2.$$

Form the product of all the primes that appear in *any* of the factorizations. Use the *largest* exponent from any factorization.

$$\text{LCM} = 2^3 \cdot 3^3 \cdot 5^2 \cdot 7 = 37{,}800$$ Use the largest exponents.

The smallest natural number divisible by 135, 280, and 300 is 37,800. ■

The least common multiple of a group of numbers can also be found by dividing by prime factors. The process is slightly different than that for finding the GCF.

Finding the Least Common Multiple (Dividing by Prime Factors Method)

Step 1 Write the numbers in a row.

Step 2 Divide each of the numbers by a common prime factor. Try 2, then try 3, and so on.

Step 3 Divide the quotients by a common prime factor. When no prime will divide all quotients, but a prime will divide some of them, divide where possible and bring any nondivisible quotients down. Continue until no prime will divide any two quotients.

Step 4 The product of all prime divisors from Steps 2 and 3 as well as all remaining quotients is the least common multiple.

EXAMPLE 7 Finding the Least Common Multiple by Dividing by Prime Factors

Find the least common multiple of 12, 18, and 60.

SOLUTION

Proceed just as in Example 4 to obtain the following.

2	12	18	60
3	6	9	30
	2	3	10

Now, even though no prime will divide 2, 3, and 10, the prime 2 will divide 2 and 10. Divide the 2 and the 10 and bring down the 3.

2	12	18	60	
3	6	9	30	
2	2	3	10	
	1	3	5	$2 \cdot 3 \cdot 2 \cdot 1 \cdot 3 \cdot 5 = 180$

The LCM of 12, 18, and 60 is 180. ■

It is shown in more advanced courses that the least common multiple of two numbers m and n can be obtained by dividing their product by their greatest common factor.

Finding the Least Common Multiple (Formula)

The least common multiple of m and n is given by

$$\textbf{LCM} = \frac{m \cdot n}{\textbf{greatest common factor of } m \textbf{ and } n}.$$

(This method works only for two numbers, not for more than two.)

This supports the result in Example 8.

EXAMPLE 8 Finding the Least Common Multiple by Formula

Use the formula to find the least common multiple of 90 and 168.

SOLUTION

In Example 5 we used the Euclidean algorithm to find that the greatest common factor of 90 and 168 is 6. Therefore, the formula gives us

$$\text{Least common multiple of 90 and 168} = \frac{90 \cdot 168}{6} = 2520.$$

PROBLEM-SOLVING HINT Problems that deal with questions such as "How many objects will there be in each group if each group contains the same number of objects?" and "When will two events occur at the same time?" can sometimes be solved using the ideas of greatest common factor and least common multiple.

EXAMPLE 9 Finding Common Starting Times of Movie Cycles

The King Theatre and the Star Theatre run movies continuously, and each starts its first feature at 1:00 P.M. If the movie shown at the King lasts 80 minutes and the movie shown at the Star lasts 2 hours, when will the two movies start again at the same time?

SOLUTION

First, convert 2 hours to 120 minutes. The question can be restated as follows: "What is the smallest number of minutes it will take for the two movies to start at the same time again?" This is equivalent to asking, "What is the least common multiple of 80 and 120?" Using any of the methods described in this section, it can be shown that the least common multiple of 80 and 120 is 240. Therefore, it will take 240 minutes, or $\frac{240}{60} = 4$ hours for the movies to start again at the same time. By adding 4 hours to 1:00 P.M., we find that they will start together again at 5:00 P.M.

EXAMPLE 10 Finding the Largest Common Size of Stacks of Cards

Joshua Hornsby has 450 football cards and 840 baseball cards. He wants to place them in stacks on a table so that each stack has the same number of cards, and no stack has different types of cards within it. What is the largest number of cards that he can have in each stack?

SOLUTION

Here, we are looking for the largest number that will divide evenly into 450 and 840. This is, of course, the greatest common factor of 450 and 840. Using any of the methods described in this section, we find that

$$\text{greatest common factor of } 450 \text{ and } 840 = 30.$$

Therefore, the largest number of cards he can have in each stack is 30.

5.3 Exercises

Decide whether each statement is true *or* false.

1. Two even natural numbers cannot be relatively prime.

2. Two different prime numbers must be relatively prime.

3. If p is a prime number, then the greatest common factor of p and p^2 is p.

4. If p is a prime number, then the least common multiple of p and p^2 is p^3.

5. There is no prime number p such that the greatest common factor of p and 2 is 2.

6. The set of all common multiples of two given natural numbers is finite.

7. Two natural numbers must have at least one common factor.

8. The least common multiple of two different primes is their product.

9. Two composite numbers may be relatively prime.

10. The set of all common factors of two given natural numbers is finite.

Use the prime factors method to find the greatest common factor of each group of numbers.

11. 70 and 120

12. 180 and 300

13. 480 and 1800

14. 168 and 504

15. 28, 35, and 56

16. 252, 308, and 504

Use the method of dividing by prime factors to find the greatest common factor of each group of numbers.

17. 60 and 84

18. 130 and 455

19. 310 and 460

20. 234 and 470

21. 12, 18, and 30

22. 450, 1500, and 432

Use the Euclidean algorithm to find the greatest common factor of each group of numbers.

23. 36 and 60

24. 25 and 70

25. 84 and 180

26. 72 and 120

27. 210 and 560

28. 150 and 480

29. Explain in your own words how to find the greatest common factor of a group of numbers.

30. Explain in your own words how to find the least common multiple of a group of numbers.

Use the prime factors method to find the least common multiple of each group of numbers.

31. 24 and 30

32. 12 and 32

33. 56 and 96

34. 28 and 70

35. 30, 40, and 70

36. 24, 36, and 48

Use the method of dividing by prime factors to find the least common multiple of each group of numbers.

37. 24 and 32

38. 35 and 56

39. 45 and 75

40. 48, 54, and 60

41. 16, 120, and 216

42. 210, 385, and 2310

Use the formula given in the text and the results of Exercises 23–28 to find the least common multiple of each group of numbers.

43. 36 and 60

44. 25 and 70

45. 84 and 180

46. 72 and 120

47. 210 and 560

48. 150 and 480

49. If p, q, and r are different primes, and a, b, and c are natural numbers such that $a > b > c$,

(a) what is the greatest common factor of $p^a q^c r^b$ and $p^b q^a r^c$?

(b) what is the least common multiple of $p^b q^a$, $q^b r^c$, and $p^a r^b$?

50. Find **(a)** the greatest common factor and **(b)** the least common multiple of

$$2^{31} \cdot 5^{17} \cdot 7^{21} \quad \text{and} \quad 2^{34} \cdot 5^{22} \cdot 7^{13}.$$

Leave your answers in prime factored form.

It is possible to extend the Euclidean algorithm in order to find the greatest common factor of more than two numbers. For example, if we wish to find the greatest common factor of 150, 210, *and* 240, *we can first use the algorithm to find the greatest common factor of two of these (say, for example,* 150 *and* 210*). Then we find the greatest common factor of that result and the third number,* 240. *The final result is the greatest common factor of the original group of numbers. Use the Euclidean algorithm as described above to find the greatest common factor of each group of numbers.*

51. 150, 210, and 240

52. 12, 75, and 120

53. 90, 105, and 315

54. 48, 315, and 450

55. 144, 180, and 192

56. 180, 210, and 630

If we allow repetitions of prime factors, we can use Venn diagrams (Chapter 2) to find the greatest common factor and the least common multiple of two numbers. For example, consider $36 = 2^2 \cdot 3^2$ *and* $45 = 3^2 \cdot 5$. *Their greatest common factor is* $3^2 = 9$, *and their least common multiple is* $2^2 \cdot 3^2 \cdot 5 = 180$.

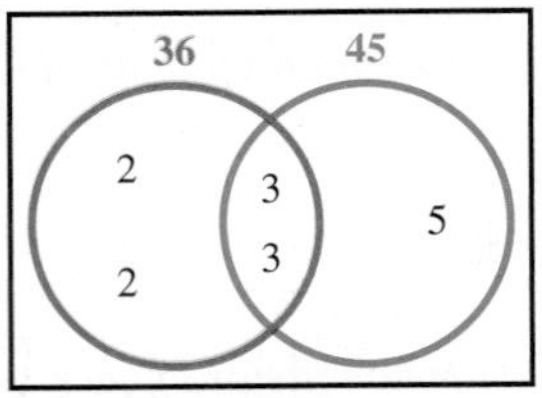

Intersection gives 3, 3.
Union gives 2, 2, 3, 3, 5.

Use this method to find **(a)** *the greatest common factor and* **(b)** *the least common multiple of the two numbers given.*

57. 12 and 18

58. 27 and 36

59. 54 and 72

60. Suppose that the least common multiple of p and q is q. What can we say about p and q?

61. Suppose that the least common multiple of p and q is pq. What can we say about p and q?

62. Suppose that the greatest common factor of p and q is p. What can we say about p and q?

63. Recall some of your early experiences in mathematics (for example, in the elementary grade classroom). What topic involving fractions required the use of the least common multiple? Give an example.

64. Recall some of your experiences in elementary algebra. What topics required the use of the greatest common factor? Give an example.

Refer to Examples 9 and 10 to solve each problem.

65. ***Inspecting Calculators*** Colleen Jones and Nancy Barre work on an assembly line, inspecting electronic calculators. Colleen inspects the electronics of every sixteenth calculator, while Nancy inspects the workmanship of every thirty-sixth calculator. If they both start working at the same time, which calculator will be the first that they both inspect?

66. ***Night Off for Security Guards*** Paul Crockett and Cindy Herring work as security guards at a publishing company. Paul has every sixth night off, and Cindy has every tenth night off. If both are off on July 1, what is the next night that they will both be off together?

67. ***Stacking Coins*** Sheila Abbruzzo has 240 pennies and 288 nickels. She wants to place the pennies and nickels in stacks so that each stack has the same number of coins, and each stack contains only one denomination of coin. What is the largest number of coins that she can place in each stack?

68. ***Bicycle Racing*** Kathryn Campbell and Tami Dreyfus are in a bicycle race, following a circular track. If they start at the same place and travel in the same direction, and Kathryn completes a revolution every 40 seconds and Tami completes a revolution every 45 seconds, how long will it take them before they reach the starting point again simultaneously?

69. ***Selling Books*** John Cross sold some books at $24 each, and used the money to buy some concert tickets at $50 each. He had no money left over after buying the tickets. What is the least amount of money he could have earned from selling the books? What is the least number of books he could have sold?

70. ***Sawing Lumber*** Jill Bos has some pieces of two-by-four lumber. Some are 60 inches long, and some are 72 inches long. She wishes to saw them so as to obtain equal-length pieces. What is the longest such piece she can saw so that no lumber is left over?

EXTENSION

Modern Cryptography

Cryptography involves secret codes, ways of disguising information in order that a "sender" can transmit it to an intended "receiver" so that an "adversary" who somehow intercepts the transmission will be unable to discern its meaning. It has become customary in discussions of cryptography to refer to the sender and receiver (in either order) as Alice and Bob (A and B) and to the adversary as Eve (E). We follow that practice here. Converting a message to disguised form is called **encryption,** and converting it back to original form is called **decryption.**

Cryptography has been employed for thousands of years. It became more crucial as the extent of military, diplomatic, then industrial, and now even personal applications expanded. As the "code makers" became more adept at designing their systems, the "code breakers" became more adept at compromising those systems. If intercepting secret information is very important to an adversary, then vast resources will likely be expended in that effort.

The basis of a cryptography system is normally some mathematical function, the "encryption algorithm," that encrypts (disguises) the message. (*Functions* are used extensively in mathematics. They are discussed in Chapter 8 of this book.). An example of a simple (and very insecure) encryption algorithm is the following:

> Replace every letter of the alphabet with the letter that *follows* it.
> (Replace z with a.)

(continued)

Then the message "zebra" would be encrypted as "afcsb." Analyzing one or more intercepted messages encrypted using this function, and trying various possibilities, would enable an adversary to quickly determine the function, and its inverse, which would be the following:

> Replace every letter of the alphabet with the letter that *precedes* it.
> (Replace a with z.)

More advanced systems also use a **key,** which is some additional information needed to perform the algorithm correctly.

By the middle of the twentieth century, state-of-the-art requirements for an effective cryptography system were the following.

Basic Requirements of a Cryptography System

1. A *secret* algorithm (or function) for encrypting and decrypting data
2. A *secret* key that provides additional information necessary for a receiver to carry out the decrypting process

The difficulty with requirement number 1 was that all known encryption functions at the time were two-way functions. Once an adversary obtained the encryption algorithm, the inverse (that is, the decryption algorithm) could be deduced mathematically. The difficulty with requirement number 2 was that the security of the key frequently dropped off after a period of use. This meant that Bob and Alice must exchange a new key fairly often so that their communications would continue to be safe. But this measure may be self-defeating, because every key exchange may be vulnerable to interception. This dilemma became known as the **key exchange problem** (or the **key distribution problem** in the case of multiple intended receivers).

The world of cryptography was revolutionized in the 1970s when researchers discovered how to construct a *one-way* function that overcame both difficulties. That *exponential function* is given by

$$\boldsymbol{C = M^k \text{ (mod } n),}$$

with the calculation carried out modulo n. The practical success of this formula, the achievement of an essentially one-way, rather than two-way, function, is made possible by the theory of large prime numbers (studied earlier in this chapter), the nature of modular arithmetic (introduced in Section 4.4), and the present state of computer hardware and algorithms.

The Diffie-Hellman-Merkle Key Exchange Scheme First the key exchange problem was solved by the **Diffie-Hellman-Merkle key exchange scheme** (announced in 1976 and named for the Stanford University team of Whitfield Diffie, Martin Hellman, and Ralph Merkle). Basically, it works as follows.

The Diffie-Hellman-Merkle Key Exchange Scheme

Alice and Bob can establish a key (a number) that they both will know, but that Eve cannot find out, even if she observes the communications between Bob and Alice as they set up their key. Alice and Bob can agree to use the function $C = M^k \pmod{n}$ with specific values for M and n. (They can agree to all this by mail, telephone, e-mail, or even casual conversation. It won't matter if Eve finds out.) Then they carry out the following sequence of individual steps.

Alice's Actions		*Bob's Actions*	
Step 1	Choose a value of a. (Keep this value secret.)	***Step 1***	Choose a value of b. (Keep this value secret.)
Step 2	Compute $\alpha = M^a \pmod{n}$.	***Step 2***	Compute $\beta = M^b \pmod{n}$.
Step 3	Send the value of α to Bob.	***Step 3***	Send the value of β to Alice.
Step 4	Receive the value of β from Bob.	***Step 4***	Receive the value of α from Alice.
Step 5	Compute the key: $\boldsymbol{K = \beta^a \pmod{n}}$.	***Step 5***	Compute the key: $\boldsymbol{K = \alpha^b \pmod{n}}$.

By this procedure, Alice and Bob will arrive at the same key value K because

$$\begin{aligned} \beta^a &= (M^b)^a && \beta = M^b \\ &= M^{ba} && \text{Rule of exponents: } (a^m)^n = a^{mn} \\ &= M^{ab} && \text{Commutative property: } ab = ba \\ &= (M^a)^b && \text{Rule of exponents: } (a^m)^n = a^{mn} \\ &= \alpha^b. && \alpha = M^a \end{aligned}$$

We illustrate the basic procedures using much smaller numbers than would be used in practice so that our computations can be done on a handheld calculator. (It is not recommended that you try working any of the following examples *without* a calculator.)

One of the essential aspects of the schemes we use here is the nature of modular arithmetic. Given a modulus n, every natural number a is "equivalent" (actually congruent) to the remainder obtained when a is divided by n. This remainder is called the **residue** of a, modulo n. To find the residue can be thought of as to "mod." In Example 1 to follow, for instance, one of the calculations will be to find the residue of 16,807, modulo 13. A quick procedure to accomplish this on a calculator is as follows.

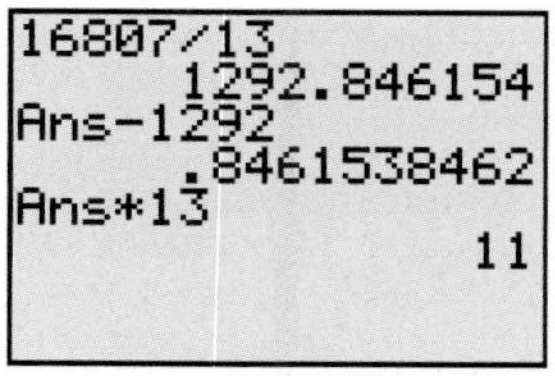

The display shows that the residue of 16,807, modulo 13, is 11.

Step 1 Divide 16,807 by 13, obtaining 1292.846154.

Step 2 Subtract the integer part of the quotient, obtaining .846154.

Step 3 Multiply by 13, obtaining 11.

So, we see that $16{,}807 \equiv 11 \pmod{13}$. We have shown that $16{,}807 = 1292 \cdot 13 + 11$.

(continued)

(*Note:* In the work that follows, we carry out some lengthy sequences of modular arithmetic. We sometimes use equals signs, $=$, rather than congruence symbols, $\equiv$, and when the modulus is understood, we sometimes omit the designation (mod n).)

This calculator routine can be summarized.

Calculator Routine for Finding the Residue of a, Modulo n

In a modular system, the residue modulo n for a number a can be found by completing these three steps, in turn.

Step 1 Divide a by the modulus n.

Step 2 Subtract the integer part of the quotient to obtain only the fractional part.

Step 3 Multiply the fractional part of the quotient by n.

The final result is the residue modulo n.

EXAMPLE 1 Using the Diffie-Hellman-Merkle Key Exchange Scheme

Establish a common key for Alice and Bob by using specific values for M, n, a, and b, and completing the steps outlined earlier for the Diffie-Hellman-Merkle key exchange scheme.

SOLUTION

Suppose Alice and Bob agree to use the values $M = 7$ and $n = 13$.

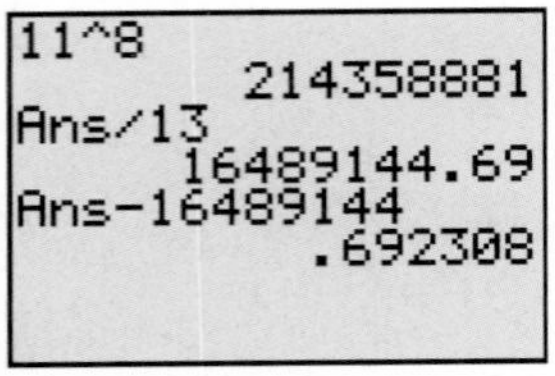

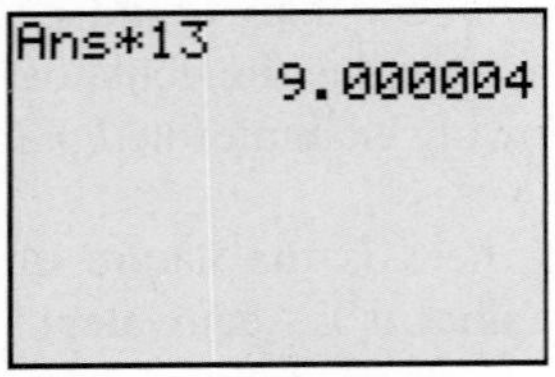

The display shows the calculation of K in the right column. (Ignore the tiny roundoff error.)

Alice's Actions

Step 1 Choose a value of a, say 5. (Alice keeps this value secret.)

Step 2
$$\alpha = M^a \pmod{n} = 7^5 \pmod{13} = 16{,}807 \pmod{13} = 11$$

Step 3 Send $\alpha = 11$ to Bob.

Step 4 Receive $\beta = 3$.

Step 5 Compute the key:
$$K = \beta^a \pmod{n} = 3^5 \pmod{13} = 243 \pmod{13} = 9.$$

Bob's Actions

Step 1 Choose a value of b, say 8. (Bob keeps this value secret.)

Step 2
$$\beta = M^b \pmod{n} = 7^8 \pmod{13} = 5{,}764{,}801 \pmod{13} = 3$$

Step 3 Send $\beta = 3$ to Alice.

Step 4 Receive $\alpha = 11$.

Step 5 Compute the key:
$$K = \alpha^b \pmod{n} = 11^8 \pmod{13} = 214{,}358{,}881 \pmod{13} = 9.$$

Both Alice and Bob arrived at the same key value, $K = 9$, which they can use for encrypting future communications to one another. ■

Suppose, at Step 3 in Example 1, Eve intercepts Bob's transmission of the value $\beta = 3$ to Alice. This will not help her, because she cannot deduce Bob's value of b that generated β. In fact it could have been any of the values

$$8, 20, 32, 44, 56, \ldots,$$

an infinite list of possibilities. Also, Eve does not know what exponent Alice will apply to 3 to obtain the key. The value $a = 5$ is Alice's secret, never communicated to anyone else, not even Bob, so Eve cannot know what key Alice will obtain. The same argument applies if Eve intercepts Alice's transmission to Bob of the value $\alpha = 11$. (She is stymied even if she intercepts both transmissions.)

RSA Public Key Cryptography At practically the same time that Diffie, Hellman, and Merkle solved the key exchange problem, another team of researchers, Ron Rivest, Adi Shamir, and Leonard Adleman, at MIT, used the same type of mathematical function to provide an even better solution that eliminated the need for key exchange. Their scheme, known as RSA (from their surnames), is called **public key cryptography.** Anyone who wants the capability of receiving encrypted data simply makes known their public key, which anyone else can then use to encrypt messages to them. The beauty of the system is that the receiver possesses another private key, necessary for decrypting but never released to anyone else.

When the **RSA code** was first introduced in 1977, Martin Gardner's "Mathematical Games" column in *Scientific American* challenged researchers to decode a message using an n with 129 digits. Some estimated it would take approximately 20,000 years to decipher without any knowledge of p or q. However, with the aid of number theory, it took 600 mathematicians in 25 different countries only 17 years to factor n into 64- and 65-digit prime factors, as shown here

114,381,625,757,888,867,669,235, 779,976,146,612,010,218,296,721, 242,362,562,561,842,935,706,935, 245,733,897,830,597,123,563,958, 705,058,989,075,147,599,290,026, 879,543,541 = 3,490,529,510,847, 650,949,147,849,619,903,898,133, 417,764,638,493,387,843,990,820, 577 × 32,769,132,993,266,709, 549,961,988,190,834,461,413,177, 642,967,992,942,539,798,288,533.

The decoded message said, "The magic words are squeamish ossifrage."

Today, RSA users select much larger values of p and q, resulting in an n of well over 300 digits. It is thought that breaking such an encryption would take all the computers in the world, working together, more time than the age of the universe. So until someone discovers new factoring techniques, or new computer designs, RSA would seem to be safe from attack.

What makes RSA successful is that we have the mathematical understanding to identify very large prime numbers, and to multiply them to obtain a product. But if the prime factors are large enough, it is impossible, given the present state of knowledge, for anyone to determine the two original factors. This is true even using very powerful computers. Large prime factors can be used so that it would take one hundred million personal computers, working together, over a thousand years to break the code. This sort of security is due to the fact that factoring large primes is mathematically much more difficult than multiplying large primes (again, even for computers).

Using RSA, Alice can receive encrypted messages from Bob in such a way that Eve cannot discern their meaning even if she intercepts them (the usual goal of cryptography). Again, we use rather small values in our examples (relative to values used in practice). Although we give examples of specific portions of the process later, we show here a complete outline of all the basic procedures, from setting up the scheme to encrypting and then decrypting a message.

RSA Basics: A Public Key Cryptography Scheme

Alice (the receiver) completes the following steps.

Step 1 Choose two prime numbers, p and q, which she keeps secret.

Step 2 Compute the *modulus* n (which is the product $p \cdot q$).

Step 3 Compute $\ell = (p - 1)(q - 1)$.

Step 4 Choose the *encryption exponent* e, which can be any integer between 1 and ℓ that is relatively prime to ℓ, that is, has no common factors with ℓ.

Step 5 Find her *decryption exponent* d, a number satisfying

$$e \cdot d = 1 \pmod{\ell}.$$

She keeps d secret.

Step 6 Provide Bob with her *public key*, which consists of the modulus n and the encryption exponent e.

(Bob's steps are on the next page.)

(continued)

Now Bob (the sender) completes the following steps. (Recall that the purpose of all this is for Bob to be able to send Alice secure messages.)

Step 7 Convert the message to be sent to Alice into a number M (sometimes called the *plaintext*).

Step 8 Encrypt M, that is, use Alice's public key (n and e) to generate the encrypted message C (sometimes called the *ciphertext*) according to the formula $C = M^e \pmod{n}$.

Step 9 Transmit C to Alice.

When Alice receives C, she completes the final step:

Step 10 Decrypt C, that is, use her private key, consisting of n (also part of her public key) and d, to reproduce the original plaintext message M according to the formula $M = C^d \pmod{n}$.

James Ellis, Clifford Cocks, and Malcolm Williamson all worked for Britain's Government Communications Headquarters in the 1970s. In a strange twist of fate, they actually discovered the mathematics of public key cryptography several years before the work at Stanford and MIT was announced (and subsequently patented). The British work was classified top secret and never came to light until some twenty years later, at approximately the same time that RSA Data Security, the company that had been built on U.S. RSA patents, was sold for $200 million.

EXAMPLE 2 Devising a Public Encryption Key

Use the values $p = 7$ and $q = 13$ (arbitrarily chosen primes) to devise Alice's public key by completing Steps 2–4 of the above outline of RSA basics.

SOLUTION

Step 2 $n = p \cdot q = 7 \cdot 13 = 91$

Step 3 $\ell = (p - 1)(q - 1) = 6 \cdot 12 = 72$

Step 4 There are many choices here, but a prime less than 72 will certainly meet the requirements. We arbitrarily choose $e = 11$.

Alice's public key is $n = 91$, $e = 11$. (Prime factors p and q must be kept secret.)

EXAMPLE 3 Finding a Private Decryption Key

Complete Step 5 of the RSA basics outline to find Alice's private decryption key.

SOLUTION

Step 5 The decryption exponent d must satisfy

$$e \cdot d = 1 \pmod{\ell} \quad \text{or} \quad 11d = 1 \pmod{72}.$$

One way to satisfy this equation is to check the powers of 11 until we find one equal (actually congruent) to 1, modulo 72:

Mod 72 congruences →

$$11^1 = 11, \quad 11^2 = 121 = 49, \quad 11^3 = 1331 = 35,$$

$$11^4 = 14{,}641 = 25, \quad 11^5 = 161{,}051 = 59, \quad 11^6 = 1{,}771{,}561 = 1.$$

(The residues were found using the calculator routine explained before Example 1.) Because we found that $11^6 = 1$, we take $d = 11^5 = 59$. This way,

$$e \cdot d = 11 \cdot 11^5 = 11^6 = 1, \qquad \text{as required.}$$

Alice's private key is $n = 91$, $d = 59$.

EXAMPLE 4 Encrypting a Message for Transmission

Complete Steps 7 and 8 of the RSA basics outline to encrypt the message "HI" for Bob to send Alice. Use Alice's public key found in Example 2: $n = 91$, $e = 11$.

SOLUTION

Step 7 A simple way to convert "HI" to a number is to note that H and I are the 8th and 9th letters of the English alphabet. Simply let the plaintext message be $M = 89$.

Step 8 Now compute the ciphertext C.

$$C = M^e \pmod{n} = 89^{11} \pmod{91}$$

This presents a new difficulty, because 89^{11} is too large to be handled as we did the powers of 11 in Example 3. But we can use a trick here, expressing 11 as $1 + 2 + 8$. (1, 2, and 8 are the unique powers of 2 that sum to 11. So we are doing something like what we did when discussing the binary system in Section 4.3.) Now we can rewrite 89^{11} in terms of smaller powers and then make use of rules of exponents (which are covered in Section 7.5).

$$\begin{aligned} 89^{11} &= 89^{1+2+8} && 1 + 2 + 8 = 11 \\ &= 89^1 \cdot 89^2 \cdot 89^8 && \text{Rule of exponents: } a^{n+m} = a^n \cdot a^m \end{aligned}$$

Now it will help to follow the maxim "mod before you multiply," in other words, compute the residue of individual factors first, then multiply those results. This keeps the numbers we must deal with smaller.

$$\begin{aligned} 89^1 &= 89 && \text{Definition of first power} \\ 89^2 &= 7921 = 4 && \text{Mod} \\ 89^8 &= 3.936588806\text{E}15 && \text{Calculator result} \end{aligned}$$

This last factor is far too large to handle like the others. But because of the way we "split up" the exponent, each subsequent power of 89 can be written as a power of an earlier one, again using rules of exponents. Specifically,

$$\begin{aligned} 89^8 &= (89^2)^4 && \text{Rule of exponents: } a^{m \cdot n} = (a^m)^n \\ &= 4^4 && 89^2 = 4 \text{ from above} \\ &= 256 && \text{Evaluate } 4^4. \\ &= 74. && \text{Mod} \end{aligned}$$

Finally we obtain

$$\begin{aligned} 89^{11} &= 89 \cdot 4 \cdot 74 && \text{Substitute.} \\ &= 26{,}344 && \text{Multiply.} \\ &= 45. && \text{Mod} \end{aligned}$$

The plaintext $M = 89$ (for the message "HI") has been converted to the ciphertext $C = 45$. ■

Now let's see if Alice can successfully decrypt the message 45 when she receives it.

(continued)

EXAMPLE 5 Decrypting a Received Message

Complete Step 10 of the RSA basics outline to decrypt the message $C = 45$ from Example 4. Use Alice's private key, found in Example 3: $d = 59$ (also, $n = 91$).

SOLUTION

Step 10 The decryption formula gives

$$\begin{aligned} M &= C^d \pmod{n} \\ &= 45^{59} \pmod{91} \\ &= 45^{1+2+8+16+32} \pmod{91} \\ &= 45 \cdot 45^2 \cdot 45^8 \cdot 45^{16} \cdot 45^{32} \pmod{91}. \end{aligned}$$

Start with the smaller powers and "mod" each factor individually.

$$\begin{aligned} 45^2 &= 2025 = 23 \\ 45^8 &= (45^2)^4 = 23^4 = 279{,}841 = 16 \\ 45^{16} &= (45^8)^2 = 16^2 = 256 = 74 \\ 45^{32} &= (45^{16})^2 = 74^2 = 5476 = 16 \end{aligned}$$

Inserting these values in the product above for M, we get

$$\begin{aligned} M &= 45 \cdot 23 \cdot 16 \cdot 74 \cdot 16 \\ &= 19{,}607{,}040 \\ &= 89. \end{aligned}$$

We have correctly decrypted $C = 45$ to obtain

$$M = 89 = \text{HI}.$$

EXTENSION EXERCISES

Find the residue in each case.

1. 45 (mod 6)

2. 67 (mod 10)

3. 225 (mod 13)

4. 418 (mod 15)

5. 5^9 (mod 12)

6. 4^{11} (mod 9)

7. 8^7 (mod 11)

8. 14^5 (mod 13)

9. 8^{27} (mod 17)

10. 45^7 (mod 23)

11. 11^{14} (mod 18)

12. 14^9 (mod 19)

Finding a Common Key *Find Alice and Bob's common key K by using the Diffie-Hellman-Merkle key exchange scheme with the given values of M, n, a, and b.*

	M	n	a	b
13.	5	13	7	6
14.	11	9	5	4
15.	5	11	6	7
16.	17	5	6	3

Apply the RSA scheme to find each missing value.

	p	q	n	ℓ
17.	5	11	____	____
18.	11	3	____	____
19.	5	13	____	____
20.	17	7	____	____

Encrypting Plaintext *Given the modulus n, the encryption exponent e, and the plaintext M, use RSA encryption to find the ciphertext C in each case.*

	n	e	M
21.	55	7	15
22.	33	7	8
23.	65	5	16
24.	119	11	12

Decrypting Ciphertext *Given the prime factors p and q, the encryption exponent e, and the ciphertext C, apply the RSA algorithm to find* **(a)** *the decryption exponent d and* **(b)** *the plaintext message M.*

	p	q	e	C
25.	5	11	3	30
26.	11	3	13	24
27.	5	13	35	17
28.	17	7	5	40

29. Describe the breakthrough represented by Diffie-Hellman-Merkle and RSA as opposed to all earlier forms of cryptography.

30. Explain why RSA would fail if mathematicians could (using computers) factor arbitrarily large numbers.

5.4 The Fibonacci Sequence and the Golden Ratio

The Fibonacci Sequence • The Golden Ratio

The solution of Fibonacci's rabbit problem is examined in Chapter 1, pages 21–22.

The Fibonacci Sequence

One of the most famous problems in elementary mathematics comes from the book *Liber Abaci*, written in 1202 by Leonardo of Pisa, a.k.a. Fibonacci. The problem is as follows:

> A man put a pair of rabbits in a cage. During the first month the rabbits produced no offspring, but each month thereafter produced one new pair of rabbits. If each new pair thus produced reproduces in the same manner, how many pairs of rabbits will there be at the end of one year?

The solution of this problem leads to a sequence of numbers known as the **Fibonacci sequence.** Here are the first fifteen terms of the Fibonacci sequence:

$$1, 1, 2, 3, 5, 8, 13, 21, 34, 55, 89, 144, 233, 377, 610.$$

Notice the pattern established in the sequence. After the first two terms (both 1), each term is obtained by adding the two previous terms. For example, the third term is obtained by adding $1 + 1$ to get 2, the fourth term is obtained by adding $1 + 2$ to get 3, and so on. This can be described by a mathematical formula known as a *recursion formula*.

If F_n represents the Fibonacci number in the nth position in the sequence, then

$$F_1 = 1$$
$$F_2 = 1$$
$$F_n = F_{n-2} + F_{n-1}, \quad \text{for } n \geq 3.$$

Using the recursion formula $F_n = F_{n-2} + F_{n-1}$, we obtain

$$F_3 = F_1 + F_2 = 1 + 1 = 2, \quad F_4 = F_2 + F_3 = 1 + 2 = 3, \quad \text{and so on.}$$

The Fibonacci sequence exhibits many interesting patterns, and by inductive reasoning we can make many conjectures about these patterns. However, as we have indicated many times earlier, simply observing a finite number of examples does not provide a proof of a statement. Proofs of the properties of the Fibonacci sequence often involve mathematical induction (covered in college algebra texts). Here we simply observe the patterns and do not attempt to provide such proofs.

The **Fibonacci Association** is a research organization dedicated to investigation into the **Fibonacci sequence** and related topics. Check your library to see if it has the journal *Fibonacci Quarterly*. The first two journals of 1963 contain a basic introduction to the Fibonacci sequence.

As an example of the many interesting properties of the Fibonacci sequence, choose any term of the sequence after the first and square it. Then multiply the terms on either side of it, and subtract the smaller result from the larger. The difference is always 1. For example, choose the sixth term in the sequence, 8. The square of 8 is 64. Now multiply the terms on either side of 8: $5 \cdot 13 = 65$. Subtract 64 from 65 to get $65 - 64 = 1$. This pattern continues throughout the sequence.

The following program for the TI-83/84 Plus utilizes the *Binet form* of the nth Fibonacci number (see Exercises 33–38) to determine its value.

```
PROGRAM: FIB
: Clr Home
: Disp "WHICH TERM"
: Disp "OF THE"
: Disp "SEQUENCE DO"
: Disp "YOU WANT?"
: Input N
: (1 + √(5))/2 → A
: (1 − √(5))/2 → B
: (A^N − B^N)/√(5) → F
: Disp F
```

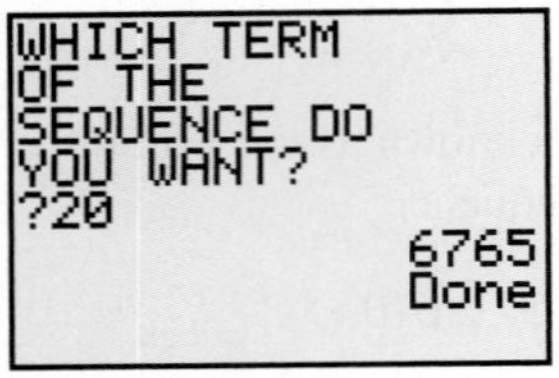

This screen indicates that the twentieth Fibonacci number is 6765.

EXAMPLE 1 Observing a Pattern of the Fibonacci Numbers

Find the sum of the squares of the first n Fibonacci numbers for $n = 1, 2, 3, 4, 5$, and examine the pattern. Generalize this relationship.

SOLUTION

$$1^2 = 1 = 1 \cdot 1 = F_1 \cdot F_2$$
$$1^2 + 1^2 = 2 = 1 \cdot 2 = F_2 \cdot F_3$$
$$1^2 + 1^2 + 2^2 = 6 = 2 \cdot 3 = F_3 \cdot F_4$$
$$1^2 + 1^2 + 2^2 + 3^2 = 15 = 3 \cdot 5 = F_4 \cdot F_5$$
$$1^2 + 1^2 + 2^2 + 3^2 + 5^2 = 40 = 5 \cdot 8 = F_5 \cdot F_6$$

The sum of the squares of the first n Fibonacci numbers seems to always be the product of F_n and F_{n+1}. This has been proven to be true, in general, using mathematical induction.

There are many other patterns similar to the one examined in Example 1, and some of them are discussed in the exercises of this section. An interesting property of the decimal value of the reciprocal of 89, the eleventh Fibonacci number, is examined in the next example.

EXAMPLE 2 Observing the Fibonacci Sequence in a Long Division Problem

Observe the steps of the long division process used to find the first few decimal places for $\frac{1}{89}$.

SOLUTION

$$
\begin{array}{r}
.011235\ldots \\
89\overline{)1.000000} \\
\underline{89} \\
110 \\
\underline{89} \\
210 \\
\underline{178} \\
320 \\
\underline{267} \\
530 \\
\underline{445} \\
850\ldots
\end{array}
$$

Notice that after the 0 in the tenths place, the next five digits are the first five terms of the Fibonacci sequence. In addition, as indicated in color in the process, the digits 1, 1, 2, 3, 5, 8 appear in the division steps. Now, look at the digits next to the ones in color, beginning with the second "1"; they, too, are 1, 1, 2, 3, 5, ...

If the division process is continued past the final step shown above, the pattern seems to stop, since to ten decimal places, $\frac{1}{89} \approx .0112359551$. (The decimal representation actually begins to repeat later in the process, since $\frac{1}{89}$ is a rational number.) However, the sum below indicates how the Fibonacci numbers are actually "hidden" in this decimal.

$$
\begin{array}{l}
.01 \\
.001 \\
.0002 \\
.00003 \\
.000005 \\
.0000008 \\
.00000013 \\
.000000021 \\
.0000000034 \\
.00000000055 \\
\underline{.000000000089} \\
\end{array}
$$

$$\frac{1}{89} = .0112359550\ldots$$

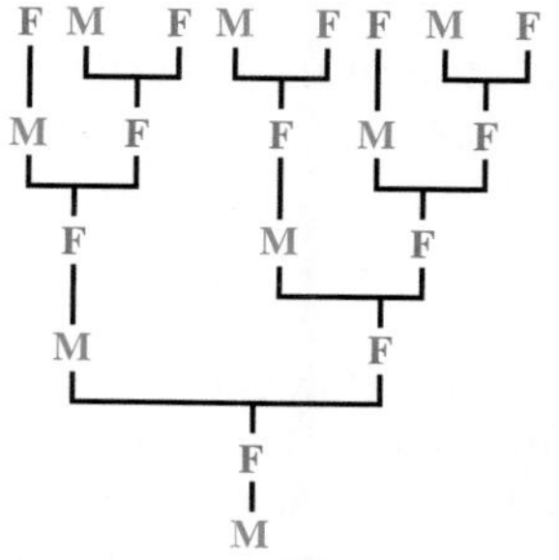

FIGURE 1

Fibonacci patterns have been found in numerous places in nature. For example, male honeybees (drones) hatch from eggs which have not been fertilized, so a male bee has only one parent, a female. On the other hand, female honeybees hatch from fertilized eggs, so a female has two parents, one male and one female. Figure 1 shows several generations of ancestors for a male honeybee.

Notice that in the first generation, starting at the bottom, there is 1 bee, in the second there is 1 bee, in the third there are 2 bees, and so on. These are the terms of the Fibonacci sequence. Furthermore, beginning with the second generation, the numbers of female bees form the sequence, and beginning with the third generation, the numbers of male bees also form the sequence.

Successive terms in the Fibonacci sequence also appear in some plants. For example, the photo (at the left on the next page) shows the double spiraling of a daisy head, with 21 clockwise spirals and 34 counterclockwise spirals. These numbers are successive terms in the sequence. Most pineapples (see the photo on the right) exhibit the Fibonacci sequence in the following way: Count the spirals formed by the "scales" of the cone, first counting from lower left to upper right. Then count the spirals from lower right to upper left. You should find that in one direction you get 8 spirals, and in the other you get 13 spirals, once again successive terms of the Fibonacci sequence. Many pinecones exhibit 5 and 8 spirals, and the cone of the giant sequoia has 3 and 5 spirals.

The Golden Ratio If we consider the quotients of successive Fibonacci numbers, a pattern emerges.

$$\frac{1}{1} = 1 \qquad \frac{13}{8} = 1.625$$

$$\frac{2}{1} = 2 \qquad \frac{21}{13} \approx 1.615384615$$

$$\frac{3}{2} = 1.5 \qquad \frac{34}{21} \approx 1.619047619$$

$$\frac{5}{3} = 1.666\ldots \qquad \frac{55}{34} \approx 1.617647059$$

$$\frac{8}{5} = 1.6 \qquad \frac{89}{55} = 1.618181818\ldots$$

These quotients seem to be approaching some "limiting value" close to 1.618. In fact, as we go farther into the sequence, these quotients approach the number

$$\frac{1 + \sqrt{5}}{2},$$

known as the **golden ratio,** and often symbolized by ϕ, the Greek letter phi.

The golden ratio appears over and over in art, architecture, music, and nature. Its origins go back to the days of the ancient Greeks, who thought that a golden rectangle exhibited the most aesthetically pleasing proportion. A **golden rectangle** is one that can be divided into a square and another (smaller) rectangle the same shape as the original rectangle. (See Figure 2 on the next page.) If we let the smaller rectangle have

---

A fraction such as

$$1 + \cfrac{1}{1 + \cfrac{1}{1 + \cfrac{1}{1 + \ddots}}}$$

is called a **continued fraction.** This continued fraction can be evaluated as follows.

Let $x = 1 + \cfrac{1}{1 + \cfrac{1}{1 + \ddots}}$

Then $x = 1 + \frac{1}{x}$

$$x^2 = x + 1$$

$$x^2 - x - 1 = 0.$$

By the quadratic formula from algebra,

$$x = \frac{1 \pm \sqrt{1 - 4(1)(-1)}}{2(1)}$$

$$x = \frac{1 \pm \sqrt{5}}{2}.$$

Notice that the positive solution

$$\frac{1 + \sqrt{5}}{2}$$

is the **golden ratio.**

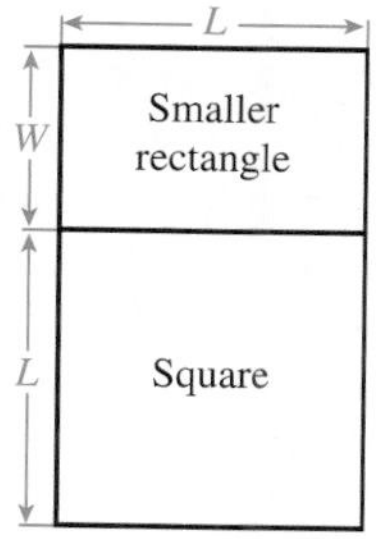

FIGURE 2

length L and width W, as shown in the figure, then we see that the original rectangle has length $L + W$ and width L. Both rectangles (being "golden") have their lengths and widths in the golden ratio, ϕ, given above, so we have

$$\frac{L}{W} = \frac{L + W}{L}$$

$$\frac{L}{W} = \frac{L}{L} + \frac{W}{L} \quad \text{Write the right side as two fractions.}$$

$$\phi = 1 + \frac{1}{\phi} \quad \text{Substitute } \tfrac{L}{W} = \phi,\ \tfrac{L}{L} = 1, \text{ and } \tfrac{W}{L} = \tfrac{1}{\phi}.$$

$$\phi^2 = \phi + 1 \quad \text{Multiply both sides by } \phi.$$

$$\phi^2 - \phi - 1 = 0. \quad \text{Write in standard quadratic form.}$$

Using the quadratic formula from algebra, the positive solution of this equation is found to be $\frac{1+\sqrt{5}}{2} \approx 1.618033989$, the golden ratio.

The Parthenon (see the photo), built on the Acropolis in ancient Athens during the fifth century B.C., is an example of architecture exhibiting many distinct golden rectangles.

A Golden Rectangle in Art The rectangle outlining the figure in *St. Jerome* by Leonardo da Vinci is an example of a golden rectangle.

To see an interesting connection between the terms of the Fibonacci sequence, the golden ratio, and a phenomenon of nature, we can start with a rectangle measuring 89 by 55 units. (See Figure 3.)

This is a very close approximation to a golden rectangle. Within this rectangle a square is then constructed, 55 units on a side. The remaining rectangle is also approximately a golden rectangle, measuring 55 units by 34 units. Each time this process is repeated, a square and an approximate golden rectangle are formed. As indicated in the figure, vertices of the square may be joined by a smooth curve known as a *spiral*. This spiral resembles the outline of a cross section of the shell of the chambered nautilus, as shown in the photograph next to Figure 3.

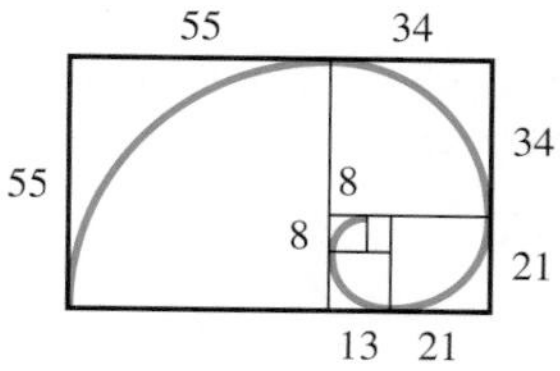

FIGURE 3

For Further Thought

Mathematical Animation

The 1959 animated film *Donald in Mathmagic Land* has endured for nearly 50 years as a classic. It provides a 25-minute trip with Donald Duck, led by the Spirit of Mathematics, through the world of mathematics. Several minutes of the film are devoted to the golden ratio (or, as it is termed there, the golden section). (*Donald in Mathmagic Land* is also discussed in the opener to Chapter 3 on page 97.)

Disney provides animation to explain the golden ratio in a way that the printed word simply cannot do. The golden ratio is seen in architecture, nature, and the human body.

For Group Discussion or Individual Investigation

1. Verify the following Fibonacci pattern in the conifer family. Obtain a pineapple, and count spirals formed by the "scales" of the cone, first counting from lower left to upper right. Then count the spirals from lower right to upper left. What do you find?
2. Two popular sizes of index cards are 3″ by 5″ and 5″ by 8″. Why do you think that these are industry-standard sizes?
3. Divide your height by the height to your navel. Find a class average. What value does this come close to?

5.4 Exercises

Answer each question concerning the Fibonacci sequence or the golden ratio.

1. The sixteenth Fibonacci number is 987 and the seventeenth Fibonacci number is 1597. What is the eighteenth Fibonacci number?

2. Recall that F_n represents the Fibonacci number in the nth position in the sequence. What are the only two values of n such that $F_n = n$?

3. $F_{23} = 28{,}657$ and $F_{25} = 75{,}025$. What is the value of F_{24}?

4. If two successive terms of the Fibonacci sequence are both odd, is the next term even or odd?

5. What is the exact value of the golden ratio?

6. What is the approximate value of the golden ratio to the nearest thousandth?

In each of Exercises 7–14, a pattern is established involving terms of the Fibonacci sequence. Use inductive reasoning to make a conjecture concerning the next equation in the pattern, and verify it. You may wish to refer to the first few terms of the sequence given in the text.

7.
$1 = 2 - 1$
$1 + 1 = 3 - 1$
$1 + 1 + 2 = 5 - 1$
$1 + 1 + 2 + 3 = 8 - 1$
$1 + 1 + 2 + 3 + 5 = 13 - 1$

8.
$1 = 2 - 1$
$1 + 3 = 5 - 1$
$1 + 3 + 8 = 13 - 1$
$1 + 3 + 8 + 21 = 34 - 1$
$1 + 3 + 8 + 21 + 55 = 89 - 1$

9. $1 = 1$
$1 + 2 = 3$
$1 + 2 + 5 = 8$
$1 + 2 + 5 + 13 = 21$
$1 + 2 + 5 + 13 + 34 = 55$

10. $1^2 + 1^2 = 2$
$1^2 + 2^2 = 5$
$2^2 + 3^2 = 13$
$3^2 + 5^2 = 34$
$5^2 + 8^2 = 89$

11. $2^2 - 1^2 = 3$
$3^2 - 1^2 = 8$
$5^2 - 2^2 = 21$
$8^2 - 3^2 = 55$

12. $2^3 + 1^3 - 1^3 = 8$
$3^3 + 2^3 - 1^3 = 34$
$5^3 + 3^3 - 2^3 = 144$
$8^3 + 5^3 - 3^3 = 610$

13. $1 = 1^2$
$1 - 2 = -1^2$
$1 - 2 + 5 = 2^2$
$1 - 2 + 5 - 13 = -3^2$
$1 - 2 + 5 - 13 + 34 = 5^2$

14. $1 - 1 = -1 + 1$
$1 - 1 + 2 = 1 + 1$
$1 - 1 + 2 - 3 = -2 + 1$
$1 - 1 + 2 - 3 + 5 = 3 + 1$
$1 - 1 + 2 - 3 + 5 - 8 = -5 + 1$

15. Every natural number can be expressed as a sum of Fibonacci numbers, where no number is used more than once. For example, $25 = 21 + 3 + 1$. Express each of the following in this way.
(a) 37 **(b)** 40 **(c)** 52

16. It has been shown that if m divides n, then F_m is a factor of F_n. Show that this is true for the following values of m and n.
(a) $m = 2, n = 6$ **(b)** $m = 3, n = 9$
(c) $m = 4, n = 8$

17. It has been shown that if the greatest common factor of m and n is r, then the greatest common factor of F_m and F_n is F_r. Show that this is true for the following values of m and n.
(a) $m = 10, n = 4$ **(b)** $m = 12, n = 6$
(c) $m = 14, n = 6$

18. For any prime number p except 2 or 5, either F_{p+1} or F_{p-1} is divisible by p. Show that this is true for the following values of p.
(a) $p = 3$ **(b)** $p = 7$ **(c)** $p = 11$

19. Earlier we saw that if a term of the Fibonacci sequence is squared and then the product of the terms on each side of the term is found, there will always be a difference of 1. Follow the steps below, choosing the seventh Fibonacci number, 13.
(a) Square 13. Multiply the terms of the sequence two positions away from 13 (i.e., 5 and 34). Subtract the smaller result from the larger, and record your answer.
(b) Square 13. Multiply the terms of the sequence three positions away from 13. Once again, subtract the smaller result from the larger, and record your answer.
(c) Repeat the process, moving four terms away from 13.
(d) Make a conjecture about what will happen when you repeat the process, moving five terms away. Verify your answer.

20. ***A Number Trick*** Here is a number trick that you can perform. Ask someone to pick any two numbers at random and to write them down. Ask the person to determine a third number by adding the first and second, a fourth number by adding the second and third, and so on, until ten numbers are determined. Then ask the person to add these ten numbers. You will be able to give the sum before the person even completes the list, because the sum will always be 11 times the seventh number in the list. Verify that this is true, by using x and y as the first two numbers arbitrarily chosen. (*Hint:* Remember the distributive property from algebra.)

Another Fibonacci-type sequence that has been studied by mathematicians is the **Lucas sequence,** *named after a French mathematician of the nineteenth century. The first ten terms of the Lucas sequence are*

$$1, 3, 4, 7, 11, 18, 29, 47, 76, 123.$$

21. What is the eleventh term of the Lucas sequence?

22. Choose any term of the Lucas sequence and square it. Then multiply the terms on either side of the one you chose. Subtract the smaller result from the larger. Repeat this for a different term of the sequence. Do you get the same result? Make a conjecture about this pattern.

23. The first term of the Lucas sequence is 1. Add the first and third terms. Record your answer. Now add the first, third, and fifth terms and record your answer. Continue this pattern, each time adding another term that is in an *odd* position in the sequence. What do you notice about all of your sums?

24. The second term of the Lucas sequence is 3. Add the second and fourth terms. Record your answer. Now add the second, fourth, and sixth terms and record your answer. Continue this pattern, each time adding another term that is in an *even* position of the sequence. What do you notice about all of your sums?

25. Many interesting patterns exist between the terms of the Fibonacci sequence and the Lucas sequence. Make a conjecture about the next equation that would appear in each of the lists and then verify it.

(a)	(b)
$1 \cdot 1 = 1$	$1 + 2 = 3$
$1 \cdot 3 = 3$	$1 + 3 = 4$
$2 \cdot 4 = 8$	$2 + 5 = 7$
$3 \cdot 7 = 21$	$3 + 8 = 11$
$5 \cdot 11 = 55$	$5 + 13 = 18$

(c) $1 + 1 = 2 \cdot 1$
$1 + 3 = 2 \cdot 2$
$2 + 4 = 2 \cdot 3$
$3 + 7 = 2 \cdot 5$
$5 + 11 = 2 \cdot 8$

26. In the text we illustrate that the quotients of successive terms of the Fibonacci sequence approach the golden ratio. Make a similar observation for the terms of the Lucas sequence; that is, find the decimal approximations for the quotients

$$\frac{3}{1}, \frac{4}{3}, \frac{7}{4}, \frac{11}{7}, \frac{18}{11}, \frac{29}{18},$$

and so on, using a calculator. Then make a conjecture about what seems to be happening.

Recall the **Pythagorean theorem** *from geometry: If a right triangle has legs of lengths a and b and hypotenuse of length c, then*

$$a^2 + b^2 = c^2.$$

Suppose that we choose any four successive terms of the Fibonacci sequence. Multiply the first and fourth. Double the product of the second and third. Add the squares of the second and third. The three results obtained form a **Pythagorean triple** *(three numbers that satisfy the equation $a^2 + b^2 = a^2$). Find the Pythagorean triple obtained this way using the four given successive terms of the Fibonacci sequence.*

27. 1, 1, 2, 3

28. 1, 2, 3, 5

29. 2, 3, 5, 8

30. Look at the values of the hypotenuse (c) in the answers to Exercises 27–29. What do you notice about each of them?

31. The following array of numbers is called **Pascal's triangle.**

1
1 1
1 2 1
1 3 3 1
1 4 6 4 1
1 5 10 10 5 1
1 6 15 20 15 6 1

This array is important in the study of counting techniques and probability (see later chapters) and appears in algebra in the binomial theorem. If the triangular array is written in a different form, as follows, and the sums along the diagonals as indicated by the dashed lines are found, there is an interesting occurrence. What do you find when the numbers are added?

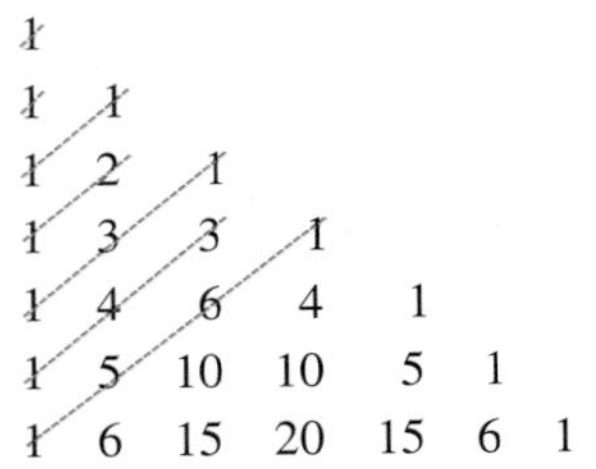

32. Write a paragraph explaining some of the occurrences of the Fibonacci sequence and the golden ratio in your everyday surroundings.

Exercises 33–38 require a scientific calculator.

33. The positive solution of the equation $x^2 - x - 1 = 0$ is $\frac{1 + \sqrt{5}}{2}$, as indicated in the text. The negative solution is $\frac{1 - \sqrt{5}}{2}$. Find the decimal approximations for both. What similarity do you notice between the two decimals?

34. In some cases, writers define the golden ratio to be the *reciprocal* of $\frac{1 + \sqrt{5}}{2}$. Find a decimal approximation for the reciprocal of $\frac{1 + \sqrt{5}}{2}$. What similarity do you notice between the decimals for $\frac{1 + \sqrt{5}}{2}$ and its reciprocal?

A remarkable relationship exists between the two solutions of $x^2 - x - 1 = 0$,

$$\phi = \frac{1 + \sqrt{5}}{2} \quad \text{and} \quad \bar{\phi} = \frac{1 - \sqrt{5}}{2},$$

and the Fibonacci numbers. To find the nth Fibonacci number without using the recursion formula, evaluate

$$\frac{\phi^n - \bar{\phi}^n}{\sqrt{5}}$$

using a calculator. For example, to find the thirteenth Fibonacci number, evaluate

$$\frac{\left(\frac{1 + \sqrt{5}}{2}\right)^{13} - \left(\frac{1 - \sqrt{5}}{2}\right)^{13}}{\sqrt{5}}$$

This form is known as the **Binet form** *of the nth Fibonacci number. Use the Binet form and a calculator to find the nth Fibonacci number for each of the following values of n.*

35. $n = 14$

36. $n = 20$

37. $n = 22$

38. $n = 25$

EXTENSION

Magic Squares

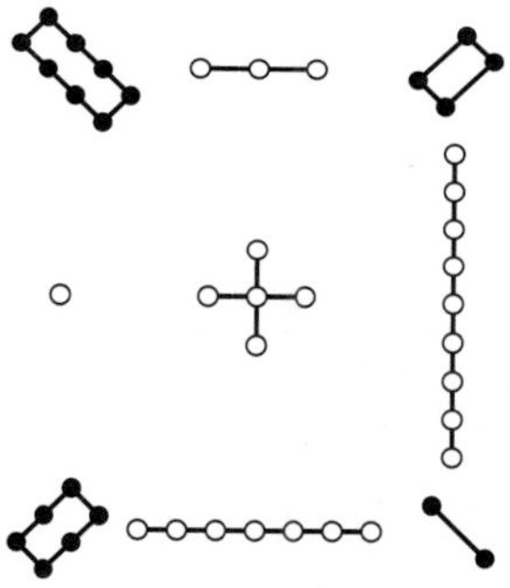

FIGURE 4

8	3	4
1	5	9
6	7	2

FIGURE 5

Legend has it that in about 2200 B.C. the Chinese Emperor Yu discovered on the bank of the Yellow River a tortoise whose shell bore the diagram in Figure 4. This so-called *lo-shu* is an early example of a **magic square.** If the numbers of dots are counted and arranged in a square fashion, the array in Figure 5 is obtained. A magic square is a square array of numbers with the property that the sum along each row, column, and diagonal is the same. This common value is called the "magic sum." The **order** of a magic square is simply the number of rows (and columns) in the square. The magic square of Figure 5 is an order 3 magic square.

By using the formula for the sum of the first n terms of an arithmetic sequence, it can be shown that if a magic square of order n has entries $1, 2, 3, \ldots, n^2$, then the sum of *all entries* in the square is

$$\frac{n^2(n^2 + 1)}{2}.$$

Because there are n rows (and columns), the magic sum of the square may be found by dividing the above expression by n. This results in the following formula for finding the magic sum.

Magic Sum Formula

If a magic square of order n has entries $1, 2, 3, \ldots, n^2$, then the magic sum MS is given by the formula

$$\text{MS} = \frac{n(n^2 + 1)}{2}.$$

(continued)

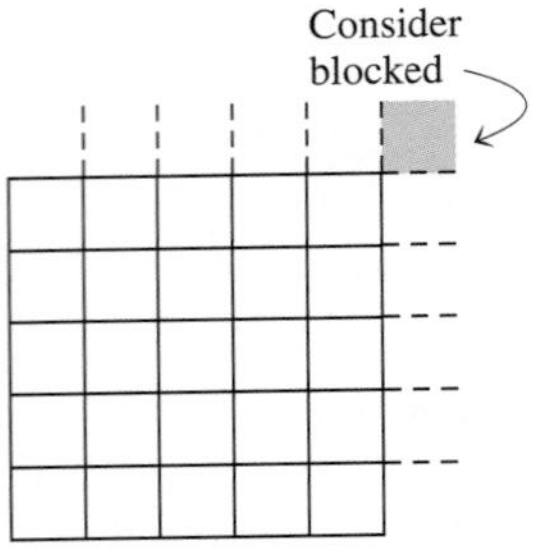

FIGURE 6

18	25	2	9	16	
17	24	1	8	15	17
23	5	7	14	16	23
4	6	13	20	22	4
10	12	19	21	3	10
11	18	25	2	9	

FIGURE 7

With $n = 3$ in this formula we find that the magic sum of the square in Figure 5, which may be verified by direct addition, is

$$\text{MS} = \frac{3(3^2 + 1)}{2} = 15.$$

There is a method of constructing an odd-order magic square which is attributed to an early French envoy, *de la Loubere,* that is sometimes referred to as the "staircase method." The method is described below for an order 5 square, with entries 1, 2, 3, . . . , 25.

Begin by sketching a square divided into 25 cells into which the numbers 1–25 are to be entered. Proceed as described below, referring to Figures 6 and 7 for clarification.

Step 1 Write 1 in the middle cell of the top row.

Step 2 Always try to enter numbers in sequence in the cells by moving diagonally from lower left to upper right. There are two exceptions to this:

(a) If you go outside of the magic square, move all the way across the row or down the column to enter the number. Then proceed to move diagonally.

(b) If you run into a cell which is already occupied (that is, you are "blocked"), drop down one cell from the last entry written and enter the next number there. Then proceed to move diagonally.

Step 3 Your last entry, 25, will be in the middle cell of the bottom row.

Figure 7 shows the completed magic square. Its magic sum is 65.

Benjamin Franklin admitted that he would amuse himself while in the Pennsylvania Assembly with magic squares or circles "or any thing to avoid Weariness." He wrote about the usefulness of mathematics in the *Gazette* in 1735, saying that no employment can be managed without arithmetic, no mechanical invention without geometry. He also thought that mathematical demonstrations are better than academic logic for training the mind to reason with exactness and distinguish truth from falsity even outside of mathematics.

The square shown here is one developed by Franklin. It has a sum of 2056 in each row and diagonal, and, in Franklin's words, has the additional property "that a four-square hole being cut in a piece of paper of such size as to take in and show through it just 16 of the little squares, when laid on the greater square, the sum of the 16 numbers so appearing through the hole, wherever it was placed on the greater square should likewise make 2056." He claimed that it was "the most magically magic square ever made by any magician."

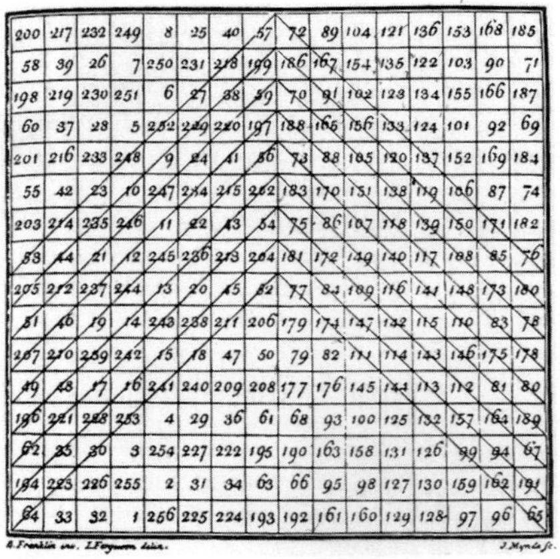

200	217	232	249	8	25	40	57	72	89	104	121	136	153	168	185
58	39	26	7	250	231	218	199	186	167	154	135	122	103	90	71
198	219	230	251	6	27	38	59	70	91	102	123	134	155	166	187
60	37	28	5	252	229	220	197	188	165	156	133	124	101	92	69
201	216	233	248	9	24	41	56	73	88	105	120	137	152	169	184
55	42	23	10	247	234	215	202	183	170	151	138	119	106	87	74
203	214	235	246	11	22	43	54	75	86	107	118	139	150	171	182
53	44	21	12	245	236	213	204	181	172	149	140	117	108	85	76
205	212	237	244	13	20	45	52	77	84	109	116	141	148	173	180
51	46	19	14	243	238	211	206	179	174	147	142	115	110	83	78
207	210	239	242	15	18	47	50	79	82	111	114	143	146	175	178
49	48	17	16	241	240	209	208	177	176	145	144	113	112	81	80
196	221	228	253	4	29	36	61	68	93	100	125	132	157	164	189
62	35	30	3	254	227	222	195	190	163	158	131	126	99	94	67
194	223	226	255	2	31	34	63	66	95	98	127	130	159	162	191
64	33	32	1	256	225	224	193	192	161	160	129	128	97	96	65

You might wish to verify the following property of this magic square: The sum of any four numbers that are opposite each other and at equal distances from the center is 514 (which is one-fourth of the magic sum).

EXTENSION EXERCISES

Given a magic square, other magic squares may be obtained by rotating the given one. For example, starting with the magic square in Figure 5, a 90° rotation in a clockwise direction gives the magic square shown here.

6	1	8
7	5	3
2	9	4

Start with Figure 5 and give the magic square obtained by each rotation described.

1. 180° in a clockwise direction

2. 90° in a counterclockwise direction

Start with Figure 7 and give the magic square obtained by each rotation described.

3. 90° in a clockwise direction

4. 180° in a clockwise direction

5. 90° in a counterclockwise direction

6. Try to construct an order 2 magic square containing the entries 1, 2, 3, 4. What happens?

Given a magic square, other magic squares may be obtained by adding or subtracting a constant value to or from each entry, multiplying each entry by a constant value, or dividing each entry by a nonzero constant value. In Exercises 7–10, start with the magic square whose figure number is indicated, and perform the operation described to find a new magic square. Give the new magic sum.

7. Figure 5, multiply by 3

8. Figure 5, add 7

9. Figure 7, divide by 2

10. Figure 7, subtract 10

According to a fanciful story by Charles Trigg in Mathematics Magazine *(September 1976, page 212), the Emperor Charlemagne (742–814) ordered a five-sided fort to be built at an important point in his kingdom. As good-luck charms, he had magic squares placed on all five sides of the fort. He had one restriction for these magic squares: all the numbers in them must be prime.*

Charlemagne's magic squares are given in Exercises 11–15, with one missing entry. Find the missing entry in each square.

11.

	71	257
47	269	491
281	467	59

12.

389		227
107	269	431
311	347	149

13.

389	227	191
71	269	
347	311	149

14.

401	227	179
47	269	491
359		137

15.

401	257	149
17		521
389	281	137

16. Compare the magic sums in Exercises 11–15. Charlemagne had stipulated that each magic sum should be the year in which the fort was built. What was that year?

Find the missing entries in each magic square.

17.

75	68	**(a)**
(b)	72	**(c)**
71	76	**(d)**

18.

1	8	13	**(a)**
(b)	14	7	2
16	9	4	**(c)**
(d)	**(e)**	**(f)**	15

19.

3	20	**(a)**	24	11
(b)	14	1	18	10
9	21	13	**(c)**	17
16	8	25	12	**(d)**
(e)	2	**(f)**	**(g)**	**(h)**

(continued)

20.

3	36	2	35	31	4
10	12	**(a)**	26	7	27
21	13	17	14	**(b)**	22
16	**(c)**	23	**(d)**	18	15
28	30	8	**(e)**	25	9
(f)	1	32	5	6	34

21. Use the "staircase method" to construct a magic square of order 7, containing the entries 1, 2, 3, . . . , 49.

The magic square shown in the photograph is from a woodcut by Albrecht Dürer entitled Melancholia.

The two bottom center numbers give 1514, *the date of the woodcut. Refer to this magic square for Exercises 22–30.*

16	3	2	13
5	10	11	8
9	6	7	12
4	15	14	1

Dürer's Magic Square

22. What is the magic sum?

23. Verify: The sum of the entries in the four corners is equal to the magic sum.

24. Verify: The sum of the entries in any 2 by 2 square at a corner of the given magic square is equal to the magic sum.

25. Verify: The sum of the entries in the diagonals is equal to the sum of the entries not in the diagonals.

26. Verify: The sum of the squares of the entries in the diagonals is equal to the sum of the squares of the entries not in the diagonals.

27. Verify: The sum of the cubes of the entries in the diagonals is equal to the sum of the cubes of the entries not in the diagonals.

28. Verify: The sum of the squares of the entries in the top two rows is equal to the sum of the squares of the entries in the bottom two rows.

29. Verify: The sum of the squares of the entries in the first and third rows is equal to the sum of the squares of the entries in the second and fourth rows.

30. Find another interesting property of Dürer's magic square and state it.

31. A magic square of order 4 may be constructed as follows. Lightly sketch in the diagonals of the blank magic square. Beginning at the upper left, move across each row from left to right, counting the cells as you go along. If the cell is on a diagonal, count it but do not enter its number. If it is not on a diagonal, enter its number. When this is completed, reverse the procedure, beginning at the bottom right and moving across from right to left. As you count the cells, enter the number if the cell is not occupied. If it is already occupied, count it but do not enter its number. You should obtain a magic square similar to the one given for Exercises 22–30. How do they differ?

With chosen values for a, b, and c, an order 3 magic square can be constructed by substituting these values in the generalized form shown here.

$a + b$	$a - b - c$	$a + c$
$a - b + c$	a	$a + b - c$
$a - c$	$a + b + c$	$a - b$

Use the given values of a, b, and c to construct an order 3 magic square, using this generalized form.

32. $a = 5, \quad b = 1, \quad c = -3$

33. $a = 16, \quad b = 2, \quad c = -6$

34. $a = 5, \quad b = 4, \quad c = -8$

35. It can be shown that if an order n magic square has least entry k, and its entries are consecutive counting numbers, then its magic sum is given by the formula

$$\text{MS} = \frac{n(2k + n^2 - 1)}{2}.$$

Construct an order 7 magic square with least entry 10 using the staircase method. Find its magic sum.

36. Use the formula of Exercise 35 to find the missing entries in the following order 4 magic square whose least entry is 24.

(a)	38	37	27
35	**(b)**	30	32
31	33	**(c)**	28
(d)	26	25	**(e)**

In a 1769 letter from Benjamin Franklin to a Mr. Peter Collinson, Franklin exhibited the following semimagic square of order 8. (Note: A square is semimagic if it is magic except that one or both diagonals fail to give the magic sum.)

52	61	4	13	20	29	36	45
14	3	62	51	46	35	30	19
53	60	5	12	21	28	37	44
11	6	59	54	43	38	27	22
55	58	7	10	23	26	39	42
9	8	57	56	41	40	25	24
50	63	2	15	18	31	34	47
16	1	64	49	48	33	32	17

37. What is the magic sum?

Verify the following properties of this semimagic square.

38. The sums in the first half of each row and the second half of each row are both equal to half the magic sum.

39. The four corner entries added to the four center entries is equal to the magic sum.

40. The "bent diagonals" consisting of eight entries, going up four entries from left to right and down four entries from left to right, give the magic sum. (For example, starting with 16, one bent diagonal sum is $16 + 63 + 57 + 10 + 23 + 40 + 34 + 17$.)

If we use a "knight's move" (up two, right one) from chess, a variation of the staircase method gives rise to the magic square shown here. (When blocked, we move to the cell just below the previous entry.)

10	18	1	14	22
11	24	7	20	3
17	5	13	21	9
23	6	19	2	15
4	12	25	8	16

Use a similar process to construct an order 5 magic square, starting with 1 in the cell described.

41. fourth row, second column (up two, right one; when blocked, move to the cell just below the previous entry)

42. third row, third column (up one, right two; when blocked, move to the cell just to the left of the previous entry)

COLLABORATIVE INVESTIGATION

Investigating an Interesting Property of Number Squares

In the Extension at the end of this chapter, we looked at magic squares. Now in this group activity we will investigate another property of squares of numbers. Begin by dividing up the class into groups of three or four students. Each student in the group should prepare a square of numbers like the one that follows:

1	2	3	4	5
6	7	8	9	10
11	12	13	14	15
16	17	18	19	20
21	22	23	24	25

Topics for Discussion

1. Each student should do the following individually:

   Choose any number in the first row. Circle it, and cross out all entries in the column below it. (For example, if you circle 4, cross out 9, 14, 19, and 24.) Now circle any remaining number in the second row, and cross out all entries in the column below it.

   Repeat this procedure for the third and fourth rows, and then circle the final remaining number in the fifth row.

   Now each student in the group should add the circled numbers and compare his or her sum with all others in the group. What do you notice?

2. How does the sum obtained in Exercise 1 compare with the magic sum for an order 5 magic square?

3. Suppose Exercise 1 was done as shown here:

1	(2)	3	4	5
6	~~7~~	(8)	9	10
11	~~12~~	~~13~~	14	(15)
(16)	~~17~~	~~18~~	19	~~20~~
~~21~~	~~22~~	~~23~~	(24)	~~25~~

   Notice that summing the circled entries is just like summing $1 + 2 + 3 + 4 + 5$, except that

   3 is replaced by $3 + 5$,
   5 is replaced by $5 + 10$,
   1 is replaced by $1 + 15$,
   4 is replaced by $4 + 20$.

   We can express this as

$$\begin{aligned} \text{sum} &= (1 + 2 + 3 + 4 + 5) \\ &\quad + (5 + 10 + 15 + 20) \\ &= 15 + 50 = 65. \end{aligned}$$

4. Explain why, whatever entries you choose to circle in the various rows, the sum is always the same.

5. Prepare a similar square of the natural numbers 1 through 36. Then repeat Exercise 1. Discuss your results. How does the sum compare with the magic sum for an order 6 magic square?

6. As a group, fill in the entries in this equation for the 6 by 6 square.

   sum = (____ + ____ + ____ + ____ + ____ + ____) + (____ + ____ + ____ + ____ + ____)

7. As a group, predict the sum of the circled numbers in a 7 by 7 square by expressing it as follows. (Do not actually construct the square.)

   sum = (____ + ____ + ____ + ____ + ____ + ____ + ____) + (____ + ____ + ____ + ____ + ____ + ____)

   How does the sum compare with the magic sum for an order 7 magic square?

8. Each individual should now prepare another 5 by 5 square and repeat Exercise 1, except this time start with a number in the first *column* and cross out remaining numbers in *rows*. In your group, discuss and explain what you observe.

CHAPTER 5 TEST

In Exercises 1–5, decide whether each statement is true *or* false.

1. No two prime numbers differ by 1.

2. There are infinitely many prime numbers.

3. If a natural number is divisible by 9, then it must also be divisible by 3.

4. If p and q are different primes, 1 is their greatest common factor and pq is their least common multiple.

5. For all natural numbers n, 1 is a factor of n and n is a multiple of n.

6. Use divisibility tests to determine whether the number

331,153,470

is divisible by each of the following.

(a) 2 **(b)** 3 **(c)** 4
(d) 5 **(e)** 6 **(f)** 8
(g) 9 **(h)** 10 **(i)** 12

7. Decide whether each number is prime, composite, or neither.

(a) 93 **(b)** 1 **(c)** 59

8. Give the prime factorization of 1440.

9. In your own words state the Fundamental Theorem of Arithmetic.

10. Decide whether each number is perfect, deficient, or abundant.

(a) 17 **(b)** 6 **(c)** 24

11. Which of the following statements is false?

A. There are no known odd perfect numbers.
B. Every even perfect number must end in 6 or 28.
C. Goldbach's Conjecture for the number 8 is verified by the equation $8 = 7 + 1$.

12. Give a pair of twin primes between 40 and 50.

13. Find the greatest common factor of 270 and 450.

14. Find the least common multiple of 24, 36, and 60.

15. ***Day Off for Fast-food Workers*** Both Sherrie Firavich and Della Daniel work at a fast-food outlet. Sherrie has every sixth day off and Della has every fourth day off. If they are both off on Wednesday of this week, what will be the day of the week that they are next off together?

16. The twenty-second Fibonacci number is 17,711 and the twenty-third Fibonacci number is 28,657. What is the twenty-fourth Fibonacci number?

17. Make a conjecture about the next equation in the following list, and verify it.

$$8 - (1 + 1 + 2 + 3) = 1$$
$$13 - (1 + 2 + 3 + 5) = 2$$
$$21 - (2 + 3 + 5 + 8) = 3$$
$$34 - (3 + 5 + 8 + 13) = 5$$
$$55 - (5 + 8 + 13 + 21) = 8$$

18. Choose the correct completion of this statement: If p is a prime number, then $2^p - 1$ is prime

A. never **B.** sometimes **C.** always.

19. (a) Give the first eight terms of a Fibonacci-type sequence with first term 1 and second term 5.

(b) Choose any term after the first in the sequence just formed. Square it. Multiply the two terms on either side of it. Subtract the smaller result from the larger. Now repeat the process with a different term. Make a conjecture about what this process will yield for any term of the sequence.

20. Which one of the following is the *exact* value of the golden ratio?

A. $\frac{1 + \sqrt{5}}{2}$ **B.** $\frac{1 - \sqrt{5}}{2}$ **C.** 1.6 **D.** 1.618

21. Briefly state what Fermat's Last Theorem says, and describe the circumstances of its proof.

22. Write a brief explanation of the acronym GIMPS. Include a definition and several examples of the term represented by the letters MP.

6

THE REAL NUMBERS AND THEIR REPRESENTATIONS

The original *Star Trek* series first aired on NBC on September 8, 1966, and spawned an entire generation of science fiction fans. In its second season, the episode "Wolf in the Fold" told the story of an alien entity that had taken over the computer of the starship *Enterprise.* In an effort to drive the entity out of the computer, Captain Kirk suggested the following to Mr. Spock:

KIRK: Spock, don't you have a compulsory scan unit built into the computer banks?

SPOCK: Yes we do, Captain, but with the entity in control. . . .

KIRK: Well aren't there certain mathematical problems which simply cannot be solved?

SPOCK: Indeed. If we can focus the attention of the computer on one of them. . .

KIRK: That ought to do it.

Later, they are able to do just that:

SPOCK: Ready?

KIRK: Implement.

SPOCK: Computer, this is a class "A" compulsory directive. Compute to the last digit the value of pi.

COMPUTER: No, no, no, no, no, . . .

SPOCK (TO KIRK): As we know, the value of pi is a transcendental figure without resolution. The computer banks will work on this problem to the exclusion of all else until we order it to stop.

KIRK: Yes, that should keep that thing busy for a while.

The alien could not comply with the compulsory directive, because pi (π) is an irrational number, and its decimal representation has no last digit. As a result, ingenuity and mathematics saved the *Enterprise.* In this chapter, we study the rational numbers and irrational numbers, which together form the real number system.

6.1 Real Numbers, Order, and Absolute Value

Sets of Real Numbers • Order in the Real Numbers • Additive Inverses and Absolute Value • Applications

The Origins of Zero The Mayan Indians of Mexico and Central America had one of the earliest numeration systems that included a symbol for zero. The very early Babylonians had a positional system, but they placed only a space between "digits" to indicate a missing power. When the Greeks absorbed Babylonian astronomy, they used the letter omicron, o, of their alphabet or $\bar{o}$ to represent "no power," or "zero." The Greek numeration system was gradually replaced by the Roman numeration system.

The Roman system was the one most commonly used in Europe from the time of Christ until perhaps 1400 A.D., when the Hindu-Arabic system began to take over. The original Hindu word for zero was *sunya,* meaning "void." The Arabs adopted this word as *sifr,* or "vacant." The word *sifr* passed into Latin as *zephirum,* which over the years became *zevero, zepiro,* and finally, *zero.*

Sets of Real Numbers

As mathematics developed, it was discovered that the *counting,* or *natural, numbers* did not satisfy all requirements of mathematicians. Consequently, new, expanded number systems were created. The mathematician Leopold Kronecker (1823–1891) once made the statement, "God made the integers, all the rest is the work of man." The *natural numbers* are those numbers with which we count discrete objects. By including 0 in the set, we obtain the set of *whole numbers.*

Natural Numbers

$\{1, 2, 3, 4, \ldots\}$ is the set of **natural numbers.**

Whole Numbers

$\{0, 1, 2, 3, \ldots\}$ is the set of **whole numbers.**

These numbers, along with many others, can be represented on **number lines** like the one pictured in Figure 1. We draw a number line by locating any point on the line and calling it 0. Choose any point to the right of 0 and call it 1. The distance between 0 and 1 gives a unit of measure used to locate other points, as shown in Figure 1. The points labeled in Figure 1 and those continuing in the same way to the right correspond to the set of whole numbers.

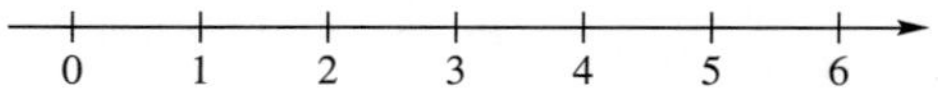

FIGURE 1

All the whole numbers starting with 1 are located to the right of 0 on the number line. But numbers may also be placed to the left of 0. These numbers, written -1, -2, -3, and so on, are shown in Figure 2. (The negative sign is used to show that the numbers are located to the *left* of 0.)

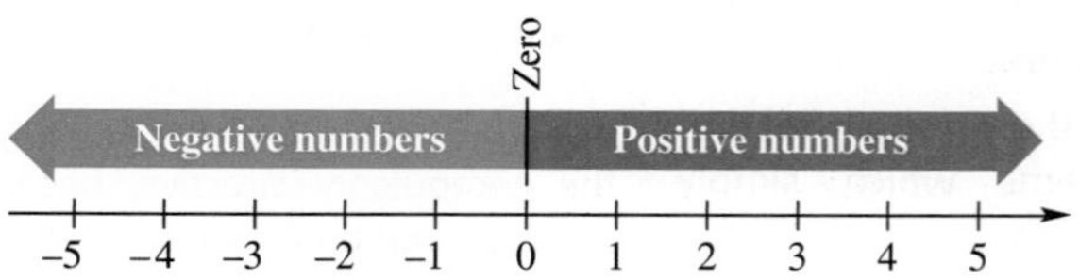

FIGURE 2

The numbers to the *left* of 0 are **negative numbers.** The numbers to the *right* of 0 are **positive numbers.** The number 0 itself is neither positive nor negative. Positive numbers and negative numbers are called **signed numbers.**

The Origins of Negative Numbers Negative numbers can be traced back to the Chinese between 200 B.C. and 220 A.D. Mathematicians at first found negative numbers ugly and unpleasant, even though they kept cropping up in the solutions of problems. For example, an Indian text of about 1150 A.D. gives the solution of an equation as -5 and then makes fun of anything so useless.

Leonardo of Pisa (Fibonacci), while working on a financial problem, was forced to conclude that the solution must be a negative number (that is, a financial loss). In 1545 A.D., the rules governing operations with negative numbers were published by **Girolamo Cardano** in his *Ars Magna* (Great Art).

There are many practical applications of negative numbers. For example, temperatures sometimes fall below zero. The lowest temperature ever recorded in meteorological records was $-128.6°$F at Vostok, Antarctica, on July 22, 1983. Altitudes below sea level can be represented by negative numbers. The shore surrounding the Dead Sea is 1312 feet below sea level; this can be represented as -1312 feet.

The set of numbers marked on the number line in Figure 2, including positive and negative numbers and zero, is part of the set of *integers.*

Integers

$\{\ldots, -3, -2, -1, 0, 1, 2, 3, \ldots\}$ is the set of **integers.**

Not all numbers are integers. For example, $\frac{1}{2}$ is not; it is a number halfway between the integers 0 and 1. Also, $3\frac{1}{4}$ is not an integer. Several numbers that are not integers are *graphed* in Figure 3. The **graph** of a number is a point on the number line. Think of the graph of a set of numbers as a picture of the set. All the numbers in Figure 3 can be written as quotients of integers. These numbers are examples of *rational numbers.*

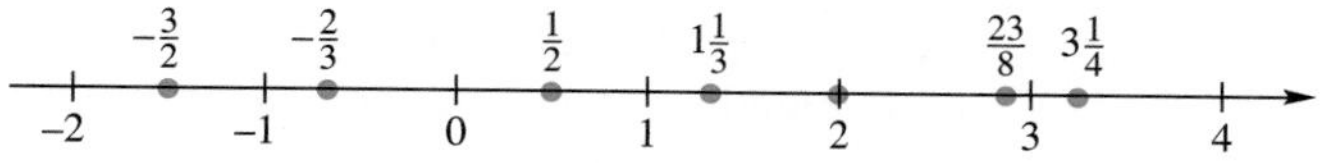

FIGURE 3

Notice that an integer, such as 2, is also a rational number; for example, $2 = \frac{2}{1}$.

Rational Numbers

$\{x \mid x$ is a quotient of two integers, with denominator not equal to $0\}$ is the set of **rational numbers.**

(Read the part in the braces as "the set of all numbers x such that x is a quotient of two integers, with denominator not equal to 0.")

The set symbolism used in the definition of rational numbers,

$$\{x \mid x \textbf{ has a certain property}\},$$

is called **set-builder notation.** This notation is convenient to use when it is not possible, or practical, to list all the elements of the set.

Although a great many numbers are rational, not all are. For example, a square that measures one unit on a side has a diagonal whose length is the square root of 2, written $\sqrt{2}$. See Figure 4. It will be shown later that $\sqrt{2}$ cannot be written as a quotient of integers. Because of this, $\sqrt{2}$ is not rational; it is *irrational.*

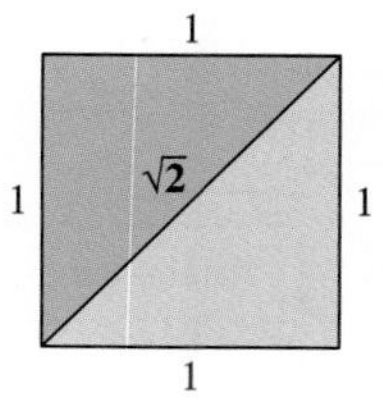

FIGURE 4

Irrational Numbers

$\{x|x$ is a number on the number line that is not rational$\}$ is the set of **irrational numbers.**

Examples of irrational numbers include $\sqrt{3}$, $\sqrt{7}$, $-\sqrt{10}$, and π, which is the ratio of the distance around a circle (its *circumference*) to the distance across it (its *diameter*).

All numbers that can be represented by points on the number line are called *real numbers.*

Real Numbers

$\{x|x$ is a number that can be represented by a point on the number line$\}$ is the set of **real numbers.**

Real numbers can be written as decimal numbers. Any rational number can be written as a decimal that will come to an end (terminate), or repeat in a fixed "block" of digits. For example, $\frac{2}{5} = .4$ and $\frac{27}{100} = .27$ are rational numbers with terminating decimals; $\frac{1}{3} = .3333\ldots$ and $\frac{3}{11} = .27272727\ldots$ are repeating decimals. The decimal representation of an irrational number will neither terminate nor repeat. Decimal representations of rational and irrational numbers will be discussed further later in this chapter.

Figure 5 illustrates two ways to represent the relationships among the various sets of real numbers.

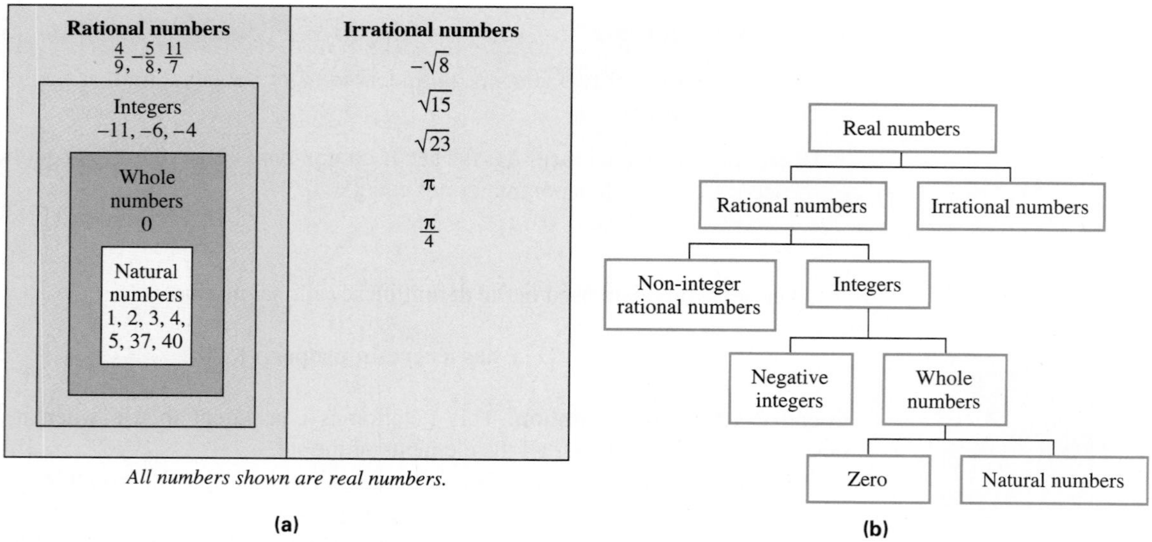

All numbers shown are real numbers.

FIGURE 5

EXAMPLE 1 Identifying Elements of a Set of Numbers

List the numbers in the set

$$\left\{-5, -\frac{2}{3}, 0, \sqrt{2}, \frac{13}{4}, 5, 5.8\right\}$$

that belong to each set of numbers.

(a) natural numbers **(b)** whole numbers **(c)** integers
(d) rational numbers **(e)** irrational numbers **(f)** real numbers

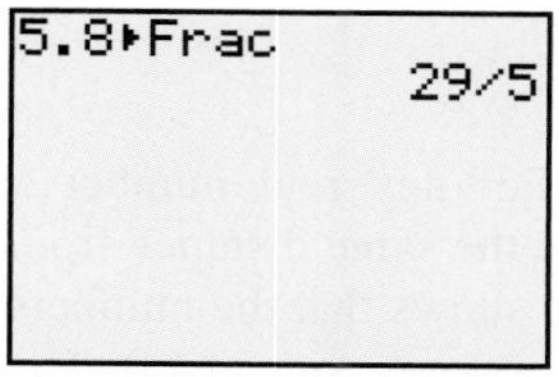

The TI-83/84 Plus calculator will convert a decimal to a fraction. See Example 1(d).

SOLUTION

(a) The only natural number in the set is 5.
(b) The whole numbers consist of the natural numbers and 0. So, the elements of the set that are whole numbers are 0 and 5.
(c) The integers in the set are -5, 0, and 5.
(d) The rational numbers are -5, $-\frac{2}{3}$, 0, $\frac{13}{4}$, 5, and 5.8, because each of these numbers *can* be written as the quotient of two integers. For example, $5.8 = \frac{58}{10} = \frac{29}{5}$.
(e) The only irrational number in the set is $\sqrt{2}$.
(f) All the numbers in the set are real numbers.

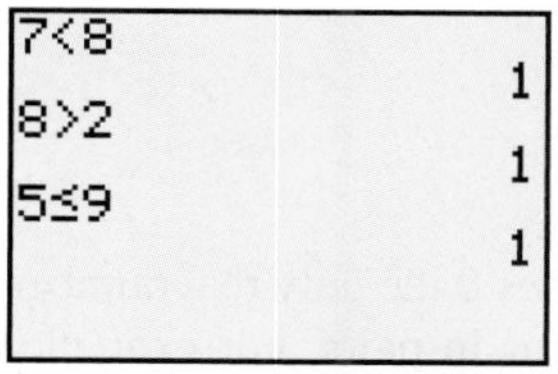

The calculator returns a 1 for these statements of inequality, signifying that each is true.

Order in the Real Numbers Two real numbers may be compared, or ordered, using the ideas of equality and inequality. Suppose that a and b represent two real numbers. If their graphs on the number line are the same point, they are **equal.** If the graph of a lies to the left of b, a **is less than** b, and if the graph of a lies to the right of b, a **is greater than** b. The **law of trichotomy** says that for two numbers a and b, one and only one of the following is true:

$$a = b, \quad a < b, \quad \text{or} \quad a > b.$$

When read from left to right, the symbol $<$ means "is less than," so

$$7 < 8. \quad \text{7 is less than 8.}$$

The symbol $>$ means "is greater than." For example,

$$8 > 2. \quad \text{8 is greater than 2.}$$

The symbol for equality, =, was first introduced by the Englishman Robert Recorde in his 1557 algebra text *The Whetstone of Witte.* He used two parallel line segments, because, he claimed, no two things can be more equal.

The symbols for order relationships, < and >, were first used by Thomas Harriot (1560–1621), another Englishman. These symbols were not immediately adopted by other mathematicians.

Notice that the symbol always points to the lesser number. For example,

$$\text{Lesser number} \longrightarrow 8 < 15.$$

The symbol $\leq$ means "is less than or equal to," so

$$5 \leq 9. \quad \text{5 is less than or equal to 9.}$$

This statement is true, since $5 < 9$ is true. ***If either the < part or the = part is true, then the inequality ≤ is true.*** Also, $8 \leq 8$ is true since $8 = 8$ is true. But it is not true that $13 \leq 9$ because neither $13 < 9$ nor $13 = 9$ is true.

The symbol $\geq$ means "is greater than or equal to." Again,

$$9 \geq 5 \quad \text{9 is greater than or equal to 5.}$$

is true because $9 > 5$ is true.

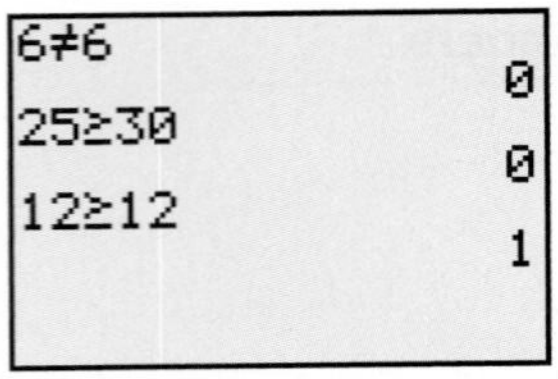

The inequalities in Example 2(a) and (d) are false, as signified by the 0. The statement in Example 2(e) is true.

EXAMPLE 2 Comparing Real Numbers

Determine whether each statement is *true* or *false*.

(a) $6 \neq 6$ **(b)** $5 < 19$ **(c)** $15 \leq 20$ **(d)** $25 \geq 30$ **(e)** $12 \geq 12$

SOLUTION

(a) The statement $6 \neq 6$ is false, because 6 *is equal to* 6.
(b) Since 5 is indeed less than 19, this statement is true.
(c) The statement $15 \leq 20$ is true, since $15 < 20$.
(d) Both $25 > 30$ and $25 = 30$ are false, so $25 \geq 30$ is false.
(e) Since $12 = 12$, the statement $12 \geq 12$ is true.

Additive Inverses and Absolute Value For any real number x (except 0), there is exactly one number on the number line the same distance from 0 as x but on the opposite side of 0. For example, Figure 6 shows that the numbers 3 and -3 are both the same distance from 0 but are on opposite sides of 0. The numbers 3 and -3 are called **additive inverses, negatives,** or **opposites,** of each other.

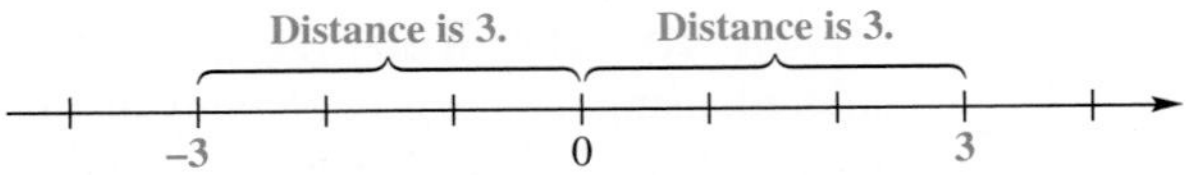

FIGURE 6

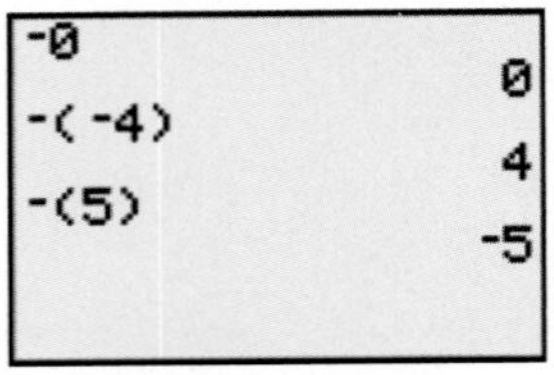

The TI-83/84 Plus distinguishes between the negative symbol (−) and the operation of subtraction.

The additive inverse of the number 0 is 0 itself. This makes 0 the only real number that is its own additive inverse. Other additive inverses occur in pairs. For example, 4 and -4, and 5 and -5, are additive inverses of each other. Several pairs of additive inverses are shown in Figure 7.

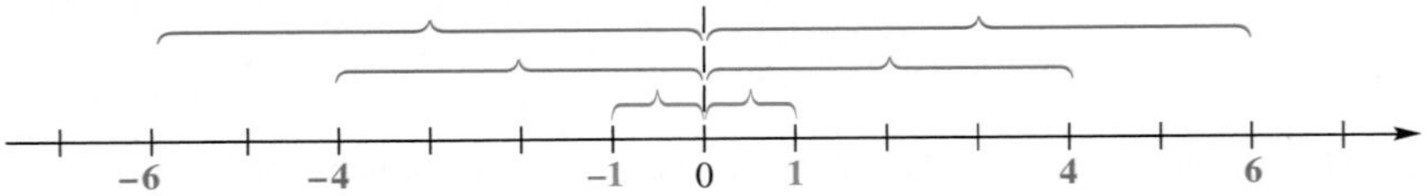

FIGURE 7

The additive inverse of a number can be indicated by writing the symbol $-$ in front of the number. With this symbol, the additive inverse of 7 is written -7. The additive inverse of -4 is written $-(-4)$, and can also be read "the opposite of -4" or "the negative of -4." Figure 7 suggests that 4 is an additive inverse of -4. Since a number can have only one additive inverse, the symbols 4 and $-(-4)$ must represent the same number, which means that

$$-(-4) = 4.$$

This idea can be generalized as follows.

Double Negative Rule

For any real number x,

$$-(-x) = x.$$

TABLE 1

Number	Additive Inverse
-4	$-(-4)$ or 4
0	0
19	-19
$-\frac{2}{3}$	$\frac{2}{3}$

Table 1 shows several numbers and their additive inverses. An important property of additive inverses will be studied later in this chapter: $a + (-a) = (-a) + a = 0$ for all real numbers a.

As mentioned above, additive inverses are numbers that are the same distance from 0 on the number line. See Figure 7. This idea can also be expressed by saying that a number and its additive inverse have the same absolute value. The **absolute value** of a real number can be defined as the distance between 0 and the number on the number line. The symbol for the absolute value of the number x is $|x|$, read **"the absolute value of *x*."** For example, the distance between 2 and 0 on the number line is 2 units, so

$$|2| = 2.$$

Because the distance between -2 and 0 on the number line is also 2 units,

$$|-2| = 2.$$

Since distance is a physical measurement, which is never negative, ***the absolute value of a number is never negative.*** For example, $|12| = 12$ and $|-12| = 12$, since both 12 and -12 lie at a distance of 12 units from 0 on the number line. Also, since 0 is a distance of 0 units from 0, $|0| = 0$.

In symbols, the absolute value of x is defined as follows.

Formal Definition of Absolute Value

For any real number x,

$$|x| = \begin{cases} x & \text{if } x \geq 0 \\ -x & \text{if } x < 0. \end{cases}$$

By this definition, if x is a positive number or 0, then its absolute value is x itself. For example, since 8 is a positive number, $|8| = 8$. However, if x is a negative number, then its absolute value is the additive inverse of x. This means that if $x = -9$, then $|-9| = -(-9) = 9$, since the additive inverse of -9 is 9.

The formal definition of absolute value can be confusing if it is not read carefully. The "$-x$" in the second part of the definition *does not* represent a negative number. Since x is negative in the second part, $-x$ represents the opposite of a negative number, that is, a positive number.

EXAMPLE 3 Using Absolute Value

Simplify by finding the absolute value.

(a) $|5|$ **(b)** $|-5|$ **(c)** $-|5|$

(d) $-|-14|$ **(e)** $|8 - 2|$ **(f)** $-|8 - 2|$

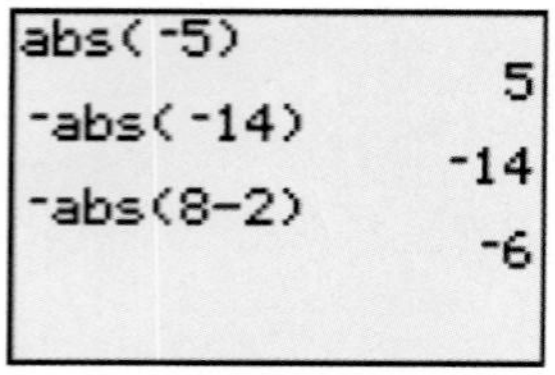

This screen supports the results of Example 3(b), (d), and (f).

SOLUTION

(a) $|5| = 5$ (b) $|-5| = -(-5) = 5$

(c) $-|5| = -(5) = -5$ (d) $-|-14| = -(14) = -14$

(e) $|8 - 2| = |6| = 6$ (f) $-|8 - 2| = -|6| = -6$

Part (e) of Example 3 shows that absolute value bars also serve as grouping symbols. You must perform any operations that appear inside absolute value symbols before finding the absolute value.

Applications

A table of data provides a concise way of relating information.

EXAMPLE 4 Interpreting Change Using a Table

The projected annual rates of employment change (in percent) in some of the fastest growing and most rapidly declining industries from 1994 through 2005 are shown in Table 2. What industry in the list is expected to see the greatest change? the least change?

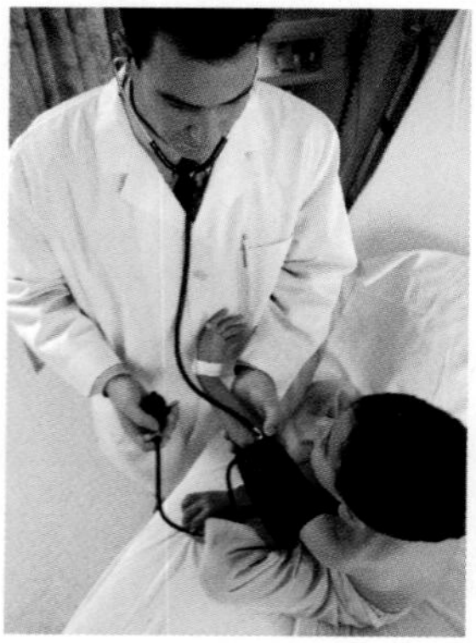

TABLE 2

Industry (1994–2005)	Percent Rate of Change
Health services	5.7
Computer and data processing services	4.9
Child day care services	4.3
Footware, except rubber and plastic	−6.7
Household audio and video equipment	−4.2
Luggage, handbags, and leather products	−3.3

Source: U.S. Bureau of Labor Statistics.

SOLUTION

We want the greatest *change,* without regard to whether the change is an increase or a decrease. Look for the number in the list with the greatest absolute value. That number is found in footware, since $|-6.7| = 6.7$. Similarly, the least change is in the luggage, handbags, and leather products industry: $|-3.3| = 3.3$.

6.1 EXERCISES

In Exercises 1– 6, give a number that satisfies the given condition.

1. An integer between 3.5 and 4.5

2. A rational number between 3.8 and 3.9

3. A whole number that is not positive and is less than 1

4. A whole number greater than 4.5

5. An irrational number that is between $\sqrt{11}$ and $\sqrt{13}$

6. A real number that is neither negative nor positive

In Exercises 7–10, decide whether each statement is true or false.

7. Every natural number is positive.

8. Every whole number is positive.

9. Every integer is a rational number.

10. Every rational number is a real number.

In Exercises 11 and 12, list all numbers from each set that are **(a)** *natural numbers;* **(b)** *whole numbers;* **(c)** *integers;* **(d)** *rational numbers;* **(e)** *irrational numbers;* **(f)** *real numbers.*

11. $\left\{-9, -\sqrt{7}, -1\frac{1}{4}, -\frac{3}{5}, 0, \sqrt{5}, 3, 5.9, 7\right\}$

12. $\left\{-5.3, -5, -\sqrt{3}, -1, -\frac{1}{9}, 0, 1.2, 1.8, 3, \sqrt{11}\right\}$

13. Explain in your own words the different sets of numbers introduced in this section, and give an example of each kind.

14. What two possible situations exist for the decimal representation of a rational number?

Use an integer to express each number representing a change or measurement in the following applications.

15. ***Height of the Sears Tower*** The Sears Tower in Chicago is 1450 feet high. (*Source:* Council on Tall Buildings and Urban Habitat.)

16. ***Population of Laredo*** Between 2000 and 2004, the population of Laredo, TX increased by 26,636. (*Source:* Estimate of the U.S. Census Bureau.

17. ***Height of Mt. Arenal*** The height of Mt. Arenal, an active volcano in Costa Rica, is 5436 feet above sea level. (*Source: The New York Times Almanac 2006.*)

18. ***Boiling Point of Chlorine*** The boiling point of chlorine is approximately 30° below 0° Fahrenheit.

19. ***Melting Point of Fluorine*** The melting point of fluorine gas is 220° below 0° Celsius.

20. ***Population of Detroit*** Between 2000 and 2004, the population of Detroit, MI decreased by 51,072. (*Source:* Estimate of the U.S. Census Bureau.)

21. ***Windchill*** When the wind speed is 20 miles per hour and the actual temperature is 10° Fahrenheit, the windchill factor is 9° below 0° Fahrenheit. (Give three responses.)

22. ***Elevation of New Orleans*** The city of New Orleans lies 8 feet below sea level. (*Source:* U.S. Geological Survey, *Elevations and Distances in the United States.*)

23. ***Depths and Heights of Seas and Mountains*** The chart gives selected depths and heights of bodies of water and mountains.

Bodies of Water	Average Depth in Feet (as a negative number)	Mountains	Altitude in Feet (as a positive number)
Pacific Ocean	−12,925	McKinley	20,320
South China Sea	−4802	Point Success	14,150
Gulf of California	−2375	Matlalcueyetl	14,636
Caribbean Sea	−8448	Ranier	14,410
Indian Ocean	−12,598	Steele	16,644

Source: The World Almanac and Book of Facts.

(a) List the bodies of water in order, starting with the deepest and ending with the shallowest.
(b) List the mountains in order, starting with the lowest and ending with the highest.
(c) *True or false:* The absolute value of the depth of the Pacific Ocean is greater than the absolute value of the depth of the Indian Ocean.
(d) *True or false:* The absolute value of the depth of the Gulf of California is greater than the absolute value of the depth of the Caribbean Sea.

24. ***Personal Savings*** The bar graph in the figure illustrates the amount of personal savings, in billions of dollars, accumulated during the years 1997 through 2001.

(a) Which year had the greatest amount of savings? Which had the least?

(b) Which years had amounts greater than $200 billion?

(c) Estimate the amounts for 1997 and 1998.

(d) Estimate the difference of the amounts for the years 1997 and 1998.

(e) How did personal savings in 1998 compare to personal savings in 1999?

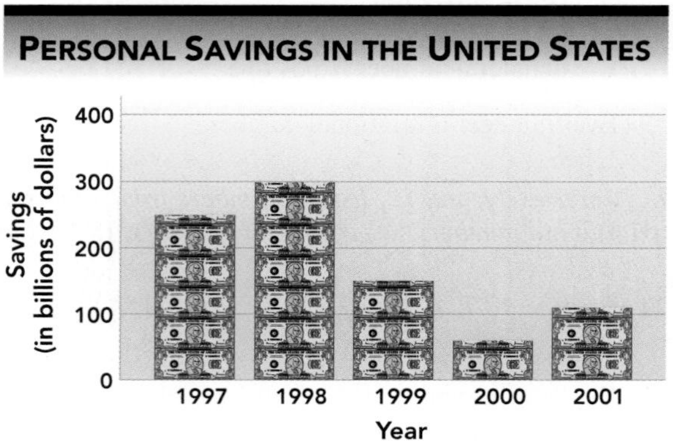

Source: U.S. Bureau of Economic Analysis.

Graph each group of numbers on a number line.

25. $-2, -6, -4, 3, 4$

26. $-5, -3, -2, 0, 4$

27. $\frac{1}{4}, 2\frac{1}{2}, -3\frac{4}{5}, -4, -1\frac{5}{8}$

28. $5\frac{1}{4}, 4\frac{5}{9}, -2\frac{1}{3}, 0, -3\frac{2}{5}$

29. Match each expression in Column I with its value in Column II. Some choices in Column II may not be used.

I	II
(a) $\lvert -7 \rvert$	A. 7
(b) $-(-7)$	B. -7
(c) $-\lvert -7 \rvert$	C. neither A nor B
(d) $-\lvert -(-7) \rvert$	D. both A and B

30. Fill in the blanks with the correct values: The opposite of -2 is _____ , while the absolute value of -2 is _____. The additive inverse of -2 is _____, while the additive inverse of the absolute value of -2 is _____ .

Find **(a)** *the additive inverse (or opposite) of each number and* **(b)** *the absolute value of each number.*

31. -2

32. -8

33. 6

34. 11

35. $7 - 4$

36. $8 - 3$

37. $7 - 7$

38. $3 - 3$

39. Use the results of Exercises 35 and 36 to complete the following: If $a - b > 0$, then the absolute value of $a - b$ in terms of a and b is _____ .

40. Look at Exercises 37 and 38 and use the results to complete the following: If $a - b = 0$, then the absolute value of $a - b$ is _____.

Select the lesser of the two given numbers.

41. $-12, -4$

42. $-9, -14$

43. $-8, -1$

44. $-15, -16$

45. $3, \lvert -4 \rvert$

46. $5, \lvert -2 \rvert$

47. $\lvert -3 \rvert, \lvert -4 \rvert$

48. $\lvert -8 \rvert, \lvert -9 \rvert$

49. $-\lvert -6 \rvert, -\lvert -4 \rvert$

50. $-\lvert -2 \rvert, -\lvert -3 \rvert$

51. $\lvert 5 - 3 \rvert, \lvert 6 - 2 \rvert$

52. $\lvert 7 - 2 \rvert, \lvert 8 - 1 \rvert$

Decide whether each statement is true or false.

53. $6 > -(-2)$

54. $-8 > -(-2)$

55. $-4 \le -(-5)$

56. $-6 \le -(-3)$

57. $\lvert -6 \rvert < \lvert -9 \rvert$

58. $\lvert -12 \rvert < \lvert -20 \rvert$

59. $-|8| > |-9|$

60. $-|12| > |-15|$

61. $-|-5| \geq -|-9|$

62. $-|-12| \leq -|-15|$

63. $|6 - 5| \geq |6 - 2|$

64. $|13 - 8| \leq |7 - 4|$

Producer Price Index *The table shows the percent change in the Producer Price Index (PPI) for selected industries from 2002 to 2003 and from 2003 to 2004. Use the table to answer Exercises 65–68.*

65. Which industry in which year represents the greatest percentage increase?

66. Which industry in which year represents the greatest percentage decrease?

67. Which industry in which year represents the least change?

68. Which industries represent a decrease for both years?

Industry	Change from 2002 to 2003	Change from 2003 to 2004
Book publishers	3.7	3.8
Telephone apparatus manufacturing	−3.5	−5.1
Construction machinery manufacturing	1.4	3.1
Petroleum refineries	25.9	25.0
Electronic computer manufacturing	−19.6	−12.3

Source: U.S. Bureau of Labor Statistics.

69. ***Comparing Employment Data*** Refer to the table in Example 4. Of the household audio/video equipment industry and computer/data processing services, which shows the greater change (without regard to sign)?

70. Students often say "Absolute value is always positive." Is this true? If not, explain why.

Give three numbers between −6 and 6 that satisfy each given condition.

71. Positive real numbers but not integers

72. Real numbers but not positive numbers

73. Real numbers but not whole numbers

74. Rational numbers but not integers

75. Real numbers but not rational numbers

76. Rational numbers but not negative numbers

6.2 Operations, Properties, and Applications of Real Numbers

Operations • Order of Operations • Properties of Addition and Multiplication of Real Numbers • Applications of Real Numbers

Operations

The result of adding two numbers is called their **sum.** The numbers being added are called **addends** (or **terms**).

Adding Real Numbers

Like Signs Add two numbers with the *same* sign by adding their absolute values. The sign of the sum (either + or −) is the same as the sign of the two numbers.

Unlike Signs Add two numbers with *different* signs by subtracting the smaller absolute value from the larger to find the absolute value of the sum. The sum is positive if the positive number has the larger absolute value. The sum is negative if the negative number has the larger absolute value.

Practical Arithmetic From the time of Egyptian and Babylonian merchants, practical aspects of arithmetic complemented mystical (or "Pythagorean") tendencies. This was certainly true in the time of **Adam Riese** (1489–1559), a "reckon master" influential when commerce was growing in Northern Europe. Riese's likeness on the stamp above comes from the title page of one of his popular books on *Rechnung* (or "reckoning"). He championed new methods of reckoning using Hindu-Arabic numerals and quill pens. (The Roman methods then in common use moved counters on a ruled board.) Riese thus fulfilled Fibonacci's efforts 300 years earlier to supplant Roman numerals and methods.

For example, to add -12 and -8, first find their absolute values:

$$|-12| = 12 \quad \text{and} \quad |-8| = 8.$$

Since -12 and -8 have the *same* sign, add their absolute values: $12 + 8 = 20$. Give the sum the sign of the two numbers. Since both numbers are negative, the sum is negative and

$$-12 + (-8) = -20.$$

Find $-17 + 11$ by subtracting the absolute values, because these numbers have different signs.

$$|-17| = 17 \quad \text{and} \quad |11| = 11$$
$$17 - 11 = 6$$

Give the result the sign of the number with the larger absolute value.

$$-17 + 11 = -6$$

Negative since $|-17| > |11|$

EXAMPLE 1 Adding Signed Numbers

Find each sum.

(a) $-6 + (-3)$ **(b)** $-12 + (-4)$ **(c)** $4 + (-1)$
(d) $-9 + 16$ **(e)** $-16 + 12$

SOLUTION

(a) $-6 + (-3) = -(6 + 3) = -9$
(b) $-12 + (-4) = -(12 + 4) = -16$
(c) $4 + (-1) = 3$ **(d)** $-9 + 16 = 7$ **(e)** $-16 + 12 = -4$

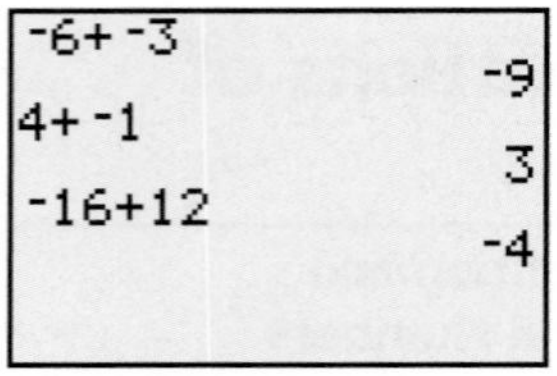

The calculator supports the results of Example 1(a), (c), and (e).

The result of subtracting two numbers is called their **difference.** In $a - b$, a is called the **minuend,** and b is called the **subtrahend.** Compare the two statements below.

$$7 - 5 = 2$$
$$7 + (-5) = 2$$

In a similar way, $9 - 3 = 9 + (-3)$. That is, to subtract 3 from 9, add the additive inverse of 3 to 9.

These examples suggest the following rule for subtraction.

Definition of Subtraction

For all real numbers a and b,

$$a - b = a + (-b).$$

(Change the sign of the subtrahend and add.)

EXAMPLE 2 Subtracting Signed Numbers

Find each difference.

(a) $6 - 8$ **(b)** $-12 - 4$

(c) $-10 - (-7)$ **(d)** $15 - (-3)$

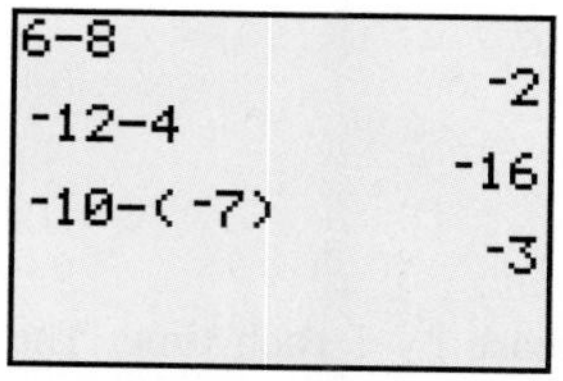

The calculator supports the results of Example 2(a), (b), and (c).

SOLUTION

Change to addition.
Change sign of the subtrahend.

(a) $6 - 8 = 6 + (-8) = -2$

Change to addition.
Sign changed.

(b) $-12 - 4 = -12 + (-4) = -16$

(c) $-10 - (-7) = -10 + [-(-7)]$ This step can be omitted.

$= -10 + 7$

$= -3$

(d) $15 - (-3) = 15 + 3 = 18$

The result of multiplying two numbers is called their **product.** The two numbers being multiplied are called **factors.** Any rules for multiplication with negative real numbers should be consistent with the usual rules for multiplication of positive real numbers and zero. To inductively obtain a rule for multiplying a positive real number and a negative real number, observe the pattern of products below.

$$4 \cdot 5 = 20$$
$$4 \cdot 4 = 16$$
$$4 \cdot 3 = 12$$
$$4 \cdot 2 = 8$$
$$4 \cdot 1 = 4$$
$$4 \cdot 0 = 0$$
$$4 \cdot (-1) = ?$$

Early ways of writing the basic operation symbols were quite different from those used today. The **addition symbol** shown below was derived from the Italian word *piú* (plus) in the sixteenth century. The + sign used today is shorthand for the Latin *et* (and).

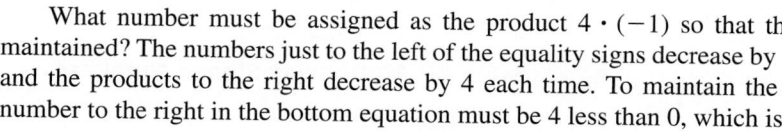

The **subtraction symbol** shown below was used by Diophantus in Greece sometime during the second or third century A.D. Our subtraction bar may be derived from a bar used by medieval traders to mark differences in weights of products.

What number must be assigned as the product $4 \cdot (-1)$ so that the pattern is maintained? The numbers just to the left of the equality signs decrease by 1 each time, and the products to the right decrease by 4 each time. To maintain the pattern, the number to the right in the bottom equation must be 4 less than 0, which is -4, so

$$4 \cdot (-1) = -4.$$

The pattern continues with

$$4 \cdot (-2) = -8$$
$$4 \cdot (-3) = -12$$
$$4 \cdot (-4) = -16,$$

and so on. In the same way,

$$-4 \cdot 2 = -8$$
$$-4 \cdot 3 = -12$$
$$-4 \cdot 4 = -16,$$

and so on. A similar observation can be made about the product of two negative real numbers. Look at the pattern that follows.

$$\begin{aligned} -5 \cdot 4 &= -20 \\ -5 \cdot 3 &= -15 \\ -5 \cdot 2 &= -10 \\ -5 \cdot 1 &= -5 \\ -5 \cdot 0 &= 0 \\ -5 \cdot (-1) &= ? \end{aligned}$$

The numbers just to the left of the equality signs decrease by 1 each time. The products on the right increase by 5 each time. To maintain the pattern, the product $-5 \cdot (-1)$ must be 5 more than 0, so

$$-5 \cdot (-1) = 5.$$

Continuing this pattern gives

$$\begin{aligned} -5 \cdot (-2) &= 10 \\ -5 \cdot (-3) &= 15 \\ -5 \cdot (-4) &= 20, \end{aligned}$$

and so on. These observations lead to the following rules for multiplication.

Early ways of writing the multiplication and division symbols were also quite different. In the seventeenth century, Leibniz used the **multiplication symbol** below to avoid × as too similar to the "unknown" *x*. The multiplication symbol × is based on St. Andrew's Cross.

The **division symbol** shown below was used by Gallimard in the eighteenth century. The familiar ÷ symbol may come from the fraction bar, embellished with the dots above and below.

Multiplying Real Numbers

Like Signs Multiply two numbers with the *same* sign by multiplying their absolute values to find the absolute value of the product. The product is positive.

Unlike Signs Multiply two numbers with *different* signs by multiplying their absolute values to find the absolute value of the product. The product is negative.

$(+) \cdot (+) = +$
$(-) \cdot (-) = +$
$(+) \cdot (-) = -$
$(-) \cdot (+) = -$

EXAMPLE 3 Multiplying Signed Numbers

Find each product.

(a) $-9 \cdot 7$ **(b)** $14 \cdot (-5)$ **(c)** $-8 \cdot (-4)$

SOLUTION

(a) $-9 \cdot 7 = -63$ **(b)** $14 \cdot (-5) = -70$ **(c)** $-8 \cdot (-4) = 32$ ■

The result of dividing two numbers is called their **quotient.** In the quotient $a \div b$ (or $\frac{a}{b}$), where $b \neq 0$, a is called the **dividend** (or numerator), and b is called the **divisor** (or denominator). For real numbers a, b, and c, if

$$\frac{a}{b} = c, \qquad \text{then} \qquad a = b \cdot c.$$

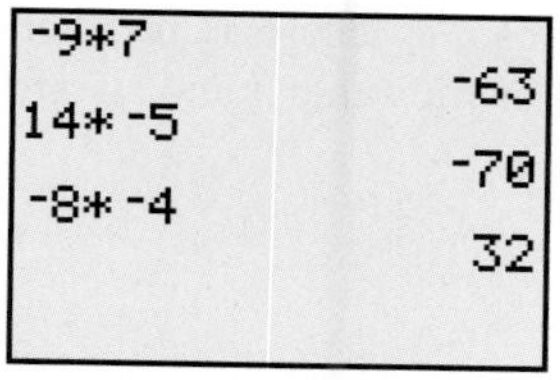

An asterisk (*) represents multiplication on this screen. The display supports the results of Example 3.

To illustrate this, consider the quotient $\frac{10}{-2}$. The value of this quotient is obtained by asking, "What number multiplied by -2 gives 10?" From our discussion of multiplication, the answer to this question must be "-5." Therefore,

$$\frac{10}{-2} = -5,$$

because $-2 \cdot (-5) = 10$. Similar reasoning leads to the following results.

$$\frac{-10}{2} = -5 \quad \text{and} \quad \frac{-10}{-2} = 5$$

These facts, along with the fact that the quotient of two positive numbers is positive, lead to the following rule for division.

$(+)/(+) = +$
$(-)/(-) = +$
$(+)/(-) = -$
$(-)/(+) = -$

Dividing Real Numbers

Like Signs Divide two numbers with the *same* sign by dividing their absolute values to find the absolute value of the quotient. The quotient is positive.

Unlike Signs Divide two numbers with *different* signs by dividing their absolute values to find the absolute value of the quotient. The quotient is negative.

EXAMPLE 4 Dividing Signed Numbers

Find each quotient.

(a) $\frac{15}{-5}$ **(b)** $\frac{-100}{-25}$ **(c)** $\frac{-60}{3}$

SOLUTION

(a) $\frac{15}{-5} = -3$ This is true because $-5 \cdot (-3) = 15$.

(b) $\frac{-100}{-25} = 4$ **(c)** $\frac{-60}{3} = -20$

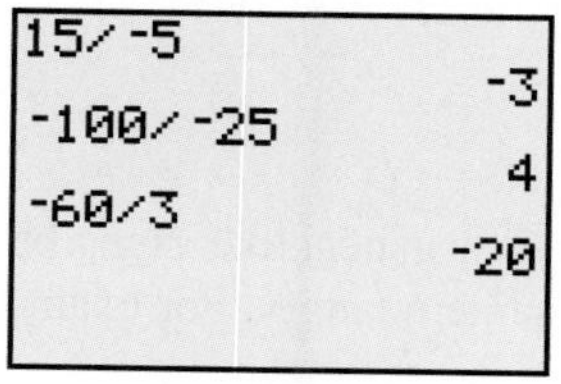

The division operation is represented by a slash (/). This screen supports the results of Example 4.

If 0 is divided by a nonzero number, the quotient is 0. That is,

$$\mathbf{\frac{0}{a} = 0, \quad \text{for } a \neq 0.}$$

This is true because $a \cdot 0 = 0$. However, we cannot divide by 0. There is a good reason for this. Whenever a division is performed, we want to obtain one and only one quotient. Now consider the division problem

$$\frac{7}{0}.$$

Dividing by zero leads to this message on the TI-83/84 Plus.

We must ask ourselves "What number multiplied by 0 gives 7?" There is no such number, since the product of 0 and any number is zero. On the other hand, if we consider the quotient

$$\frac{0}{0},$$

there are infinitely many answers to the question, "What number multiplied by 0 gives 0?" Since division by 0 does not yield a *unique* quotient, it is not permitted. To summarize these two situations, we make the following statement.

Division by Zero

Division by 0 is undefined.

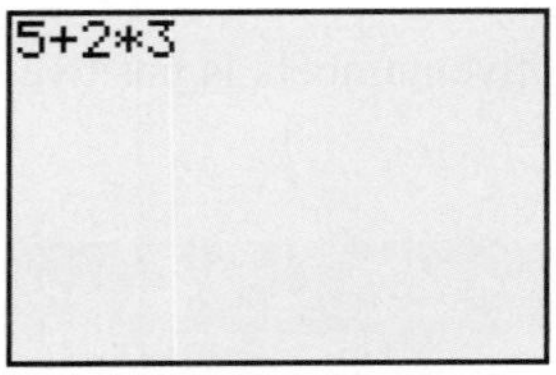

What result does the calculator give? The order of operations determines the answer. (See Example 5(a).)

Order of Operations

Given a problem such as $5 + 2 \cdot 3$, should 5 and 2 be added first or should 2 and 3 be multiplied first? When a problem involves more than one operation, we use the following **order of operations.**

Order of Operations

If parentheses or square brackets are present:

Step 1 Work separately above and below any **fraction bar.**

Step 2 Use the rules below within each set of **parentheses or square brackets.** Start with the innermost set and work outward.

If no parentheses or brackets are present:

Step 1 Apply any **exponents.**

Step 2 Do any **multiplications or divisions** in the order in which they occur, working from left to right.

Step 3 Do any **additions or subtractions** in the order in which they occur, working from left to right.

The sentence **"Please excuse my dear Aunt Sally"** is often used to help us remember the rule for order of operations. The letters **P, E, M, D, A, S** are the first letters of the words of the sentence, and they stand for ***parentheses, exponents, multiply, divide, add, subtract.*** (*Remember also that M and D have equal priority, as do A and S. Operations with equal priority are performed in order from left to right.*)

When evaluating an exponential expression that involves a negative sign, be aware that $(-a)^n$ and $-a^n$ do not necessarily represent the same quantity. For example, if $a = 2$ and $n = 6$,

$$(-2)^6 = (-2)(-2)(-2)(-2)(-2)(-2) = 64 \quad \text{The base is } -2.$$

while $$-2^6 = -(2 \cdot 2 \cdot 2 \cdot 2 \cdot 2 \cdot 2) = -64. \quad \text{The base is 2.}$$

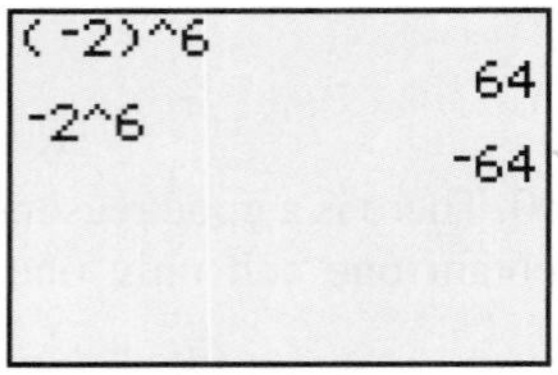

Notice the difference in the two expressions. This supports $(-2)^6 \neq -2^6$.

EXAMPLE 5 Using the Order of Operations

Use the order of operations to simplify each expression.

(a) $5 + 2 \cdot 3$

(b) $4 \cdot 3^2 + 7 - (2 + 8)$

(c) $\dfrac{2(8 - 12) - 11(4)}{5(-2) - 3}$

(d) -4^4

(e) $(-4)^4$

(f) $(-8)(-3) - [4 - (3 - 6)]$

SOLUTION

(a) $5 + 2 \cdot 3 = 5 + 6$ Multiply.

$= 11$ Add.

Be careful! Multiply first.

(b) $4 \cdot 3^2 + 7 - (2 + 8) = 4 \cdot 3^2 + 7 - 10$ Work within parentheses first.

$= 4 \cdot 9 + 7 - 10$ Apply the exponent.

$3^2 = 3 \cdot 3$, not $3 \cdot 2$.

$= 36 + 7 - 10$ Multiply.

$= 43 - 10$ Add.

$= 33$ Subtract.

(c) $\dfrac{2(8 - 12) - 11(4)}{5(-2) - 3} = \dfrac{2(-4) - 11(4)}{5(-2) - 3}$ Work separately above and below fraction bar.

$= \dfrac{-8 - 44}{-10 - 3}$ Multiply.

$= \dfrac{-52}{-13}$ Subtract.

$= 4$ Divide.

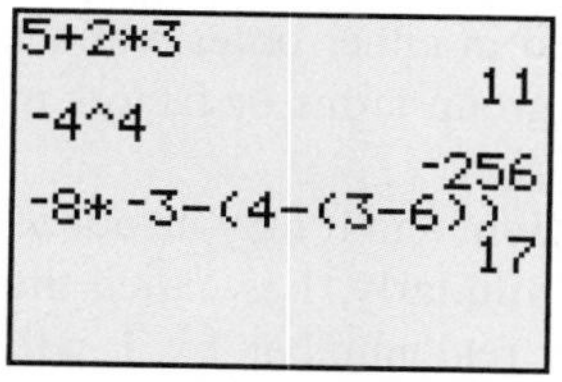

The calculator supports the results in Example 5(a), (d), and (f).

(d) $-4^4 = -(4 \cdot 4 \cdot 4 \cdot 4) = -256$

The base is 4, not -4.

(e) $(-4)^4 = (-4)(-4)(-4)(-4) = 256$

The base is -4 here.

(f) $-8(-3) - [4 - (3 - 6)] = -8(-3) - [4 - (-3)]$ Work within parentheses.

Start here.

$= -8(-3) - [4 + 3]$ Definition of subtraction

$= -8(-3) - 7$ Work within brackets.

$= 24 - 7$ Multiply.

$= 17$ Subtract. ■

Properties of Addition and Multiplication of Real Numbers

Properties of Addition and Multiplication

For real numbers a, b, and c, the following properties hold.

Closure Properties — $a + b$ **and** ab **are real numbers.**

Commutative Properties — $a + b = b + a \qquad ab = ba$

Associative Properties — $(a + b) + c = a + (b + c)$

$(ab)c = a(bc)$

Identity Properties — There is a real number 0 such that

$a + 0 = a$ and $0 + a = a$.

There is a real number 1 such that

$a \cdot 1 = a$ and $1 \cdot a = a$.

(continued)

Properties of Addition and Multiplication, Continued

Inverse Properties

For each real number a, there is a single real number $-a$ such that

$$\mathbf{a + (-a) = 0} \quad \text{and} \quad \mathbf{(-a) + a = 0.}$$

For each nonzero real number a, there is a single real number $\frac{1}{a}$ such that

$$\mathbf{a \cdot \frac{1}{a} = 1} \quad \text{and} \quad \mathbf{\frac{1}{a} \cdot a = 1.}$$

Distributive Property of Multiplication with Respect to Addition

$$\mathbf{a(b + c) = ab + ac}$$

$$\mathbf{(b + c)a = ba + ca}$$

The set of real numbers is said to be closed with respect to the operations of addition and multiplication. This means that the sum of two real numbers and the product of two real numbers are themselves real numbers. The commutative properties state that two real numbers may be added or multiplied in either order without affecting the result. The associative properties allow us to group terms or factors in any manner we wish without affecting the result.

The number 0 is called the **identity element for addition,** and it may be added to any real number to obtain that real number as a sum. Similarly, 1 is called the **identity element for multiplication,** and multiplying a real number by 1 will always yield that real number. Each real number a has an **additive inverse,** $-a$, such that the sum of a and its additive inverse is the additive identity element 0. Each nonzero real number a has a **multiplicative inverse,** or **reciprocal,** $\frac{1}{a}$, such that the product of a and its multiplicative inverse is the multiplicative identity element 1. The distributive property allows us to change certain products to sums and certain sums to products.

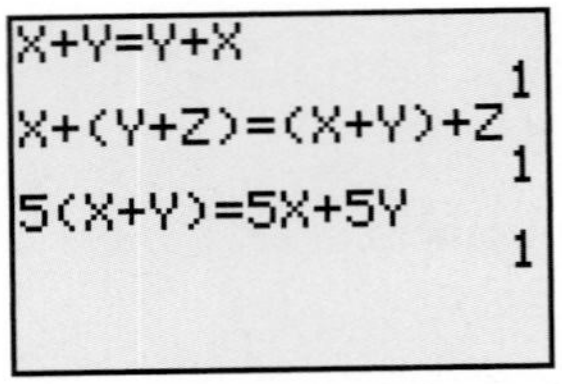

No matter what values are stored in X, Y, and Z, the commutative, associative, and distributive properties assure us that these statements are true.

EXAMPLE 6 Identifying Properties of Addition and Multiplication

Identify the property of addition or multiplication illustrated in each statement.

(a) $5 + 7$ is a real number.

(b) $5 + (6 + 8) = (5 + 6) + 8$

(c) $8 + 0 = 8$

(d) $-4\left(-\frac{1}{4}\right) = 1$

(e) $4 + (3 + 9) = 4 + (9 + 3)$

(f) $5(x + y) = 5x + 5y$

SOLUTION

(a) The statement that the sum of two real numbers is also a real number is an example of the closure property of addition.

(b) Because the grouping of the terms is different on the two sides of the equation, this illustrates the associative property of addition.

(c) Adding 0 to a number yields the number itself. This is an example of the identity property of addition.

(d) Multiplying a number by its reciprocal yields 1, and this illustrates the inverse property of multiplication.

(e) The order of the addends (terms) 3 and 9 is different, so this is justified by the commutative property of addition.

(f) The factor 5 is distributed to the terms x and y. This is an example of the distributive property of multiplication with respect to addition.

Applications of Real Numbers The usefulness of negative numbers can be seen by considering situations that arise in everyday life. For example, we need negative numbers to express the temperatures on January days in Anchorage, Alaska, where they often drop below zero. The phrases "in the red" and "in the black" mean losing money and making money, respectively. (These descriptions go back to the days when bookkeepers used red ink to represent losses and black ink to represent gains.)

PROBLEM-SOLVING HINT When problems deal with gains and losses, the gains may be interpreted as positive numbers and the losses as negative numbers. Temperatures below 0° are negative, and those above 0° are positive. Altitudes above sea level are considered positive, and those below sea level are considered negative.

EXAMPLE 7 Analyzing the Producer Price Index

The Producer Price Index is the oldest continuous statistical series published by the Bureau of Labor Statistics. It measures the average changes in prices received by producers of all commodities produced in the United States. The bar graph in Figure 8 gives the Producer Price Index (PPI) for construction materials between 1996 and 2003.

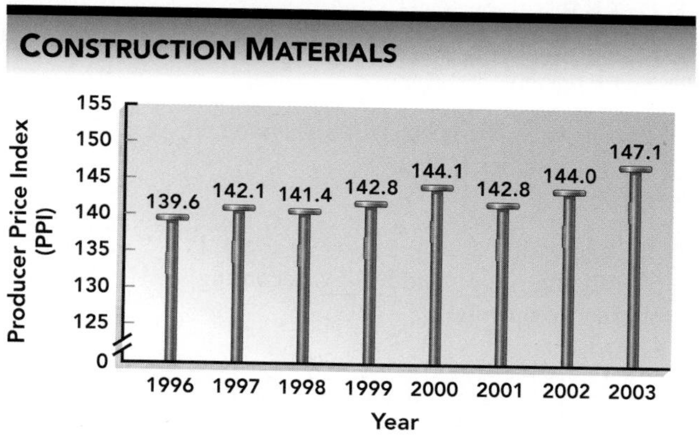

Source: U.S. Bureau of Labor Statistics, Producer Price Indexes, monthly and annual.

FIGURE 8

Use a signed number to represent the change in the PPI from

(a) 1999 to 2000. **(b)** 2000 to 2001.

SOLUTION

(a) To find this change, we start with the index number from 2000 and subtract from it the index number from 1999.

$$\underbrace{144.1}_{\text{The 2000 index}} - \underbrace{142.8}_{\text{The 1999 index}} = \underbrace{\mathbf{1.3}}_{\substack{\text{A positive number} \\ \text{indicates an increase.}}}$$

(b) Use the same procedure as in part (a).

$$\underbrace{142.8}_{\text{The 2001 index}} - \underbrace{144.1}_{\text{The 2000 index}} = 142.8 + (-144.1) = \underbrace{\mathbf{-1.3}}_{\substack{\text{A negative number} \\ \text{indicates a decrease.}}}$$

EXAMPLE 8 Determining Difference of Temperatures

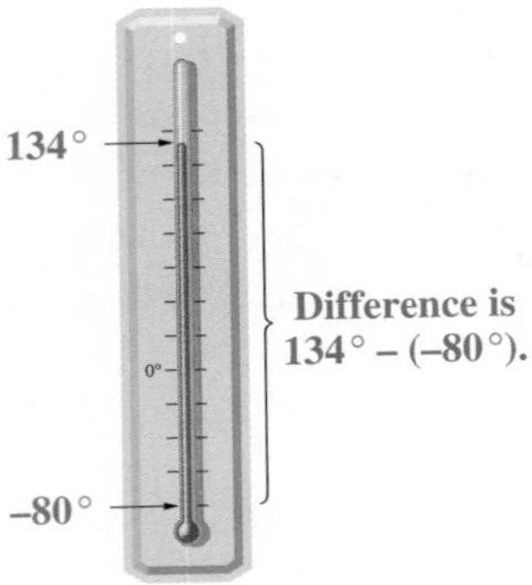

FIGURE 9

The record high temperature in the United States was 134° Fahrenheit, recorded at Death Valley, California, in 1913. The record low was −80°F, at Prospect Creek, Alaska, in 1971. See Figure 9. How much greater was the highest temperature than the lowest temperature? (*Source: The World Almanac and Book of Facts 2006.*)

SOLUTION

We must subtract the lower temperature from the higher temperature.

$$\begin{aligned} 134 - (-80) &= 134 + 80 && \text{Use the definition of subtraction.} \\ &= 214 && \text{Add.} \end{aligned}$$

The difference of the two temperatures is 214°F.

6.2 EXERCISES

Fill in each blank with the correct response.

1. The sum of two negative numbers will always be a ________________ number.
(positive/negative)

2. The sum of a number and its opposite will always be ______.

3. To simplify the expression $8 + [-2 + (-3 + 5)]$, I should begin by adding _____ and _____, according to the rule for order of operations.

4. If I am adding a positive number and a negative number, and the negative number has the larger absolute value, the sum will be a ________________ number.
(positive/negative)

5. Explain in words how to add signed numbers. Consider the various cases and give examples.

6. Explain in words how to multiply signed numbers.

Perform the indicated operations, using the order of operations as necessary.

7. $-12 + (-8)$

8. $-5 + (-2)$

9. $12 + (-16)$

10. $-6 + 17$

11. $-12 - (-1)$

12. $-3 - (-8)$

13. $-5 + 11 + 3$

14. $-9 + 16 + 5$

15. $12 - (-3) - (-5)$

16. $15 - (-6) - (-8)$

17. $-9 - (-11) - (4 - 6)$

18. $-4 - (-13) + (-5 + 10)$

19. $(-12)(-2)$

20. $(-3)(-5)$

21. $9(-12)(-4)(-1)3$

22. $-5(-17)(2)(-2)4$

23. $\frac{-18}{-3}$

24. $\frac{-100}{-50}$

25. $\frac{36}{-6}$

26. $\frac{52}{-13}$

27. $\frac{0}{12}$

28. $\frac{0}{-7}$

29. $-6 + [5 - (3 + 2)]$

30. $-8[4 + (7 - 8)]$

31. $-8(-2) - [(4^2) + (7 - 3)]$

32. $-7(-3) - [2^3 - (3 - 4)]$

33. $-4 - 3(-2) + 5^2$

34. $-6 - 5(-8) + 3^2$

35. $(-8 - 5)(-2 - 1)$

36. $\frac{(-10 + 4) \cdot (-3)}{-7 - 2}$

37. $\frac{(-6 + 3) \cdot (-4)}{-5 - 1}$

38. $\frac{2(-5 + 3)}{-2^2} - \frac{(-3^2 + 2)3}{3 - (-4)}$

39. $\frac{2(-5) + (-3)(-2^2)}{-3^2 + 9}$

40. $\frac{3(-4) + (-5)(-2)}{2^3 - 2 + (-6)}$

41. $-\frac{1}{4}[3(-5) + 7(-5) + 1(-2)]$

42. $\frac{5 - 3\left(\frac{-5 - 9}{-7}\right) - 6}{-9 - 11 + 3 \cdot 7}$

43. Which of the following expressions are undefined?

A. $\frac{8}{0}$ **B.** $\frac{9}{6 - 6}$ **C.** $\frac{4 - 4}{5 - 5}$ **D.** $\frac{0}{-1}$

44. If you have no money in your pocket and you divide it equally among your three siblings, how much does each get? Use this situation to explain division of zero by a positive integer.

Identify the property illustrated by each statement.

45. $6 + 9 = 9 + 6$

46. $8 \cdot 4 = 4 \cdot 8$

47. $7 + (2 + 5) = (7 + 2) + 5$

48. $(3 \cdot 5) \cdot 4 = 4 \cdot (3 \cdot 5)$

49. $9 + (-9) = 0$

50. $12 + 0 = 12$

51. $9 \cdot 1 = 9$

52. $\left(\frac{1}{-3}\right) \cdot (-3) = 1$

53. $0 + 283 = 283$

54. $6 \cdot (4 \cdot 2) = (6 \cdot 4) \cdot 2$

55. $2 \cdot (4 + 3) = 2 \cdot 4 + 2 \cdot 3$

56. $9 \cdot 6 + 9 \cdot 8 = 9 \cdot (6 + 8)$

57. $0 = -8 + 8$

58. $19 + 12$ is a real number.

59. $19 \cdot 12$ is a real number.

60. Work the following problem in two ways, first using the order of operations, and then using the distributive property: Evaluate $9(11 + 15)$.

Exercises 61–68 are designed to explore the properties of real numbers in further detail.

61. **(a)** Evaluate $6 - 8$ and $8 - 6$.
(b) By the results of part (a), we may conclude that subtraction is not a(n) ____________ operation.
(c) Are there *any* real numbers a and b for which $a - b = b - a$? If so, give an example.

62. **(a)** Evaluate $4 \div 8$ and $8 \div 4$.
(b) By the results of part (a), we may conclude that division is not a(n) ____________ operation.
(c) Are there *any* real numbers a and b for which $a \div b = b \div a$? If so, give an example.

63. Many everyday occurrences can be thought of as operations that have opposites or inverses. For example, the inverse operation for "going to sleep" is "waking up." For each of the given activities, specify its inverse activity.
(a) cleaning up your room
(b) earning money
(c) increasing the volume on your MP3 player

64. Many everyday activities are commutative; that is, the order in which they occur does not affect the outcome. For example, "putting on your shirt" and "putting on your pants" are commutative operations. Decide whether the given activities are commutative.
(a) putting on your shoes; putting on your socks
(b) getting dressed; taking a shower
(c) combing your hair; brushing your teeth

65. The following conversation actually took place between one of the authors of this text and his son, Jack, when Jack was four years old.

DADDY: "Jack, what is 3 + 0?"
JACK: "3"
DADDY: "Jack, what is 4 + 0?"
JACK: "4. . . and Daddy, *string* plus zero equals *string!*" What property of addition of real numbers did Jack recognize?

66. The phrase *defective merchandise counter* is an example of a phrase that can have different meanings depending upon how the words are grouped (think of the associative properties). For example, (*defective merchandise*) *counter* is a location at which we would return an item that does not work, while *defective* (*merchandise counter*) is a broken place where items are bought and sold. For each of the following phrases, determine why the associative property does not hold.
(a) difficult test question
(b) woman fearing husband
(c) man biting dog

67. The distributive property holds for multiplication with respect to addition. Does the distributive property hold for addition with respect to multiplication? That is, is $a + (b \cdot c) = (a + b) \cdot (a + c)$ true for all values of a, b, and c? (*Hint:* Let $a = 2$, $b = 3$, and $c = 4$.)

68. Suppose that a student shows you the following work.

$$-3(4 - 6) = -3(4) - 3(6) = -12 - 18 = -30$$

The student has made a very common error in applying the distributive property. Explain the student's mistake, and work the problem correctly.

Each expression in Exercises 69–76 is equal to either 81 *or* -81. *Decide which of these is the correct value.*

69. -3^4
70. $-(3^4)$
71. $(-3)^4$
72. $-(-3^4)$
73. $-(-3)^4$
74. $[-(-3)]^4$
75. $-[-(-3)]^4$
76. $-[-(-3^4)]$

77. *Federal Budget Outlays* The bar graph shows federal budget outlays for the U.S. Treasury Department for the years 2002 through 2005. Use a signed number to represent the change in outlay for each time period.
(a) 2002 to 2003
(b) 2003 to 2004
(c) 2004 to 2005
(d) 2002 to 2005

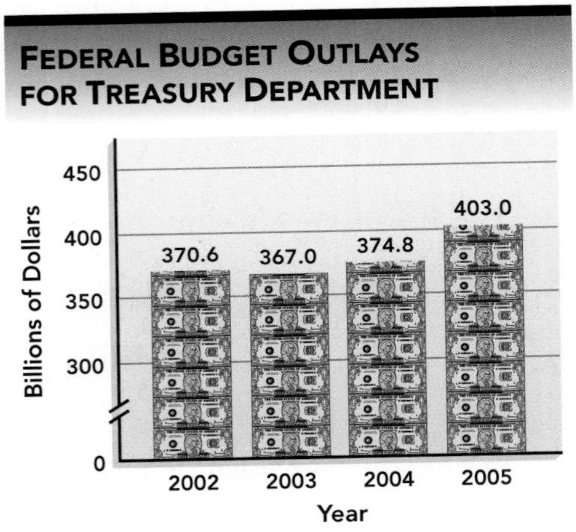

Source: U.S. Office of Management and Budget.

78. *Heights of Mountains and Depths of Trenches* The chart shows the heights in feet of some selected mountains and the depths in feet (as negative numbers) of some selected ocean trenches.

Mountain	Height	Trench	Depth
Foraker	17,400	Philippine	−32,995
Wilson	14,246	Cayman	−24,721
Pikes Peak	14,110	Java	−23,376

Source: The World Almanac and Book of Facts 2006.

(a) What is the difference between the height of Mt. Foraker and the depth of the Philippine Trench?

(b) What is the difference between the height of Pikes Peak and the depth of the Java Trench?

(c) How much deeper is the Cayman Trench than the Java Trench?

(d) How much deeper is the Philippine Trench than the Cayman Trench?

79. ***Social Security Finances*** The table shows Social Security tax revenue and cost of benefits (in billions of dollars).

Year	Tax Revenue	Cost of Benefits
2000	538	409
2010*	916	710
2020*	1479	1405
2030*	2041	2542

*Projected

Source: Social Security Board of Trustees.

(a) Find the difference between Social Security tax revenue and cost of benefits for each year shown in the table.

(b) Interpret your answer for 2030.

80. ***House of Representatives*** Based on census population projections for 2020, New York will lose 5 seats in the U.S. House of Representatives, Pennsylvania will lose 4 seats, and Ohio will lose 3. Write a signed number that represents the total projected change in the number of seats for these three states. (*Source:* Population Reference Bureau.)

81. ***House of Representatives*** Michigan is projected to lose 3 seats in the U.S. House of Representatives and Illinois 2 in 2020. The states projected to gain the most seats are California with 9, Texas with 5, Florida with 3, Georgia with 2, and Arizona with 2. Write a signed number that represents the algebraic sum of these changes. (*Source:* Population Reference Bureau.)

82. ***Checking Account Balance*** Shalita's checking account balance is \$54.00. She then takes a gamble by writing a check for \$89.00. What is her new balance? (Write the balance as a signed number.)

83. ***Checking Account Balance*** In August, Marilyn Cazayoux began with a checking account balance of \$904.89. Her checks and deposits for August are given below:

Checks	Deposits
\$35.84	\$85.00
\$26.14	\$120.76
\$3.12	

Assuming no other transactions, what was her account balance at the end of August?

84. ***Checking Account Balances*** In September, Carter Fenton began with a checking account balance of \$904.89. His checks and deposits for September are given below:

Checks	Deposits
\$41.29	\$80.59
\$13.66	\$276.13
\$84.40	

Assuming no other transactions, what was his account balance at the end of September?

85. ***Difference in Elevations*** The top of Mt. Whitney, visible from Death Valley, has an altitude of 14,494 feet above sea level. The bottom of Death Valley is 282 feet below sea level. Using 0 as sea level, find the difference of these two elevations. (*Source: World Almanac and Book of Facts 2006.*)

86. ***Altitude of Hikers*** The surface, or rim, of a canyon is at altitude 0. On a hike down into the canyon, a party of hikers stops for a rest at 130 meters below the surface. They then descend another 54 meters. What is their new altitude? (Write the altitude as a signed number.)

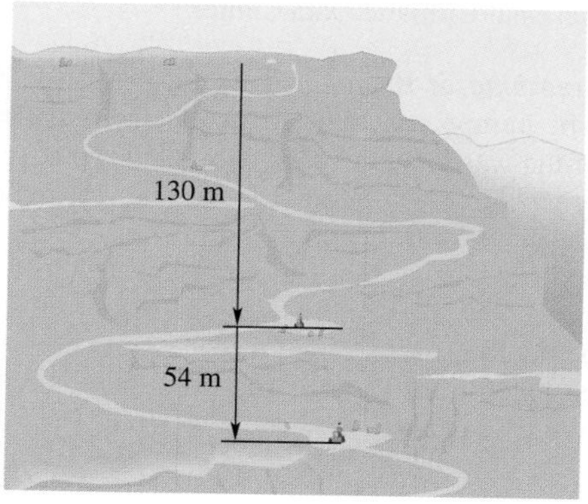

87. ***Drastic Temperature Change*** On January 23, 1943, the temperature rose 49°F in two minutes in Spearfish, South Dakota. If the starting temperature was −4°F, what was the temperature two minutes later? (*Source: Guinness World Records 2006.*)

88. ***Drastic Temperature Change*** The largest change in temperature ever recorded within a 24-hour period occurred in Browning, Montana, on January 23–24, 1916. The temperature fell 100°F from a starting temperature of 44°F. What was the low temperature during this period? (*Source: Guinness World Records, 2006.*)

89. ***Extreme Temperatures in Little Rock*** The lowest temperature ever recorded in Little Rock, Arkansas, was $-5°F$. The highest temperature ever recorded there was $117°F$ more than the lowest. What was this highest temperature? (*Source: The World Almanac and Book of Facts 2006.*)

90. ***Extreme Temperatures in Tennessee*** The lowest temperature ever recorded in Tennessee was $-32°F$. The highest temperature ever recorded there was $145°F$ more than the lowest. What was this highest temperature? (*Source:* National Climatic Data Center.)

91. ***Low Temperatures in Chicago and Huron*** The lowest temperature recorded in Chicago, Illinois, was $-27°F$ in 1985. The record low in Huron, South Dakota, was set in 1994 and was $14°F$ lower than $-27°F$. What was the record low in Huron? (*Source: The World Almanac and Book of Facts 2006.*)

92. ***Low Temperatures in Illinois and Utah*** The lowest temperature ever recorded in Illinois was $-36°F$ on January 5, 1999. The lowest temperature ever recorded in Utah was observed on February 1, 1985 and was $33°F$ lower than Illinois's record low. What is the record low temperature for Utah? (*Source:* National Climatic Data Center.)

93. ***Breaching of Humpback Whales*** No one knows just why humpback whales heave their 45-ton bodies out of the water, but leap they do. (This activity is called *breaching.*) Mark and Debbie, two researchers based on the island of Maui, noticed that one of their favorite whales, "Pineapple," leaped 15 feet above the surface of the ocean while her mate cruised 12 feet below the surface. (See the diagram at the top of the next column.) What is the difference between these two levels?

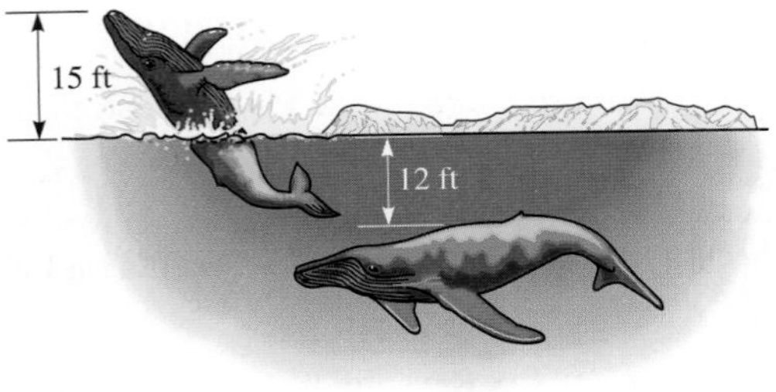

94. ***Highest Point in Louisiana*** The highest point in Louisiana is Driskill Mountain, at an altitude of 535 feet. The lowest point is at Spanish Fort, 8 feet below sea level. Using zero as sea level, find the difference of these two elevations. (*Source: The World Almanac and Book of Facts 2002.*)

95. ***Birth Date of a Greek Mathematician*** A certain Greek mathematician was born in 426 B.C. Her father was born 43 years earlier. In what year was her father born?

96. ***Federal Budget*** In 2000, the federal budget had a surplus of \$236 billion. In 2004, the federal budget had a deficit of \$413 billion. Find the difference of these amounts. (*Source:* Treasury Department.)

97. ***Credit Card Balance*** In 1998, undergraduate college students had an average credit card balance of \$1879. The average balance increased \$869 by 2000 and then dropped \$579 by 2004. What was the average credit card balance of undergraduate college students in 2004? (*Source:* Nellie Mae.)

98. ***Airline Ticket Price*** In 1999, companies paid an average of \$243 for an airline ticket. This average price had increased \$16 by 2001 and then had decreased \$40 by 2005. What was the average price companies paid for an airline ticket in 2005? (*Source:* American Express.)

6.3 Rational Numbers and Decimal Representation

Definition and the Fundamental Property • Operations with Rational Numbers • Density and the Arithmetic Mean • Decimal Form of Rational Numbers

Definition and the Fundamental Property

The set of real numbers is composed of two important mutually exclusive subsets: the rational numbers and the irrational numbers. (Two sets are *mutually exclusive* if they contain no elements in common.)

Benjamin Banneker (1731–1806) spent the first half of his life tending a farm in Maryland. He gained a reputation locally for his mechanical skills and abilities in mathematical problem solving. In 1772 he acquired astronomy books from a neighbor and devoted himself to learning astronomy, observing the skies, and making calculations. In 1789 Banneker joined the team that surveyed what is now the District of Columbia.

Banneker published almanacs yearly from 1792 to 1802. He sent a copy of his first almanac to Thomas Jefferson along with an impassioned letter against slavery. Jefferson subsequently championed the cause of this early African-American mathematician.

Recall from Section 6.1 that quotients of integers are called **rational numbers.** Think of the rational numbers as being made up of all the fractions (quotients of integers with denominator not equal to zero) and all the integers. Any integer can be written as the quotient of two integers. For example, the integer 9 can be written as the quotient $\frac{9}{1}$, or $\frac{18}{2}$, or $\frac{27}{3}$, and so on. Also, -5 can be expressed as a quotient of integers as $\frac{-5}{1}$ or $\frac{-10}{2}$, and so on. (How can the integer 0 be written as a quotient of integers?)

Rational Numbers

Rational numbers $= \{x \mid x$ is a quotient of two integers, with denominator not $0\}$

A rational number is said to be in **lowest terms** if the greatest common factor of the numerator (top number) and the denominator (bottom number) is 1. (The greatest common factor and least common multiple were discussed in Section 5.3.) Rational numbers are written in lowest terms by using the *fundamental property of rational numbers.*

Fundamental Property of Rational Numbers

If *a, b,* and *k* are integers with $b \neq 0$ and $k \neq 0$, then

$$\frac{a \cdot k}{b \cdot k} = \frac{a}{b}.$$

EXAMPLE 1 Writing a Fraction in Lowest Terms

Write $\frac{36}{54}$ in lowest terms.

SOLUTION

Since the greatest common factor of 36 and 54 is 18,

$$\frac{36}{54} = \frac{2 \cdot 18}{3 \cdot 18} = \frac{2}{3}.$$

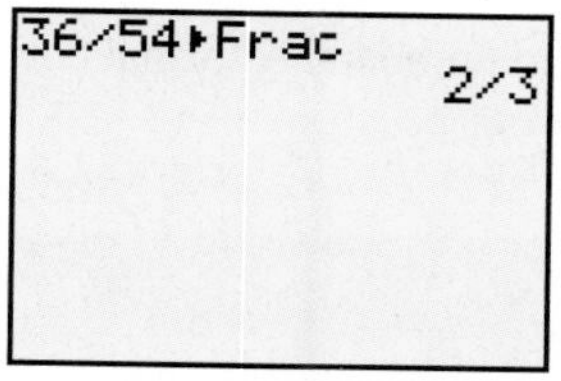

The calculator gives 36/54 in lowest terms, as illustrated in Example 1.

In Example 1, $\frac{36}{54} = \frac{2}{3}$. If we multiply the numerator of the fraction on the left by the denominator of the fraction on the right, we obtain $36 \cdot 3 = 108$. If we multiply the denominator of the fraction on the left by the numerator of the fraction on the right, we obtain $54 \cdot 2 = 108$. The result is the same in both cases.

One way of determining whether two fractions are equal is to perform this test. If the product of the **"extremes"** (36 and 3 in this case) equals the product of the **"means"** (54 and 2), the fractions are equal. This test for equality of rational numbers is called the **cross-product test.**

Cross-Product Test for Equality of Rational Numbers

For rational numbers $\frac{a}{b}$ and $\frac{c}{d}$, $b \neq 0$, $d \neq 0$,

$$\frac{a}{b} = \frac{c}{d} \quad \textbf{if and only if} \quad a \cdot d = b \cdot c.$$

Operations with Rational Numbers The operation of addition of rational numbers can be illustrated by the sketches in Figure 10. The rectangle at the top left is divided into three equal portions, with one of the portions in color. The rectangle at the top right is divided into five equal parts, with two of them in color.

The total of the areas in color is represented by the sum

$$\frac{1}{3} + \frac{2}{5}.$$

To evaluate this sum, the areas in color must be redrawn in terms of a common unit. Since the least common multiple of 3 and 5 is 15, redraw both rectangles with 15 parts. See Figure 11. In the figure, 11 of the small rectangles are in color, so

$$\frac{1}{3} + \frac{2}{5} = \frac{5}{15} + \frac{6}{15} = \frac{11}{15}.$$

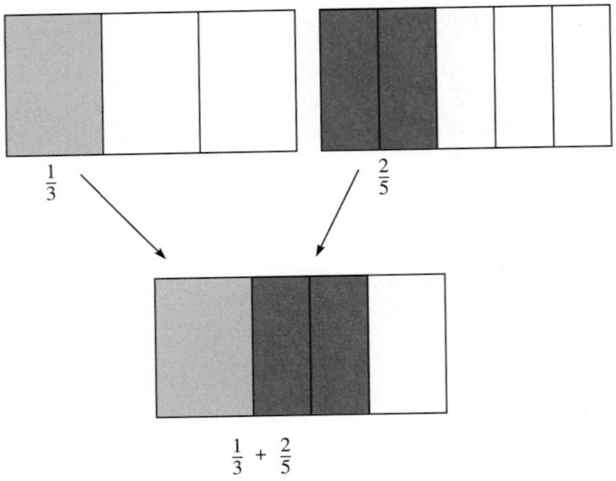

FIGURE 10

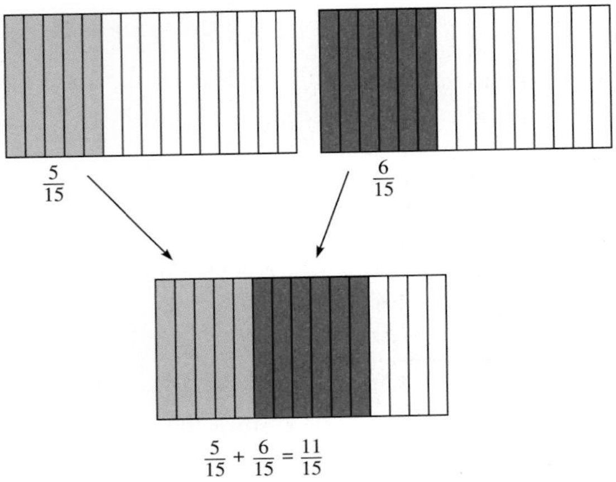

FIGURE 11

A similar example could be given for the difference of rational numbers. A formal definition of addition and subtraction of rational numbers follows.

Adding and Subtracting Rational Numbers

If $\frac{a}{b}$ and $\frac{c}{d}$ are rational numbers, then

$$\frac{a}{b} + \frac{c}{d} = \frac{ad + bc}{bd} \quad \text{and} \quad \frac{a}{b} - \frac{c}{d} = \frac{ad - bc}{bd}.$$

This formal definition is seldom used in practice. In practical problems involving addition and subtraction of rational numbers, we usually rewrite the fractions with the least common multiple of their denominators, called the **least common denominator.**

The results of Example 2 are illustrated in this screen.

EXAMPLE 2 Adding and Subtracting Rational Numbers

Perform each operation.

(a) $\frac{2}{15} + \frac{1}{10}$ **(b)** $\frac{173}{180} - \frac{69}{1200}$

SOLUTION

(a) The least common multiple of 15 and 10 is 30. Now write $\frac{2}{15}$ and $\frac{1}{10}$ with denominators of 30, and then add the numerators. Proceed as follows:

Since $30 \div 15 = 2$, $\frac{2}{15} = \frac{2 \cdot 2}{15 \cdot 2} = \frac{4}{30}$,

and since $30 \div 10 = 3$, $\frac{1}{10} = \frac{1 \cdot 3}{10 \cdot 3} = \frac{3}{30}$.

Thus, $\frac{2}{15} + \frac{1}{10} = \frac{4}{30} + \frac{3}{30} = \frac{7}{30}$.

(b) The least common multiple of 180 and 1200 is 3600.

$$\frac{173}{180} - \frac{69}{1200} = \frac{3460}{3600} - \frac{207}{3600} = \frac{3460 - 207}{3600} = \frac{3253}{3600}$$

The product of two rational numbers is defined as follows.

Multiplying Rational Numbers

If $\frac{a}{b}$ and $\frac{c}{d}$ are rational numbers, then

$$\frac{a}{b} \cdot \frac{c}{d} = \frac{ac}{bd}.$$

To illustrate the results of Example 3, we use parentheses around the fraction factors.

EXAMPLE 3 Multiplying Rational Numbers

Find each product.

(a) $\frac{3}{4} \cdot \frac{7}{10}$ **(b)** $\frac{5}{18} \cdot \frac{3}{10}$

SOLUTION

(a) $\frac{3}{4} \cdot \frac{7}{10} = \frac{3 \cdot 7}{4 \cdot 10} = \frac{21}{40}$

(b) $\frac{5}{18} \cdot \frac{3}{10} = \frac{5 \cdot 3}{18 \cdot 10} = \frac{15}{180} = \frac{1 \cdot 15}{12 \cdot 15} = \frac{1}{12}$

In practice, a multiplication problem such as this is often solved by using slash marks to indicate that common factors have been divided out of the numerator and denominator.

$$\frac{\overset{1}{\cancel{5}}}{\underset{6}{\cancel{18}}} \cdot \frac{\overset{1}{\cancel{3}}}{\underset{2}{\cancel{10}}} = \frac{1}{6} \cdot \frac{1}{2} \quad \text{3 is divided out of the terms 3 and 18; 5 is divided out of 5 and 10.}$$

$$= \frac{1}{12}$$

In a fraction, the fraction bar indicates the operation of division. Recall that, in the previous section, we defined the multiplicative inverse, or reciprocal, of the nonzero number b. The multiplicative inverse of b is $\frac{1}{b}$. We can now define division using multiplicative inverses.

For Further Thought

The Influence of Spanish Coinage on Stock Prices

Until August 28, 2000, when decimalization of the U.S. stock market began, market prices were reported with fractions having denominators with powers of 2, such as $17\frac{3}{4}$ and $112\frac{5}{8}$. Did you ever wonder why this was done?

During the early years of the United States, prior to the minting of its own coinage, the Spanish eight-reales coin, also known as the Spanish milled dollar, circulated freely in the states. Its fractional parts, the four reales, two reales, and one real, were known as **pieces of eight,** and were described as such in pirate and treasure lore. When the New York Stock Exchange was founded in 1792, it chose to use the Spanish milled dollar as its price basis, rather than the decimal base as proposed by Thomas Jefferson that same year.

In the September 1997 issue of *COINage,* Tom Delorey's article "The End of 'Pieces of Eight'" gives the following account:

> *As the Spanish dollar and its fractions continued to be legal tender in America alongside the decimal coins until 1857, there was no urgency to change the system—and by the time the Spanish-American money was withdrawn in 1857, pricing stocks in eighths of a dollar—and no less—was a tradition carved in stone. Being somewhat a conservative organization, the NYSE saw no need to fix what was not broken.*

All prices on the U.S. stock markets are now reported in decimals. (*Source:* "Stock price tables go to decimal listings," *The Times Picayune,* June 27, 2000.)

For Group Discussion or Individual Investigation

Consider this: Have you ever heard this old cheer? "Two bits, four bits, six bits, a dollar. All for the (home team), stand up and holler." The term **two bits** refers to 25 cents. Discuss how this cheer is based on the Spanish eight-reales coin.

Early U.S. cents and **half cents** used fractions to denote their denominations. The half cent used $\frac{1}{200}$ and the cent used $\frac{1}{100}$. (See Exercise 18 for a photo of an interesting error coin.)

The coins shown here were part of the collection of Louis E. Eliasberg, Sr. **Louis Eliasberg** was the only person ever to assemble a complete collection of United States coins. The Eliasberg gold coins were auctioned in 1982, while the copper, nickel, and silver coins were auctioned in two sales in 1996 and 1997. The half cent pictured sold for $506,000 and the cent sold for $27,500. The cent shown in Exercise 18 went for a mere $2970.

Definition of Division

If a and b are real numbers, $b \neq 0$, then

$$\frac{a}{b} = a \cdot \frac{1}{b}.$$

You have probably heard the rule, "To divide fractions, invert the divisor and multiply." But have you ever wondered why this rule works? To illustrate it, suppose that you have $\frac{7}{8}$ of a gallon of milk and you wish to find how many quarts you have. Since a quart is $\frac{1}{4}$ of a gallon, you must ask yourself, "How many $\frac{1}{4}$s are there in $\frac{7}{8}$?" This would be interpreted as

$$\frac{7}{8} \div \frac{1}{4} \quad \text{or} \quad \frac{\frac{7}{8}}{\frac{1}{4}}.$$

The fundamental property of rational numbers discussed earlier can be extended to rational number values of a, b, and k. With $a = \frac{7}{8}$, $b = \frac{1}{4}$, and $k = 4$ (the reciprocal of $b = \frac{1}{4}$),

$$\frac{a}{b} = \frac{a \cdot k}{b \cdot k} = \frac{\frac{7}{8} \cdot 4}{\frac{1}{4} \cdot 4} = \frac{\frac{7}{8} \cdot 4}{1} = \frac{7}{8} \cdot \frac{4}{1}.$$

Now notice that we began with the division problem $\frac{7}{8} \div \frac{1}{4}$ which, through a series of equivalent expressions, led to the multiplication problem $\left(\frac{7}{8} \cdot \frac{4}{1}\right)$. So dividing by $\frac{1}{4}$ is equivalent to multiplying by its reciprocal, $\frac{4}{1}$. By the definition of multiplication of fractions,

$$\frac{7}{8} \cdot \frac{4}{1} = \frac{28}{8} = \frac{7}{2},$$

and thus there are $\frac{7}{2}$ or $3\frac{1}{2}$ quarts in $\frac{7}{8}$ gallon.*

We now state the rule for dividing $\frac{a}{b}$ by $\frac{c}{d}$.

Dividing Rational Numbers

If $\frac{a}{b}$ and $\frac{c}{d}$ are rational numbers, where $\frac{c}{d} \neq 0$, then

$$\frac{a}{b} \div \frac{c}{d} = \frac{a}{b} \cdot \frac{d}{c} = \frac{ad}{bc}.$$

*$3\frac{1}{2}$ is a **mixed number.** Mixed numbers are covered in the exercises for this section.

This screen supports the results in Example 4(b) and (c).

EXAMPLE 4 Dividing Rational Numbers

Find each quotient.

(a) $\frac{3}{5} \div \frac{7}{15}$ **(b)** $\frac{-4}{7} \div \frac{3}{14}$ **(c)** $\frac{2}{9} \div 4$

SOLUTION

(a) $\frac{3}{5} \div \frac{7}{15} = \frac{3}{5} \cdot \frac{15}{7} = \frac{45}{35} = \frac{9 \cdot 5}{7 \cdot 5} = \frac{9}{7}$

(b) $\frac{-4}{7} \div \frac{3}{14} = \frac{-4}{7} \cdot \frac{14}{3} = \frac{-56}{21} = \frac{-8 \cdot 7}{3 \cdot 7} = \frac{-8}{3} = -\frac{8}{3}$

(c) $\frac{2}{9} \div 4 = \frac{2}{9} \div \frac{4}{1} = \frac{2}{9} \cdot \frac{1}{4} = \frac{\overset{1}{\cancel{2}}}{9} \cdot \frac{1}{\underset{2}{\cancel{4}}} = \frac{1}{18}$

$\frac{-a}{b}$, $\frac{a}{-b}$, and $-\frac{a}{b}$ are all equal.

Density and the Arithmetic Mean There is no integer between two consecutive integers, such as 3 and 4. However, a rational number can always be found between any two distinct rational numbers. For this reason, the set of rational numbers is said to be *dense.*

Density Property of the Rational Numbers

If r and t are distinct rational numbers, with $r < t$, then there exists a rational number s such that

$$r < s < t.$$

Repeated applications of the density property lead to the conclusion that there are *infinitely many* rational numbers between two distinct rational numbers.

To find the **arithmetic mean,** or **average,** of n numbers, we add the numbers and then divide the sum by n. For two numbers, the number that lies halfway between them is their average.

EXAMPLE 5 Finding the Arithmetic Mean (Average)

Find the rational number halfway between $\frac{2}{3}$ and $\frac{5}{6}$ (that is, their arithmetic mean, or average).

SOLUTION

First, find their sum.

$$\frac{2}{3} + \frac{5}{6} = \frac{4}{6} + \frac{5}{6} = \frac{9}{6} = \frac{3}{2}$$ Find a common denominator.

Now divide by 2.

$$\frac{3}{2} \div 2 = \frac{3}{2} \cdot \frac{1}{2} = \frac{3}{4}$$ To divide, multiply by the reciprocal.

The number $\frac{3}{4}$ is halfway between $\frac{2}{3}$ and $\frac{5}{6}$.

TABLE 3

Year	Number (in thousands)
1998	16,211
1999	16,477
2000	16,258
2001	16,289
2002	15,979
2003	15,776

Source: U.S. Bureau of Labor Statistics.

The computation in Example 6 is shown here.

EXAMPLE 6 Finding the Arithmetic Mean (Average)

Table 3 shows the number of labor union or employee association members, in thousands, for the years 1998–2003. What is the average number, in thousands, for this six-year period?

SOLUTION

To find this average, divide the sum by 6.

$$\frac{16{,}211 + 16{,}477 + 16{,}258 + 16{,}289 + 15{,}979 + 15{,}776}{6} = \frac{96{,}990}{6} = 16{,}165$$

The average number of workers for the six-year period is 16,165 thousand (or 16,165,000).

It is also true that between any two *real* numbers there is another *real* number. Thus, we say that the set of real numbers is dense.

Decimal Form of Rational Numbers

We have discussed rational numbers in the form of quotients of integers. Rational numbers can also be expressed as decimals. Decimal numerals have place values that are powers of 10. For example, the decimal numeral 483.039475 is read "four hundred eighty-three and thirty-nine thousand, four hundred seventy-five millionths." The place values are as shown here.

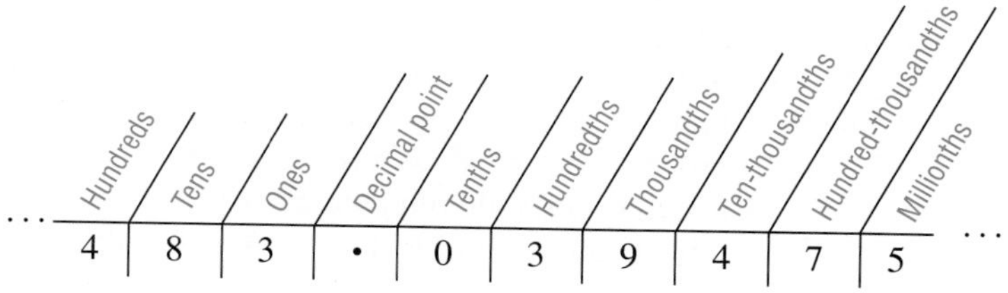

$$\begin{array}{r} .375 \\ 8\overline{)3.000} \\ \underline{24} \\ 60 \\ \underline{56} \\ 40 \\ \underline{40} \\ 0 \end{array} \qquad \begin{array}{r} .3636\ldots \\ 11\overline{)4.00000\ldots} \\ \underline{33} \\ 70 \\ \underline{66} \\ 40 \\ \underline{33} \\ 70 \\ \underline{66} \\ 40 \\ \vdots \end{array}$$

Given a rational number in the form $\frac{a}{b}$, it can be expressed as a decimal most easily by entering it into a calculator. For example, to write $\frac{3}{8}$ as a decimal, enter 3, then enter the operation of division, then enter 8. Press the equals key to find the following equivalence.

$$\frac{3}{8} = .375$$

This same result may be obtained by long division, as shown in the margin. By this result, the rational number $\frac{3}{8}$ is the same as the decimal .375. A decimal such as .375, which stops, is called a **terminating decimal.** Other examples of terminating decimals are

$$\frac{1}{4} = .25, \quad \frac{7}{10} = .7, \quad \text{and} \quad \frac{89}{1000} = .089. \qquad \text{Terminating decimals}$$

Not all rational numbers can be represented by terminating decimals. For example, convert $\frac{4}{11}$ into a decimal by dividing 11 into 4 using a calculator. The display shows

$$.3636363636, \quad \text{or perhaps} \quad .363636364.$$

However, we see that the long division process, shown in the margin, indicates that we will actually get .3636 . . . , with the digits 36 repeating over and over indefinitely.

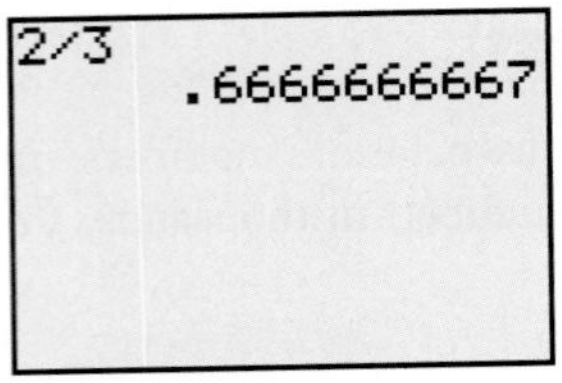

While 2/3 has a repeating decimal representation ($2/3 = .\overline{6}$), the calculator rounds off in the final decimal place displayed.

To indicate this, we write a bar (called a *vinculum*) over the "block" of digits that repeats. Therefore, we can write

$$\frac{4}{11} = .\overline{36}.$$

A decimal such as $.\overline{36}$, which continues indefinitely, is called a **repeating decimal.** Other examples of repeating decimals are

$$\frac{5}{11} = .\overline{45}, \quad \frac{1}{3} = .\overline{3}, \quad \text{and} \quad \frac{5}{6} = .8\overline{3}.$$

Repeating decimals

Because of the limitations of the display of a calculator, and because some rational numbers have repeating decimals, it is important to be able to interpret calculator results accordingly when obtaining repeating decimals.

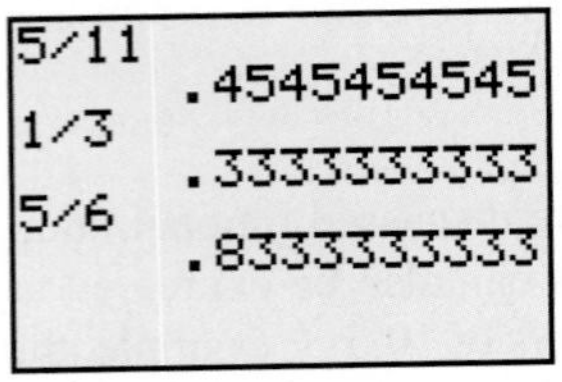

Although only ten decimal digits are shown, all three fractions have decimals that repeat endlessly.

While we shall distinguish between *terminating* and *repeating* decimals in this book, some mathematicians prefer to consider all rational numbers as repeating decimals. This can be justified by thinking this way: if the division process leads to a remainder of 0, then zeros repeat without end in the decimal form. For example, we can consider the decimal form of $\frac{3}{4}$ as follows.

$$\frac{3}{4} = .75\overline{0}$$

By considering the possible remainders that may be obtained when converting a quotient of integers to a decimal, we can draw an important conclusion about the decimal form of rational numbers. If the remainder is never zero, the division will produce a repeating decimal. This happens because each step of the division process must produce a remainder that is less than the divisor. Since the number of different possible remainders is less than the divisor, the remainders must eventually begin to repeat. This makes the digits of the quotient repeat, producing a repeating decimal.

Decimal Representation of Rational Numbers

Any rational number can be expressed as either a terminating decimal or a repeating decimal.

Simon Stevin (1548–1620) worked as a bookkeeper in Belgium and became an engineer in the Netherlands army. He is usually given credit for the development of **decimals.**

To determine whether the decimal form of a quotient of integers will terminate or repeat, we use the following rule.

Criteria for Terminating and Repeating Decimals

A rational number $\frac{a}{b}$ in lowest terms results in a **terminating decimal** if the only prime factor of the denominator is 2 or 5 (or both).

A rational number $\frac{a}{b}$ in lowest terms results in a **repeating decimal** if a prime other than 2 or 5 appears in the prime factorization of the denominator.

To find a baseball player's batting average, we divide the number of hits by the number of al-bats. A surprising paradox exists concerning averages; it is possible for Player *A* to have a higher batting average than Player *B* in each of two successive years, yet for the two-year period, Player *B* can have a higher total batting average. Look at the chart.

Year	Player *A*	Player *B*
1998	$\frac{20}{40} = .500$	$\frac{90}{200} = .450$
1999	$\frac{60}{200} = .300$	$\frac{10}{40} = .250$
Two-year total	$\frac{80}{240} = .333$	$\frac{100}{240} = .417$

In both individual years, Player *A* had a higher average, but for the two-year period, Player *B* had the higher average. This is an example of **Simpson's paradox** from statistics.

Justification of this rule is based on the fact that the prime factors of 10 are 2 and 5, and the decimal system uses ten as its base.

EXAMPLE 7 Determining Whether a Decimal Terminates or Repeats

Without actually dividing, determine whether the decimal form of the given rational number terminates or repeats.

(a) $\frac{7}{8}$ **(b)** $\frac{13}{150}$ **(c)** $\frac{6}{75}$

SOLUTION

(a) The rational number $\frac{7}{8}$ is in lowest terms. Its denominator is 8, and since 8 factors as 2^3, the decimal form will terminate. No primes other than 2 or 5 divide the denominator.

(b) The rational number $\frac{13}{150}$ is in lowest terms with denominator $150 = 2 \cdot 3 \cdot 5^2$. Since 3 appears as a prime factor of the denominator, the decimal form will repeat.

(c) First write the rational number $\frac{6}{75}$ in lowest terms.

$$\frac{6}{75} = \frac{2}{25} \quad \text{Denominator is 25.}$$

Since $25 = 5^2$, the decimal form will terminate. ■

We have seen that a rational number will be represented by either a terminating or a repeating decimal. Must a terminating decimal or a repeating decimal represent a rational number? The answer is *yes*. For example, the terminating decimal .6 represents a rational number.

$$.6 = \frac{6}{10} = \frac{3}{5}$$

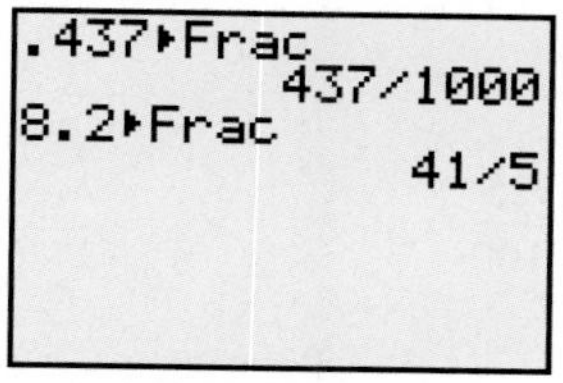

The results of Example 8 are supported in this screen.

EXAMPLE 8 Writing Terminating Decimals as Quotients of Integers

Write each terminating decimal as a quotient of integers.

(a) .437 **(b)** 8.2

SOLUTION

(a) $.437 = \frac{437}{1000}$ Read as "four hundred thirty-seven thousandths" and then write as a fraction.

(b) $8.2 = 8 + \frac{2}{10} = \frac{82}{10} = \frac{41}{5}$ Read as a decimal, write as a sum, and then add. ■

Repeating decimals cannot be converted into quotients of integers quite so quickly.

EXAMPLE 9 Writing a Repeating Decimal as a Quotient of Integers

Find a quotient of two integers equal to $.\overline{85}$.

SOLUTION

Step 1 Let $x = .\overline{85}$, so $x = .858585\ldots$.

Step 2 Multiply both sides of the equation $x = .858585\ldots$ by 100. (Use 100 since there are two digits in the part that repeats, and $100 = 10^2$.)

$$x = .858585\ldots$$
$$100x = 100(.858585\ldots)$$
$$100x = 85.858585\ldots$$

Step 3 Subtract the expressions in Step 1 from the final expressions in Step 2.

$$100x = 85.858585\ldots \quad \text{(Recall that } x = 1x \text{ and}$$
$$x = .858585\ldots \quad 100x - x = 99x.)$$
$$99x = 85 \quad \text{Subtract.}$$

Step 4 Solve the equation $99x = 85$ by dividing both sides by 99.

$$99x = 85$$
$$\frac{99x}{99} = \frac{85}{99} \quad \text{Divide by 99.}$$
$$x = \frac{85}{99} \quad \frac{99x}{99} = x$$
$$.\overline{85} = \frac{85}{99} \quad x = .\overline{85}$$

When checking with a calculator, remember that the calculator will only show a finite number of decimal places and may round off in the final decimal place shown.

$1 = .99999999999999\ldots$

Terminating or Repeating?

One of the most baffling truths of elementary mathematics is the following:

$$1 = .9999\ldots.$$

Most people believe that $.\overline{9}$ has to be less than 1, but this is not the case. The following argument shows why. Let $x = .9999\ldots$. Then

$$10x = 9.9999\ldots$$
$$x = .9999\ldots$$
$$9x = 9 \quad \text{Subtract.}$$
$$x = 1. \quad \text{Divide.}$$

Therefore, $1 = .9999\ldots$. Similarly, it can be shown that any terminating decimal can be represented as a repeating decimal with an endless string of 9s. For example, $.5 = .49999\ldots$ and $2.6 = 2.59999\ldots$. This is a way of justifying that any rational number may be represented as a repeating decimal.

6.3 Exercises

Choose the expression(s) that is (are) equivalent to the given rational number.

1. $\frac{4}{8}$

A. $\frac{1}{2}$ **B.** $\frac{8}{4}$ **C.** $.5$

D. $.5\overline{0}$ **E.** $.\overline{55}$

2. $\frac{2}{3}$

A. $.67$ **B.** $.\overline{6}$ **C.** $\frac{20}{30}$

D. $.666\ldots$ **E.** $.6$

3. $\frac{5}{9}$

A. $.56$ **B.** $.55$ **C.** $.\overline{5}$

D. $\frac{9}{5}$ **E.** $1\frac{4}{5}$

4. $\frac{1}{4}$

A. $.25$ **B.** $.24\overline{9}$ **C.** $\frac{25}{100}$

D. 4 **E.** $\frac{10}{400}$

Use the fundamental property of rational numbers to write each fraction in lowest terms.

5. $\frac{16}{48}$

6. $\frac{21}{28}$

7. $-\frac{15}{35}$

8. $-\frac{8}{48}$

Use the fundamental property to write each fraction in three other ways.

9. $\frac{3}{8}$

10. $\frac{9}{10}$

11. $-\frac{5}{7}$

12. $-\frac{7}{12}$

13. Write a fraction in lowest terms that represents the portion of each figure that is in color.

(a)

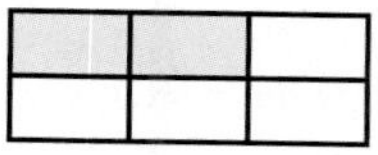

(b)

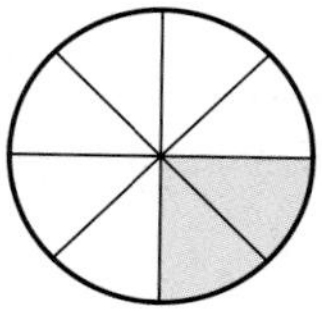

(c)

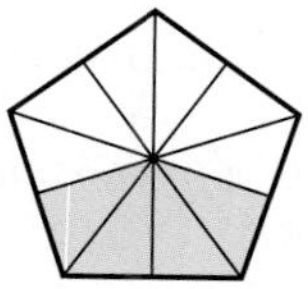

(d)

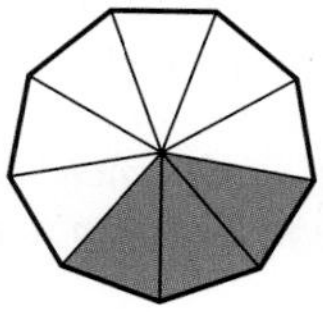

14. Write a fraction in lowest terms that represents the region described.

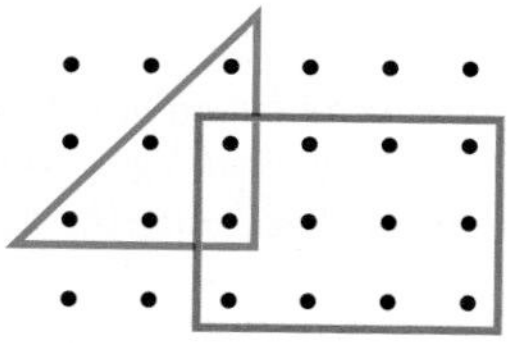

(a) the dots in the rectangle as a part of the dots in the entire figure

(b) the dots in the triangle as a part of the dots in the entire figure

(c) the dots in the rectangle as a part of the dots in the union of the triangle and the rectangle

(d) the dots in the intersection of the triangle and the rectangle as a part of the dots in the union of the triangle and the rectangle

15. Refer to the figure for Exercise 14 and write a description of the region that is represented by the fraction $\frac{1}{12}$.

16. ***Batting Averages*** In a softball league, the first six games produced the following results: Greg Tobin got 8 hits in 20 at-bats, and Jason Jordan got 12 hits in 30 at-bats. Which player (if either) had the higher batting average?

17. ***Batting Averages*** After ten games, the statistics at the top of the next column were obtained.

Player	At-bats	Hits	Home Runs
Anne Kelly	40	9	2
Christine O'Brien	36	12	3
Joanne Ha	11	5	1
Otis Taylor	16	8	0
Carol Britz	20	10	2

Answer each of the following, using estimation skills as necessary.

(a) Which player got a hit in exactly $\frac{1}{3}$ of his or her at-bats?

(b) Which player got a hit in just less than $\frac{1}{2}$ of his or her at-bats?

(c) Which player got a home run in just less than $\frac{1}{10}$ of his or her at-bats?

(d) Which player got a hit in just less than $\frac{1}{4}$ of his or her at-bats?

(e) Which two players got hits in exactly the same fractional parts of their at-bats? What was the fractional part, reduced to lowest terms?

18. Refer to the margin note discussing the use of common fractions on early U.S. copper coinage. The photo here shows an error near the bottom that occurred on an 1802 large cent. Discuss the error and how it represents a mathematical impossibility.

Perform the indicated operations and express answers in lowest terms. Use the order of operations as necessary.

19. $\frac{3}{8} + \frac{1}{8}$

20. $\frac{7}{9} + \frac{1}{9}$

21. $\frac{5}{16} + \frac{7}{12}$

22. $\frac{1}{15} + \frac{7}{18}$

23. $\frac{2}{3} - \frac{7}{8}$

24. $\frac{13}{20} - \frac{5}{12}$

25. $\frac{5}{8} - \frac{3}{14}$

26. $\frac{19}{15} - \frac{7}{12}$

27. $\frac{3}{4} \cdot \frac{9}{5}$

28. $\frac{3}{8} \cdot \frac{2}{7}$

29. $-\frac{2}{3} \cdot -\frac{5}{8}$

30. $-\frac{2}{4} \cdot \frac{3}{9}$

31. $\frac{5}{12} \div \frac{15}{4}$

32. $\frac{15}{16} \div \frac{30}{8}$

33. $-\frac{9}{16} \div -\frac{3}{8}$

34. $-\frac{3}{8} \div \frac{5}{4}$

35. $\left(\frac{1}{3} \div \frac{1}{2}\right) + \frac{5}{6}$

36. $\frac{2}{5} \div \left(-\frac{4}{5} \div \frac{3}{10}\right)$

37. ***Recipe for Grits*** The following chart appears on a package of Quaker® Quick Grits.

Microwave		Stove Top		
Servings	**1**	**1**	**4**	**6**
Water	$\frac{3}{4}$ cup	1 cup	3 cups	4 cups
Grits	3 Tbsp	3 Tbsp	$\frac{3}{4}$ cup	1 cup
Salt (optional)	dash	dash	$\frac{1}{4}$ tsp	$\frac{1}{2}$ tsp

(a) How many cups of water would be needed for 6 microwave servings?
(b) How many cups of grits would be needed for 5 stove-top servings? (*Hint:* 5 is halfway between 4 and 6.)

38. ***U.S. Immigrants*** More than 8 million immigrants were admitted to the United States during the first eight years of the 1990s. The circle graph gives the fractional number from each region of birth for these immigrants.

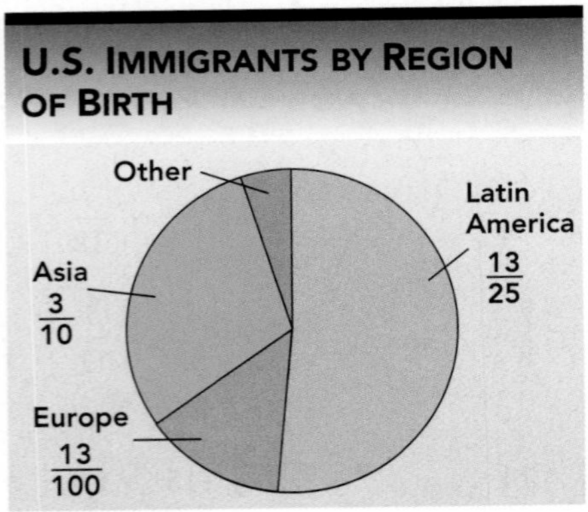

Source: U.S. Census Bureau

(a) What fractional part of the immigrants were from other regions?
(b) What fractional part of the immigrants were from Latin America or Asia?
(c) How many (in millions) were from Europe?

The **mixed number** $2\frac{5}{8}$ *represents the sum* $2 + \frac{5}{8}$. *We can convert* $2\frac{5}{8}$ *to a fraction as follows:*

$$2\frac{5}{8} = 2 + \frac{5}{8} = \frac{2}{1} + \frac{5}{8} = \frac{16}{8} + \frac{5}{8} = \frac{21}{8}.$$

The fraction $\frac{21}{8}$ *can be converted back to a mixed number by dividing 8 into 21. The quotient is 2, the remainder is 5, and the divisor is 8.*

Convert each mixed number to a fraction, and convert each fraction to a mixed number.

39. $4\frac{1}{3}$

40. $3\frac{7}{8}$

41. $2\frac{9}{10}$

42. $\frac{18}{5}$

43. $\frac{27}{4}$

44. $\frac{19}{3}$

It is possible to add mixed numbers by first converting them to fractions, adding, and then converting the sum back to a mixed number. For example,

$$2\frac{1}{3} + 3\frac{1}{2} = \frac{7}{3} + \frac{7}{2} = \frac{14}{6} + \frac{21}{6} = \frac{35}{6} = 5\frac{5}{6}.$$

The other operations with mixed numbers may be performed in a similar manner.

Perform each operation and express your answer as a mixed number.

45. $3\frac{1}{4} + 2\frac{7}{8}$

46. $6\frac{1}{5} - 2\frac{7}{15}$

47. $-4\frac{7}{8} \cdot 3\frac{2}{3}$

48. $-4\frac{1}{6} \div 1\frac{2}{3}$

Solve each problem.

49. ***Socket Wrench Measurements*** A hardware store sells a 40-piece socket wrench set. The measure of the largest socket is $\frac{3}{4}$ in., while the measure of the smallest socket is $\frac{3}{16}$ in. What is the difference between these measures?

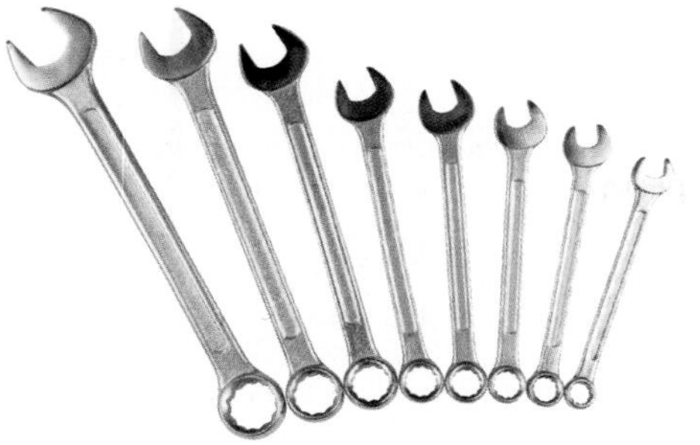

50. ***Swiss Cheese Hole Sizes*** Under existing standards, most of the holes in Swiss cheese must have diameters between $\frac{11}{16}$ and $\frac{13}{16}$ in. To accommodate new high-speed slicing machines, the USDA wants to reduce the minimum size to $\frac{3}{8}$ in. How much smaller is $\frac{3}{8}$ in. than $\frac{11}{16}$ in.? (*Source:* U.S. Department of Agriculture.)

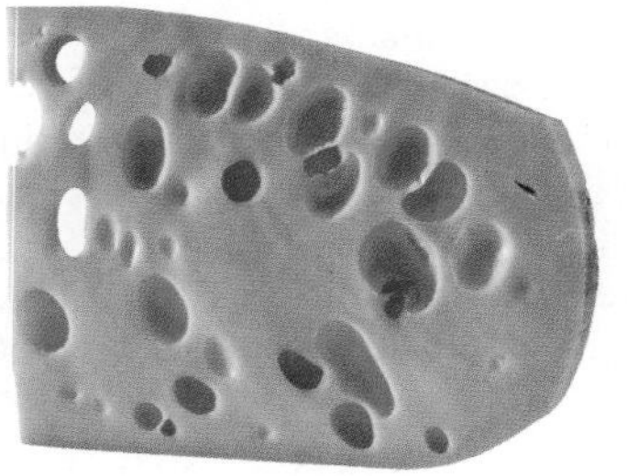

A quotient of quantities containing fractions (with denominator not zero) is called a **complex fraction.** *There are two methods that are used to simplify a complex fraction.*

Method 1 *Simplify the numerator and denominator separately. Then rewrite as a division problem, and proceed as you would when dividing fractions.*

Method 2 *Multiply both the numerator and denominator by the least common denominator of all the fractions found within the complex fraction. (This is, in effect, multiplying the fraction by 1, which does not change its value.) Apply the distributive property, if necessary, and simplify.*

Use one of the methods above to simplify each complex fraction.

51. $\dfrac{\frac{1}{2}+\frac{1}{4}}{\frac{1}{2}-\frac{1}{4}}$ **52.** $\dfrac{\frac{2}{3}+\frac{1}{6}}{\frac{2}{3}-\frac{1}{6}}$ **53.** $\dfrac{\frac{5}{8}-\frac{1}{4}}{\frac{1}{8}+\frac{3}{4}}$

54. $\dfrac{\frac{3}{16}-\frac{1}{2}}{\frac{5}{16}+\frac{1}{8}}$ **55.** $\dfrac{\frac{7}{11}+\frac{3}{10}}{\frac{1}{11}-\frac{9}{10}}$ **56.** $\dfrac{\frac{11}{15}+\frac{1}{9}}{\frac{13}{15}-\frac{2}{3}}$

The expressions in Exercises 57 and 58 are called ***continued fractions.*** *Write each in the form $\frac{p}{q}$ reduced to lowest terms. (Hint: Start at the bottom and work up.)*

57. $2+\cfrac{1}{1+\cfrac{1}{3+\cfrac{1}{2}}}$ **58.** $4+\cfrac{1}{2+\cfrac{1}{1+\cfrac{1}{3}}}$

Find the rational number halfway between the two given rational numbers.

59. $\frac{1}{2}, \frac{3}{4}$ **60.** $\frac{1}{3}, \frac{5}{12}$ **61.** $\frac{3}{5}, \frac{2}{3}$

62. $\frac{7}{12}, \frac{5}{8}$ **63.** $-\frac{2}{3}, -\frac{5}{6}$ **64.** $-3, -\frac{5}{2}$

Solve each problem.

65. ***Average Annual Salary*** The table shows the average annual salary in the eight highest-paying metropolitan areas in the United States. Find the average of these amounts to the nearest dollar.

Metropolitan Area	Average Annual Salary
San Jose, CA	\$63,056
New York, NY	\$57,708
San Francisco, CA	\$56,602
New Haven, CT, area	\$51,170
Middlesex, NJ, area	\$50,457
Jersey City, NJ	\$49,562
Newark, NJ	\$48,781
Washington, DC, area	\$48,430

Source: Bureau of Labor Statistics.

66. ***Adoption of Chinese Babies*** Since 2000, the country of China has been the most popular foreign country for U.S. adoptions. Find the average annual number of adoptions during the period 2000–2004, based on the figures in the table.

Year	Number of Adoptions
2000	4943
2001	4629
2002	6062
2003	6638
2004	7033

Source: Department of Homeland Security, Office of Immigration Statistics.

In the March 1973 issue of The Mathematics Teacher *there appeared an article by Laurence Sherzer, an eighth-grade mathematics teacher, that immortalized one of his students, Robert McKay. The class was studying the density property and Sherzer was explaining how to find a rational number between two given positive rational numbers by finding the average. McKay pointed out that there was no need to go to all that trouble. To find a number (not necessarily their average) between two positive rational numbers $\frac{a}{b}$ and $\frac{c}{d}$, he claimed, simply add the numerators and add the denominators. Much to Sherzer's surprise, this method really does work.*

For example, to find a rational number between $\frac{1}{3}$ and $\frac{1}{4}$, add $1 + 1 = 2$ to get the numerator and $3 + 4 = 7$ to get the denominator. Therefore, by **McKay's theorem,** *$\frac{2}{7}$ is between $\frac{1}{3}$ and $\frac{1}{4}$. Sherzer provided a proof of this method in the article.*

Use McKay's theorem *to find a rational number between the two given rational numbers.*

67. $\frac{5}{6}$ and $\frac{9}{13}$ **68.** $\frac{10}{11}$ and $\frac{13}{19}$

69. $\frac{4}{13}$ and $\frac{9}{16}$ **70.** $\frac{6}{11}$ and $\frac{13}{14}$

71. 2 and 3 **72.** 3 and 4

73. Apply McKay's theorem to any pair of consecutive integers, and make a conjecture about what always happens in this case.

74. Explain in your own words how to find the rational number that is one-fourth of the way between two different rational numbers.

Convert each rational number into either a repeating or a terminating decimal. Use a calculator if your instructor so allows.

75. $\frac{3}{4}$ **76.** $\frac{7}{8}$ **77.** $\frac{3}{16}$ **78.** $\frac{9}{32}$

79. $\frac{3}{11}$ **80.** $\frac{9}{11}$ **81.** $\frac{2}{7}$ **82.** $\frac{11}{15}$

Convert each terminating decimal into a quotient of integers. Write each in lowest terms.

83. .4 **84.** .9 **85.** .85

86. .105 **87.** .934 **88.** .7984

Use the method of Example 7 to decide whether each rational number would yield a repeating or a terminating decimal. (Hint: Write in lowest terms before trying to decide.)

89. $\frac{8}{15}$ **90.** $\frac{8}{35}$ **91.** $\frac{13}{125}$

92. $\frac{3}{24}$ **93.** $\frac{22}{55}$ **94.** $\frac{24}{75}$

95. Follow through on all parts of this exercise in order.
(a) Find the decimal for $\frac{1}{3}$.
(b) Find the decimal for $\frac{2}{3}$.
(c) By adding the decimal expressions obtained in parts (a) and (b), obtain a decimal expression for $\frac{1}{3} + \frac{2}{3} = \frac{3}{3} = 1$.
(d) Does your result seem bothersome? Read the margin note on terminating and repeating decimals in this section, which refers to this idea.

96. It is a fact that $\frac{1}{3} = .333\ldots$. Multiply both sides of this equation by 3. Does your answer bother you? See the margin note on terminating and repeating decimals in this section.

Use the method of Example 9 to write each rational number as a quotient of integers in lowest terms.

97. (a) .8 **(b)** $.7\overline{9}$

98. (a) .75 **(b)** $.74\overline{9}$

99. (a) .66 **(b)** $.65\overline{9}$

100. Based on your results in Exercises 97–99, predict the lowest terms form of the rational number $.4\overline{9}$.

6.4 Irrational Numbers and Decimal Representation

Definition and Basic Concepts • Irrationality of $\sqrt{2}$ and Proof by Contradiction • Operations with Square Roots • The Irrational Numbers π, ϕ, and e

Definition and Basic Concepts In the previous section, we saw that every rational number has a decimal form that terminates or repeats. Also, every repeating or terminating decimal represents a rational number. Some decimals, however, neither repeat nor terminate. For example, the decimal

Tsu Ch'ung-chih (about 500 A.D.), the Chinese mathematician honored on the above stamp, investigated the digits of π. **Aryabhata,** his Indian contemporary, gave 3.1416 as the value.

$$.102001000200001000002\ldots$$

does not terminate and does not repeat. (It is true that there is a pattern in this decimal, but no single block of digits repeats indefinitely.)*

Irrational Numbers

Irrational numbers = $\{x \mid x$ is a number represented by a nonrepeating, nonterminating decimal$\}$.

As the name implies, an irrational number cannot be represented as a quotient of integers.

The decimal number mentioned above is an irrational number. Other irrational numbers include $\sqrt{2}$, $\frac{1+\sqrt{5}}{2}$ (ϕ, from Section 5.5), π (the ratio of the circumference of a circle to its diameter), and e (a constant *approximately equal to* 2.71828). There are infinitely many irrational numbers.

The irrational number $\sqrt{2}$ was discovered by the Pythagoreans in about 500 B.C. This discovery was a great setback to their philosophy that everything is based upon the whole numbers. The Pythagoreans kept their findings secret, and legend has it that members of the group who divulged this discovery were sent out to sea, and, according to Proclus (410–485), "perished in a shipwreck, to a man."

Irrationality of $\sqrt{2}$ and Proof by Contradiction

Figure 12 illustrates how a point with coordinate $\sqrt{2}$ can be located on a number line.

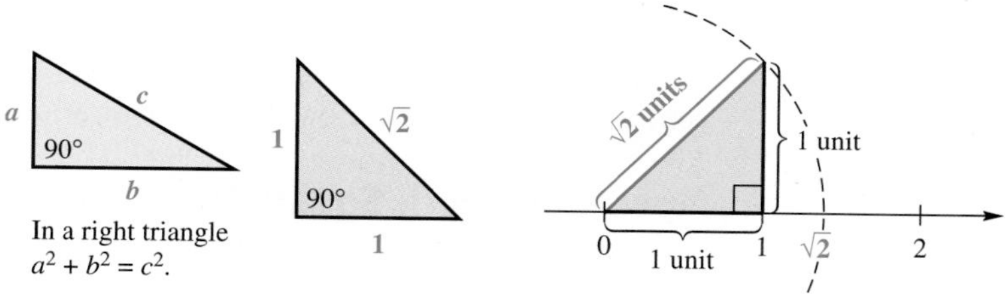

FIGURE 12

The proof that $\sqrt{2}$ is irrational is a classic example of a **proof by contradiction.** We begin by assuming that $\sqrt{2}$ is rational, which leads to a contradiction, or absurdity. The method is also called **reductio ad absurdum** (Latin for "reduce to the absurd"). In order to understand the proof, we consider three preliminary facts:

1. When a rational number is written in lowest terms, the greatest common factor of the numerator and denominator is 1.
2. If an integer is even, then it has 2 as a factor and may be written in the form $2k$, where k is an integer.
3. If a perfect square is even, then its square root is even.

*In this section, we will assume that the digits of a number such as this continue indefinitely in the pattern established. The next few digits would be 000000100000002, and so on.

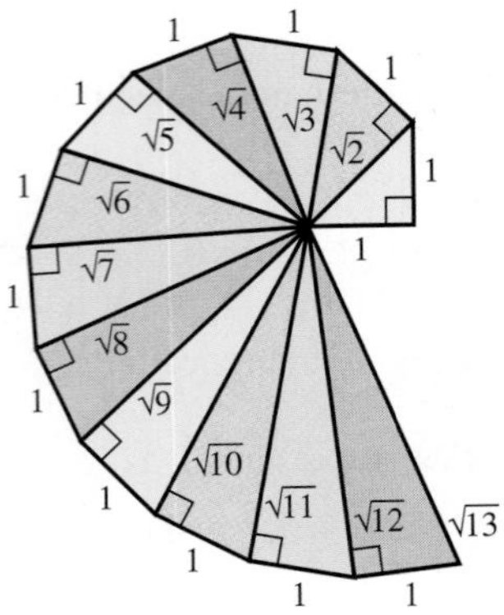

An interesting way to represent the lengths corresponding to $\sqrt{2}$, $\sqrt{3}$, $\sqrt{4}$, $\sqrt{5}$, and so on, is shown in the figure. Use the **Pythagorean theorem** to verify the lengths in the figure.

THEOREM

Statement: $\sqrt{2}$ is an irrational number.

Proof: Assume that $\sqrt{2}$ is a rational number. Then by definition,

$$\sqrt{2} = \frac{p}{q}, \quad \text{for some integers } p \text{ and } q.$$

Furthermore, assume that $\frac{p}{q}$ is the form of $\sqrt{2}$ that is written in lowest terms, so the greatest common factor of p and q is 1.

$$2 = \frac{p^2}{q^2} \quad \text{Square both sides of the equation.}$$

$$2q^2 = p^2 \quad \text{Multiply by } q^2.$$

This last equation indicates that 2 is a factor of p^2. So p^2 is even, and thus p is even. Since p is even, it may be written in the form $2k$, where k is an integer.

Now, substitute $2k$ for p in the last equation and simplify:

$$2q^2 = (2k)^2 \quad \text{Let } p = 2k.$$

$$2q^2 = 4k^2 \quad (2k)^2 = 2k \cdot 2k = 4k^2$$

$$q^2 = 2k^2. \quad \text{Divide by 2.}$$

Since 2 is a factor of q^2, q^2 must be even, and thus q must be even. This leads to a contradiction: p and q cannot both be even because they would then have a common factor of 2, although it was assumed that their greatest common factor is 1.

Therefore, since the original assumption that $\sqrt{2}$ is rational has led to a contradiction, it must follow that $\sqrt{2}$ is irrational. ■

Operations with Square Roots In everyday mathematical work, nearly all of our calculations deal with rational numbers, usually in decimal form. In our study of mathematics, however, we must sometimes perform operations with irrational numbers, and in many instances, the irrational numbers are square roots. Some examples are

$$\sqrt{2}, \quad \sqrt{3}, \quad \text{and} \quad \sqrt{13}. \quad \text{Square roots that are irrational.}$$

However, not all square roots are irrational. For example,

$$\sqrt{4} = 2, \quad \sqrt{36} = 6, \quad \text{and} \quad \sqrt{100} = 10 \quad \text{Square roots that are rational.}$$

are all rational numbers. If n is a positive integer that is not the square of an integer, then $\sqrt{n}$ is an irrational number.

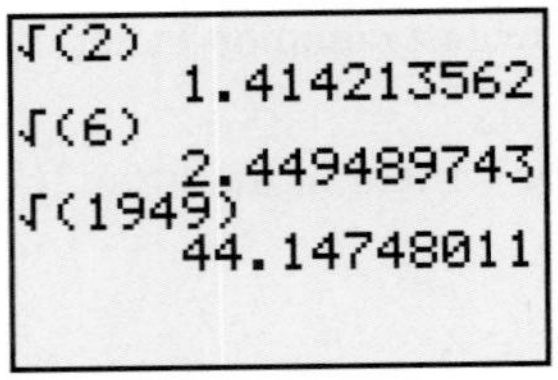

These are calculator approximations of irrational numbers.

A calculator with a square root key can give approximations of square roots of numbers that are not perfect squares. To show that they are approximations, we use the ≈ symbol to indicate "is approximately equal to." Some such calculator approximations are as follows:

$$\sqrt{2} \approx 1.414213562, \quad \sqrt{6} \approx 2.449489743, \quad \text{and} \quad \sqrt{1949} \approx 44.14748011.$$

Recall that $\sqrt{a}$, for $a \geq 0$, is the nonnegative number whose square is a; that is, $(\sqrt{a})^2 = a$. We will now look at some simple operations with square roots.

Notice that

$$\sqrt{4} \cdot \sqrt{9} = 2 \cdot 3 = 6$$

and

$$\sqrt{4 \cdot 9} = \sqrt{36} = 6.$$

Thus, $\sqrt{4} \cdot \sqrt{9} = \sqrt{4 \cdot 9}$. This is a particular case of the following product rule.

Product Rule for Square Roots

For nonnegative real numbers a and b,

$$\sqrt{a} \cdot \sqrt{b} = \sqrt{a \cdot b}.$$

Just as every rational number $\frac{a}{b}$ can be written in simplest (lowest) terms (by using the fundamental property of rational numbers), every square root radical has a simplified form.

Conditions Necessary for the Simplified Form of a Square Root Radical

A square root radical is in **simplified form** if the following three conditions are met.

1. The number under the radical **(radicand)** has no factor (except 1) that is a perfect square.
2. The radicand has no fractions.
3. No denominator contains a radical.

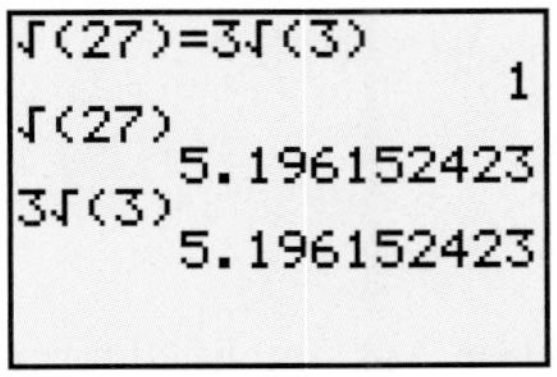

The 1 after the first line indicates that the equality is true. The calculator also shows the same approximations for $\sqrt{27}$ and $3\sqrt{3}$ in the second and third answers. (See Example 1.)

EXAMPLE 1 Simplifying a Square Root Radical (Product Rule)

Simplify $\sqrt{27}$.

SOLUTION

Since 9 is a factor of 27 and 9 is a perfect square, $\sqrt{27}$ is not in simplified form. The first condition of simplified form is not met. We simplify as follows.

$$\begin{aligned} \sqrt{27} &= \sqrt{9 \cdot 3} \\ &= \sqrt{9} \cdot \sqrt{3} && \text{Use the product rule.} \\ &= 3\sqrt{3} && \sqrt{9} = 3, \text{ since } 3^2 = 9. \end{aligned}$$

Expressions such as $\sqrt{27}$ and $3\sqrt{3}$ are *exact values* of the square root of 27. If we use the square root key of a calculator, we find

$$\sqrt{27} \approx 5.196152423.$$

If we find $\sqrt{3}$ and then multiply the result by 3, we get

$$3\sqrt{3} \approx 3(1.732050808) \approx 5.196152423.$$

Notice that these approximations are the same, as we would expect. (Due to various methods of calculating, there may be a discrepancy in the final digit of the calculation.) Understand, however, that the calculator approximations do not actually

The radical symbol above comes from the Latin word for root, *radix.* It was first used by **Leonardo of Pisa** (Fibonacci) in 1220. The sixteenth-century German symbol we use today probably is also derived from the letter r.

prove that the two numbers are equal, but only strongly suggest equality. The work done in Example 1 actually provides the mathematical justification that they are indeed equal.

A rule similar to the product rule exists for quotients.

Quotient Rule for Square Roots

For nonnegative real numbers a and positive real numbers b,

$$\frac{\sqrt{a}}{\sqrt{b}} = \sqrt{\frac{a}{b}}.$$

EXAMPLE 2 Simplifying Square Root Radicals (Quotient Rule)

Simplify each radical.

(a) $\sqrt{\frac{25}{9}}$ **(b)** $\sqrt{\frac{3}{4}}$ **(c)** $\sqrt{\frac{1}{2}}$

SOLUTION

(a) Because the radicand contains a fraction, the radical expression is not simplified. (See condition 2 of simplified form preceding Example 1.) Use the quotient rule as follows.

$$\sqrt{\frac{25}{9}} = \frac{\sqrt{25}}{\sqrt{9}} = \frac{5}{3}$$

(b) $\sqrt{\frac{3}{4}} = \frac{\sqrt{3}}{\sqrt{4}} = \frac{\sqrt{3}}{2}$

(c) $\sqrt{\frac{1}{2}} = \frac{\sqrt{1}}{\sqrt{2}} = \frac{1}{\sqrt{2}}$

This expression is not yet in simplified form, since condition 3 of simplified form is not met. To give an equivalent expression with no radical in the denominator, we use a procedure called **rationalizing the denominator.** Multiply $\frac{1}{\sqrt{2}}$ by $\frac{\sqrt{2}}{\sqrt{2}}$, which is a form of 1, the identity element for multiplication.

$$\frac{1}{\sqrt{2}} = \frac{1}{\sqrt{2}} \cdot \frac{\sqrt{2}}{\sqrt{2}} = \frac{\sqrt{2}}{2} \qquad \sqrt{2} \cdot \sqrt{2} = 2$$

The simplified form of $\sqrt{\frac{1}{2}}$ is $\frac{\sqrt{2}}{2}$.

Is $\sqrt{4} + \sqrt{9} = \sqrt{4 + 9}$ a true statement? Computation shows that the answer is *no,* since $\sqrt{4} + \sqrt{9} = 2 + 3 = 5$, while $\sqrt{4 + 9} = \sqrt{13}$, and $5 \neq \sqrt{13}$. ***Square root radicals may be combined, however, if they have the same radicand.*** Such radicals are **like radicals.** We add (and subtract) like radicals using the distributive property.

EXAMPLE 3 Adding and Subtracting Square Root Radicals

Add or subtract as indicated.

(a) $3\sqrt{6} + 4\sqrt{6}$ **(b)** $\sqrt{18} - \sqrt{32}$

SOLUTION

(a) Since both terms contain $\sqrt{6}$, they are like radicals, and may be combined.

$$\begin{aligned} 3\sqrt{6} + 4\sqrt{6} &= (3 + 4)\sqrt{6} && \text{Distributive property} \\ &= 7\sqrt{6} && \text{Add.} \end{aligned}$$

(b) If we simplify $\sqrt{18}$ and $\sqrt{32}$, then this operation can be performed.

$$\begin{aligned} \sqrt{18} - \sqrt{32} &= \sqrt{9 \cdot 2} - \sqrt{16 \cdot 2} && \text{Factor so that perfect squares are in the radicands.} \\ &= \sqrt{9} \cdot \sqrt{2} - \sqrt{16} \cdot \sqrt{2} && \text{Product rule} \\ &= 3\sqrt{2} - 4\sqrt{2} && \text{Take square roots.} \\ &= (3 - 4)\sqrt{2} && \text{Distributive property} \\ &= -1\sqrt{2} && \text{Subtract.} \\ &= -\sqrt{2} && -1 \cdot a = -a \end{aligned}$$

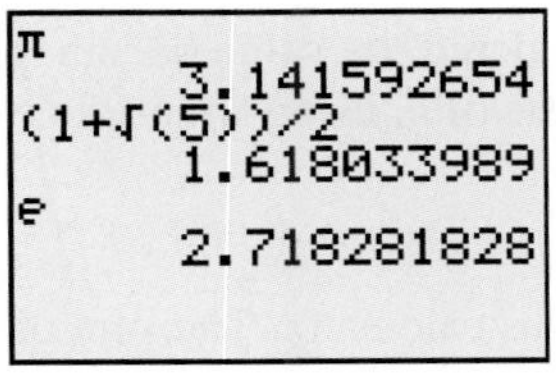

FIGURE 13

From Example 3, we see that like radicals may be added or subtracted by adding or subtracting their coefficients (the numbers by which they are multiplied) and keeping the same radical. For example,

$$9\sqrt{7} + 8\sqrt{7} = 17\sqrt{7} \quad (\text{since } 9 + 8 = 17)$$
$$4\sqrt{3} - 12\sqrt{3} = -8\sqrt{3}, \quad (\text{since } 4 - 12 = -8)$$

and so on.

In the statements of the product and quotient rules for square roots, the radicands could not be negative. While $-\sqrt{2}$ is a real number, for example, $\sqrt{-2}$ is not: there is no real number whose square is -2. The same may be said for any negative radicand. In order to handle this situation, mathematicians have extended our number system to include *complex numbers,* discussed in the Extension at the end of this chapter.

This poem, dedicated to **Archimedes** ("the immortal Syracusan"), allows us to learn the first 31 digits of the decimal representation of π. By replacing each word with the number of letters it contains, with a decimal point following the initial 3, the decimal is found. The poem was written by A. C. Orr, and appeared in the *Literary Digest* in 1906.

Now I, even I, would celebrate
In rhymes unapt, the great
Immortal Syracusan, rivaled nevermore,
Who in his wondrous lore
Passed on before,
Left men his guidance
How to circles mensurate.

The Irrational Numbers π, ϕ, and e Figure 13 shows approximations for three of the most interesting and important irrational numbers in mathematics. The first of these, π, represents the ratio of the circumference of a circle to its diameter. The second, ϕ, is the Golden Ratio, covered in detail in Section 5.5. Its exact value is $\frac{1+\sqrt{5}}{2}$. The third is e, a fundamental number in our universe. It is the base of the *natural exponential* and *natural logarithmic* functions, as seen in Section 8.6. The letter e was chosen to honor Leonhard Euler, who published extensive research on the number in 1746.

Pi (π)

$$\pi \approx 3.141592653589793238462643383279$$

The computation of the digits of π has fascinated mathematicians since ancient times. Archimedes was the first to explore it extensively, and mathematicians have

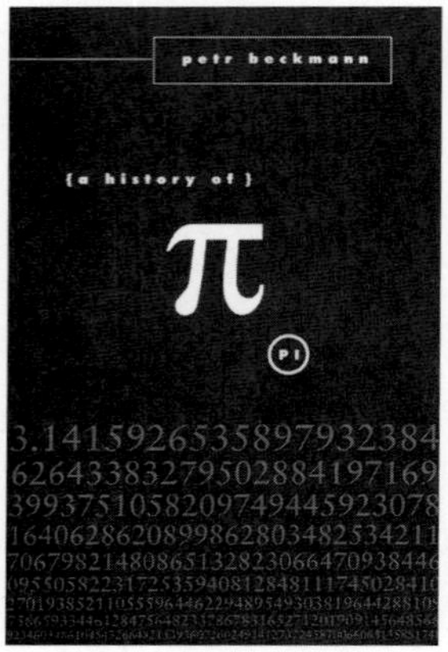

today computed its value to over 1 trillion digits. Yasumasa Kanada of the University of Tokyo and the brothers Gregory and David Chudnovsky are among the foremost of today's pi researchers. The book *A History of* π by Petr Beckmann is a classic, now in its third edition. Numerous Web sites are devoted to the history and methods of computation of pi. Some of them are as follows:

www.joyofpi.com/

www.math.utah.edu/~alfeld/Archimedes/Archimedes.html

www.super-computing.org/

www.pbs.org/wgbh/nova/sciencenow/3210/04.html

One of the methods of computing pi involves the topic of *infinite series,* as seen in Example 4.

EXAMPLE 4 Computing the Digits of Pi Using an Infinite Series

It is shown in higher mathematics that the *infinite series*

$$1 - \frac{1}{3} + \frac{1}{5} - \frac{1}{7} + \frac{1}{9} + \ldots$$

"converges" to $\frac{\pi}{4}$. That is, as more and more terms are considered, its value becomes closer and closer to $\frac{\pi}{4}$. With a calculator, approximate the value of pi using twenty-one terms of this series.

SOLUTION

Figure 14 shows the necessary calculation on the TI-83/84 Plus calculator. The sum of the first twenty-one terms is multiplied by 4, to obtain the approximation

$$3.189184782.$$

(While this is only correct to the first decimal place, better approximations are obtained using more terms of the series.)

Northern Exposure, which ran between 1990 and 1995 on the CBS network, starred Rob Morrow as Dr. Joel Fleischman, a doctor practicing in Alaska. In the episode "Nothing's Perfect" (10/12/92), he meets and falls in love with a mathematician (played by Wendel Meldrum) after accidentally running over her dog. Her area of research is **computation of the decimal digits of pi.** She mentions that a string of eight 8s appears in the decimal relatively early in the expansion. A search at The Pi Searcher (www.angio.net/pi/bigpi/cgi) confirms that this string starts at position 46,663,520 counting from the first digit after the decimal point.

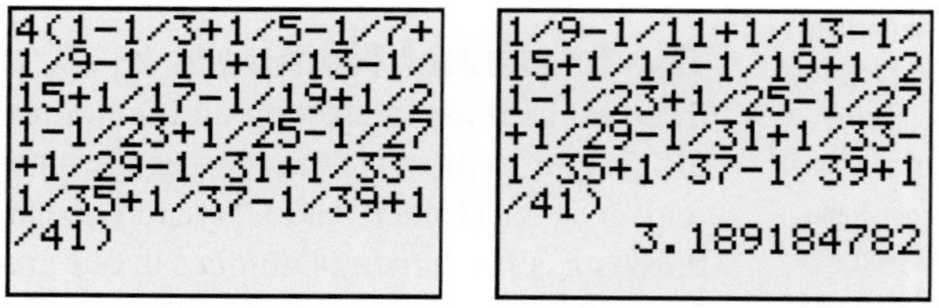

This is a continuation of the previous screen.

FIGURE 14

A rectangle that satisfies the condition that the ratio of its length to its width is equal to the ratio of the sum of its length and width to its length is called a **Golden Rectangle.** This ratio is called the **Golden Ratio.** (See Section 5.5.) The exact value of the Golden Ratio is the irrational number $\frac{1+\sqrt{5}}{2}$, and it is represented by the Greek letter ϕ (phi).

In 1767 **J. H. Lambert** proved that π is irrational (and thus its decimal will never terminate and never repeat). Nevertheless, the 1897 Indiana state legislature considered a bill that would have *legislated* the value of π. In one part of the bill, the value was stated to be 4, and in another part, 3.2. Amazingly, the bill passed the House, but the Senate postponed action on the bill indefinitely.

```
233/144
         1.618055556
377/233
         1.618025751
610/377
         1.618037135
```

FIGURE 15

Phi (ϕ)

$$\phi = \frac{1+\sqrt{5}}{2} \approx 1.6180339887498948482045868343\,65$$

Two readily accessible books on phi are *The Divine Proportion, A Study in Mathematical Beauty* by H. E. Huntley, and the more recent *The Golden Ratio* by Mario Livio. Some popular Web sites devoted to this irrational number are as follows:

www.mcs.surrey.ac.uk/Personal/R.Knott/Fibonacci/

www.goldennumber.net/

www.mathforum.org/dr.math/faq/faq.golden.ratio.html

www.geom.uiuc.edu/~demo5337/s97b/art.htm

EXAMPLE 5 Computing the Digits of Phi Using the Fibonacci Sequence

The first twelve terms of the Fibonacci sequence are

$$1, 1, 2, 3, 5, 8, 13, 21, 34, 55, 89, 144.$$

Each term after the first two terms is obtained by adding the two previous terms. Thus, the thirteenth term is $89 + 144 = 233$. As one goes farther and farther out in the sequence, the ratio of a term to its predecessor gets closer and closer to ϕ. How far out must one go in order to approximate ϕ so that the first five decimal places agree?

SOLUTION

After 144, the next three Fibonacci numbers are 233, 377, and 610. Figure 15 shows that $\frac{610}{377} \approx 1.618037135$, which agrees with ϕ to the fifth decimal place. ■

Most applications of the irrational number e are beyond the scope of this text. However, e is a fundamental constant in mathematics, and if there are intelligent beings elsewhere in the universe, they will no doubt know about this number. If you study Section 8.6, you will encounter it as a base of the important exponential and logarithmic functions.

e

$$e \approx 2.718281828459045235360287471353$$

The nature of e has made it less understood by the layman than π (or even ϕ, for that matter). The 1994 book *e: The Story of a Number* by Eli Maor has attempted to rectify this situation. These Web sites also give information on e:

www.mathforum.org/dr.math/faq/faq.e.html

www-groups.dcs.st-and.ac.uk/~history/HistTopics/e.html

http://antwrp.gsfc.nasa.gov/htmltest/gifcity/e.1mil

www.math.toronto.edu/mathnet/answers/ereal.html

Example 6 illustrates another infinite series, but this one converges to e.

EXAMPLE 6 Computing the Digits of *e* Using an Infinite Series

The infinite series

$$2 + \frac{1}{1 \cdot 2} + \frac{1}{1 \cdot 2 \cdot 3} + \frac{1}{1 \cdot 2 \cdot 3 \cdot 4} + \cdots$$

converges to e. Use a calculator to approximate e using the first seven terms of this series.

SOLUTION

Figure 16 shows the sum of the first seven terms. (The denominators have all been multiplied out.) The sum is 2.718253968, which agrees with e to four decimal places. This series converges more rapidly than the one for π in Example 4.

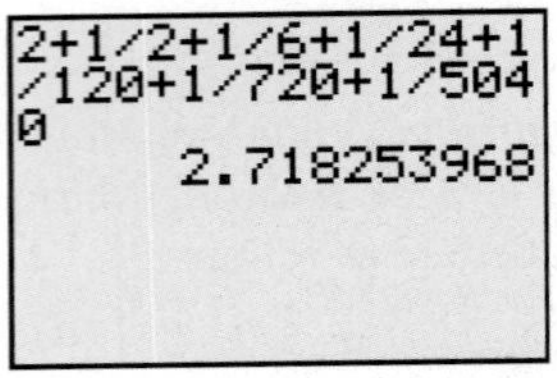

FIGURE 16

6.4 EXERCISES

Identify each number as rational *or* irrational.

1. $\frac{4}{9}$ **2.** $\frac{7}{8}$ **3.** $\sqrt{10}$ **4.** $\sqrt{14}$ **5.** 1.618

6. 2.718 **7.** $.\overline{41}$ **8.** $.\overline{32}$ **9.** π **10.** $\frac{1 + \sqrt{5}}{2}$

11. .878778777877778. . . **12.** e **13.** 3.14159 **14.** $\frac{22}{7}$

15. (a) Find the sum.

$$\begin{array}{r} .272772777277772\ldots \\ +.616116111611116\ldots \\ \hline \end{array}$$

(b) Based on the result of part (a), we can conclude that the sum of two ________ numbers may be a(n) ________ number.

16. (a) Find the sum.

$$\begin{array}{r} .010110111011110\ldots \\ +.252552555255552\ldots \\ \hline \end{array}$$

(b) Based on the result of part (a), we can conclude that the sum of two ________ numbers may be a(n) ________ number.

Use a calculator to find a rational decimal approximation for each irrational number. Give as many places as your calculator shows.

17. $\sqrt{39}$ **18.** $\sqrt{44}$ **19.** $\sqrt{15.1}$ **20.** $\sqrt{33.6}$

21. $\sqrt{884}$ **22.** $\sqrt{643}$ **23.** $\sqrt{\frac{9}{8}}$ **24.** $\sqrt{\frac{6}{5}}$

Use the methods of Examples 1 and 2 to simplify each expression. Then, use a calculator to approximate both the given expression and the simplified expression. (Both should be the same.)

25. $\sqrt{50}$ **26.** $\sqrt{32}$ **27.** $\sqrt{75}$

28. $\sqrt{150}$ **29.** $\sqrt{288}$ **30.** $\sqrt{200}$

31. $\frac{5}{\sqrt{6}}$ **32.** $\frac{3}{\sqrt{2}}$ **33.** $\sqrt{\frac{7}{4}}$

34. $\sqrt{\frac{8}{9}}$ **35.** $\sqrt{\frac{7}{3}}$ **36.** $\sqrt{\frac{14}{5}}$

Use the method of Example 3 to perform the indicated operations.

37. $\sqrt{17} + 2\sqrt{17}$

38. $3\sqrt{19} + \sqrt{19}$

39. $5\sqrt{7} - \sqrt{7}$

40. $3\sqrt{27} - \sqrt{27}$

41. $3\sqrt{18} + \sqrt{2}$

42. $2\sqrt{48} - \sqrt{3}$

43. $-\sqrt{12} + \sqrt{75}$

44. $2\sqrt{27} - \sqrt{300}$

Exercises 45–58 deal with π, ϕ, or e. Use a calculator or computer as necessary.

45. Move one matchstick to make the equation approximately true. (*Source:* www.joyofpi.com)

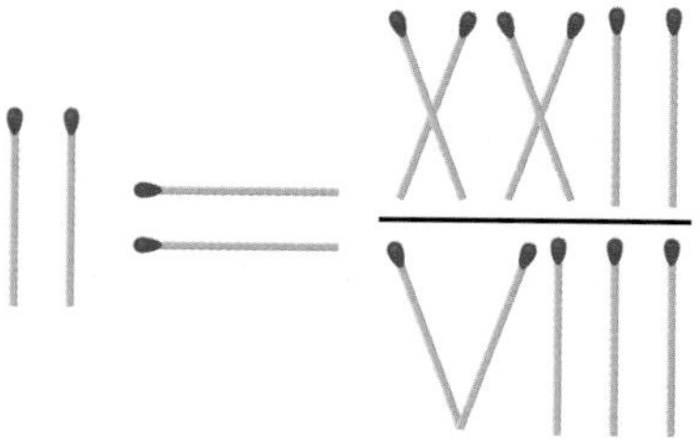

46. Find the square root of $\frac{2143}{22}$ using a calculator. Then find the square root of that result. Compare your result to the decimal given for π in the margin note. What do you notice?

47. Find the first eight digits in the decimal for $\frac{355}{113}$. Compare the result to the decimal for π given in the text. What do you notice?

48. You may have seen the statements "use $\frac{22}{7}$ for π" and "use 3.14 for π." Since $\frac{22}{7}$ is the quotient of two integers, and 3.14 is a terminating decimal, do these statements suggest that π is rational?

49. In the Bible (I Kings 7:23), a verse describes a circular pool at King Solomon's temple, about 1000 B.C. The pool is said to be ten cubits across, "and a line of 30 cubits did compass it round about." What value of π does this imply?

50. The ancient Egyptians used a method for finding the area of a circle that is equivalent to a value of 3.1605 for π. Write this decimal as a mixed number.

51. The computation of π has fascinated mathematicians and laymen for centuries. In the nineteenth century, the British mathematician William Shanks spent many years of his life calculating π to 707 decimal places. It turned out that only the first 527 were correct. Use an Internet search to find the 528th decimal digit of π (following the whole number part 3.).

52. One of the reasons for computing so many digits of π is to determine how often each digit appears and to identify any interesting patterns among the digits. Gregory and David Chudnovsky have spent a great deal of time and effort looking for patterns in the digits. For example, six 9s in a row appear relatively early in the decimal, within the first 800 decimal places. Use an Internet search to find the positions of these six 9s in a row.

53. The expression $\frac{2 \cdot 2 \cdot 4 \cdot 4 \cdot 6 \cdot 6 \cdot 8 \cdots}{1 \cdot 3 \cdot 3 \cdot 5 \cdot 5 \cdot 7 \cdot 7 \cdots}$ converges to $\frac{\pi}{2}$. Use a calculator to evaluate only the digits of the expression as shown here, and then multiply by 2. What value for an approximation for π does this give (to one decimal place)?

54. A *mnemonic device* is a scheme whereby one is able to recall facts by memorizing something completely unrelated to the facts. One way of learning the first few digits of the decimal for π is to memorize a sentence (or several sentences) and count the letters in each word of the sentence. For example, "See, I know a digit," will give the first 5 digits of π: "See" has 3 letters, "I" has 1 letter, "know" has 4 letters, "a" has 1 letter, and "digit" has 5 letters. So the first five digits are 3.1415.

Verify that the following mnemonic devices work.

(a) "May I have a large container of coffee?"

(b) "See, I have a rhyme assisting my feeble brain, its tasks ofttimes resisting."

(c) "How I want a drink, alcoholic of course, after the heavy lectures involving quantum mechanics."

55. Use a calculator to find the decimal approximations for $\phi = \frac{1+\sqrt{5}}{2}$ and its *conjugate*, $\frac{1-\sqrt{5}}{2}$. Comment on the similarities and differences in the two decimals.

56. In some literature, the Golden Ratio is defined to be the reciprocal of $\frac{1+\sqrt{5}}{2}$— that is, $\frac{2}{1+\sqrt{5}}$. Use a calculator to find a decimal approximation for $\frac{2}{1+\sqrt{5}}$ and compare it to ϕ as defined in this text. What do you observe?

57. An approximation for e is 2.718281828. A student noticed that there seems to be a repetition of four digits in this number (1, 8, 2, 8) and concluded that it is rational, because repeating decimals represent rational numbers. Was the student correct? Why or why not?

58. Use a calculator with an exponential key to find values for the following: $(1.1)^{10}$, $(1.01)^{100}$, $(1.001)^{1000}$, $(1.0001)^{10,000}$, and $(1.00001)^{100,000}$. Compare your results to the approximation given for e in this section. What do you find?

Solve each problem. Use a calculator as necessary, and give approximations to the nearest tenth unless specified otherwise.

59. ***Period of a Pendulum*** The period of a pendulum in seconds depends on its length, L, in feet, and is given by the formula

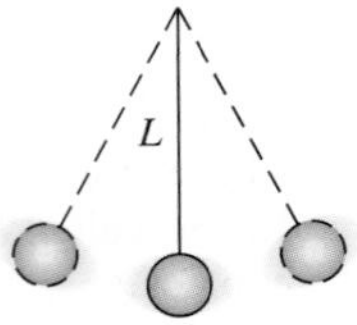

$$P = 2\pi\sqrt{\frac{L}{32}}.$$

If a pendulum is 5.1 feet long, what is its period? Use 3.14 for π.

60. ***Radius of an Aluminum Can*** The radius of the circular top or bottom of an aluminum can with surface area S and height h is given by

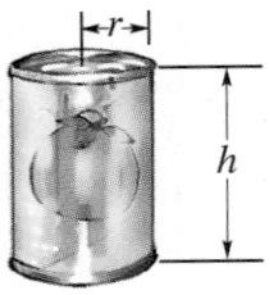

$$r = \frac{-h + \sqrt{h^2 + .64S}}{2}.$$

What radius should be used to make a can with height 12 inches and surface area 400 square inches?

61. ***Distance to the Horizon*** Jack Adrian, a friend of one of the authors of this text, has a beautiful 14th floor condo, with a stunning view, in downtown Chicago. The floor is 150 feet above the ground. Knowing that this author is a mathematics teacher, Jack emailed the author and told him that he recalled once having studied a formula for calculating the distance to the horizon, but could not remember it. He wanted to know how far he can see from his condo. The author responded:

> *To find the distance to the horizon in miles, take the square root of the height of your view and multiply that result by 1.224. That will give you the number of miles to the horizon.*

Assuming Jack's eyes are 6 feet above his floor, the total height from the ground is $150 + 6 = 156$ feet.

To the nearest tenth of a mile, how far can he see to the horizon?

62. ***Electronics Formula*** The formula

$$I = \sqrt{\frac{2P}{L}}$$

relates the coefficient of self-induction L (in henrys), the energy P stored in an electronic circuit (in joules), and the current I (in amps). Find I if $P = 120$ joules and $L = 80$ henrys.

63. ***Area of the Bermuda Triangle*** *Heron's formula* gives a method of finding the area of a triangle if the lengths of its sides are known. Suppose that a, b, and c are the lengths of the sides. Let s denote one-half of the perimeter of the triangle (called the *semiperimeter*); that is,

$$s = \frac{1}{2}(a + b + c).$$

Then the area A of the triangle is given by

$$A = \sqrt{s(s - a)(s - b)(s - c)}.$$

Find the area of the Bermuda Triangle, if the "sides" of this triangle measure approximately 850 miles, 925 miles, and 1300 miles. Give your answer to the nearest thousand square miles.

64. ***Area Enclosed by the Vietnam Veterans' Memorial*** The Vietnam Veterans' Memorial in Washington, D.C., is in the shape of an unenclosed isosceles triangle with equal sides of length 246.75 feet. If the triangle were enclosed, the third side would have length 438.14 feet. Use Heron's formula from the previous exercise to find the area of this enclosure to the nearest hundred square feet. (*Source:* Information pamphlet obtained at the Vietnam Veterans' Memorial.)

65. ***Perfect Triangles*** A *perfect triangle* is a triangle whose sides have whole number lengths and whose area is numerically equal to its perimeter. Use Heron's formula to show that the triangle with sides of length 9, 10, and 17 is perfect.

66. ***Heron Triangles*** A *Heron triangle* is a triangle having integer sides and area. Use Heron's formula to show that each of the following is a Heron triangle.
(a) $a = 11, b = 13, c = 20$
(b) $a = 13, b = 14, c = 15$
(c) $a = 7, b = 15, c = 20$

67. ***Diagonal of a Box*** The length of the diagonal of a box is given by

$$D = \sqrt{L^2 + W^2 + H^2},$$

where L, W, and H are the length, the width, and the height of the box. Find the length of the diagonal, D, of a box that is 4 feet long, 3 feet wide, and 2 feet high.

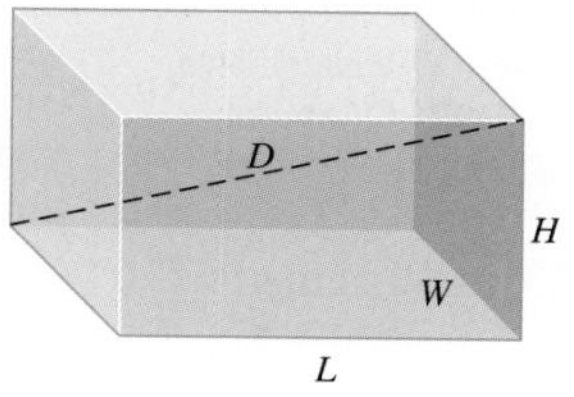

68. ***Rate of Return of an Investment*** If an investment of P dollars grows to A dollars in two years, the annual rate of return on the investment is given by

$$r = \frac{\sqrt{A} - \sqrt{P}}{\sqrt{P}}.$$

First rationalize the denominator and then find the annual rate of return (as a decimal) if \$50,000 increases to \$58,320.

69. ***Accident Reconstruction*** Police sometimes use the following procedure to estimate the speed at which a car was traveling at the time of an accident. A police officer drives the car involved in the accident under conditions similar to those during which the accident took place and then skids to a stop. If the car is driven at 30 miles per hour, then the speed at the time of the accident is given by

$$s = 30\sqrt{\frac{a}{p}},$$

where a is the length of the skid marks left at the time of the accident and p is the length of the skid marks in the police test. Find s for the following values of a and p.
(a) $a = 862$ feet; $p = 156$ feet
(b) $a = 382$ feet; $p = 96$ feet
(c) $a = 84$ feet; $p = 26$ feet

70. ***Law of Tensions*** In the study of sound, one version of the law of tensions is

$$f_1 = f_2\sqrt{\frac{F_1}{F_2}}.$$

Find f_1 to the nearest unit if $F_1 = 300$, $F_2 = 60$, and $f_2 = 260$.

The concept of square (second) root can be extended to **cube (third) root, fourth root,** *and so on. If $n \geq 2$ and a is a nonnegative number, $\sqrt[n]{a}$ represents the nonnegative number whose nth power is a. For example,*

$$\sqrt[3]{8} = 2 \text{ because } 2^3 = 8,$$
$$\sqrt[3]{1000} = 10 \text{ because } 10^3 = 1000,$$
$$\sqrt[4]{81} = 3 \text{ because } 3^4 = 81,$$

and so on. Find each root.

71. $\sqrt[3]{64}$
72. $\sqrt[3]{125}$
73. $\sqrt[3]{343}$
74. $\sqrt[3]{729}$
75. $\sqrt[3]{216}$
76. $\sqrt[3]{512}$
77. $\sqrt[4]{1}$
78. $\sqrt[4]{16}$
79. $\sqrt[4]{256}$
80. $\sqrt[4]{625}$
81. $\sqrt[4]{4096}$
82. $\sqrt[4]{2401}$

Use a calculator to approximate each root. Give as many places as your calculator shows. (Hint: To find the fourth root, find the square root of the square root.)

83. $\sqrt[3]{43}$
84. $\sqrt[3]{87}$
85. $\sqrt[3]{198}$
86. $\sqrt[4]{2107}$
87. $\sqrt[4]{10{,}265.2}$
88. $\sqrt[4]{863.5}$

6.5 Applications of Decimals and Percents

Operations with Decimals • Rounding Decimals • Percent • Applications

Operations with Decimals Because calculators have, for the most part, replaced paper-and-pencil methods for operations with decimals and percent, we will only briefly mention these latter methods. *We strongly suggest that the work in this section be done with a calculator at hand.*

```
.46+3.9+12.58
          16.94
12.1-8.723
          3.377
```

This screen supports the results in Example 1.

Addition and Subtraction of Decimals

To add or subtract decimal numbers, line up the decimal points in a column and perform the operation.

EXAMPLE 1 Adding and Subtracting Decimal Numbers

Find each of the following.

(a) .46 + 3.9 + 12.58 **(b)** 12.1 − 8.723

SOLUTION

(a) To compute the sum .46 + 3.9 + 12.58, use the following method.

$$\begin{array}{r} .46 \\ 3.90 \\ +12.58 \\ \hline 16.94 \end{array}$$

Line up decimal points.
Attach a zero as a placeholder.
← Sum

(b) To compute the difference 12.1 − 8.723, use this method.

$$\begin{array}{r} 12.100 \\ -\ 8.723 \\ \hline 3.377 \end{array}$$

Attach zeros.
← Difference

Recall that when two numbers are multiplied, the numbers are called *factors* and the answer is called the *product.* When two numbers are divided, the number being divided is called the *dividend,* the number doing the dividing is called the *divisor,* and the answer is called the *quotient.*

Technology pervades the world outside school. There is no question that students will be expected to use calculators in other settings; this technology is now part of our culture. . . students no longer have the same need to perform these (paper-and-pencil) procedures with large numbers of lengthy expressions that they might have had in the past without ready access to technology.

From *Computation, Calculators, and Common Sense (A Position of the National Council of Teachers of Mathematics).*

Multiplication and Division of Decimals

Multiplication To multiply decimals, multiply in the same manner as integers are multiplied. The number of decimal places to the right of the decimal point in the product is the *sum* of the numbers of places to the right of the decimal points in the factors.

Division To divide decimals, move the decimal point to the right the same number of places in the divisor and the dividend so as to obtain a whole number in the divisor. Divide in the same manner as integers are divided. The number of decimal places to the right of the decimal point in the quotient is the same as the number of places to the right in the dividend.

EXAMPLE 2 Multiplying and Dividing Decimal Numbers

Find each of the following.

(a) 4.613×2.52 **(b)** $65.175 \div 8.25$

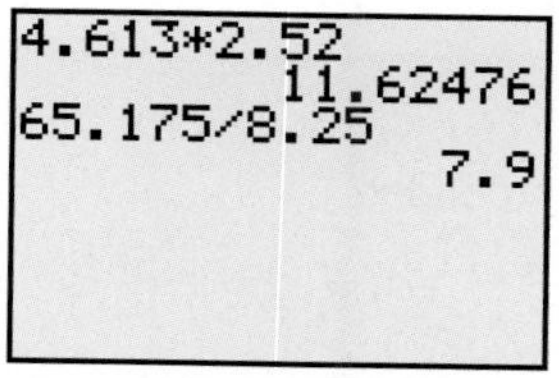

This screen supports the results in Example 2.

SOLUTION

(a) To find the product 4.613×2.52, use the following method.

$$\begin{array}{rl} 4.613 & \leftarrow \text{3 decimal places} \\ \times \quad 2.52 & \leftarrow \text{2 decimal places} \\ \hline 9226 & \\ 23065 & \\ 9226 & \\ \hline 11.62476 & \leftarrow 3 + 2 = \text{5 decimal places} \end{array}$$

(b) To find the quotient $65.175 \div 8.25$, follow these steps.

$$\begin{array}{r} 7.9 \\ \rightarrow 825\overline{)6517.5} \\ \underline{5775} \\ 742\,5 \\ \underline{742\,5} \\ 0 \end{array}$$

Bring the decimal point straight up in the answer.

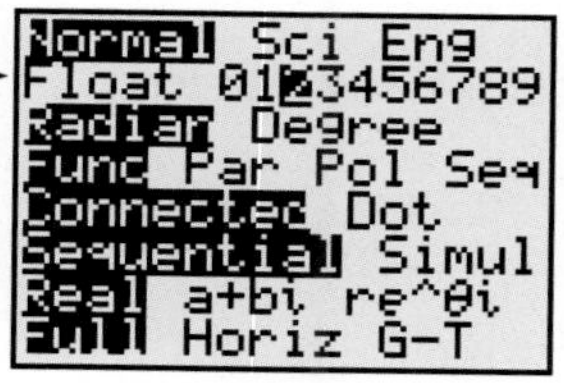

TI-83 Plus

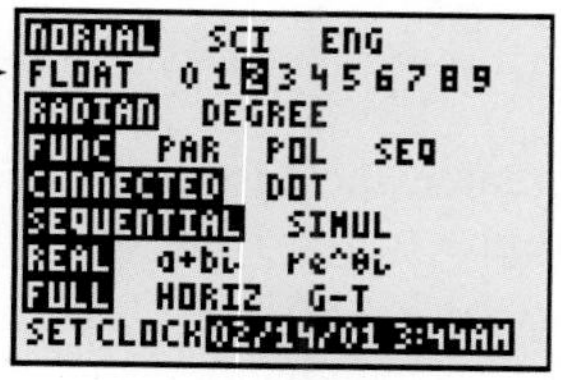

TI-84 Plus

Here the TI-83/84 Plus is set to round the answer to two decimal places.

Rounding Decimals Operations with decimals often result in long strings of digits in the decimal places. Since all these digits may not be needed in a practical problem, it is common to *round* a decimal to the necessary number of decimal places. For example, in preparing federal income tax, money amounts are rounded to the nearest dollar. Round as shown in the next example.

EXAMPLE 3 Rounding a Decimal Number

Round 3.917 to the nearest hundredth.

SOLUTION

The hundredths place in 3.917 contains the digit 1.

$$\begin{array}{c} 3.917 \\ \uparrow \text{ Hundredths place} \end{array}$$

To round this decimal, locate 3.91 and 3.92 on a number line as in Figure 17.

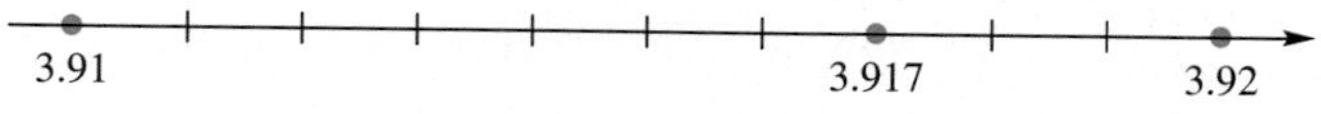

FIGURE 17

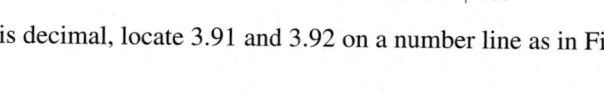

The distance from 3.91 to 3.92 is divided into ten equal parts. The seventh of these ten parts locates the number 3.917. As the number line shows, 3.917 is closer to 3.92 than it is to 3.91, so 3.917 rounded to the nearest hundredth is 3.92.

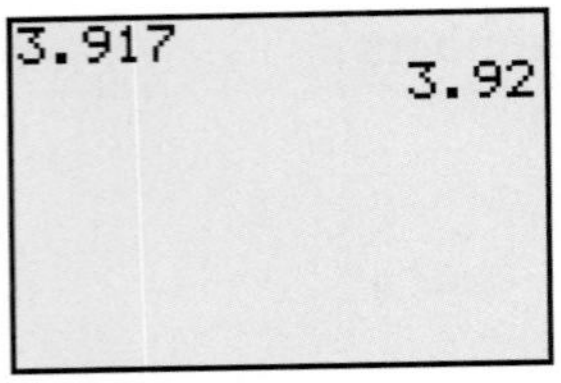

The calculator rounds 3.917 to the nearest hundredth.

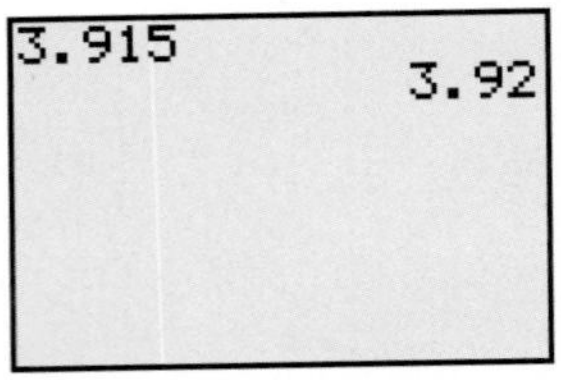

The calculator rounds 3.915 *up* to 3.92.

If the number line method of Example 3 were used to round 3.915 to the nearest hundredth, a problem would develop—the number 3.915 is exactly halfway between 3.91 and 3.92. An arbitrary decision is then made to round *up:* 3.915 rounded to the nearest hundredth is 3.92.

Rules for Rounding Decimals

Step 1 Locate the **place** to which the number is being rounded.

Step 2 Look at the next **digit to the right** of the place to which the number is being rounded.

Step 3A If this digit is **less than 5,** drop all digits to the right of the place to which the number is being rounded. Do *not change* the digit in the place to which the number is being rounded.

Step 3B If this digit is **5 or greater,** drop all digits to the right of the place to which the number is being rounded. *Add one* to the digit in the place to which the number is being rounded.

EXAMPLE 4 Rounding a Decimal Number

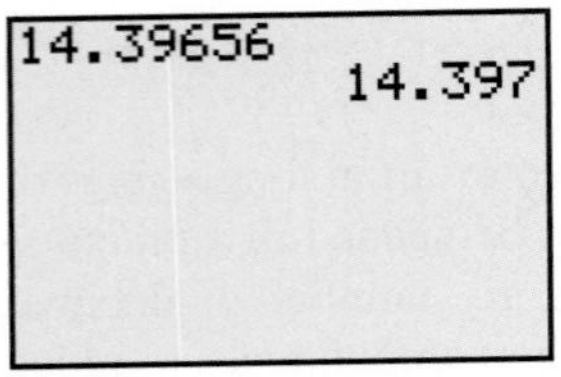

With the calculator set to round to three decimal places, the result of Example 4 is supported.

Round 14.39656 to the nearest thousandth.

SOLUTION

Step 1 Use an arrow to locate the place to which the number is being rounded.

14.39656
↑ Thousandths place

Step 2 Check to see if the first digit to the right of the arrow is 5 or greater.

14.396 5 6 Digit to the right of the arrow is 5.
↑

Step 3 Since the digit to the right of the arrow is 5 or greater, increase by 1 the digit to which the arrow is pointing. Drop all digits to the right of the arrow.

14.39656 Drop.
↑
14.397 Increase by 1.

Finally, 14.39656 rounded to the nearest thousandth is 14.397. ■

The percent sign, %, probably evolved from a symbol introduced in an Italian manuscript of 1425. Instead of "per 100," "P 100," or "P cento," which were common at that time, the author used "P$\overset{o}{c}$." By about 1650 the $\overset{o}{c}$ had become $\frac{o}{o}$, so "per $\frac{o}{o}$," was often used. Finally the "per" was dropped, leaving $\frac{0}{0}$ or %.

From *Historical Topics for the Mathematics Classroom,* the Thirty-first Yearbook of the National Council of Teachers of Mathematics, 1969.

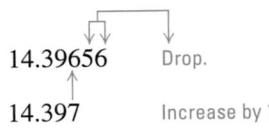

Percent One of the main applications of decimals comes from problems involving **percents.** In consumer mathematics, interest rates and discounts are often given as percents. The word *percent* means "per hundred." The symbol % represents "percent."

Gene Wilder has appeared in nearly 40 movies in his career. Among his most notable characters are accountant Leo Bloom in the 1969 version of *The Producers* and Willy Wonka in *Willy Wonka and the Chocolate Factory*, both of whom deliver interesting lines involving **percent.**

In *The Producers,* he and Max Bialystock (Zero Mostel) scheme to make a fortune by overfinancing what they think will be a Broadway flop. After enumerating the percent of profits all of Max's little old ladies have been offered in the production, reality sets in.

MAX: Leo, how much percentage of a play can there be altogether?
LEO: Max, you can only sell 100% of anything.
MAX: And how much for *Springtime for Hitler* have we sold?
LEO: 25,000%
MAX (reaching for Leo's blue security blanket): 25,000%. . . Give me that blue thing.

As Willy Wonka, upon preparing a mixture in his laboratory, Wilder delivers the following as he drinks his latest concoction.

WILLY WONKA: Invention, my dear friends, is 93% perspiration, 6% electricity, 4% evaporation, and 2% butterscotch ripple.
FEMALE VOICE: That's 105%.
MALE VOICE: Any good?
WILLY WONKA: Yes!

Percent

$$1\% = \frac{1}{100} = .01$$

EXAMPLE 5 Converting Percents to Decimals

Convert each percent to a decimal.

(a) 98% **(b)** 3.4% **(c)** .2%

SOLUTION

(a) $98\% = 98(1\%) = 98(.01) = .98$
(b) $3.4\% = 3.4(1\%) = 3.4(.01) = .034$
(c) $.2\% = .2(1\%) = .2(.01) = .002$

EXAMPLE 6 Converting Decimals to Percents

Convert each decimal to a percent.

(a) .13 **(b)** .532 **(c)** 2.3

SOLUTION

(a) $.13 = 13(.01) = 13(1\%) = 13\%$
(b) $.532 = 53.2(.01) = 53.2(1\%) = 53.2\%$
(c) $2.3 = 230(.01) = 230(1\%) = 230\%$

From Examples 5 and 6, we see that the following procedures can be used when converting between percents and decimals.

Converting Between Decimals and Percents

To convert a percent to a decimal, drop the % sign and move the decimal point two places to the left, inserting zeros as placeholders if necessary.

To convert a decimal to a percent, move the decimal point two places to the right, inserting zeros as placeholders if necessary, and attach a % sign.

EXAMPLE 7 Converting Fractions to Percents

Convert each fraction to a percent.

(a) $\frac{3}{5}$ **(b)** $\frac{14}{25}$

SOLUTION

(a) First write $\frac{3}{5}$ as a decimal. Dividing 5 into 3 gives $\frac{3}{5} = .6 = 60\%$.

The 2004 movie *Mean Girls* stars Lindsay Lohan as Cady Heron, who has been home-schooled until her senior year in high school. A scene in the school cafeteria features her sitting with The Plastics (the "mean girls" of the title). Regina George, played by Rachel McAdams, is reading a candy bar wrapper.

REGINA: 120 calories and 48 calories from fat. What **percent** is that? I'm only eating food with less than 30% calories from fat.
CADY: It's 40%. (Responding to a quizzical look from Regina.) Well, 48 over 120 equals *x* over 100, and then you cross-multiply and get the value of *x*.
REGINA: Whatever. I'm getting cheese fries.

(b) $\frac{14}{25} = .56 = 56\%$

The procedure of Example 7 is summarized as follows.

Converting a Fraction to a Percent

To convert a fraction to a percent, convert the fraction to a decimal, and then convert the decimal to a percent.

In the following examples involving percents, three methods are shown. The second method in each case involves using cross-products. The third method involves the percent key of a basic calculator.

EXAMPLE 8 Finding a Percent of a Number

Find 18% of 250.

SOLUTION

Method 1 The key word here is "of." The word "of" translates as "times," with 18% of 250 given by

$$(18\%)(250) = (.18)(250) = 45.$$

Method 2 Think "18 is to 100 as what (x) is to 250?" This translates into the equation

$$\frac{18}{100} = \frac{x}{250}$$

$$100x = 18 \cdot 250 \qquad \frac{a}{b} = \frac{c}{d} \text{ if and only if } ad = bc$$

$$x = \frac{18 \cdot 250}{100} \qquad \text{Divide by 100.}$$

$$x = 45. \qquad \text{Simplify.}$$

Method 3 Use the percent key on a calculator with the following keystrokes:

[2] [5] [0] [×] [1] [8] [%]   **45** .

With any of these methods, we find that 18% of 250 is 45.

EXAMPLE 9 Finding What Percent One Number Is of Another

What percent of 500 is 75?

SOLUTION

Method 1 Let the phrase "what percent" be represented by $x \cdot 1\%$ or $.01x$. Again the word "of" translates as "times," while "is" translates as "equals." Thus,

$$.01x \cdot 500 = 75$$

$$5x = 75 \quad \text{Multiply on the left side.}$$

$$x = 15. \quad \text{Divide by 5.}$$

Method 2 Think "What (x) is to 100 as 75 is to 500?" This translates as

$$\frac{x}{100} = \frac{75}{500}$$

$$500x = 7500 \quad \text{Cross-products}$$

$$x = 15. \quad \text{Divide by 500.}$$

Method 3 [7] [5] [÷] [5] [0] [0] [%] **15**

In each case, 15 is the percent, so we conclude that 75 is 15% of 500.

EXAMPLE 10 Finding a Number of Which a Given Number Is a Given Percent

38 is 5% of what number?

SOLUTION

Method 1

$$38 = .05x$$

$$x = \frac{38}{.05} \quad \text{Divide by .05.}$$

$$x = 760 \quad \text{Simplify.}$$

Method 2 Think "38 is to what number (x) as 5 is to 100?"

$$\frac{38}{x} = \frac{5}{100}$$

$$5x = 3800 \quad \text{Cross-products}$$

$$x = 760 \quad \text{Divide by 5.}$$

Method 3 [3] [8] [÷] [5] [%] **760**

Each method shows us that 38 is 5% of 760.

There are various shortcuts that can be used to work with percents. Suppose that you need to compute 20% of 50. Here are two such shortcuts.

1. You think "20% means $\frac{1}{5}$, and to find $\frac{1}{5}$ of something I divide by 5, so 50 divided by 5 is 10. The answer is 10."
2. You think "20% is twice 10%, and to find 10% of something I move the decimal point one place to the left. So, 10% of 50 is 5, and 20% is twice 5, or 10. The answer is 10."

Applications

PROBLEM-SOLVING HINT When applying percent it is often a good idea to restate the problem as a question similar to those found in Examples 8–10, and then answer that question. One strategy of problem solving deals with solving a simpler, similar problem.

EXAMPLE 11 Interpreting Percents from a Graph

In 2003, people in the United States spent an estimated \$29.7 billion on their pets. Use the graph in Figure 18 to determine how much of this amount was spent on pet food.

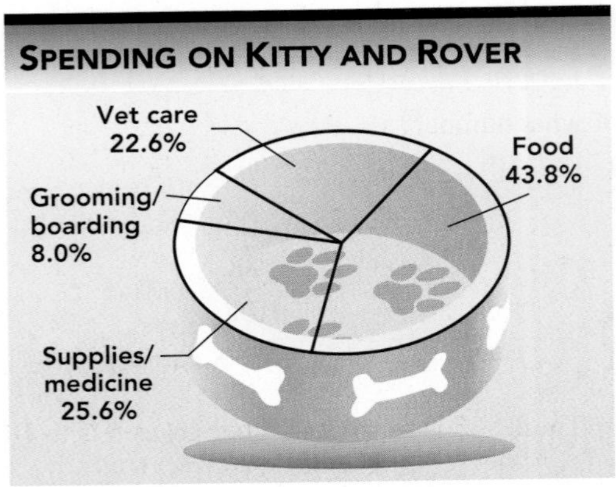

Source: American Pet Products Manufacturers Association Inc.

Dotty

FIGURE 18

SOLUTION

According to the graph, 43.8% was spent on food. We use Method 1 of Example 8 to find 43.8% of \$29.7 billion.

$$\underbrace{.438}_{43.8\%} \underbrace{\times}_{\text{of}} \underbrace{\$29.7 \text{ billion}}_{\text{Total amount}} = \underbrace{\$13.0 \text{ billion}}_{\text{Amount spent on pet food}}$$

In many applications we are asked to find the percent increase or percent decrease from one quantity to another. The following guidelines summarize how to do this.

Finding Percent Increase or Decrease

1. To find the **percent increase from a to b,** where $b > a$, subtract a from b, and divide this result by a. Convert to a percent.

Example: The percent increase from 4 to 7 is $\frac{7-4}{4} = \frac{3}{4} = 75\%$.

2. To find the **percent decrease from a to b,** where $b < a$, subtract b from a, and divide this result by a. Convert to a percent.

Example: The percent decrease from 8 to 6 is $\frac{8-6}{8} = \frac{2}{8} = \frac{1}{4} = 25\%$.

EXAMPLE 12 Finding Percent Increase of Las Vegas Population

Las Vegas, Nevada, is the fastest-growing city in the United States. In 1990, the population of Las Vegas was 258,295. By 2000, it had grown to 478,434. (*Source:* U.S. Census Bureau.)

(a) Estimate the percent increase over this period.
(b) Find the actual percent increase, to the nearest tenth of a percent, over this period.

SOLUTION

(a) For easy computation, think of the 2000 population as 470,000 and the 1990 population as 270,000. The difference is

$$470{,}000 - 270{,}000 = 200{,}000.$$

Now, think of the 1990 population as 250,000 and answer the question "What percent of 250,000 (the *original* population) is 200,000?" The fraction $\frac{200{,}000}{250{,}000}$ simplifies to $\frac{20}{25} = \frac{4}{5}$, which is 80%. Therefore, the population increased by about 80% over the ten-year period.

(b) We must find the difference between the two populations, and then determine what percent of 258,295 this difference comprises.

$$\underbrace{478{,}434}_{\text{Population in 2000}} - \underbrace{258{,}295}_{\text{Population in 1990}} = \underbrace{220{,}139}_{\text{Increase in population}}$$

Now solve the problem "What percent of 258,295 is 220,139?" This is similar to the problem in Example 9. Any of the methods explained there will indicate that the answer is approximately 85.2%. ■

For Further Thought

It's Time to End Decimal Point Abuse

Using a decimal point erroneously with a ¢ symbol is seen almost on a daily basis. Think about it. . . \$.99 represents $\frac{99}{100}$ of a dollar, or 99 cents, while 99¢ also represents 99 cents (since ¢ is the symbol for *cent*). So what does .99¢ represent? That's right, $\frac{99}{100}$ of one cent!

Look at the photos provided by one of the authors. A 20-oz single of FlavorSplash is advertised for .99¢. What do you think would happen if you gave the clerk a dime and asked for ten bottles and change? You would most likely get a dumbfounded look. A similar response would probably be forthcoming if you asked for Sierra Mist, which costs even less: .79¢. To vacuum your car, it costs .50¢, a mere half cent. At The Floor Place, fabulous floors really do cost less. . . a lot less: less than half a cent per square foot for Berber flooring. Now here's a deal: a 2 liter bottle of Coca Cola for .09¢! (No doubt, the 1 preceding the decimal point fell off. Even then, one such bottle would cost only a tiny bit more than one penny.) On the Cherokee Turnpike, you are expected to provide exact change of .25¢. Could you possibly do it?

For Group Discussion or Individual Investigation

Assume that the products shown in the photos are actually being sold for the indicated prices. Answer each of the following.

1. How many 20-oz singles of FlavorSplash should you get for \$1.00? How much change would the store owe you?
2. How much does one ounce of Sierra Mist cost?
3. If you deposit two quarters to have your car vacuumed, how many times should you be able to vacuum?
4. You want to cover your room area with 400 square feet of Berber flooring. How much will this cost?
5. How many 2 liter bottles of Coca Cola would nine cents get you?
6. How many trips on the Cherokee Turnpike should you get for a quarter?

6.5 EXERCISES

Decide whether each statement is true or false.

1. 300% of 12 is 36.

2. 25% of a quantity is the same as $\frac{1}{4}$ of that quantity.

3. When 759.367 is rounded to the nearest hundredth, the result is 759.40.

4. When 759.367 is rounded to the nearest hundred, the result is 759.37.

5. To find 50% of a quantity, we may simply divide the quantity by 2.

6. A soccer team that has won 12 games and lost 8 games has a winning percentage of 60%.

7. If 70% is the lowest passing grade on a quiz that has 50 items of equal value, then answering at least 35 items correctly will assure you of a passing grade.

8. 30 is more than 40% of 120.

9. .99¢ = 99 cents

10. If an item usually costs \$70.00 and it is discounted 10%, then the discount price is \$7.00.

Calculate each of the following using either a calculator or paper-and-pencil methods, as directed by your instructor.

11. $8.53 + 2.785$

12. $9.358 + 7.2137$

13. $8.74 - 12.955$

14. $2.41 - 3.997$

15. $25.7 \times .032$

16. 45.1×8.344

17. $1019.825 \div 21.47$

18. $-262.563 \div 125.03$

19. $\dfrac{118.5}{1.45 + 2.3}$

20. $2.45(1.2 + 3.4 - 5.6)$

Change in Population The table shows the percent change in population from 1990 through 2000 for some large cities in the United States.

City	Percent Change
New York	8.8
Los Angeles	9.7
Cleveland	2.2
Pittsburgh	−1.5
Baltimore	7.2
Buffalo	−1.6

Source: U.S. Census Bureau.

21. Which city had the greatest percent change? What was this change? Was it an increase or a decrease?

22. Which city had the least percent change? What was this change? Was it an increase or a decrease?

Postage Stamp Pricing *Refer to* For Further Thought *on decimal point abuse. At one time, the United States Postal Service sold rolls of 33-cent stamps that featured fruit berries. One such stamp is shown on the left. On the right is a photo of the pricing information found on the cellophane wrapper of such a roll.*

23. Look at the second line of the pricing information. According to the price listed *per stamp,* how many stamps should you be able to purchase for one cent?

24. The total price listed is the amount the Postal Service actually charges. If you were to multiply the listed price *per stamp* by the number of stamps, what should the total price be?

Pricing of Pie and Coffee *The photos here were taken at a flea market near Natchez, MS. The handwritten signs indicate that a piece of pie costs .10¢ and a cup of coffee ("ffee") costs .5¢. Assuming these are the actual prices, answer the questions in Exercises 25–28.*

25. How much will 10 pieces of pie and 10 cups of coffee cost?

26. How much will 20 pieces of pie and 10 cups of coffee cost?

27. How many pieces of pie can you get for \$1.00?

28. How many cups of coffee can you get for \$1.00?

Exercises 29–32 are based on formulas found in Auto Math Handbook: Mathematical Calculations, Theory, and Formulas for Automotive Enthusiasts, *by John Lawlor (1991, HP Books).*

29. *Blood Alcohol Concentration* The Blood Alcohol Concentration (BAC) of a person who has been drinking is given by the formula

$$\text{BAC} = \frac{(\text{ounces} \times \text{percent alcohol} \times .075)}{\text{body weight in lb}} - (\text{hours of drinking} \times .015).$$

Suppose a policeman stops a 190-pound man who, in two hours, has ingested four 12-ounce beers, each having a 3.2 percent alcohol content. The formula would then read

$$\text{BAC} = \frac{[(4 \times 12) \times 3.2 \times .075]}{190} - (2 \times .015).$$

(a) Find this BAC.
(b) Find the BAC for a 135-pound woman who, in three hours, has drunk three 12-ounce beers, each having a 4.0 percent alcohol content.

30. *Approximate Automobile Speed* The approximate speed of an automobile in miles per hour (MPH) can be found in terms of the engine's revolutions per minute (rpm), the tire diameter in inches, and the overall gear ratio by the formula

$$\text{MPH} = \frac{\text{rpm} \times \text{tire diameter}}{\text{gear ratio} \times 336}.$$

If a certain automobile has an rpm of 5600, a tire diameter of 26 inches, and a gear ratio of 3.12, what is its approximate speed (MPH)?

31. *Engine Horsepower* Horsepower can be found from indicated mean effective pressure (mep) in pounds per square inch, engine displacement in cubic inches, and revolutions per minute (rpm) using the formula

$$\text{Horsepower} = \frac{\text{mep} \times \text{displacement} \times \text{rpm}}{792{,}000}.$$

Suppose that an engine has displacement of 302 cubic inches, and indicated mep of 195 pounds per square inch at 4000 rpm. What is its approximate horsepower?

32. *Torque Approximation* To determine the torque at a given value of rpm, the formula below applies:

$$\text{Torque} = \frac{5252 \times \text{horsepower}}{\text{rpm}}.$$

If the horsepower of a certain vehicle is 400 at 4500 rpm, what is the approximate torque?

Round each number to the nearest **(a)** *tenth;* **(b)** *hundredth. Always round from the original number.*

33. 78.414

34. 3689.537

35. .0837

36. .0658

37. 12.68925

38. 43.99613

Convert each decimal to a percent.

39. .42

40. .87

41. .365

42. .792

43. .008

44. .0093

45. 2.1

46. 8.9

Convert each fraction to a percent.

47. $\frac{1}{5}$

48. $\frac{2}{5}$

49. $\frac{1}{100}$

50. $\frac{1}{50}$

51. $\frac{3}{8}$

52. $\frac{5}{6}$

53. $\frac{3}{2}$

54. $\frac{7}{4}$

55. Explain the difference between $\frac{1}{2}$ of a quantity and $\frac{1}{2}\%$ of the quantity.

56. On the next page Group I shows some common percents, found in many everyday situations. In Group II are fractional equivalents of these percents. Match the fractions in Group II with their equivalent percents in Group I.

I		II	
(a) 25%	(b) 10%	A. $\frac{1}{3}$	B. $\frac{1}{50}$
(c) 2%	(d) 20%	C. $\frac{3}{4}$	D. $\frac{1}{10}$
(e) 75%	(f) $33\frac{1}{3}\%$	E. $\frac{1}{4}$	F. $\frac{1}{5}$

57. Fill in each blank with the appropriate numerical response.
(a) 5% means _____ in every 100.
(b) 25% means 6 in every_____.
(c) 200% means _____ for every 4.
(d) .5% means _____ in every 100.
(e) _____% means 12 for every 2.

58. The Venn diagram shows the number of elements in the four regions formed.

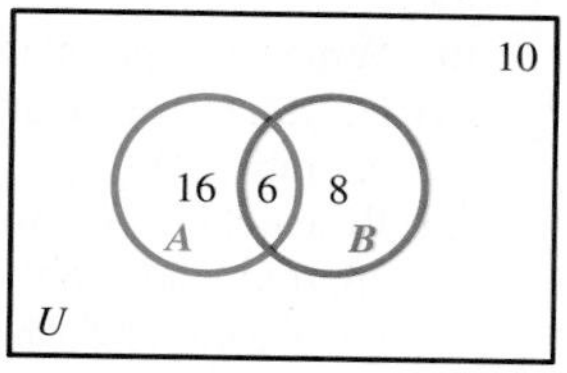

(a) What percent of the elements in the universe are in $A \cap B$?
(b) What percent of the elements in the universe are in A but not in B?
(c) What percent of the elements in $A \cup B$ are in $A \cap B$?
(d) What percent of the elements in the universe are in neither A nor B?

59. ***Discount and Markup*** Suppose that an item regularly costs \$60.00 and it is discounted 20%. If it is then marked up 20%, is the resulting price \$60.00? If not, what is it?

60. The figures in Exercise 13 of Section 6.3 are reproduced here. Express the fractional parts represented by the shaded areas as percents.

(a)
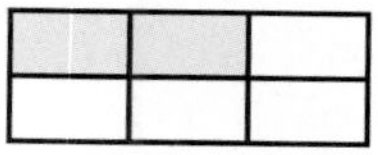

(b)
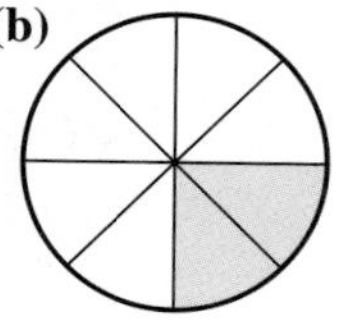

(c)
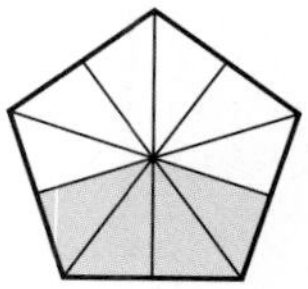

(d)
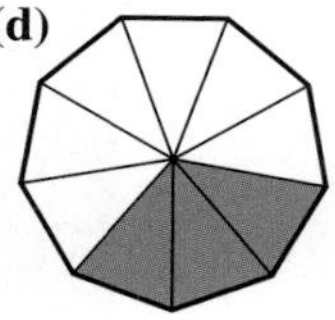

Win-Loss Record *Exercises 61 and 62 deal with winning percentage in the standings of sports teams.*

61. At the end of the regular 2005 Major League Baseball season, the standings of the Central Division of the American League were as shown. Winning percentage is commonly expressed as a decimal rounded to the nearest thousandth. To find the winning percentage of a team, divide the number of wins (W) by the total number of games played (W + L). Find the winning percentage of each team.
(a) Chicago **(b)** Cleveland **(c)** Detroit

Team	W	L
Chicago	99	63
Cleveland	93	69
Minnesota	83	79
Detroit	71	91
Kansas City	56	106

62. Repeat Exercise 61 for the following standings for the East Division of the National League.
(a) Atlanta **(b)** Philadelphia
(c) Florida and New York

Team	W	L
Atlanta	90	72
Philadelphia	88	74
Florida	83	79
New York	83	79
Washington	81	81

Work each problem involving percent.

63. What is 26% of 480?

64. Find 38% of 12.

65. Find 10.5% of 28.

66. What is 48.6% of 19?

67. What percent of 30 is 45?

68. What percent of 48 is 20?

69. 25% of what number is 150?

70. 12% of what number is 3600?

71. .392 is what percent of 28?

72. 78.84 is what percent of 292?

Use mental techniques to answer the questions in Exercises 73–76. Try to avoid using paper and pencil or a calculator.

73. ***Allowance Increase*** Dierdre Lynch's allowance was raised from \$4.00 per week to \$5.00 per week. What was the percent of the increase?
A. 25% **B.** 20% **C.** 50% **D.** 30%

74. ***Boat Purchase and Sale*** Jane Gunton bought a boat five years ago for \$5000 and sold it this year for \$2000. What percent of her original purchase did she lose on the sale?
A. 40% **B.** 50% **C.** 20% **D.** 60%

75. ***Population of Alabama*** The 2000 U.S. census showed that the population of Alabama was 4,447,000, with 26.0% represented by African Americans. What is the best estimate of the African American population in Alabama? (*Source:* U.S. Census Bureau.)
A. 500,000 **B.** 1,500,000
C. 1,100,000 **D.** 750,000

76. ***Population of Hawaii*** The 2000 U.S. census showed that the population of Hawaii was 1,212,000, with 21.4% of the population being of two or more races. What is the best estimate of this population of Hawaii? (*Source:* U.S. Census Bureau.)
A. 240,000 **B.** 300,000
C. 21,400 **D.** 24,000

Gasoline Prices *The line graph shows the average price, adjusted for inflation, that Americans have paid for a gallon of gasoline for selected years between 1970 and 2000. Use this information in Exercises 77 and 78.*

AVERAGE GASOLINE PRICES

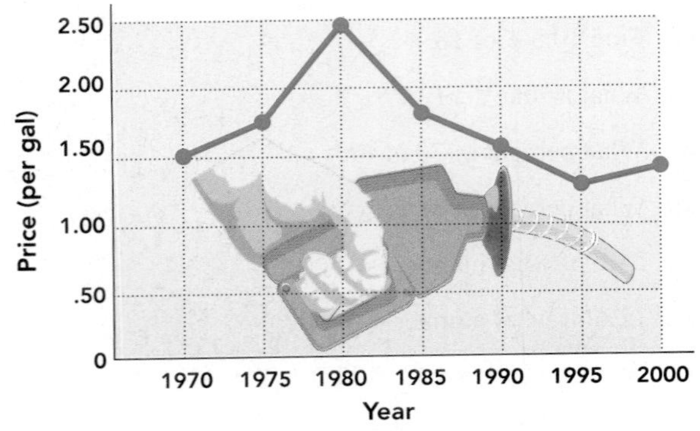

Source: American Petroleum Institute; AP research.

77. By what percent did prices increase from 1970 to 1980?

78. By what percent did prices decrease from 1980 to 1990?

Metabolic Units *One way to measure a person's cardio fitness is to calculate how many METs, or metabolic units, he or she can reach at peak exertion. One MET is the amount of energy used when sitting quietly. To calculate ideal METs, we can use one of the following expressions.*

$$14.7 - \text{age} \cdot .13 \quad \text{For women}$$

$$14.7 - \text{age} \cdot .11 \quad \text{For men}$$

(*Source: New England Journal of Medicine,* August, 2005.)

79. A 40-year-old woman wishes to calculate her ideal MET.
(a) Write the expression using her age.
(b) Calculate her ideal MET. (*Hint:* Use the order of operations.)
(c) Researchers recommend that a person reach approximately 85% of their MET when exercising. Calculate 85% of the ideal MET from part (b). Then refer to the following table. What activity can the woman do that is approximately this value?

Activity	METs	Activity	METs
Golf (with cart)	2.5	Skiing (water or downhill)	6.8
Walking (3 mph)	3.3	Swimming	7.0
Mowing lawn (power mower)	4.5	Walking (5 mph)	8.0
Ballroom or square dancing	5.5	Jogging	10.2
Cycling	5.7	Rope skipping	12.0

Source: Harvard School of Public Health.

80. Repeat parts **(a)–(c)** of Exercise 79 for a 55-year-old man.

81. ***Value of 1916-D Mercury Dime*** The 1916 Mercury dime minted in Denver is quite rare. In 1979 its value in Extremely Fine condition was \$625. The 2005 value had increased to \$6000. What was the percent increase in the value of this coin from 1979 to 2005? (*Sources: A Guide Book of United States Coins; Coin World Coin Values.*)

82. ***Value of 1903-O Morgan Dollar*** In 1963, the value of a 1903 Morgan dollar minted in New Orleans in typical Uncirculated condition was \$1500. Due to a discovery of a large hoard of these dollars late that year, the value plummeted. Its value in 2005 was \$550. What was the percent decrease in its value from 1963 to 2005? (*Sources: A Guide Book of United States Coins; Coin World Coin Values.*)

Tipping Procedure *It is customary in our society to "tip" waiters and waitresses when dining in restaurants. One common rate for tipping is 15%. A quick way of figuring a tip that will give a close approximation of 15% is as follows:*

1 *Round off the bill to the nearest dollar.*

2 *Find 10% of this amount by moving the decimal point one place to the left.*

3 *Take half of the amount obtained in Step 2 and add it to the result of Step 2.*

This will give you approximately 15% of the bill. The amount obtained in Step 3 is 5%, and

$$10\% + 5\% = 15\%.$$

Use the method above to find an approximation of 15% for each restaurant bill.

83. \$29.57 **84.** \$38.32

85. \$5.15 **86.** \$7.89

Suppose that you get extremely good service and decide to tip 20%. You can use the first two steps listed, and then in Step 3, double the amount you obtained in Step 2. Use this method to find an approximation of 20% for each restaurant bill.

87. \$59.96 **88.** \$40.24

89. \$180.43 **90.** \$199.86

91. A television reporter once asked a professional wrist-wrestler what percent of his sport was physical and what percent was mental. The athlete responded "I would say it's 50% physical and 90% mental." Comment on this response.

92. According to *The Yogi Book,* consisting of quotes by baseball Hall-of-Famer Yogi Berra, he claims that "90% of the game is half mental." Comment on this statement.

EXTENSION

Complex Numbers

Numbers such as $\sqrt{-5}$ and $\sqrt{-16}$ were called *imaginary* by the early mathematicians who would not permit these numbers to be used as solutions to problems. Gradually, however, applications were found that required the use of these numbers, making it necessary to expand the set of real numbers to form the set of **complex numbers.**

Consider the equation $x^2 + 1 = 0$. It has no real number solution, since any solution must be a number whose square is -1. In the set of real numbers all squares are nonnegative numbers, because the product of either two positive numbers or two negative numbers is positive. To provide a solution for the equation $x^2 + 1 = 0$, a new number i is defined so that

$$\mathbf{i^2 = -1.}$$

(continued)

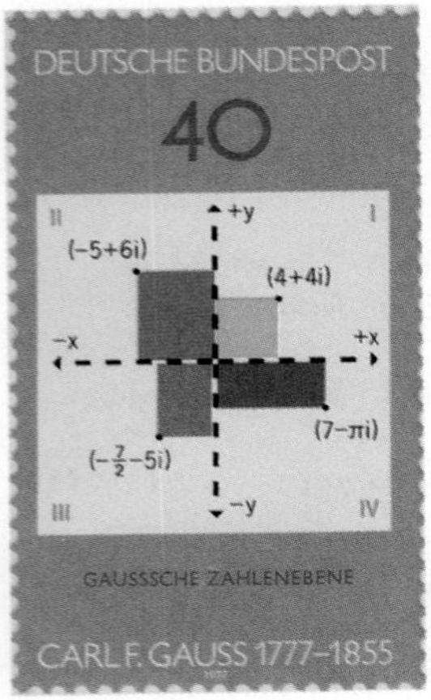

Gauss and the Complex Numbers The stamp shown above honors the many contributions made by Gauss to our understanding of complex numbers. In about 1831 he was able to show that numbers of the form $a + bi$ can be represented as points on the plane (as the stamp shows) just as real numbers are. He shared this contribution with **Robert Argand,** a bookkeeper in Paris, who wrote an essay on the geometry of the complex numbers in 1806. This went unnoticed at the time.

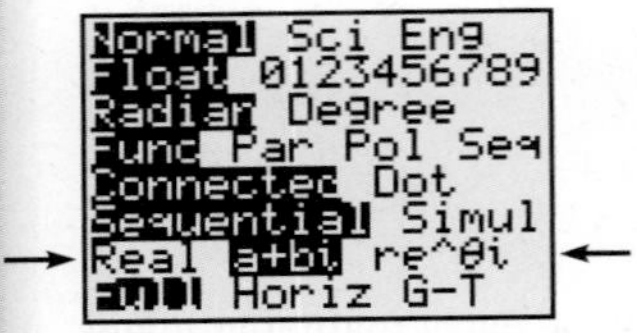

When the TI-83/84 Plus calculator is in complex mode, denoted by $a + bi$, it will perform complex number arithmetic.

That is, i is a number whose square is -1. This definition of i makes it possible to define the square root of any negative number as follows.

$\sqrt{-b}$

For any positive real number b, $\quad \boldsymbol{\sqrt{-b} = i\sqrt{b}.}$

EXAMPLE 1 Writing Square Roots Using *i*

Write each number as a product of a real number and i.

(a) $\sqrt{-100}$ **(b)** $\sqrt{-2}$

SOLUTION

(a) $\sqrt{-100} = i\sqrt{100} = 10i$

(b) $\sqrt{-2} = \sqrt{2}i = i\sqrt{2}$

It is easy to mistake $\sqrt{2}i$ for $\sqrt{2i}$, with the i under the radical. For this reason, it is common to write $\sqrt{2}i$ as $i\sqrt{2}$.

When finding a product such as $\sqrt{-4} \cdot \sqrt{-9}$, the product rule for radicals cannot be used, since that rule applies only when both radicals represent real numbers. For this reason, always change $\sqrt{-b}$ $(b > 0)$ to the form $i\sqrt{b}$ before performing any multiplications or divisions. For example,

$$\sqrt{-4} \cdot \sqrt{-9} = i\sqrt{4} \cdot i\sqrt{9} = i \cdot 2 \cdot i \cdot 3 = 6i^2.$$

Since $i^2 = -1$,

$$6i^2 = 6(-1) = -6.$$

An ***incorrect*** use of the product rule for radicals would give a wrong answer.

$$\sqrt{-4} \cdot \sqrt{-9} = \sqrt{(-4)(-9)} = \sqrt{36} = 6 \quad \text{Incorrect}$$

EXAMPLE 2 Multiplying Expressions Involving *i*

Multiply.

(a) $\sqrt{-3} \cdot \sqrt{-7}$ **(b)** $\sqrt{-2} \cdot \sqrt{-8}$ **(c)** $\sqrt{-5} \cdot \sqrt{6}$

SOLUTION

(a) $\sqrt{-3} \cdot \sqrt{-7} = i\sqrt{3} \cdot i\sqrt{7} = i^2\sqrt{3 \cdot 7} = (-1)\sqrt{21} = -\sqrt{21}$

(b) $\sqrt{-2} \cdot \sqrt{-8} = i\sqrt{2} \cdot i\sqrt{8} = i^2\sqrt{2 \cdot 8} = (-1)\sqrt{16} = (-1)4 = -4$

(c) $\sqrt{-5} \cdot \sqrt{6} = i\sqrt{5} \cdot \sqrt{6} = i\sqrt{30}$

EXAMPLE 3 Dividing Expressions Involving *i*

Divide.

(a) $\dfrac{\sqrt{-75}}{\sqrt{-3}}$ **(b)** $\dfrac{\sqrt{-32}}{\sqrt{8}}$

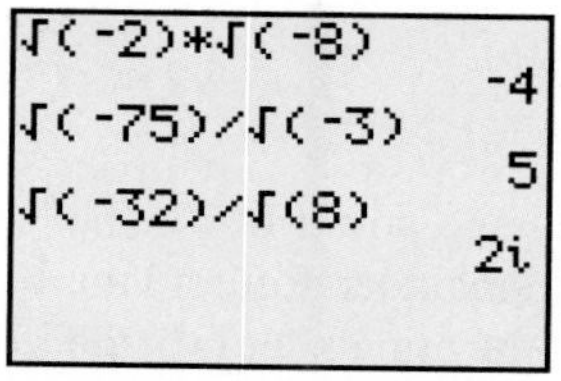

This screen supports the results of Examples 2(b), 3(a), and 3(b).

SOLUTION

(a) $\dfrac{\sqrt{-75}}{\sqrt{-3}} = \dfrac{i\sqrt{75}}{i\sqrt{3}} = \sqrt{\dfrac{75}{3}} = \sqrt{25} = 5$

(b) $\dfrac{\sqrt{-32}}{\sqrt{8}} = \dfrac{i\sqrt{32}}{\sqrt{8}} = i\sqrt{\dfrac{32}{8}} = i\sqrt{4} = 2i$

Complex numbers are defined as follows.

Complex Numbers

If a and b are real numbers, then any number of the form $\boldsymbol{a + bi}$ is called a **complex number.**

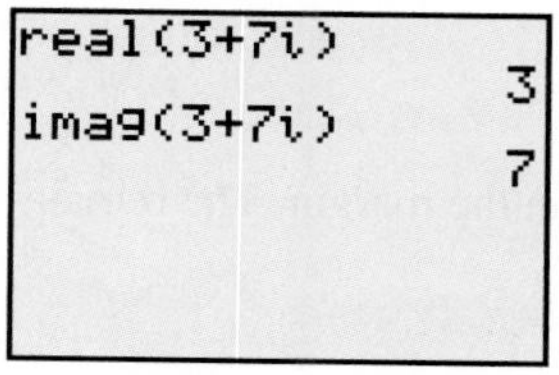

The TI-83/84 Plus calculator identifies the real and imaginary parts of $3 + 7i$.

In the complex number $a + bi$, the number a is called the **real part** and b is called the **imaginary part.*** When $b = 0$, $a + bi$ is a real number, so the real numbers are a subset of the complex numbers. Complex numbers of the form bi, where $b \neq 0$, are called **pure imaginary numbers.** In spite of their name, such numbers are very useful in applications, particularly in work with electricity.

An interesting pattern emerges when we consider various powers of i. By definition, $i^0 = 1$, and $i^1 = i$. We have seen that $i^2 = -1$, and greater powers of i can be found as shown in the following list.

$$i^3 = i \cdot i^2 = i(-1) = -i \qquad i^6 = i^2 \cdot i^4 = (-1) \cdot 1 = -1$$
$$i^4 = i^2 \cdot i^2 = (-1)(-1) = 1 \qquad i^7 = i^3 \cdot i^4 = (-i) \cdot 1 = -i$$
$$i^5 = i \cdot i^4 = i \cdot 1 = i \qquad i^8 = i^4 \cdot i^4 = 1 \cdot 1 = 1$$

A few powers of i are listed here.

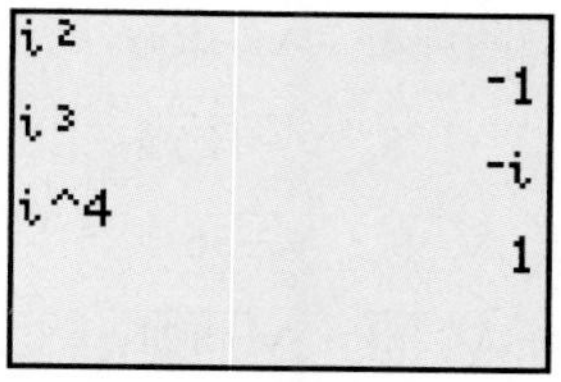

The calculator computes powers of i. Compare to the powers in the chart.

Powers of *i*

$i^1 = i$	$i^5 = i$	$i^9 = i$	$i^{13} = i$
$i^2 = -1$	$i^6 = -1$	$i^{10} = -1$	$i^{14} = -1$
$i^3 = -i$	$i^7 = -i$	$i^{11} = -i$	$i^{15} = -i$
$i^4 = 1$	$i^8 = 1$	$i^{12} = 1$	$i^{16} = 1$

As these examples suggest, the powers of i rotate through the four numbers i, -1, $-i$, and 1. Larger powers of i can be simplified by using the fact that $i^4 = 1$. For example,

$$i^{75} = (i^4)^{18} \cdot i^3 = 1^{18} \cdot i^3 = 1 \cdot i^3 = -i.$$

*In some texts, bi is called the imaginary part.

(continued)

Simplifying Large Powers of *i*

Step 1 Divide the exponent by 4.

Step 2 Observe the remainder obtained in Step 1. The large power of i is the same as i raised to the power determined by this remainder. Refer to the previous chart to complete the simplification. (If the remainder is 0, the power simplifies to $i^0 = 1$.)

EXAMPLE 4 Simplifying Powers of *i*

Simplify each power of i.

(a) i^{12} **(b)** i^{39}

SOLUTION

(a) $i^{12} = (i^4)^3 = 1^3 = 1$

(b) To find i^{39}, start by dividing 39 by 4 (Step 1), as shown in the margin. The remainder is 3. So $i^{39} = i^3 = -i$ (Step 2).

$$\begin{array}{r} 9 \\ 4\overline{)39} \\ \underline{36} \\ 3 \end{array} \leftarrow \text{Remainder}$$

Another way to simplify i^{39} is as follows.

$$i^{39} = i^{36} \cdot i^3 = (i^4)^9 \cdot i^3 = 1^9 \cdot (-i) = -i$$

EXTENSION EXERCISES

Use the method of Examples 1–3 to write each expression as a real number or a product of a real number and i.

1. $\sqrt{-144}$ **2.** $\sqrt{-196}$ **3.** $-\sqrt{-225}$ **4.** $-\sqrt{-400}$

5. $\sqrt{-3}$ **6.** $\sqrt{-19}$ **7.** $\sqrt{-75}$ **8.** $\sqrt{-125}$

9. $\sqrt{-5} \cdot \sqrt{-5}$ **10.** $\sqrt{-3} \cdot \sqrt{-3}$ **11.** $\sqrt{-9} \cdot \sqrt{-36}$

12. $\sqrt{-4} \cdot \sqrt{-81}$ **13.** $\sqrt{-16} \cdot \sqrt{-100}$ **14.** $\sqrt{-81} \cdot \sqrt{-121}$

15. $\dfrac{\sqrt{-200}}{\sqrt{-100}}$ **16.** $\dfrac{\sqrt{-50}}{\sqrt{-2}}$ **17.** $\dfrac{\sqrt{-54}}{\sqrt{6}}$

18. $\dfrac{\sqrt{-90}}{\sqrt{10}}$ **19.** $\dfrac{\sqrt{-288}}{\sqrt{-8}}$ **20.** $\dfrac{\sqrt{-48} \cdot \sqrt{-3}}{\sqrt{-2}}$

21. Why is it incorrect to use the product rule for radicals to multiply $\sqrt{-3} \cdot \sqrt{-12}$?

22. In your own words describe the relationship between complex numbers and real numbers.

Use the method of Example 4 to simplify each power of i.

23. i^8 **24.** i^{16} **25.** i^{42} **26.** i^{86}

27. i^{47} **28.** i^{63} **29.** i^{101} **30.** i^{141}

COLLABORATIVE INVESTIGATION

Budgeting to Buy a Car

You are shopping for a sports car and have put aside a certain amount of money each month for a car payment. Your instructor will assign this amount to you. After looking through a variety of resources, you have narrowed your choices to the cars listed in the table.

Year/Make/Model	Retail Price	Fuel Tank Size (in gallons)	Miles per Gallon (city)	Miles per Gallon (highway)
2006 Ford Mustang	\$26,320	16.0	18	23
2006 Ford Five Hundred	\$22,230	20.0	20	27
2006 Toyota Camry	\$25,805	18.5	20	28
2006 Mazda MX-5 Miata	\$26,700	12.7	23	30
2006 Honda CR-V XLE	\$25,450	15.3	22	27
2006 Chevrolet Tracker	\$23,900	17.0	19	25

Source: www.edmunds.com

As a group, work through the following steps to determine which car you can afford to buy.

A. Decide which cars you think are within your budget.

B. Select one of the cars you identified in part A. Have each member of the group calculate the monthly payment for this car using a different financing option. Use the formula given below, where P is principal, r is interest rate, and m is the number of monthly payments, along with the financing options table.

Financing Options

Time (in years)	Interest Rate
4	7.0%
5	8.5%
6	10.0%

$$\text{Monthly Payment} = \frac{\dfrac{Pr}{12}}{1 - \left(\dfrac{12}{12 + r}\right)^{m}}$$

C. Have each group member determine the amount of money paid in interest over the duration of the loan for his or her financing option.

D. Consider fuel expenses.

1. Assume you will travel an average of 75 miles in the city and 400 miles on the highway each week. How many gallons of gas will you need to buy each month?
2. Using typical prices for gas in your area at this time, how much money will you need to have available for buying gas?

E. Repeat parts B–D as necessary until your group can reach a consensus on the car you will buy and the financing option you will use. Write a paragraph to explain your choices.

CHAPTER 6 TEST

1. Consider $\{-4, -\sqrt{5}, -\frac{3}{2}, -.5, 0, \sqrt{3}, 4.1, 12\}$. List the elements of the set that belong to each of the following.
   (a) natural numbers
   (b) whole numbers
   (c) integers
   (d) rational numbers
   (e) irrational numbers
   (f) real numbers

2. Match each set in (a)–(d) with the correct set-builder notation description in A–D.
   (a) $\{\ldots, -4, -3, -2, -1\}$
   (b) $\{3, 4, 5, 6, \ldots\}$
   (c) $\{1, 2, 3, 4, \ldots\}$
   (d) $\{-12, \ldots, -2, -1, 0, 1, 2, \ldots, 12\}$

   A. $\{x \mid x$ is an integer with absolute value less than or equal to $12\}$
   B. $\{x \mid x$ is an integer greater than $2.5\}$
   C. $\{x \mid x$ is a negative integer$\}$
   D. $\{x \mid x$ is a positive integer$\}$

3. Decide whether each statement is true or false.
   (a) The absolute value of a number is always positive.
   (b) $|-7| = -(-7)$
   (c) $\frac{2}{5}$ is an example of a real number that is not an integer.
   (d) Every real number is either positive or negative.

Perform the indicated operations. Use the order of operations as necessary.

4. $6^2 - 4(9 - 1)$

5. $(-3)(-2) - [5 + (8 - 10)]$

6. $\dfrac{(-8 + 3) - (5 + 10)}{7 - 9}$

7. ***Changes in Car Sales*** The graph shows the percent change in car sales from January 2000 to January 2001 for various automakers. Use this graph to answer the following. (Consider absolute value.)
   (a) Which automaker had the greatest change in sales? What was that change?
   (b) Which automaker had the least change in sales? What was that change?

CAR SALES

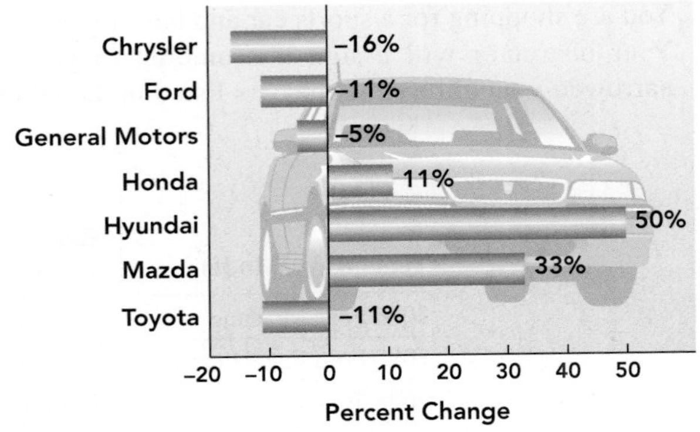

Source: Automakers.

   (c) *True* or *false:* The absolute value of the percent change for Honda was greater than the absolute value of the percent change for Toyota.
   (d) *True* or *false:* The percent change for Hyundai was more than four times greater than the percent change for Honda.

8. ***Altitude of a Plane*** The surface of the Dead Sea has altitude 1299 ft below sea level. Vangie is flying 80 ft above that surface. How much altitude must she gain to clear a 3852 ft pass by 225 ft? (*Source: The World Almanac and Book of Facts.*)

9. ***Median Home Prices*** Median pricings for existing homes in the United States for the years 1997 through 2002 are shown in the table. Complete the table, determining the change from one year to the next by subtraction.

	Year	Median-Priced Existing Homes	Change from Previous Year
	1997	\$121,800	
	1998	\$128,400	\$6600
(a)	1999	\$133,300	
(b)	2000	\$139,000	
(c)	2001	\$147,800	
(d)	2002	\$158,100	

Source: National Association of Realtors.

10. Match each statement in (a)–(f) with the property that justifies it in A–F.

(a) $7 \cdot (8 \cdot 5) = (7 \cdot 8) \cdot 5$
(b) $3x + 3y = 3(x + y)$
(c) $8 \cdot 1 = 1 \cdot 8 = 8$
(d) $7 + (6 + 9) = (6 + 9) + 7$
(e) $9 + (-9) = -9 + 9 = 0$
(f) $5 \cdot 8$ is a real number.

A. Distributive property
B. Identity property
C. Closure property
D. Commutative property
E. Associative property
F. Inverse property

11. ***Basketball Shot Statistics*** Six players on the local high school basketball team had the following shooting statistics.

Player	Field Goal Attempts	Field Goals Made
Ed Moura	40	13
Jack Pritchard	10	4
Chuck Miller	20	8
Ben Whitney	6	4
Charlie Dawkins	7	2
Jason McElwain ("J-Mac")	7	6

Answer each question, using estimation skills as necessary.

(a) Which player made more than half of his attempts?
(b) Which players made just less than $\frac{1}{3}$ of the attempts?
(c) Which player made exactly $\frac{2}{3}$ of his attempts?
(d) Which two players made the same fractional parts of their attempts? What was the fractional part, reduced to lowest terms?
(e) Which player had the greatest fractional part of shots made?

Perform each operation. Write your answer in lowest terms.

12. $\frac{3}{16} + \frac{1}{2}$

13. $\frac{9}{20} - \frac{3}{32}$

14. $\frac{3}{8} \cdot \left(-\frac{16}{15}\right)$

15. $\frac{7}{9} \div \frac{14}{27}$

16. Convert each rational number into a repeating or terminating decimal. Use a calculator if your instructor so allows.

(a) $\frac{9}{20}$ **(b)** $\frac{5}{12}$

17. Convert each decimal into a quotient of integers, reduced to lowest terms.

(a) .72 **(b)** $.\overline{58}$

18. Identify each number as rational or irrational.

(a) $\sqrt{10}$ **(b)** $\sqrt{16}$
(c) .01 **(d)** $.\overline{01}$
(e) .0101101110 . . . **(f)** π

For each of the following, **(a)** *use a calculator to find a decimal approximation and* **(b)** *simplify the radical according to the guidelines in this chapter.*

19. $\sqrt{150}$

20. $\frac{13}{\sqrt{7}}$

21. $2\sqrt{32} - 5\sqrt{128}$

22. A student using her powerful new calculator states that the *exact* value of $\sqrt{65}$ is 8.062257748. Is she correct? If not, explain.

23. Work each of the following using either a calculator or paper-and-pencil methods, as directed by your instructor.

(a) $4.6 + 9.21$ **(b)** $12 - 3.725 - 8.59$
(c) $86(.45)$ **(d)** $236.439 \div (-9.73)$

24. Round 9.0449 to the following place values:

(a) hundredths **(b)** thousandths.

25. **(a)** Find 18.5% of 90.
(b) What number is 145% of 70?

26. Consider the figure.

(a) What percent of the total number of shapes are circles?
(b) What percent of the total number of shapes are not stars?

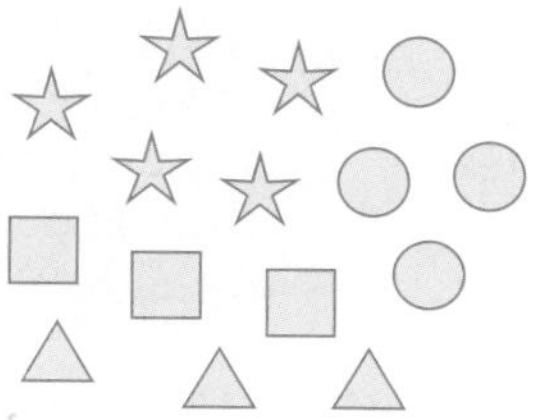

27. ***Sales of Books*** Use estimation techniques to answer the following: In 2005, Bill Schoof sold \$300,000 worth of books. In 2006, he sold \$900,000. His 2006 sales were _____ of his 2005 sales.

A. 30% **B.** $33\frac{1}{3}\%$

C. 200% **D.** 300%

28. ***Creature Comforts*** From a list of "everyday items" often taken for granted, adults were recently surveyed as to those items they wouldn't want to live without. Complete the results shown in the table if 1200 adults were surveyed.

Item	Percent That Wouldn't Want to Live Without	Number That Wouldn't Want to Live Without
Toilet paper	69%	
Zipper	42%	
Frozen Food		190
Self-stick note pads		75

(Other items included tape, hairspray, pantyhose, paper clips, and Velcro.)
Source: Market Facts for Kleenex Cottonelle.

29. ***Composition of U.S. Workforce*** The U.S. Bureau of Labor Statistics projected the composition of the U.S. workforce for the year 2006. The projected total number of people in the workforce for that year is 148,847,000. To the nearest thousand, how many of these will be in the Hispanic category?

COMPOSITION OF THE U.S. WORKFORCE IN 2006

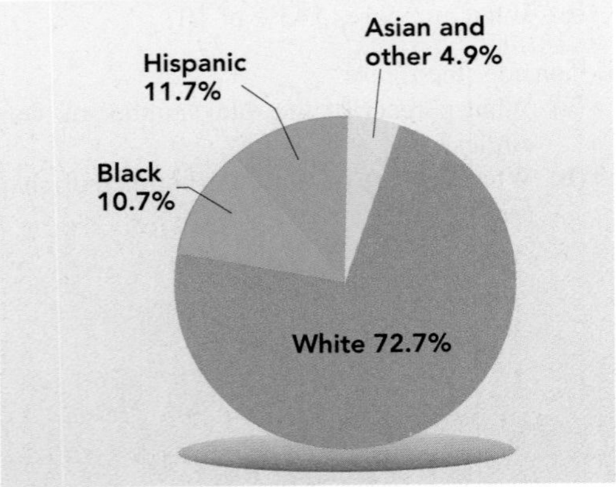

Source: U.S. Bureau of Labor Statistics.

30. ***Medicare Funding*** Current projections indicate that funding for Medicare will not cover its costs unless the program changes. The line graph shows projections for the years 2004 through 2013. What signed number represents how much the funding will have changed from 2004 to 2013?

MEDICARE FUNDS

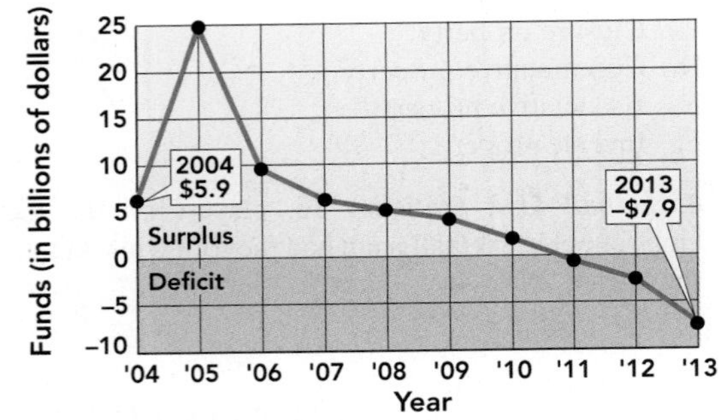

Source: Centers for Medicare and Medicaid Services.

7

THE BASIC CONCEPTS OF ALGEBRA

In the 1994 movie *Little Big League*, young Billy Heywood (Luke Edwards) inherits the Minnesota Twins baseball team and becomes manager. He leads the team to the Division Championship and then to the playoffs. But before the final playoff game, the biggest game of the year, he can't keep his mind on his job, because a homework problem is giving him trouble.

If Joe can paint a house in 3 hours, and Sam can paint the same house in 5 hours, how long does it take for them to do it together?

One of his players provides a method to solve the problem, where a and b are the individual times. He claims that the expression $\frac{a \times b}{a + b}$ gives the correct answer. With $a = 5$ and $b = 3$, the answer he gives is

$$\frac{5 \times 3}{5 + 3} = \frac{15}{8} = 1\frac{7}{8} \text{ hours.}$$

However, the scriptwriters never say whether this is truly mathematically correct. Can you determine whether this answer is indeed the correct one? See the solution on page 414.

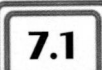

7.1 Linear Equations

Solving Linear Equations • Special Kinds of Linear Equations • Literal Equations and Formulas • Models

al-jabr, algebrista, algebra The word **algebra** comes from the title of the work *Hisâb al-jabr w'al muquâbalah*, a ninth-century treatise by the Arab Mohammed ibn Mûsâ al-Khowârizmî. The title translates as "the science of reunion and reduction," or more generally, "the science of transposition and cancellation."

In the title of Khowârizmî's book, *jabr* ("restoration") refers to transposing negative quantities across the equals symbol in solving equations. From Latin versions of Khowârizmî's text, **"al-jabr"** became the broad term covering the art of equation solving. (The prefix *al* means "the.")

Solving Linear Equations

An **algebraic expression** involves only the basic operations of addition, subtraction, multiplication, or division (except by 0), or raising to powers or taking roots on any collection of variables and numbers.

$$8x + 9, \quad \sqrt{y} + 4, \quad \text{and} \quad \frac{x^3y^8}{z} \qquad \text{Algebraic expressions}$$

An **equation** is a statement that two algebraic expressions are equal. A *linear equation in one variable* involves only real numbers and one variable.

$$x + 1 = -2, \quad y - 3 = 5, \quad \text{and} \quad 2k + 5 = 10 \qquad \text{Linear equations}$$

Linear Equation in One Variable

An equation in the variable x is **linear** if it can be written in the form

$$Ax + B = C,$$

where A, B, and C are real numbers, with $A \neq 0$.

A linear equation in one variable is also called a **first-degree equation,** because the greatest power on the variable is one.

If the variable in an equation is replaced by a real number that makes the statement true, then that number is a **solution** of the equation. For example, 8 is a solution of the equation $x - 3 = 5$, because replacing x with 8 gives a true statement. An equation is **solved** by finding its **solution set,** the set of all solutions. The solution set of the equation $x - 3 = 5$ is $\{8\}$.

Equivalent equations are equations with the same solution set. Equations generally are solved by starting with a given equation and producing a series of simpler equivalent equations. For example,

$$8x + 1 = 17, \quad 8x = 16, \quad \text{and} \quad x = 2 \qquad \text{Equivalent equations}$$

are equivalent equations because each has the same solution set, $\{2\}$. We use the addition and multiplication properties of equality to produce equivalent equations.

Addition Property of Equality

For all real numbers A, B, and C, the equations

$$A = B \quad \text{and} \quad A + C = B + C$$

are equivalent. (The same number may be added to both sides of an equation without changing the solution set.)

algebrista, algebra In Spain under Moslem rule, the word **algebrista** referred to the person who restored (reset) broken bones. Signs outside barber shops read *Algebrista y Sangrador* (bonesetter and bloodletter). Such services were part of the barber's trade. The traditional red-and-white striped barber pole symbolizes blood and bandages.

Multiplication Property of Equality

For all real numbers A, B, and C, where $C \neq 0$, the equations

$$A = B \qquad \text{and} \qquad AC = BC$$

are equivalent. (Both sides of an equation may be multiplied by the same nonzero number without changing the solution set.)

Because subtraction and division are defined in terms of addition and multiplication, respectively, the same number may be subtracted from both sides of an equation, and both sides may be divided by the same nonzero number, without affecting the solution set.

The distributive property allows us to combine *like terms*, such as $4y$ and $2y$. For example,

$$4y - 2y = (4 - 2)y = 2y.$$

EXAMPLE 1 Using the Addition and Multiplication Properties to Solve a Linear Equation

Solve $4x - 2x - 5 = 4 + 6x + 3$.

SOLUTION

The goal is to get x alone on one side of the equation.

$$4x - 2x - 5 = 4 + 6x + 3$$

$$2x - 5 = 7 + 6x \qquad \text{Combine like terms.}$$

Use the addition property to get the terms with x on the same side of the equation and the remaining terms (the numbers) on the other side.

$$2x - 5 + 5 = 7 + 6x + 5 \qquad \text{Add 5.}$$

$$2x = 12 + 6x \qquad \text{Combine like terms.}$$

$$2x - 6x = 12 + 6x - 6x \qquad \text{Subtract } 6x.$$

$$-4x = 12 \qquad \text{Combine like terms.}$$

$$\frac{-4x}{-4} = \frac{12}{-4} \qquad \text{Divide by } -4.$$

$$x = -3$$

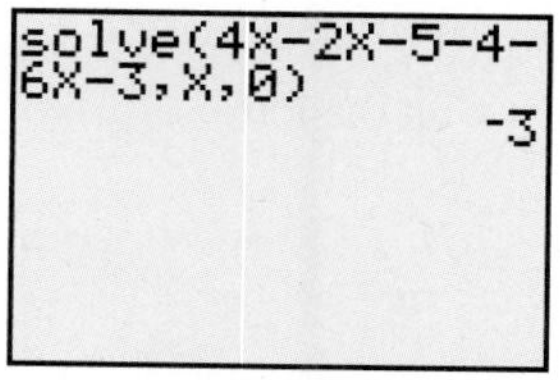

Graphing calculators can solve equations of the form

expression in X = 0.

The TI-83/84 Plus requires that you input the expression, the variable for which you are solving, and a "guess," separated by commas. The screen here shows how to solve the equation in Example 1.

Check that -3 is the solution by substituting it for x in the *original* equation.

Check:

$$4x - 2x - 5 = 4 + 6x + 3 \qquad \text{Original equation}$$

$$4(-3) - 2(-3) - 5 \stackrel{?}{=} 4 + 6(-3) + 3 \qquad \text{Let } x = -3.$$

$$-12 + 6 - 5 \stackrel{?}{=} 4 - 18 + 3 \qquad \text{Multiply.}$$

$$-11 = -11 \qquad \text{True}$$

This is not the solution. →

The true statement indicates that $\{-3\}$ is the solution set.

Solving a Linear Equation in One Variable

Step 1 **Clear fractions.** Eliminate any fractions by multiplying both sides of the equation by a common denominator.

Step 2 **Simplify each side separately.** Use the distributive property to clear parentheses, and combine like terms as needed.

Step 3 **Isolate the variable terms on one side.** Use the addition property of equality to get all terms with variables on one side of the equation and all numbers on the other.

Step 4 **Transform so that the coefficient of the variable is 1.** Use the multiplication property of equality to get an equation with just the variable (with coefficient 1) on one side.

Step 5 **Check.** Substitute the solution into the original equation.

EXAMPLE 2 Using the Distributive Property to Solve a Linear Equation

Solve $2(k - 5) + 3k = k + 6$.

SOLUTION

Step 1 Because there are no fractions in this equation, Step 1 does not apply.

Step 2 Use the distributive property to simplify and combine terms on the left side.

$$2(k - 5) + 3k = k + 6$$

$$2k - 10 + 3k = k + 6 \qquad \text{Distributive property}$$

$$5k - 10 = k + 6 \qquad \text{Combine like terms.}$$

Step 3 Next, use the addition property of equality.

$$5k - 10 + 10 = k + 6 + 10 \qquad \text{Add 10.}$$

$$5k = k + 16 \qquad \text{Combine like terms.}$$

$$5k - k = k + 16 - k \qquad \text{Subtract } k.$$

$$4k = 16 \qquad \text{Combine like terms.}$$

Step 4 Use the multiplication property of equality to get just k on the left.

$$\frac{4k}{4} = \frac{16}{4} \qquad \text{Divide by 4.}$$

$$k = 4$$

Step 5 Check that the solution set is $\{4\}$ by substituting 4 for k in the original equation. ■

Because of space limitations, we will not always show the check when solving an equation. ***You should always check your work.***

When fractions or decimals appear as numerical factors of terms in equations, our work can be made easier if we multiply each side of the equation by the least common denominator (LCD) of all the fractions. This is an application of the multiplication property of equality.

The problem-solving strategy of guessing and checking, discussed in Chapter 1, was actually used by the early Egyptians in equation solving. This method, called the **Rule of False Position,** involved making an initial guess at the solution of an equation, and then following up with an adjustment in the likely event that the guess was incorrect. For example (using our modern notation), if the equation

$$6x + 2x = 32$$

was to be solved, an initial guess might have been $x = 3$. Substituting 3 for x gives

$$6(3) + 2(3) = 32 \quad ?$$

$$18 + 6 = 32 \quad ?$$

$$24 = 32. \quad \textbf{False}$$

The guess, 3, gives a value (24) which is smaller than the desired value (32). Since 24 is $\frac{3}{4}$ of 32, the guess, 3, is $\frac{3}{4}$ of the actual solution. The actual solution, therefore, must be 4, since 3 is $\frac{3}{4}$ of 4.

Use the methods explained in this section to verify this result.

François Viète (1540–1603) was a lawyer at the court of Henry IV of France and studied equations. Viète simplified the notation of algebra and was among the first to use letters to represent numbers. For centuries, algebra and arithmetic were expressed in a cumbersome way with words and occasional symbols. Since the time of Viète, algebra has gone beyond equation solving; the abstract nature of higher algebra depends on its symbolic language.

Algebra dates back to the Babylonians of 2000 B.C. The Egyptians also worked problems in algebra, but the problems were not as complex as those of the Babylonians. In about the sixth century, the Hindus developed methods for solving problems involving interest, discounts, and partnerships.

Many Hindu and Greek works on mathematics were preserved only because Moslem scholars from about 750 to 1250 made translations of them. The Arabs took the work of the Greeks and Hindus and greatly expanded it. For example, Mohammed ibn Mûsâ al-Khowârizmî wrote books on algebra and on the Hindu numeration system (the one we use) that had tremendous influence in Western Europe; his name is remembered today in the word *algorithm*.

EXAMPLE 3 Solving a Linear Equation with Fractions

Solve $\frac{x+7}{6} + \frac{2x-8}{2} = -4.$

SOLUTION

Step 1 $6\left(\frac{x+7}{6} + \frac{2x-8}{2}\right) = 6(-4)$ — Multiply both sides by the LCD, 6, to eliminate the fractions.

Step 2 $6\left(\frac{x+7}{6}\right) + 6\left(\frac{2x-8}{2}\right) = 6(-4)$ — Distributive property

Multiply each term by 6.

$x + 7 + 3(2x - 8) = -24$ — Multiply.

$x + 7 + 6x - 24 = -24$ — Distributive property

$7x - 17 = -24$ — Combine like terms.

Step 3 $7x - 17 + 17 = -24 + 17$ — Add 17.

$7x = -7$ — Combine like terms.

Step 4 $\frac{7x}{7} = \frac{-7}{7}$ — Divide by 7.

$x = -1$

Step 5 *Check:* $\frac{x+7}{6} + \frac{2x-8}{2} = -4$ — Original equation

$\frac{-1+7}{6} + \frac{2(-1)-8}{2} = -4 \quad ?$ — Let $x = -1$.

$\frac{6}{6} + \frac{-10}{2} = -4 \quad ?$

$1 - 5 = -4 \quad ?$

$-4 = -4$ — True

The solution -1 checks, so the solution set is $\{-1\}$. ■

EXAMPLE 4 Solving a Linear Equation with Decimals

Solve $.06x + .09(15 - x) = .07(15)$.

SOLUTION

Because each decimal number is in hundredths, multiply both sides of the equation by 100. (This is done by moving the decimal points two places to the right.) To multiply the second term, $.09(15 - x)$, by 100, multiply 100(.09) first to get 9, so the product $100(.09)(15 - x)$ becomes $9(15 - x)$.

$.06x + .09(15 - x) = .07(15)$ — Original equation

$.06x + .09(15 - x) = .07(15)$ — Multiply each term by 100.

$6x + 9(15 - x) = 7(15)$

$6x + 9(15) - 9x = 105$ — Distributive property; multiply.

$-3x + 135 = 105$ — Combine like terms; multiply.

$$-3x + 135 - 135 = 105 - 135 \quad \text{Subtract 135.}$$
$$-3x = -30$$
$$\frac{-3x}{-3} = \frac{-30}{-3} \quad \text{Divide by } -3.$$
$$x = 10$$

Check to verify that the solution set is $\{10\}$.

Special Kinds of Linear Equations The preceding equations had solution sets containing one element; for example,

$$2(k - 5) + 3k = k + 6 \quad \text{has solution set } \{4\}.$$

Some equations that appear to be linear have no solutions, while others have an infinite number of solutions. Table 1 gives the names of these types of equations.

TABLE 1

Type of Equation	Number of Solutions	Indication When Solving
Conditional	One	Final line is $x =$ a number. (See Example 5(a).)
Identity	Infinite; solution set {all real numbers}	Final line is true, such as $0 = 0$. (See Example 5(b).)
Contradiction	None; solution set $\emptyset$	Final line is false, such as $0 = 1$. (See Example 5(c).)

Sofia Kovalevskaya (1850–1891) was the most widely known Russian mathematician in the late nineteenth century. She did most of her work in the theory of **differential equations**—equations invaluable for expressing rates of change. For example, in biology, the rate of growth of a population, say of microbes, can be precisely stated by differential equations.

Kovalevskaya studied privately because public lectures were not open to women. She eventually received a degree (1874) from the University of Göttingen, Germany. In 1884 she became a lecturer at the University of Stockholm and later was appointed professor of higher mathematics.

EXAMPLE 5 Recognizing Conditional Equations, Identities, and Contradictions

Solve each equation. Decide whether it is a *conditional equation*, an *identity*, or a *contradiction*.

(a) $5x - 9 = 4(x - 3)$ **(b)** $5x - 15 = 5(x - 3)$ **(c)** $5x - 15 = 5(x - 4)$

SOLUTION

(a)

$$5x - 9 = 4(x - 3)$$
$$5x - 9 = 4x - 12 \quad \text{Distributive property}$$
$$5x - 9 - 4x = 4x - 12 - 4x \quad \text{Subtract } 4x.$$
$$x - 9 = -12 \quad \text{Combine like terms.}$$
$$x - 9 + 9 = -12 + 9 \quad \text{Add 9.}$$
$$x = -3$$

The solution set, $\{-3\}$, has only one element, so $5x - 9 = 4(x - 3)$ is a conditional equation.

(b)

$$5x - 15 = 5(x - 3)$$
$$5x - 15 = 5x - 15 \quad \text{Distributive property}$$
$$0 = 0 \quad \text{Subtract } 5x \text{ and add 15.}$$

The final line, $0 = 0$, indicates that the solution set is {all real numbers}, and the equation $5x - 15 = 5(x - 3)$ is an identity. (*Note:* The first step yielded $5x - 15 = 5x - 15$, which is true for all values of x. We could have identified the equation as an identity at that point.)

(c)

$$5x - 15 = 5(x - 4)$$

$$5x - 15 = 5x - 20 \quad \text{Distributive property}$$

$$5x - 15 - 5x = 5x - 20 - 5x \quad \text{Subtract } 5x.$$

$$\mathbf{-15 = -20} \quad \text{False}$$

Because the result, $-15 = -20$, is *false*, the equation has no solution. The solution set is $\emptyset$, so the equation $5x - 15 = 5(x - 4)$ is a contradiction.

For Further Thought

The Axioms of Equality

When we solve an equation, we must make sure that it remains "balanced"—that is, any operation that is performed on one side of an equation must also be performed on the other side in order to assure that the set of solutions remains the same.

Underlying the rules for solving equations are four axioms of equality, listed below. For all real numbers a, b, and c,

1. Reflexive axiom	$a = a$
2. Symmetric axiom	If $a = b$, then $b = a$.
3. Transitive axiom	If $a = b$ and $b = c$, then $a = c$.
4. Substitution axiom	If $a = b$, then a may replace b in any statement without affecting the truth or falsity of the statement.

A relation, such as equality, which satisfies the first three of these axioms (reflexive, symmetric, and transitive), is called an equivalence relation.

For Group Discussion or Individual Investigation

1. Give an example of an everyday relation that does not satisfy the symmetric axiom.
2. Does the transitive axiom hold in sports competition, with the relation "defeats"?
3. Give an example of a relation that does not satisfy the transitive axiom.

Literal Equations and Formulas An equation involving *variables* (or letters), such as $cx + d = e$, is called a **literal equation.** The most useful examples of literal equations are *formulas*. The solution of a problem in algebra often depends on the use of a mathematical statement or **formula** in which more than one letter is used to express a relationship. Examples of formulas are

$$d = rt, \quad I = prt, \quad \text{and} \quad P = 2L + 2W. \quad \text{Formulas}$$

In some applications, the necessary formula must be solved for one of its variables. This process is called **solving for a specified variable.** The steps used are similar to those used in solving linear equations. ***When you are solving for a specified variable, the key is to treat that variable as if it were the only one; treat all other variables like numbers (constants).***

Solving for a Specified Variable

Step 1 Transform the equation so that all terms containing the specified variable are on one side of the equation and all terms without that variable are on the other side.

Step 2 If necessary, use the distributive property to combine the terms with the specified variable. The result should be the product of a sum or difference and the variable.

Step 3 Divide both sides by the factor that is multiplied by the specified variable.

EXAMPLE 6 Solving for a Specified Variable

Solve the formula $P = 2L + 2W$ for W.

SOLUTION

Solve the formula for the perimeter (distance around) of a rectangle (Figure 1) for W by isolating W on one side of the equals sign.

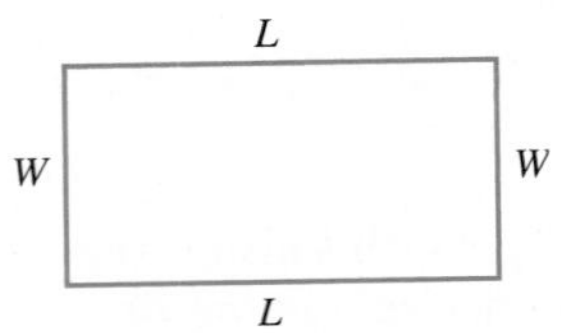

Perimeter, P, the sum of the lengths of the sides of the rectangle, is given by

$P = 2L + 2W$.

FIGURE 1

Step 1

$$P = 2L + 2W$$

$$P - 2L = 2L + 2W - 2L \quad \text{Subtract } 2L.$$

$$P - 2L = 2W$$

Step 2 Step 2 is not needed here.

Step 3

$$\frac{P - 2L}{2} = \frac{2W}{2} \quad \text{Divide both sides by 2.}$$

$$\frac{P - 2L}{2} = W \quad \text{or} \quad W = \frac{P}{2} - L$$

Models A **mathematical model** is an equation (or inequality) that describes the relationship between two quantities. A *linear model* is a linear equation.

EXAMPLE 7 Modeling the Prevention of Indoor Pollutants

One of the most effective ways of removing contaminants such as carbon monoxide and nitrogen dioxide from the air while cooking is to use a vented range hood. If a range hood removes contaminants at a flow rate of F liters of air per second, then the percent P of contaminants that are also removed from the surrounding air can be modeled by the linear equation

$$P = 1.06F + 7.18,$$

where $10 \le F \le 75$. What flow rate F must a range hood have to remove 50% of the contaminants from the air? (*Source:* Rezvan, R. L., "Effectiveness of Local Ventilation in Removing Simulated Pollutants from Point Sources," 65–75. In *Proceedings of the Third International Conference on Indoor Air Quality and Climate*, 1984.)

SOLUTION

Because $P = 50$, the equation becomes

$$\begin{aligned} 50 &= 1.06F + 7.18 \\ 5000 &= 106F + 718 && \text{Multiply by 100.} \\ 4282 &= 106F && \text{Subtract 718.} \\ F &\approx 40.40. && \text{Divide by 106.} \end{aligned}$$

Therefore, to remove 50% of the contaminants, the flow rate must be approximately 40.40 L of air per second. ■

7.1 EXERCISES

1. Which equations are linear equations in x?
   A. $3x + x - 1 = 0$ **B.** $8 = x^2$
   C. $6x + 2 = 9$ **D.** $\frac{1}{2}x - \frac{1}{x} = 0$

2. Which of the equations in Exercise 1 are not linear equations in x? Explain why.

3. Decide whether 6 is a solution of $3(x + 4) = 5x$ by substituting 6 for x. If it is not a solution, explain why.

4. Use substitution to decide whether -2 is a solution of $5(x + 4) - 3(x + 6) = 9(x + 1)$. If it is not a solution, explain why.

5. If two equations are equivalent, they have the same ________ ________.

6. The equation $4[x + (2 - 3x)] = 2(4 - 4x)$ is an identity. Let x represent the number of letters in your last name. Is this number a solution of this equation? Check your answer.

7. Which expression is equivalent to $.06(10 - x)(100)$?
   A. $.06 - .06x$ **B.** $60 - 6x$
   C. $6 - 6x$ **D.** $6 - .06x$

8. Describe in your own words the steps used to solve a linear equation.

Solve each equation.

9. $7k + 8 = 1$
10. $5m - 4 = 21$
11. $8 - 8x = -16$
12. $9 - 2r = 15$
13. $7x - 5x + 15 = x + 8$
14. $2x + 4 - x = 4x - 5$
15. $12w + 15w - 9 + 5 = -3w + 5 - 9$
16. $-4t + 5t - 8 + 4 = 6t - 4$
17. $2(x + 3) = -4(x + 1)$
18. $4(x - 9) = 8(x + 3)$
19. $3(2w + 1) - 2(w - 2) = 5$
20. $4(x - 2) + 2(x + 3) = 6$
21. $2x + 3(x - 4) = 2(x - 3)$
22. $6x - 3(5x + 2) = 4(1 - x)$
23. $6p - 4(3 - 2p) = 5(p - 4) - 10$
24. $-2k - 3(4 - 2k) = 2(k - 3) + 2$
25. $-[2z - (5z + 2)] = 2 + (2z + 7)$
26. $-[6x - (4x + 8)] = 9 + (6x + 3)$
27. $-3m + 6 - 5(m - 1) = -(2m - 4) - 5m + 5$
28. $4(k + 2) - 8k - 5 = -3k + 9 - 2(k + 6)$
29. $-[3x - (2x + 5)] = -4 - [3(2x - 4) - 3x]$
30. $2[-(x - 1) + 4] = 5 + [-(6x - 7) + 9x]$

31. $-(9 - 3a) - (4 + 2a) - 4 = -(2 - 5a) - a$

32. $(2 - 4x) - (3 - 4x) + 4 = -(-3 + 6x) + x$

33. $(2m - 6) - (3m - 4) = -(-4 + m) - 4m + 6$

34. $(3x - 4) - (5x - 8) = -(x + 12) - 6x + 1$

35. To solve the linear equation

$$.05x + .12(x + 5000) = 940,$$

we can multiply both sides by a power of 10 so that all coefficients are integers. What is the smallest power of 10 that will accomplish this goal?

36. Suppose that in solving the equation

$$\frac{1}{3}x + \frac{1}{2}x = \frac{1}{6}x,$$

you begin by multiplying both sides by 12, rather than the *least* common denominator, 6. Should you get the correct solution anyway? Explain.

Solve each equation.

37. $\frac{3x}{4} + \frac{5x}{2} = 13$

38. $\frac{8x}{3} - \frac{2x}{4} = -13$

39. $\frac{x - 8}{5} + \frac{8}{5} = -\frac{x}{3}$

40. $\frac{2r - 3}{7} + \frac{3}{7} = -\frac{r}{3}$

41. $\frac{4t + 1}{3} = \frac{t + 5}{6} + \frac{t - 3}{6}$

42. $\frac{2x + 5}{5} = \frac{3x + 1}{2} + \frac{-x + 7}{2}$

43. $.05x + .12(x + 5000) = 940$

44. $.09k + .13(k + 300) = 61$

45. $.02(50) + .08r = .04(50 + r)$

46. $.20(14{,}000) + .14t = .18(14{,}000 + t)$

47. $.05x + .10(200 - x) = .45x$

48. $.08x + .12(260 - x) = .48x$

49. The equation $x + 2 = x + 2$ is called a(n)________, because its solution set is {all real numbers}. The equation $x + 1 = x + 2$ is called a(n)____________, because its solution set is $\emptyset$.

50. Which equation is a conditional equation?

A. $2x + 1 = 3$ **B.** $x = 3x - 2x$

C. $3x + 1 = 3x$ **D.** $\frac{1}{2}x = \frac{1}{2}x$

Decide whether each equation is conditional, an identity, or a contradiction. Give the solution set.

51. $-2p + 5p - 9 = 3(p - 4) - 5$

52. $-6k + 2k - 11 = -2(2k - 3) + 4$

53. $6x + 2(x - 2) = 9x + 4$

54. $-4(x + 2) = -3(x + 5) - x$

55. $-11m + 4(m - 3) + 6m = 4m - 12$

56. $3p - 5(p + 4) + 9 = -11 + 15p$

57. $7[2 - (3 + 4r)] - 2r = -9 + 2(1 - 15r)$

58. $4[6 - (1 + 2m)] + 10m = 2(10 - 3m) + 8m$

59. When a formula is solved for a particular variable, several different equivalent forms may be possible. If we solve $A = \frac{1}{2}bh$ for h, one possible correct answer is

$$h = \frac{2A}{b}.$$

Which of the formulas is *not* equivalent to this?

A. $h = 2\left(\frac{A}{b}\right)$ **B.** $h = 2A\left(\frac{1}{b}\right)$

C. $h = \frac{A}{\frac{1}{2}b}$ **D.** $h = \frac{\frac{1}{2}A}{b}$

60. One source for geometric formulas gives the formula for the perimeter of a rectangle as

$$P = 2L + 2W,$$

while another gives it as

$$P = 2(L + W).$$

Are these equivalent? If so, what property justifies their equivalence?

Mathematical Formulas *Solve each formula for the specified variable.*

61. $d = rt$; for t (distance)

62. $I = prt$; for r (simple interest)

63. $A = bh$; for b (area of a parallelogram)

64. $P = 2L + 2W$; for L (perimeter of a rectangle)

65. $P = a + b + c$; for a (perimeter of a triangle)

66. $V = LWH$; for W (volume of a rectangular solid)

67. $A = \frac{1}{2}bh$; for b (area of a triangle)

68. $C = 2\pi r$; for r (circumference of a circle)

69. $S = 2\pi rh + 2\pi r^2$; for h (surface area of a right circular cylinder)

70. $A = \frac{1}{2}(B + b)h$; for B (area of a trapezoid)

71. $\mathrm{C} = \frac{5}{9}(\mathrm{F} - 32)$; for F (Fahrenheit to Celsius)

72. $\mathrm{F} = \frac{9}{5}\mathrm{C} + 32$; for C (Celsius to Fahrenheit)

73. $A = 2HW + 2LW + 2LH$; for H (surface area of a rectangular solid)

74. $V = \frac{1}{3}Bh$; for h (volume of a right pyramid)

Work each problem involving a linear model.

75. *College Enrollment* The linear model

$$y = .2145x + 15.69$$

provides the projected approximate enrollment, in millions, for degree-granting institutions between the years 2003 and 2012, where $x = 0$ corresponds to 2003, $x = 1$ to 2004, and so on, and y is in millions of students.

(a) Use the model to determine projected enrollment for Fall 2008.

(b) Use the model to determine the year in which enrollment is projected to reach 17 million.

76. *Mobility of Americans* The linear model

$$y = -.9x + 21.2,$$

where y represents the percent moving each year in decade x, approximates the percent of Americans moving each year since the 1950s fairly well. Here $x = 1$ represents the 1950s, $x = 2$ represents the 1960s, and so on. Use this model to answer each question.

(a) What was the approximate percent of Americans moving in the 1960s?

(b) What was the approximate percent of Americans moving in the 1980s?

77. *Indoor Air Quality and Control* The excess lifetime cancer risk R is a measure of the likelihood that an individual will develop cancer from a particular pollutant. For example, if $R = .01$ then a person has a 1% increased chance of developing cancer during a lifetime. (This would translate into 1 case of cancer for every 100 people during an average lifetime.) The value of R for formaldehyde, a highly toxic indoor air pollutant, can be calculated using the linear model $R = kd$, where k is a constant, and d is the daily dose in parts per million. The constant k for formaldehyde can be calculated using the formula

$$k = \frac{.132B}{W},$$

where B is the total number of cubic meters of air a person breathes in one day, and W is a person's weight in kilograms. (*Source*: Hines, A., T. Ghosh, S. Loyalka, and R. Warder, *Indoor Air: Quality & Control*, Prentice-Hall, 1993; Ritchie, I., and R. Lehnen, "An Analysis of Formaldehyde Concentration in Mobile and Conventional Homes," *J. Env. Health* 47: 300–305.)

(a) Find k for a person who breathes in 20 cubic meters of air per day and weighs 75 kilograms.

(b) Mobile homes in Minnesota were found to have a mean daily dose d of .42 part per million. Calculate R using the value of k found in part (a).

(c) For every 5000 people, how many cases of cancer could be expected each year from these levels of formaldehyde? Assume an average life expectancy of 72 years.

78. ***Indoor Air Quality and Control*** (See Exercise 77.) For nonsmokers exposed to environmental tobacco smoke (passive smokers), $R = .0015$. (*Source*: Hines, A., T. Ghosh, S. Loyalka, and R. Warder, *Indoor Air: Quality & Control*, Prentice-Hall, 1993.)

(a) If the average life expectancy is 72 years, what is the excess lifetime cancer risk from secondhand tobacco smoke per year?

(b) Write a linear equation that will model the expected number of cancer cases C per year if there are x passive smokers.

(c) Estimate the number of cancer cases each year per 100,000 passive smokers.

(d) The excess lifetime risk of death from smoking is $R = .44$. Currently 26% of the U.S. population smoke. If the U.S. population is 260 million, approximate the excess number of deaths caused by smoking each year.

79. ***Eye Irritation from Formaldehyde*** When concentrations of formaldehyde in the air exceed 33 μg per cubic foot (1 μg = 1 microgram = .000001 gram), a strong odor and irritation to the eyes often occurs. One square foot of hardwood plywood paneling can emit 3365 μg of formaldehyde per day. (*Source*: Hines, A., T. Ghosh, S. Loyalka, and R. Warder, *Indoor Air: Quality & Control*, Prentice-Hall, 1993.)

A 4-foot by 8-foot sheet of this paneling is attached to an 8-foot wall in a room having floor dimensions of 10 feet by 10 feet.

(a) Determine how many cubic feet of air are in the room.

(b) Find the total number of micrograms of formaldehyde that are released into the air by the paneling each day.

(c) If there is no ventilation in the room, write a linear equation that models the amount of formaldehyde F that there would be in the room after x days.

(d) How long will it take before a person's eyes become irritated in the room?

80. ***Classroom Ventilation*** According to the American Society of Heating, Refrigerating and Air-Conditioning Engineers, Inc. (ASHRAE), a nonsmoking classroom should have a ventilation rate of 15 cubic feet per minute for each person in the classroom. (*Source: ASHRAE*, 1989.)

(a) Write an equation that models the total ventilation V (in cubic feet per hour) necessary for a classroom with x students.

(b) A common unit of ventilation is an air change per hour (ach). 1 ach is equivalent to exchanging all of the air in a room every hour. If x students are in a classroom having volume 15,000 cubic feet, determine how many air exchanges per hour (A) are necessary to keep the room properly ventilated.

(c) Find the necessary number of ach A if the classroom has 40 students in it.

(d) In areas like bars and lounges that allow smoking, the ventilation rate should be increased to 50 cubic feet per minute per person. Compared to classrooms, ventilation should be increased by what factor in heavy smoking areas?

7.2 Applications of Linear Equations

Translating Words into Symbols • Guidelines for Applications • Finding Unknown Quantities • Mixture and Interest Problems • Monetary Denomination Problems • Motion Problems

Translating Words into Symbols

When algebra is used to solve practical applications, we must translate the verbal statements of the problems into mathematical statements.

PROBLEM-SOLVING HINT Usually there are key words and phrases in a verbal problem that translate into mathematical expressions involving addition, subtraction, multiplication, and division.

Translation from Words to Mathematical Expressions

Verbal Expression	Mathematical Expression (where x and y are numbers)
Addition	
The **sum** of a number and 7	$x + 7$
6 **more than** a number	$x + 6$
3 **plus** 8	$3 + 8$
24 **added to** a number	$x + 24$
A number **increased by** 5	$x + 5$
The **sum** of two numbers	$x + y$
Subtraction	
2 **less than** a number	$x - 2$
12 **minus** a number	$12 - x$
A number **decreased by** 12	$x - 12$
The **difference between** two numbers	$x - y$
A number **subtracted from** 10	$10 - x$
Multiplication	
16 **times** a number	$16x$
A number **multiplied by** 6	$6x$
$\frac{2}{3}$ **of** a number (as applied to fractions and percent)	$\frac{2}{3}x$
Twice (2 times) a number	$2x$
The **product** of two numbers	xy
Division	
The **quotient** of 8 and a number	$\frac{8}{x}$ $(x \neq 0)$
A number **divided by** 13	$\frac{x}{13}$
The **ratio** of two numbers or the **quotient** of two numbers	$\frac{x}{y}$ $(y \neq 0)$

The symbol of equality, **=**, is often indicated by the word ***is***. In fact, since equal mathematical expressions represent different names for the same number, words that indicate the idea of "sameness" translate as =. For example,

If the product of a number and 12 is decreased by 7, the result is 105

translates to the mathematical equation

$$\mathbf{12x - 7 = 105,}$$

where x represents the unknown number. (Why would $7 - 12x = 105$ be incorrect?)

Guidelines for Applications While there is no one method that allows us to solve all types of applied problems, the following six steps are helpful.

George Polya's problem-solving procedure can be adapted to applications of algebra as seen in the steps in the box. Steps 1 and 2 make up the first stage of Polya's procedure (*Understand the Problem*), Step 3 forms the second stage (*Devise a Plan*), Step 4 comprises the third stage (*Carry Out the Plan*), and Steps 5 and 6 form the last stage (*Look Back*).

Solving an Applied Problem

Step 1 **Read** the problem carefully until you understand what is given and what is to be found.

Step 2 **Assign a variable** to represent the unknown value, using diagrams or tables as needed. Write down what the variable represents. If necessary, express any other unknown values in terms of the variable.

Step 3 **Write an equation** using the variable expression(s).

Step 4 **Solve** the equation.

Step 5 **State the answer.** Does it seem reasonable?

Step 6 **Check** the answer in the words of the *original* problem.

The third step is often the hardest. To translate the problem into an equation, write the given phrases as mathematical expressions. Since equal mathematical expressions are names for the same number, translate any words that mean *equal* or *same* as the = symbol. The = symbol leads to an equation to be solved.

Finding Unknown Quantities

PROBLEM-SOLVING HINT A common type of problem involves finding two quantities when the sum of the quantities is known. Choose a variable to represent one of the unknowns and then represent the other quantity in terms of the same variable, using information from the problem. Then write an equation based on the words of the problem.

EXAMPLE 1 Finding Numbers of Strikeouts

Two outstanding major league pitchers in recent years are Randy Johnson and Pedro Martinez. In 2002, they combined for a total of 573 strikeouts. Johnson had 95 more strikeouts than Martinez. How many strikeouts did each pitcher have? (*Source: World Almanac and Book of Facts 2004.*)

SOLUTION

Step 1 **Read** the problem. We are asked to find the number of strikeouts each pitcher had.

Step 2 **Assign a variable** to represent the number of strikeouts for one of the men.

Let s = the number of strikeouts for Pedro Martinez.

We must also find the number of strikeouts for Randy Johnson. Because he had 95 more strikeouts than Martinez,

$s + 95$ = the number of strikeouts for Johnson.

Step 3 **Write an equation.** The sum of the numbers of strikeouts is 573, so

Martinez's strikeouts	+	Johnson's strikeouts	=	Total
↓		↓		↓
s	+	$(s + 95)$	=	573.

Step 4 **Solve** the equation.

$$\begin{aligned} s + (s + 95) &= 573 \\ 2s + 95 &= 573 && \text{Combine like terms.} \\ 2s + 95 - 95 &= 573 - 95 && \text{Subtract 95.} \\ 2s &= 478 && \text{Combine like terms.} \\ \frac{2s}{2} &= \frac{478}{2} && \text{Divide by 2.} \\ s &= 239 \end{aligned}$$

Here is an application of linear equations, taken from the **Greek Anthology** (about 500 A.D.), a group of 46 number problems.

Demochares has lived a fourth of his life as a boy, a fifth as a youth, a third as a man, and has spent 13 years in his dotage. How old is he?

(Answer: 60 years old)

Step 5 **State the answer.** We let s represent the number of strikeouts for Martinez, so Martinez had 239. Then the number of strikeouts for Johnson is

$s + 95 = 239 + 95 = 334.$ ← Be sure to find the second answer.

Step 6 **Check.** 334 is 95 more than 239, and the sum of 239 and 334 is 573. The conditions of the problem are satisfied, and our answer checks. ■

EXAMPLE 2 Finding Lengths of Pieces of Wood

The instructions for a woodworking project call for three pieces of wood. The longest piece must be twice the length of the middle-sized piece, and the shortest piece must be 10 inches shorter than the middle-sized piece. Jean Loeb has a board 70 inches long that she wishes to use. How long can each piece be?

SOLUTION

Step 1 **Read** the problem. There will be three answers.

Step 2 **Assign a variable.** Because the middle-sized piece appears in both pairs of comparisons, let x represent the length, in inches, of the middle-sized piece. We have

$$\begin{aligned} x &= \text{the length of the middle-sized piece,} \\ 2x &= \text{the length of the longest piece, and} \\ x - 10 &= \text{the length of the shortest piece.} \end{aligned}$$

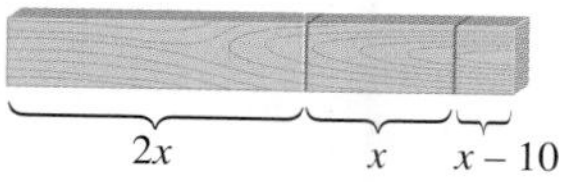

FIGURE 2

A sketch is helpful here. See Figure 2.

Problems involving age have been around since antiquity. The *Greek Anthology* gives the only information known about the life of the mathematician **Diophantus:**

Diophantus passed $\frac{1}{6}$ of his life in childhood, $\frac{1}{12}$ in youth, and $\frac{1}{7}$ more as a bachelor. Five years after his marriage was born a son who died 4 years before his father, at $\frac{1}{2}$ his father's final age.

Try to write an equation and solve it to show that Diophantus was 84 years old when he died.

In the 1941 movie *Buck Privates*, Bud Abbott and Lou Costello perform a routine that pokes fun at such problems. Slicker Smith (Abbott) tells Herbie Brown (Costello) that he is really dumb, and to prove it, he challenges Herbie to answer this question:

Suppose you're 40 years old and you're in love with a little girl that's 10 years old. You're 4 times as old as that little girl. Now, you couldn't marry that little girl, could you? So you wait 5 years. Now you're 45 and she's 15. You're three times as old as the little girl. You still can't marry her, so you wait another 15 years. Now you're twice as old as that little girl. How long will you have to wait before she catches up to you?

Watch the movie to hear Herbie's clever answer.

Step 3 **Write an equation.**

Longest		Middle-sized		Shortest	is	Total length
$2x$	$+$	x	$+$	$(x - 10)$	$=$	70

Step 4 **Solve.**

$$4x - 10 = 70 \qquad \text{Combine like terms.}$$
$$4x - 10 + 10 = 70 + 10 \qquad \text{Add 10.}$$
$$4x = 80 \qquad \text{Combine like terms.}$$
$$x = 20 \qquad \text{Divide by 4.}$$

Step 5 **State the answer.** The middle-sized piece is 20 inches long, the longest piece is $2(20) = 40$ inches long, and the shortest piece is $20 - 10 = 10$ inches long.

Step 6 **Check.** The sum of the lengths is 70 inches. All conditions of the problem are satisfied. ■

Mixture and Interest Problems

PROBLEM-SOLVING HINT Percents often are used in problems involving mixing different concentrations of a substance or different interest rates. In each case, to get the amount of pure substance or the interest, we multiply.

Mixture Problems	Interest Problems (annual)
base × rate (%) = percentage	principal × rate (%) = interest
$b \times r = p$	$P \times r = I$

In an equation, the percent should be written as a decimal. For example, 35% is written .35, not 35, and 7% is written .07, not 7.

EXAMPLE 3 Using Percents in Applications

(a) If a chemist has 40 liters of a 35% acid solution, how much pure acid is there?

(b) If \$1300 is invested for one year at 2% simple interest, how much interest is earned in one year?

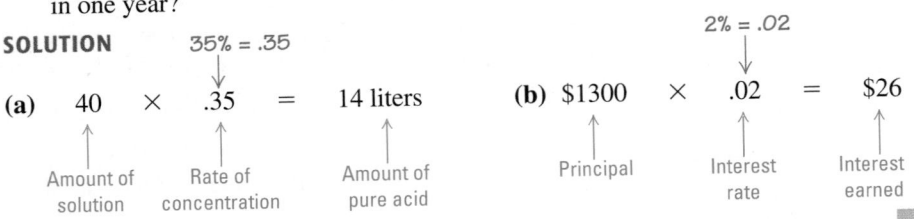

SOLUTION

(a) $40 \times .35 = 14$ liters ($35\% = .35$)

Amount of solution		Rate of concentration		Amount of pure acid
40	×	.35	=	14 liters

(b) $\$1300 \times .02 = \$26$ ($2\% = .02$)

Principal		Interest rate		Interest earned
\$1300	×	.02	=	\$26

■

PROBLEM-SOLVING HINT Sometimes we use tables to organize the information in a problem. A table enables us to more easily set up an equation for the problem, which is usually the most difficult step.

EXAMPLE 4 Solving a Mixture Problem

A chemist must mix 8 liters of a 40% acid solution with some 70% solution to obtain a 50% solution. How much of the 70% solution should be used?

SOLUTION

Step 1 **Read** the problem. The problem asks for the amount of 70% solution to be used.

Step 2 **Assign a variable.** Let $x =$ the number of liters of 70% solution to be used. The information in the problem is illustrated in Figure 3.

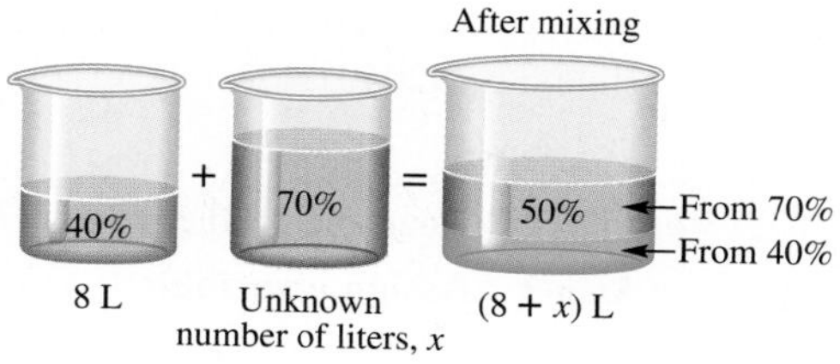

FIGURE 3

Use the given information to complete the table.

Percent (as a decimal)	Number of Liters	Liters of Pure Acid
$40\% = .40$	8	$.40(8) = 3.2$
$70\% = .70$	x	$.70x$
$50\% = .50$	$8 + x$	$.50(8 + x)$

Sum must equal (the first two entries of the right column must sum to the third)

The numbers in the right column were found by multiplying the strengths and the numbers of liters. The number of liters of pure acid in the 40% solution plus the number of liters of pure acid in the 70% solution must equal the number of liters of pure acid in the 50% solution.

Step 3 **Write an equation.**

$$3.2 + .70x = .50(8 + x)$$

Step 4 **Solve.** We will do so here without clearing the decimals.

$$3.2 + .70x = 4 + .50x \quad \text{Distributive property}$$
$$.20x = .8 \quad \text{Subtract 3.2 and } .50x.$$
$$x = 4 \quad \text{Divide by .20.}$$

Step 5 **State the answer.** The chemist should use 4 liters of the 70% solution.

The 1995 action thriller *Die Hard: With a Vengeance* features John McClane (Bruce Willis) and Zeus Carver (Samuel L. Jackson) matching wits with villain Simon Gruber (Jeremy Irons) who is planting bombs around New York. In one scene, Simon communicates with McClane and Carver on a pay telephone and tells them that in order to keep a bomb from detonating, they must dial a number that requires their solving the following riddle.

As I was going to St. Ives,
I met a man with seven wives,
Every wife had seven sacks,
Every sack had seven cats,
Every cat had seven kits.
Kits, cats, sacks, and wives,
How many were going to
St. Ives?

The rhyme is a derivation of an old application found in the **Rhind papyrus,** an Egyptian manuscript that dates back to about 1650 B.C. **Leonardo of Pisa (Fibonacci)** also included a similar problem in *Liber Abaci* in 1202.

The answer to the question is 1. Only "I" was *going* to St. Ives.

Step 6 **Check.** 8 liters of 40% solution plus 4 liters of 70% solution is

$$8(.40) + 4(.70) = \mathbf{6\ liters}$$

of acid. Similarly, 8 + 4 or 12 liters of 50% solution has

$$12(.50) = \mathbf{6\ liters}$$

of acid in the mixture. The total amount of pure acid is 6 liters both before and after mixing, so the answer checks.

The next example uses the formula for simple interest, $I = prt$. Remember that when $t = 1$, the formula becomes $I = pr$, as shown in the Problem-Solving Hint just before Example 3. Once again the idea of multiplying the total amount (principal) by the rate (rate of interest) gives the percentage (amount of interest).

EXAMPLE 5 Solving an Investment Problem

After winning the state lottery, Mark LeBeau has $40,000 to invest. He will put part of the money in an account paying 4% interest and the remainder into stocks paying 6% interest. His accountant tells him that the total annual income from these investments should be $2040. How much should he invest at each rate?

SOLUTION

Step 1 **Read** the problem again. We must find the two amounts.

Step 2 **Assign a variable.**

Let x = the amount to invest at 4%;

then $40{,}000 - x$ = the amount to invest at 6%.

The formula for interest is $I = prt$. Here the time, t, is 1 year. We organize the given information in a table.

Rate (as a decimal)	Principal	Interest	
$4\% = .04$	x	$.04x$	
$6\% = .06$	$40{,}000 - x$	$.06(40{,}000 - x)$	
	40,000	2040	← Totals

Step 3 **Write an equation.** The last column of the table gives the equation.

$$\underset{\text{Interest at 4\%}}{.04x} + \underset{\text{Interest at 6\%}}{.06(40{,}000 - x)} = \underset{\text{Total interest}}{2040}$$

Step 4 **Solve** the equation. We do so without clearing decimals.

$$.04x + .06(40{,}000) - .06x = 2040 \quad \text{Distributive property}$$
$$.04x + 2400 - .06x = 2040 \quad \text{Multiply.}$$
$$-.02x + 2400 = 2040 \quad \text{Combine like terms.}$$
$$-.02x = -360 \quad \text{Subtract 2400.}$$
$$x = 18{,}000 \quad \text{Divide by } -.02.$$

Step 5 **State the answer.** Mark should invest \$18,000 at 4%. At 6%, he should invest \$40,000 − \$18,000 = \$22,000.

Step 6 **Check** by finding the annual interest at each rate.

$$.04(\$18{,}000) = \$720 \quad \text{and} \quad .06(\$22{,}000) = \$1320$$

They total $\$720 + \$1320 = \$2040,$ as required. ■

Monetary Denomination Problems

PROBLEM-SOLVING HINT Problems that involve different denominations of money or items with different monetary values are similar to mixture and investment problems.

Money Problems

Number × Value of one = Total value

For example, if a jar contains 37 quarters, the monetary value of the coins is

$$\underset{\text{Number of coins}}{\underset{\uparrow}{37}} \times \underset{\text{Denomination}}{\underset{\uparrow}{\$.25}} = \underset{\text{Monetary value}}{\underset{\uparrow}{\$9.25}}.$$

EXAMPLE 6 Solving a Monetary Denomination Problem

For a bill totaling \$5.65, a cashier received 25 coins consisting of nickels and quarters. How many of each type of coin did the cashier receive?

SOLUTION

Step 1 **Read** the problem. The problem asks that we find the number of nickels and the number of quarters the cashier received.

Step 2 **Assign a variable.**

Let x = the number of nickels;
then $25 - x$ = the number of quarters.

We organize the information in a table.

Denomination	Number of Coins	Value
\$.05	x	$.05x$
\$.25	$25 - x$	$.25(25 - x)$
	25	5.65 ← Totals

Step 3 **Write an equation.** From the last column of the table,

$$.05x + .25(25 - x) = 5.65.$$

Can we average averages?
A car travels from A to B at 40 miles per hour and returns at 60 miles per hour. What is its rate for the entire trip?

The correct answer is not 50 miles per hour, as you might expect. Remembering the distance, rate, time relationship and letting $x =$ the distance between A and B, we can simplify a complex fraction to find the correct answer.

$$\text{Average rate for entire trip} = \frac{\text{Total distance}}{\text{Total time}} = \frac{x + x}{\frac{x}{40} + \frac{x}{60}} = \frac{2x}{\frac{3x}{120} + \frac{2x}{120}} = \frac{2x}{\frac{5x}{120}} = 2x \cdot \frac{120}{5x} = 48$$

The average rate for the entire trip is 48 miles per hour.

Step 4 **Solve.**

$$\begin{aligned} 5x + 25(25 - x) &= 565 && \text{Multiply by 100.} \\ 5x + 625 - 25x &= 565 && \text{Distributive property} \\ -20x &= -60 && \text{Subtract 625; combine terms.} \\ x &= 3 && \text{Divide by } -20. \end{aligned}$$

Step 5 **State the answer.** The cashier has 3 nickels and $25 - 3 = 22$ quarters.

Step 6 **Check.** The cashier has $3 + 22 = 25$ coins, and the value of the coins is $\$.05(3) + \$.25(22) = \$5.65$, as required.

Motion Problems

If an automobile travels at an average rate of 50 miles per hour for two hours, then it travels $50 \times 2 = 100$ miles. This is an example of the basic relationship between distance, rate, and time:

$$\text{distance} = \text{rate} \times \text{time},$$

given by the formula $d = rt$. By solving, in turn, for r and t in the formula, we obtain two other equivalent forms of the formula. The three forms are given below.

Distance, Rate, Time Relationship

$$d = rt \qquad r = \frac{d}{t} \qquad t = \frac{d}{r}$$

EXAMPLE 7 Using the Distance, Rate, Time Relationship

(a) The speed of sound is 1088 feet per second at sea level at 32°F. In 5 seconds under these conditions, how far does sound travel?

(b) The winner of the first Indianapolis 500 race (in 1911) was Ray Harroun, driving a Marmon Wasp at an average speed of 74.59 miles per hour. How long did it take for him to complete the 500-mile course? (*Source: The Universal Almanac 1997*, John W. Wright, General Editor.)

(c) At the 2004 Olympic Games in Athens, Greece, Chinese swimmer Luo Xuejuan set an Olympic record in the women's 100-m breaststroke swimming event of 66.64 seconds. What was her rate? (*Source: World Almanac and Book of Facts 2006*.)

SOLUTION

(a)

$$\underset{\text{Rate}}{1088} \times \underset{\text{Time}}{5} = \underset{\text{Distance}}{5440 \text{ feet}}$$

Here we found distance given rate and time using $d = rt$.

(b) To complete the 500 miles, it took Harroun

$$\frac{500 \leftarrow \text{Distance}}{74.59 \leftarrow \text{Rate}} = 6.70 \text{ hours} \quad \text{(rounded).} \leftarrow \text{Time}$$

Here, we found time given rate and distance, using $t = \frac{d}{r}$. To convert .70 hour to minutes, multiply by 60 to get $.70(60) = 42$ minutes. The race took him 6 hours, 42 minutes to complete.

(c) Her rate was

$$\text{Rate} = \frac{\text{Distance}}{\text{Time}} \rightarrow \frac{100}{66.64} = 1.50 \text{ meters per second (rounded).}$$

PROBLEM-SOLVING HINT Motion problems use the distance formula, $d = rt$. In this formula, ***when rate (or speed) is given in miles per hour, time must be given in hours.*** To solve such problems, ***draw a sketch*** to illustrate what is happening in the problem, and ***make a table*** to summarize the given information.

EXAMPLE 8 Solving a Motion Problem

Jeff can bike to work in $\frac{3}{4}$ hour. By bus, the trip takes $\frac{1}{4}$ hour. If the bus travels 20 mph faster than Jeff rides his bike, how far is it to his workplace?

SOLUTION

Step 1 **Read** the problem. We must find the distance between Jeff's home and his workplace.

Step 2 **Assign a variable.** Although the problem asks for a distance, it is easier here to let x be Jeff's speed when he rides his bike to work. Then the speed of the bus is $x + 20$. Thus,

$$d = rt = x \cdot \frac{3}{4} = \frac{3}{4}x, \quad \text{Trip by bike}$$

and

$$d = rt = (x + 20) \cdot \frac{1}{4} = \frac{1}{4}(x + 20). \quad \text{Trip by bus}$$

We summarize this information in a table.

	Rate	Time	Distance	
Bike	x	$\frac{3}{4}$	$\frac{3}{4}x$	Same
Bus	$x + 20$	$\frac{1}{4}$	$\frac{1}{4}(x + 20)$	

Step 3 **Write an equation.** The key to setting up the correct equation is to realize that the distance in each case is the same. See Figure 4.

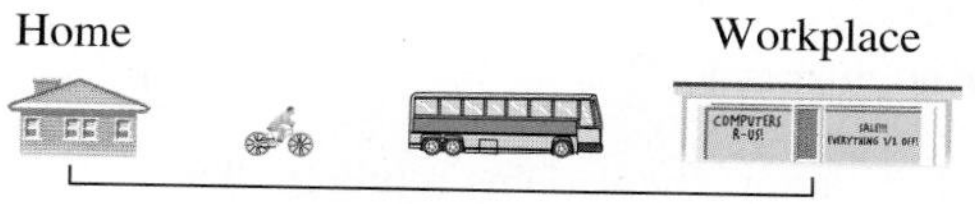

FIGURE 4

$$\frac{3}{4}x = \frac{1}{4}(x + 20) \quad \text{The distance is the same.}$$

Step 4 **Solve.** $4\left(\frac{3}{4}x\right) = 4\left(\frac{1}{4}\right)(x + 20)$ Multiply by 4.

$3x = x + 20$ Multiply.

$2x = 20$ Subtract x.

$x = 10$ Divide by 2.

Step 5 **State the answer.** The required distance is given by

$$d = \frac{3}{4}x = \frac{3}{4}(10) = \frac{30}{4} = 7.5 \text{ mi.}$$

Step 6 **Check** by finding the distance using

$$d = \frac{1}{4}(x + 20) = \frac{1}{4}(10 + 20) = \frac{30}{4} = 7.5 \text{ mi,}$$

which yields the same result. ■

PROBLEM-SOLVING HINT In motion problems such as the one in Example 8, once you have filled in two pieces of information in each row of the table, you should automatically fill in the third piece of information, using the appropriate form of the formula relating distance, rate, and time. Set up the equation based upon your sketch and the information in the table.

7.2 EXERCISES

Decide whether each of the following translates into an expression or an equation.

1. the product of a number and 5

2. 36% of a number

3. $\frac{2}{3}$ of a number is 18.

4. 9 is 4 more than a number.

5. the ratio of a number and 12

6. 48 divided by a number is 4.

7. Rework Example 6, letting the variable represent the number of quarters. Is the answer to the problem the same?

8. Explain why $19 - x$ is *not* a correct translation of "19 less than a number."

Translate each verbal phrase into a mathematical expression. Use x to represent the unknown number.

9. a number decreased by 14

10. 8 more than a number

11. the product of 7 less than a number and 5 more than the number

12. the quotient of a number and 8

13. the ratio of 15 and a nonzero number

14. $\frac{4}{7}$ of a number

15. Write a few sentences describing the six steps for problem solving.

16. Which is *not* a valid translation of "30% of a number"?

A. $.30x$ **B.** $.3x$ **C.** $\frac{3x}{10}$ **D.** $.30$

Unknown Numbers *Let x represent the number, write an equation for the sentence, and then solve.*

17. If 2 is added to five times a number, the result is equal to 5 more than four times the number. Find the number.

18. If four times a number is added to 8, the result is three times the number added to 5. Find the number.

19. If 2 is subtracted from a number and this difference is tripled, the result is 6 more than the number. Find the number.

20. If 3 is added to a number and this sum is doubled, the result is 2 more than the number. Find the number.

21. The sum of three times a number and 7 more than the number is the same as the difference between -11 and twice the number. What is the number?

22. If 4 is added to twice a number and this sum is multiplied by 2, the result is the same as if the number is multiplied by 3 and 4 is added to the product. What is the number?

Use the methods of Examples 1 and 2 or your own method to solve each problem.

23. ***Concert Revenues*** Bruce Springsteen and the E Street Band generated top revenue on the concert circuit in 2003. Springsteen and second-place Céline Dion together took in \$196.4 million from ticket sales. If Céline Dion took in \$35.4 million less than Bruce Springsteen and the E Street Band, how much revenue did each generate? (*Source: Parade*, February 15, 2004.)

24. ***Automobile Sales*** The Toyota Camry was the top-selling passenger car in the United States in 2004, followed by the Honda Accord. Honda Accord sales were 40 thousand less than Toyota Camry sales, and 814 thousand of these two cars were sold. How many of each model of car were sold? (*Source:* Ward's Communications.)

25. ***NBA Record*** In the 2004–2005 NBA regular season, the Phoenix Suns won two more than three times as many games as they lost. The Suns played 82 games. How many wins and losses did the team have? (*Source: World Almanac and Book of Facts 2006.*)

26. ***MLB Record*** In the 2005 Major League Baseball season, the Chicago White Sox won 27 fewer than twice as many games as they lost. They played 162 regular season games. How many wins and losses did the team have? (*Source: World Almanac and Book of Facts 2006.*)

27. ***U.S. Senate*** During the 109th Congress (2005–2006), the U.S. Senate had a total of 99 Democrats and Republicans. There were 11 more Republicans than Democrats. How many Democrats and Republicans were there in the Senate? (*Source: World Almanac and Book of Facts 2006.*)

28. ***U.S. House of Representatives*** The total number of Democrats and Republicans in the U.S. House of Representatives during the 109th Congress was 434. There were 30 more Republicans than Democrats. How many members of each party were there? (*Source: World Almanac and Book of Facts 2006.*)

29. ***Submarine Sandwich*** Nagaraj Nanjappa has a party-length submarine sandwich 59 inches long. He wants to cut it into three pieces so that the middle piece is 5 inches longer than the shortest piece, and the shortest piece is 9 inches shorter than the longest piece. How long should the three pieces be?

30. ***Office Manager Duties*** Tyrone Moseley, an office manager, must book airline tickets for the business trips that employees of his company need to make. In one week, he booked 55 tickets, divided among three airlines. He booked 7 more tickets on American Airlines than United Airlines. On Southwest Airlines, he booked 4 more than twice as many tickets as on United. How many tickets did he book on each airline?

31. ***U.S. Olympic Medals*** The United States earned a total of 103 medals at the 2004 Athens Olympics. The number of gold medals earned was 6 more than the number of bronze medals. The number of silver medals earned was 10 more than the number of bronze medals. How many of each kind of medal did the United States earn? (*Source:* U.S. Olympic Committee.)

32. ***Textbook Editor Duties*** In her job as a mathematics textbook editor, Joanne Ha works $7\frac{1}{2}$ hours a day. She spent a recent day making telephone calls, writing e-mails, and attending meetings. On that day, she spent twice as much time attending meetings as making telephone calls, and spent $\frac{1}{2}$ hour longer writing e-mails than making telephone calls. How many hours did she spend on each task?

Use basic formulas, as in Example 3, to solve each problem.

33. ***Acid Mixture*** How much pure acid is in 250 milliliters of a 14% acid solution?

34. ***Alcohol Mixture*** How much pure alcohol is in 150 liters of a 30% alcohol solution?

35. ***Interest Earned*** If \$10,000 is invested for one year at 3.5% simple interest, how much interest is earned?

36. ***Interest Earned*** If \$25,000 is invested at 3% simple interest for 2 years, how much interest is earned?

37. ***Monetary Value of Coins*** What is the monetary amount of 283 nickels?

38. ***Monetary Value of Coins*** What is the monetary amount of 35 half-dollars?

Use the method of Example 4 or your own method to solve each problem.

39. ***Alcohol Mixture*** In a chemistry class, 12 liters of a 12% alcohol solution must be mixed with a 20% solution to get a 14% solution. How many liters of the 20% solution are needed?

Strength	Liters of Solution	Liters of Alcohol
12%	12	
20%		
14%		

40. ***Alcohol Mixture*** How many liters of a 10% alcohol solution must be mixed with 40 liters of a 50% solution to get a 40% solution?

Strength	Liters of Solution	Liters of Alcohol
	x	
	40	
40%		

41. ***Alcohol Mixture in First Aid Spray*** A medicated first aid spray on the market is 78% alcohol by volume. If the manufacturer has 50 liters of the spray containing 70% alcohol, how much pure alcohol should be added so that the final mixture is the required 78% alcohol? (*Hint:* Pure alcohol is 100% alcohol.)

42. ***Insecticide Mixture*** How much water must be added to 3 gallons of a 4% insecticide solution to reduce the concentration to 3%? (*Hint:* Water is 0% insecticide.)

43. ***Antifreeze Mixture*** It is necessary to have a 40% antifreeze solution in the radiator of a certain car. The radiator now holds 20 liters of 20% solution. How many liters of this should be drained and replaced with 100% antifreeze to get the desired strength? (*Hint:* The number of liters drained is equal to the number of liters replaced.)

44. ***Chemical Mixture*** A tank holds 80 liters of a chemical solution. Currently, the solution has a strength of 30%. How much of this should be drained and replaced with a 70% solution to get a final strength of 40%?

Use the method of Example 5 or your own method to solve each problem. Assume all rates and amounts are annual.

45. ***Investments at Different Rates*** John Allen earned \$12,000 last year by giving tennis lessons. He invested part at 3% simple interest and the rest at 4%.

He earned a total of \$440 in interest. How much did he invest at each rate?

Rate (as a Decimal)	Principal	Interest in One Year
.03		
.04		
	12,000	440

46. ***Investments at Different Rates*** Kackie Smith won \$60,000 in a slot machine in Las Vegas. She invested part at 2% simple interest and the rest at 3%. She earned a total of \$1600 in interest. How much was invested at each rate?

Rate (as a Decimal)	Principal	Interest in One Year
.02	x	$.02x$
	$60,000 - x$	
		1600

47. ***Investments at Different Rates*** Jerome Dugas invested some money at 4.5% simple interest and \$1000 less than twice this amount at 3%. His total income from the interest was \$1020. How much was invested at each rate?

48. ***Investments at Different Rates*** Margaret Maggio invested some money at 3.5% simple interest, and \$5000 more than 3 times this amount at 4%. She earned \$1440 in interest. How much did she invest at each rate?

49. ***Investments at Different Rates*** Ed Moura has \$29,000 invested in stocks paying 5%. How much additional money should he invest in certificates of deposit paying 2% so that the average return on the two investments is 3%?

50. ***Investments at Different Rates*** Terry McGinnis placed \$15,000 in an account paying 6%. How much additional money should she deposit at 4% so that the average return on the two investments is 5.5%?

Use the method of Example 6 or your own method to solve each problem.

51. ***Coin Mixture*** Mike Easley has a box of coins that he uses when playing poker with his friends. The box currently contains 44 coins, consisting of pennies, dimes, and quarters. The number of pennies is equal to the number of dimes, and the total value is \$4.37. How many of each denomination of coin does he have in the box?

Denomination	Number of Coins	Value	
.01	x	$.01x$	
	x		
.25			
	44	4.37	Totals

52. ***Coin Mixture*** Melena Fenn found some coins while looking under her sofa pillows. There were equal numbers of nickels and quarters, and twice as many half-dollars as quarters. If she found \$2.60 in all, how many of each denomination of coin did she find?

Denomination	Number of Coins	Value	
.05	x	$.05x$	
	x		
.50	$2x$		
		2.60	Total

53. ***Attendance at a School Play*** The school production of *Hamlet* was a big success. For opening night, 410 tickets were sold. Students paid \$3 each, while nonstudents paid \$7 each. If a total of \$1650 was collected, how many students and how many nonstudents attended?

54. ***Attendance at a Concert*** A total of 550 people attended a Maynard Ferguson concert. Floor tickets cost \$40 each, while balcony tickets cost \$28 each. If a total of \$20,800 was collected, how many of each type of ticket were sold?

55. ***Attendance at a Sporting Event*** At the Sacramento Monarchs home games, Row 1 seats cost \$35 each and Row 2 seats cost \$30 each. The 105 seats in these rows were sold out for the season. The total receipts for them were \$3420. How many of each type of seat were sold? (*Source:* Sacramento Monarchs.)

56. ***Coin Mixture*** In the nineteenth century, the United States minted two-cent and three-cent pieces. Frances Steib has three times as many three-cent pieces as two-cent pieces, and the face value of these coins is \$1.21. How many of each denomination does she have?

57. ***Stamp Denominations*** In January 2006, U.S. first-class mail rates increased to 39 cents for the first ounce, plus 24 cents for each additional ounce. If Sabrina spent $15.00 for a total of 45 stamps of these two denominations, how many stamps of each denomination did she buy? (*Source:* U.S. Postal Service.)

58. ***Movie Ticket Prices*** A movie theater has two ticket prices: $8 for adults and $5 for children. If the box office took in $4116 from the sale of 600 tickets, how many tickets of each kind were sold?

From **Harry Potter and the Chamber of Secrets**

Automobile Racing *In Exercises 59–62, find the time based on the information provided. Use a calculator and round your answers to the nearest thousandth. (Source: The World Almanac and Book of Facts 2006.)*

	Event and Year	Participant	Distance	Rate
59.	Indianapolis 500, 2005	Dan Weldon (Honda)	500 miles	157.579 mph
60.	Daytona 500, 2001	Michael Waltrip (Chevrolet)	500 miles	161.794 mph
61.	Indianapolis 500, 1980	Johnny Rutherford (Hy-Gain McLaren/Goodyear)	255 miles*	148.725 mph
62.	Indianapolis 500, 1975	Bobby Unser (Jorgensen Eagle)	435 miles*	149.213 mph

*rain-shortened

Olympic Results *In Exercises 63–66, find the rate based on the information provided. Use a calculator and round your answers to the nearest hundredth. All events were at the 2004 Olympics. (Source: World Almanac and Book of Facts 2006.)*

	Event	Participant	Distance	Time
63.	100-m hurdles, Women	Joanna Hayes, USA	100 meters	12.37 seconds
64.	400-m hurdles, Women	Fani Halkia, Greece	400 meters	52.82 seconds
65.	400-m hurdles, Men	Felix Sanchez, Dominican Republic	400 meters	47.63 seconds
66.	400-m run, Men	Jeremy Wariner, USA	400 meters	44.00 seconds

Use the formula $d = rt$ in Exercises 67–70.

67. ***Distance Between Cities*** A driver averaged 53 miles per hour and took 10 hours to travel from Memphis to Chicago. What is the distance between Memphis and Chicago?

68. ***Distance Between Cities*** A small plane traveled from Warsaw to Rome, averaging 164 miles per hour. The trip took two hours. What is the distance from Warsaw to Rome?

69. Suppose that an automobile averages 45 miles per hour, and travels for 30 minutes. Is the distance traveled $45 \cdot 30 = 1350$ miles? If not, explain why not, and give the correct distance.

70. Which of the following choices is the best *estimate* for the average speed of a trip of 405 miles that lasted 8.2 hours?

A. 50 miles per hour **B.** 30 miles per hour
C. 60 miles per hour **D.** 40 miles per hour

Use the method of Example 8 or your own method to solve each problem.

71. ***Travel Times of Trains*** A train leaves Little Rock, Arkansas, and travels north at 85 kilometers per hour. Another train leaves at the same time and travels south at 95 kilometers per hour. How long will it take before they are 315 kilometers apart?

	Rate	Time	Distance
First train	85	t	
Second train			

72. ***Travel Times of Steamers*** Two steamers leave a port on a river at the same time, traveling in opposite directions. Each is traveling 22 miles per hour. How long will it take for them to be 110 miles apart?

	Rate	Time	Distance
First steamer		t	
Second steamer	22		

73. ***Travel Times of Commuters*** Nancy and Mark commute to work, traveling in opposite directions. Nancy leaves the house at 8:00 A.M. and averages 35 miles per hour. Mark leaves at 8:15 A.M. and averages 40 miles per hour. At what time will they be 140 miles apart?

74. ***Travel Times of Bicyclers*** Jeff leaves his house on his bicycle at 8:30 A.M. and averages 5 miles per hour. His wife, Joan, leaves at 9:00 A.M., following the same path and averaging 8 miles per hour. At what time will Joan catch up with Jeff?

75. ***Distance Traveled to Work*** When Wayne Pourciau drives his car to work, the trip takes 30 minutes. When he rides the bus, it takes 45 minutes. The average speed of the bus is 12 miles per hour less than his speed when driving. Find the distance he travels to work.

76. ***Distance Traveled to School*** Latoya can get to school in 15 minutes if she rides her bike. It takes her 45 minutes if she walks. Her speed when walking is 10 miles per hour slower than her speed when riding. How far does she travel to school?

77. ***Time Traveled by a Pleasure Boat*** A pleasure boat on the Mississippi River traveled from New Roads, LA, to New Orleans with a stop at White Castle. On the first part of the trip, the boat traveled at an average speed of 10 miles per hour. From White Castle to New Orleans the average speed was 15 miles per hour. The entire trip covered 100 miles. How long did the entire trip take if the two parts each took the same number of hours?

78. ***Time Traveled on a Visit*** Steve leaves Nashville to visit his cousin David in Napa, 80 miles away. He travels at an average speed of 50 miles per hour. One-half hour later David leaves to visit Steve, traveling at an average speed of 60 miles per hour. How long after David leaves will they meet?

7.3 Ratio, Proportion, and Variation

Writing Ratios • Unit Pricing • Solving Proportions • Direct Variation • Inverse Variation • Joint and Combined Variation

Writing Ratios

One of the most frequently used mathematical concepts in everyday life is *ratio*. A baseball player's batting average is actually a ratio. The slope, or pitch, of a roof on a building may be expressed as a ratio. Ratios provide a way of comparing two numbers or quantities.

During the first season (1960) of ***The Andy Griffith Show,*** the episode "Opie's Charity" featured a conversation between Opie and Andy during which Andy explained to Opie that his donation of three cents to the under-privileged children's drive at school was "a piddlin' amount."

ANDY: I was reading here just the other day where there's somewhere like 400 needy boys in this county alone, or one and a half boys per square mile.
OPIE: There is?
ANDY: Sho' is.
OPIE: I've never seen one, Pa.
ANDY: Never seen one what?
OPIE: A half a boy.
ANDY: Well it's not really a half a boy. It's **a ratio.**
OPIE: Horatio who?
ANDY: Not *Ho*ratio, *a* ratio. It's mathematics. Arithmetic. Look now Opie, just forget that part of it. Forget the part about the half a boy.
OPIE: It's pretty hard to forget a thing like that, Pa.
ANDY: Well try.
OPIE: Poor Horatio.

Ratio

A **ratio** is a quotient of two quantities. The ratio of the number a to the number b is written

$$\boldsymbol{a \text{ to } b}, \quad \boldsymbol{\frac{a}{b}}, \quad \text{or} \quad \boldsymbol{a\!:\!b}.$$

When ratios are used in comparing units of measure, the units should be the same.

EXAMPLE 1 Writing Ratios

Write a ratio for each word phrase.

(a) 5 hours to 3 hours **(b)** 6 hours to 3 days

SOLUTION

(a) The ratio of 5 hr to 3 hr is $\frac{5 \text{ hr}}{3 \text{ hr}} = \frac{5}{3}$.

(b) To find the ratio of 6 hr to 3 days, first convert 3 days to hours:

$$3 \text{ days} = 3 \text{ days} \cdot \frac{24 \text{ hr}}{1 \text{ day}} = 72 \text{ hr}.$$

The ratio of 6 hr to 3 days is thus

$$\frac{6 \text{ hr}}{3 \text{ days}} = \frac{6 \text{ hr}}{72 \text{ hr}} = \frac{6}{72} = \frac{1}{12}.$$

Unit Pricing Ratios can be applied in unit pricing, to see which size of an item offered in different sizes produces the best price per unit. To do this, set up the ratio of the price of the item to the number of units on the label. Then divide to obtain the price per unit.

EXAMPLE 2 Finding Price per Unit

The Cub Foods supermarket in Coon Rapids, Minnesota, charges the following prices for a jar of extra crunchy peanut butter

Peanut Butter

Size	Price
18-oz	$1.50
40-oz	$4.14
64-oz	$6.29

Which size is the best buy? That is, which size has the lowest unit price?

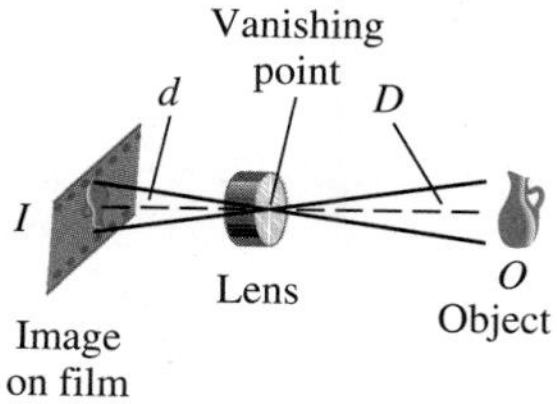

When you look a long way down a straight road or railroad track, it seems to narrow as it vanishes in the distance. The point where the sides seem to touch is called the **vanishing point.** The same thing occurs in the lens of a camera, as shown in the figure. Suppose I represents the length of the image, O the length of the object, d the distance from the lens to the film, and D the distance from the lens to the object. Then

$$\frac{\text{Image length}}{\text{Object length}} = \frac{\text{Image distance}}{\text{Object distance}}$$

or $$\frac{I}{O} = \frac{d}{D}.$$

Given the length of the image on the film and its distance from the lens, then the length of the object determines how far away the lens must be from the object to fit on the film.

SOLUTION

To find the best buy, write ratios comparing the price for each size jar to the number of units (ounces) per jar. The results in Table 2 are rounded to the nearest thousandth.

TABLE 2

Size	Unit Cost (dollars per ounce)
18-oz	$\frac{\$1.50}{18} = \$.083$ ← The best buy
40-oz	$\frac{\$4.14}{40} = \$.104$
64-oz	$\frac{\$6.29}{64} = \$.098$

Because the 18-oz size produces the lowest unit cost, it is the best buy. Buying the largest size does not always provide the best buy, although this is often true. ■

Solving Proportions We now define a special type of equation called a *proportion.*

Proportion

A **proportion** is a statement that says that two ratios are equal.

For example, $$\frac{3}{4} = \frac{15}{20} \quad \text{Proportion}$$

is a proportion that says that the ratios $\frac{3}{4}$ and $\frac{15}{20}$ are equal. In the proportion

$$\frac{a}{b} = \frac{c}{d} \quad (b, d \neq 0),$$

a, b, c, and d are the **terms** of the proportion. The a and d terms are called the **extremes,** and the b and c terms are called the **means.** We read the proportion $\frac{a}{b} = \frac{c}{d}$ as "a is to b as c is to d." Multiplying each side of this proportion by the common denominator, bd, gives

$$bd \cdot \frac{a}{b} = bd \cdot \frac{c}{d}$$

$$\frac{b}{b}(d \cdot a) = \frac{d}{d}(b \cdot c) \quad \text{Associative and commutative properties}$$

$$ad = bc. \quad \text{Commutative and identity properties}$$

We can also find the products ad and bc by multiplying diagonally.

$$\frac{a}{b} = \frac{c}{d} \qquad \begin{matrix} bc \\ ad \end{matrix}$$

For this reason, ad and bc are called **cross products.**

Cross Products

If $\frac{a}{b} = \frac{c}{d}$, then the cross products ad and bc are equal.

Also, if $ad = bc$, then $\frac{a}{b} = \frac{c}{d}$ (as long as $b \neq 0$, $d \neq 0$).

From this rule, if $\frac{a}{b} = \frac{c}{d}$ then $ad = bc$; that is, ***the product of the extremes equals the product of the means.***

If $\frac{a}{c} = \frac{b}{d}$, then $ad = cb$, or $ad = bc$. This means that the two corresponding proportions are equivalent, and

$$\text{the proportion } \frac{a}{b} = \frac{c}{d} \text{ can also be written as } \frac{a}{c} = \frac{b}{d} \quad (c \neq 0).$$

Sometimes one form is more convenient to work with than the other.

EXAMPLE 3 Solving Proportions

Solve each proportion.

(a) $\frac{63}{x} = \frac{9}{5}$ **(b)** $\frac{8}{5} = \frac{12}{r}$

SOLUTION

(a)

$$\frac{63}{x} = \frac{9}{5}$$

$$63 \cdot 5 = 9x \qquad \text{Set the cross products equal.}$$

$$315 = 9x \qquad \text{Multiply.}$$

$$35 = x \qquad \text{Divide by 9.}$$

The solution set is $\{35\}$.

(b)

$$\frac{8}{5} = \frac{12}{r}$$

$$8r = 5 \cdot 12 \qquad \text{Set the cross products equal.}$$

$$8r = 60 \qquad \text{Multiply.}$$

$$r = \frac{60}{8} = \frac{15}{2} \qquad \text{Divide by 8; express in lowest terms.}$$

The solution set is $\left\{\frac{15}{2}\right\}$. ■

EXAMPLE 4 Solving an Equation Using Cross Products

Solve the equation $\frac{m-2}{5} = \frac{m+1}{3}$.

SOLUTION

Find the cross products, and set them equal to each other.

Be sure to use parentheses.

$$\begin{aligned} 3(m-2) &= 5(m+1) && \text{Cross products} \\ 3m - 6 &= 5m + 5 && \text{Distributive property} \\ 3m &= 5m + 11 && \text{Add 6.} \\ -2m &= 11 && \text{Subtract } 5m. \\ m &= -\frac{11}{2} && \text{Divide by } -2. \end{aligned}$$

The solution set is $\{-\frac{11}{2}\}$.

EXAMPLE 5 Using a Proportion to Predict Population

Biologists use algebra to estimate the number of fish in a lake. They first catch a sample of fish and mark each specimen with a harmless tag. Some weeks later, they catch a similar sample of fish from the same areas of the lake and determine the proportion of previously tagged fish in the new sample. The total fish population is estimated by assuming that the proportion of tagged fish in the new sample is the same as the proportion of tagged fish in the entire lake.

Suppose biologists tag 300 fish on May 1. When they return on June 1 and take a new sample of 400 fish, 5 of the 400 were previously tagged. Estimate the number of fish in the lake.

SOLUTION

Let x represent the number of fish in the lake. Set up and solve a proportion.

$$\frac{300}{x} = \frac{5}{400}$$

Tagged fish on May 1 → 300; Total fish in the lake → x; 5 ← Tagged fish in the June 1 sample; 400 ← Total number in the June 1 sample

$$\begin{aligned} 5x &= 120{,}000 && \text{Cross products} \\ x &= 24{,}000 && \text{Divide by 5.} \end{aligned}$$

There are approximately 24,000 fish in the lake.

TABLE 3

Number of Rooms	Cost of the Job
1	$ 49.99
2	$ 99.98
3	$149.97
4	$199.96
5	$249.95

Direct Variation

Suppose that a carpet cleaning service charges 49.99 per room to shampoo a carpet. Table 3 shows the relationship between the number of rooms cleaned and the cost of the total job for 1 through 5 rooms.

If we divide the cost of the job by the number of rooms, in each case we obtain the quotient, or ratio, 49.99 (dollars per room). Suppose that we let x represent the number of rooms and y represent the cost for cleaning that number of rooms. Then the relationship between x and y is given by the equation

$$\frac{y}{x} = 49.99, \quad \text{or} \quad y = 49.99x.$$

This relationship between x and y is an example of *direct variation.*

Direct Variation

y varies directly as x, or **y is directly proportional to x,** if there exists a nonzero constant k such that

$$y = kx, \quad \text{or, equivalently,} \quad \frac{y}{x} = k.$$

The constant k is a numerical value called the **constant of variation.**

EXAMPLE 6 Solving a Direct Variation Problem

Suppose y varies directly as x, and $y = 50$ when $x = 20$. Find y when $x = 14$.

SOLUTION

Since y varies directly as x, there exists a constant k such that $y = kx$. Find k by replacing y with 50 and x with 20.

$$y = kx \quad \text{Variation equation}$$
$$50 = k \cdot 20 \quad \text{Substitute the given values.}$$
$$\frac{5}{2} = k \quad \text{Divide by 20; express in lowest terms.}$$

Since $y = kx$ and $k = \frac{5}{2}$,

$$y = \frac{5}{2}x.$$

Now find y when $x = 14$.

$$y = \frac{5}{2} \cdot 14 = 35$$

The value of y is 35 when $x = 14$.

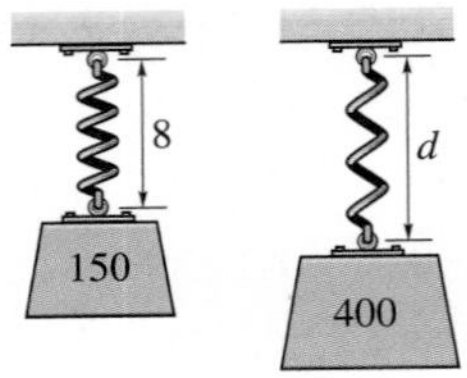

FIGURE 5

EXAMPLE 7 Solving a Direct Variation Problem

Hooke's law for an elastic spring states that the distance a spring stretches is directly proportional to the force applied. If a force of 150 pounds stretches a certain spring 8 centimeters, how much will a force of 400 pounds stretch the spring? See Figure 5.

SOLUTION

If d is the distance the spring stretches and f is the force applied, then $d = kf$ for some constant k.

$$d = kf \quad \text{Formula}$$
$$8 = k \cdot 150 \quad \text{Let } d = 8 \text{ and } f = 150.$$
$$k = \frac{8}{150} = \frac{4}{75} \quad \text{Find } k.$$

Thus $d = \frac{4}{75}f$.

For a force of 400 pounds,

$$d = \frac{4}{75}(400) = \frac{64}{3}. \quad \text{Let } f = 400.$$

The spring will stretch $\frac{64}{3}$ centimeters if a force of 400 pounds is applied. ■

In summary, follow these steps to solve a variation problem.

Solving a Variation Problem

Step 1 Write the variation equation.

Step 2 Substitute the initial values and solve for k.

Step 3 Rewrite the variation equation with the value of k from Step 2.

Step 4 Substitute the remaining values, solve for the unknown, and find the required answer.

In some cases one quantity will vary directly as a *power* of another.

Direct Variation as a Power

y* varies directly as the *n*th power of *x if there exists a nonzero real number k such that

$$y = kx^n.$$

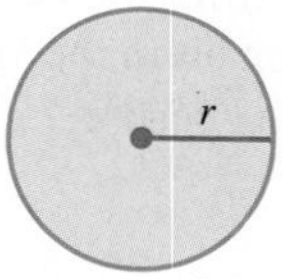

$A = \pi r^2$

FIGURE 6

An example of direct variation as a power involves the area of a circle. See Figure 6. The formula for the area of a circle is $A = \pi r^2$. Here, π is the constant of variation, and the area A varies directly as the square of the radius r.

EXAMPLE 8 Solving a Direct Variation Problem

The distance a body falls from rest varies directly as the square of the time it falls (here we disregard air resistance). If a skydiver falls 64 feet in 2 seconds, how far will she fall in 8 seconds?

SOLUTION

Step 1 If d represents the distance the skydiver falls and t the time it takes to fall, then d is a function of t, and, for some constant k, $d = kt^2$.

Step 2 To find the value of k, use the fact that the skydiver falls 64 feet in 2 seconds.

$$d = kt^2 \quad \text{Formula}$$

$$64 = k(2)^2 \quad \text{Let } d = 64 \text{ and } t = 2.$$

$$k = 16 \quad \text{Find } k.$$

Step 3 With this result, the variation equation becomes

$$d = 16t^2.$$

Step 4 Now let $t = 8$ to find the number of feet the skydiver will fall in 8 seconds.

$$d = 16t^2 = 16(8)^2 = 1024 \quad \text{Let } t = 8.$$

$8^2 = 8 \cdot 8 = 64$

The skydiver will fall 1024 feet in 8 seconds.

Inverse Variation ***In direct variation where $k > 0$, as x increases, y increases, and similarly as x decreases, y decreases.*** Another type of variation is *inverse variation.*

Inverse Variation

y* varies inversely as *x if there exists a nonzero real number k such that

$$y = \frac{k}{x}, \quad \text{or, equivalently,} \quad xy = k.$$

Also, ***y* varies inversely as the *n*th power of *x*** if there exists a nonzero real number k such that

$$y = \frac{k}{x^n}.$$

Johann Kepler (1571–1630) established the importance of the **ellipse** in 1609, when he discovered that the orbits of the planets around the sun were elliptical, not circular.

Halley's comet, which has been studied since 467 B.C., has an elliptical orbit which is long and narrow, with one axis much longer than the other. This comet was named for the British astronomer and mathematician **Edmund Halley** (1656–1742), who predicted its return after observing it in 1682. The comet appears regularly every 76 years.

EXAMPLE 9 Solving an Inverse Variation Problem

The weight of an object above Earth varies inversely as the square of its distance from the center of Earth. A space vehicle in an elliptical orbit has a maximum distance from the center of Earth (apogee) of 6700 miles. Its minimum distance from the center of Earth (perigee) is 4090 miles. See Figure 7 (not to scale). If an astronaut in the vehicle weighs 57 pounds at its apogee, what does the astronaut weigh at the perigee?

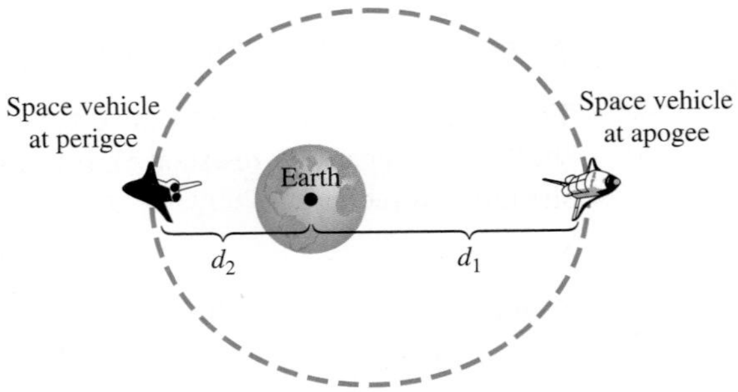

FIGURE 7

SOLUTION

If w is the weight and d is the distance from the center of Earth, then

$$w = \frac{k}{d^2}$$

for some constant k.

At the apogee the astronaut weighs 57 pounds and the distance from the center of Earth is 6700 miles. Use these values to find k.

$$57 = \frac{k}{(6700)^2} \quad \text{Let } w = 57 \text{ and } d = 6700.$$

$$k = 57(6700)^2 \quad \text{Multiply by } (6700)^2\text{; rewrite.}$$

Then the weight at the perigee with $d = 4090$ miles is

$$w = \frac{57(6700)^2}{(4090)^2} \approx 153 \text{ pounds.} \quad \text{Use a calculator.}$$

Joint and Combined Variation If one variable varies as the product of several other variables (perhaps raised to powers), the first variable is said to **vary jointly** as the others.

EXAMPLE 10 Solving a Joint Variation Problem

The strength of a rectangular beam varies jointly as its width and the square of its depth. If the strength of a beam 2 inches wide by 10 inches deep is 1000 pounds per square inch, what is the strength of a beam 4 inches wide and 8 inches deep?

SOLUTION

If S represents the strength, w the width, and d the depth, then, for some constant k

$$S = kwd^2$$

$$1000 = k(2)(10)^2 \quad \text{Let } S = 1000, w = 2, \text{ and } d = 10.$$

$10^2 = 10 \cdot 10 = 100$

$$1000 = 200k \quad \text{Apply the exponent; multiply.}$$

$$k = 5, \quad \text{Divide by 200; rewrite.}$$

so

$$S = 5wd^2.$$

Find S when $w = 4$ and $d = 8$ by substitution in $S = 5wd^2$.

$$S = 5(4)(8)^2 = 1280 \quad \text{Let } w = 4 \text{ and } d = 8.$$

The strength of the beam is 1280 pounds per square inch.

There are situations that involve combinations of direct and inverse variation. The final example shows a typical **combined variation** problem.

EXAMPLE 11 Solving a Combined Variation Problem

Body mass index, or BMI, is used by physicians to assess a person's level of fatness. A BMI from 19 through 25 is considered desirable. BMI varies directly as an individual's weight in pounds and inversely as the square of the individual's height in inches. A person who weighs 118 lb and is 64 in. tall has a BMI of 20. (The BMI is rounded to the nearest whole number.) Find the BMI of a person who weighs 165 lb with a height of 70 in.

SOLUTION

Let B represent the BMI, w the weight, and h the height. Then

$$B = \frac{kw}{h^2}$$ ← BMI varies directly as the weight. ← BMI varies inversely as the square of the height.

$$20 = \frac{k(118)}{64^2}$$ Let $B = 20$, $w = 118$, and $h = 64$.

$$k = \frac{20(64^2)}{118}$$ Multiply by 64^2; divide by 118.

$$k \approx 694.$$ Use a calculator.

Now find B when $k = 694$, $w = 165$, and $h = 70$.

$$B = \frac{694(165)}{70^2} \approx 23$$ Nearest whole number

The person's BMI is 23.

7.3 EXERCISES

Determine the ratio and write it in lowest terms.

1. 25 feet to 40 feet

2. 16 miles to 48 miles

3. 18 dollars to 72 dollars

4. 300 people to 250 people

5. 144 inches to 6 feet

6. 60 inches to 2 yards

7. 5 days to 40 hours

8. 75 minutes to 2 hours

9. Which ratio is not the same as the ratio 2 to 5?
A. .4 **B.** 4 to 10 **C.** 20 to 50 **D.** 5 to 2

10. Give three ratios that are equivalent to the ratio 4 to 3.

11. Explain the distinction between *ratio* and *proportion*. Give examples.

12. Suppose that someone told you to use cross products in order to multiply fractions. How would you explain to the person what is wrong with his or her thinking?

Decide whether each proportion is true *or* false.

13. $\frac{5}{35} = \frac{8}{56}$

14. $\frac{4}{12} = \frac{7}{21}$

15. $\frac{120}{82} = \frac{7}{10}$

16. $\frac{27}{160} = \frac{18}{110}$

17. $\frac{\frac{1}{2}}{5} = \frac{1}{10}$

18. $\frac{\frac{1}{3}}{6} = \frac{1}{18}$

Solve each equation.

19. $\frac{k}{4} = \frac{175}{20}$

20. $\frac{49}{56} = \frac{z}{8}$

21. $\frac{3x - 2}{5} = \frac{6x - 5}{11}$

22. $\frac{5 + x}{3} = \frac{x + 7}{5}$

23. $\frac{3t + 1}{7} = \frac{2t - 3}{6}$

24. $\frac{2p + 7}{3} = \frac{p - 1}{4}$

Solve each problem. In Exercises 25–31, assume all items are equally priced.

25. ***Price of Candy Bars*** If 16 candy bars cost \$20.00, how much do 24 candy bars cost?

26. ***Price of Ringtones*** If 12 ringtones cost \$30.00, how much do 8 ringtones cost?

27. ***Price of Oil*** Eight quarts of oil cost \$14.00. How much do 5 quarts of oil cost?

28. ***Price of Tires*** Four tires cost \$398.00. How much do 7 tires cost?

29. ***Price of Jeans*** If 9 pairs of jeans cost \$121.50, find the cost of 5 pairs.

30. ***Price of Shirts*** If 7 shirts cost \$87.50, find the cost of 11 shirts.

31. ***Price of Gasoline*** If 6 gallons of premium unleaded gasoline cost \$15.54, how much would it cost to completely fill a 15-gallon tank?

32. ***Sales Tax*** If sales tax on a \$16.00 DVD is \$1.32, how much would the sales tax be on a \$120.00 DVD player?

33. ***Distance Between Cities*** The distance between Kansas City, Missouri, and Denver is 600 miles. On a certain wall map, this is represented by a length of 2.4 feet. On the map, how many feet would there be between Memphis and Philadelphia, two cities that are actually 1000 miles apart?

34. ***Distance Between Cities*** The distance between Singapore and Tokyo is 3300 miles. On a certain wall map, this distance is represented by 11 inches. The actual distance between Mexico City and Cairo is 7700 miles. How far apart are they on the same map?

35. ***Distance Between Cities*** A wall map of the United States has a distance of 8.5 inches between Memphis and Denver, two cities that are actually 1040 miles apart. The actual distance between St. Louis and Des Moines is 333 miles. How far apart are St. Louis and Des Moines on the map?

36. ***Distance Between Cities*** The same map of the United States mentioned in the previous exercise has a distance of 8.0 inches between New Orleans and Chicago, two cities that are actually 912 miles apart. The actual distance between Milwaukee and Seattle is 1940 miles. How far apart are Milwaukee and Seattle on the map?

37. ***Distance Between Cities*** On a world globe, the distance between Capetown and Bangkok, two cities that are actually 10,080 kilometers apart, is 12.4 inches. The actual distance between Moscow and Berlin is 1610 kilometers. How far apart are Moscow and Berlin on this globe?

38. ***Distance Between Cities*** On a world globe, the distance between Rio de Janeiro and Hong Kong, two cities that are actually 17,615 kilometers apart, is 21.5 inches. The actual distance between Paris and Stockholm is 1605 kilometers. How far apart are Paris and Stockholm on this globe?

39. ***Cleaning Mixture*** According to the directions on a bottle of Armstrong® Concentrated Floor Cleaner, for routine cleaning, $\frac{1}{4}$ cup of cleaner should be mixed with 1 gallon of warm water. How much cleaner should be mixed with $10\frac{1}{2}$ gallons of water?

40. ***Cleaning Mixture*** The directions on the bottle mentioned in Exercise 39 also specify that for extra-strength cleaning, $\frac{1}{2}$ cup of cleaner should be used for each gallon of water. For extra-strength cleaning, how much cleaner should be mixed with $15\frac{1}{2}$ gallons of water?

41. ***Exchange Rate (Dollars and Euros)*** The euro is the common currency used by most European countries, including Italy. On January 29, 2006, the exchange rate between euros and U.S. dollars was 1 euro to \$1.2128. Ashley went to Rome and exchanged her U.S. currency for euros, receiving 300 euros. How much in U.S. dollars did she exchange? (*Source:* www.xe.com/ucc)

42. ***Exchange Rate (U.S. and Mexico)*** If 8 U.S. dollars can be exchanged for 84.30 Mexican pesos, how many pesos can be obtained for \$65? (Round to the nearest hundredth.)

43. ***Tagging Fish for a Population Estimate*** Biologists tagged 250 fish in an oxbow lake known as False River on October 5. On a later date they found 7 tagged fish in a sample of 350. Estimate the total number of fish in False River to the nearest hundred.

44. ***Tagging Fish for a Population Estimate*** On May 13 researchers at Argyle Lake tagged 420 fish. When they returned a few weeks later, their sample of 500 fish contained 9 that were tagged. Give an approximation of the fish population in Argyle Lake to the nearest hundred.

Merchandise Pricing *A supermarket was surveyed to find the prices charged for items in various sizes. Find the best buy (based on price per unit) for each particular item.*

45. Granulated Sugar

Size	Price
4-lb	$1.78
10-lb	$4.39

46. Ground Coffee

Size	Price
13-oz	$2.58
39-oz	$4.44

47. Salad Dressing

Size	Price
16-oz	$2.44
32-oz	$2.98
48-oz	$4.95

48. Black Pepper

Size	Price
2-oz	$1.79
4-oz	$2.59
8-oz	$5.59

49. Vegetable Oil

Size	Price
16-oz	$1.54
24-oz	$2.08
64-oz	$3.63
128-oz	$5.65

50. Mouthwash

Size	Price
8.5-oz	$.99
16.9-oz	$1.87
33.8-oz	$2.49
50.7-oz	$2.99

51. Tomato Ketchup

Size	Price
14-oz	$1.39
24-oz	$1.55
36-oz	$1.78
64-oz	$3.99

52. Grape Jelly

Size	Price
12-oz	$1.05
18-oz	$1.73
32-oz	$1.84
48-oz	$2.88

Two triangles are ***similar*** *if they have the same shape (but not necessarily the same size). Similar triangles have sides that are proportional. The figure shows two similar triangles.*

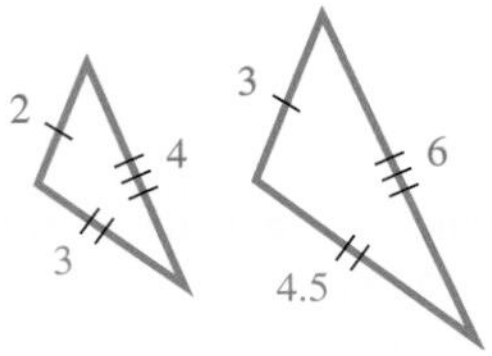

Notice that the ratios of the corresponding sides are all equal to $\frac{3}{2}$:

$$\frac{\mathbf{3}}{\mathbf{2}} = \frac{3}{2} \qquad \frac{\mathbf{4.5}}{\mathbf{3}} = \frac{3}{2} \qquad \frac{\mathbf{6}}{\mathbf{4}} = \frac{3}{2}.$$

If we know that two triangles are similar, we can set up a proportion to solve for the length of an unknown side. Use a proportion to find the lengths x and y given that the pair of triangles are similar.

53.

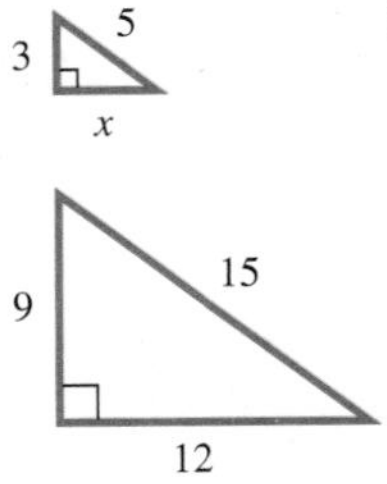

54.

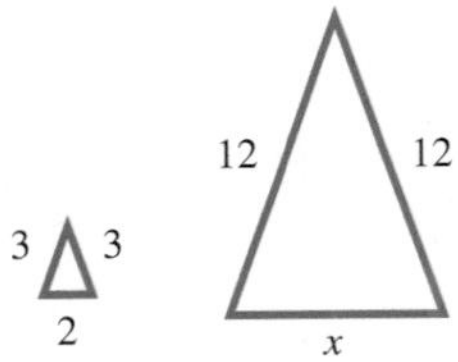

55.

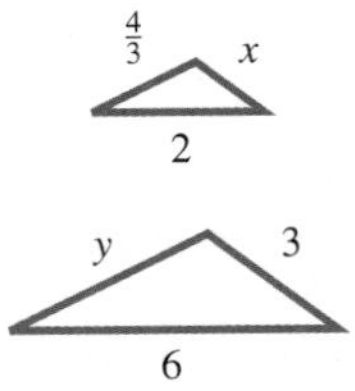

56.

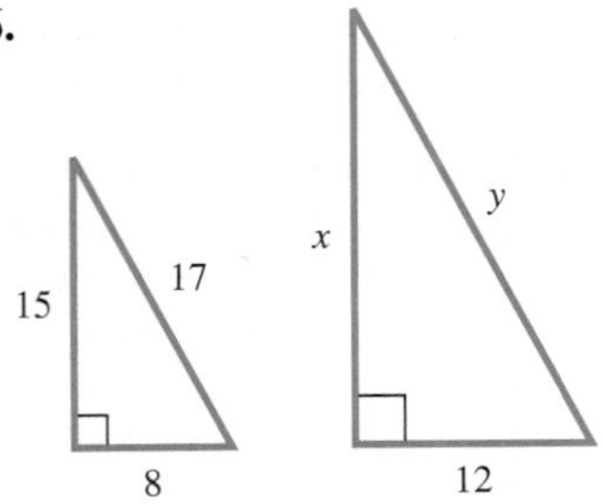

For the problems in Exercises 57 and 58, **(a)** *draw a sketch consisting of two right triangles, depicting the situation described, and* **(b)** *solve the problem. (Source: Guinness World Records.)*

57. ***George Washington's Chair*** An enlarged version of the chair used by George Washington at the Constitutional Convention casts a shadow 18 feet long at the same time a vertical pole 12 feet high casts a shadow 4 feet long. How tall is the chair?

58. ***Candle at an Exhibition*** One of the tallest candles ever constructed was exhibited at the 1897 Stockholm Exhibition. If it cast a shadow 5 feet long at the same time a vertical pole 32 feet high cast a shadow 2 feet long, how tall was the candle?

Consumer Price Index *The Consumer Price Index, issued by the U.S. Bureau of Labor Statistics, provides a means of determining the purchasing power of the U.S. dollar from one year to the next. Using the period from 1982 to 1984 as a measure of* 100.0, *the Consumer Price Index for selected years from 1990 to 2004 is shown here.*

Year	Consumer Price Index
1990	130.7
1992	140.3
1994	148.2
1996	156.9
1998	163.0
2000	172.2
2002	179.9
2004	188.9

Source: Bureau of Labor Statistics.

To use the Consumer Price Index to predict a price in a particular year, we can set up a proportion and compare it with a known price in another year, as follows:

$$\frac{\text{Price in year } A}{\text{Index in year } A} = \frac{\text{Price in year } B}{\text{Index in year } B}.$$

Use the Consumer Price Index figures in the table to find the amount that would be charged for the purchase of the same amount of groceries that cost \$120 in 1990. Give your answer to the nearest dollar.

59. in 1996

60. in 2000

61. in 2002

62. in 2004

Solve each problem involving variation.

63. If x varies directly as y, and $x = 27$ when $y = 6$, find x when $y = 2$.

64. If z varies directly as x, and $z = 30$ when $x = 8$, find z when $x = 4$.

65. If m varies directly as p^2, and $m = 20$ when $p = 2$, find m when $p = 5$.

66. If a varies directly as b^2, and $a = 48$ when $b = 4$, find a when $b = 7$.

67. If p varies inversely as q^2, and $p = 4$ when $q = \frac{1}{2}$, find p when $q = \frac{3}{2}$.

68. If z varies inversely as x^2, and $z = 9$ when $x = \frac{2}{3}$, find z when $x = \frac{5}{4}$.

69. ***Interest on an Investment*** The interest on an investment varies directly as the rate of interest. If the interest is \$48 when the interest rate is 5%, find the interest when the rate is 4.2%.

70. ***Area of a Triangle*** For a given base, the area of a triangle varies directly as its height. Find the area of a triangle with a height of 6 inches, if the area is 10 square inches when the height is 4 inches.

71. ***Speed of a Car*** Over a specified distance, speed varies inversely with time. If a car goes a certain distance in one-half hour at 30 miles per hour, what speed is needed to go the same distance in three-fourths of an hour?

72. ***Length of a Rectangle*** For a constant area, the length of a rectangle varies inversely as the width. The length of a rectangle is 27 feet when the width is 10 feet. Find the length of a rectangle with the same area if the width is 18 feet.

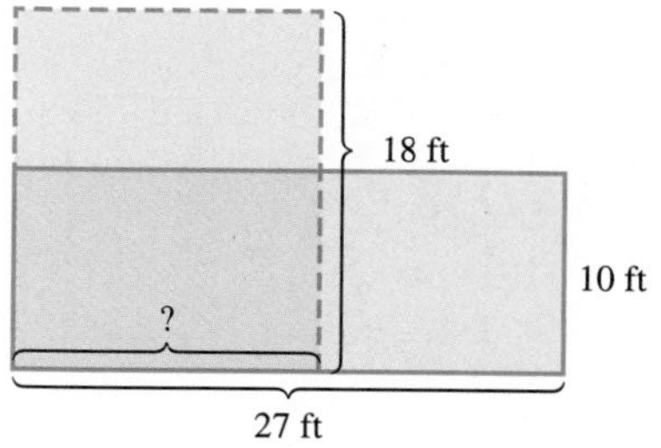

73. ***Weight of a Moose*** The weight of an object on the moon varies directly as the weight of the object on Earth. According to *Guinness World Records*, "Shad," a goat owned by a couple in California, is the largest known goat, weighing 352 pounds. Shad would weigh about 59 pounds on the moon. A bull moose weighing 1800 pounds was shot in Canada and is the largest confirmed moose. How much would the moose have weighed on the moon?

74. *Voyage in a Paddleboat* According to *Guinness World Records*, the longest recorded voyage in a paddle boat is 2226 miles in 103 days by the foot power of two boaters down the Mississippi River. Assuming a constant rate, how far would they have gone if they had traveled 120 days? (Distance varies directly as time.)

75. *Pressure Exerted by a Liquid* The pressure exerted by a certain liquid at a given point varies directly as the depth of the point beneath the surface of the liquid. The pressure at a depth of 10 feet is 50 pounds per square inch. What is the pressure at a depth of 20 feet?

76. *Pressure of a Gas in a Container* If the volume is constant, the pressure of a gas in a container varies directly as the temperature. If the pressure is 5 pounds per square inch at a temperature of 200 degrees Kelvin, what is the pressure at a temperature of 300 degrees Kelvin?

77. *Pressure of a Gas in a Container* If the temperature is constant, the pressure of a gas in a container varies inversely as the volume of the container. If the pressure is 10 pounds per square foot in a container with 3 cubic feet, what is the pressure in a container with 1.5 cubic feet?

78. *Force Required to Compress a Spring* The force required to compress a spring varies directly as the change in the length of the spring. If a force of 12 pounds is required to compress a certain spring 3 inches, how much force is required to compress the spring 5 inches?

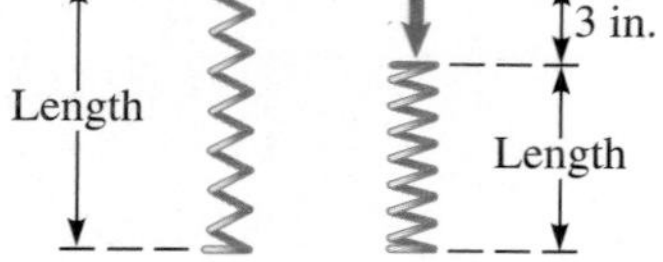

79. *Falling Body* For a body falling freely from rest (disregarding air resistance), the distance the body falls varies directly as the square of the time. If an object is dropped from the top of a tower 400 feet high and hits the ground in 5 seconds, how far did it fall in the first 3 seconds?

80. *Illumination from a Light Source* The illumination produced by a light source varies inversely as the square of the distance from the source. If the illumination produced 4 feet from a light source is 75 foot-candles, find the illumination produced 9 feet from the same source.

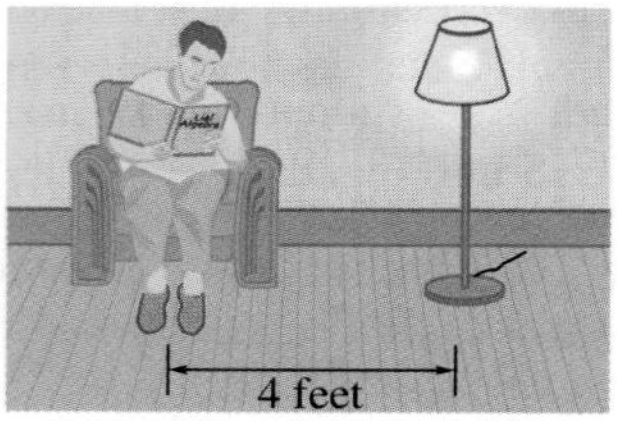

81. *Volume of Gas* Natural gas provides 35.8% of U.S. energy. (*Source*: U.S. Energy Department.) The volume of gas varies inversely as the pressure and directly as the temperature. [Temperature must be measured in *Kelvin* (K), a unit of measurement used in physics.] If a certain gas occupies a volume of 1.3 liters at 300 K and a pressure of 18 newtons, find the volume at 340 K and a pressure of 24 newtons.

82. *Skidding Car* The force needed to keep a car from skidding on a curve varies inversely as the radius of the curve and jointly as the weight of the car and the square of the speed. If 242 pounds of force keep a 2000-pound car from skidding on a curve of radius 500 feet at 30 miles per hour, what force would keep the same car from skidding on a curve of radius 750 feet at 50 miles per hour?

83. *Load Supported by a Column* The maximum load that a cylindrical column with a circular cross section can hold varies directly as the fourth power of the diameter of the cross section and inversely as the square of the height. A 9-meter column 1 meter in diameter will support 8 metric tons. How many metric tons can be supported by a column 12 meters high and $\frac{2}{3}$ meter in diameter?

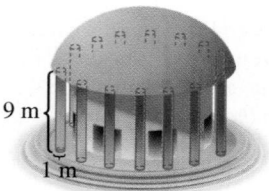

Load = 8 metric tons

Fish Weight-Estimation *Exercises 84 and 85 describe weight-estimation formulas that fishermen have used over the years. Girth is the distance around the body of the fish. (Source: Sacramento Bee, November 9, 2000.)*

84. The weight of a bass varies jointly as its girth and the square of its length. A prize-winning bass weighed in at 22.7 pounds and measured 36 inches long with a 21-inch girth. How much would a bass 28 inches long with an 18-inch girth weigh?

85. The weight of a trout varies jointly as its length and the square of its girth. One angler caught a trout that weighed 10.5 pounds and measured 26 inches long with an 18-inch girth. Find the weight of a trout that is 22 inches long with a 15-inch girth.

86. Bill Veeck was the owner of several major league baseball teams in the 1950s and 1960s. He was known to often sit in the stands and enjoy games with his paying customers. Here is a quote attributed to him:

"I have discovered in 20 years of moving around a ballpark, that the knowledge of the game is usually in inverse proportion to the price of the seats."

Explain in your own words the meaning of this statement. (To prove his point, Veeck once allowed the fans (as shown in the photo) to vote on managerial decisions.)

7.4 Linear Inequalities

Number Lines and Interval Notation • Addition Property of Inequality • Multiplication Property of Inequality • Solving Linear Inequalities • Applications • Three-Part Inequalities

Number Lines and Interval Notation Inequalities are algebraic expressions related by

$<$ "is less than,"	$\leq$ "is less than or equal to,"
$>$ is greater than,"	$\geq$ "is greater than or equal to."

We solve an inequality by finding all real number solutions for it. For example, the solution set of $x \leq 2$ includes *all real numbers* that are less than or equal to 2, not just the *integers* less than or equal to 2.

We show the solution set of an inequality by graphing. We graph all the real numbers satisfying $x \leq 2$ by placing a square bracket at 2 on a number line and drawing an arrow from the bracket to the left (to show that all numbers less than 2 are also part of the graph).* See Figure 8.

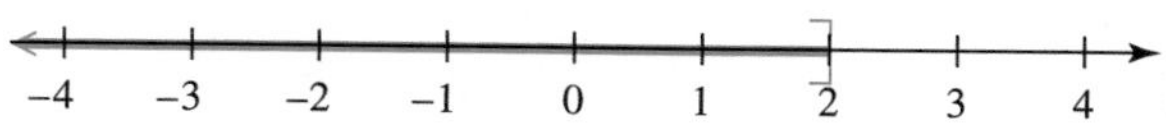

FIGURE 8

*Some texts use solid circles rather than square brackets to indicate that the end point is included on a number line graph. [Open circles are also used to indicate noninclusion rather than parentheses, as described in Example 1(a).]

Archimedes, one of the greatest mathematicians of antiquity, is shown on this Italian stamp. He was born in the Greek city of Syracuse about 287 B.C.

A colorful story about Archimedes relates his reaction to one of his discoveries. While taking a bath, he noticed that an immersed object, if heavier than a fluid, "will, if placed in it, descend to the bottom of the fluid, and the solid will, when weighed in the fluid, be lighter than its true weight by the weight of the fluid displaced." This discovery so excited him that he ran through the streets shouting "Eureka!" ("I have found it!") without bothering to clothe himself!

Archimedes met his death at age 75 during the pillage of Syracuse. He was using a sand tray to draw geometric figures when a Roman soldier came upon him. He ordered the soldier to move clear of his "circles," and the soldier obliged by killing him.

The set of numbers less than or equal to 2 is an example of an **interval** on the number line. To write intervals, we use **interval notation.** For example, using this notation, the interval of all numbers less than or equal to 2 is written as $(-\infty, 2]$. The **negative infinity** symbol $-\infty$ does not indicate a number. It is used to show that the interval includes all real numbers less than 2. As on the number line, the square bracket indicates that 2 is part of the solution. ***A parenthesis is always used next to the infinity symbol.*** The set of real numbers is written in interval notation as $(-\infty, \infty)$.

Examples of other sets written in interval notation are shown in Table 4. In these intervals, assume that $a < b$.

TABLE 4

Type of Interval	Set	Interval Notation	Graph
Open interval	$\{x \mid x > a\}$	(a, ∞)	
	$\{x \mid a < x < b\}$	(a, b)	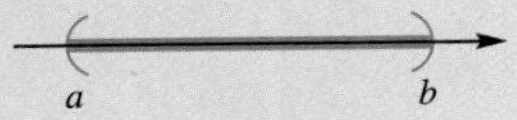
	$\{x \mid x < b\}$	$(-\infty, b)$	
Half-open interval	$\{x \mid x \geq a\}$	$[a, \infty)$	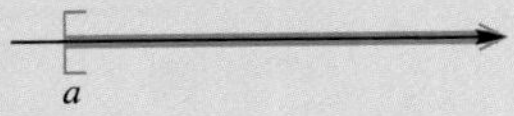
	$\{x \mid a < x \leq b\}$	$(a, b]$	
	$\{x \mid a \leq x < b\}$	$[a, b)$	
	$\{x \mid x \leq b\}$	$(-\infty, b]$	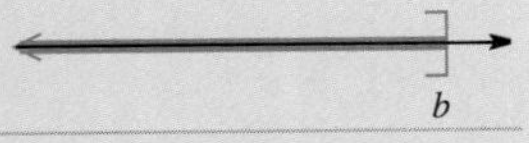
Closed interval	$\{x \mid a \leq x \leq b\}$	$[a, b]$	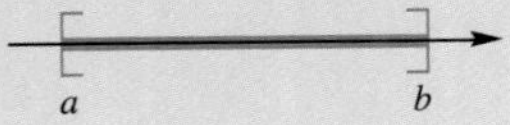

EXAMPLE 1 Graphing Intervals Written in Interval Notation on a Number Line

Write each inequality in interval notation and graph the interval.

(a) $x > -5$ **(b)** $-1 \leq x < 3$

SOLUTION

(a) The statement $x > -5$ says that x can be any number greater than -5 but cannot be -5. The interval is written $(-5, \infty)$. On a graph we place a parenthesis at -5 and draw an arrow to the right, as shown in Figure 9 on the next page. The parenthesis at -5 indicates that -5 is not part of the graph.

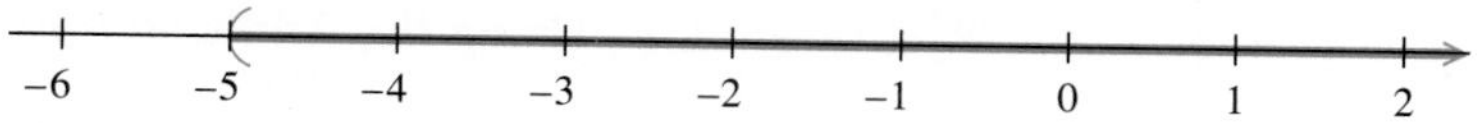

FIGURE 9

(b) The statement is read "-1 is less than or equal to x *and* x is less than 3." Thus, we want the set of numbers that are *between* -1 and 3, with -1 included and 3 excluded. In interval notation, we write $[-1, 3)$, using a square bracket at -1 because -1 is part of the graph, and a parenthesis at 3 because 3 is not part of the graph. The graph is shown in Figure 10.

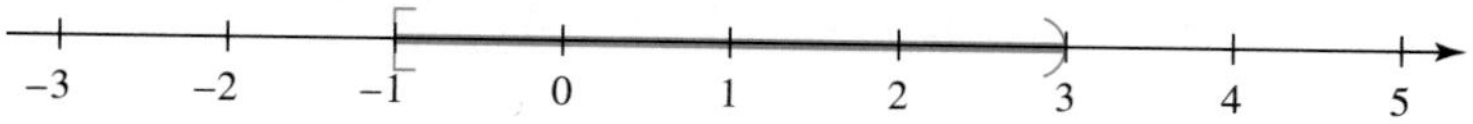

FIGURE 10

Addition Property of Inequality

Linear Inequality in One Variable

A **linear inequality in one variable** can be written in the form

$$Ax + B < C,$$

where A, B, and C are real numbers, with $A \neq 0$. (The symbol $<$ may be replaced by $>$, $\leq$, or $\geq$.)

Examples of linear inequalities in one variable include

$x + 5 < 2$, $\quad y - 3 \geq 5$, $\quad$ and $\quad 2k + 5 \leq 10$. Linear inequalities

Consider the inequality $2 < 5$. If 4 is added to each side, the result is

$$2 + 4 < 5 + 4$$
$$6 < 9. \quad \text{True}$$

Start over and subtract 8 from each side:

$$2 - 8 < 5 - 8$$
$$-6 < -3. \quad \text{True}$$

These examples suggest the **addition property of inequality.**

Addition Property of Inequality

For any real numbers A, B, and C, the inequalities

$$A < B \quad \text{and} \quad A + C < B + C$$

have exactly the same solutions.

That is, the same number may be added to each side of an inequality without changing the solutions.

The same number may also be *subtracted* from each side of an inequality.

EXAMPLE 2 Using the Addition Property of Inequality

Solve $7 + 3x > 2x - 5$.

SOLUTION

Use the addition property of inequality twice, once to isolate the terms containing x on one side of the inequality and a second time to get the integers together on the other side. (These steps can be done in either order.)

$$7 + 3x > 2x - 5$$
$$7 + 3x - 2x > 2x - 5 - 2x \quad \text{Subtract } 2x.$$
$$7 + x > -5 \quad \text{Combine like terms.}$$
$$7 + x - 7 > -5 - 7 \quad \text{Subtract 7.}$$
$$x > -12 \quad \text{Combine like terms.}$$

The solution set is $(-12, \infty)$. Its graph is shown in Figure 11.

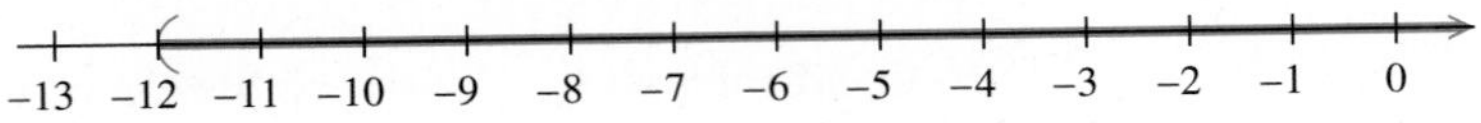

FIGURE 11

Multiplication Property of Inequality The addition property of inequality cannot be used to solve inequalities such as $4x \geq 28$. These inequalities require the *multiplication property of inequality*. To see how this property works, we look at some examples.

Multiply each side of the inequality $3 < 7$ by the positive number 2.

$$3 < 7 \quad \text{True}$$
$$2(3) < 2(7) \quad \text{Multiply each side by 2.}$$
$$6 < 14 \quad \text{True}$$

Now multiply each side of $3 < 7$ by the negative number -5.

$$3 < 7 \quad \text{True}$$
$$-5(3) < -5(7) \quad \text{Multiply each side by } -5.$$
$$-15 < -35 \quad \text{False}$$

To get a true statement when multiplying each side by -5, we must reverse the direction of the inequality symbol.

$$3 < 7 \quad \text{True}$$
$$-5(3) > -5(7) \quad \text{Multiply by } -5\text{; reverse the inequality symbol.}$$
$$-15 > -35 \quad \text{True}$$

Multiply each side of the inequality $-6 < 2$ by the positive number 4.

$$-6 < 2 \quad \text{True}$$
$$4(-6) < 4(2) \quad \text{Multiply by 4.}$$
$$-24 < 8 \quad \text{True}$$

Multiplying each side of $-6 < 2$ by -5 *and at the same time reversing the direction of the inequality symbol* gives

$$-6 < 2 \quad \text{True}$$
$$-5(-6) > -5(2) \quad \text{Multiply by } -5\text{; reverse the inequality symbol.}$$
$$30 > -10. \quad \text{True}$$

In summary, the **multiplication property of inequality** has two parts.

Multiplication Property of Inequality

For any real numbers A, B, and C, with $C \neq 0$,

1. if C is *positive,* then the inequalities

$$A < B \quad \text{and} \quad AC < BC$$

have the same solutions;

2. if C is *negative,* then the inequalities

$$A < B \quad \text{and} \quad AC > BC$$

have the same solutions.

That is, each side of an inequality may be multiplied by the same positive number without changing the solutions. If the multiplier is negative, we must reverse the direction of the inequality symbol.

The multiplication property of inequality also permits *division* of each side of an inequality by the same nonzero number.

It is important to remember the differences in the multiplication property for positive and negative numbers.

1. When each side of an inequality is multiplied or divided by a positive number, the direction of the inequality symbol *does not change*. (Also, adding or subtracting terms on each side does not change the symbol.)
2. When each side of an inequality is multiplied or divided by a negative number, the direction of the symbol *does change*. ***Reverse the direction of the inequality symbol only when multiplying or dividing each side by a negative number.***

EXAMPLE 3 Using the Multiplication Property of Inequality

Solve each inequality and graph the solution set.

(a) $3x < -18$ **(b)** $-4x \geq 8$

SOLUTION

(a)

$$3x < -18$$

$$\frac{3x}{3} < \frac{-18}{3} \quad \text{Divide by 3, a } \textit{positive} \text{ number.}$$

3 is a positive number, so the inequality symbol does not change.

$$x < -6$$

The solution set is $(-\infty, -6)$. The graph is shown in Figure 12 on the next page.

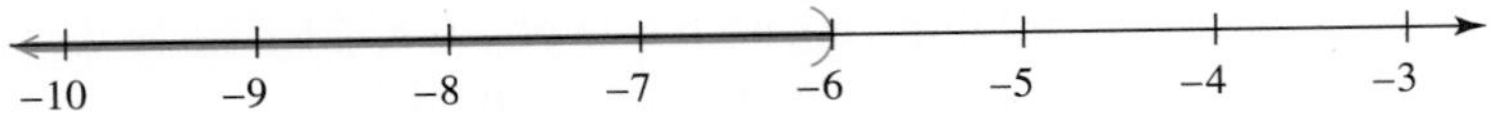

FIGURE 12

(b)

$$-4x \geq 8$$

$$\frac{-4x}{-4} \leq \frac{8}{-4} \quad \text{Divide by } -4, \text{ a } \textit{negative} \text{ number; reverse the inequality symbol.}$$

$$x \leq -2$$

Reverse the inequality when multiplying or dividing by a negative number.

The solution set $(-\infty, -2]$ is graphed in Figure 13.

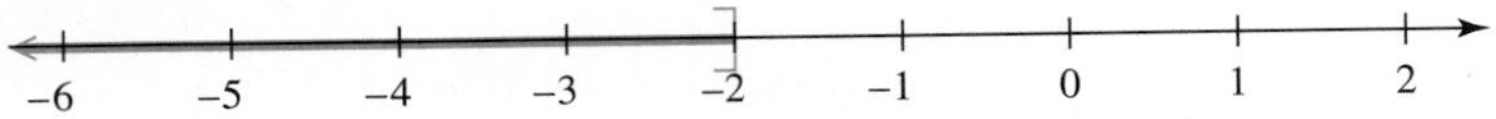

FIGURE 13

Solving Linear Inequalities

To solve a linear inequality, follow these steps.

Solving a Linear Inequality in One Variable

Step 1 **Simplify each side separately.** Use the distributive property to clear parentheses and combine like terms on each side as needed.

Step 2 **Isolate the variable terms on one side.** Use the addition property of inequality to get all terms with variables on one side of the inequality and all numbers on the other side.

Step 3 **Isolate the variable.** Use the multiplication property of inequality to change the inequality to the form $x < k$ or $x > k$, where k is a number.

Remember: Reverse the direction of the inequality symbol only when multiplying or dividing each side of an inequality by a negative number.

EXAMPLE 4 Solving a Linear Inequality

Solve $5(x - 3) - 7x \geq 4(x - 3) + 9$. Give the solution set in interval form, and then graph.

SOLUTION

Step 1 Simplify and combine like terms.

$$5(x - 3) - 7x \geq 4(x - 3) + 9$$

$$5x - 15 - 7x \geq 4x - 12 + 9 \quad \text{Distributive property}$$

$$-2x - 15 \geq 4x - 3 \quad \text{Combine like terms.}$$

Step 2 Use the addition property of inequality.

$$-2x - 15 - \mathbf{4x} \geq 4x - 3 - \mathbf{4x} \quad \text{Subtract } 4x.$$
$$-6x - 15 \geq -3$$
$$-6x - 15 + \mathbf{15} \geq -3 + \mathbf{15} \quad \text{Add 15.}$$
$$-6x \geq 12$$

Step 3 Use the multiplication property of inequality.

Remember to reverse the inequality symbol.

$$\frac{-6x}{\mathbf{-6}} \leq \frac{12}{\mathbf{-6}} \quad \text{Divide by } -6\text{, a } \textit{negative} \text{ number; reverse the symbol.}$$
$$x \leq -2$$

The solution set is $(-\infty, -2]$. Its graph is shown in Figure 14.

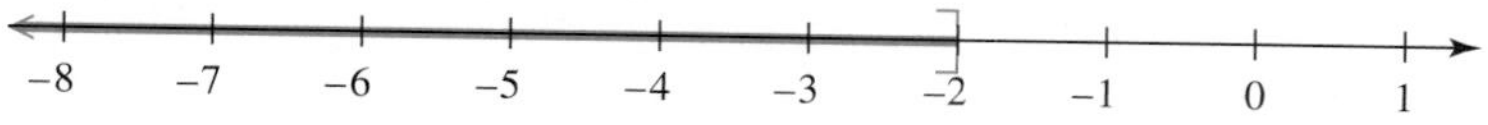

FIGURE 14

Applications

PROBLEM-SOLVING HINT Inequalities can be used to solve applied problems involving phrases that suggest inequality. The table gives some of the more common such phrases along with examples and translations.

Phrase	Example	Inequality
Is more than	A number *is more than* 4.	$x > 4$
Is less than	A number *is less than* -12.	$x < -12$
Is at least	A number *is at least* 6.	$x \geq 6$
Is at most	A number *is at most* 8.	$x \leq 8$

We use the same six problem-solving steps from Section 7.2, changing Step 3 to "Write an inequality" instead of "Write an equation."

The next example shows an application of algebra that is important to anyone who has ever asked, "What score can I make on my next test and have a (particular grade) in this course?" It uses the idea of finding the average of a number of grades. In general, to find the average of n numbers, add the numbers, then divide by n.

EXAMPLE 5 Finding an Average Test Score

Brent has test grades of 86, 88, and 78 on his first three tests in geometry. If he wants an average of at least 80 after his fourth test, what are the possible scores he can make on his fourth test?

SOLUTION

Step 1 **Read** the problem again.

Step 2 **Assign a variable.** Let x = Brent's score on his fourth test.

Step 3 **Write an inequality.** To find his average after 4 tests, add the test scores and divide by 4.

Average (is at least) 80.

$$\frac{86 + 88 + 78 + x}{4} \geq 80$$

Step 4 **Solve.**

$$\frac{252 + x}{4} \geq 80 \qquad \text{Add the known scores.}$$

$$4\left(\frac{252 + x}{4}\right) \geq 4(80) \qquad \text{Multiply by 4 to clear the fraction.}$$

$$252 + x \geq 320$$

$$252 + x - 252 \geq 320 - 252 \qquad \text{Subtract 252.}$$

$$x \geq 68 \qquad \text{Combine like terms.}$$

Step 5 **State the answer.** He must score 68 or more on the fourth test to have an average of *at least* 80.

Step 6 **Check.** Determine whether a score of 68 gives an average of 80.

$$\frac{86 + 88 + 78 + 68}{4} = \frac{320}{4} = 80$$

■

EXAMPLE 6 Using a Linear Inequality to Solve a Rental Problem

A rental company charges $15.00 to rent a chain saw, plus $2.00 per hour. Mamie Zwettler can spend no more than $35.00 to clear some logs from her yard. What is the maximum amount of time she can use the rented saw?

SOLUTION

Let h = the number of hours she can rent the saw. She must pay $15.00, plus 2.00h$, to rent the saw for h hours, and this amount must be *no more than* $35.00.

Cost of renting (is no more than) 35 dollars.

$$15 + 2h \leq 35$$

$$15 + 2h - 15 \leq 35 - 15 \qquad \text{Subtract 15.}$$

$$2h \leq 20$$

$$h \leq 10 \qquad \text{Divide by 2.}$$

Mamie can use the saw for a maximum of 10 hours. (Of course, she may use it for less time, as indicated by the inequality $h \leq 10$.) ■

Three-Part Inequalities Inequalities that say that one number is *between* two other numbers are **three-part inequalities.** For example,

$$-3 < 5 < 7 \qquad \text{Three-part inequality}$$

says that 5 is between -3 and 7.

For some applications, it is necessary to work with an inequality such as

$$3 < x + 2 < 8,$$

where $x + 2$ is between 3 and 8. To solve this inequality, we subtract 2 from each of the three parts of the inequality, giving

$$3 - 2 < x + 2 - 2 < 8 - 2$$
$$1 < x < 6.$$

The idea is to get the inequality in the form

a number $<$ x $<$ another number,

The symbols must point in the same direction and toward the lesser number.

using "is less than." The solution set (in this case the interval (1, 6)) can then easily be graphed.

EXAMPLE 7 Solving Three-Part Inequalities

Solve $4 \leq 3x - 5 < 6$. Give the solution set in interval form, and then graph.

SOLUTION

$$4 \leq 3x - 5 < 6$$
$$4 + 5 \leq 3x - 5 + 5 < 6 + 5 \quad \text{Add 5 to each part.}$$
$$9 \leq 3x < 11$$

Remember to divide all three parts by 3.

$$\frac{9}{3} \leq \frac{3x}{3} < \frac{11}{3} \quad \text{Divide each part by 3.}$$
$$3 \leq x < \frac{11}{3}$$

The solution set is $\left[3, \frac{11}{3}\right)$. Its graph is shown in Figure 15.

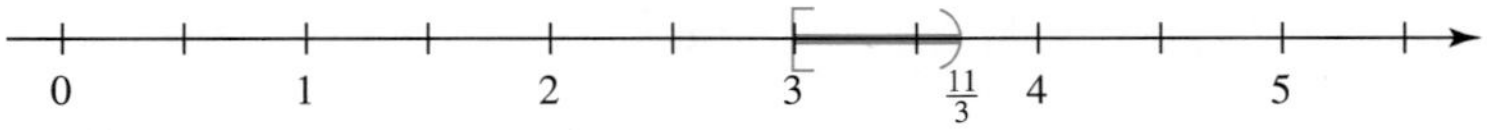

FIGURE 15

7.4 EXERCISES

In Exercises 1–6, match each set in Column I with the correct graph or interval notation in A–F in Column II. See Example 1.

I

1. $\{x \mid x \leq 3\}$ **2.** $\{x \mid x > 3\}$

3. $\{x \mid x < 3\}$ **4.** $\{x \mid x \geq 3\}$

5. $\{x \mid -3 \leq x \leq 3\}$ **6.** $\{x \mid -3 < x < 3\}$

II

A. 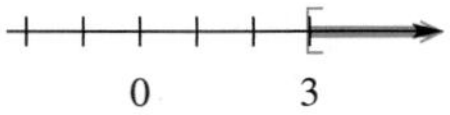

B. 

C. $(3, \infty)$ **D.** $(-\infty, 3]$

E. $(-3, 3)$ **F.** $[-3, 3]$

7. How does one determine whether to use parentheses or brackets when graphing the solution set of an inequality?

8. Describe the steps used to solve a linear inequality. Explain when it is necessary to reverse the inequality symbol.

Solve each inequality. Give the solution set in both interval and graph forms.

9. $4x + 1 \geq 21$

10. $5t + 2 \geq 52$

11. $\dfrac{3k - 1}{4} > 5$

12. $\dfrac{5z - 6}{8} < 8$

13. $-4x < 16$

14. $-2m > 10$

15. $-\dfrac{3}{4}r \geq 30$

16. $-1.5x \leq -\dfrac{9}{2}$

17. $-1.3m \geq -5.2$

18. $-2.5x \leq -1.25$

19. $\dfrac{2k - 5}{-4} > 5$

20. $\dfrac{3z - 2}{-5} < 6$

21. $x + 4(2x - 1) \geq x$

22. $m - 2(m - 4) \leq 3m$

23. $-(4 + r) + 2 - 3r < -14$

24. $-(9 + k) - 5 + 4k \geq 4$

25. $-3(z - 6) > 2z - 2$

26. $-2(x + 4) \leq 6x + 16$

27. $\dfrac{2}{3}(3k - 1) \geq \dfrac{3}{2}(2k - 3)$

28. $\dfrac{7}{5}(10m - 1) < \dfrac{2}{3}(6m + 5)$

29. $-\dfrac{1}{4}(p + 6) + \dfrac{3}{2}(2p - 5) < 10$

30. $\dfrac{3}{5}(k - 2) - \dfrac{1}{4}(2k - 7) \leq 3$

31. $3(2x - 4) - 4x < 2x + 3$

32. $7(4 - x) + 5x < 2(16 - x)$

33. $8\left(\dfrac{1}{2}x + 3\right) < 8\left(\dfrac{1}{2}x - 1\right)$

34. $10x + 2(x - 4) < 12x - 10$

35. A student solved the inequality $5x < -20$ by dividing both sides by 5 and reversing the direction of the inequality symbol. His reasoning was that since -20 is a negative number, reversing the direction of the symbol was required. Is this correct? Explain why or why not.

36. Match each set given in interval notation with its description.

(a) $(0, \infty)$ **A.** positive real numbers
(b) $[0, \infty)$ **B.** negative real numbers
(c) $(-\infty, 0]$ **C.** nonpositive real numbers
(d) $(-\infty, 0)$ **D.** nonnegative real numbers

Solve each inequality. Give the solution set in both interval and graph forms.

37. $-4 < x - 5 < 6$

38. $-1 < x + 1 < 8$

39. $-9 \leq k + 5 \leq 15$

40. $-4 \leq m + 3 \leq 10$

41. $-6 \leq 2z + 4 \leq 16$

42. $-15 < 3p + 6 < -12$

43. $-19 \leq 3x - 5 \leq 1$

44. $-16 < 3t + 2 < -10$

45. $-1 \leq \dfrac{2x - 5}{6} \leq 5$

46. $-3 \leq \dfrac{3m + 1}{4} \leq 3$

47. $4 \leq 5 - 9x < 8$

48. $4 \leq 3 - 2x < 8$

Tornado Activity *In Exercises 49–52, answer the questions based on the graph.*

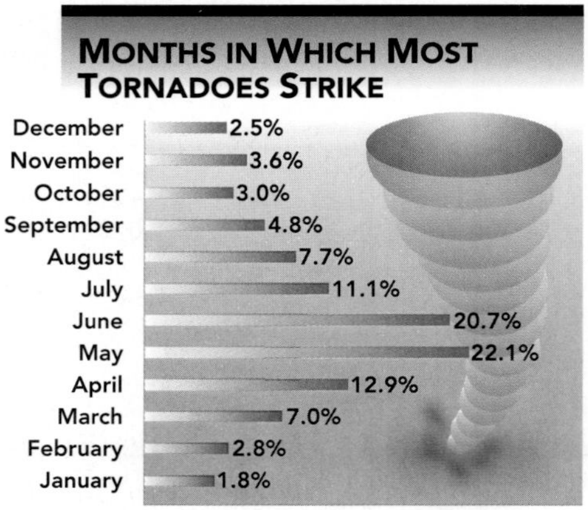

Source: The USA Today Weather Book.

49. In which months did the percent of tornadoes exceed 7.7%?

50. In which months was the percent of tornadoes at least 12.9%?

51. The data used to determine the graph were based on the number of tornadoes sighted in the United States during a twenty-year period. A total of 17,252 tornadoes were reported. In which months were fewer than 1500 reported?

52. How many more tornadoes occurred during March than October? (Use the total given in Exercise 51.)

Olympic Temperature Preferences *The weather forecast by time of day for the U.S. Olympic Track and Field Trials, held in Sacramento, California, is shown in the figure. Use this graph to work Exercises 53–56.*

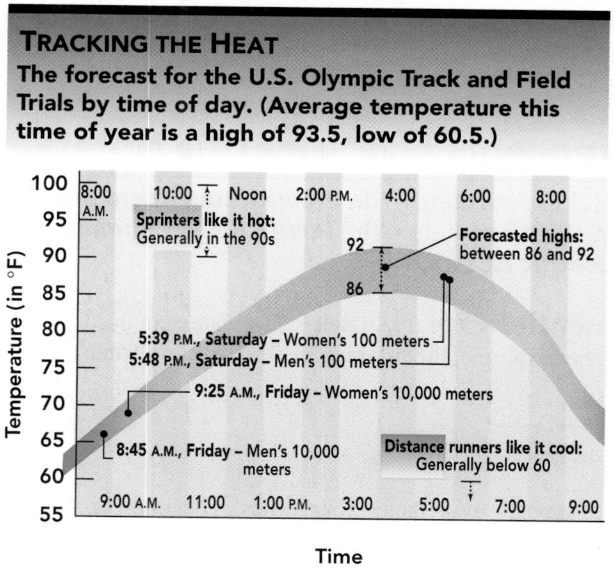

Source: Accuweather, *Sacramento Bee* research.

53. Sprinters prefer Fahrenheit temperatures in the 90s. Using the upper boundary of the forecast, in what time period is the temperature expected to be at least 90° F?

54. Distance runners prefer cool temperatures. During what time period are temperatures predicted to be no more than 70° F? Use the lower forecast boundary.

55. What range of temperatures is predicted for the Women's 100-meter event?

56. What range of temperatures is forecast for the Men's 10,000-meter event?

Solve each problem.

57. *Taxicab Fare* In a midwestern city, taxicabs charge \$3.00 for the first $\frac{1}{5}$ mile and \$.50 for each additional $\frac{1}{5}$ mile. Ed d'Hemecourt has only \$7.50 in his pocket. What is the maximum distance he can travel (not including a tip for the cabbie)?

58. *Taxicab Fare* Ten years ago taxicab fares in the city in Exercise 57 were \$.90 for the first $\frac{1}{7}$ mile and \$.10 for each additional $\frac{1}{7}$ mile. Based on the information given there and the answer you found, how much farther could Ed have traveled at that time?

59. *Grade Average* Allen Wells earned scores of 90 and 82 on his first two tests in English Literature. What score must he make on his third test to keep an average of 84 or greater?

60. *Grade Average* Beth Anderson scored 92 and 96 on her first two tests in Methods in Photography. What score must she make on her third test to keep an average of 90 or greater?

61. *Car Rental* A couple wishes to rent a car for one day while on vacation. Avis wants \$35.00 per day and 14¢ per mile, while Downtown Toyota wants \$34.00 per day and 16¢ per mile. After how many miles would the price to rent from Downtown Toyota exceed the price to rent from Avis?

62. *Car Rental* John and Sherry Gainey went to New Jersey for a week. They needed to rent a car, so they checked out two rental firms. Avis wanted \$28 per day, with no mileage fee. Downtown Toyota wanted \$108 per week and 14¢ per mile. How many miles would they have to drive before the Avis price is less than the Toyota price?

63. *Body Mass Index* A BMI (body mass index) between 19 and 25 is considered healthy. Use the formula

$$\text{BMI} = \frac{704 \times (\text{weight in pounds})}{(\text{height in inches})^2}$$

to find the weight range w, to the nearest pound, that gives a healthy BMI for each height. (*Source: Washington Post.*)

(a) 72 inches **(b)** Your height in inches

64. ***Target Heart Rate*** To achieve the maximum benefit from exercising, the heart rate in beats per minute should be in the target heart rate zone (THR). For a person aged A, the formula is

$$.7(220 - A) \leq \text{THR} \leq .85(220 - A).$$

Find the THR to the nearest whole number for each age. (*Source:* Hockey, Robert V., *Physical Fitness: The Pathway to Healthful Living,* Times Mirror/Mosby College Publishing, 1989.)

(a) 35 **(b)** Your age

Profit/Cost Analysis *A product will produce a profit only when the revenue R from selling the product exceeds the cost C of producing it (R and C in dollars). Find the least whole number of units x that must be sold for the business to show a profit for the item described.*

65. Peripheral Visions, Inc. finds that the cost to produce x studio quality DVDs is $C = 20x + 100$, while the revenue produced from them is $R = 24x$.

66. Speedy Delivery finds that the cost to make x deliveries is $C = 3x + 2300$, while the revenue produced from them is $R = 5.50x$.

7.5 Properties of Exponents and Scientific Notation

Exponents and Exponential Expressions • The Product Rule • Zero and Negative Exponents • The Quotient Rule • The Power Rules • Summary of Rules for Exponents • Scientific Notation

Exponents and Exponential Expressions

Exponents are used to write products of repeated factors. For example, the product $3 \cdot 3 \cdot 3 \cdot 3$ is written

$$\underbrace{3 \cdot 3 \cdot 3 \cdot 3}_{\text{4 factors of 3}} = 3^4. \quad \leftarrow \text{Exponent (4)}, \ \text{Base (3)}$$

The number 4 shows that 3 appears as a factor four times. The number 4 is the **exponent** and 3 is the **base.** The quantity 3^4 is called an **exponential expression.** Read 3^4 as "3 to the fourth power," or "3 to the fourth." Multiplying out the four 3s gives

$$3^4 = 3 \cdot 3 \cdot 3 \cdot 3 = 81.$$

The term **googol,** meaning 10^{100}, was coined by Professor Edward Kasner of Columbia University. A googol is made up of a 1 with one hundred zeros following it. This number exceeds the estimated number of electrons in the universe, which is 10^{79}.

The Web search engine Google is named after a googol. Sergey Brin, president and cofounder of Google, Inc., was a mathematics major. He chose the name Google to describe the vast reach of this search engine. (*Source: The Gazette*, March 2, 2001.) The term "googling" is now part of the English language.

If a googol isn't big enough for you, try a **googolplex:**

$$\text{googolplex} = 10^{\text{googol}}.$$

Exponential Expression

If a is a real number and n is a natural number, then the exponential expression a^n is defined as

$$a^n = \underbrace{a \cdot a \cdot a \cdot \ldots \cdot a}_{n \text{ factors of } a}.$$

The number a is the *base* and n is the *exponent.*

EXAMPLE 1 Evaluating Exponential Expressions

Evaluate each exponential expression.

(a) 7^2 **(b)** 5^3 **(c)** $(-2)^4$ **(d)** $(-2)^5$ **(e)** 5^1

SOLUTION

(a) $7^2 = 7 \cdot 7 = 49$ Read 7^2 as "7 squared."

$7^2 = 7 \cdot 7$, *not* $7 \cdot 2$.

(b) $5^3 = 5 \cdot 5 \cdot 5 = 125$ Read 5^3 as "5 cubed."

(c) $(-2)^4 = (-2)(-2)(-2)(-2) = 16$

(d) $(-2)^5 = (-2)(-2)(-2)(-2)(-2) = -32$

(e) $5^1 = 5$

In the exponential expression $3z^7$, the base of the exponent 7 is z, *not* $3z$. That is,

$$3z^7 = 3 \cdot z \cdot z \cdot z \cdot z \cdot z \cdot z \cdot z \quad \text{Base is } z.$$

while

$$(3z)^7 = (3z)(3z)(3z)(3z)(3z)(3z)(3z). \quad \text{Base is } 3z.$$

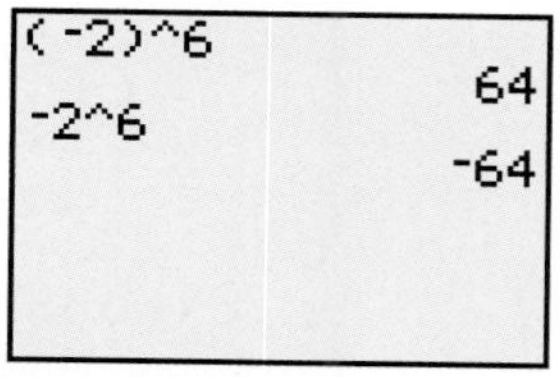

This screen supports the discussion preceding Example 2.

To evaluate $(-2)^6$, the parentheses around -2 indicate that the base is -2, so

$$(-2)^6 = (-2)(-2)(-2)(-2)(-2)(-2) = 64. \quad \text{Base is } -2.$$

In the expression -2^6, the base is 2, *not* -2. The $-$ sign tells us to find the negative, or additive inverse, of 2^6. It acts as a symbol for the factor -1.

$$-2^6 = -(2 \cdot 2 \cdot 2 \cdot 2 \cdot 2 \cdot 2) = -64 \quad \text{Base is } 2.$$

Therefore, since $64 \neq -64$, $(-2)^6 \neq -2^6$.

EXAMPLE 2 Evaluating Exponential Expressions

Evaluate each exponential expression.

(a) -4^2 **(b)** -8^4 **(c)** -2^4

SOLUTION

(a) $-4^2 = -(4 \cdot 4) = -16$ **(b)** $-8^4 = -(8 \cdot 8 \cdot 8 \cdot 8) = -4096$

(c) $-2^4 = -(2 \cdot 2 \cdot 2 \cdot 2) = -16$

The Product Rule There are several useful rules that simplify work with exponents. For example, the product $2^5 \cdot 2^3$ can be simplified as follows.

$$5 + 3 = 8$$

$$2^5 \cdot 2^3 = (2 \cdot 2 \cdot 2 \cdot 2 \cdot 2)(2 \cdot 2 \cdot 2) = 2^8$$

This result—products of exponential expressions with the same base are found by adding exponents—is generalized as the **product rule for exponents.**

Product Rule for Exponents

If m and n are natural numbers and a is any real number, then

$$a^m \cdot a^n = a^{m+n}.$$

EXAMPLE 3 Applying the Product Rule

Apply the product rule for exponents in each case.

(a) $3^4 \cdot 3^7$ **(b)** $5^3 \cdot 5$ **(c)** $y^3 \cdot y^8 \cdot y^2$
(d) $(5y^2)(-3y^4)$ **(e)** $(7p^3q)(2p^5q^2)$

SOLUTION

(a) $3^4 \cdot 3^7 = 3^{4+7} = 3^{11}$

Do not make the error of writing $3 \cdot 3 = 9$ as the base.

(b) $5^3 \cdot 5 = 5^3 \cdot 5^1 = 5^{3+1} = 5^4$

(c) $y^3 \cdot y^8 \cdot y^2 = y^{3+8+2} = y^{13}$

(d)
$$\begin{aligned}(5y^2)(-3y^4) &= 5(-3)y^2y^4 && \text{Associative and commutative properties}\\ &= -15y^{2+4} && \text{Multiply; product rule}\\ &= -15y^6\end{aligned}$$

(e)
$$\begin{aligned}(7p^3q)(2p^5q^2) &= 7(2)p^3p^5qq^2\\ &= 14p^8q^3\end{aligned}$$

Zero and Negative Exponents

We now consider 0 as an exponent. How can we define an expression such as 4^0 so that it is consistent with the product rule? By the product rule, we should have

$$4^2 \cdot 4^0 = 4^{2+0} = 4^2.$$

For the product rule to hold true, 4^0 must equal 1. This leads to the definition of a^0 for any nonzero real number a.

Zero Exponent

If a is any nonzero real number, then $\mathbf{a^0 = 1.}$

The expression $\mathbf{0^0}$ **is undefined.***

EXAMPLE 4 Applying the Definition of Zero Exponent

Evaluate each expression.

(a) 12^0 **(b)** $(-6)^0$ **(c)** -6^0
(d) $5^0 + 12^0$ **(e)** $(8k)^0$, $k \neq 0$

SOLUTION

(a) $12^0 = 1$

(b) $(-6)^0 = 1$ Base is -6.

(c) $-6^0 = -(6^0) = -1$ Base is 6.

(d) $5^0 + 12^0 = 1 + 1 = 2$

(e) $(8k)^0 = 1$, $k \neq 0$

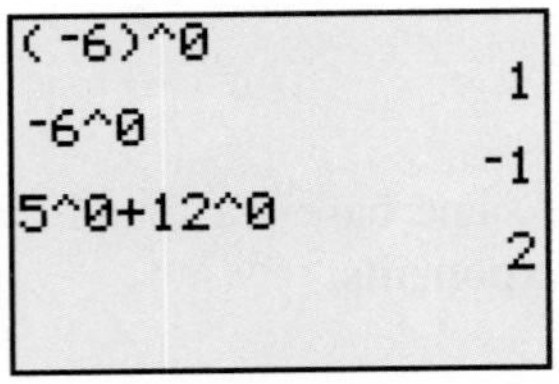

This screen supports the results in parts (b), (c), and (d) of Example 4.

*In advanced studies, 0^0 is called an *indeterminate form*.

How should we define a negative exponent? Using the product rule again,

$$8^2 \cdot 8^{-2} = 8^{2+(-2)} = 8^0 = 1.$$

This indicates that 8^{-2} is the reciprocal of 8^2. But $\frac{1}{8^2}$ is the reciprocal of 8^2, and a number can have only one reciprocal. Therefore, it is reasonable to conclude that $8^{-2} = \frac{1}{8^2}$. We can generalize and make the following definition.

Negative Exponent

For any natural number n and any nonzero real number a,

$$a^{-n} = \frac{1}{a^n}.$$

With this definition, and the ones given earlier for positive and zero exponents, the expression a^n is meaningful for any integer exponent n and any nonzero real number a.

EXAMPLE 5 Applying the Definition of Negative Exponents

Write the following expressions with only positive exponents. Assume that all variables represent nonzero real numbers.

(a) 2^{-3} **(b)** 3^{-2} **(c)** 6^{-1} **(d)** $(5z)^{-3}$
(e) $5z^{-3}$ **(f)** $(5z^2)^{-3}$ **(g)** $-m^{-2}$ **(h)** $(-m)^{-4}$

SOLUTION

(a) $2^{-3} = \frac{1}{2^3} = \frac{1}{8}$

(b) $3^{-2} = \frac{1}{3^2} = \frac{1}{9}$

(c) $6^{-1} = \frac{1}{6^1} = \frac{1}{6}$

(d) $(5z)^{-3} = \frac{1}{(5z)^3}$ Base is 5z.

(e) $5z^{-3} = 5\left(\frac{1}{z^3}\right) = \frac{5}{z^3}$ Base is z.

(f) $(5z^2)^{-3} = \frac{1}{(5z^2)^3}$

(g) $-m^{-2} = -\frac{1}{m^2}$

(h) $(-m)^{-4} = \frac{1}{(-m)^4}$ ■

EXAMPLE 6 Evaluating Exponential Expressions

Evaluate each expression.

(a) $3^{-1} + 4^{-1}$ **(b)** $5^{-1} - 2^{-1}$ **(c)** $\frac{1}{2^{-3}}$ **(d)** $\frac{2^{-3}}{3^{-2}}$

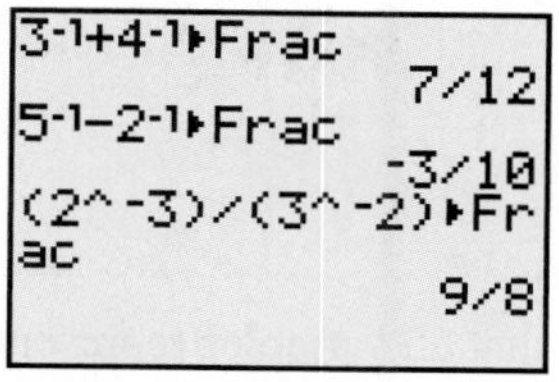

This screen supports the results in parts (a), (b), and (d) of Example 6.

SOLUTION

(a) $3^{-1} + 4^{-1} = \frac{1}{3} + \frac{1}{4} = \frac{4}{12} + \frac{3}{12} = \frac{7}{12}$ $3^{-1} = \frac{1}{3}; 4^{-1} = \frac{1}{4}$

(b) $5^{-1} - 2^{-1} = \frac{1}{5} - \frac{1}{2} = \frac{2}{10} - \frac{5}{10} = -\frac{3}{10}$

(c) $\dfrac{1}{2^{-3}} = \dfrac{1}{\frac{1}{2^3}} = 1 \div \dfrac{1}{2^3} = 1 \cdot \dfrac{2^3}{1} = 2^3 = 8$

To divide, multiply by the reciprocal.

(d) $\dfrac{2^{-3}}{3^{-2}} = \dfrac{\frac{1}{2^3}}{\frac{1}{3^2}} = \dfrac{1}{2^3} \div \dfrac{1}{3^2} = \dfrac{1}{2^3} \cdot \dfrac{3^2}{1} = \dfrac{3^2}{2^3} = \dfrac{9}{8}$

Parts (c) and (d) of Example 6 suggest the following generalizations.

Special Rules for Negative Exponents

If $a \neq 0$ and $b \neq 0$, then $\dfrac{1}{a^{-n}} = a^n$ and $\dfrac{a^{-n}}{b^{-m}} = \dfrac{b^m}{a^n}$.

The Quotient Rule A quotient, such as $\dfrac{a^8}{a^3}$, can be simplified in much the same way as a product. (In all quotients of this type, assume that the denominator is not 0.) Using the definition of an exponent,

$$\frac{a^8}{a^3} = \frac{a \cdot a \cdot a \cdot a \cdot a \cdot a \cdot a \cdot a}{a \cdot a \cdot a} = a \cdot a \cdot a \cdot a \cdot a = a^5.$$

Notice that $8 - 3 = 5$. In the same way,

$$\frac{a^3}{a^8} = \frac{a \cdot a \cdot a}{a \cdot a \cdot a \cdot a \cdot a \cdot a \cdot a \cdot a} = \frac{1}{a^5} = a^{-5}.$$

Here, $3 - 8 = -5$. These examples suggest the **quotient rule for exponents.**

Quotient Rule for Exponents

If a is any nonzero real number and m and n are integers, then

$$\frac{a^m}{a^n} = a^{m-n}.$$

EXAMPLE 7 Applying the Quotient Rule

Apply the quotient rule for exponents in each case. Assume that all variables represent nonzero real numbers.

(a) $\dfrac{3^7}{3^2}$ **(b)** $\dfrac{p^6}{p^2}$ **(c)** $\dfrac{12^{10}}{12^9}$ **(d)** $\dfrac{7^4}{7^6}$ **(e)** $\dfrac{k^7}{k^{12}}$

SOLUTION

Numerator exponent

Denominator exponent

(a) $\dfrac{3^7}{3^2} = 3^{7-2} = 3^5$

Minus sign

(b) $\dfrac{p^6}{p^2} = p^{6-2} = p^4$

(c) $\dfrac{12^{10}}{12^9} = 12^{10-9} = 12^1 = 12$

(d) $\dfrac{7^4}{7^6} = 7^{4-6} = 7^{-2} = \dfrac{1}{7^2}$

(e) $\dfrac{k^7}{k^{12}} = k^{7-12} = k^{-5} = \dfrac{1}{k^5}$

EXAMPLE 8 Applying the Quotient Rule

Write each quotient using only positive exponents. Assume that all variables represent nonzero real numbers.

(a) $\dfrac{2^7}{2^{-3}}$ (b) $\dfrac{8^{-2}}{8^5}$ (c) $\dfrac{6^{-5}}{6^{-2}}$ (d) $\dfrac{4}{4^{-1}}$ (e) $\dfrac{z^{-5}}{z^{-8}}$

SOLUTION

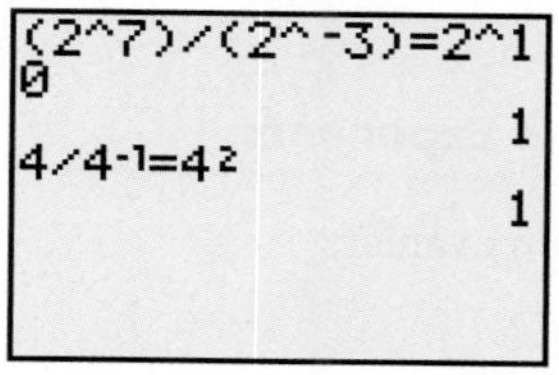

This screen supports the results in parts (a) and (d) of Example 8.

Be careful when subtracting a negative number.

(a) $\dfrac{2^7}{2^{-3}} = 2^{7-(-3)} = 2^{10}$

(b) $\dfrac{8^{-2}}{8^5} = 8^{-2-5} = 8^{-7} = \dfrac{1}{8^7}$

(c) $\dfrac{6^{-5}}{6^{-2}} = 6^{-5-(-2)} = 6^{-3} = \dfrac{1}{6^3}$

(d) $\dfrac{4}{4^{-1}} = \dfrac{4^1}{4^{-1}} = 4^{1-(-1)} = 4^2$

(e) $\dfrac{z^{-5}}{z^{-8}} = z^{-5-(-8)} = z^3$

The Power Rules The expression $(3^4)^2$ can be simplified as $(3^4)^2 = 3^4 \cdot 3^4 = 3^{4+4} = 3^8$, where $4 \cdot 2 = 8$. This example suggests the first of the **power rules for exponents.** The other two parts can be demonstrated with similar examples.

Power Rules for Exponents

If a and b are real numbers, and m and n are integers, then

$$(a^m)^n = a^{mn}, \quad (ab)^m = a^m b^m, \quad \text{and} \quad \left(\frac{a}{b}\right)^m = \frac{a^m}{b^m} \quad (b \neq 0).$$

In the statements of rules for exponents, we always assume that zero never appears to a negative power or to the power zero.

EXAMPLE 9 Applying the Power Rules

Use a power rule in each case. Assume that all variables represent nonzero real numbers.

(a) $(p^8)^3$ (b) $\left(\dfrac{2}{3}\right)^4$ (c) $(3y)^4$ (d) $(6p^7)^2$ (e) $\left(\dfrac{-2m^5}{z}\right)^3$

SOLUTION

(a) $(p^8)^3 = p^{8 \cdot 3} = p^{24}$ **(b)** $\left(\frac{2}{3}\right)^4 = \frac{2^4}{3^4} = \frac{16}{81}$

(c) $(3y)^4 = 3^4y^4 = 81y^4$ **(d)** $(6p^7)^2 = 6^2p^{7 \cdot 2} = 6^2p^{14} = 36p^{14}$

(e) $\left(\frac{-2m^5}{z}\right)^3 = \frac{(-2)^3m^{5 \cdot 3}}{z^3} = \frac{(-2)^3m^{15}}{z^3} = \frac{-8m^{15}}{z^3}$ ■

Notice that

$$6^{-3} = \left(\frac{1}{6}\right)^3 = \frac{1}{216} \quad \text{and} \quad \left(\frac{2}{3}\right)^{-2} = \left(\frac{3}{2}\right)^2 = \frac{9}{4}.$$

These are examples of two special rules for negative exponents.

Special Rules for Negative Exponents

If $a \neq 0$ and $b \neq 0$ and n is an integer, then

$$a^{-n} = \left(\frac{1}{a}\right)^n \quad \text{and} \quad \left(\frac{a}{b}\right)^{-n} = \left(\frac{b}{a}\right)^n.$$

EXAMPLE 10 Applying Special Rules for Negative Exponents

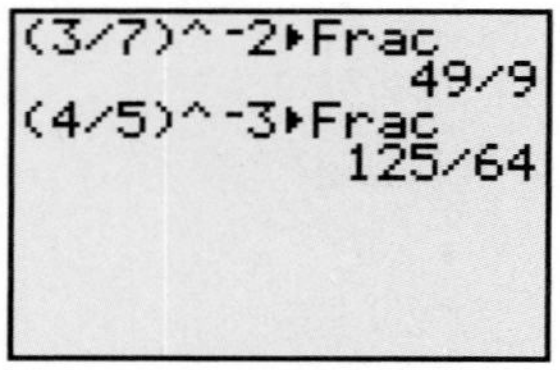

This screen supports the results of Example 10.

Write each expression with only positive exponents, and then evaluate.

(a) $\left(\frac{3}{7}\right)^{-2}$ **(b)** $\left(\frac{4}{5}\right)^{-3}$

SOLUTION

(a) $\left(\frac{3}{7}\right)^{-2} = \left(\frac{7}{3}\right)^2 = \frac{49}{9}$ **(b)** $\left(\frac{4}{5}\right)^{-3} = \left(\frac{5}{4}\right)^3 = \frac{125}{64}$ ■

Summary of Rules for Exponents The definitions and rules of this section are summarized here.

Definitions and Rules for Exponents

For all integers m and n and all real numbers a and b,

Product Rule $a^m \cdot a^n = a^{m+n}$

Quotient Rule $\frac{a^m}{a^n} = a^{m-n} \quad (a \neq 0)$

Zero Exponent $a^0 = 1 \quad (a \neq 0)$

Negative Exponent $a^{-n} = \frac{1}{a^n} \quad (a \neq 0)$

(continued)

Power Rules	$(a^m)^n = a^{mn}$ $\quad$ $(ab)^m = a^m b^m$ $\left(\frac{a}{b}\right)^m = \frac{a^m}{b^m} \quad (b \neq 0)$
Special Rules for Negative Exponents	$\frac{1}{a^{-n}} = a^n \quad (a \neq 0)$ $\quad$ $\frac{a^{-n}}{b^{-m}} = \frac{b^m}{a^n} \quad (a, b \neq 0)$ $a^{-n} = \left(\frac{1}{a}\right)^n \quad (a \neq 0)$ $\left(\frac{a}{b}\right)^{-n} = \left(\frac{b}{a}\right)^n \quad (a, b \neq 0).$

EXAMPLE 11 Writing Expressions with No Negative Exponents

Simplify each expression so that no negative exponents appear in the final result. Assume that all variables represent nonzero real numbers.

(a) $3^2 \cdot 3^{-5}$ $\quad$ **(b)** $x^{-3} \cdot x^{-4} \cdot x^2$ $\quad$ **(c)** $(4^{-2})^{-5}$

(d) $(x^{-4})^6$ $\quad$ **(e)** $\dfrac{x^{-4}y^2}{x^2y^{-5}}$ $\quad$ **(f)** $(2^3x^{-2})^{-2}$

SOLUTION

(a) $3^2 \cdot 3^{-5} = 3^{2+(-5)} = 3^{-3} = \dfrac{1}{3^3}$ or $\dfrac{1}{27}$

(b) $x^{-3} \cdot x^{-4} \cdot x^2 = x^{-3+(-4)+2} = x^{-5} = \dfrac{1}{x^5}$

(c) $(4^{-2})^{-5} = 4^{-2(-5)} = 4^{10}$

(d) $(x^{-4})^6 = x^{(-4)6} = x^{-24} = \dfrac{1}{x^{24}}$

(e)
$$\frac{x^{-4}y^2}{x^2y^{-5}} = \frac{x^{-4}}{x^2} \cdot \frac{y^2}{y^{-5}}$$
$$= x^{-4-2} \cdot y^{2-(-5)}$$
$$= x^{-6}y^7$$
$$= \frac{y^7}{x^6}$$

(f)
$$(2^3x^{-2})^{-2} = (2^3)^{-2} \cdot (x^{-2})^{-2}$$
$$= 2^{-6}x^4$$
$$= \frac{x^4}{2^6} \text{ or } \frac{x^4}{64}$$

Scientific Notation Many of the numbers that occur in science are very large, such as the number of one-celled organisms that will sustain a whale for a few hours: 400,000,000,000,000. Other numbers are very small, such as the shortest wavelength of visible light, about .0000004 meter. Writing these numbers is simplified by using *scientific notation*.

In the episode "Court-Martial" from the original *Star Trek* television series, Captain Kirk makes this statement during a scene on the bridge of the Enterprise:

Gentlemen, this computer has an auditory sensor. It can, in effect, hear sounds. By installing a booster we can increase that capability on an order of one to the fourth power. The computer should be able to bring us every sound occurring on the ship.

Can you identify the error in Kirk's statement? What do you think he might have really meant? (Think about scientific notation.)

Scientific Notation

A number is written in **scientific notation** when it is expressed in the form

$$a \times 10^n,$$

where $1 \leq |a| < 10$, and n is an integer.

As stated in the definition, scientific notation requires that the number be written as a product of a number between 1 and 10 (or -1 and -10) and some integer power of 10. (1 and -1 are allowed as values of a, but 10 and -10 are not.) For example, since

$$8000 = 8 \cdot 1000 = 8 \cdot 10^3,$$

the number 8000 is written in scientific notation as

$$8000 = 8 \times 10^3. \longleftarrow \text{Scientific notation}$$

When using scientific notation, it is customary to use × instead of a dot to show multiplication.

The steps involved in writing a number in scientific notation follow. (If the number is negative, ignore the minus sign, go through these steps, and then attach a minus sign to the result.)

Converting to Scientific Notation

Step 1 **Position the decimal point.** Place a caret, $\wedge$, to the right of the first nonzero digit, where the decimal point will be placed.

Step 2 **Determine the numeral for the exponent.** Count the number of digits from the decimal point to the caret. This number gives the absolute value of the exponent on 10.

Step 3 **Determine the sign for the exponent.** Decide whether multiplying by 10^n should make the result of Step 1 larger or smaller. The exponent should be positive to make the result larger; it should be negative to make the result smaller.

It is helpful to remember that for $n \geq 1$, $10^{-n} < 1$ and $10^n \geq 10$.

EXAMPLE 12 Converting to Scientific Notation

Convert each number from standard notation to scientific notation.

(a) 8,200,000 **(b)** .000072

SOLUTION

(a) Place a caret to the right of the 8 (the first nonzero digit) to mark the new location of the decimal point.

$$8_\wedge 200{,}000$$

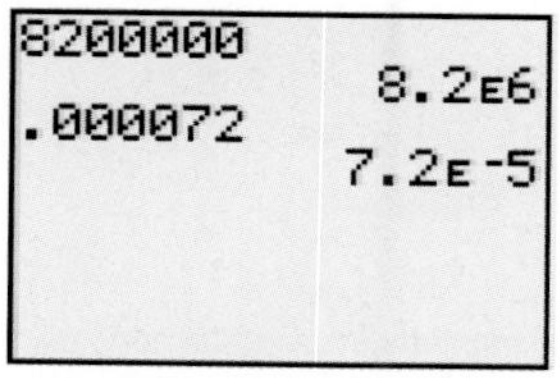

If a graphing calculator is set in scientific notation mode, it will give results as shown here. E6 means "times 10^6" and E$^-$5 means "times 10^{-5}". Compare to the results of Example 12.

Count from the decimal point, which is understood to be after the last 0, to the caret.

$$8_\wedge 200{,}000. \longleftarrow \text{Decimal point}$$

Count 6 places.

Because the number 8.2 is to be made larger, the exponent on 10 is positive.

$$8{,}200{,}000 = 8.2 \times 10^6$$

(b) $.00007_\wedge 2$ Count from left to right.

5 places

Since the number 7.2 is to be made smaller, the exponent on 10 is negative.

$$.000072 = 7.2 \times 10^{-5}$$

To convert a number written in scientific notation to standard notation, just work in reverse.

Converting from Scientific Notation to Standard Notation

Multiplying a number by a positive power of 10 makes the number larger, so move the decimal point to the right if n is positive in 10^n.

Multiplying by a negative power of 10 makes a number smaller, so move the decimal point to the left if n is negative.

If n is zero, leave the decimal point where it is.

EXAMPLE 13 Converting from Scientific Notation

Convert each number from scientific notation to standard notation.

(a) 6.93×10^5 **(b)** 4.7×10^{-6} **(c)** -1.083×10^0

SOLUTION

(a) $6.93 \times 10^5 = 6.93000$ Attach 0s as necessary.

5 places

The decimal point was moved 5 places to the right.

$$6.93 \times 10^5 = 693{,}000$$

(b) $4.7 \times 10^{-6} = 000004.7$ Attach 0s as necessary.

6 places

The decimal point was moved 6 places to the left.

$$4.7 \times 10^{-6} = .0000047$$

(c) $-1.083 \times 10^0 = -1.083$

We can use scientific notation and the rules for exponents to simplify calculations.

EXAMPLE 14 Using Scientific Notation in Computation

Evaluate $\frac{1{,}920{,}000 \times .0015}{.000032 \times 45{,}000}$ by using scientific notation.

SOLUTION

$$\frac{1{,}920{,}000 \times .0015}{.000032 \times 45{,}000} = \frac{1.92 \times 10^6 \times 1.5 \times 10^{-3}}{3.2 \times 10^{-5} \times 4.5 \times 10^4}$$ Express all numbers in scientific notation.

$$= \frac{1.92 \times 1.5 \times 10^6 \times 10^{-3}}{3.2 \times 4.5 \times 10^{-5} \times 10^4}$$ Commutative and associative properties

$$= \frac{1.92 \times 1.5}{3.2 \times 4.5} \times 10^4$$ Product and quotient rules

$$= .2 \times 10^4$$ Simplify.

$$= (2 \times 10^{-1}) \times 10^4$$ Write .2 using scientific notation.

$$= 2 \times 10^3, \text{ or } 2000$$ Product rule; multiply. ■

EXAMPLE 15 Using Scientific Notation to Solve Problems

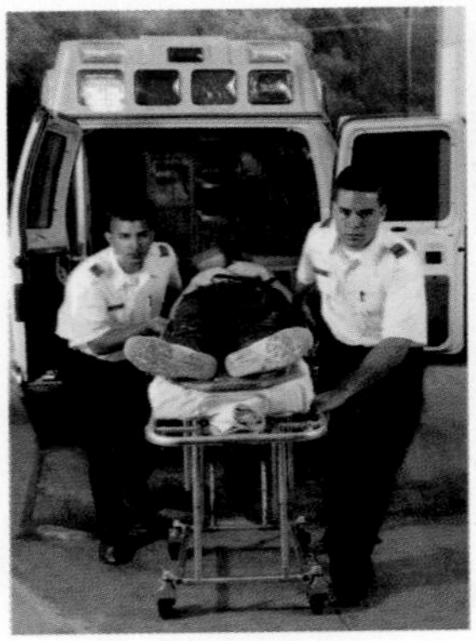

In 1990, the national health care expenditure was \$695.6 billion. By 2000, this figure had risen by a factor of 1.9; that is, it almost doubled in only 10 years. (*Source:* U.S. Centers for Medicare & Medicaid Services.)

(a) Write the 1990 health care expenditure using scientific notation.
(b) What was the expenditure in 2000?

SOLUTION

(a)
$$695.6 \text{ billion} = 695.6 \times 10^9$$ 1 billion = 10^9

$$= (6.956 \times 10^2) \times 10^9$$

$$= 6.956 \times 10^{11}$$ Product rule

In 1990, the expenditure was \$$6.956 \times 10^{11}$.

(b) Multiply the result in part (a) by 1.9.

$$(6.956 \times 10^{11}) \times 1.9 = (1.9 \times 6.956) \times 10^{11}$$ Commutative and associative properties

$$= 13.216 \times 10^{11}$$ Round to three decimal places.

The 2000 expenditure was about \$1,321,600,000,000 (over \$1 trillion). ■

7.5 EXERCISES

Match the exponential expressions in Exercises 1–6 with their equivalent expressions in Choices A–F. Choices may be used once, more than once, or not at all.

1. $\left(\frac{5}{3}\right)^2$ **2.** $\left(\frac{3}{5}\right)^2$ **3.** $\left(-\frac{3}{5}\right)^{-2}$ **4.** $\left(-\frac{5}{3}\right)^{-2}$ **5.** $-\left(-\frac{3}{5}\right)^2$ **6.** $-\left(-\frac{5}{3}\right)^2$

A. $\frac{25}{9}$ **B.** $-\frac{25}{9}$ **C.** $\frac{9}{25}$ **D.** $-\frac{9}{25}$ **E.** none of these **F.** all of these

Evaluate each exponential expression.

7. 5^4
8. 10^3
9. $(-2)^5$
10. $(-5)^4$
11. -2^3
12. -3^2
13. $-(-3)^4$
14. $-(-5)^2$
15. 7^{-2}
16. 4^{-1}
17. -7^{-2}
18. -4^{-1}
19. $\frac{2}{(-4)^{-3}}$
20. $\frac{2^{-3}}{3^{-2}}$
21. $\frac{5^{-1}}{4^{-2}}$
22. $\left(\frac{1}{2}\right)^{-3}$
23. $\left(\frac{1}{5}\right)^{-3}$
24. $\left(\frac{2}{3}\right)^{-2}$
25. $\left(\frac{4}{5}\right)^{-2}$
26. $3^{-1} + 2^{-1}$
27. $4^{-1} + 5^{-1}$
28. 8^0
29. 12^0
30. $(-23)^0$
31. $(-4)^0$
32. -2^0
33. $3^0 - 4^0$
34. $-8^0 - 7^0$

35. In order to raise a fraction to a negative power, we may change the fraction to its ________ and change the exponent to the ______ ______ of the original exponent.

36. Explain in your own words how to raise a power to a power.

37. Which one of the following is correct?
A. $-\frac{3}{4} = \left(\frac{3}{4}\right)^{-1}$ B. $\frac{3^{-1}}{4^{-1}} = \left(\frac{4}{3}\right)^{-1}$
C. $\frac{3^{-1}}{4} = \frac{3}{4^{-1}}$ D. $\frac{3^{-1}}{4^{-1}} = \left(\frac{3}{4}\right)^{-1}$

38. Which one of the following is incorrect?
A. $(3r)^{-2} = 3^{-2}r^{-2}$ B. $3r^{-2} = (3r)^{-2}$
C. $(3r)^{-2} = \frac{1}{(3r)^2}$ D. $(3r)^{-2} = \frac{r^{-2}}{9}$

Use the product, quotient, and power rules to simplify each expression. Write answers with only positive exponents. Assume that all variables represent nonzero real numbers.

39. $x^{12} \cdot x^4$
40. $\frac{x^{12}}{x^4}$
41. $\frac{5^{17}}{5^{16}}$
42. $\frac{3^{12}}{3^{13}}$
43. $\frac{3^{-5}}{3^{-2}}$
44. $\frac{2^{-4}}{2^{-3}}$
45. $\frac{9^{-1}}{9}$
46. $\frac{12}{12^{-1}}$
47. t^5t^{-12}
48. p^5p^{-6}
49. $(3x)^2$
50. $(-2x^{-2})^2$
51. $a^{-3}a^2a^{-4}$
52. $k^{-5}k^{-3}k^4$
53. $\frac{x^7}{x^{-4}}$
54. $\frac{p^{-3}}{p^5}$
55. $\frac{r^3r^{-4}}{r^{-2}r^{-5}}$
56. $\frac{z^{-4}z^{-2}}{z^3z^{-1}}$
57. $7k^2(-2k)(4k^{-5})$
58. $3a^2(-5a^{-6})(-2a)$
59. $(z^3)^{-2}z^2$
60. $(p^{-1})^3p^{-4}$
61. $-3r^{-1}(r^{-3})^2$
62. $2(y^{-3})^4(y^6)$
63. $(3a^{-2})^3(a^3)^{-4}$
64. $(m^5)^{-2}(3m^{-2})^3$
65. $(x^{-5}y^2)^{-1}$
66. $(a^{-3}b^{-5})^2$

67. Which one of the following does *not* represent the reciprocal of x $(x \neq 0)$?
A. x^{-1} B. $\frac{1}{x}$ C. $\left(\frac{1}{x^{-1}}\right)^{-1}$ D. $-x$

68. Which one of the following is *not* in scientific notation?
A. 6.02×10^{23} B. 14×10^{-6}
C. 1.4×10^{-5} D. 3.8×10^3

Convert each number from standard notation to scientific notation.

69. 230
70. 46,500
71. .02
72. .0051

Convert each number from scientific notation to standard notation.

73. 6.5×10^3

74. 2.317×10^5

75. 1.52×10^{-2}

76. 1.63×10^{-4}

Use scientific notation to perform each of the following computations. Leave the answers in scientific notation.

77. $\dfrac{.002 \times 3900}{.000013}$

78. $\dfrac{.009 \times 600}{.02}$

79. $\dfrac{.0004 \times 56{,}000}{.000112}$

80. $\dfrac{.018 \times 20{,}000}{300 \times .0004}$

81. $\dfrac{840{,}000 \times .03}{.00021 \times 600}$

82. $\dfrac{28 \times .0045}{140 \times 1500}$

Solve each problem.

83. ***U.S. Budget*** The U.S. budget first passed **$1,000,000,000** in 1917. Seventy years later in 1987 it exceeded **$1,000,000,000,000** for the first time. President George W. Bush's budget request for fiscal 2003 was **$2,128,000,000,000**. If stacked in dollar bills, this amount would stretch **144,419** mi, almost two-thirds of the distance to the moon. Write the four boldfaced numbers in scientific notation. (*Source: The Gazette*, February 5, 2002.)

84. ***Wal-Mart Employment*** In 1970, Wal-Mart had **1500** employees. In 1997, Wal-Mart became the largest private employer in the United States, with **680,000** employees. In 1999, Wal-Mart became the largest private employer in the world, with **1,100,000** employees. By 2007, the company is expected to have **2,200,000** employees. Write these four numbers in scientific notation. (*Source:* Wal-Mart.)

85. ***NASA Budget*** The budget for the Operating Plan in 2005 for the National Aeronautics and Space Administration was **$16,196.4 million.** Write this amount in scientific notation. (*Source:* www.nasa.gov)

86. ***Motor Vehicle Registrations*** In 2002, there were **229,620,000** motor vehicle registrations in the United States. Write this number in scientific notation. (*Source:* U.S. Federal Highway Administration.)

Astronomy Data *Each of the following statements (Exercises 87–90) comes from* Astronomy! A Brief Edition *by James B. Kaler (Addison-Wesley, 1997). If the number in the statement is in scientific notation, write it in standard notation without using exponents. If the number is in standard notation, write it in scientific notation.*

87. Multiplying this view over the whole sky yields a galaxy count of more than **10 billion.** (page 496)

88. The circumference of the solar orbit is . . . about **4.7 million** km . . . . (in reference to the orbit of Jupiter, page 395)

89. The solar luminosity requires that $\mathbf{2 \times 10^9}$ kg of mass be converted into energy every second. (page 327)

90. At maximum, a cosmic ray particle—a mere atomic nucleus of only $\mathbf{10^{-13}}$ cm across—can carry the energy of a professionally pitched baseball. (page 445)

Solve each problem.

91. ***Defense Budget*** The federal budget outlay for defense functions in the United States in 2003 was $404.9 billion. The estimated population of the United States in 2003 was 290,810 thousand. To the nearest dollar, what was the amount devoted to defense per person? (*Source:* U.S. Office of Management and Budget, U.S. Census Bureau.)

92. ***Powerball Lottery*** In the early years of the Powerball Lottery, a player had to choose five numbers from 1 through 49 and one number from 1 through 42. It can be shown that there are about 8.009×10^7 different ways to do this. Suppose that a group of 2000 people decided to purchase tickets for all these numbers and each ticket cost \$1.00. How much should each person have expected to pay? (*Source:* www.powerball.com)

93. ***Distance of Uranus from the Sun*** A parsec, a unit of length used in astronomy, is 1.9×10^{13} miles. The mean distance of Uranus from the sun is 1.8×10^7 miles. How many parsecs is Uranus from the sun?

94. ***Number of Inches in a Mile*** An inch is approximately 1.57828×10^{-5} mile. Find the reciprocal of this number to determine the number of inches in a mile.

95. ***Speed of Light*** The speed of light is approximately 3×10^{10} centimeters per second. How long will it take light to travel 9×10^{12} centimeters?

96. ***Rocket from Earth to the Sun*** The average distance from Earth to the sun is 9.3×10^7 miles. How long would it take a rocket, traveling at 2.9×10^3 miles per hour, to reach the sun?

97. ***Miles in a Light-Year*** A *light-year* is the distance that light travels in one year. Find the number of miles in a light-year if light travels 1.86×10^5 miles per second.

98. ***Time for Light to Travel*** Use the information given in the previous two exercises to find the number of minutes necessary for light from the sun to reach Earth.

99. ***Rocket from Venus to Mercury*** The planet Mercury has an average distance from the sun of 3.6×10^7 miles, while the mean distance of Venus from the sun is 6.7×10^7 miles. How long would it take a spacecraft traveling at 1.55×10^3 miles per hour to travel from Venus to Mercury? Assume the trip could be timed so that its start and finish would occur when the respective planets are at their average distances from the sun and the same direction from the sun. (Give your answer in hours, without scientific notation.)

100. ***Distance from an Object to the Moon*** When the distance between the centers of the moon and Earth is 4.60×10^8 meters, an object on the line joining the centers of the moon and Earth exerts the same gravitational force on each when it is 4.14×10^8 meters from the center of Earth. How far is the object from the center of the moon at that point?

7.6 Polynomials and Factoring

Basic Terminology • Addition and Subtraction • Multiplication • Special Products • Factoring • Factoring Out the Greatest Common Factor • Factoring by Grouping • Factoring Trinomials • Factoring Special Binomials

Basic Terminology A **term,** or **monomial,** is defined to be a number, a variable, or a product of numbers and variables. A **polynomial** is a term or a finite sum or difference of terms, with only nonnegative integer exponents permitted on the variables. If the terms of a polynomial contain only the variable x, then the polynomial is called a **polynomial in x.** (Polynomials in other variables are defined similarly.) Examples of polynomials include

$$5x^3 - 8x^2 + 7x - 4, \quad 9p^5 - 3, \quad 8r^2, \quad \text{and} \quad 6.$$ Polynomials

The expression $9x^2 - 4x - \frac{6}{x}$ is not a polynomial because of the presence of $-\frac{6}{x}$. The terms of a polynomial cannot have variables in a denominator.

The greatest exponent in a polynomial in one variable is the **degree** of the polynomial. A nonzero constant is said to have degree 0. (The polynomial 0 has no degree.) For example, $3x^6 - 5x^2 + 2x + 3$ is a polynomial of degree 6.

A polynomial can have more than one variable. A term containing more than one variable has degree equal to the sum of all the exponents appearing on the variables in the term. For example, $-3x^4y^3z^5$ is of degree $4 + 3 + 5 = 12$. The degree of a polynomial in more than one variable is equal to the greatest degree of any term appearing in the polynomial. By this definition, the polynomial $2x^4y^3 - 3x^5y + x^6y^2$ is of degree 8 because the x^6y^2 term has degree 8.

A polynomial containing exactly three terms is called a **trinomial** and one containing exactly two terms is a **binomial.** Table 5 shows several polynomials and gives the degree and type of each.

TABLE 5

Polynomial	Degree	Type
$9p^7 - 4p^3 + 8p^2$	7	Trinomial
$29x^{11} + 8x^{15}$	15	Binomial
$-10r^6s^8$	14	Monomial
$5a^3b^7 - 3a^5b^5 + 4a^2b^9 - a^{10}$	11	None of these

Addition and Subtraction Since the variables used in polynomials represent real numbers, a polynomial represents a real number. This means that all the properties of the real numbers mentioned in this book hold for polynomials. In particular, the distributive property holds, so

$$3m^5 - 7m^5 = (3 - 7)m^5 = -4m^5.$$

Like terms are terms that have the exact same variable factors. Thus, polynomials are added by adding coefficients of like terms; polynomials are subtracted by subtracting coefficients of like terms.

EXAMPLE 1 Adding and Subtracting Polynomials

Add or subtract, as indicated.

(a) $(2y^4 - 3y^2 + y) + (4y^4 + 7y^2 + 6y)$ **(b)** $(-3m^3 - 8m^2 + 4) - (m^3 + 7m^2 - 3)$

(c) $8m^4p^5 - 9m^3p^5 + (11m^4p^5 + 15m^3p^5)$

(d) $4(x^2 - 3x + 7) - 5(2x^2 - 8x - 4)$

SOLUTION

(a) $(2y^4 - 3y^2 + y) + (4y^4 + 7y^2 + 6y)$

$= (2 + 4)y^4 + (-3 + 7)y^2 + (1 + 6)y$

$= 6y^4 + 4y^2 + 7y$

(b) $(-3m^3 - 8m^2 + 4) - (m^3 + 7m^2 - 3)$

$= (-3 - 1)m^3 + (-8 - 7)m^2 + [4 - (-3)]$

$= -4m^3 - 15m^2 + 7$

This − symbol changes the sign of each coefficient in $m^3 + 7m^2 - 3$.

(c) $8m^4p^5 - 9m^3p^5 + (11m^4p^5 + 15m^3p^5) = 19m^4p^5 + 6m^3p^5$

(d) $4(x^2 - 3x + 7) - 5(2x^2 - 8x - 4)$

$= 4x^2 - 4(3x) + 4(7) - 5(2x^2) - 5(-8x) - 5(-4)$ Distributive property

$= 4x^2 - 12x + 28 - 10x^2 + 40x + 20$ Associative property

$= -6x^2 + 28x + 48$ Combine like terms. ■

As shown in parts (a), (b), and (d) of Example 1, polynomials in one variable are often written with their terms in **descending powers;** so the term of greatest degree is first, the one with the next greatest degree is second, and so on.

Multiplication The associative and distributive properties, together with the properties of exponents, can also be used to find the product of two polynomials. For example, to find the product of $3x - 4$ and $2x^2 - 3x + 5$, treat $3x - 4$ as a single expression and use the distributive property as follows.

$$(3x - 4)(2x^2 - 3x + 5) = (3x - 4)(2x^2) - (3x - 4)(3x) + (3x - 4)(5)$$

Now use the distributive property three separate times on the right to get

$$= (3x)(2x^2) - 4(2x^2) - (3x)(3x) - (-4)(3x) + (3x)5 - 4(5)$$
$$= 6x^3 - 8x^2 - 9x^2 + 12x + 15x - 20$$
$$= 6x^3 - 17x^2 + 27x - 20.$$

It is sometimes more convenient to write such a product vertically, as follows.

$$\begin{array}{rrrrl}
 & 2x^2 & -\ 3x & +\ 5 & \\
 & & 3x & -\ 4 & \\
\hline
 & -8x^2 & +\ 12x & -\ 20 & \leftarrow -4(2x^2 - 3x + 5) \\
6x^3 & -\ 9x^2 & +\ 15x & & \leftarrow 3x(2x^2 - 3x + 5) \\
\hline
6x^3 & -\ 17x^2 & +\ 27x & -\ 20 & \text{Add in columns.}
\end{array}$$

Be sure to place like terms in columns.

EXAMPLE 2 Multiplying Polynomials Vertically

Multiply $(3p^2 - 4p + 1)(p^3 + 2p - 8)$.

SOLUTION

Multiply each term of the second polynomial by each term of the first and add.

$$\begin{array}{rrrrrrl}
 & & & 3p^2 & -\ 4p & +\ 1 & \\
 & & & p^3 & +\ 2p & -\ 8 & \\
\hline
 & & & -\ 24p^2 & +\ 32p & -\ 8 & \text{Multiply } 3p^2 - 4p + 1 \text{ by } -8. \\
 & & 6p^3 & -\ 8p^2 & +\ 2p & & \text{Multiply } 3p^2 - 4p + 1 \text{ by } 2p. \\
3p^5 & -\ 4p^4 & +\ p^3 & & & & \text{Multiply } 3p^2 - 4p + 1 \text{ by } p^3. \\
\hline
3p^5 & -\ 4p^4 & +\ 7p^3 & -\ 32p^2 & +\ 34p & -\ 8 & \text{Add in columns.}
\end{array}$$

■

The FOIL method is a convenient way to find the product of two binomials. The memory aid FOIL (for First, Outside, Inside, Last) gives the pairs of terms to be multiplied to get the product, as shown in the next examples.

EXAMPLE 3 Using the FOIL Method

Find each product.

(a) $(6m + 1)(4m - 3)$ **(b)** $(2x + 7)(2x - 7)$

SOLUTION

(a)

$$\begin{aligned}(6m + 1)(4m - 3) &= \overset{F}{(6m)(4m)} + \overset{O}{(6m)(-3)} + \overset{I}{1(4m)} + \overset{L}{1(-3)} \\ &= 24m^2 - 18m + 4m - 3 \\ &= 24m^2 - 14m - 3 \quad \text{Combine like terms.}\end{aligned}$$

(b)

$$\begin{aligned}(2x + 7)(2x - 7) &= 4x^2 - 14x + 14x - 49 \quad \text{FOIL} \\ &= 4x^2 - 49 \quad \text{Combine like terms.}\end{aligned}$$

The **special product**

$(x + y)(x - y) = x^2 - y^2$

can be used to solve some multiplication problems. For example,

$$\begin{aligned}51 \times 49 &= (50 + 1)(50 - 1) \\ &= 50^2 - 1^2 \\ &= 2500 - 1 \\ &= 2499\end{aligned}$$

$$\begin{aligned}102 \times 98 &= (100 + 2)(100 - 2) \\ &= 100^2 - 2^2 \\ &= 10{,}000 - 4 \\ &= 9996.\end{aligned}$$

Once these patterns are recognized, multiplications of this type can be done mentally.

Special Products In part (a) of Example 3, the product of two binomials was a trinomial, while in part (b) the product of two binomials was a binomial. ***The product of two binomials of the forms $x + y$ and $x - y$ is always a binomial.*** Check by multiplying that the following is true.

Product of the Sum and Difference of Two Terms

$$(x + y)(x - y) = x^2 - y^2$$

The product $x^2 - y^2$ is called the **difference of two squares.**

EXAMPLE 4 Finding Special Products

Find each product.

(a) $(3p + 11)(3p - 11)$ **(b)** $(5m^3 - 3)(5m^3 + 3)$

(c) $(9k - 11r^3)(9k + 11r^3)$

SOLUTION

(a) Using the pattern discussed above, replace x with $3p$ and y with 11.

$$\begin{aligned}(3p + 11)(3p - 11) &= (3p)^2 - 11^2 \\ &= 9p^2 - 121\end{aligned}$$

(b)

$$\begin{aligned}(5m^3 - 3)(5m^3 + 3) &= (5m^3)^2 - 3^2 \\ &= 25m^6 - 9\end{aligned}$$

(c)

$$\begin{aligned}(9k - 11r^3)(9k + 11r^3) &= (9k)^2 - (11r^3)^2 \\ &= 81k^2 - 121r^6\end{aligned}$$

The **squares of binomials** are also special products.

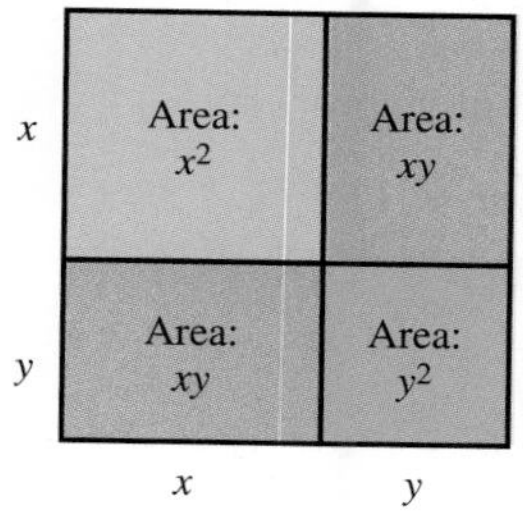

The **special product**

$$(x + y)^2 = x^2 + 2xy + y^2$$

can be illustrated geometrically using the diagram shown here. Each side of the large square has length $x + y$, so the area of the square is

$$(x + y)^2.$$

The large square is made up of two smaller squares and two congruent rectangles. The sum of the areas of these figures is

$$x^2 + 2xy + y^2.$$

Since these expressions represent the same quantity, they must be equal, thus giving us the pattern for squaring a binomial.

Squares of Binomials

$$(x + y)^2 = x^2 + 2xy + y^2$$
$$(x - y)^2 = x^2 - 2xy + y^2$$

EXAMPLE 5 Finding Special Products

Find each product.

(a) $(2m + 5)^2$ **(b)** $(3x - 7y^4)^2$

SOLUTION

(a) $(2m + 5)^2 = (2m)^2 + 2(2m)(5) + (5)^2$
$= 4m^2 + 20m + 25$

(b) $(3x - 7y^4)^2 = (3x)^2 - 2(3x)(7y^4) + (7y^4)^2$
$= 9x^2 - 42xy^4 + 49y^8$

The square of a binomial has three terms. $(x + y)^2 \neq x^2 + y^2$

Factoring The process of finding polynomials whose product equals a given polynomial is called **factoring.** For example, since

$$4x + 12 = 4(x + 3),$$

both 4 and $x + 3$ are called **factors** of $4x + 12$. Also, $4(x + 3)$ is called a **factored form** of $4x + 12$. A polynomial that cannot be written as a product of two polynomials with integer coefficients is a **prime polynomial.** A polynomial is **factored completely** when it is written as a product of prime polynomials with integer coefficients.

Factoring Out the Greatest Common Factor Some polynomials are factored by using the distributive property. We look for a monomial that is the greatest common factor (GCF) of all the terms of the polynomial. For example,

$$6x^2y^3 + 9xy^4 + 18y^5 = (3y^3)(2x^2) + (3y^3)(3xy) + (3y^3)(6y^2) \quad \text{GCF} = 3y^3$$
$$= 3y^3(2x^2 + 3xy + 6y^2).$$

EXAMPLE 6 Factoring Out the Greatest Common Factor

Factor out the greatest common factor from each polynomial.

(a) $9y^5 + y^2$ **(b)** $6x^2t + 8xt + 12t$

(c) $14m^4(m + 1) - 28m^3(m + 1) - 7m^2(m + 1)$

SOLUTION

(a) $9y^5 + y^2 = y^2 \cdot 9y^3 + y^2 \cdot 1$ The greatest common factor is y^2.
$= y^2(9y^3 + 1)$

(b) $6x^2t + 8xt + 12t = 2t(3x^2 + 4x + 6)$

(c) $14m^4(m + 1) - 28m^3(m + 1) - 7m^2(m + 1)$
The greatest common factor is $7m^2(m + 1)$. Use the distributive property.

$$\begin{aligned} &14m^4(m + 1) - 28m^3(m + 1) - 7m^2(m + 1) \\ &= [7m^2(m + 1)]\,(2m^2 - 4m - 1) \\ &= 7m^2(m + 1)\,(2m^2 - 4m - 1) \end{aligned}$$

Factoring by Grouping When a polynomial has more than three terms, it can sometimes be factored by a method called **factoring by grouping.** For example,

$$\begin{aligned} ax + ay + 6x + 6y &= (ax + ay) + (6x + 6y) && \text{Group the terms.} \\ &= a(x + y) + 6(x + y) && \text{Factor each group.} \\ &= (x + y)\,(a + 6). && \text{Factor out } (x + y). \end{aligned}$$

Experience and repeated trials are the most reliable tools for factoring by grouping.

EXAMPLE 7 Factoring by Grouping

Factor by grouping.

(a) $mp^2 + 7m + 3p^2 + 21$ **(b)** $2y^2 - 2z - ay^2 + az$

SOLUTION

(a)
$$\begin{aligned} mp^2 + 7m + 3p^2 + 21 &= (mp^2 + 7m) + (3p^2 + 21) && \text{Group the terms.} \\ &= m(p^2 + 7) + 3(p^2 + 7) && \text{Factor each group.} \\ &= (p^2 + 7)\,(m + 3) && p^2 + 7 \text{ is a common factor.} \end{aligned}$$

(b)
$$\begin{aligned} 2y^2 - 2z - ay^2 + az &= (2y^2 - 2z) + (-ay^2 + az) && \text{Group the terms.} \\ &= 2(y^2 - z) + a(-y^2 + z) && \text{Factor each group.} \end{aligned}$$

The expression $-y^2 + z$ is the negative of $y^2 - z$, so factor out $-a$ instead of a.

$$\begin{aligned} &= 2(y^2 - z) - a(y^2 - z) && \text{Factor out } -a. \\ &= (y^2 - z)\,(2 - a) && \text{Factor out } y^2 - z. \end{aligned}$$

Factoring Trinomials ***Factoring is the inverse of multiplying.*** Since the product of two binomials is usually a trinomial, we can expect factorable trinomials (that have terms with no common factor) to have two binomial factors. Thus, factoring trinomials requires using FOIL in an inverse manner.

EXAMPLE 8 Factoring Trinomials

Factor each trinomial.

(a) $4y^2 - 11y + 6$ **(b)** $6p^2 - 7p - 5$

SOLUTION

(a) To factor this polynomial, we must find integers a, b, c, and d such that

$$4y^2 - 11y + 6 = (ay + b)(cy + d).$$

By using FOIL, we see that $ac = 4$ and $bd = 6$. The positive factors of 4 are 4 and 1 or 2 and 2. Since the middle term is negative, we consider only negative factors of 6. The possibilities are -2 and -3 or -1 and -6. Now we try various arrangements of these factors until we find one that gives the correct coefficient of y.

$$(2y - 1)(2y - 6) = 4y^2 - 14y + 6 \quad \text{Incorrect}$$
$$(2y - 2)(2y - 3) = 4y^2 - 10y + 6 \quad \text{Incorrect}$$
$$(y - 2)(4y - 3) = 4y^2 - 11y + 6 \quad \text{Correct}$$

The last trial gives the correct factorization.

(b) Again, we try various possibilities. The positive factors of 6 could be 2 and 3 or 1 and 6. As factors of -5 we have only -1 and 5 or -5 and 1. Try different combinations of these factors until the correct one is found.

$$(2p - 5)(3p + 1) = 6p^2 - 13p - 5 \quad \text{Incorrect}$$
$$(3p - 5)(2p + 1) = 6p^2 - 7p - 5 \quad \text{Correct}$$

Thus, $6p^2 - 7p - 5$ factors as $(3p - 5)(2p + 1)$. ■

Each of the special patterns of multiplication given earlier can be used in reverse to get a pattern for factoring. Perfect square trinomials can be factored as follows.

Perfect Square Trinomials

$$x^2 + 2xy + y^2 = (x + y)^2$$
$$x^2 - 2xy + y^2 = (x - y)^2$$

EXAMPLE 9 Factoring Perfect Square Trinomials

Factor each polynomial.

(a) $16p^2 - 40pq + 25q^2$ **(b)** $169x^2 + 104xy^2 + 16y^4$

SOLUTION

(a) Make sure that the middle term of the trinomial being factored, $-40pq$ here, is twice the product of the two terms in the binomial $4p - 5q$.

$$-40pq = 2(4p)(-5q)$$

Since $16p^2 = (4p)^2$ and $25q^2 = (5q)^2$, use the second pattern shown above with $4p$ replacing x and $5q$ replacing y to obtain

$$16p^2 - 40pq + 25q^2 = (4p)^2 - 2(4p)(5q) + (5q)^2$$
$$= (4p - 5q)^2.$$

(b) $169x^2 + 104xy^2 + 16y^4 = (13x + 4y^2)^2$, since $2(13x)(4y^2) = 104xy^2$. ■

Factoring Special Binomials The pattern for the product of the sum and difference of two terms gives the following factorization.

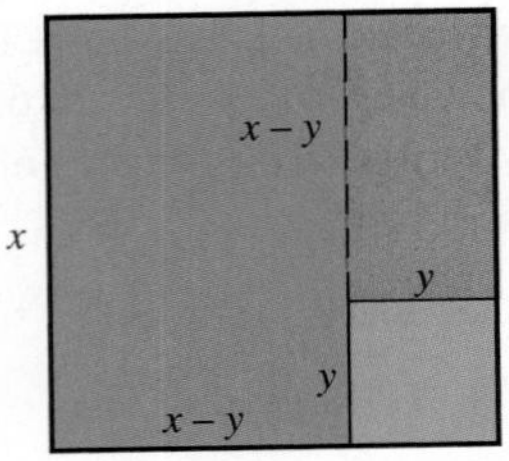

A **geometric proof** for the difference of squares property is shown above. (The proof is only valid for $x > y > 0$.)

$$x^2 - y^2 = x(x - y) + y(x - y)$$
$$= (x - y)(x + y)$$

Factor out $x - y$ in the second step.

Difference of Squares

$$x^2 - y^2 = (x + y)(x - y)$$

EXAMPLE 10 Factoring Differences of Squares

Factor each polynomial.

(a) $4m^2 - 9$ **(b)** $256k^4 - 625m^4$ **(c)** $x^2 - 6x + 9 - y^4$

SOLUTION

(a) First, recognize that $4m^2 - 9$ is the difference of squares, since $4m^2 = (2m)^2$ and $9 = 3^2$. Use the pattern for the difference of squares with $2m$ replacing x and 3 replacing y.

$$4m^2 - 9 = (2m)^2 - 3^2$$
$$= (2m + 3)(2m - 3)$$

(b) $256k^4 - 625m^4 = (16k^2)^2 - (25m^2)^2$ Difference of squares

$= (16k^2 + 25m^2)(16k^2 - 25m^2)$

$= (16k^2 + 25m^2)(4k + 5m)(4k - 5m)$ Difference of squares

(c) $x^2 - 6x + 9 - y^4 = (x^2 - 6x + 9) - y^4$ Group the first three terms.

$= (x - 3)^2 - (y^2)^2$ Perfect square trinomial

$= [(x - 3) + y^2][(x - 3) - y^2]$ Difference of squares

$= (x - 3 + y^2)(x - 3 - y^2)$ ■

Solution to the Chapter Opener The player's expression and answer are correct. Suppose a and b are the individual times. Then the hourly rates for the players are $\frac{1}{a}$ and $\frac{1}{b}$ job per hour. Multiplying rate by time worked gives the fractional part of the job performed by each player. If x represents the time they must work together to complete one whole job, then we have

$$\frac{1}{a}x + \frac{1}{b}x = 1$$
Linear equation

$$ab\left(\frac{1}{a}x + \frac{1}{b}x\right) = ab \cdot 1$$
Multiply by ab.

$$bx + ax = ab$$
Distributive property

$$x(a + b) = ab$$
Factor out x.

$$x = \frac{ab}{a + b}.$$
Divide by $a + b$.

The answer given in the movie is correct.

Two other special results of factoring are listed below. Each can be verified by multiplying on the right side of the equation.

Sum and Difference of Cubes

Sum of Cubes	$x^3 + y^3 = (x + y)(x^2 - xy + y^2)$
Difference of Cubes	$x^3 - y^3 = (x - y)(x^2 + xy + y^2)$

EXAMPLE 11 Factoring Sums and Differences of Cubes

Factor each polynomial.

(a) $x^3 + 27$ **(b)** $m^3 - 64n^3$ **(c)** $8q^6 + 125p^9$

SOLUTION

(a) $x^3 + 27 = x^3 + 3^3$ Sum of cubes

$= (x + 3)(x^2 - 3x + 9)$

(b) $m^3 - 64n^3 = m^3 - (4n)^3$ Difference of cubes

$= (m - 4n)[m^2 + m(4n) + (4n)^2]$

$= (m - 4n)(m^2 + 4mn + 16n^2)$

(c) $8q^6 + 125p^9 = (2q^2)^3 + (5p^3)^3$
$= (2q^2 + 5p^3)[(2q^2)^2 - (2q^2)(5p^3) + (5p^3)^2]$
$= (2q^2 + 5p^3)(4q^4 - 10q^2p^3 + 25p^6)$ ■

7.6 Exercises

Find each sum or difference.

1. $(3x^2 - 4x + 5) + (-2x^2 + 3x - 2)$
2. $(4m^3 - 3m^2 + 5) + (-3m^3 - m^2 + 5)$
3. $(12y^2 - 8y + 6) - (3y^2 - 4y + 2)$
4. $(8p^2 - 5p) - (3p^2 - 2p + 4)$
5. $(6m^4 - 3m^2 + m) - (2m^3 + 5m^2 + 4m) + (m^2 - m)$
6. $-(8x^3 + x - 3) + (2x^3 + x^2) - (4x^2 + 3x - 1)$
7. $5(2x^2 - 3x + 7) - 2(6x^2 - x + 12)$
8. $8x^2y - 3xy^2 + 2x^2y - 9xy^2$

Find each product.

9. $(x + 3)(x - 8)$
10. $(y - 3)(y - 9)$
11. $(4r - 1)(7r + 2)$
12. $(5m - 6)(3m + 4)$
13. $4x^2(3x^3 + 2x^2 - 5x + 1)$
14. $2b^3(b^2 - 4b + 3)$
15. $(2m + 3)(2m - 3)$
16. $(8s - 3t)(8s + 3t)$
17. $(4m + 2n)^2$
18. $(a - 6b)^2$
19. $(5r + 3t^2)^2$
20. $(2z^4 - 3y)^2$
21. $(2z - 1)(-z^2 + 3z - 4)$
22. $(k + 2)(12k^3 - 3k^2 + k + 1)$
23. $(m - n + k)(m + 2n - 3k)$
24. $(r - 3s + t)(2r - s + t)$
25. $(a - b + 2c)^2$
26. $(k - y + 3m)^2$
27. Which one of the following is a trinomial in descending powers, having degree 6?
   A. $5x^6 - 4x^5 + 12$ B. $6x^5 - x^6 + 4$
   C. $2x + 4x^2 - x^6$ D. $4x^6 - 6x^4 + 9x + 1$
28. Give an example of a polynomial of four terms in the variable x, having degree 5, written in descending powers, lacking a fourth degree term.
29. The exponent in the expression 6^3 is 3. Explain why the degree of 6^3 is not 3. What is its degree?
30. Explain in your own words how to square a binomial.

Factor the greatest common factor from each polynomial.

31. $8m^4 + 6m^3 - 12m^2$
32. $2p^5 - 10p^4 + 16p^3$
33. $4k^2m^3 + 8k^4m^3 - 12k^2m^4$
34. $28r^4s^2 + 7r^3s - 35r^4s^3$
35. $2(a + b) + 4m(a + b)$
36. $4(y - 2)^2 + 3(y - 2)$
37. $2(m - 1) - 3(m - 1)^2 + 2(m - 1)^3$
38. $5(a + 3)^3 - 2(a + 3) + (a + 3)^2$

Factor each polynomial by grouping.

39. $6st + 9t - 10s - 15$
40. $10ab - 6b + 35a - 21$
41. $rt^3 + rs^2 - pt^3 - ps^2$
42. $2m^4 + 6 - am^4 - 3a$
43. $16a^2 + 10ab - 24ab - 15b^2$
44. $15 - 5m^2 - 3r^2 + m^2r^2$
45. $20z^2 - 8zx - 45zx + 18x^2$
46. $4 - 2y - 2x + xy$
47. $1 - a + ab - b$

48. Consider the polynomial $1 - a + ab - b$ from Exercise 47. The answer given in the answer section is $(1 - a)(1 - b)$. However, there are other acceptable factored forms. Which one of A–D is *not* a factored form of this polynomial?

A. $(a - 1)(b - 1)$
B. $(-a + 1)(-b + 1)$
C. $(-1 + a)(-1 + b)$
D. $(1 - a)(b + 1)$

Factor each trinomial.

49. $x^2 - 2x - 15$
50. $r^2 + 8r + 12$
51. $y^2 + 2y - 35$
52. $x^2 - 7x + 6$
53. $6a^2 - 48a - 120$
54. $8h^2 - 24h - 320$
55. $3m^3 + 12m^2 + 9m$
56. $9y^4 - 54y^3 + 45y^2$
57. $6k^2 + 5kp - 6p^2$
58. $14m^2 + 11mr - 15r^2$
59. $5a^2 - 7ab - 6b^2$
60. $12s^2 + 11st - 5t^2$
61. $21x^2 - xy - 2y^2$
62. $30a^2 + am - m^2$
63. $24a^4 + 10a^3b - 4a^2b^2$
64. $18x^5 + 15x^4z - 75x^3z^2$

65. When a student was given the polynomial

$$4x^2 + 2x - 20$$

to factor completely on a test, she lost some credit by giving the answer $(4x + 10)(x - 2)$. She then complained to her teacher that the product $(4x + 10) \cdot (x - 2)$ is indeed $4x^2 + 2x - 20$. Do you think that the teacher was justified in not giving her full credit? Explain.

66. Write an explanation as to why most people would find it more difficult to factor

$$36x^2 - 44x - 15$$

than

$$37x^2 - 183x - 10.$$

Factor each polynomial, using the method for factoring a perfect square trinomial. It may be necessary to factor out a common factor first.

67. $9m^2 - 12m + 4$
68. $16p^2 - 40p + 25$
69. $32a^2 - 48ab + 18b^2$
70. $20p^2 - 100pq + 125q^2$
71. $4x^2y^2 + 28xy + 49$
72. $9m^2n^2 - 12mn + 4$

Factor each difference of squares.

73. $x^2 - 36$
74. $t^2 - 64$
75. $y^2 - w^2$
76. $25 - w^2$
77. $9a^2 - 16$
78. $16q^2 - 25$
79. $25s^4 - 9t^2$
80. $36z^2 - 81y^4$
81. $p^4 - 625$
82. $m^4 - 81$

Factor each sum or difference of cubes.

83. $8 - a^3$
84. $r^3 + 27$
85. $125x^3 - 27$
86. $8m^3 - 27n^3$
87. $27y^9 + 125z^6$
88. $27z^3 + 729y^3$

Each polynomial may be factored using one of the methods described in this section. Decide on the method, and then factor the polynomial completely.

89. $x^2 + xy - 5x - 5y$
90. $8r^2 - 10rs - 3s^2$
91. $p^4(m - 2n) + q(m - 2n)$
92. $36a^2 + 60a + 25$
93. $4z^2 + 28z + 49$
94. $6p^4 + 7p^2 - 3$
95. $1000x^3 + 343y^3$
96. $b^2 + 8b + 16 - a^2$
97. $125m^6 - 216$
98. $q^2 + 6q + 9 - p^2$
99. $12m^2 + 16mn - 35n^2$
100. $216p^3 + 125q^3$

101. The sum of squares usually cannot be factored. For example, $x^2 + y^2$ is prime. Notice that $x^2 + y^2 \neq (x + y)(x + y)$. By choosing $x = 4$ and $y = 2$, show that this statement is true.

102. The binomial $9x^2 + 36$ is a sum of squares. Can it be factored? If so, factor it.

7.7 Quadratic Equations and Applications

Quadratic Equations • Zero-Factor Property • Square Root Property • Quadratic Formula • Applications

Quadratic Equations

Recall that a linear equation is one that can be written in the form $ax + b = c$, where a, b, and c are real numbers, $a \neq 0$.

Quadratic Equation

An equation that can be written in the form

$$ax^2 + bx + c = 0$$

where a, b, and c are real numbers, with $a \neq 0$, is a **quadratic equation.**

A quadratic equation written in the form $ax^2 + bx + c = 0$ is in **standard form.**

Zero-Factor Property

The simplest method of solving a quadratic equation, but one that is not always easily applied, is by factoring. This method depends on the following property.

Zero-Factor Property

If $ab = 0$, then $a = 0$ or $b = 0$ or both.

When solving a quadratic equation by the zero-factor property, the equation must be in standard form before factoring.

EXAMPLE 1 Using the Zero-Factor Property

Solve $6x^2 + 7x = 3$.

SOLUTION

$$6x^2 + 7x = 3$$

$$6x^2 + 7x - 3 = 0 \quad \text{Standard form}$$

$$(3x - 1)(2x + 3) = 0 \quad \text{Factor.}$$

$$3x - 1 = 0 \quad \text{or} \quad 2x + 3 = 0 \quad \text{Zero-factor property}$$

$$3x = 1 \quad \text{or} \quad 2x = -3 \quad \text{Solve each equation.}$$

$$x = \frac{1}{3} \quad \text{or} \quad x = -\frac{3}{2}$$

Check by first substituting $\frac{1}{3}$ and then $-\frac{3}{2}$ in the original equation. The solution set is $\left\{\frac{1}{3}, -\frac{3}{2}\right\}$.

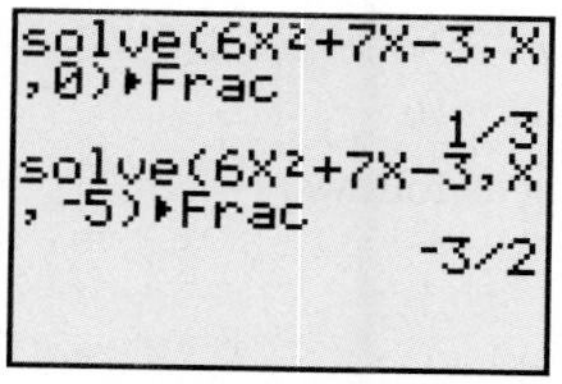

The *solve* feature gives the two solutions of the equation in Example 1. Notice that the guess 0 yields the solution 1/3, while the guess −5 yields the solution −3/2. Compare to Example 1.

Square Root Property A quadratic equation of the form $x^2 = k$, $k \geq 0$, can be solved by factoring.

$$x^2 = k$$
$$x^2 - k = 0$$
$$(x + \sqrt{k})(x - \sqrt{k}) = 0$$
$$x + \sqrt{k} = 0 \quad \text{or} \quad x - \sqrt{k} = 0$$
$$x = -\sqrt{k} \quad \text{or} \quad x = \sqrt{k}$$

This proves the square root property for solving equations.

Square Root Property

If $k \geq 0$, then the solutions of $x^2 = k$ are $\boldsymbol{x = \pm\sqrt{k}}$.

If $k > 0$, the equation $x^2 = k$ has two real solutions. If $k = 0$, there is only one solution, 0. If $k < 0$, there are no real solutions. (However, in this case, there *are* imaginary solutions. Imaginary numbers are discussed briefly in the Extension on complex numbers at the end of Chapter 6.)

EXAMPLE 2 Using the Square Root Property

Use the square root property to solve each quadratic equation for real solutions.

(a) $x^2 = 25$ **(b)** $r^2 = 18$ **(c)** $z^2 = -3$ **(d)** $(x - 4)^2 = 12$

SOLUTION

(a) Since $\sqrt{25} = 5$, the solution set of the equation $x^2 = 25$ is $\{5, -5\}$, which may be abbreviated $\{\pm 5\}$.

(b)

$$r^2 = 18$$
$$r = \pm\sqrt{18} \qquad \text{Square root property}$$
$$r = \pm\sqrt{9 \cdot 2}$$
$$r = \pm\sqrt{9} \cdot \sqrt{2} \qquad \text{Product rule for square roots}$$
$$r = \pm 3\sqrt{2} \qquad \sqrt{9} = 3$$

The solution set is $\{\pm 3\sqrt{2}\}$.

(c) Since $-3 < 0$, there are no real roots, and the solution set is $\emptyset$.

(d) Use a generalization of the square root property, working as follows.

$$(x - 4)^2 = 12$$
$$x - 4 = \pm\sqrt{12} \qquad \text{Square root property}$$
$$x = 4 \pm \sqrt{12} \qquad \text{Add 4.}$$
$$x = 4 \pm \sqrt{4 \cdot 3} \qquad \text{Simplify } \sqrt{12}.$$
$$x = 4 \pm 2\sqrt{3}$$

The solution set is $\{4 \pm 2\sqrt{3}\}$.

Completing the square, used in deriving the quadratic formula, has important applications in algebra. To transform the expression $x^2 + kx$ into the square of a binomial, we add to it the square of half the coefficient of x; that is, $\left[\left(\frac{1}{2}\right)k\right]^2 = \frac{k^2}{4}$. We then get

$$x^2 + kx + \frac{k^2}{4} = \left(x + \frac{k}{2}\right)^2.$$

For example, to make $x^2 + 6x$ the square of a binomial, we add 9, since $9 = \left[\frac{1}{2}(6)\right]^2$. This results in the trinomial $x^2 + 6x + 9$, which is equal to $(x + 3)^2$.

The Greeks had a method of completing the square geometrically. For example, to complete the square for $x^2 + 6x$, begin with a square of side x. Add three rectangles of width 1 and length x to the right side and the bottom. Each rectangle has area $1x$ or x, so the total area of the figure is now $x^2 + 6x$. To fill in the corner (that is, "complete the square"), we add 9 1-by-1 squares as shown. The new completed square has sides of length $x + 3$ and area

$$(x + 3)^2 = x^2 + 6x + 9.$$

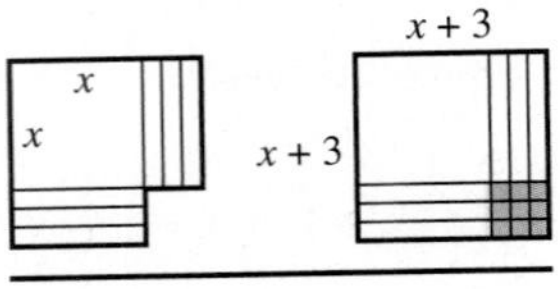

A first-season episode of *Blue Collar TV* (2004) featured Bill Engvall paying a sarcastic tribute to an underappreciated figure in his life.

To my high school algebra teacher, for teaching me that x equals minus b plus or minus the square root of b squared minus 4ac all over 2a, because Lord knows I use that information EVERY DAY!

Bill's teacher evidently did a good job, because he used the word "all" before $2a$. A common student error is to forget to write the $-b$ in the numerator with the radical expression in the **quadratic formula.** See Exercise 44 in this section.

Quadratic Formula By using a procedure called *completing the square* (see the margin note) we can derive one of the most important formulas in algebra, the *quadratic formula*. We begin with the standard quadratic equation and assume $a > 0$.

$$ax^2 + bx + c = 0$$

$$x^2 + \frac{b}{a}x + \frac{c}{a} = 0 \qquad \text{Divide by } a.$$

$$x^2 + \frac{b}{a}x = -\frac{c}{a} \qquad \text{Add } -\frac{c}{a}.$$

$$x^2 + \frac{b}{a}x + \frac{b^2}{4a^2} = \frac{b^2}{4a^2} - \frac{c}{a} \qquad \text{Add } \frac{b^2}{4a^2}.$$

$$\left(x + \frac{b}{2a}\right)^2 = \frac{b^2 - 4ac}{4a^2} \qquad \text{Factor on the left; combine terms on the right.}$$

$$x + \frac{b}{2a} = \pm\sqrt{\frac{b^2 - 4ac}{4a^2}} \qquad \text{Square root property}$$

$$x + \frac{b}{2a} = \pm\frac{\sqrt{b^2 - 4ac}}{\sqrt{4a^2}} \qquad \text{Quotient rule for square roots}$$

$$x = -\frac{b}{2a} \pm \frac{\sqrt{b^2 - 4ac}}{2a} \qquad \text{Subtract } \frac{b}{2a}.$$

$$x = \frac{-b \pm \sqrt{b^2 - 4ac}}{2a} \qquad \text{Combine terms.}$$

Be careful; $-b \pm \sqrt{b^2 - 4ac}$ is all written over $2a$.

The formula is also valid for $a < 0$.

Quadratic Formula

The solutions of $ax^2 + bx + c = 0$, $a \neq 0$, are

$$x = \frac{-b \pm \sqrt{b^2 - 4ac}}{2a}.$$

EXAMPLE 3 Using the Quadratic Formula

Solve $x^2 - 4x + 2 = 0$.

SOLUTION

Here $a = 1$, $b = -4$, and $c = 2$. Substitute these values into the quadratic formula.

$$x = \frac{-b \pm \sqrt{b^2 - 4ac}}{2a} \qquad \text{Quadratic formula}$$

$$x = \frac{-(-4) \pm \sqrt{(-4)^2 - 4(1)2}}{2(1)} \qquad a = 1,\ b = -4,\ c = 2$$

$$x = \frac{4 \pm \sqrt{16 - 8}}{2}$$

A Radical Departure from the Other Methods of Evaluating the Golden Ratio Recall from a previous chapter that the golden ratio is found in numerous places in mathematics, art, and nature. In a margin note there, we showed that

$$1 + \cfrac{1}{1 + \cfrac{1}{1 + \cfrac{1}{1 + \cdots}}}$$

is equal to the golden ratio, $\frac{1+\sqrt{5}}{2}$. Now consider this "nested" radical:

$$\sqrt{1 + \sqrt{1 + \sqrt{1 + \cdots}}}$$

Let x represent this radical. Because it appears "within itself," we can write

$$x = \sqrt{1 + x}$$
$$x^2 = 1 + x$$
$$x^2 - x - 1 = 0.$$

Using the quadratic formula, with $a = 1$, $b = -1$, and $c = -1$, it can be shown that the positive solution of this equation, and thus the value of the nested radical is . . . (you guessed it!) the golden ratio.

$$x = \frac{4 \pm 2\sqrt{2}}{2} \qquad \sqrt{16-8} = \sqrt{8} = 2\sqrt{2}$$

Factor, then divide out the common factor.

$$x = \frac{2(2 \pm \sqrt{2})}{2} \qquad \text{Factor out a 2 in the numerator.}$$

$$x = 2 \pm \sqrt{2} \qquad \text{Divide out common factor.}$$

The solution set is $\{2 + \sqrt{2}, 2 - \sqrt{2}\}$, abbreviated $\{2 \pm \sqrt{2}\}$.

EXAMPLE 4 Using the Quadratic Formula

Solve $2x^2 = x + 4$.

SOLUTION

First write the equation in standard form as $2x^2 - x - 4 = 0$.

$$x = \frac{-(-1) \pm \sqrt{(-1)^2 - 4(2)(-4)}}{2(2)} \qquad \text{Quadratic formula with } a = 2, b = -1, c = -4$$

$$x = \frac{1 \pm \sqrt{1 + 32}}{4} \qquad \text{Simplify the radicand.}$$

$$x = \frac{1 \pm \sqrt{33}}{4} \qquad \text{Add.}$$

The solution set is $\left\{\frac{1 \pm \sqrt{33}}{4}\right\}$.

Applications When solving applied problems that lead to quadratic equations, we might get a solution that does not satisfy the physical constraints of the problem. For example, if x represents a width and the two solutions of the quadratic equation are -9 and 1, the value -9 must be rejected, since a width must be a positive number.

EXAMPLE 5 Applying a Quadratic Equation

Two cars left an intersection at the same time, one heading due north, and the other due west. Some time later, they were exactly 100 miles apart. The car headed north had gone 20 miles farther than the car headed west. How far had each car traveled?

SOLUTION

Step 1 **Read** the problem carefully.

Step 2 **Assign a variable.**

Let x = the distance traveled by the car headed west;

Then $(x + 20)$ = the distance traveled by the car headed north.

See Figure 16. The cars are 100 miles apart, so the hypotenuse of the right triangle equals 100.

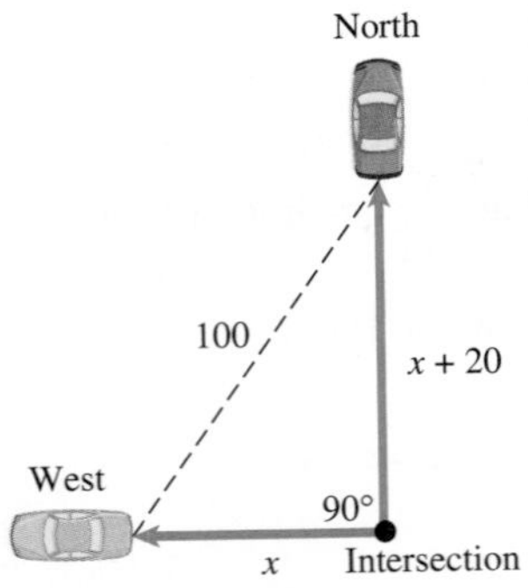

FIGURE 16

Step 3 **Write an equation.**

$$c^2 = a^2 + b^2 \quad \text{Pythagorean theorem}$$
$$100^2 = x^2 + (x + 20)^2 \quad \text{Substitute.}$$

Step 4 **Solve.**

$$10{,}000 = x^2 + x^2 + 40x + 400 \quad \text{Square the binomial.}$$
$$2x^2 + 40x - 9600 = 0 \quad \text{Standard form}$$
$$2(x^2 + 20x - 4800) = 0 \quad \text{Factor out the common factor.}$$
$$x^2 + 20x - 4800 = 0 \quad \text{Divide both sides by 2.}$$

Use the quadratic formula to find x.

$$x = \frac{-20 \pm \sqrt{400 - 4(1)(-4800)}}{2} \quad a = 1,\ b = 20,\ c = -4800$$
$$x = \frac{-20 \pm \sqrt{19{,}600}}{2}$$
$$x = 60 \quad \text{or} \quad x = -80 \quad \text{Use a calculator.}$$

Step 5 **State the answer.** Since distance cannot be negative, discard the negative solution. The required distances are 60 miles and $60 + 20 = 80$ miles.

Step 6 **Check.** Since $60^2 + 80^2 = 100^2$, the answer is correct.

EXAMPLE 6 Applying a Quadratic Equation

If a rock on Earth is projected upward from the top of a 144-foot building with an initial velocity of 112 feet per second, its position (in feet above the ground) is given by $s = -16t^2 + 112t + 144$, where t is time in seconds after it was projected. When does it hit the ground?

SOLUTION

When the rock hits the ground, its distance above the ground is 0. Find t when s is 0 by solving the following equation.

$$0 = -16t^2 + 112t + 144 \quad \text{Let } s = 0.$$
$$0 = t^2 - 7t - 9 \quad \text{Divide both sides by } -16.$$
$$t = \frac{7 \pm \sqrt{49 + 36}}{2} \quad \text{Quadratic formula}$$
$$t = \frac{7 \pm \sqrt{85}}{2} \quad \text{Add.}$$
$$t \approx 8.1 \quad \text{or} \quad t \approx -1.1 \quad \text{Use a calculator.}$$

Since time cannot be negative, discard the negative solution. The rock will hit the ground about 8.1 seconds after it is projected.

7.7 EXERCISES

Fill in each blank with the correct response.

1. For the quadratic equation $4x^2 + 5x - 9 = 0$, the values of a, b, and c are, respectively, ____, ____, and ____.

2. To solve the equation $3x^2 - 5x = -2$ by the quadratic formula, the first step is to add ____ to both sides of the equation.

3. When using the quadratic formula, if $b^2 - 4ac$ is positive, then the equation has ________ (how many?) real solution(s).

4. If a, b, and c are integers in $ax^2 + bx + c = 0$ and $b^2 - 4ac = 17$, then the equation has ________ (how many?) irrational solution(s).

Solve each equation by the zero-factor property.

5. $(x + 3)(x - 9) = 0$
6. $(m + 6)(m + 4) = 0$
7. $(2t - 7)(5t + 1) = 0$
8. $(7x - 3)(6x + 4) = 0$
9. $x^2 - x - 12 = 0$
10. $m^2 + 4m - 5 = 0$
11. $x^2 + 9x + 14 = 0$
12. $15r^2 + 7r = 2$
13. $12x^2 + 4x = 1$
14. $x(x + 3) = 4$
15. $(x + 4)(x - 6) = -16$
16. $(w - 1)(3w + 2) = 4w$

Solve each equation by using the square root property. Give only real number solutions.

17. $x^2 = 64$
18. $w^2 = 16$
19. $x^2 = 24$
20. $x^2 = 48$
21. $r^2 = -5$
22. $x^2 = -10$
23. $(x - 4)^2 = 9$
24. $(x + 3)^2 = 25$
25. $(4 - x)^2 = 3$
26. $(3 + x)^2 = 11$
27. $(2x - 5)^2 = 13$
28. $(4x + 1)^2 = 19$

Solve each equation by the quadratic formula. Give only real number solutions.

29. $4x^2 - 8x + 1 = 0$
30. $m^2 + 2m - 5 = 0$
31. $2x^2 = 2x + 1$
32. $9r^2 + 6r = 1$
33. $q^2 - 1 = q$
34. $2p^2 - 4p = 5$
35. $4k(k + 1) = 1$
36. $4r(r - 1) = 19$
37. $(g + 2)(g - 3) = 1$
38. $(x - 5)(x + 2) = 6$
39. $m^2 - 6m = -14$
40. $x^2 = 2x - 2$

41. Can the quadratic formula be used to solve the equation $2x^2 - 5 = 0$? Explain, and solve it if the answer is yes.

42. Can the quadratic formula be used to solve the equation $4x^2 + 3x = 0$? Explain, and solve it if the answer is yes.

43. Why can't the quadratic formula be used to solve the equation $2x^3 + 3x - 4 = 0$?

44. A student gave the quadratic formula incorrectly as follows: $x = -b \pm \dfrac{\sqrt{b^2 - 4ac}}{2a}$. What is wrong with this?

The expression $b^2 - 4ac$, the radicand in the quadratic formula, is called the **discriminant** *of the quadratic equation $ax^2 + bx + c = 0$, $a \neq 0$. By evaluating it we can determine, without actually solving the equation, the number and nature of the solutions of the equation. Suppose that a, b, and c are integers. Then the chart at the top of the next page shows how the discriminant can be used to analyze the solutions.*

Discriminant	Solutions
Positive, and the square of an integer	Two different rational solutions
Positive, but not the square of an integer	Two different irrational solutions
Zero	One rational solution (a double solution)
Negative	No real solutions

In Exercises 45–50, evaluate the discriminant, and then determine whether the equation has **(a)** *two different rational solutions,* **(b)** *two different irrational solutions,* **(c)** *one rational solution (a double solution), or* **(d)** *no real solutions.*

45. $x^2 + 6x + 9 = 0$

46. $4x^2 + 20x + 25 = 0$

47. $6x^2 + 7x - 3 = 0$

48. $2x^2 + x - 3 = 0$

49. $9x^2 - 30x + 15 = 0$

50. $2x^2 - x + 1 = 0$

Solve each problem by using a quadratic equation. Use a calculator as necessary, and round the answer to the nearest tenth.

51. ***Height of a Projectile*** The Mart Hotel in Dallas, Texas, is 400 feet high. Suppose that a ball is projected upward from the top of the Mart, and its position s in feet above the ground is given by the equation $s = -16t^2 + 45t + 400$, where t is the number of seconds elapsed. How long will it take for the ball to reach a height of 200 feet above the ground? (*Source: The World Almanac and Book of Facts.*)

52. ***Height of a Projectile*** The Toronto Dominion Center in Winnipeg, Manitoba, is 407 feet high. Suppose that a ball is projected upward from the top of the Center, and its position s in feet above the ground is given by the equation $s = -16t^2 + 75t + 407$, where t is the number of seconds elapsed. How long will it take for the ball to reach a height of 450 feet above the ground? (*Source: The World Almanac and Book of Facts.*)

53. ***Height of a Projectile*** Refer to the equations in Exercises 51 and 52. Suppose that the first sentence in each problem did not give the height of the building. How could you use the equation to determine the height of the building?

54. ***Position of a Searchlight Beam*** A searchlight beam moves horizontally back and forth along a wall with the distance of the light from a starting point at t minutes given by $s = 100t^2 - 300t$. How long will it take before the light returns to the starting point?

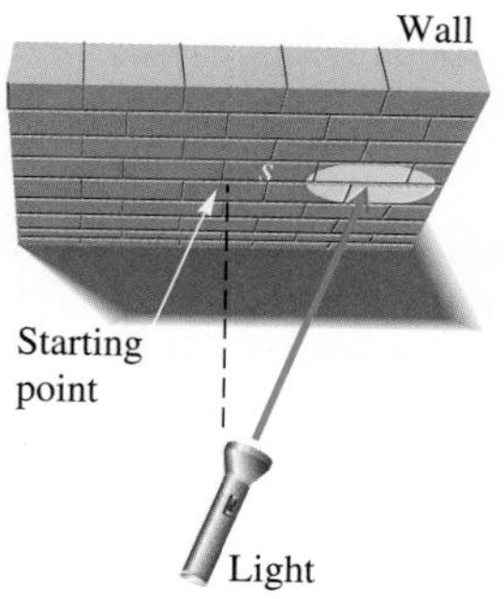

55. ***Height of a Projectile*** An object is projected directly upward from the ground. After t seconds its distance in feet above the ground is $s = 144t - 16t^2$.

(a) After how many seconds will the object be 128 feet above the ground? (*Hint:* Look for a common factor before solving the equation.)

(b) When does the object strike the ground?

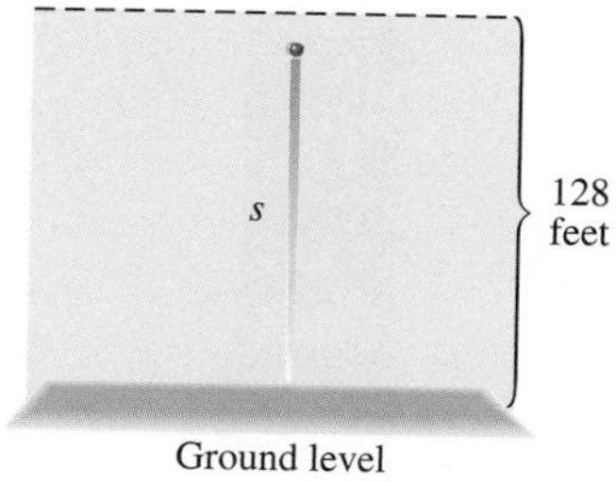

56. ***Distance of a Skid*** The formula $D = 100t - 13t^2$ gives the distance in feet a car going approximately 68 miles per hour will skid in t seconds. Find the time it would take for the car to skid 190 feet. (*Hint:* Your answer must be less than the time it takes the car to stop, which is 3.8 seconds.)

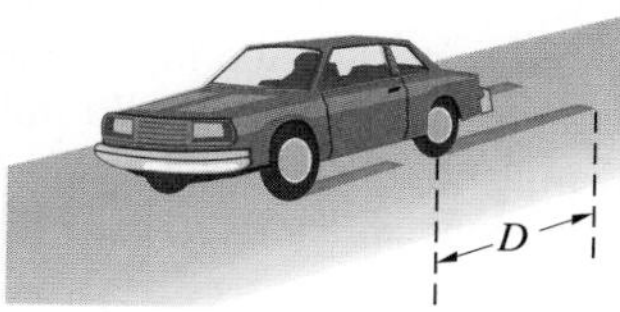

57. ***Side Lengths of a Triangle*** Find the lengths of the sides of the triangle.

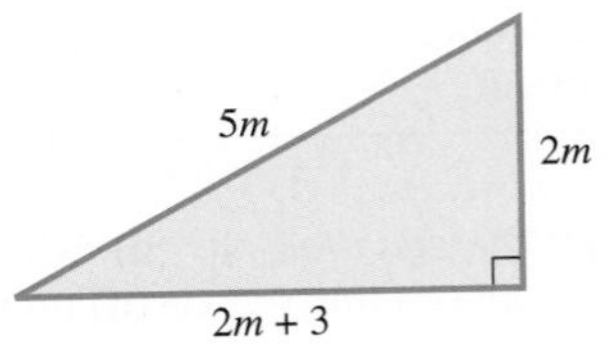

58. ***Side Lengths of a Triangle*** Find the lengths of the sides of the triangle.

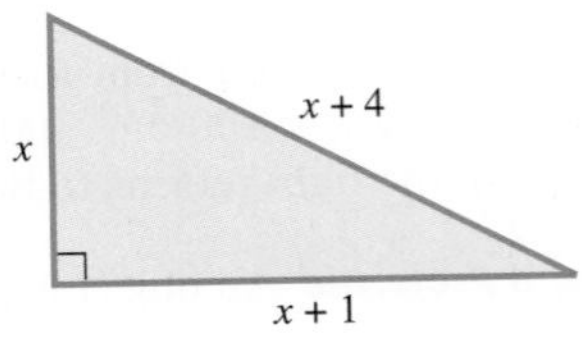

59. ***Length of a Wire*** Refer to Exercise 51. Suppose that a wire is attached to the top of the Mart and pulled tight. It is attached to the ground 100 feet from the base of the building, as shown in the figure. How long is the wire?

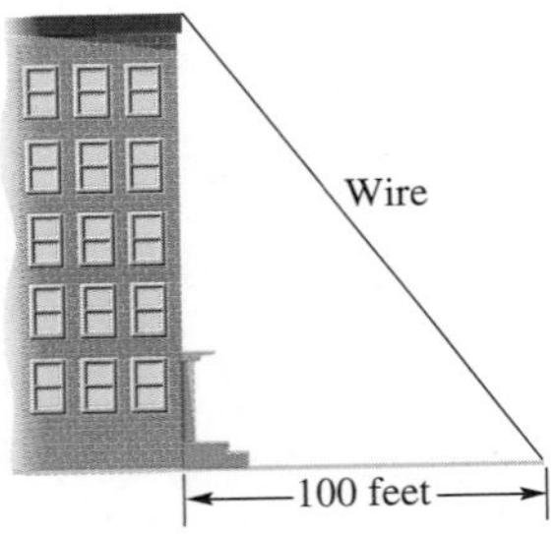

60. ***Length of a Wire*** Refer to Exercise 52. Suppose that a wire is attached to the top of the Center and pulled tight. The length of the wire is twice the distance between the base of the Center and the point on the ground where the wire is attached. How long is the wire?

61. ***Distances Traveled by Ships*** Two ships leave port at the same time, one heading due south and the other heading due east. (See the top of the next column.) Several hours later, they are 170 miles apart. If the ship traveling south travels 70 miles farther than the other, how many miles does each travel?

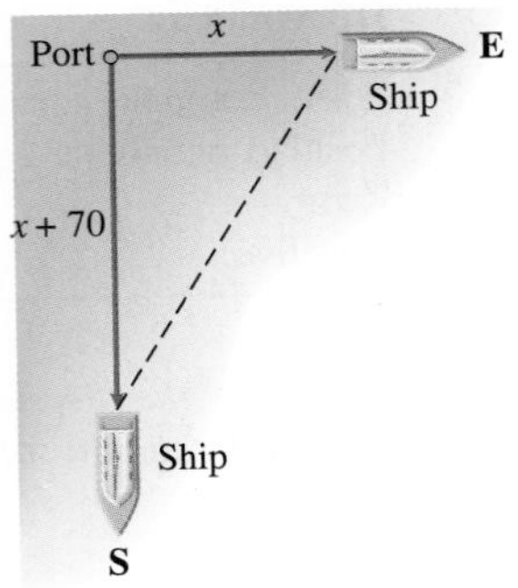

62. ***Height of a Kite*** Kim Hobbs is flying a kite that is 30 feet farther above her hand than its horizontal distance from her. The string from her hand to the kite is 150 feet long. How far is the kite above her hand?

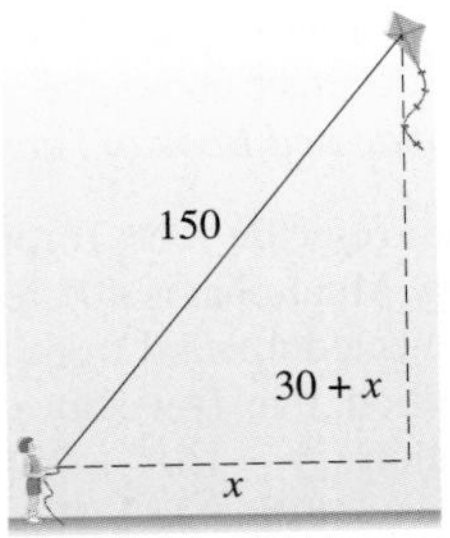

63. ***Size of a Toy Piece*** A toy manufacturer needs a piece of plastic in the shape of a right triangle with the longer leg 2 centimeters more than twice as long as the shorter leg, and the hypotenuse 1 centimeter more than the longer leg. How long should the three sides of the triangular piece be?

64. ***Size of a Developer's Property*** Michael Cardella, a developer, owns a piece of land enclosed on three sides by streets, giving it the shape of a right triangle. The hypotenuse is 8 meters longer than the longer leg, and the shorter leg is 9 meters shorter than the hypotenuse. Find the lengths of the three sides of the property.

65. ***Dimensions of Puzzle Pieces*** Two pieces of a large wooden puzzle fit together to form a rectangle with a length 1 centimeter less than twice the width. The diagonal, where the two pieces meet, is 2.5 centimeters in length. Find the length and width of the rectangle.

66. ***Leaning Ladder*** A 13-foot ladder is leaning against a house. The distance from the bottom of the ladder to the house is 7 feet less than the distance from the top of the ladder to the ground. How far is the bottom of the ladder from the house?

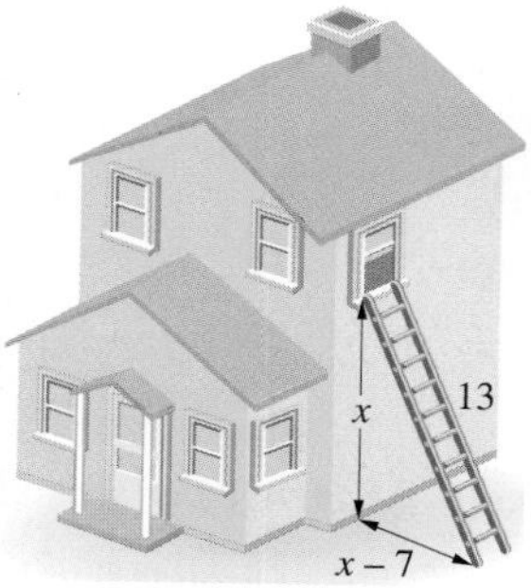

67. ***Dimensions of a Strip of Flooring Around a Rug*** Catarina and José want to buy a rug for a room that is 15 feet by 20 feet. They want to leave an even strip of flooring uncovered around the edges of the room. How wide a strip will they have if they buy a rug with an area of 234 square feet?

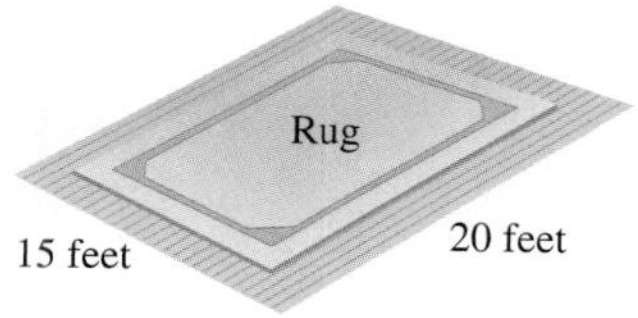

68. ***Dimensions of a Border Around a Pool*** A club swimming pool is 30 feet wide and 40 feet long. The club members want an exposed aggregate border in a strip of uniform width around the pool. They have enough material for 296 square feet. How wide can the strip be?

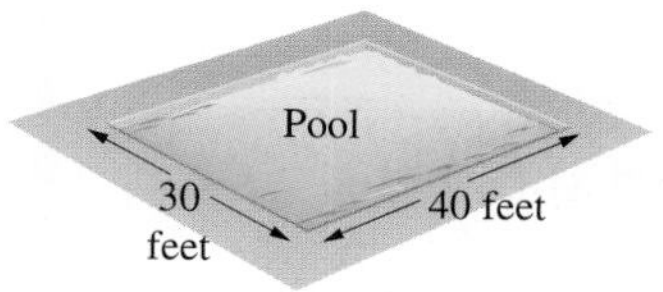

69. ***Dimensions of a Garden*** Arif's backyard is 20 meters by 30 meters. He wants to put a flower garden in the middle of the backyard, leaving a strip of grass of uniform width around the flower garden. Arif must have 184 square meters of grass. Under these conditions, what will the length and width of the garden be?

70. ***Interest Rate*** The formula $A = P(1 + r)^2$ gives the amount A in dollars that P dollars will grow to in 2 years at interest rate r (where r is given as a decimal), using compound interest. What interest rate will cause \$2000 to grow to \$2142.25 in 2 years?

71. ***Dimensions of a Piece of Sheet Metal*** A rectangular piece of sheet metal has a length that is 4 inches less than twice the width. A square piece 2 inches on a side is cut from each corner. The sides are then turned up to form an uncovered box of volume 256 cubic inches. Find the length and width of the original piece of metal.

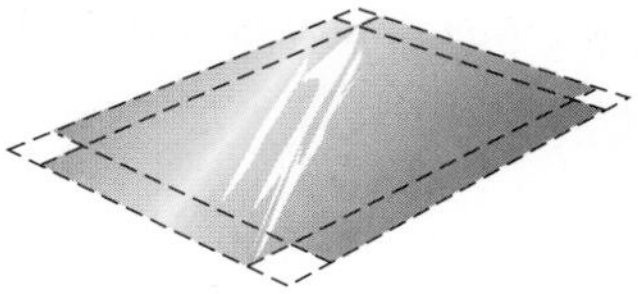

72. ***Cardboard Box Dimensions*** If a square piece of cardboard has 3-inch squares cut from its corners and then has the flaps folded up to form an open-top box, the volume of the box is given by the formula $V = 3(x - 6)^2$, where x is the length of each side of the original piece of cardboard in inches. What original length would yield a box with a volume of 432 cubic inches?

73. ***Supply and Demand for Bran Muffins*** A certain bakery has found that the daily demand for bran muffins is $\frac{3200}{p}$ where p is the price of a muffin in cents. The daily supply is $3p - 200$. Find the price at which supply and demand are equal.

74. ***Supply and Demand for Compact Discs*** In one area the demand for compact discs is $\frac{700}{P}$ per day, where P is the price in dollars per disc. The supply is $5P - 1$ per day. At what price does supply equal demand?

Froude Number *William Froude was a 19th-century naval architect who used the expression* $\frac{v^2}{g\ell}$ *in shipbuilding. This expression, known as the Froude number, was also used by R. McNeill Alexander in his research on dinosaurs. (See "How Dinosaurs Ran," in* Scientific American, *April 1991, pp. 130–136.)*

In Exercises 75 and 76, find the value of v *(in meters per second) given that* $g = 9.8$ *meters per second squared.*

75. Rhinoceros: $\ell = 1.2$; Froude number $= 2.57$

76. Triceratops: $\ell = 2.8$; Froude number $= .16$

Recall that the corresponding sides of similar triangles are proportional. (Refer to Section 7.3 Exercises 53–56.) Use this fact to find the lengths of the indicated sides of each pair of similar triangles. Check all possible solutions in both triangles. Sides of a triangle cannot be negative.

77. Side AC

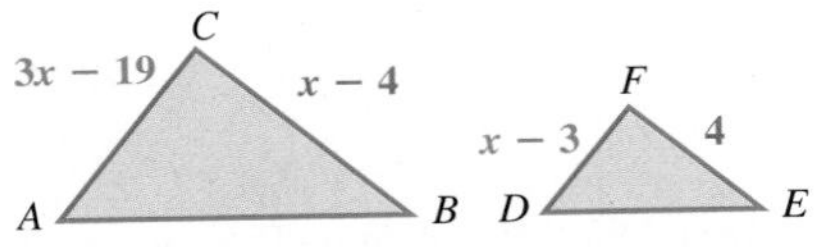

78. Side RQ

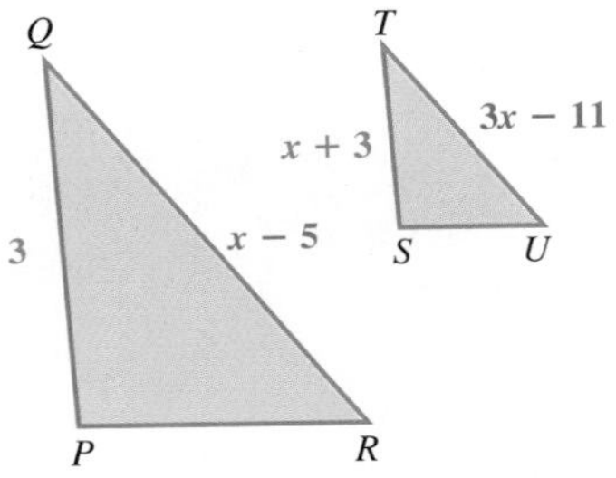

79. For centuries mathematicians wrestled with finding a formula that could solve cubic (third-degree) equations. A story from sixteenth-century Italy concerns two main characters, Girolamo Cardano and Niccolo Tartaglia. In those days, mathematicians often participated in contests. Tartaglia had developed a method of solving a cubic equation of the form $x^3 + mx = n$ and had used it in one of these contests. Cardano begged to know Tartaglia's method and after he was told, he was sworn to secrecy. Nonetheless, Cardano published Tartaglia's method in his 1545 work *Ars Magna* (although he did give Tartaglia credit).

The formula for finding one real solution of the above equation is

$$x = \sqrt[3]{\frac{n}{2} + \sqrt{\left(\frac{n}{2}\right)^2 + \left(\frac{m}{3}\right)^3}} - \sqrt[3]{-\left(\frac{n}{2}\right) + \sqrt{\left(\frac{n}{2}\right)^2 + \left(\frac{m}{3}\right)^3}}.$$

Solve $x^3 + 9x = 26$ using this formula.

COLLABORATIVE INVESTIGATION

How Does Your Walking Rate Compare to That of Olympic Race-Walkers?

Race-walking at speeds exceeding 8 miles per hour is a high fitness, long-distance competitive sport. The table below contains gold medal winners in the 2004 Olympic Games race-walking competition.

A. Complete the table by applying the proportion given below to find the race-walker's steps per minute. **Use 10 km ≈ 6.21 miles.** (Round all answers except those for steps per minute to the nearest thousandth. Round steps per minute to the nearest whole number.)

$$\frac{70 \text{ steps per minute}}{2 \text{ miles per hour}} = \frac{x \text{ steps per minute}}{y \text{ miles per hour}}$$

Event	Gold Medal Winner	Country	Time in Hours: Minutes: Seconds	Time in Minutes	Time in Hours	y Miles per Hour	x Steps per Minute
20-km Walk, Women	Athanasia Tsoumeleka	Greece	1:29:12				
20-km Walk, Men	Ivano Brugnatti	Italy	1:19:40				
50-km Walk, Men	Robert Korzeniowski	Poland	3:38:48				

Source: The World Almanac and Book of Facts 2006.

B. Using a stopwatch, take turns counting how many steps each member of the group takes in one minute while walking at a normal pace. Record the results in the table below. Then do it again at a fast pace. Record these results.

C. Use the proportion from part A to convert the numbers from part B to miles per hour and complete the chart.

1. Find the average speed for the group at a normal pace and at a fast pace.
2. What is the minimum number of steps per minute you would have to take to be a race-walker?
3. At a fast pace, did anyone in the group walk fast enough to be a race-walker? Explain how you decided.

	Normal Pace		Fast Pace	
Name	**x Steps per Minute**	**y Miles per Hour**	**x Steps per Minute**	**y Miles per Hour**

CHAPTER 7 TEST

Solve each equation.

1. $5x - 3 + 2x = 3(x - 2) + 11$

2. $\frac{2p - 1}{3} + \frac{p + 1}{4} = \frac{43}{12}$

3. Decide whether the equation

$$3x - (2 - x) + 4x = 7x - 2 - (-x)$$

is conditional, an identity, or a contradiction. Give its solution set.

4. Solve for v: $S = vt - 16t^2$.

Solve each application.

5. *Areas of Hawaiian Islands* Three islands in the Hawaiian island chain are Hawaii (the Big Island), Maui, and Kauai. Together, their areas total 5300 square miles. The island of Hawaii is 3293 square miles larger than the island of Maui, and Maui is 177 square miles larger than Kauai. What is the area of each island?

6. *Chemical Mixture* How many liters of a 20% solution of a chemical should Tippy Hurst mix with 10 liters of a 50% solution to obtain a mixture that is 40% chemical?

7. *Speeds of Trains* A passenger train and a freight train leave a town at the same time and travel in opposite directions. Their speeds are 60 mph and 75 mph, respectively. How long will it take for them to be 297 miles apart?

8. *Merchandise Pricing* Which is the better buy for processed cheese slices: 8 slices for \$2.19 or 12 slices for \$3.30?

9. *Distance Between Cities* The distance between Milwaukee and Boston is 1050 miles. On a certain map this distance is represented by 21 inches. On the same map Seattle and Cincinnati are 46 inches apart. What is the actual distance between Seattle and Cincinnati?

10. *Current in a Circuit* The current in a simple electrical circuit is inversely proportional to the resistance. If the current is 80 amps when the resistance is 30 ohms, find the current when the resistance is 12 ohms.

Solve each inequality. Give the solution set in both interval and graph forms.

11. $-4x + 2(x - 3) \geq 4x - (3 + 5x) - 7$

12. $-10 < 3k - 4 \leq 14$

13. Which one of the following inequalities is equivalent to $x < -3$?

A. $-3x < 9$ **B.** $-3x > -9$
C. $-3x > 9$ **D.** $-3x < -9$

14. *Grade Average* Paul Lorio has scores of 83, 76, and 79 on his first three tests in Math 1031 (Survey of Mathematics). If he wants an average of at least 80 after his fourth test, what are the possible scores he can make on his fourth test?

Evaluate each exponential expression.

15. $\left(\frac{4}{3}\right)^2$

16. $-(-2)^6$

17. $\left(\frac{3}{4}\right)^{-3}$

18. $-5^0 + (-5)^0$

Use the properties of exponents to simplify each expression. Write answers with positive exponents only. Assume that all variables represent nonzero real numbers.

19. $9(4p^3)(6p^{-7})$

20. $\frac{m^{-2}(m^3)^{-3}}{m^{-4}m^7}$

21. Write each number in standard notation.
(a) 6.93×10^8
(b) 1.25×10^{-7}

22. Use scientific notation to evaluate

$$\frac{(2{,}500{,}000)(.00003)}{(.05)(5{,}000{,}000)}.$$

Leave the answer in scientific notation.

23. *Time Traveled for a Radio Signal* The mean distance to Earth from the planet Pluto is 4.58×10^9 kilometers. The first U.S. space probe to Pluto, Pioneer 10, transmitted radio signals from Pluto to Earth at the speed of light, 3.00×10^5 kilometers per second. How long (in seconds) did it take for the signals to reach Earth?

Perform the indicated operations.

24. $(3k^3 - 5k^2 + 8k - 2) - (3k^3 - 9k^2 + 2k - 12)$

25. $(5x + 2)(3x - 4)$

26. $(4x^2 - 3)(4x^2 + 3)$

27. $(x + 4)(3x^2 + 8x - 9)$

28. Give an example of a polynomial in the variable t, such that it is fifth degree, in descending powers of the variable, with exactly six terms, and having a negative coefficient for its second degree term.

Factor each polynomial completely.

29. $2p^2 - 5pq + 3q^2$

30. $100x^2 - 49y^2$

31. $27y^3 - 125x^3$

32. $4x + 4y - mx - my$

Solve each quadratic equation.

33. $6x^2 + 7x - 3 = 0$

34. $x^2 - 13 = 0$

35. $x^2 - x = 7$

36. ***Time an Object Has Descended*** The equation

$$s = 16t^2 + 15t$$

gives the distance s in feet an object thrown off a building has descended in t seconds. Find the time t when the object has descended 25 feet. Use a calculator and round the answer to the nearest hundredth.

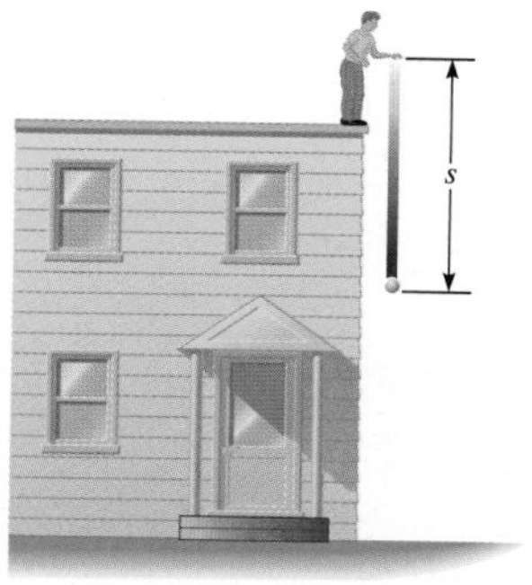

8

GRAPHS, FUNCTIONS, AND SYSTEMS OF EQUATIONS AND INEQUALITIES

During the seasons of 1969–70 and 1970–71, the NBC television network aired *The Bill Cosby Show,* in which the popular comedian played Chet Kincaid, a Los Angeles high school physical education teacher. In one of the first season episodes, Chet has to substitute for the algebra teacher one Friday. He and the entire class are stumped by the following problem:

> *How many pounds of candy that sells for \$.75 per pound must be mixed with candy that sells for \$1.25 per pound to obtain 9 pounds of a mixture that should sell for \$.96 per pound.*

Chet learns that he will have to teach the class again on Monday, and he spends his weekend trying to solve this problem. (The title of the episode is "Let *x* Equal a Lousy Weekend.") He even visits the local candy store, where the owner says that in order to solve the problem, "you have to know algebra." On Monday, the smartest student in the class, Eddie Tucker, is able to solve the problem correctly using a system of equations. Can you do this as well? See page 508 for Eddie's answer.

8.1 The Rectangular Coordinate System and Circles

Rectangular Coordinates • Distance Formula • Midpoint Formula • Circles

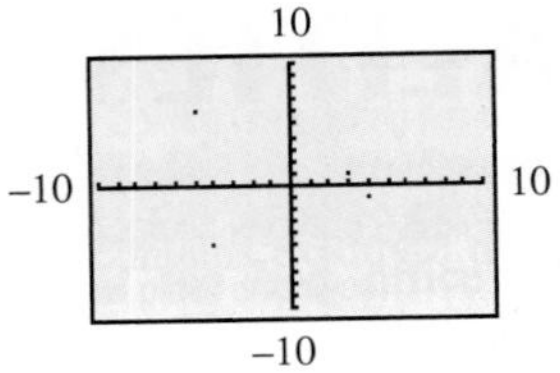

The points in Figure 2 are plotted on this calculator screen. Why is $E(-5, 0)$ not visible?

Rectangular Coordinates

Each of the pairs of numbers (1, 2), (−1, 5), and (3, 7) is an example of an **ordered pair**—a pair of numbers written within parentheses in which the order of the numbers is important. The two numbers are the **components** of the ordered pair. An ordered pair is graphed using two number lines that intersect at right angles at the zero points, as shown in Figure 1. The common zero point is called the **origin.** The horizontal line, the **x-axis,** represents the first number in an ordered pair, and the vertical line, the **y-axis,** represents the second. The *x*-axis and the *y*-axis make up a **rectangular** (or **Cartesian**) **coordinate system.** The axes form four **quadrants,** numbered I, II, III, and IV as shown in Figure 2. (A point on an axis is not considered to be in any of the four quadrants.)

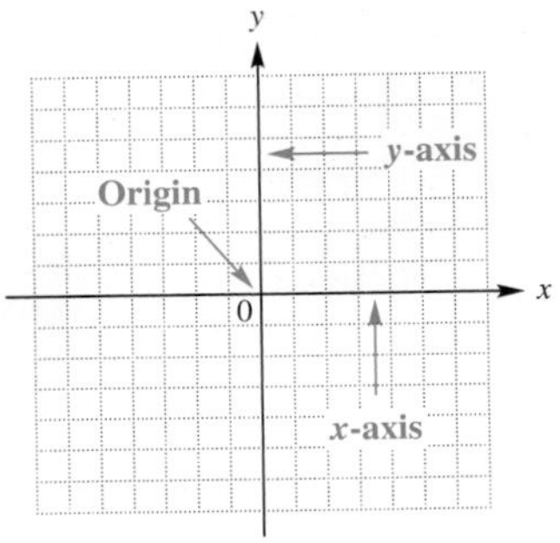

FIGURE 1

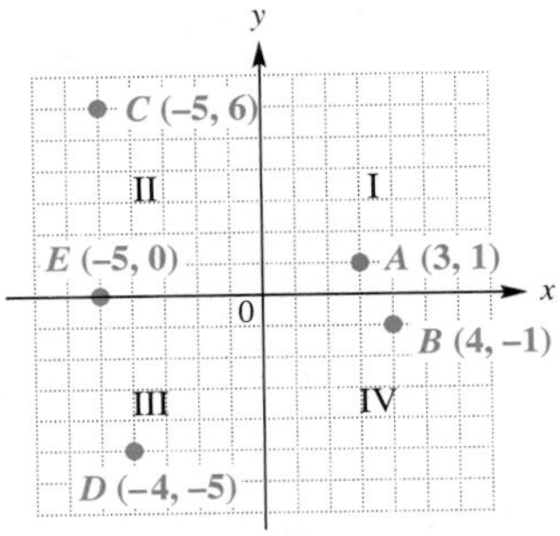

FIGURE 2

We locate, or **plot,** the point on the graph that corresponds to the ordered pair (3, 1) by going three units from zero to the right along the *x*-axis, and then one unit up parallel to the *y*-axis. The point corresponding to the ordered pair (3, 1) is labeled *A* in Figure 2. The phrase "the point corresponding to the ordered pair (3, 1)" often is abbreviated "the point (3, 1)." The numbers in an ordered pair are called the **coordinates** of the corresponding point.

The parentheses used to represent an ordered pair also are used to represent an open interval (introduced in an earlier chapter). The context of the discussion tells us whether we are discussing ordered pairs or open intervals.

Double Descartes After the French postal service issued the above stamp in honor of **René Descartes,** sharp eyes noticed that the title of Descartes's most famous book was wrong. Thus a second stamp (see facing page) was issued with the correct title. The book in question, *Discourse on Method,* appeared in 1637. In it Descartes rejected traditional Aristotelian philosophy, outlining a universal system of knowledge that was to have the certainty of mathematics. For Descartes, method was *analysis,* going from self-evident truths step-by-step to more distant and more general truths. One of these truths is his famous statement, "I think, therefore I am." (Thomas Jefferson, also a rationalist, began the *Declaration* with the words, "We hold these truths to be self-evident.")

Distance Formula

Suppose that we wish to find the distance between two points, say (3, −4) and (−5, 3). The Pythagorean theorem allows us to do this. In Figure 3 on the next page, we see that the vertical line through (−5, 3) and the horizontal line through (3, −4) intersect at the point (−5, −4). Thus, the point (−5, −4) becomes the vertex of the right angle in a right triangle. By the Pythagorean theorem, the square of the length of the hypotenuse, d, of the right triangle in Figure 3 is equal to the sum of the squares of the lengths of the two legs a and b:

$$d^2 = a^2 + b^2.$$

The length a is the distance between the endpoints of that leg. Since the x-coordinate of both points is -5, the side is vertical, and we can find a by finding the difference between the y-coordinates. Subtract -4 from 3 to get a positive value of a. Similarly, find b by subtracting -5 from 3.

$$a = 3 - (-4) = 7$$
$$b = 3 - (-5) = 8$$

Use parentheses when subtracting a negative number.

Substituting these values into the formula, we have

$$d^2 = a^2 + b^2$$
$$d^2 = 7^2 + 8^2 \quad \text{Let } a = 7 \text{ and } b = 8.$$
$$d^2 = 49 + 64$$
$$d^2 = 113$$
$$d = \sqrt{113}. \quad \text{Square root property, } d > 0$$

Therefore, the distance between $(-5, 3)$ and $(3, -4)$ is $\sqrt{113}$.

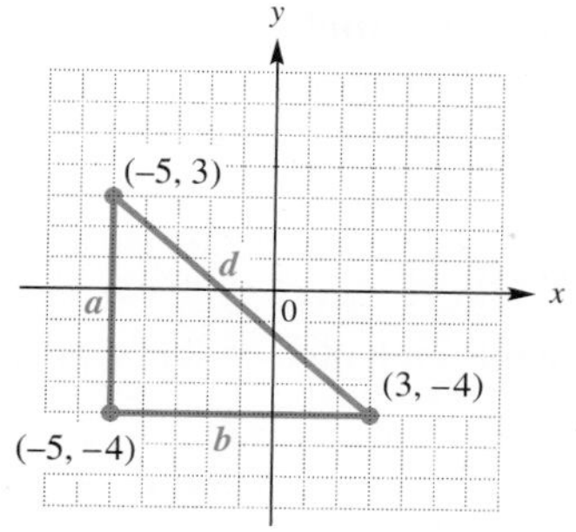

FIGURE 3

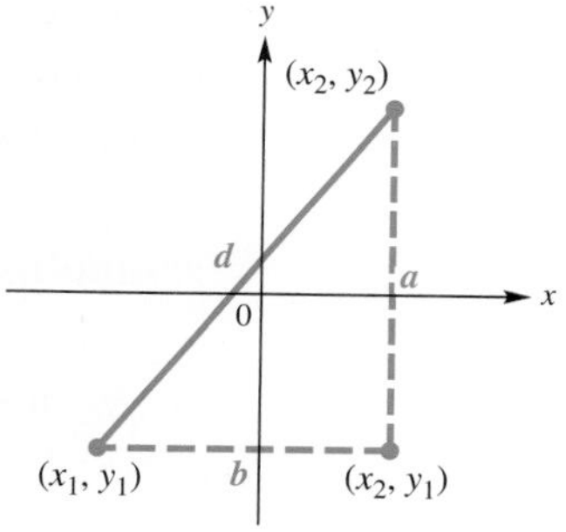

FIGURE 4

Descartes wrote his *Geometry* as an application of his method; it was published as an appendix to the *Discourse*. His attempts to unify algebra and geometry influenced the creation of what became coordinate geometry and influenced the development of calculus by Newton and Leibniz in the next generation.

In 1649 Descartes went to Sweden to tutor Queen Christina. She preferred working in the unheated castle in the early morning; Descartes was used to staying in bed until noon. The rigors of the Swedish winter proved too much for him, and he died less than a year later.

This result can be generalized. Figure 4 shows the two different points (x_1, y_1) and (x_2, y_2). To find a formula for the distance d between these two points, notice that the distance between (x_2, y_2) and (x_2, y_1) is given by $a = y_2 - y_1$, and the distance between (x_1, y_1) and (x_2, y_1) is given by $b = x_2 - x_1$. From the Pythagorean theorem,

$$d^2 = (x_2 - x_1)^2 + (y_2 - y_1)^2,$$

and by using the square root property, we obtain the distance formula.

Distance Formula

The distance between the points (x_1, y_1) and (x_2, y_2) is

$$d = \sqrt{(x_2 - x_1)^2 + (y_2 - y_1)^2}.$$

This result is called the **distance formula.**

The small numbers 1 and 2 in the ordered pairs (x_1, y_1) and (x_2, y_2) are called **subscripts.** We read x_1 as "x sub 1." Subscripts are used to distinguish between different values of a variable that have a common property. For example, in the ordered pairs $(-3, 5)$ and $(6, 4)$, -3 can be designated as x_1 and 6 as x_2. Their common property is that they are both x components of ordered pairs. This idea is used in the following example.

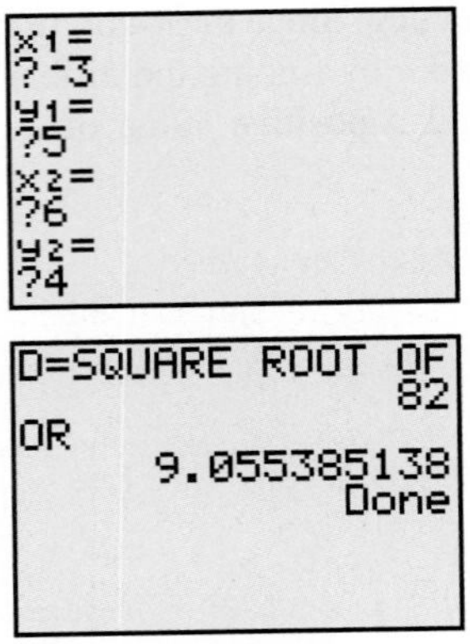

A program can be written for the distance formula. This one supports the result in Example 1, since $\sqrt{82} \approx 9.055385138$.

EXAMPLE 1 Finding the Distance Between Two Points

Find the distance between $(-3, 5)$ and $(6, 4)$.

SOLUTION

When using the distance formula to find the distance between two points, designating the points as (x_1, y_1) and (x_2, y_2) is arbitrary. Let us choose $(x_1, y_1) = (-3, 5)$ and $(x_2, y_2) = (6, 4)$.

$$d = \sqrt{(x_2 - x_1)^2 + (y_2 - y_1)^2}$$

$$= \sqrt{(6 - (-3))^2 + (4 - 5)^2} \qquad x_2 = 6,\ y_2 = 4,\ x_1 = -3,\ y_1 = 5$$

Begin with the x- and y- values of the same point.

$$= \sqrt{9^2 + (-1)^2}$$

$$= \sqrt{82}$$

Midpoint Formula The **midpoint** of a line segment is the point on the segment that is equidistant from both endpoints. Given the coordinates of the two end-points of a line segment, we can find the coordinates of the midpoint of the segment.

> **Midpoint Formula**
>
> The coordinates of the midpoint of the segment with endpoints (x_1, y_1) and (x_2, y_2) are
>
> $$\left(\frac{x_1 + x_2}{2}, \frac{y_1 + y_2}{2}\right).$$

In words, the coordinates of the midpoint of a line segment are found by calculating the averages of the x- and y-coordinates of the endpoints.

EXAMPLE 2 Finding the Midpoint of a Segment

Find the coordinates of the midpoint of the line segment with endpoints $(8, -4)$ and $(-9, 6)$.

SOLUTION

Using the midpoint formula, we find that the coordinates of the midpoint are

$$\left(\frac{8 + (-9)}{2}, \frac{-4 + 6}{2}\right) = \left(-\frac{1}{2}, 1\right).$$

EXAMPLE 3 Applying the Midpoint Formula to Data

Figure 5 on the next page depicts how the number of McDonald's restaurants worldwide increased from 1995 through 2001. Use the midpoint formula and the two given points to estimate the number of restaurants in 1998, and compare it to the actual (rounded) figure of 24,000.

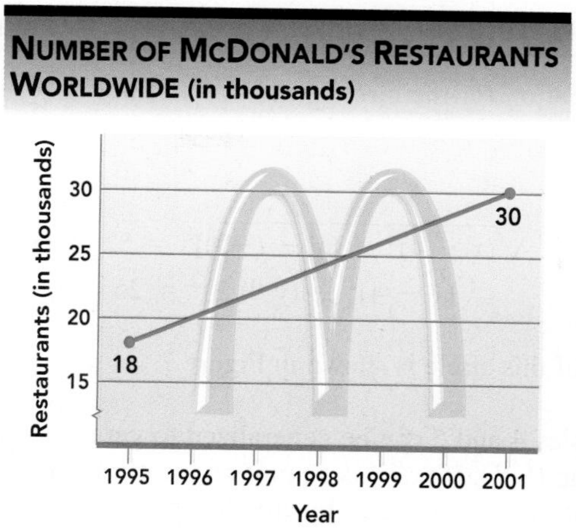

Source: McDonald's Corp.; Yahoo.com.

FIGURE 5

SOLUTION

The year 1998 lies halfway between 1995 and 2001, so we must find the coordinates of the midpoint of the segment that has endpoints (1995, 18) and (2001, 30). (Here, y is in thousands.) By the midpoint formula, this is

$$\left(\frac{1995 + 2001}{2}, \frac{18 + 30}{2}\right) = (1998, 24)$$

Thus, our estimate is 24,000 restaurants in 1998, which matches the actual (rounded) figure.

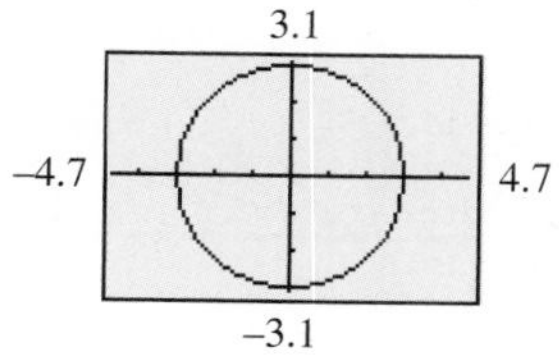

To graph $x^2 + y^2 = 9$, we solve for y to get $y_1 = \sqrt{9 - x^2}$ and $y_2 = -\sqrt{9 - x^2}$. Then we graph both in a square window.

Circles An application of the distance formula leads to one of the most familiar shapes in geometry, the circle. A **circle** is the set of all points in a plane that lie a fixed distance from a fixed point. The fixed point is called the **center** and the fixed distance is called the **radius.**

EXAMPLE 4 Finding an Equation of a Circle

Find an equation of the circle with radius 3 and center at (0, 0), and graph the circle.

SOLUTION

If the point (x, y) is on the circle, the distance from (x, y) to the center (0, 0) is 3, as shown in Figure 6.

$$\sqrt{(x_2 - x_1)^2 + (y_2 - y_1)^2} = d \quad \text{Distance formula}$$

$$\sqrt{(x - 0)^2 + (y - 0)^2} = 3 \quad x_1 = 0, y_1 = 0, x_2 = x, y_2 = y, d = 3$$

$$x^2 + y^2 = 9 \quad \text{Square both sides.}$$

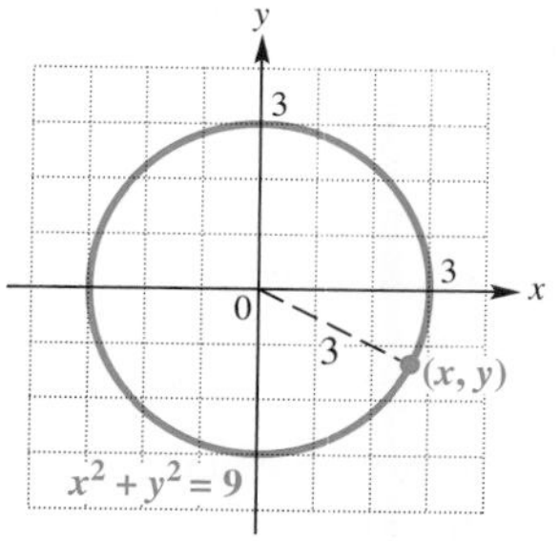

FIGURE 6

An equation of this circle is $x^2 + y^2 = 9$. It can be graphed by locating all points three units from the origin.

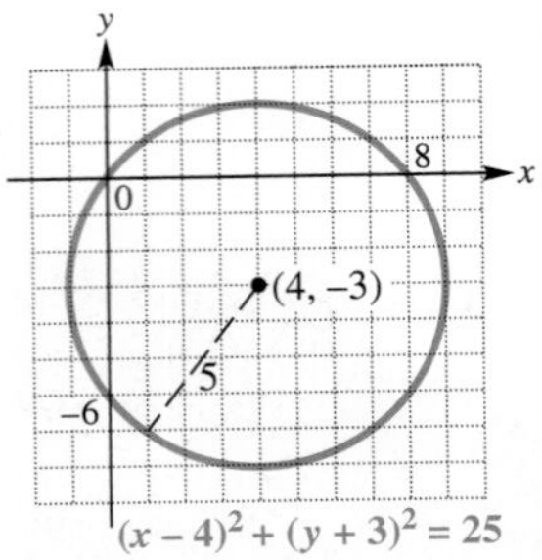

FIGURE 7

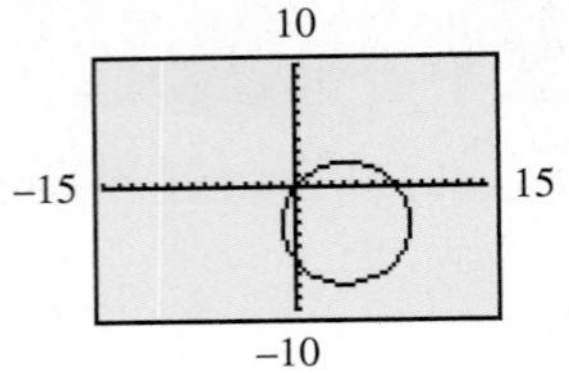

A circle can be "drawn" by using the appropriate command, entering the coordinates of the center and the radius. Compare with Example 5 and Figure 7.

EXAMPLE 5 Finding an Equation of a Circle and Graphing

Find an equation for the circle that has its center at $(4, -3)$ and radius 5, and graph the circle.

SOLUTION

$$\sqrt{(x-4)^2 + [y-(-3)]^2} = 5 \quad \text{Substitute in the distance formula.}$$
$$(x-4)^2 + (y+3)^2 = 25 \quad \text{Square both sides.}$$

The graph of this circle is shown in Figure 7. ■

Examples 4 and 5 can be generalized to get an equation of a circle with radius r and center at (h, k). If (x, y) is a point on the circle, the distance from the center (h, k) to the point (x, y) is r. Then by the distance formula, $\sqrt{(x-h)^2 + (y-k)^2} = r$. Squaring both sides gives the following equation of a circle.

Equation of a Circle

The equation of a circle of radius r with center at (h, k) is

$$(x-h)^2 + (y-k)^2 = r^2.$$

In particular, a circle of radius r with center at the origin has equation

$$x^2 + y^2 = r^2.$$

EXAMPLE 6 Finding an Equation of a Circle

Find an equation of the circle with center at $(-1, 2)$ and radius 4.

SOLUTION

$$(x-h)^2 + (y-k)^2 = r^2$$
$$[x-(-1)]^2 + (y-2)^2 = 4^2 \quad \text{Let } h = -1, k = 2, \text{ and } r = 4.$$
$$(x+1)^2 + (y-2)^2 = 16$$

■

In the equation found in Example 5, multiplying out $(x-4)^2$ and $(y+3)^2$ and then combining like terms gives

$$(x-4)^2 + (y+3)^2 = 25$$
$$x^2 - 8x + 16 + y^2 + 6y + 9 = 25$$
$$x^2 + y^2 - 8x + 6y = 0$$

This result suggests that an equation that has both x^2- and y^2-terms with equal coefficients may represent a circle. The next example shows how to tell, using the method of **completing the square.**

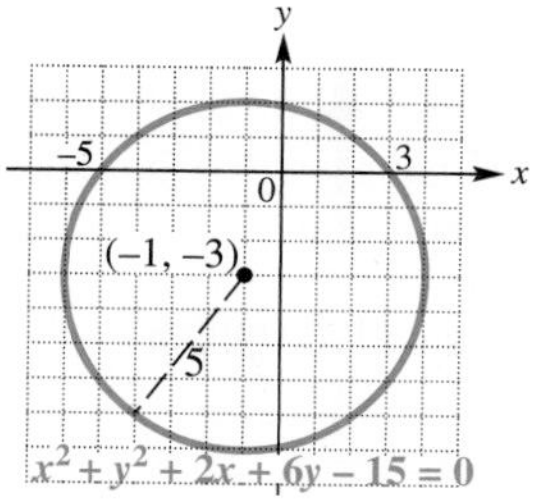

FIGURE 8

EXAMPLE 7 Completing the Square and Graphing a Circle

Graph $x^2 + y^2 + 2x + 6y - 15 = 0$.

SOLUTION

Since the equation has x^2- and y^2-terms with equal coefficients, its graph might be that of a circle. To find the center and radius, complete the squares in x and y as follows. (See page 418, where completing the square is introduced.)

$x^2 + y^2 + 2x + 6y = 15$	Add 15 to both sides.
$(x^2 + 2x \quad\quad) + (y^2 + 6y \quad\quad) = 15$	Rewrite in anticipation of completing the square.
$(x^2 + 2x + 1) + (y^2 + 6y + 9) = 15 + 1 + 9$	Complete the squares in both x and y.
$(x + 1)^2 + (y + 3)^2 = 25$	Factor on the left and add on the right.

The final equation shows that the graph is a circle with center at $(-1, -3)$ and radius 5. The graph is shown in Figure 8.

The final example in this section shows how equations of circles can be used in locating the epicenter of an earthquake.

EXAMPLE 8 Locating the Epicenter of an Earthquake

Seismologists can locate the epicenter of an earthquake by determining the intersection of three circles. The radii of these circles represent the distances from the epicenter to each of three receiving stations. The centers of the circles represent the receiving stations.

Suppose receiving stations A, B, and C are located on a coordinate plane at the points $(1, 4)$, $(-3, -1)$, and $(5, 2)$. Let the distances from the earthquake epicenter to the stations be 2 units, 5 units, and 4 units, respectively. See Figure 9. Where on the coordinate plane is the epicenter located?

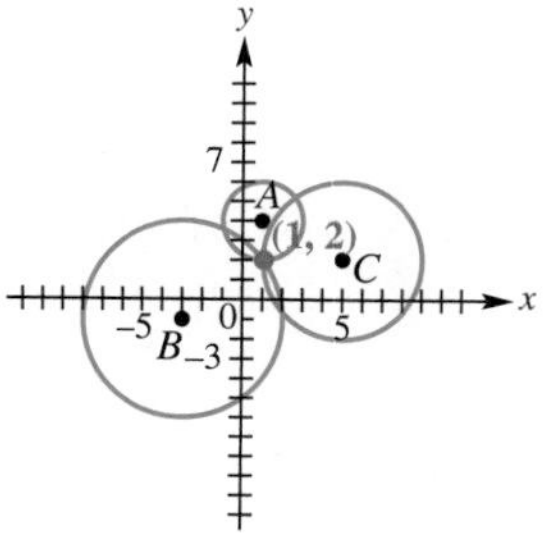

FIGURE 9

SOLUTION

Graphically, it appears that the epicenter is located at $(1, 2)$. To check this algebraically, determine the equation for each circle and substitute $x = 1$ and $y = 2$.

Station A

$$(x - 1)^2 + (y - 4)^2 = 4$$
$$(1 - 1)^2 + (2 - 4)^2 = 4$$
$$0 + 4 = 4$$
$$4 = 4$$

Station B

$$(x + 3)^2 + (y + 1)^2 = 25$$
$$(1 + 3)^2 + (2 + 1)^2 = 25$$
$$16 + 9 = 25$$
$$25 = 25$$

Station C

$$(x - 5)^2 + (y - 2)^2 = 16$$
$$(1 - 5)^2 + (2 - 2)^2 = 16$$
$$16 + 0 = 16$$
$$16 = 16$$

Thus, we can be sure that the epicenter lies at $(1, 2)$.

8.1 EXERCISES

In Exercises 1 and 2, answer each question by locating ordered pairs on the graphs.

1. *Women in Mathematics or Computer Science* The graph shows the percent of women in mathematics or computer science professions since 1970.

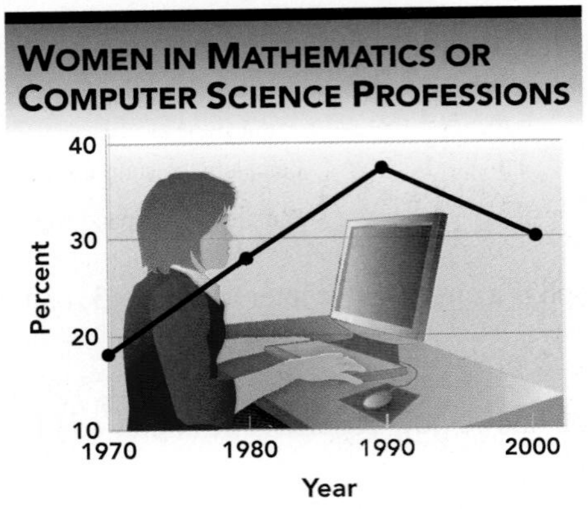

Source: U.S. Bureau of the Census and Bureau of Labor Statistics.

(a) If (x, y) represents a point on the graph, what does x represent? What does y represent?
(b) In what decade (10-year period) did the percent of women in mathematics or computer science professions decrease?
(c) When did the percent of women in mathematics or computer science professions reach a maximum?
(d) In what year was the percent of women in mathematics or computer science professions about 27%?

2. *Federal Government Tax Revenues* The graph shows federal government tax revenues in billions of dollars.

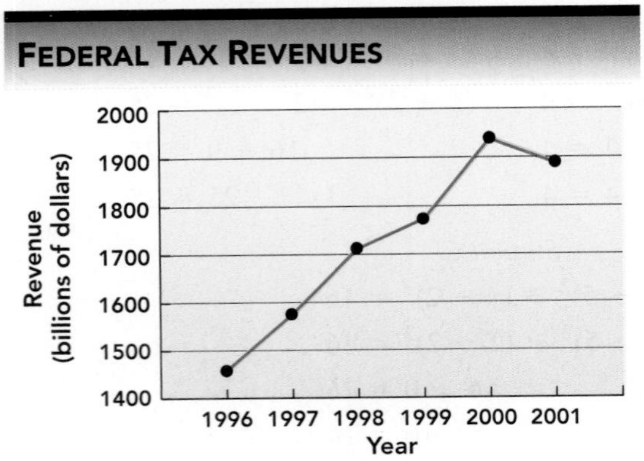

Source: U.S. Office of Management and Budget.

(a) If (x, y) represents a point on the graph, what does x represent? What does y represent?
(b) What was the revenue in 1999?
(c) In what year was revenue about \$1700 billion?

Fill in each blank with the correct response.

3. For any value of x, the point $(x, 0)$ lies on the ______ -axis.

4. For any value of y, the point $(0, y)$ lies on the ______ -axis.

5. The circle $x^2 + y^2 = 9$ has the point ____ as its center.

6. The point (___, 0) is the center of the circle

$$(x - 2)^2 + y^2 = 16.$$

Name the quadrant, if any, in which each point is located.

7. (a) $(1, 6)$
(b) $(-4, -2)$
(c) $(-3, 6)$
(d) $(7, -5)$
(e) $(-3, 0)$

8. (a) $(-2, -10)$
(b) $(4, 8)$
(c) $(-9, 12)$
(d) $(3, -9)$
(e) $(0, -8)$

9. Use the given information to determine the possible quadrants in which the point (x, y) must lie.
(a) $xy > 0$
(b) $xy < 0$
(c) $\frac{x}{y} < 0$
(d) $\frac{x}{y} > 0$

10. What must be true about one of the coordinates of any point that lies along an axis?

Locate the following points on the rectangular coordinate system, using a graph similar to Figure 2.

11. $(2, 3)$

12. $(-1, 2)$

13. $(-3, -2)$

14. $(1, -4)$

15. $(0, 5)$

16. $(-2, -4)$

17. $(-2, 4)$

18. $(3, 0)$

19. $(-2, 0)$

20. $(3, -3)$

Find
(a) *the distance between the pair of points, and*
(b) *the coordinates of the midpoint of the segment having the points as endpoints.*

21. $(3, 4)$ and $(-2, 1)$

22. $(-2, 1)$ and $(3, -2)$

23. $(-2, 4)$ and $(3, -2)$

24. $(1, -5)$ and $(6, 3)$

25. $(-3, 7)$ and $(2, -4)$

26. $(0, 5)$ and $(-3, 12)$

In Exercises 27–30, match each center-radius form of the equation of a circle with the correct graph from choices A–D.

A.

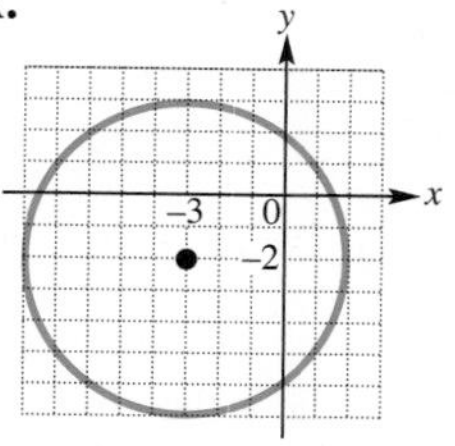

B.

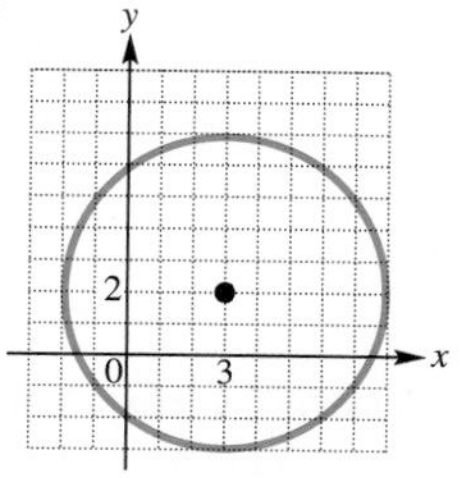

C.

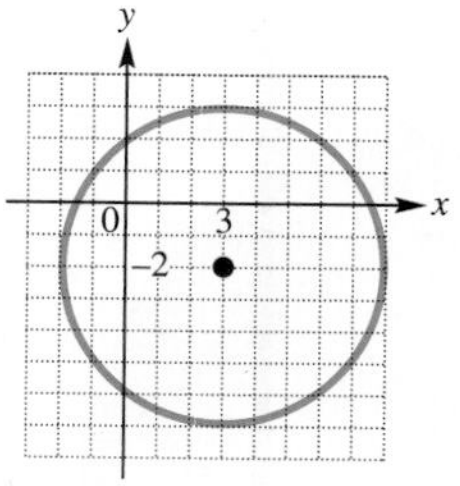

D.

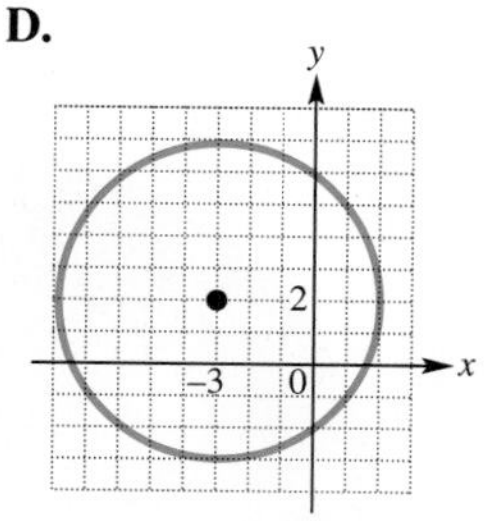

27. $(x - 3)^2 + (y - 2)^2 = 25$

28. $(x - 3)^2 + (y + 2)^2 = 25$

29. $(x + 3)^2 + (y - 2)^2 = 25$

30. $(x + 3)^2 + (y + 2)^2 = 25$

Write an equation of the circle with the given center and radius.

31. $(0, 0)$; $r = 6$

32. $(0, 0)$; $r = 5$

33. $(-1, 3)$; $r = 4$

34. $(2, -2)$; $r = 3$

35. $(0, 4)$; $r = \sqrt{3}$

36. $(-2, 0)$; $r = \sqrt{5}$

37. Suppose that a circle has an equation of the form $x^2 + y^2 = r^2$, $r > 0$. What is the center of the circle? What is the radius of the circle?

38. (a) How many points are there on the graph of $(x - 4)^2 + (y - 1)^2 = 0$? Explain your answer.
(b) How many points are there on the graph of $(x - 4)^2 + (y - 1)^2 = -1$? Explain your answer.

Find the center and the radius of each circle. (Hint: In Exercises 43 and 44 divide both sides by the greatest common factor.)

39. $x^2 + y^2 + 4x + 6y + 9 = 0$

40. $x^2 + y^2 - 8x - 12y + 3 = 0$

41. $x^2 + y^2 + 10x - 14y - 7 = 0$

42. $x^2 + y^2 - 2x + 4y - 4 = 0$

43. $3x^2 + 3y^2 - 12x - 24y + 12 = 0$

44. $2x^2 + 2y^2 + 20x + 16y + 10 = 0$

Graph each circle.

45. $x^2 + y^2 = 36$

46. $x^2 + y^2 = 81$

47. $(x - 2)^2 + y^2 = 36$

48. $x^2 + (y + 3)^2 = 49$

49. $(x + 2)^2 + (y - 5)^2 = 16$

50. $(x - 4)^2 + (y - 3)^2 = 25$

51. $(x + 3)^2 + (y + 2)^2 = 36$

52. $(x - 5)^2 + (y + 4)^2 = 49$

Find **(a)** *the distance between P and Q and* **(b)** *the coordinates of the midpoint of the segment joining P and Q.*

53.

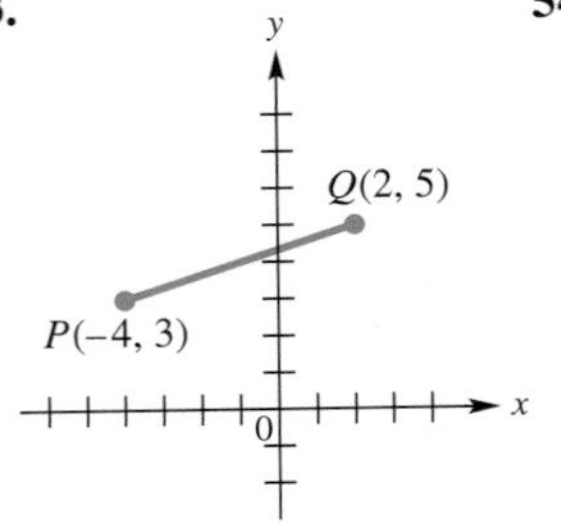

54.

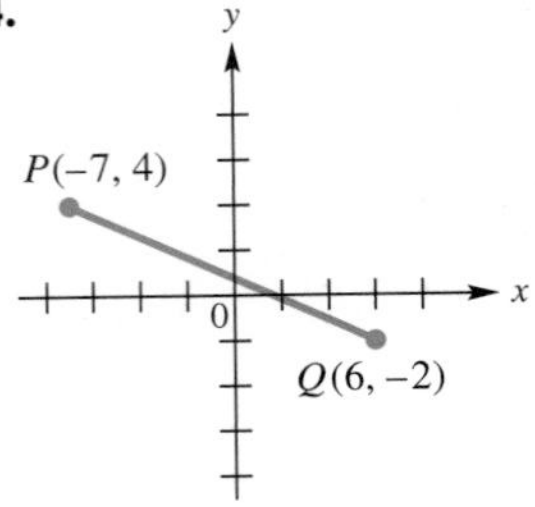

55. ***(Modeling) Tuition and Fees*** From 1998–2004, the average annual cost (in dollars) of tuition and fees at private four-year colleges rose in an approximately linear fashion. The graph depicts this growth with a line segment. Use the midpoint formula to approximate the cost during the year 2001.

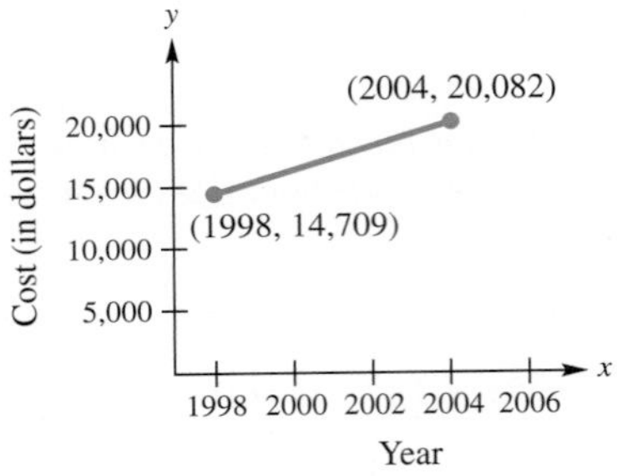

Source: The College Board.

56. ***(Modeling) Two-Year College Enrollment*** Projected enrollments in two-year colleges for 2005, 2007, and 2009 are shown in the table. Use the midpoint formula to estimate the enrollments for 2006 and 2008.

Year	Enrollment (in thousands)
2005	6038
2007	6146
2009	6257

Source: Statistical Abstract of the United States.

57. ***(Modeling) Poverty-Level Income Cutoffs*** The table at the top of the next column lists poverty-level income cutoffs for a family of four since 1983. Use the midpoint formula to estimate the poverty-level cutoffs (rounded to the nearest dollar) in 1988 and 1998.

Year	Income (in dollars)
1983	10,178
1993	14,763
2003	18,810

Source: U.S. Census Bureau.

58. An alternative form of the distance formula is

$$d = \sqrt{(x_1 - x_2)^2 + (y_1 - y_2)^2}.$$

Compare this to the form given in this section, and explain why the two forms are equivalent.

59. A student was asked to find the distance between the points (5, 8) and (2, 14), and wrote the following:

$$d = \sqrt{(5 - 8)^2 + (2 - 14)^2}.$$

Explain why this is incorrect.

60. A circle can be drawn on a piece of posterboard by fastening one end of a string, pulling the string taut with a pencil, and tracing a curve as shown in the figure. Explain why this method works.

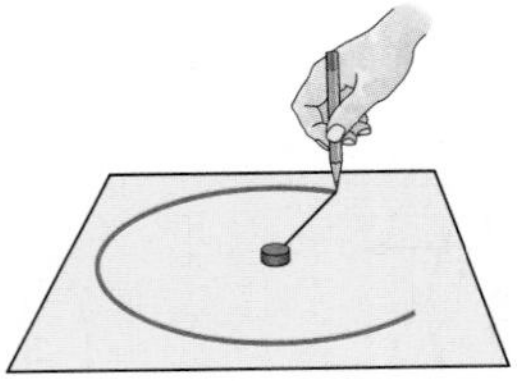

61. ***Crawfish Racing*** This figure shows how the crawfish race is held at the Crawfish Festival in Breaux Bridge, Louisiana. Explain why a circular "race-track" is appropriate for such a race.

62. ***Epicenter of an Earthquake*** Show algebraically that if three receiving stations at $(1, 4)$, $(-6, 0)$, and $(5, -2)$ record distances to an earthquake epicenter of 4 units, 5 units, and 10 units, respectively, the epicenter would lie at $(-3, 4)$.

63. ***Epicenter of an Earthquake*** Three receiving stations record the presence of an earthquake. The locations of the receiving stations and the distances to the epicenter are contained in the following three equations: $(x - 2)^2 + (y - 1)^2 = 25$, $(x + 2)^2 + (y - 2)^2 = 16$, and $(x - 1)^2 + (y + 2)^2 = 9$. Graph the circles and determine the location of the earthquake epicenter.

64. Without actually graphing, state whether the graphs of $x^2 + y^2 = 4$ and $x^2 + y^2 = 25$ will intersect. Explain your answer.

65. Can a circle have its center at $(2, 4)$ and be tangent to both axes? (*Tangent to* means touching in one point.) Explain.

66. Suppose that the endpoints of a line segment have coordinates (x_1, y_1) and (x_2, y_2).
   (a) Show that the distance between (x_1, y_1) and $\left(\frac{x_1 + x_2}{2}, \frac{y_1 + y_2}{2}\right)$ is the same as the distance between (x_2, y_2) and $\left(\frac{x_1 + x_2}{2}, \frac{y_1 + y_2}{2}\right)$.
   (b) Show that the sum of the distances between (x_1, y_1) and $\left(\frac{x_1 + x_2}{2}, \frac{y_1 + y_2}{2}\right)$, and (x_2, y_2) and $\left(\frac{x_1 + x_2}{2}, \frac{y_1 + y_2}{2}\right)$ is equal to the distance between (x_1, y_1) and (x_2, y_2).
   (c) From the results of parts (a) and (b), what conclusion can be made?

67. If the coordinates of one endpoint of a line segment $(3, -8)$ and the coordinates of the midpoint of the segment are $(6, 5)$, what are the coordinates of the other endpoint?

68. Which one of the following has a circle as its graph?
   A. $x^2 - y^2 = 9$ **B.** $x^2 = 9 - y^2$
   C. $y^2 - x^2 = 9$ **D.** $-x^2 - y^2 = 9$

69. For the three choices that are not circles in Exercise 68, explain why their equations are not those of circles.

70. An *isosceles triangle* has at least two sides of equal length. Determine whether the triangle with vertices $(0, 0)$, $(3, 4)$, and $(7, 1)$ is isosceles.

8.2 Lines, Slope, and Average Rate of Change

Linear Equations in Two Variables • Intercepts • Slope • Parallel and Perpendicular Lines • Average Rate of Change

Linear Equations in Two Variables In the previous chapter, we studied linear equations in a single variable. The solution of such an equation is a real number. A linear equation in *two* variables will have solutions written as ordered pairs. Unlike linear equations in a single variable, equations with two variables will, in general, have an infinite number of solutions.

To find ordered pairs that satisfy the equation, select any number for one of the variables, substitute it into the equation for that variable, and then solve for the other variable. For example, suppose $x = 0$ in the equation $2x + 3y = 6$. Then

$$\begin{aligned} 2x + 3y &= 6 \\ 2(0) + 3y &= 6 \quad \text{Let } x = 0. \\ 0 + 3y &= 6 \\ 3y &= 6 \\ y &= 2, \end{aligned}$$

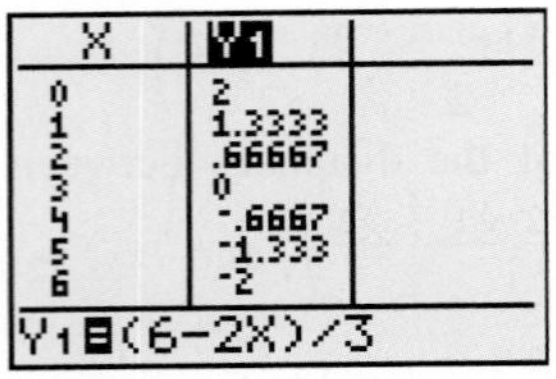

Graphing calculators can generate tables of ordered pairs. Here is an example for $2x + 3y = 6$. We must solve for y to get $Y_1 = (6 - 2X)/3$ before generating the table.

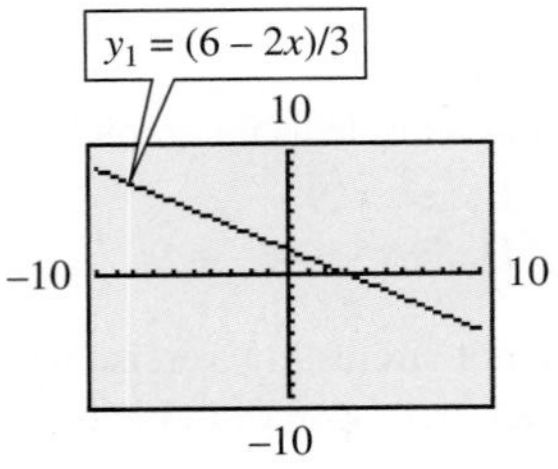

This is a calculator graph of the line shown in Figure 10(b).

giving the ordered pair $(0, 2)$. Other ordered pairs satisfying $2x + 3y = 6$ include $(6, -2)$, $(3, 0)$, $(-3, 4)$, and $(9, -4)$.

The equation $2x + 3y = 6$ is graphed by first plotting all the ordered pairs mentioned above. These are shown in Figure 10(a). The resulting points appear to lie on a straight line. If all the ordered pairs that satisfy the equation $2x + 3y = 6$ were graphed, they would form a straight line. In fact, the graph of any first-degree equation in two variables is a straight line. The graph of $2x + 3y = 6$ is the line shown in Figure 10(b).

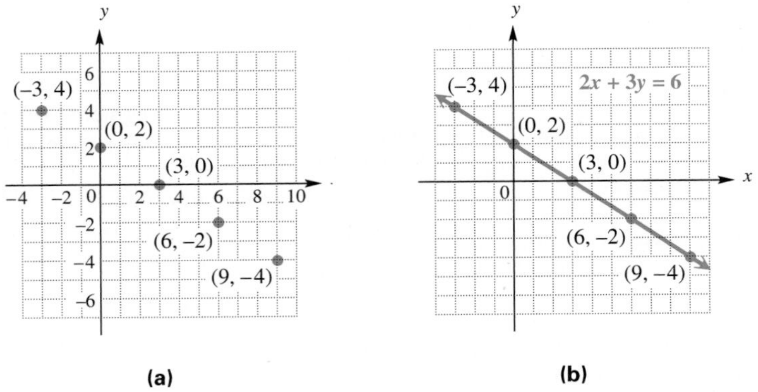

FIGURE 10

> **Linear Equation in Two Variables**
>
> An equation that can be written in the form
>
> $$Ax + By = C \quad \text{(where } A \text{ and } B \text{ are not both 0)}$$
>
> is a **linear equation in two variables.** This form is called **standard form.**

All first-degree equations with two variables have straight-line graphs. Since a straight line is determined if any two different points on the line are known, finding two different points is sufficient to graph the line.

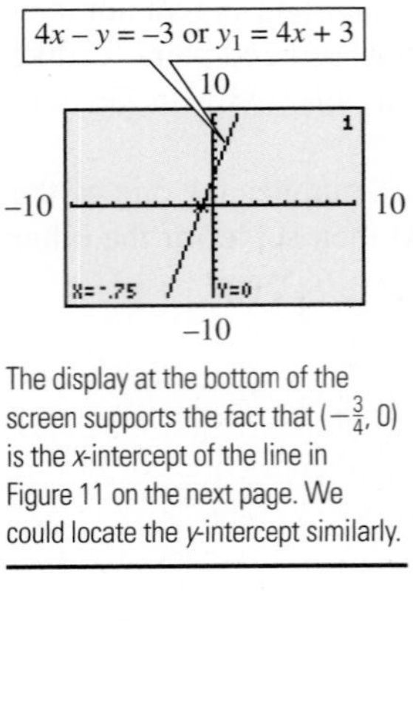

The display at the bottom of the screen supports the fact that $(-\frac{3}{4}, 0)$ is the x-intercept of the line in Figure 11 on the next page. We could locate the y-intercept similarly.

Intercepts Two points that are useful for graphing lines are the x- and y-intercepts. The ***x*-intercept** is the point (if any) where the line crosses the x-axis, and the ***y*-intercept** is the point (if any) where the line crosses the y-axis. (*Note:* In many texts, the intercepts are defined as numbers, and not points. However, in this book we will refer to intercepts as points.) Intercepts can be found as follows.

> **Intercepts**
>
> To find the x-intercept of the graph of a linear equation, let $y = 0$ and solve for x.
>
> To find the y-intercept, let $x = 0$ and solve for y.

EXAMPLE 1 Graphing an Equation Using Intercepts

Find the x- and y-intercepts of $4x - y = -3$, and graph the equation.

SOLUTION

To find the x-intercept, let $y = 0$.

$$4x - 0 = -3 \quad \text{Let } y = 0.$$
$$4x = -3$$
$$x = -\frac{3}{4} \quad x\text{-intercept is } \left(-\tfrac{3}{4}, 0\right).$$

To find the y-intercept, let $x = 0$.

$$4(0) - y = -3 \quad \text{Let } x = 0.$$
$$-y = -3$$
$$y = 3 \quad y\text{-intercept is } (0, 3).$$

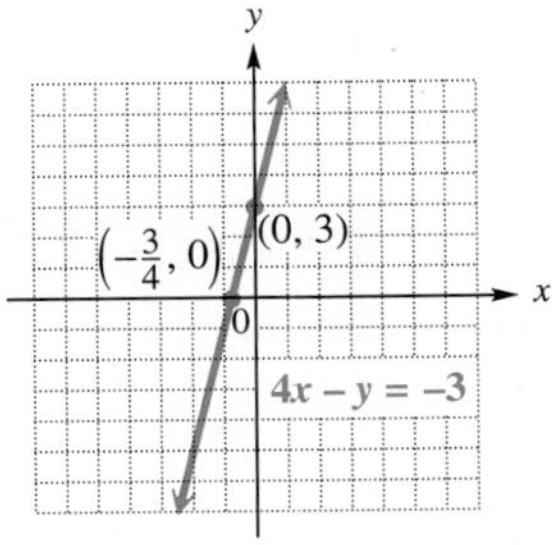

FIGURE 11

The intercepts are the two points $\left(-\frac{3}{4}, 0\right)$ and $(0, 3)$. Use these two points to draw the graph, as shown in Figure 11. ■

A line may not have an x-intercept, or it may not have a y-intercept.

EXAMPLE 2 Graphing Lines with a Single Intercept

Graph each line.

(a) $y = 2$ **(b)** $x = -1$

SOLUTION

(a) Writing $y = 2$ as $0x + 1y = 2$ shows that any value of x, including $x = 0$, gives $y = 2$, making the y-intercept $(0, 2)$. Since y is always 2, there is no value of x corresponding to $y = 0$, and so the graph has no x-intercept. The graph, shown in Figure 12(a), is a horizontal line.

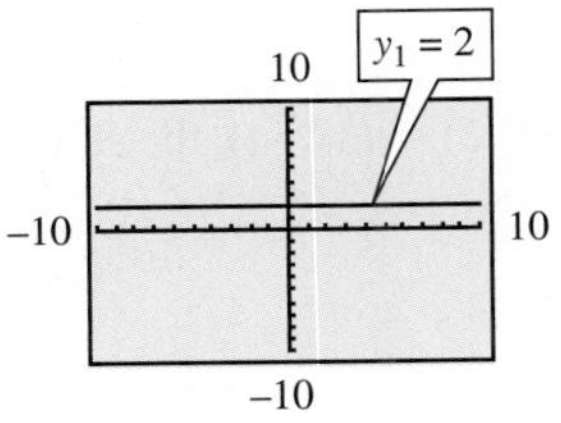

Compare this graph with the one in Figure 12(a).

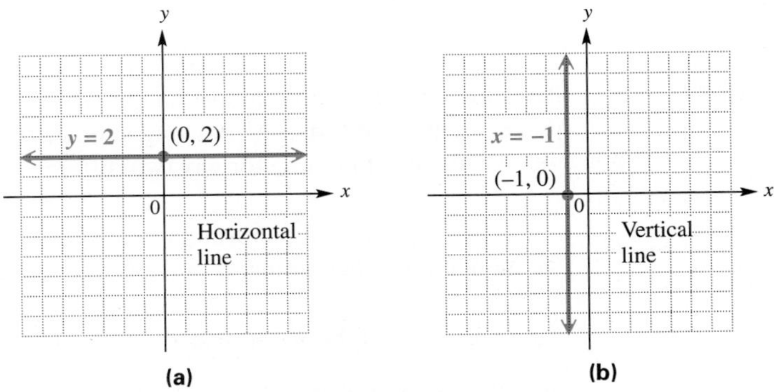

FIGURE 12

(b) In this equation, for all x, $x = -1$. No value of y makes $x = 0$. The graph has no y-intercept. The only way a straight line can have no y-intercept is to be vertical, as shown in Figure 12(b). ■

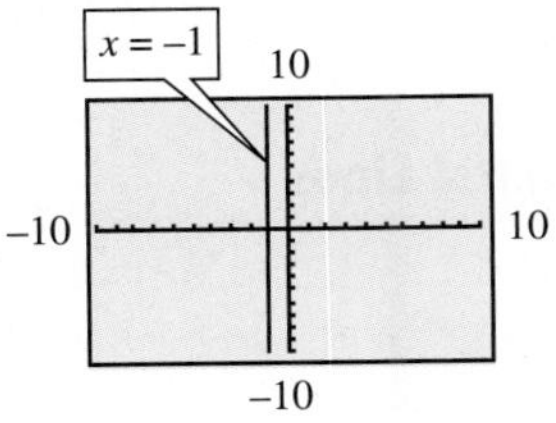

This vertical line is not an example of a *function* (see Section 8.4), so we must use a *draw* command to obtain it. Compare with Figure 12(b).

Slope Two distinct points determine a unique line. A line also can be determined by a point on the line and some measure of the "steepness" of the line. The measure of the steepness of a line is called the *slope* of the line. One way to get a measure of the steepness of a line is to compare the vertical change in the line (the *rise*) to the horizontal change (the *run*) while moving along the line from one fixed point to another.

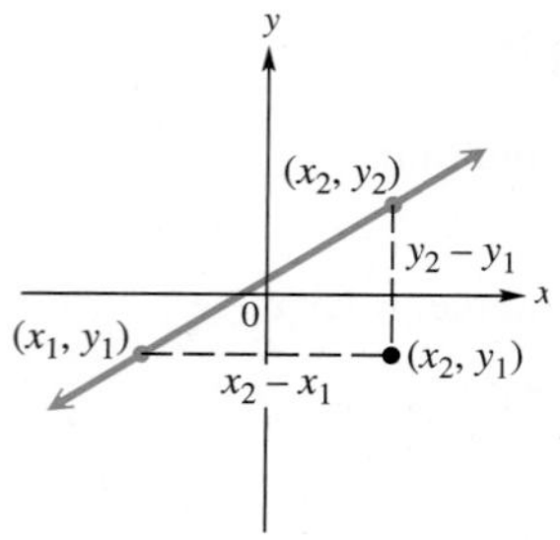

FIGURE 13

Suppose that (x_1, y_1) and (x_2, y_2) are two different points on a line. Then, going along the line from (x_1, y_1) to (x_2, y_2), the y-value changes from y_1 to y_2, an amount equal to $y_2 - y_1$. As y changes from y_1 to y_2, the value of x changes from x_1 to x_2 by the amount $x_2 - x_1$. See Figure 13. The ratio of the change in y to the change in x is called the **slope** of the line. The letter m is used to denote the slope.

Slope

If $x_1 \neq x_2$, the slope of the line through the distinct points (x_1, y_1) and (x_2, y_2) is

$$m = \frac{\text{rise}}{\text{run}} = \frac{\text{change in } y}{\text{change in } x} = \frac{y_2 - y_1}{x_2 - x_1}.$$

EXAMPLE 3 Using the Slope Formula

Find the slope of the line that passes through the points $(2, -1)$ and $(-5, 3)$.

SOLUTION

If $(2, -1) = (x_1, y_1)$ and $(-5, 3) = (x_2, y_2)$, then

$$m = \frac{y_2 - y_1}{x_2 - x_1} = \frac{3 - (-1)}{-5 - 2} = \frac{4}{-7} = -\frac{4}{7}.$$

Start with the x- and y-values of the same point.

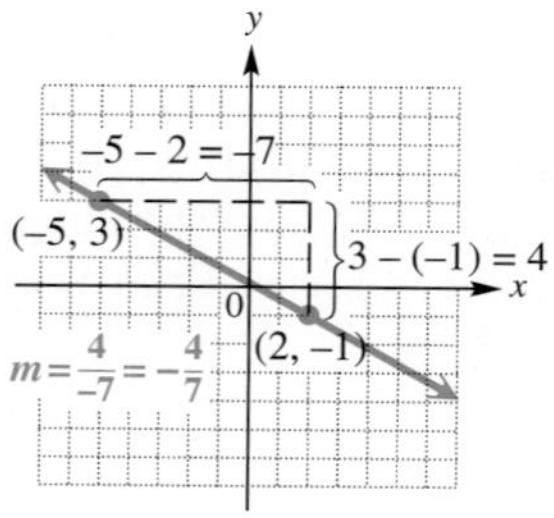

FIGURE 14

See Figure 14. On the other hand, if $(2, -1) = (x_2, y_2)$ and $(-5, 3) = (x_1, y_1)$, the slope would be

$$m = \frac{-1 - 3}{2 - (-5)} = \frac{-4}{7} = -\frac{4}{7},$$

the same answer. This example suggests that the slope is the same no matter which point is considered first. Also, using similar triangles from geometry, it can be shown that the slope is the same for *any* two different points chosen on the line.

If we apply the slope formula to a vertical or a horizontal line, we find that either the numerator or denominator in the fraction is 0.

EXAMPLE 4 Finding Slopes of Vertical and Horizontal Lines

Find the slope, if possible, of each of the following lines.

(a) $x = -3$ **(b)** $y = 5$

SOLUTION

(a) By inspection, $(-3, 5)$ and $(-3, -4)$ are two points that satisfy the equation $x = -3$. Use these two points to find the slope.

$$m = \frac{-4 - 5}{-3 - (-3)} = \frac{-9}{0} \quad \text{Undefined slope}$$

Since division by zero is undefined, the slope is undefined. This is why the definition of slope includes the restriction that $x_1 \neq x_2$.

(b) Find the slope by selecting two different points on the line, such as $(3, 5)$ and $(-1, 5)$, and by using the definition of slope.

$$m = \frac{5 - 5}{3 - (-1)} = \frac{0}{4} = 0 \quad \text{Zero slope}$$

Highway slopes are measured in percent. For example, a slope of 8% means that the road gains 8 feet in altitude for each 100 feet that the road travels horizontally. Interstate highways cannot exceed a slope of 6%. While this may not seem like much of a slope, there are probably stretches of interstate highways that would be hard work for a distance runner.

In Example 2, $x = -1$ has a graph that is a vertical line, and $y = 2$ has a graph that is a horizontal line. Generalizing from those results and the results of Example 4, we can make the following statements about vertical and horizontal lines.

Vertical and Horizontal Lines

A vertical line has an equation of the form $\boldsymbol{x = a}$, where $\boldsymbol{a}$ is a real number, and its slope is undefined. A horizontal line has an equation of the form $\boldsymbol{y = b}$, where $\boldsymbol{b}$ is a real number, and its slope is 0.

If we know the slope of a line and a point contained on the line, then we can graph the line using the method shown in the next example.

EXAMPLE 5 Graphing a Line Using Slope and a Point

Graph the line that has slope $\frac{2}{3}$ and passes through the point $(-1, 4)$.

SOLUTION

First locate the point $(-1, 4)$ on a graph as shown in Figure 15. Then,

$$m = \frac{\text{change in } y}{\text{change in } x} = \frac{2}{3}. \quad \text{Definition of slope}$$

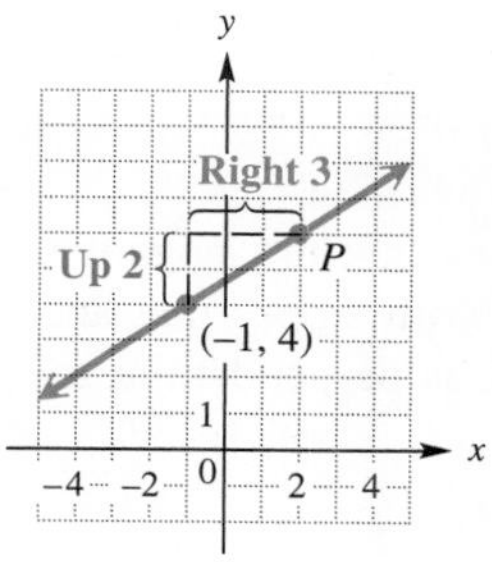

FIGURE 15

Move *up* 2 units in the y-direction and then 3 units to the *right* in the x-direction to locate another point on the graph (labeled P). The line through $(-1, 4)$ and P is the required graph.

The line graphed in Figure 14 has a negative slope, $-\frac{4}{7}$, and the line *falls* from left to right. In contrast, the line graphed in Figure 15 has a positive slope, $\frac{2}{3}$ and it *rises* from left to right. These ideas can be generalized. (Figure 16 shows lines of positive, zero, negative, and undefined slopes.)

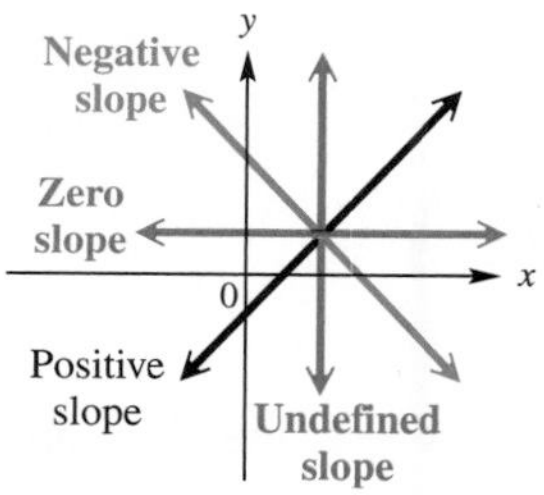

FIGURE 16

Positive and Negative Slopes

A line with a positive slope rises from left to right, while a line with a negative slope falls from left to right.

Parallel and Perpendicular Lines The slopes of a pair of parallel or perpendicular lines are related in a special way. The slope of a line measures the steepness of the line. Since parallel lines have equal steepness, their slopes also must be equal. Also, lines with the same slope are parallel.

Slopes of Parallel Lines

Two nonvertical lines with the same slope are parallel. Two nonvertical parallel lines have the same slope. Furthermore, any two vertical lines are parallel.

EXAMPLE 6 Determining Whether Two Lines Are Parallel

Determine whether the lines L_1, through $(-2, 1)$ and $(4, 5)$, and L_2, through $(3, 0)$ and $(0, -2)$, are parallel.

SOLUTION

The slope of L_1 is $m_1 = \frac{5 - 1}{4 - (-2)} = \frac{4}{6} = \frac{2}{3}.$

The slope of L_2 is $m_2 = \frac{-2 - 0}{0 - 3} = \frac{-2}{-3} = \frac{2}{3}.$

Because the slopes are equal, the lines are parallel.

Perpendicular lines are lines that meet at right angles. It can be shown that the slopes of perpendicular lines have a product of -1, provided that neither line is vertical. For example, if the slope of a line is $\frac{3}{4}$, then any line perpendicular to it has slope $-\frac{4}{3}$, because $(\frac{3}{4})(-\frac{4}{3}) = -1$.

Slopes of Perpendicular Lines

If neither is vertical, two perpendicular lines have slopes that are negative reciprocals—that is, their product is -1. Also, two lines with slopes that are negative reciprocals are perpendicular. Every vertical line is perpendicular to every horizontal line.

EXAMPLE 7 Determining Whether Two Lines Are Perpendicular

Determine whether the lines L_1, through $(0, -3)$ and $(2, 0)$, and L_2, through $(-3, 0)$ and $(0, -2)$, are perpendicular.

SOLUTION

The slope of L_1 is $m_1 = \frac{0 - (-3)}{2 - 0} = \frac{3}{2}.$

The slope of L_2 is $m_2 = \frac{-2 - 0}{0 - (-3)} = -\frac{2}{3}.$

Because the product of the slopes of the two lines is $\frac{3}{2}\left(-\frac{2}{3}\right) = -1$, the lines are perpendicular.

Average Rate of Change We have seen how the slope of a line is the ratio of the change in y (vertical change) to the change in x (horizontal change). This idea can be extended to real-life situations as follows: the slope gives the average rate of change of y per unit of change in x, where the value of y *is dependent upon the value of* x. The next example illustrates this idea of average rate of change. We assume a linear relationship between x and y.

EXAMPLE 8 Finding Average Rate of Change

Figure 17 depicts how the purchasing power of the dollar declined from 1990 to 2000, based on changes in the Consumer Price Index. In this model, the base period is 1982–1984. For example, it would have cost \$1.00 to purchase in 1990 what \$.766 would have purchased during the base period. Find the average rate of change in the purchasing power of the dollar during the decade.

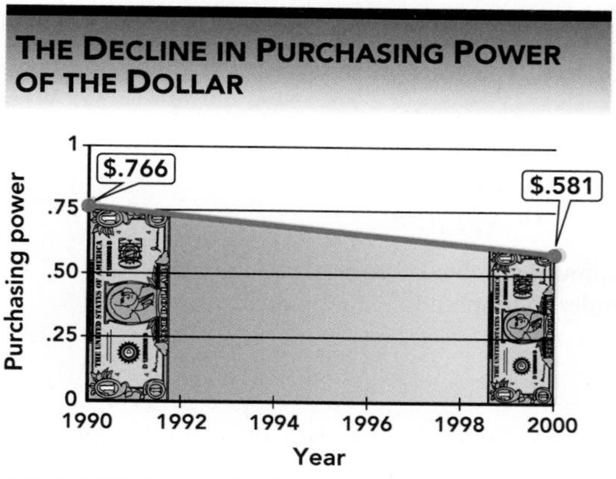

Source: U.S. Bureau of Labor Statistics.

FIGURE 17

SOLUTION

The average rate of change is found by determining the slope of the line segment. Using x for year and y for purchasing power, the two points indicated are (1990, .766) and (2000, .581). By the slope formula,

$$\text{average rate of change} = \frac{y_2 - y_1}{x_2 - x_1} = \frac{.766 - .581}{1990 - 2000} = \frac{.185}{-10} = -.0185.$$

Because $-.0185$ is approximately $-.02$, we can say that on the average, the purchasing power *dropped* by about 2 cents per year during the decade.

8.2 EXERCISES

Complete the given ordered pairs for each equation. Then graph the equation.

1. $2x + y = 5$; (0,), (, 0), (1,), (,1)

2. $3x - 4y = 24$; (0,), (, 0), (6,), (, −3)

3. $x - y = 4$; (0,), (, 0), (2,), (, −1)

4. $x + 3y = 12$; (0,), (,0), (3,), (, 6)

5. $4x + 5y = 20$; (0,), (, 0), (3,), (, 2)

6. $2x - 5y = 12$; (0,), (, 0), (, −2), (−2,)

7. $3x + 2y = 8$

x	y
0	
	0
2	
	−2

8. $5x + y = 12$

x	y
0	
	0
	−3
2	

9. Explain how to find the x-intercept of a linear equation in two variables.

10. Explain how to find the y-intercept of a linear equation in two variables.

11. Which has a horizontal line as its graph?
A. $2y = 6$ **B.** $2x = 6$
C. $x - 4 = 0$ **D.** $x + y = 0$

12. What is the minimum number of points that must be determined in order to graph a linear equation in two variables?

For each equation, give the x-intercept and the y-intercept. Then graph the equation.

13. $3x + 2y = 12$ **14.** $2x + 5y = 10$

15. $5x + 6y = 10$ **16.** $3y + x = 6$

17. $2x - y = 5$ **18.** $3x - 2y = 4$

19. $x - 3y = 2$ **20.** $y - 4x = 3$

21. $y + x = 0$ **22.** $2x - y = 0$

23. $3x = y$ **24.** $x = -4y$

25. $x = 2$ **26.** $y = -3$

27. $y = 4$ **28.** $x = -2$

In Exercises 29–36, match the equation with the figure in choices A–D in the next column that most closely resembles its graph.

29. $y + 2 = 0$ **30.** $y + 4 = 0$

31. $x + 3 = 0$ **32.** $x + 7 = 0$

33. $y - 2 = 0$ **34.** $y - 4 = 0$

35. $x - 3 = 0$ **36.** $x - 7 = 0$

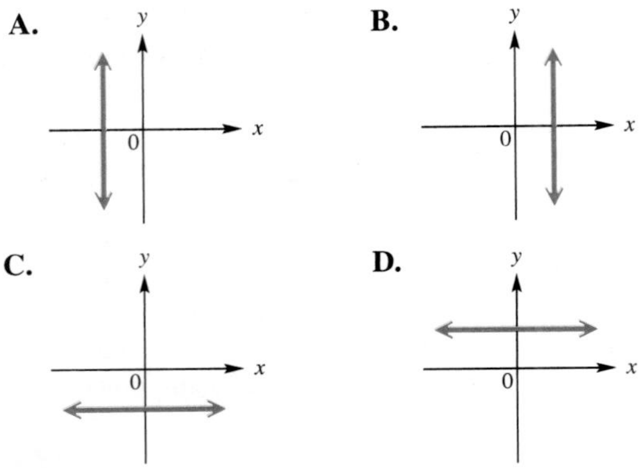

37. What is the slope (or pitch) of this roof?

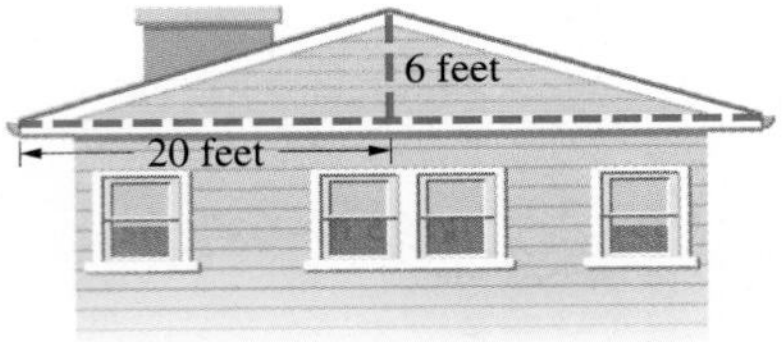

38. What is the slope (or grade) of this hill?

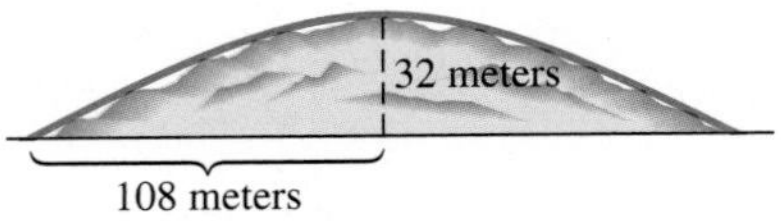

39. Use the coordinates of the indicated points to find the slope of each line.

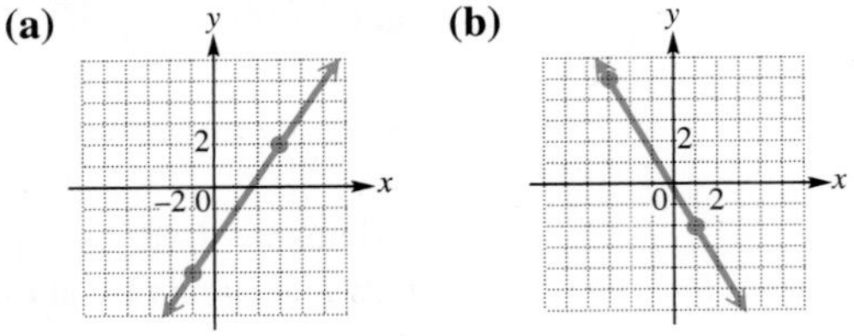

40. Tell whether the slope of the given line in (a) – (d) is positive, negative, zero, or undefined.

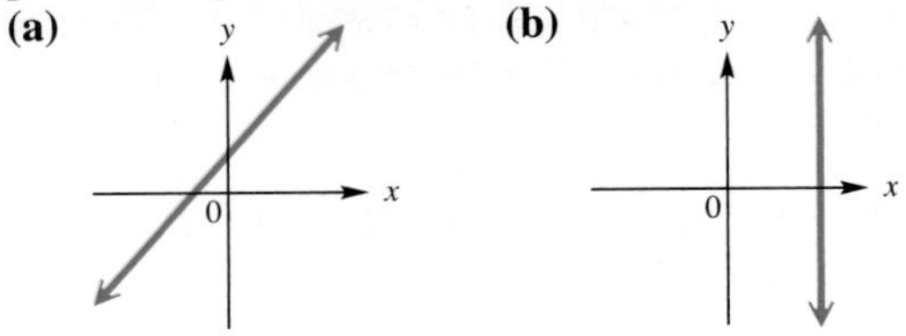

(c)

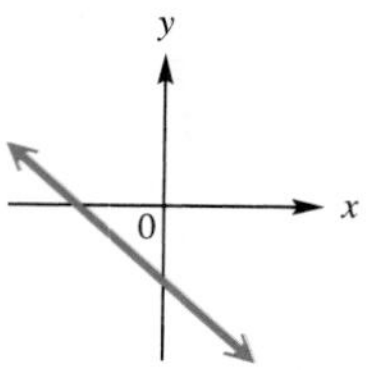

(d)

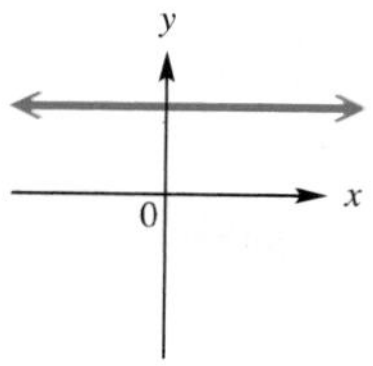

Find the slope of the line through each pair of points by using the slope formula.

41. $(-2, -3)$ and $(-1, 5)$

42. $(-4, 3)$ and $(-3, 4)$

43. $(8, 1)$ and $(2, 6)$

44. $(13, -3)$ and $(5, 6)$

45. $(2, 4)$ and $(-4, 4)$

46. $(-6, 3)$ and $(2, 3)$

47. ***Public School Data*** Figure A depicts public school enrollment (in thousands) in grades 9–12 in the United States. Figure B in the next column gives the (average) number of public school students per computer.

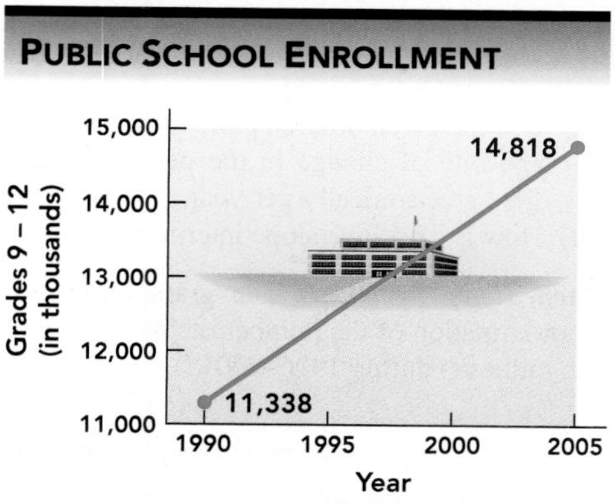

Source: *Digest of Educational Statistics*, annual; and *Projections of Educational Statistics*, annual.

FIGURE A

(a) Use the ordered pairs (1990, 11,338) and (2005, 14,818) to find the slope of the line in Figure A.

(b) The slope of the line in Figure A is __________.
(positive/negative)
This means that during the period represented, enrollment __________________.
(increased/decreased)

(c) The slope of a line represents its *rate of change*. Based on Figure A, what was the increase in students *per year* during the period shown?

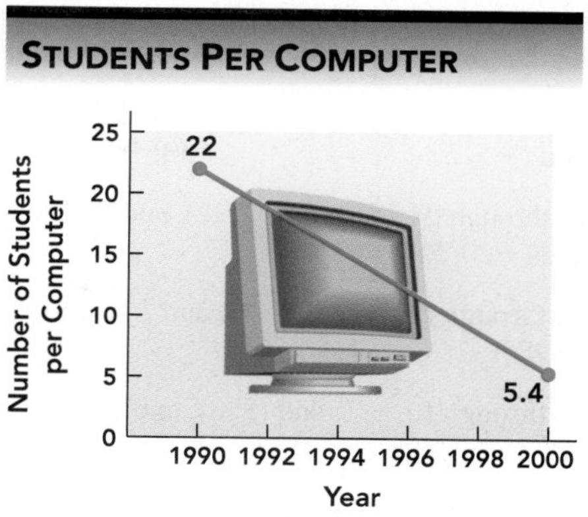

Source: Quality Education Data.

FIGURE B

(d) Use the given ordered pairs to find the slope of the line in Figure B.

(e) The slope of the line in Figure B is __________.
(positive/negative)
This means that during the period represented, the number of students per computer ____________.
(increased/decreased)

(f) Based on Figure B, what was the decrease in students per computer *per year* during the period shown?

48. Use the results of Exercise 47 to make a connection between the sign of the slope of a line and the increase or decrease in the quantity represented by y.

Use the method of Example 5 to graph each of the following lines.

49. $m = \frac{1}{2}$, through $(-3, 2)$

50. $m = \frac{2}{3}$, through $(0, 1)$

51. $m = -\frac{5}{4}$, through $(-2, -1)$

52. $m = -\frac{3}{2}$, through $(-1, -2)$

53. $m = -2$, through $(-1, -4)$

54. $m = 3$, through $(1, 2)$

55. $m = 0$, through $(2, -5)$

56. undefined slope, through $(-3, 1)$

Determine whether the lines described are parallel, perpendicular, *or* neither parallel nor perpendicular.

57. L_1 through $(4, 6)$ and $(-8, 7)$, and L_2 through $(7, 4)$ and $(-5, 5)$

58. L_1 through $(9, 15)$ and $(-7, 12)$, and L_2 through $(-4, 8)$ and $(-20, 5)$

59. L_1 through $(2, 0)$ and $(5, 4)$, and L_2 through $(6, 1)$ and $(2, 4)$

60. L_1 through $(0, -7)$ and $(2, 3)$, and L_2 through $(0, -3)$ and $(1, -2)$

61. L_1 through $(0, 1)$ and $(2, -3)$, and L_2 through $(10, 8)$ and $(5, 3)$

62. L_1 through $(1, 2)$ and $(-7, -2)$, and L_2 through $(1, -1)$ and $(5, -9)$

Use the concept of slope to solve each problem.

63. ***Steepness of an Upper Deck*** The upper deck at U.S. Cellular Field in Chicago has produced, among other complaints, displeasure with its steepness. It has been compared to a ski jump. It is 160 ft from home plate to the front of the upper deck and 250 ft from home plate to the back. The top of the upper deck is 63 ft above the bottom. What is its slope?

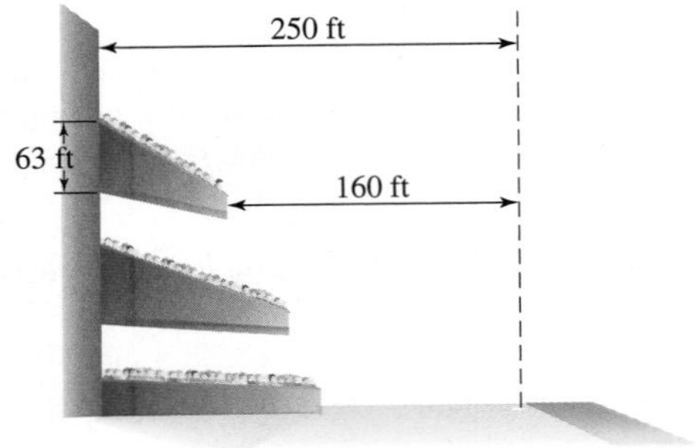

64. ***Grade (Slope) of a Ramp*** When designing the new TD Banknorth Garden arena in Boston to replace the old Boston Garden, architects were careful to design the ramps leading up to the entrances so that circus elephants would be able to walk up the ramps. The maximum grade (or slope) that an elephant will walk on is 13%. Suppose that such a ramp was constructed with a horizontal run of 150 ft. What would be the maximum vertical rise the architects could use?

Use the idea of average rate of change to solve each problem.

65. ***Electronic Filing of Tax Returns*** The percent of tax returns filed electronically for the years 1997–2002 is shown in the graph.

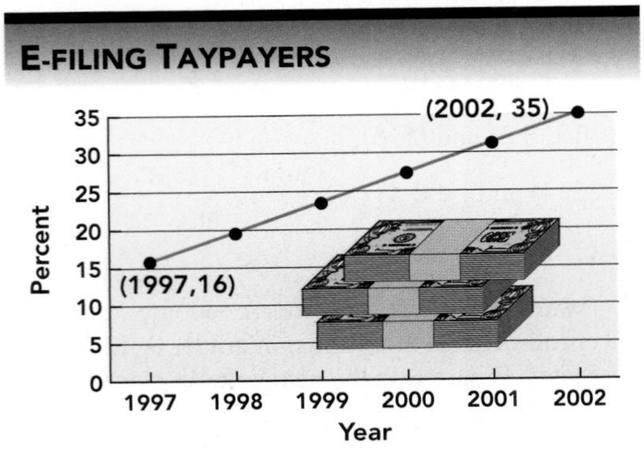

Source: Internal Revenue Service.

(a) Use the given ordered pairs to determine the average rate of change in the percent of tax returns filed electronically per year.

(b) How is a positive slope interpreted here?

66. ***Food Stamp Recipients*** The graph provides a good approximation of the number of food stamp recipients (in millions) during 1996–2001.

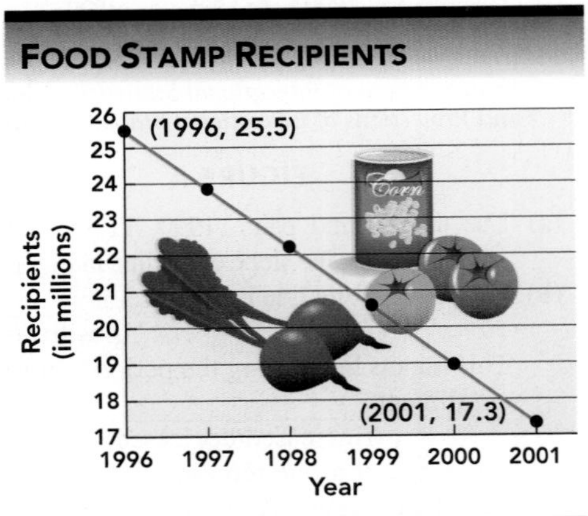

Source: U.S. Department of Agriculture.

(a) Use the given ordered pairs to find the average rate of change in food stamp recipients per year during this period.

(b) Interpret the meaning of a negative slope in this situation.

67. ***Book Sales*** The table gives book publishers' revenue for the period 1999–2002.

Book Publishers' Revenue

Year	Revenue (in millions)
1999	24,000
2000	25,000
2001	26,000
2002	27,000

Source: U.S. Census Bureau.

Find the average rate of change for 1999–2000; 2000–2001; 2001–2002. What do you notice about your answers? What does this tell you?

68. ***Cellular Telephone Subscribers*** The table in the next column gives the number of cellular telephone subscribers from 1999 through 2003.

Cellular Telephone Subscribers

Year	Subscribers (in thousands)
1999	86,047
2000	109,478
2001	128,375
2002	140,766
2003	158,722

Source: Cellular Telecommunications Industry Association, Washington, D.C. *State of the Cellular Industry* (Annual).

(a) Find the average rate of change in subscribers from 1999 to 2003.

(b) Is the average rate of change in successive years approximately the same? If the ordered pairs in the table were plotted, could an approximately straight line be drawn through them?

69. Explain the meaning of *slope*. Give examples showing cases of positive, negative, zero, and undefined slope.

70. Explain how the *grade* of a highway corresponds to the slope concept.

8.3 Equations of Lines and Linear Models

Point-Slope Form • Slope-Intercept Form • Summary of Forms of Linear Equations • Linear Models

Point-Slope Form If the slope of a line and a particular point on the line are known, it is possible to find an equation of the line. Suppose that the slope of a line is m and (x_1, y_1) is a particular point on the line. Let (x, y) be any other point on the line. Then, by the definition of slope,

$$m = \frac{y - y_1}{x - x_1}.$$

Multiplying both sides by $x - x_1$ gives the *point-slope form* of the equation of the line.

Point-Slope Form

The equation of the line through (x_1, y_1) with slope m is written in **point-slope form** as

$$y - y_1 = m(x - x_1).$$

Maria Gaetana Agnesi (1719–1799) did much of her mathematical work in coordinate geometry. She grew up in a scholarly atmosphere; her father was a mathematician on the faculty at the University of Bologna. In a larger sense she was an heir to the long tradition of Italian mathematicians.

Maria Agnesi was fluent in several languages by age 13, but she chose mathematics over literature. The curve shown below, called the **witch of Agnesi,** is studied in analytic geometry courses.

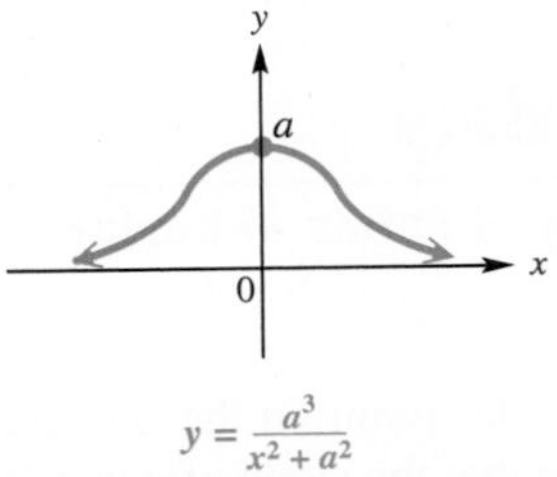

EXAMPLE 1 Finding an Equation Given the Slope and a Point

Find the standard form of an equation of the line with slope $\frac{1}{3}$, passing through the point $(-2, 5)$.

SOLUTION

$$y - y_1 = m(x - x_1) \quad \text{Point-slope form}$$

$$y - 5 = \frac{1}{3}[x - (-2)] \quad \text{Let } (x_1, y_1) = (-2, 5) \text{ and } m = \tfrac{1}{3}.$$

Substitute carefully.

$$y - 5 = \frac{1}{3}(x + 2)$$

$$3y - 15 = x + 2 \quad \text{Multiply by 3.}$$

$$x - 3y = -17 \quad \text{Standard form}$$

If two points on a line are known, it is possible to find an equation of the line. First, find the slope using the slope formula, and then use the slope with one of the given points in the point-slope form.

EXAMPLE 2 Finding an Equation Given Two Points

Find the standard form of an equation of the line passing through the points $(-4, 3)$ and $(5, -7)$.

SOLUTION

First find the slope, using the definition.

$$m = \frac{-7 - 3}{5 - (-4)} = -\frac{10}{9}$$

Either $(-4, 3)$ or $(5, -7)$ may be used as (x_1, y_1) in the point-slope form of the equation of the line. If $(-4, 3)$ is used, then $-4 = x_1$ and $3 = y_1$.

$$y - y_1 = m(x - x_1) \quad \text{Point-slope form}$$

$$y - 3 = -\frac{10}{9}[x - (-4)] \quad (x_1, y_1) = (-4, 3) \text{ and } m = -\tfrac{10}{9}$$

$$y - 3 = -\frac{10}{9}(x + 4)$$

$$9(y - 3) = -10(x + 4) \quad \text{Multiply by 9.}$$

$$9y - 27 = -10x - 40 \quad \text{Distributive property}$$

$$10x + 9y = -13 \quad \text{Standard form}$$

Slope-Intercept Form Suppose that the slope m of a line is known, and the y-intercept of the line has coordinates $(0, b)$. Then,

$$y - y_1 = m(x - x_1) \quad \text{Point-slope form}$$

$$y - b = m(x - 0) \quad \text{Let } (x_1, y_1) = (0, b).$$

$$y - b = mx$$

$$y = mx + b. \quad \text{Add } b \text{ to both sides.}$$

This last result is known as the *slope-intercept form* of the equation of the line.

Slope-Intercept Form

The equation of a line with slope m and y-intercept $(0, b)$ is written in **slope-intercept form** as

$$y = mx + b.$$

(m ↑ Slope; b ↑ y-intercept is $(0, b)$.)

The importance of the slope-intercept form of a linear equation cannot be overemphasized. First, every linear equation (of a nonvertical line) has a *unique* (one and only one) slope-intercept form. Second, in the next section we will study *linear functions,* where the slope-intercept form is necessary in specifying such functions.

EXAMPLE 3 Writing an Equation in Slope-Intercept Form

Write each of the following equations in slope-intercept form.

(a) the line described in Example 1 **(b)** the line described in Example 2

SOLUTION

(a) We determined the standard form of the equation of the line to be $x - 3y = -17$. Solve for y to obtain the slope-intercept form.

$$-3y = -x - 17 \qquad \text{Subtract } x.$$

$$y = \frac{1}{3}x + \frac{17}{3} \qquad \text{Multiply by } -\tfrac{1}{3}.$$

The slope is $\frac{1}{3}$ and the y-intercept is $(0, \frac{17}{3})$.

(b) The equation determined in Example 2 is $10x + 9y = -13$.

$$9y = -10x - 13 \qquad \text{Subtract } 10x.$$

$$y = -\frac{10}{9}x - \frac{13}{9} \qquad \text{Divide by 9.}$$

The slope is $-\frac{10}{9}$ and the y-intercept is $(0, -\frac{13}{9})$. ■

If the slope-intercept form of the equation of a line is known, the method of graphing described in Example 5 of Section 8.2 can be used to graph the line.

EXAMPLE 4 Graphing a Line Using Slope and y-Intercept

Graph the line with the equation $y = -\frac{2}{3}x + 3$.

SOLUTION

Because the equation is given in slope-intercept form, we can easily see that the slope is $-\frac{2}{3}$ and the y-intercept is $(0, 3)$. For now, interpret $-\frac{2}{3}$ as $\frac{-2}{3}$. Plot the point $(0, 3)$, and then, using the "rise over run" interpretation of slope, move *down* 2 units (because of the -2 in the numerator of the slope) and to the *right* 3 units (because of the 3 in the denominator). We arrive at the point $(3, 1)$. Plot the point $(3, 1)$, and join the two points with a line, as shown in Figure 18. (We could also have interpreted $-\frac{2}{3}$ as $\frac{2}{-3}$ and obtained a different second point; however, the line would be the same.) ■

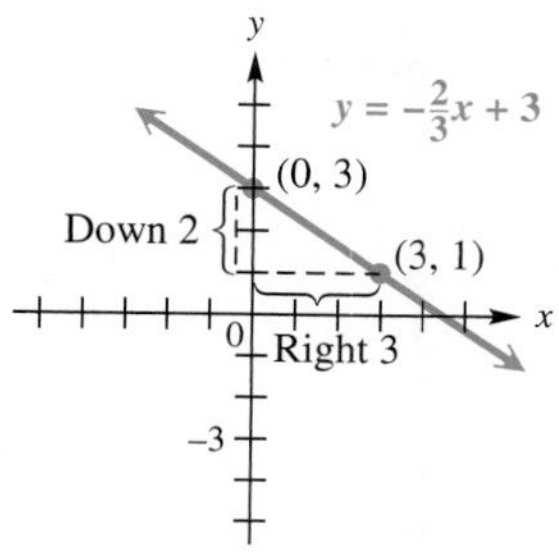

FIGURE 18

As mentioned in the previous section, parallel lines have the same slope and perpendicular lines have slopes that are negative reciprocals of each other.

EXAMPLE 5 Finding an Equation Using a Slope Relationship (Parallel Lines)

Find the slope-intercept form of the equation of the line parallel to the graph of $2x + 3y = 6$, passing through the point $(-4, 5)$.

SOLUTION

The slope of the line $2x + 3y = 6$, shown in Figure 19, can be found by solving for y.

$$2x + 3y = 6$$

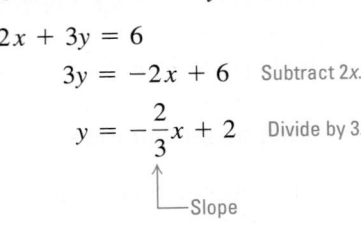

$$3y = -2x + 6 \quad \text{Subtract } 2x.$$

$$y = -\frac{2}{3}x + 2 \quad \text{Divide by 3.}$$

↑ Slope

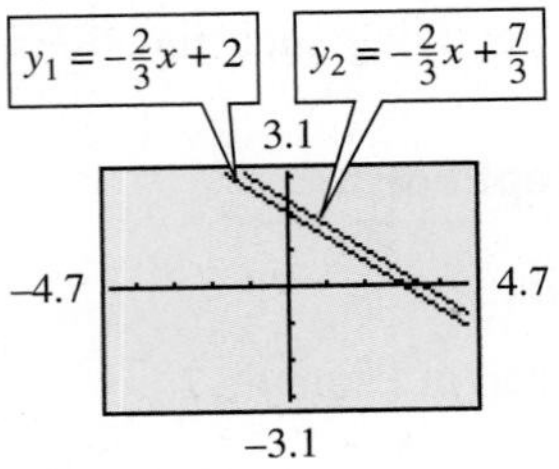

This screen gives support to the result in Example 5.

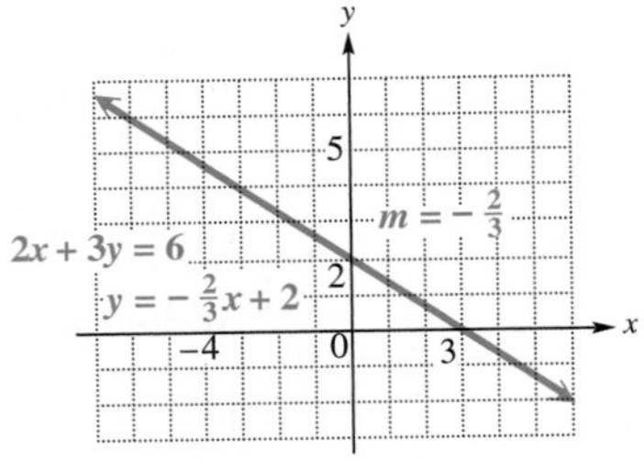

FIGURE 19

The slope is given by the coefficient of x, so $m = -\frac{2}{3}$. The required equation of the line through $(-4, 5)$ and parallel to $2x + 3y = 6$ must also have slope $-\frac{2}{3}$. To find the required equation, use the point-slope form, with $(x_1, y_1) = (-4, 5)$ and $m = -\frac{2}{3}$.

$$y - 5 = -\frac{2}{3}[x - (-4)] \quad y_1 = 5,\ m = -\tfrac{2}{3},\ x_1 = -4$$

$$y - 5 = -\frac{2}{3}(x + 4)$$

$$y - 5 = -\frac{2}{3}x - \frac{8}{3} \quad \text{Distributive property}$$

$$y = -\frac{2}{3}x - \frac{8}{3} + \frac{15}{3} \quad \text{Add } 5 = \frac{15}{3}.$$

$$y = -\frac{2}{3}x + \frac{7}{3} \quad \text{Combine like terms.}$$

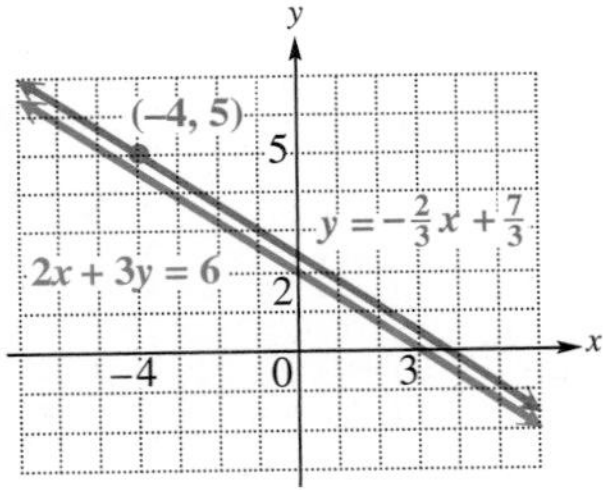

FIGURE 20

We did not clear fractions after the substitution step here because we want the equation in slope-intercept form—that is, solved for y. Both lines are shown in Figure 20. ■

EXAMPLE 6 Finding an Equation Using a Slope Relationship (Perpendicular Lines)

Find the slope-intercept form of the equation of the line perpendicular to the graph of $2x + 3y = 6$, passing through the point $(-4, 5)$.

SOLUTION

In Example 5 we found that the slope of the line $2x + 3y = 6$ is $-\frac{2}{3}$. To be perpendicular to it, a line must have a slope that is the negative reciprocal of $-\frac{2}{3}$, which is $\frac{3}{2}$. Use the point $(-4, 5)$ and slope $\frac{3}{2}$ in the point-slope form to obtain the equation of the perpendicular line shown in Figure 21 on the next page.

$$y - 5 = \frac{3}{2}[x - (-4)] \qquad y_1 = 5,\ m = \frac{3}{2},\ x_1 = -4$$

$$y - 5 = \frac{3}{2}(x + 4)$$

$$y - 5 = \frac{3}{2}x + 6 \qquad \text{Distributive property}$$

$$y = \frac{3}{2}x + 11 \qquad \text{Add 5.}$$

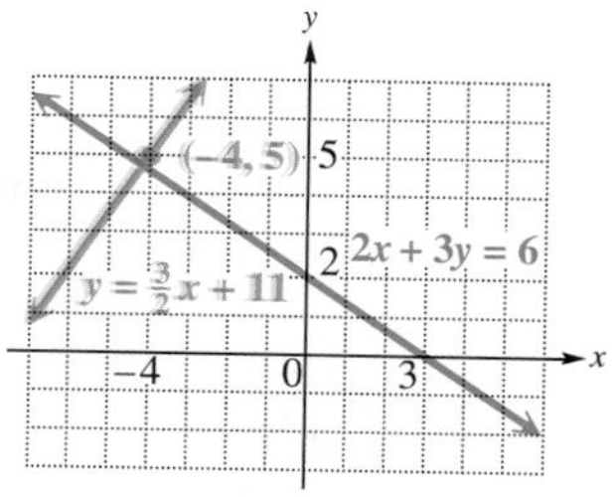

FIGURE 21

Summary of Forms of Linear Equations A summary of the various forms of linear equations from this section and the previous one follows.

Summary of Forms of Linear Equations	
$Ax + By = C$	**Standard form** (Neither A nor B is 0.)
$x = a$	**Vertical line** Undefined slope and the x-intercept is $(a, 0)$.
$y = b$	**Horizontal line** Slope is 0 and the y-intercept is $(0, b)$.
$y = mx + b$	**Slope-intercept form** Slope is m and the y-intercept is $(0, b)$.
$y - y_1 = m(x - x_1)$	**Point-slope form** Slope is m and the line passes through (x_1, y_1).

Linear Models Earlier examples and exercises gave equations that described real data. Now we show how such equations can be found. The process of writing an equation to fit a graph is called *curve-fitting*. The next example illustrates this concept for a straight line. The resulting equation is called a **linear model.**

TABLE 1

Year	Cost (in billions)
2002	264
2003	281
2004	299
2005	318
2006	336
2007	354

Source: U.S. Center for Medicare and Medicaid Services.

EXAMPLE 7 Modeling Medicare Costs

Estimates for Medicare costs (in billions of dollars) are shown in Table 1.

(a) Graph the data. Let $x = 0$ correspond to 2002, $x = 1$ to 2003, and so on. What type of equation might model the data?

(b) Find a linear equation that models the data.

(c) Use the equation from part (b) to predict Medicare costs in 2010.

SOLUTION

(a) Because $x = 0$ corresponds to 2002, $x = 1$ corresponds to 2003, and so on, the data points can be expressed as the ordered pairs

$$(0, 264),\quad (1, 281),\quad (2, 299),\quad (3, 318),\quad (4, 336),\quad \text{and} \quad (5, 354).$$

The data are graphed in Figure 22 and appear to be approximately linear, so a linear equation is appropriate.

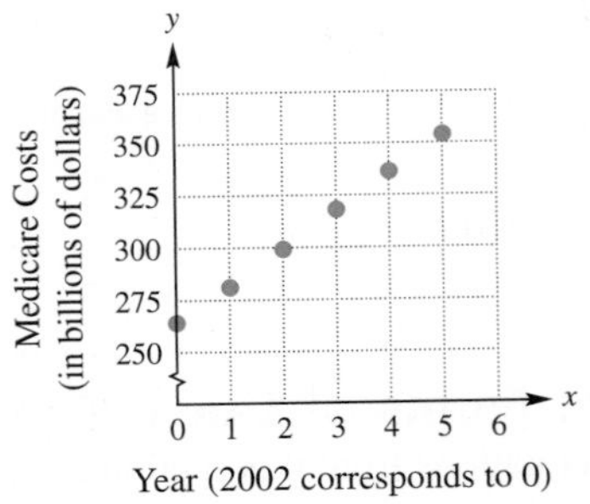

FIGURE 22

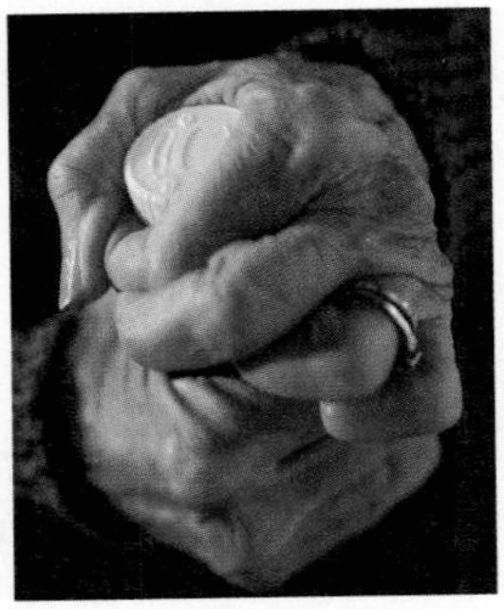

(b) We start by choosing two data points that the line should pass through. For example, if we use (0, 264) and (3, 318), then the slope of the line is

Start with the x- and y-values of the same point.

$$m = \frac{318 - 264}{3 - 0} = 18.$$

The point (0, 264) indicates that the value of b is 264. Thus,

$$y = 18x + 264.$$

The slope $m = 18$ indicates that Medicare costs might increase, on average, by \$18 billion per year.

(c) The value $x = 8$ corresponds to the year 2010. When $x = 8$,

$$y = 18(8) + 264 = 408.$$

This model predicts that Medicare costs will reach \$408 billion in 2010.

The equation $y = 18x + 264$ found in Example 7 is not unique. If two different points are chosen, a different equation may result. However, all such equations should be in approximate agreement for linear data points.

8.3 Exercises

Match each equation in Column I with the correct description given in Column II.

I	II
1. $y = 4x$	**A.** slope $= -2$, through the point $(4, 1)$
2. $y = \frac{1}{4}x$	**B.** slope $= -2$, y-intercept $(0, 1)$
3. $y = -2x + 1$	**C.** passing through the points $(0, 0)$ and $(4, 1)$
4. $y - 1 = -2(x - 4)$	**D.** passing through the points $(0, 0)$ and $(1, 4)$

Use the geometric interpretation of slope (rise *divided by* run) *to find the slope of each line. Then, by identifying the y-intercept from the graph, write the slope-intercept form of the equation of the line.*

5.

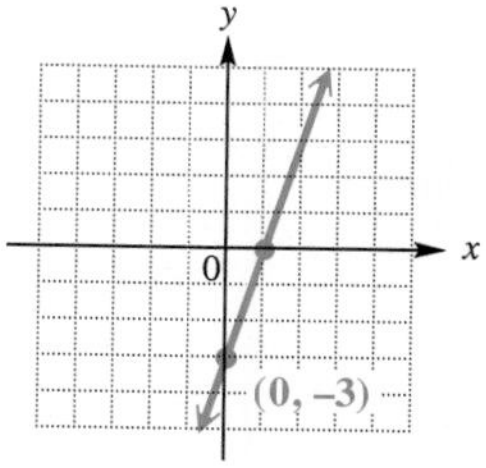

6.

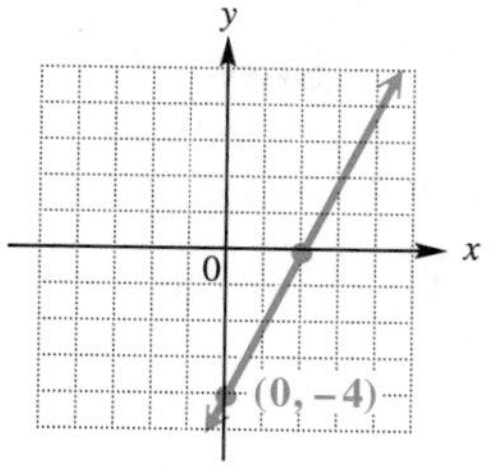

7.

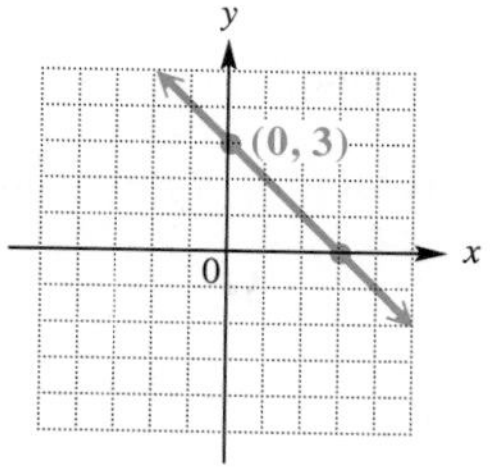

8.

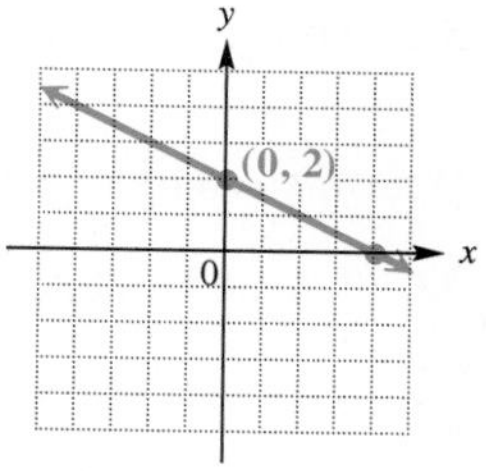

In Exercises 9–16, match each equation with the graph that it most closely resembles in Choices A–H. (Hint: Determining the signs of m and b will help you make your decision.)

9. $y = 2x + 3$

10. $y = -2x + 3$

11. $y = -2x - 3$

12. $y = 2x - 3$

13. $y = 2x$

14. $y = -2x$

15. $y = 3$

16. $y = -3$

A.

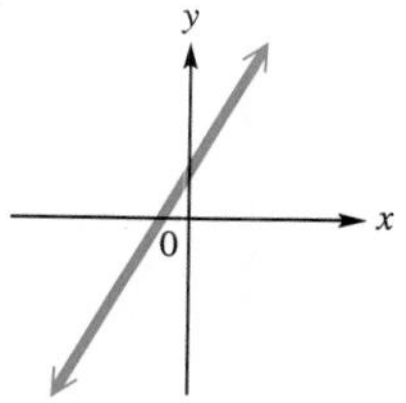

B.

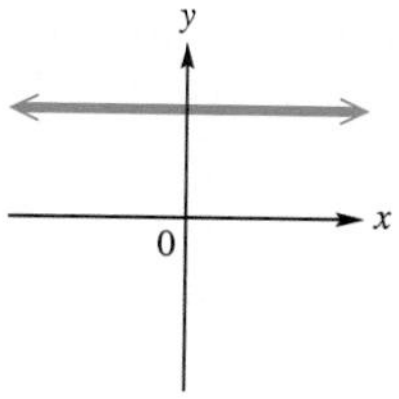

C.

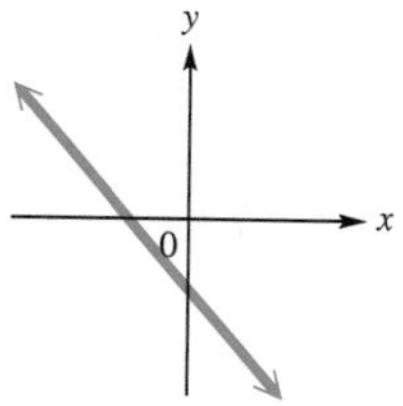

D.

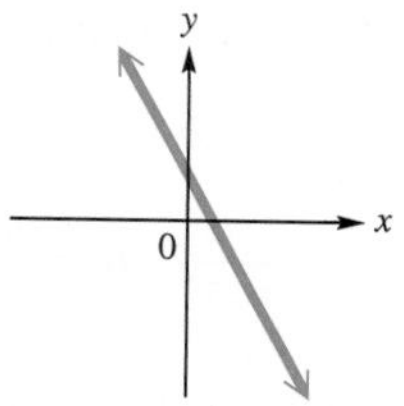

E.

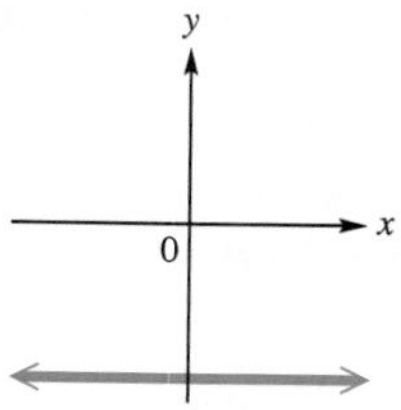

F.

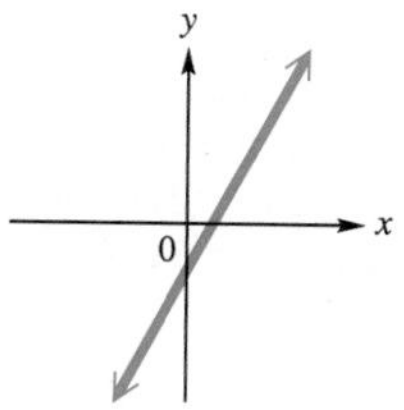

G.

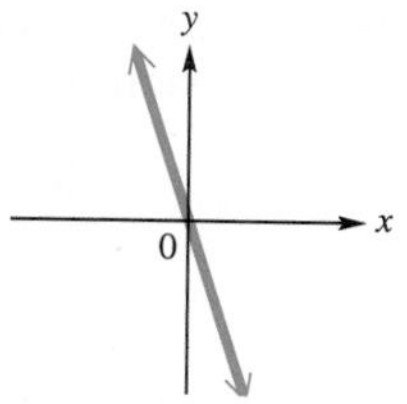

H.

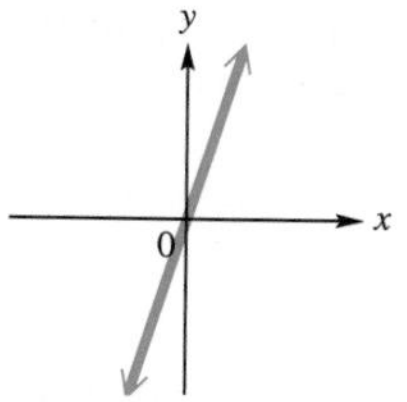

Write the slope-intercept form of the equation of the line satisfying the given conditions.

17. Through $(-2, 4)$; slope $-\frac{3}{4}$

18. Through $(-1, 6)$; slope $-\frac{5}{6}$

19. Through $(5, 8)$; slope -2

20. Through $(12, 10)$; slope 1

21. Through $(-5, 4)$; slope $\frac{1}{2}$

22. Through $(7, -2)$; slope $\frac{1}{4}$

23. x-intercept $(3, 0)$; slope 4

24. x-intercept $(-2, 0)$; slope -5

Write an equation for a line that satisfies the given conditions.

25. Through $(9, 5)$; slope 0

26. Through $(-4, -2)$; slope 0

27. Through $(9, 10)$; undefined slope

28. Through $(-2, 8)$; undefined slope

29. Through (.5, .2); vertical

30. Through $\left(\frac{5}{8}, \frac{2}{9}\right)$; vertical

31. Through $(-7, 8)$; horizontal

32. Through (2, 7); horizontal

Write the equation, in slope-intercept form if possible, of the line passing through the two points.

33. (3, 4) and (5, 8)

34. $(5, -2)$ and $(-3, 14)$

35. (6, 1) and $(-2, 5)$

36. $(-2, 5)$ and $(-8, 1)$

37. $\left(-\frac{2}{5}, \frac{2}{5}\right)$ and $\left(\frac{4}{3}, \frac{2}{3}\right)$

38. $\left(\frac{3}{4}, \frac{8}{3}\right)$ and $\left(\frac{2}{5}, \frac{2}{3}\right)$

39. (2, 5) and (1, 5)

40. $(-2, 2)$ and (4, 2)

41. (7, 6) and $(7, -8)$

42. (13, 5) and $(13, -1)$

43. $(1, -3)$ and $(-1, -3)$

44. $(-4, 6)$ and (5, 6)

Find the equation in slope-intercept form of the line satisfying the given conditions.

45. $m = 5$; $b = 15$

46. $m = -2$; $b = 12$

47. $m = -\frac{2}{3}$; $b = \frac{4}{5}$

48. $m = -\frac{5}{8}$; $b = -\frac{1}{3}$

49. Slope $\frac{2}{5}$; y-intercept (0, 5)

50. Slope $-\frac{3}{4}$; y-intercept (0, 7)

51. Explain why the point-slope form of an equation cannot be used to find the equation of a vertical line.

52. Which one of the following equations is in standard form, according to the definition of standard form given in this text?
A. $3x + 2y - 6 = 0$ **B.** $y = 5x - 12$
C. $2y = 3x + 4$ **D.** $6x - 5y = 12$

For each equation **(a)** *write in slope-intercept form,* **(b)** *give the slope of the line, and* **(c)** *give the y-intercept.*

53. $x + y = 12$

54. $x - y = 14$

55. $5x + 2y = 20$

56. $6x + 5y = 40$

57. $2x - 3y = 10$

58. $4x - 3y = 10$

Write the equation in slope-intercept form of the line satisfying the given conditions.

59. Through (7, 2); parallel to $3x - y = 8$

60. Through (4, 1); parallel to $2x + 5y = 10$

61. Through $(-2, -2)$; parallel to $-x + 2y = 10$

62. Through $(-1, 3)$; parallel to $-x + 3y = 12$

63. Through (8, 5); perpendicular to $2x - y = 7$

64. Through $(2, -7)$; perpendicular to $5x + 2y = 18$

65. Through $(-2, 7)$; perpendicular to $x = 9$

66. Through (8, 4); perpendicular to $x = -3$

Solve each problem.

67. *Private 4-year College Costs* The table lists the average annual cost (in dollars) of tuition and fees at private 4-year colleges for selected years, where year 0 represents 1990, year 5 represents 1995, and so on.

Private College Costs

Year	Cost (in dollars)
0	9391
5	12,432
9	15,380
14	20,082

Source: The College Board.

(a) Plot the four ordered pairs (year, cost). Do the points lie in approximately a straight line?
(b) Use the ordered pairs (0, 9391) and (14, 20,082) to find the equation of a line that approximates the data. Write the equation in slope-intercept form. (Round the slope to the nearest tenth.)
(c) Use the equation from part (b) to estimate the average annual cost at private 4-year colleges in 2008.

68. *Nuclear Waste* The table gives the heavy metal nuclear waste (in thousands of metric tons) from spent reactor fuel stored temporarily at reactor sites, awaiting permanent storage. (*Source*: "Burial of Radioactive Nuclear Waste under the Seabed," *Scientific American*, January 1998, p. 62.)

Heavy Metal Nuclear Waste

Year x	Waste y
1995	32
2000*	42
2010*	61
2020*	76

*Estimates by the U.S. Department of Energy.

Let $x = 0$ represent 1995, $x = 5$ represent 2000 (since $2000 - 1995 = 5$), and so on.
(a) Plot the ordered pairs (x, y). Do the points lie approximately in a line?
(b) Use the ordered pairs (0, 32) and (25, 76) to find the equation of a line that approximates the other ordered pairs. Use the form $y = mx + b$.
(c) Use the equation from part (b) to estimate the amount of nuclear waste in 2008.

69. *U.S. Post Offices* The number of post offices in the United States has been declining. Use the information given on the bar graph for the years 1990 and 2000, letting $x = 0$ represent the year 1990 and y represent the number of post offices.

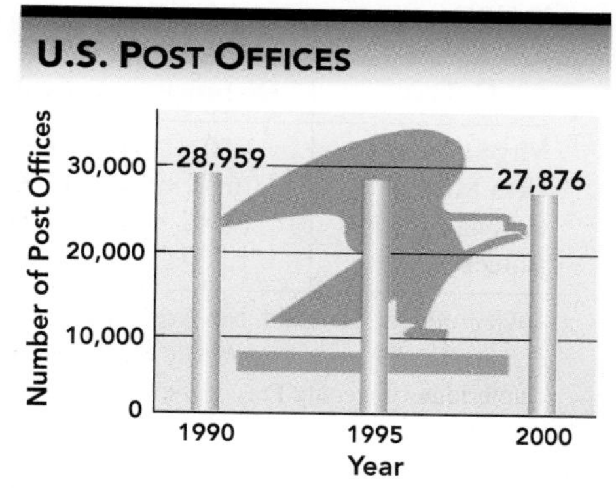

Source: U.S. Postal Service, *Annual Report of the Postmaster General.*

(a) Find a linear equation that models the data. Write it in slope-intercept form.
(b) Predict the number of post offices in 2005 using the model for part (a).

70. *Air Conditioner Choices* The graph shows the recommended air conditioner size (in British thermal units) for selected room sizes in square feet. Use the information given for rooms of 150 square feet and 1400 square feet. Let x represent the number of square feet and y represent the corresponding Btu size air conditioner. Find a linear equation that models the data. Write it in slope-intercept form.

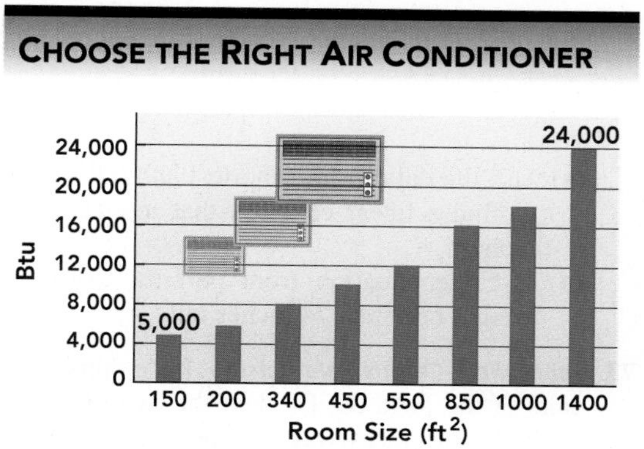

Source: Carey, Morris and James, *Home Improvement for Dummies*, IDG Books.

71. ***Distant Galaxies*** In the late 1920s, the famous observational astronomer Edwin P. Hubble (1889–1953) determined the distances to several galaxies and the velocities at which they were receding from Earth. Four galaxies with their distances in light-years and velocities in miles per second are listed in the table.

Galaxy	Distance	Velocity
Virgo	50	990
Ursa Minor	650	9,300
Corona Borealis	950	15,000
Bootes	1,700	25,000

Source: Sharov, A., and I. Novikov, *Edwin Hubble, the Discoverer of the Big Bang Universe,* Cambridge University Press, 1993.

(a) Let x represent distance and y represent velocity. Use the data for Virgo and Bootes to find an equation of a line that models the data.

(b) If the galaxy Hydra is receding at a speed of 37,000 miles per second, estimate its distance from Earth, using the equation from part (a).

72. ***Heights and Weights of Men*** A sample of 10 adult men gave the following data on their heights and weights:

Height, x (in inches)	Weight, y (in pounds)
61	120
62	140
63	130
65	150
66	142
67	130
68	135
69	175
70	149
72	168

(a) Use the data for the shortest and tallest two men to find a linear equation that models height vs. weight.

(b) Use the equation from part (a) to predict the weight of a man 74 inches tall.

73. ***Fahrenheit-Celsius Relationship*** If we think of ordered pairs of the form (C, F), then the two most common methods of measuring temperature, Celsius and Fahrenheit, can be related as follows: When C = 0, F = 32, and when C = 100, F = 212. This exercise explains how this information is used to find the formula that relates the two temperature scales.

(a) There is a linear relationship between Celsius and Fahrenheit temperatures. When C = 0°, F = ___, and when C = 100°, F = ____.

(b) Think of ordered pairs of temperatures (C, F), where C and F represent corresponding Celsius and Fahrenheit temperatures. The equation that relates the two scales has a straight-line graph that contains the two points determined in part (a). What are these two points?

(c) Find the slope of the line described in part (b).

(d) Think of the point-slope form of the equation in terms of C and F, where C replaces x and F replaces y. Use the slope from part (c) and one of the two points determined earlier to find the equation that gives F in terms of C.

(e) To obtain another form of the formula, use the equation you found in part (d) and solve for C in terms of F.

(f) The equation found in part (d) is graphed on the graphing calculator screen shown here. Observe the display at the bottom, and interpret it in the context of this exercise.

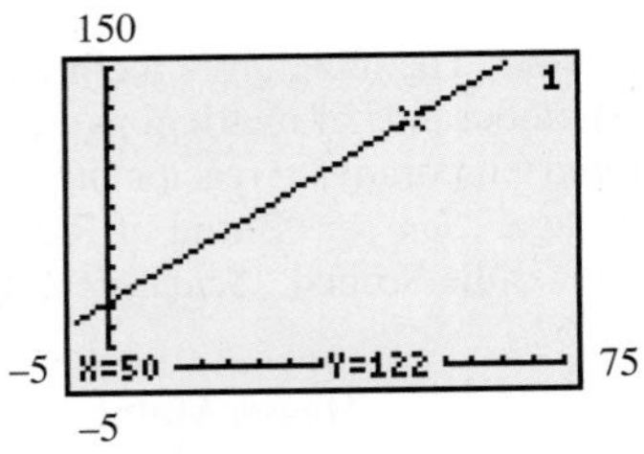

74. A table of points, generated by a graphing calculator, is shown for a line Y_1. Use any two points to find the equation of each line in slope-intercept form.

(a)

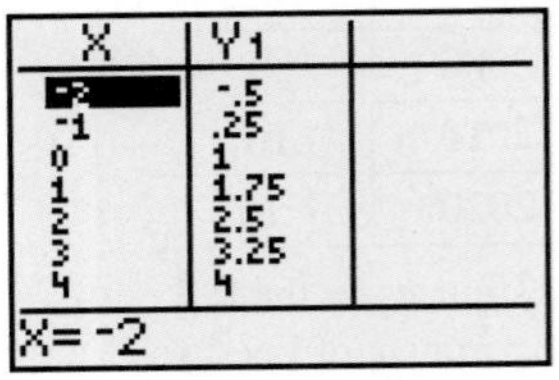

(b)

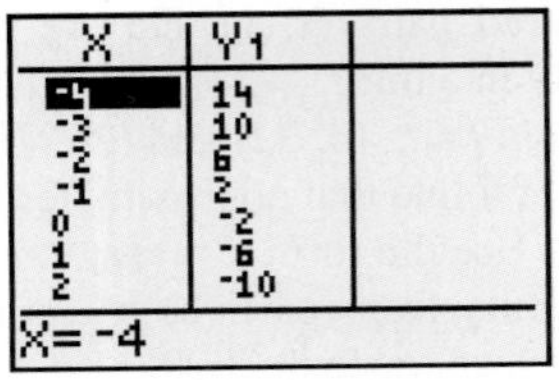

8.4 An Introduction to Functions: Linear Functions, Applications, and Models

Relations and Functions • Domain and Range • Graphs of Relations • Graphs of Functions • Function Notation • Linear Functions • Modeling with Linear Functions

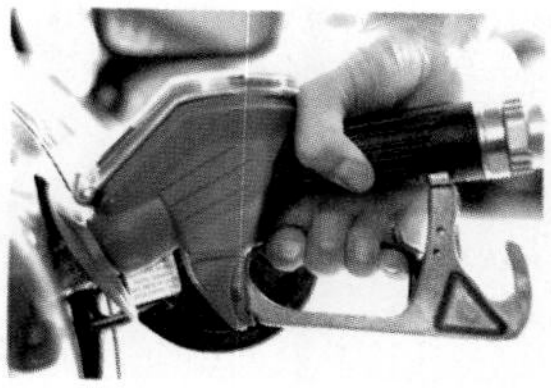

Relations and Functions We often describe one quantity in terms of another; for example, the growth of a plant is related to the amount of light it receives, the demand for a product is related to the price of the product, the cost of a trip is related to the distance traveled, and so on. To represent these corresponding quantities, we can use ordered pairs.

For example, suppose that it is time to fill up your car's tank with gasoline. At your local station, 89-octane gas is selling for \$3.10 per gallon. Experience has taught you that the final price you pay is determined by the number of gallons you buy multiplied by the price per gallon (in this case, \$3.10). As you pump the gas, two sets of numbers spin by: the number of gallons pumped and the price for that number of gallons. Table 2 uses ordered pairs to illustrate this situation.

TABLE 2

Number of Gallons Pumped	Price for This Number of Gallons
0	\$0.00 = 0 (\$3.10)
1	\$3.10 = 1 (\$3.10)
2	\$6.20 = 2 (\$3.10)
3	\$9.30 = 3 (\$3.10)
4	\$12.40 = 4 (\$3.10)

If we let x denote the number of gallons pumped, then the price y in dollars can be found by the linear equation $y = 3.10x$. Theoretically, there are infinitely many ordered pairs (x, y) that satisfy this equation, but in this application we are limited to nonnegative values for x, since we cannot have a negative number of gallons. There also is a practical maximum value for x in this situation, which varies from one car to another. What determines this maximum value?

In this example, the total price depends on the amount of gasoline pumped. For this reason, price is called the *dependent variable*, and the number of gallons is called the *independent variable*. Generalizing, if the value of the variable y depends on the value of the variable x, then y is the **dependent variable** and x the **independent variable.**

Independent variable → x, Dependent variable → y

$$(x, y)$$

Because related quantities can be written using ordered pairs, the concept of *relation* can be defined as follows.

Relation

A **relation** is a set of ordered pairs.

For example, the sets

$$F = \{(1, 2), (-2, 5), (3, -1)\} \quad \text{and} \quad G = \{(-4, 1), (-2, 1), (-2, 0)\}$$

both are relations. A special kind of relation, called a *function*, is very important in mathematics and its applications.

Function

A **function** is a relation in which for each value of the first component of the ordered pairs there is *exactly one value* of the second component.

Of the two examples of a relation just given, only set F is a function, because for each x-value, there is exactly one y-value. In set G, the last two ordered pairs have the same x-value paired with two different y-values, so G is a relation, but not a function.

$$F = \{(1, 2), (-2, 5), (3, -1)\}$$ Function

Different x-values

$$G = \{(-4, 1), (-2, 1), (-2, 0)\}$$ Not a function

Same x-values

In a function, there is exactly one value of the dependent variable, the second component, for each value of the independent variable, the first component.

Another way to think of a functional relationship is to think of the independent variable as an input and the dependent variable as an output. A calculator is an input-output machine, for example. To find 8^2, we must input 8, press the squaring key, and see that the output is 64. Inputs and outputs also can be determined from a graph or a table.

A third way to describe a function is to give a rule that tells how to determine the dependent variable for a specific value of the independent variable. Suppose the rule is given in words as "the dependent variable is twice the independent variable." As an equation, this can be written

$$y = 2x.$$

Dependent variable (y) — Independent variable (x)

EXAMPLE 1 Determining Independent and Dependent Variables

Determine the independent and dependent variables for each of the following functions. Give an example of an ordered pair belonging to the function.

(a) The years and locations of the three summer Olympic Games prior to the 2004 games by the relation {(1992, Barcelona), (1996, Atlanta), (2000, Sydney)}

(b) The procedure by which someone uses a calculator that finds square roots

TABLE 3

U.S. Refined Petroleum Product Imports

Year	Imports (millions of barrels)
1998	731
1999	775
2000	872
2001	928
2002	844

Source: American Petroleum Institute.

(c) The graph in Figure 23 that shows the relationship between the number of gallons of water in a small swimming pool and time in hours

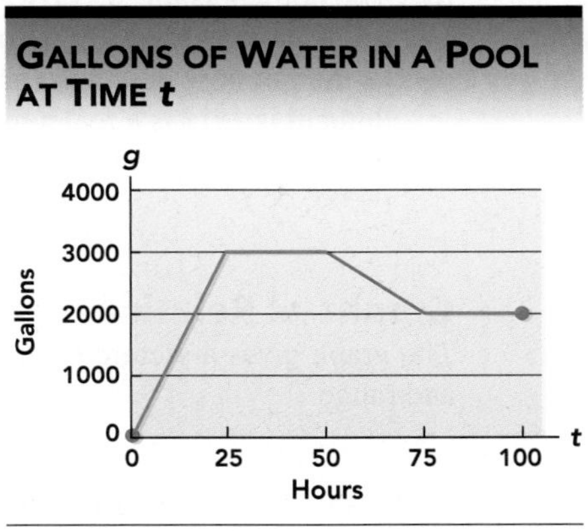

FIGURE 23

(d) The table of petroleum imports shown in Table 3

(e) $y = 3x + 4$

SOLUTION

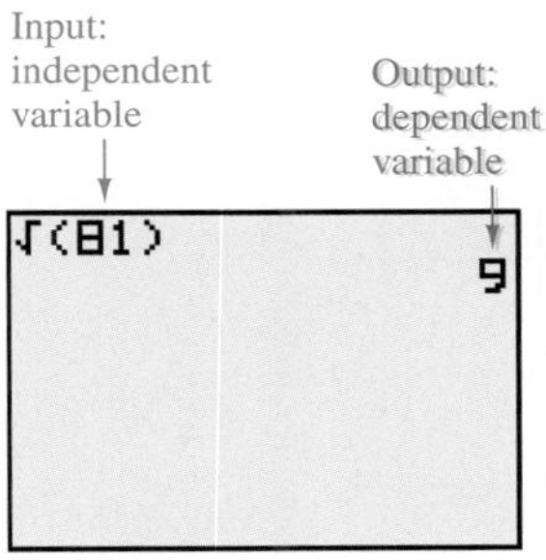

(a) The independent variable (the first component in each ordered pair) is the year. The dependent variable (the second component) is the city. For example, (1996, Atlanta) belongs to this function.

(b) The independent variable (the input) is a nonnegative real number, because the square root of a negative number is not a real number. The dependent variable (the output) is the nonnegative square root. For example, (81, 9) belongs to this function.

(c) The independent variable is time, in hours, and the dependent variable is the number of the gallons of water in the pool. One ordered pair is (25, 3000).

(d) The independent variable is the year and the dependent variable is the number of millions of barrels. An example of an ordered pair is (2000, 872).

(e) The independent variable is x, and the dependent variable is y. One ordered pair is (1, 7).

Domain and Range

Domain and Range

In a relation, the set of all values of the independent variable (x) is the **domain.** The set of all values of the dependent variable (y) is the **range.**

EXAMPLE 2 Determining Domain and Range

Give the domain and range of each function in Example 1.

SOLUTION

(a) The domain is the set of years, {1992, 1996, 2000}, and the range is the set of cities, {Barcelona, Atlanta, Sydney}.

(b) The domain is restricted to nonnegative numbers: $[0, \infty)$. The range also is $[0, \infty)$.

(c) The domain is all possible values of t, the time in hours, which is the interval [0, 100]. The range is the number of gallons at time t, the interval [0, 3000].

(d) The domain is the set of years, {1998, 1999, 2000, 2001, 2002}. The range is the set of import values (in millions of barrels), {731, 775, 872, 928, 844}.

(e) In the defining equation (or rule), $y = 3x + 4$, x can be any real number, so the domain is $\{x|x \text{ is a real number}\}$, or $(-\infty, \infty)$. Because every real number y can be produced by some value of x, the range also is the set $\{y|y \text{ is a real number}\}$, or $(-\infty, \infty)$.

Graphs of Relations

The **graph of a relation** is the graph of its ordered pairs. The graph gives a picture of the relation, which can be used to determine its domain and range.

EXAMPLE 3 Determining Domain and Range

Give the domain and range of each relation.

(a)

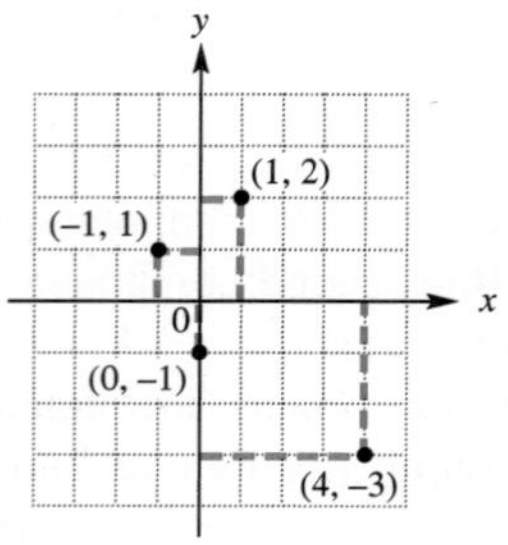

(b)

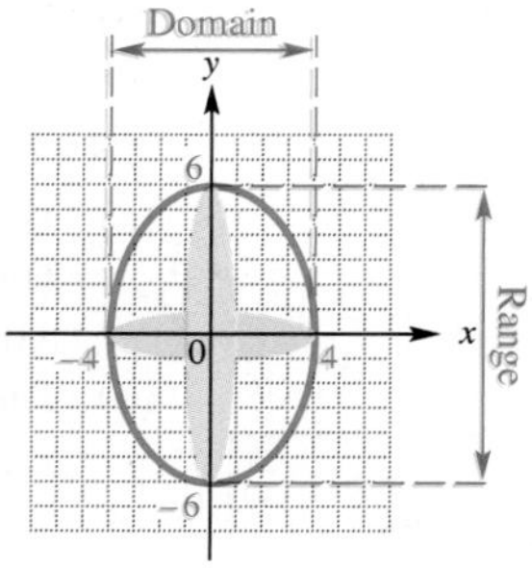

(c)

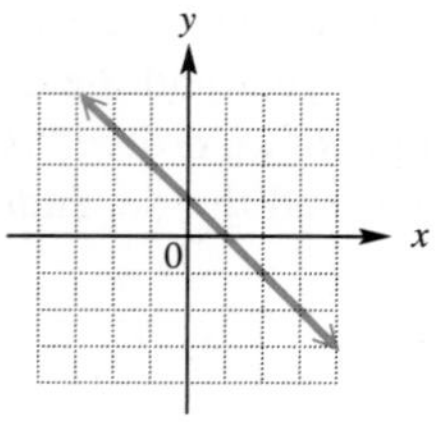

(d)

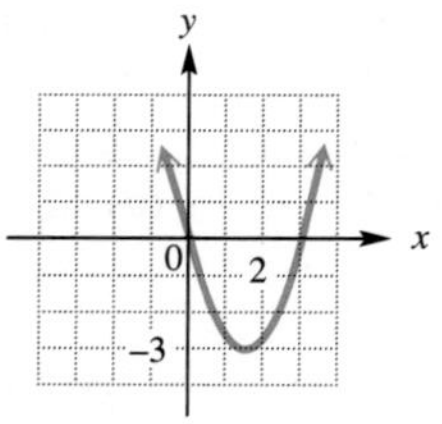

SOLUTION

(a) The domain is the set of x-values, $\{-1, 0, 1, 4\}$, and the range is the set of y-values, $\{-3, -1, 1, 2\}$.

(b) The x-values of the points on the graph include all numbers between -4 and 4, inclusive. The y-values include all numbers between -6 and 6, inclusive. Using interval notation, the domain is $[-4, 4]$ and the range is $[-6, 6]$.

(c) The arrowheads indicate that the line extends indefinitely left and right, as well as up and down. Therefore, both the domain and the range are the set of all real numbers, written $(-\infty, \infty)$.

(d) The arrowheads indicate that the graph extends indefinitely left and right, as well as upward. The domain is $(-\infty, \infty)$. Because there is a least y-value, -3, the range includes all real numbers greater than or equal to -3, written $[-3, \infty)$.

We have seen that relations can be defined by equations, such as $y = 2x + 3$ and $y^2 = x$. It is sometimes necessary to determine the domain of a relation from its equation. In this book, the following agreement on the domain of a relation is assumed.

Agreement on Domain

The domain of a relation is assumed to be all real numbers that produce real numbers when substituted for the independent variable.

To illustrate this agreement, because any real number can be used as a replacement for x in $y = 2x + 3$, the domain of this function is the set of real numbers. As another example, the function defined by $y = \frac{1}{x}$ has all real numbers except 0 as domain, because y is undefined only if $x = 0$. In general, the domain of a function defined by an algebraic expression is all real numbers, except those numbers that lead to division by 0 or an even root of a negative number.

Graphs of Functions

Most of the relations we have seen in the examples are functions—that is, each x-value corresponds to exactly one y-value. Now we look at ways to determine whether a given relation, defined algebraically, is a function.

In a function each value of x leads to only one value of y, so any vertical line drawn through the graph of a function must intersect the graph in at most one point. This is the **vertical line test for a function.**

Vertical Line Test

If a vertical line intersects the graph of a relation in more than one point, then the relation is not a function.

For example, the graph shown in Figure 24(a) is not the graph of a function, since a vertical line can intersect the graph in more than one point, while the graph in Figure 24(b) does represent a function.

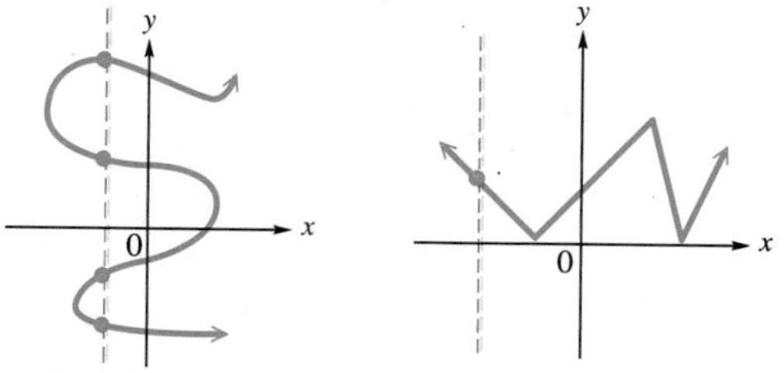

Not a function — the same x-value corresponds to four different y-values

(a)

Function — each x-value corresponds to only one y-value

(b)

FIGURE 24

The vertical line test is a simple method for identifying a function defined by a graph. It is more difficult to decide whether a relation defined by an equation is a function. The next example gives some hints that may help.

EXAMPLE 4 Determining Whether a Relation is a Function

Decide whether each equation defines a function, and give the domain.

(a) $y = \sqrt{2x - 1}$ **(b)** $y^2 = x$

(c) $y \leq x - 1$ **(d)** $y = \dfrac{5}{x - 1}$

SOLUTION

(a) In the equation $y = \sqrt{2x - 1}$, for any choice of x in the domain, there is exactly one corresponding value for y (the radical is a nonnegative number). Thus, this equation defines a function. Because the radicand cannot be negative,

$$2x - 1 \geq 0$$

$$2x \geq 1 \quad \text{Add 1.}$$

$$x \geq \frac{1}{2}. \quad \text{Divide by 2.}$$

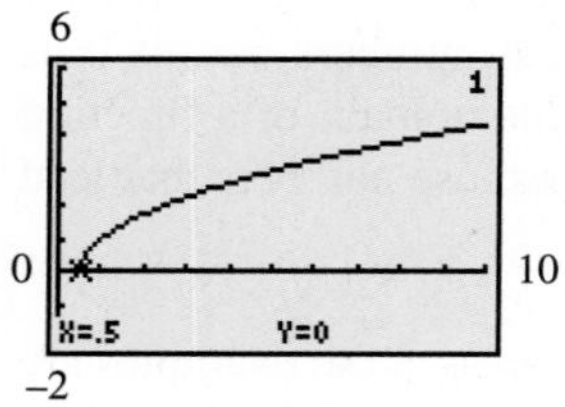

The graph of $y = \sqrt{2x - 1}$ supports the result in Example 4(a). The domain is $\left[\frac{1}{2}, \infty\right)$.

The domain is $\left[\frac{1}{2}, \infty\right)$.

(b) The ordered pairs (16, 4) and (16, −4) both satisfy the equation $y^2 = x$. Since one value of x, 16, corresponds to two values of y, 4 and −4, this equation does not define a function. Solving $y^2 = x$ for y gives

$$y = \sqrt{x} \quad \text{or} \quad y = -\sqrt{x},$$

which shows that two values of y correspond to each positive value of x. Because x is equal to the square of y, the values of x must always be nonnegative. The domain of the relation is $[0, \infty)$.

(c) By definition, y is a function of x if every value of x leads to exactly one value of y. In the inequality

$$y \leq x - 1,$$

a particular value of x, say 1, corresponds to many values of y. The ordered pairs (1, 0), (1, −1), (1, −2), (1, −3), and so on, all satisfy the inequality. For this reason, the inequality does not define a function. Any number can be used for x, so the domain is the set of real numbers $(-\infty, \infty)$.

(d) For the equation $y = \frac{5}{x - 1}$, given any value of x in the domain, we find y by subtracting 1, then dividing the result into 5. This process produces exactly one value of y for each value in the domain, so this equation defines a function. The domain includes all real numbers except those that make the denominator 0. We find these numbers by setting the denominator equal to 0 and solving for x.

$$x - 1 = 0$$

$$x = 1$$

Thus, the domain includes all real numbers except 1. In interval notation this is written as $(-\infty, 1) \cup (1, \infty)$. ■

In summary, three variations of the definition of function are given here.

Variations of the Definition of Function

1. A **function** is a relation in which for each value of the first component of the ordered pairs there is exactly one value of the second component.
2. A **function** is a set of distinct ordered pairs in which no first component is repeated.
3. A **function** is a rule or correspondence that assigns exactly one range value to each domain value.

Function Notation When a function f is defined with a rule or an equation using x and y for the independent and dependent variables, we say "y is a function of x" to emphasize that y *depends on* x. We use the notation

$$y = f(x),$$

called **function notation,** to express this and read $f(x)$ as "f of x." (In this notation the parentheses do not indicate multiplication.) For example, if $y = 2x - 7$, we write

$$f(x) = 2x - 7.$$

Do not read f(x) as "f times x."

Note that $f(x)$ is just another name for the dependent variable y. For example, if $y = f(x) = 9x - 5$, and $x = 2$, then we find y, or $f(2)$, by replacing x with 2.

$$y = f(2)$$
$$y = 9 \cdot 2 - 5 = 18 - 5 = 13.$$

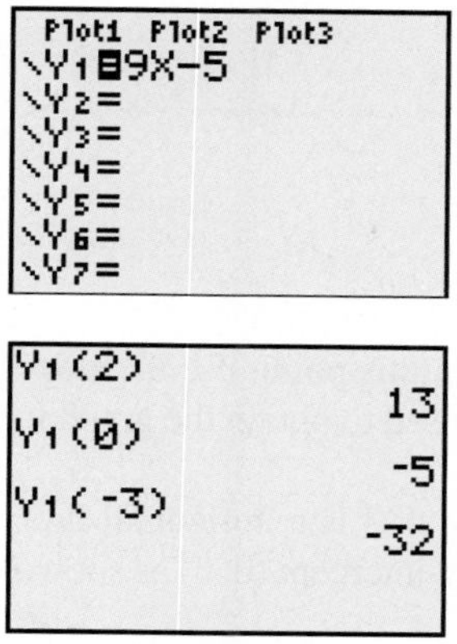

For $Y_1 = 9X - 5$, function notation capability of the TI-83/84 Plus supports the discussion here.

The statement "if $x = 2$, then $y = 13$" is abbreviated with function notation as

$$f(2) = 13.$$

Read $f(2)$ as "f of 2" or "f at 2." Also,

$$f(0) = 9 \cdot 0 - 5 = -5, \quad \text{and} \quad f(-3) = 9(-3) - 5 = -32.$$

These ideas and the symbols used to represent them can be explained as follows.

$$y = f(x) = 9x - 5$$

Name of the function

Defining expression

Value of the function

Name of the independent variable

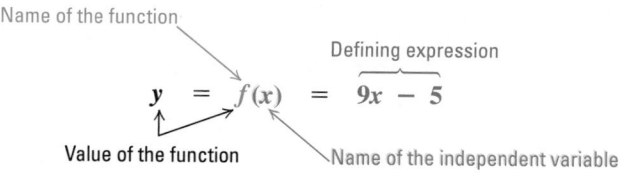

EXAMPLE 5 Using Function Notation

Let $f(x) = -x^2 + 5x - 3$. Find the following.

(a) $f(2)$ **(b)** $f(-1)$ **(c)** $f(2x)$

SOLUTION

(a)
$$f(x) = -x^2 + 5x - 3$$
$$f(2) = -2^2 + 5 \cdot 2 - 3 \quad \text{Replace } x \text{ with 2.}$$
$$= -4 + 10 - 3$$
$$= 3$$

-2^2 means $-(2^2) = -4$.

(b) $f(-1) = -(-1)^2 + 5(-1) - 3$
$= -1 - 5 - 3 = -9$

(c) $f(2x) = -(2x)^2 + 5(2x) - 3$ Replace x with $2x$.
$= -4x^2 + 10x - 3$

Linear Functions An important type of elementary function is the *linear function.*

Linear Function

A function that can be written in the form

$$f(x) = mx + b$$

for real numbers m and b is a **linear function.**

Notice that the form $f(x) = mx + b$ defining a linear function is the same as that of the slope-intercept form of the equation of a line, first seen in the previous section. We know that the graph of $f(x) = mx + b$ will be a line with slope m and y-intercept $(0, b)$.

EXAMPLE 6 Graphing Linear Functions

Graph each linear function.

(a) $f(x) = -2x + 3$ **(b)** $f(x) = 3$

SOLUTION

(a) To graph the function, locate the y-intercept, $(0, 3)$. From this point, use the slope $-2 = \frac{-2}{1}$ to go down 2 and right 1. This second point is used to obtain the graph in Figure 25(a).

(b) From the previous section, we know that the graph of $y = 3$ is a horizontal line. Therefore, the graph of $f(x) = 3$ is a horizontal line with y-intercept $(0, 3)$ as shown in Figure 25(b).

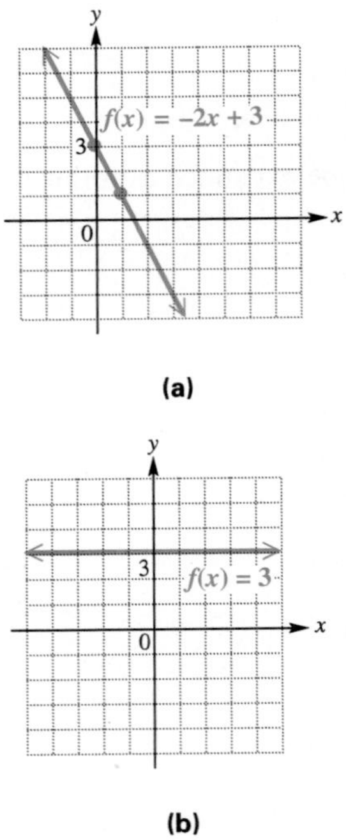

FIGURE 25

The function defined in Example 6(b) and graphed in Figure 25(b) is an example of a constant function. A **constant function** is a linear function of the form $f(x) = b$, where b is a real number. The domain of any linear function is $(-\infty, \infty)$. The range of a nonconstant linear function (like in Example 6(a)) is also $(-\infty, \infty)$, while the range of the constant function $f(x) = b$ is $\{b\}$.

Modeling with Linear Functions A company's cost of producing a product and the revenue from selling the product can be expressed as linear functions. The idea of **break-even analysis** then can be explained using the graphs of these functions. When cost is greater than revenue earned, the company loses money. When cost is less than revenue the company makes money, and when cost equals revenue the company breaks even.

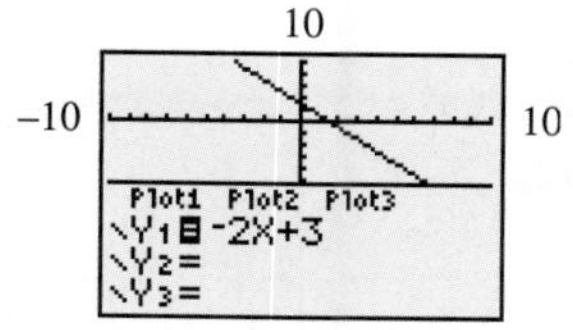

Compare with Example 6(a) and Figure 25(a).

EXAMPLE 7 Analyzing Cost, Revenue, and Profit

Peripheral Visions, Inc., produces studio quality DVDs of live concerts. The company places an ad in a trade newsletter. The cost of the ad is \$100. Each DVD costs \$20 to produce, and the company charges \$24 per disk.

(a) Express the cost C as a function of x, the number of DVDs produced.

(b) Express the revenue R as a function of x, the number of DVDs sold.

(c) When will the company break even? That is, for what value of x does revenue equal cost?

(d) Graph the cost and revenue functions on the same coordinate system, and interpret the graph.

SOLUTION

(a) The *fixed cost* is \$100, and for each DVD produced, the *variable cost* is \$20. Therefore, the cost C can be expressed as a function of x, the number of DVDs produced:

$$C(x) = 20x + 100 \quad (C \text{ in dollars}).$$

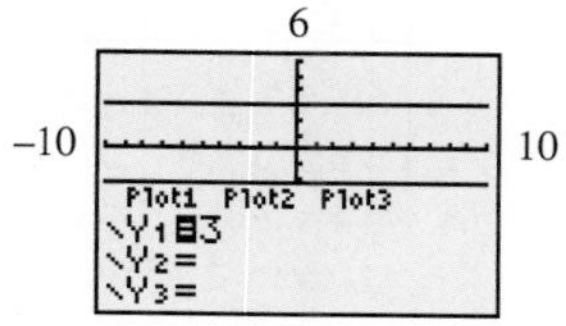

Compare with Example 6(b) and Figure 25(b).

(b) Each DVD sells for \$24, so the revenue R is given by

$$R(x) = 24x \quad (R \text{ in dollars}).$$

(c) The company will just break even (no profit and no loss) as long as revenue just equals cost, or $R(x) = C(x)$. This is true whenever

$$\begin{aligned} R(x) &= C(x) \\ 24x &= 20x + 100 && \text{Substitute for } R(x) \text{ and } C(x). \\ 4x &= 100 && \text{Subtract } 20x. \\ x &= 25. && \text{Divide by 4.} \end{aligned}$$

If 25 DVDs are produced and sold, the company will break even.

(d) Figure 26 shows the graphs of the two functions. At the break-even point, we see that when 25 DVDs are produced and sold, both the cost and the revenue are \$600. If fewer than 25 DVDs are produced and sold (that is, when $x < 25$), the company loses money. When more than 25 DVDs are produced and sold (that is, when $x > 25$), there is a profit.

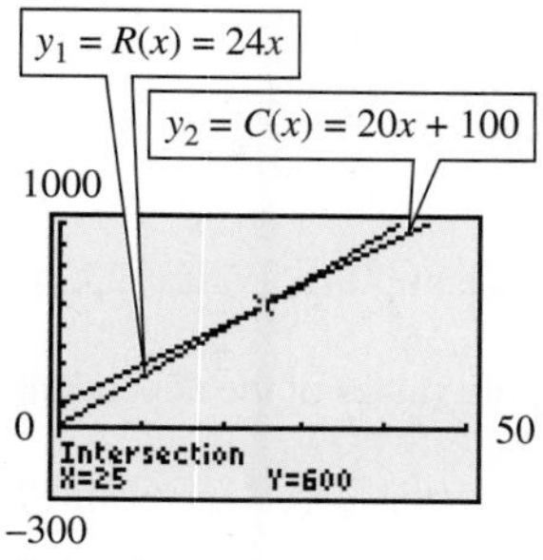

The break-even point is (25, 600), as indicated at the bottom of the screen. The calculator can find the point of intersection of the graphs. Compare with Figure 26.

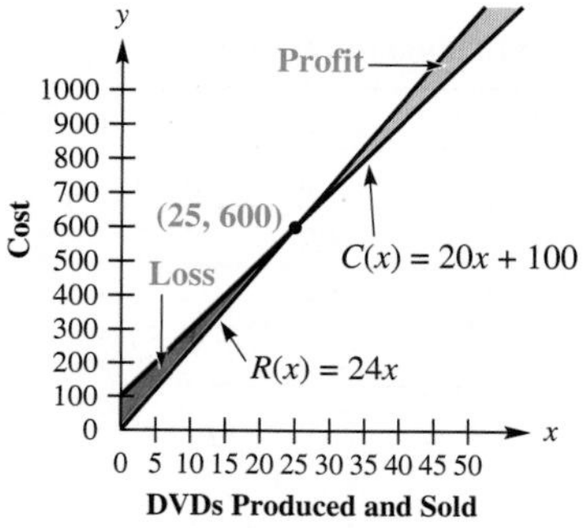

FIGURE 26

8.4 EXERCISES

1. In your own words, define *function* and give an example.

2. In your own words, define *domain of a function* and give an example.

3. In an ordered pair of a relation, is the first element a value of the independent or the dependent variable?

For each relation, decide whether it is a function, and give the domain and range.

4. $\{(1, 1), (1, -1), (2, 4), (2, -4), (3, 9), (3, -9)\}$

5. $\{(2, 5), (3, 7), (4, 9), (5, 11)\}$

6. The set containing certain countries and their predicted life expectancy estimates for persons born in 2050 is {(United States, 83.9), (Japan, 90.91), (Canada, 85.26), (Britain, 83.79), (France, 87.01), (Germany, 83.12), (Italy, 82.26)}. (*Source:* Shripad Tuljapurkar, Mountain View Research, Los Altos, California.)

7. An input–output machine accepts positive real numbers as input, and outputs both their positive and negative square roots.

8.

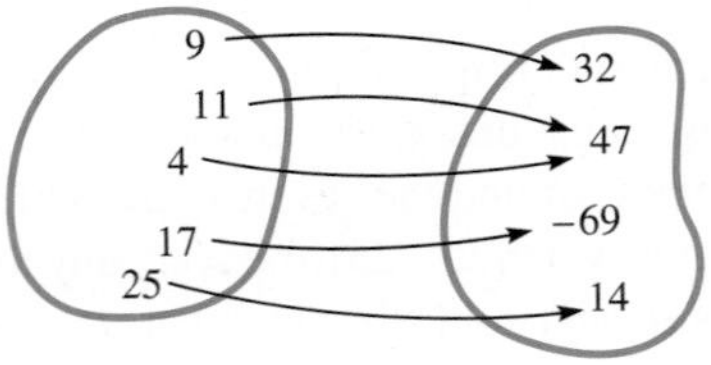

9. **U.S. Voting-Age Population in 2000 (in millions)**

Hispanic	21.3
Native American	1.6
Asian American	8.2
African American	24.6
White	152.0

Source: U.S. Bureau of the Census.

10. $\{(x, y) \mid x = |y|\}$

11.

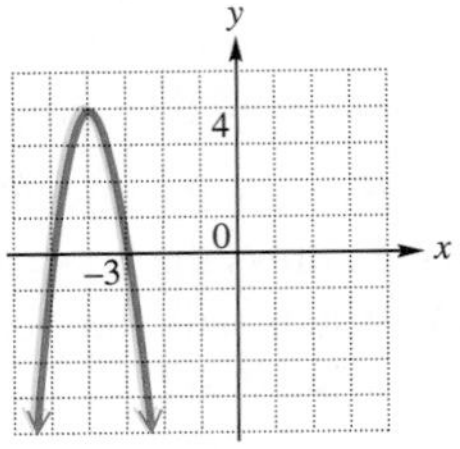

12.

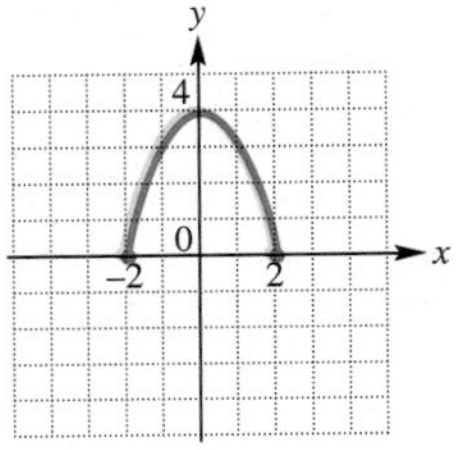

13.

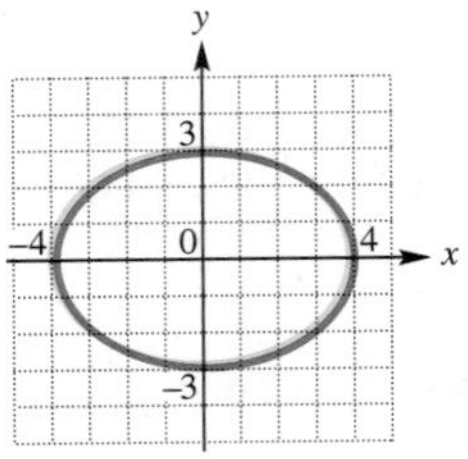

14.

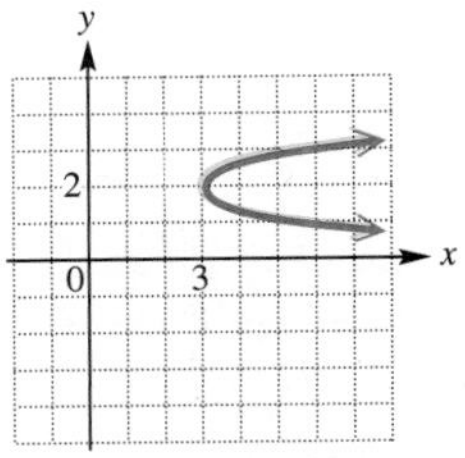

Decide whether the given relation defines y as a function of x. Give the domain.

15. $y = x^2$

16. $y = x^3$

17. $x = y^2$

18. $x = y^4$

19. $x + y < 4$

20. $x - y < 3$

21. $y = \sqrt{x}$

22. $y = -\sqrt{x}$

23. $xy = 1$

24. $xy = -3$

25. $y = \sqrt{4x + 2}$

26. $y = \sqrt{9 - 2x}$

27. $y = \dfrac{2}{x - 9}$

28. $y = \dfrac{-7}{x - 16}$

29. ***Pool Water Level*** Refer to Example 1, Figure 23, to answer the questions.
   - **(a)** What numbers are possible values of the dependent variable?
   - **(b)** For how long is the water level increasing? Decreasing?
   - **(c)** How many gallons are in the pool after 90 hours?
   - **(d)** Call this function g. What is $g(0)$? What does it mean in this example?

30. ***Electricity Consumption*** The graph shows the megawatts of electricity used on a record-breaking summer day in Sacramento, California.

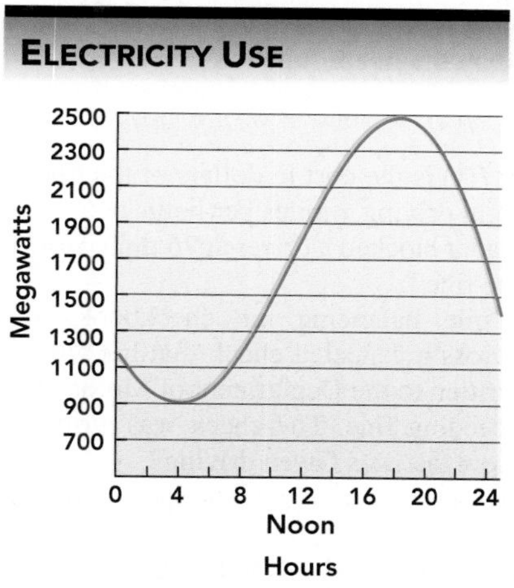

Source: Sacramento Municipal Utility District.

(a) Is this the graph of a function?
(b) What is the domain?
(c) Estimate the number of megawatts of electricity use at 8 A.M.
(d) At what time was the most electricity used? The least electricity?

31. Give an example of a function from everyday life. (*Hint:* Fill in the blanks: ______ depends on ______, so ______ is a function of ______.)

32. Choose the correct response. The notation $f(3)$ means

A. the variable f times 3, or $3f$.
B. the value of the dependent variable when the independent variable is 3.
C. the value of the independent variable when the dependent variable is 3.
D. f equals 3.

Let $f(x) = 3 + 2x$ and $g(x) = x^2 - 2$. Find each function value.

33. $f(1)$ **34.** $f(4)$ **35.** $g(2)$

36. $g(0)$ **37.** $g(-1)$ **38.** $g(-3)$

39. $f(-8)$ **40.** $f(-5)$

Sketch the graph of each linear function. Give the domain and range.

41. $f(x) = -2x + 5$ **42.** $g(x) = 4x - 1$

43. $h(x) = \frac{1}{2}x + 2$ **44.** $F(x) = -\frac{1}{4}x + 1$

45. $G(x) = 2x$ **46.** $H(x) = -3x$

47. $f(x) = 5$ **48.** $g(x) = -4$

An equation that defines y as a function of x is given. **(a)** *Solve for y in terms of x, and replace y with the function notation f(x).* **(b)** *Find f(3).*

49. $y + 2x^2 = 3$ **50.** $y - 3x^2 = 2$

51. $4x - 3y = 8$ **52.** $-2x + 5y = 9$

53. Fill in the blanks with the correct responses. The equation $2x + y = 4$ has a straight ____ as its graph. One point that lies on the line is (3, ____). If we solve the equation for y and use function notation, we have a linear function $f(x) =$ ________. For this function, $f(3) =$ ____, meaning that the point (____, ____) lies on the graph of the function.

54. Which one of the following defines a linear function?

A. $y = \frac{x - 5}{4}$ **B.** $y = \frac{1}{x}$

C. $y = x^2$ **D.** $y = \sqrt{x}$

55. ***Taxi Fares***

(a) Suppose that a taxicab driver charges \$1.50 per mile. Fill in the chart with the correct response for the price $f(x)$ she charges for a trip of x miles.

x	$f(x)$
0	
1	
2	
3	

(b) The linear function that gives a rule for the amount charged is $f(x) =$ ______.
(c) Graph this function for the domain $\{0, 1, 2, 3\}$.

56. ***Cost to Mail a Package*** Suppose that a package weighing x pounds costs $f(x)$ dollars to mail to a given location, where $f(x) = 2.75x$.

(a) What is the value of $f(3)$?

(b) Describe what 3 and the value $f(3)$ mean in part (a), using the terminology *independent variable* and *dependent variable*.

(c) How much would it cost to mail a 5-lb package? Write the answer using function notation.

57. ***Forensic Studies*** Forensic scientists use the lengths of the tibia (t), the bone from the ankle to the knee, and the femur (r), the bone from the knee to the hip socket, to calculate the height of a person. A person's height (h) is determined from the lengths of these bones using functions defined by the following formulas. All measurements are in centimeters.

For men:

$$h(r) = 69.09 + 2.24r$$

or $$h(t) = 81.69 + 2.39t$$

For women:

$$h(r) = 61.41 + 2.32r$$

or $$h(t) = 72.57 + 2.53t$$

(a) Find the height of a man with a femur measuring 56 centimeters.

(b) Find the height of a man with a tibia measuring 40 centimeters.

(c) Find the height of a woman with a femur measuring 50 centimeters.

(d) Find the height of a woman with a tibia measuring 36 centimeters.

58. ***Pool Size for Sea Otters*** Federal regulations set standards for the size of the quarters of marine mammals. A pool to house sea otters must have a volume of "the square of the sea otter's average adult length (in meters) multiplied by 3.14 and by .91 meter." If x represents the sea otter's average adult length and $f(x)$ represents the volume of the corresponding pool size, this formula can be written as $f(x) = (.91)(3.14)x^2$.

Find the volume of the pool for each of the following adult lengths (in meters). Round answers to the nearest hundredth.

(a) .8
(b) 1.0
(c) 1.2
(d) 1.5

59. ***Speeding Fines*** Suppose that speeding fines are determined by the linear function

$$f(x) = 10(x - 65) + 50, \quad x > 65,$$

where $f(x)$ is the cost in dollars of the fine if a person is caught driving x miles per hour.

(a) Radar clocked a driver at 76 mph. How much was the fine?

(b) While balancing his checkbook, Johnny ran across a canceled check that his wife Gwen had written to the Department of Motor Vehicles for a speeding fine. The check was written for \$100. How fast was Gwen driving?

(c) At what whole-number speed are tickets first given?

(d) For what speeds is the fine greater than \$200?

60. ***Expansion and Contraction of Gases*** In 1787, Jacques Charles noticed that gases expand when heated and contract when cooled. Suppose that a particular gas follows the model

$$f(x) = \frac{5}{3}x + 455,$$

where x is the temperature in Celsius and $f(x)$ is the volume in cubic centimeters. (*Source:* Bushaw, D., et al., *A Sourcebook of Applications of School Mathematics*, MAA, 1980. Reprinted with permission.)

(a) What is the volume when the temperature is 27°C?

(b) What is the temperature when the volume is 605 cubic centimeters?

(c) Determine what temperature gives a volume of 0 cubic centimeters (that is, absolute zero, or the coldest possible temperature).

Cost and Revenue Models *In each of the following,* **(a)** *express the cost C as a function of x, where x represents the quantity of items as given;* **(b)** *express the revenue R as a function of x;* **(c)** *determine the value of x for which revenue equals cost;* **(d)** *graph $y = C(x)$ and $y = R(x)$ on the same axes, and interpret the graph.*

61. Perian Herring stuffs envelopes for extra income during her spare time. Her initial cost to obtain the necessary information for the job was \$200.00. Each envelope costs \$.02 and she gets paid \$.04 per envelope stuffed. Let x represent the number of envelopes stuffed.

62. Brent Labatut runs a copying service in his home. He paid \$3500 for the copier and a lifetime service contract. Each sheet of paper he uses costs \$.01, and he gets paid \$.05 per copy he makes. Let x represent the number of copies he makes.

63. Roy Pollina operates a delivery service in a southern city. His start-up costs amounted to \$2300. He estimates that it costs him (in terms of gasoline, wear and tear on his car, etc.) \$3.00 per delivery. He charges \$5.50 per delivery. Let x represent the number of deliveries he makes.

64. Katie Simon bakes cakes and sells them at county fairs. Her initial cost for the St. Charles Parish fair this year was \$40.00. She figures that each cake costs \$2.50 to make, and she charges \$6.50 per cake. Let x represent the number of cakes sold. (Assume that there were no cakes left over.)

8.5 Quadratic Functions, Graphs, and Models

Quadratic Functions and Parabolas • Graphs of Quadratic Functions • Vertex of a Parabola • General Graphing Guidelines • A Model for Optimization

Quadratic Functions and Parabolas

In the previous section, we discussed linear functions, those that are defined by first-degree polynomials. We now look at *quadratic functions,* those defined by second-degree polynomials.

Quadratic Function

A function f is a **quadratic function** if

$$f(x) = ax^2 + bx + c,$$

where a, b, and c are real numbers, with $a \neq 0$.

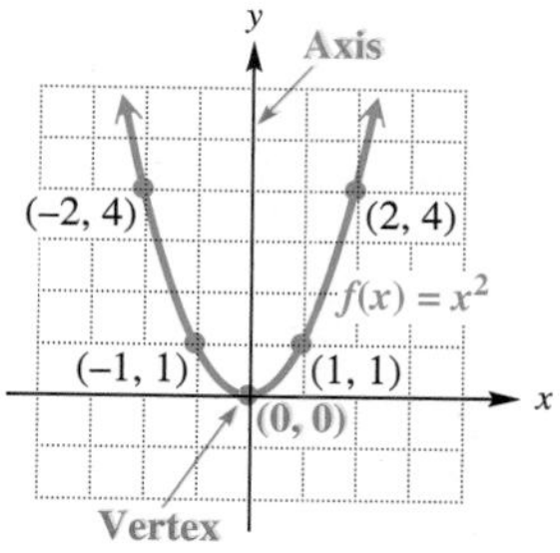

FIGURE 27

The simplest quadratic function is defined by $f(x) = x^2$. This function can be graphed by finding several ordered pairs that satisfy the equation: for example, (0, 0), (1, 1), (−1, 1), (2, 4), (−2, 4), $\left(\frac{1}{2}, \frac{1}{4}\right)$, $\left(-\frac{1}{2}, \frac{1}{4}\right)$, $\left(\frac{3}{2}, \frac{9}{4}\right)$, and $\left(-\frac{3}{2}, \frac{9}{4}\right)$. Plotting these points and drawing a smooth curve through them gives the graph shown in Figure 27. This graph is called a **parabola.** Every quadratic function has a graph that is a parabola.

Parabolas are symmetric about a line (the y-axis in Figure 27.) Intuitively, this means that if the graph were folded along the line of symmetry, the two sides would coincide. The line of symmetry for a parabola is called the **axis** of the parabola. The point where the axis intersects the parabola is the **vertex** of the parabola. The vertex is the lowest (or highest) point of a vertical parabola.

Parabolas have many practical applications. For example, the reflectors of solar ovens and flashlights are made by revolving a parabola about its axis. The **focus** of a parabola is a point on its axis that determines the curvature. See Figure 28. When the parabolic reflector of a solar oven is aimed at the sun, the light rays bounce off the reflector and collect at the focus, creating intense heat at that point. In contrast, when a lightbulb is placed at the focus of a parabolic reflector, light rays reflect out parallel to the axis.

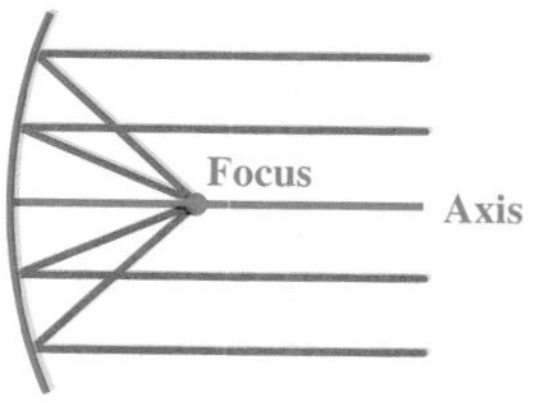

Parabolic reflector

FIGURE 28

Graphs of Quadratic Functions

The first example shows how the constant a affects the graph of a function of the form $g(x) = ax^2$.

EXAMPLE 1 Graphing Quadratic Functions ($g(x) = ax^2$)

Graph the functions defined as follows.

(a) $g(x) = -x^2$ **(b)** $g(x) = \frac{1}{2}x^2$

SOLUTION

(a) For a given value of x, the corresponding value of $g(x)$ will be the negative of what it was for $f(x) = x^2$. (See the table of values with Figure 29(a).) Because of this, the graph of $g(x) = -x^2$ is the same shape as that of $f(x) = x^2$, but opens downward. See Figure 29(a). This is generally true; the graph of $f(x) = ax^2 + bx + c$ opens downward whenever $a < 0$.

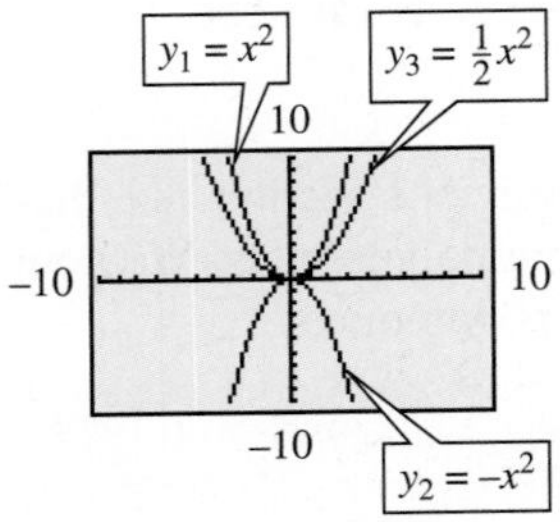

The screen illustrates the three graphs considered in Example 1 and Figures 29(a) and 29(b).

x	y
−2	−4
−1	−1
0	0
1	−1
2	−4

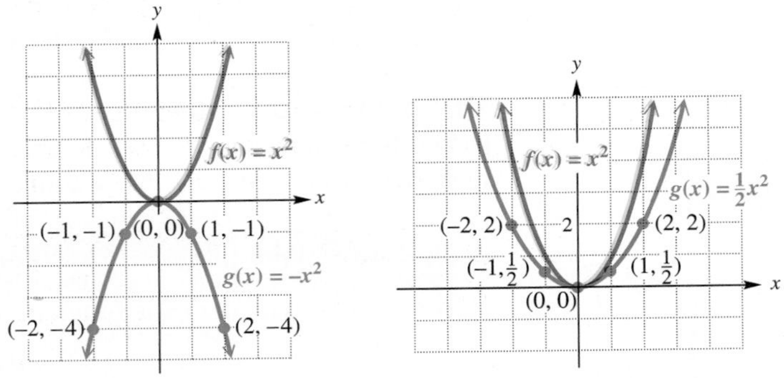

(a) (b)

FIGURE 29

(b) Choose a value of x, and then find $g(x)$. The coefficient $\frac{1}{2}$ will cause the resulting value of $g(x)$ to be less than that of $f(x) = x^2$, making the parabola wider than the graph of $f(x) = x^2$. See Figure 29(b). In both parabolas of this example, the axis is the vertical line $x = 0$ and the vertex is the origin (0, 0). ■

The next few examples show the results of horizontal and vertical shifts, called **translations,** of the graph of $f(x) = x^2$.

EXAMPLE 2 Graphing a Quadratic Function (Vertical Shift)

Graph $g(x) = x^2 - 4$.

SOLUTION

By comparing the tables of values for $g(x) = x^2 - 4$ and $f(x) = x^2$ shown with Figure 30 on the next page, we can see that for corresponding x-values, the y-values of g are each 4 less than those for f. This leads to a *vertical shift*. Thus, the graph of $g(x) = x^2 - 4$ is the same as that of $f(x) = x^2$, but translated 4 units down. See Figure 30. The vertex of this parabola (here the lowest point) is at (0, −4). The axis of the parabola is the vertical line $x = 0$.

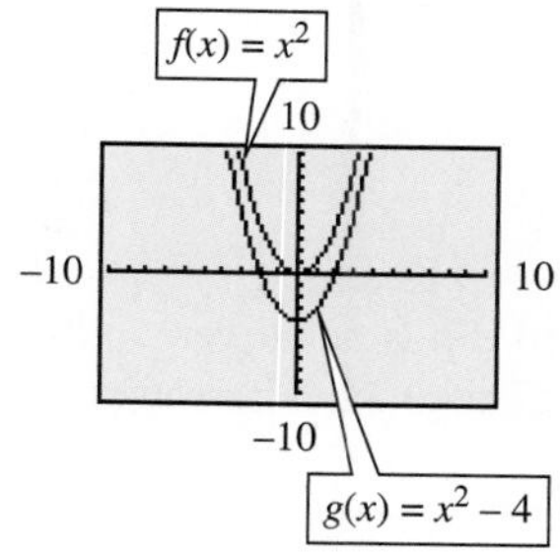

Compare with Figure 30.

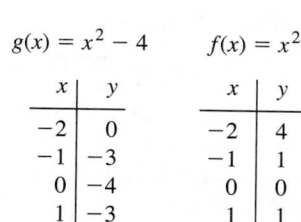

$g(x) = x^2 - 4$

x	y
−2	0
−1	−3
0	−4
1	−3
2	0

$f(x) = x^2$

x	y
−2	4
−1	1
0	0
1	1
2	4

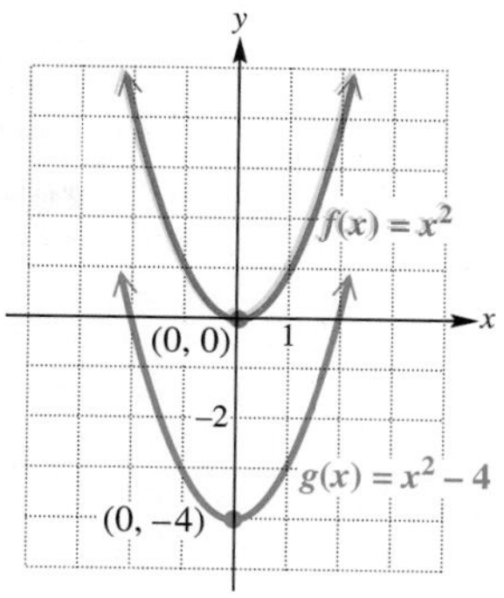

FIGURE 30

EXAMPLE 3 Graphing a Quadratic Function (Horizontal Shift)

Graph $g(x) = (x - 4)^2$.

SOLUTION

Comparing the tables of values shown with Figure 31 shows that the graph of $g(x) = (x - 4)^2$ is the same as that of $f(x) = x^2$, but translated 4 units to the right. This is a *horizontal shift.* The vertex is at (4, 0). As shown in Figure 31, the axis of this parabola is the vertical line $x = 4$.

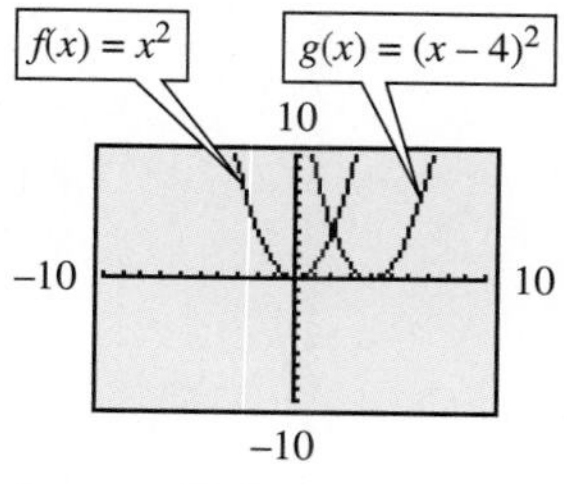

Compare with Figure 31.

$g(x) = (x - 4)^2$

x	y
2	4
3	1
4	0
5	1
6	4

$f(x) = x^2$

x	y
−2	4
−1	1
0	0
1	1
2	4

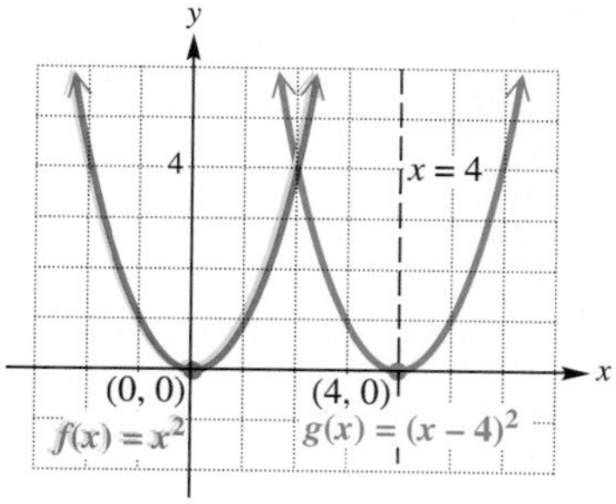

FIGURE 31

Pay close attention to this warning. → ***Errors frequently occur when horizontal shifts are involved.*** To determine the direction and magnitude of horizontal shifts, find the value of x that would cause the expression $x - h$ to equal 0. For example, the graph of

$$f(x) = (x - 5)^2$$

would be shifted 5 units to the *right*, because $x = +5$ would cause $x - 5$ to equal 0. On the other hand, the graph of

$$f(x) = (x + 4)^2$$

would be shifted 4 units to the *left*, because $x = -4$ would cause $x + 4$ to equal 0.

The following general principles apply for graphing functions of the form $f(x) = a(x - h)^2 + k$.

General Principles for Graphs of Quadratic Functions

1. The graph of the quadratic function defined by
$$f(x) = a(x - h)^2 + k, \quad a \neq 0,$$
is a parabola with vertex (h, k), and the vertical line $x = h$ as axis.
2. The graph opens upward if a is positive and downward if a is negative.
3. The graph is wider than that of $f(x) = x^2$ if $0 < |a| < 1$. The graph is narrower than that of $f(x) = x^2$ if $|a| > 1$.

EXAMPLE 4 Graphing a Quadratic Function Using General Principles

Graph $f(x) = -2(x + 3)^2 + 4$.

SOLUTION

The parabola opens downward (because $a < 0$), and is narrower than the graph of $f(x) = x^2$, since $a = -2$, and $|-2| > 1$. This parabola has vertex at $(-3, 4)$, as shown in Figure 32. To complete the graph, we plotted the additional ordered pairs $(-4, 2)$ and $(-2, 2)$.

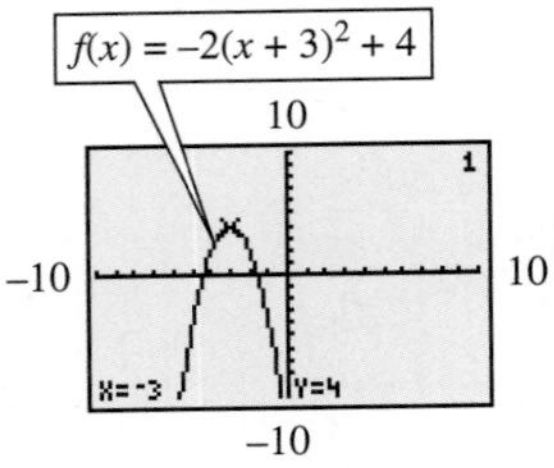

Compare with Figure 32. The vertex is $(-3, 4)$.

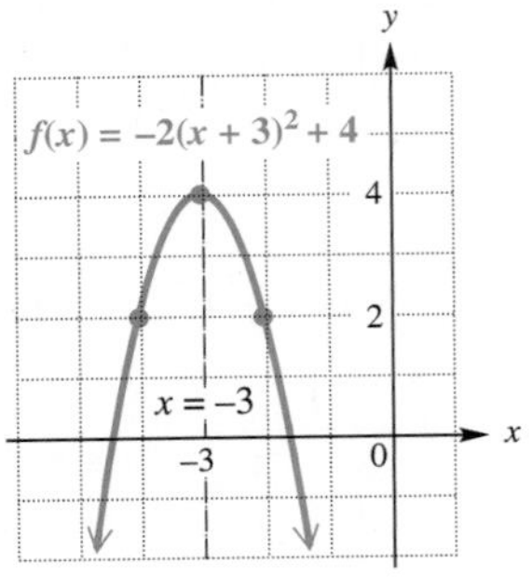

FIGURE 32

Vertex of a Parabola When the equation of a parabola is given in the form $f(x) = ax^2 + bx + c$, it is necessary to locate the vertex in order to sketch an accurate graph. This can be done in two ways. The first is by completing the square, as shown in Example 5. The second is by using a formula which can be derived by completing the square.

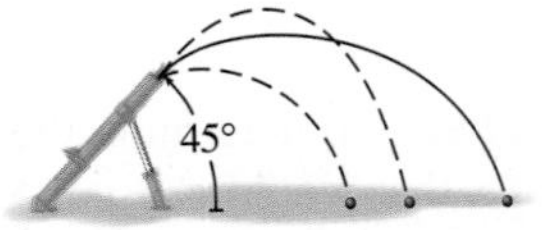

The trajectory of a shell fired from a cannon is a **parabola.** To reach the maximum range with a cannon, it is shown in calculus that the muzzle must be set at 45°. If the muzzle is elevated above 45°, the shell goes too high and falls too soon. If the muzzle is set below 45°, the shell is rapidly pulled to Earth by gravity.

EXAMPLE 5 Finding the Vertex by Completing the Square

Find the vertex of the graph of $f(x) = x^2 - 4x + 5$.

SOLUTION

To find the vertex, we need to express $x^2 - 4x + 5$ in the form $(x - h)^2 + k$. This is done by completing the square. (See pages 418 and 437.) To simplify the notation, replace $f(x)$ by y.

$$y = x^2 - 4x + 5$$

$$y - 5 = x^2 - 4x \qquad \text{Transform so that the constant term is on the left.}$$

$$y - 5 + 4 = x^2 - 4x + 4 \qquad \text{Half of } -4 \text{ is } -2;\ (-2)^2 = 4. \text{ Add 4 to both sides.}$$

$$y - 1 = (x - 2)^2 \qquad \text{Combine terms on the left and factor on the right.}$$

$$y = (x - 2)^2 + 1 \qquad \text{Add 1 to both sides.}$$

Now write the original equation as $f(x) = (x - 2)^2 + 1$. As shown earlier, the vertex of this parabola is $(2, 1)$.

Johann Kepler (1571–1630) established the importance of a curve called an **ellipse** in 1609, when he discovered that the orbits of the planets around the sun were elliptical, not circular. The orbit of Halley's comet, shown here, also is elliptical.

See For Further Thought at the end of this section for more on ellipses.

A formula for the vertex of the graph of the quadratic function $y = ax^2 + bx + c$ can be found by completing the square for the general form of the equation. In doing so, we begin by dividing by a, since the coefficient of x^2 must be 1.

$$y = ax^2 + bx + c \quad (a \neq 0)$$

$$\frac{y}{a} = x^2 + \frac{b}{a}x + \frac{c}{a} \qquad \text{Divide by } a.$$

$$\frac{y}{a} - \frac{c}{a} = x^2 + \frac{b}{a}x \qquad \text{Subtract } \frac{c}{a}.$$

$$\frac{y}{a} - \frac{c}{a} + \frac{b^2}{4a^2} = x^2 + \frac{b}{a}x + \frac{b^2}{4a^2} \qquad \text{Add } \left(\frac{1}{2} \cdot \frac{b}{a}\right)^2 = \frac{b^2}{4a^2}.$$

$$\frac{y}{a} + \frac{b^2 - 4ac}{4a^2} = \left(x + \frac{b}{2a}\right)^2 \qquad \text{Combine terms on the left and factor on the right.}$$

$$\frac{y}{a} = \left(x + \frac{b}{2a}\right)^2 - \frac{b^2 - 4ac}{4a^2} \qquad \text{Transform so that the } y\text{-term is alone on the left.}$$

$$y = a\left(x + \frac{b}{2a}\right)^2 + \frac{4ac - b^2}{4a} \qquad \text{Multiply by } a.$$

$$y = a\left[x - \underbrace{\left(-\frac{b}{2a}\right)}_{h}\right]^2 + \underbrace{\frac{4ac - b^2}{4a}}_{k}$$

The final equation shows that the vertex (h, k) can be expressed in terms of a, b, and c. However, it is not necessary to memorize the expression for k, because it can be obtained by replacing x by $-\frac{b}{2a}$.

Vertex Formula

The vertex of the graph of $f(x) = ax^2 + bx + c \ (a \neq 0)$ has coordinates

$$\left(-\frac{b}{2a},\ f\left(-\frac{b}{2a}\right)\right).$$

EXAMPLE 6 Finding the Vertex by Using the Formula

Use the vertex formula to find the vertex of the graph of the function

$$f(x) = x^2 - x - 6.$$

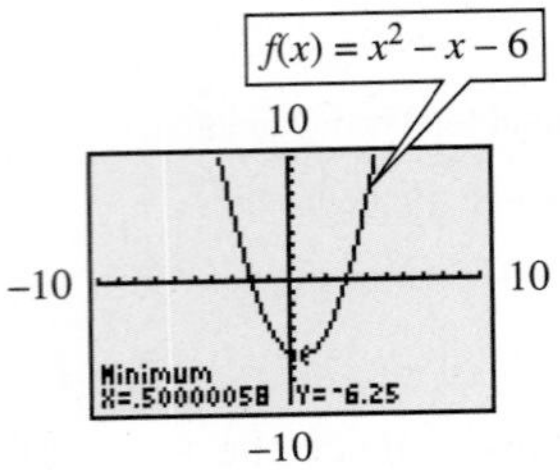

Notice the slight discrepancy when we instruct the calculator to find the vertex (a *minimum* here). This reinforces the fact that *we must understand the concepts and not totally rely on technology!*

SOLUTION

For this function, $a = 1$, $b = -1$, and $c = -6$. The x-coordinate of the vertex of the parabola is given by

$$-\frac{b}{2a} = -\frac{(-1)}{2(1)} = \frac{1}{2}.$$

The y-coordinate is $f\left(-\frac{b}{2a}\right) = f\left(\frac{1}{2}\right)$.

$$f\left(\frac{1}{2}\right) = \left(\frac{1}{2}\right)^2 - \frac{1}{2} - 6 = \frac{1}{4} - \frac{1}{2} - 6 = -\frac{25}{4}$$

Finally, the vertex is $\left(\frac{1}{2}, -\frac{25}{4}\right)$.

General Graphing Guidelines

A general approach to graphing quadratic functions using intercepts and the vertex is now given.

Graphing a Quadratic Function $f(x) = ax^2 + bx + c$

Step 1 **Decide whether the graph opens upward or downward.** Determine whether the graph opens upward (if $a > 0$) or opens downward (if $a < 0$) to aid in the graphing process.

Step 2 **Find the vertex.** Find the vertex either by using the formula or by completing the square.

Step 3 **Find the y-intercept.** Find the y-intercept by evaluating $f(0)$.

Step 4 **Find the x-intercepts.** Find the x-intercepts, if any, by solving $f(x) = 0$.

Step 5 **Complete the graph.** Find and plot additional points as needed, using the symmetry about the axis.

EXAMPLE 7 Graphing a Quadratic Function Using General Guidelines

Graph the quadratic function $f(x) = x^2 - x - 6$.

SOLUTION

Step 1 From the equation, $a = 1 > 0$, so the graph of the function opens up.

Step 2 The vertex, $\left(\frac{1}{2}, -\frac{25}{4}\right)$, was found in Example 6 using the vertex formula.

Step 3 To find the y-intercept, evaluate $f(0)$.

$$f(x) = x^2 - x - 6$$
$$f(0) = 0^2 - 0 - 6 \quad \text{Let } x = 0.$$
$$f(0) = -6$$

The y-intercept is $(0, -6)$.

Step 4 Find any x-intercepts. Because the vertex, $\left(\frac{1}{2}, -\frac{25}{4}\right)$, is in quadrant IV and the graph opens up, there will be two x-intercepts. To find them, let $f(x) = 0$ and solve.

$$f(x) = x^2 - x - 6$$
$$0 = x^2 - x - 6 \qquad \text{Let } f(x) = 0.$$
$$0 = (x - 3)(x + 2) \qquad \text{Factor.}$$
$$x - 3 = 0 \quad \text{or} \quad x + 2 = 0 \qquad \text{Zero-factor property}$$
$$x = 3 \quad \text{or} \quad x = -2$$

The x-intercepts are $(3, 0)$ and $(-2, 0)$.

Step 5 Plot the points found so far, and plot any additional points as needed. The symmetry of the graph is helpful here. The graph is shown in Figure 33.

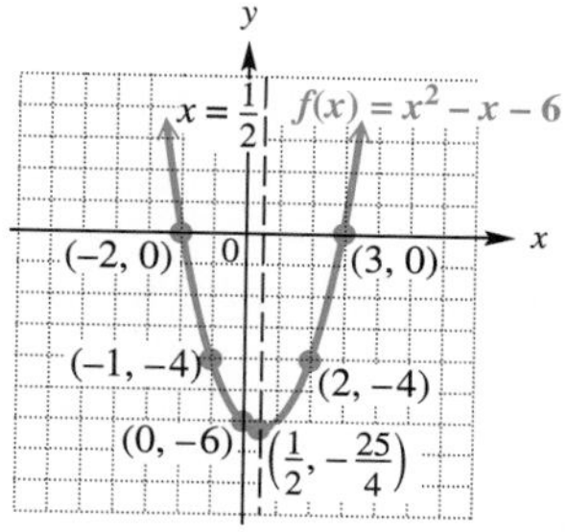

FIGURE 33

X	Y1	
-2	0	
-1	-4	
0	-6	
1	-6	
2	-4	
3	0	
4	6	

Y1■X²−X−6

This table provides other points on the graph of

$$Y_1 = X^2 - X - 6.$$

Galileo Galilei (1564–1642) died in the year Newton was born; his work was important in Newton's development of calculus. The idea of **function** is implicit in Galileo's analysis of the parabolic path of a projectile, where height and range are functions (in our terms) of the angle of elevation and the initial velocity.

According to legend, Galileo dropped objects of different weights from the tower of Pisa to disprove the Aristotelian view that heavier objects fall faster than lighter objects. He developed a formula for freely falling objects that is described by

$$d = 16t^2,$$

where d is the distance in feet that a given object falls (discounting air resistance) in a given time t, in seconds, regardless of weight.

A Model for Optimization As we have seen, the vertex of a vertical parabola is either the highest or the lowest point of the parabola. The y-value of the vertex gives the maximum or minimum value of y, while the x-value tells where that maximum or minimum occurs. Often a model can be constructed so that y can be *optimized*.

PROBLEM-SOLVING HINT In some practical problems we want to know the least or greatest value of some quantity. When that quantity can be expressed using a quadratic function $f(x) = ax^2 + bx + c$, as in the next example, the vertex can be used to find the desired value.

EXAMPLE 8 Finding a Maximum Area

A farmer has 120 feet of fencing. He wants to put a fence around three sides of a rectangular plot of land, with the side of a barn forming the fourth side. Find the maximum area he can enclose. What dimensions give this area?

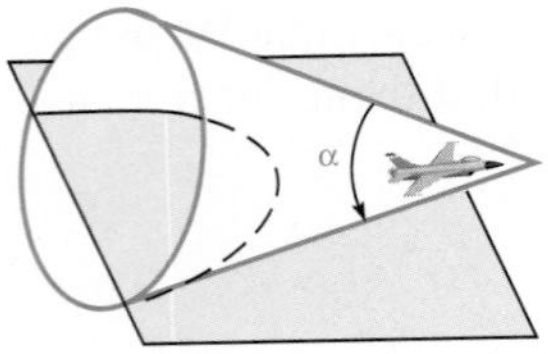

A sonic boom is a loud explosive sound caused by the shock wave that accompanies an aircraft traveling at supersonic speed. The sonic boom shock wave has the shape of a cone, and it intersects the ground in one branch of a curve known as a **hyperbola.** Everyone located along the hyperbolic curve on the ground hears the sound at the same time.

See For Further Thought on the next page for more on hyperbolas.

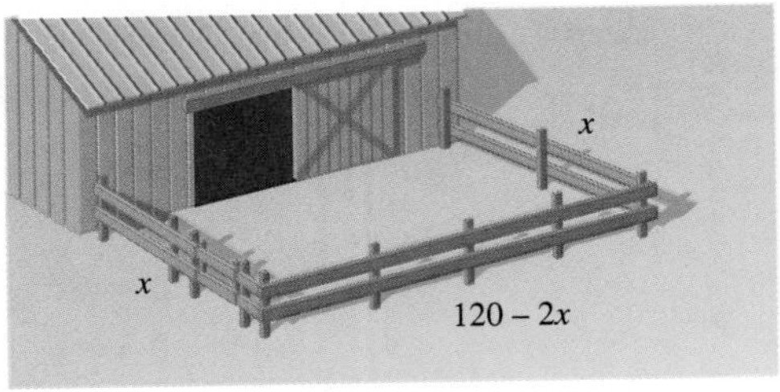

FIGURE 34

SOLUTION

Figure 34 shows the plot. Let x represent its width. Then, since there are 120 feet of fencing,

$$x + x + \text{length} = 120 \quad \text{Sum of the three fenced sides is 120 feet.}$$
$$2x + \text{length} = 120 \quad \text{Combine terms.}$$
$$\text{length} = 120 - 2x. \quad \text{Subtract } 2x.$$

The area is modeled by the product of the length and width, or

$$A(x) = (120 - 2x)x = 120x - 2x^2.$$

To make $120x - 2x^2$ (and thus the area) as large as possible, first find the vertex of the graph of the function $A(x) = 120x - 2x^2$.

$$A(x) = -2x^2 + 120x \quad \text{Standard form}$$

Here we have $a = -2$ and $b = 120$. The x-coordinate of the vertex is

$$-\frac{b}{2a} = -\frac{120}{2(-2)} = 30.$$

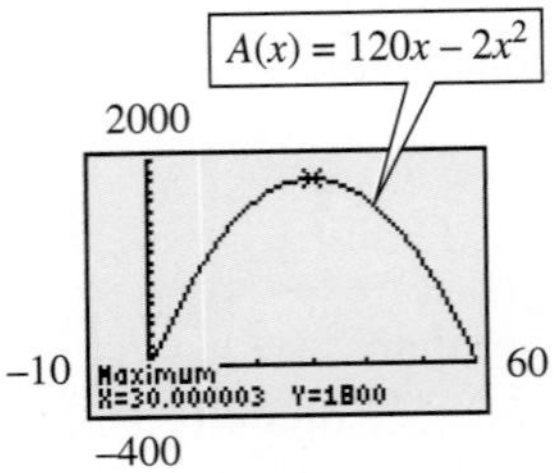

The vertex is (30, 1800), supporting the analytic result in Example 8.

The vertex is a maximum point (since $a = -2 < 0$), so the maximum area that the farmer can enclose is

$$A(30) = -2(30)^2 + 120(30) = 1800 \text{ square feet.}$$

The farmer can enclose a maximum area of 1800 square feet, when the width of the plot is 30 feet and the length is $120 - 2(30) = 60$ feet. ■

As seen in Example 8, be careful when interpreting the meanings of the coordinates of the vertex in problems involving maximum or minimum values. The first coordinate, x, gives the value for which the *function value* is a maximum or a minimum. Read the problem carefully to determine whether you are asked to find the value of the independent variable, the dependent variable (that is, the function value), or both.

For Further Thought

The Conic Sections

The circle, introduced in the first section of this chapter, the parabola, the ellipse, and the hyperbola are known as **conic sections.** As seen in the figure below, each of these geometric shapes can be obtained by intersecting a plane and an infinite cone (made up of two *nappes*).

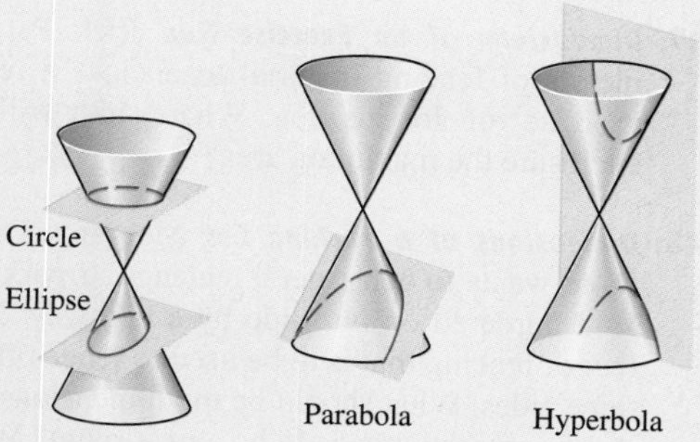

The Greek geometer Apollonius (c. 225 B.C.) was also an astronomer, and his classic work *Conic Sections* thoroughly investigated these figures. Apollonius is responsible for the names "ellipse," "parabola," and "hyperbola." The margin notes in this section show some ways that these figures appear in the world around us.

For Group Discussion or Individual Investigation

1. The terms *ellipse, parabola*, and *hyperbola* are similar to the terms *ellipsis, parable*, and *hyperbole*. What do these latter three terms mean? You might want to do some investigation as to the similarities between the mathematical terminology and these language-related terms.
2. Identify some places in the world around you where conic sections are encountered.
3. The accompanying figure shows how an ellipse can be drawn using tacks and string. Have a class member volunteer to go to the board and using string and chalk, modify the method to draw a circle. Then have two class members work together to draw an ellipse. (*Hint:* Press hard!)

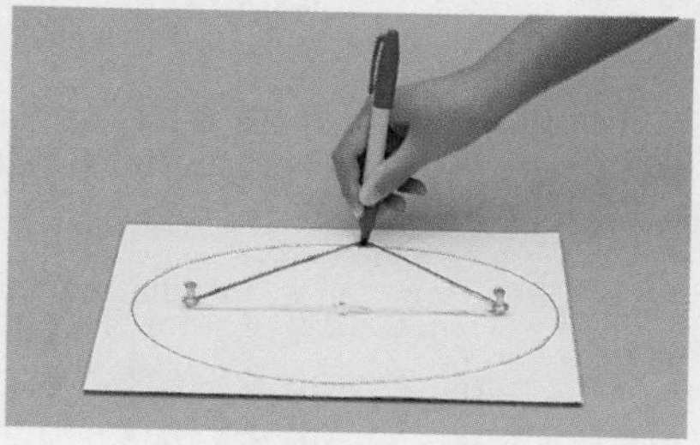

8.5 EXERCISES

In Exercises 1– 6, match each equation with the figure in A–F that most closely resembles its graph.

A.

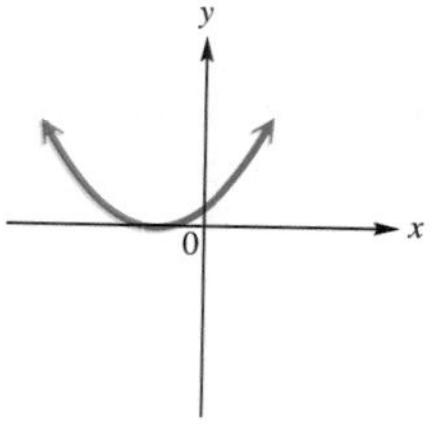

B.

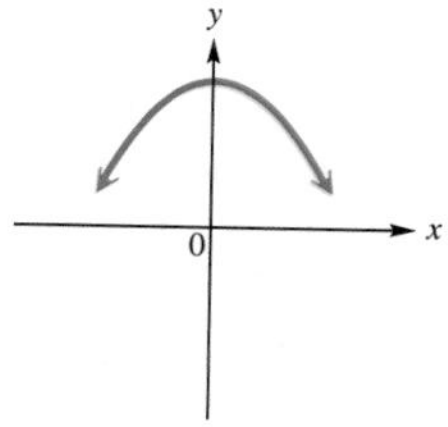

C.

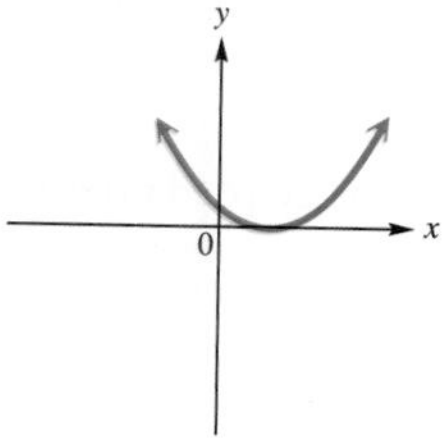

D.

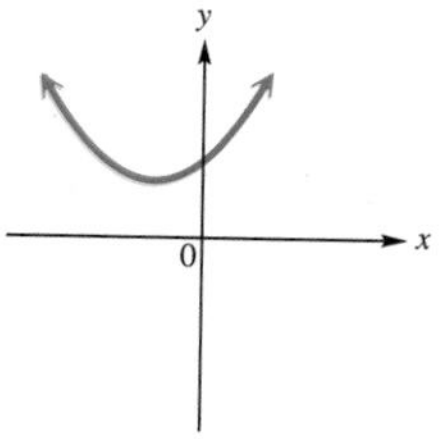

E.

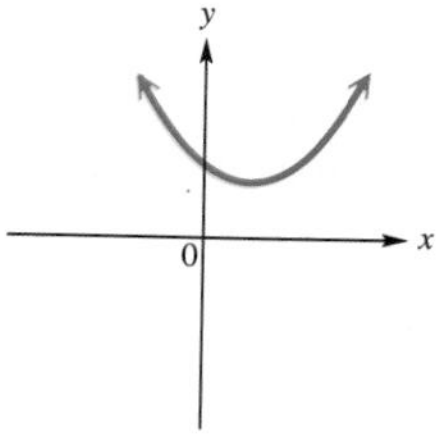

F.

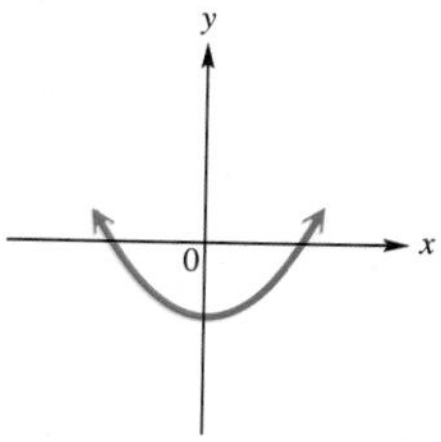

1. $g(x) = x^2 - 5$

2. $h(x) = -x^2 + 4$

3. $F(x) = (x - 1)^2$

4. $G(x) = (x + 1)^2$

5. $H(x) = (x - 1)^2 + 1$

6. $K(x) = (x + 1)^2 + 1$

7. Explain in your own words the meaning of each term.
(a) vertex of a parabola **(b)** axis of a parabola

8. Explain why the axis of the graph of a quadratic function cannot be a horizontal line.

Identify the vertex of the graph of each quadratic function.

9. $f(x) = -3x^2$
10. $f(x) = -.5x^2$
11. $f(x) = x^2 + 4$
12. $f(x) = x^2 - 4$
13. $f(x) = (x - 1)^2$
14. $f(x) = (x + 3)^2$
15. $f(x) = (x + 3)^2 - 4$
16. $f(x) = (x - 5)^2 - 8$
17. Describe how the graph of each parabola in Exercises 15 and 16 is shifted compared to the graph of $y = x^2$.

For each quadratic function, tell whether the graph opens upward or downward, and tell whether the graph is wider, narrower, or the same as the graph of $f(x) = x^2$.

18. $f(x) = -2x^2$
19. $f(x) = -3x^2 + 1$
20. $f(x) = .5x^2$
21. $f(x) = \frac{2}{3}x^2 - 4$
22. What does the value of a in $f(x) = a(x - h)^2 + k$ tell you about the graph of the function compared to the graph of $y = x^2$?
23. For $f(x) = a(x - h)^2 + k$, in what quadrant is the vertex if:
    (a) $h > 0, k > 0$;
    (b) $h > 0, k < 0$;
    (c) $h < 0, k > 0$;
    (d) $h < 0, k < 0$?
24. (a) What is the value of h if the graph of $f(x) = a(x - h)^2 + k$ has vertex on the y-axis?
    (b) What is the value of k if the graph of $f(x) = a(x - h)^2 + k$ has vertex on the x-axis?

Sketch the graph of each quadratic function using the methods described in this section. Indicate two points on each graph.

25. $f(x) = 3x^2$
26. $f(x) = -2x^2$
27. $f(x) = -\frac{1}{4}x^2$
28. $f(x) = \frac{1}{3}x^2$
29. $f(x) = x^2 - 1$
30. $f(x) = x^2 + 3$
31. $f(x) = -x^2 + 2$
32. $f(x) = -x^2 - 4$
33. $f(x) = 2x^2 - 2$
34. $f(x) = -3x^2 + 1$
35. $f(x) = (x - 4)^2$
36. $f(x) = (x - 3)^2$
37. $f(x) = 3(x + 1)^2$
38. $f(x) = -2(x + 1)^2$
39. $f(x) = (x + 1)^2 - 2$
40. $f(x) = (x - 2)^2 + 3$

Sketch the graph of each quadratic function. Indicate the coordinates of the vertex of the graph.

41. $f(x) = x^2 + 8x + 14$
42. $f(x) = x^2 + 10x + 23$
43. $f(x) = x^2 + 2x - 4$
44. $f(x) = 3x^2 - 9x + 8$
45. $f(x) = -2x^2 + 4x + 5$
46. $f(x) = -5x^2 - 10x + 2$

Solve each problem.

47. ***Dimensions of an Exercise Run*** Rick Pal has 100 meters of fencing material to enclose a rectangular exercise run for his dog. What width will give the enclosure the maximum area?

48. ***Dimensions of a Parking Lot*** Morgan's Department Store wants to construct a rectangular parking lot on land bordered on one side by a highway. It has 280 feet of fencing that is to be used to fence off the other three sides. What should be the dimensions of the lot if the enclosed area is to be a maximum? What is the maximum area?

49. ***Height of a Projected Object*** If an object on Earth is projected upward with an initial velocity of 32 feet per second, then its height after t seconds is given by

$$h(t) = 32t - 16t^2.$$

Find the maximum height attained by the object and the number of seconds it takes to hit the ground.

50. ***Height of a Projected Object*** A projectile on Earth is fired straight upward so that its distance (in feet) above the ground t seconds after firing is given by

$$s(t) = -16t^2 + 400t.$$

Find the maximum height it reaches and the number of seconds it takes to reach that height.

51. ***Height of a Projected Object*** If air resistance is neglected, a projectile on Earth shot straight upward with an initial velocity of 40 meters per second will be at a height s in meters given by the function

$$s(t) = -4.9t^2 + 40t,$$

where t is the number of seconds elapsed after projection. After how many seconds will it reach its maximum height, and what is this maximum height? Round your answers to the nearest tenth.

52. ***Height of a Projected Object*** A space robot is projected from the moon with its distance in feet given by

$$f(x) = 1.727x - .0013x^2$$

feet, where x is time in seconds. Find the maximum height the robot can reach, and the time it takes to get there. Round answers to the nearest tenth.

53. ***Carbon Monoxide Exposure*** Carbon monoxide (CO) combines with the hemoglobin of the blood to form carboxyhemoglobin (COHb), which reduces the transport of oxygen to tissues. Smokers routinely have a 4% to 6% COHb level in their blood, which can cause symptoms such as blood flow alterations, visual impairment, and poorer vigilance. The quadratic function defined by

$$T(x) = .00787x^2 - 1.528x + 75.89$$

approximates the exposure time in hours necessary to reach this 4% to 6% level, where $50 \le x \le 100$ is the amount of carbon monoxide present in the air in parts per million (ppm). (*Source: Indoor Air Quality Environmental Information Handbook: Combustion Sources*, U.S. Department of Energy, 1985.)

(a) A kerosene heater or a room full of smokers is capable of producing 50 ppm of carbon monoxide. How long would it take for a nonsmoking person to start feeling the symptoms mentioned?

(b) Find the carbon monoxide concentration necessary for a person to reach the 4% to 6% COHb level in 3 hours.

54. ***Carbon Monoxide Exposure*** Refer to Exercise 53. High concentrations of carbon monoxide (CO) can cause coma and death. The time required for a person to reach a COHb level capable of causing a coma can be approximated by

$$T(x) = .0002x^2 - .316x + 127.9,$$

where T is the exposure time in hours necessary to reach that level and $500 \le x \le 800$ is the amount of carbon monoxide in parts per million (ppm). (*Source: Indoor Air Quality Environmental Information Handbook: Combustion Sources*, U.S. Department of Energy, 1985.)

(a) What is the exposure time when $x = 600$ ppm?

(b) Estimate the concentration of CO necessary to produce a coma in 4 hours.

55. ***Automobile Stopping Distance*** Selected values of the stopping distance y in feet of a car traveling x mph are given in the table.

Speed (in mph)	Stopping Distance (in feet)
20	46
30	87
40	140
50	240
60	282
70	371

Source: National Safety Institute Student Workbook, 1993, p. 7.

The quadratic function defined by

$$f(x) = .056057x^2 + 1.06657x$$

is one model of the data. Find and interpret $f(45)$.

56. ***Investment Portfolio Mixtures*** The graph, which appears to be a portion of a parabola opening to the *right*, shows the performance of investment portfolios with different mixtures of U.S. and foreign investments for the period January 1, 1971, to December 31, 1996.

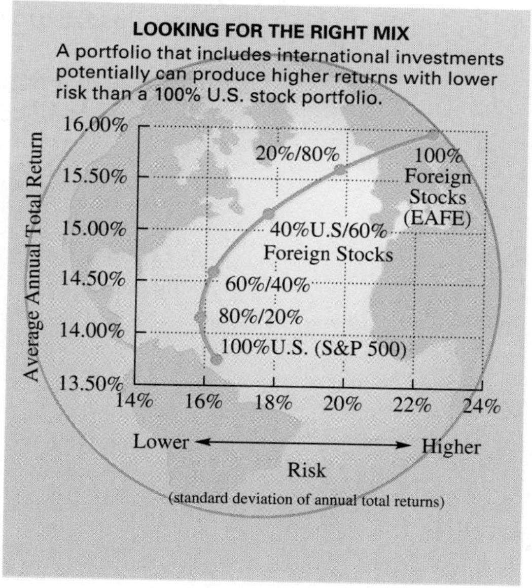

Source: Financial Ink **Newsletter, Investment Management and Research, Inc., Feb. 1998. Thanks to David Van Geffen for this information.**

Use the graph to answer the following questions.

(a) Is this the graph of a function? Explain.

(b) What investment mixture shown on the graph appears to represent the vertex? What relative amount of risk does this point represent? What return on investment does it provide?
(c) Which point on the graph represents the riskiest investment mixture? What return on investment does it provide?

57. ***Maximum Airline Revenue*** A charter flight charges a fare of \$200 per person, plus \$4 per person for each unsold seat on the plane. If the plane holds 100 passengers and if x represents the number of unsold seats, find the following.
(a) An expression for the total revenue $R(x)$ received for the flight (*Hint*: Multiply the number of people flying, $100 - x$, by the price per ticket.)
(b) The graph for the function of part (a)
(c) The number of unsold seats that will produce the maximum revenue
(d) The maximum revenue

58. ***Maximum Bus Fare Revenue*** For a trip to a resort, a charter bus company charges a fare of \$48 per person, plus \$2 per person for each unsold seat on the bus. If the bus has 42 seats and x represents the number of unsold seats, find the following.
(a) An expression that defines the total revenue, $R(x)$, from the trip (*Hint:* Multiply the total number riding, $42 - x$, by the price per ticket, $48 + 2x$.)
(b) The graph of the function from part (a)
(c) The number of unsold seats that produces the maximum revenue
(d) The maximum revenue

8.6 Exponential and Logarithmic Functions, Applications, and Models

Exponential Functions and Applications • Logarithmic Functions and Applications • Exponential Models in Nature

Exponential Functions and Applications

In this section we introduce two new types of functions.

Exponential Function

An **exponential function** with base b, where $b > 0$ and $b \neq 1$, is a function of the form

$$f(x) = b^x, \qquad \text{where } x \text{ is any real number.}$$

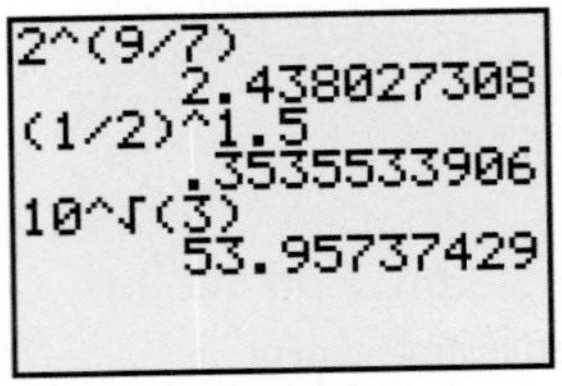

Compare with the discussion in the text.

Thus far, we have defined only integer exponents. In the definition of exponential function, we allow x to take on any real number value. By using methods not discussed in this book, expressions such as

$$2^{9/7}, \qquad \left(\frac{1}{2}\right)^{1.5}, \qquad \text{and} \qquad 10^{\sqrt{3}}$$

can be approximated. A scientific or graphing calculator is capable of determining approximations for these numbers. See the screen in the margin.

Notice that in the definition of exponential function, the base b is restricted to positive numbers, with b ≠ 1.

The graphs of $f(x) = 2^x$, $g(x) = \left(\frac{1}{2}\right)^x$, and $h(x) = 10^x$ are shown in Figure 35(a), (b), and (c). In each case, a table of selected points is given. The points are joined with a smooth curve, typical of the graphs of exponential functions. For each graph, the curve approaches but does not intersect the x-axis. For this reason, the x-axis is called the **horizontal asymptote** of the graph.

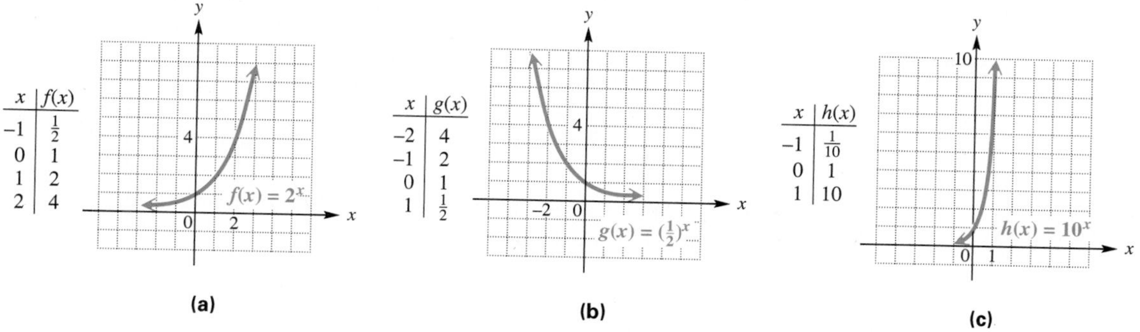

x	$f(x)$
-1	$\frac{1}{2}$
0	1
1	2
2	4

(a)

x	$g(x)$
-2	4
-1	2
0	1
1	$\frac{1}{2}$

(b)

x	$h(x)$
-1	$\frac{1}{10}$
0	1
1	10

(c)

FIGURE 35

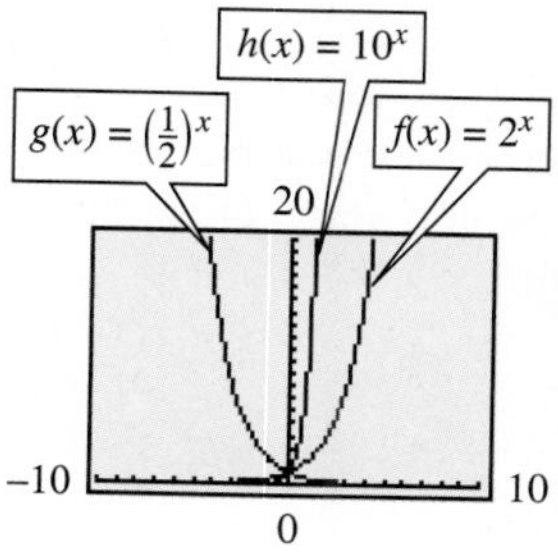

Compare with Figure 35.

This discussion illustrates the following facts about the graph of an exponential function.

Graph of $f(x) = b^x$

1. The graph always will contain the point (0, 1), because $b^0 = 1$.
2. When $b > 1$, the graph will *rise* from left to right (as in the illustrations of Figures 35(a) and (c) above, with $b = 2$ and $b = 10$). When $0 < b < 1$, the graph will *fall* from left to right (as in the illustration of Figure 35(b), with $b = \frac{1}{2}$).
3. The x-axis is the horizontal asymptote.
4. The domain is $(-\infty, \infty)$ and the range is $(0, \infty)$.

Probably the most important exponential function has the base e. (See Section 6.4.) The number e is named after Leonhard Euler (1707–1783), and is approximately 2.718281828. It is an irrational number, and its value is approached by the expression

$$\left(1 + \frac{1}{n}\right)^n$$

as n takes on larger and larger values. We write

$$\text{as } n \rightarrow \infty, \quad \left(1 + \frac{1}{n}\right)^n \rightarrow e \approx 2.718281828.$$

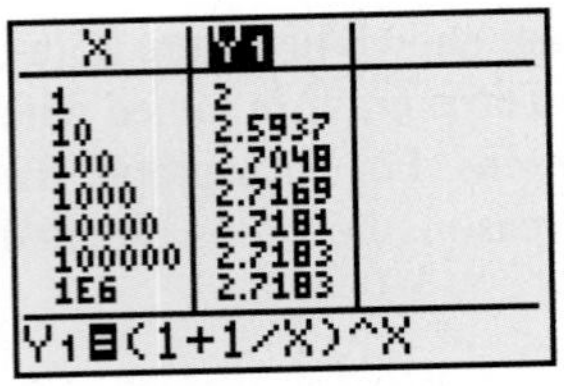

This table shows selected values for $Y_1 = \left(1 + \frac{1}{X}\right)^X$. Compare with Table 4.

See Table 4.

TABLE 4

n	Approximate Value of $\left(1 + \dfrac{1}{n}\right)^n$
1	2
100	2.70481
1000	2.71692
10,000	2.71815
1,000,000	2.71828

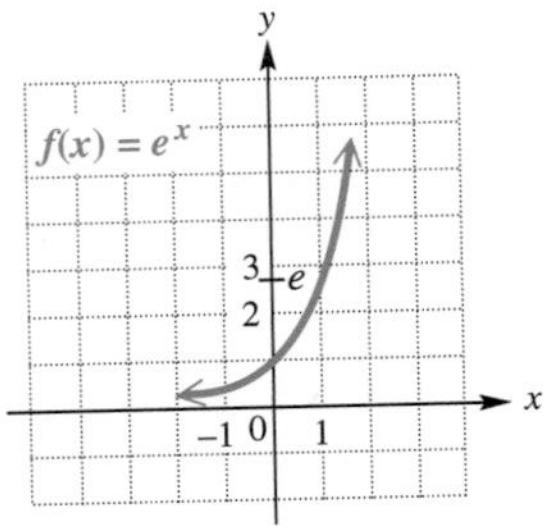

FIGURE 36

Powers of e can be approximated on a scientific or graphing calculator. Some powers of e obtained on a calculator are

$$e^{-2} \approx .1353352832, \qquad e^{1.3} \approx 3.669296668, \qquad e^{4} \approx 54.59815003.$$

The graph of the function $f(x) = e^x$ is shown in Figure 36.

A real-life application of exponential functions occurs in the computation of **compound interest.**

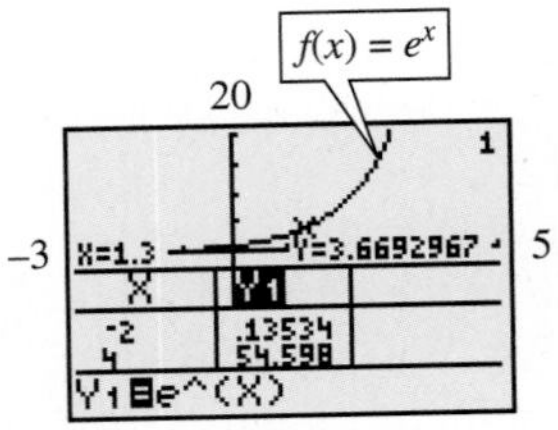

The values for e^{-2}, $e^{1.3}$, and e^{4} are approximated in this split-screen graph of $f(x) = e^x$.

Compound Interest Formula

Suppose that a principal of P dollars is invested at an annual interest rate r (in percent, expressed as a decimal), compounded n times per year. Then the amount A accumulated after t years is given by the formula

$$A = P\left(1 + \frac{r}{n}\right)^{nt}.$$

EXAMPLE 1 Applying the Compound Interest Formula

Suppose that \$1000 is invested at an annual rate of 8%, compounded quarterly (four times per year). Find the total amount in the account after ten years if no withdrawals are made.

SOLUTION

$$A = P\left(1 + \frac{r}{n}\right)^{nt} \qquad \text{Compound interest formula}$$

$$A = 1000\left(1 + \frac{.08}{4}\right)^{4 \cdot 10} \qquad P = 1000,\ r = .08,\ n = 4,\ t = 10$$

$$A = 1000\,(1.02)^{40}$$

$$A \approx 1000\,(2.20804) \qquad \text{Evaluate } 1.02^{40} \text{ with a calculator.}$$

$$A = 2208.04. \qquad \text{To the nearest cent}$$

≈ means "approximately equal to."

There would be \$2208.04 in the account at the end of ten years.

Following her success in *I Love Lucy*, **Lucille Ball** starred in *The Lucy Show* which aired for six seasons on CBS in the 1960s. She worked for Mr. Mooney (Gale Gordon), who was very careful with his money. In the September 26, 1966, show "Lucy, the Bean Queen," Lucy learned a lesson about **exponential growth.** Mr. Mooney had refused to lend her $1500 to buy furniture, because he claimed she did not know the value of money. He explained to her that if she were to save one penny on Day 1, two pennies on Day 2, four pennies on Day 3, and so on, she would have more than enough money to buy her furniture after only nineteen days. (You might want to verify this on your own.)

The compounding formula given earlier applies if the financial institution compounds interest for a finite number of compounding periods annually. Theoretically, the number of compounding periods per year can get larger and larger (quarterly, monthly, daily, etc.), and if n is allowed to approach infinity, we say that interest is compounded *continuously*. The formula for **continuous compounding** involves the number e.

Continuous Compound Interest Formula

Suppose that a principal of P dollars is invested at an annual interest rate r (in percent, expressed as a decimal), compounded continuously. Then the amount A accumulated after t years is given by the formula

$$A = Pe^{rt}.$$

EXAMPLE 2 Applying the Continuous Compound Interest Formula

Suppose that $5000 is invested at an annual rate of 6.5%, compounded continuously. Find the total amount in the account after four years if no withdrawals are made.

SOLUTION

$$A = Pe^{rt} \quad \text{Continuous compound interest formula}$$

$$A = 5000e^{.065(4)} \quad P = 5000,\ r = .065,\ t = 4$$

$$A = 5000e^{.26}$$

$$A \approx 5000(1.29693) \quad \text{Use the } e^x \text{ key on a calculator.}$$

$$A = 6484.65$$

There will be $6484.65 in the account after four years. ■

The continuous compound interest formula is an example of an **exponential growth function.** In situations involving growth or decay of a quantity, the amount or number present at time t can often be approximated by a function of the form

$$A(t) = A_0e^{kt},$$

where A_0 represents the amount or number present at time $t = 0$, and k is a constant. If $k > 0$, there is exponential growth; if $k < 0$, there is exponential decay.

Logarithmic Functions and Applications

Consider the equation

$$2^3 = 8.$$

Here 3 is the exponent (or power) to which 2 must be raised in order to obtain 8. The exponent 3 is called the *logarithm* to the base 2 of 8, and this is written

$$3 = \log_2 8.$$

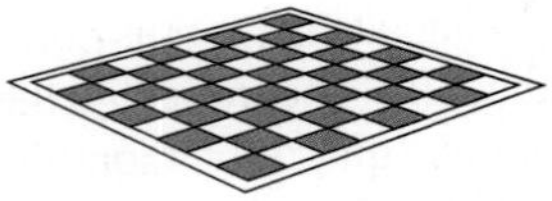

A variation of **exponential growth** is found in the legend of a Persian king, who wanted to please his executive officer, the Grand Vizier, with a gift of his choice. The Grand Vizier explained that he would like to be able to use his chessboard to accumulate wheat. A single grain of wheat would be received for the first square on the board, two grains would be received for the second square, four grains for the third, and so on, doubling the number of grains for each of the 64 squares on the board. As unlikely as it may seem, the number of grains would total 18.5 quintillion! Even with today's methods of production, this amount would take 150 years to produce. The Grand Vizier evidently knew his mathematics.

In general, we have the following relationship.

Definition of $\log_b x$

For $b > 0$, $b \neq 1$,

$$\text{if } b^y = x, \qquad \text{then } y = \log_b x.$$

Table 5 illustrates the relationship between exponential equations and logarithmic equations.

TABLE 5

Exponential Equation	Logarithmic Equation
$3^4 = 81$	$4 = \log_3 81$
$10^5 = 100{,}000$	$5 = \log_{10} 100{,}000$
$\left(\frac{1}{2}\right)^{-4} = 16$	$-4 = \log_{1/2} 16$
$10^0 = 1$	$0 = \log_{10} 1$
$4^{-3} = \frac{1}{64}$	$-3 = \log_4 \frac{1}{64}$

The concept of inverse functions (studied in more advanced algebra courses) leads us to the definition of the logarithmic function with base b.

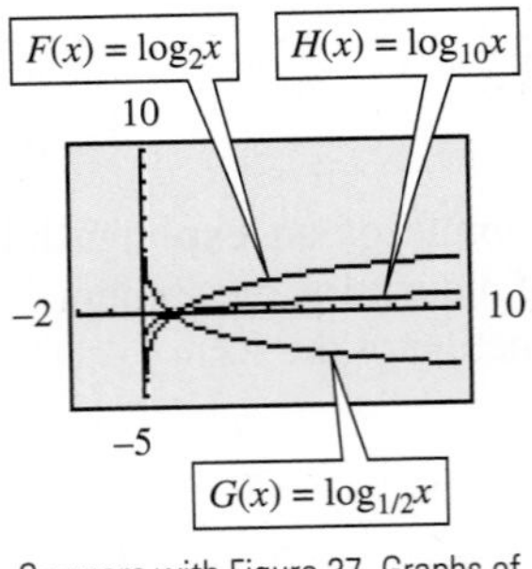

Compare with Figure 37. Graphs of logarithmic functions with bases other than 10 and e are obtained with the use of the change-of-base rule from algebra:

$$\log_a x = \frac{\log x}{\log a} = \frac{\ln x}{\ln a}.$$

Alternatively, they can be drawn by using the capability that allows the user to obtain the graph of the inverse.

Logarithmic Function

A **logarithmic function with base b,** where $b > 0$ and $b \neq 1$, is a function of the form

$$g(x) = \log_b x, \qquad \text{where } x > 0.$$

The graph of the function $g(x) = \log_b x$ can be found by interchanging the roles of x and y in the function $f(x) = b^x$. Geometrically, this is accomplished by reflecting the graph of $f(x) = b^x$ about the line $y = x$.

The graphs of

$$F(x) = \log_2 x, \quad G(x) = \log_{1/2} x, \quad \text{and} \quad H(x) = \log_{10} x$$

are shown in Figure 37(a), (b), and (c) on the next page. In each case, a table of selected points is given. These points were obtained by interchanging the roles of x and y in the tables of points given in Figure 35. The points are joined with a smooth curve, typical of the graphs of logarithmic functions. For each graph, the curve approaches but does not intersect the y-axis. Thus, the y-axis is called the **vertical asymptote** of the graph.

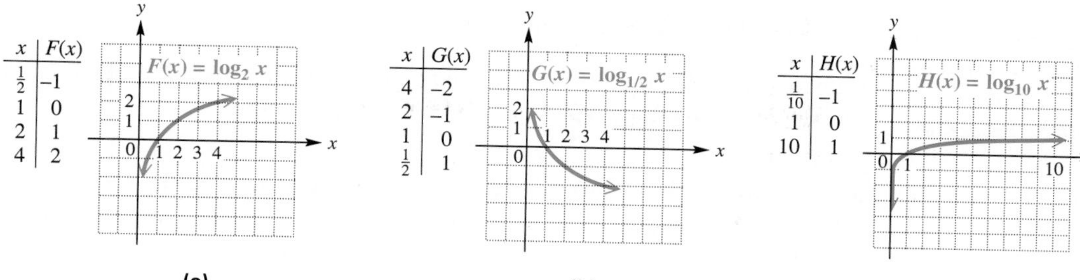

x	$F(x)$
$\frac{1}{2}$	-1
1	0
2	1
4	2

(a)

x	$G(x)$
4	-2
2	-1
1	0
$\frac{1}{2}$	1

(b)

x	$H(x)$
$\frac{1}{10}$	-1
1	0
10	1

(c)

FIGURE 37

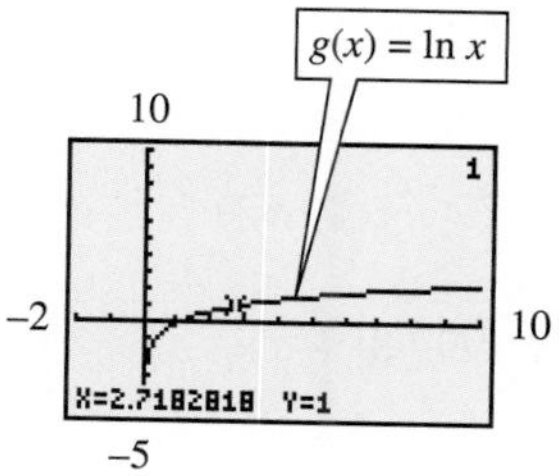

Notice that when

$x = e \approx 2.7182818,$

$y = \ln e = 1.$

Graph of $g(x) = \log_b x$

1. The graph will always contain the point (1, 0), because $\log_b 1 = 0$.
2. When $b > 1$, the graph will *rise* from left to right, from the fourth quadrant to the first (as in the illustrations of Figure 37(a) and (c) above, with $b = 2$ and $b = 10$). When $0 < b < 1$, the graph will *fall* from left to right, from the first quadrant to the fourth (as in the illustration of Figure 37(b), with $b = \frac{1}{2}$).
3. The y-axis is the vertical asymptote.
4. The domain is $(0, \infty)$ and the range is $(-\infty, \infty)$.

An important logarithmic function is the function with base e. If we interchange the roles of x and y in the graph of $f(x) = e^x$ (Figure 36), we obtain the graph of $g(x) = \log_e x$. There is a special symbol for $\log_e x$: it is $\ln x$. That is,

$$\ln x = \log_e x.$$

Figure 38 shows the graph of $g(x) = \ln x$, called the **natural logarithmic function.**

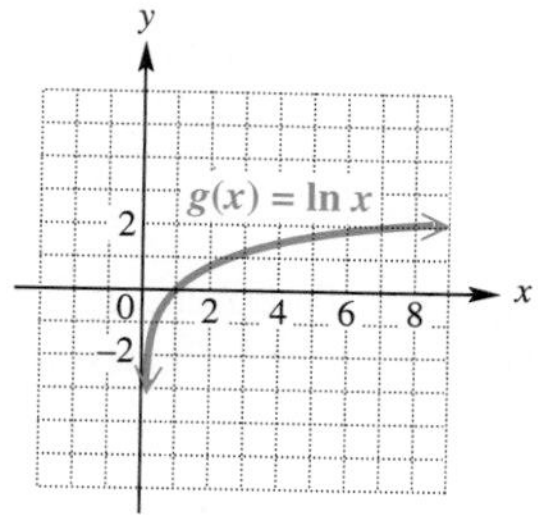

FIGURE 38

The expression $\ln e^k$ is the exponent to which the base e must be raised in order to obtain e^k. There is only one such number that will do this, and it is k. Thus, for all real numbers k,

$$\ln e^k = k.$$

The number *e* is named in honor of Leonhard Euler (1707–1783), the prolific Swiss mathematician. The value of *e* can be expressed as an **infinite series.**

$$e = 1 + \frac{1}{1} + \frac{1}{1 \cdot 2} + \frac{1}{1 \cdot 2 \cdot 3} + \frac{1}{1 \cdot 2 \cdot 3 \cdot 4} + \cdots$$

It can also be expressed in two ways using **continued fractions.**

$$e = 2 + \cfrac{1}{1 + \cfrac{1}{2 + \cfrac{1}{1 + \cfrac{1}{1 + \cfrac{1}{4 + \cdots}}}}}$$

$$e = 2 + \cfrac{1}{1 + \cfrac{1}{2 + \cfrac{2}{3 + \cfrac{3}{4 + \cfrac{4}{\cdots}}}}}$$

EXAMPLE 3 Finding the Time for an Amount to Triple

Suppose that a certain amount P is invested at an annual rate of 6.5%, compounded continuously. How long will it take for the amount to triple?

SOLUTION

We wish to find the value of t in the continuous compound interest formula that will make the amount A equal to $3P$ (since we want the initial investment, P, to triple).

$$A = Pe^{rt}$$

$$3P = Pe^{.065t} \qquad \text{Substitute } 3P \text{ for } A \text{ and } .065 \text{ for } r.$$

$$3 = e^{.065t} \qquad \text{Divide both sides by } P.$$

$$\ln 3 = \ln e^{.065t} \qquad \text{Take the natural logarithm of both sides.}$$

$$\ln 3 = .065t \qquad \text{Use the fact that } \ln e^k = k.$$

$$t = \frac{\ln 3}{.065} \qquad \text{Divide both sides by } .065.$$

A calculator shows that $\ln 3 \approx 1.098612289$. Dividing this by .065 gives

$$t \approx 16.9$$

to the nearest tenth. Therefore, it would take about 16.9 years for any initial investment P to triple under the given conditions. (The amount of time it would take for a given amount to double under given conditions is called **doubling time.** The doubling time for this example is $\frac{\ln 2}{.065} \approx 10.66$ years.)

Exponential Models in Nature

EXAMPLE 4 Modeling the Greenhouse Effect

The *greenhouse effect* refers to the phenomenon whereby emissions of gases such as carbon dioxide, methane, and chlorofluorocarbons (CFCs) have the potential to alter the climate of the earth and destroy the ozone layer. Concentrations of CFC-12, used in refrigeration technology, in parts per billion (ppb) can be modeled by the exponential function defined by

$$f(x) = .48e^{.04x},$$

where $x = 0$ represents 1990, $x = 1$ represents 1991, and so on. Use this function to approximate the concentration in 1998.

SOLUTION

Because $x = 0$ represents 1990, $x = 8$ represents 1998. Evaluate $f(8)$ using a calculator.

$$f(8) = .48e^{.04(8)} = .48e^{.32} \approx .66$$

In 1998, the concentration of CFC-12 was about .66 ppb.

Radioactive materials disintegrate according to exponential decay functions. The **half-life** of a quantity that decays exponentially is the amount of time that it takes for any initial amount to decay to half its initial value.

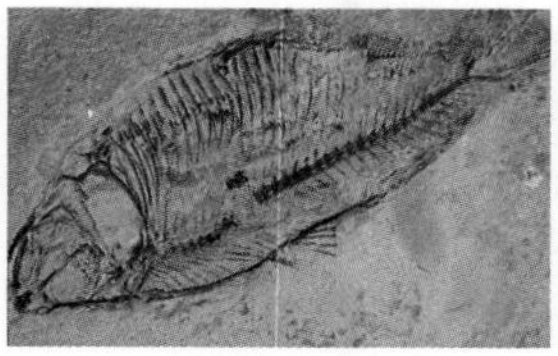

EXAMPLE 5 Carbon 14 Dating

Carbon 14 is a radioactive form of carbon that is found in all living plants and animals. After a plant or animal dies, the radiocarbon disintegrates. Scientists determine the age of the remains by comparing the amount of carbon 14 present with the amount found in living plants and animals. The amount of carbon 14 present after t years is modeled by the exponential equation $y = y_0 e^{-.0001216t}$, where y_0 represents the initial amount. **(a)** What is the half-life of carbon 14? **(b)** If an initial sample contains 1 gram of carbon 14, how much will be left after 10,000 years?

SOLUTION

(a) To find the half-life, let $y = \frac{1}{2}y_0$ in the equation.

$$\frac{1}{2}y_0 = y_0 e^{-.0001216t}$$

$$\frac{1}{2} = e^{-.0001216t} \qquad \text{Divide by } y_0.$$

$$\ln\left(\frac{1}{2}\right) = -.0001216t \qquad \text{Take the natural logarithm of both sides.}$$

$$t \approx 5700 \qquad \text{Use a calculator.}$$

The half-life of carbon 14 is about 5700 years.

(b) Evaluate y for $t = 10{,}000$ and $y_0 = 1$.

$$y = 1e^{-.0001216(10{,}000)} \approx .30 \qquad \text{Use a calculator.}$$

There will be about .30 grams remaining.

If you were to travel back in time to a period before the availability of handheld calculators, your study of logarithms would probably focus on how they can be used to perform calculations. **Tables of logarithms** were routinely included in mathematics textbooks. Properties of logarithms allow us to multiply by adding, divide by subtracting, and raise to powers and take roots by simple multiplications and divisions.

To get an idea of what computing with logarithms involved, look at the photo above. It shows a homework problem from May 15, 1930, where the student was required to calculate

$$\sqrt[5]{\frac{3.1416 \times 4771.21 \times 2.7183^{1/2}}{30.103^4 \times .4343^{1/2} \times 69.897^4}}.$$

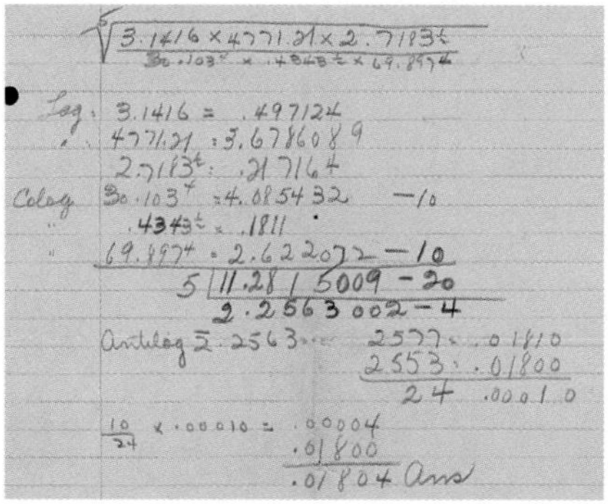

It was found on a well-preserved sheet of paper stuck inside an old mathematics text purchased in a used-book store in New Orleans. The student's answer .01804 is correct. Doesn't this make you appreciate your calculator?

A **common,** or **base ten logarithm,** of a positive number x is the exponent to which 10 must be raised in order to obtain x. Symbolically, we simply write $\log x$ to denote $\log_{10} x$. Common logarithms are still used in certain applications in science. For example, the **pH** of a substance is the negative of the common logarithm of the hydronium ion concentration in moles per liter. The pH value is a measure of the acidity or alkalinity of a solution. If pH > 7.0, the solution is alkaline; if pH < 7.0, it is acidic. The **Richter scale,** used to measure the intensity of earthquakes, is based on logarithms, and sound intensities, measured in decibels, also have a logarithmic basis.

8.6 EXERCISES

Fill in each blank with the correct response.

1. For an exponential function $f(x) = a^x$, if $a > 1$, the graph _______ (rises/falls) from left to right. If $0 < a < 1$, the graph _______ (rises/falls) from left to right.

2. The y-intercept of the graph of $y = a^x$ is _____.

3. The graph of the exponential function $f(x) = a^x$ __________ (does/does not) have an x-intercept.

4. The point (2, ____) is on the graph of $f(x) = 3^{4x-3}$.

5. For a logarithmic function $g(x) = \log_a x$, if $a > 1$, the graph _______ (rises/falls) from left to right. If $0 < a < 1$, the graph _______ (rises/falls) from left to right.

6. The x-intercept of the graph of $y = \log_a x$ is _____.

7. The graph of the exponential function $g(x) = \log_a x$ __________ (does/does not) have a y-intercept.

8. The point (98, _____) lies on the graph of $g(x) = \log_{10}(x + 2)$.

Use a calculator to find an approximation for each number. Give as many digits as the calculator displays.

9. $9^{3/7}$

10. $14^{2/7}$

11. $(.83)^{-1.2}$

12. $(.97)^{3.4}$

13. $(\sqrt{6})^{\sqrt{5}}$

14. $(\sqrt{7})^{\sqrt{3}}$

15. $\left(\frac{1}{3}\right)^{9.8}$

16. $\left(\frac{2}{5}\right)^{8.1}$

Sketch the graph of each function.

17. $f(x) = 3^x$

18. $f(x) = 5^x$

19. $f(x) = \left(\frac{1}{4}\right)^x$

20. $f(x) = \left(\frac{1}{3}\right)^x$

Use a calculator to approximate each number. Give as many digits as the calculator displays.

21. e^3

22. e^4

23. e^{-4}

24. e^{-3}

In Exercises 25–28, rewrite the exponential equation as a logarithmic equation. In Exercises 29–32, rewrite the logarithmic equation as an exponential equation.

25. $4^2 = 16$

26. $5^3 = 125$

27. $\left(\frac{2}{3}\right)^{-3} = \frac{27}{8}$

28. $\left(\frac{1}{10}\right)^{-4} = 10{,}000$

29. $5 = \log_2 32$

30. $3 = \log_4 64$

31. $1 = \log_3 3$

32. $0 = \log_{12} 1$

Use a calculator to approximate each number. Give as many digits as the calculator displays.

33. $\ln 4$

34. $\ln 6$

35. $\ln .35$

36. $\ln 2.45$

Global Warming *This figure appeared in the October 1990 issue of* National Geographic. *It shows projected temperature increases using two graphs: one an exponential-type curve, and the other linear. From the figure, approximate the increase* **(a)** *for the exponential curve, and* **(b)** *for the linear graph for each of the years in Exercises 37–40.*

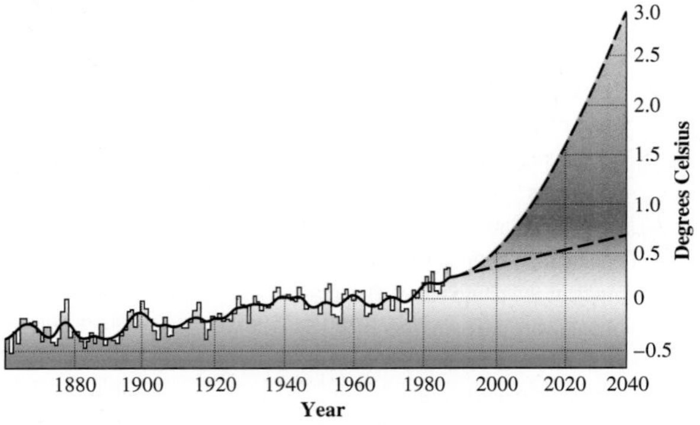

Source: "Zero Equals Average Global Temperature for the Period 1950–1979" by Dale D. Glasgow from *National Geographic,* October 1990. Reprinted by permission of the National Geographic Society.

37. 2000

38. 2010

39. 2020

40. 2040

Sketch the graph of each function. (Hint: Use the graphs of the exponential functions in Exercises 17–20 to help.)

41. $g(x) = \log_3 x$

42. $g(x) = \log_5 x$

43. $g(x) = \log_{1/4} x$

44. $g(x) = \log_{1/3} x$

Investment *Determine the amount of money that will be accumulated in an account that pays compound interest, given the initial principal in each of the following.*

45. \$20,000 invested at 3% annual interest for 4 years compounded **(a)** annually; **(b)** semiannually

46. \$35,000 invested at 4.2% annual interest for 3 years compounded **(a)** annually; **(b)** quarterly

47. \$27,500 invested at 3.95% annual interest for 5 years compounded **(a)** daily ($n = 365$); **(b)** continuously

48. \$15,800 invested at 4.6% annual interest for 6.5 years compounded **(a)** quarterly; **(b)** continuously

Comparing Investment Plans *In Exercises 49 and 50, decide which of the two plans will provide a better yield. (Interest rates stated are annual rates.)*

49. Plan A: \$40,000 invested for 3 years at 4.5%, compounded quarterly
Plan B: \$40,000 invested for 3 years at 4.4%, compounded continuously

50. Plan A: \$50,000 invested for 10 years at 4.75%, compounded daily ($n = 365$)
Plan B: \$50,000 invested for 10 years at 4.7%, compounded continuously

Solve each problem.

51. *Atmospheric Pressure* The atmospheric pressure f (in millibars) at a given altitude x (in meters) is approximated by the function

$$f(x) = 1013e^{-.0001341x}.$$

(a) Predict the pressure at 1500 meters.
(b) Predict the pressure at 11,000 meters.

52. *World Population Growth* World population in millions closely fits the exponential function defined by

$$y = 5282e^{.01405x},$$

where x is the number of years since 1990. (*Source*: U.S. Census Bureau.)

(a) The world population was about 6080 million in 2000. How closely does the function approximate this value?
(b) Use this model to approximate the population in 2005.
(c) Use the model to predict the population in 2010.

53. *Population of Pakistan* The U.S. Census Bureau projects that the approximate population of Pakistan will grow according to the function

$$f(x) = 146{,}250(2)^{.0176x},$$

where $x = 0$ represents the year 2000, $x = 25$ represents 2025, and $x = 50$ represents 2050.

(a) According to this model, what was the population of Pakistan in 2000?
(b) What will the population be in 2025?
(c) How will the population in 2025 compare to the population in 2000?

54. *Population of Brazil* The U.S. Census Bureau projects that the approximate population of Brazil will grow according to the function

$$f(x) = 176{,}000(2)^{.008x},$$

where $x = 0$ represents the year 2000, $x = 25$ represents 2025, and $x = 50$ represents 2050.

(a) According to this model, what was the population of Brazil in 2000?
(b) What will the population be in 2025?
(c) How will the population in 2025 compare to the population in 2000?

55. ***Decay of Lead*** A sample of 500 grams of radioactive lead 210 decays to polonium 210 according to the function defined by

$$A(t) = 500e^{-.032t},$$

where t is time in years. Find the amount of the sample remaining after

(a) 4 years
(b) 8 years
(c) 20 years
(d) Find the half-life.

56. ***Decay of Plutonium*** Repeat Exercise 55 for 500 grams of plutonium 241, which decays according to the function defined as follows, where t is time in years.

$$A(t) = A_0e^{-.053t}$$

57. ***Decay of Radium*** Find the half-life of radium 226, which decays according to the function defined as follows, where t is time in years.

$$A(t) = A_0e^{-.00043t}$$

58. ***Decay of Iodine*** How long will it take any quantity of iodine 131 to decay to 25% of its initial amount, knowing that it decays according to the function defined as follows, where t is time in days?

$$A(t) = A_0e^{-.087t}$$

59. ***Carbon 14 Dating*** Suppose an Egyptian mummy is discovered in which the amount of carbon 14 present is only about one-third the amount found in living human beings. About how long ago did the Egyptian die?

60. ***Carbon 14 Dating*** A sample from a refuse deposit near the Strait of Magellan had 60% of the carbon 14 of a contemporary living sample. How old was the sample?

61. ***Carbon 14 Dating*** Estimate the age of a specimen that contains 20% of the carbon 14 of a comparable living specimen.

62. ***Value of a Copier*** A small business estimates that the value $V(t)$ of a copier is decreasing according to the function defined by

$$V(t) = 5000(2)^{-.15t},$$

where t is the number of years that have elapsed since the machine was purchased, and $V(t)$ is in dollars.

(a) What was the original value of the machine?
(b) What is the value of the machine 5 years after purchase? Give your answer to the nearest dollar.
(c) What is the value of the machine 10 years after purchase? Give your answer to the nearest dollar.

Earthquake Intensity *In the United States, the intensity of an earthquake is rated using the Richter scale. The Richter scale rating of an earthquake in intensity x is given by*

$$R = \log_{10}\frac{x}{x_0},$$

where x_0 is the intensity of an earthquake of a certain (small) size. The figure shows Richter scale ratings for major Southern California earthquakes since 1920. As the figure indicates, earthquakes "come in bunches" and the 1990s were an especially busy time.

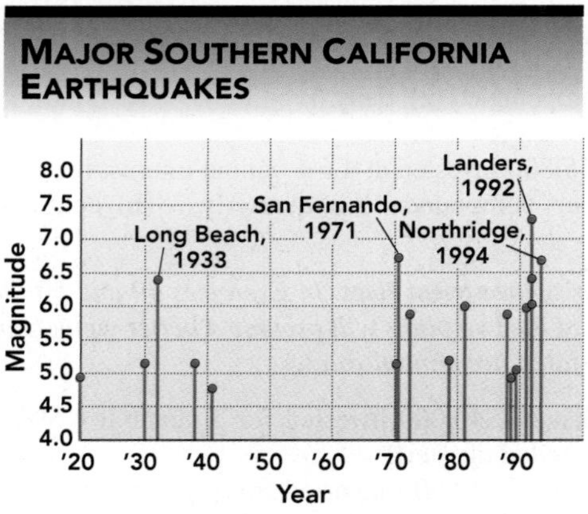

Source: Caltech; U.S. Geological Survey.

Writing the logarithmic equation just given in exponential form, we get

$$10^R = \frac{x}{x_0} \quad \text{or} \quad x = 10^R x_0.$$

The 1994 Northridge earthquake had a Richter scale rating of 6.7; the Landers earthquake had a rating of 7.3.

63. How many times as powerful was the Landers earthquake compared to the Northridge earthquake?

64. Compare the smallest rated earthquake in the graph (at 4.8) with the Landers quake. How many times as powerful was the Landers quake?

8.7 Systems of Equations and Applications

Linear Systems in Two Variables • Elimination Method • Substitution Method • Linear Systems in Three Variables • Applications of Linear Systems

Linear Systems in Two Variables

During the 1990s, the audiocassette as a medium of recorded music began its demise, while compact discs gained in popularity. Figure 39 shows these trends. With $x = 0$ representing 1990 and $x = 10$ representing 2000, the two equations

$$y = 74.523x + 289.3 \quad \text{Compact discs}$$
$$y = -34.107x + 434.49 \quad \text{Audiocassettes}$$

provide good models for the growth in sales of CDs and decline in sales of cassettes, where y is in millions of dollars. The two equations here are considered together, and such a set of equations is called a **system of equations.** The point where the graphs in Figure 39 intersect is a solution of each of the individual equations. It is also the solution of the system of equations. From the figure we see that in 1991, both CDs and cassettes had sales of about $400 million.

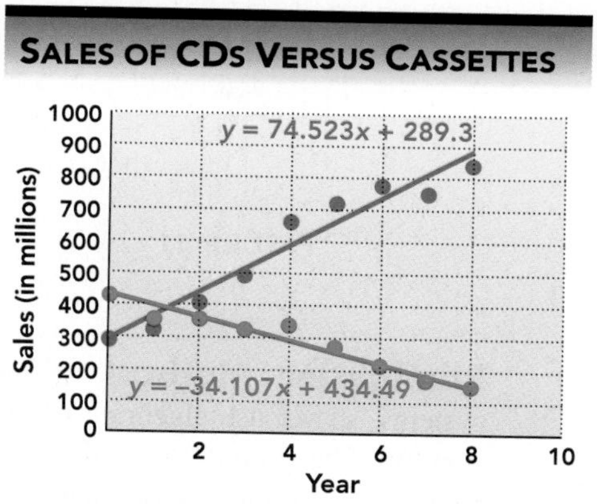

Source: World Almanac and Book of Facts.

FIGURE 39

The definition of a linear equation given earlier can be extended to more variables. Any equation of the form

$$a_1x_1 + a_2x_2 + \cdots + a_nx_n = b$$

for real numbers $a_1, a_2, \ldots, a_n$ (not all of which are 0), and b, is a **linear equation in *n* variables.** If all the equations in a system are linear, the system is a **system of linear equations,** or a **linear system.**

In Figure 40, the two linear equations $x + y = 5$ and $2x - y = 4$ are graphed in the same coordinate system. Notice that they intersect at the point (3, 2). Because (3, 2) is the only ordered pair that satisfies both equations at the same time, we say that $\{(3, 2)\}$ is the solution set of the system

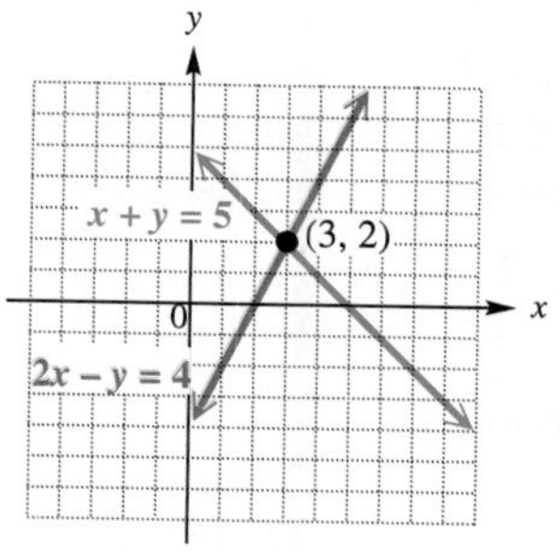

FIGURE 40

$$x + y = 5$$
$$2x - y = 4$$

Because the graph of a linear equation is a straight line, there are three possibilities for the number of solutions in the solution set of a system of two linear equations, as shown in Figures 41–43.

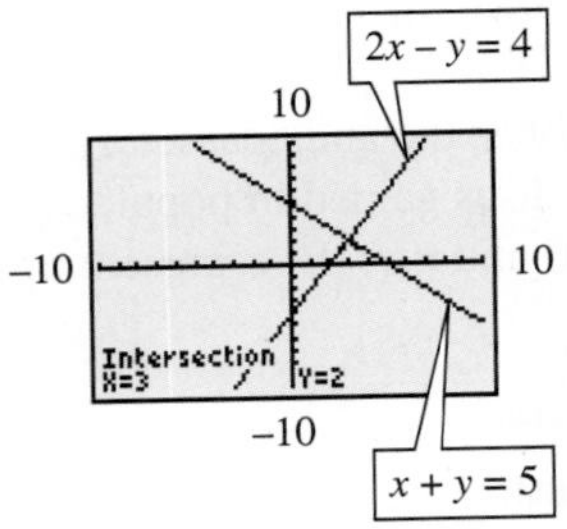

A graphing calculator supports our statement that (3, 2) is the solution of the system

$$x + y = 5$$
$$2x - y = 4.$$

Graphs of a Linear System (The Three Possibilities)

1. The two graphs intersect in a single point. The coordinates of this point give the only solution of the system. In this case, the system is **consistent** and the equations are **independent.** This is the most common case. See Figure 41.
2. The graphs are parallel lines. In this case, the system is **inconsistent** and the equations are **independent.** That is, there is no solution common to both equations of the system, and the solution set is $\emptyset$. See Figure 42.
3. The graphs are the same line. In this case, the system is **consistent** and the equations are **dependent,** because any solution of one equation of the system is also a solution of the other. The solution set is an infinite set of ordered pairs representing the points on the line. See Figure 43.

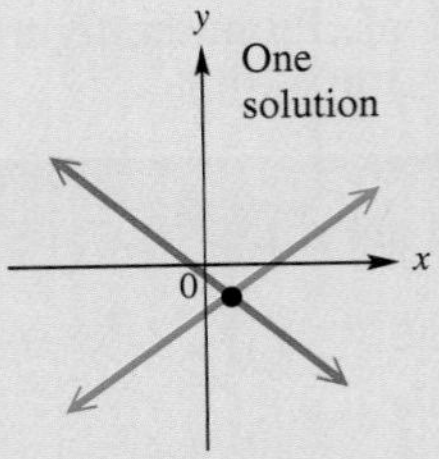

FIGURE 41

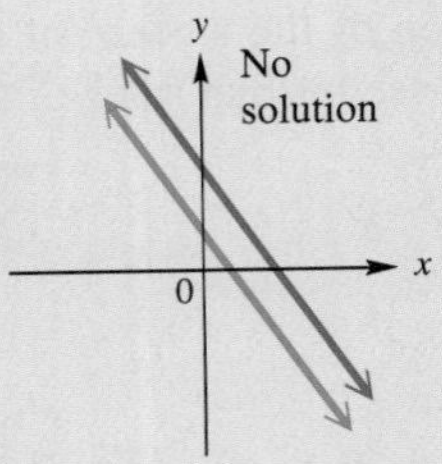

FIGURE 42

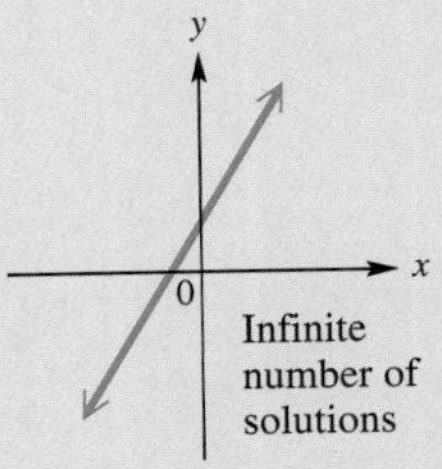

FIGURE 43

Elimination Method In most cases, we cannot rely on graphing to solve systems, so we use algebraic methods. One such method is called the **elimination method.** The elimination method involves combining the two equations of the system so that one variable is eliminated. This is done using the following fact.

$$\textbf{If } a = b \textbf{ and } c = d\textbf{, then } a + c = b + d.$$

The method of solving a system by elimination is summarized as follows.

The solution of **systems of equations** of graphs more complicated than straight lines is the principle behind the **mattang** (shown on this stamp), a stick chart used by the people of the Marshall Islands in the Pacific. A mattang is made of roots tied together with coconut fibers, and it shows the wave patterns found when approaching an island.

Solving Linear Systems by Elimination

Step 1 **Write both equations in standard form $Ax + By = C$.**

Step 2 **Make the coefficients of one pair of variable terms opposites.** Multiply one or both equations by appropriate numbers so that the sum of the coefficients of either x or y is zero.

Step 3 **Add** the new equations to eliminate a variable. The sum should be an equation with just one variable.

Step 4 **Solve** the equation from Step 3.

(continued)

Step 5 **Find the other value.** Substitute the result of Step 4 into either of the given equations and solve for the other variable.

Step 6 **Find the solution set.** Check the solution in both of the given equations. Then write the solution set.

EXAMPLE 1 Solving a System by Elimination (Two Variables)

Solve the system.

$$5x - 2y = 4 \quad (1)$$
$$2x + 3y = 13 \quad (2)$$

SOLUTION

Step 1 Both equations are already in standard form.

Step 2 Our goal is to add the two equations so that one of the variables is eliminated. Suppose we wish to eliminate the variable x. Since the coefficients of x are *not* opposites, we must first transform one or both equations so that the coefficients *are* opposites. Then, when we combine the equations, the term with x will have a coefficient of 0, and we will be able to solve for y. We begin by multiplying equation (1) by 2 and equation (2) by -5.

$$10x - 4y = 8 \quad \text{2 times each side of equation (1)}$$
$$-10x - 15y = -65 \quad \text{−5 times each side of equation (2)}$$

Step 3 Now add the two equations to eliminate x.

$$\begin{aligned} 10x - 4y &= 8 \\ -10x - 15y &= -65 \\ \hline -19y &= -57 \quad \text{Add.} \end{aligned}$$

Step 4 Solve the equation from Step 3 to get $y = 3$.

Step 5 To find x, we substitute 3 for y in either of the original equations.

$$\begin{aligned} 2x + 3y &= 13 && \text{We use equation (2).} \\ 2x + 3(3) &= 13 && \text{Let } y = 3. \\ 2x + 9 &= 13 \\ 2x &= 4 && \text{Subtract 9.} \\ x &= 2 && \text{Divide by 2.} \end{aligned}$$

Write the x-value first.

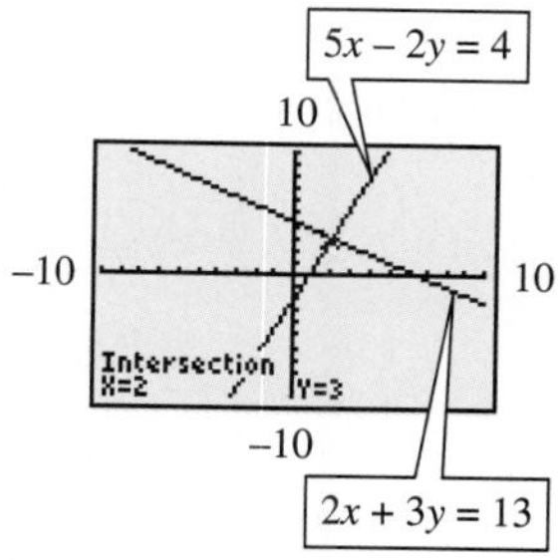

The solution in Example 1 is supported by a graphing calculator. We solve each equation for y, graph them both, and find the point of intersection of the two lines: (2, 3).

Step 6 The solution appears to be (2, 3). To check, substitute 2 for x and 3 for y in both of the original equations.

$$\begin{aligned} 5x - 2y &= 4 && (1) \\ 5(2) - 2(3) &= 4 && ? \\ 10 - 6 &= 4 && ? \\ 4 &= 4 && \text{True} \end{aligned}$$

$$\begin{aligned} 2x + 3y &= 13 && (2) \\ 2(2) + 3(3) &= 13 && ? \\ 4 + 9 &= 13 && ? \\ 13 &= 13 && \text{True} \end{aligned}$$

The solution set is $\{(2, 3)\}$.

Substitution Method Linear systems can also be solved by the **substitution method.** This method is most useful for solving linear systems in which one variable has coefficient 1 or -1. As shown in more advanced algebra courses, the substitution method also is the best choice for solving many *nonlinear* systems.

Solving Linear Systems by Substitution

Step 1 **Solve for one variable in terms of the other.** Solve one of the equations for either variable. (If one of the variables has coefficient 1 or -1, choose it, since the substitution method is usually easier this way.)

Step 2 **Substitute** for that variable in the other equation. The result should be an equation with just one variable.

Step 3 **Solve** the equation from Step 2.

Step 4 **Find the other value.** Substitute the result from Step 3 into the equation from Step 1 to find the value of the other variable.

Step 5 **Find the solution set.** Check the solution in both of the given equations. Then write the solution set.

EXAMPLE 2 Solving a System by Substitution (Two Variables)

Solve the system.

$$3x + 2y = 13 \quad (1)$$
$$4x - y = -1 \quad (2)$$

SOLUTION

Step 1 To use the substitution method, first solve one of the equations for either x or y. Since the coefficient of y in equation (2) is -1, it is easiest to solve for y in equation (2).

$$-y = -1 - 4x \quad \text{Equation (2) rearranged}$$
$$y = 1 + 4x$$

Step 2 Substitute $1 + 4x$ for y in equation (1) to obtain an equation in x.

$$3x + 2y = 13 \quad (1)$$
$$3x + 2(1 + 4x) = 13 \quad \text{Let } y = 1 + 4x.$$

Step 3 Solve for x in the equation just obtained.

$$3x + 2 + 8x = 13 \quad \text{Distributive property}$$
$$11x = 11 \quad \text{Combine terms; subtract 2.}$$
$$x = 1 \quad \text{Divide by 11.}$$

Step 4 Now solve for y. Because $y = 1 + 4x$, $y = 1 + 4(1) = 5$.

Step 5 Check to see that the ordered pair $(1, 5)$ satisfies both equations. The solution set is $\{(1, 5)\}$. ■

The next example illustrates special cases that may result when systems are solved. (We will use the elimination method, but the same conclusions will follow when the substitution method is used.)

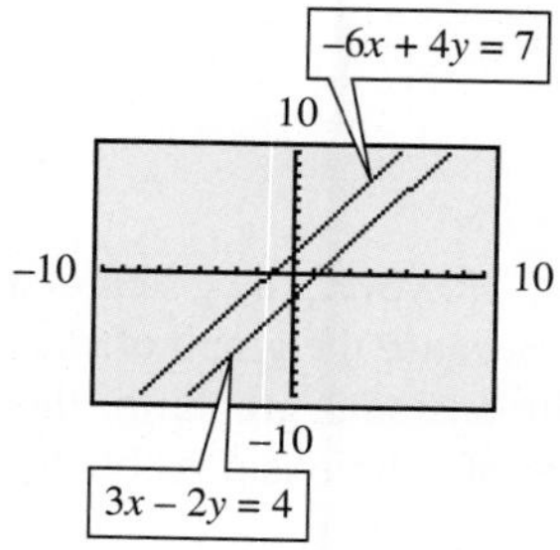

The graphs of the equations in Example 3(a) are parallel. There are no solutions.

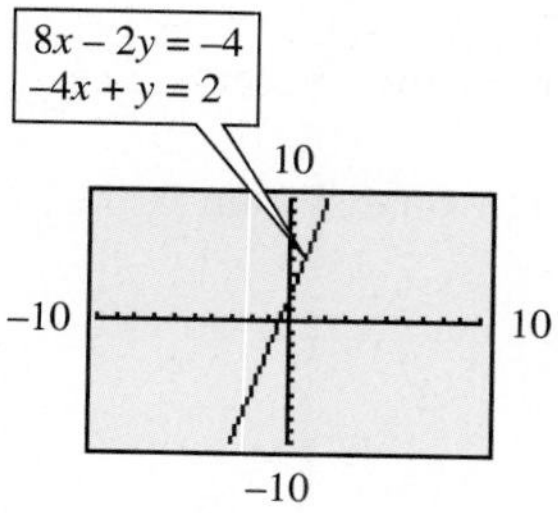

The graphs of the equations in Example 3(b) coincide. We see only one line. There are infinitely many solutions.

EXAMPLE 3 Solving Systems of Special Cases

Solve each system.

(a) $3x - 2y = 4$ (1)
$\quad -6x + 4y = 7$ (2)

(b) $-4x + y = 2$ (3)
$\quad 8x - 2y = -4$ (4)

SOLUTION

(a) Eliminate x by multiplying both sides of equation (1) by 2 and then adding.

$$\begin{aligned} 6x - 4y &= 8 && \text{2 times equation (1)} \\ -6x + 4y &= 7 && \text{(2)} \\ \hline 0 &= 15 && \text{False} \end{aligned}$$

Both variables were eliminated here, leaving the false statement $0 = 15$, indicating that these two equations have no solutions in common. The system is inconsistent, with the empty set $\emptyset$ as the solution set.

(b) Eliminate x by multiplying both sides of equation (3) by 2 and then adding the result to equation (4).

$$\begin{aligned} -8x + 2y &= 4 && \text{2 times equation (3)} \\ 8x - 2y &= -4 && \text{(4)} \\ \hline 0 &= 0 && \text{True} \end{aligned}$$

This true statement, $0 = 0$, indicates that a solution of one equation is also a solution of the other, so the solution set is an infinite set of ordered pairs. The two equations are dependent.

We write the solution set of a system of dependent equations as a set of ordered pairs by expressing x in terms of y as follows. Choose either equation and solve for x. Choosing equation (3) gives

$$-4x + y = 2 \qquad (3)$$

$$x = \frac{2 - y}{-4} = \frac{y - 2}{4}.$$

The solution set is written as

$$\left\{\left(\frac{y - 2}{4}, y\right)\right\}.$$

By selecting values for y and calculating the corresponding values for x, individual ordered pairs of the solution set can be found. For example, if $y = -2$, $x = \frac{-2-2}{4} = -1$ and the ordered pair $(-1, -2)$ is a solution. ■

Linear Systems in Three Variables A solution of an equation in three variables, such as $2x + 3y - z = 4$, is called an **ordered triple** and is written (x, y, z). For example, the ordered triples $(1, 1, 1)$ and $(10, -3, 7)$ are both solutions of the equation $2x + 3y - z = 4$, because the numbers in these ordered triples satisfy the equation when used as replacements for x, y, and z, respectively. The methods of solving systems of two equations in two variables can be extended to solving systems of equations in three variables such as

$$\begin{aligned} 4x + 8y + z &= 2 \\ x + 7y - 3z &= -14 \\ 2x - 3y + 2z &= 3. \end{aligned}$$

Theoretically, a system of this type can be solved by graphing. However, the graph of a linear equation with three variables is a *plane* and not a line. Because the graph of each equation of the system is a plane, which requires three-dimensional graphing, this method is not practical. However, it does illustrate the number of solutions possible for such systems, as Figure 44 shows.

Possibilities for Graphs of Linear Systems in Three Variables

1. The three planes may meet at a single, common point that forms the solution set of the system. See Figure 44(a).
2. The three planes may have the points of a line in common so that the set of points along that line is the solution set of the system. See Figure 44(b).
3. The three planes may coincide so that the solution set of the system is the set of all points on that plane. See Figure 44(c).
4. The planes may have no points common to all three so that there is no solution for the system. See Figure 44(d), (e), and (f).

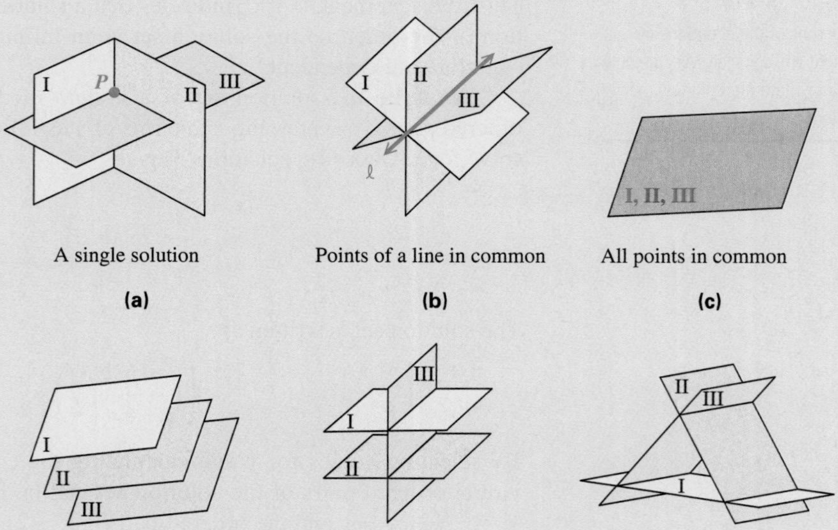

A single solution (a)

Points of a line in common (b)

All points in common (c)

No points in common (d)

No points in common (e)

No points in common (f)

FIGURE 44

Because graphing to find the solution set of a system of three equations in three variables is impractical, these systems are solved with an extension of the elimination method, summarized as follows.

Solving Linear Systems in Three Variables by Elimination

Step 1 **Eliminate a variable.** Use the elimination method to eliminate any variable using any two of the given equations. The result is an equation in two variables.

Step 2 **Eliminate the same variable again.** Eliminate the *same* variable using any *other* two equations. The result is an equation in the same two variables as in Step 1.

Step 3 **Eliminate a different variable and solve.** Use the elimination method to eliminate a second variable using the two equations in two variables that result from Steps 1 and 2. The result is an equation in one variable that gives the value of that variable.

Step 4 **Find a second value.** Substitute the value of the variable found in Step 3 into either of the equations in two variables to find the value of the second variable.

Step 5 **Find a third value.** Use the values of the two variables from Steps 3 and 4 to find the value of the third variable by substituting into any of the original equations.

Step 6 **Find the solution set.** Check the solution in all of the original equations. Then write the solution set.

EXAMPLE 4 Solving a System (Three Variables)

Solve the system.

$$\begin{aligned} 4x + 8y + z &= 2 && (1)\\ x + 7y - 3z &= -14 && (2)\\ 2x - 3y + 2z &= 3 && (3) \end{aligned}$$

SOLUTION

Step 1 The choice of which variable to eliminate is arbitrary. Suppose we decide to begin by eliminating z. To do this, multiply both sides of equation (1) by 3 and then add the result to equation (2).

$$\begin{aligned} 12x + 24y + 3z &= 6 && \text{Multiply both sides of equation (1) by 3.}\\ x + 7y - 3z &= -14 && (2)\\ \hline 13x + 31y \quad &= -8 && \text{Add.} \end{aligned}$$

Step 2 The new equation has only two variables. To get another equation without z, multiply both sides of equation (1) by -2 and add the result to equation (3). ***It is essential at this point to eliminate the same variable, z.***

$$\begin{aligned} -8x - 16y - 2z &= -4 && \text{Multiply both sides of equation (1) by } -2.\\ 2x - 3y + 2z &= 3 && (3)\\ \hline -6x - 19y &= -1 && \text{Add.} \end{aligned}$$

Step 3 Now solve the system of equations from Steps 1 and 2 for x and y. (This step is possible only if the *same* variable is eliminated in the first two steps.)

$$\begin{aligned} 78x + 186y &= -48 && \text{Multiply both sides of } 13x + 31y = -8 \text{ by 6.}\\ -78x - 247y &= -13 && \text{Multiply both sides of } -6x - 19y = -1 \text{ by 13.}\\ \hline -61y &= -61 && \text{Add.}\\ y &= 1 \end{aligned}$$

Step 4 Substitute 1 for y in either equation from Steps 1 and 2. Choosing $-6x - 19y = -1$ gives

$$\begin{aligned} -6x - 19y &= -1\\ -6x - 19(1) &= -1 && \text{Let } y = 1.\\ -6x - 19 &= -1\\ -6x &= 18\\ x &= -3. \end{aligned}$$

Step 5 Substitute -3 for x and 1 for y in any one of the three given equations to find z. Choosing equation (1) gives

$$\begin{aligned} 4x + 8y + z &= 2\\ 4(-3) + 8(1) + z &= 2 && \text{Let } x = -3 \text{ and } y = 1.\\ z &= 6. \end{aligned}$$

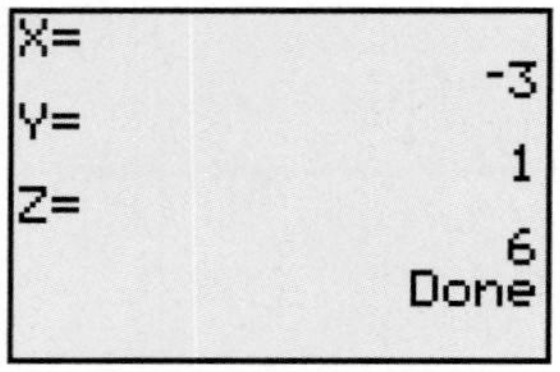

A graphing calculator can be *programmed* to solve a system such as the one in Example 4. Compare the result here to the solution in the text.

Step 6 It appears that the ordered triple $(-3, 1, 6)$ is the only solution of the system. Check that the solution satisfies all three equations of the system. We show the check here only for equation (1).

$$\begin{aligned} 4x + 8y + z &= 2 && (1)\\ 4(-3) + 8(1) + 6 &= 2 && ?\\ -12 + 8 + 6 &= 2 && ?\\ 2 &= 2 && \text{True} \end{aligned}$$

Because $(-3, 1, 6)$ also satisfies equations (2) and (3), the solution set is $\{(-3, 1, 6)\}$. ■

Applications of Linear Systems

PROBLEM-SOLVING HINT Many problems involve more than one unknown quantity. Although some problems with two unknowns can be solved using just one variable, many times it is easier to use two variables. To solve a problem with two unknowns, we write two equations that relate the unknown quantities. The system formed by the pair of equations then can be solved using the methods of this section.

Just for Fun Here is a way to find a person's age and the date of his or her birth. Ask the person to do the following.

1. Multiply the number of the month of birth by *100*.
2. Add the date of the month.
3. Multiply the sum by *2*.
4. Add *8*.
5. Multiply this result by *5*.
6. Add *4*.
7. Multiply by *10*.
8. Add *4*.
9. Add the person's age to this.

Ask the person the number obtained. You should then subtract *444* from the number you are given. The final two figures give the age, the next figure (or figures) will identify the date of the month, and the first figure (or figures) will give the number of the month.

The following steps, based on the six-step problem-solving method first introduced in Chapter 7, give a strategy for solving problems using more than one variable.

Solving an Applied Problem by Writing a System of Equations

Step 1 **Read** the problem carefully until you understand what is given and what is to be found.

Step 2 **Assign variables** to represent the unknown values, using diagrams or tables as needed. *Write down* what each variable represents.

Step 3 **Write a system of equations** that relates the unknowns.

Step 4 **Solve** the system of equations.

Step 5 **State the answer** to the problem. Does it seem reasonable?

Step 6 **Check** the answer in the words of the original problem.

Problems about the perimeter of a geometric figure often involve two unknowns and can be solved using systems of equations.

EXAMPLE 5 Solving a Perimeter Problem

A rectangular soccer field may have a width between 50 and 100 yards and a length between 50 and 100 yards. Suppose that one particular field has a perimeter of 320 yards. Its length measures 40 yards more than its width. What are the dimensions of this field? (*Source: Microsoft Encarta Encyclopedia.*)

SOLUTION

Step 1 **Read** the problem again. We are asked to find the dimensions of the field.

Step 2 **Assign variables.** Let $L =$ the length and $W =$ the width.

Step 3 **Write a system of equations.** Because the perimeter is 320 yards, we find one equation by using the perimeter formula:

$$2L + 2W = 320.$$

Because the length is 40 yards more than the width, we have

$$L = W + 40.$$

See Figure 45. The system is, therefore,

$$2L + 2W = 320 \quad (1)$$
$$L = W + 40. \quad (2)$$

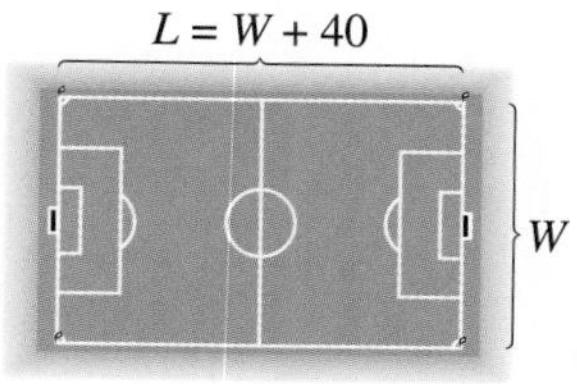

FIGURE 45

Step 4 **Solve** the system of equations. Since equation (2) is solved for L, we can substitute $W + 40$ for L in equation (1), and solve for W.

$$\begin{aligned} 2L + 2W &= 320 && (1) \\ 2(W + 40) + 2W &= 320 && L = W + 40 \\ 2W + 80 + 2W &= 320 && \text{Distributive property} \\ 4W + 80 &= 320 && \text{Combine like terms.} \\ 4W &= 240 && \text{Subtract 80.} \\ W &= 60 && \text{Divide by 4.} \end{aligned}$$

Let $W = 60$ in the equation $L = W + 40$ to find L.

$$L = 60 + 40 = 100$$

Step 5 **State the answer.** The length is 100 yards, and the width is 60 yards. Both dimensions are within the ranges given in the problem.

Step 6 **Check.** The answer is correct, because the perimeter of this soccer field is

$$2(100) + 2(60) = 320 \text{ yards},$$

and the length, 100 yards, is indeed 40 yards more than the width, because

$$100 - 40 = 60.$$

Professional sport ticket prices increase annually. Average per-ticket prices in three of the four major sports (football, basketball, and hockey) now exceed $30.00.

EXAMPLE 6 Solving a Problem About Ticket Prices

During recent National Hockey League and National Basketball Association seasons, two hockey tickets and one basketball ticket purchased at their average prices would have cost $110.40. One hockey ticket and two basketball tickets would have cost $106.32. What were the average ticket prices for the two sports? (*Source:* Team Marketing Report, Chicago.)

SOLUTION

Step 1 **Read** the problem again. There are two unknowns.

Step 2 **Assign variables.** Let h represent the average price for a hockey ticket and b represent the average price for a basketball ticket.

Step 3 **Write a system of equations.** Because two hockey tickets and one basketball ticket cost a total of $110.40, one equation for the system is

$$2h + b = 110.40.$$

By similar reasoning, the second equation is

$$h + 2b = 106.32.$$

Therefore, the system is

$$\begin{aligned} 2h + b &= 110.40 && (1) \\ h + 2b &= 106.32. && (2) \end{aligned}$$

Step 4 **Solve** the system of equations. We eliminate h.

$$\begin{aligned} 2h + b &= 110.40 && (1) \\ -2h - 4b &= -212.64 && \text{Multiply each side of (2) by } -2. \\ \hline -3b &= -102.24 && \text{Add.} \\ b &= 34.08 && \text{Divide by } -3. \end{aligned}$$

To find the value of h, let $b = 34.08$ in equation (2).

$$\begin{aligned} h + 2b &= 106.32 && (2) \\ h + 2(34.08) &= 106.32 && \text{Let } b = 34.08. \\ h + 68.16 &= 106.32 && \text{Multiply.} \\ h &= 38.16 && \text{Subtract 68.16.} \end{aligned}$$

Step 5 **State the answer.** The average price for one basketball ticket was \$34.08. For one hockey ticket, the average price was \$38.16.

Step 6 **Check** that these values satisfy the conditions stated in the problem.

We solved mixture problems earlier using one variable. Another approach is to use two variables and a system of equations.

Problems that can be solved by writing a **system of equations** have been of interest historically. The following problem first appeared in a **Hindu** work that dates back to about A.D. 850.

The mixed price of 9 citrons [a lemonlike fruit shown in the photo] and 7 fragrant wood apples is 107; again, the mixed price of 7 citrons and 9 fragrant wood apples is 101. O you arithmetician, tell me quickly the price of a citron and the price of a wood apple here, having distinctly separated those prices well.

Use a system to solve this problem. The answer can be found at the end of the exercises for this section on page 514.

EXAMPLE 7 Solving a Mixture Problem

How many ounces each of 5% hydrochloric acid and 20% hydrochloric acid must be combined to get 10 oz of solution that is 12.5% hydrochloric acid?

SOLUTION

Step 1 **Read** the problem. Two solutions of different strengths are being mixed together to get a specific amount of a solution with an "in-between" strength.

Step 2 **Assign variables.** Let x represent the number of ounces of 5% solution and y represent the number of ounces of 20% solution. Summarize the information from the problem in a table.

Percent (as a decimal)	Ounces of Solution	Ounces of Pure Acid
$5\% = .05$	x	$.05x$
$20\% = .20$	y	$.20y$
$12.5\% = .125$	10	$(.125)10$

Figure 46 also illustrates what is happening in the problem.

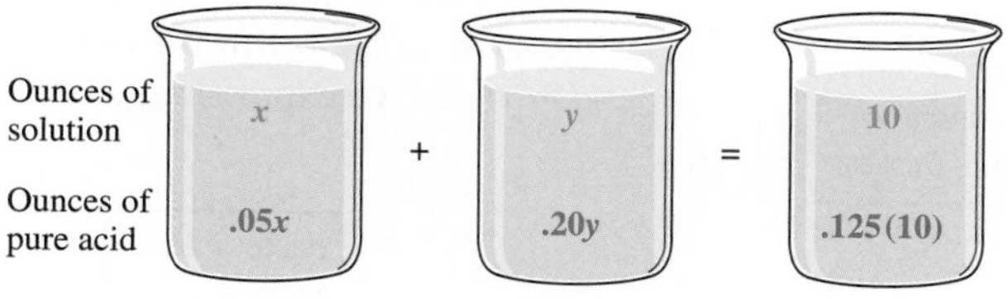

FIGURE 46

Step 3 **Write a system of equations.** When x ounces of 5% solution and y ounces of 20% solution are combined, the total number of ounces is 10, so

$$x + y = 10. \quad (1)$$

The ounces of pure acid in the 5% solution ($.05x$) plus the ounces of pure acid in the 20% solution ($.20y$) should equal the total ounces of pure acid in the mixture, which is $(.125)10$, or 1.25. That is,

$$.05x + .20y = 1.25. \quad (2)$$

Notice that these equations can be quickly determined by reading down in the table or using the labels in Figure 46.

Step 4 **Solve** the system of equations (1) and (2). We eliminate x.

$$\begin{aligned} 5x + 20y &= 125 && \text{Multiply each side of (2) by 100.} \\ -5x - 5y &= -50 && \text{Multiply each side of (1) by } -5. \\ \hline 15y &= 75 && \text{Add.} \\ y &= 5 \end{aligned}$$

Because $y = 5$ and $x + y = 10$, x is also 5.

Step 5 **State the answer.** The desired mixture will require 5 ounces of the 5% solution and 5 ounces of the 20% solution.

Step 6 **Check** that these values satisfy both equations of the system. ■

Problems that use the distance formula $d = rt$ were first introduced in Chapter 7. In many cases, these problems can be solved with systems of two linear equations. Keep in mind that setting up a table and drawing a sketch will help you solve such problems.

François Viète, a mathematician of sixteenth-century France, did much for the symbolism of mathematics. Before his time, different symbols were often used for different powers of a quantity. Viète used the same letter with a description of the power and the coefficient. According to Howard Eves in *An Introduction to the History of Mathematics*, Viète would have written

$$5BA^2 - 2CA + A^3 = D$$

as

*B*5 in *A* quad − *C* plano 2 in *A* + *A* cub aequatur *D* solido.

EXAMPLE 8 Solving a Motion Problem

Two executives in cities 400 miles apart drive to a business meeting at a location on the line between their cities. They meet after 4 hours. Find the speed of each car if one car travels 20 miles per hour faster than the other.

SOLUTION

Step 1 **Read** the problem carefully.

Step 2 **Assign variables.**

Let x = the speed of the faster car,
and y = the speed of the slower car.

We use the formula $d = rt$. Each car travels for 4 hours, so the time, t, for each car is 4, as shown in the table. The distance is found by using the formula $d = rt$ and the expressions already entered in the table.

	r	t	d
Faster car	x	4	$4x$
Slower car	y	4	$4y$

Find d from $d = rt$.

Sketch what is happening in the problem. See Figure 47.

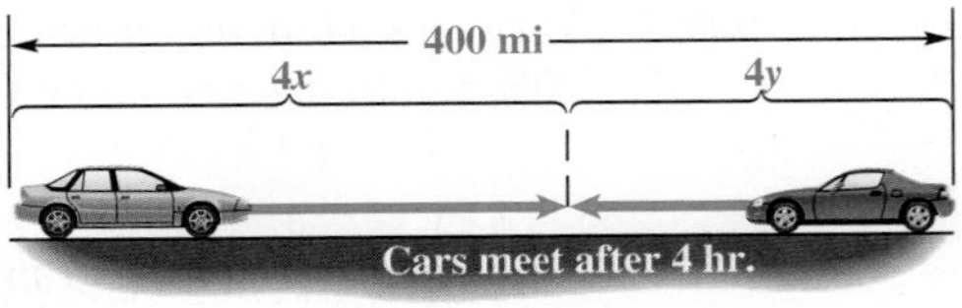

FIGURE 47

Step 3 **Write two equations.** As shown in the figure, because the total distance traveled by both cars is 400 miles, one equation is

$$4x + 4y = 400.$$

Because the faster car goes 20 miles per hour faster than the slower car, the second equation is

$$x = 20 + y.$$

Step 4 **Solve.** This system of equations,

$$4x + 4y = 400 \quad (1)$$
$$x = 20 + y, \quad (2)$$

can be solved by substitution. Replace x with $20 + y$ in equation (1) and solve for y.

$$4(20 + y) + 4y = 400 \quad \text{Let } x = 20 + y.$$
$$80 + 4y + 4y = 400 \quad \text{Distributive property}$$
$$80 + 8y = 400 \quad \text{Combine like terms.}$$
$$8y = 320 \quad \text{Subtract 80.}$$
$$y = 40 \quad \text{Divide by 8.}$$

Since $x = 20 + y$, and $y = 40$,

$$x = 20 + 40 = 60.$$

Step 5 **State the answer.** The speeds of the two cars are 40 miles per hour and 60 miles per hour.

Step 6 **Check** the answer. Because each car travels for 4 hours, the total distance traveled is

$$4(60) + 4(40) = 240 + 160 = 400 \text{ miles}, \quad \text{as required.}$$

The final example shows how a system in three variables is used to solve a problem.

EXAMPLE 9 Solving a Problem Involving Prices

At Panera Bread, a loaf of honey wheat bread costs \$2.40, a loaf of pumpernickel bread costs \$3.35, and a loaf of French bread costs \$2.10. On a recent day, three times as many loaves of honey wheat were sold as pumpernickel. The number of loaves of French bread sold was 5 less than the number of loaves of honey wheat sold. Total receipts for these breads were \$56.90. How many loaves of each type of bread were sold? (*Source:* Panera Bread menu.)

SOLUTION

Step 1 **Read** the problem again. There are three unknowns in this problem.

Step 2 **Assign variables** to represent the three unknowns.

Let x = the number of loaves of honey wheat,
y = the number of loaves of pumpernickel,
and z = the number of loaves of French bread.

Step 3 **Write a system of three equations** using the information in the problem. Because three times as many loaves of honey wheat were sold as pumpernickel,

$$x = 3y, \quad \text{or} \quad x - 3y = 0. \quad (1)$$

Solution to the Chapter Opener Problem Let x represent the number of pounds of \$.75 candy, and let y represent the number of pounds of \$1.25 candy. The system to solve is

$$x + y = 9$$
$$.75x + 1.25y = .96(9).$$

Using the methods of this section, we find that there should be 5.22 pounds at \$.75 per pound and 3.78 pounds at \$1.25 per pound.

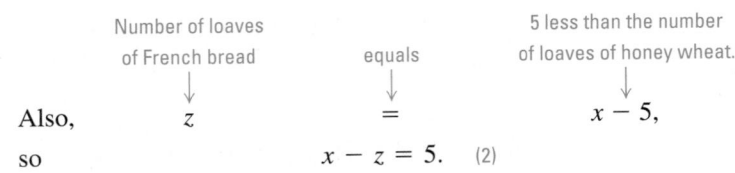

Also, z (Number of loaves of French bread) $=$ (equals) $x - 5$ (5 less than the number of loaves of honey wheat),

so $x - z = 5.$ (2)

Multiplying the cost of a loaf of each kind of bread by the number of loaves of that kind sold and adding gives the total receipts.

$$2.40x + 3.35y + 2.10z = 56.90$$

Multiply each side of this equation by 100 to clear it of decimals.

$$240x + 335y + 210z = 5690 \quad (3)$$

Step 4 **Solve** the system of three equations.

$$x - 3y = 0 \quad (1)$$
$$x - z = 5 \quad (2)$$
$$240x + 335y + 210z = 5690 \quad (3)$$

using the method of this section to find that the solution set is $\{(12, 4, 7)\}$.

Step 5 **State the answer.** The answer is 12 loaves of honey wheat, 4 loaves of pumpernickel, and 7 loaves of French bread were sold.

Step 6 **Check.** Because $12 = 3 \cdot 4$, the number of loaves of honey wheat is three times the number of loaves of pumpernickel. Also, $12 - 7 = 5$, so the number of loaves of French bread is 5 less than the number of loaves of honey wheat. Multiply the appropriate cost per loaf by the number of loaves sold and add the results to check that total receipts were \$56.90.

8.7 EXERCISES

Answer the questions in Exercises 1 and 2 by observing the graphs provided.

1. ***Music Format Sales*** The graph shows how the production of vinyl LPs, audiocassettes, and compact discs (CDs) changed over the years from 1986 through 1998.

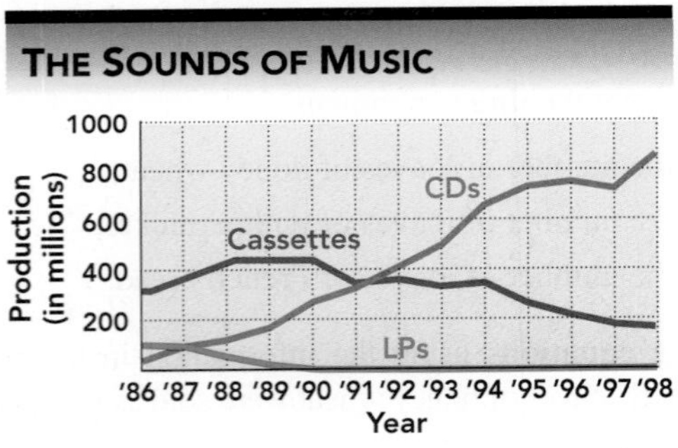

Source: Recording Industry Association of America.

(a) In what year did cassette production and CD production reach equal levels? What was that level?
(b) Express the point of intersection of the graphs of LP production and CD production as an ordered pair of the form (year, production level).
(c) Between what years did cassette production first stabilize and remain fairly constant?
(d) Describe the trend in CD production from 1986 through 1998. If a straight line were used to approximate its graph, would the line have positive, negative, or 0 slope?
(e) If a straight line were used to approximate the graph of cassette production from 1990 through 1998, would the line have positive, negative, or 0 slope? Why?

2. ***Network News Programs*** The graph on the next page shows network share (the percentage of TV sets in use) for the early evening news programs for three major broadcast networks for 1986–2000.

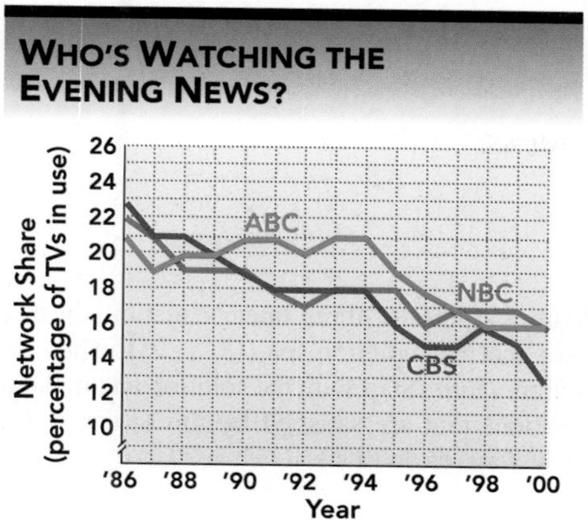

Source: Nielson Media Research.

(a) Between what years did the ABC early evening news dominate?

(b) During what year did ABC's dominance end? Which network equaled ABC's share that year? What was that share?

(c) During what years did ABC and CBS have equal network share? What was the share for each of these years?

(d) Which networks most recently had equal share? Write their share as an ordered pair of the form (year, share).

(e) Describe the general trend in viewership for the three major networks during these years.

Decide whether the ordered pair is a solution of the given system.

3. $x + y = 6$ $(5, 1)$
$x - y = 4$

4. $x - y = 17$ $(8, -9)$
$x + y = -1$

5. $2x - y = 8$ $(5, 2)$
$3x + 2y = 20$

6. $3x - 5y = -12$ $(-1, 2)$
$x - y = 1$

Solve each system by graphing.

7. $x + y = 4$
$2x - y = 2$

8. $x + y = -5$
$-2x + y = 1$

Solve each system by elimination.

9. $2x - 5y = 11$
$3x + y = 8$

10. $-2x + 3y = 1$
$-4x + y = -3$

11. $3x + 4y = -6$
$5x + 3y = 1$

12. $4x + 3y = 1$
$3x + 2y = 2$

13. $3x + 3y = 0$
$4x + 2y = 3$

14. $8x + 4y = 0$
$4x - 2y = 2$

15. $7x + 2y = 6$
$-14x - 4y = -12$

16. $x - 4y = 2$
$4x - 16y = 8$

17. $\frac{x}{2} + \frac{y}{3} = -\frac{1}{3}$
$\frac{x}{2} + 2y = -7$

18. $\frac{x}{5} + y = \frac{6}{5}$
$\frac{x}{10} + \frac{y}{3} = \frac{5}{6}$

19. $5x - 5y = 3$
$x - y = 12$

20. $2x - 3y = 7$
$-4x + 6y = 14$

Solve each system by substitution.

21. $4x + y = 6$
$y = 2x$

22. $2x - y = 6$
$y = 5x$

23. $3x - 4y = -22$
$-3x + y = 0$

24. $-3x + y = -5$
$x + 2y = 0$

25. $-x - 4y = -14$
$2x = y + 1$

26. $-3x - 5y = -17$
$4x = y - 8$

27. $5x - 4y = 9$
$3 - 2y = -x$

28. $6x - y = -9$
$4 + 7x = -y$

29. $x = 3y + 5$
$x = \frac{3}{2}y$

30. $x = 6y - 2$
$x = \frac{3}{4}y$

31. $\frac{1}{2}x + \frac{1}{3}y = 3$
$y = 3x$

32. $\frac{1}{4}x - \frac{1}{5}y = 9$
$y = 5x$

33. Explain what the following statement means: The solution set of the system

$$\begin{aligned} 2x + y + z &= 3 \\ 3x - y + z &= -2 \\ 4x - y + 2z &= 0 \end{aligned} \quad \text{is } \{(-1, 2, 3)\}.$$

34. Write a system of three linear equations in three variables that has solution set $\{(3, 1, 2)\}$. Then solve the system. (*Hint:* Start with the solution and make up three equations that are satisfied by the solution. There are many ways to do this.)

Solve each system of equations in three variables.

35. $3x + 2y + z = 8$
$2x - 3y + 2z = -16$
$x + 4y - z = 20$

36. $-3x + y - z = -10$
$-4x + 2y + 3z = -1$
$2x + 3y - 2z = -5$

37. $2x + 5y + 2z = 0$
$4x - 7y - 3z = 1$
$3x - 8y - 2z = -6$

38. $5x - 2y + 3z = -9$
$4x + 3y + 5z = 4$
$2x + 4y - 2z = 14$

39. $x + y - z = -2$
$2x - y + z = -5$
$-x + 2y - 3z = -4$

40. $x + 2y + 3z = 1$
$-x - y + 3z = 2$
$-6x + y + z = -2$

41. $2x - 3y + 2z = -1$
$x + 2y + z = 17$
$2y - z = 7$

42. $2x - y + 3z = 6$
$x + 2y - z = 8$
$2y + z = 1$

43. $4x + 2y - 3z = 6$
$x - 4y + z = -4$
$-x + 2z = 2$

44. $2x + 3y - 4z = 4$
$x - 6y + z = -16$
$-x + 3z = 8$

45. $2x + y = 6$
$3y - 2z = -4$
$3x - 5z = -7$

46. $4x - 8y = -7$
$4y + z = 7$
$-8x + z = -4$

Solve each problem involving two unknowns.

47. ***Win–Loss Record*** During the 2005 Major League Baseball regular season, the St. Louis Cardinals played 162 games. They won 38 more games than they lost. What was their win–loss record that year?

2005 MLB FINAL STANDINGS
NATIONAL LEAGUE CENTRAL

Team	W	L
St. Louis	—	—
Houston	89	73
Milwaukee	81	81
Chicago	79	83
Cincinnati	73	89
Pittsburgh	—	—

Source: mlb.com.

48. ***Win–Loss Record*** Refer to Exercise 47. During the same 162-game season, the Pittsburgh Pirates lost 28 more games than they won. What was the team's win–loss record?

49. ***Dimensions of a Basketball Court*** LeBron and Yao found that the width of their basketball court was 44 feet less than the length. If the perimeter was 288 feet, what were the length and the width of their court?

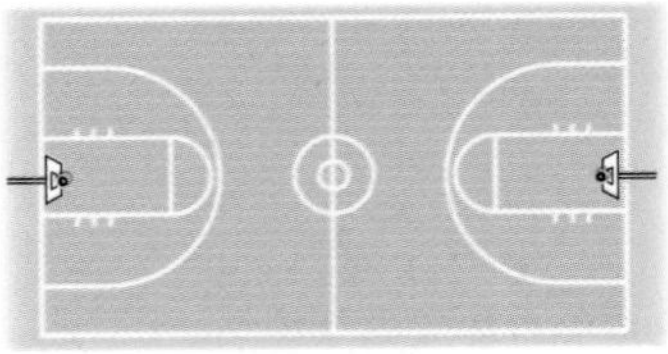

50. ***Dimensions of a Tennis Court*** Venus and Serena measured a tennis court and found that it was 42 feet longer than it was wide and had a perimeter of 228 feet. What were the length and the width of the tennis court?

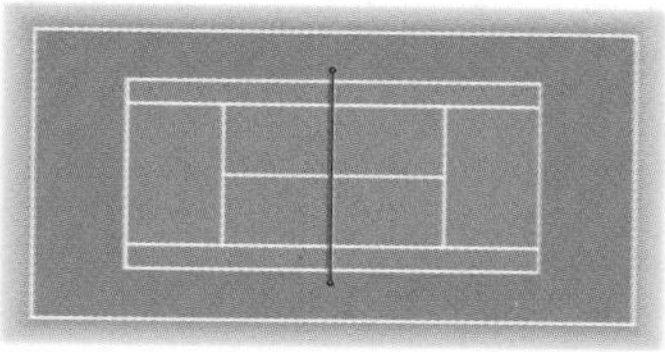

51. ***Car Rental*** On a 6-day business trip, Jerome Dugas rented a car for \$53 per day at weekday rates and \$35 per day at weekend rates. If his total rental bill was \$264, how many days did he rent at each rate? (*Source:* Enterprise.)

52. ***Popular Fiction*** A popular leisure activity of Americans is reading. In a recent year, two popular fiction titles were *How Stella Got Her Groove Back* by Terry McMillan and *The Deep End of the Ocean* by Jacquelyn Mitchard. Together, these two titles sold 1,622,962 copies. The Mitchard book sold 57,564 more copies than the McMillan book. How many copies of each title were sold? (*Source: Publishers Weekly.*)

53. ***Dimensions of a Square and a Triangle*** The side of a square is 4 centimeters longer than the side of an equilateral triangle. The perimeter of the square is 24 centimeters more than the perimeter of the triangle. Find the lengths of a side of the square and a side of the triangle.

54. ***Dimensions of a Rectangle*** The length of a rectangle is 7 feet more than the width. If the length were decreased by 3 feet and the width were increased by 2 feet, the perimeter would be 32 feet. Find the length and width of the original rectangle.

55. ***Coffee Prices*** At a business meeting at Panera Bread, the bill for two cappuccinos and three house lattes was \$10.95. At another table, the bill for one cappuccino and two house lattes was \$6.65. How much did each type of beverage cost? (*Source:* Panera Bread menu.)

56. ***Cost of Art Supplies*** For an art project Margaret Maggio bought 8 sheets of colored paper and 3 marker pens for \$6.50. She later needed 2 sheets of colored paper and 2 marker pens. These items cost \$3.00. Find the cost of 1 marker pen and 1 sheet of colored paper.

Fan Cost Index *The Fan Cost Index (FCI) represents the cost of four average-price tickets, four small soft drinks, two small beers, four hot dogs, parking for one car, two game programs, and two souvenir caps to a sporting event. For example, in 2005, the FCI for Major League Baseball was \$164.43. This was by far the least for the four major professional sports. (Source:* www.teammarketing.com.*) Use the concept of FCI in Exercises 57 and 58.*

57. The FCI prices for the National Hockey League and the National Basketball Association totaled \$514.69. The hockey FCI was \$20.05 less than that of basketball. What were the FCIs for these sports?

58. The FCI prices for Major League Baseball and the National Football League totaled \$494.25. The football FCI was \$165.39 more than that of baseball. What were the FCIs for these sports?

59. ***Travel Costs*** Tokyo and New York are among the most expensive cities worldwide for business travelers. Using average costs per day for each city (which includes room, meals, laundry, and two taxi fares), 2 days in Tokyo and 3 days in New York cost \$2015. Four days in Tokyo and 2 days in New York cost \$2490. What is the average cost per day for each city? (*Source:* ECA International.)

60. ***Prices at Wendy's*** Andrew McGinnis works at Wendy's Old Fashioned Hamburgers. During one particular lunch hour, he sold 15 single hamburgers and 10 double hamburgers, totaling \$63.25. Another lunch hour, he sold 30 singles and 5 doubles, totaling \$78.65. How much did each type of burger cost? (*Source*: Wendy's Old Fashioned Hamburgers menu.)

61. ***Cost of Clay*** For his art class, Bryce bought 2 kilograms of dark clay and 3 kilograms of light clay, paying \$22 for the clay. He later needed 1 kilogram of dark clay and 2 kilograms of light clay, costing \$13 altogether. What was the cost per kilogram for each type of clay?

62. ***Yarn and Thread Production*** A factory makes use of two basic machines, A and B, which turn out two different products, yarn and thread. Each unit of yarn requires 1 hour on machine A and 2 hours on machine B, while each unit of thread requires 1 hour on A and 1 hour on B. Machine A runs 8 hours per day, while machine B runs 14 hours per day. How many units per day of yarn and thread should the factory make to keep its machines running at capacity?

Formulas *The formulas $p = br$ (percentage = base × rate) and $I = prt$ (simple interest = principal × rate × time) are used in the applications in Exercises 67–76. To prepare to use these formulas, answer the questions in Exercises 63–66.*

63. If a container of liquid contains 120 oz of solution, what is the number of ounces of pure acid if the given solution contains the following acid concentrations?
(a) 10% **(b)** 25%
(c) 40% **(d)** 50%

64. If $50,000 is invested in an account paying simple annual interest, how much interest will be earned during the first year at the following rates?
(a) 2% **(b)** 3%
(c) 4% **(d)** 3.5%

65. If a pound of turkey costs $1.29, give an expression for the cost of x pounds.

66. If a ticket to the movie *Ice Age 2: The Meltdown* costs $9 and y tickets are sold, give an expression for the amount collected.

67. *Acid Mixture* How many liters each of 15% acid and 33% acid should be mixed to obtain 40 liters of 21% acid?

Kind of Solution	Liters of Solution	Amount of Pure Acid
.15	x	
.33	y	
.21	40	

68. *Alcohol Mixture* How many gallons each of 25% alcohol and 35% alcohol should be mixed to obtain 20 gallons of 32% alcohol?

Kind of Solution	Gallons of Solution	Amount of Pure Alcohol
.25	x	$.25x$
.35	y	$.35y$
.32	20	.32(20)

69. *Antifreeze Mixture* A truck radiator holds 18 liters of fluid. How much pure antifreeze must be added to a mixture that is 4% antifreeze in order to fill the radiator with a mixture that is 20% antifreeze?

70. *Acid Mixture* Pure acid is to be added to a 10% acid solution to obtain 27 liters of a 20% acid solution. What amounts of each should be used?

71. *Fruit Drink Mixture* A popular fruit drink is made by mixing fruit juices. Such a mixture with 50% juice is to be mixed with another mixture that is 30% juice to get 200 liters of a mixture that is 45% juice. How much of each should be used?

Kind of Juice	Number of Liters	Amount of Pure Juice
.50	x	$.50x$
.30	y	
.45		

72. *Candy Mixture* Lauren plans to mix pecan clusters that sell for $3.60 per pound with chocolate truffles that sell for $7.20 per pound to get a mixture that she can sell in Valentine boxes for $4.95 per pound. How much of the $3.60 clusters and the $7.20 truffles should she use to create 80 pounds of the mix?

	Number of Pounds	Price per Pound	Value of Candy
Clusters	x	3.60	$3.60x$
Truffles	y	7.20	$7.20y$
Mixture	80	4.95	4.95(80)

73. *Candy Mixture* A grocer plans to mix candy that sells for $1.20 per pound with candy that sells for $2.40 per pound to get a mixture that he plans to sell for $1.65 per pound. How much of the $1.20 and $2.40 candy should he use if he wants 160 pounds of the mix?

74. *Ticket Sales* Tickets to a production of *Oklahoma* at Northeastern State University cost $2.50 for general admission or $2.00 with student identification. If 184 people paid to see a performance and $406 was collected, how many of each type of admission were sold?

75. *Investment Mixture* An investor must invest a total of $15,000 in two accounts, one paying 4% simple annual interest, and the other 3%. If he wants to earn $550 annual interest, how much should he invest at each rate?

Principal	Rate	Interest
x	.04	
y	.03	
15,000		

76. *Investment Mixture* A total of $3000 is invested, part at 2% simple interest and part at 4%. If the total annual return from the two investments is $100, how much is invested at each rate?

Principal	Rate	Interest
x	.02	$.02x$
y	.04	$.04y$
3000		100

77. ***Speeds of Trains*** A train travels 150 kilometers in the same time that a plane covers 400 kilometers. If the speed of the plane is 20 kilometers per hour less than 3 times the speed of the train, find both speeds.

78. ***Speeds of Trains*** A freight train and an express train leave towns 390 kilometers apart, traveling toward one another. The freight train travels 30 kilometers per hour slower than the express train. They pass one another 3 hours later. What are their speeds?

79. ***Speeds of Boat and Current*** In his motorboat, Nguyen travels upstream at top speed to his favorite fishing spot, a distance of 36 miles, in two hours. Returning, he finds that the trip downstream, still at top speed, takes only 1.5 hours. Find the speed of Nguyen's boat and the speed of the current.

80. ***Speeds of Snow Speeder and Wind*** Braving blizzard conditions on the planet Hoth, Luke Skywalker sets out at top speed in his snow speeder for a rebel base 3600 miles away. He travels into a steady headwind, and makes the trip in 2 hours. Returning, he finds that the trip back, still at top speed but now with a tailwind, takes only 1.5 hours. Find the top speed of Luke's snow speeder and the speed of the wind.

Solve each problem involving three unknowns.

81. ***Olympic Gold Medals*** In the 2004 Olympics in Athens, Greece, the United States earned 4 fewer gold medals than silver. The number of bronze medals earned was 49 less than twice the number of silver medals. The United States earned a total of 103 medals. How many of each kind of medal did the United States earn? (*Source: World Almanac and Book of Facts.*)

82. ***Voter Affiliations*** In a random sample of 100 Americans of voting age, 10 more Americans identify themselves as Independents than Republicans. Six fewer Americans identify themselves as Republicans than Democrats. Assuming that all of those sampled are Republican, Democrat, or Independent, how many of those in the sample identify themselves with each political affiliation? (*Source:* The Gallup Organization.)

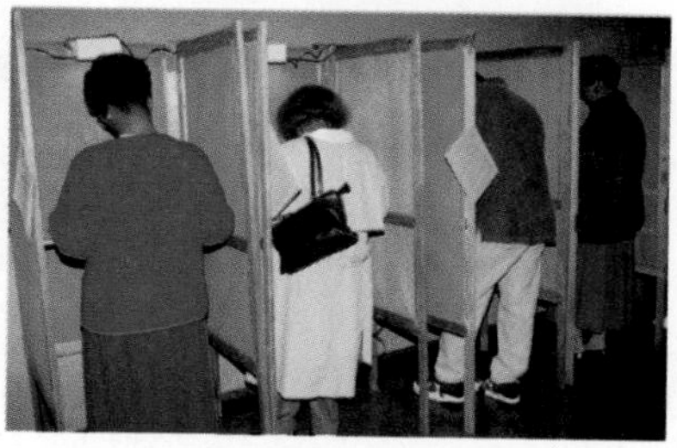

83. ***Dimensions of a Triangle*** The perimeter of a triangle is 56 inches. The longest side measures 4 inches less than the sum of the other two sides. Three times the shortest side is 4 inches more than the longest side. Find the lengths of the three sides.

84. ***Dimensions of a Triangle*** The perimeter of a triangle is 70 centimeters. The longest side is 4 centimeters less than the sum of the other two sides. Twice the shortest side is 9 centimeters less than the longest side. Find the length of each side of the triangle.

85. ***Hardware Production*** A hardware supplier manufactures three kinds of clamps, types A, B, and C. Production restrictions require it to make 10 units more type C clamps than the total of the other types and twice as many type B clamps as type A. The shop must produce a total of 490 units of clamps per day. How many units of each type can be made per day?

86. ***Television Production*** A company produces three color television sets, models X, Y, and Z. Each model X set requires 2 hr of electronics work, 2 hr of assembly time, and 1 hr of finishing time. Each model Y requires 1, 3, and 1 hr of electronics, assembly, and finishing time, respectively. Each model Z requires 3, 2, and 2 hr of the same work, respectively. There are 100 hr available for electronics, 100 hr available for assembly, and 65 hr available for finishing per week. How many of each model should be produced each week if all available time must be used?

87. ***Globetrotter Ticket Prices*** Tickets for one show on the Harlem Globetrotters' 75th Anniversary Tour cost \$10, \$18, or, for VIP seats, \$30. So far, five times as many \$18 tickets have been sold as VIP tickets. The number of \$10 tickets equals the number of \$18 tickets plus twice the number of VIP tickets. Sales of these tickets total \$9500. How many of each kind of ticket have been sold? (*Source:* www.ticketmaster.com)

88. ***Concert Ticket Prices*** Three kinds of tickets are available for a Third Day concert: "up close," "in the middle," and "far out." "Up close" tickets cost \$10 more than "in the middle" tickets, while "in the middle" tickets cost \$10 more than "far out" tickets. Twice the cost of an "up close" ticket is \$20 more than 3 times the cost of a "far out" seat. Find the price of each kind of ticket.

Answer to Margin Note problem on page 505: price of a citron: 8; price of a wood apple: 5

EXTENSION

Using Matrix Row Operations to Solve Systems

The elimination method used to solve systems introduced in the previous section can be streamlined into a systematic method by using *matrices* (singular: *matrix*). Matrices can be used to solve linear systems and matrix methods are particularly suitable for computer solutions of large systems of equations having many unknowns.

To begin, consider a system of three equations and three unknowns such as

$$\begin{aligned} a_1x + b_1y + c_1z &= d_1 \\ a_2x + b_2y + c_2z &= d_2, \\ a_3x + b_3y + c_3z &= d_3 \end{aligned} \quad \text{written in an abbreviated form as} \quad \begin{bmatrix} a_1 & b_1 & c_1 & d_1 \\ a_2 & b_2 & c_2 & d_2 \\ a_3 & b_3 & c_3 & d_3 \end{bmatrix}.$$

Such a rectangular array of numbers enclosed by brackets is called a **matrix.** Each number in the array is an **element** or **entry.** The matrix above has three **rows** (horizontal) and four **columns** (vertical) of entries, and is called a 3×4 (read "3 by 4") matrix. The constants in the last column of the matrix can be set apart from the coefficients of the variables by using a vertical line, as shown in the following **augmented matrix.**

$$\text{Rows} \rightarrow \left[\begin{array}{ccc|c} a_1 & b_1 & c_1 & d_1 \\ a_2 & b_2 & c_2 & d_3 \\ a_3 & b_3 & c_3 & d_3 \end{array}\right] \quad \text{Augmented matrix}$$

Columns

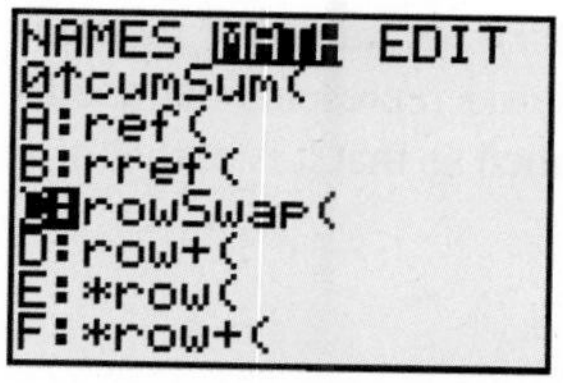

Choices C, D, E, and F provide the user of the TI-83/84 Plus calculator a means of performing row operations on matrices.

The rows of this augmented matrix can be treated the same as the equations of a system of equations, since the augmented matrix is actually a short form of the system. Any transformation of the matrix that will result in an equivalent system is permitted. The following **matrix row operations** produce such transformations.

Matrix Row Operations

For any real number k and any augmented matrix of a system of linear equations, the following operations will produce the matrix of an *equivalent system*—that is, another system with the same solution set.

1. **Interchange any two rows of a matrix.**
2. **Multiply the elements of any row of a matrix by the same nonzero number k.**
3. **Add a common multiple of the elements of one row to the corresponding elements of another row.**

If the word "row" is replaced by "equation," it can be seen that the three row operations also apply to a system of equations, so that a system of equations can be solved by transforming its corresponding matrix into the matrix of an equivalent, simpler system. The goal is a matrix in the form

$$\left[\begin{array}{cc|c} 1 & 0 & a \\ 0 & 1 & b \end{array}\right] \quad \text{or} \quad \left[\begin{array}{ccc|c} 1 & 0 & 0 & a \\ 0 & 1 & 0 & b \\ 0 & 0 & 1 & c \end{array}\right]$$

for systems with two or three equations respectively. Notice that on the left of the vertical bar there are ones down the diagonal from upper left to lower right and zeros elsewhere in the matrices. When these matrices are rewritten as systems of equations, the values of the variables are known. The **Gauss-Jordan method** is a systematic way of using the matrix row operations to change the augmented matrix of a system into the form that shows its solution. The following examples will illustrate this method.

EXAMPLE 1 Solving a Linear System Using Gauss-Jordan (Two Unknowns)

Solve the linear system.

$$\begin{aligned} 3x - 4y &= 1 \\ 5x + 2y &= 19 \end{aligned}$$

SOLUTION

The equations should all be in the same form, with the variable terms in the same order on the left, and the constant term on the right. Begin by writing the augmented matrix.

$$\left[\begin{array}{cc|c} 3 & -4 & 1 \\ 5 & 2 & 19 \end{array}\right]$$

The goal is to transform this augmented matrix into one in which the values of the variables will be easy to see. That is, since each column in the matrix represents the coefficients of one variable, the augmented matrix should be transformed so that it is of the form

$$\left[\begin{array}{cc|c} 1 & 0 & k \\ 0 & 1 & j \end{array}\right]$$

for real numbers k and j. Once the augmented matrix is in this form, the matrix can be rewritten as a linear system to get

$$\begin{aligned} x &= k \\ y &= j. \end{aligned}$$

The necessary transformations are performed as follows. It is best to work in columns beginning in each column with the element that is to become 1. In the augmented matrix,

$$\left[\begin{array}{cc|c} 3 & -4 & 1 \\ 5 & 2 & 19 \end{array}\right]$$

there is a 3 in the first row, first column position. Use row operation 2, multiplying each entry in the first row by $\frac{1}{3}$ to get a 1 in this position. (This step is abbreviated as $\frac{1}{3}$ R1.)

$$\left[\begin{array}{cc|c} 1 & -\frac{4}{3} & \frac{1}{3} \\ 5 & 2 & 19 \end{array}\right] \quad \tfrac{1}{3}\text{R1}$$

Introduce 0 in the second row, first column by multiplying each element of the first row by -5 and adding the result to the corresponding element in the second row, using row operation 3.

$$\left[\begin{array}{cc|c} 1 & -\frac{4}{3} & \frac{1}{3} \\ 0 & \frac{26}{3} & \frac{52}{3} \end{array}\right] \quad -5\text{R1} + \text{R2}$$

Obtain 1 in the second row, second column by multiplying each element of the second row by $\frac{3}{26}$, using row operation 2.

$$\left[\begin{array}{cc|c} 1 & -\frac{4}{3} & \frac{1}{3} \\ 0 & 1 & 2 \end{array}\right] \quad \tfrac{3}{26}\text{R2}$$

Finally, obtain 0 in the first row, second column by multiplying each element of the second row by $\frac{4}{3}$ and adding the result to the corresponding element in the first row.

$$\left[\begin{array}{cc|c} 1 & 0 & 3 \\ 0 & 1 & 2 \end{array}\right] \quad \tfrac{4}{3}\text{R2} + \text{R1}$$

This last matrix corresponds to the system

$$\begin{aligned} x &= 3 \\ y &= 2, \end{aligned}$$

that has the solution set $\{(3, 2)\}$. This solution could have been read directly from the third column of the final matrix. ■

A linear system with three equations is solved in a similar way. Row operations are used to get 1s down the diagonal from left to right and 0s above and below each 1.

EXAMPLE 2 Solving a System Using Gauss-Jordan (Three Unknowns)

Use the Gauss-Jordan method to solve the system.

$$\begin{aligned} x - y + 5z &= -6 \\ 3x + 3y - z &= 10 \\ x + 3y + 2z &= 5 \end{aligned}$$

SOLUTION

Because the system is in proper form, begin by writing the augmented matrix of the linear system.

$$\left[\begin{array}{ccc|c} 1 & -1 & 5 & -6 \\ 3 & 3 & -1 & 10 \\ 1 & 3 & 2 & 5 \end{array}\right]$$

The final matrix is to be of the form

$$\left[\begin{array}{ccc|c} 1 & 0 & 0 & m \\ 0 & 1 & 0 & n \\ 0 & 0 & 1 & p \end{array}\right],$$

where m, n, and p are real numbers. This final form of the matrix gives the system $x = m$, $y = n$, and $z = p$, so the solution set is $\{(m, n, p)\}$.

There is already a 1 in the first row, first column. Introduce a 0 in the second row of the first column by multiplying each element in the first row by -3 and adding the result to the corresponding element in the second row, using row operation 3.

$$\left[\begin{array}{ccc|c} 1 & -1 & 5 & -6 \\ 0 & 6 & -16 & 28 \\ 1 & 3 & 2 & 5 \end{array}\right] \quad \begin{array}{l} \\ -3\text{R1} + \text{R2} \\ \\ \end{array}$$

Now, to change the last element in the first column to 0, use row operation 3. Multiply each element of the first row by -1, then add the results to the corresponding elements of the third row.

$$\left[\begin{array}{ccc|c} 1 & -1 & 5 & -6 \\ 0 & 6 & -16 & 28 \\ 0 & 4 & -3 & 11 \end{array}\right] \quad \begin{array}{l} \\ \\ -1\text{R1} + \text{R3} \end{array}$$

The same procedure is used to transform the second and third columns. For both of these columns, first perform the step of getting 1 in the appropriate position of each column. Do this by multiplying the elements of the row by the reciprocal of the number in that position.

$$\left[\begin{array}{ccc|c} 1 & -1 & 5 & -6 \\ 0 & 1 & -\frac{8}{3} & \frac{14}{3} \\ 0 & 4 & -3 & 11 \end{array}\right] \quad \begin{array}{l} \\ \frac{1}{6}\text{R2} \\ \\ \end{array}$$

$$\left[\begin{array}{ccc|c} 1 & 0 & \frac{7}{3} & -\frac{4}{3} \\ 0 & 1 & -\frac{8}{3} & \frac{14}{3} \\ 0 & 4 & -3 & 11 \end{array}\right] \quad \text{R2 + R1}$$

$$\left[\begin{array}{ccc|c} 1 & 0 & \frac{7}{3} & -\frac{4}{3} \\ 0 & 1 & -\frac{8}{3} & \frac{14}{3} \\ 0 & 0 & \frac{23}{3} & -\frac{23}{3} \end{array}\right] \quad -4\text{R2} + \text{R3}$$

$$\left[\begin{array}{ccc|c} 1 & 0 & \frac{7}{3} & -\frac{4}{3} \\ 0 & 1 & -\frac{8}{3} & -\frac{14}{3} \\ 0 & 0 & 1 & -1 \end{array}\right] \quad \frac{3}{23}\text{R3}$$

$$\left[\begin{array}{ccc|c} 1 & 0 & 0 & 1 \\ 0 & 1 & -\frac{8}{3} & \frac{14}{3} \\ 0 & 0 & 1 & -1 \end{array}\right] \quad -\frac{7}{3}\text{R3} + \text{R1}$$

$$\left[\begin{array}{ccc|c} 1 & 0 & 0 & 1 \\ 0 & 1 & 0 & 2 \\ 0 & 0 & 1 & -1 \end{array}\right] \quad \frac{8}{3}\text{R3} + \text{R2}$$

The linear system associated with this final matrix is

$$\begin{aligned} x &= 1 \\ y &= 2 \\ z &= -1 \end{aligned}, \text{ and the solution set is } \{(1, 2, -1)\}.$$

EXTENSION EXERCISES

Use the Gauss-Jordan method to solve each system of equations.

1. $\begin{aligned} x + y &= 5 \\ x - y &= -1 \end{aligned}$

2. $\begin{aligned} x + 2y &= 5 \\ 2x + y &= -2 \end{aligned}$

3. $\begin{aligned} x + y &= -3 \\ 2x - 5y &= -6 \end{aligned}$

4. $\begin{aligned} 3x - 2y &= 4 \\ 3x + y &= -2 \end{aligned}$

5. $\begin{aligned} 2x - 3y &= 10 \\ 2x + 2y &= 5 \end{aligned}$

6. $\begin{aligned} 4x + y &= 5 \\ 2x + y &= 3 \end{aligned}$

7. $\begin{aligned} 3x - 7y &= 31 \\ 2x - 4y &= 18 \end{aligned}$

8. $\begin{aligned} 5x - y &= 14 \\ x + 8y &= 11 \end{aligned}$

9. $\begin{aligned} x + y - z &= 6 \\ 2x - y + z &= -9 \\ x - 2y + 3z &= 1 \end{aligned}$

10. $\begin{aligned} x + 3y - 6z &= 7 \\ 2x - y + 2z &= 0 \\ x + y + 2z &= -1 \end{aligned}$

11. $\begin{aligned} 2x - y + 3z &= 0 \\ x + 2y - z &= 5 \\ 2y + z &= 1 \end{aligned}$

12. $\begin{aligned} 4x + 2y - 3z &= 6 \\ x - 4y + z &= -4 \\ -x + 2z &= 2 \end{aligned}$

13. $\begin{aligned} -x + y &= -1 \\ y - z &= 6 \\ x + z &= -1 \end{aligned}$

14. $\begin{aligned} x + y &= 1 \\ 2x - z &= 0 \\ y + 2z &= -2 \end{aligned}$

15. $\begin{aligned} 2x - y + 4z &= -1 \\ -3x + 5y - z &= 5 \\ 2x + 3y + 2z &= 3 \end{aligned}$

16. $\begin{aligned} 5x - 3y + 2z &= -5 \\ 2x + 2y - z &= 4 \\ 4x - y + z &= -1 \end{aligned}$

17. $\begin{aligned} x + y - 2z &= 1 \\ 2x - y - 4z &= -4 \\ 3x - 2y + z &= -7 \end{aligned}$

18. $\begin{aligned} x + 3y - 6z &= -26 \\ 3x + y - z &= -10 \\ 2x - y - 3z &= -16 \end{aligned}$

8.8 Linear Inequalities, Systems, and Linear Programming

Linear Inequalities in Two Variables • Systems of Inequalities • Linear Programming

Linear Inequalities in Two Variables

Linear inequalities with one variable were graphed on the number line in an earlier chapter. In this section linear inequalities in two variables are graphed in a rectangular coordinate system.

Linear Inequality in Two Variables

An inequality that can be written as

$$Ax + By < C \quad \text{or} \quad Ax + By > C,$$

where A, B, and C are real numbers and A and B are not both 0, is a **linear inequality in two variables.** The symbols $\leq$ and $\geq$ may replace $<$ and $>$ in this definition.

A line divides the plane into three regions: the line itself and the two half-planes on either side of the line. Recall that the graphs of linear inequalities in one variable are intervals on the number line that sometimes include an endpoint. The graphs of linear inequalities in two variables are *regions* in the real number plane and may include a *boundary line*. The **boundary line** for the inequality $Ax + By < C$ or $Ax + By > C$ is the graph of the *equation* $Ax + By = C$. To graph a linear inequality, we follow these steps.

Graphing a Linear Inequality

Step 1 **Draw the boundary.** Draw the graph of the straight line that is the boundary. Make the line solid if the inequality involves $\leq$ or $\geq$; make the line dashed if the inequality involves $<$ or $>$.

Step 2 **Choose a test point.** Choose any point not on the line as a test point.

Step 3 **Shade the appropriate region.** Shade the region that includes the test point if it satisfies the original inequality; otherwise, shade the region on the other side of the boundary line.

EXAMPLE 1 Graphing a Linear Inequality

Graph $3x + 2y \geq 6$.

SOLUTION

First graph the straight line $3x + 2y = 6$. The graph of this line, the boundary of the graph of the inequality, is shown in Figure 48 on the next page. The graph of

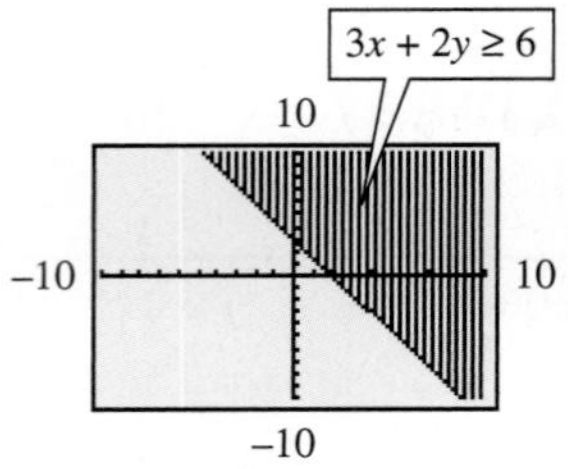

The TI-83/84 Plus allows us to shade the appropriate region for an inequality. Compare with Figure 48.

the inequality $3x + 2y \geq 6$ includes the points of the line $3x + 2y = 6$, and either the points *above* the line $3x + 2y = 6$ or the points *below* that line. To decide which, select any point not on the line $3x + 2y = 6$ as a test point. The origin, (0, 0), often is a good choice. Substitute the values from the test point (0, 0) for x and y in the inequality $3x + 2y \geq 6$.

$$3(0) + 2(0) \geq 6 \quad ?$$
$$0 \geq 6 \quad \text{False}$$

Because the result is false, (0, 0) does not satisfy the inequality, and so the solution set includes all points on the other side of the line. This region is shaded in Figure 48.

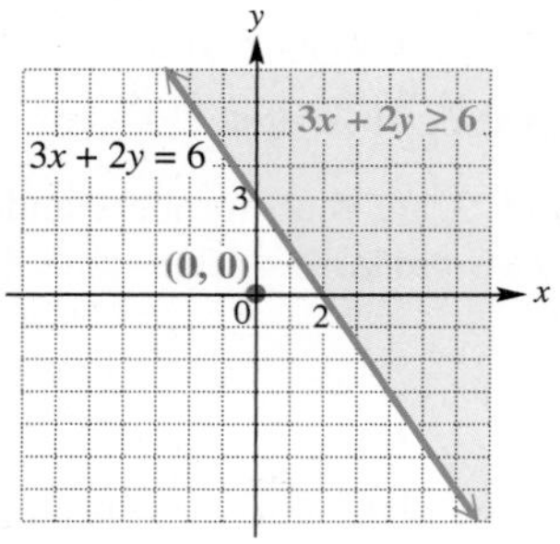

FIGURE 48

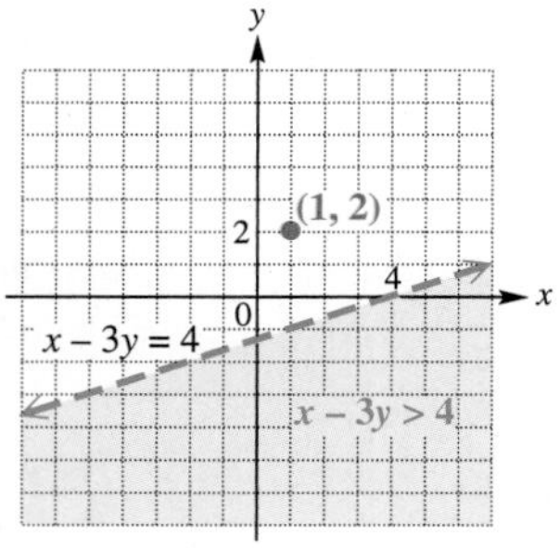

FIGURE 49

EXAMPLE 2 Graphing a Linear Inequality

Graph $x - 3y > 4$.

SOLUTION

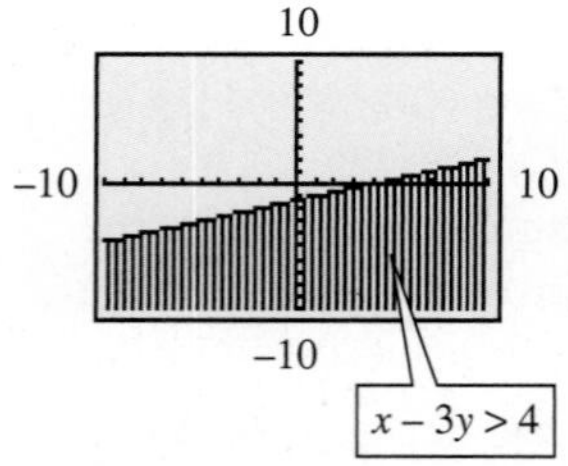

Compare with Figure 49. Can you tell from the calculator graph whether points on the boundary line are included in the solution set of the inequality?

First graph the boundary line, $x - 3y = 4$. The graph is shown in Figure 49. The points of the boundary line do not belong to the inequality $x - 3y > 4$ (since the inequality symbol is $>$ and not $\geq$). For this reason, the line is dashed. To decide which side of the line is the graph of the solution set, choose any point that is not on the line, say (1, 2). Substitute 1 for x and 2 for y in the original inequality.

$$1 - 3(2) > 4 \quad ?$$
$$-5 > 4 \quad \text{False}$$

Because of this false result, the solution set lies on the side of the boundary line that does *not* contain the test point (1, 2). The solution set, graphed in Figure 49, includes only those points in the shaded region (not those on the line).

Systems of Inequalities Methods of solving systems of *equations* were discussed in the previous section. System of inequalities with two variables may be solved by graphing. A system of linear inequalities consists of two or more such inequalities, and the solution set of such a system consists of all points that make all the inequalities true at the same time.

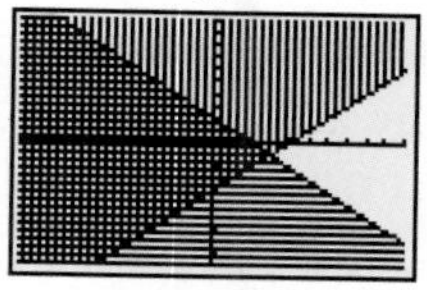

The cross-hatched region shows the solution set of the system

$y > x - 4$
$y < -x + 2.$

Graphing a System of Linear Inequalities

Step 1 **Graph each inequality in the same coordinate system.** Graph each inequality in the system, using the method described in Examples 1 and 2.

Step 2 **Find the intersection of the regions of solutions.** Indicate the intersection of the regions of solutions of the individual inequalities. This is the solution set of the system.

EXAMPLE 3 Graphing a System of Inequalities

Graph the solution set of the linear system.

$$3x + 2y \le 6$$
$$2x - 5y \ge 10$$

SOLUTION

Begin by graphing $3x + 2y \le 6$. To do this, graph $3x + 2y = 6$ as a solid line. Since (0, 0) makes the inequality true, shade the region containing (0, 0), as shown in Figure 50.

Now graph $2x - 5y \ge 10$. The solid line boundary is the graph of $2x - 5y = 10$. Since (0, 0) makes the inequality false, shade the region that does not contain (0, 0), as shown in Figure 51.

The solution set of the system is given by the intersection (overlap) of the regions of the graphs in Figures 50 and 51. The solution set is the shaded region in Figure 52, and includes portions of the two boundary lines.

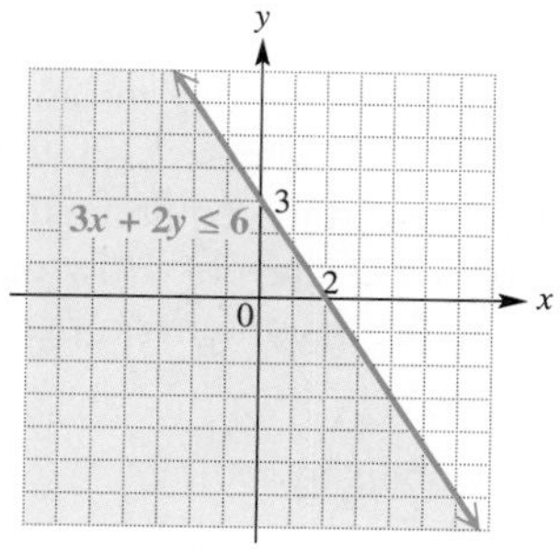

FIGURE 50

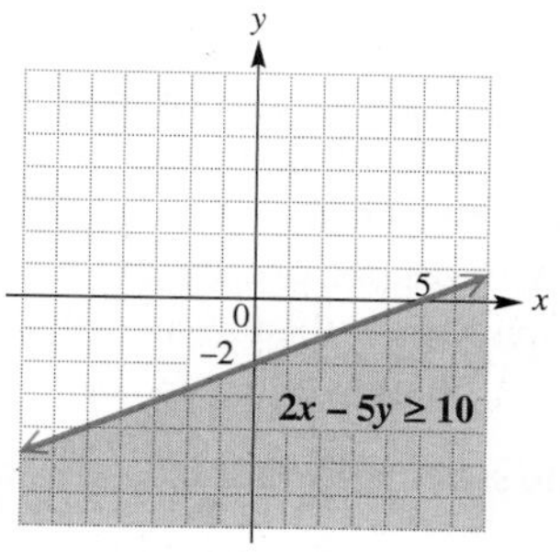

FIGURE 51

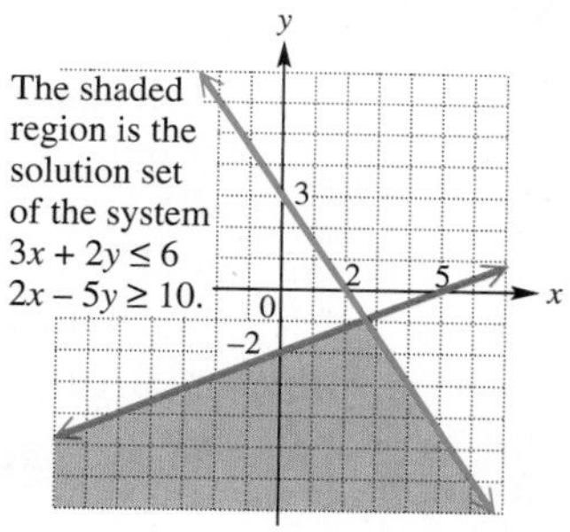

FIGURE 52

In practice, we usually do all the work in one coordinate system at the same time. In the following example, only one graph is shown.

EXAMPLE 4 Graphing a System of Inequalities

Graph the solution set of the linear system.

$$2x + 3y \ge 12$$
$$7x + 4y \ge 28$$
$$y \le 6$$
$$x \le 5$$

SOLUTION

Graph the four inequalities in one coordinate system and shade the region common to all four as shown in Figure 53. As shown, the boundary lines are all solid.

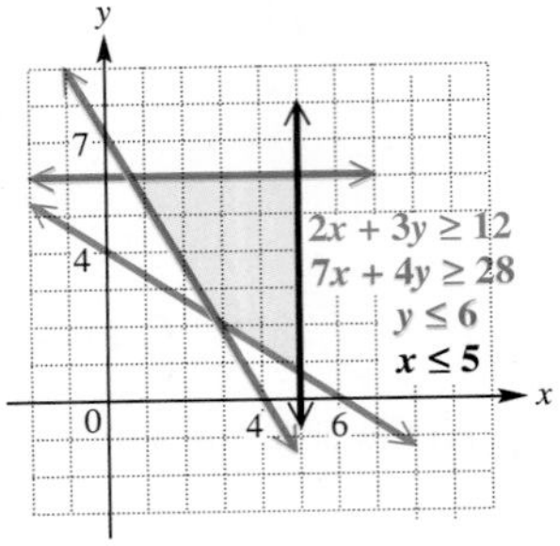

FIGURE 53

George B. Dantzig (1914–2005) of Stanford University was one of the key people behind **operations research** (OR). As a management science, OR is not a single discipline, but draws from mathematics, probability theory, statistics, and economics. The name given to this "multiplex" shows its historical origins in World War II, when operations of a military nature called forth the efforts of many scientists to research their fields for applications to the war effort and to solve tactical problems.

Operations research is an approach to problem solving and decision making. First of all, the problem has to be clarified. Quantities involved have to be designated as variables, and the objectives as functions. Use of **models** is an important aspect of OR.

Linear Programming A very important application of mathematics to business and social science is called **linear programming.** Linear programming is used to find an optimum value, for example, minimum cost or maximum profit. Procedures for solving linear programming problems were developed in 1947 by George Dantzig, while he was working on a problem of allocating supplies for the Air Force in a way that minimized total cost.

EXAMPLE 5 Maximizing Profit

The Smartski Company makes two products, DVD recorders and MP-3 players. Each DVD recorder gives a profit of \$3, while each MP-3 player gives a profit of \$7. The company must manufacture at least 1 DVD recorder per day to satisfy one of its customers, but no more than 5 because of production problems. Also, the number of MP-3 players produced cannot exceed 6 per day. As a further requirement, the number of DVD recorders cannot exceed the number of MP-3 players. How many of each should the company manufacture in order to obtain the maximum profit?

SOLUTION

We translate the statements of the problem into symbols by letting

$$x = \text{number of DVD recorders to be produced daily}$$
$$y = \text{number of MP-3 players to be produced daily.}$$

According to the statement of the problem, the company must produce at least one DVD recorder (one or more), so

$$x \geq 1.$$

No more than 5 DVD recorders may be produced:

$$x \leq 5.$$

No more than 6 MP-3 players may be made in one day:

$$y \leq 6.$$

The number of DVD recorders may not exceed the number of MP-3 players:

$$x \leq y.$$

The number of DVD recorders and of MP-3 players cannot be negative:

$$x \geq 0 \quad \text{and} \quad y \geq 0.$$

All restrictions, or **constraints,** that are placed on production can now be summarized:

$$x \geq 1, \quad x \leq 5, \quad y \leq 6, \quad x \leq y, \quad x \geq 0, \quad y \geq 0.$$

The maximum possible profit that the company can make, subject to these constraints, is found by sketching the graph of the solution set of the system. See Figure 54. The only feasible values of x and y are those that satisfy all constraints. These values correspond to points that lie on the boundary or in the shaded region, called the **region of feasible solutions.**

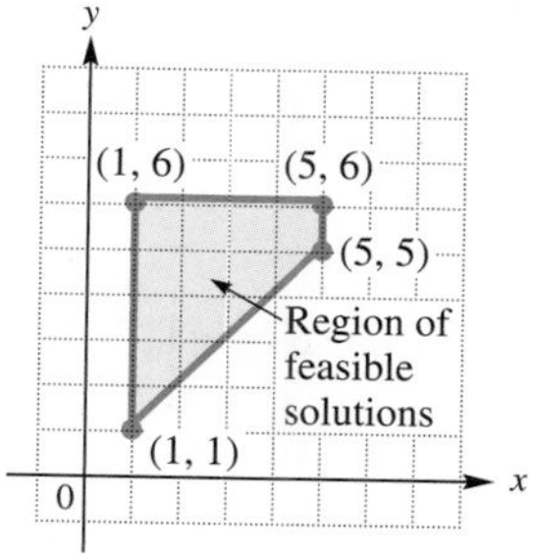

FIGURE 54

Because each DVD recorder gives a profit of \$3, the daily profit from the production of x DVD recorders is $3x$ dollars. Also, the profit from the production of y MP-3 players will be $7y$ dollars per day. The total daily profit is thus given by the following **objective function:**

$$\text{Profit} = 3x + 7y.$$

The problem may now be stated as follows: find values of x and y in the region of feasible solutions as shown in Figure 54 that will produce the maximum possible value of $3x + 7y$. It can be shown that any optimum value (maximum or minimum) will always occur at a **vertex** (or **corner point**) of the region of feasible solutions. Locate the point (x, y) that gives the maximum profit by checking the coordinates of the vertices, shown in Figure 54 and in Table 6. Find the profit that corresponds to each coordinate pair and choose the one that gives the maximum profit.

TABLE 6

Point	Profit = $3x + 7y$	
(1,1)	$3(1) + 7(1) = 10$	
(1,6)	$3(1) + 7(6) = 45$	
(5,6)	$3(5) + 7(6) = 57$	⟵ Maximum
(5,5)	$3(5) + 7(5) = 50$	

The maximum profit of \$57 is obtained when 5 DVD recorders and 6 MP-3 players are produced each day. ■

It Pays to Do Your Homework
George Dantzig, profiled on the previous page, was interviewed in the September 1986 issue of the *College Mathematics Journal.* In the interview he relates the story of how he obtained his degree without actually writing a thesis.

One day Dantzig arrived late to one of his classes and on the board were two problems. Assuming that they were homework problems, he worked on them and handed them in a few days later, apologizing to his professor for taking so long to do them. Several weeks later he received an early morning visit from his professor. The problems had not been intended as homework problems; they were actually two famous *unsolved* problems in statistics! Later, when Dantzig began to think about a thesis topic, his professor told him that the two solutions would serve as his thesis.

To solve a linear programming problem in general, use the following steps.

Solving a Linear Programming Problem

Step 1 Write all necessary constraints and the objective function.

Step 2 Graph the region of feasible solutions.

Step 3 Identify all vertices.

Step 4 Find the value of the objective function at each vertex.

Step 5 The solution is given by the vertex producing the optimum value of the objective function.

8.8 EXERCISES

In Exercises 1–4, match each system of inequalities with the correct graph from choices A–D.

A.

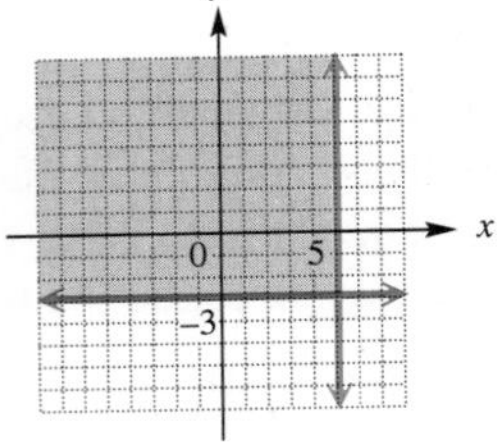

B.

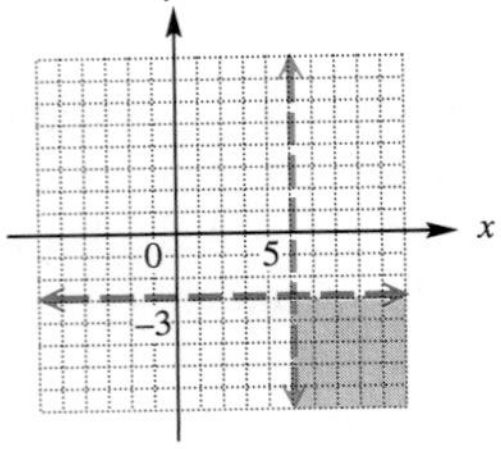

C.

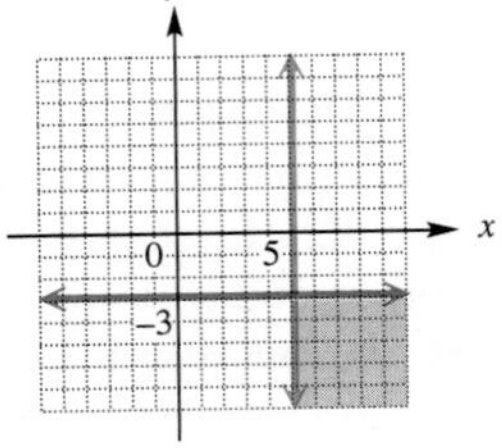

D.

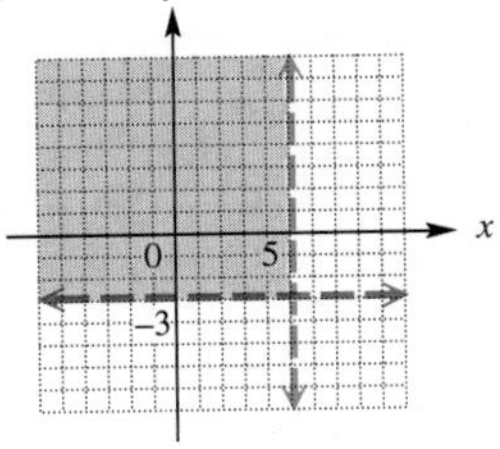

1. $x \geq 5$
$y \leq -3$

2. $x \leq 5$
$y \geq -3$

3. $x > 5$
$y < -3$

4. $x < 5$
$y > -3$

Graph each linear inequality.

5. $x + y \leq 2$

6. $x - y \geq -3$

7. $4x - y \leq 5$

8. $3x + y \geq 6$

9. $x + 3y \geq -2$

10. $4x + 6y \leq -3$

11. $x + 2y \leq -5$

12. $2x - 4y \leq 3$

13. $4x - 3y < 12$

14. $5x + 3y > 15$

15. $y > -x$

16. $y < x$

Graph each system of inequalities.

17. $x + y \leq 1$
$x \geq 0$

18. $3x - 4y \leq 6$
$y \geq 1$

19. $2x - y \geq 1$
$3x + 2y \geq 6$

20. $x + 3y \geq 6$
$3x - 4y \leq 12$

21. $-x - y < 5$
$x - y \leq 3$

22. $6x - 4y < 8$
$x + 2y \geq 4$

Exercises 23 and 24 show regions of feasible solutions. Find the maximum and minimum values of the given expressions.

23. $3x + 5y$

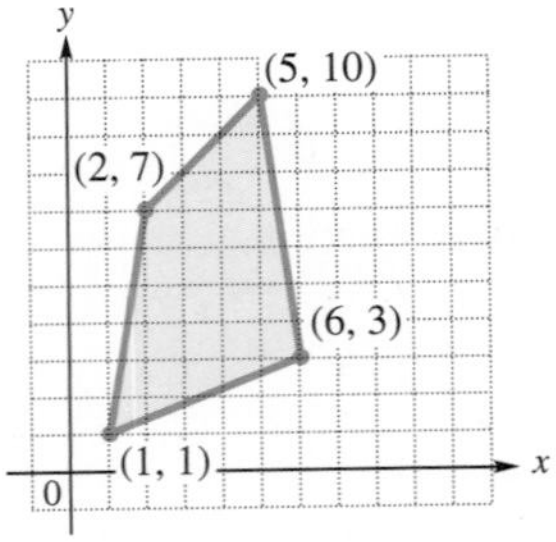

24. $40x + 75y$

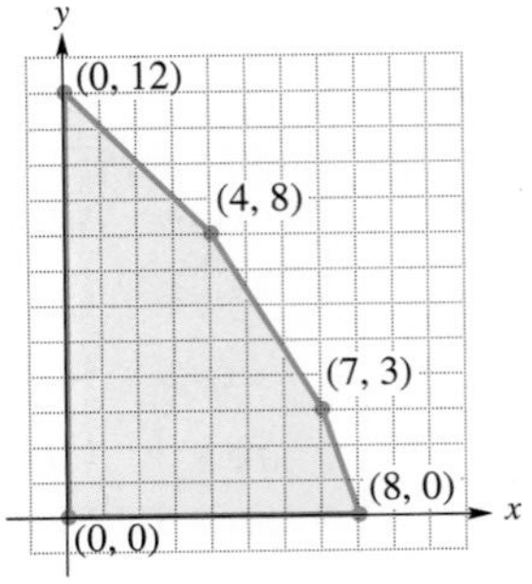

Use graphical methods to find values of x and y satisfying the given conditions. (It may be necessary to solve a system of equations in order to find vertices.) Find the value of the maximum or minimum.

25. Find $x \geq 0$ and $y \geq 0$ such that

$$\begin{aligned} 2x + 3y &\leq 6 \\ 4x + y &\leq 6 \end{aligned}$$

and $5x + 2y$ is maximized.

26. Find $x \geq 0$ and $y \geq 0$ such that

$$\begin{aligned} x + y &\leq 10 \\ 5x + 2y &\geq 20 \\ 2y &\geq x \end{aligned}$$

and $x + 3y$ is minimized.

27. Find $x \geq 2$ and $y \geq 5$ such that

$$\begin{aligned} 3x - y &\geq 12 \\ x + y &\leq 15 \end{aligned}$$

and $2x + y$ is minimized.

28. Find $x \geq 10$ and $y \geq 20$ such that

$$\begin{aligned} 2x + 3y &\leq 100 \\ 5x + 4y &\leq 200 \end{aligned}$$

and $x + 3y$ is maximized.

Solve each linear programming problem.

29. *Refrigerator Shipping Costs* A manufacturer of refrigerators must ship at least 100 refrigerators to its two West coast warehouses. Each warehouse holds a maximum of 100 refrigerators. Warehouse A holds 25 refrigerators already, while warehouse B has 20 on hand. It costs \$12 to ship a refrigerator to warehouse A and \$10 to ship one to warehouse B. How many refrigerators should be shipped to each warehouse to minimize cost? What is the minimum cost?

30. *Food Supplement Costs* Bonnie, who is dieting, requires two food supplements, I and II. She can get these supplements from two different products, A and B. Product A provides 3 grams per serving of supplement I and 2 grams per serving of supplement II. Product B provides 2 grams per serving of supplement I and 4 grams per serving of supplement II. Her dietician, Dr. Dawson, has recommended that she include at least 15 grams of each supplement in her daily diet. If product A costs 25¢ per serving and product B costs 40¢ per serving, how can she satisfy her requirements most economically?

31. *Vitamin Pill Costs* Elizabeth Lamulle takes vitamin pills. Each day, she must have at least 16 units of Vitamin A, at least 5 units of Vitamin B_1, and at least 20 units of Vitamin C. She can choose between red pills costing 10¢ each that contain 8 units of A, 1 of B_1, and 2 of C; and blue pills that cost 20¢ each and contain 2 units of A, 1 of B_1, and 7 of C. How many of each pill should she take in order to minimize her cost and yet fulfill her daily requirements?

32. *Bolt Costs* A machine shop manufactures two types of bolts. Each can be made on any of three groups of machines, but the time required on each group differs, as shown in the table in the next column.

		Machine Groups		
		I	**II**	**III**
Bolts	**Type A**	.1 min	.1 min	.1 min
	Type B	.1 min	.4 min	.5 min

Production schedules are made up one day at a time. In a day there are 240, 720, and 160 minutes available, respectively, on these machines. Type A bolts sell for 10¢ and type B bolts for 12¢. How many of each type of bolt should be manufactured per day to maximize revenue? What is the maximum revenue?

33. *Gasoline and Fuel Oil Costs* A manufacturing process requires that oil refineries manufacture at least 2 gallons of gasoline for each gallon of fuel oil. To meet the winter demand for fuel oil, at least 3 million gallons a day must be produced. The demand for gasoline is no more than 6.4 million gallons per day. If the price of gasoline is \$1.90 per gallon and the price of fuel oil is \$1.50 per gallon, how much of each should be produced to maximize revenue?

34. *Cake and Cookie Production* A bakery makes both cakes and cookies. Each batch of cakes requires two hours in the oven and three hours in the decorating room. Each batch of cookies needs one and a half hours in the oven and two-thirds of an hour in the decorating room. The oven is available no more than 15 hours a day, while the decorating room can be used no more than 13 hours a day. How many batches of cakes and cookies should the bakery make in order to maximize profits if cookies produce a profit of \$20 per batch and cakes produce a profit of \$30 per batch?

35. ***Aid to Earthquake Victims*** Earthquake victims in China need medical supplies and bottled water. Each medical kit measures 1 cubic foot and weighs 10 pounds. Each container of water is also 1 cubic foot and weighs 20 pounds. The plane can only carry 80,000 pounds with a total volume of 6000 cubic feet. Each medical kit will aid 6 people, while each container of water will serve 10 people. How many of each should be sent in order to maximize the number of people aided?

36. ***Aid to Earthquake Victims*** If each medical kit could aid 4 people instead of 6, how would the results in Exercise 35 change?

COLLABORATIVE INVESTIGATION

Living with AIDS

The graph here shows a comparison of the number of African Americans and Whites living with AIDS in the United States during 1993–2000. Form groups of 2–3 students each to work the following.

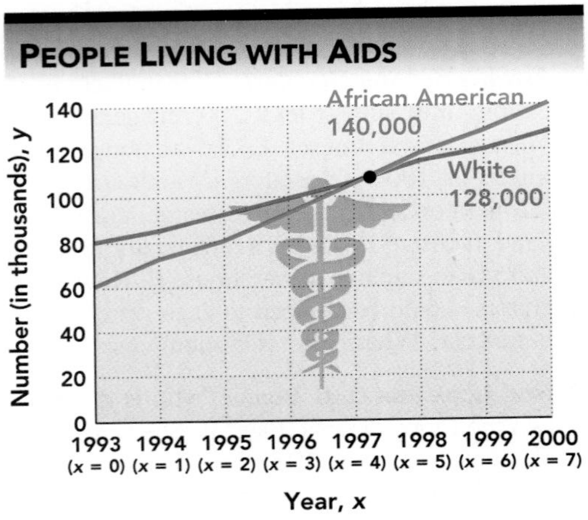

Source: U.S. Centers for Disease Control.

Topics for Discussion

1. The two lines were obtained by joining the data points that are of the form

(year, number of people in thousands).

Let $x = 0$ represent the year 1993, $x = 1$ represent 1994, and so on, and approximate the value of y for each year for African Americans. Estimate the missing values, and fill in the table. Remember that y is in thousands.

Year	Number of African Americans with AIDS (y, in thousands)
1993 ($x = 0$)	60
1994 ($x = 1$)	
1995 ($x = 2$)	
1996 ($x = 3$)	
1997 ($x = 4$)	
1998 ($x = 5$)	
1999 ($x = 6$)	
2000 ($x = 7$)	140

Now use any two data points to find an equation of the line describing this data.

2. Repeat the procedure from part 1, applying the data from the line for Whites living with AIDS.

Year	Number of Whites with AIDS (y, in thousands)
1993 ($x = 0$)	80
1994 ($x = 1$)	
1995 ($x = 2$)	
1996 ($x = 3$)	
1997 ($x = 4$)	
1998 ($x = 5$)	
1999 ($x = 6$)	
2000 ($x = 7$)	128

Now use any two data points to find an equation of the line describing this data.

3. The two equations from parts 1 and 2 form a system of two linear equations in two variables. Solve this system using any method you wish.

4. The x-coordinate of the solution of the system in part 3 should correspond to the year in which the two lines intersect. Look at the graph again. Does your x-value correspond the way it should?

5. Discuss why results of this activity might vary among groups performing it.

CHAPTER 8 TEST

1. Find the distance between the points $(-3, 5)$ and $(2, 1)$.

2. Find an equation of the circle whose center has coordinates $(-1, 2)$, with radius 3. Sketch its graph.

3. Find the x- and y-intercepts of the graph of $3x - 2y = 8$, and graph the equation.

4. Find the slope of the line passing through the points $(6, 4)$ and $(-1, 2)$.

5. Find the slope-intercept form of the equation of the line described.
   (a) passing through the point $(-1, 3)$, with slope $-\frac{2}{5}$
   (b) passing through $(-7, 2)$ and perpendicular to $y = 2x$
   (c) the line shown in the figures below (Look at the displays at the bottom.)

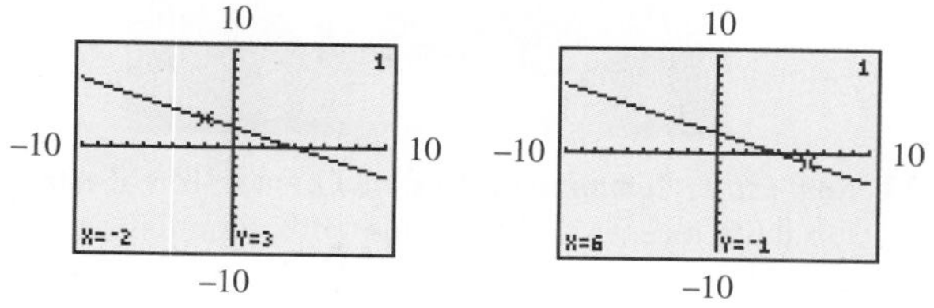

6. Which one of the following has positive slope and negative y-coordinate for its y-intercept?

A.
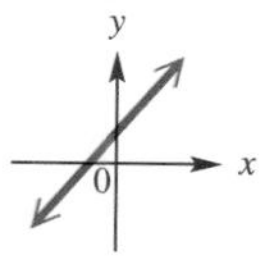

B.
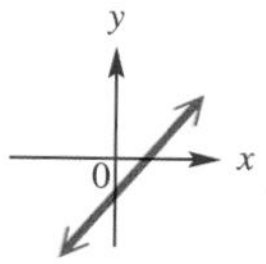

C.
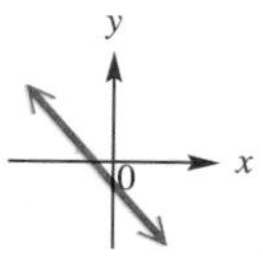

D.
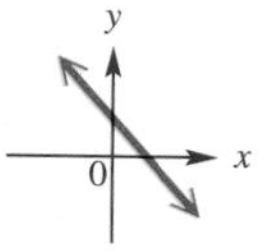

7. ***Income for Americans*** Median household income of Americans is shown in the graph in the next column.

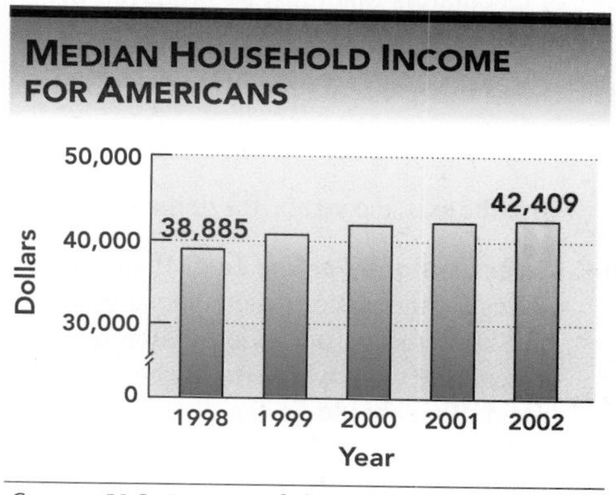

Source: U.S. Bureau of the Census.

   (a) Use the information given for the years 1998 and 2002, letting $x = 0$ represent 1998, $x = 4$ represent 2002, and y represent the median income, to write an equation that models median household income.
   (b) Use the equation to approximate the median income for 2001. How does your result compare to the actual income, \$42,228?

8. ***Library Fines*** It costs a borrower \$.05 per day for an overdue book, plus a flat \$.50 charge for all books borrowed. Let x represent the number of days the book is overdue, so y represents the total fine to the tardy user. Write an equation in the form $y = mx + b$ for this situation. Then give three ordered pairs with x-values of 1, 5, and 10 that satisfy the equation.

9. Write the slope-intercept form of the equation of the line shown.

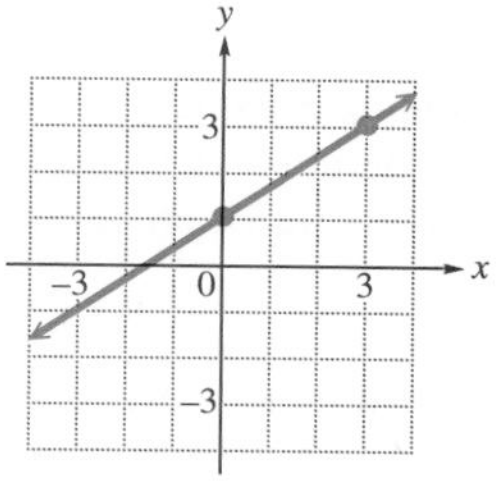

10. For the function $f(x) = x^2 - 3x + 12$,
(a) give its domain. **(b)** find $f(-2)$.

11. Give the domain of the function defined by

$$f(x) = \frac{2}{x - 3}.$$

12. ***Calculator Production*** If the cost to produce x units of calculators is $C(x) = 50x + 5000$ dollars, while the revenue is $R(x) = 60x$ dollars, find the number of units of calculators that must be produced in order to break even. What is the revenue at the break-even point?

13. Graph the quadratic function

$$f(x) = -(x + 3)^2 + 4.$$

Give the axis, the vertex, the domain, and the range.

14. ***Dimensions of a Parking Lot*** Miami-Dade Community College wants to construct a rectangular parking lot on land bordered on one side by a highway. It has 320 ft of fencing with which to fence off the other three sides. What should be the dimensions of the lot if the enclosed area is to be a maximum?

15. Use a scientific calculator to find an approximation of each of the following. Give as many digits as the calculator displays.
(a) $5.1^{4.7}$
(b) $e^{-1.85}$
(c) $\ln 23.56$

16. Which one of the following is a false statement?
A. The domain of the function $f(x) = \log_2 x$ is $(-\infty, \infty)$.
B. The graph of $F(x) = 3^x$ intersects the y-axis.
C. The graph of $G(x) = \log_3 x$ intersects the x-axis.
D. The expression $\ln x$ represents the exponent to which e must be raised in order to obtain x.

17. ***Investment*** Suppose that \$12,000 is invested in an account that pays 4% annual interest, and is left untouched for 3 years. How much will be in the account if
(a) interest is compounded quarterly (four times per year);
(b) interest is compounded continuously?

18. ***Decay of Plutonium-241*** Suppose that the amount, in grams, of plutonium-241 present in a given sample is determined by the function defined by

$$A(t) = 2.00e^{-.053t},$$

where t is measured in years. Find the amount present in the sample after the given number of years.
(a) 4 **(b)** 10
(c) 20
(d) What was the initial amount present?

Solve each system by using elimination, substitution, or a combination of the two methods.

19. $2x + 3y = 2$
$3x - 4y = 20$

20. $2x + y + z = 3$
$x + 2y - z = 3$
$3x - y + z = 5$

21. $2x + 3y - 6z = 11$
$x - y + 2z = -2$
$4x + y - 2z = 7$

Solve each problem by using a system of equations.

22. ***Julia Roberts' Box Office Hits*** Julia Roberts is one of the biggest box-office stars in Hollywood. As of April 2006, her two top-grossing domestic films, *Pretty Woman* and *Ocean's Eleven*, together earned \$914.1 million in worldwide revenues. If *Ocean's Eleven* grossed \$12.7 million less than *Pretty Woman*, how much did each film gross? (*Source:* ACNielsen EDI.)

23. ***Real Estate Commission*** Keshon Grant sells real estate. On three recent sales, he made 10% commission, 6% commission, and 5% commission. His total commissions on these sales were \$17,000, and he sold property worth \$280,000. If the 5% sale amounted to the sum of the other two, what were the three sales' prices?

24. Graph the solution set of the system of inequalities.

$$x + y \le 6$$
$$2x - y \ge 3$$

25. ***Ring Sales*** The Alessi company designs and sells two types of rings: the VIP and the SST. The company can produce up to 24 rings each day using up to 60 total hours of labor. It takes 3 hours to make one VIP ring, and 2 hours to make one SST ring. How many of each type of ring should be made daily in order to maximize profit, if profit on a VIP ring is \$30 and profit on an SST ring is \$40? What is the profit?

9

GEOMETRY

Director Robert Zemeckis' *Cast Away* was one of the top films of 2000. It stars Tom Hanks as Chuck Noland, a Federal Express employee who, as the only survivor in a plane crash, is stranded for 4 years alone on a tropical island. Not long after the crash, speaking to his "friend" Wilson, a volleyball that had washed ashore, he used geometry to assess their chances of being found. Sketching a circle and performing an arithmetic calculation on the side of a rock, Chuck realizes their futility.

So, Wilson. We were en route from Memphis for eleven and a half hours at about 475 miles an hour. They think that we are right here. But we went out of radio contact and flew around that storm for about an hour. So that's a distance of what, 400 miles? Four hundred miles squared, that's 160,000, times pi, 3.14, . . .

Chuck's calculation for the size of the search area is an application of the formula for the area of a circle, $A = \pi r^2$. He looks to Wilson, and sighs,

. . . That's twice the size of Texas. They may never find us.

What was Chuck's answer? The land area of Texas is 261,797 square miles. Was he correct? The answer can be found on page 559.

9.1 Points, Lines, Planes, and Angles

The Geometry of Euclid • Points, Lines, and Planes • Angles

Euclid's *Elements* as translated by Billingsley appeared in 1570 and was the first English language translation of the text—the most influential geometry text ever written.

Unfortunately, no copy of *Elements* exists that dates back to the time of Euclid (circa 300 B.C.), and most current translations are based upon a revision of the work prepared by Theon of Alexandria.

Although *Elements* was only one of several works of Euclid, it is, by far, the most important. It ranks second only to the Bible as the most published book in history.

The Geometry of Euclid

Let no one unversed in geometry enter here.
—Motto over the door of Plato's Academy

To the ancient Greeks, mathematics meant geometry above all—a rigid kind of geometry from a modern-day point of view. The Greeks studied the properties of figures identical in shape and size (congruent figures) as well as figures identical in shape but not necessarily in size (similar figures). They absorbed ideas about area and volume from the Egyptians and Babylonians and established general formulas. The Greeks were the first to insist that statements in geometry be given rigorous proof.

The Greek view of geometry (and other mathematical ideas) was summarized in *Elements,* written by Euclid about 300 B.C. The influence of this book has been extraordinary; it has been studied virtually unchanged to this day as a geometry textbook and as *the* model of deductive logic.

The most basic ideas of geometry are **point, line,** and **plane.** In fact, it is not really possible to define them with other words. Euclid defined a point as "that which has no part," but this definition is so vague as to be meaningless. Do you think you could decide what a point is from this definition? But from your experience in saying "this point in time" or in sharpening a pencil, you have an idea of what he was getting at. Even though we don't try to define *point,* we do agree that, intuitively, a point has no magnitude and no size.

Euclid defined a line as "that which has breadthless length." Again, this definition is vague. Based on our experience, however, we know what Euclid meant. The drawings that we use for lines have properties of no thickness and no width, and they extend indefinitely in two directions.

What do you visualize when you read Euclid's definition of a plane: "a surface which lies evenly with the straight lines on itself"? Do you think of a flat surface, such as a tabletop or a page in a book? That is what Euclid intended.

The geometry of Euclid is a model of deductive reasoning. In this chapter, we will present geometry from an inductive viewpoint, using objects and situations found in the world around us as models for study.

Points, Lines, and Planes There are certain universally accepted conventions and symbols used to represent points, lines, planes, and angles. A capital letter usually represents a point. A line may be named by two capital letters representing points that lie on the line, or by a single (usually lowercase) letter, such as ℓ. Subscripts are sometimes used to distinguish one line from another when a lowercase letter is used. For example, ℓ_1 and ℓ_2 would represent two distinct lines. A plane may be named by three capital letters representing points that lie in the plane, or by a letter of the Greek alphabet, such as α (alpha), β (beta), or γ (gamma).

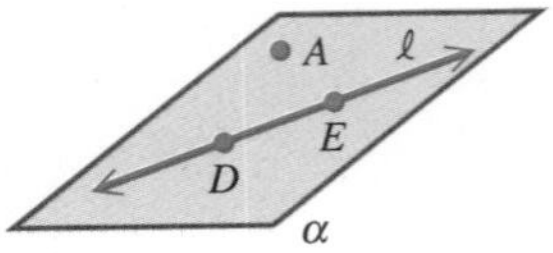

FIGURE 1

Figure 1 depicts a plane that may be represented either as α or as plane *ADE*. Contained in the plane is the line *DE* (or, equivalently, line *ED*), which is also labeled ℓ in the figure.

Selecting any point on a line divides the line into three parts: the point itself, and two **half-lines,** one on each side of the point. For example point *A* divides the

Given any three points that are not in a straight line, a plane can be passed through the points. That is why **camera tripods** have three legs—no matter how irregular the surface, the tips of the three legs determine a plane. On the other hand, a camera support with four legs would wobble unless all four legs were carefully extended just the right amount.

line shown in Figure 2 into three parts, A itself and two half-lines. Point A belongs to neither half-line. As the figure suggests, each half-line extends indefinitely in the direction opposite the other half-line.

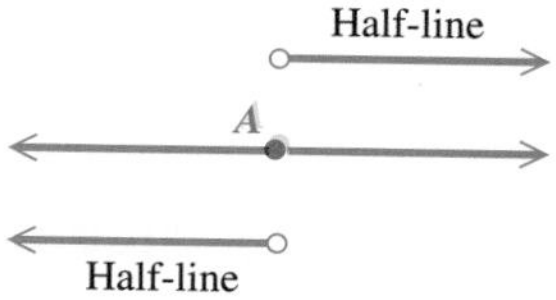

FIGURE 2

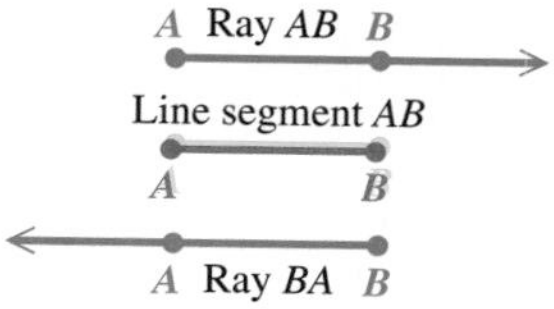

FIGURE 3

Including an initial point with a half-line gives a **ray.** A ray is named with two letters, one for the initial point of the ray, and one for another point contained in the half-line. For example, in Figure 3 ray AB has initial point A and extends in the direction of B. On the other hand, ray BA has B as its initial point and extends in the direction of A.

A **line segment** includes both endpoints and is named by its endpoints. Figure 3 shows line segment AB, which may also be designated as line segment BA.

Table 1 shows these figures along with the symbols used to represent them.

TABLE 1

Name	Figure	Symbol
Line AB or line BA		$\overleftrightarrow{AB}$ or $\overleftrightarrow{BA}$
Half-line AB		$\overset{\circ\!\!\rightarrow}{AB}$
Half-line BA		$\overset{\circ\!\!\rightarrow}{BA}$
Ray AB	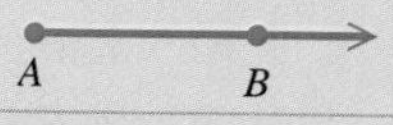	$\overset{\bullet\!\!\rightarrow}{AB}$
Ray BA		$\overset{\bullet\!\!\rightarrow}{BA}$
Segment AB or segment BA		$\overset{\bullet\!-\!\bullet}{AB}$ or $\overset{\bullet\!-\!\bullet}{BA}$

For a line, the symbol above the two letters shows two arrowheads, indicating that the line extends indefinitely in both directions. For half-lines and rays, only one arrowhead is used because these extend in only one direction. An open circle is used for a half-line to show that the endpoint is not included, while a solid circle is used for a ray to indicate the inclusion of the endpoint. Since a segment includes both endpoints and does not extend in either direction, solid circles are used to indicate endpoints of line segments.

The geometric definitions of "parallel" and "intersecting" apply to two or more lines or planes. (See Figure 4 on the next page.) **Parallel lines** lie in the same plane and never meet, no matter how far they are extended. However, **intersecting lines** do meet. If two distinct lines intersect, they intersect in one and only one point.

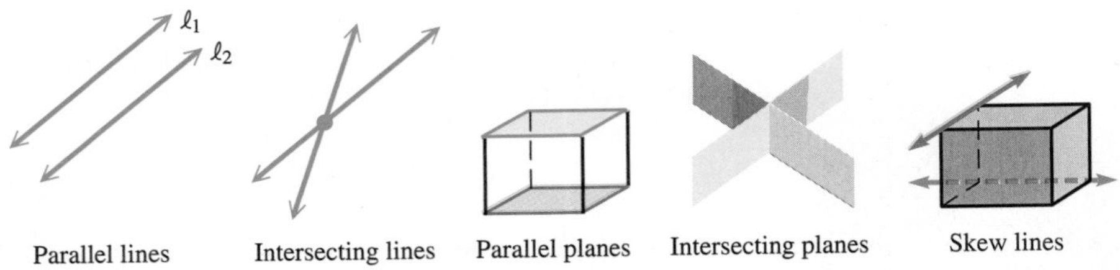

FIGURE 4

We use the symbol $\|$ to denote parallelism. If ℓ_1 and ℓ_2 are parallel lines, as in Figure 4, then this may be indicated as $\ell_1 \| \ell_2$.

Parallel planes also never meet, no matter how far they are extended. Two distinct **intersecting planes** form a straight line, the one and only line they have in common. **Skew lines** do not lie in the same plane, and they never meet, no matter how far they are extended.

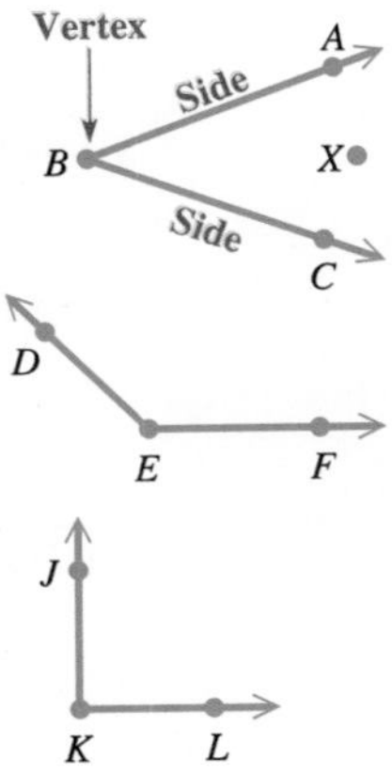

FIGURE 5

Angles

An **angle** is the union of two rays that have a common endpoint, as shown in Figure 5. It is important to remember that the angle is formed by points on the rays themselves, and no other points. In Figure 5, point X is *not* a point on the angle. (It is said to be in the *interior* of the angle.) Notice that "angle" is the first basic term in this section that is actually defined, using the undefined terms *ray* and *endpoint*.

The rays forming an angle are called its **sides.** The common endpoint of the rays is the **vertex** of the angle. There are two standard ways of naming angles using letters. If no confusion will result, an angle can be named with the letter marking its vertex. Using this method, the angles in Figure 5 can be named, respectively, angle B, angle E, and angle K. Angles also can be named with three letters: the first letter names a point on one side of the angle; the middle letter names the vertex; the third names a point on the other side of the angle. In this system, the angles in the figure can be named angle ABC, angle DEF, and angle JKL. The symbol for representing an angle is $\measuredangle$. Rather than writing "angle ABC," we may write "$\measuredangle ABC$."

An angle can be associated with an amount of rotation. For example, in Figure 6(a), we let $\overleftrightarrow{BA}$ first coincide with $\overleftrightarrow{BC}$—as though they were the same ray. We then rotate $\overleftrightarrow{BA}$ (the endpoint remains fixed) in a counterclockwise direction to form $\measuredangle ABC$.

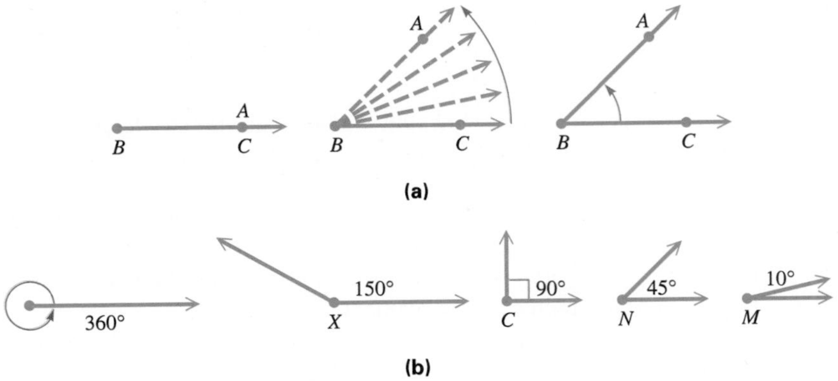

FIGURE 6

Why 360? The use of the number 360 goes back to the Babylonian culture. There are several theories regarding why 360 was chosen for the number of degrees in a complete rotation around a circle. One says that 360 was chosen because it is close to the number of days in a year, and is conveniently divisible by 2, 3, 4, 5, 6, 8, 9, 10, 12, and other numbers.

Angles are the key to the study of **geodesy,** the measurement of distances on the earth's surface.

Angles are measured by the amount of rotation, using a system that dates back to the Babylonians some two centuries before Christ. Babylonian astronomers chose the number 360 to represent the amount of rotation of a ray back onto itself. Using 360 as the amount of rotation of a ray back onto itself, **one degree,** written 1°, is defined to be $\frac{1}{360}$ of a complete rotation. Figure 6(b) shows angles of various degree measures.

Angles are classified and named with reference to their degree measures. An angle whose measure is between 0° and 90° is called an **acute angle.** Angles *M* and *N* in Figure 6(b) are acute. An angle that measures 90° is called a **right angle.** Angle *C* in the figure is a right angle. The squared symbol ⌉ at the vertex denotes a right angle. Angles that measure more than 90° but less than 180° are said to be **obtuse angles** (angle *X*, for example). An angle that measures 180° is a **straight angle.** Its sides form a straight line.

Our work in this section will be devoted primarily to angles whose measures are less than or equal to 180°. Angles whose measures are greater than 180° are discussed in Chapter 10 and are studied in more detail in trigonometry courses.

A tool called a **protractor** can be used to measure angles. Figure 7 shows a protractor measuring an angle. To use a protractor, position the hole (or dot) of the protractor on the vertex of the angle. The 0-degree measure on the protractor should be placed on one side of the angle, while the other side should extend to the degree measure of the angle. The figure indicates an angle whose measure is 135°.

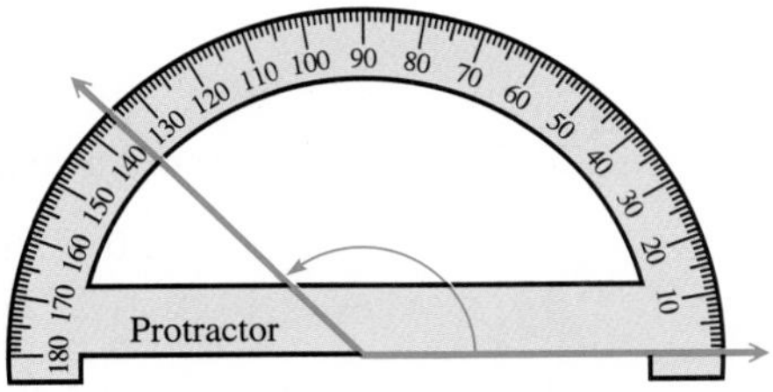

FIGURE 7

When two lines intersect to form right angles they are called **perpendicular lines.** Our sense of *vertical* and *horizontal* depends on perpendicularity.

In Figure 8, the sides of $\measuredangle NMP$ have been extended to form another angle, $\measuredangle RMQ$. The pair $\measuredangle NMP$ and $\measuredangle RMQ$ are called **vertical angles.** Another pair of vertical angles have been formed at the same time. They are $\measuredangle NMQ$ and $\measuredangle PMR$.

An important property of vertical angles follows.

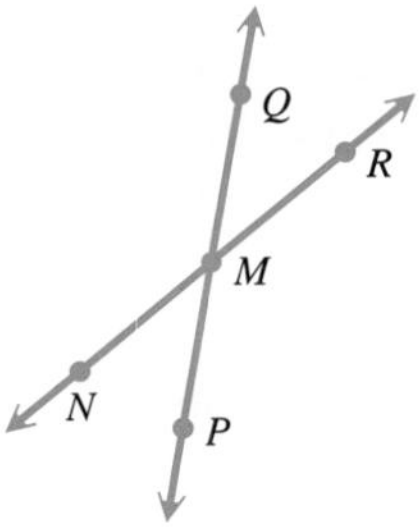

FIGURE 8

Property of Vertical Angles

Vertical angles have equal measures.

For example, $\measuredangle NMP$ and $\measuredangle RMQ$ in Figure 8 have equal measures. What other pair of angles in the figure have equal measures?

EXAMPLE 1 Finding Angle Measures

Find the measure of each marked angle in the given figure.

(a) Figure 9 **(b)** Figure 10

SOLUTION

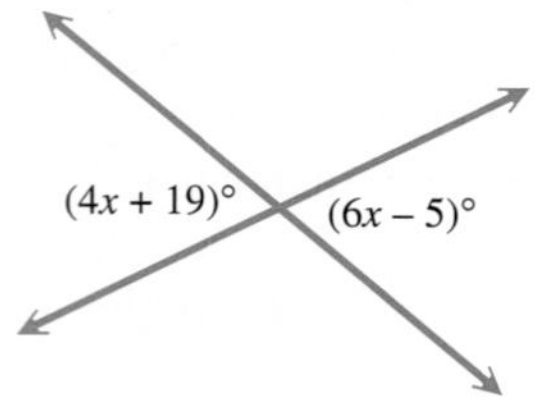

FIGURE 9

(a) Because the marked angles are vertical angles, they have the same measure. So,

$$\begin{aligned} 4x + 19 &= 6x - 5 \\ 4x + 19 - 4x &= 6x - 5 - 4x && \text{Subtract } 4x. \\ 19 &= 2x - 5 && \text{Combine like terms.} \\ 19 + 5 &= 2x - 5 + 5 && \text{Add 5.} \\ 24 &= 2x \\ 12 &= x. && \text{Divide by 2.} \end{aligned}$$

Don't stop here. → $12 = x.$

Since $x = 12$, one angle has measure $4(12) + 19 = 67$ degrees. The other has the same measure, because $6(12) - 5 = 67$ as well. Each angle measures 67°.

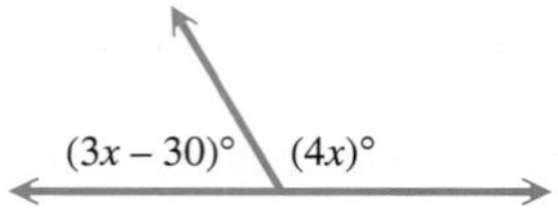

FIGURE 10

(b) The measures of the marked angles must add to 180° because together they form a straight angle.

$$\begin{aligned} (3x - 30) + 4x &= 180 && \text{The angle sum is 180.} \\ 7x - 30 &= 180 && \text{Combine like terms.} \\ 7x - 30 + 30 &= 180 + 30 && \text{Add 30.} \\ 7x &= 210 \\ x &= 30 && \text{Divide by 7.} \end{aligned}$$

Don't stop here. → $x = 30$

To find the measures of the angles, replace x with 30 in the two expressions.

$$\begin{aligned} 3x - 30 &= 3(30) - 30 = 90 - 30 = 60 \\ 4x &= 4(30) = 120 \end{aligned}$$

The two angle measures are 60° and 120°. ■

If the sum of the measures of two acute angles is 90°, the angles are said to be **complementary,** and each is called the *complement* of the other. For example, angles measuring 40° and 50° are complementary angles, because $40° + 50° = 90°$. If two angles have a sum of 180°, they are **supplementary.** The *supplement* of an angle whose measure is 40° is an angle whose measure is 140°, because $40° + 140° = 180°$.

If x represents the degree measure of an angle, 90 − x represents the measure of its complement, and 180 − x represents the measure of its supplement.

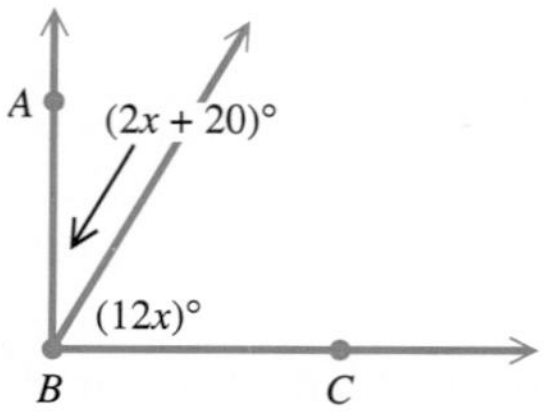

FIGURE 11

EXAMPLE 2 Finding Angle Measures

Find the measures of the angles in Figure 11, given that $\measuredangle ABC$ is a right angle.

SOLUTION

The sum of the measures of the two acute angles is 90° (that is, they are complementary), because they form a right angle. We add their measures to obtain a sum of 90 and solve the resulting equation.

$$\begin{aligned}(2x + 20) + 12x &= 90 \\ 14x + 20 &= 90 \quad \text{Combine like terms.} \\ 14x &= 70 \quad \text{Subtract 20.} \\ x &= 5 \quad \text{Divide by 14.}\end{aligned}$$

The value of x is 5. Therefore, replace x with 5 in the two expressions.

$$2x + 20 = 2(5) + 20 = 30.$$
$$12x = 12(5) = 60$$

The measures of the two angles are 30° and 60°. ■

EXAMPLE 3 Using Complementary and Supplementary Angles

The supplement of an angle measures 10° more than three times its complement. Find the measure of the angle.

SOLUTION

Let $x =$ the degree measure of the angle.
Then $180 - x =$ the degree measure of its supplement,
and $90 - x =$ the degree measure of its complement.

Now use the words of the problem to write the equation.

Supplement	measures	10 more than	three times its complement.
↓	↓	↓	↓
$180 - x$	$=$	$10 +$	$3(90 - x)$

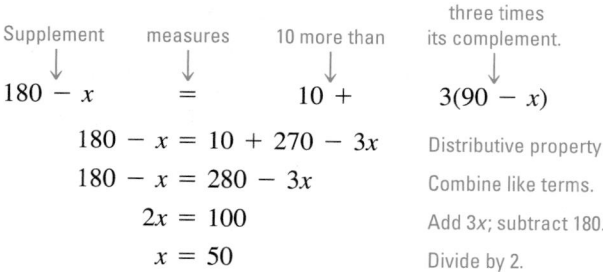

$$\begin{aligned}180 - x &= 10 + 270 - 3x \quad \text{Distributive property} \\ 180 - x &= 280 - 3x \quad \text{Combine like terms.} \\ 2x &= 100 \quad \text{Add } 3x\text{; subtract 180.} \\ x &= 50 \quad \text{Divide by 2.}\end{aligned}$$

The angle measures 50°. Because its supplement (130°) is 10° more than three times its complement (40°) (that is, $130 = 10 + 3(40)$ is true) the answer checks. ■

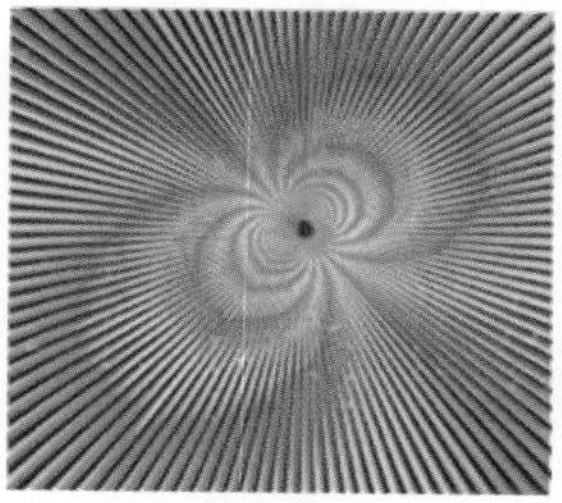

A set of parallel lines with equidistant spacing intersects an identical set, but at a small angle. The result is a **moiré pattern,** named after the fabric *moiré* ("watered") *silk*. You often see similar effects looking through window screens with bulges. Moiré patterns are related to **periodic functions,** which describe regular recurring phenomena (wave patterns such as heartbeats or business cycles). Moirés thus apply to the study of electromagnetic, sound, and water waves, to crystal structure, and to other wave phenomena.

Parallel lines are lines that lie in the same plane and do not intersect. Figure 12 shows parallel lines m and n. When a line q intersects two parallel lines, q is called a **transversal.** In Figure 12, the transversal intersecting the parallel lines forms eight angles, indicated by numbers. Angles 1 through 8 in the figure possess some special properties regarding their degree measures, as shown in Table 2 on the next page.

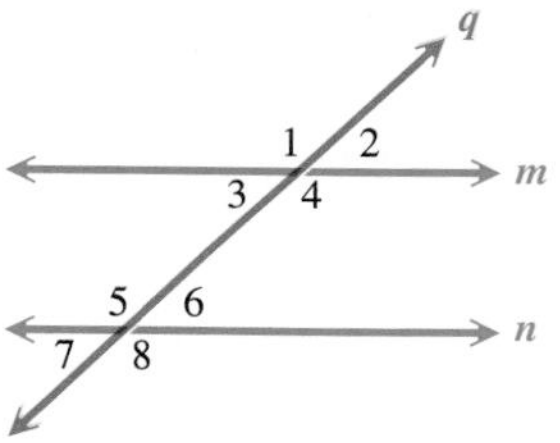

FIGURE 12

TABLE 2

Name	Figure	Rule
Alternate interior angles	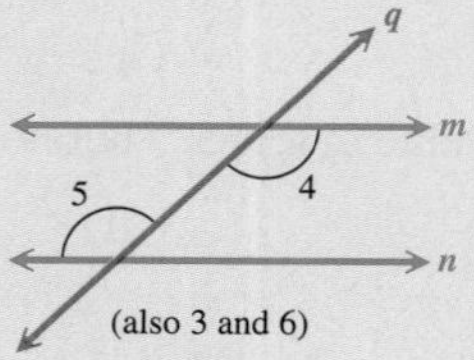	Angle measures are equal.
Alternate exterior angles	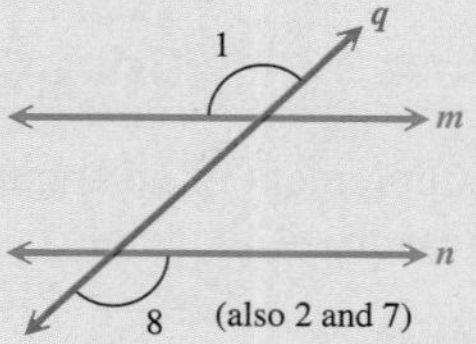	Angle measures are equal.
Interior angles on same side of transversal	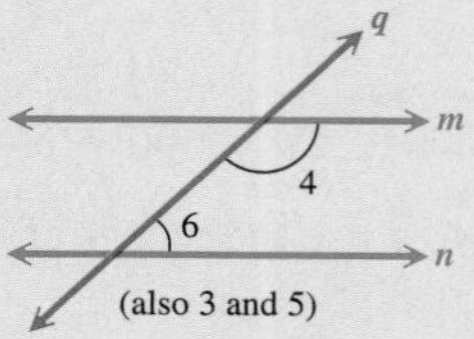	Angle measures add to 180°.
Corresponding angles	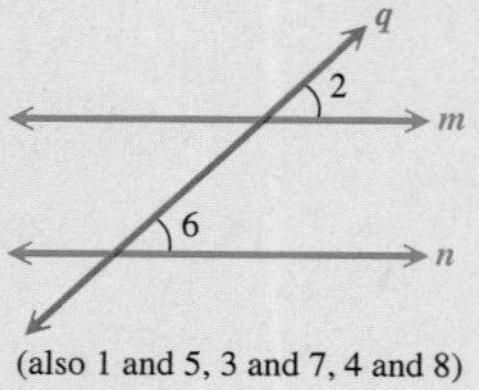	Angle measures are equal.

The converses of the above also are true. That is, if alternate interior angles are equal, then the lines are parallel, with similar results valid for alternate exterior angles, interior angles on the same side of a transversal, and corresponding angles.

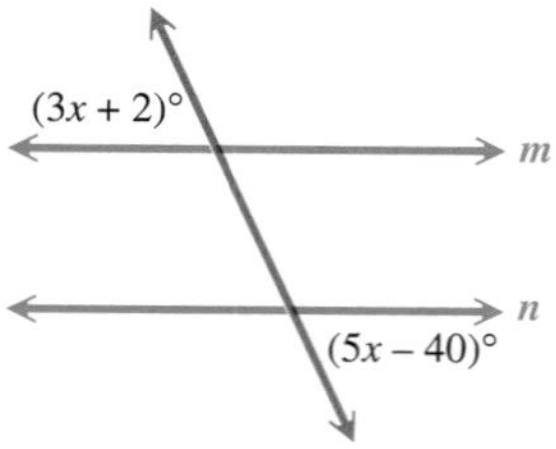

FIGURE 13

EXAMPLE 4 Finding Angle Measures

Find the measure of each marked angle in Figure 13, given that lines m and n are parallel.

SOLUTION

The marked angles are alternate exterior angles, which are equal. This gives

$$3x + 2 = 5x - 40$$
$$42 = 2x \qquad \text{Subtract } 3x\text{; add } 40.$$
$$21 = x. \qquad \text{Divide by } 2.$$

Because $3x + 2 = 3 \cdot 21 + 2 = 65$ and $5x - 40 = 5 \cdot 21 - 40 = 65$, both angles measure 65°.

9.1 Exercises

Fill in each blank with the correct response.

1. The sum of the measures of two complementary angles is _____ degrees.

2. The sum of the measures of two supplementary angles is _____ degrees.

3. The measures of two vertical angles are _____.
(equal/not equal)

4. The measures of _____ right angles add up to the measure of a straight angle.

Decide whether each statement is true *or* false.

5. A line segment has two endpoints.

6. A ray has one endpoint.

7. If *A* and *B* are distinct points on a line, then ray *AB* and ray *BA* represent the same set of points.

8. If two lines intersect, they lie in the same plane.

9. If two lines are parallel, they lie in the same plane.

10. If two lines do not intersect, they must be parallel.

11. Segment *AB* and segment *BA* represent the same set of points.

12. There is no angle that is its own complement.

13. There is no angle that is its own supplement.

14. The origin of the use of the degree as a unit of measure of an angle goes back to the Egyptians.

Exercises 15–24 name portions of the line shown. For each exercise, **(a)** *give the symbol that represents the portion of the line named, and* **(b)** *draw a figure showing just the portion named, including all labeled points.*

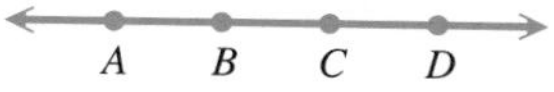

15. line segment *AB* **16.** ray *BC* **17.** ray *CB* **18.** line segment *AD*

19. half-line *BC* **20.** half-line *AD* **21.** ray *BA* **22.** ray *DA*

23. line segment *CA* **24.** line segment *DA*

Match the symbol in Column I with the symbol in Column II that names the same set of points, based on the figure.

I

25. $\overset{\bullet\!-\!\bullet}{PQ}$ **26.** $\overset{\circ\!\rightarrow}{QR}$

27. $\overset{\bullet\!\rightarrow}{QR}$ **28.** $\overleftrightarrow{PQ}$

29. $\overset{\bullet\!\rightarrow}{RP}$ **30.** $\overset{\circ\!\rightarrow}{SQ}$

31. $\overset{\bullet\!-\!\bullet}{PS}$ **32.** $\overset{\circ\!\rightarrow}{PS}$

II

A. $\overset{\circ\!\rightarrow}{QS}$ **B.** $\overset{\bullet\!\rightarrow}{RQ}$

C. $\overset{\circ\!\rightarrow}{SR}$ **D.** $\overset{\bullet\!\rightarrow}{QS}$

E. $\overset{\bullet\!-\!\bullet}{SP}$ **F.** $\overset{\bullet\!-\!\bullet}{QP}$

G. $\overleftrightarrow{RS}$ **H.** none of these

Lines, rays, half-lines, and segments may be considered sets of points. The **intersection** *(symbolized $\cap$) of two sets is composed of all elements common to both sets, while the* **union** *(symbolized $\cup$) of two sets is composed of all elements found in at least one of the two sets. Based on the figure below, specify each of the sets given in Exercises 33–40 in a simpler way.*

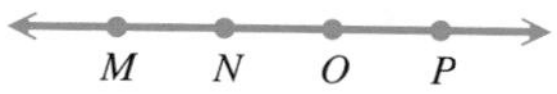

33. $\overset{\bullet\!-\!\bullet}{MN} \cup \overset{\bullet\!-\!\bullet}{NO}$ **34.** $\overset{\bullet\!-\!\bullet}{MN} \cap \overset{\bullet\!-\!\bullet}{NO}$ **35.** $\overset{\bullet\!\rightarrow}{MO} \cap \overset{\bullet\!\rightarrow}{OM}$ **36.** $\overset{\bullet\!\rightarrow}{MO} \cup \overset{\bullet\!\rightarrow}{OM}$

37. $\overset{\circ\!\rightarrow}{OP} \cap O$ **38.** $\overset{\circ\!\rightarrow}{OP} \cup O$ **39.** $\overset{\bullet\!-\!\bullet}{NP} \cap \overset{\bullet\!-\!\bullet}{OP}$ **40.** $\overset{\bullet\!-\!\bullet}{NP} \cup \overset{\bullet\!-\!\bullet}{OP}$

Give the measure of the complement of each angle.

41. 28° **42.** 32° **43.** 89° **44.** 45° **45.** $x°$ **46.** $(90 - x)°$

Give the measure of the supplement of each angle.

47. 132° **48.** 105° **49.** 26° **50.** 90° **51.** $y°$ **52.** $(180 - y)°$

Name all pairs of vertical angles in each figure.

53.

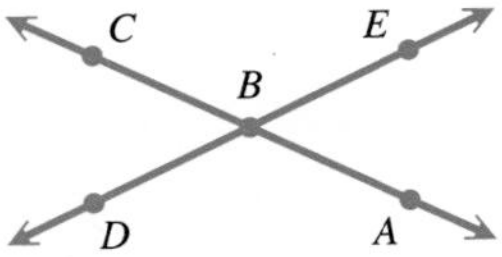

54.

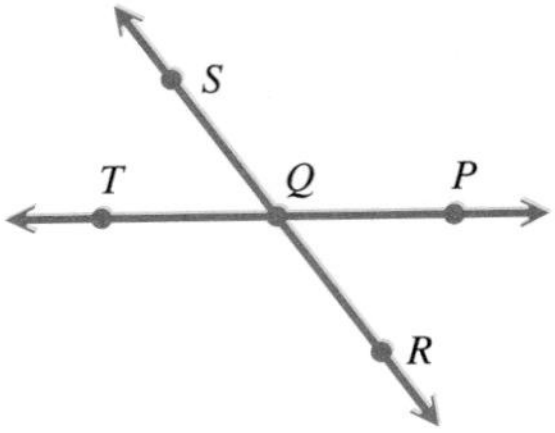

55. In Exercise 53, if $\measuredangle ABE$ has a measure of 52°, find the measures of the angles.
(a) $\measuredangle CBD$
(b) $\measuredangle CBE$

56. In Exercise 54, if $\measuredangle SQP$ has a measure of 126°, find the measures of the angles.
(a) $\measuredangle TQR$
(b) $\measuredangle PQR$

Find the measure of each marked angle.

57.

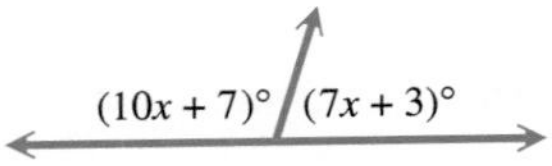

58.

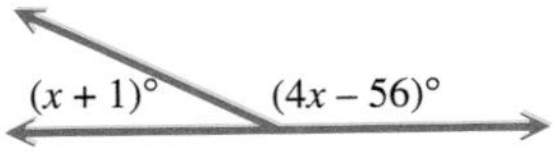

59.

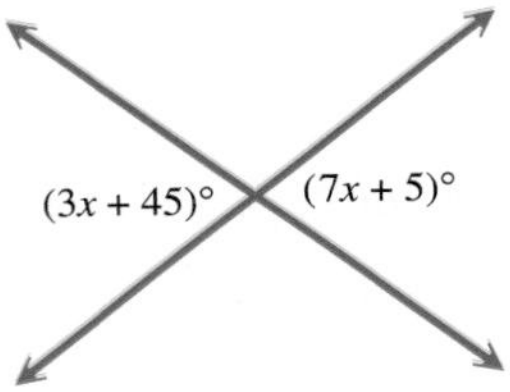

60.

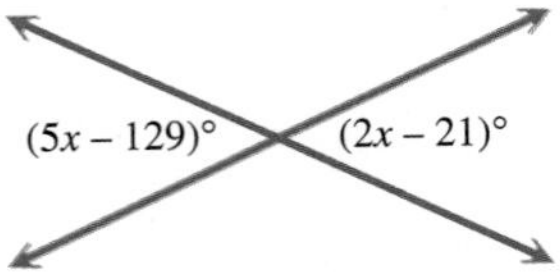

61.

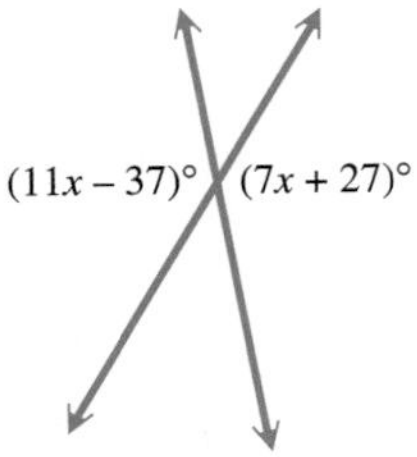

62.

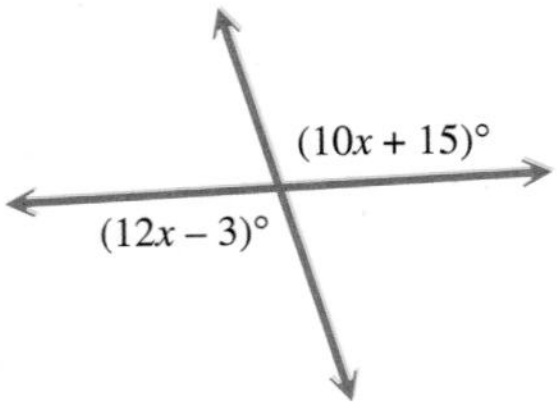

63.

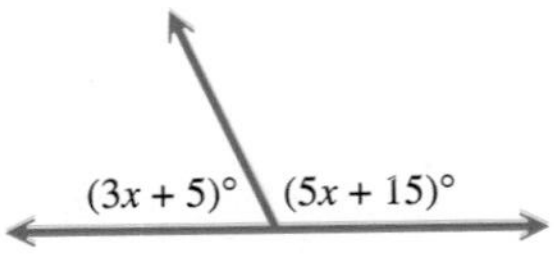

64.

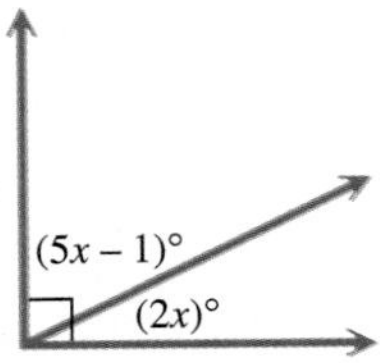

65.

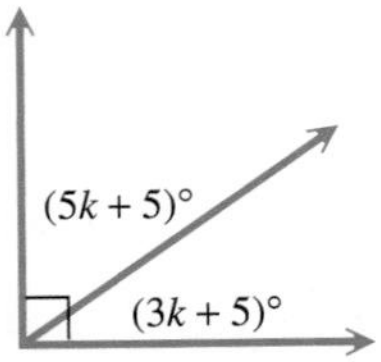

In Exercises 66–69, assume that lines m and n are parallel, and find the measure of each marked angle.

66.

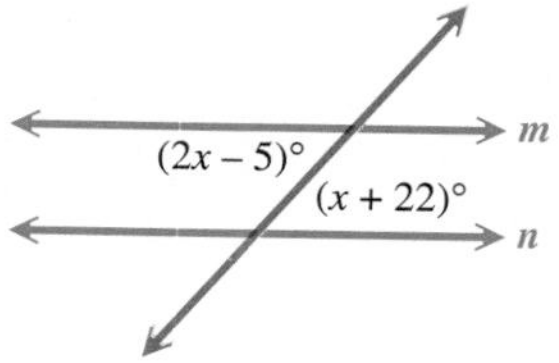

67.

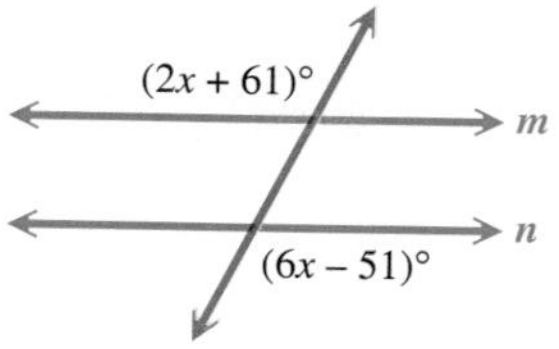

68.

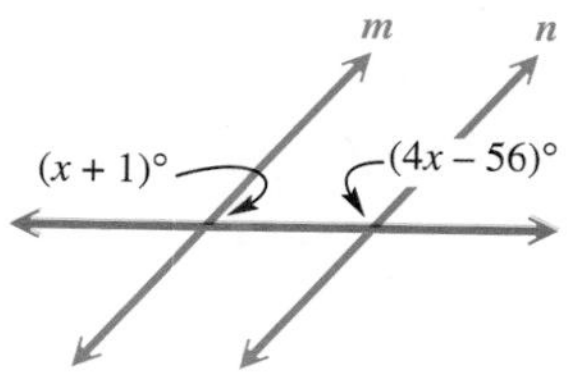

69.

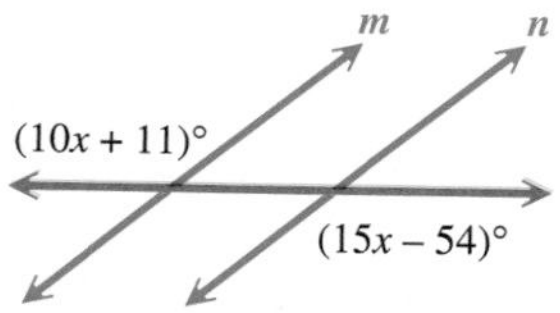

Complementary and Supplementary Angles *Solve each problem in Exercises 70–73.*

70. The supplement of an angle measures 25° more than twice its complement. Find the measure of the angle.

71. The complement of an angle measures 10° less than one-fifth of its supplement. Find the measure of the angle.

72. The supplement of an angle added to the complement of the angle gives 210°. What is the measure of the angle?

73. Half the supplement of an angle is 12° less than twice the complement of the angle. Find the measure of the angle.

74. The sketch shows parallel lines *m* and *n* cut by a transversal *q*. Using the figure, complete the steps to prove that alternate exterior angles have the same measure.

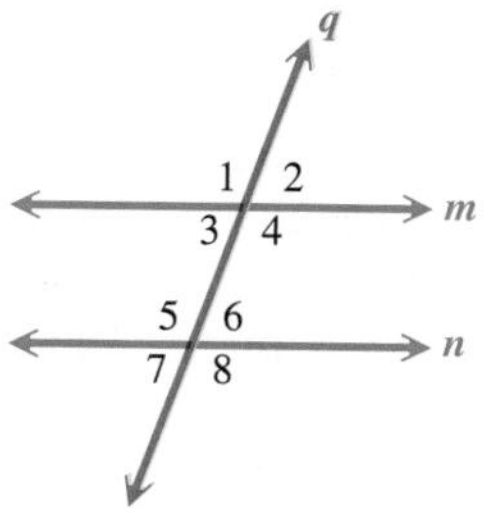

(a) Measure of $\measuredangle 2$ = measure of $\measuredangle$ ____, since they are vertical angles.

(b) Measure of $\measuredangle 3$ = measure of $\measuredangle$ ____, since they are alternate interior angles.

(c) Measure of $\measuredangle 6$ = measure of $\measuredangle$ ____, since they are vertical angles.

(d) By the results of parts (a), (b), and (c), the measure of $\measuredangle 2$ must equal the measure of $\measuredangle$ ____, showing that alternate ________ angles have equal measures.

75. Use the sketch to find the measure of each numbered angle. Assume that $m \parallel n$.

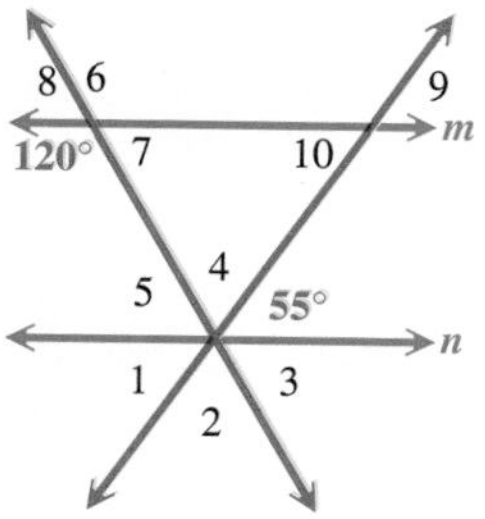

76. Complete these steps in the proof that vertical angles have equal measures. In this exercise, m($\measuredangle x$) means "the measure of the angle *x*." Use the figure at the right.

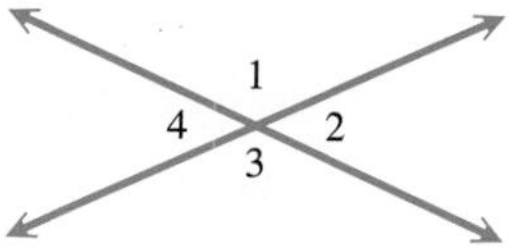

(a) m($\measuredangle 1$) + m($\measuredangle 2$) = _____°

(b) m($\measuredangle 2$) + m($\measuredangle 3$) = _____°

(c) Subtract the equation in part (b) from the equation in part (a) to get [m($\measuredangle 1$) + m($\measuredangle 2$)] − [m($\measuredangle 2$) + m($\measuredangle 3$)] = _____° − _____°.

(d) m($\measuredangle 1$) + m($\measuredangle 2$) − m($\measuredangle 2$) − m($\measuredangle 3$) = ___°

(e) m($\measuredangle 1$) − m($\measuredangle 3$) = _____°

(f) m($\measuredangle 1$) = m($\measuredangle$ _____)

77. Use the approach of Exercise 74 to prove that interior angles on the same side of a transversal are supplementary.

78. Find the values of x and y in the figure, given that $x + y = 40$.

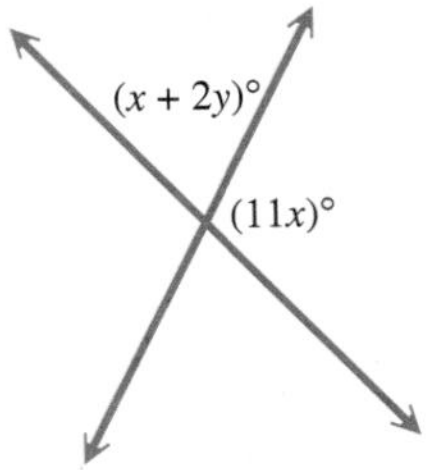

9.2 Curves, Polygons, and Circles

Curves • Triangles and Quadrilaterals • Circles

Curves The basic undefined term *curve* is used for describing figures in the plane. (See Figure 14.)

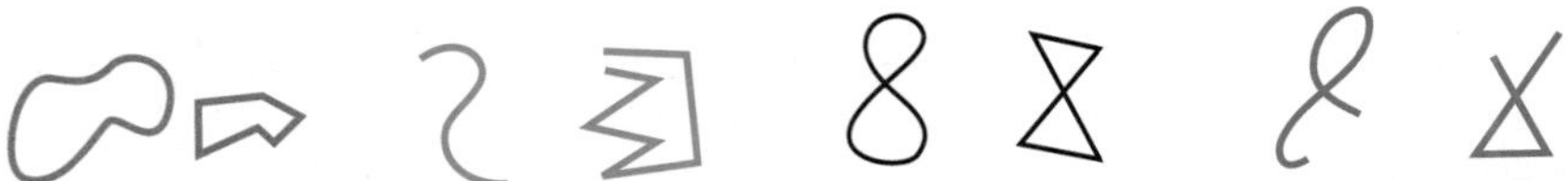

FIGURE 14

Simple Curve; Closed Curve

A **simple curve** can be drawn without lifting the pencil from the paper, and without passing through any point twice.

A **closed curve** has its starting and ending points the same, and is also drawn without lifting the pencil from the paper.

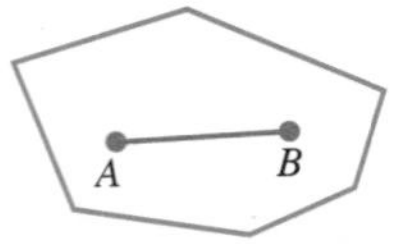

Convex

(a)

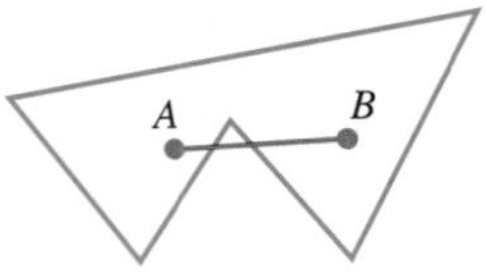

Not convex

(b)

FIGURE 15

A figure is said to be **convex** if, for any two points A and B inside the figure, the line segment AB (that is, $\overleftrightarrow{AB}$) is always completely inside the figure. Figure 15(a) shows a convex figure and (b) shows one that is not convex.

Among the most common types of curves in mathematics are those that are both simple and closed, and perhaps the most important of these are *polygons*. A **polygon** is a simple closed curve made up only of straight line segments. The line segments are called the *sides,* and the points at which the sides meet are called *vertices* (singular: *vertex*). Polygons are classified according to the number of line segments used as sides. Table 3 on the next page gives the special names. In general, if a polygon has n sides, and no particular value of n is specified, it is called an n-gon.

Some examples of polygons are shown in Figure 16. A polygon may or may not be convex. Polygons with all sides equal and all angles equal are **regular polygons.**

TABLE 3
Classification of Polygons According to Number of Sides

Number of Sides	Name
3	triangle
4	quadrilateral
5	pentagon
6	hexagon
7	heptagon
8	octagon
9	nonagon
10	decagon

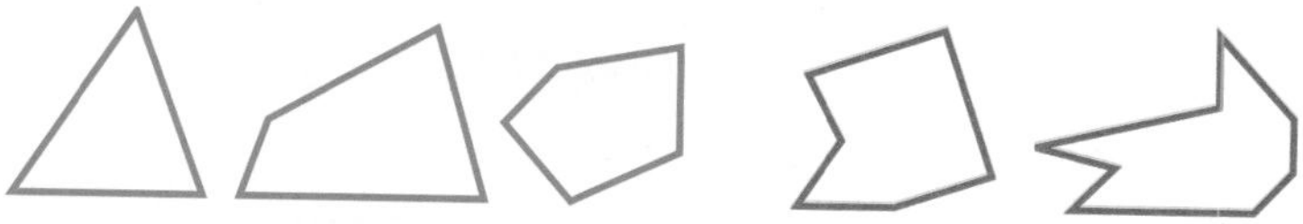

Convex   Not convex

Polygons are simple closed curves made up of straight line segments.

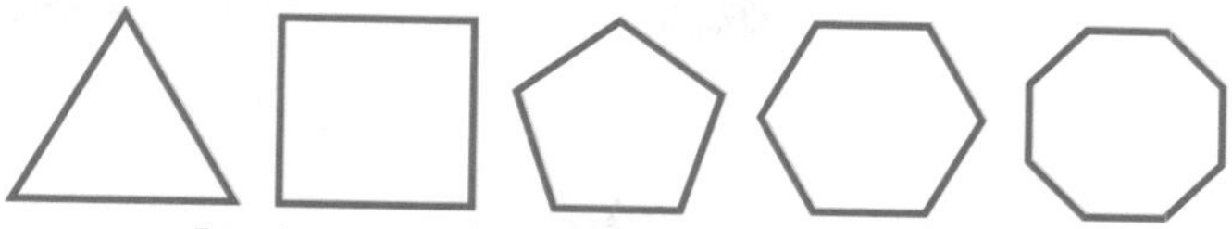

Regular polygons have equal sides and equal angles.

FIGURE 16

The puzzle-game above comes from China, where it has been a popular amusement for centuries. The figure on the left is a **tangram.** Any tangram is composed of the same set of seven tans (the pieces making up the square are shown on the right).

Mathematicians have described various properties of tangrams. While each tan is convex, only 13 convex tangrams are possible. All others, like the figure on the left, are not convex.

Triangles and Quadrilaterals

Triangles and Quadrilaterals Two of the most common types of polygons are triangles and quadrilaterals. Triangles are classified by measures of angles as well as by number of equal sides, as shown in the following box. (Notice that tick marks are used in the bottom three figures to show how side lengths are related.)

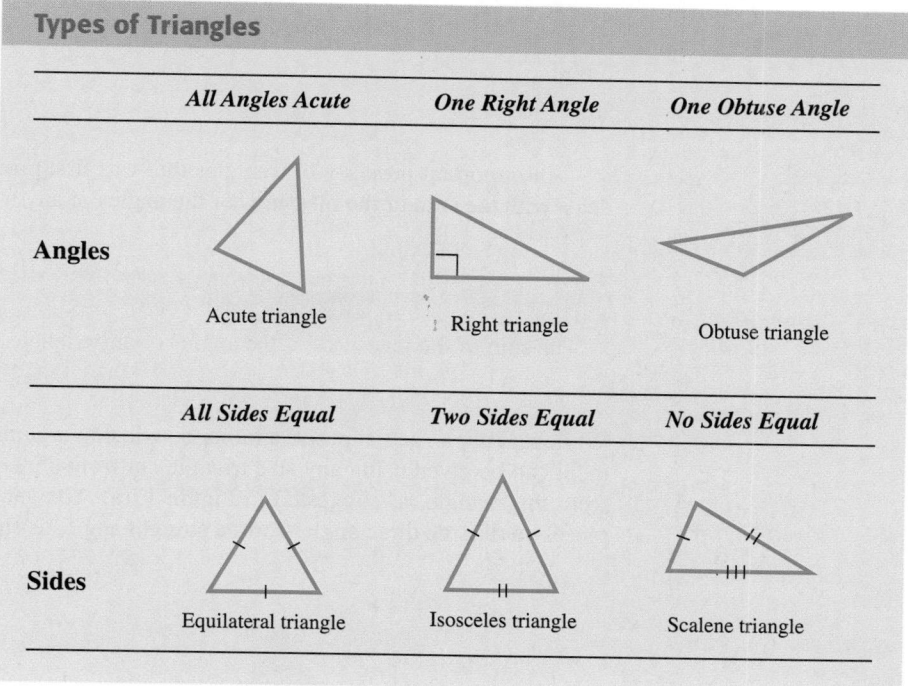

Types of Triangles

	All Angles Acute	*One Right Angle*	*One Obtuse Angle*
Angles	 Acute triangle	 Right triangle	 Obtuse triangle
	All Sides Equal	***Two Sides Equal***	***No Sides Equal***
Sides	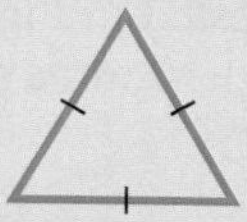 Equilateral triangle	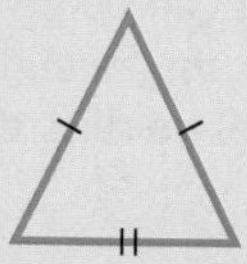 Isosceles triangle	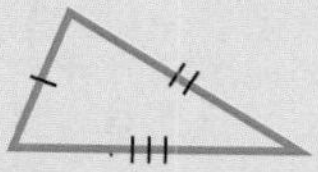 Scalene triangle

Quadrilaterals are classified by sides and angles. It can be seen in the box at the top of the next page that an important distinction involving quadrilaterals is whether one or more pairs of sides are parallel.

Types of Quadrilaterals

Sample Figure

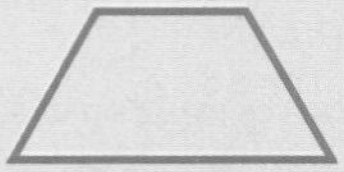

Trapezoid

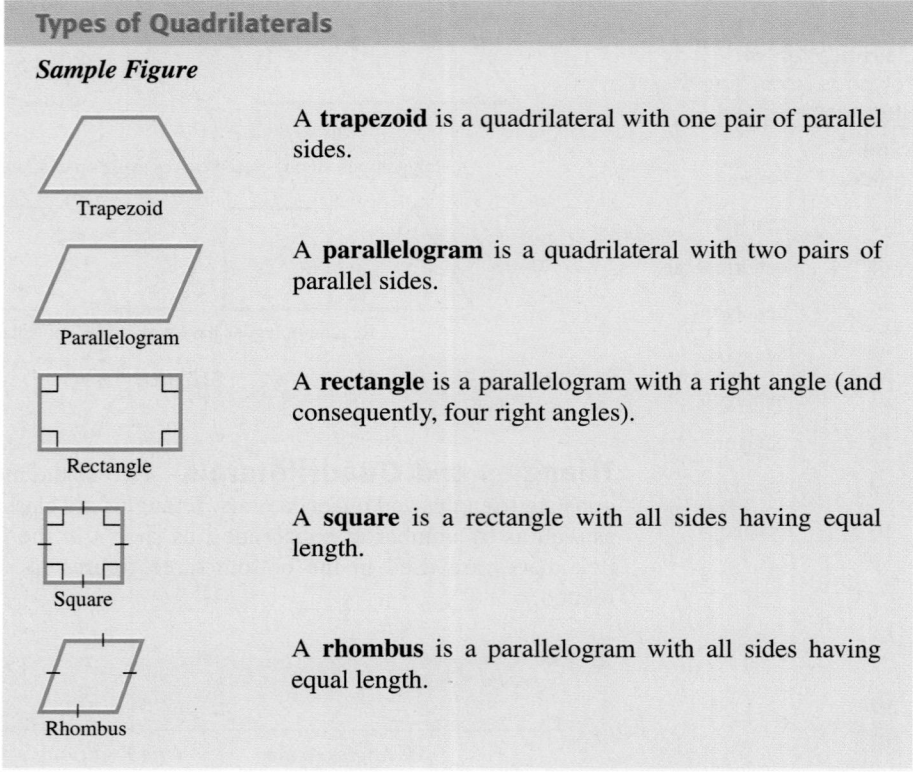

A **trapezoid** is a quadrilateral with one pair of parallel sides.

Parallelogram

A **parallelogram** is a quadrilateral with two pairs of parallel sides.

Rectangle

A **rectangle** is a parallelogram with a right angle (and consequently, four right angles).

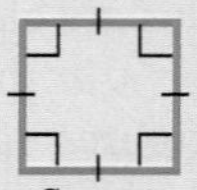

Square

A **square** is a rectangle with all sides having equal length.

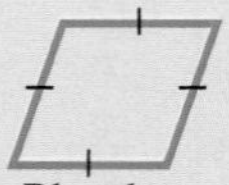

Rhombus

A **rhombus** is a parallelogram with all sides having equal length.

An important property of triangles that was first proved by the Greek geometers deals with the sum of the measures of the angles of any triangle.

Angle Sum of a Triangle

The sum of the measures of the angles of any triangle is 180°.

While it is not an actual proof, a rather convincing argument for the truth of this statement can be given using any size triangle cut from a piece of paper. Tear each corner from the triangle, as suggested in Figure 17(a). You should be able to rearrange the pieces so that the three angles form a straight angle, as shown in Figure 17(b).

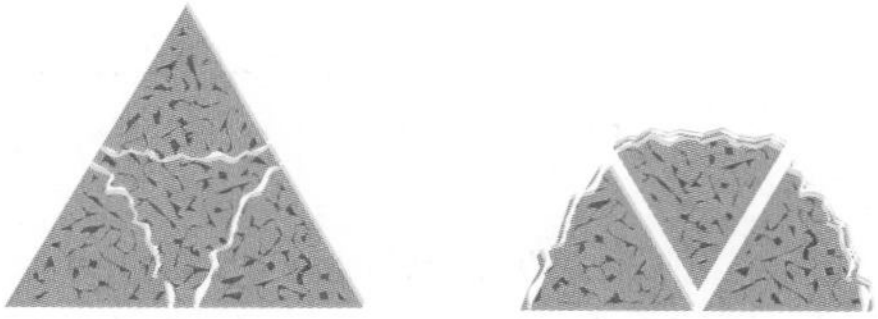

(a) **(b)**

FIGURE 17

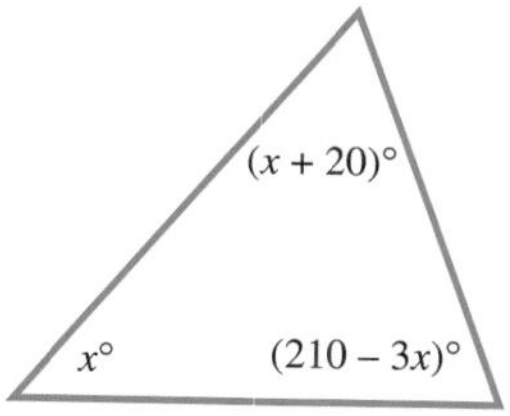

FIGURE 18

EXAMPLE 1 Finding Angle Measures in a Triangle

Find the measure of each angle in the triangle of Figure 18.

SOLUTION

By the angle sum relationship, the three angle measures must add up to 180°.

$$x + (x + 20) + (210 - 3x) = 180$$
$$-x + 230 = 180 \quad \text{Combine like terms.}$$
$$-x = -50 \quad \text{Subtract 230.}$$
There are two more values to find. ⟶ $$x = 50 \quad \text{Divide by } -1.$$

Because $x = 50$, $x + 20 = 50 + 20 = 70$ and $210 - 3x = 210 - 3(50) = 60$. Thus the measures of the three angles are 50°, 70°, and 60°. Because $50° + 70° + 60° = 180°$, the answers satisfy the angle sum relationship. ■

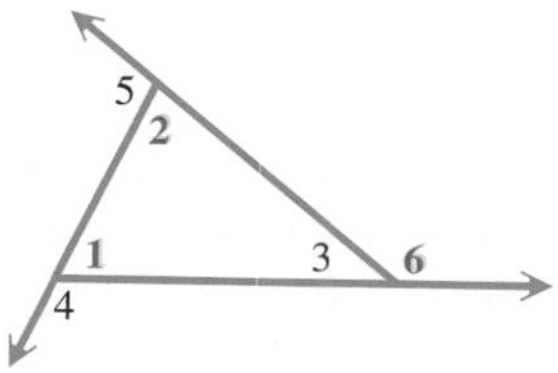

FIGURE 19

In the triangle shown in Figure 19, angles 1, 2, and 3 are called **interior angles,** while angles 4, 5, and 6 are called **exterior angles** of the triangle. Using the fact that the sum of the angle measures of any triangle is 180°, and a straight angle also measures 180°, the following property may be deduced.

Exterior Angle Measure

The measure of an exterior angle of a triangle is equal to the sum of the measures of the two opposite interior angles.

In Figure 19, the measure of angle 6 is equal to the sum of the measures of angles 1 and 2. Two other such statements can be made.

EXAMPLE 2 Finding Interior and Exterior Angle Measures

Find the measures of interior angles A, B, and C of the triangle in Figure 20, and the measure of exterior angle BCD.

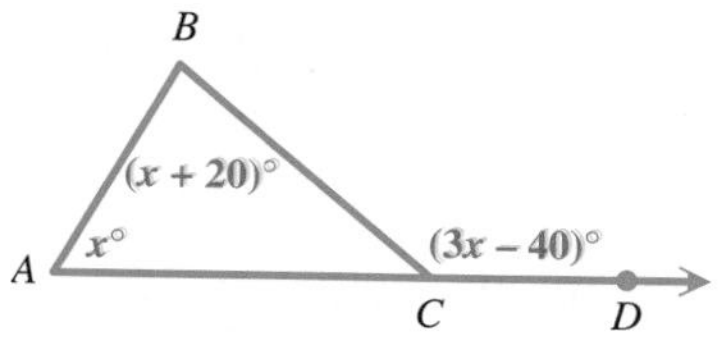

FIGURE 20

SOLUTION

By the property concerning exterior angles, the sum of the measures of interior angles A and B must equal the measure of angle BCD. Thus,

$$x + (x + 20) = 3x - 40$$
$$2x + 20 = 3x - 40 \quad \text{Combine like terms.}$$
$$-x = -60 \quad \text{Subtract } 3x\text{; subtract 20.}$$
$$x = 60. \quad \text{Divide by } -1.$$

Because the value of x is **60**,

$$\begin{aligned} \text{m(Interior angle } A) &= 60^\circ \\ \text{m(Interior angle } B) &= (60 + 20)^\circ = 80^\circ \\ \text{m(Interior angle } C) &= 180^\circ - (60^\circ + 80^\circ) = 40^\circ \\ \text{m(Exterior angle } BCD) &= [3(60) - 40]^\circ = 140^\circ. \end{aligned}$$

Circles One of the most important plane curves is the circle. It is a simple closed curve defined as follows.

Circle

A **circle** is a set of points in a plane, each of which is the same distance from a fixed point.

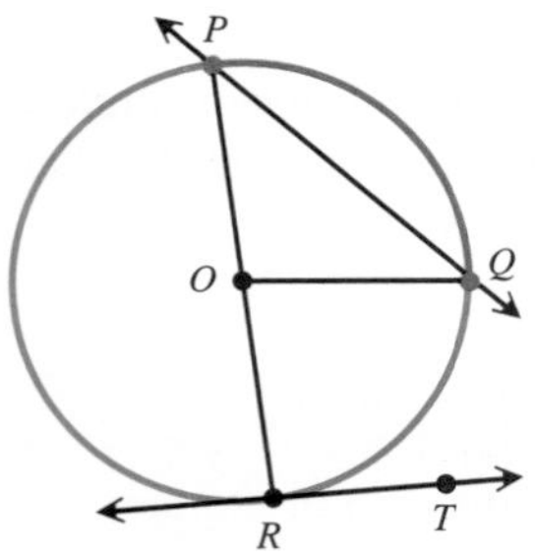

FIGURE 21

A circle may be physically constructed with compasses, where the spike leg remains fixed and the other leg swings around to construct the circle. A string may also be used to draw a circle. For example, loop a piece of chalk on one end of a piece of string. Hold the other end in a fixed position on a chalkboard, and pull the string taut. Then swing the chalk end around to draw a circle.

A circle, along with several lines and segments, is shown in Figure 21. The points P, Q, and R lie on the circle. Each lies the same distance from point O, which is called the **center** of the circle. (It is the "fixed point" referred to in the definition.) $\overline{OP}$, $\overline{OQ}$, and $\overline{OR}$ are segments whose endpoints are the center and a point on the circle. Each is called a **radius** of the circle (plural: **radii**). $\overline{PQ}$ is a segment whose endpoints both lie on the circle and is an example of a **chord.** The segment $\overline{PR}$ is a chord that passes through the center and is called a **diameter** of the circle. Notice that the measure of a diameter is twice that of a radius. A diameter such as $\overline{PR}$ in Figure 21 divides a circle into two parts of equal size, each of which is called a **semicircle.**

$\overleftrightarrow{RT}$ is a line that touches (intersects) the circle in only one point, R, and is called a **tangent** to the circle. R is the point of tangency. $\overleftrightarrow{PQ}$, which intersects the circle in two points, is called a **secant** line. (What is the distinction between a chord and a secant?)

The portion of the circle shown in red in Figure 21 is an **arc** of the circle. It consists of two endpoints (P and Q) and all points on the circle "between" these endpoints. The colored portion is called arc PQ (or QP), denoted in symbols as $\overset{\frown}{PQ}$ (or $\overset{\frown}{QP}$).

The Greeks were the first to insist that all propositions, or **theorems,** about geometry be given a rigorous proof before being accepted. According to tradition, the first theorem to receive such a proof was the following.

Thales made his fortune merely to prove how easy it is to become wealthy; he cornered all the oil presses during a year of an exceptionally large olive crop. Legend records that Thales studied for a time in Egypt and then introduced geometry to Greece, where he attempted to apply the principles of Greek logic to his newly learned subject.

Inscribed Angle

Any angle inscribed in a semicircle must be a right angle.

To be **inscribed** in a semicircle, the vertex of the angle must be on the circle with the sides of the angle going through the endpoints of the diameter at the base of the semicircle. (See Figure 22 on the next page.) This first proof was said to have been given by the Greek philosopher Thales.

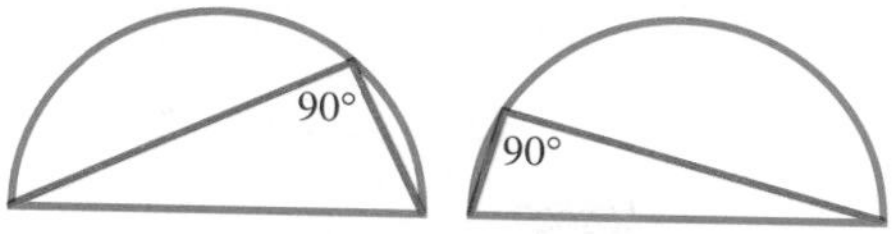

FIGURE 22

The result illustrated in Figure 22 is a special case of a more general theorem: The measure of an angle inscribed in a *circle* is one-half the measure of the intercepted arc.

9.2 EXERCISES

Fill in each blank with the correct response.

1. A segment joining two points on a circle is called a(n) ______.

2. A segment joining the center of a circle and a point on the circle is called a(n) ______ .

3. A regular triangle is called a(n) ______________ triangle.

4. A chord that contains the center of a circle is called a(n) ________.

Decide whether each statement is true *or* false.

5. A rhombus is an example of a regular polygon.

6. If a triangle is isosceles, then it is not scalene.

7. A triangle can have more than one obtuse angle.

8. A square is both a rectangle and a parallelogram.

9. A square must be a rhombus.

10. A rhombus must be a square.

11. In your own words, explain the distinction between a square and a rhombus.

12. What common traffic sign in the U.S. is in the shape of an octagon?

Identify each curve as simple, closed, both, *or* neither.

13.

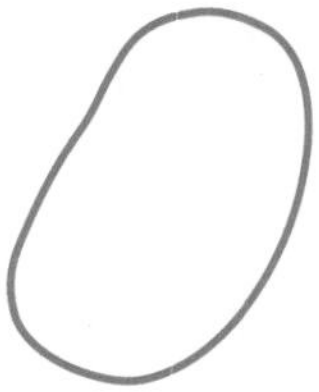

14.

15.

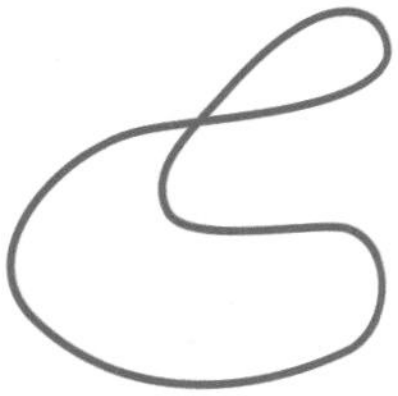

16.

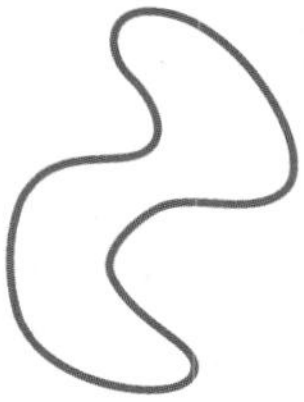

17.

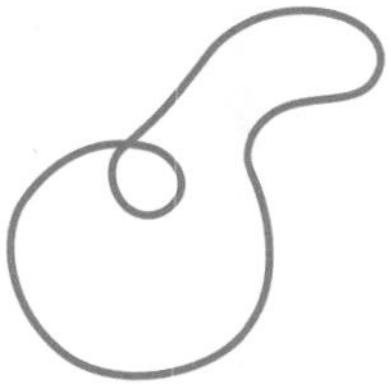

18.

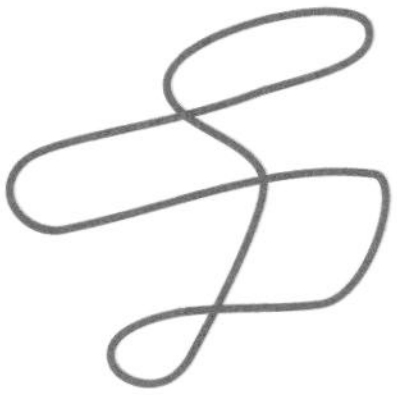

19.

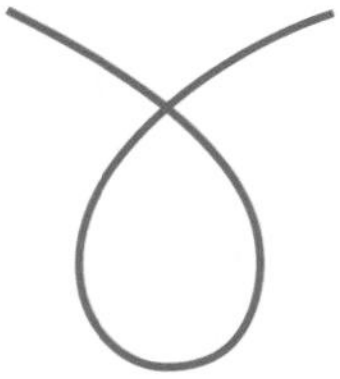

20.

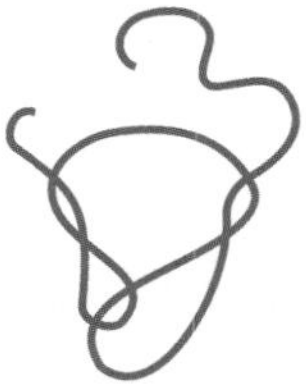

Decide whether each figure is convex *or* not convex.

21.

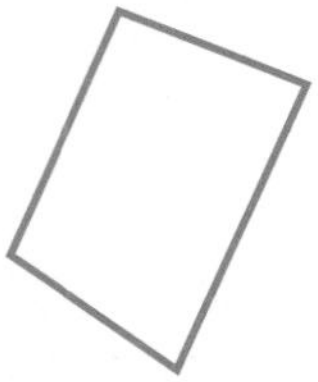

22.

23.

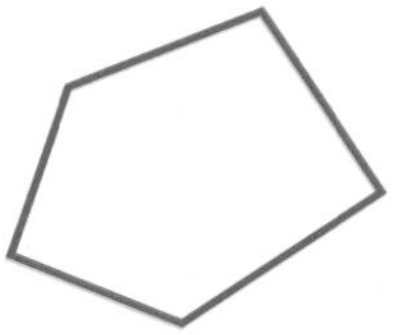

24.

25.

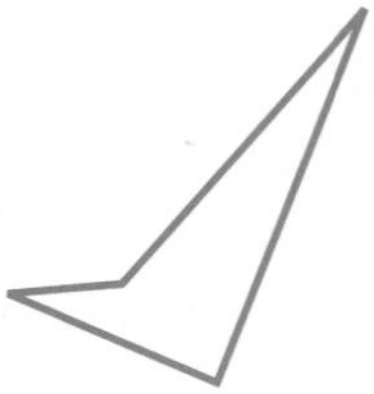

26.

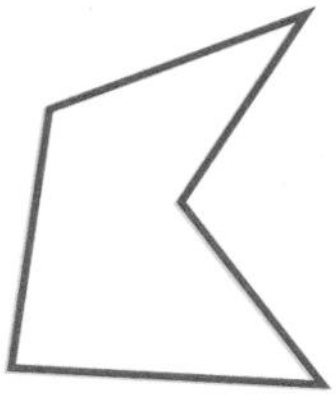

Classify each triangle as acute, right, *or* obtuse. *Also classify each as* equilateral, isosceles, *or* scalene.

27.

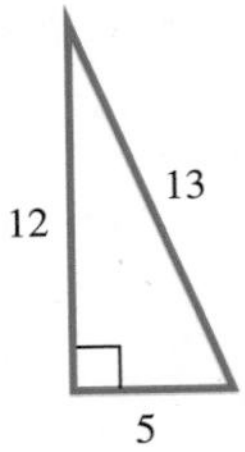

28.

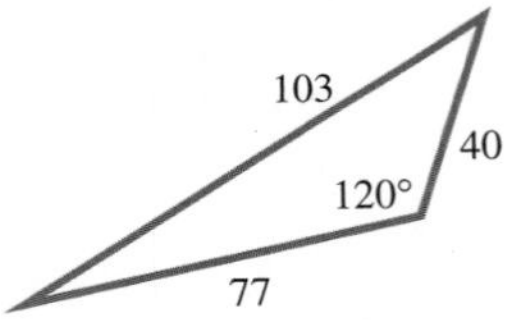

29.

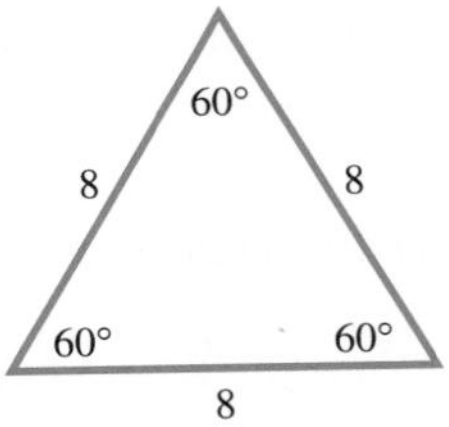

30.

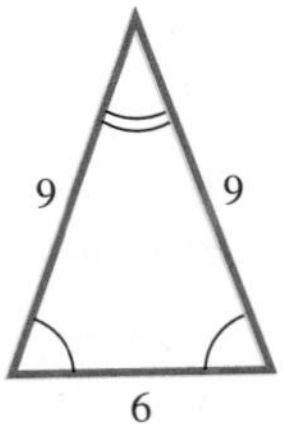

31.

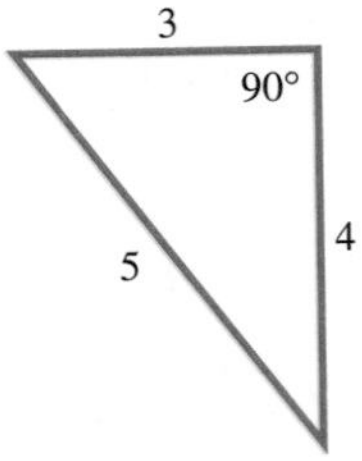

32.

33.

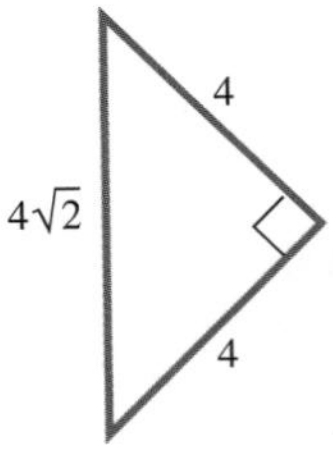

34.

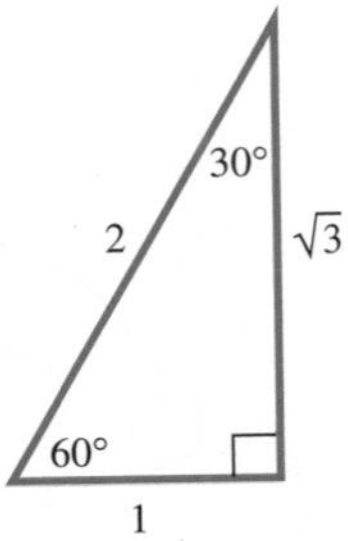

35.

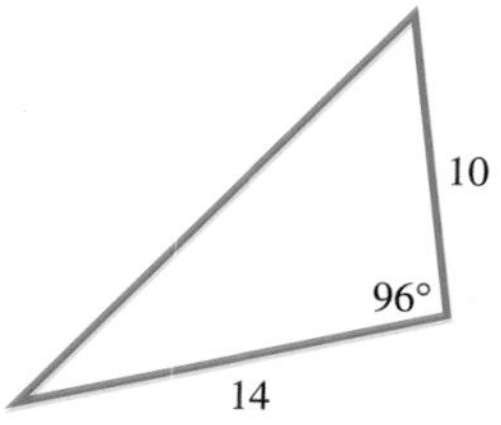

36.

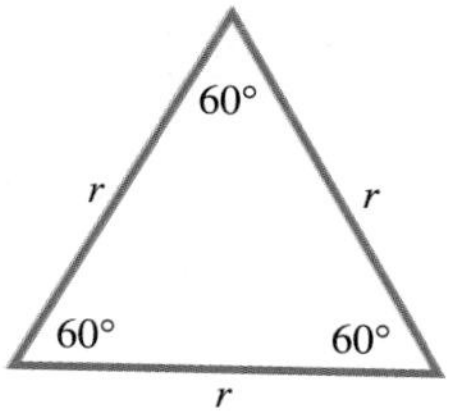

37.

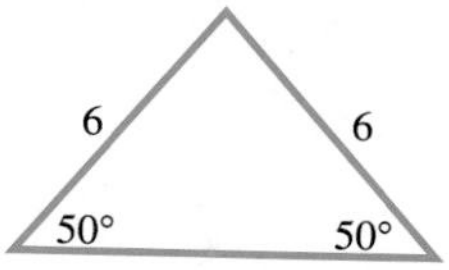

38.

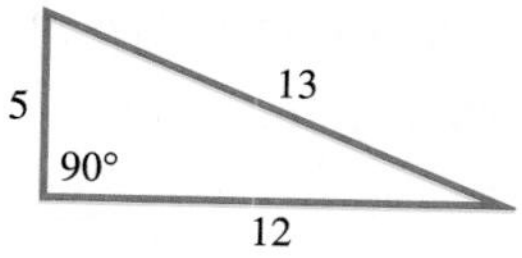

39. Write a definition of *isosceles right triangle*.

40. Explain why the sum of the lengths of any two sides of a triangle must be greater than the length of the third side.

41. Can a triangle be both right and obtuse? Explain.

42. In the classic 1939 movie *The Wizard of Oz*, the Scarecrow, upon getting a brain, says the following: "The sum of the square roots of any two sides of an isosceles triangle is equal to the square root of the remaining side." Give an example to show that his statement is incorrect.

Find the measure of each angle in triangle ABC.

43.

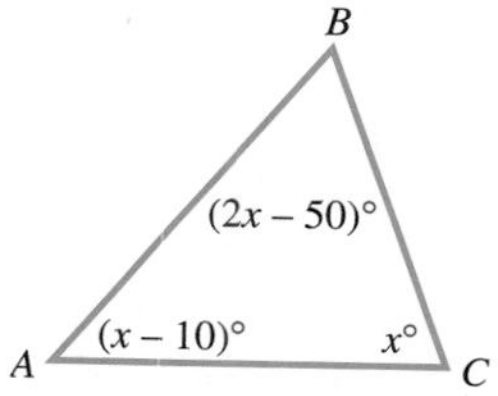

44.

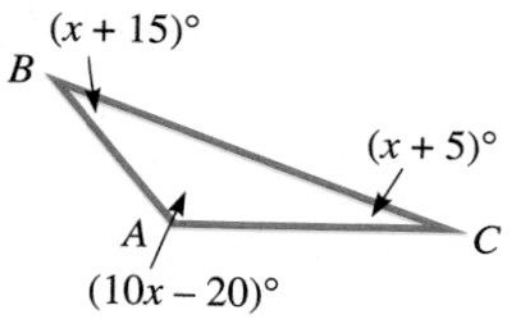

45.

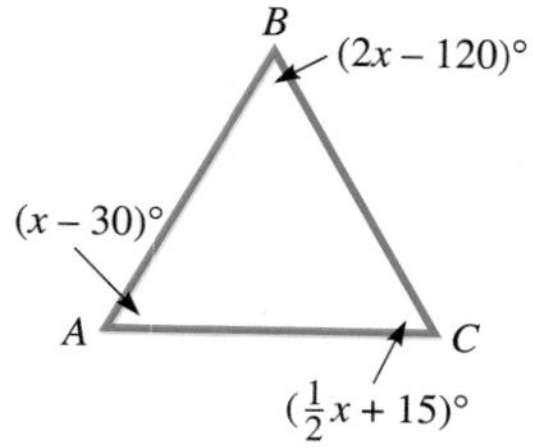

46.

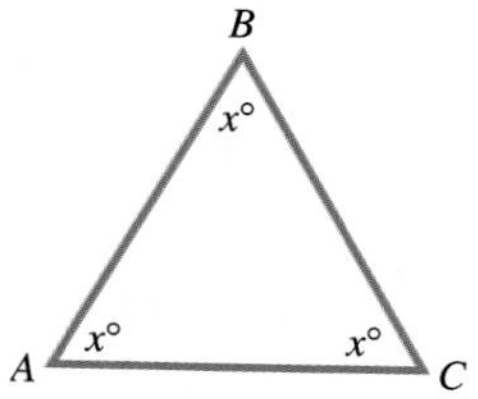

47. ***Angle Measures*** In triangle ABC, angles A and B have the same measure, while the measure of angle C is 24 degrees larger than the measure of each of A and B. What are the measures of the three angles?

48. ***Angle Measures*** In triangle ABC, the measure of angle A is 30 degrees more than the measure of angle B. The measure of angle B is the same as the measure of angle C. Find the measure of each angle.

In each triangle, find the measure of exterior angle BCD.

49.

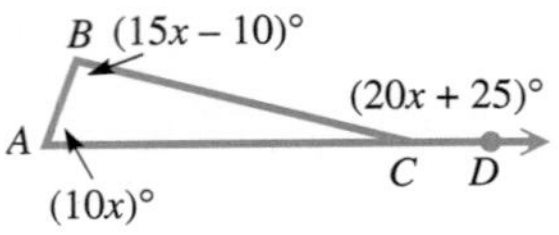

50.

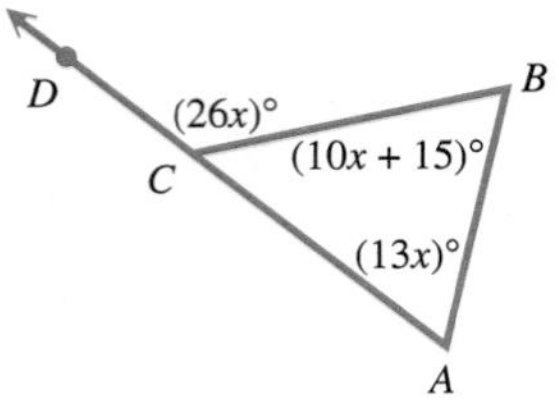

51.

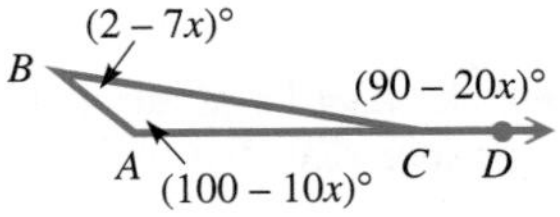

52.

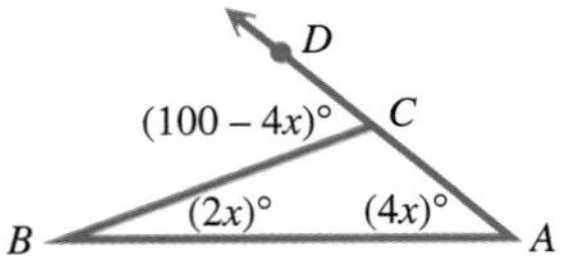

53. Using the points, segments, and lines in the figure, list all parts of the circle.

(a) center
(b) radii
(c) diameters
(d) chords
(e) secants
(f) tangents

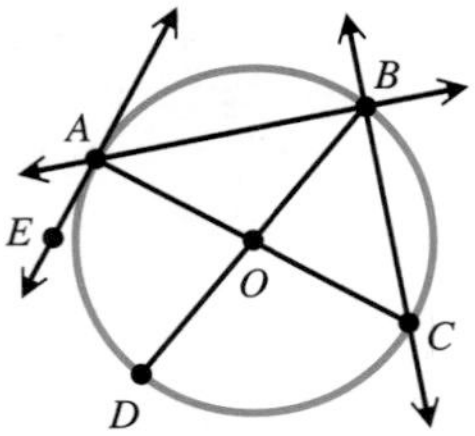

54. Refer to angles 1, 2, and 6 in the figure. Prove that the sum of the measures of angles 1 and 2 is equal to the measure of angle 6.

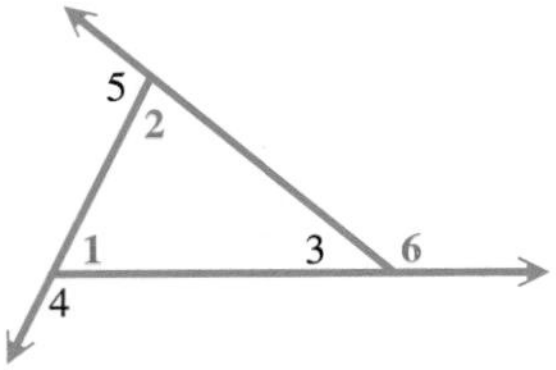

55. Go through the following argument provided by Richard Crouse in a letter to the editor of *Mathematics Teacher* in the February 1988 issue.

(a) Place the eraser end of a pencil on vertex *A* of the triangle and let the pencil coincide with side *AC* of the triangle.
(b) With the eraser fixed at *A*, rotate the pencil counterclockwise until it coincides with side *AB*.
(c) With the pencil fixed at point *B*, rotate the eraser end counterclockwise until the pencil coincides with side *BC*.
(d) With the eraser fixed at point *C* (slide the pencil to this position), rotate the point end of the pencil counterclockwise until the pencil coincides with side *AC*.
(e) Notice that the pencil is now pointing in the opposite direction. What concept from this section does this exercise reinforce?

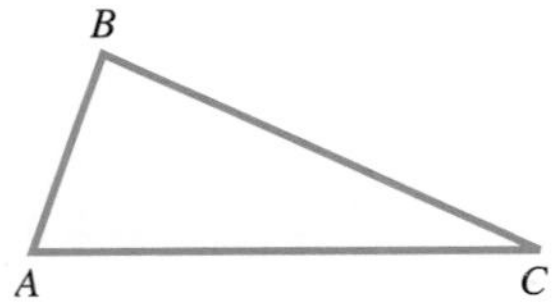

EXTENSION

Geometric Constructions

The Greeks did not study algebra as we do. To them geometry was the highest expression of mathematical science; their geometry was an abstract subject. Any practical application resulting from their work was nice but held no great importance. To the Greeks, a geometrical construction also needed abstract beauty. A construction could not be polluted with such practical instruments as a ruler. The Greeks permitted only two tools in geometrical construction: compasses for drawing circles and arcs of circles, and a straightedge for drawing straight line segments. The straightedge, unlike a ruler, could have no marks on it. It was not permitted to line up points by eye.

In his first effort as a director, Mel Gibson starred in the 1993 movie *The Man Without a Face*. As disfigured former teacher Justin McLeod, he tutors teenager Chuck Norstadt (portrayed by Nick Stahl) who has hopes of attending a boarding school. In one scene, McLeod explains to Norstadt how to find the center of a circle using any three points on the circle as he sketches the diagram on a windowpane. Although his explanation has some flaws, it conveys the general idea of how to perform this **construction.** It is based on the fact that the **perpendicular bisector** of any chord of a circle passes through the center of the circle.

Suppose the points on the circle are A, B, and C. Draw the chord AB, and construct its perpendicular bisector. Then draw BC, and construct its perpendicular bisector. The point of intersection of the two perpendicular bisectors is the center of the circle.

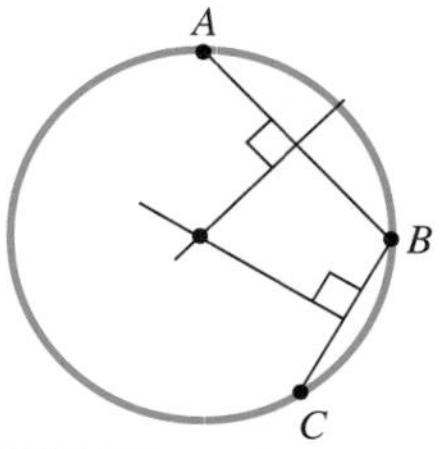

Here are four basic constructions. Their justifications are based on the *congruence properties* of Section 9.4.

Construction 1 Construct the perpendicular bisector of a given line segment.

Let the segment have endpoints A and B. Adjust the compasses for any radius greater than half the length of AB. Place the point of the compasses at A and draw an arc, then draw another arc of the same size at B. The line drawn through the points of intersection of these two arcs is the desired perpendicular bisector. See Figure 23.

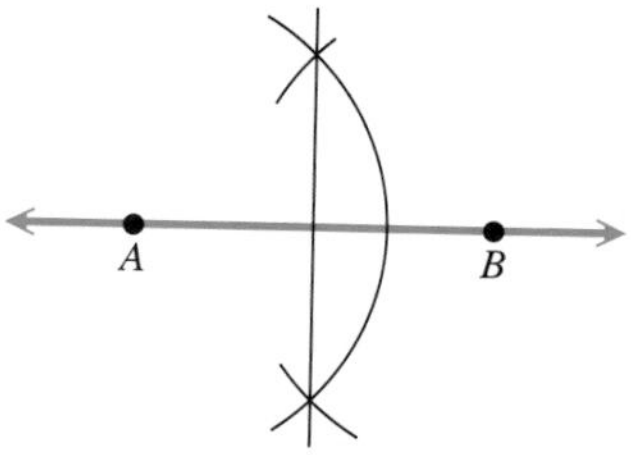

FIGURE 23

Construction 2 Construct a perpendicular from a point off a line to the line.

1. Let A be the point, r the line. Place the point of the compasses at A and draw an arc, cutting r in two points.
2. Swing arcs of equal radius from each of the two points on r which were constructed in (1). The line drawn through the intersection of the two arcs and point A is perpendicular to r. See Figure 24.

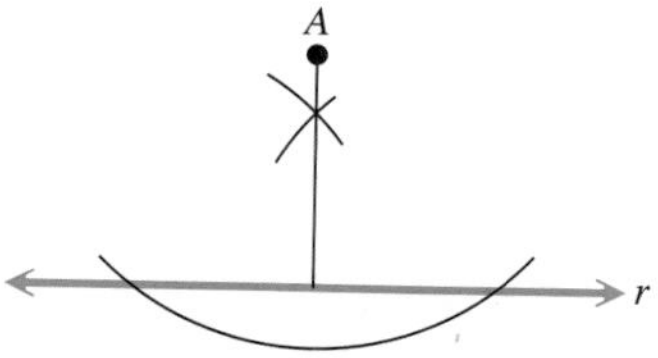

FIGURE 24

Construction 3 Construct a perpendicular to a line at some given point on the line.

1. Let r be the line and A the point. Using any convenient radius on the compasses, place the compass point at A and swing arcs that intersect r, as in Figure 25.

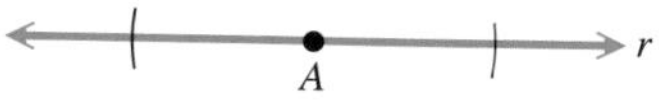

FIGURE 25

2. Increase the radius of the compasses, place the point of the compasses on the points obtained in (1) and draw arcs. A line through A and the intersection of the two arcs is perpendicular to r. See Figure 26.

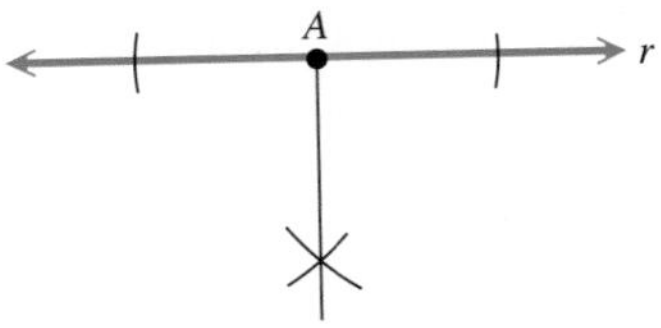

FIGURE 26

Construction 4 Copy an angle.

1. In order to copy an angle ABC on line r, place the point of the compasses at B and draw an arc. Then place the point of the compasses on r' at some point P and draw the same arc, as in Figure 27.

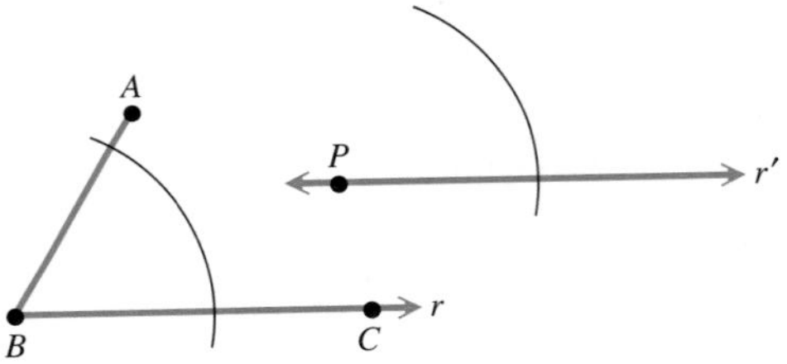

FIGURE 27

2. Measure, with your compasses, the distance between the points where the arc intersects the angle, and transfer this distance, as shown in Figure 28. Use a straightedge to join P to the point of intersection. The angle is now copied.

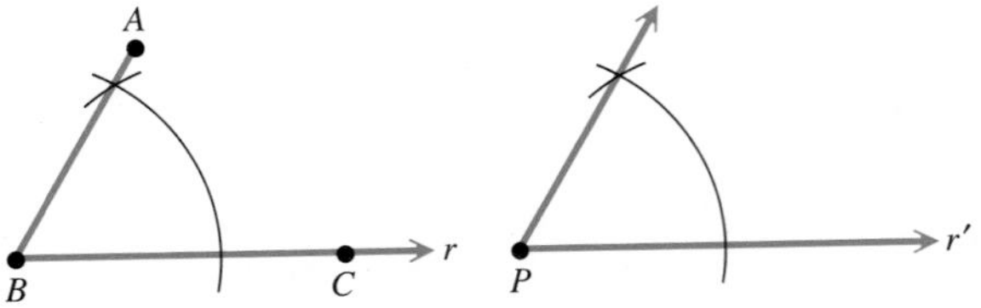

FIGURE 28

There are other basic constructions that can be found in books on plane geometry.

Euclidean tools, the compasses and unmarked straightedge, proved to be sufficient for Greek geometers to accomplish a great number of geometric constructions. Basic constructions such as copying an angle, constructing the perpendicular bisector of a segment, and bisecting an angle are easily performed and verified.

There were, however, three constructions that the Greeks were not able to accomplish with these tools. Now known as the *three famous problems of antiquity,* they are:

1. To trisect an arbitrary angle;
2. To construct the length of the edge of a cube having twice the volume of a given cube;
3. To construct a square having the same area as that of a given circle.

In the nineteenth century it was learned that these constructions are, in fact, impossible to accomplish with Euclidean tools. Over the years other methods have been devised to accomplish them. For example, trisecting an arbitrary angle can be accomplished if one allows the luxury of marking on the straightedge! But this violates the rules followed by the Greeks.

Extension Exercises

In Exercises 1 and 2, use Construction 1 to construct the perpendicular bisector of segment PQ.

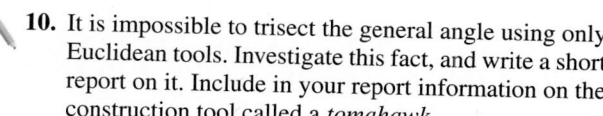

1.

2.

In Exercises 3 and 4, use Construction 2 to construct a perpendicular from P to the line r.

3.

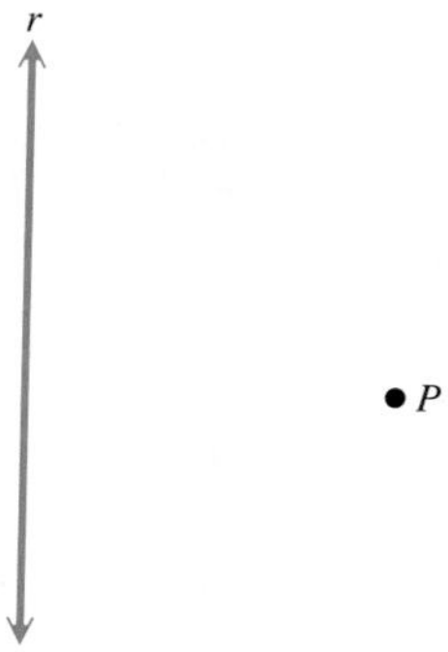

4.

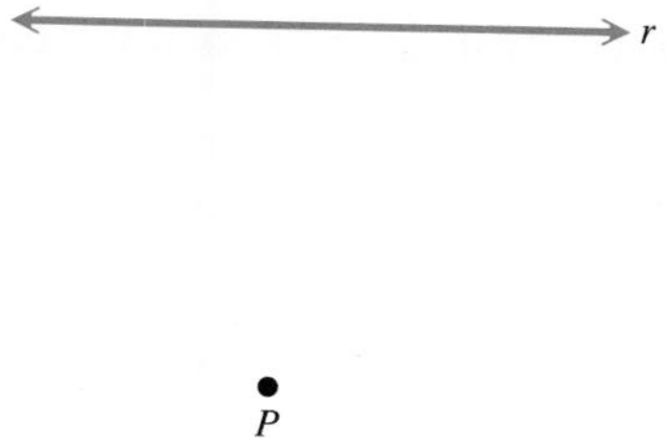

In Exercises 5 and 6, use Construction 3 to construct a perpendicular to the line r at P.

5.

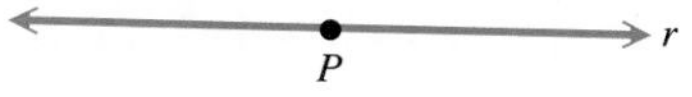

6.

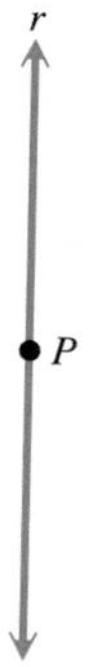

In Exercises 7 and 8, use Construction 4 to copy the given angle.

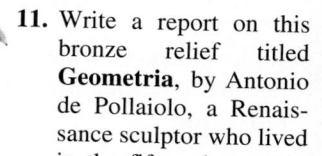

7.

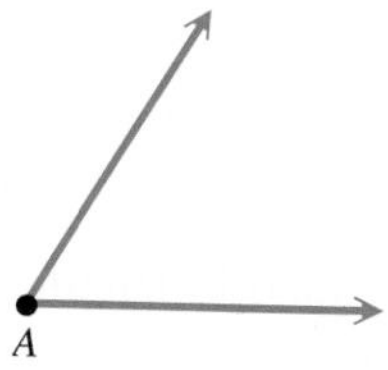

8.

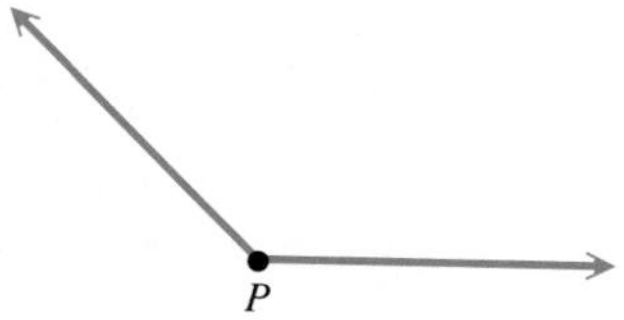

9. Construct a 45° angle.

10. It is impossible to trisect the general angle using only Euclidean tools. Investigate this fact, and write a short report on it. Include in your report information on the construction tool called a *tomahawk*.

11. Write a report on this bronze relief titled **Geometria**, by Antonio de Pollaiolo, a Renaissance sculptor who lived in the fifteenth century.

9.3 Perimeter, Area, and Circumference

Perimeter of a Polygon • Area of a Polygon • Circumference of a Circle • Area of a Circle

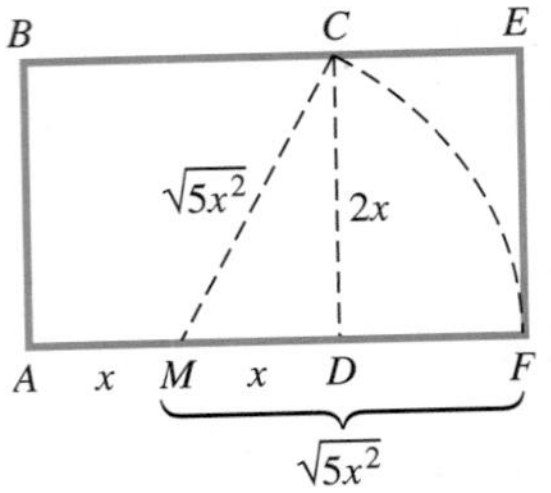

To construct a **golden rectangle,** one in which the ratio of the length to the width is equal to the ratio of the length plus the width to the length, begin with a square *ABCD*. With the point of the compasses at *M*, the midpoint of $\overleftrightarrow{AD}$, swing an arc of radius *MC* to intersect the extension of $\overleftrightarrow{AD}$ at *F*. Construct a perpendicular at *F*, and have it intersect the extension of $\overleftrightarrow{BC}$ at *E*. Then *ABEF* is a golden rectangle with ratio $(1 + \sqrt{5})/2$. (See Section 5.4 for more on the golden ratio.)

To verify this construction, let $AM = x$, so that $AD = CD = 2x$. Then, by the Pythagorean theorem,

$$MC = \sqrt{x^2 + (2x)^2} = \sqrt{x^2 + 4x^2} = \sqrt{5x^2}.$$

Because *CF* is an arc of the circle with radius *MC*, $MF = MC = \sqrt{5x^2}$. Then the ratio of length *AF* to width *EF* is

$$\frac{AF}{EF} = \frac{x + \sqrt{5x^2}}{2x} = \frac{x + x\sqrt{5}}{2x} = \frac{x(1 + \sqrt{5})}{2x} = \frac{1 + \sqrt{5}}{2}.$$

Similarly, it can be shown that

$$\frac{AF + EF}{AF} = \frac{1 + \sqrt{5}}{2}.$$

Perimeter of a Polygon

When working with a polygon, we are sometimes required to find the "distance around," or *perimeter,* of the polygon.

Perimeter

The **perimeter** of any polygon is the sum of the measures of the line segments that form its sides. Perimeter is measured in *linear units.*

The simplest polygon is a triangle. If a triangle has sides of lengths *a, b,* and *c,* then to find its perimeter we simply find the sum of *a, b,* and *c,* as shown below.

Perimeter of a Triangle

The perimeter *P* of a triangle with sides of lengths *a, b,* and *c* is given by the formula

$$P = a + b + c.$$

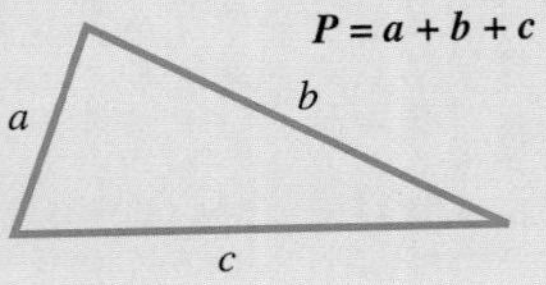

Because a rectangle is made up of two pairs of sides with the two sides in each pair equal in length, the formula for the perimeter of a rectangle may be stated as follows.

Perimeter of a Rectangle

The perimeter *P* of a rectangle with length ℓ and width *w* is given by the formula

$$P = 2\ell + 2w,$$

or equivalently,

$$P = 2(\ell + w).$$

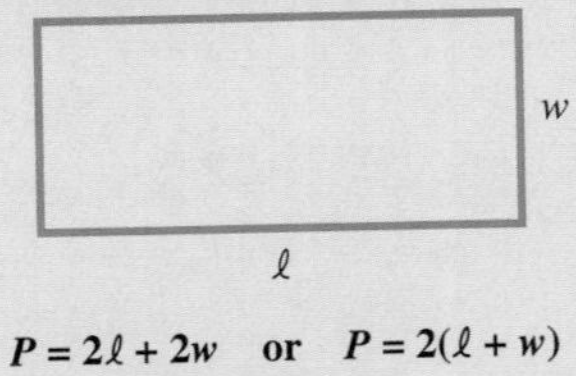

EXAMPLE 1 Using Perimeter to Determine Amount of Fencing Needed

A plot of land is in the shape of a rectangle. If it has length 50 feet and width 26 feet, how much fencing would be needed to completely enclose the plot?

SOLUTION

Since we must find the distance around the plot of land, the formula for the perimeter of a rectangle is needed.

$$P = 2\ell + 2w$$
$$P = 2(50) + 2(26) \quad \ell = 50, w = 26$$
$$P = 100 + 52 \quad \text{Multiply.}$$
$$P = 152 \quad \text{Add.}$$

The perimeter is 152 feet, so 152 feet of fencing is required.

A square is a rectangle with four sides of equal length. The formula for the perimeter of a square is a special case of the formula for the perimeter of a rectangle.

Perimeter of a Square

The perimeter P of a square with all sides of length s is given by the formula

$$P = 4s.$$

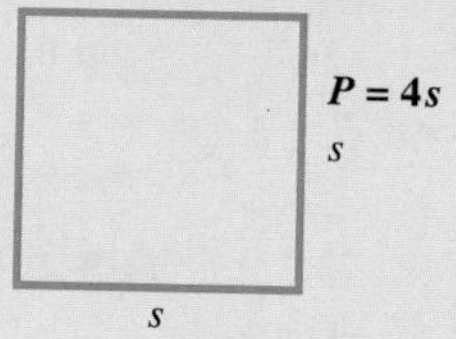

EXAMPLE 2 Using the Formula for Perimeter of a Square

A square has perimeter 54 inches. What is the length of each side?

SOLUTION

$$P = 4s$$
$$54 = 4s \quad P = 54$$
$$s = 13.5 \quad \text{Divide by 4.}$$

Each side has a measure of 13.5 inches.

PROBLEM-SOLVING HINT The six-step method of solving an applied problem from Section 7.2 can be used to solve problems involving geometric figures.

EXAMPLE 3 Finding Length and Width of a Rectangle

The length of a rectangular-shaped label is 1 centimeter more than twice the width. The perimeter is 110 centimeters. Find the length and the width.

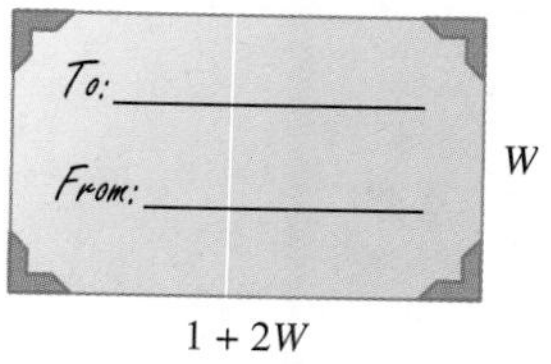

FIGURE 29

SOLUTION

Step 1 **Read the problem.** We must find the length and the width.

Step 2 **Assign a variable.** Let W represent the width. Then $1 + 2W$ can represent the length, because the length is 1 centimeter more than twice the width. Figure 29 shows a diagram of the label.

Step 3 **Write an equation.** In the formula $P = 2\ell + 2w$, replace w with W, ℓ with $1 + 2W$, and P with 110, because the perimeter is 110 centimeters.

$$110 = 2(1 + 2W) + 2W$$

Step 4 **Solve the equation.**

$$110 = 2 + 4W + 2W \quad \text{Distributive property}$$
$$110 = 2 + 6W \quad \text{Combine like terms.}$$
$$108 = 6W \quad \text{Subtract 2.}$$
$$18 = W \quad \text{Divide by 6.}$$

Step 5 **State the answer.** Because $W = 18$, the width is 18 centimeters and the length is $1 + 2W = 1 + 2(18) = 37$ centimeters.

Step 6 **Check.** Because 37 is 1 more than twice 18, and because the perimeter is $2(37) + 2(18) = 110$, the answers are correct.

Area of a Polygon

Area

The amount of plane surface covered by a polygon is called its **area.** Area is measured in *square units.*

Defining the **area** of a figure requires a basic *unit of area.* One that is commonly used is the *square centimeter,* abbreviated cm^2. One square centimeter, or 1 cm^2, is the area of a square one centimeter on a side. In place of 1 cm^2, the basic unit of area could have been 1 in.^2, 1 ft^2, 1 m^2, or any appropriate unit.

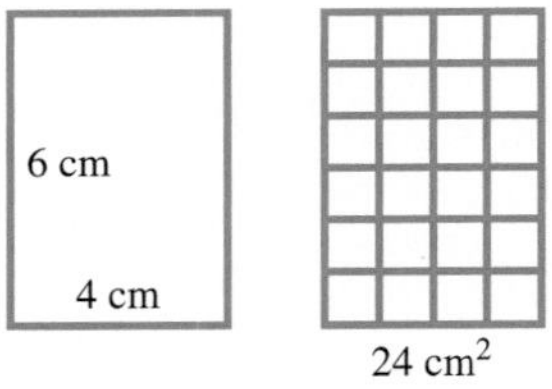

FIGURE 30

As an example, we calculate the area of the rectangle shown in Figure 30. Using the basic 1 cm^2 unit, Figure 30 shows that four squares, each 1 cm on a side, can be laid off horizontally while six such squares can be laid off vertically. A total of $24 = 4 \cdot 6$ of the small squares are needed to cover the large rectangle. Thus, the area of the large rectangle is 24 cm^2.

We can generalize the above illustration to obtain a formula for the area of a rectangle.

Area of a Rectangle

The area A of a rectangle with length ℓ and width w is given by the formula

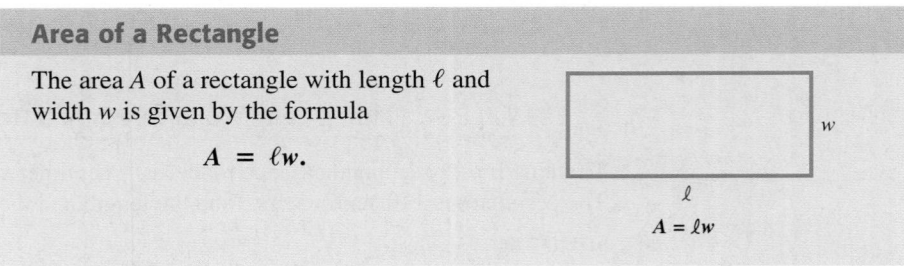

$$A = \ell w.$$

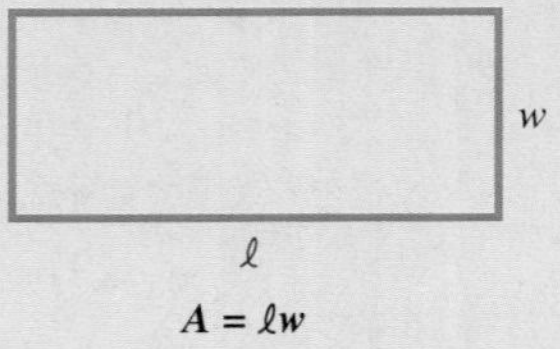

The formula for the area of a rectangle $A = \ell w$ can be used to find formulas for the areas of other figures. For example, if the letter s represents the equal lengths of the sides of a square, then $A = s \cdot s = s^2$.

Area of a Square

The area A of a square with all sides of length s is given by the formula

$$A = s^2.$$

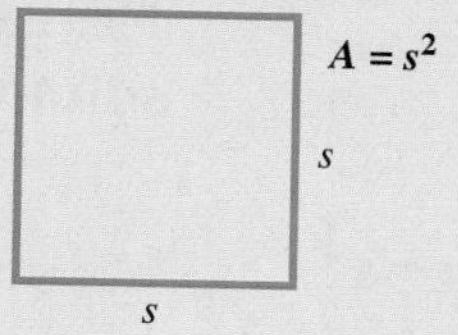

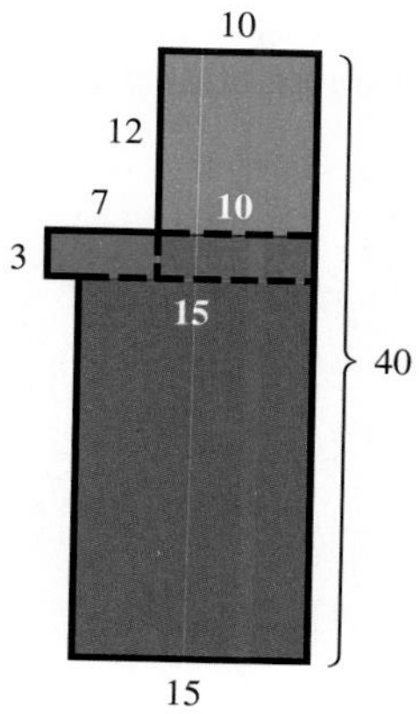

FIGURE 31

EXAMPLE 4 Using Area to Determine Amount of Carpet Needed

Figure 31 shows the floor plan of a building, made up of various rectangles. If each length given is in meters, how many square meters of carpet would be required to carpet the building?

SOLUTION

The dashed lines in the figure break up the floor area into rectangles. The areas of the various rectangles that result are

$$10 \text{ m} \cdot 12 \text{ m} = \mathbf{120 \text{ m}^2}, \qquad 3 \text{ m} \cdot 10 \text{ m} = \mathbf{30 \text{ m}^2},$$
$$3 \text{ m} \cdot 7 \text{ m} = \mathbf{21 \text{ m}^2}, \qquad 15 \text{ m} \cdot 25 \text{ m} = \mathbf{375 \text{ m}^2}.$$

$40 - 12 - 3 = 25$

Because $(120 + 30 + 21 + 375) \text{ m}^2 = 546 \text{ m}^2$, the amount of carpet needed is 546 m^2.

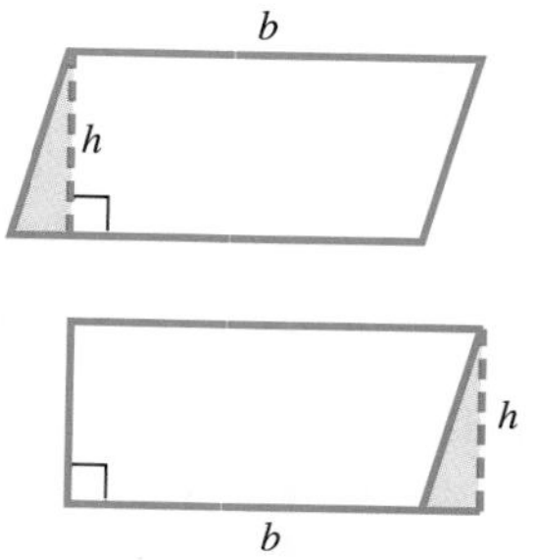

FIGURE 32

As mentioned earlier in this chapter, a **parallelogram** is a four-sided figure with both pairs of opposite sides parallel. Because a parallelogram need not be a rectangle, the formula for the area of a rectangle cannot be used directly for a parallelogram. However, this formula can be used indirectly, as shown in Figure 32. Cut off the triangle in color, and attach it at the right. The resulting figure is a rectangle with the same area as the original parallelogram.

The *height* of the parallelogram is the perpendicular distance between two of the parallel sides and is denoted by h in the figure. The width of the rectangle equals the height of the parallelogram, and the length of the rectangle is the base b of the parallelogram, so

$$A = \text{length} \cdot \text{width} \qquad \text{becomes} \qquad A = \text{base} \cdot \text{height}.$$

Area of a Parallelogram

The area A of a parallelogram with height h and base b is given by the formula

$$A = bh.$$

(*Note:* h is not the length of a side.)

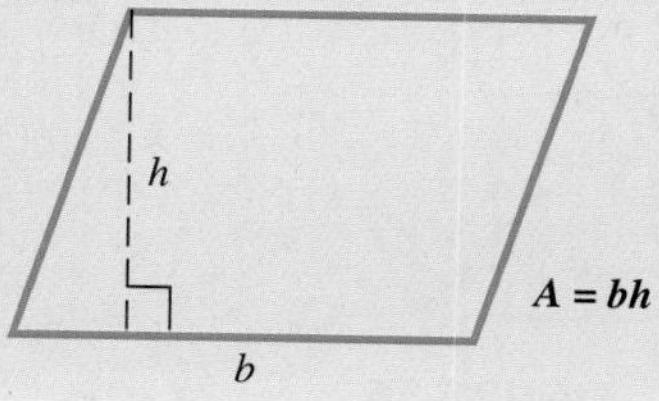

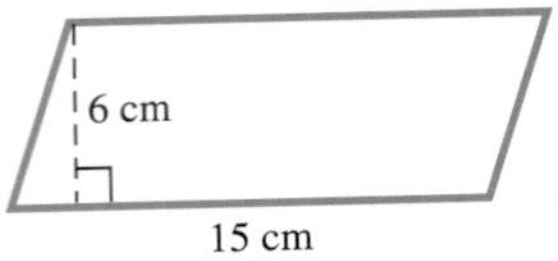

FIGURE 33

EXAMPLE 5 Using the Formula for Area of a Parallelogram

Find the area of the parallelogram in Figure 33.

SOLUTION

$$A = bh = 15 \text{ cm} \cdot 6 \text{ cm} = 90 \text{ cm}^2 \qquad b = 15 \text{ cm}, h = 6 \text{ cm}$$

The area is 90 cm^2.

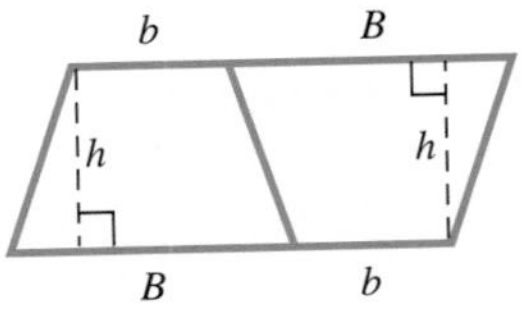

FIGURE 34

Figure 34 shows how we can find a formula for the area of a trapezoid. Notice that the figure as a whole is a parallelogram. It is made up of two trapezoids, each of which has height h, shorter base b, and longer base B. The area of the parallelogram is found by multiplying the height h by the base of the parallelogram, $b + B$, that is, $h(b + B)$. Because the area of the parallelogram is twice the area of each trapezoid, the area of each trapezoid is *half* the area of the parallelogram.

Area of a Trapezoid

The area A of a trapezoid with parallel bases b and B and height h is given by the formula

$$A = \frac{1}{2}h(b + B).$$

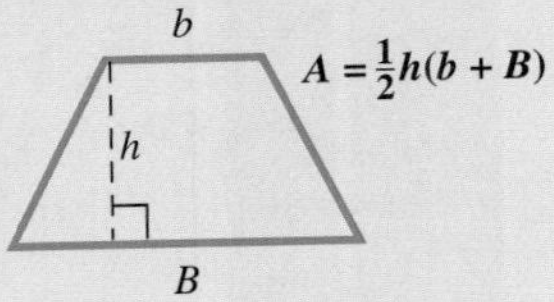

EXAMPLE 6 Using the Formula for Area of a Trapezoid

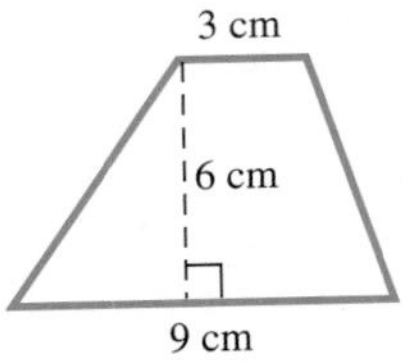

FIGURE 35

Find the area of the trapezoid in Figure 35.

SOLUTION

$$A = \frac{1}{2}h(B + b) = \frac{1}{2}(6 \text{ cm})(9 \text{ cm} + 3 \text{ cm}) \qquad h = 6 \text{ cm}, B = 9 \text{ cm}, b = 3 \text{ cm}$$

$$= \frac{1}{2}(6 \text{ cm})(12 \text{ cm}) = 36 \text{ cm}^2$$

The area of the trapezoid is 36 cm^2.

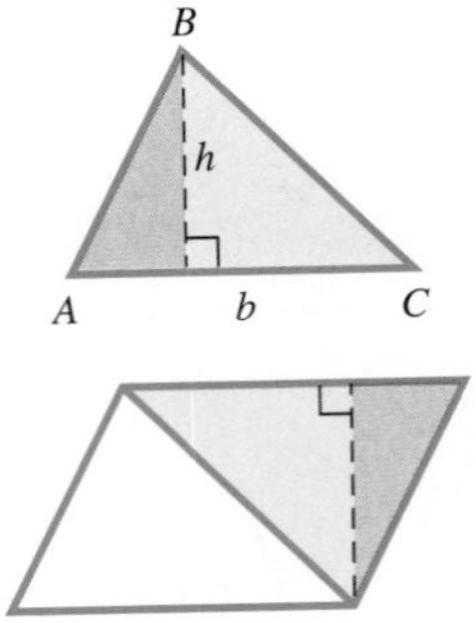

FIGURE 36

The formula for the area of a triangle can be found from the formula for the area of a parallelogram. In Figure 36 the triangle with vertices A, B, and C has been broken into two parts, one shown in color and one shown in gray. Repeating the part shown in color and the part in gray gives a parallelogram. The area of this parallelogram is $A = \text{base} \cdot \text{height}$, or $A = bh$. However, the parallelogram has *twice* the area of the triangle; in other words, the area of the triangle is *half* the area of the parallelogram.

Area of a Triangle

The area A of a triangle with height h and base b is given by the formula

$$A = \frac{1}{2}bh.$$

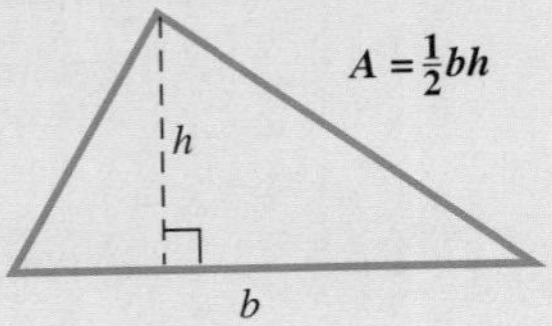

When applying the formula for the area of a triangle, remember that the height is the perpendicular distance between a vertex and the opposite side (or the extension of that side). See Figure 37.

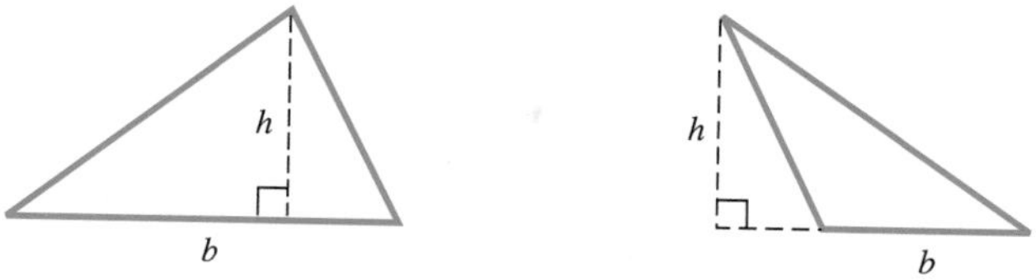

In each case, $A = \frac{1}{2}bh$.

FIGURE 37

EXAMPLE 7 Finding the Height of a Triangular Sail

The area of a triangular sail of a sailboat is 126 ft^2. The base of the sail is 12 ft. Find the height of the sail.

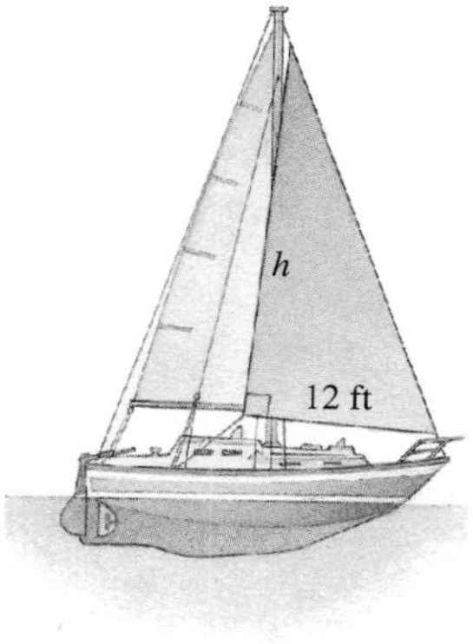

FIGURE 38

SOLUTION

Step 1 **Read.** We must find the height of the triangular sail.

Step 2 **Assign a variable.** Let $h =$ the height of the sail in feet. See Figure 38.

Step 3 **Write an equation.** Using the information given in the problem, we substitute 126 ft^2 for A and 12 ft for b in the formula for the area of a triangle.

$$A = \frac{1}{2}bh$$

$$126\text{ ft}^2 = \frac{1}{2}(12\text{ ft})h \qquad A = 126\text{ ft}^2,\ b = 12\text{ ft}$$

Step 4 **Solve.**

$$126\text{ ft}^2 = 6h\text{ ft} \qquad \text{Multiply.}$$

$$21\text{ ft} = h \qquad \text{Divide by 6 ft.}$$

Step 5 **State the answer.** The height of the sail is 21 ft.

Step 6 **Check** to see that the values $A = 126\text{ ft}^2$, $b = 12\text{ ft}$, and $h = 21\text{ ft}$ satisfy the formula for the area of a triangle.

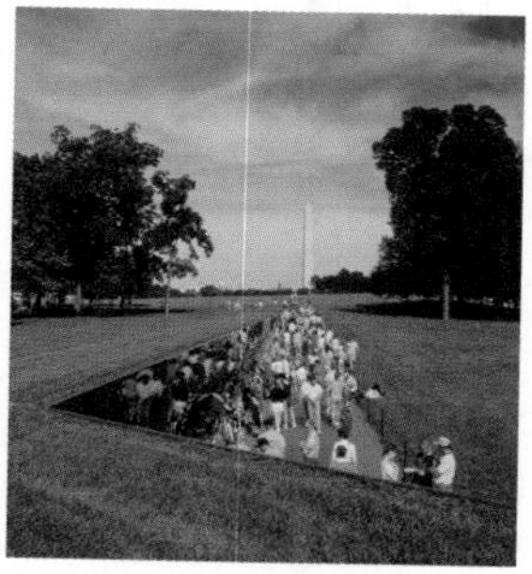

To use the formula for the area of a triangle, $A = \frac{1}{2}bh$, we must know the height from one of the sides of the triangle to the opposite vertex. Suppose that we know only the lengths of the three sides. Is there a way to determine the area from only this given information?

The answer is yes, and it leads us to the formula known as **Heron's formula.** Heron of Alexandria lived during the second half of the first century A.D., and although the formula is named after him, there is evidence that it was known to Archimedes several centuries earlier.

Let a, b, and c be lengths of the sides of any triangle. Let $s = \frac{1}{2}(a + b + c)$ represent the semiperimeter. Then the area A of the triangle is given by the formula

$$A = \sqrt{s(s-a)(s-b)(s-c)}.$$

The Vietnam Veterans' Memorial in Washington, D.C., is in the shape of an unenclosed isosceles triangle. The walls form a "V-shape," and each wall measures 246.75 feet. The distance between the ends of the walls is 438.14 feet. Use Heron's formula to find the area enclosed by the triangular shape.

Circumference of a Circle The distance around a circle is called its **circumference** (rather than "perimeter"). To understand the formula for the circumference of a circle, use a piece of string to measure the distance around a circle. Measure its diameter and then divide the circumference by the diameter. This quotient is the same, no matter what the size of the circle. The result of this measurement is an approximation for the number π. We have

$$\pi = \frac{\text{circumference}}{\text{diameter}} = \frac{C}{d}, \quad \text{or alternatively,} \quad C = \pi d.$$

Because the diameter of a circle measures twice the radius, we have $d = 2r$. These relationships allow us to state the following formulas for the circumference of a circle.

Circumference of a Circle

The circumference C of a circle of diameter d is given by the formula

$$C = \pi d.$$

Also, the circumference C of a circle of radius r is given by the formula

$$C = 2\pi r.$$

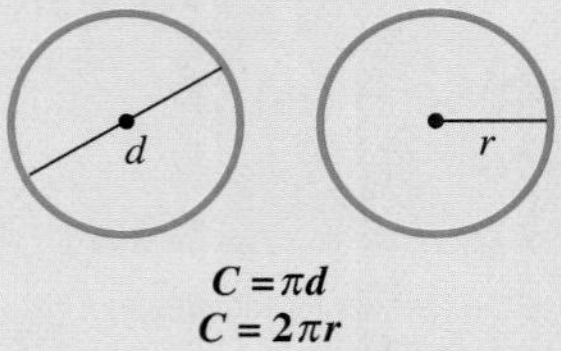

Recall that π is not a rational number. In this chapter we will use 3.14 as an approximation for π when one is required.

EXAMPLE 8 Finding the Circumference of a Circle

Find the circumference of each circle described. Use $\pi \approx 3.14$.

(a) A circle with diameter 12.6 centimeters
(b) A circle with radius 1.7 meters

SOLUTION

(a) $C = \pi d \approx 3.14(12.6\text{ cm}) = 39.564\text{ cm}$ $\quad d = 12.6\text{ cm}$

The circumference is about 39.6 centimeters, rounded to the nearest tenth.

(b) $C = 2\pi r \approx 2(3.14)(1.7\text{ m}) \approx 10.7\text{ m}$ $\quad r = 1.7\text{ m}$

The circumference is approximately 10.7 meters.

Area of a Circle Start with a circle as shown in Figure 39(a), divided into many equal pie-shaped pieces (**sectors**). Rearrange the pieces into an approximate rectangle as shown in Figure 39(b). The circle has circumference $2\pi r$, so the "length" of the approximate rectangle is one-half of the circumference, or $\frac{1}{2}(2\pi r) = \pi r$, while its "width" is r. The area of the approximate rectangle is length times width, or $(\pi r)r = \pi r^2$. By choosing smaller and smaller sectors, the figure becomes closer and closer to a rectangle, so its area becomes closer and closer to πr^2. This "limiting" procedure leads to the following formula.

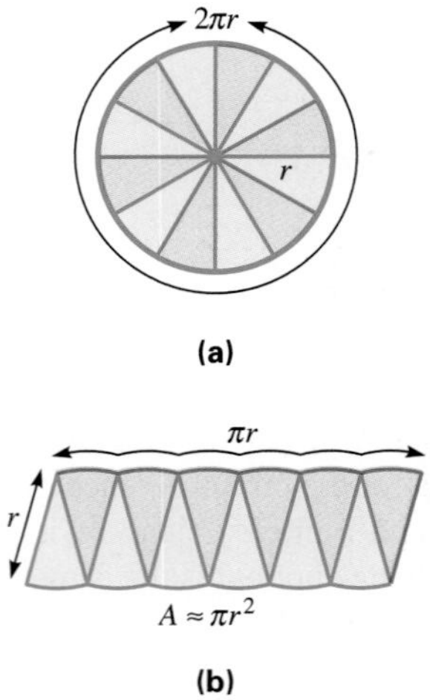

FIGURE 39

Area of a Circle

The area A of a circle with radius r is given by the formula

$$A = \pi r^2.$$

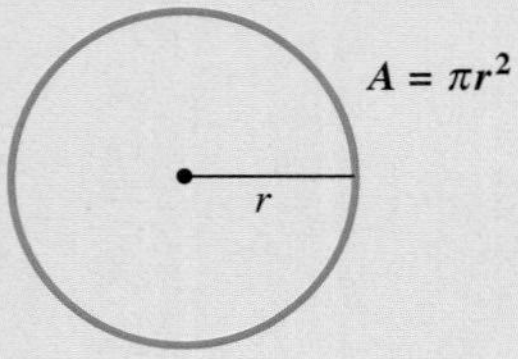

PROBLEM-SOLVING HINT The formula for the area of a circle can be used to determine the best value for the money the next time you purchase a pizza. The next example uses the idea of unit pricing.

EXAMPLE 9 Using Area to Determine Better Value for Pizza

Paw-Paw Johnny's delivers pizza. The price of an 8-inch diameter pepperoni pizza is \$6.99, while the price of a 16-inch diameter pizza is \$13.98. Which is the better buy?

SOLUTION

To determine which pizza is the better value for the money, we must first find the area of each, and divide the price by the area to determine the price per square inch.

$$\text{8-inch diameter pizza area} = \pi(4 \text{ in.})^2 \approx 50.24 \text{ in.}^2 \quad \text{Radius is } (\tfrac{1}{2})(8 \text{ in.}) = 4 \text{ in.}$$
$$\text{16-inch diameter pizza area} = \pi(8 \text{ in.})^2 \approx 200.96 \text{ in.}^2 \quad \text{Radius is } (\tfrac{1}{2})(16 \text{ in.}) = 8 \text{ in.}$$

The price per square inch for the 8-inch pizza is $\frac{\$6.99}{50.24} \approx 13.9$¢, while the price per square inch for the 16-inch pizza is $\frac{\$13.98}{200.96} \approx 7.0$¢. Therefore, the 16-inch pizza is the better buy, since it costs approximately half as much per square inch.

Solution to Chapter Opener Problem Chuck computes the approximate search area as 502,400 square miles. "Twice the size of Texas" is 2(261,797) = 523,594 square miles, so his analysis is correct.

9.3 EXERCISES

In Exercises 1–5, fill in each blank with the correct response.

1. The perimeter of an equilateral triangle with side length equal to ______ inches is the same as the perimeter of a rectangle with length 10 inches and width 8 inches.

2. A square with area 16 cm^2 has perimeter ______ cm.

3. If the area of a certain triangle is 24 square inches, and the base measures 8 inches, then the height must measure ______ inches.

4. If the radius of a circle is doubled, then its area is multiplied by a factor of ______.

5. Perimeter is to a polygon as ______________ is to a circle.

6. ***Perimeter or Area?*** *Decide whether perimeter or area would be used to solve a problem concerning the measure of the quantity.*

(a) Sod for a lawn
(b) Carpeting for a bedroom
(c) Baseboards for a living room
(d) Fencing for a yard
(e) Fertilizer for a garden
(f) Tile for a bathroom
(g) Determining the cost of planting rye grass in a lawn for the winter
(h) Determining the cost of replacing a linoleum floor with a wood floor

Use the formulas of this section to find the area of each figure. In Exercises 19–22, use 3.14 *as an approximation for* π.

7.

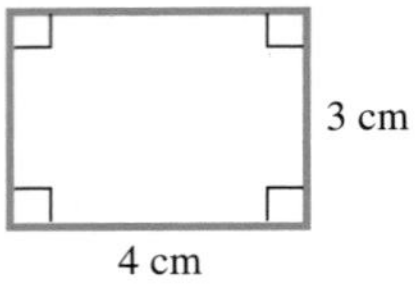

8.

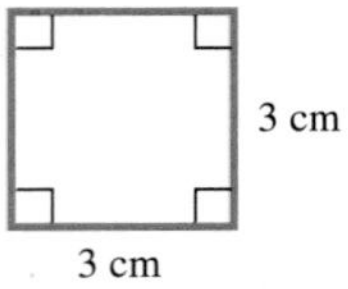

9.

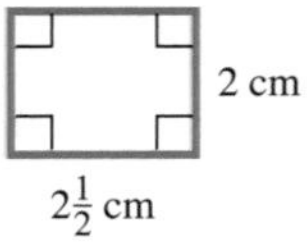

10.

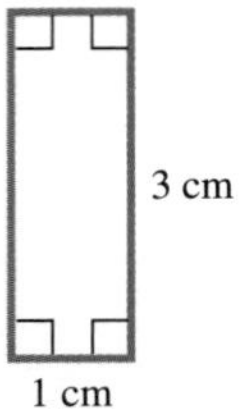

11.

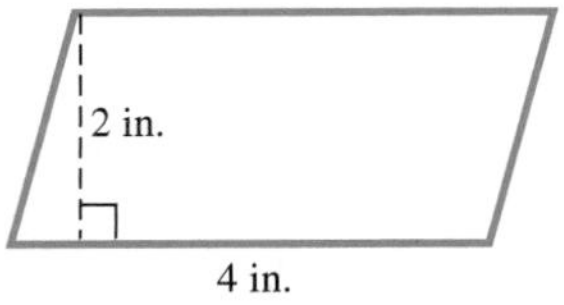

(a parallelogram)

12.

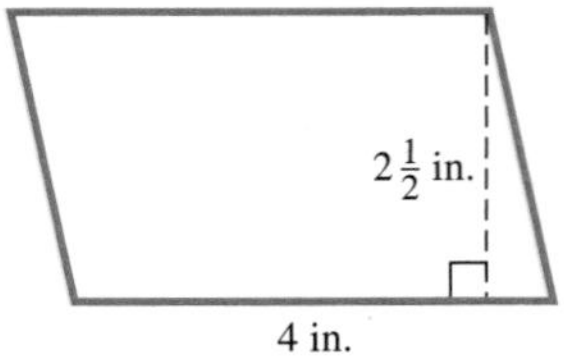

(a parallelogram)

13.

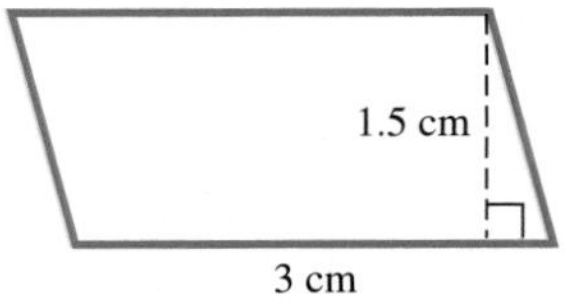

(a parallelogram)

14.

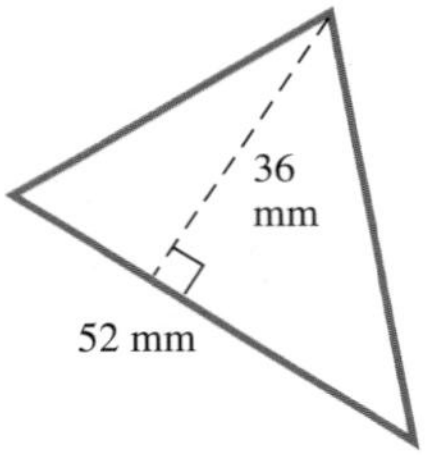

15.

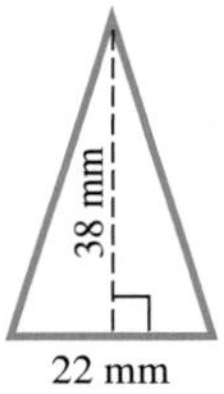

16.

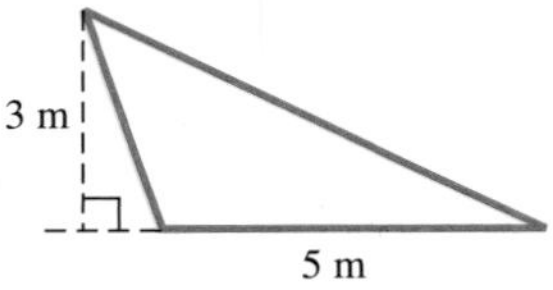

17.

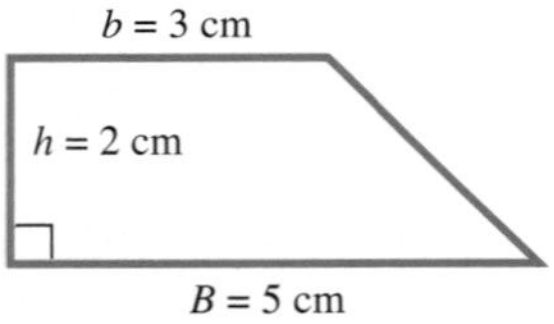

(a trapezoid)

18.

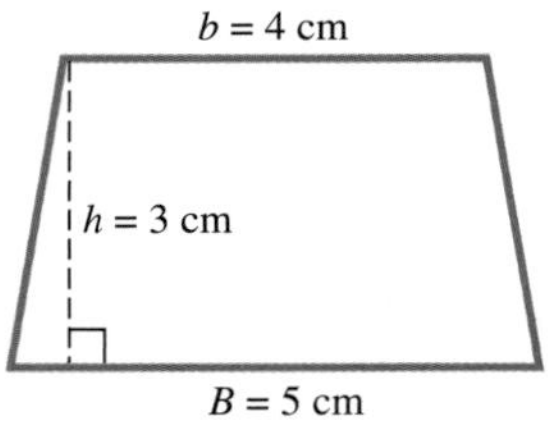

(a trapezoid)

19.

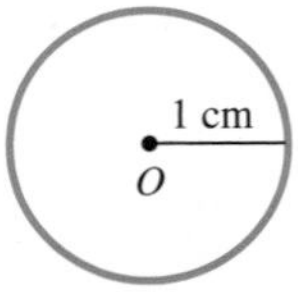

20.

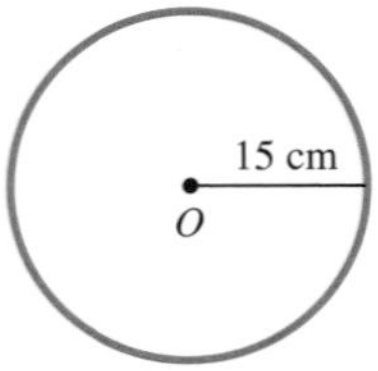

21.

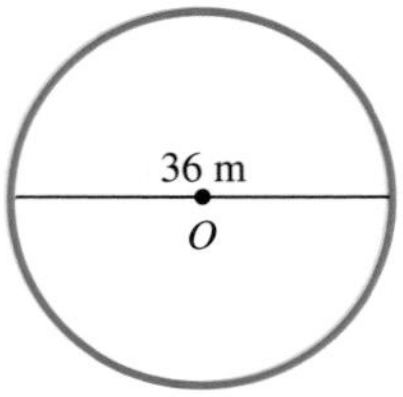

22.

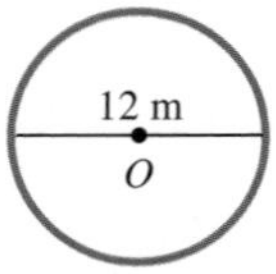

Solve each problem.

23. ***Window Side Length*** A stained-glass window in a church is in the shape of a square. The perimeter of the square is 7 times the length of a side in meters, decreased by 12. Find the length of a side of the window.

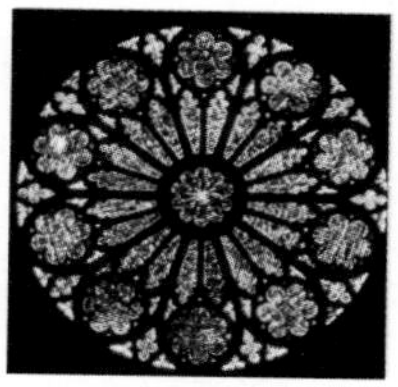

24. ***Dimensions of a Rectangle*** A video rental establishment displayed a rectangular cardboard stand-up advertisement for the movie *Failure to Launch.* The length was 20 in. more than the width, and the perimeter was 176 in. What were the dimensions of the rectangle?

25. ***Dimensions of a Lot*** A lot is in the shape of a triangle. One side is 100 ft longer than the shortest side, while the third side is 200 ft longer than the shortest side. The perimeter of the lot is 1200 ft. Find the lengths of the sides of the lot.

26. ***Pennant Side Lengths*** A wall pennant is in the shape of an isosceles triangle. Each of the two equal sides measures 18 in. more than the third side, and the perimeter of the triangle is 54 in. What are the lengths of the sides of the pennant?

27. ***Radius of a Circular Foundation*** A hotel is in the shape of a cylinder, with a circular foundation. The circumference of the foundation is 6 times the radius, increased by 12.88 ft. Find the radius of the circular foundation. (Use 3.14 as an approximation for π.)

28. ***Radius of a Circle*** If the radius of a certain circle is tripled, with 8.2 cm then added, the result is the circumference of the circle. Find the radius of the circle. (Use 3.14 as an approximation for π.)

29. ***Area of Two Lots*** The survey plat in the figure at the top of the next column shows two lots that form a trapezoid. The measures of the parallel sides are 115.80 ft and 171.00 ft. The height of the trapezoid is 165.97 ft. Find the combined area of the two lots. Round your answer to the nearest hundredth of a square foot.

30. ***Area of a Lot*** Lot A in the figure is in the shape of a trapezoid. The parallel sides measure 26.84 ft and 82.05 ft. The height of the trapezoid is 165.97 ft. Find the area of Lot A. Round your answer to the nearest hundredth of a square foot.

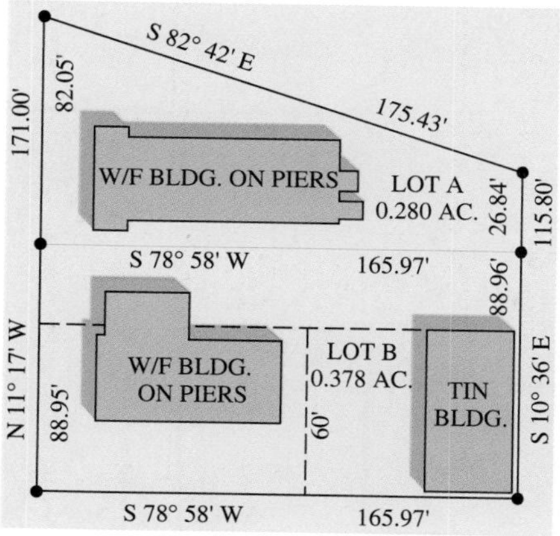

31. ***Perimeter or Area?*** In order to purchase fencing to go around a rectangular yard, would you need to use perimeter or area to decide how much to buy?

32. ***Perimeter or Area?*** In order to purchase fertilizer for the lawn of a yard, would you need to use perimeter or area to decide how much to buy?

In the chart below, one of the values r (radius), d (diameter), C (circumference), or A (area) is given for a particular circle. Find the remaining three values. Leave π in your answers.

	r	d	C	A
33.	6 in.			
34.	9 in.			
35.		10 ft		
36.		40 ft		
37.			12π cm	
38.			18π cm	
39.				100π in.^{2}
40.				256π in.^{2}

Each figure has perimeter as indicated. (Figures are not necessarily to scale.) Find the value of x.

41. $P = 58$

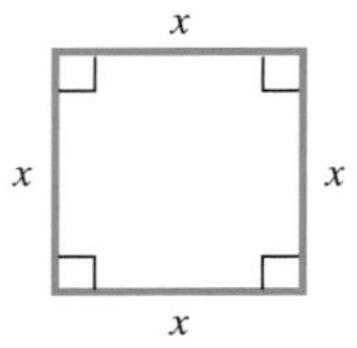

42. $P = 42$

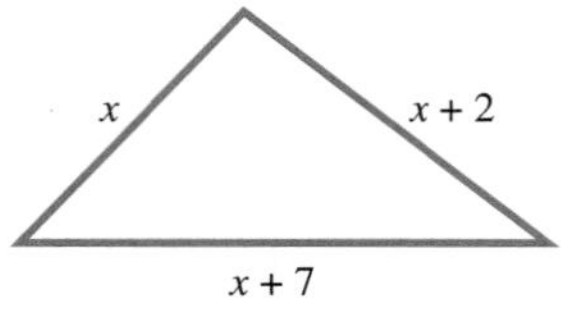

43. $P = 38$

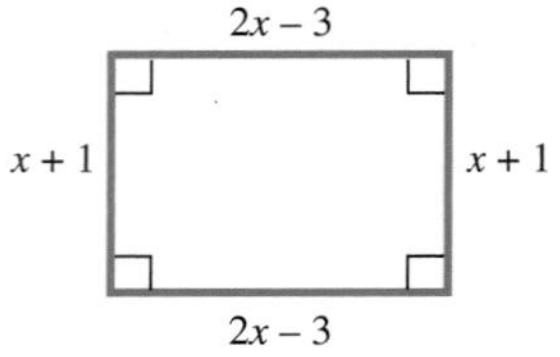

44. $P = 278$

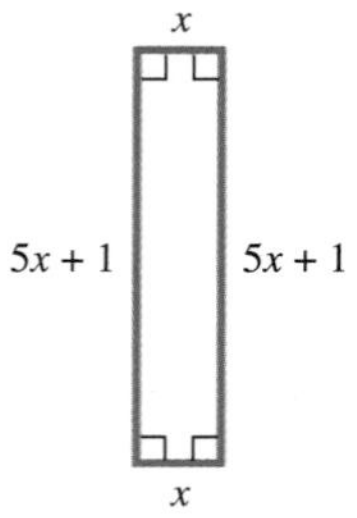

Each figure has area as indicated. Find the value of x.

45. $A = 26.01$

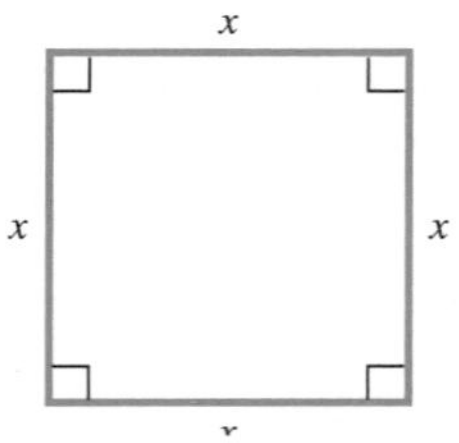

46. $A = 28$

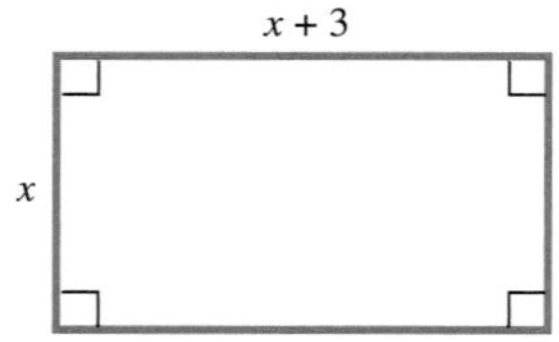

47. $A = 15$

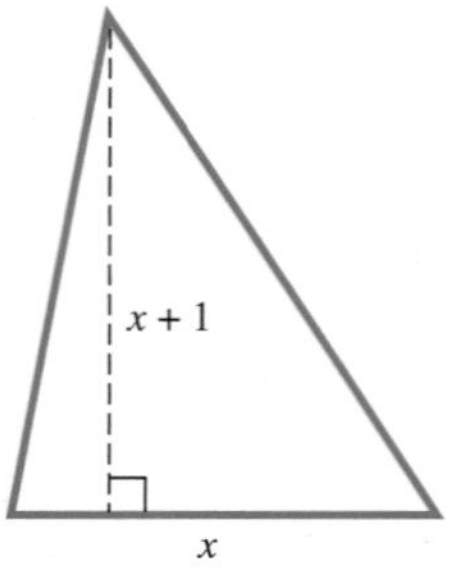

48. $A = 30$

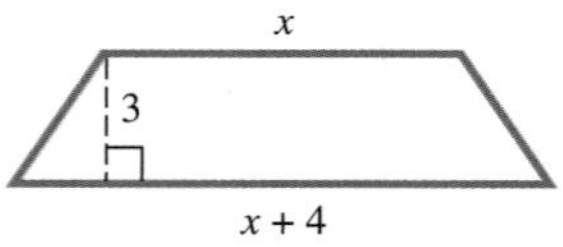

(a trapezoid)

Each circle has circumference or area as indicated. Find the value of x. Use 3.14 as an approximation for π.

49. $C = 37.68$

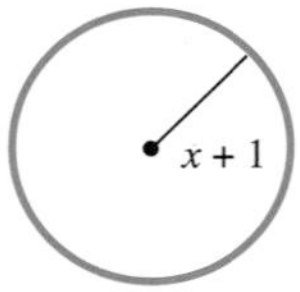

50. $C = 54.95$

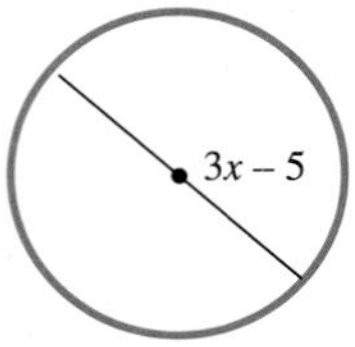

51. $A = 28.26$

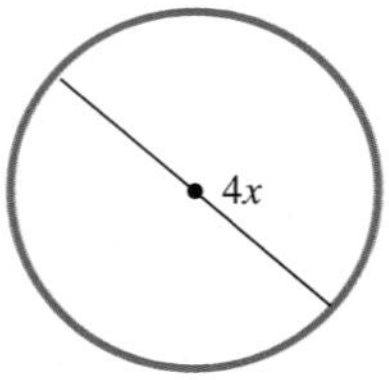

52. $A = 18.0864$

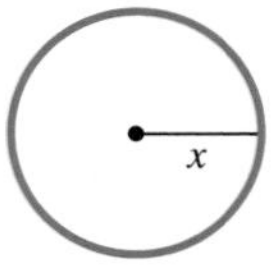

53. Work through the parts of this exercise in order, and use it to make a generalization concerning areas of rectangles.

(a) Find the area of a rectangle 4 cm by 5 cm.

(b) Find the area of a rectangle 8 cm by 10 cm.

(c) Find the area of a rectangle 12 cm by 15 cm.

(d) Find the area of a rectangle 16 cm by 20 cm.

(e) The rectangle in part (b) had sides twice as long as the sides of the rectangle in part (a). Divide the larger area by the smaller. By doubling the sides, the area increased _____ times.

(f) To get the rectangle in part (c), each side of the rectangle in part (a) was multiplied by _____. This made the larger area _____ times the size of the smaller area.

(g) To get the rectangle of part (d), each side of the rectangle of part (a) was multiplied by _____. This made the area increase to _____ times what it was originally.

(h) In general, if the length of each side of a rectangle is multiplied by n, the area is multiplied by _____.

Job Cost *Use the results of Exercise 53 to solve each problem.*

54. A ceiling measuring 9 ft by 15 ft can be painted for \$60. How much would it cost to paint a ceiling 18 ft by 30 ft?

55. Suppose carpet for a 10 ft by 12 ft room costs \$200. Find the cost to carpet a room 20 ft by 24 ft.

56. A carpet cleaner charges \$80 to shampoo an area 31 ft by 31 ft. What would be the charge for an area 93 ft by 93 ft?

57. Use the logic of Exercise 53 to answer the following: If the radius of a circle is multiplied by n, then the area of the circle is multiplied by _____.

58. Use the logic of Exercise 53 to answer the following: If the height of a triangle is multiplied by n and the base length remains the same, then the area of the triangle is multiplied by _____.

Total Area as the Sum of Areas *By considering total area as the sum of the areas of all of its parts, the area of a figure such as those in Exercises 59–62 can be determined. Find the total area of each figure. Use 3.14 as an approximation for π in Exercises 61 and 62.*

59.

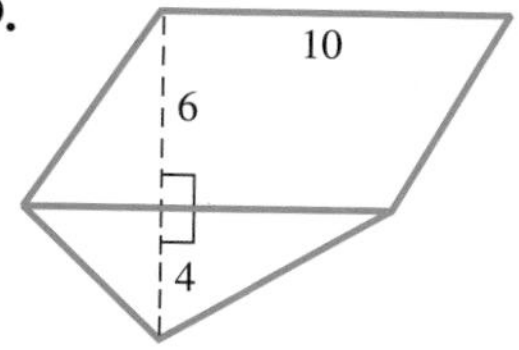

(a parallelogram and a triangle)

60.

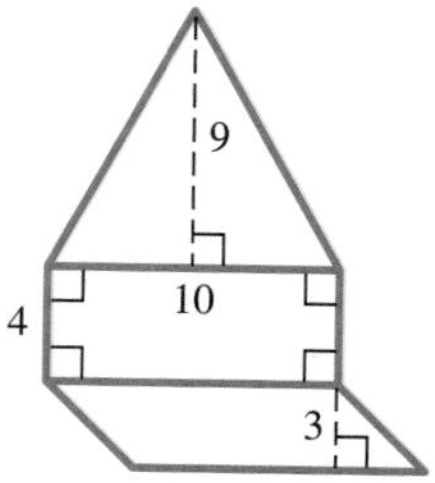

(a triangle, a rectangle, and a parallelogram)

61.

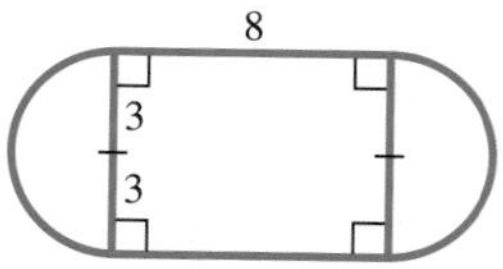

(a rectangle and two semicircles)

62.

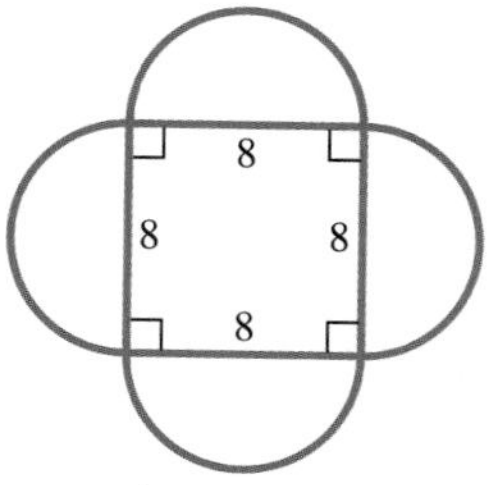

(a square and four semicircles)

Area of a Shaded Portion of a Plane Figure *The shaded areas of the figures in Exercises 63–68 may be found by subtracting the area of the unshaded portion from the total area of the figure. Use this approach to find the area of the shaded portion. Use 3.14 as an approximation for π in Exercises 66–68, and round to the nearest hundredth.*

63.

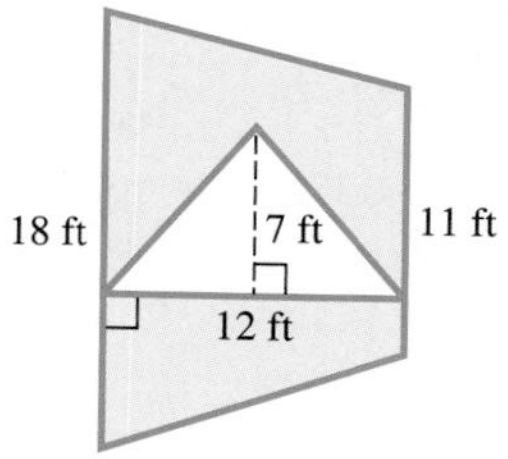

(a triangle within a trapezoid)

64.

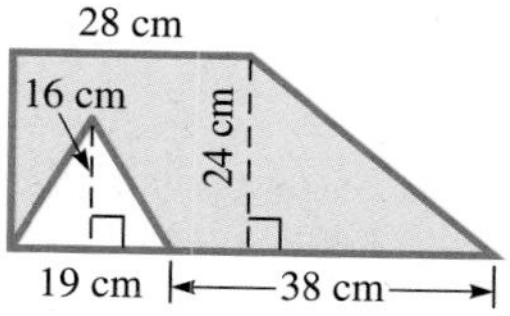

(a triangle within a trapezoid)

65.

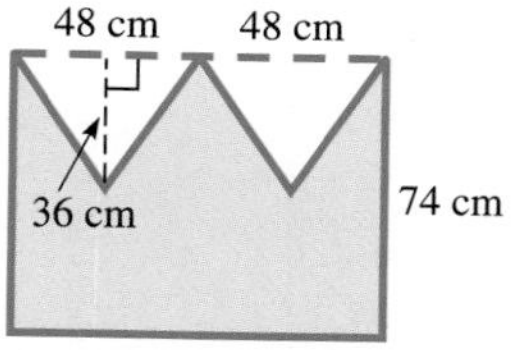

(two congruent triangles within a rectangle)

66.

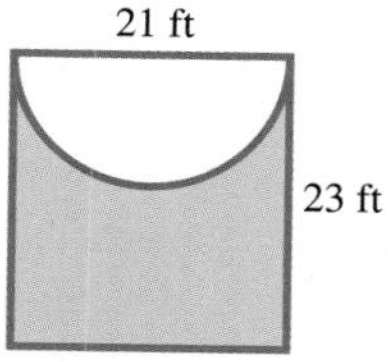

(a semicircle within a rectangle)

67.

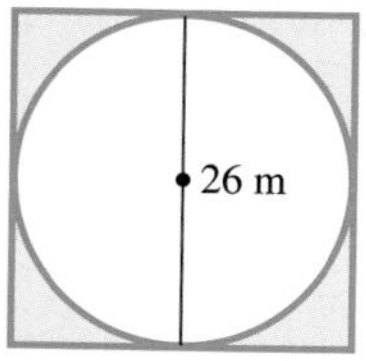

(a circle within a square)

68.

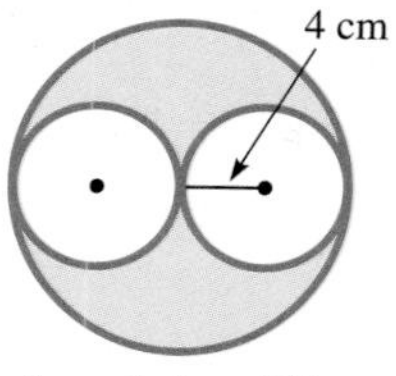

(two circles within a circle)

Pizza Pricing *The following exercises show prices actually charged by Maw-Maw Gigi's, a local pizzeria. In each case, the dimension is the diameter of the pizza. Find the best buy.*

69. Cheese pizza: 10-in. pizza sells for \$5.99, 12-in. pizza sells for \$7.99, 14-in. pizza sells for \$8.99.

70. Cheese pizza with two toppings: 10-in. pizza sells for \$7.99, 12-in. pizza sells for \$9.99, 14-in. pizza sells for \$10.99.

71. All Feasts pizza: 10-in. pizza sells for \$9.99, 12-in. pizza sells for \$11.99, 14-in. pizza sells for \$12.99.

72. Extravaganza pizza: 10-in. pizza sells for \$11.99, 12-in. pizza sells for \$13.99, 14-in. pizza sells for \$14.99.

A polygon can be inscribed within a circle or circumscribed about a circle. In the figure, triangle ABC is inscribed within the circle, while square WXYZ is circumscribed about it. These ideas will be used in some of the remaining exercises in this section and later in this chapter.

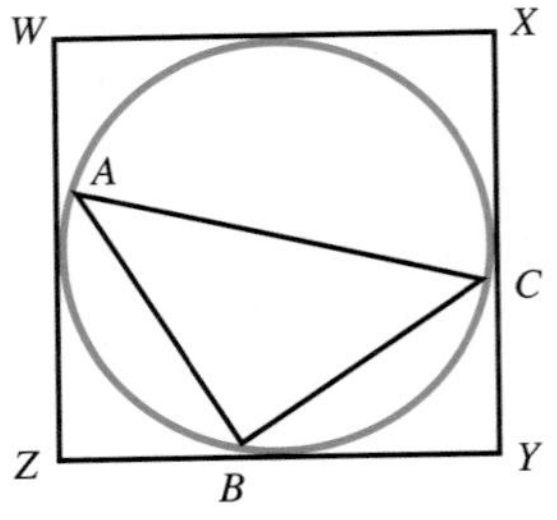

Exercises 73–80 require some ingenuity, but all may be solved using the concepts presented so far in this chapter.

73. *Diameter of a Circle* Given the circle with center O and rectangle $ABCO$, find the diameter of the circle.

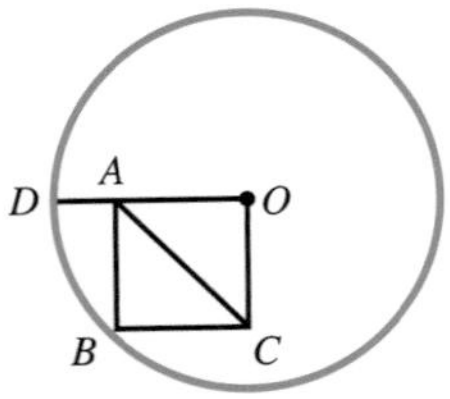

$AC = 13$ in.
$AD = 3$ in.

74. *Perimeter of a Triangle* What is the perimeter of ΔAEB, if $AD = 20$ in., $DC = 30$ in., and $AC = 34$ in.?

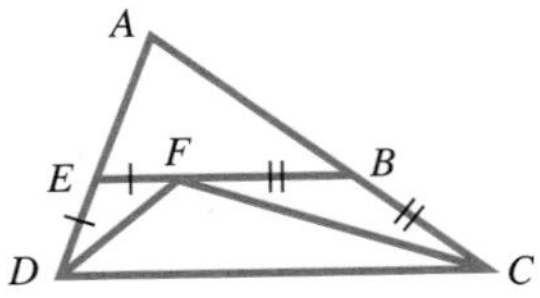

75. *Area of a Square* The area of square $PQRS$ is 1250 square feet. T, U, V, and W are the midpoints of PQ, QR, RS, and SP, respectively. What is the area of square $TUVW$?

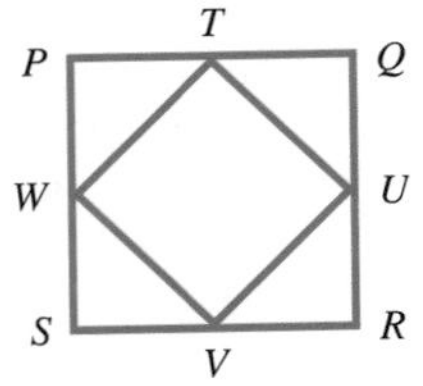

76. *Area of a Quadrilateral* The rectangle $ABCD$ has length twice the width. If P, Q, R, and S are the midpoints of the sides, and the perimeter of $ABCD$ is 96 in., what is the area of quadrilateral $PQRS$?

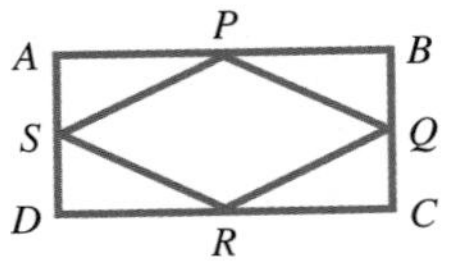

77. *Area of a Shaded Region* If $ABCD$ is a square with each side measuring 36 in., what is the area of the shaded region?

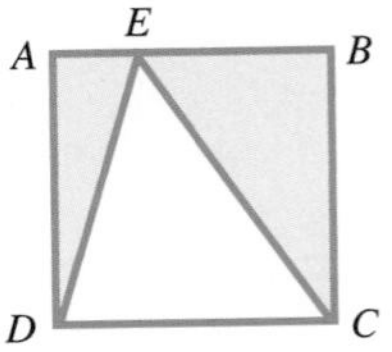

78. *Perimeter of a Polygon* Can the perimeter of the polygon shown be determined from the given information? If so, what is the perimeter?

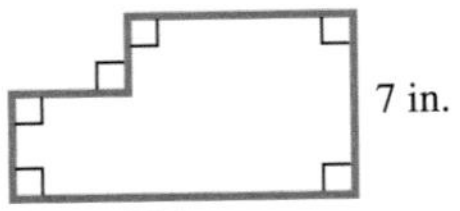

79. ***Area of a Shaded Region*** Express the area of the shaded region in terms of r, given that the circle is inscribed in the square.

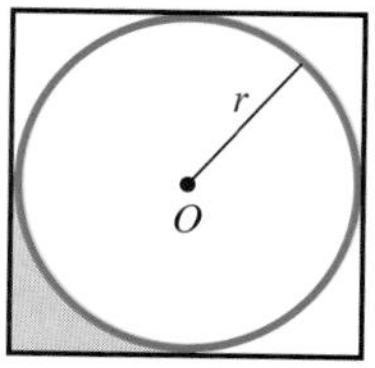

80. ***Area of a Trapezoid*** Find the area of trapezoid $ABCD$, given that the area of right triangle ABE is 30 in.^{2}.

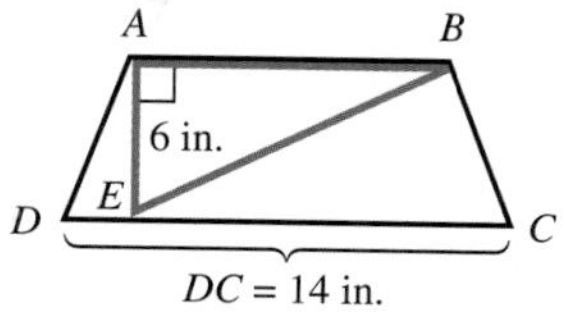

9.4 The Geometry of Triangles: Congruence, Similarity, and the Pythagorean Theorem

Congruent Triangles • Similar Triangles • The Pythagorean Theorem

Congruent Triangles

Triangles that are both the same size and the same shape are called **congruent triangles.** Informally speaking, if two triangles are congruent, then it is possible to pick up one of them and place it on top of the other so that they coincide exactly. An everyday example of congruent triangles would be the triangular supports for a child's swing set, machine-produced with exactly the same dimensions each time.

In this section we will use the "$\triangle$" symbolism to designate triangles. Figure 40 illustrates two congruent triangles, $\triangle ABC$ and $\triangle DEF$. The symbol $\cong$ denotes congruence, so $\triangle ABC \cong \triangle DEF$. Notice how the angles and sides are marked to indicate which angles are congruent and which sides are congruent. (Using precise terminology, we refer to angles or sides as being *congruent,* while the *measures* of congruent angles or congruent sides are *equal.* We will often use the terms "equal angles" or "equal sides" to describe angles of equal measure or sides of equal measure.)

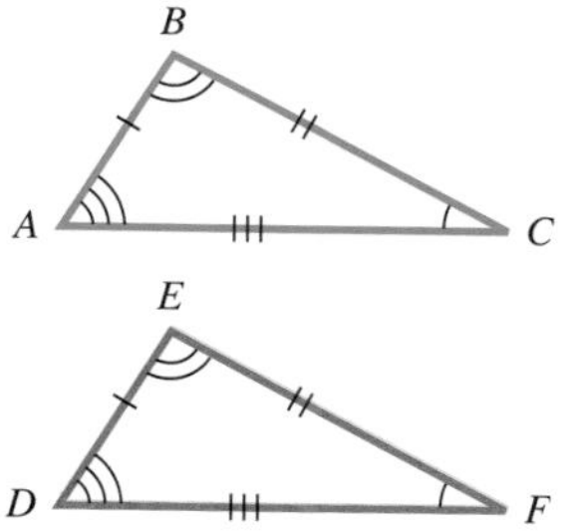

FIGURE 40

In geometry the following properties are used to prove that two triangles are congruent.

Congruence Properties

Side-Angle-Side (SAS) If two sides and the included angle of one triangle are equal, respectively, to two sides and the included angle of a second triangle, then the triangles are congruent.

Angle-Side-Angle (ASA) If two angles and the included side of one triangle are equal, respectively, to two angles and the included side of a second triangle, then the triangles are congruent.

Side-Side-Side (SSS) If three sides of one triangle are equal, respectively, to three sides of a second triangle, then the triangles are congruent.

Examples 1–3 show how to prove statements using these properties. We use a diagram with two columns, headed by STATEMENTS and REASONS.

Our knowledge of the mathematics of the Babylonians of Mesopotamia is based largely on archaeological discoveries of thousands of clay tablets. On the tablet labeled **Plimpton 322,** there are several columns of inscriptions that represent numbers. The far right column is simply one that serves to number the lines, but two other columns represent values of hypotenuses and legs of right triangles with integer-valued sides. Thus, it seems that while the famous theorem relating right-triangle side lengths is named for the Greek Pythagoras, the relationship was known more than 1000 years prior to the time of Pythagoras.

EXAMPLE 1 Proving Congruence

Refer to Figure 41.

Given: $CE = ED$

$AE = EB$

Prove: $\triangle ACE \cong \triangle BDE$

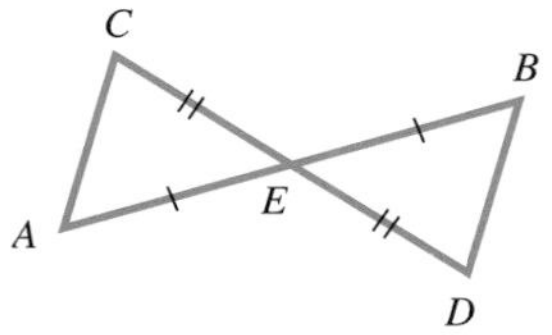

FIGURE 41

PROOF

STATEMENTS	REASONS
1. $CE = ED$	**1.** Given
2. $AE = EB$	**2.** Given
3. $\measuredangle CEA = \measuredangle DEB$	**3.** Vertical angles are equal.
4. $\triangle ACE \cong \triangle BDE$	**4.** SAS congruence property

EXAMPLE 2 Proving Congruence

Refer to Figure 42.

Given: $\measuredangle ADB = \measuredangle CBD$

$\measuredangle ABD = \measuredangle CDB$

Prove: $\triangle ADB \cong \triangle CBD$

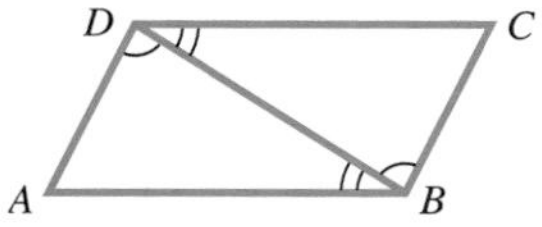

FIGURE 42

PROOF

STATEMENTS	REASONS
1. $\measuredangle ADB = \measuredangle CBD$	**1.** Given
2. $\measuredangle ABD = \measuredangle CDB$	**2.** Given
3. $DB = DB$	**3.** Reflexive property (a quantity is equal to itself)
4. $\triangle ADB \cong \triangle CBD$	**4.** ASA congruence property

EXAMPLE 3 Proving Congruence

Refer to Figure 43.

Given: $AD = CD$

$AB = CB$

Prove: $\triangle ABD \cong \triangle CBD$

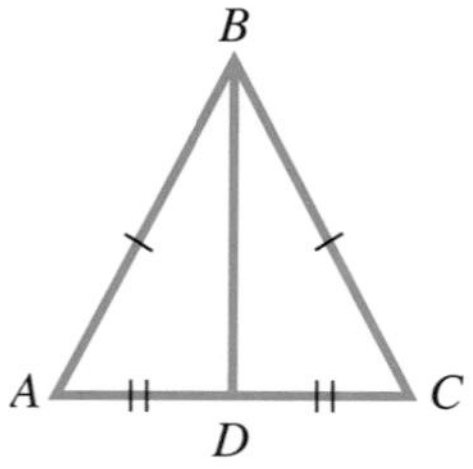

FIGURE 43

PROOF

STATEMENTS	REASONS
1. $AD = CD$	**1.** Given
2. $AB = CB$	**2.** Given
3. $BD = BD$	**3.** Reflexive property
4. $\triangle ABD \cong \triangle CBD$	**4.** SSS congruence property

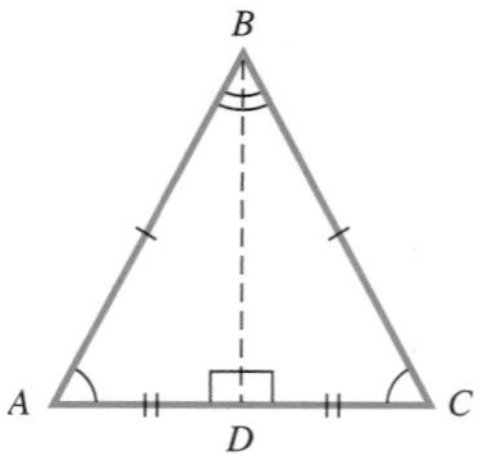

FIGURE 44

In Example 3, $\triangle ABC$ is an isosceles triangle. The results of that example allow us to make several important statements about an isosceles triangle. They are indicated symbolically in Figure 44 and stated in the following box.

Important Statements About Isosceles Triangles

If $\triangle ABC$ is an isosceles triangle with $AB = CB$, and if D is the midpoint of the base AC, then the following properties hold.

1. The base angles A and C are equal.
2. Angles ABD and CBD are equal.
3. Angles ADB and CDB are both right angles.

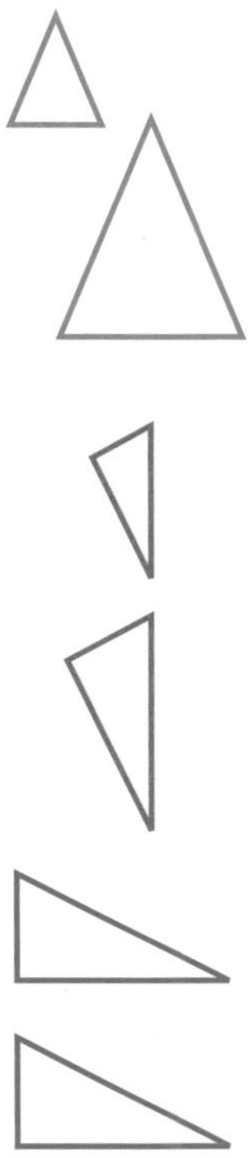

FIGURE 45

Similar Triangles Many of the key ideas of geometry depend on **similar triangles,** pairs of triangles that are exactly the same shape but not necessarily the same size. Figure 45 shows three pairs of similar triangles. (*Note:* The triangles do not need to be oriented in the same fashion in order to be similar.)

Suppose that a correspondence between two triangles ABC and DEF is set up as follows.

$\measuredangle A$ corresponds to $\measuredangle D$	side AB corresponds to side DE
$\measuredangle B$ corresponds to $\measuredangle E$	side BC corresponds to side EF
$\measuredangle C$ corresponds to $\measuredangle F$	side AC corresponds to side DF

For triangle ABC to be similar to triangle DEF, the following conditions must hold.

1. Corresponding angles must have the same measure.
2. The ratios of the corresponding sides must be constant; that is, the corresponding sides are proportional.

By showing that either of these conditions holds in a pair of triangles, we may conclude that the triangles are similar.

EXAMPLE 4 Verifying Similarity

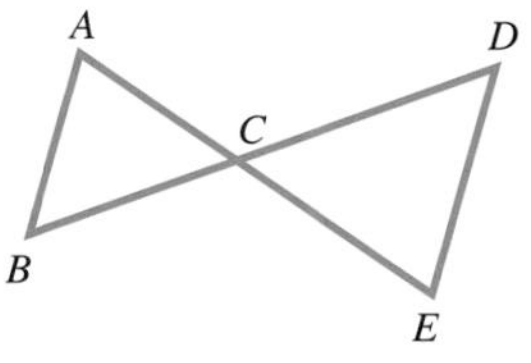

FIGURE 46

In Figure 46, $\overleftrightarrow{AB}$ is parallel to $\overleftrightarrow{ED}$. How can we verify that $\triangle ABC$ is similar to $\triangle EDC$?

SOLUTION

Because $\overleftrightarrow{AB}$ is parallel to $\overleftrightarrow{ED}$, the transversal $\overleftrightarrow{BD}$ forms equal alternate interior angles ABC and EDC. Also, transversal $\overleftrightarrow{AE}$ forms equal alternate interior angles BAC and DEC. We know that $\measuredangle ACB = \measuredangle ECD$, because they are vertical angles. Because the corresponding angles have the same measures in triangles ABC and EDC, the triangles are similar.

Once we have shown that two angles of one triangle are equal to the two corresponding angles of a second triangle, it is not necessary to show the same for the third angle. Because, in any triangle, the sum of the angles equals 180°, we may conclude that the measures of the remaining angles *must* be equal. This leads to the following Angle-Angle similarity property.

Angle-Angle (AA) Similarity Property

If the measures of two angles of one triangle are equal to those of two corresponding angles of a second triangle, then the two triangles are similar.

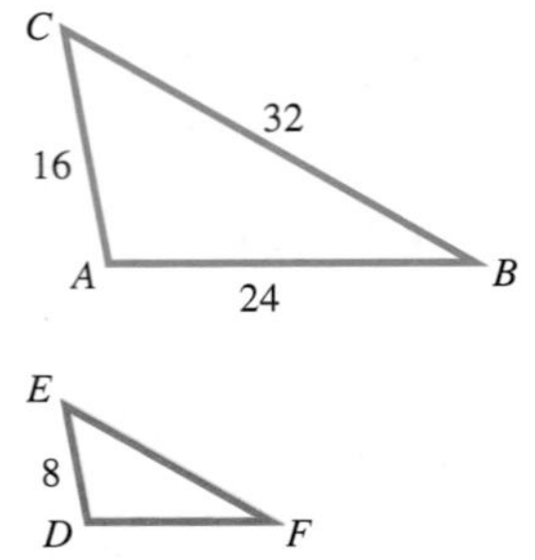

FIGURE 47

EXAMPLE 5 Finding Side Lengths in Similar Triangles

In Figure 47, $\triangle EDF$ is similar to $\triangle CAB$. Find the unknown side lengths in $\triangle EDF$.

SOLUTION

As mentioned above, similar triangles have corresponding sides in proportion. Use this fact to find the unknown sides in the smaller triangle. Side DF of the small triangle corresponds to side AB of the larger one, and sides DE and AC correspond. This leads to the proportion

$$\frac{8}{16} = \frac{DF}{24}.$$

Using a technique from algebra, set the two cross-products equal. (As an alternative method of solution, multiply both sides by 48, the least common multiple of 16 and 24.) Setting cross-products equal gives

$$\begin{aligned} 8(24) &= 16DF \\ 192 &= 16DF \\ 12 &= DF. \end{aligned}$$

Side DF has length 12.

Side EF corresponds to side CB. This leads to another proportion.

$$\begin{aligned} \frac{8}{16} &= \frac{EF}{32} \\ \frac{1}{2} &= \frac{EF}{32} \quad && \frac{8}{16} = \frac{1}{2} \\ 2EF &= 32 \quad && \text{Cross-products} \\ EF &= 16 \end{aligned}$$

Side EF has length 16. ■

EXAMPLE 6 Finding Side Lengths and Angle Measures in Similar Triangles

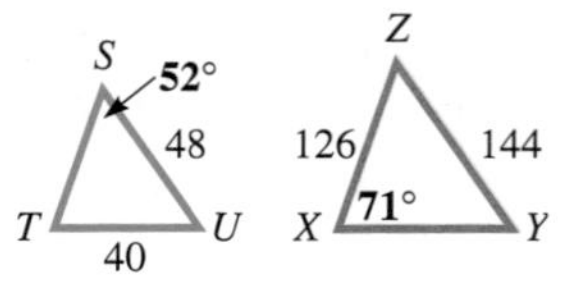

FIGURE 48

Find the measures of the unknown parts of the similar triangles STU and ZXY in Figure 48.

SOLUTION

Here angles X and T correspond, as do angles Y and U, and angles Z and S. Since angles Z and S correspond and since angle S is 52°, angle Z also must be 52°. The sum of the angles of any triangle is 180°. In the larger triangle $X = 71°$ and $Z = 52°$. To find Y, set up an equation and solve for Y.

$$\begin{aligned} X + Y + Z &= 180 \\ 71 + Y + 52 &= 180 \\ 123 + Y &= 180 \\ Y &= 57 \end{aligned}$$

Angle Y is 57°. Because angles Y and U correspond, $U = 57°$ also.

The 1984 Grenada Television Ltd. production *The Return of Sherlock Holmes: The Musgrave Ritual*, stars Jeremy Brett as Sir Arthur Conan Doyle's legendary sleuth. In solving the mystery, Holmes is faced with the problem of determining where the tip end of the shadow of a tree 64 feet tall would have been on the ground. Because the tree is no longer standing, he uses **similar triangles** to solve the problem. He finds that a vertical rod of 6 feet casts a shadow of 9 feet, so he determines that at the same time of day the tree of height 64 feet would have cast a shadow of 96 feet.

Now find the unknown sides. Sides SU and ZY correspond, as do TS and XZ, and TU and XY, leading to the following proportions.

$\frac{SU}{ZY} = \frac{TS}{XZ}$	$\frac{XY}{TU} = \frac{ZY}{SU}$
$\frac{48}{144} = \frac{TS}{126}$	$\frac{XY}{40} = \frac{144}{48}$
$\frac{1}{3} = \frac{TS}{126}$	$\frac{XY}{40} = \frac{3}{1}$
$3TS = 126$	$XY = 120$
$TS = 42$	

Side TS has length 42, and side XY has length 120.

EXAMPLE 7 Finding the Height of a Flagpole

Lucie Wanersdorfer, the Lettsworth, LA, postmaster, wants to measure the height of the office flagpole. She notices that at the instant when the shadow of the station is 18 feet long, the shadow of the flagpole is 99 feet long. The building is 10 feet high. What is the height of the flagpole?

SOLUTION

Figure 49 shows the information given in the problem. The two triangles shown there are similar, so that corresponding sides are in proportion, with

$$\frac{MN}{10} = \frac{99}{18}$$

$$\frac{MN}{10} = \frac{11}{2}$$

$$2MN = 110$$

$$MN = 55.$$

The flagpole is 55 feet high.

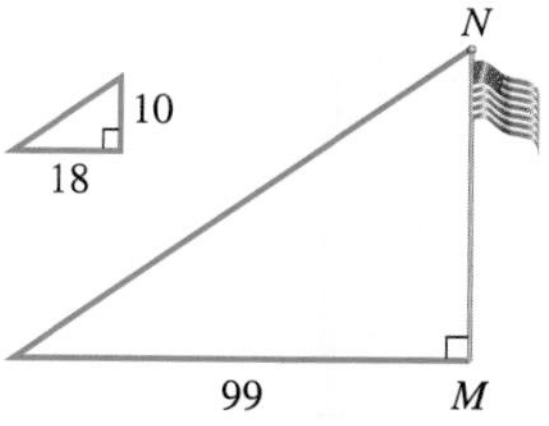

FIGURE 49

The Pythagorean Theorem

We have used the Pythagorean theorem earlier in this book, and because of its importance in mathematics, we will investigate it further in this section on the geometry of triangles. Recall that in a right triangle, the side opposite the right angle (and consequently, the longest side) is called the **hypotenuse.** The other two sides, which are perpendicular, are called the **legs.**

Pythagorean Theorem

If the two legs of a right triangle have lengths a and b, and the hypotenuse has length c, then

$$a^2 + b^2 = c^2.$$

That is, the sum of the squares of the lengths of the legs is equal to the square of the hypotenuse.

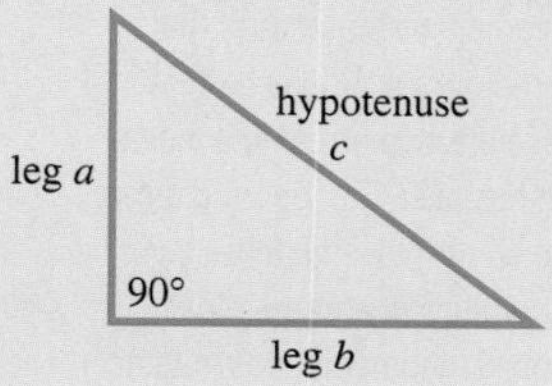

FIGURE 50

Pythagoras did not actually discover the theorem that was named after him, although legend tells that he sacrificed 100 oxen to the gods in gratitude for the discovery. There is evidence that the Babylonians knew the concept quite well. The first proof, however, may have come from Pythagoras.

Figure 50 illustrates the theorem by using a tile pattern. The side of the square along the hypotenuse measures 5 units. Those along the legs measure 3 and 4 units. If $a = 3, b = 4$, and $c = 5$, the equation of the Pythagorean theorem is satisfied.

$$a^2 + b^2 = c^2$$
$$3^2 + 4^2 = 5^2$$
$$9 + 16 = 25$$
$$25 = 25$$

The natural numbers (3, 4, 5) form a **Pythagorean triple,** because they satisfy the equation of the Pythagorean theorem. There are infinitely many such triples.

Probably the most famous mathematical statement in the history of motion pictures is heard in the 1939 classic *The Wizard of Oz.* Ray Bolger's character, the Scarecrow, wants a brain. When the Wizard grants him his "Th.D." (Doctor of Thinkology), the Scarecrow replies with a statement that has made mathematics teachers shudder for almost 70 years. (See Exercise 42 in Section 9.2.) His statement is quite impressive and sounds like the **Pythagorean theorem** but is totally incorrect. A triangle with sides of length 9, 9, and 4 provides a simple counterexample to his assertion.

If you watch this scene, also notice that between camera shots preceding his statement, the Scarecrow's position changes. These are two errors in a movie that has over 100 reported errors (see www.moviemistakes.com), yet remains one of the most beloved motion pictures in history.

EXAMPLE 8 Using the Pythogorean Theorem

Find the length a in the right triangle shown in Figure 51.

SOLUTION

$$a^2 + b^2 = c^2 \quad \text{Pythagorean theorem}$$
$$a^2 + 36^2 = 39^2 \quad b = 36, c = 39$$
$$a^2 + 1296 = 1521$$
$$a^2 = 225 \quad \text{Subtract 1296 from both sides.}$$
$$a = 15 \quad \text{Choose the positive square root, because } a > 0.$$

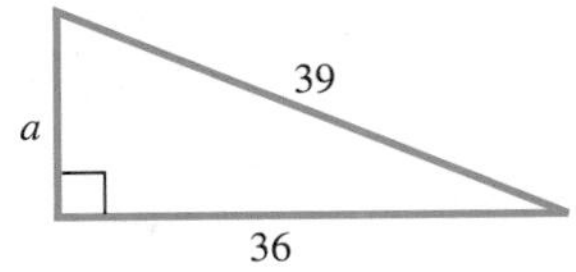

FIGURE 51

Verify that (15, 36, 39) is a Pythagorean triple as a check.

The next example comes from the Cairo Mathematical Papyrus, an Egyptian document that dates back to about 300 B.C.

EXAMPLE 9 Finding a Ladder Height by Using the Pythagorean Theorem

A ladder of length 10 cubits has its foot 6 cubits from a wall. To what height does the ladder reach?

SOLUTION

As suggested by Figure 52, the ladder forms the hypotenuse of a right triangle, and the ground and wall form the legs. Let x represent the distance from the base of the wall to the top of the ladder. Then, by the Pythagorean theorem,

$$x^2 + 6^2 = 10^2$$
$$x^2 + 36 = 100$$
$$x^2 = 64 \quad \text{Subtract 36.}$$
$$x = 8. \quad \text{Choose the positive square root of 64, because } x \text{ represents a length.}$$

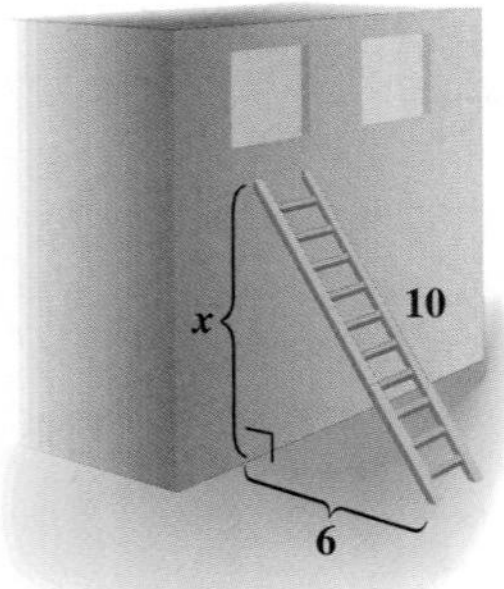

FIGURE 52

The ladder reaches a height of 8 cubits.

For Further Thought

Proving the Pythagorean Theorem

The Pythagorean theorem has probably been proved in more different ways than any theorem in mathematics. A book titled *The Pythagorean Proposition,* by Elisha Scott Loomis, was first published in 1927. It contained more than 250 different proofs of the theorem. It was reissued in 1968 by the National Council of Teachers of Mathematics as the first title in a series of "Classics in Mathematics Education."

One of the most popular proofs of the theorem follows.

For Group Discussion or Individual Investigation

Copy the accompanying figure. Keep in mind that the area of the large square must always be the same, no matter how it is determined. It is made up of four right triangles and a smaller square.

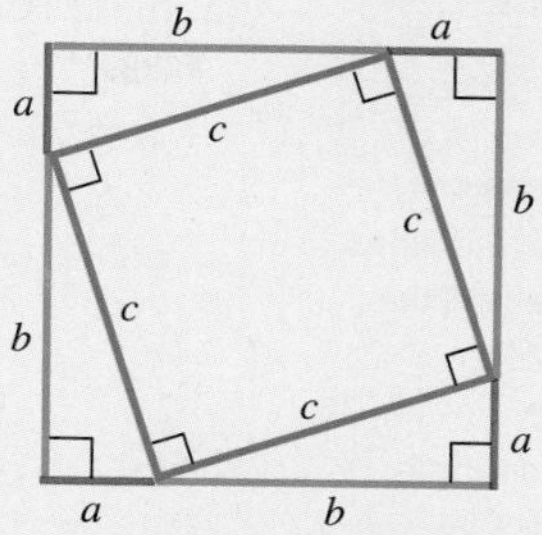

(a) The length of a side of the large square is _____ , so its area is (_____)^2 or _____.

(b) The area of the large square can also be found by obtaining the sum of the areas of the four right triangles and the smaller square. The area of each right triangle is _____ , so the sum of the areas of the four right triangles is _____. The area of the smaller square is _____.

(c) The sum of the areas of the four right triangles and the smaller square is _____.

(d) Since the areas in (a) and (c) represent the area of the same figure, the expressions there must be equal. Setting them equal to each other we obtain _____ = _____.

(e) Subtract $2ab$ from each side of the equation in (d) to obtain the desired result: _____ = _____.

Following **Hurricane Katrina** in August 2005, the pine trees of southeastern Louisiana provided thousands of examples of **right triangles.** See the photo.

Suppose the vertical distance from the base of a broken tree to the point of the break is 55 inches. The length of the broken part is 144 inches. How far along the ground is it from the base of the tree to the point where the broken part touches the ground?

The statement of the Pythagorean theorem is an *if . . . then* statement. If the antecedent (the statement following the word "if") and the consequent (the statement following the word "then") are interchanged, the new statement is called the *converse* of the original. Although the converse of a true statement may not be true, the *converse* of the Pythagorean theorem *is* also a true statement and can be used to determine if a triangle is a right triangle, given the lengths of the three sides.

Converse of the Pythagorean Theorem

If a triangle has sides of lengths a, b, and c, where c is the length of the longest side, and if $a^2 + b^2 = c^2$, then the triangle is a right triangle.

EXAMPLE 10 Applying the Converse of the Pythagorean Theorem

Lee Guidroz has been contracted to complete an unfinished 8-foot-by-12-foot laundry room on an existing house. He finds that the previous contractor built the floor so that the length of its diagonal is 14 feet, 8 inches. Is the floor "squared off" properly?

The 10 Mathematical Formulas That Changed the Face of the Earth was the theme of ten Nicaraguan stamps commemorating mathematical formulas, including this one featuring the **Pythagorean theorem,** $a^2 + b^2 = c^2$. Notice how the compasses dominate the figure.

SOLUTION

Because 14 feet, 8 inches $= 14\frac{2}{3}$ feet, he must check to see whether the following statement is true.

$$8^2 + 12^2 = \left(14\frac{2}{3}\right)^2 \quad ? \quad a^2 + b^2 = c^2$$

$$8^2 + 12^2 = \left(\frac{44}{3}\right)^2 \quad ? \quad 14\tfrac{2}{3} = \tfrac{44}{3}$$

$$208 = \frac{1936}{9} \quad ? \quad \text{Simplify.}$$

$$208 \neq 215\frac{1}{9} \quad \text{False}$$

Lee needs to fix the problem, since the diagonal, which measures 14 feet, 8 inches, should actually measure $\sqrt{208} \approx 14.4 \approx 14$ feet, 5 inches. He must correct the error to avoid major problems later.

9.4 Exercises

In Exercises 1–6, provide a STATEMENTS/REASONS proof similar to the ones in Examples 1–3.

1. Given: $AC = BD$; $AD = BC$
Prove: $\triangle ABD \cong \triangle BAC$

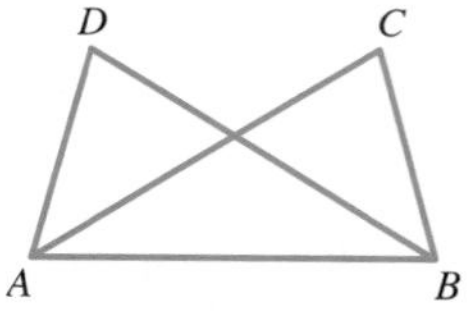

2. Given: $AC = BC$; $\measuredangle ACD = \measuredangle BCD$
Prove: $\triangle ADC \cong \triangle BDC$

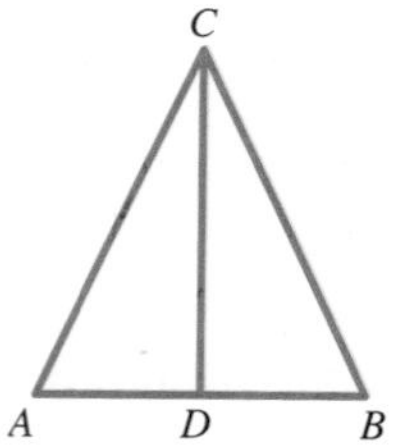

3. Given: $\overleftrightarrow{DB}$ is perpendicular to $\overleftrightarrow{AC}$; $AB = BC$
Prove: $\triangle ABD \cong \triangle CBD$

Figure for Exercise 3

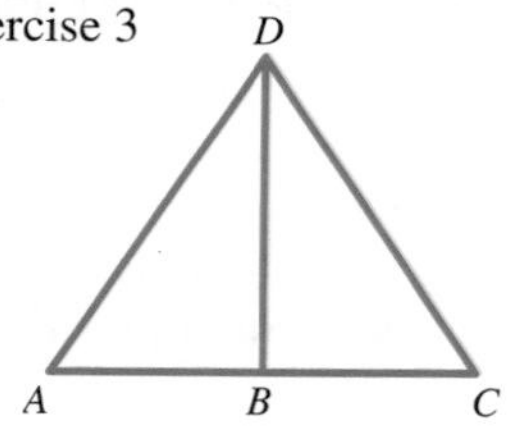

4. Given: $BC = BA$; $\measuredangle 1 = \measuredangle 2$
Prove: $\triangle DBC \cong \triangle DBA$

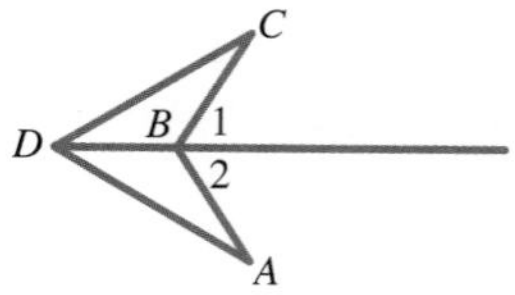

5. Given: $\measuredangle BAC = \measuredangle DAC$; $\measuredangle BCA = \measuredangle DCA$
Prove: $\triangle ABC \cong \triangle ADC$.

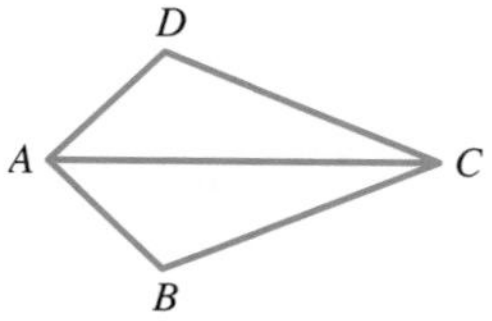

6. Given: $BO = OE$; $\overleftrightarrow{OB}$ is perpendicular to $\overleftrightarrow{AC}$; $\overleftrightarrow{OE}$ is perpendicular to $\overleftrightarrow{DF}$
Prove: $\triangle AOB \cong \triangle FOE$.

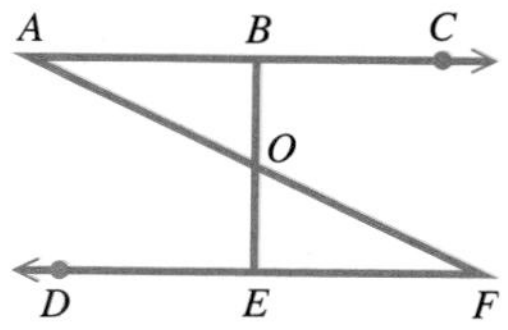

Exercises 7–10 refer to the given figure, an isosceles triangle with $AB = BC$.

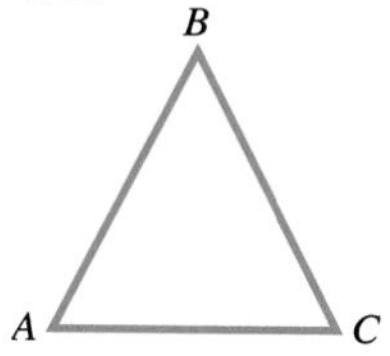

7. If $\measuredangle B$ measures 46°, then $\measuredangle A$ measures ______ and $\measuredangle C$ measures ______.

8. If $\measuredangle C$ measures 52°, what is the measure of $\measuredangle B$?

9. If $BC = 12$ in., and the perimeter of $\triangle ABC$ is 30 in., what is the length AC?

10. If the perimeter of $\triangle ABC = 40$ in., and $AC = 10$ in., what is the length AB?

11. Explain why all equilateral triangles must be similar.

12. Explain why two congruent triangles must be similar, but two similar triangles might not be congruent.

Name the corresponding angles and the corresponding sides for each of the following pairs of similar triangles.

13.

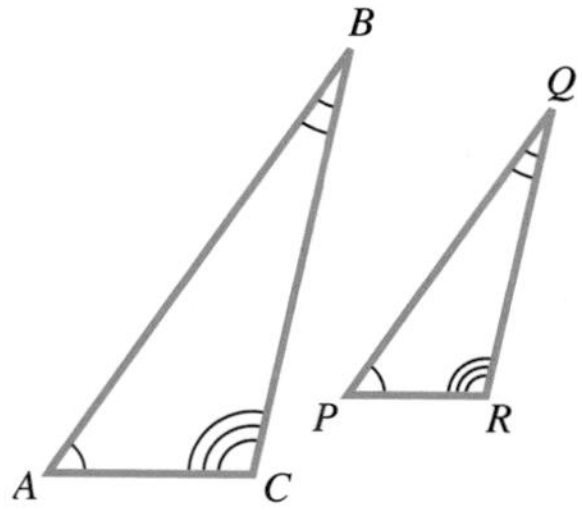

14.

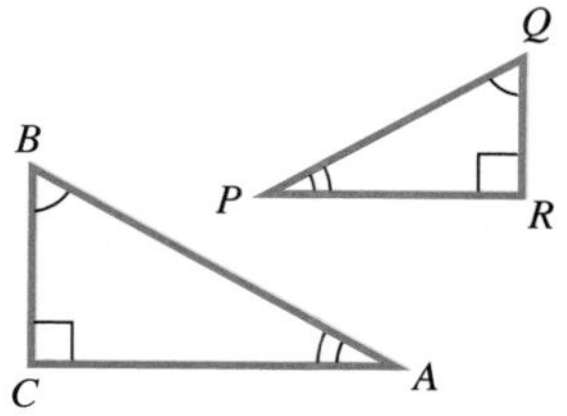

15.

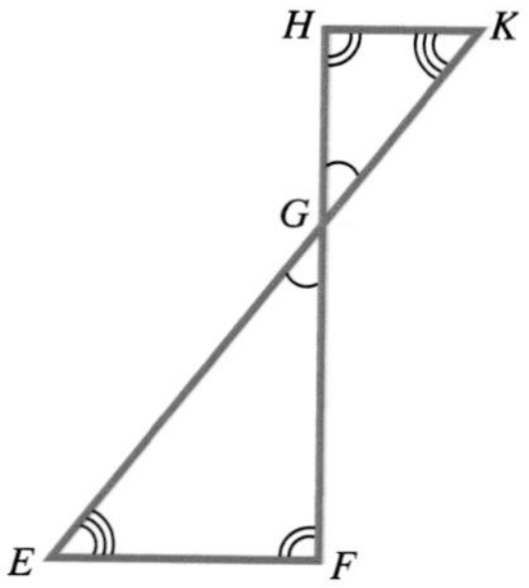

16.

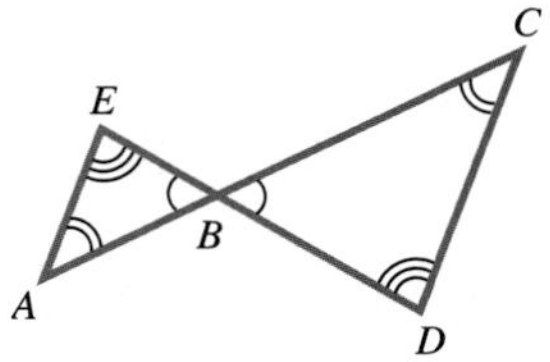

Find all unknown angle measures in each pair of similar triangles.

17.

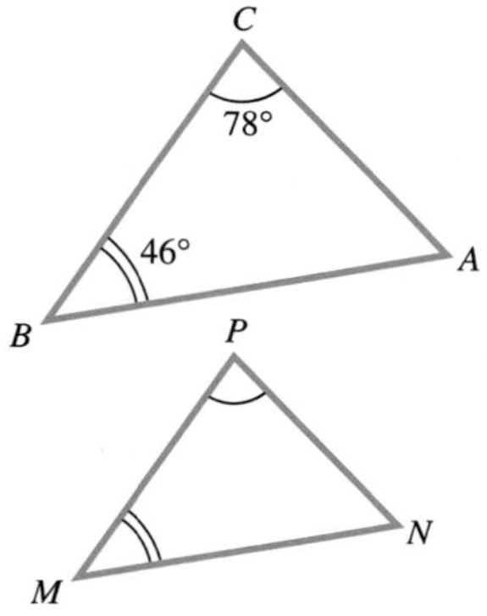

18.

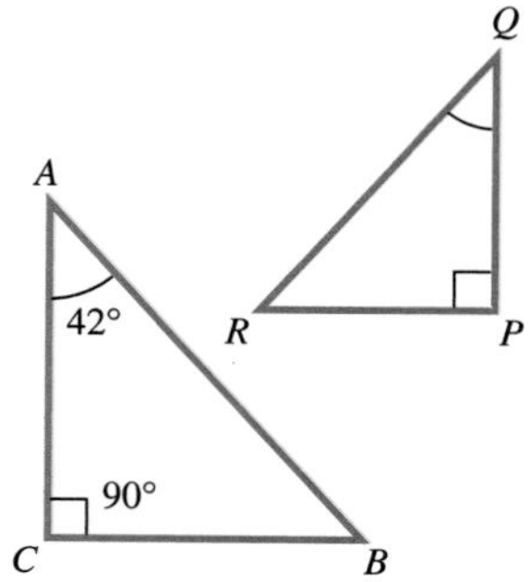

19.

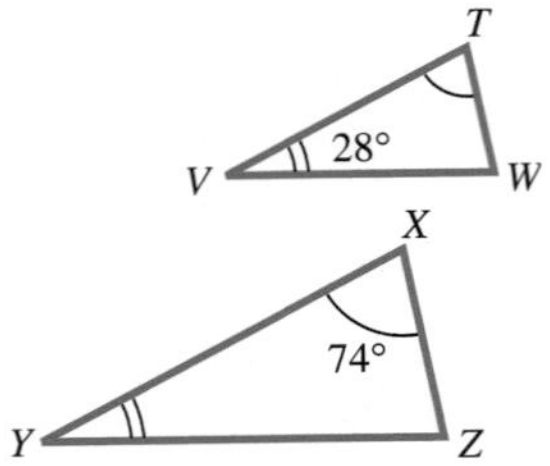

20.

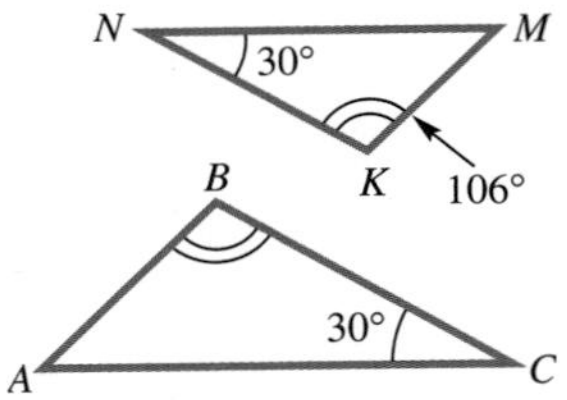

21.

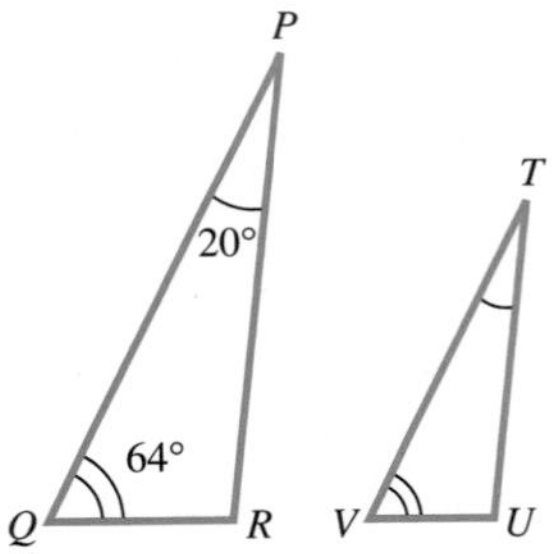

22.

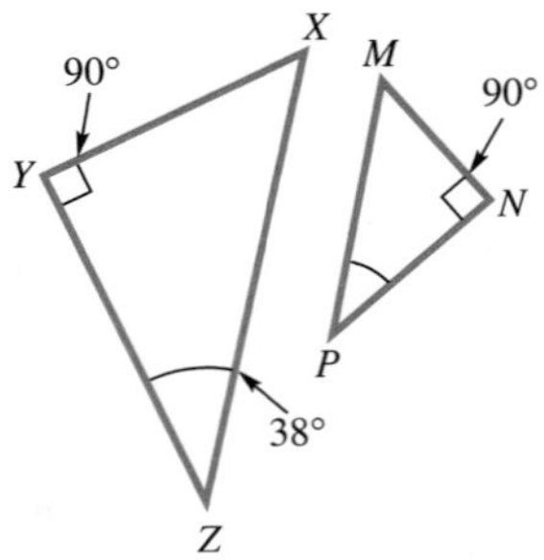

Find the unknown side lengths in each pair of similar triangles.

23.

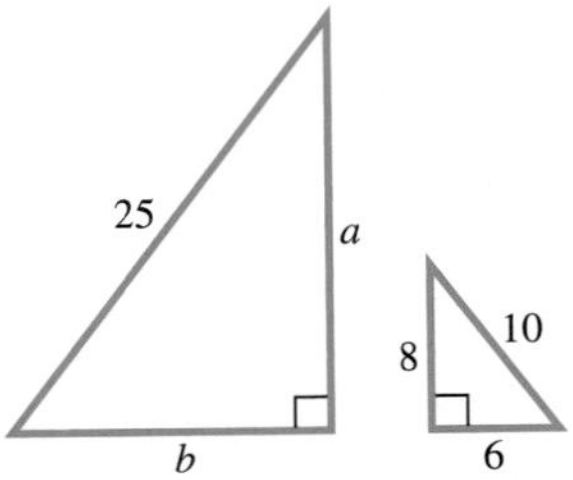

24.

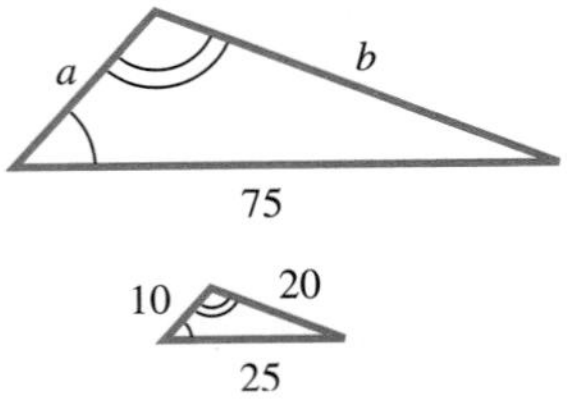

25.

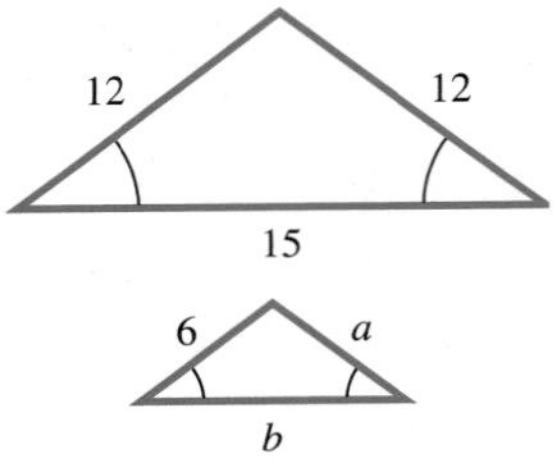

26.

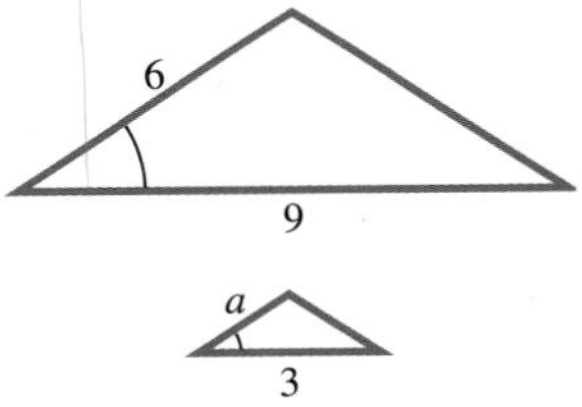

27.

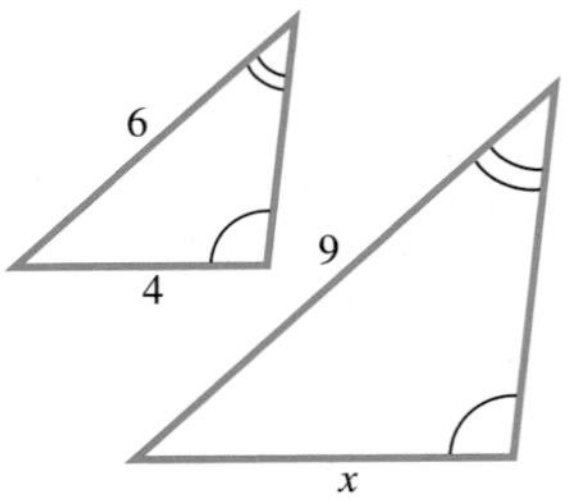

28.

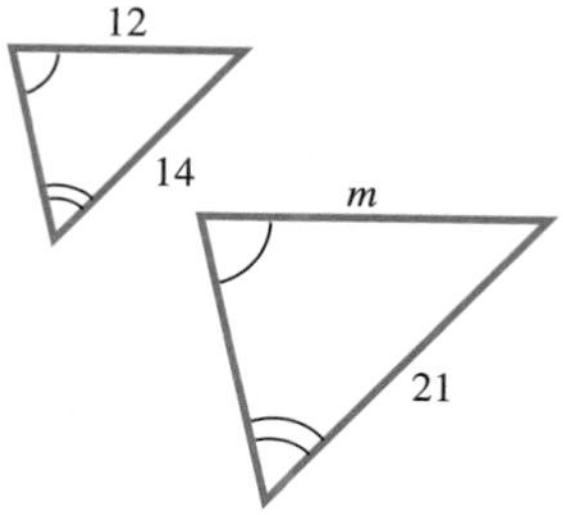

In each diagram, there are two similar triangles. Find the unknown measurement in each. (Hint: In the figure for Exercise 29, the side of length 100 *in the smaller triangle corresponds to a side of length* $100 + 120 = 220$ *in the larger triangle.)*

29.

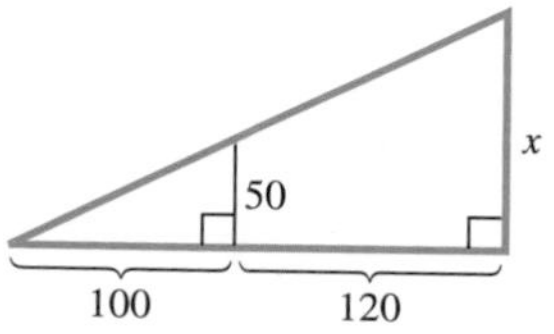

30.

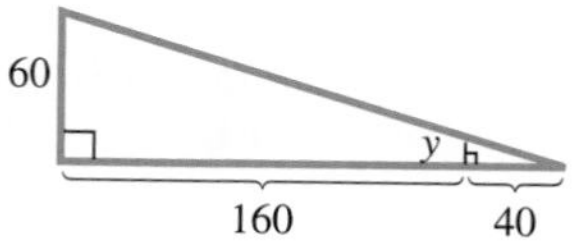

31.

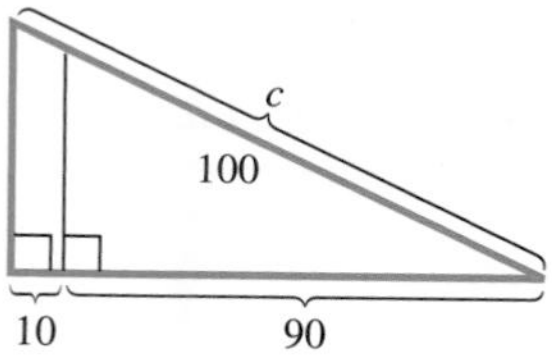

32.

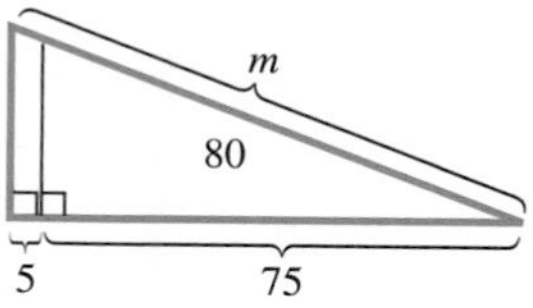

Solve each problem.

33. *Height of a Tree* A tree casts a shadow 45 m long. At the same time, the shadow cast by a vertical 2-m stick is 3 m long. Find the height of the tree.

34. *Height of a Tower* A forest fire lookout tower casts a shadow 180 ft long at the same time that the shadow of a 9-ft truck is 15 ft long. Find the height of the tower.

35. *Lengths of Sides of a Photograph* On a photograph of a triangular piece of land, the lengths of the three sides are 4 cm, 5 cm, and 7 cm, respectively. The shortest side of the actual piece of land is 400 m long. Find the lengths of the other two sides.

36. *Height of a Lighthouse Keeper* The Santa Cruz lighthouse is 14 m tall and casts a shadow 28 m long at 7 P.M. At the same time, the shadow of the lighthouse keeper is 3.5 m long. How tall is she?

37. *Height of a Building* A house is 15 ft tall. Its shadow is 40 ft long at the same time the shadow of a nearby building is 300 ft long. Find the height of the building.

38. *Distances Between Cities* By drawing lines on a map, a triangle can be formed by the cities of Phoenix, Tucson, and Yuma. On the map, the distance between Phoenix and Tucson is 8 cm, the distance between Phoenix and Yuma is 12 cm, and the distance between Tucson and Yuma is 17 cm. The actual straight-line distance from Phoenix to Yuma is 230 km. Find the distances between the other pairs of cities.

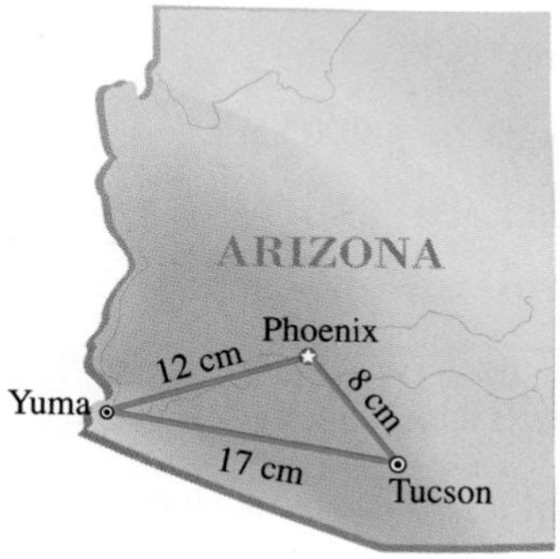

39. *Height of the World's Tallest Human* Robert Wadlow was the tallest human being ever recorded. When a 6-ft stick cast a shadow 24 in., Robert would cast a shadow 35.7 in. How tall was he?

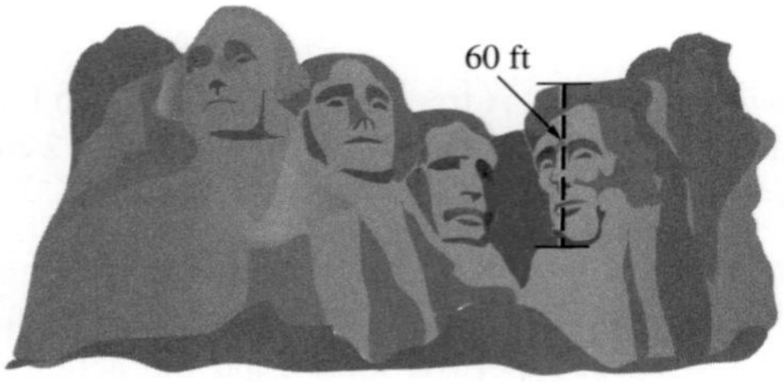

40. *Dimensions on Mount Rushmore* Assume that Lincoln was $6\frac{1}{3}$ ft tall and his head $\frac{3}{4}$ ft long. Knowing that the carved head of Lincoln at Mount Rushmore is 60 ft tall, find out how tall his entire body would be if it were carved into the mountain.

In Exercises 41–48, a and b represent the two legs of a right triangle, while c represents the hypotenuse. Find the lengths of the unknown sides.

41.

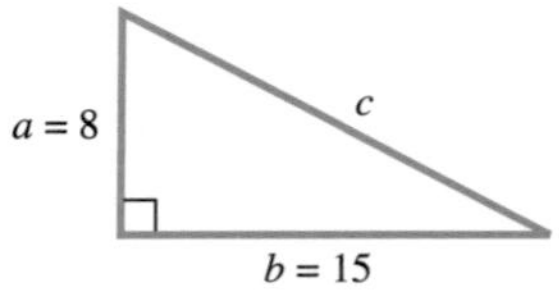

42.

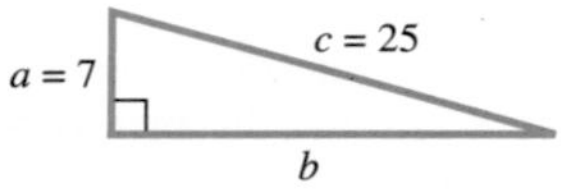

43.

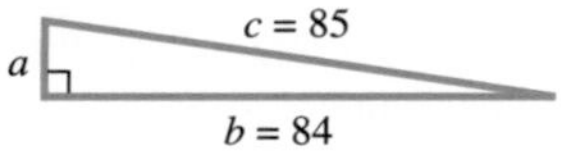

44. $a = 24$ cm; $c = 25$ cm

45. $a = 14$ m; $b = 48$ m

46. $a = 28$ km; $c = 100$ km

47. $b = 21$ in.; $c = 29$ in.

48. $b = 120$ ft; $c = 169$ ft

49. Refer to Exercise 42 in Section 9.2. Correct the Scarecrow's statement, using language similar to his.

50. Show that if $a^2 + b^2 = c^2$, then it is not necessarily true that $a + b = c$.

There are various formulas that will generate Pythagorean triples. For example, if we choose positive integers r and s, with r > s, then the set of equations

$$a = r^2 - s^2, \quad b = 2rs, \quad c = r^2 + s^2$$

generates a Pythagorean triple (a, b, c). Use the values of r and s given in each of Exercises 51–56 to generate a Pythagorean triple using this method.

51. $r = 2, s = 1$

52. $r = 3, s = 2$

53. $r = 4, s = 3$

54. $r = 3, s = 1$

55. $r = 4, s = 2$

56. $r = 4, s = 1$

57. Show that the formula given for Exercises 51–56 actually satisfies $a^2 + b^2 = c^2$.

58. It can be shown that if $(x, x + 1, y)$ is a Pythagorean triple, then so is

$$(3x + 2y + 1, \quad 3x + 2y + 2, \quad 4x + 3y + 2).$$

Use this idea to find three more Pythagorean triples, starting with 3, 4, 5. (*Hint:* Here, $x = 3$ and $y = 5$.)

If m is an odd positive integer greater than 1, then

$$\left(m, \frac{m^2 - 1}{2}, \frac{m^2 + 1}{2}\right)$$

is a Pythagorean triple. Use this to find the Pythagorean triple generated by each value of m in Exercises 59–62.

59. $m = 3$

60. $m = 5$

61. $m = 7$

62. $m = 9$

63. Show that the expressions in the directions for Exercises 59–62 actually satisfy $a^2 + b^2 = c^2$.

64. Show why (6, 8, 10) is the only Pythagorean triple consisting of consecutive even numbers.

For any integer n greater than 1,

$$(2n, n^2 - 1, n^2 + 1)$$

is a Pythagorean triple. Use this pattern to find the Pythagorean triple generated by each value of n in Exercises 65–68.

65. $n = 2$

66. $n = 3$

67. $n = 4$

68. $n = 5$

69. Show that the expressions in the directions for Exercises 65–68 actually satisfy $a^2 + b^2 = c^2$.

70. Can an isosceles right triangle have sides with integer lengths? Why or why not?

Solve each problem. (You may wish to review quadratic equations from algebra.)

71. ***Side Length of a Triangle*** If the hypotenuse of a right triangle is 1 m more than the longer leg, and the shorter leg is 7 m, find the length of the longer leg.

72. ***Side Lengths of a Triangle*** The hypotenuse of a right triangle is 1 cm more than twice the shorter leg, and the longer leg is 9 cm less than three times the shorter leg. Find the lengths of the three sides of the triangle.

73. ***Height of a Tree*** At a point on the ground 30 ft from the base of a tree, the distance to the top of the tree is 2 ft more than twice the height of the tree. Find the height of the tree.

74. ***Dimensions of a Rectangle*** The length of a rectangle is 2 in. less than twice the width. The diagonal is 5 in. Find the length and width of the rectangle.

75. ***Height of a Break in Bamboo*** (Problem of the broken bamboo, from the Chinese work *Arithmetic in Nine Sections* (1261)) There is a bamboo 10 ft high, the upper end of which, being broken, reaches the ground 3 ft from the stem. Find the height of the break.

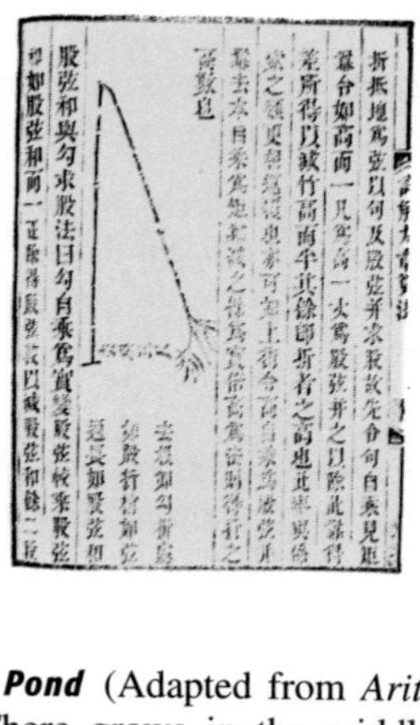

76. ***Depth of a Pond*** (Adapted from *Arithmetic in Nine Sections*) There grows in the middle of a circular pond 10 ft in diameter a reed which projects 1 ft out of the water. When it is drawn down it just reaches the edge of the pond. How deep is the water?

Squaring Off a Floor Under Construction *Imagine that you are a carpenter building the floor of a rectangular room. What must the diagonal of the room measure if your floor is to be squared off properly, given the dimensions in Exercises 77–80? Give your answer to the nearest inch.*

77. 12 ft by 15 ft

78. 14 ft by 20 ft

79. 16 ft by 24 ft

80. 20 ft by 32 ft

81. ***Garfield's Proof of the Pythagorean Theorem*** James A. Garfield, the twentieth president of the United States, provided a proof of the Pythagorean theorem using the given figure. Supply the required information in each of parts (a) through (c) in order to follow his proof.

(a) Find the area of the trapezoid *WXYZ* using the formula for the area of a trapezoid.

(b) Find the area of each of the right triangles *PWX*, *PZY*, and *PXY*.

(c) Since the sum of the areas of the three right triangles must equal the area of the trapezoid, set the expression from part (a) equal to the sum of the three expressions from part (b). Simplify the equation as much as possible.

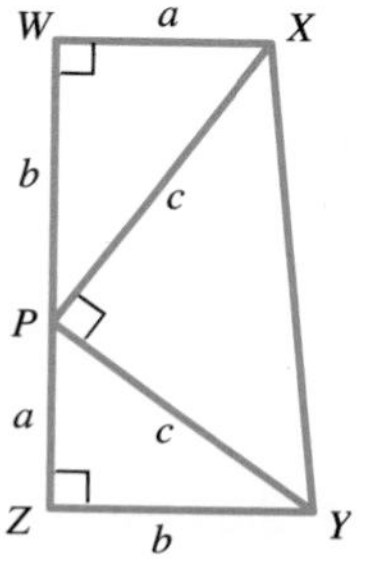

82. ***Proof of the Pythagorean Theorem by Similar Triangles*** In the figure, right triangles *ABC*, *CBD*, and *ACD* are similar. This may be used to prove the Pythagorean theorem. Fill in the blanks with the appropriate responses.

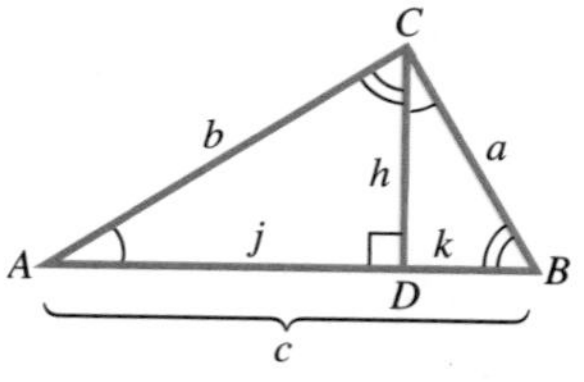

(a) By proportion, we have $\frac{c}{b} =$ ____$/j$.

(b) By proportion, we also have $\frac{c}{a} = a/$____.

(c) From part (a), $b^2 =$ _____.

(d) From part (b), $a^2 =$ _____.

(e) From the results of parts (c) and (d) and factoring, $a^2 + b^2 = c($________$)$. Since ________ $= c$, it follows that ____________.

Exercises 83–90 require some ingenuity, but all can be solved using the concepts presented so far in this chapter.

83. ***Area of a Quadrilateral*** Find the area of quadrilateral *ABCD*, if angles *A* and *C* are right angles.

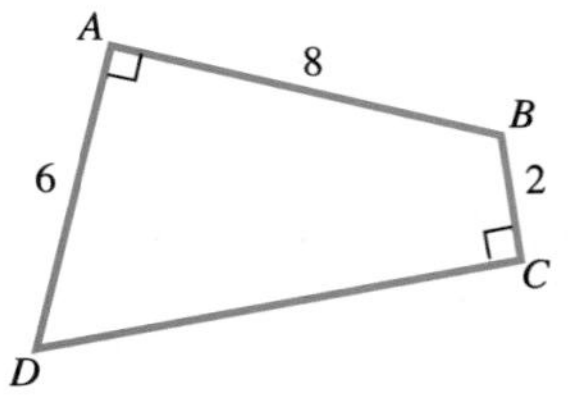

84. ***Area of a Triangle*** The perimeter of the isosceles triangle *ABC* (with *AB* = *BC*) is 128 in. The altitude *BD* is 48 in. What is the area of triangle *ABC*?

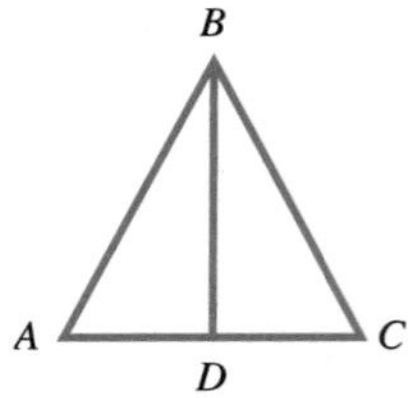

85. ***Base Measure of an Isosceles Triangle*** An isosceles triangle has a base of 24 and two sides of 13. What other base measure can an isosceles triangle with equal sides of 13 have and still have the same area as the given triangle?

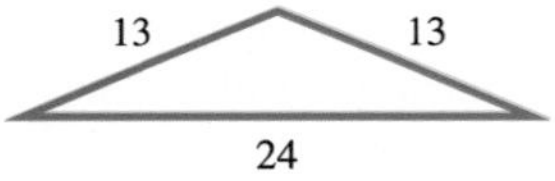

86. ***Value of a Measure in a Triangle*** In right triangle *ABC*, if *AD* = *DB* + 8, what is the value of *CD*?

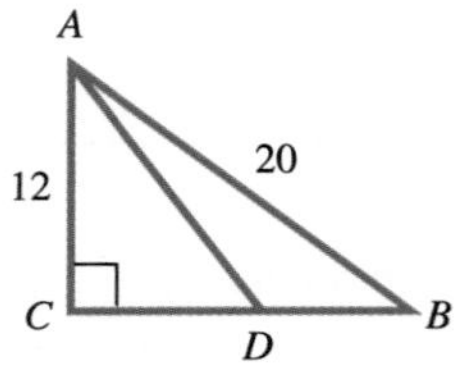

87. ***Area of a Pentagon*** In the figure at the top of the next column, pentagon *PQRST* is formed by a square and an equilateral triangle such that *PQ* = *QR* = *RS* = *ST* = *PT*. The perimeter of the pentagon is 80. Find the area of the pentagon.

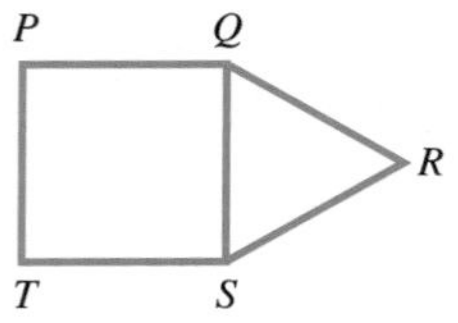

88. ***Angle Measure in a Triangle*** (A segment that *bisects* an angle divides the angle into two equal angles.) In the figure, angle *A* measures 50°. *OB* bisects angle *ABC*, and *OC* bisects angle *ACB*. What is the measure of angle *BOC*?

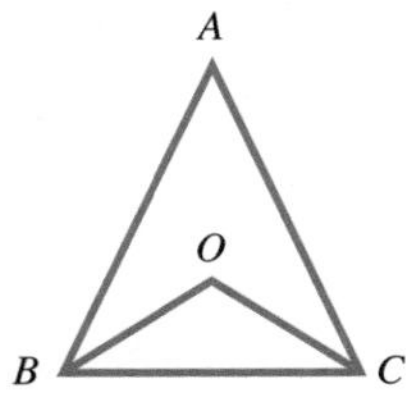

Exercises 89 and 90 refer to the given figure. The center of the circle is O.

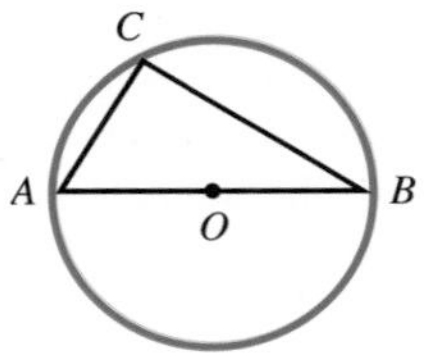

89. ***Radius of a Circle*** If $\overline{AC}$ measures 6 in. and $\overline{BC}$ measures 8 in., what is the radius of the circle?

90. ***Lengths of Chords of a Circle*** If $\overline{AB}$ measures 13 cm, and the length of $\overline{BC}$ is 7 cm more than the length of $\overline{AC}$, what are the lengths of $\overline{BC}$ and $\overline{AC}$?

Verify that the following constructions from the Extension following Section 9.2 are valid. Use a STATEMENTS/REASONS proof.

91. Construction 1

92. Construction 2

93. Construction 3

94. Construction 4

9.5 Space Figures, Volume, and Surface Area

Space Figures • Volume and Surface Area of Space Figures

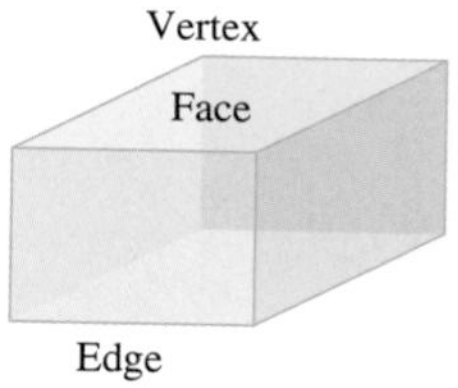

Rectangular parallelepiped (box)

FIGURE 53

Space Figures

Thus far, this chapter has discussed only **plane figures**—figures that can be drawn completely in the plane of a piece of paper. However, it takes the three dimensions of space to represent the solid world around us. For example, Figure 53 shows a "box" (a **rectangular parallelepiped** in mathematical terminology). The *faces* of a box are rectangles. The faces meet at *edges;* the "corners" are called *vertices* (plural of vertex—the same word as for the "corner" of an angle).

Boxes are one kind of space figure belonging to an important group called **polyhedra,** the faces of which are made only of polygons. Perhaps the most interesting polyhedra are the *regular polyhedra.* Recall that a *regular polygon* is a polygon with all sides equal and all angles equal. A regular polyhedron is a space figure, the faces of which are only one kind of regular polygon. It turns out that there are only five different regular polyhedra. They are shown in Figure 54. A **tetrahedron** is composed of four equilateral triangles, each three of which meet in a point. Use the figure to verify that there are four faces, four vertices, and six edges.

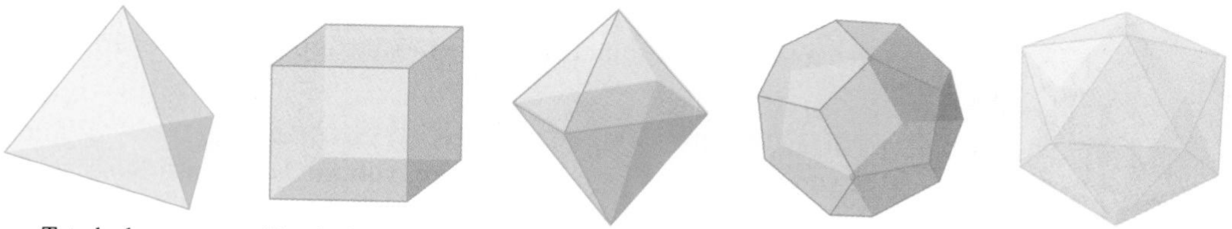

Tetrahedron Hexahedron (cube) Octahedron Dodecahedron Icosahedron

FIGURE 54

A regular quadrilateral is called a square. Six squares, each three of which meet at a point, form a **hexahedron,** or **cube**. Again, use Figure 54 to verify that a cube has 6 faces, 8 vertices, and 12 edges.

The three remaining regular polyhedra are the **octahedron,** the **dodecahedron,** and the **icosahedron.** The octahedron is composed of groups of four regular triangles (i.e., equilateral) meeting at a point. The dodecahedron is formed by groups of three regular pentagons, while the icosahedron is made up of groups of five regular triangles. The five regular polyhedra are also known as **Platonic solids,** named for the Greek philosopher Plato. He considered them as "building blocks" of nature and assigned fire to the tetrahedron, earth to the cube, air to the octahedron, and water to the icosahedron. Because the dodecahedron is different from the others because of its pentagonal faces, he assigned to it the cosmos (stars and planets). (*Source:* www.mathacademy.com)

Two other types of polyhedra are familiar space figures: pyramids and prisms. **Pyramids** are made of triangular sides and a polygonal base. **Prisms** have two faces in parallel planes; these faces are congruent polygons. The remaining faces of a prism are all parallelograms. (See Figure 55(a) and (b) on the next page.) By this definition, a box is also a prism.

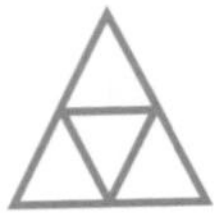

Tetrahedron

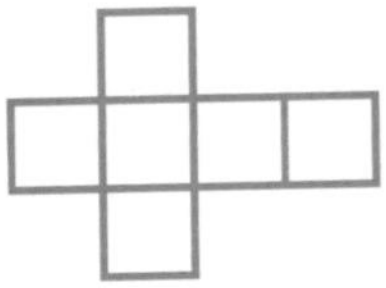

Hexahedron (cube)

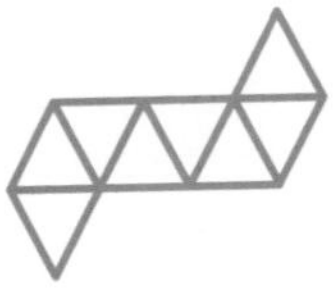

Octahedron

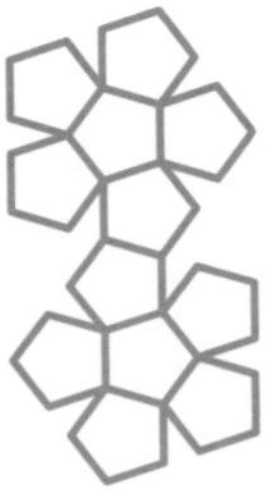

Dodecahedron

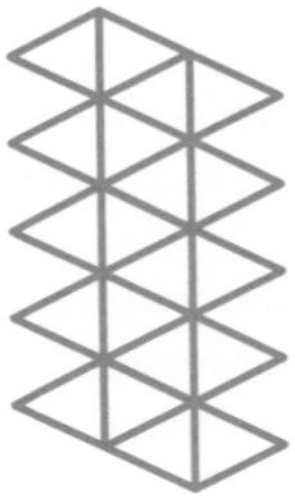

Icosahedron

Patterns such as these may be used to actually construct three-dimensional models of the **regular polyhedra**.

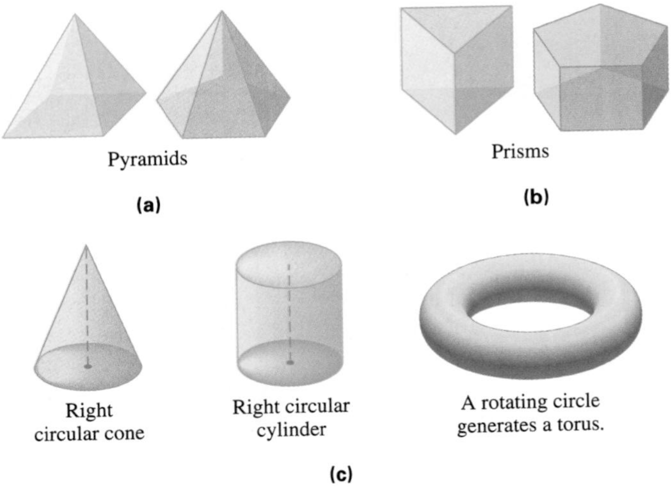

FIGURE 55

The circle, although a plane figure, is not a polygon. (Why?) Figure 55(c) shows space figures made up in part of circles, including *right circular cones* and *right circular cylinders.* The figure also shows how a circle can generate a *torus,* a doughnut-shaped solid that has interesting topological properties. See Section 9.7.

Volume and Surface Area of Space Figures

While area is a measure of surface covered by a plane figure, **volume** is a measure of capacity of a space figure. Volume is measured in *cubic* units. For example, a cube with edge measuring 1 cm has volume 1 cubic cm, which is also written as 1 cm^3, or 1 cc. The **surface area** is the total area that would be covered if the space figure were "peeled" and the peel laid flat. Surface area is measured in *square* units.

Volume and Surface Area of a Box

Suppose that a box has length ℓ, width w, and height h. Then the volume V and the surface area S are given by the formulas

$$V = \ell wh \quad \text{and} \quad S = 2\ell w + 2\ell h + 2hw.$$

In particular, if the box is a cube with edge of length s,

$$V = s^3 \quad \text{and} \quad S = 6s^2.$$

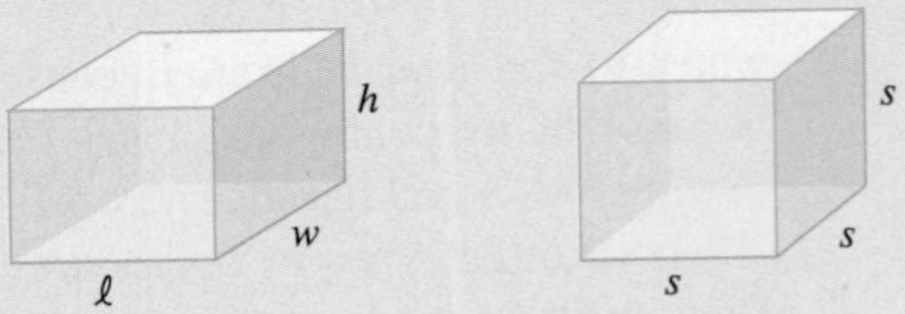

$V = \ell wh$ $\qquad$ $V = s^3$

$S = 2\ell w + 2\ell h + 2hw$ $\qquad$ $S = 6s^2$

John Conway of Princeton University offered a reward in the 1990s to anyone producing a **holyhedron,** a polyhedron with a finite number of faces and with a hole in every face. At the time, no one knew whether such an object could exist. When graduate student **Jade Vinson** arrived at Princeton, he immediately took up the challenge and in 2000 produced (at least the proof of the theoretical existence of) a holyhedron with 78,585,627 faces. The reward offered ($10,000 divided by the number of faces) earned Vinson $.0001. Subsequently, **Don Hatch** produced one with 492 faces, good for a prize of $20.33. Conway had predicted that someone will eventually find a holyhedron with fewer than 100 faces. For more details, including references, images, and an online talk on holyhedra by Vinson, start at www.hadron.org/~hatch/

EXAMPLE 1 Using the Formulas for a Box

Find the volume V and the surface area S of the box shown in Figure 56.

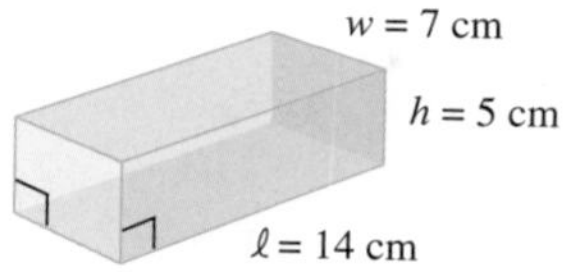

FIGURE 56

SOLUTION

$$V = \ell wh = 14 \cdot 7 \cdot 5 = 490 \quad \ell = 14, w = 7, h = 5$$

Volume is measured in cubic units, so the volume of the box is 490 cubic centimeters, or 490 cm^3.

To find the surface area, use the formula $S = 2\ell w + 2\ell h + 2hw$.

$$\begin{aligned} S &= 2(14)(7) + 2(14)(5) + 2(5)(7) \\ &= 196 + 140 + 70 \\ &= 406 \end{aligned}$$

Like areas of plane figures, surface areas of space figures are measured in square units, so the surface area of the box is 406 square centimeters, or 406 cm^2.

A typical tin can is an example of a **right circular cylinder.**

Volume and Surface Area of a Right Circular Cylinder

If a right circular cylinder has height h and radius of its base equal to r, then the volume V and the surface area S are given by the formulas

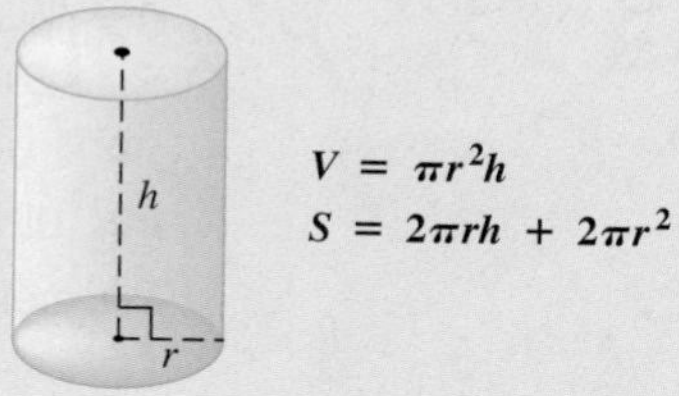

$$V = \pi r^2 h$$

and

$$S = 2\pi rh + 2\pi r^2.$$

(In the formula for S, the areas of the top and bottom are included.)

EXAMPLE 2 Using the Formulas for a Right Circular Cylinder

In Figure 57, the right circular cylinder has surface area 288π square inches, and the radius of its base is 6 inches. Find each measure.

(a) the height of the cylinder
(b) the volume of the cylinder

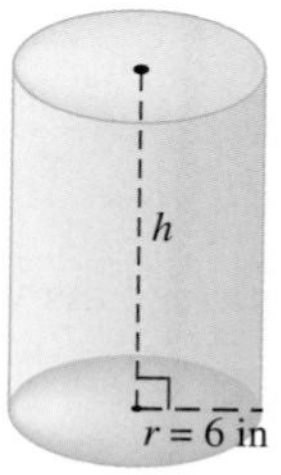

Right circular cylinder

FIGURE 57

On April 30, 2006, Episode 19 of the seventeenth season of the Fox television series *The Simpsons* dealt with the issue of gender, expectations, and curriculum in schools. In "Girls Just Want to Have Sums," Lisa Simpson's school is divided in two: she must attend the girls' school, where her mathematics class no longer poses a challenge. She disguises herself as a boy and infiltrates the boys' school and proves that she can handle the more difficult classes offered there.

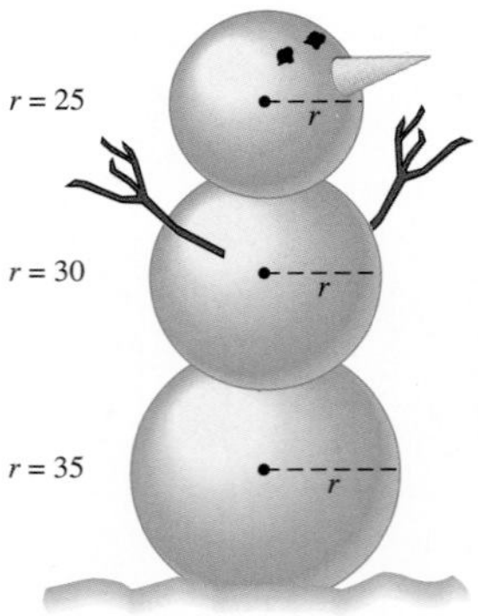

In one scene, the boys are asked by their teacher to determine the **volume of a snowman.** (See the figure above.) One boy responds to add the volumes of the spheres, because the radii are known. But Lisa corrects him, saying that he forgot the volume of the carrot nose: "one-third base times height" and gleefully follows with "Oh math, I have missed you!"

SOLUTION

(a)

$$\begin{aligned} S &= 2\pi rh + 2\pi r^2 \\ 288\pi &= 2\pi(6)h + 2\pi(6)^2 && S = 288\pi,\ r = 6 \\ 288\pi &= 12\pi h + 72\pi \\ 216\pi &= 12\pi h && \text{Subtract } 72\pi. \\ h &= 18 && \text{Divide by } 12\pi. \end{aligned}$$

The height is 18 inches.

(b)

$$V = \pi r^2 h = \pi(6)^2(18) = 648\pi \qquad r = 6,\ h = 18$$

The exact volume is 648π cubic inches, or approximately 2034.72 cubic inches, using $\pi \approx 3.14$.

The three-dimensional analogue of a circle is a **sphere.** It is defined by replacing the word "plane" with "space" in the definition of a circle (Section 9.2).

Volume and Surface Area of a Sphere

If a sphere has radius r, then the volume V and the surface area S are given by the formulas

$$V = \frac{4}{3}\pi r^3 \quad \text{and} \quad S = 4\pi r^2.$$

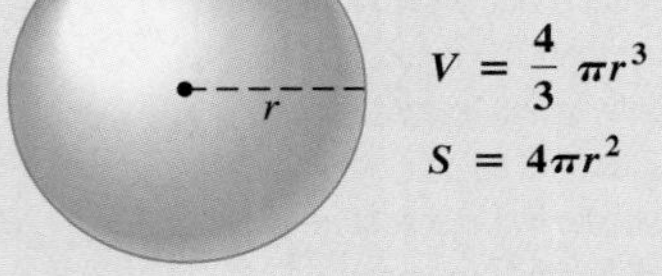

EXAMPLE 3 Using the Volume Formula for a Sphere

Suppose that a spherical tank having radius 3 meters can be filled with liquid fuel for \$200. How much will it cost to fill a spherical tank of radius 6 meters with the same fuel?

SOLUTION

We must first find the volume of the tank with radius 3 meters. Call it V_1.

$$V_1 = \frac{4}{3}\pi r^3 = \frac{4}{3}\pi(3)^3 = \frac{4}{3}\pi(27) = 36\pi \qquad r = 3$$

Now find V_2, the volume of the tank having radius 6 meters.

$$V_2 = \frac{4}{3}\pi(6)^3 = \frac{4}{3}\pi(216) = 288\pi \qquad r = 6$$

Notice that by doubling the radius of the sphere from 3 meters to 6 meters, the volume has increased 8 times, because

$$V_2 = 288\pi = 8V_1 = 8(36\pi).$$

Therefore, the cost to fill the larger tank is eight times the cost to fill the smaller one: $8(\$200) = \$1600$.

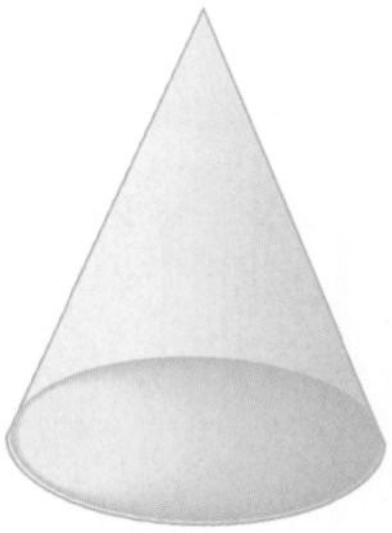

Right circular cone

FIGURE 58

The space figure shown in Figure 58 is a **right circular cone.**

Volume and Surface Area of a Right Circular Cone

If a right circular cone has height h and the radius of its circular base is r, then the volume V and the surface area S are given by the formulas

$$V = \frac{1}{3}\pi r^2 h$$

and $S = \pi r\sqrt{r^2 + h^2} + \pi r^2.$

(In the formula for S, the area of the bottom is included.)

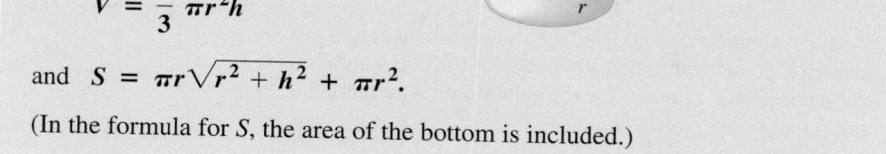

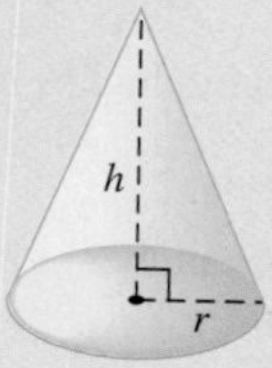

$$V = \frac{1}{3}\pi r^2 h$$
$$S = \pi r\sqrt{r^2 + h^2} + \pi r^2$$

A **pyramid** is a space figure having a polygonal base and triangular sides. Figure 59 shows a pyramid with a square base.

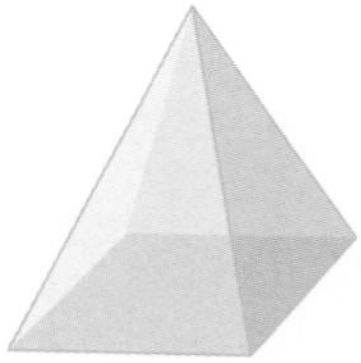

Pyramid

FIGURE 59

Volume of a Pyramid

If B represents the area of the base of a pyramid, and h represents the height (that is, the perpendicular distance from the top, or apex, to the base), then the volume V is given by the formula

$$V = \frac{1}{3}Bh.$$

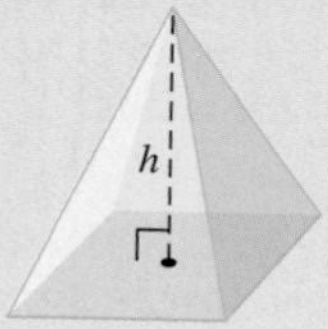

$$V = \frac{1}{3}Bh$$

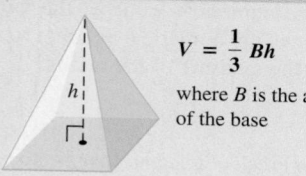

where B is the area of the base

The **Transamerica Tower** in San Francisco is a pyramid with a square base. Each side of the base has a length of 52 meters, while the height of the building is 260 meters. The formula for the volume of a pyramid indicates that the volume of the building is about 234,000 cubic meters.

EXAMPLE 4 Comparing Volumes Using Ratios

What is the ratio of the volume of a right circular cone with radius of base r and height h to the volume of a pyramid having a square base, with each side of length r, and height h?

SOLUTION

Using the formula for the volume of a cone, we have

$$V_1 = \text{Volume of the cone} = \frac{1}{3}\pi r^2 h.$$

Because the pyramid has a square base, the area B of its base is r^2. Using the formula for the volume of a pyramid, we get

$$V_2 = \text{Volume of the pyramid} = \frac{1}{3}Bh = \frac{1}{3}(r^2)h.$$

The ratio of the first volume to the second is

$$\frac{V_1}{V_2} = \frac{\frac{1}{3}\pi r^2 h}{\frac{1}{3}r^2 h} = \pi.$$

9.5 EXERCISES

Decide whether each of the following statements is true *or* false.

1. A cube with volume 64 cubic inches has surface area 96 square inches.

2. A tetrahedron has the same number of faces as vertices.

3. A dodecahedron can be used as a model for a calendar for a given year, where each face of the dodecahedron contains a calendar for a single month, and there are no faces left over.

4. Each face of an octahedron is an octagon.

5. If you double the length of the edge of a cube, the new cube will have a volume that is twice the volume of the original cube.

6. The numerical value of the volume of a sphere is $\frac{r}{3}$ times the numerical value of its surface area, where r is the measure of the radius.

Find **(a)** *the volume and* **(b)** *the surface area of each space figure. When necessary, use* 3.14 *as an approximation for* π, *and round answers to the nearest hundredth.*

7.

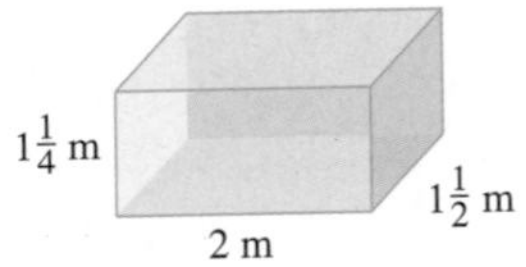

(a box)

8.

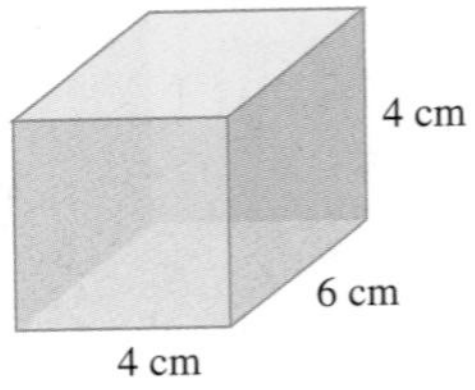

(a box)

9.

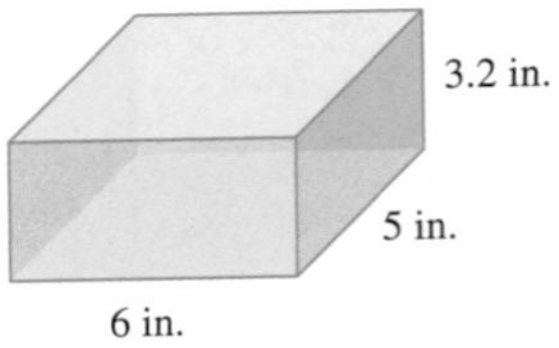

(a box)

10.

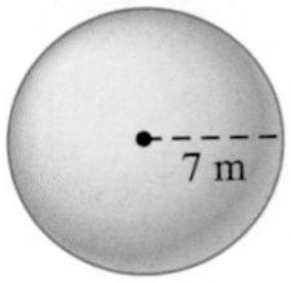

(a sphere)

11.

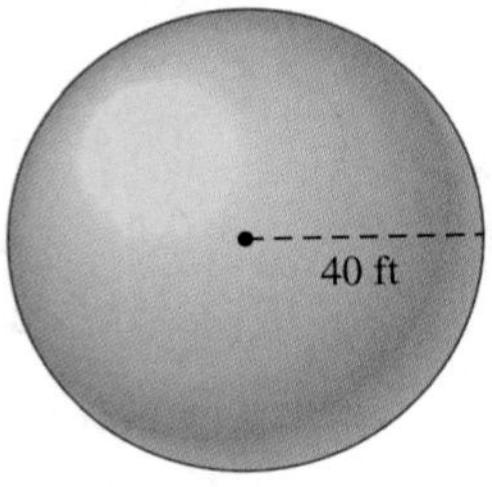

(a sphere)

12.

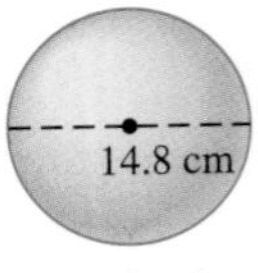

(a sphere)

13.

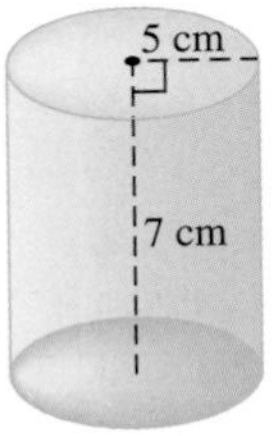

(a right circular cylinder)

14.

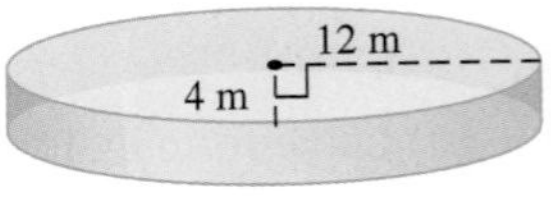

(a right circular cylinder)

15.

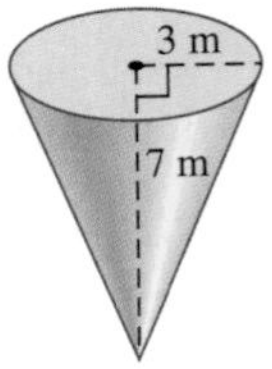

(a right circular cone)

16.

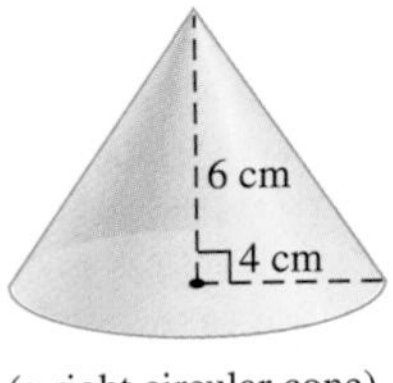

(a right circular cone)

Find the volume of each pyramid. In each case, the base is a rectangle.

17.

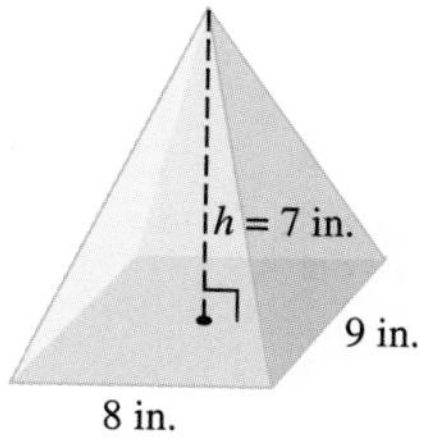

18.

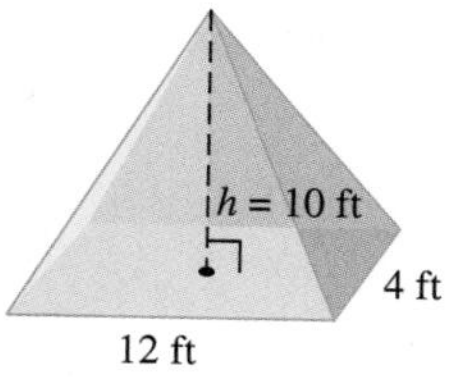

Volumes of Common Objects *Find each volume. Use 3.14 as an approximation for π when necessary.*

19. a coffee can, radius 6.3 cm and height 15.8 cm

20. a soup can, radius 3.2 cm and height 9.5 cm

21. a pork-and-beans can, diameter 7.2 cm and height 10.5 cm

22. a cardboard mailing tube, diameter 2 in. and height 40 in.

23. a coffee mug, diameter 9 cm and height 8 cm

24. a bottle of typewriter correction fluid, diameter 3 cm and height 4.3 cm

25. the Great Pyramid of Cheops, near Cairo—its base is a square 230 m on a side, while the height is 137 m

26. a hotel in the shape of a cylinder with a base radius of 46 m and a height of 220 m

27. a road construction marker, a cone with height 2 m and base radius $\frac{1}{2}$ m

28. the conical portion of a witch's hat for a Halloween costume, with height 12 in. and base radius 4 in.

In the chart at the top of the next column, one of the values r (radius), d (diameter), V (volume), or S (surface area) is given for a particular sphere. Find the remaining three values. Leave π in your answers.

	r	d	V	S
29.	6 in.			
30.	9 in.			
31.		10 ft		
32.		40 ft		
33.			$\frac{32}{3}\pi$ cm^3	
34.			$\frac{256}{3}\pi$ cm^3	
35.				4π m^2
36.				144π m^2

Solve each problem.

37. *Volume or Surface Area?* In order to determine the amount of liquid a spherical tank will hold, would you need to use volume or surface area?

38. *Volume or Surface Area?* In order to determine the amount of leather it would take to manufacture a basketball, would you need to use volume or surface area?

39. *Side Length of a Cube* One of the three famous construction problems of Greek mathematics required the construction of an edge of a cube with twice the volume of a given cube. If the length of each side of the given cube is x, what would be the length of each side of a cube with twice the original volume?

40. Work through the parts of this exercise in order, and use them to make a generalization concerning volumes of spheres. Leave answers in terms of π.

(a) Find the volume of a sphere having a radius of 1 m.

(b) Suppose the radius is doubled to 2 m. What is the volume?

(c) When the radius was doubled, by how many times did the volume increase? (To find out, divide the answer for part (b) by the answer for part (a).)

(d) Suppose the radius of the sphere from part (a) is tripled to 3 m. What is the volume?

(e) When the radius was tripled, by how many times did the volume increase?

(f) In general, if the radius of a sphere is multiplied by n, the volume is multiplied by ______.

Cost to Fill a Spherical Tank *If a spherical tank 2 m in diameter can be filled with a liquid for \$300, find the cost to fill tanks of each diameter.*

41. 6 m **42.** 8 m **43.** 10 m

44. Use the logic of Exercise 40 to answer the following: If the radius of a sphere is multiplied by n, then the surface area of the sphere is multiplied by _____.

Each of the following figures has volume as indicated. Find the value of x.

45. $V = 60$

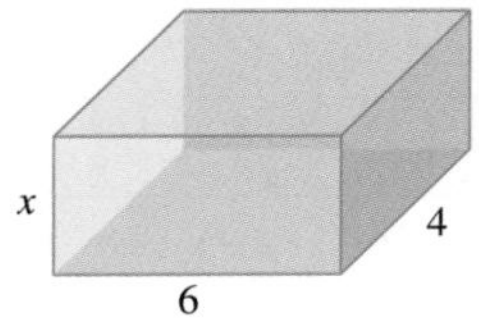

(a box)

46. $V = 450$

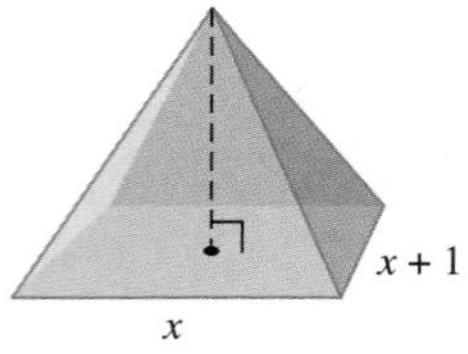

$h = 15$
Base is a rectangle.
(a pyramid)

47. $V = 36\pi$

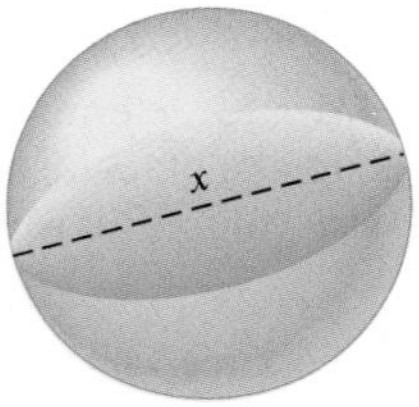

(a sphere)

48. $V = 245\pi$

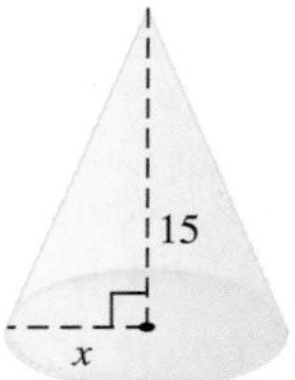

(a right circular cone)

Exercises 49–56 require some ingenuity, but all can be solved using the concepts presented so far in this chapter.

49. *Volume of a Box* The areas of the sides of a rectangular box are 30 in.^{2}, 35 in.^{2}, and 42 in.^{2}. What is the volume of the box?

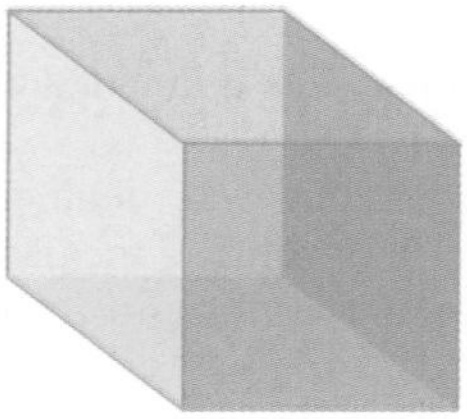

50. *Ratios of Volumes* In the figure, a right circular cone is inscribed in a hemisphere. What is the ratio of the volume of the cone to the volume of the hemisphere?

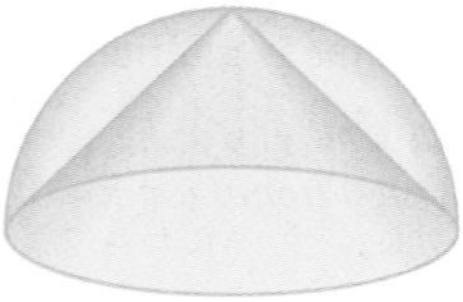

51. *Volume of a Sphere* A plane intersects a sphere to form a circle as shown in the figure. If the area of the circle formed by the intersection is 576π in.^{2}, what is the volume of the sphere?

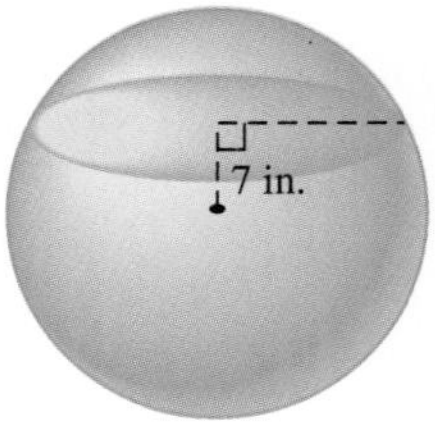

52. *Change in Volume* If the height of a right circular cylinder is halved and the diameter is tripled, how is the volume changed?

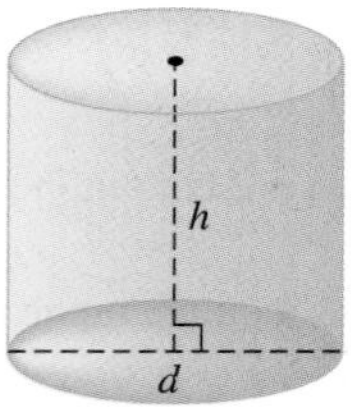

53. *Ratio of Area* What is the ratio of the area of the circumscribed square to the area of the inscribed square?

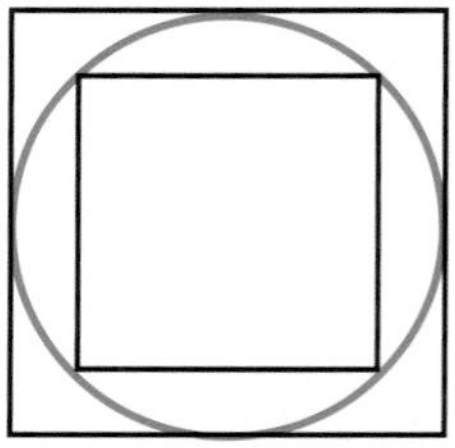

54. *Perimeter of a Square* Suppose the diameter of the circle shown is 8 in. What is the perimeter of the inscribed square $ABCD$?

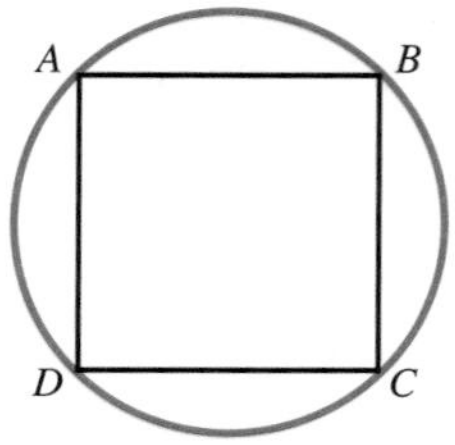

55. ***Value of a Sum*** In the circle shown with center O, the radius is 6. $QTSR$ is an inscribed square. Find the value of $PQ^2 + PT^2 + PR^2 + PS^2$.

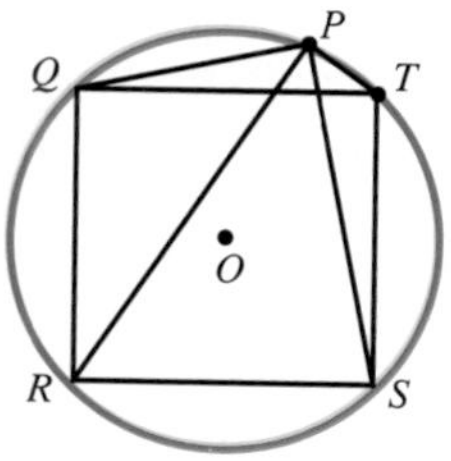

56. ***Ratio of Side Lengths*** The square $JOSH$ is inscribed in a semicircle. What is the ratio of x to y?

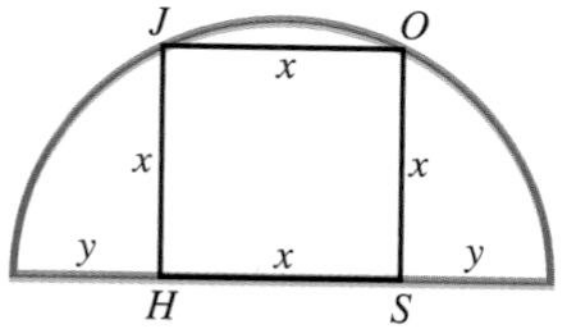

Euler's Formula *Many crystals and some viruses are constructed in the shapes of regular polyhedra. On the left in the next column is a polio virus, based on an icosahedron, and on the right is radiolara, a group of microorganisms, based on a tetrahedron.*

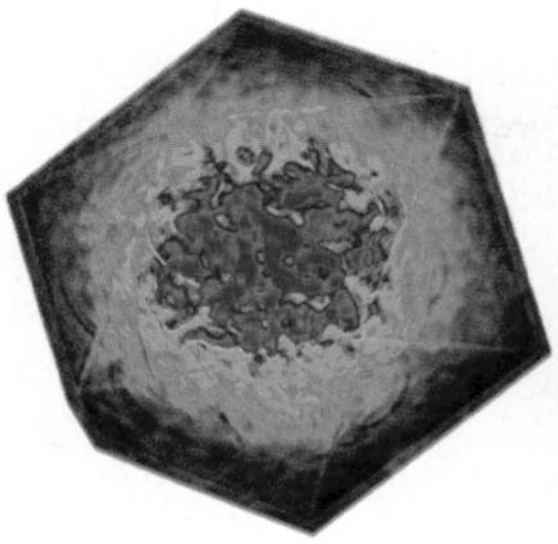

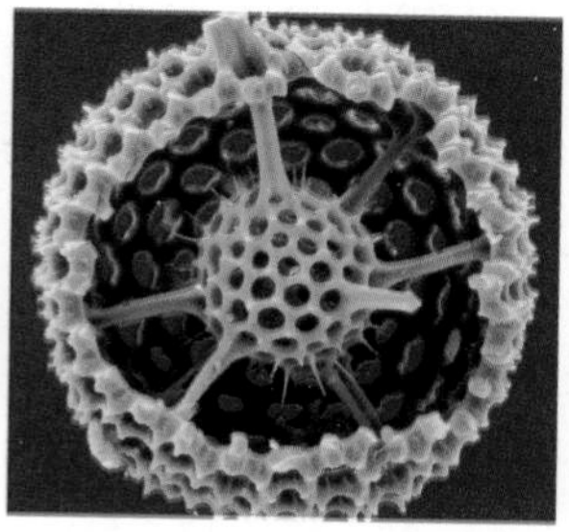

Leonhard Euler investigated a remarkable relationship among the numbers of faces (F), vertices (V) and edges (E) for the five regular polyhedra. Complete the chart in Exercises 57–61, and then draw a conclusion in Exercise 62.

	Polyhedron	Faces (F)	Vertices (V)	Edges (E)	Value of $F + V - E$
57.	Tetrahedron				
58.	Hexahedron (Cube)				
59.	Octahedron				
60.	Dodecahedron				
61.	Icosahedron				

62. Euler's formula is $F + V - E =$ ______.

9.6 Transformational Geometry

Reflections • Translations and Rotations • Size Transformations

There are many branches of geometry. In this chapter we have studied concepts of Euclidean geometry. One particular branch of geometry, known as **transformational geometry,** investigates how one geometric figure can be transformed into another. In transformational geometry we are required to reflect, rotate, and change the size of figures using concepts that we now discuss.

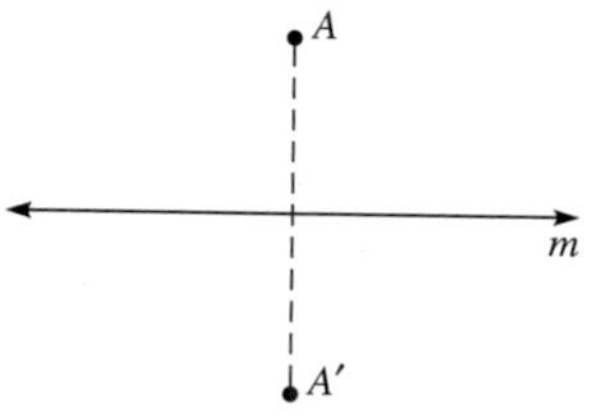

FIGURE 60

Reflections

One way to transform one geometric figure into another is by reflection. In Figure 60, line m is perpendicular to the line segment AA' and also bisects this line segment. We call point A' the **reflection image** of point A about line m. Line m is called the **line of reflection** for points A and A'. In the figure, we use a dashed line to connect points A and A' to show that these two points are images of each other under this transformation.

Point A' is the reflection image of point A only for line m; if a different line were used, A would have a different reflection image. Think of the reflection image of a point A about a line m as follows: Place a drop of ink at point A, and fold the paper along line m. The spot made by the ink on the other side of m is the reflection image of A. If A' is the image of A about line m, then A is the image of A' about the same line m.

To find the reflection image of a figure, find the reflection image of each point of the figure. The set of all reflection images of the points of the original figure is called the **reflection image** of the figure. Figure 61 shows several figures (in black) and their reflection images (in color) about the lines shown.

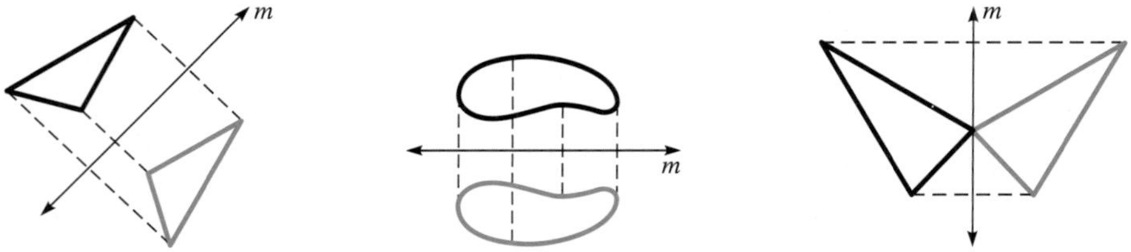

FIGURE 61

By the definition of reflection given above, each point in a plane has exactly one reflection image point with respect to a given line of reflection. Also, each reflection image point has exactly one original point. Thus, two distinct points cannot have the same reflection image. This means there is a *1-to-1 correspondence* between the set of points of the plane and the image points with respect to a given line of reflection. Any operation, such as reflection, in which there is a 1-to-1 correspondence between the points of the plane and their image points is called a **transformation;** we can call reflection about a line the **reflection transformation.**

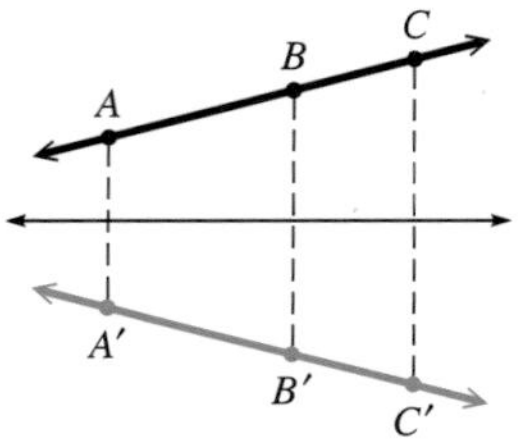

FIGURE 62

If a point A and its image, A', under a certain transformation are the same point, then point A is called an **invariant point** of the transformation. The only invariant points of the reflection transformation are the points of the line of reflection.

Three points that lie on the same straight line are called **collinear.** In Figure 62, points A, B, and C are collinear, and it can be shown that the reflection images A', B', and C', are also collinear. Thus, the reflection image of a line is also a line. We express this by saying that **reflection preserves collinearity.**

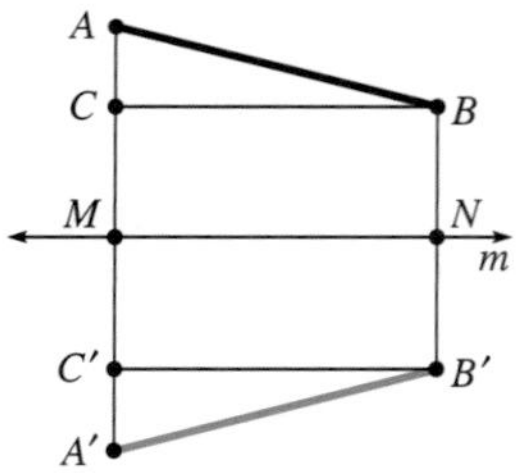

FIGURE 63

Distance is also preserved by the reflection transformation. Thus, in Figure 63, the distance between points A and B, written $|AB|$, is equal to the distance between the reflection images A' and B', or

$$|AB| = |A'B'|.$$

To prove this, we can use the definition of reflection image to verify that $|AM| = |MA'|$, and $|BN| = |NB'|$. Construct segments CB and $C'B'$, each perpendicular to BB'. Note that $CBB'C'$ is a rectangle. Because the opposite sides of a rectangle are equal and parallel, we have

$$|CB| = |C'B'|. \qquad \text{(Side)} \qquad \textbf{(1)}$$

Because $CBB'C'$ is a rectangle, we can also say

$$m\measuredangle ACB = m\measuredangle A'C'B' = 90° \qquad \text{(Angle)} \qquad \textbf{(2)}$$

where we use $m\measuredangle ACB$ to represent the measure of angle ACB.

This playful little beagle, D'Artagnan, is owned by Finley Westmoreland. Notice that the markings on his face are **symmetric with respect to a vertical line** through his cold, wet, little nose.

We know $|AM| = |MA'|$ and can show $|CM| = |MC'|$, so that

$$|AC| = |A'C'|. \qquad \text{(Side)} \qquad \textbf{(3)}$$

From statements (1), (2), and (3) above, we conclude that in triangles ABC and $A'B'C'$, two sides and the included angle of one are equal in measure to the corresponding two sides and angle of the other, and thus are congruent by SAS (Section 9.4). Because corresponding sides of congruent triangles are equal in length, we have

$$|AB| = |A'B'|,$$

which is what we wanted to show. Hence, the distance between two points equals the distance between their reflection images, and thus, reflection preserves distance. (The proof we have given is not really complete, because we have tacitly assumed that AB is not parallel to $A'B'$, and that A and B are on the same side of the line of reflection. Some modification would have to be made in the proof above to include these other cases.)

An example of a **reflection**.

The figures shown in Figure 64 are their own reflection images about the lines of reflection shown. In this case, the line of reflection is called a **line of symmetry** for the figure. Figure 64(a) has three lines of symmetry. A circle has every line through its center as a line of symmetry.

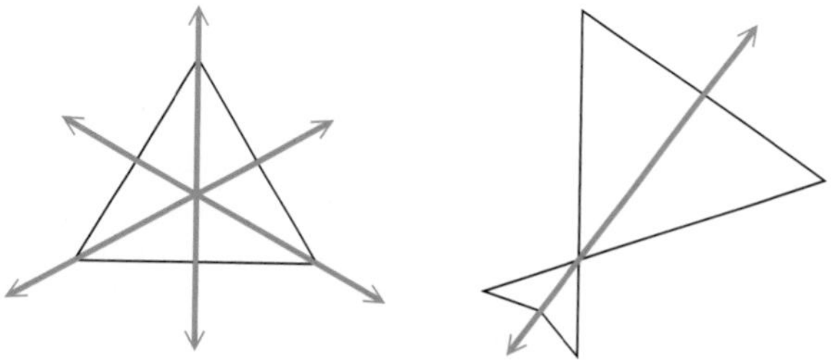

(a) **(b)**

FIGURE 64

Translations and Rotations We shall use the symbol r_m to represent a reflection about line m, and let us use $r_n \cdot r_m$ to represent a reflection about line m followed by a reflection about line n. We call $r_n \cdot r_m$ the **composition,** or **product,** of the two reflections r_n and r_m. Figure 65 on the next page shows two examples of the composition of two reflections. In Figure 65(a), lines m and n are parallel, while they intersect in Figure 65(b).

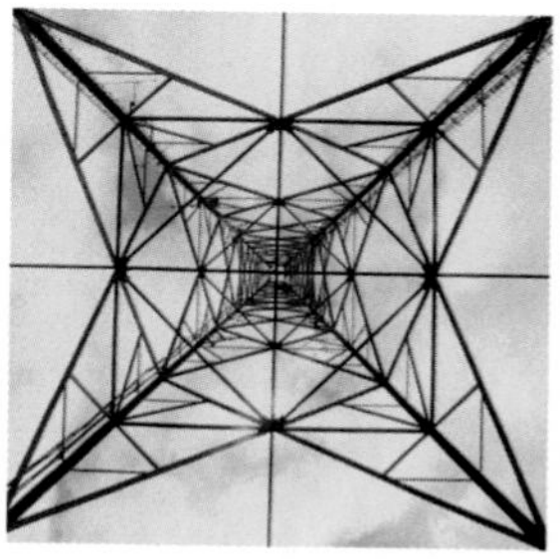

How many kinds of **symmetry** do you see here?

In Figure 65(a) both the original figure and its image under the composition of the two reflections appear to be oriented the same way and to have the same "tilt." In fact, it appears that the original figure could be slid along the dashed lines of Figure 65(a), with no rotation, so as to cover the image. This composite transformation is called a **translation.** Figure 66, also on the next page, shows a translation, and the image can be obtained as a composition of two reflections about parallel lines. Check that the distance between a point and its image under a translation is twice the distance between the two parallel lines. The distance between a point and its image under a translation is called the **magnitude** of the translation.

A translation of magnitude 0 leaves every point of the plane unchanged, and thus is called the **identity translation.** A translation of magnitude k, followed by a similar translation of magnitude k but of opposite direction, returns a point to its original

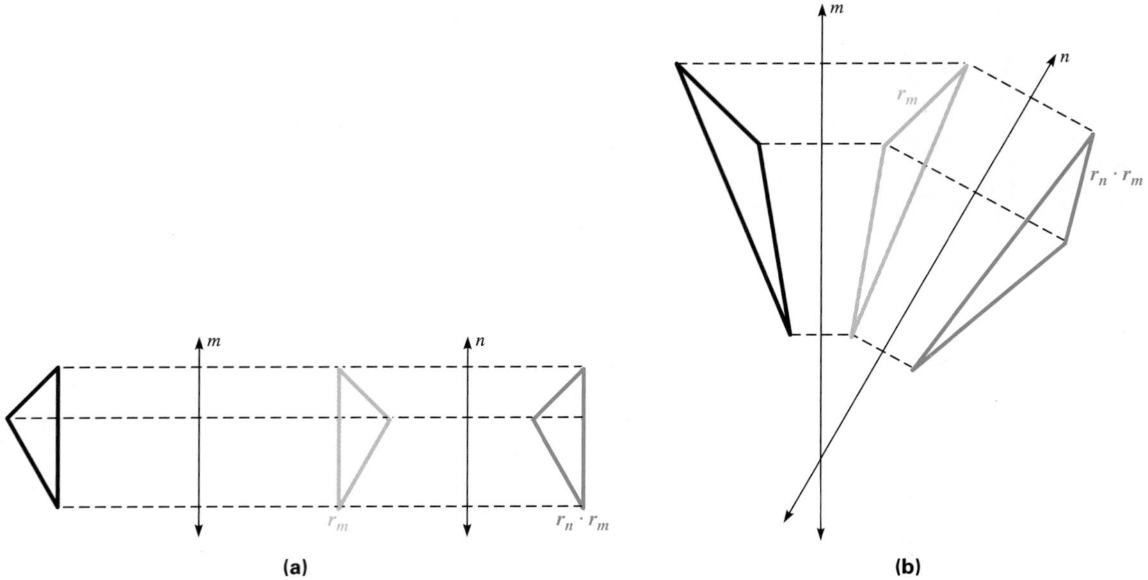

FIGURE 65

position, and thus, these two translations are called **inverses** of each other. Check that there are no invariant points in a translation of magnitude $k > 0$.

A translation preserves collinearity (three points on the same line have image points that also lie on a line) and distance (the distance between two points is the same as the distance between the images of the points).

In Figure 65(b), the original figure could be rotated so as to cover the image. Hence, we call the composition of two reflections about nonparallel lines a **rotation.** The point of intersection of these two nonparallel lines is called the **center of rotation.** The black triangle of Figure 67 was reflected about line m and then reflected about line n, resulting in a rotation with center at B. The dashed lines in color represent the paths of the vertices

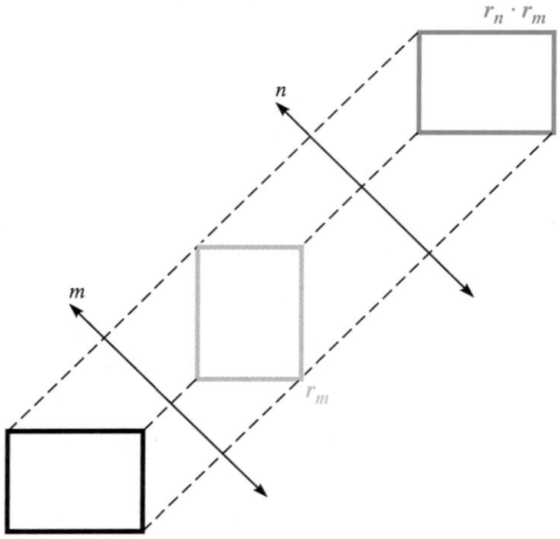

FIGURE 66

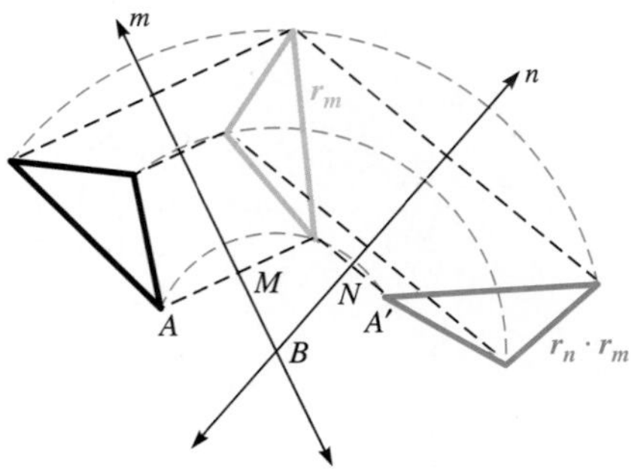

FIGURE 67

of the triangle under the rotation. It is shown in the exercises that $m\measuredangle ABA'$ is twice as large as $m\measuredangle MBN$. The measure of angle ABA' is called the **magnitude** of the rotation.

Rotations also preserve collinearity and distance. The identity transformation here is a rotation of 0° or 360°, and rotations of, say, 240° and 120° (or in general, $x°$ and $360° - x°$, $0 \leq x \leq 360°$) are inverses of each other. The center of rotation is the only invariant point of any rotation except the identity rotation.

We have defined rotations as the composition of two reflections about nonparallel lines of reflection. We could also define a rotation by specifying its center, the angle of rotation, and a direction of rotation, as shown by the following example.

EXAMPLE 1 Finding an Image Under a Rotation

Find the image of a point P under a rotation transformation having center at a point Q and magnitude 135° clockwise.

SOLUTION

To find P', the image of P, first draw angle PQM having measure 135°. Then draw an arc of a circle with center at Q and radius $|PQ|$. The point where this arc intersects side QM is P'. See Figure 68. ■

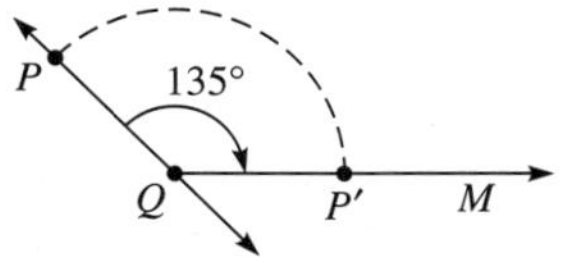

FIGURE 68

Figure 69 shows a rotation transformation having center Q and magnitude 180° clockwise. Point Q bisects the line segment from a point A to its image A', and for this reason this rotation is sometimes called a **point reflection.**

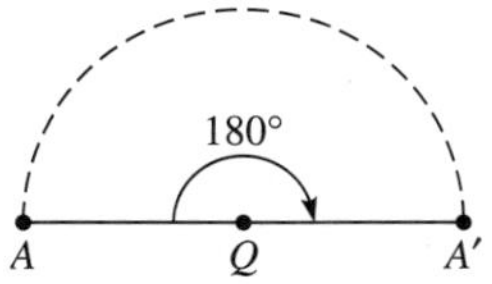

FIGURE 69

EXAMPLE 2 Finding Point Reflections

Find the point reflection images about point Q for each of the following figures.

(a)

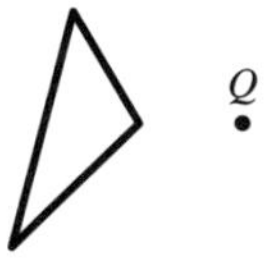

(b)

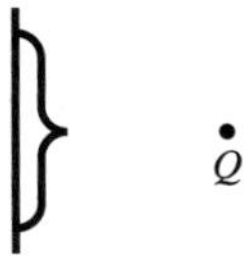

SOLUTION

The point reflection images are shown in color.

(a)

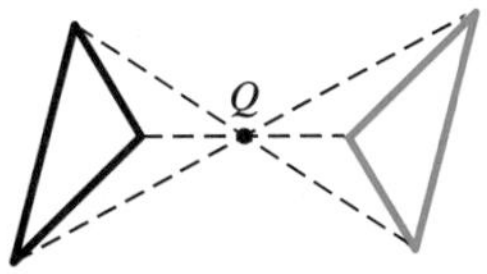

(b)

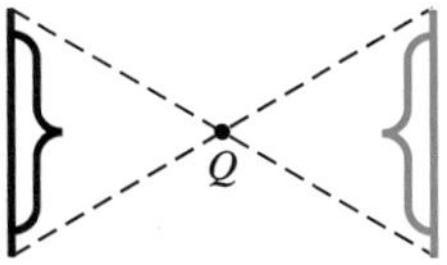

■

This creature exhibits **bilateral symmetry.** This kind of symmetry is often found in living organisms.

Let r_m be a reflection about line m, and let T be a translation having nonzero magnitude and a direction parallel to m. Then the composition of T and r_m is called a **glide reflection,** as seen in Figure 70 on the next page. Here a reflection followed by a translation is the same as a translation followed by a reflection, so that in this case

$$T \cdot r_m = r_m \cdot T.$$

Because a translation is the composition of *two* reflections, a glide reflection is the composition of *three* reflections. Because it is required that the translation have nonzero magnitude, there is no identity glide transformation.

M.C. Escher (1898–1972) was a Dutch graphic artist, most recognized for spatial illusions, impossible buildings, repeating geometric patterns (tessellations), and his incredible techniques in woodcutting and lithography.

Escher was a man studied and greatly appreciated by respected mathematicians, scientists, and crystallographers, yet he had no formal training in math or science. He was a humble man who considered himself neither an artist nor a mathematician.

Intricate repeating patterns, mathematically complex structures, and spatial perspectives all require a 'second look.' In Escher's work what you see the first time is most certainly not all there is to see." (*Sources:* M. C. Escher's Symmetry Drawing (Smaller and Smaller) and Waterfall © 2003 Cordon Art B.V., Baarn, Holland. All rights reserved; www.worldofescher.com.)

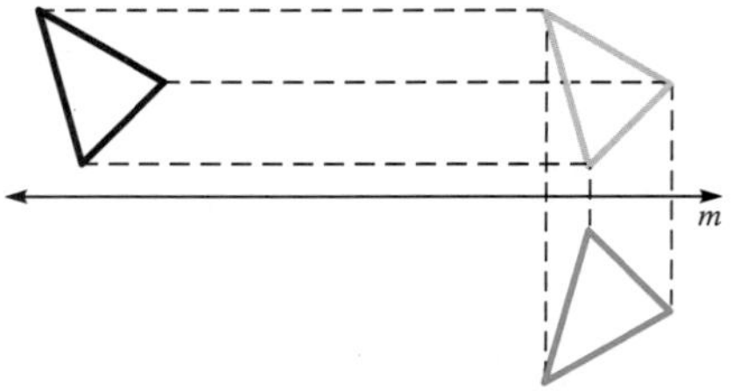

FIGURE 70

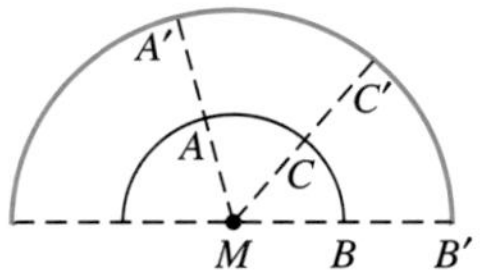

FIGURE 71

All the transformations of this section discussed so far are **isometries,** or transformations in which the image of a figure has the same size and shape as the original figure. Any isometry is either a reflection or the composition of two or more reflections.

Size Transformations

Figure 71 shows a semicircle in black, a point M, and an image semicircle in color. Distance $A'M$ is twice the distance AM and distance $B'M$ is twice the distance BM. In fact, all the points of the image semicircle, such as C', were obtained by drawing a line through M and C, and then locating C' such that $|MC'| = 2|MC|$. Such a transformation is called a **size transformation** with center M and magnitude 2. We shall assume that a size transformation can have any positive real number k as magnitude. A size transformation having magnitude $k > 1$ is called a **dilatation,** or **stretch;** while a size transformation having magnitude $k < 1$ is called a **contraction,** or **shrink.**

EXAMPLE 3 Applying Size Transformations

Apply a size transformation with center M and magnitude $\frac{1}{3}$ to the two triangles shown in black in Figure 72.

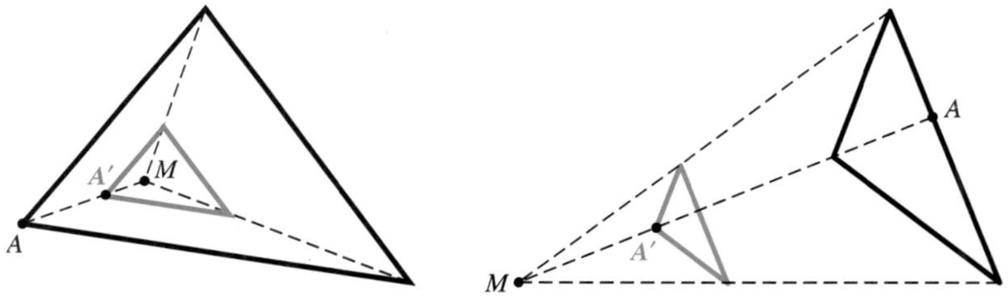

FIGURE 72

SOLUTION

To find the images of these triangles, we can find the image points of some sample points. For example, if we select point A on each of the original triangles, we can find the image points by drawing a line through A and M, and locating a point A' such that $|MA'| = \frac{1}{3}|MA|$. By doing this for all points of each of the black triangles, we get the images shown in color in Figure 72.

The identity transformation is a size transformation of magnitude 1, while size transformations of magnitude k and $\frac{1}{k}$, having the same center, are inverses of each other. The only invariant point of a size transformation of magnitude $k \neq 1$ is the center of the transformation.

EXAMPLE 4 Investigating Size Transformations

Does a size transformation **(a)** preserve collinearity? **(b)** preserve distance?

SOLUTION

(a) Figure 73 shows three collinear points, A, B, and C, and their images under two different size transformations with center at M: one of magnitude 3 and one of magnitude $\frac{1}{3}$. In each case the image points appear to be collinear, and it can be proved that they are, using similar triangles. In fact, the image of a line not through the center of the transformation is a line parallel to the original line.

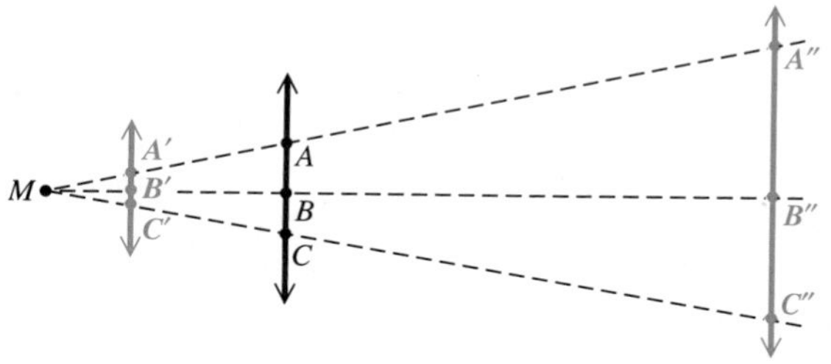

FIGURE 73

(b) As shown in Figure 73, $|AB| \neq |A'B'|$. Thus, a size transformation of magnitude $k \neq 1$ does not preserve distance and is not an isometry.

TABLE 4 Summary of Transformations

	Reflection	Translation	Rotation	Glide Reflection	Size Transformation
Example	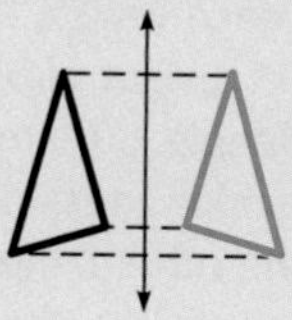	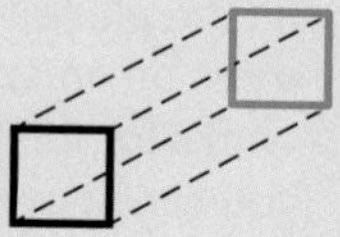	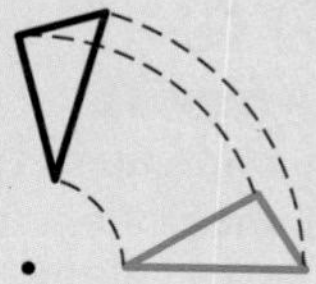	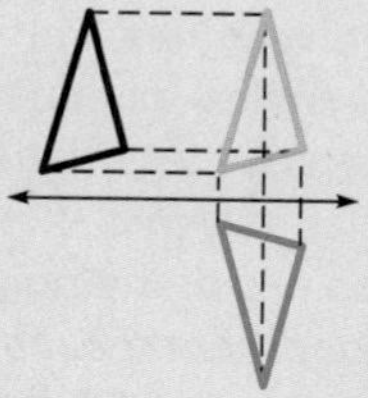	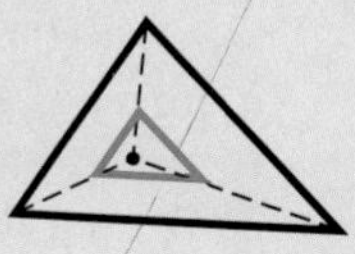
Preserve collinearity?	Yes	Yes	Yes	Yes	Yes
Preserve distance?	Yes	Yes	Yes	Yes	No
Identity transformation?	None	Magnitude 0	Magnitude 360°	None	Magnitude 1
Inverse transformation?	None	Same magnitude; opposite direction	Same center; magnitude $360° - x°$	None	Same center; magnitude $\frac{1}{k}$
Composition of n reflections?	$n = 1$	$n = 2$, parallel	$n = 2$, nonparallel	$n = 3$	No
Isometry?	Yes	Yes	Yes	Yes	No
Invariant points?	Line of reflection	None	Center of rotation	None	Center of transformation

For Further Thought

Tessellations

The authors wish to thank Suzanne Alejandre for permission to reprint this article on tessellations, which first appeared at www.mathforum.org/sum95/suzanne/whattess.html.

tessellate (verb), **tessellation** (noun): from Latin *tessera* "a square tablet" or "a die used for gambling." Latin tessera may have been borrowed from Greek *tessares,* meaning "four," since a square tile has four sides. The diminutive of *tessera* was *tessella,* a small, square piece of stone or a cubical tile used in mosaics. Since a mosaic extends over a given area without leaving any region uncovered, the geometric meaning of the word tessellate is "to cover the plane with a pattern in such a way as to leave no region uncovered." By extension, space or hyperspace may also be tessellated.

Definition

A dictionary will tell you that the word "tessellate" means to form or arrange small squares in a checkered or mosaic pattern. The word "tessellate" is derived from the Ionic version of the Greek word "tesseres," which in English means "four." The first tilings were made from square tiles.

A regular polygon has 3 or 4 or 5 or more sides and angles, all equal. A **regular tessellation** means a tessellation made up of congruent regular polygons. [Remember: *Regular* means that the sides of the polygon are all the same length. *Congruent* means that the polygons that you put together are all the same size and shape.]

Only three regular polygons tessellate in the Euclidean plane: triangles, squares, or hexagons. We can't show the entire plane, but imagine that these are pieces taken from planes that have been tiled. Here are examples of

a tessellation of triangles

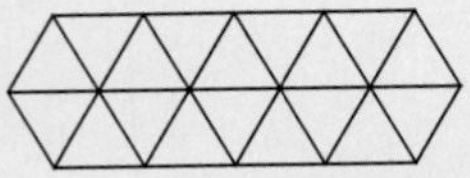

a tessellation of squares

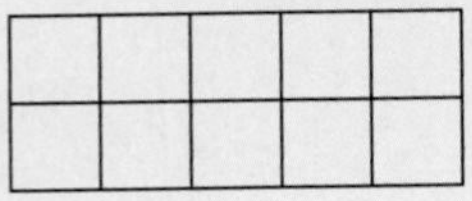

a tessellation of hexagons

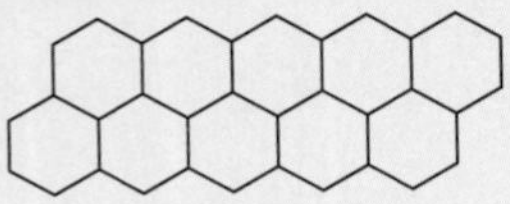

When you look at these three samples you can easily notice that the squares are lined up with each other while the triangles and hexagons are not. Also, if you look at six triangles at a time, they form a hexagon, so the tiling of triangles and the tiling of hexagons are similar and they cannot be formed by directly lining shapes up under each other—a slide (or a glide!) is involved.

You can work out the interior measure of the angles for each of these polygons:

Shape	Angle Measure in Degrees
triangle	60
square	90
pentagon	108
hexagon	120
more than six sides	more than 120 degrees

Since the regular polygons in a tessellation must fill the plane at each vertex, the interior angle must be an exact divisor of 360 degrees. This works for the triangle, square, and hexagon, and you can show working tessellations for these figures. For all the others, the interior angles are not exact divisors of 360 degrees, and therefore those figures cannot tile the plane.

Naming Conventions

A tessellation of squares is named "4.4.4.4." Here's how: choose a vertex, and then look at one of the polygons that touches that vertex. How many sides does it have?

Since it's a square, it has four sides, and that's where the first "4" comes from. Now keep going around the vertex in either direction, finding the number of sides of the polygons until you get back to the polygon you started with. How many polygons did you count?

There are four polygons, and each has four sides.

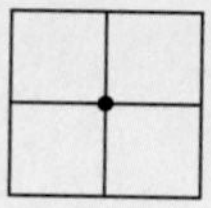

4.4.4.4

For a tessellation of regular congruent hexagons, if you choose a vertex and count the sides of the polygons that touch it, you'll see that there are three polygons and each has six sides, so this tessellation is called "6.6.6":

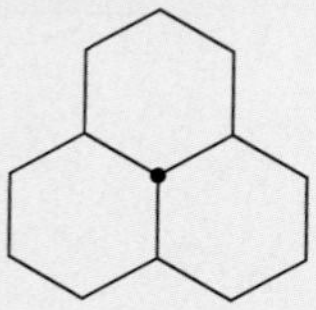

6.6.6

A tessellation of triangles has six polygons surrounding a vertex, and each of them has three sides: "3.3.3.3.3.3."

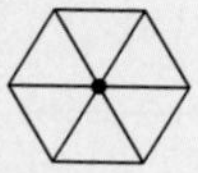

3.3.3.3.3.3

Semi-regular Tessellations

You can also use a variety of regular polygons to make **semi-regular tessellations.**

A semi-regular tessellation has two properties, which are:

1. It is formed by regular polygons.
2. The arrangement of polygons at every vertex point is identical.

Here are the **eight** semi-regular tessellations:

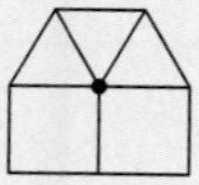

3.3.3.4.4

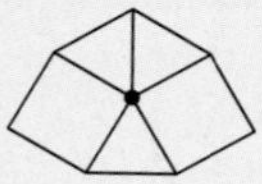

3.3.4.3.4

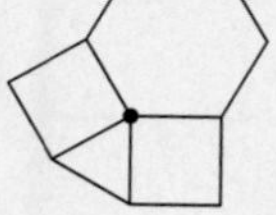

3.4.6.4

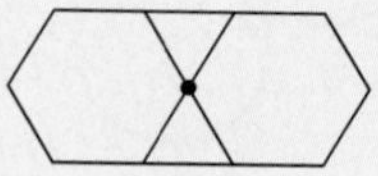

3.6.3.6

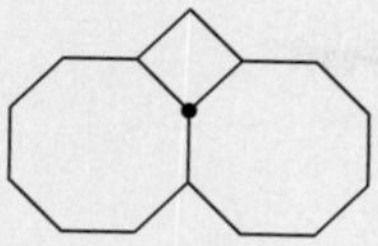

4.8.8

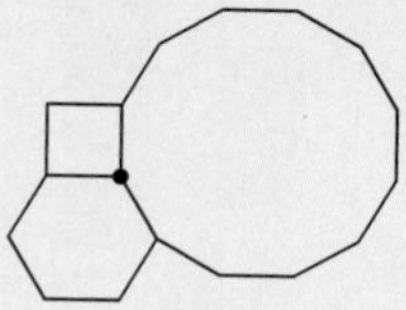

4.6.12

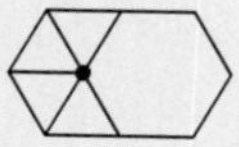

3.3.3.3.6

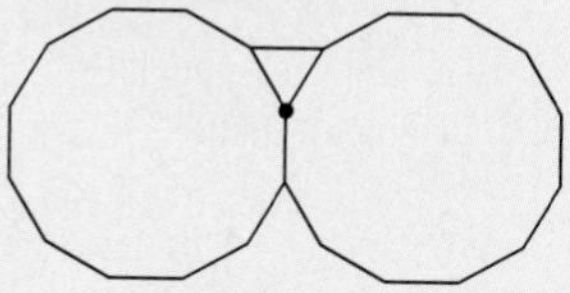

3.12.12

Interestingly there are other combinations that seem like they should tile the plane because the arrangements of the regular polygons fill the space around a point. For example:

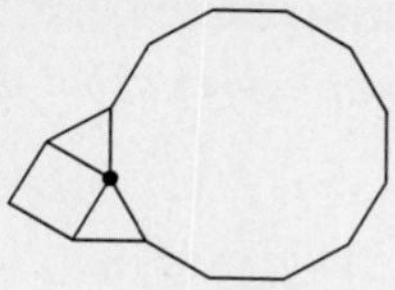

3.4.3.12

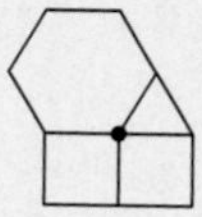

3.4.4.6

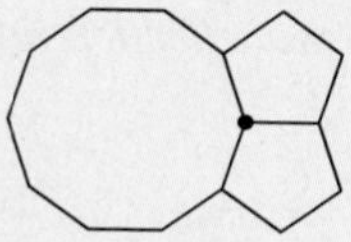

5.5.10

If you try tiling the plane with these units of tessellation you will find that they cannot be extended infinitely.

There is an infinite number of tessellations that can be made of patterns that do not have the same combination of angles at every vertex point. There are also tessellations made of polygons that do not share common edges and vertices.

For Group Discussion or Individual Investigation

1. Use the naming conventions to name each of these semi-regular tessellations.

(a)

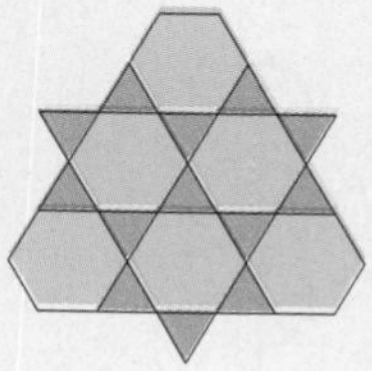

(b)

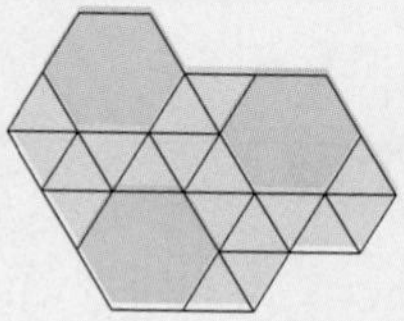

(continued)

(c)

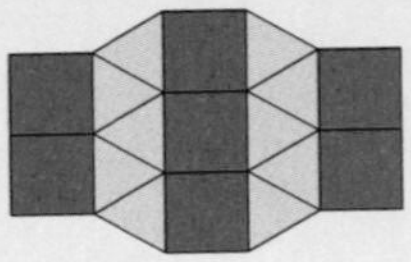

(d)

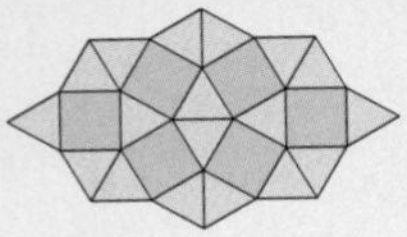

(g)

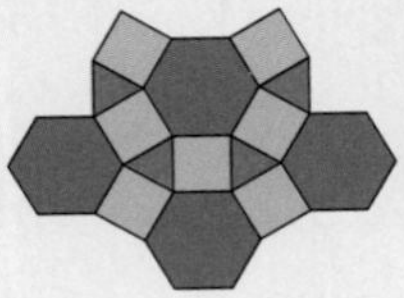

(h)

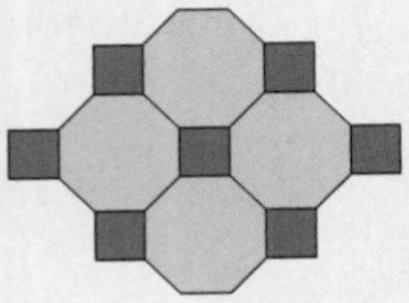

(e)

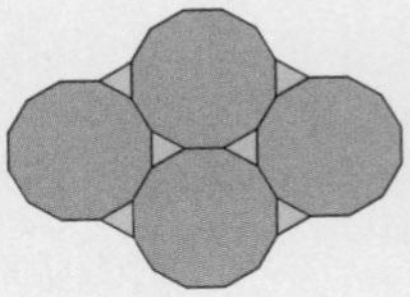

(f)

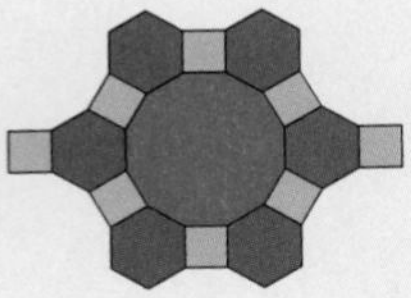

2. Why isn't this a semi-regular tessellation?

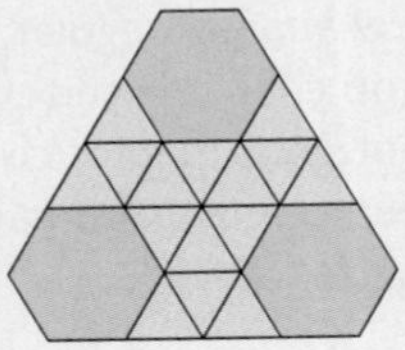

9.6 EXERCISES

Find the reflection images of the given figures about the given lines.

1.

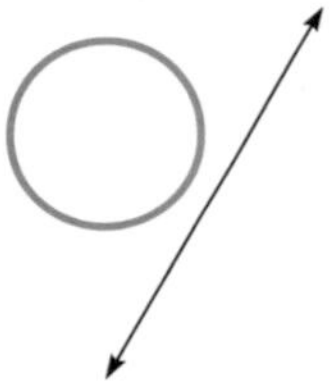

2.

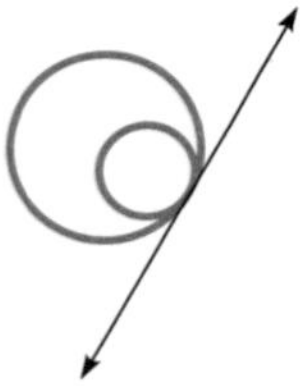

3.

4.

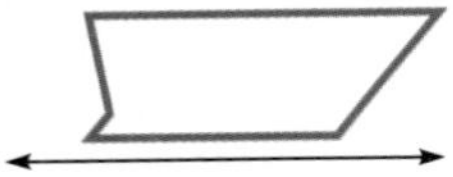

5.

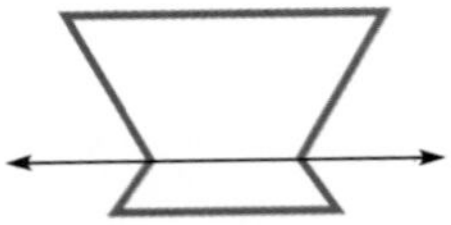

6.

7.

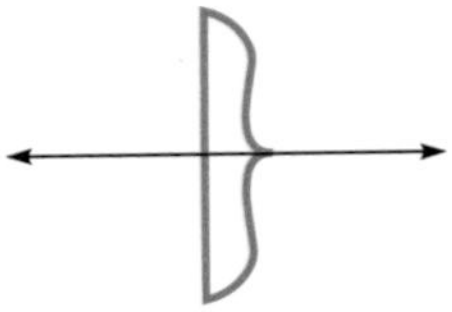

8.

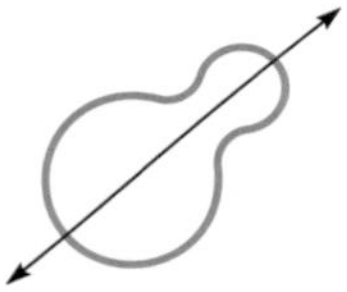

Find any lines of symmetry of the given figures.

9.

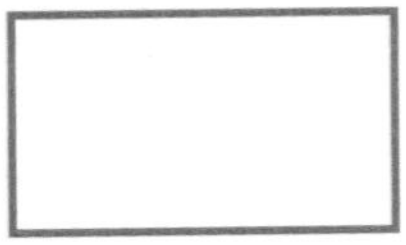

10.

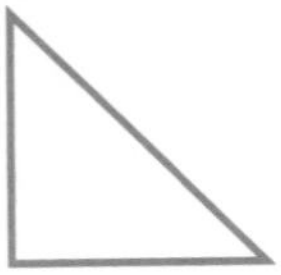

11.

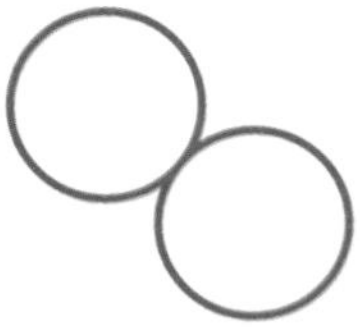

12.

First reflect the given figure about line m. Then reflect about line n.

13.

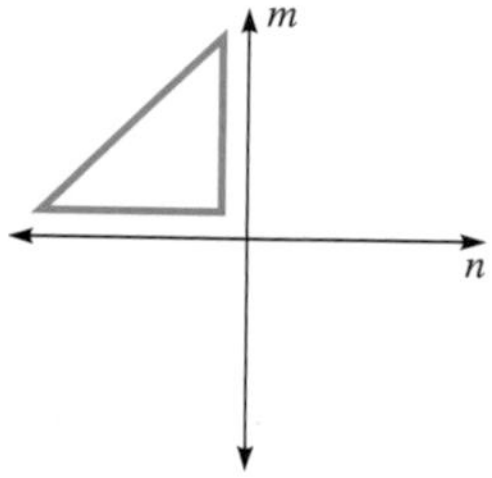

14.

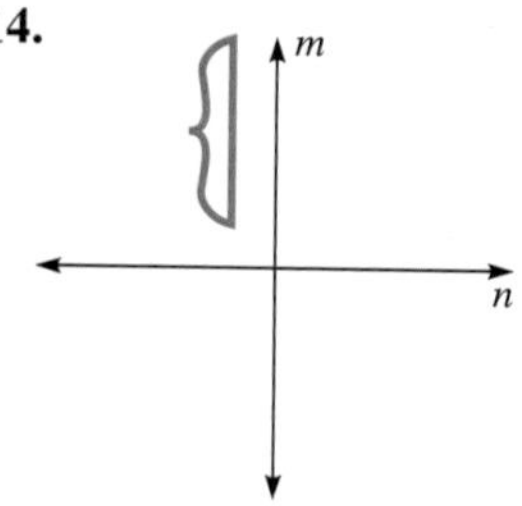

15.

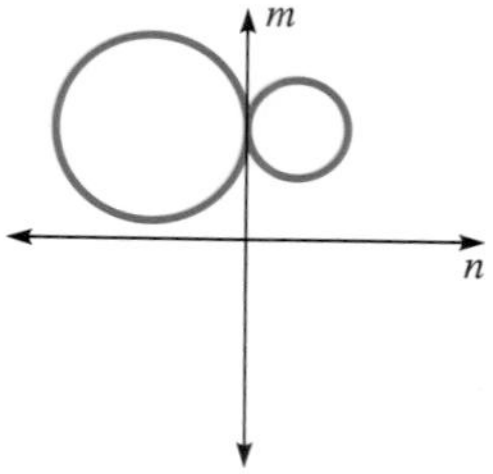

16.

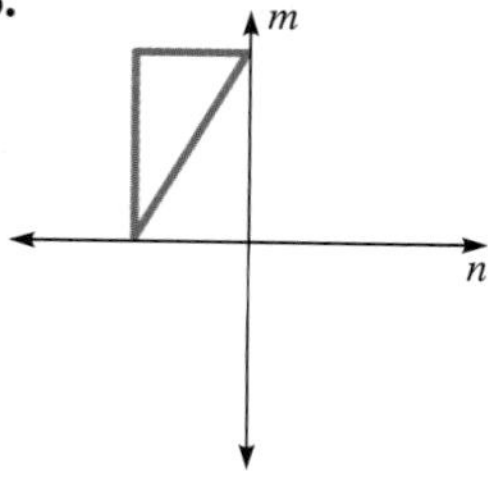

17.

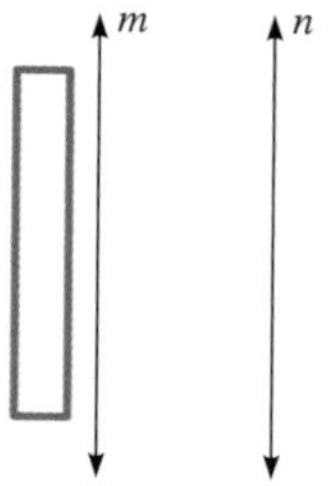

18.

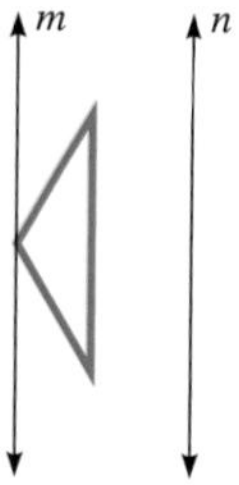

19.

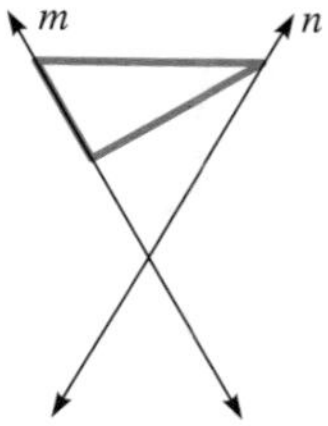

20.

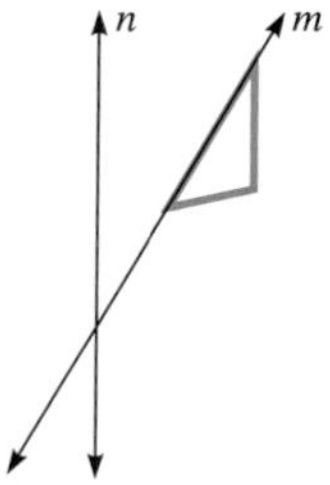

In Exercises 21–34, let T be a translation having magnitude $\frac{3}{4}$ inch to the right in a direction parallel to the bottom edge of the page. Let r_m be a reflection about line m, and let R_p be a rotation about point P having magnitude 60° clockwise.

In each of Exercises 21–32, perform the given transformations on point A of the figure below to obtain final image point A′.

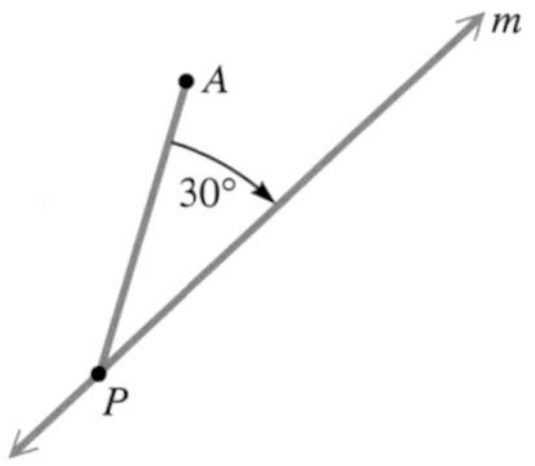

21. r_m

22. R_p

23. T

24. $r_m \cdot r_m$

25. $T \cdot T$

26. $R_p \cdot R_p$

27. $T \cdot R_p$

28. $T \cdot r_m$

29. $r_m \cdot T$

30. $R_P \cdot r_m$

31. $r_m \cdot R_P$

32. $R_P \cdot T$

33. Is $T \cdot r_m$ a glide reflection here?

34. Is $T \cdot r_m = r_m \cdot T$ true?

35. Suppose a rotation is given by $r_m \cdot r_n$, as shown in the figure below. Find the images of A, B, and C.

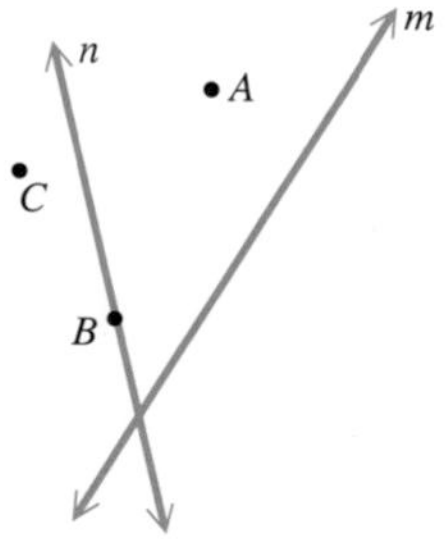

36. Does a glide reflection preserve
(a) collinearity?
(b) distance?

Find the point reflection images of each of the following figures with the given points as center.

37.

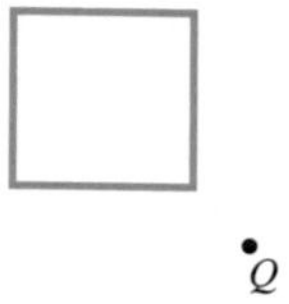

38.

39.

40.

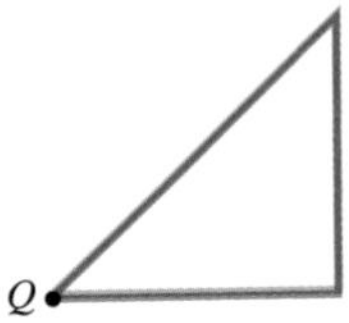

Perform the indicated size transformation.

41. magnitude 2; center M

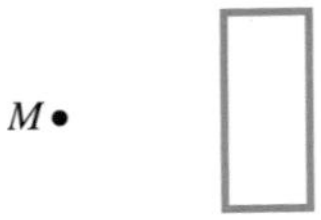

42. magnitude $\frac{1}{2}$; center M

43. magnitude $\frac{1}{2}$; center M

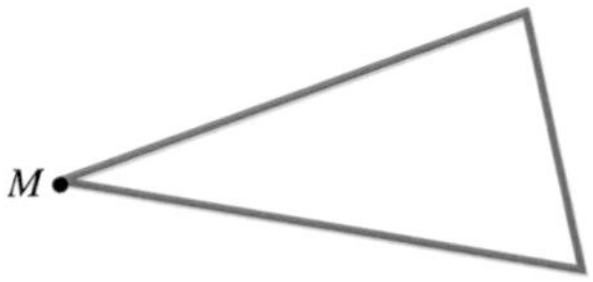

44. magnitude 2; center M

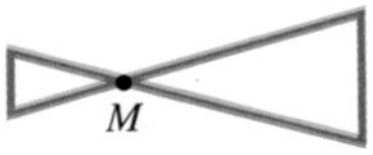

45. magnitude $\frac{1}{3}$; center M

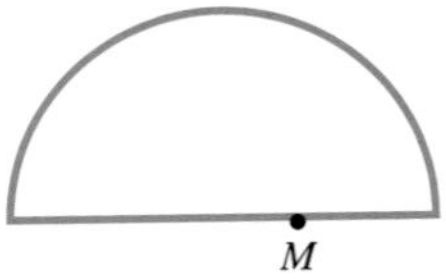

46. magnitude $\frac{1}{3}$; center M

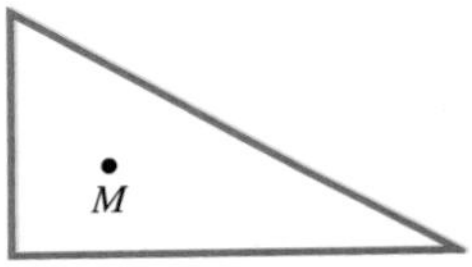

9.7 Non-Euclidean Geometry, Topology, and Networks

Euclid's Postulates and Axioms • The Parallel Postulate (Euclid's Fifth Postulate) • The Origins of Non-Euclidean Geometry • Topology • Networks

Euclid's Postulates and Axioms The *Elements* of Euclid is quite possibly the most influential mathematics book ever written. (See the margin note at the beginning of this chapter on page 530.) It begins with definitions of basic ideas such as point, line, and plane. Euclid then gives five postulates providing the foundation of all that follows.

Next, Euclid lists five axioms that he views as general truths and not just facts about geometry. See Table 5 on the next page. (To some of the Greek writers, postulates were truths about a particlar field, while axioms were general truths. Today, "axiom" is used in either case.)

Using only these ten statements and the basic rules of logic, Euclid was able to prove a large number of "propositions" about geometric figures.

John Playfair (1748–1819) wrote his *Elements of Geometry* in 1795. Playfair's Axiom is: Given a line k and a point P not on the line, there exists one and only one line m through P that is parallel to k. Playfair was a geologist who fostered "uniformitarianism," the doctrine that geological processes long ago gave Earth its features, and processes today are the same kind as those in the past.

TABLE 5

Euclid's Postulates	Euclid's Axioms
1. Two points determine one and only one straight line.	**6.** Things equal to the same thing are equal to each other.
2. A straight line extends indefinitely far in either direction.	**7.** If equals are added to equals, the sums are equal.
3. A circle may be drawn with any given center and any given radius.	**8.** If equals are subtracted from equals, the remainders are equal.
4. All right angles are equal.	**9.** Figures that can be made to coincide are equal.
5. Given a line k and a point P not on the line, there exists one and only one line m through P that is parallel to k.	**10.** The whole is greater than any of its parts.

The statement for Postulate 5 given above is actually known as Playfair's axiom on parallel lines, which is equivalent to Euclid's fifth postulate. To understand why this postulate caused trouble for so many mathematicians for so long, we must examine the original formulation.

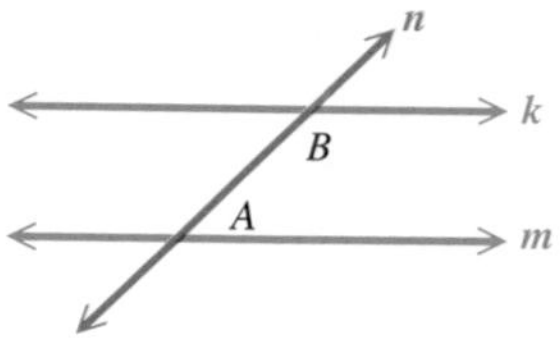

FIGURE 74

The Parallel Postulate (Euclid's Fifth Postulate) In its original form, Euclid's fifth postulate states that if two lines (k and m in Figure 74) are such that a third line, n, intersects them so that the sum of the two interior angles (A and B) on one side of line n is less than (the sum of) two right angles, then the two lines, if extended far enough, will meet on the same side of n that has the sum of the interior angles less than (the sum of) two right angles.

Euclid's parallel postulate is quite different from the other nine postulates and axioms we listed. The others are simple statements that seem in complete agreement with our experience of the world around us. But the parallel postulate is long and wordy, and difficult to understand without a sketch.

The difference between the parallel postulate and the other axioms was noted by the Greeks, as well as later mathematicians. It was commonly believed that this was not a postulate at all, but a theorem to be proved. For more than 2000 years mathematicians tried repeatedly to prove it.

The most dedicated attempt came from an Italian Jesuit, Girolamo Saccheri (1667–1733). He attempted to prove the parallel postulate in an indirect way, by so-called "reduction to absurdity." He would assume the postulate to be false and then show that the assumption leads to a contradiction of something true (an absurdity). Such a contradiction would thus prove the statement true.

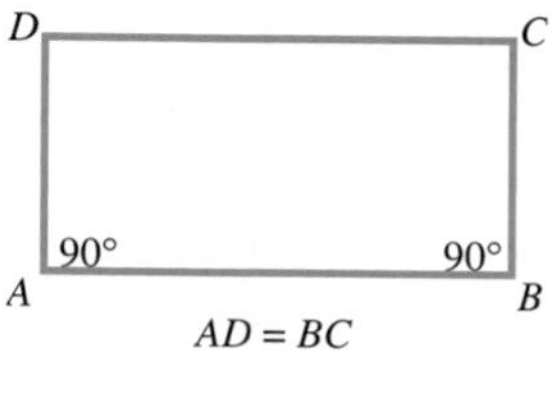

FIGURE 75

Saccheri began with a quadrilateral, as in Figure 75. He assumed angles A and B to be right angles and sides AD and BC to be equal. His plan was as follows:

1. To assume that angles C and D are obtuse angles, and to show that this leads to a contradiction.
2. To assume that angles C and D are acute angles, and to show that this also leads to a contradiction.

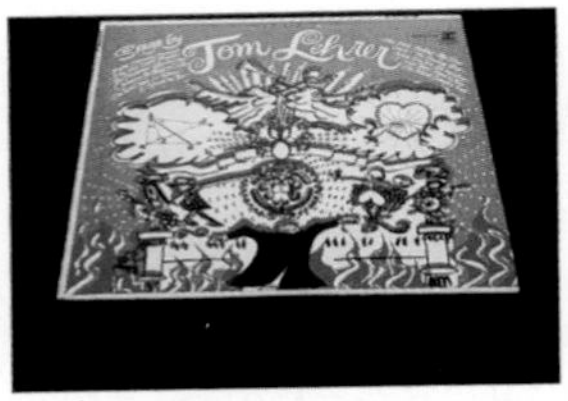

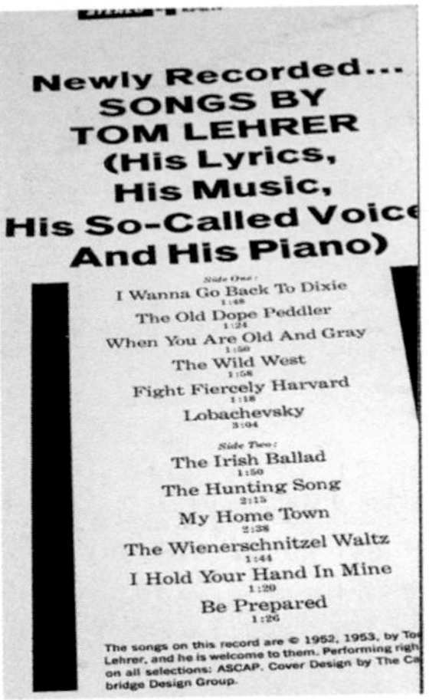

A song titled simply **Lobachevsky** first appeared on the LP *Songs by Tom Lehrer* (Reprise RS-6216). The liner notes indicate that the song "is a description of one way to get ahead in mathematics (which happens to be the author's own academic specialty) or any other academic field." Find a copy of the album, or the compact disc *The Remains of Tom Lehrer*, and see what Lehrer suggests!

3. Then if C and D can be neither acute nor obtuse angles, they must be right angles.
4. If C and D are both right angles, then it can be proved that the fifth postulate is true. It thus is a theorem rather than a postulate.

Saccheri had no trouble with part 1. However, he did not actually reach a contradiction in the second part, but produced some theorems so "repugnant" that he convinced himself he had vindicated Euclid. In fact, he published a book called in English *Euclid Freed of Every Flaw.*

Today we know that the fifth postulate is indeed an axiom, and not a theorem. It is *consistent* with Euclid's other axioms.

The ten axioms of Euclid describe the world around us with remarkable accuracy. We now realize that the fifth postulate is necessary in Euclidean geometry to establish *flatness.* That is, the axioms of Euclid describe the geometry of *plane surfaces.* By changing the fifth postulate, we can describe the geometry of other surfaces. So, other geometric systems exist as much as Euclidean geometry exists, and they can even be demonstrated in our world. They are just not as familiar. A system of geometry in which the fifth postulate is changed is called a **non-Euclidean geometry.**

The Origins of Non-Euclidean Geometry One non-Euclidean system was developed by three people working separately at about the same time. Early in the nineteenth century Carl Friedrich Gauss, one of the great mathematicians, worked out a consistent geometry replacing Euclid's fifth postulate. He never published his work, however, because he feared the ridicule of people who could not free themselves from habitual ways of thinking. Gauss first used the term "non-Euclidean." Nikolai Ivanovich Lobachevski (1793–1856) published a similar system in 1830 in the Russian language. At the same time, Janos Bolyai (1802–1860), a Hungarian army officer, worked out a similar system, which he published in 1832, not knowing about Lobachevski's work. Bolyai never recovered from the disappointment of not being the first, and did no further work in mathematics.

Lobachevski replaced Euclid's fifth postulate with:

Angles C and D in the quadrilateral of Saccheri are acute angles.

This postulate of Lobachevski can be rephrased as follows:

Through a point P off a line k (Figure 76), at least two different lines can be drawn parallel to k.

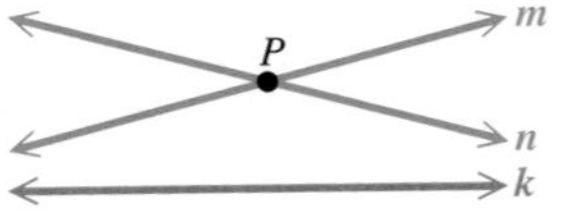

FIGURE 76

Compare this form of Lobachevski's postulate to the geometry of Euclid. How many lines can be drawn through P and parallel to k in Euclidean geometry? At first glance, the postulate of Lobachevski does not agree with what we know about the world around us. But this is only because we think of our immediate surroundings as being flat.

Many of the theorems of Euclidean geometry are valid for the geometry of Lobachevski, but many are not. For example, in Euclidean geometry, the sum of the

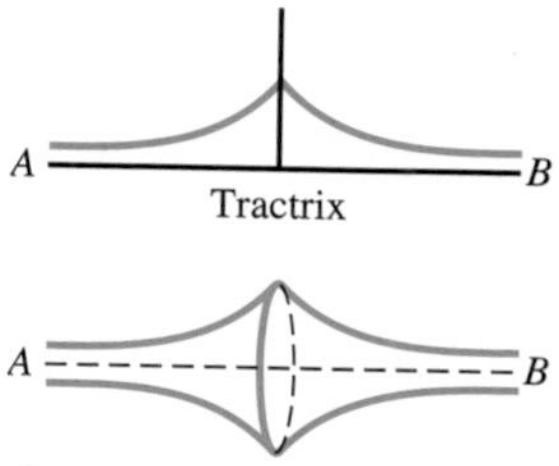

Rotating the tractrix produces the pseudosphere.

FIGURE 77

measures of the angles in any triangle is 180°. In Lobachevskian geometry, the sum of the measures of the angles in any triangle is *less* than 180°. Also, triangles of different sizes can never have equal angles, so similar triangles do not exist.

The geometry of Euclid can be represented on a plane. Since any portion of the earth that we are likely to see looks flat, Euclidean geometry is very useful for describing the everyday world around us. The non-Euclidean geometry of Lobachevski can be represented as a surface called a **pseudosphere.** This surface is formed by revolving a curve called a **tractrix** about the line AB in Figure 77.

A second non-Euclidean system was developed by Georg Riemann (1826–1866). He pointed out the difference between a line that continues indefinitely and a line having infinite length. For example, a circle on the surface of a sphere continues indefinitely but does not have infinite length. Riemann developed the idea of geometry on a sphere and replaced Euclid's fifth postulate with:

Angles C and D of the Saccheri quadrilateral are obtuse angles.

In terms of parallel lines, Riemann's postulate becomes:

Through a point P off a line k, no line can be drawn that is parallel to k.

Riemannian geometry is important in navigation. "Lines" in this geometry are really *great circles,* or circles whose centers are at the center of the sphere. The shortest distance between two points on a sphere lies along an arc of a great circle. Great circle routes on a globe don't look at all like the shortest distance when the globe is flattened out to form a map, but this is part of the distortion that occurs when the earth is represented as a flat surface. See Figure 78. The sides of a triangle drawn on a sphere would be arcs of great circles. And, in Riemannian geometry, the sum of the measures of the angles in any triangle is *more* than 180°.

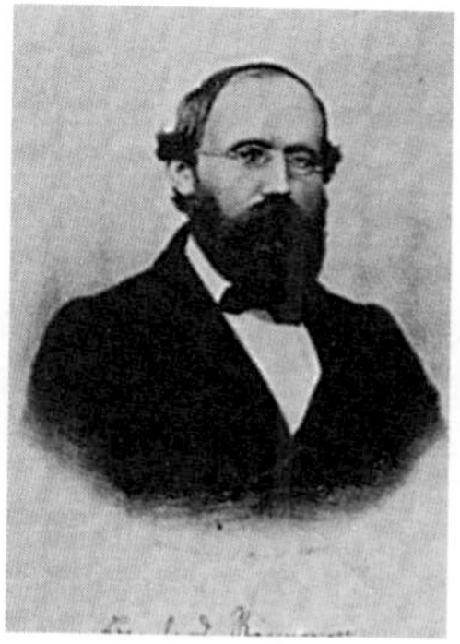

Georg Friedrich Bernhard Riemann (1826–1866) was a German mathematician. Though he lived a short time and published few papers, his work forms a basis for much modern mathematics. He made significant contributions to the theory of functions and the study of complex numbers as well as to geometry. Most calculus books today use the idea of a "Riemann sum" in defining the integral.

Riemann achieved a complete understanding of the non-Euclidean geometries of his day, expressing them on curved surfaces and showing how to extend them to higher dimensions.

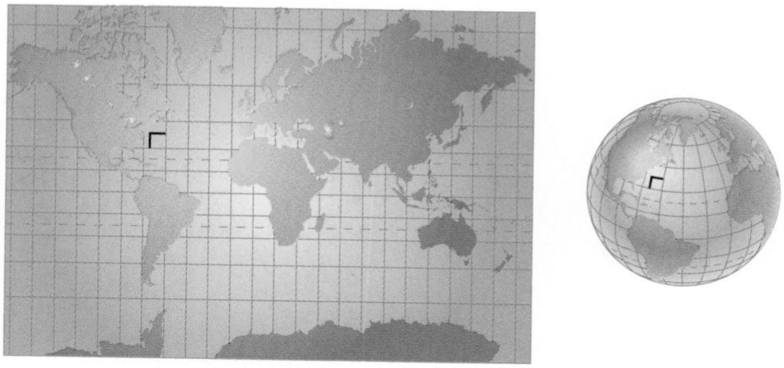

FIGURE 78

Topology This chapter began by suggesting that Euclidean geometry might seem rigid from a modern-day point of view. The plane and space figures studied in the Euclidean system are carefully distinguished by differences in size, shape, angularity, and so on. For a given figure such properties are permanent, and thus we can ask sensible questions about congruence and similarity. Suppose we studied "figures" made of rubber bands, as it were, "figures" that could be stretched, bent, or otherwise distorted without tearing or scattering. **Topology** does just that.

Topological questions concern the basic structure of objects rather than size or arrangement. For example, a typical topological question has to do with the number of holes in an object, a basic structural property that does not change during deformation. You cannot deform a rubber ball to get a rubber band without tearing it—making a hole in it. Thus the two objects are not topologically equivalent. On the other hand, a doughnut and a coffee cup are topologically equivalent, because one could be stretched so as to form the other, without changing the basic structural property.

Topology and geometry software, including games for users age 10 and up, can be found at www.geometrygames.org, a site developed by Jeff Weeks. Included are Torus and Klein Bottle Games, Kali, KaleidoTile, classroom materials for teachers of grades 6 to 10, curved spaces for software developers, and research software.

EXAMPLE 1 Determining Topological Equivalence

Decide if the figures in each pair are topologically equivalent.

(a) a football and a cereal box

(b) a doughnut and an unzipped coat

SOLUTION

(a) If we assume that a football is made of a perfectly elastic substance such as rubber or dough, it could be twisted or kneaded into the same shape as a cereal box. Thus, the two figures are topologically equivalent.

(b) A doughnut has one hole, while the coat has two (the sleeve openings). Thus, a doughnut could not be stretched and twisted into the shape of the coat without tearing another hole in it. Because of this, a doughnut and the coat are not topologically equivalent. ■

In topology, figures are classified according to their **genus**—that is, the number of cuts that can be made without cutting the figures into two pieces. The genus of an object is the number of holes in it. See Figure 79.

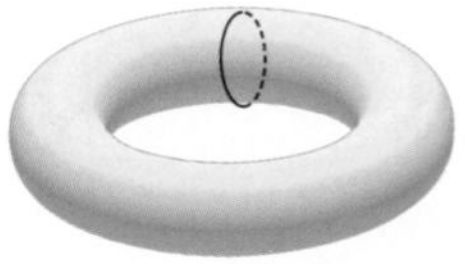

Torus One of the most useful figures in topology is the torus, a doughnut-like or tire-shaped surface. Its properties are different from those of a sphere, for example. Imagine a sphere covered with hair. You cannot comb the hairs in a completely smooth way; one fixed point remains, as you can find on your own head. In the same way, on the surface of Earth the winds are not a smooth system. There is a calm point somewhere. However, the hair on a torus could be combed smooth. The winds would be blowing smoothly on a planet shaped like a torus.

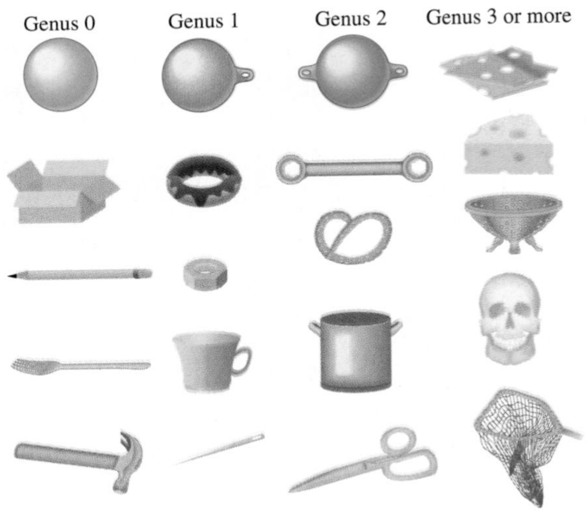

Genus of some common objects

FIGURE 79

For Further Thought

Two Interesting Topological Surfaces

Two examples of topological surfaces are the **Möbius strip** and the **Klein bottle**. The Möbius strip is a single-sided surface named after August Ferdinand Möbius (1790–1868), a pupil of Gauss.

To construct a Möbius strip, cut out a rectangular strip of paper, perhaps 3 cm by 25 cm. Paste together the two 3-cm ends after giving the paper a half-twist. To see how the strip now has only one side, mark an *x* on the strip and then mark another *x* on what appears to be the other "side." Begin at one of the *x*'s you have drawn, and trace a path along the strip. You will eventually come to the other *x* without crossing the edge of the strip.

A branch of chemistry called chemical topology studies the structures of chemical configurations. A recent advance in this area was the synthesis of the first molecular Möbius strip, which was formed by joining the ends of a double-stranded strip of carbon and oxygen atoms.

A mathematician confided
That a Möbius strip is one-sided.
And you'll get quite a laugh
If you cut one in half,
For it stays in one piece when divided.

Möbius strip

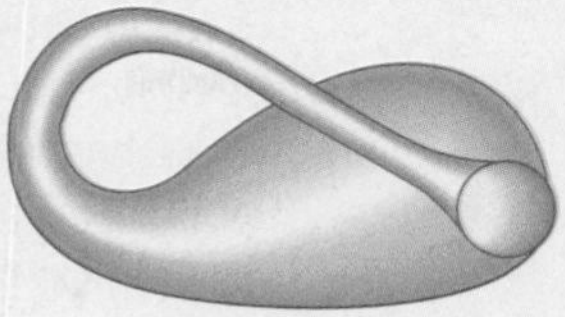

Klein bottle

Whereas a Möbius strip results from giving a paper *strip* a half-twist and then connecting it to itself, if we could do the same thing with a paper *tube* we would obtain a Klein bottle, named after Felix Klein (1849–1925). Klein produced important results in several areas, including non-Euclidean geometry and the early beginnings of group theory. (It is not possible to construct an actual Klein bottle. However, for some interesting blown glass models, see www.kleinbottle.com.)

A mathematician named Klein
Thought the Möbius strip was divine.
Said he, "If you glue
The edges of two
You'll get a weird bottle like mine."

For Group Discussion or Individual Investigation

1. The Möbius strip has other interesting properties. With a pair of scissors, cut the strip lengthwise. Do you get two strips? Repeat the process with what you have obtained from the first cut. What happens?
2. Now construct another Möbius strip, and start cutting lengthwise about $\frac{1}{3}$ of the way from one edge. What happens?
3. What would be the advantage of a conveyor belt with the configuration of a Möbius strip?

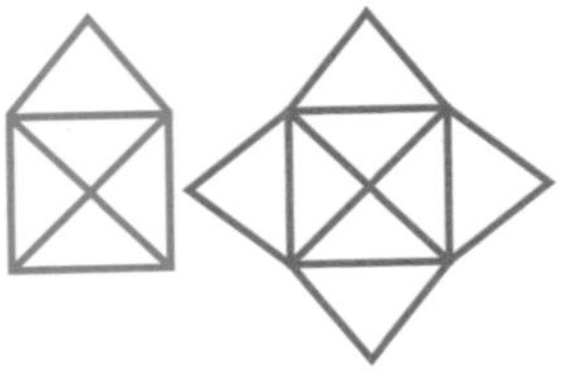

FIGURE 80

Networks* Another branch of modern geometry is *graph theory*. One topic of study in graph theory is *networks*. A **network** is a diagram showing the various paths (or **arcs**) between points (called **vertices,** or **nodes**). A network can be thought of as a set of arcs and vertices. Figure 80 shows two examples of networks. The study of networks began formally with the so-called Königsberg Bridge problem as solved by Leonhard Euler (1707–1783). In Königsberg, Germany, the River Pregel flowed through the middle of town. There were two islands in the river. During Euler's lifetime, there were seven bridges connecting the islands and the two banks of the river.

The people of the town loved Sunday strolls among the various bridges. Gradually, a competition developed to see if anyone could find a route that crossed each of the seven bridges exactly once. The problem concerns what topologists today call the *traversability* of a network. No one could find a solution. The problem became so famous that in 1735 it reached Euler, who was then at the court of the Russian empress Catherine the Great. In trying to solve the problem, Euler began by drawing a network representing the system of bridges, as in Figure 81.

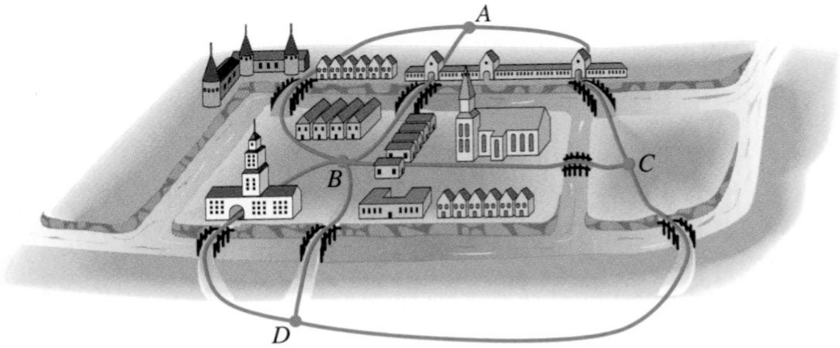

FIGURE 81

Euler first noticed that three routes meet at vertex *A*. Because 3 is an odd number, he called *A* an **odd vertex.** As three routes meet at *A*, it must be a starting or an ending point for any traverse of the network. This must be true; otherwise, when you got to *A* on your second trip there would be no way to get out. An **even vertex,** one where an even number of routes meet, need not be a starting or an ending point. (Why is this?) Three paths also meet at *C* and *D*, with five paths meeting at *B*. Thus, *B*, *C*, and *D* are also odd vertices. An odd vertex must be a starting or an ending point of a traverse. Thus, all four vertices *A*, *B*, *C*, and *D* must be starting or ending points. Because a network can have only two starting or ending points (one of each), this network cannot be traversed. The residents of Königsberg were trying to do the impossible.

Euler's result can be summarized as follows.

Results on Vertices and Traversability

1. The number of odd vertices of any network is *even*. (That is, a network must have $2n$ odd vertices, where $n = 0, 1, 2, 3, \ldots$)
2. A network with no odd vertices or exactly two odd vertices can be traversed. In the case of exactly two, start at one odd vertex and end at the other.
3. A network with more than two odd vertices cannot be traversed.

How should an artist paint a realistic view of railroad tracks going off to the horizon? In reality, the tracks are always at a constant distance apart, but they cannot be drawn that way except from overhead. The artist must make the tracks converge at a point. Only in this way will the scene look "real."

Beginning in the fifteenth century, artists led by Leone Battista Alberti, Leonardo da Vinci, and Albrecht Dürer began to study the problems of representing three dimensions in two. They found geometric methods of doing this. What artists initiated, mathematicians developed into a geometry different from that of Euclid—**projective geometry.**

Gerard Desargues (1591–1661), a French architect and engineer, published in 1636 and 1639 a treatise and proposals about perspective, and had thus invented projective geometry. However, his geometric innovations were hidden for nearly 200 years. A manuscript by Desargues turned up in 1845, about 30 years after Jean-Victor Poncelet had rediscovered projective geometry.

*An entire chapter on Graph Theory, of which networks is one topic, is available from the publisher. If this chapter is not part of this book and you are interested in more information on networks, contact Addison-Wesley.

EXAMPLE 2 Deciding Whether Networks Are Traversable

Decide whether the networks are traversable.

(a)

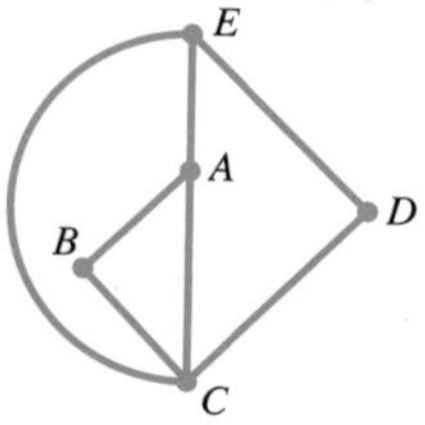

(b)

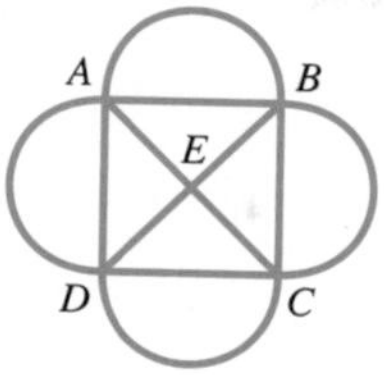

SOLUTION

(a) Because there are exactly two odd vertices (A and E) this network can be traversed. One way to traverse the network is to start at A, go through B to C, then back to A. (It is acceptable to go through a vertex as many times as needed.) Then go to E, to D, to C, and finally go back to E. It is traversable.

(b) Because vertices A, B, C, and D are all odd (five routes meet at each of them), this network is not traversable. (One of the authors of this text, while in high school, tried for hours to traverse it, not knowing he was attempting the impossible!)

EXAMPLE 3 Applying Traversability Concepts to a Floor Plan

The figure shows the floor plan of a house. Is it possible to travel through this house, going through each door exactly once?

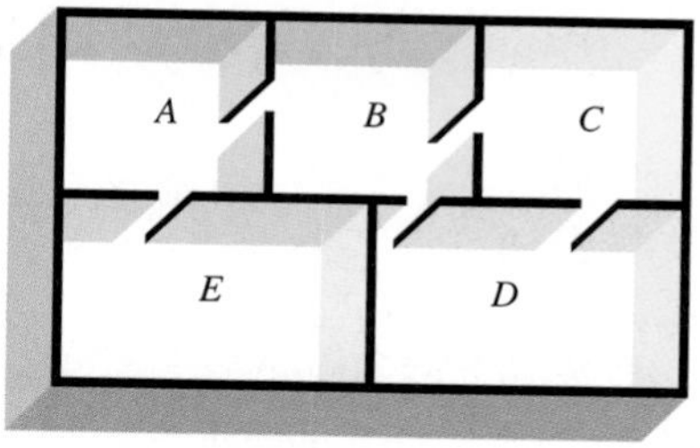

SOLUTION

Rooms A, C, and D have even numbers of doors, while rooms B and E have odd numbers of doors. If we think of the rooms as the vertices of a graph, then the fact that we have exactly two odd vertices means that it is possible to travel through each door of the house exactly once. One can either start in room B or room E. The figure here shows how this can be done starting in room E.

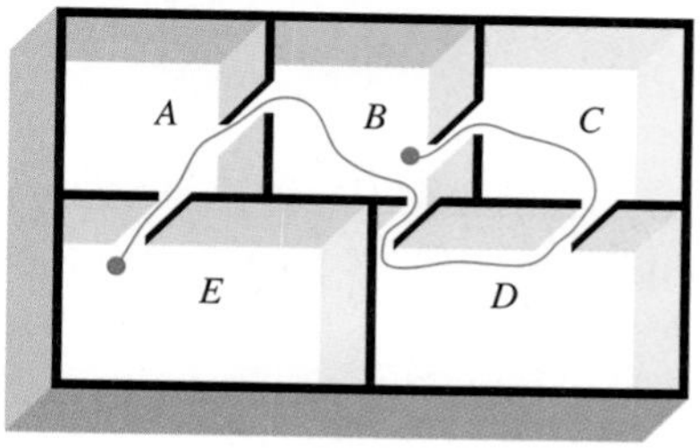

9.7 Exercises

The chart that follows characterizes certain properties of Euclidean and non-Euclidean geometries. Study it, and use it to respond to Exercises 1–10.

EUCLIDEAN	NON-EUCLIDEAN	
Dates back to about 300 B.C.	**Lobachevskian (about 1830)**	**Riemannian (about 1850)**
Lines have *infinite* length.		Lines have *finite* length.
Geometry on a plane 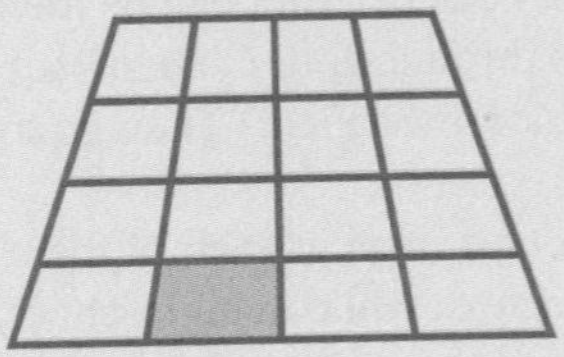	Geometry on a surface like a pseudosphere 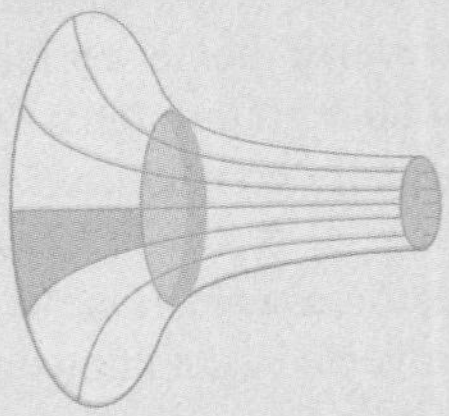	Geometry on a sphere 
Angles C and D of a Saccheri quadrilateral are *right* angles. 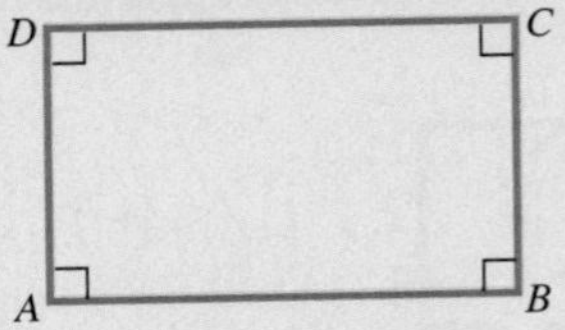	Angles C and D are *acute* angles. 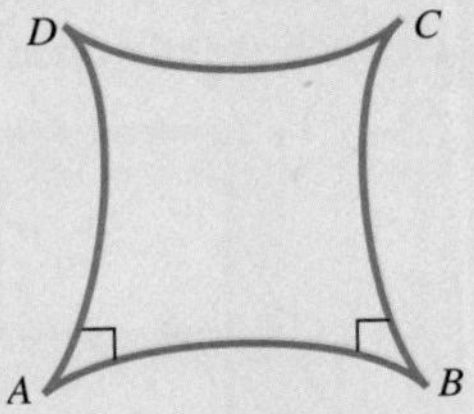	Angles C and D are *obtuse* angles. 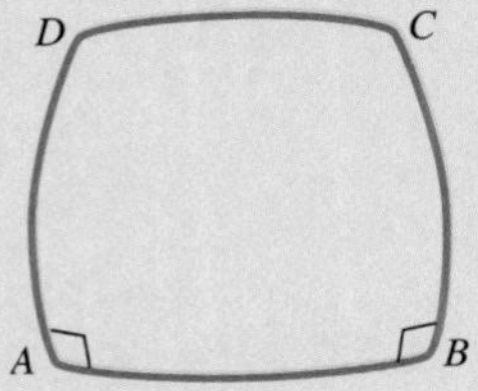
Given point P off line k, exactly *one* line can be drawn through P and parallel to k.	*More than one* line can be drawn through P and parallel to k.	*No* line can be drawn through P and parallel to k.
Typical triangle ABC 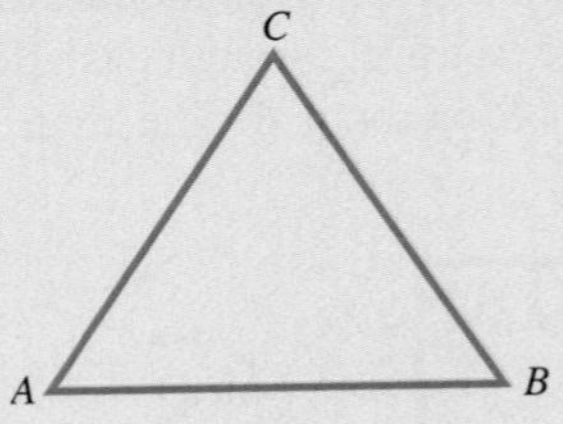	Typical triangle ABC 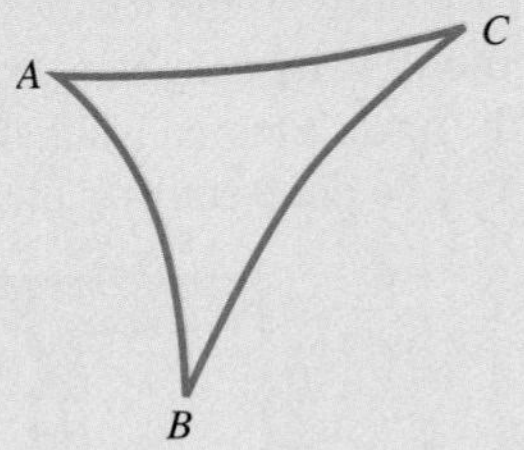	Typical triangle ABC 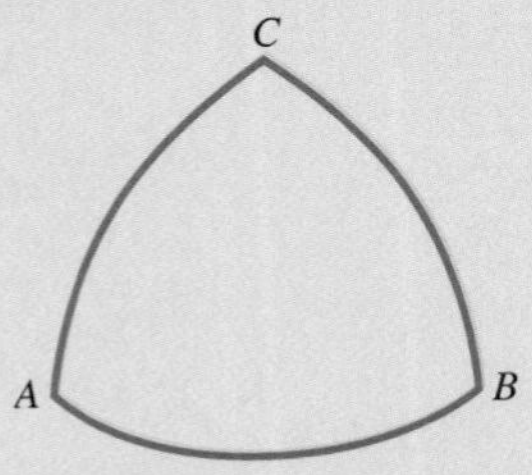
Two triangles can have the same size angles but different size sides (similarity as well as congruence).	Two triangles with the same size angles must have the same size sides (congruence only).	

1. In which geometry is the sum of the measures of the angles of a triangle equal to 180°?

2. In which geometry is the sum of the measures of the angles of a triangle greater than 180°?

3. In which geometry is the sum of the measures of the angles of a triangle less than 180°?

4. In a quadrilateral $ABCD$ in Lobachevskian geometry, the sum of the measures of the angles must be ________________ 360°.
(less than/greater than)

5. In a quadrilateral $ABCD$ in Riemannian geometry, the sum of the measures of the angles must be ________________ 360°.
(less than/greater than)

6. Suppose that m and n represent lines through P that are both parallel to k. In which geometry is this possible?

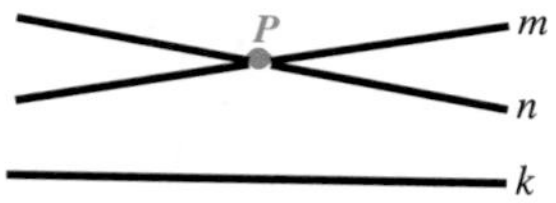

7. Suppose that m and n below *must* meet at a point. In which geometry is this possible?

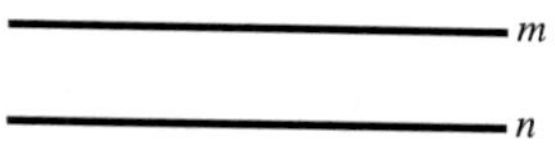

8. A globe representing the earth is a model for a surface in which geometry?

9. In which geometry is this statement possible? "Triangle ABC and triangle DEF are such that $\angle A = \angle D$, $\angle B = \angle E$, and $\angle C = \angle F$, and they have different areas."

10. Draw a figure (on a sheet of paper) as best you can showing the shape formed by the north pole N and two points A and B lying at the equator of a model of the earth.

11. Pappus, a Greek mathematician in Alexandria about A.D. 320, wrote a commentary on the geometry of the times. We will work out a theorem of his about a hexagon inscribed in two intersecting lines. First we need to define an old word in a new way: a **hexagon** consists of any six lines in a plane, no three of which meet in the same point. As the figure shows, the vertices of several hexagons are labeled with numbers. Thus 1–2 represents a line segment joining vertices 1 and 2. Segments 1–2 and 4–5 are opposite sides of a hexagon, as are 2–3 and 5–6, and 3–4 and 6–1.

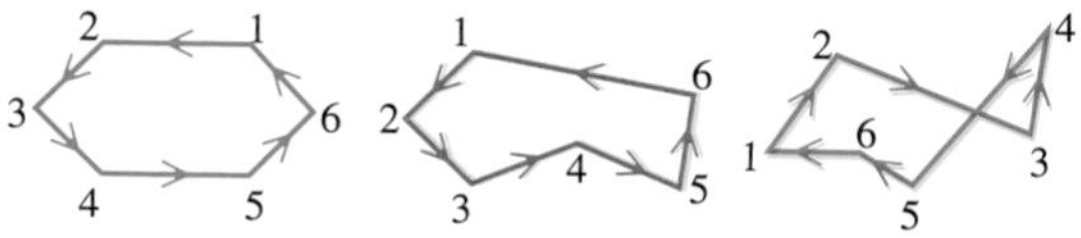

(a) Draw an angle less than 180°.
(b) Choose three points on one side of the angle. Label them 1, 5, 3 in that order, beginning with the point nearest the vertex.
(c) Choose three points on the other side of the angle. Label them 6, 2, 4 in that order, beginning with the point nearest the vertex.
(d) Draw line segments 1–6 and 3–4. Draw lines through the segments so they extend to meet in a point; call it N.
(e) Let lines through 1–2 and 4–5 meet in point M.
(f) Let lines through 2–3 and 5–6 meet in P.
(g) Draw a straight line through points M, N, and P.
(h) Write in your own words a theorem generalizing your result.

12. The following theorem comes from projective geometry:

Theorem of Desargues in a Plane Desargues's theorem states that in a plane, if two triangles are placed so that lines joining corresponding vertices meet in a point, then corresponding sides, when extended, will meet in three collinear points. (*Collinear* points are points lying on the same line.)

Draw a figure that illustrates this theorem.

In Exercises 13–20 on the next page, each figure may be topologically equivalent to none or some of the objects labeled A–E. List all topological equivalences (by letter) for each figure.

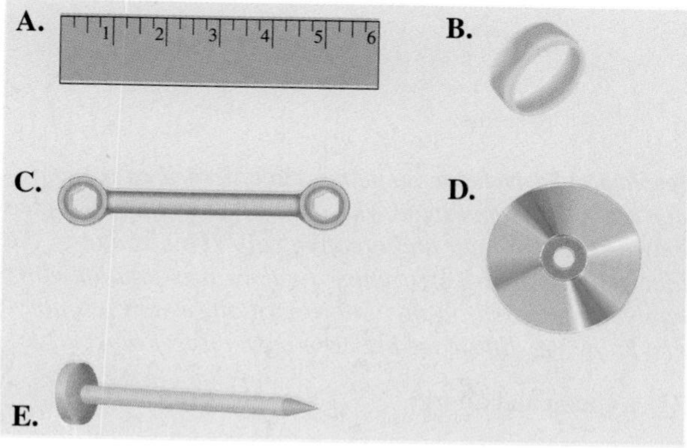

13.

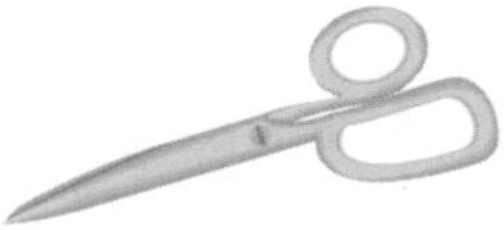

(a pair of scissors)

14.

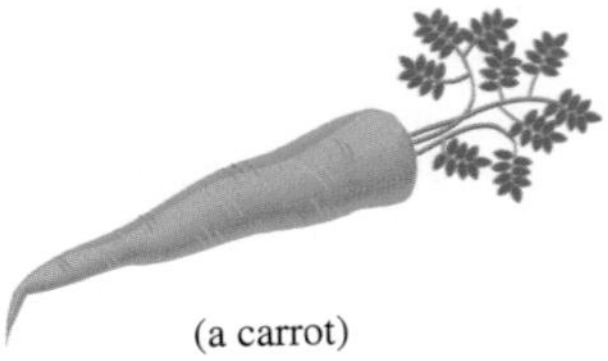

(a carrot)

15.

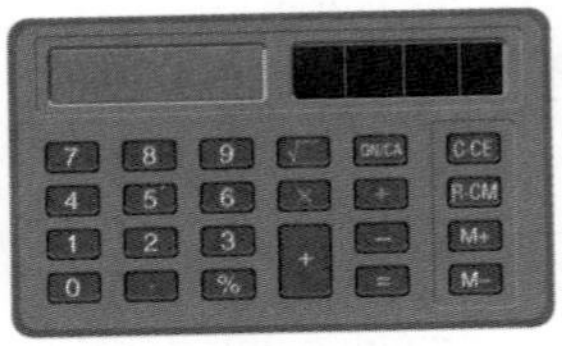

(a calculator)

16.

(a nut)

17.

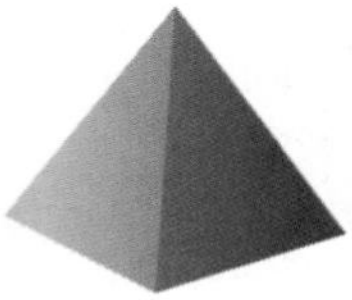

(a pyramid)

18.

(a coin)

19.

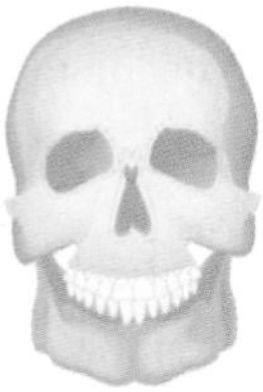

(a skull)

20.

(a needle)

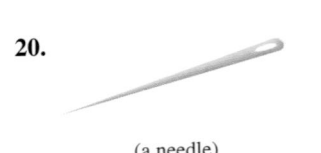

Topological Equivalence *Someone once described a topologist as "a mathematician who doesn't know the difference between a doughnut and a coffee cup." This is due to the fact that both are of genus 1—they are topologically equivalent. Based on this interpretation, would a topologist know the difference between each pair of objects?*

21. a spoon and a fork

22. a mixing bowl and a colander

23. a slice of American cheese and a slice of Swiss cheese

24. a compact disc and a phonograph record

Give the genus of each object.

25. a compact disc

26. a phonograph record

27. a sheet of loose-leaf paper made for a three-ring binder

28. a sheet of loose-leaf paper made for a two-ring binder

29. a wedding band

30. a postage stamp

For each network, decide whether each lettered vertex is even *or* odd.

31.

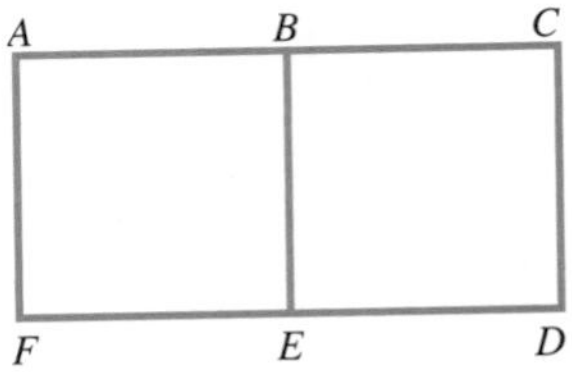

32.

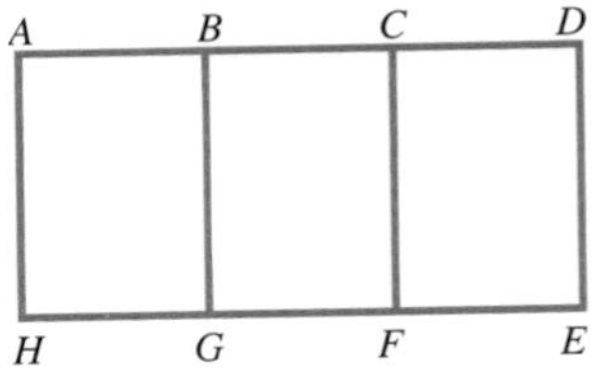

33.

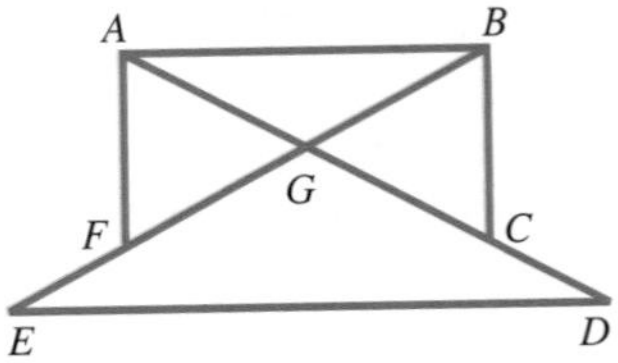

34.

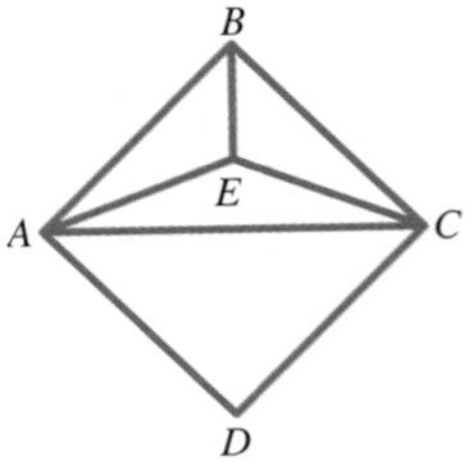

35.

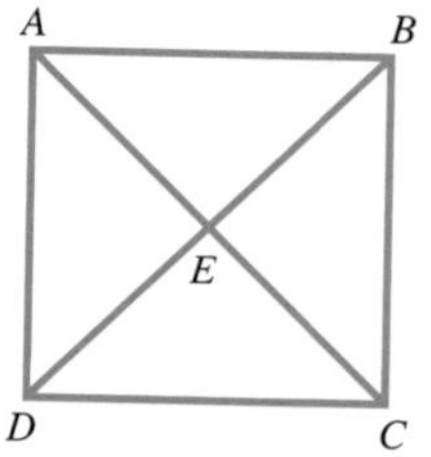

36.

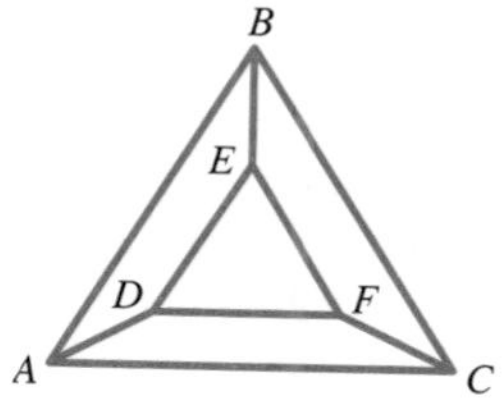

Decide whether each network is traversable. If a network is traversable, show how it can be traversed.

37.

38.

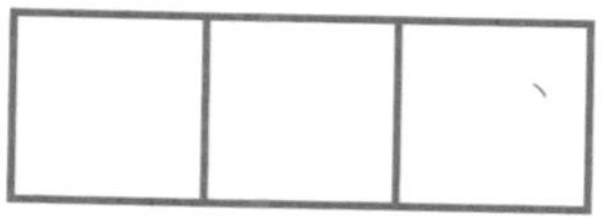

39.

40.

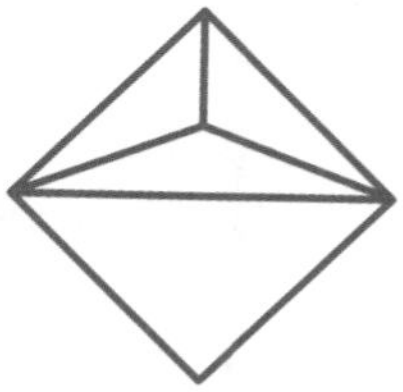

41.

42.

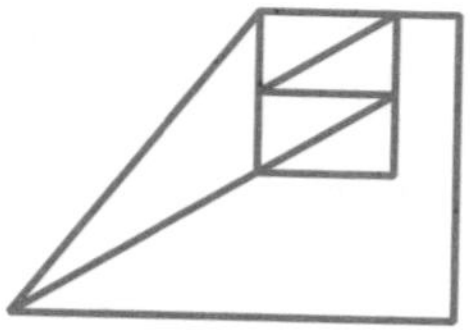

Is it possible to walk through each door of the following houses exactly once? If the answer is "yes," show how it can be done.

43.

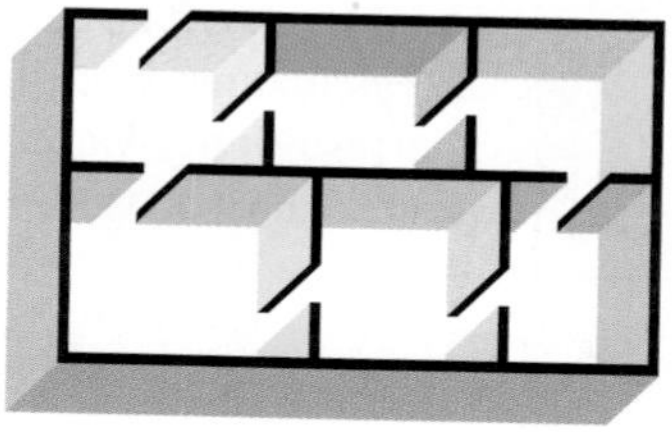

44.

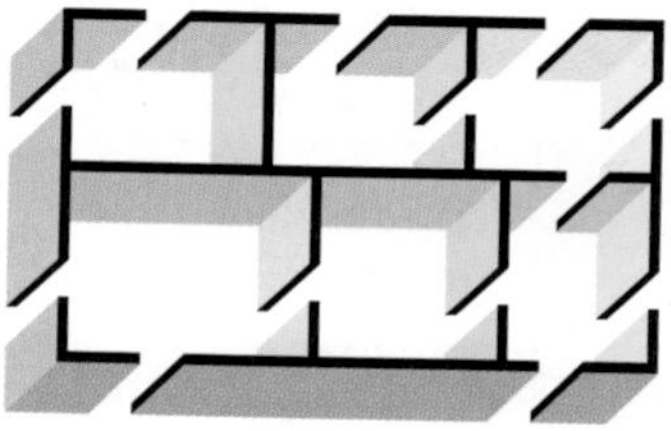

45.

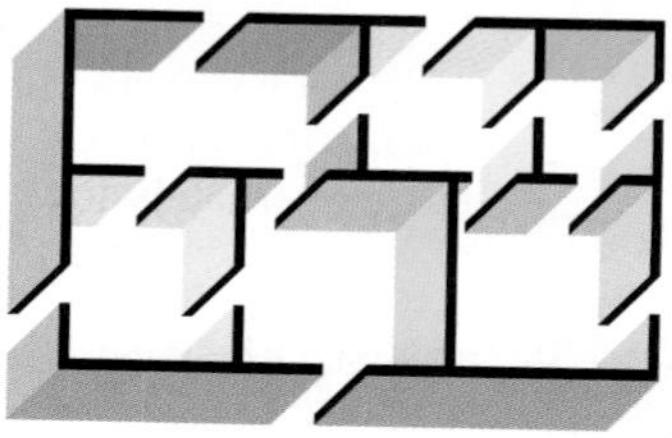

46.

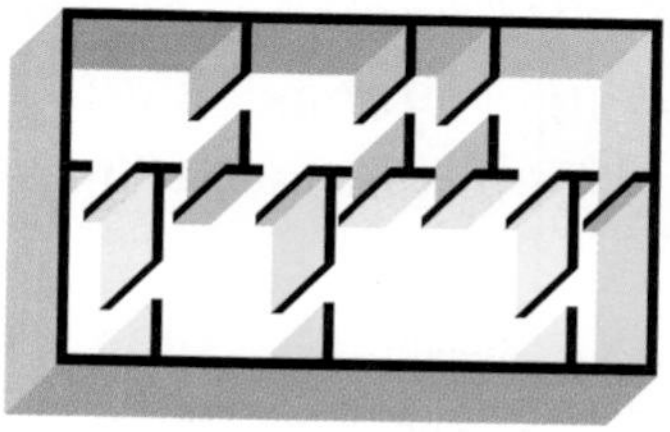

9.8 Chaos and Fractal Geometry

Chaos • Fractals

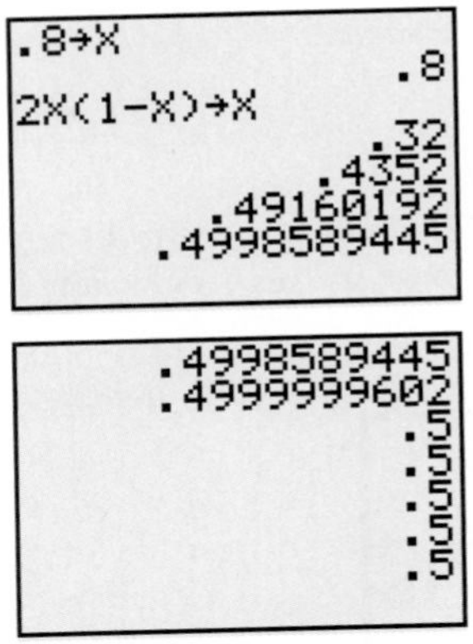

These two screens show how the TI-83/84 Plus calculator can produce the sequence described.

Chaos

Does Chaos Rule the Cosmos?

—One of the ten great unanswered questions of science, as found in the November 1992 issue of *Discover*

Consider the equation $y = kx(1 - x)$. Choosing $k = 2$ gives the equation $y = 2x(1 - x)$, which can be "iterated" by starting with an arbitrary x-value between 0 and 1, calculating the resulting y-value, substituting that y-value back in as x, calculating the resulting y-value, substituting that y-value back in as x, calculating another y-value, and so on. For example, a starting value of $x = .8$ produces the following sequence (which you can verify with a calculator):

$$.8, .32, .435, .492, .500, .500, .500, \quad \text{and so on.}$$

The sequence seems to begin randomly but quickly stabilizes at the value .500. A different initial x-value would produce another sequence which would also "converge" to .500. The value .500 can be called an *attractor* for the sequence generated by the equation $y = 2x(1 - x)$. The values of the sequence are "attracted" toward .500.

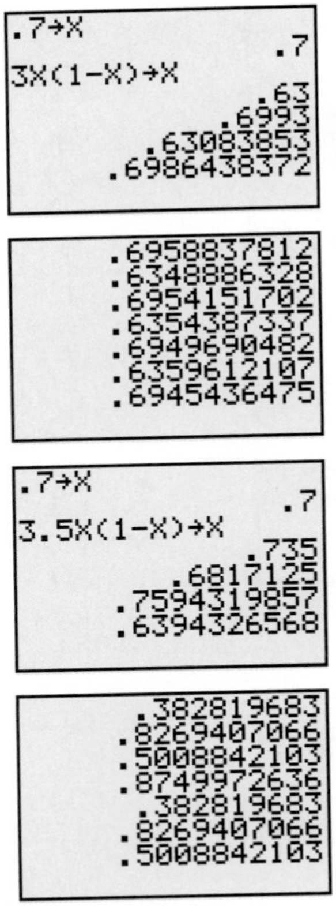

These screens support the discussion in Examples 1 and 2.

EXAMPLE 1 Finding Attractors

For the equation $y = kx(1 - x)$ with $k = 3$, begin with $x = .7$ and iterate with a calculator. What pattern emerges?

SOLUTION

Using a TI-83/84 Plus calculator, we find that the seventeenth through twentieth iterations give the sequence of terms

$$.6354387337, \quad .6949690482, \quad .6359612107, \quad .6945436475.$$

This sequence apparently converges in a manner different from the initial discussion, alternating between values near .636 and .695. Therefore, for $k = 3$, the sequence tends alternately toward two distinct attractors. ■

It happens that the equation in Example 1 exhibits the same behavior for any initial value of x between 0 and 1. You are asked to show this for several cases in the exercises.

EXAMPLE 2 Finding Attractors

In the equation of Example 1, change the multiplier k to 3.5, and find the forty-fourth through fifty-first terms. What pattern emerges?

SOLUTION

Again, using a TI-83/84 Plus calculator, and rounding to three decimal places, we get

$$.383, .827, .501, .875, .383, .827, .501, .875.$$

This sequence seems to stabilize around four alternating attractors, .383, .827, .501, and .875. ■

John Nash, a notable modern American mathematician (born in 1928), first came to the attention of the general public through his biography *A Beautiful Mind* (and the movie of the same name). In 1958 Nash narrowly lost out to René Thom (topologist and inventor of catastrophe theory, discussed on page 612) for the Fields Medal, the mathematical equivalent of the Nobel prize. Although his brilliant career was sadly interrupted by mental illness for a period of about thirty years, in 1994 Nash was awarded "the Central Bank of Sweden Prize in Economic Science in Memory of Alfred Nobel," generally regarded as equivalent to the Nobel prize. This award was for Nash's equilibrium theorem, published in his doctoral thesis in 1950. It turned out that Nash's work established a significant new way of analyzing rational conflict and cooperation in economics and other social sciences.

Notice that in our initial discussion, for $k = 2$, the sequence converged to *one* attractor. In Example 1, for $k = 3$, it converged to *two* attractors, and in Example 2, for $k = 3.5$, it converged to *four* attractors.

If k is increased further, it turns out that the number of attractors doubles over and over again, more and more often. In fact, this doubling has occurred infinitely many times before k even gets as large as 4. When we look closely at groups of these doublings we find that they are always similar to earlier groups but on a smaller scale. This is called *self-similarity,* or *scaling,* an idea that is not new, but which has taken on new significance in recent years. Somewhere before k reaches 4, the resulting sequence becomes apparently totally random, with no attractors and no stability. This type of condition is one instance of what has come to be known in the scientific community as **chaos.** This name came from an early paper by the mathematician James A. Yorke, of the University of Maryland at College Park.

The equation $y = kx(1 - x)$ does not look all that complicated, but the intricate behavior exhibited by it and similar equations has occupied some of the brightest minds (not to mention computers) in various fields—ecology, biology, physics, genetics, economics, mathematics—since about 1960. Such an equation might represent, for example, the population of some animal species where the value of k is determined by factors (such as food supply, or predators that prey on the species) that affect the increase or decrease of the population. Under certain conditions there is a long-run steady-state population (a single attractor). Under other conditions the population will eventually fluctuate between two alternating levels (two attractors), or four, or eight, and so on. But after a certain value of $k,$ the long-term population becomes totally chaotic and unpredictable.

As long as k is small enough, there will be some number of attractors and the long-term behavior of the sequence (or population) is the same regardless of the initial x-value. But once k is large enough to cause chaos, the long-term behavior of the system will change drastically when the initial x-value is changed only slightly. For example, consider the following two sequences, both generated from $y = 4x(1 - x)$.

$$.600, .960, .154, .520, .998, .006, .025, \ldots$$
$$.610, .952, .184, .601, .959, .157, .529, \ldots$$

The fact that the two sequences wander apart from one another is partly due to round-off errors along the way. But Yorke and others have shown that even "exact" calculations of the iterates would quickly produce divergent sequences just because of the slightly different initial values. This type of "sensitive dependence on initial conditions" was discovered (accidentally) back in the 1960s by Edward Lorenz when he was looking for an effective computerized model of weather patterns. He discerned the implication that any long-range weather predicting schemes might well be hopeless.

Patterns like those in the sequences above are more than just numerical oddities. Similar patterns apply to a great many phenomena in the physical, biological, and social sciences, many of them seemingly common natural systems that have been studied for hundreds of years. The measurement of a coastline, the description of the patterns in a branching tree, or a mountain range, or a cloud formation, or intergalactic cosmic dust, the prediction of weather patterns, the turbulent behavior of fluids of all kinds, the circulatory and neurological systems of the human body, fluctuations in populations and economic systems—these and many other phenomena remain mysteries, concealing their true nature somewhere beyond the reach of even our biggest and fastest computers.

Continuous phenomena are easily dealt with. A change in one quantity produces a predictable change in another. (For example, a little more pressure on the gas pedal produces a little more speed.) Mathematical functions that represent continuous events can be graphed by unbroken lines or curves, or perhaps smooth, gradually changing surfaces. The governing equations for such phenomena are "linear," and extensive mathematical methods of solving them have been developed. On the other hand, erratic events associated with certain other equations are harder to describe or predict. The science of chaos, made possible by modern computers, continues to open up new ways to deal with such events.

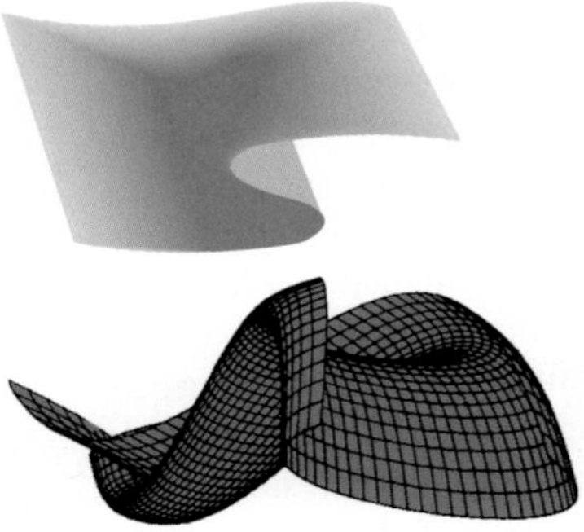

FIGURE 82

One early attempt to deal with discontinuous processes in a new way, generally acknowledged as a forerunner of chaos theory, was that of the French mathematician René Thom, who, in the 1960s, applied the methods of topology. To emphasize the feature of sudden change, Thom referred to events such as a heartbeat, a buckling beam, a stock market crash, a riot, or a tornado, as *catastrophes.* He proved that all catastrophic events (in our four-dimensional space-time) are combinations of seven elementary catastrophes. (In higher dimensions the number quickly approaches infinity.)

Each of the seven elementary catastrophes has a characteristic topological shape. Two examples are shown in Figure 82. The top figure is called a *cusp*. The bottom figure is an *elliptic umbilicus* (a belly button with an oval cross-section). Thom's work became known as **catastrophe theory.**

Computer graphics have been indispensable in the study of chaotic processes. The plotting of large numbers of points has revealed patterns that would otherwise have not been observed. (The underlying reasons for many of these patterns, however, have still not been explained.) The images shown in Figure 83 are created using chaotic processes.

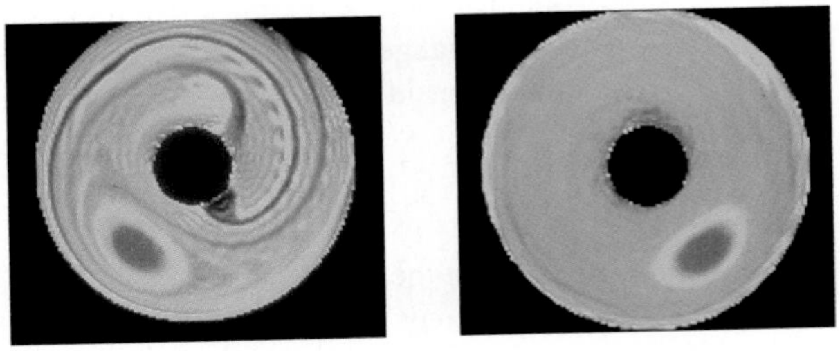

FIGURE 83

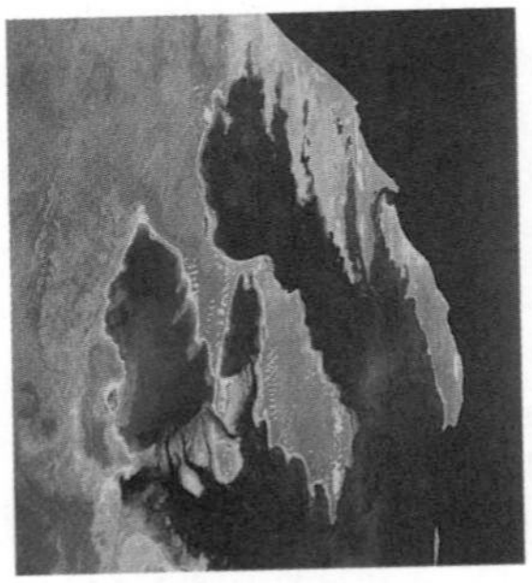

The surface of the earth, consisting of continents, mountains, oceans, valleys, and so on, has **fractal dimension** 2.2.

Fractals If there is one structure that has provided a key for the new study of nonlinear processes, it is **fractal geometry,** developed over a period of years mainly by the IBM mathematician Benoit Mandelbrot (1924–). For his work in this field, and at the recommendation of the National Science Foundation, Columbia University awarded Mandelbrot the 1985 Bernard Medal for Meritorious Service to Science.

Lines have a single dimension. Plane figures have two dimensions, and we live in a three-dimensional spatial world. In a paper published in 1967, Mandelbrot investigated the idea of measuring the length of a coastline. He concluded that such a shape defies conventional Euclidean geometry and that rather than having a natural number dimension, it has a "fractional dimension." A coastline is an example of a *self-similar shape*—a shape that repeats itself over and over on different scales. From a distance, the bays and inlets cannot be individually observed, but as one moves closer they

FIGURE 84

FIGURE 85

become more apparent. The branching of a tree, from twig to limb to trunk, also exhibits a shape that repeats itself.

In the early twentieth century, the German mathematician H. von Koch investigated the so-called Koch snowflake. It is shown in Figure 84. Starting with an equilateral triangle, each side then gives rise to another equilateral triangle. The process continues over and over, indefinitely, and a curve of infinite length is produced. The mathematics of Koch's era was not advanced enough to deal with such figures. However, using Mandelbrot's theory, it is shown that the Koch snowflake has dimension of about 1.26. This figure is obtained using a formula which involves logarithms. (Logarithms were introduced briefly in Chapter 8.)

The theory of fractals is today being applied to many areas of science and technology. It has been used to analyze the symmetry of living forms, the turbulence of liquids, the branching of rivers, and price variation in economics. Hollywood has used fractals in the special effects found in some blockbuster movies. Figure 85 shows an example of a computer-generated fractal design.

An interesting account of the science of chaos is found in the popular 1987 book *Chaos,* by James Gleick. Mandelbrot has published two books on fractals. They are *Fractals: Form, Chance, and Dimension* (1975), and *The Fractal Geometry of Nature* (1982).

Aside from providing a geometric structure for chaotic processes in nature, fractal geometry is viewed by many as a significant new art form. (To appreciate why, see the 1986 publication *The Beauty of Fractals,* by H. O. Peitgen and P. H. Richter, which contains 184 figures, many in color.) Peitgen and others have also published *Fractals for the Classroom: Strategic Activities Volume One* (Springer-Verlag, 1991).

9.8 Exercises

Exercises 1–25 are taken from an issue of Student Math Notes, *published by the National Council of Teachers of Mathematics. They were written by Dr. Tami S. Martin, Mathematics Department, Illinois State University, and the authors wish to thank N.C.T.M. and Tami Martin for permission to reproduce this activity. Because the exercises should be done in numerical order, answers to all exercises (both even- and odd-numbered) appear in the answer section of this book.*

Most of the mathematical objects you have studied have dimensions that are whole numbers. For example, such solids as cubes and icosahedrons have dimension three. Squares, triangles, and many other planar figures are two-dimensional. Lines are one-dimensional, and points have dimension zero. Consider a square with side of length one. Gather several of these squares by cutting out or using patterning blocks.

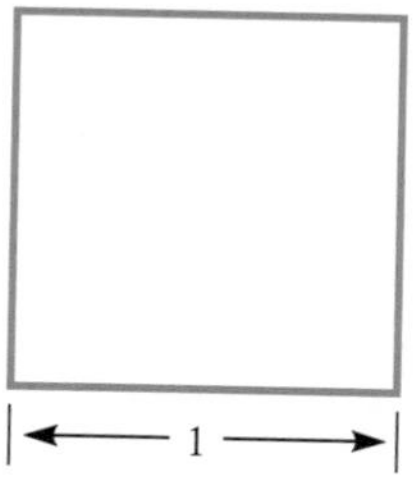

(a square)

The size of a figure is calculated by counting the number of replicas *(small pieces) that make it up. Here, a replica is the original square with edges of length one.*

1. What is the least number of these squares that can be put together edge to edge to form a larger square?

The original square is made up of one small square, so its size is one.

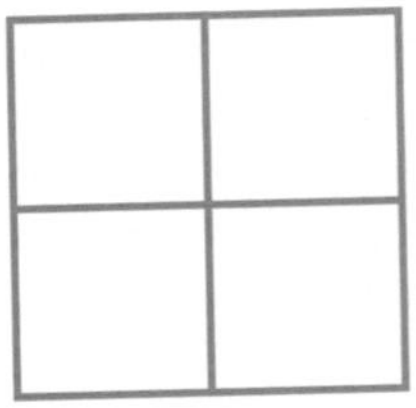

2. What is the size of the new square?

3. What is the length of each edge of the new square?

Similar figures have the same shape but are not necessarily the same size. The **scale factor** *between two similar figures can be found by calculating the ratio of corresponding edges:*

$$\frac{\text{new length}}{\text{old length}}.$$

4. What is the scale factor between the large square and the small square?

5. Find the ratio $\frac{\text{new size}}{\text{old size}}$ for the two squares.

6. Form an even larger square that is three units long on each edge. Compare this square to the small square. What is the scale factor between the two squares? What is the ratio of the new size to the old size?

7. Form an even larger square that is four units long on each edge. Compare this square to the small square. What is the scale factor between the two squares? What is the ratio of the new size to the old size?

8. Complete the table for squares.

Scale Factor	2	3	4	5	6	10
Ratio of new size to old size						

9. How are the two rows in the table related?

Consider an equilateral triangle. The length of an edge of the triangle is one unit. The size of this triangle is one.

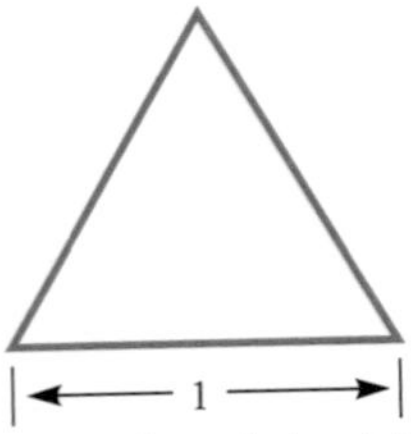

(an equilateral triangle)

10. What is the least number of equilateral triangles that can be put together edge to edge to form a similar larger triangle?

11. Complete the table for triangles.

Scale Factor	2	3	4	5	6	10
Ratio of new size to old size						

12. How does the relationship between the two rows in this table compare with the one you found in the table for squares?

One way to define the dimension, d, of a figure relates the scale factor, the new size, and the old size:

$$(\text{scale factor})^d = \frac{\text{new size}}{\text{old size}}.$$

Using a scale factor of two for squares or equilateral triangles, we can see that $2^d = \frac{4}{1}$*; that is,* $2^d = 4$*. Because* $2^2 = 4$*, the dimension, d, must be two. This definition of dimension confirms what we already know—that squares and equilateral triangles are two-dimensional figures.*

13. Use this definition and your completed tables to confirm that the square and the equilateral triangle are two-dimensional figures for scale factors other than two.

Consider a cube, with edges of length one. Let the size of the cube be one.

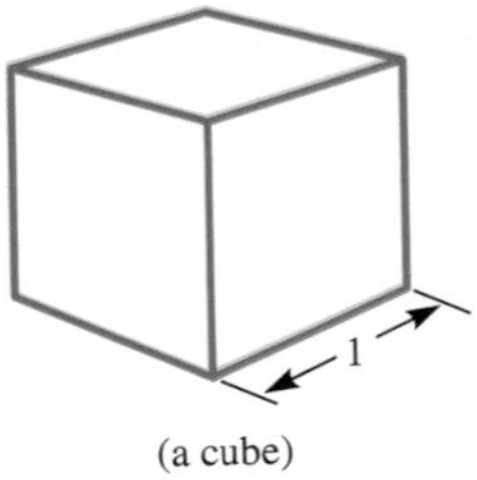

(a cube)

14. What is the least number of these cubes that can be put together face to face to form a larger cube?

15. What is the scale factor between these two cubes? What is the ratio of the new size to the old size for the two cubes?

16. Complete the table for cubes.

Scale Factor	2	3	4	5	6	10
Ratio of new size to old size						

17. How are the two rows in the table related?

18. Use the definition of dimension and a scale factor of two to verify that a cube is a three-dimensional object.

We have explored scale factors and sizes associated with two- and three-dimensional figures. Is it possible for mathematical objects to have fractional dimensions? Consider each figure formed by replacing the middle third of a line segment of length one by one upside-down V, each of whose two sides are equal in length to the segment removed. The first four stages in the development of this figure are shown.

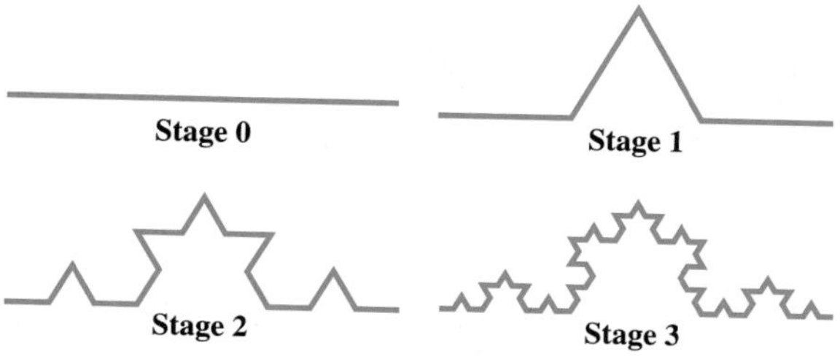

Finding the scale factor for this sequence of figures is difficult, because the overall length of a representative portion of the figure remains the same while the number of pieces increases. To simplify the procedure, follow these steps.

Step 1 *Start with any stage (e.g., Stage 1).*

Step 2 *Draw the next stage (e.g., Stage 2) of the sequence and "blow it up" so that it contains an exact copy of the preceding stage (in this example, Stage 1).*

Notice that Stage 2 contains four copies, or replicas, of Stage 1 and is three times as long as Stage 1.

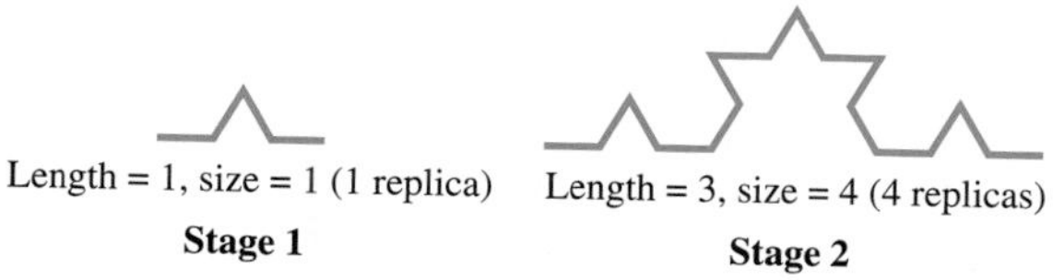

19. The scale factor is equal to the ratio $\frac{\text{new length}}{\text{old length}}$ between any two consecutive stages. The scale factor between Stage 1 and Stage 2 is _____.

20. The size can be determined by counting the number of replicas of Stage 1 found in Stage 2. Old size = 1, new size = _____.

Use the definition of dimension to compute the dimension, d, of the figure formed by this process: $3^d = \frac{4}{1}$; that is, $3^d = 4$. Since $3^1 = 3$ and $3^2 = 9$, for $3^d = 4$ the dimension of the figure must be greater than one but less than two: $1 < d < 2$.

21. Use your calculator to estimate d. Remember that d is the exponent that makes 3^d equal 4. For example, because d must be between 1 and 2, try $d = 1.5$. But $3^{1.5} = 5.196\ldots$, which is greater than 4; thus d must be smaller than 1.5. Continue until you approximate d to three decimal places. (Use logarithms for maximum accuracy.)

The original figure was a one-dimensional line segment. By iteratively adding to the line segment, an object of dimension greater than one but less than two was generated. Objects with fractional dimension are known as **fractals.** *Fractals are infinitely self-similar objects formed by repeated additions to, or removals from, a figure. The object attained at the limit of the repeated procedure is the fractal.*

Next consider a two-dimensional object with sections removed iteratively. In each stage of the fractal's development, a triangle is removed from the center of each triangular region.

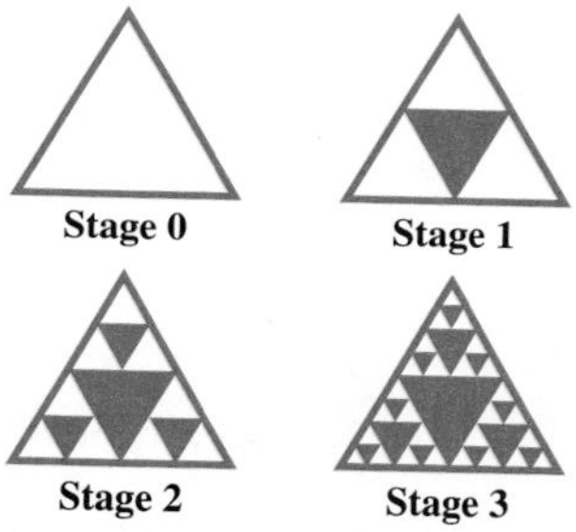

Use the process from the last example to help answer each questions.

22. What is the scale factor of the fractal?

23. Old size = 1, new size = _____.

24. The dimension of the fractal is between what two whole number values?

25. Use the definition of dimension and your calculator to approximate the dimension of this fractal to three decimal places.

Use a calculator to determine the pattern of attractors for the equation $y = kx(1 - x)$ for the given value of k and the given initial value of x.

26. $k = 3.25, x = .7$

27. $k = 3.4, x = .8$

28. $k = 3.55, x = .7$

COLLABORATIVE INVESTIGATION

Generalizing the Angle Sum Concept

In this chapter we learned that the sum of the measures of the angles of a triangle is 180°. This fact can be extended to determine a formula for the sum of the measures of the angles of any convex polygon. To begin this investigation, divide into groups of three or four students each. Prepare on a sheet of paper six figures as shown on the right.

Now we define a diagonal from vertex *A* to be a segment from *A* to a non-adjacent vertex. The triangle in Figure I has no diagonals, but the polygons in Figures II through VI do have them.

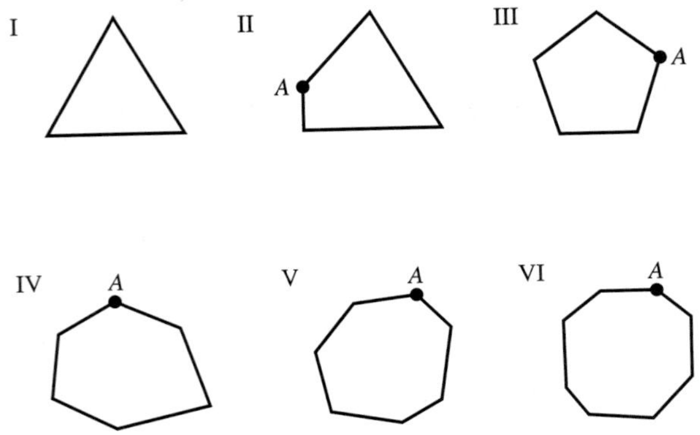

Topics for Discussion

1. Choose someone in the group to draw all possible diagonals from vertex A in each figure.
2. Now complete Table 6 as a group.

TABLE 6

Polygon	Number of Sides	Number of Tri-angles, t	Number of Degrees in Each Triangle	Sum of the Measures of All Angles of the Polygon, $t \cdot 180°$
I				
II				
III				
IV				
V				
VI				

3. Based on the table you completed, answer the following in order.
   (a) As suggested by the table, the number of triangles that a convex polygon can be divided into is _____ less than the number of sides.
   (b) Thus, if a polygon has s sides, it can be divided into _____ triangles.
   (c) From the table we see that the sum of the measures of all the angles of a polygon can be found from the expression $t \cdot 180°$. Thus, if a polygon has s sides, the sum of the measures of all the angles of the polygon is given by the expression
   (____ − _____) · ____°.
4. Use your discovery from Exercise 3(c) to find the sum of the measures of all the angles of
   (i) a nonagon **(ii)** a decagon
   (iii) a 12-sided polygon.

CHAPTER 9 TEST

1. Consider a 38° angle. Answer each of the following.
 (a) What is the measure of its complement?
 (b) What is the measure of its supplement?
 (c) Classify it as acute, obtuse, right, or straight.

Find the measure of each marked angle.

2.

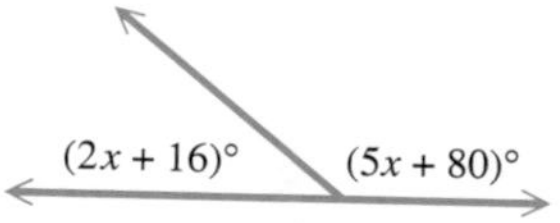

3.

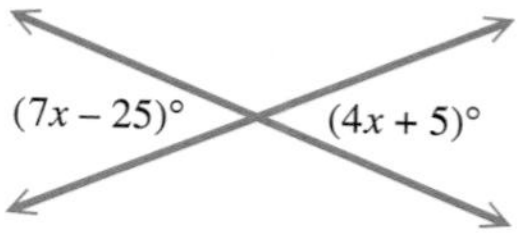

4.

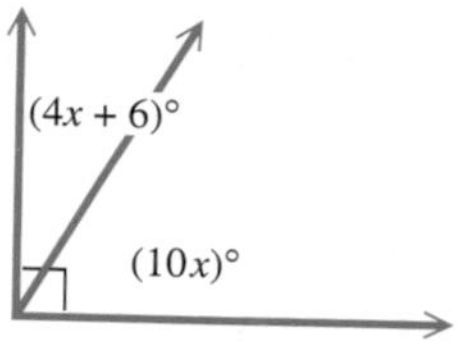

In Exercises 5 and 6, assume that lines m and n are parallel, and find the measure of each marked angle.

5.

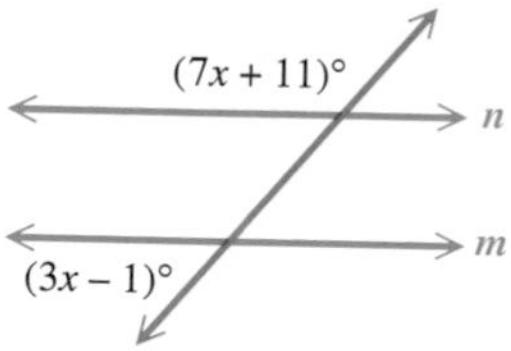

6.

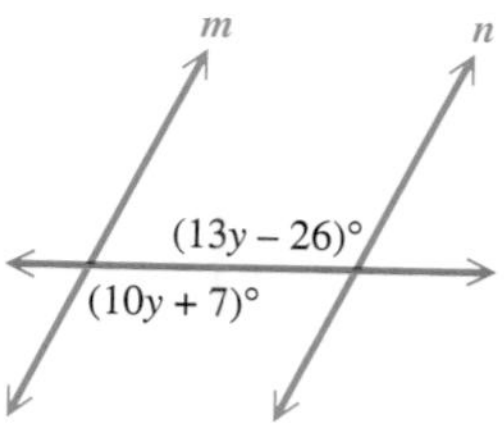

7. Explain why a rhombus must be a parallelogram, but a parallelogram might not be a rhombus.

8. Which one of the statements A–D is false?
 A. A square is a rhombus.
 B. The acute angles of a right triangle are complementary.
 C. A triangle may have both a right angle and an obtuse angle.
 D. A trapezoid may have nonparallel sides of the same length.

Identify each of the following curves as simple, closed, both, *or* neither.

9.

10.

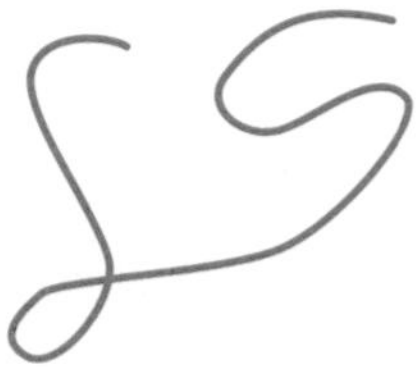

11. Find the measure of each angle in the triangle.

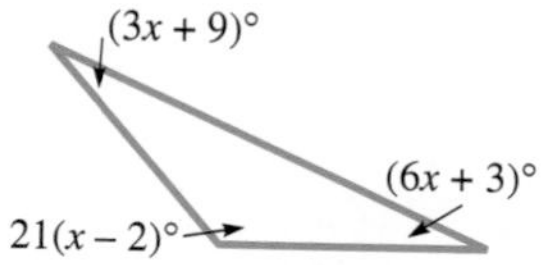

Find the area of each of the following figures.

12.

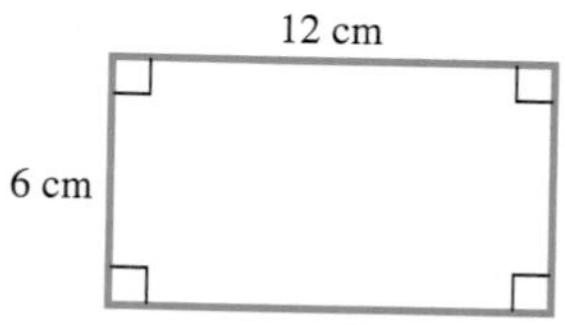

13.

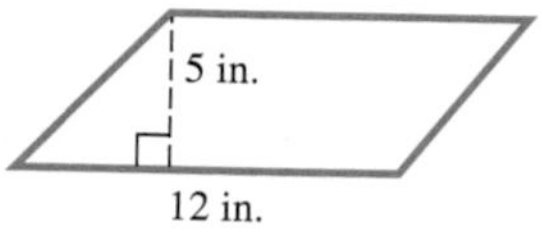

(a parallelogram)

14.

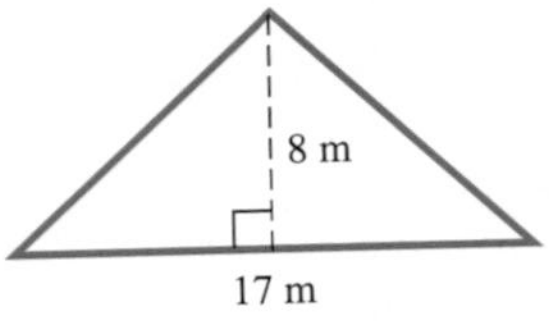

15.

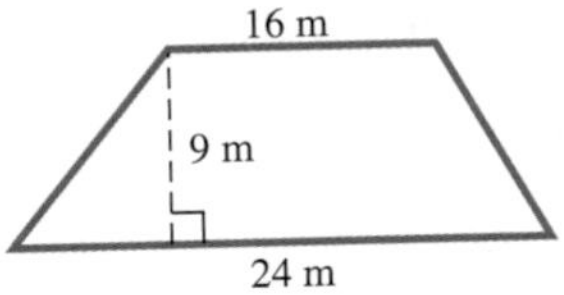

(a trapezoid)

16. If a circle has area 144π square inches, what is its circumference?

17. ***Circumference of a Dome*** The Rogers Centre in Toronto, Canada, is the first stadium with a hard-shell, retractable roof. The steel dome is 630 feet in diameter. To the nearest foot, what is the circumference of this dome?

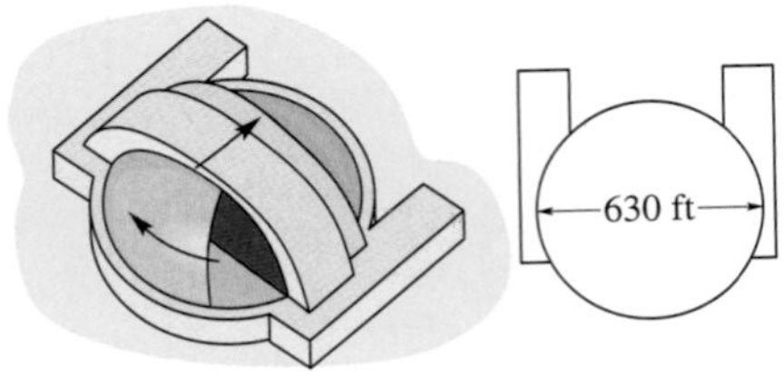

18. ***Area of a Shaded Figure*** What is the area of the shaded portion of the figure? Use 3.14 as an approximation for π.

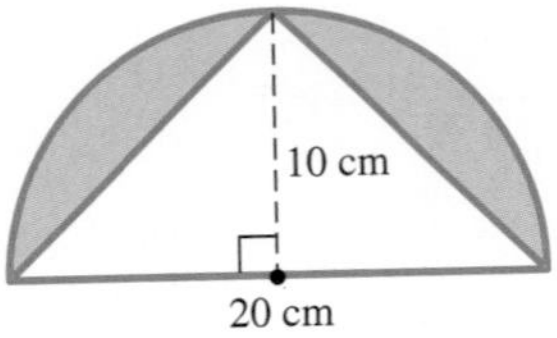

(a triangle within a semicircle)

19. Given: $\measuredangle CAB = \measuredangle DBA$; $DB = CA$
Prove: $\triangle ABD \cong \triangle BAC$

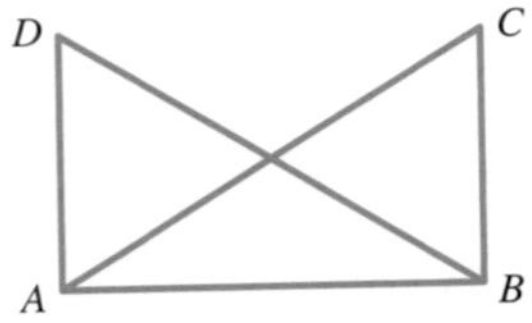

20. ***Height of a Pole*** If a 30-ft pole casts a shadow 45 ft long, how tall is a pole whose shadow is 30 ft long at the same time?

21. ***Diagonal of a Rectangle*** What is the measure of a diagonal of a rectangle that has width 20 m and length 21 m?

22. First reflect the given figure about line n, and then about line m.

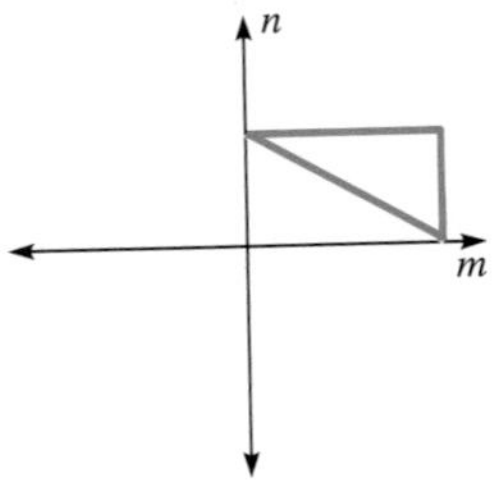

23. Find the point reflection image of the given figure with the given point as center.

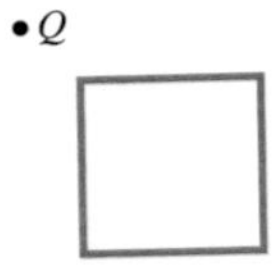

Find **(a)** *the volume and* **(b)** *the surface area of each of the following space figures. When necessary, use 3.14 as an approximation for* π.

24.

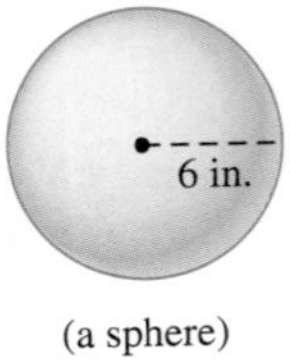

(a sphere)

25.

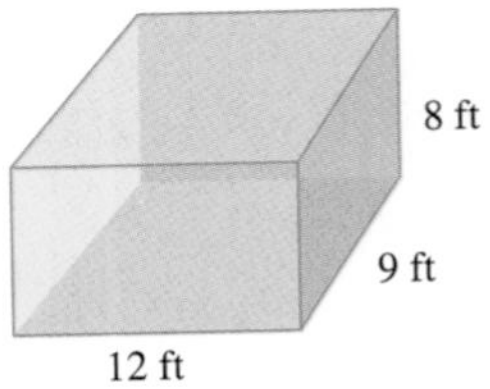

(a box)

26. 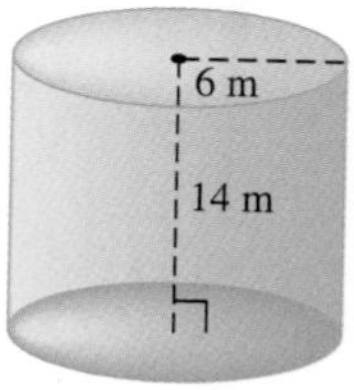

(a right circular cylinder)

27. List several main distinctions between Euclidean geometry and non-Euclidean geometry.

28. ***Topological Equivalence*** Are the following pairs of objects topologically equivalent?

(a) a page of a book and the cover of the same book

(b) a pair of glasses with the lenses removed, and the Mona Lisa

29. Decide whether it is possible to traverse the network shown. If it is possible, show how it can be done.

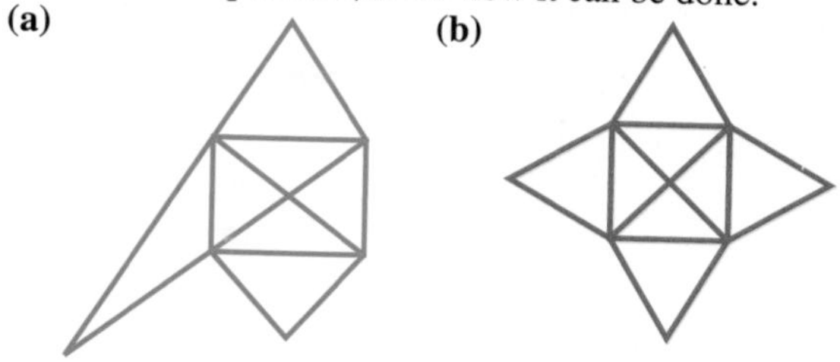

30. Use a calculator to determine the attractors for the sequence generated by the equation $y = 2.1x(1 - x)$, with initial value of $x = .6$.

10

TRIGONOMETRY

The 1958 musical *Merry Andrew* starred Danny Kaye as Andrew Larabee, a teacher with a flair for using unconventional methods in his classes. He uses a musical number to teach the Pythagorean theorem, singing and dancing to "The Square of the Hypotenuse":

> *...Parallel lines don't connect, which is just about what you might expect. Though scientific laws may change and decimals can be moved, the following is constant, and has yet to be disproved:* ***The square of the hypotenuse of a right triangle is equal to the sum of the squares of the two adjacent sides.***

The Pythagorean theorem, introduced in the previous chapter, is used extensively in the study of *trigonometry*, the topic of this chapter. The foundations of trigonometry go back at least 3000 years. The ancient Egyptians, Babylonians, and Greeks developed trigonometry to find the lengths of the sides of triangles and the measures of their angles. In Egypt, trigonometry was used to reestablish land boundaries after the annual flood of the Nile River. In Babylonia it was used in astronomy. The word *trigonometry* comes from the Greek words for triangle (*trigon*) and measurement (*metry*). Today trigonometry is used in electronics, surveying, and other engineering areas, and is necessary for further courses in mathematics, such as calculus.

10.1 Angles and Their Measures

Basic Terminology • Degree Measure • Angles in a Coordinate System

Basic Terminology

A line may be drawn through the two distinct points A and B. This line is called **line** $\boldsymbol{AB}$. The portion of the line between A and B, including points A and B themselves, is **segment** $\boldsymbol{AB}$. The portion of the line AB that starts at A and continues through B, and on past B, is called **ray** $\boldsymbol{AB}$. Point A is the endpoint of the ray. See Figure 1.

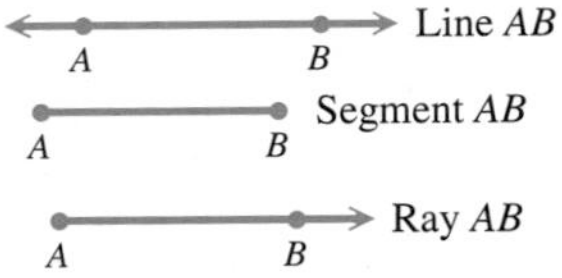

FIGURE 1

In the study of trigonometry, an **angle** is formed by rotating a ray around its endpoint. The ray in its initial position is called the **initial side** of the angle, while the ray in its location after the rotation is the **terminal side** of the angle. The endpoint of the ray is the **vertex** of the angle. Figure 2 shows the initial and terminal sides of an angle with vertex A.

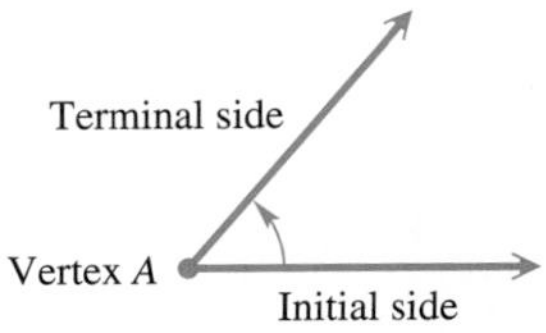

FIGURE 2

If the rotation of the terminal side is counterclockwise, the angle measure is **positive.** If the rotation is clockwise, the angle measure is **negative.** Figure 3 shows two angles, one positive and one negative.

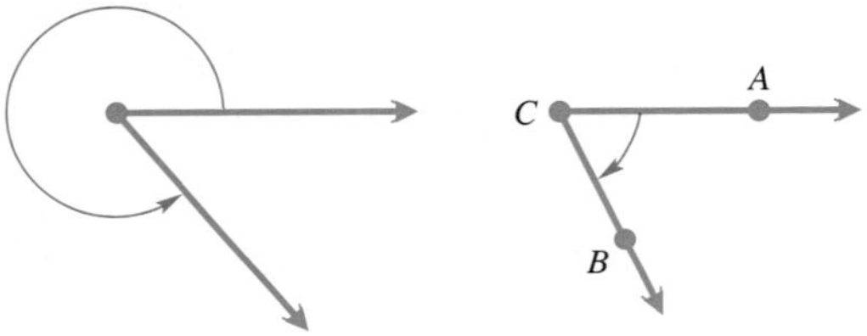

Positive angle Negative angle

FIGURE 3

An angle can be named by using the name of its vertex. For example, the angle on the right in Figure 3 can be called angle C. Alternatively, an angle can be named using three letters, with the vertex letter in the middle. Thus, the angle on the right also could be named angle ACB or angle BCA.

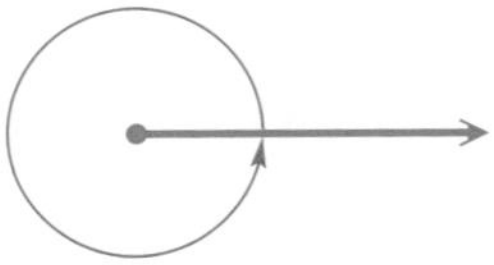

A complete rotation of a ray gives an angle whose measure is 360°.

FIGURE 4

Degree Measure

There are two systems in common use for measuring the size of angles. The most common unit of measure is the **degree.** (The other common unit of measure is called the *radian.*) Degree measure was developed by the Babylonians 4000 years ago. To use degree measure we assign 360 degrees to a complete rotation of a ray. In Figure 4, notice that the terminal side of the angle corresponds to its initial side when it makes a complete rotation.

One degree, written $1°$, represents $\frac{1}{360}$ of a rotation. Therefore, $90°$ represents $\frac{90}{360} = \frac{1}{4}$ of a complete rotation, and $180°$ represents $\frac{180}{360} = \frac{1}{2}$ of a complete rotation. Angles of measure $1°$, $90°$, and $180°$ are shown in Figure 5.

FIGURE 5

> **"Trigonometry,** perhaps more than any other branch of mathematics, developed as the result of a continual and fertile interplay of supply and demand: the supply of applicable mathematical theories and techniques available at any given time and the demands of a single applied science, astronomy. So intimate was the relation that not until the thirteenth century was it useful to regard the two subjects as separate entities."(From "The History of Trigonometry" by Edward S. Kennedy, in *Historical Topics for the Mathematics Classroom,* the Thirty-first Yearbook of N.C.T.M., 1969.)

Special angles are named as shown in the following chart.

Name	Angle Measure	Example(s)
Acute angle	Between 0° and 90°	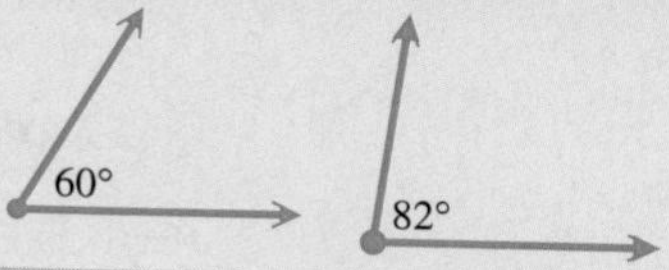
Right angle	Exactly 90°	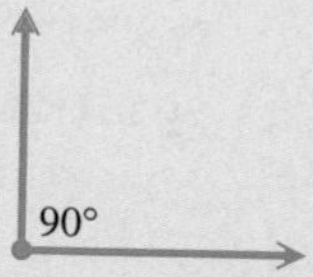
Obtuse angle	Between 90° and 180°	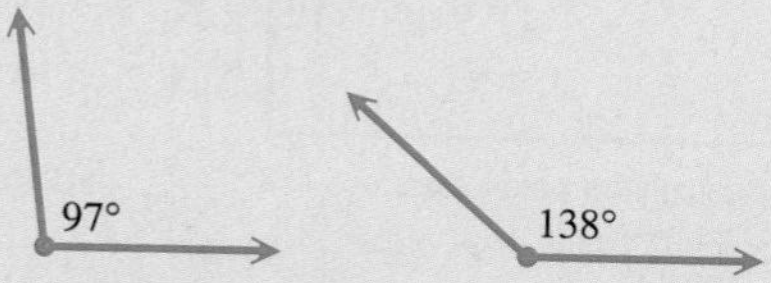
Straight angle	Exactly 180°	

If the sum of the measures of two angles is 90°, the angles are called **complementary.** Two angles with measures whose sum is 180° are **supplementary.**

EXAMPLE 1 Finding Complement and Supplement

Give the complement and the supplement of 50°.

SOLUTION

The complement of 50° is

$$90^\circ - 50^\circ = 40^\circ,$$

while the supplement of 50° is

$$180^\circ - 50^\circ = 130^\circ.$$

Do not confuse an angle with its measure. The *angle itself* consists of the vertex together with the initial and terminal sides, whereas the *measure of the angle* is the size of the rotation angle from the initial to the terminal side (commonly expressed in degrees). If angle A has a 35° rotation angle, we say that $m(\text{angle } A)$ is 35°, where $m(\text{angle } A)$ is read "the measure of angle A." It saves a lot of work, however, to abbreviate $m(\text{angle A}) = 35^\circ$ as simply angle $A = 35^\circ$.

Traditionally, portions of a degree have been measured with minutes and seconds. One **minute,** written $1'$, is $\frac{1}{60}$ of a degree.

$$1' = \left(\frac{1}{60}\right)^\circ \quad \text{or} \quad 60' = 1^\circ$$

One **second,** $1''$, is $\frac{1}{60}$ of a minute.

$$1'' = \frac{1}{60}' = \frac{1}{3600}^{\circ} \quad \text{or} \quad 60'' = 1'$$

The measure 12° 42′ 38″ represents 12 degrees, 42 minutes, 38 seconds.

EXAMPLE 2 Calculating with Degree Measure

Perform each calculation.

(a) 51° 29′ + 32° 46′ **(b)** 90° − 73° 12′

SOLUTION

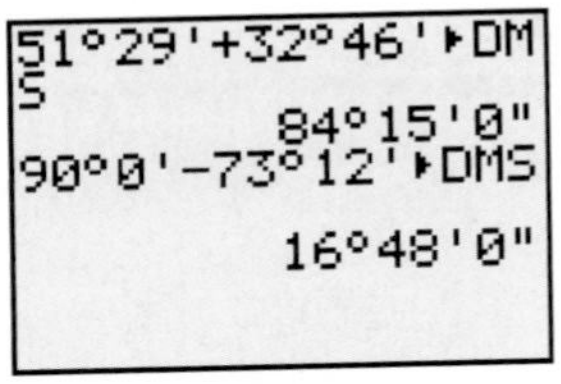

The calculations explained in Example 2 can be done with a graphing calculator capable of working with degrees, minutes, and seconds.

(a) Add the degrees and the minutes separately.

$$\begin{array}{r} 51^\circ\ 29' \\ +\ 32^\circ\ 46' \\ \hline 83^\circ\ 75' \end{array}$$

Since 75′ = 60′ + 15′ = 1° 15′, the sum is written

$$\begin{array}{r} 83^\circ \\ +\ \ 1^\circ\ 15' \\ \hline 84^\circ\ 15'. \end{array}$$

(b)

$$\begin{array}{r} 89^\circ\ 60' \\ -73^\circ\ 12' \\ \hline 16^\circ\ 48'. \end{array}$$

Write 90° as 89° 60′.

Angles can be measured in **decimal degrees.** For example, 12.4238° represents

$$12.4238^\circ = 12\frac{4238}{10{,}000}^\circ.$$

EXAMPLE 3 Converting Between Decimal Degrees and Degrees, Minutes, Seconds

(a) Convert 74° 8′ 14″ to decimal degrees. Round to the nearest thousandth of a degree.
(b) Convert 34.817° to degrees, minutes and seconds. Round to the nearest second.

SOLUTION

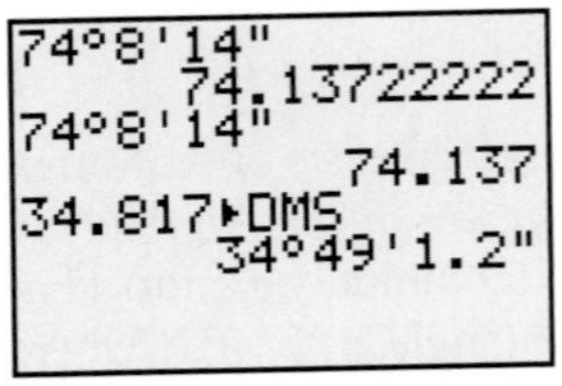

The conversions in Example 3 can be done on some graphing calculators. The second displayed result was obtained by setting the calculator to show only three places after the decimal point.

(a) $74^\circ 8' 14'' = 74^\circ + \frac{8}{60}^\circ + \frac{14}{3600}^\circ$ $1' = \frac{1}{60}^\circ$ and $1'' = \frac{1}{3600}^\circ$

$\approx 74^\circ + .1333^\circ + .0039^\circ$

$= 74.137^\circ$ Rounded to three decimal places

(b) $34.817^\circ = 34^\circ + .817^\circ$

$= 34^\circ + (.817)(60')$ $1^\circ = 60'$

$= 34^\circ + 49.02'$

$= 34^\circ + 49' + .02'$

$= 34^\circ + 49' + (.02)(60'')$ $1' = 60''$

$= 34^\circ + 49' + 1.2''$

$\approx 34^\circ\ 49'\ 1''$

Angles in a Coordinate System An angle θ (the Greek letter *theta*)* is in **standard position** if its vertex is at the origin of a rectangular coordinate system and its initial side lies along the positive x-axis. The two angles in Figure 6(a) and 6(b) are in standard position. An angle in standard position is said to lie in the quadrant in which its terminal side lies. For example, an acute angle is in quadrant I and an obtuse angle is in quadrant II. Figure 6(c) shows ranges of angle measures for each quadrant when $0° < \theta < 360°$. Angles in standard position having their terminal sides along the x-axis or y-axis, such as angles with measures 90°, 180°, 270°, and so on, are called **quadrantal angles.**

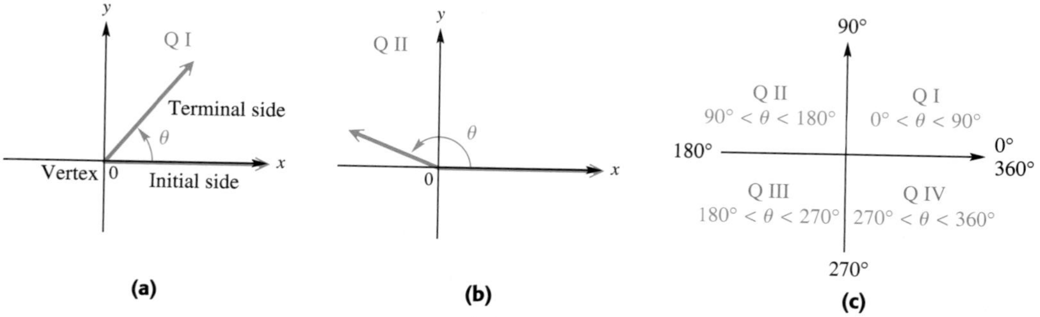

FIGURE 6

A complete rotation of a ray results in an angle of measure 360°. But there is no reason why the rotation need stop at 360°. By continuing the rotation, angles of measure larger than 360° can be produced. The angles in Figure 7(a) have measures 60° and 420°. These two angles have the same initial side and the same terminal side, but different amounts of rotation. Angles that have the same initial side and the same terminal side are called **coterminal angles.** As shown in Figure 7(b), angles with measures 110° and 830° are coterminal.

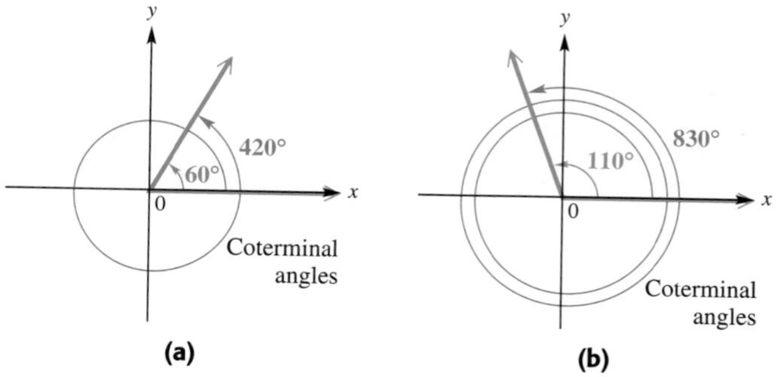

FIGURE 7

EXAMPLE 4 Finding Measures of Coterminal Angles

Find the angle of smallest possible positive measure coterminal with each angle.

(a) 908° **(b)** −75°

---

*The letters of the Greek alphabet are identified in a margin note on page 631.

SOLUTION

(a) Add or subtract 360° from 908° as many times as needed to get an angle with measure greater than 0° but less than 360°. Because

$$908° - 2 \cdot 360° = 908° - 720° = 188°,$$

an angle of 188° is coterminal with an angle of 908°. See Figure 8.

(b) Use a rotation of 360° + (−75°) = 285°. See Figure 9.

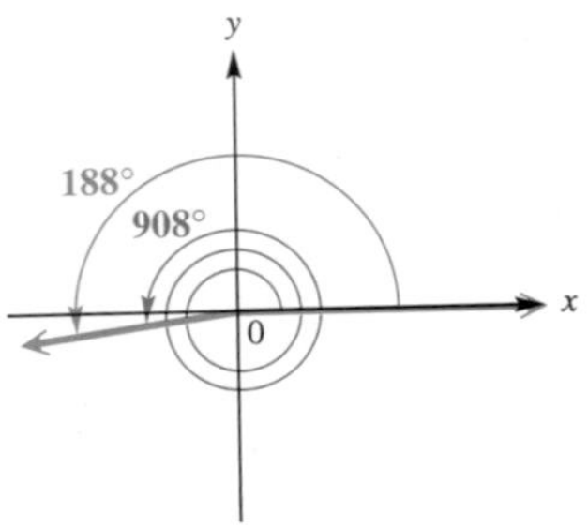

FIGURE 8

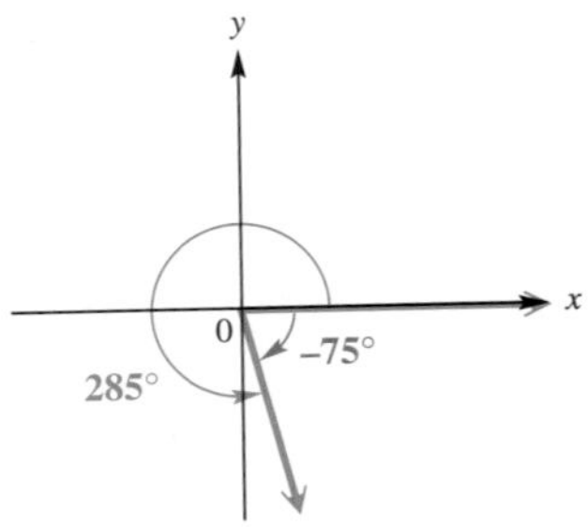

FIGURE 9

Sometimes it is necessary to find an expression that will generate all angles coterminal with a given angle. For example, suppose we wish to do this for a 60° angle. Because any angle coterminal with 60° can be obtained by adding an appropriate integer multiple of 360° to 60°, we can let n represent any integer, and the expression

$$60° + n \cdot 360°$$

will represent all such coterminal angles. Table 1 shows a few possibilities.

TABLE 1

Value of n	Angle Coterminal with 60°
2	$60° + 2 \cdot 360° = 780°$
1	$60° + 1 \cdot 360° = 420°$
0	$60° + 0 \cdot 360° = 60°$ (the angle itself)
−1	$60° + (-1) \cdot 360° = -300°$

10.1 Exercises

Give **(a)** *the complement and* **(b)** *the supplement of each angle.*

1. 30° **2.** 60° **3.** 45°

4. 55° **5.** 89° **6.** 2°

7. If an angle measures x degrees, how can we represent its complement?

8. If an angle measures x degrees, how can we represent its supplement?

Perform each calculation.

9. 62° 18′ + 21° 41′ **10.** 75° 15′ + 83° 32′ **11.** 71° 58′ + 47° 29′

12. 90° − 73° 48′ **13.** 90° − 51° 28′ **14.** 180° − 124° 51′

15. 90° − 72° 58′ 11″ **16.** 90° − 36° 18′ 47″

Convert each angle measure to decimal degrees. Use a calculator, and round to the nearest thousandth of a degree.

17. 20° 54′ **18.** 38° 42′ **19.** 91° 35′ 54″

20. 34° 51′ 35″ **21.** 274° 18′ 59″ **22.** 165° 51′ 9″

Convert each angle measure to degrees, minutes, and seconds. Use a calculator, and round to the nearest second.

23. 31.4296° **24.** 59.0854° **25.** 89.9004°

26. 102.3771° **27.** 178.5994° **28.** 122.6853°

Find the angle of smallest positive measure coterminal with each angle.

29. $-40°$ **30.** $-98°$ **31.** $-125°$ **32.** $-203°$

33. $539°$ **34.** $699°$ **35.** $850°$ **36.** $1000°$

Give an expression that generates all angles coterminal with the given angle. Let n represent any integer.

37. $30°$ **38.** $45°$ **39.** $60°$ **40.** $90°$

Sketch each angle in standard position. Draw an arrow representing the correct amount of rotation. Find the measure of two other angles, one positive and one negative, that are coterminal with the given angle. Give the quadrant of each angle.

41. $75°$ **42.** $89°$ **43.** $174°$ **44.** $234°$

45. $300°$ **46.** $512°$ **47.** $-61°$ **48.** $-159°$

10.2 Trigonometric Functions of Angles

Trigonometric Functions • Undefined Function Values

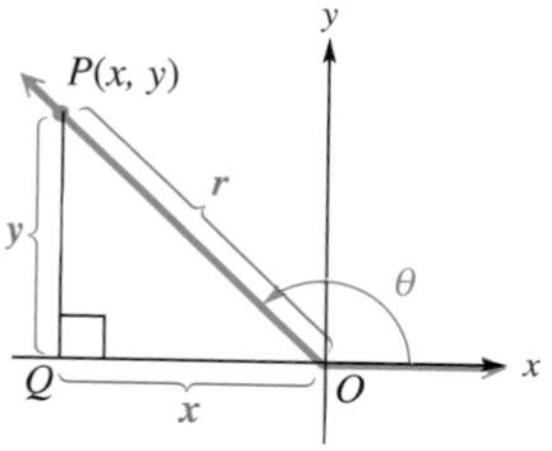

FIGURE 10

Trigonometric Functions The study of trigonometry covers the six trigonometric functions defined in this section. To define these six basic functions, start with an angle θ in standard position. Choose any point P having coordinates (x, y) on the terminal side of angle θ. (The point P must not be the vertex of the angle.) See Figure 10.

A perpendicular from P to the x-axis at point Q determines a triangle having vertices at O, P, and Q. The distance r from $P(x, y)$ to the origin, $(0, 0)$, can be found from the distance formula.

$$r = \sqrt{(x-0)^2 + (y-0)^2}$$
$$r = \sqrt{x^2 + y^2}$$

Notice that $r > 0$, because distance is never negative.

The six trigonometric functions of angle θ are called **sine, cosine, tangent, cotangent, secant,** and **cosecant.** In the following definitions, we use the customary abbreviations for the names of these functions.

"The founder of trigonometry is **Hipparchus,** who lived in Rhodes and Alexandria and died about 125 B.C. We know rather little about him. Most of what we do know comes from Ptolemy, who credits Hipparchus with a number of ideas in trigonometry and astronomy. We owe to him many astronomical observations and discoveries, the most influential astronomical theory of ancient times, and works on geography." *(Source: Mathematical Thought From Ancient to Modern Times, Volume 1, by Morris Kline.)*

Trigonometric Functions

Let (x, y) be a point other than the origin on the terminal side of an angle θ in standard position. The distance from the point to the origin is $r = \sqrt{x^2 + y^2}$. The six trigonometric functions of θ are:

$$\sin\theta = \frac{y}{r} \qquad \cos\theta = \frac{x}{r} \qquad \tan\theta = \frac{y}{x}\ (x \neq 0)$$

$$\csc\theta = \frac{r}{y}\ (y \neq 0) \qquad \sec\theta = \frac{r}{x}\ (x \neq 0) \qquad \cot\theta = \frac{x}{y}\ (y \neq 0).$$

Although Figure 10 shows a second quadrant angle, these definitions apply to any angle θ. Because of the restrictions on the denominators in the definitions of tangent, cotangent, secant, and cosecant, some angles will have undefined function values. This will be discussed in more detail later.

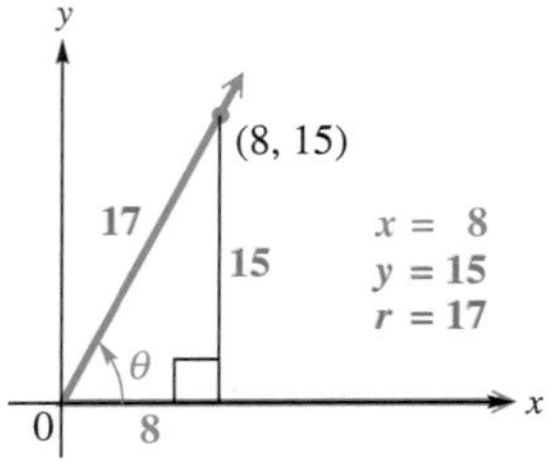

FIGURE 11

EXAMPLE 1 Finding Function Values of an Angle

The terminal side of an angle θ in standard position goes through the point (8, 15). Find the values of the six trigonometric functions of angle θ.

SOLUTION

Figure 11 shows angle θ and the triangle formed by dropping a perpendicular from the point (8, 15) to the x-axis. The point (8, 15) is 8 units to the right of the y-axis and 15 units above the x-axis, so that $x = 8$ and $y = 15$.

$$r = \sqrt{x^2 + y^2}$$
$$r = \sqrt{8^2 + 15^2} \qquad \text{Let } x = 8 \text{ and } y = 15.$$
$$r = \sqrt{64 + 225}$$
$$r = \sqrt{289}$$
$$r = 17$$

The values of the six trigonometric functions of angle θ can now be found with the definitions given above.

$$\sin\theta = \frac{y}{r} = \frac{15}{17} \qquad \cos\theta = \frac{x}{r} = \frac{8}{17} \qquad \tan\theta = \frac{y}{x} = \frac{15}{8}$$
$$\csc\theta = \frac{r}{y} = \frac{17}{15} \qquad \sec\theta = \frac{r}{x} = \frac{17}{8} \qquad \cot\theta = \frac{x}{y} = \frac{8}{15}$$

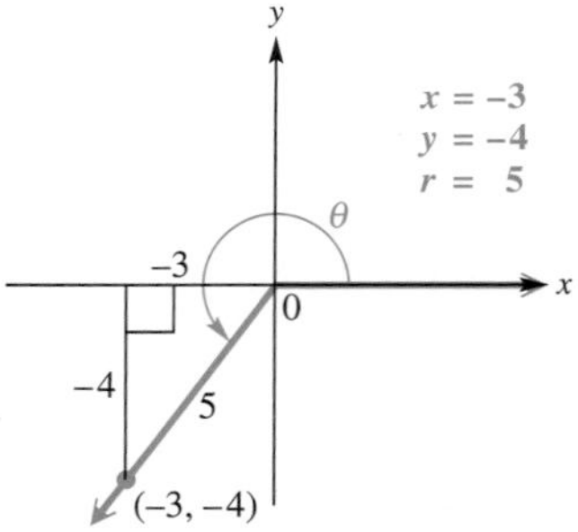

FIGURE 12

EXAMPLE 2 Finding Function Values of an Angle

The terminal side of an angle θ in standard position goes through the point $(-3, -4)$. Find the values of the six trigonometric functions of θ.

SOLUTION

As shown in Figure 12, $x = -3$ and $y = -4$. The value of r is

$$r = \sqrt{(-3)^2 + (-4)^2}$$
$$r = \sqrt{25} \longleftarrow \text{(Remember that } r > 0.)$$
$$r = 5.$$

Then by the definitions of the trigonometric functions,

$$\sin\theta = \frac{-4}{5} = -\frac{4}{5} \qquad \cos\theta = \frac{-3}{5} = -\frac{3}{5} \qquad \tan\theta = \frac{-4}{-3} = \frac{4}{3}$$
$$\csc\theta = \frac{5}{-4} = -\frac{5}{4} \qquad \sec\theta = \frac{5}{-3} = -\frac{5}{3} \qquad \cot\theta = \frac{-3}{-4} = \frac{3}{4}.$$

The six trigonometric functions can be found from *any* point on the terminal side of the angle other than the origin. To see why any point may be used, refer to

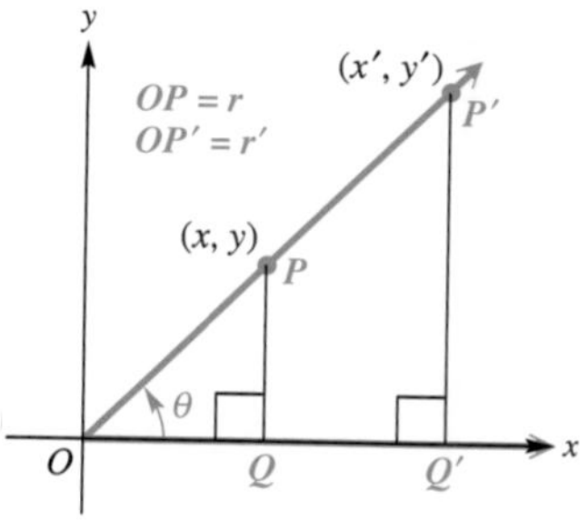

FIGURE 13

Figure 13, which shows an angle θ and two distinct points on its terminal side. Point P has coordinates (x, y) and point P' (read "P-prime") has coordinates (x', y'). Let r be the length of the hypotenuse of triangle OPQ, and let r' be the length of the hypotenuse of triangle $OP'Q'$. Because corresponding sides of similar triangles are in proportion,

$$\frac{y}{r} = \frac{y'}{r'},$$

so that $\sin\theta = \frac{y}{r}$ is the same no matter which point is used to find it. Similar results hold for the other five functions.

Undefined Function Values

If the terminal side of an angle in standard position lies along the y-axis, any point on this terminal side has x-coordinate 0. Similarly, an angle with terminal side on the x-axis has y-coordinate 0 for any point on the terminal side. Because the values of x and y appear in the denominators of some of the trigonometric functions, and because a fraction is undefined if its denominator is 0, some of the trigonometric function values of quadrantal angles (i.e., those with terminal side on an axis) will be undefined.

EXAMPLE 3 Finding Function Values and Undefined Function Values

Find values of the trigonometric functions for each angle. Identify any that are undefined.

(a) an angle of 90°

(b) an angle in standard position with terminal side through $(-3, 0)$

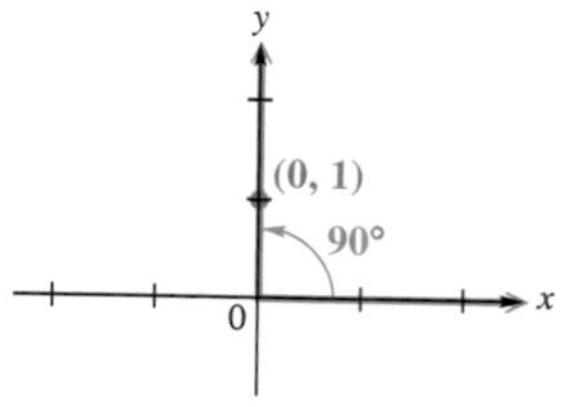

(a)

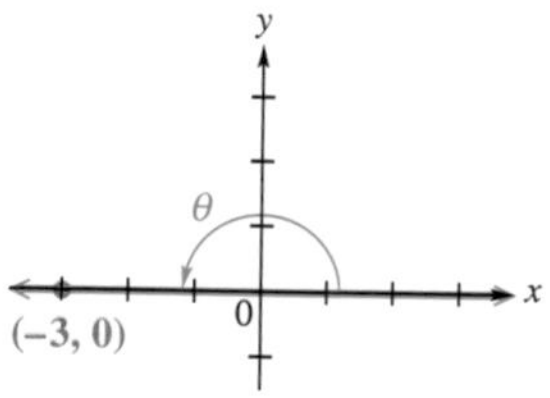

(b)

FIGURE 14

SOLUTION

(a) First, select any point on the terminal side of a 90° angle. We select the point $(0, 1)$, as shown in Figure 14(a). Here $x = 0$ and $y = 1$. Verify that $r = 1$. Then, by the definitions of the trigonometric functions,

$$\sin 90^\circ = \frac{1}{1} = 1 \qquad \cos 90^\circ = \frac{0}{1} = 0 \qquad \tan 90^\circ = \frac{1}{0} \text{ (undefined)}$$

$$\csc 90^\circ = \frac{1}{1} = 1 \qquad \sec 90^\circ = \frac{1}{0} \text{ (undefined)} \qquad \cot 90^\circ = \frac{0}{1} = 0.$$

(b) Figure 14(b) shows the angle. Here, $x = -3$, $y = 0$, and $r = 3$, so the trigonometric functions have the following values.

$$\sin\theta = \frac{0}{3} = 0 \qquad \cos\theta = \frac{-3}{3} = -1 \qquad \tan\theta = \frac{0}{-3} = 0$$

$$\csc\theta = \frac{3}{0} \text{ (undefined)} \qquad \sec\theta = \frac{3}{-3} = -1 \qquad \cot\theta = \frac{-3}{0} \text{ (undefined)}$$

The conditions under which the trigonometric function values of quadrantal angles are undefined are summarized here.

Undefined Function Values

If the terminal side of a quadrantal angle lies along the y-axis, the tangent and secant functions are undefined. If it lies along the x-axis, the cotangent and cosecant functions are undefined.

Because the most commonly used quadrantal angles are 0°, 90°, 180°, 270° and 360°, the values of the functions of these angles are summarized in Table 2.

TABLE 2 Quadrantal Angles

θ	$\sin\theta$	$\cos\theta$	$\tan\theta$	$\cot\theta$	$\sec\theta$	$\csc\theta$
0°	0	1	0	Undefined	1	Undefined
90°	1	0	Undefined	0	Undefined	1
180°	0	−1	0	Undefined	−1	Undefined
270°	−1	0	Undefined	0	Undefined	−1
360°	0	1	0	Undefined	1	Undefined

10.2 Exercises

In Exercises 1–4, sketch an angle θ in standard position such that θ has the smallest possible positive measure, and the given point is on the terminal side of θ.

1. $(-3, 4)$ **2.** $(-4, -3)$ **3.** $(5, -12)$ **4.** $(-12, -5)$

Find the values of the trigonometric functions for the angles in standard position having the following points on their terminal sides. Identify any that are undefined. Rationalize denominators when applicable.

5. $(-3, 4)$

6. $(-4, -3)$

7. $(0, 2)$

8. $(-4, 0)$

9. $(1, \sqrt{3})$

10. $(-2\sqrt{3}, -2)$

11. $(3, 5)$

12. $(-2, 7)$

13. $(-8, 0)$

14. $(0, 9)$

15. For any nonquadrantal angle θ, $\sin\theta$ and $\csc\theta$ will have the same sign. Explain why this is so.

16. If $\cot\theta$ is undefined, what is the value of $\tan\theta$?

17. How is the value of r interpreted geometrically in the definitions of the sine, cosine, secant, and cosecant functions?

18. If the terminal side of an angle θ is in quadrant III, what is the sign of each of the trigonometric function values of θ?

Suppose that the point (x, y) is in the indicated quadrant. Decide whether the given ratio is positive or negative. (Hint: It may be helpful to draw a sketch.)

19. II, $\frac{y}{r}$ **20.** II, $\frac{x}{r}$ **21.** III, $\frac{y}{r}$ **22.** III, $\frac{x}{r}$

23. IV, $\frac{x}{r}$ **24.** IV, $\frac{y}{r}$ **25.** IV, $\frac{y}{x}$ **26.** IV, $\frac{x}{y}$

Use the appropriate definition to determine each function value. If it is undefined, say so.

27. $\cos 90°$ **28.** $\sin 90°$ **29.** $\tan 90°$ **30.** $\cot 90°$

31. $\sec 90°$ **32.** $\csc 90°$ **33.** $\sin 180°$ **34.** $\sin 270°$

35. $\tan 180°$ **36.** $\cot 270°$ **37.** $\sin(-270°)$ **38.** $\cos(-270°)$

39. $\tan 0°$ **40.** $\sec(-180°)$

10.3 Trigonometric Identities

Reciprocal Identities • Signs of Function Values In Quadrants • Pythagorean Identities • Quotient Identities

The Greek Alphabet

A	α	alpha
B	β	beta
Γ	γ	gamma
Δ	δ	delta
E	ϵ	epsilon
Z	ζ	zeta
H	η	eta
Θ	θ	theta
I	ι	iota
K	κ	kappa
Λ	λ	lambda
M	μ	mu
N	ν	nu
Ξ	ξ	xi
O	o	omicron
Π	π	pi
P	ρ	rho
Σ	σ	sigma
T	τ	tau
Y	υ	upsilon
Φ	ϕ	phi
X	χ	chi
Ψ	ψ	psi
Ω	ω	omega

Reciprocal Identities The definitions of the trigonometric functions on page 627 were written so that functions above and below one another are reciprocals of each other. Because $\sin \theta = \frac{y}{r}$ and $\csc \theta = \frac{r}{y}$,

$$\sin \theta = \frac{1}{\csc \theta} \quad \text{and} \quad \csc \theta = \frac{1}{\sin \theta}.$$

Also, $\cos \theta$ and $\sec \theta$ are reciprocals, as are $\tan \theta$ and $\cot \theta$. In summary, we have the **reciprocal identities** that hold for any angle θ that does not lead to a zero denominator.

Reciprocal Identities

$$\sin \theta = \frac{1}{\csc \theta} \qquad \cos \theta = \frac{1}{\sec \theta} \qquad \tan \theta = \frac{1}{\cot \theta}$$

$$\csc \theta = \frac{1}{\sin \theta} \qquad \sec \theta = \frac{1}{\cos \theta} \qquad \cot \theta = \frac{1}{\tan \theta}$$

Identities are equations that are true for all meaningful values of the variable. For example, both $(x + y)^2 = x^2 + 2xy + y^2$ and $2(x + 3) = 2x + 6$ are identities.

When studying identities, be aware that various forms exist. For example,

$$\sin \theta = \frac{1}{\csc \theta} \quad \text{can also be written} \quad \csc \theta = \frac{1}{\sin \theta} \quad \text{and} \quad (\sin \theta)(\csc \theta) = 1.$$

You should become familiar with all forms of these identities.

EXAMPLE 1 Using the Reciprocal Identities

Find each function value.

(a) $\cos \theta$, if $\sec \theta = \frac{5}{3}$ **(b)** $\sin \theta$, if $\csc \theta = -\frac{\sqrt{12}}{2}$

SOLUTION

(a) Because $\cos \theta$ is the reciprocal of $\sec \theta$,

$$\cos \theta = \frac{1}{\sec \theta} = \frac{1}{\frac{5}{3}} = 1 \div \frac{5}{3} = 1 \cdot \frac{3}{5} = \frac{3}{5}.$$

(b) $$\sin \theta = \frac{1}{-\sqrt{12}/2} = \frac{-2}{\sqrt{12}} = \frac{-2}{2\sqrt{3}} = \frac{-1}{\sqrt{3}} = \frac{-1}{\sqrt{3}} \cdot \frac{\sqrt{3}}{\sqrt{3}} = -\frac{\sqrt{3}}{3}$$

$\sin \theta$ is the reciprocal of $\csc \theta$.

Remember to rationalize the denominator.

Signs of Function Values in Quadrants In the definitions of the trigonometric functions, r is the distance from the origin to the point (x, y). Distance is never negative, so $r > 0$. If we choose a point (x, y) in quadrant I, then both x and y will be positive. Because $r > 0$, all six of the fractions used in the definitions of the trigonometric functions will be positive, so that the values of all six functions will be positive in quadrant I.

A point (x, y) in quadrant II has $x < 0$ and $y > 0$. This makes the values of sine and cosecant positive for quadrant II angles, while the other four functions take on negative values. Similar results can be obtained for the other quadrants.

Signs of Function Values

θ in Quadrant	$\sin \theta$	$\cos \theta$	$\tan \theta$	$\cot \theta$	$\sec \theta$	$\csc \theta$
I	+	+	+	+	+	+
II	+	−	−	−	−	+
III	−	−	+	+	−	−
IV	−	+	−	−	+	−

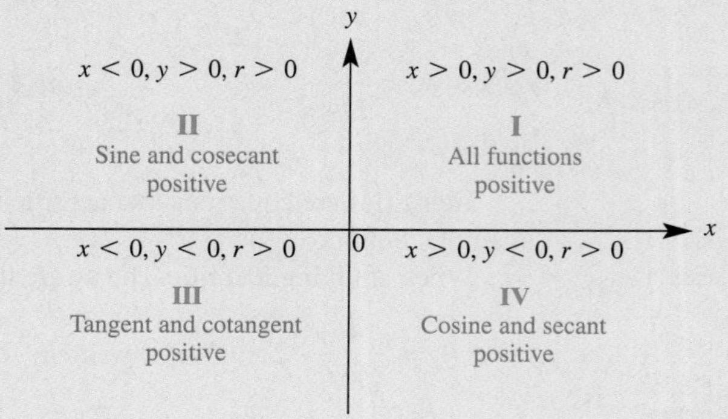

Claudius Ptolemy (c. 100–178) provided the most influential astronomical work of antiquity. Although Hipparchus is known in some circles as the founder of trigonometry, his works are lost; historians have used Ptolemy's work as their source of how the Greeks viewed trigonometry. His thirteen-book description of the Greek model of the universe, known as the *Mathematical Collection,* served as the basis for future studies both in the Islamic world and the West. Because of its importance, Islamic scientists referred to it as *al-magisti,* meaning "the greatest." Since that time, it has been called the *Almagest.* Seen here is a woodcut from one of its early printings. (*Source:* Katz, Victor J., *A History of Mathematics.*)

EXAMPLE 2 Identifying the Quadrant of an Angle

Identify the quadrant (or quadrants) of any angle θ that satisfies $\sin \theta > 0$, $\tan \theta < 0$.

SOLUTION

Because $\sin \theta > 0$ in quadrants I and II, while $\tan \theta < 0$ in quadrants II and IV, both conditions are met only in quadrant II. ■

The six trigonometric functions are defined in terms of x, y, and r, where the Pythagorean theorem shows that $r^2 = x^2 + y^2$ and $r > 0$. With these relationships, knowing the value of only one function and the quadrant in which the angle lies makes it possible to find the values of all six of the trigonometric functions.

EXAMPLE 3 Finding All Function Values Given One Value and the Quadrant

Suppose that angle α is in quadrant II and $\sin \alpha = \frac{2}{3}$. Find the values of the other five functions.

SOLUTION

We can choose any point on the terminal side of angle α. For simplicity, because $\sin \alpha = \frac{y}{r}$, we choose the point with $r = 3$. Then $y = 2$, and

$$\frac{y}{r} = \frac{2}{3}.$$

To find x, use the result $x^2 + y^2 = r^2$.

$$x^2 + y^2 = r^2$$
$$x^2 + 2^2 = 3^2 \qquad \text{Substitute.}$$
$$x^2 + 4 = 9$$
$$x^2 = 5 \qquad \text{Subtract 4.}$$
$$x = \sqrt{5} \quad \text{or} \quad x = -\sqrt{5} \qquad \text{Square root property}$$

Because α is in quadrant II, x must be negative, as shown in Figure 15, so $x = -\sqrt{5}$. This puts the point $(-\sqrt{5}, 2)$ on the terminal side of α.

Now that the values of x, y, and r are known, the values of the remaining trigonometric functions can be found.

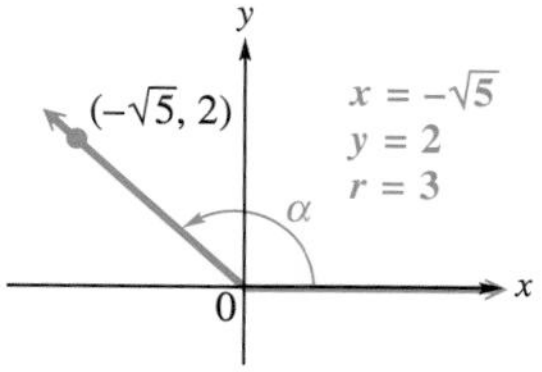

FIGURE 15

$$\cos \alpha = \frac{x}{r} = \frac{-\sqrt{5}}{3} = -\frac{\sqrt{5}}{3}$$

$$\sec \alpha = \frac{r}{x} = \frac{3}{-\sqrt{5}} = -\frac{3}{\sqrt{5}} \cdot \frac{\sqrt{5}}{\sqrt{5}} = -\frac{3\sqrt{5}}{5}$$

$$\tan \alpha = \frac{y}{x} = \frac{2}{-\sqrt{5}} = -\frac{2}{\sqrt{5}} \cdot \frac{\sqrt{5}}{\sqrt{5}} = -\frac{2\sqrt{5}}{5}$$

Remember to rationalize denominators.

$$\cot \alpha = \frac{x}{y} = \frac{-\sqrt{5}}{2} = -\frac{\sqrt{5}}{2} \qquad \csc \alpha = \frac{r}{y} = \frac{3}{2}$$

■

Pythagorean Identities We can derive three very useful new identities from the relationship $x^2 + y^2 = r^2$. Dividing both sides by r^2 gives

$$\frac{x^2}{r^2} + \frac{y^2}{r^2} = \frac{r^2}{r^2},$$

or

$$\left(\frac{x}{r}\right)^2 + \left(\frac{y}{r}\right)^2 = 1.$$

Because $\sin\theta = \frac{y}{r}$ and $\cos\theta = \frac{x}{r}$, this result becomes

$$(\cos\theta)^2 + (\sin\theta)^2 = 1,$$

or, as it is often written,

$$\boldsymbol{\sin^2\theta + \cos^2\theta = 1.}$$

Starting with $x^2 + y^2 = r^2$ and dividing through by x^2 gives

$$\frac{x^2}{x^2} + \frac{y^2}{x^2} = \frac{r^2}{x^2}$$

$$1 + \left(\frac{y}{x}\right)^2 = \left(\frac{r}{x}\right)^2$$

$$1 + (\tan\theta)^2 = (\sec\theta)^2$$

or

$$\boldsymbol{1 + \tan^2\theta = \sec^2\theta.}$$

On the other hand, dividing through by y^2 leads to

$$\boldsymbol{1 + \cot^2\theta = \csc^2\theta.}$$

These three identities are called the **Pythagorean identities** because the original equation that led to them, $x^2 + y^2 = r^2$, comes from the Pythagorean theorem.

Pythagorean Identities

$\sin^2\theta + \cos^2\theta = 1$	$1 + \tan^2\theta = \sec^2\theta$	$1 + \cot^2\theta = \csc^2\theta$

As before, we have given only one form of each identity. However, algebraic transformations can be made to get equivalent identities. For example, by subtracting $\sin^2\theta$ from both sides of $\sin^2\theta + \cos^2\theta = 1$ we get the equivalent identity

$$\cos^2\theta = 1 - \sin^2\theta.$$

Quotient Identities Recall that $\sin\theta = \frac{y}{r}$ and $\cos\theta = \frac{x}{r}$. Consider the quotient of $\sin\theta$ and $\cos\theta$, where $\cos\theta \neq 0$.

$$\frac{\sin\theta}{\cos\theta} = \frac{\frac{y}{r}}{\frac{x}{r}} = \frac{y}{r} \div \frac{x}{r} = \frac{y}{r} \cdot \frac{r}{x} = \frac{y}{x} = \tan\theta$$

Similarly, it can be shown that $\frac{\cos\theta}{\sin\theta} = \cot\theta$, for $\sin\theta \neq 0$. Thus we have two more identities, called the **quotient identities.**

Quotient Identities

$$\frac{\sin\theta}{\cos\theta} = \tan\theta \qquad \frac{\cos\theta}{\sin\theta} = \cot\theta$$

EXAMPLE 4 Finding Other Function Values Given One Value and the Quadrant

Find $\sin\alpha$ and $\tan\alpha$ if $\cos\alpha = -\frac{\sqrt{3}}{4}$ and α is in quadrant II.

SOLUTION

Start with $\sin^2\alpha + \cos^2\alpha = 1$, and replace $\cos\alpha$ with $-\frac{\sqrt{3}}{4}$.

$$\sin^2\alpha + \left(-\frac{\sqrt{3}}{4}\right)^2 = 1 \qquad \text{Replace } \cos\alpha \text{ with } -\frac{\sqrt{3}}{4}.$$

$$\sin^2\alpha + \frac{3}{16} = 1$$

$$\sin^2\alpha = \frac{13}{16} \qquad \text{Subtract } \frac{3}{16}.$$

$$\sin\alpha = \pm\frac{\sqrt{13}}{4} \qquad \text{Take square roots.}$$

$$\sin\alpha = \frac{\sqrt{13}}{4} \qquad \alpha \text{ is in quadrant II, so } \sin\alpha > 0.$$

To find $\tan\alpha$, use the quotient identity $\tan\alpha = \frac{\sin\alpha}{\cos\alpha}$.

$$\tan\alpha = \frac{\sin\alpha}{\cos\alpha} = \frac{\frac{\sqrt{13}}{4}}{-\frac{\sqrt{3}}{4}} = \frac{\sqrt{13}}{4}\left(-\frac{4}{\sqrt{3}}\right) = -\frac{\sqrt{13}}{\sqrt{3}} = -\frac{\sqrt{13}}{\sqrt{3}} \cdot \frac{\sqrt{3}}{\sqrt{3}} = -\frac{\sqrt{39}}{3}$$

10.3 Exercises

Use the appropriate reciprocal identity to find each function value. Rationalize denominators when applicable.

1. $\tan\theta$, if $\cot\theta = -3$

2. $\cot\theta$, if $\tan\theta = 5$

3. $\sin\theta$, if $\csc\theta = 3$

4. $\cos\alpha$, if $\sec\alpha = -\frac{5}{2}$

5. $\cot\beta$, if $\tan\beta = -\frac{1}{5}$

6. $\sin\alpha$, if $\csc\alpha = \sqrt{15}$

7. $\csc\alpha$, if $\sin\alpha = \frac{\sqrt{2}}{4}$

8. $\sec\beta$, if $\cos\beta = -\frac{1}{\sqrt{7}}$

9. $\tan\theta$, if $\cot\theta = -\frac{\sqrt{5}}{3}$

10. $\cot\theta$, if $\tan\theta = \frac{\sqrt{11}}{5}$

11. $\sin\theta$, if $\csc\theta = 1.5$

12. $\cos\theta$, if $\sec\theta = 7.5$

Identify the quadrant or quadrants for each angle satisfying the given conditions.

13. $\sin \alpha > 0, \cos \alpha < 0$

14. $\cos \beta > 0, \tan \beta > 0$

15. $\tan \gamma > 0, \sin \gamma > 0$

16. $\sin \beta < 0, \cos \beta > 0$

17. $\tan \omega < 0, \cos \omega < 0$

18. $\csc \theta < 0, \cos \theta < 0$

19. $\cos \beta < 0$

20. $\tan \theta > 0$

Give the signs of the six trigonometric functions for each angle.

21. 74°

22. 129°

23. 183°

24. 298°

25. 302°

26. 406°

27. 412°

28. −82°

29. −14°

30. −121°

Use identities to find the indicated value.

31. $\tan \alpha$, if $\sec \alpha = 3$, with α in quadrant IV

32. $\cos \theta$, if $\sin \theta = \frac{2}{3}$, with θ in quadrant II

33. $\sin \alpha$, if $\cos \alpha = -\frac{1}{4}$, with α in quadrant II

34. $\csc \beta$, if $\cot \beta = -\frac{1}{2}$, with β in quadrant IV

35. $\tan \theta$, if $\cos \theta = \frac{1}{3}$, with θ in quadrant IV

36. $\sec \theta$, if $\tan \theta = \frac{\sqrt{7}}{3}$, with θ in quadrant III

37. $\cos \beta$, if $\csc \beta = -4$, with β in quadrant III

38. $\sin \theta$, if $\sec \theta = 2$, with θ in quadrant IV

Find all the trigonometric function values for each angle.

39. $\tan \alpha = -\frac{15}{8}$, with α in quadrant II

40. $\cos \alpha = -\frac{3}{5}$, with α in quadrant III

41. $\cot \gamma = \frac{3}{4}$, with γ in quadrant III

42. $\sin \beta = \frac{7}{25}$, with β in quadrant II

43. $\tan \beta = \sqrt{3}$, with β in quadrant III

44. $\csc \theta = 2$, with θ in quadrant II

45. $\sin \beta = \frac{\sqrt{5}}{7}$, with $\tan \beta > 0$

46. $\cot \alpha = \frac{\sqrt{3}}{8}$, with $\sin \alpha > 0$

47. Derive the identity $1 + \cot^2 \theta = \csc^2 \theta$ by dividing $x^2 + y^2 = r^2$ by y^2.

48. Using a method similar to the one given in this section showing that $\frac{\sin \theta}{\cos \theta} = \tan \theta$, show that

$$\frac{\cos \theta}{\sin \theta} = \cot \theta.$$

10.4 Right Triangles and Function Values

Right Triangle Side Ratios • Cofunction Identities • Trigonometric Function Values of Special Angles • Reference Angles

Right Triangle Side Ratios Figure 16 on the next page shows an acute angle A in standard position. The definitions of the trigonometric function values of angle A require x, y, and r. As drawn in Figure 16, x and y are the lengths of the two legs of the right triangle ABC, and r is the length of the hypotenuse. The functions of trigonometry can be adapted to describe the ratios of these sides.

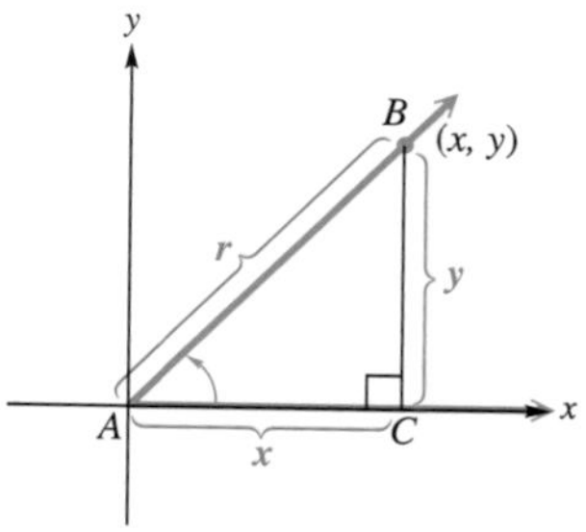

FIGURE 16

The side of length y is called the **side opposite** angle A, and the side of length x is called the **side adjacent** to angle A. The lengths of these sides can be used to replace x and y in the definitions of the trigonometric functions, with r replaced by the length of the hypotenuse, to get the following right triangle-based definitions.

Right Triangle-Based Definitions of Trigonometric Functions

For any acute angle A in standard position,

$$\sin A = \frac{y}{r} = \frac{\text{side opposite } A}{\text{hypotenuse}} \qquad \csc A = \frac{r}{y} = \frac{\text{hypotenuse}}{\text{side opposite } A}$$

$$\cos A = \frac{x}{r} = \frac{\text{side adjacent to } A}{\text{hypotenuse}} \qquad \sec A = \frac{r}{x} = \frac{\text{hypotenuse}}{\text{side adjacent to } A}$$

$$\tan A = \frac{y}{x} = \frac{\text{side opposite } A}{\text{side adjacent to } A} \qquad \cot A = \frac{x}{y} = \frac{\text{side adjacent to } A}{\text{side opposite } A}.$$

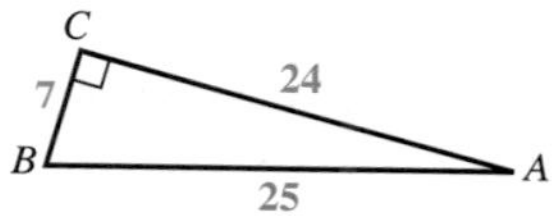

FIGURE 17

EXAMPLE 1 Finding Trigonometric Function Values of an Acute Angle

Find the values of the trigonometric functions for angles A and B in the right triangle in Figure 17.

SOLUTION

The length of the side opposite angle A is 7. The length of the side adjacent to angle A is 24, and the length of the hypotenuse is 25. Using the relationships given above,

$$\sin A = \frac{\text{side opposite}}{\text{hypotenuse}} = \frac{7}{25} \qquad \csc A = \frac{\text{hypotenuse}}{\text{side opposite}} = \frac{25}{7}$$

$$\cos A = \frac{\text{side adjacent}}{\text{hypotenuse}} = \frac{24}{25} \qquad \sec A = \frac{\text{hypotenuse}}{\text{side adjacent}} = \frac{25}{24}$$

$$\tan A = \frac{\text{side opposite}}{\text{side adjacent}} = \frac{7}{24} \qquad \cot A = \frac{\text{side adjacent}}{\text{side opposite}} = \frac{24}{7}.$$

The length of the side opposite angle B is 24, while the length of the side adjacent to B is 7, making

$$\sin B = \frac{24}{25} \qquad \cos B = \frac{7}{25} \qquad \tan B = \frac{24}{7}$$

$$\csc B = \frac{25}{24} \qquad \sec B = \frac{25}{7} \qquad \cot B = \frac{7}{24}.$$

■

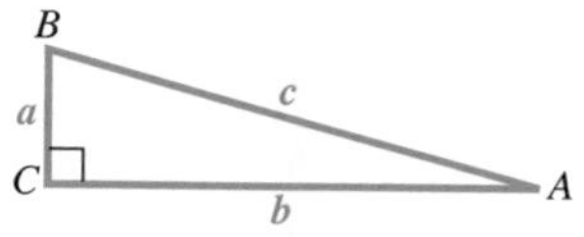

FIGURE 18

Cofunction Identities In Example 1, you may have noticed that $\sin A = \cos B$, $\cos A = \sin B$, and so on. Such relationships are always true for the two acute angles of a right triangle. Figure 18 shows a right triangle with acute angles A and B and a right angle at C. (Whenever we use A, B, and C to name the angles in a right triangle, C will be the right angle.) The length of the side opposite angle A is a, and the length of the side opposite angle B is b. The length of the hypotenuse is c.

By the definitions given above, $\sin A = \frac{a}{c}$. Because $\cos B$ is also equal to $\frac{a}{c}$,

$$\sin A = \frac{a}{c} = \cos B.$$

In a similar manner,

$$\tan A = \frac{a}{b} = \cot B \quad \text{and} \quad \sec A = \frac{c}{b} = \csc B.$$

The sum of the three angles in any triangle is 180°. Because angle C equals 90°, angles A and B must have a sum of $180° - 90° = 90°$. As mentioned in Section 10.1, angles with a sum of 90° are called *complementary angles*. Because angles A and B are complementary and $\sin A = \cos B$, the functions sine and cosine are called **cofunctions.** Also, tangent and cotangent are cofunctions, as are secant and cosecant. And because the angles A and B are complementary, $A + B = 90°$, or

$$B = 90° - A,$$

giving

$$\sin A = \cos B = \cos(90° - A).$$

Similar results are true for the other trigonometric functions. We call these results the **cofunction identities.**

Cofunction Identities

For any acute angle A,

$$\sin A = \cos(90° - A) \qquad \csc A = \sec(90° - A)$$
$$\cos A = \sin(90° - A) \qquad \sec A = \csc(90° - A)$$
$$\tan A = \cot(90° - A) \qquad \cot A = \tan(90° - A).$$

These identities can be extended to *any* angle A, and not just acute angles.

EXAMPLE 2 Writing Functions in Terms of Cofunctions

Write each of the following in terms of cofunctions.

(a) cos 52° **(b)** tan 71° **(c)** sec 24°

SOLUTION

(a) Because $\cos A = \sin(90° - A)$,

$$\cos 52° = \sin(90° - 52°) = \sin 38°.$$

(b) $\tan 71° = \cot(90° - 71°) = \cot 19°$
(c) $\sec 24° = \csc 66°$

Trigonometric Function Values of Special Angles

Certain special angles, such as 30°, 45°, and 60°, occur so often in applications of trigonometry that they deserve special study. We can find the exact trigonometric function values of these angles by using properties of geometry and the Pythagorean theorem.

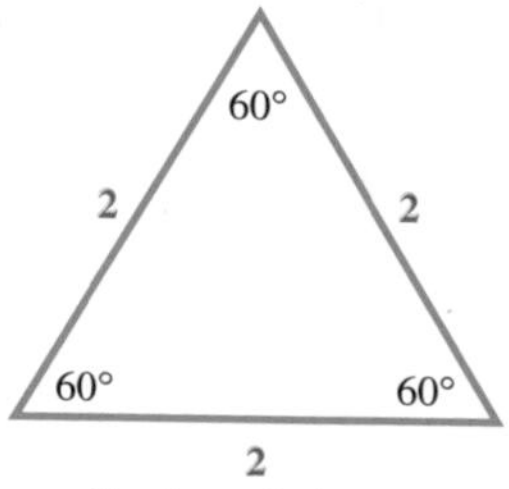

Equilateral triangle
(a)

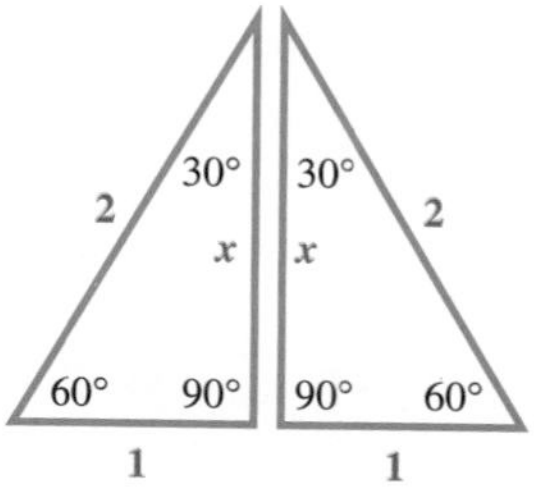

Two 30°–60° right triangles
(b)

FIGURE 19

To find the trigonometric function values for 30° and 60°, we start with an equilateral triangle, a triangle with all sides of equal length. Each angle of such a triangle has a measure of 60°. While the results we will obtain are independent of the length, for convenience, we choose the length of each side to be 2 units. See Figure 19(a).

Bisecting one angle of this equilateral triangle leads to two right triangles, each of which has angles of 30°, 60°, and 90°, as shown in Figure 19(b). Because the hypotenuse of one of these right triangles has a length of 2, the shortest side will have a length of 1. (Why?) If x represents the length of the medium side, then,

$$2^2 = 1^2 + x^2 \quad \text{Pythagorean theorem}$$
$$4 = 1 + x^2$$
$$3 = x^2 \quad \text{Subtract 1.}$$
$$\sqrt{3} = x. \quad \text{Square root property; } x > 0$$

Figure 20 summarizes our results, showing a 30°–60° right triangle. As shown in the figure, the side opposite the 30° angle has length 1; that is, for the 30° angle,

$$\text{hypotenuse} = 2, \qquad \text{side opposite} = 1, \qquad \text{side adjacent} = \sqrt{3}.$$

Using the definitions of the trigonometric functions,

$$\sin 30° = \frac{\text{side opposite}}{\text{hypotenuse}} = \frac{1}{2} \qquad \csc 30° = \frac{2}{1} = 2$$
$$\cos 30° = \frac{\text{side adjacent}}{\text{hypotenuse}} = \frac{\sqrt{3}}{2} \qquad \sec 30° = \frac{2}{\sqrt{3}} = \frac{2\sqrt{3}}{3}$$
$$\tan 30° = \frac{\text{side opposite}}{\text{side adjacent}} = \frac{1}{\sqrt{3}} = \frac{\sqrt{3}}{3} \qquad \cot 30° = \frac{\sqrt{3}}{1} = \sqrt{3}.$$

The denominator was rationalized for tan 30° and sec 30°.

In a similar manner,

$$\sin 60° = \frac{\sqrt{3}}{2} \qquad \cos 60° = \frac{1}{2} \qquad \tan 60° = \sqrt{3}$$
$$\csc 60° = \frac{2\sqrt{3}}{3} \qquad \sec 60° = 2 \qquad \cot 60° = \frac{\sqrt{3}}{3}.$$

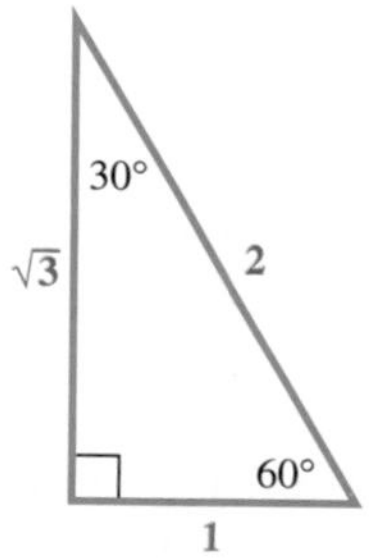

FIGURE 20

The values of the trigonometric functions for 45° can be found by starting with a 45°–45° right triangle, as shown in Figure 21. This triangle is isosceles, and, for convenience, we choose the lengths of the equal sides to be 1 unit. (As before, the results are independent of the length of the equal sides of the right triangle.) Because the shorter sides each have length 1, if r represents the length of the hypotenuse, then

$$1^2 + 1^2 = r^2$$
$$2 = r^2$$
$$\sqrt{2} = r.$$

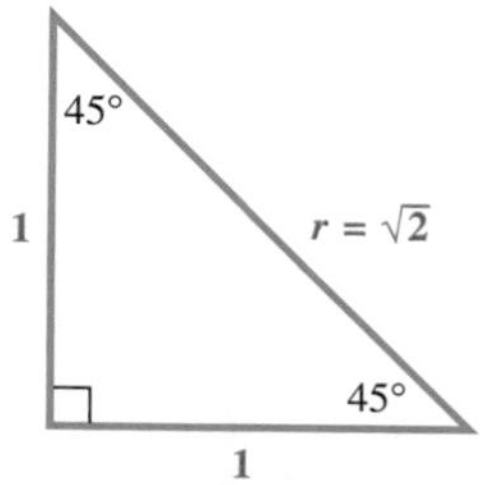

45°–45° right triangle

FIGURE 21

Using the measures indicated on the 45°–45° right triangle in Figure 21, we find

$$\sin 45^\circ = \frac{1}{\sqrt{2}} = \frac{\sqrt{2}}{2} \qquad \cos 45^\circ = \frac{1}{\sqrt{2}} = \frac{\sqrt{2}}{2} \qquad \tan 45^\circ = \frac{1}{1} = 1$$

$$\csc 45^\circ = \frac{\sqrt{2}}{1} = \sqrt{2} \qquad \sec 45^\circ = \frac{\sqrt{2}}{1} = \sqrt{2} \qquad \cot 45^\circ = \frac{1}{1} = 1.$$

The importance of these exact trigonometric function values of 30°, 60°, and 45° angles cannot be overemphasized. They are summarized in Table 3. You should be able to reproduce the function values in the table if you remember the values of sin 30°, cos 30°, and sin 45°. Then complete the rest of the chart using the reciprocal, quotient, and cofunction identities. Another option is to visualize the appropriate triangles and use the ratios of the definitions.

TABLE 3 Function Values of Special Angles

θ	$\sin\theta$	$\cos\theta$	$\tan\theta$	$\cot\theta$	$\sec\theta$	$\csc\theta$
30°	$\frac{1}{2}$	$\frac{\sqrt{3}}{2}$	$\frac{\sqrt{3}}{3}$	$\sqrt{3}$	$\frac{2\sqrt{3}}{3}$	2
45°	$\frac{\sqrt{2}}{2}$	$\frac{\sqrt{2}}{2}$	1	1	$\sqrt{2}$	$\sqrt{2}$
60°	$\frac{\sqrt{3}}{2}$	$\frac{1}{2}$	$\sqrt{3}$	$\frac{\sqrt{3}}{3}$	2	$\frac{2\sqrt{3}}{3}$

Reference Angles Now that we have discussed trigonometric function values for acute angles, we can use those results to find trigonometric function values for other types of angles. Associated with every nonquadrantal angle in standard position is a positive acute angle called its *reference angle*. A **reference angle** for an angle θ, written θ', is the positive acute angle made by the terminal side of angle θ and the x-axis. Figure 22 shows several angles θ (each less than one complete counterclockwise revolution) in quadrants II, III, and IV, respectively, with the reference angle θ' also shown. In quadrant I, θ and θ' are the same. If an angle θ is negative or has measure greater than 360°, its reference angle is found by first finding its coterminal angle that is between 0° and 360°, and then using the diagram in Figure 22.

A common error is to find the reference angle by using the terminal side of θ and the y-axis. ***The reference angle is always found with reference to the x-axis.***

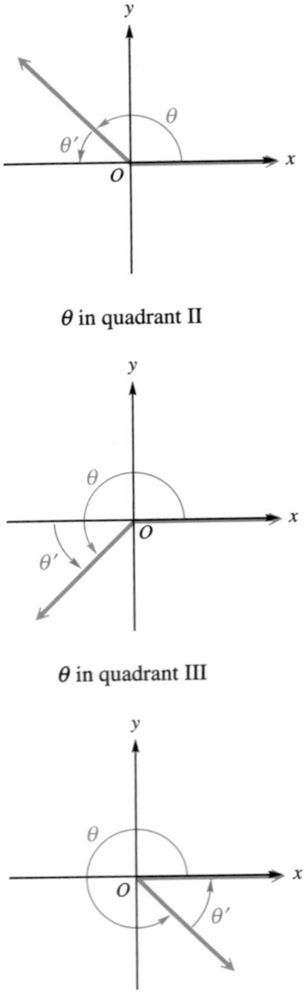

θ in quadrant II

θ in quadrant III

θ in quadrant IV

FIGURE 22

EXAMPLE 3 Finding Reference Angles

Find the reference angles for each angle.

(a) 218° **(b)** 1387°

SOLUTION

(a) As shown in Figure 23 on the next page, the positive acute angle made by the terminal side of this angle and the x-axis is $218^\circ - 180^\circ = 38^\circ$. For $\theta = 218^\circ$, the reference angle is $\theta' = 38^\circ$.

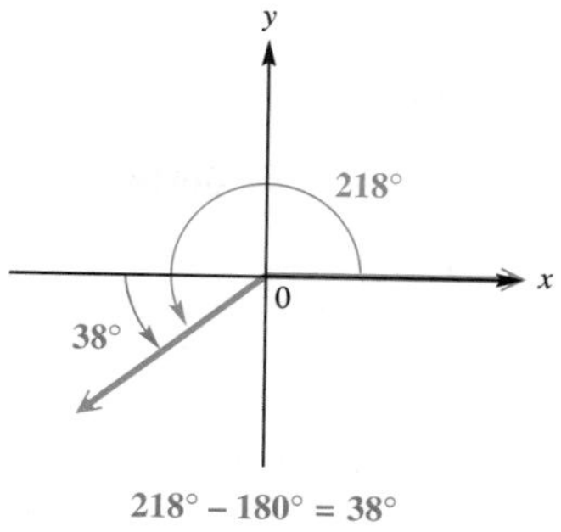

FIGURE 23

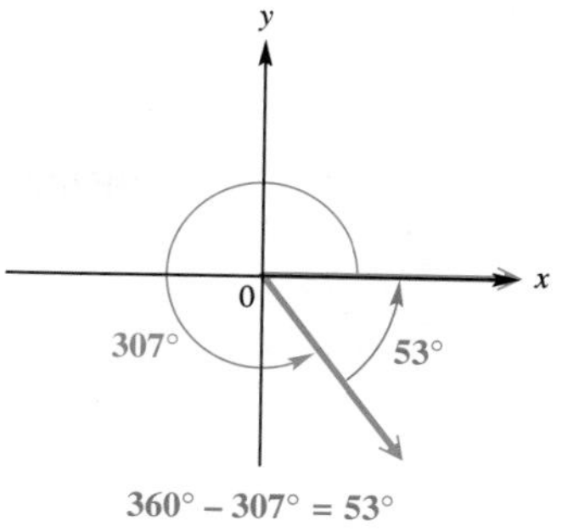

FIGURE 24

(b) First find a coterminal angle between 0° and 360°. Divide 1387° by 360° to get a quotient of about 3.9. Begin by subtracting 360° three times (because of the whole number 3 in 3.9):

$$1387° - 3 \cdot 360° = 307°.$$

The reference angle for 307° (and thus for 1387°) is $360° - 307° = 53°$. See Figure 24.

The preceding example suggests the following results for finding the reference angle θ' for any angle θ between 0° and 360°.

Reference Angle θ' for θ, where $0° < \theta < 360°$

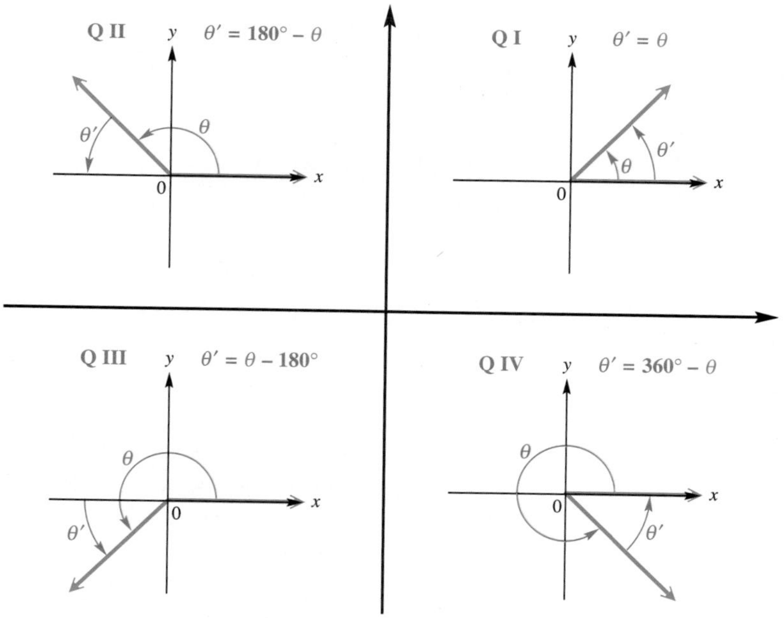

We can now find exact trigonometric function values of all angles with reference angles of 30°, 60°, or 45°. In Example 4 we show how to use these function values to find the trigonometric function values for 210°.

EXAMPLE 4 Finding Trigonometric Function Values of a Quadrant III Angle

Find the exact values of the trigonometric functions for 210°.

SOLUTION

Even though a 210° angle is not an angle of a right triangle, the ideas mentioned earlier can still be used to find the trigonometric function values for this angle. To do so, we draw an angle of 210° in standard position, as shown in Figure 25. We choose point P on the terminal side of the angle so that the distance from the origin O to P is 2. By the results from 30°–60° right triangles, the coordinates of point P become $(-\sqrt{3},-1)$, with $x = -\sqrt{3}$, $y = -1$, and $r = 2$. Then, by the definitions of the trigonometric functions,

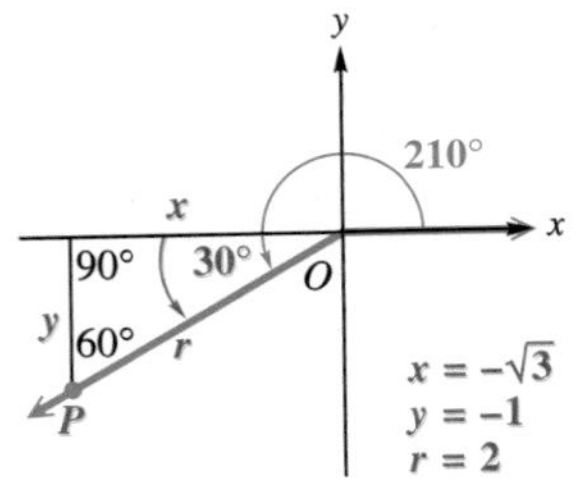

FIGURE 25

$$\sin 210° = -\frac{1}{2} \qquad \cos 210° = -\frac{\sqrt{3}}{2} \qquad \tan 210° = \frac{\sqrt{3}}{3}$$

$$\csc 210° = -2 \qquad \sec 210° = -\frac{2\sqrt{3}}{3} \qquad \cot 210° = \sqrt{3}.$$

Historian Reviel Netz of Stanford University has been studying an ancient manuscript by the Greek mathematician Archimedes ("Unveiling the work of Archimedes," *Science News,* Vol. 157, p. 77, January 29, 2000). It is known as *Archimedes Palimpsest,* and dates from the 10th century A.D., surviving on a parchment that was later cut, scraped, and overwritten with a description of a church ritual. The use of ultraviolet photography and digital imaging makes it possible to read beneath the lines of the ritual and see Archimedes' text and diagrams. According to Netz, the diagrams suggest that Greek mathematicians emphasized qualitative relationships over quantitative accuracy.

Notice in Example 4 that the trigonometric function values of 210° correspond in absolute value to those of its reference angle 30°. The signs are different for the sine, cosine, secant, and cosecant functions, because 210° is a quadrant III angle. These results suggest a method for finding the trigonometric function values of a nonacute angle, using the reference angle. In Example 4, the reference angle for 210° is 30°, as shown in Figure 25. Simply by using the trigonometric function values of the reference angle, 30°, and choosing the correct signs for a quadrant III angle, we obtain the same results as found in Example 4.

The values of the trigonometric functions for any nonquadrantal angle θ can be determined by finding the function values for an angle between 0° and 90°. To do this, perform the following steps.

Finding Trigonometric Function Values for Any Nonquadrantal Angle

Step 1 If $\theta > 360°$, or if $\theta < 0°$, find a coterminal angle by adding or subtracting 360° as many times as needed to obtain an angle greater than 0° but less than 360°.

Step 2 Find the reference angle θ'.

Step 3 Find the necessary values of the trigonometric functions for the reference angle θ'.

Step 4 Determine the correct signs for the values found in Step 3. (Use the table of signs in Section 10.3.) This result gives the values of the trigonometric functions for angle θ.

EXAMPLE 5 Finding Trigonometric Function Values Using Reference Angles

Use reference angles to find each exact value.

(a) $\cos(-240°)$ **(b)** $\tan 675°$

SOLUTION

(a) The reference angle for $-240°$ is $60°$, as shown in Figure 26. Because the cosine is negative in quadrant II,

$$\cos(-240°) = -\cos 60° = -\frac{1}{2}.$$

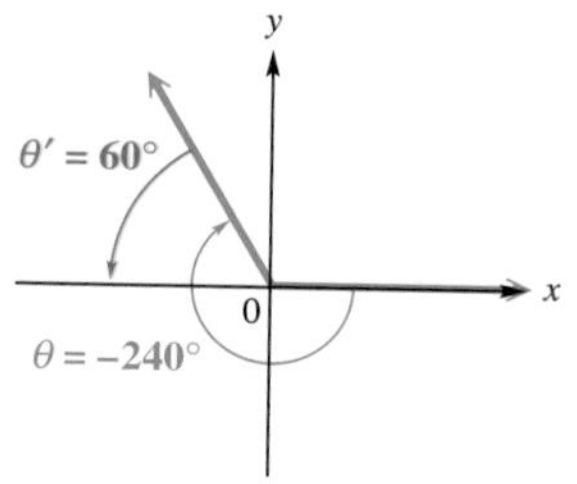

FIGURE 26

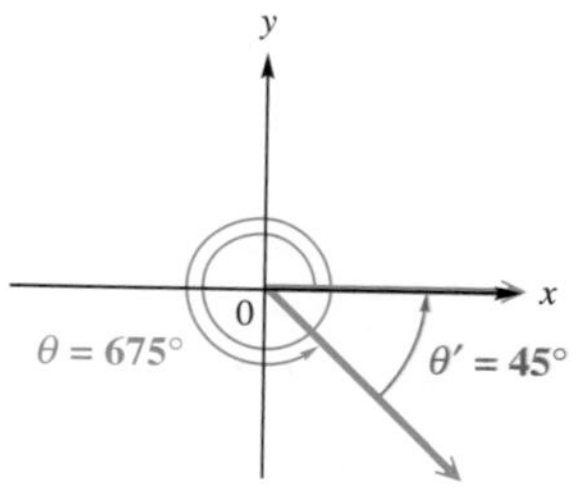

FIGURE 27

(b) Begin by subtracting $360°$ to obtain a coterminal angle between $0°$ and $360°$.

$$675° - 360° = 315°$$

As shown in Figure 27, the reference angle is $360° - 315° = 45°$. An angle of $315°$ is in quadrant IV, so the tangent will be negative, and

$$\tan 675° = \tan 315° = -\tan 45° = -1.$$

The ideas discussed in this section can be used inversely to find the measures of certain angles, given a trigonometric function value and an interval in which an angle θ must lie. We are most often interested in the interval $0° \le \theta < 360°$.

EXAMPLE 6 Finding Angle Measures Given an Interval and a Function Value

Find all values of θ, if $0° \le \theta < 360°$ and $\cos\theta = -\frac{\sqrt{2}}{2}$.

SOLUTION

Because cosine here is negative, θ must lie in either quadrant II or III. The absolute value of $\cos\theta$ is $\frac{\sqrt{2}}{2}$, so the reference angle θ' must be $45°$. The two possible angles θ are sketched in Figure 28. The quadrant II angle θ must equal $180° - 45° = 135°$, and the quadrant III angle θ must equal $180° + 45° = 225°$.

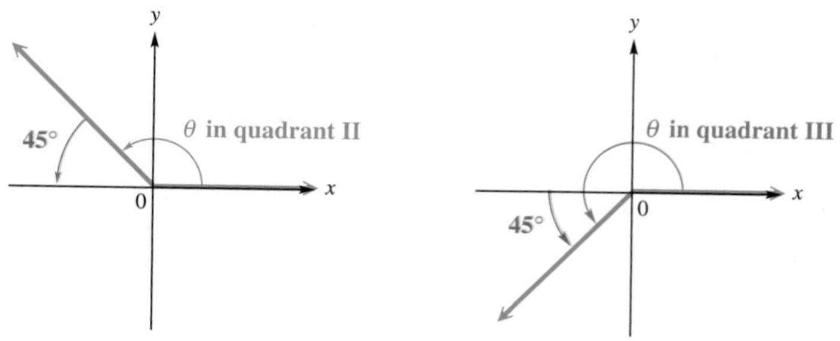

FIGURE 28

10.4 Exercises

Find the values of the six trigonometric functions for angle A. (In Exercises 5 and 6, answers will be in terms of variables.) Leave answers as fractions.

1.

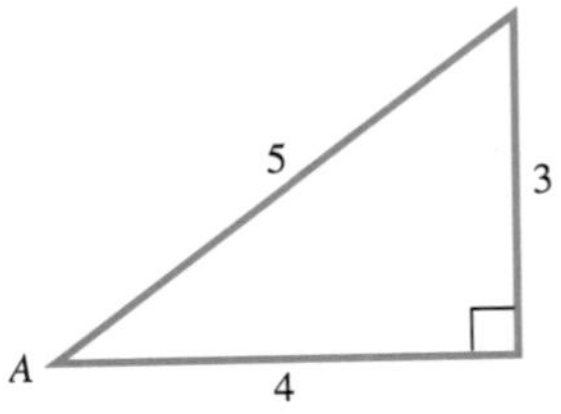

2.

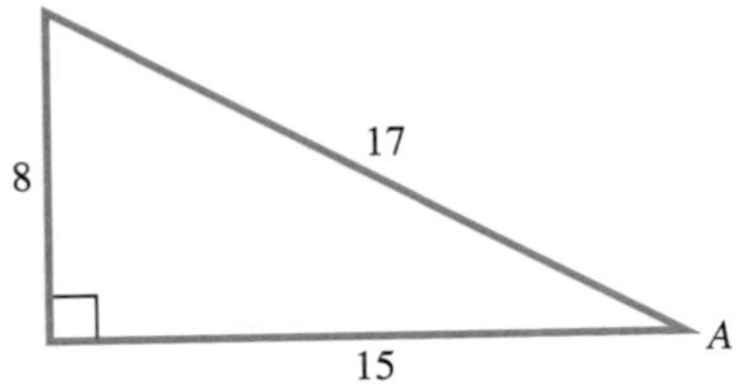

3.

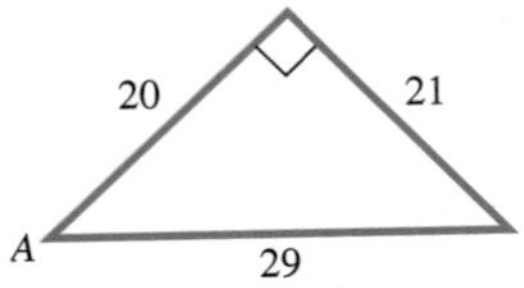

4.

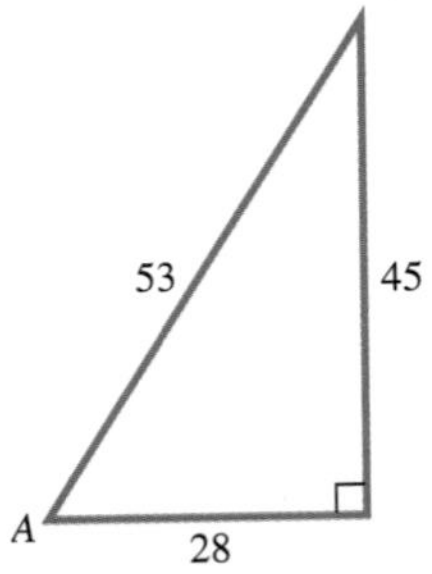

5.

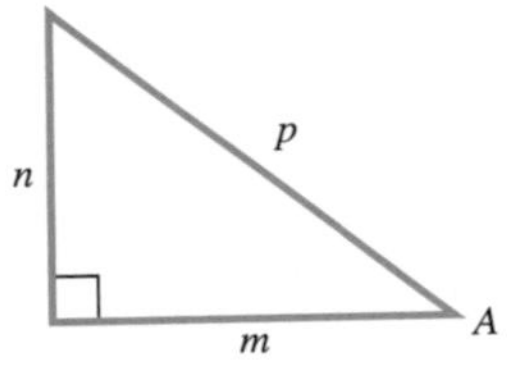

6.

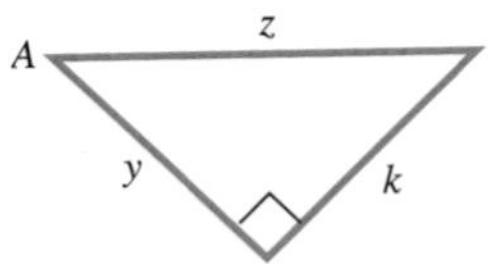

Suppose ABC is a right triangle with sides of lengths a, b, and c, with right angle C (see Figure 18 on page 637). Find the unknown side length using the Pythagorean theorem, and then find the values of the six trigonometric functions for angle B. Rationalize denominators when applicable.

7. $a = 5, b = 12$ **8.** $a = 3, b = 5$ **9.** $a = 6, c = 7$ **10.** $b = 7, c = 12$

Write each of the following in terms of the cofunction.

11. tan 50° **12.** cot 73° **13.** csc 47° **14.** sec 39°

15. tan 25.4° **16.** sin 38.7° **17.** cos 13° 30′ **18.** tan 26° 10′

Give the exact *trigonometric function value. Do not use a calculator.*

19. tan 30° **20.** cot 30° **21.** sin 30° **22.** cos 30°

23. csc 45° **24.** sec 45° **25.** cos 45° **26.** sin 45°

27. sin 60° **28.** cos 60° **29.** tan 60° **30.** cot 60°

Find the reference angle for each angle.

31. 98° **32.** 212° **33.** −135° **34.** −60° **35.** 750° **36.** 480°

Find the exact values *of the six trigonometric functions for each angle. Rationalize denominators when applicable.*

37. 120° **38.** 135° **39.** 150° **40.** 225° **41.** 240°

42. 300° **43.** 315° **44.** 405° **45.** 420° **46.** 480°

47. 495° **48.** 570° **49.** 750° **50.** 1305° **51.** 1500°

52. 2670° **53.** −390° **54.** −510° **55.** −1020° **56.** −1290°

Complete the table with exact *trigonometric function values using the methods of this section.*

	θ	$\sin\theta$	$\cos\theta$	$\tan\theta$	$\cot\theta$	$\sec\theta$	$\csc\theta$
57.	**30°**	$\frac{1}{2}$	$\frac{\sqrt{3}}{2}$			$\frac{2\sqrt{3}}{3}$	2
58.	**45°**			1	1		
59.	**60°**		$\frac{1}{2}$	$\sqrt{3}$		2	
60.	**120°**	$\frac{\sqrt{3}}{2}$		$-\sqrt{3}$			$\frac{2\sqrt{3}}{3}$
61.	**135°**	$\frac{\sqrt{2}}{2}$	$-\frac{\sqrt{2}}{2}$			$-\sqrt{2}$	$\sqrt{2}$
62.	**150°**	–	$-\frac{\sqrt{3}}{2}$	$-\frac{\sqrt{3}}{3}$			2
63.	**210°**	$-\frac{1}{2}$		$\frac{\sqrt{3}}{3}$	$\sqrt{3}$		-2
64.	**240°**	$-\frac{\sqrt{3}}{2}$	$-\frac{1}{2}$			-2	$-\frac{2\sqrt{3}}{3}$

Find all values of θ, if $0° \leq \theta < 360°$, and the given condition is true.

65. $\sin\theta = -\frac{1}{2}$ **66.** $\cos\theta = -\frac{1}{2}$ **67.** $\tan\theta = 1$

68. $\cot\theta = \sqrt{3}$ **69.** $\sin\theta = \frac{\sqrt{3}}{2}$ **70.** $\cos\theta = \frac{\sqrt{3}}{2}$

71. $\sec\theta = -2$ **72.** $\csc\theta = -2$ **73.** $\sin\theta = -\frac{\sqrt{2}}{2}$

74. $\cos\theta = -\frac{\sqrt{2}}{2}$ **75.** $\tan\theta = -\sqrt{3}$ **76.** $\cot\theta = -1$

10.5 Applications of Right Triangles

Calculator Approximations for Function Values • Finding Angles Using Inverse Functions • Significant Digits • Solving Triangles • Applications

Calculator Approximations for Function Values With the technological advances of this era in mind, the examples and exercises that follow in this chapter assume that all students have access to scientific calculators. However, because calculators differ among makes and models, students should always consult their owner's manual for specific information if questions arise concerning their use.

Thus far in this book, we have studied only one type of measure for angles—degree measure; another type of measure, *radian measure*, is studied in more theoretical work. ***When evaluating trigonometric functions of angles given in degrees, it is a common error to use the incorrect mode; remember that the calculator must be set in the degree mode.***

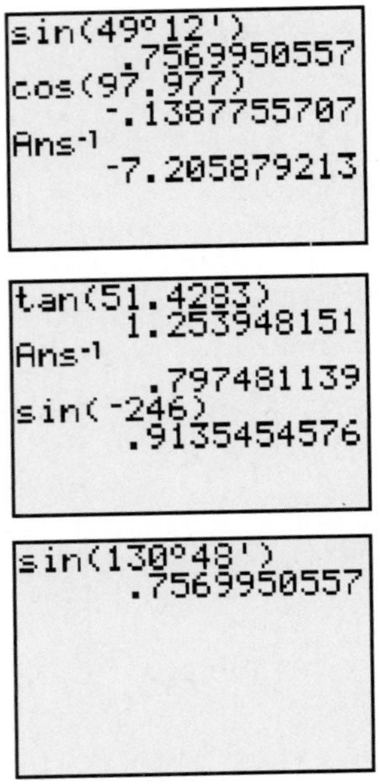

The three screens above show the results for parts (a)—(e) in Example 1. Notice that the calculator permits entering the angle measure in degrees and minutes in parts (a) and (e). In the fifth line of the first screen, Ans^{-1} tells the calculator to find the reciprocal of the answer given in the previous line.

EXAMPLE 1 Finding Function Values with a Calculator

Use a calculator to approximate the value of each trigonometric function.

(a) $\sin 49°\,12'$ **(b)** $\sec 97.977°$ **(c)** $\cot 51.4283°$
(d) $\sin(-246°)$ **(e)** $\sin 130°\,48'$

SOLUTION

(a) Convert $49°\,12'$ to decimal degrees.

$$49°\,12' = 49\frac{12}{60}^{\circ} = 49.2°$$

$$\sin 49°\,12' = \sin 49.2° \approx .75699506 \quad \text{To eight decimal places}$$

(b) Calculators do not have secant keys. However, $\sec\theta = \frac{1}{\cos\theta}$ for all angles θ where $\cos\theta \neq 0$. So find $\sec 97.977°$ by first finding $\cos 97.977°$ and then taking the reciprocal to get

$$\sec 97.977° \approx -7.20587921.$$

(c) Use the identity $\cot\theta = \frac{1}{\tan\theta}$.

$$\cot 51.4283° \approx .79748114$$

(d) $\sin(-246°) \approx .91354546$
(e) $130°\,48'$ is equal to $130.8°$, so

$$\sin 130°\,48' = \sin 130.8° \approx .75699506.$$

Notice that the values found in parts (a) and (e) of Example 1 are the same. The reason for this is that $49°\,12'$ is the reference angle for $130°\,48'$ and the sine function is positive for a quadrant II angle.

Finding Angles Using Inverse Functions So far in this section we have used a calculator to find trigonometric function values of angles. This process can be reversed using *inverse functions*. Inverse functions are denoted by using -1 as a superscript. For example, the inverse of f is denoted f^{-1} (read "f inverse"). For now we restrict our attention to angles θ in the interval $0° \le \theta \le 90°$. The measure of an angle can be found from one of its trigonometric function values using inverse functions as shown in the next example.

EXAMPLE 2 Using Inverse Functions to Find Angles

Use a calculator to find a value of θ such that $0° \le \theta \le 90°$, and θ satisfies each of the following. Leave answers in decimal degrees.

(a) $\sin\theta = .81815000$ **(b)** $\sec\theta = 1.0545829$

This screen supports the results of Example 2.

SOLUTION

(a) We find θ using a key labeled [arc] or [INV] together with the [sin] key. Some calculators may require a key labeled [$\sin^{-1}$] instead. Check your owner's manual to see how your calculator handles this. Again, make sure the calculator is set for degree measure. You should get

$$\theta \approx 54.900028°.$$

(b) Use the identity $\cos\theta = \frac{1}{\sec\theta}$. Enter 1.0545829 and find the reciprocal. This gives $\cos\theta \approx .9482421913$. Now find θ as described in part (a) using inverse cosine. The result is

$$\theta \approx 18.514704°.$$

Compare Examples 1(b) and 2(b). Note that the reciprocal key is used *before* the inverse cosine key when finding the angle, but *after* the cosine key when finding the trigonometric function value.

Significant Digits

Supppose we quickly measure a room as 15 feet by 18 feet. To calculate the length of a diagonal of the room, we can use the Pythagorean theorem.

$$d^2 = 15^2 + 18^2$$
$$d^2 = 549$$
$$d = \sqrt{549} \approx 23.430749$$

Should this answer be given as the length of the diagonal of the room? Of course not. The number 23.430749 contains 6 decimal places, while the original data of 15 feet and 18 feet are only accurate to the nearest foot. Because the results of a problem can be no more accurate than the least accurate number in any calculation, we really should say that the diagonal of the 15-by-18-foot room is 23 feet.

If a wall measured to the nearest foot is 18 feet long, this actually means that the wall has length between 17.5 feet and 18.5 feet. If the wall is measured more accurately as 18.3 feet long, then its length is really between 18.25 feet and 18.35 feet. A measurement of 18.00 feet would indicate that the length of the wall is between 17.995 feet and 18.005 feet. The measurement 18 feet is said to have two *significant digits* of accuracy; 18.0 has three significant digits, and 18.00 has four.

Consider the measurement 900 meters. We cannot tell whether this represents a measurement to the nearest meter, ten meters, or hundred meters. To avoid this problem, we write the number in scientific notation as 9.00×10^2 to the nearest meter, 9.0×10^2 to the nearest ten meters, or 9×10^2 to the nearest hundred meters. These three cases have three, two, and one significant digits, respectively.

A **significant digit** is a digit obtained by actual measurement. A number that represents the result of counting, or a number that results from theoretical work and is not the result of a measurement, is an **exact number.**

Most values of trigonometric functions are approximations, and virtually all measurements are approximations. To perform calculations on such approximate numbers, follow the rules given below. (The rules for rounding are found in Section 6.5.)

Calculation with Significant Digits

For *adding* and *subtracting*, round the answer so that the last digit you keep is in the rightmost column in which all the numbers have significant digits.

For *multiplying* or *dividing*, round the answer to the least number of significant digits found in any of the given numbers.

For *powers* and *roots*, round the answer so that it has the same number of significant digits as the numbers whose power or root you are finding.

When solving for angles, use Table 4 to determine the significant digits in angle measure.

TABLE 4 Significant Digits for Angles

Number of Significant Digits	Angle Measure to Nearest:
2	Degree
3	Ten minutes, or nearest tenth of a degree
4	Minute, or nearest hundredth of a degree
5	Tenth of a minute, or nearest thousandth of a degree

For example, an angle measuring $52° \, 30'$ has three significant digits (assuming that $30'$ is measured to the nearest ten minutes).

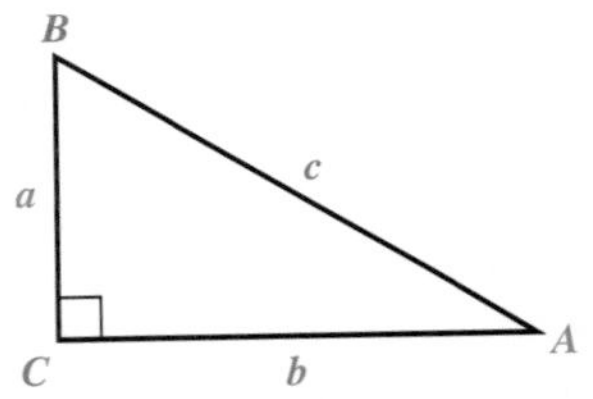

FIGURE 29

Solving Triangles

To **solve a triangle** means to find the measures of all the angles and all the sides of the triangle. In using trigonometry to solve triangles, a labeled sketch is an important aid. It is conventional to use a to represent the length of the side opposite angle A, b for the length of the side opposite angle B, and so on. As mentioned earlier, in a right triangle the letter c is reserved for the hypotenuse. Figure 29 shows the labeling of a typical right triangle.

In Examples 3–6, note that for convenience we use equality symbols (=), even though the side lengths and angle measures are often approximations.

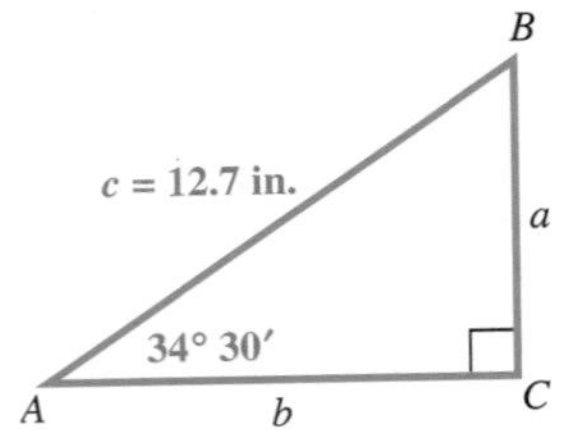

FIGURE 30

EXAMPLE 3 Solving a Right Triangle Given an Angle and a Side

Solve right triangle ABC, if $A = 34° \, 30'$ and $c = 12.7$ in. See Figure 30.

SOLUTION

To solve the triangle, find the measures of the remaining sides and angles. The value of a can be found with a trigonometric function involving the known values of angle A and side c. Because the sine of angle A is given by the quotient of the side opposite A and the hypotenuse, use sin A.

$$\sin A = \frac{a}{c}$$

$$\sin 34° \, 30' = \frac{a}{12.7} \quad \text{Substitute known values.}$$

$$a = 12.7 \sin 34° 30' \quad \text{Multiply both sides by 12.7; rewrite.}$$
$$a = 12.7(.56640624) \quad \text{Use a calculator.}$$
$$a = 7.19 \text{ in.}$$

The value of b could be found with the Pythagorean theorem. It is better, however, to use the information given in the problem rather than a result just calculated. If a mistake were to be made in finding a, then b also would be incorrect. Also, rounding more than once may cause the result to be less accurate. Using $\cos A$ gives

$$\cos A = \frac{\text{side adjacent}}{\text{hypotenuse}} = \frac{b}{c}$$
$$\cos 34° 30 = \frac{b}{12.7}$$
$$b = 12.7 \cos 34° 30'$$
$$b = 10.5 \text{ in.}$$

Once b has been found, the Pythagorean theorem could be used as a check. All that remains to solve triangle ABC is to find the measure of angle B.

$$A + B = 90°$$
$$B = 90° - A$$
$$B = 89° 60' - 34° 30'$$
$$B = 55° 30'$$

In Example 3 we could have started by finding the measure of angle B and then used the trigonometric function values of B to find the unknown sides. The process of solving a right triangle (like many problems in mathematics) can usually be done in several ways, each resulting in the correct answer. However, in order to retain as much accuracy as can be expected, always use the given information as much as possible, and avoid rounding off in intermediate steps.

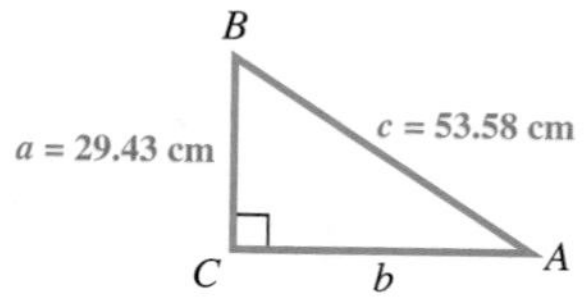

FIGURE 31

EXAMPLE 4 Solving a Right Triangle Given Two Sides

Solve right triangle ABC if $a = 29.43$ cm and $c = 53.58$ cm.

SOLUTION

Draw a sketch showing the given information, as in Figure 31. One way to begin is to find the sine of angle A, and then use inverse sine.

$$\sin A = \frac{\text{side opposite}}{\text{hypotenuse}} = \frac{29.43}{53.58}$$

Using [INV] [sin] or [$\sin^{-1}$] on a calculator, we find that $A = 33.32°$. The measure of B is $90° - 33.32° = 56.68°$.

We now find b from the Pythagorean theorem, $a^2 + b^2 = c^2$, or $b^2 = c^2 - a^2$.

$$b^2 = 53.58^2 - 29.43^2 \quad c = 53.58 \text{ and } a = 29.43$$
$$b = 44.77 \text{ cm}$$

Applications

PROBLEM-SOLVING HINT The process of solving right triangles is easily adapted to solving applied problems. ***A crucial step in such applications involves sketching the triangle and labeling the given parts correctly.*** Then we can use the methods described in the earlier examples to find the unknown value or values.

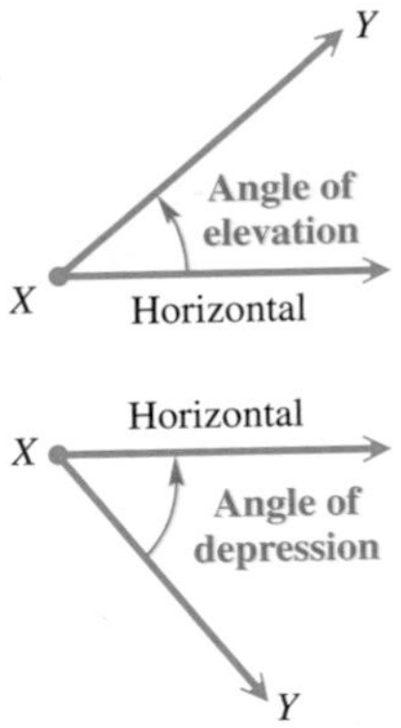

FIGURE 32

Many applications of right triangles involve the *angle of elevation* or the *angle of depression.* The **angle of elevation** from point X to point Y (above X) is the acute angle formed by ray XY and a horizontal ray with endpoint at X. The angle of elevation is always measured from the horizontal. See the angle at the top of Figure 32. The **angle of depression** from point X to point Y (below X) is the acute angle formed by ray XY and a horizontal ray with endpoint X, as shown at the bottom of Figure 32.

Errors are often made in interpreting the angle of depression. ***Remember that both the angle of elevation and the angle of depression are measured between the line of sight and the horizontal.***

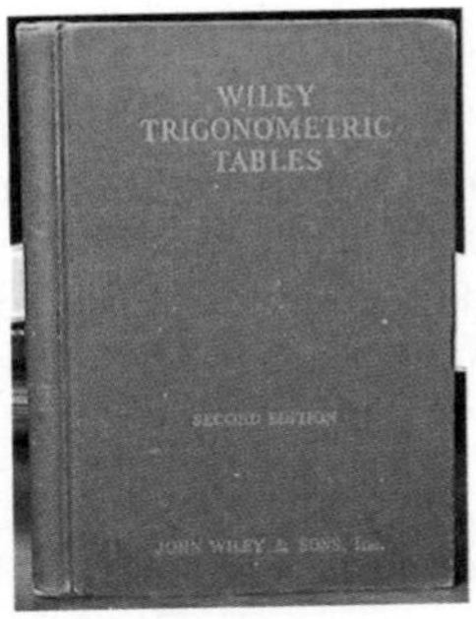

Prior to the advent of the scientific calculator in the 1970s, trigonometry students were required to use **tables of trigonometric function values.** While most trigonometry texts included short versions of these tables, there were books like the one shown here that gave function values for angles, as well as information on their logarithms to make computations easier. (*Author's note:* Today's students don't know how fortunate they are not to have to use such tables.)

EXAMPLE 5 Finding a Length When the Angle of Elevation Is Known

Ana deArmas knows that when she stands 123 feet from the base of a flagpole, the angle of elevation to the top is $26° 40'$. If her eyes are 5.30 feet above the ground, find the height of the flagpole.

SOLUTION

The length of the side adjacent to Ana is known and the length of the side opposite her is to be found. (See Figure 33.) The ratio that involves these two values is the tangent.

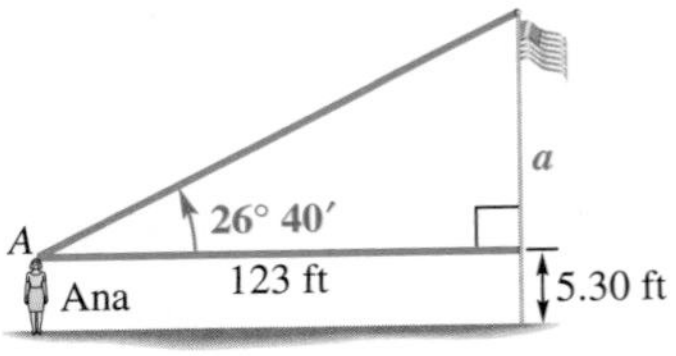

FIGURE 33

$$\tan A = \frac{\text{side opposite}}{\text{side adjacent}}$$

$$\tan 26° 40' = \frac{a}{123}$$

$$a = 123 \tan 26° 40'$$

$$a = 61.8 \text{ feet}$$

Because Ana's eyes are 5.30 feet above the ground, the height of the flagpole is

$$61.8 + 5.30 = 67.1 \text{ feet.}$$

EXAMPLE 6 Finding the Angle of Elevation When Lengths Are Known

The length of the shadow of a building 34.09 meters tall is 37.62 meters. Find the angle of elevation of the sun.

SOLUTION

As shown in Figure 34, the angle of elevation of the sun is angle B. Because the side opposite B and the side adjacent to B are known, use the tangent ratio to find B.

$$\tan B = \frac{34.09}{37.62}$$

$$B = 42.18° \quad \text{Use the inverse tangent function of a calculator.}$$

The angle of elevation of the sun is 42.18°.

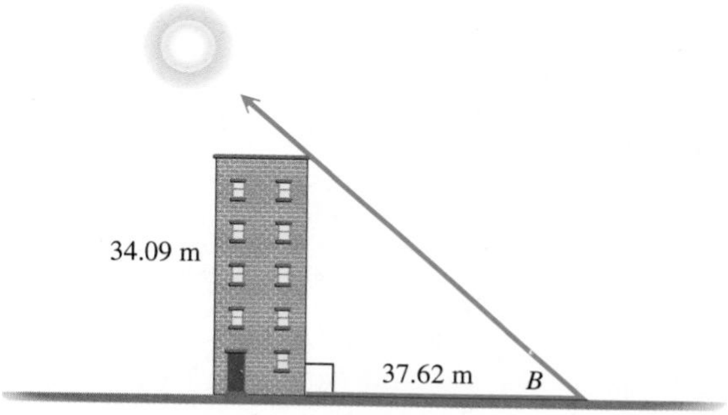

FIGURE 34

10.5 Exercises

Use a calculator to find a decimal approximation for each value. Give as many digits as your calculator displays.

1. $\tan 29°\,30'$

2. $\sin 38°\,42'$

3. $\cot 41°\,24'$

4. $\cos 27°\,10'$

5. $\sec 13°\,15'$

6. $\csc 44°\,30'$

7. $\sin 39°\,40'$

8. $\tan 17°\,12'$

9. $\csc 145°\,45'$

10. $\cot 183°\,48'$

11. $\cos 421°\,30'$

12. $\sec 312°\,12'$

13. $\tan(-80°\,6')$

14. $\sin(-317°\,36')$

15. $\cot(-512°\,20')$

16. $\cos(-15')$

Find a value of θ such that $0° \le \theta \le 90°$, and θ satisfies the statement. Leave your answer in decimal degrees.

17. $\sin\theta = .84802194$

18. $\tan\theta = 1.4739716$

19. $\sec\theta = 1.1606249$

20. $\cot\theta = 1.2575516$

21. $\sin\theta = .72144101$

22. $\sec\theta = 2.7496222$

23. $\tan\theta = 6.4358841$

24. $\sin\theta = .27843196$

In the remaining exercises in this set, use a calculator as necessary.

Solve each right triangle.

25.

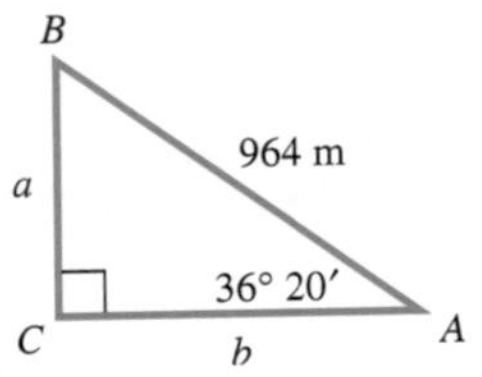

26.

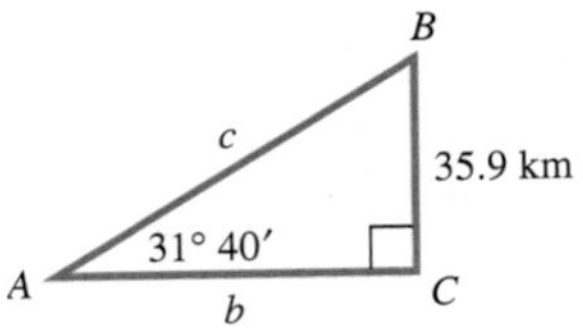

27.

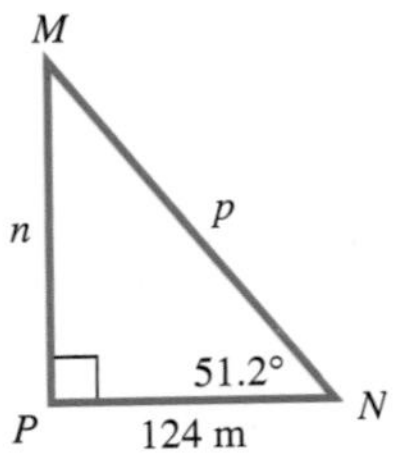

28.

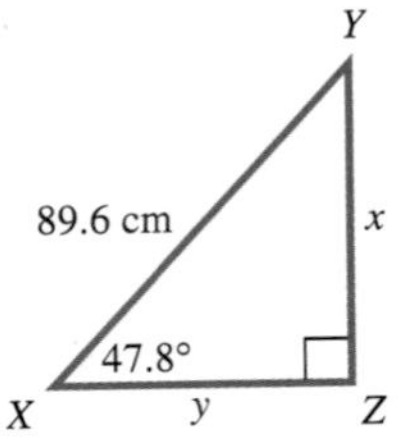

29.

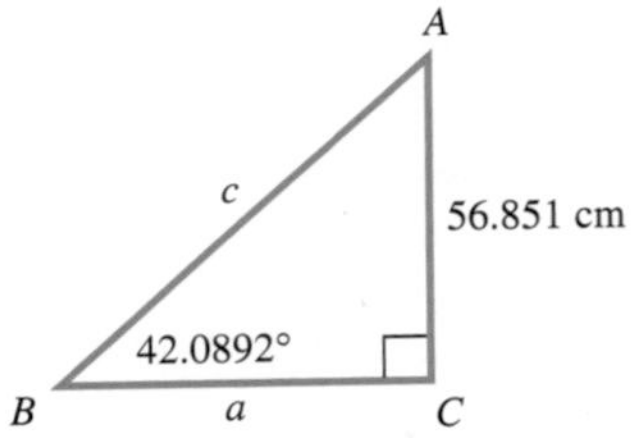

30.

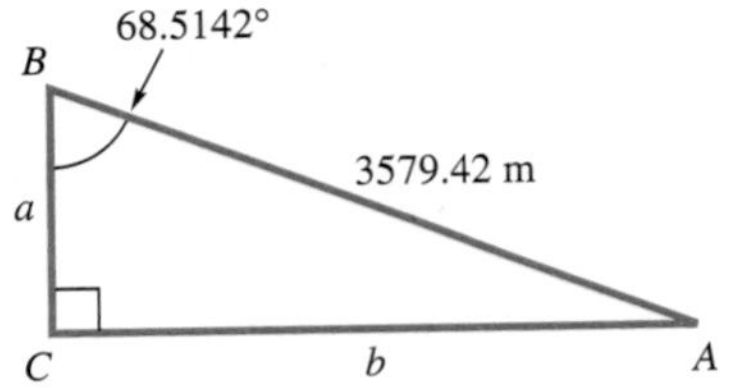

Solve each right triangle. In each case, $C = 90°$. If the angle information is given in degrees and minutes, give the answers in the same way. If given in decimal degrees, do likewise in your answers. When two sides are given, give answers in degrees and minutes.

31. $A = 28.00°$, $c = 17.4$ ft

32. $B = 46.00°$, $c = 29.7$ m

33. $B = 73.00°$, $b = 128$ in.

34. $A = 61°\ 00'$, $b = 39.2$ cm

35. $a = 76.4$ yd, $b = 39.3$ yd

36. $a = 958$ m, $b = 489$ m

37. $a = 18.9$ cm, $c = 46.3$ cm

38. $b = 219$ m, $c = 647$ m

39. $A = 53°\ 24'$, $c = 387.1$ ft

40. $A = 13°\ 47'$, $c = 1285$ m

41. $B = 39°\ 9'$, $c = .6231$ m

42. $B = 82°\ 51'$, $c = 4.825$ cm

Solve each problem.

43. ***Ladder Leaning Against a Wall*** A 13.5-meter fire-truck ladder is leaning against a wall. Find the distance the ladder goes up the wall if it makes an angle of 43° 50′ with the ground.

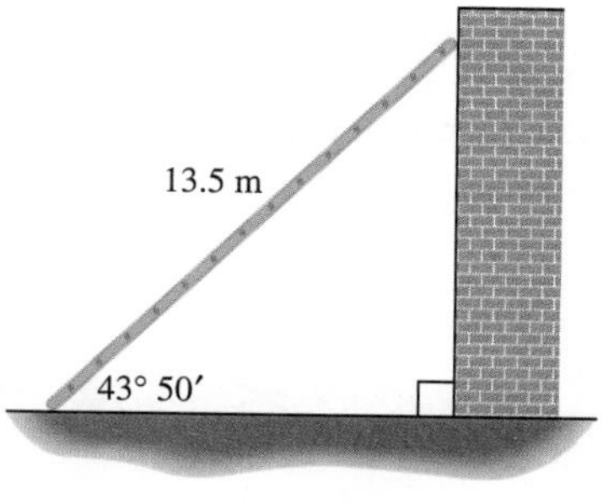

44. ***Antenna Mast Guy Wire*** A guy wire 77.4 meters long is attached to the top of an antenna mast that is 71.3 meters high. Find the angle that the wire makes with the ground.

45. ***Guy Wire to a Tower*** Find the length of a guy wire that makes an angle of 45° 30′ with the ground if the wire is attached to the top of a tower 63.0 meters high.

46. ***Distance Across a Lake*** To find the distance *RS* across a lake, a surveyor lays off $RT = 53.1$ meters, with angle $T = 32°\ 10'$ and angle $S = 57°\ 50'$. Find length *RS*.

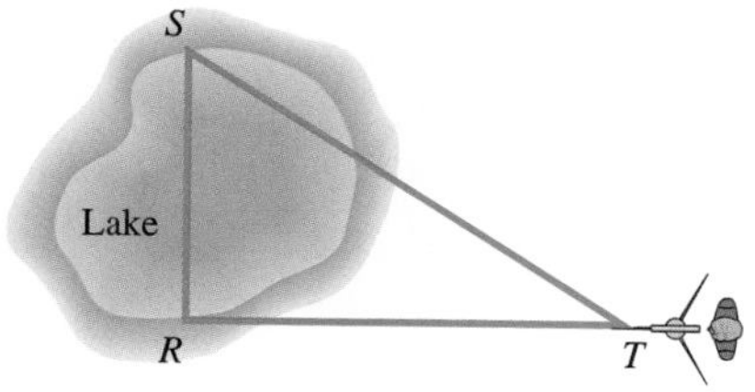

47. ***Side Lengths of a Triangle*** The length of the base of an isosceles triangle is 42.36 inches. Each base angle is 38.12°. Find the length of each of the two equal sides of the triangle. (*Hint:* Divide the triangle into two right triangles.)

48. ***Altitude of a Triangle*** Find the altitude of an isosceles triangle having a base of 184.2 cm if the angle opposite the base is 68° 44′.

49. ***Cloud Ceiling*** The U.S. Weather Bureau defines a *cloud ceiling* as the altitude of the lowest clouds that cover more than half the sky. To determine a cloud ceiling, a powerful searchlight projects a circle of light vertically on the bottom of the cloud. An observer sights the circle of light in the crosshairs of a tube called a *clinometer.* A pendant hanging vertically from the tube and resting on a protractor gives the angle of elevation. Find the cloud ceiling if the searchlight is located 1000 feet from the observer and the angle of elevation is 30.0° as measured with a clinometer at eye-height 6 feet. (Assume three significant digits.)

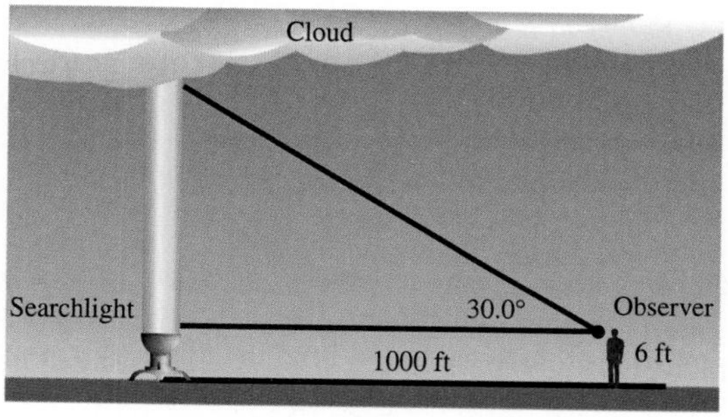

50. ***Length of a Shadow*** Suppose the angle of elevation of the sun is 23.4°. Find the length of the shadow cast by Cindy Newman, who is 5.75 feet tall.

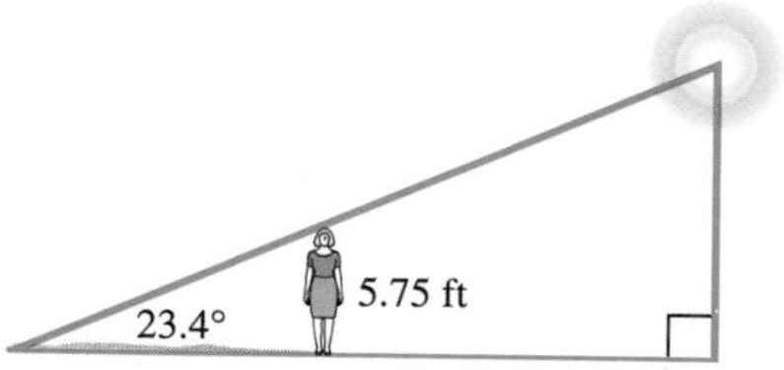

51. ***Height of a Tower*** The shadow of a vertical tower is 40.6 meters long when the angle of elevation of the sun is 34.6°. Find the height of the tower.

52. ***Angle of Elevation of the Sun*** Find the angle of elevation of the sun if a 48.6-foot flagpole casts a shadow 63.1 feet long.

53. ***Distance from the Ground to the Top of a Building*** The angle of depression from the top of a building to a point on the ground is 32° 30′. How far is the point on the ground from the top of the building if the building is 252 meters high?

54. ***Airplane Distance*** An airplane is flying 10,500 feet above the level ground. The angle of depression from the plane to the base of a tree is 13° 50′. How far horizontally must the plane fly to be directly over the tree?

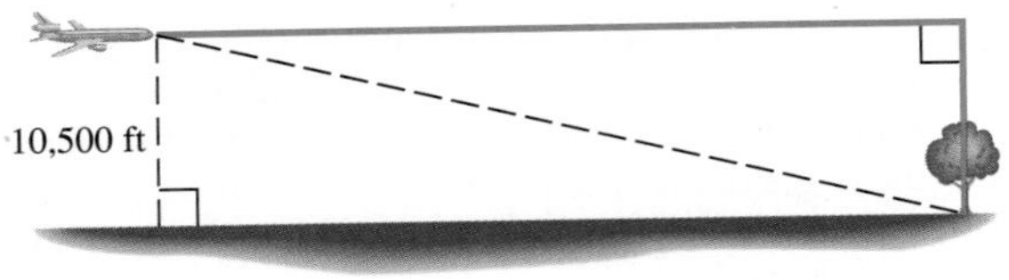

55. ***Height of a Building*** The angle of elevation from the top of a small building to the top of a nearby taller building is 46° 40′, while the angle of depression to the bottom is 14° 10′. If the smaller building is 28.0 meters high, find the height of the taller building.

56. ***Mounting a Video Camera*** A video camera is to be mounted on a bank wall so as to have a good view of the head teller. Find the angle of depression that the lens should make with the horizontal.

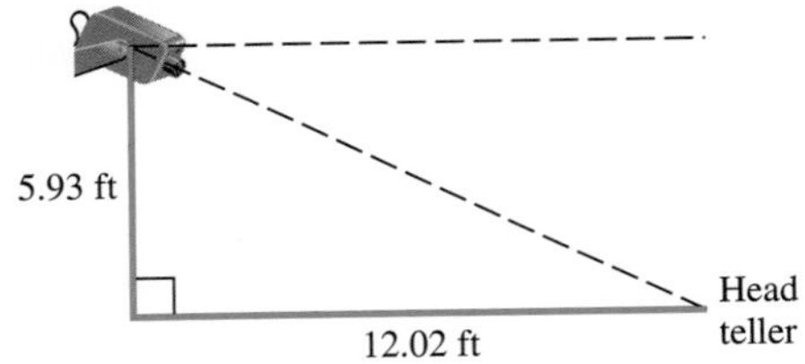

57. ***Error in Measurement*** A degree may seem like a very small unit, but an error of one degree in measuring an angle may be significant. For example, suppose a laser beam directed toward the visible center of the moon misses its assigned target by 30 seconds. How far is it (in miles) from its assigned target? Take the distance from the surface of Earth to that of the moon to be 234,000 miles. (*Source: A Sourcebook of Applications of School Mathematics* by Donald Bushaw et al. Copyright © 1980 by The Mathematical Association of America.)

58. ***Height of Mt. Everest*** The highest mountain peak in the world is Mt. Everest, located in the Himalayas. The height of this enormous mountain was determined in 1856 by surveyors using trigonometry long before it was first climbed in 1953. This difficult measurement had to be done from a great distance. At an altitude of 14,545 feet on a different mountain, the straight line distance to the peak of Mt. Everest is 27.0134 miles and its angle of elevation is $\theta = 5.82°$. (*Source:* Dunham, W., *The Mathematical Universe,* John Wiley & Sons, 1994.)

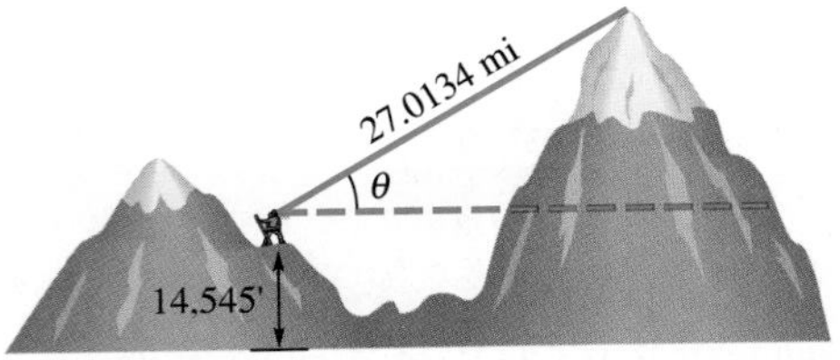

(a) Approximate the height (in feet) of Mt. Everest.

(b) In the actual measurement, Mt. Everest was over 100 miles away and the curvature of Earth had to be taken into account. Would the curvature of Earth make the peak appear taller or shorter than it actually is?

Find the exact value of each labeled part in each figure.

59.

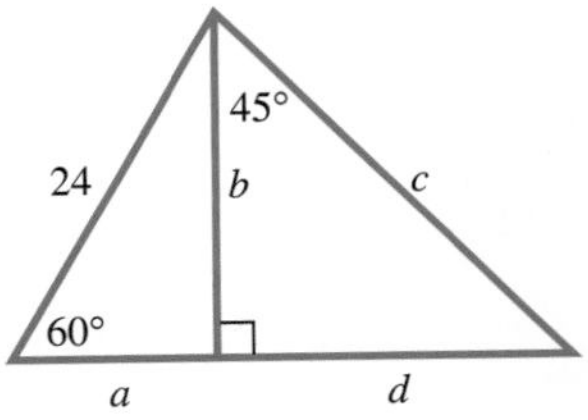

60.

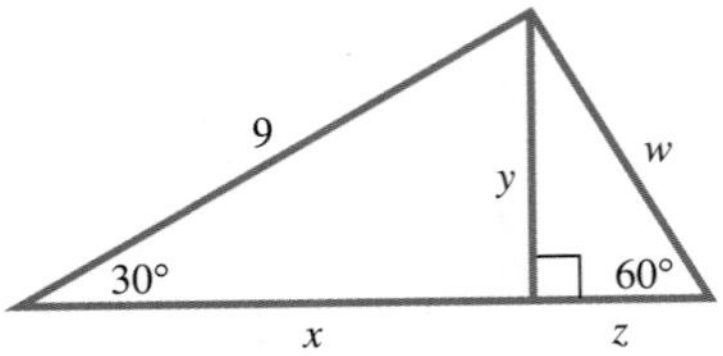

10.6 The Laws of Sines and Cosines; Area Formulas

Oblique Triangles • Law of Sines • Law of Cosines • Area Formulas

Oblique Triangles

A triangle that is not a right triangle is called an **oblique triangle.** The measures of the three sides and the three angles of a triangle can be found if at least one side and any other two measures are known. There are four possible cases.

1. One side and two angles are known. (SAA)
2. Two sides and one angle not included between the two sides are known. (SSA; this case may lead to more than one triangle.)
3. Two sides and the angle included between the two sides are known. (SAS)
4. Three sides are known. (SSS)

If we know three angles of a triangle, we cannot find unique side lengths, because AAA assures us only of similarity, not congruence. For example, there are infinitely many triangles ABC with $A = 35°$, $B = 65°$, and $C = 80°$.

To solve oblique triangles, we need to use the laws of sines and cosines.

Law of Sines

To derive the law of sines, we start with an oblique triangle, such as the acute triangle in Figure 35(a) or the obtuse triangle in Figure 35(b). The following discussion applies to both triangles. First, construct the perpendicular from B to side AC or side AC extended, meeting that side at point D. Let h be the length of this perpendicular. Then c is the hypotenuse of right triangle ADB, and a is the hypotenuse of right triangle BDC. Therefore,

$$\text{in triangle } ADB, \quad \sin A = \frac{h}{c} \quad \text{or} \quad h = c \sin A,$$

$$\text{in triangle } BDC, \quad \sin C = \frac{h}{a} \quad \text{or} \quad h = a \sin C.$$

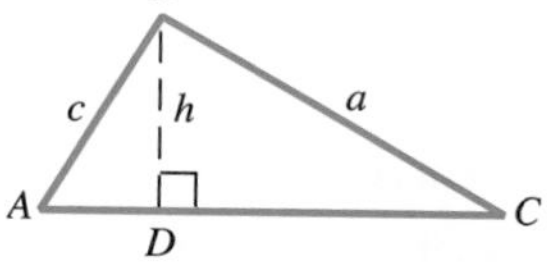

Acute triangle ABC

(a)

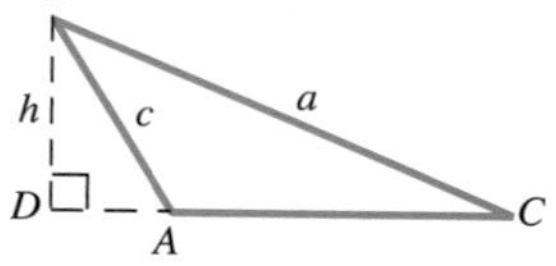

Obtuse triangle ABC

(b)

FIGURE 35

Because $h = c \sin A$ and $h = a \sin C$,

$$a \sin C = c \sin A$$

$$\frac{a}{\sin A} = \frac{c}{\sin C}. \quad \text{Divide both sides by } \sin A \sin C.$$

By constructing the perpendiculars from other vertices, it can also be shown that

$$\frac{a}{\sin A} = \frac{b}{\sin B} \quad \text{and} \quad \frac{b}{\sin B} = \frac{c}{\sin C}.$$

This discussion proves the following theorem.

Law of Sines

In any triangle ABC with sides a, b, and c,

$$\frac{a}{\sin A} = \frac{b}{\sin B}, \qquad \frac{a}{\sin A} = \frac{c}{\sin C}, \qquad \text{and} \qquad \frac{b}{\sin B} = \frac{c}{\sin C}.$$

This can be written in compact form as

$$\frac{a}{\sin A} = \frac{b}{\sin B} = \frac{c}{\sin C}.$$

Sometimes an alternative form of the law of sines is more convenient to use.

$$\frac{\sin A}{a} = \frac{\sin B}{b} = \frac{\sin C}{c}$$

If two angles and the side opposite one of the angles are known, the law of sines can be used directly to solve for the side opposite the other known angle. The triangle can then be solved completely.

EXAMPLE 1 Using the Law of Sines to Solve a Triangle (SAA)

Solve triangle ABC if $A = 32.0°$, $B = 81.8°$, and $a = 42.9$ centimeters.

SOLUTION

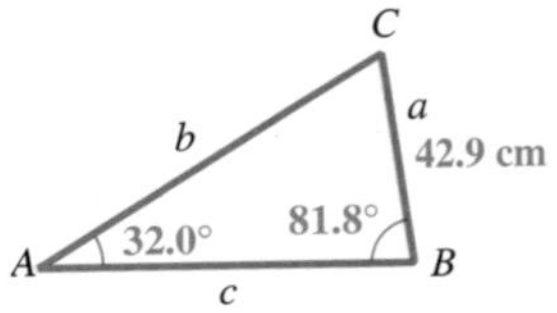

FIGURE 36

Start by drawing a triangle, roughly to scale, and labeling the given parts as in Figure 36. Because the values of A, B, and a are known, use the part of the law of sines that involves these variables.

$$\frac{a}{\sin A} = \frac{b}{\sin B} \qquad \text{Law of sines}$$

$$\frac{42.9}{\sin} = \frac{b}{\sin 81.8°}. \qquad \text{Substitute given values.}$$

$$b = \frac{42.9 \sin 81.8°}{\sin 32.0°} \qquad \text{Multiply by } \sin 81.8°\text{; rewrite.}$$

When using a calculator to find b, keep intermediate answers in the calculator until the final result is found. Then round to the proper number of significant digits. In this case, find sin 81.8°, and then multiply that number by 42.9. Keep the result in the calculator while you find sin 32.0°, and then divide. Because the given information is accurate to three significant digits, round the value of b to get

$$b = 80.1 \text{ centimeters.}$$

Find C from the fact that the sum of the angles of any triangle is 180°.

$$A + B + C = 180^\circ$$
$$C = 180^\circ - A - B$$
$$C = 180^\circ - 32.0^\circ - 81.8^\circ$$
$$C = 66.2^\circ$$

Now use the law of sines again to find c. (Note that the Pythagorean theorem does not apply here, because triangle ABC is oblique, and the Pythagorean theorem applies only to right triangles.)

$$\frac{a}{\sin A} = \frac{c}{\sin C}$$
$$\frac{42.9}{\sin 32.0^\circ} = \frac{c}{\sin 66.2^\circ}$$
$$c = \frac{42.9 \sin 66.2^\circ}{\sin 32.0^\circ}$$
$$c = 74.1 \text{ centimeters}$$

PROBLEM-SOLVING HINT In applications of oblique triangles such as the one in Example 1, a correctly labeled sketch is essential in order to set up the correct equation.

EXAMPLE 2 Using the Law of Sines in an Application (ASA)

Roosevelt Brown wishes to measure the distance across the Big Muddy River. See Figure 37. He finds that $C = 112.90^\circ$, $A = 31.10^\circ$, and $b = 347.6$ feet. Find the required distance.

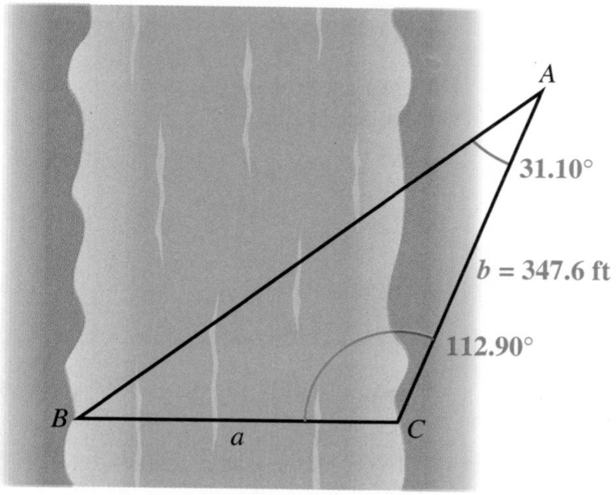

FIGURE 37

SOLUTION

To use the law of sines, one side and the angle opposite it must be known. Because the only side whose length is given is b, angle B must be found before the law of sines can be used.

$$\begin{aligned} B &= 180° - A - C \\ &= 180° - 31.10° - 112.90° \\ &= 36.00° \end{aligned}$$

Now use the form of the law of sines involving A, B, and b to find a, the distance across the river.

$$\frac{a}{\sin A} = \frac{b}{\sin B}$$

$$\frac{a}{\sin 31.10°} = \frac{347.6}{\sin 36.00°} \qquad \text{Substitute.}$$

$$a = \frac{347.6 \sin 31.10°}{\sin 36.00°} \qquad \text{Multiply by } \sin 31.10°.$$

$$a = 305.5 \text{ feet} \qquad \text{Use a calculator.}$$

Law of Cosines If we are given two sides and the included angle (SAS) or three sides of a triangle (SSS), a unique triangle is formed. In these cases, however, we cannot begin the solution of the triangle by using the law of sines because we are not given a side and the angle opposite it. Both cases require the use of the *law of cosines.*

Remember the following property of triangles when applying the law of cosines.

Property of Triangle Side Lengths

In any triangle, the sum of the lengths of any two sides must be greater than the length of the remaining side.

For example, it would be impossible to construct a triangle with sides of lengths 3, 4, and 10. See Figure 38.

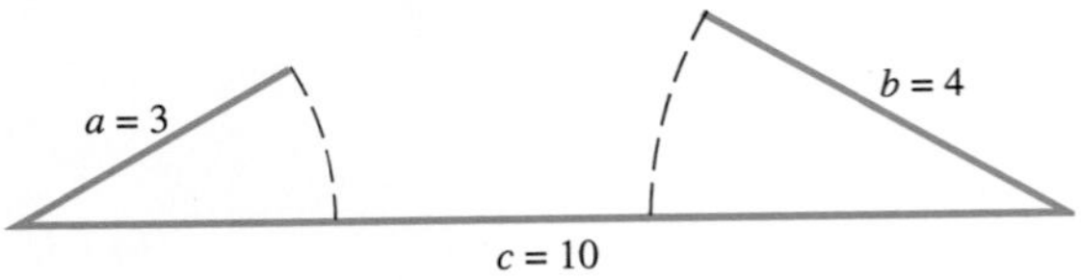

No triangle is formed.

FIGURE 38

To derive the law of cosines, let ABC be any oblique triangle. Choose a coordinate system so that vertex B is at the origin and side BC is along the positive x-axis. See Figure 39 on the next page.

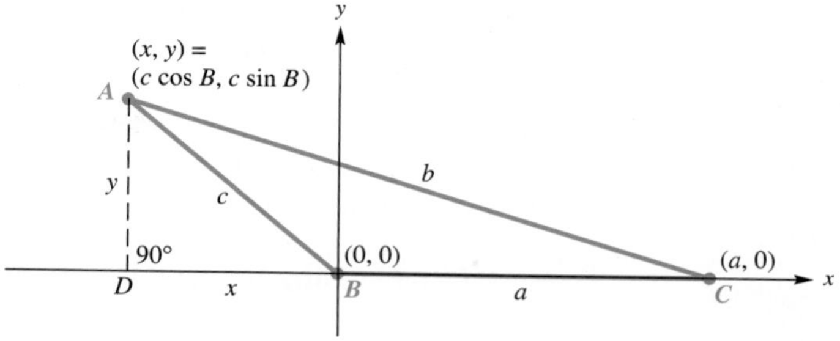

FIGURE 39

Let (x, y) be the coordinates of vertex A of the triangle. Verify that for angle B, whether obtuse or acute,

$$\sin B = \frac{y}{c} \quad \text{and} \quad \cos B = \frac{x}{c}.$$

(Here x is negative if B is obtuse.) From these results

$$y = c \sin B \quad \text{and} \quad x = c \cos B,$$

so that the coordinates of point A become

$$(c \cos B, c \sin B).$$

Point C has coordinates $(a, 0)$, and AC has length b, so

$$b = \sqrt{(c \cos B - a)^2 + (c \sin B - 0)^2} \qquad \text{Distance formula}$$

$$b^2 = (c \cos B - a)^2 + (c \sin B)^2 \qquad \text{Square both sides.}$$

$$= c^2 \cos^2 B - 2ac \cos B + a^2 + c^2 \sin^2 B$$

$$= a^2 + c^2 (\cos^2 B + \sin^2 B) - 2ac \cos B$$

$$= a^2 + c^2(1) - 2ac \cos B \qquad \text{Pythagorean identity}$$

$$= a^2 + c^2 - 2ac \cos B.$$

This result is one form of the law of cosines. In the work above, we could just as easily have placed A or C at the origin. This would have given the same result, but with the variables rearranged. These various forms of the law of cosines are summarized in the following theorem.

Law of Cosines

In any triangle ABC with sides a, b, and c,

$$a^2 = b^2 + c^2 - 2bc \cos A$$
$$b^2 = a^2 + c^2 - 2ac \cos B$$
$$c^2 = a^2 + b^2 - 2ab \cos C.$$

The law of cosines says that the square of a side of a triangle is equal to the sum of the squares of the other two sides, minus twice the product of those two sides and the cosine of the angle included between them.

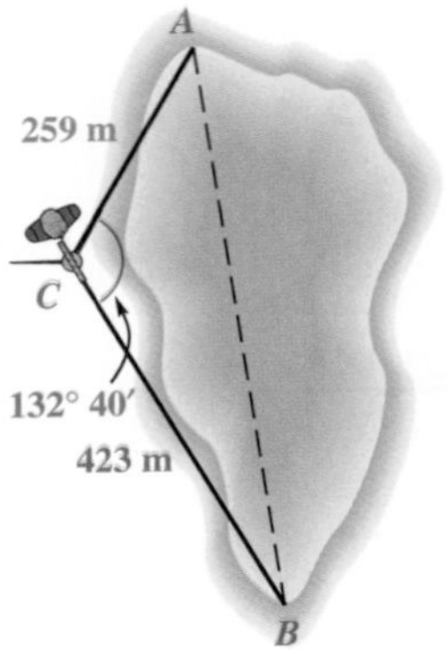

FIGURE 40

If we let $C = 90°$ in the third form of the law of cosines given above, we have $\cos C = \cos 90° = 0$, and the formula becomes

$$c^2 = a^2 + b^2,$$

the familiar equation of the Pythagorean theorem. Thus, the Pythagorean theorem is a special case of the law of cosines.

EXAMPLE 3 Using the Law of Cosines in an Application (SAS)

A surveyor wishes to find the distance between two inaccessible points A and B on opposite sides of a lake. See Figure 40. While standing at point C, she finds that $AC = 259$ meters, $BC = 423$ meters, and angle ACB measures $132°\,40'$. Find the distance AB.

SOLUTION

The law of cosines can be used here, since we know the lengths of two sides of the triangle and the measure of the included angle.

$$AB^2 = 259^2 + 423^2 - 2(259)(423)\cos 132°\,40'$$

$$AB^2 = 394{,}510.6 \quad \text{Use a calculator.}$$

$$AB \approx 628 \quad \text{Take the square root and round to 3 significant digits.}$$

The distance AB is approximately 628 meters.

This photo shows the cover of a short text (only 60 pages) on **spherical trigonometry,** published in 1928. In the historical sketch provided at the end of the book, author Pauline Sperry states that Albattani, Geber of Seville, and Nassir Eddin continued the work of Ptolemy in the Arab world, while Johanne Müller (better known as **Regiomontanus**) did the same in Europe. Regiomontanus's work *De triangulis omnimodis* was "the first systematic treatment of trigonometry independent of astronomy in the western world and dominated the latter middle ages to a remarkable degree."

EXAMPLE 4 Using the Law of Cosines to Solve a Triangle (SSS)

Solve triangle ABC if $A = 42.3°$, $b = 12.9$ meters, and $c = 15.4$ meters. See Figure 41.

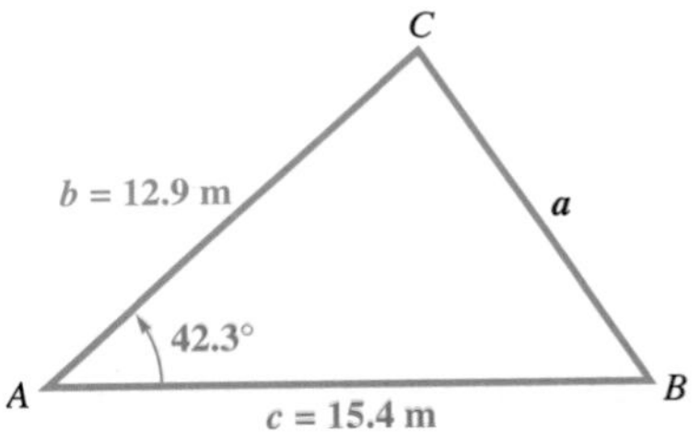

FIGURE 41

SOLUTION

$$a^2 = b^2 + c^2 - 2bc\cos A \quad \text{Law of cosines.}$$

$$a^2 = 12.9^2 + 15.4^2 - 2(12.9)(15.4)\cos 42.3° \quad \text{Substitute.}$$

$$a^2 = 109.7 \quad \text{Use a calculator.}$$

$$a = 10.5 \text{ meters} \quad \sqrt{109.7} \approx 10.47 \approx 10.5$$

We now must find the measures of angles B and C. There are several approaches that can be used at this point. We shall use the law of sines to find one of these angles. Of the two remaining angles, B must be the smaller, because it is opposite the shorter of the two sides b and c. Therefore, it cannot be obtuse, and we will avoid any ambiguity when we find its sine.

When two sides and an angle opposite one of them are given as in Example 4, under certain conditions *two* triangles satisfying those conditions may exist. This is called the **ambiguous** case. See a standard trigonometry text for more information.

$$\frac{\sin 42.3^\circ}{10.47} = \frac{\sin B}{12.9}$$ Use 10.47, the approximation for *a before* rounding to 10.5.

$$\sin B = \frac{12.9 \sin 42.3^\circ}{10.47}$$ Multiply by 12.9; rewrite.

$$B = 56.0^\circ$$ Use the inverse sine function.

The easiest way to find C is to subtract the measures of A and B from 180°.

$$C = 180^\circ - 42.3^\circ - 56.0^\circ$$
$$= 81.7^\circ$$

Had we chosen to use the law of sines to find C rather than B in Example 4, we would not have known whether C equals 81.7° or its supplement, 98.3°.

EXAMPLE 5 Using the Law of Cosines to Solve a Triangle (SSS)

Solve triangle ABC if $a = 9.47$ feet, $b = 15.9$ feet, and $c = 21.1$ feet.

SOLUTION

We may use the law of cosines to solve for any angle of the triangle. We solve for C, the largest angle, using the law of cosines. We will be able to tell if C is obtuse if $\cos C < 0$.

$$c^2 = a^2 + b^2 - 2ab \cos C$$ Law of cosines

$$\cos C = \frac{a^2 + b^2 - c^2}{2ab}$$ Solve for cos *C*.

$$\cos C = \frac{(9.47)^2 + (15.9)^2 - (21.1)^2}{2(9.47)(15.9)}$$ Substitute.

$$\cos C = -.34109402$$ Use a calculator.

$$C = 109.9^\circ$$ Use the inverse cosine function.

We can use either the law of sines or the law of cosines to find $B = 45.1^\circ$. (Verify this.) Because $A = 180^\circ - B - C$,

$$A = 25.0^\circ.$$

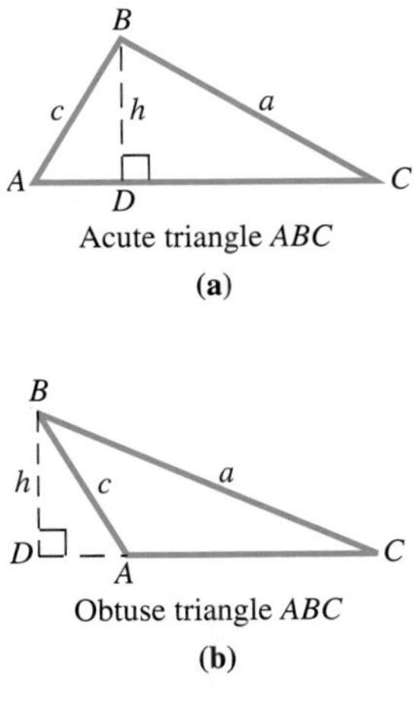

FIGURE 42

Area Formulas

The method used to derive the law of sines can also be used to derive a useful formula to find the area of a triangle. A familiar formula for the area of a triangle is $\mathcal{A} = \frac{1}{2}bh$, where $\mathcal{A}$ represents the area, b the base, and h the height. This formula cannot always be used easily, because in practice h is often unknown. To find an alternative formula, refer to acute triangle ABC in Figure 42(a) or obtuse triangle ABC in Figure 42(b).

A perpendicular has been drawn from B to the base of the triangle (or the extension of the base). This perpendicular forms two right triangles. Using triangle ABD,

$$\sin A = \frac{h}{c},$$

or

$$h = c \sin A.$$

Substituting into the formula $\mathscr{A} = \frac{1}{2}bh$,

$$\mathscr{A} = \frac{1}{2}b(c \sin A)$$

or

$$\mathscr{A} = \frac{1}{2}bc \sin A.$$

Any other pair of sides and the angle between them could have been used, as stated in the next theorem.

Area of a Triangle

In any triangle ABC, the area $\mathscr{A}$ is given by any of the following formulas:

$$\mathscr{A} = \frac{1}{2}bc \sin A, \qquad \mathscr{A} = \frac{1}{2}ab \sin C, \qquad \mathscr{A} = \frac{1}{2}ac \sin B.$$

That is, the area is given by half the product of the lengths of two sides and the sine of the angle included between them.

EXAMPLE 6 Finding the Area of a Triangle Using $\mathscr{A} = \frac{1}{2}ab \sin C$

Find the area of triangle ABC given $A = 24° 40'$, $b = 27.3$ centimeters, and $C = 52° 40'$.

SOLUTION

Before we can use the formula given above, we must use the law of sines to find either a or c. Because the sum of the measures of the angles of any triangle is 180°,

$$B = 180° - 24° 40' - 52° 40' = 102° 40'.$$

Now use the form of the law of sines that relates a, b, A, and B to find a.

$$\frac{a}{\sin A} = \frac{b}{\sin B}$$

$$\frac{a}{\sin 24° 40'} = \frac{27.3}{\sin 102° 40'}$$

Solve for a to verify that $a = 11.7$ centimeters. Now find the area.

$$\begin{aligned}\mathscr{A} &= \frac{1}{2}ab \sin C \\ &= \frac{1}{2}(11.7)(27.3) \sin 52° 40' \\ &= 127\end{aligned}$$

The area of triangle ABC is 127 square centimeters to three significant digits. ■

The law of cosines can be used to derive a formula for the area of a triangle when only the lengths of the three sides are known. This formula is known as Heron's formula, named after the Greek mathematician Heron of Alexandria.

Heron of Alexandria lived during approximately the same period as Ptolemy and Hipparchus. He wrote extensively on both mathematical and physical subjects. The most importmant of his mathematical works is *Metrica*, which was discovered in Constantinople in the late 19th century. The Arabian scholar al-Biruni claims that Archimedes actually discovered the triangle area formula which bears Heron's name. (*Source:* Eves, Howard, *An Introduction to the History of Mathematics*, 6th edition.)

Heron's Area Formula

If a triangle has sides of lengths a, b, and c, and if the **semiperimeter** is

$$s = \frac{1}{2}(a + b + c),$$

then the area of the triangle is

$$\mathscr{A} = \sqrt{s(s - a)(s - b)(s - c)}.$$

EXAMPLE 7 Finding the Area of a Triangle Using Heron's Formula

Determine the area of the triangle having sides of lengths $a = 29.7$ feet, $b = 42.3$ feet, and $c = 38.4$ feet.

SOLUTION

To use Heron's area formula, first find s.

$$s = \frac{1}{2}(a + b + c) \qquad \text{Semiperimeter formula}$$

$$s = \frac{1}{2}(29.7 + 42.3 + 38.4)$$

$$s = 55.2$$

The area is

$$\mathscr{A} = \sqrt{s(s - a)(s - b)(s - c)}$$

$$= \sqrt{55.2(55.2 - 29.7)(55.2 - 42.3)(55.2 - 38.4)} \qquad \text{Heron's area formula}$$

$$= \sqrt{55.2(25.5)(12.9)(16.8)}$$

$$\mathscr{A} = 552 \text{ square feet.} \qquad \text{Three significant digits}$$

10.6 Exercises

Use the law of sines to solve each triangle.

1.

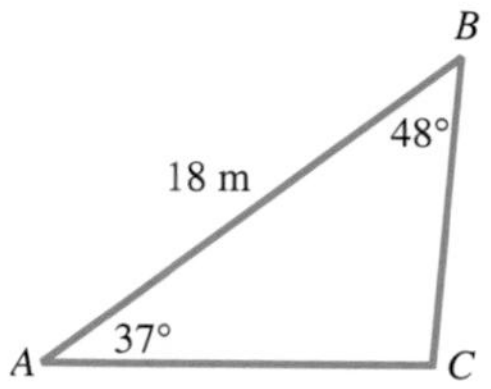

2.

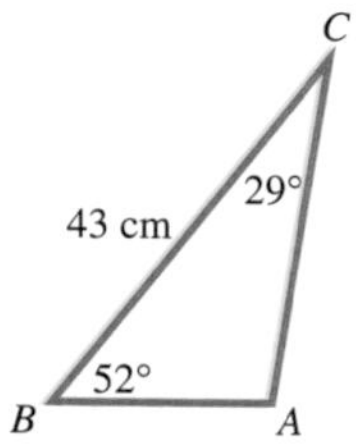

3.

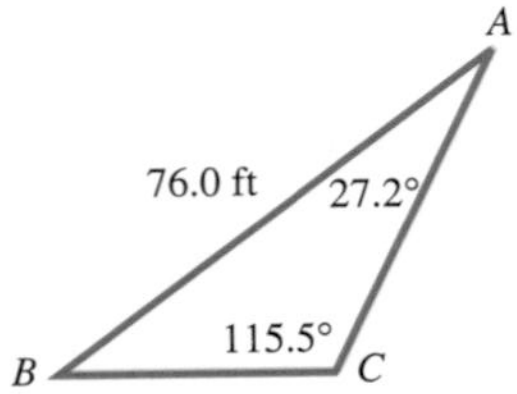

4.

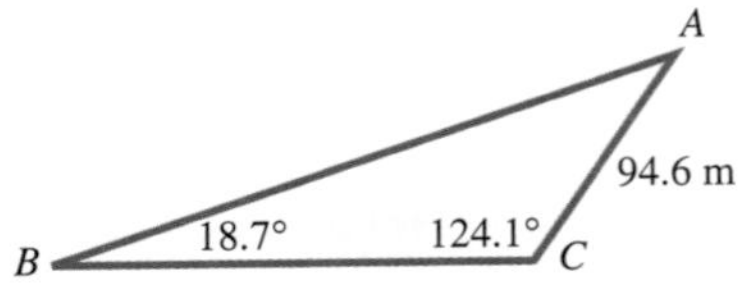

5. $A = 68.41°$, $B = 54.23°$, $a = 12.75$ ft

6. $C = 74.08°$, $B = 69.38°$, $c = 45.38$ m

7. $A = 87.2°$, $b = 75.9$ yd, $C = 74.3°$

8. $B = 38° 40'$, $a = 19.7$ cm, $C = 91° 40'$

9. $B = 20° 50'$, $C = 103° 10'$, $AC = 132$ ft

10. $A = 35.3°$, $B = 52.8°$, $AC = 675$ ft

11. $A = 39.70°$, $C = 30.35°$, $b = 39.74$ m

12. $C = 71.83°$, $B = 42.57°$, $a = 2.614$ cm

13. $B = 42.88°$, $C = 102.40°$, $b = 3974$ ft

14. $A = 18.75°$, $B = 51.53°$, $c = 2798$ yd

15. $A = 39° 54'$, $a = 268.7$ m, $B = 42° 32'$

16. $C = 79° 18'$, $c = 39.81$ mm, $A = 32° 57'$

Use the law of sines to solve each problem.

17. ***Distance Across a River*** To find the distance AB across a river, a distance $BC = 354$ meters is laid off on one side of the river. It is found that $B = 112° 10'$ and $C = 15° 20'$. Find AB.

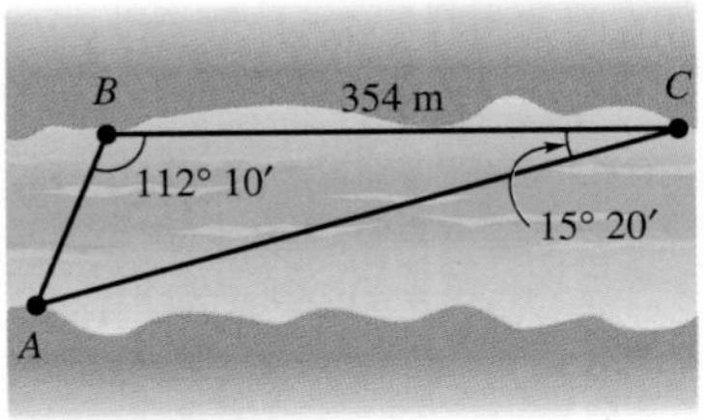

18. ***Distance Across a Canyon*** To determine the distance RS across a deep canyon, Anne Tomlin lays off a distance $TR = 582$ yards. She then finds that $T = 32° 50'$ and $R = 102° 20'$. Find RS.

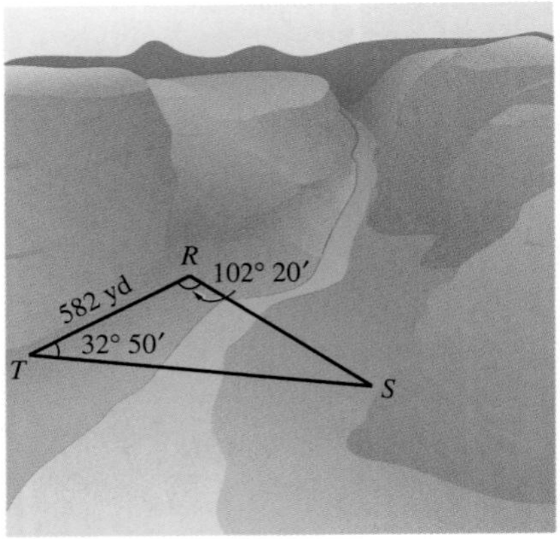

19. ***Measurement of a Folding Chair*** A folding chair is to have a seat 12.0 inches deep with angles as shown in the figure. How far down from the seat should the crossing legs be joined? (Find x in the figure.)

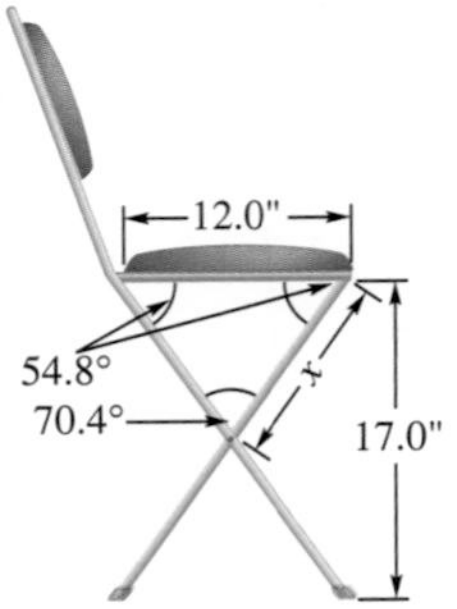

20. ***Angle Formed by Radii of Gears*** Three gears are arranged as shown in the figure. Find the measure of angle θ.

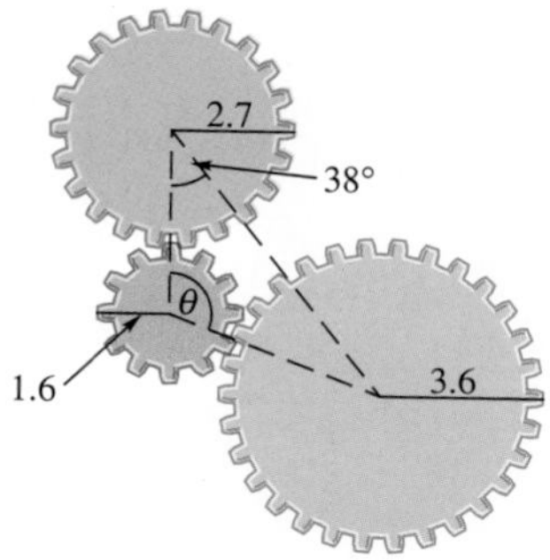

21. ***Distance Between Atoms*** Three atoms with atomic radii of 2.0, 3.0, and 4.5 are arranged as in the figure. Find the distance between the centers of atoms A and C.

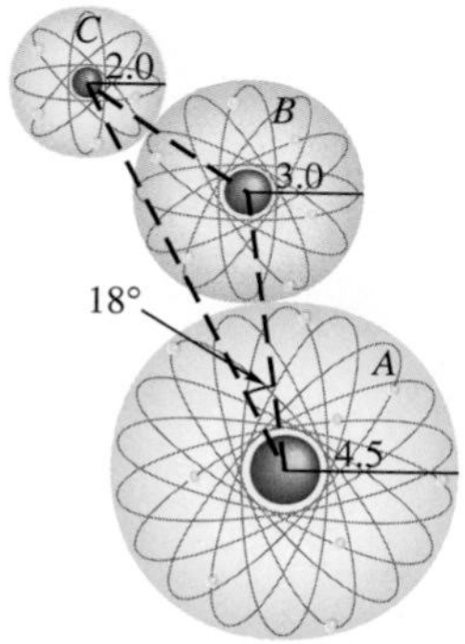

22. ***Height of a Balloon*** A balloonist is directly above a straight road 1.5 miles long that joins two villages. She finds that the town closer to her is at an angle of depression of 35°, and the farther town is at an angle of depression of 31°. How high above the ground is the balloon?

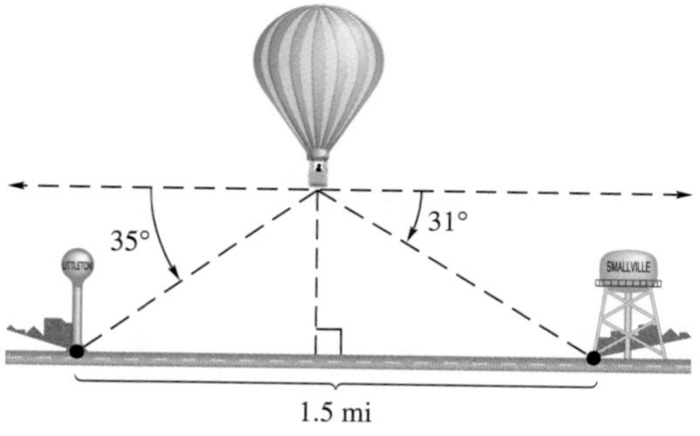

Use the law of cosines to find the length of the remaining sides of each triangle. Do not use a calculator.

23.

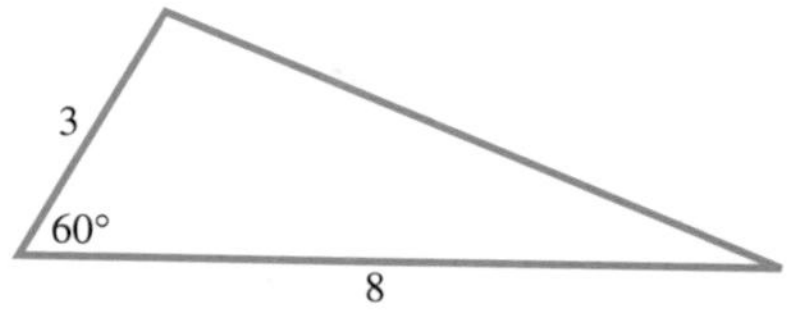

24.

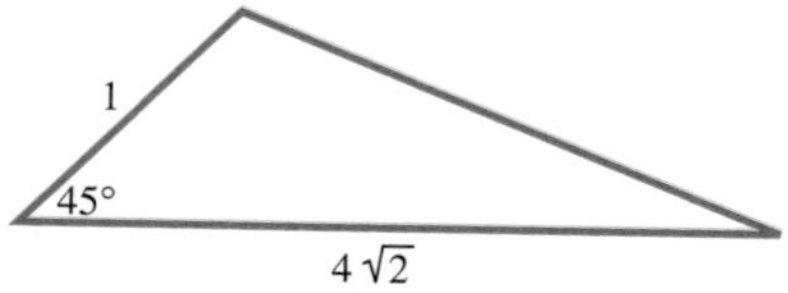

Use the law of cosines to find the value of θ in each triangle. Do not use a calculator.

25.

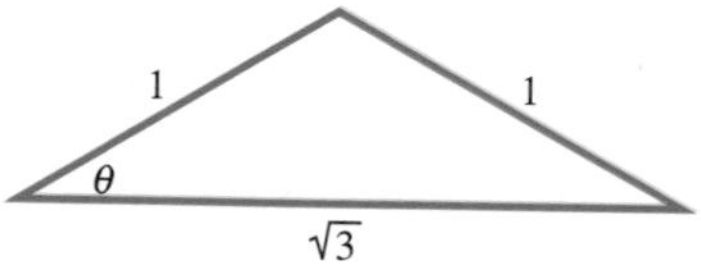

26.

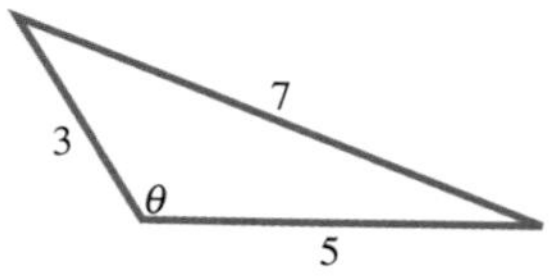

Use the law of cosines to solve each triangle.

27. $C = 28.3°$, $b = 5.71$ in., $a = 4.21$ in.

28. $A = 41.4°$, $b = 2.78$ yd, $c = 3.92$ yd

29. $C = 45.6°$, $b = 8.94$ m, $a = 7.23$ m

30. $A = 67.3°$, $b = 37.9$ km, $c = 40.8$ km

31. $A = 80° \, 40'$, $b = 143$ cm, $c = 89.6$ cm

32. $C = 72° \, 40'$, $a = 327$ ft, $b = 251$ ft

33. $B = 74.80°$, $a = 8.919$ in., $c = 6.427$ in.

34. $C = 59.70°$, $a = 3.725$ mi, $b = 4.698$ mi

35. $A = 112.8°$, $b = 6.28$ m, $c = 12.2$ m

36. $B = 168.2°$, $a = 15.1$ cm, $c = 19.2$ cm

37. $a = 3.0$ ft, $b = 5.0$ ft, $c = 6.0$ ft

38. $a = 4.0$ ft, $b = 5.0$ ft, $c = 8.0$ ft

39. $a = 9.3$ cm, $b = 5.7$ cm, $c = 8.2$ cm

40. $a = 28$ ft, $b = 47$ ft, $c = 58$ ft

41. $a = 42.9$ m, $b = 37.6$ m, $c = 62.7$ m

42. $a = 187$ yd, $b = 214$ yd, $c = 325$ yd

Solve each problem, using the law of sines or the law of cosines as needed.

43. ***Distance Across a Lake*** Points A and B are on opposite sides of Lake Yankee. From a third point, C, the angle between the lines of sight to A and B is 46.3°. If AC is 350 meters long and BC is 286 meters long, find the length AB.

44. ***Diagonals of a Parallelogram*** The sides of a parallelogram are 4.0 cm and 6.0 cm. One angle is 58° while another is 122°. Find the lengths of the diagonals of the parallelogram.

45. ***Playhouse Layout*** The layout for a child's playhouse in her backyard shows the dimensions given in the figure. Find the value of x.

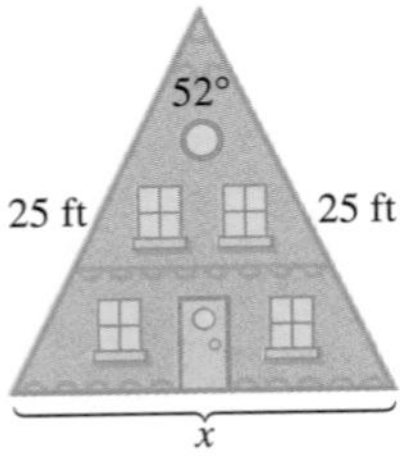

46. ***Distance Between Points on a Crane*** A crane with a counterweight is shown in the figure. Find the horizontal distance between points A and B.

47. ***Angles Between a Beam and Cables*** A weight is supported by cables attached to both ends of a horizontal beam, as shown in the figure. What angles are formed between the beam and the cables?

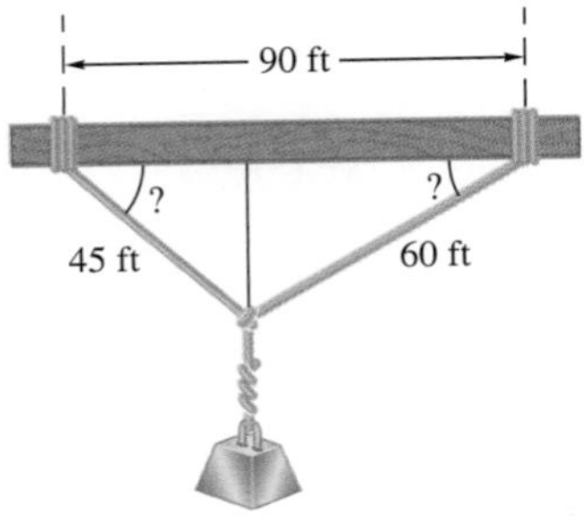

48. ***Measurement Using Triangulation*** Surveyors are often confronted with obstacles, such as trees, when measuring the boundary of a lot. One technique used to obtain an accurate measurement is the so-called *triangulation method.* In this technique, a triangle is

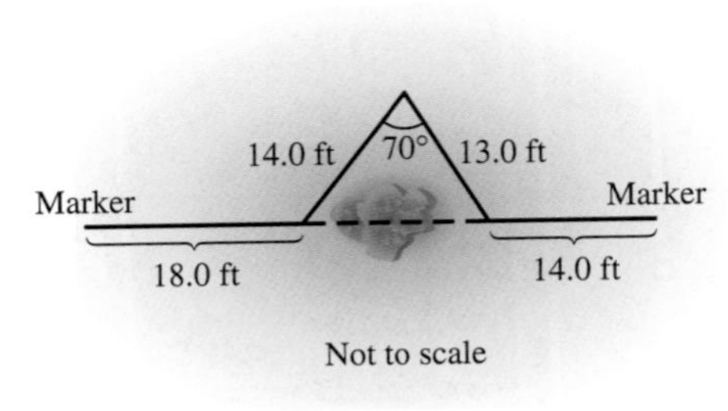

constructed around the obstacle and one angle and two sides of the triangle are measured. Use this technique to find the length of the property line (the straight line between the two markers) in the figure. (*Source:* Kavanagh, B. and S. Bird, *Surveying Principles and Applications,* Fifth Edition, Prentice Hall, 2000.)

Find the exact area of each triangle using the formula $\mathcal{A} = \frac{1}{2}bh$, *and then verify that the formula* $\mathcal{A} = \frac{1}{2}ab \sin C$ *gives the same result.*

49.

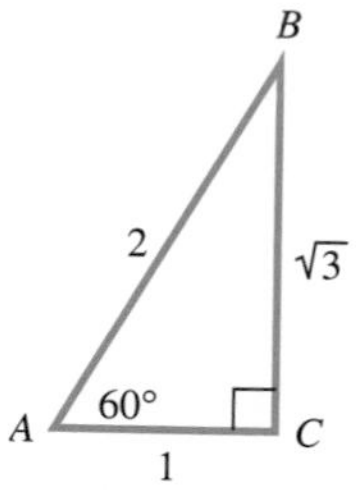

50.

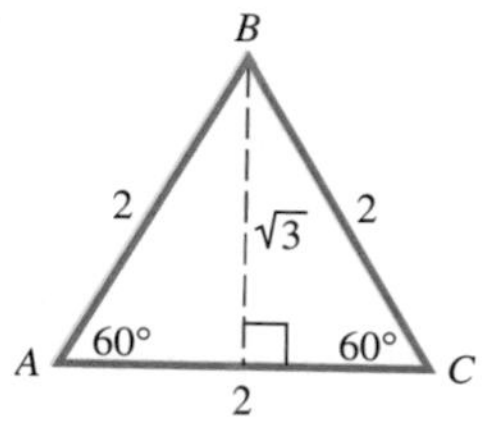

51.

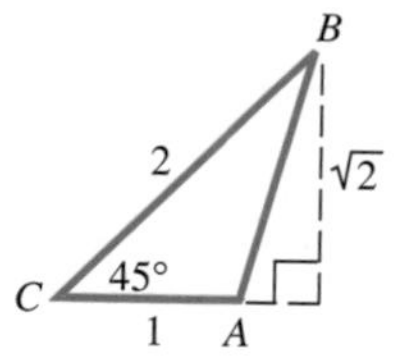

52.

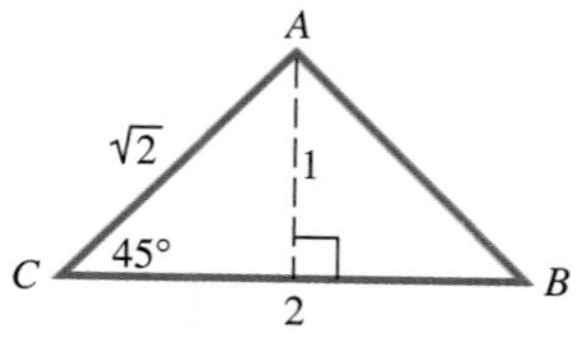

Find the area of each triangle using the formula involving the sine function (page 662).

53. $A = 42.5°, b = 13.6$ m, $c = 10.1$ m

54. $C = 72.2°, b = 43.8$ ft, $a = 35.1$ ft

55. $B = 124.5°, a = 30.4$ cm, $c = 28.4$ cm

56. $C = 142.7°, a = 21.9$ km, $b = 24.6$ km

57. $A = 56.80°, b = 32.67$ in., $c = 52.89$ in.

58. $A = 34.97°, b = 35.29$ m, $c = 28.67$ m

Find the exact area of each triangle using the formula $\mathcal{A} = \frac{1}{2}bh$, *and then verify that Heron's formula gives the same result.*

59.

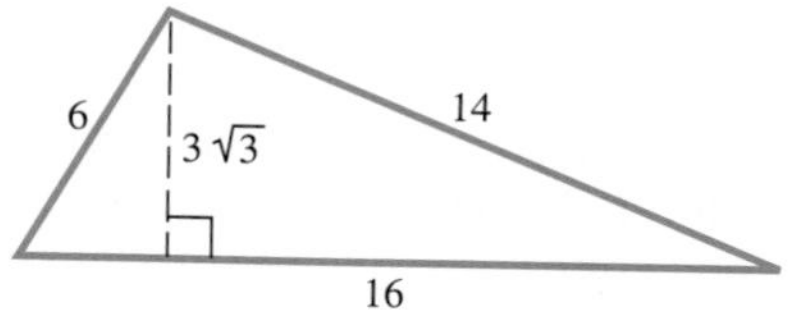

60.

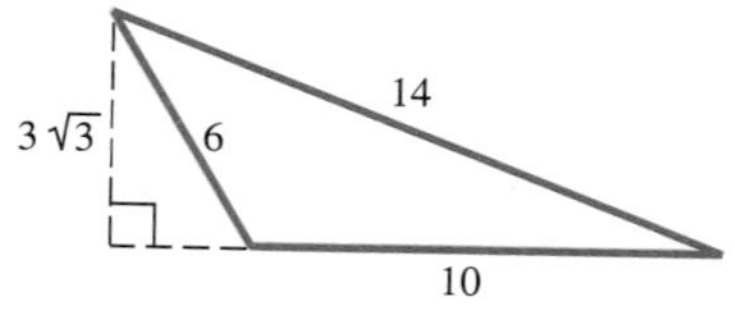

Find the area of each triangle using Heron's formula.

61. $a = 12$ m, $b = 16$ m, $c = 25$ m

62. $a = 22$ in., $b = 45$ in., $c = 31$ in.

63. $a = 154$ cm, $b = 179$ cm, $c = 183$ cm

64. $a = 25.4$ yd, $b = 38.2$ yd, $c = 19.8$ yd

65. $a = 76.3$ ft, $b = 109$ ft, $c = 98.8$ ft

66. $a = 15.89$ in., $b = 21.74$ in., $c = 10.92$ in.

Solve each problem.

67. *Area of a Metal Plate* A painter is going to apply a special coating to a triangular metal plate on a new building. Two sides measure 16.1 meters and 15.2 meters. She knows that the angle between these sides is 125°. What is the area of the surface she plans to cover with the coating?

68. *Area of a Triangular Lot* A real estate agent wants to find the area of a triangular lot. A surveyor takes measurements and finds the two sides are 52.1 meters and 21.3 meters, and the angle between them is 42.2°. What is the area of the triangular lot?

69. ***Required Amount of Paint*** A painter needs to cover a triangular region 75 meters by 68 meters by 85 meters. A can of paint covers 75 square meters of area. How many cans (to the next higher number of cans) will be needed?

70. ***Area of the Bermuda Triangle*** Find the area of the Bermuda Triangle if the sides of the triangle have approximate lengths 850 miles, 925 miles, and 1300 miles.

COLLABORATIVE INVESTIGATION

Making a *Point* About Trigonometric Function Values

An equation of the form $Ax + By = 0$ represents a line passing through the origin in the plane. By restricting x to take on only nonpositive or only nonnegative values, we obtain a ray with endpoint at the origin. This ray can be considered the terminal side of an angle in standard position. For example, the figure shows an angle θ in standard position whose terminal side has the equation

$$2x - y = 0, x \geq 0.$$

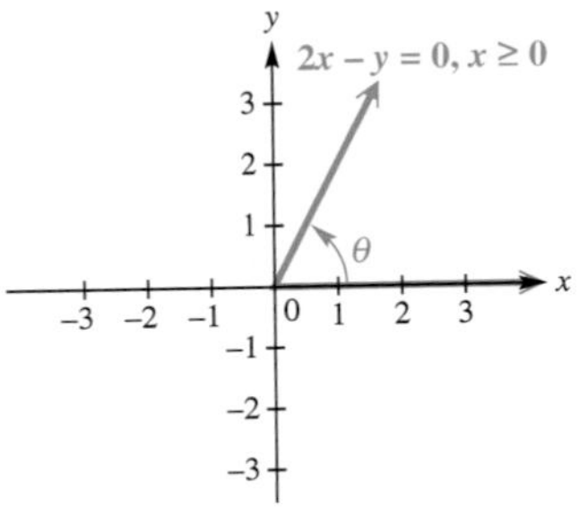

To find the trigonometric function values of θ, we can use *any* point on its terminal side except the origin. Similarly, we can find the value of θ using any point and the appropriate inverse trigonometric function. This investigation supports these statements.

Topics for Discussion

1. Divide the class into four groups: I, II, III, and IV. Each group will have a different value of x assigned to it:
   Group I: $x = 2$ Group II: $x = 4$
   Group III: $x = 6$ Group IV: $x = 8$
2. Find the coordinates of the point on the terminal side of θ corresponding to your value of x.
3. Find the corresponding value of r using the equation $r = \sqrt{x^2 + y^2}$.
4. Find the exact values of the trigonometric functions of θ. Simplify any radical expressions completely.
5. With your calculator in degree mode, find each: **(a)** $\sin^{-1}\left(\frac{y}{r}\right)$, **(b)** $\cos^{-1}\left(\frac{x}{r}\right)$, and **(c)** $\tan^{-1}\left(\frac{y}{x}\right)$.
6. Compare the results of Topic 4 among the different groups. Did the point chosen make any difference when finding the trigonometric function values?
7. Compare the results of Topic 5 among the different groups. Every value should be the same. What is this value? What does it represent?
8. Similar triangles have sides that are proportional, as discussed in Section 9.4. Discuss how this property justifies the results of Topic 4.

CHAPTER 10 TEST

1. Convert 74° 17′ 54″ to decimal degrees.
2. Find the angle of smallest positive measure coterminal with −157°.
3. If $(2, -5)$ is on the terminal side of angle θ in standard position, find $\sin\theta$, $\cos\theta$, and $\tan\theta$.
4. If $\cos\theta < 0$ and $\cot\theta > 0$, in what quadrant does θ terminate?
5. If $\cos\theta = \frac{4}{5}$ and θ is in quadrant IV, find the values of the other trigonometric functions of θ.
6. Give the six trigonometric function values of angle A in the triangle at the top of the next page.

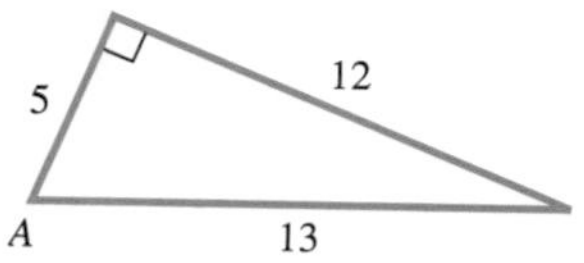

7. Give the *exact* value of each expression. If it is undefined, say so.
   (a) $\cos 60°$
   (b) $\tan 45°$
   (c) $\tan(-270°)$
   (d) $\sec 210°$
   (e) $\csc(-180°)$
   (f) $\sec 135°$

8. Use a calculator to approximate the following.
   (a) $\sin 78° 21'$
   (b) $\tan 11.7689°$
   (c) $\sec 58.9041°$
   (d) $\cot 13.5°$

9. Find a value of θ in the interval $0° \leq \theta \leq 90°$ in decimal degrees, if $\sin \theta = .27843196$.

10. Find two values of θ in the interval $0° \leq \theta < 360°$ that satisfy

$$\cos \theta = -\frac{\sqrt{2}}{2}.$$

11. Solve the triangle.

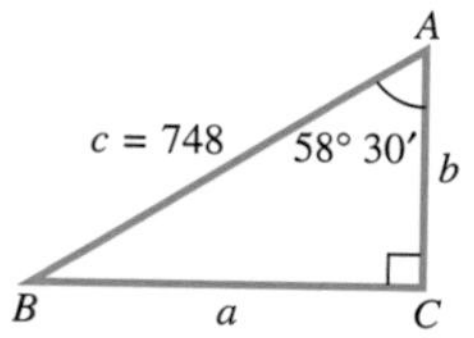

12. ***Height of a Flagpole*** To measure the height of a flagpole, Mike de la Hoz found that the angle of elevation from a point 24.7 feet from the base to the top is $32° 10'$. What is the height of the flagpole?

Find the indicated part of each triangle.

13. $A = 25.2°$, $a = 6.92$ yd, $b = 4.82$ yd; find C

14. $C = 118°$, $b = 130$ km, $a = 75$ km; find c

15. $a = 17.3$ ft, $b = 22.6$ ft, $c = 29.8$ ft; find B

16. Find the area of triangle ABC in Exercise 14.

17. Find the area of a triangle having sides of lengths 22, 26, and 40.

Solve each problem.

18. ***Height of a Balloon*** The angles of elevation of a balloon from two points A and B on level ground are $24° 50'$ and $47° 20'$, respectively. As shown in the figure, points A, B, and C are in the same vertical plane and A and B are 8.4 miles apart. Approximate the height of the balloon above the ground to the nearest tenth of a mile.

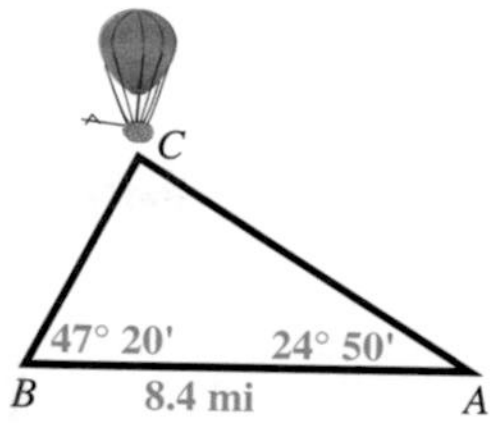

19. ***Length of a Tunnel*** To measure the distance through a mountain for a proposed tunnel, a point C is chosen that can be reached from each end of the tunnel. If $AC = 3800$ meters, $BC = 2900$ meters, and angle $C = 110°$, find the length of the tunnel.

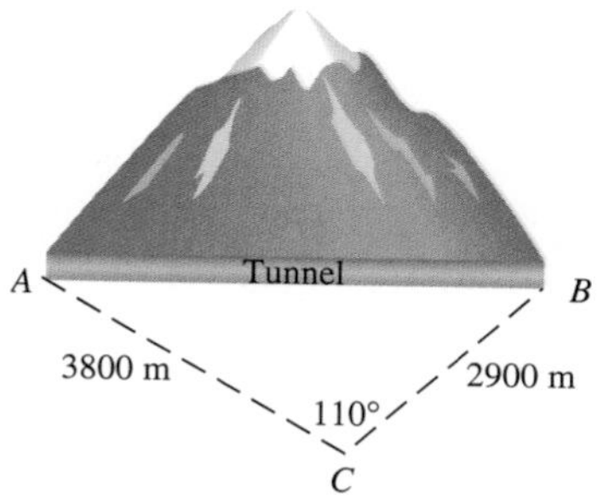

20. ***Distances on a Baseball Diamond*** A baseball diamond is a square 90.0 feet on a side, with home plate and the three bases as vertices. The pitcher's position is located 60.5 feet from home plate. Find the distance from the pitcher's position to each of the bases.

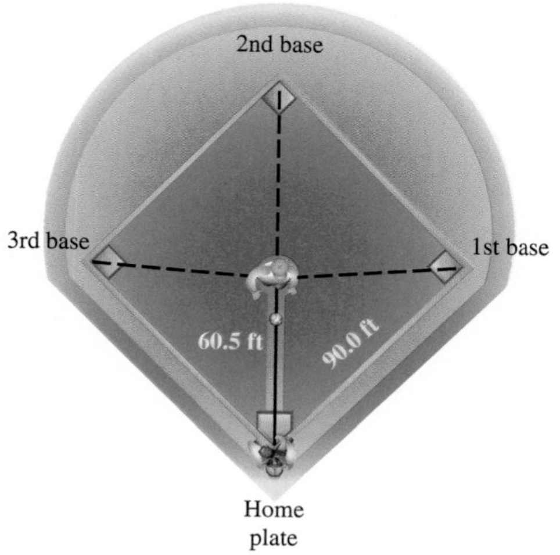

11

COUNTING METHODS

When two dice, each with six possible faces, are tossed, there are $6 \cdot 6 = 36$ ways the dice can land. This is an application of the *fundamental counting principle* introduced in this chapter.

The game of craps, which involves finding the *sum* of two faces, is a mainstay in gaming casinos. A win occurs if the first roll is a natural (7, 11), and a loss occurs if it is craps (2, 3, 12). If a point is rolled (4, 5, 6, 8, 9, 10), it must be repeated before a 7 is tossed in order to win. If 7 is tossed before the point, you lose. (*Source*: www.ildado.com/craps_rules.html)

Craps was popular among U.S. servicemen during World War II. In the Oscar-nominated 1941 movie *Buck Privates*, Bud Abbott and Lou Costello perform one of their many comedy routines on the silver screen, with Costello getting the better of Abbott by twisting around the rules of the game. Later mathematics-related scenes appear in the movie as well, one involving a money-changing scheme and the other a word problem dealing with the ages of Costello and a little girl. (See the margin note in Chapter 7 on page 358.)

Most questions addressed in this chapter will contain a phrase such as "How many . . . ?" or "In how many ways . . . ?" or "Find the number of. . . ." In effect, such a question asks for the cardinal number of some particular set. (See Chapter 2.) There are many possible reasons for knowing the number of elements in a particular set. One major reason is to be able to calculate the likelihood that some event may occur, that is, the *probability* of the event. (Probability is the subject of Chapter 12.)

For example, the genetic code of an individual human consists of approximately six billion DNA bases stored in a linear sequence. The nucleus of every cell contains a copy of this code, tightly coiled in the shape of a double helix. Because each base in the sequence can be any one of four types, called adenine (A), guanine (G), thymine (T), and cytosine (C), there are approximately $4^{6,000,000,000}$ different sequences possible. Of this huge number, only .1% to 1% is unique to the individual, with 99% to 99.9% being common to all humans. The probabilities used in DNA profiling are based on the tiny .1% to 1% that is unique.

The cardinal numbers of some sets are easy to find. Others are more difficult. Some are so large, or so involved, that even state-of-the-art computers cannot determine them. A number of methods are developed in this chapter that are useful in answering "How many . . ." questions.

11.1 Counting by Systematic Listing

One-Part Tasks • Product Tables for Two-Part Tasks • Tree Diagrams for Multiple-Part Tasks • Other Systematic Listing Methods

Counting methods can be used to find the number of moves required to solve a Rubik's Cube. The scrambled cube must be modified so that each face is a solid color. Rubik's royalties from sales of the cube in Western countries made him Hungary's richest man.

The methods of counting presented in this section involve listing the possible results for a given task. This approach is practical only for fairly short lists. When listing possible results, it is extremely important to use a *systematic* approach, or we are likely to miss some results.

One-Part Tasks

The results for simple, one-part tasks can often be listed easily. For the task of tossing a single fair coin, for example, the list is *heads, tails,* with two possible results. If the task is to roll a single fair die (a cube with faces numbered 1 through 6), the different results are 1, 2, 3, 4, 5, 6, a total of six possibilities.

EXAMPLE 1 Selecting a Club President

Consider a club N with five members:

$$N = \{\text{Alan, Bill, Cathy, David, Evelyn}\},$$

or, in abbreviated form, $N = \{A, B, C, D, E\}$.

In how many ways can this group select a president (assuming all members are eligible)?

SOLUTION

The task in this case is to select one of the five members as president. There are five possible results: A, B, C, D, and E. ■

Product Tables for Two-Part Tasks

EXAMPLE 2 Building Numbers from a Set of Digits

Determine the number of two-digit numbers that can be written using digits from the set $\{1, 2, 3\}$.

SOLUTION

This task consists of two parts:

1. Choose a first digit.
2. Choose a second digit.

The results for a two-part task can be pictured in a **product table** such as Table 1. From the table we obtain our list of possible results:

$$11, 12, 13, 21, 22, 23, 31, 32, 33.$$

There are nine possibilities.

TABLE 1

		Second Digit		
		1	2	3
First Digit	1	11	12	13
	2	21	22	23
	3	31	32	33

EXAMPLE 3 Rolling a Pair of Dice

Determine the number of different possible results when two ordinary dice are rolled.

SOLUTION

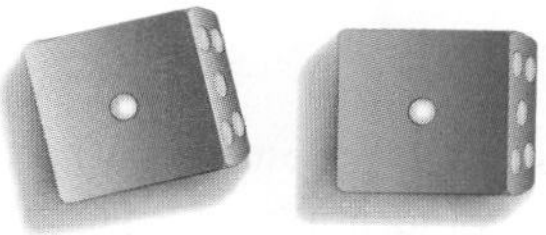

Assume the dice are easily distinguishable. Perhaps one is red and the other green. Then the task consists of two parts:

1. Roll the red die.
2. Roll the green die.

The product table in Table 2 shows that there are thirty-six possible results.

TABLE 2 Rolling Two Fair Dice

		Green Die					
		1	2	3	4	5	6
Red Die	1	(1, 1)	(1, 2)	(1, 3)	(1, 4)	(1, 5)	(1, 6)
	2	(2, 1)	(2, 2)	(2, 3)	(2, 4)	(2, 5)	(2, 6)
	3	(3, 1)	(3, 2)	(3, 3)	(3, 4)	(3, 5)	(3, 6)
	4	(4, 1)	(4, 2)	(4, 3)	(4, 4)	(4, 5)	(4, 6)
	5	(5, 1)	(5, 2)	(5, 3)	(5, 4)	(5, 5)	(5, 6)
	6	(6, 1)	(6, 2)	(6, 3)	(6, 4)	(6, 5)	(6, 6)

You will want to refer to Table 2 when various dice-rolling problems occur in the remainder of this chapter and the next.

EXAMPLE 4 Electing Two Club Officers

Find the number of ways that club *N* of Example 1 can elect both a president and a secretary. Assume that all members are eligible, but that no one can hold both offices.

SOLUTION

Again, the required task has two parts:

1. Determine the president.
2. Determine the secretary.

Constructing Table 3 gives us the possibilities (where, for example, *AB* denotes president *A* and secretary *B*, while *BA* denotes president *B* and secretary *A*).

TABLE 3 Electing Two Officers

		Secretary				
		A	*B*	*C*	*D*	*E*
President	*A*		*AB*	*AC*	*AD*	*AE*
	B	*BA*		*BC*	*BD*	*BE*
	C	*CA*	*CB*		*CD*	*CE*
	D	*DA*	*DB*	*DC*		*DE*
	E	*EA*	*EB*	*EC*	*ED*	

Notice that certain entries (down the main diagonal, from upper left to lower right) are omitted from the table, since the cases *AA*, *BB*, and so on would imply one person holding both offices. Altogether, there are twenty possibilities.

Bone dice were unearthed in the remains of a Roman garrison, Vindolanda, near the border between England and Scotland. Life on the Roman frontier was occupied with gaming as well as fighting. Some of the Roman dice were loaded in favor of 6 and 1.

Life on the American frontier was reflected in cattle brands that were devised to keep alive the memories of hardships, feuds, and romances. A rancher named Ellis from Paradise Valley in Arizona designed his cattle brand in the shape of a pair of dice. You can guess that the pips were 6 and 1.

EXAMPLE 5 Selecting Committees for a Club

Find the number of ways that club *N* can appoint a committee of two members to represent them at an association conference.

SOLUTION

The required task again has two parts. In fact, we can refer to Table 3 again, but this time, the order of the two letters (people) in a given pair really makes no difference. For example, *BD* and *DB* are the same committee. (In Example 4, *BD* and *DB* were different results since the two people would be holding different offices.) In the case of committees, we eliminate not only the main diagonal entries, but also all entries below the main diagonal. The resulting list contains ten possibilities:

AB, *AC*, *AD*, *AE*, *BC*, *BD*, *BE*, *CD*, *CE*, *DE*.

Tree Diagrams for Multiple-Part Tasks

PROBLEM-SOLVING HINT A task that has more than two parts is not easy to analyze with a product table. Another helpful device is the **tree diagram,** which we use in the following examples.

EXAMPLE 6 Building Numbers from a Set of Digits

Find the number of three-digit numbers that can be written using digits from the set $\{1, 2, 3\}$, assuming that **(a)** repeated digits are allowed and **(b)** repeated digits are not allowed.

SOLUTION

(a) The task of constructing such a number has three parts:

1. Select the first digit.
2. Select the second digit.
3. Select the third digit.

As we move from left to right through the tree diagram in Figure 1, the tree branches at the first stage to all possibilities for the first digit. Then each first-stage branch again branches, or splits, at the second stage, to all possibilities for the second digit. Finally, the third-stage branching shows the third-digit possibilities. The list of possible results (twenty-seven of them) is shown in Figure 1.

(b) For the case of nonrepeating digits, we could construct a whole new tree diagram, as in Figure 2, or we could simply go down the list of numbers from the first tree diagram and strike out any that contain repeated digits. In either case we obtain only six possibilities.

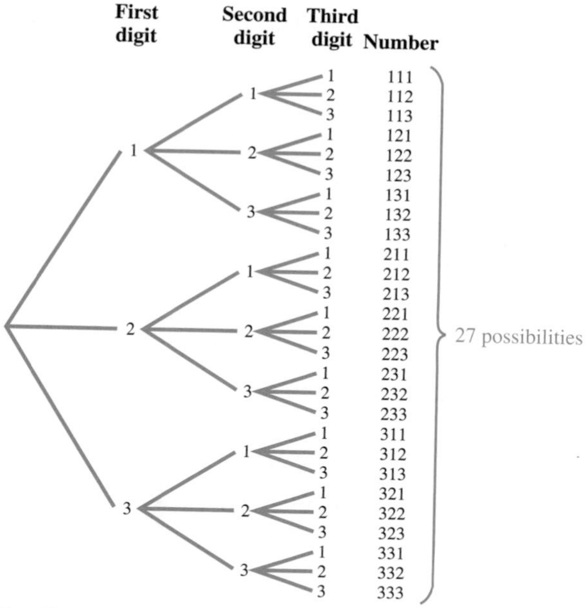

Tree diagram for three-digit numbers with digits from the set $\{1, 2, 3\}$

FIGURE 1

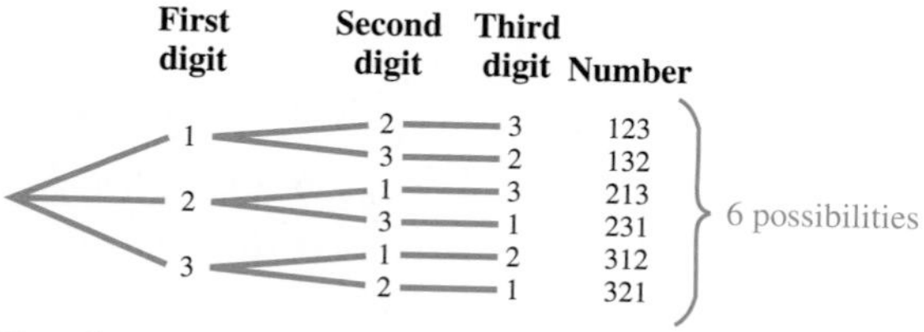

Tree diagram for nonrepeating three-digit numbers with digits from the set $\{1, 2, 3\}$

FIGURE 2

Notice the distinction between parts (a) and (b) of Example 6. There are twenty-seven possibilities when "repetitions (of digits) are allowed," but only six possibilities when "repetitions are not allowed."

Here is another way to phrase the problem of Example 6:

> A three-digit number is to be determined by placing three slips of paper (marked 1, 2, and 3) into a hat and drawing out three slips in succession. Find the number of possible results if the drawing is done **(a)** *with replacement* and **(b)** *without replacement*.

Drawing "with replacement" means drawing a slip, recording its digit, and replacing the slip into the hat so that it is again available for subsequent draws. Drawing "with replacement" has the effect of "allowing repetitions," while drawing "without replacement" has the effect of "not allowing repetitions."

The words "repetitions" and "replacement" are important in the statement of a problem. In Example 2, since no restrictions were stated, we assume that *repetitions* (of digits) *are allowed,* or equivalently that digits are selected *with replacement.*

EXAMPLE 7 Selecting Switch Settings on a Printer

Michelle Clayton's computer printer allows for optional settings with a panel of four on-off switches in a row. How many different settings can she select if no two adjacent switches can both be off?

SOLUTION

This situation is typical of user-selectable options on various devices, including computer equipment, garage door openers, and other appliances. In Figure 3 we denote "on" and "off" with 1 and 0, respectively (a common practice). The number of possible settings is seen to be eight. Notice that each time on the tree diagram that a switch is indicated as off (0), the next switch can only be on (1). This is to satisfy the restriction that no two adjacent switches can both be off.

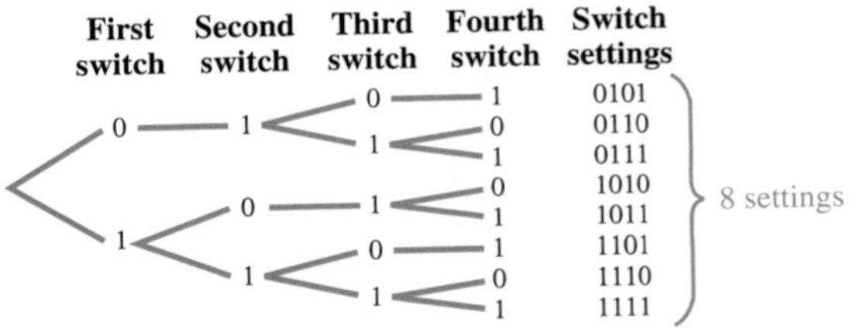

Tree diagram for printer settings

FIGURE 3

EXAMPLE 8 Seating Attendees at a Concert

Arne, Bobbette, Chuck, and Deirdre have tickets for four reserved seats in a row at a concert. In how many different ways can they seat themselves so that Arne and Bobbette will sit next to each other?

SOLUTION

Here we have a four-part task: assign people to the first, second, third, and fourth seats. The tree diagram in Figure 4 on the next page avoids repetitions, because no person can occupy more than one seat. Also, once *A* or *B* appears in the tree, the other one must occur at the next stage. (Why is this?) No splitting occurs from stage three to stage four because by that time there is only one person left unassigned. The right column in the figure shows the twelve possible seating arrangements.

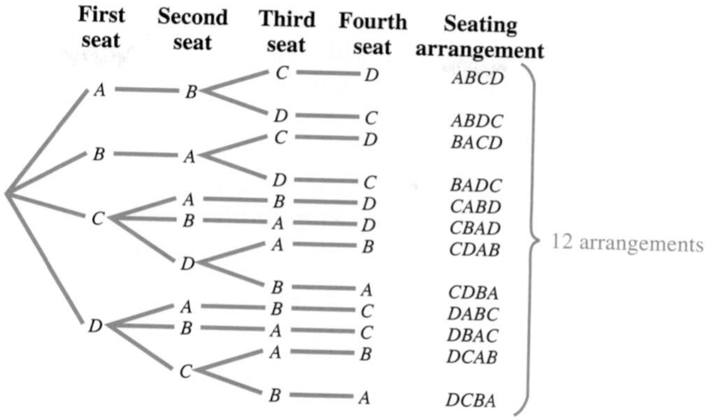

Tree diagram for concert seating

FIGURE 4

Although we have applied tree diagrams only to tasks with three or more parts, they can also be used for two-part or even simple, one-part tasks. Product tables, on the other hand, are practical only for two-part tasks.

Other Systematic Listing Methods There are additional systematic ways to produce complete listings of possible results besides product tables and tree diagrams.

In Example 4, where we used a product table (Table 3) to list all possible president-secretary pairs for the club $N = \{A, B, C, D, E\}$, we could systematically construct the same list using a sort of alphabetical or left-to-right approach. First, consider the results where *A* is president. Any of the remaining members (*B, C, D,* or *E*) could then be secretary. That gives us the pairs *AB, AC, AD,* and *AE*. Next, assume *B* is president. The secretary could then be *A, C, D,* or *E*. We get the pairs *BA, BC, BD,* and *BE*. Continuing in order, we get the complete list just as in Example 4:

AB, AC, AD, AE, BA, BC, BD, BE, CA, CB,
CD, CE, DA, DB, DC, DE, EA, EB, EC, ED.

EXAMPLE 9 Counting Triangles in a Figure

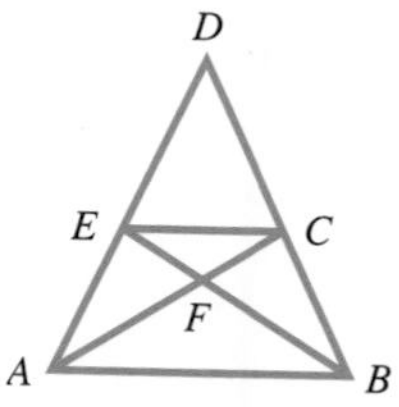

FIGURE 5

How many different triangles (of any size) are included in Figure 5?

SOLUTION

One systematic approach is to label the points as shown, begin with *A*, and proceed in alphabetical order to write all three-letter combinations, then cross out the ones that are not triangles in the figure.

ABC, ABD, ABE, ABF, ACD, ACE, ~~ACF~~, ~~ADE~~, ~~ADF~~, AEF,
~~BCD~~, BCE, BCF, BDE, ~~BDF~~, ~~BEF~~, CDE, ~~CDF~~, CEF, ~~DEF~~

Finally, there are twelve different triangles in the figure. Why are *ACB* and *CBF* (and many others) not included in the list?

Another method might be first to identify the triangles consisting of a single region each: *DEC, ECF, AEF, BCF, ABF*. Then list those consisting of two regions each: *AEC, BEC, ABE, ABC;* and those with three regions each: *ACD, BED*. There are no triangles with four regions, but there is one with five: *ABD*. The total is again twelve. Can you think of other systematic ways of getting the same list? ■

Notice that in the first method shown in Example 9, the labeled points were considered in alphabetical order. In the second method, the single-region triangles were listed by using a top-to-bottom and left-to-right order. Using a definite system helps to ensure that we get a complete list.

The **"tree diagram"** on the map came from research on the feasibility of using motor-sailers (motor-driven ships with wind-sail auxiliary power) on the North Atlantic run. At the beginning of a run, weather forecasts and computer analysis are used to choose the best of the 45 million possible routes.

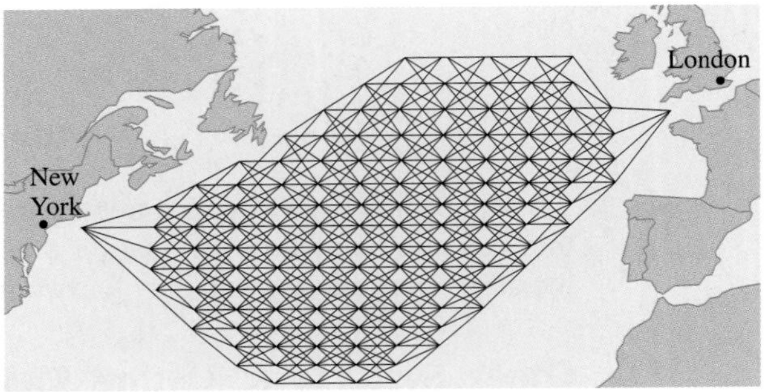

11.1 EXERCISES

Electing Officers of a Club *Refer to Examples 1 and 4, involving the club*

$$N = \{\text{Alan, Bill, Cathy, David, Evelyn}\}.$$

Assuming all members are eligible, but that no one can hold more than one office, list and count the different ways the club could elect each group of officers.

1. a president and a treasurer
2. a president and a treasurer if the president must be a female
3. a president and a treasurer if the two officers must not be the same sex
4. a president, a secretary, and a treasurer, if the president and treasurer must be women
5. a president, a secretary, and a treasurer, if the president must be a man and the other two must be women
6. a president, a secretary, and a treasurer, if the secretary must be a woman and the other two must be men

Appointing Committees *List and count the ways club N could appoint a committee of three members under each condition.*

7. There are no restrictions.
8. The committee must include more men than women.

Refer to Table 2 (the product table for rolling two dice). Of the 36 possibilities, determine the number for which the sum (for both dice) is the following.

9. 2 **10.** 3 **11.** 4

12. 5 **13.** 6 **14.** 7

15. 8 **16.** 9 **17.** 10

18. 11 **19.** 12 **20.** odd

21. even

22. from 6 through 8 inclusive

23. between 6 and 10

24. less than 5

25. Construct a product table showing all possible two-digit numbers using digits from the set

$$\{1, 2, 3, 4, 5, 6\}.$$

Of the thirty-six numbers in the product table for Exercise 25, list the ones that belong to each category.

26. odd numbers

27. numbers with repeating digits

28. multiples of 6

29. prime numbers

30. triangular numbers

31. square numbers

32. Fibonacci numbers

33. powers of 2

34. Construct a tree diagram showing all possible results when three fair coins are tossed. Then list the ways of getting each result.
- **(a)** at least two heads
- **(b)** more than two heads
- **(c)** no more than two heads
- **(d)** fewer than two heads

35. Extend the tree diagram of Exercise 34 for four fair coins. Then list the ways of getting each result.
- **(a)** more than three tails
- **(b)** fewer than three tails
- **(c)** at least three tails
- **(d)** no more than three tails

Determine the number of triangles (of any size) in each figure.

36.

37.

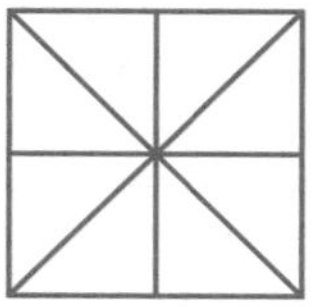

38.

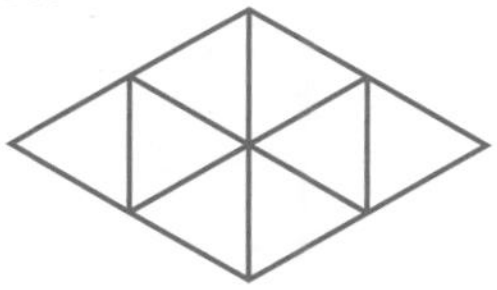

39.

Determine the number of squares (of any size) in each figure.

40.

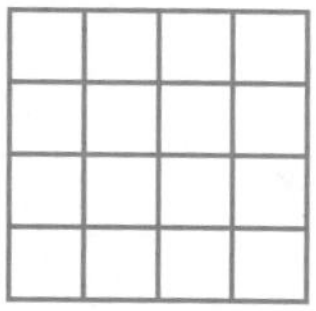

41.

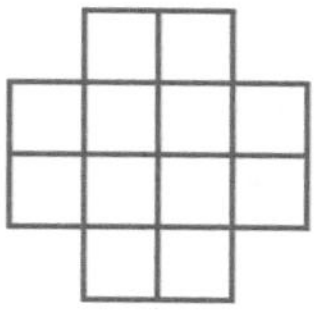

42.

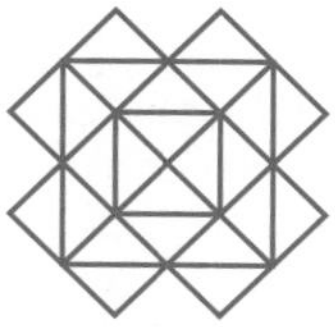

43.

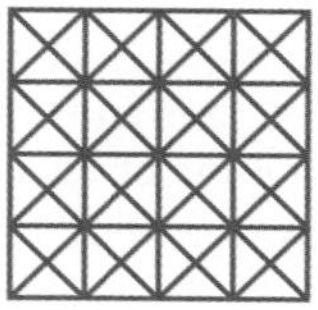

Consider only the smallest individual cubes and assume solid stacks (no gaps). Determine the number of cubes in each stack that are not visible from the perspective shown.

44.

45.

46.

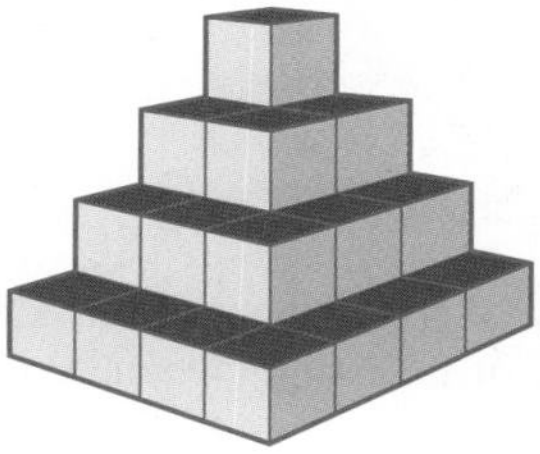

47.

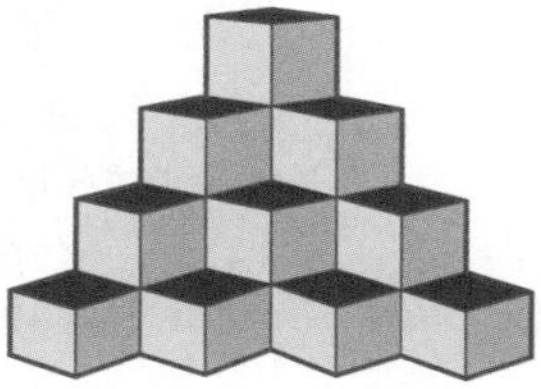

48. In the plane figure shown here, only movement that tends downward is allowed. Find the total number of paths from *A* to *B*.

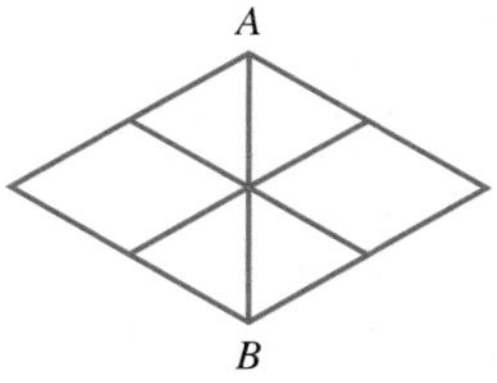

49. Find the number of paths from A to B in the figure shown here if the directions on various segments are restricted as shown.

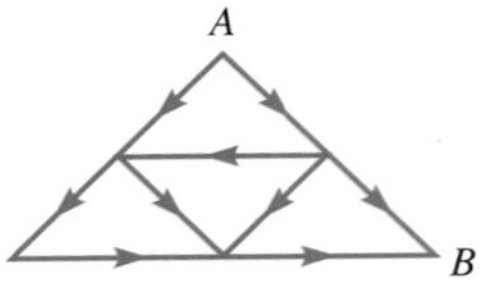

In each of Exercises 50–52, determine the number of different ways the given number can be written as the sum of two primes.

50. 30 **51.** 40 **52.** 95

53. *Shaking Hands in a Group* A group of twelve strangers sat in a circle, and each one got acquainted only with the person to the left and the person to the right. Then all twelve people stood up and each one shook hands (once) with each of the others who was still a stranger. How many handshakes occurred?

54. *Number of Games in a Chess Tournament* Fifty people enter a single-elimination chess tournament. (If you lose one game, you're out.) Assuming no ties occur, what is the number of games required to determine the tournament champion?

55. *Sums of Digits* How many of the numbers from 10 through 100 have the sum of their digits equal to a perfect square?

56. *Sums of Digits* How many three-digit numbers have the sum of their digits equal to 22?

57. *Integers Containing the Digit 2* How many integers between 100 and 400 contain the digit 2?

58. *Selecting Dinner Items* Frank Capek and friends are dining at the Bay Steamer Restaurant this evening, where a complete dinner consists of three items:

(1) soup (clam chowder or minestrone) or salad (fresh spinach or shrimp),
(2) sourdough rolls or bran muffin, and
(3) entree (lasagna, lobster, or roast turkey).

Frank selects his meal subject to the following restrictions. He cannot stomach more than one kind of seafood at a sitting. Also, whenever he tastes minestrone, he cannot resist having lasagna as well. And he cannot face the teasing he would receive from his companions if he were to order both spinach and bran. Use a tree diagram to determine the number of different choices Frank has.

Setting Options on a Computer Printer *For Exercises 59–61, refer to Example 7. How many different settings could Michelle choose in each case?*

59. No restrictions apply to adjacent switches.

60. No two adjacent switches can be off *and* no two adjacent switches can be on.

61. There are five switches rather than four, and no two adjacent switches can be on.

62. *Building Numbers from Sets of Digits* Determine the number of odd, nonrepeating three-digit numbers that can be written using digits from the set $\{0, 1, 2, 3\}$.

63. *Lattice Points on a Line Segment* A line segment joins the points (8, 12) and (53, 234) in the Cartesian plane. Including its endpoints, how many lattice points does this line segment contain? (A *lattice point* is a point with integer coordinates.)

64. *Lengths of Segments Joining Lattice Points* In the pattern that follows, dots are one unit apart horizontally and vertically. If a segment can join any two dots, how many segments can be drawn with each length?

(a) 1 **(b)** 2 **(c)** 3
(d) 4 **(e)** 5

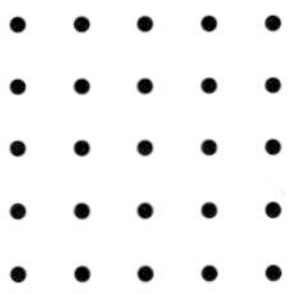

65. *Counting Matchsticks in a Grid* Uniform-length matchsticks are used to build a rectangular grid as shown here. If the grid is 15 matchsticks high and 28 matchsticks wide, how many matchsticks are used?

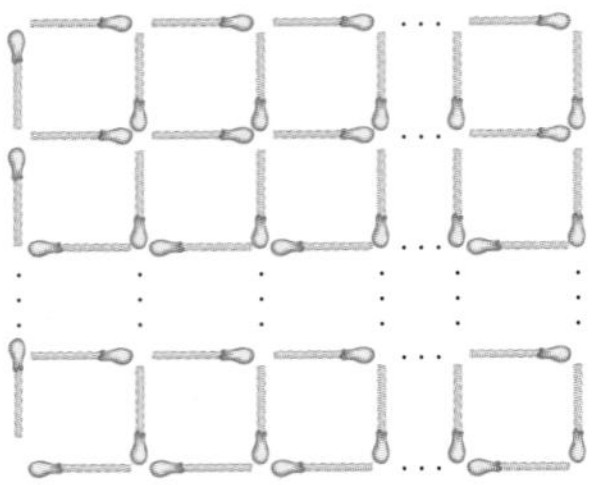

66. ***Patterns in Floor Tiling*** A square floor is to be tiled with square tiles as shown, with blue tiles on the main diagonals and red tiles everywhere else. (In all cases, both blue and red tiles must be used and the two diagonals must have a common blue tile at the center of the floor.)

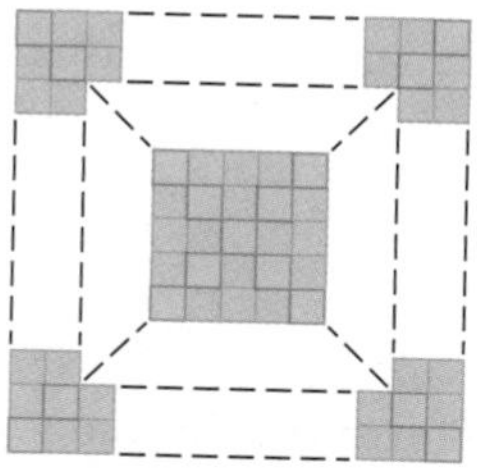

(a) If 81 blue tiles will be used, how many red tiles will be needed?

(b) For what numbers in place of 81 would this problem still be solvable?

(c) Find a formula expressing the number of red tiles required in general.

67. ***Shaking Hands in a Group*** Jeff Howard and his son were among four father-and-son pairs who gathered to trade baseball cards. As each person arrived, he shook hands with anyone he had not known previously. Each person ended up making a different number of new acquaintances (0–6), except Jeff and his son, who each met the same number of people. How many hands did Jeff shake?

In Exercises 68–71, restate the given counting problem in two ways, first **(a)** *using the word* repetition, *and then* **(b)** *using the word* replacement.

68. Example 2

69. Example 3

70. Example 4

71. Exercise 7

11.2 Using the Fundamental Counting Principle

Uniformity and the Fundamental Counting Principle • Factorials • Arrangements of Objects

Uniformity and the Fundamental Counting Principle

In Section 11.1, we obtained complete lists of all possible results for various tasks. However, if the total number of possibilities is all we need to know, then an actual listing usually is unnecessary and often is difficult or tedious to obtain, especially when the list is long. In this section, we develop ways to calculate "how many" using the *fundamental counting principle.*

Figure 6 repeats Figure 2 of Section 11.1 (for Example 6(b)) which shows all possible nonrepeating three-digit numbers with digits from the set {1, 2, 3}.

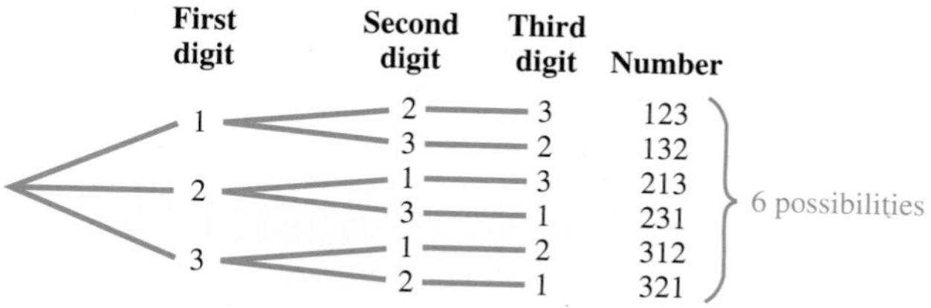

FIGURE 6

The tree diagram in Figure 6 is "uniform" in the sense that a given part of the task can be done in the same number of ways no matter which choices were selected for previous parts. For example, there are always two choices for the second digit. (If the first digit is 1, the second can be 2 or 3; if the first is 2, the second can be 1 or 3; if the first is 3, the second can be 1 or 2.)

Example 6(a) of Section 11.1 addressed the same basic situation: *Find the number of three-digit numbers that can be written using the digits* 1, 2, *and* 3. In that case repetitions were allowed. With repetitions allowed, there were many more possibilities (27 rather than 6—see Figure 1 of Section 11.1). But the uniformity criterion mentioned above still applied. No matter what the first digit is, there are three choices for the second (1, 2, 3). And no matter what the first and second digits are, there are three choices for the third. This uniformity criterion can be stated in general as follows.

Uniformity Criterion for Multiple-Part Tasks

A multiple-part task is said to satisfy the **uniformity criterion** if the number of choices for any particular part is the same *no matter which choices were selected for previous parts.*

The uniformity criterion is not satisfied by all multiple-part counting problems. Refer to Example 7 (and Figure 3) of Section 11.1. After the first switch (two possibilities), other switches had either one or two possible settings depending on how previous switches were set. (This "nonuniformity" arose, in that case, from the requirement that no two adjacent switches could both be off.)

In the many cases where uniformity does hold, we can avoid having to construct a tree diagram by using the **fundamental counting principle,** stated as follows.

Fundamental Counting Principle

When a task consists of k separate parts and satisfies the uniformity criterion, if the first part can be done in n_1 ways, the second part can then be done in n_2 ways, and so on through the kth part, which can be done in n_k ways, then the total number of ways to complete the task is given by the product

$$n_1 \cdot n_2 \cdot n_3 \cdot \ldots \cdot n_k.$$

PROBLEM-SOLVING HINT A problem-solving strategy suggested in Chapter 1 was: "If a formula applies, use it." The fundamental counting principle provides a formula that applies to a variety of problems. The trick is to visualize the "task" at hand as being accomplished in a sequence of two or more separate parts. A helpful technique when applying the fundamental counting principle is to write out all the separate parts of the task, with a blank for each one. Reason out how many ways each part can be done and enter these numbers in the blanks. Finally, multiply these numbers together.

Richard Dedekind (1831–1916) studied at the University of Göttingen, where he was Gauss's last student. His work was not recognized during his lifetime, but his treatment of the infinite and of what constitutes a real number are influential even today.

While on vacation in Switzerland, Dedekind met Georg Cantor (profiled on page 52). Dedekind was interested in Cantor's work on infinite sets. Perhaps because both were working in new and unusual fields of mathematics, such as number theory, and because neither received the professional attention he deserved during his lifetime, the two struck up a lasting friendship.

EXAMPLE 1 Counting the Two-Digit Numbers

How many two-digit numbers are there in our (base-ten) system of counting numbers? (While 40 is a two-digit number, 04 is not.)

SOLUTION

Our "task" here is to select, or construct, a two-digit number. We can set up the work as follows, showing the two parts to be done.

Part of task	Select first digit	Select second digit
Number of ways	___	___

There are nine choices for the first digit (1 through 9). Since there were no stated or implied restrictions, we assume that repetition of digits is allowed. Therefore, no matter which nonzero digit is used as the first digit, all nine choices are available for the second digit. Also, unlike the first digit, the second digit may be zero, so we have ten choices for the second digit. We can now fill in the blanks and multiply.

Part of task	Select first digit		Select second digit	
Number of ways	9	$\cdot$	10	$= 90$

There are 90 two-digit numbers. (As a check, notice that they are the numbers from 10 through 99, a total of $99 - 10 + 1 = 90$.)

EXAMPLE 2 Building Two-Digit Numbers with Restrictions

Find the number of two-digit numbers that do not contain repeated digits.

SOLUTION

The basic task is again to select a two-digit number, and there are two parts:

1. Select the first digit.
2. Select the second digit.

But a new restriction applies—no repetition of digits. There are nine choices for the first digit (1 through 9). Then nine choices remain for the second digit, since one nonzero digit has been used and cannot be repeated, but zero is now available. The total number is $9 \cdot 9 = 81$.

EXAMPLE 3 Electing Club Officers with Restrictions

In how many ways can Club *N* of the previous section elect a president and a secretary if no one may hold more than one office and the secretary must be a man?

SOLUTION

Recall that $N = \{A, B, C, D, E\} = \{\text{Alan, Bill, Cathy, David, Evelyn}\}$. Considering president first, there are five choices (no restrictions). But now we have a problem with finding the number of choices for secretary. If a woman was selected president (*C* or *E*), there are three choices for secretary (*A*, *B*, and *D*). If a man was selected president, only two choices (the other two men) remain for secretary. In other words, the uniformity criterion is not met and our attempt to apply the fundamental counting principle has failed.

All is not lost, however. In finding the total number of ways, there is no reason we cannot consider secretary first. There are three choices (A, B, and D). Now, no matter which man was chosen secretary, both of the other men, and both women, are available for president (four choices in every case). In this order, we satisfy the uniformity criterion and can use the fundamental counting principle. The total number of ways to elect a president and a secretary is $3 \cdot 4 = 12$.

PROBLEM-SOLVING HINT Example 3 suggests a useful problem-solving strategy: Whenever one or more parts of a task have special restrictions, try considering that part (or those parts) before other parts.

EXAMPLE 4 Counting Three-Digit Numbers with Restrictions

How many nonrepeating odd three-digit counting numbers are there?

SOLUTION

The most restricted digit is the third, since it must be odd. There are five choices (1, 3, 5, 7, and 9). Next, consider the first digit. It can be any nonzero digit except the one already chosen as the third digit. There are eight choices. Finally, the second digit can be any digit (including 0) except for the two (nonzero) digits already used. There are eight choices. We can summarize this reasoning as follows.

Part of task	Select third digit		Select first digit		Select second digit	
Number of ways	5	·	8	·	8	= 320

There are 320 nonrepeating odd three-digit counting numbers.

EXAMPLE 5 Counting License Plates

In some states, auto license plates have contained three letters followed by three digits. How many such licenses are possible?

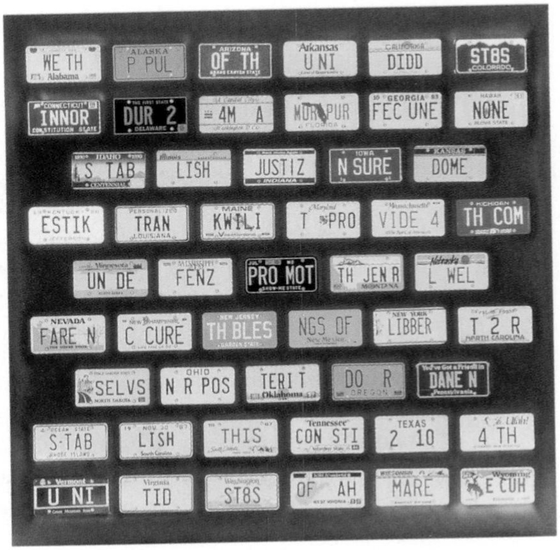

SOLUTION

The basic task is to design a license number with three letters followed by three digits. There are six component parts to this task. Since there are no restrictions on letters or digits, the fundamental counting principle gives

$$26 \cdot 26 \cdot 26 \cdot 10 \cdot 10 \cdot 10 = 26^3 \cdot 10^3 = 17{,}576{,}000$$

possible licenses. (In practice, a few of the possible sequences of letters are considered undesirable and are not used.)

EXAMPLE 6 Building Numbers from a Set of Digits

A four-digit number is to be constructed using only digits from the set $\{1, 2, 3\}$.

(a) How many such numbers are possible?

(b) How many of these numbers are odd and less than 2000?

SOLUTION

(a) To construct such a number, we must select four digits, in succession, from the given set of three digits, where the selection is done with replacement (since repetition of digits is apparently allowed). By the fundamental counting principle, the number of possibilities is

$$3 \cdot 3 \cdot 3 \cdot 3 = 3^4 = 81.$$

(b) The number is less than 2000 only if the first digit is 1 (just one choice) and is odd only if the fourth digit is 1 or 3 (two choices). The second and third digits are unrestricted (three choices for each). The answer is

$$1 \cdot 3 \cdot 3 \cdot 2 = 18.$$

As a check, can you list the eighteen possibilities?

PROBLEM-SOLVING HINT Two of the problem-solving strategies of Chapter 1 were to "first solve a similar simpler problem," and to "look for a pattern." In fact, a problem at hand may sometimes prove to be essentially the same, or at least fit the same pattern, as another problem already solved.

EXAMPLE 7 Distributing Golf Clubs

Vern has four antique wood head golf clubs that he wants to give to his three sons, Mark, Chris, and Scott.

(a) How many ways can the clubs be distributed?

(b) How many choices are there if the power driver must go to Mark and the number 3 wood must go to either Chris or Scott?

SOLUTION

(a) The task is to distribute four clubs among three sons. Consider the clubs in succession and, for each one, ask how many sons could receive it. In effect, we must select four sons, in succession, from the set {Mark, Chris, Scott}, selecting with replacement. Compare this with Example 6(a), in which we selected four digits, in succession, from the set $\{1, 2, 3\}$, selecting with replacement. In this case, we are selecting sons rather than digits, but the pattern is the same and the numbers are the same. Again our answer is $3^4 = 81$.

(b) Just as in Example 6(b), one part of the task is now restricted to a single choice and another part is restricted to two choices. As in that example, the number of possibilities is $1 \cdot 3 \cdot 3 \cdot 2 = 18$.

EXAMPLE 8 Seating Attendees at a Concert

Rework Example 8 of Section 11.1, this time using the fundamental counting principle.

SOLUTION

Recall that Arne, Bobbette, Chuck, and Deirdre (A, B, C, and D) are to seat themselves in four adjacent seats (say 1, 2, 3, and 4) so that A and B are side-by-side. One approach to accomplish this task is to make three successive decisions as follows.

1	2	3	4
X	X	_	_
_	X	X	_
_	_	X	X

Seats available to A and B

1. Which pair of seats should A and B occupy? There are *three* choices (1 and 2, 2 and 3, 3 and 4, as illustrated in the margin).
2. Which order should A and B take? There are *two* choices (A left of B, or B left of A).
3. Which order should C and D take? There are *two* choices (C left of D, or D left of C, not necessarily right next to each other).

(Why did we not ask which two seats C and D should occupy?) The fundamental counting principle now gives the total number of choices:

$$3 \cdot 2 \cdot 2 = 12 \quad \text{Same result as in Section 11.1}$$

Short Table of Factorials

Factorial values increase rapidly. The value of 100! is a number with 158 digits.

$0! = 1$
$1! = 1$
$2! = 2$
$3! = 6$
$4! = 24$
$5! = 120$
$6! = 720$
$7! = 5040$
$8! = 40{,}320$
$9! = 362{,}880$
$10! = 3{,}628{,}800$

Factorials This section began with a discussion of nonrepeating three-digit numbers with digits from the set $\{1, 2, 3\}$. The number of possibilities was

$$3 \cdot 2 \cdot 1 = 6,$$

in keeping with the fundamental counting principle. That product can also be thought of as the total number of distinct *arrangements* of the three digits 1, 2, and 3.

Similarly, the number of distinct arrangements of four objects, say A, B, C, and D, is, by the fundamental counting principle,

$$4 \cdot 3 \cdot 2 \cdot 1 = 24.$$

Since this type of product occurs so commonly in applications, we give it a special name and symbol as follows. For any counting number n, the product of *all* counting numbers from n down through 1 is called ***n* factorial,** and is denoted $n!$.

Factorial Formula

For any counting number n, the quantity ***n* factorial** is given by

$$n! = n(n - 1)(n - 2)\ldots 2 \cdot 1.$$

The first few factorial values are easily found by simple multiplication, but they rapidly become very large, as indicated in the margin. The use of a calculator is advised in most cases. (See the margin notes that follow.)

PROBLEM-SOLVING HINT Sometimes expressions with factorials can be evaluated easily by observing that, in general, $n! = n(n-1)!$, $n! = n(n-1)(n-2)!$, and so on. For example,

$$8! = 8 \cdot 7!, \quad 12! = 12 \cdot 11 \cdot 10 \cdot 9!, \quad \text{and so on.}$$

This pattern is especially helpful in evaluating quotients of factorials, such as

$$\frac{10!}{8!} = \frac{10 \cdot 9 \cdot 8!}{8!} = 10 \cdot 9 = 90.$$

EXAMPLE 9 Evaluating Expressions Containing Factorials

Evaluate each expression.

(a) $3!$ **(b)** $6!$ **(c)** $(6-3)!$ **(d)** $6! - 3!$

(e) $\frac{6!}{3!}$ **(f)** $\left(\frac{6}{3}\right)!$ **(g)** $15!$ **(h)** $100!$

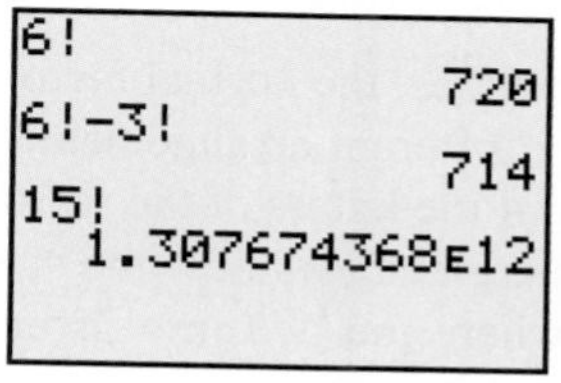

The results of Example 9(b), (d), and (g) are illustrated in this calculator screen.

SOLUTION

(a) $3! = 3 \cdot 2 \cdot 1 = 6$

(b) $6! = 6 \cdot 5 \cdot 4 \cdot 3 \cdot 2 \cdot 1 = 720$

(c) $(6-3)! = 3! = 6$

(d) $6! - 3! = 720 - 6 = 714$

(e) $\frac{6!}{3!} = \frac{6 \cdot 5 \cdot 4 \cdot 3!}{3!} = 6 \cdot 5 \cdot 4 = 120$ Note application of problem-solving hint

(f) $\left(\frac{6}{3}\right)! = 2! = 2 \cdot 1 = 2$

(g) $15! = 1.307674368000 \times 10^{12}$

(h) $100! = 9.332621544 \times 10^{157}$

Notice the distinction between parts (c) and (d) and between parts (e) and (f) above. Parts (g) and (h) were performed on an advanced scientific calculator. (Part (h) is beyond the capability of most scientific calculators.)

So that factorials will be defined for all whole numbers, including zero, it is common to define 0! as follows.

The definition $0! = 1$ is illustrated here.

Definition of Zero Factorial

$$0! = 1$$

(We will see later that this special definition makes other results easier to state.)

Arrangements of Objects When finding the total number of ways to *arrange* a given number of distinct objects, we can use a factorial. The fundamental counting principle would do, but factorials provide a shortcut.

Arrangements of n Distinct Objects

The total number of different ways to arrange n distinct objects is $\boldsymbol{n!}$.

EXAMPLE 10 Arranging Essays

Erika Berg has seven essays to include in her English 1A folder. In how many different orders can she arrange them?

SOLUTION

The number of ways to arrange seven distinct objects is $7! = 5040$. ■

EXAMPLE 11 Arranging Preschoolers

Lynn Damme is taking thirteen preschoolers to the park. How many ways can the children line up, in single file, to board the van?

SOLUTION

Thirteen children can be arranged in $13! = 6{,}227{,}020{,}800$ different ways. ■

D_1AD_2
D_2AD_1

D_1D_2A
D_2D_1A

AD_1D_2
AD_2D_1

In counting arrangements of objects that contain look-alikes, the normal factorial formula must be modified to find the number of truly different arrangements. For example, the number of distinguishable arrangements of the letters of the word DAD is not $3! = 6$ but rather $\frac{3!}{2!} = 3$. The listing in the margin shows how the six total arrangements consist of just three groups of two, where the two in a given group look alike. In general, the distinguishable arrangements can be counted as follows.

Arrangements of n Objects Containing Look-Alikes

The number of **distinguishable arrangements** of n objects, where one or more subsets consist of look-alikes (say n_1 are of one kind, n_2 are of another kind, . . . , and n_k are of yet another kind), is given by

$$\frac{n!}{n_1!n_2!\ldots n_k!}.$$

EXAMPLE 12 Counting Distinguishable Arrangements

Determine the number of distinguishable arrangements of the letters in each word.

(a) HEEDLESS **(b)** NOMINEE

SOLUTION

(a) For the letters of HEEDLESS, the number of distinguishable arrangements is

8 letters total ⟶
3 E's, 2 S's ⟶

$$\frac{8!}{3!\,2!} = 3360.$$

(b) For the letters of NOMINEE, the number of distinguishable arrangements is

7 letters total →
2 N's, 2 E's →
$$\frac{7!}{2!\,2!} = 1260.$$

For Further Thought

Stirling's Approximation for *n*!

Although all factorial values are counting numbers, they can be approximated using **Stirling's formula,**

$$n! \approx \sqrt{2\pi n} \cdot n^n \cdot e^{-n},$$

which involves two famous irrational numbers, π and e. For example, while the exact value of 5! is $5 \cdot 4 \cdot 3 \cdot 2 \cdot 1 = 120$, the corresponding approximation is

$$5! \approx \sqrt{2\pi 5} \cdot 5^5 \cdot e^{-5} \approx 118.019168,$$

which is off by less than 2, an error of only 1.65%.

For Group Discussion or Individual Investigation

Use a calculator to fill in all values in the table. The column values are defined as follows.

$C = n!$ (exact value, by calculator)

$S \approx n!$ (Stirling's approximation, by calculator)

D = Difference $(C - S)$

P = Percentage difference $\left(\frac{D}{C} \cdot 100\%\right)$

Try to obtain percentage differences accurate to two decimal places.

n	*C*	*S*	*D*	*P*
1				
2				
3				
4				
5				
6				
7				
8				
9				
10				

Based on your calculations, answer each question.

1. In general, is Stirling's approximation too low or too high?
2. Observe the values in the table as *n* grows larger.
   (a) Do the differences (*D*) get larger or smaller?
   (b) Do the percentage differences (*P*) get larger or smaller?
   (c) Does Stirling's formula become more accurate or less accurate?

11.2 Exercises

1. Explain the fundamental counting principle in your own words.

2. Describe how factorials can be used in counting problems.

For Exercises 3–6, n and m are counting numbers. Do the following: **(a)** *Tell whether the given statement is true in general, and* **(b)** *explain your answer, using specific examples.*

3. $(n + m)! = n! + m!$

4. $(n \cdot m)! = n! \cdot m!$

5. $(n - m)! = n! - m!$

6. $n! = n(n - 1)!$

Evaluate each expression without using a calculator.

7. $4!$

8. $7!$

9. $\frac{8!}{5!}$

10. $\frac{16!}{14!}$

11. $\frac{5!}{(5-2)!}$

12. $\frac{6!}{(6-4)!}$

13. $\frac{9!}{6!(6-3)!}$

14. $\frac{10!}{4!(10-4)!}$

15. $\frac{n!}{(n-r)!}$, where $n = 7$ and $r = 4$

16. $\frac{n!}{r!(n-r)!}$, where $n = 12$ and $r = 4$

Evaluate each expression using a calculator. (Some answers may not be exact.)

17. $11!$

18. $17!$

19. $\frac{12!}{7!}$

20. $\frac{15!}{9!}$

21. $\frac{13!}{(13-3)!}$

22. $\frac{16!}{(16-6)!}$

23. $\frac{20!}{10! \cdot 10!}$

24. $\frac{18!}{6! \cdot 12!}$

25. $\frac{n!}{(n-r)!}$, where $n = 23$ and $r = 10$

26. $\frac{n!}{r!(n-r)!}$, where $n = 28$ and $r = 15$

Arranging Letters *Find the number of distinguishable arrangements of the letters of each word.*

27. SYNDICATE

28. GOOGOL

29. NONSENSE

30. HEEBIE-JEEBIES

Settings on a Switch Panel *A panel containing three on–off switches in a row is to be set.*

31. Assuming no restrictions on individual switches, use the fundamental counting principle to find the total number of possible panel settings.

32. Assuming no restrictions, construct a tree diagram to list all the possible panel settings of Exercise 31.

33. Now assume that no two adjacent switches can both be off. Explain why the fundamental counting principle does not apply.

34. Construct a tree diagram to list all possible panel settings under the restriction of Exercise 33.

35. ***Rolling Dice*** Table 2 in the previous section shows that there are 36 possible outcomes when two fair dice are rolled. How many would there be if three fair dice were rolled?

36. ***Counting Five-Digit Numbers*** How many five-digit numbers are there in our system of counting numbers?

Matching Club Members with Tasks *Recall the club*

$$N = \{\text{Alan, Bill, Cathy, David, Evelyn}\}.$$

In how many ways could they do each of the following?

37. line up all five members for a photograph

38. schedule one member to work in the office on each of five different days, assuming members may work more than one day

39. select a male and a female to decorate for a party

40. select two members, one to open their next meeting and another to close it, given that Bill will not be present

Building Numbers from Sets of Digits *In Exercises 41–44, counting numbers are to be formed using only digits from the set* $\{3, 4, 5\}$. *Determine the number of different possibilities for each type of number described.*

41. two-digit numbers

42. odd three-digit numbers

43. four-digit numbers with one pair of adjacent 4s and no other repeated digits (*Hint:* You may want to split the task of designing such a number into three parts, such as *(1)* position the pair of 4s, *(2)* position the 3, and *(3)* position the 5.)

44. five-digit numbers beginning and ending with 3 and with unlimited repetitions allowed

Selecting Dinner Items *The Casa Loma Restaurant offers four choices in the soup and salad category (two soups and two salads), two choices in the bread category, and three choices in the entree category. Find the number of dinners available in each case.*

45. One item is to be included from each of the three categories.

46. Only soup and entree are to be included.

Selecting Answers on a Test *Determine the number of possible ways to mark your answer sheet (with an answer for each question) for each test.*

47. a six-question true-or-false test

48. a twenty-question multiple-choice test with five answer choices for each question

Selecting a College Class Schedule *Tiffany Connolly's class schedule for next semester must consist of exactly one class from each of the four categories shown.*

Category	Choices	Number of Choices
English	Medieval Literature Composition Modern Poetry	3
Mathematics	History of Mathematics College Algebra Finite Mathematics	3
Computer Information Science	Introduction to Spreadsheets Advanced Word Processing C Programming BASIC Programming	4
Sociology	Social Problems Sociology of the Middle East Aging in America Minorities in America Women in American Culture	5

For each situation in Exercises 49–54, use the table at the bottom of the left column to determine the number of different sets of classes Tiffany can take.

49. All classes shown are available.

50. She is not eligible for Modern Poetry or for C Programming.

51. All sections of Minorities in America and Women in American Culture already are filled.

52. She does not have the prerequisites for Medieval Literature, Finite Mathematics, or C Programming.

53. Funding has been withdrawn for three of the computer courses and for two of the Sociology courses.

54. She must complete English Composition and Aging in America next semester to fulfill her degree requirements.

55. ***Selecting Clothing*** Sean took two pairs of shoes, five pairs of pants, and six shirts on a trip. Assuming all items are compatible, how many different outfits can he wear?

56. ***Selecting Music Equipment*** A music equipment outlet stocks ten different guitars, four guitar cases, six amplifiers, and three effects processors, with all items mutually compatible and all suitable for beginners. How many different complete setups could Lionel choose to start his musical career?

57. ***Counting ZIP Codes*** Tonya's ZIP code is 85726. How many ZIP codes, altogether, could be formed, each one using those same five digits?

58. ***Listing Phone Numbers*** John Cross keeps the phone numbers for his seven closest friends (three men and four women) in his digital phone memory. How many ways can he list them if

(a) men are listed before women?

(b) men are all listed together?

(c) no two men are listed next to each other?

Seating Arrangements at a Theater *Arne, Bobbette, Chuck, Deirdre, Ed, and Fran have reserved six seats in a row at the theater, starting at an aisle seat. (Refer to Example 8 in this section.)*

59. In how many ways can they arrange themselves? (*Hint:* Divide the task into the series of six parts shown below, performed in order.)

(a) If A is seated first, how many seats are available for him?

(b) Now, how many are available for B?

(c) Now, how many for C?
(d) Now, how many for D?
(e) Now, how many for E?
(f) Now, how many for F?
Now multiply together your six answers above.

60. In how many ways can they arrange themselves so that Arne and Bobbette will be next to each other?

1	2	3	4	5	6
X	X	_	_	_	_
_	X	X	_	_	_
_	_	X	X	_	_
_	_	_	X	X	_
_	_	_	_	X	X

Seats available to A and B

(*Hint:* First answer the following series of questions, assuming these parts are to be accomplished in order.)
(a) How many pairs of adjacent seats can A and B occupy?
(b) Now, given the two seats for A and B, in how many orders can they be seated?
(c) Now, how many seats are available for C?
(d) Now, how many for D?
(e) Now, how many for E?
(f) Now, how many for F?
Now multiply your six answers above.

61. In how many ways can they arrange themselves if the men and women are to alternate seats and a man must sit on the aisle? (*Hint:* First answer the following series of questions.)
(a) How many choices are there for the person to occupy the first seat, next to the aisle? (It must be a man.)
(b) Now, how many choices of people may occupy the second seat from the aisle? (It must be a woman.)
(c) Now, how many for the third seat? (one of the remaining men)
(d) Now, how many for the fourth seat? (a woman)
(e) Now, how many for the fifth seat? (a man)
(f) Now, how many for the sixth seat? (a woman)
Now multiply your six answers above.

62. In how many ways can they arrange themselves if the men and women are to alternate with either a man or a woman on the aisle? (*Hint:* First answer the following series of questions.)
(a) How many choices of people are there for the aisle seat?
(b) Now, how many are there for the second seat? (This person may not be of the same sex as the person on the aisle.)
(c) Now, how many choices are there for the third seat?
(d) Now, how many for the fourth seat?
(e) Now, how many for the fifth seat?
(f) Now, how many for the sixth seat?
Now multiply your six answers above.

63. Try working Example 4 by considering digits in the order first, then second, then third. Explain what goes wrong.

64. Try working Example 4 by considering digits in the order third, then second, then first. Explain what goes wrong.

65. Repeat Example 4 but this time allow repeated digits. Does the order in which digits are considered matter in this case?

11.3 Using Permutations and Combinations

Permutations • Combinations • Guidelines on Which Method to Use

Permutations In Section 11.2 we introduced factorials as a way of counting the number of *arrangements* of a given set of objects. For example, the members of the club

$$N = \{\text{Alan, Bill, Cathy, David, Evelyn}\}$$

can arrange themselves in a row for a photograph in $5! = 120$ different ways. We have also used previous methods, like tree diagrams and the fundamental counting principle, to answer questions such as: How many ways can club N elect a president, a secretary, and a treasurer if no one person can hold more than one office? This again is a matter of *arrangements*. The difference is that only three, rather than all five, of the members are involved in each arrangement. A common way to rephrase the basic question here is as follows:

How many arrangements are there of five things taken three at a time?

The answer, by the fundamental counting principle, is $5 \cdot 4 \cdot 3 = 60$. The factors begin with 5 and proceed downward, just as in a factorial product, but do not go all the way to 1. (In this example the product stops when there are three factors.) We now generalize this idea.

In the context of counting problems, arrangements are often called **permutations;** the number of permutations of n distinct things taken r at a time is denoted $_nP_r$.* Since the number of objects being arranged cannot exceed the total number available, we assume for our purposes here that $r \leq n$. Applying the fundamental counting principle to arrangements of this type gives

$$_nP_r = n(n - 1)(n - 2) \ldots [n - (r - 1)].$$

Notice that the first factor is $n - 0$, the second is $n - 1$, the third is $n - 2$, and so on. The rth factor, the last one in the product, will be the one with $r - 1$ subtracted from n, as shown above. We can express permutations, in general, in terms of factorials, and obtain a practical formula for calculation as follows.

$$
\begin{aligned}
_nP_r &= n(n-1)(n-2)\ldots[n-(r-1)] \\
&= n(n-1)(n-2)\ldots(n-r+1) && \text{Simplify the last factor.} \\
&= \frac{n(n-1)(n-2)\ldots(n-r+1)\,(n-r)\,(n-r-1)\ldots 2\cdot 1}{(n-r)(n-r-1)\ldots 2\cdot 1} && \text{Multiply and divide by } (n-r)\,(n-r-1)\ldots 2\cdot 1. \\
&= \frac{n!}{(n-r)!} && \text{Definition of factorial}
\end{aligned}
$$

We summarize as follows.

Factorial Formula for Permutations

The number of **permutations,** or *arrangements,* of n distinct things taken r at a time, where $r \leq n$, can be calculated as

$$_nP_r = \frac{n!}{(n - r)!}.$$

Note that although we sometimes refer to a symbol such as $_4P_2$ as "a permutation"(see Examples 1 and 2), the symbol actually represents "the number of permutations of 4 distinct things taken 2 at a time."

*Alternative notations are $P(n, r)$ and P_r^n.

EXAMPLE 1 Using the Factorial Formula for Permutations

Evaluate each permutation.

(a) ${}_4P_2$ **(b)** ${}_8P_5$ **(c)** ${}_5P_5$

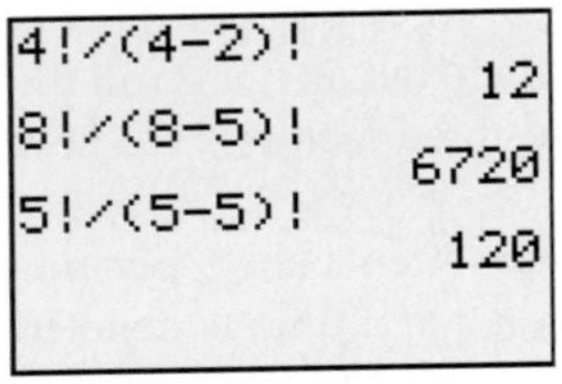

This screen uses factorials to support the results of Example 1.

SOLUTION

(a) $${}_4P_2 = \frac{4!}{(4-2)!} = \frac{4!}{2!} = \frac{24}{2} = 12$$

(b) $${}_8P_5 = \frac{8!}{(8-5)!} = \frac{8!}{3!} = \frac{40{,}320}{6} = 6720$$

(c) $${}_5P_5 = \frac{5!}{(5-5)!} = \frac{5!}{0!} = \frac{120}{1} = 120$$

Notice that ${}_5P_5$ is equal to 5!. It is true for all whole numbers n that

$${}_nP_n = n!.$$

(This is the number of possible arrangements of n distinct objects taken all n at a time.)

Most graphing and scientific calculators allow direct calculation of permutations, in which case the factorial formula is not needed.

This screen uses the permutations feature to support the results of Example 2.

EXAMPLE 2 Calculating Permutations Directly

Evaluate each permutation.

(a) ${}_{10}P_6$ **(b)** ${}_{28}P_0$ **(c)** ${}_{18}P_{12}$

SOLUTION

(a) ${}_{10}P_6 = 151{,}200$ **(b)** ${}_{28}P_0 = 1$
(c) ${}_{18}P_{12} = 8{,}892{,}185{,}702{,}400$

Concerning part (c), many calculators will not display this many digits, so you may obtain an answer such as 8.8921857×10^{12}.

PROBLEM-SOLVING HINT Permutations can be used any time we need to know the number of size-r arrangements that can be selected from a size-n set. The word *arrangement* implies an ordering, so we use permutations only in cases when

1. repetitions are not allowed, and
2. **order is important.**

Change ringing, the English way of ringing church bells, combines mathematics and music. Bells are rung first in sequence, 1, 2, 3, ... Then the sequence is permuted ("changed"). On six bells, 720 different "changes" (different permutations of tone) can be rung: ${}_6P_6 = 6!$.

Composers work out changes so that musically interesting and harmonious sequences occur regularly.

The church bells are swung by means of ropes attached to the wheels beside them. One ringer swings each bell, listening intently and watching the other ringers closely. If one ringer gets lost and stays lost, the rhythm of the ringing cannot be maintained; all the ringers have to stop.

A ringer can spend weeks just learning to keep a bell going and months learning to make the bell ring in exactly the right place. Errors of $\frac{1}{4}$ second mean that two bells are ringing at the same time. Even errors of $\frac{1}{10}$ second can be heard.

EXAMPLE 3 Building Numbers from a Set of Digits

How many nonrepeating three-digit numbers can be written using digits from the set $\{3, 4, 5, 6, 7, 8\}$?

SOLUTION

Repetitions are not allowed since the numbers are to be "nonrepeating." (For example, 448 is not acceptable.) Also, order is important. (For example, 476 and 746 are *distinct* cases.) So we use permutations:

$$_6P_3 = 6 \cdot 5 \cdot 4 = 120.$$

The next example involves multiple parts, and hence calls for the fundamental counting principle, but the individual parts can be handled with permutations.

EXAMPLE 4 Designing Account Numbers

Suppose certain account numbers are to consist of two letters followed by four digits and then three more letters, where repetitions of letters or digits are not allowed *within* any of the three groups, but the last group of letters may contain one or both of those used in the first group. How many such accounts are possible?

SOLUTION

The task of designing such a number consists of three parts:

1. Determine the first set of two letters.
2. Determine the set of four digits.
3. Determine the final set of three letters.

Each part requires an arrangement without repetitions, which is a permutation. Multiply together the results of the three parts.

$$_{26}P_2 \cdot {}_{10}P_4 \cdot {}_{26}P_3 = \underbrace{650}_{\text{Part 1}} \cdot \underbrace{5040}_{\text{Part 2}} \cdot \underbrace{15{,}600}_{\text{Part 3}}$$
$$= 51{,}105{,}600{,}000$$

Combinations

We introduced permutations to evaluate the number of arrangements of n things taken r at a time, where repetitions are not allowed. The order of the items was important. Recall that club

$$N = \{\text{Alan, Bill, Cathy, David, Evelyn}\}$$

could elect three officers in $_5P_3 = 60$ different ways. With three-member committees, on the other hand, order is not important. The committees B, D, E and E, B, D are not different. The possible number of committees is not the number of arrangements of size 3. Rather, it is the number of *subsets* of size 3 (since the order of elements in a set makes no difference).

Subsets in this new context are called **combinations.** The number of combinations of n things taken r at a time (that is, the number of size r subsets, given a set of size n) is written $_nC_r$.* Since there are n things available and we are choosing r of them, we can read $_nC_r$ as "n choose r."

The size-3 committees (subsets) of the club (set) $N = \{A, B, C, D, E\}$ are:

$$\{A, B, C\}, \quad \{A, B, D\}, \quad \{A, B, E\}, \quad \{A, C, D\}, \quad \{A, C, E\},$$
$$\{A, D, E\}, \quad \{B, C, D\}, \quad \{B, C, E\}, \quad \{B, D, E\}, \quad \{C, D, E\}.$$

There are ten subsets of size 3, so ten is the number of three-member committees possible. Just as with permutations, repetitions are not allowed. For example, $\{E, E, B\}$ is not a valid three-member subset, just as EEB is not a valid three-member arrangement.

To see how to find the number of such subsets without listing them all, notice that each size-3 subset (combination) gives rise to six size-3 arrangements (permutations). For example, the single combination ADE yields these six permutations:

$$A, D, E \quad A, E, D \quad D, A, E \quad D, E, A \quad E, A, D \quad E, D, A.$$

There must be six times as many size-3 permutations as there are size-3 combinations, or, in other words, one-sixth as many combinations as permutations. Therefore,

$$_5C_3 = \frac{_5P_3}{6} = \frac{60}{6} = 10.$$

"**Bilateral cipher**" (above) was invented by **Francis Bacon** early in the seventeenth century to code political secrets. This binary code, *a* and *b* in combinations of five, has 32 permutations. Bacon's "biformed alphabet" (bottom four rows) uses two type fonts to conceal a message in some straight text. The decoder deciphers a string of *a*s and *b*s, groups them by fives, then deciphers letters and words. This code was applied to Shakespeare's plays in efforts to prove Bacon the rightful author.

Again, the 6 appears in the denominator because there are six different ways to arrange a set of three things (since $3! = 3 \cdot 2 \cdot 1 = 6$). Generalizing from this example, we can obtain a formula for evaluating numbers of combinations.

$$_nC_r = \frac{_nP_r}{r!} \qquad r \text{ things can be arranged in } r! \text{ ways.}$$

$$= \frac{\dfrac{n!}{(n-r)!}}{r!} \qquad \text{Substitute the factorial formula for } _nP_r.$$

$$= \frac{n!}{r!(n-r)!}. \qquad \text{Simplify algebraically.}$$

We summarize as follows.

Factorial Formula for Combinations

The number of **combinations,** or *subsets,* of n distinct things taken r at a time, where $r \leq n$, can be calculated as

$$_nC_r = \frac{_nP_r}{r!} = \frac{n!}{r!(n-r)!}.$$

*Alternative notations are $C(n, r)$, C_r^n, and $\binom{n}{r}$.

In Examples 5 and 6, you will again note the "shorthand" terminology whereby we refer to ${}_nC_r$ as "a combination" even though it actually represents "the number of combinations of n distinct things taken r at a time."

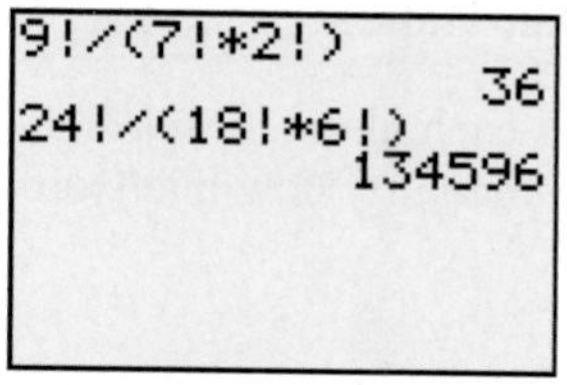

This screen uses factorials to support the results of Example 5.

EXAMPLE 5 Using the Factorial Formula for Combinations

Evaluate each combination.

(a) ${}_9C_7$ **(b)** ${}_{24}C_{18}$

SOLUTION

(a) $${}_9C_7 = \frac{9!}{7!(9-7)!} = \frac{9!}{7!2!} = \frac{362{,}880}{5040 \cdot 2} = 36$$

(b) $${}_{24}C_{18} = \frac{24!}{18!(24-18)!} = \frac{24!}{18!6!} = 134{,}596$$

A calculator that does permutations directly likely will do combinations directly as well. We illustrate in Example 6.

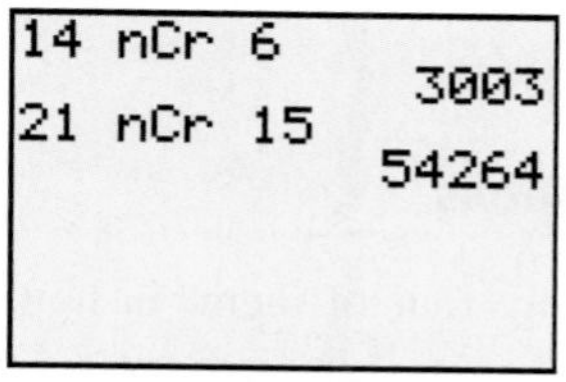

This screen uses the combinations feature to support the results of Example 6.

EXAMPLE 6 Calculating Combinations Directly

Evaluate each combination.

(a) ${}_{14}C_6$ **(b)** ${}_{21}C_{15}$

SOLUTION

(a) ${}_{14}C_6 = 3003$ **(b)** ${}_{21}C_{15} = 54{,}264$

PROBLEM-SOLVING HINT Combinations have an important common property with permutations (repetitions are not allowed) and also have an important distinction (order is *not* important with combinations). Combinations are applied only when

1. repetitions are not allowed, and
2. **order is *not* important.**

EXAMPLE 7 Finding the Number of Subsets

Find the number of different subsets of size 2 in the set $\{a, b, c, d\}$. List them to check the answer.

SOLUTION

A subset of size 2 must have two distinct elements, so repetitions are not allowed. And since the order in which the elements of a set are listed makes no difference, we see that order is not important. Use the combinations formula with $n = 4$ and $r = 2$.

$$_4C_2 = \frac{4!}{2!(4-2)!} = \frac{4!}{2!2!} = 6$$

The six subsets of size 2 are $\{a, b\}$, $\{a, c\}$, $\{a, d\}$, $\{b, c\}$, $\{b, d\}$, $\{c, d\}$. ■

The set of 52 playing cards in the standard deck has four suits.

♠ spades ♦ diamonds
♥ hearts ♣ clubs

Ace is the unit card. Jacks, queens, and kings are "face cards." Each suit contains thirteen denominations: ace, 2, 3, . . . , 10, jack, queen, king. (In some games, ace rates above king, instead of counting as 1.)

EXAMPLE 8 Finding the Number of Possible Poker Hands

A common form of poker involves hands (sets) of five cards each, dealt from a standard deck consisting of 52 different cards (illustrated in the margin). How many different 5-card hands are possible?

SOLUTION

A 5-card hand must contain five distinct cards, so repetitions are not allowed. Also, the order is not important since a given hand depends only on the cards it contains, and not on the order in which they were dealt or the order in which they are displayed. Since order does not matter, use combinations (and a calculator):

$$_{52}C_5 = \frac{52!}{5!(52-5)!} = \frac{52!}{5!\,47!} = 2{,}598{,}960.$$ ■

EXAMPLE 9 Finding the Number of Subsets of Books

Melvin wants to buy ten different books but can afford only four of them. In how many ways can he make his selections?

SOLUTION

The four books selected must be distinct (repetitions are not allowed), and also the order of the four chosen has no bearing in this case, so we use combinations:

$$_{10}C_4 = \frac{10!}{4!(10-4)!} = \frac{10!}{4!\,6!} = 210 \text{ ways.}$$ ■

Notice that, according to our formula for combinations,

$$_{10}C_6 = \frac{10!}{6!(10-6)!} = \frac{10!}{6!4!} = 210,$$

which is the same as $_{10}C_4$. In fact, Exercise 58 asks you to prove the fact that, in general, for all whole numbers n and r, with $r \leq n$,

$$_nC_r = {_nC_{n-r}}.$$

See the margin note as well. It indicates that $_{10}C_6$ is equal to $_{10}C_4$.

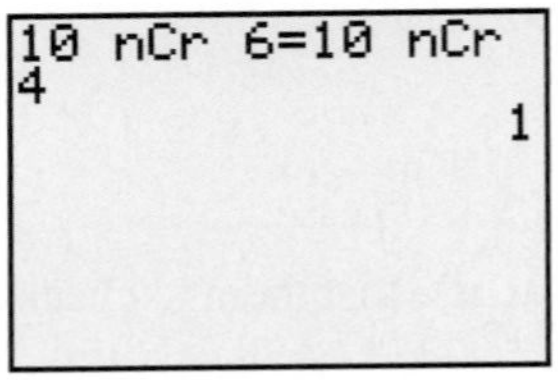

The "1" indicates that the statement

$$_{10}C_6 = {_{10}C_4}$$

is true.

Guidelines on Which Method to Use Both permutations and combinations produce the number of ways of selecting r items from n items where repetitions are not allowed. Permutations apply to arrangements (where order is important), while combinations apply to subsets (where order is not important). These similarities and differences, as well as the appropriate formulas, are summarized in the following table.

Permutations	Combinations
Number of ways of selecting r items out of n items	
Repetitions are not allowed.	
Order is important.	Order is not important.
Arrangements of n items taken r at a time	Subsets of n items taken r at a time
${}_nP_r = \frac{n!}{(n-r)!}$	${}_nC_r = \frac{n!}{r!(n-r)!}$
Clue words: arrangement, schedule, order	Clue words: set, group, sample, selection

In cases where r items are to be selected from n items and repetitions *are* allowed, it is usually best to make direct use of the fundamental counting principle.

PROBLEM-SOLVING HINT Many counting problems involve selecting some of the items from a given set of items. The particular conditions of the problem will determine which specific technique to use. The following guidelines may help.

1. **If selected items can be repeated, use the fundamental counting principle.**
   Example: How many four-digit numbers are there?
   $$9 \cdot 10^3 = 9000$$
2. **If selected items cannot be repeated, and order is important, use permutations.**
   Example: How many ways can three of eight people line up at a ticket counter?
   $${}_8P_3 = \frac{8!}{(8-3)!} = 336$$
3. **If selected items cannot be repeated, and order is *not* important, use combinations.**
   Example: How many ways can a committee of three be selected from a group of twelve people?
   $${}_{12}C_3 = \frac{12!}{3!(12-3)!} = 220$$

EXAMPLE 10 Distributing Toys to Children

In how many ways can a mother distribute three different toys among her seven children if a child may receive anywhere from none to all three toys?

SOLUTION

Because a given child can be a repeat recipient, repetitions are allowed here, so we use the fundamental counting principle. Each of the three toys can go to any of the seven children. The number of possible distributions is $7 \cdot 7 \cdot 7 = 343$.

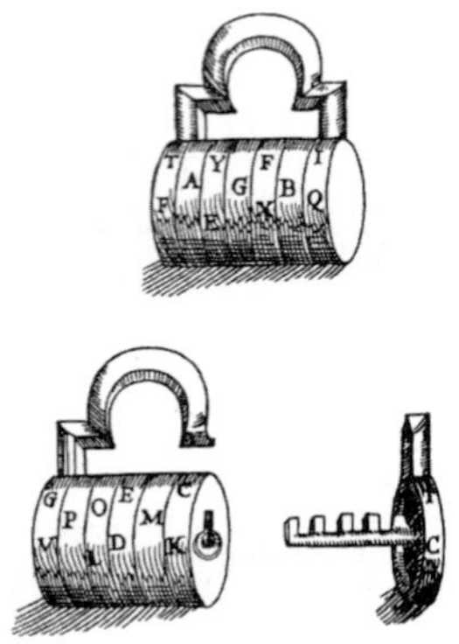

The illustration above is from the 1560s text ***Logistica***, by the mathematician J. Buteo. Among other topics, the book discusses the number of possible throws of four dice and the number of arrangements of the cylinders of a combination lock. Note that "combination" is a misleading name for these locks since repetitions are allowed, and, also, order makes a difference.

EXAMPLE 11 Selecting Committees

How many different three-member committees could club N appoint so that exactly one woman is on the committee?

SOLUTION

Recall that $N = \{\text{Alan, Bill, Cathy, David, Evelyn}\}$. Two members are women; three are men. Although the question mentioned only that the committee must include exactly one woman, in order to complete the committee two men must be selected as well. Therefore the task of selecting the committee members consists of two parts:

1. Choose one woman.
2. Choose two men.

Because order is not important for committees, use combinations for the two parts. One woman can be chosen in ${}_2C_1 = \frac{2!}{1!\,1!} = 2$ ways, and two men can be chosen in ${}_3C_2 = \frac{3!}{2!\,1!} = 3$ ways. Finally, use the fundamental counting principle to obtain $2 \cdot 3 = 6$ different committees. This small number can be checked by listing.

$$\{C, A, B\}, \{C, A, D\}, \{C, B, D\}, \{E, A, B\}, \{E, A, D\}, \{E, B, D\}.$$

EXAMPLE 12 Selecting Attendees for an Event

Every member of the Alpha Beta Gamma fraternity would like to attend a special event this weekend, but only ten members will be allowed to attend. How many ways could the lucky ten be selected if there are a total of forty-eight members?

SOLUTION

In this case, ten distinct men are required (repetitions are not allowed), and the order of selection makes no difference, so we use combinations.

$${}_{48}C_{10} = \frac{48!}{10!\,38!} = 6{,}540{,}715{,}896 \quad \text{Use a calculator.}$$

EXAMPLE 13 Selecting Escorts

When the ten fraternity men of Example 12 arrive at the event, four of them are selected to escort the four homecoming queen candidates. In how many ways can this selection be made?

SOLUTION

Of the ten, four distinct men are required, and order is important here because different orders will pair the men with different women. Use permutations to obtain

$${}_{10}P_4 = \frac{10!}{6!} = 5040 \text{ possible selections.}$$

Most, if not all, of the exercises in this section will call for permutations and/or combinations. And in the case of multiple-part tasks, the fundamental counting principle may also be required.

For Further Thought

Poker Hands

In 5-card poker, played with a standard 52-card deck, 2,598,960 different hands are possible. (See Example 8.) The desirability of the various hands depends upon their relative chance of occurrence, which, in turn, depends on the number of different ways they can occur, as shown in Table 4. Note that an ace can generally be positioned either below 2 (as a 1) or above king (as a 14). This is important in counting straight flush hands and straight hands.

TABLE 4 Categories of Hands in 5-Card Poker

Event E	Description of Event E	Number of Outcomes Favorable to E
Royal flush	Ace, king, queen, jack, and 10, all of the same suit	4
Straight flush	5 cards of consecutive denominations, all in the same suit (excluding royal flush)	36
Four of a kind	4 cards of the same denomination, plus 1 additional card	______
Full house	3 cards of one denomination, plus 2 cards of a second denomination	3744
Flush	Any 5 cards all of the same suit (excluding royal flush and straight flush)	______
Straight	5 cards of consecutive denominations (not all the same suit)	10,200
Three of a kind	3 cards of one denomination, plus 2 cards of two additional denominations	54,912
Two pairs	2 cards of one denomination, plus 2 cards of a second denomination, plus 1 card of a third denomination	______
One pair	2 cards of one denomination, plus 3 additional cards of three different denominations	1,098,240
No pair	No two cards of the same denomination (and excluding any sort of flush or straight)	1,302,540
Total		**2,598,960**

For Group Discussion or Individual Investigation

As the table shows, a full house is a relatively rare occurrence. (Only four of a kind, straight flush, and royal flush are less likely.) To verify that there are 3744 different full house hands possible, carry out the following steps.

1. Explain why there are ${}_4C_3$ different ways to select three aces from the deck.
2. Explain why there are ${}_4C_2$ different ways to select two 8s from the deck.
3. If "aces and 8s" (three aces and two 8s) is one kind of full house, show that there are ${}_{13}P_2$ different kinds of full house altogether.
4. Multiply the expressions from Steps 1, 2, and 3 together. Explain why this product should give the total number of full house hands possible.
5. Find the three missing values in the right column of Table 4. (Answers are on page 727.)

11.3 EXERCISES

Evaluate each expression.

1. ${}_7P_4$

2. ${}_{14}P_5$

3. ${}_{12}C_4$

4. ${}_{15}C_9$

Determine the number of permutations (arrangements) of each of the following.

5. 18 things taken 5 at a time

6. 12 things taken 7 at a time

Determine the number of combinations (subsets) of each of the following.

7. 8 things taken 4 at a time

8. 14 things taken 5 at a time

Use a calculator to evaluate each expression.

9. ${}_{22}P_9$

10. ${}_{33}C_{11}$

11. Is it possible to evaluate ${}_6P_{10}$? Explain.

12. Is it possible to evaluate ${}_9C_{12}$? Explain.

13. Explain how permutations and combinations differ.

14. Explain how factorials are related to permutations.

15. Decide whether each object is a permutation or a combination.
(a) a telephone number
(b) a Social Security number
(c) a hand of cards in poker
(d) a committee of politicians
(e) the "combination" on a student gym locker combination lock
(f) a lottery choice of six numbers where the order does not matter
(g) an automobile license plate number

Exercises 16–23 can be solved with permutations even though the problem statements will not always include a form of the word "permutation," or "arrangement," or "ordering."

16. ***Placing in a Race*** How many different ways could first-, second-, and third-place finishers occur in a race with six runners competing?

17. ***Arranging New Home Models*** Jeff Hubbard, a contractor, builds homes of eight different models and presently has five lots to build on. In how many different ways can he arrange homes on these lots? Assume five different models will be built.

18. ***ATM PIN Numbers*** An automated teller machine (ATM) requires a four-digit personal identification number (PIN), using the digits 0–9. (The first digit may be 0.) How many such PINs have no repeated digits?

19. ***Electing Officers of a Club*** How many ways can president and vice president be determined in a club with twelve members?

20. ***Counting Prize Winners*** First, second, and third prizes are to be awarded to three different people. If there are ten eligible candidates, how many outcomes are possible?

21. ***Counting Prize Winners*** How many ways can a teacher give five different prizes to five of her 25 students?

22. ***Scheduling Security Team Visits*** A security team visits 12 offices each night. How many different ways can the team order its visits?

23. ***Sums of Digits*** How many counting numbers have four distinct nonzero digits such that the sum of the four digits is
(a) 10? **(b)** 11?

Exercises 24–31 can be solved with combinations even though the problem statements will not always include the word "combination" or "subset."

24. ***Sampling Cell Phones*** How many ways can a sample of five cell phones be selected from a shipment of twenty-four cell phones?

25. ***Detecting Defective Cell Phones*** If the shipment of Exercise 24 contains six defective phones, how many of the size-five samples would not include any of the defective ones?

26. ***Committees of U.S. Senators*** How many different five-member committees could be formed from the 100 U.S. senators?

27. ***Selecting Hands of Cards*** Refer to the standard 52-card deck pictured on page 698 and notice that the deck contains four aces, twelve face cards, thirteen hearts (all red), thirteen diamonds (all red), thirteen spades (all black), and thirteen clubs (all black). Of the 2,598,960 different five-card hands possible, decide how many would consist of the following cards.
   (a) all diamonds
   (b) all black cards
   (c) all aces

28. ***Selecting Lottery Entries*** In a $\frac{7}{39}$ lottery, you select seven distinct numbers from the set 1 through 39, where order makes no difference. How many different ways can you make your selection?

29. ***Arranging New Home Models*** Jeff Hubbard (the contractor) is to build six homes on a block in a new subdivision. Overhead expenses have forced him to limit his line to two different models, standard and deluxe. (All standard model homes are the same and all deluxe model homes are the same.)
   (a) How many different choices does Jeff have in positioning the six houses if he decides to build three standard and three deluxe models?
   (b) If Jeff builds only two deluxes and four standards, how many different positionings can he use?

30. ***Choosing a Monogram*** Judy Zahrndt wants to name her new baby so that his monogram (first, middle, and last initials) will be distinct letters in alphabetical order and he will share her last name. How many different monograms could she select?

31. ***Number of Paths from Point to Point*** How many paths are possible from A to B if all motion must be to the right or downward? (*Hint:* It takes ten unit steps to get from A to B and three of the ten must be downward.)

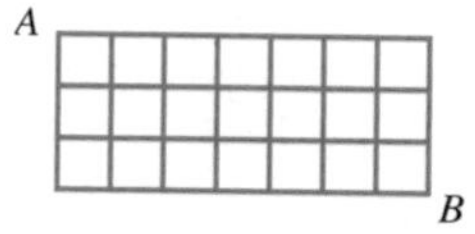

For Exercises 32–56, you may use permutations, combinations, the fundamental counting principle, or other counting methods as appropriate. (Some problems may require using more than one method.)

32. ***Selecting Lottery Entries*** In SuperLotto Plus, a California state lottery game, you select five distinct numbers from 1 to 47, and one MEGA number from 1 to 27, hoping that your selection will match a random list selected by lottery officials.
   (a) How many different sets of six numbers can you select?
   (b) Eileen Burke always includes her age and her husband's age as two of the first five numbers in her SuperLotto Plus selections. How many ways can she complete her list of six numbers?

33. ***Drawing Cards*** How many cards must be drawn (without replacement) from a standard deck of 52 to guarantee the following?
   (a) Two of the cards will be of the same suit.
   (b) Three of the cards will be of the same suit.

34. ***Flush Hands in Poker*** How many different 5-card poker hands would contain only cards of a single suit?

35. ***Identification Numbers in Research*** Subject identification numbers in a certain scientific research project consist of three letters followed by three digits and then three more letters. Assume repetitions are not allowed within any of the three groups, but letters in the first group of three may occur also in the last group of three. How many distinct identification numbers are possible?

36. ***Radio Station Call Letters*** Radio stations in the United States have call letters that begin with K or W (for west or east of the Mississippi River, respectively). Some have three call letters, such as WBZ in Boston, WLS in Chicago, and KGO in San Francisco. Assuming no repetition of letters, how many three-letter sets of call letters are possible?

37. ***Radio Station Call Letters*** Most stations that were licensed after 1927 have four call letters starting with K or W, such as WXYZ in Detroit or KRLD in Dallas. Assuming no repetitions, how many four-letter sets are possible?

38. ***Scheduling Games in a Basketball League*** Each team in an eight-team basketball league is scheduled to play each other team three times. How many games will be played altogether?

39. ***Scheduling Batting Orders in Baseball*** The Coyotes, a youth league baseball team, have seven pitchers, who only pitch, and twelve other players, all of whom can play any position other than pitcher. For Saturday's game, the coach has not yet determined which nine players to use nor what the batting order will be, except that the pitcher will bat last. How many different batting orders may occur?

40. ***Ordering Performers in a Music Recital*** A music class of eight girls and seven boys is having a recital. If each member is to perform once, how many ways can the program be arranged in each of the following cases?

(a) All girls must perform first.
(b) A girl must perform first and a boy must perform last.
(c) Elisa and Doug will perform first and last, respectively.
(d) The entire program will alternate between girls and boys.
(e) The first, eighth, and fifteenth performers must be girls.

41. ***Scheduling Daily Reading*** Carole begins each day by reading from one of seven inspirational books. How many ways can she choose the books for one week if the selection is done
(a) with replacement?
(b) without replacement?

42. ***Counting Card Hands*** How many of the possible 5-card hands from a standard 52-card deck would consist of the following cards?
(a) four clubs and one non-club
(b) two face cards and three non-face cards
(c) two red cards, two clubs, and a spade

43. ***Dividing People into Groups*** In how many ways could twenty-five people be divided into five groups containing, respectively, three, four, five, six, and seven people?

44. ***Points and Lines in a Plane*** If any two points determine a line, how many lines are determined by seven points in a plane, no three of which are collinear?

45. ***Points and Triangles in a Plane*** How many triangles are determined by twenty points in a plane, no three of which are collinear?

46. ***Counting Possibilities on a Combination Lock*** How many different three-number "combinations" are possible on a combination lock having 40 numbers on its dial? (*Hint:* "Combination" is a misleading name for these locks since repetitions are allowed and also order makes a difference.)

47. ***Selecting Drivers and Passengers for a Trip*** Michael Grant, his wife and son, and four additional friends are driving, in two vehicles, to the seashore.
(a) If all seven people can drive, how many ways can the two drivers be selected? (Everyone wants to drive the sports car, so it is important which driver gets which car.)
(b) If the sports car must be driven by Michael, his wife, or their son, how many ways can the drivers now be determined?
(c) If the sports car will accommodate only two people, and there are no other restrictions, how many ways can both drivers and passengers be assigned to both cars?

48. ***Winning the Daily Double in Horse Racing*** At the race track, you win the "daily double" by purchasing a ticket and selecting the winners of both of two specified races. If there are six and eight horses running in the first and second races, respectively, how many tickets must you purchase to guarantee a winning selection?

49. ***Winning the Trifecta in Horse Racing*** Many race tracks offer a "trifecta" race. You win by selecting the correct first-, second-, and third-place finishers. If eight horses are entered, how many tickets must you purchase to guarantee that one of them will be a trifecta winner?

50. ***Selecting Committee Members*** Nine people are to be distributed among three committees of two, three, and four members and a chairperson is to be selected for each committee. How many ways can this be done? (*Hint:* Break the task into the following sequence of parts.)
    (a) Select the members of the two-person committee.
    (b) Select the members of the three-person committee.
    (c) Select the chair of the two-person committee.
    (d) Select the chair of the three-person committee.
    (e) Select the chair of the four-person committee.

51. ***Arranging New Home Models*** (See Exercise 29.) Because of his good work, Jeff Hubbard gets a contract to build homes on three additional blocks in the subdivision, with six homes on each block. He decides to build nine deluxe homes on these three blocks: two on the first block, three on the second, and four on the third. The remaining nine homes will be standard.
    (a) Altogether on the three-block stretch, how many different choices does Jeff have for positioning the eighteen homes? (*Hint:* Consider the three blocks separately and use the fundamental counting principle.)
    (b) How many choices would he have if he built 2, 3, and 4 deluxe models on the three different blocks as before, but not necessarily on the first, second, and third blocks in that order?

52. ***Building Numbers from Sets of Digits***
    (a) How many six-digit counting numbers use all six digits 4, 5, 6, 7, 8, and 9?
    (b) Suppose all these numbers were arranged in increasing order: 456,789; 456,798; and so on. Which number would be 364th in the list?

53. ***Arranging a Wedding Reception Line*** At a wedding reception, the bride and groom, the maid of honor and best man, two bridesmaids, and two ushers will form a reception line. How many ways can they be arranged in each of the following cases?

    (a) Any order will do.
    (b) The bride and groom must be the last two in line.
    (c) The groom must be last in line with the bride next to him.

54. ***Assigning Student Grades*** A professor teaches a class of 60 students and another class of 40 students. Five percent of the students in each class are to receive a grade of A. How many different ways can the A grades be distributed?

55. ***Sums of Digits*** How many counting numbers have four distinct nonzero digits such that the sum of the four digits is
    (a) 12? **(b)** 13?

56. ***Screening Computer Processors*** A computer company will screen a shipment of 30 processors by testing a random sample of five of them. How many different samples are possible?

57. Verify that ${}_{12}C_9 = {}_{12}C_3$.

58. Use the factorial formula for combinations to prove that in general, ${}_nC_r = {}_nC_{n-r}$.

59. **(a)** Use the factorial formula for permutations to evaluate, for any whole number n, ${}_nP_0$.
    (b) Explain the meaning of the result in part (a).

60. **(a)** Use the factorial formula for combinations and the definition of 0! to evaluate, for any whole number n, ${}_nC_0$.
    (b) Explain the meaning of the result in part (a).

11.4 Using Pascal's Triangle

Pascal's Triangle • Applications

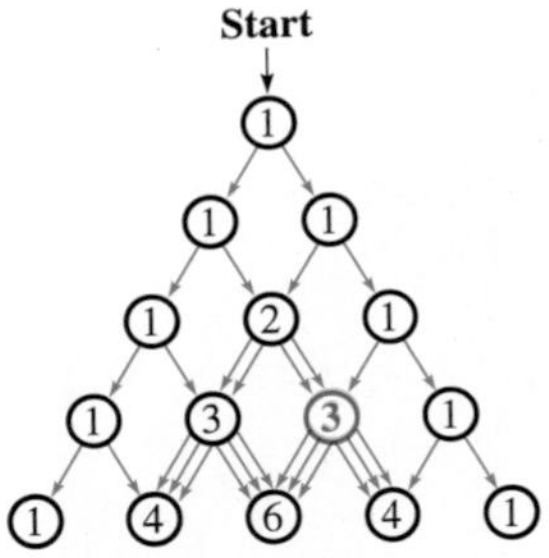

FIGURE 7

Pascal's Triangle The triangular array in Figure 7 represents what we can call "random walks" that begin at START and proceed downward according to the following rule. At each circle (branch point), a coin is tossed. If it lands heads, we go downward to the left. If it lands tails, we go downward to the right. At each point, left and right are equally likely. In each circle we have recorded the number of different routes that could bring us to that point. For example, the colored 3 can be reached as the result of three different coin-tossing sequences: htt, tht, and tth.

Another way to generate the same pattern of numbers is to begin with 1s down both diagonals and then fill in the interior entries by adding the two numbers just above a given position (to the left and right). For example, the colored 28 in Table 5 is the result of adding 7 and 21 in the row above it.

TABLE 5 Pascal's Triangle

Row Number												Row Sum
0	1											1
1	1	1										2
2	1	2	1									4
3	1	3	3	1								8
4	1	4	6	4	1							16
5	1	5	10	10	5	1						32
6	1	6	15	20	15	6	1					64
7	1	7	21	35	35	21	7	1				128
8	1	8	28	56	70	56	28	8	1			256
9	1	9	36	84	126	126	84	36	9	1		512
10	1	10	45	120	210	252	210	120	45	10	1	1024

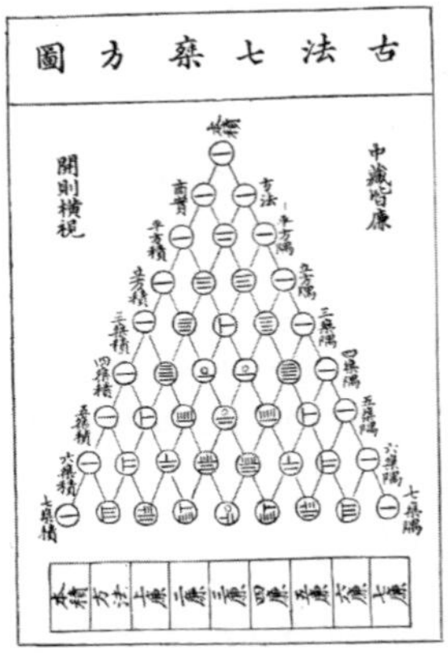

"Pascal's" triangle shown in the 1303 text ***Szu-yuen Yu-chien*** (*The Precious Mirror of the Four Elements*) by the Chinese mathematician Chu Shih-chieh.

By continuing to add pairs of numbers, we extend the array indefinitely downward, always beginning and ending each row with 1s. (The table shows just rows 0 through 10.) This unending "triangular" array of numbers is called **Pascal's triangle,** since Blaise Pascal wrote a treatise about it in 1653. There is evidence, though, that it was known as early as around 1100 and may have been studied in China or India still earlier.

At any rate, the "triangle" possesses many interesting properties. In counting applications, the most useful property is that, in general, entry number r in row number n is equal to ${}_nC_r$—the number of *combinations* of n things taken r at a time. This correspondence is shown (through row 7) in Table 6 on the next page.

TABLE 6 Combination Values in Pascal's Triangle

Row Number	
0	$_0C_0$
1	$_1C_0$ $_1C_1$
2	$_2C_0$ $_2C_1$ $_2C_2$
3	$_3C_0$ $_3C_1$ $_3C_2$ $_3C_3$
4	$_4C_0$ $_4C_1$ $_4C_2$ $_4C_3$ $_4C_4$
5	$_5C_0$ $_5C_1$ $_5C_2$ $_5C_3$ $_5C_4$ $_5C_5$
6	$_6C_0$ $_6C_1$ $_6C_2$ $_6C_3$ $_6C_4$ $_6C_5$ $_6C_6$
7	$_7C_0$ $_7C_1$ $_7C_2$ $_7C_3$ $_7C_4$ $_7C_5$ $_7C_6$ $_7C_7$
	and so on

Having a copy of Pascal's triangle handy gives us another option for evaluating combinations. Any time we need to know the number of combinations of n things taken r at a time (that is, the number of subsets of size r in a set of size n), we can simply read entry number r of row number n. Keep in mind that the *first* row shown is *row number 0*. Also, the first entry of each row can be called entry number 0. This entry gives the number of subsets of size 0 (which is always 1 since there is only one empty set).

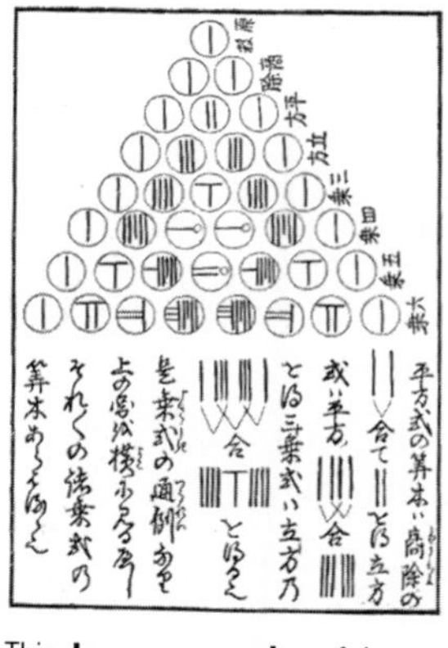

This **Japanese version** of the triangle dates from the eighteenth century. The "stick numerals" evolved from bamboo counting pieces used on a ruled board. Possibly Omar Khayyam, twelfth-century Persian mathematician and poet, may also have divined its patterns in pursuit of algebraic solutions. (The triangle lists the coefficients of the binomial expansion, explained in For Further Thought on page 709.)

Applications

EXAMPLE 1 Applying Pascal's Triangle to Counting People

A group of ten people includes six women and four men. If five of these people are randomly selected to fill out a questionnaire, how many different samples of five people are possible?

SOLUTION

Since this is simply a matter of selecting a subset of five from a set of ten (or combinations of ten things taken five at a time), we can read $_{10}C_5$ from row 10 of Pascal's triangle in Table 5. The answer is 252. ■

EXAMPLE 2 Applying Pascal's Triangle to Counting People

Among the 252 possible samples of five people in Example 1, how many of them would consist of exactly two women and three men?

SOLUTION

Two women can be selected from six women in $_6C_2$ different ways, and three men can be selected from four men in $_4C_3$ different ways. These combination values can be read from Pascal's triangle. Then, since the task of obtaining two women and three men requires both individual parts, the fundamental counting principle tells us to multiply the two values:

$$_6C_2 \cdot {}_4C_3 = 15 \cdot 4 = 60. \quad ■$$

EXAMPLE 3 Applying Pascal's Triangle to Coin Tossing

If five fair coins are tossed, in how many different ways could exactly three heads be obtained?

SOLUTION

There are various "ways" of obtaining exactly three heads because the three heads can occur on different subsets of the coins. For example, hhtht and thhth are just two of many possibilities. When such a possibility is written down, exactly three positions are occupied by an h, the other two by a t. Each distinct way of choosing three positions from a set of five positions gives a different possibility. (Once the three positions for h are determined, each of the other two positions automatically receives a t.) So our answer is just the number of size-three subsets of a size-five set, that is, the number of combinations of five things taken three at a time. We read this answer from row 5 of Pascal's triangle:

$$_5C_3 = 10.$$

Notice that row 5 of Pascal's triangle also provides answers to several other questions about tossing five fair coins. They are summarized in Table 7.

TABLE 7 Tossing Five Fair Coins

Number of Heads n	Ways of Obtaining Exactly n Heads	Listing
0	$_5C_0 = 1$	ttttt
1	$_5C_1 = 5$	htttt, thttt, tthtt, tttht, tttth
2	$_5C_2 = 10$	hhttt, hthtt, httht, hftth, thhtt, ththt, thtth, tthht, tthth, ttthh
3	$_5C_3 = 10$	hhhtt, hhtht, hhtth, hthht, hthth, htthh, thhht, thhth, ththh, tthhh
4	$_5C_4 = 5$	hhhht, hhhth, hhthh, hthhh, thhhh
5	$_5C_5 = 1$	hhhhh

To analyze the tossing of a different number of fair coins, we can simply take the pertinent numbers from a different row of Pascal's triangle. Repeated coin tossing is an example of a "binomial" experiment (because each toss has *two* possible outcomes, heads and tails). The likelihoods of various occurrences in such a situation will be addressed in Section 12.4 on binomial probability.

For Further Thought

The Binomial Theorem

The combination values, which comprise Pascal's triangle, also arise in a totally different mathematical context. In algebra, "binomial" refers to a two-term expression such as $x + y$, or $a + 2b$, or $w^3 - 4$. The first few powers of the binomial $x + y$ are shown here.

$$(x + y)^0 = 1$$
$$(x + y)^1 = x + y$$
$$(x + y)^2 = x^2 + 2xy + y^2$$
$$(x + y)^3 = x^3 + 3x^2y + 3xy^2 + y^3$$
$$(x + y)^4 = x^4 + 4x^3y + 6x^2y^2 + 4xy^3 + y^4$$
$$(x + y)^5 = x^5 + 5x^4y + 10x^3y^2 + 10x^2y^3 + 5xy^4 + y^5$$

Notice that the numerical coefficients of these expansions form the first six rows of Pascal's triangle. In our study of counting, we have called these numbers *combinations*, but in the study of algebra, they are called *binomial coefficients* and are usually denoted $\binom{n}{r}$ rather than ${}_nC_r$.

Generalizing the pattern of the powers shown above yields the important result known as the **binomial theorem.**

Binomial Theorem

For any positive integer n,

$$(x + y)^n = \binom{n}{0} \cdot x^n + \binom{n}{1} \cdot x^{n-1}y + \binom{n}{2} \cdot x^{n-2}y^2 + \binom{n}{3} \cdot x^{n-3}y^3 + \cdots + \binom{n}{n-1} \cdot xy^{n-1} + \binom{n}{n} \cdot y^n,$$

where each binomial coefficient can be calculated by the formula

$$\binom{n}{r} = \frac{n!}{r!(n-r)!}.$$

EXAMPLE Applying the Binomial Theorem

Write out the binomial expansion for $(2a + 5)^4$.

SOLUTION

We take the initial coefficients from row 4 of Pascal's triangle and then simplify algebraically.

$$(2a + 5)^4$$
$$= \binom{4}{0} \cdot (2a)^4 + \binom{4}{1} \cdot (2a)^3 \cdot 5 + \binom{4}{2} \cdot (2a)^2 \cdot 5^2 + \binom{4}{3} \cdot (2a) \cdot 5^3 + \binom{4}{4} \cdot 5^4$$

Recall that $(xy)^n = x^n \cdot y^n$.

$$= 1 \cdot 2^4 \cdot a^4 + 4 \cdot 2^3 \cdot a^3 \cdot 5 + 6 \cdot 2^2 \cdot a^2 \cdot 5^2 + 4 \cdot 2 \cdot a \cdot 5^3 + 1 \cdot 5^4$$
$$= 16a^4 + 160a^3 + 600a^2 + 1000a + 625$$

For Group Discussion or Individual Investigation

Write out the binomial expansion for each of the following powers.

1. $(x + y)^6$
2. $(x + y)^7$
3. $(w + 4)^5$
4. $(4x + 2y)^4$
5. $(u - v)^6$
   (*Hint:* First change $u - v$ to $u + (-v)$.)
6. $(5m - 2n)^3$
7. How many terms are in the binomial expansion for $(x + y)^n$?
8. Identify the 15th term only of the expansion for $(a + b)^{18}$.

11.4 EXERCISES

Read each combination value directly from Pascal's triangle.

1. ${}_4C_2$ **2.** ${}_5C_3$ **3.** ${}_6C_3$

4. ${}_7C_5$ **5.** ${}_8C_5$ **6.** ${}_9C_6$

7. ${}_9C_2$ **8.** ${}_{10}C_7$

Selecting Committees of Congressmen *A committee of four Congressmen will be selected from a group of seven Democrats and three Republicans. Find the number of ways of obtaining each result.*

9. exactly one Democrat

10. exactly two Democrats

11. exactly three Democrats

12. exactly four Democrats

Tossing Coins *Suppose eight fair coins are tossed. Find the number of ways of obtaining each result.*

13. exactly three heads

14. exactly four heads

15. exactly five heads

16. exactly six heads

Selecting Classrooms *Peg Cheever, searching for an Economics class, knows that it must be in one of nine classrooms. Since the professor does not allow people to enter after the class has begun, and there is very little time left, Peg decides to try just four of the rooms at random.*

17. How many different selections of four rooms are possible?

18. How many of the selections of Exercise 17 will fail to locate the class?

19. How many of the selections of Exercise 17 will succeed in locating the class?

20. What fraction of the possible selections will lead to "success"? (Give three decimal places.)

For a set of five elements, find the number of different subsets of each size. (Use row 5 of Pascal's triangle to find the answers.)

21. 0 **22.** 1 **23.** 2

24. 3 **25.** 4 **26.** 5

27. How many subsets (of any size) are there for a set of five elements?

28. Find and explain the relationship between the row number and row sum in Pascal's triangle.

Over the years, many interesting patterns have been discovered in Pascal's triangle. We explore a few of them in Exercises 29–35.*

29. Refer to Table 5.

(a) Choose a row whose row number is prime. Except for the 1s in this row, what is true of all the other entries?

(b) Choose a second prime row number and see if the same pattern holds.

(c) Use the usual method to construct row 11 in Table 5, and verify that the same pattern holds in that row.

30. Name the next five numbers of the diagonal sequence in the figure. What are these numbers called? (See Section 1.2.)

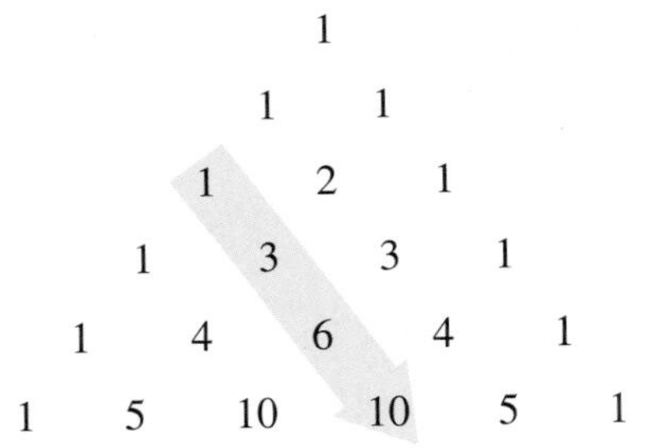

31. Complete the sequence of sums on the diagonals shown in the figure at the top of the next page. What pattern do these sums make? What is the name of this important sequence of numbers? The presence of this sequence in the triangle apparently was not recognized by Pascal.

*For example, see the article "Serendipitous Discovery of Pascal's Triangle" by Francis W. Stanley in *The Mathematics Teacher,* February 1975.

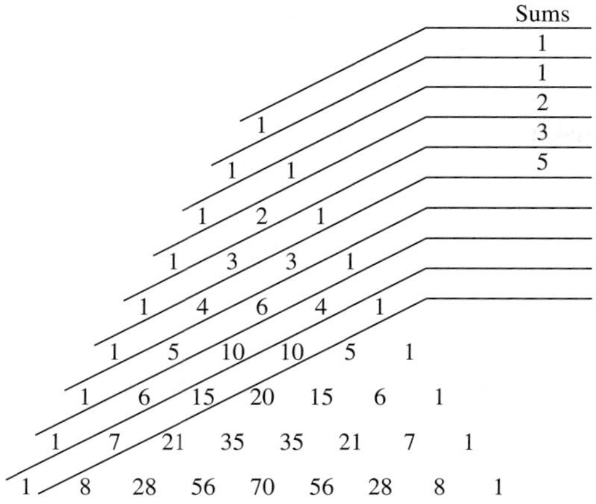

32. Construct another "triangle" by replacing every number in Pascal's triangle (rows **0** through **5**) by its remainder when divided by 2. What special property is shared by rows **2** and **4** of this new triangle?

33. What is the next row that would have the same property as rows **2** and **4** in Exercise 32?

34. How many even numbers are there in row **256** of Pascal's triangle? (Work Exercises 32 and 33 first.)

35. The figure shows a portion of Pascal's triangle with several inverted triangular regions outlined. For any one of these regions, what can be said of the sum of the squares of the entries across its top row?

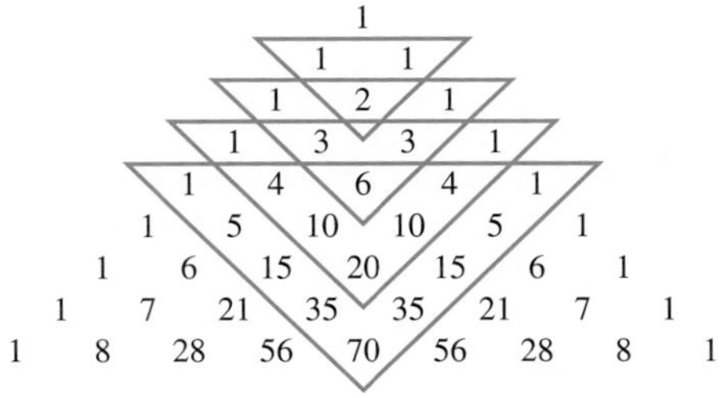

36. More than a century before Pascal's treatise on the "triangle" appeared, another work by the Italian mathematician Niccolo Tartaglia (1506–1559) came out and included the table of numbers shown here.

1	1	1	1	1	1
1	2	3	4	5	6
1	3	6	10	15	21
1	4	10	20	35	56
1	5	15	35	70	126
1	6	21	56	126	252
1	7	28	84	210	462
1	8	36	120	330	792

Explain the connection between Pascal's triangle and Tartaglia's "rectangle."

37. It was stated in the text that each interior entry in Pascal's triangle can be obtained by adding the two numbers just above it (to the left and right). This fact, known as the "Pascal identity," can be written as ${}_nC_r = {}_{n-1}C_{r-1} + {}_{n-1}C_r$. Use the factorial formula for combinations (along with some algebra) to prove the Pascal identity.

The "triangle" that Pascal studied and published in his treatise was actually more like a truncated corner of Tartaglia's rectangle, as shown here.

1	1	1	1	1	1	1	1	1	1
1	2	3	4	5	6	7	8	9	
1	3	6	10	15	21	28	36		
1	4	10	20	35	56	84			
1	5	15	35	70	126				
1	6	21	56	126					
1	7	28	84						
1	8	36							
1	9								
1									

Each number in the array can be calculated in various ways. In each of Exercises 38–41, consider the number N to be located anywhere in the array. By checking several locations in the given array, determine how N is related to the sum of all entries in the shaded cells. Describe the relationship in words.

38.

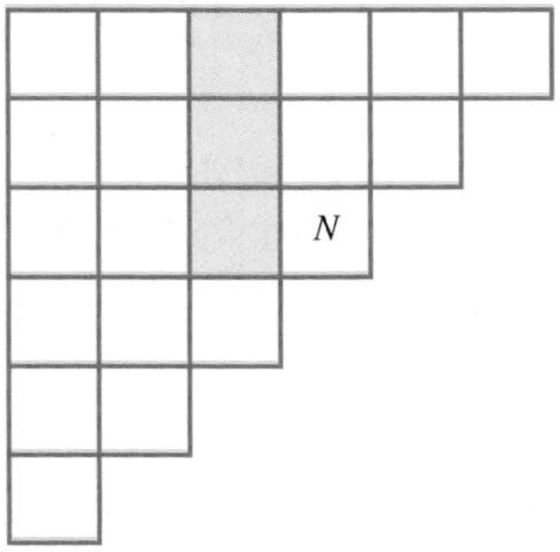

39.

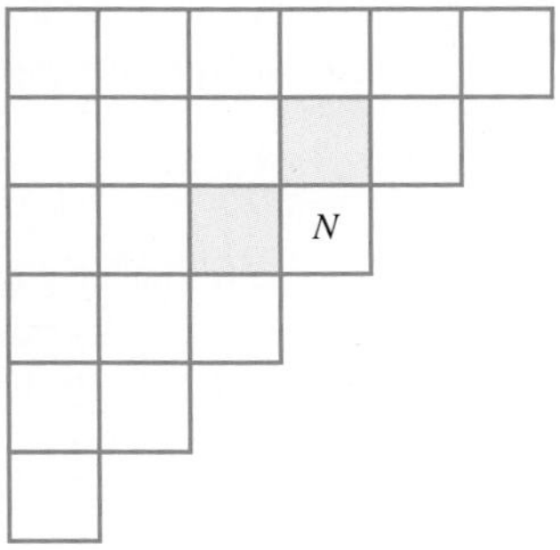

40.

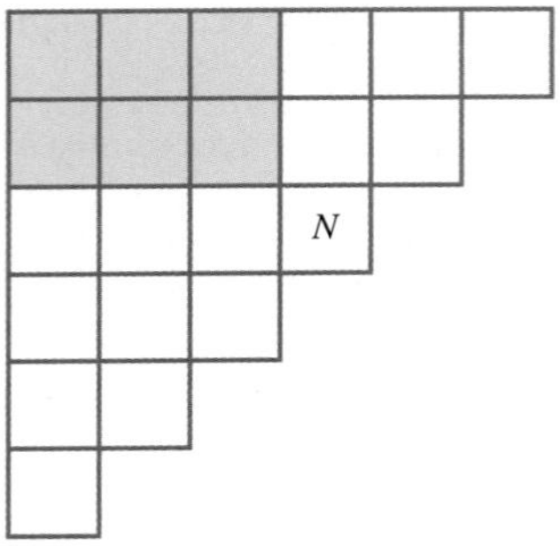

41.

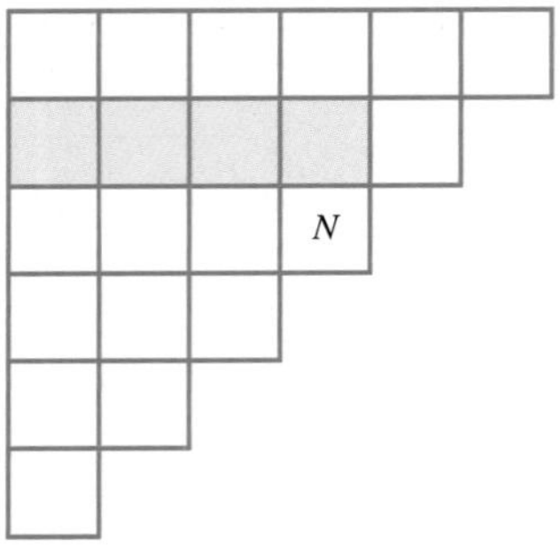

11.5 Counting Problems Involving "Not" and "Or"

Problems Involving "Not" • Problems Involving "Or"

The counting techniques in this section, which can be thought of as *indirect techniques*, are based on some useful correspondences (from Chapters 2 and 3) between set theory, logic, and arithmetic, as shown in Table 8.

TABLE 8 Set Theory/Logic/Arithmetic Correspondences

	Set Theory	Logic	Arithmetic
Operation or Connective (Symbol)	Complement (′)	Not (~)	Subtraction (−)
Operation or Connective (Symbol)	Union (∪)	Or (∨)	Addition (+)

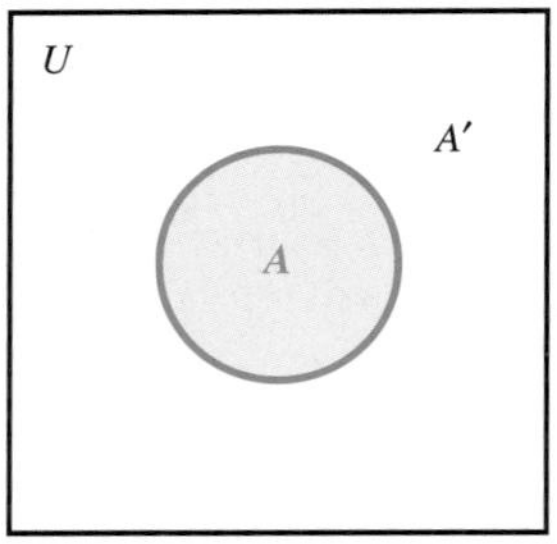

The complement of a set

FIGURE 8

Problems Involving "Not" Suppose U is the set of all possible results of some type. (All possibilities comprise the "universal set U," as discussed in Chapter 2.) Let A be the set of all those results that satisfy a given condition. For any set S, its cardinal number is written $n(S)$, and its complement is written S'. Figure 8 suggests that

$$n(A) + n(A') = n(U).$$

Also, $n(A) = n(U) - n(A')$ and $n(A') = n(U) - n(A)$.

We focus here on the form that expresses the following indirect counting principle (based on the complement/not/subtraction correspondence from Table 8).

Complements Principle of Counting

The number of ways a certain condition can be satisfied is the total number of possible results minus the number of ways the condition would **not** be satisfied. Symbolically, if A is any set within the universal set U, then

$$n(A) = n(U) - n(A').$$

EXAMPLE 1 Counting the Proper Subsets of a Set

For the set $S = \{a, b, c, d, e, f\}$, find the number of proper subsets.

SOLUTION

Recall that a proper subset of S is any subset with fewer than all six elements. Subsets of several different sizes would satisfy this condition. However, it is easier to consider the one subset that is not proper, namely S itself. From set theory, we know that set S has a total of $2^6 = 64$ subsets. Thus, from the complements principle, the number of proper subsets is $64 - 1 = 63$. In words, the number of subsets that *are* proper is the total number of subsets minus the number of subsets that are *not* proper. ■

Consider the tossing of three fair coins. Since each coin will land either heads (h) or tails (t), the possible results can be listed as follows.

hhh, hht, hth, thh, htt, tht, tth, ttt   Results of tossing three fair coins

(Even without the listing, we could have concluded that there would be eight possibilities. There are two possible outcomes for each coin, so the fundamental counting principle gives $2 \cdot 2 \cdot 2 = 2^3 = 8$.)

Suppose we wanted the number of ways of obtaining *at least* one head. In this case, "at least one" means one or two or three. Rather than dealing with all three cases, we can note that "at least one" is the opposite (or complement) of "fewer than one" (which is zero). Because there is only one way to get zero heads (ttt), and there are a total of eight possibilities, the complements principle gives the number of ways of getting at least one head: $8 - 1 = 7$. (The number of outcomes that include at least one head is the total number of outcomes minus the number of outcomes that do *not* include at least one head.) We find that indirect counting methods can often be applied to problems involving "at least," or "at most," or "less than," or "more than."

EXAMPLE 2 Counting Coin-Tossing Results

If four fair coins are tossed, in how many ways can at least one tail be obtained?

SOLUTION

By the fundamental counting principle, $2^4 = 16$ different results are possible. Exactly one of these fails to satisfy the condition of "at least one tail" (namely, no tails, or hhhh). So our answer (from the complements principle) is $16 - 1 = 15$. ■

EXAMPLE 3 Counting Selections of Airliner Seats

Carol Britz and three friends are boarding an airliner just before departure time. There are only ten seats left, three of which are aisle seats. How many ways can the four people arrange themselves in available seats so that at least one of them sits on the aisle?

SOLUTION

The word "arrange" implies that order is important, so we shall use permutations. "At least one aisle seat" is the opposite (complement) of "no aisle seats." The total number of ways to arrange four people among ten seats is ${}_{10}P_4 = 5040$. The number of ways to arrange four people among seven (non-aisle) seats is ${}_7P_4 = 840$. Therefore, by the complements principle, the number of arrangements with at least one aisle seat is $5040 - 840 = 4200$. ■

Problems Involving "Or" The complements principle is one way of counting indirectly. Another technique is to count the elements of a set by breaking that set into simpler component parts. If $S = A \cup B$, the cardinal number formula (from Section 2.4) says to find the number of elements in S by adding the number in A to the number in B. We must then subtract the number in the intersection $A \cap B$ if A and B are not disjoint, as in Figure 9. But if A and B are disjoint, as in Figure 10, the subtraction is not necessary.

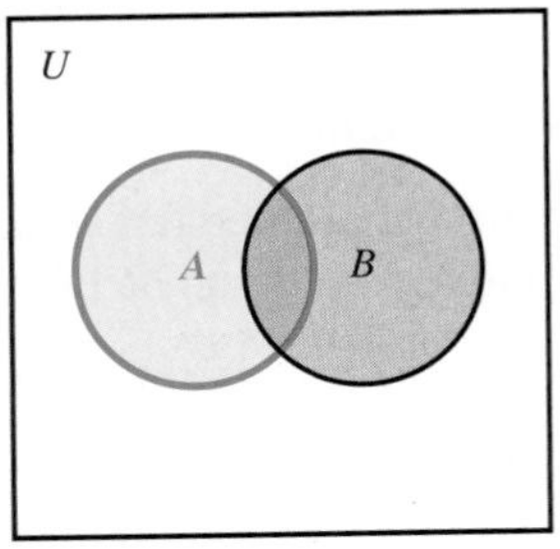

Nondisjoint sets

FIGURE 9

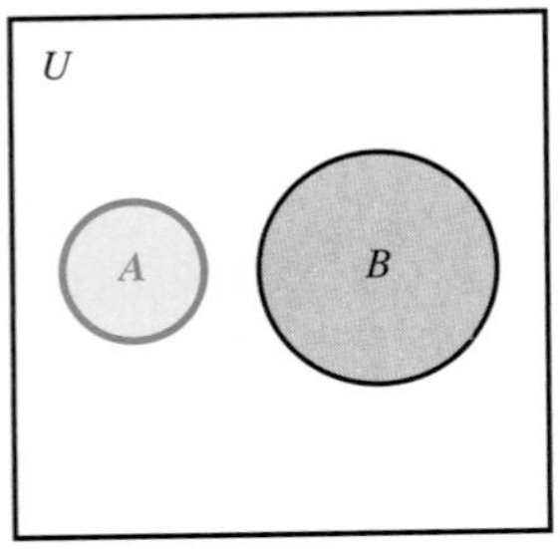

Disjoint sets

FIGURE 10

The following principle reflects the union/or/addition correspondence from Table 8.

Additive Principle of Counting

The number of ways that one **or** the other of two conditions could be satisfied is the number of ways one of them could be satisfied plus the number of ways the other could be satisfied minus the number of ways they could both be satisfied together.

If A and B are any two sets, then

$$n(A \cup B) = n(A) + n(B) - n(A \cap B).$$

If sets A and B are disjoint, then

$$n(A \cup B) = n(A) + n(B).$$

EXAMPLE 4 Counting Card Hands

How many five-card hands consist of either all clubs or all red cards?

SOLUTION

No hand that satisfies one of these conditions could also satisfy the other, so the two sets of possibilities (all clubs, all red cards) are disjoint. Therefore the second formula of the additive principle applies, and we obtain

$n(\text{all clubs } or \text{ all red cards})$	$= n(\text{all clubs}) + n(\text{all red cards})$	Additive counting principle
	$= {}_{13}C_5 + {}_{26}C_5$	13 clubs, 26 red cards
	$= 1287 + 65{,}780$	Substitute values.
	$= 67{,}067.$	

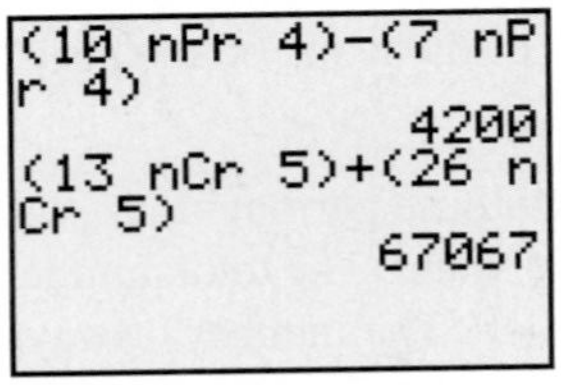

Results in Examples 3 and 4 are supported in this screen.

EXAMPLE 5 Counting Selections from a Diplomatic Delegation

Table 9 categorizes a diplomatic delegation of 18 congressional members as to political party and gender. If one of the members is chosen randomly to be spokesperson for the group, in how many ways could that person be a Democrat or a woman?

TABLE 9

	Men (M)	Women (W)	Totals
Republican (R)	5	3	8
Democrat (D)	4	6	10
Totals	9	9	18

SOLUTION

Since D and W are not disjoint (6 delegates are both Democrats and women), the first formula of the additive principle is required.

$$\begin{aligned} n(D \text{ or } W) &= n(D \cup W) && \text{Union/or correspondence} \\ &= n(D) + n(W) - n(D \cap W) && \text{Additive principle} \\ &= 10 + 9 - 6 && \text{Substitute values.} \\ &= 13. \end{aligned}$$

EXAMPLE 6 Counting Course Selections for a Degree Program

Peggy Jenders needs to take twelve more specific courses for a bachelors degree, including four in math, three in physics, three in computer science, and two in business. If five courses are randomly chosen from these twelve for next semester's program, how many of the possible selections would include at least two math courses?

SOLUTION

Of all the information given here, what is important is that there are four math courses and eight other courses to choose from, and that five of them are being selected for next semester. If T denotes the set of selections that include at least two math courses, then we can write

$$T = A \cup B \cup C$$

where A = the set of selections with exactly two math courses,

B = the set of selections with exactly three math courses,

and C = the set of selections with exactly four math courses.

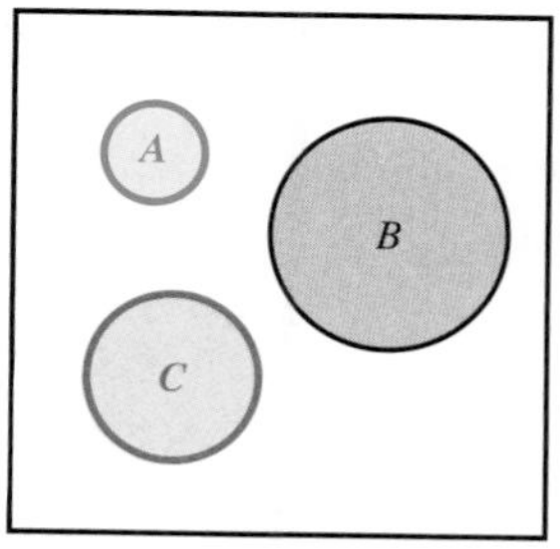

$T = A \cup B \cup C$

FIGURE 11

(In this case, *at least two* means exactly two **or** exactly three **or** exactly four.) The situation is illustrated in Figure 11. By previous methods, we know that

$$n(A) = {}_4C_2 \cdot {}_8C_3 = 6 \cdot 56 = 336,$$

$$n(B) = {}_4C_3 \cdot {}_8C_2 = 4 \cdot 28 = 112,$$

and

$$n(C) = {}_4C_4 \cdot {}_8C_1 = 1 \cdot 8 = 8,$$

so that, by the additive principle,

$$n(T) = 336 + 112 + 8 = 456.$$

EXAMPLE 7 Counting Three-Digit Numbers with Conditions

How many three-digit counting numbers are multiples of 2 or multiples of 5?

SOLUTION

A multiple of 2 must end in an even digit (0, 2, 4, 6, or 8), so there are $9 \cdot 10 \cdot 5 = 450$ three-digit multiples of 2. A multiple of 5 must end in either 0 or 5, so there are $9 \cdot 10 \cdot 2 = 180$ of those. A multiple of both 2 and 5 is a multiple of 10 and must end in 0. There are $9 \cdot 10 \cdot 1 = 90$ of those. By the additive principle there are

$$450 + 180 - 90 = 540$$

possible three-digit numbers that are multiples of 2 or multiples of 5.

EXAMPLE 8 Counting Card-Drawing Results

A single card is drawn from a standard 52-card deck.

(a) In how many ways could it be a heart or a king?
(b) In how many ways could it be a club or a face card?

SOLUTION

(a) A single card can be both a heart and a king (the king of hearts), so use the first additive formula. There are thirteen hearts, four kings, and one card that is both a heart and a king:

$$13 + 4 - 1 = 16.$$

(b) There are 13 clubs, 12 face cards, and 3 cards that are both clubs and face cards, giving

$$13 + 12 - 3 = 22.$$

EXAMPLE 9 Counting Subsets of a Set with Conditions

How many subsets of a 25-element set have more than three elements?

SOLUTION

It would be a real job to count directly all subsets of size 4, 5, 6, . . . , 25. It is much easier to count those with three or fewer elements and apply the complements principle.

There is	${}_{25}C_0 = 1$	size-0 subset.
There are	${}_{25}C_1 = 25$	size-1 subsets.
There are	${}_{25}C_2 = 300$	size-2 subsets.
There are	${}_{25}C_3 = 2300$	size-3 subsets.

The total number of subsets (of all sizes, 0 through 25) is $2^{25} = 33{,}554{,}432$ (use a calculator). So the number with more than three elements must be

$$\begin{aligned} 33{,}554{,}432 - (1 + 25 + 300 + 2300) &= 33{,}554{,}432 - 2626 \\ &= 33{,}551{,}806. \end{aligned}$$

In Example 9, we used both the additive principle (to get the number of subsets with no more than three elements) and the complements principle.

PROBLEM-SOLVING HINT As you work the exercises of this section, keep in mind that indirect methods may be best, and that you may also be able to use permutations, combinations, the fundamental counting principle, or listing procedures such as product tables or tree diagrams. Also, you may want to obtain combination values, when needed, from Pascal's triangle, or find combination and permutation values on a calculator.

11.5 EXERCISES

How many proper subsets are there of each set?

1. {A, B, C, D}

2. {u, v, w, x, y, z}

Tossing Coins *If you toss seven fair coins, in how many ways can you obtain each result?*

3. at least one head ("At least one" is the complement of "none.")

4. at least two heads ("At least two" is the complement of "zero or one.")

5. at least two tails

6. at least one of each (a head and a tail)

Rolling Dice *If you roll two fair dice (say red and green), in how many ways can you obtain each result? (Refer to Table 2 in Section 11.1.)*

7. at least 2 on the green die

8. a sum of at least 3

9. a 4 on at least one of the dice

10. a different number on each die

Drawing Cards *If you draw a single card from a standard 52-card deck, in how many ways can you obtain each result?*

11. a card other than the ace of spades

12. a nonface card

Identifying Properties of Counting Numbers *How many two-digit counting numbers meet each requirement?*

13. not a multiple of 10

14. greater than 70 or a multiple of 10

15. ***Choosing Country Music Albums*** Michelle Cook's collection of eight country music albums includes *When the Sun Goes Down* by Kenny Chesney. Michelle will choose three of her albums to play on a drive to Nashville. (Assume order is not important.)

(a) How many different sets of three albums could she choose?

(b) How many of these sets would not include *When the Sun Goes Down*?

(c) How many of them would include *When the Sun Goes Down*?

16. ***Choosing Broadway Hits*** The ten longest Broadway runs include *The Phantom of the Opera* and *Les Misérables.* Four of the ten are chosen randomly. (Assume order is not important.)

(a) How many ways can the four be chosen?

(b) How many of those groups of four would include neither of the two productions mentioned?

(c) How many of them would include at least one of the two productions mentioned?

17. ***Choosing Days of the Week*** How many different ways could three distinct days of the week be chosen so that at least one of them begins with the letter S? (Assume order of selection is not important.)

18. ***Choosing School Assignments for Completion*** Chalon Bridges has nine major assignments to complete for school this week. Two of them involve writing essays. Chalon decides to work on two of the nine assignments tonight. How many different choices of two would include at least one essay assignment? (Assume order is not important.)

Selecting Restaurants *Byron Hopkins wants to dine at three different restaurants during a visit to a mountain resort. If two of eight available restaurants serve seafood, find the* *number of ways that at least one of the selected restaurants will serve seafood given the following conditions.*

19. The order of selection is important.

20. The order of selection is not important.

21. ***Seating Arrangements on an Airliner*** Refer to Example 3. If one of the group decided at the last minute not to fly, then how many ways could the remaining three arrange themselves among the ten available seats so that at least one of them will sit on the aisle?

22. ***Identifying Properties of Counting Numbers*** Find the number of four-digit counting numbers containing at least one zero, under each of the following conditions.
(a) Repeated digits are allowed.
(b) Repeated digits are not allowed.

23. ***Selecting Faculty Committees*** A committee of four faculty members will be selected from a department of twenty-five which includes professors Fontana and Spradley. In how many ways could the committee include at least one of these two professors?

24. ***Selecting Search and Rescue Teams*** A Civil Air Patrol unit of twelve members includes four officers. In how many ways can four members be selected for a search and rescue mission such that at least one officer is included?

Drawing Cards *If a single card is drawn from a standard 52-card deck, in how many ways could it be the following? (Use the additive principle.)*

25. a club or a jack

26. a face card or a black card

Counting Students Who Enjoy Music and Literature *Of a group of 50 students, 30 enjoy music, 15 enjoy literature, and 10 enjoy both music and literature. How many of them enjoy the following?*

27. at least one of these two subjects (Use the additive principle.)

28. neither of these two subjects (complement of "at least one")

Counting Hands of Cards *Among the 2,598,960 possible 5-card poker hands from a standard 52-card deck, how many contain the following cards?*

29. at least one card that is not a club (complement of "all clubs")

30. cards of more than one suit (complement of "all the same suit")

31. at least one face card (complement of "no face cards")

32. at least one diamond, but not all diamonds (complement of "no diamonds or all diamonds")

The Size of Subsets of a Set *If a given set has twelve elements, how many of its subsets have the given numbers of elements?*

33. at most two elements

34. at least ten elements

35. more than two elements

36. from three through nine elements

37. ***Counting License Numbers*** If license numbers consist of three letters followed by three digits, how many different licenses could be created having at least one letter or digit repeated? (*Hint:* Use the complements principle of counting.)

38. ***Drawing Cards*** If two cards are drawn from a 52-card deck without replacement (that is, the first card is not replaced in the deck before the second card is drawn), in how many different ways is it possible to obtain a king on the first draw and a heart on the second? (*Hint:* Split this event into the two disjoint components "king of hearts and then another heart" and "non-heart king and then heart." Use the fundamental counting principle on each component, then apply the additive principle.)

39. Extend the additive counting principle to three overlapping sets (as in the figure) to show that

$$n(A \cup B \cup C) = n(A) + n(B) + n(C) - n(A \cap B) - n(A \cap C) - n(B \cap C) + n(A \cap B \cap C).$$

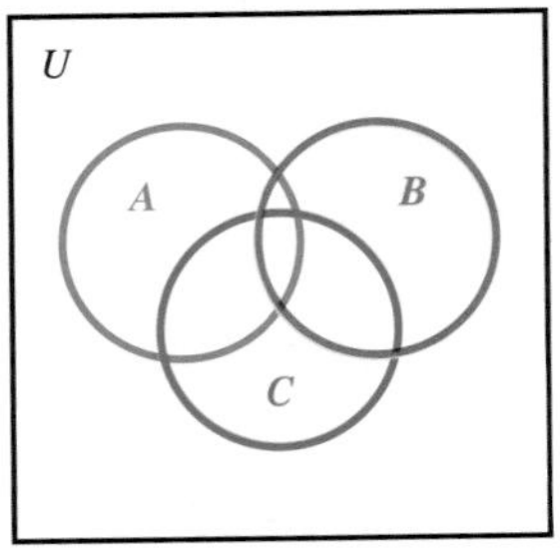

40. How many of the counting numbers 1 through 300 are *not* divisible by 2, 3, or 5? (*Hint:* Use the complements principle and the result of Exercise 39.)

Selecting National Monuments to Visit *Edward Roberts is planning a long-awaited driving tour, which will take him and his family on the southern route to the West Coast. Ed is interested in seeing the twelve national monuments listed here, but will have to settle for seeing just three of them because some family members are anxious to get to Disneyland.*

New Mexico	Arizona	California
Gila Cliff Dwellings	Canyon de Chelly	Devils Postpile
Petroglyph	Organ Pipe Cactus	Joshua Tree
White Sands	Saguaro	Lava Beds
Aztec Ruins		Muir Woods
		Pinnacles

In how many ways could the three monuments chosen include the following?

41. sites in only one state

42. at least one site not in California

43. sites in fewer than all three states

44. sites in exactly two of the three states

Counting Categories of Poker Hands *Table 4 in this chapter (For Further Thought in Section 11.3) described the various kinds of hands in 5-card poker. Verify each statement in Exercises 45–48. (Explain all steps of your argument.)*

45. There are four ways to get a royal flush.

46. There are 36 ways to get a straight flush.

47. There are 10,200 ways to get a straight.

48. There are 54,912 ways to get three of a kind.

49. Explain why the complements principle of counting is called an "indirect" method.

50. Explain the difference between the two formulas of the additive principle of counting.

COLLABORATIVE INVESTIGATION

Solving a Traveling Salesman Problem

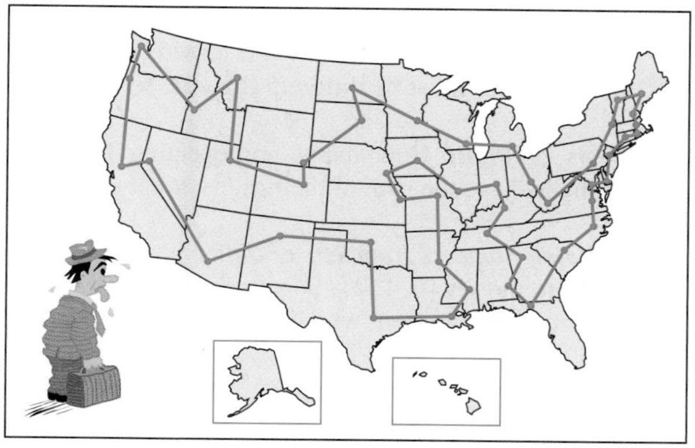

In 1985, Shen Lin came up with the route shown above for a salesman wanting to visit all capital cities in the forty-eight contiguous states, starting and ending at the same capital and traveling the shortest possible total distance. He could not prove that his 10,628-mile route was the shortest possible, but he offered $100 to anyone who could find a shorter one.

This is an example of a classic problem, the so-called **traveling salesman problem** (or **TSP**), which has many practical applications in business and industry but has baffled mathematicians for years. In the case above, there are 47! possible routes, although many of them can be quickly eliminated, leaving $\frac{24!}{3}$ possibilities to consider. This is still a 24-digit number, far too large for even state-of-the-art computers to analyze directly.

Although computer scientists have so far failed to find an "efficient algorithm" to solve the general traveling salesman problem, successes are periodically achieved for particular cases. In 1998, Rice University researchers David Applegate, Robert Bixby, and William Cook, along with Vasek Chvatal of Rutgers University, announced a breakthrough solution to the traveling salesman problem for 13,509 U.S. cities with populations of at least 500 people. In 2004, the solution was found for a case with 24,978 points to visit.

A much smaller set (of seven cities, A through G), which can be completely analyzed using a calculator, is shown here.

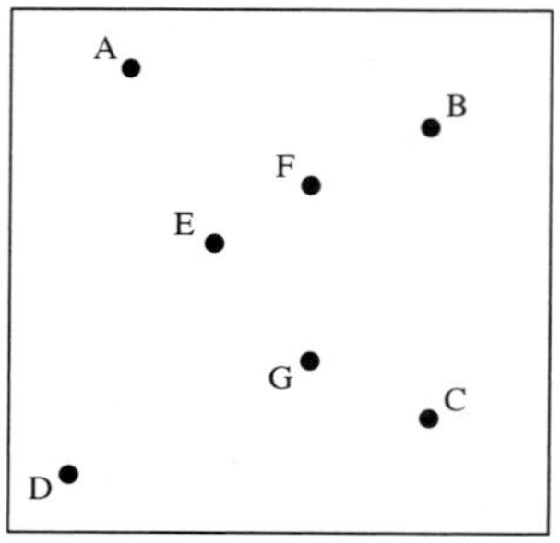

Notice that certain routes clearly are *not* the shortest. For example, it is apparent that the route ACEDFGBA involves too much jumping back and forth across the diagram to result in the least possible total distance. (In fact, the total distance for this route is 360 miles, considerably more than necessary.) The fifteen distances given here (in miles) between pairs of cities should be sufficient data for computing the shortest possible route.

AB = 51	AF = 36	BF = 22	CG = 22	EF = 22
AD = 71	BC = 50	BG = 45	DE = 45	EG = 28
AE = 32	BE = 45	CD = 61	DG = 45	FG = 30

Topics for Discussion

Divide the class into groups of 3 or 4 students each. Each group is to do the following.

1. Study the drawing, and make a list of all routes that you think may be the shortest.
2. For each candidate route, add the appropriate seven terms to get a total distance.
3. Arrive at a group consensus as to which route is shortest.

Now bring the whole class back together, and do the following.

1. Make a list of routes, with total distances, that the various groups thought were shortest.
2. Observe whether the different groups all agreed on which route was shortest.
3. As a class, try to achieve a consensus on the shortest route. Do you think that someone else may be able to find a shorter one?

CHAPTER 11 TEST

If digits may be used from the set {0, 1, 2, 3, 4, 5, 6}, *find the number of possibilities in each category.*

1. three-digit numbers
2. even three-digit numbers
3. three-digit numbers without repeated digits
4. three-digit multiples of five without repeated digits
5. ***Counting Triangles in a Figure*** Determine the number of triangles (of any size) in the figure shown here.

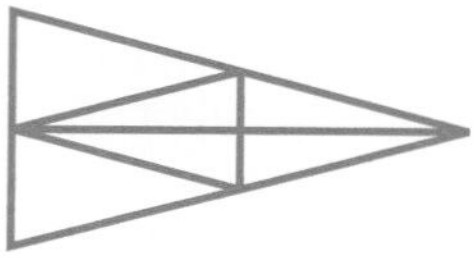

6. ***Tossing Coins*** Construct a tree diagram showing all possible results when a fair coin is tossed four times, if the third toss must be different than the second.
7. ***Sums of Digits*** How many nonrepeating four-digit numbers have the sum of their digits equal to 30?
8. ***Building Numbers from Sets of Digits*** Using only digits from the set {0, 1, 2}, how many three-digit numbers can be written which have no repeated odd digits?

Evaluate each expression.

9. 5!
10. $\frac{8!}{5!}$
11. ${}_{12}P_4$
12. ${}_{7}C_3$

13. ***Building Words from Sets of Letters*** How many five-letter "words" without repeated letters are possible using the English alphabet? (Assume that any five letters make a "word.")

14. ***Building Words from Sets of Letters*** Using the Russian alphabet (which has 32 letters), and allowing repeated letters, how many five-letter "words" are possible?

Scheduling Assignments *Andrea has seven homework assignments to complete. She wants to do four of them on Tuesday and the other three on Wednesday.*

15. In how many ways can she order Tuesday's work?

16. Assuming she finishes Tuesday's work successfully, in how many ways can she order Wednesday's work?

Arranging Letters *Find the number of distinguishable arrangements of the letters of each word.*

17. TATTLE

18. OLIGOPOLY

Selecting Groups of Basketball Players *If there are twelve players on a basketball team, find the number of choices the coach has in selecting each of the following.*

19. four players to carry the team equipment

20. two players for guard positions and two for forward positions

21. five starters and five subs

22. a set of three or more of the players

Choosing Switch Settings *Determine the number of possible settings for a row of four on–off switches under each condition.*

23. There are no restrictions.

24. The first and fourth switches must be on.

25. The first and fourth switches must be set the same.

26. No two adjacent switches can both be off.

27. No two adjacent switches can be set the same.

28. At least two switches must be on.

Choosing Subsets of Letters *Four distinct letters are to be chosen from the set*

$$\{A, B, C, D, E, F, G\}.$$

Determine the number of ways to obtain a subset that includes each of the following.

29. the letter D

30. both A and E

31. either A or E, but not both

32. equal numbers of vowels and consonants

33. more consonants than vowels

34. State the fundamental counting principle in your own words.

35. If ${}_nC_r = 495$ and ${}_nC_{r+1} = 220$, find the value of ${}_{n+1}C_{r+1}$.

36. If you write down the second entry of each row of Pascal's triangle (starting with row 1), what sequence of numbers do you obtain?

37. Explain why there are $r!$ permutations of n things taken r at a time corresponding to each combination of n things taken r at a time.

12

PROBABILITY

Suppose you're on a game show, and you're given the choice of three doors: Behind one of the doors is a car, and behind the other doors, goats. Of course, you want to win the car. You pick one of the doors, say Door 1, and the host, who knows what's behind the other doors, opens another door, say Door 3, to reveal a goat. He then says to you, "Do you want to change your choice?" Is it to your advantage to switch to Door 2?

This question appeared in *Parade* magazine in a column written by Marilyn vos Savant in the early 1990s. This probability problem, known as the Monty Hall Problem, was named after the host of the popular game show *Let's Make a Deal*. Marilyn's answer caused an incredible amount of discussion and argument among the general public at that time. The problem and an explanation of its answer were presented by Charlie Eppes in his "Math for Non-Mathematicians" class in the May 13, 2005, episode "Man Hunt" of the CBS series *NUMB3RS*.

The answers and its justification can also be found at the interactive Web site www.math.ucsd.edu/~crypto/Monty/monty.html. *Would YOU switch doors*? (See page 763 for the answer.)

12.1 Basic Concepts

Historical Background • Probability • The Law of Large Numbers • Probability in Genetics • Odds

Christiaan Huygens (1629–1695), a brilliant Dutch mathematician and scientist, was the first to write a formal treatise on **probability.** It appeared in 1657 and was based on the Pascal–Fermat correspondence.

Historical Background

Nearly all of the limited work in probability from the fifteenth through the eighteenth century concerned games and gambling. For example, in 1654, two French mathematicians, Pierre de Fermat (about 1601–1665) and Blaise Pascal (1623–1662), corresponded with each other regarding a problem posed by the Chevalier de Méré, a gambler and member of the aristocracy. *If the two players of a game are forced to quit before the game is finished, how should the pot be divided?* Pascal and Fermat solved the problem by developing basic methods of determining each player's chance, or probability, of winning.

One of the first to apply probability to matters other than gambling was the French mathematician Pierre Simon de Laplace (1749–1827), who is usually credited with being the "father" of probability theory.

But it was not until the twentieth century that a coherent mathematical theory of probability had been developed. It came mainly through a line of remarkable scholars in Russia, including P. L. Chebyshev (1821–1922), his student A. A. Markov (1856–1922), and finally Andrei Nikolaevich Kolmogorov (1903–1987), whose *Foundations of the Theory of Probability* was published in 1933.

Probability

If you go to a supermarket and select five pounds of peaches at 89¢ per pound, you can easily predict the amount you will be charged at the checkout counter: $5 \cdot \$.89 = \$4.45$. The amount charged for such purchases is a **deterministic phenomenon.** It can be predicted exactly on the basis of obtainable information, namely, in this case, number of pounds and cost per pound.

On the other hand, consider the problem faced by the produce manager of the market, who must order peaches to have on hand each day without knowing exactly how many pounds customers will buy during the day. The customer demand is an example of a **random phenomenon.** It fluctuates in such a way that its value (on a given day) cannot be predicted exactly with obtainable information.

The study of probability is concerned with such random phenomena. Even though we cannot be certain whether a given result will occur, we often can obtain a good measure of its *likelihood,* or **probability.** This chapter discusses various ways of finding and using probabilities.

In the study of probability, we say that any observation, or measurement, of a random phenomenon is an **experiment.** The possible results of the experiment are called **outcomes,** and the set of all possible outcomes is called the **sample space.**

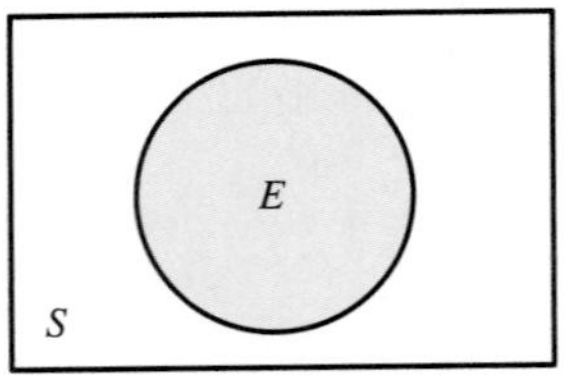

Every event is a subset of the sample space.

Usually we are interested in some particular collection of the possible outcomes. Any such subset of the sample space is called an **event.** Outcomes that belong to the event are commonly referred to as "favorable outcomes," or "successes." Any time a success is observed, we say that the event has "occurred." The probability of an event, being a numerical measure of the event's likelihood, is determined in one of two ways, either *empirically* (experimentally) or *theoretically* (mathematically).

"But is it probable," asked Pascal, "that probability gives assurance? Nothing gives certainty but truth; nothing gives rest but the sincere search for truth." When Pascal wrote that, he had gone to live at the Jansenist convent of Port-Royal after a carriage accident in 1654.

Pascal's notes on Christianity were collected after his death in the *Pensées* (thoughts). The above quotation is included. Another develops Pascal's "rule of the wager": If you bet God exists and live accordingly, you will have gained much even if God does not exist; if you bet the opposite and God does exist, you will have lost the reason for living right—hence everything.

EXAMPLE 1 Finding Probability When Tossing a Coin

If a single coin is tossed, find the probability that it will land heads up.

SOLUTION

There is no apparent reason for one side of a coin to land up any more often than the other (in the long run), so we would normally assume that heads and tails are equally likely. We express this assumption by saying that the coin is "fair."

Now the experiment here is the tossing of a single fair coin, the sample space is $S = \{\text{h}, \text{t}\}$, and the event whose probability we seek is $E = \{\text{h}\}$. Since one of two possible outcomes is a head, the probability of heads is the quotient of 1 and 2.

$$\text{Probability (heads)} = \frac{1}{2}, \quad \text{written} \quad P(\text{h}) = \frac{1}{2} \quad \text{or} \quad P(E) = \frac{1}{2}.$$

EXAMPLE 2 Finding Probability When Tossing a Cup

If a Styrofoam cup is tossed, find the probability that it will land on its top.

SOLUTION

Intuitively, it seems that a cup will land on its side much more often than on its top or its bottom. But just how much more often is not clear. To get an idea, we performed the experiment of tossing such a cup 50 times. It landed on its side 44 times, on its top 5 times, and on its bottom just 1 time. By the frequency of "success" in this experiment, we concluded that

$$P(\text{top}) \approx \frac{5}{50} = \frac{1}{10}. \quad \longleftarrow \text{Write in lowest terms.}$$

In Example 1 involving the tossing of a fair coin, the number of possible outcomes was obviously two, both were equally likely, and one of the outcomes was a head. No actual experiment was required. The desired probability was obtained *theoretically.* Theoretical probabilities apply to all kinds of games of chance (dice rolling, card games, roulette, lotteries, and so on), and also apparently to many phenomena in nature.

Laplace, in his famous *Analytic Theory of Probability,* published in 1812, gave a formula that applies to any such theoretical probability, as long as the sample space S is finite and all outcomes are equally likely. (It is sometimes referred to as the *classical definition of probability.*)

Theoretical Probability Formula

If all outcomes in a sample space S are equally likely, and E is an event within that sample space, then the **theoretical probability** of event E is given by

$$P(E) = \frac{\textbf{number of favorable outcomes}}{\textbf{total number of outcomes}} = \frac{n(E)}{n(S)}.$$

On the other hand, Example 2 involved the tossing of a cup, where the likelihoods of the various outcomes were not intuitively clear. It took an actual experiment to arrive at a probability value of $\frac{1}{10}$, and that value, based on a portion of all possible tosses of the cup, should be regarded as an approximation of the true theoretical probability. The value was found according to the *experimental,* or *empirical,* probability formula.

In 1827, **Robert Brown** (1773–1858), a Scottish physician and botanist, described the irregular motion of microscopic pollen grains suspended in water. Such "Brownian motion," as it came to be called, was not understood until 1905 when Albert Einstein explained it by treating molecular motion as a random phenomenon.

Empirical Probability Formula

If E is an event that may happen when an experiment is performed, then the **empirical probability** of event E is given by

$$P(E) \approx \frac{\textbf{number of times event } E \textbf{ occurred}}{\textbf{number of times the experiment was performed}}.$$

Usually it is clear in applications which of the two probability formulas should be used.

EXAMPLE 3 Finding the Probability of Having Daughters

Michelle Brown wants to have exactly two daughters. Assuming that boy and girl babies are equally likely, find her probability of success if

(a) she has a total of two children,
(b) she has a total of three children.

SOLUTION

(a) The equal likelihood assumption here allows the use of theoretical probability. But how can we determine the number of favorable outcomes and the total number of possible outcomes?

One way is to use a tree diagram (Section 11.1) to enumerate the possibilities, as shown in Figure 1. From the outcome column we obtain the sample space $S = \{\text{gg, gb, bg, bb}\}$. Only one outcome, marked with an arrow, is favorable to the event of exactly two daughters: $E = \{\text{gg}\}$. By the theoretical probability formula,

$$P(E) = \frac{n(E)}{n(S)} = \frac{1}{4}.$$

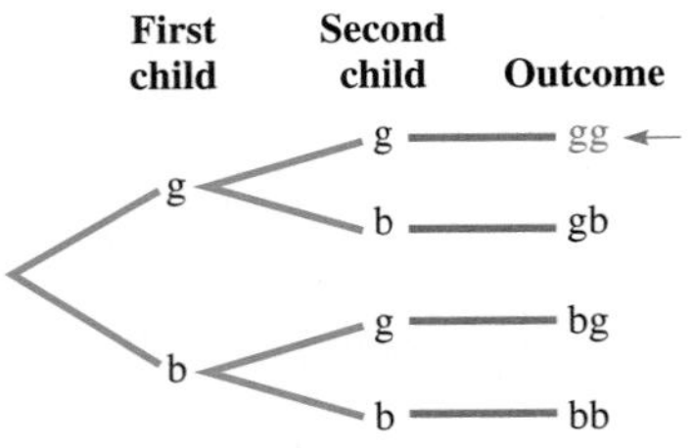

Exactly two girls among two children

FIGURE 1

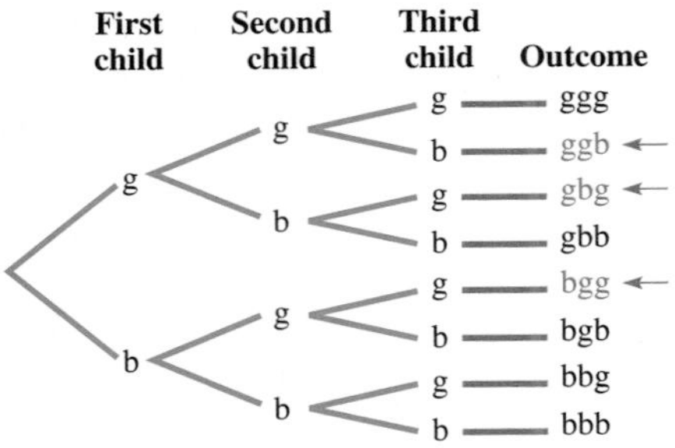

Exactly two girls among three children

FIGURE 2

(b) For three children altogether, we construct another tree diagram, as shown in Figure 2. In this case, we see that $S = \{\text{ggg, ggb, gbg, gbb, bgg, bgb, bbg, bbb}\}$ and $E = \{\text{ggb, gbg, bgg}\}$, so $P(E) = \frac{3}{8}$. ■

TABLE 1
Number of Poker Hands in 5-Card Poker; Nothing Wild

Event E	Number of Outcomes Favorable to E
Royal flush	4
Straight flush	36
Four of a kind	624
Full house	3744
Flush	5108
Straight	10,200
Three of a kind	54,912
Two pairs	123,552
One pair	1,098,240
No pair	1,302,540
Total	2,598,960

EXAMPLE 4 Finding Probability When Dealing Cards

Find the probability of being dealt each of the following hands in five-card poker. Use a calculator to obtain answers to eight decimal places.

(a) a full house (three of one denomination and two of another)
(b) a royal flush (the five highest cards—ace, king, queen, jack, ten—of a single suit)

SOLUTION

(a) Table 1 summarizes the various possible kinds of five-card hands. Since the 2,598,960 possible individual hands all are equally likely, we can enter the appropriate numbers from the table into the theoretical probability formula.

$$P(\text{full house}) = \frac{3744}{2{,}598{,}960} = \frac{6}{4165} \approx .00144058$$

(b) The table shows that there are four royal flushes, one for each suit, so that

$$P(\text{royal flush}) = \frac{4}{2{,}598{,}960} = \frac{1}{649{,}740} \approx .00000154.$$

Examples 3 and 4 both utilized the theoretical probability formula because we were able to enumerate all possible outcomes and all were equally likely. In Example 3, however, the equal likelihood of girl and boy babies was *assumed.* In fact, male births typically occur a little more frequently. (At the same time, there usually are more females living at any given time, due to higher infant mortality rates among males and longer female life expectancy in general.) Example 5 shows a way of incorporating such empirical information.

EXAMPLE 5 Finding the Probability of the Gender of a Resident

In the year 2003, the U.S. populace included 143,037,000 males and 147,773,000 females. If a person was selected randomly from the population in that year, what is the probability that the person would be a female?

SOLUTION

In this case, we calculate the empirical probability from the given experimental data.

$$\begin{aligned} P(\text{female}) &= \frac{\text{number of female residents}}{\text{total number of residents}} \\ &= \frac{147{,}773{,}000}{143{,}037{,}000 + 147{,}773{,}000} \\ &\approx .508 \end{aligned}$$

The Law of Large Numbers Now think again about the cup of Example 2. If we tossed it 50 more times, we would have 100 total tosses upon which to base an empirical probability of the cup landing on its top. The new value would likely be (at least slightly) different from what we obtained before. It would still be an empirical probability, but it would be "better" in the sense that it is based upon a larger set of outcomes.

The **law of large numbers** also can be stated as follows.

A theoretical probability really says nothing about one, or even a few, repetitions of an experiment, but only about the proportion of successes we would expect over the long run.

As we increase the number of tosses, the resulting empirical probability values may approach some particular number. If so, that number can be defined as the theoretical probability of that particular cup landing on its top. This "limiting" value can occur only as the actual number of observed tosses approaches the total number of possible tosses of the cup. Since there are potentially an infinite number of possible tosses, we could never actually find the theoretical probability we want. But we can still assume such a number exists. And as the number of actual observed tosses increases, the resulting empirical probabilities should tend ever closer to the theoretical value. This very important principle is known as the **law of large numbers** (or sometimes as the "law of averages").

Law of Large Numbers

As an experiment is repeated more and more times, the proportion of outcomes favorable to any particular event will tend to come closer and closer to the theoretical probability of that event.

EXAMPLE 6 Graphing a Sequence of Proportions

A fair coin was tossed 35 times, producing the following sequence of outcomes.

tthhh ttthh hthtt hhthh ttthh thttt hhthh

Calculate the ratio of heads to total tosses after the first toss, the second toss, and so on through all 35 tosses, and plot these ratios on a graph.

SOLUTION

After the first toss, we have 0 heads out of 1 toss, for a ratio of $\frac{0}{1} = .00$. After two tosses, we have $\frac{0}{2} = .00$. After three tosses, we have $\frac{1}{3} = .33$. Verify that the first six ratios are

$$.00, .00, .33, .50, .60, .50.$$

The thirty-five ratios are plotted as points in Figure 3. Notice that the fluctuations away from .50 become smaller as the number of tosses increases, and the ratios appear to approach .50 toward the right side of the graph, in keeping with the law of large numbers.

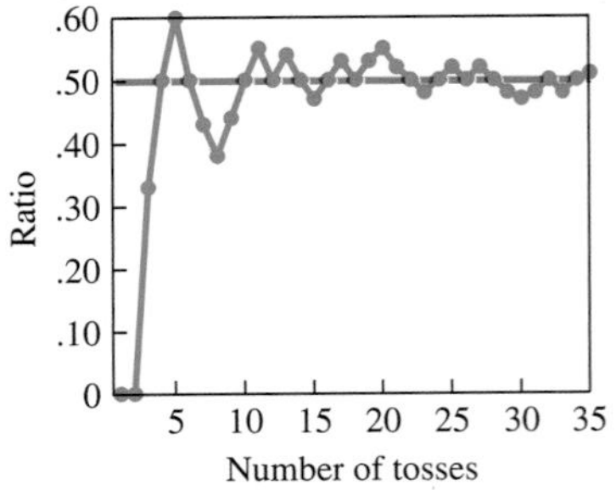

Ratio of heads to total tosses

FIGURE 3

Gregor Johann Mendel (1822–1884) came from a peasant family who managed to send him to school. By 1847 he had been ordained and was teaching at the Abbey of St. Thomas. He finished his education at the University of Vienna and returned to the abbey to teach mathematics and natural science. Mendel began to carry out experiments on plants in the abbey garden, notably pea plants, whose distinct traits (unit characters) he had puzzled over. In 1865 he published his results. (The Czech stamp above commemorates the centennial.) His work was not appreciated at the time even though he had laid the foundation of **classical genetics.** Mendel had established the basic laws of heredity: laws of unit characters, of dominance, and of segregation.

The law of large numbers provides an important connection between empirical and theoretical probabilities. Having obtained, by experiment, an empirical probability for an event, we can then, by inductive reasoning, *estimate* that event's theoretical probability. The more repetitions the estimate is based upon, the more reliable it is.

Likewise, if we know the theoretical probability of an event, we can then, by deductive reasoning, *predict* (estimate) the fraction of times the event will occur in a series of repeated experiments. The prediction should be more accurate for larger numbers of repetitions.

Probability in Genetics Probabilities, both empirical and theoretical, have been valuable tools in many areas of science. An important early example was the work of the Austrian monk Gregor Mendel, who used the idea of randomness to help establish the study of genetics. Mendel published his results in 1865, but they were largely ignored until 1900 when others rediscovered and recognized the importance of his work.

In an effort to understand the mechanism of character transmittal from one generation to the next in plants, Mendel counted the number of occurrences of various characteristics. For example, he found that the flower color in certain pea plants obeyed this scheme:

Pure red crossed with pure white produces red.

Mendel theorized that red is "dominant" (symbolized with the capital letter R), while white is "recessive" (symbolized with the lowercase letter r). The pure red parent carried only genes for red (R), and the pure white parent carried only genes for white (r). The offspring would receive one gene from each parent, hence one of the four combinations shown in the body of Table 2. Because every offspring receives one gene for red, that characteristic dominates and the offspring exhibits the color red.

TABLE 2 First to Second Generation

		Second Parent	
		r	**r**
First	**R**	Rr	Rr
Parent	**R**	Rr	Rr

Now each of these second-generation offspring, though exhibiting the color red, still carries one of each gene. So when two of them are crossed, each third-generation offspring will receive one of the gene combinations shown in Table 3. Mendel theorized that each of these four possibilities would be equally likely, and produced experimental counts that were close enough to support this hypothesis. (In more recent years, some have accused Mendel, or his assistants, of fudging the experimental data, but his conclusions have not been disputed.)

TABLE 3 Second to Third Generation

		Second Parent	
		R	**r**
First	**R**	RR	Rr
Parent	**r**	rR	rr

Smoking 1.4 cigarettes
Spending 1 hour in a coal mine
Living 2 days in New York or Boston
Eating 40 teaspoons of peanut butter
Living 2 months with a cigarette smoker
Flying 1000 miles in a jet
Traveling 300 miles in a car
Riding 10 miles on a bicycle

Risk is the probability that a harmful event will occur. Almost every action or substance exposes a person to some risk, and the assessment and reduction of risk accounts for a great deal of study and effort in our world. The list above, from *Calculated Risk*, by J. Rodricks, contains activities that carry an annual increased risk of death by one chance in a million.

EXAMPLE 7 Finding Probabilities of Flower Colors

Referring to Table 3 on the previous page, determine the probability that a third-generation offspring will exhibit each of the following flower colors. Base the probabilities on the following sample space of equally likely outcomes: $S = \{\text{RR, Rr, rR, rr}\}$.

(a) red **(b)** white

SOLUTION

(a) Since red dominates white, any combination with at least one gene for red (R) will result in red flowers. Since three of the four possibilities meet this criterion, $P(\text{red}) = \frac{3}{4}$.

(b) Only the combination rr has no gene for red, so $P(\text{white}) = \frac{1}{4}$.

Due to the probabilistic laws of genetics, vast improvements have resulted in food supply technology, through hybrid animal and crop development. Also, the understanding and control of human diseases have been advanced by related studies.

Odds

Whereas probability compares the number of favourable outcomes to the total number of outcomes, **odds** compare the number of favorable outcomes to the number of unfavorable outcomes. Odds are commonly quoted, rather than probabilities, in horse racing, lotteries, and most other gambling situations. And the odds quoted normally are odds "against" rather than odds "in favor."

Odds

If all outcomes in a sample space are equally likely, a of them are favorable to the event E, and the remaining b outcomes are unfavorable to E, then the **odds in favor** of E are a to b, and the **odds against** E are b to a.

EXAMPLE 8 Finding the Odds of Getting a Resort Job

Jennifer has been promised one of six summer jobs, three of which would be at a nearby beach resort. If she has equal chances for all six jobs, find the odds that she will land one at the resort.

SOLUTION

Since three possibilities are favorable and three are not, the odds of working at the resort are 3 to 3, or 1 to 1. (The common factor of 3 has been divided out.) Odds of 1 to 1 are often termed "even odds," or a "50–50 chance."

EXAMPLE 9 Finding the Odds of Winning a Raffle

Donn Demaree has purchased six tickets for an office raffle where the winner will receive an iPod unit. If 51 tickets were sold altogether and each has an equal chance of winning, what are the odds against Donn's winning the iPod?

SOLUTION

Donn has six chances to win and 45 chances to lose, so the odds against winning are 45 to 6, or 15 to 2. We simplify an odds ratio, just as a fraction.

EXAMPLE 10 Converting from Probability to Odds

Suppose the probability of rain tomorrow is .13. Give this information in terms of odds.

SOLUTION

We can say that

$$P(\text{rain}) = .13 = \frac{13}{100}.$$ ← Convert the decimal fraction to a quotient of integers.

This fraction does not mean there are necessarily 13 favorable outcomes and 100 possible outcomes, only that they would occur in this *ratio.* The corresponding number of *unfavorable* outcomes would be

$$100 - 13 = 87. \quad \text{(Total − favorable = unfavorable)}$$

So the odds are 13 to 87 in favor; we can say the odds are 87 to 13 against rain tomorrow.

EXAMPLE 11 Converting from Odds to Probability

In a certain sweepstakes, your odds of winning are 1 to 99,999,999. What is the probability you will win?

SOLUTION

The given odds provide a ratio of 1 favorable and 99,999,999 unfavorable outcomes, therefore 1 + 99,999,999 = 100,000,000 total outcomes. So

$$P(\text{win}) = \frac{\text{favorable}}{\text{total}} = \frac{1}{100{,}000{,}000} = .00000001.$$

12.1 EXERCISES

In each of Exercises 1–4, give the probability that the spinner shown would land on **(a)** *red,* **(b)** *yellow,* **(c)** *blue.*

1. 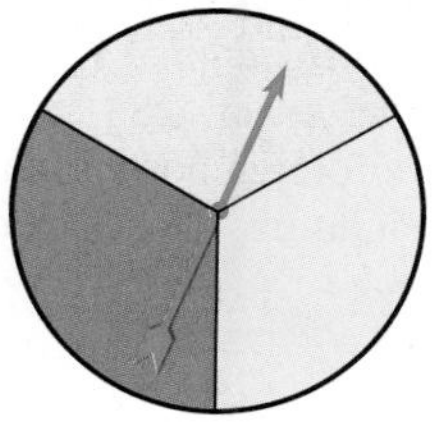

2. 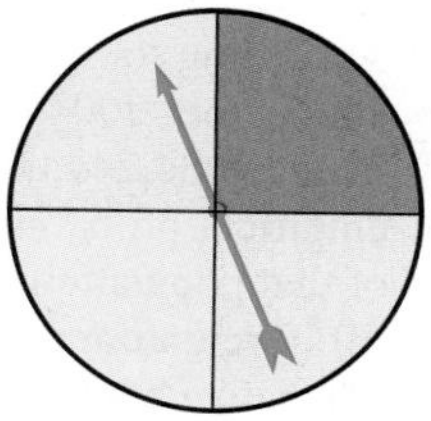

3. 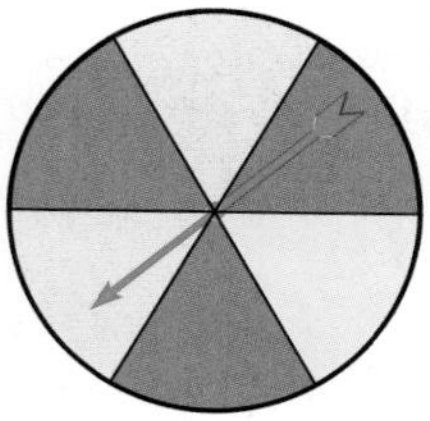

4. 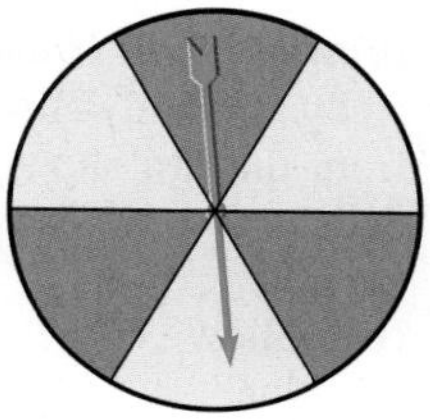

Solve each probability problem.

5. ***Using Spinners to Generate Numbers*** Suppose the spinner shown here is spun once, to determine a single-digit number, and we are interested in the event E that the resulting number is odd. Give each of the following.

(a) the sample space
(b) the number of favorable outcomes
(c) the number of unfavorable outcomes
(d) the total number of possible outcomes
(e) the probability of an odd number
(f) the odds in favor of an odd number

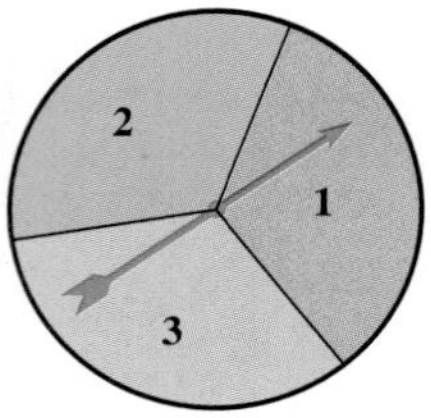

6. ***Lining Up Preschool Children*** Lynn Damme's group of preschool children includes eight girls and five boys. If Lynn randomly selects one child to be first in line, with E being the event that the one selected is a girl, give each of the following.

(a) the total number of possible outcomes
(b) the number of favorable outcomes
(c) the number of unfavorable outcomes
(d) the probability of event E
(e) the odds in favor of event E

7. ***Using Spinners to Generate Numbers*** The spinner of Exercise 5 is spun twice in succession to determine a two-digit number. Give each of the following.

(a) the sample space
(b) the probability of an odd number
(c) the probability of a number with repeated digits
(d) the probability of a number greater than 30
(e) the probability of a prime number

8. ***Probabilities in Coin Tossing*** Two fair coins are tossed (say a dime and a quarter). Give each of the following.

(a) the sample space
(b) the probability of heads on the dime
(c) the probability of heads on the quarter
(d) the probability of getting both heads
(e) the probability of getting the same outcome on both coins

9. ***Drawing Balls from an Urn*** Anne Kelly randomly chooses a single ball from the urn shown here. Find the odds in favor of each event.

(a) red
(b) yellow
(c) blue

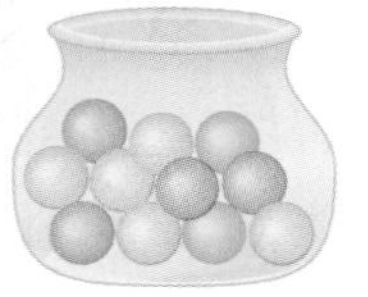

10. ***Random Selection of Club Officers*** Five people (Alan, Bill, Cathy, David, and Evelyn) form a club $N = \{A, B, C, D, E\}$. If they choose a president randomly, find the odds against each result.

(a) Cathy **(b)** a woman
(c) a person whose name begins with a consonant

11. ***Random Selection of Fifties Music*** "Jukebox" Joe has fifty hit singles from the fifties, including exactly one by Buddy Holly, two by The Drifters, three by Bobby Darin, four by The Coasters, and five by Fats Domino. If Joe randomly selects one hit from his collection of fifty, find the probability it will be by each of the following.

(a) Buddy Holly **(b)** The Drifters
(c) Bobby Darin **(d)** The Coasters
(e) Fats Domino

12. ***Probabilities in Coin Tossing*** Three fair coins are tossed.

(a) Write out the sample space.

Determine the probability of each event.

(b) no heads **(c)** exactly one head
(d) exactly two heads **(e)** three heads

13. ***Number Sums for Rolling Two Dice*** The sample space for the rolling of two fair dice appeared in Table 2 of Section 11.1. Reproduce that table, but replace each of the 36 equally likely ordered pairs with its corresponding sum (for the two dice). Then find the probability of rolling each sum.

(a) 2 **(b)** 3 **(c)** 4
(d) 5 **(e)** 6 **(f)** 7
(g) 8 **(h)** 9 **(i)** 10
(j) 11 **(k)** 12

In Exercises 14 and 15, compute answers to three decimal places.

14. ***Probability of Seed Germination*** In a hybrid corn research project, 200 seeds were planted, and 170 of them germinated. Find the empirical probability that any particular seed of this type will germinate.

15. ***Probabilities of Native- and Foreign-Born Persons*** According to the *Encyclopedia Britannica Almanac 2006*, the 2004 U.S. population of 288,280,000 included 34,244,000 who were foreign born. Find the empirical probability that a randomly chosen person of that population would be

(a) foreign born
(b) native born.

16. ***Probabilities of Two Daughters Among Four Children*** In Example 3, what would be Michelle's probability of having exactly two daughters if she were to have four children altogether? (You may want to use a tree diagram to construct the sample space.)

17. Explain the difference between theoretical and empirical probability.

18. Explain the difference between probability and odds.

Genetics in Snapdragons *Mendel found no dominance in snapdragons (in contrast to peas) with respect to red and white flower color. When pure red and pure white parents are crossed (see Table 2), the resulting* Rr *combination (one of each gene) produces second-generation offspring with* pink *flowers. These second-generation pinks, however, still carry one red and one white gene, so when they are crossed the third generation is still governed by Table 3.*

Find each probability for third-generation snapdragons.

19. $P(\text{red})$ 20. $P(\text{pink})$ 21. $P(\text{white})$

Genetics in Pea Plants *Mendel also investigated various characteristics besides flower color. For example, round peas are dominant over recessive wrinkled peas. First, second, and third generations can again be analyzed using Tables 2 and 3, where* R *represents round and* r *represents wrinkled.*

22. Explain why crossing pure round and pure wrinkled first-generation parents will always produce round peas in the second-generation offspring.

23. When second-generation round pea plants (each of which carries both R and r genes) are crossed, find the probability that a third-generation offspring will have **(a)** round peas **(b)** wrinkled peas.

Genetics of Cystic Fibrosis **Cystic fibrosis** *is one of the most common inherited diseases in North America (including the United States), occurring in about 1 of every 2000 Caucasian births and about 1 of every 250,000 non-Caucasian births. Even with modern treatment, victims usually die from lung damage by their early twenties.*

If we denote a cystic fibrosis gene with a c *and a disease-free gene with a* C *(since the disease is recessive), then only a* cc *person will actually have the disease. Such persons would ordinarily die before parenting children, but a child can also inherit the disease from two* Cc *parents (who themselves are healthy, that is, have no symptoms but are "carriers" of the disease). This is like a pea plant inheriting white flowers from two red-flowered parents which both carry genes for white.*

24. Find the empirical probability (to four decimal places) that cystic fibrosis will occur in a randomly selected infant birth among U.S. Caucasians.

25. Find the empirical probability (to six decimal places) that cystic fibrosis will occur in a randomly selected infant birth among U.S. non-Caucasians.

26. Among 150,000 North American Caucasian births, about how many occurrences of cystic fibrosis would you expect?

Suppose that both partners in a marriage are cystic fibrosis carriers (a rare occurrence). Construct a chart similar to Table 3 and determine the probability of each of the following events.

27. Their first child will have the disease.

28. Their first child will be a carrier.

29. Their first child will neither have nor carry the disease.

Suppose a child is born to one cystic fibrosis carrier parent and one non-carrier parent. Find the probability of each of the following events.

30. The child will have cystic fibrosis.

31. The child will be a healthy cystic fibrosis carrier.

32. The child will neither have nor carry the disease.

Genetics of Sickle-Cell Anemia **Sickle-cell anemia** *occurs in about 1 of every 500 black baby births and about 1 of every 160,000 non-black baby births. It is ordinarily fatal in early childhood. There is a test to identify carriers. Unlike cystic fibrosis, which is recessive, sickle-cell anemia is* **codominant.** *This means that inheriting two sickle-cell genes causes the disease, while inheriting just one sickle-cell gene causes a mild (non-fatal) version (which is called* **sickle-cell trait***). This is similar to a snapdragon plant manifesting pink flowers by inheriting one red gene and one white gene.*

In Exercises 33 and 34, find the empirical probabilities of the given events.

33. A randomly selected black baby will have sickle-cell anemia. (Give your answer to three decimal places.)

34. A randomly selected non-black baby will have sickle-cell anemia. (Give your answer to six decimal places.)

35. Among 80,000 births of black babies, about how many occurrences of sickle-cell anemia would you expect?

Find the theoretical probability of each condition in a child both of whose parents have sickle-cell trait.

36. The child will have sickle-cell anemia.

37. The child will have sickle-cell trait.

38. The child will be healthy.

39. ***Women's 100-Meter Run*** In the history of track and field, no woman has broken the 10-second barrier in the 100-meter run.

(a) From the statement above, find the empirical probability that a woman runner will break the 10-second barrier next year.
(b) Can you find the theoretical probability for the event of part (a)?
(c) Is it possible that the event of part (a) will occur?

40. Is there any way a coin could fail to be "fair"? Explain.

41. On page 27 of their book *Descartes' Dream,* Philip Davis and Reuben Hersh ask the question, "Is probability real or is it just a cover-up for ignorance?" What do you think? Are some things truly random, or is everything potentially deterministic?

42. If $P(E) = .37$, find
(a) the odds in favor of E,
(b) the odds against E.

43. If the odds in favor of event E are 12 to 19, find $P(E)$.

44. If the odds against event E are 10 to 3, find $P(E)$.

Probabilities of Poker Hands *In 5-card poker, find the probability of being dealt each of the following. Give each answer to eight decimal places. (Refer to Table 1.)*

45. a straight flush

46. two pairs

47. four of a kind

48. four queens

49. a hearts flush (*not* a royal flush or a straight flush)

50. ***Probabilities in Dart Throwing*** If a dart hits the square target shown here at random, what is the probability that it will hit in a colored region? (*Hint:* Compare the area of the colored regions to the total area of the target.)

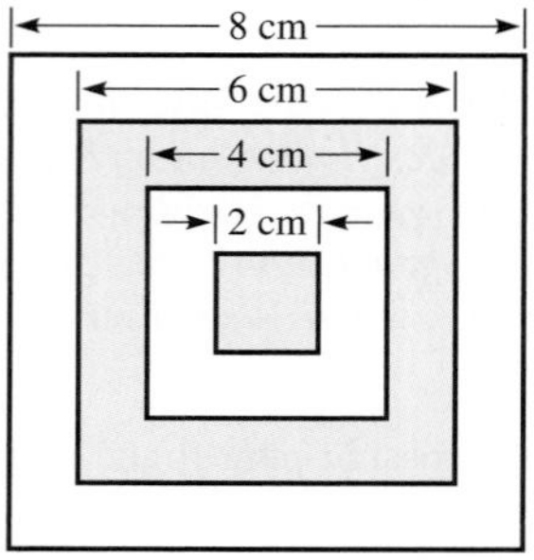

51. ***Probabilities in Olympic Curling*** In the Olympic event of curling, the scoring area (shown here) consists of four concentric circles on the ice with radii of 6 inches, 2 feet, 4 feet, and 6 feet.

If a team member lands a (43-pound) stone *randomly* within the scoring area, find the probability that it ends up centered on
(a) red **(b)** white **(c)** blue.

52. ***Drawing Cards*** When drawing cards without replacement from a standard 52-card deck, find the maximum number of cards you could possibly draw and still get
   (a) fewer than three black cards,
   (b) fewer than six spades,
   (c) fewer than four face cards,
   (d) fewer than two kings.

The remaining exercises require careful thought to determine n(E) and n(S). (In some cases, you may want to employ counting methods from Chapter 11, such as the fundamental counting principle, permutations, or combinations.)

Probabilities of Seating Arrangements *Three married couples arrange themselves randomly in six consecutive seats in a row. Find the probability of each event in Exercises 53–56. (Hint: In each case the denominator of the probability fraction will be* $6! = 720$*, the total number of ways to arrange six items.)*

53. Each man will sit immediately to the left of his wife.

54. Each man will sit immediately to the left of a woman.

55. The women will be in three adjacent seats.

56. The women will be in three adjacent seats, as will the men.

57. ***Selecting Slopes*** If two distinct numbers are chosen randomly from the set $\{-2, -\frac{4}{3}, -\frac{1}{2}, 0, \frac{1}{2}, \frac{3}{4}, 3\}$, find the probability that they will be the slopes of two perpendicular lines.

58. ***Racing Bets*** At most horse-racing tracks, the "trifecta" is a particular race where you win if you correctly pick the "win," "place," and "show" horses (the first-, second-, and third-place winners), in their proper order. If five horses of equal ability are entered in today's trifecta race, and you select an entry, what is the probability that you will be a winner?

59. ***Probabilities of Student Course Schedules*** Suppose you plan to take three courses next term. If you select them randomly from a listing of twelve courses, five of which are science courses, what is the probability that all three courses you select will be science courses?

60. ***Selecting Symphony Performances*** Cheryl Chechvala randomly selects three symphony performances to attend this season, choosing from a schedule of ten performances, three of which will feature works by Beethoven. Find the probability that Cheryl will select all of the Beethoven programs.

Selecting Class Reports *Assuming that Ben, Jill, and Pam are three of the 36 members of the class, and that three of the class members will be chosen randomly to deliver their reports during the next class meeting, find the probability (to six decimal places) of each event.*

61. Ben, Jill, and Pam are selected, in that order.

62. Ben, Jill, and Pam are selected, in any order.

63. ***Random Selection of Prime Numbers*** If two distinct prime numbers are randomly selected from among the first eight prime numbers, what is the probability that their sum will be 24?

64. ***Building Numbers from Sets of Digits*** The digits 1, 2, 3, 4, and 5 are randomly arranged to form a five-digit number. Find the probability of each of the following events.
   (a) The number is even.
   (b) The first and last digits of the number both are even.

65. ***Random Sums*** Two integers are randomly selected from the set $\{1, 2, 3, 4, 5, 6, 7, 8, 9\}$ and are added together. Find the probability that their sum is 11 if they are selected
   (a) with replacement
   (b) without replacement.

Finding Palindromic Numbers *Numbers that are* **palindromes** *read the same forward and backward. For example, 30203 is a five-digit palindrome. If a single number is chosen randomly from each of the following sets, find the probability that it will be palindromic.*

66. the set of all two-digit numbers

67. the set of all three-digit numbers

12.2 Events Involving "Not" and "Or"

Properties of Probability • Events Involving "Not" • Events Involving "Or"

Pierre Simon de Laplace (1749–1827) began in 1773 to solve the problem of why Jupiter's orbit seems to shrink and Saturn's orbit seems to expand. Eventually Laplace worked out a complete theory of the solar system. *Celestial Mechanics* resulted from almost a lifetime of work. In five volumes, it was published between 1799 and 1825 and gained for Laplace the reputation "Newton of France."

Laplace's work on probability was actually an adjunct to his celestial mechanics. He needed to demonstrate that probability is useful in interpreting scientific data. He also wrote a popular exposition of the system, which contains (in a footnote!) the "nebular hypothesis" that the sun and planets originated together in a cloud of matter, which then cooled and condensed into separate bodies.

Properties of Probability

Remember from the previous section that an empirical probability, based upon experimental observation, may be the best value available but still is only an approximation to the ("true") theoretical probability. For example, no human has ever been known to jump higher than 8.5 feet vertically, so the empirical probability of such an event is zero. However, observing the rate at which high jump records have been broken, we suspect that the event is, in fact, possible and may one day occur. Hence it must have some nonzero theoretical probability, even though we have no way of assessing its exact value.

Recall also that the theoretical probability formula,

$$P(E) = \frac{n(E)}{n(S)},$$

is valid only when all outcomes in the sample space S are equally likely. For the experiment of tossing two fair coins, we can write $S = \{\text{hh, ht, th, tt}\}$ and compute correctly that

$$P(\text{both heads}) = \frac{1}{4},$$

whereas if we define the sample space with non-equally likely outcomes as $S = \{\text{both heads, both tails, one of each}\}$, we are led to

$$P(\text{both heads}) = \frac{1}{3}, \qquad \text{which is } \textit{wrong}.$$

(To convince yourself that $\frac{1}{4}$ is a better value than $\frac{1}{3}$, toss two fair coins 100 times or so to see what the empirical fraction seems to approach.)

Since any event E is a subset of the sample space S, we know that $0 \le n(E) \le n(S)$. Dividing all members of this inequality by $n(S)$ gives

$$\frac{0}{n(S)} \le \frac{n(E)}{n(S)} \le \frac{n(S)}{n(S)}, \quad \text{or} \quad \mathbf{0 \le P(E) \le 1.}$$

In words, the probability of any event is a number from 0 through 1, inclusive.

If event E is *impossible* (cannot happen), then $n(E)$ must be 0 (E is the empty set), so $P(E) = 0$. If event E is *certain* (cannot help but happen), then $n(E) = n(S)$, so $P(E) = \frac{n(E)}{n(S)} = \frac{n(S)}{n(S)} = 1$. These properties are summarized below.

Properties of Probability

Let E be an event from the sample space S. That is, E is a subset of S. Then the following properties hold.

1. $\mathbf{0 \le P(E) \le 1}$ (The probability of an event is a number from 0 through 1, inclusive.)
2. $\mathbf{P(\emptyset) = 0}$ (The probability of an impossible event is 0.)
3. $\mathbf{P(S) = 1}$ (The probability of a certain event is 1.)

EXAMPLE 1 Finding Probability When Rolling a Die

When a single fair die is rolled, find the probability of each event.

(a) the number 2 is rolled **(b)** a number other than 2 is rolled
(c) the number 7 is rolled **(d)** a number less than 7 is rolled

SOLUTION

(a) Since one of the six possibilities is a 2, $P(2) = \frac{1}{6}$.
(b) There are five such numbers, 1, 3, 4, 5, and 6, so $P(\text{a number other than 2}) = \frac{5}{6}$.
(c) None of the possible outcomes is 7. Thus, $P(7) = \frac{0}{6} = 0$.
(d) Since all six of the possible outcomes are less than 7, $P(\text{a number less than } 7) = \frac{6}{6} = 1$.

Notice that no probability in Example 1 was less than 0 or greater than 1, which illustrates probability property 1. The "impossible" event of part (c) had probability 0, illustrating property 2. And the "certain" event of part (d) had probability 1, illustrating property 3.

Events Involving "Not"

Table 4 repeats the information of Table 8 of Section 11.5, with a third correspondence added in row 3. These correspondences are the basis for the probability rules developed in this section and the next. For example, the probability of an event *not* happening involves the *complement* and *subtraction*, according to row 1 of the table.

TABLE 4 Set Theory/Logic/Arithmetic Correspondences

	Set Theory	Logic	Arithmetic
1. Operation or Connective (Symbol)	Complement (′)	Not (~)	Subtraction (−)
2. Operation or Connective (Symbol)	Union (∪)	Or (∨)	Addition (+)
3. Operation or Connective (Symbol)	Intersection (∩)	And (∧)	Multiplication (·)

The rule for the probability of a complement is stated as follows. It is illustrated in Figure 4.

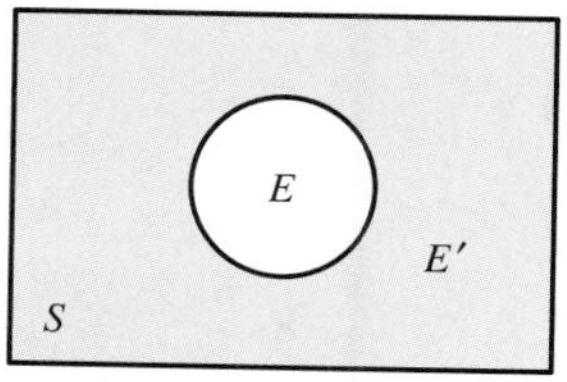

The logical connective "not" corresponds to "complement" in set theory.

$P(\text{not } E) = P(S) - P(E)$
$= 1 - P(E)$

FIGURE 4

Probability of a Complement (for Not *E*)

The probability that an event *E* will *not* occur is equal to one minus the probability that it *will* occur.

$$P(\text{not } E) = 1 - P(E)$$

Notice that the events of Examples 1(a) and (b), namely "2" and "not 2," are complements of one another, and that their probabilities add up to 1. This illustrates the above probability rule. The equation

$$P(E) + P(E') = 1$$

is a rearrangement of the formula for the probability of a complement. Another form of the equation, also useful at times, is

$$P(E) = 1 - P(E').$$

EXAMPLE 2 Finding the Probability of a Complement

When a single card is drawn from a standard 52-card deck, what is the probability that it will not be a king?

SOLUTION

$$P(\text{not a king}) = 1 - P(\text{king}) = 1 - \frac{4}{52} = \frac{48}{52} = \frac{12}{13}. \leftarrow$$ Remember to write in lowest terms. ■

Mary Somerville (1780–1872) is associated with Laplace because of her brilliant exposition of his *Celestial Mechanics.* She combined a deep understanding of science with the ability to communicate its concepts to the general public.

Somerville studied Euclid thoroughly and perfected her Latin so she could read Newton's *Principia.* In about 1816 she went to London and soon became part of its literary and scientific circles. She also corresponded with Laplace and other Continental scientists.

Somerville's book on Laplace's theories came out in 1831 with great acclaim. Then followed a panoramic book, *Connection of the Physical Sciences* (1834). A statement in one of its editions suggested that irregularities in the orbit of Uranus might indicate that a more remote planet, not yet seen, existed. This caught the eye of the scientists who worked out the calculations for Neptune's orbit.

EXAMPLE 3 Finding the Probability of a Complement

If five fair coins are tossed, find the probability of obtaining at least one head.

SOLUTION

There are $2^5 = 32$ possible outcomes for the experiment of tossing five fair coins. Most include at least one head. In fact, only the outcome ttttt does not include at least one head. If E denotes the event "at least one head," then E' is the event "not at least one head," and

$$P(E) = 1 - P(E') = 1 - \frac{1}{32} = \frac{31}{32}.$$ ■

Events Involving "Or" Examples 2 and 3 showed how the probability of an event can be approached indirectly, by first considering the complement of the event. Another indirect approach is to break the event into simpler component events. Row 2 of Table 4 indicates that the probability of one event *or* another should involve the *union* and *addition.*

EXAMPLE 4 Selecting From a Set of Numbers

If one number is selected randomly from the set $\{1, 2, 3, 4, 5, 6, 7, 8, 9, 10\}$, find the probability that it will be

(a) odd or a multiple of 4
(b) odd or a multiple of 3.

SOLUTION

Define the following events:

$S = \{1, 2, 3, 4, 5, 6, 7, 8, 9, 10\}$ Sample space
$A = \{1, 3, 5, 7, 9\}$ Odd outcomes
$B = \{4, 8\}$ Multiples of 4
$C = \{3, 6, 9\}$ Multiples of 3

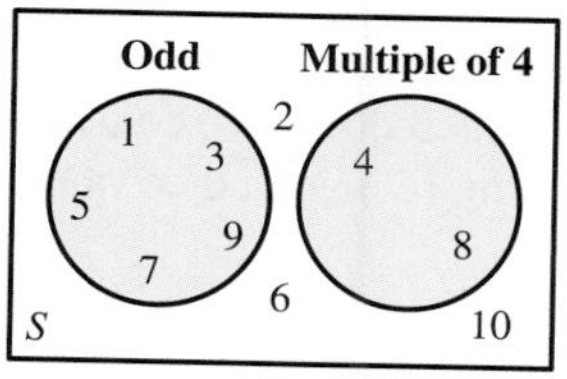

FIGURE 5

(a) Figure 5 shows the positioning of the 10 integers within the sample space and within the pertinent sets A and B. The composite event "A or B" corresponds to the set $A \cup B = \{1, 3, 4, 5, 7, 8, 9\}$. By the theoretical probability formula,

Of 10 total outcomes, 7 are favorable.

$$P(A \text{ or } B) = \frac{7}{10}.$$

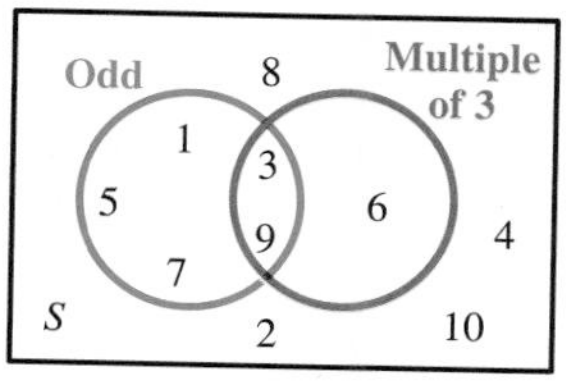

FIGURE 6

(b) Figure 6 shows the situation. We see that

Of 10 total outcomes, 6 are favorable.

$$P(A \text{ or } C) = \frac{6}{10} = \frac{3}{5}.$$

Would an addition formula have worked in Example 4? Let's check. In part (a),

$$P(A \text{ or } B) = P(A) + P(B) = \frac{5}{10} + \frac{2}{10} = \frac{7}{10}, \quad \text{Correct}$$

which is correct. In part (b),

$$P(A \text{ or } C) = P(A) + P(C) = \frac{5}{10} + \frac{3}{10} = \frac{8}{10} = \frac{4}{5}, \quad \text{Incorrect}$$

which is incorrect. The trouble in part (b) is that A and C are not disjoint sets. They have outcomes in common. Just as with the additive counting principle in Chapter 11, an adjustment must be made here to compensate for counting the common outcomes twice. The correct calculation is

$$P(A \text{ or } C) = P(A) + P(C) - P(A \text{ and } C)$$
$$= \frac{5}{10} + \frac{3}{10} - \frac{2}{10} = \frac{6}{10} = \frac{3}{5}. \quad \text{Correct}$$

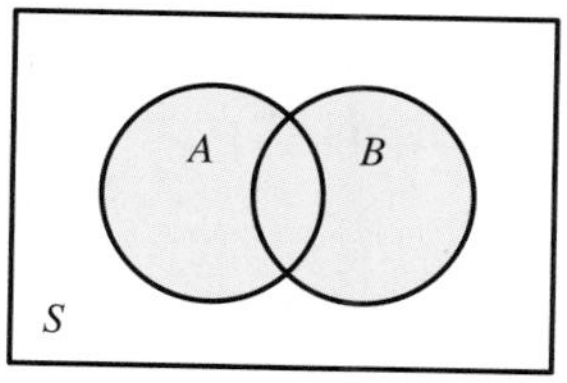

The logical connective "or" corresponds to "union" in set theory.

$P(A \text{ or } B)$
$= P(A) + P(B) - P(A \text{ and } B)$

FIGURE 7

In probability theory, events that are disjoint sets are called *mutually exclusive events,* which are defined as follows.

Mutually Exclusive Events

Two events A and B are **mutually exclusive events** if they have no outcomes in common. (Mutually exclusive events cannot occur simultaneously.)

The results observed in Example 4 are generalized as follows. The two possibilities are illustrated in Figures 7 and 8.

Addition Rule of Probability (for *A* or *B*)

If A and B are any two events, then

$$\mathbf{P(A \text{ or } B) = P(A) + P(B) - P(A \text{ and } B).}$$

If *A* and *B* are mutually exclusive, then

$$\mathbf{P(A \text{ or } B) = P(A) + P(B).}$$

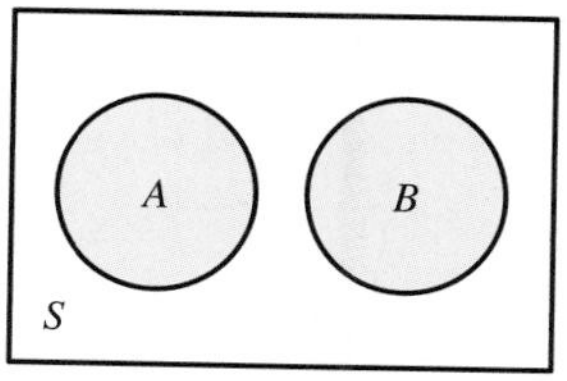

When A and B are mutually exclusive,

$P(A \text{ or } B) = P(A) + P(B)$.

FIGURE 8

Actually, the first formula in the addition rule applies in all cases. (The third term on the right drops out when A and B are mutually exclusive, because $P(A \text{ and } B) = 0$).

Still it is good to remember the second formula in the preceding box for the many cases where the component events are mutually exclusive. In this section, we consider only cases where the event "A and B" is simple. We deal with more involved composites involving "and" in the next section.

EXAMPLE 5 Finding the Probability of an Event Involving "Or"

If a single card is drawn from a standard 52-card deck, what is the probability that it will be a spade or a red card?

SOLUTION

First note that "spade" and "red" cannot both occur, because there are no red spades. (All spades are black.) Therefore, we can use the formula for mutually exclusive events. With 13 spades and 26 red cards in the deck, we obtain

$$P(\text{spade or red}) = P(\text{spade}) + P(\text{red}) = \frac{13}{52} + \frac{26}{52} = \frac{39}{52} = \frac{3}{4}.$$

We often need to consider composites of more than two events. When each event involved is mutually exclusive of all the others, we extend the addition rule to the appropriate number of components.

EXAMPLE 6 Treating Unions of Several Components

Jacob will spend from 1 to 6 hours on his homework. If x represents the number of hours to be spent, then the probabilities of the various values of x, rounded to the nearest hour, are shown in Table 5. Find the probabilities that he will spend

(a) fewer than 3 hours
(b) more than 2 hours
(c) more than 1 but no more than 5 hours
(d) fewer than 5 hours.

SOLUTION

Because the time periods in Table 5 are mutually exclusive of one another, we can simply add the appropriate component probabilities.

(a)
$$\begin{aligned} P(\text{fewer than } 3) &= P(1 \text{ or } 2) && \text{Fewer than 3 means 1 or 2.}\\ &= P(1) + P(2) && \text{Addition rule}\\ &= .05 + .10 && \text{Substitute values from Table 5.}\\ &= .15 \end{aligned}$$

(b)
$$\begin{aligned} P(\text{more than } 2) &= P(3 \text{ or } 4 \text{ or } 5 \text{ or } 6) && \text{More than 2 means 3, 4, 5, or 6.}\\ &= P(3) + P(4) + P(5) + P(6) && \text{Addition rule}\\ &= .20 + .40 + .10 + .15 && \text{Substitute values from Table 5.}\\ &= .85 \end{aligned}$$

(c)
$$\begin{aligned} &P(\text{more than 1 but no more than 5})\\ &= P(2 \text{ or } 3 \text{ or } 4 \text{ or } 5) && \text{2, 3, 4, and 5 are more than 1 and no more than 5.}\\ &= P(2) + P(3) + P(4) + P(5) && \text{Addition rule}\\ &= .10 + .20 + .40 + .10 && \text{Substitute values from Table 5.}\\ &= .80 \end{aligned}$$

TABLE 5

x	$P(x)$
1	.05
2	.10
3	.20
4	.40
5	.10
6	.15

(d) Although we could take a direct approach here, as in parts (a), (b), and (c), we will combine the complement rule with the addition rule.

$P(\text{fewer than } 5)$	$= 1 - P(\text{not fewer than } 5)$	Complement rule
	$= 1 - P(5 \text{ or more})$	5 or more is equivalent to not fewer than 5.
	$= 1 - P(5 \text{ or } 6)$	5 or more means 5 or 6.
	$= 1 - [P(5) + P(6)]$	Addition rule
	$= 1 - (.10 + .15)$	Substitute values from Table 5.
	$= 1 - .25$	Add inside the parentheses first.
	$= .75$	

Table 5, in Example 6, lists all possible time intervals so the corresponding probabilities add up to 1, a necessary condition for the way part (d) was done. The time spent on homework here is an example of a **random variable.** (It is "random" since we cannot predict which of its possible values will occur.) A listing like Table 5, which shows all possible values of a random variable, along with the probabilities that those values will occur, is called a **probability distribution** for that random variable. Since *all* possible values are listed, they make up the entire sample space, and so the listed probabilities must add up to 1 (by probability property 3). Probability distributions will occur in Exercises 32 and 33 of this section and will be discussed further in later sections.

EXAMPLE 7 Finding the Probability of an Event Involving "Or"

Find the probability that a single card drawn from a standard 52-card deck will be a diamond or a face card.

SOLUTION

In this case, the component events "diamond" and "face card" can both occur. (The jack, queen, and king of diamonds belong to both.) So, we must use the first formula of the addition rule. If D denotes "diamond" and F denotes "face card," we obtain

$P(D \text{ or } F) = P(D) + P(F) - P(D \text{ and } F)$	Addition rule
$= \frac{13}{52} + \frac{12}{52} - \frac{3}{52}$	There are 13 diamonds, 12 face cards, and 3 that are both.
$= \frac{22}{52}$	Add and subtract.
$= \frac{11}{26}.$	Write in lowest terms.

EXAMPLE 8 Finding the Probability of an Event Involving "Or"

Of 20 elective courses to be offered this term, Juanita plans to enroll in one, which she will choose by throwing a dart at the schedule of courses. If 8 of the courses are recreational, 9 are interesting, and 3 are both recreational and interesting, find the probability that the course she chooses will have at least one of these two attributes.

SOLUTION

If R denotes "recreational" and I denotes "interesting," then $P(R) = \frac{8}{20}$, $P(I) = \frac{9}{20}$, and $P(R \text{ and } I) = \frac{3}{20}$. Because R and I are not mutually exclusive, we use the formula for that case.

$$P(R \text{ or } I) = \frac{8}{20} + \frac{9}{20} - \frac{3}{20} = \frac{14}{20} = \frac{7}{10}.$$

Remember to write in lowest terms.

12.2 EXERCISES

1. ***Determining Whether Two Events Are Mutually Exclusive*** Julie Davis has three office assistants. If A is the event that at least two of them are men and B is the event that at least two of them are women, are A and B mutually exclusive?

2. ***Attending Different Colleges*** Nancy Hart earned her college degree several years ago. Consider the following four events.

   Her alma mater is in the East.
   Her alma mater is a private college.
   Her alma mater is in the Northwest.
   Her alma mater is in the South.

   Are these events all mutually exclusive of one another?

3. Explain the difference between the two formulas in the addition rule of probability on page 739, illustrating each one with an appropriate example.

Probabilities for Rolling a Die *For the experiment of rolling a single fair die, find the probability of each event.*

4. not less than 2
5. not prime
6. odd or less than 5
7. even or prime
8. odd or even
9. less than 3 or greater than 4

Probability and Odds for Drawing a Card *For the experiment of drawing a single card from a standard 52-card deck, find* **(a)** *the probability, and* **(b)** *the odds in favor, of each event.*

10. not an ace
11. king or queen
12. club or heart
13. spade or face card
14. not a heart, or a 7
15. neither a heart nor a 7

Number Sums for Rolling a Pair of Dice *For the experiment of rolling an ordinary pair of dice, find the probability that the sum will be each of the following. (You may want to use a table showing the sum for each of the 36 equally likely outcomes.)*

16. 11 or 12
17. even or a multiple of 3
18. odd or greater than 9
19. less than 3 or greater than 9
20. Find the probability of getting a prime number in each case.
    (a) A number is chosen randomly from the set $\{1, 2, 3, 4, \ldots, 12\}$.
    (b) Two dice are rolled and the sum is observed.
21. Suppose, for a given experiment, A, B, C, and D are events, all mutually exclusive of one another, such that $A \cup B \cup C \cup D = S$ (the sample space). By extending the addition rule of probability on page 739 to this case, and utilizing probability property 3, what statement can you make?

Probabilities of Poker Hands *If you are dealt a 5-card hand (this implies without replacement) from a standard 52-card deck, find the probability of getting each of the following. Refer to Table 1 of Section 12.1, and give answers to six decimal places.*

22. a flush or three of a kind
23. a full house or a straight
24. a black flush or two pairs
25. nothing any better than two pairs

Probabilities in Golf Scoring *The table gives golfer Amy Donlin's probabilities of scoring in various ranges on a par-70 course. In a given round, find the probability of each event in Exercises 26–30.*

x	$P(x)$
Below 60	.04
60–64	.06
65–69	.14
70–74	.30
75–79	.23
80–84	.09
85–89	.06
90–94	.04
95–99	.03
100 or above	.01

26. 95 or higher

27. par or above

28. in the 80s

29. less than 90

30. not in the 70s, 80s, or 90s

31. What are the odds of Amy's shooting below par?

32. ***Drawing Balls from an Urn*** Anne Kelly randomly chooses a single ball from the urn shown here, and x represents the color of the ball chosen. Construct a complete probability distribution for the random variable x.

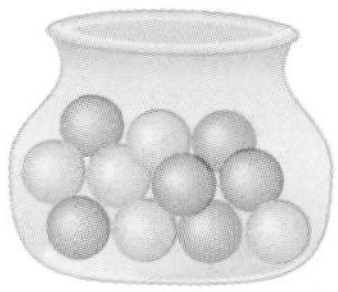

33. Let x denote the sum of two distinct numbers selected randomly from the set $\{1, 2, 3, 4, 5\}$. Construct the probability distribution for the random variable x.

34. ***Comparing Empirical and Theoretical Probabilities for Rolling Dice*** Roll a pair of dice 50 times, keeping track of the number of times the sum is "less than 3 or greater than 9" (that is 2, 10, 11, or 12).

(a) From your results, calculate an empirical probability for the event "less than 3 or greater than 9."

(b) By how much does your answer differ from the *theoretical* probability of Exercise 19?

For Exercises 35–38, let A be an event within the sample space S, and let $n(A) = a$ *and* $n(S) = s$.

35. Use the complements principle of counting to find an expression for $n(A')$.

36. Use the theoretical probability formula to express $P(A)$ and $P(A')$.

37. Evaluate, and simplify, $P(A) + P(A')$.

38. What rule have you proved?

The remaining exercises require careful thought for the determination of n(E) and n(S). (In some cases, you may want to employ counting methods from Chapter 11, such as the fundamental counting principle, permutations, or combinations.)

Building Numbers from Sets of Digits *Suppose we want to form three-digit numbers using the set of digits* $\{0, 1, 2, 3, 4, 5\}$. *For example,* 501 *and* 224 *are such numbers but* 035 *is not.*

39. How many such numbers are possible?

40. How many of these numbers are multiples of 5?

41. If one three-digit number is chosen at random from all those that can be made from the above set of digits, find the probability that the one chosen is not a multiple of 5.

42. ***Multiplying Numbers Generated by Spinners*** An experiment consists of spinning both spinners shown here and multiplying the resulting numbers together. Find the probability that the resulting product will be even.

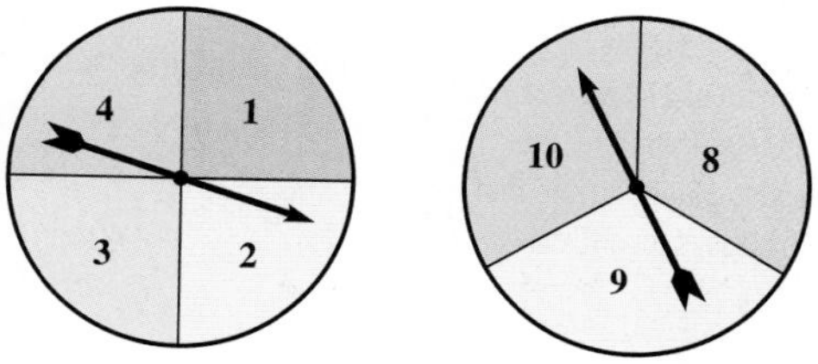

43. ***Drawing Colored Marbles from Boxes*** A bag contains fifty blue and fifty green marbles. Two marbles at a time are randomly selected. If both are green, they are placed in box A; if both are blue, in box B; if one is green and the other is blue, in box C. After all marbles are drawn, what is the probability that the numbers of marbles in box A and box B are the same? (This problem is borrowed with permission from the December 1992 issue of *Mathematics Teacher,* page 736.)

12.3 Conditional Probability; Events Involving "And"

Conditional Probability • Events Involving "And"

Even **a rare occurrence** can sometimes cause widespread controversy. When Mattel Toys marketed a new talking Barbie doll a few years ago, some of the Barbies were programmed to say "Math class is tough." The National Council of Teachers of Mathematics (NCTM), the American Association of University Women (AAUW), and numerous consumers voiced complaints about the damage such a message could do to the self-confidence of children and to their attitudes toward school and mathematics. Mattel subsequently agreed to erase the phrase from the microchip to be used in future doll production.

Incidentally, each Barbie was programmed to say four different statements, randomly selected from a pool of 270 prerecorded statements. Therefore, the probability of getting one that said "Math class is tough" was only

$$\frac{1 \cdot {}_{269}C_3}{{}_{270}C_4} \approx .015.$$

Other messages included in the pool were "I love school, don't you?," "I'm studying to be a doctor," and "Let's study for the quiz."

Conditional Probability

Sometimes the probability of an event must be computed using the knowledge that some other event has happened (or is happening, or will happen—the timing is not important). This type of probability is called *conditional probability.*

Conditional Probability

The probability of event B, computed on the assumption that event A has happened, is called the **conditional probability of B, given A,** and is denoted **$P(B \mid A)$.**

EXAMPLE 1 Selecting from a Set of Numbers

From the sample space $S = \{1, 2, 3, 4, 5, 6, 7, 8, 9, 10\}$, a single number is to be selected randomly. Given the events

A: The selected number is odd, and B: The selected number is a multiple of 3,

find each probability.

(a) $P(B)$ **(b)** $P(A \text{ and } B)$ **(c)** $P(B \mid A)$

SOLUTION

(a) $B = \{3, 6, 9\}$, so $P(B) = \frac{n(B)}{n(S)} = \frac{3}{10}$.

(b) A and B is the set $A \cap B = \{1, 3, 5, 7, 9\} \cap \{3, 6, 9\} = \{3, 9\}$, so

$$P(A \text{ and } B) = \frac{n(A \cap B)}{n(S)} = \frac{2}{10} = \frac{1}{5}.$$

(c) The given condition, that A occurs, effectively reduces the sample space from S to A, and the elements of the new sample space A, that are also in B, are the elements of $A \cap B$. Therefore,

$$P(B \mid A) = \frac{n(A \cap B)}{n(A)} = \frac{2}{5}.$$

Example 1 illustrates some important points. First, because

$$\frac{n(A \cap B)}{n(A)} = \frac{\frac{n(A \cap B)}{n(S)}}{\frac{n(A)}{n(S)}} \qquad \text{Multiply numerator and denominator by } \tfrac{1}{n(S)}.$$

$$= \frac{P(A \cap B)}{P(A)}, \qquad \text{Theoretical probability formula}$$

the final line of the example gives the following convenient formula for computing conditional probability, which works in all cases.

Natural disasters, such as tornadoes and cyclones, earthquakes, tsunamis, volcanic eruptions, firestorms, and floods, can kill thousands of people. But a **cosmic impact,** the collision of a meteor, comet, or asteroid with Earth, could be at least as catastrophic as full-scale nuclear war, killing a billion or more people. Reported at the Web site www.impact.arc.nasa.gov is that a large enough object (1 kilometer or more in diameter) could even put the human species at risk of annihilation by causing drastic climate changes and destroying food crops. By the end of 2004, the Spaceguard Survey had already discovered more than half of the near-Earth asteroids (NEAs) in this size range and had plans to locate 90% of them by the end of 2008.

Although the risk of finding one on a collision course with the Earth is slight, it is anticipated that, if we did, we would be able to deflect it before impact.

The photo above shows damage caused by a cosmic fragment that disintegrated in the atmosphere over Tunguska, Siberia, in 1908, with an explosive energy of more than 10 megatons of TNT. Nearly 1000 square miles of uninhabited forest were flattened.

Conditional Probability Formula

The **conditional probability of B, given A,** is given by

$$P(B \mid A) = \frac{P(A \cap B)}{P(A)} = \frac{P(A \text{ and } B)}{P(A)}.$$

A second observation from Example 1 is that the conditional probability of B, given A, was $\frac{2}{5}$, whereas the "unconditional" probability of B (with no condition given) was $\frac{3}{10}$, so the condition did make a difference.

EXAMPLE 2 Finding Probabilities of Boys and Girls in a Family

Given a family with two children, find the probabilities that

(a) both are girls, given that at least one is a girl, and
(b) both are girls, given that the older child is a girl.

(Assume boys and girls are equally likely.)

SOLUTION

We define the following events.

$S = \{\text{gg, gb, bg, bb}\}$	The sample space
$A = \{\text{gg}\}$	Both are girls.
$B = \{\text{gg, gb, bg}\}$	At least one is a girl.
$C = \{\text{gg, gb}\}$	The older one is a girl.

Note that $A \cap B = \{\text{gg}\}$.

(a) $P(A \mid B) = \dfrac{P(A \text{ and } B)}{P(B)} = \dfrac{\frac{1}{4}}{\frac{3}{4}} = \dfrac{1}{4} \div \dfrac{3}{4} = \dfrac{1}{4} \cdot \dfrac{4}{3} = \dfrac{1}{3}$

(b) $P(A \mid C) = \dfrac{P(A \text{ and } C)}{P(C)} = \dfrac{\frac{1}{4}}{\frac{2}{4}} = \dfrac{1}{4} \div \dfrac{2}{4} = \dfrac{1}{4} \cdot \dfrac{4}{2} = \dfrac{1}{2}$ ■

We noted earlier that in Example 1, the condition A did affect the value of $P(B)$. However, sometimes a conditional probability is no different than the corresponding unconditional probability, in which case we call the two events *independent*. Independent events are defined generally as follows.

Independent Events

Two events A and B are called **independent events** if knowledge about the occurrence of one of them has no effect on the probability of the other one, that is, if

$$P(B \mid A) = P(B), \quad \text{or, equivalently,} \quad P(A \mid B) = P(A).$$

EXAMPLE 3 Checking Events for Independence

A single card is to be drawn from a standard 52-card deck. (The sample space S has 52 elements.) Given the events

A: The selected card is a face card, and B: The selected card is black,

(a) Find $P(B)$. **(b)** Find $P(B \mid A)$.
(c) Determine whether events A and B are independent.

SOLUTION

(a) There are 26 black cards in the 52-card deck, so

$$P(B) = \frac{26}{52} = \frac{1}{2}. \quad \text{Theoretical probability formula}$$

(b) $$P(B \mid A) = \frac{P(B \text{ and } A)}{P(A)} \quad \text{Conditional probability formula}$$

$$= \frac{\frac{6}{52}}{\frac{12}{52}} \quad \text{Of 52 cards, 12 are face cards and 6 are black face cards.}$$

$$= \frac{6}{52} \cdot \frac{52}{12} = \frac{1}{2} \quad \text{Calculate and write in lowest terms.}$$

(c) Because $P(B \mid A) = P(B)$, events A and B are independent. ■

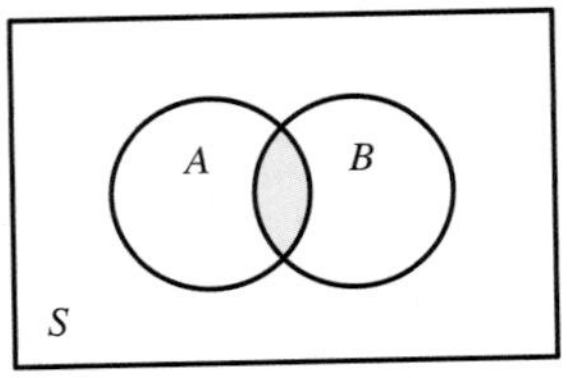

The logical connective "and" corresponds to "intersection" in set theory.

$P(A \text{ and } B) = P(A) \cdot P(B|A)$

FIGURE 9

Events Involving "And" If we multiply both sides of the conditional probability formula by $P(A)$, we obtain an expression for $P(A \cap B)$, which applies to events of the form "A and B." The resulting formula is related to the fundamental counting principle of Chapter 11. It is illustrated in Figure 9. Just as the calculation of $P(A$ or $B)$ is simpler when A and B are mutually exclusive, the calculation of $P(A$ and $B)$ is simpler when A and B are independent.

Multiplication Rule of Probability (for A and B)

If A and B are any two events, then

$$P(A \text{ and } B) = P(A) \cdot P(B \mid A).$$

If A and B are independent, then

$$P(A \text{ and } B) = P(A) \cdot P(B).$$

The first formula in the multiplication rule actually applies in all cases. ($P(B \mid A) = P(B)$ when A and B are independent.) Still, the independence of the component events is clear in many cases, so it is good to remember the second formula as well.

EXAMPLE 4 Selecting from a Set of Books

Each year, Diane Carr adds to her book collection a number of new publications that she believes will be of lasting value and interest. She has categorized each of her twenty acquisitions for 2007 as hardcover or paperback and as fiction or nonfiction. The numbers of books in the various categories are shown in Table 6.

TABLE 6 Year 2007 Books

	Fiction (F)	Nonfiction (N)	Totals
Hardcover (H)	3	5	8
Paperback (P)	8	4	12
Totals	11	9	20

If Diane randomly chooses one of these 20 books, find the probability it will be

(a) hardcover, **(b)** fiction, given it is hardcover, **(c)** hardcover and fiction.

SOLUTION

(a) Eight of the 20 books are hardcover, so $P(H) = \frac{8}{20} = \frac{2}{5}$.

(b) The given condition that the book is hardcover reduces the sample space to eight books. Of those eight, just three are fiction, so $P(F \mid H) = \frac{3}{8}$.

(c) $P(H \text{ and } F) = P(H) \cdot P(F \mid H) = \frac{2}{5} \cdot \frac{3}{8} = \frac{3}{20}$ Multiplication rule

It is easier here if we simply notice, directly from Table 6, that 3 of the 20 books are "hardcover and fiction." This verifies that the general multiplication rule of probability did give us the correct answer.

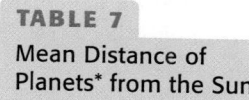

EXAMPLE 5 Selecting from a Set of Planets

TABLE 7
Mean Distance of Planets* from the Sun

Mercury	57.9
Venus	108.2
Earth	149.6
Mars	227.9
Jupiter	778.3
Saturn	1427
Uranus	2870
Neptune	4497
Pluto	5900

Table 7 lists the nine generally acknowledged planets of our solar system (prior to 2006)* together with their mean distances from the sun, in millions of kilometers. (Data is from the *Time Almanac 1999.*) Selina must choose two distinct planets to cover in her astronomy report. If she selects randomly, find the probability that the first one selected is closer to the sun than Mars and the second is closer than Saturn.

SOLUTION

We define the events

A: The first is closer than Mars, and B: The second is closer than Saturn.

Then $P(A) = \frac{3}{9} = \frac{1}{3}$. (Three of the original nine choices are favorable.) If the planet selected first is closer than Mars, it is also closer than Saturn, and since that planet is no longer available, $P(B \mid A) = \frac{4}{8} = \frac{1}{2}$. (Four of the remaining eight are favorable.) The desired probability is

$$P(A \text{ and } B) = P(A) \cdot P(B \mid A) = \frac{1}{3} \cdot \frac{1}{2} = \frac{1}{6} \approx .167. \quad \text{Multiplication rule}$$

In Example 5, the condition that A had occurred changed the probability of B, since the selection was done, in effect, without replacement. (Repetitions were not allowed.) Events A and B were not independent. On the other hand, in the next example, the same events, A and B, will be independent.

*In 2006, Pluto was officially downgraded from "planet" status.

EXAMPLE 6 Selecting from a Set of Planets

Selina must again select two planets, but this time one is for an oral report, the other is for a written report, and they need not be distinct. (The same planet may be selected for both.) Again find the probability that, if she selects randomly, the first is closer than Mars and the second is closer than Saturn.

SOLUTION

Defining events A and B as in Example 5, we have $P(A) = \frac{3}{9} = \frac{1}{3}$, just as before. But the selection is now done *with* replacement. Repetitions *are* allowed. In this case event B is independent of event A, so we can use the second form of the multiplication rule, obtaining a different answer than in Example 5.

$$P(A \text{ and } B) = P(A) \cdot P(B) = \frac{1}{3} \cdot \frac{5}{9} = \frac{5}{27} \approx .185$$

EXAMPLE 7 Selecting from a Deck of Cards

A single card is drawn from a standard 52-card deck. Let B denote the event that the card is black, and let D denote the event that it is a diamond. Answer each question.

(a) Are events B and D independent? **(b)** Are events B and D mutually exclusive?

SOLUTION

(a) For the unconditional probability of D, we get $P(D) = \frac{13}{52} = \frac{1}{4}$. (Thirteen of the 52 cards are diamonds.) But for the conditional probability of D, given B, we have $P(D \mid B) = \frac{0}{26} = 0$. (None of the 26 black cards are diamonds.) Since the conditional probability $P(D \mid B)$ is different than the unconditional probability $P(D)$, B and D are not independent.

(b) Mutually exclusive events, defined in the previous section, are events that cannot both occur for a given performance of an experiment. Since no card in the deck is both black and a diamond, B and D are mutually exclusive.

(People sometimes get the idea that "mutually exclusive" and "independent" mean the same thing. This example shows that this is not so.)

The multiplication rule of probability, can be extended to cases where more than two events are involved.

EXAMPLE 8 Selecting from an Urn of Balls

Anne is still drawing balls from the same urn (shown at the side). This time she draws three balls, without replacement. Find the probability that she gets red, yellow, and blue balls, in that order.

SOLUTION

Using appropriate letters to denote the colors, and subscripts to indicate first, second, and third draws, the event can be symbolized R_1 and Y_2 and B_3, so

$$\begin{aligned} P(R_1 \text{ and } Y_2 \text{ and } B_3) &= P(R_1) \cdot P(Y_2 \mid R_1) \cdot P(B_3 \mid R_1 \text{ and } Y_2) \\ &= \frac{4}{11} \cdot \frac{5}{10} \cdot \frac{2}{9} = \frac{4}{99} \approx .0404. \end{aligned}$$

The **search for extraterrestrial intelligence (SETI)** may have begun in earnest as early as 1961 when Dr. Frank Drake presented an equation for estimating the number of possible civilizations in the Milky Way galaxy whose communications we might detect. Over the years, the effort has been advanced by many scientists, including the late astronomer and exobiologist Carl Sagan, who popularized the issue in TV appearances and in his book *The Cosmic Connection: An Extraterrestrial Perspective* (Dell Paperback). "There must be other starfolk," said Sagan. In fact, some astronomers have estimated the odds against life on Earth being the only life in the universe at one hundred billion billion to one.

Other experts disagree. Freeman Dyson, a noted mathematical physicist and astronomer, says in his book *Disturbing the Universe* that after considering the same evidence and arguments, he believes it is just as likely as not (even odds) that there never was any other intelligent life out there.

EXAMPLE 9 Selecting from a Deck of Cards

If five cards are drawn without replacement from a standard 52-card deck, find the probability that they all are hearts.

SOLUTION

Each time a heart is drawn, the number of available cards decreases by one and the number of hearts decreases by one. The probability of drawing only hearts is

$$\frac{13}{52} \cdot \frac{12}{51} \cdot \frac{11}{50} \cdot \frac{10}{49} \cdot \frac{9}{48} = \frac{33}{66{,}640} \approx .000495.$$

If you studied counting methods (Chapter 11), you may prefer to solve the problem of Example 9 by using the theoretical probability formula and combinations. The total possible number of 5-card hands, drawn without replacement, is ${}_{52}C_5$, and the number of those containing only hearts is ${}_{13}C_5$, so the required probability is

$$\frac{{}_{13}C_5}{{}_{52}C_5} = \frac{\frac{13!}{5!8!}}{\frac{52!}{5!47!}} \approx .000495. \quad \text{Use a calculator.}$$

EXAMPLE 10 Using Both Addition and Multiplication Rules

The local garage employs two mechanics, Alex and Ben. Your consumer club has found that Alex does twice as many jobs as Ben, Alex does a good job three out of four times, and Ben does a good job only two out of five times. If you plan to take your car in for repairs, find the probability that a good job will be done.

SOLUTION

We define the events

A: work done by Alex; B: work done by Ben; G: good job done.

Since Alex does twice as many jobs as Ben, the (unconditional) probabilities of events A and B are, respectively, $\frac{2}{3}$ and $\frac{1}{3}$. Since Alex does a good job three out of four times, the probability of a good job, given that Alex did the work, is $\frac{3}{4}$. And since Ben does well two out of five times, the probability of a good job, given that Ben did the work, is $\frac{2}{5}$. (These last two probabilities are conditional.) These four values can be summarized as

$$P(A) = \frac{2}{3}, \quad P(B) = \frac{1}{3}, \quad P(G|A) = \frac{3}{4}, \quad \text{and} \quad P(G|B) = \frac{2}{5}.$$

Event G can occur in two mutually exclusive ways: Alex could do the work and do a good job ($A \cap G$), or Ben could do the work and do a good job ($B \cap G$). Thus,

$$\begin{aligned} P(G) &= P(A \cap G) + P(B \cap G) && \text{Addition rule} \\ &= P(A) \cdot P(G|A) + P(B) \cdot P(G|B) && \text{Multiplication rule} \\ &= \frac{2}{3} \cdot \frac{3}{4} + \frac{1}{3} \cdot \frac{2}{5} && \text{Substitute the values.} \\ &= \frac{1}{2} + \frac{2}{15} = \frac{19}{30} \approx .633. \end{aligned}$$

Multiply first, then add.

The tree diagram in Figure 10 shows a graphical way to organize the work of Example 10. Use the given information to draw the tree diagram, then find the probability of a good job by adding the probabilities from the indicated branches of the tree.

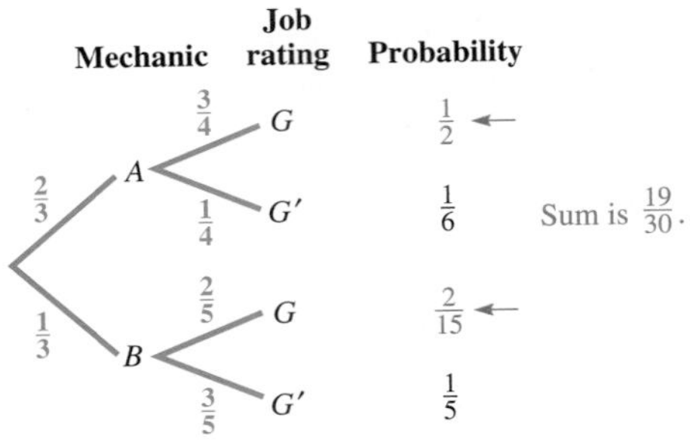

Garage mechanics experiment

FIGURE 10

EXAMPLE 11 Selecting Door Prizes

Since 1999, the **search for extraterrestrial intelligence (SETI)** has been carried out mainly through **SETI@HOME,** the largest distributed computing project on Earth. Some of the world's largest radio telescopes, like the one built into a 20-acre natural bowl in Aricebo, Puerto Rico (pictured above), collect data from the most likely regions of our galaxy. The data are analyzed by personal computer users around the world. The four millionth participant registered on October 1, 2002. To learn more, or if you would like a chance at being the first to "contact" an extraterrestrial civilization, check out the website www.setiathome.ssl.berkeley.edu.

Michael Bolinder is among five door prize winners at a Christmas party. The five winners are asked to choose, without looking, from a bag which, they are told, contains five tokens, four of them redeemable for candy canes and one specific token redeemable for a \$100 gift certificate. Can Michael improve his chance of getting the gift certificate by drawing first among the five people?

SOLUTION

We denote candy cane by C, gift certificate by G, and first draw, second draw, and so on by subscripts 1, 2, ... Then if Michael draws first, his probability of getting the gift certificate is

$$P(G_1) = \frac{1}{5}.$$

If he draws second, his probability of getting the gift certificate is

$$\begin{aligned} P(G_2) &= P(C_1 \text{ and } G_2) \\ &= P(C_1) \cdot P(G_2 \mid C_1) \\ &= \frac{4}{5} \cdot \frac{1}{4} = \frac{1}{5}. \quad \text{Same result as above} \end{aligned}$$

For the third draw,

$$\begin{aligned} P(G_3) &= P(C_1 \text{ and } C_2 \text{ and } G_3) \\ &= P(C_1) \cdot P(C_2 \mid C_1) \cdot P(G_3 \mid C_1 \text{ and } C_2) \\ &= \frac{4}{5} \cdot \frac{3}{4} \cdot \frac{1}{3} = \frac{1}{5}. \quad \text{Same result as above} \end{aligned}$$

Likewise, the probability of getting the gift certificate is $\frac{1}{5}$ when drawing fourth or when drawing fifth. Therefore, the order in which the five winners draw does not affect Michael's chances. ■

For Further Thought

The Birthday Problem

A classic problem (with a surprising result) involves the probability that a given group of people will include at least one pair of people with the same birthday (the same day of the year, not necessarily the same year). This problem can be analyzed using the probability of a complement formula (Section 12.2) and the multiplication rule of probability from this section. Suppose there are three people in the group. Then

P(at least one duplication of birthdays)

$= 1 - P(\text{no duplications})$ Complement formula

$= 1 - P$(2nd is different than 1st and 3rd is different than 1st and 2nd)

$= 1 - \frac{364}{365} \cdot \frac{363}{365}$ Multiplication rule

$\approx 1 - .992$

$= .008$

(To simplify the calculations, we have assumed 365 possible birth dates, ignoring February 29.)

By doing more calculations like the one above, we find that the smaller the group, the smaller the probability of a duplication; the larger the group, the larger the probability of a duplication. The table below shows the probability of at least one duplication for numbers of people through 50.

For Group Discussion or Individual Investigation

1. Based on the data shown in the table, what are the odds in favor of a duplication in a group of 30 people?
2. Estimate from the table the least number of people for which the probability of duplication is at least $\frac{1}{2}$.
3. How small a group is required for the probability of a duplication to be *exactly* 0?
4. How large a group is required for the probability of a duplication to be *exactly* 1?

Number of People	Probability of at Least One Duplication	Number of People	Probability of at Least One Duplication	Number of People	Probability of at Least One Duplication
2	.003	19	.379	36	.832
3	.008	20	.411	37	.849
4	.016	21	.444	38	.864
5	.027	22	.476	39	.878
6	.040	23	.507	40	.891
7	.056	24	.538	41	.903
8	.074	25	.569	42	.914
9	.095	26	.598	43	.924
10	.117	27	.627	44	.933
11	.141	28	.654	45	.941
12	.167	29	.681	46	.948
13	.194	30	.706	47	.955
14	.223	31	.730	48	.961
15	.253	32	.753	49	.966
16	.284	33	.775	50	.970
17	.315	34	.795		
18	.347	35	.814		

12.3 EXERCISES

For each experiment, determine whether the two given events are independent.

1. ***Tossing Coins*** A fair coin is tossed twice. The events are "head on the first" and "head on the second."

2. ***Rolling Dice*** A pair of dice are rolled. The events are "even on the first" and "odd on the second."

3. ***Comparing Planets' Mean Distances from the Sun*** Two planets are selected, without replacement, from the list in Table 7. The events are "first is closer than Jupiter" and "second is farther than Neptune."

4. ***Comparing Mean Distances from the Sun*** Two celestial bodies are selected, with replacement, from the list in Table 7. The events are "first is closer than Earth" and "second is farther than Uranus."

5. ***Guessing Answers on a Multiple-choice Test*** The answers are all guessed on a twenty-question multiple-choice test. The events are "first answer is correct" and "last answer is correct."

6. ***Selecting Committees of U.S. Senators*** A committee of five is randomly selected from the 100 U.S. Senators. The events are "first member selected is a Republican" and "second member selected is a Republican." (Assume that there are both Republicans and non-Republicans in the Senate.)

Comparing Gender and Career Motivation of University Students *One hundred college seniors attending a career fair at a major northeastern university were categorized according to gender and according to primary career motivation, as summarized here.*

		Primary Career Motivation			
		Money	**Allowed to be Creative**	**Sense of Giving to Society**	**Total**
Gender	**Male**	18	21	19	58
	Female	14	13	15	42
	Total	32	34	34	100

If one of these students is to be selected at random, find the probability that the student selected will satisfy each condition.

7. female

8. motivated primarily by creativity

9. not motivated primarily by money

10. male and motivated primarily by money

11. male, given that primary motivation is a sense of giving to society

12. motivated primarily by money or creativity, given that the student is female

Selecting Pets *A pet store has seven puppies, including four poodles, two terriers, and one retriever. If Rebecka and Aaron, in that order, each select one puppy at random,* with replacement *(they may both select the same one), find the probability of each event.*

13. both select a poodle

14. Rebecka selects a retriever, Aaron selects a terrier

15. Rebecka selects a terrier, Aaron selects a retriever

16. both select a retriever

Selecting Pets *Suppose two puppies are selected as earlier, but this time* without replacement *(Rebecka and Aaron cannot both select the same puppy). Find the probability of each event.*

17. both select a poodle

18. Aaron selects a terrier, given Rebecka selects a poodle

19. Aaron selects a retriever, given Rebecka selects a poodle

20. Rebecka selects a retriever

21. Aaron selects a retriever, given Rebecka selects a retriever

22. both select a retriever

Drawing Cards *Let two cards be dealt successively,* without replacement, *from a standard 52-card deck. Find the probability of each event in Exercises 23–27.*

23. spade second, given spade first

24. club second, given diamond first

25. two face cards

26. no face cards

27. The first card is a jack and the second is a face card.

28. Given events A and B within the sample space S, the following sequence of steps establishes formulas that can be used to compute conditional probabilities. Justify each statement.

(a) $P(A \text{ and } B) = P(A) \cdot P(B \mid A)$

(b) Therefore, $P(B \mid A) = \dfrac{P(A \text{ and } B)}{P(A)}$.

(c) Therefore, $P(B \mid A) = \dfrac{n(A \text{ and } B)/n(S)}{n(A)/n(S)}$.

(d) Therefore, $P(B \mid A) = \dfrac{n(A \text{ and } B)}{n(A)}$.

Considering Conditions in Card Drawing *Use the results of Exercise 28 to find each probability when a single card is drawn from a standard 52-card deck.*

29. $P(\text{queen} \mid \text{face card})$

30. $P(\text{face card} \mid \text{queen})$

31. $P(\text{red} \mid \text{diamond})$

32. $P(\text{diamond} \mid \text{red})$

33. If one number is chosen randomly from the integers 1 through 10, the probability of getting a number that is *odd and prime,* by the multiplication rule, is

$$P(\text{odd}) \cdot P(\text{prime} \mid \text{odd}) = \frac{5}{10} \cdot \frac{3}{5} = \frac{3}{10}.$$

Compute the product $P(\text{prime}) \cdot P(\text{odd} \mid \text{prime})$, and compare to the product above.

34. What does Exercise 33 imply, in general, about the probability of an event of the form A and B?

35. ***Gender in Sequences of Babies*** One of the authors of this book has three sons and no daughters. Assuming boy and girl babies are equally likely, what is the probability of this event?

36. ***Gender in Sequences of Babies*** Under the assumptions of Exercise 35, what is the probability that three successive births will be a boy, then a girl, then a boy?

The remaining exercises, and groups of exercises, may require concepts from earlier sections, such as the complements principle of counting and addition rules, as well as the multiplication rule of this section.

Probabilities in Warehouse Grocery Shopping *Emily Falzon manages a grocery warehouse which encourages volume shopping on the part of its customers. Emily has discovered that, on any given weekday,* 80 *percent of the customer sales amount to more than* \$100. *That is, any given sale on such a day has a probability of* .80 *of being for more than* \$100. *(Actually, the conditional probabilities throughout the day would change slightly, depending on earlier sales, but this effect would be negligible for the first several sales of the day, so we can treat them as independent.)*

Find the probability of each event. (Give answers to three decimal places.)

37. The first two sales on Wednesday are both for more than \$100.

38. The first three sales on Wednesday are all for more than \$100.

39. None of the first three sales on Wednesday is for more than \$100.

40. Exactly one of the first three sales on Wednesday is for more than \$100.

Pollution from the Space Shuttle Launch Site *One problem encountered by developers of the space shuttle program is air pollution in the area surrounding the launch site. A certain direction from the launch site is considered critical in terms of hydrogen chloride pollution from the exhaust cloud. It has been determined that weather conditions would cause emission cloud movement in the critical direction only 5% of the time.*

In Exercises 41–44 on the next page, find the probability for each event. Assume that probabilities for a particular launch in no way depend on the probabilities for other launches. (Give answers to two decimal places.)

41. A given launch will not result in cloud movement in the critical direction.

42. No cloud movement in the critical direction will occur during any of 5 launches.

43. Any 5 launches will result in at least one cloud movement in the critical direction.

44. Any 10 launches will result in at least one cloud movement in the critical direction.

Ordering Job Interviews *Four men and three women are waiting to be interviewed for jobs. If they are all selected in random order, find the probability of each event in Exercises 45–49.*

45. All the women will be interviewed first.

46. All the men will be interviewed first.

47. The first person interviewed will be a woman.

48. The second person interviewed will be a woman.

49. The last person interviewed will be a woman.

50. In Example 8, where Anne draws three balls without replacement, what would be her probability of getting one of each color, where the order does not matter?

51. ***Gender in Sequences of Babies*** Assuming boy and girl babies are equally likely, find the probability that it would take
(a) at least three births to obtain two girls,
(b) at least four births to obtain two girls,
(c) at least five births to obtain two girls.

52. ***Drawing Cards*** Cards are drawn, without replacement, from an ordinary 52-card deck.
(a) How many must be drawn before the probability of obtaining at least one face card is greater than $\frac{1}{2}$?
(b) How many must be drawn before the probability of obtaining at least one king is greater than $\frac{1}{2}$?

Fair Decisions from Biased Coins *Many everyday decisions, like who will drive to lunch, or who will pay for the coffee, are made by the toss of a (presumably fair) coin and using the criterion "heads, you will; tails, I will." This criterion is not quite fair, however, if the coin is biased (perhaps due to slightly irregular construction or wear). John von Neumann suggested a way to make perfectly fair decisions even with a possibly biased coin. If a coin, biased so that*

$$P(\text{h}) = .5200 \quad \text{and} \quad P(\text{t}) = .4800,$$

is tossed twice, find each probability. (Give answers to four decimal places.)

53. $P(\text{hh})$

54. $P(\text{ht})$

55. $P(\text{th})$

56. $P(\text{tt})$

57. Having completed Exercises 53–56, can you suggest what von Neumann's scheme may have been?

Programming a Garage Door Opener *A certain brand of automatic garage door opener utilizes a transmitter control with six independent switches, each one set on or off. The receiver (wired to the door) must be set with the same pattern as the transmitter.* Exercises 58–61 are based on ideas similar to those of the "birthday problem" in the "For Further Thought" feature in this section.*

58. How many different ways can the owner of one of these garage door openers set the switches?

59. If two residents in the same neighborhood each have one of this brand of opener, and both set the switches randomly, what is the probability to four decimal places that they are able to open each other's garage doors?

60. If five neighbors with the same type of opener set their switches independently, what is the probability of at least one pair of neighbors using the same settings? (Give your answer to four decimal places.)

61. What is the minimum number of neighbors who must use this brand of opener before the probability of at least one duplication of settings is greater than $\frac{1}{2}$?

62. ***Choosing Cards*** There are three cards, one that is green on both sides, one that is red on both sides, and one that is green on one side and red on the other. One of the three cards is selected randomly and laid on the table. If it happens that the card on the table has a red side up, what is the probability that it is also red on the other side?

---

*For more information, see "Matching Garage-Door Openers," by Bonnie H. Litwiller and David R. Duncan in the March 1992 issue of *Mathematics Teacher*, p. 217.

Weather Conditions on Successive Days *In November, the rain in a certain valley tends to fall in storms of several days' duration. The unconditional probability of rain on any given day of the month is .500. But the probability of rain on a day that follows a rainy day is .800, and the probability of rain on a day following a nonrainy day is .300. Find the probability of each event. Give answers to three decimal places.*

63. rain on two randomly selected consecutive days in November

64. rain on three randomly selected consecutive days in November

65. rain on November 1 and 2, but not on November 3

66. rain on the first four days of November, given that October 31 was clear all day

Engine Failures in a Vintage Aircraft *In a certain four-engine vintage aircraft, now quite unreliable, each engine has a 10% chance of failure on any flight, as long as it is carrying its one-fourth share of the load. But if one engine fails, then the chance of failure increases to 20% for each of the other three engines. And if a second engine fails, each of the remaining two has a 30% chance of failure. Assuming that no two engines ever fail simultaneously, and that the aircraft can continue flying with as few as two operating engines, find each probability for a given flight of this aircraft. (Give answers to four decimal places.)*

67. no engine failures

68. exactly one engine failure (any one of four engines)

69. exactly two engine failures (any two of four engines)

70. a failed flight

One-and-one Foul Shooting in Basketball *In basketball, "one-and-one" foul shooting is done as follows: if the player makes the first shot (1 point), he is given a second shot. If he misses the first shot, he is not given a second shot (see the tree diagram).*

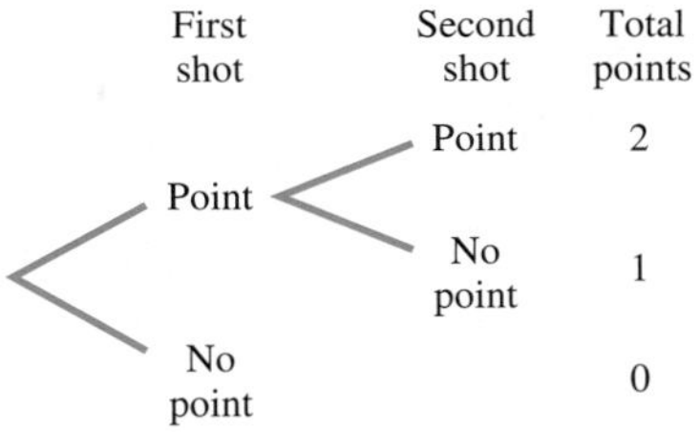

Susan Dratch, a basketball player, has a 70% foul shot record. (She makes 70% of her foul shots.) Find the probability that, on a given one-and-one foul shooting opportunity, Susan will score each number of points.

71. no points

72. one point

73. two points

74. ***Comparing Empirical and Theoretical Probabilities in Dice Rolling*** Roll a pair of dice until a sum of seven appears, keeping track of how many rolls it took. Repeat the process a total of 50 times, each time recording the number of rolls it took to get a sum of seven.

(a) Use your experimental data to compute an em-pirical probability (to two decimal places) that it would take at least three rolls to get a sum of seven.

(b) Find the theoretical probability (to two decimal places) that it would take at least three rolls to obtain a sum of seven.

75. Go to the Web site mentioned in the *natural disasters* margin note in this section and write a report on the threat to humanity of cosmic impacts. Include an explanation of the abbreviation *NEO*.

12.4 Binomial Probability

Binomial Probability Distribution • Binomial Probability Formula

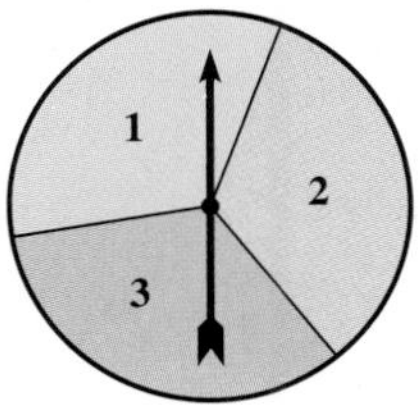

Binomial Probability Distribution

Suppose the spinner in the margin is spun twice, where we are interested in the number of times a 2 is obtained. (Assume each of the three sectors contains a 120-degree arc, so that each sector is equally likely on a given spin.) We can think of the outcome 2 as a "success," while outcomes 1 and 3 would be "failures." When the outcomes of an experiment are divided into just two categories, success and failure, the associated probabilities are called "binomial" (the prefix *bi* meaning *two*). Repeated performances of such an experiment, where the probability of success remains constant throughout all repetitions, are also known as repeated **Bernoulli trials** (after James Bernoulli). If we use an ordered pair to represent the result of each pair of spins, then the sample space for this experiment is

$$S = \{(1, 1), (1, 2), (1, 3), (2, 1), (2, 2), (2, 3), (3, 1), (3, 2), (3, 3)\}.$$

The nine outcomes in S are all equally likely. (This follows from the numbers 1, 2, and 3 being equally likely on a particular spin.)

If x denotes the number of 2s occurring on each pair of spins, then x is an example of a *random variable.* Although we cannot predict the result of any particular pair of spins, we can compute likelihoods, or probabilities, of various events from the sample space listing. In S, the number of 2s is 0 in four cases, 1 in four cases, and 2 in one case, as reflected in Table 8. Because the table includes all possible values of x, together with their probabilities, it is an example of a *probability distribution.* In this case, we have a **binomial probability distribution.** Notice that the probability column in Table 8 has a sum of 1, in agreement with property 3 of probability (Section 12.2).

TABLE 8

Probability Distribution for the Number of 2s in Two Spins

x	$P(x)$
0	$\frac{4}{9}$
1	$\frac{4}{9}$
2	$\frac{1}{9}$
	Sum $= \frac{9}{9} = 1$

In order to develop a general formula for binomial probabilities, we can consider another way to obtain the probability values in Table 8. The various spins of the spinner are independent of one another, and on each spin the probability of success (S) is $\frac{1}{3}$ and the probability of failure (F) is $\frac{2}{3}$. We will denote success on the first spin by S_1, failure on the second by F_2, and so on. Then, using the rules of probability, we have

$$\begin{aligned} P(x = 0) &= P(F_1 \text{ and } F_2) \\ &= P(F_1) \cdot P(F_2) && \text{Multiplication rule} \\ &= \frac{2}{3} \cdot \frac{2}{3} && \text{Substitute values.} \\ &= \frac{4}{9}, \end{aligned}$$

$$\begin{aligned} P(x = 1) &= P[(S_1 \text{ and } F_2) \text{ or } (F_1 \text{ and } S_2)] && \text{2 ways to get } x = 1 \\ &= P(S_1 \text{ and } F_2) + P(F_1 \text{ and } S_2) && \text{Addition rule} \\ &= P(S_1) \cdot P(F_2) + P(F_1) \cdot P(S_2) && \text{Multiplication rule} \\ &= \frac{1}{3} \cdot \frac{2}{3} + \frac{2}{3} \cdot \frac{1}{3} && \text{Substitute values.} \\ &= \frac{2}{9} + \frac{2}{9} \\ &= \frac{4}{9}, \end{aligned}$$

and

$$\begin{aligned} P(x = 2) &= P(S_1 \text{ and } S_2) \\ &= P(S_1) \cdot P(S_2) && \text{Multiplication rule} \\ &= \frac{1}{3} \cdot \frac{1}{3} && \text{Substitute values.} \\ &= \frac{1}{9}. \end{aligned}$$

James Bernoulli (1654–1705) is also known as Jacob or Jacques. He was charmed away from theology by the writings of Leibniz, became his pupil, and later headed the mathematics faculty at the University of Basel. His results in probability are contained in the *Art of Conjecture,* which was published in 1713, after his death, and which also included a reprint of the earlier Huygens paper. Bernoulli also made many contributions to calculus and analytic geometry.

Notice the following pattern in the above calculations. There is only one way to get $x = 0$ (namely, F_1 and F_2). And there is only one way to get $x = 2$ (namely, S_1 and S_2). But there are two ways to get $x = 1$. One way is S_1 and F_2; the other is F_1 and S_2. There are two ways because the one success required can occur on the first spin or on the second spin. How many ways can exactly one success occur in two repeated trials? This question is equivalent to:

How many size-one subsets are there of the set of two trials?

The answer is ${}_2C_1 = 2$. (The expression ${}_2C_1$ denotes "combinations of 2 things taken 1 at a time." Combinations were discussed in Section 11.3.) Each of the two ways to get exactly one success has a probability equal to $\frac{1}{3} \cdot \frac{2}{3}$, the probability of success times the probability of failure.

If the same spinner is spun three times rather than two, then x, the number of successes (2s) could have values of 0, 1, 2, or 3. Then the number of ways to get exactly 1 success is ${}_3C_1 = 3$. They are: S_1 and F_2 and F_3, F_1 and S_2 and F_3, F_1 and F_2 and S_3. The probability of each of these three ways is $\frac{1}{3} \cdot \frac{2}{3} \cdot \frac{2}{3} = \frac{4}{27}$. So

$$P(x = 1) = 3 \cdot \frac{4}{27} = \frac{12}{27} = \frac{4}{9}.$$

Figure 11 shows all possibilities for three spins, and Table 9 gives the associated probability distribution. In the tree diagram, the number of ways of getting two successes in three trials is 3, in agreement with the fact that ${}_3C_2 = 3$. Also the sum of the $P(x)$ column in Table 9 is again 1.

TABLE 9

Probability Distribution for the Number of 2s in Three Spins

x	$P(x)$
0	$\frac{8}{27}$
1	$\frac{12}{27}$
2	$\frac{6}{27}$
3	$\frac{1}{27}$
	Sum $= \frac{27}{27} = 1$

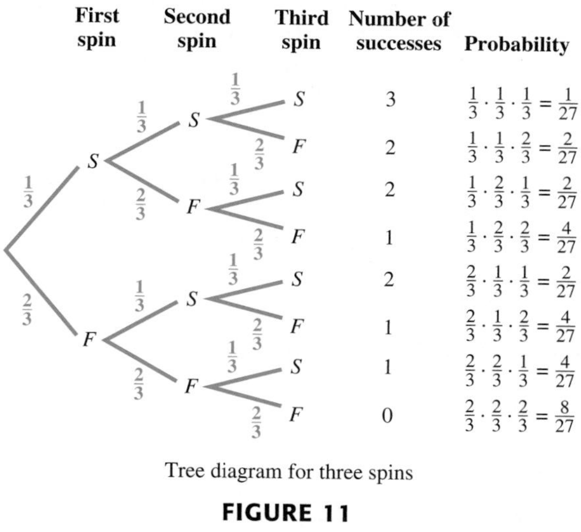

Tree diagram for three spins

FIGURE 11

PROBLEM-SOLVING HINT One of the problem-solving strategies from Chapter 1 was "Look for a pattern." Having constructed complete probability distributions for binomial experiments with 2 and 3 repeated trials (and probability of success $\frac{1}{3}$), we can now generalize the observed pattern to any binomial experiment, as shown next.

Binomial Probability Formula In general, let

n = the number of repeated trials,

p = the probability of success on any given trial,

$q = 1 - p$ = the probability of failure on any given trial,

and x = the number of successes that occur.

Note that p remains fixed throughout all n trials. This means that all trials are independent of one another. The random variable x (number of successes) can have any integer value from 0 through n. In general, x successes can be assigned among n repeated trials in ${}_nC_x$ different ways, since this is the number of different subsets of x positions among a set of n positions. Also, regardless of which x of the trials result in successes, there will always be x successes and $n-x$ failures, so we multiply x factors of p and $n-x$ factors of q together.

Binomial Probability Formula

When n independent repeated trials occur, where

p = probability of success and q = probability of failure

with p and q (where $q = 1 - p$) remaining constant throughout all n trials, the probability of exactly x successes is given by

$$P(x) = {}_nC_x p^x q^{n-x} = \frac{n!}{x!(n-x)!} p^x q^{n-x}.$$

Tables of binomial probability values are commonly available in statistics texts, for various values of p, often for n ranging up to about 20. Also, computer software packages for statistics will usually do these calculations for you automatically, as will some handheld calculators. In the following examples, we use the formula.

From the DISTR menu

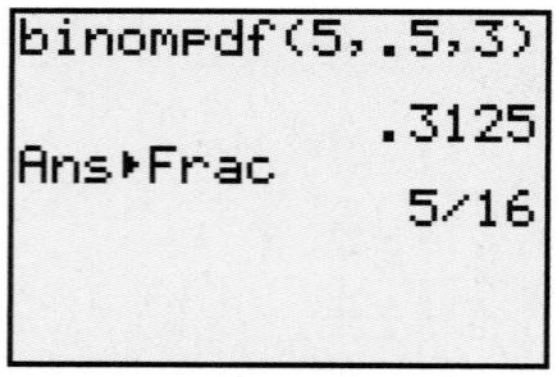

The TI-83/84 Plus calculator will find the probability discussed in Example 1.

EXAMPLE 1 Finding Probability in Coin Tossing

Find the probability of obtaining exactly three heads in five tosses of a fair coin.

SOLUTION

Let heads be "success." Then this is a binomial experiment with $n = 5$, $p = \frac{1}{2}$, $q = \frac{1}{2}$, and $x = 3$. By the binomial probability formula,

$$P(3) = {}_5C_3\left(\frac{1}{2}\right)^3\left(\frac{1}{2}\right)^2 = 10 \cdot \frac{1}{8} \cdot \frac{1}{4} = \frac{5}{16}.$$

This screen supports the answer in Example 2.

EXAMPLE 2 Finding Probability in Dice Rolling

Find the probability of obtaining exactly two 5s in six rolls of a fair die.

SOLUTION

Let 5 be "success." Then $n = 6$, $p = \frac{1}{6}$, $q = \frac{5}{6}$, and $x = 2$.

$$P(2) = {}_6C_2\left(\frac{1}{6}\right)^2\left(\frac{5}{6}\right)^4 = 15 \cdot \frac{1}{36} \cdot \frac{625}{1296} = \frac{3125}{15{,}552} \approx .201$$

In the case of repeated independent trials, when an event involves more than one specific number of successes, we can employ the binomial probability formula along with the complement or addition rules.

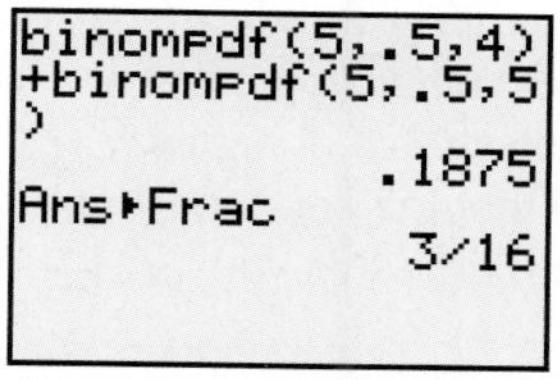

This screen supports the answer in Example 3.

EXAMPLE 3 Finding Probability of Female Children

A couple plans to have 5 children. Find the probability they will have more than 3 girls. (Assume girl and boy babies are equally likely.)

SOLUTION

Let a girl be "success." Then $n = 5$, $p = q = \frac{1}{2}$, and $x > 3$.

$$P(x > 3) = P(x = 4 \text{ or } 5) \qquad \text{More than 3 means 4 or 5.}$$

$$= P(4) + P(5) \qquad \text{Addition rule}$$

$$= {}_5C_4\left(\frac{1}{2}\right)^4\left(\frac{1}{2}\right)^1 + {}_5C_5\left(\frac{1}{2}\right)^5\left(\frac{1}{2}\right)^0 \qquad \text{Binomial probability formula}$$

$$= 5 \cdot \frac{1}{16} \cdot \frac{1}{2} + 1 \cdot \frac{1}{32} \cdot 1 \qquad \text{Simplify.}$$

$$= \frac{5}{32} + \frac{1}{32} = \frac{6}{32} = \frac{3}{16} = .1875$$

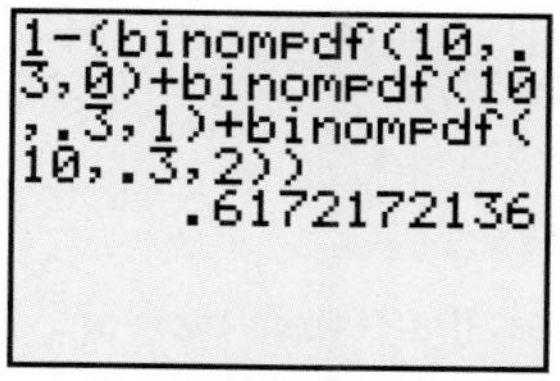

This screen supports the answer in Example 4.

EXAMPLE 4 Finding Probability of Hits in Baseball

Scott Davidson, a baseball player, has a well-established career batting average of .300. In a brief series with a rival team, Scott will bat 10 times. Find the probability that he will get more than two hits in the series.

SOLUTION

This "experiment" involves $n = 10$ repeated Bernoulli trials, with probability of success (a hit) given by $p = .3$ (which implies $q = 1 - .3 = .7$). Since, in this case, "more than 2" means

"3 or 4 or 5 or 6 or 7 or 8 or 9 or 10"

(which is eight different possibilities), it will be less work to apply the complement rule.

$$P(x > 2) = 1 - P(x \leq 2) \qquad \text{Complement rule}$$

$$= 1 - P(x = 0 \text{ or } 1 \text{ or } 2) \qquad \text{Only three different possibilities}$$

$$= 1 - [P(0) + P(1) + P(2)] \qquad \text{Addition rule}$$

$$= 1 - [{}_{10}C_0(.3)^0(.7)^{10} + {}_{10}C_1(.3)^1(.7)^9 + {}_{10}C_2(.3)^2(.7)^8] \qquad \text{Binomial probability formula}$$

$$\approx 1 - [.0282 + .1211 + .2335] \qquad \text{Simplify.}$$

$$= 1 - .3828$$

$$= .6172$$

For Further Thought

First Success on Trial *x*

In the case of *n* independent repeated Bernoulli trials, the formula developed in this section gives the probability of exactly *x* successes. Sometimes, however, we are interested not in the event that exactly *x* successes will occur in *n* trials, but rather the event that the first success will occur on the *x*th trial. Consider the probability that, in a series of coin tosses, the first success (head) will occur on the fourth toss. This implies a failure first, then a second failure, then a third failure, and finally a success. Symbolically, the event is F_1 and F_2 and F_3 and S_4. The probability of this sequence of outcomes is $q \cdot q \cdot q \cdot p$, or $q^3 \cdot p$. In general, if the probability of success stays constant at p (which implies a probability of failure of $q = 1 - p$), then the probability that the first success will occur on the *x*th trial can be computed as follows.

$$P(F_1 \text{ and } F_2 \ldots \text{ and } F_{x-1} \text{ and } S_x) = q^{x-1} \cdot p$$

For Group Discussion or Individual Investigation

1. Explain why, in the formula above, there is no combination factor, such as the ${}_nC_x$ in the binomial probability formula.
2. ***Union Members in an Industry*** If 30 percent of all workers in a certain industry are union members, and workers in this industry are selected successively at random, find the probability that the first union member to occur will be on the third selection.
3. ***Getting Caught when Speeding*** If the probability of getting caught when you exceed the speed limit on a certain stretch of highway is .38, find the probability that the first time you will get caught is the fourth time that you exceed the speed limit.
4. ***Gender in Sequences of Babies*** Assuming male and female babies are equally likely, find the probability that a family's fourth child will be their first daughter.

5. ***Aborted Rocket Launches*** If a certain type of rocket always has a four percent chance of an aborted launching, find the probability that the first launch to be aborted is the 20th launch.

12.4 EXERCISES

For Exercises 1–24, give all numerical answers as common fractions reduced to lowest terms. For Exercises 25–54, give all numerical answers to three decimal places.

Coin Tossing *If three fair coins are tossed, find the probability of each number of heads.*

1. 0 **2.** 1 **3.** 2

4. 3 **5.** 1 or 2 **6.** at least 1

7. no more than 1 **8.** fewer than 3

9. ***Gender in Sequences of Babies*** Assuming boy and girl babies are equally likely, find the probability that a family with three children will have exactly two boys.

10. Pascal's triangle was shown in Table 5 of Section 11.4. Explain how the probabilities in Exercises 1–4 here relate to row 3 of the "triangle." (Recall that we referred to the topmost row of the triangle as "row number 0," and to the leftmost entry of each row as "entry number 0.")

11. Generalize the pattern in Exercise 10 to complete the following statement. If n fair coins are tossed, the probability of exactly x heads is the fraction whose numerator is entry number ___ of row number ___ in Pascal's triangle, and whose denominator is the sum of the entries in row number ___.

Binomial Probability Applied to Tossing Coins *Use the pattern noted in Exercises 10 and 11 to find the probabilities of each number of heads when seven fair coins are tossed.*

12. 0 **13.** 1 **14.** 2 **15.** 3

16. 4 **17.** 5 **18.** 6 **19.** 7

Binomial Probability Applied to Rolling Dice *A fair die is rolled three times. A 4 is considered "success," while all other outcomes are "failures." Find the probability of each number of successes.*

20. 0 **21.** 1

22. 2 **23.** 3

24. Exercises 10 and 11 established a way of using Pascal's triangle rather than the binomial probability formula to find probabilities of different numbers of successes in coin-tossing experiments. Explain why the same process would not work for Exercises 20–23.

For n repeated independent trials, with constant probability of success p for all trials, find the probability of exactly x successes in each case.

25. $n = 5, \quad p = \frac{1}{3}, \quad x = 4$

26. $n = 10, \quad p = .7, \quad x = 5$

27. $n = 20, \quad p = \frac{1}{8}, \quad x = 2$

28. $n = 30, \quad p = .6, \quad x = 22$

For Exercises 29–31, refer to Example 4.

29. ***Batting Averages in Baseball*** Does Scott's probability of a hit really remain constant at exactly .300 through all ten times at bat? Explain your reasoning.

30. ***Batting Averages in Baseball*** If Scott's batting average is exactly .300 going into the series, and that value is based on exactly 1200 career hits out of 4000 previous times at bat, what is the greatest his average could possibly be (to three decimal places) when he goes up to bat the tenth time of the series? What is the least his average could possibly be when he goes up to bat the tenth time of the series?

31. Do you think the use of the binomial probability formula was justified in Example 4, even though p is not strictly constant? Explain your reasoning.

Random Selection of Answers on a Multiple-choice Test *Yesha Brill is taking a ten-question multiple-choice test for which each question has three answer choices, only one of which is correct. Yesha decides on answers by rolling a fair die and marking the first answer choice if the die shows 1 or 2, the second if it shows 3 or 4, and the third if it shows 5 or 6. Find the probability of each event.*

32. exactly four correct answers

33. exactly seven correct answers

34. fewer than three correct answers

35. at least seven correct answers

Side Effects of Prescription Drugs *It is known that a certain prescription drug produces undesirable side effects in 30% of all patients who use it. Among a random sample of eight patients using the drug, find the probability of each event.*

36. None have undesirable side effects.

37. Exactly one has undesirable side effects.

38. Exactly two have undesirable side effects.

39. More than two have undesirable side effects.

Likelihood of Capable Students Attending College *In a certain state, it has been shown that only 50% of the high school graduates who are capable of college work actually enroll in colleges. Find the probability that, among nine capable high school graduates in this state, each number will enroll in college.*

40. exactly 4 **41.** from 4 through 6

42. none **43.** all 9

44. ***Student Ownership of Personal Computers*** At a large midwestern university, 80% of all students have their own personal computers. If five students at that university are selected at random, find the probability that exactly three of them have their own computers.

45. ***Frost Survival Among Orange Trees*** If it is known that 65% of all orange trees will survive a hard frost, then what is the probability that at least half of a group of six trees will survive such a frost?

46. ***Rate of Favorable Media Coverage of an Incumbent President*** An extensive survey revealed that, during a certain presidential election campaign, 64% of the political columns in a certain group of major newspapers were favorable to the incumbent president. If a sample of fifteen of these columns is selected at random, what is the probability that exactly ten of them will be favorable?

Samantha stands on the street corner tossing a coin. She decides she will toss it 10 times, each time walking 1 block north if it lands heads up, and 1 block south if it lands tails up. In each of the following exercises, find the probability that she will end up in the indicated location. (In each case, ask how many successes, say heads, would be required and use the binomial formula. Some ending positions may not be possible with 10 tosses.) The random process involved here illustrates what we call a **random walk.** *It is a simplified model of Brownian motion, mentioned on page 726. Further applications of the idea of a random walk are found in the Extension at the end of this chapter.*

47. 10 blocks north of her corner

48. 6 blocks north of her corner

49. 6 blocks south of her corner

50. 5 blocks south of her corner

51. 2 blocks north of her corner

52. at least 2 blocks north of her corner

53. at least 2 blocks from her corner

54. on her corner

12.5 Expected Value

Expected Value • Games and Gambling • Investments • Business and Insurance

TABLE 10

x	$P(x)$
1	.05
2	.10
3	.20
4	.40
5	.10
6	.15

Expected Value

The probability distribution in Table 10, from Example 6 of Section 12.2, shows the probabilities assigned by Jacob to the various lengths of time his homework may take on a given night. If Jacob's friend Omer asks him how many hours his studies will take, what would be his best guess? Six different time values are possible, with some more likely than others. One thing Jacob could do is calculate a "weighted average" by multiplying each possible time value by its probability and then adding the six products.

$$1(.05) + 2(.10) + 3(.20) + 4(.40) + 5(.10) + 6(.15)$$
$$= .05 + .20 + .60 + 1.60 + .50 + .90 = 3.85$$

Thus 3.85 hours is the **expected value** (or the **mathematical expectation**) of the quantity of time to be spent. Since the original time values in the table were rounded to the nearest hour, the expected value also should be rounded, to 4 hours.

Expected Value

If a random variable x can have any of the values $x_1, x_2, x_3, \ldots, x_n$, and the corresponding probabilities of these values occurring are $P(x_1), P(x_2), P(x_3), \ldots, P(x_n)$, then the **expected value of x** is given by

$$E(x) = x_1 \cdot P(x_1) + x_2 \cdot P(x_2) + x_3 \cdot P(x_3) + \cdots + x_n \cdot P(x_n).$$

Solution to the Chapter Opener Problem One way to look at the problem, given that the car is *not* behind Door 3, is that Doors 1 and 2 are now equally likely to contain the car. Thus, switching doors will neither help nor hurt your chances of winning the car.

However, there is another way to look at the problem. When you picked Door 1, the probability was $\frac{1}{3}$ that it contained the car. Being shown the goat behind Door 3 doesn't really give you any new information; after all, you knew that there was a goat behind at least one of the other doors. So seeing the goat behind Door 3 does nothing to change your assessment of the probability that Door 1 has the car. It remains $\frac{1}{3}$. But because Door 3 has been ruled out, the probability that Door 2 has the car is now $\frac{2}{3}$. Thus, you should switch.

Analysis of this problem depends on the psychology of the host. If we suppose that the host must *always* show you a losing door and then give you an option to switch, then you should switch. This was not specifically stated in the problem as posed above but was pointed out by many mathematicians who became involved in the discussion.

(The authors wish to thank David Berman of the University of New Orleans for his assistance with this explanation.)

EXAMPLE 1 Finding the Expected Number of Boys

Find the expected number of boys for a three-child family (that is, the expected value of the number of boys). Assume girls and boys are equally likely.

SOLUTION

The sample space for this experiment is

$$S = \{ggg, ggb, gbg, bgg, gbb, bgb, bbg, bbb\}.$$

The probability distribution is shown in Table 11, along with the products and their sum, which gives the expected value.

TABLE 11

Number of Boys x	Probability $P(x)$	Product $x \cdot P(x)$
0	$\frac{1}{8}$	0
1	$\frac{3}{8}$	$\frac{3}{8}$
2	$\frac{3}{8}$	$\frac{6}{8}$
3	$\frac{1}{8}$	$\frac{3}{8}$

Expected value: $E(x) = \frac{12}{8} = \frac{3}{2}$

The expected number of boys is $\frac{3}{2}$, or 1.5. This result seems reasonable. Since boys and girls are equally likely, "half" the children are expected to be boys. ■

Notice that the expected value for the number of boys in the family is itself an impossible value. So, the expected value itself could never occur. Many times the expected value actually cannot occur; it is only a kind of long run average of the various values that could occur. (For more information on "averages," see Section 13.2.) If we recorded the number of boys in many different three-child families, then by the law of large numbers, as the number of observed families increased, the observed average number of boys should approach the expected value.

Example 1 did not involve money, but many uses of expected value do involve monetary expectations. Common applications are in gambling, investments, and business decisions.

Games and Gambling

EXAMPLE 2 Finding Expected Winnings

A player pays \$3 to play the following game: He tosses three fair coins and receives back "payoffs" of \$1 if he tosses no heads, \$2 for one head, \$3 for two heads, and \$4 for three heads. Find the player's expected net winnings for this game.

SOLUTION

Display the information as in Table 12 on the next page. (Notice that, for each possible event, "net winnings" are "gross winnings" (payoff) minus cost to play.) Probabilities are derived from the sample space

$$S = \{ttt, htt, tht, tth, hht, hth, thh, hhh\}.$$

TABLE 12

Number of Heads	Payoff	Net Winnings x	Probability $P(x)$	Product $x \cdot P(x)$
0	\$1	−\$2	$\frac{1}{8}$	$-\$\frac{2}{8}$
1	2	−1	$\frac{3}{8}$	$-\frac{3}{8}$
2	3	0	$\frac{3}{8}$	0
3	4	1	$\frac{1}{8}$	$\frac{1}{8}$

Expected value: $E(x) = -\$\frac{1}{2} = -\$.50$

The expected net loss of 50 cents is a long-run average only. On any particular play of this game, the player would lose \$2 or lose \$1 or break even or win \$1. Over a long series of plays, say 100, there would be some wins and some losses, but the total net result would likely be around a $100 \cdot (\$.50) = \$50$ *loss*.

A game in which the expected net winnings are zero is called a **fair game.** The game in Example 2 has negative expected net winnings, so it is unfair against the player. A game with positive expected net winnings is unfair in favor of the player.

EXAMPLE 3 Finding the Cost for a Fair Game

The \$3 cost to play the game of Example 2 makes the game unfair against the player (since the player's expected net winnings are negative). What cost would make this a fair game?

SOLUTION

We already computed, in Example 2, that the \$3 cost to play resulted in an expected net loss of \$.50. Therefore we can conclude that the \$3 cost was 50 cents too high. A fair cost to play the game would then be $\$3 - \$.50 = \$2.50$.

The result in Example 3 can be verified as follows. First disregard the cost to play and find the expected *gross* winnings (by summing the products of payoff times probability).

$$E(\text{gross winnings}) = \$1 \cdot \frac{1}{8} + \$2 \cdot \frac{3}{8} + \$3 \cdot \frac{3}{8} + \$4 \cdot \frac{1}{8} = \frac{\$20}{8} = \$2.50$$

Since the long-run expected gross winnings (payoff) are \$2.50, this amount is a fair cost to play.

EXAMPLE 4 Finding the Cost for a Fair Game

In a certain state lottery, a player chooses three digits, in a specific order. (Leading digits may be 0, so numbers such as 028 and 003 are legitimate entries.) The lottery operators randomly select a three-digit sequence, and any player matching their selection receives a payoff of \$600. What is a fair cost to play this game?

SOLUTION

In this case, no cost has been proposed, so we have no choice but to compute expected gross winnings. The probability of selecting all three digits correctly is $\frac{1}{10} \cdot \frac{1}{10} \cdot \frac{1}{10} = \frac{1}{1000}$, and the probability of not selecting all three correctly is $1 - \frac{1}{1000} = \frac{999}{1000}$. The expected gross winnings are

$$E(\text{gross winnings}) = \$600 \cdot \frac{1}{1000} + \$0 \cdot \frac{999}{1000} = \$.60.$$

Thus the fair cost to play this game is 60 cents. (In fact, the lottery charges \$1 to play, so players should expect to lose 40 cents per play *on the average*.)

Roulette ("little wheel") was invented in France in the seventeenth or early eighteenth century. It has been a featured game of chance in the gambling casino of Monte Carlo.

The disk is divided into red and black alternating compartments, numbered 1 to 36 (but not in that order). There is a compartment also for 0 (and for 00 in the United States). In roulette, the wheel is set in motion, and an ivory ball is thrown into the bowl opposite to the direction of the wheel. When the wheel stops, the ball comes to rest in one of the compartments—the number and color determine who wins.

The players bet against the banker (person in charge of the pool of money) by placing money or equivalent chips in spaces on the roulette table corresponding to the wheel's colors or numbers. Bets can be made on one number or several, on odd or even, on red or black, or on combinations. The banker pays off according to the odds against the particular bet(s). For example, the classic payoff for a winning single number is \$36 for each \$1 bet.

Of course, state lotteries must be unfair against players because they are designed to help fund benefits (such as the state's school system) as well as to cover administrative costs and certain other expenses. Among people's reasons for playing may be a willingness to support such causes, but most people undoubtedly play for the chance to "beat the odds" and be one of the few net winners.

Gaming casinos are major business enterprises, by no means designed to break even; the games they offer are always unfair in favor of the house. The bias does not need to be great, however, since even relatively small average losses per player multiplied by large numbers of players can result in huge profits for the house.

EXAMPLE 5 Finding Expected Winnings in Roulette

One simple type of *roulette* is played with an ivory ball and a wheel set in motion. The wheel contains thirty-eight compartments. Eighteen of the compartments are black, eighteen are red, one is labeled "zero," and one is labeled "double zero." (These last two are neither black nor red.) In this case, assume the player places \$1 on either red or black. If the player picks the correct color of the compartment in which the ball finally lands, the payoff is \$2; otherwise the payoff is zero. Find the expected net winnings.

SOLUTION

By the expected value formula, expected net winnings are

$$E(\text{net winnings}) = (\$1)\frac{18}{38} + (-\$1)\frac{20}{38} = -\$\frac{1}{19}.$$

The expected net loss here is $\$\frac{1}{19}$, or about 5.3¢, per play.

Investments

Expected value can be a useful tool for evaluating investment opportunities.

EXAMPLE 6 Finding Expected Investment Profits

Todd Hall has \$5000 to invest and will commit the whole amount, for six months, to one of three technology stocks. A number of uncertainties could affect the prices of these stocks, but Todd is confident, based on his research, that one of only several possible profit scenarios will prove true of each one at the end of the six-month period.

His complete analysis is shown in Table 13. (For example, stock *ABC* could lose \$400, gain \$800, or gain \$1500.)

TABLE 13

Company *ABC*		Company *RST*		Company *XYZ*	
Profit or Loss x	Probability $P(x)$	Profit or Loss x	Probability $P(x)$	Profit or Loss x	Probability $P(x)$
−\$400	.2	\$500	.8	\$0	.4
800	.5	1000	.2	700	.3
1500	.3			1200	.1
				2000	.2

Find the expected profit (or loss) for each of the three stocks and select Todd's optimum choice based on these calculations.

SOLUTION

Apply the expected value formula.

$$ABC\text{:}\quad -\$400 \cdot (.2) + \$800 \cdot (.5) + \$1500 \cdot (.3) = \$770$$

$$RST\text{:}\quad \$500 \cdot (.8) + \$1000 \cdot (.2) = \$600$$

$$XYZ\text{:}\quad \$0 \cdot (.4) + \$700 \cdot (.3) + \$1200 \cdot (.1) + \$2000 \cdot (.2) = \$730$$

The largest expected profit is \$770. By this analysis, Todd should invest the money in stock *ABC*. ■

Of course, by investing in stock *ABC*, Todd may in fact *lose* \$400 over the six months. The "expected" return of \$770 is only a long-run average over many identical situations. Since this particular investment situation may never occur again, you may argue that using expected values is not the best approach for Todd to use.

One possible alternative would be to adopt the view of an optimist, who might ignore the various probabilities and just hope for the best possibility associated with each choice and make a decision accordingly. Or, on the other hand, a pessimist may assume the worst case probably will occur, and make a decision in accordance with that view.

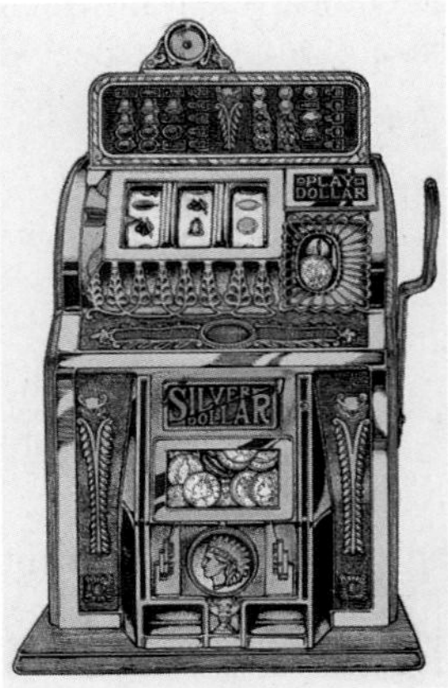

The first **Silver Dollar Slot Machine** was fashioned in 1929 by the Fey Manufacturing Company, San Francisco, inventors of the 3-reel, automatic payout machine (1895).

EXAMPLE 7 Choosing Stock Investments

Decide which stock of Example 6 Todd would pick in each case.

(a) He is an optimist. **(b)** He is a pessimist.

SOLUTION

(a) Disregarding the probabilities, he would focus on the best that could possibly happen with each stock. Since *ABC* could return as much as \$1500, *RST* as much as \$1000, and *XYZ* as much as \$2000, the optimum is \$2000. He would buy stock *XYZ*.

(b) In this case, he would focus on the worst possible cases. Since *ABC* might return as little as −\$400 (a \$400 loss), *RST* as little as \$500, and *XYZ* as little as \$0, he would buy stock *RST* (the best of the three worst cases).

Business and Insurance Expected value can be used to help make decisions in various areas of business, including insurance.

EXAMPLE 8 Finding Expected Lumber Revenue

Michael Crenshaw, a lumber wholesaler, is considering the purchase of a (railroad) carload of varied dimensional lumber. Michael calculates that the probabilities of reselling the load for \$10,000, \$9000, or \$8000 are .22, .33, and .45, respectively. In order to ensure an *expected* profit of at least \$3000, how much can Michael afford to pay for the load?

SOLUTION

The expected revenue (or income) from resales can be found in Table 14.

TABLE 14

Income x	Probability $P(x)$	Product $x \cdot P(x)$
\$10,000	.22	\$2200
9000	.33	2970
8000	.45	3600

Expected revenue: \$8770

In general, we have the relationship

$$\text{profit} = \text{revenue} - \text{cost}.$$

Therefore, in terms of expectations,

$$\text{expected profit} = \text{expected revenue} - \text{cost}.$$

So \$3000 = \$8770 − cost, or equivalently, cost = \$8770 − \$3000 = \$5770. Michael can pay up to \$5770 and still maintain an expected profit of at least \$3000.

EXAMPLE 9 Analyzing an Insurance Decision

A farmer will realize a profit of \$150,000 on his wheat crop, unless there is rain before harvest, in which case he will realize only \$40,000. The long-term weather forecast assigns rain a probability of .16. (So the probability of no rain is 1 − .16 = .84.) An insurance company offers crop insurance of \$150,000 against rain for a premium of \$20,000. Should the farmer buy the insurance?

SOLUTION

In order to make a wise decision, the farmer computes his expected profit under both options: to insure and not to insure. The complete calculations are summarized in the two "expectation" Tables 15 and 16.

For example, if insurance is purchased and it rains, the farmer's net profit is

$$\begin{bmatrix}\text{Insurance}\\\text{proceeds}\end{bmatrix} + \begin{bmatrix}\text{Reduced}\\\text{crop profit}\end{bmatrix} - \begin{bmatrix}\text{Insurance}\\\text{premium}\end{bmatrix} \leftarrow \text{Net profit}$$

$$\$150{,}000 \quad + \quad \$40{,}000 \quad - \quad \$20{,}000 \quad = \quad \$170{,}000.$$

TABLE 15

Expectation for Insuring

	Net Profit x	Probability $P(x)$	Product $x \cdot P(x)$
Rain	\$170,000	.16	\$27,200
No rain	130,000	.84	109,200

Expected profit: **\$136,400**

TABLE 16

Expectation for Not Insuring

	Net Profit x	Probability $P(x)$	Product $x \cdot P(x)$
Rain	\$40,000	.16	\$6400
No rain	150,000	.84	126,000

Expected profit: **\$132,400**

By comparing expected profits (**136,400** > **132,400**), we conclude that the farmer is better off buying the insurance.

For Further Thought

Expected Value of Games of Chance

Slot machines are a popular game for those who want to lose their money with very little mental effort. We cannot calculate an expected value applicable to all slot machines since payoffs vary from machine to machine. But we can calculate the "typical expected value."

A player operates a slot machine by pulling a handle after inserting a coin or coins. Reels inside

This Cleveland Indians fan hit four 7s in a row on a progressive nickel slot machine at the Sands Casino in Las Vegas in 1988.

(continued)

For Further Thought

the machine then rotate, and come to rest in some random order. Assume that three reels show the pictures listed in Table 17. For example, of the 20 pictures on the first reel, 2 are cherries, 5 are oranges, 5 are plums, 2 are bells, 2 are melons, 3 are bars, and 1 is the number 7.

TABLE 17

	Reels		
Pictures	**1**	**2**	**3**
Cherries	2	5	4
Oranges	5	4	5
Plums	5	3	3
Bells	2	4	4
Melons	2	1	2
Bars	3	2	1
7s	1	1	1
Totals	20	20	20

A picture of cherries on the first reel, but not on the second, leads to a payoff of 3 coins (*net* winnings: 2 coins); a picture of cherries on the first two reels, but not the third, leads to a payoff of 5 coins (*net* winnings: 4 coins). All other winning combinations are as listed in Table 18.

Since, according to Table 17, there are 2 ways of getting cherries on the first reel, 15 ways of *not* getting cherries on the second reel, and 20 ways of getting anything on the third reel, we have a total of $2 \cdot 15 \cdot 20 = 600$ ways of getting a net payoff of 2. Since there are 20 pictures per reel, there are a total of $20 \cdot 20 \cdot 20 = 8000$ possible outcomes. Hence, the probability of receiving a net payoff of 2 coins is 600/8000. Table 18 takes into account all *winning* outcomes, with the necessary products for finding expectation added in the last column.

TABLE 18 Calculating Expected Loss on a Three-Reel Slot Machine

Winning Combinations	Number of Ways	Probability	Number of Coins Received	Net Winnings (in coins)	Probability Times Winnings
1 cherry (on first reel)	$2 \cdot 15 \cdot 20 = 600$	600/8000	3	2	1200/8000
2 cherries (on first two reels)	$2 \cdot 5 \cdot 16 = 160$	160/8000	5	4	640/8000
3 cherries	$2 \cdot 5 \cdot 4 = 40$	40/8000	10	9	360/8000
3 oranges	$5 \cdot 4 \cdot 5 = 100$	100/8000	10	9	900/8000
3 plums	$5 \cdot 3 \cdot 3 = 45$	45/8000	14	13	585/8000
3 bells	__ · __ · __ = __	__/8000	18	___	____/8000
3 melons (jackpot)	__ · __ · __ = __	__/8000	100	___	____/8000
3 bars (jackpot)	__ · __ · __ = __	__/8000	200	___	____/8000
3 7s (jackpot)	__ · __ · __ = __	__/8000	500	___	____/8000
Totals	___				6318/8000

(continued)

For Further Thought

However, since a *nonwinning* outcome can occur in 8000 − 988 = 7012 ways (with winnings of −1 coin), the product $(-1) \cdot 7012/8000$ must also be included. Hence, the expected value of this particular slot machine is

$$\frac{6318}{8000} + (-1) \cdot \frac{7012}{8000} \approx -.087 \text{ coin.}$$

On a machine costing one dollar per play, the expected *loss* (per play) is about

$$(.087)\ (1 \text{ dollar}) = 8.7 \text{ cents.}$$

Actual slot machines vary in expected loss per dollar of play. But author Hornsby was able to beat a Las Vegas slot machine in 1988. (See the photo on page 768.)

Table 19 comes from an article by Andrew Sterrett in *The Mathematics Teacher* (March 1967), in which he discusses rules for various games of chance and calculates their expected values. He uses expected values to find expected times it would take to lose $1000 if you played continually at the rate of $1 per play and one play per minute.

For Group Discussion or Individual Investigation

1. Explain why the entries of the "Net Winnings" column of Table 18 are all one fewer than the corresponding entries of the "Number of Coins Received" column.
2. Find the 29 missing values in Table 18. (Refer to Table 17 for the values in the "Number of Ways" column.)
3. In order to make your money last as long as possible in a casino, which game should you play?

TABLE 19 Expected Time to Lose $1000

Game	Expected Value	Days	Hours	Minutes
Roulette (with one 0)	−$.027	25	16	40
Roulette (with 0 and 00)	−$.053	13	4	40
Chuck-a-luck	−$.079	8	19	46
Keno (one number)	−$.200	3	11	20
Numbers	−$.300	2	7	33
Football pool (4 winners)	−$.375	1	20	27
Football pool (10 winners)	−$.658	1	1	19

12.5 EXERCISES

1. Explain in words what is meant by "expected value of a random variable."

2. Explain what a couple means by the statement, "We expect to have 1.5 sons."

3. ***Tossing Coins*** Five fair coins are tossed. Find the expected number of heads.

4. ***Drawing Cards*** Two cards are drawn, with replacement, from a standard 52-card deck. Find the expected number of diamonds.

Expected Winnings in a Die-rolling Game *For Exercises 5 and 6, a game consists of rolling a single fair die and pays off as follows:* $3 *for a* 6, $2 *for a* 5, $1 *for a* 4, *and no payoff otherwise.*

5. Find the expected winnings for this game.

6. What is a fair price to pay to play this game?

Expected Winnings in a Die-rolling Game *For Exercises 7 and 8, consider a game consisting of rolling a single fair die, with payoffs as follows. If an even number of spots turns up, you receive that many dollars. But if an odd number of spots turns up, you must pay that many dollars.*

7. Find the expected net winnings of this game.

8. Is this game fair, or unfair against the player, or unfair in favor of the player?

9. ***Expected Winnings in a Coin-tossing Game*** A certain game involves tossing 3 fair coins, and it pays 10¢ for 3 heads, 5¢ for 2 heads, and 3¢ for 1 head. Is 5¢ a fair price to pay to play this game? (That is, does the 5¢ cost to play make the game fair?)

10. ***Expected Winnings in Roulette*** In a form of roulette slightly different from that in Example 5, a more generous management supplies a wheel having only thirty-seven compartments, with eighteen red, eighteen black, and one zero. Find the expected net winnings if you bet on red in this game.

11. ***Expected Number of Absences in a Math Class*** In a certain mathematics class, the probabilities have been empirically determined for various numbers of absences on any given day. These values are shown in the table below. Find the expected number of absences on a given day. (Give the answer to two decimal places.)

Number absent	0	1	2	3	4
Probability	.12	.32	.35	.14	.07

12. ***Expected Profit of an Insurance Company*** An insurance company will insure a $100,000 home for its total value for an annual premium of $330. If the company spends $20 per year to service such a policy, the probability of total loss for such a home in a given year is .002, and you assume that either total loss or no loss will occur, what is the company's expected annual gain (or profit) on each such policy?

Profits from a College Foundation Raffle *A college foundation raises funds by selling raffle tickets for a new car worth* $36,000.

13. If 600 tickets are sold for $120 each, determine
    (a) the expected *net* winnings of a person buying one of the tickets,
    (b) the total profit for the foundation, assuming they had to purchase the car,
    (c) the total profit for the foundation, assuming the car was donated.

14. For the raffle described in Exercise 13, if 720 tickets are sold for $120 each, determine
    (a) the expected *net* winnings of a person buying one of the tickets,
    (b) the total profit for the foundation, assuming they had to purchase the car,
    (c) the total profit for the foundation, assuming the car was donated.

Winnings and Profits of a Raffle *Five thousand raffle tickets are sold. One first prize of* $1000, *two second prizes of* $500 *each, and five third prizes of* $100 *each are to be awarded, with all winners selected randomly.*

15. If you purchased one ticket, what are your expected gross winnings?

16. If you purchased two tickets, what are your expected gross winnings?

17. If the tickets were sold for $1 each, how much profit goes to the raffle sponsor?

18. ***Expected Sales at a Theater Snack Bar*** A children's theater found in a random survey that 65 customers bought one snack bar item, 40 bought two items, 26 bought three items, 14 bought four items, and 18 avoided the snack bar altogether. Use this information to find the expected number of snack bar items per customer. (Round your answer to the nearest tenth.)

19. ***Expected Number of Children to Attend an Amusement Park*** An amusement park, considering adding some new attractions, conducted a study over several typical days and found that, of 10,000 families entering the park, 1020 brought just one child (defined as younger than age twelve), 3370 brought two children, 3510 brought three children, 1340 brought four children, 510 brought five children, 80 brought six children, and 170 brought no children at all. Find the expected number of children per family attending this park. (Round your answer to the nearest tenth.)

20. ***Expected Sums of Randomly Selected Numbers*** Five cards are numbered 1 through 5. Two of these cards are chosen randomly (without replacement), and the numbers on them are added. Find the expected value of this sum.

21. ***Prospects for Electronics Jobs in a City*** In a certain California city, projections for the next year are that there is a 20% chance that electronics jobs will increase by 1200, a 50% chance that they will increase by 500, and a 30% chance that they will decrease by 800. What is the expected change in the number of electronics jobs in that city in the next year?

22. ***Expected Winnings in Keno*** In one version of the game *keno,* the house has a pot containing 80 balls, numbered 1 through 80. A player buys a ticket for $1 and marks one number on it (from 1 to 80). The house then selects 20 of the 80 numbers at random. If the number selected by the player is among the 20 selected by the management, the player is paid $3.20. Find the expected net winnings for this game.

23. Refer to Examples 6 and 7. Considering the three different approaches (expected values, optimist, and pessimist), which one seems most reasonable to you, and why?

Contractor Decisions Based on Expected Profits *Dawn Casselberry, a commercial building contractor, will commit her company to one of three projects depending on her analysis of potential profits or losses as shown here.*

Project *A*		Project *B*		Project *C*	
Profit or Loss x	**Probability** $P(x)$	**Profit or Loss** x	**Probability** $P(x)$	**Profit or Loss** x	**Probability** $P(x)$
$40,000	.10	$0	.20	$60,000	.65
180,000	.60	210,000	.35	340,000	.35
250,000	.30	290,000	.45		

Determine which project Dawn should choose according to each approach.

24. expected values

25. the optimist viewpoint

26. the pessimist viewpoint

Expected Winnings in a Game Show *A game show contestant is offered the option of receiving a computer system worth* $2300, *or accepting a chance to win either a luxury vacation worth* $5000 *or a boat worth* $8000. *If the second option is chosen the contestant's probabilities of winning the vacation or the boat are* .20 *and* .15, *respectively.*

27. If the contestant were to turn down the computer system and go for one of the other prizes, what would be the expected winnings?

28. Purely in terms of monetary value, what is the contestant's wiser choice?

Evaluating an Insurance Purchase *The promoter of an outdoor concert expects a gate profit of* $100,000, *unless it rains, which would reduce gate profit to* $30,000. *The probability of rain is* .28. *The promoter can purchase insurance coverage of* $100,000 *against rain losses for a premium of* $25,000. *Use this information for Exercises 29–32.*

29. Find the expected net profit when the insurance is purchased.

30. Find the expected net profit when the insurance is not purchased.

31. Based on expected values, which is the promoter's wiser choice?

32. If you were the promoter, would you base your decision on expected values? Explain your reasoning.

Expected Values in Business Accounts *The table below illustrates how a salesman for Levi Strauss & Co. rates his accounts by considering the existing volume of each account plus potential additional volume.**

1	2	3	4	5	6	7
Account Number	**Existing Volume**	**Potential Additional Volume**	**Probability of Getting Additional Volume**	**Expected Value of Additional Volume**	**Existing Volume plus Expected Value of Additional Volume**	**Classification**
1	$15,000	$10,000	.25	$2500	$17,500	
2	40,000	0	—	—	40,000	
3	20,000	10,000	.20	2000		
4	50,000	10,000	.10	1000		
5	5000	50,000	.50			
6	0	100,000	.60			
7	30,000	20,000	.80			

Use the table to work Exercises 33–36.

33. Compute the missing expected values in column 5.

34. Compute the missing amounts in column 6.

35. In column 7, classify each account according to this scheme: Class A if the column 6 value is $55,000 or more; Class B if the column 6 value is at least $45,000 but less than $55,000; Class C if the column 6 value is less than $45,000.

36. Considering all seven of this salesman's accounts, compute the total additional volume he can "expect" to get.

37. ***Expected Winnings in Keno*** Recall that in the game keno of Exercise 22, the house randomly selects 20 numbers from the counting numbers 1–80. In the variation called 6-spot keno, the player pays 60¢ for his ticket and marks 6 numbers of his choice. If the 20 numbers selected by the house contain at least 3 of those chosen by the player, he gets a payoff according to this scheme.

3 of the player's numbers among the 20	$.35
4 of the player's numbers among the 20	2.00
5 of the player's numbers among the 20	60.00
6 of the player's numbers among the 20	1250.00

Find the player's expected net winnings in this game. [*Hint:* The four probabilities required here can be found using combinations (Section 11.3), the fundamental counting principle (Section 11.2), and the theoretical probability formula (Section 12.1).]

*This information was provided by James McDonald of Levi Strauss & Co., San Francisco.

EXTENSION

Estimating Probabilities by Simulation

		Second Parent	
		R	r
First Parent	R	RR	Rr
	r	rR	rr

Simulation methods, also called **"Monte Carlo" methods,** have been successfully used in many areas of scientific study for nearly a century. Most practical applications require huge numbers of random digits, so computers are used to produce them. A computer, however, cannot toss coins. It must use an algorithmic process, programmed into the computer, which is called a **random number generator.** It is very difficult to avoid all nonrandom patterns in the results, so the digits produced are called "pseudorandom" numbers. They must pass a battery of tests of randomness before being "approved for use."

Computer scientists and physicists have been encountering unexpected difficulties with even the most sophisticated random number generators. In recent years, researchers have discovered that a random number generator can pass all the tests, and work just fine for some simulation applications, but then produce faulty answers when used with a different simulation. Therefore, random number generators apparently cannot be approved for all uses in advance, but must be carefully checked along with each new simulation application proposed.

An important area within probability theory is the process called **simulation.** It is possible to study a complicated, or unclear, phenomenon by *simulating,* or imitating, it with a simpler phenomenon involving the same basic probabilities.

For example, recall from Section 12.1 Mendel's discovery that when two Rr pea plants (red-flowered but carrying both red and white genes) are crossed, the offspring will have red flowers if an R gene is received from either parent, or from both. This is because red is dominant and white is recessive. Table 3, reproduced here in the margin, shows that three of the four equally likely possibilities result in red-flowered offspring.

Now suppose we want to know (or at least approximate) the probability that three offspring in a row will have red flowers. It is much easier (and quicker) to toss coins than to cross pea plants. And the equally likely outcomes, heads and tails, can be used to simulate the transfer of the equally likely genes, R and r. If we toss two coins, say a nickel and a penny, then we can interpret the results as follows.

hh ⇒ RR ⇒ red gene from first parent and red gene from second parent
⇒ red flowers
ht ⇒ Rr ⇒ red gene from first parent and white gene from second parent
⇒ red flowers
th ⇒ rR ⇒ white gene from first parent and red gene from second parent
⇒ red flowers
tt ⇒ rr ⇒ white gene from first parent and white gene from second parent
⇒ white flowers

Although nothing is certain for a few tosses, the law of large numbers indicates that larger and larger numbers of tosses should become better and better indicators of general trends in the genetic process.

EXAMPLE 1 Simulating Genetic Processes

Toss two coins 50 times and use the results to approximate the probability that the crossing of Rr pea plants will produce three successive red-flowered offspring.

SOLUTION

We actually tossed two coins 50 times and got the following sequence.

th, hh, th, tt, th, hh, ht, th, ht, th, hh, hh,
tt, th, hh, ht, ht, ht, ht, th, hh, hh, hh, tt,
ht, tt, hh, ht, ht, hh, tt, tt, tt, th, tt, tt, hh,
ht, ht, ht, hh, tt, th, hh, tt, hh, ht, tt, tt, tt

By the color interpretation described above, this gives the following sequence of flower colors in the offspring.

Only "both tails" gives white. →

red–red–red–white–red–red–red–red–red–red–red–red–
white–red–red–red–red–red–red–red–red–red–red–white–
red–white–red–red–red–red–white–white–white–red–white–white–red–
red–red–red–red–white–red–red–white–red–red–white–white–white

We now have an experimental list of 48 sets of three successive plants, the 1st, 2nd, and 3rd entries, then the 2nd, 3rd, and 4th entries, and so on. Do you see why there are 48 in all?

Now we just count up the number of these sets of three that are "red-red-red." Since there are 20 of those, our empirical probability of three successive red offspring, obtained through simulation, is $\frac{20}{48} = \frac{5}{12}$, or about .417. By applying the multiplication rule of probability (with all outcomes independent of one another), we find that the theoretical value is $(\frac{3}{4})^3 = \frac{27}{64}$, or about .422, so our approximation obtained by simulation is very close. ■

Pilots, astronauts, race car drivers, and others often receive a portion of their training in **simulators.** Some of these devices, which may be viewed as very technical, high-cost versions of video games, mimic, or imitate conditions to be encountered later in the "real world." A simulator session allows estimation of the likelihood, or probability, of different responses that the learner would display under actual conditions. Repeated sessions help the learner to develop more successful responses before actual equipment and lives are put at risk.

In human births boys and girls are (essentially) equally likely. Therefore, an individual birth can be simulated by tossing a fair coin, letting a head correspond to a girl and a tail to a boy.

EXAMPLE 2 Simulating Births with Coin Tossing

A sequence of 40 actual coin tosses produced the results below.

bbggb, gbbbg, gbgbb, bggbg, bbbbg, gbbgg, gbbgg, bgbbg

(For every head we have written g, for girl; for every tail, b, for boy.) Refer to this sequence to answer the following questions.

(a) How many pairs of two successive births are represented by the above sequence?
(b) How many of those pairs consist of both boys?
(c) Find the empirical probability, based on this simulation, that two successive births both will be boys. Give your answer to three decimal places.

SOLUTION

(a) Beginning with the 1st–2nd pair and ending with the 39th–40th pair, there are 39 pairs.
(b) Observing the sequence of boys and girls, we count 11 pairs of two consecutive boys.
(c) Utilizing parts (a) and (b), we have $\frac{11}{39} \approx .282$. ■

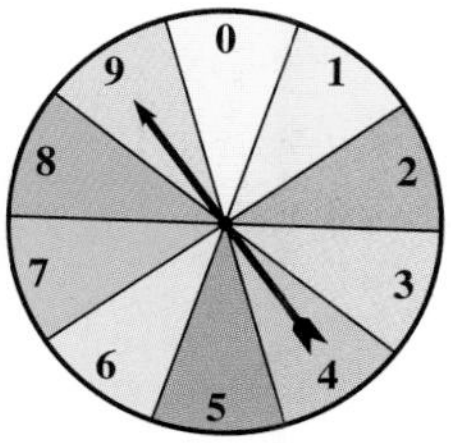

FIGURE 12

Another way to simulate births is to generate a random sequence of digits, perhaps interpreting even digits as girls and odd digits as boys. The digits might be generated by spinning the spinner in Figure 12. It turns out that many kinds of phenomena can be simulated using random digits, so we can save lots of effort by using the spinner to obtain a table of random digits, like in Table 20 on the next page, and then use that table to carry out any simulation experiment that is needed. Notice that the 250 random digits in Table 20 have been grouped conveniently so that we can easily follow down a column or across a row.

TABLE 20

→ 51592	73219 ←
77876	55707 ←
36500	48007
40571	65191 ←
04822	06772
→ 53033	94928
92080	15709 ←
01587	39922
36006	96365
63698	14655
→ 17297	65587
22841	76905
→ 91979	12369
96480	54219
74949	89329
76896	90060
47588	06975
45521	05050
02472	69774
55184	78351 ←
40177	11464
84861	84086
86937	51497 ←
20931	12307
22454	68009

EXAMPLE 3 Simulating Births with Random Numbers

A couple plans to have five children. Use random number simulation to estimate the probability they will have more than three boys.

SOLUTION

Let each sequence of five digits, as they appear in Table 20, represent a family with five children, and (arbitrarily) associate odd digits with boys, even digits with girls. (Recall that 0 is even.) Verify that, of the fifty families simulated, only the ten marked with arrows have more than 3 boys (4 boys or 5 boys). Therefore, the estimated (empirical) probability is

$$P(\text{more than 3 boys}) = \frac{10}{50} = .20.$$

The theoretical value for the probability estimated in Example 3, above, would be the same as that obtained in Example 3 of Section 12.4. It was .1875. So our estimate above was fairly close. In light of the law of large numbers, a larger sampling of random digits (more than 50 simulated families) would most likely yield a closer approximation. Extensive tables of random digits are available in statistical research publications. Computers can also be programmed to generate sequences of "pseudorandom" digits, which serve the same purposes. In most simulation experiments, much larger samples than we are using here are necessary to obtain reliable results.

EXAMPLE 4 Simulating Card Drawing with Random Numbers

Use random number simulation to estimate the probability that two cards drawn from a standard deck with replacement will both be of the same suit.

SOLUTION

Use this correspondence: 0 and 1 mean clubs, 2 and 3 mean diamonds, 4 and 5 mean hearts, 6 and 7 mean spades, 8 and 9 are disregarded. Now read down the columns of Table 20. Suppose we (arbitrarily) use the first digit of each five-digit group. The first time from top to bottom gives the sequence

$$5\text{–}7\text{–}3\text{–}4\text{–}0\text{–}5\text{–}0\text{–}3\text{–}6\text{–}1\text{–}2\text{–}7\text{–}7\text{–}4\text{–}4\text{–}0\text{–}5\text{–}4\text{–}2\text{–}2.$$

First digits of the left groups

(Five 8s and 9s were omitted.) Starting again at the top, we obtain

$$7\text{–}5\text{–}4\text{–}6\text{–}0\text{–}1\text{–}3\text{–}1\text{–}6\text{–}7\text{–}1\text{–}5\text{–}0\text{–}0\text{–}6\text{–}7\text{–}1\text{–}5\text{–}1\text{–}6.$$

First digits of the right groups

(Again, there happened to be five 8s and 9s.) This 40-digit sequence of digits yields the sequence of suits shown next.

5 gives hearts, 7 gives spades, 3 gives diamonds, and so on.

hearts–spades–diamonds–hearts–clubs–hearts–clubs–diamonds–spades–
clubs–diamonds–spades–spades–hearts–hearts–clubs–hearts–hearts–
diamonds–diamonds–spades–hearts–hearts–spades–clubs–clubs–
diamonds–clubs–spades–spades–clubs–hearts–clubs–clubs–spades–
spades–clubs–hearts–clubs–spades

Verify that, of the 39 successive pairs of suits (hearts–spades, spades–diamonds, diamonds–hearts, etc.), 9 of them are pairs of the same suit. This makes the estimated probability $\frac{9}{39} \approx .23$. (For comparison, the theoretical value is .25.)

Extension Exercises

1. ***Simulating Pea Plant Reproduction with Coin Tossing*** Explain why, in Example 1, fifty tosses of the coins produced only 48 sets of three successive offspring.

2. ***Simulating Pea Plant Reproduction with Coin Tossing*** Use the sequence of flower colors of Example 1 to approximate the probability that *four* successive offspring all will have red flowers.

3. ***Comparing the Likelihoods of Girl and Boy Births*** Should the probability of two successive girl births be any different from that of two successive boy births?

4. ***Finding Empirical Probability*** Simulate 40 births by tossing coins yourself, and obtain an empirical probability for two successive girls.

5. ***Simulating Boy and Girl Children with Random Numbers*** Use Table 20 to simulate fifty families with three children. Let 0–4 correspond to boys and 5–9 to girls, and use the middle three digits of the left hand groupings (159, 787, 650, and so on). Estimate the probability of exactly two boys in a family of three children. Compare the estimation with the theoretical probability, which is $\frac{3}{8} = .375$.

Simulating One-and-One Foul Shooting with Random Numbers *Exercises 71–73 of Section 12.3 involved one-and-one foul shooting in basketball. Susan, who had a 70% foul-shooting record, had probabilities of scoring* 0, 1, *or* 2 *points of* .30, .21, *and* .49, *respectively.*

Use Table 20 (with digits 0–6 representing hit and 7–9 representing miss) to simulate 50 *one-and-one shooting opportunities for Susan. Begin at the top of the first column (*5, 7, 3, *etc., to the bottom), then move to the second column (*1, 7, 6, *etc.), going until* 50 *one-and-one opportunities are obtained. (Some "opportunities" involve one shot while others involve two shots.) Keep a tally of the numbers of times* 0, 1, *and* 2 *points are scored.*

Number of Points	Tally
0	
1	
2	

From the tally, find the empirical probability that, on a given opportunity, Susan will score as follows. (Round to two decimal places.)

6. no points

7. 1 point

8. 2 points

Determining the Path of a Random Walk Using a Die and a Coin *Exercises 47–54 of Section 12.4 illustrated a simple version of the idea of a "random walk." Atomic particles released in nuclear fission also move in a random fashion. During World War II, John von Neumann and Stanislaw Ulam used simulation with random numbers to study particle motion in nuclear reactions. Von Neumann coined the name "Monte Carlo" for the methods used, and since then the terms "Monte Carlo methods" and "simulation methods" have often been used with very little distinction.*

The figure suggests a model for random motion in two dimensions. Assume that a particle moves in a series of 1-unit "jumps," each one in a random direction, any one of 12 *equally likely possibilities. One way to choose directions is to roll a fair die and toss a fair coin. The die determines one of the directions 1–6, coupled with heads on the coin. Tails on the coin reverses the direction of the die, so that the die coupled with tails gives directions 7–12. Thus 3h (meaning* 3 *with the die and heads with the coin) gives direction* 3*; 3t gives direction* 9 *(opposite to* 3*); and so on.*

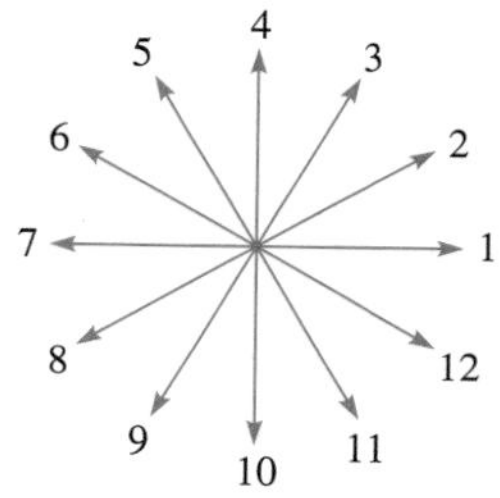

9. Simulate the motion described above with 10 rolls of a die (and tosses of a coin). Draw the 10-jump path you get. Make your drawing accurate enough so you can estimate (by measuring) how far from its starting point the particle ends up.

10. Repeat the experiment of Exercise 9 four more times. Measure distance from start to finish for each of the 5 "random trips." Add these 5 distances and divide the sum by 5, to arrive at an "expected net distance" for such a trip.

For Exercises 11 and 12, consider another two-dimensional random walk governed by the following conditions.

1. *Start out from a given street corner, and travel one block north. At each intersection:*
2. *Turn left with probability* $\frac{1}{6}$.

3. *Go straight with probability* $\frac{2}{6}$ $(= \frac{1}{3})$.
4. *Turn right with probability* $\frac{3}{6}$ $(= \frac{1}{2})$.

(Never turn around.)

11. *A Random Walk Using a Fair Die* Explain how a fair die could be used to simulate this random walk.

12. *A Random Walk Using a Random Number Table* Use Table 20 to simulate this random walk. For every 1 encountered in the table, turn left and proceed for another block. For every 2 or 3, go straight and proceed for another block. For every 4, 5, or 6, turn right and proceed for another block. Disregard all other digits, that is, 0s, 7s, 8s, and 9s. (Do you see how this scheme satisfies the probabilities given before Exercise 11?) This time begin at the upper right corner of the table, running down the column 9, 7, 7, and so on, to the bottom. Then start at the top of the next column to the left, 1, 0, 0, and so on, to the bottom. When these two columns of digits are used up, stop the "walk." Describe, in terms of distance and direction, where you have ended up relative to your starting point.

COLLABORATIVE INVESTIGATION

Finding Empirical Values of π

The information in this investigation was obtained from Burton's History of Mathematics: An Introduction, *Third Edition, by David M. Burton, published by Wm. C. Brown, 1995, page 440.*

The following problem was posed by Georges Louis Leclerc, Comte de Buffon (1707–1788) in his *Histoire Naturelle* in 1777. A large plane area is ruled with equidistant parallel lines, the distance between two consecutive lines of the series being a. A thin needle of length $\ell < a$ is tossed randomly onto the plane. What is the probability that the needle will intersect one of these lines?

The answer to this problem is found using integral calculus, and the probability p is shown to be $p = \frac{2\ell}{\pi a}$. Solving for π gives us the formula

$$\pi = \frac{2\ell}{pa}, \qquad (1)$$

which can be used to approximate the value of π experimentally. This was first observed by Pierre Simon de Laplace, and such an experiment was carried out by Johann Wolf, a professor of astronomy at Bern, in about 1850. In this investigation, we will perform a similar experiment.

See www.angelfire.com/wa/hurben/buff.html for a dynamic illustration of this Buffon Needle Problem.

Topics for Discussion

Divide the class into groups of 3 or 4 students each. Each group will need the following materials.

1. a sheet of paper with a series of parallel lines evenly spaced across it
2. a thin needle, or needlelike object, with a length less than the distance between adjacent parallel lines on the paper

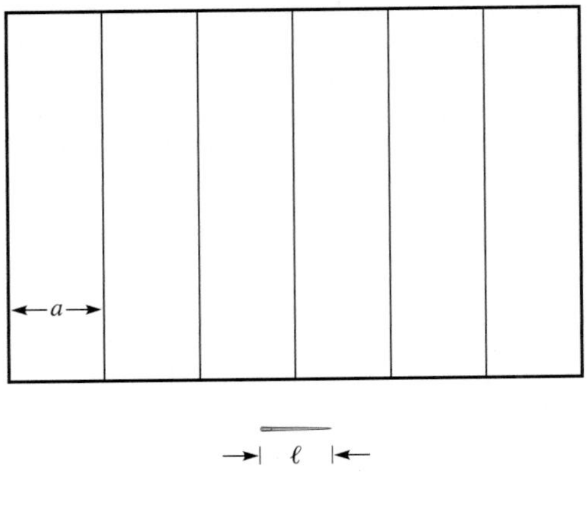

Each group should carry out these steps:

1. Measure and record the distance between lines (a) and the length of the needle (ℓ), using the same units for both.
2. Assign one member to drop the needle onto the paper, another to determine whether the needle "hits" a line or not, and another to keep a tally of hits and misses.
3. Discuss ways to minimize bias so that the position and orientation of the dropped needle will be as random as possible.
4. Drop the needle 100 times, and record the number of hits.
5. Calculate the probability $p =$ (number of hits)/100. Is this probability value theoretical or empirical?
6. Enter the calculated value of p and the measured values of a and ℓ into formula (1) to obtain a value of π. Round this value to four decimal places.

Now come back together as a class and record the various values obtained for π. Discuss the following questions.

1. The correct value of π, to four decimal places, is 3.1416. Which value of π, reported by the various groups, is most accurate? How far off is it?
2. Was it necessary to drop the needle 100 times, or could more or fewer tosses have been used?
3. Wolf tossed his needle 5000 times and it hit a line 2532 times, leading to an experimental value of π equal to 3.1596. How far off was Wolf's value?
4. How could the experiment be modified to produce "better" values for π?
5. Why could different groups use different ℓ to a ratios and still all obtain legitimate approximations for π?

CHAPTER 12 TEST

1. Explain the difference between *empirical* and *theoretical* probabilities.
2. State the *law of large numbers,* and use coin tossing to illustrate it.

Drawing Cards *A single card is chosen at random from a standard 52-card deck. Find the odds against its being each of the following.*

3. a heart
4. a red queen
5. a king or a black face card

Genetics of Cystic Fibrosis *The chart represents genetic transmission of cystic fibrosis.* C *denotes a normal gene while* c *denotes a cystic fibrosis gene. (Normal is dominant.) Both parents in this case are* Cc, *which means that they inherited one of each gene, and are therefore carriers but do not have the disease.*

		Second Parent	
		C	c
First Parent	C		Cc
	c		

6. Complete the chart, showing all four equally likely gene combinations.
7. Find the probability that a child of these parents will also be a carrier without the disease.
8. What are the odds that a child of these parents actually will have cystic fibrosis?

Days Off for Pizza Parlor Workers *The manager of a pizza parlor (which operates seven days a week) allows each of three employees to select one day off next week. Assuming the selection is done randomly and independently, find the probability of each event.*

9. All three select different days.
10. All three select the same day, given that all three select a day beginning with the same letter.
11. Exactly two of them select the same day.

Building Numbers from Sets of Digits *Two numbers are randomly selected without replacement from the set* $\{1, 2, 3, 4, 5\}$. *Find the probability of each event.*

12. Both numbers are even.
13. Both numbers are prime.

14. The sum of the two numbers is odd.

15. The product of the two numbers is odd.

Selecting Committees *A three-member committee is selected randomly from a group consisting of three men and two women.*

16. Let x denote the number of men on the committee, and complete the probability distribution table.

x	$P(x)$
0	0
1	
2	
3	

17. Find the probability that the committee members are not all men.

18. Find the expected number of men on the committee.

Rolling Dice *A pair of dice are rolled. Find the following.*

19. the probability of "doubles" (the same number on both dice)

20. the odds in favor of a sum greater than 2

21. the odds against a sum of "7 or 11"

22. the probability of a sum that is even and less than 5

Making Par in Golf *Greg Brueck has a .78 chance of making par on each hole of golf that he plays. Today he plans to play just three holes. Find the probability of each event. Round answers to three decimal places.*

23. He makes par on all three holes.

24. He makes par on exactly two of the three holes.

25. He makes par on at least one of the three holes.

26. He makes par on the first and third holes but not on the second.

Drawing Cards *Two cards are drawn, without replacement, from a standard 52-card deck. Find the probability of each event.*

27. Both cards are red.

28. Both cards are the same color.

29. The second card is a queen, given that the first card is an ace.

30. The first card is a face card and the second is black.

13

STATISTICS

In 2001, the Academy of Motion Picture Arts and Sciences presented four Oscars, including its annual Best Picture Award, to *A Beautiful Mind*, the story of mathematician John Forbes Nash, Jr. Nash was the recipient of the Nobel Memorial Prize in Economic Sciences for his work, and the movie related the story of how he and his wife dealt with the schizophrenia that afflicted him. You may wish to visit the Web page www.abeautifulmind.com/main.html to learn more about this movie and participate in an interactive session based on code-breaking, one of the branches of mathematics in which Nash did his research.

A note in the *Wall Street Journal* once claimed that during a television presentation of the Academy Awards, the average winner took 1 minute, 39 seconds to thank 7.8 friends, relatives, and loyal supporters "for making it all possible." These averages are examples of arithmetic means. *Means*, *medians*, and *modes* are all measures of central tendency, a topic in the branch of mathematics known as *statistics*.

13.1 Visual Displays of Data

Basic Concepts • Frequency Distributions • Grouped Frequency Distributions • Stem-and-Leaf Displays • Bar Graphs, Circle Graphs, and Line Graphs

Basic Concepts Governments collect and analyze an amazing quantity of "statistics"; the census, for example, is a vast project of gathering data. The census is not a new idea; two thousand years ago Mary and Joseph traveled to Bethlehem to be counted in a census. Long before, the Egyptians had recorded numerical information that is still being studied. For a long time, in fact, "statistics" referred to information about the government. The word itself comes from the Latin *statisticus,* meaning "of the state."

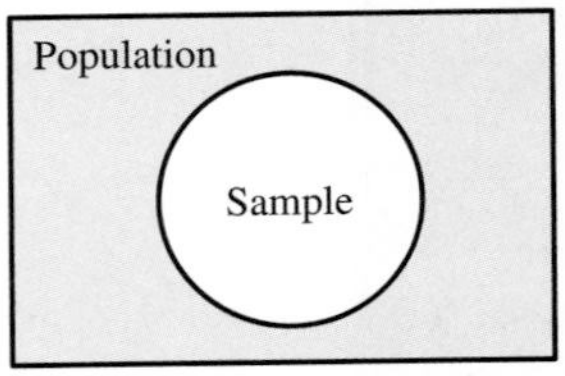

It is often important in statistics to distinguish between a **population,** which includes *all* items of interest, and a **sample,** which includes *some* (but ordinarily not all) of the items in the population. For example, to predict the outcome of an approaching presidential election, we may be interested in a population of many millions of voter preferences (those of all potential voters in the country). As a practical matter, however, even national polling organizations with considerable resources will obtain only a relatively small sample, say 2000, of those preferences. In general, a sample can be any subset of a population. See the Venn diagram in the margin.

The study of statistics can be divided into two main areas. The first, **descriptive statistics,** has to do with collecting, organizing, summarizing, and presenting data (information). The main tools of descriptive statistics are tables of numbers, various kinds of graphs, and various calculated quantities, such as averages.

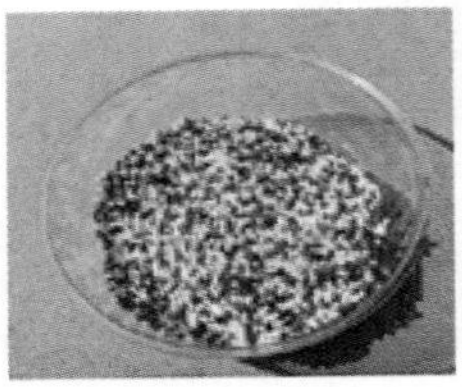

A population of 10,000

The second main area of statistics, **inferential statistics,** has to do with drawing inferences or conclusions (making conjectures) about populations based on information from samples. It is in this area that we can best see the relationship between probability and statistics.

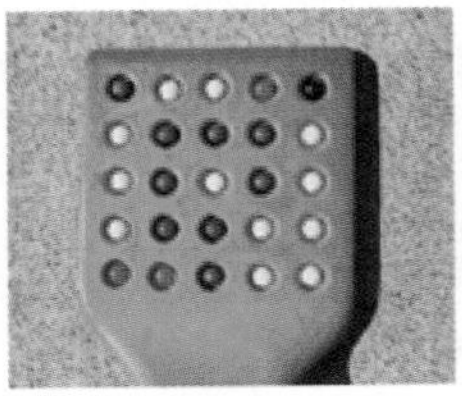

A random sample of 25

The Lansford Publishing Company, of San Jose, California, has produced a sampling demonstrator which consists of a large bowl containing 10,000 beads of various colors, along with several paddles for easily drawing samples of various sizes. Beneath the bowl, at the side, we show two random samples from the bowl, the first of which consists of 25 beads, 9 of which are green. From this sample we can "infer" that the bowl (the population) must contain about $\frac{9}{25}$, or 36%, green beads, that is, about 3600. This is a matter of *inductive reasoning.* From a particular sample we have made a conjecture about the population in general. If we increase the size of our sample, then the proportion of green beads in the larger sample will most likely give a better estimate of the proportion in the population. The second sample shown contains 100 beads, 28 of which are green. So the new estimate would be about $\frac{28}{100}$, or 28%, that is, 2800 green beads in the bowl.

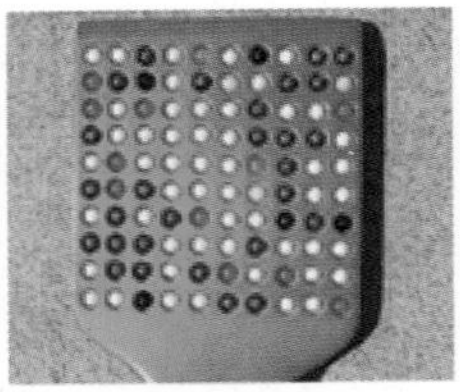

A random sample of 100

In fact, the bowl is known to contain 30%, or 3000 green beads. So our larger sample estimate, 2800, was considerably more accurate than the smaller sample estimate, 3600. Knowing the true proportion of green beads in the population, we can turn the situation around, using *deductive reasoning* to predict the proportion of green beads in a particular sample. Our prediction for a sample of 100 beads would be 30% of 100, or 30. The sample, being random, may or may not end up containing exactly 30 green beads.

To summarize this discussion: If we know what a population is like, then probability theory enables us to conclude what is likely to happen in a sample (deductive reasoning); if we know what a sample is like, then inferential statistics enables us to draw inferences about the population (inductive reasoning).

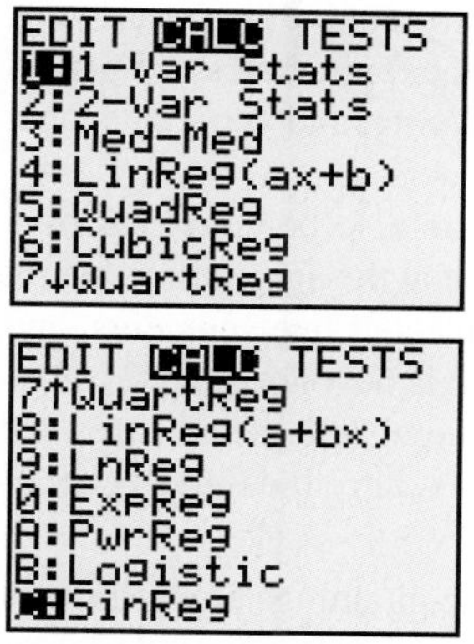

Various statistical options on the TI-83/84 Plus.

Most scientific and graphing calculators will perform all computations we describe in this chapter, including sorting of data (once they are entered). So your purpose in studying this material should be to understand the meanings of the various computed quantities. It is recommended that the actual computations, especially for sizable collections of data, be done on a calculator (or a computer with suitable statistical software).

Information that has been collected but not yet organized or processed is called **raw data.** It is often **quantitative** (or **numerical**), but can also be **qualitative** (or **nonnumerical**), as illustrated in Table 1.

TABLE 1 Examples of Raw Data

Quantitative data: The number of siblings in ten different families: 3, 1, 2, 1, 5, 4, 3, 3, 8, 2

Qualitative data: The makes of six different automobiles: Toyota, Ford, Nissan, Toyota, Chevrolet, Honda

Quantitative data are generally more useful when they are **sorted,** or arranged in numerical order. In sorted form, the first list in Table 1 appears as

1, 1, 2, 2, 3, 3, 3, 4, 5, 8.

Frequency Distributions When a data set includes many repeated items, it can be organized into a **frequency distribution,** which lists the distinct data values (x) along with their frequencies (f). The frequency designates the number of times the corresponding item occurred in the data set. It is also helpful to show the **relative frequency** of each distinct item. This is the fraction, or percentage, of the data set represented by the item. If n denotes the total number of items, and a given item, x, occurred f times, then the relative frequency of x is $\frac{f}{n}$. Example 1 illustrates these ideas.

EXAMPLE 1 Constructing Frequency and Relative Frequency Distributions

The 25 members of a psychology class were polled as to the number of siblings in their individual families. Construct a frequency distribution and a relative frequency distribution for their responses, which are shown here.

2, 3, 1, 3, 3, 5, 2, 3, 3, 1, 1, 4, 2, 4, 2, 5, 4, 3, 6, 5, 1, 6, 2, 2, 2

SOLUTION

The data range from a low of 1 to a high of 6. The frequencies (obtained by inspection) and relative frequencies are shown in Table 2.

TABLE 2 Frequency and Relative Frequency Distributions for Numbers of Siblings

Number x	Frequency f	Relative Frequency $\frac{f}{n}$
1	4	$\frac{4}{25} = 16\%$
2	7	$\frac{7}{25} = 28\%$
3	6	$\frac{6}{25} = 24\%$
4	3	$\frac{3}{25} = 12\%$
5	3	$\frac{3}{25} = 12\%$
6	2	$\frac{2}{25} = 8\%$

■

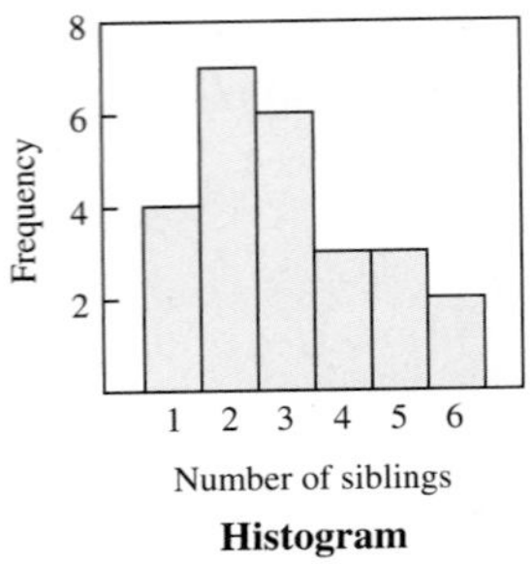

FIGURE 1

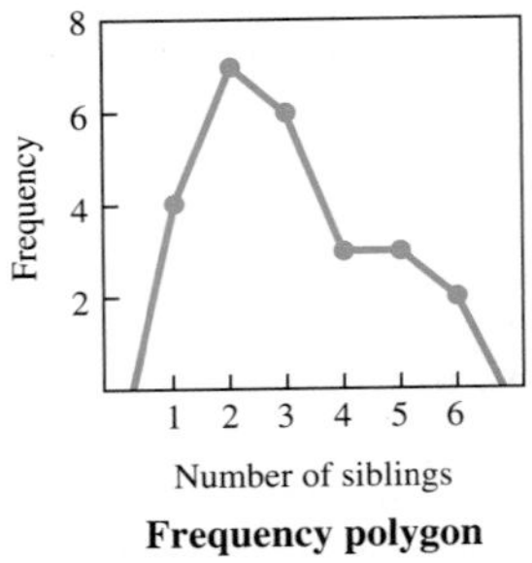

FIGURE 2

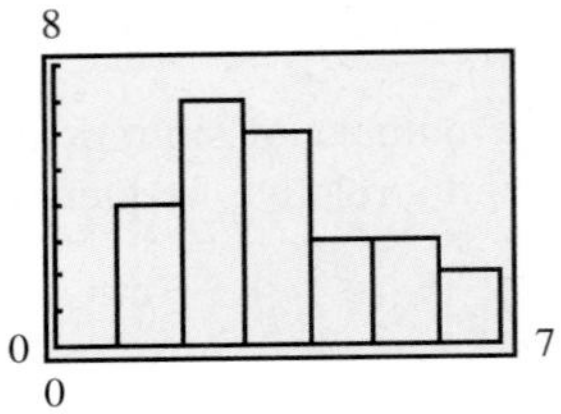

This histogram was generated using the data in Table 2. Compare with Figure 1.

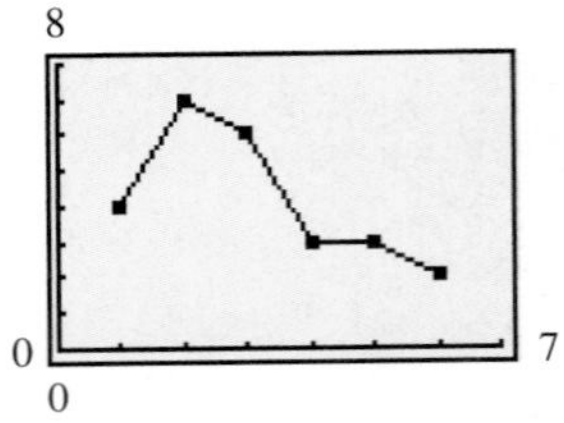

This line graph resembles the frequency polygon in Figure 2. It was generated using the data in Table 2.

The numerical data of Table 2 can be more easily interpreted with the aid of a **histogram.** A series of rectangles, whose lengths represent the frequencies, are placed next to one another as shown in Figure 1. On each axis, horizontal and vertical, a label and the numerical scale should be shown.

The information shown in the histogram in Figure 1 can also be conveyed by a **frequency polygon,** as in Figure 2. Simply plot a single point at the appropriate height for each frequency, connect the points with a series of connected line segments, and complete the polygon with segments that trail down to the axis beyond 1 and 6.

The frequency polygon is an instance of the more general *line graph,* used for many kinds of data, not just frequencies. Line graphs were first introdued in Chapter 1.

Grouped Frequency Distributions

Data sets containing large numbers of items are often arranged into groups, or *classes.* All data items are assigned to their appropriate classes, and then a **grouped frequency distribution** can be set up and a graph displayed. Although there are no fixed rules for establishing the classes, most statisticians agree on a few general guidelines.

Guidelines for the Classes of a Grouped Frequency Distribution

1. Make sure each data item will fit into one, and only one, class.
2. Try to make all classes the same width.
3. Make sure the classes do not overlap.
4. Use from 5 to 12 classes. (Too few or too many classes can obscure the tendencies in the data.)

EXAMPLE 2 Constructing a Histogram and a Frequency Polygon

Forty students, selected randomly in the school cafeteria on a Monday morning, were asked to estimate the number of hours they had spent studying in the past week (including both in-class and out-of-class time). Their responses are recorded here.

18	60	72	58	20	15	12	26	16	29
26	41	45	25	32	24	22	55	30	31
55	39	29	44	29	14	40	31	45	62
36	52	47	38	36	23	33	44	17	24

Tabulate a grouped frequency distribution and a grouped relative frequency distribution and construct a histogram and a frequency polygon for the given data.

SOLUTION

Scanning the data, we see that they range from a low of 12 to a high of 72 (that is, over a range of $72 - 12 = 60$ units.). The widths of the classes should be uniform (by Guideline 2), and there should be from 5 to 12 classes (by Guideline 4). Five classes would imply a class width of about $\frac{60}{5} = 12$, while twelve classes would imply a class width of about $\frac{60}{12} = 5$. We arbitrarily choose a class width of 10, a good round number between 5 and 12. We (arbitrarily again) let our classes run from 10 through 19, from 20 through 29, and so on up to 70 through 79, for a total of seven classes. All four guidelines are met.

We go through the data set, tallying each item into the appropriate class. The tally totals produce class frequencies, which in turn produce relative frequencies, as shown in Table 3. The histogram is displayed in Figure 3.

TABLE 3 Grouped Frequency and Relative Frequency Distributions for Weekly Study Times

Class Limits	Tally	Frequency f	Relative Frequency $\frac{f}{n}$
10–19	𝍸 |	6	$\frac{6}{40} = 15.0\%$
20–29	𝍸 𝍸 |	11	$\frac{11}{40} = 27.5\%$
30–39	𝍸 ||||	9	$\frac{9}{40} = 22.5\%$
40–49	𝍸 ||	7	$\frac{7}{40} = 17.5\%$
50–59	||||	4	$\frac{4}{40} = 10.0\%$
60–69	||	2	$\frac{2}{40} = 5.0\%$
70–79	|	1	$\frac{1}{40} = 2.5\%$
	Total:	$n = 40$	

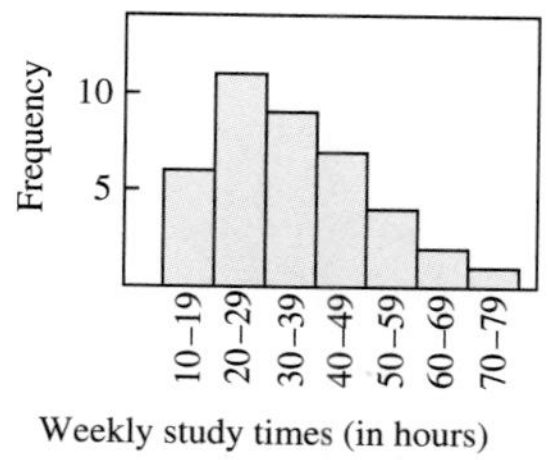

Grouped frequency histogram

FIGURE 3

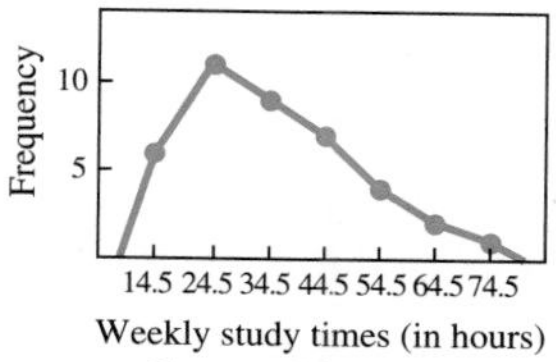

Grouped frequency polygon

FIGURE 4

In Table 3 (and Figure 3) the numbers 10, 20, 30, and so on are called the **lower class limits.** They are the smallest possible data values within the respective classes. The numbers 19, 29, 39, and so on are called the **upper class limits.** The common **class width** for the distribution is the difference of any two successive lower class limits (such as 30–20), or of any two successive upper class limits (such as 59–49). The class width for this distribution is 10, as noted earlier.

To construct a frequency polygon, notice that, in a *grouped* frequency distribution, the data items in a given class are generally not all the same. We can obtain the "middle" value, or **class mark,** by adding the lower and upper class limits and dividing this sum by 2. We locate all the class marks along the horizontal axis and plot points above the class marks. The heights of the plotted points represent the class frequencies. The resulting points are connected just as for an ordinary (nongrouped) frequency distribution. The result, in this case, is shown in Figure 4. ■

TABLE 4 Grouped Frequency Distribution for Weekly Study Times

Class Limits	Frequency
10–19	6
20–29	11
30–39	9
40–49	7
50–59	4
60–69	2
70–79	1

Stem-and-Leaf Displays

In Table 3, the tally marks give a good visual impression of how the data are distributed. (It is clear, for example, that the highest frequency occurs in the class 20–29 and that the items are concentrated more heavily in the lower classes generally.) In fact, the tally marks are almost like a histogram turned on its side. Nevertheless, once the tallying is done, the tally marks are usually dropped, and the grouped frequency distribution is presented as in Table 4.

The pictorial advantage of the tally marks is now lost. Furthermore, we cannot tell, from the grouped frequency distribution itself (or from the tally marks either, for that matter), what any of the original items were. We only know, for example, that there were seven items in the class 40–49. We do not know specifically what any of them were.

To avoid the difficulties mentioned previously, we can employ a tool of exploratory data analysis, the **stem-and-leaf display,** as shown in Example 3.

EXAMPLE 3 Constructing a Stem-and-Leaf Display

Present the study times data of Example 2 in a stem-and-leaf display.

SOLUTION

Going back to Example 2 for the original raw data, we arrange the numbers in Table 5. The tens digits, to the left of the vertical line, are the "stems," while the corresponding ones digits are the "leaves." We have entered all items from the first row of the original data, from left to right, then the items from the second row, and so on through the fourth row.

TABLE 5 Stem-and-Leaf Display for Weekly Study Times

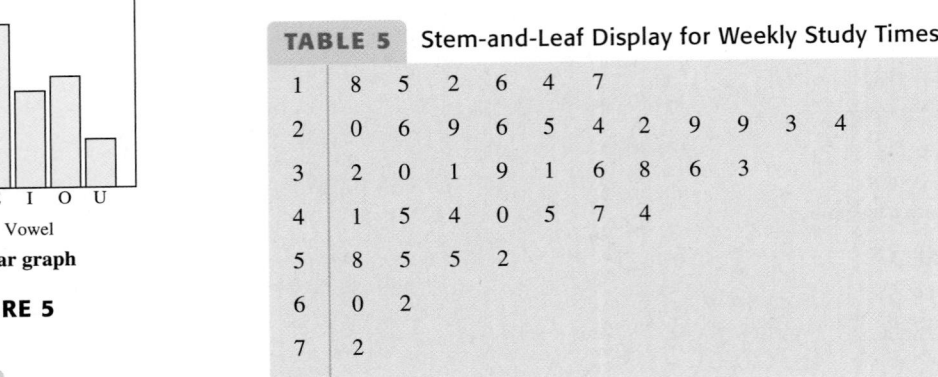

1	8	5	2	6	4	7					
2	0	6	9	6	5	4	2	9	9	3	4
3	2	0	1	9	1	6	8	6	3		
4	1	5	4	0	5	7	4				
5	8	5	5	2							
6	0	2									
7	2										

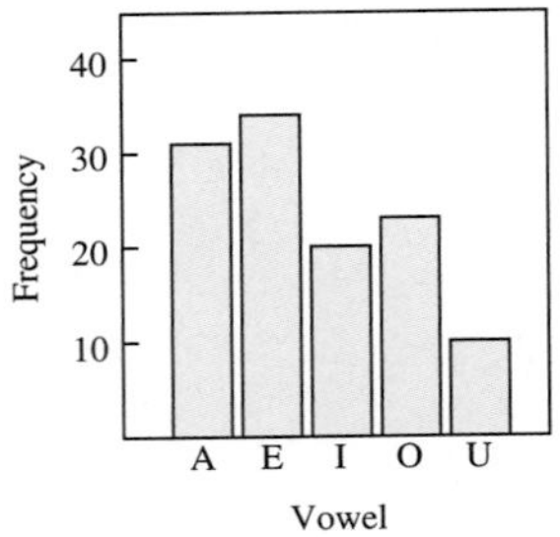

Bar graph

FIGURE 5

Notice that the stem-and-leaf display of Example 3 conveys at a glance the same pictorial impressions that a histogram would convey without the need for constructing the drawing. It also preserves the exact data values.

TABLE 6 Student Expenses

Expense	Percent of Total
Food	30%
Rent	25%
Entertainment	15%
Clothing	10%
Books	10%
Other	10%

Bar Graphs, Circle Graphs, and Line Graphs

A frequency distribution of nonnumerical observations can be presented in the form of a **bar graph,** which is similar to a histogram except that the rectangles (bars) usually are not touching one another and sometimes are arranged horizontally rather than vertically. The bar graph of Figure 5 shows the frequencies of occurrence of the vowels A, E, I, O, and U in this paragraph.

A graphical alternative to the bar graph is the **circle graph,** or **pie chart,** which uses a circle to represent the total of all the categories and divides the circle into sectors, or wedges (like pieces of pie), whose sizes show the relative magnitudes of the categories. The angle around the entire circle measures 360°. For example, a category representing 20% of the whole should correspond to a sector whose central angle is 20% of 360°, that is, $.20(360°) = 72°$.

EXAMPLE 4 Constructing a Circle Graph

Nola Akala found that, during her first semester of college, her expenses fell into categories as shown in Table 6. Present this information in a circle graph.

SOLUTION

The central angle of the sector for food is $.30(360°) = 108°$. Rent is $.25(360°) = 90°$. Calculate the other four angles similarly. Then draw a circle and mark off the angles with a protractor. The completed circle graph appears in Figure 6.

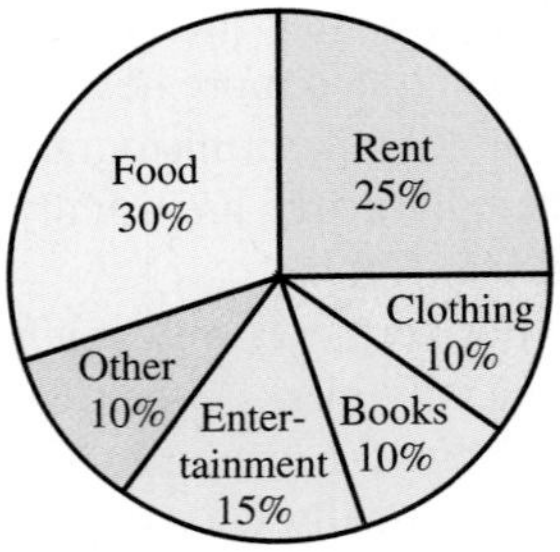

Expense categories

FIGURE 6

A circle graph shows, at a glance, the relative magnitudes of various categories. If we are interested in demonstrating how a quantity *changes,* say with respect to time, we use a **line graph.** We connect a series of line segments that rise and fall with time, according to the magnitude of the quantity being illustrated. To compare the patterns of change for two or more quantities, we can even plot multiple line graphs together. (A line graph looks somewhat like a frequency polygon, but the quantities graphed are not necessarily frequencies.)

EXAMPLE 5 Constructing and Interpreting a Line Graph

Nola, from Example 4, wanted to keep track of her major expenses, food and rent, over the course of four years of college (eight semesters), in order to see how each one's budget percentage changed with time and also how the two compared. Use the data she collected (Table 7) to show this information in a line graph, and state any significant conclusions that are apparent from the graph.

TABLE 7 Food and Rent Expense Percentages

Semester	Food	Rent
First	30%	25%
Second	31	26
Third	30	28
Fourth	29	29
Fifth	28	34
Sixth	31	34
Seventh	30	37
Eighth	29	38

SOLUTION

A comparison line graph for the given data (Figure 7) shows that the food percentage stayed fairly constant over the four years (at close to 30%), while the rent percentage, starting several points below food, rose steadily, surpassing food after the fourth semester and finishing significantly higher than food.

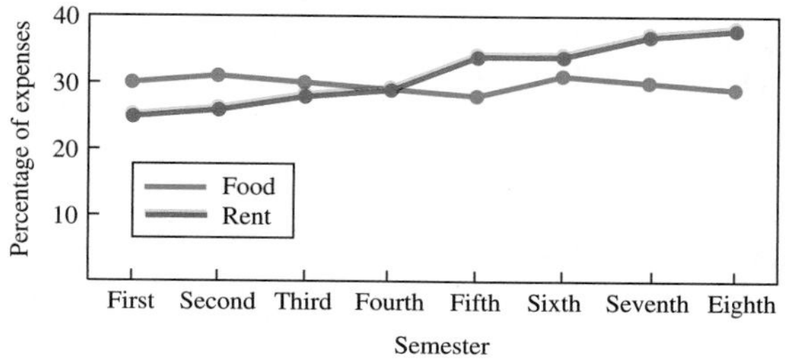

Comparison line graph

FIGURE 7

For Further Thought

Expected and Observed Frequencies

When fair coins are tossed, the results on particular tosses cannot be reliably predicted. As more and more coins are tossed, however, the proportions of heads and tails become more predictable. This is a consequence of the "law of large numbers." For example, if five coins are tossed, then the resulting number of heads, denoted x, is a "random variable," whose possible values are 0, 1, 2, 3, 4, and 5. If the five coins are tossed repeatedly, say 64 separate times, then the binomial probability formula can be used to get **expected frequencies** (or **theoretical frequencies**), as shown in the table to the right. The first two columns of the table comprise the **expected frequency distribution** for 64 tosses of five fair coins. In an actual experiment, we could obtain **observed frequencies** (or **empirical frequencies**), which would most likely differ somewhat from the expected frequencies. But 64 repetitions of the experiment should be enough to provide fair consistency between expected and observed values.

Number of Heads x	Expected Frequency e	Observed Frequency o
0	2	
1	10	
2	20	
3	20	
4	10	
5	2	

For Group Discussion or Individual Investigation

Toss five coins a total of 64 times, keeping a record of the results.

1. Enter your experimental results in the third column of the table above, producing an **observed frequency distribution.**
2. Construct two histograms, one from the expected frequency distribution and one from your observed frequency distribution.
3. Compare the two histograms, and explain why they are different.

13.1 EXERCISES

In Exercises 1 and 2, use the given data to do the following:

(a) *Construct frequency and relative frequency distributions, in a table similar to Table 2.*

(b) *Construct a histogram.*

(c) *Construct a frequency polygon.*

1. *Questions Omitted on a Math Exam* The following data are the numbers of questions omitted on a math exam by the 30 members of the class.

1	0	3	0	0	2	1	2	2	0	0	5	1	1	3
4	2	0	2	0	1	0	1	2	3	3	4	0	1	0

2. *Quiz Scores in an Economics Class* The following data are quiz scores for the members of an economics class.

8	5	6	10	4	7	2	7	6	3	1	7	4	9
5	9	2	6	5	4	6	6	8	4	10	8	9	7

In each of Exercises 3–6, use the given data to do the following:

(a) *Construct grouped frequency and relative frequency distributions, in a table similar to Table 3. (In each case, follow the suggested guidelines for class limits and class width.)*

(b) *Construct a histogram.*

(c) *Construct a frequency polygon.*

3. *Heights of Baseball Players* The heights (in inches) of the 54 starting players in a baseball tournament were as follows.

53	51	65	62	61	55	59	52	62
64	48	54	64	57	51	67	60	49
49	59	54	52	53	60	58	60	64
52	56	56	58	66	59	62	50	58
60	63	64	52	60	58	63	53	56
58	61	55	50	65	56	61	55	54

Use five classes with a uniform class width of 5 inches, and use a lower limit of 45 inches for the first class.

4. ***Charge Card Account Balances*** The following raw data represent the monthly account balances (to the nearest dollar) for a sample of 50 brand-new charge card users.

138	78	175	46	79	118	90	163	88	107
126	154	85	60	42	54	62	128	114	73
129	130	81	67	119	116	145	105	96	71
100	145	117	60	125	130	94	88	136	112
118	84	74	62	81	110	108	71	85	165

Use seven classes with a uniform width of 20 dollars, where the lower limit of the first class is 40 dollars.

5. ***Daily High Temperatures*** The following data represent the daily high temperatures (in degrees Fahrenheit) for the month of June in a southwestern U.S. city.

79	84	88	96	102	104	99	97	92	94
85	92	100	99	101	104	110	108	106	106
90	82	74	72	83	107	111	102	97	94

Use nine classes with a uniform width of 5 degrees, where the lower limit of the first class is 70 degrees.

6. ***IQ Scores of Tenth Graders*** The following data represent IQ scores of a group of 50 tenth graders.

113	109	118	92	130	112	114	117	122	115
127	107	108	113	124	112	111	106	116	118
121	107	118	118	110	124	115	103	100	114
104	124	116	123	104	135	121	126	116	111
96	134	98	129	102	103	107	113	117	112

Use nine classes with a uniform width of 5, where the lower limit of the first class is 91.

In each of Exercises 7–10, construct a stem-and-leaf display for the given data. In each case, treat the ones digits as the leaves. For any single-digit data, use a stem of 0.

7. ***Games Won in the National Basketball Association*** On a certain date approaching midseason, the teams in the National Basketball Association had won the following numbers of games.

20	29	11	26	11	12	7	26	18
19	14	13	22	9	25	11	10	15
10	22	23	31	8	24	15	24	15

8. ***Accumulated College Units*** The students in a calculus class were asked how many college units they had accumulated to date. Their responses are shown at the top of the next column.

12	4	13	12	21	22	15	17	33	24
32	42	26	11	53	62	42	25	13	8
54	18	21	14	19	17	38	17	20	10

9. ***Distances to School*** Following are the daily round-trip distances to school (in miles) for 30 randomly chosen students at a community college in California.

12	30	10	11	18	26	34	18	8	12
26	14	5	22	4	25	9	10	6	21
18	18	9	16	44	23	4	13	36	8

10. ***Yards Gained in the National Football League*** The following data represent net yards gained per game by National Football League running backs who played during a given week of the season.

28	19	36	73	37	88	67	33	54	123	79
12	39	45	22	58	7	73	30	43	24	36
51	43	33	55	40	29	112	60	94	86	62
42	29	18	25	41	3	49	102	16	32	46

Network Television News Ratings *Graphs of all kinds often are artistically embellished in order to catch attention or to emphasize meaning. The comparison line graph here, presented in the form of a television screen, shows ratings trends for evening news broadcasts at three major U.S. television networks. Refer to the graph for Exercises 11–15.*

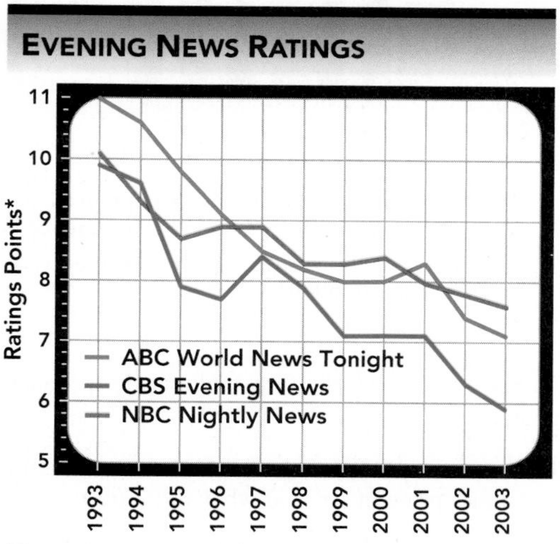

*One rating point represents 1 percent of the total U.S. households with a television.

Source: Nielsen Media Research unpublished data, www.nielsenmedia.com

Ratings taken for month of November.

11. Of all U.S. households with televisions, about what total percentage (to the nearest whole percentage) watched the three networks' news broadcasts in 1993?

12. Of all U.S. households with televisions, about what total percentage (to the nearest whole percentage) watched the three networks' news broadcasts in 2003?

13. Which network's ratings dropped the most over the given period, and how much did they drop?

14. In which one-year period did none of the three networks experience a ratings drop?

15. Over the entire period represented, what was the least spread between the highest and lowest rated networks? In what year did it occur?

Reading Bar Graphs of Economic Indicators *The bar graphs here show trends in several economic indicators over the period 2000–2005. Refer to these graphs for Exercises 16–20.*

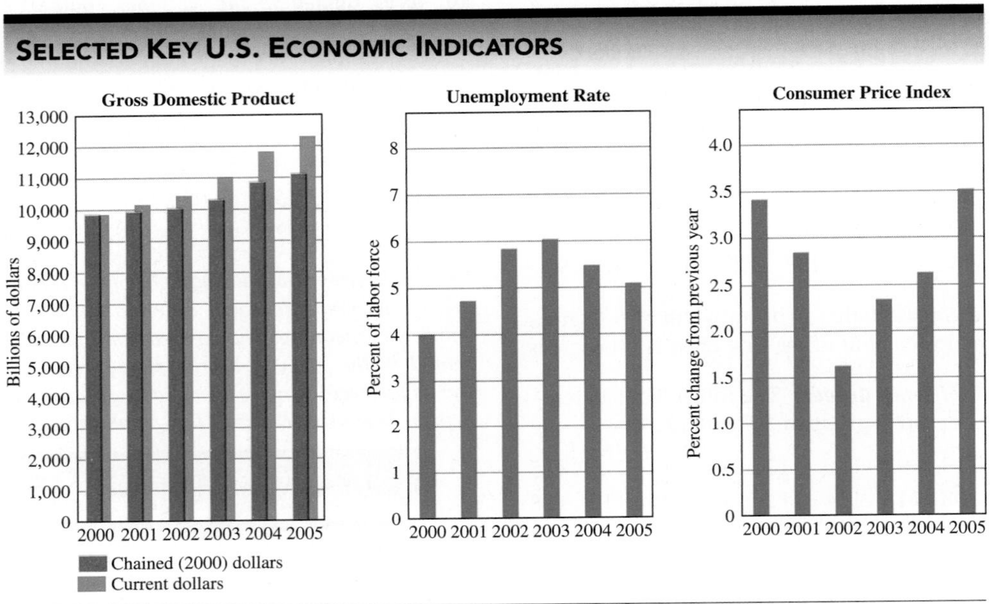

Source: U.S. Department of Commerce and U.S. Department of labor; except 2005 figures, which are estimates from The Conference Board.

Source: 2006 World Book Year Book, page 169. World Book, Inc., 233 N. Michigan Ave., Chicago, IL 60601.

16. About what was the gross domestic product, in chained (2000) dollars, in 2005?

17. Over the six-year period, about what was the highest consumer price index, and when did it occur?

18. What was the greatest year-to-year change in the unemployment rate, and when did it occur?

19. Describe the apparent trends in both chained dollar and current dollar gross domestic product. Would it be possible for one of these to actually decrease the same year that the other increased? Why or why not?

20. Explain why the gross domestic product would generally increase when the unemployment rate decreases.

Reading Circle Graphs of Primary Energy Sources *The circle graph on the next page shows the primary energy sources used to generate electricity in the U.S. in 2003. Refer to the graph for Exercises 21 and 22.*

21. To the nearest degree, what is the central angle of the "Coal" sector?

22. Assuming the "Other" category includes five primary sources, what percentage (to the nearest tenth of a percent) did each of the "others" contribute on the average?

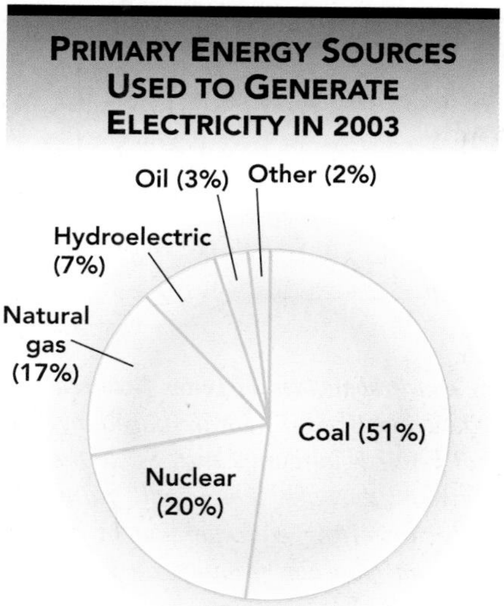

Source: www.eia.doe.gov

23. ***Sources of Job Training*** A survey by the Bureau of Labor Statistics asked American workers how they were trained for their jobs. The percentages who responded in various categories are shown in the table below. Use the information in the table to draw a circle graph.

Principal Source of Training	Approximate Percentage of Workers
Trained in school	33%
Informal on-the-job training	25
Formal training from employers	12
Trained in military, or correspondence or other courses	10
No particular training, or could not identify any	20

24. ***Correspondence Between Education and Earnings*** Bureau of Labor Statistics data for a recent year showed that the average annual earnings of American workers corresponded to educational level as shown in the table below. Draw a bar graph that shows this information.

Educational Level	Average Annual Earnings
Less than 4 years of high school	$18,990
High school graduate	28,763
Associate degree	39,015
Bachelor's degree	50,916
Master's degree	61,698

Net Worth of Retirement Savings *Dave Chwalik, wishing to retire at age 60, is studying the comparison line graph here, which shows (under certain assumptions) how the net worth of his retirement savings (initially $200,000 at age 60) will change as he gets older and as he withdraws living expenses from savings. Refer to the graph for Exercises 25–28.*

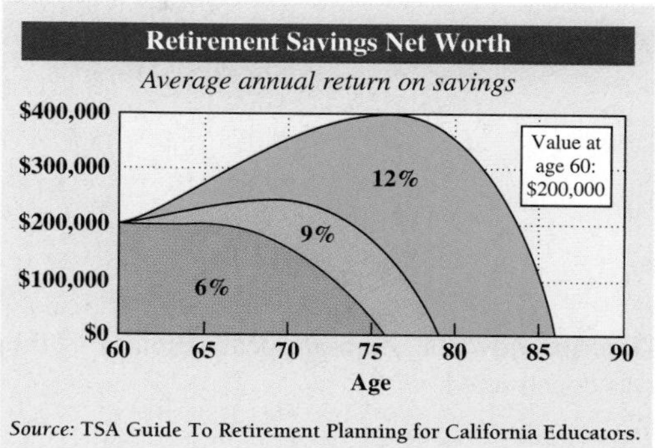

Source: TSA Guide To Retirement Planning for California Educators.

25. Assuming Dave can maintain an average annual return of 9%, how old will he be when his money runs out?

26. If he could earn an average of 12% annually, what maximum net worth would Dave achieve? At about what age would the maximum occur?

27. Suppose Dave reaches age 70, in good health, and the average annual return has proved to be 6%.
    (a) About how much longer can he expect his money to last?
    (b) What options might he consider in order to extend that time?

28. At age 77, about how many times more will Dave's net worth be if he averages a 12% return than if he averages a 9% return?

Sample Masses in a Geology Laboratory *Stem-and-leaf displays can be modified in various ways in order to obtain a reasonable number of stems. The following data, representing the measured masses (in grams) of thirty mineral samples in a geology lab, are shown in a* **double-stem** *display in Table 8 on the next page.*

60.7	41.4	50.6	39.5	46.4
58.1	49.7	38.8	61.6	55.2
47.3	52.7	62.4	59.0	44.9
35.6	36.2	40.6	56.9	42.6
34.7	48.3	55.8	54.2	33.8
51.3	50.1	57.0	42.8	43.7

TABLE 8 Stem-and-Leaf Display for Mineral Sample Masses

(30–34)	3	4.7	3.8				
(35–39)	3	9.5	8.8	5.6	6.2		
(40–44)	4	1.4	4.9	0.6	2.6	2.8	3.7
(45–49)	4	6.4	9.7	7.3	8.3		
(50–54)	5	0.6	2.7	4.2	1.3	0.1	
(55–59)	5	8.1	5.2	9.0	6.9	5.8	7.0
(60–64)	6	0.7	1.6	2.4			

29. Describe how the stem-and-leaf display of Table 8 was constructed.

30. Explain why Table 8 is called a "double-stem" display.

31. In general, how many stems (total) are appropriate for a stem-and-leaf display? Explain your reasoning.

32. According to the U.S. National Climatic Data Center, the highest temperatures (in degrees Fahrenheit) ever recorded in the 50 states were as follows.

112	100	128	120	134	118	106	110	109	112
100	118	117	116	118	121	114	114	105	109
107	112	114	115	118	117	118	125	106	110
122	108	110	121	113	120	119	111	104	111
120	113	120	117	105	110	118	112	114	116

Present these data in a double-stem display.

33. ***Letter Occurrence Frequencies in the English Language*** The table at the top of the next column shows commonly accepted percentages of occurrence for the various letters in English language usage. (Code breakers have carefully analyzed these percentages as an aid in deciphering secret codes.) For example, notice that E is the most commonly occurring letter, followed by T, A, O, N, and so on. The letters Q and Z occur least often. Referring to Figure 5 in the text, would you say that the relative frequencies of occurrence of the vowels in the associated paragraph were typical or unusual? Explain your reasoning.

Letter	Percent
E	13
T	9
A, O	8
N	7
I, R	$6\frac{1}{2}$
S, H	6
D	4

Letter	Percent
L	$3\frac{1}{2}$
C, M, U	3
F, P, Y	2
W, G, B	$1\frac{1}{2}$
V	1
K, X, J	$\frac{1}{2}$
Q, Z	$\frac{1}{5}$

Frequencies and Probabilities of Letter Occurrence *The percentages shown in Exercise 33 are based on a very large sampling of English language text. Since they are based upon experiment, they are "empirical" rather than "theoretical." By converting each percent in that table to a decimal fraction, you can produce an* ***empirical probability distribution.***

For example, if a single letter is randomly selected from a randomly selected passage of text, the probability that it will be an E *is* .13. *The probability that a randomly selected letter would be a vowel* (A, E, I, O, *or* U) *is*

$$(.08 + .13 + .065 + .08 + .03) = .385.$$

34. Rewrite the distribution shown in Exercise 33 as an empirical probability distribution. Give values to three decimal places. Note that the 26 probabilities in this distribution—one for each letter of the alphabet—should add up to 1 (except for, perhaps, a slight round-off error).

35. **(a)** From your distribution of Exercise 34, construct an empirical probability distribution just for the vowels A, E, I, O, and U. (*Hint:* Divide each vowel's probability, from Exercise 34, by .385 to obtain a distribution whose five values add up to 1.) Give values to three decimal places.

(b) Construct an appropriately labeled bar chart from your distribution of part (a).

36. Based on the occurrences of vowels in the paragraph represented by Figure 5, construct a probability distribution for the vowels. The frequencies are:

A–31, E–34, I–20, O–23, U–10.

Give probabilities to three decimal places.

37. Is the probability distribution of Exercise 36 theoretical or empirical? Is it different from the distribution of Exercise 35? Which one is more accurate? Explain your reasoning.

38. ***Frequencies and Probabilities of Study Times*** Convert the grouped frequency distribution of Table 3 to an empirical probability distribution, using the same classes and giving probability values to three decimal places.

39. ***Probabilities of Study Times*** Recall that the distribution of Exercise 38 was based on weekly study times for a sample of 40 students. Suppose one of those students was chosen randomly. Using your distribution, find the probability that the study time in the past week for the student selected would have been in each of the following ranges.
   (a) 30–39 hours
   (b) 40–59 hours
   (c) fewer than 30 hours
   (d) at least 50 hours

Favorite Sports Among Recreation Students *The 40 members of a recreation class were asked to name their favorite sports. The table in the next column shows the numbers who responded in various ways.*

Sport	Number of Class Members
Sailing	9
Hang gliding	5
Bungee jumping	7
Sky diving	3
Canoeing	12
Rafting	4

Use this information in Exercises 40–42.

40. If a member of this class is selected at random, what is the probability that the favorite sport of the person selected is bungee jumping?

41. **(a)** Based on the data in the table, construct a probability distribution, giving probabilities to three decimal places.
   (b) Is the distribution of part (a) theoretical or is it empirical?
   (c) Explain your answer to part (b).

42. Explain why a frequency polygon trails down to the axis at both ends while a line graph ordinarily does not.

13.2 Measures of Central Tendency

Mean • Median • Mode • Central Tendency from Stem-and-Leaf Displays • Symmetry in Data Sets • Summary

Video Recyclers, a local business that sells "previously viewed" movies, had the following daily sales over a one-week period:

$$\$305, \quad \$285, \quad \$240, \quad \$376, \quad \$198, \quad \$264.$$

It would be desirable to have a single number to serve as a kind of representative value for this whole set of numbers—that is, some value around which all the numbers in the set tend to cluster, a kind of "middle" number or a **measure of central tendency.** Three such measures are discussed in this section.

Many calculators find the mean (as well as other statistical measures) automatically when a set of data items are entered. To recognize these calculators, look for a key marked $\boxed{\bar{x}}$, or perhaps $\boxed{\mu}$, or look in a menu such as "LIST" for a listing of mathematical measures.

Mean

The most common measure of central tendency is the **mean** (more properly called the **arithmetic mean**). The mean of a sample is denoted $\bar{x}$ (read "x bar"), while the mean of a complete population is denoted μ (the lower case Greek letter *mu*). Inferential statistics often involves both sample means and the population mean in the same discussion, but for our purposes here, data sets are considered to be samples, so we use $\bar{x}$.

The mean of a set of data items is found by adding up all the items and then dividing the sum by the number of items. (The mean is what most people associate with the word "average.") Since adding up, or summing, a list of items is a common procedure in statistics, we use the symbol for "summation," Σ (the capital Greek letter **sigma**). Therefore, the sum of n items, say $x_1, x_2, \ldots, x_n$, can be denoted $\Sigma x = x_1 + x_2 + \cdots + x_n$, and the mean is found as follows.

Mean

The **mean** of n data items $x_1, x_2, \ldots, x_n$, is given by the formula

$$\bar{x} = \frac{\Sigma x}{n}.$$

Now use this formula to find the central tendency of the daily sales figures for Video Recyclers:

A calculator can find the mean of items in a list. This screen supports the text discussion of daily sales figures.

$$\begin{aligned}\text{Mean} = \bar{x} &= \frac{\Sigma x}{n}\\ &= \frac{305 + 285 + 240 + 376 + 198 + 264}{6}\\ &= \frac{1668}{6}\\ &= 278.\end{aligned}$$

The mean value (the “average daily sales”) for the week is \$278.

EXAMPLE 1 Finding the Mean of a List of Sales Figures

Last year’s annual sales for eight different flower shops were

\$374,910	\$321,872	\$242,943	\$351,147
\$382,740	\$412,111	\$334,089	\$262,900.

Find the mean annual sales for the eight shops.

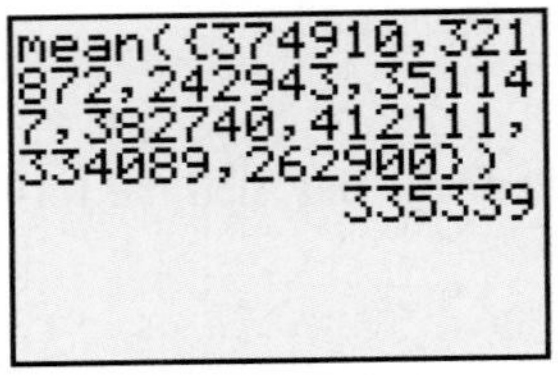

This screen supports the result in Example 1.

SOLUTION

$$\bar{x} = \frac{\Sigma x}{n} = \frac{2,682,712}{8} = 335,339$$

The mean annual sales amount is \$335,339.

The following table shows the units and grades earned by one student last term.

Course	Grade	Units
Mathematics	A	3
History	C	3
Chemistry	B	5
Art	B	2
PE	A	1

In one common system of finding a **grade-point average,** an A grade is assigned 4 points, with 3 points for B, 2 for C, and 1 for D. Find the grade-point average by multiplying the number of units for a course and the number assigned to each grade, and then adding these products. Finally, divide this sum by the total number of units.

Course	Grade	Grade Points	Units	(Grade Points) · (Units)
Mathematics	A	4	3	12
History	C	2	3	6
Chemistry	B	3	5	15
Art	B	3	2	6
PE	A	4	1	4
		Totals:	14	43

$$\text{Grade-point average} = \frac{43}{14} = 3.07 \text{ (rounded)}$$

The calculation of a grade-point average is an example of a *weighted mean,* because the grade points for each course grade must be weighted according to the number of units of the course. (For example, five units of A is better than two units of A.) The number of units is called the **weighting factor.**

Weighted Mean

The **weighted mean** of n numbers, $x_1, x_2, \ldots, x_n$, that are weighted by the respective factors $f_1, f_2, \ldots, f_n$ is given by the formula

$$\overline{w} = \frac{\Sigma(x \cdot f)}{\Sigma f}.$$

In words, the weighted mean of a group of (weighted) items is the sum of all products of items times weighting factors, divided by the sum of all weighting factors.

The weighted mean formula is commonly used to find the mean for a frequency distribution. In this case, the weighting factors are the frequencies.

Salary x	Number of Employees f
$12,000	8
$16,000	11
$18,500	14
$21,000	9
$34,000	2
$50,000	1

EXAMPLE 2 Finding the Mean of a Frequency Distribution of Salaries

Find the mean salary for a small company that pays annual salaries to its employees as shown in the frequency distribution in the margin.

SOLUTION

According to the weighted mean formula, we can set up the work as follows.

Salary x	Number of Employees f	Salary · Number $x \cdot f$
$12,000	8	$ 96,000
$16,000	11	$176,000
$18,500	14	$259,000
$21,000	9	$189,000
$34,000	2	$ 68,000
$50,000	1	$ 50,000
Totals:	45	$838,000

$$\text{Mean salary} = \frac{\$838{,}000}{45} = \$18{,}622 \quad \text{(rounded)}$$

In this screen supporting Example 2, the first list contains the salaries and the second list contains their frequencies.

For some data sets the mean can be a misleading indicator of average. Consider Shady Sam who runs a small business that employs five workers at the following annual salaries.

$$\$16{,}500, \quad \$16{,}950, \quad \$17{,}800, \quad \$19{,}750, \quad \$20{,}000$$

The employees, knowing that Sam accrues vast profits to himself, decide to go on strike and demand a raise. To get public support, they go on television and tell about their miserable salaries, pointing out that the mean salary in the company is only

$$\bar{x} = \frac{\$16{,}500 + \$16{,}950 + \$17{,}800 + \$19{,}750 + \$20{,}000}{5}$$
$$= \frac{\$91{,}000}{5} = \$18{,}200.$$

The local television station schedules an interview with Sam to investigate. In preparation, Sam calculates the mean salary of *all* workers (including his own salary of \$188,000):

$$\bar{x} = \frac{\$16{,}500 + \$16{,}950 + \$17{,}800 + \$19{,}750 + \$20{,}000 + \$188{,}000}{6}$$
$$= \frac{\$279{,}000}{6} = \$46{,}500.$$

When the TV crew arrives, Sam calmly assures them that there is no reason for his employees to complain since the company pays a generous mean salary of \$46,500.

The employees, of course, would argue that when Sam included his own salary in the calculation, it caused the mean to be a misleading indicator of average. This was so because Sam's salary is not typical. It lies a good distance away from the general grouping of the items (salaries). An extreme value like this is referred to as an **outlier.** Since a single outlier can have a significant effect on the value of the mean, we say that the mean is "highly sensitive to extreme values."

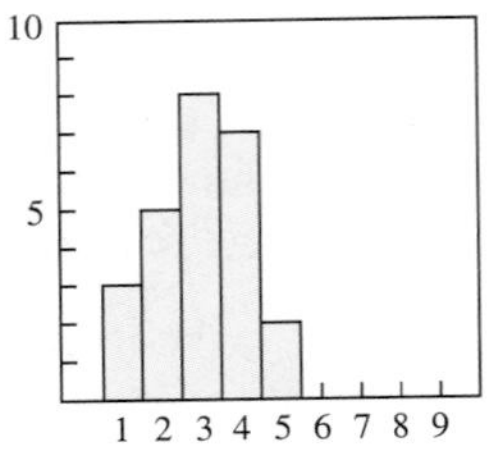

Mean = 3.00
Median = 3.00

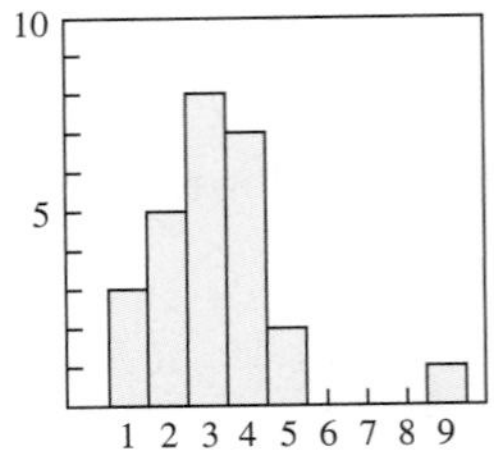

Mean = 3.23
Median = 3.00

The introduction of a single "outlier" above increased the mean by 8 percent but left the median unaffected.

Outliers should usually be considered as *possible* errors in the data.

Median

Another measure of central tendency, which is not so sensitive to extreme values, is the **median.** This measure divides a group of numbers into two parts, with half the numbers below the median and half above it.

Median

To find the **median** of a group of items:

Step 1 Rank the items (that is, arrange them in numerical order from least to greatest).

Step 2 If the number of items is *odd,* the median is the middle item in the list.

Step 3 If the number of items is *even,* the median is the mean of the two middle items.

For Shady Sam's business, all salaries (including Sam's), arranged in numerical order, are

$$\$16{,}500, \quad \$16{,}950, \quad \$17{,}800, \quad \$19{,}750, \quad \$20{,}000, \quad \$188{,}000,$$

so $$\text{median} = \frac{\$17{,}800 + \$19{,}750}{2} = \frac{\$37{,}550}{2} = \$18{,}775.$$

This figure is a representative average, based on all six salaries, that the employees would probably agree is reasonable.

EXAMPLE 3 Finding Medians of Lists of Numbers

Find the median of each list of numbers.

(a) 6, 7, 12, 13, 18, 23, 24

(b) 17, 15, 9, 13, 21, 32, 41, 7, 12

(c) 147, 159, 132, 181, 174, 253

SOLUTION

(a) This list is already in numerical order. The number of values in the list, 7, is odd, so the median is the middle value, or 13.

(b) First, place the numbers in numerical order from least to greatest.

7, 9, 12, 13, 15, 17, 21, 32, 41

↑ Median (15)

The middle number can now be picked out; the median is 15.

(c) First write the numbers in numerical order.

132, 147, 159, 174, 181, 253

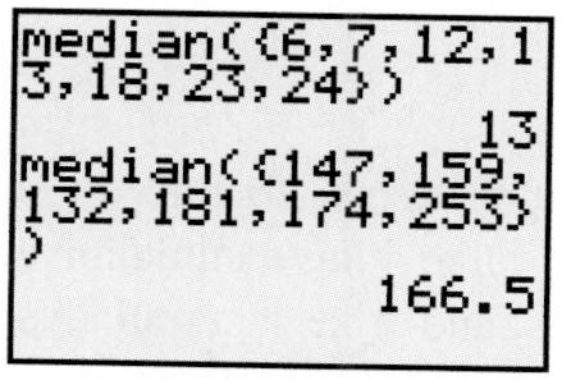

The calculator can find the median of the entries in a list. This screen supports the results in parts (a) and (c) of Example 3.

Since the list contains an even number of items, namely 6, there is no single middle item. Find the median by taking the mean of the two middle items, 159 and 174.

$$\frac{159 + 174}{2} = \frac{333}{2} = 166.5 \leftarrow \text{Median}$$

In the case of a frequency distribution, locating the middle item (the median) is a bit different. First find the total number of items in the set by adding the frequencies ($n = \Sigma f$). Then the median is the item whose *position* is given by the following formula.

Position of the Median in a Frequency Distribution

$$\textbf{Position of median} = \frac{n + 1}{2} = \frac{\Sigma f + 1}{2}$$

Notice that this formula gives only the position, and not the actual value, of the median. The next example shows how the formula is used to find the median.

EXAMPLE 4 Finding Medians for Frequency Distributions

Find the medians for the following distributions.

(a)

Value	1	2	3	4	5	6
Frequency	1	3	2	4	8	2

(b)

Value	2	4	6	8	10
Frequency	5	8	10	6	6

SOLUTION

(a) Arrange the work as follows. Tabulate the values and frequencies, and also the **cumulative frequencies,** which tell, for each different value, how many items have that value or a lesser value.

Value	Frequency	Cumulative Frequency	
1	1	1	1 item 1 or less
2	3	4	1 + 3 = 4 items 2 or less
3	2	6	4 + 2 = 6 items 3 or less
4	4	10	6 + 4 = 10 items 4 or less
5	8	18	10 + 8 = 18 items 5 or less
6	2	20	18 + 2 = 20 items 6 or less

Total: 20

Adding the frequencies shows that there are 20 items total, so

$$\text{position of median} = \frac{20 + 1}{2} = \frac{21}{2} = 10.5.$$

The median, then, is the average of the tenth and eleventh items. To find these items, make use of the cumulative frequencies. Since the value 4 has a cumulative frequency of 10, that is, 10 items have a value of 4 or less, and 5 has a cumulative frequency of 18, the tenth item is 4 and the eleventh item is 5, making the median $\frac{4+5}{2} = \frac{9}{2} = 4.5$.

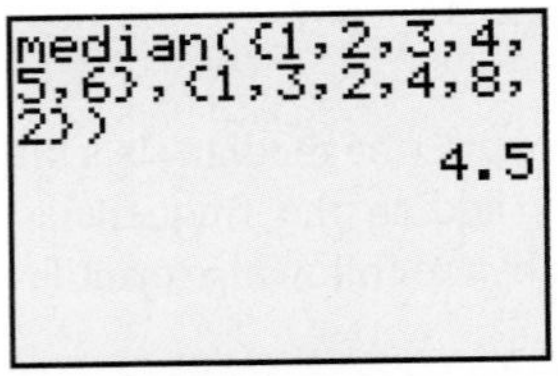

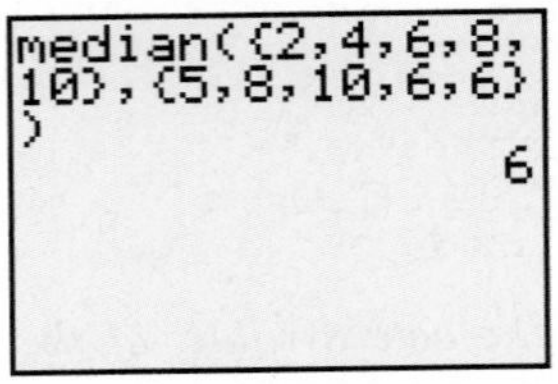

These two screens support the results in Example 4.

(b)

Value	Frequency	Cumulative Frequency
2	5	5
4	8	13
6	10	23
8	6	29
10	6	35

Total: 35

There are 35 items total, so

$$\text{position of median} = \frac{35 + 1}{2} = \frac{36}{2} = 18.$$

From the cumulative frequency column, the fourteenth through the twenty-third items are all 6s. This means the eighteenth item is a 6, so the median is 6. ■

Mode The third important measure of central tendency is the **mode.** If ten students earned scores on a business law examination of

74, 81, 39, 74, 82, 80, 100, 92, 74, 85,

then we notice that more students earned the score 74 than any other score. This fact makes 74 the mode of this list.

Mode

The **mode** of a data set is the value that occurs most often.

EXAMPLE 5 Finding Modes for Sets of Data

Find the mode for each set of data.

(a) 51, 32, 49, 49, 74, 81, 92 **(b)** 482, 485, 483, 485, 487, 487, 489

(c) 10,708, 11,519, 10,972, 17,546, 13,905, 12,182

(d)

Value	19	20	22	25	26	28
Frequency	1	3	8	7	4	2

SOLUTION

(a) 51, 32, 49, 49, 74, 81, 92
The number 49 occurs more often than any other. Therefore, 49 is the mode. The numbers do not need to be in numerical order when looking for the mode.

(b) 482, 485, 483, 485, 487, 487, 489
Both 485 and 487 occur twice. This list is said to have *two* modes, or to be *bimodal.*

(c) No number here occurs more than once. This list has no mode.

(d)

Value	**Frequency**	
19	1	
20	3	
22	8	← Greatest frequency
25	7	
26	4	
28	2	

The frequency distribution shows that the most frequently occurring value (and thus the mode) is 22.

That we have included the mode as a measure of *central tendency* is traditional, probably because many important kinds of data sets do have their most frequently occurring values "centrally" located. However, there is no reason the mode cannot be one of the least values in the set or one of the greatest. In such a case, the mode really is not a good measure of "central tendency."

When the data items being studied are nonnumeric, the mode may be the only usable measure of central tendency. For example, the bar graph of Figure 5 in Section 13.1 showed frequencies of occurrence of vowels in a sample paragraph. Since the vowels A, E, I, O, and U are not numbers, they cannot be added, so their mean does not exist. Furthermore, the vowels cannot be arranged in any meaningful numerical order, so their median does not exist either. The mode, however, does exist. As the bar graph shows, the mode is the letter E.

Sometimes, a distribution is **bimodal** (literally, "two modes"), as in Example 5(b). In a large distribution, this term is commonly applied even when the two modes do not have exactly the same frequency. When a distribution has three or more different items sharing the highest frequency of occurrence, that information is not often useful. (Too many modes tend to obscure the significance of these "most frequent" items.) We say that such a distribution has *no* mode.

Central Tendency from Stem-and-Leaf Displays As shown in Section 13.1, data are sometimes presented in a stem-and-leaf display in order to give a graphical impression of their distribution. We can also calculate measures of central tendency from a stem-and-leaf display. The median and mode are more easily identified when the "leaves" are **ranked** (arranged in numerical order) on their "stems." In Table 9, we have rearranged the leaves of Table 5 in Section 13.1 (which showed the weekly study times from Example 2 of that section).

TABLE 9 Stem-and-Leaf Display for Weekly Study Times, with Leaves Ranked

1	2	4	5	6	7	8					
2	0	2	3	4	4	5	6	6	9	9	9
3	0	1	1	2	3	6	6	8	9		
4	0	1	4	4	5	5	7				
5	2	5	5	8							
6	0	2									
7	2										

EXAMPLE 6 Finding the Mean, Median, and Mode from a Stem-and-Leaf Display

Find the mean, median, and mode for the data in Table 9.

SOLUTION

If you enter the data into a calculator with statistical capabilities, the mean will be automatically computed for you. Otherwise, add all items (reading from the stem-and-leaf display) and divide by $n = 40$.

$$\text{mean} = \frac{12 + 14 + 15 + \cdots + 60 + 62 + 72}{40} = \frac{1395}{40} = 34.875$$

In this case, $n = 40$ (an even number), so the median is the average of the twentieth and twenty-first items, in order. Counting leaves, we see that these will be the third and fourth items on the stem 3. So

$$\text{median} = \frac{31 + 32}{2} = 31.5.$$

By inspection, we see that 29 occurred three times and no other value occurred that often, so

$$\text{mode} = 29.$$

Symmetry in Data Sets The most useful way to analyze a data set often depends on whether the distribution is **symmetric** or **nonsymmetric.** In a "symmetric" distribution, as we move out from the central point, the pattern of frequencies is the same (or nearly so) to the left and to the right. In a "nonsymmetric" distribution, the patterns to the left and right are different. Figure 8 shows several types of symmetric distributions, while Figure 9 shows some nonsymmetric distributions. A nonsymmetric distribution with a tail extending out to the left, shaped like a J, is called **skewed to the left.** If the tail extends out to the right, the distribution is **skewed to the right.** Notice that a bimodal distribution may be either symmetric or nonsymmetric.

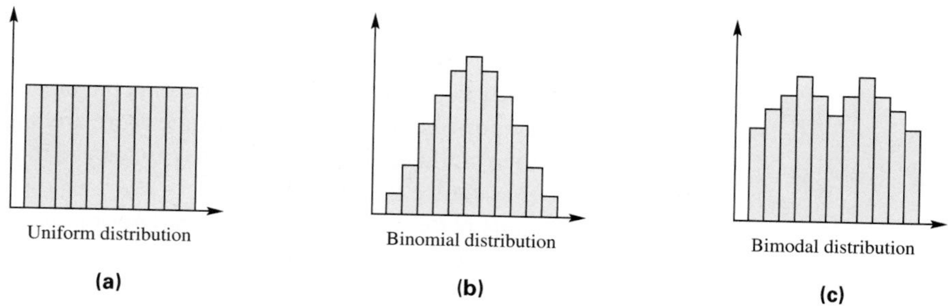

Some symmetric distributions

FIGURE 8

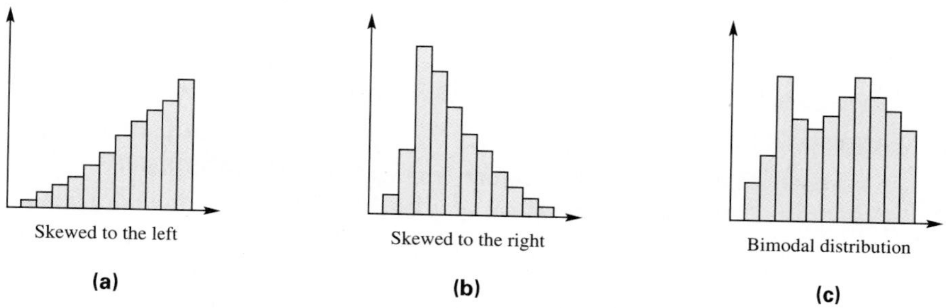

Some non-symmetric distributions

FIGURE 9

Summary We conclude this section with a summary of the measures presented and a brief discussion of their relative advantages and disadvantages.

Summary of the Common Measures of Central Tendency

The **mean** of a set of numbers is found by adding all the values in the set and dividing by the number of values.

The **median** is a kind of "middle" number. To find the median, first arrange the values in numerical order. For an *odd* number of values, the median is the middle value in the list. For an *even* number of values, the median is the mean of the two middle values.

The **mode** is the value that occurs most often. Some sets of numbers have two most frequently occurring values and are **bimodal.** Other sets have no mode at all (if no value occurs more often than the others or if more than two values occur most often).

For distributions of numeric data, the mean and median will always exist, while the mode may not exist. On the other hand, for nonnumeric data, it may be that none of the three measures exists, or that only the mode exists.

Because even a single change in the data may cause the mean to change, while the median and mode may not be affected at all, ***the mean is the most "sensitive" measure***. In a symmetric distribution, the mean, median, and mode (if a single mode exists) will all be equal. In a nonsymmetric distribution, the mean is often unduly affected by relatively few extreme values, and therefore may not be a good representative measure of central tendency. For example, distributions of salaries, or of family incomes, often include a few values that are much higher than the bulk of the items. In such cases, the median is a more useful measure. Any time there are many more items on one side of the mean than on the other, extreme nonsymmetry is indicated and the median is most likely a better choice than the mean.

The mode is the only measure covered here that must always be equal to one of the data items of the distribution. In fact, more of the data items are equal to the mode than to any other number. A fashion shop planning to stock only one hat size for next season would want to know the mode (the most common) of all hat sizes among their potential customers. Likewise, a designer of family automobiles would be interested in the most common family size. In examples like these, designing for the mean or the median might not be right for anyone.

For Further Thought

Simpson's Paradox

In baseball statistics, a player's "batting average" gives the average number of hits per time at bat. For example, a player who has gotten 84 hits in 250 times at bat has a batting average of $\frac{84}{250} = .336$. This "average" can be interpreted as the empirical probability of that player's getting a hit the next time at bat.

The following are actual comparisons of hits and at-bats for two major league players in the 1989 and 1990 seasons. The numbers illustrate a puzzling statistical occurrence known as **Simpson's paradox.** (This information was reported by Richard J. Friedlander on page 845 of the November 1992 issue of the *MAA Journal.*)

For Group Discussion or Individual Investigation

1. Fill in the ten blanks in the table, giving batting averages to three decimal places.
2. Which player had a better average in 1989?
3. Which player had a better average in 1990?
4. Which player had a better average in 1989 and 1990 combined?
5. Did the results above surprise you? How can it be that one player's batting average leads another's for each of two years, and yet trails the other's for the combined years?

	Dave Justice			Andy Van Slyke		
	Hits	**At-bats**	**Batting Average**	**Hits**	**At-bats**	**Batting Average**
1989	12	51	____	113	476	____
1990	124	439	____	140	493	____
Combined (1989–1990)	____	____	____	____	____	____

13.2 EXERCISES

For each list of data, calculate **(a)** *the mean,* **(b)** *the median, and* **(c)** *the mode or modes (if any). Round mean values to the nearest tenth.*

1. 5, 9, 10, 14, 32

2. 21, 25, 32, 48, 53, 62, 62, 64

3. 128, 230, 196, 224, 196, 233

4. 26, 31, 46, 31, 26, 29, 31

5. 3.1, 4.5, 6.2, 7.1, 4.5, 3.8, 6.2, 6.3

6. 14,320, 16,950, 17,330, 15,470

7. .78, .93, .66, .94, .87, .62, .74, .81

8. .53, .03, .28, .18, .39, .28, .14, .22, .04

9. 128, 131, 136, 125, 132, 128, 125, 127

10. 8.97, 5.64, 2.31, 1.02, 4.35, 7.68

Overseas Visitors to the United States *The table shows the top 10 overseas countries (this does not include Canada or Mexico) whose citizens visited the United States in 2004. Use this information for Exercises 11–16. (Source of data:* Encyclopaedia Britannica Almanac 2006, *page 833.)*

Top Countries of Origin for Visitors to the United States (2004)

Rank	Country	Number of Visitors	% Change from 2003
1.	United Kingdom	4,302,737	+ 9
2.	Japan	3,747,620	+18
3.	Germany	1,319,904	+12
4.	France	775,274	+13
5.	South Korea	626,595	+ 1
6.	Australia	519,955	+28
7.	Italy	470,805	+15
8.	The Netherlands	424,872	+14
9.	Brazil	384,734	+10
10.	Ireland	345,119	+36

For each set of countries, find **(a)** *the mean number of 2004 visitors and* **(b)** *the median number of 2004 visitors. (Give answers to the nearest thousand.)*

11. the countries ranked 1 through 5

12. the countries ranked 6 through 10

13. the countries ranked 1 through 10

Estimate (to the nearest thousand) the number of 2003 visitors from each of the following countries.

14. United Kingdom

15. Australia

16. Ireland

Forest Loss and Atmospheric Pollution *Amid concerns about global warming and rain forest destruction, the World Resources Institute reported the five countries with the greatest average yearly forest loss for a recent five-year period, as well as the five greatest emitters of chlorofluorocarbons in a one-year period. The estimates are shown here.*

Country	Average Yearly Forest Loss (in square miles)
Brazil	9700
Colombia	3400
Indonesia	2400
Mexico	2375
Ivory Coast	2000

Country	Chlorofluorocarbon Emissions (in metric tons)
United States	130,000,000
Japan	95,000,000
C.I.S.	67,000,000
Germany	34,000,000
United Kingdom	25,000,000

17. Find the mean value of yearly forest loss per country for the five countries listed. Give your answer to the nearest 100 square miles.

18. Find the mean value of chlorofluorocarbon emissions per country for the five countries listed. Give your answer to the nearest million metric tons.

Measuring Elapsed Times *While doing an experiment, a physics student recorded the following sequence of elapsed times (in seconds) in a lab notebook.*

2.16, 22.2, 2.96, 2.20, 2.73, 2.28, 2.39

19. Find the mean.

20. Find the median.

The student from Exercises 19 and 20, when reviewing the calculations later, decided that the entry 22.2 should have been recorded as 2.22, and made that change in the listing.

21. Find the mean for the new list.

22. Find the median for the new list.

23. Which measure, the mean or the median, was affected more by correcting the error?

24. In general, which measure, mean or median, is affected less by the presence of an extreme value in the data?

Scores on Statistics Examinations *Jay Beckenstein earned the following scores on his six statistics exams last semester.*

$$79, 81, 44, 89, 79, 90$$

25. Find the mean, the median, and the mode for Jay's scores.

26. Which of the three averages probably is the best indicator of Jay's ability?

27. If Jay's instructor gives him a chance to replace his score of 44 by taking a "make-up" exam, what must he score on the make-up to get an overall average (mean) of 85?

Exercises 28 and 29 give frequency distributions for sets of data values. For each set find the **(a)** *mean (to the nearest hundredth),* **(b)** *median, and* **(c)** *mode or modes (if any).*

28.

Value	Frequency
2	5
4	1
6	8
8	4

29.

Value	Frequency
603	13
597	8
589	9
598	12
601	6
592	4

30. ***Average Employee Salaries*** A company has six employees with a salary of $19,500, eight with a salary of $23,000, four with a salary of $28,300, two with a salary of $34,500, seven with a salary of $36,900, and one with a salary of $145,500. Find the mean salary for the employees (to the nearest hundred dollars).

Grade-point Averages *Find the grade-point average for each of the following students. Assume* A = 4, B = 3, C = 2, D = 1, F = 0. *Round to the nearest hundredth.*

31.

Units	Grade
4	C
7	B
3	A
3	F

32.

Units	Grade
2	A
6	B
5	C

Area and Population in the Commonwealth of Independent States *The table gives the land area and population of the twelve members of the Commonwealth of Independent States. (All were former republics of the Soviet Union.) Use this information for Exercises 33–36. (Source of data:* Encyclopaedia Britannica Almanac 2006, *pages 316–615.)*

State	Area (square miles)	Population
Armenia	11,500	2,991,000
Azerbaijan	33,400	8,343,000
Belarus	80,200	9,828,000
Georgia	27,100	4,694,000
Kazakhstan	1,052,100	15,144,000
Kyrgystan	77,200	5,081,000
Moldova	13,068	4,216,000
Russia	6,592,800	144,315,000
Tajikistan	55,300	6,606,000
Turkmenistan	188,500	4,940,000
Ukraine	233,100	47,470,000
Uzbekistan	172,700	26,009,000

33. Find the mean area (to the nearest 100 square miles) for these 12 states.

34. Discuss the meaningfulness of the average calculated in Exercise 33.

35. Find the mean population (to the nearest 1000) for the 12 states.

36. Discuss the meaningfulness of the average calculated in Exercise 35.

Languages Spoken by Americans *The table lists the ten most common languages (other than English) spoken at home by Americans. The numbers in the right column indicate the change (increase or decrease), over a recent ten-year period, in the number of Americans speaking the different languages. (Source of data:* 1995 Information Please Almanac, *page 835.) Use the table for Exercises 37–40. (The rankings are given for information only. They will not enter into your calculations.)*

Most Common Non-English Languages Among Americans

Rank	Language	Change in Number of People
1	Spanish	5,789,839
2	French	129,901
3	German	−59,644
4	Italian	−324,631
5	Chinese	617,476
6	Tagalog	391,289
7	Polish	−102,667
8	Korean	350,766
9	Vietnamese	303,801
10	Portuguese	68,759

37. Find the mean change per language for all ten languages.

38. Find the mean change per language for the seven categories with increases.

39. Find the mean change per language for the three categories with decreases.

40. Use your answers for Exercises 38 and 39 to find the weighted mean for the change per language for all ten categories. Compare with the value calculated in Exercise 37.

41. ***Triple Crown Horse Races*** The table in the next column shows the winning horse and the "value to winner" for each of the so-called Triple Crown races in 2005. No horse has won all three in the same year since Affirmed did so in 1978. Find the average (mean) value to winner for the three races. (*Source:* 2006 World Book Year Book, *page 199.*)

Race	Winning Horse	Value to Winner
Kentucky Derby	Giacomo	$1,639,600
Preakness Stakes	Afleet Alex	600,000
Belmont Stakes	Afleet Alex	600,000

Olympic Medal Standings *The top eleven medal-winning nations in the 2006 Winter Olympics at Torino, Italy, are shown in the table. Use the given information for Exercises 42–45.*

Medal Standings for the 2006 Winter Olympics

Nation	Gold	Silver	Bronze	Total
Germany	11	12	6	29
United States	9	9	7	25
Austria	9	7	7	23
Russian Fed.	8	6	8	22
Canada	7	10	7	24
Sweden	7	2	5	14
Korea	6	3	2	11
Switzerland	5	4	5	14
Italy	5	0	6	11
France	3	2	4	9
Netherlands	3	2	4	9

Source: www.torino2006.org

Calculate each of the following for all nations shown.

42. the mean number of gold medals

43. the median number of silver medals

44. the mode, or modes, for the number of bronze medals

45. each of the following for the total number of medals
(a) mean
(b) median
(c) mode or modes

In Exercises 46 and 47, use the given stem-and-leaf display to identify **(a)** *the mean,* **(b)** *the median, and* **(c)** *the mode (if any) for the data represented.*

46. *Auto Repair Charges* The display here represents prices (to the nearest dollar) charged by 23 different auto repair shops for a new alternator (installed). Give answers to the nearest cent.

9	7
10	2 4
10	5 7 9
11	1 3 4 4
11	5 5 8 8 9
12	0 4 4
12	5 7 7 9
13	8

47. *Scores on a Biology Exam* The display here represents scores achieved on a 100-point biology exam by the 34 members of the class.

4	7
5	1 3 6
6	2 5 5 6 7 8 8
7	0 4 5 6 7 7 8 8 8 8 9
8	0 1 1 3 4 5 5
9	0 0 0 1 6

48. *Calculating a Missing Test Score* Dinya Floyd's Business professor lost his grade book, which contained Dinya's five test scores for the course. A summary of the scores (each of which was an integer from 0 to 100) indicates that the mean was 88, the median was 87, and the mode was 92. (The data set was not bimodal.) What is the least possible number among the missing scores?

49. Explain what an "outlier" is and how it affects measures of central tendency.

50. *Consumer Preferences in Food Packaging* A food processing company that packages individual cups of instant soup wishes to find out the best number of cups to include in a package. In a survey of 22 consumers, they found that five prefer a package of 1, five prefer a package of 2, three prefer a package of 3, six prefer a package of 4, and three prefer a package of 6.

(a) Calculate the mean, median, and mode values for preferred package size.

(b) Which measure in part (a) should the food processing company use?

(c) Explain your answer to part (b).

51. *Scores on a Math Quiz* The following are scores earned by 15 college students on a 20-point math quiz.

0, 1, 3, 14, 14, 15, 16, 16, 17, 17, 18, 18, 18, 19, 20

(a) Calculate the mean, median, and mode values for these scores.

(b) Which measure in part (a) is most representative of the data?

(c) Explain your answer to part (b).

In Exercises 52–55, begin a list of the given numbers, in order, starting with the least one. Continue the list only until the median of the listed numbers is a multiple of 4. Stop at that point and find **(a)** *the number of numbers listed, and* **(b)** *the mean of the listed numbers (to two decimal places).*

52. counting numbers

53. prime numbers

54. Fibonacci numbers (see page 257)

55. triangular numbers (see pages 14–15)

56. Seven consecutive whole numbers add up to 147. What is the result when their mean is subtracted from their median?

57. If the mean, median, and mode are all equal for the set $\{70, 110, 80, 60, x\}$, find the value of x.

58. Michael Coons wants to include a fifth counting number, n, along with the numbers 2, 5, 8, and 9 so that the mean and median of the five numbers will be equal. How many choices does Michael have for the number n, and what are those choices?

For Exercises 59–61, refer to the grouped frequency distribution shown here.

Class Limits	Frequency f
21–25	5
26–30	3
31–35	8
36–40	12
41–45	21
46–50	38
51–55	35
56–60	20

59. Is it possible to identify any specific data items that occurred in this sample?

60. Is it possible to compute the actual mean for this sample?

61. Describe how you might approximate the mean for this sample. Justify your procedure.

62. *Average Employee Salaries* Refer to the salary data of Example 2. Explain what is wrong with simply calculating the mean salary as follows.

$$\bar{x} = \frac{\Sigma x}{n} = \frac{\$12{,}000 + \$16{,}000 + \$18{,}500 + \$21{,}000 + \$34{,}000 + \$50{,}000}{6} = \$25{,}250$$

13.3 Measures of Dispersion

Range • Standard Deviation • Interpreting Measures of Dispersion • Coefficient of Variation

TABLE 10

	A	B
	5	1
	6	2
	7	7
	8	12
	9	13
Mean	7	7
Median	7	7

The mean is a good indicator of the central tendency of a set of data values, but it does not give the whole story about the data. To see why, compare distribution A with distribution B in Table 10.

Both distributions of numbers have the same mean (and the same median also), but beyond that, they are quite different. In the first, 7 is a fairly typical value; but in the second, most of the values differ considerably from 7. What is needed here is some measure of the **dispersion,** or *spread,* of the data. Two of the most common measures of dispersion, the *range* and the *standard deviation,* are discussed in this section.

Range

The **range** of a data set is a straightforward measure of dispersion.

Range

For any set of data, the **range** of the set is given by

Range = (greatest value in the set) − (least value in the set).

For a short list of data, calculation of the range is simple. For a more extensive list, it is more difficult to be sure you have accurately identified the greatest and least values.

EXAMPLE 1 Finding and Comparing Range Values

Find the range for each of distributions A and B given in Table 10, and describe what they imply.

SOLUTION

In distribution A, the greatest value is 9 and the least is 5. Thus,

$$\text{Range} = \text{greatest} - \text{least} = 9 - 5 = 4.$$

In distribution B,

$$\text{Range} = 13 - 1 = 12.$$

We can say that even though the two distributions have identical averages, distribution B exhibits three times more dispersion, or *spread*, than distribution A. ■

Once the data are entered, a calculator with statistical functions may actually show the range (among other things), or at least sort the data and identify the minimum and maximum items. (The associated symbols may be something like MIN Σ and MAX Σ, or min$X$ and max$X$.) Given these two values, a simple subtraction produces the range.

TABLE 11

Dive	Mark	Myrna
1	28	27
2	22	27
3	21	28
4	26	6
5	18	27
Mean	23	23
Median	22	27
Range	10	22

The range can be misleading if it is interpreted unwisely. For example, suppose three judges for a diving contest assign points to Mark and Myrna on five different dives, as shown in Table 11. The ranges for the divers make it tempting to conclude that Mark is a more consistent diver than Myrna. However, Myrna is actually more consistent, with the exception of one very poor score. That score, 6, is an outlier which, if not actually recorded in error, must surely be due to some special circumstance. (Notice that the outlier does not seriously affect Myrna's median score, which is more typical of her overall performance than is her mean score.)

Standard Deviation One of the most useful measures of dispersion, the *standard deviation,* is based on *deviations from the mean* of the data values.

EXAMPLE 2 Finding Deviations from the Mean

Find the deviations from the mean for all data values of the sample

32, 41, 47, 53, 57.

SOLUTION

Add these values and divide by the total number of values, 5. The mean is 46. To find the deviations from the mean, subtract 46 from each data value.

Data value	32	41	47	53	57
Deviation	**−14**	**−5**	**1**	**7**	**11**

$32 - 46 = -14$ (first deviation); $57 - 46 = 11$ (last deviation)

To check your work, add the deviations. ***The sum of the deviations for a set of data is always 0.***

It is perhaps tempting now to find a measure of dispersion by finding the mean of the deviations. However, this number always turns out to be 0 no matter how much dispersion is shown by the data; this is because the positive deviations just cancel out the negative ones. This problem of positive and negative numbers canceling each other can be avoided by *squaring* each deviation. (The square of a negative number is positive.) The following chart shows the squares of the deviations for the data above.

Data value	32	41	47	53	57
Deviation	−14	−5	1	7	11
Square of deviation	**196**	**25**	**1**	**49**	**121**

$(-14) \cdot (-14) = 196$; $11 \cdot 11 = 121$

An average of the squared deviations could now be found by dividing their sum by the number of data values n (5 in this case), which we would do if our data values comprised a population. However, since we are considering the data to be a sample, we divide by $n - 1$ instead. The average that results is itself a measure of dispersion, called the **variance,** but a more common measure is obtained by taking the square root of the variance. This makes up, in a way, for squaring the deviations earlier, and gives a kind of average of the deviations from the mean, which is called the sample **standard deviation.** It is denoted by the letter s. (The standard deviation of a population is denoted σ, the lowercase Greek letter sigma.)

Most calculators find square roots, such as $\sqrt{98}$, to as many digits as you need using a key like $\boxed{\sqrt{x}}$. In this book, we normally give from two to four significant figures for such calculations.

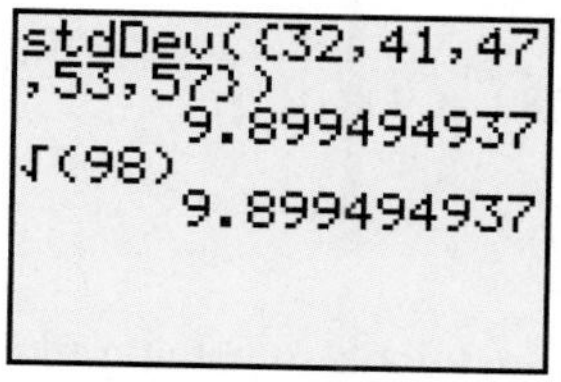

This screen supports the text discussion. Note that the standard deviation reported agrees with the approximation for $\sqrt{98}$.

Continuing our calculations from the chart on page 808, we obtain

$$s = \sqrt{\frac{196 + 25 + 1 + 49 + 121}{4}} = \sqrt{\frac{392}{4}} = \sqrt{98} \approx 9.90.$$

The algorithm (process) described above for finding the sample standard deviation can be summarized as follows.*

Calculation of Standard Deviation

Let a sample of n numbers $x_1, x_2, \ldots, x_n$ have mean $\bar{x}$. Then the **sample standard deviation, *s*,** of the numbers is given by

$$s = \sqrt{\frac{\Sigma(x - \bar{x})^2}{n - 1}}.$$

The individual steps involved in this calculation are as follows.

Step 1 Calculate $\bar{x}$, the mean of the numbers.

Step 2 Find the deviations from the mean.

Step 3 Square each deviation.

Step 4 Sum the squared deviations.

Step 5 Divide the sum in Step 4 by $n - 1$.

Step 6 Take the square root of the quotient in Step 5.

The preceding description helps show why standard deviation measures the amount of spread in a data set. For actual calculation purposes, we recommend the use of a scientific calculator, or a statistical calculator, that does all the detailed steps automatically. We illustrate both methods in the following example.

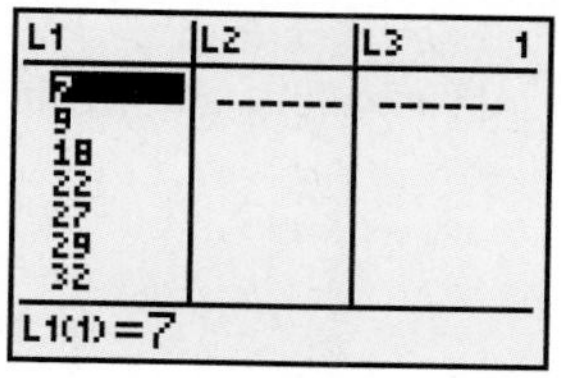

The sample in Example 3 is stored in a list. (The last entry, 40, is not shown here.)

EXAMPLE 3 Finding a Sample Standard Deviation

Find the standard deviation of the sample

$$7, 9, 18, 22, 27, 29, 32, 40$$

by using **(a)** the step-by-step process, and **(b)** the statistical functions of a calculator.

SOLUTION

(a) Carry out the six steps summarized above.

Step 1 Find the mean of the values.

$$\frac{7 + 9 + 18 + 22 + 27 + 29 + 32 + 40}{8} = 23$$

*Although the reasons cannot be explained at this level, dividing by $n - 1$ rather than n produces a sample measure that is more accurate for purposes of inference. In most cases, the results using the two different divisors differ only slightly.

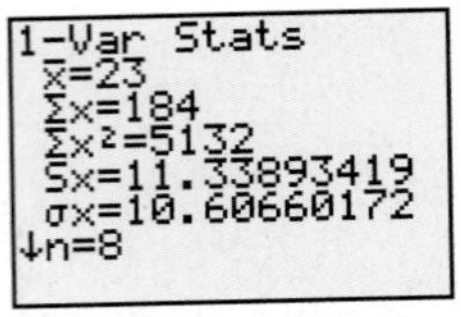

The arrows point to the mean and the sample and population standard deviations. See Example 3.

Step 2 Find the deviations from the mean.

Data value	7	9	18	22	27	29	32	40
Deviation	−16	−14	−5	−1	4	6	9	17

Step 3 Square each deviation.

Squares of deviations: 256 196 25 1 16 36 81 289

Step 4 Sum the squared deviations.

$$256 + 196 + 25 + 1 + 16 + 36 + 81 + 289 = 900$$

Step 5 Divide by $n - 1 = 8 - 1 = 7$: $\frac{900}{7} \approx 128.57$.

Step 6 Take the square root: $\sqrt{128.57} \approx 11.3$.

(b) Enter the eight data values. (The key for entering data may look something like this: $\boxed{\Sigma+}$. Find out which key it is on your calculator.) Then press the key for standard deviation. (This one may look like

STDEV or SD or S_{n-1} or σ_{n-1}.

If your calculator also has a key that looks like σ_n, it is probably for *population* standard deviation, which involves dividing by n rather than by $n - 1$, as mentioned earlier.) The result should again be 11.3. (If you *mistakenly* used the population standard deviation key, the result is 10.6 instead.)

For data given in the form of a frequency distribution, some calculators allow entry of both values and frequencies, or each value can be entered separately the number of times indicated by its frequency. Then press the standard deviation key.

The following example is included only to strengthen your understanding of frequency distributions and standard deviation, not as a practical algorithm for calculating.

EXAMPLE 4 Finding the Standard Deviation of a Frequency Distribution

Find the sample standard deviation for the frequency distribution shown in Table 12.

TABLE 12

Value	Frequency
2	5
3	8
4	10
5	2

SOLUTION

Complete the calculations as shown in Table 13 on next page. To find the numbers in the "Deviation" column, first find the mean, and then subtract the mean from the numbers in the "Value" column.

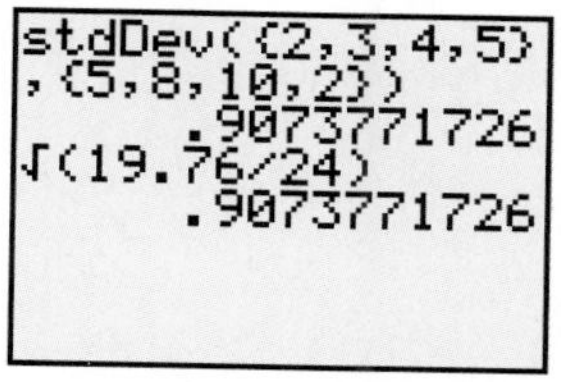

The screen supports the result in Example 4.

TABLE 13

Value	Frequency	Value Times Frequency	Deviation	Squared Deviation	Squared Deviation Times Frequency
2	5	10	−1.36	1.8496	9.2480
3	8	24	−.36	.1296	1.0368
4	10	40	.64	.4096	4.0960
5	2	10	1.64	2.6896	5.3792
Sums	25	84			19.76

$$\bar{x} = \frac{84}{25} = 3.36 \qquad s = \sqrt{\frac{19.76}{24}} \approx \sqrt{.8233} \approx .91$$

Central tendency and dispersion (or "spread tendency") are different and independent aspects of a set of data. Which one is more critical can depend on the specific situation. For example, suppose tomatoes sell by the basket. Each basket costs the same, and each contains one dozen tomatoes. If you want the most fruit possible per dollar spent, you would look for the basket with the highest average weight per tomato (regardless of the dispersion of the weights). On the other hand, if the tomatoes are to be served on an hors d'oeuvre tray where "presentation" is important, you would look for a basket with uniform-sized tomatoes, that is a basket with the lowest weight dispersion (regardless of the average of the weights). See the illustration at the side.

Higher average

Lower dispersion

The more desirable basket depends on your objective.

Another situation involves target shooting (also illustrated at the side). The five hits on the top target are, *on average*, very close to the bulls eye, but the large dispersion (spread) implies that improvement will require much effort. On the other hand, the bottom target exhibits a poorer average, but the smaller dispersion means that improvement will require only a minor adjustment of the gun sights. (In general, consistent errors can be dealt with and corrected more easily than more dispersed errors.)

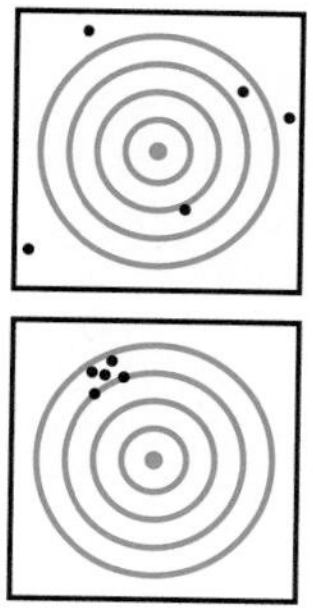

Good average, poor consistency

Good consistency, poor average

In this case, good consistency (lesser dispersion) is more desirable than a good average (central tendency).

Interpreting Measures of Dispersion A main use of dispersion measures is to compare the amounts of spread in two (or more) data sets as we did with distributions A and B at the beginning of this section. A common technique in inferential statistics is to draw comparisons between populations by analyzing samples that come from those populations. (See the discussion of inferential statistics on page 782.)

EXAMPLE 5 Comparing Populations Based on Samples

Two companies, A and B, sell 12-ounce jars of instant coffee. Five jars of each were randomly selected from markets, and the contents were carefully weighed, with the following results.

A: 12.02, 12.08, 11.99, 11.96, 11.99

B: 12.40, 12.21, 12.36, 12.22, 12.27

Find **(a)** which company provides more coffee in their jars, and **(b)** which company fills its jars more consistently.

SOLUTION

From the given data, we calculate the mean and standard deviation values for both samples. These values are shown in Table 14.

TABLE 14

Sample *A*	Sample *B*
$\bar{x}_A = 12.008$	$\bar{x}_B = 12.292$
$s_A = .0455$	$s_B = .0847$

(a) Since $\bar{x}_B$ is greater than $\bar{x}_A$, we *infer* that Company *B* most likely provides more coffee (greater mean).

(b) Since s_A is less than s_B, we *infer* that Company *A* seems more consistent (smaller standard deviation).

The conclusions drawn in Example 5 are tentative, because the samples were small. We could place more confidence in our inferences if we used larger samples, for then it would be more likely that the samples were accurate representations of their respective populations. In a more detailed study of inferential statistics, you would learn techniques best suited to small samples as well as those best suited to larger samples. You would also learn to state your inferences more precisely so that the degree of uncertainty is conveyed along with the basic conclusion.

It is clear that a larger dispersion value means more "spread" than a smaller one. But it is difficult to say exactly what a single dispersion value says about a data set. *It is impossible* (though it would be nice) to make a general statements like: "Exactly half of the items of any distribution lie within one standard deviation of the mean of the distribution." Such statements can be made only of specialized kinds of distributions. (See, for example, Section 13.5 on the normal distribution.) There is, however, one useful result that does apply to all data sets, no matter what their distributions are like. This result is named for the Russian mathematician Pafnuty Lvovich Chebyshev.

Pafnuty Lvovich Chebyshev (1821–1894) was a Russian mathematician known mainly for his work on the theory of prime numbers. Chebyshev and French mathematician and statistician **Jules Bienaymé** (1796–1878) independently developed an important inequality of probability now known as the Bienaymé–Chebyshev inequality.

Chebyshev's Theorem

For any set of numbers, regardless of how they are distributed, the fraction of them that lie within k standard deviations of their mean (where $k > 1$) is *at least*

$$1 - \frac{1}{k^2}.$$

Be sure to notice the words *at least* in the theorem. In certain distributions the fraction of items within k standard deviations of the mean may be more than $1 - \frac{1}{k^2}$, but in no case will it ever be less. The theorem is meaningful for any value of k greater than 1 (integer or noninteger).

EXAMPLE 6 Applying Chebyshev's Theorem

What is the minimum percentage of the items in a data set which lie within 3 standard deviations of the mean?

SOLUTION

With $k = 3$, we calculate

$$1 - \frac{1}{3^2} = 1 - \frac{1}{9} = \frac{8}{9} \approx .889 = 88.9\%. \leftarrow \text{Minimum percentage}$$

$3^2 = 3 \cdot 3$, not $3 \cdot 2$

Coefficient of Variation Look again at the top target pictured on page 811. The dispersion, or spread, among the five bullet holes may not be especially impressive if the shots were fired from 100 yards, but would be much more so at, say, 300 yards. There is another measure, the *coefficient of variation*, which takes this distinction into account. It is not strictly a measure of dispersion, as it combines central tendency and dispersion. It expresses the standard deviation as a percentage of the mean. ***Often this is a more meaningful measure than a straight measure of dispersion, especially when comparing distributions whose means are appreciably different.***

Coefficient of Variation

For any set of data, the **coefficient of variation** is given by

$$V = \frac{s}{\bar{x}} \cdot 100 \quad \text{for a sample} \qquad \text{or} \qquad V = \frac{\sigma}{\mu} \cdot 100 \quad \text{for a population.}$$

EXAMPLE 7 Comparing Samples

Compare the dispersions in the two samples A and B.

A: 12, 13, 16, 18, 18, 20 $\quad$ B: 125, 131, 144, 158, 168, 193

SOLUTION

Using a calculator, we obtain the values shown in Table 15. The values of V_A and V_B were found using the formula above.

TABLE 15

Sample A	Sample B
$\bar{x}_A = 16.167$	$\bar{x}_B = 153.167$
$s_A = 3.125$	$s_B = 25.294$
$V_A = 19.3$	$V_B = 16.5$

From the calculated values, we see that sample B has a much larger dispersion (standard deviation) than sample A. But sample A actually has the larger relative dispersion (coefficient of variation). The dispersion within sample A is larger as a percentage of the sample mean.

FOR FURTHER THOUGHT

Measuring Skewness in a Distribution

Section 13.2 included a discussion of "symmetry in data sets." Here we present a common method of measuring the amount of "skewness," or nonsymmetry, inherent in a distribution.

In a skewed distribution, the mean will be farther out toward the tail than the median, as shown in the sketch.

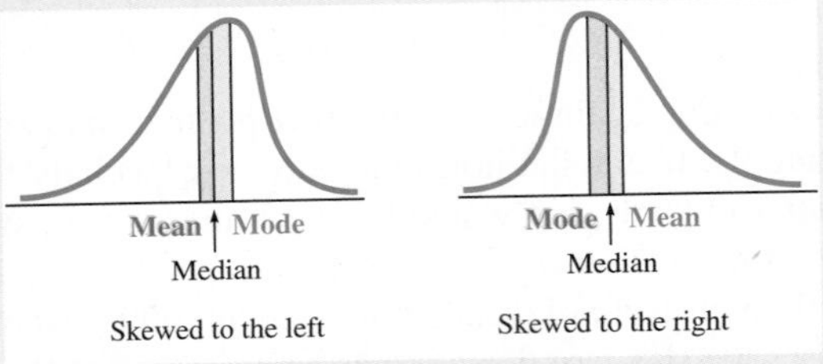

The degree of skewness can be measured by the **skewness coefficient,** which involves both central tendency and dispersion, and is calculated as follows.

$$SK = \frac{3 \cdot (\text{mean} - \text{median})}{\text{standard deviation}}$$

For Group Discussion or Individual Investigation

1. Under what conditions would the skewness coefficient be each of the following?
   (a) positive **(b)** negative
2. Explain why the mean of a skewed distribution is always farther out toward the tail than the median.
3. Suppose that the mean length of a patient stay (in days) in U.S. hospitals is 2.7 days, with a standard deviation of 7.1 days. Make a sketch of how this distribution may look.

13.3 EXERCISES

1. If your calculator finds both kinds of standard deviation, the sample standard deviation and the population standard deviation, which of the two will be a larger number for a given set of data? (*Hint:* Recall the difference between how the two standard deviations are calculated.)

2. If your calculator finds only one kind of standard deviation, explain how you would determine whether it is sample or population standard deviation (assuming your calculator manual is not available).

Find **(a)** *the range, and* **(b)** *the standard deviation for each sample in Exercises 3–12. Round fractional answers to the nearest hundredth.*

3. 2, 5, 6, 8, 9, 11, 15, 19

4. 8, 5, 12, 8, 9, 15, 21, 16, 3

5. 25, 34, 22, 41, 30, 27, 31

6. 67, 83, 55, 68, 77, 63, 84, 72, 65

7. 318, 326, 331, 308, 316, 322, 310, 319, 324, 330

8. 5.7, 8.3, 7.4, 6.6, 7.4, 6.8, 7.1, 8.0, 8.5, 7.9, 7.1, 7.4, 6.9, 8.2

9. 84.53, 84.60, 84.58, 84.48, 84.72, 85.62, 85.03, 85.10, 84.96

10. 206.3, 210.4, 209.3, 211.1, 210.8, 213.5, 212.6, 210.5, 211.0, 214.2

11.

Value	Frequency
9	3
7	4
5	7
3	5
1	2

12.

Value	Frequency
14	8
16	12
18	15
20	14
22	10
24	6
26	3

Use Chebyshev's theorem for Exercises 13–28.

Find the least possible fraction of the numbers in a data set lying within the given number of standard deviations of the mean. Give answers as standard fractions reduced to lowest terms.

13. 2

14. 4

15. $\frac{7}{2}$

16. $\frac{11}{4}$

Find the least possible percentage (to the nearest tenth of a percent) of the items in a distribution lying within the given number of standard deviations of the mean.

17. 3

18. 5

19. $\frac{5}{3}$

20. $\frac{5}{2}$

In a certain distribution of numbers, the mean is 70 and the standard deviation is 8. At least what fraction of the numbers are between the following pairs of numbers? Give answers as common fractions reduced to lowest terms.

21. 54 and 86

22. 46 and 94

23. 38 and 102

24. 30 and 110

In the same distribution (mean 70 and standard deviation 8), find the largest fraction of the numbers that could meet the following requirements. Give answers as common fractions reduced to lowest terms.

25. less than 54 or more than 86

26. less than 50 or more than 90

27. less than 42 or more than 98

28. less than 52 or more than 88

Bonus Pay for a Baseball Team *Leigh Jacka owns a minor league baseball team. Each time the team wins a game, Leigh pays the nine starting players, the manager, and two coaches bonuses, which are certain percentages of their regular salaries. The amounts paid are listed here.*

\$80,	\$105,	\$120,	\$175,	\$185,	\$190,
\$205,	\$210,	\$215,	\$300,	\$320,	\$325

Use this distribution of bonuses for Exercises 29–34.

29. Find the mean of the distribution.

30. Find the standard deviation of the distribution.

31. How many of the bonus amounts are within one standard deviation of the mean?

32. How many of the bonus amounts are within two standard deviations of the mean?

33. What does Chebyshev's theorem say about the number of the amounts that are within two standard deviations of the mean?

34. Explain any discrepancy between your answers for Exercises 32 and 33.

In Exercises 35 and 36, two samples are given. In each case, **(a)** *find both sample standard deviations,* **(b)** *find both sample coefficients of variation,* **(c)** *decide which sample has the higher dispersion, and* **(d)** *decide which sample has the higher relative dispersion.*

35. *A:* 3, 7, 4, 3, 8 *B:* 10, 8, 10, 6, 7, 3, 5

36. *A:* 68, 72, 69, 65, 71, 72, 68, 71, 67, 67
B: 26, 35, 30, 28, 31, 36, 38, 29, 34, 33

37. ***Comparing Battery Lifetimes*** Two brands of car batteries, both carrying 6-year warranties, were sampled and tested under controlled conditions. Five of each brand failed after the numbers of months shown here.

Brand A: 75, 65, 70, 64, 71

Brand B: 69, 70, 62, 72, 60

(a) Calculate both sample means.
(b) Calculate both sample standard deviations.
(c) Which brand battery apparently lasts longer?
(d) Which brand battery has the more consistent lifetime?

Lifetimes of Engine Control Modules *Chris Christensen manages the service department of a trucking company. Each truck in the fleet utilizes an electronic engine control module, which must be replaced when it fails. Long-lasting modules are desirable, of course, but a preventive replacement program can also avoid costly breakdowns on the highway. For this purpose it is desirable that the modules be fairly consistent in their lifetimes; that is, they should all last about the same number of miles before failure, so that the timing of preventive replacements can be done accurately.*

Chris tested a sample of 20 *Brand A modules, and they lasted* 43,560 *highway miles on the average (mean), with a standard deviation of* 2116 *miles. The listing below shows how long each of another sample of* 20 *Brand B modules lasted. Use the data for Exercises 38–40.*

50,660, 41,300, 45,680, 48,840, 47,300,
51,220, 49,100, 48,660, 47,790, 47,210,
50,050, 49,920, 47,420, 45,880, 50,110,
49,910, 47,930, 48,800, 46,690, 49,040

38. According to the sampling, which brand of module has the longer average life (in highway miles)?

39. Which brand of module apparently has a more consistent (or uniform) length of life (in highway miles)?

40. If Brands A and B are the only modules available, which one should Chris purchase for the maintenance program? Explain your reasoning.

Utilize the following sample for Exercises 41–46.

13, 14, 16, 18, 20, 22, 25

41. Compute the mean and standard deviation for the sample (each to the nearest hundredth).

42. Now add 5 to each item of the given sample and compute the mean and standard deviation for the new sample.

43. Go back to the original sample. This time subtract 10 from each item, and compute the mean and standard deviation of the new sample.

44. Based on your answers for Exercises 41–43, make conjectures about what happens to the mean and standard deviation when all items of the sample have the same constant k added or subtracted.

45. Go back to the original sample again. This time multiply each item by 3, and compute the mean and standard deviation of the new sample.

46. Based on your answers for Exercises 41 and 45, make conjectures about what happens to the mean and standard deviation when all items of the sample are multiplied by the same constant k.

47. In Section 13.2 we showed that the mean, as a measure of central tendency, is highly sensitive to extreme values. Which measure of dispersion, covered in this section, would be more sensitive to extreme values? Illustrate your answer with one or more examples.

48. ***A Cereal Marketing Survey*** A food distribution company conducted a survey to determine whether a proposed premium to be included in boxes of their cereal was appealing enough to generate new sales. Four cities were used as test markets, where the cereal was distributed with the premium, and four cities as control markets, where the cereal was distributed without the premium. The eight cities were chosen on the basis of their similarity in terms of population, per capita income, and total cereal purchase volume. The results follow.

		Percent Change in Average Market Share per Month
Test cities	1	+18
	2	+15
	3	+7
	4	+10
Control cities	1	+1
	2	−8
	3	−5
	4	0

(a) Find the mean of the change in market share for the four test cities.
(b) Find the mean of the change in market share for the four control cities.
(c) Find the standard deviation of the change in market share for the test cities.
(d) Find the standard deviation of the change in market share for the control cities.
(e) Find the difference between the means of (a) and (b). This difference represents the estimate of the percent change in sales due to the premium.
(f) The two standard deviations from (c) and (d) were used to calculate an "error" of ± 7.95 for the estimate in (e). With this amount of error, what are the least and greatest estimates of the increase in sales?

On the basis of the interval estimate of part (f) the company decided to mass produce the premium and distribute it nationally.

For Exercises 49–51, refer to the grouped frequency distribution shown on the next page. (Also refer to Exercises 59–61 in Section 13.2.)

49. Is it possible to identify any specific data items that occurred in this sample?

50. Is it possible to compute the actual standard deviation for this sample?

Class Limits	Frequency f
21–25	5
26–30	3
31–35	8
36–40	12
41–45	21
46–50	38
51–55	35
56–60	20

51. Describe how you might approximate the standard deviation for this sample. Justify your procedure.

52. Suppose the frequency distribution of Example 4 involved 50 or 100 (or even more) distinct data values, rather than just four. Explain why the procedure of that example would then be very inefficient.

53. A "J-shaped" distribution can be skewed either to the right or to the left. (When skewed right, it is sometimes called a "reverse J" distribution.)

(a) In a J-shaped distribution skewed to the right, which data item would be the mode, the greatest or the least item?

(b) In a J-shaped distribution skewed to the left, which data item would be the mode, the greatest or the least item?

(c) Explain why the mode is a weak measure of central tendency for a J-shaped distribution.

13.4 Measures of Position

The *z*-Score • Percentiles • Deciles and Quartiles • The Box Plot

Measures of central tendency and measures of dispersion give us an effective way of characterizing an overall set of data. Central tendency has to do with where, along a number scale, the overall data set is centered. Dispersion has to do with how much the data set is spread out from the center point. And Chebyshev's theorem, stated in the previous section, tells us in a general sense what portions of the data set may be dispersed different amounts from the center point.

In many cases, we are especially interested in certain individual items in the data set, rather than in the set as a whole. We still need a way of measuring how an item fits into the collection, how it compares to other items in the collection, or even how it compares to another item in another collection. There are several common ways of creating such measures. Since they measure an item's position within the data set, they usually are called **measures of position.**

The *z*-Score

Each individual item in a sample can be assigned a ***z*-score,** which is defined as follows.

The *z*-score

If x is a data item in a sample with mean $\bar{x}$ and standard deviation s, then the ***z*-score** of x is given by

$$z = \frac{x - \bar{x}}{s}.$$

Because $x - \bar{x}$ gives the amount by which x differs (or deviates) from the mean $\bar{x}$, $\frac{x - \bar{x}}{s}$ gives the number of standard deviations by which x differs from $\bar{x}$. Notice that z will be positive if x is greater than $\bar{x}$ but negative if x is less than $\bar{x}$. Chebyshev's theorem assures us that, in any distribution whatsoever, at least 89% (roughly) of the items will lie within three standard deviations of the mean. That is, at least 89% of the items will have z-scores between -3 and 3. In fact, many common distributions, especially symmetric ones, have considerably more than 89% of their items within three standard deviations of the mean (as we will see in the next section). Hence, a z-score greater than 3 or less than -3 is rare.

EXAMPLE 1 Comparing Positions Using *z*-Scores

Two friends, Reza and Weihua, who take different history classes, had midterm exams on the same day. Reza's score was 86 while Weihua's was only 78. Which student did relatively better, given the class data shown here?

	Reza	**Weihua**
Class mean	73	69
Class standard deviation	8	5

SOLUTION

Calculate as follows.

$$\text{Reza: } z = \frac{86 - 73}{8} = 1.625 \qquad \text{Weihua: } z = \frac{78 - 69}{5} = 1.8$$

Since Weihua's z-score is higher, she was positioned relatively higher within her class than Reza was within his class.

Percentiles When you take the Scholastic Aptitude Test (SAT), or any other standardized test taken by large numbers of students, your raw score usually is converted to a **percentile** score, which is defined as follows.

Percentile

If approximately n percent of the items in a distribution are less than the number x, then x is the ***n*th percentile** of the distribution, denoted P_n.

For example, if you scored at the eighty-third percentile on the SAT, it means that you outscored approximately 83% of all those who took the test. (It does *not* mean that you got 83% of the answers correct.) Since the percentile score positions an item within the data set, it is another "measure of position." The following example approximates percentiles for a fairly small collection of data.

EXAMPLE 2 Finding Percentiles

The following are the numbers of dinner customers served by a restaurant on 40 consecutive days. (The numbers have been ranked least to greatest.)

46	51	52	55	56	56	58	59	59	59
61	61	62	62	63	63	64	64	64	65
66	66	66	67	67	67	68	68	69	69
70	70	71	71	72	75	79	79	83	88

For this data set, find **(a)** the thirty-fifth percentile, and **(b)** the eighty-sixth percentile.

SOLUTION

(a) The thirty-fifth percentile can be taken as the item below which 35 percent of the items are ranked. Since 35 percent of 40 is $.35(40) = 14$, we take the fifteenth item, or 63, as the thirty-fifth percentile.

(b) Since 86 percent of 40 is $.86(40) = 34.4$, we round *up* and take the eighty-sixth percentile to be the thirty-fifth item, or 72.

Technically, percentiles originally were conceived as a set of 99 values $P_1, P_2, P_3, \ldots, P_{99}$ (not necessarily data items) along the scale that would divide the data set into 100 equal parts. They were computed only for very large data sets. With smaller data sets, as in Example 2, dividing the data into 100 parts would necessarily leave many of those parts empty. However, the modern techniques of exploratory data analysis seek to apply the percentile concept to even small data sets. Thus, we use approximation techniques as in Example 2. Another option is to divide the data into a lesser number of equal (or nearly equal) parts.

Deciles and Quartiles **Deciles** are the nine values (denoted $D_1, D_2, \ldots, D_9$) along the scale that divide a data set into ten (approximately) equal parts, and **quartiles** are the three values (Q_1, Q_2, and Q_3) that divide a data set into four (approximately) equal parts. Since deciles and quartiles serve to position particular items within portions of a distribution, they also are "measures of position." We can evaluate deciles by finding their equivalent percentiles:

$$D_1 = P_{10}, \quad D_2 = P_{20}, \quad D_3 = P_{30}, \quad \ldots, \quad D_9 = P_{90}.$$

EXAMPLE 3 Finding Deciles

Find the fourth decile for the dinner customer data of Example 2.

SOLUTION

Refer to the ranked data table. The fourth decile is the fortieth percentile, and 40% of 40 is $.40\,(40) = 16$. We take the fourth decile to be the seventeenth item, or 64.

Although the three quartiles also are equivalent to corresponding percentiles, notice that the second quartile, Q_2, also is equivalent to the median, a measure of central tendency introduced in Section 13.2. A common convention for computing quartiles goes back to the way we computed the median.

Finding Quartiles

For any set of data (ranked in order from least to greatest):

The **second quartile, Q_2,** is just the median, the middle item when the number of items is odd, or the mean of the two middle items when the number of items is even.

The **first quartile, Q_1,** is the median of all items below Q_2.

The **third quartile, Q_3,** is the median of all items above Q_2.

EXAMPLE 4 Finding Quartiles

Find the three quartiles for the data of Example 2.

SOLUTION

Refer to the ranked data. The two middle data items are 65 and 66, so

$$Q_2 = \frac{65 + 66}{2} = 65.5.$$

The least 20 items (an even number) are all below Q_2, and the two middle items in that set are 59 and 61.

$$Q_1 = \frac{59 + 61}{2} = 60$$

The greatest 20 items are above Q_2.

$$Q_3 = \frac{69 + 70}{2} = 69.5$$

The Box Plot A **box plot,** or **box-and-whisker plot,** involves the median (a measure of central tendency), the range (a measure of dispersion), and the first and third quartiles (measures of position), all incorporated into a simple visual display.

Box Plot

For a given set of data, a **box plot** (or **box-and-whisker plot**) consists of a rectangular box positioned above a numerical scale, extending from Q_1 to Q_3, with the value of Q_2 (the median) indicated within the box, and with "whiskers" (line segments) extending to the left and right from the box out to the minimum and maximum data items.

EXAMPLE 5 Constructing a Box Plot

Construct a box plot for the weekly study times data of Example 2 in Section 13.1.

SOLUTION

To determine the quartiles and the minimum and maximum values more easily, we use the stem-and-leaf display (with ranked leaves), given in Table 9 of Section 13.2:

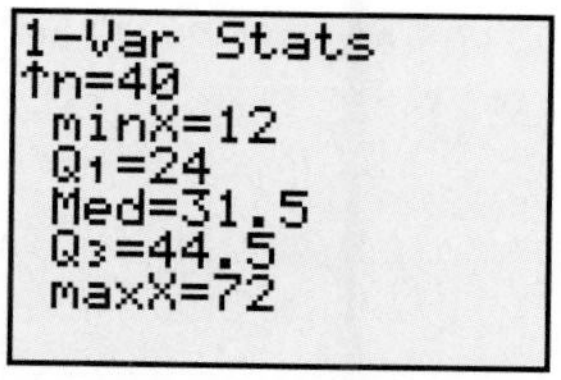

This screen supports the results of Example 5.

1	2 4 5 6 7 8
2	0 2 3 4 4 5 6 6 9 9 9
3	0 1 1 2 3 6 6 8 9
4	0 1 4 4 5 5 7
5	2 5 5 8
6	0 2
7	2

The median (determined earlier in Example 6 of Section 13.2) is

$$\frac{31 + 32}{2} = 31.5.$$

From the stem-and-leaf display,

$$Q_1 = \frac{24 + 24}{2} = 24 \quad \text{and} \quad Q_3 = \frac{44 + 45}{2} = 44.5.$$

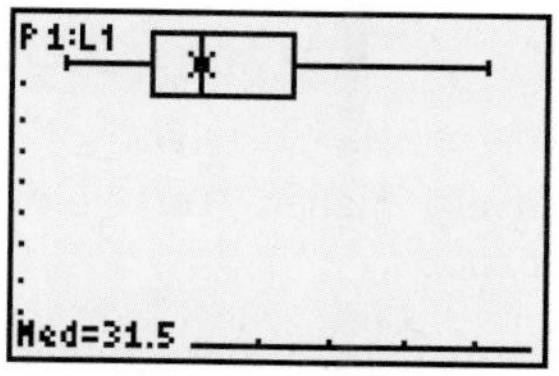

This box plot corresponds to the results of Example 5. It indicates the median in the display at the bottom. The TRACE function of the TI-83/84 Plus will locate the minimum, maximum, and quartile values as well.

The minimum and maximum items are 12 and 72. The box plot is shown in Figure 10.

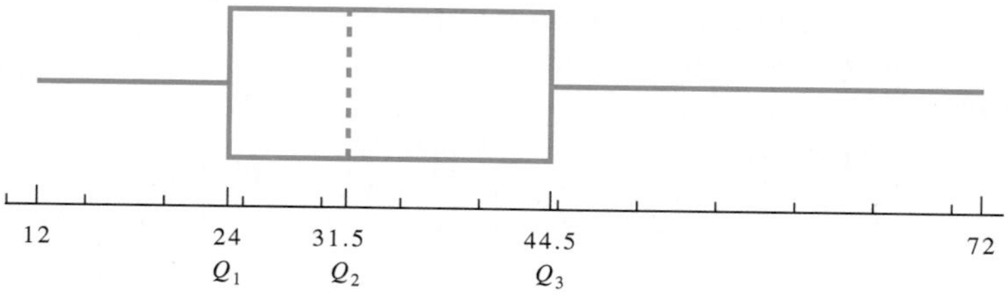

Weekly study times (in hours)

Box plot

FIGURE 10

Notice in Figure 10 that the box plot clearly conveys the following:

1. central tendency (the location of the median);
2. the location of the middle half of the data (the extent of the box);
3. dispersion (the range is the extent of the whiskers);
4. skewness (the nonsymmetry of both the box and the whiskers).

13.4 EXERCISES

In Exercises 1–4, make use of z-scores.

1. ***Relative Positions on Calculus Quizzes*** In a calculus class, Chris Doran scored 5 on a quiz for which the class mean and standard deviation were 4.6 and 2.1, respectively. Lynn Colgin scored 6 on another quiz for which the class mean and standard deviation were 4.9 and 2.3, respectively. Relatively speaking, which student did better?

2. ***Relative Performances in Track Events*** In Saturday's track meet, Ryan Flahive, a high jumper, jumped

6 feet 3 inches. Conference high jump marks for the past season had a mean of 6 feet even and a standard deviation of 3.5 inches. Michael Drago, Ryan's teammate, achieved 18 feet 4 inches in the long jump. In that event the conference season average (mean) and standard deviation were 16 feet 6 inches and 1 foot 10 inches, respectively. Relative to this past season in this conference, which athlete had a better performance on Saturday?

3. ***Relative Lifetimes of Tires*** The lifetimes of Brand A tires are distributed with mean 45,000 miles and standard deviation 4500 miles, while Brand B tires last for only 38,000 miles on the average (mean) with standard deviation 2080 miles. Jutta's Brand A tires lasted 37,000 miles and Arvind's Brand B tires lasted 35,000 miles. Relatively speaking, within their own brands, which driver got the better wear?

4. ***Relative Ratings of Fish Caught*** In a certain lake, the trout average 12 inches in length with a standard deviation of 2.75 inches. The bass average 4 pounds in weight with a standard deviation of .8 pound. If Imelda caught an 18-inch trout and Timothy caught a 6-pound bass, then relatively speaking, which catch was the better trophy?

Numbers of Restaurant Customers *Refer to the dinner customers data of Example 2. Approximate each of the following. (Use the methods illustrated in this section.)*

5. the fifteenth percentile

6. the seventy-fifth percentile

7. the third decile

8. the eighth decile

Leading Foreign Trade Countries of the Middle East *In Exercises 9–20, refer to the data on the top 10 exporting countries of the Middle East for 2005. (Source:* The 2006 World Book Year Book, *page 256.)*

Country	Population (millions)	Foreign Trade (billion U.S. $)	
		Exports	Imports
Saudi Arabia	26.3	113	36
United Arab Emirates	4.3	69	46
Turkey	74.2	69	95
Iran	68.9	39	31
Israel	6.8	34	37
Kuwait	2.7	27	11
Qatar	.6	15	6
Oman	2.8	13	6
Egypt	76.3	11	19
Iraq	27.3	10	10

Compute z-scores (accurate to one decimal place) for Exercises 9–12.

9. Iran's population

10. Kuwait's exports

11. Turkey's imports

12. Israel's population

In each of Exercises 13–16, determine which country occupied the given position.

13. the fifteenth percentile in population

14. the third quartile in exports

15. the eighth decile in imports

16. the first quartile in exports

17. Determine who was relatively higher: Turkey in imports or Saudi Arabia in exports.

18. Construct box plots for the both exports and imports, one above the other in the same drawing.

19. What does your box plot of Exercise 18 *for exports* indicate about the following characteristics of the exports data?
(a) the central tendency **(b)** the dispersion
(c) the location of the middle half of the data items

20. Comparing your two box plots of Exercise 18, what can you say about the 2005 trade balance of the Middle East region?

21. The text stated that, for *any* distribution of data, at least 89% of the items will be within three standard deviations of the mean. Why couldn't we just move some items farther out from the mean to obtain a new distribution that would violate this condition?

22. Describe the basic difference between a measure of central tendency and a measure of position.

This chapter has introduced three major characteristics: central tendency, dispersion, and position, and has developed various ways of measuring them in numerical data. In each of Exercises 23–26, a new measure is described. Explain in each case which of the three characteristics you think it would measure and why.

23. Midrange $= \dfrac{\text{minimum item} + \text{maximum item}}{2}$

24. Midquartile $= \dfrac{Q_1 + Q_3}{2}$

25. Interquartile range $= Q_3 - Q_1$

26. Semi-interquartile range $= \dfrac{Q_3 - Q_1}{2}$

27. The "skewness coefficient" was defined in For Further Thought of the previous section, and it is equivalent to

$$\frac{3 \cdot (\bar{x} - Q_2)}{s}.$$

Is this a measure of individual data items or of the overall distribution?

28. In a national standardized test, Jennifer scored at the ninety-second percentile. If 67,500 individuals took the test, about how many scored higher than Jennifer did?

29. Let the three quartiles (from least to greatest) for a large population of scores be denoted Q_1, Q_2, and Q_3.

(a) Is it necessarily true that

$$Q_2 - Q_1 = Q_3 - Q_2?$$

(b) Explain your answer to part (a).

In Exercises 30–33, answer yes *or* no *and explain your answer. (Consult Exercises 23–26 for definitions of some of these measures.)*

30. Is the midquartile necessarily the same as the median?

31. Is the midquartile necessarily the same as the midrange?

32. Is the interquartile range necessarily half the range?

33. Is the semi-interquartile range necessarily half the interquartile range?

34. ***Relative Positions on a Standardized Chemistry Test*** Omer and Alessandro participated in the standardization process for a new statewide chemistry test. Within the large group participating, their raw scores and corresponding z-scores were as shown here.

	Raw Score	z-score
Omer	60	.69
Alessandro	72	1.67

Find the overall mean and standard deviation of the distribution of scores. (Give answers to two decimal places.)

Rating Passers in the National Football League *Since the National Football League began keeping official statistics in 1932, one important aspect of the performance of quarterbacks, namely their passing effectiveness, has been rated by several different methods. The current system, adopted in 1973, is based on four performance components considered most important: completions, touchdowns, yards gained, and interceptions, as percentages of the number of passes attempted. The computation can be accomplished using the following formula.**

$$\text{Rating} = \frac{\left(250 \cdot \frac{C}{A}\right) + \left(1000 \cdot \frac{T}{A}\right) + \left(12.5 \cdot \frac{Y}{A}\right) + 6.25 - \left(1250 \cdot \frac{I}{A}\right)}{3},$$

where $A =$ attempted passes, $C =$ completed passes, $T =$ touchdown passes, $Y =$ yards gained passing, and $I =$ interceptions.

In addition to the weighting factors (coefficients) appearing in the formula, the four category ratios are limited to nonnegative values with the following maximums.

$$.775 \text{ for } \frac{C}{A}, \quad .11875 \text{ for } \frac{T}{A}, \quad 12.5 \text{ for } \frac{Y}{A}, \quad .095 \text{ for } \frac{I}{A}$$

*This version of the formula was derived by Joseph Farabee of American River College after he sifted through a variety of descriptions in sports columns and NFL publications.

These limitations are intended to prevent any one component of performance from having an undue effect on the overall rating. They are not often invoked, but in special cases can have a significant effect. (See Exercises 43–46.)

The preceding formula rates all passers against the same performance standard and is applied, for example, after a single game, an entire season, or a career. The ratings for the ten leading passers in the league for 2005 postseason play are ranked in the table at the right. Find the requested measures (to one decimal place) in Exercises 35–40.

35. the three quartiles

36. the third decile

37. the sixty-fifth percentile

38. the midrange (See Exercise 23.)

39. the midquartile (see Exercise 24.)

40. the interquartile range (See Exercise 25.)

41. Construct a box plot for the rating points data.

42. The eleventh-ranked passer in the 2005 postseason was Rex Grossman of the Chicago Bears. Rex attempted 41 passes, completed 17, passed for 1 touchdown, gained 192 yards passing, and was intercepted once. Compute his postseason rating.

NFL Passer	Rating Points
Roethlisberger, Pittsburgh	101.7
Brady, New England	92.2
Manning, Indianapolis	90.9
Hasselbeck, Seattle	89.7
Delhomme, Carolina	82.4
Plummer, Denver	72.0
Brunell, Washington	69.6
Leftwich, Jacksonville	61.1
Kitna, Cincinnati	60.1
Simms, Tampa Bay	56.7

Source: www.nfl.com.

Sid Luckman of the Chicago Bears, the passing champion of the 1943 (full) season, had the following pertinent statistics in that year.

202 attempts, 110 completions, 28 touchdowns, 2194 yards, 12 interceptions

Use these numbers in the rating formula for the following exercises.

43. Compute Luckman's 1943 rating, being careful to replace each ratio in the formula with its allowed maximum if necessary.

44. Compute Luckman's rating assuming no restrictions on the ratios.

45. Which component of Luckman's passing game that year was apparently very unusual, even among high-caliber passers?

46. Compare the two ratings of Exercises 43 and 44 with Steve Young's all-time record season rating of 112.8 (achieved in 1994).

13.5 The Normal Distribution

Discrete and Continuous Random Variables • Definition and Properties of a Normal Curve • A Table of Standard Normal Curve Areas • Interpreting Normal Curve Areas

Discrete and Continuous Random Variables

A random variable that can take on only certain fixed values is called a **discrete random variable.** For example, the number of heads in 5 tosses of a coin is discrete since its only possible values are 0, 1, 2, 3, 4, and 5. A variable whose values are not restricted in this way is a **continuous random variable.** For example, the diameter of camellia blossoms would be a continuous variable, spread over a scale perhaps from 5 to 25 centimeters. The values would not be restricted to whole numbers, or even to tenths, or hundredths, etc. Although in practice we may not measure the diameters to more accuracy than, say, tenths of centimeters, they could theoretically take on any of the infinitely many values

TABLE 16

Probability Distribution

x	$P(x)$
0	.03125
1	.15625
2	.31250
3	.31250
4	.15625
5	.03125

from 5 to 25. In terms of set theory, we can say that a discrete random variable can take on only a countable number of values, whereas a continuous random variable can take on an uncountable number of values.

In contrast to most distributions discussed earlier in this chapter, which were *empirical* (based on observation), the distributions covered in this section and the next are *theoretical* (based on theoretical probabilities). A knowledge of theoretical distributions enables us to identify when actual observations are inconsistent with stated assumptions, which is the key to inferential statistics.

The theoretical probability distribution for the discrete random variable "number of heads" when 5 fair coins are tossed is shown in Table 16, and Figure 11 shows the corresponding histogram. The probability values can be found using the binomial probability formula (Section 12.4) or using Pascal's triangle (Section 11.4).

Since each rectangle in Figure 11 is 1 unit wide, the *area* of the rectangle is also equal to the probability of the corresponding number of heads. The area, and thus the probability, for the event "1 head or 2 heads" is shaded in the figure. The graph consists of 6 distinct rectangles since "number of heads" is a *discrete* variable with 6 possible values. The sum of the 6 rectangular areas is exactly 1 square unit.

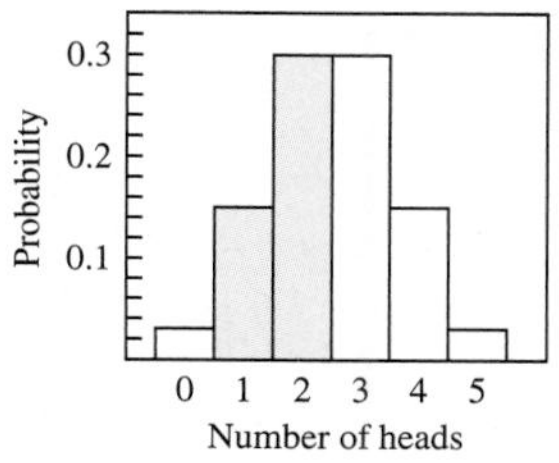

FIGURE 11

In contrast to the discrete "number of heads" distribution in Table 16, a probability distribution for camellia blossom diameters cannot be tabulated or graphed in quite the same way, since this variable is *continuous*. The graph would be smeared out into a "continuous" bell-shaped curve (rather than a set of rectangles) as shown in Figure 12. The vertical scale on the graph in this case shows what we call "probability density," the probability per unit along the horizontal axis.

Definition and Properties of a Normal Curve The curve just described is highest at a diameter value of 15 cm, its center point, and drops off rapidly and equally toward a zero level in both directions. A symmetric, bell-shaped curve of this type is called a **normal curve.** Any random variable whose graph has this characteristic shape is said to have a **normal distribution.** A great many continuous random variables have this type of normal distribution, and even the distributions of discrete variables can often be approximated closely by normal curves. Thus, normal curves are well worth studying. (Of course, not all distributions are normal or approximately so—for example, income distributions usually are not.)

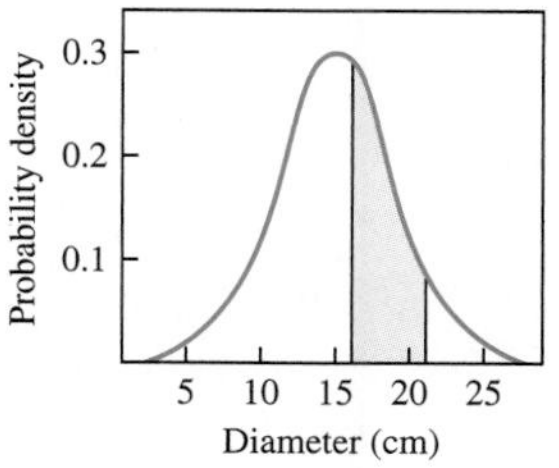

FIGURE 12

On a normal curve, if the quantity shown on the horizontal axis is the number of standard deviations from the mean, rather than values of the random variable itself, then we call the curve the **standard normal curve.** In that case, the horizontal dimension measures numbers of standard deviations, with the value being 0 at the mean of the distribution. The number of standard deviations, associated with a point on the horizontal axis, is the "standard score" for that point. It is the same as the z-score defined in the previous section.

The area under the curve along a certain interval is numerically equal to the probability that the random variable will have a value in the corresponding interval. The total area under the standard normal curve is exactly 1 square unit.

The area of the shaded region in Figure 12 is equal to the probability of a randomly chosen blossom having a diameter in the interval from exactly 16.4 cm to exactly 21.2 cm. To find that area, we first convert 16.4 and 21.2 to standard deviation units. With this approach, we can deal with any normal curve by comparing it to the standard normal curve. A single table of areas for the standard normal curve then provides probabilities for all normal curves, as will be shown later in this section.

The normal curve was first developed by **Abraham De Moivre** (1667–1754), but his work went unnoticed for many years. It was independently redeveloped by Pierre de Laplace (1749–1827) and Carl Friedrich Gauss (1777–1855). Gauss found so many uses for this curve that it is sometimes called the *Gaussian curve.*

Figure 13 shows several of infinitely many possible normal curves. Each is completely characterized by its mean and standard deviation. Only one of these, the one marked S, is the *standard* normal curve. That one has mean 0 and standard deviation 1.

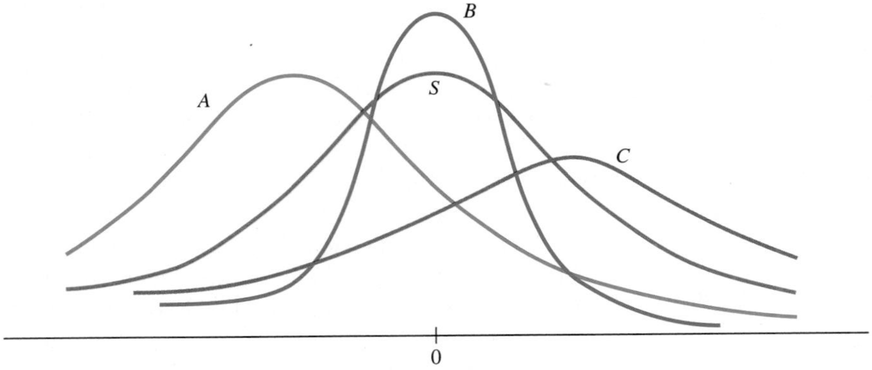

Normal curve S is standard, with mean = 0 and standard deviation = 1.
Normal curve A has mean < 0 and standard deviation = 1.
Normal curve B has mean = 0 and standard deviation < 1.
Normal curve C has mean > 0 and standard deviation > 1.

FIGURE 13

Several properties of normal curves are summarized below and are illustrated in Figure 14.

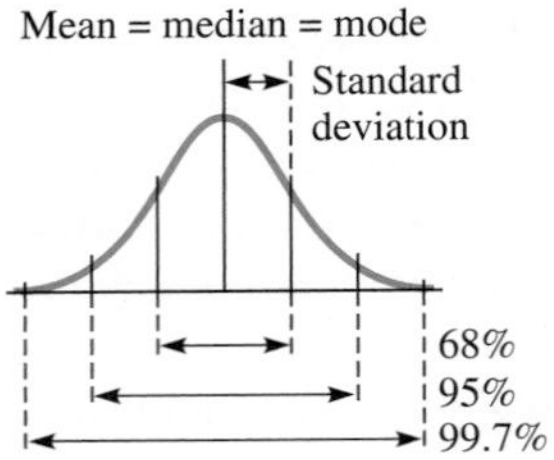

FIGURE 14

Properties of Normal Curves

The graph of a normal curve is bell-shaped and symmetric about a vertical line through its center.

The mean, median, and mode of a normal curve are all equal and occur at the center of the distribution.

Empirical Rule About 68% of all data values of a normal curve lie within 1 standard deviation of the mean (in both directions), about 95% within 2 standard deviations, and about 99.7% within 3 standard deviations.

The empirical rule indicates that a very small percentage of the items in a normal distribution will lie more than 3 standard deviations from the mean (approximately .3%, divided equally between the upper and lower tails of the distribution). As we move away from the center, the curve falls rapidly, and then more and more gradually, toward the axis. But it *never* actually reaches the axis. No matter how far out we go, there is always a chance (though very small) of an item occurring even farther out. Theoretically then, the range of a true normal distribution is infinite.

It is important to realize that the percentage of items within a certain interval is equivalent to the probability that a randomly chosen item will lie in that interval. Thus, one result of the empirical rule is that if we choose a single number at random from a normal distribution, the probability that it will lie within 1 standard deviation of the mean is about .68.

Close but Never Touching
When a curve approaches closer and closer to a line, without ever actually meeting it (as a normal curve approaches the horizontal axis), the line is called an **asymptote,** and the curve approaches the line **asymptotically.**

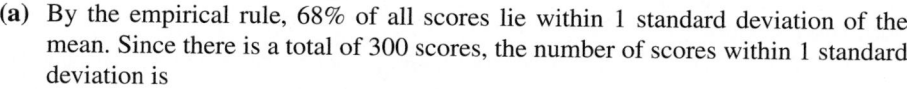

EXAMPLE 1 Applying the Empirical Rule

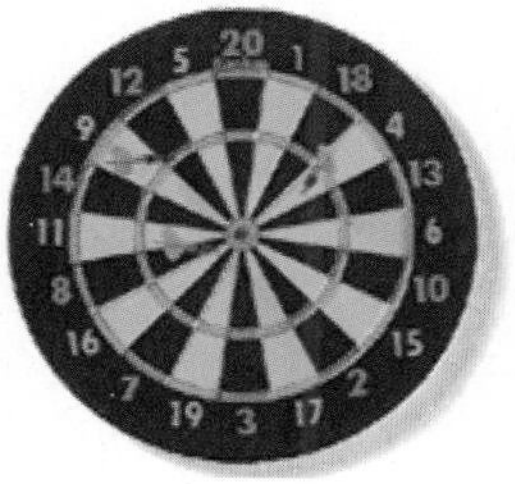

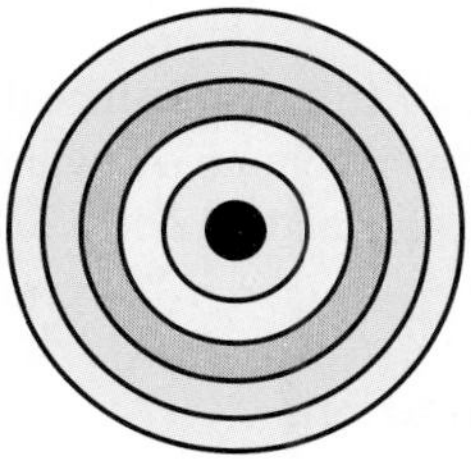

Suppose 300 chemistry students take a midterm exam and that the distribution of their scores can be treated as normal. Find the number of scores falling into each of the following intervals.

(a) Within 1 standard deviation of the mean
(b) Within 2 standard deviations of the mean

SOLUTION

(a) By the empirical rule, 68% of all scores lie within 1 standard deviation of the mean. Since there is a total of 300 scores, the number of scores within 1 standard deviation is

$$.68(300) = 204. \quad 68\% = .68$$

(b) A total of 95% of all scores lie within 2 standard deviations of the mean.

$$.95(300) = 285 \quad 95\% = .95$$

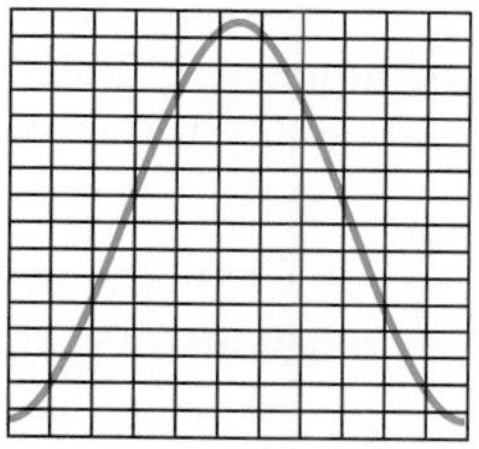

A normal distribution occurs in darts if the player, always aiming at the bull's-eye, tosses a fairly large number of times, and the aim on each toss is affected by independent random errors.

A Table of Standard Normal Curve Areas

Most questions we need to answer about normal distributions involve regions other than those within 1, 2, or 3 standard deviations of the mean. For example, we might need the percentage of items within $1\frac{1}{2}$ or $2\frac{1}{5}$ standard deviations of the mean, or perhaps the area under the curve from .8 to 1.3 standard deviations above the mean. In such cases, we need more than the empirical rule. The traditional approach is to refer to a table of area values, such as Table 17, which appears on page 829. Computer software packages designed for statistical uses usually will produce the required values on command and some advanced calculators also have this capability. Those tools are recommended. As an optional approach, we illustrate the use of Table 17 here.

The table gives the fraction of all scores in a normal distribution that lie between the mean and z standard deviations from the mean. *Because of the symmetry of the normal curve, the table can be used for values above the mean or below the mean.* All of the items in the table can be thought of as corresponding to the area under the curve. The total area is arranged to be 1.000 square unit, with .500 square unit on each side of the mean. The table shows that at 3.30 standard deviations from the mean, essentially all of the area is accounted for. Whatever remains beyond is so small that it does not appear in the first three decimal places.

EXAMPLE 2 Applying the Normal Curve Table

Use Table 17 to find the percent of all scores that lie between the mean and the following values.

(a) One standard deviation above the mean
(b) 2.45 standard deviations below the mean

SOLUTION

(a) Here $z = 1.00$ (the number of standard deviations, written as a decimal to the nearest hundredth). Refer to Table 17. Find 1.00 in the z column. The table entry is .341, so 34.1% of all values lie between the mean and one standard deviation above the mean.

Carl Friedrich Gauss (1777–1855) was one of the greatest mathematical thinkers of history. In his *Disquisitiones Arithmeticae,* published in 1798, he pulled together work by predecessors and enriched and blended it with his own into a unified whole. The book is regarded by many as the true beginning of the theory of numbers.

Of his many contributions to science, the statistical method of least squares is the most widely used today in astronomy, biology, geodesy, physics, and the social sciences. Gauss took special pride in his contributions to developing the method. Despite an aversion to teaching, he taught an annual course in the method for the last twenty years of his life.

It has been said that Gauss was the last person to have mastered all of the mathematics known in his day.

Another way of looking at this is to say that the area in color in Figure 15 represents 34.1% of the total area under the normal curve.

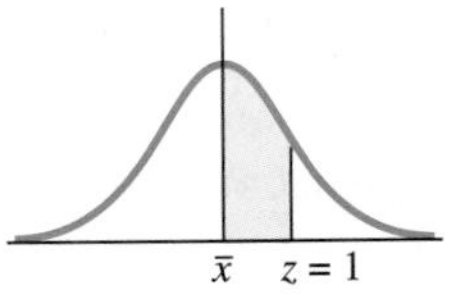

FIGURE 15

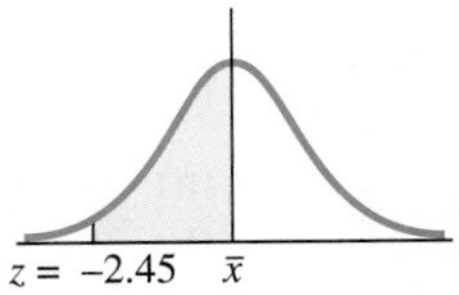

FIGURE 16

(b) Even though we go *below* the mean here (to the left), Table 17 still works since the normal curve is symmetrical about its mean. Find 2.45 in the z column. A total of .493, or 49.3%, of all values lie between the mean and 2.45 standard deviations below the mean. This region is colored in Figure 16. ■

EXAMPLE 3 Finding Probabilities of Phone Call Durations

The lengths of long-distance phone calls placed through a certain company are distributed normally with mean 6 minutes and standard deviation 2 minutes. If 1 call is randomly selected from phone company records, what is the probability that it will have lasted more than 10 minutes?

SOLUTION

Here 10 minutes is two standard deviations above the mean. The probability of such a call is equal to the area of the colored region in Figure 17.

From Table 17, the area between the mean and two standard deviations above is .477 ($z = 2.00$). The total area to the right of the mean is .500. Find the area from $z = 2.00$ to the right by subtracting: $.500 - .477 = .023$. So the probability of a call exceeding 10 minutes is .023, or 2.3%. ■

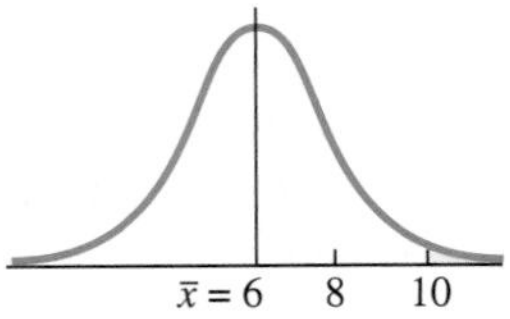

FIGURE 17

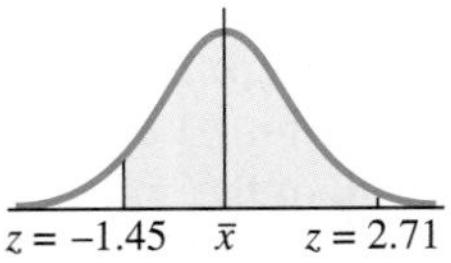

FIGURE 18

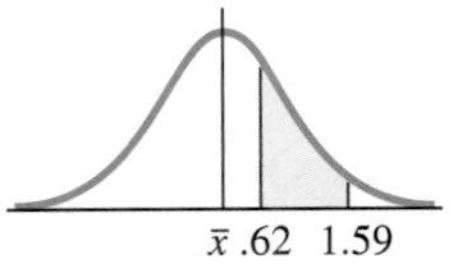

FIGURE 19

EXAMPLE 4 Finding Areas Under the Normal Curve

Find the total areas indicated in the regions in color in each of Figures 18 and 19.

SOLUTION

For Figure 18, find the area from 1.45 standard deviations below the mean to 2.71 standard deviations above the mean. From Table 17, $z = 1.45$ leads to an area of .426, while $z = 2.71$ leads to .497. The total area is the sum of these, or $.426 + .497 = .923$.

To find the indicated area in Figure 19, refer again to Table 17. From the table, $z = .62$ leads to an area of .232, while $z = 1.59$ gives .444. To get the area between these two values of z, subtract the areas: $.444 - .232 = .212$. ■

The column under *A* gives the proportion of the area under the entire curve that is between $z = 0$ and a positive value of z.

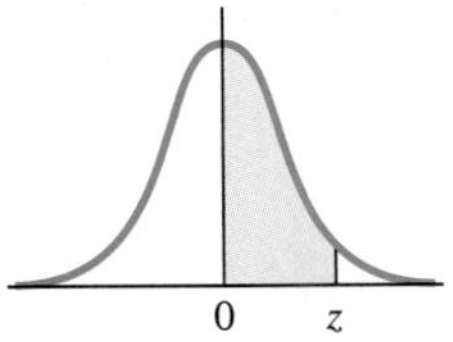

Because the curve is symmetric about the 0-value, the area between $z = 0$ and a *negative* value of z can be found by using the corresponding positive value of z.

TABLE 17 Areas Under the Standard Normal Curve

z	*A*	*z*	*A*	*z*	*A*	*z*	*A*	*z*	*A*	*z*	*A*
.00	.000	.56	.212	1.12	.369	1.68	.454	2.24	.487	2.80	.497
.01	.004	.57	.216	1.13	.371	1.69	.454	2.25	.488	2.81	.498
.02	.008	.58	.219	1.14	.373	1.70	.455	2.26	.488	2.82	.498
.03	.012	.59	.222	1.15	.375	1.71	.456	2.27	.488	2.83	.498
.04	.016	.60	.226	1.16	.377	1.72	.457	2.28	.489	2.84	.498
.05	.020	.61	.229	1.17	.379	1.73	.458	2.29	.489	2.85	.498
.06	.024	.62	.232	1.18	.381	1.74	.459	2.30	.489	2.86	.498
.07	.028	.63	.236	1.19	.383	1.75	.460	2.31	.490	2.87	.498
.08	.032	.64	.239	1.20	.385	1.76	.461	2.32	.490	2.88	.498
.09	.036	.65	.242	1.21	.387	1.77	.462	2.33	.490	2.89	.498
.10	.040	.66	.245	1.22	.389	1.78	.462	2.34	.490	2.90	.498
.11	.044	.67	.249	1.23	.391	1.79	.463	2.35	.491	2.91	.498
.12	.048	.68	.252	1.24	.393	1.80	.464	2.36	.491	2.92	.498
.13	.052	.69	.255	1.25	.394	1.81	.465	2.37	.491	2.93	.498
.14	.056	.70	.258	1.26	.396	1.82	.466	2.38	.491	2.94	.498
.15	.060	.71	.261	1.27	.398	1.83	.466	2.39	.492	2.95	.498
.16	.064	.72	.264	1.28	.400	1.84	.467	2.40	.492	2.96	.498
.17	.067	.73	.267	1.29	.401	1.85	.468	2.41	.492	2.97	.499
.18	.071	.74	.270	1.30	.403	1.86	.469	2.42	.492	2.98	.499
.19	.075	.75	.273	1.31	.405	1.87	.469	2.43	.492	2.99	.499
.20	.079	.76	.276	1.32	.407	1.88	.470	2.44	.493	3.00	.499
.21	.083	.77	.279	1.33	.408	1.89	.471	2.45	.493	3.01	.499
.22	.087	.78	.282	1.34	.410	1.90	.471	2.46	.493	3.02	.499
.23	.091	.79	.285	1.35	.411	1.91	.472	2.47	.493	3.03	.499
.24	.095	.80	.288	1.36	.413	1.92	.473	2.48	.493	3.04	.499
.25	.099	.81	.291	1.37	.415	1.93	.473	2.49	.494	3.05	.499
.26	.103	.82	.294	1.38	.416	1.94	.474	2.50	.494	3.06	.499
.27	.106	.83	.297	1.39	.418	1.95	.474	2.51	.494	3.07	.499
.28	.110	.84	.300	1.40	.419	1.96	.475	2.52	.494	3.08	.499
.29	.114	.85	.302	1.41	.421	1.97	.476	2.53	.494	3.09	.499
.30	.118	.86	.305	1.42	.422	1.98	.476	2.54	.494	3.10	.499
.31	.122	.87	.308	1.43	.424	1.99	.477	2.55	.495	3.11	.499
.32	.126	.88	.311	1.44	.425	2.00	.477	2.56	.495	3.12	.499
.33	.129	.89	.313	1.45	.426	2.01	.478	2.57	.495	3.13	.499
.34	.133	.90	.316	1.46	.428	2.02	.478	2.58	.495	3.14	.499
.35	.137	.91	.319	1.47	.429	2.03	.479	2.59	.495	3.15	.499
.36	.141	.92	.321	1.48	.431	2.04	.479	2.60	.495	3.16	.499
.37	.144	.93	.324	1.49	.432	2.05	.480	2.61	.495	3.17	.499
.38	.148	.94	.326	1.50	.433	2.06	.480	2.62	.496	3.18	.499
.39	.152	.95	.329	1.51	.434	2.07	.481	2.63	.496	3.19	.499
.40	.155	.96	.331	1.52	.436	2.08	.481	2.64	.496	3.20	.499
.41	.159	.97	.334	1.53	.437	2.09	.482	2.65	.496	3.21	.499
.42	.163	.98	.336	1.54	.438	2.10	.482	2.66	.496	3.22	.499
.43	.166	.99	.339	1.55	.439	2.11	.483	2.67	.496	3.23	.499
.44	.170	1.00	.341	1.56	.441	2.12	.483	2.68	.496	3.24	.499
.45	.174	1.01	.344	1.57	.442	2.13	.483	2.69	.496	3.25	.499
.46	.177	1.02	.346	1.58	.443	2.14	.484	2.70	.497	3.26	.499
.47	.181	1.03	.348	1.59	.444	2.15	.484	2.71	.497	3.27	.499
.48	.184	1.04	.351	1.60	.445	2.16	.485	2.72	.497	3.28	.499
.49	.188	1.05	.353	1.61	.446	2.17	.485	2.73	.497	3.29	.499
.50	.191	1.06	.355	1.62	.447	2.18	.485	2.74	.497	3.30	.500
.51	.195	1.07	.358	1.63	.448	2.19	.486	2.75	.497	3.31	.500
.52	.198	1.08	.360	1.64	.449	2.20	.486	2.76	.497	3.32	.500
.53	.202	1.09	.362	1.65	.451	2.21	.486	2.77	.497	3.33	.500
.54	.205	1.10	.364	1.66	.452	2.22	.487	2.78	.497	3.34	.500
.55	.209	1.11	.367	1.67	.453	2.23	.487	2.79	.497	3.35	.500

A Basic Consumer (T)issue It all started when a reporter on consumer issues for a Midwest TV station received a complaint that rolls of Brand X toilet paper manufactured by Company Y did not have the number of sheets claimed on the wrapper. Brand X is supposed to have 375 sheets, but three rolls of it were found by reporters to have 360, 361, and 363. Shocked Company Y executives said that the **odds against** six rolls having fewer than 375 sheets each are 1 billion to 1. They counted sheets and found that several rolls of Brand X actually had 380 sheets each (machines count the sheets only in 10s). TV reporters made an independent count, and their results agreed with Company Y.

What happened the first time? Well, the reporters hadn't actually counted sheets, but had measured rolls and divided the length of a roll by the length of one sheet. Small variations in length can be expected, which add up over a roll, giving false results.

This true story perhaps points up the distinction in probability and statistics between **discrete values** and **continuous values.**

Interpreting Normal Curve Areas Examples 2–4 emphasize the *equivalence* of three quantities, as follows.

Meaning of Normal Curve Areas

In the standard normal curve, the following three quantities are equivalent.

1. **Percentage** (of total items that lie in an interval)
2. **Probability** (of a randomly chosen item lying in an interval)
3. **Area** (under the normal curve along an interval)

Which quantity we think of depends upon how a particular question is formulated. They are all evaluated by using A-values from Table 17.

In general, when we use Table 17, z is the z-score of a particular data item x. Recall from the previous section that the z-score is related to the mean $\bar{x}$ and the standard deviation s of the distribution by the z-score formula

$$z = \frac{x - \bar{x}}{s}.$$

For example, suppose a normal curve has mean 220 and standard deviation 12. To find the number of standard deviations that a given value is from the mean, use the formula above. For the value 247,

$$z = \frac{247 - 220}{12} = \frac{27}{12} = 2.25,$$

so 247 is 2.25 standard deviations above the mean. For 204,

$$z = \frac{204 - 220}{12} = \frac{-16}{12} \approx -1.33,$$

so 204 is 1.33 standard deviations *below* the mean. (We know it is below rather than above the mean since the z-score is negative rather than positive.)

EXAMPLE 5 Applying the Normal Curve to Driving Distances

In one area, the average motorist drives about 1200 miles per month, with standard deviation 150 miles. Assume that the number of miles is closely approximated by a normal curve, and find the percent of all motorists driving the following distances.

(a) Between 1200 and 1600 miles per month
(b) Between 1000 and 1500 miles per month

SOLUTION

(a) Start by finding how many standard deviations 1600 miles is above the mean. Use the formula for z.

$$z = \frac{1600 - 1200}{150} = \frac{400}{150} \approx 2.67$$

From Table 17, .496, or 49.6%, of all motorists drive between 1200 and 1600 miles per month.

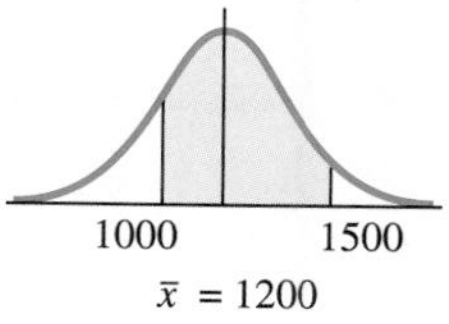

FIGURE 20

(b) As shown in Figure 20, values of z must be found for both 1000 and 1500.

$$\text{For } 1000\text{:}\quad z = \frac{1000 - 1200}{150} = \frac{-200}{150} \approx -1.33.$$

$$\text{For } 1500\text{:}\quad z = \frac{1500 - 1200}{150} = \frac{300}{150} = 2.00.$$

From Table 17, $z = -1.33$ leads to an area of .408, while $z = 2.00$ gives .477. This means a total of

$$.408 + .477 = .885, \quad \text{or} \quad 88.5\%,$$

of all motorists drive between 1000 and 1500 miles per month. ■

Each example thus far has given a data value x and then required that z be found. The next example gives z and asks for the data value.

EXAMPLE 6 Identifying a Data Value Within a Normal Distribution

A particular normal distribution has mean $\bar{x} = 81.7$ and standard deviation $s = 5.21$. What data value from the distribution would correspond to $z = -1.35$?

SOLUTION

Substitute the given values for z, $\bar{x}$, and s into the z-score formula, and solve for x.

$$z = \frac{x - \bar{x}}{s}$$

$$-1.35 = \frac{x - 81.7}{5.21}$$

$$-1.35(5.21) = \frac{x - 81.7}{5.21}(5.21)$$

$$-7.0335 = x - 81.7$$

$$74.6665 = x$$

Rounding to the nearest tenth, the required data value is 74.7. ■

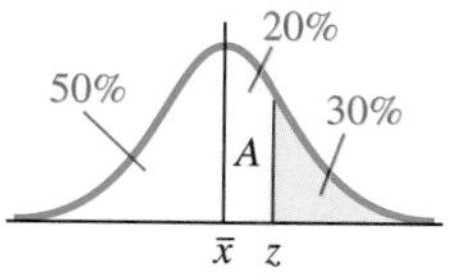

FIGURE 21

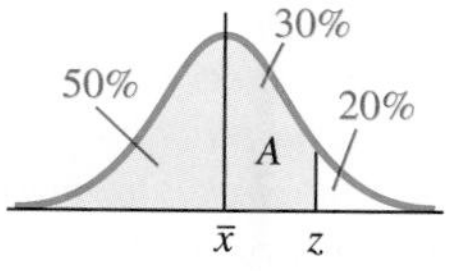

FIGURE 22

EXAMPLE 7 Finding z-Values for Given Areas Under the Normal Curve

Assuming a normal distribution, find the z-value meeting each condition.

(a) 30% of the total area is to the right of z.
(b) 80% of the total area is to the left of z.

SOLUTION

(a) Because 50% of the area lies to the right of the mean, there must be 20% between the mean and z. (See Figure 21.) In Table 17, $A = .200$ corresponds to $z = .52$ or .53, or we could average the two: $z = .525$.

(b) This situation is shown in Figure 22. The 50% to the left of the mean plus 30% additional makes up the 80%. From Table 17, $A = .300$ implies $z = .84$. ■

13.5 EXERCISES

Note: For problems requiring the calculation of z-scores or A-values, our answers are based on Table 17. By using a calculator or computer package, you will sometimes obtain a slightly more accurate answer.

Identify each variable quantity as discrete *or* continuous.

1. the number of heads in 50 tossed coins

2. the number of babies born in one day at a certain hospital

3. the average weight of babies born in a week

4. the heights of seedling pine trees at six months of age

5. the time as shown on a digital watch

6. the time as shown on a watch with a sweep hand

Measuring the Mass of Ore Samples *Suppose 100 geology students measure the mass of an ore sample. Due to human error and limitations in the reliability of the balance, not all the readings are equal. The results are found to closely approximate a normal curve, with mean 86 g and standard deviation 1 g.*

Use the symmetry of the normal curve and the empirical rule to estimate the number of students reporting readings in the following ranges.

7. more than 86 g

8. more than 85 g

9. between 85 and 87 g

10. between 84 and 87 g

Distribution of IQ Scores *On standard IQ tests, the mean is 100, with a standard deviation of 15. The results come very close to fitting a normal curve. Suppose an IQ test is given to a very large group of people. Find the percent of people whose IQ scores fall into each category.*

11. less than 100

12. greater than 115

13. between 70 and 130

14. more than 145

Find the percent of area under a normal curve between the mean and the given number of standard deviations from the mean. (Note that positive indicates above the mean, while negative indicates below the mean.)

15. 1.50

16. .92

17. −1.08

18. −2.25

Find the percent of the total area under a normal curve between the given values of z.

19. $z = 1.41$ and $z = 1.83$

20. $z = -1.74$ and $z = -1.14$

21. $z = -3.11$ and $z = 2.06$

22. $z = -1.98$ and $z = 1.02$

Find a value of z such that each condition is met.

23. 10% of the total area is to the right of z.

24. 4% of the total area is to the left of z.

25. 9% of the total area is to the left of z.

26. 23% of the total area is to the right of z.

Lifetimes of Lightbulbs *The Better lightbulb has an average life of 600 hr, with a standard deviation of 50 hr. The length of life of the bulb can be closely approximated by a normal curve. A warehouse manager buys and installs 10,000 such bulbs. Find the total number that can be expected to last each amount of time.*

27. at least 600 hr

28. between 600 and 675 hr

29. between 675 and 740 hr

30. between 490 and 720 hr

31. less than 740 hr

32. less than 510 hr

Weights of Chickens *The chickens at Ben and Ann Rice's farm have a mean weight of 1850 g with a standard deviation of 150 g. The weights of the chickens are closely approximated by a normal curve. Find the percent of all chickens having each weight.*

33. more than 1700 g

34. less than 1800 g

35. between 1750 and 1900 g

36. between 1600 and 2000 g

Filling Cereal Boxes *A certain dry cereal is packaged in 24-oz boxes. The machine that fills the boxes is set so that, on the average, a box contains 24.5 oz. The machine-filled boxes have contents weights that can be closely approximated by a normal curve. What percentage of the boxes will be underweight if the standard deviation is as follows?*

37. .5 oz

38. .4 oz

39. .3 oz

40. .2 oz

41. ***Recommended Daily Vitamin Allowances*** In nutrition, the recommended daily allowance of vitamins is a number set by the government to guide an individual's daily vitamin intake. Actually, vitamin needs vary drastically from person to person, but the needs are closely approximated by a normal curve. To calculate the recommended daily allowance, the government first finds the average need for vitamins among people in the population and the standard deviation. The **recommended daily allowance** is then defined as the mean plus 2.5 times the standard deviation. What fraction of the population will receive adequate amounts of vitamins under this plan?

Recommended Daily Vitamin Allowances *Find the recommended daily allowance for each vitamin if the mean need and standard deviation are as follows. (See Exercise 41.)*

42. mean need = 1800 units;
standard deviation = 140 units

43. mean need = 159 units;
standard deviation = 12 units

Assume the following distributions are all normal, and use the areas under the normal curve given in Table 17 to find the appropriate areas.

44. ***Filling Cartons with Milk*** A machine that fills quart milk cartons is set up to average 32.2 oz per carton, with a standard deviation of 1.2 oz. What is the probability that a filled carton will contain less than 32 oz of milk?

45. ***Finding Blood Clotting Times*** The mean clotting time of blood is 7.47 sec, with a standard deviation of 3.6 sec. What is the probability that an individual's blood-clotting time will be less than 7 sec or greater than 8 sec?

46. ***Sizes of Fish*** The average length of the fish caught in Lake Amotan is 12.3 in., with a standard deviation of 4.1 in. Find the probability that a fish caught there will be longer than 18 in.

47. ***Size Grading of Eggs*** To be graded extra large, an egg must weigh at least 2.2 oz. If the average weight for an egg is 1.5 oz, with a standard deviation of .4 oz, how many of five dozen randomly chosen eggs would you expect to be extra large?

Distribution of Student Grades *Anessa Davis teaches a course in marketing. She uses the following system for assigning grades to her students.*

Grade	Score in Class
A	Greater than $\bar{x} + 1.5s$
B	$\bar{x} + .5s$ to $\bar{x} + 1.5s$
C	$\bar{x} - .5s$ to $\bar{x} + .5s$
D	$\bar{x} - 1.5s$ to $\bar{x} - .5s$
F	Below $\bar{x} - 1.5s$

From the information in the table, what percent of the students receive the following grades?

48. A

49. B

50. C

51. Do you think this system would be more likely to be fair in a large freshman class in psychology or in a graduate seminar of five students? Why?

Normal Distribution of Student Grades *A teacher gives a test to a large group of students. The results are closely approximated by a normal curve. The mean is 75 with a standard deviation of 5. The teacher wishes to give As to the top 8% of the students and Fs to the bottom 8%. A grade of B is given to the next 15%, with Ds given similarly. All other students get Cs. Find the bottom cutoff (rounded to the nearest whole number) for the following grades. (Hint: Use Table 17 to find z-scores from known A-values.)*

52. A

53. B

54. C

55. D

A normal distribution has mean 76.8 and standard deviation 9.42. Follow the method of Example 6 and find data values corresponding to the following values of z. Round to the nearest tenth.

56. $z = .72$

57. $z = 1.44$

58. $z = -2.39$

59. $z = -3.87$

60. What percentage of the items lie within 1.25 standard deviations of the mean

(a) in any distribution (using the results of Chebyshev's theorem)?

(b) in a normal distribution (by Table 17)?

61. Explain the difference between the answers to parts (a) and (b) in Exercise 60.

EXTENSION

How to Lie with Statistics

The statement that there are "lies, damned lies, and statistics" is attributed to Benjamin Disraeli, Queen Victoria's prime minister. Other people have made even stronger comments about statistics. This often intense distrust of statistics has come about because of a belief that "you can prove anything with numbers." It must be admitted that there is often a conscious or unconscious distortion in many published statistics. The classic book on distortion in statistics is *How to Lie with Statistics*, by Darrell Huff (first published in 1972). This extension quotes some common methods of distortion that Huff gives in his book.

The Sample with the Built-in Bias

A house-to-house survey purporting to study magazine readership was once made in which a key question was: What magazines does your household read? When the results were tabulated and analyzed it appeared that a great many people loved Harper's and not very many read True Story. Now there were publishers' figures around at the time that showed very clearly that True Story had more millions of circulation than Harper's had hundreds of thousands. Perhaps we asked the wrong kind of people, the designers of the survey said to themselves. But no, the questions had been asked in all sorts of neighborhoods all around the country. The only reasonable conclusion then was that a good many of the respondents, as people are called when they answer such questions, had not told the truth. About all the survey had uncovered was snobbery.

In the end it was found that if you wanted to know what certain people read it was no use asking them. You could learn a good deal more by going to their houses and saying you wanted to buy old magazines and what could be had? Then all you had to do was count the Yale Reviews and the Love Romances. Even that dubious device, of course, does not tell you what people read, only what they have been exposed to.

Similarly, the next time you learn from your reading that the average American (you hear a good deal about him these days, most of it faintly improbable) brushes his teeth 1.02 times a day–a figure pulled out of the air, but it may be as good as anyone else's–ask yourself a question. How can anyone have found out such a thing? Is a woman who has read in countless advertisements that non-brushers are social offenders going to confess to a stranger that she does not brush her teeth regularly? The statistic may have meaning to one who wants to know only what people say about tooth brushing but it does not tell a great deal about the frequency with which bristle is applied to incisor.

The Well-Chosen Average

A common trick is to use a different kind of average each time, the word "average" having a very loose meaning. It is a trick commonly used, sometimes in innocence but often in guilt, by people wishing to influence public opinion or sell advertising space. When you are told that something is an average you still don't know very much about it unless you can find out which of the common kinds of average it is–mean, median or mode.

Try your skepticism on some items from "A letter from the Publisher" in Time magazine. Of new subscribers it said, "Their median age is 34 years and their average family income is $7270 a year." An earlier survey of "old TIMErs" had found that their "median age was 41 years . . . . Average income was $9535 . . . ." The natural question is why, when median is given for ages both times, the kind of average for incomes is carefully unspecified. Could it be that the mean was used

(continued)

Stand Up and Be Counted
According to an Associated Press release in 1988 (*The Sacramento Union,* September 28), the U.S. Census Bureau anticipated that the 1990 census would cost up to $3 billion (the cost in 1980 was $1.1 billion). Even so, significant problems were foreseen in finding the necessary workers to conduct the census and getting response from the populace. Census Director John G. Keane cited a deteriorating climate in the nation for taking censuses and surveys, due to Americans' attitude toward being counted.

instead because it is bigger, thus seeming to dangle a richer readership before advertisers?

Be careful when reading charts or graphs—often there is no numerical scale, or no units are given. This makes the chart pretty much meaningless. Huff has a good example of this:

The Little Figures That Are Not There

Before me are wrappers from two boxes of Grape-Nuts Flakes. They are slightly different editions, as indicated by their testimonials: one cites Two-Gun Pete and the other says "If you want to be like Hoppy . . . you've got to eat like Hoppy!" Both offer charts to show ("Scientists proved it's true!") that these flakes "start giving you energy in 2 minutes!" In one case the chart hidden in these forests of exclamation points has numbers up the side; in the other case the numbers have been omitted. This is just as well, since there is no hint of what the numbers mean. Both show a steeply climbing brown line ("energy release"), but one has it starting one minute after eating Grape-Nuts Flakes, the other 2 minutes later. One line climbs about twice as fast as the other, suggesting that even the draftsman didn't think these graphs meant anything.

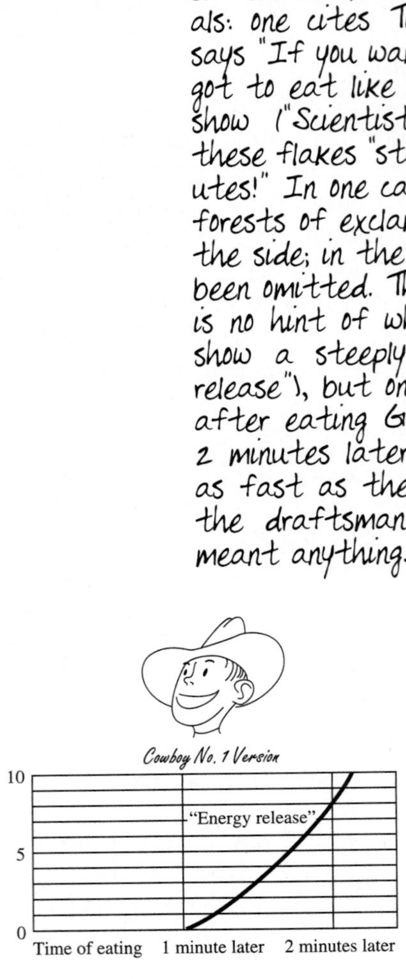

The Gee-Whiz Graph

About the simplest kind of statistical picture, or graph, is the line variety. It is very useful for showing trends, something practically everybody is interested in showing or knowing about or spotting or deploring or forecasting. We'll let our graph show how national income increased ten percent in a year.

Begin with paper ruled into squares. Name the months along the bottom. Indicate billions of dollars up the side. Plot your points and draw your line, and your graph will look like this:

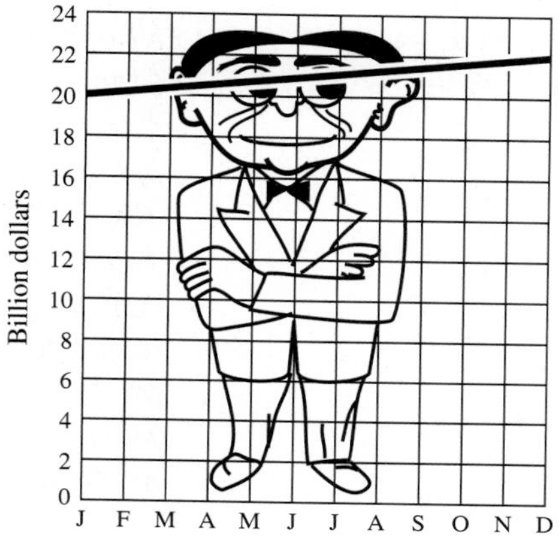

Now that's clear enough. It shows what happened during the year and it shows it month by month. He who runs may see and understand, because the whole graph is in proportion and there is a zero line at the bottom for comparison. Your ten percent looks like ten percent—an upward trend that is substantial but perhaps not overwhelming.

That is very well if all you want to do is convey information. But suppose you wish to win an argument, shock a reader, move him into action, sell him something. For that, this chart lacks schmaltz. Chop off the bottom.

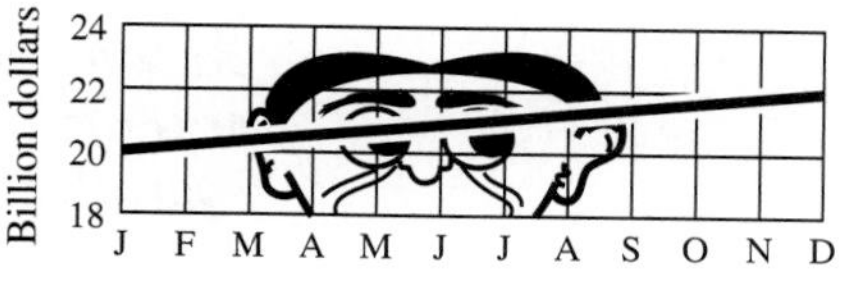

Now that's more like it. (You've saved paper too, something to point out if any carping person objects to your misleading graphics.) The figures are the same and so is the curve. It

(continued)

is the same graph. Nothing has been falsified—except the impression that it gives. But what the hasty reader sees now is a national income line that has climbed halfway up the paper in twelve months, all because most of the chart isn't there any more. Like the missing parts of speech in sentences that you met in grammar classes, it is "understood." Of course, the eye doesn't "understand" what isn't there, and a small rise has become, visually, a big one.

Now that you have practiced to deceive, why stop with truncating? You have a further trick available that's worth a dozen of that. It will make your modest rise of ten percent look livelier than one hundred percent is entitled to look. Simply change the proportion between the side and the bottom. There's no rule against it, and it does give your graph a prettier shape. All you have to do is let each mark up the side stand for only one-tenth as many dollars as before.

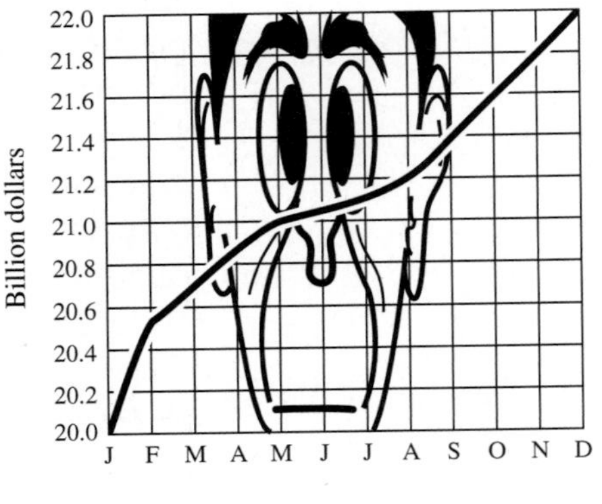

That is impressive, isn't it? Anyone looking at it can just feel prosperity throbbing in the arteries of the country. It is a subtler equivalent of editing "National income rose ten percent" into ". . . climbed a whopping ten percent." It is vastly more effective, however, because it contains no adjectives or adverbs to spoil the illusion of objectivity. There's nothing anyone can pin on you.

Suppose Diana makes twice as much money as Mike. One way to show this is with a graph using silver dollars to represent the income of each. If we used one silver dollar for Mike and two for Diana, we would be fine. But it is more common to use proportional dollars. It is common to find a dollar of one size used for Mike, with one twice as wide for Diana. This is wrong—the larger dollar actually has four times the area, giving the impression that Diana earns four times as much as Mike. Huff gives another example of this:

The One-Dimensional Picture

> Newsweek once showed how "U.S. Old Folks Grow Older" by means of a chart on which appeared two male figures, one representing the 68.2-year life expectancy of today, the other the 34-year life expectancy of 1879–1889. It was the same old story. One figure was twice as tall as the other and so would have had eight times the bulk or weight. This picture sensationalized facts in order to make a better story. It would be called a form of yellow journalism.

Cause and Effect Many people often assume that just because two things changed together one caused the other. The classic example of this is the fact that teachers' salaries and liquor consumption increased together over the last few decades. Neither of these caused the other; rather, both were caused by the same underlying growth in national prosperity. Another case of faulty cause and effect reasoning is in the claim that going to college raises your income. See Huff's comments below.

> Reams of pages of figures have been collected to show the value in dollars of a college education, and stacks of pamphlets have been published to bring these figures-and conclusions more or less based on them-to the attention of potential students. I am not quarreling with the intention. I am in favor of education myself, particularly if it includes a course in elementary statistics. Now these figures have pretty conclusively demonstrated that people who have gone to college make more money than people who have not. The exceptions are numerous, of course, but the tendency is strong and clear.
>
> The only thing wrong is that along with the figures and facts goes a totally unwarranted conclusion . . . . It says that these figures show that if you (your son, your daughter) attend college you will probably earn more

(continued)

> money than if you decide to spend the next four years in some other manner. This unwarranted conclusion has for its basis the equally unwarranted assumption that since college-trained folks make more money, they make it because they went to college. Actually we don't know but that these are the people who would have made more money even if they had not gone to college. There are a couple of things that indicate rather strongly that this is so. Colleges get a disproportionate number of two groups of people: the bright and the rich. The bright might show good earning power without college knowledge. And as for the rich ones . . . well, money breeds money in several obvious ways. Few children of rich parents are found in low-income brackets whether they go to college or not.

Extrapolation This refers to predicting the future, based only on what has happened in the past. However, it is very rare for the future to be just like the past—something will be different. One of the best examples of extrapolating incorrectly is given by Mark Twain:

> In the space of one hundred and seventy-six years the Lower Mississippi has shortened itself two hundred and forty-two miles. That is an average of a trifle over one mile and a third per year. Therefore, any calm person, who is not blind or idiotic, can see that in the Old Oölitic Silurian Period, just a million years ago next November, the Lower Mississippi River was upward of one million three hundred thousand miles long, and stuck out over the Gulf of Mexico like a fishing rod. And by the same token any person can see that seven hundred and forty-two years from now the Lower Mississippi will be only a mile and three-quarters long, and Cairo and New Orleans will have joined their streets together, and be plodding comfortably along under a single mayor and a mutual board of aldermen. There is something fascinating about science. One gets such wholesale returns of conjecture out of such a trifling investment of fact.

Extension Exercises

Reading Line Graphs *The Norwegian stamp at the side features two graphs.*

1. What does the solid line represent?

2. Has it increased much?

3. What can you tell about the dashed line?

4. Does the graph represent a long period of time or a brief period of time?

Changing Value of the British Pound *The illustrations below show the decline in the value of the British pound for a certain period.*

$2.40

$1.72

5. Calculate the percent of decrease in the value by using the formula

$$\text{Percent of decrease} = \frac{\text{old value} - \text{new value}}{\text{old value}}.$$

6. We estimate that the smaller banknote shown has about 50% less area than the larger one. Do you think this is close enough?

Deciphering Advertising Claims *Several advertising claims are given below. Decide what further information you might need before deciding to accept the claim.*

7. 98% of all Toyotas ever sold in the United States are still on the road.

8. Sir Walter Raleigh pipe tobacco is 44% fresher.

9. Eight of 10 dentists responding to a survey preferred Trident Sugarless Gum.

10. A Volvo has $\frac{2}{3}$ the turning radius of a Continental.

11. A Ford LTD is as quiet as a glider.

12. Circulation has increased for *The Wall Street Journal*, as shown by the graph below.

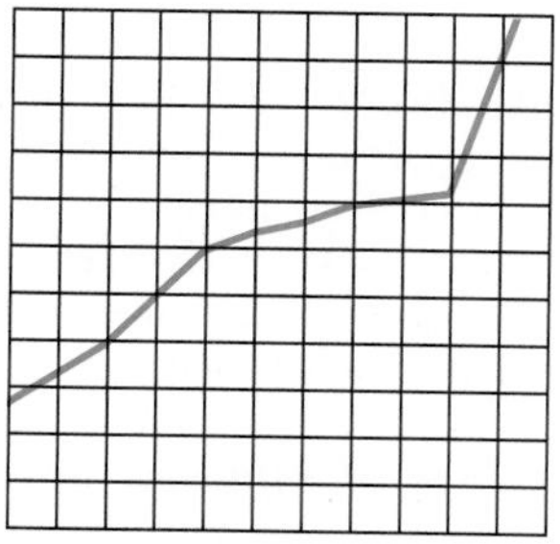

Exercises 13–16 come from Huff's book. Decide how these exercises describe possibly misleading uses of numbers.

13. ***The Extent of Federal Taxes*** Each map below shows what portion of our national income is now being taken and spent by the federal government. The top map does this by shading the areas of most of the states *west* of the Mississippi to indicate that federal spending has become equal to the total incomes of the people of those states. The bottom map does this for states *east* of the Mississippi.

The Darkening Shadow
(Western Style)

(Eastern Style)

To show we aren't cheating, we added MD, DE, and RI for good measure

14. ***Faculty-Student Marriages*** Long ago, when Johns Hopkins University had just begun to admit women students, someone not particularly enamored of co-education reported a real shocker: Thirty three and one-third percent of the women at Hopkins had married faculty members!

15. ***Comparative Death Rates*** The death rate in the Navy during the Spanish-American War was nine per thousand. For civilians in New York City during the same period it was sixteen per thousand. Navy recruiters later used these figures to show that it was safer to be in the Navy than out of it.

16. ***Reported Cases of Influenza and Pneumonia*** If you should look up the latest available figures on influenza and pneumonia, you might come to the strange conclusion that these ailments are practically confined to three southern states, which account for about eighty percent of the reported cases.

Questioning the Reasonableness of Samples *When a sample is selected from a population, it is important to decide whether the sample is reasonably representative of the entire population. For example, a sample of the general population that is only 20% women should make us suspicious. The same is true of a questionnaire asking, "Do you like to answer questionnaires?"*

In each case, pick the most representative sample.

17. A factory has 10% management employees, 30% clerical employees, and 60% assembly-line workers. A sample of 50 is chosen to discuss parking.

A. 4 management, 21 clerical, 25 assembly-line
B. 6 management, 15 clerical, 29 assembly-line
C. 8 management, 9 clerical, 33 assembly-line

18. A college has 35% freshmen, 28% sophomores, 21% juniors, and 16% seniors. A sample of 80 is chosen to discuss methods of electing student officers.

A. 22 freshmen, 22 sophomores, 24 juniors, 12 seniors
B. 24 freshmen, 20 sophomores, 22 juniors, 14 seniors
C. 28 freshmen, 23 sophomores, 16 juniors, 13 seniors

19. A computer company with plants in Boca Raton, Jacksonville, and Tampa produces 42% of its output in Boca Raton, with 27% and 31% coming from Jacksonville and Tampa, respectively. A sample of 120 parts is chosen for quality testing.

A. 38 from Boca Raton, 39 from Jacksonville, 43 from Tampa
B. 43 from Boca Raton, 37 from Jacksonville, 40 from Tampa
C. 50 from Boca Raton, 31 from Jacksonville, 39 from Tampa

20. At one resort, 56% of all guests come from the Northeast, 29% from the Midwest, and 15% from Texas. A sample of 75 guests is chosen to discuss the dinner menu.

A. 41 from the Northeast, 21 from the Midwest, 13 from Texas
B. 45 from the Northeast, 18 from the Midwest, 12 from Texas
C. 47 from the Northeast, 20 from the Midwest, 8 from Texas

Use the information supplied in each problem to solve for the given variables.

21. ***Employee Categories at an Insurance Agency*** An insurance agency has 7 managers, 25 agents, and 18 clerical employees. A sample of 10 is chosen.

(a) Let m be the number of managers in the sample, a the number of agents, and c the number of clerical employees. Find m, a, and c, if

$$c = m + 2$$
$$a = 2c.$$

(b) To check that your answer is reasonable, calculate the numbers of the office staff that *should* be in the sample, if all groups are represented proportionately.

22. ***Faculty Categories at a Small College*** A small college has 12 deans, 24 full professors, 39 associate professors, and 45 assistant professors. A sample of 20 employees is chosen to discuss the graduation speaker.

(a) Let d be the number of deans in the sample, f the number of full professors, a the number of associate professors, and s the number of assistant professors. Find d, f, a, and s if

$$f = 2d$$
$$a = f + d + 1$$
$$a = s.$$

(*Hint:* There are 2 deans.)

(b) To check that your answer is reasonable, calculate the number of each type of employee that *should* be in the sample, if all groups are represented proportionately.

13.6 Regression and Correlation

Linear Regression • Correlation

One very important branch of inferential statistics, called **regression analysis,** is used to compare quantities or variables, to discover relationships that exist between them, and to formulate those relationships in useful ways.

Linear Regression Suppose a sociologist gathers data on a few (say ten) of the residents of a small village in a remote region in order to get an idea of how annual income (in dollars) relates to age in that village. The data are shown in Table 18.

TABLE 18
Age Versus Income

Resident	Age	Annual Income
A	19	2150
B	23	2550
C	27	3250
D	31	3150
E	36	4250
F	40	4200
G	44	4350
H	49	5000
I	52	4950
J	54	5650

The first step in analyzing these data is to graph the results, as shown in the **scatter diagram** of Figure 23. (Graphing calculators will plot scatter diagrams.)

Once a scatter diagram has been produced, we can draw a curve that best fits the pattern exhibited by the sample data points. This curve can have any one of many characteristic shapes, depending on how the quantities involved are related. We would like to infer a graph for the entire population of points for the related quantities, but we have available only a sample of those points, so the best-fitting curve for the sample points is called an **estimated regression curve.** If, as in the present discussion, the points in the scatter diagram seem to lie approximately along a straight line, the relation is assumed to be linear, and the line that best fits the data points is called the **estimated regression line.**

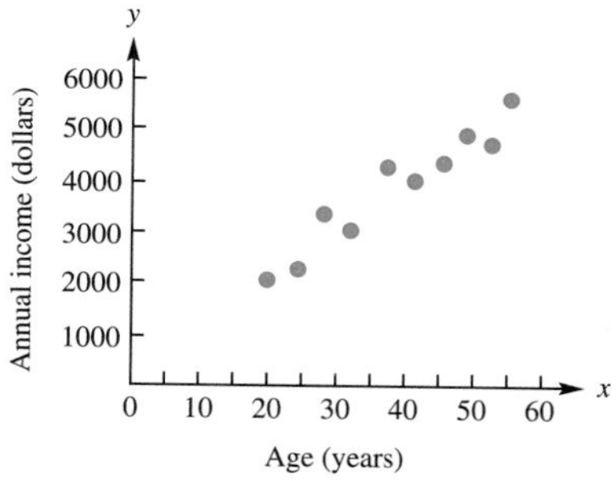

Scatter diagram

FIGURE 23

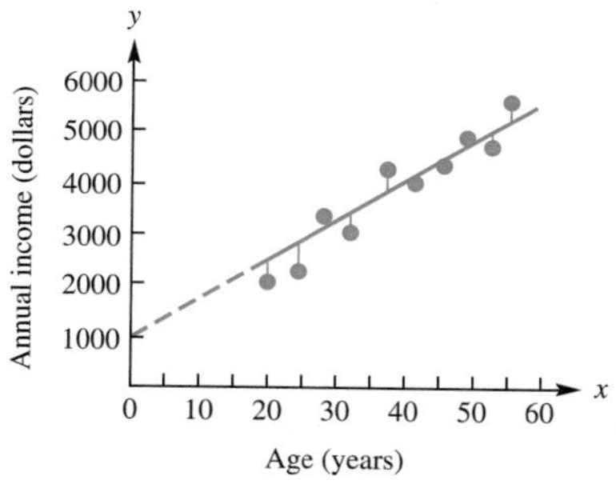

A tentative estimated regression line

FIGURE 24

If we let x denote age and y denote income in the data of Table 18 and assume that the best-fitting curve is a line, then the equation of that line will take the form

$$y = ax + b,$$

where a is the slope of the line and b is the y-coordinate of the y-intercept (the y-value at which the line, if extended, would intersect the y-axis).

To completely identify the estimated regression line, we must find the values of the "regression coefficients" a and b, which requires some calculation. In Figure 24, a tentative line has been drawn through the scatter diagram.

For each x-value in the data set, the corresponding y-value usually differs from the value it would have if the data point were exactly on the line. These differences are shown in the figure by vertical segments. Choosing another line would make some of these differences greater and some lesser. The most common procedure is to choose the line where the sum of the squares of all these differences is minimized. This is called the **method of least squares,** and the resulting line is called the **least squares line.**

In the equation of the least squares line, the variable y' can be used to distinguish the *predicted* values (which would give points on the least squares line) from the *observed* values y (those occurring in the data set).

The least squares criterion mentioned above leads to specific values of a and b. We shall not give the details, which involve differential calculus, but the results are given here. (Σ—the Greek letter *sigma*—represents summation just as in earlier sections.)

Francis Galton (1822–1911) learned to read at age three, was interested in mathematics and machines, but was an indifferent mathematics student at Trinity College, Cambridge. After several years as a traveling gentleman of leisure, he became interested in researching methods of predicting weather. It was during this research on weather that Galton developed early intuitive notions of **correlation** and **regression** and posed the problem of multiple regression.

Galton's key statistical work is *Natural Inheritance.* In it, he set forth his ideas on regression and correlation. He discovered the correlation coefficient while pondering Alphonse Bertillon's scheme for classifying criminals by physical characteristics. It was a major contribution to statistical method.

Regression Coefficient Formulas

The **least squares line** $y' = ax + b$ that provides the best fit to the data points $(x_1, y_1), (x_2, y_2), \ldots, (x_n, y_n)$ has

$$a = \frac{n(\Sigma xy) - (\Sigma x)(\Sigma y)}{n(\Sigma x^2) - (\Sigma x)^2} \quad \text{and} \quad b = \frac{\Sigma y - a(\Sigma x)}{n}.$$

EXAMPLE 1 Computing and Graphing a Least Squares Line

Find the equation of the least squares line for the age and income data given in Table 18. Graph the line.

SOLUTION

Start with the two columns on the left in Table 19 (which just repeat the original data). Then find the products $x \cdot y$, and the squares x^2.

TABLE 19

x	y	$x \cdot y$	x^2
19	2150	40,850	361
23	2550	58,650	529
27	3250	87,750	729
31	3150	97,650	961
36	4250	153,000	1296
40	4200	168,000	1600
44	4350	191,400	1936
49	5000	245,000	2401
52	4950	257,400	2704
54	5650	305,100	2916
Sums: 375	39,500	1,604,800	15,433

From the table, $\Sigma x = 375$, $\Sigma y = 39{,}500$, $\Sigma xy = 1{,}604{,}800$, and $\Sigma x^2 = 15{,}433$. There are 10 pairs of values, so $n = 10$. Now find a with the formula given above.

$$a = \frac{10(1{,}604{,}800) - 375(39{,}500)}{10(15{,}433) - (375)^2} = \frac{1{,}235{,}500}{13{,}705} \approx 90.15$$

Finally, use this value of a to find b.

$$b = \frac{39{,}500 - 90.15(375)}{10} \approx 569.4$$

The equation of the least squares line is

$$y' = 90x + 569. \quad \text{Coefficients are rounded.}$$

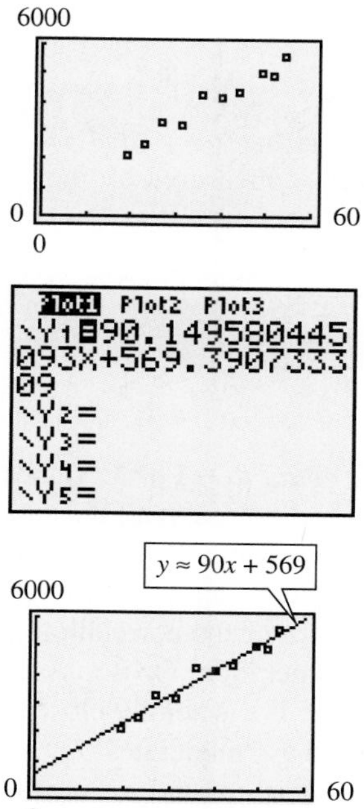

The information in Figure 25 and the accompanying discussion is supported in these screens.

Letting $x = 20$ in this equation gives $y' = 2369$, and $x = 50$ implies $y' = 5069$. The two points (20, 2369) and (50, 5069) are used to graph the regression line in Figure 25. Notice that the intercept coordinates (0, 569) also fit the extended line.

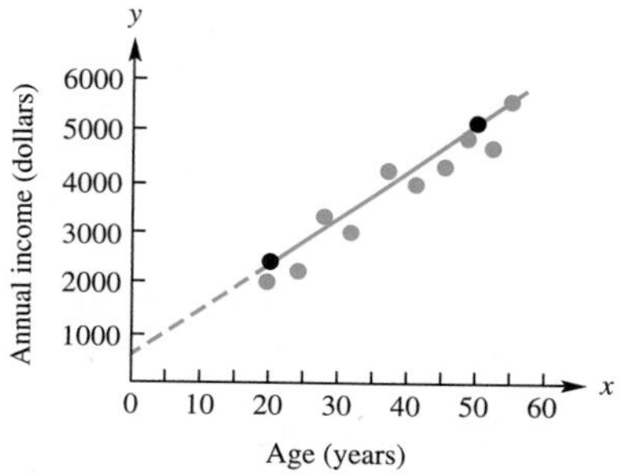

FIGURE 25

A computer, or a scientific, statistical, or graphing calculator is recommended for finding regression coefficients. (See the margin notes and the statistical options on page 783.) That way tedious calculations, such as in Example 1, can be avoided and the regression line produced automatically.

EXAMPLE 2 Predicting from a Least Squares Line

Use the result of Example 1 to predict the income of a village resident who is 35 years old.

SOLUTION

Use the equation $y' = 90x + 569$ and replace x with 35.

$$y' = 90(35) + 569 \quad \text{Let } x = 35.$$
$$= 3719$$

Based on the given data, a 35-year-old will make about \$3719 per year.

Correlation Once an equation for the line of best fit (the least squares line) has been found, it is reasonable to ask, "Just how good is this line for predictive purposes?" If the points already observed fit the line quite closely, then future pairs of scores can be expected to do so. If the points are widely scattered about even the "best-fitting" line, then predictions are not likely to be accurate. In general, the closer the *sample* data points lie to the least squares line, the more likely it is that the entire *population* of (x, y) points really do form a line, that is, that x and y really are related linearly. Also, the better the fit, the more confidence we can have that our least squares line (based on the sample) is a good estimator of the true population line.

One common measure of the strength of the linear relationship in the sample is called the **sample correlation coefficient,** denoted r. It is calculated from the sample data according to the following formula.

Sample Correlation Coefficient Formula

In linear regression, the strength of the linear relationship is measured by the correlation coefficient

$$r = \frac{n(\Sigma xy) - (\Sigma x)(\Sigma y)}{\sqrt{n(\Sigma x^2) - (\Sigma x)^2} \cdot \sqrt{n(\Sigma y^2) - (\Sigma y)^2}}.$$

The value of r is always between −1 and 1, or perhaps equal to −1 or 1. Values of exactly 1 or −1 indicate that the least squares line goes exactly through all the data points. If r is close to 1 or −1, but not exactly equal, then the line comes "close" to fitting through all the data points, and the linear correlation between x and y is "strong." If r is equal, or nearly equal, to 0, there is no linear correlation, or the correlation is weak, meaning that the points are a totally disordered conglomeration. Or it may be that the points form an ordered pattern but one that is not linear. If r is neither close to 0 nor close to 1 or −1, we might describe the linear correlation as "moderate."

A positive value of r indicates a regression line with positive slope and we say the linear relationship between x and y is direct; as x increases, y also increases. A negative r-value means the line has negative slope, so there is an inverse relationship between x and y; as x increases, y decreases.

The slope a and y-intercept b of the regression equation, along with r^2 and r, are given. Compare with Examples 1 and 3.

EXAMPLE 3 Finding a Correlation Coefficient

Find r for the age and income data of Table 19.

SOLUTION

Almost all values needed to find r were computed in Example 1.

$$n = 10 \qquad \Sigma x = 375 \qquad \Sigma y = 39{,}500$$

$$\Sigma xy = 1{,}604{,}800 \qquad \Sigma x^2 = 15{,}433$$

The only missing value is Σy^2. Squaring each y in the original data and adding the squares gives

$$\Sigma y^2 = 167{,}660{,}000.$$

Now use the formula to find that $r = .98$ (to two decimal places). This value of r, very close to 1, shows that age and income in this village are highly correlated. The fact that r is positive indicates that the linear relationship is direct; as age increases, income also increases.

EXAMPLE 4 Analyzing the Aging Trend in the U.S. Population

The World Almanac and Book of Facts 2002 (page 385) reported the following U.S. Census Bureau data concerning the aging U.S. population over the 20th century.

Year	1900	1910	1920	1930	1940	1950	1960	1970	1980	1990	2000
Percent 65 and over	4.1	4.3	4.7	5.4	6.8	8.1	9.2	9.8	11.3	12.5	12.4

Let x represent time, in decades, from 1900, so $x = 0$ in 1900, $x = 1$ in 1910, $x = 2$ in 1920, and so on. Let y represent percent 65 and over in the population. Based on the data table, carry out the following.

(a) Plot a scatter diagram.
(b) Compute and graph the least squares regression line.
(c) Compute the correlation coefficient.
(d) Use the regression line to predict the percent 65 and over in 2020, and discuss the validity of the prediction.

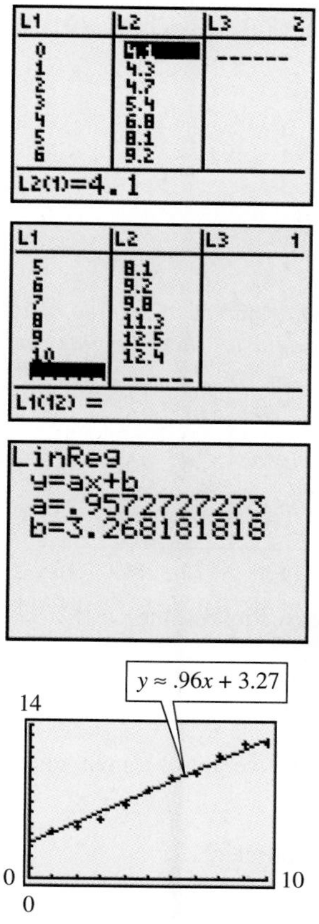

These screens support Example 4(b).

SOLUTION

(a) The data points are plotted in Figure 26.
(b) We entered the x- and y-values into lists L1 and L2, respectively, in a calculator to obtain the line

$$y' = .96x + 3.27 \quad \text{Coefficients are rounded.}$$

This line is shown in Figure 26 as a dashed line.

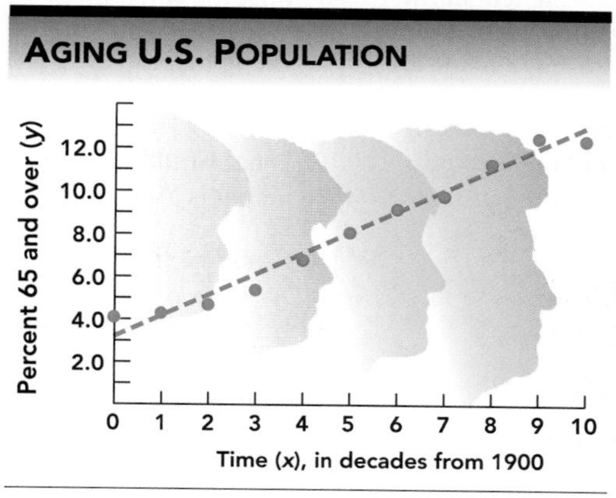

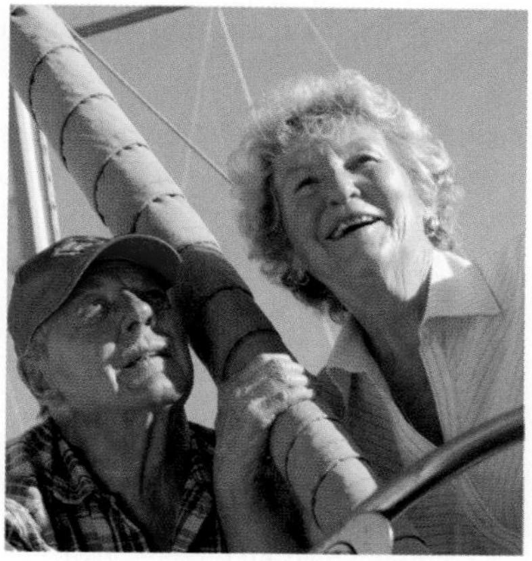

FIGURE 26

(c) $\Sigma x = 55,\ \Sigma x^2 = 385,\ n = 11,\ \Sigma y = 88.6,$ All values are from the calculator, using two-variable statistics.

$\Sigma y^2 = 816.78,\ \Sigma(xy) = 548.3$

$$r = \frac{n(\Sigma xy) - (\Sigma x)(\Sigma y)}{\sqrt{n(\Sigma x^2) - (\Sigma x)^2} \cdot \sqrt{n(\Sigma y^2) - (\Sigma y)^2}}$$ Correlation coefficient formula

$$= \frac{11 \cdot 548.3 - 55 \cdot 88.6}{\sqrt{11 \cdot 385 - 55^2} \cdot \sqrt{11 \cdot 816.78 - 88.6^2}}$$

$$= .988560\ldots$$

$$r \approx .99$$

(d) $y = .96x + 3.27$ Computed regression line

$= .96 \cdot 12 + 3.27$ The year 2020 corresponds to $x = 12$.

$y \approx 14.8$ 14.79 has been rounded here.

Although the correlation was strong ($r \approx .99$) for the data points we had, it is risky to extrapolate a regression line too far out. There may be factors (such as declining numbers of baby boomers in the population) that may slow the aging phenomenon.

13.6 EXERCISES

Correlating Fertilizer and Corn Ear Size *In a study to determine the linear relationship between the length (in decimeters) of an ear of corn (y) and the amount (in tons per acre) of fertilizer used (x), the following values were determined.*

$n = 10 \qquad \Sigma xy = 75$

$\Sigma x = 30 \qquad \Sigma x^2 = 100$

$\Sigma y = 24 \qquad \Sigma y^2 = 80$

1. Find an equation for the least squares line.

2. Find the correlation coefficient.

3. If 3 tons per acre of fertilizer are used, what length (in decimeters) would the regression equation predict for an ear of corn?

Correlating Celsius and Fahrenheit Temperatures *In an experiment to determine the linear relationship between temperatures on the Celsius scale (y) and on the Fahrenheit scale (x), a student got the following results.*

$n = 5 \qquad \Sigma xy = 28{,}050$

$\Sigma x = 376 \qquad \Sigma x^2 = 62{,}522$

$\Sigma y = 120 \qquad \Sigma y^2 = 13{,}450$

4. Find an equation for the least squares line.

5. Find the reading on the Celsius scale that corresponds to a reading of 120° Fahrenheit, using the equation of Exercise 4.

6. Find the correlation coefficient.

Correlating Heights and Weights of Adult Men *A sample of 10 adult men gave the following data on their heights and weights.*

Height (inches) (x)	62	62	63	65	66
Weight (pounds) (y)	120	140	130	150	142

Height (inches) (x)	67	68	68	70	72
Weight (pounds) (y)	130	135	175	149	168

7. Find the equation of the least squares line.

8. Using the results of Exercise 7, predict the weight of a man whose height is 60 inches.

9. What would be the predicted weight of a man whose height is 70 inches?

10. Compute the correlation coefficient.

Correlating Reading Ability and IQs *The table below gives reading ability scores and IQs for a group of 10 individuals.*

Reading (x)	83	76	75	85	74
IQ (y)	120	104	98	115	87

Reading (x)	90	75	78	95	80
IQ (y)	127	90	110	134	119

11. Plot a scatter diagram with reading on the horizontal axis.

12. Find the equation of a regression line.

13. Use your regression line equation to estimate the IQ of a person with a reading score of 65.

Correlating Yearly Sales of a Company *Sales, in thousands of dollars, of a certain company are shown here.*

Year (x)	0	1	2	3	4	5
Sales (y)	48	59	66	75	80	90

14. Find the equation of the least squares line.

15. Find the correlation coefficient.

16. If the linear trend displayed by this data were to continue beyond year 5, what sales amount would you predict in year 7?

Comparing the Ages of Dogs and Humans *It often is said that a dog's age can be multiplied by 7 to obtain the equivalent human age. A more accurate correspondence (through the first 14 years) is shown in this table from* The Old Farmer's Almanac, *2000 edition, page 180.*

Dog age (x)	$\frac{1}{2}$	1	2	3	4	5	6	7
Equivalent human age (y)	10	15	24	28	32	36	40	44

Dog age (x)	8	9	10	11	12	13	14
Equivalent human age (y)	48	52	56	60	64	68	70.5

17. Plot a scatter diagram for the given data.

18. Find the equation of the regression line, and graph the line on the scatter diagram of Exercise 17.

19. Describe where the data points show the most pronounced departure from the regression line, and explain why this might be so.

20. Compute the correlation coefficient.

Statistics on the Westward Population Movement *The data here (from* The World Almanac and Book of Facts 2000, *page 532) show the increase in the percentage of U.S. population in the West since about the time of the California Gold Rush.*

Census Year	Time, in Decades from 1850 (x)	Percentage in West (y)
1850	0	.8%
1870	2	2.6
1890	4	5.0
1910	6	7.7
1930	8	10.0
1950	10	13.3
1970	12	17.1
1990	14	21.2

21. Taking x and y as indicated in the table, find the equation of the regression line.

22. Compute the correlation coefficient.

23. Describe the degree of correlation (for example, as strong, moderate, or weak).

24. What would be the value of x in the year 2010? Use the regression line to predict the percentage in the West for that year.

Comparing State Populations with Governors' Salaries *The table shows the ten most populous states (as of the 2000 census) and the salaries of their governors (as of October 2001). (Source:* The World Almanac and Book of Facts 2002, *page 380 and page 99).*

Rank	State	Population, in Thousands (x)	Governor's Salary, in Thousands of Dollars (y)
1	California	33,872	175
2	Texas	20,852	115
3	New York	18,976	179
4	Florida	15,982	123
5	Illinois	12,419	151
6	Pennsylvania	12,281	105
7	Ohio	11,353	126
8	Michigan	9938	177
9	New Jersey	8414	83
10	Georgia	8186	123

25. Find the equation of the regression line.

26. Compute the correlation coefficient.

27. Describe the degree of correlation (for example, as strong, moderate, or weak).

28. What governor's salary would this linear model predict for a state with a population of 15 million citizens?

COLLABORATIVE INVESTIGATION

Combining Sets of Data

Divide your class into two separate groups, one consisting of the women and the other consisting of the men. Each group is to select a recorder to write the group's results. As a group, carry out the following tasks. (You may want to devise a way to allow the members of the groups to provide personal data anonymously.)

1. Record the number of members (n) in your group.
2. Collect shoe sizes (x) and heights in inches (y) for all members of the group.
3. Compute the mean, median, and mode(s), if any, for each of the two sets of data.
4. Compute the standard deviation of each of the two sets of data.
5. Construct a box plot for each of the two sets of data.
6. Plot a scatter diagram for the x-y data collected.
7. Find the equation of the least squares regression line ($y' = ax + b$).
8. Evaluate the correlation coefficient (r).
9. Evaluate the strength of the linear relationship between shoe size and height for your group.

Now re-combine your two groups into one. Discuss and carry out the following tasks.

1. If possible, compute the mean of the heights for the combined group, using only the means for the two individual groups and the number of members in each of the two groups. If this is not possible, explain why and describe how you *could* find the combined mean. Obtain the combined mean.
2. Do the same as in item 1 above for the median of the heights for the combined group.
3. Do the same for the mode of the heights for the combined group.
4. Fill in the table below, pertaining to heights, and discuss any apparent relationships among the computed statistics.

	Number of Members	Mean	Median	Mode
Women				
Men				
Combined				

CHAPTER 13 TEST

Statistical Trends in U.S. Farm Workers and Farms *Refer to the bar graphs and line graph shown here and on the next page for Exercises 1–6 which follow.*

Decline in Farm Workers, 1820–1994*

Source: U.S. Department of Agriculture, Economic Research Service.

Of the approximately 2.9 million workers in the U.S. in 1820, 71.8%, or about 2.1 million, were employed in farm occupations. The percentage of U.S. workers in farm occupations had declined drastically by the turn of the century, and by 1994 only 2.5% of all U.S. workers were employed in farm occupations.

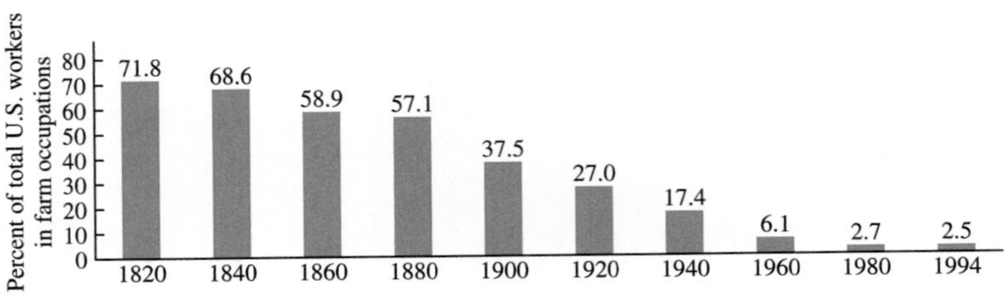

*Figures not compiled for years after 1994. Total workers for 1994 are employed workers age 15 and older; total workers for 1900 to 1960 are members of the experienced civilian labor force 14 and older; total workers for 1820 to 1880 are gainfully employed workers 10 and older.

U.S. Farms, 1940–2000

Source: National Agricultural Statistics Service, U.S. Department of Agriculture.

Since 1940, the number of farms in the U.S. has sharply declined, while the size of the average farm has grown.

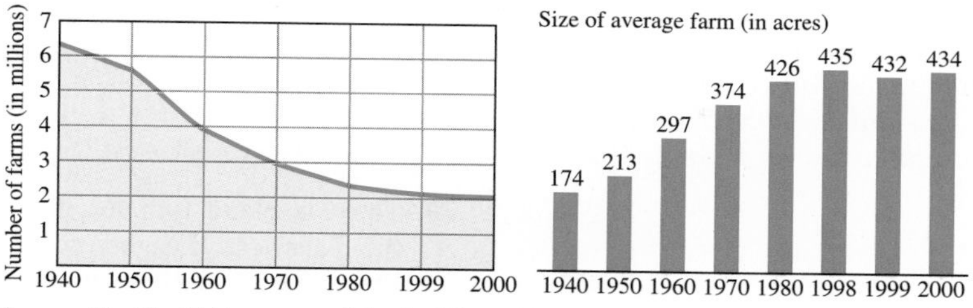

Source: The World Almanac and Book of Facts 2002, page 130.

1. From the given information, approximate when half of all U.S. workers were working in farm occupations?

2. Is it true that there were about 4 times as many farm workers in 1840 as in 1940? Explain your answer.

3. About how many farms were there in 1970?

4. In which year was the greater amount of land being farmed, 1940 or 2000? By about what percentage did it exceed the lesser amount?

Use the following sample for Exercises 5 and 6.

8, 12, 6, 10, 7, 6, 7

5. Compute the following.
   (a) the mean (b) the median
   (c) the mode or modes, if any

6. Compute the following.
   (a) the range (b) the standard deviation

Client Contacts of a Publisher's Representative *Laura Griffin Hiller, a sales representative for a publishing company, recorded the following numbers of client contacts for the twenty-two days that she was on the road in the month of March. Use the given data for Exercises 7–9.*

12	8	15	11	20	18	14	22	13	26	17
19	16	25	19	10	7	18	24	15	30	24

7. Construct grouped frequency and relative frequency distributions. Use five uniform classes of width 5 where the first class has a lower limit of 6. (Round relative frequencies to two decimal places.)

8. From your frequency distribution of Exercise 7, construct **(a)** a histogram and **(b)** a frequency polygon. Use appropriate scales and labels.

9. For the data above, how many uniform classes would be required if the first class had limits 7–9?

In Exercises 10–13, find the indicated measures for the following frequency distribution.

Value	8	10	12	14	16	18
Frequency	3	8	10	8	5	1

10. the mean
11. the median
12. the mode
13. the range

14. ***Exam Scores in a Physics Class*** The following data are exam scores achieved by the students in a physics class. Arrange the data into a stem-and-leaf display with leaves ranked.

79	43	65	84	77	70	52	61	80	66
68	48	55	78	71	38	45	64	67	73
77	50	67	91	84	33	49	61	79	72

Use the stem-and-leaf display shown here for Exercises 15–20.

2	3	3	4					
2	6	7	8	9	9			
3	0	1	1	2	3	3	3	4
3	5	6	7	8	8	9		
4	1	2	2	4				
4	5	7	9					
5	2	4						
5	8							
6	0							

Compute the measures required in Exercises 15–19.

15. the median
16. the mode(s), if any
17. the range
18. the third decile
19. the eighty-fifth percentile

20. Construct a box plot for the given data, showing values for the five important quantities on the numerical scale.

Test Scores in a Training Institute *A certain training institute gives a standardized test to large numbers of applicants nationwide. The resulting scores form a normal distribution with mean 80 and standard deviation 5. Find the percent of all applicants with scores as follows. (Use the empirical rule.)*

21. between 70 and 90

22. greater than 95 or less than 65

23. less than 75

24. between 85 and 90

Heights of Spruce Trees *In a certain young forest, the heights of the spruce trees are normally distributed with mean 5.5 meters and standard deviation 2.1 meters. If a single tree is selected randomly, find the probability (to the nearest thousandth) that its height will fall in each of the following intervals.*

25. less than 6.5 meters

26. between 6.2 and 9.4 meters

Season Statistics in Major League Baseball *During the 2005 Major League Baseball season, each team played a total of 162 regular season games. The tables below show the statistics on games won for all three divisions of both leagues. In each case, n = number of teams in the division, $\bar{x}$ = average (mean) number of games won, and s = standard deviation of number of games won.*

American League

East Division	Central Division	West Division
$n = 5$ $\bar{x} = 82.2$ $s = 12.6$	$n = 5$ $\bar{x} = 80.4$ $s = 17.3$	$n = 4$ $\bar{x} = 82.8$ $s = 11.3$

National League

East Division	Central Division	West Division
$n = 5$ $\bar{x} = 85.0$ $s = 3.8$	$n = 6$ $\bar{x} = 81.5$ $s = 11.7$	$n = 5$ $\bar{x} = 74.4$ $s = 5.7$

Refer to the preceding tables for Exercises 27–29.

27. Overall, who had the greatest winning average, the East teams, the Central teams, or the West teams?

28. Overall, where were the teams the least "consistent" in number of games won, East, Central, or West?

29. Find (to the nearest tenth) the average number of games won for all West Division teams.

30. The Cleveland Indians, in the Central Division of the American League, won 93 games, while the Philadelphia Phillies, in the East Division of the National League, won 88 games. Use z-scores to determine which of these two teams did relatively better within its own division of 5 teams.

Carry out the following for the paired data values shown here. (In Exercises 32–34, give all calculated values to two decimal places.)

x	1	4	6	7
y	9	7	8	1

31. Plot a scatter diagram.

32. Find the equation for the least squares regression line.

33. Use your equation from Exercise 32 to predict y when $x = 3$.

34. Find the sample correlation coefficient for the given data.

35. Evaluate the strength of the linear relationship for the above data. How confident are you in your evaluation? Explain.

36. Relate the concepts of *inferential statistics* and *inductive reasoning*.

14

PERSONAL FINANCIAL MANAGEMENT

The second season of *The Andy Griffith Show* provided an episode that beautifully illustrated an application of the mathematics of finance. In "Mayberry Goes Bankrupt," Sheriff Andy Taylor was reluctantly forced to evict kindly old gentlemen Frank Myers from his home due to nonpayment of taxes. But Frank then produced a bond given to him by his grandfather. It was purchased for $100 in 1861 and paid 8.5% interest compounded annually. The bond had been stored away for 100 years. When the town banker told the Mayor he could not pay Frank, Andy explained why this was so.

Well, Mayor ... according to the computation machines down at the bank ... and they're good machines ... we, that is the town of Mayberry, owe Frank Myers $349,119.27. (To which Frank responds: *I'll take it in cash.*)

By applying the formula for *compound interest* found in Section 14.1, you will find that those were indeed good machines at the Bank of Mayberry, for the figure quoted is correct to the penny.

14.1 The Time Value of Money

Interest • Simple Interest • Future Value and Present Value • Compound Interest • Effective Annual Yield • Inflation

Interest To determine the value of money, we consider not only the amount (number of dollars), but also the particular point in time that the value is to be determined. If we borrow an amount of money today, we will repay a larger amount later. This increase in value is known as **interest.** The money *gains value over time.*

The amount of a loan or a deposit is called the **principal.** The interest is usually computed as a percent of the principal. This percent is called the **rate of interest** (or the **interest rate,** or simply the **rate**). The rate of interest is always assumed to be an annual rate unless otherwise stated.

Interest calculated only on principal is called **simple interest.** Interest calculated on principal plus any previously earned interest is called **compound interest.**

Simple Interest

Simple interest is calculated according to the following formula.

Simple Interest

If P = principal, r = annual interest rate, and t = time (in years), then the **simple interest** I is given by

$$I = Prt.$$

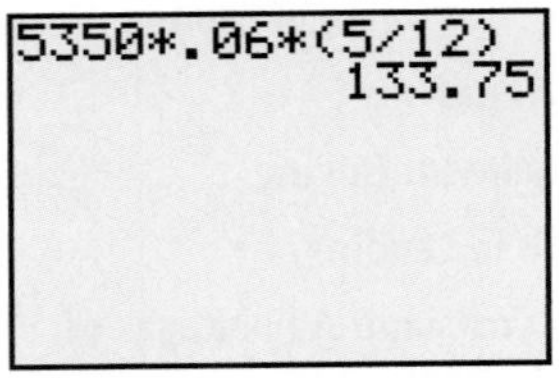

This is the computation required to solve Example 1.

EXAMPLE 1 Finding Simple Interest

Find the simple interest paid to borrow \$5350 for 5 months at 6%.

SOLUTION

5 months is $\frac{5}{12}$ of a year.

$$I = Prt = \$5350(.06)\left(\frac{5}{12}\right) = \$133.75 \qquad P = \$5350,\ r = 6\% = .06,\ t = \tfrac{5}{12}$$

Future Value and Present Value

In Example 1, at the end of 5 months the borrower would have to repay

$$\underset{\text{Principal}}{\$5350} + \underset{\text{Interest}}{\$133.75} = \$5483.75.$$

The total amount repaid is sometimes called the **maturity value** (or simply the **value**) of the loan. We will generally refer to it as the **future value,** or **future amount,** since when a loan is being set up, repayment will be occurring in the future. We will use A to denote future amount (or value). The original principal, denoted P, can also be thought of as **present value.** The future value depends on the principal (present value) and the interest as follows.

$$A = P + I = P + Prt = P(1 + rt)$$

We summarize as follows.

Future Value for Simple Interest

If a principal P is borrowed at simple interest for t years at an annual interest rate of r, then the **future value** of the loan, denoted A, is given by

$$A = P(1 + rt).$$

EXAMPLE 2 Finding Future Value for Simple Interest

Alex Ortega took out a simple interest loan for \$210 to purchase textbooks and school supplies. If the annual interest rate is 7% and he must repay the loan after 8 months, find the future value (the maturity value) of the loan.

This is the computation required to solve Example 2.

SOLUTION

$$A = P(1 + rt) = \$210\left[1 + .07\left(\frac{8}{12}\right)\right] = \$219.80.$$

At the end of 8 months, Alex will need to repay \$219.80.

Sometimes the future value is known, and we need to compute the present value. For this purpose, we solve the future value formula for P.

EXAMPLE 3 Finding Present Value for Simple Interest

Suppose that Alex (Example 2) is granted an 8-month *deferral* of the \$210 payment rather than a loan. That is, instead of incurring interest for 8 months, he will have to pay just the \$210 at the end of that period. If he has extra money right now, and if he can earn 4% simple interest on savings, what lump sum must he deposit now so that its value will be \$210 after 8 months?

This is the computation required to solve Example 3.

SOLUTION

$$A = P(1 + rt) \quad \text{Future value formula for simple interest}$$

$$P = \frac{A}{1 + rt} \quad \text{Solve the formula for } P.$$

$$P = \frac{\$210}{1 + (.04)(\frac{8}{12})} = \$204.55 \quad \text{Substitute known values and simplify.}$$

A deposit of \$204.55 now, growing at 4% simple interest, will grow to \$210 over an 8-month period.

Compound Interest As mentioned earlier, interest paid on principal plus interest is called *compound interest*. After a certain period, the interest earned so far is *credited* (added) to the account, and the sum (principal plus interest) then earns interest during the next period.

Example 4 shows the dramatic effect that compound interest can have over time.

To borrow from the title of the Clint Eastwood classic *The Good, the Bad, and the Ugly,* the 1994 movie *Blank Check* includes a scene that qualifies as **ugly mathematics**. Twelve-year-old Preston Waters receives a check for \$11.00 and uses his computer to determine how long it will take for this amount to grow to \$1,000,000 at 3.45% annual interest. While there is no information on the number of compounding periods, the answer given in the movie is incorrect for any number. The computer determines that it would take 342,506 years. Even with interest compounded just once a year, the time would "only" be 337 years.

EXAMPLE 4 Comparing Retirement Plans

Compare the results at age 65 for the following two retirement plans. Both plans earn 8% annual interest throughout the account building period.

Plan A: A person begins saving at age 20, deposits \$2000 on every birthday from age 21 to age 30 (10 deposits, or \$20,000 total), and thereafter makes no additional contributions.

Plan B: A person waits until age 30 to start saving, makes deposits of \$2000 on every birthday from age 31 to age 65 (35 deposits, or \$70,000 total).

SOLUTION

Table 1 shows how both accounts build over the years. \$20,000, deposited earlier, produces \$83,744 more than \$70,000, deposited later.

TABLE 1

Age	Plan A	Plan B
20	0	0
25	11,733	0
30	28,973	0
35	42,571	11,733
40	62,551	28,973
45	91,908	54,304
50	135,042	91,524
55	198,422	146,212
60	291,547	226,566
65	428,378	344,634

EXAMPLE 5 Comparing Simple and Compound Interest

Compare simple and compound interest for a \$1000 deposit at 6% interest for 3 years.

SOLUTION

$$A = P(1 + rt) = \$1000(1 + .06 \cdot 3) = \$1180$$

After 3 years, \$1000 grows to \$1180 subject to 6% simple interest.

The result of compounding (annually) for 3 years is shown in Table 2.

TABLE 2 A \$1000 Deposit at 6% Interest Compounded Annually

Year	Beginning Balance	Interest Earned $I = Prt$	Ending Balance
1	\$1000.00	\$1000.00(.06)(1) = \$60.00	\$1060.00
2	\$1060.00	\$1060.00(.06)(1) = \$63.60	\$1123.60
3	\$1123.60	\$1123.60(.06)(1) = \$67.42	\$1191.02

The following item shows how one community plans to use the **power of compound interest** to help it plan for the future. (News broadcast © CBS Inc. 1977. All rights reserved. Originally broadcast over the CBS Radio Network as part of the *Hourly News* by Charles Osgood on March 2, 1977.)

"Eli Hooker, of Union City, Michigan, doesn't want his community to go broke when the tricentennial rolls around in 2076. Hooker says the local Bicentennial Committee last year didn't have enough money. He was chairman. So, to plan ahead, he has collected twenty-five dollars apiece from 42 residents of Union City, squirreling it away in a local bank. At seven percent interest, compounded, the fund would have a million dollars by 2076.

Hooker hopes that will be enough for the parades and fireworks a tricentennial celebration will require. But Eli Hooker isn't being an entirely selfless patriot about all this: he hopes one of the big events on July 4, 2076, will be the unveiling of a monument to the 42 people of Union City who contributed 99 years in advance."

Unfortunately, that million dollars won't be worth as much then as we might think. If the community decides to hire people to parade around in historical costumes, it might have paid them \$4 per hour in 1976. The going wage in 2076, assuming 7% annual inflation for a hundred years, would be well over \$3000 per hour.

Under annual compounding for 3 years, \$1000 grows to \$1191.02, which is \$11.02 more than under simple interest. ■

Based on the compounding pattern of Example 5, we now develop a future value formula for compound interest. In practice, earned interest can be credited to an account at time intervals other than 1 year (usually more often). For example, it can be done semiannually, quarterly, monthly, or daily. (Daily compounding is quite common.) This time interval is called the **compounding period** (or simply the **period**). Start with the following definitions:

$$P = \text{original principal deposited,} \qquad r = \text{annual interest rate,}$$
$$m = \text{number of periods per year,} \qquad n = \text{total number of periods.}$$

During each individual compounding period, interest is earned according to the simple interest formula, and as interest is added, the beginning principal increases from one period to the next. During the first period, the interest earned is given by

$$\text{Interest} = P(r)\left(\frac{1}{m}\right) \qquad \text{Interest} = \text{(Principal)(rate)(time); one period} = \tfrac{1}{m}\text{ year}$$
$$= P\left(\frac{r}{m}\right). \qquad \text{Rewrite } (r)(\tfrac{1}{m}) \text{ as } (\tfrac{r}{m}).$$

At the end of the first period, the account then contains

$$\text{Ending amount} = \underset{\text{Beginning amount}}{P} + \underset{\text{Interest}}{P\left(\frac{r}{m}\right)}$$
$$= P\left(1 + \frac{r}{m}\right). \qquad \text{Factor } P \text{ from both terms.}$$

Now during the second period, the interest earned is given by

$$\text{Interest} = \left[P\left(1 + \frac{r}{m}\right)\right](r)\left(\frac{1}{m}\right) \qquad \text{Interest} = \text{[Principal](rate)(time)}$$
$$= P\left(1 + \frac{r}{m}\right)\left(\frac{r}{m}\right),$$

so that the account ends the second period containing

$$\text{Ending amount} = \underset{\text{Beginning amount}}{P\left(1 + \frac{r}{m}\right)} + \underset{\text{Interest}}{P\left(1 + \frac{r}{m}\right)\left(\frac{r}{m}\right)}$$
$$= P\left(1 + \frac{r}{m}\right)\left[1 + \frac{r}{m}\right] \qquad \text{Factor } P(1 + \tfrac{r}{m}) \text{ from both terms.}$$
$$= P\left(1 + \frac{r}{m}\right)^2.$$

King Hammurabi tried to hold interest rates at 20 percent for both silver and gold, but moneylenders ignored his decrees.

Consider one more period, namely, the third. The interest earned is

$$\text{Interest} = \left[P\left(1+\frac{r}{m}\right)^2\right](r)\left(\frac{1}{m}\right) \qquad \text{Interest} = [\text{Principal}]\,(\text{rate})\,(\text{time})$$
$$= P\left(1+\frac{r}{m}\right)^2\left(\frac{r}{m}\right),$$

so the account ends the third period containing

$$\text{Ending amount} = \underbrace{P\left(1+\frac{r}{m}\right)^2}_{\text{Beginning amount}} + \underbrace{P\left(1+\frac{r}{m}\right)^2\left(\frac{r}{m}\right)}_{\text{Interest}}$$
$$= P\left(1+\frac{r}{m}\right)^2\left[1+\frac{r}{m}\right] \qquad \text{Factor } P(1+\tfrac{r}{m})^2 \text{ from both terms.}$$
$$= P\left(1+\frac{r}{m}\right)^3.$$

Table 3 summarizes the preceding results. Inductive reasoning gives the bottom row of expressions for period number n.

TABLE 3 Compound Amount

Period Number	Beginning Amount	Interest Earned During Period	Ending Amount
1	P	$P\left(\frac{r}{m}\right)$	$P\left(1+\frac{r}{m}\right)$
2	$P\left(1+\frac{r}{m}\right)$	$P\left(1+\frac{r}{m}\right)\left(\frac{r}{m}\right)$	$P\left(1+\frac{r}{m}\right)^2$
3	$P\left(1+\frac{r}{m}\right)^2$	$P\left(1+\frac{r}{m}\right)^2\left(\frac{r}{m}\right)$	$P\left(1+\frac{r}{m}\right)^3$
.	.	.	.
.	.	.	.
.	.	.	.
n	$P\left(1+\frac{r}{m}\right)^{n-1}$	$P\left(1+\frac{r}{m}\right)^{n-1}\left(\frac{r}{m}\right)$	$P\left(1+\frac{r}{m}\right)^n$

The lower right entry of the table provides the following formula.

The quantity

$$\left(1+\frac{r}{m}\right)^n$$

can be evaluated using a key such as $\boxed{y^x}$ on a scientific calculator or $\boxed{\wedge}$ on a graphing calculator.

Future Value for Compound Interest

If P dollars are deposited at an annual interest rate of r, compounded m times per year, and the money is left on deposit for a total of n periods, then the **future value, A** (the final amount on deposit), is given by

$$A = P\left(1+\frac{r}{m}\right)^n.$$

EXAMPLE 6 Finding Future Value for Compound Interest

Find the future value (final amount on deposit) and the amount of interest earned for the following deposits.

(a) \$18,950 at 6% compounded quarterly for 5 years
(b) \$6429 at 2.7% compounded monthly for 30 months

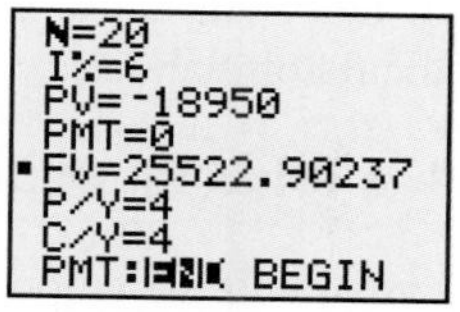

The financial functions of the TI-83/84 Plus calculator allow the user to solve for a missing quantity regarding the time value of money. The arrow indicates the display that supports the answer in Example 6(a).

SOLUTION

(a) We have $P = \$18{,}950$, $r = 6\% = .06$, $m = 4$. Over 5 years, we obtain $n = 5m = (5)(4) = 20$. Using the future value formula and a calculator, we get

$$A = \$18{,}950\left(1 + \frac{.06}{4}\right)^{20} = \$25{,}522.90.$$

To find the interest earned, subtract the amount originally deposited (present value) from the future value:

$$\text{Interest earned} = \underset{\text{Future value}}{\$25{,}522.90} - \underset{\text{Present value}}{\$18{,}950} = \$6572.90.$$

(b) Here $P = \$6429$, $r = 2.7\% = .027$, $m = 12$, and $n = 30$, so

$$A = \$6429\left(1 + \frac{.027}{12}\right)^{30} = \$6877.42.$$

Thus,

$$\text{Interest earned} = \underset{\text{Future value}}{\$6877.42} - \underset{\text{Present value}}{\$6429} = \$448.42.$$

Just as with simple interest, compound interest problems sometimes require us to solve the formula for P in order to compute the present value when the future value is known.

EXAMPLE 7 Finding Present Value for Compound Interest

The Cebelinskis' daughter will need \$17,000 in 5 years to help pay for her college education. What lump sum, deposited today at 7% compounded quarterly, will produce the necessary amount?

SOLUTION

This question requires that we find the present value P based on the following.

Future value:	$A = \$17{,}000$
Annual rate:	$r = 7\% = .07$
Periods per year:	$m = 4$
Total number of periods:	$n = (5)(4) = 20$

Thus,

$$A = P\left(1 + \frac{r}{m}\right)^n, \quad \text{Future value for compound interest}$$

so

$$P = \frac{A}{(1 + \frac{r}{m})^n} \quad \text{Solve the formula for } P.$$

The 1957 movie *The Pajama Game* was inspired by the Broadway musical of the same name. It was recently revived on Broadway and starred Harry Connick, Jr. The female lead in the movie, "Babe" Williams, was played by Doris Day. She and her coworkers at the Sleeptite Pajama Factory were attempting to get a $7\frac{1}{2}$ cent per hour raise. The musical number **"Seven and a Half Cents"** features three computations, determining how much this seemingly small raise will earn in three time periods. Based on pencil and paper calculations, the lyrics state that in 5 years, the raise will amount to \$852.74, in 10 years \$1705.48, and in 20 years \$3411.96. Assuming that the figure for 5 years is correct, is the one for 10 years also correct? Now, how about the one for 20 years? Oops!

and
$$P = \frac{\$17{,}000}{(1 + \frac{.07}{4})^{20}} \qquad \text{Substitute known values.}$$
$$P = \$12{,}016.02.$$

Assuming interest of 7% compounded quarterly can be maintained, \$17,000 can be attained 5 years in the future by depositing \$12,016.02 today. ■

The next example will involve the use of logarithms. For help with those steps, see the discussion of logarithms in Section 8.6 (or see your calculator manual).

EXAMPLE 8 Finding the Time Required to Double a Principal Deposit

In a savings account paying 3% interest, compounded daily, when will the amount in the account be twice the original principal?

SOLUTION

The future value must equal two times the present value.

$$2P = P\left(1 + \frac{r}{m}\right)^n \qquad \text{Substitute } 2P \text{ for } A \text{ in the future value formula.}$$
$$2 = \left(1 + \frac{r}{m}\right)^n \qquad \text{Divide both sides by } P.$$
$$2 = \left(1 + \frac{.03}{365}\right)^n \qquad \text{Substitute values of } r \text{ and } m.$$
$$\log 2 = \log\left(1 + \frac{.03}{365}\right)^n \qquad \text{Take the logarithm of both sides.}$$
$$\log 2 = n \log\left(1 + \frac{.03}{365}\right) \qquad \text{Use the power property of logarithms.}$$
$$n = \frac{\log 2}{\log(1 + \frac{.03}{365})} \qquad \text{Solve for } n.$$
$$n = 8434 \qquad \text{Round to the nearest whole number.}$$

Because n denotes the number of periods, which is days in this case, the required amount of time is 8434 days, or 23 years, 39 days (ignoring leap years). ■

Effective Annual Yield Banks, credit unions, and other savings institutions often give two quantities when advertising the rates they will pay. The first, the actual annualized interest rate, is the **nominal rate** (the "named" or "stated" rate). The second quantity is the equivalent rate that would produce the same final amount, or future value, at the end of 1 year if the interest being paid were simple rather than compound. This is called the "effective rate," or more commonly the **effective annual yield.** (You may also see this rate denoted *APY* for "annual percentage yield.") Because the interest is normally compounded multiple times per year, the yield will usually be somewhat higher than the nominal rate.

The book, *You Can Do the Math*, by Ron Lipsman of the University of Maryland, is a practical resource for most aspects of **personal financial management.** The associated Web site, www.math.umd.edu/~rll/cgi-bin/finance.html, provides "calculators," with which you can easily input your own values to get the results of many different financial computations.

EXAMPLE 9 Finding Effective Annual Yield

What is the effective annual yield of an account paying a nominal rate of 3.50%, compounded quarterly?

Financial calculators are programmed to compute effective interest rate. Compare to the result in Example 9.

SOLUTION

From the given data, $r = .035$ and $m = 4$. Suppose we deposited $P = \$1$ and left it for 1 year ($n = 4$). Then the compound future value formula gives

$$A = \left(1 + \frac{.035}{4}\right)^4 \approx 1.0355.$$

The initial deposit of \$1, after 1 year, has grown to \$1.0355. As usual, interest earned = future value − present value = 1.0355 − 1 = .0355 = 3.55%. A nominal rate of 3.50% results in an effective annual yield of 3.55%.

Generalizing the procedure of Example 9 gives the following formula.

Effective Annual Yield

A nominal interest rate of r, compounded m times per year, is equivalent to an **effective annual yield** of

$$\mathbf{Y = \left(1 + \frac{r}{m}\right)^m - 1.}$$

Consumers can make practical use of the effective annual yield concept when shopping for loans or savings opportunities. A borrower should seek the least yield available, while a depositor should look for the greatest.

EXAMPLE 10 Comparing Savings Rates

Tamara Johnson-Draper wants to deposit \$2800 into a savings account and has narrowed her choices to the three institutions represented here. Which is the best choice?

Institution	Rate on Deposits of \$1000 to \$5000
Friendly Credit Union	2.08% annual rate, compounded monthly
Premier Savings	2.09% annual yield
Neighborhood Bank	2.05% compounded daily

SOLUTION

Compare the effective annual yields for the three institutions:

For Friendly, $$Y = \left(1 + \frac{.0208}{12}\right)^{12} - 1 = .0210 = 2.10\%.$$

For Premier, $$Y = 2.09\%.$$

For Neighborhood, $$Y = \left(1 + \frac{.0205}{365}\right)^{365} - 1 = .0207 = 2.07\%.$$

The best of the three yields, 2.10%, is offered by Friendly Credit Union.

Inflation As we have shown, interest reflects how money *gains value over time* when it is borrowed or lent. On the other hand, in terms of the equivalent number of goods or services that a given amount of money will buy, it is normally more today than it will be later. In this sense, the money *loses value over time.* This periodic increase in the cost of living is called **inflation.**

TABLE 4

Consumer Price Index (CPI-U) 1975 to 2005 [Period from 1982 to 1984: 100]

Year	Average CPI-U	Percent Change in CPI-U
1975	53.8	9.1
1976	56.9	5.8
1977	60.6	6.5
1978	65.2	7.6
1979	72.6	11.3
1980	82.4	13.5
1981	90.9	10.3
1982	96.5	6.2
1983	99.6	3.2
1984	103.9	4.3
1985	107.6	3.6
1986	109.6	1.9
1987	113.6	3.6
1988	118.3	4.1
1989	124.0	4.8
1990	130.7	5.4
1991	136.2	4.2
1992	140.3	3.0
1993	144.5	3.0
1994	148.2	2.6
1995	152.4	2.8
1996	156.9	3.0
1997	160.5	2.3
1998	163.0	1.6
1999	166.6	2.2
2000	172.2	3.4
2001	177.1	2.8
2002	179.9	1.6
2003	184.0	2.3
2004	188.9	2.7
2005	195.3	3.4

Source: Bureau of Labor Statistics.

Inflation in an economy usually is expressed as a monthly or annual rate of price increases, estimated by government agencies in a systematic way. In the United States, the Bureau of Labor Statistics publishes **consumer price index (CPI)** figures, which reflect the prices of certain items purchased by large numbers of people. The items include such things as food, housing, automobiles, fuel, and clothing. Items such as yachts or expensive jewelry are not included.

CPI figures are reported on a monthly basis, but annual figures give adequate information for most consumers. Table 4 gives values of the primary index representing "all urban consumers" (the CPI-U). The value shown for a given year is actually the average (arithmetic mean) of the twelve monthly figures for that year. The percent change value for a given year shows the change in the CPI-U since the previous year. For example, the average CPI-U for 2005 (195.3) was 3.4% greater than the average value for 2004 (188.9).

From 1913 to the present, the record inflation rate occurred in the year 1918 and was 18.0%. The record **deflation** rate (with price levels actually decreasing from year to year) occurred in 1921 and was −10.5%. Severe deflation has not occurred since the 1930s. A period of minor deflation, normally brief and often accompanying an overall slowdown in the general economy, usually is referred to as a **recession** (rather than deflation). A business recession *can* coincide with price inflation.

Unlike account values under interest compounding, which make sudden jumps at just certain points in time (such as quarterly, monthly, or daily), price levels tend to fluctuate gradually over time. Thus it is appropriate, for inflationary estimates, to use the formula for continuous compounding (introduced in Section 8.6.)

Future Value for Continuous Compounding

If an initial deposit of P dollars earns continuously compounded interest at an annual rate r for a period of t years, then the **future value, A,** is given by

$$A = Pe^{rt}.$$

EXAMPLE 11 Predicting Inflated Salary Levels

Suppose you earn a salary of $24,000 per year. About what salary would you need 20 years from now to maintain your purchasing power in case the inflation rate were to persist at each of the following levels?

(a) 2% (approximately the 1999 level)
(b) 13% (approximately the 1980 level)

SOLUTION

(a) In this case we can use the continuous compounding future value formula with $P = \$24{,}000$, $r = .02$, and $t = 20$. Using the e^x key on a calculator, we find

$$A = Pe^{rt} = (\$24{,}000)e^{(.02)(20)} = \$35{,}803.79.$$

The required salary 20 years from now would be about $36,000.

(b) For this level of inflation, we would have

$$A = Pe^{rt} = (\$24{,}000)e^{(.13)(20)} = \$323{,}129.71.$$

The required salary 20 years from now would be about $323,000.

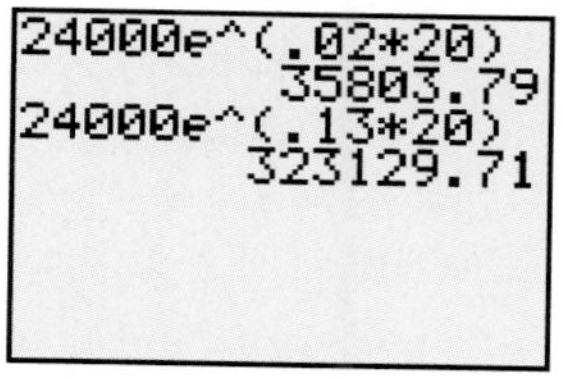

These are the computations required to solve the two parts of Example 11.

Notice that in Example 11, we stated the required salaries as rounded numbers, \$36,000 and \$323,000. In the first place, the given inflation rates were rounded. Also, even the more precise CPI figures published by the government are only an attempt to reflect cost of living on an average basis. Finally, the assumption that any given rate will persist for 20 years is itself a wild speculation. (Consider the year-to-year fluctuations in Table 4.) For all these reasons (and more), you should regard all inflationary projections into the future as only approximate forecasts.

To compare equivalent general price levels in any 2 years, we can use the following proportion. (A proportion is a statement that says that two ratios are equal. See Section 7.3.)

Inflation Proportion

For a given consumer product or service subject to average inflation,

$$\frac{\textbf{Price in year A}}{\textbf{Price in year B}} = \frac{\textbf{CPI in year A}}{\textbf{CPI in year B}}.$$

Inflation depends on the prices of many goods and services—the so-called "market basket" upon which CPI calculations are based. Selecting market basket items and gathering and analyzing price and expenditure data is a huge task for the Bureau of Labor Statistics (BLS), but is crucial because many millions of workers, retirees, and program recipients are directly affected by CPI figures each year. Also pegged to the CPI are income tax exemptions and deductions and the break points between tax brackets.

The BLS constantly attempts to refine and improve the accuracy of the computed indexes. An example was the introduction, in 2002, of a supplemental index, the **Chained Consumer Price Index** for all Urban Consumers, denoted C-CPI-U. This index tracks changes in relative prices within categories, and seeks to reflect the degree to which consumers shift to alternative items as prices fluctuate. It was hoped that these alternative approaches would lead to more accurate measures of the true inflation rate. Current data are published regularly online at the CPI home page, www.bls.gov/cpi.

EXAMPLE 12 Comparing a Tuition Increase to Average Inflation

Ricardo's college tuition in 2005 was \$9410. His uncle attended the same school in 1986 and paid \$3990 in tuition. Compare the school's tuition increase to average inflation over the same period.

SOLUTION

Let x represent what we would expect the tuition to be in 2005 if it had increased at the average rate since 1986.

$$\frac{\text{Price in 2005}}{\text{Price in 1986}} = \frac{\text{CPI in 2005}}{\text{CPI in 1986}} \qquad \text{Inflation proportion}$$

$$\frac{x}{\$3990} = \frac{195.3}{109.6} \qquad \text{Substitute; CPI values are from Table 4.}$$

$$x = \frac{195.3}{109.6} \cdot \$3990 \qquad \text{Solve for } x.$$

$$x \approx \$7110$$

Now compare the actual 2005 tuition, \$9410, with the expected figure, \$7110.

$$\frac{\$9410}{\$7110} = 1.32$$

Over the period from 1986 to 2005, tuition at Ricardo's college increased approximately 32% more than the average CPI-U rate.

Especially when working with quantities such as inflation, where continual fluctuations and inexactness prevail, we often develop rough "rules of thumb" for obtaining quick estimates. One example is the estimation of the **years to double,** which is the number of years it takes for the general level of prices to double for a given annual rate of inflation. We can derive an estimation rule as follows.

$$A = Pe^{rt}$$ Future value formula

$$2P = Pe^{rt}$$ Prices are to double.

$$2 = e^{rt}$$ Divide both sides by P.

$$\ln 2 = rt$$ Take the natural logarithm of both sides.

$$t = \frac{\ln 2}{r}$$ Solve for t.

$$t = \frac{100 \ln 2}{100r}$$ Multiply numerator and denominator by 100.

$$\textbf{years to double} \approx \frac{\textbf{70}}{\textbf{annual inflation rate}}$$ $100 \ln 2 \approx 70$

Because $100 \ln 2 \approx 70$, the two results shown here are approximately equal. See Example 13 and the discussion preceding it.

(Because r is the inflation rate as a *decimal,* $100r$ is the inflation rate as a *percent.*)

The result above usually is called the **rule of 70.** The value it produces, if not a whole number, should be rounded *up* to the next whole number of years.

EXAMPLE 13 Estimating Years to Double by the Rule of 70

Estimate the years to double for an annual inflation rate of 2.3%.

SOLUTION

$$\text{Years to double} \approx \frac{70}{2.3} = 30.43$$ Rule of 70

With a sustained inflation rate of 2.3%, prices would double in about 31 years.

14.1 EXERCISES

In the following exercises, assume whenever appropriate that, unless otherwise known, there are 12 months per year, 30 days per month, and 365 days per year.

Find the simple interest owed for each loan.

1. \$1400 at 8% for 1 year

2. \$8000 at 5% for 1 year

3. \$650 at 6% for 9 months

4. \$6000 at 7% for 4 months

5. \$2675 at 8.2% for $2\frac{1}{2}$ years

6. \$1460 at 7.82% for 22 months

Find the future value of each deposit if the account pays **(a)** *simple interest, and* **(b)** *interest compounded annually.*

7. \$700 at 3% for 6 years

8. \$2000 at 4% for 5 years

9. \$2500 at 2% for 3 years

10. \$5000 at 8% for 5 years

Solve each interest-related problem.

11. ***Simple Interest on a Late Property Tax Payment*** George Atkins was late on his property tax payment to the county. He owed \$7500 and paid the tax 4 months late. The county charges a penalty of 8% simple interest. Find the amount of the penalty.

12. ***Simple Interest on a Loan for Work Uniforms*** Michael Bolinder bought a new supply of delivery uniforms. He paid \$815 for the uniforms and agreed to pay for them in 5 months at 9% simple interest. Find the amount of interest that he will owe.

13. ***Simple Interest on a Small Business Loan*** Marvin Tidwell opened a security service on March 1. To pay for office furniture and guard dogs, he borrowed \$14,800 at the bank and agreed to pay the loan back in 10 months at 9% simple interest. Find the *total amount* required to repay the loan.

14. ***Simple Interest on a Tax Overpayment*** David Fontana is owed \$260 by the Internal Revenue Service for overpayment of last year's taxes. The IRS will repay the amount at 6% simple interest. Find the *total amount* Fontana will receive if the interest is paid for 8 months.

Find the missing final amount (future value) and/or interest earned.

	Principal	Rate	Compounded	Time	Final Amount	Compound Interest
15.	\$ 975	4%	quarterly	4 years	\$1143.26	______
16.	\$1150	7%	semiannually	6 years	\$1737.73	______
17.	\$ 480	6%	semiannually	9 years	______	\$337.17
18.	\$2370	10%	quarterly	5 years	______	______
19.	\$7500	$3\frac{1}{2}$ %	annually	25 years	______	______
20.	\$3450	2.4%	semiannually	10 years	______	______

For each deposit, find the future value (final amount on deposit) when compounding occurs **(a)** *annually,* **(b)** *semiannually, and* **(c)** *quarterly.*

	Principal	Rate	Time
21.	\$1000	10%	3 years
22.	\$3000	5%	7 years
23.	\$12,000	8%	5 years
24.	\$15,000	7%	9 years

Occasionally a savings account may actually pay interest compounded continuously. For each deposit, find the interest earned *if interest is compounded* **(a)** *semiannually,* **(b)** *quarterly,* **(c)** *monthly,* **(d)** *daily, and* **(e)** *continuously.*

	Principal	Rate	Time
25.	\$1040	7.6%	4 years
26.	\$1550	2.8%	33 months (Assume 1003 days in parts (d) and (e).)

27. Describe the effect of interest being compounded more and more often. In particular, how good is continuous compounding?

Solve each interest-related problem.

28. ***Finding the Amount Borrowed in a Simple Interest Loan*** Chris Siragusa takes out a 7% simple interest loan today which will be repaid 15 months from now with a payoff amount of \$815.63. What amount is Chris borrowing?

29. ***Finding the Amount Borrowed in a Simple Interest Loan*** What is the maximum amount you can borrow today if it must be repaid in 4 months with simple interest at 8% and you know that at that time you will be able to repay no more than \$1500?

30. In the development of the future value formula for compound interest in the text, at least four specific problem-solving strategies were employed. Identify (name) as many of them as you can and describe their use in this case.

Find the present value for each future amount.

31. \$1000 (6% compounded annually for 5 years)

32. \$14,000 (4% compounded quarterly for 3 years)

33. \$9860 (8% compounded semiannually for 10 years)

34. \$15,080 (5% compounded monthly for 4 years)

Finding the Present Value of a Compound Interest Retirement Account *Tom and Louise want to establish an account that will supplement their retirement income beginning 30 years from now. For each interest rate find the lump sum they must deposit today so that $500,000 will be available at time of retirement.*

35. 8% compounded quarterly

36. 6% compounded quarterly

37. 8% compounded daily

38. 6% compounded daily

Finding the Effective Annual Yield in a Savings Account *Suppose a savings and loan pays a nominal rate of 4% on savings deposits. Find the effective annual yield if interest is compounded as stated in Exercises 39–45. (Give answers to the nearest thousandth of a percent.)*

39. annually

40. semiannually

41. quarterly

42. monthly

43. daily

44. 1000 times per year

45. 10,000 times per year

46. Judging from Exercises 39–45, what do you suppose is the effective annual yield if a nominal rate of 4% is compounded continuously? Explain your reasoning.

Comparing Savings Rates and Yields *In April 2006, Schools Financial Credit Union advertised the following savings rates for balances from $20,000 to $49,999.99 in two types of accounts. Use this information in Exercises 47 and 48.*

	Rate	Yield
Super$hare[SM] Account	2.00%	2.02%
Premier Money Market Account	3.00%	3.04%

47. If you deposit $30,000 in Super$hares, how much would you have in 1 year?

48. How often does compounding occur in the Premier Money Market Account, daily, monthly, quarterly, or semiannually?

Solve each interest-related problem.

49. ***Finding Years to Double*** How long would it take to double your money in an account paying 6% compounded quarterly? (Give your answer in years plus days, ignoring leap years.)

50. ***Comparing Principal and Interest Amounts*** After what time period would the interest earned equal the original principal in an account paying 6% compounded daily? (Give your answer in years plus days, ignoring leap years.)

51. Solve the effective annual yield formula for r to obtain a general formula for nominal rate in terms of yield and the number of compounding periods per year.

52. ***Finding the Nominal Rate of a Savings Account*** Ridgeway Savings compounds interest monthly, and the effective annual yield is 2.95%. What is the nominal rate?

53. ***Comparing Bank Savings Rates*** Bank A pays a nominal rate of 3.800% compounded daily on deposits. Bank B produces the same annual yield as the first but compounds interest only quarterly and pays no interest on funds deposited for less than an entire quarter.

(a) What nominal rate does Bank B pay (to the nearest thousandth of a percent)?

(b) Which bank should Patty Demko choose if she has $2000 to deposit for 10 months? How much more interest will she earn than in the other bank?

(c) Which bank should Sashaya Davis choose for a deposit of $6000 for one year? How much interest will be earned?

Estimating the Years to Double by the Rule of 70 *Use the rule of 70 to estimate the years to double for each annual inflation rate.*

54. 1%

55. 2%

56. 8%

57. 9%

Estimating the Inflation Rate by the Rule of 70 *Use the rule of 70 to estimate the annual inflation rate (to the nearest tenth of a percent) that would cause the general level of prices to double in each time period.*

58. 5 years

59. 7 years

60. 16 years

61. 22 years

Solve each inflation-related problem.

62. Derive a rule for estimating the "years to triple," that is, the number of years it would take for the general levels of prices to triple for a given annual inflation rate.

63. What would you call your rule of Exercise 62? Explain your reasoning.

Estimating Future Prices for Constant Annual Inflation *The year 2006 prices of several items are given below. Find the estimated future prices required to fill the blanks in the chart. (Give a number of significant figures consistent with the 2006 price figures provided.)*

Item	2006 Price	2015 Price 2% Inflation	2025 Price 2% Inflation	2015 Price 10% Inflation	2025 Price 10% Inflation
64. House	$265,000	______	______	______	______
65. Fast food meal	$ 5.89	______	______	______	______
66. Gallon of gasoline	$ 2.65	______	______	______	______
67. Small car	$ 18,500	______	______	______	______

Estimating Future Prices for Variable Annual Inflation *As seen in Table 4, inflation rates do not often stay constant over a period of years. For example, from 1996 to 1997 the index increased from* 156.9 *to* 160.5, *or* 2.3%, *while the following year, 1997 to 1998, it increased from* 160.5 *to* 163.0, *only* 1.6%. *Assume that prices for the items below increased at the average annual rates shown in Table 4. Use the inflation proportion to find the missing prices in the last column of the chart. Round to the nearest dollar.*

Item	Price	Year Purchased	Price in 2005
68. Desk	$ 450	2002	____
69. Evening dress	$ 175	2000	____
70. Designer puppy	$ 250	1998	____
71. Lawn tractor	$1099	1996	____

Solve each interest-related problem.

72. ***Finding the Present Value of a Future Real Estate Purchase*** A California couple are selling their small dairy farm to a developer, but they wish to defer receipt of the money until 2 years from now, when they will be in a lower tax bracket. Find the lump sum that the developer can deposit today, at 5% compounded quarterly, so that enough will be available to pay the couple $1,450,000 in 2 years.

73. ***Finding the Present Value of a Future Equipment Purchase*** Human gene sequencing is a major research area of biotechnology. One company (PE Biosystems) leased 300 of its sequencing machines to a sibling company (Celera). If Celera was able to earn 7% compounded quarterly on invested money, what lump sum did they need to invest in order to purchase those machines 18 months later at a price of $300,000 each? (*Source: Forbes,* February 21, 2000, p. 102.)

74. Research how the Bureau of Labor Statistics measures changes in consumer prices, and explain some of the effects those measurements end up having on citizens.

75. CPI represents consumer price index. Investigate and explain the related concept PPI.

EXTENSION

Annuities

A series of deposits, periodically made into an account where deposited funds earn compound interest, is called an **annuity.** The mathematics of annuities is based on summing the terms of a **geometric sequence.**

Geometric Sequence In a geometric sequence, each term, after the first, is generated by multiplying the previous term by the **common ratio,** a number remaining constant throughout the sequence. For example, a geometric sequence with first term 3 and common ratio 2 starts out as follows:

$$3, 6, 12, 24, 48, 96, \ldots.$$

In general, if the first term is denoted a, the common ratio is denoted r, and there are n terms altogether, then the complete sequence is

$$a, ar, ar^2, ar^3, \ldots, ar^{n-1}.$$

(Verify by inductive reasoning that the nth term really is ar^{n-1}.)

The sum of all n terms of the geometric sequence above can be written

$$S = a + ar + ar^2 + \cdots + ar^{n-2} + ar^{n-1}.$$

Multiply both sides of this equation by r:

$$Sr = ar + ar^2 + ar^3 + \cdots + ar^{n-1} + ar^n.$$

Now position these two equations, one below the other, and subtract as follows.

$$\begin{aligned} S &= a \quad + ar \quad + ar^2 \quad + \cdots + ar^{n-2} \quad + ar^{n-1} \\ Sr &= ar \quad + ar^2 \quad + ar^3 \quad + \cdots + ar^{n-1} \quad + ar^n \\ \hline S - Sr &= (a - ar) + (ar - ar^2) + (ar^2 - ar^3) + \cdots + (ar^{n-2} - ar^{n-1}) + (ar^{n-1} - ar^n) \end{aligned}$$

Now we can rearrange and regroup the terms on the right to obtain the following.

$$\begin{aligned} S - Sr &= a + (ar - ar) + (ar^2 - ar^2) + \cdots + (ar^{n-1} - ar^{n-1}) - ar^n \\ &= a + 0 + 0 + \cdots + 0 - ar^n \end{aligned}$$

Notice that all terms on the right, except a and $-ar^n$, were arranged in pairs to cancel out. So we get the following.

$$\begin{aligned} S - Sr &= a - ar^n \\ S(1 - r) &= a(1 - r^n) && \text{Factor both sides.} \\ S &= \frac{a(1 - r^n)}{1 - r} && \text{Solve for the sum } S. \end{aligned}$$

The Sum of a Geometric Sequence

If a geometric sequence has first term a and common ratio r, and has n terms altogether, then the sum of all n terms is given by

$$\mathbf{S = \frac{a(1 - r^n)}{1 - r}}.$$

Ordinary Annuities Now let's see how this summation formula can be applied to systematic savings or investment plans. Consider the following situation.

> ***When I deposit money into a certain account, I can earn interest at an annual rate r, compounded m times per year. Assume I will make a regular deposit of R (in dollars)* at the end *of each compounding period. What will be the total value of my account after n periods?***

This type of account, which receives regular periodic deposits *at the end of* each compounding period, is called an **ordinary annuity** account. The total term (time period) over which deposits are made, consisting of the n compounding periods, is called the **accumulation period** of the annuity.

As each deposit R goes into the account, it begins to earn interest and grows to some amount according to the future value formula for compound interest. In Table 5 we list the n deposits, starting with the last one, which is made at the very end of the nth compounding period. Because the last deposit (deposit number n) has no time to accumulate interest, its final value is just R.

TABLE 5 Value of an Annuity Account

Deposit Number	Future Value of Deposit
n	R
$n - 1$	$R\left(1 + \frac{r}{m}\right)$
$n - 2$	$R\left(1 + \frac{r}{m}\right)^2$
.	.
.	.
.	.
3	$R\left(1 + \frac{r}{m}\right)^{n-3}$
2	$R\left(1 + \frac{r}{m}\right)^{n-2}$
1	$R\left(1 + \frac{r}{m}\right)^{n-1}$

(Why is the exponent in the future value of the first deposit $n - 1$ rather than n?)

The future value of the account must be the sum of the future values of all the individual deposits, which is the sum of the entries of the right-hand column of the table. But those entries comprise a geometric sequence. Its first term is R, the common ratio is $(1 + \frac{r}{m})$, and the number of terms is n. So, in this case, the sum of the sequence (the **future value of the annuity**) is given by

$$\text{Sum} = \frac{R[1 - (1 + \frac{r}{m})^n]}{1 - (1 + \frac{r}{m})}.$$

By reversing the differences in the numerator and denominator and simplifying the denominator, we obtain the following formula.

Future Value of an Ordinary Annuity

If money in an annuity account earns interest at an annual rate r, compounded m times per year, and regular deposits of amount R are made into this account at the end of each compounding period, then the final value (or **future value**) of the account at the end of n periods is given by

$$V = \frac{R\left[\left(1 + \frac{r}{m}\right)^n - 1\right]}{\frac{r}{m}}.$$

EXAMPLE 1 Saving with an Ordinary Annuity Account

Alisher Abdullayev sets up an ordinary annuity account with his credit union. Starting a month from today he will deposit \$100 per month, and deposits will earn 5.1% per year, compounded monthly. Find

(a) the value of Alisher's account 10 years from today,
(b) the total of all his deposits over the 10 years, and
(c) the total interest earned.

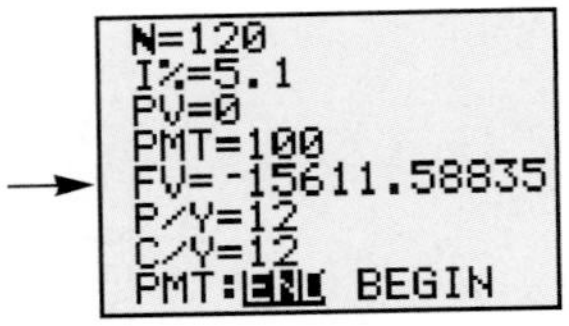

Compare the future value indicated by the arrow with the result in Example 1(a).

SOLUTION

(a) In this case, $R = \$100$, $r = 5.1\% = .051$, $m = 12$, and $n = 10 \cdot 12 = 120$. The formula for future value of an ordinary annuity gives

$$V = \frac{\$100\left[\left(1 + \frac{.051}{12}\right)^{120} - 1\right]}{\frac{.051}{12}} = \$15{,}611.59.$$

(b) Over 10 years, monthly deposits of \$100 will total

$$\$100 \cdot 10 \cdot 12 = \$12{,}000.$$

(c) Interest earned = \$15,611.59 (Future value) − \$12,000 (Total deposits)
= \$3611.59

The future value of Alisher's account, at the end of 10 years, will be \$15,611.59. This value results from a total of \$12,000 in deposits which, over those 10 years, earn an additional \$3611.59 in interest.

It often happens that we would like to have a known amount available at some future date but do not have the necessary present amount to deposit as a lump sum in order to accomplish that goal. A practical option is to make smaller periodic deposits into an interest-bearing account so that the sum of the various deposits, with their accrued compound interest, equals the desired future amount. Such an account is sometimes called a **sinking fund.** It is actually an ordinary annuity where the unknown is the regular deposit rather than the future value.

EXAMPLE 2 Finding the Regular Deposit for a Sinking Fund

Victor Tran wishes to accumulate \$14,000 to replace equipment for his business 2 years from now. His sinking fund will earn 4%, compounded monthly. Find

(a) the regular deposit required at the end of each month,
(b) the total of all deposits, and
(c) the total interest earned.

SOLUTION

(a) $$V = \frac{R[(1 + \frac{r}{m})^n - 1]}{\frac{r}{m}}$$ Future value of an ordinary annuity

$$R = \frac{V(\frac{r}{m})}{(1 + \frac{r}{m})^n - 1}$$ Solve for R.

$$R = \frac{\$14{,}000(\frac{.04}{12})}{(1 + \frac{.04}{12})^{24} - 1}$$ Substitute known values.

$$R = \$561.28$$ There are 24 months in 2 years.

Victor must deposit \$561.28 monthly to have the \$14,000 available in 2 years.

(b) Over the term of the account,

Total amount deposited = 24 (Number of monthly deposits) · \$561.28 (Amount of each deposit)
= \$13,470.72.

(c) Interest earned = \$14,000 (Accumulated sum) − \$13,470.72 (Total amount deposited)
= \$529.28

Tax-Deferred Annuities The U.S. tax code encourages individuals to save for retirement by contributing a portion of current income to various kinds of tax-deferred retirement plans, with restrictions based on employment status, income levels, and other factors. One type of retirement plan is the **tax-deferred annuity (TDA),** or **tax-sheltered annuity (TSA).** Contributions to such a plan reduce "take-home" pay in the current year, but they also reduce income tax paid currently because they are deducted from current taxable income.

EXAMPLE 3 Assessing the Value of Tax Deferral

Jim Walker contributes \$500 per month to his TDA, where money earns 6% compounded monthly. If these dollars were not deferred, they would incur income tax of 30% (federal and state combined). Compare Jim's end-of-year results with the results of not making these contributions.

SOLUTION

At the end of the year, we have

$$\text{Annuity value} = \frac{R[(1 + \frac{r}{m})^n - 1]}{\frac{r}{m}} \qquad \text{Future value formula}$$

$$= \frac{\$500[(1 + \frac{.06}{12})^{12} - 1]}{\frac{.06}{12}} \qquad \text{Substitute known values.}$$

$$= \$6167.78.$$

$$\text{Total deposits} = 12 \cdot \$500 = \$6000 \longleftarrow \text{This amount is tax-deferred.}$$

Now, what if Jim had not made the contributions?

$$\text{Tax on } \$6000 = \$6000 \cdot 30\% = \$1800 \qquad (30\% \text{ is the tax rate})$$

$$\text{Remainder} = \$6000 - \$1800 = \$4200$$

$$\text{Advantage of contributing} = \$6167.78 - \$4200 = \$1967.78$$

It is true that, had Jim not contributed, he would have had $4200 more to spend this year, but for retirement purposes, it is better to have $6167.78 tucked away in an account earning interest than to have $4200 gone forever.

EXAMPLE 4 Comparing Deferred and Nondeferred Saving

Suppose Jim (see Example 3) is now 25 and hopes to retire at age 55. Compare his retirement account balance at retirement under the deferred plan with the corresponding balance if he were to set aside the same $500 per month on a nondeferred basis.

SOLUTION

Tax-deferred plan:

$$V = \frac{R[(1 + \frac{r}{m})^n - 1]}{\frac{r}{m}} \qquad \text{Future value formula}$$

There are 360 months in 30 years.

$$= \frac{\$500[(1 + \frac{.06}{12})^{360} - 1]}{\frac{.06}{12}} \qquad \text{Substitute known values.}$$

$$= \$502{,}257.52$$

Non-tax-deferred plan: In this case, each $500 set aside is currently taxed at 30%, which leaves only a $350 deposit (70% of $500) each month.

$$V = \frac{\$350[(1 + \frac{.06}{12})^{360} - 1]}{\frac{.06}{12}} \qquad \text{Substitute values in the future value formula.}$$

$$= \$351{,}580.26$$

During the entire 30 years, Jim reduced his take-home pay by $500 per month under both plans, but he ends up with

$$\$502{,}257.52 - \$351{,}580.26 = \$150{,}677.26$$

more in the tax-deferred account.

Tax-deferred is not the same as "tax-free," which makes the result of Example 4 less impressive than it may seem. This type of retirement income is taxed at the time it is withdrawn. People who sell tax-deferred plans stress that you will probably be in a lower tax bracket after retirement because of reduced income. But many find this is not so, especially compared to early career income, and especially if your retirement account investments have done very well, which could make your required withdrawals quite large. Besides, who can tell what general tax rates will be 30 years from now?

EXTENSION EXERCISES

Finding the Earnings of an Ordinary Annuity Account *For each of the following ordinary annuity accounts, find* **(a)** *the future value of the account (at the end of the accumulation period),* **(b)** *the total of all deposits, and* **(c)** *the total interest earned.*

	Regular Deposit	Compounded	Annual Interest Rate	Accumulation Period
1.	\$1000	yearly	8.5%	10 years
2.	\$2000	yearly	4.5%	20 years
3.	\$50	monthly	6.0%	5 years
4.	\$75	monthly	7.2%	10 years
5.	\$20	weekly	5.2%	3 years
6.	\$30	weekly	7.8%	5 years

An annuity account into which regular deposits are made at the beginning of each compounding period, rather than at the end, is called an **annuity due.** *(It is usually assumed that a deposit is made at the end of compounding period number n so that the future value of an annuity due is the accumulation of $n + 1$ deposits rather than just n.)*

7. If the interest rate, regular deposit amount, compounding frequency, and accumulation period all are the same in both cases, would the future value of an annuity due be less or more than that of an ordinary annuity? Explain.

8. Modify the derivation of the formula for future value of an ordinary annuity (presented in the text) to show that the **future value of an annuity due** is given by

$$V = \frac{R\left[\left(1 + \frac{r}{m}\right)^{n+1} - 1\right]}{\frac{r}{m}}.$$

Finding the Earnings of an Annuity Due *Use the formula of Exercise 8 to find the future value of each annuity due.*

	Regular Deposit	Compounded	Annual Interest Rate	Accumulation Period
9.	\$50	monthly	6.0%	5 years
10.	\$75	monthly	7.2%	10 years
11.	\$20	weekly	5.2%	3 years
12.	\$30	weekly	7.8%	5 years

Finding Regular Deposits for a Sinking Fund *For each sinking fund account, find* **(a)** *the required periodic deposit (using the future value formula, solved for R),* **(b)** *the total amount deposited, and* **(c)** *the total interest earned.*

	Future Value	Compounded	Annual Interest Rate	Accumulation Period
13.	\$4000	yearly	5.4%	3 years
14.	\$100,000	monthly	7.2%	10 years
15.	\$500	weekly	2.6%	5 years
16.	\$250,000	quarterly	4.8%	30 years

17. In the tax-deferred plan of Example 4, consider the entire 30-year accumulation period.
(a) What amount did Jim contribute?
(b) What amount of interest was earned?

19. An **IRA (Individual Retirement Account)** is another option for building retirement funds. A traditional IRA defers tax on current contributions but taxes later withdrawals, like a tax-deferred annuity. The newer **Roth IRA,** available since 1998, does *not* defer current contributions (which are taxed in the year they are made) but exempts all later withdrawals from taxation (including contributions and gains).
(a) Explain why a Roth IRA may be more advantageous for a person who starts it at an early age.
(b) What might have been motivating factors for Congress to authorize the Roth IRA in 1998?

18. Do you think it is practical for a financial institution (bank, brokerage, insurance company) to guarantee to pay you a set interest rate for years into the future on your annuity account? Why or why not?

20. The tax code generally limits deductible contributions to a tax-deferred retirement account to a percentage of your adjusted gross income (AGI), and also prohibits early withdrawals (before age $59\frac{1}{2}$); and monetary penalties are imposed for violations of these rules. Why do you suppose this is true?

14.2 Installment Buying

Closed-End Credit • Open-End Credit

Closed-End Credit When you borrow your neighbor's lawnmower, you (they hope) return the whole thing all at once when you are finished with it. However, when *money* is borrowed (in any sizable amount), timely repayment is much more likely if the debt is repaid over a period of time in a series of smaller payments.

Borrowing to finance purchases, and repaying with periodic payments is called **installment buying.** In this section we discuss the two general types of consumer installment credit. The first type, **closed-end** credit, involves borrowing a set amount up front and paying a series of equal installments (payments) until the loan is paid off. Furniture, appliances, and cars commonly are financed through closed-end credit (sometimes called **fixed installment loans**).

With the second type of consumer installment credit, **open-end** credit, there is no fixed number of installments—the consumer continues paying until no balance is owed. With open-end credit, additional credit often is extended before the initial amount is paid off. Examples of open-end credit include most department store charge accounts and bank charge cards such as MasterCard and VISA.

Installment loans set up under closed-end credit often are based on **add-on interest.** This means that if an amount P is borrowed, the annual interest rate is to be r, and payments will extend over t years, then the required interest comes from the simple interest formula $I = Prt$. You simply "add on" this amount of interest to the principal borrowed to arrive at the total debt (or amount to be repaid).

$$\text{Amount to be repaid} = \underset{\substack{\uparrow \\ \text{Amount borrowed}}}{P} + \underset{\substack{\uparrow \\ \text{Interest due}}}{Prt}$$
$$= P(1 + rt)$$

This is just the future value of a simple interest loan as shown in Section 14.1. The total debt is then equally divided among the payments (usually monthly) to be made over the t years.

EXAMPLE 1 Repaying an Add-On Loan

Chris and Heather are a newlywed couple. They buy \$3500 worth of furniture and appliances to furnish their first apartment. They pay \$700 down and agree to pay the balance at a 7% add-on rate for 2 years. Find

(a) the total amount to be repaid,
(b) the monthly payment, and
(c) the total cost of the purchases, including finance charges.

SOLUTION

(a)
$$\begin{aligned} \text{Amount to be repaid} &= P(1 + rt) && \text{Formula} \\ &= \$2800(1 + .07 \cdot 2) && \text{Substitute known values.} \\ &= \$3192 \end{aligned}$$

(b)
$$\text{Monthly payment} = \frac{\$3192}{24} \quad \begin{array}{l} \leftarrow \text{Amount to be repaid} \\ \leftarrow \text{Number of payments} \end{array}$$
$$= \$133$$

(c)
$$\text{Total cost of purchases} = \underset{\text{Down payment}}{\$700} + \underset{\text{Loan repayment}}{\$3192}$$
$$= \$3892$$

Chris and Heather will end up paying \$3892, which is 11.2% more than the price tag total of their purchases. ■

Credit card debt among students has raised increasing concern in recent years among lawmakers, college officials, and consumer advocacy groups. One survey found that undergraduates carried an average credit card balance of \$2,748, while graduate students carried an average balance of \$4,776. Ineffective management of such debt can set students up for a variety of financial and other difficulties. (*Source:* www.ericdigests.org)

Notice that the repayment amount in Example 1, $P(1 + rt)$, was the same as the final amount A in a savings account paying a simple interest rate r for t years. (See Section 14.1.) But in the case of savings, the bank keeps all of your money for the entire time period. In Example 1, Chris and Heather did not keep the full principal amount for the full time period. They repaid it in 24 monthly installments. The 7% add-on rate turns out to be equivalent to a much higher "true annual interest" rate, as we shall see in the next section.

Another type of fixed installment loan, the **real estate mortgage,** usually involves much larger amounts and longer repayment periods than are involved in the consumer loans covered in this section. We will discuss mortgages in Section 14.4.

Open-End Credit With a typical department store account, or bank card, a credit limit is established initially and the consumer can make many purchases during a month (up to the credit limit). The required monthly payment can vary from a set minimum (which may depend on the account balance) up to the full balance.

For purchases made in a store or other retail location, the customer may authorize adding the purchase price to his or her charge account by signing on a paper "charge slip," or perhaps on an electronic signature pad. Authorization also can be given by telephone for mail orders, services, contributions, and other expenditures. Such transactions can be accomplished directly over the Internet as well, though many consumers have been hesitant to do so.

At the end of each billing period (normally once a month), the customer receives an **itemized billing,** a statement listing purchases and cash advances, the total balance owed, the minimum payment required, and perhaps other account information. Any charges beyond cash advanced and cash prices of items purchased are called **finance charges.** Finance charges may include interest, an annual fee, credit insurance coverage, a time payment differential, or carrying charges.

A few open-end lenders still use one of the simpler methods of calculating finance charges, such as the **unpaid balance method** which is illustrated in Example 2. Typically, you would begin with the unpaid account balance at the *end of the previous month* (or *billing period*) and apply to it the current monthly interest rate (or the daily rate multiplied by the number of days in the current billing period). Any purchases or returns in the current period do not affect the finance charge calculation.

EXAMPLE 2 Using the Unpaid Balance Method for Credit Card Charges

The table below shows Eric Kanemoto's VISA account activity for a 2-month period. If the bank charges interest of 1.1% per month on the unpaid balance, and there are no other finance charges, find the missing quantities in the table, and the total finance charges for the 2 months.

Month	Unpaid Balance at Beginning of Month	Finance Charge	Purchases During Month	Returns	Payment	Unpaid Balance at End of Month
July	\$179.82	______	\$101.58	\$ 0	\$30	______
August	______	______	\$ 19.84	\$42.95	\$85	______

SOLUTION

The July finance charge is 1.1% of \$179.82, which is \$1.98. Then, adding any finance charge and purchases and subtracting any returns and payments, we arrive at the unpaid balance at the end of July (and the beginning of August):

$$\$179.82 + \$1.98 + \$101.58 - \$30 = \$253.38.$$

A similar calculation gives an August finance charge of

$$1.1\% \text{ of } \$253.38 = .011 \cdot \$253.38 = \$2.79,$$

and an unpaid balance at the end of August of

$$\$253.38 + \$2.79 + \$19.84 - \$42.95 - \$85 = \$148.06.$$

The total finance charges for the 2 months were

$$\$1.98 + \$2.79 = \$4.77.$$

Most open-end credit plans now calculate finance charges by the **average daily balance method.** From the bank's point of view, this seems fairer since it considers balances on all days of the billing period and thus comes closer to charging card holders for the credit they actually utilize. The average daily balance method is illustrated in Example 3.

Learning the following rhyme will help you in problems like the one found in Example 3.

Thirty days hath September,
April, June, and November.
All the rest have thirty-one,
Save February which has
twenty-eight days clear,
And twenty-nine in each
leap year.

EXAMPLE 3 Using the Average Daily Balance Method for Credit Card Charges

The activity in Paige Dunbar's MasterCard account for one billing period is shown below. If the previous balance (on March 3) was $209.46, and the bank charges 1.3% per month on the average daily balance, find **(a)** the average daily balance for the next billing (April 3), **(b)** the finance charge to appear on the April 3 billing, and **(c)** the account balance on April 3.

March 3	Billing date	
March 12	Payment	$50.00
March 17	Clothes	$28.46
March 20	Mail order	$31.22
April 1	Auto parts	$59.10

SOLUTION

(a) First make a table that shows the beginning date of the billing period and the dates of all transactions in the billing period. Along with each of these, compute the running balance on that date.

Date	Running Balance
March 3	$209.46
March 12	$209.46 − $50 = $159.46
March 17	$159.46 + $28.46 = $187.92
March 20	$187.92 + $31.22 = $219.14
April 1	$219.14 + $59.10 = $278.24

Next, tabulate the running balance figures, along with the number of days until the balance changed. Multiply each balance amount by the number of days. The sum of these products gives the "sum of the daily balances."

Date	Running Balance	Number of Days Until Balance Changed	(Running Balance) · (Number of Days)
March 3	$209.46	9	$1885.14
March 12	$159.46	5	$ 797.30
March 17	$187.92	3	$ 563.76
March 20	$219.14	12	$2629.68
April 1	$278.24	2	$ 556.48
		Totals: 31	$6432.36

$$\text{Average daily balance} = \frac{\text{Sum of daily balances}}{\text{Days in billing period}} = \frac{\$6432.36}{31} = \$207.50.$$

Dunbar will pay a finance charge based on the average daily balance of $207.50.

E-commerce, or electronic commerce, refers to the use of the Internet as a distribution channel for goods and services. It can be as simple as traditional companies selling online. But the dynamics of the Internet also facilitate the formation of entirely new kinds of companies and company partnerships, often marketing previously undreamed of products and services.

(b) The finance charge for the April 3 billing will be

$$1.3\% \text{ of } \$207.50 = .013 \cdot \$207.50 = \$2.70.$$

(c) The account balance on the April 3 billing will be the latest running balance plus the finance charge: $\$278.24 + \$2.70 = \$280.94.$

Bank card and other open-end credit accounts normally require some minimum payment each month, which depends in some way on your balance. But if you need to delay payment, you can carry most of your balance from month to month, and pay the resulting finance charges. Whatever method is used for calculating finance charges, you can avoid finance charges altogether in many cases by paying your entire new balance by the due date each month. In fact, this is a wise practice, because these types of credit accounts, although convenient, are relatively expensive. Examples 1 and 2 involved typical monthly interest rates of 1.1% and 1.3%, respectively, which are equivalent to annual rates of 13.2% and 15.6%. When selecting a bank card, pay careful attention to the interest rates and other finance charges involved. Purchases charged to your account during the month will not be billed to you until your next billing date. And even after that, most lenders allow a grace period (perhaps 20 days or more) between the billing date and the payment due date. With careful timing, you may be able to purchase an item and delay having to pay for it for up to nearly 2 months without paying interest.

Although the open-end credit examples in this section have mainly addressed the actual interest charges, there are other features of these accounts that can be at least as important as interest. For example:

1. Is an annual fee charged? If so, how much is it?
2. Is a special "introductory" rate offered? If so, how long will it last?
3. Are there other incentives, such as rebates, credits toward certain purchases, or return of interest charges for long-time use?

14.2 EXERCISES

Round all monetary answers to the nearest cent unless directed otherwise. Assume, whenever appropriate, that unless otherwise known, there are 12 months per year, 30 days per month, and 365 days per year.

Financing an Appliance Purchase *Suppose you buy appliances costing* \$2150 *at a store charging* 12% *add-on interest, you make a* \$500 *down payment, and you agree to monthly payments over two years.*

1. Find the total amount you will be financing.
2. Find the total interest you will pay.
3. Find the total amount to be repaid.
4. Find the monthly payment.
5. Find your total cost, for appliances plus interest.

Financing a New Car Purchase *Suppose you want to buy a new car that costs* \$16,500. *You have no cash—only your old car, which is worth* \$3000 *as a trade-in.*

6. How much do you need to finance to buy the new car?
7. The dealer says the interest rate is 9% add-on for 3 years. Find the total interest.
8. Find the total amount to be repaid.
9. Find the monthly payment.
10. Find your total cost, for the new car plus interest.

In Exercises 11–16 on the next page, use the add-on method of calculating interest to find the total interest and the monthly payment.

	Amount of Loan	Length of Loan	Interest Rate
11.	$4500	3 years	9%
12.	$2700	2 years	8%
13.	$ 750	18 months	7.4%
14.	$2450	30 months	9.2%
15.	$ 535	16 months	11.1%
16.	$ 798	29 months	10.3%

Work each problem. Give monetary answers to the nearest cent unless directed otherwise.

17. *Finding the Monthly Payment for an Add-On Interest Furniture Loan* The Giordanos buy $8500 worth of furniture for their new home. They pay $3000 down. The store charges 10% add-on interest. The Giordanos will pay off the furniture in 30 monthly payments ($2\frac{1}{2}$ years). Find the monthly payment.

18. *Finding the Monthly Payment for an Add-On Interest Auto Loan* Find the monthly payment required to pay off an auto loan of $9780 over 3 years if the add-on interest rate is 9.3%.

19. *Finding the Monthly Payment for an Add-On Interest Home Electronics Loan* The total purchase price of a new home entertainment system is $14,240. If the down payment is $2900 and the balance is to be financed over 48 months at 10% add-on interest, what is the monthly payment?

20. *Finding the Monthly Payment for an Add-On Interest Loan* What are the monthly payments Donna De Simone pays on a loan of $1680 for a period of 10 months if 9% add-on interest is charged?

21. *Finding the Amount Borrowed in an Add-On Interest Car Loan* Joshua Eurich has misplaced the sales contract for his car and cannot remember the amount he originally financed. He does know that the add-on interest rate was 9.8% and the loan required a total of 48 monthly payments of $314.65 each. How much did Joshua borrow (to the nearest dollar)?

22. *Finding an Add-On Interest Rate* Susan Dratch is making monthly payments of $207.31 to pay off a $3\frac{1}{2}$ year loan for $6400. What is her add-on interest rate (to the nearest tenth of a percent)?

23. *Finding the Term of an Add-On Interest Loan* How long (in years) will it take Michael Garbin to pay off an $8000 loan with monthly payments of $172.44 if the add-on interest rate is 9.2%?

24. *Finding the Number of Payments of an Add-On Interest Loan* How many monthly payments must Jawann make on a $10,000 loan if he pays $417.92 a month and the add-on interest rate is 10.15%?

Finding the Finance Charge of an Open-End Charge Account *Find the finance charge on each open-end charge account. Assume interest is calculated on the unpaid balance of the account.*

	Unpaid Balance	Monthly Interest Rate
25.	$325.50	1.0%
26.	$450.25	1.2%
27.	$242.88	1.12%
28.	$655.33	1.21%

Finding the Unpaid Balance of an Open-End Charge Account *Complete each table, showing the unpaid balance at the end of each month. Assume a monthly interest rate of* 1.1% *on the unpaid balance.*

29.

Month	Unpaid Balance at Beginning of Month	Finance Charge	Purchases During Month	Returns	Payment	Unpaid Balance at End of Month
February	$319.10	____	$86.14	0	$50	____
March	____	____	109.83	$15.75	60	____
April	____	____	39.74	0	72	____
May	____	____	56.29	18.09	50	____

30.

Month	Unpaid Balance at Beginning of Month	Finance Charge	Purchases During Month	Returns	Payment	Unpaid Balance at End of Month
October	$828.63	____	$128.72	$23.15	$125	____
November	____	____	291.64	0	170	____
December	____	____	147.11	17.15	150	____
January	____	____	27.84	139.82	200	____

31.

Month	Unpaid Balance at Beginning of Month	Finance Charge	Purchases During Month	Returns	Payment	Unpaid Balance at End of Month
August	$684.17	____	$155.01	$38.11	$100	____
September	____	____	208.75	0	75	____
October	____	____	56.30	0	90	____
November	____	____	190.00	83.57	150	____

32.

Month	Unpaid Balance at Beginning of Month	Finance Charge	Purchases During Month	Returns	Payment	Unpaid Balance at End of Month
March	$1230.30	____	$308.13	$74.88	$250	____
April	____	____	488.35	0	350	____
May	____	____	134.99	18.12	175	____
June	____	____	157.72	0	190	____

Finding Finance Charges *Find the finance charge for each charge account. Assume interest is calculated on the average daily balance of the account.*

	Average Daily Balance	Monthly Interest Rate
33.	$ 249.94	1.0%
34.	$ 450.21	1.06%
35.	$1073.40	1.125%
36.	$1320.42	1.375%

Finding Finance Charges and Account Balances Using the Average Daily Balance Method *For each credit card account, assume one month between billing dates (with the appropriate number of days) and interest of 1.2% per month on the average daily balance. Find* **(a)** *the average daily balance,* **(b)** *the monthly finance charge, and* **(c)** *the account balance for the next billing.*

37. Previous balance: $728.36

May 9	Billing date	
May 17	Payment	$200
May 30	Dinner	$ 46.11
June 3	Theater tickets	$ 64.50

38. Previous balance: $514.79

January 27	Billing date	
February 9	Candy	$11.08
February 13	Returns	$26.54
February 20	Payment	$59
February 25	Repairs	$71.19

39. Previous balance: $462.42

June 11	Billing date	
June 15	Returns	$106.45
June 20	Jewelry	$115.73
June 24	Car rental	$ 74.19
July 3	Payment	$115
July 6	Flowers	$ 68.49

40. Previous balance: $983.25

August 17	Billing date	
August 21	Mail order	$ 14.92
August 23	Returns	$ 25.41
August 27	Beverages	$ 31.82
August 31	Payment	$108
September 9	Returns	$ 71.14
September 11	Concert tickets	$110
September 14	Cash advance	$100

Finding Finance Charges *Assume no purchases or returns are made in Exercises 41 and 42.*

41. At the beginning of a 31-day billing period, Margaret Dent-Dzierzanowski has an unpaid balance of $720 on her credit card. Three days before the end of the billing period, she pays $600. Find her finance charge at 1.0% per month using the following methods.
(a) unpaid balance method
(b) average daily balance method

42. Christine Grexa's VISA bill dated April 14 shows an unpaid balance of $1070. Five days before May 14, the end of the billing period, Grexa makes a payment of $900. Find her finance charge at 1.1% per month using the following methods.
(a) unpaid balance method
(b) average daily balance method

Analyzing a "90 Days Same as Cash" Offer *One version of the "90 Days Same as Cash" promotion was offered by a "major purchase card," which established an account charging 1.3167% interest per month on the account balance. Interest charges are added to the balance each month, becoming part of the balance on which interest is computed the next month. If you pay off the original purchase charge within 3 months, all interest charges are cancelled. Otherwise you are liable for all the interest. Suppose you purchase $2900 worth of carpeting under this plan.*

43. Find the interest charge added to the account balance at the end of
(a) the first month,
(b) the second month,
(c) the third month.

44. Suppose you pay off the account 1 day late (3 months plus 1 day). What total interest amount must you pay? (Do not include interest for the one extra day.)

45. Treating the 3 months as $\frac{1}{4}$ year, find the equivalent simple interest rate for this purchase (to the nearest tenth of a percent).

Various Charges of a Bank Card Account *Carole's bank card account charges 1.1% per month on the average daily balance as well as the following special fees:*

Cash advance fee:	*2% (not less than $2 nor more than $10)*
Late payment fee:	*$15*
Over-the-credit-limit fee:	*$5*

In the month of June, Carole's average daily balance was $1846. She was on vacation during the month and did not get her account payment in on time, which resulted in a late payment and also resulted in charges accumulating to a sum above her credit limit. She also used her card for six $100 cash advances while on vacation. Find the following based on account transactions in that month.

46. interest charges to the account

47. special fees charged to the account

Write out your response to each of the following.

48. Is it possible to use a bank credit card for your purchases without paying anything for credit? If so, explain how.

49. Obtain applications or descriptive brochures for several different bank card programs, compare their features (including those in fine print), and explain which deal would be best for you, and why.

50. Research and explain the difference, if any, between a "credit" card and a "debit" card.

51. Many charge card offers include the option of purchasing credit insurance coverage, which would make your monthly payments if you became disabled and could not work and/or would pay off the account balance if you died. Find out the details on at least one such offer, and discuss why you would or would not accept it.

52. Make a list of "special incentives" offered by bank cards you are familiar with, and briefly describe the pros and cons of each one.

53. One bank offered a card with a "low introductory rate" of 5.9%, good through the end of the year. And furthermore, you could receive back a percentage (up to 100%!) of all interest you pay, as shown in the table.

Use your card for:	2 years	5 years	10 years	15 years	20 years
Get back:	10%	25%	50%	75%	100%

(As soon as you take a refund, the time clock starts over.) Because you can eventually claim all your interest payments back, is this card a good deal? Explain why or why not.

54. Either recall a car-buying experience you have had, or visit a new-car dealer and interview a salesperson. Write a description of the procedure involved in purchasing a car on credit.

Comparing Bank Card Accounts *Daniel and Nora Onishi are considering two bank card offers that are the same in all respects except that Bank A charges no annual fee and charges monthly interest of* 1.18% *on the unpaid balance while Bank B charges a* $30 *annual fee and monthly interest of* 1.01% *on the unpaid balance. From their records, the Onishis have found that the unpaid balance they tend to carry from month to month is quite consistent and averages* $900.

55. Estimate their total yearly cost to use the card if they choose the card from
(a) Bank A **(b)** Bank B.

56. Which card is their better choice?

14.3 Truth in Lending

Annual Percentage Rate (APR) • Unearned Interest

Free consumer information is available in abundance from the Federal Trade Commission at www.ftc.gov. For example, you can find out about "Abusive Lending," "Credit," "Identity Theft," "Investments," "Privacy," and many other consumer topics. Consumers can register fraud complaints at this site using an online complaint form.

Annual Percentage Rate (APR)

The Consumer Credit Protection Act, which was passed in 1968, has commonly been known as the **Truth in Lending Act.** In this section we discuss two major issues addressed in the law:

1. How can I tell the true annual interest rate a lender is charging?
2. How much of the finance charge am I entitled to save if I decide to pay off a loan sooner than originally scheduled?

Question 1 above arose because lenders were computing and describing the interest they charged in several different ways. For example, how does 1.5% per month at Sears compare to 9% per year add-on interest at a furniture store? Truth in Lending standardized the so-called true annual interest rate, or **annual percentage rate,** commonly denoted **APR.** All sellers (car dealers, stores, banks, insurance agents, credit card companies, and the like) must disclose the APR when you ask, and the contract must state the APR whether or not you ask. This enables a borrower to more easily compare the true costs of different loans.

Theoretically, as a borrower you should not need to calculate APR yourself, but it is possible to verify the value stated by the lender if you wish. Since the formulas for finding APR are quite involved, it is easiest to use a table provided by the Federal Reserve Bank. We show an abbreviated version in Table 6 on the next page. It will identify APR values to the nearest half percent from 8.0% to 14.0%, which should do in most cases. (You should be able to obtain a more complete table from your local bank.) Our table is designed to apply to loans requiring *monthly* payments and extending over the most common lengths for consumer loans from 6 to 60 months. Apart from home mortgages, which are discussed in the next section, these conditions characterize most closed-end consumer loans.

Table 6 relates the following three quantities.

APR = true annual interest rate (shown across the top)

n = total number of scheduled monthly payments (shown down the left side)

h = finance charge per $100 of amount financed (shown in the body of the table)

TABLE 6 Annual Percentage Rate (APR) for Monthly Payment Loans

Number of Monthly Payments (n)	Annual Percentage Rate (APR)												
	8.0%	8.5%	9.0%	9.5%	10.0%	10.5%	11.0%	11.5%	12.0%	12.5%	13.0%	13.5%	14.0%
	(Finance charge per \$100 of amount financed) (h)												
6	\$2.35	\$2.49	\$2.64	\$2.79	\$2.94	\$3.08	\$3.23	\$3.38	\$3.53	\$3.68	\$3.83	\$3.97	\$4.12
12	4.39	4.66	4.94	5.22	5.50	5.78	6.06	6.34	6.62	6.90	7.18	7.46	7.74
18	6.45	6.86	7.28	7.69	8.10	8.52	8.93	9.35	9.77	10.19	10.61	11.03	11.45
24	8.55	9.09	9.64	10.19	10.75	11.30	11.86	12.42	12.98	13.54	14.10	14.66	15.23
30	10.66	11.35	12.04	12.74	13.43	14.13	14.83	15.54	16.24	16.95	17.66	18.38	19.10
36	12.81	13.64	14.48	15.32	16.16	17.01	17.86	18.71	19.57	20.43	21.30	22.17	23.04
48	17.18	18.31	19.45	20.59	21.74	22.90	24.06	25.23	26.40	27.58	28.77	29.97	31.17
60	21.66	23.10	24.55	26.01	27.48	28.96	30.45	31.96	33.47	34.99	36.52	38.06	39.61

Later in this section, we will give the formula connected with this table. As you will see, though, it expresses h in terms of n and APR and is *not* easily solved to express APR in terms of n and h. Hence we have provided the table for determining APR values.

EXAMPLE 1 Finding the APR for an Add-On Loan

Recall that Chris and Heather (in Example 1 of Section 14.2) paid \$700 down on a \$3500 purchase and agreed to pay the balance at a 7% add-on rate for 2 years. Find the APR for their loan.

SOLUTION

As shown previously, the total amount financed was

$$\text{Purchase price} - \text{down payment} = \$3500 - \$700 = \$2800.$$

The finance charge (interest) was

$$I = Prt = \$2800 \cdot .07 \cdot 2 = \$392.$$

Next find the finance charge per \$100 of the amount financed. To do this, divide the finance charge by the amount financed, then multiply by \$100.

$$\left(\begin{array}{c}\text{Finance charge per}\\ \text{\$100 financed}\end{array}\right) = \frac{\text{Finance charge}}{\text{Amount financed}} \cdot \$100$$

$$= \frac{\$392}{\$2800} \cdot \$100 = \$14$$

```
3500-700
                2800
2800*.07*2
                 392
(392/2800)*100
                  14
```

These are the computations required in the solution of Example 1.

This amount, \$14, represents h, the finance charge per \$100 of the amount financed. Because the loan was to be paid over 24 months, look down to the "24 monthly payments" row of Table 6 ($n = 24$). Then look across the table for the h-value closest to \$14.00, which is \$14.10. From that point, read up the column to find the APR, 13.0% (to the nearest half percent). In this case, a 7% add-on rate is equivalent to an APR of 13.0%.

EXAMPLE 2 Finding the APR for a Car Loan

After a down payment on your new car, you still owe \$7454. You agree to repay the balance in 48 monthly payments of \$185 each. What is the APR on your loan?

SOLUTION

First find the finance charge.

Total payments (48 · \$185); Amount financed (\$7454)

$$\text{Finance charge} = 48 \cdot \$185 - \$7454$$
$$= \$1426$$

Multiply first. Then subtract.

```
48*185-7454
        1426.00
(1426/7454)*100
          19.13
```

These are the computations required in the solution of Example 2.

Now find the finance charge per \$100 financed as in Example 1.

$$\begin{pmatrix}\text{Finance charge per} \\ \text{\$100 financed}\end{pmatrix} = \frac{\text{Finance charge}}{\text{Amount financed}} \cdot \$100$$
$$= \frac{\$1426}{\$7454} \cdot \$100 = \$19.13$$

Find the "48 payments" row of Table 6, read across to find the number closest to 19.13, which is 19.45. From there read up to find the APR, which is 9.0%. ■

Rights and responsibilities apply to all credit accounts, and the consumer should read all disclosures provided by the lender. *The Fair Credit Billing Act* and *The Fair Credit Reporting Act* regulate, among other things, procedures for billing and for disputing bills, and for providing and disputing personal information on consumers.

If you have ever applied for a charge account, a personal loan, insurance, or a job, then information about where you work and live, how you pay your bills, and whether you've been sued, arrested, or have filed for bankruptcy appears in the files of Consumer Reporting Agencies (CRAs), which sell that information to creditors, employers, insurers, and other businesses.

For more detailed information on these and many other consumer issues, you may want to consult the Web site www.consumeraction.gov.

Unearned Interest Question 2 at the beginning of the section arises when a borrower decides, for one reason or another, to pay off a closed-end loan earlier than originally scheduled. In such a case, it turns out that the lender has not loaned as much money for as long as planned and so he has not really "earned" the full finance charge originally disclosed.

If a loan is paid off early, the amount by which the original finance charge is reduced is called the **unearned interest.** We will discuss two common methods of calculating unearned interest, the **actuarial method** and the **rule of 78.** The Truth in Lending Act requires that the method for calculating this refund (or reduction) of finance charge be disclosed at the time the loan is initiated. Whichever method is used, the borrower may not, in fact, save all the unearned interest, since the lender is entitled to impose an **early payment penalty** to recover certain costs. A lender's intention to impose such a penalty in case of early payment also must be disclosed at initiation of the loan.

We will now state and illustrate both methods of computing unearned interest.

Unearned Interest—Actuarial Method

For a closed-end loan requiring *monthly* payments, which is paid off earlier than originally scheduled, let

R = regular monthly payment

k = remaining number of scheduled payments (*after* current payment)

h = finance charge per \$100, corresponding to a loan with the same APR and k monthly payments.

Then the **unearned interest, u,** is given by

$$u = kR\left(\frac{h}{\$100 + h}\right).$$

Once the unearned interest u is calculated (by any method), the amount required to actually pay off the loan early is easily found. It consists of the present regular payment due, plus k additional future payments, minus the unearned interest.

Payoff Amount

A closed-end loan requiring regular monthly payments R can be paid off early, along with the current payment. If the original loan had k additional payments scheduled (after the current payment), and the unearned interest is u, then, disregarding any possible prepayment penalty, the **payoff amount** is given by

$$\textbf{Payoff amount} = (k + 1)R - u.$$

EXAMPLE 3 Finding the Early Payoff Amount by the Actuarial Method

Suppose you get an unexpected pay raise and want to pay off your car loan of Example 2 at the end of 3 years rather than paying for 4 years as originally agreed. Find **(a)** the unearned interest (the amount you will save by retiring the loan early) and **(b)** the "payoff amount" (the amount required to pay off the loan at the end of 3 years).

SOLUTION

(a) From Example 2, recall that $R = \$185$ and APR $= 9.0\%$. The current payment is payment number 36, so $k = 48 - 36 = 12$. Use Table 6, with 12 payments and APR 9.0%, to obtain $h = \$4.94$. Finally, by the actuarial method formula,

```
12*185(4.94/(100
+4.94))
             104.51
(12+1)*185-104.5
1
            2300.49
```

These are the computations required in the solution of Example 3.

$$u = 12 \cdot \$185\left(\frac{\$4.94}{\$100 + \$4.94}\right) = \$104.51.$$

By the actuarial method, you will save \$104.51 in interest by retiring the loan early.

(b) The payoff amount is found by using the appropriate formula.

$$\text{Payoff amount} = (12 + 1)\$185 - \$104.51 = \$2300.49$$

The required payoff amount at the end of 3 years is \$2300.49.

The actuarial method, because it is based on the APR value, probably is the better method. However, some lenders still use the second method, the **rule of 78.**

Unearned Interest—Rule of 78

For a closed-end loan requiring *monthly* payments, which is paid off earlier than originally scheduled, let

F = original finance charge

n = number of payments originally scheduled

k = remaining number of scheduled payments (*after* current payment).

Then the **unearned interest,** u, is given by

$$u = \frac{k(k + 1)}{n(n + 1)} \cdot F.$$

EXAMPLE 4 Finding the Early Payoff Amount by the Rule of 78

Again assume that the loan in Example 2 is paid off at the time of the thirty-sixth monthly payment. This time use the rule of 78 to find **(a)** the unearned interest, and **(b)** the payoff amount.

SOLUTION

(a) From Example 2, the original finance charge is $F = \$1426$. Also, $n = 48$ and $k = 12$. By the rule of 78,

$$u = \frac{12(12 + 1)}{48(48 + 1)} \cdot \$1426 = \$94.58.$$

By the rule of 78, you will save \$94.58 in interest, which is \$9.93 *less* than your savings by the actuarial method.

(b) $$\text{Payoff amount} = (12 + 1)\$185 - \$94.58 = \$2310.42$$

The payoff amount is \$2310.42, which is \$9.93 *more* than the payoff amount calculated by the actuarial method in Example 3.

```
(12(12+1))/(48(4
8+1))*1426
           94.58
(12+1)*185-94.58

         2310.42
```

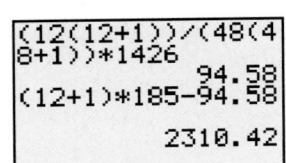

These are the computations required in the solution of Example 4.

When the rule of 78 was first introduced into financial law (by the Indiana legislature in 1935), loans were ordinarily written for 1 year or less, interest rates were relatively low, and loan amounts were less than they tend to be today. For these reasons the rule of 78 was acceptably accurate then. Today, however, with very accurate tables and/or calculators readily available, the rule of 78 is used much less often than previously.*

Suppose you want to compute unearned interest accurately (so you don't trust the rule of 78), but the APR value, or the number of scheduled payments, or the number of remaining payments (or at least one of the three) is not included in Table 6. Then what? Actually, in the actuarial method, you can evaluate h (the finance charge per \$100 financed) using the same formula that was used to generate Table 6.

Finance Charge per \$100 Financed

If a closed-end loan requires n equal monthly payments and APR denotes the true annual interest rate for the loan (as a decimal), then $\boldsymbol{h}$, the **finance charge per \$100 financed,** is given by

$$h = \frac{n \cdot \frac{\text{APR}}{12} \cdot \$100}{1 - \left(1 + \frac{\text{APR}}{12}\right)^{-n}} - \$100.$$

EXAMPLE 5 Finding Unearned Interest and Early Payoff Amount

Sergei borrowed \$4000 to pay for music equipment for his band. His loan contract states an APR of 9.8% and stipulates 28 monthly payments of \$160.39 each. Sergei decides to pay the loan in full at the time of his nineteenth payment. Find **(a)** the unearned interest, and **(b)** the payoff amount.

*For an excellent discussion of the shortcomings of the rule of 78, see Alonzo F. Johnson's article, "The Rule of 78: A Rule that Outlived Its Useful Life," in *Mathematics Teacher,* September 1988.

SOLUTION

(a) First find h from the finance charge formula just given.

$$h = \frac{9\left(\frac{.098}{12}\right)\$100}{1 - \left(1 + \frac{.098}{12}\right)^{-9}} - \$100 = \$4.13$$

Remember to use the remaining number of payments, 28 − 19 = 9, as the value of n.

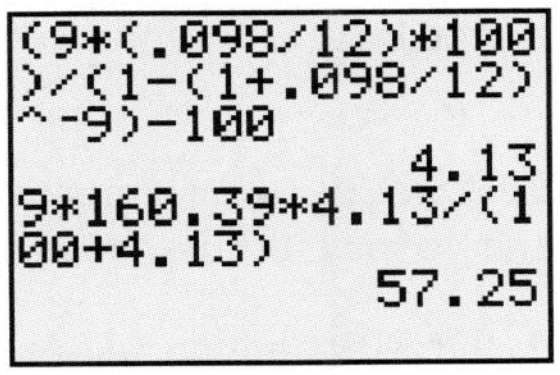

These are the computations required in the solution of Example 5(a).

Next use the actuarial formula for unearned interest.

Regular monthly payment:	$R = \$160.39$
Remaining number of payments:	$k = 28 - 19 = 9$
Finance charge per \$100:	$h = \$4.13$

$$u = 9 \cdot \$160.39 \cdot \frac{\$4.13}{\$100 + \$4.13} = \$57.25$$

The amount of interest Sergei will save is \$57.25.

(b) Payoff amount $= (9 + 1)(\$160.39) - \$57.25 = \$1546.65.$

To pay off the loan, Sergei must pay \$1546.65. ■

14.3 EXERCISES

Round all monetary answers to the nearest cent unless otherwise directed.

Finding True Annual Interest Rate *Find the APR (true annual interest rate), to the nearest half percent, for each loan.*

	Amount Financed	Finance Charge	Number of Monthly Payments
1.	\$1000	\$75	12
2.	\$1700	\$202	24
3.	\$6600	\$750	30
4.	\$5900	\$1150	48

Finding the Monthly Payment *Find the monthly payment for each loan.*

	Purchase Price	Down Payment	Finance Charge	Number of Monthly Payments
5.	\$3000	\$500	\$250	24
6.	\$4280	\$450	\$700	36
7.	\$3950	\$300	\$800	48
8.	\$8400	\$2500	\$1300	60

Finding True Annual Interest Rate *Find the APR (true annual interest rate), to the nearest half percent, for each loan.*

	Purchase Price	Down Payment	Add-on Interest Rate	Number of Payments
9.	\$4190	\$390	6%	12
10.	\$3250	\$750	7%	36
11.	\$7480	\$2200	5%	18
12.	\$12,800	\$4500	6%	48

Unearned Interest by the Actuarial Method *Each loan was paid in full before its due date.* **(a)** *Obtain the value of h from Table 6. Then* **(b)** *use the actuarial method to find the amount of unearned interest, and* **(c)** *find the payoff amount.*

	Regular Monthly Payment	APR	Remaining Number of Scheduled Payments After Payoff
13.	\$346.70	11.0%	18
14.	\$783.50	8.5%	12
15.	\$595.80	9.5%	6
16.	\$314.50	10.0%	24

Finding Finance Charge and True Annual Interest Rate *For each loan, find* **(a)** *the finance charge, and* **(b)** *the APR.*

17. Frank Barry financed a $1990 computer with 24 monthly payments of $91.50 each.

18. Pat Peterson bought a horse trailer for $5090. She paid $1240 down and paid the remainder at $152.70 per month for $2\frac{1}{2}$ years.

19. Mike Karelius still owed $2000 on his new garden tractor after the down payment. He agreed to pay monthly payments for 18 months at 6% add-on interest.

20. Phil Givant paid off a $15,000 car loan over 3 years with monthly payments of $487.54 each.

Comparing the Actuarial Method and the Rule of 78 for Unearned Interest *Each loan was paid off early. Find the unearned interest by* **(a)** *the actuarial method, and* **(b)** *the rule of 78.*

	Amount Financed	Regular Monthly Payments	Total Number of Payments Scheduled	Remaining Number of Scheduled Payments After Payoff
21.	$3310	$201.85	18	6
22.	$10,230	$277.00	48	12
23.	$29,850	$641.58	60	12
24.	$16,730	$539.82	36	18

Unearned Interest by the Actuarial Method *Each loan was paid in full before its due date.* **(a)** *Obtain the value of h from the appropriate formula. Then* **(b)** *use the actuarial method to find the amount of unearned interest, and* **(c)** *find the payoff amount.*

	Regular Monthly Payment	APR	Remaining Number of Scheduled Payments After Payoff
25.	$212	8.6%	4
26.	$575	9.33%	8

Comparing Loan Choices *Wei-Jen Luan needs to borrow* $5000 *to pay for Kings season tickets for her family. She can borrow the amount from a finance company (at* 6.5% *add-on interest for* 3 *years) or from the credit union (*36 *monthly payments of* $164.50 *each). Use this information for Exercises 27–30.*

27. Find the APR (to the nearest half percent) for each loan and decide which one is Wei-Jen's better choice.

28. Suppose Wei-Jen takes the credit union loan. At the time of her thirtieth payment she decides that this loan is a nuisance and decides to pay it off. If the credit union uses the rule of 78 for computing unearned interest, how much will she save by paying in full now?

29. What would Wei-Jen save in interest if she paid in full at the time of the thirtieth payment and the credit union used the actuarial method for computing unearned interest?

30. Under the conditions of Exercise 29, what amount must Wei-Jen come up with to pay off her loan?

31. Describe why, in Example 1, the APR and the add-on rate differ. Which one is more legitimate? Why?

Approximating the APR of an Add-On Rate *To convert an add-on interest rate to its corresponding APR, some people recommend using the formula*

$$APR = \frac{2n}{n+1} \cdot r,$$

where r is the add-on rate and n is the total number of payments.

32. Apply the given formula to calculate the APR (to the nearest half percent) for the loan of Example 1 ($r = .07$, $n = 24$).

33. Compare your APR value in Exercise 32 to the value of Example 1. What do you conclude?

The Rule of 78 with Prepayment Penalty *A certain retailer's credit contract designates the rule of 78 for computing unearned interest, and also imposes a "prepayment penalty." In case of any payoff earlier than the due date, they will charge an additional 10% of the original finance charge. Find the least value of k (remaining payments after payoff) that would result in any net savings in each case.*

34. 24 payments originally scheduled

35. 36 payments originally scheduled

36. 48 payments originally scheduled

The actuarial method of computing unearned interest assumes that, throughout the life of the loan, the borrower is paying interest at the rate given by APR for money actually being used by the borrower. When contemplating complete payoff along with the current payment, think of k future payments as applying to a separate loan with the same APR, and h being the finance charge per \$100 *of* that *loan. Refer to the following formula.*

$$u = kR\left(\frac{h}{\$100 + h}\right)$$

37. Describe in words the quantity represented by

$$\frac{h}{\$100 + h}.$$

38. Describe in words the quantity represented by kR.

39. Explain why the product of the two quantities above represents unearned interest.

Write out your response to each exercise.

40. Why might a lender be justified in imposing a prepayment penalty?

41. Discuss reasons that a borrower may want to pay off a loan early.

42. Find out what federal agency you can contact if you have questions about compliance with the Truth in Lending Act. (Any bank, or retailer's credit department, should be able to help you with this, or you could try a Web search.)

43. Study the following table, which pertains to a 12-month loan. The column 3 entries are designed so that they are in the same ratios as the column 2 entries but will add up to 1 because their denominators are all equal to

$$1 + 2 + 3 + 4 + 5 + \ldots + 12 = \frac{12 \cdot 13}{2} = 78.$$

(This is the origin of the term "rule of 78.")

Month	Fraction of Loan Principal Used by Borrower	Fraction of Finance Charge Owed
1	12/12	12/78
2	11/12	11/78
3	10/12	10/78
4	9/12	9/78
5	8/12	8/78
6	7/12	7/78
7	6/12	6/78
8	5/12	5/78
9	4/12	4/78
10	3/12	3/78
11	2/12	2/78
12	1/12	1/78
		78/78 = 1

Suppose the loan is paid in full after eight months. Use the table to determine the unearned fraction of the total finance charge.

44. Find the fraction of unearned interest of Exercise 43 by using the rule of 78 formula.

14.4 The Costs and Advantages of Home Ownership

Fixed-Rate Mortgages • Adjustable-Rate Mortgages • Closing Costs • Taxes, Insurance, and Maintenance

Heating a house is another cost that may get you involved with banks and interest rates after you finally get a roof over your head. The roof you see above does more than keep of the rain. It holds glass tube solar collectors, part of the solar heating system in the building.

Fixed-Rate Mortgages For many decades, home ownership has been considered a centerpiece of the "American dream." And since homes and condominiums have generally kept pace over the years with the rate of inflation, they have also been sound investments. For most people, a home represents the largest purchase of their lifetime, and it is certainly worth careful consideration.

A loan for a substantial amount, extending over a lengthy time interval (typically up to 30 years), for the purpose of buying a home or other property or real estate, and for which the property is pledged as security for the loan, is called a **mortgage.** (In some areas, a mortgage may also be called a **deed of trust** or a **security deed.**) The time until final payoff is called the **term** of the mortgage. The portion of the purchase price of the home which the buyer pays initially is called the **down payment.** The **principal amount of the mortgage** (the amount borrowed) is found by subtracting the down payment from the purchase price.

With a **fixed-rate mortgage,** the interest rate will remain constant throughout the term, and the initial principal balance, together with interest due on the loan, is repaid to the lender through regular (constant) periodic (we assume monthly) payments. This is called **amortizing** the loan. The regular monthly payment needed to amortize a loan depends on the amount financed, the term of the loan, and the interest rate, according to the following formula.

Regular Monthly Payment

The **regular monthly payment** required to repay a loan of P dollars, together with interest at an annual rate r, over a term of t years, is given by

$$R = \frac{P\left(\frac{r}{12}\right)}{1 - \left(\frac{12}{12 + r}\right)^{12t}}.$$

EXAMPLE 1 Using a Formula to Find a Monthly Mortgage Payment

The arrow indicates a payment of $612.81, supporting the result of Example 1.

Find the monthly payment necessary to amortize a $75,000 mortgage at 5.5% annual interest for 15 years.

SOLUTION

$$R = \frac{\$75{,}000\left(\frac{.055}{12}\right)}{1 - \left(\frac{12}{12 + .055}\right)^{(12)(15)}} = \$612.81$$

With a programmable or financial calculator, you can store the formula above and minimize the work. Another approach is to use a tool such as Table 7 on the next page, which gives payment values (per $1000 principal) for typical ranges of mortgage terms and interest rates. The entries in the table are given to five decimal places so that accuracy to the nearest cent can be obtained for most normal mortgage amounts.

EXAMPLE 2 Using a Table to Find a Monthly Mortgage Payment

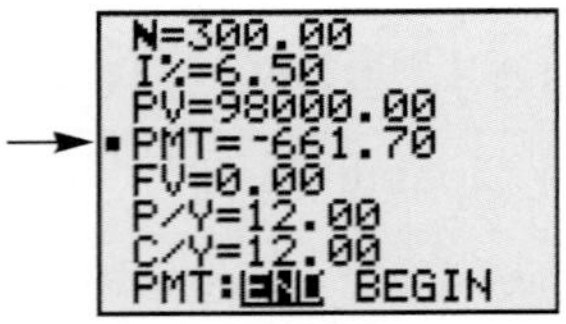

Under the conditions of Example 2, the monthly payment is $661.70. Compare with the table method.

Find the monthly payment necessary to amortize a $98,000 mortgage at 6.5% for 25 years.

SOLUTION

In Table 7, read down to the 6.5% row and across to the column for 25 years, to find the entry 6.75207. As this is the monthly payment amount needed to amortize a loan of $1000, and our loan is for $98,000, our required monthly payment is

$$98 \cdot \$6.75207 = \$661.70.$$

So that the borrower pays interest only on the money actually owed in a month, interest on real-estate loans is computed on the decreasing balance of the loan. Each equal monthly payment is first applied toward interest for the previous month. The remainder of the payment is then applied toward reduction of the principal amount owed. Payments in the early years of a real-estate loan are mostly interest (typically 80% or more); only a small amount goes toward reducing the principal. The amount of interest decreases each month, so that larger and larger amounts of the payment will apply to the principal. During the last years of the loan, most of the monthly payment is applied toward the principal. (See Table 9 on page 893.)

TABLE 7 Monthly Payments to Repay Principal and Interest on a $1000 Mortgage

Annual rate (r)	Term of Mortgage (Years) (t) 5	10	15	20	25	30
4.0%	$18.41652	$10.12451	$7.39688	$6.05980	$5.27837	$4.77415
4.5%	18.64302	10.36384	7.64993	6.32649	5.55832	5.06685
5.0%	18.87123	10.60655	7.90794	6.59956	5.84590	5.36822
5.5%	19.10116	10.85263	8.17083	6.87887	6.14087	5.67789
6.0%	19.33280	11.10205	8.43857	7.16431	6.44301	5.99551
6.5%	19.56615	11.35480	8.71107	7.45573	6.75207	6.32068
7.0%	19.80120	11.61085	8.98828	7.75299	7.06779	6.65302
7.5%	20.03795	11.87018	9.27012	8.05593	7.38991	6.99215
8.0%	20.27639	12.13276	9.55652	8.36440	7.71816	7.33765
8.5%	20.51653	12.39857	9.84740	8.67823	8.05227	7.68913
9.0%	20.75836	12.66758	10.14267	8.99726	8.39196	8.04623
9.5%	21.00186	12.93976	10.44225	9.32131	8.73697	8.40854
10.0%	21.24704	13.21507	10.74605	9.65022	9.08701	8.77572
10.5%	21.49390	13.49350	11.05399	9.98380	9.44182	9.14739
11.0%	21.74242	13.77500	11.36597	10.32188	9.80113	9.52323
11.5%	21.99261	14.05954	11.68190	10.66430	10.16469	9.90291
12.0%	22.24445	14.34709	12.00168	11.01086	10.53224	10.28613

Bay Area borrowers are embracing nontraditional loans. This shows the percentage of Bay Area loans that were interest-only or option ARMs.

Year	Interest-only	Option ARM*
2005	42.6%	29.1%
2004	43.7	9.6
2003	20.3	0.8
2002	12.0	1.7
2001	2.9	1.6
2000	1.5	5.5

*Also called negative amortization
Source: LoanPerformance

The table above appeared in the April 13, 2006, *San Francisco Chronicle*. The data are for the San Francisco Bay Area, but nationwide trends are similar. Reasons that borrowers are increasingly choosing these nontraditional mortgages, according to the article, include recent "soaring home prices, more flexible lending standards and a greater willingness by consumers to finance their lifestyle by borrowing against their home equity." (Your "equity" is the difference between what your house is worth and what you owe on it.)

In the long run, these trends may or may not hurt borrowers (as well as lenders), but such loans are clearly more likely to cause payment shock when principal payments begin or interest rates adjust upward.

Once the regular monthly payment has been determined, as in Examples 1 and 2, an **amortization schedule** (or **repayment schedule**) can be generated. It will show the allotment of payments for interest and principal, and the principal balance, for one or more months during the life of the loan. Tables showing these breakdowns are available from lenders, or can be produced on a computer spreadsheet. The following steps demonstrate how the computations work.

Step 1 $\text{Interest for the month} = \begin{pmatrix}\text{Old balance}\\ \text{of principal}\end{pmatrix}\begin{pmatrix}\text{Annual}\\ \text{interest rate}\end{pmatrix}\left(\frac{1}{12}\text{ year}\right)$

Step 2 $\text{Payment on principal} = \begin{pmatrix}\text{Monthly}\\ \text{payment}\end{pmatrix} - \begin{pmatrix}\text{Interest for}\\ \text{the month}\end{pmatrix}$

Step 3 $\text{New balance of principal} = \begin{pmatrix}\text{Old balance}\\ \text{of principal}\end{pmatrix} - \begin{pmatrix}\text{Payment on}\\ \text{principal}\end{pmatrix}$

This sequence of steps is done for the end of each month. The new balance obtained in Step 3 becomes the Step 1 old balance for the next month.

Homes and condominiums are sometimes available for a very low down payment from the inventory of HUD (the Department of Housing and Urban Development). For current information on these opportunities, consult your local real estate companies.

EXAMPLE 3 Preparing an Amortization Schedule

The Petersons paid for many years on a \$60,000 mortgage with a term of 30 years and a 4.5% interest rate. Prepare an amortization schedule for the first 2 months of their mortgage.

SOLUTION

First get the monthly payment. We use Table 7. (You could also use the formula.)

Mortgage amount in \$1000s ↓ ; Intersection of 4.5% row with 30-year column in Table 7 ↓

$$R = 60 \cdot \$5.06685 = \$304.01$$

Now apply Steps 1 through 3.

Step 1 Interest for the month $= \$60{,}000(.045)(\frac{1}{12}) = \$225$

Step 2 Payment on principal $= \$304.01 - \$225 = \$79.01$

Step 3 New balance of principal $= \$60{,}000 - \$79.01 = \$59{,}920.99$

Starting with an old balance of \$59,920.99, repeat the steps for the second month.

Step 1 Interest for the month $= \$59{,}920.99(.045)(\frac{1}{12}) = \$224.70$

Step 2 Payment on principal $= \$304.01 - \$224.70 = \$79.31$

Step 3 New balance of principal $= \$59{,}920.99 - \$79.31 = \$59{,}841.68$

These calculations are summarized in Table 8.

TABLE 8 Amortization Schedule

Payment Number	Interest Payment	Principal Payment	Balance of Principal
			\$60,000.00
1	\$225.00	\$79.01	\$59,920.99
2	\$224.70	\$79.31	\$59,841.68

Prevailing mortgage interest rates have varied considerably over the years. Table 9 on the next page shows portions of the amortization schedule for the Petersons' loan of Example 3, and also shows what the corresponding values would have been had their interest rate been 14.5%. (Rates that high have not been seen for many years.) Notice how much interest is involved in this home mortgage. At the (low) 4.5% rate, \$49,444.03 in interest was paid along with the \$60,000 principal. At a rate of 14.5%, the interest alone would total the huge sum of \$204,504.88, which is about 3.4 times greater than the mortgage principal.

TABLE 9 Amortization Schedules for a $60,000, 30-Year Mortgage

4.5% Interest Monthly Payment: $304.01				14.5% Interest Monthly Payment: $734.73			
Payment Number	**Interest Payment**	**Principal Payment**	**Balance of Principal**	**Payment Number**	**Interest Payment**	**Principal Payment**	**Balance of Principal**
Initially →			60,000.00	Initially →			60,000.00
1	225.00	79.01	59,920.99	1	725.00	9.73	59,990.27
2	224.70	79.31	59,841.68	2	724.88	9.85	59,980.42
3	224.41	79.60	59,762.08	3	724.76	9.97	59,970.45
12	221.68	82.33	59,032.06	12	723.63	11.11	59,875.11
60	205.47	98.54	54,694.75	60	714.96	19.77	59,149.53
175	152.47	151.54	40,506.65	175	656.05	78.69	54,214.82
176	151.90	152.11	40,354.54	176	655.10	79.64	54,135.18
236	113.60	190.41	30,102.65	236	571.02	163.72	47,092.76
237	112.88	191.13	29,911.52	237	569.04	165.70	46,927.07
240	110.73	193.28	29,333.83	240	562.96	171.78	46,417.87
303	59.33	244.68	15,575.66	303	368.65	366.09	30,142.47
304	58.41	245.60	15,330.05	304	364.22	370.51	29,771.96
359	2.27	301.74	302.88	359	17.44	717.29	725.96
360	1.14	302.88	0.00	360	8.77	725.96	0.00
Totals:	49,444.03	60,000.00		**Totals:**	204,504.88	60,000.00	

Adjustable-Rate Mortgages The lending industry uses many variations on the basic fixed-rate mortgage to cut down on the risk to lenders and also to make home loans accessible to more people. An **adjustable-rate mortgage (ARM),** also known as a **variable-rate mortgage (VRM),** generally starts out with a lower rate than similar fixed-rate loans, but the rate changes periodically, reflecting changes in prevailing rates. To understand how ARMs work, you must know about *indexes, margins, discounts, caps, negative amortization,* and *convertibility.*

Your ARM interest rate may change every 1, 3, or 5 years (occasionally more frequently). The frequency of change in rate is called the **adjustment period.** For example, an ARM with an adjustment period of 1 year is called a 1-year ARM. When the rate changes, your payment normally changes also. These adjustments are caused by fluctuations in an **index** upon which your rate is based. A variety of indexes are used, including the 1-, 3-, and 5-year U.S. Treasury security rates. Also used are national and regional "cost of funds" indexes.

To determine your interest rate, the lender will add to the applicable index a few percentage points called the **margin.** The index and the margin are both important in determining the cost of the loan.

EXAMPLE 4 Comparing ARM Payments Before and After a Rate Adjustment

Suppose you pay \$20,000 down on a \$180,000 house and take out a 1-year ARM for a 30-year term. The lender uses the 1-year Treasury index (presently at 4%) and a 2% margin.

(a) Find your monthly payment for the first year.
(b) Suppose that after a year the 1-year Treasury index has increased to 5.1%. Find your monthly payment for the second year.

SOLUTION

(a) First find the amount of your mortgage:

$$\text{Mortgage amount} = \underset{\text{Cost of house}}{\$180{,}000} - \underset{\text{Down payment}}{\$20{,}000} = \$160{,}000.$$

Your first-year interest rate will be

$$\text{ARM interest rate} = \text{Index rate} + \text{Margin} = 4\% + 2\% = 6\%.$$

Now from Table 7 (using 6% over 30 years) we obtain 5.99551, so

$$\text{First-year monthly payment} = 160 \cdot \$5.99551 = \$959.28.$$

(b) During the first year, a small amount of the mortgage principal has been paid, so in effect we will now have a new "mortgage amount." (Also, the term will now be 1 year less than the original term.) The amortization schedule for the first year (not shown here) yields a loan balance, after the twelfth monthly payment, of \$158,035.19.

Also, for the second year,

$$\text{ARM interest rate} = \text{Index rate} + \text{Margin} = 5.1\% + 2\% = 7.1\%.$$

Because 7.1% is not included in Table 7, we resort to the regular monthly payment formula, using the new mortgage balance and also using 29 years for the remaining term.

$$\text{Second-year monthly payment} = \frac{P\left(\frac{r}{12}\right)}{1 - \left(\frac{12}{12 + r}\right)^{12t}} \quad \text{Regular payment formula}$$

$$= \frac{\$158{,}035.19\left(\frac{.071}{12}\right)}{1 - \left(\frac{12}{12 + .071}\right)^{(12)(29)}} \quad \text{Substitute known values.}$$

$$= \$1072.74$$

The first ARM interest rate adjustment has caused the second-year monthly payment to rise to \$1072.74, which is an increase of \$113.46 over the initial monthly payment. ■

Sometimes initial rates offered on an ARM are less than the sum of the index and the margin. Such a **discount** often is the result of the seller (a new home builder, for example) paying the lender an amount in order to reduce the buyer's initial interest rate. This arrangement is called a "seller buydown." Discounts result in lower initial monthly payments, but often are combined with higher initial fees, or even an increase in the price of the house. Also, when the discount expires, you may experience sizable increases in your monthly payments.

EXAMPLE 5 Discounting a Mortgage Rate

In Example 4, suppose that a seller buydown discounts your initial (first-year) rate by 1.5%. Find your first-year and second-year monthly payments.

SOLUTION

The 4% index rate is discounted to 2.5%. Adding the 2% margin yields a net first-year rate of 4.5% (rather than the 6% of Example 4), so the Table 7 entry is 5.06685, and

$$\text{First-year monthly payment} = 160(\$5.06685) = \$810.70.$$

The amortization schedule shows a balance at the end of the first year of \$157,418.79. The discount now expires, and the index has increased to 5.1%, so for the second year,

$$\text{ARM interest rate} = \text{Index rate} + \text{Margin} = 5.1\% + 2\% = 7.1\%.$$

(This is just as in Example 4.) Using the monthly payment formula, with $r = 7.1\%$ and $t = 29$,

$$\text{Second-year monthly payment} = \frac{\$157{,}418.79\left(\frac{.071}{12}\right)}{1 - \left(\frac{12}{12 + .071}\right)^{(12)(29)}} = \$1068.55.$$

Substitute values in the monthly payment formula.

The initial monthly payment of \$810.70 looks considerably better than the \$959.28 of Example 4, but at the start of year two, monthly payments jump by \$257.85. Such an increase may cause you to experience "payment shock."

To reduce the risk of payment shock, be sure the ARM you obtain includes **cap** features. An **interest rate cap** limits the amount your interest rate can increase. A **periodic cap** limits how much the rate can increase from one adjustment period to the next (typically about 1% per 6 months or 2% per year). An **overall cap** limits how much the rate can increase over the life of the loan (typically about 5% total). Most all ARMs must have overall rate caps by law and many also have periodic caps. Some ARMs also have **payment caps,** which limit how much the *payment* can increase at each adjustment. However, since such a cap limits the payment but not the interest rate, it is possible that your payment may not cover even the interest owed. In that case you could find yourself owing more principal at the end of an adjustment period than you did at the beginning. This situation is called **negative amortization.** To make up the lost interest, the lender will increase your payments more in the future, or may extend the term of the loan.

Another way to reduce risk is with a **convertibility** feature. This allows you, at certain points in time, to elect to convert the mortgage to a fixed-rate mortgage. If you elect to convert, your fixed rate will be determined by prevailing rates at the time of conversion. If you think you may want to completely, or partially, pay off the principal of the loan ahead of schedule, try to negotiate an ARM with no **prepayment penalty.** This means that you will be able to make early principal payments without paying any special fees, and your early payment(s) will go totally toward reducing the principal.

Refinancing means initiating a new home mortgage and paying off the old one. In times of relatively low rates, lenders often encourage homeowners to refinance. Because setting up a new loan will involve costs, make sure you have a good reason before refinancing. Some possible reasons:

1. The new loan may be comparable to the current loan with a rate low enough to recoup the costs in a few years.
2. You may want to pay down your balance faster by switching to a shorter term loan.
3. You may want to borrow out equity to cover other major expenses, such as education for your children.
4. Your present loan may have a large balloon feature which necessitates refinancing.
5. You may be uncomfortable with your variable (hence uncertain) ARM and want to switch to a fixed-rate loan.

(*Source:* www.business-to-business-resources.com)

Closing Costs Besides the down payment and the periodic costs of home ownership, including principal and interest payments, there are some significant one-time expenses that apply to both fixed-rate and adjustable-rate mortgages and are paid when the mortgage is originally set up. Lumped together, these charges are called **closing costs** (or **settlement charges**). The **closing,** or **settlement,** occurs when all details of the transaction have been determined and the final contracts are signed. It is the formal process by which the ownership title is transferred from seller to buyer. Within 3 days after your loan application, the lender is required to deliver or mail to you certain materials that include a "good faith estimate" of your closing costs. That is the time to ask questions, clarify all proposed costs, and, if you wish, shop around for alternative providers of settlement services. At the time of the closing, it will likely be too late to avoid paying for services already provided. We illustrate typical closing costs in the following example.

EXAMPLE 6 Computing Total Closing Costs

For a $58,000 mortgage, the borrower was charged the following closing costs.

Loan origination fee (1% of mortgage amount)	$____
Broker loan fee	1455
Lender document and underwriting fees	375
Lender tax and wire fees	205
Fee to title company	200
Title insurance fee	302
Title reconveyance fee	65
Document recording fees	35

Compute the total closing costs for this mortgage.

SOLUTION

"Loan origination fees" are commonly referred to as **points.** Each "point" amounts to 1% of the mortgage amount. By imposing points, the lender can effectively raise the interest rate without raising monthly payments (because points are normally paid at closing rather than over the life of the loan). In this case, "one point" translates to $580. Adding this to the other amounts listed gives total closing costs of $3217. ■

Not all mortgage loans will involve the same closing costs. You should seek to know just what services are being provided and whether their costs are reasonable. There is considerable potential for abuse in the area of closing costs. Therefore, in 1974 Congress passed the **Real Estate Settlement Procedures Act (RESPA)** to standardize Truth in Lending practices in the area of consumer mortgages. The lender is required to provide you with clarifying information, including the APR for your loan. The APR, the cost of your loan as a yearly rate, will probably be higher than the stated interest rate since it must reflect all points and other fees paid directly for credit as well as the actual interest.

Taxes, Insurance, and Maintenance The primary financial considerations for most new homeowners are the following.

1. Accumulating the down payment
2. Having sufficient cash and income to qualify for the loan
3. Making the mortgage payments

Here we consider three additional ongoing expenses, taxes, insurance, and maintenance, all of which should be anticipated realistically. They can be significant, especially if you have stretched your resources just to purchase the house.

Property taxes are collected by your county or other local government. Depending on your location and the value of your home, they can range up to several thousand dollars annually. A more positive fact is that property taxes, and also mortgage interest, are deductible on your income taxes. Therefore, money expended for those items will decrease what you pay in income taxes. This is how the government, through the tax code, encourages home ownership, just as saving for retirement is encouraged. (See the Extension on *Annuities* earlier in this chapter.)

EXAMPLE 7 Taking Taxes into Account in Home Ownership

Mr. and Mrs. Waite, early in their working careers, are in a 30% combined state and federal income tax bracket. They have calculated that they can afford a net average monthly expenditure of $1400 for a home. (They have already taken into account their current condominium rent that they would save, and utilities would be about the same in a home as what they pay presently.) Can they afford the home of their dreams, which would require a 20-year, $205,000 fixed-rate mortgage at 6% plus $1920 in annual property taxes?

SOLUTION

Let's "do the math."

$$\text{Regular monthly mortgage payment} = \underset{\substack{\uparrow \\ \text{Mortgage amount in \$1000s}}}{205} \cdot \underset{\substack{\uparrow \\ \text{Value from Table 7}}}{\$7.16431} \approx \$1469$$

$$\text{Monthly property taxes} = \frac{\$1920 \leftarrow \text{Annual taxes}}{12 \leftarrow \text{Months per year}} = \$160$$

$$\text{Total monthly expense} = \underset{\substack{\uparrow \\ \text{Mortgage payment}}}{\$1469} + \underset{\substack{\uparrow \\ \text{Property taxes}}}{\$160} = \$1629$$

Because $\$1629 > \$1400$, it seems that the Waites cannot afford this home. But wait—remember that mortgage interest and property taxes are both income tax deductible. Let's consider further.

$$\text{Monthly interest} = \$205{,}000\,(.06)\left(\frac{1}{12}\right) \qquad \text{Interest} = Prt$$

$$= \$1025$$

Of the $1469 mortgage payment, $1025 would be interest (initially).

$$\text{Monthly deductible expenses} = \underset{\substack{\uparrow \\ \text{Mortgage interest}}}{\$1025} + \underset{\substack{\uparrow \\ \text{Property tax}}}{\$160} = \$1185$$

Government-backed mortgages, including FHA (Federal Housing Administration) and VA (Veterans Administration) loans, carry a government guarantee to protect the lender in case the borrower fails to repay the loan. Those who do not qualify for these loans, and obtain a conventional loan instead, usually are required to buy private mortgage insurance (PMI) as part of the loan package.

This feature can be a surprise to first-time home purchasers. It was introduced to protect lenders but indirectly protects the buyers, who may lose a bundle if some catastrophe makes it impossible to make the payments.

Over the second half of the twentieth century the amount of credit life insurance in force (for the purpose of covering consumer debt) increased by over 500 times, while group life insurance increased by only about 200 times, ordinary life insurance increased by about 60 times, and industrial life insurance actually decreased. (*Source:* American Council of Life Insurance.)

$$\text{Monthly income tax savings} = \underset{\text{Deductible expense}}{\$1185} \cdot \underset{\text{Tax bracket}}{30\%} \approx \$356$$

$$\text{Net monthly cost of home} = \underset{\text{Gross cost}}{\$1629} - \underset{\text{Tax savings}}{\$356} = \$1273$$

By considering the effect of taxes, we see that the net monthly cost, \$1273, is indeed within the Waites' \$1400 affordability limit. ■

Homeowner's insurance usually covers losses due to fire, storm damage, and other casualties. Some types, such as earthquake or hurricane coverage, could be unavailable or very expensive, depending on your location. Also, all homes require **maintenance**, but these costs can vary greatly, depending mainly on the size, construction type, age, and condition of the home. These expenses are necessary in order to protect the investment you made in your home, and to keep it comfortable and attractive. They are not generally tax deductible.

EXAMPLE 8 Including Insurance and Maintenance in Home Costs

In moving ahead with their home purchase, the Waites (Example 7) estimate that homeowner's insurance will be about \$550 per year, and that maintenance will be about \$650 per year. Will these additional expenses mean that they cannot afford the home?

SOLUTION

These items carry no tax advantage (they are not deductible), so the added monthly expense is simply

$$\text{Monthly insurance and maintenance expense} = \frac{\$550 + \$650}{12} = \$100.$$

With insurance and maintenance included, we obtain

$$\text{Adjusted monthly expense} = \$1273 + \$100 = \$1373,$$

which is still less than \$1400. The Waites can still afford the purchase. ■

Payments for property taxes and homeowner's insurance are commonly made from a **reserve account** (also called an **escrow** or an **impound account**) maintained by the mortgage lender. The borrower must pay enough each month, along with amortization costs, so that the reserve account will be sufficient to make the payments when they come due. (In some cases, the homeowner pays taxes directly to the taxing authority and insurance premiums directly to an insurance company, totally separate from mortgage payments to the lender.)

Some reasonable questions may have occurred to you concerning Examples 7 and 8.

1. Is it wise to buy a house that puts you so close to your spending limit, in light of future uncertainties?

2. What about the fact that the interest portion of mortgage payments, and therefore the tax savings, will decrease over time?
3. Won't taxes, insurance, and maintenance costs likely increase over time, making it more difficult to keep up the payments?

Some possible responses follow.

1. The Waites may have built in sufficient leeway when they decided on their $1400 per month allowance.
2. (a) Look again at Table 9 to see how slowly the interest portion drops.
   (b) Most people find that their income over time increases faster than expenses for a home that carries a fixed-rate mortgage. (A variable-rate mortgage is more risky and should have its initial rate locked in for as long as possible.)
   (c) Prevailing interest rates rise and fall over time. While rising rates will not affect your fixed-rate mortgage, falling rates may offer the opportunity to reduce your mortgage expenses by refinancing. (See the margin note on refinancing for guidelines.)
3. Here again, increases in income will probably keep pace. While most things, including personal income, tend to follow inflation, the fixed-rate mortgage insures that mortgage amortization, a major item, will stay constant.

14.4 Exercises

Round all monetary answers to the nearest cent unless directed otherwise.

Monthly Payment on a Fixed-Rate Mortgage *Find the monthly payment needed to amortize principal and interest for each fixed-rate mortgage. You can use either the regular monthly payment formula or Table 7, as appropriate.*

	Loan Amount	Interest Rate	Term
1.	$70,000	10.0%	20 years
2.	$50,000	11.0%	15 years
3.	$57,300	8.7%	25 years
4.	$85,000	7.9%	30 years

	Loan Amount	Interest Rate	Term
5.	$227,750	12.5%	25 years
6.	$95,450	15.5%	5 years
7.	$132,500	7.6%	22 years
8.	$205,000	5.5%	10 years

Amortization of a Fixed-Rate Mortgage *Complete the first one or two months (as required) of each amortization schedule for a fixed-rate mortgage.*

9. Mortgage: $58,500
   Interest rate: 10.0%
   Term of loan: 30 years

Amortization Schedule

Payment Number	Total Payment	Interest Payment	Principal Payment	Balance of Principal
1	(a) ______	(b) ______	(c) ______	(d) ______

10. Mortgage: $87,000
    Interest rate: 8.5%
    Term of loan: 20 years

Amortization Schedule

Payment Number	Total Payment	Interest Payment	Principal Payment	Balance of Principal
1	(a) ______	(b) ______	(c) ______	(d) ______

11. Mortgage: $143,200
Interest rate: 6.5%
Term of loan: 15 years

Amortization Schedule

Payment Number	Total Payment	Interest Payment	Principal Payment	Balance of Principal
1	(a) ________	(b) ________	(c) ________	(d) ________
2	(e) ________	(f) ________	(g) ________	(h) ________

12. Mortgage: $124,750
Interest rate: 9%
Term of loan: 25 years

Amortization Schedule

Payment Number	Total Payment	Interest Payment	Principal Payment	Balance of Principal
1	(a) ________	(b) ________	(c) ________	(d) ________
2	(e) ________	(f) ________	(g) ________	(h) ________

13. Mortgage: $113,650
Interest rate: 8.2%
Term of loan: 10 years

Amortization Schedule

Payment Number	Total Payment	Interest Payment	Principal Payment	Balance of Principal
1	(a) ________	(b) ________	(c) ________	(d) ________
2	(e) ________	(f) ________	(g) ________	(h) ________

14. Mortgage: $150,000
Interest rate: 6.25%
Term of loan: 16 years

Amortization Schedule

Payment Number	Total Payment	Interest Payment	Principal Payment	Balance of Principal
1	(a) ________	(b) ________	(c) ________	(d) ________
2	(e) ________	(f) ________	(g) ________	(h) ________

Finding Monthly Mortgage Payments *Find the total monthly payment, including taxes and insurance, on each mortgage.*

	Mortgage	Interest Rate	Term of Loan	Annual Taxes	Annual Insurance
15.	$ 62,300	7%	20 years	$610	$220
16.	$ 51,800	10%	25 years	$570	$145
17.	$ 89,560	6.5%	10 years	$915	$409
18.	$ 72,890	5.5%	15 years	$1850	$545
19.	$115,400	8.8%	20 years	$1295.16	$444.22
20.	$128,100	11.3%	30 years	$1476.53	$565.77

Comparing Total Principal and Interest on a Mortgage *Suppose* $140,000 *is owed on a house after the down payment is made. The monthly payment for principal and interest at* 8.5% *for* 30 *years is* $140 \cdot \$7.68913 = \$1076.48$.

21. How many monthly payments will be made?

22. What is the total amount that will be paid for principal and interest?

23. The total interest charged is the total amount paid minus the amount financed. What is the total interest?

24. Which is more—the amount financed or the total interest paid? By how much?

Long-Term Effect of Interest Rates *You may remember seeing home mortgage interest rates fluctuate widely in a period of not too many years. The following exercises show the effect of changing rates. Refer to Table 9, which compared the amortization of a* $60,000, *30-year mortgage for rates of* 4.5% *and* 14.5%. *Give values of each of the following for* **(a)** *a* 4.5% *rate, and* **(b)** *a* 14.5% *rate.*

25. monthly payments

26. percentage of first monthly payment that goes toward principal

27. balance of principal after 1 year

28. balance of principal after 20 years

29. the first monthly payment that includes more toward principal than toward interest

30. amount of interest included in final payment

The Effect of the Term on Total Amount Paid *Suppose a* $60,000 *mortgage is to be amortized at* 7.5% *interest. Find the total amount of interest that would be paid for each term.*

31. 10 years

32. 20 years

33. 30 years

34. 40 years

The Effect of Adjustable Rates on the Monthly Payment *For each adjustable-rate mortgage, find* **(a)** *the initial monthly payment,* **(b)** *the monthly payment for the second adjustment period, and* **(c)** *the change in monthly payment at the first adjustment. (The "adjusted balance" is the principal balance at the time of the first rate adjustment. Assume no caps apply.)*

	Beginning Balance	Term	Initial Index Rate	Margin	Adjustment Period	Adjusted Index Rate	Adjusted Balance
35.	$75,000	20 years	6.5%	2.5%	1 year	8.0%	$73,595.52
36.	$44,500	30 years	7.2%	2.75%	3 years	6.6%	$43,669.14

The Effect of Rate Caps on Adjustable-Rate Mortgages *Jeff Walenski has a 1-year ARM for* $50,000 *over a 20-year term. The margin is* 2% *and the index rate starts out at* 7.5% *and increases to* 10.0% *at the first adjustment. The balance of principal at the end of the first year is* $49,119.48. *The ARM includes a periodic rate cap of* 2% *per adjustment period. (Use this information for Exercises 37–40.)*

37. Find **(a)** the interest owed and **(b)** the monthly payment due for the first month of the first year.

38. Find **(a)** the interest owed and **(b)** the monthly payment due for the first month of the second year.

39. What is the monthly payment adjustment at the end of the first year?

40. If the index rate has dropped slightly at the end of the second year, will the third-year monthly payments necessarily drop? Why or why not?

Closing Costs of a Mortgage *For Exercises 41–44, refer to the following list of closing costs for the purchase of a* $175,000 *house requiring a* 20% *down payment, and find each requested amount.*

Title insurance premium	$240
Document recording fee	30
Loan fee (two points)	____
Appraisal fee	225
Prorated property taxes	685
Prorated fire insurance premium	295

41. the mortgage amount

42. the loan fee

43. the total closing costs

44. the total amount of cash required of the buyer at closing (including down payment)

Consider the scenario of Example 7. Recalling that mortgage interest is income tax deductible, find (to the nearest dollar) the additional initial net monthly savings resulting from each strategy. (In each case, only the designated item changes. All other features remain the same.)

45. Change the mortgage term from 20 years to 30 years.

46. Change the mortgage from fixed at 6% to an ARM with an initial rate of 5%.

In each case below, find all of the following quantities for a \$200,000 *fixed-rate mortgage. (Give answers to the nearest dollar.)*
(a) *Monthly mortgage payment (principal and interest)*
(b) *Monthly house payment (including property taxes and insurance)*
(c) *Initial monthly interest*
(d) *Income tax deductible portion of initial house payment*
(e) *Net initial monthly cost for the home (considering tax savings)*

	Term of Mortgage	Interest Rate	Annual Property Tax	Annual Insurance	Owner's Income Tax Bracket
47.	15 years	5.5%	\$960	\$480	20%
48.	20 years	6.0%	\$840	\$420	25%
49.	10 years	6.5%	\$1092	\$540	30%
50.	30 years	7.5%	\$1260	\$600	40%

On the basis of material in this section, or your own research, give brief written responses to each problem.

51. Give other ways the Waites (Examples 7 and 8) could possibly decrease the initial net monthly payments for their home.

52. Suppose your ARM allows conversion to a fixed-rate loan at each of the first five adjustment dates. Describe circumstances under which you would want to convert.

53. Describe each type of mortgage.

Graduated payment
Balloon payment
Interest-only
Option ARM ("neg-am")

54. Should a home buyer always pay the smallest down payment that will be accepted? Explain.

55. Should a borrower always choose the shortest term available in order to minimize the total interest expense? Explain.

56. Under what conditions would an ARM probably be a better choice than a fixed-rate mortgage?

57. Why are second-year monthly payments (slightly) less in Example 5 than in Example 4 even though the term, 29 years, and the interest rate, 7.1%, are the same in both cases?

58. Do you think that the discount in Example 5 actually makes the overall cost of the mortgage less? Explain.

59. Discuss the term "payment shock" mentioned at the end of Example 5.

60. Find out what is meant by each term and describe some of the features of each.

FHA-backed mortgage
VA-backed mortgage
Conventional mortgage

14.5 Financial Investments

Stocks • Bonds • Mutual Funds • Evaluating Investment Returns • Building a Nest Egg

Stocks In a general sense, an *investment* is a way of putting resources to work so that (hopefully) they will grow. A variety of "consumer" investments, from cash to real estate to collectibles, or even your own business, will be compared later in this section (in For Further Thought: Summary of Investments). But our main emphasis will be on the basic mathematical aspects of a restricted class of financial investments, namely, *stocks, bonds,* and *mutual funds.*

Buying stock in a corporation makes you a part owner of the corporation. You then share in any profits the company makes, and your share of profits is called a **dividend.** If the company prospers (or if increasing numbers of investors believe it will prosper in the future), your stock will be attractive to others, so that you may, if you wish, sell your shares at a profit. The profit you make by selling for more than you paid is called a **capital gain.** A negative gain, or **capital loss,** results if you sell for less than you paid. By **return on investment,** we mean the net difference between what you receive (including your sale price and any dividends received) and what you paid (your purchase price plus any other expenses of buying and selling the stock).

EXAMPLE 1 Finding the Return on a Stock Ownership

Brenda Cochrane bought 100 shares of stock in Company A on January 15, 2006, paying \$30 per share. On January 15, 2007, she received a dividend of 50¢ per share, and the stock price had risen to \$30.85 per share. (Ignore any costs other than the purchase price.) Find the following.

(a) Brenda's total cost for the stock
(b) The total dividend amount
(c) Brenda's capital gain if she sold the stock on January 15, 2007
(d) Brenda's total return on her one year of ownership of this stock
(e) The percentage return

SOLUTION

(a) $\text{Cost} = (\$30 \text{ per share}) \cdot (100 \text{ shares}) = \$3000$

She paid \$3000 total.

(b) $\text{Dividend} = (\$.50 \text{ per share}) \cdot (100 \text{ shares}) = \$50$

The dividend amount was \$50.

(c) $\text{Capital gain} = \underbrace{(\$30.85 - \$30.00)}_{\text{Change in price per share}} \cdot \underbrace{100}_{\text{Number of shares}} = \$.85 \cdot 100 = \$85$

The capital gain was \$85.

(d) $\text{Total return} = \underbrace{\$50}_{\text{Dividend}} + \underbrace{\$85}_{\text{Capital gain}} = \$135$

Total return on the investment was \$135.

(e) $$\text{Percentage return} = \frac{\text{Total return}}{\text{Total cost}} \cdot 100\% = \frac{\$135}{\$3000} \cdot 100\% = 4.5\%$$

The percentage return on the investment was 4.5%.

The price of a share of stock is determined by the law of supply and demand at institutions called **stock exchanges.** In the United States, the oldest and largest exchange is the New York Stock Exchange (NYSE), established in 1792 and located on Wall Street in New York City. Members of the public buy their stock through stockbrokers, people who have access to the exchange. Stockbrokers charge a fee for buying or selling stock.

The **NASDAQ (National Association of Securities Dealers Advanced Quotations),** unlike the NYSE and other exchanges which actually carry out trading at specific locations, is an electronic network of brokerages established in 1971 to facilitate the trading of over-the-counter (unlisted) stocks. Most of the high-tech company stocks, which grew very rapidly as a group in the 1980s and especially the 1990s (and many of which declined as rapidly in 2000–2002), are (or were) listed on the Nasdaq market.

Current stock prices can be found in many daily newspapers or by consulting an online computer service that provides this information. Table 10 on the next page shows only a small portion of the activities on the New York Stock Exchange for a single day of trading.

The headings at the tops of the columns of Table 10 indicate the meanings of the numbers across the row. For example, the information for Albertsons (with stock symbol ABS) appears as follows.

YTD %CHG	52-WEEK HI	LO	STOCK (SYM)	DIV	YLD %	PE	VOL 100s	CLOSE	NET CHG
6.8	36.99	26.88	Albertsons **ABS**	.76	2.3	27	13617	33.62	0.01

The number 6.8 shows that the share value of Albertsons has increased 6.8% so far this calendar year. The next two numbers show that \$36.99 is the highest price that the stock has reached in the last 52 weeks, while \$26.88 is the lowest. Following the company symbol is .76, showing that the company is currently paying an annual dividend of 76¢ per share of stock. As mentioned earlier, a dividend is money paid by a company to owners of its stock. Dividends generally go up when the company is doing well and go down when business is bad. Then comes 2.3, the **current yield** on the company's stock, in percent. The dividend of 76¢ per share represents 2.3% of the current purchase price of the stock.

The 27 after the 2.3 is the **price-to-earnings ratio,** the current purchase price per share divided by the earnings per share over the past 12 months. After 27 is 13617, the number of shares sold that day, in hundreds. On the day in question, a total of $13{,}617 \cdot 100 = 1{,}361{,}700$ shares of Albertsons stock were sold. The next number in the listings, 33.62, shows that the stock closed for the day at \$33.62 per share. That is, the final trade in Albertsons stock, prior to the closing of the market at 4 P.M. Eastern time, was executed at that price. The final number shown, 0.01, indicates that this day's closing price was 1¢ higher than that of the previous trading day.

EXAMPLE 2 Reading the Stock Table

Use the stock table (Table 10) to find the required amounts below.

(a) The highest price for the last 52 weeks for Alcoa (AA)
(b) The dividend for Pioneer-Standard Electronics (PIOS)

SOLUTION

(a) Find the correct line from the stock table.

1.5	45.71	27.36	Alcoa **AA**	.60	1.7	46	33587	36.07	−0.18

The highest price for this stock for the last 52 weeks (52-WEEK HI) is \$45.71.

(b) The dividend appears just after the company symbol. From the table, find the dividend for Pioneer-Standard Electronics, .12, or 12¢ per share. (Notice the special symbol ♣ preceding the company name in the listing. This means that Pioneer-Standard Electronics is one of a number of companies whose annual reports can be obtained free by readers of the *Wall Street Journal.*)

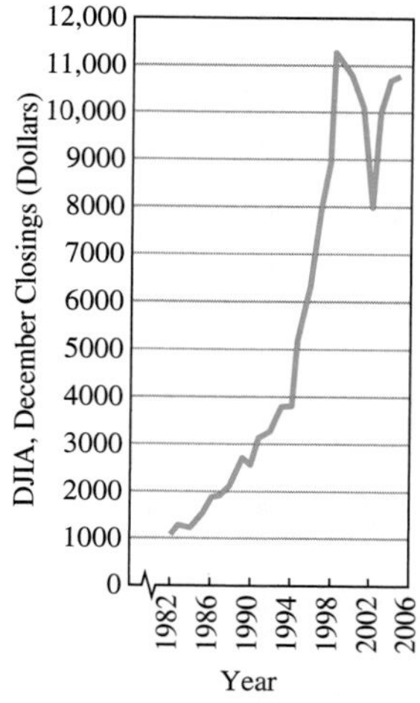

The **Dow Jones Industrial Average** (**DJIA,** or simply the "**Dow**") has been the most popular measure of general stock price trends since its inception in 1896. A significant time of increasing prices is a **bull market,** while an extended period of decreasing prices is a **bear market.** The Dow is a composite of 30 stocks from various sectors of the economy, which are changed occasionally to keep the list representative of the overall economy.

Stock price trends have historically been upward but with lots of variation within that general pattern. A strong bull market that began in August of 1982 saw the Dow triple over 5 years, until October 19, 1987 (known as Black Monday), when the Dow lost 22.6% of its value in a single day. The bull soon regained strength and mostly continued through the 1990s. Then another bear market set in. Starting in 2000, the Dow declined for 3 years in a row, a very rare occurrence historically.

(Source: *The World Book Year Books,* 1983–2006)

TABLE 10 Selected Stock Quotes

YTD % CHG	52-WEEK HI	52-WEEK LO	STOCK (SYM)	DIV	YLD %	PE	VOL 100s	CLOSE	NET CHG
4.3	39.80	18	AgilentTch A		...	dd	24333	29.75	-1.30
46.8	15.80	7.70	♣AgnicoEgl AEM	.02g	.1	dd	4609	14.49	-0.20
0.1	20.60	14.40	AgreeRlty ADC	1.84	9.9	10	115	18.50	-0.06
-7.0	12.20	8.72	Agrium AGU	.11g	1.1	dd	4229	9.86	0.09
-23.4	31.95	21.75	Ahold AHO	.65e	2.9	...	1969	22.52	0.02
4.7	11.05	4.70	AirNetSys ANS		...	18	73	8.63	-0.02
11.3	53.52	32.25	AirProduct APD	.80	1.5	22	6726	52.19	-0.04
▲ 55.4	23.10	7	AirborneInc ABF	.16	.7	cc	4386	23.05	0.15
11.3	20.74	9.20	Airgas ARG		...	26	3201	16.83	0.11
▼ -36.4	11.20	4.01	Airlease FLY	.44a	10.6	69	502	4.16	-0.09
-9.5	12.25	2.60	AirTranHldg AAI		...	dd	1732	5.97	0.23
10.9	17.15	12.60	AlamoGp ALG	.24	1.5	15	113	15.80	0.05
-52.6	20	3	Alamosa APS		...	dd	584	5.65	0.17
-0.5	33.90	17.40	AlaskaAir ALK		...	dd	2160	28.94	0.09
18.4	30.65	14.18	♣AlbanyInt AIN	.15e	.6	27	1650	25.70	0.05
30.7	32.60	16.50	Albemarle ALB	.52	1.7	23	1304	31.36	-1.16
26.8	57.91	37.35	AlbertoCl ACV	.36	.6	27	2872	56.72	-0.47
30.5	51.95	31.70	AlbertoCl A ACVA	.36	.7	24	517	51	-0.48
6.8	36.99	26.88	Albertsons ABS	.76	2.3	27	13617	33.62	0.01
10.6	48.75	28	Alcan AL	.60g	1.5	dd	32069	39.74	0.96
-22.7	33.20	10.53	Alcatel ADS ALA	.14e	1.1	...	7869	12.80	0.07
1.5	45.71	27.36	Alcoa AA	.60	1.7	46	33587	36.07	-0.18
12.9	39.30	29.90	Alcon ACL n		...	...	7187	38.40	1.40
19.1	68	55.75	Alexanders ALX		...	55	1	67.75	0.17
9.4	46.98	35.86	AlexREEq ARE	2.00f	4.4	26	300	44.98	-0.01
0.9	219.84	176.47	Allghny Y s	stk	...	dd	83	190.30	0.05
-2.3	55.09	31.89	AllghnyEngy AYE	1.72	4.9	10	13036	35.40	0.13
4.0	21.07	12.50	AllghnyTch ATI	.80	4.6	dd	2639	17.42	0.26
-26.6	15	5.66	AllenTele ALN		...	dd	664	6.24	-0.06
-20.2	93.30	56.06	♣Allergan AGN x	.36	.6	37	7596	59.87	0.06
19.4	31.10	21.14	Allete ALE	1.10	3.7	17	1850	30.08	-0.13
-6.7	15.06	13.33	AllianceCal. AKP n	.91	6.5	...	34	14	...
-8.9	54.32	38.60	AllncCapMgt AC	2.64e	6.0	16	1326	44.03	0.17
21.5	26.20	11.05	AllianceData ADS n		...	...	549	23.27	0.52
4.9	14.40	9.85	Alliance AIQ n		...	31	109	12.80	...
-9.3	15.05	12.90	AllianceNa AFB n	.16e	1.2	...	432	13.61	0.01
-6.9	15.11	13.24	AllianceNY AYN n	.15e	1.1	...	87	13.96	0.12
-9.1	32.29	27.50	AlliantEngy LNT	2.00	7.2	...	2241	27.61	-0.24
38.4	115.40	56.07	AlliantTech ATK s		...	31	4258	106.85	-1.15
-0.8	30.20	18	Allianz ADS AZ	.13e	.6	...	704	23.42	-0.18
0.0	29	20	♣AlldCap ALD	2.20f	8.5	12	8286	25.99	-0.25
▲ 22.9	27.70	16.80	AlldIrishBk ADS AIB	.77e	2.7	...	1485	28.40	0.74
-18.1	19.90	8.90	AlldWaste AW		...	...	10393	11.52	-0.03
6.4	57.50	36.70	AllmericaFnl AFC	.25	.5	99	1376	47.42	0.02
16.0	45.90	30	Allstate ALL	.84f	2.1	26	18794	39.08	-0.32
-18.8	65.15	46.74	Alltel AT	1.36	2.7	18	17160	50.12	1.06
▲ 1.0	50	47.95	Alltel un n		...	...	9441	50.25	0.75
134.0	37.50	10.70	Alltrista ALC		...	dd	366	36.74	0.74
-28.2	**32.47**	**13.75**	**Alpharma A ALO**	**.18**	**.9**	**dd**	**4863**	**18.98**	**1.29**
-34.1	2.15	1.10	AlpineGp AGI		...	dd	2	1.12	-0.03
10.3	30.30	10.35	Alstom ADS ALS	.47e	3.8	...	140	12.52	0.05
16.1	24.70	16.67	AluCpChina ACH n	.21p	...	...	5	20.30	-0.05
13.5	66.50	42.20	♣AmbacFnl ABK	.36	.5	16	5436	65.70	0.22
-0.7	25.95	24.26	♣Ambac AKB	1.77	7.1	...	385	25.07	-0.03
0.6	26.10	23.80	♣Ambac 7% wi AFK n	.80e	3.1	...	195	25.40	0.01
-4.8	25.80	13.20	AmBev ADS pfC ABV	.32e	1.7	...	6013	19.32	-0.30
-8.9	9.25	4.75	AmcastInd AIZ	.56	11.4	dd	257	4.90	-0.05
-41.3	57.34	16.35	AmdocTr AAE	1.51	8.6	...	1040	17.62	0.27
-47.5	66.50	16.11	Amdocs DOX		...	57	20114	17.82	-0.03
30.2	90.40	53.75	AmerHess AHC	1.20	1.5	10	10071	81.38	-0.06
2.7	45.48	36.53	Ameren AEE	2.54	5.8	13	4478	43.43	-0.12
3.6	27.45	25.50	Ameren ACES AEEE n		...	...	119	27	-0.04
4.5	9.59	7	♣AmFstMtg MFA	1.20	13.1	...	647	9.14	0.01
-3.5	23	11.75	AmMovil ADS AMX	.09e	.5	...	9682	18.80	0.26
1.7	10.71	1.45	AmWstHldg B AWA		...	dd	2536	3.56	0.05
69.3	36.20	10.15	AmerAxle AXL		...	14	2476	36.19	0.54
3.1	51.20	39.70	♣AEP AEP	2.40	5.4	16	7929	44.86	-0.85
25.7	46.55	24.20	AmExpress AXP	.32	.7	44	44846	44.87	0.38
13.0	30.75	18.35	AmFnl AFG	.50	1.8	79	1270	27.74	-0.26
64.8	23.80	9.95	AmGreetgs AM	.40	1.8	dd	9438	22.71	0.18
▼ -17.0	87.10	66	AmIntGp AIG	.17	.3	...	161049	65.90	-2.17
7.4	14.14	11.68	AmLandLse ANL	1.00	7.1	29	109	14.07	0.02
61.0	20.60	4.80	AmMedSecGp AMZ		...	20	1308	20.05	-0.05
12.8	10.30	8.29	AmRE Ptnrs ACP s		...	8	75	9.70	0.30
-23.9	13	6.40	AmRltyInv ARL	j	...	6	34	7.51	0.11
138.1	6	1.60	AmResdntInv INV	j	...	dd	227	5	-0.10
15.1	4.15	1	AmRetire ACR		...	...	192	2.75	0.10
4.2	10.35	8.08	AmSftyIns ASI	.24e	2.6	9	7	9.25	0.02
▲ 13.2	77.73	51.24	AmStandard ASD		...	20	5376	77.24	-0.31
8.7	42	29.70	AmStateWtr AWR	1.30	3.4	18	306	38	0.50
-58.0	27.48	3.50	AmTower A AMT		...	dd	16194	3.98	-0.10
3.5	44.33	28.81	♣AmWtrWks AWK	.98	2.3	28	3276	43.21	-0.04
23.6	64.90	14	♣AmeriCredit ACF		...	11	14745	39.01	-0.99
3.0	25.35	18.05	Amerigas APU	2.20	9.5	31	318	23.05	0.12
19.4	79.70	51.06	♣AmeriSrcBrg ABC	.08e	.1	30	11825	75.86	0.63
5.7	80	47.50	AmeronInt AMN	1.28	1.8	11	61	73.11	0.56

YTD % CHG	52-WEEK HI	52-WEEK LO	STOCK (SYM)	DIV	YLD %	PE	VOL 100s	CLOSE	NET CHG
-52.8	14	3.51	ParametTch PMTC		...	dd	63445	3.69	0.01
65.1	4.36	1.12	Paravant PVAT		...	39	906	3.88	-0.07
2.0	20	10.14	Parexel PRXL		...	70	740	14.63	0.50
12.0	21.48	17.06	ParkBcp PFED	.48	2.4	...	8	19.90	-0.30
▲ 80.8	5.90	1.65	ParkOH PKOH		...	dd	223	5.75	0.10
11.7	29.51	13.18	ParkerVision PRKR		...	dd	118	23.46	-0.09
-19.4	16.80	8.55	Parlex PRLX		...	dd	43	12.70	-0.05
2.4	3.53	1.56	ParluxFrag PARL		...	dd	161	1.90	0.07
-20.0	15.35	2.70	PrthusTchi		...	dd	32	4	...
-27.0	7.67	3.85	♣PrtnrComm PTNR		...	...	878	5	0.05
13.4	16.41	13.50	PtnrsTr PRTR n		...	...	252	15.90	0.12
46.6	1.79	0.25	Partsbase PRTS		...	dd	71	1.07	-0.05
124.8	18.30	4.80	PartyCity PCTY n		...	23	593	16.88	-0.07
-5.8	25.96	18.90	Pathmark PTMK		...	dd	546	23.22	0.33
-12.8	10.27	5.50	Pathmark wt		...	...	2	7.41	-0.27
26.8	9.94	5.45	PatrickInd PATK	.16	1.8	dd	3	9	-0.12
32.4	15	8.46	PatriotBkCp PBIX	.40f	2.8	13	116	14.10	-0.19
49.7	35	15	PatriotTrns PATR		...	35	1	29.37	1.12
12.5	47.48	30.26	PattrsnDentl PDCO		...	35	1123	46.06	0.06
37.5	**35.65**	**11.06**	**PattersnUtiEngy PTEN**		**...**	**19**	**14575**	**32.04**	**-1.98**
-32.6	5.30	0.14	PaulaFnl PFCO	.12j	...	dd	285	0.91	0.11
5.3	43.49	28.27	Paychex PAYX	.44	1.2	50	33503	36.69	1.13
15.5	**28.70**	**12**	**PayPal PYPL n**		**...**	**...**	**6058**	**23.20**	**-1.40**
-4.7	**5.10**	**1.38**	**PCMall MALL**		**...**	**11**	**802**	**3.87**	**-0.23**
-46.7	**22.61**	**9.65**	**PDF Sol PDFS n**		**...**	**dd**	**1088**	**11.20**	**-0.65**
4.0	8.50	5	PeakInt PEAK		...	dd	71	7.80	-0.02
13.4	14.10	5.63	PediatricSvc PSAI n		...	...	344	9.48	0.18
-5.8	23.50	11.40	PeerlsMfg PMFG s	.06	.4	9	87	17	...
56.0	2.27	0.78	PeerlessSys PRLS		...	dd	759	1.95	0.10
38.3	16.70	6.01	PeetsCof&Tea PEET		...	71	861	15.60	...
-81.0	23.04	1.65	PegasComm A PGTV		...	...	12579	1.98	-0.04
32.0	21.22	7.67	PegasusSol PEGS		...	dd	1393	18.75	-0.43
93.1	9.50	2.11	Pegasys PEGA		...	19	910	8.40	0.24
48.3	30.71	9.50	PemcoAviatn PAGI		...	7	91	23.50	-0.94
-78.0	**18.55**	**1.31**	**Pemstar PMTR**		**...**	**dd**	**14316**	**2.64**	**0.18**
34.1	17.60	8.94	Penford PENX	.24	1.4	cc	47	16.70	-0.25
18.9	41.77	12.95	PennNtl PENN		...	21	2973	36.08	-0.66
85.3	11	3.50	PennTraf PNFT		...	dd	303	9.82	0.22
-0.9	26.10	24.55	PennFed pf PFSBP	2.23	8.9	...	4	25.15	-0.05
10.6	27.70	18.25	PennFedFnl PFSB	.24	.9	15	117	27.45	-0.05
12.6	32.40	19.50	Pennichuck PNNW s	.78	2.6	19	115	30.40	-0.05
▲ 36.0	29.66	16.81	♣PennRkFnl PRFS	.84	2.8	15	75	30.24	0.59
-3.5	35.50	31.15	PnsWdsBcp PWOD n	1.01e	3.0	...	9	33.77	0.47
-0.7	21.51	12.07	PenwstPharm PPCO		...	dd	540	19.91	0.66
11.2	18.93	14.70	PeopBcp IN PFDC	.60	3.4	12	15	17.90	0.15
48.8	28.33	15.20	PeopBcp OH PEBO s	.60b	2.2	14	97	27.30	0.06
23.5	20	13.20	PeopBcpNC PEBK	.40b	2.3	14	15	17.72	0.22
▲ 31.2	27.82	20.66	PeopBkCT PBCT	1.44f	5.2	21	1465	27.90	0.32
▲ 13.7	11.25	9.10	PeopSdnyFnl PSFC	.36	3.0	26	25	11.82	0.72
-40.1	51	15.78	Peoplesoft PSFT		...	...	72883	24.08	-0.30
13.2	13.45	5.83	PerSeTch PSTI		...	cc	639	12.17	0.16
40.9	1.98	0.87	Perceptron PRCP		...	dd	76	1.86	0.16
-89.1	33.55	0.73	PeregSys PRGN		...	dd	103056	1.61	0.16
8.0	39.21	21.81	PerfrmFdGp PFGC		...	32	2409	38	0.28
-37.7	17.65	6.25	PerfrmncTech PTIX		...	24	787	8.30	-0.18
-1.6	18.88	10.80	PericmSemi PSEM		...	dd	1464	14.27	0.18
2.7	**18.30**	**10.56**	**Perrigo PRGO**		**...**	**17**	**4926**	**12.14**	**-0.75**
▲ 34.9	13.05	5.10	PerryEllis PERY		...	13	155	12.81	-0.19
-35.5	2.08	0.15	PrstncSftwr PRSW		...	dd	100	0.80	0.02
53.8	**5.24**	**1.08**	**PervasvSftwr PVSW**		**...**	**20**	**6572**	**4.49**	**0.93**
6.7	25.75	19	Petco PETC		...	...	356	23.40	-0.10
3.2	8.85	4.38	♣PetrlDev PETD		...	9	101	6.37	-0.01
22.2	8.99	3.95	PetrqstEngy PQUE		...	33	584	6.50	0.23
57.0	16.60	4.09	PETsMART PETM		...	50	7448	15.45	0.20
▲ 10.3	14.96	12	PFS Bcp Inc PBNC n	.05p	...	...	7	15	0.04
-20.7	38.36	19.40	PharmPdtDev PPDI		...	69	9466	25.61	0.10
-17.4	25.25	9.70	♣Phrmacopia PCOP		...	dd	1822	11.48	0.18
-46.7	34.25	5.01	Phrmacyclcs PCYC		...	dd	569	5.30	0.08
5.9	11.57	5.05	Pharmanetics PHAR		...	dd	373	7.68	0.18
-52.6	3.74	0.16	PharmChem PCHM		...	dd	44	0.45	-0.15
12.4	44.29	24.35	♣PhlConsHldg PHLY		...	24	435	42.37	1.41
-29.5	3.56	0.63	PhlpSvcs PSCD		...	dd	85	1.03	-0.02
15.6	15.63	8.35	PhnxTch PTEC		...	...	521	13.46	-0.04
-19.5	5.40	0.49	PhotoMedex PHMD		...	dd	417	1.49	-0.06
-4.4	54.50	20.06	♣PhotonDyn PHTN		...	dd	6779	43.65	0.90
-9.1	**35.57**	**16.85**	**Photronics PLAB**		**...**	**cc**	**56238**	**28.50**	**-4.74**
19.7	18.25	11.85	PHSB Fin PHSB	.32	2.2	17	14	14.30	0.02
-45.4	4.90	0.50	Physiomtrx PHYX		...	dd	210	1.19	0.09
-87.9	8.39	0.03	PinaclHldg BIGT		...	dd	16510	0.04	-0.02
▲ **33.1**	**10.17**	**2.15**	**PinnacleSys PCLE**		**...**	**dd**	**30907**	**10.57**	**0.92**
-2.1	15.50	7.40	♣PionrStdElec PIOS	.12	1.0	dd	1719	12.43	0.53
-21.9	24.70	2.36	♣Pivotal PVTL		...	dd	1381	4.80	0.06
20.9	46.49	30.38	Pixar PIXR		...	61	2156	43.49	0.22
-29.1	35.74	7.81	Pixelwrks PXLW		...	dd	4084	11.38	-0.07
14.2	32	13.80	♣PlanarSys PLNR		...	26	2324	24.10	-0.35
-22.0	27.30	11.17	PLATO Learn TUTR s		...	45	2226	12.95	-0.03

EXAMPLE 3 Finding the Cost of a Stock Purchase

Find the cost for 100 shares of Allstate (ALL) stock, purchased at the closing price for the day.

SOLUTION

From Table 10, the closing price for the day for Allstate was \$39.08 per share.

$$\text{Total cost} = \underset{\text{Price per share}}{\$39.08} \cdot \underset{\text{Number of shares}}{100} = \$3908.00$$

One hundred shares of this stock would cost \$3908.00, plus any broker's fees.

On a given day's listings, individual stocks may carry a variety of special features. These are indicated with designated symbols which are explained where the listings are published. A few of the more prominent features are shown here.

Some Special Features in Stock Listings

1. A line listed in boldface print distinguishes a stock whose price changed by 5% or more (as long as the previous day's closing price was at least \$2).
2. An underlined row in the listings indicates a stock with an especially high volume of trading for the day relative to its usual volume. In the case of the NYSE listings and the Nasdaq National Market, the 40 largest volume percentage leaders are indicated in this way.
3. A "pf" between the company name and symbol indicates a "preferred" stock, a special category of stock issued by some companies in addition to their ordinary, or "common," stock. A company must pay dividends to holders of its preferred stock before it can pay dividends to holders of its common stock. Additional advantages of preferred stocks include attractive yields, stable dividend income, and tax advantages for corporate investors.
4. The symbol ▲ at the left indicates a new 52-week high, while the symbol ▼ indicates a new 52-week low.

To buy or sell a stock, it is generally necessary to use a broker (although a growing number of companies will sell their shares directly to investors, which saves brokerage expenses). A broker has representatives at the exchange who will execute a buy or sell order. The broker will charge a **commission** (the broker's fee) for executing an order. Before 1974, commission rates were set by stock exchange rules and were uniform from broker to broker. Since then, however, rates are competitive and vary considerably among brokers.

Full-service brokers, who offer research, professional opinions on buying and selling individual issues, and various other services, tend to charge the highest commissions on transactions they execute. Many **discount brokers,** on the other hand, merely buy and sell stock for their clients, offering little in the way of additional services.

During the 1990s, when it often seemed that almost any reasonable choice of stock would pay off for almost any investor, a number of online companies appeared, offering much cheaper transactions than were available through conventional brokers.

Subsequently, most conventional brokerage firms, as well as the major discount brokerages, introduced their own automated (online) services. Investors today should compare a number of options before deciding which brokerage can best provide the stock investment services they need.

Commissions will normally be basically some percentage of the value of a purchase or sale (called the **principal**). Some firms will charge additional amounts in certain cases. For example, rates may depend on whether the order is for a **round lot** of shares (a multiple of 100) or an **odd lot** (any portions of an order for fewer than 100 shares). On the odd-lot portion, you may be charged an **odd-lot differential** (say, 10¢ per share). Also, it is possible to place a **limit order,** where the broker is instructed to execute a buy or sell if and when a stock reaches a predesignated price. An extra fee may be added to the commision on a limit order. One prominent discount brokerage does not charge these special fees but uses a tiered commission structure depending on the principal amount of the purchase or sale and also distinguishes between **broker-assisted** and **automated** trades, as shown here.

Typical Discount Commission Structure

Broker-Assisted Trade

Principal Amount	Commission
Up to \$2499.99	\$35 + 1.7% of principal
\$2500.00–\$6249.99	\$65 + .66% of principal
\$6250.00–\$19,999.99	\$76 + .34% of principal
\$20,000.00–\$49,999.99	\$100 + .22% of principal
\$50,000.00–\$499,999.99	\$155 + .11% of principal
\$500,000.00 or more	\$255 + .09% of principal

Automated Trade

Number of Shares	Commission
Up to 1000	\$29.95
More than 1000	3¢ per share

Also, the Securities and Exchange Commission (SEC), a federal agency that regulates stock markets, supports its own activities by charging the exchanges, based on volume of transactions. Typically this charge is passed on to investors, through brokers, in the form of an **SEC fee,** assessed on stock sales only (not purchases), say 3.01¢ per \$1000 of principal (rounded *up* to the next cent). For example, to find the fee for a sale of \$1600, first divide \$1600 by \$1000, then multiply by 3.01¢.

$$\frac{\$1600}{\$1000} \cdot 3.01¢ = 4.816¢, \quad \text{which is rounded to } 5¢.$$

EXAMPLE 4 Finding Total Cost for a Broker-Assisted Stock Purchase

Find the total cost (including expenses) for a broker-assisted purchase of 765 shares of Air Lease stock at \$4.22 per share. Use the typical discount commission structure outlined above.

SOLUTION

$$\text{Basic cost} = \underset{\text{Price per share}}{\$4.22} \cdot \underset{\text{Number of shares}}{765} = \$3228.30$$

Because this principal amount falls in the second tier of the commission structure, the broker's commission is \$65 plus .66% of this amount, or

$$\text{Broker's commission} = \$65 + .0066 \cdot \$3228.30 = \$65 + \$21.31 = \$86.31.$$

Multiply first, then add.

$$\text{Total cost} = \underset{\text{Basic cost}}{\$3228.30} + \underset{\text{Commission}}{86.31} = \$3314.61$$

The total cost for this purchase is \$3314.61.

EXAMPLE 5 Finding Proceeds for an Automated Stock Sale

Find the amount received by a person executing an automated sale of 1500 shares of ParkerVision (PRKR) at \$23.46 per share.

SOLUTION

The basic price of the stock sold is

$$(\$23.46 \text{ per share}) \cdot (1500 \text{ shares}) = \$35,190.00.$$

Because more than 1000 shares are sold, the commission is

$$(3¢ \text{ per share}) \cdot (1500 \text{ shares}) = \$45.00.$$

Find the SEC fee as described before Example 4.

$$\frac{\$35,190.00}{\$1000} \cdot 3.01¢ = 105.9219¢$$

Round up to obtain an SEC fee of \$1.06. Then

$$\text{Amount received} = \underset{\text{Basic price}}{\$35,190.00} - \underset{\text{Commission}}{\$45.00} - \underset{\text{SEC fee}}{\$1.06} = \$35,143.94.$$

The seller receives \$35,143.94.

Bonds Rather than contributing your capital (money) to a company, you may prefer merely to *lend* money to the company, receiving an agreed-upon rate of interest for the use of your money. In this case you would buy a bond from the company rather than purchase stock. The bond (loan) is issued with a stated term (life span), after which the bond "matures" and the principal (or **face value**) is paid back to you. Over the term, the company pays you a fixed rate of interest, which depends upon prevailing interest rates at the time of issue (and to some extent on the company's credit rating), rather than on the underlying value of the company.

A corporate bondholder has no stake in company profits, but is (quite) certain to receive timely interest payments. The potential return is generally less, but so is the risk of loss. If a company is unable to pay both bond interest and stock dividends, the bondholders must be paid first.

EXAMPLE 6 Finding the Return on a Bond Investment

Chris Campbell invests \$10,000 in a 5-year corporate bond paying 6% annual interest, paid semiannually. Find the total return on this investment, assuming Chris holds the bond to maturity (for the entire 5 years). Ignore any broker's fees.

SOLUTION

By the simple interest formula (see Section 14.1),

$$I = Prt = \$10{,}000(.06)(5) = \$3000.$$

Notice that this amount, \$3000, is the *total* return, over the 5-year period, not the annual return. Also, the fact that interest is paid twice per year (3% of the face value each time) has no bearing here since compounding does not occur. ■

Historically the overall rate of return over most time periods has been greater for stocks than for bonds. This fact (along with the general appeal of ownership) has led some investors to avoid bonds altogether. On the other hand, there are fewer uncertainties involved with bonds, and a bond investor's capital is subject to less risk. These facts can make bonds more attractive than stocks, especially for retired investors who depend on a steady stream of income from their investments. Many investors, however, seek to build a balanced portfolio that includes both stocks and bonds.

An abundance of information about **mutual funds** is available on the Internet. For example, the Investment Company Institute (ICI), the national association of the American investment company industry, provides material at the site www.ici.org/.

Mutual Funds A **mutual fund** is a pool of money collected by an investment company from many individuals and/or institutions and invested in many stocks, bonds, or money market instruments (perhaps various combinations of all of these). The mutual fund industry has expanded tremendously. By the late 1990s approximately \$5 trillion was invested in more than 7000 separate funds. By 2000 over 50 million U.S. households held mutual funds, and the largest category of these holdings was within retirement accounts.

Mutual Funds Versus Individual Stocks and Bonds

Advantages of Mutual Funds

1. *Simplicity*

   Let someone else do the work.

2. *Diversification*

   Being part of a large pool makes it easier to own interests in many different stocks, which decreases vulnerability to large losses in one particular stock or one particular sector of the economy.

3. *Access to New Issues*

   Initial public offerings (IPOs) can sometimes be very profitable. Whereas individual investors find it difficult to get in on these, the large overall assets of a mutual fund give its managers much more clout.

4. *Economies of Scale*

   Large stock purchases usually incur smaller expenses per dollar invested. In addition, the expenses are shared by large numbers of investors.

(*continued*)

5. *Professional Management*

   Professional managers may have the expertise to pick stocks and time purchases to better achieve the stated objectives of the fund.

6. *Indexing*

   With minimal management, an index fund can maintain a portfolio that mimics a popular index (such as the S&P 500). This makes it easy for individual fund investors to achieve returns at least close to those of the index.

Disadvantages of Mutual Funds

1. *Impact of One-Time Charges and Recurring Fees*

   Sales charges, management fees, "12b-1" fees (used to pay sales representatives), and fund expenses can mount and can be difficult to identify. And paying management fees does not guarantee getting *quality* management.

2. *Hidden Cost of Brokerage*

   Recurring fees and expenses are added to give the *expense ratio* of a fund. But commissions paid by the fund for stock purchases and sales are in addition to the expense ratio. (Since 1995, funds have been required to disclose their average commission costs in their annual reports.)

3. *Some Hidden Risks of Fund Ownership*
   (a) In the event of a market crash, getting out of the fund may mean accepting securities rather than cash, and these may be difficult to redeem for a fair price.
   (b) Managers may stray from the stated strategies of the fund.
   (c) Since tax liability is passed on to fund investors, and depends on purchases and sales made by fund managers, investors may be unable to avoid inheriting unwanted tax basis.

Source: Forbes Guide to the Markets, John Wiley & Sons, Inc., 1999.

New ways of providing financial services have emerged in the last several years. As of the turn of the century, American consumers were spending somewhere around $350 billion annually on fees and commissions for banking, insurance, and securities brokerage services. And sweeping legislation in late 1999 repealed many legal and regulatory barriers between companies operating in those three areas.

One result of the new laws was an increased number of company mergers as businesses set themselves up to offer one-stop shopping for insurance, investments, and banking. (*Source: The Sacramento Bee,* November 13, 1999.)

Because a mutual fund owns many different stocks, each share of the fund owns a fractional interest in each of those companies. In an "open-end" fund (by far the most common type), new shares are created and issued to buyers while the fund company absorbs the shares of sellers. Each day, the value of a share in the fund (called the *net asset value,* or *NAV*) is determined as follows.

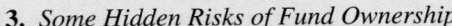

Net Asset Value of a Mutual Fund

If A = Total fund assets, L = Total fund liabilities, and N = Number of shares outstanding, then the **net asset value** is given by

$$NAV = \frac{A - L}{N}.$$

EXAMPLE 7 Finding the Number of Shares in a Mutual Fund Purchase

Suppose, on a given day, a mutual fund has \$500 million worth of stock, \$500,000 in cash (not invested), and \$300,000 in other assets. Total liabilities amount to \$4 million, and there are 25 million shares outstanding. If Laura Griffin Hiller invests \$50,000 in this fund, how many shares will she obtain?

SOLUTION

$$A = \$500{,}000{,}000 + \$500{,}000 + \$300{,}000 = \$500{,}800{,}000$$
$$L = \$4{,}000{,}000$$
$$N = 25{,}000{,}000$$

So,
$$NAV = \frac{\$500{,}800{,}000 - \$4{,}000{,}000}{25{,}000{,}000}$$ Substitute values in the NAV formula.
$$= \$19.872.$$

Because $\frac{\$50{,}000}{\$19.872} \approx 2516$, Laura's \$50,000 investment will purchase (to the nearest whole number) 2516 shares.

Evaluating Investment Returns Regardless of the type of investment you have, or are considering (stocks, bonds, mutual funds, or others), it is important to be able to accurately evaluate and compare returns (profits or losses). Even though past performance is never a guarantee of future returns, it is crucial information. A wealth of information on stock and bond markets and mutual funds is available in many publications (for example, the *Wall Street Journal* and *Barron's Weekly*) and on numerous Internet sources (for example, Nasdaq.com).

The most important measure of performance of an investment is the **annual rate of return.** This is not *necessarily* the same as percentage return, as calculated in Example 1, because annual rate of return depends on the time period involved. Some commonly reported periods are daily, seven-day, monthly, month-to-date, quarterly, quarter-to-date, annual, year-to-date, 2-year, 3-year, 5-year, 10-year, 20-year, and "since inception" (since the fund or other investment vehicle was begun).

Examples 8 and 9 illustrate some of the many complications that arise when one tries to evaluate and compare different investments. Yet it is important to compare different performances on an equal basis.

EXAMPLE 8 Analyzing a Stock's Annual Rate of Return

Todd Hall owns 50 shares of stock. His brokerage statement for the end of August showed that the stock closed that month at a price of \$40 (per share) and the statement for the end of September showed a closing price of \$40.12. For purposes of illustration, disregard any possible dividends, and assume that the money gained in a given month has no opportunity to earn additional returns. (Those earnings cannot be "reinvested.") Find the following.

(a) The value of these shares at the end of August
(b) Todd's monthly return on this stock
(c) The monthly percentage return
(d) The annual rate of return

SOLUTION

(a) $$\text{Value} = \underbrace{\$40}_{\text{Price per share}} \cdot \underbrace{50}_{\text{Number of shares}} = \$2000$$

(b) $$\begin{aligned}\text{Return} &= \underbrace{(\$40.12 - \$40)}_{\text{Change in price per share}} \cdot \underbrace{50}_{\text{Number of shares}} \\ &= \$6\end{aligned}$$

(c) $$\begin{aligned}\text{Percentage return} &= \frac{\text{Total return}}{\text{Total value}} \cdot 100\% = \frac{\$6}{\$2000} \cdot 100\% \\ &= .3\%\end{aligned}$$

(d) Because monthly returns do not earn more returns, we use an "arithmetic" return here. Simply add .3% twelve times (or multiply 12 times .3%), to obtain 3.6%. The annual rate of return is 3.6%.

EXAMPLE 9 Analyzing a Mutual Fund's Annual Rate of Return

Erika Gutierrez owns 80 shares of a mutual fund with a net asset value of $10 on October 1. Assume, for the sake of simplicity, that the only return that the fund earns is dividends of .4% per month, which are automatically reinvested. Find the following.

(a) The value of Erika's holdings in this fund on October 1
(b) The monthly return
(c) The annual rate of return

SOLUTION

(a) Value $=$ NAV $\cdot$ number of shares $= \$10 \cdot 80 = \$800$
(b) Return $= .4\%$ of $\$800 = \$3.20$
(c) In this case, monthly returns get reinvested. Compounding occurs. Therefore we use a "geometric" return rather than arithmetic. To make this calculation, first find the "return relative," which is 1 plus the monthly rate of return: $1 + .4\% = 1 + .004 = 1.004$. Then multiply this return relative 12 times (raise it to the 12th power) to get an annual return relative: $1.004^{12} \approx 1.049$. Finally, subtract 1 and multiply by 100 to convert back to a percentage rate. The annual percentage rate is 4.9%.

A geometric return such as the one found in Example 9 above often is called the "effective annual rate of return." (Notice that an arithmetic return calculation would have ignored the reinvestment compounding and would have understated the effective annual rate of return by .1%, because $12(.4\%) = 4.8\%$.)

Building a Nest Egg The most important function of investing, for most people, is to build an account for some future use, probably for retirement living.

In Example 4 of Section 14.1, we compared two retirement programs, Plan A and Plan B, to emphasize the importance of starting your retirement investment program early in life. The numbers shown there in Table 1 can be verified by applying the formula for the future value of an ordinary annuity (page 870) for periods when regular deposits are being made, and the compound interest formula (page 858) for periods when deposits are no longer made.

There are two major barriers to building your retirement nest egg (or college tuition fund, or legacy to leave to your heirs or to charity, or whatever your goal is for accumulating wealth). These barriers are *inflation* and (*income*) *taxes*.

Table 4 (page 862) shows actual historical inflation rates over 30 years (1975–2005). This reflects an average of about 4.4% per year. If you had invested for retirement over that time span, the inflation proportion (see page 863) indicates that you would have ended up in 2005 paying about $\frac{195.3}{53.8} = 3.6$ times as much for goods and services as when you started, in 1975.

This inflationary effect, if unanticipated, could make a sizable dent in retirement living, but there is a fairly painless way to make provision for it. Recall that the CPI reflects the (usually) rising trend in pricing, and that salaries and wages more or less follow along. Periodically (maybe once a year), you can simply increase your contributions by enough to keep pace with inflation.

The theory of geometric series, explained in the Annuities extension following Section 14.1, allows us to derive a formula to compute the future values of an account where regular deposits are systematically adjusted for inflation. If the inflation rate is i, then the amount R is deposited at the end of the year 1, $R(1+i)$ is deposited at the end of year 2, and so on, with the deposit at the end of the year n, in general, being $R(1 + i)^{n-1}$.

Future Value of an Inflation-Adjusted Retirement Account

Assume annual deposits into a retirement account, adjusted for inflation. If

i = annual rate of inflation,

R = initial deposit (at the end of year 1),

r = annual rate of return on money in the account,

and n = number of years deposits are made,

then the value V of the account at the end of n years is given by

$$V = R\left[\frac{(1 + r)^n - (1 + i)^n}{r - i}\right].$$

In Example 4 of Section 14.1, the person executing Plan A deposited \$2000 on each birthday from age 21 to age 30, then stopped depositing, and at age 65 had \$428,378. The person with Plan B waited until age 31 to make the first \$2000 deposit, then deposited \$2000 every year to age 65 and came out with only \$344,634.

EXAMPLE 10 Adjusting a Retirement Account for Inflation

Compute the final account value, at age 65, for the person under Plan B, assuming that the 35 annual deposits had been adjusted for a 4% annual inflation rate.

SOLUTION

Use the "inflation-adjusted" formula given above with $R = \$2000$, $r = .08$, $i = .04$, and $n = 35$. The final account value is

$$V = \$2000\left[\frac{(1 + .08)^{35} - (1 + .04)^{35}}{.08 - .04}\right] \approx \$541,963.$$

With inflation adjustment, Plan B, by age 65, accumulates the amount \$541,963. ■

Putting off paying taxes for as long as possible, though often urged by financial advisors, is not always the best plan. The **Roth IRA**, available since 1998, is funded with taxed dollars, but all funds, both principal and earnings, are withdrawn later tax-free. For a young person, with many years to "grow" the investment, the prospect of avoiding taxes on that growth is attractive, especially if withdrawals will be made when the account holder is in a higher tax bracket than when contributions were made. Advantages also include less restriction on who qualifies, no forced withdrawals, and no restrictions on when withdrawals can be made.

The second barrier to building wealth—taxes—is not so easily dealt with. You will find that the more you invest and the more you earn, the higher the percentage that our progressive tax system will claim of your earnings. Probably your most effective tool in softening (though not overcoming) this effect is to take full advantage of tax-deferred accounts, such as a TSA (Tax-Sheltered Annuity, or 403(b) Plan); an IRA (Individual Retirement Account, or 401(k) plan), especially if your employer will contribute matching funds; or, if self-employed, a SEP (Simplified Employee Pension) or a Keogh plan.

With most tax-deferred retirement accounts, the accumulated money is all taxed when it is withdrawn (presumably during retirement years). Therefore, it may seem that tax deferral is not an advantage but simply puts off when the tax is paid, assuming your tax rate remains the same. To demonstrate the decided advantage of deferral, we will compare the two situations using the following formulas, which are based on the ordinary annuity formula. For a fair comparison, we look at both account values *after all taxes have been paid.*

Tax-Deferred Versus Taxable Retirement Accounts

In both cases,

R = amount withheld annually from current salary to build the retirement account,

n = number of years contributions are made,

r = annual rate of return on money in the account,

t = marginal tax rate of account holder,

V = final value of the account, **following all accumulations and payment of all taxes.**

Tax-Deferred Account

The entire amount R is contributed each year, all contributions, plus earnings, earn a return over the n years, at which time tax is paid on all money in the account. The final account value is given by

$$V = \frac{(1 - t)R[(1 + r)^n - 1]}{r}.$$

Taxable Account

Each amount R withheld from salary is taxed up front, decreasing the amount of the annual deposit to $(1 - t)R$. Furthermore, annual earnings are also taxed at the end of each year. But at the end of the accumulation period, no more tax will be due. The final account value is given by

$$V = \frac{R[(1 + r(1 - t))^n - 1]}{r}.$$

EXAMPLE 11 Comparing Tax-Deferred and Taxable Retirement Accounts

Alaina and Marin, both in a 20% marginal tax bracket, will each contribute \$2000 annually for 20 years to retirement accounts that return 5% annually. Alaina chooses a tax-deferred account, while Marin chooses a taxable account. Compare their final account values at the end of the accumulation period, after payment of all taxes.

SOLUTION

We have $R = \$2000, n = 20, r = .05, \text{ and } t = .20 = .2.$
For Alaina:

$$V = \frac{(1 - .2)\$2000[(1 + .05)^{20} - 1]}{.05} \approx \$52{,}906.$$

Substitute values in the "tax-deferred" formula.

For Marin:

$$V = \frac{\$2000[(1 + .05(1 - .2))^{20} - 1]}{.05} \approx \$47{,}645.$$

Substitute values in the "taxable" formula.

By deferring taxes, Alaina ends up with \$5261, or about 11%, more.

Greater contributions, a longer accumulation period, a higher rate of return, and a higher tax bracket will accentuate the effect of tax deferral.

EXAMPLE 12 Comparing Tax-Deferred and Taxable Retirement Accounts

Repeat Example 11, but this time let $R = \$5000, n = 40, r = .10, \text{ and } t = .40 = .4.$

SOLUTION

For Alaina:

$$V = \frac{(1 - .4)\ \$5000\ [(1 + .10)^{40} - 1]}{.10} \approx \$1{,}327{,}778.$$

Substitute values in the "tax-deferred" formula.

For Marin:

$$V = \frac{\$5000[(1 + .10\ (1 - .4))^{40} - 1]}{.10} \approx \$464{,}286.$$

Substitute values in the "taxable" formula.

With these higher parameters, Alaina's advantage is dramatic: \$863,492, or about 186%, more.

For Further Thought

Summary of Investments

The summary of various types of investments in Table 11 is based on average cases and is intended only as a general comparison of investment opportunities. There are numerous exceptions to the characteristics shown. For example, though the chart shows no selling fees for mutual funds, this is really true only for the so-called no-load funds. The "loaded" funds do charge "early redemption fees" or other kinds of sales charges. That is not to say that a no-load fund is necessarily better. For example, the absence of sales fees may be offset by higher "management fees."

These various costs, as well as other important factors, will be clearly specified in a fund's prospectus and should be studied carefully before

TABLE 11 Summary of Investments

Investment	Protection of Principal	Protection Against Inflation	Rate of Return (%)	Certainty of Return	Selling Fees	Liquidity	Long-Term Growth	Requires Work from You	Expert Knowledge Required
Cash	excellent	none	0			excellent	no	no	no
Ordinary life insurance	good	poor	2–4	excellent	high	excellent	no	no	no
Savings bonds	excellent	poor	3–5	excellent	none	excellent	no	no	no
Bank savings	excellent	poor	2–4	excellent	none	excellent	no	no	no
Credit union	excellent	poor	3–4	excellent	none	excellent	no	no	no
Corporation bonds	good	poor	5–7	excellent	low	good	no	some	some
Tax-free municipal bonds	excellent	poor	2–5	excellent	medium	good	no	no	some
Corporation stock	good	good	3–12	good	medium	good	yes	some	some
Preferred stock	good	poor	4–8	excellent	medium	good	no	some	some
Mutual funds	good	good	3–12	good	none	good	yes	no	no
Your own home	good	good	4–10	good	medium	poor	yes	yes	no
Mortgages on the homes of others	fair	poor	5–8	fair	high	poor	no	yes	some
Raw land	fair	good	?	fair	medium	poor	yes	no	yes
Rental properties	good	good	4–8	fair	medium	poor	yes	yes	some
Stamps, coins, antiques	fair	good	0–10	fair	high	poor	yes	yes	yes
Your own business	fair	good	0–20	poor	high	poor	yes	yes	yes

a purchase decision is made. Some expertise is not entirely unnecessary for mutual fund investing, just less so than with certain other options, like direct stock purchases, raw land, or collectibles.

To use the chart, read across the columns to find the general characteristics of the investment. "Liquidity" refers to the ability to get cash from the investment quickly.

For Group Discussion or Individual Investigation

List (a) some investment types you would recommend considering and (b) some you would recommend avoiding for a friend whose main investment objective is as follows.

1. Avoid losing money.
2. Avoid losing purchasing power.
3. Avoid having to learn about investments.
4. Avoid having to work on investments.
5. Be sure to realize a gain.
6. Be able to "cash in" at any time.
7. Get as high a return as possible.
8. Be able to "cash in" without paying fees.
9. Realize quick profits.
10. Realize growth over the long term.

14.5 EXERCISES

Refer to the stock table (Table 10 on page 905) for Exercises 1–36.

Reading Stock Charts *Find each of the following.*

1. closing price for Agilent Technologies (A)
2. sales for the day for Airborne (ABF)
3. change from the previous day for Party City (PCTY)
4. closing price for Peet's Coffee & Tea (PEET)
5. 52-week high for Patterson Dental (PDCO)
6. 52-week low for Alaska Air Group (ALK)
7. dividend for Allegheny Energy (AYE)
8. dividend for PennFed Financial Services, preferred (PFSBP)
9. sales for the day for Perceptron (PRCP)
10. change from the previous day for Alltrista (ALC)
11. year-to-date percentage change for Alcan (AL)
12. year-to-date percentage change for PeopleSoft (PSFT)
13. price-to-earnings ratio for Patriot Transportation (PATR)
14. price-to-earnings ratio for AirNet Systems (ANS)

Finding Stock Costs *Find the basic cost (ignoring any broker fees) for each stock purchase, at the day's closing prices.*

15. 200 shares of Albany International (AIN)
16. 400 shares of Ameren (AEE)
17. 300 shares of PC Mall (MALL)
18. 700 shares of PetsMart (PETM)

Finding Stock Costs *Find the cost, at the day's closing price, for each stock purchase. Include typical discount broker commissions as described in the text.*

Stock	Symbol	Number of Shares	Transaction Type
19. Paychex	PAYX	60	broker-assisted
20. Amerada Hess	AHC	70	broker-assisted
21. Alltel	AT	355	automated
22. Performance Food Group	PFGC	585	automated
23. Agrium	AGU	2500	broker-assisted
24. Pixelworks	PXLW	1500	automated
25. Pathmark Stores	PTMK	2400	automated
26. Alamosa	APS	10,000	broker-assisted

Finding Receipts for Stock Sales *Find the amount received by the sellers of each stock (at the day's closing prices). Deduct sales expenses as described in the text.*

	Stock	Symbol	Number of Shares	Transaction Type
27.	Albemarle	ALB	400	broker-assisted
28.	PartsBase	PRTS	600	broker-assisted
29.	Pharmacopeia	PCOP	500	automated
30.	Alberto Culver	ACV	700	automated
31.	PLATO Learning	TUTRs	1350	automated
32.	American Greetings	AM	2740	automated
33.	American States Water	AWR	1480	broker-assisted
34.	Photronics	PLAB	1270	broker-assisted

Finding Net Results of Combined Transactions *For each combined transaction (executed at the closing price of the day), find the* net *amount paid out or taken in. Assume typical expenses as outlined in the text.*

35. Paul Borge bought 100 shares of Alpine Group (AGI) and sold 20 shares of Peak International (PEAK), both broker-assisted trades.

36. Nara Lee bought 800 shares of PETCO Animal Supplies (PETC) and sold 1200 shares of Alaska Air Group (ALK), both automated trades.

Costs and Returns of Stock Investments *For each of the following stock investments, find* **(a)** *the total purchase price,* **(b)** *the total dividend amount,* **(c)** *the capital gain or loss,* **(d)** *the total return, and* **(e)** *the percentage return.*

	Number of Shares	Purchase Price per Share	Dividend per Share	Sale Price per Share
37.	40	$20	$2	$44
38.	20	$25	$1	$22
39.	100	$12.50	$1.08	$10.15
40.	200	$ 8.80	$1.12	$11.30

Total Return on Bond Investments *Find the total return earned by each bond in Exercises 41–44.*

	Face Value	Annual Interest Rate	Term to Maturity
41.	$ 1000	5.5%	5 years
42.	$ 5000	6.4%	10 years
43.	$10,000	7.11%	3 months
44.	$50,000	4.88%	6 months

Net Asset Value of a Mutual Fund *For each investment, find* **(a)** *the net asset value, and* **(b)** *the number of shares purchased.*

	Amount Invested	Total Fund Assets	Total Fund Liabilities	Total Shares Outstanding
45.	$ 3500	$875 million	$ 36 million	80 million
46.	$ 1800	$643 million	$102 million	50 million
47.	$25,470	$2.31 billion	$135 million	263 million
48.	$83,250	$1.48 billion	$ 84 million	112 million

Finding Monthly and Annual Investment Returns *For each investment, assume that there is no opportunity for reinvestment of returns. In each case find* **(a)** *the monthly return,* **(b)** *the annual return, and* **(c)** *the annual percentage return.*

	Amount Invested	Monthly Percentage Return
49.	$ 645	1.3%
50.	$ 895	.9%
51.	$2498	2.3%
52.	$4983	1.8%

Effective Annual Rate of Return of Mutual Fund Investments *Assume that each mutual fund investment earns monthly returns that are reinvested and subsequently earn at the same rate. In each case find* **(a)** *the beginning value of the investment,* **(b)** *the first monthly return, and* **(c)** *the effective annual rate of return.*

	Beginning NAV	Number of Shares Purchased	Monthly Percentage Return
53.	$ 9.63	125	1.5%
54.	$12.40	185	2.3%
55.	$11.94	350	1.83%
56.	$18.54	548	2.22%

Finding Commissions *For each of the following trades, determine the missing amounts in columns (a), (b), and (c).*

	Number of Shares	Price per Share	(a) Principal Amount	(b) Broker-assisted Commission	(c) Automated Commission
57.	10	\$1.00	______	______	______
58.	10	\$100.00	______	______	______
59.	400	\$1.00	______	______	______
60.	400	\$100.00	______	______	______
61.	4000	\$1.00	______	______	______
62.	4000	\$100.00	______	______	______

63. Referring to Exercises 57–62, fill in the blanks in the following statement.

A broker-assisted purchase is cheaper than an automated purchase only when a relatively ________ number of shares are purchased for a relatively _______ price per share.

Inflation-Adjusted Retirement Accounts *Find the future value (to the nearest dollar) of each inflation-adjusted retirement account. Deposits are made at the end of each year.*

	Annual Inflation Rate	Initial Deposit	Annual Rate of Return	Number of Years
64.	2%	\$1000	6%	20
65.	3%	\$2000	5%	25
66.	4%	\$2500	7%	30
67.	1%	\$5000	6.5%	40

The Effect of Tax Deferral on Retirement Accounts *Find the final value, after all taxes are paid, for each account if* **(a)** *taxes are deferred, or* **(b)** *taxes are not deferred. In both cases, deposits are made at the end of each year.*

	Marginal Tax Rate	Regular Deferred Contribution	Annual Rate of Return	Number of Years
68.	10%	\$500	7%	15
69.	15%	\$1000	5%	30
70.	28%	\$3000	10%	25
71.	35%	\$1500	6%	10

72. ***Monthly Deposits Adjusted for Inflation*** In an inflation-adjusted retirement account, let i denote annual inflation, let r denote annual return, and let n denote number of years, as in the text, but suppose deposits are made monthly rather than yearly with initial deposit R.

(a) What expression represents the amount that would be deposited at the end of the second month?

(b) Modify the future value formula for an inflation-adjusted account to accommodate the more frequent deposits.

(c) Find (to the nearest dollar) the future value for monthly deposits of \$100 (initially) for 20 years if $i = .03$ and $r = .06$.

73. ***Finding an Unknown Rate*** Given the compound interest formula

$$A = P\left(1 + \frac{r}{m}\right)^n,$$

do the following.

(a) Solve the formula for r.

(b) Find the annual rate r if \$50 grows to \$85.49 in 10 years with interest compounded quarterly. Give r to the nearest tenth of a percent.

74. ***Comparing Continuous with Annual Compounding*** It was suggested in the text that inflation is more accurately reflected by the continuous compounding formula than by periodic compounding. Compute the ratio

$$\frac{Pe^{rt}}{P(1 + r)^n} = \frac{e^{rt}}{(1 + r)^n}$$

to find how much (to the nearest tenth of a percent) continuous compounding exceeds annual compounding. Use a rate of 3% over a period of 30 years.

Spreading Mutual Fund Investments Among Asset Classes *Mutual funds (as well as other kinds of investments) normally are categorized into one of several "asset classes," according to the kinds of stocks or other securities they* *hold. Small capitalization funds are most aggressive, while cash is most conservative. Many investors, often with the help of an advisor, try to construct their portfolios in accordance with their stages of life. Basically, the idea is that a younger person can afford to be more aggressive (and assume more risk) while an older investor should be more conservative (assuming less risk). The investment pyramids here show percentage ranges recommended by Edward Jones, one of the largest investment firms in the country. In Exercises 75–78, divide the given investor's money into the five categories so as to position them right at the middle of the recommended ranges.*

Asset Classes	
Aggressive Growth (small cap)	■
Growth	▲
Growth & Income	◆
Income	●
Cash	✚

75. Michelle Clayton, in her early investing years, with $20,000 to invest

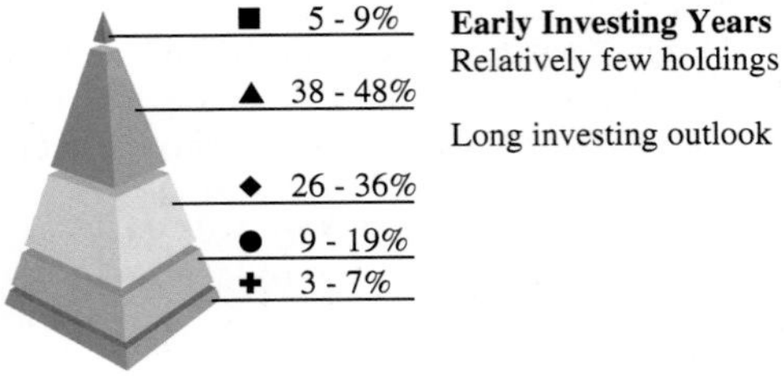

76. Jeff Hubbard, in his good earning years, with $250,000 to invest

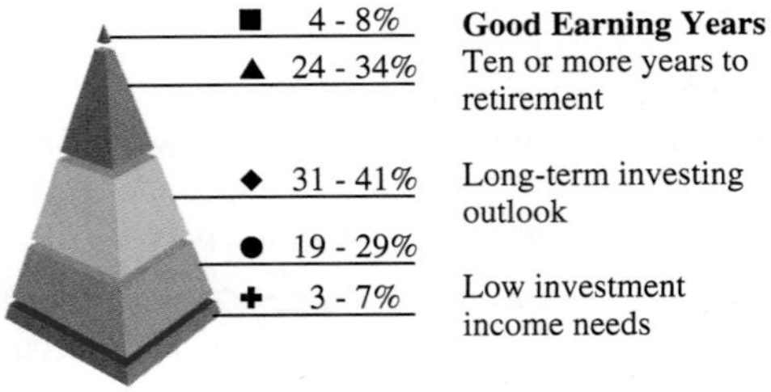

77. Paul Crockett, in his high income/saving years, with $400,000 to invest

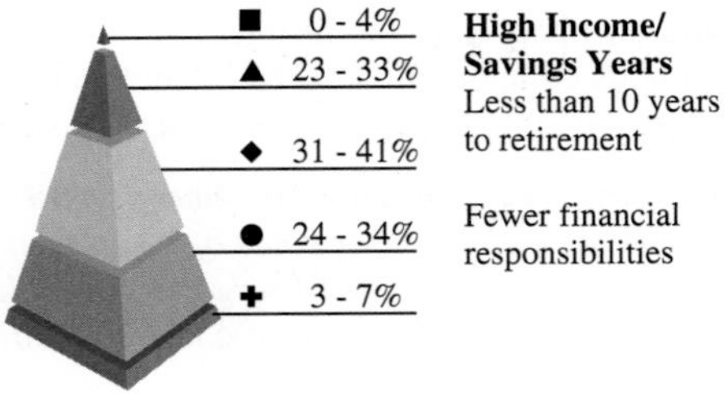

78. Amy Donlin, retired, with $845,000 to invest

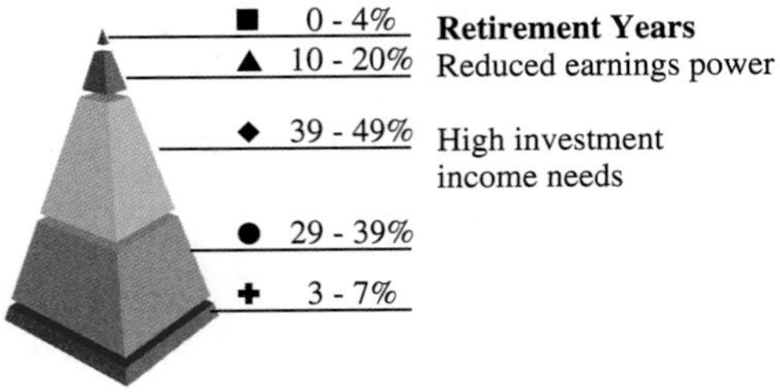

The Effect of Taxes on Investment Returns *Income from municipal bonds generally is exempt from federal, and sometimes state, income tax. For an investor with a marginal combined state and federal tax rate of 35%, a taxable return of 6% would yield only 3.9% (65% of 6%). So at that tax rate, a tax-exempt return of 3.9% is equivalent to a taxable return of 6%. Find the tax-exempt rate of return that is equivalent to the given taxable rate of return for each investor.*

Investor	Marginal Combined Tax Rate	Taxable Rate of Return
79. Paul Altier	25%	5%
80. Eileen Burke	30%	7%
81. Carol Britz	35%	8%
82. Greg Brueck	40%	10%

The Real Rate of Return *When selecting long-term bonds, income investors should pay attention to the real rate of return (RRR), defined as interest rate minus inflation rate. Find the RRR in each case. (Source: Edward Jones and Company: Income Advantage, September/October 2002, p. 11.)*

Year	Long-term Bond Rate	Inflation Rate
83. 1981	14.2%	8.9%
84. 2002	6.6%	1.3%

Reading Stock Charts *Refer to the stock table in the text (Table 10 on page 905) to answer each question.*

85. What special feature was true of Allied Capital (ALD)?

86. Give an example of a stock with all three of the following special features:

(1) The price on this day was a new 52-week high.

(2) The price changed by 5% or more from the previous day.

(3) The stock was among the 40 largest volume percentage leaders on this day.

Write a response to each of the following. Some research may be required.

87. Describe the concept of "dollar-cost averaging," and relate it to advantage number 1 of mutual funds as listed in the text.

88. Discuss advantage number 3 of mutual funds as listed in the text.

89. In light of Exercises 83 and 84, describe the importance to bond investors of the real rate of return.

90. Log onto www.napfa.org to find out about financial planners. Then report on what you would look for in a planner.

91. A great deal of research has been done in attempts to understand and predict trends in the stock market. Discuss the term *financial engineering*. Within this area of study, what is a *rocket scientist* (or *quant*)?

92. In connection with mutual funds, discuss the difference between a *front-end load* and a *contingent deferred sales charge*.

93. What is meant by the terms *growth stock* and *income stock*?

94. With respect to investing in corporate America, describe the difference between being an owner and being a lender.

95. Describe the meaning of the graph shown here.

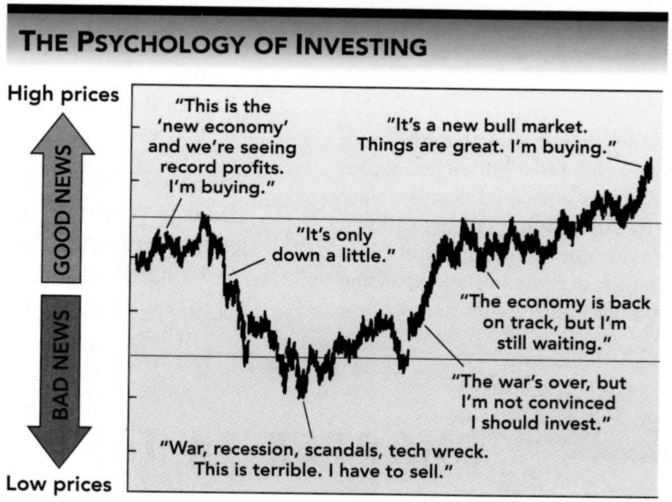
THE PSYCHOLOGY OF INVESTING
High prices
Low prices
GOOD NEWS
BAD NEWS
"This is the 'new economy' and we're seeing record profits. I'm buying."
"It's a new bull market. Things are great. I'm buying."
"It's only down a little."
"The economy is back on track, but I'm still waiting."
"The war's over, but I'm not convinced I should invest."
"War, recession, scandals, tech wreck. This is terrible. I have to sell."

Source: Edward Jones and Company, 2002.

COLLABORATIVE INVESTIGATION

To Buy or to Rent?

Divide your class into groups of at least four students each. Every group is to first read the following.

> Scott and Aimee, a young couple with a child, Alaina, live in a rented apartment. After taxes, Scott earns \$36,000 per year and Aimee's part-time job brings in \$6000 per year. In the foreseeable future, their earnings probably will just keep pace with inflation. They are presently operating on the following monthly budget.

Rent	\$1050
Food	600
Day care	525
Clothing	420
Utilities	120
Entertainment	300
Savings	450
Other	300

They have accumulated \$29,000 in savings.

Having found a house they would like to buy, and having consulted several lenders, they have discovered that, at best, purchasing the house would involve costs as shown here.

Price of house	$127,500
Down payment required (20%)	______
Closing costs (required at closing):	
Loan fee (1 point)	______
Appraisal fee	300
Title insurance premium	375
Document and recording fees	70
Prorated property taxes	650
Prorated fire insurance premium	270
Other	230
Total immediate costs:	______
Fixed-rate 30-year mortgage at 7.5%	
Annual taxes:	$1350
Annual fire insurance:	$540

In order to own their own home, Scott and Aimee are willing to cut their clothing and entertainment expenditures by 30% each, and could decrease their savings allotments by 20%. (Some savings still will be necessary to provide for additional furnishings they would need and for expenses of an expanding family in the future.) They also figure that the new house and yard would require $130 monthly for maintenance and that utilities will be twice what they have been in the apartment. Other than that, their present budget allotments would remain the same.

Now divide your group into two "subgroups." The first subgroup is to answer these questions:

1. What amount would the home purchase cost Scott and Aimee in immediate expenditures?
2. Do they have enough cash on hand?

The second subgroup is to answer these questions:

3. What amount would the house cost Scott and Aimee on an ongoing monthly basis?
4. What amount will their monthly budget allow toward this cost?
5. Can they meet the monthly expenses of owning the house?

Within your group, decide whether Scott and Aimee can afford to purchase the house. Select a representative to report your findings to the class.

Compare the evaluations of the various groups, and try to resolve any discrepancies.

CHAPTER 14 TEST

Find all monetary answers to the nearest cent. Use tables and formulas from the chapter as necessary.

Finding the Future Value of a Deposit *Find the future value of each deposit.*

1. $100 for 5 years at 6% simple interest
2. $50 for 2 years at 8% compounded quarterly

Solve each problem.

3. ***Effective Annual Yield of an Account*** Find the effective annual yield to the nearest hundredth of a percent for an account paying 3% compounded monthly.
4. ***Years to Double by the Rule of 70*** Use the rule of 70 to estimate the years to double at an inflation rate of 5%.
5. ***Finding the Present Value of a Deposit*** What amount deposited today in an account paying 4% compounded semiannually would grow to $100,000 in 10 years?
6. ***Finding Bank Card Interest by the Average Daily Balance Method*** Peg Cheever's MasterCard statement shows an average daily balance of $680. Find the interest due if the rate is 1.6% per month.

Analyzing a Consumer Loan *Jeff Howard buys a turquoise necklace for his wife on their anniversary. He pays* $4000 *for the necklace with* $1000 *down. The dealer charges add-on interest of* 7.5% *per year. Jeff agrees to make payments for* 24 *months. Use this information for Exercises 7–10.*

7. Find the total amount of interest he will pay.
8. Find the monthly payment.
9. Find the APR value for this loan (to the nearest half percent).
10. Find the unearned interest if he repays the loan in full with six payments remaining. Use the most accurate method available.

11. ***True Annual Interest Rate in Consumer Financing*** Newark Hardware wants to include financing terms in their advertising. If the price of a floor waxer is $150 and the finance charge with no down payment is $5 over a 6-month period (six equal monthly payments), find the true annual interest rate (APR).

12. Explain what a mutual fund is and discuss several of its advantages and disadvantages.

Finding the Monthly Payment on a Home Mortgage *Find the monthly payment required for each home loan.*

13. The mortgage is for $150,000 at a fixed rate of 7% for 20 years. Amortize principal and interest only.

14. The purchase price of the home is $218,000. The down payment is 20%. Interest is at 8.5% fixed for a term of 30 years. Annual taxes are $1500 and annual insurance is $750.

Solve each problem.

15. ***The Cost of Points in a Home Loan*** If the lender in Exercise 14 charges two *points,* how much does that add to the cost of the loan?

16. Explain in general what *closing costs* are. Are they different from *settlement charges*?

17. ***Adjusting the Rate in an Adjustable-Rate Mortgage*** To buy your home you obtain a 1-year ARM with 2.25% margin and a 2% periodic rate cap. The index starts at 7.85% but has increased to 10.05% by the first adjustment date. What interest rate will you pay during the second year?

18. ***Reading Stock Charts*** According to the stock table (Table 10 on page 905), how many shares of Patterson Dental (PDCO) were traded on the day represented?

Finding the Return on a Stock Investment *Kathryn Campbell bought* 1000 *shares of stock at* $12.75 *per share. She received a dividend of* $1.38 *per share shortly after the purchase and another dividend of* $1.02 *per share one year later. Eighteen months after buying the stock she sold it for* $10.36 *per share.*

19. Find Kathryn's total return on this stock.

20. Find her percentage return. Is this the *annual rate of return* on this stock transaction? Why or why not?

21. ***The Final Value of a Retirement Account*** $1800 is deposited at the end of each year in a tax-deferred retirement account. The account earns a 6% annual return, the marginal tax rate is 25%, and taxes are paid at the end of 30 years. Find the final value of the account.

22. What is meant by saying that a retirement account is "adjusted for inflation"?

15

GRAPH THEORY

Have you ever played the popular game *The Six Degrees of Kevin Bacon?* It was developed by three Albright College friends a number of years ago. The object of the game is to link actor Kevin Bacon to another actor or actress using movies, with six or fewer titles in the link. Thanks to the International Movie Data Base (www.imdb.com) this goal can be reached almost instantaneously. The minimum number of links necessary is called the *Bacon number* for the other performer. For example, the silent film star Charles Chaplin has a Bacon number of three, as he can be linked to Kevin Bacon in three steps as follows:

- Charles Chaplin was in *The Great Dictator* (1940) with Don Brodie.
- Don Brodie was in *Murphy's Law* (1986) with David Hayman.
- David Hayman was in *Where the Truth Lies* (2005) with Kevin Bacon.

One of the finest movies about mathematics ever to come from Hollywood is *Stand and Deliver* (1988), starring Edward James Olmos as Jaime Escalante. The movie is based on the true story of an inner-city high school teacher who defies all odds and prepares his students for the Advanced Placement calculus exam. Edward James Olmos' Bacon number is two. Can you justify this result? (The answer is on page 937).

The linking process used in *The Six Degrees of Kevin Bacon* can be applied to other situations (Web pages, for example) and can be modeled using concepts from *graph theory*, the topic of this chapter. (*Sources:* www.imdb.com, www.oracleofbacon.org, www-distance.syr.edu/bacon.html.)

Unlike graphs in the rectangular coordinate plane, studied earlier in the text, the "graphs" in this chapter are convenient diagrams that show relations or connections between objects in a collection. These graphs enable us to communicate and analyze complex information in a simple way. As a result, *graph theory* has many real-world applications, with particularly important ones related to the Internet and telecommunications.

15.1 Basic Concepts

Graphs • Walks, Paths, and Circuits • Complete Graphs and Subgraphs • Graph Coloring

Graphs

Consider the following. A preschool teacher has ten children in her class: Andy, Claire, Dave, Erin, Glen, Katy, Joe, Mike, Sam, and Tim. The teacher wants to analyze the social interactions among these children. She observes with whom the children play during recess over a 2-week period. Here are her observations:

Child	*Played with*
Andy	no one
Claire	Dave, Erin, Glen, Katy, Sam
Dave	Claire, Erin, Glen, Katy
Erin	Claire, Dave, Katy
Glen	Claire, Dave, Sam
Katy	Claire, Dave, Erin
Joe	Mike, Tim
Mike	Joe
Sam	Claire, Glen
Tim	Joe

Here it is not easy to see patterns in the children's friendships. Instead, the information may be shown in a diagram, as in Figure 1. Each letter represents the first letter of a child's name. Such a diagram is called a *graph*.* Each dot is called a **vertex** (plural **vertices**). In Figure 1, each vertex represents one of the preschool children. The lines between vertices are called the **edges** of the graph. In Figure 1, the edges show the relationship "play with each other." An edge must always begin and end at a vertex. We may have a vertex with no edges joined to it, such as A in Figure 1.

The **World Wide Web** can be studied as a very large graph. The vertices are Web pages, with the edges links between Web pages. In 2005, it was estimated that there were about 11.5 billion Web pages. Because of the way the Web is connected, you can get from any randomly chosen Web page to any other using hyperlinks, in an average of 19 clicks. Even if the Web becomes 10 or 100 times larger, this will still be possible in only 20 or 21 clicks. Because of this, search engines such as Google can find keywords very quickly. (*Sources*: http://jmm.aaa.net.au/articles/15080.htm; Antonio Gulli Universit di Pisa, Informatica; Alessio Signorini, University of Iowa, Computer Science; R. Albert, H. Jeong, and A.-L. Barabsi, *Nature* 401, 130–131, 1999.)

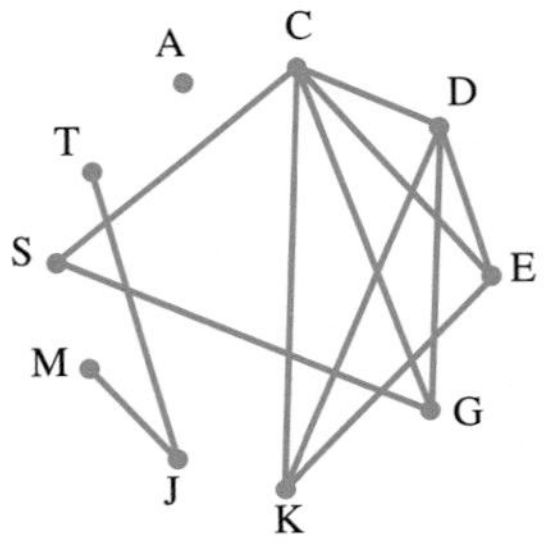

FIGURE 1

Graph

A **graph** is a collection of vertices (at least one) and edges. Each edge goes from a vertex to a vertex.

*Here *graph* will mean finite graph. A finite graph is a graph with finitely many edges and vertices.

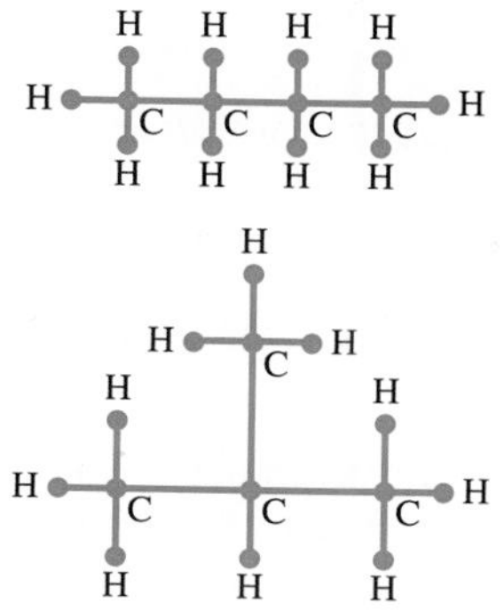

Why are they called graphs? The name was first used by chemist **A. Crum Brown.** He made a major breakthrough in chemistry in 1864 by inventing the now familiar diagrams to show different molecules with the same chemical composition. For example, both molecules illustrated above have chemical formula C_4H_{10}. Brown called his sketches **graphic formulae.** Mathematicians got involved in helping to classify all possible structures with a particular chemical composition and adopted the term "graph."

The graph in Figure 1 has ten vertices and twelve edges. The only vertices of the graph are the dots; there is *no* vertex where edges TJ and SG cross, since the crossing point does not represent one of the children. The positions of the vertices and the lengths of the edges have no significance; the edges do not even have to be drawn as straight lines. ***Only the relation between the vertices has significance, indicated by the presence or absence of an edge between them.***

Why not draw two edges for each friendship pair? For example, why not draw one edge showing that Tim plays with Joe and another edge showing that Joe plays with Tim? Since Tim and Joe play with each other, the extra edge would show no additional information. Graphs in which there is no more than one edge between any two vertices and in which no edge goes from a vertex to the same vertex are called **simple graphs.** See Figure 2. ***In this chapter, the word* graph *means simple graph, unless indicated otherwise.*** (The meaning of "simple" here is different from its meaning in the phrase "simple curves" discussed with geometry in Chapter 9.)

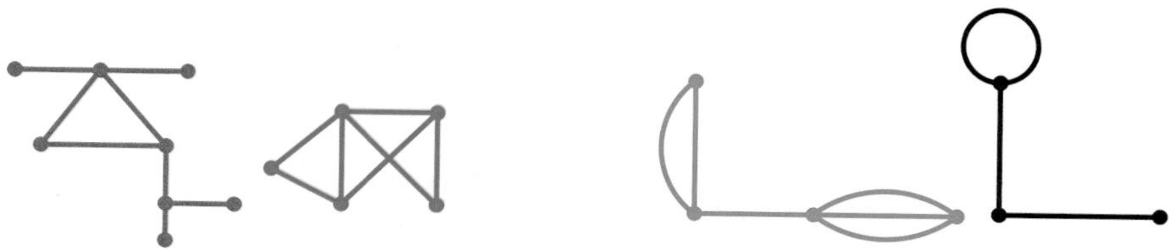

These are simple graphs. These are not simple graphs.

FIGURE 2

From the graph in Figure 1, we can determine the size of a child's friendship circle by counting the number of edges coming from that child's vertex. There are five edges coming from the vertex labeled C, indicating that Claire plays with five different children. The number of edges joined to a vertex is called the **degree of the vertex.** In Figure 1, the degree of vertex C is 5, the degree of vertex M is 1, and the degree of vertex A is 0, as no edges are joined to A. (Here, the use of *degree* has nothing to do with its use in geometry in measuring the size of an angle.)

While Figure 1 clarifies the friendship patterns of the preschoolers, the graph can be drawn in a different, more informative way as in Figure 3. The graph still has edges between the same pairs of vertices. We think of Figure 3 as one graph, even though it has three pieces, since it represents a single situation. By drawing the graph in this way, we make the friendship patterns more apparent. Although the graph in Figure 3 looks different from that in Figure 1, it shows the same relationships between the vertices; the two graphs are said to be *isomorphic.*

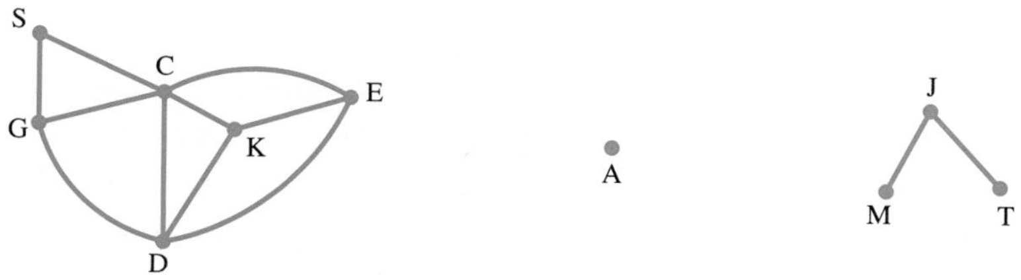

FIGURE 3

The word **isomorphism** comes from the Greek words *isos,* meaning "same," and *morphe,* meaning "form." Other words that come from the Greek root *isos* include *isobars*: lines drawn on a map connecting places with the same barometric pressure; *isosceles*: a triangle with two legs (*scelos*) of equal length; and *isomers*: chemical compounds with the same chemical composition, but different chemical structures.

Isomorphic Graphs

Two graphs are **isomorphic** if there is a one-to-one matching between vertices of the two graphs with the property that whenever there is an edge between two vertices of either one of the graphs, there is an edge between the corresponding vertices of the other graph.

A useful way to think about isomorphic graphs is to imagine a graph drawn with computer software that allows you to drag vertices without detaching any of the edges from their vertices. Any graph you can obtain by simply dragging vertices in this way will be isomorphic to the original graph.

As demonstrated with the friendship graph, replacing a graph with an isomorphic graph sometimes can convey more about the relationships being examined. For example it is clear from Figure 3 that there are no friendship links between the group consisting of Joe, Tim, and Mike, and the other children. We say that the graph is *disconnected.*

Connected and Disconnected Graphs

A graph is **connected** if one can move from each vertex of the graph to every other vertex of the graph *along edges of the graph.* If not, the graph is **disconnected.** The connected pieces of a graph are called the **components** of the graph.

Figure 3 shows clearly that the friendship relationships among the children in the preschool class have three components. Although the graph in Figure 1 also is disconnected and includes three components, the way in which the graph is drawn does not make this as clear.

Using the graph in Figure 3 to analyze the relationships in her class, the teacher might feel concerned that Tim, Mike, and Joe do not interact with the other children. Certainly she would feel concerned about Andy's isolation. She might observe that Claire seems to get along well with other children and encourage interaction between Claire and Tim, Mike, and Joe, and she might make a special attempt to encourage the children to include Andy in their games. In summary, showing the friendship relations as a graph like that in Figure 3 made it much easier to analyze those relationships.

PROBLEM-SOLVING HINT Using different colors for different components can help you determine how many components a graph has. Choose a color, begin at any vertex, and color all edges and vertices that you can get to from your starting vertex along edges of the graph. If the whole graph is not colored, then choose a different color, choose an uncolored vertex, and repeat the process. Continue until the whole graph is colored. It will then be easy to see the different components of the graph.

A **call graph** has telephone numbers as vertices, while edges represent a call placed from one number to another. A call graph for one company for just one day contained nearly 54 million vertices and 170 million edges. It had 3.7 million components. One of these was a giant component containing nearly 80% of the vertices.

EXAMPLE 1 Determining Number of Components

Is the graph in Figure 4 connected or disconnected? How many components does the graph have?

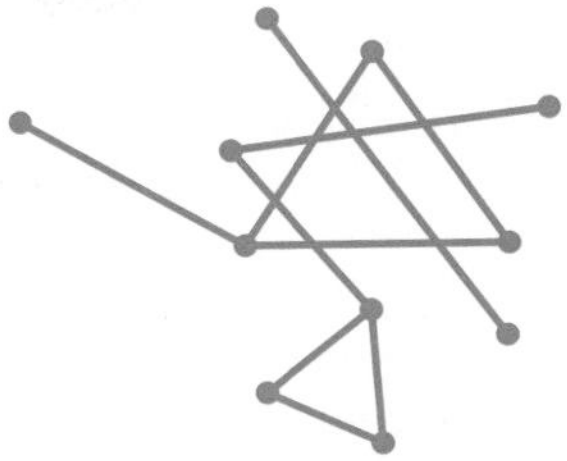

FIGURE 4

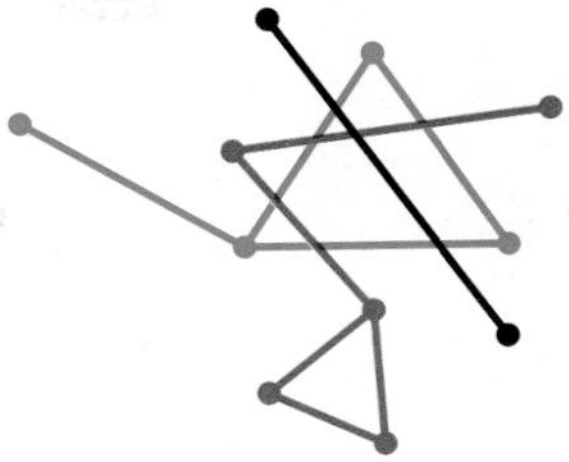

FIGURE 5

SOLUTION

Coloring the graph helps to solve the problem. The graph with color (Figure 5) shows three connected components. The original graph is disconnected. ■

EXAMPLE 2 Deciding Whether Graphs Are Isomorphic

Are the two graphs in Figure 6 isomorphic? Justify your answer.

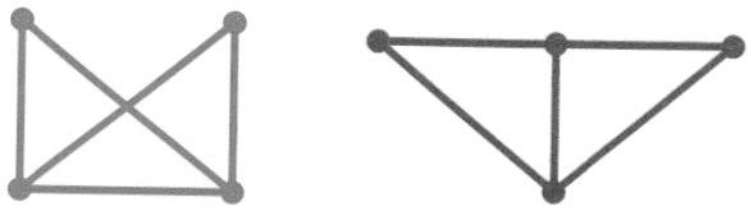

(a) **(b)**

FIGURE 6

SOLUTION

These graphs are isomorphic. To show this, we label corresponding vertices with the same letters and color-code the matching edges in graphs (a) and (b) of Figure 7.

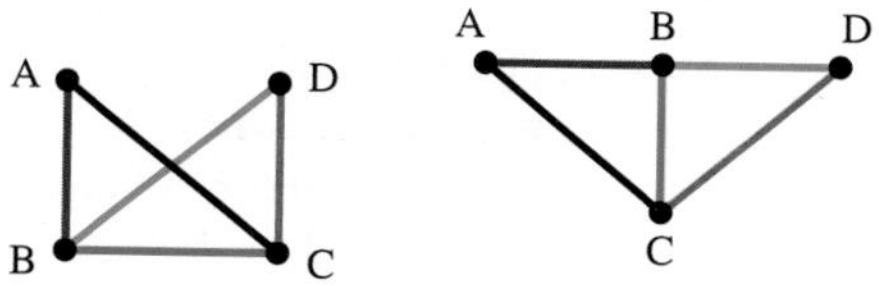

(a) **(b)**

FIGURE 7

■

James Joseph Sylvester (1814–1897) was the first mathematician to use the word *graph* for these diagrams in the context of mathematics. Sylvester spent most of his life in Britain, but had an important influence on American mathematics from 1876 to 1884 as a professor at the newly founded Johns Hopkins University in Baltimore, Maryland. As a young man he worked as an actuary and lawyer in London, tutoring math on the side; one of his students was Florence Nightingale. His only published book, *The Laws of Verse,* was about poetry.

EXAMPLE 3 Deciding Whether Graphs Are Isomorphic

Are the two graphs in Figure 8 isomorphic?

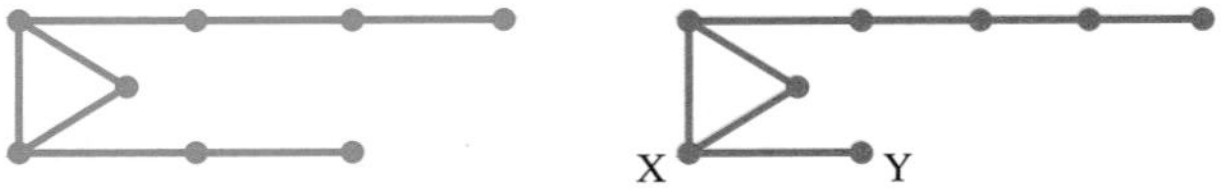

(a) (b)

FIGURE 8

SOLUTION

These graphs are not isomorphic. If we imagine graph (a) drawn with our special computer software, we can see that no matter how we drag the vertices we can not get this graph to look exactly like graph (b). For example, graph (b) has vertex X of degree 3 joined to vertex Y of degree one; neither of the vertices of degree 3 in graph (a) is joined to a vertex of degree one. ■

EXAMPLE 4 Summing Degrees

For each graph in Figure 9, determine the number of edges and the sum of the degrees of the vertices. (Note that graph (b) is not a simple graph.)

(a) (b)

FIGURE 9

SOLUTION

Start by writing the degree of each vertex next to the vertex, as in Figure 10.

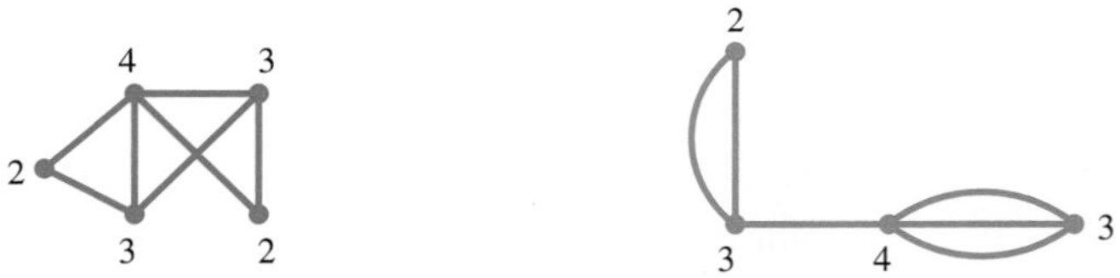

(a) (b)

FIGURE 10

For graph (a):
Number of edges $= 7$
Sum of degrees of vertices
$= 2 + 3 + 2 + 3 + 4 = 14.$

For graph (b):
Number of edges $= 6$
Sum of degrees of vertices
$= 2 + 3 + 4 + 3 = 12.$ ■

For each graph in Example 4, the sum of the degrees of the vertices is twice the number of edges. This is true for all graphs. Check that it is true for graphs in previous examples.

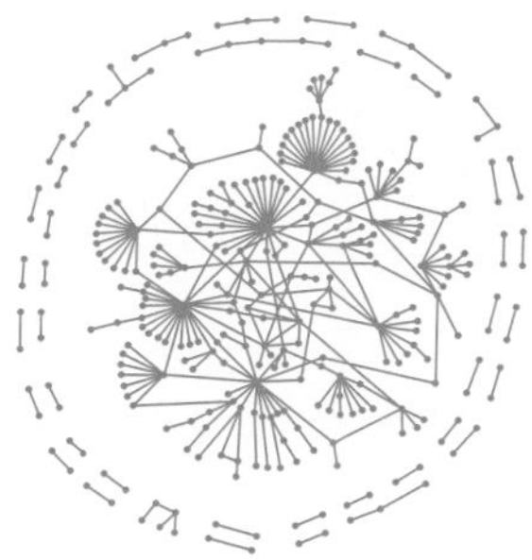

Networks are at the forefront of research in biology. Biologists have made enormous strides in understanding the structure of molecules in living organisms. It is now important to understand how these molecules interact with each other. This can be viewed as a network, or graph. For example, current research aims to identify how proteins in cells interact with other proteins. Insights into such networks may help explain the growth of cancerous tumors; this may lead to better treatments for cancer. It was discovered that as graphs, these biological networks are surprisingly similar to the graphs representing the World Wide Web.

Sum of the Degrees Theorem

In any graph, the sum of the degrees of the vertices equals twice the number of edges.

To understand why this theorem is true, think of any graph. Imagine cutting each edge at its midpoint. (Don't add any vertices to your graph, however.) Now you have precisely twice as many half edges as you had edges in your original graph. For each vertex, count the number of half edges joined to the vertex; this is, of course, the degree of the vertex. Add these answers to determine the sum of the degrees of the vertices. You know that the number of half edges is twice the number of edges in your original graph. So, the sum of the degrees of the vertices in the graph is twice the number of edges in the graph.

EXAMPLE 5 Using the Sum of the Degrees Theorem

A graph has precisely six vertices, each of degree 3. How many edges does this graph have?

SOLUTION

The sum of the degrees of the vertices of the graph is

$$3 + 3 + 3 + 3 + 3 + 3 = 18.$$

By the theorem above, this number is twice the number of edges, so the number of edges in the graph is $\frac{18}{2}$, or 9.

EXAMPLE 6 Planning a Tour

Suppose you decide to explore the upper Midwest next summer. You plan to travel by Greyhound bus and would like to visit Grand Forks, Fargo, Bemidji, St. Cloud, Duluth, Minneapolis, Escanaba, and Green Bay. A check of the Greyhound website shows direct bus links between destinations as indicated.

City	*Direct bus links with*
Grand Forks	Bemidji, Fargo
Fargo	Grand Forks, St. Cloud, Minneapolis
Bemidji	Grand Forks, St. Cloud
St. Cloud	Bemidji, Fargo, Minneapolis
Duluth	Escanaba, Minneapolis
Minneapolis	Fargo, St. Cloud, Duluth, Green Bay
Escanaba	Duluth, Green Bay
Green Bay	Escanaba, Minneapolis

Draw a graph with vertices representing the destinations and edges representing the relation "there is a direct bus link." Is this a connected graph? Explain the significance of the graph.

SOLUTION

Begin with eight vertices representing the eight destinations as in Figure 11. Use the bus link information to fill in the edges of the graph.

The first letter of the city name is shown, except Y stands for Green Bay.

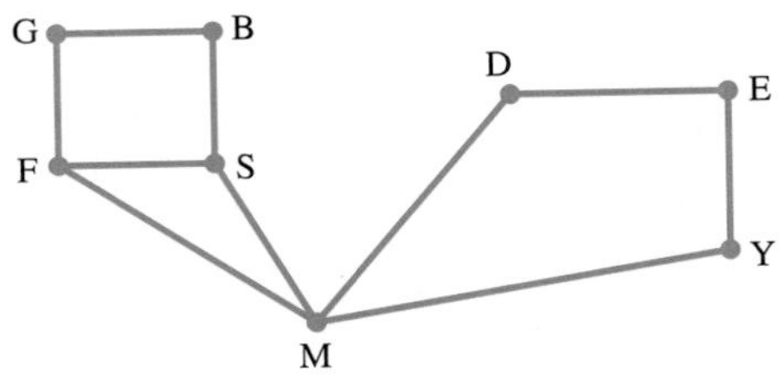

FIGURE 11

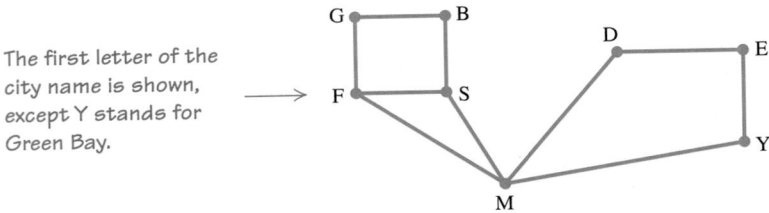

Figure 11 is a connected graph. It shows that you can travel by bus from any of your chosen destinations to any other. ■

Maps of subway systems in large cities usually do not faithfully represent the positions of stations, nor the distances between them. They are essentially graphs, with vertices representing stations, and edges showing subway links.

Walks, Paths, and Circuits

Example 6 suggests several important ideas. For example, what trips could you take among the eight destinations, using only direct bus links? If you do not mind riding the same bus route more than once, you could take this route: M → S → F → S → B. Note that you used the St. Cloud to Fargo link twice. You could not take the M → D → S route, however, since there is no direct bus link from Duluth to St. Cloud. Each possible route on the graph is called a *walk*.

Walk

A **walk** in a graph is a sequence of vertices, each linked to the next vertex by a specified edge of the graph.

We can think of a walk as a route we can trace with a pencil without lifting the pencil from the graph. M → S → F → S → B and F → M → D → E are walks; M → D → S is not a walk, however, since there is no edge between D and S.

Suppose that the travel time for your trip is restricted, and you do not want to use any bus link more than once. A *path* is a special kind of walk.

Path

A **path** in a graph is a walk that uses no edge more than once.*

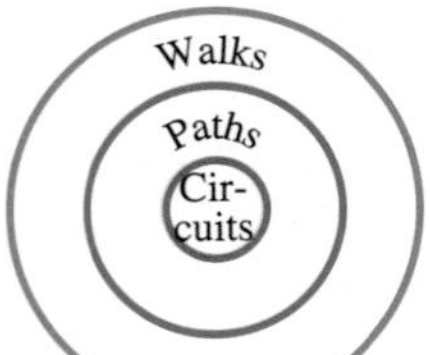

This diagram shows a **Venn diagram** relationship among walks, paths, and circuits.

M → S → F → M, F → M → D → E, and M → S → F → M → D are paths. Trace these paths with your finger on the graph in Figure 11. Note that a path may use a *vertex* more than once. B → S → F → M → S → B is not a path, because it uses the edge BS more than once. M → D → S is not a path, as it is not even a walk.

Perhaps you want to begin and end your tour in the same city (and still not use any bus route more than once). A path such as this is known as a *circuit.* Two circuits you could take are: M → S → F → M and M → S → F → M → D → E → Y → M.

*Some books call this a *trail* and use the term *path* for a walk that visits no vertex more than once.

Circuit

A **circuit** in a graph is a path that begins and ends at the same vertex.

Notice that a circuit is a kind of path, and therefore is also a kind of walk.

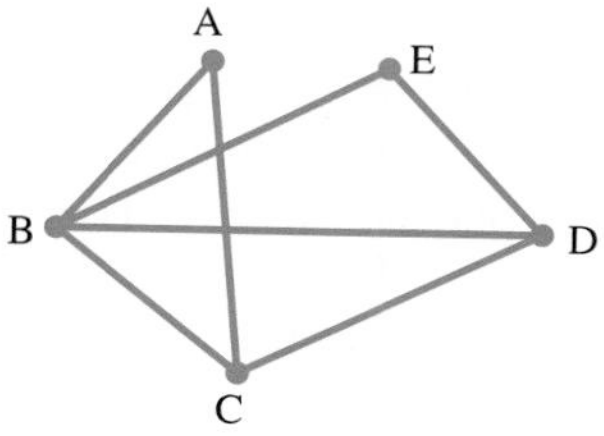

FIGURE 12

EXAMPLE 7 Classifying Walks

Using the graph in Figure 12, classify each sequence as a walk, a path, or a circuit.

(a) $E \to C \to D \to E$ **(b)** $A \to C \to D \to E \to B \to A$

(c) $B \to D \to E \to B \to C$ **(d)** $A \to B \to C \to D \to B \to A$

SOLUTION

	Walk	Path	Circuit
(a)	No (no edge E to C)	No*	No*
(b)	Yes	Yes	Yes
(c)	Yes	Yes	No
(d)	Yes	No (edge AB is used twice)	No**

* If a sequence of vertices is not a walk, it cannot possibly be either a path or a circuit.

** If a sequence of vertices is not a path, it cannot possibly be a circuit, since a circuit is defined as a special kind of path.

When planning your trip around the Upper Midwest, it would be useful to know the length of time the bus ride takes. We write the time for each bus ride on the graph, as shown in Figure 13.

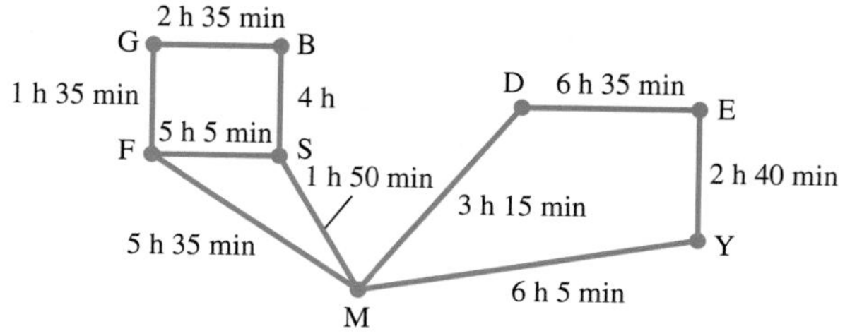

FIGURE 13

A graph with numbers on the edges as in Figure 13 is known as a **weighted graph.** The numbers on the edges are called **weights.** Weighted graphs will be important in later sections of this chapter.

Mazes have always fascinated people. Mazes are found on coins from ancient Knossos (Greece), in the sand drawings at Nazca in Peru, on Roman pavements, on the floors of renaissance cathedrals in Europe, and, of course, as hedge mazes in gardens. Graphs can be used to clarify the structure of mazes.

PROBLEM-SOLVING HINT Drawing a sketch is a useful problem-solving strategy even when the original problem does not specifically refer to a geometric shape. Example 8 illustrates how drawing a sketch can simplify a problem that at first seems complicated.

EXAMPLE 8 Analyzing a Round-Robin Tournament

In a round-robin tournament, every contestant plays every other contestant. The winner is the contestant who wins the most games. Suppose six tennis players compete in a round-robin tournament. How many matches will be played in the tournament?

SOLUTION

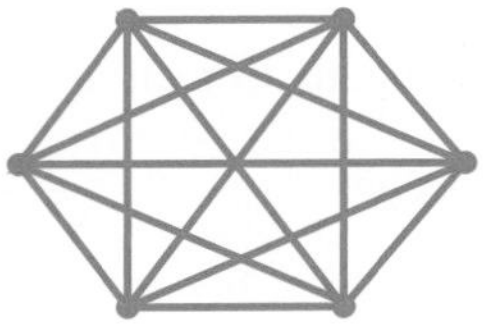

FIGURE 14

To help solve the problem, draw a graph with six vertices to represent the six players. Then, as in Figure 14, draw an edge for each match between a pair of contestants. We can count the edges to find out how many matches must be played. Or, we can reason as follows: There are 6 vertices each of degree 5, so the sum of the degrees of the vertices in this graph is $6 \cdot 5$, which is 30. We know that the sum of the degrees is twice the number of edges. Therefore, there are 15 edges in this graph and, thus, 15 matches in the tournament.

Complete Graphs and Subgraphs

The graph in Figure 14 is an example of a *complete graph.*

Complete Graph

A **complete graph** is a graph in which there is exactly one edge going from each vertex to each other vertex in the graph.

EXAMPLE 9 Deciding Whether a Graph Is Complete

Decide whether each of the graphs in Figure 15 is complete. If a graph is not complete, explain why it is not.

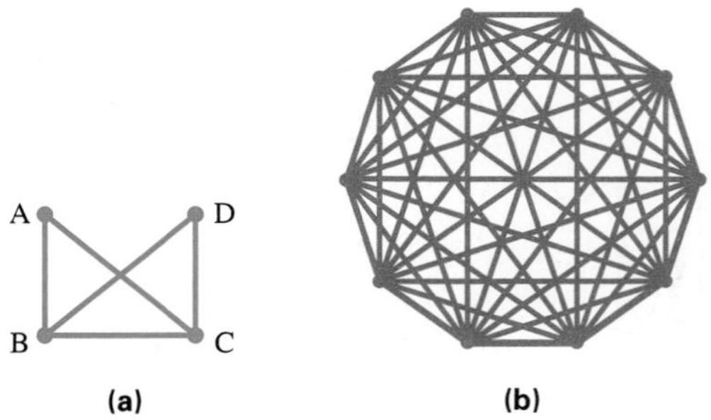

FIGURE 15

SOLUTION

Graph (a) is not complete, since there is no edge from vertex A to vertex D. Graph (b) is complete. (Check that there is an edge from each of the ten vertices to each of the other nine vertices in the graph.)

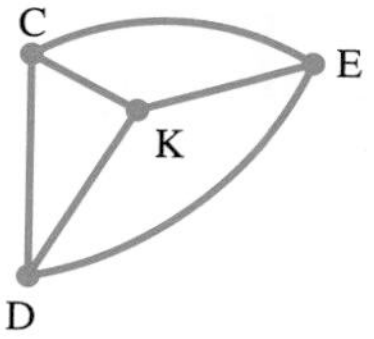

FIGURE 16

EXAMPLE 10 Finding a Complete Graph

Find a complete graph with four vertices in the preschoolers' friendship pattern depicted in Figure 3.

SOLUTION

Figure 16 shows a portion of the friendship graph from Figure 3. This is a complete graph with four vertices.

Do you see some complete graphs with three vertices in Figure 16? In general, a graph consisting of some of the vertices of the original graph and some of the original edges between those vertices is called a **subgraph.** Vertices or edges not included in the original graph cannot be in the subgraph. (Notice the similarity to the idea of subset, discussed earlier in the text.) A subgraph may include anywhere from one to all the vertices of the original graph and anywhere from none to all the edges of the original graph. Notice that a subgraph is a graph, and recall that in a graph every edge goes from a vertex to a vertex. Therefore, no edge can be included in a subgraph without the vertices at both of its ends also being included.

The subgraph shown in Figure 16 is a complete graph, but a subgraph does not have to be complete. For example, graphs (a) and (b) in Figure 17 are both subgraphs of the graph in Figure 3.

Graph (b) in Figure 17 is a subgraph even though it has no edges. (However, we could not form a subgraph by taking edges without vertices. In a graph every edge goes from a vertex to a vertex.)

Graph (c) in Figure 17 is *not* a subgraph of the graph in Figure 3, since there was no edge between M and T in the original graph.

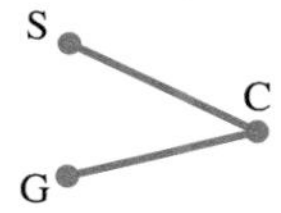

(a)

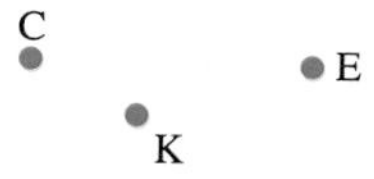

(b)

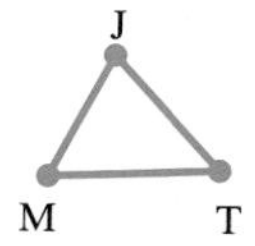

(c)

FIGURE 17

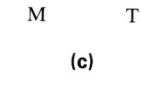

Graph Coloring Consider another application of graph theory. Tabatha works at the college learning center, and one of her tasks is to schedule study groups for the courses listed at the left below. Study groups for different courses must be scheduled at different times if there are students taking both courses. The column on the right shows, for each course, which of the other courses have students in common with it. For example, the first row shows that each of the courses statistics, chemistry, history, and physics has one or more students also taking algebra.

Course	***Other courses taken by students in this course***
Algebra	Statistics, Chemistry, History, Physics
History	Statistics, Chemistry, Biology, Algebra
Chemistry	Physics, Biology, Writing, Algebra, History
Biology	Physics, Statistics, Writing, History, Chemistry
Physics	Statistics, Algebra, Chemistry, Biology
Statistics	Writing, Algebra, History, Biology, Physics
Writing	Chemistry, Biology, Statistics

Tabatha's task is to figure out the least number of time slots needed for the seven study groups and to decide which (if any) can be scheduled at the same time.

We can draw a graph to show the information, with a vertex for each course and an edge between two vertices *if there are students taking both courses*. See Figure 18.

Each letter is the first letter of a course name.

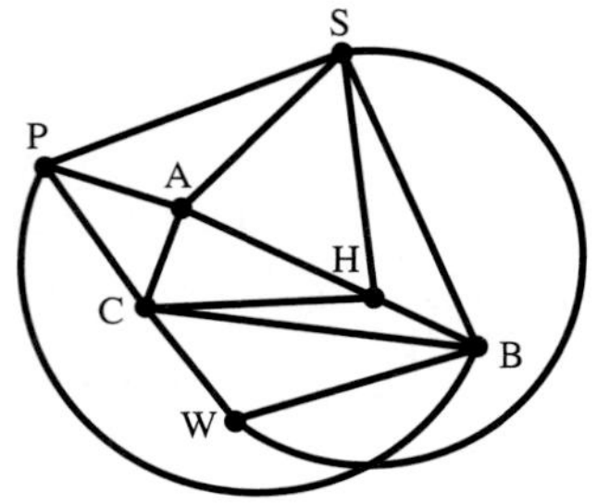

FIGURE 18

If there is an edge between two vertices, the corresponding courses have students in common, and the study groups must *not* meet at the same time. To show this we use a different color for each time slot for study groups. We will color the vertices with the different colors to show when the study groups could be scheduled. To make sure that study groups are at different times if the courses have students in common, we must make sure that any two vertices joined by an edge have different colors. To ensure the least number of time slots, we use as few colors as possible. In Figure 19, we color the vertices of the graph as required using three colors. We cannot color the graph as required with fewer than three colors.

Vertices that are joined by an edge must have different colors.

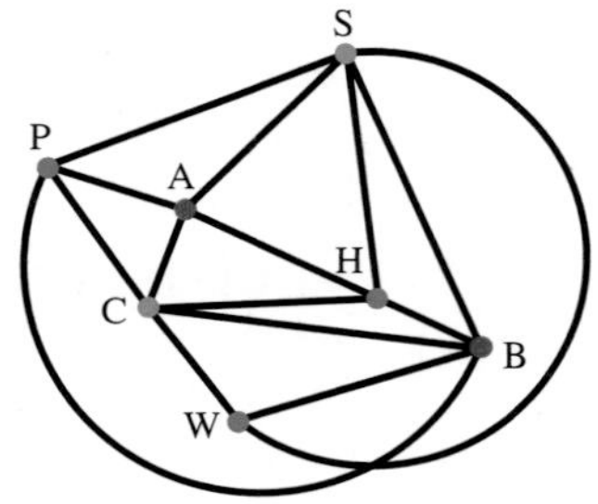

FIGURE 19

Tabatha can schedule the required study groups in no fewer than 3 time slots. She can schedule statistics and chemistry at one time, physics, history, and writing at a second time, and biology and algebra at a third time.

The method we used is called a *coloring** for the graph.

Coloring and Chromatic Number

A **coloring** for a graph is a coloring of the vertices in such a way that vertices joined by an edge have different colors. The **chromatic number** of a graph is the least number of colors needed to make a coloring.

---

*This is sometimes called a *vertex* coloring for the graph.

Have you ever had bad reception on your cell phone? This often occurs because signals are being sent at roughly the same frequency to too many cell phones located close together. Graph coloring can be used to help analyze this problem. However, with about one billion cell phone users in the world at present and a fairly narrow frequency range available for cell phone transmission, many believe that governments should make more frequencies available for cell phone usage.

Using this terminology, Tabatha had to determine the chromatic number for the graph in Figure 18. The chromatic number was 3, which equaled the least number of time slots needed for the study groups.

The idea of using vertex coloring to solve problems like this is straightforward. Determining the chromatic number for a graph with many vertices or producing a coloring using the least possible number of colors can be difficult. No one has yet found an efficient method for finding the exact chromatic number for arbitrarily large graphs. (The term *efficient algorithm* has a specific, technical meaning for computer scientists. See Section 15.3.) While the following method for coloring a graph may not give a coloring with the least number of colors, it may be helpful.

Coloring a Graph

Step 1 Choose a vertex with greatest degree, and color it. Use the same color to color as many vertices as you can without coloring two vertices the same color if they are joined by an edge.

Step 2 Choose a new color, and repeat what you did in Step 1 for vertices not already colored.

Step 3 Repeat Step 1 until all vertices are colored.

Graph coloring is used to solve many practical problems. It is useful in management science for solving scheduling problems, such as the study groups example above. Graph coloring is also related to allocating transmission frequencies to TV and radio stations and cell phone companies.

The coloring of graphs has an interesting history involving maps. Maps need to be colored so that territories with common boundaries have different colors. How many colors are needed? For map makers, four colors have always sufficed. In about 1850, mathematicians started pondering this. For over a hundred years, they were neither able to prove that four colors are always enough, nor find a map that needed more than four colors. This was called the **four-color problem.** Finally, in 1976, two mathematicians, Kenneth Appel and Wolfgang Haken, provided a proof that four colors are always enough. This proof created a controversy in mathematics, as it relied on computer calculations so lengthy (fifty 24-hour days of computer time) that no person could check the entire proof. Because of this, some mathematicians still feel that the four-color theorem has not yet been proved, and many keep looking for a shorter proof.

Answer to Chapter Opener Question

Olmos was not in a movie with Bacon, so Olmos' Bacon number must be greater than one.

- Edward James Olmos was in *Talent for the Game* (1991) with Terry Kinney.
- Terry Kinney was in *Sleepers* (1996) with Kevin Bacon.

Thus, Olmos has a Bacon number of two.

15.1 Exercises

For Exercises 1–6, determine how many vertices and how many edges each graph has.

1.

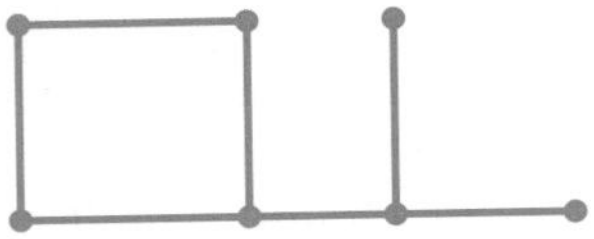

2.

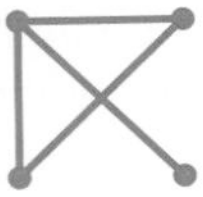

3.

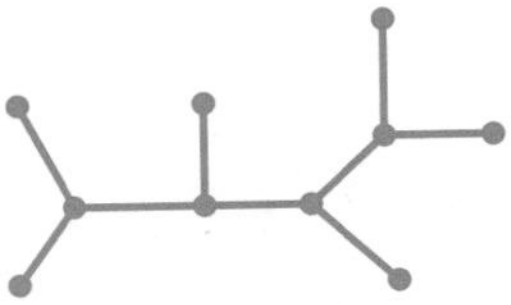

4.

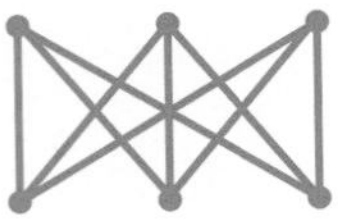

5.

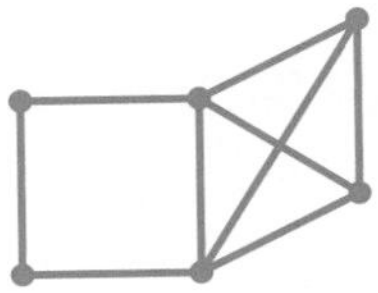

6.

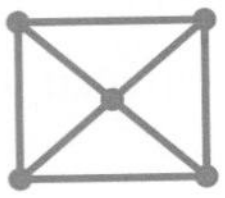

7.–10. For Exercises 7–10, refer to the graphs shown in Exercises 1–4. For each of these graphs, find the degree of each vertex in the graph. Then add the degrees to get the sum of the degrees of the vertices of the graph. What relationship do you notice between the sum of degrees and the number of edges?

In Exercises 11–16, determine whether the two graphs are isomorphic. If so, label corresponding vertices of the two graphs with the same letters and color-code corresponding edges, as in Example 2. (Note that there is more than one correct answer for many of these exercises.)

11.

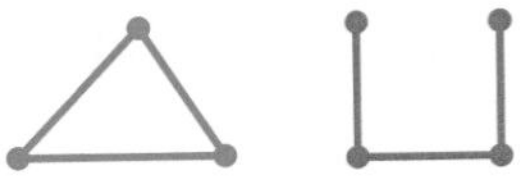

(a) (b)

12.

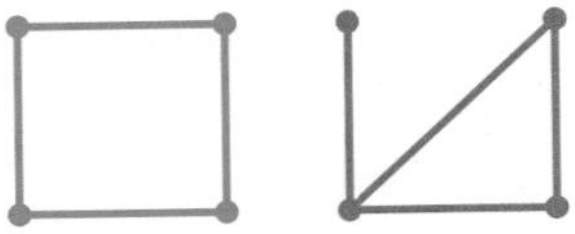

(a) (b)

13.

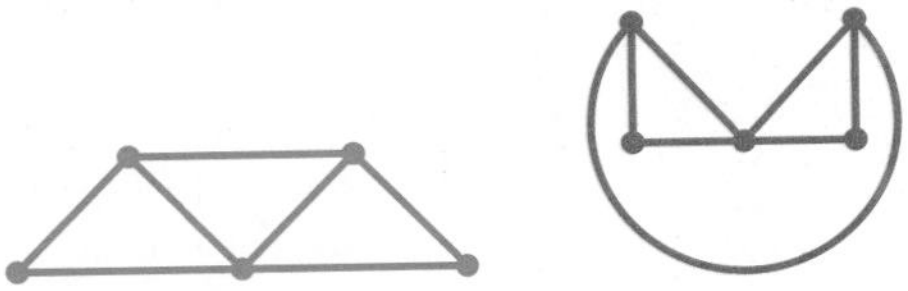

(a) (b)

14.

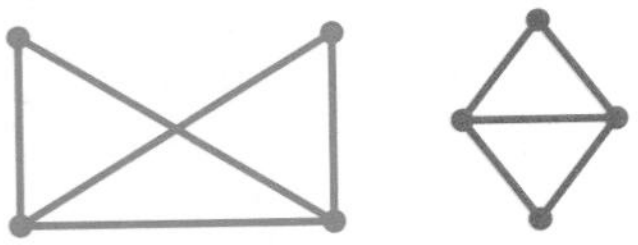

(a) (b)

15.

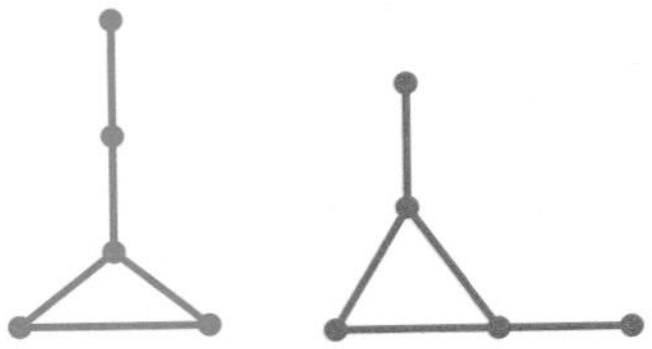

(a) (b)

16.

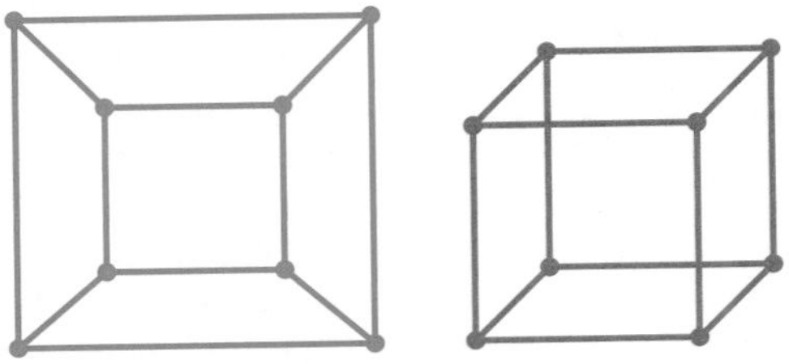

(a) (b)

In Exercises 17–22, determine whether the graph is connected or disconnected. Then determine how many components the graph has.

17.

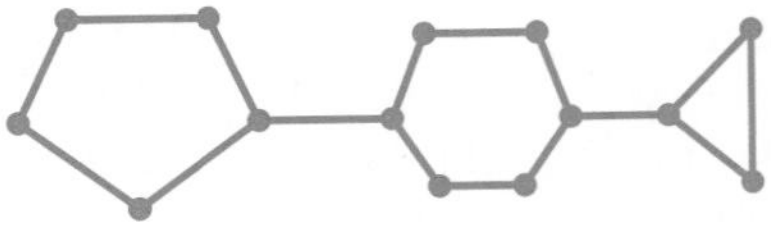

18.

19.

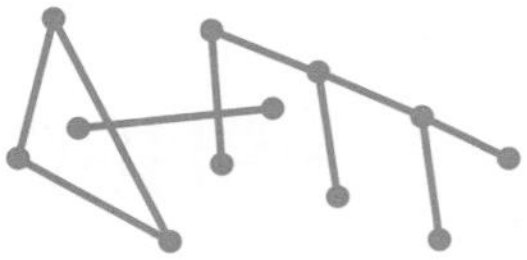

20.

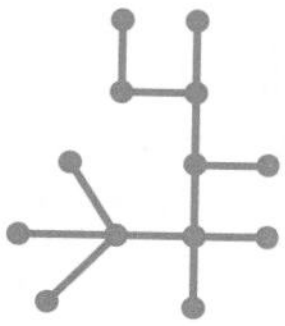

21.

22.

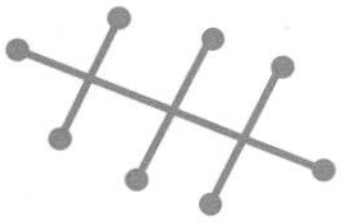

In Exercises 23–26, use the theorem that relates the sum of degrees to the number of edges to determine the number of edges in the graph (without drawing the graph).

23. A graph with 5 vertices, each of degree 4

24. A graph with 7 vertices, each of degree 4

25. A graph with 5 vertices, three of degree 1, one of degree 2, and one of degree 3

26. A graph with 8 vertices, two of degree 1, three of degree 2, one of degree 3, one of degree 5, and one of degree 6

In Exercises 27–29, refer to the following graph.

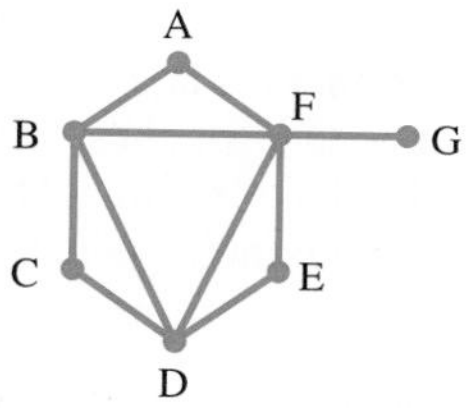

27. Which are walks in the graph? If not, why not?
(a) $A \rightarrow B \rightarrow C$
(b) $B \rightarrow A \rightarrow D$
(c) $E \rightarrow F \rightarrow A \rightarrow E$
(d) $B \rightarrow D \rightarrow F \rightarrow B \rightarrow D$
(e) $D \rightarrow E$
(f) $C \rightarrow B \rightarrow C \rightarrow B$

28. Which are paths in the graph? If not, why not?
(a) $B \rightarrow D \rightarrow E \rightarrow F$
(b) $D \rightarrow F \rightarrow B \rightarrow D$
(c) $B \rightarrow D \rightarrow F \rightarrow B \rightarrow D$
(d) $D \rightarrow E \rightarrow F \rightarrow G \rightarrow F \rightarrow D$
(e) $B \rightarrow C \rightarrow D \rightarrow B \rightarrow A$
(f) $A \rightarrow B \rightarrow E \rightarrow F \rightarrow A$

29. Which are circuits in the graph? If not, why not?
(a) $A \rightarrow B \rightarrow C \rightarrow D \rightarrow E \rightarrow F$
(b) $A \rightarrow B \rightarrow D \rightarrow E \rightarrow F \rightarrow A$
(c) $C \rightarrow F \rightarrow E \rightarrow D \rightarrow C$
(d) $G \rightarrow F \rightarrow D \rightarrow E \rightarrow F$
(e) $F \rightarrow D \rightarrow F \rightarrow E \rightarrow D \rightarrow F$

In Exercises 30 and 31, refer to the following graph.

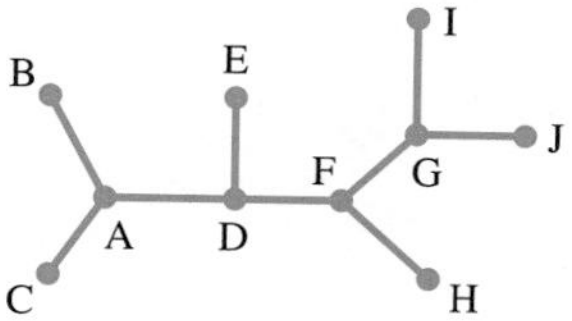

30. Which are walks in the graph? If not, why not?
(a) $F \rightarrow G \rightarrow J \rightarrow H \rightarrow F$
(b) $D \rightarrow F$
(c) $B \rightarrow A \rightarrow D \rightarrow F \rightarrow H$
(d) $B \rightarrow A \rightarrow D \rightarrow E \rightarrow D \rightarrow F \rightarrow H$
(e) $I \rightarrow G \rightarrow J$
(f) $I \rightarrow G \rightarrow J \rightarrow I$

31. Which are paths in the graph? If not, why not?
(a) $A \rightarrow B \rightarrow C$
(b) $J \rightarrow G \rightarrow I \rightarrow G \rightarrow F$
(c) $D \rightarrow E \rightarrow I \rightarrow G \rightarrow F$
(d) $C \rightarrow A$
(e) $C \rightarrow A \rightarrow D \rightarrow E$
(f) $C \rightarrow A \rightarrow D \rightarrow E \rightarrow D \rightarrow A \rightarrow B$

In Exercises 32–37, refer to the following graph. In each case, determine whether the sequence of vertices is **(i)** *a walk,* **(ii)** *a path,* **(iii)** *a circuit in the graph.*

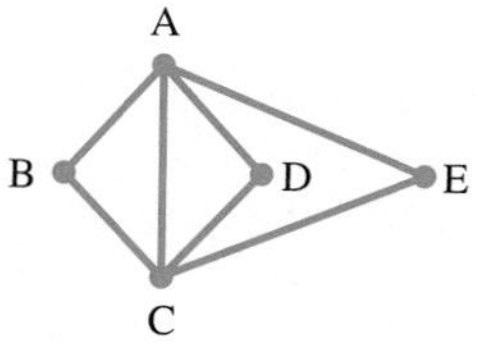

32. $A \rightarrow B \rightarrow C \rightarrow D \rightarrow E$

33. $A \rightarrow B \rightarrow C$

34. $A \rightarrow B \rightarrow C \rightarrow D \rightarrow A$

35. $A \rightarrow B \rightarrow A \rightarrow C \rightarrow D \rightarrow A$

36. $A \rightarrow B \rightarrow C \rightarrow A \rightarrow D \rightarrow C \rightarrow E \rightarrow A$

37. $C \rightarrow A \rightarrow B \rightarrow C \rightarrow D \rightarrow A \rightarrow E$

In Exercises 38–43, determine whether the graph is a complete graph. If not, explain why it is not complete.

38.

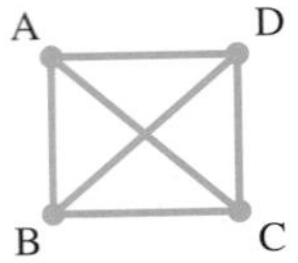

39.

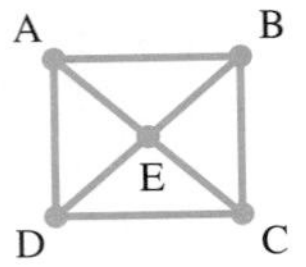

40.

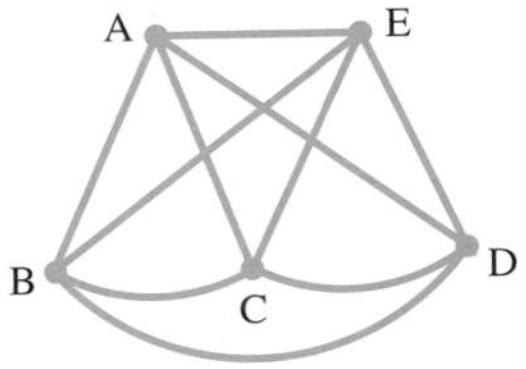

41.

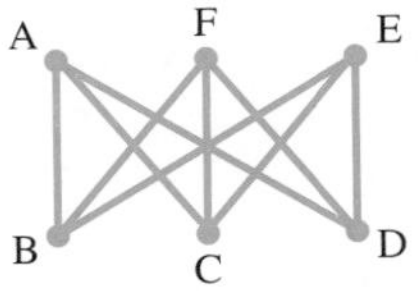

42.

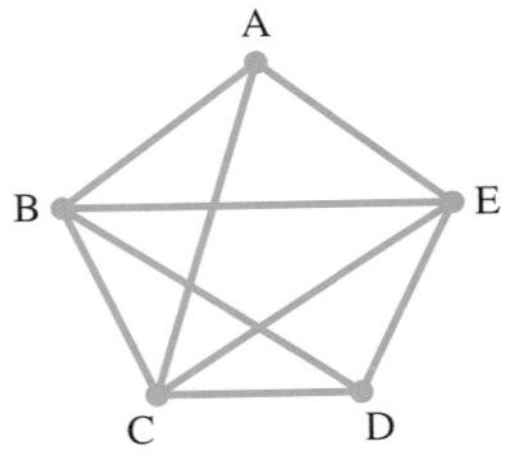

43.

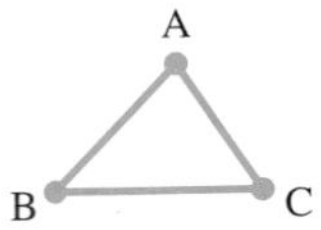

Solve each problem.

44. *Chess Competition* Students from two schools compete in chess. Each school has a team of four students. Each student must play one game against each student on the opposing team. Draw a graph with vertices representing the students, and edges representing the chess games. How many games must be played in the competition?

45. *Chess Competition* A chess master plays seven simultaneous games with seven other players. Draw a graph with vertices representing the players and edges representing the chess games. How many games are being played?

46. *Dancing Partners* At a party there were four males and five females. During the party each male danced with each female (and no female pair or male pair danced together). Draw a graph with vertices representing the people at the party and edges showing the relationship "danced with." How many edges are there in the graph?

47. *Number of Handshakes* There are six members on a hockey team (including the goalie). At the end of a hockey game, each member of the team shakes hands with each member of the opposing team. How many handshakes occur?

In Exercises 48 and 49, use the theorem that the sum of the degrees of the vertices in a graph is twice the number of edges.

48. *Number of Handshakes* There are seven people at a business meeting. One of these shakes hands with four people, four shake hands with two people, and two shake hands with three people. How many handshakes occur?

49. *Spread of a Rumor* A lawyer is preparing his argument in a libel case. He has evidence that a libelous rumor about his client was discussed in various telephone conversations among eight people. Two of the people involved had four telephone conversations in which the rumor was discussed, one person had three, four had two, and one had one such telephone conversation. How many telephone conversations were there in which the rumor was discussed among the eight people?

50. *Vertices and Edges of a Cube* Draw a graph with vertices representing the vertices (the corners) of a cube and edges representing the edges of the cube. In your graph, find a circuit that visits four different vertices. What figure does your circuit form on the actual cube?

51. ***Vertices and Edges of a Tetrahedron*** Draw a graph with vertices representing the vertices (or corners) of a tetrahedron and edges representing the edges of the tetrahedron. In your graph, identify a circuit that visits three different vertices. What figure does your circuit form on the actual tetrahedron?

52. ***Students in the Same Class*** Mary, Erin, Sue, Jane, Katy, and Brenda are friends at college. Mary, Erin, Sue, and Jane are in the same math class. Sue, Jane, and Katy take the same English composition class.

(a) Draw a graph with vertices representing the six students and edges representing the relation "take a common class."

(b) Is the graph connected or disconnected? How many components does the graph have?

(c) In your graph, identify a subgraph that is a complete graph with four vertices.

(d) In your graph, identify three different subgraphs that are complete graphs with three vertices. (There are several correct answers.)

53. Here is another theorem about graphs: *In any graph, the number of vertices with odd degree must be even.* Explain why this theorem is true. (*Hint:* Use the theorem about the relationship between the number of edges and the sum of degrees.)

54. Draw two nonisomorphic (simple) graphs with 6 vertices, with each vertex having degree 3.

55. Explain why the graphs in Exercise 54 are not isomorphic.

56. ***Analyzing a Cube with a Graph*** Draw a graph whose vertices represent the *faces* of a cube and in which an edge between two vertices shows that the corresponding faces of the actual cube share a common boundary. What is the degree of each vertex in your graph? What does the degree of any vertex in your graph tell you about the actual cube?

57. Graphs may be used to clarify the structure of mazes. Vertices represent entrances and points in the maze where there is a dead end or a choice of two or more edges by which to proceed. For example, the 1690 design for the hedge maze at Hampton Court in England and a corresponding graph appear in the next column. Write a paper on mazes. Use graphs to analyze the structures of the mazes you discuss. You may want to focus on a particular kind of maze, such as mazes in gardens, mazes in art and architecture, or mazes in ancient cultures.

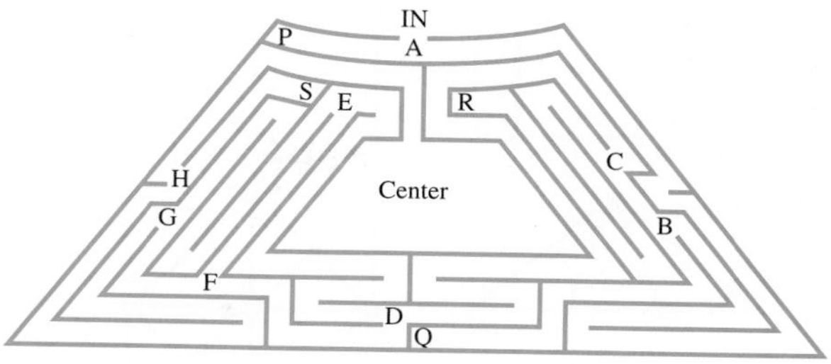

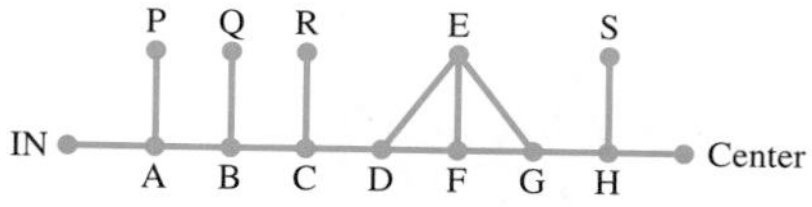

58. Graphs may be used to clarify the rhyme schemes in poetry. Vertices represent words at the end of a line, edges are drawn, and a path is used to indicate the rhyme scheme. For example, here is the first stanza of *Ode to Autumn* by John Keats, with the rhyme scheme analyzed using a graph:

Season of mists and mellow fruitfulness, A
Close bosom-friend of the maturing sun; B
Conspiring with him how to load and bless A
With fruit the vines that round the thatch-eves run; B
To bend with apples the moss'd cottage-trees C
And fill all fruit with ripeness to the core; D
To swell the gourd, and plump the hazel shells E
With a sweet kernel; to set budding more, D
And still more, later flowers for the bees, C
Until they think that warm days will never cease, C
For Summer has o'er brimm'd their clammy cells. E

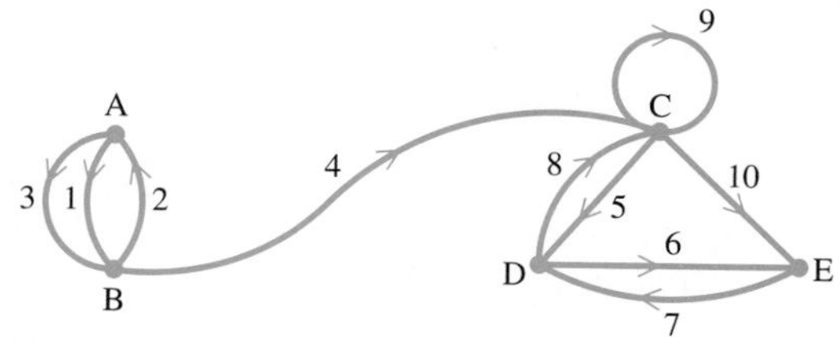

Write a paper on rhyme schemes in poetry, using graphs to illustrate. Focus either on a particular poet or on a particular type of poem.

In Exercises 59 through 64 on the next page, color the graph using as few colors as possible. Determine the chromatic number of the graph. (Hint: the chromatic number is fixed, but there may be more than one correct coloring.)

59. 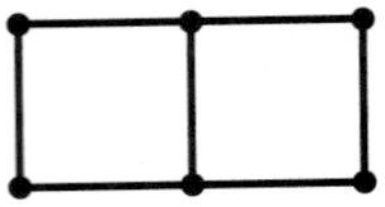

60. 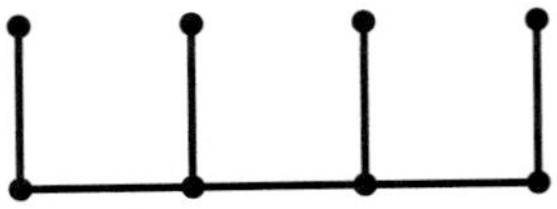

61. 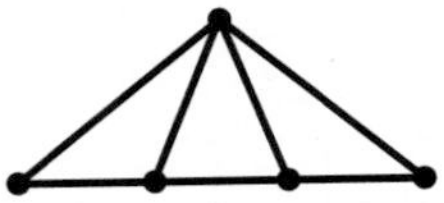

62. 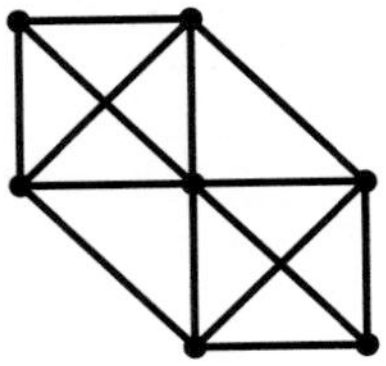

63. 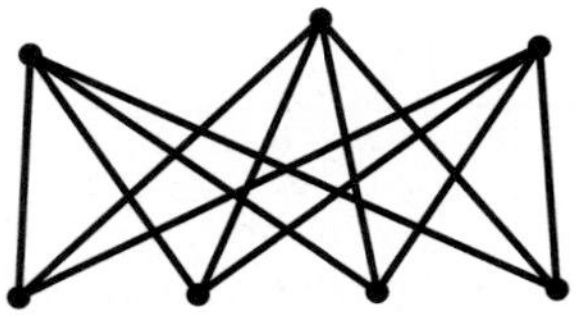

64. 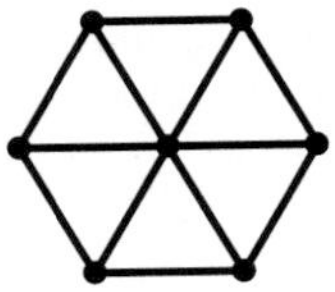

65. Color each graph using as few colors as possible. Use this to determine the chromatic number of the graph. (Graphs like these are called **cycles.**)

(a)  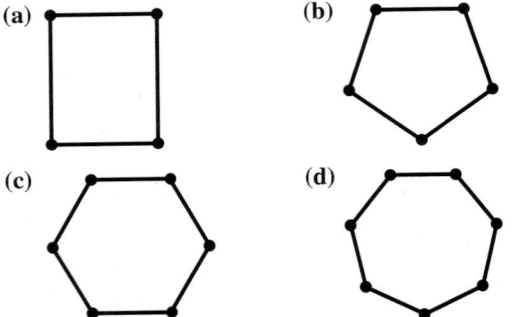

(b) 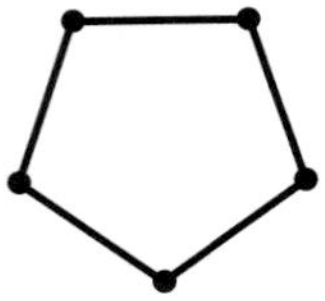

(c) 

(d) 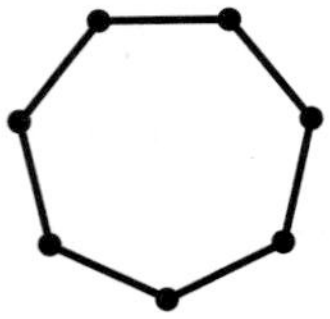

(e) Use your results from parts (a)–(d) to make a prediction about the chromatic number of a cycle. (*Hint:* consider two cases.)

66. Sketch a complete graph with the specified number of vertices, color the graph with as few colors as possible, and use this to determine the chromatic number of the graph.

(a) 3 vertices

(b) 4 vertices

(c) 5 vertices

(d) Write a general principle by completing this statement: A complete graph with n vertices has chromatic number ____.

(e) Why is the statement in part (d) true?

(f) Generalize further by completing this statement: If a graph has a *subgraph* that is a complete graph with n vertices, then the chromatic number of the graph must be at least ____.

In Exercises 67–70, color the vertices using as few colors as possible. Then state the chromatic number of the graph. (Hint: it might help to first identify the largest subgraph that is a complete graph.)

67. 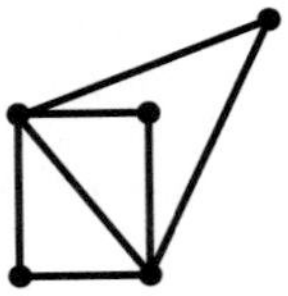

68. 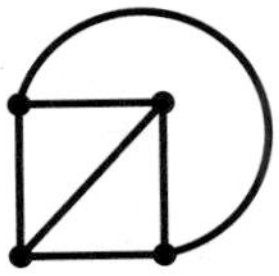

69. 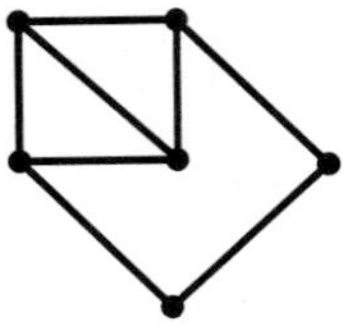

70. 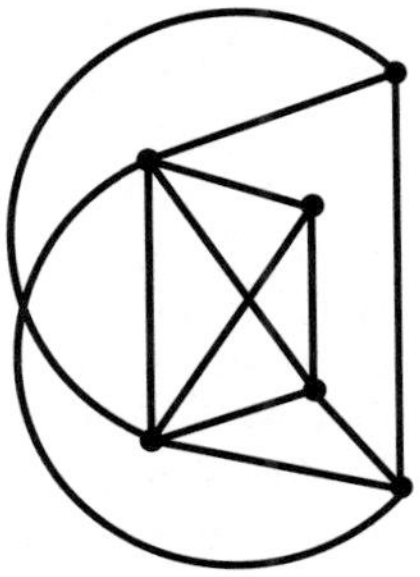

71. ***Scheduling Meeting Times*** At a college, Campus Life must schedule weekly meeting times for the six organizations listed below in such a way that organizations with members in common meet at different times. Use graph coloring to determine the least number of different meeting times and to decide which organizations should meet at the same time.

Organization	*Members also belong to*
Choir	Caribbean Club, Dance Club, Theater, Service Club
Caribbean Club	Dance Club, Service Club, Choir
Service Club	Forensics, Choir, Theater, Caribbean Club
Forensics	Dance Club, Service Club
Theater	Choir, Service Club
Dance Club	Choir, Caribbean Club, Forensics

72. ***Assigning Frequencies to Transmitters*** Interference can occur between radio stations. To avoid this, transmitters that are less than 60 miles apart must be assigned different broadcast frequencies. The transmitters for 7 radio stations are labeled A through G.

The information below shows, for each transmitter, which of the others are within 60 miles. Use graph coloring to determine the least number of different broadcast frequencies that must be assigned to the 7 transmitters. Include a plan for which transmitters could broadcast on the same frequency to use this least number of frequencies.

Transmitter	***Transmitters within 60 miles of this***
A	B, E, F
B	A, C, E, F, G
C	B, D, G
D	C, F, G
E	A, B, F
F	A, B, D, E, G
G	B, C, D, F

73. *Inviting Colleagues to a Gathering* Tiffany wants to invite her work colleagues to her home. However, several of her colleagues do not get along with each other, as summarized below. Tiffany plans to organize several gatherings, so that colleagues who do not get along are not there at the same time. Use graph coloring to determine the least number of gatherings needed to achieve this, and decide which colleagues should be invited each time.

- Joe does not get along with Brad and Phil.
- Caitlin does not get along with Lindsay and Brad.
- Lindsay does not get along with Phil.
- Mary does not get along with Lindsay.
- Eva gets along with everyone.

Graph Coloring *In Exercises 74–76, the graph shows the vertices and edges of the figure specified. Color the vertices of each figure in such a way that vertices with an edge between them have different colors. Use graph coloring to find a way to do this, and specify the least number of colors needed.*

74. Cube

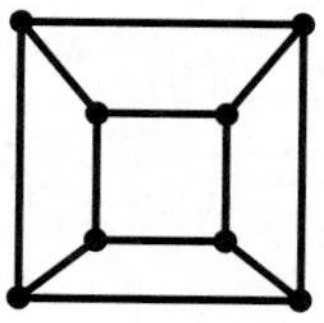

75. Octahedron

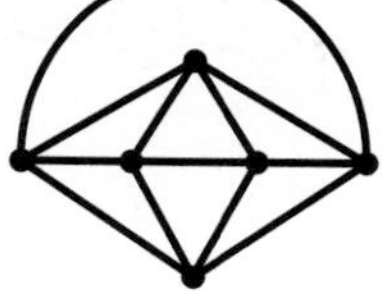

76. Dodecahedron

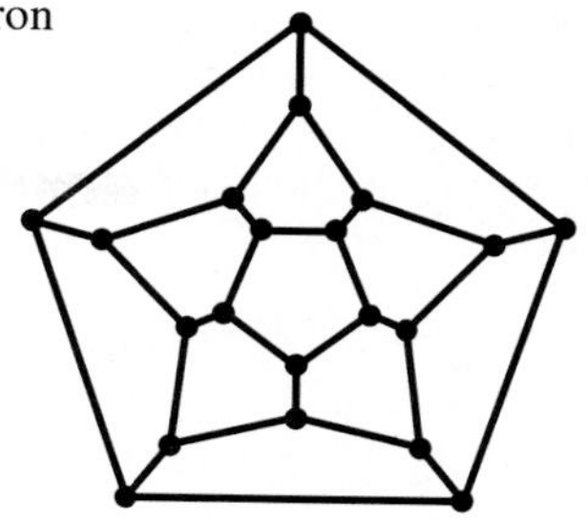

Graph Coloring *In Exercises 77 and 78, draw a vertex for each region shown in the map. Draw an edge between two vertices if the two regions have a common boundary. (Do not draw an edge if the regions meet at just one point.) Then find a coloring for the graph, using as few colors as possible.*

77.

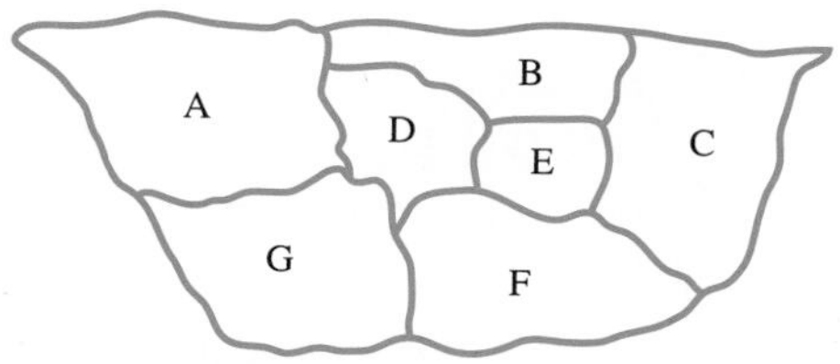

78.

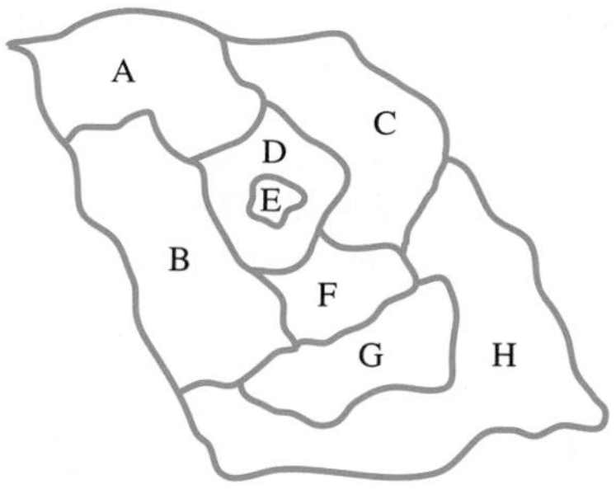

79. *Map Coloring* The map shows areas of upstate New York. Use graph coloring to determine the least number of colors that can be used to color the map so that areas with common boundaries have different colors.

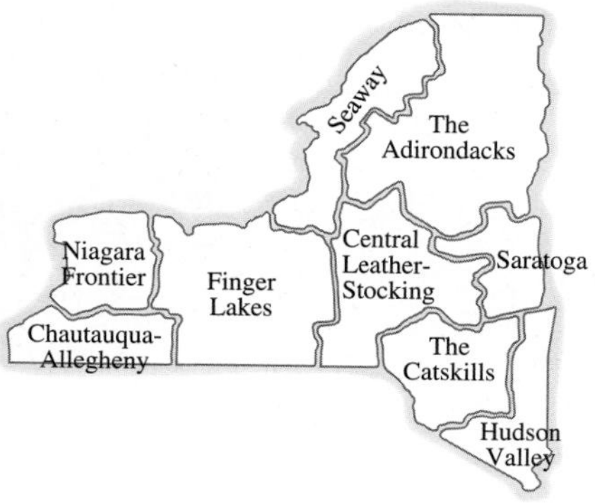

80. ***Map Coloring*** The map shows countries of Southern Africa. Use graph coloring to determine the least number of colors that can be used to color the map so that countries with common boundaries have different colors.

Map Drawing *In Exercises 81–83, draw a map with the specified number of countries that can be colored using the stated number of colors, and no fewer. (Countries may not consist of disconnected pieces.) If it is not possible to draw such a map, say why not.*

81. 6 countries, 3 colors

82. 6 countries, 4 colors

83. 6 countries, 5 colors

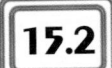

84. Write a short paper on the history of the four-color problem. Include discussion of the proof by Appel and Haken.

15.2 Euler Circuits

Königsberg Bridge Problem • Fleury's Algorithm

The ideas developed in this section are important for planning efficient routes for snowplows and mail delivery, and for many other practical applications.

Königsberg Bridge Problem In the early 1700s, the city of Königsberg was the capital of East Prussia. (Königsberg is now called Kaliningrad and is in Russia.) The river Pregel ran through the city in two branches with an island between the branches. Figure 20 shows an engraving of the city as it looked in the early 1700s.

A university was founded in Königsberg in 1544. It was established as a Lutheran center of learning. Its most famous professor was the philosopher **Immanual Kant,** who was born in Königsberg in 1724. The university was completely destroyed during World War II.

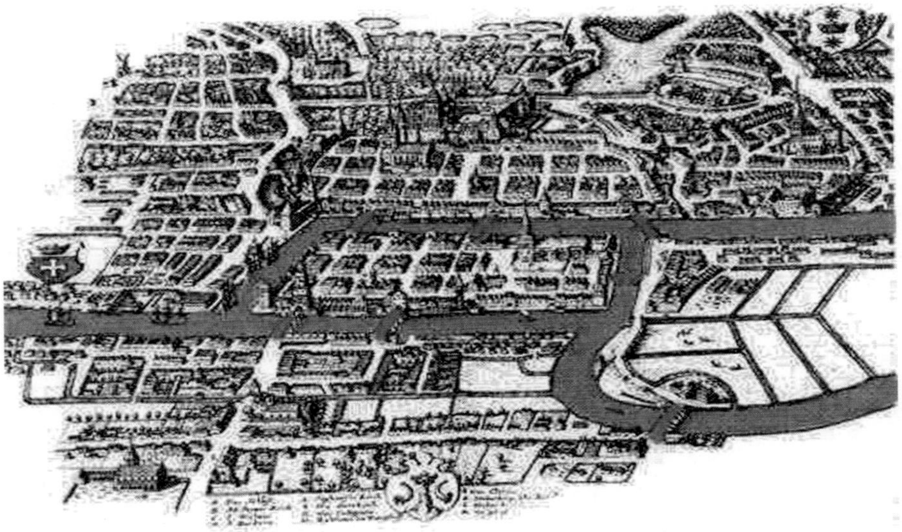

FIGURE 20

Leonhard Euler (1707–1783) was a devoted father and grandfather. He had thirteen children and frequently worked on mathematics with children playing around him. He lost the sight in his right eye at age 31, a couple of years after writing the Königsberg bridge paper. He became completely blind at age 58, but produced more than half his mathematical work after that. Euler wrote nearly a thousand books and papers.

There were seven bridges joining various parts of the city. Figure 21 shows a map of the river and bridges in the city. According to Leonhard Euler (pronounced "oiler"), the following problem was well known in his time (around 1730): Is it possible for a citizen of Königsberg to take a stroll through the city, crossing each bridge exactly once, and beginning and ending at the same place?

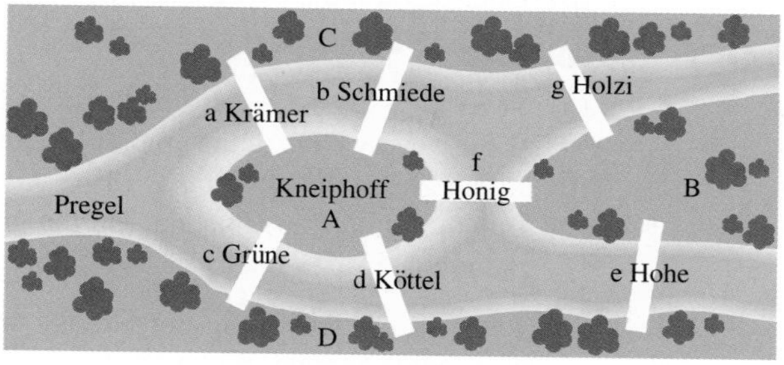

FIGURE 21

See if you can find such a route on the map in Figure 21. (You cannot cross the river anywhere except at the bridges shown on the map.)

We can simplify the problem by drawing the map as a graph. See Figure 22. The graph focuses on the relations between the vertices; namely the relation "there is a bridge." In this graph we have *two* edges between vertices A and C and also between vertices A and D; thus this graph is not a simple graph. We have labeled the edges with lowercase letters. Can you find a route as required in the puzzle?

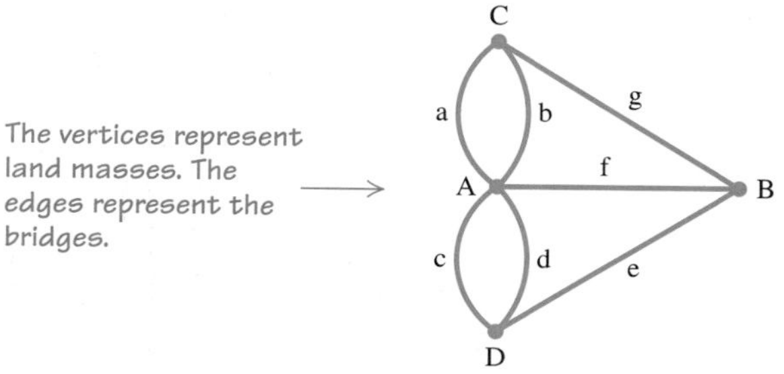

FIGURE 22

The Königsberg bridge problem requires more than a path, and more than a circuit. (Recall that a path is a walk that uses no edge more than once, and a circuit is a path that begins and ends at the same vertex.) The problem requires a circuit that uses *every* edge, *exactly* once. This is called an *Euler circuit.*

Euler Path and Euler Circuit

An **Euler path** in a graph is a path that uses every edge of the graph exactly once. An **Euler circuit** in a graph is a circuit that uses every edge of the graph exactly once.

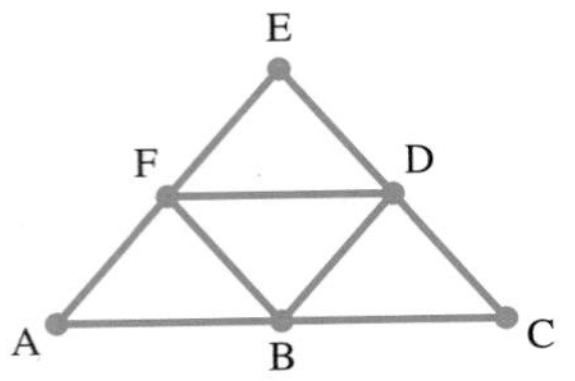

FIGURE 23

EXAMPLE 1 Recognizing Euler Circuits

Consider the graph in Figure 23.

(a) Is A → B → C → D → E → F → A an Euler circuit for this graph? Justify your answer.

(b) Does the graph have an Euler circuit?

SOLUTION

(a) A → B → C → D → E → F → A is a circuit, but not an Euler circuit, since it does not use every edge of the graph. For example, the edge BD is not used in this circuit.

(b) In the graph, the circuit A → B → C → D → E → F → D → B → F → A is an Euler circuit. Trace this path on the graph to check that it does use every edge of the graph exactly once. ■

Returning to the Königsberg bridge problem, it seemed that no route of the type required could be found. Why was this so? Euler published a paper in 1736 that explained why it is impossible to find a walk of the required kind, and he provided a simple way for deciding whether a given graph has an Euler circuit. Euler did not use the terms *graph* and *circuit*, but his paper was, in fact, the first paper on graph theory. In modern terms, Euler proved the first part of the theorem below. Part 2 was not proved until 1873.

The following is from Euler's paper on the **Königsberg bridge problem:** "I was told that while some denied the possibility of doing this and others were in doubt, no one maintained that it was actually possible. On the basis of the above, I formulated the following very general problem for myself: Given any configuration of the river and the branches into which it may divide, as well as any number of bridges, to determine whether or not it is possible to cross each bridge exactly once." (*Source:* Newman, James R. (ed.), "The Konigsberg Bridges," *Scientific American*, July 1953, in *Readings from Scientific American: Mathematics in the Modern World*, 1968.)

Euler's Theorem

Suppose we have a connected graph.

1. If the graph has an Euler circuit, then each vertex of the graph has even degree.
2. If each vertex of the graph has even degree, then the graph has an Euler circuit.

What does Euler's theorem predict about the connected graph in Figure 23? The degrees of the vertices are as follows:

A: 2 B: 4 C: 2 D: 4 E: 2 F: 4.

Each vertex has even degree, so it follows from part 2 of the theorem that the graph has an Euler circuit. (We already knew this, since we found an Euler circuit for this graph in Example 1.)

What does Euler's theorem suggest about the Königsberg bridge problem? Note that the degree of vertex C in Figure 22 is 3, which is an odd number, and the graph is connected. So, by part 1 of the theorem, the graph cannot have an Euler circuit. (In fact, the graph has many vertices with odd degree, but the presence of *any* vertex of odd degree indicates that the graph does not have an Euler circuit.)

Why is Euler's theorem true? Consider the first part of the theorem:

If a connected graph has an Euler circuit, then each vertex has even degree.

Suppose we have a connected graph, and an Euler circuit in the graph. Suppose the circuit begins and ends at a vertex A. Imagine traveling along the Euler circuit, from A all the way back to A, putting an arrow on each edge in your direction of travel. (Of course, you travel along each edge exactly once.) Now consider any vertex B in the graph. It has edges joined to it, each with an arrow; some of the arrows point toward B, some point away from B. The number of arrows pointing toward B is the same as the number of arrows pointing away from B, for every time you visit B, you come in on one unused edge and leave via a

By 1875 a new bridge had been built in Königsberg, joining the land areas we labeled C and D. It was now possible for a citizen of Königsberg to take a walk that used each bridge exactly once, provided he or she started at A and ended at B, or vice versa.

different unused edge. In other words, you can pair off the edges joined to B, with each pair having one arrow pointing toward B and one arrow pointing away from B. Thus, the total number of edges joined to B must be an even number and, therefore, the degree of B is even. A similar argument shows that our starting vertex, A, also has even degree.

The second part of Euler's theorem states the following:

If each vertex of a connected graph has even degree,
then the graph has an Euler circuit.

To show that this is true, we will provide a recipe (Fleury's algorithm) for finding an Euler circuit in any such graph. First, however, we consider more examples of how Euler's theorem can be applied.

EXAMPLE 2 Using Euler's Theorem

Decide which of the graphs in Figure 24 has an Euler circuit. Justify your answers.

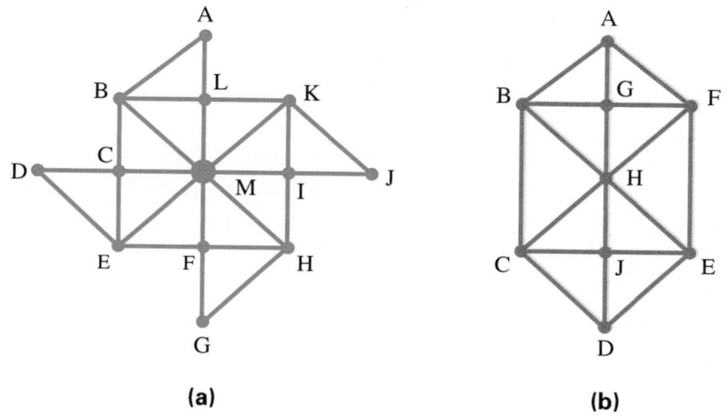

FIGURE 24

SOLUTION

Graph (a): The graph is connected. If we write the degree next to each vertex on the graph, as in Figure 25, we see that each vertex has even degree. Since the graph is connected, and each vertex has even degree, it follows from Euler's theorem that the graph has an Euler circuit.

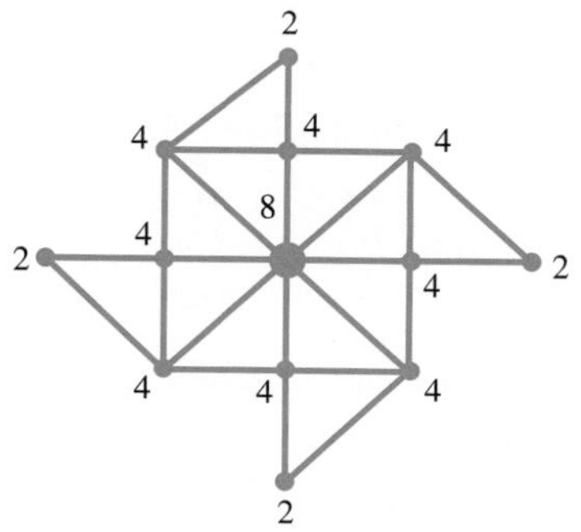

FIGURE 25

Graph (b): Check the degrees of the vertices. Note that vertex A has degree three, an odd degree. We need look no further. It follows from Euler's theorem that the connected graph (b) in Figure 24 does not have an Euler circuit. (All the other vertices in this graph except for D have even degree; however, if just one vertex has odd degree, then the graph does not have an Euler circuit.)

PROBLEM-SOLVING HINT When solving a challenging problem, it is useful to ask yourself whether the problem is related to another problem that you do know how to solve. Example 3 illustrates this strategy.

EXAMPLE 3 Tracing Patterns

Beginning and ending at the same place, is it possible to trace the pattern shown in Figure 26(a) without lifting the pencil off the page and without tracing over any part of the pattern more than once?

(a)

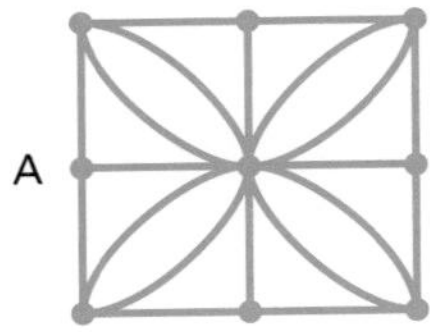

(b)

FIGURE 26

SOLUTION

The pattern in (a) is not a graph. (It has no vertices.) But what we are being asked to do is very much like being asked to find an Euler circuit. So we imagine that there are vertices at points where lines or curves of the pattern meet, obtaining the graph in Figure 26(b).

Now we can use Euler's theorem to decide whether we can trace the pattern. Notice that vertex A has odd degree. Thus, by Euler's theorem, this graph does not have an Euler circuit. It follows that we cannot trace the original pattern in the manner required.

The tracing problem in Example 3 illustrates a practical problem that occurs when designing robotic arms to trace patterns. Automated machines that engrave identification tags for pets require such a mechanical arm.

Fleury's Algorithm

Fleury's algorithm can be used to find an Euler circuit in any connected graph in which each vertex has even degree. An algorithm is like a recipe; follow the steps and you achieve what you need. Before introducing Fleury's algorithm, we need a definition.

Cut Edge

A **cut edge** in a graph is an edge whose removal disconnects a component of the graph.

We call such an edge a cut edge, since removing the edge *cuts* a connected piece of a graph into two pieces.*

*Some mathematicians use "bridge" instead of "cut edge" to describe such an edge.

The word **algorithm** comes from the name of the Persian mathematician, **Abu Ja'far Muhammad ibn Musa Al-Khowârizmî,** who lived from about A.D. 780 to A.D. 850. One of the books he wrote was on the Hindu-Arabic system of numerals (the system we use today), and calculations using them. The Latin translation of this book was called *Algoritmi de numero Indorum,* meaning "Al-Khowârizmî on the Hindu Art of Reckoning."

EXAMPLE 4 Identifying Cut Edges

Identify the cut edges in the graph in Figure 27.

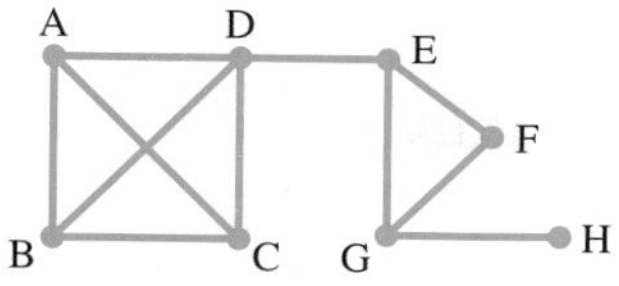

FIGURE 27

SOLUTION

This graph has only one component, so we must look for edges whose removal would disconnect the graph.

DE is a cut edge; if we removed this edge, we would disconnect the graph into two components, obtaining the graph of Figure 28(a). HG is also a cut edge; its removal would disconnect the graph as shown in Figure 28(b).

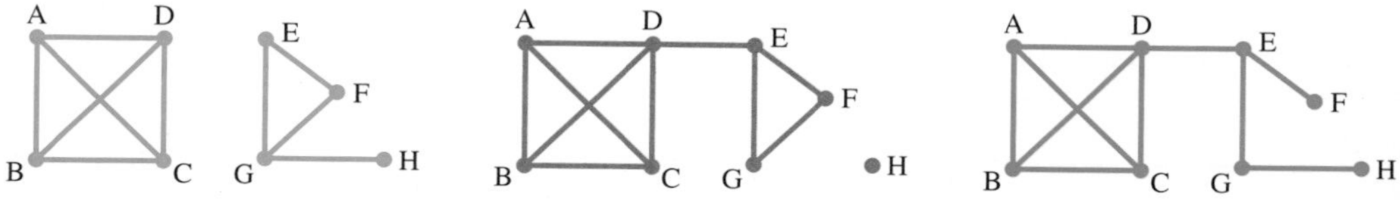

(a) (b) (c)

FIGURE 28

None of the other edges is a cut edge; we could remove any *one* of the other edges and still have a connected graph (that is, we would still have a path from each vertex of the graph to each other vertex). For example, if we remove edge GF, we end up with a graph that still is connected, as shown in Figure 28(c).

Recall that Fleury's algorithm is for finding an Euler circuit in a connected graph in which each vertex has even degree.

Fleury's Algorithm

Step 1 Start at any vertex. Go along any edge from this vertex to another vertex. *Remove this edge from the graph.*

Step 2 You are now on a vertex of the revised graph. Choose any edge from this vertex, subject to only one condition: do not use a cut edge (*of the revised graph*) unless you have no other option. Go along your chosen edge. *Remove this edge from the graph.*

Step 3 Repeat Step 2 until you have used all the edges and gotten back to the vertex at which you started.

The **Tshokwe people of northeastern Angola** have a tradition involving continuous patterns in sand. The tracings are called *sona* and they have ritual significance. They are used as mnemonics for stories about the gods and ancestors. Elders trace the drawings while telling the stories. The sona shown here is called *skin of a leopard*. The men draw the grid of dots first, as an aid tc recalling the sona. *Source:* Gerdes, Paul. *Geometry from Africa – Mathematical and Educational Explorations.* Mathematical Association of America, 1999, p.171.

EXAMPLE 5 Using Fleury's Algorithm

Find an Euler circuit for the graph in Figure 29.

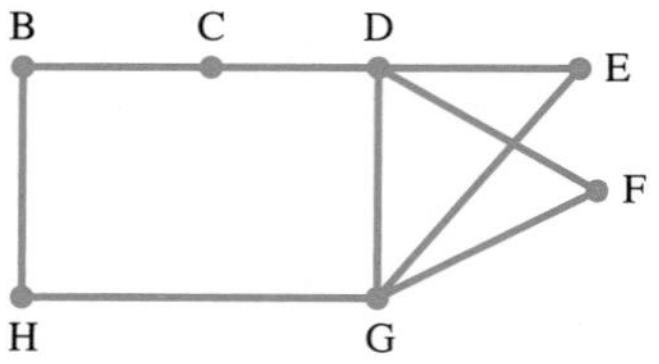

FIGURE 29

SOLUTION

First we must check that this graph does indeed have an Euler circuit. Check that the graph is connected, and that each vertex has even degree.

We can start at any vertex; let's start at C. We could go from C to B or from C to D. Let's go to D, removing the edge CD. (We simply show it scratched out, but you must think of this edge as gone.) Our path begins: $C \rightarrow D$. We show the revised graph in Figure 30(a).

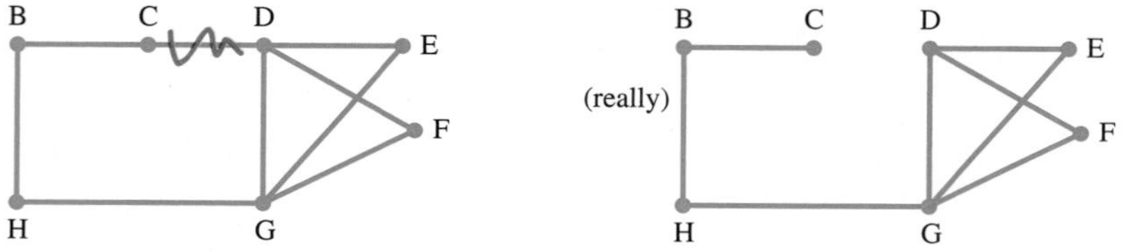

FIGURE 30(a)

Note that none of the edges DE, DF, or DG are cut edges for our current graph, so we can choose to go along any one of these. Let's go from D to G, removing edge DG. So far, our path is $C \rightarrow D \rightarrow G$. (Be sure to keep a record of your path as you go.) Figure 30(b) shows the revised graph.

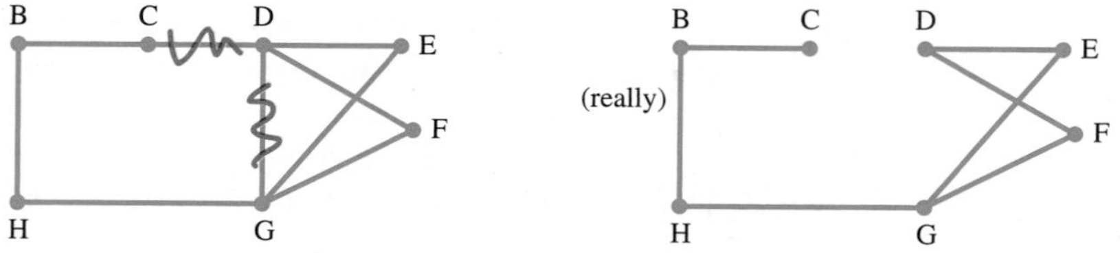

FIGURE 30(b)

The remaining edges at G are GE, GF, and GH. Note, however, that GH is a cut edge for our revised graph, and we have other options. Thus, according to Fleury's algorithm, *we must not use GH at this stage.* (If you used GH now, you would never be able to get back to use the edges GF, GE, and so on.) Let's go from G to F. Our path is $C \rightarrow D \rightarrow G \rightarrow F$, and we show our revised graph in Figure 30(c).

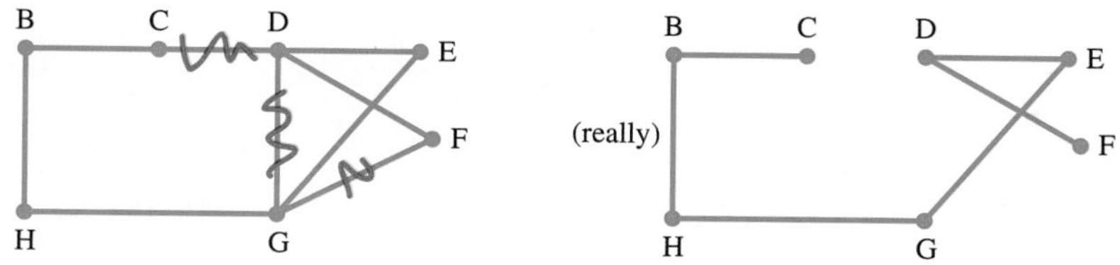

FIGURE 30(c)

Now we have only one edge from F, namely, FD. This is a cut edge of the revised graph, but *since we have no other option* we may use this edge. Our path is $C \rightarrow D \rightarrow G \rightarrow F \rightarrow D$, and our revised graph is shown in Figure 30(d).

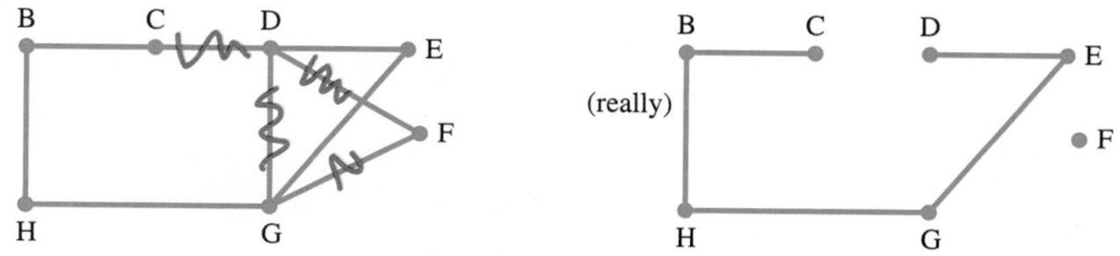

FIGURE 30(d)

We have only one edge left at D, namely DE. It is a cut edge of the remaining graph, but since we have no other option, we may use it.

It is now clear how to complete the Euler circuit. (Be sure to return to the starting vertex.) The complete Euler circuit is: $C \rightarrow D \rightarrow G \rightarrow F \rightarrow D \rightarrow E \rightarrow G \rightarrow H \rightarrow B \rightarrow C$. ■

If we start with a connected graph with all vertices having even degree, then Fleury's algorithm will always produce an Euler circuit for the graph. ***Note that a graph that has an Euler circuit always has more than one Euler circuit.***

When we have an algorithm for performing some task, we can program a computer to do it for us. Even for a large graph, a computer can quickly apply Fleury's algorithm. This is not the case for all algorithms. In the next section, we shall encounter an algorithm that, when used for large graphs, requires too much time for even the fastest computers.

Euler circuits have many practical applications. Consider planning a route for mail delivery. Figure 31 on the next page shows a map of a rural district. In rural areas, mailboxes are placed along the same side of a road, so the mail delivery vehicle needs to travel along the road in one direction only. Suppose the roads off the main road in Figure 31 show a mail delivery region.

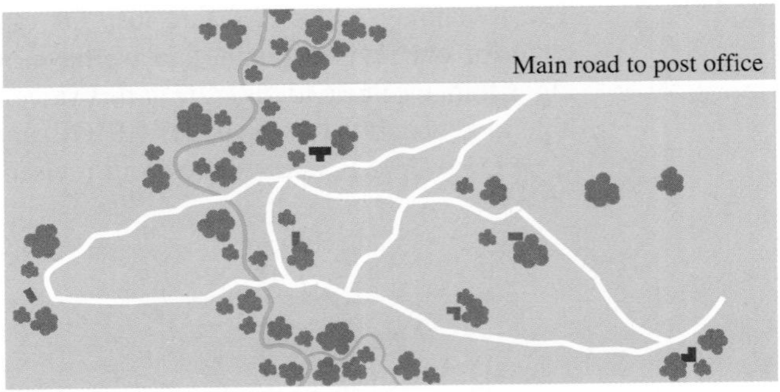

FIGURE 31

It would be ideal if the mail carrier could find an Euler circuit covering the delivery route. If such a circuit existed, it would minimize the distance traveled for the delivery. A glance at the map (thinking of road intersections as vertices) reveals that there is no Euler circuit for this route. As the next best thing, the delivery vehicle should retrace its path as little as possible. For simplicity, we assume that the time to travel any of the stretches of road when not delivering mail is roughly the same. In Figure 32, we added dotted edges to indicate stretches of road to be covered twice. For this delivery route, it is easy to see that we effectively inserted as few edges as possible to obtain a graph that has an Euler circuit.

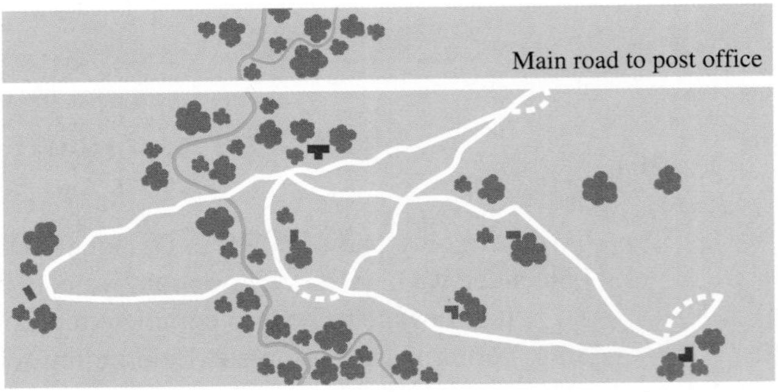

FIGURE 32

There are additional complications for mechanical brooms for street sweeping:

1. The mechanical brooms have to travel along each curb, in the direction of traffic flow (and some streets are one-way).
2. It takes time to switch the brooms from one side of the vehicle to the other.
3. Sweeping has to be coordinated with city parking so that there are no parked cars when sweeping is to be done.

Now we can use Fleury's algorithm to find an Euler circuit on the graph in Figure 32, and this will give the most efficient route for mail delivery.

For a simple delivery area such as this, it is not hard to choose an efficient route intuitively. For route planning in large cities, however, the techniques of this section offer significant savings on services such as mail delivery, mechanized street sweeping, snowplowing, and electric meter checking. In the 1980s a computerized system for planning street sweeping in Washington, D.C., resulted in a 20% savings. In large cities the street sweeping budget is tens of millions of dollars, so savings of 20% easily justify the initial cost of setting up a computerized system.

The study of Euler circuits began in the eighteenth century with a little mathematical puzzle, but the ideas developed provide efficient management techniques for our complex modern cities.

For Further Thought

Route Planning

When planning routes for mail delivery, street sweeping, or snowplowing, the roads in the region to be covered usually do not have an Euler circuit; unavoidably, certain stretches of road must be traveled more than once. (This is referred to as **deadheading.**) Planners then try to find a route on which the driver will spend as little time as possible deadheading. Here we explore how this can be done.

For simplicity, we use street grids for which the distance from corner to corner in any direction is the same. For example, suppose our street grid looks like this:

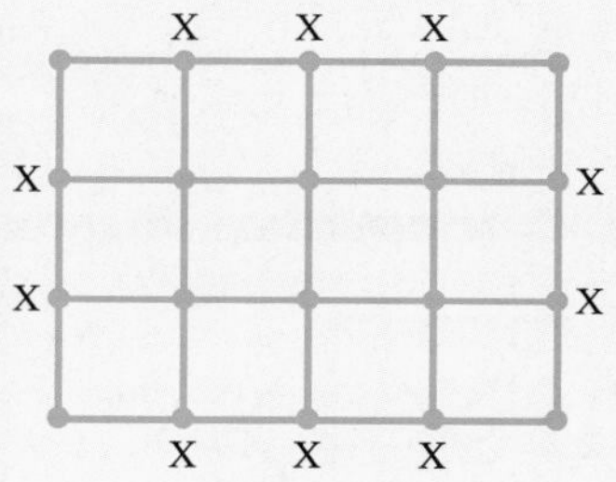

We have marked with an X every vertex with odd degree. Now we must insert edges *coinciding with existing roads* to change this into a graph that has an Euler circuit. If we insert edges as shown in graph (a), we obtain a graph that has an Euler circuit. (Check that all vertices now have even degree.) This solution introduces 11 edges—more deadheading than is necessary.

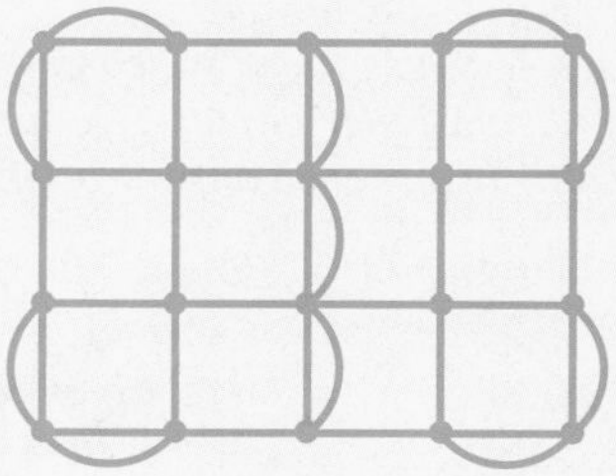

(a)

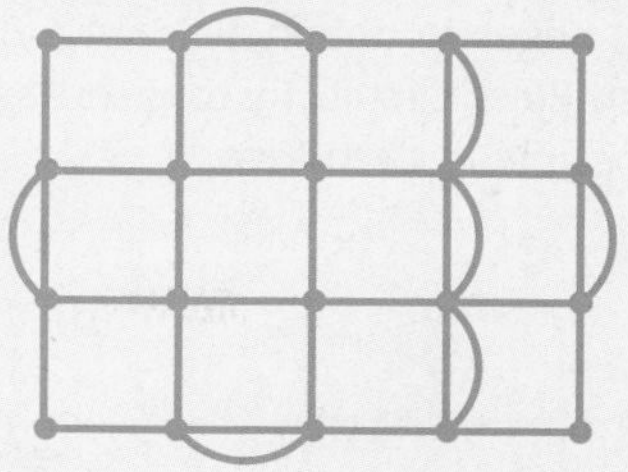

(b)

If we insert edges as shown in graph (b), with just 7 additional edges we obtain a graph with an Euler circuit. This is, in fact, the least number of edges that must be inserted (keeping to the street grid) to change the original street grid graph into a graph that has an Euler circuit. (There are different ways to insert the seven edges.)

A certain amount of trial and error is needed to find the least number of edges that must be inserted to ensure that the street grid graph has an Euler circuit.

For Group Discussion or Individual Investigation

1. In the graph below, we have inserted just six edges in our original street grid. Why is this not a better solution than the one in (b) above?

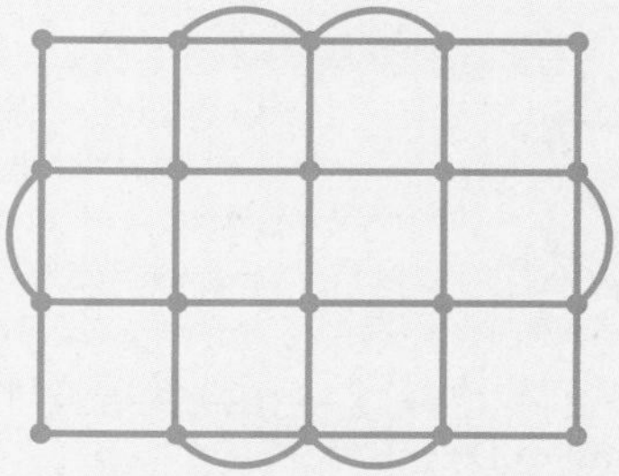

(continued)

2. For each of the street grids (c), (d), and (e), insert edges to obtain a street grid graph that has an Euler circuit. Try to insert as few edges as possible in each case.

(c)
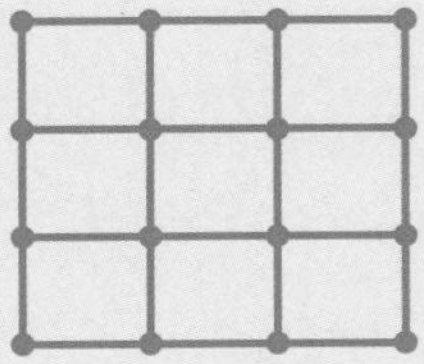
(d)
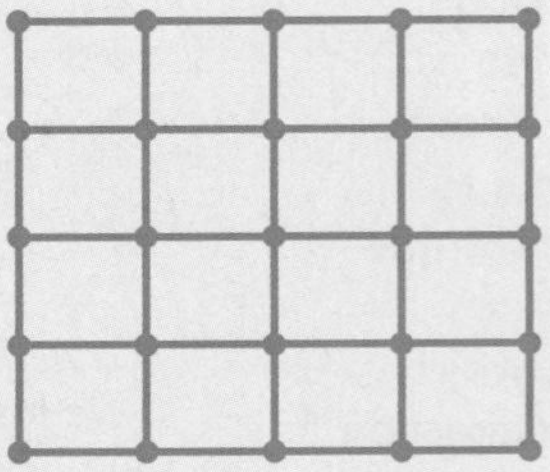
(e)
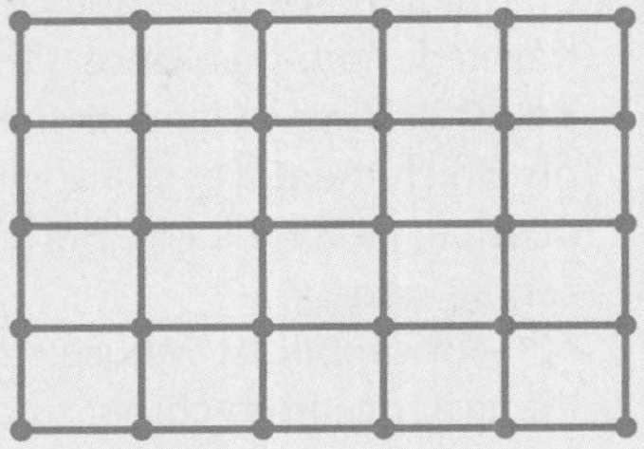

15.2 EXERCISES

In Exercises 1–3, a graph is shown and some sequences of vertices are specified. Determine which of these sequences show Euler circuits. If not, explain why not.

1.
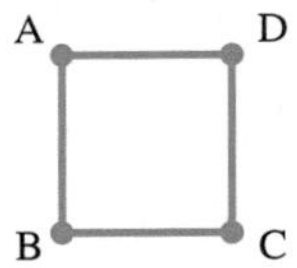

(a) A → B → C → D → A → B → C → D → A
(b) C → B → A → D → C
(c) A → C → D → B → A
(d) A → B → C → D

2.
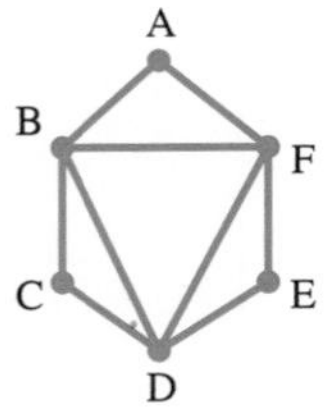

(a) A → B → C → D → E → F → A
(b) F → B → D → B
(c) A → B → C → D → E → F → B → D → F → A
(d) A → B → F → D → B → C → D → E → F → A

3.
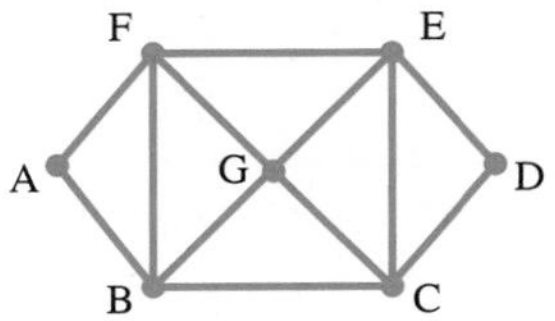

(a) A → B → C → D → E → F → A
(b) A → B → C → D → E → G → C → E → F → G → B → F → A
(c) A → B → C → D → E → C → G → E → F → G → E → F → A
(d) A → B → G → E → D → C → G → F → B → C → E → F → A

In Exercises 4–8, use Euler's theorem to decide whether the graph has an Euler circuit. (Do not actually find an Euler circuit.) Justify each answer briefly.

4.
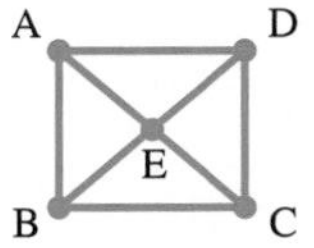

5.
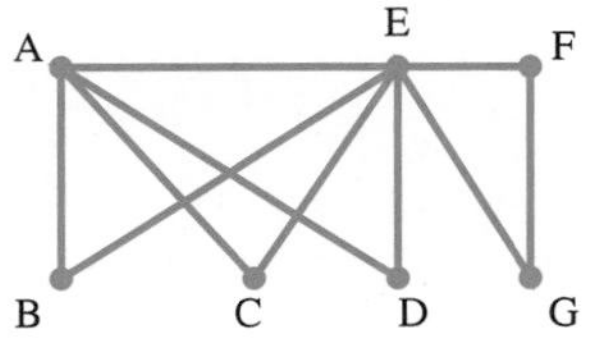

6.

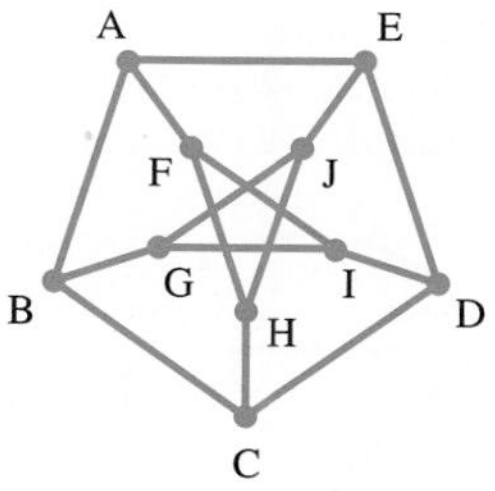

7.

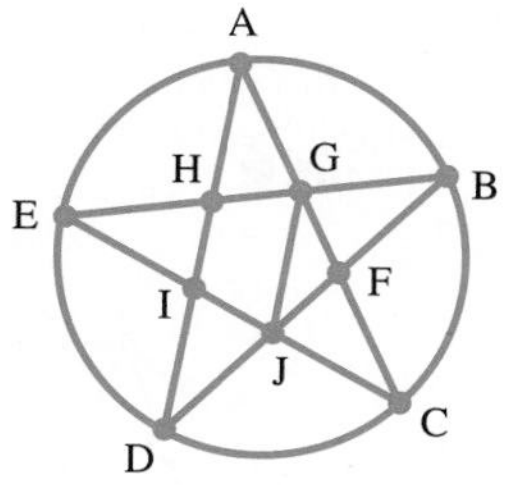

8.

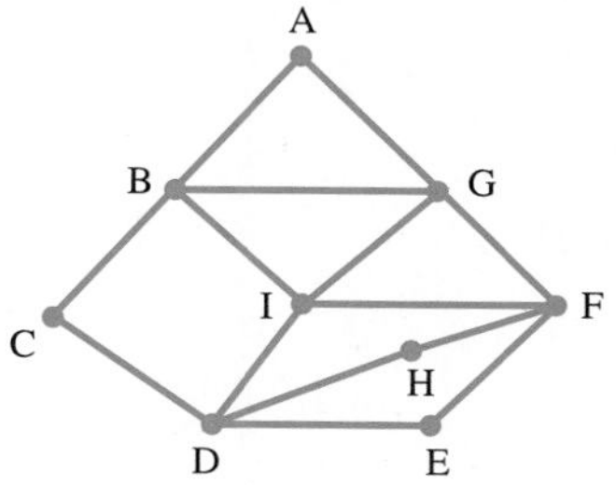

In Exercises 9 and 10, use Euler's theorem to determine whether it is possible to begin and end at the same place, trace the pattern without lifting your pencil, and trace over no line in the pattern more than once.

9.

10.

In Exercises 11–15, use Euler's theorem to determine whether the graph has an Euler circuit, justifying each answer. Then determine whether the graph has a circuit that visits each vertex exactly once, except that it returns to its starting vertex. If so, write down the circuit. (There is more than one correct answer for some of these.)

11.

12.

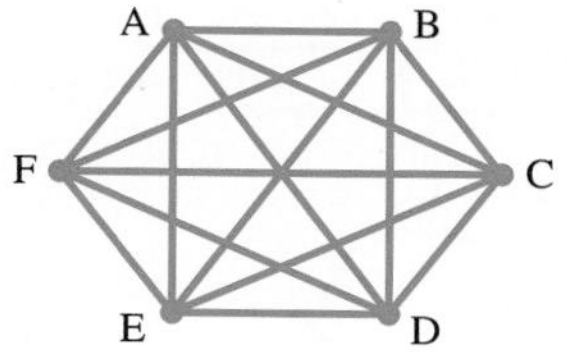

13.

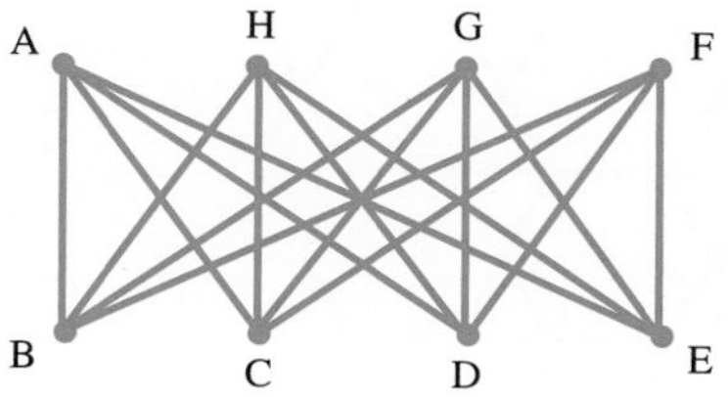

14.

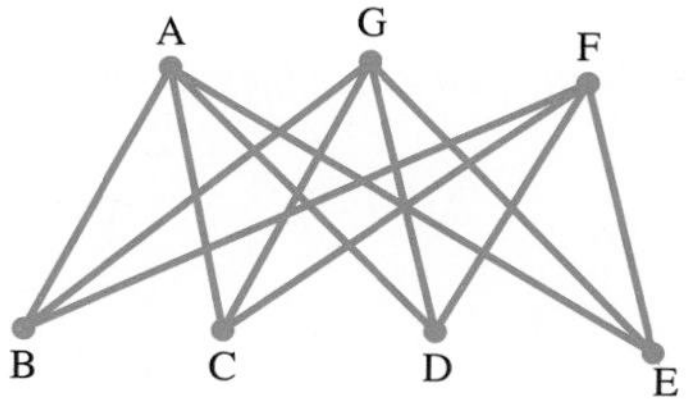

15.

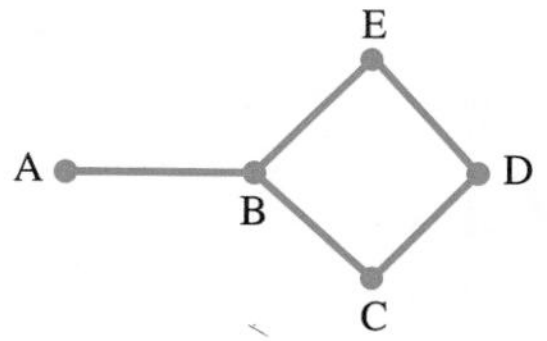

In Exercises 16–19 different floor tilings are shown. The material applied between tiles is called grout. For which of these floor tilings could the grout be applied beginning and ending at the same place, without going over any section twice, and without lifting the tool? Justify answers.

16.

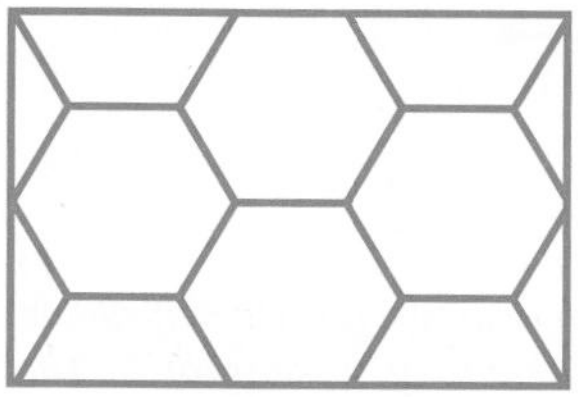

17.

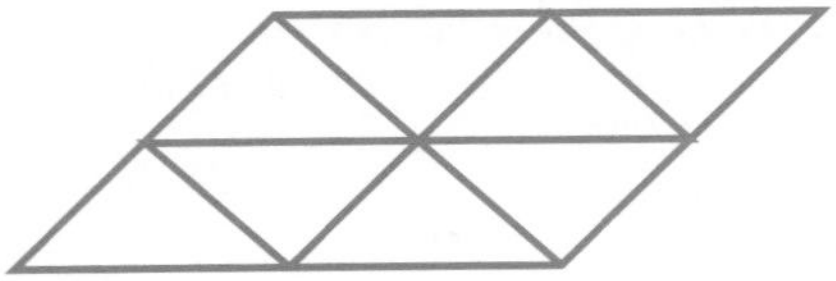

18.

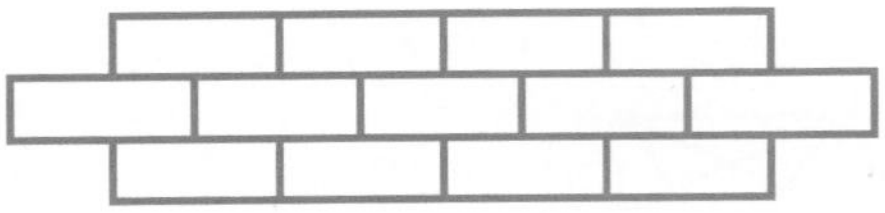

19.

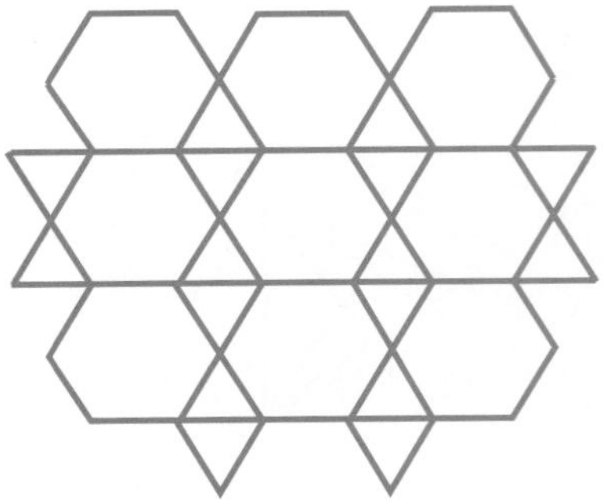

In Exercises 20–22, identify all cut edges in the graph. If there are none, say so.

20.

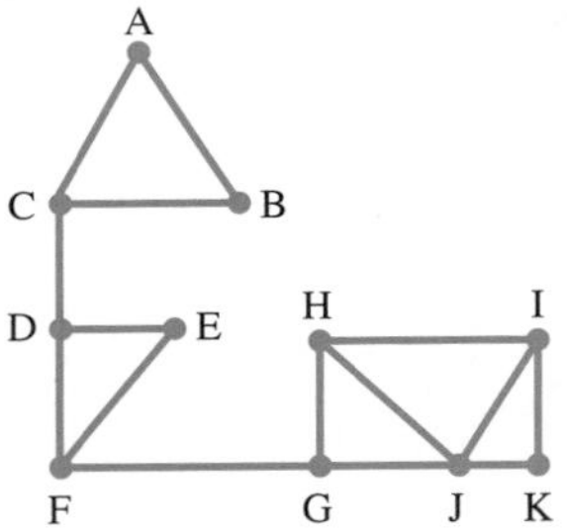

21.

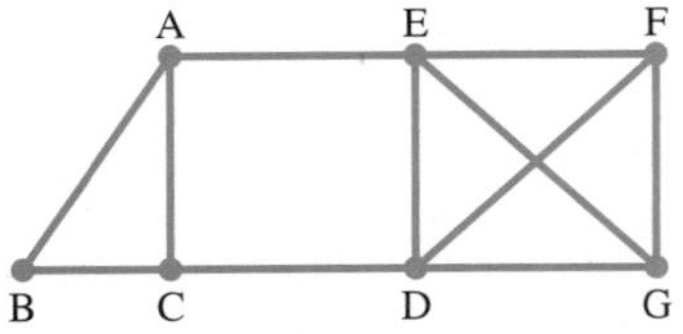

22.

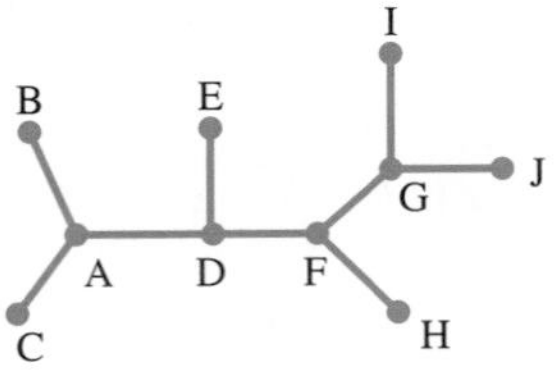

In Exercises 23–25, a graph is shown for which a student has been asked to find an Euler circuit starting at A. The student's revisions of the graph after the first few steps of Fleury's algorithm are shown, and in each case the student is now at B. For each graph, determine all edges that Fleury's algorithm permits the student to use for the next step.

23.

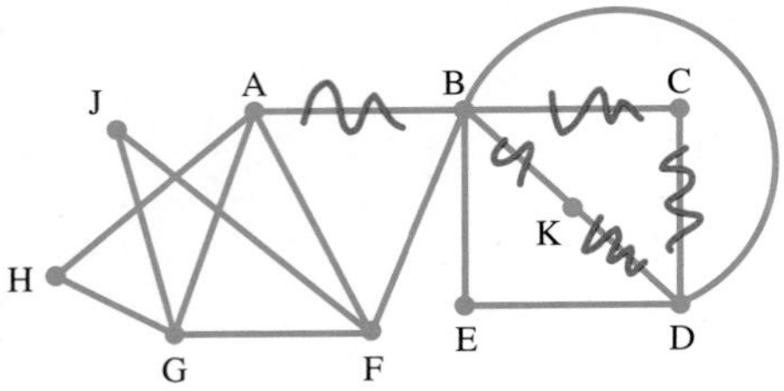

24.

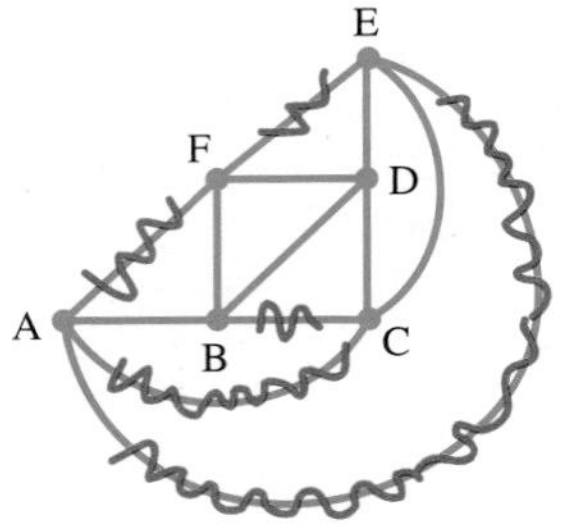

25.

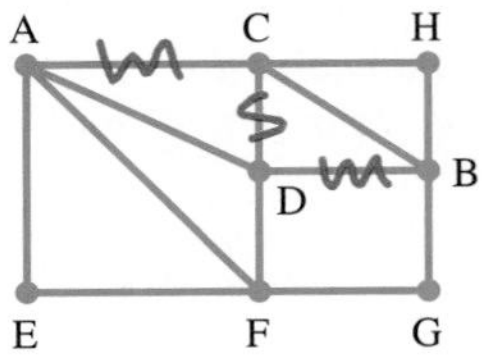

In Exercises 26–28, use Fleury's algorithm to find an Euler circuit for the graph, beginning and ending at A. (There are many different correct answers.)

26.

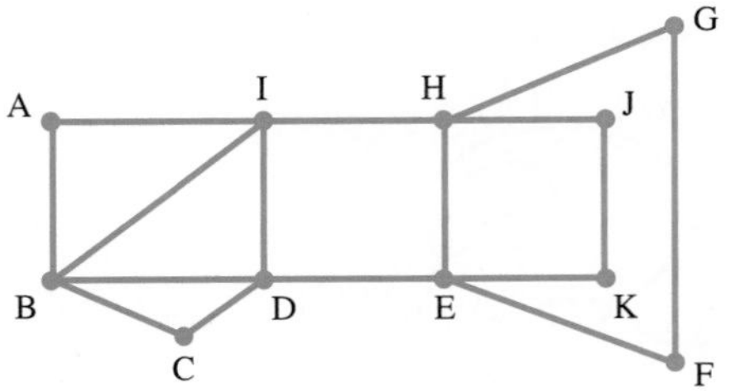

27.

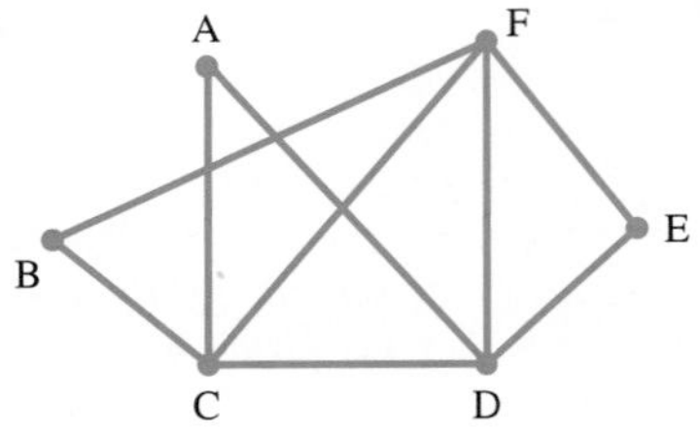

28.

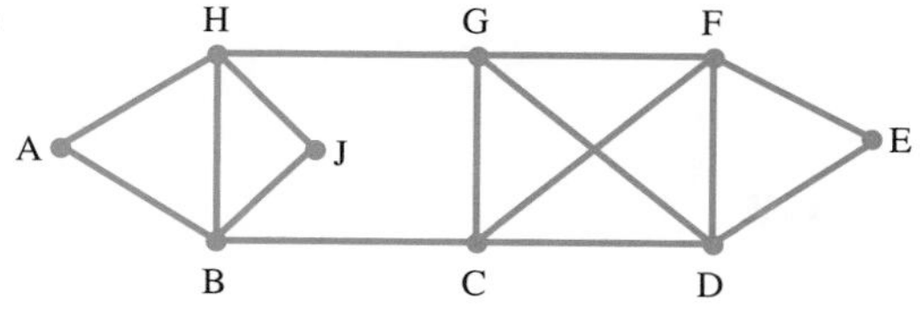

In Exercises 29–31, use Euler's theorem to determine whether the graph has an Euler circuit. If not, explain why not. If the graph does have an Euler circuit, use Fleury's algorithm to find an Euler circuit for the graph. (There are many different correct answers.)

29.

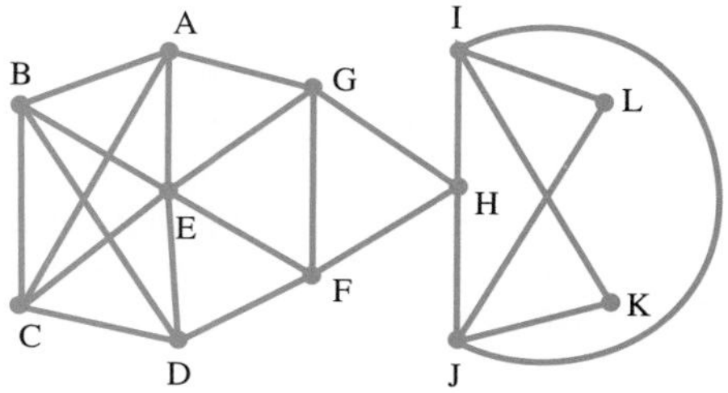

30.

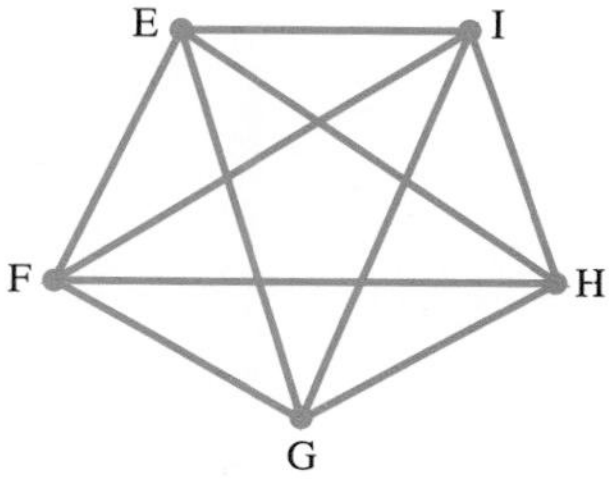

31.

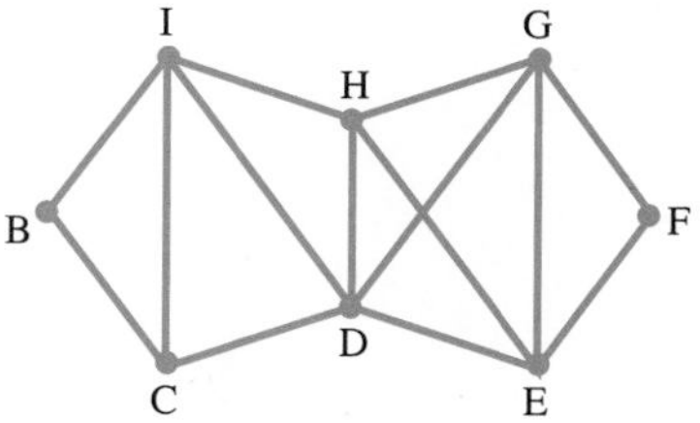

32. *Garden Design* The graph below shows the layout of the paths in a botanical garden. The edges represent the paths. Has the garden been designed in such a way that it is possible for a visitor to find a route that begins and ends at the entrance to the garden (represented by the vertex A) and that goes along each path exactly once? If so, use Fleury's algorithm to find such a route.

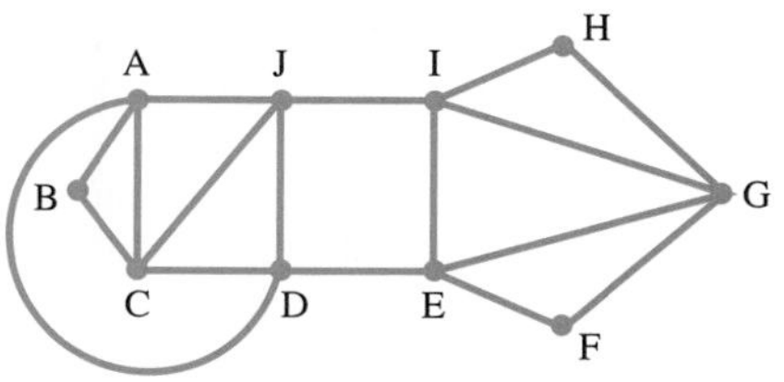

33. *Parking Pattern* The map shows the roads on which parking is permitted at a national monument. This is a pay and display facility. A security guard has the task of periodically checking that all parked vehicles have a valid parking ticket displayed. He is based at the central complex, labeled A. Is there a route that he can take to walk along each of the roads exactly once, beginning and ending at A? If so, use Fleury's algorithm to find such a route.

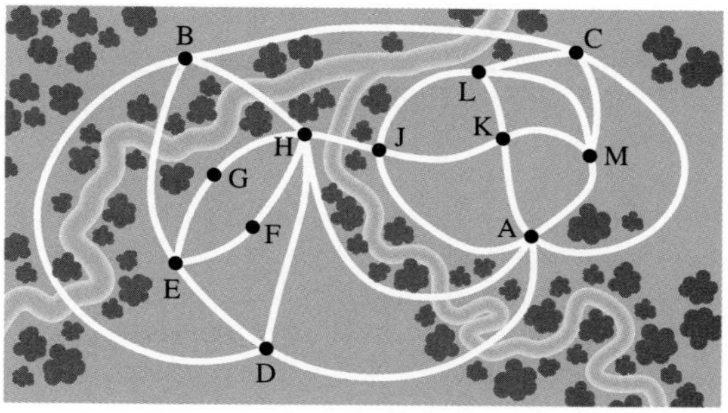

Building Floor Plans *In Exercises 34–36, the floor plan of a building is shown. For which of these is it possible to start outside, walk through each door exactly once, and end up back outside? Justify each answer. (Hint: Think of the rooms and "outside" as the vertices of a graph, and the doors as the edges of the graph.)*

34.

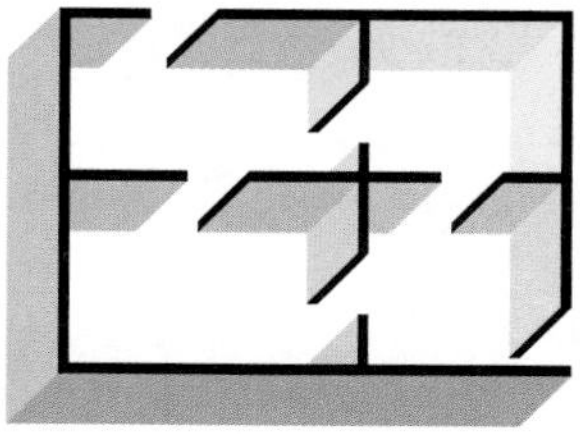

35.

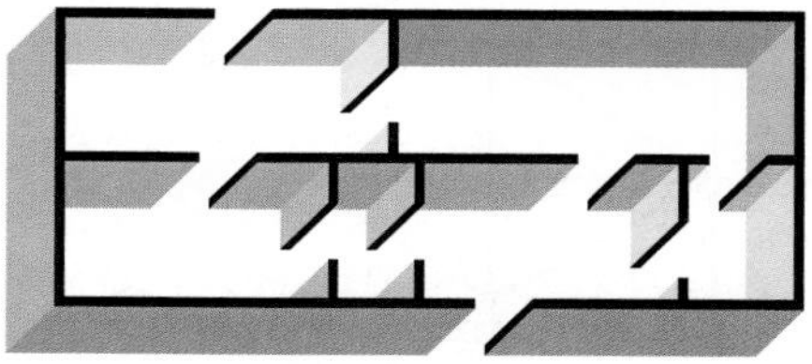

36.

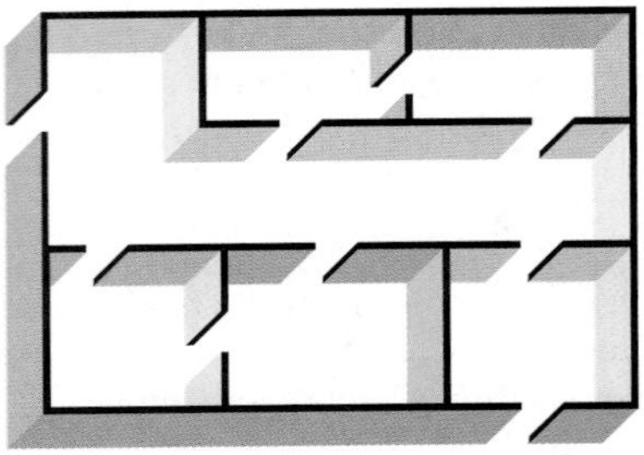

Exercises 37–44 are based on the following theorem:

1. *If a graph has an Euler path that begins and ends at different vertices, then these two vertices are the only vertices with odd degree. (All the rest have even degree.)*
2. *If exactly two vertices in a connected graph have odd degree, then the graph has an Euler path beginning at one of these vertices and ending at the other.*

In Exercises 37–40, determine whether the graph has an Euler path that begins and ends at different vertices. Justify your answer. If the graph has such a path, say at which vertices the path must begin and end.

37.

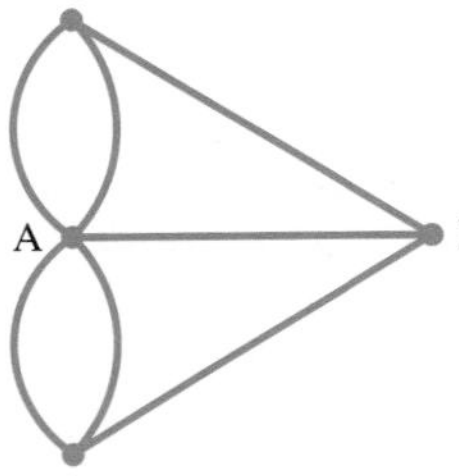

38.

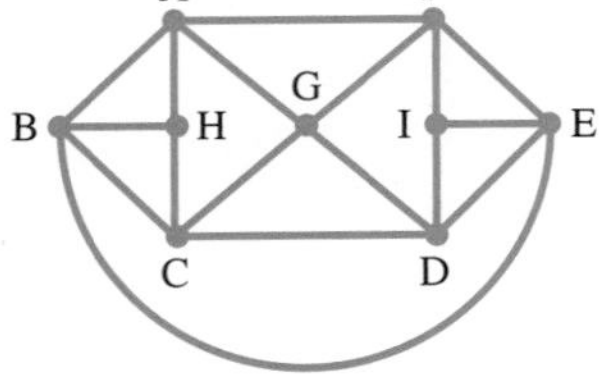

39.

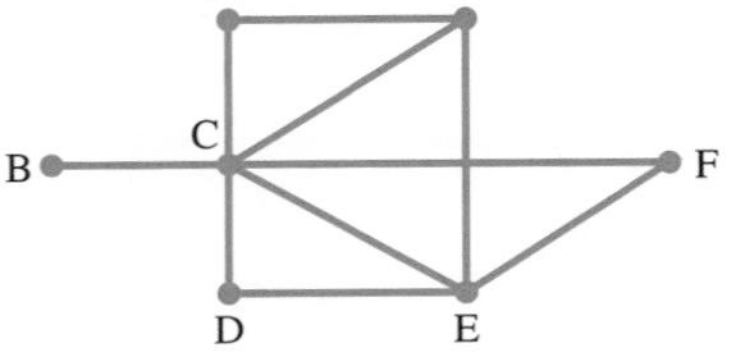

40.

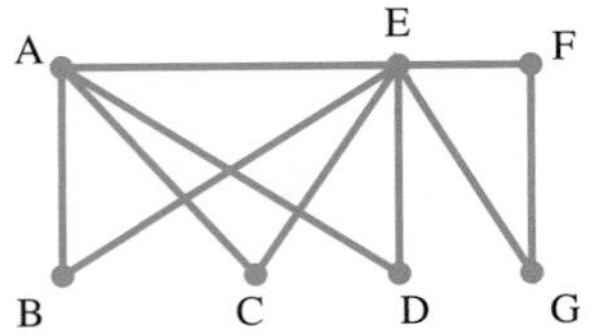

Building Floor Plans *In Exercises 41–43, refer to the floor plan indicated, and determine whether it is possible to start in one of the rooms of the building, walk through each door exactly once, and end up in a different room from the one you started in. Justify your answer.*

41. Refer to the floor plan shown in Exercise 34.

42. Refer to the floor plan shown in Exercise 35.

43. Refer to the floor plan shown in Exercise 36.

44. ***New York City Bridges*** The accompanying schematic map shows a portion of the New York City area, including tunnels and bridges.

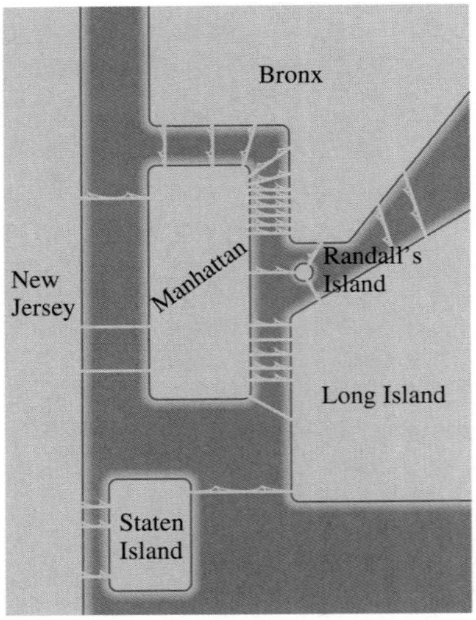

(a) Is it possible to take a drive around the New York City area using each tunnel and bridge exactly once, beginning and ending in the same place?

(b) Is it possible to take a drive around the New York City area using each tunnel and bridge exactly once, beginning and ending in different places? If so, where must your drive begin and end?

In Exercises 45–47, the graph does not have an Euler circuit. For each graph find a circuit that uses as many edges as possible. (There is more than one correct answer in each case.) How many edges did you use in the circuit?

45. 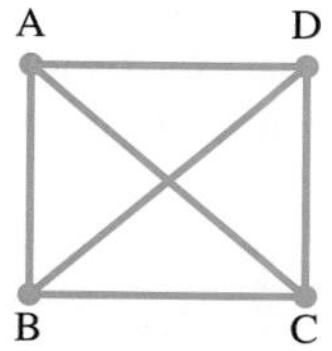

46. 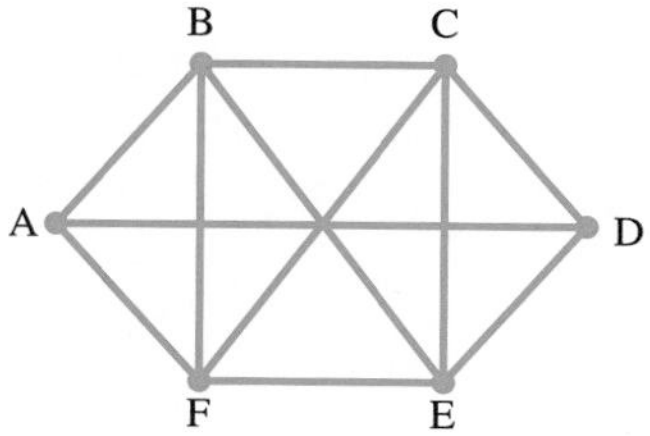

47. 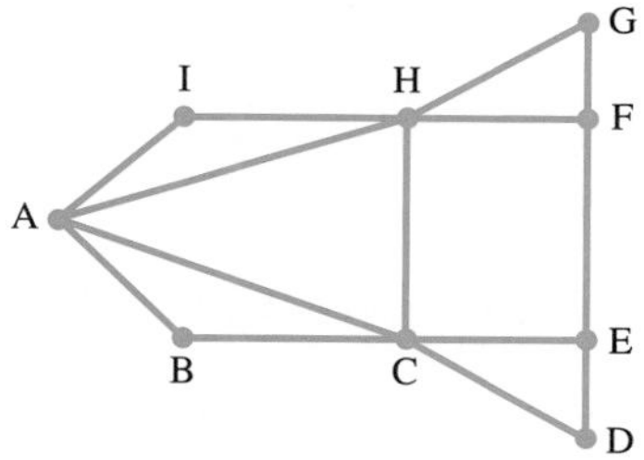

48. There are only five regular polyhedra: the cube, the tetrahedron, the octahedron, the dodecahedron, and the icosahedron. (See Chapter 9 on Geometry.) For which of these can you trace your finger along each edge exactly once, beginning and ending at the same vertex? Justify your answer carefully, using the ideas developed in this section.

49. Which *complete graphs* have Euler circuits? Justify your answer carefully.

50. Write a paper on the sand tracings of the Bushoong or Tshokwe people. Include discussion of the role these drawings play in the culture.

51. Write a paper on continuous tracings in Danish folk culture.

52. Write a paper on the life of Leonhard Euler.

15.3 Hamilton Circuits and Algorithms

Hamilton Circuits • Minimum Hamilton Circuits • Brute Force Algorithm • Nearest Neighbor Algorithm

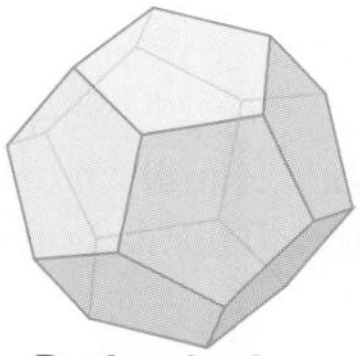

Dodecahedron

Hamilton sold the **Icosian game** idea to a games dealer for 25 British pounds. It went on the market in 1859, but it was not a huge success. A later version of the game, *A Voyage Around the World,* consisted of a regular dodecahedron with pegs at each of the 20 vertices. The vertices had names such as Brussels, Delhi, and Zanzibar. The aim was to travel to all vertices along the edges of the dodecahedron, visiting each vertex exactly once. We can show the vertices and edges of a dodecahedron precisely as in Figure 33.

Hamilton Circuits In this section we examine **Hamilton circuits** in graphs. The story of Hamilton circuits is a story of very large numbers and of unsolved problems in both mathematics and computer science.

We start with a game, called the Icosian game, invented by Irish mathematician William Hamilton in the mid-nineteenth century. It uses a wooden board marked with the graph shown in Figure 33.

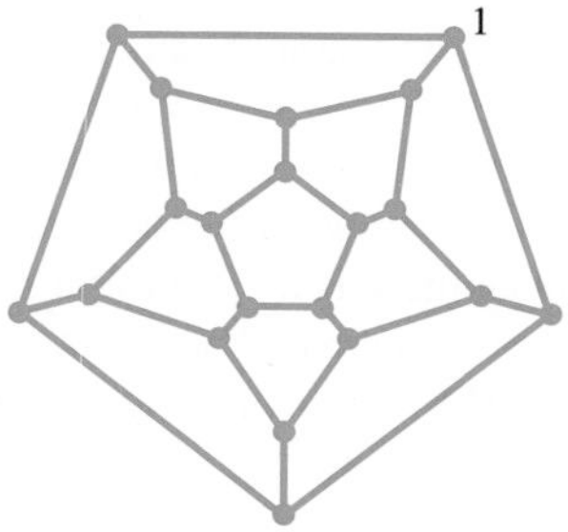

The Icosian Game

FIGURE 33

The game includes 20 pegs numbered 1 through 20. (The word *Icosian* comes from the Greek word for 20.) At each vertex of the graph there is a hole for a peg.

William Rowan Hamilton (1805–1865) spent most of his life in Dublin, Ireland. He was good friends with the poet William Wordsworth, whom he met while touring England and Scotland. Hamilton tried writing poetry, but Wordsworth tactfully suggested that his talents were in mathematics rather than poetry. Catherine Disney was the first great love of his life, but under pressure from her parents, she married another man, much wealthier than Hamilton. Hamilton never seemed to quite get over this.

The simplest version of the game is as follows: put the pegs into the holes in order, following along the edges of the graph, in such a way that peg 20 ends up in a hole that is joined by an edge to the hole of peg 1. Try this now, numbering the vertices in the graph in Figure 33, starting at vertex 1.

The Icosian game asks you to find a circuit in the graph, but the circuit need not be an Euler circuit. The circuit must visit each *vertex* exactly once, except for returning to the starting vertex to complete the circuit. (The circuit may or may not travel all edges of the graph.) Circuits such as this are called *Hamilton circuits.*

Hamilton Circuit

A **Hamilton circuit** in a graph is a circuit that visits each vertex exactly once (returning to the starting vertex to complete the circuit).

EXAMPLE 1 Identifying Hamilton Circuits

Which of the following are Hamilton circuits for the graph in Figure 34? Justify your answers briefly.

(a) $A \rightarrow B \rightarrow E \rightarrow D \rightarrow C \rightarrow F \rightarrow A$

(b) $A \rightarrow B \rightarrow C \rightarrow D \rightarrow E \rightarrow F \rightarrow C \rightarrow E \rightarrow B \rightarrow F \rightarrow A$

(c) $B \rightarrow C \rightarrow D \rightarrow E \rightarrow F \rightarrow B$

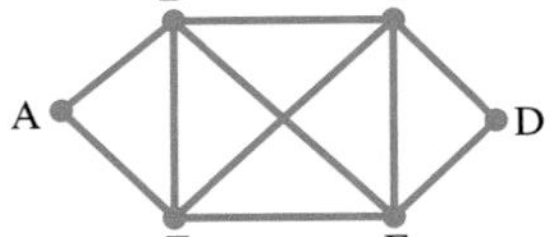

FIGURE 34

SOLUTION

(a) $A \rightarrow B \rightarrow E \rightarrow D \rightarrow C \rightarrow F \rightarrow A$ is a Hamilton circuit for the graph. It visits each vertex of the graph exactly once, and then returns to the starting vertex. (Trace this circuit on the graph to check it.)

(b) $A \rightarrow B \rightarrow C \rightarrow D \rightarrow E \rightarrow F \rightarrow C \rightarrow E \rightarrow B \rightarrow F \rightarrow A$ is not a Hamilton circuit, since it visits vertex B (and vertices C, E, and F) more than once. (This is, however, an Euler circuit for the graph.)

(c) $B \rightarrow C \rightarrow D \rightarrow E \rightarrow F \rightarrow B$ is not a Hamilton circuit since it does not visit all vertices in the graph. ■

Some graphs, such as the graph in Figure 35, do not have a Hamilton circuit. There is no way to visit all the vertices and return to the starting vertex without visiting vertex I more times than allowed.

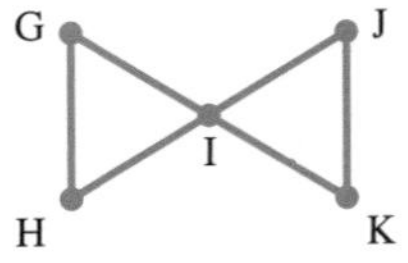

FIGURE 35

It can be difficult to determine whether a particular graph has a Hamilton circuit, since there is no theorem that gives necessary and sufficient conditions for a Hamilton circuit to exist. This remains an unsolved problem in Hamilton circuits. (Contrast the simple method provided by Euler's theorem for checking whether a graph has an Euler circuit.)

Fortunately, in many real-world applications of Hamilton circuits we are dealing with complete graphs, and any complete graph with three or more vertices does have a Hamilton circuit. (Recall our discussion of complete graphs in Section 15.1.) For example, $A \rightarrow B \rightarrow C \rightarrow D \rightarrow E \rightarrow F \rightarrow A$ is a Hamilton circuit for the complete graph shown in Figure 36. We could form a Hamilton circuit for any complete graph with three or more vertices in a similar way.

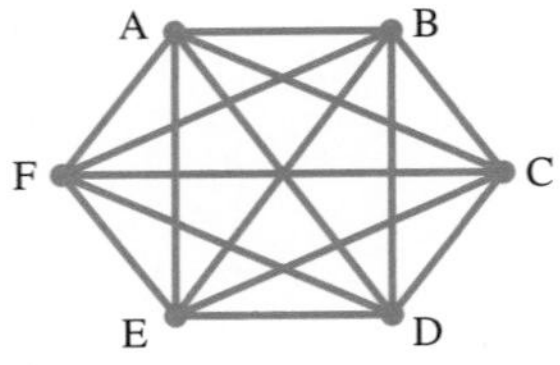

FIGURE 36

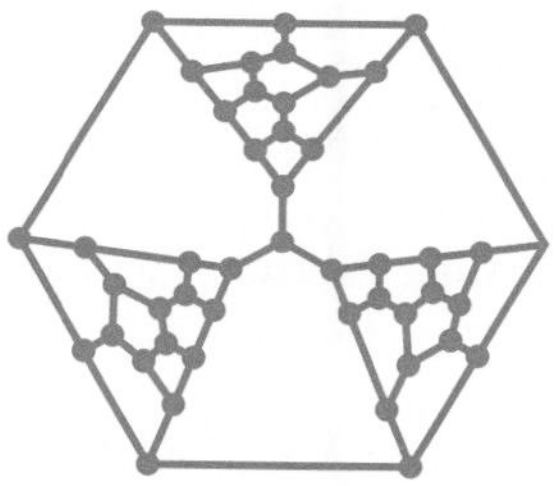

The Tutte graph, shown here, has no Hamilton circuit. This graph has a history tied up with the four-color problem for map coloring (Section 15.1). In 1880 Peter Tate provided a "proof" for the four-color theorem. It was based on the assumption that every connected graph, with all vertices of degree three, and which can be drawn with edges crossing nowhere except at vertices, has a Hamilton circuit. But in 1946, Tutte produced the graph shown here, showing that Tate's basic assumption was wrong, and his proof was, therefore, invalid.

Hamilton Circuits for Complete Graphs

Any complete graph with three or more vertices has a Hamilton circuit.

The graph in Figure 36 has many Hamilton circuits. Try finding some that are different from the one given earlier.

How many Hamilton circuits does a complete graph have? Before we count, we need an agreement about when two lists of vertices will be considered to be *different* Hamilton circuits. For example, the Hamilton circuits $B \rightarrow C \rightarrow D \rightarrow E \rightarrow F \rightarrow A \rightarrow B$ and $A \rightarrow B \rightarrow C \rightarrow D \rightarrow E \rightarrow F \rightarrow A$ visit the vertices in essentially the same order in the graph in Figure 36, although the lists start with different vertices. If you mark the circuit on the graph, you can describe it starting at any vertex. For our purposes, it is convenient to consider the two lists above as representing the same Hamilton circuit.

When Hamilton Circuits Are the Same

Hamilton circuits that differ *only* in their starting points will be considered to be the same circuit.

PROBLEM-SOLVING HINT Tree diagrams are often useful for counting.

We begin by counting the number of Hamilton circuits in a complete graph with four vertices. (See Figure 37.) We can use the same starting point for all the circuits we count. We choose A as the starting point. From A we can go to any of the three remaining vertices (B, C, or D). No matter which vertex we choose, we then have two unvisited vertices for our next choice. Then there is only one way to complete the Hamilton circuit. We illustrate this counting procedure in the tree diagram shown in Figure 38.

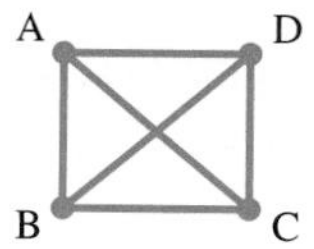

FIGURE 37

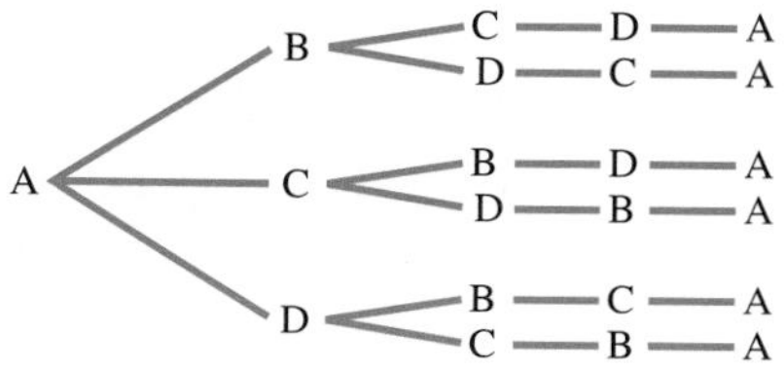

FIGURE 38

This tree diagram shows that we have $3 \cdot 2 \cdot 1$ Hamilton circuits in the graph. We write this as 3! (read "3 factorial"). Refer to Chapter 11 on counting methods for more on factorial notation.

Returning to Figure 36, we can count the number of Hamilton circuits in the complete graph with 6 vertices. Start at vertex A. From A we can proceed to any of the five remaining vertices. No matter which we choose, we then have four choices for the next vertex along the circuit, three choices for the next, and two for the next.

Large Numbers 70! is the first factorial greater than 10^{100}. Most calculators will calculate numbers up to only about 10^{99}, so if you try calculating 70! you probably will get an error message. (Try it.) 10^{100} is called one googol, a name invented by a nine-year-old child when asked to name a very, very large number.

These numbers are unimaginably large. Consider:

$$25! \approx 2 \times 10^{25}$$
$$50! \approx 3 \times 10^{64}$$
$$100! \approx 9 \times 10^{157}$$

Comparison:

Approximate number of *meters* from Saturn to the sun: 10^{12}

Estimate of the number of *inches* to the farthest object in the universe: 10^{27}.

Approximate number of molecules we would have if we filled a sphere the size of the earth with water: 10^{49}.

(We could draw a counting tree for this, but it would be rather large.) This means that we have $5 \cdot 4 \cdot 3 \cdot 2 \cdot 1 = 5!$ different Hamilton circuits in all.

Likewise, we would find that a complete graph with 10 vertices has 9! Hamilton circuits. (A nice pattern is emerging here.) In general, for a complete graph the number of Hamilton circuits in the graph can be obtained by calculating the factorial of the number that is one less than the number of vertices.

Number of Hamilton Circuits in a Complete Graph

A complete graph with n vertices has $(n - 1)!$ Hamilton circuits.

As the number of vertices in a complete graph increases, the number of Hamilton circuits for that graph increases very quickly. Previously you considered how quickly exponential functions increase; factorials increase even faster. For example, $25! = 15{,}511{,}210{,}043{,}330{,}985{,}984{,}000{,}000$, or approximately 1.6×10^{25}—more than a trillion trillion.

Minimum Hamilton Circuits

Where would we come across real-life problems involving Hamilton circuits for such large graphs? Consider, for example, that on a typical day, a UPS van might have to make deliveries to 100 different locations. For simplicity, assume that none are priority deliveries, so they can be made in any order. UPS wants to minimize the time to make these deliveries. Think of the 100 locations and the UPS distribution center as the vertices of a complete graph with 101 vertices. In principle, the van can go from any of these locations directly to any other, shown by having an edge between each pair of vertices. Suppose we estimate the travel time between each pair of locations (of course, this would depend on distance and traffic conditions along the route). These estimates provide weights on the edges of the graph. The objective is to visit each location exactly once, to begin and end at the same place, and to take as little time as possible. In graph theory terms, we need a Hamilton circuit for the graph that has *least possible total weight.* The **total weight** of a circuit is the sum of the weights on the edges in the circuit. We call such a circuit a *minimum Hamilton circuit* for the graph.

Minimum Hamilton Circuit

In a weighted graph, a **minimum Hamilton circuit** is a Hamilton circuit with least possible total weight.

A problem whose solution requires us to find a minimum Hamilton circuit for a complete, weighted graph often is called a **traveling salesman problem** (or TSP). Think of the vertices of the complete graph as the cities that a salesperson must visit, and the weights on the edges as the cost of traveling directly between the cities. To minimize costs, the salesperson needs a minimum Hamilton circuit for the graph.

Traveling salesman problems arise in business and industry. They are relevant to efficient routing of telephone calls and Internet connections. As another example, to manufacture integrated circuit silicon chips, many lines have to be etched on a silicon wafer. Minimizing production time involves deciding the order in which to etch the lines, a traveling salesman problem. Likewise, to manufacture circuit boards for integrated circuits, laser-drilled holes must be made for connections. Again, the order for drilling the holes is a critical factor in production time.

Traveling Salesman Problems and Archaeology Archaeologists often excavate sites in which there is no clear evidence to show which deposits were made earlier, and which later. Some archaeologists have used minimum Hamilton circuits to solve the problem. For example, if a site consists of a number of burials, they consider a complete graph with vertices representing the various burials, and weights on the edges corresponding roughly to how dissimilar the burials are. A minimum Hamilton circuit in this graph gives the best guess for the order in which the deposits were made.

Brute Force Algorithm Suppose we are given a complete, weighted graph. How can we find a minimum Hamilton circuit for the graph? One way is to systematically list all the Hamilton circuits in the graph, find the total weight of each, and choose a circuit with least total weight. (In fact, it is sufficient to add up the weights on the edges for just half of the Hamilton circuits, since the total weight of a circuit is the same as the total weight of the circuit that uses the same edges in reverse order.) The method just described is sometimes called the *brute force* algorithm since we find the solution by checking *all* the Hamilton circuits.

Brute Force Algorithm

Step 1 Choose a starting point.

Step 2 List all the Hamilton circuits with that starting point.

Step 3 Find the total weight of each circuit.

Step 4 Choose a Hamilton circuit with least total weight.

EXAMPLE 2 Using the Brute Force Algorithm

Find a minimum Hamilton circuit for the complete, weighted graph shown in Figure 39.

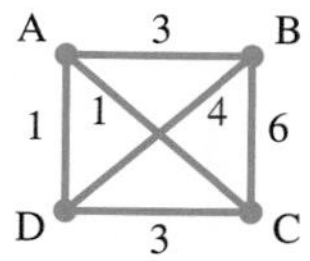

FIGURE 39

SOLUTION

Choose a starting point, A. List all the Hamilton circuits starting at A. Since this is a complete graph with 4 vertices, there are $3! = 3 \cdot 2 \cdot 1 = 6$ circuits. Thus, we must find 6 Hamilton circuits.

We need a systematic way of writing down all the Hamilton circuits. The counting tree of Figure 38 provides a guide. We start by finding all the Hamilton circuits that begin $A \rightarrow B$, then include those that begin $A \rightarrow C$, and finally include those that begin $A \rightarrow D$, as shown below. Once all the Hamilton circuits are listed, pair those that visit the vertices in precisely opposite orders. (Circuits in these pairs have the same total weight. This will save some adding.) Finally, determine the sum of the weights in each circuit.

Circuit	Total weight of the circuit
1. $A \rightarrow B \rightarrow C \rightarrow D \rightarrow A$	$3 + 6 + 3 + 1 = 13$
2. $A \rightarrow B \rightarrow D \rightarrow C \rightarrow A$	$3 + 4 + 3 + 1 = 11$
3. $A \rightarrow C \rightarrow B \rightarrow D \rightarrow A$	$1 + 6 + 4 + 1 = 12$
4. $A \rightarrow C \rightarrow D \rightarrow B \rightarrow A$ (opposite of 2)	11
5. $A \rightarrow D \rightarrow B \rightarrow C \rightarrow A$ (opposite of 3)	12
6. $A \rightarrow D \rightarrow C \rightarrow B \rightarrow A$ (opposite of 1)	13

We can now see that $A \rightarrow B \rightarrow D \rightarrow C \rightarrow A$ is a minimum Hamilton circuit for the graph. The weight of this circuit is 11. ■

Just how fast are the fastest computers? Computer speeds are sometimes measured in **flops** (floating point operations per second). A speed of one gigaflop means that the computer can perform about one billion (10^9) calculations per second. A modern desktop computer can perform at a few gigaflops. In 2005, the fastest computer was the IBM Blue Gene/L, which could perform at almost 300 teraflops. One teraflop is about one trillion (10^{12}) calculations per second. (*Source*: www.top500.org/lists.)

In principle, the brute force algorithm provides a way to find a minimum Hamilton circuit in any complete, weighted graph. In practice, it takes far too long to obtain a complete, weighted graph with 7 vertices. There are 6! or 720 Hamilton circuits in the graph, and we would have to calculate the total weight for half of these. That means we would have to do 360 separate calculations, in addition to listing the Hamilton circuits.

Most real-world problems involve a lot more than 7 vertices. Manufacturing a large integrated circuit can involve a graph with almost one million vertices; likewise, telecommunications companies routinely deal with graphs with millions of vertices. As the number of vertices in the graph increases, the task of finding a minimum Hamilton circuit using the brute force algorithm soon becomes too time-consuming for even our fastest computers. For example, if one of today's super-computers had started using the brute force algorithm on a 100-vertex traveling salesman problem when the universe was created, it would still be far from done. In contrast, Fleury's algorithm for finding an Euler circuit in a graph does not take too long for our computers, even for rather large graphs. Algorithms that do not take too much computer time are called **efficient algorithms.** Computer scientists have so far been unable to find an efficient algorithm for the traveling salesman problem, and they suspect that it is simply impossible to create such an algorithm.

The traveling salesman problem is just one of a collection of problems for which there is no known efficient algorithm. Many of these unsolved problems are related in such a way that an efficient algorithm for solving any one of them could be adapted to solve all of them. Probably the most important unsolved problem in computer science is this: either to find an efficient algorithm for the traveling salesman problem, or to explain why no one could create such an algorithm. (You might find this referred to as the P = NP problem.)

For the traveling salesman problem there are some algorithms that do not take too much computer time, and that give *reasonably good* solutions *most of the time.* Such algorithms are called **approximate algorithms,** since they give an approximate solution to the problem. We shall consider one such algorithm for the traveling salesman problem. The underlying idea is that from each vertex we proceed to a "nearest" available vertex.

DNA Computers Researchers are developing computers based on DNA instead of silicon chips. Chemical reactions replace electric signals, and since these happen simultaneously and very quickly, computation is much faster. Programming a DNA computer is more difficult than programming a conventional computer, however, since it involves synthesizing DNA molecules to represent the specific problem. DNA computers have so far solved only small problems that are easily solved by a conventional computer (including a 15 city TSP). It seems likely, however, that they will soon be sufficiently developed to solve in a few months problems that would take 10,000 years on a conventional computer.

Nearest Neighbor Algorithm

Nearest Neighbor Algorithm

Step 1 Choose a starting point for the circuit. Call this vertex A.

Step 2 Check all the edges joined to A, and choose one that has least weight. Proceed along this edge to the next vertex.

Step 3 At each vertex you reach, check the edges from there *to vertices not yet visited.* Choose one with least weight. Proceed along this edge to the next vertex.

Step 4 Repeat Step 3 until you have visited all the vertices.

Step 5 Return to the starting vertex.

NP complete is a technical term in computer science describing problems for which all known algorithms take way too much time for large graphs for even the fastest imaginable computers. The traveling salesman problem and determining the chromatic number of a graph are NP complete. The popular puzzle SUDOKU also is in the class of NP complete problems.

EXAMPLE 3 Using the Nearest Neighbor Algorithm

A courier is based at the head office (A), and must deliver documents to four other offices (B, C, D, and E). The estimated time of travel (in minutes) between each of these offices is shown on the graph in Figure 40. The courier wants to visit the locations in an order that takes the least time. Use the nearest neighbor algorithm to find an approximate solution to this problem. Calculate the total time required to cover the chosen route.

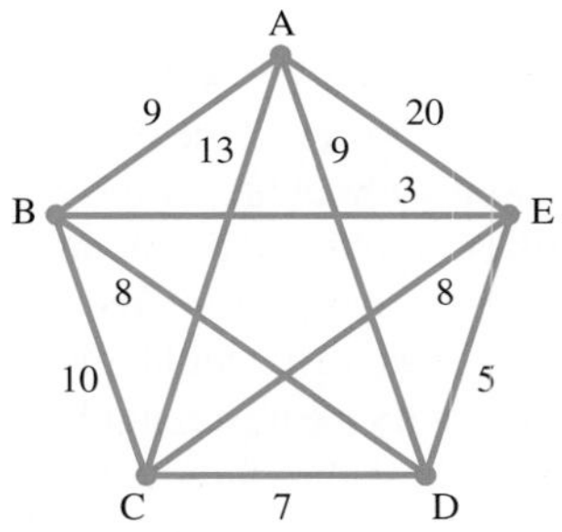

FIGURE 40

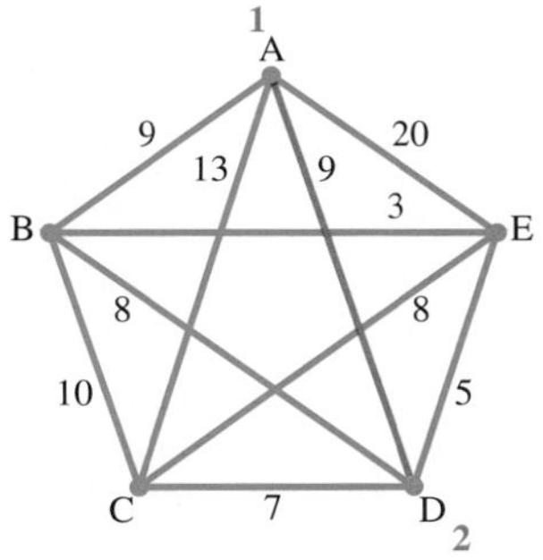

FIGURE 41

SOLUTION

Step 1 Choose a starting point. Let's start at A.

Step 2 Choose an edge with least weight joined to A. Both AB and AD have weight 9, which is less than the weights of the other two edges joined to A. Let's choose AD.* We keep a record of the circuit as we form it:

$$A \xrightarrow{9} D.$$

To help ensure that we do not visit a vertex twice, we number the vertices as we visit them and color the edges used. Our diagram now appears as in Figure 41.

Continue to number the vertices as we work through the rest of the solution.

Step 3 Now check edges joined to D, excluding DA (since we have already been to A). DC has weight 7, DB has weight 8, and DE has weight 5. DE has smallest weight, so proceed along DE to E. Our circuit begins:

$$A \xrightarrow{9} D \xrightarrow{5} E.$$

Step 4 Now repeat the process at E. Check all edges from E that go to a vertex not yet visited. EC has weight 8 and EB has weight 3. Proceed along EB to B. Our circuit so far:

$$A \xrightarrow{9} D \xrightarrow{5} E \xrightarrow{3} B.$$

*If using a computer to do this, we would instruct the computer to make a random choice between AB and AD.

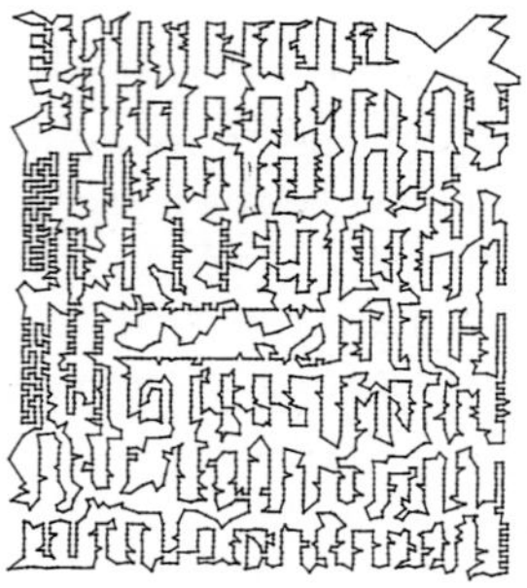

Records for Traveling Salesman Problems There is a race on to set records for exact solutions to traveling salesman problems. Computer scientists do this by using parallel processing and more sophisticated algorithms (all still exponential time algorithms—see For Further Thought on the next page). The illustration shows the exact solution to a traveling salesman problem with 3038 vertices on a printed circuit board, obtained in 1993.

Here are some of the other records, showing the year each was set and the number of vertices in the graph.

1980: 318
1987: 666
1994: 7397
1998: 13,509
2001: 15,112
2004: 24,978

Source: www.tsp.gatech.edu/history/milestone.html

Step 5 We have one vertex not yet visited, C. We go next to C and now have visited all the vertices. We return to our starting point, A. Our Hamilton circuit is

$$A \xrightarrow{9} D \xrightarrow{5} E \xrightarrow{3} B \xrightarrow{10} C \xrightarrow{13} A.$$

Its total weight is $9 + 5 + 3 + 10 + 13 = 40$.

Our advice to the courier would thus be to visit the offices in the order shown in this circuit. This route will take about 40 minutes. ■

You might think that you could get a quicker route for the courier just by looking at the graph, rather than following the rules of the algorithm. You would be right. The circuit we found in Example 3 is not the minimum Hamilton circuit for the graph in Figure 41. (See below.) However, the point of an approximate algorithm is that a computer can implement it without taking too much time, and most of the time it will give a reasonably good solution to the problem.

If we had a computer performing the nearest neighbor algorithm for us, we could make a small adjustment that would give better results without taking too much longer. We could repeat the nearest neighbor algorithm for all possible starting points, and then choose from these the Hamilton circuit with least weight. If we do this for the example above, we obtain the results in Table 1.

TABLE 1

Starting Vertex	Circuit Using Nearest Neighbor	Total Weight
A	$A \to D \to E \to B \to C \to A$	$9 + 5 + 3 + 10 + 13 = 40$
B	$B \to E \to D \to C \to A \to B$	$3 + 5 + 7 + 13 + 9 = 37$
C	$C \to D \to E \to B \to A \to C$	$7 + 5 + 3 + 9 + 13 = 37$
D	$D \to E \to B \to A \to C \to D$	$5 + 3 + 9 + 13 + 7 = 37$
E	$E \to B \to D \to C \to A \to E$	$3 + 8 + 7 + 13 + 20 = 51$

With this information, we would recommend to the courier that he use a circuit with total weight 37, for example the circuit $D \to E \to B \to A \to C \to D$. This does not force him to start his journey at D. If he followed the route $A \to C \to D \to E \to B \to A$, he would be using the same edges as in $D \to E \to B \to A \to C \to D$, and his traveling time still would be 37 minutes.

Can we now be sure that we have found the minimum Hamilton circuit for the graph in Figure 41? No. This is still an *approximate* solution. If we check the total weights of all possible Hamilton circuits in the graph, we find the Hamilton circuit

$$A \xrightarrow{9} B \xrightarrow{3} E \xrightarrow{8} C \xrightarrow{7} D \xrightarrow{9} A$$

with total weight just 36 minutes. The nearest neighbor algorithm simply will not find this minimum Hamilton circuit for the graph. However, all we expect of an approximate algorithm is that it give a reasonably good solution for the problem in a reasonable amount of time.

For Further Thought

The Speed of Algorithms

How do computer scientists classify the speed of algorithms? First, they write the number of steps a computer using the algorithm takes to solve a problem as a function of the "size" of the problem. We can think of the size of the problem as the number of vertices in the graph. Let's suppose that our graph has n vertices. For some algorithms the number of steps is a *polynomial function of n*, for example $n^4 + 2n$. For other algorithms, the number of steps is an *exponential function of n*, for example 2^n. (See Chapters 7 and 8 for more on polynomial and exponential functions.) As n increases, the number of steps increases much faster for exponential functions than for polynomial functions. Functions that involve factorials increase even faster. In the table below we show approximate values for functions of the three types for different values of n. (We use scientific notation to make it easier to compare the sizes of the numbers.)

n	n^4	2^n	$n!$
5	6.3×10^2	3.2×10	1.2×10^2
10	1.0×10^4	1.0×10^3	3.6×10^6
15	5.1×10^4	3.3×10^4	1.3×10^{12}
20	1.6×10^5	1.0×10^6	2.4×10^{18}
25	3.9×10^5	3.4×10^7	1.6×10^{25}
30	8.1×10^5	1.1×10^9	2.7×10^{32}
40	2.6×10^6	1.1×10^{12}	8.2×10^{47}
50	6.3×10^6	1.1×10^{15}	3.0×10^{64}

In the first line of the table (where $n = 5$), n^4 gives the largest value. But as n gets larger, 2^n grows much faster than n^4, while $n!$ grows extremely fast. For example, if we double the size of n from 15 to 30, n^4 becomes a little more than 10 times as large, 2^n becomes about 3×10^4 or 30,000 times as large, while $n!$ becomes approximately 2×10^{20} times as large. 10^{20} is more than a billion billion.

Algorithms whose time functions grow no faster than a polynomial function are called *polynomial time algorithms.* These are *efficient* algorithms; they do not take too much computer time.

Algorithms whose time functions grow faster than any polynomial function are called *exponential time algorithms.* These are *not efficient* algorithms. They are by nature too time-consuming for our computers. Our silicon chip computers are getting faster each year, but even this increase in speed hardly puts a dent in the time required for a computer to implement an exponential time algorithm for a very large graph.

Because the brute force algorithm for the traveling salesman problem has a time function involving $n!$, which grows faster than an exponential function of n, this algorithm is an exponential time algorithm.

For Group Discussion or Individual Investigation

Use the table to help answer these questions:

1. By approximately what factor does n^4 grow if we double n from 10 to 20?
2. By approximately what factor does 2^n grow if we double n from 10 to 20?
3. By approximately what factor does $n!$ grow if we double n from 10 to 20?
4. Repeat Exercises 1–3 if we double n from 25 to 50.
5. Suppose it would take 2^{30} years using computers at their present speeds to solve a certain problem. (2^{30} is a little over one billion.) If we assume that computers double in speed each year, how long would we have to wait before we could solve the problem in 1 year?

15.3 EXERCISES

In Exercises 1 and 2, a graph is shown, and some paths in the graph are specified. Determine which paths are Hamilton circuits for the graph. If not, say why not.

1.

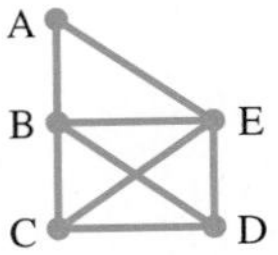

(a) A → E → C → D → E → B → A
(b) A → E → C → D → B → A
(c) D → B → E → A → B
(d) E → D → C → B → E

2.

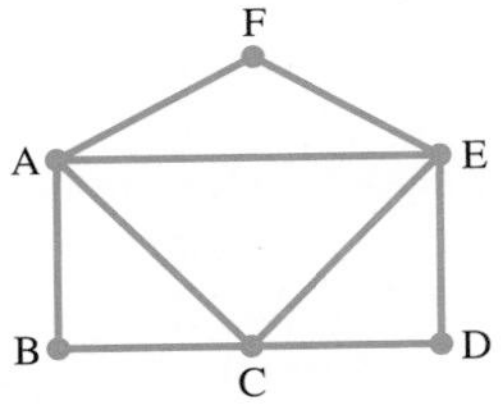

(a) A → B → C → D → E → C → A → E → F → A
(b) A → C → D → E → F → A
(c) F → A → C → E → F
(d) C → D → E → F → A → B

In Exercises 3 and 4, determine whether the string of vertices is a circuit, whether it is an Euler circuit, and whether it is a Hamilton circuit. Justify your answers.

3.

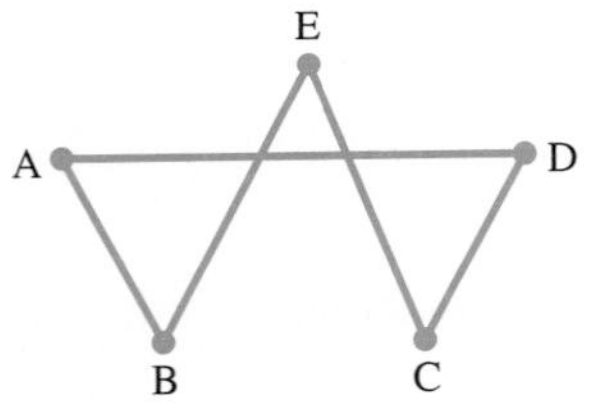

(a) A → B → C → D → E → A
(b) B → E → C → D → A → B
(c) E → B → A → D → A → D → C → E

4.

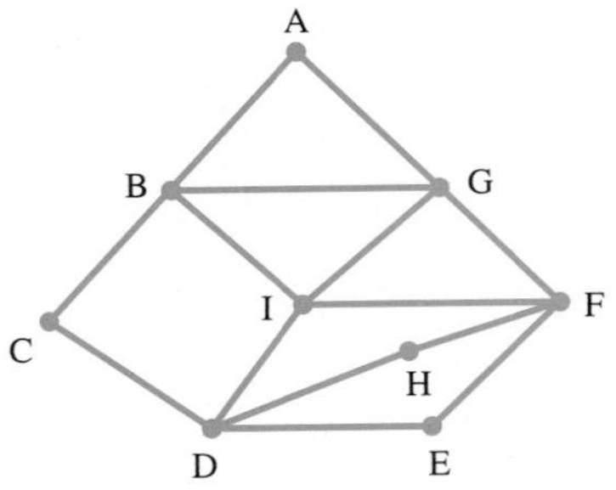

(a) A → B → C → D → E → F → G → A
(b) B → I → G → F → E → D → H → F → I → D → C → B → G → A → B
(c) A → B → C → D → E → F → G → H → I → A

In Exercises 5–10, determine whether the graph has a Hamilton circuit. If so, find one. (There are many different correct answers.)

5.

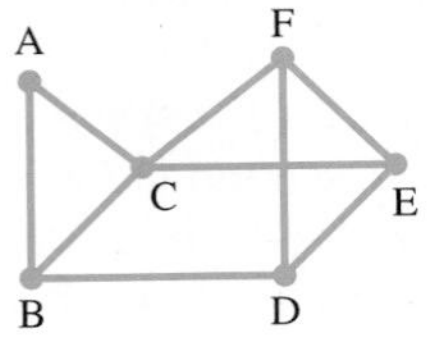

6.

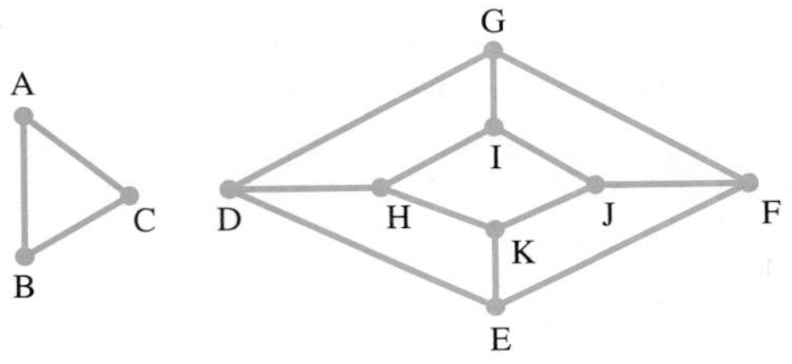

7.

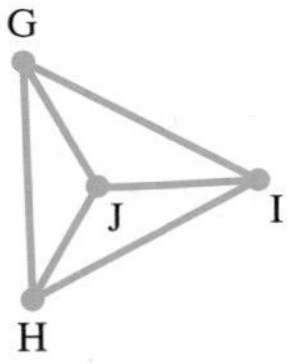

8.

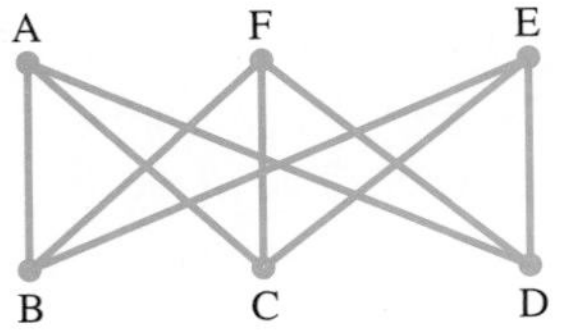

9.

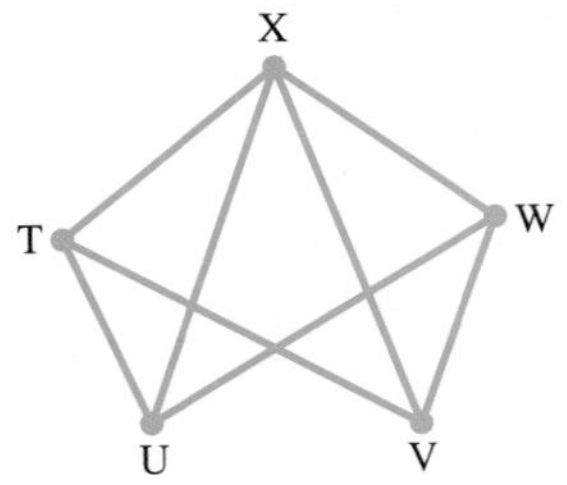

10.

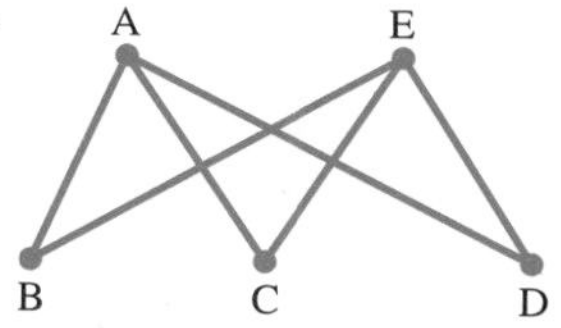

11. Draw a graph that has a Hamilton circuit, but no Euler circuit. Specify the Hamilton circuit, and explain why the graph has no Euler circuit. (There are many different correct answers.)

12. Draw a graph that has an Euler circuit, but no Hamilton circuit. Specify an Euler circuit in your graph. (There are many different correct answers.)

13. Draw a graph that has both an Euler circuit and a Hamilton circuit. Specify these circuits. (There are many different correct answers.)

14. Decide whether each statement is true or false. If the statement is false, give an example to show that it is false.

(a) A Hamilton circuit for a graph must visit each vertex in the graph.

(b) An Euler circuit for a graph must visit each vertex in the graph.

(c) A Hamilton circuit for a graph must use each edge in the graph.

(d) An Euler circuit for a graph must use each edge in the graph.

(e) A circuit cannot be both a Hamilton circuit and an Euler circuit.

(f) An Euler circuit must visit no vertex more than once, except the vertex where the circuit begins and ends.

In Exercises 15–20, determine whether an Euler circuit, a Hamilton circuit, or neither would solve the problem.

15. ***Bandstands at a Festival*** The vertices of a graph represent bandstands at a festival and the edges represent paths between the bandstands. A visitor wants to visit each bandstand exactly once, returning to her starting point when she is finished.

16. ***Relay Team Running Order*** The vertices of a complete graph represent the members of a five-person relay team. The team manager wants a circuit that will show the order in which the team members will run. (He will decide later who will start.)

17. ***Paths in a Botanical Garden*** The vertices of a graph represent places where paths in a botanical garden cross and the edges represent the paths. A visitor wants to walk along each path in the garden exactly once, returning to his starting point when finished.

18. ***Traveling in Western Europe*** The vertices of a graph represent the countries on the continent (Western Europe), with edges representing border crossings between the countries. A traveler wants to travel over each border crossing exactly once, returning to the first country visited for his flight home to the United States.

19. ***Traveling in Africa*** Vertices represent countries in sub-Saharan Africa, with an edge between two vertices if those countries have a common border. A traveler wants to visit each country exactly once, returning to the first country visited for her flight home to the United States.

20. ***Reading X-Rays*** In using X-rays to analyze the structure of crystals, an X-ray diffractometer measures the intensity of reflected radiation from the crystal in thousands of different positions. Consider the complete graph with vertices representing the positions where measurements must be taken. The researcher must decide the order in which to take these readings, with the diffractometer returning to its starting point when finished.

In Exercises 21–24, use your calculator, if necessary, to find the value.

21. 4! **22.** 6!

23. 9! **24.** 14!

In Exercises 25–28, determine how many Hamilton circuits there are in a complete graph with this number of vertices. (Leave answers in factorial notation.)

25. 10 vertices **26.** 15 vertices

27. 18 vertices **28.** 60 vertices

29. List all Hamilton circuits in the graph which start at P.

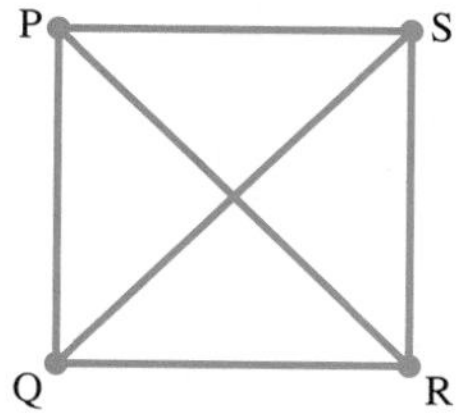

In Exercises 30–36, refer to the following graph. List all Hamilton circuits in the graph that start with the indicated vertices.

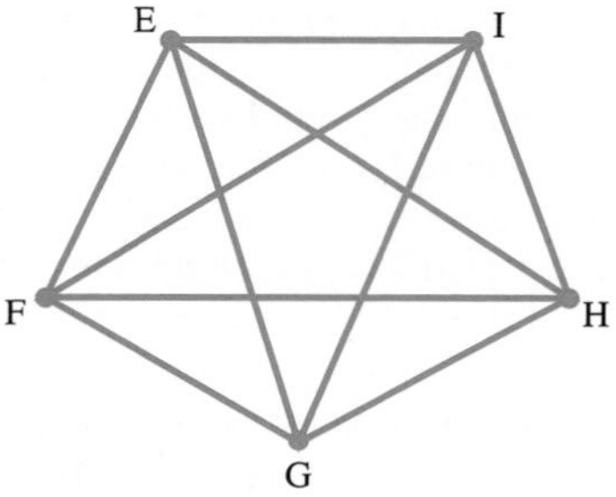

30. Starting E → F → G

31. Starting E → H → I

32. Starting E → I → H

33. Starting E → F

34. Starting E → I

35. Starting E → G

36. Starting at E

37. List all Hamilton circuits in the graph which start at A.

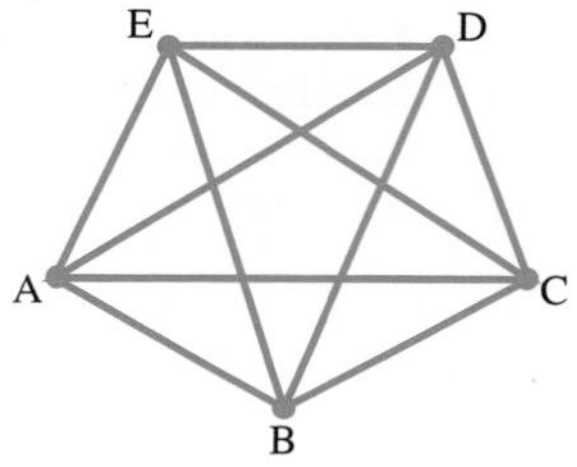

In Exercises 38–41, use the brute force algorithm to find a minimum Hamilton circuit for the graph. In each case determine the total weight of the minimum Hamilton circuit.

38.

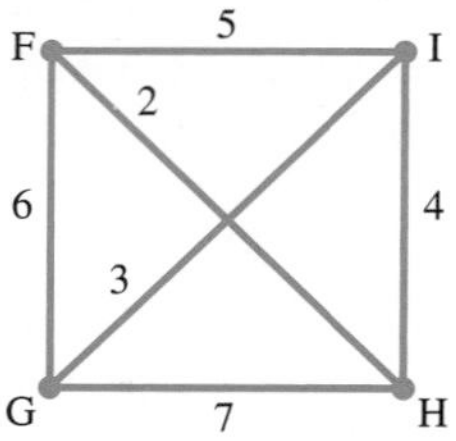

39.

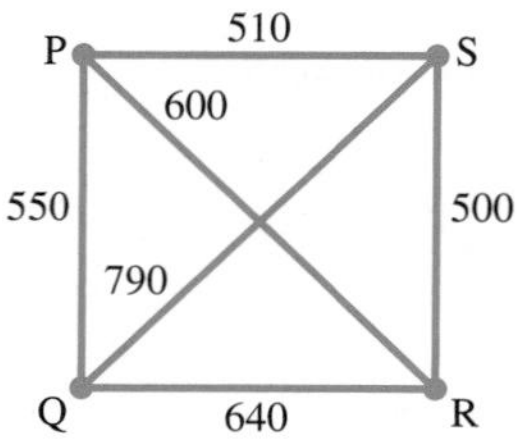

40.

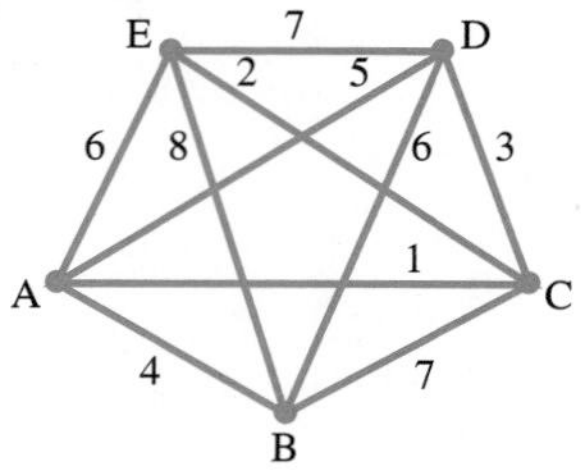

41.

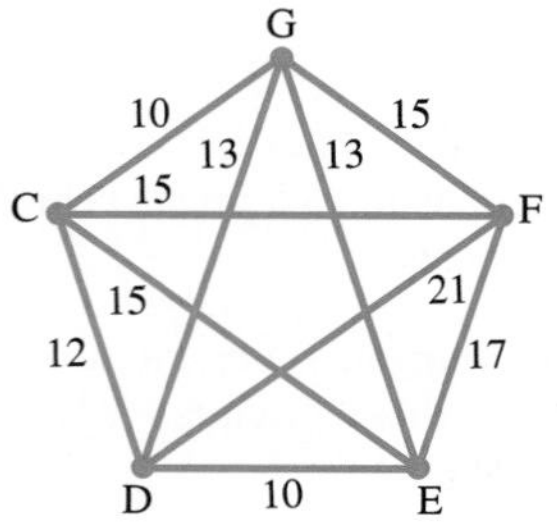

In Exercises 42–44, use the nearest neighbor algorithm starting at each of the indicated vertices to determine an approximate solution to the problem of finding a minimum Hamilton circuit for the graph. In each case, find the total weight of the circuit you have found.

42.

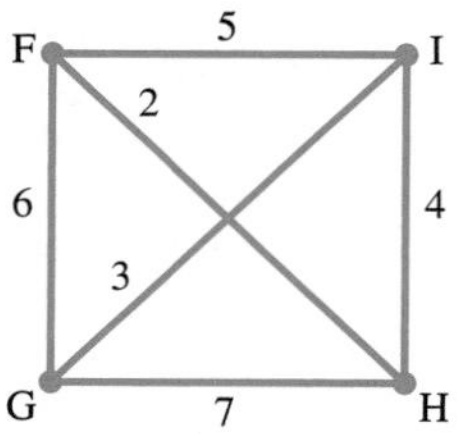

(a) Starting at F
(b) Starting at G
(c) Starting at H
(d) Starting at I

43.

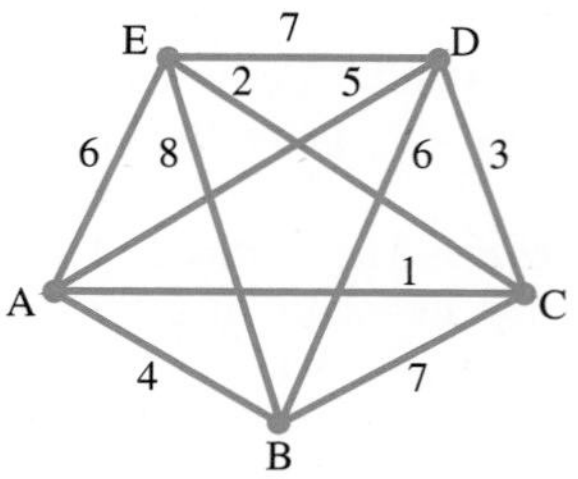

(a) Starting at A
(b) Starting at C
(c) Starting at D
(d) Starting at E

44.

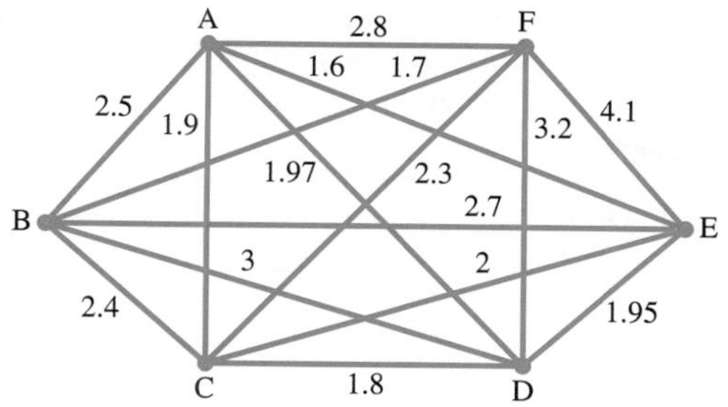

(a) Starting at A
(b) Starting at B
(c) Starting at C
(d) Starting at D
(e) Starting at E
(f) Starting at F

45. Refer to the accompanying graph. Complete parts (a)–(c) in order.

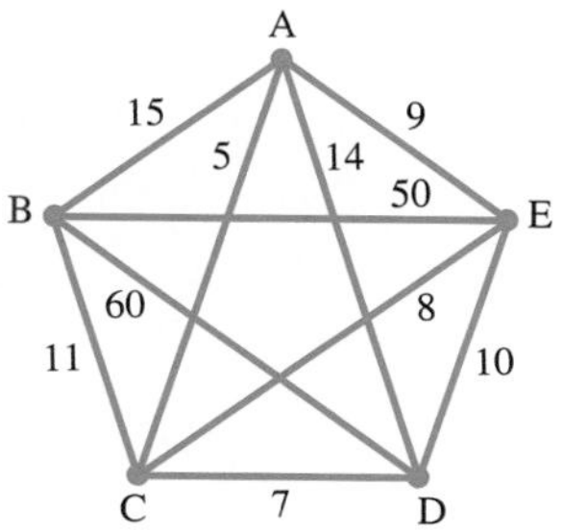

(a) Use the nearest neighbor algorithm starting at each of the vertices in turn to determine an approximate solution to the problem of finding a minimum Hamilton circuit for the graph. In each case, find the total weight of the circuit you have found.
(b) Which of the circuits that you found in part (a) gives the best solution to the problem of finding a minimum Hamilton circuit for the graph?
(c) Just by looking carefully at the graph, find a Hamilton circuit in the graph that has lower total weight than any of the circuits you found in part (a).

46. A graph is called a *complete bipartite graph* if the vertices can be separated into two groups in such a way that there are no edges between vertices in the same group, and there is an edge between each vertex in the first group and each vertex in the second group. An example is shown below.

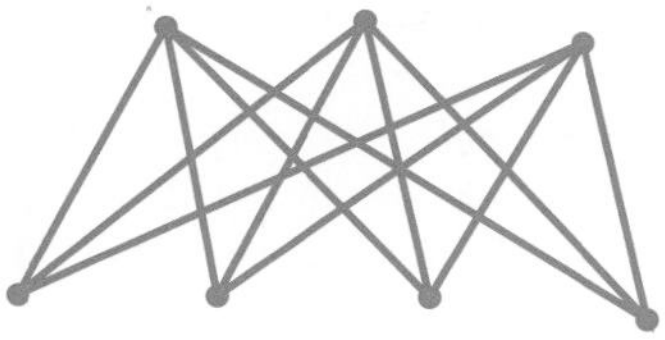

The notation we use for these graphs is $K_{m,n}$ where m and n are the numbers of vertices in the two groups. For example, the graph shown in this exercise is $K_{4,3}$ (or $K_{3,4}$). Answer the following in order:
(a) Draw $K_{2,2}$, $K_{2,3}$, $K_{2,4}$, $K_{3,3}$, and $K_{4,4}$.
(b) For each of the graphs you have drawn in part (a), find a Hamilton circuit, or say that the graph has no Hamilton circuit.
(c) Make a conjecture: What must be true about m and n for $K_{m,n}$ to have a Hamilton circuit?
(d) What must be true about m and n for $K_{m,n}$ to have an Euler circuit? Justify your answer.

In Exercises 47–50, find all Hamilton circuits in the graph which start at A. (Hint: Use counting trees.)

47.

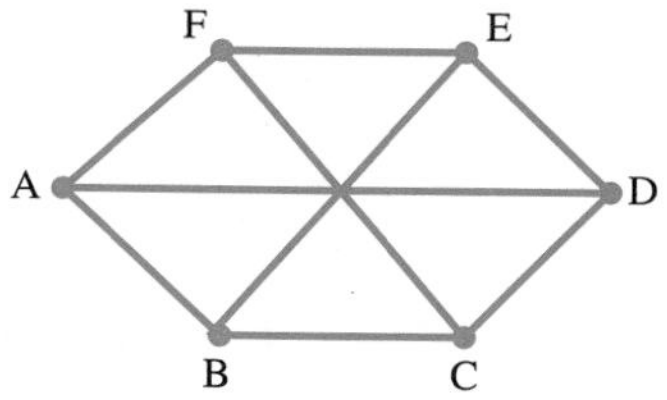

48.

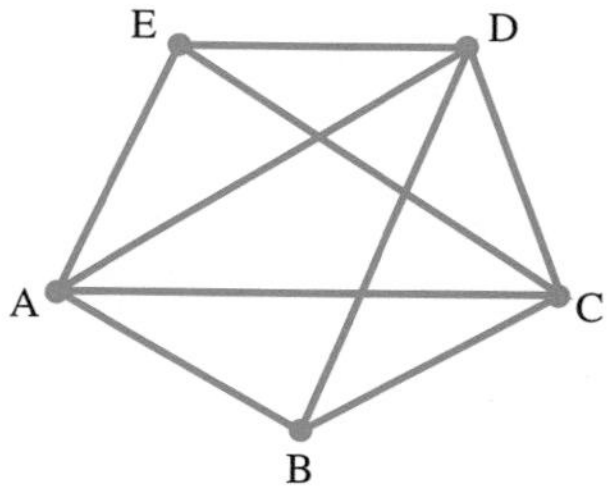

49.

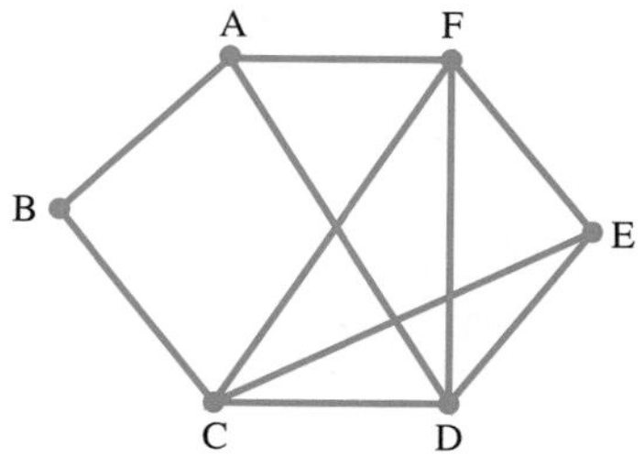

50.

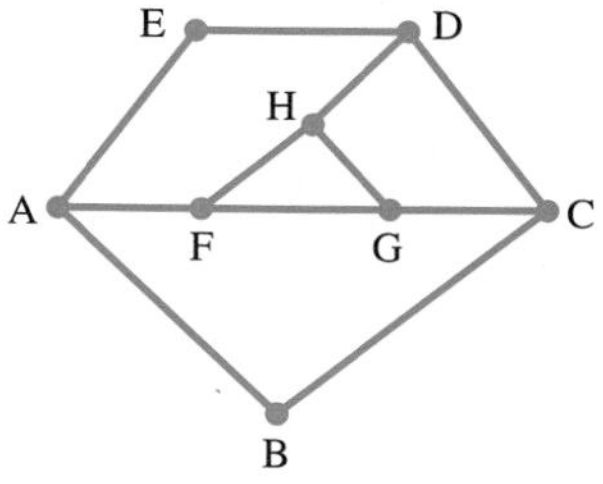

51. Paul A. M. Dirac proved the following theorem in 1952: Suppose G is a (simple) graph with n vertices, $n \geq 3$. If the degree of each vertex is greater than or equal to $\frac{n}{2}$, then the graph has a Hamilton circuit. Refer to graphs (1)–(5) as you answer parts (a)–(e) in order.

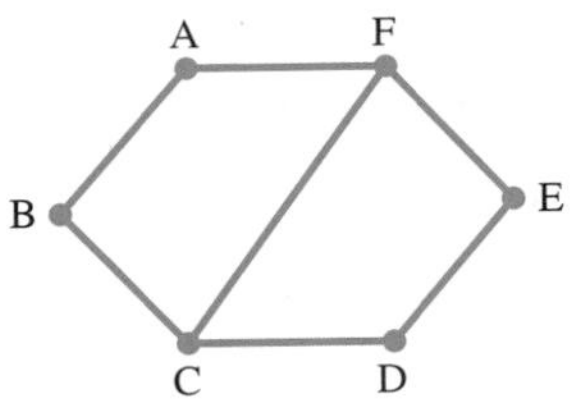

(1)

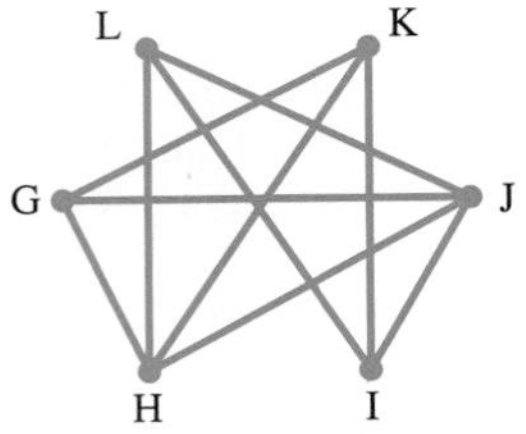

(2)

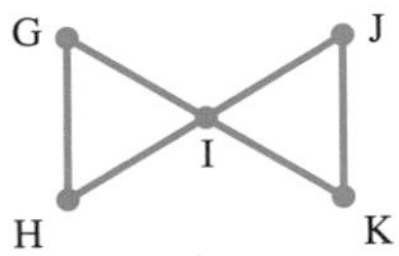

(3)

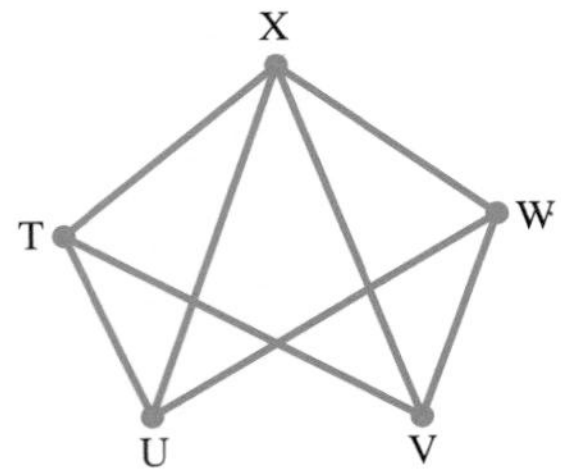

(4)

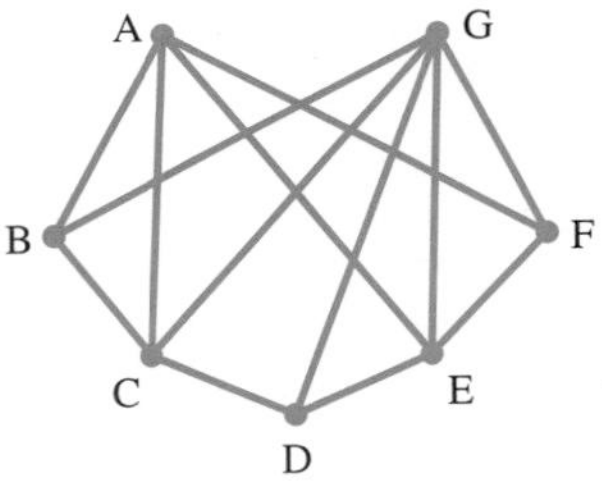

(5)

(a) Which of the graphs satisfy the condition that the degree of each vertex is greater than or equal to $\frac{n}{2}$?

(b) For which of the graphs can we conclude *from Dirac's theorem* that the graph has a Hamilton circuit?

(c) If a graph does *not* satisfy the condition that the degree of each vertex is greater than or equal to $\frac{n}{2}$, can we be sure that the graph does *not* have a Hamilton circuit? Justify your answer. (*Hint:* Study the accompanying graphs.)

(d) Is Dirac's theorem still true if $n < 3$? Justify your answer.

(e) Use Dirac's theorem to write a convincing argument that any complete graph with 3 or more vertices has a Hamilton circuit.

The graph below shows the Icosian game (described in the text) with the vertices labeled. In Exercises 52–54, find a Hamilton circuit for the graph in the specified version of the game. (We suggest you write numbers on the graph in the order in which you visit the vertices. There are different correct answers for these exercises.)

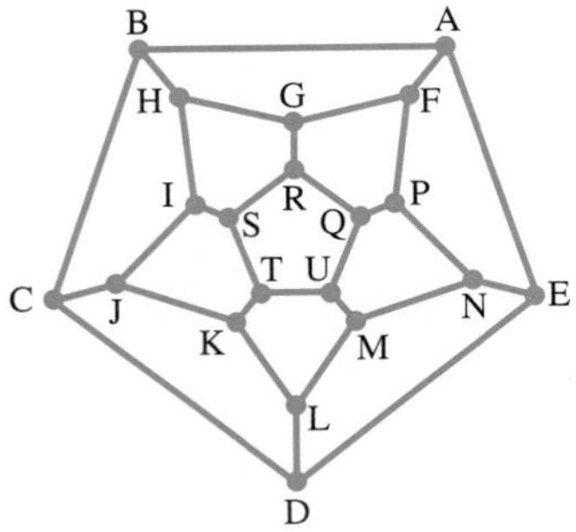

52. The circuit begins at A.

53. The circuit begins A → F.

54. The circuit begins A → B → H.

55. Refer to Exercise 58 in Section 15.1 for how one can analyze the rhyme scheme in poetry using graphs. Use this idea to write a paper on sestinas. Be sure to discuss what Hamilton circuits have to do with sestinas.

56. Find out what NP complete problems are, and write a paper on them. Include a careful description of at least one NP complete problem that is different from the traveling salesman problem.

57. Write a paper on very large numbers. Be sure to include some good examples of actual situations that involve very, very large numbers.

15.4 Trees and Minimum Spanning Trees

Connected Graphs and Trees • Spanning Trees • Minimum Spanning Trees • Kruskal's Algorithm • Number of Vertices and Edges

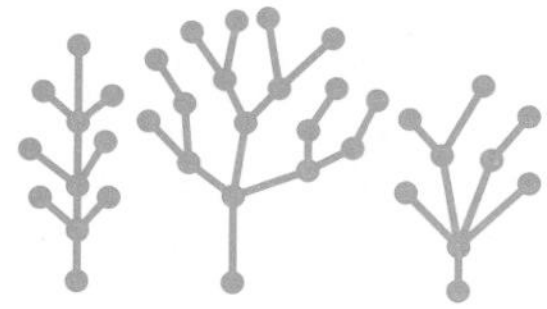

In graph theory, the botanical words go beyond trees. A **forest** is a graph with all components trees, as shown here. Later in this section you will find the words **root** and **leaf** applied to graphs. The terms **pruning** and **separating** arise in applications of trees to computer analysis of images.

Connected Graphs and Trees

In the previous two sections, we explored graphs with special kinds of circuits. In this section we examine graphs (called *trees*) that have no circuits.

Consider a problem that Maria has in her garden. She wants to install an underground irrigation system. The system must connect the faucet (F) to outlets at various points in the garden, specifically the rose bed (R), perennial bed (P), daffodil bed (D), annuals bed (A), berry patch (B), vegetable patch (V), cut flower bed (C), and shade bed (S). These are shown as the vertices of the graph in Figure 42. The irrigation system will have pipes connecting these points.

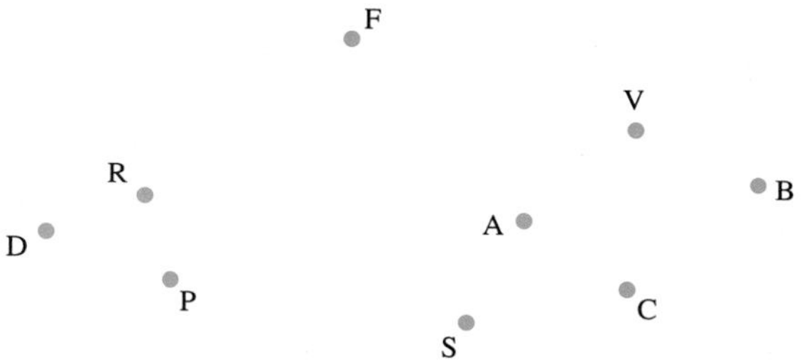

FIGURE 42

For the irrigation system to work well, there should be as few pipes as possible connecting all the outlets.

Think about this in terms of a graph with edges representing pipes. Maria wants water to flow from the faucet to each of the outlets; this means that we need to create a *connected* graph that includes all the vertices in Figure 42. We discussed connected graphs in Section 15.1. Now that you are familiar with paths, we can give a definition more useful for our purposes here:

Connected Graph

A **connected graph** is one in which there is *at least one path* between each pair of vertices.

To see that this definition agrees with our earlier definition (on page 928), observe that if we select any pair of vertices in the connected graph in Figure 43 (for example, A and C) then there is at least one path between them. (In fact, there are many paths between A and C; for example $A \rightarrow B \rightarrow E \rightarrow C$ and $A \rightarrow F \rightarrow C$.

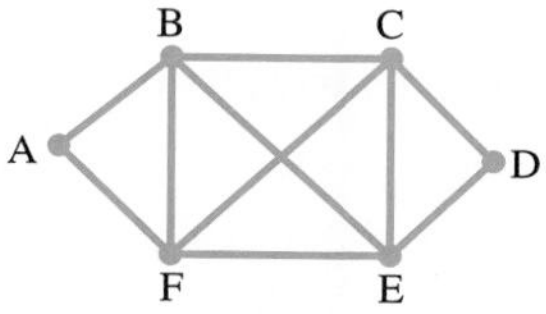

FIGURE 43

Maria needs a *connected* graph. Also she must use as few pipes as possible to connect all the outlets to the system. This means that the graph *must not contain any circuits*. If the graph had a circuit, we could remove one of the pipes in the circuit, and still have a connected system of pipes.

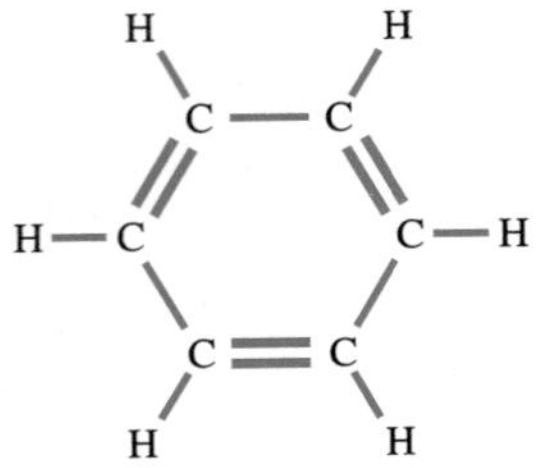

Benzene

It seems mathematician **Arthur Cayley** (see page 976) invented the term "tree" in graph theory in about 1857 in connection with problems related to counting all trees of certain types. Later he realized the relevancy of his work to nineteenth-century organic chemistry, and in the 1870s published a note on this. Of course, not all molecules have treelike structures. Friedrich Kekulé's realization that the structure of benzene is *not* a tree is considered one of the most brilliant breakthroughs in organic chemistry.

Figure 44 shows one possible solution to Maria's problem.

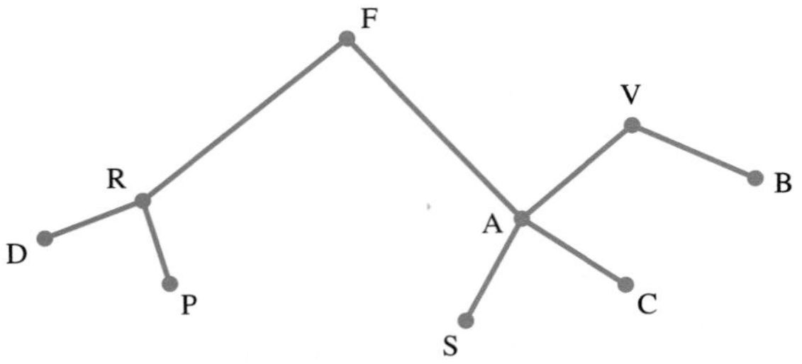

FIGURE 44

There is a name for graphs that are connected and contain no circuits.

Tree

We call a graph a **tree** if the graph is *connected* and contains *no circuits.*

All five of the graphs shown in Figure 45 are trees.

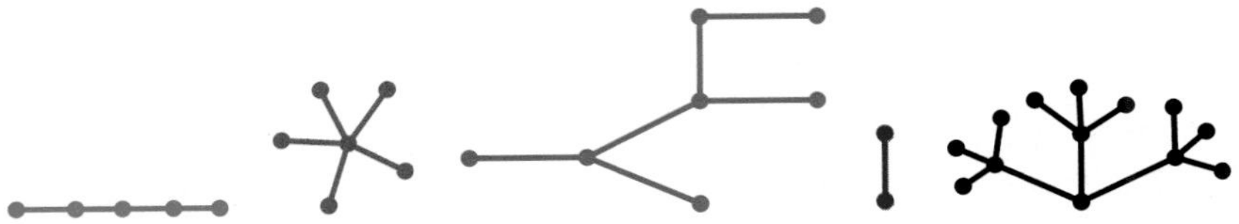

FIGURE 45

In contrast, none of the graphs shown in Figure 46 is a tree. Graph (a) is not a tree, because it is not connected. Graphs (b) and (c) are not trees, because each contains at least one circuit.

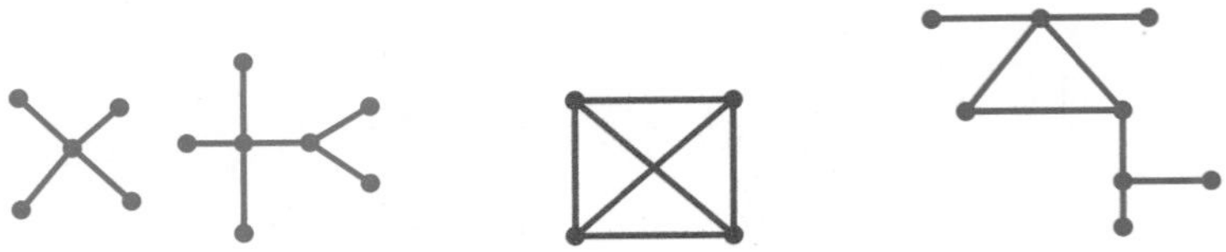

(a) **(b)** **(c)**

FIGURE 46

To better understand the concept of a tree, we consider an important property of trees.

Unique Path Property of Trees

In a tree there is always **exactly one path** from each vertex in the graph to any other vertex in the graph.

For example, starting with vertices S and B in the tree in Figure 44, there is exactly one path from vertex S to vertex B, namely, S → A → V → B.

This property follows from the definition of a tree. For as a tree is connected, there is always at least one path between each pair of vertices. Also, if the graph is a tree, there cannot be two different paths between a pair of vertices; if there were, these together would be a circuit.

For Further Thought

Binary Coding

The unique path property of a tree can be used to make a code. We illustrate with a **binary code,** that is, one that uses 0s and 1s. (Recall from Chapter 4 that when we represent numbers in binary form we use only the symbols 0 and 1.)

To set up the code we use a special kind of tree like that shown in the figure below. This tree is a **directed graph** (there are arrows on the edges). The vertex at the top (with no arrows pointing toward it) is called the **root** of the directed tree; the vertices with no arrows pointing away from them are called **leaves** of the directed tree. We label each leaf with a letter we want to encode. The diagram shown encodes only 8 letters, but we could easily draw a bigger tree with more leaves to represent more letters.

We now write a 0 on each branch that goes left and a 1 on each branch that goes right.

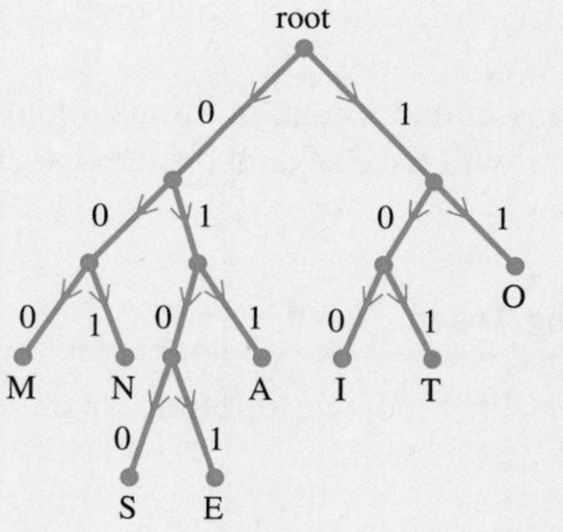

To show how the encoding works, we write the word MAT using the code. Follow the unique path from the root of the tree down to the appropriate leaf, noting in order the labels on the edges.

M is written 000.
A is written 011.
T is written 101.

So MAT is written 000011101.

We can easily translate this code using our tree. Let us see how we could decode 000011101. Referring to the tree, there is only one path from the root to a leaf that can give rise to those first three 0s, and that is the path leading to M. So we can begin to separate the code word into letters: 000–011101. Again following down from the root, the path 011 leads us unambiguously to A. So we have 000-011-101. The path 101 leads unambiguously to T.

The reason we can translate the string of 0s and 1s back to letters without ambiguity is that no letter has a code the same as the first part of the code for a different letter.

For Group Discussion

1. Use the binary encoding tree to write the binary code for each of the following words: ANT, SEAT
2. Use the encoding tree to find the word represented by each of the following codes: 0001111001 0100011101
3. Decode the following messages (commas are inserted to show separation of words).
   (a) 0110011010100, 100001, 1010101001101
   (b) 0101011101, 0001000011010100, 100001, 101100001
   (c) 011, 0000101011001, 1010101011000
   (d) 00001010101101, 0000101, 011101, 1010101011

It is the unique path property of trees that makes them so important in real-world applications. This also is the reason trees are not a very useful model for telephone networks; each edge of a tree is a cut edge, so failure along one link would disrupt the service. Telephone companies rely on networks with many circuits to provide alternative routes for calls.

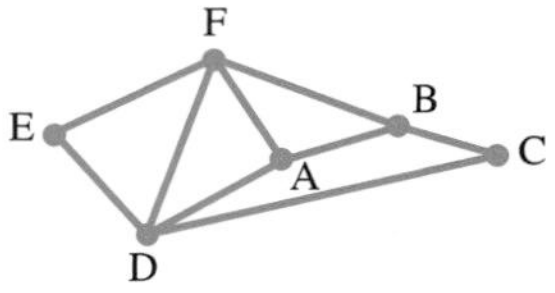

FIGURE 47

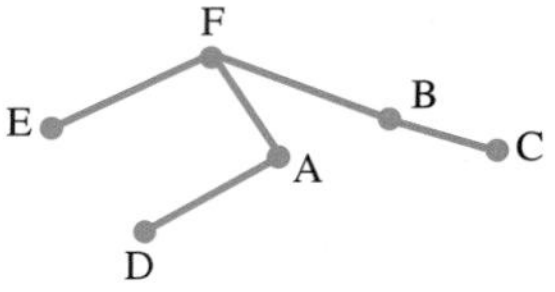

FIGURE 48

EXAMPLE 1 Renovating an Irrigation System

Joe has an old underground irrigation system in his garden, with connections at the faucet (F), outlets A through E in the garden, and existing underground pipes as shown in the graph in Figure 47. He wants to renovate this system, keeping only some of the pipes. Water must still be able to flow to each of the original outlets and the graph of the irrigation system must be a tree. Design an irrigation system that meets Joe's objectives.

SOLUTION

In graph theory terms, we need a *subgraph* of the graph in Figure 47. Since water still must be able to flow to each of the original outlets, we need a connected subgraph that *includes all the vertices of the original graph.* Finally, this subgraph must be a *tree*. Thus, we must remove edges (but no vertices) from the graph, without disconnecting the graph, to obtain a subgraph that is a tree. One way to achieve this is shown in Figure 48. ■

Arthur Cayley (1821–1895) was a close friend of James Sylvester. (See pages 930 and 980.) They worked together as lawyers at the courts of Lincolns Inn in London and discussed mathematics with each other during work hours. Cayley considered his work as a lawyer simply a way to support himself so that he could do mathematics. In the fourteen years he worked as a lawyer, he published 250 mathematical papers. At age 42, he took a huge paycut to become a mathematics professor at Cambridge University.

Spanning Trees

A subgraph of the type shown in Figure 48 in Example 1 is called a *spanning tree* for the graph. This kind of subgraph "spans" the graph in the sense that it connects all the vertices of the original graph into a subgraph.

Spanning Tree

A **spanning tree** for a graph is a subgraph that includes every vertex of the original, and is a tree.

We can find a spanning tree for any connected graph (think about that) and, if the original graph has at least one circuit, it will have several different spanning trees. The following example illustrates this.

EXAMPLE 2 Finding Spanning Trees

Consider the graph shown in Figure 49. Find two different spanning trees for the graph.

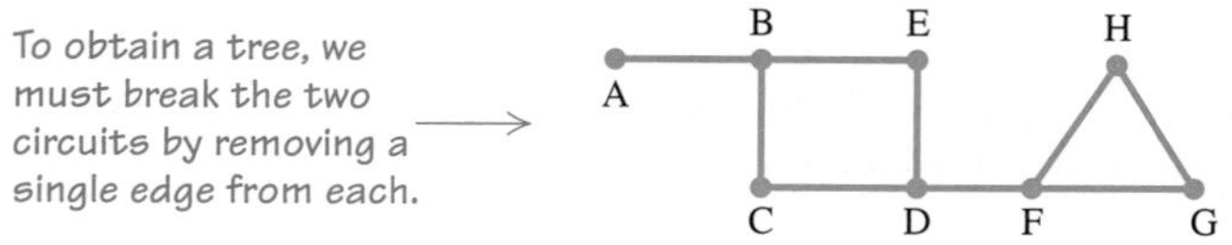

FIGURE 49

SOLUTION

There are several choices for which edges to remove. We could remove edges CD and FH, leaving the subgraph in Figure 50(a). Alternatively, we could remove edges ED and FG, leaving the subgraph in Figure 50(b). There are other correct solutions.

These are two different spanning trees.

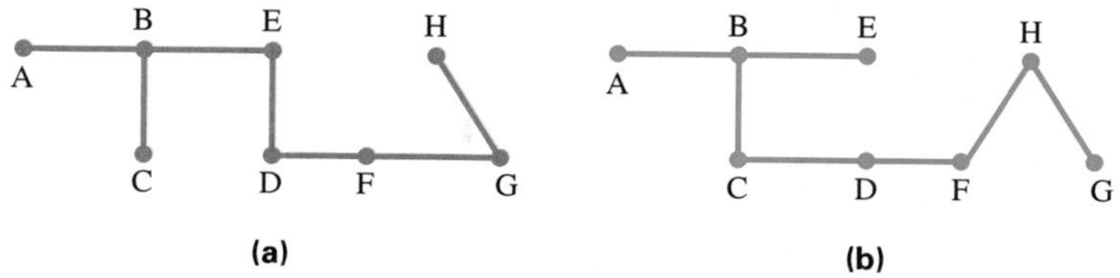

FIGURE 50

Note that if two spanning trees use different edges from the original graph, we consider them to be different spanning trees even when they are isomorphic graphs. For example, the subgraphs shown in Figure 50 are *different* spanning trees, even though they happen to be isomorphic graphs. (Did you notice they were isomorphic? To see this, change the labels in graph (a) as follows: interchange C and E, and interchange G and H. By dragging vertices without detaching edges, we can obtain graphs that look alike.)

Minimum Spanning Trees Consider this situation. There are six villages in a rural district. At present all the roads in the district are gravel roads. There is not enough funding to pave all the roads, but there is a pressing need to have good paved roads so that emergency vehicles can travel easily between the villages. The vertices in the graph in Figure 51 represent the villages in the district, and the edges represent the existing gravel roads. Estimated costs for paving the different roads are shown, in millions of dollars. (These estimates take into account distances, as well as features of the terrain.)

Computer analysis of images is a field of active research, particularly in medicine. (Computer analysis of images refers to having a computer detect relevant features of an image.) Some of this research uses minimum spanning trees. They have been used to identify protein fibers in photographs of cells. The computer detects some points lying on the fibers and then determines a minimum spanning tree to deduce the overall structure of the fiber. Some studies have used minimum spanning trees to analyze cancerous tissue.

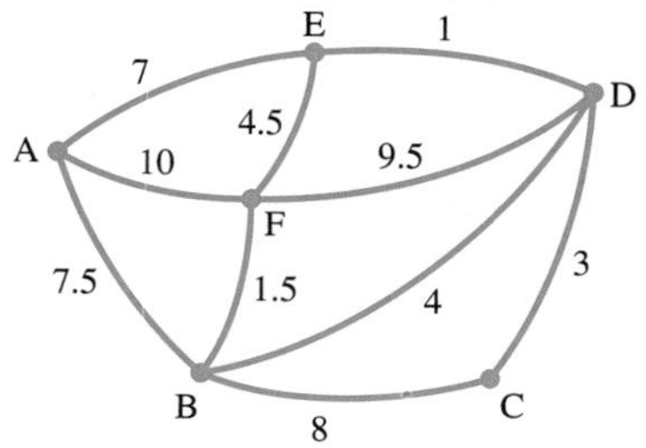

FIGURE 51

The district council wants to pave just enough roads so that emergency vehicles will be able to travel from any village to any other along paved roads, though possibly by a roundabout route. Of course, the council would like to achieve this at *lowest possible cost.*

The problem calls for exactly one path from each vertex in the graph to any other, so it requires a spanning tree. Also, this spanning tree should have minimum total weight. (The total weight of a spanning tree is the sum of the weights of the edges in the tree.)

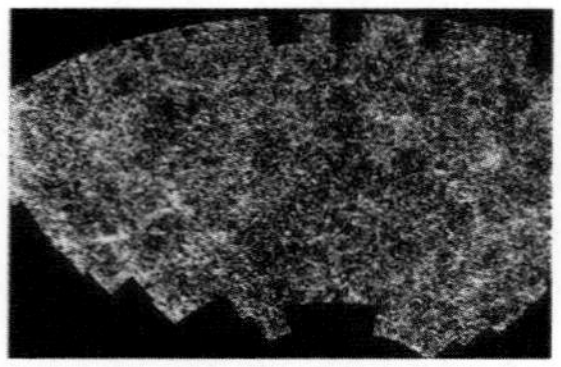

Minimum Spanning Trees in Astronomy The structure of the universe may hold clues about the formation of the universe. The image shows the large-scale structure of the universe in a region comprising approximately 10% of the sky. It shows the distribution of about 2 million galaxies. Galaxies are not uniformly distributed throughout the universe; instead, they seem to be arranged in clusters, which are grouped together into superclusters, which themselves seem to be arranged in vast chains called filaments. As the image shows, there are voids between the filaments. Some astronomers are using spanning trees to help map the filamentary structure of the universe.

Minimum Spanning Tree

A spanning tree that has minimum total weight is called a **minimum spanning tree** for the graph.

Kruskal's Algorithm Fortunately for the district planners, there is a good algorithm for finding a minimum spanning tree for any connected, weighted graph. It is called *Kruskal's algorithm.* (The algorithm is an efficient algorithm; that is, it does not take too much computer time for even very large graphs.)

Kruskal's Algorithm for Finding a Minimum Spanning Tree for Any Connected, Weighted Graph

Choose edges for the spanning tree as follows.

Step 1 First edge: choose any edge with minimum weight.

Step 2 Next edge: choose any edge with minimum weight from *those not yet selected.* (At this stage, the subgraph may look disconnected.)

Step 3 Continue to choose edges of minimum weight from those not yet selected, except *do not select any edge that creates a circuit* in the subgraph.

Step 4 Repeat Step 3 until the subgraph connects all vertices of the original graph.

Algorithms such as Kruskal's often are called *greedy* algorithms; can you think of a reason?

Because Kruskal's algorithm gives a subgraph that is connected, includes all vertices of the original graph, and has no circuits, it certainly provides a spanning tree for the graph. Also, the algorithm always will give us a *minimum* spanning tree for the graph. The explanation is beyond the scope of this text.

EXAMPLE 3 Applying Kruskal's Algorithm

Use Kruskal's algorithm to find a minimum spanning tree for the graph in Figure 51.

SOLUTION

Note that the graph is connected, so Kruskal's algorithm applies. Choose an edge with minimum weight; ED has the least weight of all the edges, so it is selected first. We color the edge a different color to show that it has been selected. See Figure 52(a).

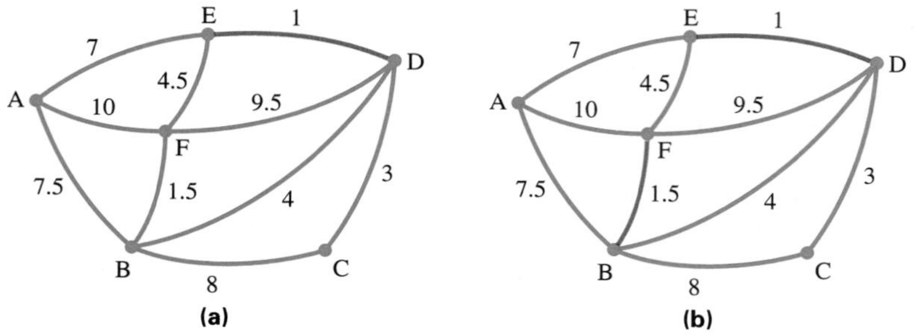

FIGURE 52

Weather data are complex, involving numerous readings of temperature, pressure, and wind velocity over huge areas, making accurate weather prediction difficult at best. Some current research uses minimum spanning trees to help computers interpret these data; for example, minimum spanning trees are used to identify developing cold fronts.

Of the remaining edges, BF has the least weight, so it is the second edge chosen, as shown in Figure 52(b). Note that it does not matter that at this stage the subgraph is not connected.

Continuing, we choose CD, then BD, as shown in Figure 52(c). Of the edges remaining, EF has the least weight; however, EF *would create a circuit* in the subgraph. So, ignore EF and note that, of the edges remaining, AE has the least weight. Thus, AE is included in the spanning tree, as shown in Figure 52(d).

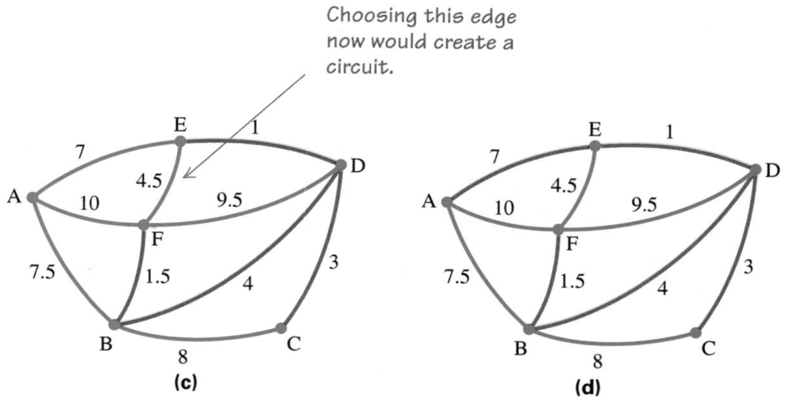

FIGURE 52

We now have a subgraph of the original that connects all vertices of the original graph into a tree. This subgraph is the required minimum spanning tree for the original graph. Its total weight is $1 + 1.5 + 3 + 4 + 7$, which is 16.5. Thus, the district council can achieve its objective at a cost of 16.5 million dollars, by paving only the roads represented by edges in the minimum spanning tree. ■

Number of Vertices and Edges

There is an interesting relation between the number of vertices and the number of edges in a tree. Consider the graphs shown in Figure 53. Information about these graphs is shown in Table 2 on the next page.

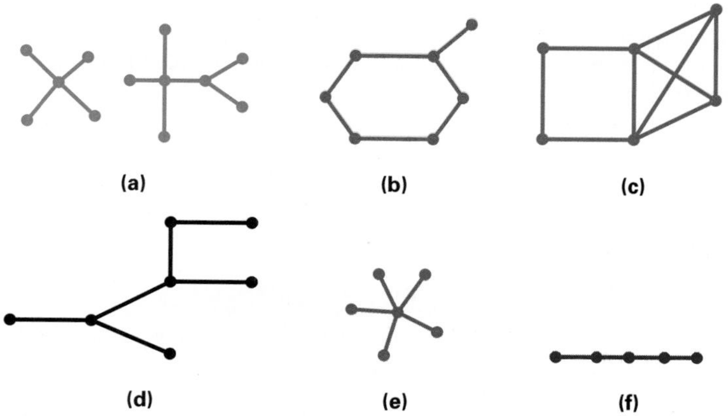

FIGURE 53

TABLE 2

Graph	Tree?	Number of Vertices	Number of Edges
(a)	No	12	10
(b)	No	7	7
(c)	No	6	9
(d)	Yes	7	6
(e)	Yes	6	5
(f)	Yes	5	4

The first three graphs listed are not trees; there is no uniform relation between the number of edges and the number of vertices for these graphs. Such graphs may have more or fewer vertices than edges, or an equal number of both. In contrast, the last three graphs listed are trees. What do you notice about the number of edges compared to the number of vertices for these trees? The same relation holds for any tree.

Number of Vertices and Edges in a Tree

If a graph is a tree, then the number of edges in the graph is one less than the number of vertices.

A tree with n vertices has $n - 1$ edges.

It is not hard to understand why this should be true. (See Exercise 56 of this section.) If we consider only connected graphs, we have also a converse for this theorem: For a connected graph, if the number of edges is one less than the number of vertices, then the graph is a tree.

James Sylvester was known for his flowery language. He described the connection between graphs and chemistry as follows: "Chemistry has the same quickening and suggestive influence upon the algebraist as a visit to the Royal Academy, or the old masters may be supposed to have had on a Browning or a Tennyson."

EXAMPLE 4 Using the Vertex/Edge Relation

A chemist has synthesized a new chemical compound. She knows from her analyses that a molecule of the compound contains 54 atoms and that the molecule has a tree-like structure. How many chemical bonds are there in the molecule?

SOLUTION

Think of the atoms in the molecule as the vertices of a tree, and the bonds as edges of the tree. This tree has 54 vertices. Since the number of edges in a tree is one less than the number of vertices, the molecule must have 53 bonds. ■

PROBLEM-SOLVING HINT Sometimes trial and error is a good problem-solving strategy. Choose an arbitrary proposed solution to the problem and check whether it works. If the arbitrary proposed solution does not work, we may gain insight into what to try next.

EXAMPLE 5 Finding a Graph with Specified Properties

Suppose a graph is a tree and has 15 vertices. What is the greatest number of vertices of degree 5 that this graph could have? Draw such a tree.

SOLUTION

Because the graph is a tree with 15 vertices, it must have 14 edges. Do we know anything about graphs that will help us relate the number of edges to the degrees of vertices? Recall that the sum of the degrees of the vertices of a graph is twice the number of edges. Thus, the sum of the degrees of the vertices in the tree must be $2 \cdot 14$, which is 28.

Now we use trial and error. We want the greatest possible number of vertices with degree 5. Four vertices with degree 5 would contribute 20 to the total degree sum. This would leave $28 - 20 = 8$ as the degree sum of the remaining 11 vertices, but this would mean that some of those vertices would have no edges joined to them. The graph would not be connected and would not be a tree.

Let's try 3 vertices with degree 5. That would contribute 15 to the total degree sum; this would leave $28 - 15 = 13$ as the degree sum for the remaining 12 vertices. After a little thought we see that we can draw a graph with these specifications. See Figure 54. Thus, the greatest possible number of vertices of degree 5 in a tree with 15 vertices is 3.

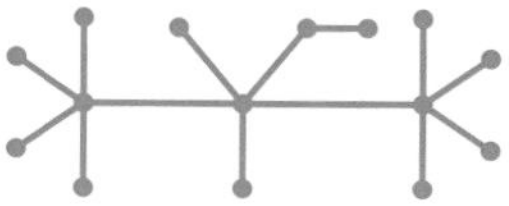

FIGURE 54

15.4 EXERCISES

In Exercises 1–7, determine whether the graph is a tree. If not, explain why it is not.

1.

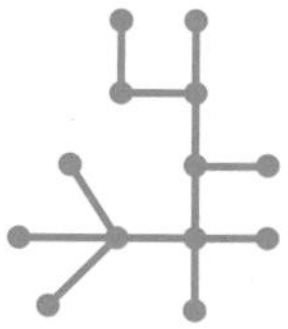

2.

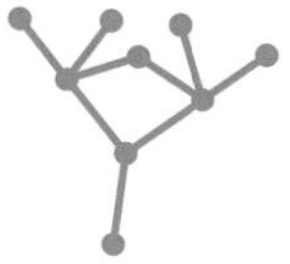

3.

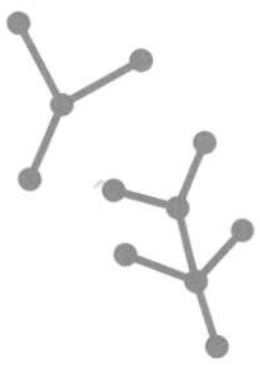

4.

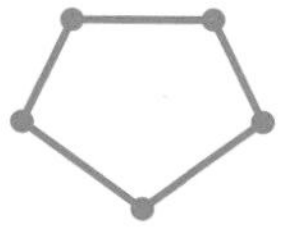

5.

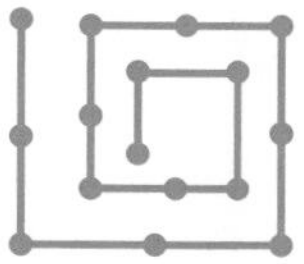

6. a complete graph with 6 vertices

7. a connected graph with all vertices having even degree

In Exercises 8–10, add edges (no vertices) to the graph to change the graph into a tree, or explain why this is not possible. (Note: There may be more than one correct answer.)

8.

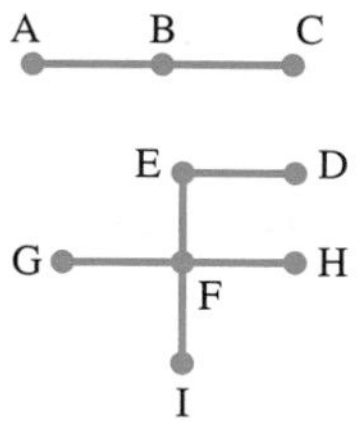

9.

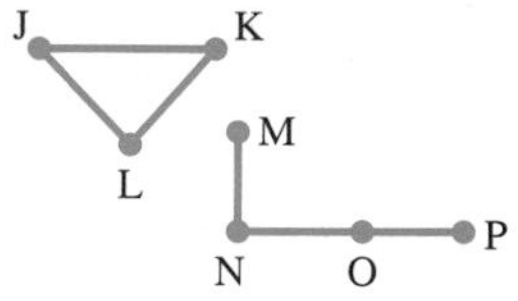

10.

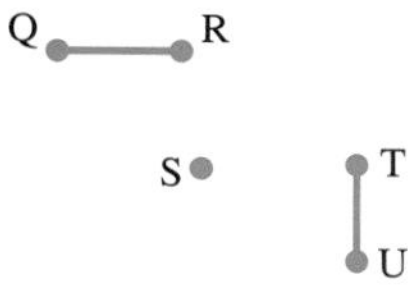

In Exercises 11–13, determine whether the graph described must be a tree.

11. ***Spread of a Rumor*** A sociologist is investigating the spread of a rumor. He finds that 18 people know of the rumor. One of these people must have started the rumor, and each must have heard the rumor for the first time from one of the others. He draws a graph with vertices representing the 18 people and edges showing from whom each person first heard the rumor.

12. ***Internet Search Engine*** You are using the Google search engine to search the Web for information on a topic for a paper. You start at the Google page and follow links you find, sometimes going back to a site you've already visited. To keep a record of the sites you visit, you show each site as a vertex of a graph; you draw an edge each time you connect for the *first* time to a *new* site.

13. ***Tracking Infectious Disease*** A patient has a highly infectious disease. An employee of the Centers for Disease Control is trying to quarantine all people who have had contact over the past week with either the patient or with someone already included in the contact network. The employee draws a graph with vertices representing the patient and all people who have had contact as described. The edges of the graph represent the relation "the two people have had contact."

In Exercises 14–17, determine whether the statement is true *or* false. *If the statement is false, draw an example to show that it is false.*

14. Every graph with no circuits is a tree.

15. Every connected graph in which each edge is a cut edge is a tree.

16. Every graph in which there is a path between each pair of vertices is a tree.

17. Every graph in which each edge is a cut edge is a tree.

In Exercises 18–20, find three different spanning trees for each graph. (There are many different correct answers.)

18.
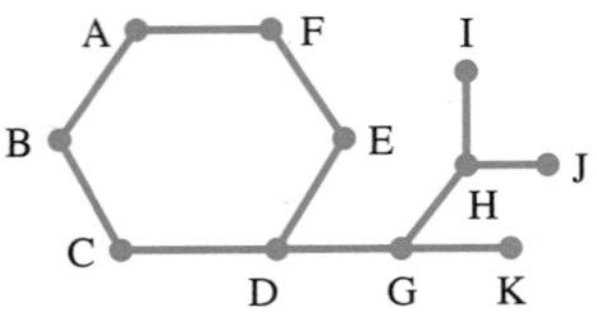

19.
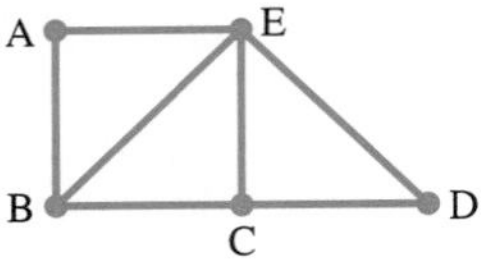

20.
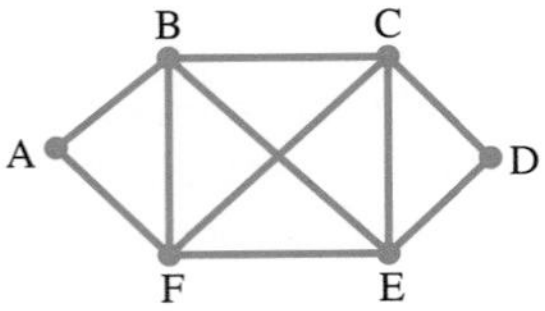

In Exercises 21–23, find all spanning trees for the graph.

21.
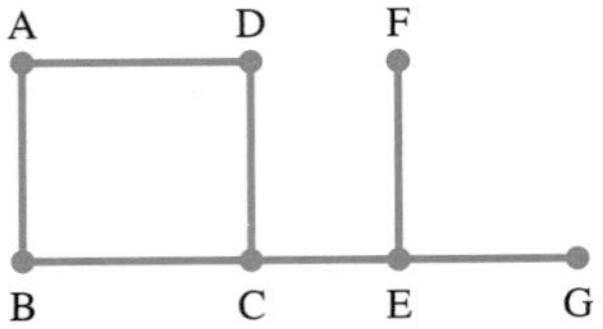

22.
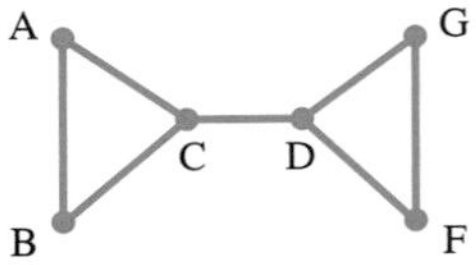

23.
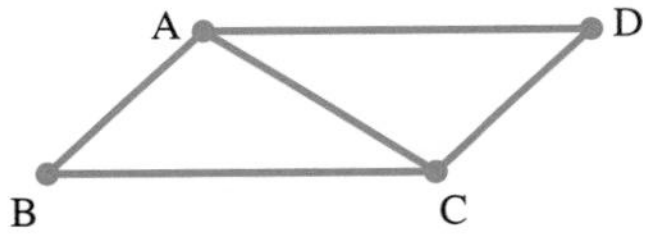

In Exercises 24–26, determine how many spanning trees the graph has.

24.
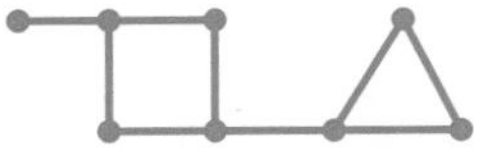

25.

26.
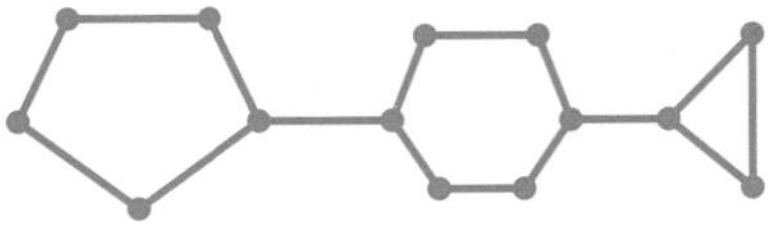

27. What is a general principle about the number of spanning trees in graphs such as those in Exercises 24–26?

28. Complete the parts of this exercise in order.

(a) Find all the spanning trees of each of the following graphs.

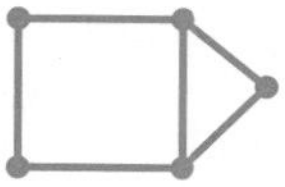

(i)

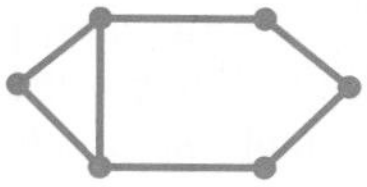

(ii)

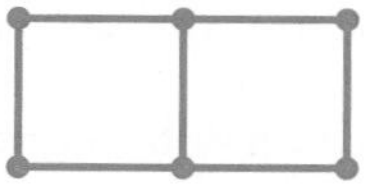

(iii)

(b) For each of the following graphs, determine how many spanning trees the graph has.

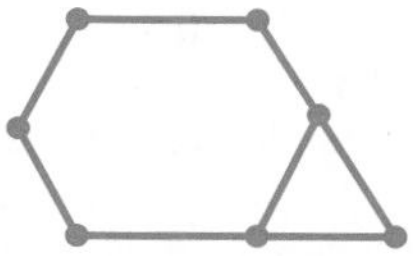

(i)

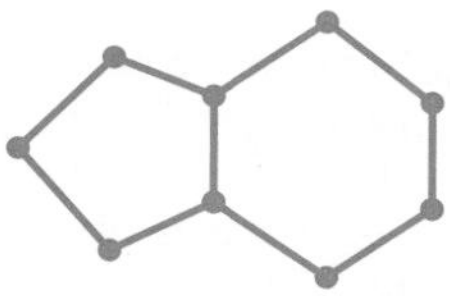

(ii)

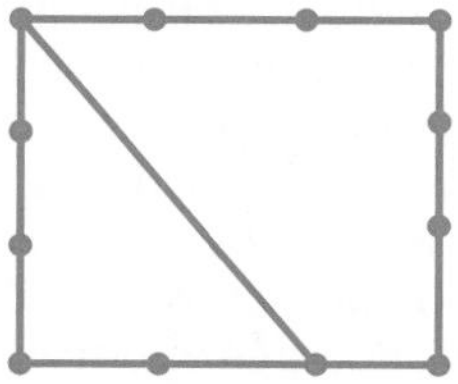

(iii)

(c) What is a general principle about the number of spanning trees in graphs of the kind shown in (a) and (b)?

In Exercises 29–32, use Kruskal's algorithm to find a minimum spanning tree for the graph. Find the total weight of this minimum spanning tree.

29.

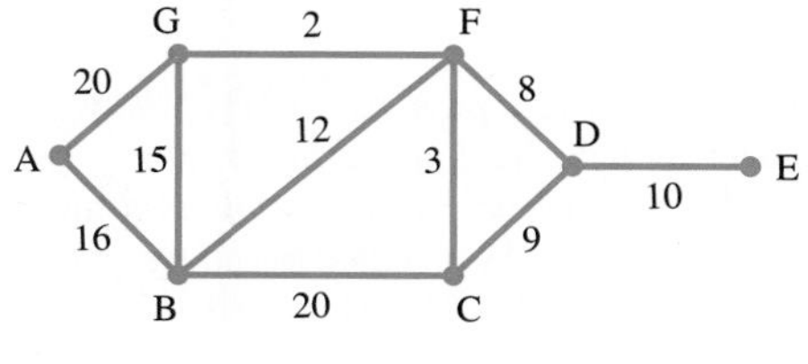

30.

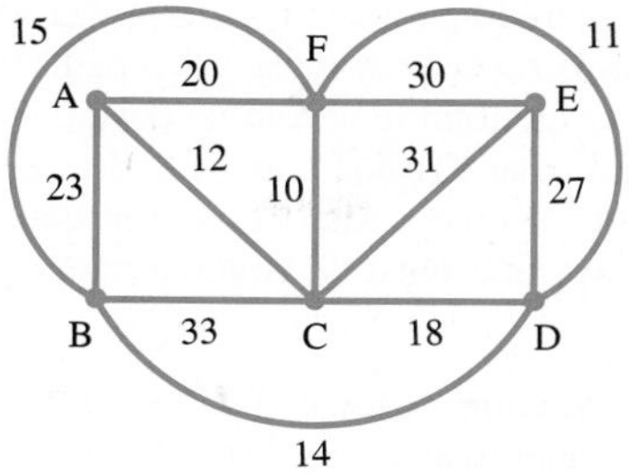

31.

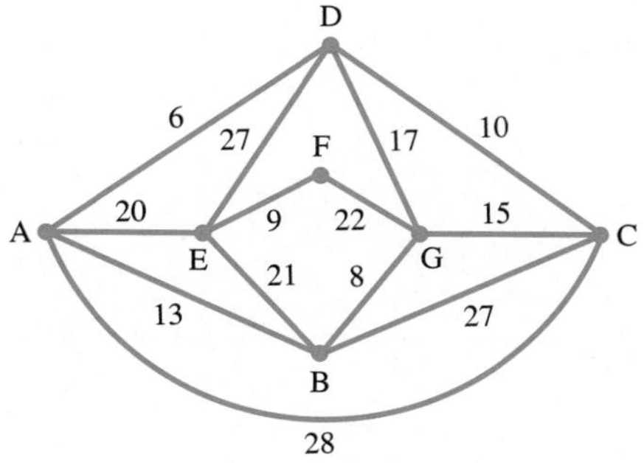

32.

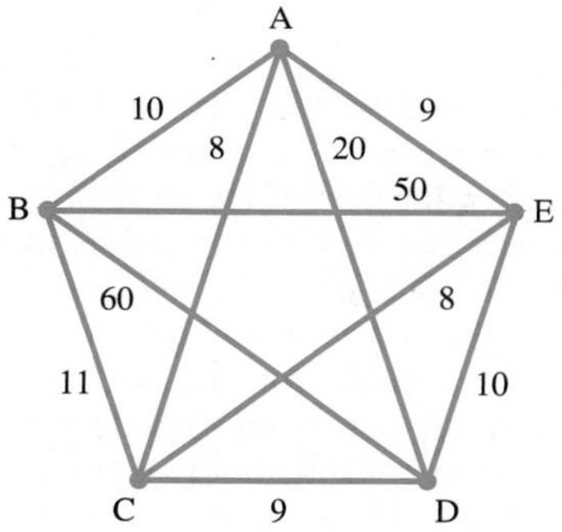

33. ***School Building Layout*** A school consists of 6 separate buildings, represented by the vertices in the following graph.

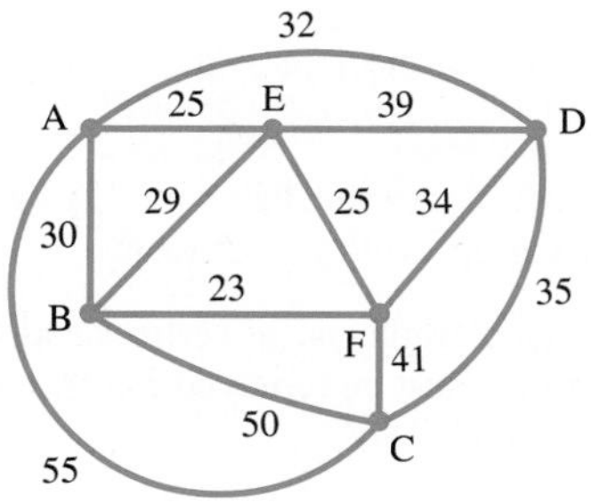

There are paths between some of the buildings as shown. The graph also shows the length in feet of each path. School administrators want to cover some of these paths with roofs so that students will be able to walk between buildings without getting wet when it rains. To minimize cost, they must select paths to be covered such that the total length to be covered is as small as possible. Use Kruskal's algorithm to determine which paths to cover. Also determine the total length of pathways that must be covered under your plan.

34. ***Town Water Distribution*** A town council is planning to provide town water to an area that previously relied on private wells. Water will be fed into the area at the point represented by the vertex labeled A in the accompanying graph.

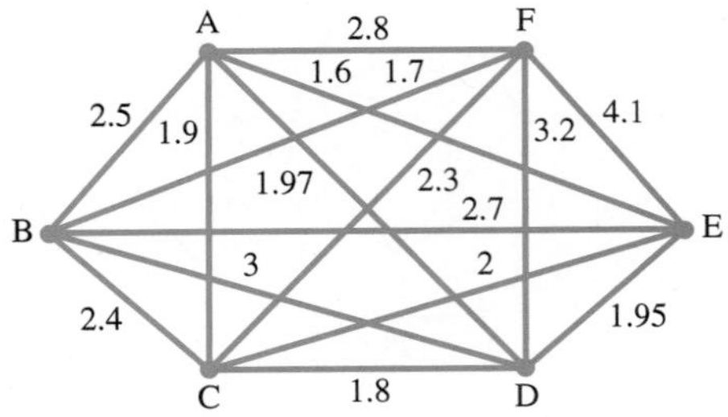

Water must be piped to each of five main distribution points, represented by the vertices labeled B through F in the accompanying graph. Town engineers have estimated the cost of laying the pipes to carry the water between each pair of points in millions of dollars, as indicated in the graph. They must now select which pipes should be laid, so that there is exactly one route for the water to be pumped from A to any one of the five distribution points (possibly via another distribution point), and they want to achieve this at minimum cost. Use Kruskal's algorithm to decide which pipes should be laid. Find the total cost of laying the pipes you select.

35. How many edges are there in a tree with 34 vertices?

36. How many vertices are there in a tree with 40 edges?

37. How many edges are there in a spanning tree for a complete graph with 63 vertices?

38. A connected graph has 27 vertices and 43 edges. How many edges must be removed to form a spanning tree for the graph?

39. We have said that we can always find at least one spanning tree for a connected graph, and usually more than one. Could it happen that different spanning trees for the same graph have different numbers of edges? Justify your answer.

40. Suppose we have a tree with 9 vertices.
- **(a)** Determine the number of edges in the graph.
- **(b)** Determine the sum of the degrees of the vertices in the graph.
- **(c)** Determine the least possible number of vertices of degree 1 in this graph.
- **(d)** Determine the greatest possible number of vertices of degree 1 in this graph.
- **(e)** Answer parts (a) through (d) for a tree with n vertices.

41. Suppose we have a tree with 10 vertices.
- **(a)** Determine the number of edges in the graph.
- **(b)** Determine the sum of the degrees of the vertices in the graph.
- **(c)** Determine the least possible number of vertices of degree 4 in this graph.
- **(d)** Determine the greatest possible number of vertices of degree 4 in this graph. Draw a graph to illustrate your answer.

42. Suppose we have a tree with 17 vertices.
- **(a)** Determine the number of edges in the graph.
- **(b)** Determine the sum of the degrees of the vertices in the graph.
- **(c)** Determine the greatest possible number of vertices of degree 5 in this graph. Draw a graph to illustrate your answer.

43. ***Computer Network Layout*** A business has 23 employees, each with his or her own desk computer, all working in the same office. The managers want to network the computers. To do this they need to install cables between individual computers so that every computer is linked into the network. Because of the way the office is laid out, it is not convenient to simply connect all the computers in one long line. Determine the least number of cables the managers need to install to achieve their objective.

44. ***Design of a Garden*** Maria Jimenez has 12 vegetable and flower beds in her garden and wants to build flagstone paths between the beds so that she can get from each bed along flagstone paths to every other bed. She also wants a path linking her front door to one of the beds. Determine the minimum number of paths she must build to achieve this.

Exercises 45–47 on the next page refer to the following situation: We start with a tree and then draw in extra edges and vertices. For each, say whether it is possible to draw in the number of edges and vertices specified to end up with a connected graph. If it is possible, determine whether the resulting graph must be a tree, may be a tree, or cannot possibly be a tree. Justify each answer briefly.

45. Draw in the same number of vertices as edges.

46. Draw in more vertices than edges.

47. Draw in more edges than vertices.

48. Starting with a tree, is it possible to draw in the same number of vertices as edges and end up with a disconnected graph? If so, give an example. If not, justify briefly.

Exercises 49–51 require the following theorem, proved by Cayley in 1889: A complete graph with n vertices has n^{n-2} spanning trees.

49. How many spanning trees are there for a complete graph with 3 vertices? Draw a complete graph with 3 vertices and find all the spanning trees.

50. How many spanning trees are there for a complete graph with 4 vertices? Draw a complete graph with 4 vertices and find all the spanning trees.

51. How many spanning trees are there for a complete graph with 5 vertices?

52. Find all nonisomorphic trees with 4 vertices. How many are there?

53. Find all nonisomorphic trees with 5 vertices. How many are there?

54. Find all nonisomorphic trees with 6 vertices. How many are there?

55. Find all nonisomorphic trees with 7 vertices. How many are there?

56. In this exercise, we explore why the number of edges in a tree is one less than the number of vertices. Since the statement is clearly true for a tree with only one vertex, we will consider a tree *with more than one vertex.* Answer parts (a)–(h) in order:

(a) How many components does the tree have?

(b) Why must the tree have at least one edge?

(c) Remove one edge from the tree. How many components does the resulting graph have?

(d) You have not created any new circuits by removing the edge, so each of the components of the resulting graph is a tree. If the remaining graph still has edges, choose any edge and remove it. (You have now removed 2 edges from the original tree.) Altogether, how many components remain?

(e) Repeat the procedure described in (d). If you remove 3 edges from the original tree, how many components remain? If you remove 4 edges from your original tree, how many components remain?

(f) Repeat the procedure in (d) until you have removed all the edges from the tree. If you have to remove n edges to achieve this, determine an expression involving n for the number of components remaining.

(g) What *are* the components that remain when you have removed all the edges from the tree?

(h) What can you conclude about the number of vertices in a tree with n edges?

57. Write a paper on the arrangement of galaxies and the filamentary structure of the universe. Include discussion of how minimum spanning trees are used to analyze this.

58. Write a paper on James Sylvester and Arthur Cayley. Include discussion of their work in graph theory.

COLLABORATIVE INVESTIGATION

Finding the Number of Edges in a Complete Graph

In this investigation we explore two short methods for finding the number of edges in a complete graph with a specified number of vertices. By comparing the methods, we obtain a special sum formula that you encountered in Chapter 1.

Divide into groups of 4 or 5 students. Assign to each group member a different complete graph, with 4, 5, 6, or 7 vertices. Each member should sketch his or her complete graph and count the number of edges. Next, make a table showing the number of vertices and the number of edges in complete graphs with 3 through 7 vertices.

For Group Discussion

Do you notice a pattern in the number of edges for these complete graphs? Guess how many edges there would be in a complete graph with 8 vertices. How many are in a complete graph with 10 vertices?

If you spotted a pattern in the table, you could find the number of vertices in a complete graph with, for example, 100 vertices. Of course, completing the table all the way up to 100 would take a long time! A short way of determining the number of edges in a complete graph with any specified number of vertices can be used instead. Divide your group into two groups, A and B, and complete the following tasks. When both groups have finished, come back together to share what you have learned.

Group A

1. Start by studying the number of edges in a complete graph with eight vertices. Proceed as follows.

   Have someone in the group draw the eight vertices (no edges yet).

   (a) Choose any one of the vertices and draw in all necessary edges *from this vertex.* (Remember, there must be an edge to every other vertex in the graph.) Record the number of edges you drew from this first vertex: **(a)** ____

   (b) Now choose a different vertex—it is already joined to your first vertex by an edge. Draw in all necessary additional edges *from this vertex.* Record the number of *additional* edges you drew from this vertex: **(b)** ____

   Continue this process, determining how many *additional* edges you must draw from a third vertex, from a fourth vertex, and so on until the graph is complete. Record the number of additional edges in succession:

   (c) ____ **(d)** ____ **(e)** ____ **(f)** ____ **(g)** ____

   Now add your answers from (a) through (g) to find the total number of edges:

   Total number of edges = ____ + ____ + ____ + ____ + ____ + ____ + ____ = ____

2. Repeat the procedure described above to find the number of edges in a complete graph with 10 vertices.

3. As in Step 1 above, write down the numbers you must add to find the number of edges in a complete graph with 100 vertices. (Since there are so many numbers, write only the first three numbers, ellipsis points to indicate the omission, and then the last three numbers.)

   Total number of edges = ____ + ____ + ____ + . . . + ____ + ____ + ____

   To add these numbers by hand (or even with a calculator) would take a long time. Perhaps you remember a formula from a previous chapter? You will discuss this further when you rejoin Group B.

Group B

1. Start by studying the number of edges in a complete graph with eight vertices. Proceed as follows.

   Have someone draw a graph with 8 vertices. Then answer the following:

   (a) Each vertex in your graph has the same degree. What is it? **(a)** ____

   (b) Use your answer from (a) to find the sum of the degrees of the vertices in your graph. **(b)** ____

   (c) Find the number of edges in the complete graph, by using the theorem from Section 15.1 that relates the sum of degrees to the number of edges in the graph. **(c)** ____

2. Repeat the procedure described above to find the number of edges in a complete graph with 10 vertices.

3. Repeat the procedure described above to find the number of edges in a complete graph with 100 vertices.

Group B should now rejoin Group A. The members of Group A must explain their method to the whole group. Then the members of Group B must explain their method to the whole group. Use your combined knowledge to answer the following questions.

1. By comparing the work on Step 3 of Group A and Group B, find the sum: $99 + 98 + 97 + \ldots + 3 + 2 + 1 =$ ____

2. Determine the number of edges in a complete graph with 210 vertices.

3. Write a formula for the number of edges in a complete graph with m vertices.

4. Determine the sum of the following numbers (*Hint:* Regard the sum as the number of edges in a complete graph.):

   $155 + 154 + 153 + \ldots + 3 + 2 + 1 =$ ____

CHAPTER 15 TEST

In Exercises 1–4, refer to the following graph.

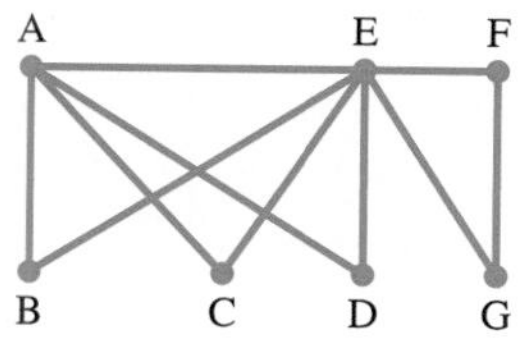

1. Determine how many vertices the graph has.

2. Determine the sum of the degrees of the vertices.

3. Determine how many edges the graph has.

4. Which of the following are paths in the graph? If not, why not?
   (a) B → A → C → E → B → A
   (b) A → B → E → A
   (c) A → C → D → E

5. Which of the following are circuits in the graph? If not, why not?
   (a) A → B → E → D → A
   (b) A → B → C → D → E → F → G → A
   (c) A → B → E → F → G → E → D → A → E → C → A

6. Draw a graph that has 2 components.

7. A graph has 10 vertices, 3 of degree 4, and the rest of degree 2. Use the theorem that relates the sum of degrees to the number of edges to determine the number of edges in the graph (without drawing the graph).

8. Determine whether graphs (a) and (b) are isomorphic. If they are, justify this by labeling corresponding vertices of the two graphs with the same letters, and color-coding the corresponding edges.

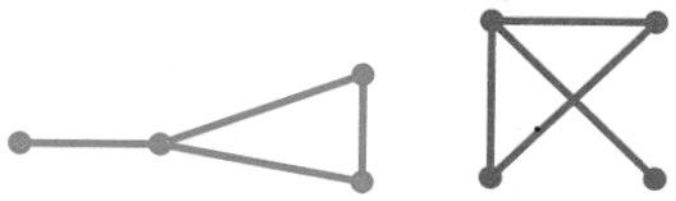

(a) **(b)**

9. ***Planning for Dinner*** Julia is planning to invite some friends for dinner. She plans to invite John, Adam, Bill, Tina, Nicole, and Rita. John, Nicole, and Tina all know each other. Adam knows John and Tina. Bill also knows Tina, and he knows Rita. Draw a graph with vertices representing the six friends whom Julia plans to invite to dinner and edges representing the relationship "know each other." Is your graph connected or disconnected? Which of the guests knows the greatest number of other guests?

10. ***Chess Competition*** There are 8 contestants in a chess competition. Each contestant plays one chess game against every other contestant. The winner is the contestant who wins the most games. How many chess games will be played in the competition?

11. Is the accompanying graph a complete graph? Justify your answer.

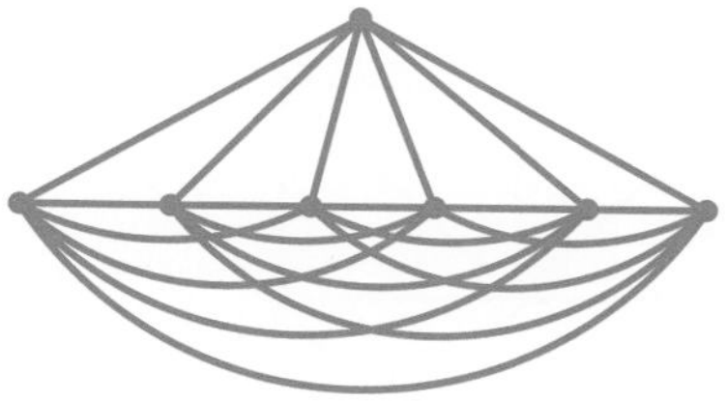

12. Color each graph using as few colors as possible. Use the result to determine the chromatic number of the graph.

(a) **(b)**

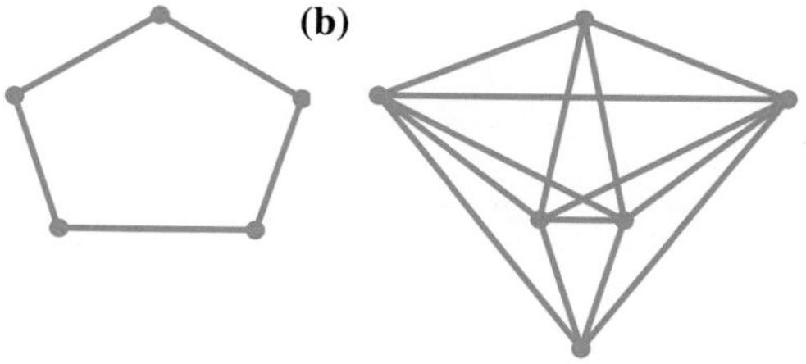

13. ***Scheduling Exams*** A teacher at a high school must schedule exams for the senior class. Subjects in which there are exams are listed on the left. Several students must take more than one of the exams, as shown on the right. Exams in subjects with students in common must not be scheduled at the same time. Use graph coloring to determine the least number of exam times needed, and to identify exams that can be given at the same times.

Exam	*Students taking this must also take*
History	English, Mathematics, Biology, Geography, Psychology
English	Mathematics, Psychology, History
Mathematics	Chemistry, Biology, History, English
Chemistry	Biology, Mathematics
Biology	Mathematics, Chemistry, History
Psychology	English, History, Geography
Geography	Psychology, History

14. Refer to the graph for Exercises 1–4. Which of the following are Euler circuits for the graph? If not, why not?
(a) $A \rightarrow B \rightarrow E \rightarrow D \rightarrow A$
(b) $A \rightarrow B \rightarrow C \rightarrow D \rightarrow E \rightarrow F \rightarrow G \rightarrow A$
(c) $A \rightarrow B \rightarrow E \rightarrow F \rightarrow G \rightarrow E \rightarrow D \rightarrow A \rightarrow E \rightarrow C \rightarrow A$

In Exercises 15 and 16, use Euler's theorem to decide whether the graph has an Euler circuit. (Do not actually find an Euler circuit.) Justify your answer briefly.

15.

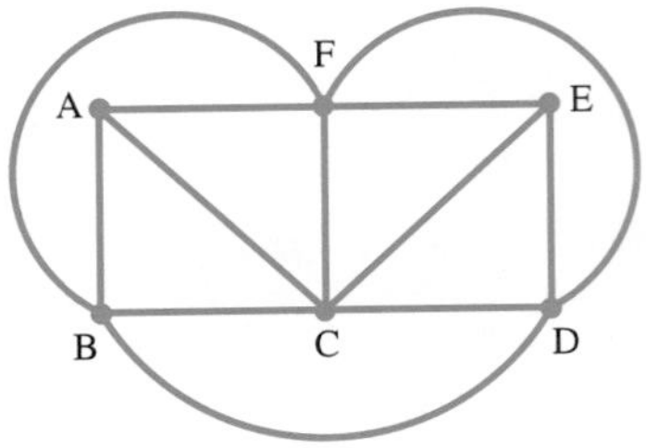

16.

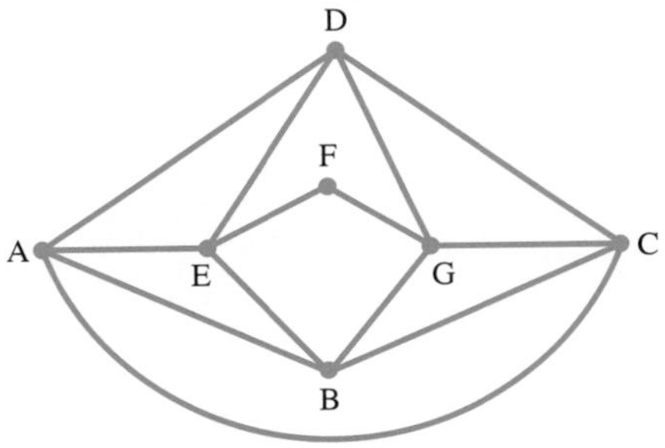

17. ***Building Floor Plan*** The floor plan of a building is shown. Is it possible to start outside, walk through each door exactly once, and end up back outside? Justify your answer.

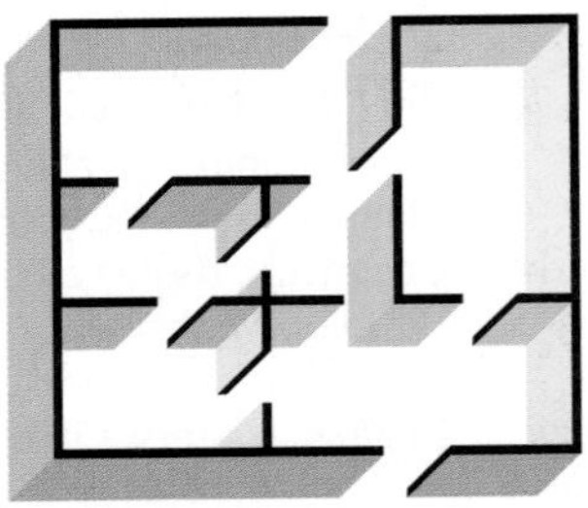

18. Use Fleury's algorithm to find an Euler circuit for the accompanying graph, beginning $F \rightarrow B$.

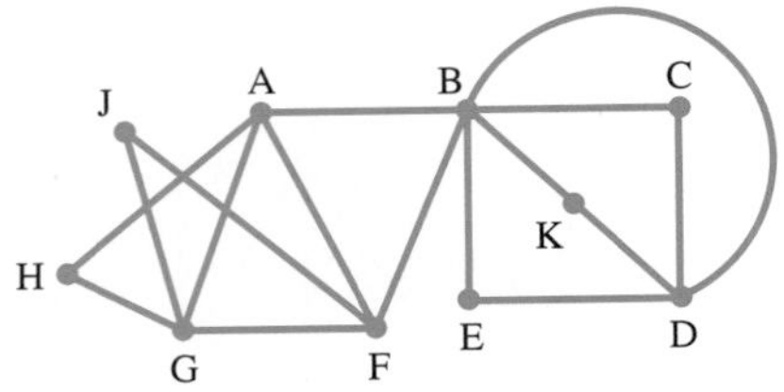

19. Refer to the graph for Exercises 1–4. Which of the following are Hamilton circuits for the graph? If not, why not?
(a) $A \rightarrow B \rightarrow E \rightarrow D \rightarrow A$
(b) $A \rightarrow B \rightarrow C \rightarrow D \rightarrow E \rightarrow F \rightarrow G \rightarrow A$
(c) $A \rightarrow B \rightarrow E \rightarrow F \rightarrow G \rightarrow E \rightarrow D \rightarrow A \rightarrow E \rightarrow C \rightarrow A$

20. List all Hamilton circuits in the graph that start $F \rightarrow G$. How many are there?

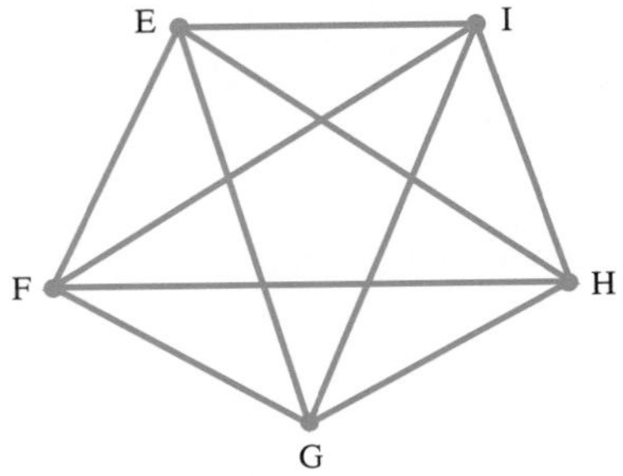

21. Use the brute force algorithm to find a minimum Hamilton circuit for the accompanying graph. Determine the total weight of the minimum Hamilton circuit.

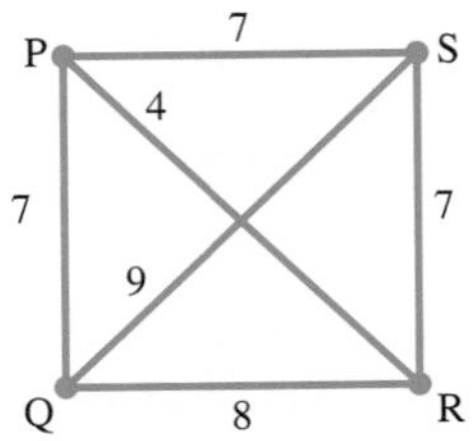

22. Use the nearest neighbor algorithm starting at A to find an approximate solution to the problem of finding a minimum Hamilton circuit for the graph on the next page. Find the total weight of this circuit.

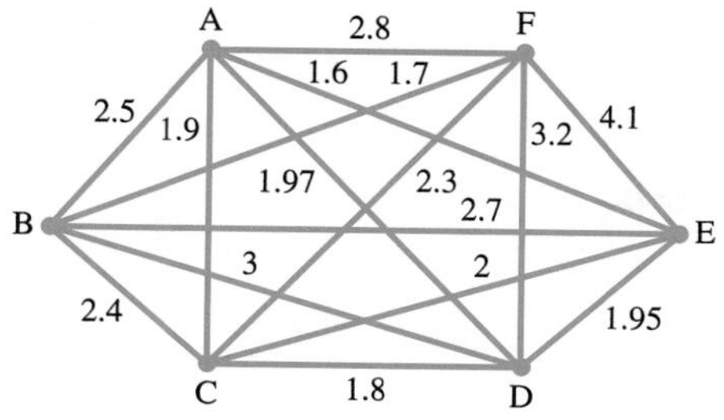

23. How many Hamilton circuits are there in a complete graph with 25 vertices? (Leave your answer as a factorial.)

24. ***Rock Band Tour Plan*** The agent for a rock band based in Milwaukee is planning a tour for the band. He plans for the band to visit Minneapolis, Santa Barbara, Orlando, Phoenix, and St. Louis. Since the band members want to minimize the time they are away from home, they do not want to go to any of these cities more than once on the tour. Consider the complete graph with vertices representing the 6 cities mentioned. Does the problem of planning the tour require an Euler circuit, a Hamilton circuit, or neither for its solution?

25. Draw three nonisomorphic trees with 7 vertices.

In Exercises 26–28, decide whether the statement is true *or* false.

26. Every tree has a Hamilton circuit.

27. In a tree each edge is a cut edge.

28. Every tree is connected.

29. Find all the different spanning trees for the following graph. How many different spanning trees are there?

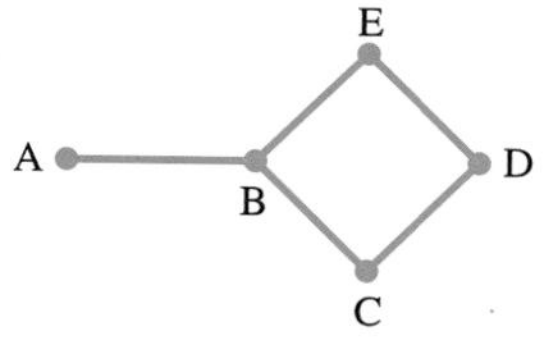

30. Use Kruskal's algorithm to find a minimum spanning tree for the following graph.

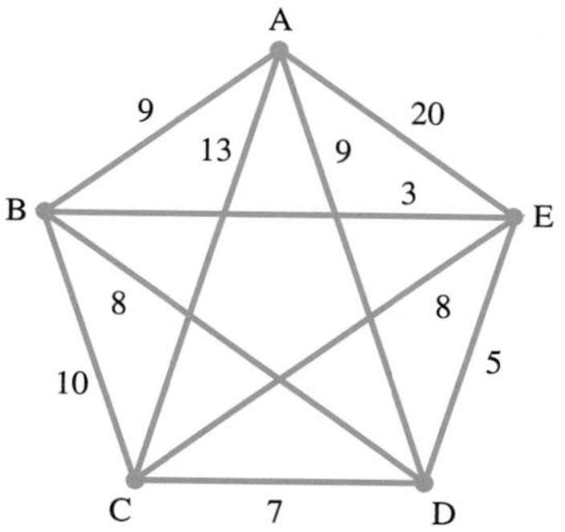

31. Determine the number of edges in a tree with 50 vertices.

16

VOTING AND APPORTIONMENT

On May 14, 2006, the NBC television network aired the final episode of the popular series *The West Wing*, which for seven seasons depicted Martin Sheen as U.S. president Josiah Bartlett. The series ended with the election of a new president, but the outcome of that election was in question for a short time based on the results from Nevada, a state with only five electoral votes.

During the actual presidential election of 2000, the country received a true-to-life lesson in the electoral process. Out of 105 million votes cast, George W. Bush recorded about 550,000 fewer popular votes than Al Gore, but Bush won the election by collecting 271 electoral votes to Gore's 266.

The Electoral College was designed to discourage party politics, yet protect individual states from excessive federal power. Each state has as many electors as its House of Representatives delegation, modified by adding two votes corresponding to its two Senate seats. This apportionment actually gives smaller states electoral power disproportionate to their populations.

In this chapter, we study different systems of voting and different ways of apportioning representatives in a legislative body. As you will learn, the design of a completely fair apportionment method or voting method is an impossible dream: every solution can give undesirable results in certain circumstances.

Source: U.S. National Archives & Records Administration; William C. Kimberling, "The Electoral College."

16.1 The Possibilities of Voting

Introduction • Plurality Method • Pairwise Comparison Method • Borda Method • Hare Method

Harold Washington won the 1983 Chicago Democratic mayoral primary with only 36% of the vote. He eventually became Chicago's first black mayor. He died of a heart attack on Nov. 15, 1987, while in office.

Introduction

If a group of people want to make one choice from two alternatives, then each member of the group indicates which of the two alternatives he or she prefers. The alternative awarded the majority of the votes is the choice of the group. Selecting one alternative from two is a straightforward process. The outcome, ignoring ties, is uncontestable. The essentially two-party political system of the United States means nearly all elections for public office at the state and federal level are straightforward. Most other democratic nations have three or more major political parties. Elections in these multiple-party democracies are not straightforward.

Suppose a group of 100 people want to select one alternative from among three. The group decides each member should indicate his or her first preference, and the alternative with the most votes will be selected as the choice of the group. It is possible that none of the alternatives gets a **majority** (more than half) of the votes. For example, the outcome of the vote may be

33 votes for Alternative a,

31 votes for Alternative b,

and 36 votes for Alternative c.

While Alternative c does *not* have a majority of the votes, it does have a **plurality** (the greatest number) of the votes. Alternative c is declared the people's choice, even though 64 of 100 voters chose different alternatives.

Such a situation is not rare. Whenever a group of people are required to select one alternative from three or more, an alternative favored by far fewer than a majority of the people may be selected.

Most voting methods require each voter to give a complete ranking of the candidates/alternatives. In other words, it is necessary to know which candidate each voter ranks first, second, third, and so on. The number of possible rankings (or arrangements) of n candidates is calculated by the factorial formula $\boldsymbol{n!}$.

PROBLEM-SOLVING HINT Many techniques covered in Chapter 11 on Counting Methods have direct application in the discussion of voting.

Suppose, for example, that there are twelve voters and three candidates/alternatives. Let the three candidates be a, b, and c. There are

$$3! = 3 \cdot 2 \cdot 1 = 6 \text{ possible ways}$$

to rank the three candidates. The notation $a > b > c$ indicates the ranking in which Candidate a is the voter's first choice, Candidate b is the second choice, and Candidate c is the third choice. Suppose 3 of the 12 voters chose the ranking $a > b > c$, 3 voters chose the ranking $c > a > b$, 4 voters chose the ranking $b > a > c$, and 2 voters chose the ranking c > b > a. This **profile of voters** is displayed in Table 1. Although there are six possible rankings, only four are reflected in this voter profile (because no voters chose the other two possible rankings).

In our discussion of voting methods, we do not consider final election outcomes that result in ties.

TABLE 1

Number of Voters	Ranking
3	a > b > c
3	c > a > b
4	b > a > c
2	c > b > a

Plurality Method

The *plurality method* is the most common voting method. It is the process used by the one hundred voters in the opening discussion, and in nearly all state and federal elections in the United States.

Plurality Method

In the **plurality method** of voting, each voter gives one vote to his or her top-ranked candidate. The candidate with the most votes, a plurality of the votes, wins the election.

The plurality method does not require a voter to rank the candidates completely. Only the voter's first choice is important in the tally of votes. When an election involves only two alternatives, the plurality of the votes is the majority of votes.

EXAMPLE 1 Selecting a Banquet Entrée by the Plurality Method

TABLE 2

Number of Voters	Ranking
6	b > p > c
5	c > p > b
4	p > c > b

The Culinary Arts Club is planning its annual banquet. Entrée choices are beef, chicken, or pork. The 15 members of the club rank the choices according to the profile in Table 2, where b represents beef, c chicken, and p pork. Determine the plurality winner among the three entrée choices.

SOLUTION

Each voter's complete ranking is shown, but *only the top-ranked candidate matters in a plurality method election.* Beef receives 6 votes, chicken receives 5, and pork receives 4. The winner is beef because it receives a plurality of the votes. ■

EXAMPLE 2 Electing a Board Chairperson by the Plurality Method

TABLE 3

Number of Voters	Ranking
3	a > b > c
3	c > a > b
4	b > a > c
2	c > b > a

Twelve voters rank the candidates for chairperson in their school board election according to the profile, in Table 3, where a represents Alice Adams, b represents Bobby Brown, and c represents Cathy Coutts. Determine the plurality winner of this election.

SOLUTION

The top-ranked candidate will be elected. Candidate Alice Adams gets 3 votes and Bobby Brown gets 4 votes. Cathy Coutts gets 3 votes plus 2 more votes, for a total of 5 votes. Thus, Cathy Coutts, with a plurality of the votes, is elected chairperson. ■

Pairwise Comparison Method

The *pairwise comparison method* requires voters to make a choice between every pair of candidates, and all possible pairs of candidates must be compared to find the outcome of the election. Pairwise comparison method elections are similar to round-robin tournaments in sports.

Recall from Section 11.3 that, in general,

$${}_nC_r = \frac{n!}{r!(n-r)!}.$$

For example,

$${}_3C_2 = \frac{3!}{2!1!} = \frac{(3 \cdot 2 \cdot 1)}{(2 \cdot 1)(1)} = 3.$$

Pairwise Comparison Method

In the **pairwise comparison method,** each voter gives a complete ranking of the candidates. For *each pair* of candidates a and b, the number of voters preferring Candidate a is compared with the number of voters preferring Candidate b. The candidate receiving more votes is awarded one point. If the two candidates receive an equal number of votes, each is awarded a half point. After all pairs have been compared, the candidate with the most points wins the election.

If there are n candidates, then the number of pairwise comparisons needed for a pairwise comparison method election is calculated by the combinations formula ${}_nC_2$.

A voter's complete ranking of the candidates allows the direct comparison of any pair of candidates. All candidate positions in the rankings are ignored, except the positions of the two candidates being considered. The voter's preferred candidate of the pair is the candidate ranking higher in the order. For example, a voter ranking the six candidates s > u > n > d > a > y prefers Candidate s to Candidate a, in a direct pairwise comparison. The voter prefers Candidate u in direct comparisons with any of Candidates n, d, a, or y.

EXAMPLE 3 Selecting a Banquet Entrée by the Pairwise Comparison Method

TABLE 4

Number of Voters	Ranking
6	b > p > c
5	c > p > b
4	p > c > b

Some members of the Culinary Arts Club are disappointed with the beef outcome of the plurality method election of Example 1. They request that the election be redone, using the pairwise comparison method. Determine the winning entrée by the pairwise comparison method.

SOLUTION

The profile of how the 15 club members rank beef, pork, and chicken is repeated in Table 4. There are $n = 3$ candidates, so

$${}_nC_2 = {}_3C_2 = 3 \text{ pairs}$$

must be examined: p and c, p and b, c and b.

p and c:

10 voters prefer p to c. 6 with the 1st ranking plus 4 with the 3rd ranking

5 voters prefer c to p. The 5 with the 2nd ranking

Pork beats chicken (10 to 5) and receives **1 point.**

p and b:

9 voters prefer p to b. 5 with the 2nd ranking plus 4 with the 3rd ranking

6 voters prefer b to p. The 6 with the 1st ranking

Pork beats beef (9 to 6) and receives **1 point.**

c and b:

9 voters prefer c to b. 5 with the 2nd ranking plus 4 with the 3rd ranking

6 voters prefer b to c. The 6 with the 1st ranking

Chicken beats beef (9 to 6) and receives **1 point.**

All three pairs have been considered and the results are pork with 2 points, chicken with 1 point, and beef with 0 points. The winner of the entrée election, using the pairwise comparison method, is pork.

TABLE 5

Number of Voters	Ranking
3	a > b > c > d
2	c > a > b > d
5	d > a > b > c
6	c > b > d > a
4	b > a > d > c

EXAMPLE 4 Selecting a Meeting Site by the Pairwise Comparison Method

The twenty-member executive board of the College Art Association is meeting to decide the location of the next yearly meeting. They are considering four cities:

Atlanta a, Boston b, Chicago c, and Dallas d.

Each executive board member gives a complete ranking of the four cities. The profile is in Table 5 (Only 5 of the $4! = 24$ possibilities actually appear as voter rankings.) Determine the winning city by the pairwise comparison method.

SOLUTION

There are four candidates, so a total of ${}_nC_2 = {}_4C_2 = 6$ pairs of cities must be compared:

a and b, a and c, a and d, b and c, b and d, c and d.

Compare a and b. Atlanta is preferred to Boston by the $3 + 2 + 5 = 10$ voters with the first three rankings. The remaining $6 + 4 = 10$ voters with the last two rankings prefer Boston to Atlanta. The tie for this pair means **Atlanta** and **Boston** each receive $\frac{1}{2}$ **point**.

Atlanta wins the comparison with Chicago by a margin of

$$3 + 5 + 4 = 12 \quad \text{to} \quad 2 + 6 = 8.$$ Atlanta, 1 point

Dallas wins the comparison with Atlanta by a margin of

$$5 + 6 = 11 \quad \text{to} \quad 3 + 2 + 4 = 9.$$ Dallas, 1 point

Boston wins when compared with Chicago,

$$3 + 5 + 4 = 12 \quad \text{to} \quad 2 + 6 = 8.$$ Boston, 1 point

Boston also wins when compared with Dallas,

$$3 + 2 + 6 + 4 = 15 \quad \text{to} \quad 5.$$ Boston, 1 point

Chicago wins the comparison with Dallas,

$$3 + 2 + 6 = 11 \quad \text{to} \quad 5 + 4 = 9.$$ Chicago, 1 point

Counting points, we see that Atlanta has 1.5, Boston has 2.5, and Chicago and Dallas each have 1. The host city selected by the pairwise comparison method is Boston.

Notice that the choice of voting method can lead to different outcomes. Beef is the plurality method selection of the Culinary Arts Club in Example 1, but Example 3 shows that pork wins a pairwise comparison method election. If the plurality method is used to decide the outcome of the election in Example 4, then Chicago, with a total of eight first-place votes, is selected instead of Boston. (Atlanta has 3 first-place votes, Boston has 4, and Dallas has 5.)

Jean-Charles de Borda (1733–1799), a Frenchman of diverse talents, was a military officer with skills on land and sea, serving as both a cavalry officer and a naval captain. In addition to voting theory, he studied and wrote about other topics in mathematics and physics. Borda was fully aware that his voting method is easy for an insincere voter to manipulate. Such a voter could misrepresent his real ranking in order to help a favorite candidate. In reply to his detractors, Borda said he assumed his method would be used by honest men.

Borda Method What is it about pork in the Culinary Arts Club profile that allows it to fare so well in the pairwise comparison method, while losing to beef in a plurality method election? Every club member ranks pork first or second, and never last. The profile for the executive board of the College Art Association shows Boston also is highly ranked by many voters. This allows Boston to win the pairwise comparison method election, although Boston loses a plurality method election.

The *Borda method* was first proposed in 1781 by French military officer Jean-Charles de Borda (1733–1799). Winners of many popular music awards and the winner of the annual Heisman Trophy are decided with versions of the Borda method. Unlike the plurality method, which considers only the candidate a voter ranks first, the Borda method uses a voter's complete ranking of the candidates. The pairwise comparison method also uses a voter's complete ranking, but the outcome depends on many individual comparisons. Indeed, the number of comparisons necessary makes the pairwise comparison method difficult to use for contests or elections in which there are more than four or five candidates. In contrast, the Borda method computes a weighted sum for each candidate to decide the outcome of the election.

Borda Method

In the **Borda method,** each voter must give a complete ranking of the candidates. Let n be the number of candidates. Each first-place vote a candidate receives is worth $n - 1$ points. Each second-place vote earns $n - 2$ points. A third-place vote earns $n - 3$ points, fourth place earns $n - 4$ points, . . . , and last place earns $n - n = 0$ points. Points are tallied separately for each candidate. The candidate with the highest tally of points is the winner.

EXAMPLE 5 Selecting a Banquet Entrée by the Borda Method

Determine the entrée selection for the Culinary Arts Club's annual banquet if they use a Borda method election.

SOLUTION

Recall, from Examples 1 and 3, that there are $n = 3$ choices, beef, pork, and chicken. A first-place vote earns $n - 1 = 2$ points, a second-place vote earns 1 point, and a third-place vote earns 0 points. See Table 6.

TABLE 6

Number of Voters	Ranking	First Place Earns 2 Points	Second Place Earns 1 Point	Third Place Earns 0 Points
6	b > p > c	b	p	c
5	c > p > b	c	p	b
4	p > c > b	p	c	b

Beef earns $6 \cdot 2 = 12$ points from the six voters with the first ranking, $5 \cdot 0 = 0$ points from the five voters with the second ranking, and $4 \cdot 0 = 0$ points from the four voters with the third ranking.

The Jesus Seminar, a collection of biblical scholars, uses a Borda-like voting method to decide on the credibility of the Gospels. Did Jesus really deliver the Sermon on the Mount? Following a proper scholarly debate, members vote by dropping a colored bead in the voting box. Depositing a red bead means the scholar believes the Sermon on the Mount is definitely a true historical event, a pink bead means it is likely true, a gray bead means maybe it is true, a black bead means the scholar rejects any possibility of the event being factual. A weighted mathematical formula is used to compile the results of the vote and render a decision on the validity of the event under discussion.

The tally for beef is $12 + 0 + 0 = 12$ points.

Chicken earns $6 \cdot 0 = 0$ points from the six voters with the first ranking, $5 \cdot 2 = 10$ points from the five voters with the second ranking, and $4 \cdot 1 = 4$ points from the four voters with the last ranking.

The tally for chicken is $0 + 10 + 4 = 14$ points.

Pork earns $6 \cdot 1 = 6$ points from the six voters with the first ranking, $5 \cdot 1 = 5$ points from the five voters with the second ranking, and $4 \cdot 2 = 8$ points from the four voters with the third ranking.

The tally for **pork** is $6 + 5 + 8 =$ **19 points.** ← Winner

Pork is the winner of the Borda method and the pairwise comparison method entrée elections. Both methods take into account the overall popularity of pork. ■

EXAMPLE 6 Selecting a Meeting Site by the Borda Method

Midwestern members of the executive board (see Example 4) are disappointed that Chicago was not selected as the site of the next yearly meeting, so they suggest a Borda method election. Determine the winning city by the Borda method.

SOLUTION

The modified voter profile is given in Table 7. The $n = 4$ choices, Atlanta, Boston, Chicago, and Dallas, mean each first-place vote earns 3 points, second place earns 2 points, third place earns 1 point, and fourth place earns 0 points.

TABLE 7

Number of Voters	Ranking	First Place (3 points)	Second Place (2 points)	Third Place (1 point)	Fourth Place (0 points)
3	a > b > c > d	a	b	c	d
2	c > a > b > d	c	a	b	d
5	d > a > b > c	d	a	b	c
6	c > b > d > a	c	b	d	a
4	b > a > d > c	b	a	d	c

Atlanta earns $(3 \cdot 3) + (2 \cdot 2) + (5 \cdot 2) + (6 \cdot 0) + (4 \cdot 2) = 31$ points.

Boston earns $(3 \cdot 2) + (2 \cdot 1) + (5 \cdot 1) + (6 \cdot 2) + (4 \cdot 3) =$ **37 points.** ← Winner

Chicago earns $(3 \cdot 1) + (2 \cdot 3) + (5 \cdot 0) + (6 \cdot 3) + (4 \cdot 0) = 27$ points.

Dallas earns $(3 \cdot 0) + (2 \cdot 0) + (5 \cdot 3) + (6 \cdot 1) + (4 \cdot 1) = 25$ points.

The winner of the Borda method election is Boston. ■

In a traditional n-candidate **Borda method** election, a first-place vote earns $n - 1$ points, second place earns $n - 2$ points, third place earns $n - 3$ points, and so on, with last place earning 0 points. The important defining feature is that the points for the various places in a voter's ranking must be equally spaced. In other words, the difference between the points given for any two consecutive rankings must be equal.

Instead of using the traditional 2-, 1-, 0-point scheme, try finding the outcome of the Culinary Arts Club Borda method vote using 1 point for first place, 0 points for second, and −1 points for third place. Try again using a 17-, 10-, 3-point scheme.

Pork, the Borda method selection in Example 5, also is the selection using the pairwise comparison method for the same profile in Example 3. Boston, the Borda method selection in Example 6, also is the pairwise comparison method selection for the same profile in Example 4. The next example shows that the Borda and pairwise comparison methods do not *always* select the same winner.

EXAMPLE 7 Selecting a Class Pet by the Borda and Pairwise Comparison Methods

Fifteen members of Miss Green's kindergarten class are voting on a class pet. The choices are a hamster h, a fish f, a turtle t, or a canary c. The profile in Table 8 shows remarkable agreement among the students. It uses only two of the $4! = 24$ possible rankings. Compare the Borda and pairwise comparison results for this profile.

TABLE 8

Number of Voters	Ranking
10	$h > f > t > c$
5	$f > t > c > h$

SOLUTION

The number of class pet candidates is $n = 4$. If the Borda method is used for the class selection, first place earns 3 points, second place 2 points, third place 1 point, and fourth place 0 points.

A hamster earns $(10 \cdot 3) + (5 \cdot 0) = 30$ points.
A **fish** earns $(10 \cdot 2) + (5 \cdot 3) =$ **35 points.** ← Borda winner
A turtle earns $(10 \cdot 1) + (5 \cdot 2) = 20$ points.
A canary earns $(10 \cdot 0) + (5 \cdot 1) = 5$ points.

The winner of the Borda method election is a fish, with 35 Borda points.

If the pairwise comparison method is used, a hamster wins each pairwise comparison with a fish, a turtle, and a canary, by a margin of 10 to 5. A fish wins each comparison with a turtle and a canary, by a margin of 15 to 0. A turtle wins in the comparison with a canary, by a margin of 15 to 0.

A **hamster** earns **3 points.** ← Pairwise comparison winner
A fish earns 2 points.
A turtle earns 1 point.
A canary earns 0 points.

All pairwise comparisons are complete. The winner using the pairwise comparison method is a hamster, with 3 pairwise comparison points. The Borda and pairwise comparison methods selected different winners in this case.

Hare Method

The *Hare method* was first proposed in 1861 by British political theorist Thomas Hare (1806–1891). It is a variation of the plurality method in which candidates are eliminated in sequential rounds of voting. Votes are transferred from eliminated candidates to remaining candidates. The president of France and public officials in Australia are selected by the Hare method. It also is known as the *plurality with elimination method,* or the *single transferable vote system.*

Hare elections do not require the voters initially to rank the complete list of candidates. Instead, voters need to decide only which candidate they prefer most in each round. This is considered an advantage of the method.

Thomas Hare (1806–1891) was a British lawyer. His reason for devising his voting method was to secure proportional representation of all classes in the United Kingdom, including minorities, in the House of Commons and other elected groups. The **Hare method** was considered to be "among the greatest improvements yet made in the theory and practice of government" by Hare's contemporary, John Stuart Mill.

Hare Method

In a **Hare method** election, each voter gives one vote to his or her favorite candidate in the first round. If a candidate receives a *majority* of the votes, he or she is declared the winner. If no candidate receives a majority of the votes, the candidate (or candidates) with the fewest votes is (are) eliminated, and a second election is conducted, in the same way, on the remaining candidates.

The rounds of voting continue to eliminate candidates until one candidate receives a *majority* of the votes. That candidate wins the election.

It may take many rounds of voting to learn the winner of a Hare method election. To simulate Hare elections in the following examples, it is assumed each voter has given a complete, initial ranking of the candidates. It also is assumed that when a candidate is eliminated from a voter's ranking, the voter votes for the next candidate in the ranking. For example, if the voter initially ranks the candidates s > u > n > d > a > y and Candidate s is eliminated in the first round of voting, the voter will vote for u in the second round of voting. If s, u, and n are eliminated, the voter will vote for d in the next round.

EXAMPLE 8 Selecting a Banquet Entrée by the Hare Method

Chicken entrée supporters of the Culinary Arts Club (see Examples 1, 3, 5), still hoping for a way to prevail, suggest a Hare method election. Does this method give them victory?

SOLUTION

The 15-member club profile is repeated in Table 9. Club members vote for their favorite of the three entrées in the first round, with the following results.

TABLE 9

Number of Voters	Ranking
6	b > p > c
5	c > p > b
4	p > c > b

Beef gets 6 votes.

Chicken gets 5 votes.

Pork gets 4 votes.

No entrée receives a majority (8 or more) of the votes in the first round. Pork receives the fewest votes and is eliminated.

Club members vote for their favorite of the remaining entrées, beef or chicken, in the second round of Hare method voting. The 4 club members with the third ranking vote for chicken, because pork is eliminated in the first round and they prefer chicken to beef. The other 11 voters vote as they did in the first round.

Beef gets 6 votes.

Chicken gets $5 + 4 = 9$ votes. ← Hare winner

Chicken receives a majority of the votes in the second round and indeed is declared the entrée selection by the Hare method.

EXAMPLE 9 Selecting a Meeting Site by the Hare Method

Chicago supporters still are grumbling about being able to win a plurality method election, but lose both the pairwise comparison and the Borda method elections to Boston. They convince the College Art Association to evaluate their vote yet again using the Hare method. Determine the winning city by the Hare method.

TABLE 10

Number of Voters	Ranking
3	a > b > c > d
2	c > a > b > d
5	d > a > b > c
6	c > b > d > a
4	b > a > d > c

SOLUTION

The twenty-member executive board profile is repeated in Table 10. First round voting gives the following results.

Atlanta gets 3 votes.

Boston gets 4 votes.

Chicago gets $2 + 6 = 8$ votes.

Dallas gets 5 votes.

No city receives a majority (11 or more) of the votes in the first round. Atlanta, with the fewest votes, is eliminated.

Boston, Chicago, and Dallas are voted on in the second round of the Hare election. The 3 voters with the first ranking vote for Boston, because Atlanta was eliminated. Other voters vote as they did in the first round.

Boston gets $4 + 3 = 7$ votes.

Chicago gets 8 votes.

Dallas gets 5 votes.

No city receives a majority. Dallas has the fewest votes and is eliminated.

The third-round vote is between Boston and Chicago. The five voters with the middle ranking prefer Boston to Chicago; with both southern cities eliminated, they transfer their votes to Boston. The other 15 voters vote as they did in the second round.

Boston gets $7 + 5 = 12$ votes. ← Hare winner

Chicago gets 8 votes.

Boston is the winner of the Hare method election. Chicago loses out again. ■

EXAMPLE 10 Selecting a Theme Movie by the Hare Method

TABLE 11

Number of Voters	Ranking
1	e > a > r > p
6	a > e > p > r
7	r > p > e > a
3	p > e > r > a
2	p > e > a > r

Nineteen members of the Outer Banks Film Confederation meet to decide which of four *Star Wars* movies will be the theme of a charity costume ball. They must choose from *The Empire Strikes Back* e, *Revenge of the Sith* r, *The Phantom Menace* p, and *Attack of the Clones* a. The profile of complete rankings for the 19 members of the confederation is shown in Table 11. Determine the winning film by the Hare method.

SOLUTION

We use the profile to carry out a simulated Hare method election. In the first round, each member votes for their favorite of the four movies, with the following results.

The Empire Strikes Back gets 1 vote.

Attack of the Clones gets 6 votes.

Revenge of the Sith gets 7 votes.

The Phantom Menace gets $3 + 2 = 5$ votes.

No movie gets a majority (ten or more) of the votes. *The Empire Strikes Back* has the fewest votes and is eliminated.

In the second round, each member votes for their favorite among r, p, and a. The voter with the first ranking now votes for a, because e is eliminated. The other 18 voters vote as they did in the first round. Second round results are as follows.

Attack of the Clones gets 6 + 1 = 7 votes.

Revenge of the Sith gets 7 votes.

The Phantom Menace gets 5 votes.

Again no movie gets a majority of the votes. *The Phantom Menace* has the fewest votes and is eliminated.

The final round of voting is between a and r. The five voters with p as their first choice do not transfer their five votes in a single block. The three members with the fourth ranking prefer r to a. The two members with the fifth ranking prefer a to r, so

Attack of the Clones gets 7 + 2 = 9 votes.

Revenge of the Sith gets 7 + 3 = 10 votes. ← Hare winner

Revenge of the Sith wins a majority of the votes in the final round and is selected by the Hare method as the theme of the Outer Banks Film Confederation costume ball.

The voting methods introduced above are summarized in Table 12.

TABLE 12 Summary of Voting Methods

Voting Method	Mechanics of the Election Procedure
Plurality Method	Each voter gives one vote to his or her top-ranked candidate. The candidate with the most votes, a *plurality*, wins the election.
Pairwise Comparison Method	Each voter gives a complete ranking of the candidates. Each pair of candidates, a and b, are compared, and the one ranked higher by more voters receives one point. (If a and b are ranked higher by equal numbers of voters, each receives a half point.) After all pairs have been compared, the candidate with the most points wins the election.
Borda Method	Each voter gives a complete ranking of the candidates. If there are n candidates, then a candidate receives $n - 1$ points for each first-place ranking by a voter, $n - 2$ points for each second-place ranking,..., and $n - n = 0$ points for each nth place (last-place) ranking. The candidate with the most points wins the election.
Hare Method	Each voter gives one vote to his or her top-ranked candidate. If a candidate receives a *majority* of the votes, that candidate wins the election. If no candidate receives a majority of the votes, the candidate (or candidates) with the fewest votes is (are) eliminated, and a second-round election is conducted, in the same way, on the remaining candidates. Rounds continue to eliminate candidates until a candidate receives a majority of the votes. That candidate wins the election. *Note*: An option in the Hare method is to initially obtain complete rankings from the voters. This way, only one actual vote is required. If no candidate receives a majority of first-place rankings, subsequent rounds of voting, as required, can be simulated from the voter profile.

For Further Thought

Approval Voting Method

The *approval voting method* was introduced in the 1970s. Like the Hare method, it is a variation of the plurality method. It is used to elect the secretary-general of the United Nations. Approval voting is useful for elections involving numerous candidates. The strength of the approval voting method is that voters do not need to rank the list of candidates completely. Voters need to decide only which candidates they like and which they dislike.

Approval Voting Method

In the **approval voting method,** each voter is allowed to give one vote to as many candidates as he or she wishes. The candidate with the most approval votes wins the election.

Recent work on voting theory includes the following examples. Steven J. Brams of New York University and Peter C. Fishburn of AT&T are working hard to educate voters about the advantages of approval voting. In spite of its somewhat sensitive nature, the method has many good qualities such as encouraging voter turnout by allowing voters to more easily express their feelings.

Donald G. Saari of Northwestern University is using highly sophisticated mathematical methods to shed light on the paradoxical nature of voting methods. Saari's studies offer political and social scientists a solid mathematical foundation for their work in the new millennium.

Many emerging democracies around the globe need a voting system that is easy for their citizens to use. Approval voting is a good alternative, but it is very temperamental, as you will see. Although voters are not required to rank the candidates completely for an approval election, we will include a complete ranking profile here, from which a voting simulation can be carried out.

If three candidates are in an approval method election, then each voter can decide to vote for his or her top-ranked candidate only, or for both of his or her top-ranked candidates. (If the voter voted for all three candidates, the vote would be meaningless, because it would fail to make a distinction between the candidates.) The familiar profile for the Culinary Arts Club is repeated in the table with an additional column of information necessary for an approval voting method election simulation.

Number of Voters	Ranking	Number of Voters Voting for Two of the Three Candidates
6	b > p > c	$x,\quad 0 \le x \le 6$
5	c > p > b	$y,\quad 0 \le y \le 5$
4	p > c > b	$z,\quad 0 \le z \le 4$

The variable x in $0 \le x \le 6$ in the first row represents how many of the six voters with this ranking vote for both b, their top-ranked candidate, and p, their second-ranked candidate. The variable y in $0 \le y \le 5$ in the second row represents how many of the five voters with the second ranking vote for both top-ranked c and second-ranked p. The variable z in $0 \leq z \leq 4$ represents how many of the four voters with the third ranking vote for both top-ranked p and second-ranked c.

Approval voting method results of this profile can be generalized. Beef always receives six approval votes from the voters with the top ranking. Beef cannot get any additional votes, however, because it is not ranked second by anyone. Chicken always receives five votes from the middle ranking. Chicken also can get up to $z = 4$ additional votes from the voters with the third ranking who rank it second. Pork always receives four votes from the third ranking. Pork also can get up to $x = 6$ and $y = 5$ additional votes from the 11 voters' with the first and second rankings who rank pork second.

In summary:

Beef gets 6 approval votes.

Chicken gets $5 + z$ approval votes, $0 \leq z \leq 4$.

Pork gets $4 + x + y$ approval votes, $0 \leq x \leq 6, 0 \leq y \leq 5$.

The outcome of the election depends on the values of the variables x, y, and z, which, in turn, depend on the individual voters' decisions whether to approve second-place candidates.

For Group Discussion or Individual Investigation

1. Determine the winner of this approval election under each of the following assumptions.
   (a) All the voters distrust the new system and vote only for their single top-ranked candidate.
   (b) Three of the six voters with the first ranking decide to approve of both pork and beef. All other voters approve only their top-ranked candidate.
   (c) Two of the six voters with the first ranking decide to approve of both beef and pork, one of the five voters with the second ranking decides to approve of both chicken and pork, and three of the four voters with the third ranking decide to approve of both pork and chicken.
2. What other voting method is equivalent to the approval method under assumption 1(a) above?
3. Compare approval voting with the other four methods discussed in this section. Do you "approve" of the approval voting method?
4. The members of a soccer team are electing their captain using the approval method. Candidates include team members Joan, Lori, Mary, and Alison. The 13 ballots have been marked as shown; an x indicates an approval vote for the candidate named in the first column.
   (a) Which candidate is selected as team captain by the approval method?
   (b) If the soccer team decides to name the two candidates with the most approval votes as their co-captains, which two candidates will be selected?

	Voter												
Candidate	**#1**	**#2**	**#3**	**#4**	**#5**	**#6**	**#7**	**#8**	**#9**	**#10**	**#11**	**#12**	**#13**
Joan	X	X	X		X	X	X	X	X				
Lori	X	X		X				X	X		X		X
Mary			X		X		X		X	X	X		
Alison	X		X	X		X						X	

16.1 EXERCISES

A local animal shelter is choosing a poster dog. The choices are an Australian shepherd a, a boxer b, a cocker spaniel c, or a dalmatian d. Thirteen staff members completely rank the four dogs on their ballots. The thirteen individual ballots are the columns in the table. In each case, complete the following.

(a) *How many different ways can a staff member complete his or her ballot?*

(b) *Write a voter profile for this election.*

(c) *Use the plurality method to determine the shelter's poster dog.*

1.

Voter	#1	#2	#3	#4	#5	#6	#7	#8	#9	#10	#11	#12	#13
1st place	b	b	a	c	d	d	c	c	b	a	a	a	a
2nd place	c	c	d	d	c	a	a	a	c	c	d	b	b
3rd place	a	a	c	b	b	b	d	d	a	d	c	c	c
4th place	d	d	b	a	a	c	b	b	d	b	b	d	d

2.

Voter	#1	#2	#3	#4	#5	#6	#7	#8	#9	#10	#11	#12	#13
1st place	a	b	d	c	a	a	b	c	d	a	a	a	a
2nd place	b	c	b	b	d	d	c	b	b	b	b	b	b
3rd place	c	a	c	a	c	c	d	d	c	c	d	d	c
4th place	d	d	a	d	b	b	a	a	a	d	c	c	d

3. If there are $n = 5$ candidates, then how many different rankings of the candidates are possible? What if there are $n = 7$ candidates?

4. If there are $n = 6$ candidates, how many different rankings of the candidates are possible? What if there are $n = 8$ candidates?

5. Comment on the implications of the number of different rankings for elections involving more than $n = 4$ candidates.

6. How many pairwise comparisons are needed to learn the outcome of an election involving $n = 5$ candidates? What if there are 7 candidates?

7. How many pairwise comparisons are needed to learn the outcome of an election involving $n = 6$ candidates? What if there are 8 candidates?

8. Comment on the implications of the number of pairwise comparisons needed to learn the outcome of an election involving more than $n = 4$ candidates.

9. Comment on the logistics of using the pairwise comparison method to determine the outcome of an election involving $n = 8$ candidates.

Applying Four Voting Methods to a Voter Profile A 13-*member committee is selecting a chairperson. The 3 candidates are Albert Werner a, Barbara Hightower b, and Charles Smith c. Each committee member completely ranked the candidates on a separate ballot. For each voter profile determine the chairperson using the*

(a) *plurality method*

(b) *pairwise comparison method*

(c) *Borda method*

(d) *Hare method.*

10.

Number of voters	Ranking
4	a > b > c
2	b > c > a
4	b > a > c
3	c > a > b

11.

Number of Voters	Ranking
3	a > c > b
4	c > b > a
2	b > a > c
4	b > c > a

A 13-member committee is selecting a new company logo from 3 alternatives a, b, and c. Each committee member completely ranked the possible logos on a separate ballot. For each voter profile, determine the new company logo using the
(a) *plurality method*
(b) *pairwise comparison method*
(c) *Borda method*
(d) *Hare method.*

12.

Number of Voters	Ranking
2	a > c > b
3	c > b > a
4	b > a > c
4	b > c > a

13.

Number of Voters	Ranking
6	a > b > c
1	b > c > a
3	b > a > c
3	c > a > b

A senator invited one person from each of the 21 counties in her state to a weekend workshop. The senator asked the attendees to rank the issues of job creation j, education e, health care h, and gun control g in order of importance to themselves and their counties. The 21 members each gave a complete ranking of the issues on a separate ballot. For each voter profile, determine which issue workshop members felt was the highest priority using the
(a) *plurality method*
(b) *pairwise comparison method*
(c) *Borda method*
(d) *Hare method.*

14.

Number of Voters	Ranking
3	e > h > g > j
6	h > e > g > j
5	j > g > e > h
4	g > e > h > j
3	j > e > h > g

15.

Number of Voters	Ranking
6	h > j > g > e
5	e > g > j > h
4	g > j > h > e
3	j > h > g > e
3	e > j > h > g

16. For the profile of voters found in Exercise 1(b), determine the animal shelter's poster dog using the following voting methods.
(a) pairwise comparison method
(b) Borda method 6.
(c) Hare method 6.

17. Repeat Exercise 16 using the profile of voters in Exercise 2(b).

Members of the Adventure Club are selecting an activity for their yearly vacation. Choices are hiking in the desert h, rock climbing c, white-water kayaking k, bungee jumping b, or just sitting by a mountain lake and tanning t. For each voter profile of the 55 members, determine the selected activity using the
(a) *plurality method*
(b) *pairwise comparison method*
(c) *Borda method*
(d) *Hare method.*

18.

Number of Voters	Ranking
18	t > k > h > b > c
12	c > h > k > b > t
10	b > c > h > k > t
9	k > b > h > c > t
4	h > c > k > b > t
2	h > b > k > c > t

19.

Number of Voters	Ranking
18	t > b > h > k > c
12	c > h > b > k > t
10	k > c > h > b > t
9	b > k > h > c > t
4	h > c > b > k > t
2	h > k > b > c > t

20. ***Holding a Runoff Election*** One common solution to an election in which no candidate receives a majority of first-place votes is to have a runoff between the two candidates with the most first-place votes. For the Adventure Club voter profile in Exercise 18, what activity is selected if a runoff is held between the two candidates with the most first-place votes?

21. For the workshop voter profile in Exercise 15, what issue is the highest priority if a runoff is held between the two candidates with the most first-place votes?

22. For the workshop voter profile in Exercise 14, what issue is the highest priority if a runoff is held between the two candidates with the most first-place votes?

23. Repeat Exercise 20 using the voter profile in Exercise 19.

A Different Kind of Runoff Election *Another solution to an election in which no candidate receives a majority of first-place votes is to have a runoff between the candidates that rank second and third in first-place votes. The winner of that election faces the candidate with the most first-place votes to decide the final outcome of the election. Use this process to select a winner in each case.*

24. The voter profile in Exercise 18

25. The voter profile in Exercise 19

26. The voter profile in Exercise 12

27. The voter profile in Exercise 13

Each table represents a pairwise comparisons method election.
(a) *Find the missing number of pairwise comparisons won.*
(b) *Identify the winning candidate and the number of pairwise points for the winner.*

28.

Candidate	a	b	c	d	e	f	g
Number of pairwise comparisons won	4	6	3	1	5	1	?

29.

Candidate	a	b	c	d	e	f	g
Number of pairwise comparisons won	3	5	7	1	2	?	1

30.

Candidate	a	b	c	d	e	f	g	h
Number of pairwise comparisons won	4	5	2	?	3	1	2	3

31.

Candidate	a	b	c	d	e	f	g	h
Number of pairwise comparisons won	2	6	3	4	?	2	2	2

Each table represents a Borda method election.
(a) *Find the missing number of Borda points.* **(b)** *Identify the winning candidate.*

32. Total number of voters: 15

Candidate	a	b	c
Number of Borda points	17	14	?

33. Total number of voters: 15

Candidate	a	b	c
Number of Borda points	15	14	?

34. Total number of voters: 20

Candidate	a	b	c	d	e
Number of Borda points	?	50	25	40	55

35. Total number of voters: 20

Candidate	a	b	c	d	e
Number of Borda points	35	40	?	40	30

The Coombs Method *The Coombs method of voting is a variation of the Hare method. Voting takes place in rounds, but instead of eliminating the candidate with the fewest first-place votes, the candidate with the most last-place votes is eliminated. In each case use the Coombs method to determine which issue the senator's workshop members felt was the highest priority.*

36. The profile in Exercise 14

37. The profile in Exercise 15

38. Which voting method—plurality, pairwise comparison, Borda, or Hare—do you personally think is the best way to select one alternative from among three or more alternatives? Why do you think it is the best method?

39. Suppose only two candidates are running for office. Show that the pairwise comparison method, the Borda method, and the Hare method all select the same winner. Explain why all three methods reduce to a simple plurality method vote of giving one vote to your top-ranked candidate.

40. What is the least possible number of rounds of voting required in a Hare method election? How would it occur?

41. Discuss possible advantages and disadvantages of using the Hare method option of having voters give a complete ranking of candidates.

42. Which voting method do you think would be best in an election with a very large number of candidates? Why?

43. Research and report on the method used for processing new bills (laws), including proposed amendments, in the U.S. Congress.

Devising a Profile for Consistency Among Voting Methods *Given each voter profile for a 4-candidate election, arrange a total of 21 voters in such a way that the plurality method, the pairwise comparison method, the Borda method, and the Hare method would all select candidate d.*

44.

Number of Voters	Ranking
?	a > b > d > c
?	b > c > d > a
?	c > d > a > b
?	d > a > b > c
?	a > d > c > b

45.

Number of Voters	Ranking
?	d > b > a > c
?	b > c > a > d
?	c > a > d > b
?	a > d > b > c
?	d > a > c > b

16.2 The Impossibilities of Voting

Introduction • Majority Criterion • Condorcet Criterion • Monotonicity Criterion • Irrelevant Alternatives Criterion • Arrow's Impossibility Theorem

Introduction Different voting methods applied to the same voter profile can show remarkable agreement. If the pairwise comparison, Borda, or Hare method is used, then Boston wins the College Art Association (CAA) city selection in the previous section. Boston does lose a plurality method CAA election, because Chicago has a plurality of first-place rankings.

The Culinary Arts Club entrée selection shows how different voting methods applied to the same voter profile can result in totally different outcomes. Beef is selected by the plurality method. Pork is selected by the pairwise comparison method and the Borda method. The Hare method selects chicken.

One argument offered against the **electoral college system** is that it is unstable under very small, even accidental changes in the vote. The 2000 presidential election illustrated this risk. In Palm Beach County, Florida, as many as 2000 voters confused by the controversial "butterfly" ballot may have voted for Reform Party candidate Patrick Buchanan instead of their actual choice, Al Gore. Since George W. Bush won the Florida popular vote (and thus all 25 of Florida's electoral votes) by a margin of only 537, the effect of this voting accident was magnified to a national scale.

Of course, votes were misrecorded or miscounted in other counties; indeed any voting process is subject to this kind of error. And nationwide only about 51% of the voting-age population actually cast votes in the 2000 presidential election. Determining the will of the people is an elusive goal. *Source:* Wand, Jonathan N., Kenneth W. Shotts, Jasjeet S. Sekhon, Walter R. Mebane, Jr., Michael C. Herron, Henry E. Brady, 2001. "The Butterfly Did It: The Aberrant Vote for Buchanan in Palm Beach County Florida." (http://elections.berkeley.edu/election2000/butterfly.pdf)

The voting methods presented in the previous section are all used in real-world decision making, but none is perfect. Defects are revealed by considering the following four desirable attributes for any voting method.

1. the *majority criterion*
2. the *Condorcet criterion*
3. the *monotonicity criterion*
4. the *irrelevant alternatives* (IA) *criterion*

The first two criteria concern desirable qualities for a voting method when it is used a single time to determine a winner. The third and fourth criteria concern desirable qualities of a voting method when the method is to be used twice in the election procedure.

Majority Criterion

When a voter profile includes a candidate with a majority of first-place rankings, that is, a **majority candidate,** it seems reasonable to expect that candidate to be elected. It is an attribute many people consider a standard or criterion of election fairness.

Majority Criterion

If a candidate has a majority of first-place rankings in a voter profile, then that candidate should be the winner of the election.

If there are only two candidates, then a majority candidate wins the election no matter which method from Section 16.1 is used. However, Example 1 shows that a majority candidate in an election involving three or more candidates may not win the election if the Borda method is used.

EXAMPLE 1 Showing That the Borda Method Fails to Satisfy the Majority Criterion

The College Art Association has a new executive board. Its members are meeting to select a city to host the yearly meeting. The sites under consideration are Portland p, Boston b, Chicago c, and Oakland o. The voter profile for the new executive board shows how the 20 members rank the cities. (See Table 13.)

TABLE 13

Number of Votes	Ranking	First Place (3 points)	Second Place (2 points)	Third Place (1 point)	Fourth Place (0 points)
8	c > p > b > o	c	p	b	o
4	c > b > o > p	c	b	o	p
4	b > p > o > c	b	p	o	c
4	b > o > p > c	b	o	p	c

Board members agree to use the Borda method to decide the outcome of the vote. Determine the Borda winner.

SOLUTION

Chicago supporters are delighted to discover they have a *majority* of first-place rankings. They are confident that, with 12 of 20 first-place votes, Chicago will be selected as the meeting site. The Borda tallies are calculated.

Portland earns $(8 \cdot 2) + (4 \cdot 0) + (4 \cdot 2) + (4 \cdot 1) = 28$ points.
Oakland earns $(8 \cdot 0) + (4 \cdot 1) + (4 \cdot 1) + (4 \cdot 2) = 16$ points.
Chicago earns $(8 \cdot 3) + (4 \cdot 3) + (4 \cdot 0) + (4 \cdot 0) = 36$ points.
Boston earns $(8 \cdot 1) + (4 \cdot 2) + (4 \cdot 3) + (4 \cdot 3) = 40$ **points.** ←Winner

The Borda method selects Boston as the site for next year's meeting, even though Chicago had a majority of first-place rankings.

This failure to select the majority candidate is a defect in the Borda method. No voting method that can result in a violation of the majority criterion can be considered completely satisfactory.

Do not misunderstand what it means when a voting method fails to satisfy a criterion. For some voter profiles, if a candidate has a majority of first-place rankings, the Borda method *does* select that candidate; however, it takes only one contradictory profile to show that a voting system violates a criterion. Because the Borda method *does not* select the majority candidate for the CAA voter profile in Example 1, the Borda method does not satisfy the majority criterion.

The plurality method, the Hare method, and the pairwise comparison method do satisfy the majority criterion. Whatever the voter profile, if a majority candidate exists, these methods always select it. We can see this by reasoning as follows.

1. *Plurality method* Any candidate with a majority of first-place rankings automatically also has a plurality of first-place rankings and will, therefore, always be selected by the plurality method.
2. *Pairwise comparison method* Suppose Candidate x has a majority of first-place rankings in a field of n candidates. Every candidate is involved in $n - 1$ comparisons (once with each of the other candidates). Candidate x wins all $n - 1$ comparisons with the other candidates and receives $n - 1$ comparison points. Every other candidate loses the comparison with x and, therefore, receives at most $n - 2$ points. So, the majority candidate will always be selected by the pairwise comparison method.
3. *Hare method* Any candidate with a majority of first-place rankings will win the election in the first round of voting and will, therefore, always be selected by the Hare method.

The Marquis de Condorcet (1743–1794) was a French aristocrat. His interests included mathematics, social science, economics, and politics. Sadly, Condorcet's political passions at the end of the French Revolution sent him to prison, where he died of questionable causes. It is hard to understand how France's counterpart to America's Thomas Jefferson could have come to such a tragic end. Imagine how different the government of the United States would be today if Jefferson's great mind had been eliminated from American history at the end of the American Revolution.

Condorcet Criterion A candidate who can win a pairwise comparison with every other candidate is called a **Condorcet candidate,** in honor of French aristocrat Marquis de Condorcet (1743–1794). Not every voter profile yields a candidate who can win each pairwise comparison, however. It is not unusual for a voter profile to include a Condorcet candidate, but neither is it commonplace. Another standard of election fairness calls for a Condorcet candidate, if one exists, to be selected as the choice of the group.

The Condorcet Criterion

If a Condorcet candidate exists for a voter profile, then the Condorcet candidate should be the winner of the election.

The Condorcet criterion is satisfied by all the methods of Section 16.1 for any election involving only two candidates. However, it is not always satisfied when there are more than two candidates.

EXAMPLE 2 Showing That the Plurality Method Fails to Satisfy the Condorcet Criterion

Show that the plurality method of voting fails to satisfy the Condorcet criterion by applying it to the voter profile of the Culinary Arts Club of Section 16.1.

SOLUTION

TABLE 14

Number of Voters	Ranking
6	$b > p > c$
5	$c > p > b$
4	$p > c > b$

The voter profile is repeated in Table 14 and shows that pork wins pairwise comparisons with both beef and chicken. The nine voters of the second and third rankings prefer pork to beef, while only six voters of the first ranking prefer beef to pork. Also, the ten voters of the first and third rankings prefer pork to chicken, while only five voters of the second ranking prefer chicken to pork.

If the plurality method is used by the club to select the entrée, then beef wins, because it is ranked first by six (a plurality) of the members. Beef is selected in spite of the popularity of pork displayed by the pairwise comparisons.

Because the plurality method has failed to select pork, the Condorcet candidate, this method does not satisfy the Condorcet criterion.

Example 2 established that the plurality method fails to satisfy the Condorcet criterion. What about the other methods of Section 16.1?

1. *Pairwise comparison method* A Condorcet candidate, by definition, wins pairwise comparisons with every other candidate, earning $n - 1$ pairwise points, while no other candidate can possibly earn more than $n - 2$ points. So, the Condorcet candidate is always selected by the pairwise comparison method. (In fact, this method was created to satisfy the Condorcet criterion.)
2. *Borda method* First observe that a majority candidate (having a majority of first-place rankings) automatically wins all pairwise comparisons with other candidates and is, therefore, also a Condorcet candidate. In Example 1, Chicago was a majority candidate, and therefore a Condorcet candidate, but was *not* selected by the Borda method. (Boston was the winner.) This counterexample establishes that the Borda method fails to satisfy the Condorcet criterion.
3. *Hare method* We saw in Example 2 that pork was a Condorcet candidate. However, in Example 8 of Section 16.1, the Hare method selected chicken rather than pork for the same profile. The Hare method also fails to satisfy the Condorcet criterion.

Again recall that other voter profiles may not violate these criteria. But, a single counterexample is enough to establish that a given voting method fails to satisfy a criterion.

The majority criterion and the Condorcet criterion are desirable attributes of a voting method when it is used in an election. The next two criteria involve a voting method's desirable attributes when a second election is held using the same method.

Monotonicity Criterion Suppose the outcome of a first election is not binding, as in a straw poll. Voters may change their rankings before the next election. Suppose those who rearrange their rankings move the winner of the first election to the top of their rankings. It is reasonable to expect the winner of the first election, who now enjoys even more support, to win the second election also. This desirable attribute is known as the *monotonicity criterion.*

Monotonicity Criterion

If Candidate x wins an election and, before a second election, the voters who rearrange their rankings move Candidate x to the top of their rankings, then Candidate x should win the second election.

The plurality method respects the monotonicity criterion. If Candidate x has a plurality of votes in the first election and only x can get more first-place votes, then x must have a plurality in the second vote. The Borda method and the pairwise comparison method can fail the monotonicity criterion as it is stated. They can fail because, while a rearranging voter is required to move winning Candidate x to the top of his or her ranking, how that voter shifts the *other* candidates is not restricted. The unrestricted movement of the other candidates is not obvious in the statement of the criterion, but it is implied. It allows the Borda point tallies and the pairwise comparison points of all candidates to increase, which may prevent the original winner from winning again in the second election. The Hare method, too, can violate the monotonicity criterion.

TABLE 15

Number of Voters	Ranking
7	m > b > s > c
8	b > c > m > s
10	c > s > m > b
2	m > c > s > b
2	m > s > b > c

EXAMPLE 3 Showing That the Hare Method Fails to Satisfy the Monotonicity Criterion

The College Art Association increased its executive board to 29 members. Voting for the site of the yearly meeting is among Miami m, Boston b, Chicago c, and Seattle s. The board members agree to use the Hare method to make their selection. They decide to hold a preliminary, nonbinding vote Friday afternoon and to meet again Saturday morning for the official vote. The Friday afternoon voter profile of the 29 members is given in Table 15. Show that the monotonicity criterion is violated in this case.

SOLUTION

Round One results are as follows:

Miami gets $7 + 2 + 2 = 11$ votes.

Boston gets 8 votes.

Chicago gets 10 votes.

Seattle gets 0 votes.

No city has a majority (15 or more) of the 29 votes. Seattle has the fewest votes and is eliminated.

Round Two results are as follows:

Miami gets 11 votes.

Boston gets 8 votes.

Chicago gets 10 votes.

Again no city has a majority of the votes. Boston has the fewest votes and is eliminated. The eight voters with the second ranking vote for Chicago in the next round, because Boston is eliminated.

Round Three results are as follows:

Miami gets 11 votes.

Chicago gets $10 + 8 = 18$ votes.

Friday afternoon, Chicago is selected by the preliminary Hare method election.

Friday evening, Chicago supporters wine and dine the four voters with the two bottom rankings in Table 15. They convince the four voters to rearrange their rankings, placing Chicago first. Table 16 shows the voter profile for the official vote on Saturday morning. The four voters with the bottom rankings have moved Chicago into first place. (The two voters with the bottom ranking were seriously swayed.) The other 25 voters do not rearrange their rankings. Chicago supporters believe the four additional first-place rankings will work in their favor.

TABLE 16

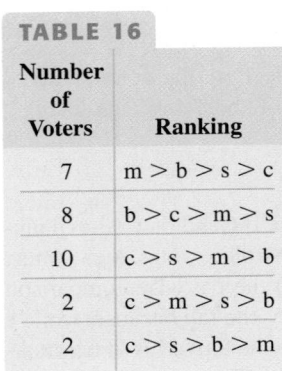

Number of Voters	Ranking
7	m > b > s > c
8	b > c > m > s
10	c > s > m > b
2	c > m > s > b
2	c > s > b > m

Round One results on Saturday are as follows:

Miami gets 7 votes.

Boston gets 8 votes.

Chicago gets $10 + 2 + 2 = 14$ votes.

Seattle gets 0 votes.

No city has a majority of the 29 votes. Seattle has the fewest votes, and is eliminated.

Round Two results are as follows:

Miami gets 7 votes.

Boston gets 8 votes.

Chicago gets 14 votes.

No city has a majority of the votes. Miami has the fewest votes and is eliminated. The seven Miami supporters with the top ranking will vote for Boston in the next round, because Miami is eliminated.

Round Three results are as follows:

Boston gets $8 + 7 = $ **15 votes.** ← Winner

Chicago gets 14 votes.

Boston wins the official Saturday morning Hare method election by a single vote in the final round. Chicago loses the official vote, although Chicago has more first-place votes than it did in the preliminary round. Because the winner of the first election, Chicago, did not win the second election, this example violates the monotonicity criterion. Therefore, the Hare method does not satisfy the monotonicity criterion. ■

The conditions of the monotonicity criterion require the four voters who rearrange their rankings in Example 3 to move Chicago into first place. The voters also are permitted to reshuffle the other candidates in their rankings. To make the example obvious and dramatic, Chicago was simply exchanged with the top-ranked candidate in the two bottom rankings and the other candidates were not moved.

An alternative version of the monotonicity criterion exists. *If a voter rearranges his or her ranking before the second vote, the alternative version allows only the exchange of the winning Candidate x and the candidate immediately above it in a voter's ranking.*

The pairwise comparison method respects the alternative version of the criterion. A voter's ranking can be rearranged only by exchanging winning Candidate x and the candidate immediately above x in his or her ranking. Even if the two are exchanged, the initial order of the set of candidates ranked below original winner Candidate x and the candidate immediately above x must remain the same. The initial order of the set of candidates ranked above the pair also must remain the same. This means the outcomes of all the pairwise comparisons that do not involve x do not change. The outcomes of the pairwise comparisons involving x might change, but the outcomes must change in favor of x because only the original winner x is allowed to move up in the rankings. The number of pairwise comparisons x wins in the second pairwise comparison election cannot decrease, and the points for the other candidates cannot increase, so the original winner x also wins a second pairwise comparison method election.

The Borda method, the Hare method, and the pairwise comparison method can fail to satisfy the monotonicity criterion (as originally stated). The criterion is upheld by the plurality method.

Irrelevant Alternatives Criterion Suppose, that for some reason, the outcome of an original election is not considered the final outcome. If a candidate who lost the first election drops out before a second vote, it is reasonable to expect the winner of the first election also to win the second. This voting method attribute is known as the *irrelevant alternatives criterion.*

Irrelevant Alternatives (IA) Criterion

If Candidate x wins a first election and one (or more) of the losing alternatives drops out before a second vote, the winner x of the first election should win the second election.

TABLE 17

Number of Voters	Ranking
6	b > p > c
5	c > p > b
4	p > c > b

The Culinary Arts Club profile, repeated in Table 17, shows that the plurality method and the Hare method can fail to satisfy the irrelevant alternatives (IA) criterion. Beef, with six first-place votes, is selected by the plurality method as the entrée for the banquet. If pork is removed from the ballot, then, in a second election between beef and chicken, the former pork supporters vote for chicken. Chicken receives a majority of the votes and wins the second plurality election. (If chicken is removed from the ballot after one plurality vote, then pork wins the second decision between it and beef.)

The Hare method selects chicken as the Culinary Arts Club entree. When beef is removed from the ballot, the second Hare election is reduced to a simple majority election between pork and chicken. The former beef supporters vote for pork, causing it to win a second Hare election.

The Borda method can fail to respect the IA criterion. When an alternative is removed from a voter's ranking, the candidates formerly ranked below it move up in the ranking. Moving up in the rankings can allow an original Borda loser to gain points and defeat the original winner in a second election. Suppose a voter has the ranking, p > l > a > c > e, and p is the original Borda winner. If l drops out before a second vote, p still is ranked first, a and c move up, and e remains ranked last: p > a > c > e. Relative Borda tallies of a and c increase for this voter's ranking; the relative tally for the winner p does not change. Candidate p might lose a second Borda election, if enough rankings are shifted.

The pairwise comparison method also can fail to satisfy the IA criterion. The outcome of a pairwise comparison election depends on how many individual comparisons each candidate wins. When a candidate is dropped from an election, the pairwise comparisons involving the candidate are no longer considered in the voting process. The next example shows how the decreased number of comparisons in a second vote can change a pairwise comparison method election outcome.

EXAMPLE 4 Showing That the Pairwise Comparison Method Fails to Satisfy the IA Criterion

The conductor, along with the music director and a small group of trustees, are selecting a percussionist to replace their beloved timpani player who is retiring after 40 years with the world-famous American Orchestra. Five hopeful percussionists, v, w, x, y, and z, perform audition pieces. The committee agrees to use the pairwise comparison method for their selection, because it allows direct, one-on-one comparisons of the five musicians. Only the results of the ten possible pairwise comparisons are shown; the actual voter profile is not given.

v ties with w; v gets $\frac{1}{2}$ point and w gets $\frac{1}{2}$ point.
v beats x; v gets 1 point.
y beats v; y gets 1 point.
v beats z; v gets 1 point.
x beats w; x gets 1 point.
y beats w; y gets 1 point.
z ties with w; z gets $\frac{1}{2}$ point and w gets $\frac{1}{2}$ point.
x beats y; x gets 1 point.
x beats z; x gets 1 point.
z beats y; z gets 1 point.

The conductor is about to call and congratulate the selected percussionist when he gets an e-mail from y. Sensing she had not played well enough to win the audition, percussionist y accepted an offer to join a small, yet popular, European orchestra. The conductor and the selection committee feel they should revisit their decision, now that y no longer is a candidate. They agree to vote again on the four remaining candidates v, w, x, and z, using the pairwise comparison method.

Show that the first and second elections yield different winners.

SOLUTION

The pairwise comparison point total for each candidate is gathered from the ten individual results in the first election.

x gets **3 points.** ← Winner of first election
v gets $2\frac{1}{2}$ points.
y gets 2 points.
z gets $1\frac{1}{2}$ points.
w gets 1 point.

Percussionist x wins the most comparisons and wins the first election by the pairwise comparison method.

The only change in the voter profile (not shown) for the second election is that the committee members remove y from their rankings for the second vote. Consequently, the results of the second round of comparisons agree with the previous results, but there are no longer any comparisons with percussionist y. Only six pairwise comparisons are needed for four candidates.

v ties with w; v gets $\frac{1}{2}$ point and w gets $\frac{1}{2}$ point.
v beats x; v gets 1 point.
v beats z; v gets 1 point.
x beats w; x gets 1 point.
z ties with w; z gets $\frac{1}{2}$ point and w gets $\frac{1}{2}$ point.
x beats z; x gets 1 point.

The results gathered from the six individual comparisons surprise the voters.

v gets $2\frac{1}{2}$ points. ⟵ Winner of second election
x gets 2 points.
w gets 1 point.
z gets $\frac{1}{2}$ point.

The second pairwise comparison method vote selects percussionist v as the next timpani player of the American Orchestra. The first election selected percussionist x, while the second election selected percussionist v, violating the IA criterion. Therefore, the pairwise comparison method does not satisfy the IA criterion. ■

The IA criterion seems like a reasonable expectation of a voting method. The wording of the monotonicity criterion disguises the ability to reshuffle candidates. The IA criterion does not have any hidden features. It simply says removing a loser from the ballot should not make the original winner unable to win a second election. Surprisingly, the plurality method, the Borda method, the Hare method, and the pairwise comparison method can all violate the IA criterion for certain voter profiles.

Arrow's Impossibility Theorem Section 16.1 introduced five ways to select one alternative from a set of alternatives (including the approval method in For Further Thought); however, each of the five methods is inherently flawed. Approval voting permits multiple outcomes for the same profile of voters. Each of the remaining four voting methods fails to satisfy at least one of the four criteria considered desirable attributes of a selection process. Table 18 on the next page summarizes which criteria are satisfied and which criteria are not satisfied by the plurality, pairwise comparison, Hare, and Borda voting methods.

Plurality voting considers only top-ranked candidates, so it manages to satisfy the majority criterion and monotonicity criterion. The Hare method uses a series of plurality elections, but only satisfies the majority criterion. The Borda method considers how a voter ranks a complete list of candidates, and although using it is easy, the Borda method fails to satisfy any of the criteria.

The pairwise comparison method also considers a voter's complete ranking of the candidates and satisfies the majority criterion and Condorcet criterion. While this method does not satisfy either of the desired criteria for a second election, it is the only method to meet both desired attributes of a single election. This should make the pairwise comparison method the ideal choice if only one vote is taken.

TABLE 18 Summary of Desirable Criteria and Which Voting Methods Satisfy Them

	Voting Method			
Criterion	**Plurality Method**	**Pairwise Comparison Method**	**Borda Method**	**Hare Method**
Majority criterion	satisfied	satisfied	not satisfied (Example 1)	satisfied
Condorcet criterion	not satisfied (Example 2)	satisfied	not satisfied	not satisfied
Monotonicity criterion	satisfied	not satisfied	not satisfied	not satisfied (Example 3)
Irrelevant alternatives criterion	not satisfied	not satisfied (Example 4)	not satisfied	not satisfied

Unfortunately, the pairwise comparison method is difficult to use for elections with six or more candidates because of the large number of comparisons required to determine the outcome. The next example reveals another difficulty with the pairwise comparison method—it often fails to select a winner.

EXAMPLE 5 Showing That the Pairwise Comparison Method Does Not Always Select a Winner

A family of opera enthusiasts wants to name their new parrot after a female opera character. They narrow the choices down to Tosca, Aida, Carmen, and Lulu. The family agrees to use the pairwise comparison method to select the parrot's name. Results of the six pairwise comparisons are given. (The family voting profile is not shown.)

Tosca beats Aida; Tosca gets 1 point.
Tosca beats Carmen; Tosca gets 1 point.
Lulu beats Tosca; Lulu gets 1 point.
Aida beats Carmen; Aida gets 1 point.
Aida beats Lulu; Aida gets 1 point.
Lulu beats Carmen; Lulu gets 1 point.

Conduct a pairwise comparison of the four alternatives Tosca, Aida, Carmen, and Lulu.

SOLUTION

The pairwise comparison points for each name are calculated.

Tosca gets 2 points.
Aida gets 2 points.
Lulu gets 2 points.
Carmen gets 0 points.

Kenneth J. Arrow (1921–) is an American economist. In 1972, he was awarded the Nobel Memorial Prize in Economic Sciences for his work in the theory of general economic equilibrium. His famous impossibility theorem about voting (1951) resulted from discussions he had while working at the RAND Corporation. Arrow and his colleagues were applying the then-new mathematical science of game theory to the age-old science of military and diplomatic affairs. Arrow's discovery of unavoidable imperfection ended the first era of voting theory. The focus has shifted from the question of what voting method is perfect to much more mathematically challenging questions such as what can be expected of a voting method, and what can be done to help minimize the possibility that a particular paradox may occur. *Source:* From *Nobel Lectures, Economics 1969–1980,* Editor Assar Lindbeck, World Scientific Publishing Co., Singapore, 1992

The outcome is inconclusive, because no alternative has more points than all the others. Tosca, Aida, and Lulu each get 2 points. The pairwise comparison method fails to select an opera name for the family's new parrot. ■

All of the voting methods presented in Section 16.1 have shortcomings. Why wasn't a better voting method presented? Keep in mind that the four fairness criteria are reasonable expectations. Wanting to elect a majority or Condorcet candidate is reasonable. Wanting the original winner to win a second election is reasonable. It is especially reasonable if the winning candidate has moved up in the voters' rankings (monotonicity) or if an original loser drops out (IA) before the second election. These criteria may be reasonable individually, but together they are impossible to satisfy.

Economist Kenneth Arrow (see the margin note) set out to invent a better voting system. He discovered his goal was mathematically impossible. A perfect voting system cannot exist. This famous result is known as **Arrow's Impossibility Theorem.** It is said that the theorem took Arrow less than a week to prove and ultimately ended the search for a perfect voting system that had started nearly 200 years earlier.

Arrow's Impossibility Theorem

For an election with more than two alternatives, there does not exist and never will exist any voting method that simultaneously satisfies the majority criterion, the Condorcet criterion, the monotonicity criterion, and the irrelevant alternatives criterion.

Arrow proved that for any voting method there will be some voter profile for which one or more of the criteria are not satisfied. The CAA voter profile in Example 1 shows that the Borda method does not satisfy the majority criterion. In Example 2, the Culinary Arts Club profile shows that the plurality method does not satisfy the Condorcet criterion. The monotonicity criterion is violated by the Hare method for a different CAA voter profile in Example 3. The pairwise comparison method does not satisfy the irrelevant alternatives criterion, as shown by the orchestra selection committee profile in Example 4. Arrow's theorem also says that a contradictory profile of voters will exist for any new voting system yet to be invented.

16.2 Exercises

Identifying Violations of the Majority Criterion *Answer each question for each voter profile.*

(a) *Which alternative is a majority candidate?*

(b) *Which alternative is the Borda method winner?*

(c) *Does the Borda method violate the majority criterion for the voter profile?*

1.

Number of Voters	Ranking
6	a > b > c
3	b > c > a
2	c > b > a

2.

Number of Voters	Ranking
6	b > a > c
3	a > c > b
2	c > a > b

3.

Number of Voters	Ranking
20	a > b > c > d
6	b > c > d > a
5	c > b > d > a
5	d > b > a > c

4.

Number of Voters	Ranking
4	b > c > a > d
9	c > a > d > b
4	a > c > b > d
19	d > c > b > a

5.

Number of Voters	Ranking
16	a > b > c > d > e
3	b > c > d > e > a
5	c > d > b > e > a
3	d > b > c > a > e
3	e > c > d > a > b

6.

Number of Voters	Ranking
16	b > e > c > d > a
3	a > c > b > d > e
6	c > a > e > b > d
2	d > a > c > e > b
3	e > c > a > d > b

Identifying Violations of the Condorcet Criterion *Answer each question for each voting situation.*

(a) *Which candidate is a Condorcet candidate?*
(b) *Which candidate is selected by the plurality method?*
(c) *Which candidate is selected by the Borda method?*
(d) *Which candidate is selected by the Hare method?*
(e) *Which voting method(s)—plurality, Hare, or Borda—violate(s) the Condorcet criterion for this profile of voters?*

7. In Exercise 10 of the previous section, a 13-member committee is selecting a chairperson. The three candidates are Albert Werner a, Barbara Hightower b, and Charles Smith c. The committee members ranked the candidates according to the following voter profile.

Number of Voters	Ranking
4	a > b > c
2	b > c > a
4	b > a > c
3	c > a > b

8. Repeat Exercise 7 using the following voter profile (from Exercise 11 of the previous section).

Number of Voters	Ranking
3	a > c > b
4	c > b > a
2	b > a > c
4	b > c > a

9. In Exercise 14 of the previous section, a senator is holding a workshop. The senator asked the 21 workshop members to rank the issues of job creation j, education e, health care h, and gun control g in the order of importance to themselves and the counties they represent. The workshop members rank the issues according to the following voter profile.

Number of Voters	Ranking
3	e > h > g > j
6	h > e > g > j
5	j > g > e > h
4	g > e > h > j
3	j > e > h > g

10. Repeat Exercise 9 using the following voter profile (from Exercise 15 of the previous section).

Number of Voters	Ranking
6	h > j > g > e
5	e > g > j > h
4	g > j > h > e
3	j > h > g > e
3	e > j > h > g

11. In Exercise 18 of the previous section, members of the Adventure Club are selecting an activity for their yearly vacation. Choices are hiking in the desert h, rock climbing c, white-water kayaking k, bungee jumping b, or just sitting by a mountain lake and tanning t. The 55 Adventure Club members ranked the choices according to the following voter profile.

Number of Voters	Ranking
18	t > k > h > b > c
12	c > h > k > b > t
10	b > c > h > k > t
9	k > b > h > c > t
4	h > c > k > b > t
2	h > b > k > c > t

12. Repeat Exercise 11 using the following voter profile (from Exercise 19 of the previous section).

Number of Voters	Ranking
18	t > b > h > k > c
12	c > h > b > k > t
10	k > c > h > b > t
9	b > k > h > c > t
4	h > c > b > k > t
2	h > k > b > c > t

13. ***Identifying Violations of the Monotonicity Criterion*** A 14-member committee is selecting a site for its next meeting. The choices are Montreal m, Chicago c, San Francisco s, and Boston b.

(a) The committee members decide to use the pairwise comparison method to select a site in a nonbinding decision. Prior to any discussion, the 14 members rank the choices according to the following voter profile.

Number of Voters	Ranking
5	m > c > s > b
4	b > s > c > m
3	b > s > m > c
2	c > m > s > b

Show that Montreal is selected by the pairwise comparison method in the preliminary nonbinding decision.

(b) The 2 committee members with the bottom ranking in the table rearrange their ranking after listening to the discussions. The other 12 committee members stick with their original rankings of the cities. For the official vote, the 14 members rank the choices according to the following voter profile.

Number of Voters	Ranking
5	m > c > s > b
4	b > s > c > m
3	b > s > m > c
2	m > b > c > s

Use the pairwise comparison method to determine the site selection of the committee.

(c) Does the pairwise comparison method violate the monotonicity criterion in this selection process?

14. A 19-member committee is selecting a site for the next meeting. The choices are Dallas d, Chicago c, San Francisco s, and Boston b.

(a) The committee decides to use the Borda method to select a site in a nonbinding decision. Prior to any discussion, the 19 members rank the choices according to the following voter profile.

Number of Voters	Ranking
7	s > b > c > d
5	b > d > c > s
3	d > s > c > b
4	c > s > d > b

Show that the Borda method selects San Francisco in the preliminary nonbinding decision.

(b) The 7 committee members with the bottom two rankings rearrange their rankings after listening to the discussions. The other 12 committee members stick with their original rankings of the cities. For the official vote, the 19 members rank the choices according to the following voter profile.

Number of Voters	Ranking
7	s > b > c > d
5	b > d > c > s
3	s > b > d > c
4	s > b > c > d

Use the Borda method to determine the site selection of the committee.

(c) Does the Borda method violate the monotonicity criterion in this selection process?

15. A 17-member committee is selecting a site for the next meeting. The choices are Dallas d, Chicago c, Atlanta a, and Boston b.

(a) The committee decides to use the Hare method to select a site in a nonbinding decision. The 17 members rank the choices as follows.

Number of Voters	Ranking
6	a > c > b > d
5	b > d > a > c
4	c > d > b > a
2	c > a > b > d

Show that Atlanta is selected by the Hare method in the preliminary nonbinding decision.

(b) The 2 committee members with the bottom ranking rearrange their rankings after listening to the discussions. The other 15 committee members stick with their original rankings of the cities. For the official vote, the 17 members rank the choices according to the following voter profile.

Number of Voters	Ranking
6	a > c > b > d
5	b > d > a > c
4	c > d > b > a
2	a > c > b > d

Use the Hare method to determine the site selection of the committee.

(c) Does the Hare method violate the monotonicity criterion in this selection process?

16. ***Identifying Violations of the Irrelevant Alternatives Criterion*** Thirteen voters ranked three candidates a, b, and c according to the following voter profile.

Number of Voters	Ranking
6	a > b > c
5	b > c > a
2	c > b > a

(a) Show that Candidate a is selected if the plurality method is used to determine the outcome of the election.

(b) If losing Candidate c drops out, which candidate is selected by a second plurality method election?

(c) Does the plurality method violate the irrelevant alternatives criterion in this election process?

17. Four candidates, a, b, c, and d, are ranked by 175 voters according to the following voter profile.

Number of Voters	Ranking
75	a > c > d > b
50	c > a > b > d
30	b > c > d > a
20	d > b > c > a

(a) Show that Candidate a is selected if the plurality method is used to determine the outcome of the election.

(b) If losing Candidate b drops out of the election, which candidate is selected by a second plurality vote?

(c) Does the plurality method violate the irrelevant alternatives criterion in this election process?

18. A subcommittee of the senator's workshop is asked to include the issue of military spending m in its breakout session discussion of the issues job creation j, education e, health care h, and gun control g. The 11 subcommittee members rank the five issues as shown in the voter profile on the next page.

Number of Voters	Ranking
5	e > j > h > g > m
2	m > j > e > h > g
2	h > m > e > g > j
2	g > m > e > j > h

(a) If the pairwise comparison method is used, show that the subcommittee felt the issue of education was the highest priority.

(b) The workshop subcommittee decided to delete the losing issues of health care and gun control from further discussion. Use the pairwise comparison method to determine which of the remaining issues of education, job creation, or military spending the subcommittee felt was the highest priority.

(c) Does the pairwise comparison method violate the irrelevant alternatives criterion in this selection process?

19. The conductor, music director, and a small group of trustees select a new percussionist using the pairwise comparison method in Example 4. When Percussionist y drops out of the audition process, the pairwise comparison method violates the irrelevant alternatives criterion by selecting Percussionist v instead of original winner Percussionist x in a second pairwise comparison. If w, rather than y, had dropped out, would the IA criterion have been violated in a second pairwise comparison?

20. Twenty-five voters rank three candidates a, b, and c according to the following voter profile.

Number of Voters	Ranking
13	c > b > a
8	b > a > c
4	b > c > a

(a) Show that Candidate b wins a Borda method election.

(b) Which candidate wins a second Borda method election, if losing Candidate a drops out of the election?

(c) Does the Borda method violate the irrelevant alternatives criterion in this election process?

21. Thirty-four voters rank three candidates a, b, and c according to the following voter profile.

Number of Voters	Ranking
12	a > c > b
10	b > a > c
8	c > b > a
4	c > a > b

(a) Show that Candidate a wins a Hare method election.

(b) If losing Candidate c drops out of the election, which of the remaining candidates wins a Hare method election?

(c) Does the Hare method violate the irrelevant alternatives criterion in this election process?

22. For the profile of voters in Exercise 15(a), Atlanta is selected by the Hare method in the preliminary nonbinding decision.

(a) If the losing city, Chicago, withdraws after the nonbinding vote, which city wins the official vote? Use the voter profile in Exercise 15(a) with Chicago deleted for the official Hare method election.

(b) Does the Hare method violate the irrelevant alternatives criterion in this selection process?

23. Explain why a violation of the majority criterion is an automatic violation of the Condorcet criterion.

24. What is it about the departure for Europe of Percussionist y in Example 4 that hurts Percussionist x, the original winner, and benefits Percussionist v in the second pairwise comparison method selection?

25. Construct a voter profile for 19 voters and 6 candidates that has a majority candidate. Show that the majority candidate must win 5 pairwise comparison points, but that another candidate can win 4 pairwise points. (*Hint:* Only two of the 6! possible rankings are needed for the profile.)

26. Construct a voter profile for 40 voters and 4 candidates that shows the Borda method violates the majority criterion. Do this by assigning the remaining 19 voters to the rankings in the given incomplete voter profile in such a way that the majority candidate does not win a Borda method election.

Number of Voters	Ranking
21	d > b > c > a
?	b > c > a > d
?	c > b > d > a
?	a > c > b > d

27. Construct a voter profile for 13 voters and 4 candidates that has a Condorcet candidate that fails to be elected by the Borda method, but is selected by both the Hare method and the plurality method.

28. Construct a voter profile for 13 voters and 4 candidates that has a Condorcet candidate that fails to be elected by the Borda method and the plurality method, but is selected by the Hare method.

29. Construct a voter profile for 18 voters and 4 candidates that shows the pairwise comparison method violates the monotonicity criterion. The original voter profile for the 18 voters is given.

Number of Voters	Ranking
6	a > x > y > z
5	z > y > x > a
4	z > y > a > x
3	x > a > y > z

(a) Show that Candidate a wins the pairwise comparison method election for the original given profile.

(b) Rearrange the ranking of the 3 voters in the bottom row in such a way that for the altered voter profile Candidate z wins a second pairwise comparison method election. Candidate a must be moved to first place in the new ranking.

30. Construct a voter profile for 14 voters and 4 candidates that shows the Borda method violates the monotonicity criterion. The original voter profile for the 14 voters is given.

Number of Voters	Ranking
5	m > b > s > c
4	b > s > c > m
3	s > m > c > b
2	c > m > s > b

(a) Show that Candidate m wins the Borda method election for the original given profile.

(b) Rearrange the rankings of the 5 voters in the bottom two rows in such a way that for the altered voter profile, Candidate b wins a second Borda method election. Candidate m must be moved to first place in the new ranking.

31. Delete one of the losing candidates from the given voter profile in such a way that the plurality method violates the irrelevant alternatives criterion in a second election.

Number of Voters	Ranking
15	a > b > c > d
8	b > a > c > d
9	c > b > a > d
6	d > b > a > c

32. Delete 2 or 3 of the losing candidates from the given voter profile in such a way that the pairwise comparison method violates the irrelevant alternatives criterion in a second election.

Number of Voters	Ranking
15	a > c > b > d > e
6	e > c > a > b > d
6	b > e > a > d > c
6	d > e > a > c > b

33. Construct a voter profile for 41 voters and 3 candidates g, j, and e in such a way that the Borda method violates the irrelevant alternatives criterion in a second election. Arrange the voter profile so that j wins the first Borda election with the most Borda points, and e has the fewest Borda points. Delete e from the second Borda method vote.

34. The discussion on the monotonicity criterion explains why the pairwise comparison method satisfies the alternative monotonicity criterion.

(a) Explain why the plurality method satisfies the alternative monotonicity criterion.

(b) Explain why the Borda method satisfies the alternative monotonicity criterion.

35. Now that you know the pros and cons of the plurality, pairwise comparison, Borda, and Hare methods, have you changed your mind about which one you personally think is the best method? Explain why you changed your mind or why you are staying with your original choice.

36. If you can have only two of the four criteria discussed in this section satisfied by a voting method, which would you choose? Why do you think the two criteria you selected are important?

16.3 The Possibilities of Apportionment

Introduction • Hamilton Method • Jefferson Method • Webster Method

Introduction The story of apportionment began as the dust of the American Revolution settled at the Constitutional Convention in 1787 in Philadelphia. America's founding fathers invented a system of government with three branches—executive, judicial, and legislative. The *Great Compromise,* found in Article I, Sections 2 and 3, created a legislative branch consisting of the Senate and the House of Representatives. Each state in the union would be represented by two senators in the Senate, so that, within the Senate, all states have equal voting influence. The number of representatives a state has in the House of Representatives would be determined by the size of the state's population. Therefore, within the House more populous states have greater voting influence than less populous states.

The founding fathers did not specify exactly how the number of representatives for each state was to be determined—only that the number must be based on state population. Whatever the decision method, this allotment of House of Representative seats to the various states is called *apportionment.* In general, **apportionment** is a division or partition of identical, indivisible things according to some plan or proportion.

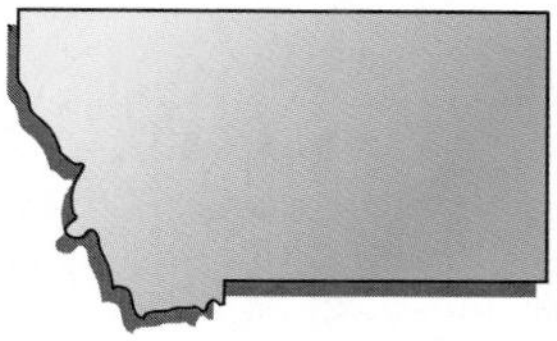

The Hill-Huntington method has been challenged on Constitutional grounds as recently as 1991. The state of Montana challenged the reapportionment of the House after the 1990 census. Montana lost 50% of its seats when its apportionment decreased from 2 seats to 1 seat. Montana asked the Supreme Court to require the Census Bureau to recompute the apportionments using an apportionment method that minimizes the differences in district populations. If successful, Montana would have regained its lost seat at the expense of Washington state. The Court unanimously rejected Montana's challenge and ruled that, since congressional districts cannot cross state lines, the districts are bound to be lopsided occasionally.

The U.S. House of Representatives is not the only place where apportionment is an issue. Many countries have Parliamentary governments in which legislative seats are apportioned based on the percent of the total vote received by a political party in a general election. Transportation boards decide, on the basis of ridership, how many trains are put on each rail line and how many buses are assigned to each route. Based on student enrollments, faculty positions at a college or university are divided among various departments and teaching assistants are meted out to professors.

In the case of U.S. House of Representatives apportionment, the country's rapidly growing population made it impossible to fix the number of seats when the Constitution was written. However, the founding fathers did specify that each congressional district represented by a seat in the House should have a population of at least 30,000. The increasing and shifting population of the United States, and the addition of each new state to the union, meant that the apportionment issue would be revisited often.

The Constitution requires a census of the population of the United States every ten years and that the House of Representatives be reapportioned according to the current census data. The first census was conducted in 1790, but a debate about what apportionment method to use postponed the first House apportionment until 1794. The House has been reapportioned two years after each subsequent census, except in the 1920s.

Since the initial apportionment of House seats in 1794, only four different methods have ever been used. One of these, the **Hill-Huntington method,** was introduced in 1942 and is still used currently. As it is more involved, mathematically, we cover it in the exercises at the end of this section. The **Adams method,** proposed by John Quincy Adams, the sixth president of the United States, was never actually put into practice and is also covered in the exercises. The three methods that have been used historically, the **Hamilton method,** the **Jefferson method,** and the **Webster method,** are presented in this section.

Alexander Hamilton (1755–1804) was born in the British West Indies and came to New York in 1773. He was General Washington's aide-de-camp in the American Revolutionary War, an author of the influential *Federalist Papers,* and the first secretary of the treasury under President Washington. He is among the least understood of the founding fathers because of his aristocratic views and often is regarded as the patron saint of capitalism. Hamilton distrusted Aaron Burr, and the two argued publicly for years before their fateful duel in 1804 that took Hamilton's life.

Hamilton Method The Hamilton method was among the earliest methods considered for apportioning the original House of Representatives after the first census in 1790. This method is sensible and easy to calculate. Congress sent a bill to President Washington adopting the Hamilton method, but he rejected it with the very first presidential veto. Congress could not override the veto and eventually adopted the Jefferson method. Hamilton's method was eventually adopted, in 1852, and was used through 1892.

The Hamilton method has four steps. The first step determines how many people in the overall population are to be represented by a single seat in the House. This is called the *standard divisor.* The next step produces what is called a *standard quota* for each state, which is the number of seats (not necessarily an integer) to which a state is entitled, based on its population. Fractional parts of seats do not exist, so the third step is a practical way to ensure that the total number of seats promised is equal to or smaller than the total number of seats available. The fourth step offers a pragmatic scheme for distributing any seats that still are available to the states Hamilton considered most deserving. It is this fourth step in the Hamilton method that caused President Washington to reject it.

Hamilton Method

1. Compute the **standard divisor,**
$$d = \frac{\textbf{total population}}{\textbf{total number of seats}}.$$
2. Compute the **standard quota** for each state,
$$Q = \frac{\textbf{state's population}}{d}.$$
3. Round each state's standard quota Q *down* to the nearest integer. Each state will get at least this many seats, but must get at least one.
4. Give any additional seats one at a time (until no seats are left) to the states with the largest fractional parts of their standard quotas.

EXAMPLE 1 Apportioning Computers to Schools Using the Hamilton Method

The Highwood School District received a generous gift of 109 computers from a local manufacturer. The benefactor stipulated that the division of the identical machines must be based on the individual enrollments at the various schools in the district. In other words, the larger the school's enrollment, the more machines it is entitled to receive. Naturally, each school must receive an integer number of computers because fractional parts of computers are worthless. The district decides to apportion the computers by using the Hamilton method. In legislative lingo, the computers are congressional seats, the schools are states, and the school enrollments are state populations. The district has five schools: Applegate, enrollment 335; Bayshore, enrollment 456; Claypool, enrollment 298; Delmar, enrollment 567; and Edgewater, enrollment 607. Apportion the 109 computers to the five schools using the Hamilton method.

SOLUTION

The total enrollment for the district is

$$335 + 456 + 298 + 567 + 607 = 2263.$$

1. The standard divisor is

$$d = \frac{\text{total enrollment}}{\text{number of computers}} = \frac{2263}{109} = 20.761.$$

The standard divisor represents the number of students there are for each of the 109 computers. In other words, there is one computer for every 20.761 students in the district.

2. The standard quotas are as follows.

$$\text{Applegate School: } Q = \frac{335}{20.761}; \quad \text{Bayshore School: } Q = \frac{456}{20.761};$$

$$\text{Claypool School: } Q = \frac{298}{20.761}; \quad \text{Delmar School: } Q = \frac{567}{20.761};$$

$$\text{Edgewater School: } Q = \frac{607}{20.761}.$$

Each standard quota Q is given to two decimal places in the third column of Table 19. Usually, two decimal places are sufficient.

TABLE 19

School	Enrollment	Standard Q Quota	Rounded-Down Q	Computers Apportioned
Applegate	335	16.14	16	16
Bayshore	456	21.96	21	**22**
Claypool	298	14.35	14	**15**
Delmar	567	27.31	27	27
Edgewater	607	29.24	29	29
Totals	2263	109	107	109

3. The standard quotas are rounded down to the nearest integer in the fourth column of the table. Each school receives at least this number of computers.
4. The total number of promised computers is only 107; thus, two additional computers must be distributed. The Hamilton method distributes these two computers one at a time to the schools with the largest fractional parts of their standard quotas, hence to Bayshore School, with fractional part .96, and to Claypool School with fractional part .35.

The final apportionment of the 109 computers is shown in the final column of the table. The bold entries of **22** and **15** for the Bayshore and Claypool Schools, respectively, indicate that these schools received more computers than their rounded-down standard quota Q. ■

Consider Delmar School in Example 1. The fractional part of its standard quota is .31, which is not drastically smaller than the fraction (.35) that gained Claypool School an additional computer. In practice, the fractional parts of two states' standard quotas might need to be calculated to four or five decimal places to decide which is larger. Indeed, a state with a quota of $Q = 14.8678$ could get the last additional seat available (for a total of 15 seats), and a state with a quota of $Q = 14.8677$, if there was one, would get only its rounded-down quota of 14 seats. The Hamilton method has much more serious and subtle problems than the obvious one discussed here.

It may seem that a more mathematical and direct approach to apportionment would be to replace the practical rounding-down scheme in Step 3 of the Hamilton method with a traditional rounding scheme. If the fractional part of Q is greater than or equal to .5, then Q is rounded up to the nearest integer. If the fractional part of Q is less than .5, then Q is rounded down to the nearest integer. If a traditional rounding scheme is used on the values of Q from the Highwood School District example, only 108 of the 109 computers are apportioned as shown in Table 20. In some cases, traditional rounding will result in too many objects being apportioned rather than too few.

TABLE 20

School	Standard Quota Q	Q Rounded to Nearest Integer
Applegate	16.14	16
Bayshore	21.96	22
Claypool	14.35	14
Delmar	27.31	27
Edgewater	29.24	29
Totals	109	108

EXAMPLE 2 Applying the Hamilton Method to the 15 States of 1794

The 1790 census determined that the population of the United States was 3,615,920. The Union had 15 states, and it was decided that the original House would have 105 seats. Use the Hamilton method to apportion the 105 seats.

SOLUTION

1. Compute the standard divisor: $d = \dfrac{3{,}615{,}920}{105} = 34{,}437.33.$
2. Virginia had the largest population with 630,560 residents. The standard quota for Virginia is $Q = \frac{630{,}560}{34{,}437.33} = 18.310$. Using Q to three places is necessary for this example. The populations of the other states and their standard quotas are shown in Table 21 on the next page.
3. The values of Q are rounded down in the fourth column of the table. The total of this column, the number of seats promised, is 97. Eight more seats must be assigned.
4. The apportionment of the seats by the Hamilton method is in the final column. The bold entries indicate the states that receive more seats than their rounded-down values of Q. They are the eight states with the largest fractional parts of their standard quotas.

TABLE 21

State	Population	Standard Quota Q	Rounded-Down Q	Number of Seats
Virginia	630,560	18.310	18	18
Massachusetts	475,327	13.803	13	**14**
Pennsylvania	432,879	12.570	12	**13**
North Carolina	353,523	10.266	10	10
New York	331,589	9.629	9	**10**
Maryland	278,514	8.088	8	8
Connecticut	236,841	6.877	6	**7**
South Carolina	206,236	5.989	5	**6**
New Jersey	179,570	5.214	5	5
New Hampshire	141,822	4.118	4	4
Vermont	85,533	2.484	2	2
Georgia	70,835	2.057	2	2
Kentucky	68,705	1.995	1	**2**
Rhode Island	68,446	1.988	1	**2**
Delaware	55,540	1.613	1	**2**
Totals	3,615,920	105	97	105

The Hamilton apportionment of the original House gives Delaware two seats. This actually is not a Constitutionally correct apportionment. The Constitution specifically requires that each seat represent at least 30,000 people. Delaware's population is 4460 people short of deserving two House seats.

The state of Virginia, where Thomas Jefferson made his home, would have received 18 seats in the first House if the Hamilton method of apportionment had been used. The actual 1794 House apportionment used the Jefferson method shown later in Example 4. ■

TABLE 22

State	Standard Quota Q
Kentucky	1.**995**
South Carolina	5.**989**
Rhode Island	1.**988**
Connecticut	6.**877**
Massachusetts	13.**803**
New York	9.**629**
Delaware	1.**613**
Pennsylvania	12.**570**

The only specific mandates given by the Constitution for apportioning seats in the House of Representatives are that apportionment be based on state population and that each seat represent at least 30,000 people. The theoretical Hamilton apportionment of the House in Example 2 can be used to illustrate one of the more serious and subtle problems with the method. This is a problem of which President Washington was certainly aware, when he vetoed the bill that would have made the Hamilton method the law of the land.

In Example 2, the additional eight seats are assigned based on fractional parts of the standard quotas, in the order shown in Table 22. It may be argued that these eight additional seats are not awarded based on population when the fourth step of the Hamilton method is executed. The state with the next largest fractional part of its standard quota is Vermont, with $Q = 2.\mathbf{484}$. The fraction .570 for Pennsylvania, the last state to receive an additional seat, is unarguably larger than Vermont's fraction of .484. Simple division shows that the fractional part .484 of Vermont's value of Q is approximately 19% ($\frac{.484}{2.484}$) of its total value of $Q = 2.484$, while the fractional part .570 of Pennsylvania's value of Q is only about 5% ($\frac{.570}{12.570}$) of its total value of $Q = 12.570$. Yet, it is Pennsylvania, not Vermont, that gets an extra seat when Step 4 of Hamilton's method is used to assign the remaining eight seats. The original detractors of the Hamilton method argued that this kind of situation showed that the method did not follow the Constitutional mandate of apportionment by population. President Washington evidently felt their argument was valid.

Jefferson Method Unable to override Washington's veto of the Hamilton method, Congress adopted the method Thomas Jefferson was promoting. It was used to apportion the original House in 1794, based on the 1790 census. The Jefferson method eliminates the pragmatic, yet problematic and perhaps unconstitutional, Step 4 of the Hamilton method. It replaces the Hamilton method's straightforward standard divisor d with a modified, slightly magical divisor, md. Each state's population is divided by the modified divisor md to produce its modified quota mQ. Each state receives as many seats as the integer part of its modified quota mQ. The beauty of Jefferson's method is that the sum of the integer parts of all the values of mQ is equal to the total number of seats to be apportioned. No additional seats are left to be assigned. How is that possible? The modified, slightly magical divisor md is carefully and painstakingly chosen to force it to happen.

Thomas Jefferson (1743–1826) is perhaps the most important man in American history. He shaped the very foundation of the country, helping to write both the Declaration of Independence and the Constitution of the United States. Many of the ideas in these important documents are the fruits of Jefferson's brilliant philosophical mind. Jefferson was secretary of state under President Washington and later was elected president himself. Some of his other accomplishments include devising the decimal monetary system and nearly doubling the land area of the United States with the Louisiana Purchase in 1803.

Jefferson Method

1. Compute ***md***, the **modified divisor.**
2. Compute ***mQ***, the **modified quota** for each state.

$$mQ = \frac{\textbf{state's population}}{md}$$

3. Round each state's modified quota ***mQ*** *down* to the nearest integer.
4. Give each state this integer number of seats.

A standard quota Q is computed by dividing a state's population by the standard divisor d. The values of Q must, by definition, sum to the total number of seats to be apportioned. Rounded down, the values of Q may sum to less than the total number of seats to be apportioned. If the rounded-down values of Q actually sum to the number of things to be apportioned, the Jefferson apportionment simply uses $md = d$. Otherwise, for the Jefferson method to work, the values of the modified quotas must be greater than the values of Q so that when the values of mQ are rounded down they sum to exactly the number of seats to be apportioned, not less. If the values of mQ must be slightly greater than the values of Q, then md, the divisor used to compute the values of mQ, must be slightly less than the standard divisor d used to compute the values of Q. In other words, because lesser divisors produce greater quotients, the Jefferson method produces greater modified quotas for each state by dividing the state populations by a number less than the standard divisor d.

Algorithms exist to find the values of *md* needed for Jefferson method apportionments. The algorithms are technical and beyond the range of this discussion. The values of *md* in the following examples were found by trial and error, slowly decreasing the value of the standard divisor *d*, until the rounded values of *mQ* summed to the exact number of objects to be apportioned.

EXAMPLE 3 Apportioning New Sailboats to Caribbean Resorts

Sea Isle Vacations operates six secluded resorts in the Caribbean. They recently purchased thirty new sailboats. Executives at the company headquarters in Virginia decide to use the Jefferson method to apportion the new boats, basing the apportionment on the number of rooms at each resort. The six individual resorts are named after hurricanes that spared them. The number of rooms at each location is in the second column of Table 23. The total number of rooms at the six resorts is 2013. Complete the Jefferson apportionment.

SOLUTION

1. The standard divisor is

$$d = \frac{\text{total number of rooms}}{\text{number of new boats}} = \frac{2013}{30} = 67.1.$$

The value of the standard divisor means that, theoretically, for every 67.1 rooms at a resort, that resort is entitled to one of the new boats. The modified divisor *md* was found by slowly decreasing the value of *d*. For this example, when $md = 60$ the Jefferson method works. That is, when *d* is reduced to $md = 60$, the resulting rounded-down modified quotients sum to exactly 30. (In fact, values of *md* greater than 58.5 but less than or equal to 60 also work.)

TABLE 23

Resort	Number of Rooms	Standard Quota *Q*	Rounded-Down *Q*	Modified Quota *mQ*	Rounded-Down *mQ*
Anna	345	5.14	5	5.75	5
Bob	234	3.49	3	3.90	3
Cathy	420	6.26	6	7.00	7
David	330	4.92	4	5.50	5
Ellen	289	4.31	4	4.82	4
Floyd	395	5.89	5	6.58	6
Totals	2013	30.00	27	33.55	30

2. Using $md = 60$, we obtain modified quota values.

$$\text{Resort Anna: } mQ = \frac{345}{60} = 5.75; \qquad \text{Resort Bob: } mQ = \frac{234}{60} = 3.90$$

Values for the remaining four resorts are calculated similarly. All six values are shown, to two decimal places, in the fifth column of Table 23. For comparison, the third column shows standard quota values, to two decimal places.

3. The modified quotas are rounded down in the final column of Table 23. The rounded-down values of mQ for the six resorts sum to exactly 30. This is guaranteed by the choice of $md = 60$. The rounded-down values of Q, given for comparison in the fourth column, do not sum to 30.
4. Each resort gets the number of new sailboats in the final column. The Jefferson method has automatically apportioned the exact number of new sailboats available. No boats remain unallocated.

If the Hamilton method is used to apportion the sailboats, the results are different from the results in Example 3. Rounding down the values of Q leaves three sailboats unassigned, as the total in the fourth column shows. The three boats go to resorts David, Floyd, and Bob, the resorts with the largest fractional portions of their values of Q. So, David and Floyd end up with five and six boats, respectively, no different than by the Jefferson method. But Bob ends up with four boats, one more than the three alloted by the Jefferson method, and Cathy gets six, one fewer than the seven alloted by the Jefferson method.

The difference between the Hamilton and Jefferson methods, in this situation, can be summarized as follows: The Hamilton method increases the apportionment for the smallest secluded resort, Bob, while the Jefferson method increases the apportionment for the largest resort, Cathy.

It is not a coincidence that the Jefferson method favored the largest resort. The Jefferson method always shows a preference for larger states. This bias helped Jefferson's large home state of Virginia in the actual apportionment of the House in 1794.

EXAMPLE 4 Applying the Jefferson Method to the 15 States of 1794

Use the Jefferson method to apportion the 1794 House of Representatives based on the 1790 census. (Census workers calculated the total population of the existing fifteen states as 3,615,920. The number of seats had been set at 105. The standard divisor $d = 34{,}437.33$ was computed in Example 2.)

The U.S. Department of the Treasury began issuing commemorative quarters for each state starting in 1999. The quarters are being phased into circulation in the same order that the states they represent joined the Union. The next time you find a new state quarter in your pocket, spend a minute or two reflecting on how the addition of that state changed the way the seats in the House of Representatives were apportioned and the mathematics involved in the process.

SOLUTION

1. The modified divisor used by Congress for the actual Jefferson method apportionment was $md = 33{,}000$. Considering the hours of manual calculations it probably required to test each guess of the value of md, it must have been a welcome surprise that when $md = 33{,}000$, the Jefferson method works.
2. The modified quota for Virginia is $\frac{630{,}560}{33{,}000} = 19.108$. The modified quotas for the other states are in Table 24 on the next page.
3. The rounded-down modified quotas mQ are in the final column of the table. The rounded-down values sum to the required number of 105 seats, because of the choice of the value of md.
4. Each state gets the number of seats shown in the final column of the table, that is, the integer part of its modified quota mQ.

The numbers in the final column represent the way the House of Representatives was actually apportioned in 1794. The largest state, Jefferson's home state of Virginia, received 19 seats in the 1794 apportionment of the House of Representatives. Delaware, the smallest state in the Union, received only one seat when the Jefferson method was used.

TABLE 24

State	Modified Population	Rounded-Down Quota mQ	mQ
Virginia	630,560	19.108	19
Massachusetts	475,327	14.404	14
Pennsylvania	432,879	13.118	13
North Carolina	353,523	10.713	10
New York	331,589	10.048	10
Maryland	278,514	8.440	8
Connecticut	236,841	7.177	7
South Carolina	206,236	6.250	6
New Jersey	179,570	5.442	5
New Hampshire	141,822	4.298	4
Vermont	85,533	2.592	2
Georgia	70,835	2.147	2
Kentucky	68,705	2.082	2
Rhode Island	68,446	2.074	2
Delaware	55,540	1.683	1
Totals	3,615,920	109.576	105

Daniel Webster (1782–1852) was an important constitutional lawyer, a brilliant speaker, and a folk hero. He served in the House and the Senate and was secretary of state under Presidents Harrison, Tyler, and Fillmore. Webster spent his life keeping a watchful eye on the power of individual states and was an ardent supporter of the national government. He disliked the institution of slavery, but he supported the Compromise of 1850 on constitutional grounds and his support helped it pass. The Compromise of 1850 delayed the American Civil War for ten years. In that time, the free states gained enough economic power to abolish slavery in America.

The Jefferson method is considered a **divisor method,** because it uses a modified divisor md to produce modified quotas. Values of mQ are rounded down in the Jefferson method. Other divisor methods, such as the Webster method discussed next, use different rounding schemes. When President Washington rejected the Hamilton method on constitutional grounds, supporters of other apportionment methods questioned whether the modified divisor in the Jefferson method was constitutionally sound. Jefferson argued that the Constitution required only that the House apportionment be based on population, and not on a standard divisor. Jefferson method supporters convinced skeptics that the modified divisor md still produced quotients that reflect state populations and, since the same divisor is used for all states, the method was on firm constitutional ground.

Webster Method The Jefferson method was used for decades without incident, until the 1820s. The apportionment of the House based on the 1820 census revealed a major flaw in this method that is discussed later. The flaw might have been dismissed as a freak incident, if the same anomaly had not occurred again following the 1830 census. The Jefferson method was replaced by a method proposed and championed by Daniel Webster. No one suspected it was flawed in the same way. The Webster method was used to apportion the House following the 1840 census, then again from 1900 until it was replaced by the Hill-Huntington method in 1941.

The appeal of the Webster method is simple. Standard quotas are not rounded down to the nearest integer. Instead, they are rounded using a traditional rounding scheme. A quota with a fractional part greater than or equal to .5 is rounded up to the next highest integer, and a quota with a fractional part less than .5 is rounded down to the next lowest integer. Discussion following Example 1 concerning the Highwood School District computer apportionment shows apportionment does not necessarily work when a traditional rounding scheme is used on the standard quota values of Q. However, the Webster method, based on a traditional rounding scheme, does work, because, like the Jefferson method, it is a divisor method. The Webster method uses a modified divisor *md*. The value of *md* is selected so that when the modified quotas ($mQ = \frac{\text{state population}}{md}$) are traditionally rounded, they sum to exactly the number of seats to be apportioned.

Webster Method

1. Compute ***md***, the **modified divisor.**
2. Compute ***mQ***, the **modified quota** for each state.

$$mQ = \frac{\text{state's population}}{md}$$

3. Round each state's modified quota ***mQ*** *up* to the nearest integer if its fractional part is greater than or equal to .5 and *down* to the nearest integer if its fractional part is less than .5.
4. Give each state this integer number of seats.

The Webster-modified divisor *md* is a bit harder to find than the modified divisor for the Jefferson method because the rounding of mQ values can go either way, up or down, in the traditional rounding scheme of the Webster method. The Jefferson method always rounds the values of mQ down to the nearest integer. Rounding the values of mQ down allows the modified divisor for the Jefferson method to be found by slowly decreasing the value of the standard divisor d, to produce slightly greater modified quotas with a greater sum. Traditional rounding of the standard quota values of Q can cause their sum to be less than, equal to, or even greater than the actual number of seats to be apportioned.

If the rare case of equality occurs, then simply let $md = d$ for the Webster apportionment. If the sum of the traditionally rounded values of Q is less than the number of objects to be apportioned, then slowly decrease the value of d to find the modified Webster divisor *md*, as in the Jefferson method. If the values of Q sum to more than the number of objects to be apportioned when traditionally rounded, then the value of *md* for the Webster apportionment is found by slowly increasing the value of d, the standard divisor. This is because greater divisors make lesser quotients, and therefore lesser sums.

EXAMPLE 5 Apportioning Computers to Schools Using the Webster Method

Use the Webster method to apportion the 109 computers given to the Highwood School District, and compare the results with those of Example 1, in which the Hamilton method was used. The total enrollment in the district is 2263, as shown in Example 1.

SOLUTION

1. The value of the standard divisor $d = 20.761$ and the standard quotas for the five schools were computed in Example 1. Discussion following Example 1 shows that the values of Q when traditionally rounded sum to 108, and 109 computers are to be apportioned. Here the value of md is found by slowly *decreasing* the value of d by trial and error to produce a slightly greater sum of mQ values. When $md = 20.6$, the Webster method works.

2. The modified quota for Applegate School is $mQ = \frac{335}{20.6} = 16.26$. The modified quotas for the other schools are in the third column of Table 25.

TABLE 25

School	Enrollment	Modified Quota mQ	Traditionally Rounded mQ
Applegate	335	16.26	16
Bayshore	456	22.14	22
Claypool	298	14.47	14
Delmar	567	27.52	28
Edgewater	607	29.47	29
Totals	2263		109

3. The five values of mQ in the fourth column are traditionally rounded. The rounded values sum to exactly 109. This is guaranteed by the choice of $md = 20.6$.
4. Each school gets the number of computers in the final column. The Webster method automatically apportioned exactly 109 computers.

The Webster apportionment agrees with the Hamilton apportionment (Example 1) for the Applegate, Bayshore, and Edgewater schools, but not for the other two schools. The Webster method gives Delmar School 28 computers and Claypool School 14. The Hamilton method apportions 27 computers to Delmar and 15 to Claypool. ■

EXAMPLE 6 Apportioning Legislative Seats Using the Webster Method

A newly formed nation on a small island in the Pacific adopted a constitution modeled on the Constitution of the United States. Unlike the United States, the new nation, Timmu, decided to have just one legislative assembly with 131 seats. Like the U.S. House of Representatives, the seats of Timmu's legislature are to be apportioned based on the populations of the nation's individual states. The Timmu Constitution does not specify an apportionment method, but the leadership of the island nation voted to use the Webster method.

A census found that the population of Timmu is currently 47,841. A house size of 131 seats means that the standard divisor $d = \frac{47{,}841}{131} = 365.20$. (We retain five significant figures in this value, which is consistent with the total population figure.) The small nation is divided into six states. Table 26 on the next page shows the state populations, the values of the standard quotas to three places, and the traditionally rounded values of the standard quotas.

TABLE 26

State	Population	Standard Quota Q	Traditionally Rounded Q
Abo	5672	15.531	16
Boa	8008	21.928	22
Cio	2400	6.572	7
Dao	6789	18.590	19
Effo	4972	13.614	14
Foti	20,000	54.765	55
Totals	47,841		133

Complete the Webster apportionment to find an appropriate modified divisor and modified quotients, and to allocate exactly 131 seats among the six states.

SOLUTION

1. The sum of the traditionally rounded values of Q is 133, which is greater than the number of seats in the Timmu house. The value of the modified divisor md needed for the Webster method is found by slowly *increasing* the value of the standard divisor $d = 365.20$, which produces smaller quotas with a smaller sum. We find that when $md = 366.98$, the Webster method works.
2. The modified quota for Abo is found by dividing its population by the value of the modified divisor.

$$mQ = \frac{5672}{366.98} = 15.456, \quad \text{to three places.}$$

   Modified quotas to three places are given in the third column of Table 27.
3. The six values of mQ are traditionally rounded in the fourth column of the table. The value of $md = 366.98$ guarantees the traditionally rounded values of mQ sum to 131 exactly.
4. Each state receives the number of seats shown in the final column of the table.

TABLE 27

State	Population	Modified Quota mQ	Traditionally Rounded mQ
Abo	5672	15.456	15
Boa	8008	21.821	22
Cio	2400	6.540	7
Dao	6789	18.500	19
Effo	4972	13.548	14
Foti	20,000	54.499	54
Totals	47,841		131

16.3 EXERCISES

1. ***New Trees for Wisconsin Parks*** The schoolchildren of Wisconsin began a statewide effort last year to raise money to purchase new trees for five state parks. Altogether, the students raised enough money to purchase 239 trees. A committee decided that the trees should be apportioned based on the amount of land in each of the five parks as shown in the table.

State park	a	b	c	d	e
Acres	1429	8639	7608	6660	5157

   (a) Find the total number of acres in the five state parks and compute the standard divisor for the apportionment of the 239 trees.
   (b) Use the Hamilton method to apportion the trees.
   (c) Use the Jefferson method to apportion the trees. As always, the modified divisor needed for the Jefferson apportionment is found by slowly decreasing the value of the standard divisor computed in part (a).
   (d) Round the values of all of the parks' standard quotas, using a traditional rounding scheme. Is the sum of the traditionally rounded values less than, equal to, or greater than the number of trees to be apportioned?
   (e) Should the value of the modified divisor for the Webster method apportionment be less than, equal to, or greater than the standard divisor computed in part (a)? Why?
   (f) Use the Webster method to apportion the trees.
   (g) Which of the three tree apportionments are the same? Which is different?

2. School enrollments for the Highwood School District, the standard divisor, and the Hamilton apportionment of its 109 new computers are in Example 1. The Webster apportionment is in Example 5.
   (a) Use the Jefferson method to apportion the 109 computers the Highwood School District received.
   (b) Which of the apportionments of the 109 computers are the same? Which is different?
   (c) The students at Claypool School studied the various apportionment methods and launched a massive letter-writing campaign to convince the Highwood School District to use the Hamilton method, rather than the Jefferson or Webster method, to apportion the new computers. Why?

3. ***Assigning Faculty to Courses*** The English department at Oaks College has the faculty to offer 11 sections in any combination of the four courses: Fiction Writing, Poetry, Short Story, or Multicultural Literature. The number of sections apportioned to each course is based on enrollments in the courses as shown.

Course	Fiction	Poetry	Short Story	Multicultural
Enrollment	56	35	78	100

   (a) Find the total enrollment for the four courses and the standard divisor for the apportionment of the 11 sections.
   (b) Use the Hamilton method to apportion the sections.
   (c) Use the Jefferson method to apportion the sections.
   (d) Round the values of all of the courses' standard quotas, using a traditional rounding scheme. Is the sum of the traditionally rounded values less than, equal to, or greater than the number of sections to be apportioned?
   (e) Should the value of the modified divisor for the Webster method apportionment be less than, equal to, or greater than the standard divisor computed in part (a)? Why?
   (f) Use the Webster method to apportion the sections.
   (g) Which of the three section apportionments are the same? Which is different?
   (h) If you were a student enrolled in the poetry course at Oaks College, which apportionment method would you hope was used to apportion the sections? Why?
   (i) If you were one of the 100 students enrolled in the multicultural literature course, which apportionment method would you hope was used to apportion the sections? Why?

4. The number of rooms at each Sea Isle resort and the standard divisor are in Example 3.
   (a) Use the Webster method to apportion the 30 new sailboats. Follow the process in Exercise 1 parts (d) and (e) to find *md*.
   (b) The Jefferson apportionment of the sailboats is in Example 3. The Hamilton apportionment follows Example 3. Which sailboat apportionments are the same? Which is different?

5. The information for apportioning the Timmu House of Representatives and the Webster apportionment of the 131 seats are in Example 6.
   (a) Use the Hamilton method to apportion the seats.
   (b) Use the Jefferson method to apportion the seats.
   (c) Which Timmu House of Representatives apportionments are the same? Which are different?

6. If the leadership of Timmu uses $md = 366.97$ for the Webster method apportionment of its legislative seats, show that 132 seats, not 131, are apportioned.

7. If the leadership of Timmu uses $md = 366.99$ for the Webster method apportionment of its legislative seats, show that 130 seats, not 131, are apportioned. Use three decimal places for mQ.

8. Show that the Webster method apportionment of the original 1794 United States House of Representatives agrees with the Hamilton method apportionment in Example 2.

9. ***Assigning Nurses to Hospitals*** Forty nurses from North Carolina have volunteered to help at 5 hospitals treating the victims of an earthquake in a remote mountainous region of Mexico. The Mexican government decides to apportion the nurses based on the number of beds at each of the hospitals.

Hospital	A	B	C	D	E
Number of beds	137	237	337	455	555

(a) Find the total number of beds at the 5 hospitals and compute the standard divisor for the apportionment of the 40 nurses.
(b) Use the Hamilton method to apportion the nurses.
(c) Use the Jefferson method to apportion the nurses.
(d) Round the values of the hospitals' standard quotas, using a traditional rounding scheme. Is the sum less than, equal to, or greater than the number of nurses to be apportioned?
(e) Should the value of the modified divisor for the Webster method apportionment of the nurses be less than, equal to, or greater than the standard divisor computed in part (a)? Why?
(f) Use the Webster method to apportion the nurses.
(g) What do you notice about the three apportionments of the 40 nurses?

10. The governor of Wisconsin challenged the schoolchildren of his state to raise enough money for an additional 200 trees. He also decided to postpone the purchase and apportionment of the trees until the children met his challenge, which took them only six months. The trees were then apportioned based on the number of acres at each of the five parks, as given in Exercise 1.
(a) Find the total number of acres in the parks and compute the standard divisor for the apportionment of 439 trees.
(b) Use the Hamilton method to apportion the trees.
(c) Use the Jefferson method to apportion the trees.
(d) Round the values of the parks' standard quotas using a traditional rounding scheme. Is the sum of the traditionally rounded values less than, equal to, or greater than the number of trees to be apportioned?
(e) Should the value of the modified divisor for the Webster method apportionment be less than, equal to, or greater than the value of the standard divisor computed in part (a)? Why?
(f) Use the Webster method to apportion the trees.
(g) What do you notice about the three apportionments of 439 trees? Comment on the political wisdom of Wisconsin's governor.

11. Create a population profile for 5 states with a total population of 1000 for which both of the following conditions are true: the Hamilton method, the Jefferson method, and the Webster method apportionments of 100 legislative seats all agree; the modified divisors for the Jefferson and Webster apportionments equal the standard divisor.

12. Create an enrollment profile for 4 science courses with a total enrollment of 198 students for which both of the following conditions are true: the Hamilton method and Webster method apportionments of 9 teaching assistants to the 4 courses are the same, and the Jefferson method apportionment is different; the modified divisor for the Webster apportionment equals the standard divisor.

13. Create a ridership profile for 5 school bus routes that must service 949 students for which the following condition is true: the Hamilton method, the Jefferson method, and the Webster method apportionments of 16 school buses to the 5 routes all are different.

Adams Method *An apportionment method considered by Congress but never adopted for use was proposed by John Quincy Adams, the sixth president of the United States. The Adams Method is a divisor method. The method uses a modified divisor selected to produce modified quotas that when rounded up to the nearest integer produce a sum that is equal to the number of seats being apportioned.*

Adams Method

1. Compute md, the **modified divisor.**
2. Compute mQ, the **modified quota** for each state.
3. Round each state's modified quota mQ *up* to the nearest integer.
4. Give each state this integer number of seats.

The standard quotas rounded up to the nearest integer can automatically sum to the number of seats to be apportioned. In these cases, the Adams modified divisor md equals d, the standard divisor, and the apportionment is done. Otherwise, the modified divisor needed for an Adams method apportionment is always found by slowly increasing the value of the standard divisor.

14. Determine why the modified divisor needed for an Adams method apportionment is always found by slowly increasing the value of the standard divisor.

15. **(a)** Use the Adams method to apportion the 11 sections the English department at Oaks College is offering. See Exercise 3 for the enrollment numbers. The standard divisor is computed in Exercise 3(a).
(b) Compare the Adams method apportionment of the 11 sections with the Hamilton, Jefferson, and Webster apportionments.

16. **(a)** Use the Adams method to apportion the 40 volunteer nurses to the five hospitals in Mexico. See Exercise 9. The standard divisor is computed in Exercise 9(a).
(b) What do you notice about the Adams method apportionment of the nurses?

17. **(a)** Use the Adams method to apportion the 131 seats in Timmu's House of Representatives. See Example 6.
(b) What do you notice about the four different apportionments of Timmu's 131 legislative seats? See Exercise 5 and Example 6.

Hill-Huntington Method *Congress currently uses the Hill-Huntington method to apportion the U.S. House of Representatives. The Hill-Huntington method is also a divisor method. The method uses a modified divisor to produce modified quotas that, when rounded by the unique Hill-Huntington rounding scheme, sum to exactly the number of seats being apportioned.*

Executing the Hill-Huntington method of apportionment is very similar to executing the Webster method. First, the standard quotas Q are rounded by the Hill-Huntington rounding scheme and summed. This sum can be less than, equal to, or greater than the number of seats being apportioned. Based on the sum of the rounded values of Q, a modified divisor for the Hill-Huntington method is selected. The modified divisor is used to produce modified quotas that are rounded according to the Hill-Huntington rounding scheme and summed. If the sum of modified quotas is equal to the number of seats being apportioned, then the apportionment is complete and each state receives its rounded integer number of seats. If the sum is not equal to the number of seats being apportioned, the modified divisor is tweaked until the sum is equal to the number of seats being apportioned.

Hill-Huntington Method

1. Compute ***md***, the **modified divisor.**
2. Compute ***mQ***, the **modified quota** for each state.
3. Round each state's modified quota ***mQ*** to the nearest integer using the Hill-Huntington rounding scheme.
4. Give each state this integer number of seats.

The Hill-Huntington unique rounding scheme uses the geometric mean of two numbers. The geometric mean of two non-negative numbers a and b is the square root of the product of the two numbers, $\sqrt{a \cdot b}$. *The Hill-Huntington method uses the geometric mean of* $a =$ *the integer part of a modified quota and* $b =$ *the integer part of the modified quota* $+1$ *as the cutoff point for rounding the modified quota up to the nearest integer, instead of down to the nearest integer.*

For example, suppose a state has a modified quota of $mQ = 6.49$. *The Hill-Huntington method cutoff point for rounding this modified quota up to* 7, *the nearest integer, is the geometric mean of* 6 *and* $6 + 1 = 7$. *The geometric mean of* 6 *and* 7 *is equal to* $\sqrt{6 \cdot 7} = \sqrt{42} = 6.481$. *The value* $mQ = 6.49$ *is greater than* 6.481, *the geometric mean of* 6 *and* 7, *so the modified quota is rounded up to* 7. *Notice the modified quota* $mQ = 6.49$ *would not have been rounded up to* 7 *using a traditional rounding scheme, but is rounded up to* 7 *by the Hill-Huntington rounding scheme.*

18. Find the Hill-Huntington method cutoff point for rounding a modified quota of $mQ = 5.470$ up to 6, the nearest integer.

19. Find the Hill-Huntington cutoff point for rounding up $mQ = 56.498$.

20. Find the Hill-Huntington cutoff point for rounding up $mQ = 11.71$.

21. Find the Hill-Huntington cutoff point for rounding up $mQ = 32.497$.

22. If initially the Hill-Huntington rounded values of Q sum to less than the number of objects being apportioned, then how is the standard divisor d modified to produce the modified divisor md necessary for the Hill-Huntington method? Why?

23. If initially the Hill-Huntington rounded values of Q sum to more than the number of objects being apportioned, then how is the standard divisor d modified to produce the modified divisor md necessary for the Hill-Huntington method? Why?

24. If initially the Hill-Huntington rounded values of Q sum to exactly the number of objects being apportioned, then how is the standard divisor d modified to produce the modified divisor md necessary for the Hill-Huntington method? Why?

25. ***Comparing Different Apportionment Methods***
(a) Use the Hill-Huntington method to apportion the 11 sections the English department at Oaks College is offering. See Exercise 3. The standard divisor is computed in Exercise 3(a).
(b) Compare the Hill-Huntington apportionment of the 11 sections with the Hamilton, Jefferson, Webster, and Adams apportionments.

26. (a) Use the Hill-Huntington Method to apportion the 40 volunteer nurses to the five hospitals in Mexico. See Exercise 9. The standard divisor is computed in Exercise 9(a).
(b) Compare the Hill-Huntington apportionment of the nurses with the Hamilton, Jefferson, Webster, and Adams apportionments.

27. (a) Use the Hill-Huntington method to apportion the 131 seats in Timmu's House of Representatives. See Example 6.
(b) Compare the Hill-Huntington apportionment of the 131 seats with the Hamilton, Jefferson, Webster, and Adams apportionments.

28. Which apportionment method do you think is best? Give three reasons to support your choice.

29. Suppose the total population is 1000 and there are 100 seats to be apportioned to five states. What simple condition guarantees that the Hamilton, Jefferson, and Webster apportionments are all the same and the modified divisors for the Jefferson and Webster methods equal the standard divisor? Explain why the condition guarantees what it does.

30. What do you think is the problem with the final step of the Hamilton method of apportionment? Explain.

16.4 The Impossibilities of Apportionment

Introduction • Quota Rule • Alabama Paradox • Population Paradox • New States Paradox • Balinski and Young Impossibility Theorem

Introduction The Hamilton, Jefferson, and Webster methods of apportionment have all been used to apportion the seats in the U.S. House of Representatives. Over time, all three of these methods fell from favor. What happened in the 1822 and 1832 apportionments that made Congress abandon the Jefferson method and replace it with the Webster method? Why was the Webster method immediately replaced by the Hamilton method? How did the Hamilton method fall from favor at the end of the 1800s? Why did the Webster method make a comeback, only to be replaced by the Hill-Huntington method that is currently in favor?

These questions can be answered by considering potential pitfalls of each apportionment method. For example, the modified divisors in both the Jefferson and Webster methods can produce apportionments that violate a basic measure of fairness called the *quota rule*. This is what happened in the 1822 and 1832 Jefferson apportionments. Also, three quantities vary in the apportionment process: the number of objects being apportioned; the population on which the apportionment is based; and the number of parties getting a part of the apportionment. A change in any of the three quantities can unpredictably affect the Hamilton method apportionment. In legislative lingo, if the number of seats, the state populations, or the number of states changes, then the Hamilton method can produce apportionments that are paradoxical. All three of these quantities were dramatically changing during the last half of the 1800s. The changes naturally revealed the serious problems of the Hamilton method.

In 1842, the **U.S. House of Representatives** actually decreased in size. The apportionment for 1832 had divided 240 seats; the apportionment of 1842 divided only 223 seats. The House has had 435 seats since 1912.

Notwithstanding a direct Constitutional mandate, the **House** was not reapportioned following the 1920 census. Technically, this means that none of the business conducted by the House in the 1920s was completely honorable.

Quota Rule The Jefferson method apportionment of the House of Representatives following the 1820 census caused concern. New York had a standard quota of $Q = 32.503$, which rounds (up) to 33. However, when the seats were apportioned, New York received 34 seats, because the modified divisor used for the Jefferson method gave New York a modified quota mQ that rounded (down) to 34. This caused some alarm (because New York received more seats than its value of Q rounded up to the nearest integer), but not enough to cause rejection of the Jefferson method. According to the 1830 census, New York had a standard quota of $Q = 38.593$, which, when rounded up to the nearest integer, is 39. The Jefferson apportionment calculated a modified quota mQ for New York that rounded (down) to 40. Thus, two consecutive House apportionments using the Jefferson method had awarded the state of New York more House seats than many people believed was fair.

The Jefferson method was known to naturally favor large states such as New York. The method's partiality for large states alone had not been enough to condemn it. However, the second time the Jefferson method apportioned more seats to a state than the state's value of Q rounded up to the nearest integer, it was rejected and replaced by the Webster method. The Jefferson method failed to pass the basic test of fairness known as the *quota rule*.

Quota Rule

The integer number of objects apportioned to each state must always be either the standard quota Q rounded down to the nearest integer, or the standard quota Q rounded up to the nearest integer.

Note that an apportionment method satisfies the **quota rule** only if there is no possible circumstance under which the required condition would be violated.

A **quota method** of apportionment uses the standard divisor d and apportions seats to each state equal to the standard quota Q either rounded up or rounded down to the nearest integer. The quota rule is satisfied by any quota method of apportionment by definition.

The Hamilton method is a quota method. It computes the standard divisor and uses it to produce standard quotas. The values of Q are then rounded down and each state is guaranteed at least this number of seats. This means that no state can receive fewer seats than its value of Q rounded down to the nearest integer. Any additional seats are awarded one at a time to the states with the largest fractional parts of their values of Q. A state cannot receive more than one additional seat in the fourth step of the Hamilton method because there are never more additional seats than states. If a state does receive an additional seat, then its apportionment of seats equals its rounded-down value of Q plus one, which equals its rounded up value of Q. This means no state receives more seats than its value of Q rounded up to the nearest integer.

Historical House of Representatives apportionments of the 1820s and 1830s reveal that the Jefferson method does not satisfy the quota rule, because a state can receive more seats than its value of Q rounded up to the nearest integer. A state can also receive fewer seats than its value of Q rounded down to the nearest integer. The Webster method also fails to satisfy the quota rule. Both the Jefferson and Webster methods are **divisor methods** of apportionment; they use modified divisors, not the standard divisor. Although the modified divisors of the Jefferson and Webster methods have withstood Constitutional debate, the divisors are the reason the two methods and all divisor methods of apportionment fail to satisfy the quota rule.

TABLE 28

Factory	Cars Produced per Week
A	1429
B	3000
C	1642
D	9074
E	4382
F	1111
Total	20,638

EXAMPLE 1 Checking a Jefferson Apportionment for Quota Rule Violation

The Nexus Motor Company currently produces 20,638 automobiles per week at six different factories, as shown in Table 28. The CEO and the board want to apportion 500 new employees, using the Jefferson method, according to the weekly automobile production at each facility. Legislatively speaking, the newly hired employees are seats, the six factories are states, and the production at each factory is a state population. Complete the apportionment, and observe whether the quota rule is satisfied.

SOLUTION

1. The standard divisor is

$$d = \frac{\text{total weekly production}}{\text{number of new employees}} = \frac{20{,}638}{500} = 41.276.$$

The value of the modified divisor *md* is found by trial and error, slowly decreasing the value of the standard divisor *d*. Here, $md = 41.055$ works in the Jefferson method.

2. The modified quotas for the six factories are in Table 29. The mQ values are calculated by dividing the weekly production of each factory by the modified divisor *md*. The standard quota values are given for comparison.

TABLE 29

Factory	Cars Produced per Week	Standard Quota Q	Rounded-Down Q	Modified Quota mQ	Rounded-Down mQ
A	1429	34.621	34	34.807	34
B	3000	72.681	72	73.073	73
C	1642	39.781	39	39.995	39
D	9074	**219.837**	**219**	221.021	**221**
E	4382	106.163	106	106.735	106
F	1111	26.916	26	27.061	27
Totals	20,638	500	496		500

3. The rounded-down modified quotas are in the final column of the table. They sum to exactly 500. The rounded-down values of the standard quotas are given for comparison.

4. Each factory gets the number of new employees shown in the final column of the table.

The Jefferson method violates the quota rule in this apportionment, and therefore, the Jefferson method does *not* satisfy the quota rule. The standard quota of the largest factory, Factory D, is $Q = 219.837$. The quota rule states that the number of employees apportioned to Factory D must be either 219 or 220. The apportionment of 221 new employees to Factory D violates the quota rule.

The Hill-Huntington method, adopted in 1941 and currently used to apportion the House, is named for Joseph Hill and Edward V. Huntington. Hill was chief statistician of the Bureau of the Census. He proposed the method in 1911. Huntington, a professor of mathematics at Harvard, gave it his endorsement.

The method was studied by the National Academy of Sciences. The Academy concluded in 1929 that the Hill-Huntington method was more neutral than the other method under consideration in its treatment of both large and small states. In 1980, Balinski and Young (see the margin note on p. 1047) showed that the Academy's conclusion was false. The method does have a slight bias for small states, because of its liberal rounding scheme. Nonetheless, Hill-Huntington remains the law.

The large weekly production of Factory D in Example 1 is naturally favored by Jefferson's method. Decreasing the value of the standard divisor to find the value of *md* allows the modified quota of the large production factory to grow more rapidly than the modified quotas of the other factories.

The same kind of paradoxical apportionment is also possible when the Webster method is used. Therefore, the Webster method also fails to satisfy the quota rule. The standard divisor is always decreased to find the value of *md* required for the Jefferson method. This means the Jefferson method violates the upper bound of the quota rule if it fails to satisfy the rule. The Webster method can require the value of the standard divisor to be increased to find the value of the modified divisor. Increasing the value of the standard divisor can cause the modified quota of a small state to decrease more rapidly than the modified quotas of other states. The rapid decrease can mean the small state receives an apportionment that is smaller than its value of Q rounded down. It is this possibility of a quota rule violation that taints the Jefferson and Webster methods.

Alabama Paradox The required tweaking of the standard divisor in any divisor method is an open invitation for quota violations, especially if one state's population is large (or small) in relation to other state populations. The Hamilton method and all quota methods automatically satisfy the quota rule by their definition. The Hamilton method would be the method of choice if the quota rule was the sole test of fairness for an apportionment method, but it is not the only such benchmark.

As the population of the United States grew and additional states joined the union, the size of the U.S. House of Representatives slowly increased from the original 105 seats to the current 435 seats. Preparations for increasing the size of the House following the 1880 census revealed a surprising flaw in the Hamilton method. Government workers calculated the different Hamilton method apportionments for all the different House sizes from 275 to 350 seats. This was quite an amazing feat, considering spreadsheet programs were not available and all calculations were done manually. The sample apportionments showed that if the House had 299 seats, then Alabama would receive 8 seats. However, if the House size was 300, then Alabama would receive 7 seats. This odd result is known as the *Alabama paradox*.

Alabama Paradox

The situation in which an increase in the number of objects being apportioned actually forces a state to lose one of those objects is known as the **Alabama paradox.**

TABLE 30

School	Enrollment
Applegate	335
Bayshore	456
Claypool	298
Delmar	567
Edgewater	607
Total	2263

EXAMPLE 2 Checking a Hamilton Apportionment for Alabama Paradox Occurrence

A longtime resident of Highwood has given one additional computer to the school district. The gift brings the total number of computers to be apportioned by the Hamilton method to 110. School enrollments are shown in Table 30. Complete the apportionment and observe whether the Alabama paradox has occurred.

SOLUTION

1. Total enrollment in the school district is 2263, and now 110 computers are to be apportioned. The standard divisor is

$$d = \frac{\text{total enrollment}}{\text{number of computers}} = \frac{2263}{110} = 20.573.$$

2. Standard quotas for each school are computed by dividing the enrollment for the school by d. Enrollments and the standard quotas for the five schools are shown in Table 31.

TABLE 31

School	Enrollment	Standard Quota Q	Rounded-Down Q	110 Computers Apportioned	109 Computers Apportioned
Applegate	335	16.28	16	16	16
Bayshore	456	22.16	22	22	22
Claypool	298	14.49	14	14	15
Delmar	567	27.56	27	28	27
Edgewater	607	29.50	29	30	29
Totals	2263	110	108	110	109

3. The standard quotas are rounded down to the nearest integer in the fourth column. Each school receives at least this integer number of computers. The values of Q sum to 108, so two additional computers are to be apportioned in the fourth step of Hamilton's method.
4. The two additional computers are apportioned one at a time to the two schools with the largest fractional parts of their standard quotas. Delmar School gets the first additional computer, and Edgewater School gets the other. The final apportionment of the 110 computers is in the fifth column of the table. For comparison, the last column of the table shows the previous Hamilton apportionment of 109 computers of Example 1 of Section 16.3. (See that example for details of that apportionment.)

Comparing the last two columns of the table shows that the gift of the additional computer has caused the Alabama paradox to occur. With the Hamilton method, Claypool School receives 15 computers when 109 computers are apportioned and only 14 when 110 are apportioned. The increase in the number of computers being apportioned actually forces Claypool School to forfeit one of its machines. ■

The paradoxical apportionment in Example 2 is caused by the pragmatic fourth step in the Hamilton method. Additional seats are distributed to the states with the greatest fractional parts of their standard quotas. And if the number of items to be apportioned changes, these fractional parts can shift erratically, possibly causing strange phenomena such as the Alabama paradox.

Population Paradox The quota rule is an extremely important measure of fairness for an apportionment method. The Hamilton method always satisfies the quota rule, even if it does allow the possibility of the Alabama paradox. The Alabama paradox can occur only when the number of objects being apportioned *increases.* If the number of objects being apportioned is permanently *fixed,* then the Hamilton method appears to shine again. The size of the U.S. House of Representatives has been fixed at 435 seats for many years. Why are the seats still apportioned with a divisor method that invites quota rule violations? What keeps the Hamilton method from really shining as the apportionment method of choice? The Hamilton method allows for paradoxical outcomes other than the Alabama paradox. One of them, discovered around 1900, is known as the *population paradox.*

Population Paradox

The **population paradox** occurs when, based on updated population figures, a reapportionment of a fixed number of seats causes a state to lose a seat to another state, although the percent increase in the population of the state that loses the seat is greater than the percent increase in the population of the state that gains the seat.

EXAMPLE 3 Checking a Hamilton Apportionment for Population Paradox Occurrence

TABLE 32

School	Number of Interviews
Lincoln	55
St. Francis	125
Marshall	190
Total	370

TABLE 33

School	Revised Number of Interviews
Lincoln	60
St. Francis	145
Marshall	210
Total	415

Eleven students in a graduate school education class plan to interview 370 educators at three different grammar schools, as shown in Table 32. They are interested in how art and music education impact scores on standardized mathematics exams. The students decide to apportion themselves into three interview teams using the Hamilton method, based on the number of educators to be interviewed at each school. In legislative lingo, the interviewers are comparable to House seats; they are what is being apportioned. The three schools are states and the number of interviews to be conducted at each school is a state population.

A month before the interviews are scheduled, the graduate students decide to include additional interviews with retired educators from the three schools. Five of the retired educators are from Lincoln, 20 are from St. Francis, and 20 are from Marshall. The interviewers must now talk with $370 + 45 = 415$ people. They will now re-apportion themselves using the Hamilton method, based on the revised numbers shown in Table 33.

Complete both apportionments and observe whether the population paradox has occurred.

SOLUTION

The initial Hamilton apportionment proceeds as follows.

1. Eleven graduate students are to interview 370 educators. The standard divisor is

$$d = \frac{\text{total number of interviews}}{\text{number of interviewers}} = \frac{370}{11} = 33.64.$$

2. Standard quotas are computed by dividing the number of interviews at each school by $d = 33.64$. The values of Q are shown in the third column of Table 34 on the next page.

3. The values of Q are rounded down in the fourth column. At least this many graduate students will conduct the interviews at each school. The sum of the rounded Q values is 9, so two additional interviewers are to be assigned in the next step of the Hamilton apportionment.

TABLE 34

School	Number of Interviews	Standard Quota Q	Rounded-Down Q	Interviewers Apportioned
Lincoln	55	1.635	1	1
St. Francis	125	3.716	3	**4**
Marshall	190	5.648	5	**6**
Totals	370	11	9	11

4. The two schools with the largest fractional parts of their values of Q are St. Francis (.716) and Marshall (.648). The final apportionment of the interviewers is in the last column of the table. Bold numbers in the last column indicate that St. Francis and Marshall get more interviewers than their value of Q rounded down.

With additional interviews, the Hamilton reapportionment proceeds as follows.

1. The new standard divisor is

$$d = \frac{415}{11} = 37.73.$$

2. The values of Q are recalculated by dividing the new number of interviews at each school by the new standard divisor $d = 37.73$. See the third column of Table 35.

TABLE 35

School	Number of Interviews	Standard Quota Q	Rounded-Down Q	Interviewers Apportioned
Lincoln	60	1.59	1	**2**
St. Francis	145	3.84	3	**4**
Marshall	210	5.57	5	5
Totals	415	11	9	11

3. The values of Q are rounded down in the fourth column. They sum to 9, leaving two additional interviewers to be apportioned in the final step.
4. The two schools with the largest fractional parts of their Q values are St. Francis (.84) and Lincoln (.59). Bold entries in the last column indicate these two schools get more interviewers than their values of Q rounded down. The revised apportionment of the interviewers is in the last column.

The original Hamilton apportionment gives one additional interviewer to each of the St. Francis and Marshall interview teams. An additional interviewer still is assigned to the St. Francis team by the revised apportionment, but the second additional interviewer is assigned to the Lincoln interview team, rather than Marshall.

Comparing the growth rates of the number of interviews at Lincoln and Marshall reveals that the team of graduate student interviewers has suffered the population paradox. The addition of the retired educators to the interview list causes the number of interviews at Marshall to increase by $\frac{210 - 190}{190} = 10.53\%$, and the number of interviews at Lincoln to increase by $\frac{60 - 55}{55} = 9.09\%$. The percent increase in the number of interviews to be conducted at Marshall is greater than the percent increase in the number of interviews at Lincoln, yet in the final Hamilton method apportionment the Marshall interview team loses a member to the Lincoln interview team.

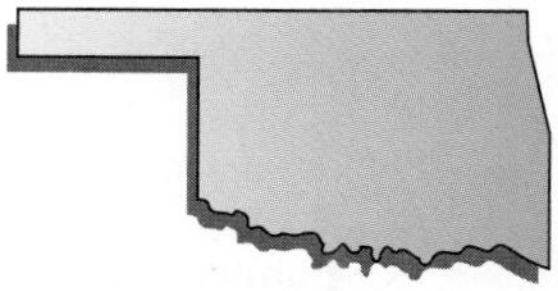

The **new states paradox** occurred in 1907, when Oklahoma joined the Union as the 46th state. Remember this fact when you find a new commemorative quarter for Oklahoma (issued in 2008) among the coins in your pocket.

New States Paradox The Hamilton method of apportionment allows the population paradox for the same reason it allows the Alabama paradox. Both paradoxes are triggered by the pragmatic fourth step of the method, which awards the unapportioned seats based on the fractional parts of the values of Q, the standard quotas. Paradoxes occur because the changes in the fractional parts of Q, due to an increased number of seats or revised populations, are erratic at best. If the surplus seats are distributed based on erratic values, then unsettling results are bound to occur. Another paradox triggered by the pragmatic fourth step of the Hamilton method is called the *new states paradox.* The new states paradox was discovered by Balinski and Young. (See p. 1047.)

If a new state is added to the existing states, then the number of seats or objects being apportioned is usually increased to prevent a decrease in the existing apportionments. The purpose of the increase is to keep all existing apportionments the same, while adding the new state and its fair share of seats to the mix. The new states paradox may result from such an addition.

New States Paradox

The **new states paradox** occurs when a reapportionment of an increased number of seats, necessary due to the addition of a new state, causes a shift in the apportionment of the original states.

TABLE 36

Community	Population
Original	8500
Annexed	1671
Total	10,171

EXAMPLE 4 Checking a Hamilton Apportionment for New States Paradox Occurrence

Several years ago the town of Lake Bluff annexed a small unincorporated subdivision. The population of the original town was 8500 and the annexed subdivision had a population of 1671. (See Table 36.) The new town was governed by 100 council members. The Hamilton method was used to apportion council seats to the new town.

During the following years, council members from the annexed subdivision lobbied intensely and tirelessly. They finally convinced the town to annex a second small unincorporated subdivision with a population of 545. Populations of the original town and the first annexed subdivision did not change. The town council voted to increase the number of council seats from 100 to 105. They reasoned that if the established standard divisor of $d = 101.71$ was used, then a population of 545 would have a standard quota of $Q = \frac{545}{101.71} = 5.36$, which rounds down to 5.

TABLE 37

Community	Population
Original	8500
1st Annexed	1671
2nd Annexed	545
Total	10,716

The new situation is reflected in Table 37. Complete the Hamilton apportionment after the first annexation, and then again after the second annexation. Observe whether the new states paradox has occurred.

SOLUTION

The apportionment following the first annexation proceeds as follows.

1. The total population of the new community was 8500 + 1671 = 10,171, and 100 council seats were available. This means the standard divisor is

$$d = \frac{\text{total population}}{\text{number of seats}} = \frac{10{,}171}{100} = 101.71.$$

2. The standard quotas for the original and annexed populations are in the third column of Table 38.

TABLE 38

Community	Population	Standard Quota Q	Rounded-Down Q	Council Seats Apportioned
Original	8500	83.57	83	84
Annexed	1671	16.43	16	16
Totals	10,171	100	99	100

Hayes or Tilden? Congress decided to divide 283 seats following the 1870 census. They selected this number of seats because with it the official apportionment done by the Hamilton method agreed with the theoretical outcome of the Webster method. A power grab on the floor of the House resulted in the apportionment of 292 seats, without using the lawful Hamilton method. It was this unconstitutional House that gave Rutherford B. Hayes enough electoral votes to become president. If the Hamilton method had been used, Samuel J. Tilden, who had won a majority of the national votes, would have had enough electoral votes to become president.

3. The rounded-down values of Q sum to 99, leaving one seat to be apportioned.
4. The additional council seat is awarded to the original town. Its fractional part of Q is .57, which is greater than .43, the annexed population's fractional part of Q. The apportionment of the 100 seats is in the final column of the table.

An updated Hamilton method apportionment of the 105 council seats shows why the first annexed subdivision lobbied so intensely for the annexation of another subdivision. The updated apportionment was done with a revised value of the standard divisor d, reflecting the increased total population of Lake Bluff and the increased number of council seats.

1. The updated total population was 8500 + 1671 + 545 = 10,716. The revised standard divisor is

$$d = \frac{10{,}716}{105} = 102.057.$$

2. The updated values of Q, the standard quotas, are in the third column of Table 39.

TABLE 39

Community	Population	Standard Quota Q	Rounded-Down Q	Council Seats Apportioned
Original	8500	83.29	83	83
1st Annexed	1671	16.37	16	17
2nd Annexed	545	5.34	5	5
Totals	10,716	105	104	105

3. The rounded-down values of Q sum to 104, leaving one additional council seat to be apportioned.
4. The fractional part of Q for the first annexed subdivision is .37, which is greater than the fractional parts of the values of Q for the other two populations. It receives the additional available seat. The reapportionment of the 105 council seats is in the final column of the table.

Comparing the two apportionments shows that annexing a second subdivision with a population of 545 and adding new seats to the council causes the new states paradox to occur. The shift in the apportionment for the original two communities favors the first annexed subdivision. The first apportionment splits 100 council seats, giving 84 seats to the original town and 16 seats to the first annexed subdivision. The second apportionment again splits 100 seats between the original town and the first annexed subdivision, but the apportionment shifts, giving only 83 seats to the original town and 17 seats to the first annexed subdivision. ■

Balinski and Young Impossibility Theorem As mentioned previously, the Jefferson and Webster methods are examples of divisor methods. All divisor methods are afflicted by the possibility of a quota rule violation, since they depend on a modified standard divisor—not the standard divisor. Modifying the standard divisor is necessary to force the modified quotas, rounded according to the prescribed scheme, to sum to exactly the number of seats being apportioned.

The Hamilton method is an example of a quota method. The Hamilton method uses the standard divisor, so it avoids the possibility of a quota rule violation. However, it allows puzzling apportionments. The pragmatic final step of the Hamilton method permits the possibility of the Alabama paradox, the population paradox, or the new states paradox. In other words, increases in the number of seats to be apportioned, changes in population figures, or increases in the number of states can cause the Hamilton method to go awry. An unexpected 1980 discovery by Michel Balinski and Peyton Young is hinted at in the following Table 40, which is a summary of fairness standards relative to the three apportionment methods discussed above.

Michel L. Balinski is a mathematician at CNRS and Ecole Polytechnique (Paris). **H. Peyton Young** is a mathematical economist at Johns Hopkins University. In addition to their important impossibility theorem and results about the Hill-Huntington method, they have shown that the Webster method is the only divisor method of apportionment that is not biased to either large or small states.

TABLE 40 Summary of Fairness Standards and Which Apportionment Methods Meet Them

	Fairness Standard			
Method	**Quota Rule**	**Alabama Paradox**	**Population Paradox**	**New States Paradox**
Hamilton method	Satisfied	Can occur	Can occur	Can occur
Jefferson method	Not satisfied	Cannot occur	Cannot occur	Cannot occur
Webster method	Not satisfied	Cannot occur	Cannot occur	Cannot occur

Scholars and politicians have always expected that mathematicians would eventually find the perfect apportionment method. Instead, mathematicians Balinski and Young showed that a perfectly satisfactory method was impossible.

Balinski and Young Impossibility Theorem

Any apportionment method that satisfies the quota rule will, by its nature, permit the possibility of a paradoxical apportionment. Likewise, any apportionment method that does not permit the possibility of a paradoxical apportionment will, by its nature, fail to satisfy the quota rule.

Balinski and Young, like Kenneth Arrow (see the margin note on p. 1017), answered an age-old question of great social importance. Unfortunately, even very talented mathematicians cannot find solutions that do not exist. Fortunately, they can tell the world when perfection does not exist and can begin developing ways of coping with unavoidable imperfection.

16.4 EXERCISES

Quota Rule Violations with the Jefferson Method *In each case, show that if the Jefferson method of apportionment is used, then a violation of the quota rule occurs.*

1. 132 seats are apportioned.

State	a	b	c	d
Population	17,179	7500	49,400	5824

2. 200 seats are apportioned.

State	a	b	c	d
Population	67,000	35,000	15,000	9900

3. 290 seats are apportioned.

State	a	b	c	d	e
Population	2567	1500	8045	950	1099

4. 150 seats are apportioned.

State	a	b	c	d	e
Population	1720	3363	6960	24,223	8800

Alabama Paradox with the Hamilton Method *In each case, show that if the Hamilton method of apportionment is used, then the Alabama paradox occurs.*

5. Seats increase from 204 to 205.

State	a	b	c	d
Population	3462	7470	4265	5300

6. Seats increase from 71 to 72.

State	a	b	c	d
Population	1050	2040	3060	4050

7. Seats increase from 126 to 127.

State	a	b	c	d	e
Population	263	808	931	781	676

8. Seats increase from 45 to 46.

State	a	b	c	d	e
Population	309	289	333	615	465

Population Paradox with the Hamilton Method *In each case, show that the population paradox occurs when the Hamilton method is used for both initial and revised populations. (Populations are given in thousands.)*

9. 11 seats are apportioned.

State	a	b	c
Initial population	55	125	190
Revised population	61	148	215

10. 11 seats are apportioned.

State	a	b	c
Initial population	55	125	190
Revised population	62	150	218

11. 13 seats are apportioned.

State	a	b	c
Initial population	930	738	415
Revised population	975	750	421

12. 13 seats are apportioned.

State	a	b	c
Initial population	89	125	225
Revised population	97	145	247

New States Paradox with the Hamilton Method *In each case, use the Hamilton method to apportion legislative seats to two states with the given populations. Show that the new states paradox occurs if a new state with the indicated population and the appropriate number of additional seats is included in a second Hamilton method apportionment. The appropriate number of seats to add for the second apportionment is the state's standard quota of the new state, computed using the original two-state standard divisor, rounded down to the nearest integer. The populations are given in hundreds.*

13. 75 seats are apportioned.

State	Original State a	Original State b	New State c
Population	3184	8475	330

14. 75 seats are apportioned.

State	Original State a	Original State b	New State c
Population	3184	8475	350

15. 83 seats are apportioned.

State	Original State a	Original State b	New State c
Population	7500	9560	1500

16. 83 seats are apportioned.

State	Original State a	Original State b	New State c
Population	7500	9560	1510

17. ***Playing Politics with Apportionment*** In Example 4, council members from the first subdivision annexed to the town of Lake Bluff lobbied for the annexation of a second unincorporated subdivision with a population of 545 because they had realized it would cause the new states paradox to occur. The first annexed subdivision took one of the original 100 council seats away from the original town when 5 new seats were included in an updated Hamilton method apportionment. If the population of the second subdivision had been only 531, why would it have been prudent for the first subdivision to wait until a baby was born and the population increased to 532?

18. Consider the situation described in Exercise 17. If, by strange coincidence, the population of the second subdivision had been 548, why would it have been urgent for the first subdivision to act before a baby was born and the population increased to 549?

19. Show that if the Adams method is used to apportion 220 legislative seats to the five states with the population profile given in Exercise 4, then a violation of the quota rule occurs.

20. Show that if the Adams method is used to apportion 219 legislative seats to the five states with the population profile given in Exercise 4, then a violation of the quota rule occurs.

21. Knowing the pros and cons of the Hamilton, Jefferson, and Webster methods, would you change your mind about which apportionment method you think is the best? Explain.

22. Which impossibility of apportionment—a quota rule violation, the Alabama paradox, the population paradox, or the new states paradox—do you find most disturbing? Explain.

23. Refer to Exercise 22. Which impossibility of apportionment do you find least disturbing? Explain.

24. In your own words, describe the problem with the final step of the Hamilton method of apportionment.

25. Why is the Adams method of apportionment tainted by the possibility of a quota rule violation? What kind of a quota rule violation can occur if the Adams method is used? Why?

26. Is the Hill-Huntington method of apportionment tainted by the possibility of a quota rule violation? Explain. If quota rule violations can occur, then describe them.

COLLABORATIVE INVESTIGATION

Class Favorites, an Election Exploration

Part One: To be completed following Section 16.1: The Possibilities of Voting

Divide the class into groups of six students each. Within each group designate three teams A, B, and C of two students each. Do the following activities in order.

1. Each group should select a different collection of four or five candidates/alternatives, from which they want to learn the class favorite. Suggestions include candidates for a student government post, movies nominated for "Best Picture," songs nominated in any of the Grammy categories, the best *Lord of the Rings* movie, or candidates in a presidential primary.

2. Groups should prepare ballots for a class vote on their selected collection of candidates. The ballots must provide a list of all of the $n!$ possible rankings of the $n = 4$ or 5 candidates. Groups should distribute a ballot to each member of the class.

3. Each member of the class should mark the complete ranking representing his or her arrangement of the particular collection of candidates. Ballots should then be returned to the appropriate groups.

4. Groups should create a voter profile of the class based on the returned ballots. To do this, simply tally how many members of the class selected each of the $n!$ rankings. Organize the profile into a table showing how many voters preferred each ranking. The voter profile will be needed in Part Two of this Collaborative Investigation.

5. Teams should determine the outcome of the election using different voting methods.
   Team A: Find the outcome of the election using the plurality method. Also, find the outcome of the election using the Borda method.
   Team B: Find the outcome of the election using the pairwise comparison method. Keep a careful record of the pairwise results; it will be useful for Part Two of the Collaborative Investigation.
   Team C: Find the outcome of the election using the Hare method.

6. When the three teams have completed their individual tasks in Item 5, reassemble as a six-student group.
   - Team A should report the winner of the plurality election and the winner of the Borda election. Team B should report the winner of the pairwise comparison method election. Team C should report the outcome of the Hare election.
   - Prepare a chart showing which candidate was selected by each method: plurality, pairwise comparison, Borda, and Hare.
   - If the outcomes of the various elections show agreement, study the voter profile and try to explain why the methods agree. If the outcomes of the various elections do not agree, try to decide what it is about the profile of voters that causes the disagreement.

7. Choose a representative from each group to report to the entire class on the election the group conducted.

Part Two: To be completed following Section 16.2: The Impossibilities of Voting

Reassemble the original groups of six students (divided into three teams). Do the following activities in order.

1. Study the voter profile to find out if there is a majority candidate. Use the notes from Team B to decide if a Condorcet candidate exists for the voter profile.

2. Determine which of the plurality, pairwise comparison, Borda, and Hare methods satisfy the majority criterion and Condorcet criterion for the group's voter profile. In other words, which methods elect the majority candidate, if one exists, and which do not? Do the same for the Condorcet candidate, if one exists.

3. Now investigate the outcomes of a second election, after a losing candidate has been removed from the group's collection. Team members should consider the irrelevant alternatives criterion.
   Team A: Remove a candidate who did not win the plurality election from the collection. Determine the winner of the remaining candidates, using the plurality method. Remove a candidate who did not win the

Borda election from the collection. Determine the winner of the remaining candidates using the Borda method. If there is extra time, try removing a different loser from an election or remove more than one loser.
Team B: Remove a candidate who did not win the pairwise comparison election from the collection. Determine the winner of the remaining candidates using the pairwise comparison method. Notes from Part One are helpful. If time permits, remove a different loser and recalculate the pairwise winner.
Team C: Remove a candidate who did not win the Hare election from the collection. Determine the winner of the remaining candidates using the Hare method. If time permits, remove a different loser and conduct another Hare election.

4. When all the team members have completed their individual tasks in Item 3, return to a group discussion.
- Team A should report whether it can demonstrate that the plurality method does not satisfy the irrelevant alternatives (IA) criterion. It also should report whether it can show that the Borda method does not satisfy the IA criterion.
- Team B should report whether it can show that the pairwise comparison method does not satisfy the IA criterion.
- Team C should report whether it can show that the Hare method does not satisfy the IA criterion.

5. Choose a representative from each group to report to the class what the group showed about the majority criterion, the Condorcet criterion, and the IA criterion.

CHAPTER 16 TEST

Twenty-seven members of a sorority are debating where to spend spring break. They have narrowed the list of possibilities to Aruba a, the Bahamas b, Cancún c, or the Dominican Republic d. Each of the 27 sisters gave a complete ranking of the four choices on a separate ballot. The voter profile of the sorority members is summarized in the table.

Number of Voters	Ranking
5	a > b > d > c
6	b > c > a > d
5	c > d > b > a
7	d > a > b > c
4	c > a > d > b

Determine the spring break destination of the sorority using the

1. plurality method.

2. Borda method.

3. Hare method.

4. Why would the pairwise comparison method be a poor choice of voting method for the sorority destination decision?

5. How many different complete rankings are possible in an election with 7 candidates?

6. How many comparisons must be considered in a pairwise comparison election with 10 alternatives?

7. What is the majority criterion?

8. What is the Condorcet criterion?

9. Explain the difference between the monotonicity criterion and the alternative version of the monotonicity criterion. In your opinion, which is a stricter criterion?

10. Why is the irrelevant alternatives criterion an important measure of fairness in a two-stage election process?

11. Show that the Borda method violates the majority criterion for the following voter profile.

Number of Voters	Ranking
16	a > b > c
8	b > c > a
7	c > b > a

12. Use the pairwise comparison method to determine which alternative is preferred by the committee with the voter profile on the next page.

Number of Voters	Ranking
5	a > c > b
6	c > b > a
4	b > a > c
6	b > c > a

13. Consider the voter profile of Exercise 12. Which alternative is a Condorcet candidate? Find the preferred alternative of the committee by using each of the plurality, Borda, and Hare methods. Which voting methods violate the Condorcet criterion and which uphold it?

14. Use the pairwise comparison method to determine the preferred alternative for the following profile of voters.

Number of Voters	Ranking
8	c > m > b > s
7	s > b > m > c
6	s > b > c > m
5	m > c > b > s

Use the pairwise comparison method to redetermine the preferred alternative for the voter profile, if the five voters with the last ranking rearrange their ranking to be c > s > m > b. What do the outcomes of the two pairwise comparison method selections show about the voting method?

15. Use the Hare method to determine the preferred alternative for the following profile of voters.

Number of Voters	Ranking
9	a > c > b > d
8	b > d > a > c
7	c > d > b > a
5	c > a > b > d

Use the Hare method to redetermine the preferred alternative for the voter profile, if the five voters with the last ranking rearrange their ranking to be a > c > b > d. What do the outcomes of the two Hare method selections show about the voting method?

16. Use the plurality method to determine the preferred alternative for the following profile of voters.

Number of Voters	Ranking
10	a > b > c
7	b > c > a
5	c > b > a

Use the plurality method to redetermine the preferred alternative for the voter profile, if alternative c is dropped from the selection process. What do the outcomes of the two plurality method selections show about the voting method?

17. Use the Borda method to determine the preferred alternative for the following profile of voters.

Number of Voters	Ranking
7	a > b > c > d
5	b > c > d > a
4	d > c > a > b

Use the Borda method to redetermine the preferred alternative for the voter profile, if alternative d is dropped from the selection process. What do the outcomes of the two Borda method selections show about the voting method?

18. Use the Hare method to determine the preferred alternative for the profile of voters in Exercise 17. Use the Hare method to redetermine the preferred alternative for the voter profile, if alternative b is dropped from the selection process. What do the outcomes of the two Hare method selections show about the voting method?

19. Who is Kenneth J. Arrow? Explain in your own words what Arrow's impossibility theorem says about voting methods.

Apportioning Representatives to the Wards of a City *The city of Smithapolis, population 2,600,700, is divided into five wards and governed by a city council of 195 representatives. The 195 council seats are apportioned to the five wards based on their populations. A recent city census reported the following populations (in hundreds) for the five wards.*

Ward	1st	2nd	3rd	4th	5th
Population	1429	8639	7608	6660	1671

Apportion the 195 council seats of Smithapolis using the

20. Hamilton method.

21. Jefferson method.

22. Webster method.

23. What is the quota rule?

24. Explain the Alabama paradox.

25. Explain the population paradox.

26. Explain the new states paradox.

27. ***Quota Rule Violation with the Jefferson Method*** Show that if the Jefferson method is used to apportion 100 legislative seats to four states with the given populations, then a violation of the quota rule occurs. The populations are given in hundreds.

State	a	b	c	d
Population	2354	4500	5598	23,000

28. ***Alabama Paradox with the Hamilton Method*** Show that if the Hamilton method is used to apportion legislative seats to five states with the given populations, then the Alabama paradox occurs when the number of seats being apportioned increases from $n = 126$ to $n = 127$. The populations are given in thousands.

State	a	b	c	d	e
Population	263	809	931	781	676

29. ***Population Paradox with the Hamilton Method*** Use the Hamilton method to apportion 11 legislative seats to three states with the given initial populations. Show that the population paradox occurs when the Hamilton method is used to apportion the 11 legislative seats again, after the population figures are revised by a census. The populations are given in thousands.

State	a	b	c
Initial population	55	125	190
Revised population	63	150	220

30. ***New States Paradox with the Hamilton Method*** Use the Hamilton method to apportion 100 legislative seats to two states with the given populations. Show that the new states paradox occurs if a new state with the indicated population and the appropriate number of additional seats is included in the second Hamilton method apportionment. The appropriate number of seats to add for the second apportionment is the state's standard quota of the new state, computed using the original two-state standard divisor, rounded down to the nearest integer. The populations are given in thousands.

State	Original State a	Original State b	New State c
Population	49	160	32

31. Who are Balinski and Young? In your own words, explain what the Balinski and Young impossibility theorem says about apportionment methods.

APPENDIX

THE METRIC SYSTEM

Joseph Louis Lagrange (1736–1813) was born in Turin, Italy, and became a professor at age 19. In 1776 he came to Berlin at the request of Frederick the Great to take the position Euler left. A decade later Lagrange settled permanently in Paris. Napoleon was among many who admired and honored him.

Lagrange's greatest work was in the theory and application of **calculus.** He carried forward Euler's work of putting calculus on firm algebraic ground in his theory of functions. His *Analytic Mechanics* (1788) applied calculus to the motion of objects. Lagrange's contributions to algebra had great influence on Galois and hence the theory of groups. He also wrote on number theory; he proved, for example, that every integer is the sum of at most four squares. His study of the moon led to methods for finding longitude.

The metric system was developed by a committee of the French Academy just after the French Revolution of 1789. The president of the committee was the mathematician Joseph Louis Lagrange.

The advantages of the metric system can be seen when compared to our English system. In the English system, one inch is one-twelfth of a foot, while one foot is one-third of a yard. One mile is equivalent to 5280 feet, or 1760 yards. Obviously, there is no consistency in subdivisions. In the metric system, prefixes are used to indicate multiplications or divisions by powers of ten. For example, the basic unit of length in the metric system is the *meter* (which is a little longer than one yard). To indicate one thousand meters, attach the prefix **"kilo-"** to get **kilo**meter. To indicate one one-hundredth of a meter, use the prefix **"centi-"** to obtain **centi**meter. A complete list of the prefixes of the metric system is shown in Table 1 below, with the most commonly used prefixes appearing in heavy type.

TABLE 1 Metric Prefixes

Prefix	Multiple	Prefix	Multiple
exa	1,000,000,000,000,000,000	deci	.1
peta	1,000,000,000,000,000	**centi**	.01
tera	1,000,000,000,000	**milli**	.001
giga	1,000,000,000	micro	.000001
mega	1,000,000	nano	.000000001
kilo	1000	pico	.000000000001
hecto	100	femto	.000000000000001
deka	10	atto	.000000000000000001

Length and Area

Lagrange urged the committee devising the metric system to find some natural measure for length from which weight and volume measures could be derived. It was decided that one **meter (m)** would be the basic unit of length, with a meter being defined as one ten-millionth of the distance from the equator to the North Pole.

To obtain measures longer than one meter, Greek prefixes were added. For measures smaller than a meter, Latin prefixes were used. A meter is a little longer than a yard (about 39.37 inches). A **centimeter (cm)** is one one-hundredth of a meter and is about $\frac{2}{5}$ of an inch. See Figure 1.

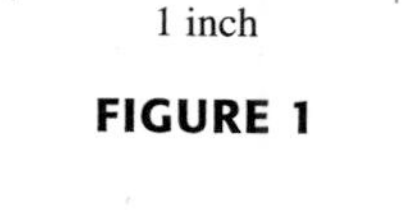

FIGURE 1

A Comparison of Distances

Length in Meters	Approximate Related Distances
10^{19}	Distance to the North Star
10^{12}	Distance of Saturn from the sun
10^{11}	Distance of Venus from the sun
10^{9}	Diameter of the sun
10^{8}	Diameter of Jupiter
10^{7}	Diameter of Earth; distance from Washington, D.C. to Tokyo
10^{6}	Distance from Chicago to Wichita, Kansas
10^{5}	Average distance across Lake Michigan
10^{4}	Average width of the Grand Canyon
10^{3}	Length of the Golden Gate Bridge
10^{2}	Length of a football field
10^{1}	Average height of a two-story house
10^{0}	Width of a door
10^{-1}	Width of your hand
10^{-2}	Diameter of a piece of chalk
10^{-3}	Thickness of a dime
10^{-4}	Thickness of a piece of paper
10^{-5}	Diameter of a red blood cell
10^{-7}	Thickness of a soap bubble
10^{-8}	Average distance between molecules of air in a room
10^{-9}	Diameter of a molecule of oil
10^{-14}	Diameter of an atomic nucleus
10^{-15}	Diameter of a proton

Because the metric system is based on decimals and powers of ten, conversions within the system involve multiplying and dividing by powers of ten. For example, to convert 2.5 m to centimeters, multiply 2.5 by 100 (since 100 cm = 1 m) to obtain 250 cm. On the other hand, to convert 18.6 cm to meters, divide by 100 to obtain .186 m. Other conversions are made in the same manner, using the meanings of the prefixes. Why is 42 m equal to 42,000 millimeters (mm)?

Long distances usually are measured in kilometers. A **kilometer (km)** is 1000 meters. (According to a popular dictionary, the word *kilometer* may be pronounced with the accent on either the first or the second syllable. Scientists usually stress the second syllable.) A kilometer is equal to about .6 mile. Figure 2 shows the ratio of 1 kilometer to 1 mile.

Conversions from meters to kilometers, and vice versa, are made by multiplying or dividing by 1000 as necessary. For example, 37 kilometers equals 37,000 meters, while 583 meters equals .583 km.

The area of a figure can be measured in square metric units. Figure 3 shows a square that is 1 cm on each side; thus, it is a **square centimeter (cm^2).** One square meter (m^2) is the area of a square with sides one meter long. How many cm^2 are in one m^2?

According to Paul G. Hewitt, in *Conceptual Physics*, 7th edition (HarperCollins, 1993):

> The distance from the equator to the North Pole was thought at the time to be close to 10,000 kilometers. One ten-millionth of this, the meter, was carefully determined and marked off by means of scratches on a bar of platinum-iridium alloy. This bar is kept at the International Bureau of Weights and Measures in France. The standard meter in France has since been calibrated in terms of the wavelength of light—it is 1,650,763.73 times the wavelength of orange light emitted by the atoms of the gas krypton-86. The meter is now defined as being the length of the path traveled by light in a vacuum during a time interval of 1/299,792,458 of a second.

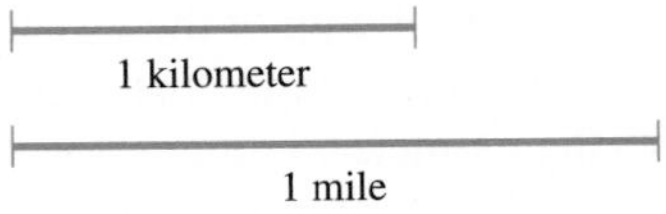

Observe the ratio.

FIGURE 2

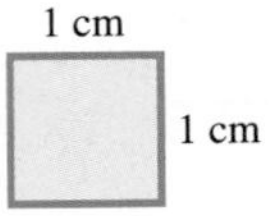

Area is 1 cm^2.

FIGURE 3

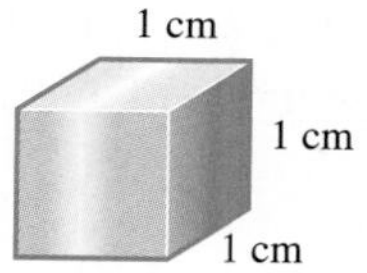

Volume is 1 cm^3.

FIGURE 4

In September 1999, the ***Mars Climate Orbiter*** was lost due to "a failure to use metric units in the coding of a ground software file. . . ." According to Edward Weiler, associate administrator for NASA's Office of Space Science, the metric conversion error that led to the loss "should have been caught five ways to Sunday," but it wasn't. (*Source:* "NASA's Mars Losses Spark Anger and Opportunity" by Leonard David, Washington Contributing Editor to www.space.com.)

Volume, Mass, and Weight

The volume of a three-dimensional figure is measured in cubic units. If, for example, the dimensions are given in centimeters, the volume may be determined by the appropriate formula from geometry, and it will be in **cubic centimeters (cm^3).** See Figure 4 for a sketch of a box whose volume is one cm^3.

In the metric system, one **liter (l)** is the quantity assigned to the volume of a box that is 10 cm on a side. (See Figure 5.) A liter is a little more than a quart. (See Figure 6.) Notice the advantage of this definition over the equivalent one in the English system—using a ruler marked in centimeters, a volume of 1 liter (symbolized 1 L) can be constructed. On the other hand, given a ruler marked in inches, it would be difficult to construct a volume of 1 quart.

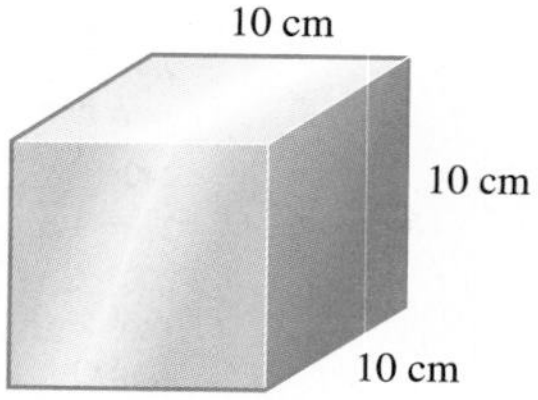

The volume of this box is 1 liter.

FIGURE 5

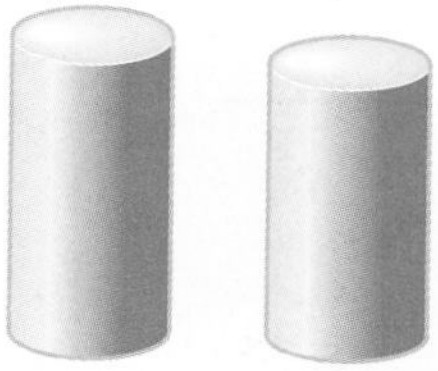

FIGURE 6

The prefixes mentioned earlier are used throughout the metric system, so one **milliliter (ml)** is one one-thousandth of a liter, one **centiliter (cl)** is one one-hundredth of a liter, one **kiloliter (kl)** is 1000 liters, and so on. Milliliters are used extensively in science and medicine. Many beverages now are sold by milliliters and by liters. For example, 750 ml is a common size for wine bottles, and many soft drinks now are sold in 1- and 2-liter bottles.

Because of the way a liter is defined as the volume of a box 10 cm on a side,

$$1 \text{ L} = 10 \text{ cm} \times 10 \text{ cm} \times 10 \text{ cm} = 1000 \text{ cm}^3 \quad \text{or} \quad \frac{1}{1000} \text{ L} = 1 \text{ cm}^3.$$

Since $\frac{1}{1000}$ L $= 1$ ml, we have the following relationship:

$$\mathbf{1 \text{ ml} = 1 \text{ cm}^3.}$$

For example, the volume of a box which is 8 cm by 6 cm by 5 cm may be given as 240 cm^3 or as 240 ml.

The box in Figure 4 is 1 cm by 1 cm by 1 cm. The volume of this box is 1 cm^3, or 1 ml. By definition, the mass of the water that fills such a box is **1 gram (g).** A nickel five-cent piece has a mass close to 5 grams, or 5 g. The volume of water used to define a gram is very small, so a gram is a very small mass. For everyday use, a **kilogram (kg),** or one thousand grams, is more practical. A kilogram weighs about 2.2 pounds. A common abbreviation for kilogram is the word **kilo.**

Extremely small masses can be measured with **milligrams (mg)** and **centigrams (cg).** These measures are so small that they, like centiliters and milliliters, are used mainly in science and medicine.

Anders Celsius and his scale for measuring temperature are honored on this Swedish stamp. The original scale had the freezing point of water at 100° and the boiling point at 0°, but biologist Carl von Linne inverted the scale, giving us the familiar Celsius scale of today.

Temperature

In the metric system temperature is measured in **degrees Celsius.** On the Celsius temperature scale, water freezes at 0° and boils at 100°. These two numbers are easier to remember than the corresponding numbers on the Fahrenheit scale, 32° and 212°. The thermometer in Figure 7 shows some typical temperatures in both Fahrenheit and Celsius.

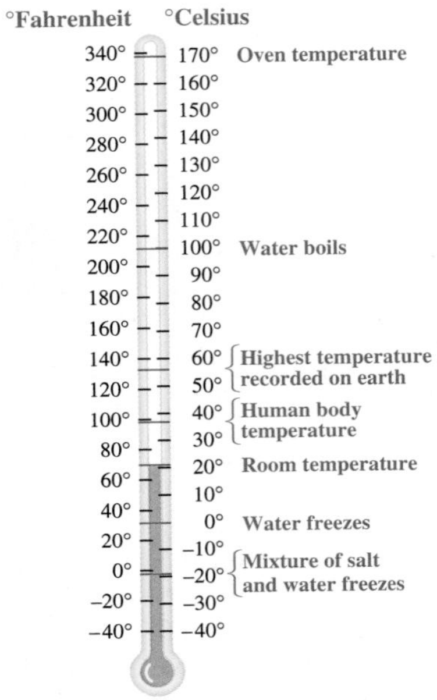

FIGURE 7

Switching a country to the metric system requires a tremendous educational campaign. The stamps shown here were issued as part of one such campaign. They show the various metric units and items measured in those units.

The formulas given below can be used to convert between Celsius and Fahrenheit temperatures.

Celsius-Fahrenheit Conversion Formulas

To convert a reading from Fahrenheit to Celsius, use $C = \frac{5}{9}(F - 32)$.

To convert from Celsius to Fahrenheit, use $F = \frac{9}{5}C + 32$.

Metric Conversions

Due to legislation enacted by Congress, the metric system is used in the United States, and an ultimate goal is for the two systems to be in use, side-by-side, with public acceptance of both systems. Many industries now use the metric system. In particular,

Further information on the **metric system** can be obtained from the U.S. Metric Association, 10245 Andasol Avenue, Northridge, CA 91325-1504.

industries that export a great many goods are using the metric system, since this is compatible with most of the countries with which they trade.

Some scientific calculators are programmed to do conversions between the English and metric systems. Approximate conversions can be made with the aid of Tables 2 and 3 below.

TABLE 2 Metric to English

To Convert from	To	Multiply by
meters	yards	1.0936
meters	feet	3.2808
meters	inches	39.37
kilometers	miles	.6214
grams	pounds	.0022
kilograms	pounds	2.20
liters	quarts	1.0567
liters	gallons	.2642

TABLE 3 English to Metric

To Convert from	To	Multiply by
yards	meters	.9144
feet	meters	.3048
inches	meters	.0254
miles	kilometers	1.609
pounds	grams	454
pounds	kilograms	.454
quarts	liters	.9464
gallons	liters	3.785

APPENDIX EXERCISES

Perform each conversion by multiplying or dividing by the appropriate powers of 10.

1. 8 m to millimeters

2. 14.76 m to centimeters

3. 8500 cm to meters

4. 250 mm to meters

5. 68.9 cm to millimeters

6. 3.25 cm to millimeters

7. 59.8 mm to centimeters

8. 3.542 mm to centimeters

9. 5.3 km to meters

10. 9.24 km to meters

11. 27,500 m to kilometers

12. 14,592 m to kilometers

Use a metric ruler to perform each measurement, first in centimeters, then in millimeters.

13.

14.

15.

16. Based on your measurement of the line segment in Exercise 13, one inch is about how many centimeters? how many millimeters?

Perform each conversion by multiplying or dividing by the appropriate powers of 10.

17. 6 L to centiliters

18. 4.1 L to milliliters

19. 8.7 L to milliliters

20. 12.5 L to centiliters

21. 925 cl to liters

22. 412 ml to liters

23. 8974 ml to liters

24. 5639 cl to liters

25. 8000 g to kilograms

26. 25,000 g to kilograms

27. 5.2 kg to grams

28. 12.42 kg to grams

29. 4.2 g to milligrams

30. 3.89 g to centigrams

31. 598 mg to grams

32. 7634 cg to grams

Use the formulas given in the text to perform each conversion. If necessary, round to the nearest degree.

33. 86°F to Celsius

34. 536°F to Celsius

35. −114°F to Celsius

36. −40°F to Celsius

37. 10°C to Fahrenheit

38. 25°C to Fahrenheit

39. −40°C to Fahrenheit

40. −15°C to Fahrenheit

Solve each problem. Refer to geometry formulas as necessary.

41. One nickel weighs 5 g. How many nickels are in 1 kg of nickels?

42. Sea water contains about 3.5 g salt per 1000 ml of water. How many grams of salt would be in one liter of sea water?

43. Helium weighs about .0002 g per milliliter. How much would one liter of helium weigh?

44. About 1500 g sugar can be dissolved in a liter of warm water. How much sugar could be dissolved in one milliliter of warm water?

45. Northside Foundry needed seven metal strips, each 67 cm long. Find the total cost of the strips, if they sell for $8.74 per meter.

46. Uptown Dressmakers bought fifteen pieces of lace, each 384 mm long. The lace sold for $54.20 per meter. Find the cost of the fifteen pieces.

47. Imported marble for desktops costs $174.20 per square meter. Find the cost of a piece of marble 128 cm by 174 cm.

48. A special photographic paper sells for $63.79 per square meter. Find the cost to buy 80 pieces of the paper, each 9 cm by 14 cm.

49. An importer received some special coffee beans in a box measuring 82 cm by 1.1 m by 1.2 m. Give the volume of the box, both in cubic centimeters and cubic meters.

50. A fabric center receives bolts of woolen cloth in crates measuring 1.5 m by 74 cm by 97 cm. Find the volume of a crate, both in cubic centimeters and cubic meters.

51. A medicine is sold in small bottles holding 800 ml each. How many of these bottles can be filled from a vat holding 160 L of the medicine?

52. How many 2-liter bottles of soda pop would be needed for a wedding reception if 80 people are expected, and each drinks 400 ml of soda?

Perform each conversion. Use a calculator and/or the table in the text as necessary.

53. 982 yd to meters

54. 12.2 km to miles

55. 125 mi to kilometers

56. 1000 mi to kilometers

57. 1816 g to pounds

58. 1.42 lb to grams

59. 47.2 lb to grams

60. 7.68 kg to pounds

61. 28.6 L to quarts

62. 59.4 L to quarts

63. 28.2 gal to liters

64. 16 qt to liters

Metric measures are very common in medicine. Since we convert among metric measures by moving the decimal point, errors in locating the decimal point in medical doses are not unknown. Decide whether each dose of medicine seems reasonable *or* unreasonable.

65. Take 2 kg of aspirin three times a day.

66. Take 4 L of liquid Mylanta every evening just before bedtime.

67. Take 25 ml of cough syrup daily.

68. Soak your feet in 6 L of hot water.

69. Inject $\frac{1}{2}$ L of insulin every morning.

70. Apply 40 g of salve to a cut on your finger.

Select the most reasonable choice for each of the following.

71. length of an adult cow
A. 1 m **B.** 3 m **C.** 5 m

72. length of a Cadillac
A. 1 m **B.** 3 m **C.** 5 m

73. distance from Seattle to Miami
A. 500 km **B.** 5000 km **C.** 50,000 km

74. length across an average nose
A. 3 cm **B.** 30 cm **C.** 300 cm

75. distance across a page of a book
A. 1.93 mm **B.** 19.3 mm **C.** 193 mm

76. weight of a book
A. 1 kg **B.** 10 kg **C.** 1000 kg

77. weight of a large automobile
A. 1300 kg **B.** 130 kg **C.** 13 kg

78. volume of a 12-ounce bottle of beverage
A. 35 ml **B.** 355 ml **C.** 3550 ml

79. height of a person
A. 180 cm **B.** 1800 cm **C.** 18 cm

80. diameter of the earth
A. 130 km **B.** 1300 km **C.** 13,000 km

81. length of a long freight train
A. 8 m **B.** 80 m **C.** 800 m

82. volume of a grapefruit
A. 1 L **B.** 4 L **C.** 8 L

83. the length of a pair of Levis
A. 70 cm **B.** 700 cm **C.** 7 cm

84. a person's weight
A. 700 kg **B.** 7 kg **C.** 70 kg

85. diagonal measure of the picture tube of a table model TV set
A. 5 cm **B.** 50 cm **C.** 500 cm

86. width of a standard bedroom door
A. 1 m **B.** 3 m **C.** 5 m

87. thickness of a marking pen
A. .9 mm **B.** 9 mm **C.** 90 mm

88. length around the rim of a coffee mug
A. 300 mm **B.** 30 mm **C.** 3000 mm

89. the temperature at the surface of a frozen lake
A. 0°C **B.** 10°C **C.** 32°C

90. the temperature in the middle of Death Valley on a July afternoon
A. 25°C **B.** 40°C **C.** 65°C

91. surface temperature of desert sand on a hot summer day
A. 30°C **B.** 60°C **C.** 90°C

92. temperature of boiling water
A. 100°C **B.** 120°C **C.** 150°C

93. air temperature on a day when you need a sweater
A. 30°C **B.** 20°C **C.** 10°C

94. air temperature when you go swimming
A. 30°C **B.** 15°C **C.** 10°C

95. temperature when baking a cake
A. 120°C **B.** 170°C **C.** 300°C

96. temperature of bath water
A. 35°C **B.** 50°C **C.** 65°C

ANSWERS TO SELECTED EXERCISES

CHAPTER 1 The Art of Problem Solving

1.1 Exercises (Pages 7–10)

1. deductive **3.** inductive **5.** deductive **7.** deductive **9.** inductive **11.** inductive **13.** Answers will vary. **15.** 21 **17.** 3072 **19.** 63 **21.** $\frac{11}{12}$ **23.** 216 **25.** 52 **27.** 5 **29.** One such list is 10, 20, 30, 40, 50, ... **31.** $(98{,}765 \times 9) + 3 = 888{,}888$ **33.** $3367 \times 15 = 50{,}505$ **35.** $33{,}334 \times 33{,}334 = 1{,}111{,}155{,}556$ **37.** $3 + 6 + 9 + 12 + 15 = \frac{15(6)}{2}$ **39.** $5(6) + 5(36) + 5(216) + 5(1296) + 5(7776) = 6(7776 - 1)$ **41.** $\frac{1}{2} + \frac{1}{4} + \frac{1}{8} + \frac{1}{16} + \frac{1}{32} = 1 - \frac{1}{32}$ **43.** 20,100 **45.** 320,400 **47.** 15,400 **49.** 2550 **51.**

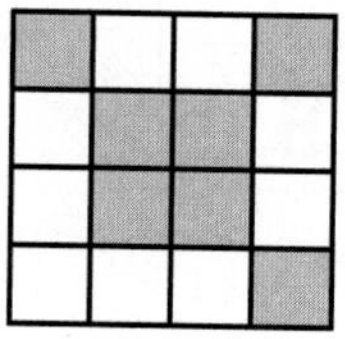

53. 1 (These are the numbers of chimes a clock rings, starting with 12 o'clock, if it rings the number of hours on the hour, and 1 chime on the half-hour.) **55. (a)** The middle digit is always 9, and the sum of the first and third digits is always 9 (considering 0 as the first digit if the difference has only two digits). **(b)** Answers will vary. **57.** 142,857; 285,714; 428,571; 571,428; 714,285; 857,142. Each result consists of the same six digits, but in a different order. $142{,}857 \times 7 = 999{,}999$ **59.** 21 **61.** Answers will vary.

1.2 Exercises (Pages 17–19)

1. 79 **3.** 450 **5.** 4032 **7.** 32,758 **9.** 57; 99 **11.** $(4321 \times 9) - 1 = 38{,}888$ **13.** $999{,}999 \times 4 = 3{,}999{,}996$ **15.** $21^2 - 15^2 = 6^3$ **17.** $5^2 - 4^2 = 5 + 4$ **19.** $1 + 5 + 9 + 13 = 4 \times 7$ **21.** 45,150 **23.** 228,150 **25.** 2601 **27.** 250,000 **29.** $S = n(n + 1)$ **31.** Answers will vary. **33.** row 1: 28, 36; row 2: 36, 49, 64; row 3: 35, 51, 70, 92; row 4: 28, 45, 66, 91, 120; row 5: 18, 34, 55, 81, 112, 148; row 6: 8, 21, 40, 65, 96, 133, 176 **35.** $8(1) + 1 = 9 = 3^2$; $8(3) + 1 = 25 = 5^2$; $8(6) + 1 = 49 = 7^2$; $8(10) + 1 = 81 = 9^2$ **37.** The pattern is 1, 0, 1, 0, 1, 0, ... **39.**

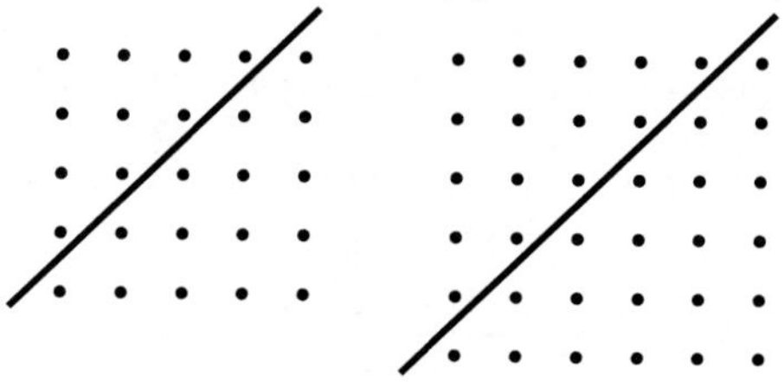

41. (a) a triangular number **(b)** a triangular number **43.** 256 **45.** 117 **47.** 235 **49.** $N_n = \frac{n(7n - 5)}{2}$ **51.** a square number **53.** a perfect cube

1.3 Exercises (Pages 27–33)

1. 42 **3.** 6 **5.** If you multiply the two digits in the numbers in the first row, you will get the second row of numbers. The second row of numbers is a pattern of two numbers (8 and 24) repeating. **7.** I put the ring in the box and put my lock on the box. I send you the box. You put your lock on, as well, and send it back to me. I then remove my lock with my key and send the box (with your lock still on) back to you, so you can remove your lock with your key and get the ring. **9.** 59 **11.** You should choose a sock from the box labeled *red and green socks*. Because it is mislabeled, it contains only red socks or only green socks, determined by the

56

sock you choose. If the sock is green, relabel this box *green socks*. Since the other two boxes were mislabeled, switch the remaining label to the other box and place the label that says *red and green socks* on the unlabeled box. No other choice guarantees a correct relabeling because you can remove only one sock.

13. D **15.** A

17. One example of a solution follows:

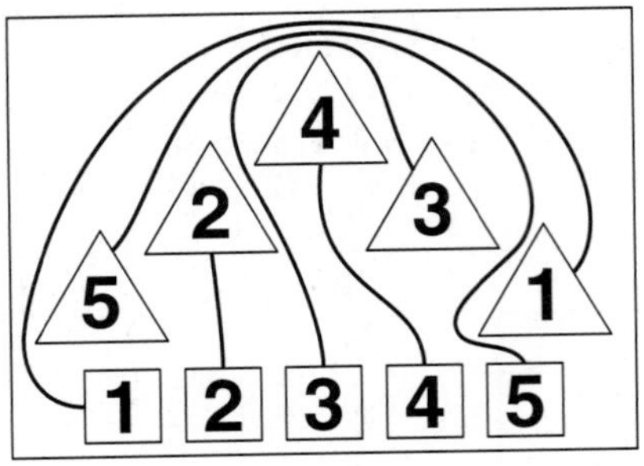

19.

9	7	2	14	11	5	4	12	13	3	6	10	15	1	8

(or the same arrangement reading right to left)

21. Here is one solution.

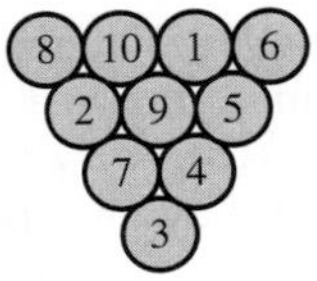

23. D **25.** $\frac{1}{3}$

27. One possible sequence is shown here. The numbers represent the number of gallons in each bucket in each successive step.

Big	7	4	4	1	1	0	7	5	5
Small	0	3	0	3	0	1	1	3	0

29. 90 **31.** 55 miles per hour **33.** 07

35. A kilogram of \$10 gold pieces is worth twice as much as half a kilogram of \$20 gold pieces. (The denomination has nothing to do with the answer!) **37.** 3 **39.** 5 **41.** 3 socks **43.** 35 **45.** 6 **47.** the nineteenth day **49.** 1967 **51.** Eve has \$5, and Adam has \$7.

53.

$$\begin{array}{r} 4\ 0\ 2 \\ \times \quad 3\ 9 \\ \hline 1\ 5,6\ 7\ 8 \end{array}$$

55.

6	**12**	**7**	9
1	15	**4**	14
11	**5**	10	**8**
16	**2**	13	**3**

57. 25 pitches (The visiting team's pitcher retires 24 consecutive batters through the first eight innings, using only one pitch per batter. His team does not score either. Going into the bottom of the ninth tied 0–0, the first batter for the home team hits his first pitch for a home run. The pitcher threw 25 pitches and loses the game by a score of 1–0.)

59. Q **61.** Here is one solution. **63.** 8 **65.** The CEO is a woman.

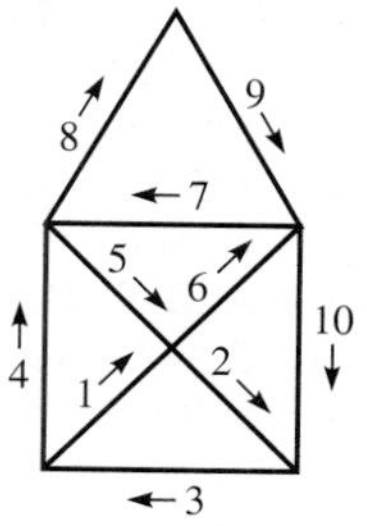

67. Dan (36) is married to Jessica (29); James (30) is married to Cathy (31). **69.** 12; All months have 28 days.
71. The products always differ by 1. **73.** 6

	X	X
X		X
X	X	

One of several possibilities

1.4 Exercises (Pages 38–41)

1. 43.8 **3.** 2.3589 **5.** 7.48 **7.** 7.1289 **9.** 6340.338097 **11.** 1 **13.** 1.061858759 **15.** 2.221441469 **17.** 3.141592653 **19.** .7782717162 **21.** yes **23.** positive **25.** 1 **27.** the same as **29.** 0 **31.** positive **33.** Answers will vary. **35.** Answers will vary. **37.** ShELLOIL **39.** BOSE **41.** Answers will vary. **43.** 63 **45.** 14 **47.** B **49.** A **51.** D **53.** 5% **55.** 260,000 **57.** 1998; 2000 **59.** \$250 billion; \$300 billion **61.** from 2004 to 2005 **63.** They are approximately the same.

Chapter 1 Test (Pages 48–49)

1. inductive **2.** deductive **3.** 256, 3125 (The nth term in the sequence is n^n.) **4.** $65{,}359{,}477{,}124{,}183 \times 68 = 4{,}444{,}444{,}444{,}444{,}444$ **5.** 351 **6.** 31,375 **7.** 65; $65 = 1 + 7 + 13 + 19 + 25$ **8.** 1, 8, 21, 40, 65, 96, 133, 176; The pattern is 1, 0, 1, 0, 1, 0, 1, 0, ... **9.** The first two terms are both 1. Each term after the second is found by adding the two previous terms. **10.** $\frac{1}{4}$ **11.** 9 **12.** 35 **13.** $629 + 154 = 783$ is one of several solutions. **14.** 8, 53, and 54 **15.** 3 **16.** The sum of the digits is always 9. **17.** 9.907572861 (Answers may vary due to the model of calculator used.) **18.** 34.328125 **19.** B **20.** **(a)** 1990–2000 **(b)** 1990 **(c)** 1980

CHAPTER 2 The Basic Concepts of Set Theory

2.1 Exercises (Pages 56–58)

1. C **3.** E **5.** B **7.** H **9.** {1, 2, 3, 4, 5, 6} **11.** {0, 1, 2, 3, 4} **13.** {6, 7, 8, 9, 10, 11, 12, 13, 14} **15.** {−15, −13, −11, −9, −7, −5, −3, −1} **17.** {2, 4, 8, 16, 32, 64, 128, 256} **19.** {0, 2, 4, 6, 8, 10} **21.** {21, 22, 23, . . .} **23.** {Lake Erie, Lake Huron, Lake Michigan, Lake Ontario, Lake Superior} **25.** {5, 10, 15, 20, 25, . . .} **27.** $\left\{1, \frac{1}{2}, \frac{1}{3}, \frac{1}{4}, \frac{1}{5}, \ldots\right\}$ **In Exercises 29 and 31, there are other ways to describe the sets.** **29.** $\{x \mid x$ is a rational number$\}$ **31.** $\{x \mid x$ is an odd natural number less than $76\}$ **33.** finite **35.** infinite **37.** infinite **39.** infinite **41.** 8 **43.** 500 **45.** 26 **47.** 39 **49.** 28 **51.** Answers will vary. **53.** well defined **55.** not well defined **57.** not well defined **59.** $\in$ **61.** $\notin$ **63.** $\in$ **65.** $\notin$ **67.** $\in$ **69.** false **71.** true **73.** true **75.** true **77.** false **79.** true **81.** true **83.** true **85.** false **87.** true **89.** Answers will vary. **91.** {2} and {3, 4} (Other examples are possible.) **93.** {a, b} and {a, c} (Other examples are possible.) **95.** **(a)** {LU, NT, PFE, GE} **(b)** {GE, MOT, TWX, C, TXN, EMC, AWE}

2.2 Exercises (Pages 63–65)

1. F **3.** C **5.** A **7.** $\not\subseteq$ **9.** $\subseteq$ **11.** $\subseteq$ **13.** $\not\subseteq$ **15.** both **17.** $\subseteq$ **19.** both **21.** neither **23.** true **25.** false **27.** true **29.** true **31.** true **33.** true **35.** false **37.** false **39.** true **41.** false **43.** **(a)** 8 **(b)** 7 **45.** **(a)** 64 **(b)** 63 **47.** **(a)** 32 **(b)** 31 **49.** {2, 3, 5, 7, 9, 10} **51.** {2} **53.** {1, 2, 3, 4, 5, 6, 7, 8, 9, 10} or U **55.** {Higher cost, Lower cost, Educational, More time to see the sights, Less time to see the sights, Cannot visit relatives along the way, Can visit relatives along the way} **57.** {Higher cost, More time to see the sights, Cannot visit relatives along the way} **59.** $\emptyset$ **61.** {A, B, C, D, E} (All are present.) **63.** {A, B, C}, {A, B, D}, {A, B, E}, {A, C, D}, {A, C, E}, {A, D, E}, {B, C, D}, {B, C, E}, {B, D, E}, {C, D, E} **65.** {A}, {B}, {C}, {D}, {E} **67.** 32 **69.** **(a)** 15 **(b)** 16; It is now possible to select *no* bills. **71.** **(a)** s **(b)** s **(c)** $2s$ **(d)** Adding one more element will always double the number of subsets, so the expression 2^n is true in general.

2.3 Exercises (Pages 75–79)

1. B **3.** A **5.** E **7.** {a, c} **9.** {a, b, c, d, e, f} **11.** {a, b, c, d, e, f, g} **13.** {b, d, f} **15.** {d, f} **17.** {a, b, c, e, g} **19.** {a, c, e, g} **21.** {a} **23.** {e, g} **25.** {e, g} **27.** {d, f} **29.** {e, b, g}

In Exercises 31–35, there may be other acceptable descriptions. **31.** the set of all elements that either are in A, or are not in B and not in C **33.** the set of all elements that are in C but not in B, or are in A **35.** the set of all elements that are in A but not in C, or in B but not in C **37.** $\{e, h, c, l, b\}$ **39.** $\{l, b\}$ **41.** $\{e, h, c, l, b\}$ **43.** the set of all tax returns showing business income or filed in 2005 **45.** the set of all tax returns filed in 2005 without itemized deductions **47.** the set of all tax returns with itemized deductions or showing business income, but not selected for audit **49.** always true **51.** always true **53.** not always true **55.** **(a)** {1, 3, 5, 2} **(b)** {1, 2, 3, 5} **(c)** For any sets X and Y, $X \cup Y = Y \cup X$. **57.** **(a)** {1, 3, 5, 2, 4} **(b)** {1, 3, 5, 2, 4} **(c)** For any sets X, Y, and Z, $X \cup (Y \cup Z) = (X \cup Y) \cup Z$. **59.** **(a)** {4} **(b)** {4} **(c)** For any sets X and Y, $(X \cup Y)' = X' \cap Y'$. **61.** $X \cup \emptyset = X$; For any set X, $X \cup \emptyset = X$. **63.** true **65.** false **67.** true **69.** true **71.** $A \times B = \{(2, 4), (2, 9), (8, 4), (8, 9), (12, 4), (12, 9)\}$; $B \times A = \{(4, 2), (4, 8), (4, 12), (9, 2), (9, 8), (9, 12)\}$ **73.** $A \times B = \{$(d, p), (d, i), (d, g), (o, p), (o, i), (o, g), (g, p), (g, i), (g, g)$\}$; $B \times A = \{$(p, d), (p, o), (p, g), (i, d), (i, o), (i, g), (g, d), (g, o), (g, g)$\}$ **75.** $n(A \times B) = 6$; $n(B \times A) = 6$ **77.** $n(A \times B) = 210$; $n(B \times A) = 210$ **79.** 6

81.

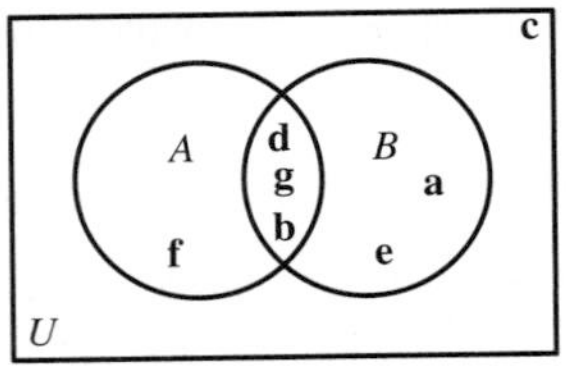

83.

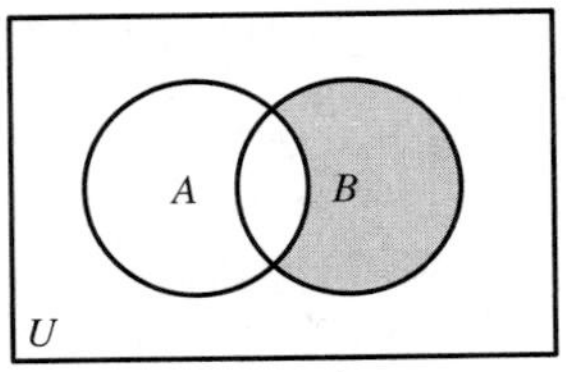

$B \cap A'$

85.

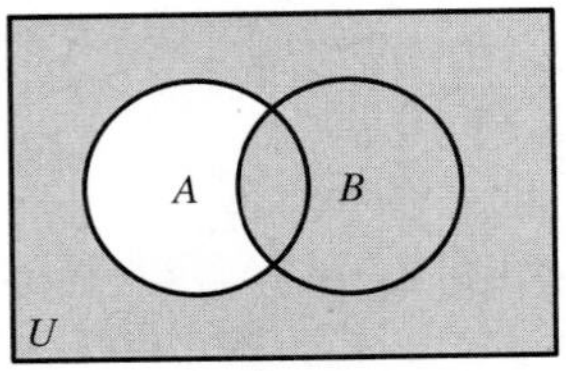

$A' \cup B$

87.

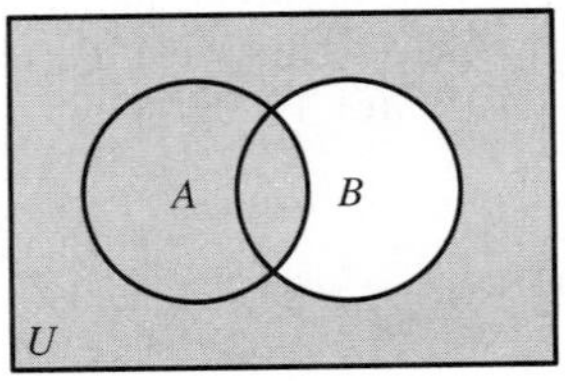

$B' \cup A$

89.

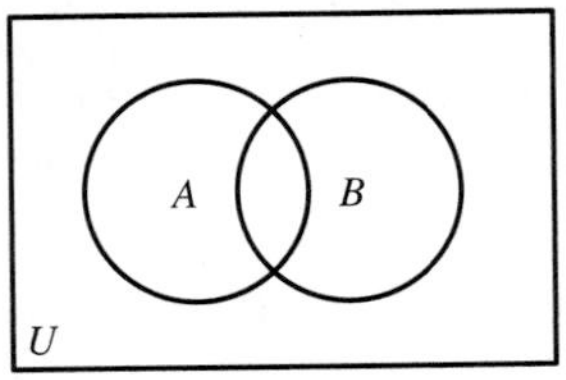

$B' \cap B = \emptyset$

91.

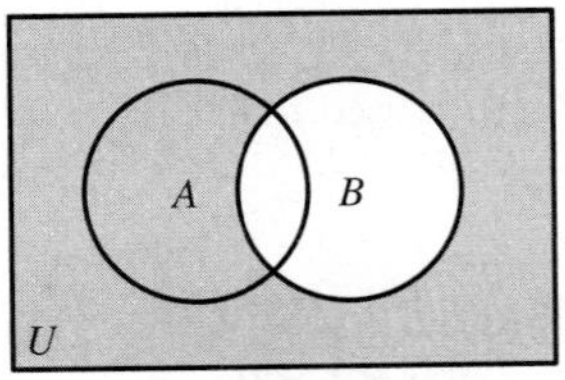

$B' \cup (A' \cap B')$

93.

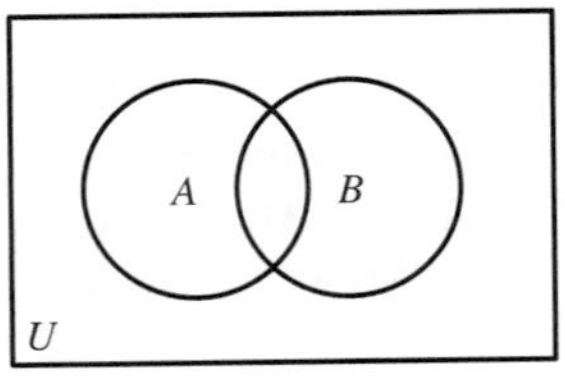

$U' = \emptyset$

95.

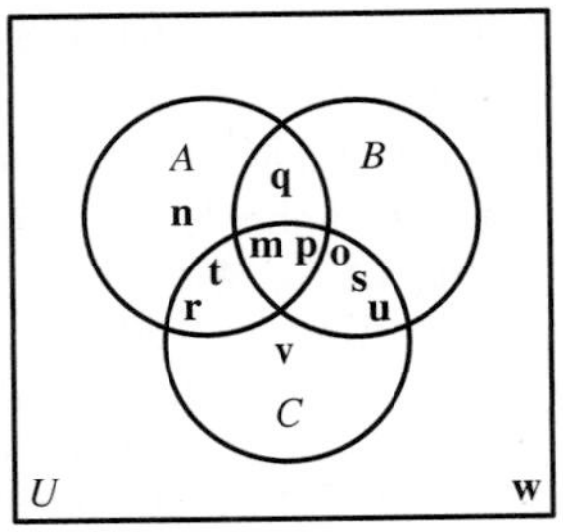

97.

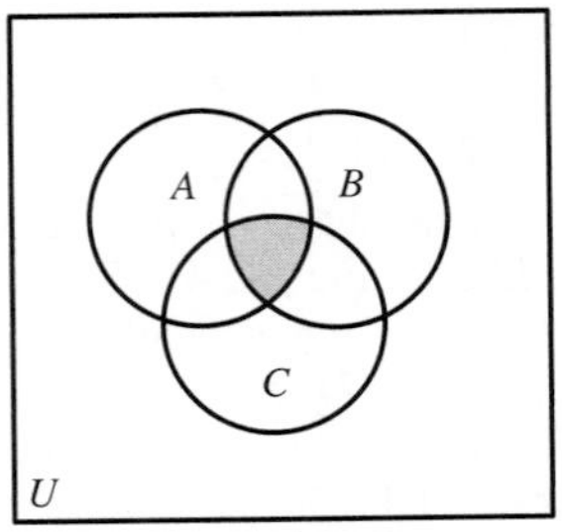

$(A \cap B) \cap C$

99.

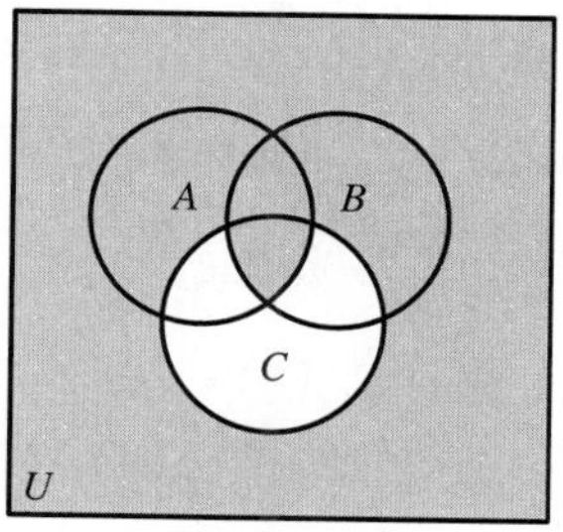

$(A \cap B) \cup C'$

101.

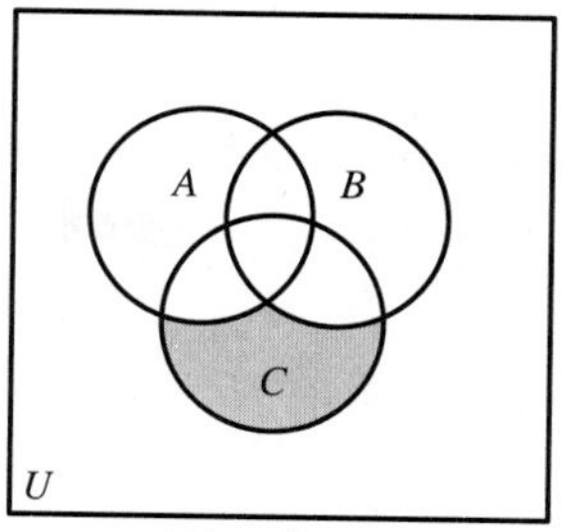

$(A' \cap B') \cap C$

103.

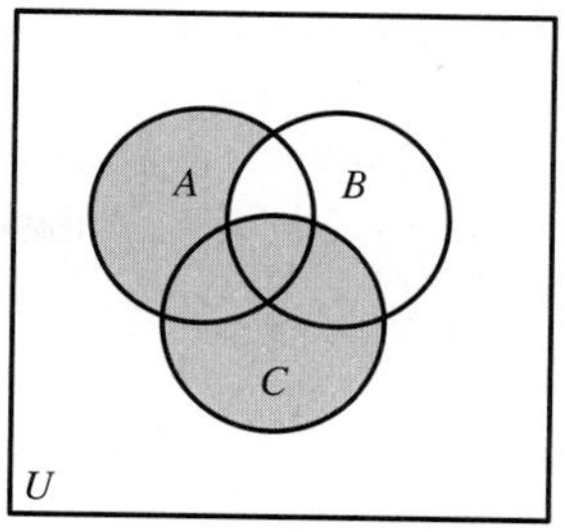

$(A \cap B') \cup C$

105.

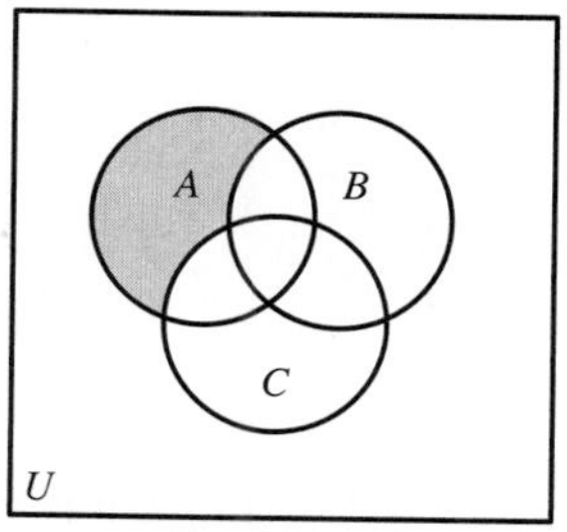

$(A \cap B') \cap C'$

107.

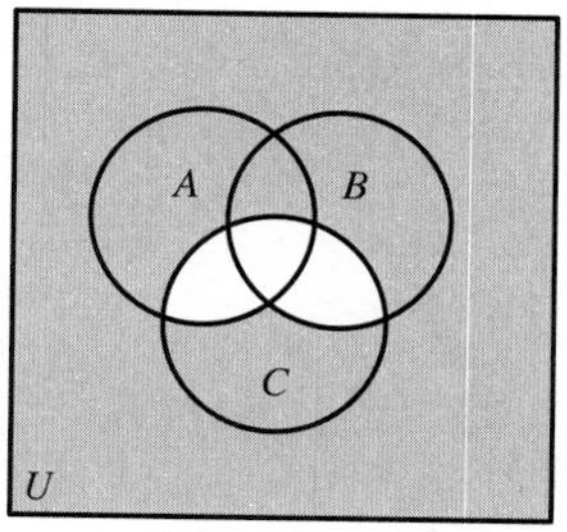

$(A' \cap B') \cup C'$

109. $A' \cap B'$ or $(A \cup B)'$ **111.** $(A \cup B) \cap (A \cap B)'$ or $(A \cup B) - (A \cap B)$ **113.** $(A \cap B) \cup (A \cap C)$ or $A \cap (B \cup C)$ **115.** $(A \cap B) \cap C'$ or $(A \cap B) - C$ **117.** $A \cap B = \emptyset$ **119.** This statement is true for any set A. **121.** $A = \emptyset$ **123.** $A = \emptyset$ **125.** $A = \emptyset$ **127.** $B \subseteq A$ **129.** always true **131.** always true **133.** not always true **135.** always true **137.** no

2.4 Exercises (Pages 82–86)

1. (a) 6 **(b)** 8 **(c)** 0 **(d)** 2 **(e)** 9 **3. (a)** 1 **(b)** 3 **(c)** 4 **(d)** 0 **(e)** 2 **(f)** 10 **(g)** 2 **(h)** 5 **5.** 17 **7.** 2 **9.** 35

11.

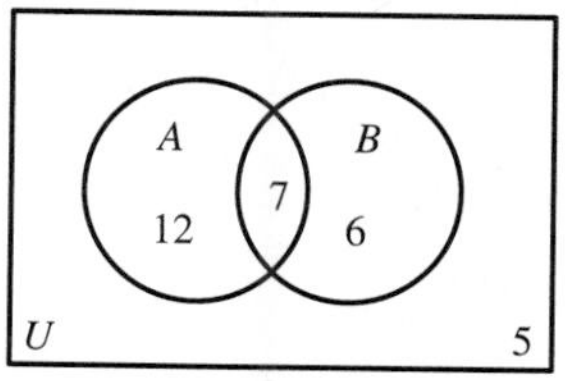

13.

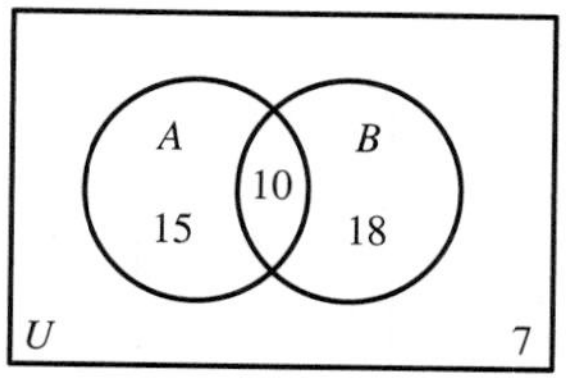

15.

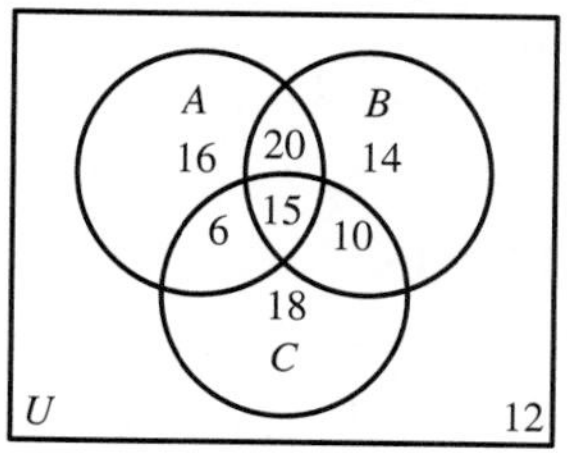

17.

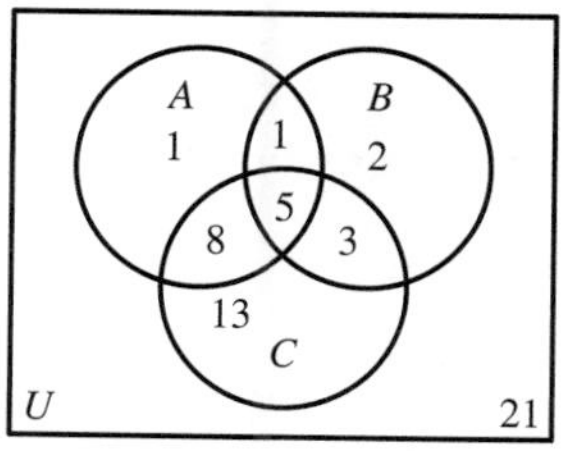

19. (a) 3 **(b)** 5 **21. (a)** 18 **(b)** 15 **(c)** 20 **(d)** 5 **23.** Yes, he should be reassigned to the job in Siberia. His figures add up to 142 people.

25. (a) 37 **(b)** 22 **(c)** 50 **(d)** 11 **(e)** 25 **(f)** 11 **27. (a)** 31 **(b)** 24 **(c)** 11 **(d)** 45 **29. (a)** 1 **(b)** 1, 2, 3, 4, 5, 6, 7, 8, 9, 10, 11, 12, 13, 14, 15 **(c)** 1, 2, 3, 4, 5, 9, 11 **(d)** 5, 8, 13 **31. (a)** 9 **(b)** 9 **(c)** 20 **(d)** 20 **(e)** 27 **(f)** 15 **33.** Answers will vary.

2.5 Exercises (Pages 92–94)

1. B; 1 **3.** A; $\aleph_0$ **5.** F; 0

7. (Other correspondences are possible.)

$$\begin{array}{ccc} \{\text{I}, & \text{II}, & \text{III}\} \\ \updownarrow & \updownarrow & \updownarrow \\ \{\text{x}, & \text{y}, & \text{z}\} \end{array}$$

9. (Other correspondences are possible.)

$$\begin{array}{cccccc} \{\text{a}, & \text{d}, & \text{i}, & \text{t}, & \text{o}, & \text{n}\} \\ \updownarrow & \updownarrow & \updownarrow & \updownarrow & \updownarrow & \updownarrow \\ \{\text{a}, & \text{n}, & \text{s}, & \text{w}, & \text{e}, & \text{r}\} \end{array}$$

11. 11 **13.** 0 **15.** $\aleph_0$ **17.** $\aleph_0$ **19.** $\aleph_0$ **21.** 12 **23.** $\aleph_0$ **25.** both **27.** equivalent **29.** equivalent

31.
$$\begin{array}{ccccccc} \{2, & 4, & 6, & 8, & \ldots, & 2n, & \ldots\} \\ \updownarrow & \updownarrow & \updownarrow & \updownarrow & & \updownarrow & \\ \{1, & 2, & 3, & 4, & \ldots, & n, & \ldots\} \end{array}$$

33.
$$\begin{array}{cccccc} \{1{,}000{,}000, & 2{,}000{,}000, & 3{,}000{,}000, & \ldots, & 1{,}000{,}000n, & \ldots\} \\ \updownarrow & \updownarrow & \updownarrow & & \updownarrow & \\ \{\ 1, & 2, & 3, & \ldots, & n, & \ldots\} \end{array}$$

35.
$$\begin{array}{cccccccc} \{2, & 4, & 8, & 16, & 32, & \ldots, & 2^n, & \ldots\} \\ \updownarrow & \updownarrow & \updownarrow & \updownarrow & \updownarrow & & \updownarrow & \\ \{1, & 2, & 3, & 4, & 5, & \ldots, & n, & \ldots\} \end{array}$$

37. This statement is not always true. For example, let A = the set of counting numbers, B = the set of real numbers. **39.** This statement is not always true. For example, A could be the set of all subsets of the set of reals. Then $n(A)$ would be an infinite number *greater* than c. **41. (a)** Rays emanating from point P will establish a geometric pairing of the points on the semicircle with the points on the line.

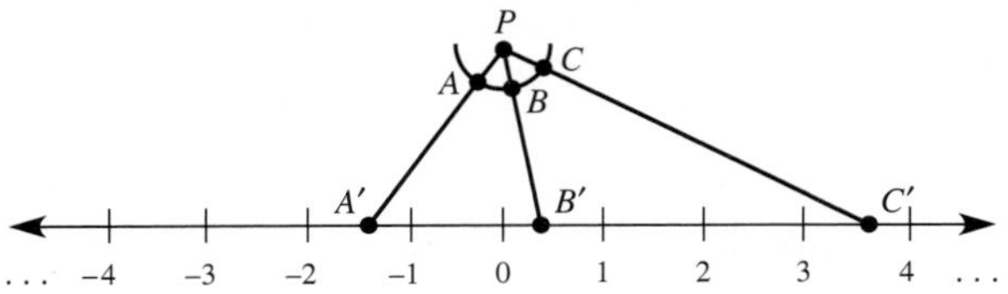

(b) The set of real numbers is infinite, having been placed in a one-to-one correspondence with a proper subset of itself.

43.
$$\begin{array}{ccccccc} \{3, & 6, & 9, & 12, & \ldots, & 3n, & \ldots\} \\ \updownarrow & \updownarrow & \updownarrow & \updownarrow & & \updownarrow & \\ \{6, & 9, & 12, & 15, & \ldots, & 3n+3, & \ldots\} \end{array}$$

45.
$$\begin{array}{ccccccc} \left\{\frac{3}{4},\right. & \frac{3}{8}, & \frac{3}{12}, & \frac{3}{16}, & \ldots, & \frac{3}{4n}, & \left.\ldots\right\} \\ \updownarrow & \updownarrow & \updownarrow & \updownarrow & & \updownarrow & \\ \left\{\frac{3}{8},\right. & \frac{3}{12}, & \frac{3}{16}, & \frac{3}{20}, & \ldots, & \frac{3}{4n+4}, & \left.\ldots\right\} \end{array}$$

47.
$$\begin{array}{cccccc} \left\{\frac{1}{9},\right. & \frac{1}{18}, & \frac{1}{27}, & \ldots, & \frac{1}{9n}, & \left.\ldots\right\} \\ \updownarrow & \updownarrow & \updownarrow & & \updownarrow & \\ \left\{\frac{1}{18},\right. & \frac{1}{27}, & \frac{1}{36}, & \ldots, & \frac{1}{9n+9}, & \left.\ldots\right\} \end{array}$$

49. Answers will vary. **51.** Answers will vary.

Chapter 2 Test (Pages 95–96)

1. {a, b, c, d, e} **2.** {a, b, d} **3.** {c, f, g, h} **4.** {a, c} **5.** true **6.** false **7.** true **8.** true **9.** false **10.** true **11.** true **12.** true **13.** 8 **14.** 15 **Answers may vary for Exercises 15–18.** **15.** the set of odd integers between −4 and 10 **16.** the set of months of the year **17.** $\{x \mid x$ is a negative integer$\}$ **18.** $\{x \mid x$ is a multiple of 8 between 20 and 90$\}$ **19.** $\subseteq$ **20.** neither

21.

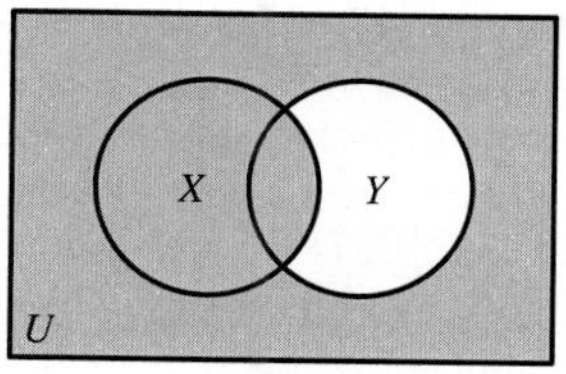

$X \cup Y'$

22.

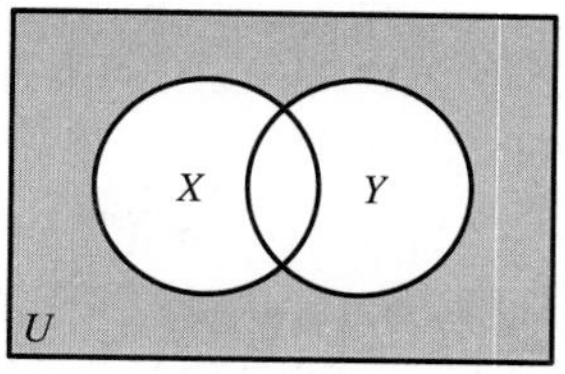

$X' \cap Y'$

23.

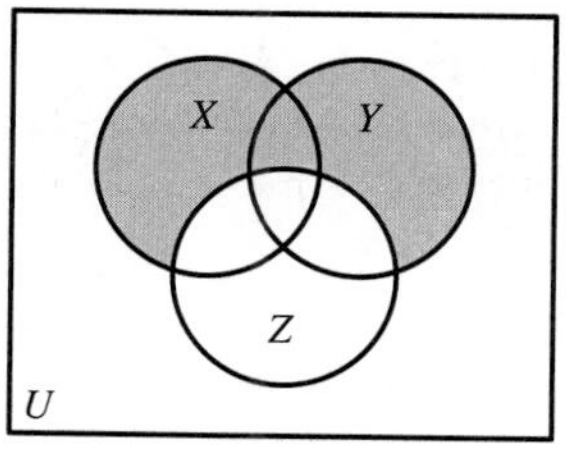

$(X \cup Y) - Z$

24.

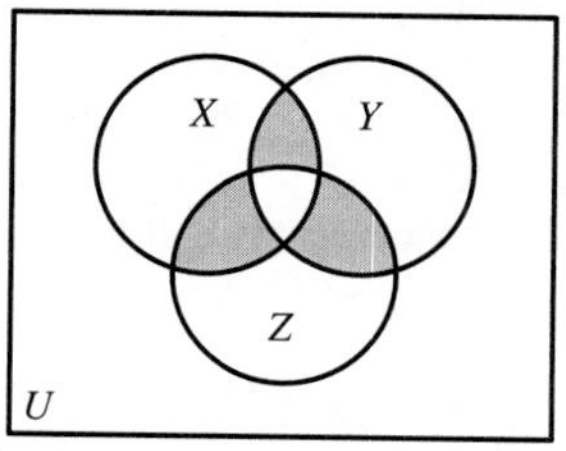

$[(X \cap Y) \cup (Y \cap Z) \cup (X \cap Z)] - (X \cap Y \cap Z)$

25. {Electric razor} **26.** {Adding machine, Barometer, Pendulum clock, Thermometer} **27.** {Electric razor} **28.** Answers will vary. **29.** **(a)** 22 **(b)** 12 **(c)** 28 **30.** **(a)** 16 **(b)** 32 **(c)** 33 **(d)** 45 **(e)** 14 **(f)** 26

CHAPTER 3 Introduction to Logic

3.1 Exercises (Pages 103–104)

1. statement **3.** not a statement **5.** statement **7.** statement **9.** statement **11.** not a statement **13.** statement **15.** compound **17.** not compound **19.** not compound **21.** compound **23.** Her aunt's name is not Hildegard. **25.** At least one dog does not have its day. **27.** No book is longer than this book. **29.** At least one computer repairman can play poker. **31.** Someone does not love somebody sometime. **33.** $x \leq 12$ **35.** $x < 5$ **37.** Answers will vary. **39.** She does not have green eyes. **41.** She has green eyes and he is 56 years old. **43.** She does not have green eyes or he is 56 years old. **45.** She does not have green eyes or he is not 56 years old. **47.** It is not the case that she does not have green eyes and he is 56 years old. **49.** $p \wedge \sim q$ **51.** $\sim p \vee q$ **53.** $\sim(p \vee q)$ or, equivalently, $\sim p \wedge \sim q$ **55.** Answers will vary. **57.** C **59.** A, B **61.** A, C **63.** B **65.** true **67.** true **69.** true **71.** true **73.** false **75.** Answers will vary. **77.** Everyone here has done that at one time or another.

3.2 Exercises (Pages 115–117)

1. false **3.** true **5.** They must both be false. **7.** T **9.** T **11.** F **13.** T **15.** T **17.** T **19.** It is a disjunction, because it means "$5 > 2$ or $5 = 2$." **21.** T **23.** F **25.** T **27.** T **29.** F **31.** F **33.** T **35.** T **37.** T **39.** 4 **41.** 16 **43.** 128 **45.** seven **47.** FFTF **49.** FTTT **51.** TTTT **53.** FFFT **55.** TFFF **57.** FFFFTFFF **59.** FTFTTTTT **61.** TTTTTTTTTTTTFTTT **63.** You can't pay me now and you can't pay me later. **65.** It is not summer or there is snow. **67.** I did not say yes or she did not say no. **69.** $5 - 1 \neq 4$ or $9 + 12 = 7$ **71.** Neither Dasher nor Blitzen will lead Santa's sleigh next Christmas. **73.** T **75.** T **77.**

p	q	$p \vee q$
T	T	F
T	F	T
F	T	T
F	F	F

79. F **81.** T

3.3 Exercises (Pages 124–127)

1. If you see it on the Internet, then you can believe it. **3.** If the person is Garrett Olinde, then his area code is 225. **5.** If the soldier is a marine, then the soldier loves boot camp. **7.** If it is a koala, then it does not live in Iowa.

9. If it is an opium-eater, then it has no self-command. **11.** true **13.** true **15.** true **17.** false **19.** Answers will vary. **21.** F **23.** T **25.** T **27.** If they do not raise alpacas, then he trains dogs. **29.** If she has a ferret for a pet, then they raise alpacas and he trains dogs. **31.** If he does not train dogs, then they do not raise alpacas or she has a ferret for a pet. **33.** $b \rightarrow p$ **35.** $p \rightarrow \sim r$ **37.** $p \wedge (r \rightarrow \sim b)$ **39.** $p \rightarrow r$ **41.** T **43.** F **45.** T **47.** F **49.** T **51.** T **53.** Answers will vary. **55.** TTTF **57.** TTFT **59.** TTTT; tautology **61.** TFTF **63.** TTTTTTFT **65.** TTTFTTTTTTTTTTTT **67.** one **69.** That is an authentic Persian rug and I am not surprised. **71.** The English measures are not converted to metric measures and the spacecraft does not crash on the surface of Saturn. **73.** You want to be happy for the rest of your life and you make a pretty woman your wife. **75.** You do not give your plants tender, loving care or they flourish. **77.** She does or he will. **79.** The person is not a resident of Oregon City or is a resident of Oregon. **81.** equivalent **83.** equivalent **85.** equivalent **87.** not equivalent **89.** equivalent **91.** $(p \wedge q) \vee (p \wedge \sim q)$; The statement simplifies to p. **93.** $p \vee (\sim q \wedge r)$ **95.** $\sim p \vee (p \vee q)$; The statement simplifies to T. **97.** The statement simplifies to $p \wedge q$. **99.** The statement simplifies to F.

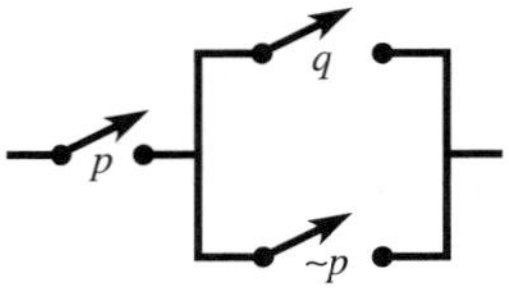

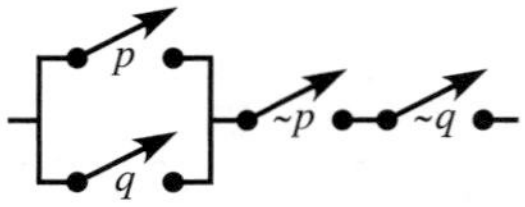

101. The statement simplifies to $(r \wedge \sim p) \wedge q$. **103.** The statement simplifies to $p \vee q$.

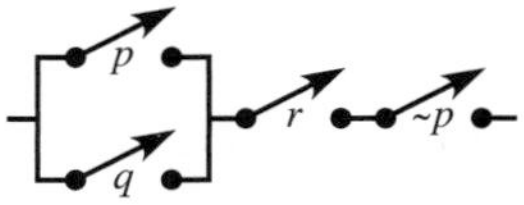

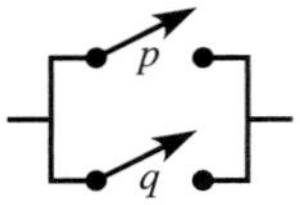

3.4 Exercises (Pages 132–134)

1. **(a)** If you were an hour, then beauty would be a minute. **(b)** If beauty were not a minute, then you would not be an hour. **(c)** If you were not an hour, then beauty would not be a minute. **3.** **(a)** If you don't fix it, then it ain't broke. **(b)** If it's broke, then fix it. **(c)** If you fix it, then it's broke. **5.** **(a)** If it is dangerous to your health, then you walk in front of a moving car. **(b)** If you do not walk in front of a moving car, then it is not dangerous to your health. **(c)** If it is not dangerous to your health, then you do not walk in front of a moving car. **7.** **(a)** If they flock together, then they are birds of a feather. **(b)** If they are not birds of a feather, then they do not flock together. **(c)** If they do not flock together, then they are not birds of a feather. **9.** **(a)** If he comes, then you built it. **(b)** If you don't build it, then he won't come. **(c)** If he doesn't come, then you didn't build it. **11.** **(a)** $\sim q \rightarrow p$ **(b)** $\sim p \rightarrow q$ **(c)** $q \rightarrow \sim p$ **13.** **(a)** $\sim q \rightarrow \sim p$ **(b)** $p \rightarrow q$ **(c)** $q \rightarrow p$ **15.** **(a)** $(q \vee r) \rightarrow p$ **(b)** $\sim p \rightarrow (\sim q \wedge \sim r)$ **(c)** $(\sim q \wedge \sim r) \rightarrow \sim p$ **17.** Answers will vary. **19.** If it is muddy, then I'll wear my galoshes. **21.** If 18 is positive, then $18 + 1$ is positive. **23.** If a number is an integer, then it is a rational number. **25.** If I do crossword puzzles, then I am driven crazy. **27.** If Gerald Guidroz is to shave, then he must have a day's growth of beard. **29.** If I go from Park Place to Baltic Avenue, then I pass GO. **31.** If a number is a whole number, then it is an integer. **33.** If their pitching improves, then the Orioles will win the pennant. **35.** If the figure is a rectangle, then it is a parallelogram with a right angle. **37.** If a triangle has two sides of the same length, then it is isosceles. **39.** If a two-digit number whose units digit is 5 is squared, then it will end in 25. **41.** D **43.** Answers will vary. **45.** true **47.** false **49.** false **51.** contrary **53.** consistent **55.** contrary **57.** Answers will vary. One example is: That man is Carter Fenton. That man sells books.

3.5 Exercises (Pages 137–138)

1. valid **3.** invalid **5.** valid **7.** invalid **9.** invalid **11.** invalid **13.** yes

15. All people with blue eyes have blond hair.
Dinya Norris does not have blond hair.
____
Dinya Norris does not have blue eyes.

17. invalid **19.** valid **21.** invalid **23.** valid **25.** invalid

27. invalid **29.** valid **31.** Answers will vary.

Extension Exercises (Pages 141–144)

1. First, Piotr Knightovich, Yorki; Second, Ivan Rookov, Porki; Third, Boris Bishopnik, Gorki; Fourth, Yuri Pawnchev, Corki. **3.** Ben Ashby, Jane Kenny, Dirk and Daisy, American; Hans Gruber, Sue Rogers, Merlyns, English; Peter Owen, Carol Dodds, Starr Twins, own composition; Steven Thorp, Nancy O'Hara, Rose and Thorn, Irish.

5.

7	2	6	4	9	1	5	3	8
9	4	5	8	6	3	2	1	7
8	1	3	7	5	2	4	9	6
6	8	1	3	2	4	7	5	9
2	9	7	6	1	5	3	8	4
3	5	4	9	8	7	1	6	2
4	7	9	1	3	8	6	2	5
5	3	8	2	4	6	9	7	1
1	6	2	5	7	9	8	4	3

7.

2	7	5	3	8	4	6	1	9
8	3	6	5	9	1	4	2	7
4	1	9	7	6	2	8	3	5
3	4	7	2	5	8	1	9	6
6	2	8	1	4	9	7	5	3
5	9	1	6	7	3	2	4	8
9	5	4	8	1	6	3	7	2
7	8	2	4	3	5	9	6	1
1	6	3	9	2	7	5	8	4

9.

1	8	2	5	7	9	3	4	6
5	3	4	8	1	6	2	9	7
7	6	9	2	3	4	8	1	5
4	1	8	7	9	2	6	5	3
3	7	6	1	8	5	4	2	9
2	9	5	4	6	3	7	8	1
9	2	1	3	4	7	5	6	8
8	5	7	6	2	1	9	3	4
6	4	3	9	5	8	1	7	2

11.

9	8	3	2	4	5	7	1	6
2	6	1	3	9	7	8	4	5
7	5	4	8	6	1	9	3	2
1	7	6	4	5	3	2	9	8
4	9	5	7	2	8	3	6	1
3	2	8	6	1	9	4	5	7
5	4	7	9	8	6	1	2	3
8	1	2	5	3	4	6	7	9
6	3	9	1	7	2	5	8	4

3.6 Exercises (Pages 151–154)

1. valid by reasoning by transitivity **3.** valid by modus ponens **5.** fallacy by fallacy of the converse **7.** valid by modus tollens **9.** fallacy by fallacy of the inverse **11.** valid by disjunctive syllogism **13.** invalid **15.** valid **17.** invalid **19.** valid **21.** invalid **23.** invalid

25. Every time something squeaks, I use WD-40.
Every time I use WD-40, I go to the hardware store.
―――――――――――――――――――――――――――
Every time something squeaks, I go to the hardware store.

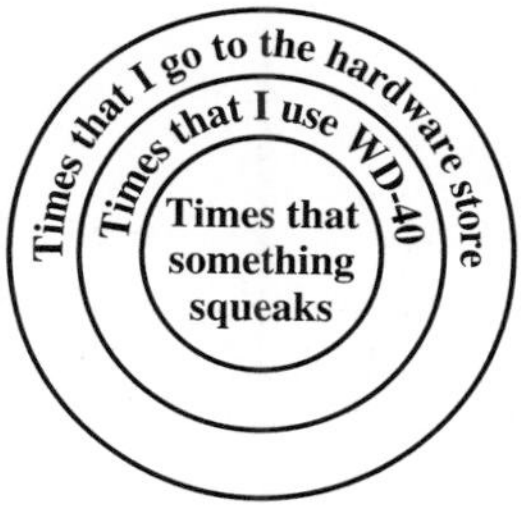

27. valid

29. invalid **31.** invalid **33.** valid **35.** valid **37.** If tell you the time, then my life will be miserable. **39.** If it is my poultry, then it is a duck. **41.** If it is a guinea pig, then it is hopelessly ignorant of music. **43.** If it is a teachable kitten, then it does not have green eyes. **45.** If I can read it, then I have not filed it. **47. (a)** $p \to \sim s$ **(b)** $r \to s$ **(c)** $q \to p$ **(d)** None of my poultry are officers. **49. (a)** $r \to \sim s$ **(b)** $u \to t$ **(c)** $\sim r \to p$ **(d)** $\sim u \to \sim q$ **(e)** $t \to s$ **(f)** All pawnbrokers are honest. **51. (a)** $r \to w$ **(b)** $\sim u \to \sim t$ **(c)** $v \to \sim s$ **(d)** $x \to r$ **(e)** $\sim q \to t$ **(f)** $y \to p$ **(g)** $w \to s$ **(h)** $\sim x \to \sim q$ **(i)** $p \to \sim u$ **(j)** I can't read any of Brown's letters.

Collaborative Investigation (Page 155)

1. 1, *Death in Beijing*, John Gunn, red; 2, *A Killer Abroad*, Mary Hemlock, brown; 3, *Murder in the Sun*, Geoffrey Stringer, green; 4, *The Final Case*, Sandra Bludgeon, blue; 5, *Mayhem in Madagascar*, Dahlia Dagger, yellow; 6, *Lurking in the Shadows*, Philip G Rott, black.

Chapter 3 Test (Pages 156–157)

1. $6 - 3 \neq 3$ **2.** Some men are not created equal. **3.** No members of the class went on the field trip. **4.** That's the way you feel and I won't accept it. **5.** She did not apply or did not get a FEMA trailer. **6.** $\sim p \to q$ **7.** $p \to q$ **8.** $\sim q \leftrightarrow \sim p$ **9.** You won't love me and I will love you. **10.** It is not the case that you will love me or I will not love you. (Equivalently: You won't love me and I will love you.) **11.** T **12.** T **13.** T **14.** F **15.** Answers will vary. **16. (a)** The antecedent must be true and the consequent must be false. **(b)** Both component statements must be true. **(c)** Both component statements must be false. **17.** TFFF **18.** TTTT (tautology) **19.** false **20.** true

Wording may vary in the answers for Exercises 21–25. **21.** If the number is an integer, then it is a rational number. **22.** If a polygon is a rhombus, then it is a quadrilateral. **23.** If a number is divisible by 9, then it is divisible by 3. **24.** If she digs dinosaur bones, then she is a paleontologist. **25. (a)** If the graph helps me understand it, then a picture paints a thousand words. **(b)** If a picture doesn't paint a thousand words, then the graph won't help me understand it. **(c)** If the graph doesn't help me understand it, then a picture doesn't paint a thousand words. **26. (a)** $(q \wedge r) \rightarrow {\sim}p$ **(b)** $p \rightarrow ({\sim}q \vee {\sim}r)$ **(c)** $({\sim}q \vee {\sim}r) \rightarrow p$ **27.** valid **28. (a)** A **(b)** F **(c)** C **(d)** D **29.** valid **30.** invalid

CHAPTER 4 Numeration and Mathematical Systems

4.1 Exercises (Pages 166–168)

1. 13,036 **3.** 7,630,729 **5.** 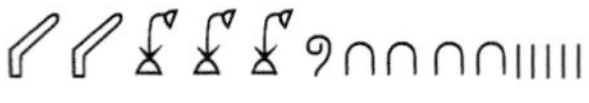 **7.** 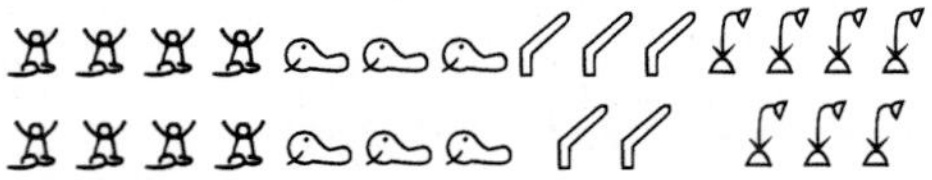

9. 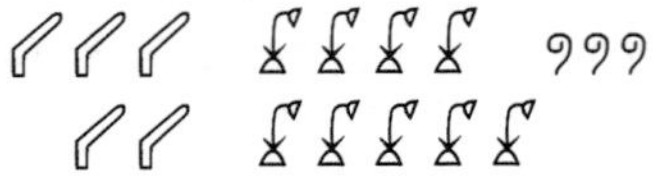 **11.** 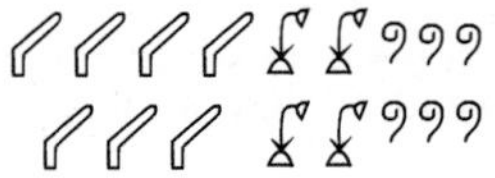 **13.** 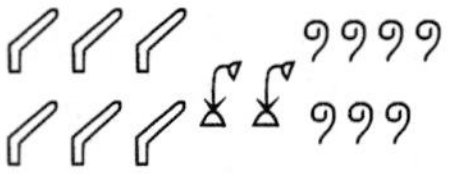 **15.** 935

17. 3007 九 **19.** 百六十 **21.** 七千零十二 **23.** 一千三百六十八 to 一千六百四十四 **25.** 六百十八 to 九百零七 **27.** 216 **29.** 53,601 **31.** 113

33. 7598 **35.** 1378 **37.** 5974 **39.** 622,500 shekels **41.** Answers will vary. **43.** Answers will vary. **45.** 99,999 **47.** 3124 **49.** $10^d - 1$ **51.** $7^d - 1$ **53.** Answers will vary.

4.2 Exercises (Pages 177–178)

1. $(7 \cdot 10^1) + (3 \cdot 10^0)$ **3.** $(3 \cdot 10^3) + (7 \cdot 10^2) + (7 \cdot 10^1) + (4 \cdot 10^0)$ **5.** $(4 \cdot 10^3) + (9 \cdot 10^2) + (2 \cdot 10^1) + (4 \cdot 10^0)$ **7.** $(1 \cdot 10^7) + (4 \cdot 10^6) + (2 \cdot 10^5) + (0 \cdot 10^4) + (6 \cdot 10^3) + (0 \cdot 10^2) + (4 \cdot 10^1) + (0 \cdot 10^0)$ **9.** 42 **11.** 6209 **13.** 70,401,009 **15.** 89 **17.** 32 **19.** 109 **21.** 733 **23.** 6 **25.** 206 **27.** 256 **29.** 63,259 **31.** 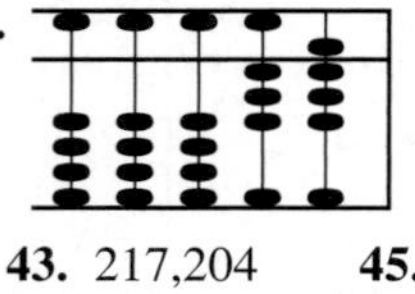 **33.** 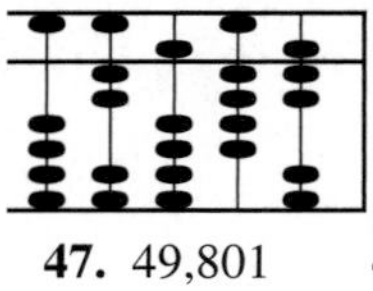 **35.** 1885 **37.** 38,325

39. 3,035,154 **41.** 496 **43.** 217,204 **45.** 242 **47.** 49,801 **49.** 460 **51.** 32,798

4.3 Exercises (Pages 187–189)

1. 1, 2, 3, 4, 5, 6, 10, 11, 12, 13, 14, 15, 16, 20, 21, 22, 23, 24, 25, 26 **3.** 1, 2, 3, 4, 5, 6, 7, 8, 10, 11, 12, 13, 14, 15, 16, 17, 18, 20, 21, 22 **5.** 13_{five}; 20_{five} **7.** $\text{B6E}_{\text{sixteen}}$; $\text{B70}_{\text{sixteen}}$ **9.** 3 **11.** 11 **13.** least: $1000_{\text{three}} = 27$; greatest: $2222_{\text{three}} = 80$ **15.** 14 **17.** 11 **19.** 956 **21.** 881 **23.** 28,854 **25.** 139 **27.** 5601 **29.** 321_{five} **31.** 10011_{two} **33.** 93_{sixteen} **35.** 2131101_{five} **37.** 1001001010_{two} **39.** 102112101_{three} **41.** 111134_{six} **43.** 32_{seven} **45.** 1031321_{four} **47.** 11110111_{two} **49.** 467_{eight} **51.** 11011100_{two} **53.** $\text{2D}_{\text{sixteen}}$ **55.** 37_{eight} **57.** 1427 **59.** 1000011_{two} **61.** 1101011_{two} **63.** HELP **65.** $10011101100101111110111_{\text{two}}$ **67.** Answers will vary. **69. (a)** The binary ones digit is 1. **(b)** The binary twos digit is 1. **(c)** The binary fours digit is 1. **(d)** The binary eights digit is 1. **(e)** The binary sixteens digit is 1. **71.** 6 **73.** yes **75.** yes **77.** yes **79.** yes **81.** Answers will vary. **83.** no **85.** yes **87.** Answers will vary. **89.** 20120011_{three} **91.** 25657_{nine}

4.4 Exercises (Pages 199–203)

1. 5 **3.** 6 **5.** row 2: 0, 6, 10; row 3: 9, 0, 9; row 4: 0, 4, 0, 8, 0; row 5: 1, 9, 2, 7; row 6: 6, 0, 0; row 7: 4, 11, 6, 8, 3, 5; row 8: 8, 4, 0, 0; row 9: 6, 3, 9, 3, 9, 6, 3; row 10: 6, 4, 0, 10, 8, 6, 4; row 11: 10, 9, 8, 7, 6, 5, 4, 3, 2 **7.** yes **9.** row 1: 0; row 2: 0; row 3: 0, 1, 2 row 4: 0, 1, 2 **11.** yes **13.** yes (0 is its own inverse, 1 and 4 are inverses of each other, and 2 and 3 are inverses of each other.) **15.** yes **17.** yes (1 is the identity element.) **19.** 3 **21.** 4 **23.** Answers will vary. **25.** 0700 **27.** 0000 **29.** false **31.** true **33.** 3 **35.** 3 **37.** 1 **39.** 10 **41.** Answers will vary. **43.** 5 **45.** 4 **47.** row 1: 0; row 2: 2, 3, 4, 5, 6, 0, 1; row 3: 3, 4, 5, 6, 0, 1, 2; row 4: 4, 5, 6, 0, 1, 2, 3; row 5: 5, 6, 0, 1, 2, 3, 4; row 6: 6, 0, 1, 2, 3, 4, 5 **49.** row 2: 1 **51.** row 2: 1, 3, 7; row 3: 3, 0; row 4: 3, 2, 1; row 5: 6, 7, 4; row 7: 3, 1, 6, 4; row 8: 6, 5, 2 **53.** {3, 10, 17, 24, 31, 38, . . . } **55.** {1, 2, 3, 4, 5, 6, . . . } **57.** 100,000 **59.** **(a)** 365 **(b)** Friday **61.** 62 **63.** Chicago: July 23 and 29; New Orleans: July 5 and August 16; San Francisco: August 9 **65.** Sunday **67.** Wednesday **69.** June **71.** June **73.** yes **75.** 6 **77.** 2

4.5 Exercises (Pages 208–210)

1. all properties; 1 is the identity element; 1 is its own inverse, as is 2. **3.** closure, commutative, associative, and identity properties; 1 is the identity element; 2, 4, and 6 have no inverses. **5.** all properties except the inverse; 1 is the identity element; 5 has no inverse. **7.** all properties; *F* is the identity element; *A* and *B* are inverses; *F* is its own inverse. **9.** all properties; *t* is the identity element; *s* and *r* are inverses; *t* and *u* are their own inverses. **11.** *a* **13.** *a* **15.** row *b*: *d*; row *c*: *d*, *b*; row *d*: *b*, *c* **17.** associative, commutative, identity (U), closure **19.** Answers may vary. One possibility is shown.

	a	*b*	*c*	*d*
a	*a*	*b*	*c*	*d*
b	*b*	*a*	*d*	*c*
c	*c*	*d*	*a*	*b*
d	*d*	*c*	*b*	*a*

21. no **23.** **(a)** true **(b)** true **(c)** true **(d)** true

25. $a + b + c = 1$ or $a = 0$ **27.** **(a)** $a = 0$ **(b)** $a = 0$ **29.** Each side simplifies to *e*. **31.** Each side simplifies to *d*. **33.**

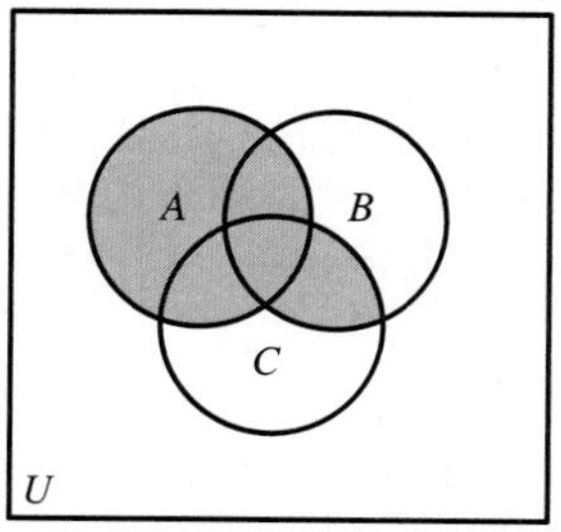

$A \cup (B \cap C) = (A \cup B) \cap (A \cup C)$

35. Both final columns read TTTTTFFF, when set up in the manner described in Chapter 3.

4.6 Exercises (Pages 216–218)

1. No operation is specified. **3.** yes **5.** no; closure, inverse **7.** yes **9.** no; inverse **11.** no; associative, identity, inverse **13.** no; inverse **15.** yes **17.** no; closure, identity, inverse **19.** Answers will vary. **21.** *S* **23.** *N* **25.** Each side is equal to *M*. **27.** Each side is equal to *Q*. **29.** *N* **31.** *R* **33.** *T* **35.** no **37.** yes **39.** Answers will vary. **41.** Answers will vary. **43.** row *B*: *D*; row *C*: *D*, *B*; row *D*: *C*, *B* **45.** **(a)** yes **(b)** Answers will vary. **47.** **(a)** no **(b)** Answers will vary.

Chapter 4 Test (Pages 219–220)

1. ancient Egyptian; 2426 **2.** 7561 **3.** $(6 \cdot 10^4) + (0 \cdot 10^3) + (9 \cdot 10^2) + (2 \cdot 10^1) + (3 \cdot 10^0)$ **4.** 1998 **5.** 22,184 **6.** 12,827 **7.** 89 **8.** 50 **9.** 57,007 **10.** 110001_{two} **11.** 43210_{five} **12.** 256_{eight} **13.** {2, 7, 12, 17, 22, 27, . . . } **14.** **(a)** 5 **(b)** 4 **(c)** 9 **(d)** 8 **15.** **(a)** 3 **(b)** 1 **16.** 48 **17.** There is less repetition of symbols. **18.** There are fewer symbols to learn. **19.** There are fewer digits in the numerals. **20.** 0 **21.** $\frac{1}{3}$ **22.** commutative **23.** row 3: 1, 7; row 5: 7, 3; row 7: 5, 1 **24.** **(a)** yes **(b)** 1 **25.** **(a)** yes **(b)** Answers will vary. **26.** **(a)** yes **(b)** Answers will vary. **27.** **(a)** no **(b)** Answers will vary. **28.** **(a)** yes **(b)** Answers will vary.

CHAPTER 5 Number Theory

5.1 Exercises (Pages 230–233)

1. true **3.** false **5.** false **7.** true **9.** false **11.** false **13.** 1, 2, 3, 4, 6, 12 **15.** 1, 2, 4, 5, 10, 20 **17.** 1, 2, 3, 4, 5, 6, 8, 10, 12, 15, 20, 24, 30, 40, 60, 120 **19. (a)** no **(b)** yes **(c)** no **(d)** yes **(e)** no **(f)** no **(g)** yes **(h)** no **(i)** no **21. (a)** no **(b)** no **(c)** no **(d)** yes **(e)** no **(f)** no **(g)** no **(h)** no **(i)** no **23. (a)** no **(b)** yes **(c)** no **(d)** no **(e)** no **(f)** no **(g)** yes **(h)** no **(i)** no **25. (a)** Answers will vary. **(b)** 13 **(c)** square root; square root; square root **(d)** prime **27.** 2, 3; no **29.** It must be 0.

31.

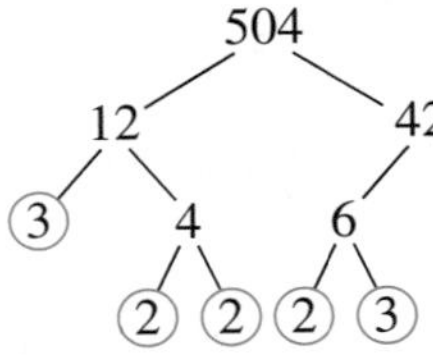

$504 = 3 \cdot 2^3 \cdot 3 \cdot 7$
$= 2^3 \cdot 3^2 \cdot 7$

33. $2^4 \cdot 3 \cdot 5$ **35.** $2^3 \cdot 3^2 \cdot 5$ **37.** $3 \cdot 13 \cdot 17$

39. yes **41.** no **43.** yes **45.** no **47.** The number must be divisible by both 3 and 5. That is, the sum of the digits must be divisible by 3, and the last digit must be 5 or 0. **49.** $75 = 3 \cdot 25$, $75 = 5 \cdot 15$ **51.** 0, 2, 4, 6, 8 **53.** 0, 4, 8 **55.** 0, 6 **57.** 6 **59.** 10 **61.** 27 **63.** leap year **65.** leap year **67.** Answers will vary. **69.** Answers will vary. **71.** $41^2 - 41 + 41 = 41^2$, and 41^2 is not a prime. **72. (a)** 1763 **(b)** 1847 **73.** B **75. (a)** 65,537 **(b)** 251 **77.** Answers will vary. **79.** Answers will vary. **81.** composite: $30{,}031 = 59 \cdot 509$ **83.** 63 **85.** $2^p - 1$ **87.** 3 and 31

5.2 Exercises (Pages 238–240)

1. true **3.** true **5.** true **7.** false **9.** true **11.** The sum of the proper divisors is 496: $1 + 2 + 4 + 8 + 16 + 31 + 62 + 124 + 248 = 496$. **13.** 8191 is prime; 33,550,336 **15.** $1 + \frac{1}{2} + \frac{1}{3} + \frac{1}{6} = 2$ **17.** abundant **19.** deficient **21.** 12, 18, 20, 24 **23.** $1 + 3 + 5 + 7 + 9 + 15 + 21 + 27 + 35 + 45 + 63 + 105 + 135 + 189 + 315 = 975$, and $975 > 945$, so 945 is abundant. **25.** $1 + 2 + 4 + 8 + 16 + 32 + 37 + 74 + 148 + 296 + 592 = 1210$ and $1 + 2 + 5 + 10 + 11 + 22 + 55 + 110 + 121 + 242 + 605 = 1184$ **27.** $3 + 11$ **29.** $3 + 23$ **31.** Let $a = 5$ and $b = 3$; $11 = 5 + 2 \cdot 3$ **33.** 71 and 73 **35.** 137 and 139 **37. (a)** $5 = 9 - 4 = 3^2 - 2^2$ **(b)** $11 = 36 - 25 = 6^2 - 5^2$ **39.** $5^2 + 2 = 27 = 3^3$ **41.** Answers will vary. **43.** No; for the first six, the sequence is 6, 8, 6, 8, 6, 6. **45.** one; not happy **47.** both; happy **49.** Answers will vary. **51.** B **53.** 7; yes **55.** 15; no **57.** 27; no **59.** 24; 23; 25; yes; no **61.** Answers will vary. **63.** B

5.3 Exercises (Pages 247–249)

1. true **3.** true **5.** false **7.** true **9.** true **11.** 10 **13.** 120 **15.** 7 **17.** 12 **19.** 10 **21.** 6 **23.** 12 **25.** 12 **27.** 70 **29.** Answers will vary. **31.** 120 **33.** 672 **35.** 840 **37.** 96 **39.** 225 **41.** 2160 **43.** 180 **45.** 1260 **47.** 1680 **49. (a)** $p^b q^c r^c$ **(b)** $p^a q^a r^b$ **51.** 30 **53.** 15 **55.** 12 **57. (a)** 6 **(b)** 36 **59. (a)** 18 **(b)** 216 **61.** p and q are relatively prime. **63.** Answers will vary. **65.** 144th **67.** 48 **69.** \$600; 25 books

Extension Exercises (Pages 256–257)

1. 3 **3.** 4 **5.** 5 **7.** 2 **9.** 2 **11.** 13 **13.** 12 **15.** 3 **17.** 55; 40 **19.** 65; 48 **21.** 5 **23.** 61 **25. (a)** 27 **(b)** 35 **27. (a)** 11 **(b)** 23 **29.** Answers will vary.

5.4 Exercises (Pages 262–265)

1. 2584 **3.** 46,368 **5.** $\frac{1+\sqrt{5}}{2}$ **7.** $1 + 1 + 2 + 3 + 5 + 8 = 21 - 1$; Each expression is equal to 20. **9.** $1 + 2 + 5 + 13 + 34 + 89 = 144$; Each expression is equal to 144. **11.** $13^2 - 5^2 = 144$; Each expression is equal to 144. **13.** $1 - 2 + 5 - 13 + 34 - 89 = -8^2$; Each expression is equal to -64. **15.** (There are other ways to do this.) **(a)** $37 = 34 + 3$ **(b)** $40 = 34 + 5 + 1$ **(c)** $52 = 34 + 13 + 5$ **17. (a)** The greatest common factor of 10 and 4 is 2, and the greatest common factor of $F_{10} = 55$ and $F_4 = 3$ is $F_2 = 1$. **(b)** The greatest common factor of 12 and 6 is 6, and the greatest common factor of $F_{12} = 144$ and $F_6 = 8$ is $F_6 = 8$. **(c)** The greatest common factor of 14 and 6 is 2, and the greatest common factor of $F_{14} = 377$ and $F_6 = 8$ is $F_2 = 1$. **19. (a)** $5 \cdot 34 - 13^2 = 1$ **(b)** $13^2 - 3 \cdot 55 = 4$ **(c)** $2 \cdot 89 - 13^2 = 9$ **(d)** The difference will be 25, because we are obtaining the squares of the terms of the Fibonacci sequence. $13^2 - 1 \cdot 144 = 25 = 5^2$. **21.** 199 **23.** Each sum is 2 less than a Lucas number. **25. (a)** $8 \cdot 18 = 144$; Each expression is equal to 144. **(b)** $8 + 21 = 29$; Each expression is equal to 29.

(c) $8 + 18 = 2 \cdot 13$; Each expression is equal to 26. **27.** 3, 4, 5 **29.** 16, 30, 34 **31.** The sums are 1, 1, 2, 3, 5, 8, 13. They are terms of the Fibonacci sequence. **33.** $\frac{1+\sqrt{5}}{2} \approx 1.618033989$ and $\frac{1-\sqrt{5}}{2} \approx -.618033989$. After the decimal point, the digits are the same. **35.** 377 **37.** 17,711

Extension Exercises (Pages 267–269)

1.

2	7	6
9	5	1
4	3	8

3.

11	10	4	23	17
18	12	6	5	24
25	19	13	7	1
2	21	20	14	8
9	3	22	16	15

5.

15	16	22	3	9
8	14	20	21	2
1	7	13	19	25
24	5	6	12	18
17	23	4	10	11

7.

24	9	12
3	15	27
18	21	6

Magic sum is 45.

9.

$\frac{17}{2}$	12	$\frac{1}{2}$	4	$\frac{15}{2}$
$\frac{23}{2}$	$\frac{5}{2}$	$\frac{7}{2}$	7	8
2	3	$\frac{13}{2}$	10	11
5	6	$\frac{19}{2}$	$\frac{21}{2}$	$\frac{3}{2}$
$\frac{11}{2}$	9	$\frac{25}{2}$	1	$\frac{9}{2}$

Magic sum is $32\frac{1}{2}$.

11. 479 **13.** 467 **15.** 269 **17.** **(a)** 73 **(b)** 70 **(c)** 74 **(d)** 69 **19.** **(a)** 7 **(b** 22 **(c)** 5 **(d)** 4 **(e)** 15 **(f)** 19 **(g)** 6 **(h)** 23

21.

30	39	48	1	10	19	28
38	47	7	9	18	27	29
46	6	8	17	26	35	37
5	14	16	25	34	36	45
13	15	24	33	42	44	4
21	23	32	41	43	3	12
22	31	40	49	2	11	20

23. Each sum is equal to 34. **25.** Each sum is equal to 68. **27.** Each sum is equal to 9248. **29.** Each sum is equal to 748. **31.**

16	2	3	13
5	11	10	8
9	7	6	12
4	14	15	1

The second and third columns are interchanged.

33.

18	20	10
8	16	24
22	12	14

35.

39	48	57	10	19	28	37
47	56	16	18	27	36	38
55	15	17	26	35	44	46
14	23	25	34	43	45	54
22	24	33	42	51	53	13
30	32	41	50	52	12	21
31	40	49	58	11	20	29

Magic sum is 238.

37. 260 **39.** $52 + 45 + 16 + 17 + 54 + 43 + 10 + 23 = 260$

41.

5	13	21	9	17
6	19	2	15	23
12	25	8	16	4
18	1	14	22	10
24	7	20	3	11

Chapter 5 Test (Pages 270–271)

1. false **2.** true **3.** true **4.** true **5.** true **6.** **(a)** yes **(b)** yes **(c)** no **(d)** yes **(e)** yes **(f)** no **(g)** yes **(h)** yes **(i)** no **7.** **(a)** composite **(b)** neither **(c)** prime **8.** $2^5 \cdot 3^2 \cdot 5$ **9.** Answers will vary. **10.** **(a)** deficient **(b)** perfect **(c)** abundant **11.** C **12.** 41 and 43 **13.** 90 **14.** 360

15. Monday **16.** 46,368 **17.** $89 - (8 + 13 + 21 + 34) = 13$; Each expression is equal to 13. **18.** B **19. (a)** 1, 5, 6, 11, 17, 28, 45, 73 **(b)** The process will yield 19 for any term chosen. **20.** A **21.** Answers will vary. **22.** Answers will vary.

CHAPTER 6 The Real Numbers and Their Representations

6.1 Exercises (Pages 280–283)

1. 4 **3.** 0 **5.** $\sqrt{12}$ (There are others.) **7.** true **9.** true **11. (a)** 3, 7 **(b)** 0, 3, 7 **(c)** $-9, 0, 3, 7$ **(d)** $-9, -1\frac{1}{4}, -\frac{3}{5}, 0, 3, 5.9, 7$ **(e)** $-\sqrt{7}, \sqrt{5}$ **(f)** All are real numbers. **13.** Answers will vary. **15.** 1450 **17.** 5436 **19.** $-220°$ **21.** 20; 10°; $-9°$ **23. (a)** Pacific Ocean, Indian Ocean, Caribbean Sea, South China Sea, Gulf of California **(b)** Point Success, Ranier, Matlalcueyetl, Steele, McKinley **(c)** true **(d)** false

25. 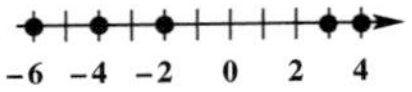

27. 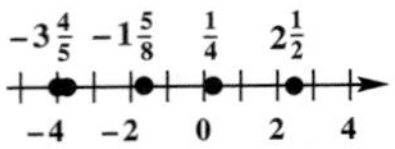

29. (a) A **(b)** A **(c)** B **(d)** B **31. (a)** 2 **(b)** 2

33. (a) -6 **(b)** 6 **35. (a)** -3 **(b)** 3 **37. (a)** 0 **(b)** 0 **39.** $a - b$ **41.** -12 **43.** -8 **45.** 3 **47.** $|-3|$ or 3 **49.** $-|-6|$ or -6 **51.** $|5 - 3|$ or 2 **53.** true **55.** true **57.** true **59.** false **61.** true **63.** false **65.** petroleum refineries, 2002 to 2003 **67.** construction machinery manufacturing, 2002 to 2003 **69.** computer/data processing services **Answers will vary in Exercises 71–75.** **71.** $\frac{1}{2}, \frac{5}{8}, 1\frac{3}{4}$ **73.** $-3\frac{1}{2}, -\frac{2}{3}, \frac{3}{7}$ **75.** $\sqrt{5}, \pi, -\sqrt{3}$

6.2 Exercises (Pages 292–296)

1. negative **3.** -3; 5 **5.** Answers will vary. **7.** -20 **9.** -4 **11.** -11 **13.** 9 **15.** 20 **17.** 4 **19.** 24 **21.** -1296 **23.** 6 **25.** -6 **27.** 0 **29.** -6 **31.** -4 **33.** 27 **35.** 39 **37.** -2 **39.** not a real number **41.** 13 **43.** A, B, C **45.** commutative property of addition **47.** associative property of addition **49.** inverse property of addition **51.** identity property of multiplication **53.** identity property of addition **55.** distributive property **57.** inverse property of addition **59.** closure property of multiplication **61. (c)** Yes; choose $a = b$. For example, $a = b = 2: 2 - 2 = 2 - 2$. **63. (a)** messing up your room **(b)** spending money **(c)** decreasing the volume on your MP3 player **65.** identity **67.** No, it does not hold true. **69.** -81 **71.** 81 **73.** -81 **75.** -81 **77. (a)** -3.6 (billion dollars) **(b)** 7.8 (billion dollars) **(c)** 28.2 (billion dollars) **(d)** 32.4 (billion dollars) **79. (a)** 2000: \$129 billion; 2010: \$206 billion; 2020: \$74 billion; 2030: $-$\$501 billion **(b)** The cost of Social Security will exceed revenue in 2030 by \$501 billion. **81.** 16 **83.** \$1045.55 **85.** 14,776 feet **87.** 45°F **89.** 112°F **91.** -41°F **93.** 27 feet **95.** 469 B.C. **97.** \$2169

6.3 Exercises (Pages 306–310)

1. A, C, D **3.** C **5.** $\frac{1}{3}$ **7.** $-\frac{3}{7}$ **Answers will vary in Exercises 9 and 11.** **9.** $\frac{6}{16}, \frac{9}{24}, \frac{12}{32}$ **11.** $-\frac{10}{14}, -\frac{15}{21}, -\frac{20}{28}$ **13. (a)** $\frac{1}{3}$ **(b)** $\frac{1}{4}$ **(c)** $\frac{2}{5}$ **(d)** $\frac{1}{3}$ **15.** the dots in the intersection of the triangle and the rectangle as a part of the dots in the entire figure **17. (a)** O'Brien **(b)** Ha **(c)** Ha **(d)** Kelly **(e)** Taylor and Britz; $\frac{1}{2}$ **19.** $\frac{1}{2}$ **21.** $\frac{43}{48}$ **23.** $-\frac{5}{24}$ **25.** $\frac{23}{56}$ **27.** $\frac{27}{20}$ **29.** $\frac{5}{12}$ **31.** $\frac{1}{9}$ **33.** $\frac{3}{2}$ **35.** $\frac{3}{2}$ **37. (a)** $4\frac{1}{2}$ cups **(b)** $\frac{7}{8}$ cup **39.** $\frac{13}{3}$ **41.** $\frac{29}{10}$ **43.** $6\frac{3}{4}$ **45.** $6\frac{1}{8}$ **47.** $-17\frac{7}{8}$ **49.** $\frac{9}{16}$ inch **51.** 3 **53.** $\frac{3}{7}$ **55.** $-\frac{103}{89}$ **57.** $\frac{25}{9}$ **59.** $\frac{5}{8}$ **61.** $\frac{19}{30}$ **63.** $-\frac{3}{4}$ **65.** \$53,221 **67.** $\frac{14}{19}$ **69.** $\frac{13}{29}$ **71.** $\frac{5}{2}$ **73.** It gives the rational number halfway between the two integers (their average). **75.** .75 **77.** .1875 **79.** $.\overline{27}$ **81.** $.\overline{285714}$ **83.** $\frac{2}{5}$ **85.** $\frac{17}{20}$ **87.** $\frac{467}{500}$ **89.** repeating **91.** terminating **93.** terminating **95. (a)** $.\overline{3}$ or .333 . . . **(b)** $.\overline{6}$ or .666 . . . **(c)** $.\overline{9}$ or .999 . . . **(d)** $1 = .\overline{9}$ **97. (a)** $\frac{4}{5}$ **(b)** $\frac{4}{5}$ **99. (a)** $\frac{33}{50}$ **(b)** $\frac{33}{50}$

6.4 Exercises (Pages 318–321)

1. rational **3.** irrational **5.** rational **7.** rational **9.** irrational **11.** irrational **13.** rational **15. (a)** $.\overline{8}$ **(b)** irrational; rational **17.** 6.244997998 **19.** 3.885871846 **21.** 29.73213749 **23.** 1.060660172 **25.** $5\sqrt{2}$; 7.071067812 **27.** $5\sqrt{3}$; 8.660254038 **29.** $12\sqrt{2}$; 16.97056275 **31.** $\frac{5\sqrt{6}}{6}$; 2.041241452 **33.** $\frac{\sqrt{7}}{2}$; 1.322875656 **35.** $\frac{\sqrt{21}}{3}$; 1.527525232 **37.** $3\sqrt{17}$ **39.** $4\sqrt{7}$ **41.** $10\sqrt{2}$

43. $3\sqrt{3}$ **45.**

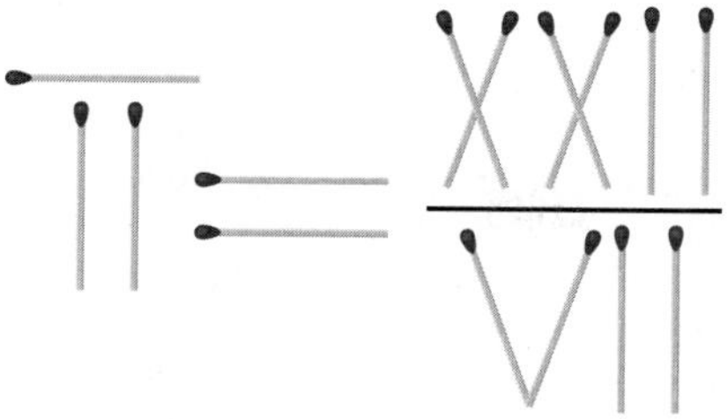

47. The result is 3.1415929, which agrees with the first seven digits in the decimal for π. **49.** 3 **51.** 4 **53.** 3.3 **55.** ϕ is positive, while its conjugate is negative. The units digit of ϕ is 1, and the units digit of its conjugate is 0. The decimal digits agree. **57.** It is just a coincidence that 1828 appears back-to-back early in the decimal. There is no repetition indefinitely, which would be indicative of a rational number. **59.** 2.5 seconds **61.** 15.3 miles **63.** 392,000 square miles **65.** The area and the perimeter are both numerically equal to 36. **67.** 5.4 feet **69. (a)** 70.5 mph **(b)** 59.8 mph **(c)** 53.9 mph **71.** 4 **73.** 7 **75.** 6 **77.** 1 **79.** 4 **81.** 8 **The number of decimal digits shown will vary among caculator models in Exercises 83–87.** **83.** 3.50339806 **85.** 5.828476683 **87.** 10.06565066

6.5 Exercises (Pages 331–335)

1. true **3.** false **5.** true **7.** true **9.** false **11.** 11.315 **13.** -4.215 **15.** .8224 **17.** 47.5 **19.** 31.6 **21.** Los Angeles; The population increased, as represented by 9.7%. **23.** three (and you would have .01¢ left over) **25.** \$.06 or 6¢ **27.** 1000 **29. (a)** .031 **(b)** .035 **31.** 297 **33. (a)** 78.4 **(b)** 78.41 **35. (a)** .1 **(b)** .08 **37. (a)** 12.7 **(b)** 12.69 **39.** 42% **41.** 36.5% **43.** .8% **45.** 210% **47.** 20% **49.** 1% **51.** $37\frac{1}{2}\%$ **53.** 150% **55.** Answers will vary. **57. (a)** 5 **(b)** 24 **(c)** 8 **(d)** .5 or $\frac{1}{2}$ **(e)** 600 **59.** No, the price is \$57.60. **61. (a)** .611 **(b)** .574 **(c)** .438 **63.** 124.8 **65.** 2.94 **67.** 150% **69.** 600 **71.** 1.4% **73.** A **75.** C **77.** about 67% **79. (a)** $14.7 - 40 \cdot .13$ **(b)** 9.5 **(c)** 8.075; walking (5 mph) **81.** 860% **83.** \$4.50 **85.** \$.75 **87.** \$12.00 **89.** \$36.00 **91.** Answers will vary.

Extension Exercises (Page 338)

1. $12i$ **3.** $-15i$ **5.** $i\sqrt{3}$ **7.** $5i\sqrt{3}$ **9.** -5 **11.** -18 **13.** -40 **15.** $\sqrt{2}$ **17.** $3i$ **19.** 6 **21.** The product rule requires that a and b be nonnegative. **23.** 1 **25.** -1 **27.** $-i$ **29.** i

Chapter 6 Test (Pages 340–342)

1. (a) 12 **(b)** 0, 12 **(c)** $-4, 0, 12$ **(d)** $-4, -\frac{3}{2}, -.5, 0, 4.1, 12$ **(e)** $-\sqrt{5}, \sqrt{3}$ **(f)** $-4, -\sqrt{5}, -\frac{3}{2}, -.5, 0, \sqrt{3}, 4.1, 12$ **2. (a)** C **(b)** B **(c)** D **(d)** A **3. (a)** false **(b)** true **(c)** true **(d)** false **4.** 4 **5.** 3 **6.** 10 **7. (a)** Hyundai; 50% **(b)** General Motors; -5% **(c)** false **(d)** true **8.** 5296 ft **9. (a)** \$4900 **(b)** \$5700 **(c)** \$8800 **(d)** \$10,300 **10. (a)** E **(b)** A **(c)** B **(d)** D **(e)** F **(f)** C **11. (a)** Whitney **(b)** Moura and Dawkins **(c)** Whitney **(d)** Pritchard and Miller; $\frac{2}{5}$ **(e)** McElwain ("J-Mac") **12.** $\frac{11}{16}$ **13.** $\frac{57}{160}$ **14.** $-\frac{2}{5}$ **15.** $\frac{3}{2}$ **16. (a)** .45 **(b)** $.41\overline{6}$ **17. (a)** $\frac{18}{25}$ **(b)** $\frac{58}{99}$ **18. (a)** irrational **(b)** rational **(c)** rational **(d)** rational **(e)** irrational **(f)** irrational **19. (a)** 12.247448714 **(b)** $5\sqrt{6}$ **20. (a)** 4.913538149 **(b)** $\frac{13\sqrt{7}}{7}$ **21. (a)** -45.254834 **(b)** $-32\sqrt{2}$ **22.** Answers will vary. **23. (a)** 13.81 **(b)** $-.315$ **(c)** 38.7 **(d)** -24.3 **24. (a)** 9.04 **(b)** 9.045 **25. (a)** 16.65 **(b)** 101.5 **26. (a)** $26\frac{2}{3}\%$ **(b)** $66\frac{2}{3}\%$ **27.** D **28.** 828; 504; 16%; 6% **29.** 17,415,000 **30.** -13.8 (billion dollars)

CHAPTER 7 The Basic Concepts of Algebra

7.1 Exercises (Pages 351–354)

1. A and C **3.** Both sides are evaluated as 30, so 6 is a solution. **5.** solution set **7.** B **9.** $\{-1\}$ **11.** $\{3\}$ **13.** $\{-7\}$ **15.** $\{0\}$ **17.** $\{-\frac{5}{3}\}$ **19.** $\{-\frac{1}{2}\}$ **21.** $\{2\}$ **23.** $\{-2\}$ **25.** $\{7\}$ **27.** $\{2\}$ **29.** $\{\frac{3}{2}\}$ **31.** $\{-5\}$ **33.** $\{3\}$ **35.** 2 (that is, 10^2, or 100) **37.** $\{4\}$ **39.** $\{0\}$ **41.** $\{0\}$ **43.** $\{2000\}$ **45.** $\{25\}$ **47.** $\{40\}$ **49.** identity, contradiction **51.** contradiction; $\emptyset$ **53.** conditional; $\{-8\}$ **55.** conditional; $\{0\}$ **57.** identity; {all real numbers} **59.** D **61.** $t = \frac{d}{r}$ **63.** $b = \frac{A}{h}$ **65.** $a = P - b - c$ **67.** $b = \frac{2A}{h}$ **69.** $h = \frac{S - 2\pi r^2}{2\pi r}$ or $h = \frac{S}{2\pi r} - r$ **71.** $F = \frac{9}{5}C + 32$ **73.** $H = \frac{A - 2LW}{2W + 2L}$ **75. (a)** 16.8 million **(b)** 2009 **77. (a)** .0352 **(b)** approximately .015 or 1.5% **(c)** approximately 1 case **79. (a)** 800 cubic feet **(b)** 107,680 μg **(c)** $F = 107{,}680x$ **(d)** approximately .25 day, or 6 hr

7.2 Exercises (Pages 364–369)

1. expression **3.** equation **5.** expression **7.** yes **9.** $x - 14$ **11.** $(x - 7)(x + 5)$ **13.** $\frac{15}{x}$ $(x \neq 0)$ **15.** Answers will vary. **17.** 3 **19.** 6 **21.** -3 **23.** Springsteen: $115.9 million; Dion: $80.5 million **25.** wins: 62; losses: 20 **27.** Democrats: 44; Republicans: 55 **29.** shortest piece: 15 inches; middle piece: 20 inches; longest piece: 24 inches **31.** gold: 35; silver: 39; bronze: 29 **33.** 35 milliliters **35.** $350 **37.** $14.15 **39.** 4 liters **41.** $18\frac{2}{11}$ liters **43.** 5 liters **45.** $4000 at 3%; $8000 at 4% **47.** $10,000 at 4.5%; $19,000 at 3% **49.** $58,000 **51.** 17 pennies, 17 dimes, 10 quarters **53.** 305 students, 105 nonstudents **55.** 54 seats on Row 1; 51 seats on Row 2 **57.** 39-cent stamps: 28; 24-cent stamps: 17 **59.** 3.173 hours **61.** 1.715 hours **63.** 8.08 meters per second **65.** 8.40 meters per second **67.** 530 miles **69.** No, it is not correct. The distance is $45\left(\frac{1}{2}\right) = 22.5$ miles. **71.** $1\frac{3}{4}$ hours **73.** 10:00 A.M. **75.** 18 miles **77.** 8 hours

7.3 Exercises (Pages 378–383)

1. $\frac{5}{8}$ **3.** $\frac{1}{4}$ **5.** $\frac{2}{1}$ **7.** $\frac{3}{1}$ **9.** D **11.** Answers will vary. **13.** true **15.** false **17.** true **19.** $\{35\}$ **21.** $\{-1\}$ **23.** $\{-\frac{27}{4}\}$ **25.** $30.00 **27.** $8.75 **29.** $67.50 **31.** $38.85 **33.** 4 feet **35.** 2.7 inches **37.** 2.0 inches **39.** $2\frac{5}{8}$ cups **41.** $363.84 **43.** 12,500 fish **45.** 10-lb size; $.439 **47.** 32-oz size; $.093 **49.** 128-oz size; $.044 **51.** 36-oz size; $.049 **53.** $x = 4$ **55.** $x = 1$; $y = 4$ **57.** **(a)**

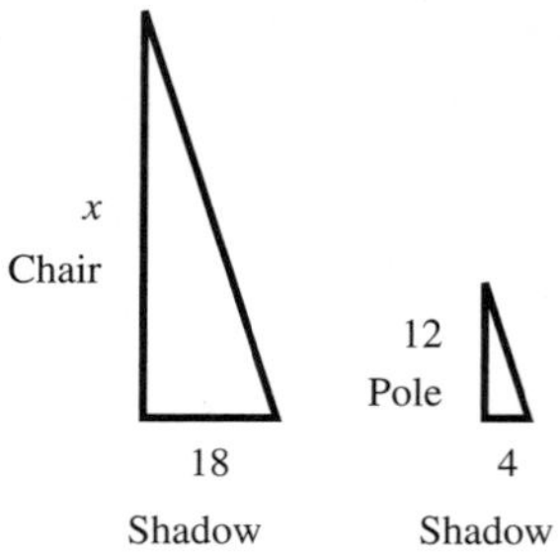

(b) 54 feet **59.** $144 **61.** $165 **63.** 9 **65.** 125 **67.** $\frac{4}{9}$ **69.** $40.32 **71.** 20 miles per hour **73.** about 302 pounds **75.** 100 pounds per square inch **77.** 20 pounds per square foot **79.** 144 feet **81.** 1.105 liters **83.** $\frac{8}{9}$ metric ton **85.** 6.2 pounds

7.4 Exercises (Pages 391–394)

1. D **3.** B **5.** F **7.** Use parentheses when the symbol is $<$ or $>$. Use brackets when the symbol is $\leq$ or $\geq$.

9. $[5, \infty)$ 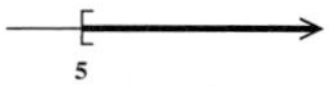

11. $(7, \infty)$ 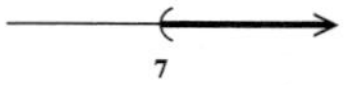

13. $(-4, \infty)$ 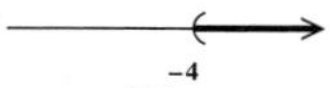

15. $(-\infty, -40]$ 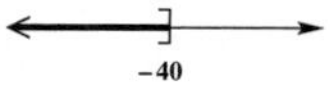

17. $(-\infty, 4]$ 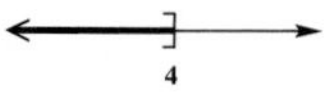

19. $\left(-\infty, -\frac{15}{2}\right)$ 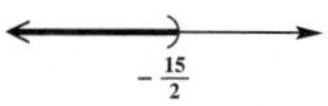

21. $\left[\frac{1}{2}, \infty\right)$ 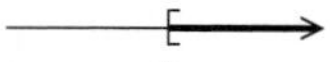

23. $(3, \infty)$ 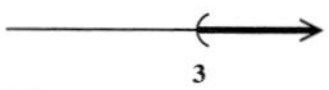

25. $(-\infty, 4)$ 

27. $\left(-\infty, \frac{23}{6}\right]$ 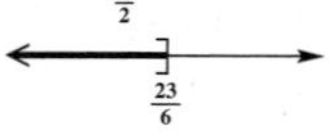

29. $\left(-\infty, \frac{76}{11}\right)$ 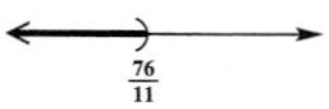

31. $(-\infty, \infty)$ 

33. $\emptyset$
35. Answers will vary. **37.** $(1, 11)$ 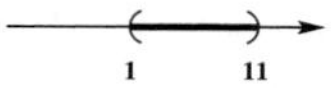

39. $[-14, 10]$ 

41. $[-5, 6]$ 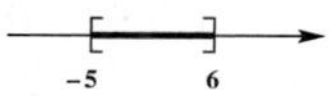

43. $\left[-\frac{14}{3}, 2\right]$ 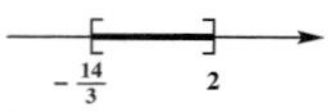

45. $\left[-\frac{1}{2}, \frac{35}{2}\right]$ 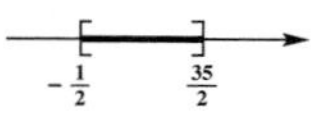

47. $\left(-\frac{1}{3}, \frac{1}{9}\right]$ 

49. April, May, June, July **51.** January, February, March, August, September, October, November, December **53.** from about 2:30 P.M. to 6:00 P.M. **55.** about 84°F–91°F **57.** 2 miles **59.** at least 80 **61.** 50 miles **63.** **(a)** 140 to 184 pounds **(b)** Answers will vary. **65.** 26 DVDs

7.5 Exercises (Pages 404–407)

1. A **3.** A **5.** D **7.** 625 **9.** -32 **11.** -8 **13.** -81 **15.** $\frac{1}{49}$ **17.** $-\frac{1}{49}$ **19.** -128 **21.** $\frac{16}{5}$ **23.** 125 **25.** $\frac{25}{16}$ **27.** $\frac{9}{20}$ **29.** 1 **31.** 1 **33.** 0 **35.** reciprocal, additive inverse **37.** D **39.** x^{16} **41.** 5 **43.** $\frac{1}{27}$ **45.** $\frac{1}{81}$ **47.** $\frac{1}{t^7}$ **49.** $9x^2$ **51.** $\frac{1}{a^5}$ **53.** x^{11} **55.** r^6 **57.** $-\frac{56}{k^2}$ **59.** $\frac{1}{z^4}$ **61.** $-\frac{3}{r^7}$ **63.** $\frac{27}{a^{18}}$ **65.** $\frac{x^5}{y^2}$ **67.** D **69.** 2.3×10^2 **71.** 2×10^{-2} **73.** 6500

75. .0152 **77.** 6×10^5 **79.** 2×10^5 **81.** 2×10^5 **83.** 1×10^9; 1×10^{12}; 2.128×10^{12}; 1.44419×10^5 **85.** $\$1.61964 \times 10^{10}$ **87.** 1×10^{10} **89.** 2,000,000,000 **91.** \$1392 **93.** approximately 9.474×10^{-7} parsec **95.** 300 seconds **97.** approximately 5.87×10^{12} miles **99.** 20,000 hours

7.6 Exercises (Pages 415–417)

1. $x^2 - x + 3$ **3.** $9y^2 - 4y + 4$ **5.** $6m^4 - 2m^3 - 7m^2 - 4m$ **7.** $-2x^2 - 13x + 11$ **9.** $x^2 - 5x - 24$ **11.** $28r^2 + r - 2$ **13.** $12x^5 + 8x^4 - 20x^3 + 4x^2$ **15.** $4m^2 - 9$ **17.** $16m^2 + 16mn + 4n^2$ **19.** $25r^2 + 30rt^2 + 9t^4$ **21.** $-2z^3 + 7z^2 - 11z + 4$ **23.** $m^2 + mn - 2n^2 - 2km + 5kn - 3k^2$ **25.** $a^2 - 2ab + b^2 + 4ac - 4bc + 4c^2$ **27.** A **29.** Answers will vary. **31.** $2m^2(4m^2 + 3m - 6)$ **33.** $4k^2m^3(1 + 2k^2 - 3m)$ **35.** $2(a + b)(1 + 2m)$ **37.** $(m - 1)(2m^2 - 7m + 7)$ **39.** $(2s + 3)(3t - 5)$ **41.** $(t^3 + s^2)(r - p)$ **43.** $(8a + 5b)(2a - 3b)$ **45.** $(5z - 2x)(4z - 9x)$ **47.** $(1 - a)(1 - b)$ **49.** $(x - 5)(x + 3)$ **51.** $(y + 7)(y - 5)$ **53.** $6(a - 10)(a + 2)$ **55.** $3m(m + 1)(m + 3)$ **57.** $(3k - 2p)(2k + 3p)$ **59.** $(5a + 3b)(a - 2b)$ **61.** $(7x + 2y)(3x - y)$ **63.** $2a^2(4a - b)(3a + 2b)$ **65.** Answers will vary. **67.** $(3m - 2)^2$ **69.** $2(4a - 3b)^2$ **71.** $(2xy + 7)^2$ **73.** $(x + 6)(x - 6)$ **75.** $(y + w)(y - w)$ **77.** $(3a + 4)(3a - 4)$ **79.** $(5s^2 + 3t)(5s^2 - 3t)$ **81.** $(p^2 + 25)(p + 5)(p - 5)$ **83.** $(2 - a)(4 + 2a + a^2)$ **85.** $(5x - 3)(25x^2 + 15x + 9)$ **87.** $(3y^3 + 5z^2)(9y^6 - 15y^3z^2 + 25z^4)$ **89.** $(x + y)(x - 5)$ **91.** $(m - 2n)(p^4 + q)$ **93.** $(2z + 7)^2$ **95.** $(10x + 7y)(100x^2 - 70xy + 49y^2)$ **97.** $(5m^2 - 6)(25m^4 + 30m^2 + 36)$ **99.** $(6m - 7n)(2m + 5n)$

7.7 Exercises (Pages 422–426)

1. 4, 5, −9 **3.** two **5.** $\{-3, 9\}$ **7.** $\left\{\frac{7}{2}, -\frac{1}{5}\right\}$ **9.** $\{-3, 4\}$ **11.** $\{-7, -2\}$ **13.** $\left\{-\frac{1}{2}, \frac{1}{6}\right\}$ **15.** $\{-2, 4\}$ **17.** $\{\pm 8\}$ **19.** $\{\pm 2\sqrt{6}\}$ **21.** $\emptyset$ **23.** $\{1, 7\}$ **25.** $\{4 \pm \sqrt{3}\}$ **27.** $\left\{\frac{5 \pm \sqrt{13}}{2}\right\}$ **29.** $\left\{\frac{2 \pm \sqrt{3}}{2}\right\}$ **31.** $\left\{\frac{1 \pm \sqrt{3}}{2}\right\}$ **33.** $\left\{\frac{1 \pm \sqrt{5}}{2}\right\}$ **35.** $\left\{\frac{-1 \pm \sqrt{2}}{2}\right\}$ **37.** $\left\{\frac{1 \pm \sqrt{29}}{2}\right\}$ **39.** $\emptyset$ **41.** Answers will vary. $\left\{\pm\frac{\sqrt{10}}{2}\right\}$ **43.** The presence of $2x^3$ makes it a *cubic* equation (degree 3). **45.** 0; (c) **47.** 121; (a) **49.** 360; (b) **51.** 5.2 seconds **53.** Find s when $t = 0$. **55.** **(a)** 1 second and 8 seconds **(b)** 9 seconds after it is projected **57.** 2.3, 5.3, 5.8 **59.** 412.3 feet **61.** eastbound ship: 80 miles; southbound ship: 150 miles **63.** 5 centimeters, 12 centimeters, 13 centimeters **65.** length: 2 centimeters; width: 1.5 centimeters **67.** 1 foot **69.** length: 26 meters; width: 16 meters **71.** length: 20 inches; width: 12 inches **73.** \$.80 **75.** 5.5 meters per second **77.** 5 or 14 **79.** $\{2\}$

Chapter 7 Test (Pages 428–429)

1. $\{2\}$ **2.** $\{4\}$ **3.** identity; {all real numbers} **4.** $v = \frac{S + 16t^2}{t}$ **5.** Hawaii: 4021 square miles; Maui: 728 square miles; Kauai: 551 square miles **6.** 5 liters **7.** 2.2 hours **8.** 8 slices for \$2.19 **9.** 2300 miles **10.** 200 amps **11.** $(-\infty, 4]$ 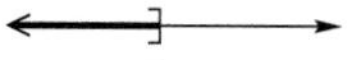

12. $(-2, 6]$ 

13. C **14.** at least 82 **15.** $\frac{16}{9}$ **16.** −64 **17.** $\frac{64}{27}$ **18.** 0 **19.** $\frac{216}{p^4}$ **20.** $\frac{1}{m^{14}}$ **21.** **(a)** 693,000,000 **(b)** .000000125 **22.** 3×10^{-4} **23.** about 15,300 seconds **24.** $4k^2 + 6k + 10$ **25.** $15x^2 - 14x - 8$ **26.** $16x^4 - 9$ **27.** $3x^3 + 20x^2 + 23x - 36$ **28.** One example is $t^5 + 2t^4 + 3t^3 - 4t^2 + 5t + 6$. **29.** $(2p - 3q)(p - q)$ **30.** $(10x + 7y)(10x - 7y)$ **31.** $(3y - 5x)(9y^2 + 15yx + 25x^2)$ **32.** $(4 - m)(x + y)$ **33.** $\left\{-\frac{3}{2}, \frac{1}{3}\right\}$ **34.** $\{\pm\sqrt{13}\}$ **35.** $\left\{\frac{1 \pm \sqrt{29}}{2}\right\}$ **36.** .87 second

CHAPTER 8 Graphs, Functions, and Systems of Equations and Inequalities

8.1 Exercises (Pages 438–441)

1. **(a)** x represents the year; y represents the percent of women in mathematics or computer science professions. **(b)** 1990–2000 **(c)** 1990 **(d)** 1980 **3.** x **5.** (0, 0) **7.** **(a)** I **(b)** III **(c)** II **(d)** IV **(e)** none

9. **(a)** I or III **(b)** II or IV **(c)** II or IV **(d)** I or III **11.–20.**

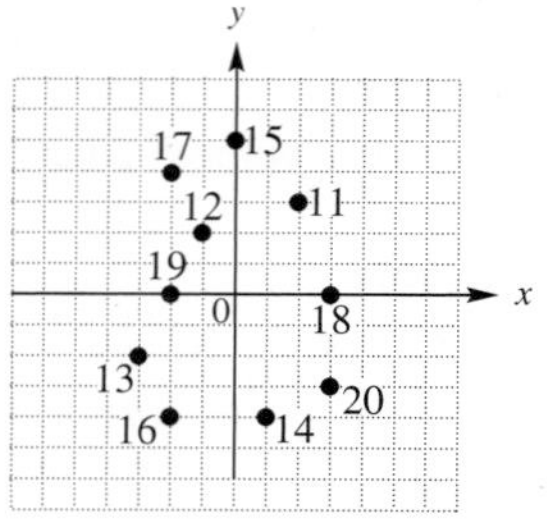

21. **(a)** $\sqrt{34}$ **(b)** $\left(\frac{1}{2}, \frac{5}{2}\right)$ **23.** **(a)** $\sqrt{61}$ **(b)** $\left(\frac{1}{2}, 1\right)$ **25.** **(a)** $\sqrt{146}$ **(b)** $\left(-\frac{1}{2}, \frac{3}{2}\right)$ **27.** B **29.** D
31. $x^2 + y^2 = 36$ **33.** $(x + 1)^2 + (y - 3)^2 = 16$ **35.** $x^2 + (y - 4)^2 = 3$ **37.** center: (0, 0); radius: r
39. $(-2, -3)$; 2 **41.** $(-5, 7)$; 9 **43.** (2, 4); 4 **45.**

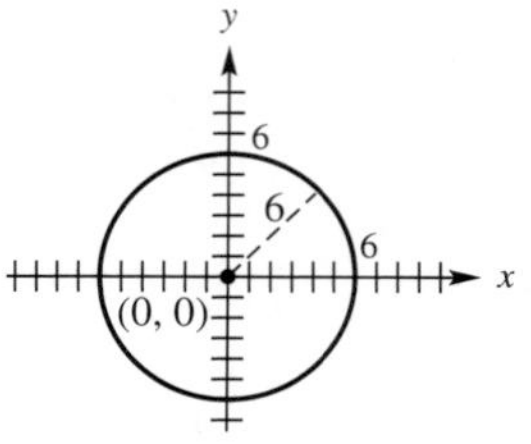

$x^2 + y^2 = 36$

47.

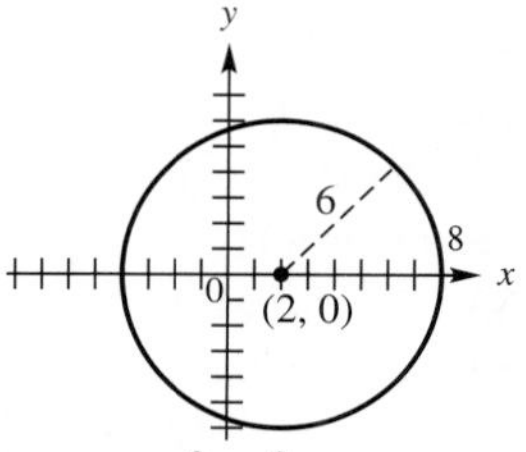

$(x - 2)^2 + y^2 = 36$

49.

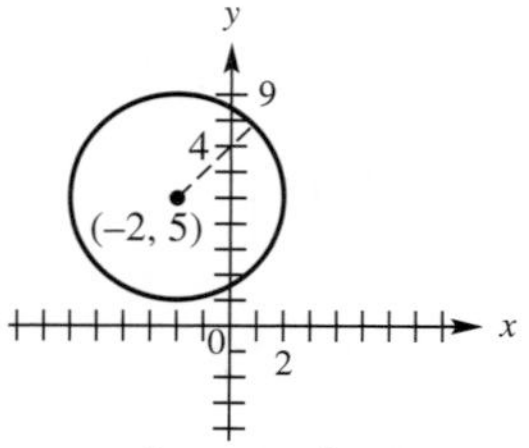

$(x + 2)^2 + (y - 5)^2 = 16$

51.

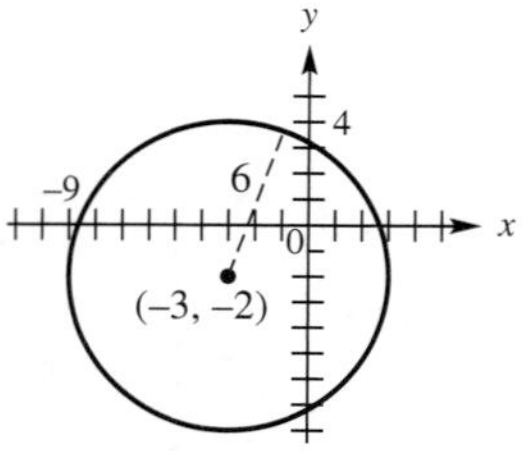

$(x + 3)^2 + (y + 2)^2 = 36$

53. **(a)** $\sqrt{40} = 2\sqrt{10}$ **(b)** $(-1, 4)$

55. \$17,396 **57.** 1988: \$12,471; 1998: \$16,787 **59.** Answers will vary. **61.** Answers will vary.
63. The epicenter is $(-2, -2)$.

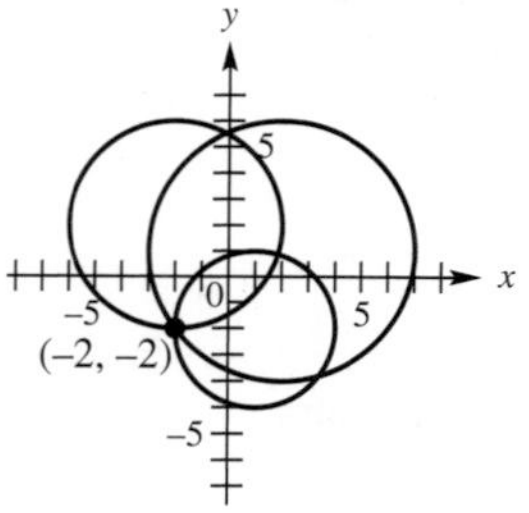

65. Answers will vary. **67.** (9, 18)
69. Answers will vary.

8.2 Exercises (Pages 447–451)

1. $(0, 5), \left(\frac{5}{2}, 0\right), (1, 3), (2, 1)$ **3.** $(0, -4), (4, 0), (2, -2), (3, -1)$ **5.** $(0, 4), (5, 0), \left(3, \frac{8}{5}\right), \left(\frac{5}{2}, 2\right)$

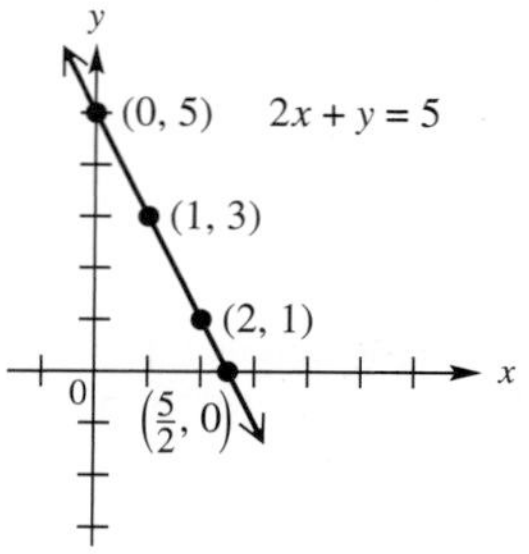

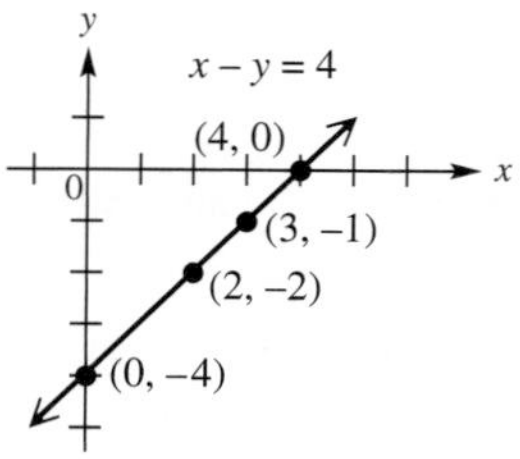

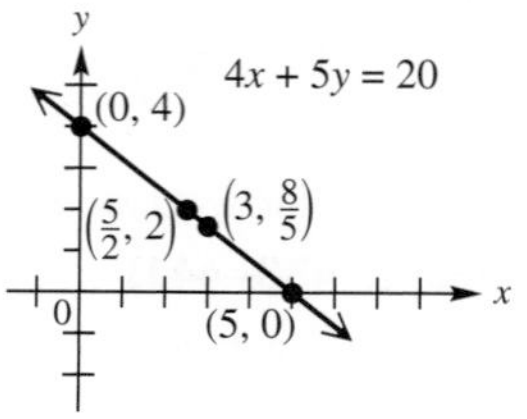

7. (0, 4), $\left(\frac{8}{3}, 0\right)$, (2, 1), (4, −2) **9.** Answers will vary. **11.** A **13.** (4, 0); (0, 6)

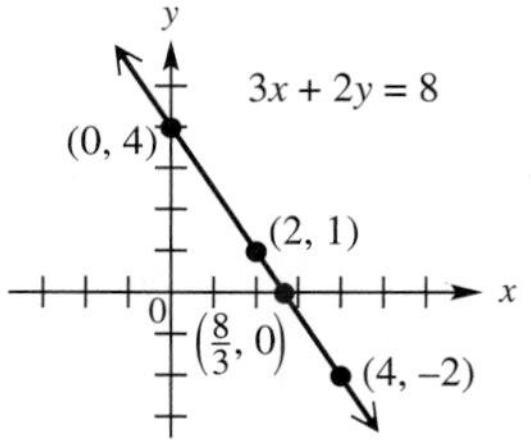

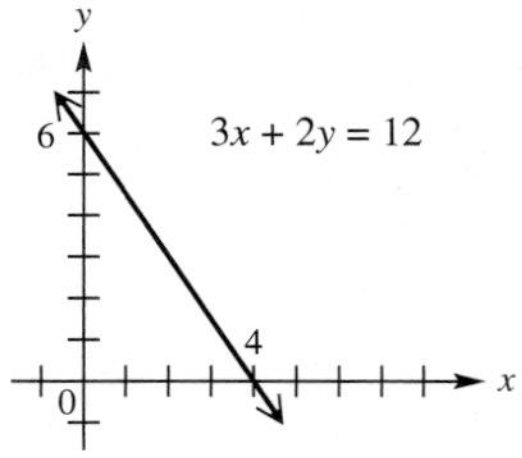

15. (2, 0); $\left(0, \frac{5}{3}\right)$ **17.** $\left(\frac{5}{2}, 0\right)$; (0, −5) **19.** (2, 0); $\left(0, -\frac{2}{3}\right)$

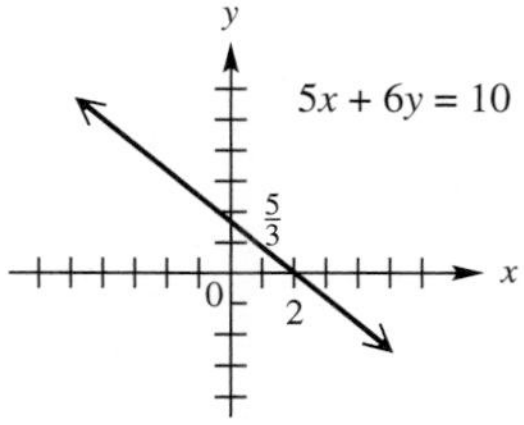

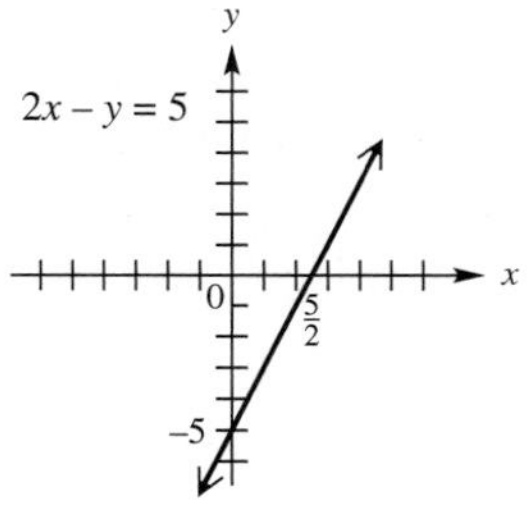

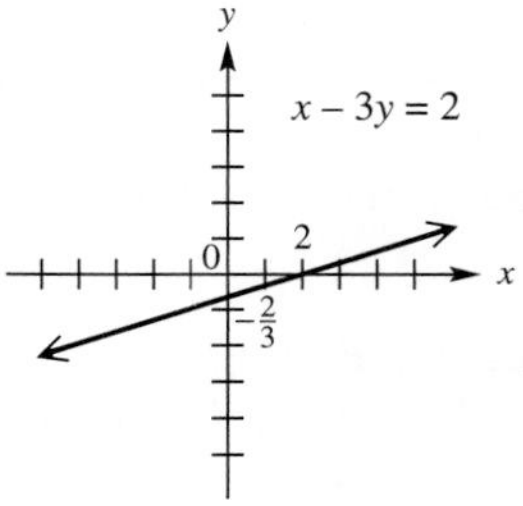

21. (0, 0); (0, 0) **23.** (0, 0); (0, 0) **25.** (2, 0); none

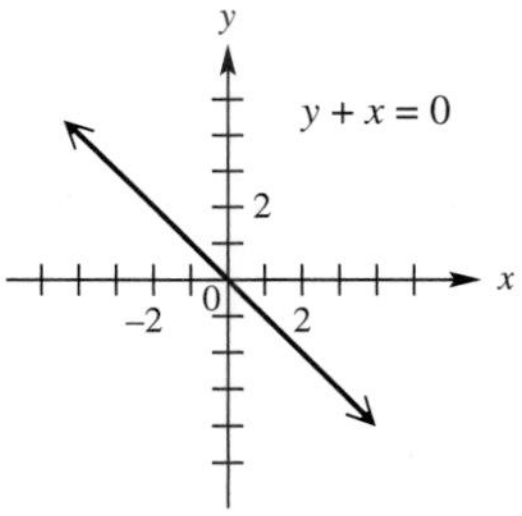

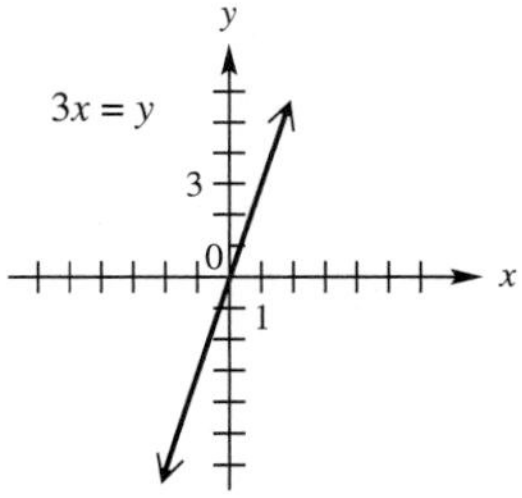

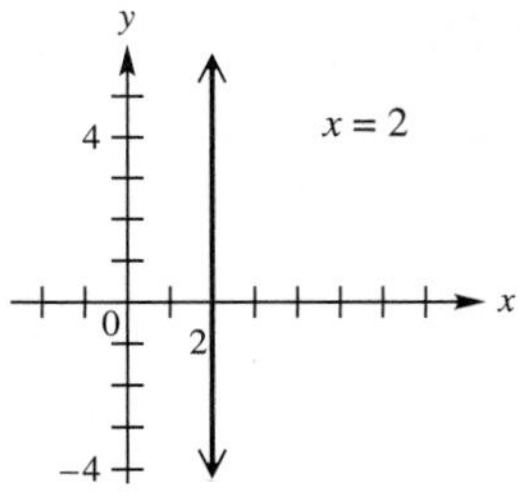

27. none; (0, 4) **29.** C **31.** A **33.** D **35.** B **37.** $\frac{3}{10}$

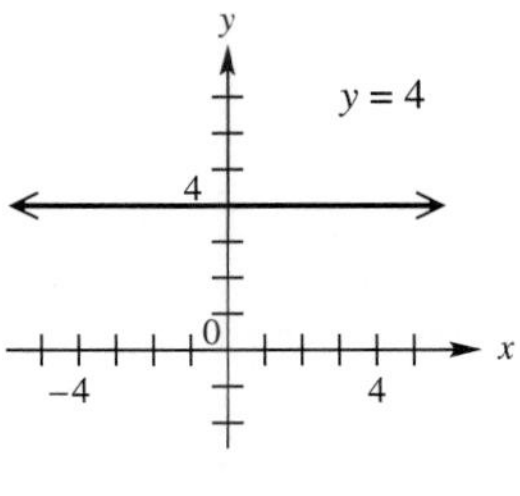

39. **(a)** $\frac{3}{2}$ **(b)** $-\frac{7}{4}$ **41.** 8 **43.** $-\frac{5}{6}$ **45.** 0 **47.** **(a)** slope = 232 (This represents enrollment in thousands.) **(b)** positive; increased **(c)** 232,000 students **(d)** −1.66 **(e)** negative; decreased **(f)** 1.66 students per computer

49.

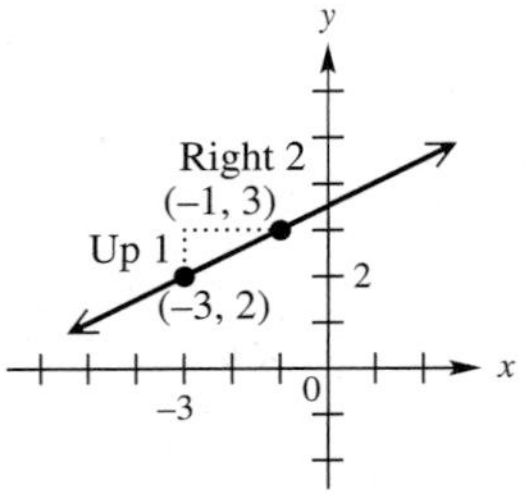

51.

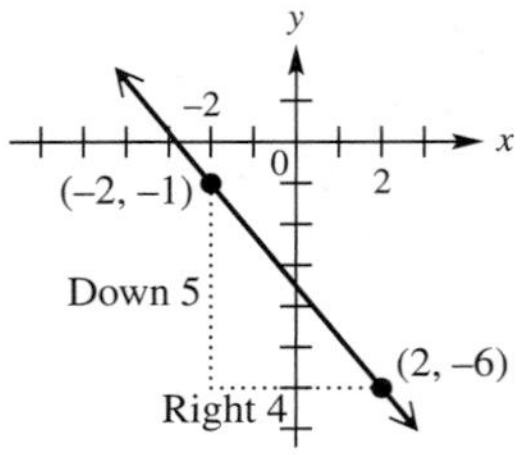

53.

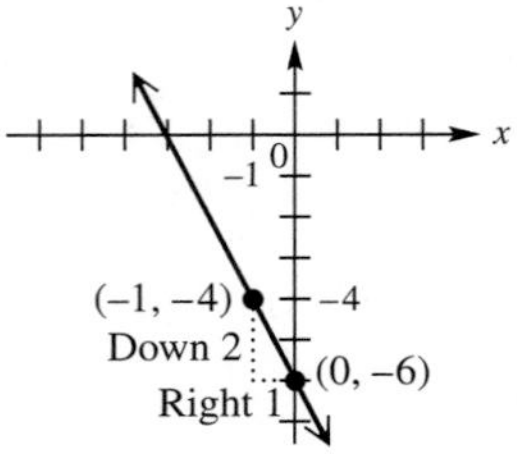

55.

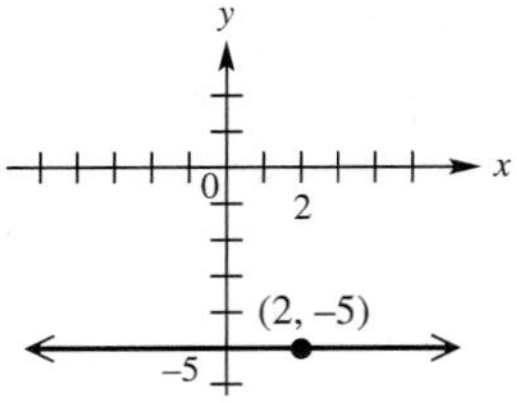

57. parallel **59.** perpendicular **61.** neither parallel nor perpendicular

63. $\frac{7}{10}$ **65.** **(a)** 3.8% per year **(b)** As the years increase, the number of electronic filings also increases. **67.** In each case, 1000 million per year; The average rate of change is the same. The data points lie on the same straight line. **69.** Answers will vary.

8.3 Exercises (Pages 456–460)

1. D **3.** B **5.** $y = 3x - 3$ **7.** $y = -x + 3$ **9.** A **11.** C **13.** H **15.** B **17.** $y = -\frac{3}{4}x + \frac{5}{2}$ **19.** $y = -2x + 18$ **21.** $y = \frac{1}{2}x + \frac{13}{2}$ **23.** $y = 4x - 12$ **25.** $y = 5$ **27.** $x = 9$ **29.** $x = .5$ **31.** $y = 8$ **33.** $y = 2x - 2$ **35.** $y = -\frac{1}{2}x + 4$ **37.** $y = \frac{2}{13}x + \frac{6}{13}$ **39.** $y = 5$ **41.** $x = 7$ **43.** $y = -3$ **45.** $y = 5x + 15$ **47.** $y = -\frac{2}{3}x + \frac{4}{5}$ **49.** $y = \frac{2}{5}x + 5$ **51.** Answers will vary. **53.** **(a)** $y = -x + 12$ **(b)** -1 **(c)** $(0, 12)$ **55.** **(a)** $y = -\frac{5}{2}x + 10$ **(b)** $-\frac{5}{2}$ **(c)** $(0, 10)$ **57.** **(a)** $y = \frac{2}{3}x - \frac{10}{3}$ **(b)** $\frac{2}{3}$ **(c)** $\left(0, -\frac{10}{3}\right)$ **59.** $y = 3x - 19$ **61.** $y = \frac{1}{2}x - 1$ **63.** $y = -\frac{1}{2}x + 9$ **65.** $y = 7$
67. **(a)** yes

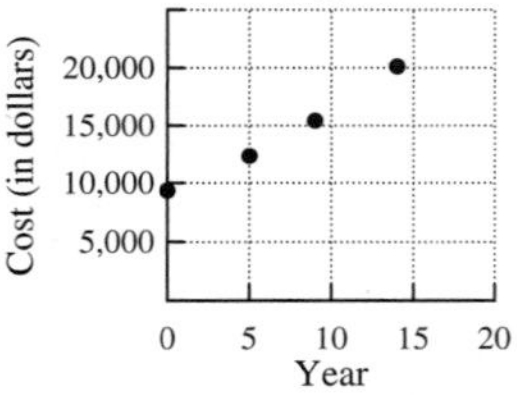

(b) $y = 763.6x + 9391$ **(c)** $23,136

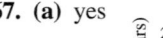

69. **(a)** $y = -108.3x + 28{,}959$ **(b)** 27,335 **71.** **(a)** $y = 14.55x + 262.42$ **(b)** 2525 light-years **73.** **(a)** 32°; 212° **(b)** (0, 32); (100, 212) **(c)** $\frac{9}{5}$ **(d)** $F = \frac{9}{5}C + 32$ **(e)** $C = \frac{5}{9}(F - 32)$
(f) When Celsius temperature is 50°, Fahrenheit temperature is 122°.

8.4 Exercises (Pages 470–473)

1. Answers will vary. **3.** It is the independent variable. **5.** function; domain: $\{2, 3, 4, 5\}$; range: $\{5, 7, 9, 11\}$ **7.** not a function; domain: $(0, \infty)$; range: $(-\infty, 0) \cup (0, \infty)$ **9.** function; domain: {Hispanic, Native American, Asian American, African American, White}; range in millions: {21.3, 1.6, 8.2, 24.6, 152.0} **11.** function; domain: $(-\infty, \infty)$; range: $(-\infty, 4]$ **13.** not a function; domain: $[-4, 4]$; range: $[-3, 3]$ **15.** function; domain: $(-\infty, \infty)$ **17.** not a function; domain: $[0, \infty)$ **19.** not a function; domain: $(-\infty, \infty)$ **21.** function; domain: $[0, \infty)$ **23.** function; domain: $(-\infty, 0) \cup (0, \infty)$ **25.** function; domain: $\left[-\frac{1}{2}, \infty\right)$ **27.** function; domain: $(-\infty, 9) \cup (9, \infty)$
29. **(a)** $[0, 3000]$ **(b)** 25 hours; 25 hours **(c)** 2000 gallons **(d)** $g(0) = 0$; The pool is empty at time zero.

31. Here is one example: The cost of gasoline; number of gallons purchased; cost; number of gallons **33.** 5 **35.** 2 **37.** -1 **39.** -13

41.

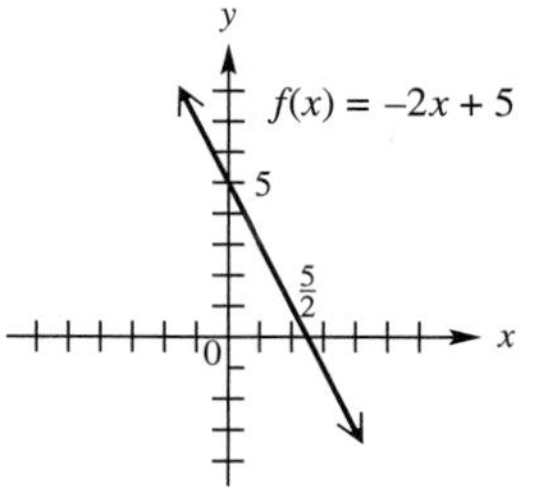

domain and range: $(-\infty, \infty)$

43.

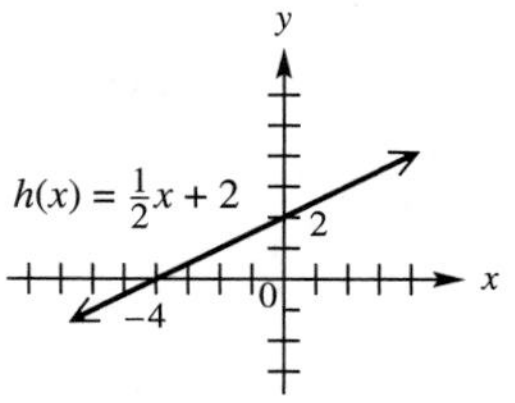

domain and range: $(-\infty, \infty)$

45.

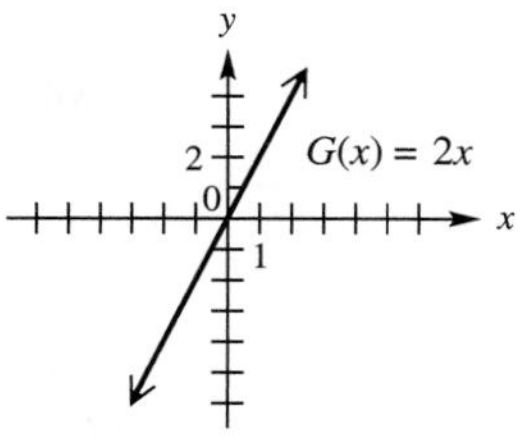

domain and range: $(-\infty, \infty)$

47.

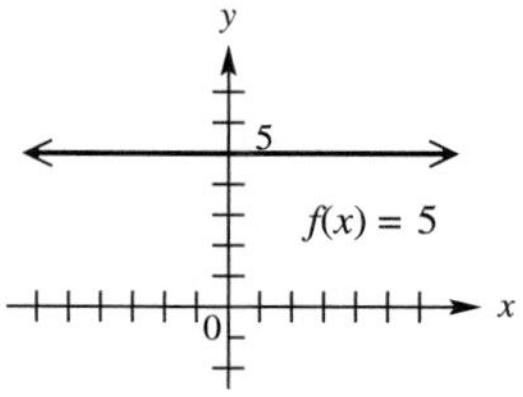

domain: $(-\infty, \infty)$; range: $\{5\}$

49. (a) $f(x) = 3 - 2x^2$ **(b)** -15 **51. (a)** $f(x) = \dfrac{8 - 4x}{-3}$ **(b)** $\dfrac{4}{3}$

53. line; -2; $-2x + 4$; -2; 3, -2 **55. (a)** \$0; \$1.50; \$3.00; \$4.50 **(b)** $1.50x$ **(c)**

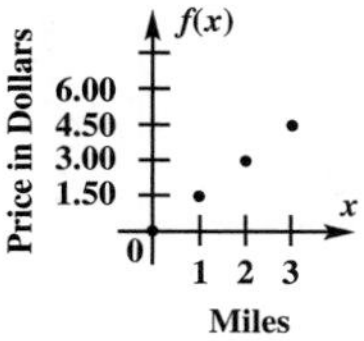

57. (a) 194.53 centimeters **(b)** 177.29 centimeters **(c)** 177.41 centimeters **(d)** 163.65 centimeters **59. (a)** \$160 **(b)** 70 mph **(c)** 66 mph **(d)** for speeds more than 80 mph

61. (a) $C(x) = .02x + 200$
(b) $R(x) = .04x$
(c) 10,000
(d)

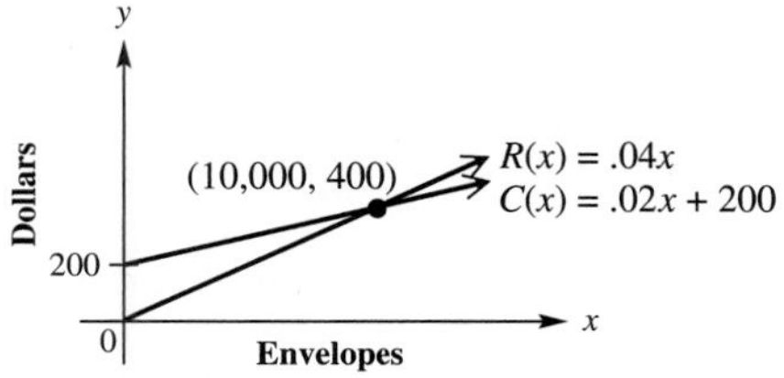

For $x < 10{,}000$, a loss
For $x > 10{,}000$, a profit

63. (a) $C(x) = 3.00x + 2300$
(b) $R(x) = 5.50x$
(c) 920
(d)

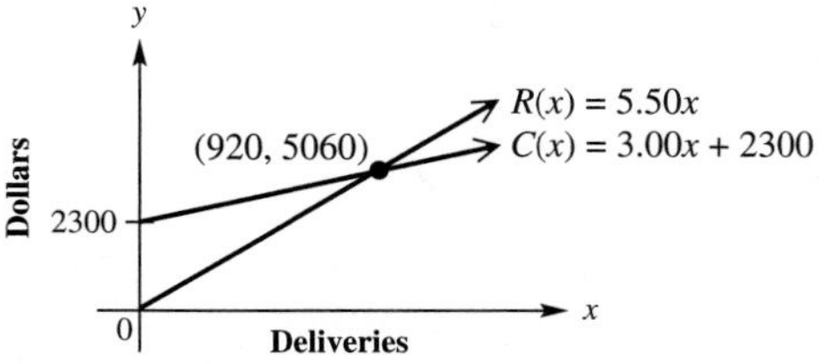

For $x < 920$, a loss
For $x > 920$, a profit

8.5 Exercises (Pages 481–484)

1. F **3.** C **5.** E **7.** Answers will vary **9.** (0, 0) **11.** (0, 4) **13.** (1, 0) **15.** $(-3, -4)$ **17.** Answers will vary **19.** downward; narrower **21.** upward; wider **23. (a)** I **(b)** IV **(c)** II **(d)** III

25.

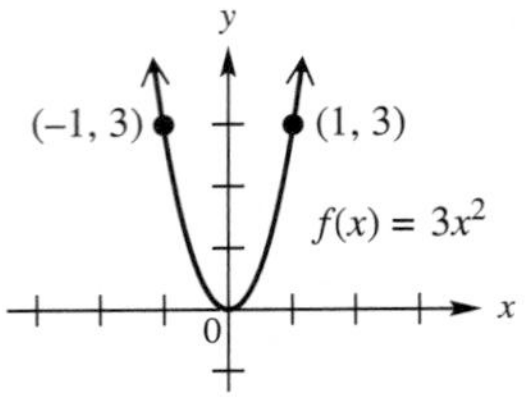

27.

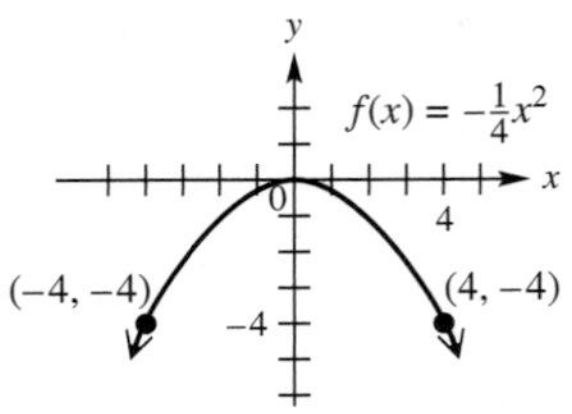

29.

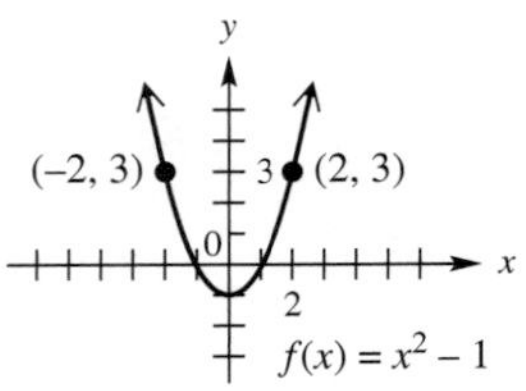

31.

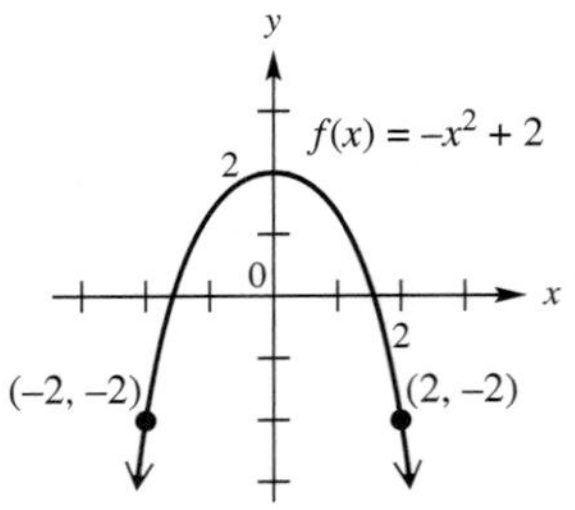

33.

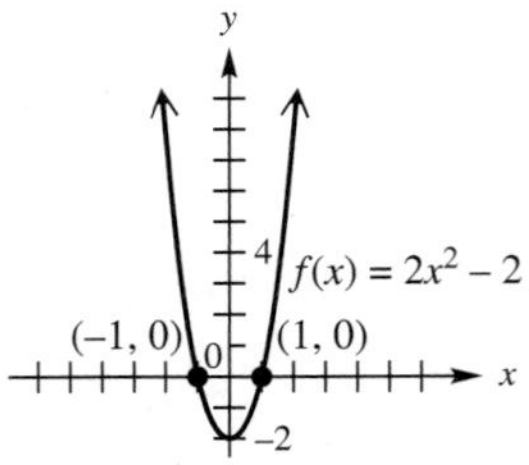

35.

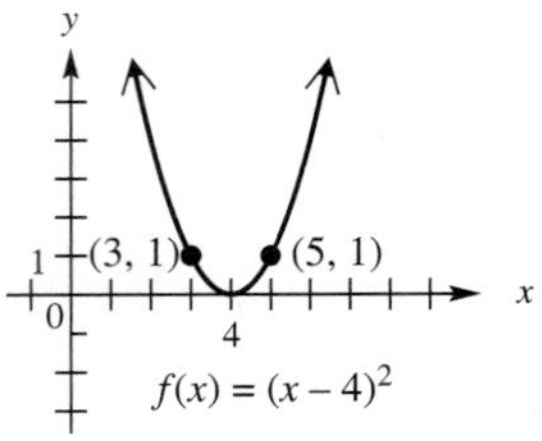

37.

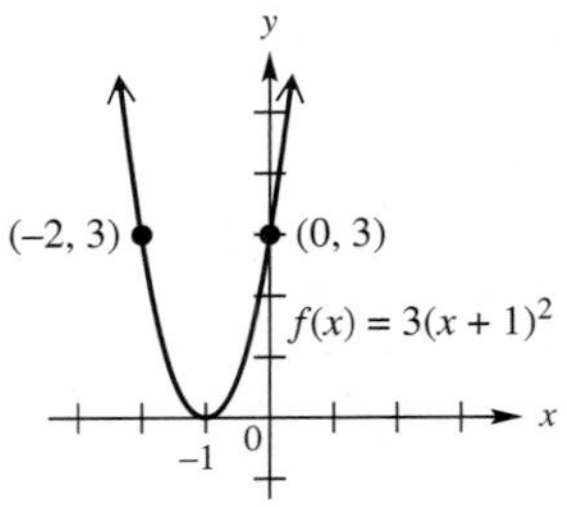

39.

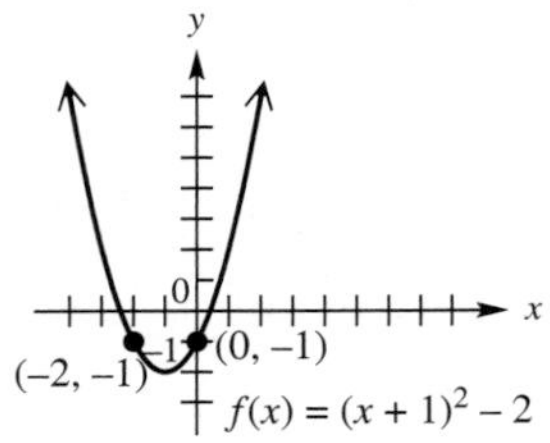

41.

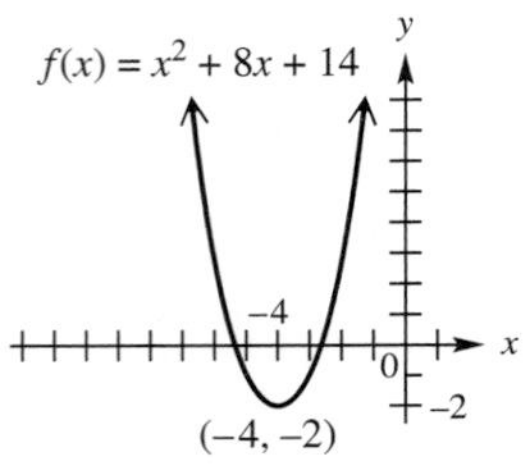

43.

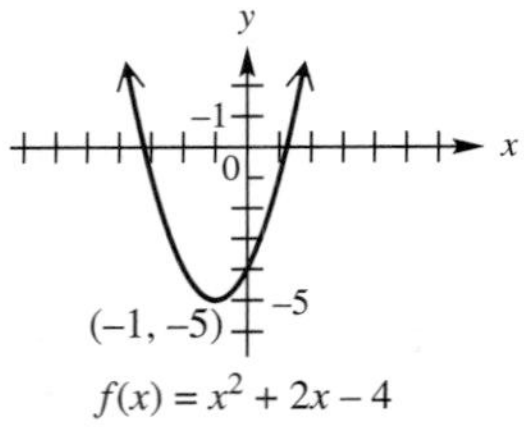

45.

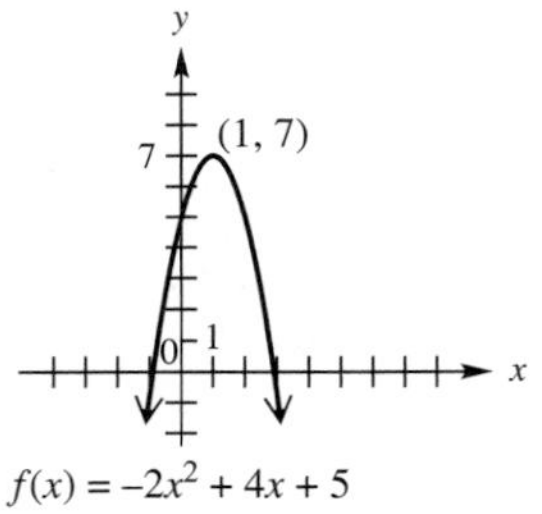

47. 25 meters **49.** 16 feet; 2 seconds

51. 4.1 seconds; 81.6 meters **53.** **(a)** 19.2 hours **(b)** 84.3 ppm **55.** $f(45) = 161.5$; This means that when the speed is 45 mph, the stopping distance is 161.5 feet.

57. **(a)** $R(x) = (100 - x)(200 + 4x) = 20{,}000 + 200x - 4x^2$ **(b)**

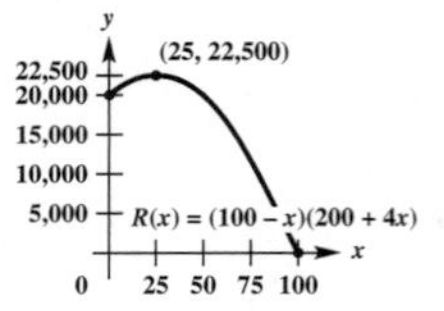

(c) 25 **(d)** $22,500

8.6 Exercises (Pages 492–494)

1. rises; falls **3.** does not **5.** rises; falls **7.** does not **9.** 2.56425419972 **11.** 1.25056505582 **13.** 7.41309466897 **15.** .0000210965628481

17.

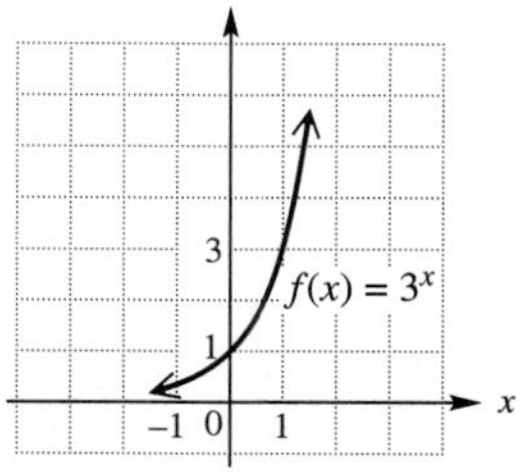

19.

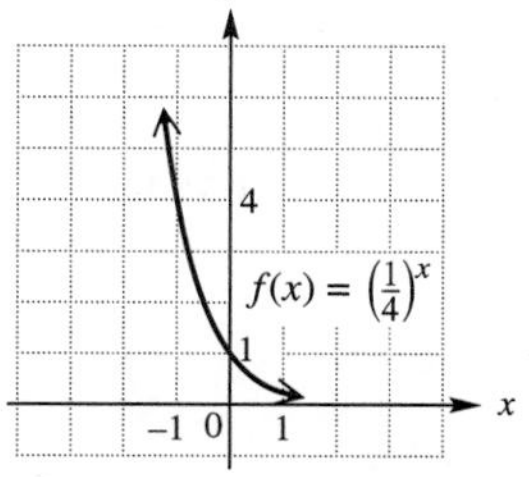

21. 20.0855369232 **23.** .018315638889 **25.** $2 = \log_4 16$ **27.** $-3 = \log_{2/3}\left(\frac{27}{8}\right)$ **29.** $2^5 = 32$ **31.** $3^1 = 3$ **33.** 1.38629436112 **35.** -1.0498221245 **37.** **(a)** .5°C **(b)** .35°C **39.** **(a)** 1.6°C **(b)** .5°C

41.

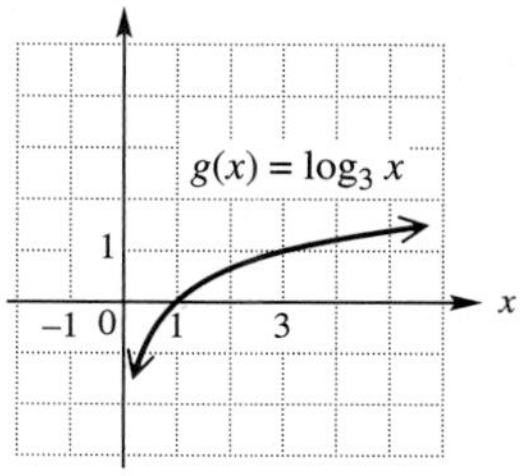

43.

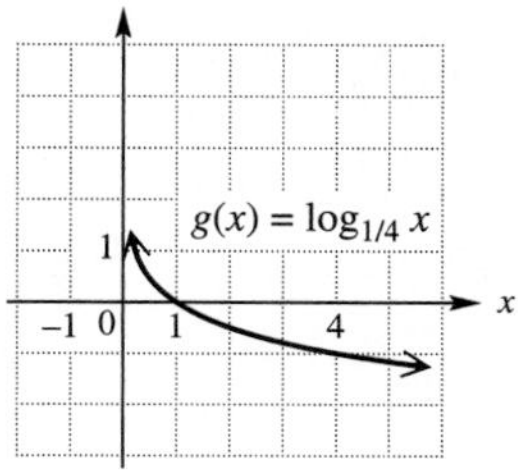

45. **(a)** \$22,510.18 **(b)** \$22,529.85

47. **(a)** \$33,504.35 **(b)** \$33,504.71 **49.** Plan A is better by \$102.65 **51.** **(a)** 828 millibars **(b)** 232 millibars **53.** **(a)** 146,250 **(b)** 198,403 **(c)** It will have increased by almost 36%. **55.** **(a)** 440 grams **(b)** 387 grams **(c)** 264 grams **(d)** 21.66 years **57.** 1611.97 years **59.** about 9000 years **61.** about 13,000 years **63.** almost 4 times as powerful

8.7 Exercises (Pages 508–514)

1. **(a)** 1991; about 350 million **(b)** (1987, 100 million) **(c)** 1988–1990 **(d)** CD production generally increased during these years; positive **(e)** The slope would be negative, because the line falls from left to right. **3.** yes **5.** no

7. $\{(2, 2)\}$

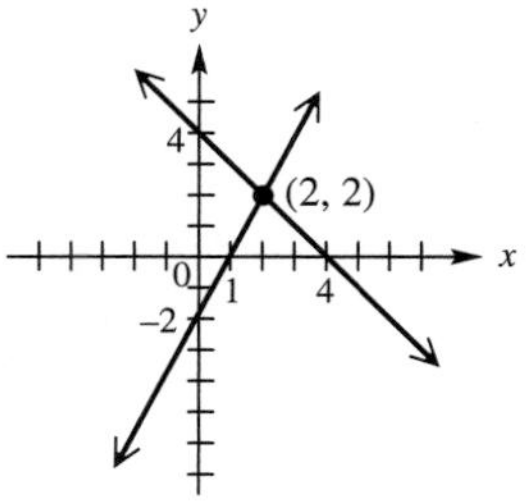

9. $\{(3, -1)\}$ **11.** $\{(2, -3)\}$ **13.** $\left\{\left(\frac{3}{2}, -\frac{3}{2}\right)\right\}$ **15.** $\left\{\left(\frac{6-2y}{7}, y\right)\right\}$

17. $\{(2, -4)\}$ **19.** $\emptyset$ **21.** $\{(1, 2)\}$ **23.** $\left\{\left(\frac{22}{9}, \frac{22}{3}\right)\right\}$ **25.** $\{(2, 3)\}$ **27.** $\{(5, 4)\}$ **29.** $\left\{\left(-5, -\frac{10}{3}\right)\right\}$ **31.** $\{(2, 6)\}$ **33.** Answers will vary. **35.** $\{(1, 4, -3)\}$ **37.** $\{(0, 2, -5)\}$ **39.** $\left\{\left(-\frac{7}{3}, \frac{22}{3}, 7\right)\right\}$ **41.** $\{(4, 5, 3)\}$ **43.** $\{(2, 2, 2)\}$ **45.** $\left\{\left(\frac{8}{3}, \frac{2}{3}, 3\right)\right\}$ **47.** wins: 100; losses: 62 **49.** length: 94 feet; width: 50 feet **51.** weekend days: 3; weekdays: 3 **53.** square: 12 centimeters; triangle: 8 centimeters **55.** cappuccino: \$1.95; house latte: \$2.35 **57.** NHL: \$247.32; NBA: \$267.37 **59.** Tokyo: \$430; New York: \$385 **61.** dark clay: \$5 per kilogram; light clay: \$4 per kilogram **63.** **(a)** 12 ounces **(b)** 30 ounces **(c)** 48 ounces **(d)** 60 ounces **65.** $\$1.29x$ **67.** 15% solution: $26\frac{2}{3}$ liters; 33% solution: $13\frac{1}{3}$ liters **69.** 3 liters **71.** 50% juice: 150 liters; 30% juice: 50 liters **73.** \$1.20 candy: 100 pounds; \$2.40 candy: 60 pounds **75.** at 4%: \$10,000; at 3%: \$5000 **77.** train: 60 kilometers per hour; plane: 160 kilometers per hour **79.** boat: 21 mph; current: 3 mph **81.** gold: 35; silver: 39; bronze: 29 **83.** shortest: 10 inches; middle: 20 inches; longest: 26 inches **85.** type A: 80; type B: 160; type C: 250 **87.** \$10 tickets: 350; \$18 tickets: 250; \$30 tickets: 50

Extension Exercises (Page 518)

1. $\{(2, 3)\}$ **3.** $\{(-3, 0)\}$ **5.** $\left\{\left(-\frac{7}{2}, -1\right)\right\}$ **7.** $\{(1, -4)\}$ **9.** $\{(-1, 23, 16)\}$ **11.** $\{(2, 1, -1)\}$ **13.** $\{(3, 2, -4)\}$ **15.** $\{(0, 1, 0)\}$ **17.** $\{(-1, 2, 0)\}$

8.8 Exercises (Pages 524–526)

1. C **3.** B **5.**

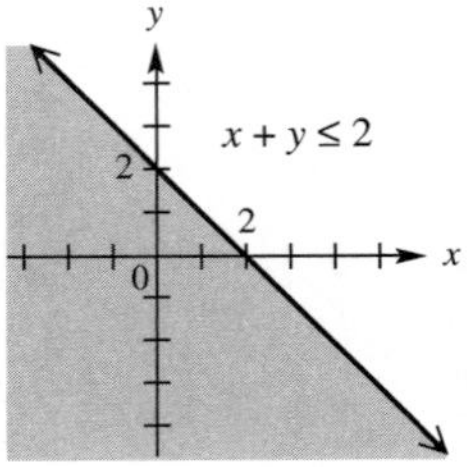

7.

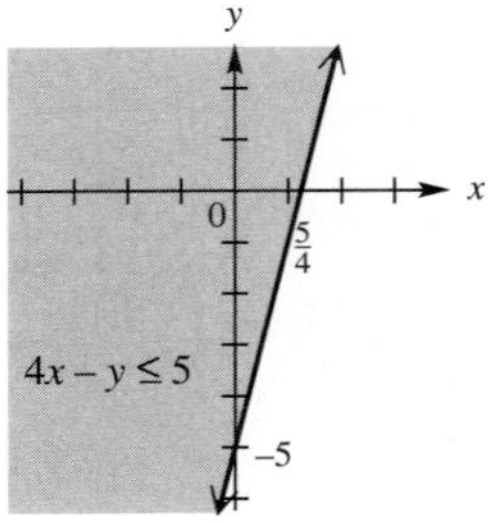

9.

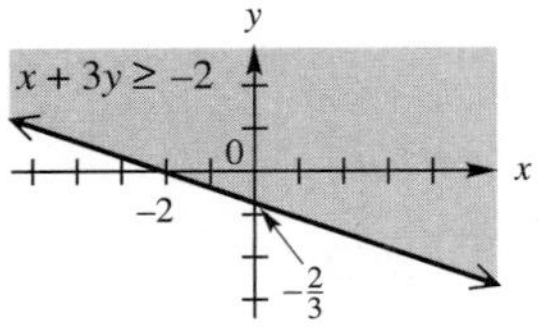

11.

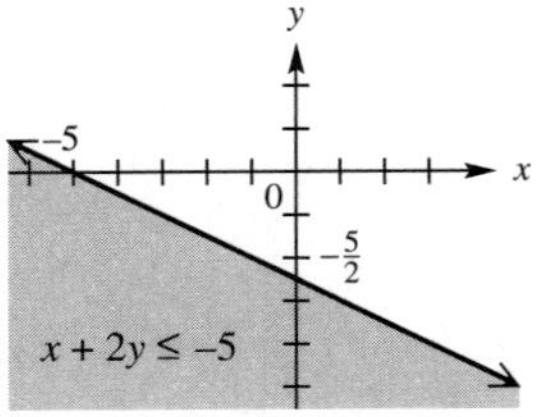

13.

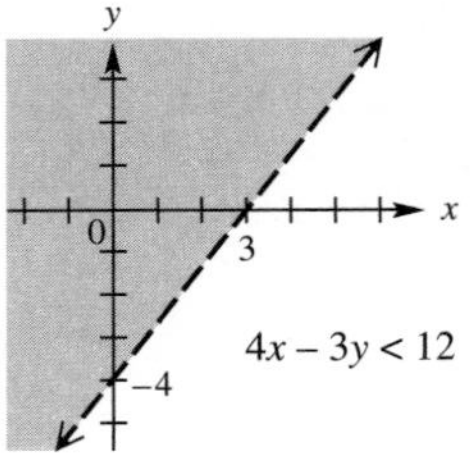

15.

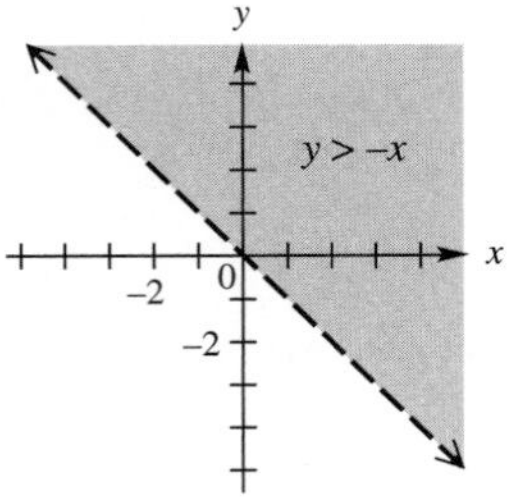

17.

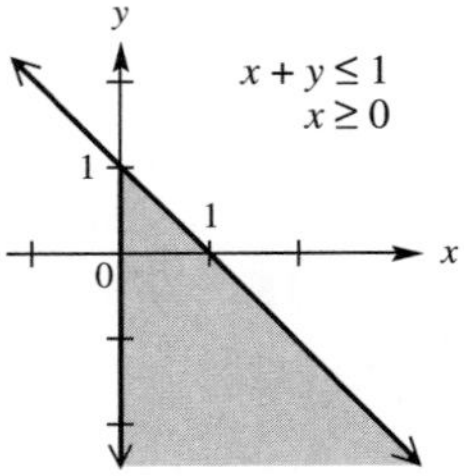

19.

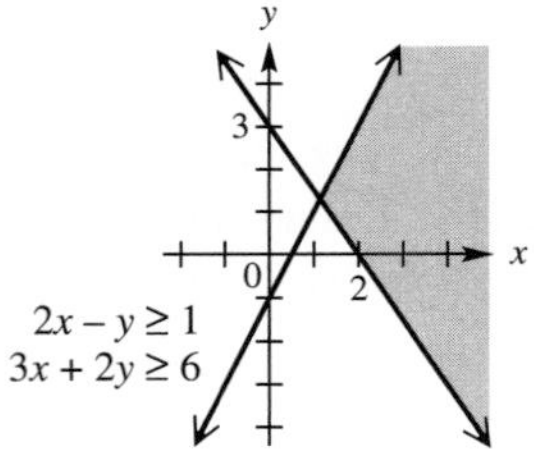

21.

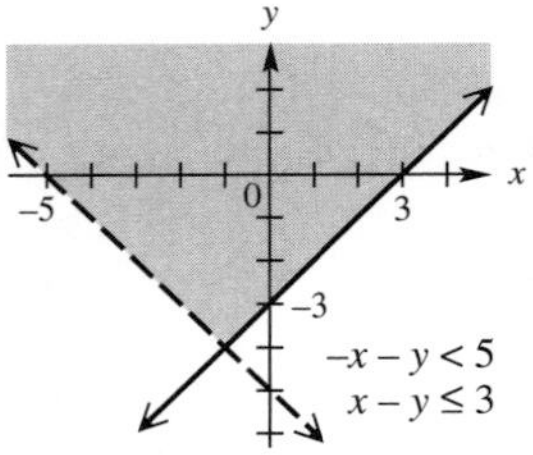

23. maximum of 65 at (5, 10); minimum of 8 at (1, 1) **25.** $\left(\frac{6}{5}, \frac{6}{5}\right); \frac{42}{5}$ **27.** $\left(\frac{17}{3}, 5\right); \frac{49}{3}$ **29.** Ship 20 to A and 80 to B, for a minimum cost of \$1040. **31.** Take 3 red pills and 2 blue pills, for a minimum cost of 70¢ per day. **33.** Produce 6.4 million gallons of gasoline and 3.2 million gallons of fuel oil, for a maximum revenue of \$16,960,000. **35.** Ship 4000 medical kits and 2000 containers of water.

Chapter 8 Test (Pages 527–528)

1. $\sqrt{41}$ **2.** $(x + 1)^2 + (y - 2)^2 = 9$

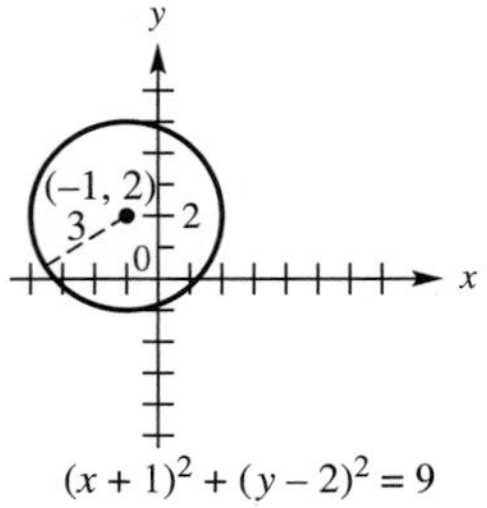

3. x-intercept: $\left(\dfrac{8}{3}, 0\right)$; y-intercept: $(0, -4)$

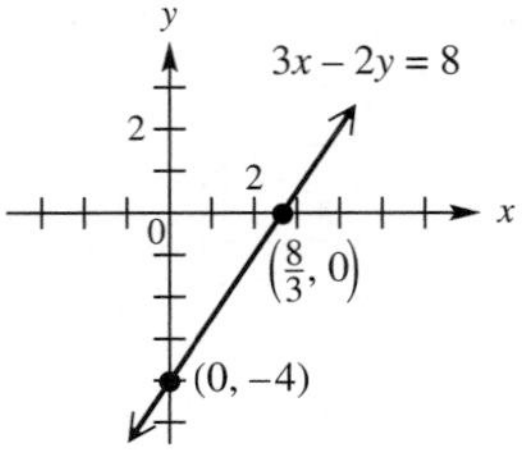

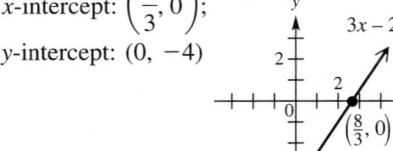

4. $\frac{2}{7}$ **5. (a)** $y = -\frac{2}{5}x + \frac{13}{5}$ **(b)** $y = -\frac{1}{2}x - \frac{3}{2}$ **(c)** $y = -\frac{1}{2}x + 2$ **6.** B **7. (a)** $y = 881x + 38{,}885$ **(b)** \$41,528; The model value is less than the actual income. **8.** $y = .05x + .50$; (1, .55), (5, .75), (10, 1.00) **9.** $y = \frac{2}{3}x + 1$ **10. (a)** $(-\infty, \infty)$ **(b)** 22 **11.** $(-\infty, 3) \cup (3, \infty)$ **12.** 500 calculators; \$30,000

13. axis: $x = -3$; vertex: $(-3, 4)$; domain: $(-\infty, \infty)$; range: $(-\infty, 4]$

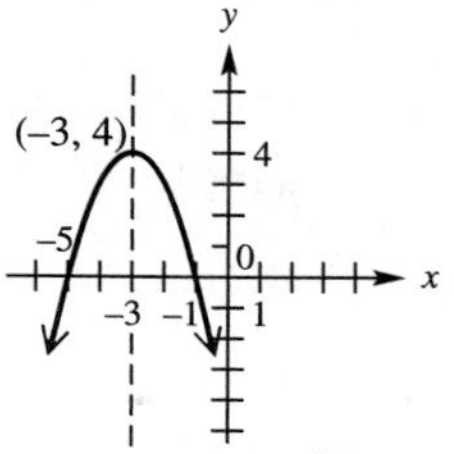

$f(x) = -(x + 3)^2 + 4$

14. 80 feet by 160 feet

15. (a) 2116.31264888 **(b)** .157237166314 **(c)** 3.15955035878 **16.** A **17. (a)** \$13,521.90 **(b)** \$13,529.96 **18. (a)** 1.62 grams **(b)** 1.18 grams **(c)** .69 gram **(d)** 2.00 grams **19.** $\{(4, -2)\}$ **20.** $\{(2, 0, -1)\}$ **21.** $\{(1, 2z + 3, z)\}$ **22.** *Pretty Woman*: \$463.4 million; *Ocean's Eleven*: \$450.7 million **23.** \$40,000 at 10%; \$100,000 at 6%; \$140,000 at 5% **24.**

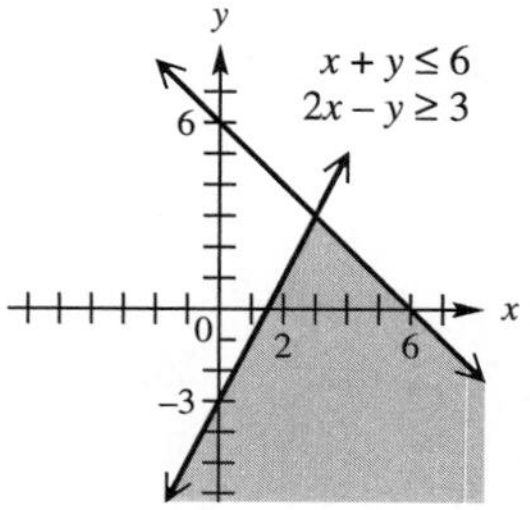

25. Manufacture 0 VIP rings and 24 SST rings, for a maximum profit of \$960.

CHAPTER 9 GEOMETRY

9.1 Exercises (Pages 537–540)

1. 90 **3.** equal **5.** true **7.** false **9.** true **11.** true **13.** false

15. (a) $\overline{AB}$ **(b)**

17. (a) $\overrightarrow{CB}$ **(b)**

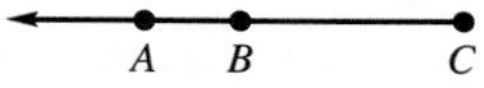

19. (a) $\overrightarrow{BC}$ **(b)**

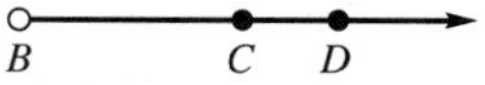

21. (a) $\overrightarrow{BA}$ **(b)**

23. (a) $\overline{CA}$ **(b)**

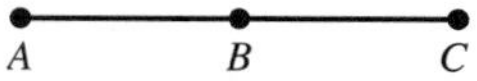

25. F **27.** D **29.** B **31.** E **There may be other correct forms of the answers in Exercises 33–39.** **33.** $\overleftrightarrow{MO}$ **35.** $\overline{MO}$ **37.** $\emptyset$ **39.** $\overrightarrow{OP}$ **41.** 62° **43.** 1° **45.** $(90 - x)°$ **47.** 48° **49.** 154° **51.** $(180 - y)°$ **53.** $\measuredangle CBD$ and $\measuredangle ABE$; $\measuredangle CBE$ and $\measuredangle DBA$ **55. (a)** 52° **(b)** 128° **57.** 107° and 73° **59.** 75° and 75° **61.** 139° and 139° **63.** 65° and 115° **65.** 35° and 55° **67.** 117° and 117° **69.** 141° and 141° **71.** 80° **73.** 52° **75.** Measures are given in numerical order, starting with angle 1: 55°, 65°, 60°, 65°, 60°, 120°, 60°, 60°, 55°, 55°. **77.** Answers will vary.

9.2 Exercises (Pages 545–548)

1. chord **3.** equilateral (or equiangular) **5.** false **7.** false **9.** true **11.** Answers will vary. **13.** both **15.** closed **17.** closed **19.** neither **21.** convex **23.** convex **25.** not convex **27.** right, scalene **29.** acute, equilateral **31.** right, scalene **33.** right, isosceles **35.** obtuse, scalene **37.** acute, isosceles **39.** Answers will vary. **41.** Answers will vary. **43.** $A = 50°$; $B = 70°$; $C = 60°$; **45.** $A = B = C = 60°$ **47.** $A = B = 52°$; $C = 76°$ **49.** 165° **51.** 170° **53. (a)** O **(b)** $\overline{OA}$, $\overline{OC}$, $\overline{OB}$, $\overline{OD}$ **(c)** $\overline{AC}$, $\overline{BD}$ **(d)** $\overline{AC}$, $\overline{BD}$, $\overline{BC}$, $\overline{AB}$ **(e)** $\overleftrightarrow{BC}$, $\overleftrightarrow{AB}$ **(f)** $\overleftrightarrow{AE}$ **55. (e)** The sum of the measures of the angles of a triangle is 180° (because the pencil has gone through one-half of a complete rotation).

Extension Exercises (Page 551)

1. With radius of the compasses greater than one-half the length PQ, place the point of the compasses at P and swing arcs above and below line r. Then with the same radius and the point of the compasses at Q, swing two more arcs above and below line r. Locate the two points of intersections of the arcs above and below, and call them A and B. With a straightedge, join A and B. AB is the perpendicular bisector of PQ. **3.** With the radius of the compasses greater than the distance from P to r, place the point of the compasses at P and swing an arc intersecting line r in two points. Call these points A and B. Swing arcs of equal radius to the left of line r, with the point of the compasses at A and at B, intersecting at points Q. With a straightedge, join P and Q. PQ is the perpendicular from P to line r. **5.** With any radius, place the point of the compasses at P and swing arcs to the left and right, intersecting line r in two points. Call these points A and B. With an arc of sufficient length, place the point of the compasses first at A and then at B, and swing arcs either both above or both below line r, intersecting at point Q. With a straightedge, join P and Q. PQ is perpendicular to line r at P. **7.** With any radius, place the point of the compasses at A and swing an arc intersecting the sides of angle A at two points. Call the point of intersection on the horizontal side B and call the other point of intersection C. Draw a horizontal working line, and locate any point A' on this line. With the same radius used earlier, place the point of the compasses at A' and swing an arc intersecting the working line at B'. Return to angle A, and set the radius of the compasses equal to BC. On the working line, place the point of the compasses at B' and swing an arc intersecting the first arc at C'. Now draw line $A'C'$. Angle A' is equal to angle A. **9.** Use Construction 3 to construct a perpendicular to a line at a point. Then use Construction 4 to bisect one of the right angles formed. This yields a 45° angle. **11.** Answers will vary.

9.3 Exercises (Pages 559–565)

1. 12 **3.** 6 **5.** circumference **7.** 12 cm^2 **9.** 5 cm^2 **11.** 8 $in.^2$ **13.** 4.5 cm^2 **15.** 418 mm^2 **17.** 8 cm^2 **19.** 3.14 cm^2 **21.** 1017.36 m^2 **23.** 4 m **25.** 300 ft, 400 ft, 500 ft **27.** 46 ft **29.** 23,800.10 ft^2 **31.** perimeter **33.** 12 in., 12π in., 36π $in.^2$ **35.** 5 ft, 10π ft, 25π ft^2 **37.** 6 cm, 12 cm, 36π cm^2 **39.** 10 in., 20 in., 20π in. **41.** 14.5 **43.** 7 **45.** 5.1 **47.** 5 **49.** 5 **51.** 1.5 **53. (a)** 20 cm^2 **(b)** 80 cm^2 **(c)** 180 cm^2 **(d)** 320 cm^2 **(e)** 4 **(f)** 3; 9 **(g)** 4; 16 **(h)** n^2 **55.** $800 **57.** n^2 **59.** 80 **61.** 76.26 **63.** 132 ft^2 **65.** 5376 cm^2 **67.** 145.34 m^2 **69.** 14-in. pizza **71.** 14-in. pizza **73.** 26 in. **75.** 625 ft^2 **77.** 648 $in.^2$ **79.** $\frac{(4-\pi)r^2}{4}$

9.4 Exercises (Pages 572–578)

1.

STATEMENTS	REASONS
1. $AC = BD$	1. Given
2. $AD = BC$	2. Given
3. $AB = AB$	3. Reflexive property
4. $\triangle ABD \cong \triangle BAC$	4. SSS congruence property

3.

STATEMENTS	REASONS
1. $\overleftrightarrow{DB}$ is perpendicular to $\overleftrightarrow{AC}$.	1. Given
2. $AB = BC$	2. Given
3. $\measuredangle ABD = \measuredangle CBD$	3. Both are right angles by definition of perpendicularity.
4. $DB = DB$	4. Reflexive property
5. $\triangle ABD \cong \triangle CBD$	5. SAS congruence property

5.

STATEMENTS	REASONS
1. $\measuredangle BAC = \measuredangle DAC$	1. Given
2. $\measuredangle BCA = \measuredangle DCA$	2. Given
3. $AC = AC$	3. Reflexive property
4. $\triangle ABD \cong \triangle ADC$	4. ASA congruence property

7. 67°, 67° **9.** 6 in. **11.** Answers will vary. **13.** $\measuredangle A$ and $\measuredangle P$; $\measuredangle C$ and $\measuredangle R$; $\measuredangle B$ and $\measuredangle Q$; $\overleftrightarrow{AC}$ and $\overleftrightarrow{PR}$; $\overleftrightarrow{CB}$ and $\overleftrightarrow{RQ}$; $\overleftrightarrow{AB}$ and $\overleftrightarrow{PQ}$; **15.** $\measuredangle H$ and $\measuredangle F$; $\measuredangle K$ and $\measuredangle E$; $\measuredangle HGK$ and $\measuredangle FGE$; $\overleftrightarrow{HK}$ and $\overleftrightarrow{FE}$; $\overleftrightarrow{GK}$ and $\overleftrightarrow{GE}$; $\overleftrightarrow{HG}$ and $\overleftrightarrow{FG}$; **17.** $\measuredangle P = 78°$; $\measuredangle M = 46°$; $\measuredangle A = \measuredangle N = 56°$ **19.** $\measuredangle T = 74°$; $\measuredangle Y = 28°$; $\measuredangle Z = \measuredangle W = 78°$ **21.** $\measuredangle T = 20°$; $\measuredangle V = 64°$; $\measuredangle R = \measuredangle U = 96°$ **23.** $a = 20$; $b = 15$ **25.** $a = 6$; $b = \frac{15}{2}$ **27.** $x = 6$

29. $x = 110$ **31.** $c = 111\frac{1}{9}$ **33.** 30 m **35.** 500 m, 700 m **37.** 112.5 ft **39.** 8 ft, 11 in. **41.** $c = 17$ **43.** $a = 13$ **45.** $c = 50$ m **47.** $a = 20$ in. **49.** The sum of the squares of the two shorter sides of a right triangle is equal to the square of the longest side. **51.** (3, 4, 5) **53.** (7, 24, 25) **55.** (12, 16, 20) **57.** Answers will vary. **59.** (3, 4, 5) **61.** (7, 24, 25) **63.** Answers will vary. **65.** (4, 3, 5) **67.** (8, 15, 17) **69.** Answers will vary. **71.** 24 m **73.** 16 ft **75.** 4.55 ft **77.** 19 ft, 3 in. **79.** 28 ft, 10 in. **81. (a)** $\frac{1}{2}(a + b)(a + b)$ **(b)** PWX: $\frac{1}{2}ab$; PZY: $\frac{1}{2}ab$; PXY: $\frac{1}{2}c^2$

(c) $\frac{1}{2}(a + b)(a + b) = \frac{1}{2}ab + \frac{1}{2}ab + \frac{1}{2}c^2$. When simplified, this gives $a^2 + b^2 = c^2$.

83. $24 + 4\sqrt{6}$ **85.** 10 **87.** $256 + 64\sqrt{3}$ **89.** 5 in. **91.** Answers will vary. **93.** Answers will vary.

9.5 Exercises (Pages 584–587)

1. true **3.** true **5.** false **7. (a)** $3\frac{3}{4}$ m^3 **(b)** $14\frac{3}{4}$ m^2 **9. (a)** 96 in.^{3} **(b)** 130.4 in.^{2} **11. (a)** 267,946.67 ft^3 **(b)** 20,096 ft^2 **13. (a)** 549.5 cm^3 **(b)** 376.8 cm^2 **15. (a)** 65.94 m^3 **(b)** 100.00 m^2 **17.** 168 in.^{3} **19.** 1969.10 cm^3 **21.** 427.29 cm^3 **23.** 508.68 cm^3 **25.** 2,415,766.67 m^3 **27.** .52 m^3 **29.** 12 in., 288π in.^{3}, 144π in.^{2} **31.** 5 ft, $\frac{500}{3}\pi$ ft^3, 100π ft^2 **33.** 2 cm, 4 cm, 16π cm^2 **35.** 1 m, 2 m, $\frac{4}{3}\pi$ m^3 **37.** volume **39.** $\sqrt[3]{2}\,x$ **41.** \$8100 **43.** \$37,500 **45.** 2.5 **47.** 6 **49.** 210 in.^{3} **51.** $\frac{62{,}500}{3}\pi$ in.^{3} **53.** 2 to 1 **55.** 288 **57.** 4, 4, 6, 2 **59.** 8, 6, 12, 2 **61.** 20, 12, 30, 2

9.6 Exercises (Pages 596–598)

The answers are given in blue for this section.

1.

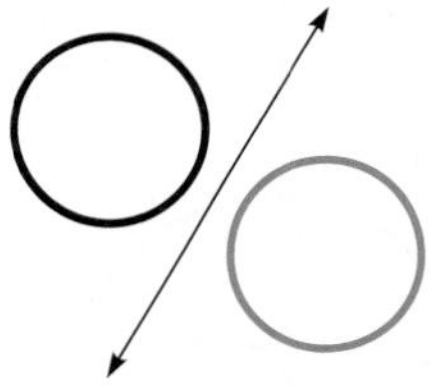

3.

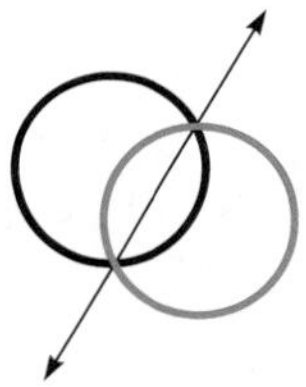

5.

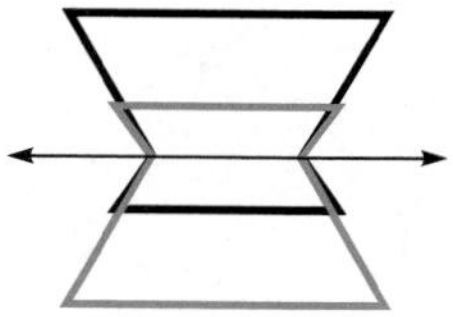

7. The figure is its own reflection image.

9.

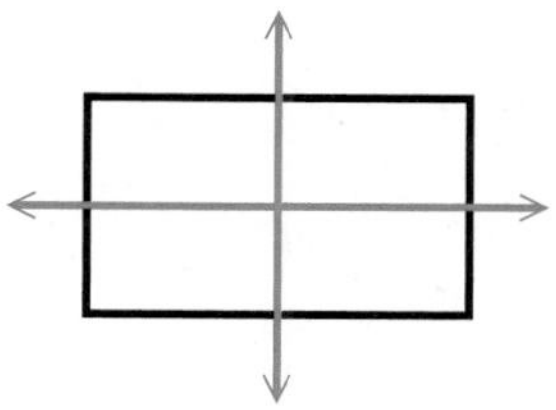

11.

13.

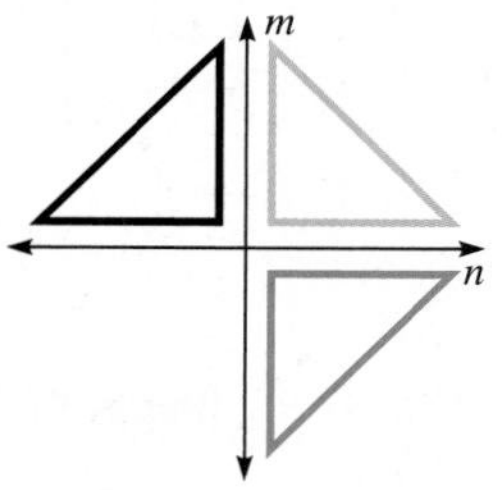

15.

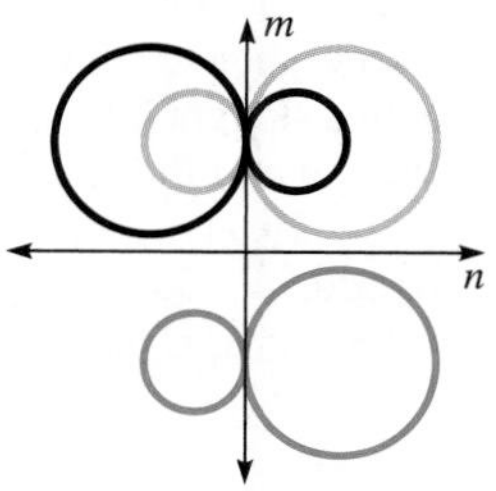

17.

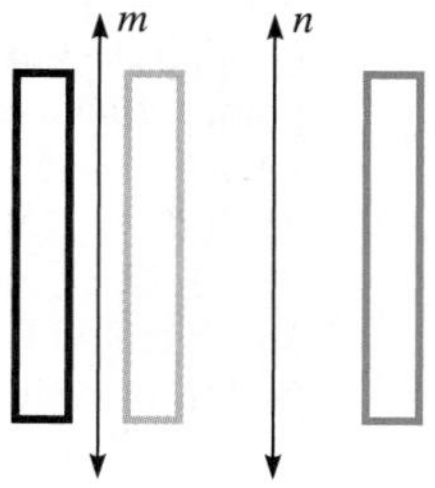

19.

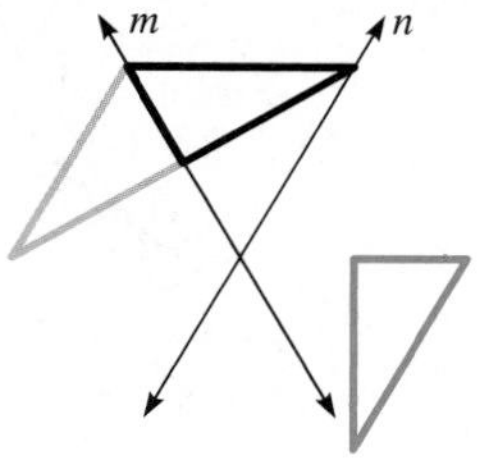

21.

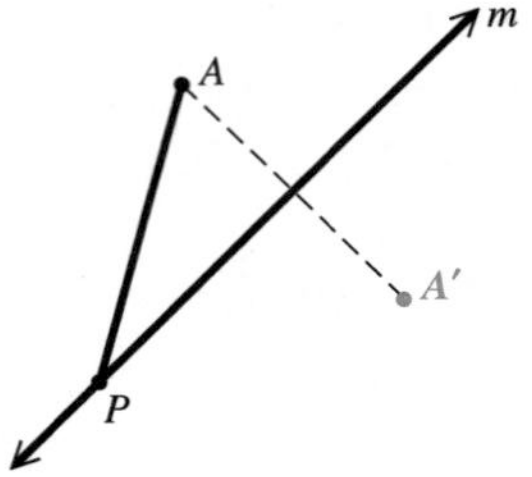

23.

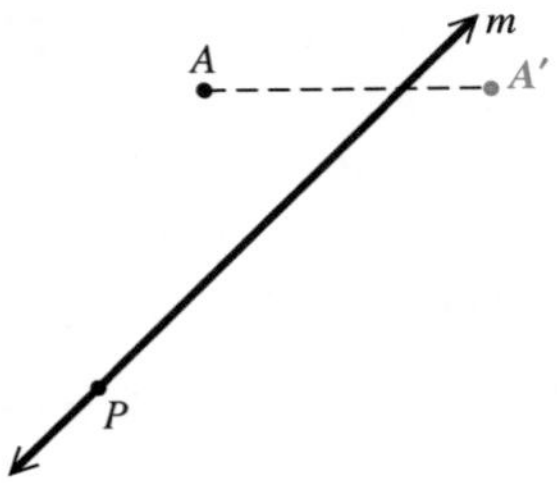

25.

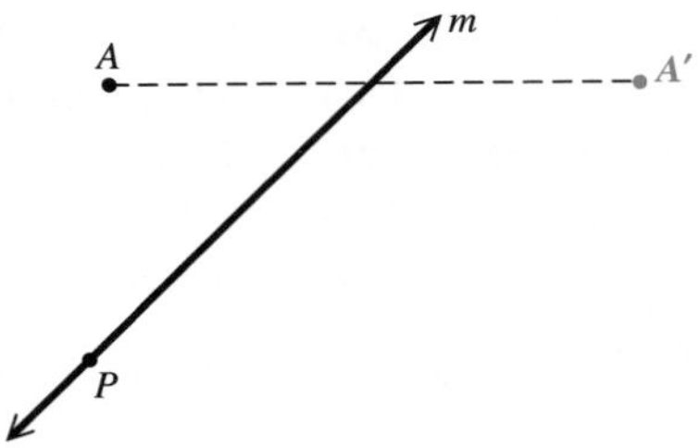

27.

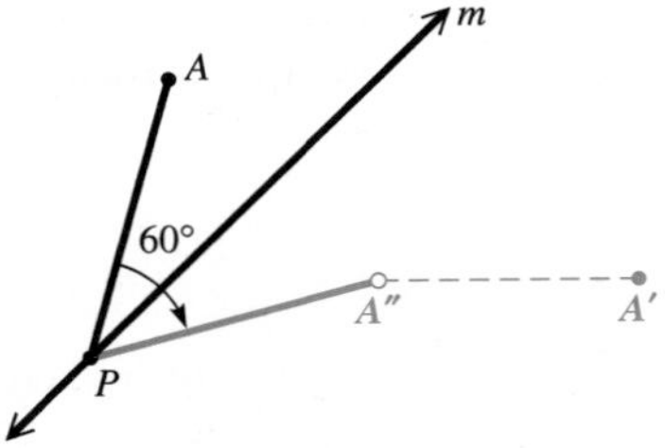

29.

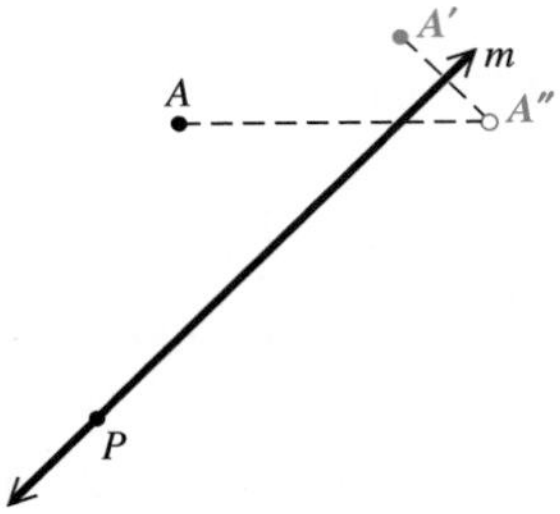

31.

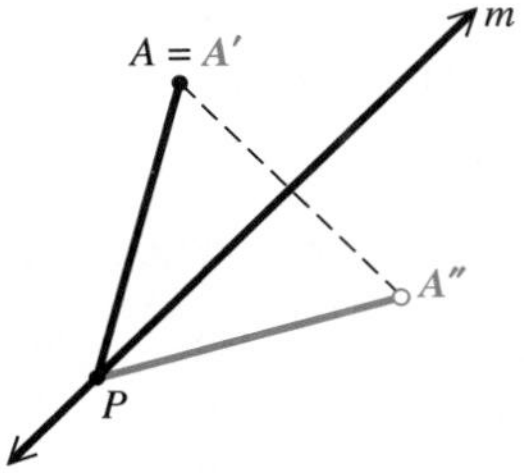

33. no **35.**

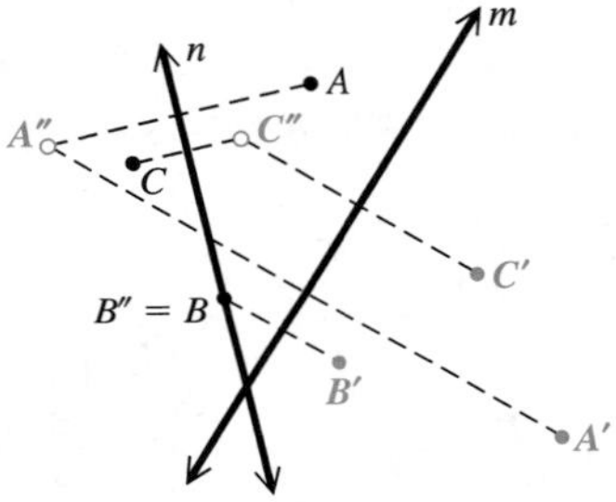

37.

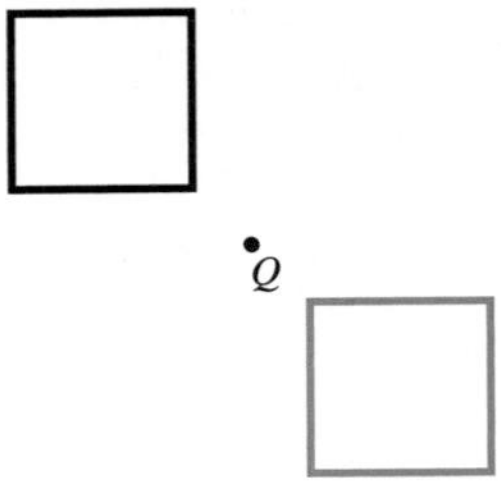

39.

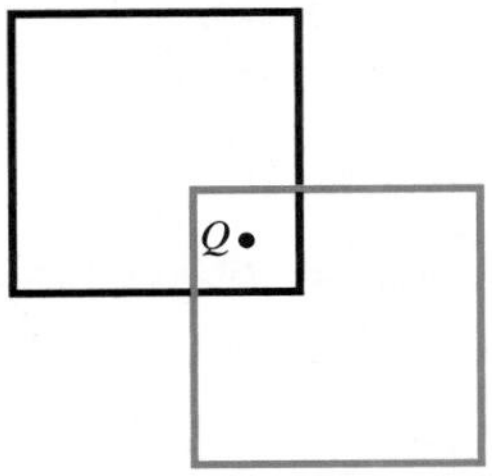

41.

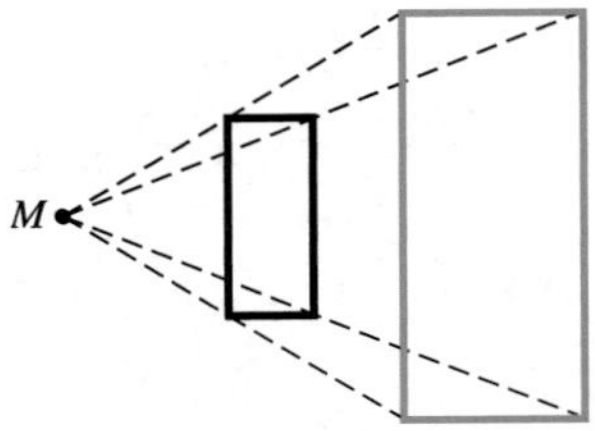

43.

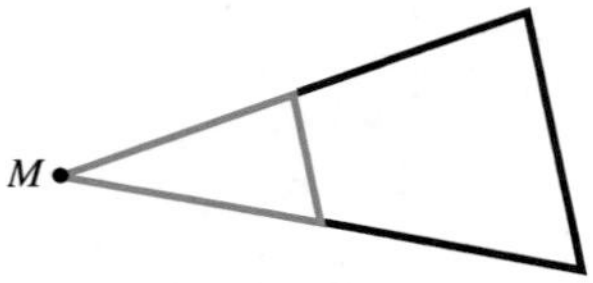

45.

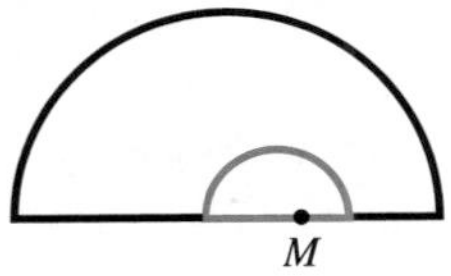

9.7 Exercises (Pages 606–609)

1. Euclidean **3.** Lobachevskian **5.** greater than **7.** Riemannian **9.** Euclidean

11. (a)–(g)

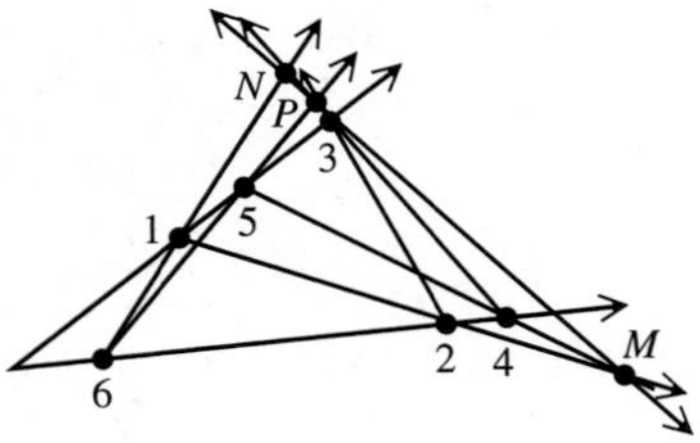

(h) Suppose that a hexagon is inscribed in an angle. Let each pair of opposite sides be extended so as to intersect. Then the three points of intersection thus obtained will lie in a straight line. **13.** C **15.** A, E **17.** A, E **19.** none of them **21.** no **23.** yes **25.** 1 **27.** 3 **29.** 1 **31.** *A*, *C*, *D*, and *F* are even; *B* and *E* are odd. **33.** *A*, *B*, *C*, and *F* are odd; *D*, *E*, and *G* are even. **35.** *A*, *B*, *C*, and *D* are odd; *E* is even

37. traversable

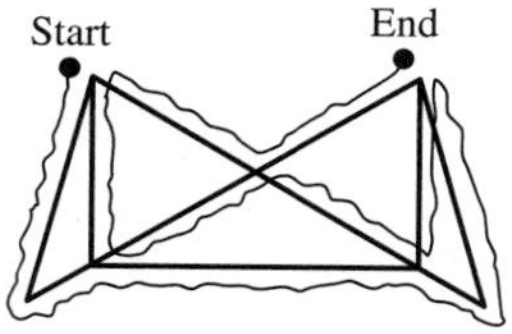

39. not traversable

41. traversable

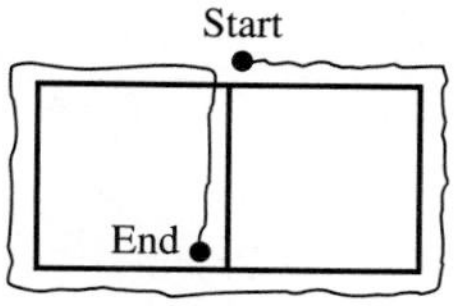

43. yes

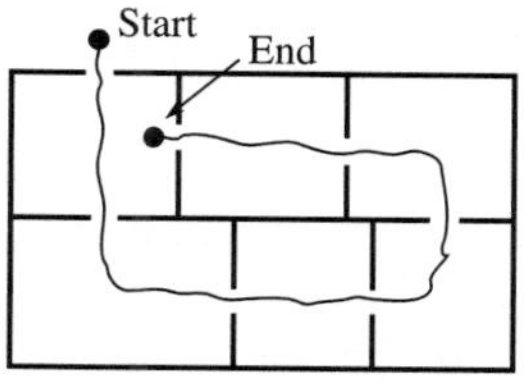

45. no

9.8 Exercises (Pages 613–615)

1. 4 **2.** 4 **3.** 2 **4.** $\frac{2}{1} = 2$ **5.** $\frac{4}{1} = 4$ **6.** $\frac{3}{1} = 3$; $\frac{9}{1} = 9$ **7.** $\frac{4}{1} = 4$; $\frac{16}{1} = 16$ **8.** 4, 9, 16, 25, 36, 100 **9.** Each ratio in the bottom row is the square of the scale factor in the top row. **10.** 4 **11.** 4, 9, 16, 25, 36, 100 **12.** Each ratio in the bottom row is again the square of the scale factor in the top row. **13.** Answers will vary. Some examples are: $3^d = 9$, thus $d = 2$; $5^d = 25$, thus $d = 2$; $4^d = 16$, thus $d = 2$. **14.** 8 **15.** $\frac{2}{1} = 2$; $\frac{8}{1} = 8$ **16.** 8, 27, 64, 125, 216, 1000 **17.** Each ratio in the bottom row is the cube of the scale factor in the top row. **18.** Since $2^3 = 8$, the value of d in $2^d = 8$ must be 3. **19.** $\frac{3}{1} = 3$. **20.** 4 **21.** 1.262 or $\frac{\ln 4}{\ln 3}$ **22.** $\frac{2}{1} = 2$ **23.** 3 **24.** It is between 1 and 2. **25.** 1.585 or $\frac{\ln 3}{\ln 2}$ **27.** .842, .452, .842, .452, ... The two attractors are .842 and .452.

Chapter 9 Test (Pages 617–619)

1. (a) 52° **(b)** 142° **(c)** acute **2.** 40°, 140° **3.** 45°, 45° **4.** 30°, 60° **5.** 130°, 50° **6.** 117°, 117° **7.** Answers will vary. **8.** C **9.** both **10.** neither **11.** 30°, 45°, 105° **12.** 72 cm^2 **13.** 60 $in.^2$ **14.** 68 m^2 **15.** 180 m^2 **16.** 24 π in. **17.** 1978 ft **18.** 57 cm^2

19.

STATEMENTS	REASONS
1. $\measuredangle CAB = \measuredangle DBA$	1. Given
2. $DB = CA$	2. Given
3. $AB = AB$	3. Reflexive property
4. $\triangle ABD \cong \triangle BAC$	4. SAS congruence property

20. 20 ft **21.** 29 m **22.**

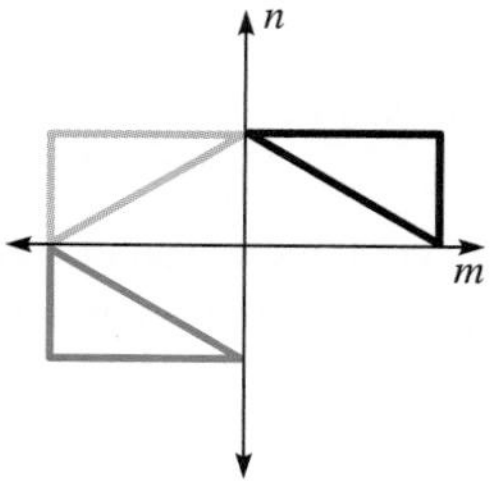

23.

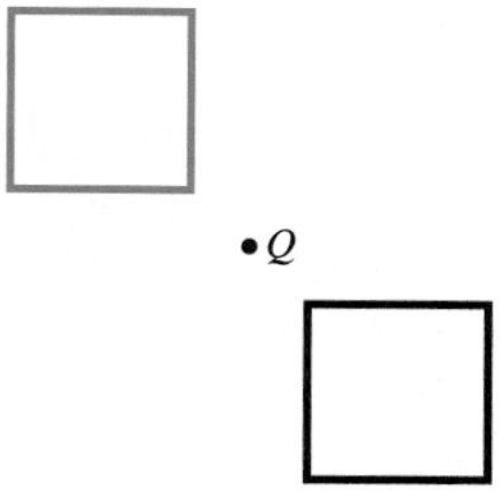

24. (a) 904.32 $in.^3$ **(b)** 452.16 $in.^2$ **25. (a)** 864 ft^2 **(b)** 552 ft^2 **26. (a)** 1582.56 m^3 **(b)** 753.60 m^2 **27.** Answers will vary. **28. (a)** yes **(b)** no

29. **(a)** yes

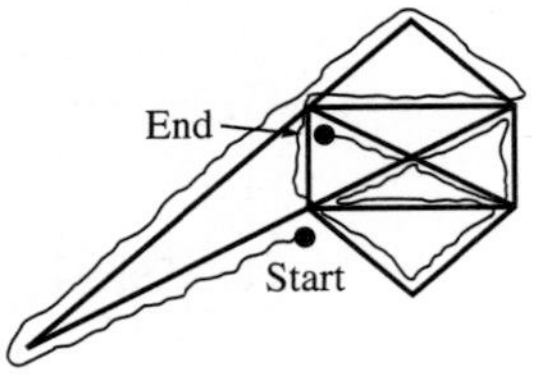

(b) no **30.** The only attractor is .5238095238.

CHAPTER 10 Trigonometry

10.1 Exercises (Pages 626–627)

1. **(a)** 60° **(b)** 150° **3.** **(a)** 45° **(b)** 135° **5.** **(a)** 1° **(b)** 91° **7.** $(90 - x)$ degrees **9.** 83°59′ **11.** 119°27′ **13.** 38°32′ **15.** 17°1′49″ **17.** 20.900° **19.** 91.598° **21.** 274.316° **23.** 31°25′47″ **25.** 89°54′1″ **27.** 178°35′58″ **29.** 320° **31.** 235° **33.** 179° **35.** 130° **37.** $30° + n \cdot 360°$ **39.** $60° + n \cdot 360°$

Angles other than those given are possible in Exercises 41–47.

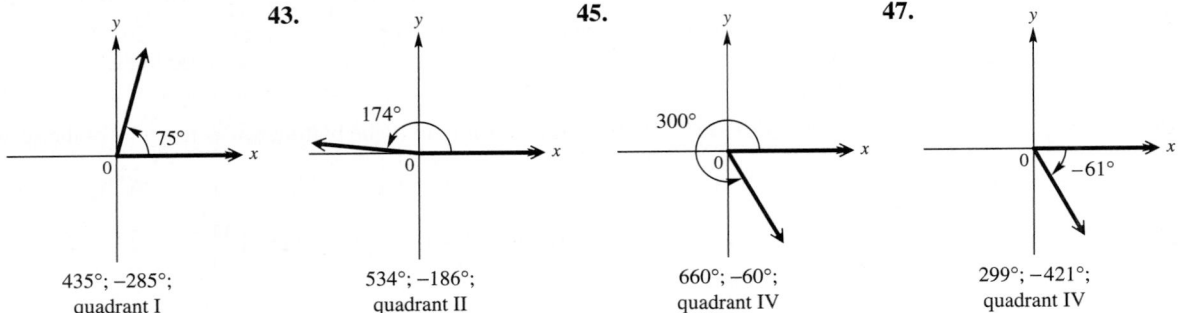

41.

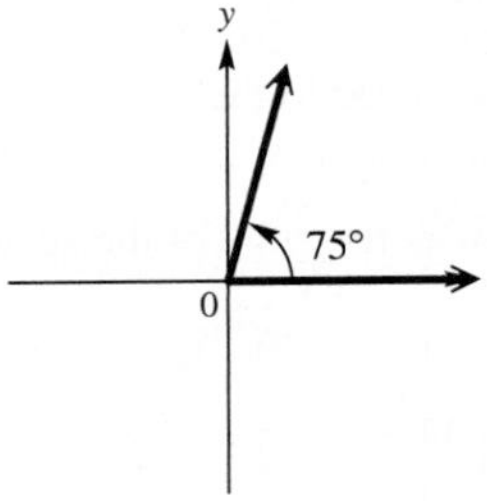

435°; –285°; quadrant I

43.

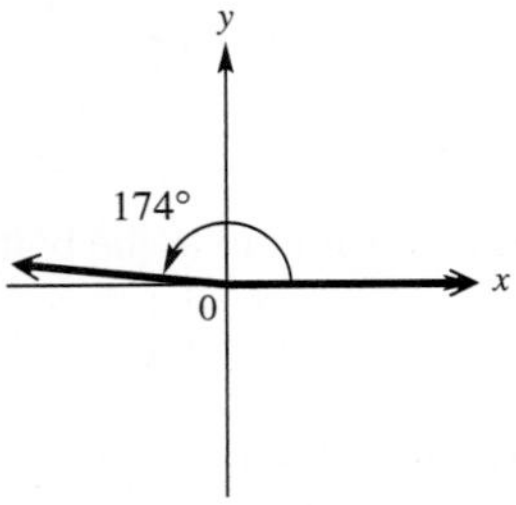

534°; –186°; quadrant II

45.

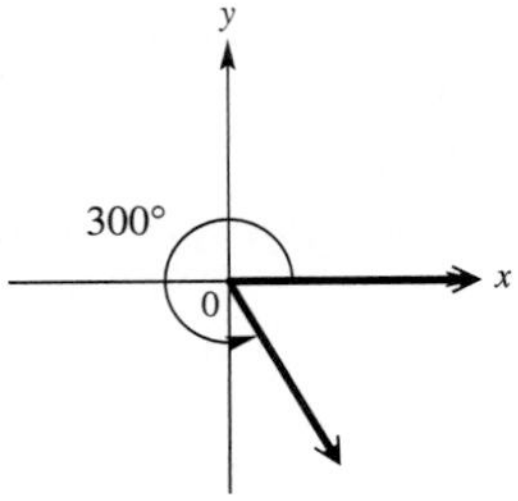

660°; –60°; quadrant IV

47.

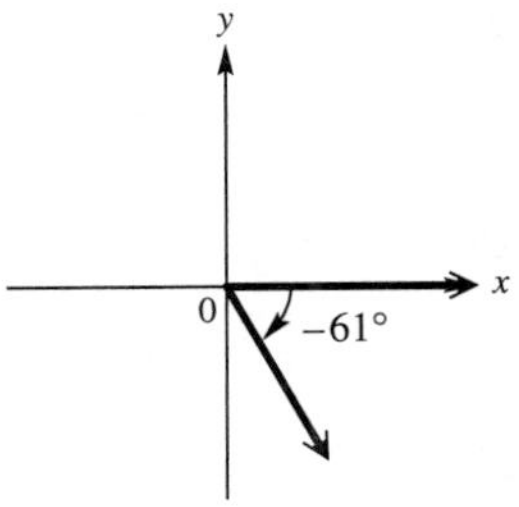

299°; –421°; quadrant IV

10.2 Exercises (Pages 630–631)

1.

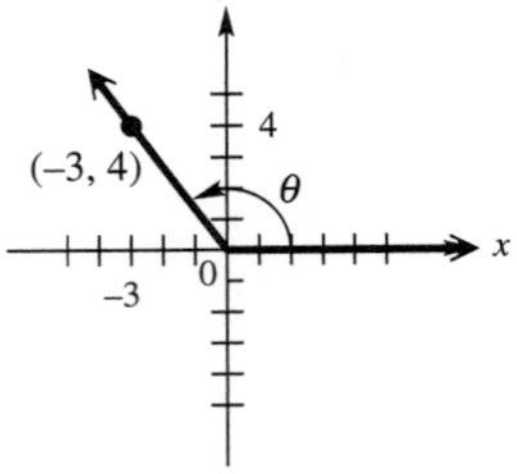

3.

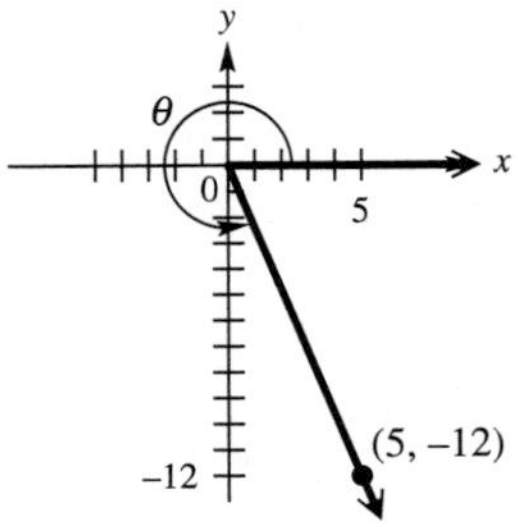

5. $\frac{4}{5}; -\frac{3}{5}; \frac{4}{3}; -\frac{3}{4}; -\frac{5}{3}; \frac{5}{4}$

7. 1; 0; undefined; 0; undefined; 1

9. $\frac{\sqrt{3}}{2}; \frac{1}{2}; \sqrt{3}; \frac{\sqrt{3}}{3}; 2; \frac{2\sqrt{3}}{3}$ **11.** $\frac{5\sqrt{34}}{34}; \frac{3\sqrt{34}}{34}; \frac{5}{3}; \frac{3}{5}; \frac{\sqrt{34}}{3}; \frac{\sqrt{34}}{5}$ **13.** 0; −1; 0; undefined; −1; undefined **15.** Answers will vary. **17.** It is the distance from a point (x, y) on the terminal side of the angle to the origin. **19.** positive **21.** negative **23.** positive **25.** negative **27.** 0 **29.** undefined **31.** undefined **33.** 0 **35.** 0 **37.** 1 **39.** 0

10.3 Exercises (Pages 635–636)

1. $-\frac{1}{3}$ **3.** $\frac{1}{3}$ **5.** −5 **7.** $2\sqrt{2}$ **9.** $-\frac{3\sqrt{5}}{5}$ **11.** $\frac{2}{3}$ **13.** II **15.** I **17.** II **19.** II or III

21. +; +; + **23.** −; −; + **25.** −; +; − **27.** +; +; + **29.** −; +; − **31.** $-2\sqrt{2}$ **33.** $\frac{\sqrt{15}}{4}$ **35.** $-2\sqrt{2}$ **37.** $-\frac{\sqrt{15}}{4}$

In Exercises 39–46, we give, in order, sine, cosine, tangent, cotangent, secant, and cosecant.

39. $\frac{15}{17}; -\frac{8}{17}; -\frac{15}{8}; -\frac{8}{15}; -\frac{17}{8}; \frac{17}{15}$ **41.** $-\frac{4}{5}; -\frac{3}{5}; \frac{4}{3}; \frac{3}{4}; -\frac{5}{3}; -\frac{5}{4}$

43. $-\frac{\sqrt{3}}{2}; -\frac{1}{2}; \sqrt{3}; \frac{\sqrt{3}}{3}; -2; \frac{2\sqrt{3}}{3}$ **45.** $\frac{\sqrt{5}}{7}; \frac{2\sqrt{11}}{7}; \frac{\sqrt{55}}{22}; \frac{2\sqrt{55}}{5}; \frac{7\sqrt{11}}{22}; \frac{7\sqrt{5}}{5}$ **47.** Answers will vary.

10.4 Exercises (Pages 644–645)

1. $\frac{3}{5}; \frac{4}{5}; \frac{3}{4}; \frac{4}{3}; \frac{5}{4}; \frac{5}{3}$ **3.** $\frac{21}{29}; \frac{20}{29}; \frac{21}{20}; \frac{20}{21}; \frac{29}{20}; \frac{29}{21}$ **5.** $\frac{n}{p}; \frac{m}{p}; \frac{n}{m}; \frac{m}{n}; \frac{p}{m}; \frac{p}{n}$

In Exercises 7 and 9, we give, in order, the unknown side, sine, cosine, tangent, cotangent, secant, and cosecant.

7. $c = 13; \frac{12}{13}; \frac{5}{13}; \frac{12}{5}; \frac{5}{12}; \frac{13}{5}; \frac{13}{12}$ **9.** $b = \sqrt{13}; \frac{\sqrt{13}}{7}; \frac{6}{7}; \frac{\sqrt{13}}{6}; \frac{6\sqrt{13}}{13}; \frac{7}{6}; \frac{7\sqrt{13}}{13}$ **11.** cot 40°

13. sec 43° **15.** cot 64.6° **17.** sin 76°30′ **19.** $\frac{\sqrt{3}}{3}$ **21.** $\frac{1}{2}$ **23.** $\sqrt{2}$ **25.** $\frac{\sqrt{2}}{2}$ **27.** $\frac{\sqrt{3}}{2}$

29. $\sqrt{3}$ **31.** 82° **33.** 45° **35.** 30°

In Exercises 37–55, we give, in order, sine, cosine, tangent, cotangent, secant, and cosecant.

37. $\frac{\sqrt{3}}{2}; -\frac{1}{2}; -\sqrt{3}; -\frac{\sqrt{3}}{3}; -2; \frac{2\sqrt{3}}{3}$ **39.** $\frac{1}{2}; -\frac{\sqrt{3}}{2}; -\frac{\sqrt{3}}{3}; -\sqrt{3}; -\frac{2\sqrt{3}}{3}; 2$

41. $-\frac{\sqrt{3}}{2}; -\frac{1}{2}; \sqrt{3}; \frac{\sqrt{3}}{3}; -2; -\frac{2\sqrt{3}}{3}$ **43.** $-\frac{\sqrt{2}}{2}; \frac{\sqrt{2}}{2}; -1; -1; \sqrt{2}; -\sqrt{2}$ **45.** $\frac{\sqrt{3}}{2}; \frac{1}{2}; \sqrt{3}; \frac{\sqrt{3}}{3}; 2; \frac{2\sqrt{3}}{3}$

47. $\frac{\sqrt{2}}{2}; -\frac{\sqrt{2}}{2}; -1; -1; -\sqrt{2}; \sqrt{2}$ **49.** $\frac{1}{2}; \frac{\sqrt{3}}{2}; \frac{\sqrt{3}}{3}; \sqrt{3}; \frac{2\sqrt{3}}{3}; 2$ **51.** $\frac{\sqrt{3}}{2}; \frac{1}{2}; \sqrt{3}; \frac{\sqrt{3}}{3}; 2; \frac{2\sqrt{3}}{3}$

53. $-\frac{1}{2}; \frac{\sqrt{3}}{2}; -\frac{\sqrt{3}}{3}; -\sqrt{3}; \frac{2\sqrt{3}}{3}; -2$ **55.** $\frac{\sqrt{3}}{2}; \frac{1}{2}; \sqrt{3}; \frac{\sqrt{3}}{3}; 2; \frac{2\sqrt{3}}{3}$ **57.** $\frac{\sqrt{3}}{3}, \sqrt{3}$

59. $\frac{\sqrt{3}}{2}, \frac{\sqrt{3}}{3}, \frac{2\sqrt{3}}{3}$ **61.** −1, −1 **63.** $-\frac{\sqrt{3}}{2}, -\frac{2\sqrt{3}}{3}$ **65.** 210°, 330° **67.** 45°, 225° **69.** 60°, 120°

71. 120°, 240° **73.** 225°, 315° **75.** 120°, 300°

10.5 Exercises (Pages 651–654)

1. .5657728 **3.** 1.1342773 **5.** 1.0273488 **7.** .6383201 **9.** 1.7768146 **11.** .4771588
13. −5.7297416 **15.** 1.9074147 **17.** 57.997172° **19.** 30.502748° **21.** 46.173581°
23. 81.168073° **25.** $B = 53°40'$; $a = 571$ m; $b = 777$ m **27.** $M = 38.8°$; $n = 154$ m; $p = 198$ m
29. $A = 47.9108°$; $c = 84.816$ cm; $a = 62.942$ cm **31.** $B = 62.00°$; $a = 8.17$ ft; $b = 15.4$ ft
33. $A = 17.00°$; $a = 39.1$ in.; $c = 134$ in. **35.** $c = 85.9$ yd; $A = 62°50'$; $B = 27°10'$ **37.** $b = 42.3$ cm; $A = 24°10'$; $B = 65°50'$ **39.** $B = 36°36'$; $a = 310.8$ ft; $b = 230.8$ ft **41.** $A = 50°51'$; $a = .4832$ m; $b = .3934$ m **43.** 9.35 meters **45.** 88.3 meters **47.** 26.92 inches **49.** 583 feet **51.** 28.0 meters
53. 469 meters **55.** 146 meters **57.** 34.0 miles **59.** $a = 12, b = 12\sqrt{3}, d = 12\sqrt{3}, c = 12\sqrt{6}$

10.6 Exercises (Pages 663–668)

1. $C = 95°, b = 13$ m, $a = 11$ m **3.** $B = 37.3°, a = 38.5$ ft, $b = 51.0$ ft **5.** $C = 57.36°, b = 11.13$ ft, $c = 11.55$ ft **7.** $B = 18.5°, a = 239$ yd, $c = 230$ yd **9.** $A = 56°00', AB = 361$ ft, $BC = 308$ ft
11. $B = 110.0°, a = 27.01$ m, $c = 21.36$ m **13.** $A = 34.72°, a = 3326$ ft, $c = 5704$ ft
15. $C = 97°34', b = 283.2$ m, $c = 415.2$ m **17.** 118 meters **19.** 10.4 inches **21.** 12 **23.** 7 **25.** 30°

27. $c = 2.83$ in., $A = 44.9°$, $B = 106.8°$ **29.** $c = 6.46$ m, $A = 53.1°$, $B = 81.3°$ **31.** $a = 156$ cm, $B = 64°50'$, $C = 34°30'$ **33.** $b = 9.529$ in., $A = 64.59°$, $C = 40.61°$ **35.** $a = 15.7$ m, $B = 21.6°$, $C = 45.6°$ **37.** $A = 30°$, $B = 56°$, $C = 94°$ **39.** $A = 82°$, $B = 37°$, $C = 61°$ **41.** $A = 42.0°$, $B = 35.9°$, $C = 102.1°$ **43.** 257 meters **45.** 22 feet **47.** 36° with the 45-foot cable, 26° with the 60-foot cable **49.** $\frac{\sqrt{3}}{2}$ **51.** $\frac{\sqrt{2}}{2}$ **53.** 46.4 m^2 **55.** 356 cm^2 **57.** 722.9 $in.^2$ **59.** $24\sqrt{3}$ **61.** 78 m^2 **63.** 12,600 cm^2 **65.** 3650 ft^2 **67.** 100 m^2 **69.** 33 cans

Chapter 10 Test (Pages 668–669)

1. 74.2983° **2.** 203° **3.** $\sin\theta = -\frac{5\sqrt{29}}{29}$; $\cos\theta = \frac{2\sqrt{29}}{29}$; $\tan\theta = -\frac{5}{2}$ **4.** III **5.** $\sin\theta = -\frac{3}{5}$; $\tan\theta = -\frac{3}{4}$; $\cot\theta = -\frac{4}{3}$; $\sec\theta = \frac{5}{4}$; $\csc\theta = -\frac{5}{3}$ **6.** $\sin A = \frac{12}{13}$; $\cos A = \frac{5}{13}$; $\tan A = \frac{12}{5}$; $\cot A = \frac{5}{12}$; $\sec A = \frac{13}{5}$; $\csc A = \frac{13}{12}$ **7. (a)** $\frac{1}{2}$ **(b)** 1 **(c)** undefined **(d)** $-\frac{2\sqrt{3}}{3}$ **(e)** undefined **(f)** $-\sqrt{2}$

8. (a) .97939940 **(b)** .20834446 **(c)** 1.9362132 **(d)** 4.16529977 **9.** 16.16664145° **10.** 135°, 225° **11.** $B = 31°30'$, $a = 638$, $b = 391$ **12.** 15.5 ft **13.** 137.5° **14.** 180 km **15.** 49.0° **16.** 4300 km^2 **17.** 264 square units **18.** 2.7 miles **19.** 5500 meters **20.** distance to both first and third bases: 63.7 feet; distance to second base: 66.8 feet

CHAPTER 11 Counting Methods

11.1 Exercises (Pages 678–681)

1. *AB, AC, AD, AE, BA, BC, BD, BE, CA, CB, CD, CE, DA, DB, DC, DE, EA, EB, EC, ED*; 20 ways **3.** *AC, AE, BC, BE, CA, CB, CD, DC, DE, EA, EB, ED*; 12 ways **5.** *ACE, AEC, BCE, BEC, DCE, DEC*; 6 ways **7.** *ABC, ABD, ABE, ACD, ACE, ADE, BCD, BCE, BDE, CDE*; 10 ways **9.** 1 **11.** 3 **13.** 5 **15.** 5 **17.** 3 **19.** 1 **21.** 18 **23.** 15

25.

	1	2	3	4	5	6
1	11	12	13	14	15	16
2	21	22	23	24	25	26
3	31	32	33	34	35	36
4	41	42	43	44	45	46
5	51	52	53	54	55	56
6	61	62	63	64	65	66

27. 11, 22, 33, 44, 55, 66 **29.** 11, 13, 23, 31, 41, 43, 53, 61 **31.** 16, 25, 36, 64 **33.** 16, 32, 64

35. (a) tttt **(b)** hhhh, hhht, hhth, hhtt, hthh, htht, htth, thhh, thht, thth, tthh **(c)** httt, thtt, ttht, ttth, tttt **(d)** hhhh, hhht, hhth, hhtt, hthh, htht, htth, httt, thhh, thht, thth, thtt, tthh, ttht, ttth

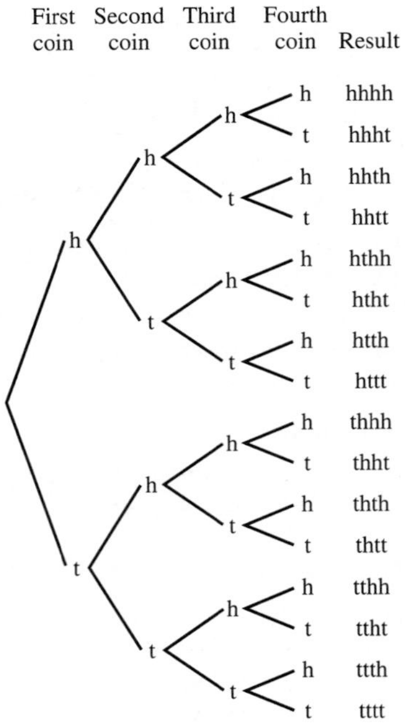

37. 16 **39.** 36 **41.** 17 **43.** 72 **45.** 12 **47.** 10 **49.** 6 **51.** 3 **53.** 54 **55.** 18 **57.** 138 **59.** 16 **61.** 13 **63.** 4 **65.** 883 **67.** 3 **69.** **(a)** Determine the number of ordered pairs of digits that can be selected from the set {1, 2, 3, 4, 5, 6} if repetition of digits is allowed. **(b)** Determine the number of ordered pairs of digits that can be selected from the set {1, 2, 3, 4, 5, 6} if the selection is done with replacement. **71.** **(a)** Find the number of ways to select three letters from the set {A, B, C, D, E} if repetition of letters is not allowed. **(b)** Find the number of ways to select three letters from the set {A, B, C, D, E} if the selection is done without replacement.

11.2 Exercises (Pages 689–692)

1. Answers will vary. **3.** **(a)** no **(b)** Answers will vary. **5.** **(a)** no **(b)** Answers will vary. **7.** 24 **9.** 336 **11.** 20 **13.** 84 **15.** 840 **17.** 39,916,800 **19.** 95,040 **21.** 1716 **23.** 184,756 **25.** $4.151586701 \times 10^{12}$ **27.** 362,880 **29.** 1680 **31.** $2^3 = 8$ **33.** Answers will vary. **35.** $6^3 = 216$ **37.** $5! = 120$ **39.** $3 \cdot 2 = 6$ **41.** $3 \cdot 3 = 9$ **43.** $3 \cdot 2 \cdot 1 = 6$ **45.** $4 \cdot 2 \cdot 3 = 24$ **47.** $2^6 = 64$ **49.** $3 \cdot 3 \cdot 4 \cdot 5 = 180$ **51.** $3 \cdot 3 \cdot 4 \cdot 3 = 108$ **53.** $3 \cdot 3 \cdot 1 \cdot 3 = 27$ **55.** $2 \cdot 5 \cdot 6 = 60$ **57.** $5! = 120$ **59.** **(a)** 6 **(b)** 5 **(c)** 4 **(d)** 3 **(e)** 2 **(f)** 1; 720 **61.** **(a)** 3 **(b)** 3 **(c)** 2 **(d)** 2 **(e)** 1 **(f)** 1; 36 **63.** Answers will vary. **65.** 450; no

11.3 Exercises (Pages 702–705)

1. 840 **3.** 495 **5.** 1,028,160 **7.** 70 **9.** $1.805037696 \times 10^{11}$ **11.** Answers will vary. **13.** Answers will vary. **15.** **(a)** permutation **(b)** permutation **(c)** combination **(d)** combination **(e)** permutation **(f)** combination **(g)** permutation **17.** ${}_8P_5 = 6720$ **19.** ${}_{12}P_2 = 132$ **21.** ${}_{25}P_5 = 6{,}375{,}600$ **23.** **(a)** ${}_4P_4 = 24$ **(b)** ${}_4P_4 = 24$ **25.** ${}_{18}C_5 = 8568$ **27.** **(a)** ${}_{13}C_5 = 1287$ **(b)** ${}_{26}C_5 = 65{,}780$ **(c)** 0 (impossible) **29.** **(a)** ${}_6C_3 = 20$ **(b)** ${}_6C_2 = 15$ **31.** ${}_{10}C_3 = 120$ **33.** **(a)** 5 **(b)** 9 **35.** ${}_{26}P_3 \cdot {}_{10}P_3 \cdot {}_{26}P_3 = 175{,}219{,}200{,}000$ **37.** $2 \cdot {}_{25}P_3 = 27{,}600$ **39.** $7 \cdot {}_{12}P_8 = 139{,}708{,}800$ **41.** **(a)** $7^7 = 823{,}543$ **(b)** $7! = 5040$ **43.** ${}_{25}C_3 \cdot {}_{22}C_4 \cdot {}_{18}C_5 \cdot {}_{13}C_6 \approx 2.473653743 \times 10^{14}$ **45.** ${}_{20}C_3 = 1140$ **47.** **(a)** ${}_7P_2 = 42$ **(b)** $3 \cdot 6 = 18$ **(c)** ${}_7P_2 \cdot 5 = 210$ **49.** ${}_8P_3 = 336$ **51.** **(a)** ${}_6C_2 \cdot {}_6C_3 \cdot {}_6C_4 = 4500$ **(b)** $3! \cdot {}_6C_2 \cdot {}_6C_3 \cdot {}_6C_4 = 27{,}000$ **53.** **(a)** $8! = 40{,}320$ **(b)** $2 \cdot 6! = 1440$ **(c)** $6! = 720$ **55.** **(a)** $2 \cdot 4! = 48$ **(b)** $3 \cdot 4! = 72$ **57.** Each equals 220. **59.** **(a)** 1 **(b)** Answers will vary.

11.4 Exercises (Pages 710–712)

1. 6 **3.** 20 **5.** 56 **7.** 36 **9.** ${}_7C_1 \cdot {}_3C_3 = 7$ **11.** ${}_7C_3 \cdot {}_3C_1 = 105$ **13.** ${}_8C_3 = 56$ **15.** ${}_8C_5 = 56$ **17.** ${}_9C_4 = 126$ **19.** $1 \cdot {}_8C_3 = 56$ **21.** 1 **23.** 10 **25.** 5 **27.** 32 **29.** **(a)** All are multiples of the row number. **(b)** The same pattern holds. **(c)** The same pattern holds. **31.** . . . 8, 13, 21, 34, . . . ; A number in this sequence is the sum of the two preceding terms. This is the Fibonacci sequence. **33.** row 8 **35.** The sum of the squares of the entries across the top row equals the entry at the bottom vertex. **37.** Answers will vary. **Wording may vary for Exercises 39 and 41.** **39.** sum = N; Any entry in the array equals the sum of the two entries immediately above it and immediately to its left. **41.** sum = N; Any entry in the array equals the sum of the row of entries from the cell immediately above it to the left boundary of the array.

11.5 Exercises (Pages 717–719)

1. $2^4 - 1 = 15$ **3.** $2^7 - 1 = 127$ **5.** 120 **7.** $36 - 6 = 30$ **9.** $6 + 6 - 1 = 11$ **11.** 51 **13.** $90 - 9 = 81$ **15.** **(a)** ${}_8C_3 = 56$ **(b)** ${}_7C_3 = 35$ **(c)** $56 - 35 = 21$ **17.** ${}_7C_3 - {}_5C_3 = 25$ **19.** ${}_8P_3 - {}_6P_3 = 216$ **21.** ${}_{10}P_3 - {}_7P_3 = 510$ **23.** ${}_{25}C_4 - {}_{23}C_4 = 3795$ **25.** $13 + 4 - 1 = 16$ **27.** $30 + 15 - 10 = 35$ **29.** $2{,}598{,}960 - {}_{13}C_5 = 2{,}597{,}673$ **31.** $2{,}598{,}960 - {}_{40}C_5 = 1{,}940{,}952$ **33.** ${}_{12}C_0 + {}_{12}C_1 + {}_{12}C_2 = 79$ **35.** $2^{12} - 79 = 4017$ **37.** $26^3 \cdot 10^3 - {}_{26}P_3 \cdot {}_{10}P_3 = 6{,}344{,}000$ **39.** Answers will vary. **41.** ${}_4C_3 + {}_3C_3 + {}_5C_3 = 15$ **43.** ${}_{12}C_3 - 4 \cdot 3 \cdot 5 = 160$ **45.** Answers will vary. **47.** Answers will vary. **49.** Answers will vary.

Chapter 11 Test (Pages 720–721)

1. $6 \cdot 7 \cdot 7 = 294$ **2.** $6 \cdot 7 \cdot 4 = 168$ **3.** $6 \cdot 6 \cdot 5 = 180$ **4.** $6 \cdot 5 \cdot 1 = 30$ end in 0; $5 \cdot 5 \cdot 1 = 25$ end in 5; $30 + 25 = 55$ **5.** 13

6.

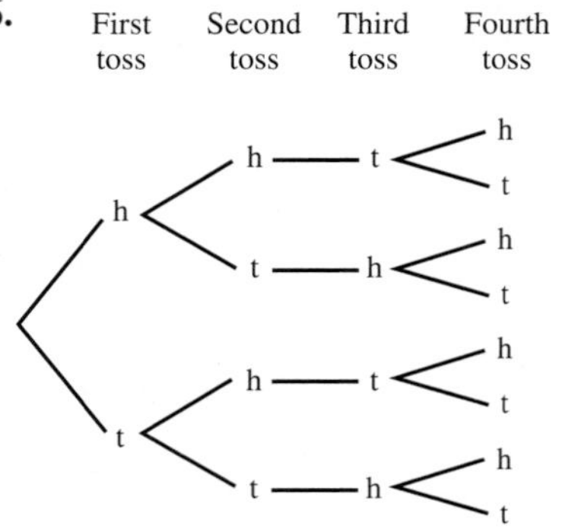

7. $4! = 24$ **8.** 12 **9.** 120 **10.** 336 **11.** 11,880 **12.** 35 **13.** ${}_{26}P_5 = 7{,}893{,}600$ **14.** $32^5 = 33.554{,}432$ **15.** ${}_7P_4 = 840$ **16.** $3! = 6$ **17.** 120 **18.** 30,240 **19.** ${}_{12}C_4 = 495$ **20.** ${}_{12}C_2 \cdot {}_{10}C_2 = 2970$ **21.** ${}_{12}C_5 \cdot {}_7C_5 = 16{,}632$ **22.** $2^{12} - [{}_{12}C_0 + {}_{12}C_1 + {}_{12}C_2] = 4017$ **23.** $2^4 = 16$ **24.** $2^2 = 4$ **25.** $2 \cdot 2^2 = 8$ **26.** 8 **27.** 2 **28.** $16 - (1 + 4) = 11$ **29.** ${}_6C_3 = 20$ **30.** ${}_5C_2 = 10$ **31.** $2 \cdot {}_5C_3 = 20$ **32.** ${}_5C_2 = 10$ **33.** ${}_5C_4 + {}_2C_1 \cdot {}_5C_3 = 25$ **34.** Answers will vary. **35.** $495 + 220 = 715$ **36.** the counting numbers **37.** Answers will vary.

CHAPTER 12 Probability

12.1 Exercises (Pages 731–735)

1. (a) $\frac{1}{3}$ **(b)** $\frac{1}{3}$ **(c)** $\frac{1}{3}$ **3. (a)** $\frac{1}{2}$ **(b)** $\frac{1}{3}$ **(c)** $\frac{1}{6}$ **5. (a)** $\{1, 2, 3\}$ **(b)** 2 **(c)** 1 **(d)** 3 **(e)** $\frac{2}{3}$ **(f)** 2 to 1 **7. (a)** $\{11, 12, 13, 21, 22, 23, 31, 32, 33\}$ **(b)** $\frac{2}{3}$ **(c)** $\frac{1}{3}$ **(d)** $\frac{1}{3}$ **(e)** $\frac{4}{9}$ **9. (a)** 4 to 7 **(b)** 5 to 6 **(c)** 2 to 9 **11. (a)** $\frac{1}{50}$ **(b)** $\frac{2}{50} = \frac{1}{25}$ **(c)** $\frac{3}{50}$ **(d)** $\frac{4}{50} = \frac{2}{25}$ **(e)** $\frac{5}{50} = \frac{1}{10}$ **13. (a)** $\frac{1}{36}$ **(b)** $\frac{2}{36} = \frac{1}{18}$ **(c)** $\frac{3}{36} = \frac{1}{12}$ **(d)** $\frac{4}{36} = \frac{1}{9}$ **(e)** $\frac{5}{36}$ **(f)** $\frac{6}{36} = \frac{1}{6}$ **(g)** $\frac{5}{36}$ **(h)** $\frac{4}{36} = \frac{1}{9}$ **(i)** $\frac{3}{36} = \frac{1}{12}$ **(j)** $\frac{2}{36} = \frac{1}{18}$ **(k)** $\frac{1}{36}$ **15. (a)** $\frac{34{,}244{,}000}{288{,}280{,}000} \approx .119$ **(b)** $\frac{288{,}280{,}000 - 34{,}244{,}000}{288{,}280{,}000} = \frac{254{,}036{,}000}{288{,}280{,}000} \approx .881$ **17.** Answers will vary. **19.** $\frac{1}{4}$ **21.** $\frac{1}{4}$ **23. (a)** $\frac{3}{4}$ **(b)** $\frac{1}{4}$ **25.** $\frac{1}{250{,}000} = .000004$ **27.** $\frac{1}{4}$ **29.** $\frac{1}{4}$ **31.** $\frac{2}{4} = \frac{1}{2}$ **33.** $\frac{1}{500} = .002$ **35.** about 160 **37.** $\frac{2}{4} = \frac{1}{2}$ **39. (a)** 0 **(b)** no **(c)** yes **41.** Answers will vary. **43.** $\frac{12}{31}$ **45.** $\frac{36}{2{,}598{,}960} \approx .00001385$ **47.** $\frac{624}{2{,}598{,}960} \approx .00024010$ **49.** $\frac{1}{4} \cdot \frac{5108}{2{,}598{,}960} \approx .00049135$ **51. (a)** $\frac{5}{9}$ **(b)** $\frac{49}{144}$ **(c)** $\frac{5}{48}$ **53.** $3 \cdot 1 \cdot 2 \cdot 1 \cdot 1 \cdot 1 = 6$; $\frac{6}{720} = \frac{1}{120} \approx .0083$ **55.** $4 \cdot 3! \cdot 3! = 144$; $\frac{144}{720} = \frac{1}{5} = .2$ **57.** $\frac{2}{{}_7C_2} = \frac{2}{21} \approx .095$ **59.** $\frac{{}_5C_3}{{}_{12}C_3} = \frac{1}{22} \approx .045$ **61.** $\frac{1}{{}_{36}P_3} \approx .000023$ **63.** $\frac{3}{28} \approx .107$ **65. (a)** $\frac{8}{9^2} = \frac{8}{81} \approx .099$ **(b)** $\frac{4}{{}_9C_2} = \frac{1}{9} \approx .111$ **67.** $\frac{9 \cdot 10}{9 \cdot 10^2} = \frac{1}{10}$

12.2 Exercises (Pages 742–743)

1. yes **3.** Answers will vary. **5.** $\frac{1}{2}$ **7.** $\frac{5}{6}$ **9.** $\frac{2}{3}$ **11. (a)** $\frac{2}{13}$ **(b)** 2 to 11 **13. (a)** $\frac{11}{26}$ **(b)** 11 to 15

15. (a) $\frac{9}{13}$ **(b)** 9 to 4 **17.** $\frac{2}{3}$ **19.** $\frac{7}{36}$ **21.** $P(A) + P(B) + P(C) + P(D) = 1$ **23.** .005365

25. .971285 **27.** .76 **29.** .92 **31.** 6 to 19

33.

x	$P(x)$
3	.1
4	.1
5	.2
6	.2
7	.2
8	.1
9	.1

35. $n(A') = s - a$ **37.** $P(A) + P(A') = 1$ **39.** 180 **41.** $\frac{2}{3}$ **43.** 1

12.3 Exercises (Pages 752–755)

1. independent **3.** not independent **5.** independent **7.** $\frac{42}{100} = \frac{21}{50}$ **9.** $\frac{68}{100} = \frac{17}{25}$ **11.** $\frac{19}{34}$

13. $\frac{4}{7} \cdot \frac{4}{7} = \frac{16}{49}$ **15.** $\frac{2}{7} \cdot \frac{1}{7} = \frac{2}{49}$ **17.** $\frac{4}{7} \cdot \frac{3}{6} = \frac{2}{7}$ **19.** $\frac{1}{6}$ **21.** 0 **23.** $\frac{12}{51} = \frac{4}{17}$ **25.** $\frac{12}{52} \cdot \frac{11}{51} = \frac{11}{221}$

27. $\frac{4}{52} \cdot \frac{11}{51} = \frac{11}{663}$ **29.** $\frac{1}{3}$ **31.** 1 **33.** $\frac{3}{10}$ (the same) **35.** $\frac{1}{2} \cdot \frac{1}{2} \cdot \frac{1}{2} = \frac{1}{8}$ **37.** .640 **39.** .008

41. .95 **43.** .23 **45.** $\frac{1}{35}$ **47.** $\frac{3}{7}$ **49.** $\frac{3}{7}$ **51. (a)** $\frac{3}{4}$ **(b)** $\frac{1}{2}$ **(c)** $\frac{5}{16}$ **53.** .2704 **55.** .2496

57. Answers will vary. **59.** $\frac{1}{64} \approx .0156$ **61.** 10 **63.** .400 **65.** .080 **67.** $(.90)^4 = .6561$

69. ${}_4C_2 \cdot (.10) \cdot (.20) \cdot (.70)^2 = .0588$ **71.** .30 **73.** .49 **75.** Answers will vary.

12.4 Exercises (Pages 760–762)

1. $\frac{1}{8}$ **3.** $\frac{3}{8}$ **5.** $\frac{3}{4}$ **7.** $\frac{1}{2}$ **9.** $\frac{3}{8}$ **11.** $x; n; n$ **13.** $\frac{7}{128}$ **15.** $\frac{35}{128}$ **17.** $\frac{21}{128}$ **19.** $\frac{1}{128}$

21. $\frac{25}{72}$ **23.** $\frac{1}{216}$ **25.** .041 **27.** .268 **29.** Answers will vary. **31.** Answers will vary. **33.** .016

35. .020 **37.** .198 **39.** .448 **41.** .656 **43.** .002 **45.** .883 **47.** $\frac{1}{1024} \approx .001$ **49.** $\frac{45}{1024} \approx .044$

51. $\frac{210}{1024} = \frac{105}{512} \approx .205$ **53.** $\frac{772}{1024} \approx .754$

12.5 Exercises (Pages 770–773)

1. Answers will vary. **3.** $\frac{5}{2}$ **5.** \$1 **7.** 50¢ **9.** no (expected net winnings: $-\frac{3}{4}$¢) **11.** 1.72

13. (a) −\$60 **(b)** \$36,000 **(c)** \$72,000 **15.** 50¢ **17.** \$2500 **19.** 2.7 **21.** an increase of 250 **23.** Answers will vary. **25.** Project *C* **27.** \$2200 **29.** \$83,400 **31.** Purchase the insurance (because \$83,400 > \$80,400). **33.** 25,000; 60,000; 16,000 **35.** C; C; C; B; C; A; B **37.** about −15¢

Extension Exercises (Pages 777–778)

1. Answers will vary. **3.** no **5.** $\frac{18}{50} = .36$ (This is quite close to .375, the theoretical value.) **7.** $\frac{11}{50} = .22$
9. Answers will vary. **11.** Answers will vary.

Chapter 12 Test (Pages 779–780)

1. Answers will vary. **2.** Answers will vary. **3.** 3 to 1 **4.** 25 to 1 **5.** 11 to 2 **6.** row 1: CC; row 2: cC, cc
7. $\frac{1}{2}$ **8.** 1 to 3 **9.** $\frac{7}{7} \cdot \frac{6}{7} \cdot \frac{5}{7} = \frac{30}{49}$ **10.** $\frac{7}{19}$ **11.** $1 - \left(\frac{30}{49} + \frac{1}{49}\right) = \frac{18}{49}$ **12.** $\frac{{}_2C_2}{{}_5C_2} = \frac{1}{10}$
13. $\frac{{}_3C_2}{{}_5C_2} = \frac{3}{10}$ **14.** $\frac{6}{10} = \frac{3}{5}$ **15.** $\frac{3}{10}$ **16.** $\frac{3}{10}; \frac{6}{10}; \frac{1}{10}$, **17.** $\frac{9}{10}$ **18.** $\frac{18}{10} = \frac{9}{5}$ **19.** $\frac{6}{36} = \frac{1}{6}$ **20.** 35 to 1
21. 7 to 2 **22.** $\frac{4}{36} = \frac{1}{9}$ **23.** $(.78)^3 \approx .475$ **24.** ${}_3C_2\,(.78)^2(.22) \approx .402$ **25.** $1 - (.22)^3 \approx .989$
26. $(.78)(.22)(.78) \approx .134$ **27.** $\frac{25}{102}$ **28.** $\frac{25}{51}$ **29.** $\frac{4}{51}$ **30.** $\frac{3}{26}$

CHAPTER 13 Statistics

13.1 Exercises (Pages 788–793)

1. (a)

x	f	f/n
0	10	$\frac{10}{30} \approx 33\%$
1	7	$\frac{7}{30} \approx 23\%$
2	6	$\frac{6}{30} = 20\%$
3	4	$\frac{4}{30} \approx 13\%$
4	2	$\frac{2}{30} \approx 7\%$
5	1	$\frac{1}{30} \approx 3\%$

(b)

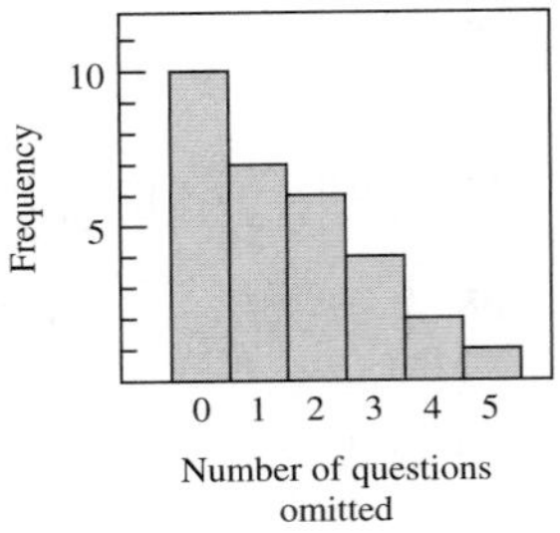

(c)

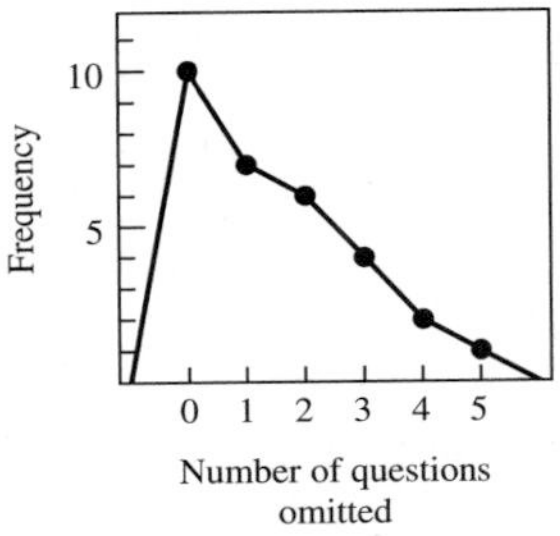

3. (a)

Class Limits	Tally	Frequency f	Relative Frequency f/n
45–49	|||	3	$\frac{3}{54} \approx 5.6\%$
50–54	~~||||~~ ~~||||~~ ||||	14	$\frac{14}{54} \approx 25.9\%$
55–59	~~||||~~ ~~||||~~ ~~||||~~ |	16	$\frac{16}{54} \approx 29.6\%$
60–64	~~||||~~ ~~||||~~ ~~||||~~ ||	17	$\frac{17}{54} \approx 31.5\%$
65–69	||||	4	$\frac{4}{54} \approx 7.4\%$

Total: $n = 54$

(b)

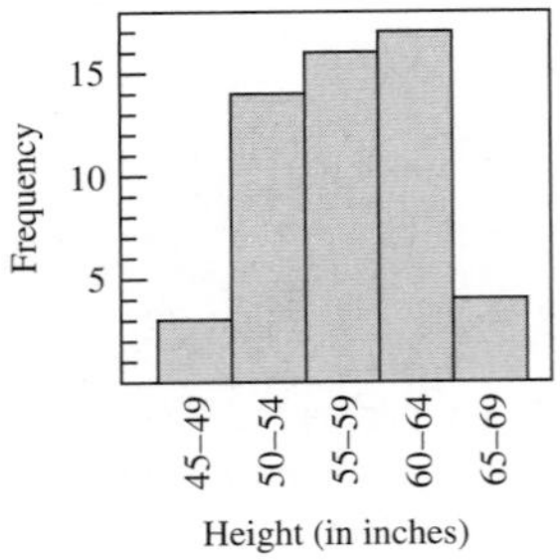

(c)

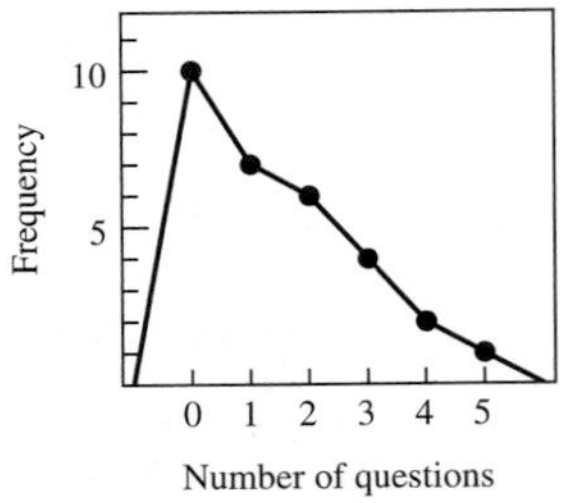

5. (a)

Class Limits	Tally	Frequency f	Relative Frequency f/n
70–74	||	2	$\frac{2}{30} \approx 6.7\%$
75–79	|	1	$\frac{1}{30} \approx 3.3\%$
80–84	|||	3	$\frac{3}{30} = 10.0\%$
85–89	||	2	$\frac{2}{30} \approx 6.7\%$
90–94	卌	5	$\frac{5}{30} \approx 16.7\%$
95–99	卌	5	$\frac{5}{30} \approx 16.7\%$
100–104	卌 |	6	$\frac{6}{30} = 20.0\%$
105–109	||||	4	$\frac{4}{30} \approx 13.3\%$
110–114	||	2	$\frac{2}{30} \approx 6.7\%$

Total: $n = 30$

(b)

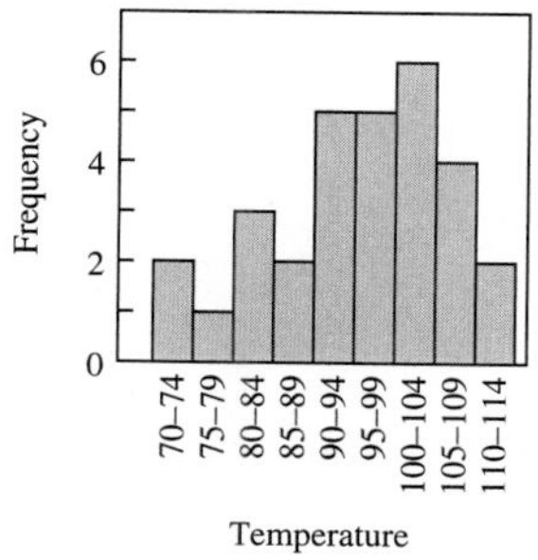

(c)

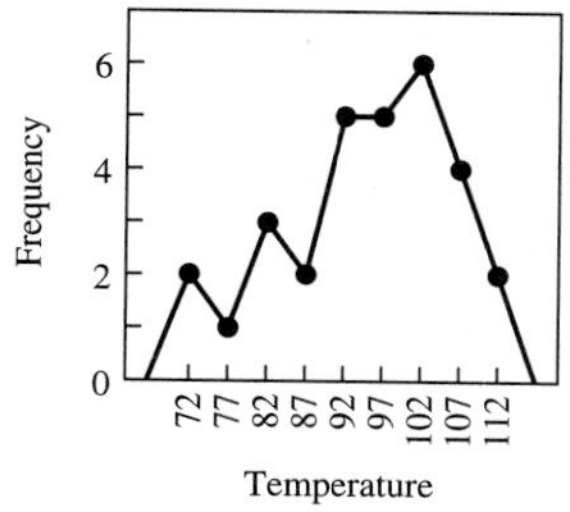

7.

0	7	9	8										
1	1	1	2	8	9	4	3	1	0	5	0	5	5
2	0	9	6	6	2	5	2	3	4	4			
3	1												

9.

0	8	5	4	9	6	9	4	8				
1	2	0	1	8	8	2	4	0	8	8	6	3
2	6	6	2	5	1	3						
3	0	4	6									
4	4											

11. about 31% **13.** CBS; about 4% **15.** about .4%; 1998 **17.** 3.5% in 2005 **19.** Answers will vary. **21.** 184°

23.

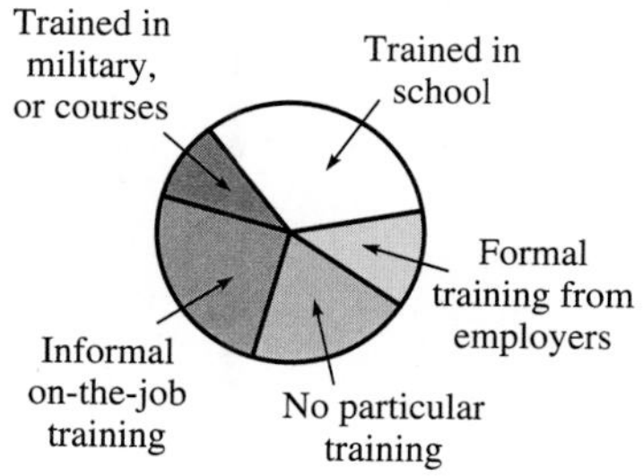

25. about 79 years **27.** **(a)** about 6 years **(b)** Answers will vary. **29.** Answers will vary. **31.** Answers will vary. **33.** Answers will vary.

35. **(a)**

Letter	Probability
A	.208
E	.338
I	.169
O	.208
U	.078

(b)

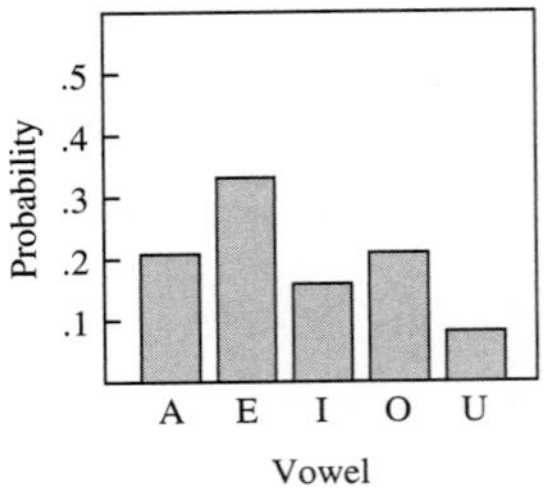

37. Answers will vary. **39.** **(a)** .225 **(b)** .275 **(c)** .425 **(d)** .175

41. **(a)**

Sport	Probability
Sailing	.225
Hang gliding	.125
Bungee jumping	.175
Sky diving	.075
Canoeing	.300
Rafting	.100

(b) empirical **(c)** Answers will vary.

13.2 Exercises (Pages 803–807)

1. **(a)** 14 **(b)** 10 **(c)** none **3.** **(a)** 201.2 **(b)** 210 **(c)** 196 **5.** **(a)** 5.2 **(b)** 5.35 **(c)** 4.5 and 6.2 **7.** **(a)** .8 **(b)** .795 **(c)** none **9.** **(a)** 129 **(b)** 128 **(c)** 125 and 128 **11.** **(a)** 2,154,000 **(b)** 1,320,000 **13.** **(a)** 1,292,000 **(b)** 573,000 **15.** 406,000 **17.** 4000 square miles **19.** 5.27 seconds **21.** 2.42 seconds **23.** the mean **25.** mean = 77; median = 80; mode = 79 **27.** 92 **29.** **(a)** 597.42 **(b)** 598 **(c)** 603 **31.** 2.41 **33.** 711,400 square miles **35.** 23,303,000 **37.** 716,489 **39.** −162,314 **41.** \$946,533.33 **43.** 4 **45.** **(a)** 17.4 **(b)** 14 **(c)** none **47.** **(a)** 74.8 **(b)** 77.5 **(c)** 78 **49.** Answers will vary. **51.** **(a)** mean = 13.7; median = 16; mode = 18 **(b)** median **(c)** Answers will vary. **53.** **(a)** 4 **(b)** 4.25 **55.** **(a)** 6 **(b)** 9.33 **57.** 80 **59.** no **61.** Answers will vary.

13.3 Exercises (Pages 814–817)

1. the sample standard deviation **3.** **(a)** 17 **(b)** 5.53 **5.** **(a)** 19 **(b)** 6.27 **7.** **(a)** 23 **(b)** 7.75 **9.** **(a)** 1.14 **(b)** .37 **11.** **(a)** 8 **(b)** 2.41 **13.** $\frac{3}{4}$ **15.** $\frac{45}{49}$ **17.** 88.9% **19.** 64% **21.** $\frac{3}{4}$ **23.** $\frac{15}{16}$ **25.** $\frac{1}{4}$ **27.** $\frac{4}{49}$ **29.** \$202.50 **31.** six **33.** There are at least nine. **35.** **(a)** $s_A = 2.35; s_B = 2.58$ **(b)** $V_A = 46.9; V_B = 36.9$ **(c)** sample B **(d)** sample A **37.** **(a)** $\overline{x}_A = 69.0; \overline{x}_B = 66.6$ **(b)** $s_A = 4.53; s_B = 5.27$ **(c)** brand A, since $\overline{x}_A > \overline{x}_B$ **(d)** brand A, since $s_A < s_B$ **39.** Brand A ($s_B = 2235 > 2116$) **41.** 18.29; 4.35 **43.** 8.29; 4.35 **45.** 54.86; 13.04 **47.** Answers will vary. **49.** no **51.** Answers will vary. **53.** **(a)** least **(b)** greatest **(c)** Answers will vary.

13.4 Exercises (Pages 821–824)

1. Lynn (since $z = .48 > .19$) **3.** Arvind (since $z = -1.44 > -1.78$) **5.** 58 **7.** 62 **9.** 1.3 **11.** 2.4 **13.** Kuwait **15.** United Arab Emirates **17.** Turkey in imports (Turkey's imports z-score was 2.4, Saudi Arabia's exports z-score was 2.2, and $2.4 > 2.2$.) **19.** **(a)** The median is \$30.5 billion. **(b)** The range is

113 − 10 = 103. **(c)** The middle half of the items extend from \$13 billion to \$69 billion. **21.** Answers will vary. **23.** Answers will vary. **25.** Answers will vary. **27.** the overall distribution **29. (a)** no **(b)** Answers will vary. **31.** Answers will vary. **33.** Answers will vary. **35.** $Q_1 = 61.1$, $Q_2 = 77.2$, $Q_3 = 90.9$ **37.** $P_{65} = 89.7$ **39.** 76.0

41.

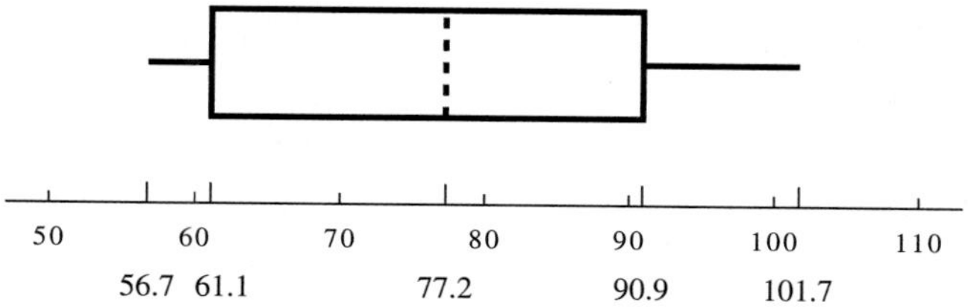

43. 107.5 **45.** high percentage of touchdown passes (about 13.9%)

13.5 Exercises (Pages 832–834)

1. discrete **3.** continuous **5.** discrete **7.** 50 **9.** 68 **11.** 50% **13.** 95% **15.** 43.3% **17.** 36.0% **19.** 4.5% **21.** 97.9% **23.** 1.28 **25.** −1.34 **27.** 5000 **29.** 640 **31.** 9970 **33.** 84.1% **35.** 37.8% **37.** 15.9% **39.** 4.7% **41.** .994 or 99.4% **43.** 189 units **45.** .888 **47.** about 2 eggs **49.** 24.2% **51.** Answers will vary. **53.** 79 **55.** 68 **57.** 90.4 **59.** 40.3 **61.** Answers will vary.

Extension Exercises (Pages 841–842)

1. We have no way of telling. **3.** We can tell only that it rises and then falls **5.** 28% **7.** How long have Toyotas been sold in the United States? How do other makes compare? **9.** The dentists preferred Trident Sugarless Gum to what? Which and how many dentists were surveyed? What percentage responded? **11.** Just how quiet *is* a glider, really? **13.** The maps convey their impressions in terms of *area* distribution, whereas personal income distribution may be quite different. The top map probably implies too high a level of government spending, while the bottom map implies too low a level. **15.** By the time the figures were used, circumstances may have changed greatly. (The Navy was much larger.) Also, New York City was most likely not typical of the nation as a whole. **17.** B **19.** C **21. (a)** $m = 1, a = 6, c = 3$ **(b)** There should be 1.4 managers, 5 agents, and 3.6 clerical employees.

13.6 Exercises (Pages 848–849)

1. $y' = .3x + 1.5$ **3.** 2.4 decimeters **5.** 48.9° **7.** $y' = 3.35x - 78.4$ **9.** 156 lb

11.

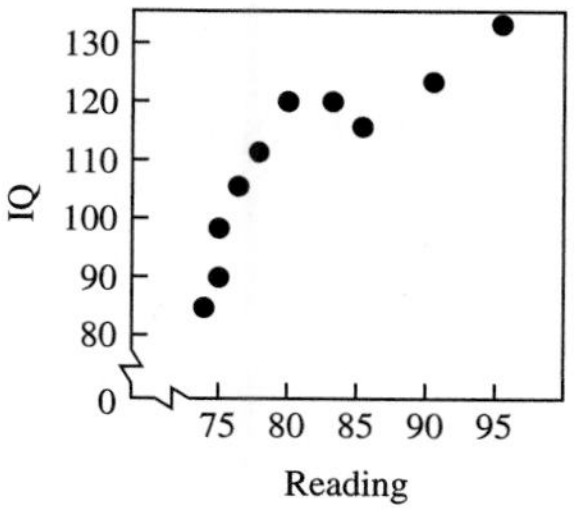

13. 79 **15.** $r = .996$ **17.**

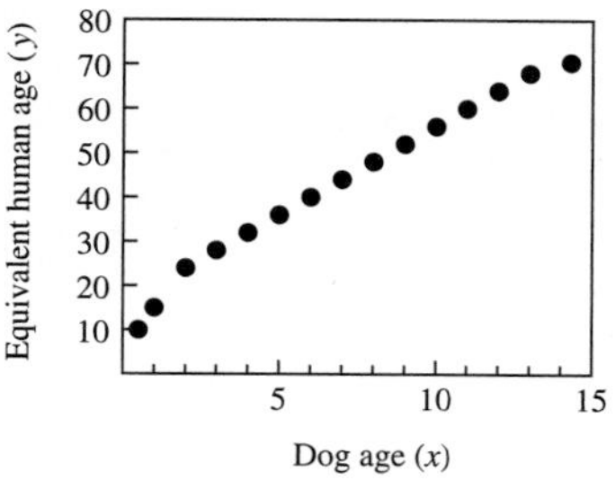

19. Answers will vary. **21.** $y' = 1.44x - .39$ **23.** The linear correlation is strong. **25.** $y' = .00197x + 105.7$ **27.** The linear correlation is moderate.

Chapter 13 Test (Pages 850–852)

1. about 1890 **2.** No. The *percentage* was 4 times as great, not the number of workers **3.** about 3 million **4.** The amount of land farmed in 1940 was about 20% to 28% greater than the amount farmed in 2000 **5. (a)** 8 **(b)** 7 **(c)** 6 and 7 **6. (a)** 6 **(b)** 2.24

7.

Class Limits	Frequency f	Relative Frequency f/n
6–10	3	$\frac{3}{22} \approx .14$
11–15	6	$\frac{6}{22} \approx .27$
16–20	7	$\frac{7}{22} \approx .32$
21–25	4	$\frac{4}{22} \approx .18$
26–30	2	$\frac{2}{22} \approx .09$

8. (a)

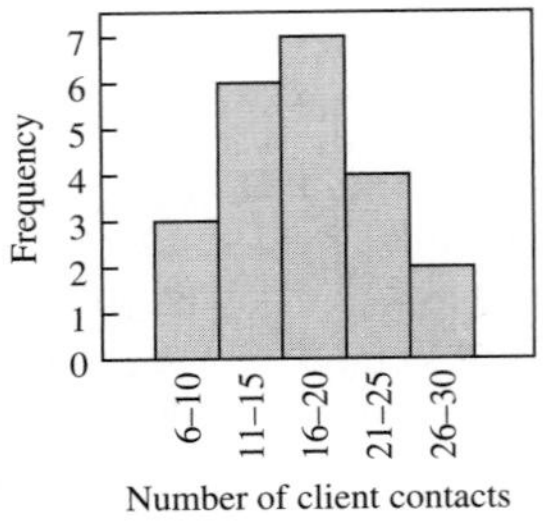

(b)

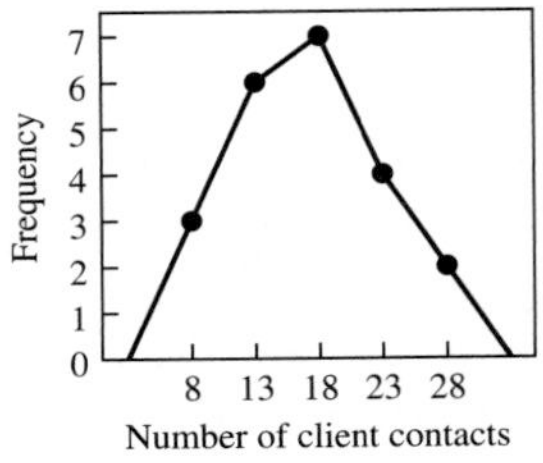

9. 8 **10.** 12.4 **11.** 12 **12.** 12 **13.** 10

14.

3	3 8
4	3 5 8 9
5	0 2 5
6	1 1 4 5 6 7 7 8
7	0 1 2 3 7 7 8 9 9
8	0 4 4
9	1

15. 35 **16.** 33 **17.** 37 **18.** 31 **19.** 49

20.

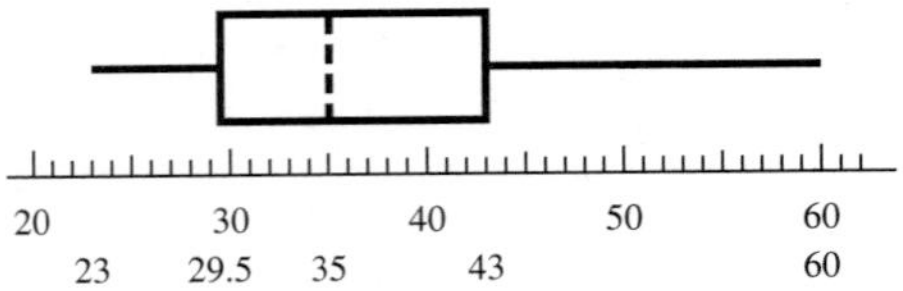

21. about 95% **22.** about .3% **23.** about 16%

24. about 13.5% **25.** .684 **26.** .340 **27.** East **28.** Central **29.** 78.1 **30.** Phillies (.79 > .73)

31.

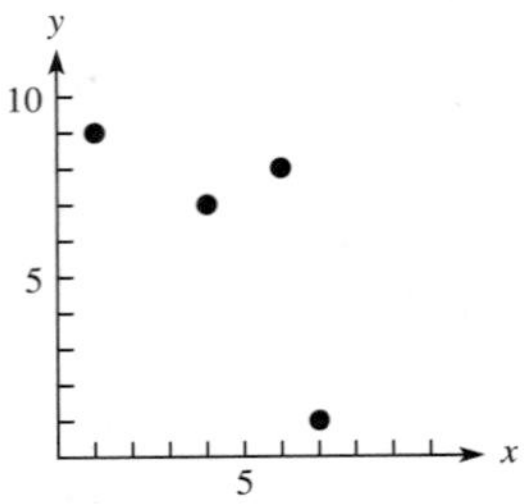

32. $y' = -.98x + 10.64$ **33.** 7.70 **34.** $-.72$
35. Answers will vary. **36.** Answers will vary.

CHAPTER 14 Personal Financial Management

14.1 Exercises (Pages 864–867)

1. \$112 **3.** \$29.25 **5.** \$548.38 **7.** **(a)** \$826 **(b)** \$835.84 **9.** **(a)** \$2650 **(b)** \$2653.02 **11.** \$200 **13.** \$15,910 **15.** \$168.26 **17.** \$817.17 **19.** \$17,724.34; \$10,224.34 **21.** **(a)** \$1331 **(b)** \$1340.10 **(c)** \$1344.89 **23.** **(a)** \$17,631.94 **(b)** \$17,762.93 **(c)** \$17,831.37 **25.** **(a)** \$361.56 **(b)** \$365.47 **(c)** \$368.13 **(d)** \$369.44 **(e)** \$369.48 **27.** Answers will vary. **29.** \$1461.04 **31.** \$747.26 **33.** \$4499.98 **35.** \$46,446.11 **37.** \$45,370.91 **39.** 4.000% **41.** 4.060% **43.** 4.081% **45.** 4.081% **47.** \$30,606 **49.** 11 years, 233 days **51.** $r = m[(Y + 1)^{1/m} - 1]$ **53.** **(a)** 3.818% **(b)** Bank A; \$6.52 **(c)** no difference; \$232.38 **55.** 35 years **57.** 8 years **59.** 10.0% **61.** 3.2% **63.** Answers will vary. **65.** \$7.05; \$8.61; \$14.50; \$39.40 **67.** \$22,100; \$27,100; \$45,500; \$124,000 **69.** \$198 **71.** \$1368 **73.** \$81,102,828.75 **75.** Answers will vary.

Extension Exercises (Pages 873–874)

1. **(a)** \$14,835.10 **(b)** \$10,000 **(c)** \$4835.10 **3.** **(a)** \$3488.50 **(b)** \$3000 **(c)** \$488.50 **5.** **(a)** \$3374.70 **(b)** \$3120 **(c)** \$254.70 **7.** Answers will vary. **9.** \$3555.94 **11.** \$3398.08 **13.** **(a)** \$1263.86 **(b)** \$3791.58 **(c)** \$208.42 **15.** **(a)** \$1.80 **(b)** \$468.00 **(c)** \$32.00 **17.** **(a)** \$180,000.00 **(b)** \$322,257.52 **19.** Answers will vary.

14.2 Exercises (Pages 878–882)

1. \$1650 **3.** \$2046 **5.** \$2546 **7.** \$3645 **9.** \$476.25 **11.** \$1215; \$158.75 **13.** \$83.25; \$46.29 **15.** \$79.18; \$38.39 **17.** \$229.17 **19.** \$330.75 **21.** \$10,850 **23.** 6 years **25.** \$3.26 **27.** \$2.72 **29.** \$3.51, \$358.75; \$358.75, \$3.95, \$396.78; \$396.78, \$4.36, \$368.88; \$368.88, \$4.06, \$361.14 **31.** \$7.53, \$708.60; \$708.60, \$7.79, \$850.14; \$850.14, \$9.35, \$825.79; \$825.79, \$9.08, \$791.30 **33.** \$2.50 **35.** \$12.08 **37.** **(a)** \$607.33 **(b)** \$7.29 **(c)** \$646.26 **39.** **(a)** \$473.96 **(b)** \$5.69 **(c)** \$505.07 **41.** **(a)** \$7.20 **(b)** \$6.62 **43.** **(a)** \$38.18 **(b)** \$38.69 **(c)** \$39.20 **45.** 16.0% **47.** \$32 **49.** Answers will vary. **51.** Answers will vary. **53.** Answers will vary. **55.** **(a)** \$127.44 **(b)** \$139.08

14.3 Exercises (Pages 887–889)

1. 13.5% **3.** 8.5% **5.** \$114.58 **7.** \$92.71 **9.** 11.0% **11.** 9.5% **13.** **(a)** \$8.93 **(b)** \$511.60 **(c)** \$6075.70 **15.** **(a)** \$2.79 **(b)** \$97.03 **(c)** \$4073.57 **17.** **(a)** \$206 **(b)** 9.5% **19.** **(a)** \$180 **(b)** 11.0% **21.** **(a)** \$41.29 **(b)** \$39.70 **23.** **(a)** \$420.68 **(b)** \$368.47 **25.** **(a)** \$1.80 **(b)** \$14.99 **(c)** \$1045.01 **27.** finance company APR: 12.0%; credit union APR: 11.5%; choose credit union **29.** \$32.27 **31.** Answers will vary. **33.** Answers will vary. **35.** 12 **37.** Answers will vary. **39.** Answers will vary. **41.** Answers will vary. **43.** $\frac{5}{39}$

14.4 Exercises (Pages 899–902)

1. \$675.52 **3.** \$469.14 **5.** \$2483.28 **7.** \$1034.56 **9.** **(a)** \$513.38 **(b)** \$487.50 **(c)** \$25.88 **(d)** \$58,474.12 **11.** **(a)** \$1247.43 **(b)** \$775.67 **(c)** \$471.76 **(d)** \$142,728.24 **(e)** \$1247.43 **(f)** \$773.11 **(g)** \$474.32 **(h)** \$142,253.92 **13.** **(a)** \$1390.93 **(b)** \$776.61 **(c)** \$614.32 **(d)** \$113,035.68 **(e)** \$1390.93 **(f)** \$772.41 **(g)** \$618.52 **(h)** \$112,417.16 **15.** \$552.18 **17.** \$1127.27 **19.** \$1168.44 **21.** 360 **23.** \$247,532.80 **25.** **(a)** \$304.01 **(b)** \$734.73 **27.** **(a)** \$59,032.06 **(b)** \$59,875.11 **29.** **(a)** payment 176 **(b)** payment 304 **31.** \$25,465.20 **33.** \$91,030.80 **35.** **(a)** \$674.79 **(b)** \$746.36 **(c)** an increase of \$71.57 **37.** **(a)** \$395.83 **(b)** \$466.07 **39.** \$65.02 **41.** \$140,000 **43.** \$4275 **45.** \$240 **47.** **(a)** \$1634 **(b)** \$1754 **(c)** \$917 **(d)** \$997 **(e)** \$1555 **49.** **(a)** \$2271 **(b)** \$2407 **(c)** \$1083 **(d)** \$1174 **(e)** \$2055 **51.** Answers will vary. **53.** Answers will vary. **55.** Answers will vary. **57.** Answers will vary. **59.** Answers will vary.

14.5 Exercises (Pages 917–921)

1. \$29.75 **3.** 7¢ per share lower **5.** \$47.48 **7.** \$1.72 per share **9.** 7600 shares **11.** 10.6% higher **13.** 35 **15.** \$5140.00 **17.** \$1161.00 **19.** \$2273.82 **21.** \$17,822.55 **23.** \$24,804.23 **25.** \$55,800.00 **27.** \$12,424.96 **29.** \$5709.88 **31.** \$17,441.47 **33.** \$56,021.45 **35.** \$30.56 net paid out **37. (a)** \$800 **(b)** \$80 **(c)** \$960 **(d)** \$1040 **(e)** 130% **39. (a)** \$1250 **(b)** \$108 **(c)** −\$235 **(d)** −\$127 **(e)** −10.16% **41.** \$275 **43.** \$177.75 **45. (a)** \$10.49 **(b)** 334 **47. (a)** \$8.27 **(b)** 3080 **49. (a)** \$8.39 **(b)** \$100.68 **(c)** 15.6% **51. (a)** \$57.45 **(b)** \$689.40 **(c)** 27.6% **53. (a)** \$1203.75 **(b)** \$18.06 **(c)** 19.56% **55. (a)** \$4179 **(b)** \$76.48 **(c)** 24.31% **57.** \$10.00, \$35.17, \$29.95 **59.** \$400.00, \$41.80, \$29.95 **61.** \$4000.00, \$91.40, \$120.00 **63.** large; low **65.** \$129,258 **67.** \$993,383 **69. (a)** \$56,473.02 **(b)** \$49,712.70 **71. (a)** \$12,851.28 **(b)** \$11,651.81 **73. (a)** $r = m[(\frac{A}{P})^{1/n} - 1]$ **(b)** 5.4%

	Aggressive Growth	Growth	Growth & Income	Income	Cash
75.	\$1400	\$ 8600	\$ 6200	\$ 2800	\$ 1000
77.	\$8000	\$112,000	\$144,000	\$116,000	\$20,000

79. 3.75% **81.** 5.2% **83.** 5.3% **85.** The annual report was available free to *Wall Street Journal* readers. **87.** Answers will vary. **89.** Answers will vary. **91.** Answers will vary. **93.** Answers will vary. **95.** Answers will vary.

Chapter 14 Test (Pages 922–923)

1. \$130 **2.** \$58.58 **3.** 3.04% **4.** 14 years **5.** \$67,297.13 **6.** \$10.88 **7.** \$450 **8.** \$143.75 **9.** 14.0% **10.** \$34.13 **11.** 11.5% **12.** Answers will vary. **13.** \$1162.95 **14.** \$1528.48 **15.** \$3488 **16.** Answers will vary. **17.** 12.1% **18.** 112,300 shares **19.** \$10 **20.** .078%; No; the given return was not for a 1-year period. **21.** \$106,728.55 **22.** Answers will vary.

Chapter 15 Graph Theory

15.1 Exercises (Pages 938–944)

1. 7 vertices, 7 edges **3.** 10 vertices, 9 edges **5.** 6 vertices, 9 edges **7.** Two have degree 3. Three have degree 2. Two have degree 1. Sum of degrees is 14. This is twice the number of edges.
9. Six have degree 1. Four have degree 3. Sum of degrees is 18. This is twice the number of edges.
11. not isomorphic
13. Yes. Corresponding edges should be the same color. AB should match AB, etc.

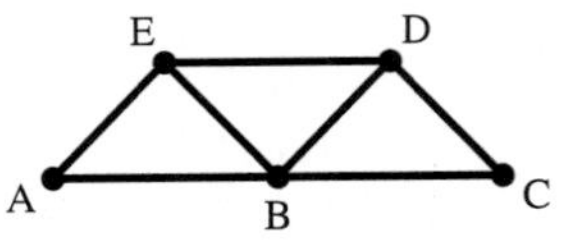

(a)

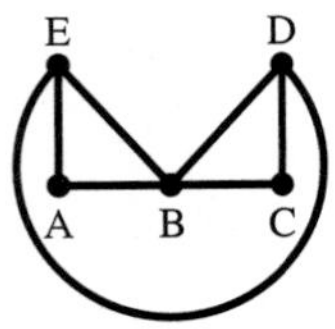

(b)

15. not isomorphic **17.** connected, 1 component **19.** disconnected, 3 components **21.** disconnected, 2 components **23.** 10 **25.** 4 **27. (a)** yes **(b)** No, because there is no edge A to D. **(c)** No, because there is no edge A to E. **(d)** yes **(e)** yes **(f)** yes **29. (a)** No, because it does not return to starting vertex. **(b)** yes **(c)** No, because there is no edge C to F. **(d)** No, because it does not return to starting vertex. **(e)** No, because edge F to D is used more than once. **31. (a)** No, because there is no edge B to C. **(b)** No, because edge I to G is used more than once. **(c)** No, because there is no edge E to I. **(d)** yes **(e)** yes **(f)** No, because edge A to D is used more than once. **33.** It is a walk and a path, not a circuit. **35.** It is a walk, not a path, not a circuit.

37. It is a walk and a path, not a circuit. **39.** No, because there is no edge A to C, for example.
41. No, because there is no edge A to F, for example. **43.** yes
45.

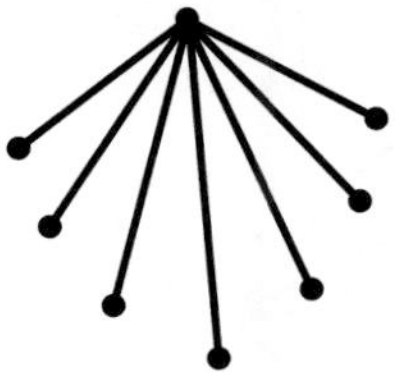

7 games **47.** 36 handshakes **49.** 10 telephone conversations

51. A → B → D → A corresponds to tracing around the edges of a single face. (The circuit is a triangle.)

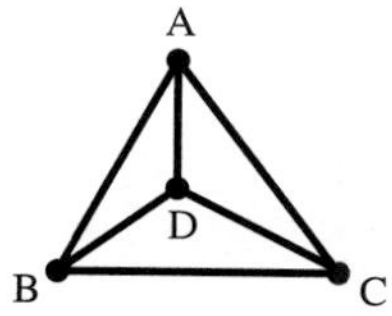

53. Answers will vary. **55.** Answers will vary. **57.** Answers will vary.

59.

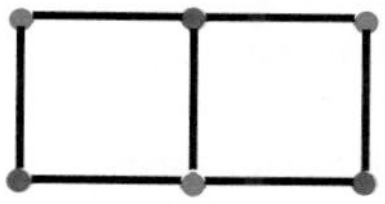

2

61.

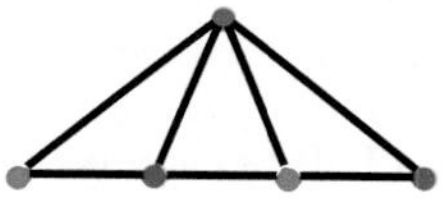

3

63.

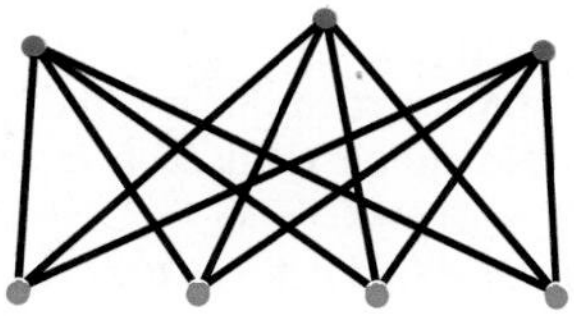

2

65. **(a)**

(b)

(c)

(d)

(e) A cycle with an odd number of vertices has chromatic number 3. A cycle with an even number of vertices has chromatic number 2.

67.

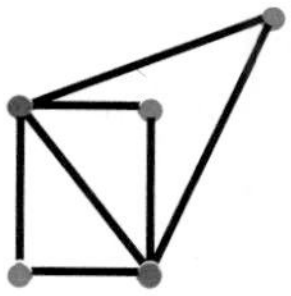

3

69.

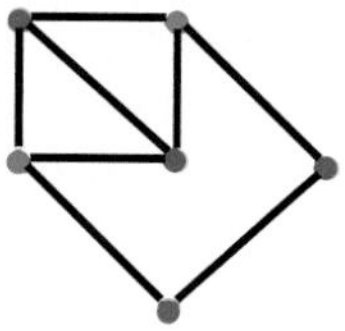

3

71.

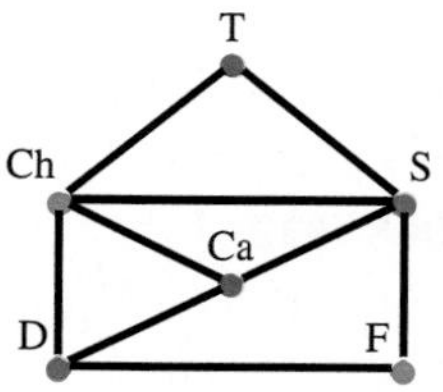

Three times: Choir and Forensics, Service Club and Dance Club, and Theater and Caribbean Club. (There are other possible groupings.)

73.

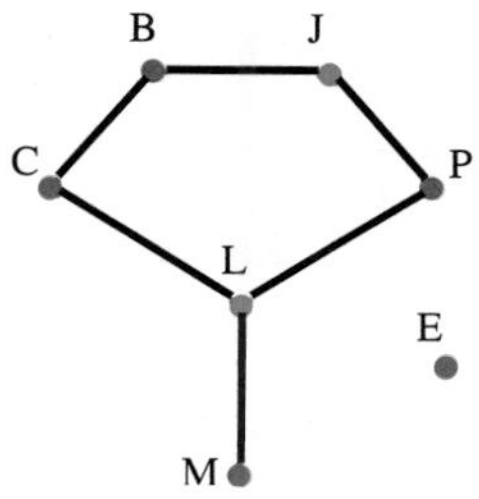

Three gatherings: Brad and Phil and Mary, Joe and Lindsay, and Caitlin and Eva. (There are other possible groupings.)

75.

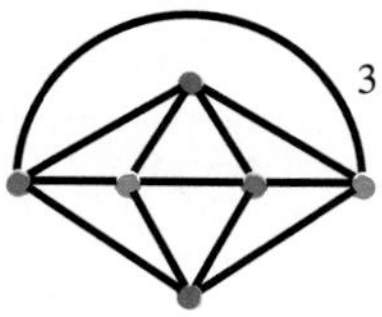

3 colors

77.

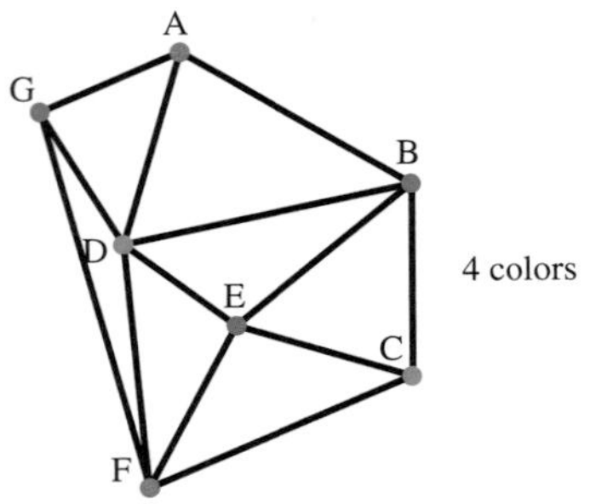

79.

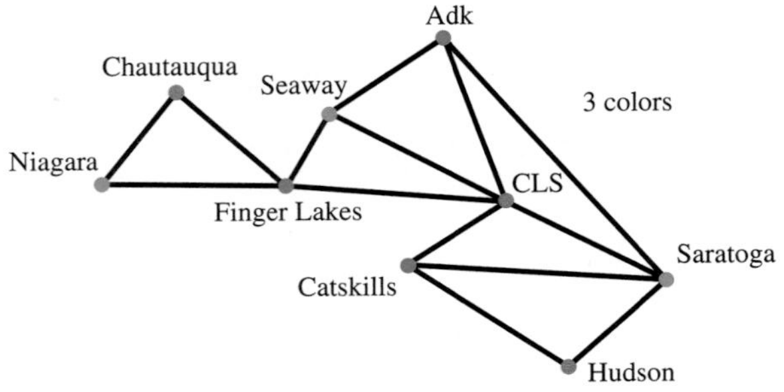

In Exercises 81 and 82, there are many different ways to draw the map.

81.

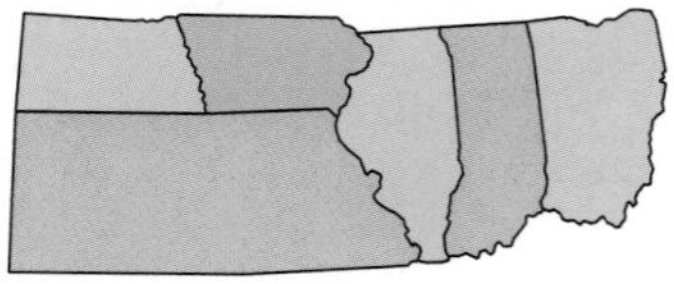

83. By the four-color theorem, this is not possible.

15.2 Exercises (Pages 954–959)

1. **(a)** No, it is not a path. **(b)** yes **(c)** No, it is not a walk. **(d)** No, it is not a circuit. **3.** **(a)** No, some edges are not used (e.g., B → F). **(b)** yes **(c)** No, it is not a path. **(d)** yes **5.** Yes, all vertices have even degree. **7.** No, some vertices have odd degree (e.g., G). **9.** yes **11.** It has an Euler circuit; all vertices have even degree. No circuit visits each vertex exactly once. **13.** It has an Euler circuit; all vertices have even degree. A → B → H → C → G → D → F → E → A visits each vertex exactly once. **15.** It has no Euler circuit; some vertices have odd degree (e.g., B). It has no circuit that visits each vertex exactly once. **17.** No, some vertices have odd degree. **19.** Yes, all vertices have even degree. **21.** none **23.** B → E or B → D **25.** B → C or B → H **27.** A → C → B → F → E → D → C → F → D → A **29.** Graph has an Euler circuit. A → G → H → J → I → L → J → K → I → H → F → G → E → F → D → E → C → D → B → C → A → E → B → A **31.** Graph does not have an Euler circuit; some vertices have odd degree (e.g., C). **33.** There is such a route: A → D → B → C → A → H → D → E → B → H → G → E → F → H → J → L → C → M → A → K → M → L → K → J → A **35.** It is not possible; some vertices have odd degree (e.g., room at upper left). **37.** no; There are more than two vertices with odd degree. **39.** yes; B and G **41.** possible; Exactly two of the rooms have an odd number of doors. **43.** not possible; All rooms have an even number of doors.

There are other correct answers in Exercises 45–47.

45. A → B → D → C → A; There are 4 edges in any such circuit. **47.** A → B → C → D → E → F → G → H → C → A → H → I → A; There are 12 edges in any such circuit. **49.** A complete graph has an Euler circuit if the number of vertices is an odd number greater than or equal to 3; In a complete graph with n vertices, the degree of each vertex is $n - 1$. "$n - 1$ is even" is equivalent to "n is odd." **51.** Answers will vary.

15.3 Exercises (Pages 968–972)

1. **(a)** No, because it visits vertex E twice. **(b)** yes **(c)** No, because it does not visit C. **(d)** No, because it does not visit A. **3.** **(a)** None, because there is no edge B to C. **(b)** all three **(c)** none; Edge AD is used twice. **5.** A → B → D → E → F → C → A **7.** G → H → J → I → G **9.** X → T → U → W → V → X

11. Hamilton circuit: A → B → C → D → A. The graph has no Euler circuit, because at least one of the vertices has odd degree. (In fact, all have odd degree.)

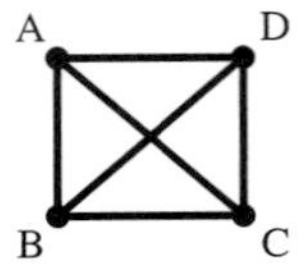

13. A → B → C → D → E → F → A is both a Hamilton and an Euler circuit.

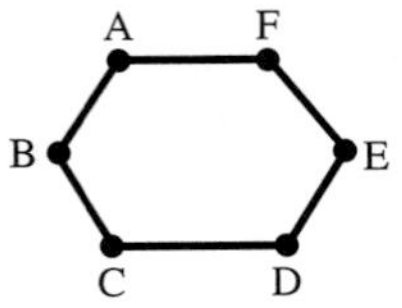

15. Hamilton circuit **17.** Euler circuit **19.** Hamilton circuit **21.** 24 **23.** 362,880 **25.** 9!
27. 17! **29.** P → Q → R → S → P P → Q → S → R → P P → R → Q → S → P P → R → S → Q → P
P → S → Q → R → P P → S → R → Q → P
31. E → H → I → F → G → E E → H → I → G → F → E
33. E → F → G → H → I → E E → F → G → I → H → E E → F → H → G → I → E
E → F → H → I → G → E E → F → I → G → H → E E → F → I → H → G → E
35. E → G → F → H → I → E E → G → F → I → H → E E → G → H → F → I → E
E → G → H → I → F → E E → G → I → F → H → E E → G → I → H → F → E
37. A → B → C → D → E → A A → C → B → D → E → A A → B → C → E → D → A
A → C → B → E → D → A A → B → D → C → E → A A → C → D → B → E → A
A → B → D → E → C → A A → C → D → E → B → A A → B → E → C → D → A
A → C → E → B → D → A A → B → E → D → C → A A → C → E → D → B → A
A → D → B → C → E → A A → E → B → C → D → A A → D → B → E → C → A
A → E → B → D → C → A A → D → C → B → E → A A → E → C → B → D → A
A → D → C → E → B → A A → E → C → D → B → A A → D → E → B → C → A
A → E → D → B → C → A A → D → E → C → B → A A → E → D → C → B → A
39. Minimum Hamilton circuit is P → Q → R → S → P; Weight is 2200. **41.** Minimum Hamilton circuit is C → D → E → F → G → C; Weight is 64. **43.** **(a)** A → C → E → D → B → A; Total weight is 20. **(b)** C → A → B → D → E → C; Total weight is 20. **(c)** D → C → A → B → E → D; Total weight is 23. **(d)** E → C → A → B → D → E; Total weight is 20. **45.** **(a)** A → C → D → E → B → A; Total weight is 87. B → C → A → E → D → B; Total weight is 95. C → A → E → D → B → C; Total weight is 95. D → C → A → E → B → D; Total weight is 131. E → C → A → D → B → E; Total weight is 137. **(b)** A → C → D → E → B → A with total weight 87 **(c)** For example, A → B → C → D → E → A has total weight 52.
47. A → B → C → D → E → F → A A → B → C → F → E → D → A A → B → E → D → C → F → A
A → B → E → F → C → D → A A → D → E → F → C → B → A A → D → E → B → C → F → A
A → D → C → B → E → F → A A → D → C → F → E → B → A A → F → E → B → C → D → A
A → F → E → D → C → B → A A → F → C → D → E → B → A A → F → C → B → E → D → A
49. A → B → C → D → E → F → A A → B → C → E → D → F → A A → B → C → E → F → D → A
A → B → C → F → E → D → A A → D → E → F → C → B → A A → D → F → E → C → B → A
A → F → D → E → C → B → A A → F → E → D → C → B → A
51. **(a)** graphs (2) and (4) **(b)** graphs (2) and (4) **(c)** No. Graph (1) provides a counterexample. **(d)** No; if $n < 3$ the graph will have no circuits at all. **(e)** Degree of each vertex in a complete graph with n vertices is $(n - 1)$. If $n \geq 3$, then $(n - 1) > n/2$. So we can conclude from Dirac's theorem that the graph has a Hamilton circuit.
53. A → F → G → R → S → T → U → Q → P → N → M → L → K → J → I → H → B → C → D → E → A
55. Answers will vary. **57.** Answers will vary.

15.4 Exercises (Pages 981–985)

1. tree **3.** No, because it is not connected. **5.** tree **7.** No, because it has a circuit.
9. It is not possible, since the graph has a circuit. **11.** tree **13.** not necessarily a tree **15.** true

17. false

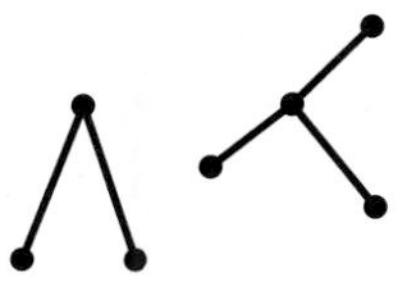

19.

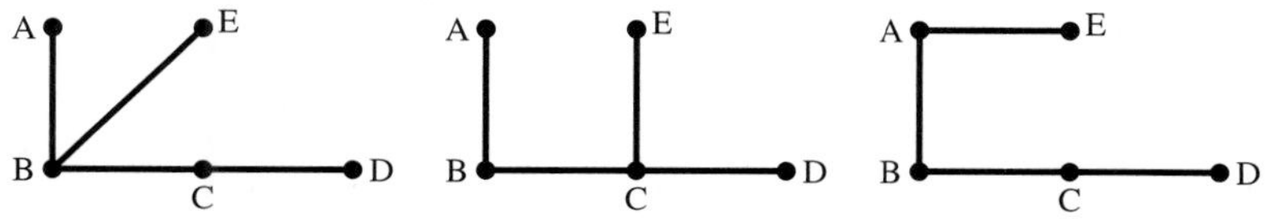

21.

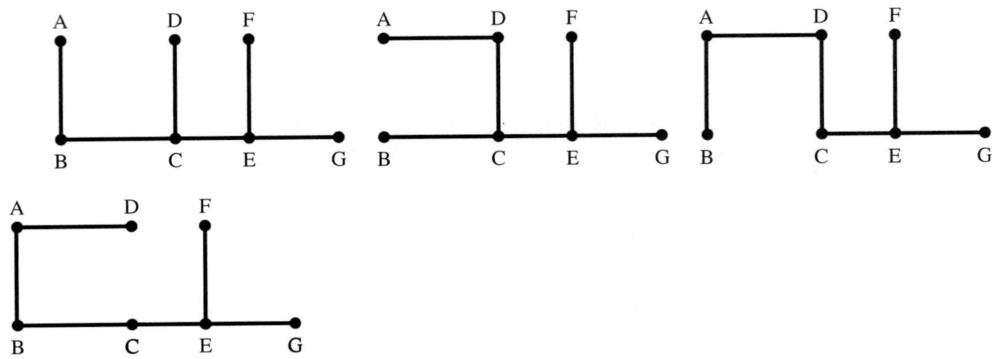

23.

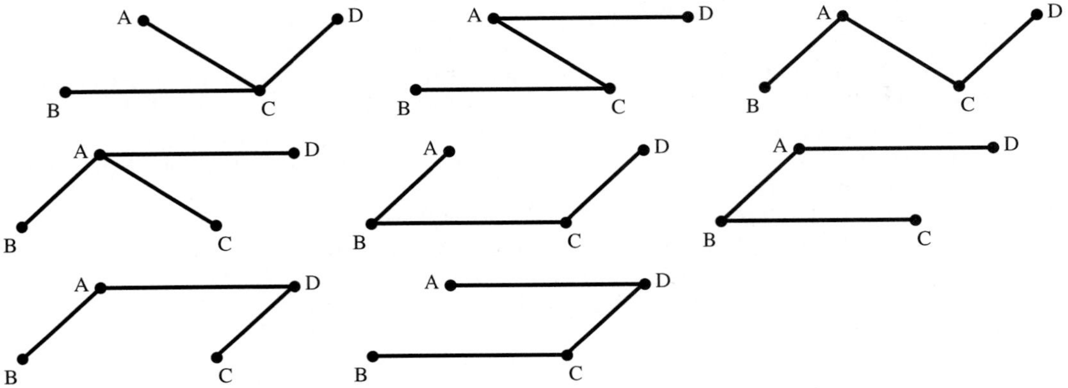

25. 20 **27.** If a connected graph has circuits, none of which have common edges, then the number of spanning trees for the graph is the product of the numbers of edges in all the circuits.

29. Total weight is 51.

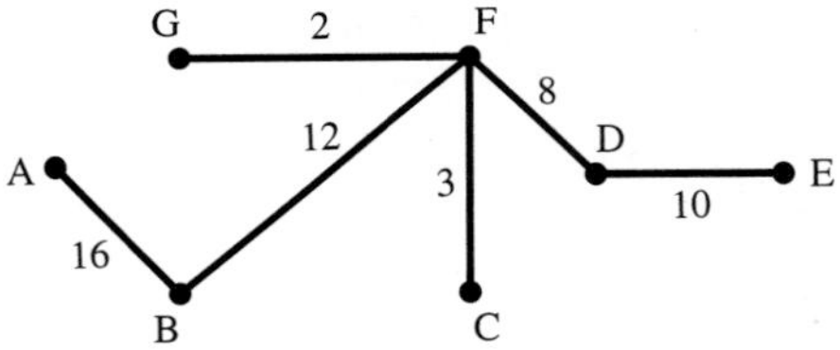

31. Total weight is 66.

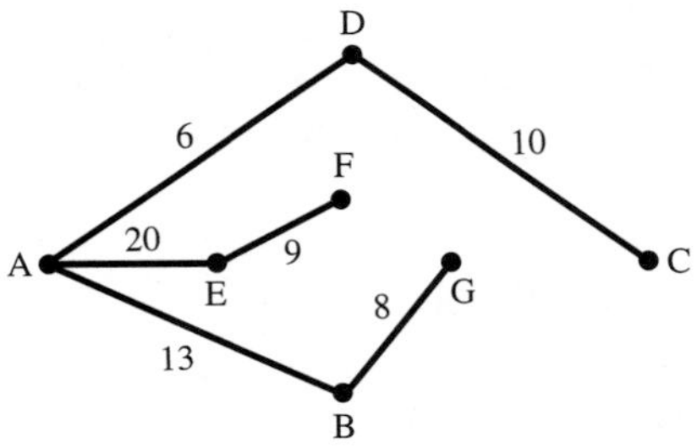

33. Total length to be covered is 140 ft.

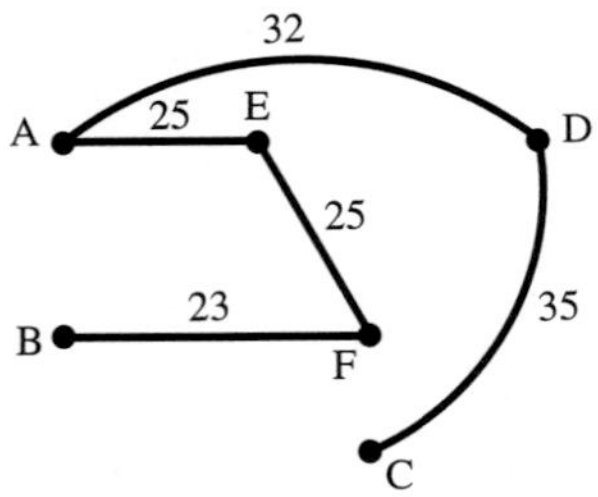

35. 33 **37.** 62 **39.** Different spanning trees must have the same number of edges, because the number of vertices in the tree is the number of vertices in the original graph, and the number of edges has to be one less than this.
41. **(a)** 9 **(b)** 18 **(c)** 0 **(d)** 2

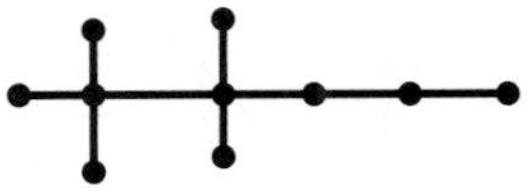

43. 22 cables **45.** This is possible. Graph must be a tree because it has one fewer edge than vertices. **47.** This is possible. Graph cannot be a tree, because it would have at least as many edges as vertices.

49. 3:

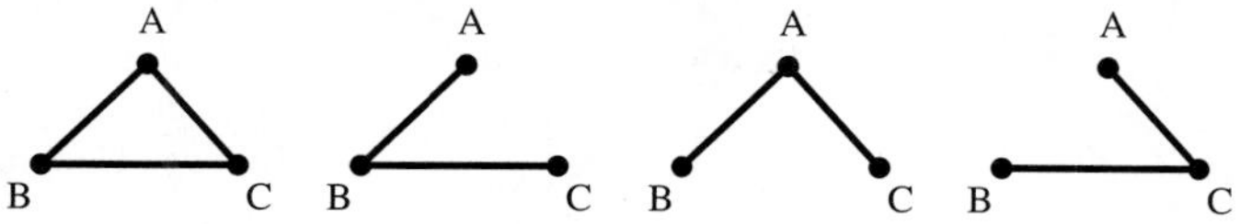

51. 125

53. 3 nonisomorphic trees:

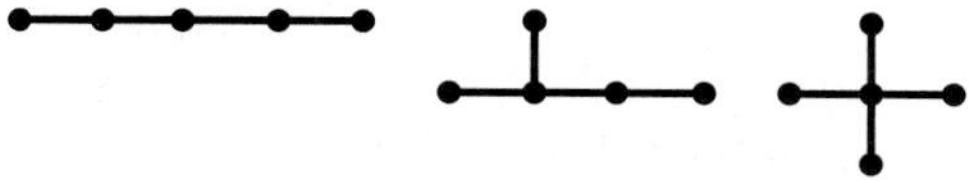

55. 11 nonisomorphic trees:

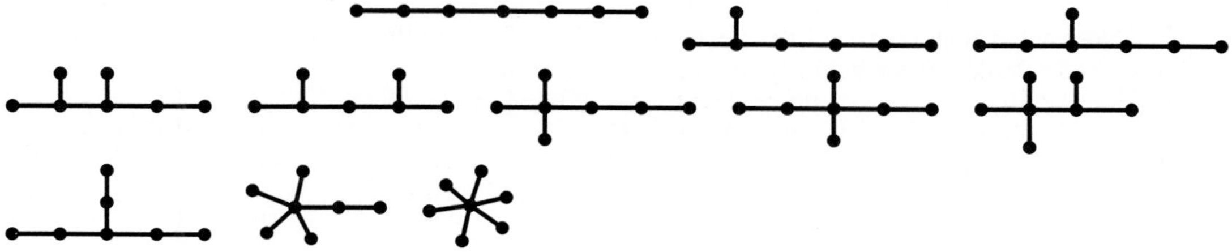

57. Answers will vary.

Chapter 15 Test **(Pages 987–989)**

1. 7 **2.** 20 **3.** 10 **4.** **(a)** No, because edge AB is used twice. **(b)** yes **(c)** No, because there is no edge C to D. **5.** **(a)** yes **(b)** No, because for example, there is no edge B to C. **(c)** yes

6. For example:

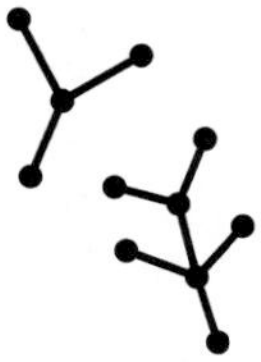

7. 13 edges

8. The graphs are isomorphic. Corresponding edges should be the same color. AB should match AB, etc.

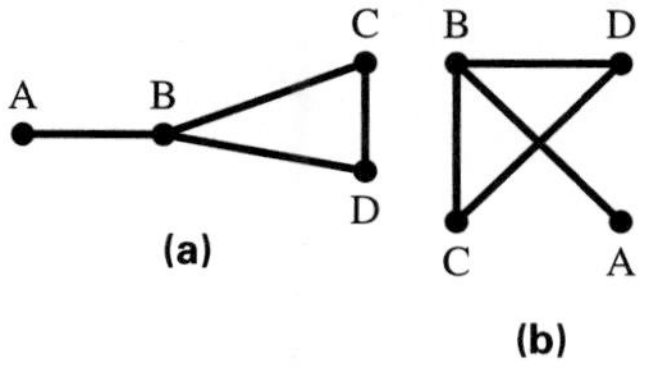

9.

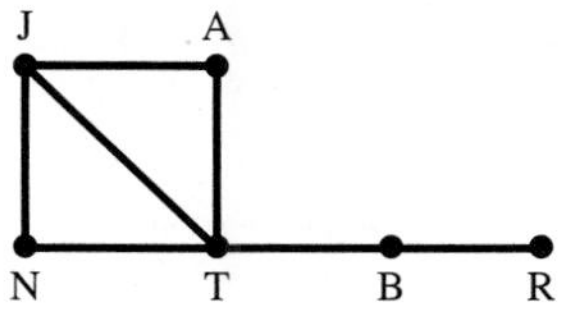

The graph is connected. Tina knows the greatest number of other guests.

10. 28 games **11.** Yes, because there is an edge from each vertex to each of the remaining 6 vertices.

12. **(a)**

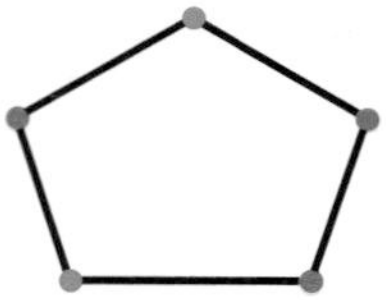

Chromatic number: 3

(b)

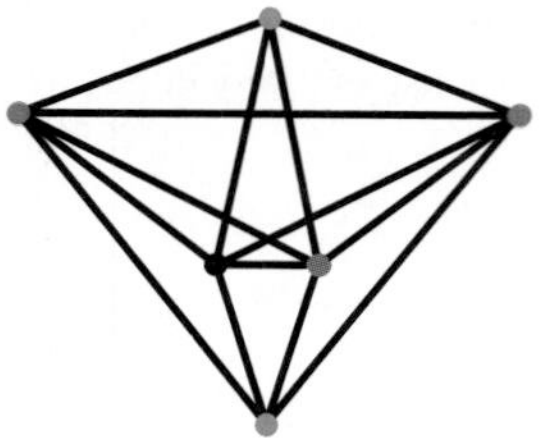

Chromatic number: 5

13.

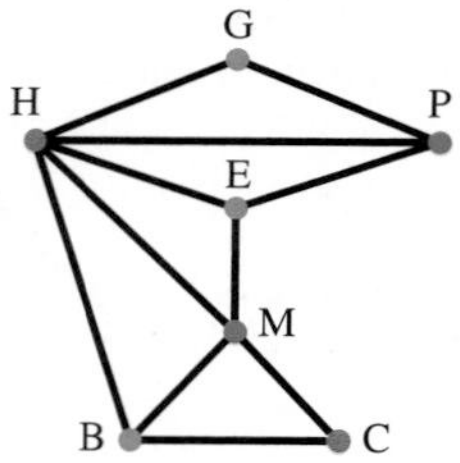

The Chromatic number is 3, so three separate exam times are needed. Exams could be at the same times as follows: Geography, Biology, and English; History and Chemistry; Mathematics and Psychology.

14. **(a)** No, because it does not use all the edges. **(b)** No, because it is not a circuit (for example, no edge B to C). **(c)** yes **15.** No, because some vertices have odd degree. **16.** Yes, because all vertices have even degree. **17.** No, because two of the rooms have an odd number of doors.
18. F → B → E → D → B → C → D → K → B → A → H → G → F → A → G → J → F
19. **(a)** No, because it does not visit all vertices. **(b)** No, because it is not a circuit (for example, no edge B to C). **(c)** No, because it visits some vertices twice before returning to starting vertex.
20. F → G → H → I → E → F F → G → H → E → I → F F → G → I → H → E → F F → G → I → E → H → F F → G → E → H → I → F F → G → E → I → H → F. There are 6 such Hamilton circuits. **21.** P → Q → S → R → P; Total weight is 27. **22.** A → E → D → C → F → B → A; Total weight is 11.85. **23.** 24! **24.** Hamilton circuit
25. Any three of these:

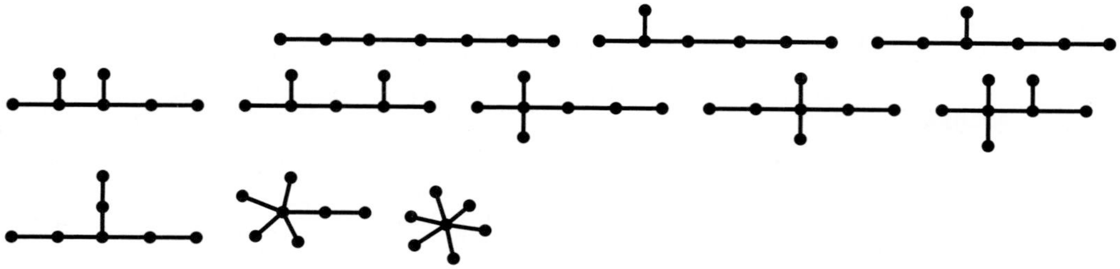

26. false **27.** true **28.** true
29. There are 4 spanning trees.

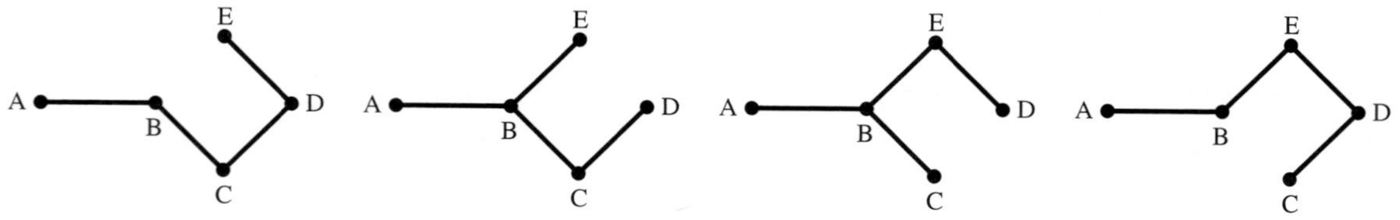

30. Weight is 24.

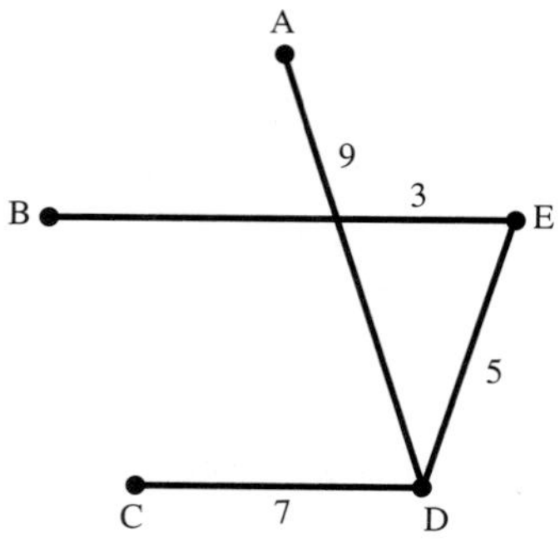

31. 49

CHAPTER 16 Voting and Apportionment

16.1 Exercises (Pages 1004–1007)

1. (a) 4! = 24
(b)

Number of Voters	Ranking
3	b > c > a > d
2	a > d > c > b
1	c > d > b > a
1	d > c > b > a
1	d > a > b > c
2	c > a > d > b
1	a > c > d > b
2	a > b > c > d

(c) Australian shepherd

3. 5! = 120; 7! = 5040 **5.** For any value of $n \geq 5$ there are at least 120 possible rankings of the candidates. This means a voter must select one ranking from a huge number of possibilities. It also means the mechanics of any election method, except the plurality method, are difficult to manage. **7.** ${}_6C_2 = 15$; ${}_8C_2 = 28$ **9.** With 40,320 possible rankings to be examined, finding the outcomes of 28 different pairwise comparisons seems a nearly impossible task. **11. (a)** b, with 6 first-place votes **(b)** c, with 2 (pairwise) points **(c)** b, with 16 (Borda) points **(d)** c **13.** Logo a is selected by all the methods. **15. (a)** e, with 8 first-place votes **(b)** j, with 3 (pairwise) points **(c)** j, with 40 (Borda) points **(d)** h **17. (a)** a, with 3 (pairwise) points **(b)** b, with 24 (Borda) points **(c)** a, Australian shepherd has 7 of 13 first-place votes—a majority—at the first round of voting. **19. (a)** t, with 18 first-place votes **(b)** h, with 4 (pairwise) points **(c)** b, with 136 (Borda) points **(d)** k **21.** h beats e. **23.** c beats t. **25.** In a runoff between k and c, activity k is selected. Activity k faces activity t, and k is selected. **27.** Logo a **29. (a)** 2 **(b)** c, with 7 pairwise points **31. (a)** 7 **(b)** e, with 7 pairwise points **33. (a)** 16 **(b)** c **35. (a)** 55 **(b)** c **37.** j **39.** Answers will vary. **41.** Answers will vary. **43.** Answers will vary. **45.** Answers will vary; one possible arrangement of the voters is 2, 4, 5, 7, 3.

16.2 Exercises (Pages 1017–1022)

1. (a) Alternative a (6 of 11 first-place votes) **(b)** b **(c)** yes **3. (a)** Alternative a (20 of 36 first-place votes) **(b)** b **(c)** yes **5. (a)** Alternative a (16 of 30 first-place votes) **(b)** b **(c)** yes **7. (a)** a **(b)** b **(c)** b **(d)** a **(e)** plurality and Borda methods **9. (a)** e **(b)** j **(c)** e **(d)** h **(e)** plurality and Hare methods **11. (a)** h **(b)** t **(c)** k **(d)** b **(e)** all three methods **13. (a)** m has 2 pairwise points, b and s have $1\frac{1}{2}$ pairwise points each, and c has 1 pairwise point. **(b)** b has $2\frac{1}{2}$ pairwise points, m has 2 pairwise points, s has 1 pairwise point, c has $\frac{1}{2}$ pairwise point, so b is selected. **(c)** Yes, the rearranging voters moved m, the winner of the nonbinding election, to the top of their ranking, but b wins the official selection process. **15. (a)** d drops out after one round of votes; b drops out after Round two; in the third vote a is preferred to c by a margin of 11 to 6. **(b)** d drops out after one round of votes; c drops out after Round two; in the third vote b is preferred to a by a margin of 9 to 8. **(c)** Yes, the rearranging voters moved a, the winner of the nonbinding election, to the top of their ranking, but b wins the official selection process. **17. (a)** Candidate a has 75 first-place votes. **(b)** Candidate c has 80 first-place votes. **(c)** yes **19.** No; however, the second pairwise comparison results in a tie. **21. (a)** Round one eliminates b; a is preferred to c in Round two, by a margin of 22 to 12. **(b)** b **(c)** yes **23.** Answers will vary.
25. Answers will vary. One possible voter profile is given.

Number of Voters	Ranking
10	a > b > c > d > e > f
9	b > f > e > c > d > a

Candidate a is a majority candidate and wins all of its pairwise comparisons by a margin of 10 to 9, earning 5 pairwise points. Candidate b wins all of its comparisons, except the one with a, earning 4 pairwise points.

27. Answers will vary. One possible profile is the voter profile for the animal shelter poster dog contest in Exercises 2 and 17 of Section 16.1. **29. (a)** a has 2 pair-wise points. **(b)** Answers will vary. The 3 voters in the bottom row all switch to the ranking a > z > x > y. **31.** Answers will vary. Delete Candidate c. **33.** Answers will vary. One possible profile is given.

Number of voters	Ranking
21	g > j > e
12	j > e > g
8	j > g > e

35. Answers will vary.

16.3 Exercises (Pages 1035–1038)

1. (a) 29,493; 123.40

(b)

State park	a	b	c	d	e
Number of trees	11	70	62	54	42

(c) $md = 122$

State park	a	b	c	d	e
Number of trees	11	70	62	54	42

(d) The traditionally rounded values of Q sum to 240, which is greater than the number of trees to be apportioned.

State park	a	b	c	d	e
Traditionally rounded Q	12	70	62	54	42

(e) The value of md for the Webster apportionment should be greater than $d = 123.40$, because greater divisors make lesser modified quotas with a lesser total sum.

(f) $md = 124$

State park	a	b	c	d	e
Number of trees	12	70	61	54	42

(g) The Hamilton and Jefferson apportionments are the same. The Webster apportionment is different from the other two apportionments.

3. (a) 269; 24.45

(b)

Course	Fiction	Poetry	Short Story	Multicultural
Number of sections	2	2	3	4

(c) $md = 20$

Course	Fiction	Poetry	Short Story	Multicultural
Number of sections	2	1	3	5

(d) The traditionally rounded values of Q sum to 10, which is less than the number of sections to be apportioned.

Course	Fiction	Poetry	Short Story	Multicultural
Traditionally rounded Q	2	1	3	4

(e) The value of md for the Webster apportionment should be less than $d = 24.45$ because lesser divisors make greater modified quotas with a greater total sum.

(f) $md = 23$

Course	Fiction	Poetry	Short Story	Multicultural
Number of sections	2	2	3	4

(g) The Hamilton and Webster apportionments are the same. The Jefferson apportionment is different from the other two apportionments. **(h)** The Hamilton and Webster methods both apportion two sections of poetry; the Jefferson method apportions only one. If there are two sections of poetry, then the 35 enrolled students can be divided into two small sections with 17 and 18 students each instead of all 35 students being forced into one large section. **(i)** The Jefferson method apportions 5 sections of multicultural literature instead of 4. This means that the average class size will be 20 students, rather than 25 students.

5. **(a)**

State	Abo	Boa	Cio	Dao	Effo	Foti
Number of seats	15	22	6	19	14	55

(b) $md = 356$

State	Abo	Boa	Cio	Dao	Effo	Foti
Number of seats	15	22	6	19	13	56

(c) The Hamilton, Jefferson, and Webster apportionments all are different.

7.

State	Abo	Boa	Cio	Dao	Effo	Foti	**Total**
Number of seats	15	22	7	18	14	54	130

9. **(a)** 1721; 43.025

(b)

Hospital	A	B	C	D	E
Number of nurses	3	5	8	11	13

(c) $md = 40$

Hospital	A	B	C	D	E
Number of nurses	3	5	8	11	13

(d) The traditionally rounded values of Q sum to 41, which is greater than the number of nurses to be apportioned.

Hospital	A	B	C	D	E
Traditionally rounded Q	3	6	8	11	13

(e) The value of *md* for the Webster apportionment should be greater than $d = 43.025$, because greater divisors make lesser modified quotas with a lesser total sum.

(f) $md = 43.1$

Hospital	A	B	C	D	E
Number of nurses	3	5	8	11	13

(g) All three apportionments are the same.

11. Answers will vary. One possible population profile is given.

State	a	b	c	d	e	**Total**
Population	50	230	280	320	120	1000

13. Answers will vary. One possible ridership profile is given.

Bus route	a	b	c	d	e	**Total**
Number of riders	131	140	303	178	197	949

15. **(a)** $md = 29$

Course	Fiction	Poetry	Short Story	Multicultural
Number of sections	2	2	3	4

(b) The Adams apportionment is the same as the Hamilton and Webster apportionments. It is different from the Jefferson apportionment

17. **(a)** $md = 377.3$

State	Abo	Boa	Cio	Dao	Effo	Foti
Number of seats	16	22	7	18	14	54

(b) All four methods produce different apportionments of the 131 seats in Timmu's legislature.

19. $\sqrt{56 \cdot 57} = 56.498$ **21.** $\sqrt{32 \cdot 33} = 32.496$ **23.** If the sum is greater, then *md* is found by slowly increasing the value of *d*, because a greater divisor produces lesser modified quotas with a lesser sum.

25. **(a)** $md = 24$

Course	Fiction	Poetry	Short Story	Multicultural
Number of sections	2	2	3	4

(b) The Hill-Huntington apportionment is the same as the Hamilton, Webster, and Adams apportionments. It is different from the Jefferson apportionment.

27. **(a)** $md = 367$

State	Abo	Boa	Cio	Dao	Effo	Foti
Number of seats	15	22	7	19	14	54

(b) The Hill-Huntington apportionment is the same as the Webster apportionment. The other three apportionments differ.

29. Answers will vary.

16.4 Exercises (Pages 1048–1049)

1. $md = 595$

State	a	b	c	d
***Q* rounded down/up**	28/29	12/13	**81/82**	9/10
Number of seats	28	12	**83**	9

3. $md = 48.4$

State	a	b	c	d	e
***Q* rounded down/up**	52/53	30/31	**164/165**	19/20	22/23
Number of seats	53	30	**166**	19	22

5.

State	a	b	c	d
Number of seats if $n = 204$	**35**	74	42	53
Number of seats if $n = 205$	**34**	75	43	53

7.

State	a	b	c	d	e
Number of seats if $n = 126$	**10**	29	34	28	25
Number of seats if $n = 127$	**9**	30	34	29	25

9.

State	a	b	c
Initial number of seats	**1**	4	**6**
Percent growth	**10.91%**	18.40%	**13.16%**
Revised number of seats	**2**	4	**5**

11.

State	a	b	c
Initial number of seats	6	**5**	**2**
Percent growth	4.84%	**1.63%**	**1.45%**
Revised number of seats	6	**4**	**3**

13. Two additional seats are added for the second apportionment.

State	Original State a	Original State b	New State c
Initial number of seats	20	55	*****
Revised number of seats	21	54	2

15. Seven additional seats are added for the second apportionment.

State	Original State a	Original State b	New State c
Initial number of seats	36	47	*****
Revised number of seats	37	46	7

17. The new states paradox does not occur if the new population is 531, but it does occur if the new population is 532.
19. $md = 208$

State	a	b	c	d	e
Q rounded down	8	16	33	**118**	42
Number of seats	9	17	34	**117**	43

21. Answers will vary. 23. Answers will vary. 25. Answers will vary.

Chapter 16 Test (Pages 1051–1053)

1. Cancún 2. Aruba 3. the Bahamas 4. Aruba and Cancún each have two pairwise points and the Bahamas and Dominican Republic each have one pairwise point, so the pairwise comparison method vote results in a tie. 5. $7! = 5040$ 6. ${}_{10}C_2 = 45$ 7. Answers will vary. 8. Answers will vary. 9. Answers will vary. 10. Answers will vary. 11. Alternative a has a majority of the first-place votes, 16 of 31. The Borda method selects alternative b. 12. c 13. Alternative c is the Condorcet candidate. Plurality and Borda violate the Condorcet criterion and select b. Hare does not violate the criterion because it selects c. 14. Alternative c is selected before the five voters rearrange their ranking. Alternative s is selected after they rearrange their ranking. The rearranging voters moved alternative c, the previous pairwise selection, to the top of their ranking, but c is not selected a second time. This shows that the pairwise comparison method can violate the monotonicity criterion. 15. Alternative a is selected before the five voters rearrange their ranking. Alternative b is selected after they rearrange their ranking. The rearranging voters moved alternative a, the previous Hare selection, to the top of their ranking, but a is not selected a second time. This shows that the Hare method can violate the monotonicity criterion. 16. Alternative a is selected before losing alternative c is dropped. Alternative b is selected after c is dropped. The two outcomes show that the plurality method can violate the irrelevant alternatives criterion. A losing alternative was dropped from the selection process, but the original preferred alternative is not selected a second time. 17. Alternative b is selected before losing alternative d is dropped. Alternative a is selected after d is dropped. The two outcomes show that the Borda method can violate the irrelevant alternatives criterion. A losing alternative was dropped from the selection process, but the original preferred alternative is not selected a second time. 18. Alternative a is selected before losing alternative b is dropped. Alternative c is selected after b is dropped. The two outcomes show that the Hare method can violate the irrelevant alternatives criterion. A losing alternative was dropped from the selection process, but the original preferred alternative is not selected a second time. 19. Answers will vary.
20. standard divisor $d = 133.3692$ (That is, each seat represents 13,337 Smithapolis citizens.)

Ward	1st	2nd	3rd	4th	5th
Number of seats	11	65	57	50	12

21. $md = 131$

Ward	1st	2nd	3rd	4th	5th
Number of seats	10	65	58	50	12

22. $md = 133.7$

Ward	1st	2nd	3rd	4th	5th
Number of seats	11	65	57	50	12

23. Answers will vary. **24.** Answers will vary. **25.** Answers will vary. **26.** Answers will vary.
27. $md = 347$

State	a	b	c	d
Q rounded down	6	12	15	**64**
Number of seats	6	12	16	**66**

28.

State	a	b	c	d	e
Number of seats if $n = 126$	**10**	29	34	28	25
Number of seats if $n = 127$	**9**	30	34	29	25

29.

State	a	b	c
Initial number of seats	**1**	4	**6**
Percent growth	**14.55%**	20.00%	**15.79%**
Revised number of seats	**2**	4	**5**

30. Fifteen additional seats are added for the second apportionment.

State	Original State a	Original State b	New State c
Initial number of seats	23	77	*****
Revised number of seats	24	76	15

31. Answers will vary.

Appendix Exercises (Pages A-5–A-7)

1. 8000 mm **3.** 85 m **5.** 689 mm **7.** 5.98 cm **9.** 5300 m **11.** 27.5 km **13.** 2.54 cm; 25.4 mm **15.** 5 cm; 50 mm **17.** 600 cl **19.** 8700 ml **21.** 9.25 L **23.** 8.974 L **25.** 8 kg **27.** 5200 g **29.** 4200 mg **31.** .598 g **33.** 30°C **35.** −81°C **37.** 50°F **39.** −40°F **41.** 200 nickels **43.** .2 g **45.** $40.99 **47.** $387.98 **49.** 1,082,400 cm^3; 1.0824 m^3 **51.** 200 bottles **53.** 897.9 m **55.** 201.1 km **57.** 3.995 lb **59.** 21,428.8 g **61.** 30.22 qt **63.** 106.7 L **65.** unreasonable **67.** reasonable **69.** unreasonable **71.** B **73.** B **75.** C **77.** A **79.** A **81.** C **83.** A **85.** B **87.** B **89.** A **91.** B **93.** C **95.** B

CREDITS

1 Cinergi Pictures/The Kobal Collection **4** ABC/Courtesy Everett Collection **6** Beth Anderson **6** Excerpted from *In Mathematical Circles, Volume 1* by Howard Eves. Published by the Mathematical Association of America. **7** Chris Pizzello/Reuters/Corbis **14** Digital Vision **20** Chuck Painter/Stanford News Service **20** Copyright 1945, © 1973 renewed by Princeton University Press; reprinted with permission of Princeton University Press. **22** CBS Broadcasting, Inc./Courtesy Everett Collection; Image used with permission. **22** Texas Instruments/NCTM. Images used with permission. **25** PhotoDisc **26** PhotoDisc **27** © 2004 by David Deutsch and Benjamin Goldman. All Rights Reserved. Reprinted with permission from *Mathematics Teacher*, Nov 2004. © 2004 by the National Council of Teachers of Mathematics. All Rights Reserved. **27–28** Reprinted with permission from *Mathematics Teacher*. © 2003–2005 by the National Council of Teachers of Mathematics. All Rights Reserved. **27** Warner Bros./The Kobal Collection **27** Corbis **29** PhotoDisc **30** Digital Vision **31** Beth Anderson **33** Beth Anderson **34** Beth Anderson **34** Corbis **34** William Perlman/Star Ledger/Corbis **39** Beth Anderson **39** Beth Anderson **40** Pier Paolo Cito/Pool/Reuters/Corbis **42–43** Reprinted with permission from *Mathematics Teacher*, Sept. 1991. © 1991 by the National Council of Teachers of Mathematics. All Rights Reserved. **48–49** Reprinted with permission from *Mathematics Teacher*. © 1994, 1997, 1998, 2000, 2002, 2004 by the National Council of Teachers of Mathematics. All Rights Reserved. **51** Sandollar/The Kobal Collection **52** Bavaria-Verlag **54** Beth Anderson **58** *Information Please* ® Database. © 2006 Pearson Education, Inc. All Rights Reserved. Data from NY Stock Exchange. **65** Beth Anderson **65** Beth Anderson **80** Reuters/Corbis **80** Reuters/Corbis **83** The Everett Collection **85** Neal Preston/CORBIS **86** Corbis **87** Silvio Fiore/Superstock **90** Image Source Getty **91** Digital Vision **92** PhotoDisc **95** PhotoDisc **96** PhotoDisc **97** Disney Enterprises, Inc. **98** Corbis/Bettmann **100** Erich Lessing/Art Resource, NY **102** Constantin Film/The Kobal Collection **103** Dream Works Photo/Entertainment Pictures/ZUMA/Corbis **104** Mathematical People, Copyright © 1985 by Berkhauser Boston. Reprinted by permission. **109** Library of Congress **111** Library of Congress **116** Digital Vision **117** PhotoFest **124** PhotoDisc **124** PhotoDisc **126** Digital Vision **126** Poirier/Roger Viollet/Getty Image (Getty Editorial) **128** Corbis/Bettmann **132** PhotoDisc **134** Corbis/Bettmann **134** John Hornsby **138** 2006 Penny Press, Inc. Used with permission **139–142, 155** "Copyright © 2004, Penny Press, Inc. England's Best Logic Problems, May 2004 used with permission of the publisher. All Rights Reserved." **140, 143–144, 155** Courtesy International Sudoku Authority **148** PhotoDisc **148** Python Pictures. Courtesy The Everett Collection **150** Courtesy Airmont Publishing Co., Inc. All Rights Reserved. **151** Reuters/Corbis **152** Jeff J Mitchell/Reuters/Corbis **154** Courtesy Airmont Publishing Co., Inc. All Rights Reserved. **159** Photofest NYC **160** PhotoDisc **160** Courtesy Prindle, Weber & Schmidt, Inc. **161** Bridgeman Art Library **161** The Metropolitan Museum, Egyptian Exhibition. Rogers Fund 1930 **165** American Museum of Natural History **171** © Trinity College. Reprinted by permission of the Master and Fellows of Trinity College Cambridge **172** Pearson Asset Library **173** Courtesy, IBM Corp. **181** Corbis/Bettmann **186** ® Kellogg Company **193** Tom Wurt/Stock Boston **212** Courtesy, Professor Noether **216** St. Andrew's University **221** Pearson Asset Library/Photofest **223** Warner Bros./Southside Amusement Co./The Kobal Collection **225** Hugh C. Williams, University of Manitoba **228** Roger Viollet/Liaison **236** Tri-Star/Phoenix Pictures/The Kobal Collection **236** Roger Viollet/Liaison **241** Giraudon/Art Resource NY **260** Getty **260** PhotoDisc **261** Scala/Art Resource NY **261** PhotoDisc **262** Disney Enterprises, Inc. **266** Francis G. Mayer/CORBIS **266** The Benjamin Franklin Collection/Sterling Memorial Library at Yale University **267** StadtmuseumAachen/ET Archive, London/SuperStock **268** Metropolitan Museum of Art, Harris Brisbane Dick Fund, 1943. **273** PhotoFest **280** PhotoDisc **291** © 2006 Pearson Education, Inc. All Rights Reserved. **295** © 2006 Pearson Education, Inc. All Rights Reserved. **300** Spanish Coins courtesy of Larry Stevens. Used with permission **301** Doug Plasencia/Courtesy of Bowers and Merena Galleries **307** Doug Plasencia/Courtesy of Bowers and Merena Galleries **308** © 2002 Pearson Education, Inc. All Rights Reserved. **309** PhotoDisc **309** PhotoDisc **316** Image courtesy of St. Martins Press. © 1976 St. Martin's Press. Used with permission. **316** The Everett Collection **317** Jacket cover copyright © 2002, from *The Golden Ratio: The Story of Phi, The World's Most Astonishing Number* by Mario Livio. Used by permission of Broadway Books, a division of Random House, Inc. **317** Cover copyright © 1994 from *e: The Story of a Number* by Eli Maor. Used by permission of Princeton University Press. **319** Courtesy of www.joyofpi.com. **320** PhotoDisc **325** Wolper/Warner Bros./The Kobal Collection **326** Paramount/The Kobal Collection **328** © 2006 Pearson Education, Inc. All Rights Reserved. **329** PhotoDisc **330** Photo provided by John C.D. Diamantopoulos, Ph.D. Mathematics and Computer Science Dept., Northeastern State University © 2006 John C.D. Diamantopoulos. All Rights Reserved. **334** © 2002 Pearson Education, Inc. All Rights Reserved. **334** PhotoDisc **335** Comstock **342** © 2006 Pearson Education, Inc. All Rights Reserved. **342** © 2006 Pearson Education, Inc. All Rights Reserved. **343** Columbia Pictures/courtesy The Everett Collection **345** Getty RF Blend Images **350** PhotoDisc **353** Comstock **353** Getty RF MedioImages **353–354** © 2004 Pearson Education, Inc. All Rights Reserved. **357** Corbis **358** The Everett Collection **359** © 2006 Pearson Education, Inc. All Rights Reserved. **360** 20th Century Fox Film Corp. Courtesy The Everett Collection **362** Getty Editorial **365** Albert Ferreira/Reuters/Corbis **365** Beth Anderson **366** MedioImages Getty **368** Warner Bros./Photofest **370** CBS-TV/The Kobal Collection **370** Beth Anderson **373** Corbis **377** PhotoDisc **379** Corbis **383** Bettmann/Corbis **386** © 2004 Pearson Education, Inc. All Rights Reserved. **387** © 2004 Pearson Education, Inc. All Rights Reserved. **390** Stockdisc Classic **393** Corbis **402** NBC/Photofest **404** PhotoDisc **406** PhotoDisc **407** NASA **419** WB Television/Photofest **431** Everett Collection

435 © 2005 Pearson Education, Inc. All Rights Reserved. **435** Beth Anderson **437** National Information Service for Earthquake Engineering **450** AP/Wideworld Photos **450** © 2006 Pearson Education, Inc. All Rights Reserved. **456** © 2007 Pearson Education, Inc. All Rights Reserved. **456** PhotoDisc **459** © 2002 Pearson Education, Inc. All Rights Reserved **460** Data from Sahrov and Novikov, *Edwin Hubble: the Discoverer of the Big Bang Universe,* Cambridge University Press, 1993. Lial, et al., © 2006 Pearson Education, Inc. All Rights Reserved. **461** Digital Vision **462** Digital Vision **472** Corbis **477** NASA **483** Investment Management and Research, Inc. **487** The Everett Collection **490** Digital Vision **491** PhotoDisc **492** Reprinted by permission of the National Geographic Society **493** Beth Anderson **494** © 2002 Pearson Education, Inc. All Rights Reserved. **495** © 2002 Pearson Education, Inc. All Rights Reserved. **500** © 2005 Pearson Education, Inc. All Rights Reserved. **504** Getty Editorial **505** Nigel Cattlin/Photo Researchers, Inc. **507** PhotoDisc **508–509** © 2002 Pearson Education, Inc. All Rights Reserved. **511** AP/Wideworld Photos **513** PhotoDisc **514** Getty Editorial **514** Nigel Cattlin/Photo Researchers, Inc. **522** Stanford University News Service **528** The Kobal Collection **529** The Kobal Collection **531** PhotoDisc **544** Corbis/Bettmann **549** The Everett Collection **550** Reprinted with permission from *The Trisection Problem*, copyright © 1971 by the National Council of Teachers of Mathematics. All rights reserved. **551** Alinari/Art Resource, N.Y. **557** Corbis **559** PhotoDisc **566** George Arthur Plympton Coll, Columbia Univ **569** The Everett Collection **570** Photofest **571** Reprinted with permission from *The Pythagorean Theorem*, copyright © 1968 by the National Council of Teachers of Mathematics. All rights reserved. **571** John Hornsby **582** Photofest **583** Digital Vision **587** Custom Medical Stock Photo **587** Photo Researchers **589** courtesy of Gwen Hornsby **589** BrandX CD MagLands **589** The Image Bank Getty Images **591** PhotoDisc **592** The M.C. Escher Company - Holland **592** The M.C. Escher Company - Holland **594–596** The authors would like to thank The Math Forum at Drexel and Suzanne Alejandre for permission to reprint this article. **595–596** "Tessellations courtesy of Coolmath.com. All Rights Reserved." **599** Corbis-Hulton/UPI Bettmann **600** Reprise Records/Rhino Entertainment Company, a Warner Music Group Company **601** Corbis/Bettmann **604** Courtesy B. K. Tomlin **611** AP/Wideworld Photos **612** Courtesy of Philip Marcus **612** NASA **613–615** Pages 1–4 from National Council of Teachers of Mathematics Student Math Notes, November 1991. Reprinted by permission of the National Council of Teachers of Mathematics. **621** The Everett Collection **633** Smithsonian Institution **650** Courtesy of John Hornsby and John Wiley and Sons, Inc.. **657** PhotoDisc **671** Photofest **674** British Museum **680** Corbis **684** National Museum of American Art, Washington, DC/Art Resource, N.Y. **685** Corbis **695** Washington National Cathedral **699** Digital Vision **702** Stockbyte Platinum **704** Image Source **704** PhotoDisc **703** Corbis **705** PhotoDisc **723** The Everett Collection **724** Corbis **726** Lithograph by Rudolph Hoffman, 1859, after photoprint by Maull and Polyblanck/Library of Congress Prints and Photographs Division **734** Stockbyte Platinum/Getty **735** PhotoDisc **738** Stock Montage, Inc./Historical Pictures Archive **744** Beth Anderson **745** Bettmann/Corbis **749** PhotoDisc **750** Philippe Psaila/Liaison Agency **753** Courtesy of NASA **754** Jonathan Nourok/Photo Edit **756** Museum fur Volkerlunde un Schwerizerischches **760** PhotoDisc **761** PhotoDisc **765** PhotoDisc **766** From David C. Christensen, *Slot Machines: A Pictorial History.* Vestal Press, 1977. **768** Courtesy John Hornsby **771** PhotoDisc **772** Atlantide Phototravel/Corbis **775** David Loh/Reuters/Corbis **781** Dreamworks/Universal/The Kobal Collection/Reed, Eli **782** Courtesy Vern Heeren **787** PhotoDisc **790** From *The 2006 World Book Yearbook, Inc.* p. 169 **805** Doug Pensinger/Getty Images **812** Sovfoto **822** MedioImages/Getty **826** North Wind Picture Archives **830** Superstock **833** Corbis **833** StockDisc **834–840** From *How to Lie with Statistics* by Darrell Huff, illustrated by Irving Geis. Copyright © 1954 and renewed © 1982 by Darrell Huff and Irving Geis. Reprinted by permission of W.W. Norton and Company, Inc. **844** Library of Congress **847** PhotoDisc **849** Sara Piaseczynski **852** Mike Cassese/Reuters/Corbis **853** The Everett Collection **856** Walt Disney/The Kobal Collection **857** Originally broadcast over the CBS News Radio Network as part of the Hourly News by Charles Osgood, March 2, 1977. ©1977 by CBS, Inc. All rights reserved. Reprinted by permission. **858** Oriental Institute Museum, University of Chicago **860** Warner Bros/The Kobal Collection **875** Wells Fargo Bank **878** Digital Vision **889** Cameramann International, Ltd. **891** © 2006 Hearst Communications, Inc. **896** www.business.to.business.resources.com **904** Jean Miele/Corbis **910** PhotoDisc **921** Edward Jones and Company, 2002. Used by permission. **925** Warner Bros./courtesy The Everett Collection **925** Dan Herrick/DMI/Rex Features, courtesy The Everett Collection **926** PhotoDisc **929** PhotoDisc **930** Bettmann/Corbis **931** Reprinted with permission of the author. © 2002 Kim Sneppen, Associate Professor Niels Bohr Institute, Copenhagen, AND Reprinted with permission from Sneppen and Maslov, "Specificity and Stability in Topology of Protein Networks." SCIENCE 3, May 2002 **931** PhotoDisc **933** MedioImages Getty **937** Digital Vision **940** Digital Vision **941** John Keats **944** Courtesy of the School of Mathematics and Statistics, University of St. Andrew's, Fife, Scotland **945** Bettmann/Corbis **946** Readings from Scientific American: Mathematics in the Modern World, 1968 **950** www.turning.mi.sanu.ac.yu/~jablans/tch.htm. **960** Hulton-Deutch Collections/Corbis **963** AFP/Corbis **966** What's Happening in Mathematical Sciences, Vol. 2, 1994, p. 46, American Mathematical Society. Reprinted with permission. Figure supplied by Bill Cooke of Bellcore. **966** http://www.tsp.gatech.edu/history/milestone.html **976** Bettmann/Corbis **978** Maddow, Sutherland, Efstathiou & Loveday/Science Photo Library/Photo Researchers **991** Warner Bros TV/The Kobal Collection **992** AP/Wideworld Photos **996** Northwind Pictures Archive **999** National Portrait Gallery, London **1001** LUCASFILM/20TH Century Fox/The Kobal Collection/Hamshere, Keith **1009** Bettmann/Corbis **1014** PhotoDisc **1017** Corbis **1024** Bettmann/Corbis **1028** Bettmann/Corbis **1029** Beth Anderson **1030** AP/Wideworld Photos **1031** Bettmann/Corbis **1039** AP/Wideworld Photos **1046** Library of Congress **1047** Courtesy Michel Balinski **1047** Courtesy Michel Balinski

INDEX OF APPLICATIONS

General Interest

Geometry

INDEX

D

E

F

G

H

M

O

P

T

U

V

W

X

Y

Z

Modern Period (Early) 1450 A.D. to 1800 A.D.
Logarithms; modern number theory; analytic geometry; calculus; the exploitation of the calculus

1700 A.D. to 1750 A.D.

Leonhard Euler pioneers work in topology, organizing calculus and using it to describe motion of objects and forces acting on them.

Age of French Enlightenment is ushered in by thinkers such as Diderot, Montesquieu, Rousseau, and Voltaire.

First suspension bridge is completed.

James Watt creates steam engine.

1750 A.D. to 1800 A.D.

Benjamin Banneker, a self-taught mathematician and astronomer, compiles a yearly almanac.

Joseph Lagrange develops theory of functions; studies of moon lead to methods for finding longitude.

Celsius thermometer is invented.

Benjamin Franklin does kite experiment.

American Revolution and French Revolution take place.

1800 A.D. to 1850 A.D.

Carl Gauss publishes masterpiece on theory of numbers (important to development of statistics and geometry).

Non-Euclidean geometry is developed.

Pierre Simon de Laplace works out mathematical formulas describing interacting gravitation forces in the solar system.

Augustin Louis Cauchy does important work in complex analysis.

Evariste Galois develops theory of groups.

Georg Riemann founds second non-Euclidean system.

Arthur Cayley and James Sylvester develop matrix theory.

Joseph Antoine Ferdinand Plateau, a Belgian scientist, develops the phenakistoscope—the first apparatus that allowed "moving pictures".

Napoleon Bonaparte attempts domination of Europe.

Georg Ohm describes principles of electric resistance.

Joseph Jacquard improves mechanical loom, allowing mass production of fabric.

Michael Faraday discovers electromagnetic induction.

Charles Babbage develops analytic engine (forerunner of computers).

Ada Augusta, Lady Lovelace, devises the concept of computer programming.

First telegraph is used.

Mathematical Events
Cultural Events